Developments in Mathematics

Volume 74

Aims and Scope

The **Developments in Mathematics** (DEVM) book series is devoted to publishing well-written monographs within the broad spectrum of pure and applied mathematics. Ideally, each book should be self-contained and fairly comprehensive in treating a particular subject. Topics in the forefront of mathematical research that present new results and/or a unique and engaging approach with a potential relationship to other fields are most welcome. High-quality edited volumes conveying current state-of-the-art research will occasionally also be considered for publication. The DEVM series appeals to a variety of audiences including researchers, postdocs, and advanced graduate students.

Dorina Mitrea · Irina Mitrea · Marius Mitrea

Geometric Harmonic Analysis III

Integral Representations, Calderón-Zygmund Theory, Fatou Theorems, and Applications to Scattering

Springer

Dorina Mitrea
Department of Mathematics
Baylor University
Waco, TX, USA

Irina Mitrea
Department of Mathematics
Temple University
Philadelphia, PA, USA

Marius Mitrea
Department of Mathematics
Baylor University
Waco, TX, USA

ISSN 1389-2177 ISSN 2197-795X (electronic)
Developments in Mathematics
ISBN 978-3-031-22734-9 ISBN 978-3-031-22735-6 (eBook)
https://doi.org/10.1007/978-3-031-22735-6

Mathematics Subject Classification: 32A, 26B20, 31B, 35J, 42B

This Springer imprint is published by the registered company Springer Nature Switzerland AG
The registered company address is: Gewerbestrasse 11, 6330 Cham, Switzerland

Dedicated with love to our parents

Prefacing the Full Series

The current work is part of a series, comprised of five volumes, [112], [113], [114], [115], [116]. In broad terms, the principal aim is to develop tools in Real and Harmonic Analysis, of geometric measure theoretic flavor, capable of treating a broad spectrum of boundary value problems formulated in rather general geometric and analytic settings.

In Volume I ([112]), we establish a sharp version of Divergence Theorem (aka Fundamental Theorem of Calculus) which allows for an inclusive class of vector fields whose boundary trace is only assumed to exist in a nontangential pointwise sense.

Volume II ([113]) is concerned with function spaces measuring size and/or smoothness, such as Hardy spaces, Besov spaces, Triebel-Lizorkin spaces, Sobolev spaces, Morrey spaces, Morrey-Campanato spaces, spaces of functions of Bounded Mean Oscillations, etc., in general, geometric settings. Work here also highlights the close interplay between the differentiability properties of functions and singular integral operators.

The topic of singular integral operators is properly considered in Volume III ([114]), where we develop a versatile Calderón-Zygmund theory for singular integral operators of convolution type (and with variable coefficient kernels) on uniformly rectifiable sets in the Euclidean ambient, and the setting of Riemannian manifolds. Applications to scattering by rough obstacles are also discussed in this volume.

In Volume IV ([115]), we focus on singular integral operators of boundary layer type which enjoy more specialized properties (compared with generic, garden variety singular integral operators treated earlier in Volume III). Applications to Complex Analysis in several variables are subsequently presented, starting from the realizations that many natural integral operators in this setting, such as the Bochner-Martinelli operator, are actual particular cases of double layer potential operators associated with the complex Laplacian.

In Volume V ([116]), where everything comes together, finer estimates for a certain class of singular integral operators (of chord-dot-normal type) are produced in a manner which indicates how their size is affected by the (infinitesimal and global) flatness of the "surfaces" on which they are defined. Among the library of double layer potential operators associated with a given second-order system, we then identify those double layers which fall under this category of singular integral operators. It is precisely for this subclass of double layer potentials that Fredholm theory may then be implemented assuming the underlying domain has a compact boundary, which is sufficiently flat at infinitesimal scales. For domains with unbounded boundaries, this very category of double layer potentials may be outright inverted, using a Neumann series argument, assuming the "surface" in question is sufficiently flat globally. In turn, this opens the door for solving a large variety of boundary value problems for second-order systems (involving boundary data from Muckenhoupt weighted Lebesgue spaces, Lorentz spaces, Hardy spaces, Sobolev spaces, BMO, VMO, Morrey spaces, Hölder spaces, etc.) in a large class of domains which, for example, are allowed to have spiral singularities (hence more general than domains locally described as upper-graphs of functions). In the opposite direction, we show that the boundary value problems formulated for systems lacking such special layer potentials may fail to be Fredholm solvable even for really tame domains, like the upper half-space, or the unit disk. Save for the announcement [111], all principal results appear here in print for the first time.

We close with a short epilogue, attempting to place the work undertaken in this series into a broader picture. The main goal is to develop machinery of geometric harmonic analysis flavor capable of ultimately dealing with boundary value problems of a very general nature. One of the principal tools (indeed, the piecè de résistance) in this regard is a new and powerful version of the Divergence Theorem, devised in Volume I, whose very formulation has been motivated and shaped from the outset by its eventual applications to Harmonic Analysis, Partial Differential Equations, Potential Theory, and Complex Analysis. The fact that its footprints may be clearly recognized in the makeup of such a diverse body of results, as presented in Volumes II-V, serves as a testament to the versatility and potency of our brand of Divergence Theorem. Alas, our enterprise is multifaceted, so its success is crucially dependent on many other factors. For one thing, it is necessary to develop a robust Calderón-Zygmund theory for singular integrals of boundary layer type (as we do in Volumes III-IV), associated with generic weakly elliptic systems, capable of accommodating a large variety of function spaces of interest considered in rather inclusive geometric settings (of the sort discussed in Volume II). This renders these (boundary-to-domain) layer potentials useful mechanisms for generating lots of null-solutions for the given system of partial differential operators, whose format is compatible with the demands in the very formulation of the boundary value problem we seek to solve. Next, in order to be able to solve the boundary integral equation to which matters are reduced in this fashion, the success of employing Fredholm theory hinges on the ability to suitably estimate the essential norms of the (boundary-to-boundary) layer potentials. In this vein, we succeed in relating the distance from such layer potentials to the space of compact operators to the flatness of the boundary of the domain in question (measured in terms of infinitesimal mean oscillations of the unit normal) in a desirable manner which shows that, in a precise quantitative fashion, *the flatter the domain the smaller the proximity to compact operators*. This subtle and powerful result, bridging between analysis and geometry, may be regarded as a far-reaching extension of the pioneering work of Radon and Carleman in the early 1900's.

Ultimately, our work aligns itself with the program stemming from A.P. Calderón's 1978 ICM plenary address in which he advocates the use of layer potentials *"for much more general elliptic systems* [*than the Laplacian*]" – see [15, p. 90], and may be regarded as an optimal extension of the pioneering work of E.B. Fabes, M. Jodeit, and N.M. Rivière in [41] (where layer potential methods have been first used to solve boundary value problems for the Laplacian in bounded $\mathscr{C}^1$ domains). In this endeavor, we have been also motivated by the problem[1] posed by A.P. Calderón on [15, p. 95], asking to identify the function spaces on which singular integral operators (of boundary layer type) are well defined and continuous. This is relevant since, as Calderón mentions, "*A clarification of this question would be very important in the study of boundary value problems for elliptic equations* [*in rough domains*]. *The methods employed so far seem to be insufficient for the treatment of these problems.*" We also wish to mention that our work is also in line with the issue raised as an open problem by C. Kenig in [79, Problem 3.2.2, pp. 116–117], where he asked whether operators of layer potential type may be inverted on appropriate Lebesgue and Sobolev spaces in suitable subclasses on NTA domains with compact Ahlfors regular boundaries.

The task of making geometry and analysis work in unison is fraught with difficulties, and only seldom can a two-way street be built on which to move between these two worlds without loss of information. Given this, it is actually surprising that, in many instances, we come very close to having optimal hypotheses, almost an accurate embodiment of the slogan *if it makes sense to write it, then it's true.*

Acknowledgments: The authors gratefully acknowledge partial support from the Simons Foundation (through grants # 426669, # 958374, # 318658, # 616050, and # 637481), as well as NSF (grant # 1900938). Portions of this work have been completed at Baylor University in Waco, Temple University in Philadelphia, the Institute for Advanced Study in Princeton, MSRI in Berkeley, and the American Institute of Mathematics in San Jose. We wish to thank these institutions for their generous hospitality. Last, but not least, we are grateful to Michael E. Taylor for gently yet persistently encouraging us over the years to complete this project.

Waco, USA — Dorina Mitrea
Philadelphia, USA — Irina Mitrea
Waco, USA — Marius Mitrea

[1] In the last section of [15], simply titled "*Problems*," Calderón singles two directions for further study. The first one is the famous question whether the smallness condition on $\|a'\|_{L^\infty}$ (the Lipschitz constant of the curve $\{(x, a(x)) : x \in \mathbb{R}\}$ on which he proved the L^2-boundedness of the Cauchy operator) may be removed (as is well known, this has been solved in the affirmative by Coifman, McIntosh, and Meyer in [20]). We are referring here to the second (and final) problem formulated by Calderón on [15, p. 95].

Description of Volume III

From the outset, the very formulation of our versions of the Divergence Theorem from Volume I (cf. [112, §1.2–§1.12]) has been motivated and shaped by potential applications to Harmonic Analysis, Partial Differential Equations, Function Space Theory, and Complex Analysis. We have envisioned these versions of the Divergence Theorem not as end-products, in and of themselves, but as effective tools to further progress in these areas of mathematics. This has already become apparent in Volume II ([113]), when dealing with function spaces measuring smoothness of Sobolev type on the geometric measure theoretic boundaries of sets of locally finite perimeter.

In the opening chapter of the present volume (Chapter 1, titled "Integral Representations and Integral Identities"), we further elaborate on this vision. We begin in §1.1 by revisiting the classical Cauchy-Pompeiu integral representation formula in open sets $\Omega \subseteq \mathbb{C}$ with a lower Ahlfors regular boundary and whose arc-length $\sigma := \mathcal{H}^1 \lfloor \partial\Omega$ is a doubling measure. Our Divergence Theorem specialized to this setting then permits us to identify a very general class of functions for which the Cauchy-Pompeiu integral representation formula is valid. By means of counterexamples, we show that the analytic conditions imposed are in the nature of best possible. A very general version of the Cauchy integral representation formula, allowing one to recover a holomorphic function from its (nontangential) boundary trace via the (boundary-to-domain) Cauchy integral operator, is then obtained as a corollary. In the same spirit, generalizations of the classical Morera Theorem and Residue Theorem are established. Variants with no explicit lower Ahlfors regularity assumptions made on the topological boundary are also discussed.

This line of work continues in §1.2 where higher-dimensional versions of some of the main results from §1.1 are extended to open subsets of $\mathbb{R}^n$ with $n \in \mathbb{N}$, $n \geq 2$ arbitrary, now involving the Clifford algebra $\mathcal{C}\ell_n$ (cf. the discussion in [112, §6.4]) in place of the field of complex numbers $\mathbb{C}$, and the Dirac operator D (from (A.0.37)) in lieu of the Cauchy-Riemann operator $\overline{\partial}$. Once again, all integral representation formulas in Clifford Analysis derived here make essential use of our brand of the Divergence Theorem from Volume I ([112]). A more general point of view is adopted in §1.6 where integral representation formulas are derived, under rather inclusive geometric and analytic assumptions, for injectively elliptic first-order systems.

Granted the availability of a fundamental solution for second-order, homogeneous, constant (complex) coefficient, weakly elliptic systems, which exhibits a number of desirable properties (as discussed in [109, Theorem 11.1, pp. 393–395]), in §1.5 and §1.7, we derive boundary layer potential representations and Green-type formulas for such systems, under minimal geometric and analytic assumptions. Chapter 1 ends with a section (§1.8) on Rellich-type identities in open sets with Ahlfors regular boundaries.

Chapter 2 contains a powerful and nuanced Calderón-Zygmund theory for singular integral operators on uniformly rectifiable sets. The results here, which build on the work of many predecessors, including Calderón, Zygmund, Mikhlin, Coifman, McIntosh, Meyer, David, Jerison, Kenig, Semmes, and Stein, among many others, enhance, refine, and sharpen those in [63] by placing more economical demands on the ambient geometry. As a preamble, in §2.1, we consider integral operators on Hölder spaces on upper Ahlfors regular sets, and in §2.2, we deal with singular integrals on Lebesgue spaces on Ahlfors regular quasi-metric spaces. The latter section contains a number of results in the spirit of Calderón-Zygmund theory, and a rather versatile version of Cotlar's Lemma. Principal-value singular integral operators on uniformly rectifiable sets are treated in earnest in §2.3, while, in §2.4, we consider boundary-to-domain integral operators on open sets with uniformly rectifiable boundaries.

The results established so far are combined in §2.5to produce a very versatile jump-formula, identifying the nontangential trace of a boundary-to-domain integral operator in an open set with a uniformly rectifiable boundary as a jump-term (involving the Fourier transform of the integral kernel function) and a principal-value singular integral operator on the boundary (of the sort alluded to earlier). In §2.9, we present some applications to integral operators whose kernels are fashioned by applying a suitable number of derivatives to the fundamental solutions of certain elliptic differential operators. In §2.8, the last section in Chapter 2, we extend the scope of the Calderón-Zygmund theory developed so far by considering singular integral operators on uniformly rectifiable sets, with variable coefficient kernels that are actually the Schwartz kernels of classical pseudodifferential operators of order -1 with odd principal symbol, acting between vector bundles over a Riemannian manifold.

The main goal in Chapter 3 is to derive quantitative Fatou-type theorems in uniformly rectifiable domains. The distinctive format of a Fatou-type theorem is that size/integrability properties of the nontangential maximal operator for a null-solution of an elliptic equation in a certain domain implies the a.e. existence of the pointwise nontangential boundary trace of the function in question. The quantitative aspect is reflected in the fact that not only the boundary trace exists but also contains considerable information concerning the size and regularity of the original function. In §3.1, one of the central results is a quantitative Fatou-type theorem for null-solutions of injectively elliptic first-order (homogeneous, constant complex coefficient) systems of differential operators in arbitrary uniformly rectifiable domains. In turn, this theorem has a large spectrum of applications, including the theory of Hardy spaces associated with injectively elliptic first-order systems in UR domains discussed in §3.2. Fatou-type theorem for second-order systems is subsequently established in §3.3 where, among other things, we prove a quantitative Fatou-type theorem for the gradient of null-solutions of second-order systems in UR domains, along with a version in which both the trace of the function and its gradient are shown to exist. Lastly, in §3.4, we show that for a null-solution of a weakly elliptic system, having sufficient regularity on the Sobolev/Besov/Triebel-Lizorkin scales in a given bounded two-sided NTA domain with an Ahlfors regular boundary guarantees the existence of its nontangential pointwise trace on the boundary of the domain in question.

Chapter 4 is primarily reserved for examining the role of Green functions in establishing uniqueness for the Dirichlet Problem for weakly elliptic homogeneous, constant (complex) coefficient, second-order systems, formulated in broad geometric settings. In this vein, see Theorem 4.1.1 and Theorem 4.1.6, which may be regarded as sharp embodiments of the heuristic principle asserting that uniqueness holds in the Dirichlet/Regularity Problem for a given system L and domain Ω granted the existence of Green functions which "pair well" with the null-solutions of the boundary value problem in question. A sensible feature of these results is the absence of concrete function spaces in their formulation, which makes them directly applicable to a wide range of specific cases of interest. Basically, all results in this chapter make essential use of the brand of Divergence Theorem developed earlier, in Volume I ([112]).

Chapter 5 continues the discussion initiated in Chapter 4 by further specializing matters to the case of the Laplace operator. Here, we are concerned with issues of a potential theoretic nature such as: When is the Poisson kernel associated with a domain $\Omega \subseteq \mathbb{R}^n$ (as the Radon-Nikodym derivative of the harmonic measure with respect to the surface measure) expressible as the normal derivative of the Green function? When is said Poisson kernel bounded from below away from zero? When can one represent the solution of the Dirichlet Problem with L^p data as an integral involving the Poisson kernel? Throughout, we are interested in minimal regularity conditions, a perspective through which such questions have been raised, for example, by J. Garnett and D.E. Marshall in [49].

In Chapter 6, the last chapter in this volume, we treat certain fundamental aspects of scattering theory. Concretely, in §6.1, we produce integral representations in exterior domains of a very general geometric nature for null-solutions of the Helmholtz operator which satisfy Sommerfeld's radiation condition at infinity. In §6.2–§6.7, we develop a unified approach to scattering theory, which identifies the broadest possible family of radiation conditions for null-solutions of the vector Helmholtz operator. This family contains, as particular cases, the Sommerfeld, Silver-Müller, and McIntosh-Mitrea radiation conditions corresponding to scattering by acoustic waves, electromagnetic waves, and null-solutions of perturbed Dirac operators, respectively. The new idea is that one can associate a radiation condition with each factorization of the (vector) Helmholtz operator (as a product of two first-order systems).

To close, we wish to remark that all these applications make essential use of our earlier versions of the Divergence Theorem.

Contents

Chapter 1
Integral Representations and Integral Identities

Working in the complex plane, in §1.1 we prove versions of the classical Cauchy-Pompeiu integral representation formula, Morera's Theorem, the Residue Theorem, and the Schwarz-Pompeiu formula under minimal smoothness assumptions. This line of work continues in §1.2 where higher-dimensional Euclidean spaces are considered, now working in the context of Clifford algebras. Integral representation formulas are then derived in domains of a very general geometric nature, first involving boundary layer potentials associated with weakly elliptic second-order systems in §1.5, then in relation to injectively elliptic first-order systems in §1.6. Next, Green-type formulas for second-order systems under optimal assumptions are presented in §1.7, while the last section in this chapter (§1.8) concerns Rellich-type identities in a rather inclusive setting.

Throughout, we seek to work in geometric and analytic settings which are fairly optimal, in light of the conclusions we have in mind. Indeed, the background hypotheses we typically adopt are not too far off from the kind of assumptions one would have to impose to simply have a meaningfully formulated conclusion. In this sense, our results are an almost accurate embodiment of the slogan *if it makes sense to write it, then it's true.*

1.1 One Variable Complex Analysis

We begin with a brief historical survey of the Cauchy integral operator and related topics, designed to highlight a number of major landmarks and breakthroughs. Cauchy's integral reproducing formula for holomorphic functions apparently first appeared in 1831 (cf. [17]). Subsequently, in his 1873 dissertation [163], Y.V. Sokhotski studied the boundary behavior of the Cauchy integral operator and derived jump-formulas under Hölder regularity assumptions on the density function. Another significant achievement of Sokhotski's work was pursuing the study of the Cauchy integral operator as a topic of independent interest. In 1885, A. Harnack re-derived Sokhotski's jump-formulas in [57] by decomposing the Cauchy integral operator

D. Mitrea et al., *Geometric Harmonic Analysis III*, Developments in Mathematics 74,
https://doi.org/10.1007/978-3-031-22735-6_1

as the two-dimensional harmonic double layer plus the tangential derivative of the two-dimensional harmonic single layer, under rather restrictive hypotheses imposed both on the underlying domain and the density function. However, it is J. Plemelj's work in 1908 which has been particularly influential on this topic. In [145] and [146] a new, more rigorous, derivation of Sokhotski jump-formulas is presented (under Hölder regularity assumptions on the density function), and the Cauchy integral operator is used for the first time as a key tool in solving a boundary value problem (of transmission type). In 1912 D. Pompeiu has discovered an integral representation formula involving the Cauchy operator along with a double integral taking into account the failure of the function to be holomorphic. First published in a series of papers [148, 149, 150] in 1912, and then revisited in 1913 in [151], this basic integral representation formula has become known as the Cauchy-Pompeiu formula. Other significant contributions are due to I.I. Privalov who, in 1918 has proved Sokhotski's jump-formulas rigorously by today's standards and established the boundedness of Cauchy's operator on Hölder spaces on Lyapunov domains, and N.I. Muskhelishvili who, starting in 1922, has systematically employed Cauchy type integral operators to solve boundary value problems in the theory of elasticity. Significantly, it was in S.G. Mikhlin's 1948 paper [106] where the action of the Cauchy integral operator has been first considered on square-integrable functions. Assuming the underlying curve Σ to be of class $\mathscr{C}^{1+\varepsilon}$, for some $\varepsilon > 0$, Mikhlin has shown that the principal value Cauchy operator is bounded on $L^2(\Sigma, \mathcal{H}^1)$. Under the same regularity assumption on Σ, this result was subsequently extended to a bound on $L^p(\Sigma, \mathcal{H}^1)$ for each $p \in (1, \infty)$ by B.V. Khvedelidze in [83] (see also the more timely exposition in [52, Vol. I, Theorem 2.1, p. 19]). The year 1977 marks A.P. Calderón's breakthrough, pertaining to L^p-bounds, $1 < p < \infty$, for the principal-value Cauchy integral operator on Lipschitz curves in the plane with a sufficiently small Lipschitz constant, in [14]. Subsequently, in 1982, R. Coifman, A. McIntosh, and Y. Meyer have refined in [20] Calderón's result for the Cauchy integral operator by allowing arbitrary Lipschitz curves. Finally, in 1984 G. David has succeeded in characterized in [30] the class of $\mathcal{H}^1$-measurable, connected, sets $\Sigma \subset \mathbb{C}$ for which the Cauchy operator is bounded on $L^p(\Sigma, \mathcal{H}^1)$ as those for which[1,2]

$$\sup_{z \in \mathbb{C},\, r>0} \frac{\mathcal{H}^1\big(B(z,r) \cap \Sigma\big)}{r} < \infty. \tag{1.1.1}$$

Our first application is a very general version of the Cauchy-Pompeiu representation formula, alluded to above, for functions in the complex plane. To state it, recall first that the Cauchy-Riemann operator $\overline{\partial}$ in the plane and its conjugate are, respectively, defined as

[1] the necessity of condition (1.1.1) is elementary and has been known a little earlier; it appeared in the paper [143] where the conjecture was made that this is also sufficient for the boundedness of the Cauchy operator (see also the exposition in [52, Vol. I, Theorem 3.3, pp. 27-28])

[2] of course, condition (1.1.1) expresses the fact that Σ is upper Ahlfors regular (also occasionally referred to as the Carleson condition, particularly in the Russian literature)

$$\overline{\partial} := \partial_{\bar{z}} := \tfrac{1}{2}(\partial_x + \mathrm{i}\partial_y) \text{ and } \partial := \partial_z := \tfrac{1}{2}(\partial_x - \mathrm{i}\partial_y), \tag{1.1.2}$$

where $\mathrm{i} := \sqrt{-1} \in \mathbb{C}$. Given any Lebesgue measurable set $\Omega \subseteq \mathbb{R}^2 \equiv \mathbb{C}$ of locally finite perimeter, introduce $\sigma := \mathcal{H}^1 \lfloor \partial\Omega$ and denote by ν the geometric measure theoretic outward unit normal to Ω. We then define the `complex arc-length measure` on $\partial\Omega$ as

$$\mathrm{d}\zeta := \mathrm{i}\nu(\zeta)\,\mathrm{d}\sigma(\zeta) \text{ on } \partial_*\Omega. \tag{1.1.3}$$

When $\mathrm{d}\zeta$ regarded as a measure in $\mathbb{R}^2$ supported on $\partial_*\Omega$, which is further identified with a distribution in $\mathbb{R}^2$, from (1.1.3), (1.1.2), [112, (5.6.10)], and the fact that $\sigma_* = \mathcal{H}^{n-1}\lfloor \partial_*\Omega$, we conclude that the complex arc-length measure on $\partial\Omega$ satisfies [3]

$$\mathrm{d}\zeta = -2\mathrm{i}\overline{\partial}\mathbf{1}_\Omega \tag{1.1.4}$$

in the sense of distributions in $\mathbb{R}^2$ (this should be compared with the result described in [112, Proposition 5.6.3]).

Theorem 1.1.1 *Let $\Omega \subseteq \mathbb{R}^2 \equiv \mathbb{C}$ be an open set with a lower Ahlfors regular boundary, such that $\sigma := \mathcal{H}^1\lfloor\partial\Omega$ is a doubling measure on $\partial\Omega$. In particular, Ω is a set of locally finite perimeter, and its geometric measure theoretic outward unit normal ν is defined σ-a.e. on $\partial_*\Omega$.*

In this context, suppose $u : \Omega \to \mathbb{C}$ is an $\mathcal{L}^2$-measurable complex-valued function which, for some $\kappa > 0$, satisfies

$$\int_{\partial\Omega} \frac{(\mathcal{N}_\kappa u)(\zeta)}{1+|\zeta|}\,\mathrm{d}\sigma(\zeta) < +\infty \text{ and } u\big|_{\partial\Omega}^{\kappa-\mathrm{n.t.}} \text{ exists } \sigma\text{-a.e. on } \partial_{\mathrm{nta}}\Omega. \tag{1.1.5}$$

Also, with the Cauchy-Riemann operator $\overline{\partial}$ taken in the sense of distributions in Ω, assume that

$$\overline{\partial}u \in L^1\Big(\Omega, \frac{\mathrm{d}\mathcal{L}^2(\zeta)}{1+|\zeta|}\Big). \tag{1.1.6}$$

Then for any $\kappa' > 0$ the nontangential trace $u\big|_{\partial\Omega}^{\kappa'-\mathrm{n.t.}}$ exists σ-a.e. on $\partial_{\mathrm{nta}}\Omega$ and is actually independent of κ'. Moreover, with the dependence on the parameter κ' dropped,

for each fixed $z \in \Omega$, the function defined σ-a.e. on $\partial_*\Omega$ as

$$\partial_*\Omega \ni \zeta \longmapsto \frac{\big(u\big|_{\partial\Omega}^{\mathrm{n.t.}}\big)(\zeta)}{\zeta - z} \text{ belongs to } L^1(\partial_*\Omega, \sigma), \tag{1.1.7}$$

and for $\mathcal{L}^2$-a.e. point $z \in \Omega$ one has (with $\mathrm{d}\zeta$ denoting the complex arc-length measure on $\partial\Omega$; cf. (1.1.3))

---

[3] if Ω is a domain of class $\mathscr{C}^1$ then one may check without difficulty that (1.1.4) holds in the sense of distributions in the plane

$$u(z) = \frac{1}{2\pi \mathrm{i}} \int_{\partial_* \Omega} \frac{\big(u\big|^{\mathrm{n.t.}}_{\partial\Omega}\big)(\zeta)}{\zeta - z}\, \mathrm{d}\zeta - \frac{1}{\pi} \int_{\Omega} \frac{(\overline{\partial} u)(\zeta)}{\zeta - z}\, \mathrm{d}\mathcal{L}^2(\zeta), \tag{1.1.8}$$

provided Ω is bounded, or $\partial\Omega$ is unbounded. In the remaining case, i.e., when Ω is unbounded and $\partial\Omega$ is bounded (that is, when Ω is an exterior domain), formula (1.1.8) *holds under the additional assumption that there exists $\lambda \in (1, \infty)$ such that*

$$⨍_{B(0,\lambda R)\setminus B(0,R)} |u|\, \mathrm{d}\mathcal{L}^2 = o(1) \text{ as } R \to \infty. \tag{1.1.9}$$

As a corollary, if actually $\overline{\partial} u = 0$ in Ω then one has the following integral representation formula

$$u(z) = \frac{1}{2\pi \mathrm{i}} \int_{\partial_* \Omega} \frac{\big(u\big|^{\mathrm{n.t.}}_{\partial\Omega}\big)(\zeta)}{\zeta - z}\, \mathrm{d}\zeta, \qquad \forall z \in \Omega, \tag{1.1.10}$$

provided Ω is bounded, or $\partial\Omega$ is unbounded. In addition, when Ω is unbounded and $\partial\Omega$ is bounded, formula (1.1.10) *continues to be valid provided* (1.1.9) *is also assumed.*

Before presenting the proof of this theorem we make a series of remarks.

Remark 1. Together, [112, Lemma 8.3.1] and the first property in (1.1.5) imply

$$u \in L^{\infty}_{\mathrm{loc}}(\Omega, \mathcal{L}^2). \tag{1.1.11}$$

In particular, $u \in L^1_{\mathrm{loc}}(\Omega, \mathcal{L}^2)$ which ensures that it is meaningful to consider the distribution $\overline{\partial} u$ in (1.1.6), and to speak of Lebesgue points of u (as done in relation to (1.1.15)).

Remark 2. Regarding the first condition in (1.1.5) observe that

$$\left.\begin{array}{r} \partial\Omega \text{ upper Ahlfors regular,} \\ \text{and } \mathcal{N}_\kappa u \in L^p(\partial\Omega, \sigma) \\ \text{with } 1 \le p < \infty \end{array}\right\} \Longrightarrow \int_{\partial\Omega} \frac{(\mathcal{N}_\kappa u)(\zeta)}{1 + |\zeta|}\, \mathrm{d}\sigma(\zeta) < +\infty, \tag{1.1.12}$$

thanks to Hölder's inequality and [112, Lemma 7.2.1]. Also, obviously,

whenever Ω has a bounded topological boundary, we have

$$\int_{\partial\Omega} \frac{(\mathcal{N}_\kappa u)(\zeta)}{1 + |\zeta|}\, \mathrm{d}\sigma(\zeta) < +\infty \Longleftrightarrow \mathcal{N}_\kappa u \in L^1(\partial\Omega, \sigma). \tag{1.1.13}$$

We also wish to note that it is possible to further relax the first condition in (1.1.5) and replace it by the demand that, for some $\varepsilon > 0$,

$$\mathcal{N}_\kappa^\varepsilon u \in L^1_{\mathrm{loc}}(\partial\Omega, \sigma) \text{ and } \int_{\partial_*\Omega} \frac{\left|\left(u\Big|_{\partial\Omega}^{\kappa-\mathrm{n.t.}}\right)(\zeta)\right|}{1+|\zeta|}\,\mathrm{d}\sigma(\zeta) < +\infty. \tag{1.1.14}$$

Such a version would require using [112, Corollary 1.5.2] (in the version recorded in [112, (1.5.23)]) in place of [112, Theorem 1.2.1], as we do in the proof of Theorem 1.1.1 given further below.

Remark 3. Concerning the integrability condition on $\overline{\partial}u$, [112, Lemma 3.5.7] ensures that (1.1.6) is equivalent with having, for $\mathcal{L}^2$-a.e. point $z \in \mathbb{C}$,

$$\int_\Omega \left|\frac{(\overline{\partial}u)(\zeta)}{\zeta - z}\right| \mathrm{d}\mathcal{L}^2(\zeta) < +\infty. \tag{1.1.15}$$

Let us also note here that under the assumption

$$\overline{\partial}u \in L^p_{\mathrm{loc}}(\Omega, \mathcal{L}^2) \text{ for some } p \in (1, \infty), \tag{1.1.16}$$

elliptic regularity implies, bearing in mind (1.1.11), that $u \in W^{1,p}_{\mathrm{loc}}(\Omega)$ (for general results of this nature see [153], [177]). In particular, u is locally Hölder continuous in Ω whenever (1.1.16) holds with $p > 2$.

Remark 4. In the case when Ω is an exterior domain, there is a version of the integral representation formula (1.1.8) which brings to the forefront the behavior of u at infinity in a manner different than the decay condition imposed in (1.1.9). Specifically, from [112, (1.4.5)], [112, Proposition 4.7.1] and (1.1.56) it follows that if Ω is an exterior domain in $\mathbb{C}$ and one assumes that

$$\begin{gathered} u \text{ is continuous in a neighborhood of infinity} \\ \text{and } \fint_{|z|=R} |u(z)|\,\mathrm{d}\mathcal{H}^1(z) = o(R) \text{ as } R \to \infty, \end{gathered} \tag{1.1.17}$$

(hence, in particular, if u is continuous and bounded in a neighborhood of infinity) then the limit

$$u_\infty := \lim_{R\to\infty} \fint_{|z|=R} u(z)\,\mathrm{d}\mathcal{H}^1(z) \text{ exists in } \mathbb{C} \tag{1.1.18}$$

and for each Lebesgue point $z \in \Omega$ of u with the property that (1.1.15) holds one has

$$u(z) = \frac{1}{2\pi\mathrm{i}} \int_{\partial_*\Omega} \frac{\left(u\big|_{\partial\Omega}^{\mathrm{n.t.}}\right)(\zeta)}{\zeta - z}\,\mathrm{d}\zeta - \frac{1}{\pi}\int_\Omega \frac{(\overline{\partial}u)(\zeta)}{\zeta - z}\,\mathrm{d}\mathcal{L}^2(\zeta) + u_\infty. \tag{1.1.19}$$

Remark 5. In the case when Ω is an exterior domain, (1.1.9) is satisfied if we impose the pointwise decay condition

$$u(\zeta) = o(1) \text{ as } |\zeta| \to \infty. \tag{1.1.20}$$

`Remark 6.` It turns out that if [112, Theorem 1.5.1] is used in place of [112, Theorem 1.4.1] in the proof of Theorem 1.1.1, then

$$\text{the doubling assumption on } \sigma := \mathcal{H}^1 \lfloor \partial\Omega \text{ may be relaxed to simply asking that } \sigma \text{ is a locally finite measure on } \partial\Omega. \tag{1.1.21}$$

The price to pay is having to demand that the aperture parameter κ is sufficiently large (depending on Ω) to begin with, and we may lose the flexibility of changing it when considering nontangential boundary traces. Modulo these nuances, the format of the main result (i.e., the integral representation formula (1.1.8)) remains the same.

This being said, *the lower Ahlfors regularity condition for $\partial\Omega$ may not be simply dropped.* To see this, consider the open subset $\Omega := B(0,1)\setminus\{0\}$ of $\mathbb{C}$ and the function $u : \Omega \to \mathbb{C}$ defined as $u(\zeta) := \frac{1}{\zeta}$ for each $\zeta \in \Omega$. It is clear that $\sigma := \mathcal{H}^1 \lfloor \partial\Omega$ is a locally finite measure. Also, the function u is holomorphic in Ω, and the conditions in (1.1.5) are satisfied for any $\kappa > 0$. In addition, $\big(u\big|^{\text{n.t.}}_{\partial\Omega}\big)(\zeta) = \frac{1}{\zeta}$ for every $\zeta \in \partial B(0,1)$. If we now fix $z \in \Omega$ and define $f(\zeta) := \frac{1}{\zeta(\zeta - z)}$ for $\zeta \in \mathbb{C} \setminus \{z, 0\}$, we have that f is meromorphic with poles of order one at 0 and z. Hence, the boundary integral in the right-hand side of (1.1.8) may be computed using the Residue Theorem as

$$\int_{\partial B(0,1)} f(\zeta)\,\mathrm{d}\zeta = 2\pi\mathrm{i}\big[\mathrm{Res}(f,z) + \mathrm{Res}(f,0)\big] = 2\pi\mathrm{i}\Big[\frac{1}{z} - \frac{1}{z}\Big] = 0. \tag{1.1.22}$$

Given that the solid integral in (1.1.8) is zero, if (1.1.8) were to hold in this case, we would obtain $u(z) = 0$ for each $z \in \Omega$, a contradiction. This shows that the lower Ahlfors regularity condition for $\partial\Omega$ may not be simply discarded.

`Remark 7.` Both assumptions in (1.1.5) are necessary. To see that this is the case, consider the open subset of $\mathbb{C}$ described as $\Omega := \big\{z \in B(0,1) : z \notin [0,1)\big\}$ and bring in the holomorphic function $u : \Omega \to \mathbb{C}$ given by $u(\zeta) := \frac{1}{\zeta}$ for each $\zeta \in \Omega$. Then the boundary integral in (1.1.8) is just as in (1.1.22), hence zero. Thus, (1.1.8) becomes $u(z) = 0$ for each $z \in \Omega$, a contradiction.

In this scenario, Ω satisfies all geometric hypotheses stipulated in Theorem 1.1.1, and u satisfies all but the first condition in (1.1.5). The latter presently fails. Specifically, since for each fixed $\kappa > 0$ we have $\big(\mathcal{N}_\kappa u\big)(x) \approx x^{-1}$ uniformly for $x \in (0,1) \subseteq \partial\Omega$, it follows that

$$\mathcal{N}_\kappa u \in L^{1,\infty}(\partial\Omega, \sigma) \text{ but } \mathcal{N}_\kappa u \notin L^1(\partial\Omega, \sigma). \tag{1.1.23}$$

This shows that the first condition in (1.1.5) is indeed necessary.

As regards the necessity of the second condition in (1.1.5), consider the open subset of $\mathbb{C}$ described as

$$\Omega := \Big\{z \in B(0,1) : z \notin \big[-\tfrac{1}{2}, +\tfrac{1}{2}\big] \times \{0\}\Big\}, \tag{1.1.24}$$

pick some $f \in \mathscr{C}^0_c\big((-\frac{1}{2}, +\frac{1}{2})\big)$ which is not identically zero, then define the function

$$u(z) := \frac{1}{2\pi \mathrm{i}} \int_{-1/2}^{1/2} \frac{f(t)}{z-t}\,\mathrm{d}t, \qquad \forall z \in \Omega. \tag{1.1.25}$$

Then u is holomorphic in Ω, and

$$\big(u\big|_{\partial\Omega}^{\mathrm{n.t.}}\big)(\zeta) = \frac{1}{2\pi \mathrm{i}} \int_{-1/2}^{1/2} \frac{f(t)}{\zeta-t}\,\mathrm{d}t \;\text{ for each } \zeta \in \partial_*\Omega = \partial B(0,1). \tag{1.1.26}$$

Consequently, Fubini's Theorem gives that for each point $z \in \Omega$ we have

$$\begin{aligned}
\int_{\partial_*\Omega} \frac{\big(u\big|_{\partial\Omega}^{\mathrm{n.t.}}\big)(\zeta)}{\zeta - z}\,\mathrm{d}\zeta &= \frac{1}{2\pi \mathrm{i}} \int_{\partial B(0,1)} \Big(\int_{-1/2}^{1/2} \frac{f(t)}{\zeta-t}\,\mathrm{d}t \Big) \frac{\mathrm{d}\zeta}{\zeta - z} \\
&= \int_{-1/2}^{1/2} f(t) \Big(\frac{1}{2\pi \mathrm{i}} \int_{\partial B(0,1)} \frac{\mathrm{d}\zeta}{(\zeta - t)(\zeta - z)} \Big)\,\mathrm{d}t \\
&= 0,
\end{aligned} \tag{1.1.27}$$

since the last integral on the unit circle vanishes, by the Residue Theorem. As such, if the integral representation formula (1.1.8) were to hold, it would presently imply that $u(z) = 0$ for each $z \in \Omega$. However, this is not the case. For example, for $\mathcal{L}^1$-a.e. point $x \in \big(-\frac{1}{2}, +\frac{1}{2}\big)$ we have

$$\lim_{y \to 0^+} u(x \pm \mathrm{i}y) = \pm\frac{1}{2} f(x) + \lim_{\varepsilon \to 0^+} \frac{1}{2\pi \mathrm{i}} \Big(\int_{-1/2}^{x-\varepsilon} + \int_{x+\varepsilon}^{+1/2} \Big) \frac{f(t)}{z-t}\,\mathrm{d}t, \tag{1.1.28}$$

which implies that

$$\begin{aligned}
&\text{there exists a Lebesgue measurable set } A \subseteq \big(-\tfrac{1}{2}, +\tfrac{1}{2}\big) \text{ with } \mathcal{L}^1(A) > 0 \\
&\text{such that } \lim_{y \to 0^+} u(x + \mathrm{i}y) - \lim_{y \to 0^+} u(x - \mathrm{i}y) = f(x) \neq 0 \text{ for each } x \in A.
\end{aligned} \tag{1.1.29}$$

In the current setting, Ω satisfies all geometric hypotheses stipulated in Theorem 1.1.1, and u satisfies all but the second condition in (1.1.5). However, (1.1.29) proves that the latter fails and, even though

$$u\big|_{\partial\Omega}^{\kappa-\mathrm{n.t.}} \text{ does exist at } \sigma\text{-a.e. point on } \partial_*\Omega, \tag{1.1.30}$$

the failure of the second condition in (1.1.5) ultimately invalidates (1.1.8).

Remark 8. Another example which highlights the necessity of the first condition in (1.1.5) is as follows. Consider

$$\Omega := B(0,1), \quad u : \Omega \to \mathbb{C}, \quad u(z) := \mathrm{e}^{\frac{1+z}{1-z}}, \quad \forall z \in \Omega. \tag{1.1.31}$$

Note that u extends to a holomorphic function in $\mathbb{C} \setminus \{1\}$ (which actually satisfies $|u(z)| \leq 1$ for $|z| > 1$), and a simple calculation shows that

$$u(e^{i\theta}) = e^{i\left(\frac{\sin\theta}{1-\cos\theta}\right)} \in \partial B(0,1), \qquad \forall\theta \in (0,2\pi). \tag{1.1.32}$$

Thus, for each $\kappa > 0$ we trivially have

$$u \text{ as in } (1.1.31) \Longrightarrow \begin{cases} u\big|_{\partial\Omega}^{\kappa-\mathrm{n.t.}} \text{ exists } \sigma\text{-a.e. on } \partial\Omega \text{ and} \\ \text{belongs to } L^\infty(\partial\Omega,\sigma) \hookrightarrow L^1(\partial\Omega,\sigma). \end{cases} \tag{1.1.33}$$

On the other hand, given any $\kappa > \sqrt{2}$, it follows that $|e^{i\theta} - (1-\theta)| < (1+\kappa)\theta$ provided $\theta \in (0,1)$ is sufficiently small. Hence, $1-\theta \in \Gamma_\kappa(e^{i\theta})$ for every $\theta \in (0,1)$ small, so

$$\big(\mathcal{N}_\kappa u\big)(e^{i\theta}) \geq |u(1-\theta)| \approx e^{2/\theta}, \quad \text{uniformly for } \theta \in (0,1) \text{ small}, \tag{1.1.34}$$

which goes to show that

$$u \text{ as in } (1.1.31) \Longrightarrow \mathcal{N}_\kappa u \notin L^1(\partial\Omega,\sigma). \tag{1.1.35}$$

As such, the Cauchy-Pompeiu representation formula (1.1.8), which in this case simply becomes $u = \mathcal{C}\big(u\big|_{\partial\Omega}^{\kappa-\mathrm{n.t.}}\big)$ where $\mathcal{C}$ is the Cauchy operator for the unit disk, necessarily fails since otherwise one would have (for any choice of $p \in (1,\infty)$)

$$\begin{aligned} \big\|\mathcal{N}_\kappa u\big\|_{L^1(\partial\Omega,\sigma)} &\leq c_p\big\|\mathcal{N}_\kappa u\big\|_{L^p(\partial\Omega,\sigma)} = c_p\big\|\mathcal{N}_\kappa\big(\mathcal{C}\big(u\big|_{\partial\Omega}^{\kappa-\mathrm{n.t.}}\big)\big)\big\|_{L^p(\partial\Omega,\sigma)} \\ &\leq C_p\big\|u\big|_{\partial\Omega}^{\kappa-\mathrm{n.t.}}\big\|_{L^p(\partial\Omega,\sigma)} \leq C_p\big\|u\big|_{\partial\Omega}^{\kappa-\mathrm{n.t.}}\big\|_{L^\infty(\partial\Omega,\sigma)} < +\infty, \end{aligned} \tag{1.1.36}$$

contradicting (1.1.35).

Remark 9. Consider the case when Ω is a UR domain in $\mathbb{C} \equiv \mathbb{R}^2$. In this situation, we have $\partial_*\Omega = \partial\Omega$, hence $\partial_{\mathrm{nta}}\Omega$ has full $\mathcal{H}^1$ measure in $\partial\Omega$. As noted later, in Comment 4 following the statement of Theorem 3.1.6, if u is holomorphic in Ω and satisfies

$$\mathcal{N}_\kappa u \in L^p(\partial\Omega,\sigma) \text{ for some } p \in (1,\infty), \tag{1.1.37}$$

then thanks to [112, (7.2.5)] and the Fatou-type result from Theorem 3.1.6 both conditions in (1.1.5) are satisfied. If actually $\Omega \subset \mathbb{C} \equiv \mathbb{R}^2$ is a bounded NTA domain with an Ahlfors regular boundary, then the second condition in (1.1.5) is guaranteed by the Fatou-type result from Proposition 5.5.2.

These observations permit streamlining the assumptions made in relation to the version of Cauchy's reproducing formula recorded in (1.1.10) in the geometric settings mentioned above.

In the proof of Theorem 1.1.1, presented a little further below, we shall need the following extension of [109, Exercise 7.47, p. 292].

Lemma 1.1.2 *Let Ω be an arbitrary open set in $\mathbb{R}^2 \equiv \mathbb{C}$, and consider a complex-valued function $u \in L^1_{\mathrm{loc}}(\Omega,\mathcal{L}^2)$ with the property that $\overline{\partial}u \in L^1_{\mathrm{loc}}(\Omega,\mathcal{L}^2)$ (where the Cauchy-Riemann operator $\overline{\partial}$ is applied in the sense of distributions in Ω).*

Then for each Lebesgue point $z \in \mathbb{C}$ of u with the additional property that the complex-valued functions

$$\Omega \ni \zeta \mapsto \frac{u(\zeta)}{\zeta - z} \text{ and } \Omega \ni \zeta \mapsto \frac{(\overline{\partial} u)(\zeta)}{\zeta - z} \text{ belong to } L^1_{\mathrm{loc}}(\Omega, \mathcal{L}^2), \tag{1.1.38}$$

one has

$$\overline{\partial}_\zeta \Big[\frac{u(\zeta)}{\zeta - z}\Big] = \pi\, u(z)\delta_z + \frac{(\overline{\partial} u)(\zeta)}{\zeta - z} \text{ in } \mathcal{D}'(\Omega), \tag{1.1.39}$$

where δ_z denotes the Dirac distribution with mass at z in Ω.

In particular, formula (1.1.39) *holds for $\mathcal{L}^2$-a.e. point $z \in \Omega$.*

Proof Let us check (1.1.39). For this, without loss of generality we may assume that $0 \in \Omega$ and $z = 0$, the origin in $\mathbb{C}$. Select a real-valued function $\theta \in \mathscr{C}^\infty(\mathbb{C})$ with the property that $\theta = 0$ on $B(0, 1)$ and $\theta = 1$ on $\mathbb{C} \setminus B(0, 2)$. For each parameter $\varepsilon \in \big(0, \frac{1}{2}\,\mathrm{dist}(0, \partial\Omega)\big)$ define $\theta_\varepsilon : \mathbb{C} \to \mathbb{R}$ by setting $\theta_\varepsilon(\zeta) := \theta(\zeta/\varepsilon)$ for every $\zeta \in \mathbb{C}$. Then

$$\begin{aligned} 1 - \theta_\varepsilon \in \mathscr{C}^\infty_c(\Omega), \quad & \mathrm{supp}\,(\nabla\theta_\varepsilon) \subseteq \overline{B(0, 2\varepsilon)} \setminus B(0, \varepsilon), \\ \text{and } \lim_{\varepsilon \to 0^+} & \theta_\varepsilon(\zeta) = 1 \text{ for every } \zeta \in \mathbb{C} \setminus \{0\}. \end{aligned} \tag{1.1.40}$$

Also, there exists a constant $C \in (0, \infty)$ such that

$$\sup_{\zeta \in \mathbb{C}} \big|(\nabla^j \theta_\varepsilon)(\zeta)\big| \leq C\varepsilon^{-j} \text{ for every } j \in \mathbb{N}_0. \tag{1.1.41}$$

To proceed, fix $\varphi \in \mathscr{C}^\infty_c(\Omega)$ and write

$$\begin{aligned} {}_{\mathcal{D}'(\Omega)}\Big\langle \overline{\partial}_\zeta \Big[\frac{u(\zeta)}{\zeta}\Big], \varphi \Big\rangle_{\mathcal{D}(\Omega)} &= -{}_{\mathcal{D}'(\Omega)}\Big\langle \frac{u(\zeta)}{\zeta}, \overline{\partial}\varphi \Big\rangle_{\mathcal{D}(\Omega)} \\ &= -\int_\Omega \frac{u(\zeta)}{\zeta} (\overline{\partial}\varphi)(\zeta)\, \mathrm{d}\mathcal{L}^2(\zeta) \\ &= -\lim_{\varepsilon \to 0^+} \int_\Omega \frac{u(\zeta)}{\zeta} (\overline{\partial}\varphi)(\zeta)\theta_\varepsilon(\zeta)\, \mathrm{d}\mathcal{L}^2(\zeta) \end{aligned} \tag{1.1.42}$$

where for the last equality in (1.1.42) we have used (1.1.40)-(1.1.41) and Lebesgue's Dominated Convergence Theorem. Next, observe that $\theta_\varepsilon\varphi \in \mathscr{C}^\infty_c(\Omega)$ and 0 is not contained in $\mathrm{supp}\,(\theta_\varepsilon\varphi)$. Consequently, the complex-valued function

$$\Omega \ni \zeta \longmapsto \frac{1}{\zeta}\varphi(\zeta)\theta_\varepsilon(\zeta) \text{ belongs to } \mathscr{C}^\infty_c(\Omega), \tag{1.1.43}$$

hence

$$\Omega \ni \zeta \mapsto \frac{u(\zeta)}{\zeta}\varphi(\zeta)\theta_\varepsilon(\zeta) \text{ belongs to } L^1_{\mathrm{comp}}(\Omega, \mathcal{L}^2) \hookrightarrow \mathscr{E}'(\Omega). \tag{1.1.44}$$

Keeping these in mind, in the sense of distributions in Ω we may compute

$$\overline{\partial}_\zeta\Big[\frac{u(\zeta)}{\zeta}\varphi(\zeta)\theta_\varepsilon(\zeta)\Big] = \frac{(\overline{\partial}u)(\zeta)}{\zeta}\varphi(\zeta)\theta_\varepsilon(\zeta) + \frac{u(\zeta)}{\zeta}(\overline{\partial}\varphi)(\zeta)\theta_\varepsilon(\zeta) + \frac{u(\zeta)}{\zeta}\varphi(\zeta)(\overline{\partial}\theta_\varepsilon)(\zeta) \tag{1.1.45}$$

and, further,

$$\begin{aligned} 0 &= {}_{\mathscr{E}'(\Omega)}\Big\langle \overline{\partial}_\zeta\Big[\frac{u(\zeta)}{\zeta}\varphi(\zeta)\theta_\varepsilon(\zeta)\Big], 1\Big\rangle_{\mathscr{E}(\Omega)} \\ &= \int_\Omega \frac{(\overline{\partial}u)(\zeta)}{\zeta}\varphi(\zeta)\theta_\varepsilon(\zeta)\,\mathrm{d}\mathcal{L}^2(\zeta) + \int_\Omega \frac{u(\zeta)}{\zeta}(\overline{\partial}\varphi)(\zeta)\theta_\varepsilon(\zeta)\,\mathrm{d}\mathcal{L}^2(\zeta) \\ &\quad + \int_\Omega \frac{u(\zeta)}{\zeta}\varphi(\zeta)(\overline{\partial}\theta_\varepsilon)(\zeta)\,\mathrm{d}\mathcal{L}^2(\zeta). \end{aligned} \tag{1.1.46}$$

In turn, this implies

$$\begin{aligned} \int_\Omega \frac{u(\zeta)}{\zeta}(\overline{\partial}\varphi)(\zeta)\theta_\varepsilon(\zeta)\,\mathrm{d}\mathcal{L}^2(\zeta) &= -\int_\Omega \frac{(\overline{\partial}u)(\zeta)}{\zeta}\varphi(\zeta)\theta_\varepsilon(\zeta)\,\mathrm{d}\mathcal{L}^2(\zeta) \\ &\quad - \int_\Omega \frac{u(\zeta)}{\zeta}\varphi(\zeta)(\overline{\partial}\theta_\varepsilon)(\zeta)\,\mathrm{d}\mathcal{L}^2(\zeta). \end{aligned} \tag{1.1.47}$$

Thanks to assumptions and Lebesgue's Dominated Convergence Theorem, we may compute

$$\begin{aligned} \lim_{\varepsilon\to 0^+} \int_\Omega \frac{(\overline{\partial}u)(\zeta)}{\zeta}\varphi(\zeta)\theta_\varepsilon(\zeta)\,\mathrm{d}\mathcal{L}^2(\zeta) &= \int_\Omega \frac{(\overline{\partial}u)(\zeta)}{\zeta}\varphi(\zeta)\,\mathrm{d}\mathcal{L}^2(\zeta) \\ &= {}_{\mathcal{D}'(\Omega)}\Big\langle \frac{(\overline{\partial}u)(\zeta)}{\zeta}, \varphi\Big\rangle_{\mathcal{D}(\Omega)}. \end{aligned} \tag{1.1.48}$$

To deal with the last integral in (1.1.47), split it as I + II where

$$\begin{aligned} \mathrm{I} &:= -\frac{1}{\varepsilon}\int_\Omega \frac{1}{\zeta}[(\varphi u)(\zeta) - (\varphi u)(0)](\overline{\partial}\theta)(\zeta/\varepsilon)\,\mathrm{d}\mathcal{L}^2(\zeta), \\ \mathrm{II} &:= -\Big(\int_\Omega \frac{1}{\zeta}(\overline{\partial}\theta_\varepsilon)(\zeta)\,\mathrm{d}\mathcal{L}^2(\zeta)\Big)(\varphi u)(0). \end{aligned} \tag{1.1.49}$$

Using the support condition and estimate from (1.1.40)-(1.1.41), the fact that 0 is a Lebesgue point of u (hence also of φu), and the properties of φ, term I may be estimated by

$$|\mathrm{I}| \le C\Big(\sup_{\mathbb{C}} |\nabla\theta|\Big) \fint_{B(0,2\varepsilon)} \big|(\varphi u)(\zeta) - (\varphi u)(0)\big| \, d\mathcal{L}^2(\zeta) \xrightarrow[\varepsilon\to 0^+]{} 0. \tag{1.1.50}$$

Concerning II, observe that for each $\varepsilon > 0$ we have

$$\begin{aligned} -\int_\Omega \frac{1}{\zeta}(\bar\partial\theta_\varepsilon)(\zeta)\, d\mathcal{L}^2(\zeta) &= -\int_\Omega \frac{1}{\zeta}\bar\partial[\theta_\varepsilon - 1](\zeta)\, d\mathcal{L}^2(\zeta) \\ &= {}_{\mathcal{D}'(\Omega)}\Big\langle \bar\partial_\zeta\Big[\frac{1}{\zeta}\Big],\ \theta_\varepsilon - 1\Big\rangle_{\mathcal{D}(\Omega)} \\ &= \pi\big\langle \delta, \theta_\varepsilon - 1\big\rangle = \pi(\theta_\varepsilon - 1)(0) = -\pi, \end{aligned} \tag{1.1.51}$$

by virtue of the fact that $1/(\pi\zeta)$ is a fundamental solution for $\bar\partial_\zeta$ (cf., e.g., [109, Theorem 7.43, p. 289]), and the properties of θ_ε. Finally, by combining (1.1.42) with (1.1.47)-(1.1.51) we arrive at the version of (1.1.39) written for $z = 0$. This concludes the proof of the first claim in the statement of the lemma.

There remains to deal with the very last claim in the statement of the lemma. In this regard, observe that since for every compact set $K \subset \Omega$ and every ball $B \subset \mathbb{C}$ we have

$$\begin{aligned} &\int_B \Big(\int_K \Big|\frac{u(\zeta)}{\zeta - z}\Big| \, d\mathcal{L}^2(\zeta)\Big)\, d\mathcal{L}^2(z) \\ &\qquad \le C\Big(\int_K |u(\zeta)|\, d\mathcal{L}^2(\zeta)\Big)\Big(\sup_{z\in K}\int_B \frac{d\mathcal{L}^2(z)}{|\zeta - z|}\Big) < +\infty, \end{aligned} \tag{1.1.52}$$

since $u \in L^1_{\rm loc}(\Omega, \mathcal{L}^2)$. From this we then deduce that for $\mathcal{L}^2$-a.e. point $z \in \mathbb{C}$ we have

$$\int_K \Big|\frac{u(\zeta)}{\zeta - z}\Big| \, d\mathcal{L}^2(\zeta) < +\infty \text{ for every compact set } K \subseteq \Omega. \tag{1.1.53}$$

Likewise, using the fact that $\bar\partial u \in L^1_{\rm loc}(\Omega, \mathcal{L}^2)$ we conclude that for $\mathcal{L}^2$-a.e. point $z \in \mathbb{C}$ we have

$$\int_K \Big|\frac{(\bar\partial u)(\zeta)}{\zeta - z}\Big| \, d\mathcal{L}^2(\zeta) < +\infty \text{ for every compact set } K \subseteq \Omega. \tag{1.1.54}$$

Together, (1.1.53)-(1.1.54) imply that

$$\text{(1.1.38) holds at } \mathcal{L}^2\text{-a.e. point } z \in \mathbb{C}. \tag{1.1.55}$$

Since Lebesgue's Differentiation Theorem also ensures that $\mathcal{L}^2$-a.e. $z \in \Omega$ is a Lebesgue point for u, from what we have already established in the first part of the proof we conclude that formula (1.1.39) is valid at $\mathcal{L}^2$-a.e. point $z \in \Omega$. □

At last, we are ready to present the proof of Theorem 1.1.1.

Proof of Theorem 1.1.1 From assumptions and [112, Corollary 8.9.9] it follows that for any $\kappa' > 0$ the nontangential trace $u\big|_{\partial\Omega}^{\kappa'-\mathrm{n.t.}}$ exists σ-a.e. on $\partial_{\mathrm{nta}}\Omega$ and is actually independent of κ'. Next, the claim in (1.1.7) is seen from (1.1.5), [112, (8.9.8)], [112, (8.9.44)], and [112, (8.8.52)].

Fix now a Lebesgue point $z \in \Omega$ of u with the property that (1.1.15) holds, and define the vector field

$$\vec{F}_z(\zeta) := \Big(\frac{u(\zeta)}{\zeta - z}, \mathrm{i}\frac{u(\zeta)}{\zeta - z}\Big) \text{ for } \mathcal{L}^2\text{-a.e. } \zeta \in \Omega. \tag{1.1.56}$$

Then from (1.1.56) and (1.1.11) we see that

$$\vec{F}_z \in \big[L^1_{\mathrm{loc}}(\Omega, \mathcal{L}^2)\big]^2. \tag{1.1.57}$$

Also, with δ_z denoting the Dirac distribution with mass at z, we have

$$\mathrm{div}\vec{F}_z(\zeta) = 2\overline{\partial}_\zeta\Big[\frac{u(\zeta)}{\zeta - z}\Big] = 2\pi\, u(z)\delta_z + 2\,\frac{(\overline{\partial}u)(\zeta)}{\zeta - z} \text{ in } \mathcal{D}'(\Omega), \tag{1.1.58}$$

where the last equality, making use of the fact that z is a Lebesgue point for u and that (1.1.15) holds, comes from Lemma 1.1.2. In particular, if we now consider $K := \overline{B\big(z, \mathrm{dist}(z, \partial\Omega)/2\big)}$, then K is a compact set contained in Ω, and (1.1.58) together with the assumption made in (1.1.15) imply

$$\mathrm{div}\vec{F}_z \text{ belongs to } \mathscr{E}'_K(\Omega) + L^1(\Omega, \mathcal{L}^2). \tag{1.1.59}$$

Going further, (1.1.56) and [112, (8.3.45)] allow us to estimate

$$\begin{aligned}(\mathcal{N}_\kappa^{\Omega\setminus K}\vec{F}_z)(\zeta) &\le \sqrt{2}(\mathcal{N}_\kappa u)(\zeta) \cdot \sup_{\eta\in\Gamma_\kappa(\zeta)\setminus K}\big[|\eta - z|^{-1}\big] \\ &\le C_K\,\frac{(\mathcal{N}_\kappa u)(\zeta)}{|\zeta - z|} \text{ for all } \zeta \in \partial\Omega. \end{aligned}\tag{1.1.60}$$

In turn, from [112, (8.2.26)], (1.1.60), and the first condition in (1.1.5) we conclude that

$$\mathcal{N}_\kappa^{\Omega\setminus K}\vec{F}_z \in L^1(\partial\Omega, \sigma). \tag{1.1.61}$$

Moreover, (1.1.56) and the assumptions on u imply that $\vec{F}_z\big|_{\partial\Omega}^{\kappa-\mathrm{n.t.}}$ exists σ-a.e. in $\partial_{\mathrm{nta}}\Omega$ and in fact

$$\Big(\vec{F}_z\Big|_{\partial\Omega}^{\kappa-\mathrm{n.t.}}\Big)(\zeta) = \left(\frac{\big(u\big|_{\partial\Omega}^{\kappa-\mathrm{n.t.}}\big)(\zeta)}{\zeta - z}, \mathrm{i}\frac{\big(u\big|_{\partial\Omega}^{\kappa-\mathrm{n.t.}}\big)(\zeta)}{\zeta - z}\right) \text{ for } \sigma\text{-a.e. } \zeta \in \partial_{\mathrm{nta}}\Omega. \tag{1.1.62}$$

Finally, observe that in the case when Ω is an exterior domain, condition (1.1.9) implies that

$$\int_{B(0,\lambda R)\setminus B(0,R)} |\vec{F}_z|\, d\mathcal{L}^2 = o(R) \text{ as } R \to \infty. \tag{1.1.63}$$

In summary, we have proved that $\vec{F}_z$ satisfies all hypotheses in [112, Theorem 1.4.1]. In combination with (1.1.58) and (1.1.62) (and also assuming (1.1.9) in the case when Ω is an exterior domain), this permits us to write

$$\begin{aligned}
2\pi\, u(z) + 2\int_\Omega \frac{(\overline{\partial} u)(\zeta)}{\zeta - z}\, d\mathcal{L}^2(\zeta) &= {}_{(\mathscr{C}_b^\infty(\Omega))^*}\big(\mathrm{div}\vec{F}_z, 1\big)_{\mathscr{C}_b^\infty(\Omega)} \\
&= \int_{\partial_*\Omega} \big\langle \nu, \vec{F}_z\big|^{\text{n.t.}}_{\partial\Omega}\big\rangle\, d\sigma \\
&= \int_{\partial_*\Omega} \left\langle \nu(\zeta), \left(\frac{\big(u\big|^{\text{n.t.}}_{\partial\Omega}\big)(\zeta)}{\zeta - z}, \mathrm{i}\frac{\big(u\big|^{\text{n.t.}}_{\partial\Omega}\big)(\zeta)}{\zeta - z} \right) \right\rangle d\sigma(\zeta) \\
&= \int_{\partial_*\Omega} \frac{\big(u\big|^{\text{n.t.}}_{\partial\Omega}\big)(\zeta)}{\zeta - z}\nu(\zeta)\, d\sigma(\zeta) \\
&= \frac{1}{\mathrm{i}}\int_{\partial_*\Omega} \frac{\big(u\big|^{\text{n.t.}}_{\partial\Omega}\big)(\zeta)}{\zeta - z}\, d\zeta,
\end{aligned} \tag{1.1.64}$$

where the last equality uses (1.1.3). From this (1.1.8) follows, at each Lebesgue point $z \in \Omega$ of u such that (1.1.15) holds, hence at $\mathcal{L}^2$-a.e. point $z \in \Omega$. □

We continue by presenting an integration by parts formula for the Cauchy-Riemann operator involving singular functions. This result is going to be a key ingredient in the treatment of the Residue Theorem for rough domains subsequently discussed in Theorem 1.1.4.

Theorem 1.1.3 *Suppose $\Omega \subseteq \mathbb{R}^2 \equiv \mathbb{C}$ is an open set with a lower Ahlfors regular boundary, and with the property that $\sigma := \mathcal{H}^1\lfloor\partial\Omega$ is a doubling measure on $\partial\Omega$. In this setting, assume $f \in \mathcal{D}'(\Omega)$ has the property that there exists some compact set $K \subset \Omega$ and some aperture parameter $\kappa > 0$ for which*

$$\begin{gathered}
f\big|_{\Omega\setminus K} \text{ is holomorphic in } \Omega\setminus K, \quad \mathcal{N}_\kappa^{\Omega\setminus K}\big(f\big|_{\Omega\setminus K}\big) \in L^1(\partial\Omega, \sigma), \\
\text{and } f\big|^{\kappa-\text{n.t.}}_{\partial\Omega} \text{ exists (in } \mathbb{C}) \text{ at } \sigma\text{-a.e. point on } \partial_{\text{nta}}\Omega.
\end{gathered} \tag{1.1.65}$$

Then $\overline{\partial} f \in \mathscr{E}'(\Omega)$, and for any $\kappa' > 0$ the nontangential trace $f\big|^{\kappa'-\text{n.t.}}_{\partial\Omega}$ exists σ-a.e. on $\partial_{\text{nta}}\Omega$ and is actually independent of κ'. When regarding the latter function as being defined σ-a.e. on $\partial_*\Omega$, this belongs to $L^1(\partial_*\Omega, \sigma)$ and, with the dependence on the parameter κ' dropped,

$$ {}_{\mathscr{E}'(\Omega)}\big\langle \overline{\partial} f, 1\big\rangle_{\mathscr{E}(\Omega)} = \frac{1}{2\mathrm{i}}\int_{\partial_*\Omega} \big(f\big|^{\text{n.t.}}_{\partial\Omega}\big)(\zeta)\, d\zeta \tag{1.1.66}$$

in the case when either Ω is bounded, or $\partial\Omega$ is unbounded. Moreover, if Ω is unbounded and $\partial\Omega$ is bounded, formula (1.1.66) *continues to hold under the additional assumption that there exists some $\lambda \in (1, \infty)$ such that*

$$\int_{B(0,\lambda R)\setminus B(0,R)} |f|\,\mathrm{d}\mathcal{L}^2 = o(R) \text{ as } R \to \infty. \tag{1.1.67}$$

In the particular case corresponding to having $K = \varnothing$, the above result yields the following version of Cauchy's vanishing formula. Let Ω and σ be as before, and assume f is a holomorphic function in Ω with the property that there exists $\kappa > 0$ such that

$$\mathcal{N}_\kappa f \in L^1(\partial\Omega, \sigma) \text{ and } f\big|_{\partial\Omega}^{\kappa-\mathrm{n.t.}} \text{ exists (in } \mathbb{C}) \text{ at } \sigma\text{-a.e. point on } \partial_{\mathrm{nta}}\Omega. \tag{1.1.68}$$

Then for any parameter $\kappa' > 0$ the nontangential trace $f\big|_{\partial\Omega}^{\kappa'-\mathrm{n.t.}}$ exists σ-a.e. on $\partial_{\mathrm{nta}}\Omega$ and is actually independent of κ'. The latter function, when regarded as being defined σ-a.e. on $\partial_*\Omega$, belongs to $L^1(\partial_*\Omega, \sigma)$ and, with the dependence on the parameter κ' dropped, Cauchy's vanishing formula

$$\int_{\partial_*\Omega} \big(f\big|_{\partial\Omega}^{\mathrm{n.t.}}\big)(\zeta)\,\mathrm{d}\zeta = 0 \tag{1.1.69}$$

holds in the case when either Ω is bounded, or $\partial\Omega$ is unbounded. Finally, if Ω is unbounded and $\partial\Omega$ is bounded, formula (1.1.69) *continues to hold under the additional assumption made in* (1.1.67).

In relation to Theorem 1.1.3 it is worth observing that it is possible to further relax the condition that $\mathcal{N}_\kappa^{\Omega\setminus K}\big(f\big|_{\Omega\setminus K}\big) \in L^1(\partial\Omega, \sigma)$ in (1.1.68) and replace it by the demand that, for some sufficiently small $\varepsilon > 0$,

$$\mathcal{N}_\kappa^{\varepsilon} f \in L^1_{\mathrm{loc}}(\partial\Omega, \sigma) \text{ and } \int_{\partial_*\Omega} \big|\big(f\big|_{\partial\Omega}^{\kappa-\mathrm{n.t.}}\big)(\zeta)\big|\,\mathrm{d}\sigma(\zeta) < +\infty. \tag{1.1.70}$$

Such a version would require using [112, Corollary 1.5.2] (in the version recorded in [112, (1.5.23)]) in place of [112, Theorem 1.4.1] (as we do in the proof of Theorem 1.1.3 given below).

We also wish to note that in Comment 4, following the statement of Theorem 3.1.6, a Fatou-type result has been presented according to which, in the case when Ω is actually a UR domain, assuming the σ-a.e. existence of the nontangential pointwise trace $f\big|_{\partial\Omega}^{\kappa-\mathrm{n.t.}}$ (i.e., assuming the second condition in (1.1.68)) becomes superfluous. This observation permits us to streamline the statement of Theorem 1.1.3 in the class of UR domains.

Proof of Theorem 1.1.3 The assumptions on Ω imply that this is a set of locally finite perimeter, hence its geometric measure theoretic outward unit normal $\nu = (\nu_1, \nu_2)$ is defined σ-a.e. on $\partial_*\Omega$. The fact that $f\big|_{\Omega\setminus K}$ is holomorphic in $\Omega \setminus K$ forces $\mathrm{supp}(\overline{\partial} f) \subseteq K$, hence $\overline{\partial} f \in \mathscr{E}'_K(\Omega)$. If we now consider the vector field

$$\vec{F} := (f, if) \in [\mathcal{D}'(\Omega)]^2 \tag{1.1.71}$$

then

$$\begin{aligned} &\text{the nontangential trace } \vec{F}\big|_{\partial\Omega}^{\kappa-\text{n.t.}} \text{ exists (in } \mathbb{C}^2) \ \sigma\text{-a.e. on } \partial_{\text{nta}}\Omega, \\ &\vec{F}\big|_{\Omega\setminus K} \in \big[L^1_{\text{loc}}(\Omega \setminus K, \mathcal{L}^2)\big]^2, \text{ and } \mathcal{N}_\kappa^{\Omega\setminus K}\big(\vec{F}\big|_{\Omega\setminus K}\big) \in L^1(\partial\Omega, \sigma). \end{aligned} \tag{1.1.72}$$

Also,

$$\operatorname{div}\vec{F} = \partial_x f + \mathrm{i}\partial_y f = 2\overline{\partial} f \text{ in } \mathcal{D}'(\Omega), \tag{1.1.73}$$

hence

$$\operatorname{div}\vec{F} \in \mathscr{E}'_K(\Omega). \tag{1.1.74}$$

Moreover, in the case when Ω is an exterior domain, assumption (1.1.67) implies that $\vec{F}$ satisfies [112, (1.4.8)]. Finally, we note that at σ-a.e. point on $\partial_*\Omega$ we have

$$\big\langle \nu, \vec{F}\big|_{\partial\Omega}^{\kappa-\text{n.t.}} \big\rangle = \nu_1\big(f\big|_{\partial\Omega}^{\kappa-\text{n.t.}}\big) + \mathrm{i}\nu_2\big(f\big|_{\partial\Omega}^{\kappa-\text{n.t.}}\big) = \nu\big(f\big|_{\partial\Omega}^{\kappa-\text{n.t.}}\big). \tag{1.1.75}$$

Granted all these, formula (1.1.66) now follows from [112, Theorem 1.4.1], bearing in mind [112, (4.6.21)] and (1.1.3). This finishes the proof of Theorem 1.1.3. □

Here is the statement of our version of the Residue Theorem for rough domains advertised earlier.

Theorem 1.1.4 *Let $\Omega \subseteq \mathbb{R}^2 \equiv \mathbb{C}$ be an open set with a lower Ahlfors regular boundary, and with the property that $\sigma := \mathcal{H}^1 \lfloor \partial\Omega$ is a doubling measure on $\partial\Omega$. Suppose f is a meromorphic function in Ω whose poles are contained in some compact set $K \subset \Omega$. Assume that, for some aperture parameter $\kappa > 0$,*

$$\begin{aligned} &\mathcal{N}_\kappa^{\Omega\setminus K}\big(f\big|_{\Omega\setminus K}\big) \in L^1(\partial\Omega, \sigma) \text{ and } f\big|_{\partial\Omega}^{\kappa-\text{n.t.}} \\ &\text{exists (in } \mathbb{C}) \text{ at } \sigma\text{-a.e. point on } \partial_{\text{nta}}\Omega. \end{aligned} \tag{1.1.76}$$

Then for any $\kappa' > 0$ the nontangential trace $f\big|_{\partial\Omega}^{\kappa'-\text{n.t.}}$ exists σ-a.e. on $\partial_{\text{nta}}\Omega$ and is actually independent of κ'. When regarding the latter function as defined σ-a.e. on $\partial_*\Omega$, this belongs to $L^1(\partial_*\Omega, \sigma)$ and, with the dependence on the parameter κ' dropped,

$$2\pi\mathrm{i} \cdot \sum_{z \text{ pole of } f} \operatorname{Res}(f, z) = \int_{\partial_*\Omega} \big(f\big|_{\partial\Omega}^{\text{n.t.}}\big)(\zeta)\, \mathrm{d}\zeta \tag{1.1.77}$$

in the case when either Ω is bounded, or $\partial\Omega$ is unbounded. Finally, when Ω is an exterior domain, formula (1.1.77) *continues to hold under the additional assumption that there exists some number $\lambda \in (1, \infty)$ such that $\int_{B(0,\lambda R)\setminus B(0,R)} |f|\, \mathrm{d}\mathcal{L}^2 = o(R)$ as $R \to \infty$.*

Proof Let $\{z_j\}_{1\le j\le N} \subseteq K$ be the distinct poles of the meromorphic function f. More specifically, assume that for each index $j \in \{1, \dots, N\}$ there exists some open set $\mathcal{O}_j \subseteq \Omega$ with $z_j \in \mathcal{O}_j$, a holomorphic function $g_j : \mathcal{O}_j \to \mathbb{C}$ with $g_j(z_j) \neq 0$, and an integer $m_j \in \mathbb{N}$ such that

$$f(z) = \frac{g_j(z)}{(z - z_j)^{m_j}} \quad \text{for each } z \in \mathcal{O}_j \setminus \{z_j\}. \tag{1.1.78}$$

Then for each $j \in \{1, \dots, N\}$ it follows that z_j is a pole of order m_j for f and

$$\begin{aligned}\operatorname{Res}(f, z_j) &= \frac{1}{(m_j - 1)!} \lim_{z \to z_j} \Big(\frac{\partial}{\partial z}\Big)^{m_j - 1} \big[(z - z_j)^{m_j} f(z)\big] \\ &= \frac{1}{(m_j - 1)!} g_j^{(m_j - 1)}(z_j).\end{aligned} \tag{1.1.79}$$

Without loss of generality we may assume that the sets $\{\mathcal{O}_j\}_{1 \leq j \leq N}$ are mutually disjoint. Let us also introduce $\mathcal{O}_0 := \Omega \setminus \{z_1, \dots, z_N\}$. Hence,

$$\Omega = \bigcup_{j=0}^{N} \mathcal{O}_j \ \text{ and } \ \mathcal{O}_{j_1} \cap \mathcal{O}_{j_2} = \varnothing \ \text{ if } 1 \leq j_1 \neq j_2 \leq N. \tag{1.1.80}$$

We wish to extend the function $f \in \mathscr{C}^\infty\big(\Omega \setminus \{z_1, \dots, z_N\}\big)$ to a distribution u in Ω. To do so, set

$$u_0 := f\big|_{\mathcal{O}_0} \in L^1_{\text{loc}}(\mathcal{O}_0, \mathcal{L}^2) \subseteq \mathcal{D}'(\mathcal{O}_0), \tag{1.1.81}$$

and for each $j \in \{1, \dots, N\}$ introduce

$$u_j := \frac{(-1)^{m_j - 1}}{(m_j - 1)!} g_j \Big(\frac{\partial}{\partial z}\Big)^{m_j - 1} \Big[\frac{1}{z - z_j}\Big] \in \mathcal{D}'(\mathcal{O}_j), \tag{1.1.82}$$

where the expression in the brackets is regarded as a locally integrable function (in the variable z) and the subsequent iterated derivatives (in z) are taken in the sense of distributions in $\mathcal{O}_j$. Since, as seen from (1.1.82) and (1.1.78), we have

$$u_j\big|_{\mathcal{O}_j \setminus \{z_j\}} = f\big|_{\mathcal{O}_j \setminus \{z_j\}} \ \text{ for each } j \in \{1, \dots, N\}, \tag{1.1.83}$$

it follows that

$$u_{j_1}\big|_{\mathcal{O}_{j_1} \cap \mathcal{O}_{j_2}} = u_{j_2}\big|_{\mathcal{O}_{j_1} \cap \mathcal{O}_{j_2}} \ \text{ whenever } 0 \leq j_1 \neq j_2 \leq N. \tag{1.1.84}$$

As such, there exists a unique distribution $u \in \mathcal{D}'(\Omega)$ with the property that

$$u\big|_{\mathcal{O}_j} = u_j \ \text{ in } \mathcal{O}_j \ \text{ for each } j \in \{0, 1, \dots, N\}. \tag{1.1.85}$$

See, e.g., [109, Exercise 2.156, p. 93]. Thus,

$$u \in \mathcal{D}'(\Omega) \ \text{ satisfies } \ u\big|_{\Omega \setminus \{z_1, \dots, z_N\}} = f\big|_{\Omega \setminus \{z_1, \dots, z_N\}} \tag{1.1.86}$$

as wanted. In particular, (1.1.76) implies that

$$\mathcal{N}_\kappa^{\Omega\setminus K}\big(u\big|_{\Omega\setminus K}\big) \in L^1(\partial\Omega,\sigma) \text{ and } u\Big|_{\partial\Omega}^{\kappa-\mathrm{n.t.}} \text{ exists (in } \mathbb{C}) \text{ at } \sigma\text{-a.e. point on } \partial_{\mathrm{nta}}\Omega. \tag{1.1.87}$$

To proceed, observe that since f is holomorphic in $\Omega \setminus \{z_1, \dots, z_N\}$ we have

$$\overline{\partial} u_0 = 0 \text{ in } \mathcal{D}'(\mathcal{O}_0). \tag{1.1.88}$$

Also, since for each $j \in \{1, \dots, N\}$ we have $\overline{\partial}[1/(z - z_j)] = \pi\delta_{z_j}$ (see, e.g., [109, Theorem 7.43, p. 289]), for each $j \in \{1, \dots, N\}$ we may write

$$\begin{aligned}
\overline{\partial} u_j &= \frac{\pi(-1)^{m_j-1}}{(m_j-1)!}\, g_j \Big(\frac{\partial}{\partial z}\Big)^{m_j-1} \delta_{z_j} \\
&= \frac{\pi(-1)^{m_j-1}}{(m_j-1)!} \sum_{k=0}^{m_j-1} \binom{m_j-1}{k} (-1)^{m_j-1+k} g_j^{(m_j-1-k)}(z_j) \Big(\frac{\partial}{\partial z}\Big)^k \delta_{z_j} \\
&= \sum_{k=0}^{m_j-1} \binom{m_j-1}{k} \frac{\pi(-1)^k}{(m_j-1)!} g_j^{(m_j-1-k)}(z_j) \Big(\frac{\partial}{\partial z}\Big)^k \delta_{z_j} \in \mathscr{E}'(\mathcal{O}_j)
\end{aligned} \tag{1.1.89}$$

where the second equality uses [109, Exercise 2.45, p. 34]. Consider now the distribution

$$w := \sum_{j=1}^{N} \sum_{k=0}^{m_j-1} \frac{\pi(-1)^k}{(m_j-1)!} \binom{m_j-1}{k} g_j^{(m_j-1-k)}(z_j) \Big(\frac{\partial}{\partial z}\Big)^k \delta_{z_j} \in \mathscr{E}'(\Omega) \tag{1.1.90}$$

and note that, as seen from (1.1.88)-(1.1.90), we have

$$w\big|_{\mathcal{O}_j} = \overline{\partial} u_j \text{ in } \mathcal{D}'(\mathcal{O}_j) \text{ for each } j \in \{0, 1, \dots, N\}. \tag{1.1.91}$$

From (1.1.80), (1.1.91), and (1.1.85) we then conclude that

$$\overline{\partial} u = w \text{ in } \mathcal{D}'(\Omega). \tag{1.1.92}$$

Based on (1.1.92), (1.1.90), and (1.1.79) we may now compute

$$
\begin{aligned}
{}_{\mathcal{E}'(\Omega)}\big\langle \overline{\partial} u, 1 \big\rangle_{\mathcal{E}(\Omega)} &= {}_{\mathcal{E}'(\Omega)}\big\langle w, 1 \big\rangle_{\mathcal{E}(\Omega)} \\
&= \sum_{j=1}^{N}\sum_{k=0}^{m_j-1} \frac{\pi(-1)^k}{(m_j-1)!}\binom{m_j-1}{k} g_j^{(m_j-1-k)}(z_j)\, {}_{\mathcal{E}'(\Omega)}\Big\langle \Big(\frac{\partial}{\partial z}\Big)^k \delta_{z_j}, 1 \Big\rangle_{\mathcal{E}(\Omega)} \\
&= \sum_{j=1}^{N} \frac{\pi}{(m_j-1)!} g_j^{(m_j-1)}(z_j) \\
&= \pi \cdot \sum_{z \text{ pole of } f} \operatorname{Res}(f, z).
\end{aligned}
\tag{1.1.93}
$$

Having established these, Theorem 1.1.3 applies to the complex distribution u (playing the role of f) and (1.1.66) yields (1.1.77) on account of (1.1.93). □

We wish to remark that, using the version of Theorem 1.1.3 mentioned just after its statement, it is possible to relax the condition $\mathcal{N}_\kappa^{\Omega\setminus K}\big(f\big|_{\Omega\setminus K}\big) \in L^1(\partial\Omega, \sigma)$ appearing in (1.1.76) and replace it by the demand that

$$
\mathcal{N}_\kappa^{\varepsilon} f \in L^1_{\text{loc}}(\partial\Omega, \sigma) \quad \text{and} \quad \int_{\partial_*\Omega} \Big| \big(f\big|_{\partial\Omega}^{\kappa-\text{n.t.}}\big)(\zeta) \Big| \, d\sigma(\zeta) < +\infty. \tag{1.1.94}
$$

In particular, the first condition above is satisfied if $f\big|_{\Omega\setminus K}$ has a continuous extension to $\overline{\Omega}\setminus K$ (a hypothesis which also takes care of the second condition in (1.1.76)). Hence, in such a scenario, if it is meaningful to write it[4] then the residue formula (1.1.77) is true.

This version of the Residue Theorem is actually more efficient than the standard technology (based on choosing a suitable contour of integration, evaluating various integrals, and passing to limit) even in such mundane scenarios as the task of showing that

$$
\int_{-\infty}^{+\infty} \frac{e^{ix}}{x^2+1}\, dx = \frac{\pi}{e}. \tag{1.1.95}
$$

Specifically, choosing $\Omega := \mathbb{R}^2_+$ and considering the meromorphic function

$$
f(z) := \frac{e^{iz}}{z^2+1} \quad \text{for } z \in \Omega, \tag{1.1.96}
$$

then f has a simple pole at $z = i$, with residue $(2ie)^{-1}$, the function f may be extended continuously to a neighborhood of $\partial\Omega$, and $\int_{\partial\Omega} |f|\, d\sigma = \int_{\mathbb{R}} \frac{dx}{x^2+1} < +\infty$. Thus, (1.1.77) holds and this gives (1.1.95).

Incidentally, it is not much harder to see that the first condition in (1.1.76) is true in this case. Indeed, if $K := \overline{B(i, 1/2)}$ then for each $\kappa > 0$ fixed the last part in [112, Lemma 8.3.7] shows (bearing in mind that $|e^{iz}| \leq 1$ for each $z \in \Omega$ and $|e^{iz}| = 1$ for each $z \in \partial\Omega$) that

[4] i.e., if the integral in the right side of (1.1.77) is absolutely convergent

$$\Big(\mathcal{N}_\kappa^{\Omega\setminus K}\big(f\big|_{\Omega\setminus K}\big)\Big)(x) \approx \frac{1}{x^2+1}, \text{ uniformly for } x \in \mathbb{R} \equiv \partial\Omega, \tag{1.1.97}$$

from which the desired conclusion follows.

Compared with Theorem 1.1.1, one remarkable aspect of the theorem below is that no explicit lower Ahlfors regularity assumptions are made on the boundary of $\partial\Omega$.

Theorem 1.1.5 *Let Ω be a nonempty bounded open subset of $\mathbb{R}^2$ with the property that $\partial\Omega$ has finitely many connected components and $\mathcal{H}^1(\partial\Omega) < +\infty$. Abbreviate $\sigma := \mathcal{H}^1 \lfloor \partial\Omega$ and denote by ν the geometric measure theoretic outward unit normal to Ω. In this context, suppose*

$$u \in L^2(\Omega, \mathcal{L}^2) \cap L^p_{\text{loc}}(\Omega, \mathcal{L}^2) \text{ with } p > 2, \tag{1.1.98}$$

is a complex-valued function which, for some sufficiently large aperture parameter $\kappa = \kappa_\Omega \in (0, \infty)$ and some small truncation parameter $\varepsilon > 0$, satisfies

$$\int_{\partial\Omega} \mathcal{N}_\kappa^\varepsilon u \, d\sigma < +\infty \text{ and } u\big|_{\partial\Omega}^{\kappa-\text{n.t.}} \text{ exists } \sigma\text{-a.e. on } A_\kappa(\partial\Omega). \tag{1.1.99}$$

Also, with the Cauchy-Riemann operator $\overline{\partial}$ taken in the sense of distributions in Ω, assume that

$$\overline{\partial} u \in L^1_{\text{loc}}(\Omega, \mathcal{L}^2). \tag{1.1.100}$$

Then the nontangential trace $u\big|_{\partial\Omega}^{\kappa-\text{n.t.}}$ exists σ-a.e. on $\partial_*\Omega$ and, as a function, belongs to $L^1(\partial_*\Omega, \sigma)$. Moreover, for each Lebesgue point $z \in \Omega$ of u with the property that

$$\int_\Omega \left| \frac{(\overline{\partial} u)(\zeta)}{\zeta - z} \right| d\mathcal{L}^2(\zeta) < +\infty \tag{1.1.101}$$

one has (with $d\zeta$ denoting the complex arc-length measure on $\partial\Omega$; cf. (1.1.3))

$$u(z) = \frac{1}{2\pi i} \int_{\partial_*\Omega} \frac{\big(u\big|_{\partial\Omega}^{\kappa-\text{n.t.}}\big)(\zeta)}{\zeta - z} \, d\zeta - \frac{1}{\pi} \int_\Omega \frac{(\overline{\partial} u)(\zeta)}{\zeta - z} \, d\mathcal{L}^2(\zeta). \tag{1.1.102}$$

Proof Pick a Lebesgue point $z \in \Omega$ of u with the property that (1.1.101) holds, and define the vector field

$$\vec{F}_z(\zeta) := \Big(\frac{u(\zeta)}{\zeta - z}, i\frac{u(\zeta)}{\zeta - z} \Big) \text{ for } \mathcal{L}^2\text{-a.e. } \zeta \in \Omega. \tag{1.1.103}$$

Choose a compact neighborhood $K_o \subseteq \Omega$ of z, select a cutoff function $\psi \in \mathscr{C}_c^\infty(\Omega)$ with $\psi \equiv 1$ in K_o, and decompose $\vec{F}_z = (1-\psi)\vec{F}_z + \psi\vec{F}_z$. Since from (1.1.103) and (1.1.98) we have

$$(1-\psi)\vec{F}_z \in \big[L^2(\Omega, \mathcal{L}^2)\big]^2 \tag{1.1.104}$$

and (keeping in mind that $p > 2$)

$$\psi\vec{F}_z \in \big[L^1(\Omega,\mathcal{L}^2)\big]^2 \text{ and } \operatorname{supp}(\psi\vec{F}_z) \subseteq K_o, \tag{1.1.105}$$

we conclude that

$$\vec{F}_z \in \big[L^2(\Omega,\mathcal{L}^2) + \mathscr{E}'_{K_o}(\Omega)\big]^2. \tag{1.1.106}$$

In addition, much as in (1.1.58), we presently have

$$\operatorname{div}\vec{F}_z(\zeta) = 2\pi\, u(z)\delta_z + 2\,\frac{(\bar{\partial}u)(\zeta)}{\zeta - z} \text{ in } \mathcal{D}'(\Omega), \tag{1.1.107}$$

where δ_z denotes the Dirac distribution with mass at z in Ω, hence

$$\operatorname{div}\vec{F}_z \in L^1(\Omega,\mathcal{L}^2) + \mathscr{E}'_{K_o}(\Omega), \tag{1.1.108}$$

thanks to (1.1.101). Finally, as is apparent from (1.1.103) and (1.1.99), for each sufficiently small truncation parameter $\varepsilon \in \big(0, \operatorname{dist}(K_o, \partial\Omega)\big)$ we have

$$\int_{\partial\Omega} \mathcal{N}_\kappa^\varepsilon \vec{F}_z \,\mathrm{d}\sigma < +\infty, \tag{1.1.109}$$

and the nontangential trace $\vec{F}_z\big|_{\partial\Omega}^{\kappa-\mathrm{n.t.}}$ exists σ-a.e. on $A_\kappa(\partial\Omega)$.

Granted these properties, [112, Corollary 1.6.5] applies. In particular, this implies that the nontangential trace $u\big|_{\partial\Omega}^{\kappa-\mathrm{n.t.}}$ exists σ-a.e. on $\partial_*\Omega$ and, as a function, belongs to $L^1(\partial_*\Omega,\sigma)$. Moreover, [112, (1.6.17)] yields (1.1.102) by reasoning as in (1.1.64).□

Here is a version of Cauchy's reproducing formula in open subsets of the plane for which no explicit lower Ahlfors regularity assumptions are made on theirs boundaries.

Corollary 1.1.6 *Let Ω be a nonempty bounded open subset of $\mathbb{R}^2$ with the property that $\partial\Omega$ has finitely many connected components and $\mathcal{H}^1(\partial\Omega) < +\infty$. Suppose*

$$u \in L^2(\Omega,\mathcal{L}^2) \tag{1.1.110}$$

is a holomorphic function which, for some sufficiently large aperture parameter $\kappa = \kappa_\Omega \in (0,\infty)$, satisfies

$$\int_{\partial\Omega} \mathcal{N}_\kappa u \,\mathrm{d}\mathcal{H}^1 < +\infty \text{ and } u\big|_{\partial\Omega}^{\kappa-\mathrm{n.t.}} \text{ exists } \mathcal{H}^1\text{-a.e. on } A_\kappa(\partial\Omega). \tag{1.1.111}$$

Then the nontangential trace $u\big|_{\partial\Omega}^{\kappa-\mathrm{n.t.}}$ exists at $\mathcal{H}^1$-a.e. point on $\partial_*\Omega$ and, as a function, belongs to $L^1(\partial_*\Omega,\mathcal{H}^1)$. Moreover,

$$u(z) = \frac{1}{2\pi\mathrm{i}}\int_{\partial_*\Omega} \frac{\big(u\big|_{\partial\Omega}^{\kappa-\mathrm{n.t.}}\big)(\zeta)}{\zeta - z}\,\mathrm{d}\zeta \text{ for each } z \in \Omega. \tag{1.1.112}$$

Proof This is a direct consequence of Theorem 1.1.5. □

We continue with a general version of Morera's theorem, of the sort described in the corollary below. Once again, a remarkable feature is the lack of an explicit lower Ahlfors regularity assumption.

Corollary 1.1.7 *Let Ω be a nonempty bounded open subset of $\mathbb{R}^2$ with the property that $\partial\Omega$ has finitely many connected components and $\mathcal{H}^1(\partial\Omega) < +\infty$. Assume*

$$\begin{aligned} &f \in L^\infty(\Omega, \mathcal{L}^2) \text{ is such that } \bar{\partial} f \in L^1(\Omega, \mathcal{L}^2) \text{ and} \\ &f_b(\zeta) := \lim_{\Omega \ni z \to \zeta} f(z) \text{ exists at } \mathcal{H}^1\text{-a.e. } \zeta \in \partial\Omega. \end{aligned} \tag{1.1.113}$$

Then f_b belongs to $L^1(\partial_*\Omega, \mathcal{H}^1)$ and

$$\int_{\partial_*\Omega} f_b(\zeta)\,\mathrm{d}\zeta = 2\mathrm{i}\int_\Omega \bar{\partial} f\,\mathrm{d}\mathcal{L}^2. \tag{1.1.114}$$

Proof This follows by applying [112, Corollary 1.6.4] to $\vec{F} := (f, if)$. □

The subsequent discussion serves a preamble and motivation for the statement of a sharp version of the Schwarz-Pompeiu formula in the unit disk in the complex plane, formulated in Theorem 1.1.8.

Given a simply connected UR domain $\Omega \subseteq \mathbb{R}^2 \equiv \mathbb{C}$, abbreviate $\sigma := \mathcal{H}^1 \lfloor \partial\Omega$. In addition, fix an aperture parameter $\kappa > 0$ together with an integrability exponent $p \in (1, \infty)$. Recall that $L^p(\partial\Omega, \sigma)$ stands for the space of σ-measurable p-th power integrable complex-valued functions defined on $\partial\Omega$, and denote by $L^p_{\mathbb{R}}(\partial\Omega, \sigma)$ the subspace of $L^p(\partial\Omega, \sigma)$ consisting of real-valued functions. Next, bring in the Hardy space in Ω associated with the Cauchy-Riemann operator, i.e.,

$$\begin{aligned} &\mathcal{H}^p(\Omega), \text{ the collection of all holomorphic functions in } \Omega \text{ satisfying} \\ &\mathcal{N}_\kappa u \in L^p(\partial\Omega, \sigma), \text{ and also } u(\infty) = 0 \text{ when } \Omega \text{ is an exterior domain.} \end{aligned} \tag{1.1.115}$$

(See §3.2 for a more general point of view.) Then Fatou's theorem (cf. Theorem 3.1.6) guarantees that the boundary Hardy space

$$\mathcal{H}^p(\partial\Omega, \sigma) := \Big\{ u\big|^{\kappa-\mathrm{n.t.}}_{\partial\Omega} : u \in \mathcal{H}^p(\Omega) \Big\} \tag{1.1.116}$$

is a well-defined subspace of $L^p(\partial\Omega, \sigma)$, which is independent of the aperture parameter $\kappa \in (0, \infty)$.

In this context, consider the question of whether

$$\operatorname{Re} \mathcal{H}^p(\partial\Omega, \sigma) = L^p_{\mathbb{R}}(\partial\Omega, \sigma), \tag{1.1.117}$$

i.e., if the collection of all real parts of functions in the boundary Hardy space $\mathcal{H}^p(\partial\Omega, \sigma)$ is precisely the "real" Lebesgue space $L^p_{\mathbb{R}}(\partial\Omega, \sigma)$. Of course, the difficult direction is to checking whether for any given $f \in L^p_{\mathbb{R}}(\partial\Omega, \sigma)$ there exists

$$u \in \mathcal{H}^p(\Omega) \text{ with } \operatorname{Re} u\big|^{\kappa-\mathrm{n.t.}}_{\partial\Omega} = f \text{ at } \sigma\text{-a.e. point on } \partial\Omega. \tag{1.1.118}$$

In this regard, a couple of lines of attack present themselves. One approach relies on the following two ingredients[5]:

(i) the solvability of the L^p Dirichlet Problem (with nontangential maximal function control, boundary trace taken in a nontangential pointwise sense, and decay at infinity in exterior domains) for the Laplacian in the set Ω;

(ii) the L^p-comparability of the nontangential maximal function with the Lusin area-function, in a uniform fashion, for harmonic functions in Ω (normalized by requiring that they vanish at a fixed point, or at infinity).

More specifically, use item *(i)* to find a solution w_1 of the aforementioned Dirichlet Problem with datum $f \in L^p_{\mathbb{R}}(\partial\Omega, \sigma)$, Without loss of generality, this solution may be assumed itself to be real-valued. In view of the fact that Ω is simply connected, there exists a real-valued harmonic function w_2 in Ω with the property that $u := w_1 + iw_2$ is holomorphic in Ω (cf. [112, Proposition 5.8.1]). Thanks to the Cauchy-Riemann equations, $|\nabla w_1| = |\nabla w_2|$ hence, in particular, w_1 and w_2 have the same Lusin area-function (cf. (2.4.31)). Since the nontangential maximal function of w_1 belongs to $L^p(\partial\Omega, \sigma)$, we may then invoke item *(ii)* to conclude that the nontangential maximal function of w_2 also belongs to $L^p(\partial\Omega, \sigma)$. These properties eventually place u in the Hardy space $\mathcal{H}^p(\Omega)$ (cf. (1.1.115)) and $\operatorname{Re} u\big|^{\text{n.t.}}_{\partial\Omega} = f$, as wanted.

To describe another approach to finding a function u as in (1.1.118), denote by K_Δ the principal-value harmonic double layer on $\partial\Omega$ (cf. (A.0.62)), and denote by $\mathscr{C}$ the boundary-to-domain Cauchy integral operator in Ω. The approach in question requires that the operator

$$\tfrac{1}{2}I + K_\Delta \text{ is invertible on } L^p_{\mathbb{R}}(\partial\Omega, \sigma) \text{ or, equivalently, on } L^p(\partial\Omega, \sigma), \tag{1.1.119}$$

a scenario in which (the theory developed in [115, Chapter 1] ensures that) a solution of (1.1.118) is given by

$$u := \mathscr{C}\Big(\big(\tfrac{1}{2}I + K_\Delta\big)^{-1} f\Big) \text{ in } \Omega. \tag{1.1.120}$$

If a uniqueness result, modulo additive (purely imaginary) constants, accompanies (1.1.118), then we may recast (1.1.120) as the integral representation formula

$$u = \mathscr{C}\Big(\big(\tfrac{1}{2}I + K_\Delta\big)^{-1}\big(\operatorname{Re} u\big|^{\kappa-\text{n.t.}}_{\partial\Omega}\big)\Big) + ic \text{ in } \Omega, \tag{1.1.121}$$

valid for any function $u \in \mathcal{H}^p(\Omega)$, where $c \in \mathbb{R}$ is a constant which depends on u. This raises the prospect of recovering (up to additive purely imaginary constants) any function in $\mathcal{H}^p(\Omega)$ from only knowing the nontangential pointwise boundary trace of its real part (compare with the Cauchy reproducing formula (1.1.10) which uses the entire boundary trace). One obstacle in the way of making (1.1.121) a reality is accommodating the demand formulated in (1.1.119). As shown in [116, §4.2, §7.2], this may be arranged provided Ω is sufficiently flat, globally or infinitesimally. Ideal

[5] We shall return to item *(i)*, in a more general, higher-dimensional context, in [116, §8.1]. For item *(ii)*, see [27]

prototypes of such domains include: the unit disk $\mathbb{D}$ in the plane, the complement of its closure, and the upper/lower half-planes $\mathbb{C}_{\pm}$. For example, what is special about the unit disk $\mathbb{D}$ is that not only $\big(\frac{1}{2}I + K_\Delta\big)^{-1}$ exists, but it has a very simple form, namely

$$\big(\tfrac{1}{2}I + K_\Delta\big)^{-1} f = 2f - \fint_{\partial\mathbb{D}} f \, d\mathcal{H}^1 \quad \text{for each } f \in L^p(\partial\mathbb{D}, \mathcal{H}^1). \tag{1.1.122}$$

As such, for $\Omega := \mathbb{D}$, formula (1.1.121) becomes (bearing in mind that the Cauchy operator maps constants to constants)

$$\begin{aligned} u(z) &= 2\Big(\mathscr{C}\big(\mathrm{Re}\, u\big|_{\partial\mathbb{D}}^{\kappa-\mathrm{n.t.}}\big)\Big)(z) - \fint_{\partial\mathbb{D}} \big(\mathrm{Re}\, u\big|_{\partial\mathbb{D}}^{\kappa-\mathrm{n.t.}}\big)(\zeta)\, d\mathcal{H}^1(\zeta) + \mathrm{i}c \\ &= \frac{1}{2\pi\mathrm{i}} \int_{\partial\mathbb{D}} \big(\mathrm{Re}\, u\big|_{\partial\mathbb{D}}^{\kappa-\mathrm{n.t.}}\big)(\zeta) \frac{2}{\zeta - z}\, d\zeta - \frac{1}{2\pi\mathrm{i}} \int_{\partial\mathbb{D}} \big(\mathrm{Re}\, u\big|_{\partial\mathbb{D}}^{\kappa-\mathrm{n.t.}}\big)(\zeta) \frac{1}{\zeta}\, d\zeta + \mathrm{i}c \\ &= \frac{1}{2\pi\mathrm{i}} \int_{\partial\mathbb{D}} \big(\mathrm{Re}\, u\big|_{\partial\mathbb{D}}^{\kappa-\mathrm{n.t.}}\big)(\zeta) \frac{\zeta + z}{(\zeta - z)\zeta}\, d\zeta + \mathrm{i}c \quad \text{for all } z \in \mathbb{D}, \end{aligned} \tag{1.1.123}$$

where we have used $d\zeta = \mathrm{i}\zeta\, d\mathcal{H}^1(\zeta)$ and $\frac{2}{\zeta - z} - \frac{1}{\zeta} = \frac{\zeta+z}{(\zeta-z)\zeta}$. Specializing (1.1.123) to the case when $z = 0$ forces

$$\begin{aligned} u(0) &= \frac{1}{2\pi\mathrm{i}} \int_{\partial\mathbb{D}} \big(\mathrm{Re}\, u\big|_{\partial\mathbb{D}}^{\kappa-\mathrm{n.t.}}\big)(\zeta) \frac{1}{\zeta}\, d\zeta + \mathrm{i}c \\ &= \frac{1}{2\pi} \int_{\partial\mathbb{D}} \big(\mathrm{Re}\, u\big|_{\partial\mathbb{D}}^{\kappa-\mathrm{n.t.}}\big)(\zeta)\, d\mathcal{H}^1(\zeta) + \mathrm{i}c \\ &= \fint_{\partial\mathbb{D}} \big(\mathrm{Re}\, u\big|_{\partial\mathbb{D}}^{\kappa-\mathrm{n.t.}}\big)(\zeta)\, d\mathcal{H}^1(\zeta) + \mathrm{i}c = (\mathrm{Re}\, u)(0) + \mathrm{i}c, \end{aligned} \tag{1.1.124}$$

where the last step uses (a slight variant of) the classical Mean Value Formula for the harmonic function $\mathrm{Re}\, u$ in $\mathbb{D}$. In turn, (1.1.124) forces $c = (\mathrm{Im}\, u)(0)$ which, when used back in (1.1.123), leads to the conclusion that for each function $u \in \mathcal{H}^p(\mathbb{D})$ we have

$$u(z) = \frac{1}{2\pi\mathrm{i}} \int_{\partial\mathbb{D}} \frac{\big(\mathrm{Re}\, u\big|_{\partial\mathbb{D}}^{\kappa-\mathrm{n.t.}}\big)(\zeta)}{\zeta} \frac{\zeta + z}{\zeta - z}\, d\zeta + \mathrm{i}(\mathrm{Im}\, u)(0) \quad \text{for all } z \in \mathbb{D}. \tag{1.1.125}$$

In particular, this implies the classical Schwarz formula stating that for each function $u \in \mathscr{C}^0(\overline{\mathbb{D}})$ which is holomorphic in $\mathbb{D}$ we have

$$u(z) = \frac{1}{2\pi\mathrm{i}} \int_{\partial\mathbb{D}} \frac{(\mathrm{Re}\, u)(\zeta)}{\zeta} \frac{\zeta + z}{\zeta - z}\, d\zeta + \mathrm{i}(\mathrm{Im}\, u)(0) \quad \text{at each } z \in \mathbb{D}. \tag{1.1.126}$$

The integral representation formula (1.1.125) is naturally associated with solutions of the boundary value problem (formulated using the Cauchy-Riemann operator $\overline{\partial}$

in the complex plane)

$$\begin{cases} u \in \mathscr{C}^\infty(\Omega), \\ \overline{\partial} u = 0 \text{ in } \Omega, \\ \mathcal{N}_\kappa u \in L^p(\partial\Omega, \sigma), \\ \operatorname{Re} u\big|_{\partial\Omega}^{\kappa-\mathrm{n.t.}} = f \text{ at } \sigma\text{-a.e. point on } \partial\Omega, \end{cases} \tag{1.1.127}$$

in the case when $\Omega := \mathbb{D}$, the unit disk in the complex plane. With stronger demands, such as $u \in \mathscr{C}^0(\overline{\Omega})$, replacing the nontangential maximal function condition in the third line above, this boundary value problem[6] this problem has been treated by Hermann Amandus Schwarz in [157]. It is therefore appropriate that we refer to (1.1.126) as the L^p Schwarz Problem[7].

Here our goal is to extend the class of functions for which integral representation formulas in the spirit of (1.1.125) are valid, by going beyond Hardy spaces. In particular, we are interested in allowing functions u which are not necessarily holomorphic, a scenario in which (1.1.125) should be augmented through the consideration of a solid integral involving $\partial_{\bar z} u$, much the manner in which Cauchy's integral representation formula (1.1.10) has been generalized by the Cauchy-Pompeiu formula (1.1.8). In view of this, it is natural to call resulting extension the *Schwarz-Pompeiu formula*. In the case of the unit disk, this is formally stated in the theorem below.

Theorem 1.1.8 *Denote by $\mathbb{D} := \{z \in \mathbb{C} : |z| < 1\}$ the open unit disk in the complex plane, and abbreviate $\sigma := \mathcal{H}^1 \lfloor \partial\mathbb{D}$. Also, recall that $\overline{\partial} = \partial_{\bar z} = \frac{1}{2}(\partial_x + i\partial_y)$ stands for the Cauchy-Riemann operator*[8]*. Having fixed an aperture parameter $\kappa > 0$, consider a Lebesgue-measurable function $u : \mathbb{D} \to \mathbb{C}$ satisfying*

$$\begin{gathered} \mathcal{N}_\kappa u \in L^1(\partial\mathbb{D}, \sigma), \quad \overline{\partial} u \in L^1\Big(\mathbb{D}, \frac{\mathcal{L}^2(\zeta)}{|\zeta|}\Big), \\ \text{the nontangential trace } (\operatorname{Re} u)\big|_{\partial\mathbb{D}}^{\kappa-\mathrm{n.t.}} \text{ exists at } \sigma\text{-a.e. point on } \partial\mathbb{D}, \end{gathered} \tag{1.1.128}$$

and $0 \in \mathbb{D}$ is a Lebesgue point for u, i.e.,

$$u(0) = \lim_{r\to 0^+} \fint_{B(0,r)} u \, d\mathcal{L}^2. \tag{1.1.129}$$

Then for any other aperture parameter $\kappa' > 0$ the nontangential boundary trace $(\operatorname{Re} u)\big|_{\partial\mathbb{D}}^{\kappa'-\mathrm{n.t.}}$ exists at σ-a.e. point on $\partial\mathbb{D}$, is actually independent of the parameter κ',

---

[6] itself a precursor of the more general Riemann-Hilbert problem, in which a linear combination of the real and imaginary parts of a holomorphic function is prescribed on the boundary

[7] As noted earlier, if (1.1.119) holds (something considered in [116, Chapters 4 and 7]) then a solution of the L^p Schwarz Problem (1.1.127) is given by (1.1.120). Uniqueness, modulo some additive purely imaginary constant, follows from uniqueness in the L^p Dirichlet Problem for the Laplacian (a topic considered later, in Theorem 5.7.6; see also §4.3 and [116, Chapter 8]).

[8] for this, we shall freely use a variety of alternative pieces of notation; e.g., we agree that $\overline{\partial} = \partial_{\bar z} = \overline{\partial}_z$ stand for the Cauchy-Riemann operator in the variable z

and for $\mathcal{L}^2$-a.e. point $z \in \mathbb{D}$ one has (with the dependence on the aperture parameter dropped)

$$u(z) = \frac{1}{2\pi \mathrm{i}} \int_{\partial\mathbb{D}} \frac{\big((\mathrm{Re}\, u)\big|^{\text{n.t.}}_{\partial\mathbb{D}}\big)(\zeta)}{\zeta} \frac{\zeta + z}{\zeta - z}\, \mathrm{d}\zeta + \mathrm{i}\mathrm{Im}\, u(0)$$
$$- \frac{1}{2\pi} \int_{\mathbb{D}} \left\{ \frac{(\bar\partial u)(\zeta)}{\zeta} \frac{\zeta + z}{\zeta - z} + \frac{\overline{(\bar\partial u)(\zeta)}}{\bar\zeta} \frac{1 + z\bar\zeta}{1 - z\bar\zeta} \right\} \mathrm{d}\mathcal{L}^2(\zeta). \tag{1.1.130}$$

Before presenting the proof of this theorem, few comments are in order. First, the fact that u is Lebesgue-measurable permits us to meaningfully define the nontangential maximal function $\mathcal{N}_\kappa u$. Second, from [112, Lemma 8.3.1] and the first property in (1.1.128) we see that

$$u \in L^\infty_{\mathrm{loc}}(\mathbb{D}, \mathcal{L}^2). \tag{1.1.131}$$

As a consequence, $u \in L^1_{\mathrm{loc}}(\mathbb{D}, \mathcal{L}^2)$ hence it is meaningful to speak of Lebesgue points of u and to consider the distribution $\bar\partial u$. The latter is further assumed to be of function-type[9], and to actually belong to the weighted Lebesgue space $L^1\Big(\mathbb{D}, \frac{\mathcal{L}^2(\zeta)}{|\zeta|}\Big)$. Third, as is apparent from the proof given below, formula (1.1.130) is actually valid at each $z \in \mathbb{D}$ which is a Lebesgue point for u and has the property that $\int_{\mathbb{D}} \frac{|(\bar\partial u)(\zeta)|}{|\zeta - z|}\, \mathrm{d}\mathcal{L}^2(\zeta) < +\infty$.

As a corollary of (1.1.130) and Fatou's theorem, corresponding to the particular case in which the function in question is annihilated by the Cauchy-Riemann operator, one has

$$u(z) = \frac{1}{2\pi \mathrm{i}} \int_{\partial\mathbb{D}} \frac{\big((\mathrm{Re}\, u)\big|^{\text{n.t.}}_{\partial\mathbb{D}}\big)(\zeta)}{\zeta} \frac{\zeta + z}{\zeta - z}\, \mathrm{d}\zeta + \mathrm{i}\mathrm{Im}\, u(0) \quad \text{for each } z \in \mathbb{D}, \tag{1.1.132}$$

whenever u is a holomorphic function in $\mathbb{D}$ with $\mathcal{N}_\kappa u \in L^1(\partial\mathbb{D}, \sigma)$.

We also want to remark that, in contrast with the Cauchy-Pompeiu formula (1.1.8), the integral representation formula (1.1.130) reconstructs a function u from knowing $\bar\partial u$ and only the boundary trace of $\mathrm{Re}\, u$ plus the value of $\mathrm{Im}\, u$ at one point[10]. It is also possible to utilize the nontangential boundary trace of $\mathrm{Im}\, u$, instead of that of $\mathrm{Re}\, u$. Specifically, if in place of the second line of (1.1.128) we now assume that $(\mathrm{Im}\, u)\big|^{\kappa-\text{n.t.}}_{\partial\mathbb{D}}$ exists at σ-a.e. point on $\partial\mathbb{D}$, then (1.1.130) written for $\mathrm{i}u$ in place of u yields the integral representation formula

[9] i.e., the distribution naturally induced by a locally integrable function, via integration

[10] This being said, the present considerations are restricted to the unit disk, whereas the Cauchy-Pompeiu formula (1.1.8) has been established in large classes of domains.

$$u(z) = \frac{1}{2\pi}\int_{\partial\mathbb{D}} \frac{\big((\operatorname{Im} u)\big|^{\text{n.t.}}_{\partial\mathbb{D}}\big)(\zeta)}{\zeta}\,\frac{\zeta+z}{\zeta-z}\,\mathrm{d}\zeta + \operatorname{Re} u(0)$$

$$-\frac{1}{2\pi}\int_{\mathbb{D}}\left\{\frac{(\overline{\partial}u)(\zeta)}{\zeta}\,\frac{\zeta+z}{\zeta-z} - \frac{\overline{(\overline{\partial}u)(\zeta)}}{\overline{\zeta}}\,\frac{1+z\overline{\zeta}}{1-z\overline{\zeta}}\right\}\mathrm{d}\mathcal{L}^2(\zeta) \tag{1.1.133}$$

valid at $\mathcal{L}^2$-a.e. point $z \in \mathbb{D}$.

Lastly, we wish to remark that our version of the Schwarz-Pompeiu formula in the above theorem is a generalization of the classical Schwarz formula (1.1.126) which is fairly optimal. Indeed, the hypotheses in Theorem 1.1.8 are not too far off from the kind of assumptions one would have to impose simply to make sense of the integral representation formula (1.1.130). Thus, in this sense, this result is almost an accurate embodiment of the slogan *if you can write it, then it's true.*

To illustrate this, we note that[11]

the Schwarz-Pompeiu formula recorded in (1.1.130) may fail even in the class of holomorphic functions in the unit disk if in the first membership in (1.1.128) the space $L^1(\partial\mathbb{D},\sigma)$ is replaced by $L^{1,\infty}(\partial\mathbb{D},\sigma)$. (1.1.134)

In a nutshell, *the margin between* L^1 *and weak-*L^1 *delineates the natural range of validity for the Schwarz-Pompeiu formula* (1.1.130). An example of the sort alluded to in (1.1.134) is offered by the function

$$u : \mathbb{D} \longrightarrow \mathbb{C}, \quad u(z) := \frac{1-z}{1+z} \quad \text{for each } z \in \mathbb{D}, \tag{1.1.135}$$

which satisfies

$$u \in \mathscr{C}^\infty\big(\overline{\mathbb{D}}\setminus\{-1\}\big), \quad \overline{\partial}u = 0 \text{ in } \mathbb{D}, \quad \operatorname{Im} u(0) = 0, \tag{1.1.136}$$

and

$$\begin{gathered} u(\zeta) = -2\mathrm{i}\frac{\operatorname{Im}\zeta}{|1+\zeta|^2} \quad \text{for each } \zeta \in \partial\mathbb{D}\setminus\{-1\}, \\ \text{hence } (\operatorname{Re} u)\big|^{\text{n.t.}}_{\partial\mathbb{D}} = 0 \text{ at } \mathcal{L}^1\text{-a.e. point on } \partial\mathbb{D}. \end{gathered} \tag{1.1.137}$$

Also, [112, Lemma 8.3.7] tells us that, for any fixed aperture parameter $\kappa > 0$,

$$\big(\mathcal{N}_\kappa u\big)(\zeta) \approx \frac{1}{|1-\zeta|}, \quad \text{uniformly for } \zeta \in \partial\mathbb{D}\setminus\{-1\}, \tag{1.1.138}$$

from which we conclude (see the discussion in [112, Example 6.2.2])

$$\mathcal{N}_\kappa u \in L^{1,\infty}(\partial\mathbb{D},\sigma) \text{ but } \mathcal{N}_\kappa u \notin L^1(\partial\mathbb{D},\sigma). \tag{1.1.139}$$

[11] in the scenario described in (1.1.134) the local version of Fatou theorem for holomorphic (even harmonic) functions in the unit disk guarantees the existence of the nontangential boundary trace almost everywhere

Clearly, the Schwarz-Pompeiu formula (1.1.130) does not hold for this function u (as the right side is identically zero), and the only hypothesis in Theorem 1.1.8 which is presently violated is the failure of $\mathcal{N}_\kappa u$ to be in $L^1(\partial\mathbb{D}, \sigma)$.

We next take up the task of providing the proof of Theorem 1.1.8.

Proof of Theorem 1.1.8 First, from assumptions and [112, Proposition 8.9.8] we see that for any other aperture parameter $\kappa' > 0$ the nontangential trace $(\operatorname{Re} u)\big|_{\partial\mathbb{D}}^{\kappa'-\mathrm{n.t.}}$ exists at σ-a.e. point on $\partial\mathbb{D}$, and is actually independent of the parameter κ'. As regards the integral representation formula (1.1.130), the idea is to put ourselves in a position where we can invoke [112, Corollary 1.4.3]. This requires some preparations. First, from assumptions and [112, Lemma 3.5.7] it follows that for $\mathcal{L}^2$-a.e. point $z \in \mathbb{C}$ we have

$$\int_{\mathbb{D}} \frac{|(\bar{\partial} u)(\zeta)|}{|\zeta - z|}\, d\mathcal{L}^2(\zeta) < +\infty. \tag{1.1.140}$$

Fix some $z \in \mathbb{D}$ which is a Lebesgue point for the function u and such that (1.1.140) holds (we know that $\mathcal{L}^2$-a.e. point in $\mathbb{D}$ enjoys these properties). For each point $\zeta \in \mathbb{D} \setminus \{0, z\}$ introduce

$$\phi(\zeta) := \frac{1}{2\pi}\left\{\frac{u(\zeta)}{\zeta}\frac{\zeta + z}{\zeta - z} + \frac{\overline{u(\zeta)}}{\bar{\zeta}}\frac{1 + z\bar{\zeta}}{1 - z\bar{\zeta}}\right\} \tag{1.1.141}$$

and

$$\psi(\zeta) := \frac{1}{2\pi}\left\{\frac{u(\zeta)}{\zeta}\frac{\zeta + z}{\zeta - z} - \frac{\overline{u(\zeta)}}{\bar{\zeta}}\frac{1 + z\bar{\zeta}}{1 - z\bar{\zeta}}\right\}, \tag{1.1.142}$$

then use these scalar-valued functions to define the vector field (with complex-valued components)

$$\vec{F}(\zeta) := \big(\phi(\zeta), i\psi(\zeta)\big) \text{ for each } \zeta \in \mathbb{D} \setminus \{0, z\}. \tag{1.1.143}$$

Based on (1.1.131), (1.1.141)-(1.1.142), and (1.1.143) we see that

$$\vec{F} \in \big[L^1_{\mathrm{loc}}(\mathbb{D}, \mathcal{L}^2)\big]^2. \tag{1.1.144}$$

Next, bring in $K := \overline{B(0, R)}$ for some fixed $R \in \big(|z|, 1\big)$, and note that K is a compact subset of the unit disk such that both 0 and z are contained in its interior. Then (1.1.141)-(1.1.143) allow us to estimate

$$\big(\mathcal{N}_\kappa^{\mathbb{D}\setminus K}\vec{F}\big)(\zeta) \le C_K\big(\mathcal{N}_\kappa u\big)(\zeta) \text{ for all } \zeta \in \partial\mathbb{D}. \tag{1.1.145}$$

In turn, from [112, (8.2.26)], (1.1.145), and the first condition in (1.1.128) we conclude that

$$\mathcal{N}_\kappa^{\mathbb{D}\setminus K}\vec{F} \in L^1(\partial\mathbb{D}, \sigma). \tag{1.1.146}$$

Denote by ∂ the conjugate of the Cauchy-Riemann operator, and agree to let ∂_ζ indicate that this acts in the variable ζ. The vector field $\vec{F}$ has been designed so that

$$
\begin{aligned}
\mathrm{div}\vec{F} &= \partial_x\phi + \mathrm{i}\partial_y\psi \\
&= \frac{1}{\pi}\overline{\partial}_\zeta\Big[\frac{u(\zeta)}{\zeta}\frac{\zeta+z}{\zeta-z}\Big] + \frac{1}{\pi}\partial_\zeta\Big[\frac{\overline{u(\zeta)}}{\overline{\zeta}}\frac{1+z\overline{\zeta}}{1-z\overline{\zeta}}\Big] \quad \text{in } \mathcal{D}'(\mathbb{D}).
\end{aligned}
\tag{1.1.147}
$$

To be able to continue, write

$$
\frac{\zeta+z}{\zeta(\zeta-z)} = \frac{2}{\zeta-z} - \frac{1}{\zeta}
\tag{1.1.148}
$$

and use this together with Lemma 1.1.2 twice (mindful of the fact that both 0 and z are Lebesgue points for u) to compute

$$
\begin{aligned}
\overline{\partial}_\zeta\Big[\frac{u(\zeta)}{\zeta}\frac{\zeta+z}{\zeta-z}\Big] &= \overline{\partial}_\zeta\Big[u(\zeta)\Big(\frac{2}{\zeta-z}-\frac{1}{\zeta}\Big)\Big] = \overline{\partial}_\zeta\Big[\frac{2u(\zeta)}{\zeta-z}\Big] - \overline{\partial}_\zeta\Big[\frac{u(\zeta)}{\zeta}\Big] \\
&= \frac{(\overline{\partial}u)(\zeta)}{\zeta}\frac{\zeta+z}{\zeta-z} + 2\pi u(z)\delta_z - \pi u(0)\delta_0 \quad \text{in } \mathcal{D}'(\mathbb{D}).
\end{aligned}
\tag{1.1.149}
$$

Also, keeping in mind that the function

$$
\mathbb{D} \ni \zeta \longmapsto \frac{1+\overline{z}\zeta}{1-\overline{z}\zeta} \in \mathbb{C}
\tag{1.1.150}
$$

is holomorphic, we may once again invoke Lemma 1.1.2 (once more using (1.1.129)) to write

$$
\begin{aligned}
\partial_\zeta\Big[\frac{\overline{u(\zeta)}}{\overline{\zeta}}\frac{1+z\overline{\zeta}}{1-z\overline{\zeta}}\Big] &= \overline{\overline{\partial}_\zeta\Big[\frac{u(\zeta)}{\zeta}\frac{1+\overline{z}\zeta}{1-\overline{z}\zeta}\Big]} = \overline{\frac{(\overline{\partial}u)(\zeta)}{\zeta}\frac{1+\overline{z}\zeta}{1-\overline{z}\zeta} + \pi u(0)\delta_0} \\
&= \frac{\overline{(\overline{\partial}u)(\zeta)}}{\overline{\zeta}}\frac{1+z\overline{\zeta}}{1-z\overline{\zeta}} + \pi\overline{u(0)}\delta_0 \quad \text{in } \mathcal{D}'(\mathbb{D}).
\end{aligned}
\tag{1.1.151}
$$

Upon noting that $-\pi u(0)\delta_0 + \pi\overline{u(0)}\delta_0 = -2\pi\mathrm{i}\big(\mathrm{Im}\,u(0)\big)\delta_0$, from (1.1.147)-(1.1.151) we therefore obtain

$$
(\mathrm{div}\vec{F})(\zeta) = \frac{1}{\pi}\frac{(\overline{\partial}u)(\zeta)}{\zeta}\frac{\zeta+z}{\zeta-z} + \frac{1}{\pi}\frac{\overline{(\overline{\partial}u)(\zeta)}}{\overline{\zeta}}\frac{1+z\overline{\zeta}}{1-z\overline{\zeta}} + 2u(z)\delta_z - 2\mathrm{i}\big(\mathrm{Im}\,u(0)\big)\delta_0
\tag{1.1.152}
$$

in the sense of distributions in $\mathbb{D}$, in the variable ζ. As such, from (1.1.152), (1.1.140), and the last assumption in the first line of (1.1.128) we see that

$$\operatorname{div}\vec{F} \text{ belongs to } \mathscr{E}'(\mathbb{D}) + L^1(\mathbb{D}, \mathcal{L}^2). \tag{1.1.153}$$

Towards the goal of eventually employing [112, Corollary 1.4.3], the next step is to check the hypothesis [112, (1.4.36)]. Guided by [112, (1.4.35)] and (1.1.143), for each $\zeta = x + \mathrm{i}y \in \mathbb{D} \setminus K$ we compute

$$\begin{aligned} f(\zeta) := x\phi(\zeta) + \mathrm{i}y\psi(\zeta) &= \frac{1}{2\pi}\left\{\frac{u(\zeta)}{\zeta}\frac{\zeta+z}{\zeta-z}\right\}\zeta + \frac{1}{2\pi}\left\{\frac{\overline{u(\zeta)}}{\overline{\zeta}}\frac{1+z\overline{\zeta}}{1-z\overline{\zeta}}\right\}\overline{\zeta} \\ &= \frac{1}{2\pi}u(\zeta)\frac{\zeta+z}{\zeta-z} + \frac{1}{2\pi}\overline{u(\zeta)}\frac{1+z\overline{\zeta}}{1-z\overline{\zeta}} \\ &= \frac{1}{2\pi}\big(u(\zeta)+\overline{u(\zeta)}\big)\frac{\zeta+z}{\zeta-z} + \frac{1}{2\pi}\overline{u(\zeta)}\left\{\frac{1+z\overline{\zeta}}{1-z\overline{\zeta}} - \frac{\zeta+z}{\zeta-z}\right\} \\ &= \frac{1}{\pi}(\operatorname{Re} u)(\zeta)\frac{\zeta+z}{\zeta-z} - \frac{1}{\pi}\overline{u(\zeta)}\frac{z(1-|\zeta|^2)}{(1-z\overline{\zeta})(\zeta-z)}, \end{aligned} \tag{1.1.154}$$

where the first equality defines $f(\zeta)$, and where we have used (1.1.141)-(1.1.142) in the second equality. From the second line of (1.1.128) we know that the nontangential boundary trace $(\operatorname{Re} u)\big|_{\partial\mathbb{D}}^{\kappa-\mathrm{n.t.}}$ exists at σ-a.e. point on $\partial\mathbb{D}$. Also, the first property in the first line of (1.1.128) implies that for σ-a.e. point $\zeta_* \in \partial\mathbb{D}$ the function u is essentially bounded in the nontangential approach region $\Gamma_\kappa(\zeta_*)$. Mindful of the fact that $1 - |\zeta|^2$ vanishes for $\zeta \in \partial\mathbb{D}$, we conclude from this that

$$\begin{aligned} &\text{the nontangential pointwise boundary trace of the function} \\ &(\mathbb{D}\setminus K) \ni \zeta \longmapsto \overline{u(\zeta)}\tfrac{z(1-|\zeta|^2)}{(1-z\overline{\zeta})(\zeta-z)} \in \mathbb{C} \text{ is zero } \sigma\text{-a.e. on } \partial\mathbb{D}. \end{aligned} \tag{1.1.155}$$

All things considered, the conclusion is that for $f : \mathbb{D}\setminus K \to \mathbb{C}$ defined as in (1.1.154) we have

$$\Big(f\big|_{\partial\mathbb{D}}^{\kappa-\mathrm{n.t.}}\Big)(\zeta) = \frac{1}{\pi}\big((\operatorname{Re} u)\big|_{\partial\mathbb{D}}^{\kappa-\mathrm{n.t.}}\big)(\zeta)\frac{\zeta+z}{\zeta-z} \quad \text{at } \sigma\text{-a.e. } \zeta \in \partial\mathbb{D}. \tag{1.1.156}$$

Consequently,

$$\begin{aligned} \int_{\partial\mathbb{D}} \Big(f\big|_{\partial\mathbb{D}}^{\kappa-\mathrm{n.t.}}\Big)(\zeta)\,\mathrm{d}\sigma(\zeta) &= \frac{1}{\mathrm{i}}\int_{\partial\mathbb{D}} \Big(f\big|_{\partial\mathbb{D}}^{\kappa-\mathrm{n.t.}}\Big)(\zeta)\frac{1}{\zeta}\,\mathrm{d}\zeta \\ &= \frac{1}{\pi\mathrm{i}}\int_{\partial\mathbb{D}} \frac{\big(\operatorname{Re} u\big|_{\partial\mathbb{D}}^{\kappa-\mathrm{n.t.}}\big)(\zeta)}{\zeta}\frac{\zeta+z}{\zeta-z}\,\mathrm{d}\zeta, \end{aligned} \tag{1.1.157}$$

since $\mathrm{d}\zeta = \mathrm{i}\nu(\zeta)\,\mathrm{d}\sigma(\zeta) = \mathrm{i}\zeta\,\mathrm{d}\sigma(\zeta)$.

At this stage, we have checked that $\vec{F}$ satisfies all hypotheses in [112, Corollary 1.4.3]. Together with (1.1.152) and (1.1.157), the Divergence Formula [112,

(1.4.38)] then allows us to write

$$2u(z) - 2\mathrm{i}\operatorname{Im} u(0) + \frac{1}{\pi}\int_{\mathbb{D}}\left\{\frac{(\bar\partial u)(\zeta)}{\zeta}\frac{\zeta+z}{\zeta-z} + \frac{\overline{(\bar\partial u)(\zeta)}}{\bar\zeta}\frac{1+z\bar\zeta}{1-z\bar\zeta}\right\}\mathrm{d}\mathcal{L}^2(\zeta)$$

$$= {}_{(\mathscr{C}_b^\infty(\mathbb{D}))^*}\big(\operatorname{div}\vec F, 1\big)_{\mathscr{C}_b^\infty(\mathbb{D})} = \int_{\partial\mathbb{D}}\Big(f\big|_{\partial\mathbb{D}}^{\kappa-\mathrm{n.t.}}\Big)(\zeta)\,\mathrm{d}\sigma(\zeta)$$

$$= \frac{1}{\pi\mathrm{i}}\int_{\partial\mathbb{D}}\frac{\big((\operatorname{Re} u)\big|_{\partial\mathbb{D}}^{\kappa-\mathrm{n.t.}}\big)(\zeta)}{\zeta}\frac{\zeta+z}{\zeta-z}\,\mathrm{d}\zeta. \tag{1.1.158}$$

From this (1.1.130) readily follows, at each Lebesgue point $z \in \mathbb{D}$ of u such that (1.1.140) holds, hence also at $\mathcal{L}^2$-a.e. point $z \in \mathbb{D}$. □

There is also a version of the Schwarz-Pompeiu formula in the complement of the closed unit disk in the plane, of the sort described in our next theorem.

Theorem 1.1.9 *Denote by $\mathbb{D}$ the open unit disk in the complex plane, and abbreviate $\sigma := \mathcal{H}^1\lfloor\partial\mathbb{D}$. Also, let $\bar\partial$ stand for the Cauchy-Riemann operator in the complex plane. Consider a Lebesgue-measurable function $u : \mathbb{C}\setminus\overline{\mathbb{D}} \to \mathbb{C}$ satisfying, for some aperture parameter $\kappa > 0$,*

$$\begin{gathered}\mathcal{N}_\kappa u \in L^1(\partial\mathbb{D},\sigma), \quad \bar\partial u \in L^1\Big(\mathbb{C}\setminus\overline{\mathbb{D}}, \frac{\mathcal{L}^2(\zeta)}{|\zeta|}\Big),\\ \text{the nontangential trace } (\operatorname{Re} u)\big|_{\partial\mathbb{D}}^{\kappa-\mathrm{n.t.}} \text{ exists at } \sigma\text{-a.e. point on } \partial\mathbb{D},\end{gathered} \tag{1.1.159}$$

and such that, for some $\lambda \in (1,\infty)$,

$$\fint_{B(0,\lambda R)\setminus B(0,R)} |u|\,\mathrm{d}\mathcal{L}^2 = o(1) \text{ as } R \to \infty. \tag{1.1.160}$$

Then for any other aperture parameter $\kappa' > 0$ the nontangential boundary trace $(\operatorname{Re} u)\big|_{\partial\mathbb{D}}^{\kappa'-\mathrm{n.t.}}$ exists at σ-a.e. point on $\partial\mathbb{D}$, is actually independent of κ', and for $\mathcal{L}^2$-a.e. point $z \in \mathbb{C}\setminus\overline{\mathbb{D}}$ one has (with the dependence on the aperture parameter dropped)

$$\begin{aligned} u(z) = &-\frac{1}{2\pi\mathrm{i}}\int_{\partial\mathbb{D}}\frac{\big((\operatorname{Re} u)\big|_{\partial\mathbb{D}}^{\mathrm{n.t.}}\big)(\zeta)}{\zeta}\frac{\zeta+z}{\zeta-z}\,\mathrm{d}\zeta\\ &-\frac{1}{2\pi}\int_{\mathbb{C}\setminus\overline{\mathbb{D}}}\left\{\frac{(\bar\partial u)(\zeta)}{\zeta}\frac{\zeta+z}{\zeta-z} + \frac{\overline{(\bar\partial u)(\zeta)}}{\bar\zeta}\frac{1+z\bar\zeta}{1-z\bar\zeta}\right\}\mathrm{d}\mathcal{L}^2(\zeta). \end{aligned} \tag{1.1.161}$$

As with (1.1.134), this result is sharp in the sense that *the Schwarz-Pompeiu formula* (1.1.161) *may fail even in the class of holomorphic functions in the complement of the closed unit disk in the plane if in the first membership in* (1.1.159) *the space*

$L^1(\partial\mathbb{D},\sigma)$ is replaced by its weak version, $L^{1,\infty}(\partial\mathbb{D},\sigma)$. To see this, consider the function

$$u:\mathbb{C}\setminus\overline{\mathbb{D}}\to\mathbb{C} \text{ defined by } u(z):=\frac{1}{1+z} \text{ for each } z\in\mathbb{C}\setminus\overline{\mathbb{D}}. \tag{1.1.162}$$

Then

$$\begin{gathered} u\in\mathscr{C}^\infty(\mathbb{C}\setminus\overline{\mathbb{D}}),\ \overline{\partial}u=0 \text{ in } \mathbb{C}\setminus\overline{\mathbb{D}}, \\ \Big(u\big|^{\rm n.t.}_{\partial(\mathbb{C}\setminus\overline{\mathbb{D}})}\Big)(z) \text{ exists for all } z\in\partial\mathbb{D}\setminus\{-1\}, \\ u(z)=O(|z|^{-1}) \text{ as } |z|\to\infty, \text{ and } \mathcal{N}_\kappa u\in L^{1,\infty}(\partial\mathbb{D},\sigma), \end{gathered} \tag{1.1.163}$$

with the last property a consequence of [112, Lemma 8.3.7] (and the discussion in [112, Example 6.2.2]). In addition, for each $z=x+iy\in\partial\mathbb{D}\setminus\{-1\}$ we have

$$\big((\mathrm{Re}\,u)\big|^{\rm n.t.}_{\partial\mathbb{D}}\big)(z)=\mathrm{Re}\Big(\frac{1+\overline{z}}{|1+z|^2}\Big)=\frac{1+x}{(1+x)^2+y^2}=\frac{1+x}{2+2x}=\frac{1}{2}. \tag{1.1.164}$$

Hence, if $z\in\mathbb{C}\setminus\overline{\mathbb{D}}$, then

$$\begin{aligned} \frac{1}{2\pi i}\int_{\partial\mathbb{D}}\frac{\big((\mathrm{Re}\,u)\big|^{\rm n.t.}_{\partial\mathbb{D}}\big)(z)(\zeta+z)}{\zeta(\zeta-z)}\,d\zeta &= \frac{1}{4\pi i}\int_{\partial\mathbb{D}}\frac{2}{\zeta-z}\,d\zeta-\frac{1}{4\pi i}\int_{\partial\mathbb{D}}\frac{1}{\zeta}\,d\zeta \\ &= 0-\frac{1}{2}=-\frac{1}{2}, \end{aligned} \tag{1.1.165}$$

where the penultimate equality uses the Residue Theorem. As such, formula (1.1.161) fails for this function u (as the right side is a constant), and the only hypothesis in Theorem 1.1.9 which is presently violated is the membership of $\mathcal{N}_\kappa u$ to $L^1(\partial\mathbb{D},\sigma)$.

We now give the proof of Theorem 1.1.9.

Proof of Theorem 1.1.9 The claims pertaining to the nontangential boundary trace of $\mathrm{Re}\,u$ follow from assumptions and [112, Proposition 8.9.8]. To justify (1.1.161), we reason as in the proof of Theorem 1.1.9, now making use of [112, Corollary 1.4.4] in place of [112, Corollary 1.4.3], and otherwise natural adjustments. Specifically, we now fix a Lebesgue point $z\in\mathbb{C}\setminus\overline{\mathbb{D}}$ for the function u, with the additional property that

$$\int_{\mathbb{C}\setminus\overline{\mathbb{D}}}\frac{|(\overline{\partial}u)(\zeta)|}{|\zeta-z|}\,d\mathcal{L}^2(\zeta)<+\infty, \tag{1.1.166}$$

and once again consider a vector field $\vec{F}$ defined as in (1.1.141)-(1.1.143) but for $\zeta\in\mathbb{C}\setminus\overline{\mathbb{D}}$. The fact that we now work in the complement of the closed unit disk affects the nature of $\mathrm{div}\vec{F}$. Specifically, since $\mathbb{C}\setminus\overline{\mathbb{D}}\ni\zeta\mapsto\frac{1}{\zeta}\in\mathbb{C}$ is a holomorphic function, in place of (1.1.149) we now have

$$\bar{\partial}_\zeta\Big[\frac{u(\zeta)}{\zeta}\frac{\zeta+z}{\zeta-z}\Big]=\bar{\partial}_\zeta\Big[u(\zeta)\Big(\frac{2}{\zeta-z}-\frac{1}{\zeta}\Big)\Big]=\bar{\partial}_\zeta\Big[\frac{2u(\zeta)}{\zeta-z}\Big]-\bar{\partial}_\zeta\Big[\frac{u(\zeta)}{\zeta}\Big]$$

$$=\frac{(\bar{\partial}u)(\zeta)}{\zeta}\frac{\zeta+z}{\zeta-z}+2\pi u(z)\delta_z \text{ in } \mathcal{D}'\big(\mathbb{C}\setminus\overline{\mathbb{D}}\big). \tag{1.1.167}$$

Moreover, in view of the fact that the functions

$$\mathbb{C}\setminus\overline{\mathbb{D}}\ni\zeta\longmapsto\frac{1}{\zeta}\in\mathbb{C} \text{ and } \mathbb{C}\setminus\overline{\mathbb{D}}\ni\zeta\longmapsto\frac{1+\bar{z}\zeta}{1-\bar{z}\zeta}\in\mathbb{C} \tag{1.1.168}$$

are holomorphic, in place of (1.1.151) we now obtain

$$\partial_\zeta\Big[\frac{\overline{u(\zeta)}}{\bar{\zeta}}\frac{1+z\bar{\zeta}}{1-z\bar{\zeta}}\Big]=\overline{\bar{\partial}_\zeta\Big[\frac{u(\zeta)}{\zeta}\frac{1+\bar{z}\zeta}{1-\bar{z}\zeta}\Big]}=\overline{\frac{(\bar{\partial}u)(\zeta)}{\zeta}\frac{1+\bar{z}\zeta}{1-\bar{z}\zeta}}$$

$$=\frac{\overline{(\bar{\partial}u)(\zeta)}}{\bar{\zeta}}\frac{1+z\bar{\zeta}}{1-z\bar{\zeta}} \text{ in } \mathcal{D}'\big(\mathbb{C}\setminus\overline{\mathbb{D}}\big). \tag{1.1.169}$$

Ultimately, these are responsible for the absence of the term $\mathrm{i}\,\mathrm{Im}\,u(0)$ in (1.1.161), compared to (1.1.130). Finally, if $f(\zeta)$ is defined as in (1.1.154) but now for $\zeta\in\mathbb{C}\setminus\overline{\mathbb{D}}$ with $\zeta\neq z$, we deduce from (1.1.160) and the final formula in (1.1.154) that

$$\int_{B(0,\lambda R)\setminus B(0,R)}|f|\,\mathrm{d}\mathcal{L}^2=o(R^2) \text{ as } R\to\infty. \tag{1.1.170}$$

Granted these properties, we may now invoke [112, Corollary 1.4.4] and conclude from [112, (1.4.54)] (which is responsible for the change in sign for the boundary integral in (1.1.161), compared to (1.1.130)) that (1.1.161) holds. □

We conclude by presenting a companion result to Theorem 1.1.8, dealing with a sharp form of the Schwarz-Pompeiu formula for $\mathbb{C}_+ := \{z\in\mathbb{C}: \mathrm{Im}\,z>0\}$. The goal is now to recover a given complex-valued function defined in $\mathbb{C}_+$ from the knowledge of the nontangential boundary trace of its real part on $\mathbb{R}\equiv\partial\mathbb{C}_+$, and the action of the Cauchy-Riemann operator $\bar{\partial}$ on said function in $\mathbb{C}_+$. This is made precise in the theorem below, where an optimal class of functions for which such a goal may be be accomplish is identified.

Theorem 1.1.10 *Consider a Lebesgue-measurable function $u:\mathbb{C}_+\to\mathbb{C}$ which, for some aperture parameter $\kappa\in(0,\infty)$, satisfies*

$$\mathcal{N}_\kappa u\in L^1\Big(\mathbb{R},\frac{\mathcal{L}^1(t)}{1+|t|}\Big),\quad \bar{\partial}u\in L^1(\mathbb{C}_+,\mathcal{L}^2),\ \text{and}$$
$$\text{the nontangential trace } (\mathrm{Re}\,u)\Big|_{\partial\mathbb{C}_+}^{\kappa-\mathrm{n.t.}} \text{ exists } \mathcal{L}^1\text{-a.e. on } \mathbb{R}\equiv\partial\mathbb{C}_+. \tag{1.1.171}$$

Then for any other aperture parameter $\kappa'>0$ the nontangential boundary trace $(\mathrm{Re}\,u)\Big|_{\partial\mathbb{C}_+}^{\kappa'-\mathrm{n.t.}}$ exists at $\mathcal{L}^1$-a.e. point on $\mathbb{R}\equiv\partial\mathbb{C}_+$, is actually independent of κ', and *for $\mathcal{L}^2$-a.e. point $z \in \mathbb{C}_+$ one has (with the dependence on the aperture parameter dropped)*

$$u(z) = \frac{1}{2\pi \mathrm{i}} \int_{\mathbb{R}} \frac{\big((\operatorname{Re} u)\big|^{\text{n.t.}}_{\partial \mathbb{C}_+}\big)(t)}{t - z}\, \mathrm{d}t - \frac{1}{\pi} \int_{\mathbb{C}_+} \left\{ \frac{(\overline{\partial} u)(\zeta)}{\zeta - z} - \frac{\overline{(\overline{\partial} u)(\zeta)}}{\overline{\zeta} - z} \right\} \mathrm{d}\mathcal{L}^2(\zeta). \tag{1.1.172}$$

Before presenting its proof, we wish to make a number of comments designed to shed further light on the nature and scope of Theorem 1.1.10. First, the hypothesis that u is Lebesgue-measurable allows us to meaningfully define the nontangential maximal function $\mathcal{N}_\kappa u$. In turn, from [112, Lemma 8.3.1] and the first property listed in (1.1.171) we deduce that

$$u \in L^\infty_{\text{loc}}(\mathbb{C}_+, \mathcal{L}^2). \tag{1.1.173}$$

In particular, u belongs to $L^1_{\text{loc}}(\mathbb{C}_+, \mathcal{L}^2)$, which makes it possible consider the distribution $\overline{\partial} u$ in $\mathbb{C}_+$. This distribution is further assumed to be of function-type, and to actually belong to the Lebesgue space $L^1(\mathbb{C}_+, \mathcal{L}^2)$. Also, as may be seen from the proof given below, formula (1.1.172) is in fact true for each Lebesgue point $z \in \mathbb{C}_+$ of u satisfying $\int_{\mathbb{C}_+} \frac{|(\overline{\partial} u)(\zeta)|}{|\zeta - z|}\, \mathrm{d}\mathcal{L}^2(\zeta) < +\infty$.

A special case of (1.1.172), corresponding to the scenario in which the function in question is annihilated by the Cauchy-Riemann operator, reads[12]

$$u(z) = \frac{1}{2\pi \mathrm{i}} \int_{\mathbb{R}} \frac{\big((\operatorname{Re} u)\big|^{\text{n.t.}}_{\partial \mathbb{C}_+}\big)(t)}{t - \zeta}\, \mathrm{d}t \quad \text{at each point } z \in \mathbb{C}_+, \tag{1.1.174}$$

$$\text{provided } u \text{ is holomorphic in } \mathbb{C}_+ \text{ and } \mathcal{N}_\kappa u \in L^1\Big(\mathbb{R}, \tfrac{\mathcal{L}^1(t)}{1+|t|}\Big).$$

In contrast with what the Cauchy-Pompeiu formula (1.1.8) would give when $\Omega := \mathbb{C}_+$, the integral representation formula (1.1.172) reconstructs a function u from knowing $\overline{\partial} u$ and only the boundary trace of $\operatorname{Re} u$. There is also a version of (1.1.172) emphasizing the use of the imaginary part in place of the real part. Concretely, if in place of the second line of (1.1.171) it is now assumed that $(\operatorname{Im} u)\big|^{\kappa-\text{n.t.}}_{\partial \mathbb{C}_+}$ exists at $\mathcal{L}^1$-a.e. point on $\mathbb{R} \equiv \partial \mathbb{C}_+$, then (1.1.172) written for $\mathrm{i}u$ in place of u yields the integral representation formula

[12] here the classical Fatou theorem for holomorphic functions in the upper half-plane is used to ensure that the nontangential boundary trace exists

$$u(z) = \frac{1}{2\pi} \int_{\mathbb{R}} \frac{\big((\operatorname{Im} u)\big|^{\text{n.t.}}_{\partial\mathbb{C}_+}\big)(t)}{t - \zeta}\, dt$$

$$- \frac{1}{\pi} \int_{\mathbb{C}_+} \left\{ \frac{(\bar\partial u)(\zeta)}{\zeta - z} + \frac{\overline{(\bar\partial u)(\zeta)}}{\bar\zeta - z} \right\} d\mathcal{L}^2(\zeta) \text{ at } \mathcal{L}^2\text{-a.e. point } z \in \mathbb{C}_+. \quad (1.1.175)$$

It is of significance to note that our version of the Schwarz-Pompeiu formula in the above theorem is in the nature of best possible. Indeed,[13]

the Schwarz-Pompeiu formula recorded in (1.1.172) may fail even in the class of holomorphic functions in the upper half-plane if in the first membership in (1.1.171) the space $L^1\Big(\mathbb{R}, \frac{\mathcal{L}^1(t)}{1+|t|}\Big)$ is replaced by its weak version, $L^{1,\infty}\Big(\mathbb{R}, \frac{\mathcal{L}^1(t)}{1+|t|}\Big)$. (1.1.176)

To see that this is the case, consider the function

$$u : \mathbb{C}_+ \longrightarrow \mathbb{C}, \quad u(z) := \frac{i}{z} \text{ for each } z \in \mathbb{C}_+. \quad (1.1.177)$$

This satisfies

$$u \in \mathscr{C}^\infty\big(\overline{\mathbb{C}_+} \setminus \{0\}\big), \quad \bar\partial u = 0 \text{ in } \mathbb{C}_+, \quad (1.1.178)$$

and

$$\begin{aligned} & u(t) = i/t \text{ for each } t \in \mathbb{R} \setminus \{0\}, \text{ hence} \\ & (\operatorname{Re} u)\big|^{\text{n.t.}}_{\partial\mathbb{C}_+} = 0 \text{ at } \mathcal{L}^1\text{-a.e. point on } \mathbb{R} \equiv \partial\mathbb{C}_+. \end{aligned} \quad (1.1.179)$$

In addition, from [112, Lemma 8.3.7] we see that, for any fixed aperture parameter $\kappa > 0$,

$$\big(\mathcal{N}_\kappa u\big)(t) \approx \frac{1}{|t|}, \text{ uniformly for } t \in \mathbb{R} \setminus \{0\}, \quad (1.1.180)$$

from which we deduce, based on straightforward calculus, that

$$\mathcal{N}_\kappa u \in L^{1,\infty}\Big(\mathbb{R}, \frac{\mathcal{L}^1(t)}{1+|t|}\Big) \text{ but } \mathcal{N}_\kappa u \notin L^1\Big(\mathbb{R}, \frac{\mathcal{L}^1(t)}{1+|t|}\Big). \quad (1.1.181)$$

The bottom line is that the Schwarz-Pompeiu formula (1.1.172) does not hold for this function u (as the right side is identically zero), and the only hypothesis in Theorem 1.1.10 which is presently violated is the failure of $\mathcal{N}_\kappa u$ to be in $L^1\Big(\mathbb{R}, \frac{\mathcal{L}^1(t)}{1+|t|}\Big)$.

The proof of Theorem 1.1.10 is given next.

[13] in the scenario described in (1.1.176) the local version of Fatou theorem for holomorphic (even harmonic) functions in the upper half-plane guarantees the existence of the nontangential boundary trace almost everywhere

Proof of Theorem 1.1.10 The claims regarding the nontangential boundary trace of $\mathrm{Re}\,u$ are seen from assumptions and [112, Corollary 8.9.9]. The bulk of the argument concerns the integral representation formula (1.1.172). The strategy is to arrange matters so that we can eventually employ the two-dimensional version of [112, Corollary 1.4.2]. To set the stage, from assumptions and [112, Lemma 3.5.7] (with $m := 0$) we see that for $\mathcal{L}^2$-a.e. point $z \in \mathbb{C}_+$ we have

$$\int_{\mathbb{C}_+} \frac{|(\overline{\partial} u)(\zeta)|}{|\zeta - z|}\, d\mathcal{L}^2(\zeta) < +\infty. \tag{1.1.182}$$

Henceforth, we fix some $z \in \mathbb{C}_+$ which is a Lebesgue point for the function u and which satisfies (1.1.182) (these properties are enjoyed by $\mathcal{L}^2$-a.e. point in $\mathbb{C}_+$). Define the vector field

$$\vec{F} = (F_1, F_2) : \mathbb{C}_+ \setminus \{z\} \longrightarrow \mathbb{C}^2 \tag{1.1.183}$$

by setting

$$\vec{F}(\zeta) := \left(\frac{u(\zeta)}{\zeta - z} - \frac{\overline{u(\zeta)}}{\overline{\zeta} - z}, \mathrm{i}\frac{u(\zeta)}{\zeta - z} + \mathrm{i}\frac{\overline{u(\zeta)}}{\overline{\zeta} - z} \right) \text{ at each } \zeta \in \mathbb{C}_+ \setminus \{z\}. \tag{1.1.184}$$

In particular, (1.1.173) ensures that

$$\vec{F} \in \big[L^1_{\mathrm{loc}}(\mathbb{C}_+, \mathcal{L}^2)\big]^2. \tag{1.1.185}$$

Bring in $K := \overline{B\big(z, 2^{-1}\,\mathrm{Im}\, z\big)}$, which a compact neighborhood of z, contained in $\mathbb{C}_+$. Then (1.1.184) together with [112, Lemma 8.3.7] imply

$$\big(\mathcal{N}_\kappa^{\mathbb{C}_+\setminus K} \vec{F}\big)(t) \leq C_z \frac{(\mathcal{N}_\kappa u)(t)}{1 + |t|} \text{ for all } t \in \mathbb{R} \equiv \partial\mathbb{C}_+. \tag{1.1.186}$$

In turn, from [112, (8.2.26)], (1.1.186), and the first condition in (1.1.171) we conclude that

$$\mathcal{N}_\kappa^{\mathbb{C}_+\setminus K} \vec{F} \in L^1(\mathbb{R}, \mathcal{L}^1). \tag{1.1.187}$$

To continue, note that the last scalar component of $\vec{F}$ is given by

$$\begin{aligned} F_2(\zeta) &= \mathrm{i}\frac{u(\zeta)}{\zeta - z} + \mathrm{i}\frac{\overline{u(\zeta)}}{\overline{\zeta} - z} = \mathrm{i}\frac{u(\zeta) + \overline{u(\zeta)}}{\zeta - z} + \overline{u(\zeta)}\left(\frac{\mathrm{i}}{\overline{\zeta} - z} - \frac{\mathrm{i}}{\zeta - z} \right) \\ &= 2\mathrm{i}\frac{(\mathrm{Re}\,u)(\zeta)}{\zeta - z} - \overline{u(\zeta)}\frac{2\,\mathrm{Im}\,\zeta}{(\overline{\zeta} - z)(\zeta - z)} \text{ at each } \zeta \in \mathbb{C}_+ \setminus \{z\}. \end{aligned} \tag{1.1.188}$$

The second line of (1.1.171) gives that the nontangential trace $\big((\mathrm{Re}\,u)\big|^{\kappa-\mathrm{n.t.}}_{\partial\mathbb{C}_+}\big)(t)$ exists at $\mathcal{L}^1$-a.e. point $t \in \mathbb{R} \equiv \partial\mathbb{C}_+$. Also, the first property in the first line of (1.1.171) guarantees that the function u is essentially bounded in the nontangential approach region $\Gamma_\kappa(t)$ for $\mathcal{L}^1$-a.e. point $t \in \mathbb{R} \equiv \partial\mathbb{C}_+$. In view of the fact that $\operatorname{Im}\zeta$ vanishes for $\zeta \in \partial\mathbb{C}_+ \equiv \mathbb{R}$, this ultimately leads to the conclusion that

$$\text{the nontangential pointwise boundary trace of the function} \\ (\mathbb{C}_+ \setminus \{z\}) \ni \zeta \mapsto \overline{u(\zeta)}\frac{2\operatorname{Im}\zeta}{(\overline{\zeta}-z)(\zeta-z)} \in \mathbb{C} \text{ is zero } \mathcal{L}^1\text{-a.e. on } \mathbb{R}. \tag{1.1.189}$$

As a consequence, we obtain

$$\Big(F_2\Big|^{\kappa-\text{n.t.}}_{\partial\mathbb{C}_+}\Big)(t) = 2\mathrm{i}\frac{\big((\operatorname{Re} u)\big|^{\kappa-\text{n.t.}}_{\partial\mathbb{C}_+}\big)(t)}{t-z} \quad \text{at } \mathcal{L}^1\text{-a.e. point } t \in \mathbb{R} \equiv \partial\mathbb{C}_+. \tag{1.1.190}$$

Going further, if δ_z denotes the Dirac distribution with mass at z in $\mathbb{C}_+$, then Lemma 1.1.2 gives

$$\overline{\partial}_\zeta\Big[\frac{u(\zeta)}{\zeta - z}\Big] = \pi\, u(z)\delta_z + \frac{(\overline{\partial}u)(\zeta)}{\zeta - z} \quad \text{in } \mathcal{D}'(\mathbb{C}_+). \tag{1.1.191}$$

Denote by ∂ the conjugate of the Cauchy-Riemann operator, and agree to let ∂_ζ indicate that this acts in the variable ζ. Then, keeping in mind that the function

$$\mathbb{C}_+ \ni \zeta \longmapsto \frac{1}{\zeta - \overline{z}} \in \mathbb{C} \tag{1.1.192}$$

is holomorphic, and that $\overline{\partial}u \in L^1_{\text{loc}}(\mathbb{C}_+, \mathcal{L}^2)$, we see that

$$\partial_\zeta\Big[\frac{\overline{u(\zeta)}}{\overline{\zeta} - z}\Big] = \overline{\overline{\partial}_\zeta\Big[\frac{u(\zeta)}{\zeta - \overline{z}}\Big]} = \overline{\frac{(\overline{\partial}u)(\zeta)}{\zeta - z}} = \frac{\overline{(\overline{\partial}u)(\zeta)}}{\overline{\zeta} - z} \quad \text{in } \mathcal{D}'(\mathbb{C}_+). \tag{1.1.193}$$

Relying on (1.1.184), (1.1.191), and (1.1.193) we may now compute (with the divergence considered in the sense of distributions in the variable $\zeta \in \mathbb{C}_+$)

$$\begin{aligned}(\operatorname{div}\vec{F})(\zeta) &= 2\overline{\partial}_\zeta\Big[\frac{u(\zeta)}{\zeta - z}\Big] - 2\partial_\zeta\Big[\frac{\overline{u(\zeta)}}{\overline{\zeta} - z}\Big] \\ &= 2\pi\, u(z)\delta_z + \frac{2(\overline{\partial}u)(\zeta)}{\zeta - z} - \frac{2\overline{(\overline{\partial}u)(\zeta)}}{\overline{\zeta} - z} \quad \text{in } \mathcal{D}'(\mathbb{C}_+).\end{aligned} \tag{1.1.194}$$

As such, from (1.1.194), (1.1.182), and the last assumption in the first line of (1.1.171) we see that

$$\operatorname{div}\vec{F} \text{ belongs to } \mathscr{E}'(\mathbb{C}_+) + L^1(\mathbb{C}_+, \mathcal{L}^2). \tag{1.1.195}$$

At this stage, all hypotheses of [112, Corollary 1.4.2] (with $n := 2$) have been verified, including [112, (1.4.26)]. We may therefore invoke the version of the Divergence Formula recorded in [112, (1.4.24)] which presently gives

$$
\begin{aligned}
2\pi\, u(z)+2 \int_{\mathbb{C}_+} & \left\{ \frac{(\overline{\partial} u)(\zeta)}{\zeta - z} - \frac{\overline{(\overline{\partial} u)(\zeta)}}{\overline{\zeta} - z} \right\} \mathrm{d}\mathcal{L}^2(\zeta) \\
&= {}_{(\mathscr{C}_b^\infty(\mathbb{C}_+))^*}\big(\mathrm{div}\vec{F}, 1\big)_{\mathscr{C}_b^\infty(\mathbb{C}_+)} = - \int_{\mathbb{R}} \Big(F_2\Big|_{\partial\mathbb{C}_+}^{\kappa-\mathrm{n.t.}} \Big)(t)\, \mathrm{d}\mathcal{L}^1(t) \\
&= -2\mathrm{i} \int_{\mathbb{R}} \frac{\big((\mathrm{Re}\, u)\big|_{\partial\mathbb{C}_+}^{\kappa-\mathrm{n.t.}}\big)(t)}{t - z}\, \mathrm{d}t, \qquad (1.1.196)
\end{aligned}
$$

thanks also to (1.1.194) and (1.1.190). Having established this, we conclude that (1.1.172) holds at each Lebesgue point $z \in \mathbb{C}_+$ of u satisfying (1.1.182) holds, hence also at $\mathcal{L}^2$-a.e. point $z \in \mathbb{C}_+$. □

1.2 Integral Representation Formulas in Clifford Analysis

There is a higher-dimensional version of Theorem 1.1.1 for subsets Ω of $\mathbb{R}^n$ with $n \in \mathbb{N}$, $n \geq 2$ arbitrary, involving the Clifford algebra $\mathcal{C}\ell_n$ (cf. the discussion in [112, §6.4]) in place of the field of complex numbers $\mathbb{C}$, and the Dirac operator D from (A.0.37) in lieu of the Cauchy-Riemann operator $\overline{\partial}$ from (1.1.2), presented a little later in Theorem 1.2.2. To facilitate the proof of the latter theorem, we first take care of the technical result described in the lemma below.

Lemma 1.2.1 *Assume Ω is an open set in $\mathbb{R}^n$ and suppose $u \in L^1_{\mathrm{loc}}(\Omega, \mathcal{L}^n) \otimes \mathcal{C}\ell_n$ is such that $Du \in L^1_{\mathrm{loc}}(\Omega, \mathcal{L}^n) \otimes \mathcal{C}\ell_n$ (with the Dirac operator D taken in the sense of distributions in Ω). For each fixed point $x \in \mathbb{R}^n$ consider the Clifford algebra-valued functions*

$$
\Phi_x(y) := \frac{1}{\omega_{n-1}} \frac{x - y}{|x - y|^n} \odot (Du)(y) \ \text{for } \mathcal{L}^n\text{-a.e. } y \in \Omega, \qquad (1.2.1)
$$

and, for $j \in \{1, \dots, n\}$,

$$
\Psi_x^{(j)}(y) := \frac{1}{\omega_{n-1}} \frac{x - y}{|x - y|^n} \odot \mathbf{e}_j \odot u(y) \ \text{for } \mathcal{L}^n\text{-a.e. } y \in \Omega. \qquad (1.2.2)
$$

Then for each Lebesgue point $x \in \Omega$ of the function u with the additional property that

$$
\begin{aligned}
&\Phi_x \text{ belongs to the space } L^1_{\mathrm{loc}}(\Omega, \mathcal{L}^n) \otimes \mathcal{C}\ell_n \text{ and} \\
&\Psi_x^{(j)} \in L^1_{\mathrm{loc}}(\Omega, \mathcal{L}^n) \otimes \mathcal{C}\ell_n \text{ for each } j \in \{1, \dots, n\},
\end{aligned} \qquad (1.2.3)
$$

one has

$$
\sum_{j=1}^n \partial_j \Psi_x^{(j)} = u(x)\delta_x + \Phi_x \ \text{in } \mathcal{D}'(\Omega) \otimes \mathcal{C}\ell_n, \qquad (1.2.4)
$$

where δ_x denotes the Dirac distribution with mass at x in Ω.

Furthermore, for every $x \in \mathbb{R}^n \setminus \Omega$ *the conditions in* (1.2.3) *are satisfied and, in place of* (1.2.4), *one now has*

$$\sum_{j=1}^{n} \partial_j \Psi_x^{(j)} = \Phi_x \ \textit{in}\ \mathcal{D}'(\Omega) \otimes \mathcal{C}\ell_n. \tag{1.2.5}$$

Of course, thanks to Lebesgue's Differentiation Theorem, we see that

$$\begin{array}{l}\text{formula (1.2.4) is valid for } \mathcal{L}^n\text{-a.e. point } x \in \Omega \\ \text{such that the memberships listed in (1.2.3) hold.}\end{array} \tag{1.2.6}$$

We also wish to comment on the nature of the very last assumption made in (1.2.3). Our first observation in this regard is that since for every compact set $K \subset \Omega$ and every ball $B \subset \mathbb{R}^n$ we have

$$\int_B \Big(\int_K \sum_{j=1}^{n} |\Psi_x^{(j)}(y)|\,\mathrm{d}y \Big)\,\mathrm{d}x$$
$$\leq C_n \Big(\int_K |u(y)|\,\mathrm{d}y \Big) \Big(\sup_{y\in K} \int_B \frac{\mathrm{d}x}{|x-y|^{n-1}} \Big) < +\infty, \tag{1.2.7}$$

it follows that

$$\begin{array}{l}\mathcal{L}^n\text{-a.e. point } x \in \Omega \text{ has the property that} \\ \Psi_x^{(j)} \in L^1_{\mathrm{loc}}(\Omega, \mathcal{L}^n) \otimes \mathcal{C}\ell_n \text{ for } 1 \leq j \leq n.\end{array} \tag{1.2.8}$$

Our second observation is that, as seen from Hölder's inequality,

$$\begin{array}{l}\text{if the function } u \text{ actually belongs to the space } L^p_{\mathrm{loc}}(\Omega, \mathcal{L}^n) \otimes \mathcal{C}\ell_n \text{ for} \\ \text{some exponent } p \in (n, \infty] \text{ then for } \textit{every} \text{ point } x \in \Omega \text{ we have that} \\ \Psi_x^{(j)} \in L^1_{\mathrm{loc}}(\Omega, \mathcal{L}^n) \otimes \mathcal{C}\ell_n \text{ for each exponent } j \in \{1, \dots, n\}.\end{array} \tag{1.2.9}$$

We now turn to the proof of Lemma 1.2.1.

Proof of Lemma 1.2.1 Consider the claim made in (1.2.4). Without loss of generality it may be assumed that $0 \in \Omega$ and $x = 0$, the origin in $\mathbb{R}^n$. Pick a scalar-valued function $\theta \in \mathscr{C}^\infty(\mathbb{R}^n)$ with the property that $\theta = 0$ on $B(0,1)$ and $\theta = 1$ on $\mathbb{R}^n \setminus B(0,2)$. For each $\varepsilon \in \big(0, \frac{1}{2}\operatorname{dist}(0, \partial\Omega)\big)$ define $\theta_\varepsilon : \mathbb{R}^n \to \mathbb{R}$ by setting $\theta_\varepsilon(y) := \theta(y/\varepsilon)$ for every $y \in \mathbb{R}^n$. Then

$$\begin{array}{c} 1 - \theta_\varepsilon \in \mathscr{C}^\infty_c(\Omega), \quad \operatorname{supp}(\nabla\theta_\varepsilon) \subseteq \overline{B(0,2\varepsilon)} \setminus B(0,\varepsilon), \\ \text{and } \lim\limits_{\varepsilon\to 0^+} \theta_\varepsilon(y) = 1 \text{ for every } y \in \mathbb{R}^n \setminus \{0\}, \end{array} \tag{1.2.10}$$

and there exists a constant $C \in (0, \infty)$ such that

$$\big|(\nabla^j \theta_\varepsilon)(y)\big| \leq C\varepsilon^{-j} \ \text{for every } y \in \mathbb{R}^n \text{ and every } j \in \mathbb{N}_0. \tag{1.2.11}$$

Next, fix $\varphi \in \mathscr{C}^\infty_c(\Omega)$ and write

$$
\begin{aligned}
{}_{\mathcal{D}'(\Omega)\otimes\mathcal{C}\ell_n}\Big\langle \sum_{j=1}^{n} \partial_j \Psi_0^{(j)}, \varphi \Big\rangle_{\mathcal{D}(\Omega)}
&= -\sum_{j=1}^{n} {}_{\mathcal{D}'(\Omega)\otimes\mathcal{C}\ell_n}\big\langle \Psi_0^{(j)}, \partial_j\varphi \big\rangle_{\mathcal{D}(\Omega)} \\
&= \sum_{j=1}^{n} \frac{1}{\omega_{n-1}} \int_{\Omega} \frac{y}{|y|^n} \odot \mathbf{e}_j \odot u(y)(\partial_j\varphi)(y)\, dy \\
&= \lim_{\varepsilon\to 0^+} \frac{1}{\omega_{n-1}} \int_{\Omega} \frac{y}{|y|^n} \odot (D\varphi)(y) \odot u(y)\theta_\varepsilon(y)\, dy,
\end{aligned}
\tag{1.2.12}
$$

where for the last equality in (1.2.12) we have used (1.2.10)-(1.2.11) and Lebesgue's Dominated Convergence Theorem. To proceed, observe that $\theta_\varepsilon\varphi \in \mathscr{C}_c^\infty(\Omega)$ and $0 \notin \operatorname{supp}(\theta_\varepsilon\varphi)$. Consequently,

$$
\frac{1}{\omega_{n-1}} \frac{y}{|y|^n} \odot \mathbf{e}_j \odot u(y)(\theta_\varepsilon\varphi)(y) \in L^1_{\rm comp}(\Omega, \mathcal{L}^n) \otimes \mathcal{C}\ell_n \hookrightarrow \mathscr{E}'(\Omega) \otimes \mathcal{C}\ell_n. \tag{1.2.13}
$$

Recall that

$$
\frac{1}{\omega_{n-1}} \sum_{j=1}^{n} \partial_j \Big(\frac{y}{|y|^n} \odot \mathbf{e}_j \Big) = -\delta \ \text{ in } \mathcal{D}'(\Omega) \otimes \mathcal{C}\ell_n. \tag{1.2.14}
$$

In particular,

$$
\frac{1}{\omega_{n-1}} \sum_{j=1}^{n} \partial_j \Big(\frac{y}{|y|^n} \odot \mathbf{e}_j \Big) = 0 \ \text{ for each } y \in \Omega \setminus \{0\}. \tag{1.2.15}
$$

Bearing this in mind, in the sense of distributions in Ω we may compute

$$
\begin{aligned}
\frac{1}{\omega_{n-1}} \sum_{j=1}^{n} \partial_j \Big(\frac{y}{|y|^n} \odot \mathbf{e}_j \odot u\,\theta_\varepsilon\varphi \Big) &= -\Phi_0\theta_\varepsilon\varphi + \frac{1}{\omega_{n-1}} \frac{y}{|y|^n} \odot (D\varphi) \odot u\,\theta_\varepsilon \\
&\quad + \frac{1}{\omega_{n-1}} \frac{y}{|y|^n} \odot (D\theta_\varepsilon) \odot u\,\varphi.
\end{aligned}
\tag{1.2.16}
$$

Together with (1.2.13) this implies

$$
\begin{aligned}
0 &= {}_{\mathscr{E}'(\Omega)\otimes\mathcal{C}\ell_n}\Big\langle \frac{1}{\omega_{n-1}} \sum_{j=1}^{n} \partial_j\Big(\frac{y}{|y|^n} \odot \mathbf{e}_j \odot u\,\theta_\varepsilon\varphi\Big), 1\Big\rangle_{\mathscr{E}(\Omega)} \\
&= -\frac{1}{\omega_{n-1}} \int_\Omega \Phi_0\,\theta_\varepsilon\varphi \,\mathrm{d}\mathcal{L}^n + \frac{1}{\omega_{n-1}} \int_\Omega \frac{y}{|y|^n} \odot (D\,\varphi)(y) \odot u(y)\theta_\varepsilon(y)\,\mathrm{d}y \\
&\quad + \frac{1}{\omega_{n-1}} \int_\Omega \frac{y}{|y|^n} \odot (D\,\theta_\varepsilon)(y) \odot (\varphi u)(y)\,\mathrm{d}y,
\end{aligned} \tag{1.2.17}
$$

hence

$$
\begin{aligned}
&\frac{1}{\omega_{n-1}} \int_\Omega \frac{y}{|y|^n} \odot (D\,\varphi)(y) \odot u(y)\theta_\varepsilon(y)\,\mathrm{d}y \\
&\qquad = \int_\Omega \Phi_0\,\theta_\varepsilon\varphi\,\mathrm{d}\mathcal{L}^n - \frac{1}{\omega_{n-1}} \int_\Omega \frac{y}{|y|^n} \odot (D\,\theta_\varepsilon)(y) \odot (\varphi u)(y)\,\mathrm{d}y.
\end{aligned} \tag{1.2.18}
$$

Thanks to assumptions and Lebesgue's Dominated Convergence Theorem, we may compute

$$
\lim_{\varepsilon\to 0^+} \int_\Omega \Phi_0\,\theta_\varepsilon\varphi\,\mathrm{d}\mathcal{L}^n = \int_\Omega \Phi_0\,\varphi\,\mathrm{d}\mathcal{L}^n = {}_{\mathcal{D}'(\Omega)\otimes\mathcal{C}\ell_n}\big\langle \Phi_0, \varphi\big\rangle_{\mathcal{D}(\Omega)}. \tag{1.2.19}
$$

To handle the last integral in (1.2.18), split

$$
-\frac{1}{\omega_{n-1}} \int_\Omega \frac{y}{|y|^n} \odot (D\,\theta_\varepsilon)(y) \odot u(y)\varphi(y)\,\mathrm{d}y = I + II, \tag{1.2.20}
$$

where

$$
\begin{aligned}
I &:= -\frac{1}{\omega_{n-1}\varepsilon} \int_\Omega \frac{y}{|y|^n} \odot (D\,\theta)(y/\varepsilon) \odot [(\varphi u)(y) - (\varphi u)(0)]\,\mathrm{d}y \\
II &:= -\frac{1}{\omega_{n-1}} \Big(\int_\Omega \frac{y}{|y|^n} \odot (D\,\theta_\varepsilon)(y)\,\mathrm{d}y \Big) \odot (\varphi u)(0).
\end{aligned} \tag{1.2.21}
$$

Using the support condition and estimate from (1.2.10)-(1.2.11), the fact that 0 is a Lebesgue point of u (hence also of φu), and the properties of φ, term I may be estimated by

$$
|I| \le C_n \Big(\sup_{\mathbb{R}^n} |\nabla\theta| \Big) \fint_{B(0,2\varepsilon)} \big|(\varphi u)(y) - (\varphi u)(0)\big|\,\mathrm{d}y \xrightarrow[\varepsilon\to 0^+]{} 0. \tag{1.2.22}
$$

As for II, observe that for each $\varepsilon > 0$ we have

$$
\begin{aligned}
-\frac{1}{\omega_{n-1}}\int_\Omega \frac{y}{|y|^n}\odot(D\,\theta_\varepsilon)(y)\,\mathrm{d}y \\
&= -\frac{1}{\omega_{n-1}}\int_\Omega \frac{y}{|y|^n}\odot D\,[\theta_\varepsilon - 1](y)\,\mathrm{d}y \\
&= {}_{\mathcal{D}'(\Omega)\otimes\mathcal{C}\ell_n}\Big\langle \frac{1}{\omega_{n-1}}\sum_{j=1}^n \partial_j\Big(\frac{y}{|y|^n}\odot\mathbf{e}_j\Big), \theta_\varepsilon - 1\Big\rangle_{\mathcal{D}(\Omega)} \\
&= -\big\langle \delta, \theta_\varepsilon - 1\big\rangle = -(\theta_\varepsilon - 1)(0) = 1,
\end{aligned}
\tag{1.2.23}
$$

by virtue of (1.2.14) and the properties of θ_ε. Finally, by combining (1.2.12) with (1.2.18)-(1.2.23) we arrive at the version of (1.2.4) written for $x = 0$.

Finally, consider the claims in the very last portion of the statement of the lemma. The new aspect is that having fixed an arbitrary point $x \in \mathbb{R}^n \setminus \Omega$ now renders the function $\Omega \ni y \mapsto (x-y)/|x-y|^n$ of class $\mathscr{C}^\infty$ in Ω. Granted this, and keeping in mind the hypotheses on u, it follows that the conditions in (1.2.3) are satisfied. Moreover, (1.2.5) is seen by a direct calculation making use of the product formula for distributional derivatives (in a product involving a smooth function and a distribution), since now we have

$$
\sum_{j=1}^n \partial_{y_j}\Big(\frac{x-y}{|x-y|^n}\Big)\odot\mathbf{e}_j = 0 \text{ for each } y \in \Omega. \tag{1.2.24}
$$

This finishes the proof of the lemma. □

Here is the higher-dimensional version of Theorem 1.1.1 promised earlier.

Theorem 1.2.2 *Assume $\Omega \subseteq \mathbb{R}^n$ is an open set with a lower Ahlfors regular boundary, and with the property that $\sigma := \mathcal{H}^{n-1}\lfloor\partial\Omega$ is a doubling measure on $\partial\Omega$. In particular, Ω is a set of locally finite perimeter, and its geometric measure theoretic outward unit normal ν is defined σ-a.e. on $\partial_*\Omega$.*

In this context, suppose $u : \Omega \to \mathcal{C}\ell_n$ is an $\mathcal{L}^n$-measurable Clifford algebra-valued function which, for some $\kappa > 0$, satisfies

$$
\int_{\partial\Omega}\frac{(\mathcal{N}_\kappa u)(x)}{1+|x|^{n-1}}\,\mathrm{d}\sigma(x) < +\infty \text{ and } u\Big|_{\partial\Omega}^{\kappa-\mathrm{n.t.}} \text{ exists } \sigma\text{-a.e. on } \partial_{\mathrm{nta}}\Omega. \tag{1.2.25}
$$

Furthermore, with the Dirac operator D taken in the sense of distributions in Ω, assume that

$$
Du \in L^1\Big(\Omega, \frac{\mathrm{d}y}{1+|y|^{n-1}}\Big). \tag{1.2.26}
$$

Then for any $\kappa' > 0$ the nontangential trace $u\big|_{\partial\Omega}^{\kappa'-\mathrm{n.t.}}$ exists σ-a.e. on $\partial_{\mathrm{nta}}\Omega$ and is actually independent of κ'. Moreover, with the dependence on the parameter κ' dropped, *for each fixed $x \in \Omega$, the $\mathcal{C}\ell_n$-valued function defined σ-a.e. on $\partial_*\Omega$ as*

$$\partial_*\Omega \ni y \mapsto \frac{x-y}{|x-y|^n} \odot \nu(y) \odot \Big(u\big|^{\text{n.t.}}_{\partial\Omega}\Big)(y) \text{ belongs to } L^1(\partial_*\Omega, \sigma) \otimes \mathcal{C}\ell_n, \tag{1.2.27}$$

and for $\mathcal{L}^n$-a.e. point $x \in \Omega$ one has

$$\begin{aligned} u(x) = {} & \frac{1}{\omega_{n-1}} \int_{\partial_*\Omega} \frac{x-y}{|x-y|^n} \odot \nu(y) \odot \Big(u\big|^{\text{n.t.}}_{\partial\Omega}\Big)(y)\, d\sigma(y) \\ & - \frac{1}{\omega_{n-1}} \int_{\Omega} \frac{x-y}{|x-y|^n} \odot (Du)(y)\, dy, \end{aligned} \tag{1.2.28}$$

when either Ω is bounded, or $\partial\Omega$ is unbounded. In the remaining case, i.e., when Ω is unbounded and $\partial\Omega$ is bounded, formula (1.2.28) *continues to hold under the additional assumption that there exists $\lambda \in (1, \infty)$ such that*

$$\int_{[B(0,\lambda R)\setminus B(0,R)]\cap\Omega} |u(x)|\, d\mathcal{L}^n = o(R^n) \text{ as } R \to \infty. \tag{1.2.29}$$

As a corollary, if actually $Du = 0$ in Ω then one has the following integral representation formula

$$u(x) = \frac{1}{\omega_{n-1}} \int_{\partial_*\Omega} \frac{x-y}{|x-y|^n} \odot \nu(y) \odot \Big(u\big|^{\text{n.t.}}_{\partial\Omega}\Big)(y)\, d\sigma(y), \qquad \forall\, x \in \Omega, \tag{1.2.30}$$

provided either Ω is bounded, or $\partial\Omega$ is unbounded. In the remaining case, i.e., when Ω is unbounded and $\partial\Omega$ is bounded, formula (1.2.30) *continues to be true under the additional assumption that there exists $\lambda \in (1, \infty)$ such that* (1.2.29) *holds.*

Before proving this theorem we wish to remark that [112, Lemma 8.3.1] and the first property in (1.2.25) imply

$$u \in L^\infty_{\text{loc}}(\Omega, \mathcal{L}^n) \otimes \mathcal{C}\ell_n. \tag{1.2.31}$$

In particular, $u \in L^1_{\text{loc}}(\Omega, \mathcal{L}^n) \otimes \mathcal{C}\ell_n$ which ensures that it is meaningful to speak of the Clifford algebra-valued distribution Du in (1.2.26).

Concerning the first condition in (1.2.25) it is worth noting that, thanks to Hölder's inequality and [112, Lemma 7.2.1],

$$\left.\begin{array}{r} \partial\Omega \text{ upper Ahlfors regular, and} \\ \mathcal{N}_\kappa u \in L^p(\partial\Omega, \sigma) \text{ with } 1 \le p < \infty \end{array}\right\} \Longrightarrow \int_{\partial\Omega} \frac{(\mathcal{N}_\kappa u)(x)}{1+|x|^{n-1}}\, d\sigma(x) < +\infty, \tag{1.2.32}$$

and, obviously,

$$\begin{gathered} \text{whenever } \partial\Omega \text{ is bounded, we have} \\ \int_{\partial\Omega} \frac{(\mathcal{N}_\kappa u)(x)}{1+|x|^{n-1}}\, d\sigma(x) < +\infty \Longleftrightarrow \mathcal{N}_\kappa u \in L^1(\partial\Omega, \sigma). \end{gathered} \tag{1.2.33}$$

Let us also remark that it is possible to further relax the first condition in (1.2.25) to demanding instead that

$$\mathcal{N}_\kappa u \in L^1_{\mathrm{loc}}(\partial\Omega, \sigma) \text{ and } \int_{\partial_*\Omega} \frac{\big|\big(u\big|^{\kappa-\mathrm{n.t.}}_{\partial\Omega}\big)(x)\big|}{1+|x|^{n-1}}\,\mathrm{d}\sigma(y) < +\infty. \tag{1.2.34}$$

This would require using [112, Corollary 1.5.2] (in the version recorded in [112, (1.5.23)]) in place of [112, Theorem 1.4.1] (as we do in the proof of Theorem 1.2.2 given further below). We also wish to point out that, if [112, Theorem 1.5.1] is employed in place of [112, Theorem 1.4.1] in the proof of Theorem 1.2.2, then the doubling assumption on $\sigma := \mathcal{H}^{n-1}\lfloor\partial\Omega$ may be relaxed to simply demanding that this is a locally finite measure. In such a scenario, we need to impose the condition that the aperture parameter κ is sufficiently large (depending on Ω), and the ability of changing this parameter when considering nontangential boundary traces is lost. However, modulo these adjustments, the format of the main result (i.e., the integral representation formula (1.2.28)) stays the same.

As regards the integrability condition imposed on Du, [112, Lemma 3.5.7] guarantees that (1.2.26) is equivalent with having, for $\mathcal{L}^n$-a.e. point $x \in \mathbb{R}^n$,

$$\int_{\Omega} \frac{|(Du)(y)|}{|x-y|^{n-1}}\,\mathrm{d}y < +\infty. \tag{1.2.35}$$

In the case when Ω is an exterior domain, there is a version of the integral representation formula (1.2.28) which brings to the forefront the behavior of u at infinity in a manner different than the growth condition imposed in (1.2.29). Specifically, [112, (1.4.5)] and [112, Proposition 4.7.1] imply (on account of (1.2.43) and (1.2.48)) that if Ω is an exterior domain in $\mathbb{R}^n$ and one assumes that

$$\begin{gathered} u \text{ is continuous in a neighborhood of infinity and} \\ \fint_{|y|=R} |u(y)|\,\mathrm{d}\mathcal{H}^{n-1}(y) = o(R) \text{ as } R \to \infty, \end{gathered} \tag{1.2.36}$$

(hence, in particular, if u is continuous and bounded in a neighborhood of infinity) then the limit

$$u_\infty := \lim_{R\to\infty} \fint_{|y|=R} u(y)\,\mathrm{d}\mathcal{H}^{n-1}(y) \text{ exists in } \mathcal{C}\ell_n \tag{1.2.37}$$

and for $\mathcal{L}^n$-a.e. point $x \in \Omega$ one has

$$\begin{aligned} u(x) = {}& \frac{1}{\omega_{n-1}} \int_{\partial_*\Omega} \frac{x-y}{|x-y|^n} \odot \nu(y) \odot \Big(u\big|^{\mathrm{n.t.}}_{\partial\Omega}\Big)(y)\,\mathrm{d}\sigma(y) \\ & - \frac{1}{\omega_{n-1}} \int_{\Omega} \frac{x-y}{|x-y|^n} \odot (Du)(y)\,\mathrm{d}y + u_\infty. \end{aligned} \tag{1.2.38}$$

In light of the last part in Lemma 1.2.1, our proof of Theorem 1.2.2 gives more, namely that for each $x \in \mathbb{R}^n \setminus \overline{\Omega}$ one has

$$0 = \frac{1}{\omega_{n-1}} \int_{\partial_* \Omega} \frac{x-y}{|x-y|^n} \odot \nu(y) \odot \Big(u\big|^{\text{n.t.}}_{\partial\Omega}\Big)(y)\, d\sigma(y) - \frac{1}{\omega_{n-1}} \int_{\Omega} \frac{x-y}{|x-y|^n} \odot (Du)(y)\, dy, \tag{1.2.39}$$

provided either Ω is bounded, or $\partial\Omega$ is unbounded. Moreover, formula (1.2.39) continues to be valid in the remaining case, i.e., when Ω is unbounded and $\partial\Omega$ is bounded, under the additional assumption that there exists $\lambda \in (1, \infty)$ such that (1.2.29) holds.

After this digression, we are ready to present the proof of Theorem 1.2.2.

Proof of Theorem 1.2.2 That for any $\kappa' > 0$ the nontangential trace $u\big|^{\kappa'-\text{n.t.}}_{\partial\Omega}$ exists σ-a.e. on $\partial_{\text{nta}}\Omega$ and is actually independent of κ' follows from assumptions and [112, Corollary 8.9.9]. Also, the claim in (1.2.27) is a consequence of (1.2.25) and [112, (6.4.34), (8.8.52), (8.9.8), (8.9.44)].

To proceed, for each point $x \in \Omega$ with the property that (1.2.35) holds consider the Clifford algebra-valued function defined as

$$\Phi_x(y) := \frac{1}{\omega_{n-1}} \frac{x-y}{|x-y|^n} \odot (Du)(y) \ \text{ for } \mathcal{L}^n\text{-a.e. } y \in \Omega. \tag{1.2.40}$$

In view of (1.2.26), the function Φ_x is $\mathcal{L}^n$-measurable and satisfies

$$\int_{\Omega} |\Phi_x(y)|\, dy \le \frac{1}{\omega_{n-1}} \int\limits_{\Omega} \frac{|(Du)(y)|}{|x-y|^{n-1}}\, dy < +\infty, \tag{1.2.41}$$

thanks to [112, (6.4.35)] and (1.2.35). Hence, for each $x \in \Omega$ such that (1.2.35) holds we have

$$\Phi_x \in L^1(\Omega, \mathcal{L}^n) \otimes C\ell_n. \tag{1.2.42}$$

Moving on, if for each fixed $x \in \Omega$ we define

$$\vec{F}_x(y) := \left\{ \frac{1}{\omega_{n-1}} \frac{x-y}{|x-y|^n} \odot \mathbf{e}_j \odot u(y) \right\}_{1 \le j \le n} \quad \text{for } \mathcal{L}^n\text{-a.e. } y \in \Omega, \tag{1.2.43}$$

then thanks to (1.2.31) we have

$$\vec{F}_x \in L^1_{\text{loc}}(\Omega, \mathcal{L}^n) \otimes C\ell_n, \qquad \forall x \in \Omega. \tag{1.2.44}$$

In addition, from (1.2.43) and the last property in (1.2.25) we conclude that, for each fixed $x \in \Omega$, the nontangential trace $\vec{F}_x\big|^{\kappa-\text{n.t.}}_{\partial\Omega}$ exists at σ-a.e. point on $\partial_{\text{nta}}\Omega$ and, in fact,

$$\Big(\vec{F}_x\Big|_{\partial\Omega}^{\kappa-\mathrm{n.t.}}\Big)(y) = \left\{\frac{1}{\omega_{n-1}}\frac{x-y}{|x-y|^n} \odot \mathbf{e}_j \odot \Big(u\Big|_{\partial\Omega}^{\kappa-\mathrm{n.t.}}\Big)(y)\right\}_{1\le j\le n} \tag{1.2.45}$$

for σ-a.e. $y \in \partial_{\mathrm{nta}}\Omega$.

Going further, introduce $K := \overline{B\big(x, \mathrm{dist}(x,\partial\Omega)/2\big)}$ which is a compact subset of Ω. Then (1.2.43) and [112, (6.4.35), (8.3.45)] allow us to estimate

$$\begin{aligned}\big(\mathcal{N}_\kappa^{\Omega\setminus K}\vec{F}_x\big)(y) &\le C_n\big(\mathcal{N}_\kappa^{\Omega\setminus K}u\big)(y)\cdot \sup_{z\in\Gamma_\kappa(y)\setminus K}|x-z|^{-(n-1)}\\ &\le C_{x,n,\kappa}\frac{(\mathcal{N}_\kappa u)(y)}{|x-y|^{n-1}} \quad \text{for all } y \in \partial\Omega.\end{aligned} \tag{1.2.46}$$

Collectively, [112, (8.2.26)], (1.2.46), and the first condition in (1.2.25) prove that for each fixed $x \in \Omega$ we have

$$\mathcal{N}_\kappa^{\Omega\setminus K}\vec{F}_x \in L^1(\partial\Omega,\sigma). \tag{1.2.47}$$

Granted (1.2.42) and (1.2.44), we may also invoke Lemma 1.2.1. This gives that for each Lebesgue point $x \in \Omega$ of u such that (1.2.35) holds we have

$$\mathrm{div}\vec{F}_x = u(x)\delta_x + \Phi_x \in \mathscr{E}'(\Omega)\otimes \mathcal{C}\ell_n + L^1(\Omega,\mathcal{L}^n)\otimes\mathcal{C}\ell_n, \tag{1.2.48}$$

where the divergence has been taken in the sense of distributions in Ω. Finally, we wish to note that in the case when Ω is an exterior domain, condition (1.2.29) implies that for each fixed $x \in \Omega$ we have

$$\text{the vector field } \vec{F}_x \text{ satisfies the growth condition [112, (1.4.8)].} \tag{1.2.49}$$

Having established (1.2.44), (1.2.45), (1.2.47), (1.2.48), (1.2.49), [112, Theorem 1.4.1] applies and, with the dependence on the parameter κ dropped, formula [112, (1.4.6)] presently gives that for each Lebesgue point $x \in \Omega$ of u such that (1.2.35) holds we have

$$\begin{aligned}u(x) + \int_\Omega \Phi_x(y)\,\mathrm{d}y &= {}_{(\mathscr{C}_b^\infty(\Omega))^*}\big(\mathrm{div}\vec{F}_x, 1\big)_{\mathscr{C}_b^\infty(\Omega)}\\ &= \int_{\partial_*\Omega}\sum_{j=1}^n \nu_j(y)\big(\vec{F}_x\big|_{\partial\Omega}^{\mathrm{n.t.}}\big)_j(y)\,\mathrm{d}\sigma(y)\\ &= \frac{1}{\omega_{n-1}}\int_{\partial_*\Omega}\frac{x-y}{|x-y|^n}\odot\Big(\sum_{j=1}^n \nu_j(y)\mathbf{e}_j\Big)\odot\Big(u\Big|_{\partial\Omega}^{\mathrm{n.t.}}\Big)(y)\,\mathrm{d}\sigma(y)\\ &= \frac{1}{\omega_{n-1}}\int_{\partial_*\Omega}\frac{x-y}{|x-y|^n}\odot\nu(y)\odot\Big(u\Big|_{\partial\Omega}^{\mathrm{n.t.}}\Big)(y)\,\mathrm{d}\sigma(y).\end{aligned} \tag{1.2.50}$$

In view of (1.2.40), this proves (1.2.28) (also assuming (1.2.29) in the case when Ω is an exterior domain) at each Lebesgue point $x \in \Omega$ of u such that (1.2.35) holds.□

In the last part of this section we discuss an application to vector fields satisfying the so called generalized Cauchy-Riemann system in the sense of Fefferman-Stein (cf. [45, (8.1), p. 168]). Given an arbitrary open set $\Omega \subseteq \mathbb{R}^n$, recall that a vector field $u = (u_j)_{1 \leq j \leq n} \in \left[\mathscr{C}^\infty(\Omega)\right]^n$ is said to satisfy the Moisil-Teodorescu system (cf. [136], [137], [138], [178]), or generalized Cauchy-Riemann equations[14], if

$$\text{(MT/GCR)} \quad \begin{cases} \dfrac{\partial u_j}{\partial x_k} = \dfrac{\partial u_k}{\partial x_j} \quad \text{in } \Omega \text{ for } 1 \leq j, k \leq n, \\[2ex] \displaystyle\sum_{j=1}^n \frac{\partial u_j}{\partial x_j} = 0 \text{ in } \Omega. \end{cases} \tag{1.2.51}$$

If this is the case, then identifying the given vector field u with the Clifford algebra-valued function

$$u := \sum_{j=1}^n u_j \mathbf{e}_j : \Omega \to \mathcal{C}\!\ell_n \tag{1.2.52}$$

permits us to express (1.2.51) concisely as

$$Du = 0 \text{ in } \Omega. \tag{1.2.53}$$

Let us now strengthen the hypotheses on Ω by also assuming that $\partial\Omega$ is a lower Ahlfors regular set and that $\sigma := \mathcal{H}^{n-1}\lfloor\partial\Omega$ is a doubling measure on $\partial\Omega$. Also, suppose that either Ω is bounded, or $\partial\Omega$ is unbounded and $n \geq 2$. In this setting, suppose the vector field $u = (u_j)_{1 \leq j \leq n} \in \left[\mathscr{C}^\infty(\Omega)\right]^n$ satisfies the Moisil-Teodorescu system (aka the generalized Cauchy-Riemann equations) (1.2.51) in Ω and, additionally, for each index $j \in \{1, \dots, n\}$ we have

$$\int_{\partial\Omega} \frac{(\mathcal{N}_\kappa u_j)(x)}{1 + |x|^{n-1}} \, d\sigma(x) < +\infty \quad \text{and} \quad u_j\Big|_{\partial\Omega}^{\text{n.t.}} \quad \text{exists } \sigma\text{-a.e. on } \partial_{\text{nta}}\Omega. \tag{1.2.54}$$

Then Theorem 1.2.2 implies that

$$u(x) = \frac{1}{\omega_{n-1}} \int_{\partial_*\Omega} \frac{x - y}{|x - y|^n} \odot \nu(y) \odot \left(u\Big|_{\partial\Omega}^{\text{n.t.}}\right)(y) \, d\sigma(y), \qquad \forall x \in \Omega. \tag{1.2.55}$$

Taking the vector part of both sides yields, on account of [112, (6.4.87)], that for each $x \in \Omega$ we have

[14] a piece of terminology used in [165], [170] (both of these papers cite [138])

$$
\begin{aligned}
u(x) = {} & \frac{1}{\omega_{n-1}} \int_{\partial_* \Omega} \frac{\langle y - x, \nu(y) \rangle}{|x-y|^n} \Big(u\big|_{\partial\Omega}^{\text{n.t.}} \Big)(y) \, d\sigma(y) \\
& + \frac{1}{\omega_{n-1}} \int_{\partial_* \Omega} \frac{\Big\langle x - y, \Big(u\big|_{\partial\Omega}^{\text{n.t.}} \Big)(y) \Big\rangle}{|x-y|^n} \nu(y) \, d\sigma(y) \\
& - \frac{1}{\omega_{n-1}} \int_{\partial_* \Omega} \Big\langle \nu(y), \Big(u\big|_{\partial\Omega}^{\text{n.t.}} \Big)(y) \Big\rangle \frac{x-y}{|x-y|^n} \, d\sigma(y). \qquad (1.2.56)
\end{aligned}
$$

Anticipating notation to be introduced later, in terms of the harmonic double layer $\mathcal{D}$ from (2.5.202) and the operators $\mathcal{R}_{jk}$ defined in (2.5.256), for each $j \in \{1, \dots, n\}$ this further implies the following integral representation formula for the j-th component of the vector field u satisfying the generalized Cauchy-Riemann equations in Ω as well as (1.2.54):

$$
u_j = \mathcal{D}\Big(u_j\big|_{\partial\Omega}^{\text{n.t.}} \Big) - \sum_{k=1}^{n} \mathcal{R}_{jk}\Big(u_k\big|_{\partial\Omega}^{\text{n.t.}} \Big) \text{ in } \Omega. \qquad (1.2.57)
$$

Finally, we wish to note that formula (1.2.57) also holds if Ω is unbounded and either $\partial\Omega$ is bounded or $n = 1$, provided the decay condition (1.2.29) is satisfied.

1.3 First and Second-Order Elliptic Systems of Partial Differential Operators

We begin by introducing the notion of weak ellipticity, in relation to second-order systems, which plays a basic role in this work.

Definition 1.3.1 *Fix $n, M \in \mathbb{N}$, with $n \geq 2$, and let*

$$
L = \big(a_{rs}^{\alpha\beta} \partial_r \partial_s \big)_{1 \leq \alpha, \beta \leq M} \qquad (1.3.1)
$$

be a homogeneous $M \times M$ second-order system in $\mathbb{R}^n$, with complex constant coefficients. Call L `weakly elliptic` *provided its $M \times M$ symbol (or characteristic) matrix*[15]

$$
L(\xi) := -\big(a_{rs}^{\alpha\beta} \xi_r \xi_s \big)_{1 \leq \alpha, \beta \leq M}, \qquad \forall \xi = (\xi_r)_{1 \leq r \leq n} \in \mathbb{R}^n, \qquad (1.3.2)
$$

is an invertible linear mapping for each $\xi \in \mathbb{R}^n \setminus \{0\}$, i.e., it satisfies

$$
\det\big[L(\xi) \big] \neq 0, \qquad \forall \xi \in \mathbb{R}^n \setminus \{0\}. \qquad (1.3.3)
$$

[15] The reader is alerted to the fact that our present definition of $L(\xi)$ differs by a minus sign from what [109, (11.3.2), p. 391] would give for second-order systems

In certain segments of the literature on partial differential equations, a system satisfying (1.3.3) is referred to as being elliptic in the sense of Petrowskiĭ (cf., e.g., [107, p. 275]). The notion of weak ellipticity for a second-order system brought forth in Definition 1.3.1 is ubiquitous in this work. The present terminology is suggested by the fact that the demand imposed in (1.3.3) is weaker than ellipticity considered in a strong sense (à la Legendre-Hadamard) which requires the characteristic matrix (evaluated at points in $\mathbb{R}^n \setminus \{0\}$) to be a strictly positive definite matrix. Here is a formal definition of the latter brand of ellipticity:

Definition 1.3.2 *Fix $M, n \in \mathbb{N}$ and let L be a homogeneous, second-order, constant complex coefficient, $M \times M$ system in $\mathbb{R}^n$. Say that L satisfies the* `Legendre-Hadamard (strong) ellipticity condition` *provided*

$$\begin{gathered}\text{there exists a constant } c \in (0,\infty) \text{ with the property that}\\ \mathrm{Re}\big\langle -L(\xi)\zeta, \overline{\zeta}\big\rangle \geq c\,|\xi|^2\,|\zeta|^2 \ \text{ for all } \xi \in \mathbb{R}^n \text{ and } \zeta \in \mathbb{C}^M.\end{gathered} \tag{1.3.4}$$

It is then clear that, in the context of the above definition,

$$\begin{gathered}L \text{ is Legendre-Hadamard (strongly) elliptic if and only if}\\ L^\top \text{ is Legendre-Hadamard (strongly) elliptic.}\end{gathered} \tag{1.3.5}$$

Of course, any Legendre-Hadamard (strongly) elliptic second-order system is weakly elliptic, but the latter label applies to a larger class. In this vein, it is instructive to contrast the two notions of ellipticity discussed above by consider the case of the complex Lamé system, defined in relation to any two complex parameters $\lambda, \mu \in \mathbb{C}$ as

$$L_{\lambda,\mu}u := \mu\Delta u + (\lambda+\mu)\nabla\mathrm{div}u, \qquad \forall u = (u_1,\dots,u_n) \in \big[\mathcal{D}'(\mathbb{R}^n)\big]^n, \tag{1.3.6}$$

where $\Delta := \partial_1^2 + \cdots + \partial_n^2$, the Laplacian in $\mathbb{R}^n$, acts on the individual components of u. If for any $a = (a_1,\dots,a_N) \in \mathbb{C}^N$ and $b = (b_1,\dots,b_M) \in \mathbb{C}^M$, where $N, M \in \mathbb{N}$, we agree to define $a \otimes b$ to be the $N \times M$ matrix

$$a \otimes b := \big(a_j b_k\big)_{\substack{1\leq j\leq N\\ 1\leq k\leq M}} \tag{1.3.7}$$

then the characteristic matrix for the complex Lamé system (1.3.6) may be expressed as

$$L_{\lambda,\mu}(\xi) = -\mu|\xi|^2 I_{n\times n} - (\lambda+\mu)\xi\otimes\xi, \qquad \forall \xi \in \mathbb{R}^n. \tag{1.3.8}$$

In relation to (1.3.8), the following result appears in [109, Proposition 10.14, p. 366].

Proposition 1.3.3 *Given any complex numbers $\mu, \lambda \in \mathbb{C}$, it follows that $L_{\lambda,\mu}(\xi)$ is invertible for some (or every) $\xi \in \mathbb{R}^n \setminus \{0\}$ if and only if $\mu \neq 0$ and $\lambda + 2\mu \neq 0$. In particular,*

$$\begin{gathered}\text{the complex Lamé system (1.3.6) is weakly elliptic}\\ \text{if and only if one has } \mu \neq 0 \text{ and } \lambda + 2\mu \neq 0.\end{gathered} \tag{1.3.9}$$

Moreover, if $\mu \neq 0$ and $\lambda + 2\mu \neq 0$, then for each $\xi \in \mathbb{R}^n \setminus \{0\}$ one has

$$\big[L_{\lambda,\mu}(\xi)\big]^{-1} = \frac{-1}{\mu|\xi|^2}\Big[I_{n\times n} - \frac{\lambda+\mu}{\lambda+2\mu}\frac{\xi}{|\xi|}\otimes\frac{\xi}{|\xi|}\Big]. \tag{1.3.10}$$

As far as strong ellipticity is concerned, we claim that

> the Lamé system $L_{\lambda,\mu} = \mu\Delta + (\lambda+\mu)\nabla\mathrm{div}$, with moduli $\lambda, \mu \in \mathbb{C}$, is Legendre-Hadamard strongly elliptic if and only if $\mathrm{Re}\,\mu > 0$ and $\mathrm{Re}\,(2\mu+\lambda) > 0$. (1.3.11)

Indeed, given any $\lambda, \mu \in \mathbb{C}$, this a consequence of the fact that

$$\inf_{\substack{\xi\in\mathbb{R}^n,\,\zeta\in\mathbb{C}^n\\ |\xi|=1,\,|\zeta|=1}} \mathrm{Re}\,\big\langle -L_{\lambda,\mu}(\xi)\zeta, \overline{\zeta}\big\rangle = \min\Big\{\mathrm{Re}\,\mu, \mathrm{Re}\,(\lambda+2\mu)\Big\} \tag{1.3.12}$$

plus rescaling. To justify (1.3.12), we proceed by "double inequality." In one direction, for each pair of vectors $\xi \in \mathbb{R}^n$, $\zeta \in \mathbb{C}^n$ with $|\xi| = 1$ and $|\zeta| = 1$ write

$$\begin{aligned}
\mathrm{Re}\,\big\langle -L_{\lambda,\mu}(\xi)\zeta, \overline{\zeta}\big\rangle &= \mathrm{Re}\,\mu + \mathrm{Re}\,(\lambda+\mu)|\langle\xi,\zeta\rangle|^2 \\
&= \mathrm{Re}\,\mu\big(1 - |\langle\xi,\zeta\rangle|^2\big) + \mathrm{Re}\,(\lambda+2\mu)|\langle\xi,\zeta\rangle|^2 \\
&\geq \min\Big\{\mathrm{Re}\,\mu, \mathrm{Re}\,(\lambda+2\mu)\Big\}\cdot\big(1 - |\langle\xi,\zeta\rangle|^2\big) \\
&\quad + \min\Big\{\mathrm{Re}\,\mu, \mathrm{Re}\,(\lambda+2\mu)\Big\}\cdot|\langle\xi,\zeta\rangle|^2 \\
&= \min\Big\{\mathrm{Re}\,\mu, \mathrm{Re}\,(\lambda+2\mu)\Big\},
\end{aligned} \tag{1.3.13}$$

where we have used the fact that $1 - |\langle\xi,\zeta\rangle|^2 \geq 0$, by Cauchy-Schwarz's inequality. Taking infimum establishes one of the desired inequalities. The opposite inequality is obtained by simply specializing the first line in (1.3.13) to the case when $\zeta \in \mathbb{C}^n$ is a unit vector orthogonal to $\xi \in S^{n-1}$ and to the case when $\zeta := \xi \in S^{n-1}$. This proves (1.3.12) from which (1.3.11) follows.

Moving on, we note that for future endeavors is going to be important to adopt an alternative point of view, namely start with a given coefficient tensor and associate it to it a homogeneous second-order system. Specifically, assume $n, M \in \mathbb{N}$ and consider a coefficient tensor

$$A = \Big(a_{rs}^{\alpha\beta}\Big)_{\substack{1\leq r,s\leq n\\ 1\leq\alpha,\beta\leq M}} \tag{1.3.14}$$

with complex entries. To this, we agree to associate the $M\times M$ second-order system (as usual, the summation convention over repeated indices is in effect)

$$L = L_A := \Big(a_{rs}^{\alpha\beta}\partial_r\partial_s\Big)_{\substack{1\leq\alpha\leq M\\ 1\leq\beta\leq M}}. \tag{1.3.15}$$

We continue by introducing the class of injectively elliptic first-order systems in $\mathbb{R}^n$.

Definition 1.3.4 *Consider an arbitrary $N \times M$ homogeneous first-order system with constant complex coefficients in $\mathbb{R}^n$ (where $N, M \in \mathbb{N}$)*

$$D = \Big(\sum_{j=1}^{n} a_j^{\alpha\beta} \partial_j\Big)_{\substack{1\leq\alpha\leq N\\ 1\leq\beta\leq M}} \tag{1.3.16}$$

and recall that its principal symbol is defined as the $N \times M$ matrix

$$\mathrm{Sym}(D;\xi) := \mathrm{i}\Big(\sum_{j=1}^{n} a_j^{\alpha\beta} \xi_j\Big)_{\substack{1\leq\alpha\leq N\\ 1\leq\beta\leq M}} \quad \textit{for each } \xi = (\xi_1, \dots, \xi_n) \in \mathbb{R}^n. \tag{1.3.17}$$

Then D is `injectively elliptic`, *i.e., it has the property that the mapping*

$$\mathrm{Sym}(D;\xi) : \mathbb{C}^M \longrightarrow \mathbb{C}^N \textit{ is injective whenever } \xi \in \mathbb{R}^n \setminus \{0\}. \tag{1.3.18}$$

Various equivalent characterizations of the class of injectively elliptic first-order systems in $\mathbb{R}^n$ are contained in the lemma below.

Lemma 1.3.5 *Fix $N, M \in \mathbb{N}$ and let*

$$D = \Big(\sum_{j=1}^{n} a_j^{\alpha\beta} \partial_j\Big)_{\substack{1\leq\alpha\leq N\\ 1\leq\beta\leq M}} \tag{1.3.19}$$

be some $N \times M$ homogeneous first-order system with constant complex coefficients in $\mathbb{R}^n$. Then the following statements are equivalent:

(i) The first-order system D is injectively elliptic (in the sense of Definition 1.3.4).

(ii) With 'bar' denoting complex conjugation, the homogeneous, constant (complex) coefficient second-order $M \times M$ system in $\mathbb{R}^n$ given by

$$L := -D^*D = -(\overline{D})^\top D \tag{1.3.20}$$

is Legendre-Hadamard strongly elliptic (hence also weakly elliptic, in the sense of Definition 1.3.1).

(iii) There exists some $M \times N$ homogeneous first-order system $\widetilde{D}$ with constant complex coefficients in $\mathbb{R}^n$ with the property that the composition

$L := \widetilde{D}D$ is a weakly elliptic homogeneous, constant (complex) coefficient, second-order $M \times M$ system in $\mathbb{R}^n$. (1.3.21)

Proof Assume first that D is injectively elliptic. Since from (1.3.19) and (A.0.38) we have

$$D^{\top} = -\Big(\sum_{j=1}^{n} a_j^{\alpha\beta}\partial_j\Big)_{\substack{1\leq\beta\leq M,\\ 1\leq\alpha\leq N}} \tag{1.3.22}$$

it follows that the system $L := D^* D$ (cf. (1.3.20)) may be written explicitly as

$$L = \Big(-\overline{a_j^{\beta\alpha}}\, a_k^{\beta\gamma}\partial_j\partial_k\Big)_{1\leq\alpha,\gamma\leq M}. \tag{1.3.23}$$

Then for each two vectors $\xi \in \mathbb{R}^n$ and $w \in \mathbb{C}^M$ the characteristic matrix of L (cf. (1.3.2)) satisfies

$$\langle L(\xi)w, \overline{w}\rangle = \overline{a_j^{\beta\alpha}}\, a_k^{\beta\gamma}\xi_j\xi_k w_\alpha \overline{w_\gamma} = \big|\mathrm{Sym}(D;\xi)w\big|^2, \tag{1.3.24}$$

so, in particular,

$$\begin{aligned} &\langle L(\xi)w, \overline{w}\rangle \in (0,\infty) \ \text{ for all vectors} \\ &\xi \in \mathbb{R}^n \ \text{ and } \ w \in \mathbb{C}^M \ \text{ with } \ |\xi| = |w| = 1. \end{aligned} \tag{1.3.25}$$

From this, the Extreme Value Theorem (for continuous functions on compact sets), and homogeneity we then conclude that there exists some $c \in (0,\infty)$ with the property that

$$\mathrm{Re}\,\langle L(\xi)w, \overline{w}\rangle \geq c|\xi|^2|w|^2 \ \text{ for all } \ \xi \in \mathbb{R}^n \ \text{ and } \ w \in \mathbb{C}^M. \tag{1.3.26}$$

In light of Definition 1.3.2, this shows that the system $-L$ satisfies the Legendre-Hadamard (strong) ellipticity condition. The veracity of the implication *(i)* ⇒ *(ii)* is therefore established.

Next, choosing $\widetilde{D} := -D^*$ proves that we also have *(ii)* ⇒ *(iii)*. To verify that *(iii)* ⇒ *(i)*, assume that there exists some $M \times N$ homogeneous first-order system $\widetilde{D}$ with constant complex coefficients in $\mathbb{R}^n$ such that (1.3.21) holds. Then the characteristic matrix $L(\xi)$ of L is invertible for each vector $\xi \in \mathbb{R}^n \setminus \{0\}$, and since

$$L(\xi) = \mathrm{Sym}(\widetilde{D};\xi)\mathrm{Sym}(D;\xi) \ \text{ for each } \ \xi \in \mathbb{R}^n, \tag{1.3.27}$$

it follows that (1.3.18) holds. The proof of the lemma is therefore complete. □

This lemma suggests making the following definition.

Definition 1.3.6 *Let D be a homogeneous first-order $N \times M$ system with constant complex coefficients in $\mathbb{R}^n$ (where $N, M \in \mathbb{N}$). Say that a homogeneous first-order $M \times N$ system $\widetilde{D}$ with constant complex coefficients in $\mathbb{R}^n$* `complements` *D provided* (1.3.21) *holds.*

In this piece of terminology, Lemma 1.3.5 guarantees that

> a given homogeneous constant coefficient first-order system D has a complement if and only if D is injectively elliptic, and if D is injectively elliptic then D^* complements D. (1.3.28)

It is worth mentioning that there could be infinitely many homogeneous first-order systems $\widetilde{D}$ complementing a given homogeneous first-order injectively elliptic system D in $\mathbb{R}^n$. For example any operator of the form $\operatorname{div}A$, where $A \in \mathbb{R}^{n\times n}$ is a strictly positive-definite matrix, complements the gradient operator in $\mathbb{R}^n$.

Several notable examples of injectively elliptic first-order systems are discussed below.

Example 1.3.7 *The Cauchy-Riemann operator* $\overline{\partial}$ *from* (1.1.2) *and the Dirac operator* D *from* (A.0.37) *are injectively elliptic first-order systems. A moment's reflection shows that the gradient operator* ∇ *is also an injectively elliptic first-order system in* $\mathbb{R}^n$. *More generally, the Jacobian* (A.0.48), *i.e., the operator acting on each vector-valued distribution* $u = (u_1, \dots, u_M) \in \big[\mathcal{D}'(\mathbb{R}^n)\big]^M$ *(for some fixed* $M \in \mathbb{N}$*) according to*

$$\nabla u := \big(\partial_k u_j\big)_{\substack{1\leq j\leq M\\ 1\leq k\leq n}} = \begin{pmatrix} \partial_1 u_1 & \cdots & \partial_n u_1 \\ \vdots & \vdots & \vdots \\ \partial_1 u_M & \cdots & \partial_n u_M \end{pmatrix}. \tag{1.3.29}$$

is injectively elliptic. To justify the latter, observe that for each $\xi = (\xi_k)_{1\leq k\leq n} \in \mathbb{R}^n$ and each $u = (u_j)_{1\leq j\leq M} \in \mathbb{C}^M$ we have

$$(-\mathrm{i})\mathrm{Sym}(\nabla; \xi)u = \big(\xi_k u_j\big)_{\substack{1\leq j\leq M\\ 1\leq k\leq n}} = u \otimes \xi. \tag{1.3.30}$$

This further implies

$$\Big((-\mathrm{i})\mathrm{Sym}(\nabla; \xi)u\Big)\xi = (u \otimes \xi)\xi = |\xi|^2 u, \tag{1.3.31}$$

from which it is then clear that the Jacobian operator (1.3.29) is indeed injectively elliptic.

Example 1.3.8 *The Moisil-Teodorescu system (aka the generalized Cauchy-Riemann equations) acting on a vector-valued distribution* $u = (u_1, \dots, u_n)$ *in* $\mathbb{R}^n$ *according to*

$$Du := \Big(\big(\partial_j u_k - \partial_k u_j\big)_{1\leq j,k\leq n}, \sum_{j=1}^{n} \partial_j u_j\Big) \tag{1.3.32}$$

is injectively elliptic. To justify this, observe first that

$$\begin{aligned} &\mathrm{Sym}(D; \xi)u := \mathrm{i}\Big(\big(\xi_j u_k - \xi_k u_j\big)_{1\leq j,k\leq n}, \textstyle\sum_{j=1}^{n} \xi_j u_j\Big) \\ &\text{for each } u = (u_j)_{1\leq j\leq n} \text{ and } \xi := (\xi_1, \dots, \xi_n) \in \mathbb{R}^n. \end{aligned} \tag{1.3.33}$$

If $\xi \neq 0$ and yet $\mathrm{Sym}(D; \xi)u = 0$, multiplying each $\xi_j u_k - \xi_k u_j$ by ξ_j, summing in j, then using $\sum_{j=1}^{n} \xi_j u_j = 0$, leads to the conclusion that $|\xi|^2 u_k = 0$ for each $k \in \{1, \dots, n\}$. Hence, $u = 0$ as desired.

Example 1.3.9 *Consider a matrix* $A = (a_{jk})_{1\leq j,k\leq n} \in \mathbb{C}^{n\times n}$ *which is weakly elliptic, i.e., satisfies*

$$\langle A\xi, \xi\rangle = a_{jk}\xi_j\xi_k \neq 0 \textit{ for all } \xi \in \mathbb{R}^n \setminus \{0\}. \tag{1.3.34}$$

Then the modified Moisil-Teodorescu system, associated with A, acting on a vector-valued distribution $u = (u_1, \dots, u_n)$ in $\mathbb{R}^n$ according to

$$Du := \Big((\partial_j u_k - \partial_k u_j)_{1\leq j,k\leq n}, \operatorname{div} Au\Big) \tag{1.3.35}$$

is injectively elliptic. Indeed,

$$\begin{aligned} \operatorname{Sym}(D;\xi)u &:= \mathrm{i}\Big((\xi_j u_k - \xi_k u_j)_{1\leq j,k\leq n}, a^n_{jk}\xi_j u_k\Big) \\ \text{for each } u &= (u_j)_{1\leq j\leq n} \text{ and } \xi := (\xi_1, \dots, \xi_n) \in \mathbb{R}^n. \end{aligned} \tag{1.3.36}$$

Assume $\xi \neq 0$ yet $\operatorname{Sym}(D;\xi)u = 0$. For an arbitrary $\ell \in \{1, \dots, n\}$, multiply the equality $a^n_{jk}\xi_j u_k = 0$ by ξ_ℓ then use the fact that $\xi_\ell u_k = \xi_k u_\ell$ to obtain $a_{jk}\xi_j\xi_k u_\ell = 0$. In view of the fact that (1.3.34) ensures $a_{jk}\xi_j\xi_k \neq 0$, this forces $u_\ell = 0$ for each $\ell \in \{1, \dots, n\}$. Thus, $u = 0$ as wanted.

Example 1.3.10 *The first-order system $D := (\partial_{\bar z_j})_{1\leq j\leq n}$ in $\mathbb{C}^n \equiv \mathbb{R}^{2n}$ is injectively elliptic.* Indeed, since for each $\xi \in \mathbb{C}^n \equiv \mathbb{R}^{2n}$ we have $|\operatorname{Sym}(D;\xi)u| = \frac{1}{2}|\xi||u|$, the desired conclusion readily follows.

Example 1.3.11 *With the summation convention over repeated indices in effect, suppose*

$$L = \Big(a^{\alpha\beta}_{rs}\partial_r\partial_s\Big)_{\substack{1\leq\alpha\leq M\\ 1\leq\beta\leq M}}. \tag{1.3.37}$$

is a weakly elliptic homogeneous constant (complex) coefficient $M\times M$ second-order system in $\mathbb{R}^n$. Then the first-order, homogeneous, constant coefficient, $(n\cdot M)\times M$ system

$$Du := A\nabla u := (a^{\alpha\beta}_{rs}\partial_s u_\beta)_{\substack{1\leq\alpha\leq M\\ 1\leq r\leq n}} \tag{1.3.38}$$

is injectively elliptic. Indeed, we have $\mathrm{i}\xi_r\big(\operatorname{Sym}(D;\xi)u\big)_{\alpha,r} = \big(L(\xi)u\big)_\alpha$ for each $\xi \in \mathbb{R}^n$, $u \in \mathbb{C}^M$, and $\alpha \in \{1, \dots, M\}$. Since the characteristic matrix $L(\xi)$ is invertible whenever $\xi \in \mathbb{R}^n \setminus \{0\}$, it follows that $\operatorname{Sym}(D;\xi)$ is, as claimed, injective whenever $\xi \in \mathbb{R}^n \setminus \{0\}$.

Example 1.3.12 *Suppose L is a weakly elliptic, second-order, homogeneous, constant (complex) coefficient $M\times M$ system associated with some coefficient tensor $A = \big(a^{\alpha\beta}_{rs}\big)_{\substack{1\leq r,s\leq n\\ 1\leq\alpha,\beta\leq M}}$ as in* (1.3.15). *In relation to this, introduce the homogeneous constant coefficient first-order differential operator $\mathbb{D}_A$, mapping distributions in $\mathbb{R}^n$ with $n\times M$ components $w = \big(w^\beta_s\big)_{\substack{1\leq s\leq n\\ 1\leq\beta\leq M}} \in \big[\mathcal{D}'(\mathbb{R}^n)\big]^{n\times M}$ into*

$$\mathbb{D}_A w := \begin{pmatrix} (\partial_r w^\beta_s - \partial_s w^\beta_r)_{\substack{1\leq r,s\leq n\\ 1\leq\beta\leq M}} \\ (a^{\alpha\beta}_{rs}\partial_r w^\beta_s)_{1\leq\alpha\leq M} \end{pmatrix}. \tag{1.3.39}$$

Then

as a linear operator, the matrix $\mathrm{Sym}(\mathbb{D}_A;\xi)$ *is injective for each vector* $\xi \in \mathbb{R}^n \setminus \{0\}$*, hence the homogeneous constant coefficient first-order system* $\mathbb{D}_A$ *is injectively elliptic,* (1.3.40)

and

$$\mathrm{Ker}\, L \ni u = (u_\beta)_{1\leq\beta\leq M} \longmapsto \nabla u = \big(\partial_s u_\beta\big)_{\substack{1\leq s\leq n\\ 1\leq\beta\leq M}} \in \mathrm{Ker}\,\mathbb{D}_A \text{ is a well-defined linear mapping.} \tag{1.3.41}$$

To see that this is the case, observe that the principal symbol of $\mathbb{D}_A$ at each point $\xi = (\xi_r)_{1\leq r\leq n} \in \mathbb{R}^n$ is the linear mapping acting on $w = \big(w_s^\beta\big)_{\substack{1\leq s\leq n\\ 1\leq\beta\leq M}} \in \mathbb{C}^{n\times M}$ according to

$$\mathrm{Sym}(\mathbb{D}_A;\xi)w = \mathrm{i}\begin{pmatrix} \big(\xi_r w_s^\beta - \xi_s w_r^\beta\big)_{\substack{1\leq r,s\leq n\\ 1\leq\beta\leq M}} \\ \big(a_{rs}^{\alpha\beta}\xi_r w_s^\beta\big)_{1\leq\alpha\leq M} \end{pmatrix}. \tag{1.3.42}$$

If $\xi = (\xi_r)_{1\leq r\leq n} \in \mathbb{R}^n \setminus \{0\}$ and $w = \big(w_s^\beta\big)_{\substack{1\leq s\leq n\\ 1\leq\beta\leq M}} \in \mathbb{C}^{n\times M}$ are such that $\mathrm{Sym}(\mathbb{D}_A;\xi)w = 0$, then

$$\begin{aligned} \xi_r w_s^\beta - \xi_s w_r^\beta &= 0 \text{ for } 1\leq r,s\leq n \text{ and } 1\leq\beta\leq M, \\ a_{rs}^{\alpha\beta}\xi_r w_s^\beta &= 0 \text{ whenever } 1\leq\alpha\leq M. \end{aligned} \tag{1.3.43}$$

Since $\xi \neq 0$, there exists some $r_* \in \{1,\dots,n\}$ such that $\xi_{r_*} \neq 0$. If we now define

$$\eta_\beta := \frac{w_{r_*}^\beta}{\xi_{r_*}} \in \mathbb{C} \text{ for each } \beta \in \{1,\dots,M\}, \tag{1.3.44}$$

then the first line in (1.3.43) implies that, for each $\beta \in \{1,\dots,M\}$ and for each $s \in \{1,\dots,n\}$, we have

$$\xi_{r_*} w_s^\beta = \xi_s w_{r_*}^\beta = \xi_s\xi_{r_*}\frac{w_{r_*}^\beta}{\xi_{r_*}} = \xi_s\xi_{r_*}\eta_\beta. \tag{1.3.45}$$

Thus,

$$w_s^\beta = \xi_s\eta_\beta \text{ for each } \beta \in \{1,\dots,M\} \text{ and } s \in \{1,\dots,n\}. \tag{1.3.46}$$

In concert with the second line in (1.3.43), this permits us to write

$$a_{rs}^{\alpha\beta}\xi_r\xi_s\eta_\beta = 0 \text{ for each } \alpha \in \{1,\dots,M\}. \tag{1.3.47}$$

In view of the weak ellipticity condition (1.3.3), we conclude from (1.3.47) that each η_β is zero which, in concert with (1.3.46), ultimately proves that $w_s^\beta = 0$ whenever $1\leq s\leq n$ and $1\leq\beta\leq M$. Bearing (1.3.18) in mind, the claims in (1.3.40) follow. Finally, (1.3.41) is clear from definitions.

In relation to Example 1.3.12, it is worth noting that

if $M := 1$, $L := \Delta$, and $a_{rs}^{\alpha\beta} = \delta_{\alpha\beta}\delta_{rs}$ then the first-order differential operator $\mathbb{D}_A$ from (1.3.39) reduces to the Moisil-Teodorescu system (aka the generalized Cauchy-Riemann equations) discussed in Example 1.3.8. (1.3.48)

Example 1.3.13 *The deformation tensor in* $\mathbb{R}^n$*, given by*

$$\begin{aligned} &\operatorname{Def}\vec{u} := \tfrac{1}{2}\big(\partial_j u_k + \partial_k u_j\big)_{1\leq j,k\leq n} \\ &\textit{for each } \vec{u} = (u_1, \dots, u_n) \in \big[\mathcal{D}'(\mathbb{R}^n)\big]^n, \end{aligned} \tag{1.3.49}$$

is an injectively elliptic, first-order homogeneous, constant (real) coefficient, $n^2 \times n$ *system in* $\mathbb{R}^n$.

To justify this, observe that for every $\xi = (\xi_1, \dots, \xi_n) \in \mathbb{R}^n$ we have

$$\begin{aligned} (-\mathrm{i})\mathrm{Sym}(\mathrm{Def};\xi)\,\vec{u} &= \tfrac{1}{2}\big(\xi_j u_k + \xi_k u_j\big)_{1\leq j,k\leq n} \\ \text{for each } \vec{u} &= (u_1, \dots, u_n) \in \mathbb{R}^n. \end{aligned} \tag{1.3.50}$$

As a consequence, if $\xi = (\xi_1, \dots, \xi_n) \in \mathbb{R}^n \setminus \{0\}$ is such that $(-\mathrm{i})\mathrm{Sym}(\mathrm{Def};\xi)\,\vec{u} = 0$ for some vector $\vec{u} = (u_1, \dots, u_n) \in \mathbb{R}^n$ then $\xi_j u_k + \xi_k u_j = 0$ for each integers $j, k \in \{1, \dots, n\}$. In particular, for each fixed index $k \in \{1, \dots, n\}$ we have the equality $0 = \xi_j(\xi_j u_k + \xi_k u_j) = |\xi|^2 u_k + \xi_k\langle\xi, \vec{u}\rangle$, hence

$$\vec{u} = -\frac{\langle\xi, \vec{u}\rangle}{|\xi|^2}\xi. \tag{1.3.51}$$

After taking the dot product with ξ this further implies

$$\langle\xi, \vec{u}\rangle = -\frac{\langle\xi, \vec{u}\rangle}{|\xi|^2}|\xi|^2 = -\langle\xi, \vec{u}\rangle, \tag{1.3.52}$$

from which we conclude that $\langle\xi, \vec{u}\rangle = 0$. In light of (1.3.51), this forces $\vec{u} = 0$, which ultimately shows that the symbol mapping $\mathrm{Sym}(\mathrm{Def};\xi)$ is indeed injective.

Example 1.3.14 *Employing the formalism associated with differential forms in* $\mathbb{R}^n$*, with* d *and* δ *denoting the exterior derivative operator and its formal transpose, respectively, consider the first-order homogeneous, constant (complex) coefficient differential operator*

$$D_{\alpha,\beta} := \alpha\mathrm{d} + \beta\delta \ \textit{ where } \alpha, \beta \in \mathbb{C}, \tag{1.3.53}$$

acting on differential forms (of mixed, arbitrary degrees). Since, in view of the fact $\mathrm{d}^2 = \delta^2 = 0$*, we have*

$$D^2_{\alpha,\beta} = (\alpha\beta)(\mathrm{d}\delta + \delta\mathrm{d}) = -(\alpha\beta)\Delta, \tag{1.3.54}$$

where Δ is the Laplacian in $\mathbb{R}^n$, we conclude from Lemma 1.3.5 that $D_{\alpha,\beta}$ is an injectively elliptic first-order operator for each $\alpha, \beta \in \mathbb{C} \setminus \{0\}$.

As far as weak ellipticity for second-order systems is concerned, the first example that comes to mind is that of the Laplace operator Δ in $\mathbb{R}^n$. In fact, we have already noted that any Legendre-Hadamard strongly elliptic second-order system is also weakly elliptic, and Lemma 1.3.5 tells us that

the second-order system $L := -D^*D$ is Legendre-Hadamard strongly elliptic, hence also weakly elliptic, whenever D is a homogeneous (complex) constant coefficient injectively elliptic first-order system. (1.3.55)

Consequently,

the second-order system $L := -D^*D$ is Legendre-Hadamard strongly elliptic, hence also weakly elliptic, whenever D is any of the first-order systems from Examples 1.3.7-1.3.14. (1.3.56)

Thus, among other things,

if D is the Jacobian operator (1.3.29), then its transpose $D^\top$ is given by the formula $D^\top w = \big(- \partial_k w_{jk} \big)_{1 \le j \le M}$ for each matrix-valued distribution $w = (w_{jk})_{\substack{1 \le j \le M \\ 1 \le k \le n}} \in \big[\mathcal{D}'(\mathbb{R}^n)\big]^{M \times n}$, a scenario in which the second-order system $L := -D^*D$ becomes $-\nabla^\top \nabla = \Delta$, the Laplacian in $\mathbb{R}^n$ acting on the M components of vector-valued distributions, (1.3.57)

and

if D is the deformation tensor (1.3.49), then its transpose $D^\top$ is given by the formula $D^\top w = -\frac{1}{2}\big(\partial_k w_{jk} + \partial_k w_{kj}\big)_{1 \le j \le n}$ for each matrix-valued distribution $w = (w_{jk})_{1 \le j,k \le n} \in \big[\mathcal{D}'(\mathbb{R}^n)\big]^{n \times n}$, so in this case the system $L := -D^*D$ takes the form $-\mathrm{Def}^\top \mathrm{Def} = \frac{1}{2}\big(\Delta + \nabla \mathrm{div}\big)$ on vector-valued distributions in $\mathbb{R}^n$. (1.3.58)

The above discussion suggests the consideration of a family of second-order systems, indexed by triplets of complex numbers $a, b, c \in \mathbb{C}$, acting on vector-valued distributions in $\mathbb{R}^n$ according to

$$L_{a,b,c} := -2a\, \mathrm{Def}^\top \mathrm{Def} + b \nabla \mathrm{div} - c \nabla^\top \nabla. \tag{1.3.59}$$

In relation to this, we have the following results:

$$\text{the system } L_{a,b,c} \text{ is weakly elliptic} \\ \text{if and only if } a + c \neq 0 \text{ and } 2a + b + c \neq 0, \tag{1.3.60}$$

and

$$\begin{aligned}&\text{the system } L_{a,b,c} \text{ is Legendre-Hadamard strongly elliptic}\\&\quad\text{if and only if } \operatorname{Re}(a+c) > 0 \text{ and } \operatorname{Re}(2a+b+c) > 0.\end{aligned} \tag{1.3.61}$$

Indeed, from (1.3.57), (1.3.58), and (1.3.6) we see that

$$\begin{aligned}&\text{the system } L_{a,b,c} \text{ defined in (1.3.59) coincides with the complex Lamé}\\&\text{system } L_{\lambda,\mu} = \mu\Delta+(\lambda+\mu)\nabla\mathrm{div} \text{ for the moduli } \lambda := b-c \text{ and } \mu := a+c,\end{aligned} \tag{1.3.62}$$

so the claims in (1.3.60)-(1.3.61) become consequences of (1.3.9) and (1.3.11).

Other relevant examples of weakly elliptic and strongly elliptic systems are discussed below.

Example 1.3.15 *Work in $\mathbb{R}^2 \equiv \mathbb{C}$. Bring in the Cauchy-Riemann operator $\partial_{\bar z}$ and its complex conjugate ∂_z from* (1.1.2). *For each parameter $\lambda \in \mathbb{C}$ then define the scalar, second-order, homogeneous, differential operator*

$$L_\lambda := \partial_{\bar z}^2 - \lambda^2 \partial_z^2. \tag{1.3.63}$$

Then for each $\xi = (\xi_1, \xi_2) \in \mathbb{R}^2$ we have

$$L_\lambda(\xi) = -\tfrac{1}{4}\big[(\xi_1 + i\xi_2)^2 - \lambda^2(\xi_1 - i\xi_2)^2\big] \tag{1.3.64}$$

If $\xi \in \mathbb{R}^2 \setminus \{0\}$ is such that $L_\lambda(\xi) = 0$, then (1.3.64) *implies*

$$\lambda = \pm\frac{\xi_1 + i\xi_2}{\xi_1 - i\xi_2} = \pm\Big(\frac{\xi_1 + i\xi_2}{|\xi|}\Big)^2, \tag{1.3.65}$$

which ultimately forces $|\lambda| = 1$. This analysis ultimately shows that

$$\begin{aligned}&\textit{the differential operator } L_\lambda \textit{ is weakly elliptic}\\&\qquad\textit{if and only if } |\lambda| \neq 1.\end{aligned} \tag{1.3.66}$$

For the scalar operator L_λ defined in (1.3.63), *the Legendre-Hadamard* (*strong*) *ellipticity condition* (1.3.4) *amounts to the demand that*

$$\operatorname{Re}[-L_\lambda(\xi)] > 0 \text{ for each } \xi \in S^1. \tag{1.3.67}$$

Testing this for the choices $\xi = (1, \pm 1)/\sqrt{2}$ forces, in light of (1.3.64),

$$0 < \operatorname{Re}\big[(1 \pm i)^2 - \lambda^2(1 \mp i)^2\big] = \operatorname{Re}\big[\pm 2i \pm 2i\lambda^2\big] = \pm\operatorname{Re}\big[2i + 2i\lambda^2\big] \tag{1.3.68}$$

an obvious impossibility, irrespective of $\lambda \in \mathbb{C}$. This proves that the differential operator

$$\begin{aligned}&L_\lambda \textit{ is not Legendre-Hadamard (strongly) elliptic}\\&\qquad\textit{for any choice of } \lambda \in \mathbb{C}.\end{aligned} \tag{1.3.69}$$

Our next example pertains to a family of operators acting on differential forms of arbitrary degrees, which contains the complex Lamé system (1.3.6) as a particular case.

Example 1.3.16 *We shall employ the formalism associated with differential forms in $\mathbb{R}^n$. In particular,* d *and* δ *denote, respectively, the exterior derivative operator and its formal transpose. For $\alpha, \beta \in \mathbb{C}$, consider the second-order homogeneous, constant (complex) coefficient differential operator*

$$L_{\alpha,\beta} := \alpha \mathrm{d}\delta + \beta\delta\mathrm{d} \tag{1.3.70}$$

acting on differential forms (of mixed, arbitrary degrees). Then

$$\text{the differential operator } L_{\alpha,\beta} \text{ is weakly elliptic for each } \alpha, \beta \in \mathbb{C} \setminus \{0\}, \tag{1.3.71}$$

and

$$L_{\alpha,\beta} \text{ is Legendre-Hadamard (strongly) elliptic if and only if } \mathrm{Re}\,\alpha > 0 \text{ and } \mathrm{Re}\,\beta > 0. \tag{1.3.72}$$

Let us justify (1.3.71). In view of the fact

$$\mathrm{d}^2 = \delta^2 = 0 \text{ and } \mathrm{d}\delta + \delta\mathrm{d} = -\Delta \tag{1.3.73}$$

where Δ is the Laplacian in $\mathbb{R}^n$ (acting on differential forms componentwise), we have

$$L_{\beta,\alpha} L_{\alpha,\beta} = (\alpha\beta)\Delta^2 = L_{\alpha,\beta} L_{\beta,\alpha} \tag{1.3.74}$$

At the level of principal symbols, this translates into[16]

$$L_{\beta,\alpha}(\xi) L_{\alpha,\beta}(\xi) = (\alpha\beta)|\xi|^4 = L_{\alpha,\beta}(\xi) L_{\beta,\alpha}(\xi) \text{ for all } \xi \in \mathbb{R}^n. \tag{1.3.75}$$

Hence, if $\alpha\beta \neq 0$ we conclude that $L_{\alpha,\beta}(\xi)$ is invertible for each $\xi \in \mathbb{R}^n \setminus \{0\}$, which ultimately proves (1.3.71).

As far as (1.3.72) is concerned, via homogeneity this becomes a consequence of

$$\inf_{|\xi|=1,\,|u|=1} \mathrm{Re}\big\langle -L_{\alpha,\beta}(\xi)u, \overline{u}\big\rangle = \min\big\{\mathrm{Re}\,\alpha, \mathrm{Re}\,\beta\big\} \tag{1.3.76}$$

which, in turn, is justified by "double inequality." To set the stage, observe that for each $\xi \in \mathbb{R}^n$ and each differential form u, formula (1.3.70) entails

$$L_{\alpha,\beta}(\xi)u = -\Big[\alpha\xi \wedge (\xi \vee u) + \beta\xi \vee (\xi \wedge u)\Big]. \tag{1.3.77}$$

Next, if ξ is a vector in $\mathbb{R}^n$ and u is a differential form in $\mathbb{R}^n$ with $|\xi| = 1$ and $|u| = 1$, we may write (based on (1.3.77), item *(3)* in [112, Lemma 6.4.7], and the second formula in [112, (6.4.133)])

[16] suppressing the identity operator (acting on differential forms) in the middle term

$$\begin{aligned}
\mathrm{Re}\,\big\langle -L_{\alpha,\beta}(\xi)u,\overline{u}\big\rangle &= \mathrm{Re}\Big[\alpha\big\langle \xi\wedge(\xi\vee u),\overline{u}\big\rangle+\beta\big\langle \xi\vee(\xi\wedge u),\overline{u}\big\rangle\Big]\\
&= \mathrm{Re}\Big[\alpha|\xi\vee u|^2+\beta|\xi\wedge u|^2\Big]=(\mathrm{Re}\,\alpha)|\xi\vee u|^2+(\mathrm{Re}\,\beta)|\xi\wedge u|^2\\
&\geq \min\big\{\mathrm{Re}\,\alpha,\mathrm{Re}\,\beta\big\}\cdot\big(|\xi\vee u|^2+|\xi\wedge u|^2\big)\\
&= \min\big\{\mathrm{Re}\,\alpha,\mathrm{Re}\,\beta\big\}, \qquad (1.3.78)
\end{aligned}$$

which is one of the desired inequalities. The inequality in the opposite direction may be seen by specializing

$$\mathrm{Re}\,\big\langle -L_{\alpha,\beta}(\xi)u,\overline{u}\big\rangle = (\mathrm{Re}\,\alpha)|\xi\vee u|^2+(\mathrm{Re}\,\beta)|\xi\wedge u|^2\Big] \qquad (1.3.79)$$

(itself implicit in (1.3.78)) to the case when $u := \xi \in S^{n-1}$ and to the case when $u := *\xi$ with $\xi \in S^{n-1}$ (where 'star' denotes the Hodge isomorphism). Thus, (1.3.72) is established.

The manner in which the operator $L_{\alpha,\beta}$ defined in (1.3.70) ties up with the complex Lamé system from (1.3.6) is as follows. From (1.3.70) and the last formula in (1.3.73) we have

$$L_{\alpha,\beta} = \alpha\mathrm{d}\delta+\beta(-\Delta-\mathrm{d}\delta) = -\beta\Delta+(\alpha-\beta)\mathrm{d}\delta \qquad (1.3.80)$$

hence,

when acting on vector fields, the operator $L_{\alpha,\beta}$ agrees with the complex Lamé system (1.3.6) written for the Lamé moduli λ,μ chosen such that $\mu=-\beta$ and $\lambda+\mu=-(\alpha-\beta)$, i.e., for $\lambda := -\alpha+2\beta$ and $\mu := -\beta$. (1.3.81)

In this vein, it is reassuring to note that the demands on α,β in (1.3.71) agree with the conditions imposed on λ,μ in (1.3.9), while the demands on α,β in (1.3.72) agree with the conditions imposed on λ,μ in (1.3.11).

1.4 Fundamental Solutions for Weakly Elliptic Second-Order Systems of Partial Differential Operators

Consider a function satisfying

$$\begin{gathered}
\Theta\in\mathscr{C}^0(\mathbb{R}^n\setminus\{0\}),\quad \int_{S^{n-1}}\Theta(\xi)\,\mathrm{d}\mathcal{H}^{n-1}(\xi)=0,\\
\text{and }\Theta\text{ is positive homogeneous of degree }-n.
\end{gathered} \qquad (1.4.1)$$

Typically, such a function fails to belong to $L^1_{\mathrm{loc}}(\mathbb{R}^n,\mathcal{L}^n)$, so associating a distribution with Θ requires some care. Specifically, with Θ from (1.4.1) we associate the principal value distribution P.V. $\Theta \in \mathcal{S}'(\mathbb{R}^n)$ defined as

$$\langle \text{P.V.}\,\Theta, \varphi\rangle := \lim_{\varepsilon\to 0^+} \int_{x\in\mathbb{R}^n,\ |x|>\varepsilon} \Theta(x)\varphi(x)\,dx, \qquad \forall \varphi \in \mathcal{S}(\mathbb{R}^n). \tag{1.4.2}$$

Then this is a well-defined tempered distribution in $\mathbb{R}^n$ with the property that

$$(\text{P.V.}\,\Theta)\big|_{\mathbb{R}^n\setminus\{0\}} = \Theta\big|_{\mathbb{R}^n\setminus\{0\}} \text{ in } \mathcal{D}'(\mathbb{R}^n \setminus \{0\}). \tag{1.4.3}$$

In addition, for each $\varphi \in \mathcal{S}'(\mathbb{R}^n)$ one has

$$(\text{P.V.}\,\Theta) * \varphi \in \mathcal{S}'(\mathbb{R}^n) \cap \mathscr{C}^\infty(\mathbb{R}^n) \tag{1.4.4}$$

and

$$\Big((\text{P.V.}\,\Theta) * \varphi\Big)(x) = \lim_{\varepsilon\to 0^+} \int_{y\in\mathbb{R}^n, |x-y|>\varepsilon} \Theta(x-y)\varphi(y)\,dy, \qquad \forall x \in \mathbb{R}^n. \tag{1.4.5}$$

Functions satisfying (1.4.1) arise naturally in applications. To give one basic example, suppose some function $\Phi \in \mathscr{C}^1(\mathbb{R}^n \setminus \{0\})$ which is positive homogeneous of degree $1-n$ has been given, and some index $j \in \{1,\dots,n\}$ has been fixed. Then $\partial_j\Phi$ satisfies the conditions in (1.4.1) and, as such, P.V.$(\partial_j\Phi)$ is a well-defined tempered distribution. Moreover, $\Phi \in L^1_{\text{loc}}(\mathbb{R}^n, \mathcal{L}^n)$, hence Φ may be canonically regarded as a distribution in $\mathbb{R}^n$, and for each $j \in \{1,\dots,n\}$ the distributional derivative $\partial_j\Phi$ may be computed as follows:

$$\partial_j\Phi = \Big(\int_{S^{n-1}} \Phi(\xi)\xi_j\,d\mathcal{H}^{n-1}(\xi)\Big)\delta + \text{P.V.}(\partial_j\Phi) \text{ in } \mathcal{S}'(\mathbb{R}^n), \tag{1.4.6}$$

where δ is Dirac's distribution in $\mathbb{R}^n$ with mass at the origin.

One remarkable aspect of the proposition below is that it provides a description of a fundamental solution for a second-order system purely in terms of ordinary calculus (i.e., without any reference to the theory of distributions).

Proposition 1.4.1 *Pick $n, M \in \mathbb{N}$ with $n \geq 2$, and consider the homogeneous, second-order, $M \times M$ system, with constant complex coefficients,*

$$L = \big(a^{\alpha\beta}_{rs}\partial_r\partial_s\big)_{1\leq\alpha,\beta\leq M} \tag{1.4.7}$$

(with the summation convention over repeated indices in effect). Also, consider a matrix-valued function $E : \mathbb{R}^n \setminus \{0\} \to \mathbb{C}^{M\times M}$ whose entries belong to the intersection $\mathscr{C}^2(\mathbb{R}^n\setminus\{0\})\cap L^{n/(n-1)}_{\text{loc}}(\mathbb{R}^n, \mathcal{L}^n)$ and have first-order derivatives positive homogeneous of degree $1-n$ in $\mathbb{R}^n \setminus \{0\}$.

Then the following statements are equivalent:

(1) When viewed in $\big[L^1_{\text{loc}}(\mathbb{R}^n, \mathcal{L}^n)\big]^{M\times M}$, the function E is a fundamental solution for the system L in $\mathbb{R}^n$.

(2) Each column of E is a (pointwise) null-solution of the system L in $\mathbb{R}^n \setminus \{0\}$, and

$$\int_{S^{n-1}} \xi_r a_{rs}^{\alpha\gamma}(\partial_s E_{\gamma\beta})(\xi)\,\mathrm{d}\mathcal{H}^{n-1}(\xi) = \delta_{\alpha\beta} \quad \text{for every } \alpha, \beta \in \{1, \dots, M\}. \tag{1.4.8}$$

Let us organize the complex numbers $a_{rs}^{\alpha\beta}$ used in the writing of L in (1.4.7) as a "tensor" of coefficients $A = \big(a_{rs}^{\alpha\beta}\big)_{\substack{1\le r,s\le n\\ 1\le\alpha,\beta\le M}}$. Then associate with coefficient tensor A the (co-)directional derivative operator along a vector $\xi = (\xi_1, \dots, \xi_n) \in \mathbb{R}^n$ by setting

$$\partial_\xi^A u := \Big(\xi_r a_{rs}^{\alpha\beta}\,\partial_s u_\beta\Big)_{1\le\alpha\le M}, \tag{1.4.9}$$

for every vector-valued function $u = (u_\beta)_{1\le\beta\le M}$. The upshot is that this piece of notation allows us to recast (1.4.8) succinctly as

$$\int_{S^{n-1}} \partial_\xi^A E(\xi)\,\mathrm{d}\mathcal{H}^{n-1}(\xi) = I_{M\times M}, \tag{1.4.10}$$

where the conormal derivative operator ∂_ξ^A acts on the columns of E.

Proof of Proposition 1.4.1 Start by writing $(E_{\gamma\beta})_{1\le\gamma,\beta\le M}$ for the scalar components of the given matrix-valued function E. Fix an arbitrary vector-valued test function $\varphi = (\varphi_\beta)_{1\le\beta\le M}$, with components in $\mathscr{C}_c^\infty(\mathbb{R}^n)$. Then, with L acting on the columns of E, making use of (1.4.6), [112, (4.5.8)], and (1.4.5) for each $\alpha \in \{1, \dots, M\}$ we may write

$$\begin{aligned}
\big((LE) * \varphi\big)_\alpha &= a_{rs}^{\alpha\gamma}\big(\partial_r(\partial_s E_{\gamma\beta})\big) * \varphi_\beta \\
&= a_{rs}^{\alpha\gamma}\Big(\int_{S^{n-1}} (\partial_s E_{\gamma\beta})(\xi)\xi_r\,\mathrm{d}\mathcal{H}^{n-1}(\xi)\Big)(\delta * \varphi_\beta) \\
&\quad + a_{rs}^{\alpha\gamma}\Big(\mathrm{P.V.}(\partial_r\partial_s E_{\gamma\beta})\Big) * \varphi_\beta \\
&= \Big(\int_{S^{n-1}} \xi_r a_{rs}^{\alpha\gamma}(\partial_s E_{\gamma\beta})(\xi)\,\mathrm{d}\mathcal{H}^{n-1}(\xi)\Big)\varphi_\beta \\
&\quad + \lim_{\varepsilon\to 0^+}\int_{|y-\cdot|>\varepsilon} a_{rs}^{\alpha\gamma}(\partial_r\partial_s E_{\gamma\beta})(\cdot - y)\varphi_\beta(y)\,\mathrm{d}y \ \text{ in } \mathcal{D}'(\mathbb{R}^n),
\end{aligned} \tag{1.4.11}$$

where we have also used the fact that $\delta * \varphi_\beta = \varphi_\beta$ for each $\beta \in \{1, \dots, M\}$. Having established this, we now turn in earnest to the proof of the equivalence claimed in the statement of proposition.

First, assume that E is a fundamental solution for L in $\mathbb{R}^n$. Then $LE = \delta I_{M\times M}$ in $\mathcal{D}'(\mathbb{R}^n)$ implies that $L\big(E\big|_{\mathbb{R}^n\setminus\{0\}}\big) = 0$ in $\mathcal{D}'(\mathbb{R}^n)$. Since by assumption the entries in $E\big|_{\mathbb{R}^n\setminus\{0\}}$ belong to $\mathscr{C}^2(\mathbb{R}^n \setminus \{0\})$, we arrive at the conclusion that $LE = 0$ pointwise in $\mathbb{R}^n \setminus \{0\}$. Explicitly, for each $\alpha, \beta \in \{1, \dots, M\}$,

$$a_{rs}^{\alpha\gamma}(\partial_r\partial_s E_{\gamma\beta})(x) = 0, \qquad \forall x \in \mathbb{R}^n \setminus \{0\}. \tag{1.4.12}$$

This proves the first claim in item *(2)*. Next, for each arbitrary $\mathbb{C}^M$-valued test function $\varphi = (\varphi_\beta)_{1\leq\beta\leq M}$, with components in the space $\mathscr{C}_c^\infty(\mathbb{R}^n)$, we may write $\varphi = (\delta I_{M\times M}) * \varphi = (LE) * \varphi$ in $\mathcal{D}'(\mathbb{R}^n)$. In light of (1.4.11) and (1.4.12), this forces

$$\varphi_\alpha = \Big(\int_{S^{n-1}} \xi_r a_{rs}^{\alpha\gamma}(\partial_s E_{\gamma\beta})(\xi)\,\mathrm{d}\mathcal{H}^{n-1}(\xi)\Big)\varphi_\beta, \quad \forall \alpha \in \{1,\dots,M\}. \tag{1.4.13}$$

Since $\varphi_\beta \in \mathscr{C}_c^\infty(\mathbb{R}^n)$ has been arbitrarily chosen for each $\beta \in \{1,\dots,M\}$, identity (1.4.8) follows from (1.4.13). This finishes the proof of *(1)*$\Rightarrow$*(2)*.

Conversely, suppose $LE = 0 \in \mathbb{C}^{M\times M}$ pointwise in $\mathbb{R}^n \setminus \{0\}$, and assume (1.4.8) holds. Granted these, formula (1.4.11) simply reduces to $(LE) * \varphi = \varphi$ for each $\varphi = (\varphi_\beta)_{1\leq\beta\leq M}$ with components in $\mathscr{C}_c^\infty(\mathbb{R}^n)$. In turn, this readily implies that $LE = \delta I_{M\times M}$ in $\mathcal{D}'(\mathbb{R}^n)$, as wanted. □

Our next theorem provides a canonical way of associating a suitable fundamental solution for any given second-order, homogeneous, weakly elliptic system, with constant (complex) complex coefficients. In various degrees of generality, this issue has been considered in a number of sources, including [74, pp. 72-76], [66, p. 169], [142, Proposition 5.2.3, p. 349], [122], [108], [139, p. 104], [161]. The theorem below is a special case of [109, Theorem 11.1, pp. 393-395], which treats weakly elliptic, homogeneous, constant coefficient systems of arbitrary (even) order. To state it, we make two conventions. The first convention regards the normalization of the Fourier transform in $\mathbb{R}^n$; the reader is alerted to the fact that the normalization we shall consistently employ throughout is *different* from those employed in [175], or [166]. Specifically, we define the Fourier transform $\widehat{\phi}$ of any given complex-valued Schwartz function ϕ in $\mathbb{R}^n$ as

$$\widehat{\phi}(\xi) := \int_{\mathbb{R}^n} \mathrm{e}^{-\mathrm{i}\langle x,\xi\rangle}\phi(x)\,\mathrm{d}x, \qquad \xi \in \mathbb{R}^n. \tag{1.4.14}$$

The action of the Fourier transform then extends to tempered distributions via duality, in the usual fashion (cf., e.g., [66], [108], [109]). The second convention is that, unless explicitly stated otherwise, the summation convention over repeated indices is in effect.

Theorem 1.4.2 *Fix $n, M \in \mathbb{N}$, with $n \geq 2$, and let*

$$L = \big(a_{rs}^{\alpha\beta}\partial_r\partial_s\big)_{1\leq\alpha,\beta\leq M} \tag{1.4.15}$$

be a second-order homogeneous $M \times M$ system in $\mathbb{R}^n$, with complex constant coefficients, which is weakly elliptic in the sense of Definition 1.3.1. That is, the characteristic matrix[17] *of L, given by*

$$L(\xi) := -\big(a_{rs}^{\alpha\beta}\xi_r\xi_s\big)_{1\leq\alpha,\beta\leq M}, \qquad \forall \xi = (\xi_r)_{1\leq r\leq n} \in \mathbb{R}^n, \tag{1.4.16}$$

[17] Once again, the reader is alerted to the fact that our present definition of $L(\xi)$ differs by a minus sign from what [109, (11.3.2), p. 391] would give for second-order systems

satisfies

$$\det\big[L(\xi)\big] \neq 0 \text{ for each } \xi \in \mathbb{R}^n \setminus \{0\}. \tag{1.4.17}$$

Consider the $M \times M$ *matrix-valued function* $E = \big(E_{\alpha\beta}\big)_{1 \le \alpha,\beta \le M}$ *defined at each* $x \in \mathbb{R}^n \setminus \{0\}$ *by*[18]

$$E(x) := \begin{cases} -\dfrac{\Delta_x^{(n-1)/2}}{4(2\pi \mathrm{i})^{n-1}} \left\{ \displaystyle\int_{S^{n-1}} |\langle x, \xi\rangle| \big[L(\xi)\big]^{-1} \, d\mathcal{H}^{n-1}(\xi) \right\} & \text{if } n \text{ is odd,} \\[2ex] \dfrac{\Delta_x^{(n-2)/2}}{(2\pi \mathrm{i})^{n}} \left\{ \displaystyle\int_{S^{n-1}} \ln|\langle x, \xi\rangle| \, \big[L(\xi)\big]^{-1} \, d\mathcal{H}^{n-1}(\xi) \right\} & \text{if } n \text{ is even.} \end{cases} \tag{1.4.18}$$

In relation to the $\mathbb{C}^{M\times M}$*-valued function* (1.4.18), *the following properties hold:*
(1) For each $\alpha, \beta \in \{1, \dots, M\}$ *one has*

$$\begin{aligned} &E_{\alpha\beta} \in \mathscr{C}^\infty(\mathbb{R}^n \setminus \{0\}) \cap L^1_{\mathrm{loc}}(\mathbb{R}^n, \mathcal{L}^n), \\ &E_{\alpha\beta}(-x) = E_{\alpha\beta}(x) \text{ for all } x \in \mathbb{R}^n \setminus \{0\}. \end{aligned} \tag{1.4.19}$$

In fact, each entry $E_{\alpha\beta}$ *is a real-analytic function in* $\mathbb{R}^n \setminus \{0\}$*. Moreover, each* $E_{\alpha\beta}$ *is an even tempered distribution in* $\mathbb{R}^n$ *(induced via integration against Schwartz functions).*

(2) If for each $y \in \mathbb{R}^n$ *one denotes by* δ_y *Dirac's delta distribution with mass at* y *in* $\mathbb{R}^n$*, then in the sense of tempered distributions in* $\mathbb{R}^n$ *one has*

$$L_x\big[E(x-y)\big] = \delta_y(x)\, I_{M\times M}, \qquad \forall y \in \mathbb{R}^n, \tag{1.4.20}$$

where $I_{M\times M}$ *is the* $M \times M$ *identity matrix, and the subscript* x *indicates that the operator* L *in* (1.4.20) *is applied to each column of the matrix* $E(x-y)$ *in the variable* x*. Componentwise, for each* $\alpha, \beta \in \{1, \dots, M\}$ *one therefore has*

$$a^{\alpha\gamma}_{rs} \partial_{x_r} \partial_{x_s} \big[E_{\gamma\beta}(x-y)\big] = \delta_{\alpha\beta}\delta_y(x) = \begin{cases} 0 & \text{if } \alpha \neq \beta, \\ \delta_y(x) & \text{if } \alpha = \beta, \end{cases} \qquad x, y \in \mathbb{R}^n. \tag{1.4.21}$$

(3) For each multi-index $\gamma \in \mathbb{N}_0^n$ *with* $|\gamma| > 0$*, the tempered distribution* $\partial^\gamma E$ *is positive homogeneous of degree* $2 - n - |\gamma|$ *in* $\mathbb{R}^n$*. This is also true for* $|\gamma| = 0$ *provided* $n \ge 3$*, i.e., the tempered distribution* E *is positive homogeneous of degree* $2 - n$ *in* $\mathbb{R}^n$ *if* $n \ge 3$*.*

Finally, corresponding to $n = 2$ *and* $|\gamma| = 0$*, one may express*

$$E(x) = \Phi(x) - \frac{\ln|x|}{4\pi^2} \int_{S^1} \big[L(\xi)\big]^{-1} \, d\mathcal{H}^1(\xi), \qquad \forall x \in \mathbb{R}^2 \setminus \{0\}, \tag{1.4.22}$$

[18] for each $m \in \mathbb{N}$, we let Δ_x^m denote the m-fold application of the Laplacian in the variable x

where $\Phi : \mathbb{R}^2 \setminus \{0\} \to \mathbb{C}^{M\times M}$*, given by*

$$\Phi(x) := -\frac{1}{4\pi^2}\int_{S^1} \ln\left|\left\langle \frac{x}{|x|}, \xi\right\rangle\right| \big[L(\xi)\big]^{-1}\, d\mathcal{H}^1(\xi), \qquad \forall x \in \mathbb{R}^2 \setminus \{0\}, \tag{1.4.23}$$

is a function of class $\mathscr{C}^\infty$ *and positive homogeneous of degree* 0 *in* $\mathbb{R}^2 \setminus \{0\}$.

(4) For each $\gamma \in \mathbb{N}_0^n$ *there exists a finite constant* $C_\gamma > 0$ *such that for each* $x \in \mathbb{R}^n \setminus \{0\}$ *one has*

$$\big|(\partial^\gamma E)(x)\big| \leq \begin{cases} \dfrac{C_\gamma}{|x|^{n+|\gamma|-2}} & \text{if either } n \geq 3, \text{ or } n = 2 \text{ and } |\gamma| > 0, \\ & \text{or } n = 2 \text{ and } \int_{S^1} \big[L(\xi)\big]^{-1}\, d\mathcal{H}^1(\xi) = 0, \\ C_0\big(1 + \big|\ln|x|\big|\big) & \text{whenever } n = 2 \text{ and } |\gamma| = 0. \end{cases} \tag{1.4.24}$$

(5) For each indexes $\alpha, \beta \in \{1, \ldots, M\}$ *one has*[19]

$$\int_{S^{n-1}} \xi_r a_{rs}^{\alpha\gamma} (\partial_s E_{\gamma\beta})(\xi)\, d\mathcal{H}^{n-1}(\xi) = \int_{S^{n-1}} \xi_r a_{sr}^{\gamma\alpha} (\partial_s E_{\beta\gamma})(\xi)\, d\mathcal{H}^{n-1}(\xi) = \delta_{\alpha\beta}. \tag{1.4.25}$$

In fact, given any vector $\omega \in S^{n-1}$*, for each* $\alpha, \beta \in \{1, \ldots, M\}$ *one has*

$$\int_{\substack{\xi\in S^{n-1}\\ \langle\xi,\omega\rangle>0}} \xi_r a_{rs}^{\alpha\gamma} (\partial_s E_{\gamma\beta})(\xi)\, d\mathcal{H}^{n-1}(\xi) = \int_{\substack{\xi\in S^{n-1}\\ \langle\xi,\omega\rangle>0}} \xi_r a_{sr}^{\gamma\alpha} (\partial_s E_{\beta\gamma})(\xi)\, d\mathcal{H}^{n-1}(\xi) = \frac{1}{2}\delta_{\alpha\beta}. \tag{1.4.26}$$

(6) Let 'hat' denote the Fourier transform in $\mathbb{R}^n$ *(originally defined on Schwartz functions as in* (1.4.14)*, then extended to tempered distributions via duality). Then* $\widehat{E}$ *is a tempered distribution in* $\mathbb{R}^n$ *(which is positive homogeneous of degree* -2 *if* $n \geq 3$*), whose restriction to* $\mathbb{R}^n \setminus \{0\}$ *is a (matrix-valued) function of class* $\mathscr{C}^\infty$*. In fact,*

[19] The format of the identities in (1.4.25) strongly depends on the choice of a coefficient tensor to represent the system L as in (1.3.1). Later on, in Definition 1.7.1, we shall recognize the integrands in (1.4.25) as being conormal derivatives of columns and rows of E on the surface S^{n-1}. Bearing this in mind, it follows from Proposition 1.7.4 that any two coefficient tensors used to write the system L as in (1.3.1) produce conormal derivatives which differ by a linear combination of tangential derivatives. Granted this, the independence of (1.4.25) on the choice of the coefficient tensor used to represent the original system L is then implied by the integration by parts on the boundary formula from [113, (11.1.7)].

$$\widehat{E}(\xi) = \big[L(\xi)\big]^{-1} \text{ for every } \xi \in \mathbb{R}^n \setminus \{0\}, \tag{1.4.27}$$

and

$$\text{if } n \geq 3 \text{ then } \widehat{E} = \big[L(\cdot)\big]^{-1} \text{ as (tempered) distributions in } \mathbb{R}^n, \tag{1.4.28}$$

which also implies

$$\text{if } n \geq 3 \text{ then } E = (2\pi)^{-n}\overline{\big[L(\cdot)\big]^{-1}} \text{ as (tempered) distributions in } \mathbb{R}^n. \tag{1.4.29}$$

More generally, given any $\gamma \in \mathbb{N}_0^n$ it follows that the tempered distribution $\widehat{\partial^\gamma E}$ is a function of class $\mathscr{C}^\infty$ when restricted to $\mathbb{R}^n \setminus \{0\}$ which, regarded as such, satisfies

$$\widehat{\partial^\gamma E}(\xi) = \mathrm{i}^{|\gamma|}\,\xi^\gamma\big[L(\xi)\big]^{-1} \text{ for every } \xi \in \mathbb{R}^n \setminus \{0\}, \tag{1.4.30}$$

and

$$\begin{array}{l}\text{if } \gamma \in \mathbb{N}_0^n \text{ then } \widehat{\partial^\gamma E} = \mathrm{i}^{|\gamma|}\,\xi^\gamma\big[L(\xi)\big]^{-1} \text{ as tempered}\\ \text{distributions in } \mathbb{R}^n \text{ when either } |\gamma| > 0 \text{ or } n \geq 3.\end{array} \tag{1.4.31}$$

(7) Writing E_L in place of E to emphasize the dependence on L, the fundamental solution E_L defined in (1.4.18) *satisfies*

$$\begin{array}{c}(E_L)^\top = E_{L^\top}, \quad \overline{(E_L)} = E_{\overline{L}}, \quad (E_L)^* = E_{L^*},\\ \text{as well as } E_{\lambda L} = \lambda^{-1}E_L \text{ for each } \lambda \in \mathbb{C} \setminus \{0\}.\end{array} \tag{1.4.32}$$

where $\top$, $\bar{\cdot}$, $*$ denote, respectively, transposition, complex conjugation, and complex (or Hermitian) adjunction. In concert with (1.4.21)*, the first formula in* (1.4.32) *also implies that for each indexes $\alpha, \beta \in \{1, \dots, M\}$ one has*

$$a_{rs}^{\gamma\alpha}\partial_{x_r}\partial_{x_s}\big[E_{\beta\gamma}(x-y)\big] = \delta_{\alpha\beta}\delta_y(x) = \begin{cases} 0 & \text{if } \alpha \neq \beta,\\ \delta_y(x) & \text{if } \alpha = \beta,\end{cases} \qquad x, y \in \mathbb{R}^n. \tag{1.4.33}$$

A more general version of the last formula in (1.4.32)*, which is visible from* (1.4.18)*, is that for each invertible matrix $C \in \mathbb{C}^{M\times M}$ one has*[20]

$$E_{LC} = C^{-1}E_L \text{ and } E_{CL} = E_L C^{-1}. \tag{1.4.34}$$

Finally, pick an arbitrary unitary transformation R in $\mathbb{R}^n$ (i.e., a matrix $R \in \mathbb{R}^{n\times n}$ with the property that $R^\top R = I_{n\times n}$), and consider the second-order homogeneous complex constant coefficient $M \times M$ system in $\mathbb{R}^n$ formally written as

$$L \circ R := \Big(a_{rs}^{\alpha\beta}(R\nabla)_r(R\nabla)_s\Big)_{1\leq\alpha,\beta\leq M}. \tag{1.4.35}$$

Then the characteristic matrix of $L \circ R$ satisfies

---

[20] with the systems LC and CL interpreted in the sense of multiplication of $M \times M$ matrices

$$(L \circ R)(\xi) = L(R\xi) \text{ for each } \xi \in \mathbb{R}^n, \tag{1.4.36}$$

so the system $L \circ R$ *is weakly elliptic, and from* (1.4.18), *the invariance of integration on* S^{n-1} *under unitary transformations (cf., e.g.,* [109, (14.9.11), p. 578]), *and the invariance of the action of the Laplacian under unitary transformations (cf., e.g.,* [109, Exercise 7.77, p. 319]) *one obtains*

$$E_{L \circ R} = E_L \circ R \text{ in } \mathbb{R}^n \setminus \{0\}. \tag{1.4.37}$$

(8) Any $M \times M$ *matrix with tempered distribution entries* $\widetilde{E}$ *which is a fundamental solution of the system* L *in* $\mathbb{R}^n$ *is of the form* $\widetilde{E} = E + Q$ *where* E *is as in* (1.4.18) *and* Q *is an* $M \times M$ *matrix whose entries are polynomials in* $\mathbb{R}^n$ *and whose columns,* Q_k *with* $k \in \{1, \ldots, M\}$, *satisfy the pointwise equations* $LQ_k = 0 \in \mathbb{C}^M$ *in* $\mathbb{R}^n$ *for each* $k \in \{1, \ldots, M\}$.

We wish to make five comments in relation to Theorem 1.4.2. Our first comment concerns the proof of the integral identities recorded in item *(5)*, since these are not explicitly stated in [109, Theorem 11.1, pp. 393-395]. Specifically, having fixed two arbitrary indexes $\alpha, \beta \in \{1, \ldots, M\}$, consider the vector field

$$\vec{F} := \big(a_{rs}^{\alpha\gamma} \partial_s E_{\gamma\beta}\big)_{1 \le r \le n} \in \big[L^1_{\mathrm{loc}}\big(B(0,1), \mathcal{L}^n\big) \cap \mathscr{C}^\infty\big(\overline{B(0,1)} \setminus \{0\}\big)\big]^n, \tag{1.4.38}$$

and notice that, thanks to (1.4.21), we have

$$\operatorname{div}\vec{F} = a_{rs}^{\alpha\gamma} \partial_r \partial_s E_{\gamma\beta} = \delta_{\alpha\beta}\delta_0 \in \mathscr{E}'\big(B(0,1)\big). \tag{1.4.39}$$

Consequently, the Divergence Formula in the version stated in [112, (1.4.17)] applies and presently gives

$$\begin{aligned} \int_{S^{n-1}} \xi_r a_{rs}^{\alpha\gamma} (\partial_s E_{\gamma\beta})(\xi) \, d\mathcal{H}^{n-1}(\xi) &= \int_{S^{n-1}} \xi \cdot \vec{F}(\xi) \, d\mathcal{H}^{n-1}(\xi) \\ &= {}_{\mathscr{E}'(B(0,1))}\big\langle \operatorname{div}\vec{F}, 1 \big\rangle_{\mathscr{E}(B(0,1))} \\ &= {}_{\mathscr{E}'(B(0,1))}\big\langle \delta_{\alpha\beta}\delta_0, 1 \big\rangle_{\mathscr{E}(B(0,1))} \\ &= \delta_{\alpha\beta}. \end{aligned} \tag{1.4.40}$$

Likewise, since the vector field

$$\vec{G} := \big(a_{sr}^{\gamma\alpha} \partial_s E_{\beta\gamma}\big)_{1 \le r \le n} \in \big[L^1_{\mathrm{loc}}\big(B(0,1), \mathcal{L}^n\big) \cap \mathscr{C}^\infty\big(\overline{B(0,1)} \setminus \{0\}\big)\big]^n \tag{1.4.41}$$

satisfies (cf. (1.4.33))

$$\operatorname{div}\vec{G} = a_{sr}^{\gamma\alpha} \partial_r \partial_s E_{\beta\gamma} = \delta_{\alpha\beta}\delta_0 \in \mathscr{E}'\big(B(0,1)\big), \tag{1.4.42}$$

we may once again invoke the Divergence Formula in the version stated in [112, (1.4.17)] to conclude that

$$\begin{aligned}\int_{S^{n-1}} \xi_r a^{\gamma\alpha}_{sr}(\partial_s E_{\beta\gamma})(\xi)\,d\mathcal{H}^{n-1}(\xi) &= \int_{S^{n-1}} \xi\cdot\vec{G}(\xi)\,d\mathcal{H}^{n-1}(\xi)\\ &= {}_{\mathscr{E}'(B(0,1))}\big\langle \mathrm{div}\vec{G}, 1\big\rangle_{\mathscr{E}(B(0,1))}\\ &= {}_{\mathscr{E}'(B(0,1))}\big\langle \delta_{\alpha\beta}\delta_0, 1\big\rangle_{\mathscr{E}(B(0,1))}\\ &= \delta_{\alpha\beta}. \end{aligned}\tag{1.4.43}$$

Together, (1.4.40) and (1.4.43) prove (1.4.25). Once this has been established, the formulas claimed in (1.4.26) become consequences of (1.4.25) taking into account the fact that, as seen from (1.4.19), the integrands are even functions.

Our second comment is that for each $\varepsilon > 0$ we have

$$E \in \bigcap_{0<\alpha<1}\Big[\dot{\mathscr{C}}^{\alpha}\big(\mathbb{R}^n\setminus B(0,\varepsilon)\big)\Big]^{M\times M}. \tag{1.4.44}$$

Indeed, when $n \geq 3$, this is a consequence of the fact that, as seen from (1.4.24), the matrix-valued function E is both bounded and Lipschitz in $\mathbb{R}^n \setminus B(0,\varepsilon)$. When $n = 2$, the function Φ continues to be both bounded and Lipschitz in $\mathbb{R}^n \setminus B(0,\varepsilon)$ (since it is smooth and positive homogeneous of degree 0 in $\mathbb{R}^2 \setminus \{0\}$), but this is not the case for the logarithmic term in (1.4.22). This being said, for each given pair of points $x, y \in \mathbb{R}^n \setminus B(0,\varepsilon)$ with $|x| > |y|$ we have

$$\begin{aligned}\big|\ln|x| - \ln|y|\big| &= \Big|\ln\Big(\frac{|x|}{|y|}\Big)\Big| = \ln\Big(1+\frac{|x|-|y|}{|y|}\Big)\\ &\leq C_\alpha\Big(\frac{|x|-|y|}{|y|}\Big)^{\alpha} \leq C_\alpha\varepsilon^{-\alpha}|x-y|^{\alpha}.\end{aligned}\tag{1.4.45}$$

Above, the first inequality is a consequence of the elementary estimate

$$\ln(1+t) \leq C_\alpha t^{\alpha} \ \text{ for each } t\in(0,\infty) \text{ and } \alpha\in(0,1), \tag{1.4.46}$$

and the final inequality in (1.4.45) is implied by the triangle inequality. Now, (1.4.44) readily follows from (1.4.45) and (1.4.22).

Our third comment is that

if $n = 2$ then, for any given vector $c \in \mathbb{C}^M$, the limit $\lim\limits_{x\to\infty} E(x)c$ exists (in $\mathbb{C}^M$) if and only if $c = 0 \in \mathbb{C}^M$, (1.4.47)

and

if $n = 2$ then, for any given vector $c \in \mathbb{C}^M$, having $E(x)c = o(\ln|x|)$ as $|x| \to \infty$ is equivalent to having $E(x)c = O(1)$ as $|x| \to \infty$, which is further equivalent to having $\Big(\int_{S^1}\big[L(\xi)\big]^{-1}\,d\mathcal{H}^1(\xi)\Big)c = 0$. (1.4.48)

To justify (1.4.47), assume $n = 2$ and $c \in \mathbb{C}^M$ is such that $a := \lim_{x\to\infty} E(x)c$ exists. Then (1.4.22) implies that

$$a = \lim_{x\to\infty} \left\{ \Phi(x)c - \frac{\ln|x|}{4\pi^2}\Big(\int_{S^1} \big[L(\xi)\big]^{-1}\, d\mathcal{H}^1(\xi)\Big)c \right\}$$

$$= \lim_{x\to\infty} \ln|x| \left\{ \frac{\Phi(x)c}{\ln|x|} - \frac{1}{4\pi^2}\Big(\int_{S^1} \big[L(\xi)\big]^{-1}\, d\mathcal{H}^1(\xi)\Big)c \right\}. \tag{1.4.49}$$

Since $\lim_{x\to\infty} \frac{\Phi(x)c}{\ln|x|} = 0$ given that the matrix-valued function Φ is continuous and positive homogeneous of degree 0 in $\mathbb{R}^2 \setminus \{0\}$ (hence bounded), we see that (1.4.49) forces

$$\Big(\int_{S^1} \big[L(\xi)\big]^{-1}\, d\mathcal{H}^1(\xi)\Big)c = 0. \tag{1.4.50}$$

Re-writing the first equality in (1.4.49) in light of this piece of information leads to the conclusion that

$$a = \lim_{x\to\infty} \big(\Phi(x)c\big). \tag{1.4.51}$$

In particular, for each fixed $x \in \mathbb{R}^2 \setminus \{0\}$ we have, using again the fact that Φ is positive homogeneous of degree 0 in $\mathbb{R}^2 \setminus \{0\}$,

$$\Phi(x)c = \lim_{t\to\infty} \big(\Phi(tx)c\big) = a. \tag{1.4.52}$$

This goes to show that $\Phi(x)c = a$ for each $x \in \mathbb{R}^2 \setminus \{0\}$, hence $E(x)c = a$ for each $x \in \mathbb{R}^2 \setminus \{0\}$, thanks to this, (1.4.22), and (1.4.50). In particular, $E(\cdot)c$ regarded as a locally integrable vector-valued function in $\mathbb{R}^2$ is actually constant. Applying the system L and recalling (1.4.20) then yields $\delta c = 0$ in the sense of distributions, which ultimately shows that $c = 0 \in \mathbb{C}^M$, as wanted. This establishes the direct implication in (1.4.47), and the opposite implication is trivial. The justification of (1.4.48) uses similar ideas, the crux of the matter being the sequence of implications (regarding the behavior as $|x| \to \infty$)

$$E(x)c = o(\ln|x|) \Longrightarrow \Big(\int_{S^1} \big[L(\xi)\big]^{-1}\, d\mathcal{H}^1(\xi)\Big)c = 0$$

$$\Longrightarrow E(x)c = \Phi(x)c = O(1). \tag{1.4.53}$$

Our fourth comment is that, in the two-dimensional setting, it is clear from (1.4.22) that

$$E \text{ is positive homogeneous of degree 0 in } \mathbb{R}^2 \setminus \{0\}$$
$$\text{if and only if } \int_{S^1} \big[L(\xi)\big]^{-1}\, d\mathcal{H}^1(\xi) = 0. \tag{1.4.54}$$

Our fifth (and final) comment is that a suitable version of (1.4.29) continues to be true in the case when $n = 2$. This requires extending the matrix-valued function $[L(\cdot)]^{-1} \in \big[\mathscr{C}^\infty(\mathbb{R}^2 \setminus \{0\})\big]^{M\times M}$, which is homogeneous of degree -2, to a matrix-valued tempered distribution in $\mathbb{R}^2$, through the consideration of its finite part. See, e.g., [103, Theorem 6.8, p. 198] for details in the scalar case.

Moving on, we wish to further elaborate on the manner in which a linear change of variables in the Euclidean space affects a given weakly elliptic system and its fundamental solution constructed as in Theorem 1.4.2.

Proposition 1.4.3 *Assume that two integers $n, M \in \mathbb{N}$, with $n \geq 2$, have been fixed. Given a (complex) coefficient tensor $A = \big(a_{jk}^{\alpha\beta}\big)_{\substack{1\leq j,k\leq n\\ 1\leq\alpha,\beta\leq M}}$, associate the second-order, homogeneous, complex constant coefficient, $M \times M$ system in $\mathbb{R}^n$*

$$L = L_A := \big(a_{rs}^{\alpha\beta}\partial_r\partial_s\big)_{1\leq\alpha,\beta\leq M}. \tag{1.4.55}$$

Also, for any matrix $W = (w_{jk})_{1\leq j,k\leq n} \in \mathbb{C}^{n\times n}$ define two new coefficient tensors, namely

$$W \circ A := \big(w_{j\ell}a_{\ell k}^{\alpha\beta}\big)_{\substack{1\leq j,k\leq n\\ 1\leq\alpha,\beta\leq M}} \tag{1.4.56}$$

and

$$A \circ W := \big(a_{j\ell}^{\alpha\beta}w_{\ell k}\big)_{\substack{1\leq j,k\leq n\\ 1\leq\alpha,\beta\leq M}}. \tag{1.4.57}$$

In addition, generalize (1.4.35) *by defining the second-order, homogeneous, complex constant coefficient, $M \times M$ system in $\mathbb{R}^n$, formally written as*

$$L \circ W := \Big(a_{jk}^{\alpha\beta}(W\nabla)_j(W\nabla)_k\Big)_{1\leq\alpha,\beta\leq M}. \tag{1.4.58}$$

Then, in relation to these, the following statements are true:

(i) For any matrix $W \in \mathbb{R}^{n\times n}$ one has

$$(L_A) \circ W = L_{W^\top\circ A\circ W} \tag{1.4.59}$$

and

$$L_{W^\top\circ A\circ W}(\xi) = L_A(W\xi) \text{ for all } \xi \in \mathbb{R}^n. \tag{1.4.60}$$

In particular,

$$\begin{gathered}\text{if the real } n\times n \text{ matrix } W \text{ is invertible then}\\ L_{W^\top\circ A\circ W} \text{ is weakly elliptic if and only if } L_A \text{ is weakly elliptic.}\end{gathered} \tag{1.4.61}$$

Also,

$$\begin{gathered}\text{if the real } n\times n \text{ matrix } W \text{ is invertible then}\\ L_{W^\top\circ A\circ W} \text{ is Legendre-Hadamard (strongly) elliptic}\\ \text{if and only if } L_A \text{ is Legendre-Hadamard (strongly) elliptic.}\end{gathered} \tag{1.4.62}$$

(ii) For any invertible matrix $W \in \mathbb{R}^{n\times n}$ one has

$$\begin{gathered}\big(L_A \circ (W^{-1})^\top\big)(u\circ W) = L_{W^{-1}\circ A\circ (W^{-1})^\top}(u\circ W) = (L_A u)\circ W\\ \textit{for any vector distribution } u \in \big[\mathcal{D}'(\mathbb{R}^n)\big]^M,\end{gathered} \tag{1.4.63}$$

where the composition of a distribution with an invertible real matrix is understood in the usual sense[21] *(cf., e.g.,* [109, Proposition 4.33, p. 140]*).*

(iii) If L_A is weakly elliptic then for any invertible matrix $W \in \mathbb{R}^{n\times n}$ one has

$$\begin{gathered}E_{L\circ W} = |\det W|^{-1} E_L \circ (W^{-1})^\top \ \textit{ if } n \geq 3,\\ \textit{as (matrix) tempered distributions,}\end{gathered} \tag{1.4.64}$$

and

$$\begin{gathered}\textit{the (matrix) tempered distributions}\\ E_{L\circ W} \textit{ and } |\det W|^{-1} E_L \circ (W^{-1})^\top\\ \textit{differ by a constant (in } \mathbb{C}^{M\times M}) \textit{ if } n = 2.\end{gathered} \tag{1.4.65}$$

Proof To justify (1.4.59), use (1.4.56)-(1.4.58) to write

$$\begin{aligned}(L_A)\circ W &= \Big(a^{\alpha\beta}_{jk}(W\nabla)_j(W\nabla)_k\Big)_{1\leq\alpha,\beta\leq M}\\ &= \Big(a^{\alpha\beta}_{jk} w_{jr}\partial_r w_{ks}\partial_s\Big)_{1\leq\alpha,\beta\leq M} = L_{W^\top\circ A\circ W},\end{aligned} \tag{1.4.66}$$

as wanted. Next, for each $\xi = (\xi_1,\dots,\xi_n) \in \mathbb{R}^n$ we have

$$\begin{aligned}L_{W^\top\circ A\circ W}(\xi) &= -\Big(a^{\alpha\beta}_{jk} w_{jr}\xi_r w_{ks}\xi_s\Big)_{1\leq\alpha,\beta\leq M}\\ &= -\Big(a^{\alpha\beta}_{jk}(W\xi)_j(W\xi)_k\Big)_{1\leq\alpha,\beta\leq M} = L_A(W\xi),\end{aligned} \tag{1.4.67}$$

proving (1.4.60). The claim in (1.4.61) is then a consequence of (1.4.60) and Definition 1.3.1. Also, if L_A is Legendre-Hadamard (strongly) elliptic (cf. Definition 1.3.2) then for all $\xi \in \mathbb{R}^n$ and $\zeta \in \mathbb{C}^M$ we may write (bearing in mind that W is invertible)

$$\begin{aligned}\operatorname{Re}\big\langle -L_{W^\top\circ A\circ W}(\xi)\zeta,\overline{\zeta}\big\rangle &= \operatorname{Re}\big\langle -L_A(W\xi)\zeta,\overline{\zeta}\big\rangle\\ &\geq c\,|W\xi|^2\,|\zeta|^2 \geq \widetilde{c}\,|\xi|^2\,|\zeta|^2\end{aligned} \tag{1.4.68}$$

for some $c, \widetilde{c} \in (0,\infty)$ independent of ξ and ζ. Ultimately, this establishes (1.4.62).

Going further, the first equality in (1.4.63) is seen directly from (1.4.59). To show the second equality in (1.4.63), assume first that $u \in \big[\mathscr{C}^\infty(\mathbb{R}^n)\big]^M$. Then

---

[21] i.e., $\langle u\circ W, \varphi\rangle := |\det W|^{-1}\langle u, \varphi\circ W^{-1}\rangle$ for each test function φ in $\mathbb{R}^n$

$$
\begin{aligned}
L_{W^{-1}\circ A\circ(W^{-1})^{\top}}(u\circ W) &= \Big(a_{jk}^{\alpha\beta}(W^{-1})_{rj}\big((W^{-1})^{\top}\big)_{ks}\partial_r\partial_s(u_\beta\circ W)\Big)_{1\le\alpha\le M}\\
&= \Big(a_{jk}^{\alpha\beta}(W^{-1})_{rj}(W^{-1})_{sk}\big((\partial_\ell\partial_m u_\beta)\circ W\big)w_{\ell r}w_{ms}\Big)_{1\le\alpha\le M}\\
&= \Big(a_{jk}^{\alpha\beta}\delta_{\ell j}\delta_{mk}\big((\partial_\ell\partial_m u_\beta)\circ W\big)\Big)_{1\le\alpha\le M}\\
&= \Big(a_{jk}^{\alpha\beta}\partial_j\partial_k u_\beta\Big)_{1\le\alpha\le M}\circ W=(L_A u)\circ W,
\end{aligned}
\tag{1.4.69}
$$

by Chain Rule and definitions. Having proved (1.4.69), the second equality in (1.4.63) follows from the sequential density of $\mathscr{C}_c^\infty(\mathbb{R}^n)$ in $\mathcal{D}'(\mathbb{R}^n)$ (cf., e.g., [109, Theorem 2.106, p. 75]) and the continuity of partial differentiation on the space of distributions.

Consider now the claims made in item *(iii)*. In all dimensions $n\ge 2$ we may compute, in the sense of distributions in $\mathbb{R}^n$,

$$
\begin{aligned}
(L\circ W)\big(E_L\circ(W^{-1})^{\top}\big) &= L_{W^{\top}\circ A\circ W}\big(E_L\circ(W^{-1})^{\top}\big)=(LE_L)\circ(W^{-1})^{\top}\\
&= \big(\delta\circ(W^{-1})^{\top}\big)I_{M\times M}=|\det W|\delta\, I_{M\times M},
\end{aligned}
\tag{1.4.70}
$$

with the first equality coming from (1.4.59), the second equality implied by (1.4.63), the third equality a consequence of (1.4.20), and the last equality seen from [109, Proposition 4.33, p. 140]. This shows that in all dimensions $n\ge 2$ the tempered distribution $|\det W|^{-1}E_L\circ(W^{-1})^{\top}$ is a fundamental solution for the weakly elliptic operator $L\circ W$. Based on this and items *(8)* and *(4)* in Theorem 1.4.2, we then conclude that the claims made in (1.4.64)-(1.4.65) are true. □

It is also of interest to specialize Theorem 1.4.2 to the case of the complex Lamé system $L_{\lambda,\mu}$ in the regime of Lamé molduli λ, μ, ensuring its weak ellipticity.

Proposition 1.4.4 *Let $L_{\lambda,\mu}$ be the complex Lamé system* (1.3.6) *associated to two Lamé moduli $\mu,\lambda\in\mathbb{C}$ satisfying*

$$
\mu\neq 0 \text{ and } \lambda+2\mu\neq 0 \tag{1.4.71}
$$

(thus ensuring the weak ellipticity of $L_{\lambda,\mu}$; cf. (1.3.9)*). Then the fundamental solution E of $L_{\lambda,\mu}$ from* (1.4.18) *takes the explicit form $E=(E_{jk})_{1\le j,k\le n}$, a matrix whose (j,k) entry is defined at each point $x=(x_1,\dots,x_n)\in\mathbb{R}^n\setminus\{0\}$ according to*[22]

[22] at least in lower dimensions and for real Lamé moduli, this goes back to the work of Lord Kelvin (aka William Thomson)

$$E_{jk}(x) = \begin{cases} \dfrac{-1}{2\mu(2\mu+\lambda)\omega_{n-1}}\left[\dfrac{(3\mu+\lambda)\delta_{jk}}{(n-2)|x|^{n-2}} + (\mu+\lambda)\dfrac{x_j x_k}{|x|^n}\right] & \textit{if } n \geq 3, \\[2ex] \dfrac{1}{4\pi\mu(2\mu+\lambda)}\left[(3\mu+\lambda)\delta_{jk}\ln|x| - (\mu+\lambda)\dfrac{x_j x_k}{|x|^2}\right] + c_{\mu,\lambda}\delta_{jk} & \textit{if } n = 2, \end{cases} \tag{1.4.72}$$

for every $j, k \in \{1, \dots, n\}$, *where* $c_{\mu,\lambda} \in \mathbb{C}$ *is the constant given by*

$$c_{\mu,\lambda} := \frac{(1+\ln 4)(\mu+\lambda)}{8\pi\mu(2\mu+\lambda)} - \frac{\ln 2}{2\pi\mu} = -\frac{(1+\ln 4)(3\mu+\lambda)}{8\pi\mu(2\mu+\lambda)} + \frac{1}{4\pi\mu}. \tag{1.4.73}$$

Proof Assume first that $n \geq 3$. Then the $n \times n$ matrix whose entries are as in the top line of (1.4.72) is known to be a matrix-valued tempered distribution which is a fundamental solution for the Lamé system (1.3.6) in $\mathbb{R}^n$ (see, e.g., [108, Theorem 10.15 on p. 323, Exercise 10.32 on p. 339], [87, (9.2)], [86, Chapter 9]). Denote by Q the difference between this fundamental solution and what E from (1.4.18) becomes when $L := L_{\lambda,\mu}$. Then the uniqueness result in item *(8)* of Theorem 1.4.2 ensures that Q is a polynomial in $\mathbb{R}^n$. In view of (1.4.24) (with $\gamma = 0$), the fact that we are presently assuming $n \geq 3$, and the first line in (1.4.72), said polynomial decays at infinity, hence actually $Q = 0$. The desired conclusion is therefore established when $n \geq 3$.

To treat the two-dimensional case, fix $\lambda, \mu \in \mathbb{C}$ such that $\mu \neq 0$ and $\lambda + 2\mu \neq 0$. Corresponding to $n = 2$, in light of (1.3.10) formula (1.4.18) becomes

$$\begin{aligned} E(x) = \frac{1}{4\pi^2\mu}\left\{\int_{S^1} \ln\big|\langle x, \xi\rangle\big|\, d\mathcal{H}^1(\xi)\right\} I_{2\times 2} \\ - \frac{\lambda+\mu}{4\pi^2\mu(\lambda+2\mu)}\int_{S^1} \big(\ln\big|\langle x, \xi\rangle\big|\big)\xi \otimes \xi\, d\mathcal{H}^1(\xi), \end{aligned} \tag{1.4.74}$$

for each $x \in \mathbb{R}^2 \setminus \{0\}$. Using the rotation invariance of integrals over the unit sphere, for each fixed $x \in \mathbb{R}^2 \setminus \{0\}$ we may compute

$$\begin{aligned} \int_{S^1} \ln\big|\langle x, \xi\rangle\big|\, d\mathcal{H}^1(\xi) &= \int_{S^1} \ln\big|\langle |x|\mathbf{e}_1, \xi\rangle\big|\, d\mathcal{H}^1(\xi) \\ &= 2\pi \ln|x| + \int_{S^1} \ln|\xi_1|\, d\mathcal{H}^1(\xi) \\ &= 2\pi \ln|x| + \int_0^{2\pi} \ln|\cos\theta|\, d\theta \\ &= 2\pi \ln|x| - 2\pi \ln 2, \end{aligned} \tag{1.4.75}$$

where the last equality uses the fact that

$$\int_0^{2\pi} \ln|\cos\theta|\, d\theta = -2\pi\ln 2. \tag{1.4.76}$$

We shall establish this identity later, in the very last portion of the proof, where we will also show that for each fixed $x \in \mathbb{R}^2 \setminus \{0\}$ we have

$$\int_{S^1} \big(\ln|\langle x,\xi\rangle|\big)\xi\otimes\xi\, d\mathcal{H}^1(\xi) = \Big\{-\frac{\pi}{2} - \pi\ln 2 + \pi\ln|x|\Big\}I_{2\times 2} + \pi\frac{x\otimes x}{|x|^2}. \tag{1.4.77}$$

Granted these, it is apparent from (1.4.75)-(1.4.77) that formula (1.4.74) reduces precisely to the $n = 2$ version of (1.4.72) with $c_{\mu,\lambda} \in \mathbb{C}$ as in (1.4.73).

At this stage, there remains to prove (1.4.76) and (1.4.77). As regards the first identity, we note that breaking up the interval of integration at π, then making a natural change of variables in the integral over $[\pi, 2\pi]$ yields

$$\int_0^{2\pi} \ln|\cos\theta|\, d\theta = 2\int_0^{\pi} \ln|\cos\theta|\, d\theta = \int_0^{\pi} \ln\big(\cos^2\theta\big)\, d\theta. \tag{1.4.78}$$

To proceed, we compute

$$\begin{aligned}
\int_0^{\pi} \ln\big(\sin^2\theta\big)\, d\theta &= \int_0^{\pi/2} \ln\big(\sin^2\theta\big)\, d\theta + \int_{\pi/2}^{\pi} \ln\big(\sin^2\theta\big)\, d\theta \\
&= \int_0^{\pi/2} \ln\big(\sin^2\theta\big)\, d\theta + \int_0^{\pi/2} \ln\big(\cos^2\theta\big)\, d\theta \\
&= \int_0^{\pi/2} \ln\Big[\big(2^{-1}\sin(2\theta)\big)^2\Big]\, d\theta \\
&= -\pi\ln 2 + \int_0^{\pi/2} \ln\big(\sin^2(2\theta)\big)\, d\theta \\
&= -\pi\ln 2 + \frac{1}{2}\int_0^{\pi} \ln\big(\sin^2\theta\big)\, d\theta,
\end{aligned} \tag{1.4.79}$$

which implies

$$\int_0^{\pi} \ln\big(\sin^2\theta\big)\, d\theta = -2\pi\ln 2. \tag{1.4.80}$$

A similar computation further gives

$$
\begin{aligned}
\int_0^{\pi} \ln\big(\cos^2\theta\big)\,\mathrm{d}\theta &= \int_0^{\pi/2} \ln\big(\cos^2\theta\big)\,\mathrm{d}\theta + \int_0^{\pi/2} \ln\big(\sin^2\theta\big)\,\mathrm{d}\theta \\
&= \int_0^{\pi/2} \ln\Big[\big(2^{-1}\sin(2\theta)\big)^2\Big]\,\mathrm{d}\theta \\
&= -\pi\ln 2 + \int_0^{\pi/2} \ln\big(\sin^2(2\theta)\big)\,\mathrm{d}\theta \\
&= -\pi\ln 2 + \frac{1}{2}\int_0^{\pi} \ln\big(\sin^2\theta\big)\,\mathrm{d}\theta = -2\pi\ln 2, \qquad (1.4.81)
\end{aligned}
$$

where the last step uses (1.4.80). Then (1.4.76) follows from (1.4.78) and (1.4.81).

Turning our attention to (1.4.77), we first note that a variant of [108, Proposition 13.47, p. 440] gives that if $x \in \mathbb{R}^n \setminus \{0\}$ is fixed, then for each $j, k \in \{1, \dots, n\}$ one has

$$
\int_{S^{n-1}} \big(\ln|\langle x, \xi\rangle|\big)\xi_j\xi_k\,\mathrm{d}\mathcal{H}^{n-1}(\xi) = a_n\delta_{jk} + b_n\frac{x_j x_k}{|x|^2} + \frac{\omega_{n-1}}{n}\ln|x|\delta_{jk}, \qquad (1.4.82)
$$

where

$$
\begin{aligned}
a_n &:= \frac{1}{n-1}\int_{S^{n-1}} \big(\ln|\xi_1|\big)(1-\xi_1^2)\,\mathrm{d}\mathcal{H}^{n-1}(\xi), \\
b_n &:= \frac{1}{n-1}\int_{S^{n-1}} \big(\ln|\xi_1|\big)(n\xi_1^2 - 1)\,\mathrm{d}\mathcal{H}^{n-1}(\xi).
\end{aligned}
\qquad (1.4.83)
$$

Specializing this to the case $n = 2$ gives that for each $x \in \mathbb{R}^2 \setminus \{0\}$ we have

$$
\int_{S^1} \big(\ln\big|\langle x, \xi\rangle\big|\big)\xi \otimes \xi\,\mathrm{d}\mathcal{H}^1(\xi) = \Big\{a_2 + \pi\ln|x|\Big\}I_{2\times 2} + b_2\frac{x \otimes x}{|x|^2}, \qquad (1.4.84)
$$

and we claim that

$$
a_2 = -\frac{\pi}{2} - \pi\ln 2 \ \text{ and } \ b_2 = \pi. \qquad (1.4.85)
$$

In this regard, we first note that

$$
a_2 = \frac{1}{2}\int_0^{2\pi} (\sin^2\theta)\ln\big(\cos^2\theta\big)\,\mathrm{d}\theta = \int_0^{\pi} (\sin^2\theta)\ln\big(\cos^2\theta\big)\,\mathrm{d}\theta, \qquad (1.4.86)
$$

and we find it useful to also consider

$$
c_2 := \int_0^{\pi} (\cos^2\theta)\ln\big(\cos^2\theta\big)\,\mathrm{d}\theta. \qquad (1.4.87)
$$

Then an integration by parts permits us to compute

$$c_2 = \int_0^{\pi} (\sin\theta)'(\cos\theta)\ln\big(\cos^2\theta\big)\,d\theta$$

$$= \Big[(\sin\theta)(\cos\theta)\ln\big(\cos^2\theta\big)\Big]\Big|_0^{\pi} + a_2 + 2\int_0^{\pi}\sin^2\theta\,d\theta, \tag{1.4.88}$$

hence $c_2 = a_2 + \pi$. Since, by (1.4.86), (1.4.87), and (1.4.81) we also have,

$$a_2 + c_2 = \int_0^{\pi} \ln\big(\cos^2\theta\big)\,d\theta = -2\pi\ln 2, \tag{1.4.89}$$

we ultimately conclude that

$$a_2 = -\frac{\pi}{2} - \pi\ln 2 \ \text{ and } \ c_2 = \frac{\pi}{2} - \pi\ln 2. \tag{1.4.90}$$

Consequently, from (1.4.83) with $n = 2$, (1.4.87), (1.4.90), and (1.4.81), we deduce that

$$b_2 = \frac{1}{2}\int_0^{2\pi} \ln\big(\cos^2\theta\big)(2\cos^2\theta - 1)\,d\theta = \int_0^{\pi} \ln\big(\cos^2\theta\big)(2\cos^2\theta - 1)\,d\theta$$

$$= 2c_2 - \int_0^{\pi} \ln\big(\cos^2\theta\big)\,d\theta = (\pi - 2\pi\ln 2) + 2\pi\ln 2 = \pi. \tag{1.4.91}$$

Together with (1.4.90), this finishes the proof of (1.4.85).

Having established this, formula (1.4.77) follows by combining (1.4.84) with (1.4.85). The proof of Proposition 1.4.4 is therefore complete. □

Here is what Theorem 1.4.2 yields when specialized to the case of the operator $\mathrm{Def}^{\top}\mathrm{Def}$.

Corollary 1.4.5 *Fix $n \in \mathbb{N}$ with $n \geq 2$ are recall the deformation tensor* Def *in $\mathbb{R}^n$ defined in* (1.3.49). *Then the fundamental solution associated as in Theorem 1.4.2 with the homogeneous, constant coefficient, weakly elliptic, second-order, $n \times n$ system $L := \mathrm{Def}^{\top}\mathrm{Def}$ in $\mathbb{R}^n$ is the matrix-valued function $E = (E_{jk})_{1\leq j,k\leq n}$ with entries defined at each point $x = (x_1, \dots, x_n) \in \mathbb{R}^n \setminus \{0\}$ as*

$$E_{jk}(x) := \begin{cases} \dfrac{1}{2\omega_{n-1}}\left[\dfrac{3\delta_{jk}}{(n-2)|x|^{n-2}} + \dfrac{x_j x_k}{|x|^n}\right] & \text{if } n \geq 3, \\[2ex] \dfrac{1}{4\pi}\left[-3\delta_{jk}\ln|x| + \dfrac{x_j x_k}{|x|^2}\right] + \dfrac{3\ln 4 - 1}{8\pi}\delta_{jk} & \text{if } n = 2, \end{cases} \tag{1.4.92}$$

for every $j, k \in \{1, \dots, n\}$.

Proof Since from (1.3.58) we know that, when acting on vector-valued distributions in $\mathbb{R}^n$, the operator $L = \mathrm{Def}^{\top}\mathrm{Def} = -\frac{1}{2}(\Delta + \nabla\mathrm{div})$ agrees with the Lamé system $L_{\mu,\lambda}$ for $\mu := -1/2$ and $\lambda := 0$, formula (1.4.92) is implied by (1.4.72) specialized to these values of the Lamé moduli (which satisfy (1.4.71)). □

Our next result explicitly identifies the fundamental solution produced by Theorem 1.4.2 for the family of "generalized Lamé" operators $L_{\alpha,\beta}$ introduced earlier, in Example 1.3.16. See also the discussion in (1.3.81), according to which the aforementioned result contains Proposition 1.4.4 as a particular case. To state it, the reader is reminded that the standard fundamental solution for the Laplacian Δ in $\mathbb{R}^n$ has been defined in (1.5.56). Also, the standard fundamental solution for the bi-Laplacian Δ^2 in $\mathbb{R}^n$ (with $\mathbb{N}$ satisfying $n \geq 2$) is given by

$$E_{\Delta^2}(x) := \begin{cases} \dfrac{1}{2(n-2)(n-4)\omega_{n-1}} \dfrac{1}{|x|^{n-4}} & \text{if } n \geq 3 \text{ and } n \neq 4, \\ \dfrac{1}{8\pi^2} \ln |x| & \text{if } n = 4, \\ \dfrac{1}{8\pi} |x|^2 \ln |x| & \text{if } n = 2, \end{cases} \tag{1.4.93}$$

for each $x \in \mathbb{R}^n \setminus \{0\}$ (see, e.g., [109, Theorem 7.21, pp. 251–252]).

Proposition 1.4.6 *Fix $n \in \mathbb{N}$ with $n \geq 2$. Select two complex numbers $\alpha, \beta \in \mathbb{C} \setminus \{0\}$ along with a degree $\ell \in \{0, 1, \dots, n\}$ and consider the second-order homogeneous, constant (complex) coefficient weakly elliptic*[23] *differential operator*

$$L_{\alpha,\beta} := \alpha \mathrm{d}\delta + \beta\delta\mathrm{d} \tag{1.4.94}$$

acting on differential forms of degree ℓ in $\mathbb{R}^n$. Then, if E_Δ and E_{Δ^2} are as in (1.5.56) *and* (1.4.93)*, respectively, it follows that the fundamental solution of $L_{\alpha,\beta}$ from Theorem 1.4.2 is the double form*[24] *defined at each $z \in \mathbb{R}^n \setminus \{0\}$ according to*[25]

$$E^{\ell}_{\alpha,\beta}(z) = \sum_{|I|=|J|=\ell}{}' \Big\{ -\alpha^{-1} E_\Delta(z)\delta_{IJ} \tag{1.4.95}$$

$$- (\beta^{-1} - \alpha^{-1}) \sum_{i,j=1}^{n} \partial_{z_i}\partial_{z_j} \big[E_{\Delta^2}(z)\big] \varepsilon^{iI}_{jJ} + c^{\ell}_{IJ} \Big\} \mathrm{d}x^I \otimes \mathrm{d}y^J$$

where c^{ℓ}_{IJ} is a complex constant if $n = 2$, and actually $c_{IJ} = 0$ if $n \geq 3$. More explicitly, if $n \geq 3$ then for each $z \in \mathbb{R}^n \setminus \{0\}$ one has

---

[23] see (1.3.71)

[24] a double form is the algebraic equivalent of a matrix-valued function in the present context

[25] as in [112, §6.4], the symbol $\sum'$ indicates that the sum is performed only over strictly increasing multi-indices

$$E^{\ell}_{\alpha,\beta}(z) = \sum_{|I|=|J|=\ell}{}' \Big\{ - \Big(\alpha^{-1} + (\beta^{-1} - \alpha^{-1})\frac{n-\ell}{2}\Big) \frac{1}{(2-n)\omega_{n-1}} \frac{1}{|z|^{n-2}} \delta_{IJ}$$

$$- (\beta^{-1} - \alpha^{-1}) \sum_{i,j=1}^{n} \frac{1}{2\omega_{n-1}} \frac{z_i z_j}{|z|^n} \varepsilon^{iI}_{jJ} \Big\} \, dx^I \otimes dy^J \tag{1.4.96}$$

whereas if $n = 2$ *then corresponding to* $\ell = 0$ *one has*

$$E^{0}_{\alpha,\beta}(z) = -\frac{\ln |z|}{2\pi}\beta^{-1} + \frac{\ln 4}{4\pi}\beta^{-1} \text{ for each } z \in \mathbb{R}^2 \setminus \{0\}, \tag{1.4.97}$$

corresponding to $\ell = 1$ *one has*

$$E^{1}_{\alpha,\beta}(z) = \sum_{i,j=1}^{2} \Big\{ - \frac{1}{4\pi}(\alpha^{-1} + \beta^{-1})\delta_{ij} \ln |z|$$

$$- \frac{1}{4\pi}(\alpha^{-1} - \beta^{-1})\frac{z_i z_j}{|z|^2} + c^1_{ij} \Big\} \, dx_i \otimes dy_j \tag{1.4.98}$$

for each $z \in \mathbb{R}^2 \setminus \{0\}$ *with*

$$c^1_{ij} := \frac{1 + \ln 4}{8\pi}(\alpha^{-1} + \beta^{-1})\delta_{ij} - \frac{1}{4\pi}\beta^{-1}\delta_{ij}, \tag{1.4.99}$$

and corresponding to $\ell = 2$ *one has*

$$E^{2}_{\alpha,\beta}(z) = \Big\{ - \frac{\ln |z|}{2\pi}\alpha^{-1} + \frac{\ln 4}{4\pi}\alpha^{-1} \Big\} (dx_1 \wedge dx_2) \otimes (dy_1 \wedge dy_2) \tag{1.4.100}$$

for each $z \in \mathbb{R}^2 \setminus \{0\}$.

Proof By virtue of (1.3.74) it follows that the double form

$$\widetilde{E}^{\ell}_{\alpha,\beta} := L_{\alpha^{-1},\beta^{-1}} \Big(\sum_{|I|=\ell}{}' E_{\Delta^2} \, dx^I \otimes dy^I \Big) \tag{1.4.101}$$

is a fundamental solution for the operator $L_{\alpha,\beta}$ acting on differential forms of degree ℓ in $\mathbb{R}^n$. Use the last formula in (1.3.73) to express

$$L_{\alpha^{-1},\beta^{-1}} = \alpha^{-1} d\delta + \beta^{-1}\delta d = \alpha^{-1}(d\delta + \delta d) + (\beta^{-1} - \alpha^{-1})\delta d$$

$$= -\alpha^{-1}\Delta + (\beta^{-1} - \alpha^{-1})\delta d. \tag{1.4.102}$$

Keeping in mind that $\Delta E_{\Delta^2} = E_\Delta$, we first see that

$$-\alpha^{-1}\Delta\Big(\sum_{|I|=\ell}{}' E_{\Delta^2} \, dx^I \otimes dy^I\Big) = -\alpha^{-1}\Big(\sum_{|I|=|J|=\ell}{}' E_\Delta \delta_{IJ} dx^I \otimes dy^J \Big). \tag{1.4.103}$$

Second, based on definitions (cf. [112, (6.6.141), (6.4.142)]) and item *(ii)* in [112, Lemma 6.4.6] we compute

$$\begin{aligned}\delta\mathrm{d}\Big(\sum_{|I|=\ell}{}' E_{\Delta^2}\,\mathrm{d}x^I\otimes\mathrm{d}y^I\Big) &= \delta\Big(\sum_{|K|=\ell+1}{}'\sum_{j=1}^{n}\sum_{|I|=\ell}{}'\partial_j E_{\Delta^2}\varepsilon^{jI}_K\,\mathrm{d}x^K\otimes\mathrm{d}y^I\Big)\\
&= -\sum_{|J|=\ell}{}'\sum_{i,j=1}^{n}\sum_{|I|=\ell}{}'\partial_i\partial_j E_{\Delta^2}\Big(\sum_{|K|=\ell+1}{}'\varepsilon^{jI}_K\varepsilon^K_{iJ}\Big)\,\mathrm{d}x^J\otimes\mathrm{d}y^I\\
&= -\sum_{|I|=\ell}{}'\sum_{|J|=\ell}{}'\sum_{i,j=1}^{n}\partial_i\partial_j E_{\Delta^2}\varepsilon^{jI}_{iJ}\,\mathrm{d}x^J\otimes\mathrm{d}y^I\\
&= -\sum_{|I|=|J|=\ell}{}'\sum_{i,j=1}^{n}\partial_i\partial_j E_{\Delta^2}\varepsilon^{iI}_{jJ}\,\mathrm{d}x^I\otimes\mathrm{d}y^J,\end{aligned} \tag{1.4.104}$$

where in the last equality we have simply intertwined I and J. We can then deduce from (1.4.101) and (1.4.103)-(1.4.104) that the double form defined at each point $z\in\mathbb{R}^n\setminus\{0\}$ by

$$\begin{aligned}\widetilde{E}^{\ell}_{\alpha,\beta}(z) = \sum_{|I|=|J|=\ell}{}'\Big\{&-\alpha^{-1}E_{\Delta}(z)\delta_{IJ}\\
&-(\beta^{-1}-\alpha^{-1})\sum_{i,j=1}^{n}\partial_{z_i}\partial_{z_j}\big[E_{\Delta^2}(z)\big]\varepsilon^{iI}_{jJ}\Big\}\mathrm{d}x^I\otimes\mathrm{d}y^J\end{aligned} \tag{1.4.105}$$

is a fundamental solution for the operator $L_{\alpha,\beta}$ acting on ℓ-forms in $\mathbb{R}^n$ with $n\geq 2$. Using (1.5.56) and (1.4.93), we may write $\widetilde{E}^{\ell}_{\alpha,\beta}$ explicitly. Concretely, if $n\geq 3$ then for each $z\in\mathbb{R}^n\setminus\{0\}$ we have

$$\begin{aligned}\widetilde{E}^{\ell}_{\alpha,\beta}(z) = \sum_{|I|=|J|=\ell}{}'\Big\{&-\Big(\alpha^{-1}+(\beta^{-1}-\alpha^{-1})\frac{n-\ell}{2}\Big)\frac{1}{(2-n)\omega_{n-1}}\frac{1}{|z|^{n-2}}\delta_{IJ}\\
&-(\beta^{-1}-\alpha^{-1})\sum_{i,j=1}^{n}\frac{1}{2\omega_{n-1}}\frac{z_i z_j}{|z|^n}\varepsilon^{iI}_{jJ}\Big\}\mathrm{d}x^I\otimes\mathrm{d}y^J\end{aligned} \tag{1.4.106}$$

while if $n=2$ then for each $z\in\mathbb{R}^2\setminus\{0\}$ we have

$$\widetilde{E}^{\ell}_{\alpha,\beta}(z) = \sum_{|I|=|J|=\ell}{}' \left\{ -\left(\alpha^{-1} + (\beta^{-1} - \alpha^{-1})\frac{2-\ell}{2}\right)\frac{1}{2\pi} \ln |z| \delta_{IJ} - (\beta^{-1} - \alpha^{-1})\frac{1}{4\pi} \sum_{i,j=1}^{2} \frac{z_i z_j}{|z|^2} \varepsilon^{iI}_{jJ} + \widetilde{c}^{\ell}_{IJ} \right\} \mathrm{d}x^I \otimes \mathrm{d}y^J \tag{1.4.107}$$

where

$$\widetilde{c}^{\ell}_{IJ} = -\frac{1}{8\pi}(\beta^{-1} - \alpha^{-1}) \sum_{i,j=1}^{2} \delta_{ij} \varepsilon^{iI}_{jJ} = -\frac{2-\ell}{8\pi}(\beta^{-1} - \alpha^{-1})\delta_{IJ}. \tag{1.4.108}$$

Next, if $E^{\ell}_{L_{\alpha,\beta}}$ denotes the fundamental solution of $L_{\alpha,\beta}$ acting on ℓ-forms from Theorem 1.4.2, then the uniqueness result from item *(8)* in Theorem 1.4.2 guarantees the existence of a polynomial Q in $\mathbb{R}^n$ with the property that

$$E^{\ell}_{L_{\alpha,\beta}} = \widetilde{E}^{\ell}_{\alpha,\beta} + Q. \tag{1.4.109}$$

From this, (1.4.24), and (1.4.106)-(1.4.108) we then deduce that Q is constant if $n = 2$, and vanishes identically if $n \geq 3$. Having established that, the claims pertaining to (1.4.95) and (1.4.96) follow.

There remains to identify $E^{\ell}_{L_{\alpha,\beta}}$ in the manner specified in (1.4.97)-(1.4.100), in the two-dimensional setting. To set the stage, we first make a couple of general observations. From definitions we see that

$$L_{\alpha,\beta}(\xi) = \alpha\xi \wedge (\xi \vee \cdot) + \beta\xi \vee (\xi \wedge \cdot) \text{ for all } \xi \in \mathbb{R}^n. \tag{1.4.110}$$

In concert with (1.3.75), this shows that for each $\xi \in \mathbb{R}^n \setminus \{0\}$ we have

$$\begin{aligned}\left[L_{\alpha,\beta}(\xi)\right]^{-1} &= (\alpha\beta)^{-1}|\xi|^{-4} L_{\beta,\alpha}(\xi) = |\xi|^{-4} L_{\alpha^{-1},\beta^{-1}}(\xi) \\ &= \alpha^{-1}|\xi|^{-4}\xi \wedge (\xi \vee \cdot) + \beta^{-1}|\xi|^{-4}\xi \vee (\xi \wedge \cdot)\end{aligned} \tag{1.4.111}$$

which, for $\xi = (\xi_1, \ldots, \xi_n) \in S^{n-1}$, reduces to

$$\begin{aligned}\left[L_{\alpha,\beta}(\xi)\right]^{-1} &= \alpha^{-1}\xi \wedge (\xi \vee \cdot) + \beta^{-1}\xi \vee (\xi \wedge \cdot) \\ &= \sum_{i,j=1}^{n} \xi_i\xi_j \Big(\alpha^{-1}\mathbf{e}_i \wedge (\mathbf{e}_j \vee \cdot) + \beta^{-1}\mathbf{e}_i \vee (\mathbf{e}_j \wedge \cdot)\Big).\end{aligned} \tag{1.4.112}$$

Let us now work in the two-dimensional setting. If $E^{\ell}_{L_{\alpha,\beta}}(x)$ is regarded as a linear map on the space of differential forms of degree ℓ, then from (1.4.18) (used with $n = 2$), (1.4.112), and (1.4.77) we deduce that

$$
\begin{aligned}
E^{\ell}_{L_{\alpha,\beta}}(z) &= -\frac{1}{4\pi^2}\int_{S^1} \ln\big|\langle z,\xi\rangle\big|\big[L_{\alpha,\beta}(\xi)\big]^{-1}\,d\mathcal{H}^1(\xi) \qquad (1.4.113)\\
&= -\frac{1}{4\pi^2}\sum_{i,j=1}^{2}\Big(\int_{S^1}\big(\ln\big|\langle z,\xi\rangle\big|\big)\xi_i\xi_j\,d\mathcal{H}^1(\xi)\Big)\times\\
&\qquad\qquad \times\Big(\alpha^{-1}\mathbf{e}_i\wedge(\mathbf{e}_j\vee\cdot)+\beta^{-1}\mathbf{e}_i\vee(\mathbf{e}_j\wedge\cdot)\Big)\\
&= \frac{1+\ln 4-2\ln|z|}{8\pi}\sum_{i,j=1}^{2}\delta_{ij}\Big(\alpha^{-1}\mathbf{e}_i\wedge(\mathbf{e}_j\vee\cdot)+\beta^{-1}\mathbf{e}_i\vee(\mathbf{e}_j\wedge\cdot)\Big)\\
&\quad -\frac{1}{4\pi}\sum_{i,j=1}^{2}\frac{z_iz_j}{|z|^2}\Big(\alpha^{-1}\mathbf{e}_i\wedge(\mathbf{e}_j\vee\cdot)+\beta^{-1}\mathbf{e}_i\vee(\mathbf{e}_j\wedge\cdot)\Big)
\end{aligned}
$$

for each $z=(z_1,z_2)\in\mathbb{R}^2\setminus\{0\}$. Next, a computation based on [112, (6.4.126)] and [112, Lemma 6.4.7] shows that, as linear maps on differential forms of degree $\ell\in\{0,1,2\}$ in $\mathbb{R}^2$,

$$
\sum_{i,j=1}^{2}\delta_{ij}\Big(\alpha^{-1}\mathbf{e}_i\wedge(\mathbf{e}_j\vee\cdot)+\beta^{-1}\mathbf{e}_i\vee(\mathbf{e}_j\wedge\cdot)\Big)=\begin{cases}2\beta^{-1}\mathrm{Id} & \text{if } \ell=0,\\ (\alpha^{-1}+\beta^{-1})\mathrm{Id} & \text{if } \ell=1,\\ 2\alpha^{-1}\mathrm{Id} & \text{if } \ell=2,\end{cases} \tag{1.4.114}
$$

where Id is the identity. In a similar fashion, the linear map

$$
\sum_{i,j=1}^{2}\frac{z_iz_j}{|z|^2}\Big(\alpha^{-1}\mathbf{e}_i\wedge(\mathbf{e}_j\vee\cdot)+\beta^{-1}\mathbf{e}_i\vee(\mathbf{e}_j\wedge\cdot)\Big) \tag{1.4.115}
$$

acting on ℓ-forms may be canonically identified with

$$
\begin{cases}\beta^{-1}\mathrm{Id} & \text{if } \ell=0,\\ (\alpha^{-1}-\beta^{-1})\dfrac{z\otimes z}{|z|^2}+\beta^{-1}\mathrm{Id} & \text{if } \ell=1,\\ \alpha^{-1}\mathrm{Id} & \text{if } \ell=2,\end{cases} \tag{1.4.116}
$$

where Id is once again the identity, and the middle case entails a matrix regarded as a linear map on 1-forms (themselves canonically identified with vectors). Finally, from (1.4.113)-(1.4.116) we conclude that (1.4.97)-(1.4.100) holds. □

It has been noted earlier that Proposition 1.4.6 contains Proposition 1.4.4 as a particular case. While this is a byproduct of (1.3.81), it is instructive to check this directly. Specifically, fix two numbers $\alpha,\beta\in\mathbb{C}\setminus\{0\}$ then define

$$\lambda := -\alpha + 2\beta \text{ and } \mu := -\beta. \tag{1.4.117}$$

As a consequence, $\alpha = -2\mu - \lambda$ and $\beta = -\mu$, so

$$\alpha^{-1} + \beta^{-1} = -\frac{3\mu + \lambda}{\mu(2\mu + \lambda)} \text{ and } \alpha^{-1} - \beta^{-1} = \frac{\mu + \lambda}{\mu(2\mu + \lambda)}. \tag{1.4.118}$$

Keeping these in mind and also observing that for any two indices $I, J \in \{1, \dots, n\}$ we have

$$\sum_{i,j=1}^{n} \frac{z_i z_j}{|z|^n} \varepsilon_{jJ}^{iI} = \frac{\delta_{IJ}}{|z|^{n-2}} - \frac{z_I z_J}{|z|^n} \text{ for all } z \in \mathbb{R}^n \setminus \{0\}, \tag{1.4.119}$$

one can verify without difficulty that if $n \geq 3$ then the fundamental solution $E^1_{\alpha,\beta}$ defined as in (1.4.96) with $\ell := 1$ may be naturally identified with the matrix-valued fundamental solution for the Lamé system $L_{\mu,\lambda}$ whose entries are defined as in (1.4.72) with $n \geq 3$.

In addition, it is apparent from (1.4.118) and definitions that if $n = 2$ then the fundamental solution $E^1_{\alpha,\beta}$ from (1.4.98)-(1.4.99) reduces precisely to the matrix-valued fundamental solution for the Lamé system $L_{\mu,\lambda}$ whose entries are defined as in (1.4.72) with $n = 2$.

Scalar Weakly Elliptic Operators

In the remaining portion of this section we consider in greater detail the *scalar* case, with the goal of providing *fully explicit* formulas for the fundamental solutions of weakly elliptic scalar operators (originally constructed in Theorem 1.4.2 via integral expressions). Henceforth we take $M = 1$ in (1.3.1). In this scenario, we agree to drop the dependence on M and simply write $L = a_{rs}\partial_r\partial_s$ and introduce the $n \times n$ matrix of coefficients $A := (a_{rs})_{1 \leq r,s \leq n} \in \mathbb{C}^{n \times n}$.

Let us re-visit the two notions of ellipticity, discussed earlier for general systems, and re-adjust them as to better reflect the fact that we are now considering scalar operators. First, the weak ellipticity condition for L from (1.3.3) may be equivalently expressed in the current case as

$$\langle A\xi, \xi \rangle \neq 0 \text{ for every } \xi \in \mathbb{R}^n \setminus \{0\}, \tag{1.4.120}$$

which may be further recast as

$$\langle A\xi, \xi \rangle \in \mathbb{C} \setminus \{0\} \text{ for every } \xi \in S^{n-1}, \tag{1.4.121}$$

and also equivalently rephrased as

$$\inf_{\xi \in S^{n-1}} \big|\langle A\xi, \xi \rangle\big| > 0. \tag{1.4.122}$$

Whenever either of these equivalent descriptions happens, we shall call $A \in \mathbb{C}^{n \times n}$ a weakly elliptic matrix.

Second, the Legendre-Hadamard ellipticity condition from Definition 1.3.2 for the operator $L = a_{rs}\partial_r\partial_s$ becomes equivalent to the demand that the coefficient matrix $A = (a_{rs})_{1\leq r,s\leq n}$ has the property that there exists some $\kappa > 0$ such that

$$\mathrm{Re}\langle A\xi, \xi\rangle = \mathrm{Re}\big[a_{rs}\xi_r\xi_s\big] \geq \kappa|\xi|^2, \qquad \forall\xi = (\xi_1, \ldots, \xi_n) \in \mathbb{R}^n, \tag{1.4.123}$$

which, in turn, may be rephrased as

$$\langle A\xi, \xi\rangle \in \big\{z \in \mathbb{C} : \mathrm{Re}\, z \geq \kappa\big\} \text{ for every } \xi \in S^{n-1}. \tag{1.4.124}$$

Whenever this happens, we shall say that the matrix $A \in \mathbb{C}^{n\times n}$ is Legendre-Hadamard elliptic. Another equivalent way of express the Legendre-Hadamard ellipticity of a given complex matrix $A = (a_{rs})_{1\leq r,s\leq n}$ is to simply demand that

$$\inf_{\xi\in S^{n-1}} \mathrm{Re}\,\langle A\xi, \xi\rangle = \inf_{\xi\in S^{n-1}} \mathrm{Re}\,\big[a_{rs}\xi_r\xi_s\big] > 0. \tag{1.4.125}$$

As is apparent from definitions,

$$\begin{aligned}&\text{if } A \in \mathbb{C}^{n\times n} \text{ is Legendre-Hadamard elliptic,}\\ &\text{then the matrix } A \text{ is also weakly elliptic.}\end{aligned} \tag{1.4.126}$$

On a related note, recall that the (strict) positive definiteness of a complex matrix $A = (a_{rs})_{1\leq r,s\leq n}$ amounts to the existence of a real number $\kappa > 0$ such that

$$\mathrm{Re}\,\langle A\zeta, \overline{\zeta}\,\rangle = \mathrm{Re}\big(a_{rs}\zeta_r\overline{\zeta_s}\big) \geq \kappa|\zeta|^2, \qquad \forall\zeta = (\zeta_r)_{1\leq r\leq n} \in \mathbb{C}^n. \tag{1.4.127}$$

With $A^* := (A^c)^\top$ and $I_{n\times n} \in \mathbb{C}^{n\times n}$ denoting the identity matrix, this is further equivalent to the demand that

$$\begin{aligned}&\tfrac{1}{2}(A + A^*) \geq \kappa I_{n\times n} \text{ in } \mathbb{C}^n, \text{ i.e.,}\\ &\big\langle \tfrac{1}{2}(A + A^*)\zeta, \overline{\zeta}\,\big\rangle \geq \kappa|\zeta|^2, \quad \forall\zeta \in \mathbb{C}^n.\end{aligned} \tag{1.4.128}$$

Obviously,

$$\begin{aligned}&\text{if } A \in \mathbb{C}^{n\times n} \text{ is (strictly) positive definite then the}\\ &\text{matrix } A \text{ is also Legendre-Hadamard elliptic,}\end{aligned} \tag{1.4.129}$$

and

$$\begin{aligned}&\text{if } A \in \mathbb{C}^{n\times n} \text{ is (strictly) positive definite and}\\ &\text{invertible, then so is the inverse matrix } A^{-1}.\end{aligned} \tag{1.4.130}$$

Complex matrices that are symmetric and Legendre-Hadamard elliptic have a number of other useful properties. Among other things, they turn out to be invertible, as indicated in our next lemma.

Lemma 1.4.7 *Fix a matrix $A \in \mathbb{C}^{n\times n}$ which is symmetric and Legendre-Hadamard elliptic, i.e., satisfies*

$$A = A^\top \text{ and } \kappa := \inf_{\xi \in S^{n-1}} \operatorname{Re}\langle A\xi, \xi\rangle > 0. \tag{1.4.131}$$

Then both A and Re A *are positive definite. In fact,*

$$\operatorname{Re}\langle A\zeta, \overline{\zeta}\rangle \geq \kappa|\zeta|^2, \qquad \forall \zeta \in \mathbb{C}^n, \tag{1.4.132}$$

and

$$\big\langle(\operatorname{Re} A)\zeta, \overline{\zeta}\big\rangle \geq \kappa|\zeta|^2, \qquad \forall \zeta \in \mathbb{C}^n. \tag{1.4.133}$$

Also,

$$|A\zeta| \geq \kappa|\zeta| \text{ and } |(\operatorname{Re} A)\zeta| \geq \kappa|\zeta| \text{ for every } \zeta \in \mathbb{C}^n. \tag{1.4.134}$$

Finally,

$$\det A \neq 0 \text{ and } \det(\operatorname{Re} A) \neq 0, \tag{1.4.135}$$

hence both A and Re A *are invertible.*

Proof For each $\zeta \in \mathbb{C}^n$ we may write

$$\begin{aligned}\operatorname{Re}\langle A\zeta, \overline{\zeta}\rangle &= \operatorname{Re}\big\langle A(\operatorname{Re}\zeta + i\operatorname{Im}\zeta), \operatorname{Re}\zeta - i\operatorname{Im}\zeta\big\rangle \\ &= \operatorname{Re}\Big\{\big\langle A(\operatorname{Re}\zeta), \operatorname{Re}\zeta\big\rangle + \big\langle A(\operatorname{Im}\zeta), \operatorname{Im}\zeta\big\rangle\Big\} \\ &\geq \kappa\,|\operatorname{Re}\zeta|^2 + \kappa\,|\operatorname{Im}\zeta|^2 = \kappa\,|\zeta|^2. \end{aligned}\tag{1.4.136}$$

The second equality in (1.4.136) uses the fact that A is symmetric, while the second property in (1.4.131) is used for the inequality in (1.4.136). This establishes (1.4.132). Combining (1.4.132) with (1.4.127) then shows that A is positive definite. The symmetry of A implies $\operatorname{Re} A = (\operatorname{Re} A)^\top$ and, further,

$$\big\langle(\operatorname{Re} A)\zeta, \overline{\zeta}\big\rangle \in \mathbb{R} \text{ for each } \zeta \in \mathbb{C}^n. \tag{1.4.137}$$

Hence, in a similar fashion to (1.4.136), for each $\zeta \in \mathbb{C}^n$ we have

$$\begin{aligned}\langle(\operatorname{Re} A)\zeta, \overline{\zeta}\rangle &= \big\langle(\operatorname{Re} A)(\operatorname{Re}\zeta + i\operatorname{Im}\zeta), \operatorname{Re}\zeta - i\operatorname{Im}\zeta\big\rangle \\ &= \big\langle(\operatorname{Re} A)(\operatorname{Re}\zeta), \operatorname{Re}\zeta\big\rangle + \big\langle(\operatorname{Re} A)(\operatorname{Im}\zeta), \operatorname{Im}\zeta\big\rangle \\ &= \operatorname{Re}\big\langle A(\operatorname{Re}\zeta), \operatorname{Re}\zeta\big\rangle + \operatorname{Re}\big\langle A(\operatorname{Im}\zeta), \operatorname{Im}\zeta\big\rangle \\ &\geq \kappa\,|\operatorname{Re}\zeta|^2 + \kappa\,|\operatorname{Im}\zeta|^2 = \kappa\,|\zeta|^2, \end{aligned}\tag{1.4.138}$$

proving (1.4.133). Collectively, (1.4.133), (1.4.127), and (1.4.137) then prove that Re A is also positive definite. Next, combining (1.4.132) with the Cauchy-Schwarz inequality justifies (1.4.134). In turn, the estimates in (1.4.134) prove that the linear maps $A : \mathbb{C}^n \to \mathbb{C}^n$ and $\operatorname{Re} A : \mathbb{C}^n \to \mathbb{C}^n$ are injective, thus invertible. In particular, (1.4.135) holds. □

Remark 1.4.8 *As a corollary of Lemma 1.4.7 we have:*

$$\begin{aligned}&\operatorname{sym} A := \tfrac{1}{2}(A + A^{\top}) \textit{ is invertible whenever the matrix}\\ &A \in \mathbb{C}^{n\times n} \textit{ happens to be Legendre-Hadamard elliptic.}\end{aligned} \tag{1.4.139}$$

Other significant consequences of Lemma 1.4.7 are noted below.

Corollary 1.4.9 *For any symmetric matrix $A \in \mathbb{C}^{n\times n}$, being Legendre-Hadamard elliptic is actually equivalent to being (strictly) positive definite.*

Moreover, any symmetric Legendre-Hadamard elliptic matrix $A \in \mathbb{C}^{n\times n}$ is invertible, and A^{-1} is symmetric and strictly positive definite (hence also Legendre-Hadamard elliptic and weakly elliptic).

Proof In the first claim, the left-to-right implication is contained in Lemma 1.4.7 (cf. (1.4.132)), while the right-to-left implication has been noted in (1.4.129). The second claim in the statement is seen from Lemma 1.4.7 (cf. (1.4.135)), the fact that the transpose commutes with the inverse, the first claim in the statement of the current corollary, (1.4.130), and (1.4.126). □

As remarked earlier, strict positive definiteness (cf. (1.4.127)) implies Legendre-Hadamard ellipticity (cf. (1.4.123)) which, in turn, implies weak ellipticity (cf. (1.4.120)). Regarding this hierarchy, we wish to propose yet another brand of ellipticity, sitting in between weak ellipticity and Legendre-Hadamard ellipticity, as defined below.

Definition 1.4.10 *Call a matrix $A \in \mathbb{C}^{n\times n}$* `mildly elliptic` *provided there exists some $z \in \mathbb{C} \setminus \{0\}$ such that*[26]

$$\langle Ax, x\rangle \notin \{zt : t \in [0, \infty)\} \textit{ for every } x \in \mathbb{R}^n \setminus \{0\}. \tag{1.4.140}$$

Equivalently,

$$\begin{aligned}&\text{a matrix } A \in \mathbb{C}^{n\times n} \text{ is mildly elliptic if } A \text{ is weakly elliptic and}\\ &\text{the set } \{\langle Ax, x\rangle : x \in \mathbb{R}^n \setminus \{0\}\} \text{ does not coincide with } \mathbb{C} \setminus \{0\}.\end{aligned} \tag{1.4.141}$$

As is clear from (1.4.124), (1.4.140), and (1.4.121), for every $A \in \mathbb{C}^{n\times n}$ we have

$$\begin{aligned}A \text{ is Legendre-Hadamard elliptic } &\Longrightarrow A \text{ is mildly elliptic}\\ &\Longrightarrow A \text{ is weakly elliptic.}\end{aligned} \tag{1.4.142}$$

Later on we shall need the following lemma:

Lemma 1.4.11 *Fix $n \in \mathbb{N}$, $n \geq 2$, and let $B = (b_{jk})_{1\leq j,k\leq n}$ be an $n \times n$ matrix with complex entries which is symmetric and mildly elliptic. Then the $n \times n$ matrix*

$$\int_{S^{n-1}} \frac{(B\xi)\otimes\xi}{\langle B\xi, \xi\rangle}\, d\mathcal{H}^{n-1}(\xi) \textit{ is symmetric.} \tag{1.4.143}$$

[26] recall that the numerical range of the matrix A is the set $\{\langle A\zeta, \zeta^c\rangle : \zeta \in \mathbb{C}^n,\ |\zeta| = 1\}$, where the superscript 'c' indicates complex conjugation

Proof From Definition 1.4.10 it follows that there exists a simply connected open set $\mathcal{O} \subseteq \mathbb{C} \setminus \{0\}$ such that

$$\langle B\xi, \xi\rangle \in \mathcal{O} \text{ for every } \xi \in S^{n-1}. \tag{1.4.144}$$

The simple connectivity of the set $\mathcal{O}$ implies (see [112, Proposition 5.8.1]) the existence of

$$\begin{aligned} &h : \mathcal{O} \to \mathbb{C} \setminus \{0\} \text{ holomorphic function, with the} \\ &\text{property that } h'(z) = 1/z \text{ at every point } z \in \mathcal{O}. \end{aligned} \tag{1.4.145}$$

In particular, for each differentiable function $g : \mathbb{R}^n \to \mathcal{O}$ and each $j \in \{1, \dots, n\}$, the Cauchy-Riemann equations and Chain Rule give

$$\partial_{\xi_j}\big[h(g(\xi))\big] = h'(g(\xi))(\partial_j g)(\xi) = \frac{(\partial_j g)(\xi)}{g(\xi)}, \qquad \forall \xi \in \mathbb{R}^n. \tag{1.4.146}$$

To proceed, define $f(\xi) := h(\langle B\xi, \xi\rangle)$ for each ξ near S^{n-1}. Thanks to (1.4.144), f is a well-defined, smooth, complex-valued function. Having fixed two arbitrary indices $j, k \in \{1, \dots, n\}$, we may then compute

$$\begin{aligned} 0 &= \int_{S^{n-1}} \partial_{\tau_{jk}}\big[f(\langle B\xi, \xi\rangle)\big]\, d\mathcal{H}^{n-1}(\xi) = \int_{S^{n-1}} \frac{\partial_{\tau_{jk}}\big[\langle B\xi, \xi\rangle\big]}{\langle B\xi, \xi\rangle}\, d\mathcal{H}^{n-1}(\xi) \\ &= \int_{S^{n-1}} \frac{\xi_j \partial_{\xi_k}\langle B\xi, \xi\rangle - \xi_k \partial_{\xi_j}\langle B\xi, \xi\rangle}{\langle B\xi, \xi\rangle}\, d\mathcal{H}^{n-1}(\xi) \\ &= \int_{S^{n-1}} \frac{2\xi_j (B\xi)_k - 2\xi_k (B\xi)_j}{\langle B\xi, \xi\rangle}\, d\mathcal{H}^{n-1}(\xi) \\ &= 2\left(\int_{S^{n-1}} \frac{(B\xi)\otimes\xi}{\langle B\xi, \xi\rangle}\, d\mathcal{H}^{n-1}(\xi)\right)_{kj} \\ &\quad - 2\left(\int_{S^{n-1}} \frac{(B\xi)\otimes\xi}{\langle B\xi, \xi\rangle}\, d\mathcal{H}^{n-1}(\xi)\right)_{jk}. \end{aligned} \tag{1.4.147}$$

Above, the first equality is a very particular case of the integration by parts formula on the boundary of the unit ball. See [113, Lemma 11.1.1], used here with $\Omega := B(0, 1)$, $\varphi := f$, and some $\psi \in \mathscr{C}_c^1(\mathbb{R}^n)$ which is identically one near S^{n-1}. The second equality in (1.4.147) is simply a consequence of (1.4.146), while the third equality follows from (A.0.115). The fourth equality in (1.4.147) uses the fact that B is symmetric, and the last equality in (1.4.147) is seen from (1.3.7). In turn, (1.4.147) readily implies (1.4.143), finishing the proof of the lemma. □

Next we elaborate on the nature of the square-root determinant function considered on the set of symmetric, Legendre-Hadamard elliptic, complex matrices. Our main result in this regard is stated in Proposition 1.4.12 below. To set the stage, consider

$$\mathfrak{M}_n := \{A \in \mathbb{C}^{n\times n} : A \text{ is Legendre-Hadamard elliptic and symmetric}\}$$
$$= \{A \in \mathbb{C}^{n\times n} : A \text{ is strictly positive definite and symmetric}\}, \tag{1.4.148}$$

where the second equality comes from Corollary 1.4.9. Observe (see (1.4.139)) that

$$\det A \neq 0 \text{ for every } A \in \mathfrak{M}_n. \tag{1.4.149}$$

Since any $n \times n$ symmetric matrix with complex entries is uniquely determined by its entries lying on or below the main diagonal, we may naturally identify $\mathfrak{M}_n$ with an open convex subset of $\mathbb{C}^{n(n+1)/2}$ (in fact, with an open convex conical[27] set with vertex at the origin).

Proposition 1.4.12 *Under the above identification of $\mathfrak{M}_n$ with an open subset of $\mathbb{C}^{n(n+1)/2}$, there exists a unique separately holomorphic function (i.e., holomorphic in each scalar complex component in $\mathbb{C}^{n(n+1)/2}$)*

$$\mathfrak{M}_n \ni A \longmapsto (\det A)^{\frac{1}{2}} \in \mathbb{C} \tag{1.4.150}$$

with the property that

$$(\det A)^{\frac{1}{2}} \in (0, \infty) \textit{ whenever } A \in \mathfrak{M}_n \textit{ has real entries.} \tag{1.4.151}$$

Thus $(\det A)^{-\frac{1}{2}}$ *is unambiguously defined as a number in $\mathbb{C} \setminus \{0\}$ for each $A \in \mathfrak{M}_n$ and, in fact,*

$$(\det A)^{-\frac{1}{2}} = \pi^{-n/2} \int_{\mathbb{R}^n} e^{-\langle A\xi,\xi\rangle}\, d\xi, \qquad \forall A \in \mathfrak{M}_n. \tag{1.4.152}$$

As a preamble, we record the following unique continuation result which appears in [108, Lemma 7.53, p. 268].

Lemma 1.4.13 *Let $N \in \mathbb{N}$ and assume O is an open and convex subset of $\mathbb{C}^N$ with the property that $O \cap \mathbb{R}^N \neq \varnothing$ (where $\mathbb{R}^N$ is canonically embedded into $\mathbb{C}^N$). Also, suppose $g_1, g_2 : O \to \mathbb{C}$ are two functions which are separately holomorphic (i.e., in each scalar complex component in $\mathbb{C}^N$) such that*

$$g_1\big|_{O\cap\mathbb{R}^N} = g_2\big|_{O\cap\mathbb{R}^N}. \tag{1.4.153}$$

Then $g_1 = g_2$ everywhere in O.

We are ready to present the proof of Proposition 1.4.12.

Proof of Proposition 1.4.12 For starters, we claim that

$$(\det A)\left(\int_{\mathbb{R}^n} e^{-\langle A\xi,\xi\rangle}\, d\xi\right)^2 = \pi^n \text{ for each } A \in \mathfrak{M}_n. \tag{1.4.154}$$

[27] i.e., a set stable under multiplication by strictly positive numbers

Before proving (1.4.154) we wish to note that the integral appearing above is absolutely convergent. Indeed, if $A \in \mathfrak{M}_n$ then there exists $\kappa \in (0, \infty)$ such that $\operatorname{Re}\langle A\xi, \xi\rangle \geq \kappa|\xi|^2$ for every $\xi \in \mathbb{R}^n$, hence

$$\int_{\mathbb{R}^n} \left| e^{-\langle A\xi,\xi\rangle} \right| d\xi = \int_{\mathbb{R}^n} e^{-\operatorname{Re}\langle A\xi,\xi\rangle} d\xi \leq \int_{\mathbb{R}^n} e^{-\kappa|\xi|^2} d\xi = (\pi/\kappa)^{n/2}, \tag{1.4.155}$$

in view of the well-known formula

$$\int_{-\infty}^{+\infty} e^{-\kappa t^2} dt = (\pi/\kappa)^{1/2}. \tag{1.4.156}$$

In particular, the mapping $g : \mathfrak{M}_n \to \mathbb{C}$ given by

$$g(A) := (\det A)\left(\int_{\mathbb{R}^n} e^{-\langle A\xi,\xi\rangle} d\xi \right)^2, \qquad \forall A \in \mathfrak{M}_n, \tag{1.4.157}$$

is well defined. Moreover, a moment's reflection shows that g is holomorphic in each variable separately. In concert with (1.4.156)-(1.4.157), a diagonalization process (for real symmetric matrices, involving an orthogonal transformation in $\mathbb{R}^n$) then shows that

$$g(A) = \pi^n \text{ whenever } A \in \mathfrak{M}_n \text{ has real entries.} \tag{1.4.158}$$

With this in hand, Lemma 1.4.13 applies (here is where the convexity of $\mathfrak{M}_n$ is first relevant) and yields $g(A) = \pi^n$ for every $A \in \mathfrak{M}_n$. This finishes the proof of the claim made in (1.4.154).

We next claim that

$$f(A) := \left(\int_{\mathbb{R}^n} e^{-\langle A\xi,\xi\rangle} d\xi \right)^{-1}, \qquad \forall A \in \mathfrak{M}_n, \tag{1.4.159}$$

is the unique mapping $f : \mathfrak{M}_n \to \mathbb{C}$ satisfying

$$\begin{aligned} &f \text{ is separately holomorphic,} \\ &f(A)^2 = \det A \text{ for every matrix } A \in \mathfrak{M}_n, \\ &f(A) \in (0, \infty) \text{ whenever } A \in \mathfrak{M}_n \text{ has real entries.} \end{aligned} \tag{1.4.160}$$

Indeed, from (1.4.154), (1.4.155), and (1.4.149), we conclude that the mapping (1.4.159) is well-defined and enjoys all properties stipulated in (1.4.160). To establish the uniqueness of such a mapping, it suffices to show that if $f_1, f_2 : \mathfrak{M}_n \to \mathbb{C}$ are two separately holomorphic functions satisfying

$$\begin{aligned} &f_1(A)^2 = f_2(A)^2 \text{ for every matrix } A \in \mathfrak{M}_n, \\ &f_1(A), f_2(A) \in (0, \infty) \text{ whenever } A \in \mathfrak{M}_n \text{ has real entries,} \end{aligned} \tag{1.4.161}$$

then they necessarily coincide on $\mathfrak{M}_n$. To see that this is the case, observe first that the properties in (1.4.161) imply

$$f_1(A) = f_2(A) \text{ whenever } A \in \mathfrak{M}_n \text{ has real entries} \tag{1.4.162}$$

which, thanks to Lemma 1.4.13 (whose applicability is ensured by the convexity of $\mathfrak{M}_n$), ultimately forces $f_1 = f_2$ in $\mathfrak{M}_n$. This finishes the proof of the fact that (1.4.159) is the unique complex-valued map defined on $\mathfrak{M}_n$ satisfying (1.4.160).

At this stage, defining

$$\mathfrak{M}_n \ni A \longmapsto (\det A)^{\frac{1}{2}} := f(A) \in \mathbb{C} \tag{1.4.163}$$

all desired conclusions follow. □

Remark 1.4.14 *For any matrix $A \in \mathbb{C}^{n\times n}$ satisfying* (1.4.131), *formula* (1.4.152) *actually implies* (*in view of* (1.4.155)) *the following stronger version of* (1.4.135)*:*

$$|\det A| \geq \kappa^n \text{ and } |\det(\operatorname{Re} A)| \geq \kappa^n. \tag{1.4.164}$$

Remark 1.4.15 *A related formula to* (1.4.152)*, to the effect that*

$$(\det A)^{-\frac{1}{2}} = \frac{1}{\omega_{n-1}} \int_{S^{n-1}} \frac{1}{\langle A\xi, \xi\rangle^{\frac{n}{2}}} \, d\mathcal{H}^{n-1}(\xi), \qquad \forall A \in \mathfrak{M}_n, \tag{1.4.165}$$

has been proved in [108, (7.8.50), p. 270] (*see also* [109, (7.12.47), p. 314] *and* [109, §7.12, p. 305] *for related matters*). *Here we wish to note that this may be used with the same effectiveness for the purpose of establishing* (1.4.150)-(1.4.151).

Remark 1.4.16 *In view of the fact that for each $A \in \mathbb{C}^{n\times n}$ we have*

$$\langle A\xi, \xi\rangle = \langle(\operatorname{sym} A)\xi, \xi\rangle, \qquad \forall \xi \in \mathbb{R}^n, \tag{1.4.166}$$

it follows from (1.4.152) *that*

$$\big[\det(\operatorname{sym} A)\big]^{-\frac{1}{2}} = \pi^{-n/2} \int_{\mathbb{R}^n} e^{-\langle A\xi,\xi\rangle} \, d\xi, \tag{1.4.167}$$

for every $A \in \mathbb{C}^{n\times n}$ Legendre-Hadamard elliptic.

Formula (1.4.165) is useful in dealing with the last integral in (1.4.22) for scalar Legendre-Hadamard elliptic operators. Specifically, for every Legendre-Hadamard elliptic matrix $A \in \mathbb{C}^{2\times 2}$ the case $n = 2$ of (1.4.165) gives (in light of (1.4.166), and the fact that, when $M = 1$ and $n = 2$, we have $L(\xi) = -\langle A\xi, \xi\rangle$ for each $\xi \in \mathbb{R}^2$)

$$-\int_{S^1}[L(\xi)]^{-1}\,d\mathcal{H}^1(\xi)=\int_{S^1}\frac{1}{\langle A\xi,\xi\rangle}\,d\mathcal{H}^1(\xi)$$
$$=\int_{S^1}\frac{1}{\langle(\mathrm{sym}\,A)\xi,\xi\rangle}\,d\mathcal{H}^1(\xi)$$
$$=2\pi\big[\det(\mathrm{sym}\,A)\big]^{-\frac{1}{2}}. \tag{1.4.168}$$

Consequently, when $n=2$ and $M=1$ formula (1.4.22) becomes

$$E(x)=\frac{1}{4\pi^2}\int_{S^1}\frac{\ln\big|\langle\frac{x}{|x|},\xi\rangle\big|}{\langle A\xi,\xi\rangle}\,d\mathcal{H}^1(\xi)+\frac{\ln|x|}{2\pi\sqrt{\det(\mathrm{sym}\,A)}} \quad \text{for all } x\in\mathbb{R}^2\setminus\{0\}. \tag{1.4.169}$$

In particular, this raises the prospect of finding a fully explicit formula[28] for the fundamental solution E associated with any second-order, scalar, constant (complex) coefficient, Legendre-Hadamard elliptic operator L in $\mathbb{R}^n$ with $n\geq 2$, something accomplished in the proposition below.

Proposition 1.4.17 *Consider the special case of Legendre-Hadamard elliptic scalar operators in $\mathbb{R}^n$ with $n\geq 2$. That is, assume $M=1$, in which scenario $L=a_{rs}\partial_r\partial_s$, and suppose the coefficient matrix $A=(a_{rs})_{1\leq r,s\leq n}\in\mathbb{C}^{n\times n}$ is Legendre-Hadamard elliptic; as noted in* (1.4.123)*, this amounts to the existence of some $\kappa>0$ such that*

$$\mathrm{Re}\big[a_{rs}\xi_r\xi_s\big]\geq\kappa|\xi|^2, \qquad \forall\,\xi=(\xi_1,\dots,\xi_n)\in\mathbb{R}^n. \tag{1.4.170}$$

Then the fundamental solution E of L from (1.4.18) *takes the explicit form*

$$E(x)=\begin{cases}-\dfrac{1}{(n-2)\omega_{n-1}\sqrt{\det(\mathrm{sym}\,A)}}\big\langle(\mathrm{sym}\,A)^{-1}x,x\big\rangle^{\frac{2-n}{2}} & \text{if } n\geq 3,\\[2ex] \dfrac{1}{4\pi\sqrt{\det(\mathrm{sym}\,A)}}\log\big\langle(\mathrm{sym}\,A)^{-1}x,x\big\rangle+c_A & \text{if } n=2,\end{cases} \tag{1.4.171}$$

at every $x\in\mathbb{R}^n\setminus\{0\}$. Here, $\mathrm{sym}\,A:=\frac{1}{2}(A+A^\top)$ is the symmetric part of the coefficient matrix A, and the square-root $\sqrt{\det(\mathrm{sym}\,A)}$ is understood in the sense of Proposition 1.4.12. Also, c_A is a complex constant which depends solely on A, while log *denotes the principal branch of the complex logarithm function (defined by the requirement that* $\log z=\ln z$ *if $z\in(0,\infty)$; see the discussion in* [112, Remark 5.8.2]). *Finally, ω_{n-1} denotes the surface measure of the unit sphere in $\mathbb{R}^n$, and*

[28] i.e., a formula lacking integrals

$$\langle a,b\rangle := \sum_{j=1}^{n} a_j b_j \text{ for any generic} \\ a=(a_j)_{1\le j\le n}\in\mathbb{C}^n \text{ and } b=(b_j)_{1\le j\le n}\in\mathbb{C}^n. \tag{1.4.172}$$

The presence of the additive constant c_A in (1.4.171) when $n = 2$ is expected given that assignment (acting from the set of all complex 2×2 Legendre-Hadamard elliptic matrices)

$$A \longmapsto E_A(x) := \frac{1}{4\pi\sqrt{\det(\operatorname{sym} A)}} \log\big\langle(\operatorname{sym} A)^{-1}x, x\big\rangle \tag{1.4.173}$$

does *not* satisfy the dilation law $E_{\lambda A} = \lambda^{-1}E_A$ for each $\lambda\in(0,\infty)$ predicted by (1.4.32).

Proof of Proposition 1.4.17 Let us temporarily denote by E_0 the function defined as in (1.4.171) for the choice $c_A := 0$. From [108, Theorem 7.54, p. 270] we know that

$$E_0 \in L^1_{\rm loc}(\mathbb{R}^n,\mathcal{L}^n)\cap\mathscr{C}^\infty(\mathbb{R}^n\setminus\{0\}) \tag{1.4.174}$$

is a tempered distribution in $\mathbb{R}^n$ which is a fundamental solution for the weakly elliptic differential operator $L = a_{rs}\partial_r\partial_s$. Moreover, a cursory inspection reveals that there exists a finite constant $C > 0$ such that for each $x\in\mathbb{R}^n\setminus\{0\}$ we have

$$|E_0(x)| \le \begin{cases} \dfrac{C}{|x|^{n-2}} & \text{if } n\ge 3,\\[2ex] C\big(1+\big|\ln|x|\big|\big) & \text{if } n=2.\end{cases} \tag{1.4.175}$$

With E denoting the fundamental solution of $L = a_{rs}\partial_r\partial_s$ from (1.4.18), the uniqueness result in item *(8)* of Theorem 1.4.2 applies and gives that $Q := E - E_0$ is a polynomial in $\mathbb{R}^n$. In view of (1.4.24) (with $\gamma = 0$) and (1.4.175), said polynomial decays at infinity if $n \ge 3$, and grows at most logarithmically if $n = 2$. This forces Q to vanish identically when $n \ge 3$, and be constant (say, $Q(x) = c_A$ for every $x\in\mathbb{R}^2$) when $n = 2$. The desired conclusion now follows. □

We continue pursuing the goal of producing *fully explicit* formulas for the fundamental solutions of scalar weakly elliptic operators (originally constructed in Theorem 1.4.2 via integral expressions). Henceforth, with each given a matrix of complex coefficients $A = (a_{rs})_{1\le r,s\le n}\in\mathbb{C}^{n\times n}$, we agree to associate the second-order scalar operator in $\mathbb{R}^n$ given by

$$L_A := \nabla\cdot A\nabla = a_{rs}\partial_r\partial_s. \tag{1.4.176}$$

In particular, the "characteristic 1×1 matrix" of L_A is given by

$$L_A(\xi) = -a_{rs}\xi_r\xi_s = -\langle A\xi,\xi\rangle \text{ for each } \xi=(\xi_1,\dots,\xi_n)\in\mathbb{R}^n. \tag{1.4.177}$$

Fundamental Solutions for Weakly Elliptic Scalar Operators when $n = 2$

To set the stage for the subsequent discussion, let us define

$$M_2 := \{A \in \mathbb{C}^{2\times 2} : A^\top = A \text{ and } A \text{ is weakly elliptic}\}. \tag{1.4.178}$$

In relation to this, we have the following structural result.

Lemma 1.4.18 *A matrix $A \in \mathbb{C}^{2\times 2}$ belongs to M_2 if and only if A has the form*

$$A = a\begin{pmatrix} 1 & -\frac{\lambda_1+\lambda_2}{2} \\ -\frac{\lambda_1+\lambda_2}{2} & \lambda_1\lambda_2 \end{pmatrix} \quad \textit{with } a \in \mathbb{C}\setminus\{0\} \textit{ and } \lambda_1, \lambda_2 \in \mathbb{C}\setminus\mathbb{R}, \tag{1.4.179}$$

or, equivalently, A is symmetric and

$$\langle A\xi, \xi\rangle = a(\xi_1 - \lambda_1\xi_2)(\xi_1 - \lambda_2\xi_2), \quad \forall\,\xi = (\xi_1, \xi_2) \in \mathbb{R}^2. \tag{1.4.180}$$

Proof Simple algebra shows that (1.4.179) is equivalent with A being a symmetric matrix satisfying (1.4.180). To proceed, suppose A is as in (1.4.179). Then clearly A is symmetric. Also, thanks to (1.4.180), having $\langle A\xi, \xi\rangle = 0$ for some $\xi \in \mathbb{R}^2 \setminus \{0\}$ forces either $a = 0$, or $\xi_1 = \lambda_1\xi_2$, or $\xi_1 = \lambda_2\xi_2$. However, none of these possibilities materialize since, by assumption, $a \in \mathbb{C}\setminus\{0\}$ and $\lambda_1, \lambda_2 \in \mathbb{C}\setminus\mathbb{R}$. In view of (1.4.120) and (1.4.178) this proves that $A \in M_2$.

To prove the converse implication, fix an arbitrary matrix $A \in M_2$ and write

$$A = \begin{pmatrix} a & b \\ b & c \end{pmatrix}, \tag{1.4.181}$$

where $a, b, c \in \mathbb{C}$. The fact that A is weakly elliptic entails (cf. (1.4.120))

$$a = \langle A\mathbf{e}_1, \mathbf{e}_1\rangle \in \mathbb{C}\setminus\{0\}. \tag{1.4.182}$$

Denote by $\lambda_1, \lambda_2 \in \mathbb{C}$ the roots of the quadratic equation $az^2 + 2bz + c = 0$. Then

$$az^2 + 2bz + c = a(z - \lambda_1)(z - \lambda_2), \tag{1.4.183}$$

which, in view of the equivalence between (1.4.179) and (1.4.180), further implies

$$A = \begin{pmatrix} a & b \\ b & c \end{pmatrix} = a\begin{pmatrix} 1 & -\frac{\lambda_1+\lambda_2}{2} \\ -\frac{\lambda_1+\lambda_2}{2} & \lambda_1\lambda_2 \end{pmatrix}. \tag{1.4.184}$$

To complete the proof we only need to see that $\lambda_1, \lambda_2 \in \mathbb{C}\setminus\mathbb{R}$. Otherwise, if for instance $\lambda_1 \in \mathbb{R}$, we could take $\xi = (\lambda_1, 1) \in \mathbb{R}^2 \setminus \{0\}$ and obtain that

$$\langle A\xi, \xi\rangle = a\lambda_1^2 + 2b\lambda_1 + c = 0, \tag{1.4.185}$$

which contradicts the assumption that A is weakly elliptic. The desired conclusion follows. □

In view of Lemma 1.4.18, from now on we shall assume that every $A \in M_2$ is expressed as in (1.4.179) and decompose

$$M_2 = M_2^0 \sqcup M_2^+ \sqcup M_2^-, \tag{1.4.186}$$

where

$$M_2^0 := \{A \in M_2 : \operatorname{Im}\lambda_1 \cdot \operatorname{Im}\lambda_2 < 0\}, \tag{1.4.187}$$

$$M_2^+ := \{A \in M_2 : \operatorname{Im}\lambda_1 > 0 \text{ and } \operatorname{Im}\lambda_2 > 0\}, \tag{1.4.188}$$

$$M_2^- := \{A \in M_2 : \operatorname{Im}\lambda_1 < 0 \text{ and } \operatorname{Im}\lambda_2 < 0\}. \tag{1.4.189}$$

For further use we note that, since $I_{2\times 2}$ can be written as in (1.4.179) with $a := 1$, $\lambda_1 := \mathrm{i}$, and $\lambda_2 := -\mathrm{i}$,

$$\text{the identity matrix } I_{2\times 2} \text{ belongs to } M_2^0. \tag{1.4.190}$$

Lemma 1.4.19 *Let $A \in \mathbb{C}^{2\times 2}$ be an arbitrary weakly elliptic matrix. Then its symmetric part* $\operatorname{sym} A := \frac{1}{2}(A + A^\top)$ *belongs to M_2. Moreover, if* $\operatorname{sym} A \in M_2^\pm$ *then*

$$\int_{S^1} \frac{1}{\langle A\xi, \xi\rangle}\, d\mathcal{H}^1(\xi) = 0. \tag{1.4.191}$$

Finally, if $\operatorname{sym} A \in M_2^0$*, then expressing* $\operatorname{sym} A$ *as in* (1.4.179) *one has*

$$\int_{S^1} \frac{1}{\langle A\xi, \xi\rangle}\, d\mathcal{H}^1(\xi) = \frac{4\pi\mathrm{i} \cdot \operatorname{sgn}(\operatorname{Im}\lambda_1)}{a(\lambda_1 - \lambda_2)}. \tag{1.4.192}$$

Proof That $\operatorname{sym} A$ belongs to M_2 is clear from assumptions, definitions, and (1.4.166). Express $\operatorname{sym} A$ in its canonical form as in (1.4.179). Then based on (1.4.166) and (1.4.180) we may write

$$\int_{S^1} \frac{1}{\langle A\xi, \xi\rangle}\, d\mathcal{H}^1(\xi) = \frac{1}{a}\int_{S^1} \frac{d\mathcal{H}^1(\xi)}{(\xi_1 - \lambda_1\xi_2)(\xi_1 - \lambda_2\xi_2)}. \tag{1.4.193}$$

Observe that for each $f \in L^1(S^1, \mathcal{H}^1)$ we have

$$\int_{S^1} f(\xi)\, d\mathcal{H}^1(\xi) = 2\int_{-\infty}^{\infty} f_{\text{even}}\Big(\frac{t}{\sqrt{1+t^2}}, \frac{1}{\sqrt{1+t^2}}\Big)\frac{dt}{1+t^2}, \tag{1.4.194}$$

where f_{even} is the even part of f, defined as

$$f_{\text{even}}(\xi) := \tfrac{1}{2}\big(f(\xi) + f(-\xi)\big), \qquad \xi \in S^1. \tag{1.4.195}$$

Indeed, (1.4.194) follows by making the changes of variables $\xi_1 = \pm t/\sqrt{1+t^2}$ and $\xi_2 = \pm 1/\sqrt{1+t^2}$ with $t \in \mathbb{R}$ in (1.4.194). See also [116, §1.6] for a more general version of (1.4.194).

Note that the last integrand in (1.4.193) is even. Bearing this in mind and applying (1.4.194), we may further recast the right-hand side of (1.4.193) as

$$\frac{2}{a}\int_{-\infty}^{\infty}\frac{\mathrm{d}t}{(t-\lambda_1)(t-\lambda_2)} = \int_{\mathbb{R}} g(t)\,\mathrm{d}t, \tag{1.4.196}$$

where g is the meromorphic function

$$g(z) := \frac{2}{a(z-\lambda_1)(z-\lambda_2)} \quad \text{for each } z \in \mathbb{C}\setminus\{\lambda_1, \lambda_2\}. \tag{1.4.197}$$

If $\mathrm{sym}\,A \in M_2^{\pm}$ then either $\lambda_1, \lambda_2 \in \mathbb{C}_+$, or $\lambda_1, \lambda_2 \in \mathbb{C}_-$, and in either scenario the Residue Theorem gives $\int_{\mathbb{R}} g(t)\,\mathrm{d}t = 0$. This justifies (1.4.191). On the other hand, if $\mathrm{sym}\,A \in M_2^0$ then either $\lambda_1 \in \mathbb{C}_+$ and $\lambda_2 \in \mathbb{C}_-$, or $\lambda_1 \in \mathbb{C}_-$ and $\lambda_2 \in \mathbb{C}_+$, so another application of the Residue Theorem then justifies (1.4.192). □

With the piece of terminology introduced in Definition 1.4.10, the lemma below (which complements Lemma 1.4.24) implies that any matrix in M_2^0 is mildly elliptic.

Lemma 1.4.20 *Let $A = (a_{rs})_{1\le r,s\le 2} \in M_2^0$ be arbitrary. Then*

$$\langle Ax, x\rangle \notin a_{11}\,(-\infty, 0] \text{ for every } x \in \mathbb{R}^2\setminus\{0\}. \tag{1.4.198}$$

Proof In view of Lemma 1.4.18, we may assume that the matrix A is written as in (1.4.179). In particular, $a_{11} = a \in \mathbb{C}\setminus\{0\}$. Seeking a contradiction, assume that there exist some $t \in (-\infty, 0]$ along with some $x = (x_1, x_2) \in \mathbb{R}^2\setminus\{0\}$ such that

$$at = \langle Ax, x\rangle = a(x_1 - \lambda_1 x_2)(x_1 - \lambda_2 x_2), \tag{1.4.199}$$

where the second equality comes from (1.4.180). In view of the fact that $a \neq 0$, this implies

$$t = (x_1 - \lambda_1 x_2)(x_1 - \lambda_2 x_2). \tag{1.4.200}$$

If $x_2 = 0$ then $x_1 \neq 0$ and (1.4.200) forces $t = x_1^2$, contradicting the fact that $t \le 0$. If $x_2 \neq 0$, using (1.4.199) we may write

$$\begin{aligned}\frac{tx_2\mathrm{Im}\,\lambda_2}{|x_1-\lambda_2x_2|^2} &= \frac{\mathrm{Im}\,(tx_1 - tx_2\overline{\lambda_2})}{|x_1-\lambda_2x_2|^2} = \mathrm{Im}\left(\frac{t}{x_1-\lambda_2x_2}\right)\\ &= \mathrm{Im}\,(x_1-\lambda_1x_2) = -x_2\mathrm{Im}\,\lambda_1,\end{aligned} \tag{1.4.201}$$

hence

$$\frac{t\,\mathrm{Im}\,\lambda_2}{|x_1-\lambda_2x_2|^2} = -\mathrm{Im}\,\lambda_1. \tag{1.4.202}$$

Using that $t \geq 0$ and bearing in mind that $A \in M_2^0$, it follows that

$$0 \leq \frac{t\mathrm{Im}\,\lambda_1 \cdot \mathrm{Im}\,\lambda_2}{|x_1 - \lambda_2 x_2|^2} = -(\mathrm{Im}\,\lambda_1)^2 < 0, \tag{1.4.203}$$

which is a contradiction. □

Recall the notion of mildly elliptic matrix introduced in Definition 1.4.10 (cf. also (1.4.141)).

Lemma 1.4.21 *Consider an arbitrary symmetric matrix $A \in \mathbb{C}^{2\times 2}$. Then A is mildly elliptic if and only if $A \in M_2^0$.*

Proof That any matrix $A \in M_2^0$ is mildly elliptic follows from Lemma 1.4.20 and Definition 1.4.10. Suppose next that $A \in \mathbb{C}^{2\times 2}$ is a symmetric mildly elliptic matrix. From (1.4.142) and (1.4.178) we then see that $A \in M_2$. Bearing this in mind together with (1.4.186) and (1.4.141), the goal is to prove that

$$\big\{\langle Ax, x\rangle : x \in \mathbb{R}^2 \setminus \{0\}\big\} = \mathbb{C} \setminus \{0\} \text{ whenever } A \in M_2^{\pm}. \tag{1.4.204}$$

Fix an arbitrary $A \in M_2^{\pm}$, and express it as in (1.4.179). Consider the semi-circles

$$\begin{aligned} S_+^1 &:= \big\{\xi = (\xi_1, \xi_2) \in S^1 : \xi_2 \geq 0\big\}, \\ S_-^1 &:= \big\{\xi = (\xi_1, \xi_2) \in S^1 : \xi_2 \leq 0\big\}. \end{aligned} \tag{1.4.205}$$

and note that since $-\xi \in S_-^1$ for each $\xi \in S_+^1$, we have

$$\big\{\langle A\xi, \xi\rangle : \xi \in S^1\big\} = \big\{\langle A\xi, \xi\rangle : \xi \in S_+^1\big\}. \tag{1.4.206}$$

Bearing this in mind, via dilation, it suffices to show that the closed curve[29]

$$\gamma := \big\{\langle A\xi, \xi\rangle : \xi \in S_+^1\big\} \subseteq \mathbb{C} \setminus \{0\} \tag{1.4.207}$$

goes around the origin $0 \in \mathbb{C}$. This, in turn, follows if we succeed in showing that the winding number of the curve γ around the point $0 \in \mathbb{C}$ is nonzero, i.e.,

$$\mathrm{Ind}_\gamma(0) := \frac{1}{2\pi \mathrm{i}} \int_\gamma \frac{dz}{z} \neq 0. \tag{1.4.208}$$

To this end, use the parametrization

[29] indeed, the fact that the end-points of S_+^1 are diametrically opposite implies that the end-points of γ coincide

$$
\begin{aligned}
\gamma(t) &:= \Big\langle A\Big(\frac{-t}{\sqrt{1+t^2}}, \frac{1}{\sqrt{1+t^2}}\Big), \Big(\frac{-t}{\sqrt{1+t^2}}, \frac{1}{\sqrt{1+t^2}}\Big)\Big\rangle \\
&= a\Big(\frac{-t}{\sqrt{1+t^2}} - \frac{\lambda_1}{\sqrt{1+t^2}}\Big)\Big(\frac{-t}{\sqrt{1+t^2}} - \frac{\lambda_2}{\sqrt{1+t^2}}\Big) \\
&= a\frac{(t+\lambda_1)(t+\lambda_2)}{1+t^2}, \qquad -\infty < t < +\infty, \qquad (1.4.209)
\end{aligned}
$$

the second equality being a consequence of (1.4.180). Since

$$
\begin{aligned}
\gamma'(t) &= a\Big(\frac{t^2 + (\lambda_1+\lambda_2)t + \lambda_1\lambda_2}{1+t^2}\Big)' \\
&= a\frac{-(\lambda_1+\lambda_2)(t^2-1) + 2(1-\lambda_1\lambda_2)t}{(1+t^2)^2} \qquad (1.4.210)
\end{aligned}
$$

for each $t \in \mathbb{R}$, we have

$$
\begin{aligned}
\frac{1}{2\pi \mathrm{i}}\int_\gamma \frac{\mathrm{d}z}{z} &= \pm\frac{1}{2\pi \mathrm{i}}\int_{-\infty}^{+\infty} \frac{-(\lambda_1+\lambda_2)(t^2-1) + 2(1-\lambda_1\lambda_2)t}{(t+\lambda_1)(t+\lambda_2)(1+t^2)}\,\mathrm{d}t \\
&= \pm\frac{1}{2\pi \mathrm{i}}\int_{\mathbb{R}} f(t)\,\mathrm{d}t, \qquad (1.4.211)
\end{aligned}
$$

where the $\pm$ is the sign of the parametrization (1.4.209) ("plus" if the parametrization covers the curve γ counterclockwise, and "minus" if the parametrization covers the curve γ clockwise), and where f is the meromorphic function

$$
f(z) := \frac{-(\lambda_1+\lambda_2)(z^2-1) + 2(1-\lambda_1\lambda_2)z}{(z+\lambda_1)(z+\lambda_2)(z+\mathrm{i})(z-\mathrm{i})} \qquad (1.4.212)
$$
$$
\text{defined for each } z \in \mathbb{C} \setminus \{\lambda_1, \lambda_2, \pm\mathrm{i}\}.
$$

Given that $A \in M_2^{\pm}$, we either have $\lambda_1, \lambda_2 \in \mathbb{C}_+$, or $\lambda_1, \lambda_2 \in \mathbb{C}_-$. In the first scenario, the Residue Theorem gives (treating $\mathbb{R}$ as the boundary of $\mathbb{C}_+$)

$$
\begin{aligned}
\frac{1}{2\pi \mathrm{i}}\int_{\mathbb{R}} f(t)\,\mathrm{d}t &= \mathrm{Res}(f, \mathrm{i}) = \frac{2(\lambda_1+\lambda_2) + 2(1-\lambda_1\lambda_2)\mathrm{i}}{(\mathrm{i}+\lambda_1)(\mathrm{i}+\lambda_2)(2\mathrm{i})} \\
&= \frac{-(\lambda_1+\lambda_2)\mathrm{i} + (1-\lambda_1\lambda_2)}{(\mathrm{i}+\lambda_1)(\mathrm{i}+\lambda_2)} = \frac{-(\mathrm{i}+\lambda_1)(\mathrm{i}+\lambda_2)}{(\mathrm{i}+\lambda_1)(i+\lambda_2)} \\
&= -1, \qquad (1.4.213)
\end{aligned}
$$

where the last step uses the fact that $(\mathrm{i}+\lambda_1)(\mathrm{i}+\lambda_2) \neq 0$ since we are presently assuming $\lambda_1, \lambda_2 \in \mathbb{C}_+$.

Finally, if $\lambda_1, \lambda_2 \in \mathbb{C}_-$ then another application of the Residue Theorem gives (now treating $\mathbb{R}$ as the boundary of $\mathbb{C}_-$, with a negative orientation)

$$\begin{aligned}\frac{1}{2\pi \mathrm{i}}\int_{\mathbb{R}} f(t)\,\mathrm{d}t &= -\mathrm{Res}(f,-\mathrm{i}) = -\frac{2(\lambda_1+\lambda_2)-2(1-\lambda_1\lambda_2)\mathrm{i}}{(-\mathrm{i}+\lambda_1)(-\mathrm{i}+\lambda_2)(-2\mathrm{i})}\\ &= -\frac{(\lambda_1+\lambda_2)\mathrm{i}+(1-\lambda_1\lambda_2)}{(-\mathrm{i}+\lambda_1)(-\mathrm{i}+\lambda_2)} = -\frac{-(-\mathrm{i}+\lambda_1)(-\mathrm{i}+\lambda_2)}{(-\mathrm{i}+\lambda_1)(-\mathrm{i}+\lambda_2)}\\ &= 1, \end{aligned} \tag{1.4.214}$$

where the last step relies on the observation that $(-\mathrm{i}+\lambda_1)(-\mathrm{i}+\lambda_2) \neq 0$, given that we now assume $\lambda_1, \lambda_2 \in \mathbb{C}_-$. Thus, (1.4.208) holds in all cases, finishing the proof.□

Going further, we briefly digress for the purpose of introducing notation and terminology which will become useful shortly. Fix an integer $m \in \mathbb{N}$ and denote by $\mathcal{P}_m$ the set of all permutation of $\{1,\dots,m\}$. Given an arbitrary set E, denote by E^m the m-fold Cartesian product of E with itself. On this set, consider the equivalence relation $\sim$ stipulating, for each $(a_1,\dots,a_m) \in E^m$ and $(b_1,\dots,b_m) \in E^m$, that

$$\begin{aligned}&(a_1,\dots,a_m) \sim (b_1,\dots,b_m) \text{ if and only if there exists}\\ &\sigma \in \mathcal{P}_m \text{ such that } b_j = a_{\sigma(j)} \text{ for each } j \in \{1,\dots,m\}.\end{aligned} \tag{1.4.215}$$

For each $(a_1,\dots,a_m) \in E^m$ denote by $\big[(a_1,\dots,a_m)\big]$ its equivalence class modulo the equivalence relation $\sim$. The symbol $E^m/\!\sim$ stands for the set of all such equivalence classes. We shall refer to elements in $E^m/\!\sim$ as being m-families (with elements in E), and use the notation

$$\begin{aligned}&\{a_1,\dots,a_m\}^* := \big[(a_1,\dots,a_m)\big] \in E^m/\!\sim\\ &\quad\text{for each } \ (a_1,\dots,a_m) \in E^m.\end{aligned} \tag{1.4.216}$$

Thus, informally speaking, m-families are "sets of cardinality m, in which repetitions are allowed." If $|\cdot - \cdot|$ is a metric on E, then we equip $E^m/\!\sim$ with the distance[30]

$$D_m\Big(\{a_1,\dots,a_m\}^*, \{b_1,\dots,b_m\}^*\Big) := \min_{\sigma\in\mathcal{P}_m} \max_{1\le j\le m} |a_j - b_{\sigma(j)}|. \tag{1.4.217}$$

For example, corresponding to $m = 2$, we have the following distance

$$\begin{aligned}&D_2\Big(\{\lambda_1,\lambda_2\}^*,\{\widetilde{\lambda}_1,\widetilde{\lambda}_2\}^*\Big)\\ &\qquad := \min\Big\{\max\big\{|\lambda_1-\widetilde{\lambda}_1|, |\lambda_2-\widetilde{\lambda}_2|\big\}, \max\big\{|\lambda_1-\widetilde{\lambda}_2|, |\lambda_2-\widetilde{\lambda}_1|\big\}\Big\}\\ &\qquad \text{for all } \lambda_1,\lambda_2,\widetilde{\lambda}_1,\widetilde{\lambda}_2 \in \mathbb{C},\end{aligned} \tag{1.4.218}$$

on the set of all 2-families with elements in $\mathbb{C}$. As is apparent from the above definition, D_2 is translation invariant, and commutes with positive dilations.

[30] It may be checked, based on definitions, that D_m is indeed a distance function on $E^m/\!\sim$.

Informally speaking, our next result asserts that the location of the two (possibly identical) roots of a quadratic equation, considered together as a 2-family, changes continuously with respect to the coefficients of the quadratic polynomial.

Lemma 1.4.22 *For each $a, b, c \in \mathbb{C}$ with $a \neq 0$, denote by $\lambda_1(a,b,c) \in \mathbb{C}$ and $\lambda_2(a,b,c) \in \mathbb{C}$ the two roots of the quadratic equation*

$$az^2 + 2bz + c = 0. \tag{1.4.219}$$

Then for each fixed triplet $a_o, b_o, c_o \in \mathbb{C}$ with $a_o \neq 0$ it follows that

$$\begin{aligned} &D_2\Big(\big\{\lambda_1(a,b,c), \lambda_2(a,b,c)\big\}^*, \big\{\lambda_1(a_o,b_o,c_o), \lambda_2(a_o,b_o,c_o)\big\}^*\Big) \\ &\qquad \textit{converges to } 0^+ \textit{ as } (a,b,c) \longrightarrow (a_o,b_o,c_o). \end{aligned} \tag{1.4.220}$$

The conclusion in (1.4.220) may be rephrased as saying that the function assigning to each triplet $a, b, c \in \mathbb{C}$ with $a \neq 0$ the 2-family $\big\{\lambda_1(a,b,c), \lambda_2(a,b,c)\big\}^*$ is continuous when regarded as a mapping defined from an open subset of $\mathbb{C}^3$ taking values in the metric space $\big(\mathbb{C}^2/\sim, D_2\big)$.

Proof of Lemma 1.4.22 Since $a \neq 0$, equation (1.4.219) has the same roots as

$$z^2 + 2(b/a)z + (c/a) = 0. \tag{1.4.221}$$

In turn, the roots of (1.4.221) are given by the quadratic formula which, in view of (1.4.216), implies

$$\big\{\lambda_1(a,b,c), \lambda_2(a,b,c)\big\}^* = \left\{ -\tfrac{b}{a} + \sqrt{\tfrac{b^2}{a^2} - \tfrac{c}{a}}, -\tfrac{b}{a} - \sqrt{\tfrac{b^2}{a^2} - \tfrac{c}{a}} \right\}^*. \tag{1.4.222}$$

Above, we agree to define the square-root of $z \in \mathbb{C}$ as $\sqrt{z} := |z|^{1/2} e^{i\theta/2}$ where $\theta \in [0, 2\pi)$ is such that $z = |z| e^{i\theta}$ (which determines θ uniquely if $z \neq 0$). The square-root function thus defined $\mathbb{C} \ni z \to \sqrt{z} \in \mathbb{C}$ is continuous in $\mathbb{C} \setminus (0, \infty)$ and at each real number $z_o \in (0, \infty)$ we have

$$\lim_{\mathbb{C}_+ \ni z \to z_o} \sqrt{z} = \sqrt{z_o} \text{ whereas } \lim_{\mathbb{C}_- \ni z \to z_o} \sqrt{z} = -\sqrt{z_o}. \tag{1.4.223}$$

While this change in sign renders the square-root function discontinuous at each point on the real semi-axis $(0, \infty)$, it does not affect the continuity of the 2-family valued function $\mathbb{C} \ni z \to \big\{\sqrt{z}, -\sqrt{z}\big\}^*$, in the sense that for each $z_o \in \mathbb{C}$ we have

$$D\Big(\big\{\sqrt{z}, -\sqrt{z}\big\}^*, \big\{\sqrt{z_o}, -\sqrt{z_o}\big\}^*\Big) \longrightarrow 0^+ \text{ as } z \to z_o. \tag{1.4.224}$$

Then the claim made in (1.4.220) follows from this and (1.4.222), bearing in mind that D_2 is translation invariant. □

Moving on, since any 2×2 symmetric matrix with complex entries is uniquely determined by its entries lying on or below the main diagonal, we may naturally identify M_2 with a subset of $\mathbb{C}^3$ (stable under multiplication by real numbers). With this identification in mind, we have the following result pertaining to the topology of M_2.

Lemma 1.4.23 *When viewed as subsets in $\mathbb{C}^3$ (as indicated above), the sets M_2^0, M_2^+, and M_2^- are open, pathwise connected, and mutually disjoint. As such, these sets are the connected components of the open set M_2 (regarded as a subset of $\mathbb{C}^3$).*

Proof That M_2 is an open subset of $\mathbb{C}^3$ (under the identification described above) is clear from (1.4.122). Note that M_2^0 is canonically identified with

$$\Big\{(a,b,c)\in\mathbb{C}^3 : \text{ one root of } az^2+2bz+c=0 \text{ is in } \mathbb{C}_+ \text{ and one in } \mathbb{C}_-\Big\}. \quad (1.4.225)$$

From this and Lemma 1.4.22 we then see that M_2^0 is an open subset of $\mathbb{C}^3$. Likewise, since $M_2^{\pm}$ are canonically identified with

$$\Big\{(a,b,c)\in\mathbb{C}^3 : \text{ both roots of } az^2+2bz+c=0 \text{ are in } \mathbb{C}_{\pm}\Big\}, \quad (1.4.226)$$

from Lemma 1.4.22 we conclude that they are open subsets of $\mathbb{C}^3$.

Next, fix two arbitrary matrices $A, B\in M_2^0$. From (1.4.187) and Lemma 1.4.18 we know that we can write

$$A = a\begin{pmatrix} 1 & -\frac{\lambda_1+\lambda_2}{2} \\ -\frac{\lambda_1+\lambda_2}{2} & \lambda_1\lambda_2 \end{pmatrix}, \qquad B = b\begin{pmatrix} 1 & -\frac{\gamma_1+\gamma_2}{2} \\ -\frac{\gamma_1+\gamma_2}{2} & \gamma_1\gamma_2 \end{pmatrix}, \quad (1.4.227)$$

with $a, b\in\mathbb{C}\setminus\{0\}$, $\operatorname{Im}\lambda_1\cdot\operatorname{Im}\lambda_2<0$ and $\operatorname{Im}\gamma_1\cdot\operatorname{Im}\gamma_2<0$. Relabeling if needed we may assume that $\operatorname{Im}\lambda_1, \operatorname{Im}\gamma_1>0$ and $\operatorname{Im}\lambda_1, \operatorname{Im}\gamma_2<0$. It is useful to express a, b as

$$a=|a|e^{i\theta_a} \text{ and } b=|b|e^{i\theta_b} \text{ for some } \theta_a,\theta_b\in[0,2\pi]. \quad (1.4.228)$$

For each $t\in[0,1]$ we then define

$$\begin{aligned} c(t) &:= |a|^{1-t}|b|^t e^{i(1-t)\theta_a+it\theta_b}, \\ \rho_1(t) &:= (1-t)\lambda_1+t\gamma_1, \\ \rho_2(t) &:= (1-t)\lambda_2+t\gamma_2. \end{aligned} \quad (1.4.229)$$

Clearly, $c(t)\in\mathbb{C}\setminus\{0\}$, $\operatorname{Im}\rho_1(t)>0$, $\operatorname{Im}\rho_2(t)<0$ for every $t\in[0,1]$. Hence, if we set

$$C(t) := c(t)\begin{pmatrix} 1 & -\frac{\rho_1(t)+\rho_2(t)}{2} \\ -\frac{\rho_1(t)+\rho_2(t)}{2} & \rho_1(t)\rho_2(t) \end{pmatrix} \quad (1.4.230)$$

it follows that the matrix $C(t)$ belongs to M_2^0 for every $t\in[0,1]$. We have therefore constructed a function $[0,1]\ni t\mapsto C(t)\in\mathbb{C}^{2\times 2}$ which is continuous, its image is contained in M_2^0, and satisfies $C(0) = A$ and $C(1) = B$. Hence, M_2^0 is pathwise connected.

In the case when $A, B \in M_2^+$ (respectively, M_2^-) the same type of construction works. This time, (1.4.229) ensures that if $\operatorname{Im}\lambda_1$, $\operatorname{Im}\lambda_2$, $\operatorname{Im}\gamma_1$, $\operatorname{Im}\gamma_2$ are all strictly positive (respectively, strictly negative) then $\operatorname{Im}\rho_1(t)$, $\operatorname{Im}\rho_2(t)$ are strictly positive (respectively, strictly negative). Consequently, $C(t)$ as define in (1.4.230) belongs to M_2^+ (respectively, M_2^-).

In summary, the above argument shows that M_2^0, M_2^+, M_2^- are all pathwise connected. Since, by design, M_2^0, M_2^+, and M_2^- are also mutually disjoint open subsets of $\mathbb{C}^3$, whose union is M_2, it follows that these sets are precisely the connected components of the open set M_2. □

The prototypes of all weakly elliptic scalar second-order differential operators in the two dimensional setting, namely the Laplacian $\Delta = \partial_{x_1}^2 + \partial_{x_2}^2$, Bitsadze's operator $\frac{1}{4}(\partial_{x_1} - \mathrm{i}\partial_{x_2})^2$, and the square of the Cauchy-Riemann operator $\frac{1}{4}(\partial_{x_1} + \mathrm{i}\partial_{x_2})^2$, may be expressed using matrices of coefficients belonging to M_2^0, M_2^+, and M_2^-, respectively. Of course, Δ may be written using the identity as its matrix of coefficients, which is actually Legendre-Hadamard elliptic. Our next lemma sheds further light on this phenomenon.

Lemma 1.4.24 *The subset of M_2 described as*

$$M_2^{\mathrm{LH}} := \big\{A \in \mathbb{C}^{2\times 2} : A^\top = A \text{ and } A \text{ is Legendre-Hadamard elliptic}\big\} \tag{1.4.231}$$

is included in M_2^0.

It should be noted that the above inclusion is strict. For example,

$$A := \begin{pmatrix} 1 & -1 \\ -1 & 1 + 2\mathrm{i} \end{pmatrix} \tag{1.4.232}$$

is a complex 2×2 matrix which may be written as in (1.4.179) with

$$a := 1 \in \mathbb{C} \setminus \{0\} \text{ and } \lambda_1 := \mathrm{i} \in \mathbb{C} \setminus \mathbb{R}, \quad \lambda_2 := 2 + \mathrm{i} \in \mathbb{C} \setminus \mathbb{R}. \tag{1.4.233}$$

As such, Lemma 1.4.18 ensures that the matrix A belongs to M_2. In addition, for each $\xi = (\xi_1, \xi_2) \in \mathbb{R}^2$ we have

$$\operatorname{Re}\langle A\xi, \xi\rangle = \operatorname{Re}\big(\xi_1^2 - 2\xi_1\xi_2 + (1 + 2\mathrm{i})\xi_2^2\big) = \xi_1^2 - 2\xi_1\xi_2 + \xi_2^2 = (\xi_1 - \xi_2)^2. \tag{1.4.234}$$

In particular, $\operatorname{Re}\langle A\xi, \xi\rangle$ vanishes whenever ξ belongs to the diagonal in $\mathbb{R} \times \mathbb{R}$, so A fails to be Legendre-Hadamard elliptic.

Proof of Lemma 1.4.24 Recall from (1.4.190) that $I_{2\times 2} \in M_2^0$. Fix $A \in M_2^{\mathrm{LH}}$ and consider the continuous path

$$[0, 1] \ni t \mapsto C(t) = (1 - t)I_{2\times 2} + tA \in \mathbb{C}^{2\times 2}. \tag{1.4.235}$$

Then $C(0) = I_{2\times2}$, $C(1) = A$, and $C(t)^{\top} = C(t)$ for every $t \in [0, 1]$. Moreover for each $\xi \in \mathbb{R}^2$

$$\mathrm{Re}\langle C(t)\xi, \xi\rangle = (1-t)|\xi|^2 + t\mathrm{Re}\langle A\xi, \xi\rangle \geq ((1-t) + \kappa t)|\xi|^2 \tag{1.4.236}$$

where we have used that A is Legendre-Hadamard elliptic with constant $\kappa > 0$ (cf. (1.4.123)). Thus, for every $t \in [0, 1]$ the matrix $C(t)$ is Legendre-Hadamard elliptic with constant $(1-t) + \kappa t > 0$. Consequently, $C(t) \in M_2^{\mathrm{LH}} \subset M_2$ for every $t \in [0, 1]$. Moreover, since M_2^0 is a connected component of M_2 and $I_{2\times2} \in M_2^0$ it follows that $C(t) \in M_2^0$ for every $t \in [0, 1]$. Therefore, $A = C(1) \in M_2^0$ as desired. □

Here is another example, of a different nature than (1.4.232), which will play a role later on, in [115, §8.2], in the treatment of the operators L_λ defined in (1.3.63).

Example 1.4.25 *For any $a, b \in \mathbb{C}$, a matrix of the form*

$$A = a\begin{pmatrix} 1 & b \\ b & -1 \end{pmatrix} \tag{1.4.237}$$

belongs to M_2 if and only if

$$a \in \mathbb{C} \setminus \{0\} \ \textit{and}\ b \in \mathbb{C} \setminus \mathbb{R}. \tag{1.4.238}$$

In addition,

$$\begin{gathered} \textit{whenever } (1.4.238) \textit{ holds then} \\ A \in M_2^+ \textit{ if } \mathrm{Im}\, b < 0, \textit{ and } A \in M_2^- \textit{ if } \mathrm{Im}\, b > 0. \end{gathered} \tag{1.4.239}$$

As a consequence of this, (1.4.186), *and Lemma 1.4.24, any matrix A as in* (1.4.237)-(1.4.238) *is not Legendre-Hadamard elliptic.*

Indeed, if $\lambda_1, \lambda_2 \in \mathbb{C}$ solve $-\frac{\lambda_1+\lambda_2}{2} = b$ and $\lambda_1\lambda_2 = -1$, then having $\lambda_1, \lambda_2 \in \mathbb{C} \setminus \mathbb{R}$ becomes equivalent to having $b \in \mathbb{C} \setminus \mathbb{R}$. According to Lemma 1.4.18, the conditions in (1.4.238) are then equivalent to the membership of A to M_2. In addition, having $\lambda_1\lambda_2 = -1$ guarantees that $\lambda_1, \lambda_2 \neq 0$, so we may express $\lambda_2 = \frac{-1}{\lambda_1} = -\frac{\overline{\lambda_1}}{|\lambda_1|^2}$, which further implies $\mathrm{Im}\,\lambda_2 = \frac{\mathrm{Im}\,\lambda_1}{|\lambda_1|^2}$. As a consequence, $\mathrm{Im}\,\lambda_1$ and $\mathrm{Im}\,\lambda_2$ have the same sign, which is actually the opposite of $\mathrm{Im}\, b$. The claims in (1.4.239) now follow from this and (1.4.186)-(1.4.189).

Let us move on. The following residue computation is going to be relevant shortly (see the proof of Lemma 1.4.28).

Lemma 1.4.26 *For every $\lambda \in \mathbb{C} \setminus \mathbb{R}$ one has*

$$I_\lambda := \int_{S^1} \frac{\xi_1 - \lambda\xi_2}{\xi_2 + \lambda\xi_1}\, d\mathcal{H}^1(\xi) = -2\pi \mathrm{i} \cdot \mathrm{sign}\,(\mathrm{Im}\,\lambda). \tag{1.4.240}$$

Proof Upon noting that the integrand in I_λ is even, it follows from (1.4.194) that

$$I_\lambda = 2\int_{-\infty}^{\infty} \frac{t-\lambda}{1+\lambda t}\,\frac{dt}{1+t^2}. \tag{1.4.241}$$

Assume now that $\operatorname{Im}\lambda > 0$ and consider the meromorphic function

$$f(z) := \frac{(z-\lambda)}{(1+\lambda z)(z-i)(z+i)} \quad \text{for each } z \in \mathbb{C}_-. \tag{1.4.242}$$

Then, according to the Residue Theorem,

$$I_\lambda = -4\pi i \cdot \operatorname{Res}(f, -i) = -4\pi i \frac{-i-\lambda}{(1-i\lambda)(-2i)} = -2\pi i, \tag{1.4.243}$$

which is in agreement with (1.4.240). When $\operatorname{Im}\lambda < 0$, we considering the meromorphic function

$$f(z) := \frac{(z-\lambda)}{(1+\lambda z)(z-i)(z+i)} \quad \text{for each } z \in \mathbb{C}_+. \tag{1.4.244}$$

and use the Residue Theorem to compute

$$I_\lambda = 4\pi i \cdot \operatorname{Res}(f, i) = 4\pi i \frac{i-\lambda}{(1+i\lambda)2i} = 2\pi i, \tag{1.4.245}$$

as wanted. □

The subsequent discussion branches out into a number of cases, focused on special classes of matrices.

Case I: *Matrices in* M_2^0.

Fix $A \in M_2^0$, that is,

$$A = a\begin{pmatrix} 1 & -\frac{\lambda_1+\lambda_2}{2} \\ -\frac{\lambda_1+\lambda_2}{2} & \lambda_1\lambda_2 \end{pmatrix} \quad \text{with } a \in \mathbb{C}\setminus\{0\} \text{ and } \operatorname{Im}\lambda_1 \cdot \operatorname{Im}\lambda_2 < 0. \tag{1.4.246}$$

Let $L = L_A := \nabla \cdot A\nabla$ be the second-order differential operator associated with A as in (1.4.176). In the present case this takes the form

$$L = a\Big[\partial_1^2 - \frac{\lambda_1+\lambda_2}{2}\partial_1\partial_2 - \frac{\lambda_1+\lambda_2}{2}\partial_2\partial_1 + \lambda_1\lambda_2\partial_2^2\Big], \tag{1.4.247}$$

hence also

$$L = a(\partial_1 - \lambda_1\partial_2)(\partial_1 - \lambda_2\partial_2). \tag{1.4.248}$$

Recall the piece of notation introduced in (1.4.9) and note that, in the current setting, this becomes

$$\partial_\xi^A = a\Big[\xi_1\partial_1 - \frac{\lambda_1+\lambda_2}{2}\xi_1\partial_2 - \frac{\lambda_1+\lambda_2}{2}\xi_2\partial_1 + \lambda_1\lambda_2\xi_2\partial_2\Big] \tag{1.4.249}$$

for each $\xi = (\xi_1, \xi_2) \in \mathbb{R}^2$. Bring in

$$A^{adj} := a\begin{pmatrix} \lambda_1\lambda_2 & \frac{\lambda_1+\lambda_2}{2} \\ \frac{\lambda_1+\lambda_2}{2} & 1 \end{pmatrix}. \tag{1.4.250}$$

Finally, define an auxiliary function $\varphi_A : \mathbb{R}^2 \to \mathbb{C}$ by setting

$$\begin{aligned}\varphi_A(x) &:= \langle A^{adj}x, x\rangle = a[\lambda_1\lambda_2 x_1^2 + (\lambda_1+\lambda_2)x_1x_2 + x_2^2] \\ &= a(x_2+\lambda_1 x_1)(x_2+\lambda_2 x_1) \text{ for each } x = (x_1,x_2)\in\mathbb{R}^2.\end{aligned} \tag{1.4.251}$$

Since, generally speaking,

$$\begin{gathered}\text{for each fixed number } \lambda \in \mathbb{C}\setminus\mathbb{R} \text{ we have} \\ |x_1+\lambda x_2| \approx |x| \text{ uniformly for } x=(x_1,x_2)\in\mathbb{R}^2,\end{gathered} \tag{1.4.252}$$

it follows that $\varphi_A(x) = 0$ if and only if $x = 0$.

In the next two lemmas we take a closer look at the function $\log\varphi_A$ in $\mathbb{R}^2\setminus\{0\}$.

Lemma 1.4.27 *Employing notation introduced above in relation to a matrix $A \in M_2^0$, one has $\varphi_A(x) \neq at$ for each $t \in (-\infty, 0]$ and $x \in \mathbb{R}^2\setminus\{0\}$. Hence, if* log *is a holomorphic branch of the complex logarithm associated with the simply connected domain $\mathbb{C}\setminus\{at : t\in(-\infty,0]\}$ (cf.* [112, Remark 5.8.2]), *it follows that* $\log\varphi_A$ *is a well-defined function in $\mathscr{C}^\infty(\mathbb{R}^2\setminus\{0\})$.*

Proof Note that by (1.4.250) we may express

$$A^{adj} = a\lambda_1\lambda_2\begin{pmatrix} 1 & \frac{\lambda_1^{-1}+\lambda_2^{-1}}{2} \\ \frac{\lambda_1^{-1}+\lambda_2^{-1}}{2} & \lambda_1^{-1}\lambda_2^{-1} \end{pmatrix}, \tag{1.4.253}$$

with $(a\lambda_1\lambda_2) \in \mathbb{C}\setminus\{0\}$ and

$$\operatorname{Im}(\lambda_1^{-1})\cdot\operatorname{Im}(\lambda_2^{-1}) = \frac{\operatorname{Im}\lambda_1\cdot\operatorname{Im}\lambda_2}{|\lambda_1\lambda_2|^2} < 0. \tag{1.4.254}$$

Hence A^{adj} belongs to M_2^0, so Lemma 1.4.20 then guarantees that $\varphi_A(x) \neq at$ whenever $t \leq 0$ and $x \in \mathbb{R}^2\setminus\{0\}$. □

Here is the second lemma alluded to above, designed to facilitate eventually invoking Proposition 1.4.1 later on.

Lemma 1.4.28 *Using previous notation introduced above in relation to the matrix $A \in M_2^0$ (cf.* (1.4.246) *and* (1.4.251)), *one has*

$$\int_{S^1} \partial_\xi^A[\log \varphi_A](\xi)\, \mathrm{d}\mathcal{H}^1(\xi) = -2\pi i a(\lambda_1 - \lambda_2) \cdot \operatorname{sign}(\operatorname{Im}\lambda_1). \tag{1.4.255}$$

Proof Fix an arbitrary $\xi = (\xi_1, \xi_2) \in \mathbb{R}^2 \setminus \{0\}$. Then (1.4.251) and [112, (5.8.6)] imply

$$\nabla[\log \varphi_A](\xi) = \Big(\frac{\lambda_1}{\xi_2 + \lambda_1\xi_1} + \frac{\lambda_2}{\xi_2 + \lambda_2\xi_1}, \frac{1}{\xi_2 + \lambda_1\xi_1} + \frac{1}{\xi_2 + \lambda_2\xi_1}\Big), \tag{1.4.256}$$

so on account of (1.4.249) we may compute

$$\begin{aligned}
&\partial_\xi^A[\log\varphi_A](\xi) \\
&\quad = a\Big[\frac{\lambda_1\xi_1}{\xi_2 + \lambda_1\xi_1} + \frac{\lambda_2\xi_1}{\xi_2 + \lambda_2\xi_1} - \frac{\lambda_1 + \lambda_2}{2}\frac{\xi_1}{\xi_2 + \lambda_1\xi_1} - \frac{\lambda_1 + \lambda_2}{2}\frac{\xi_1}{\xi_2 + \lambda_2\xi_1} \\
&\qquad - \frac{\lambda_1 + \lambda_2}{2}\frac{\lambda_1\xi_2}{\xi_2 + \lambda_1\xi_1} - \frac{\lambda_1 + \lambda_2}{2}\frac{\lambda_2\xi_2}{\xi_2 + \lambda_2\xi_1} + \frac{\lambda_1\lambda_2\xi_2}{\xi_2 + \lambda_1\xi_1} + \frac{\lambda_1\lambda_2\xi_2}{\xi_2 + \lambda_2\xi_1}\Big] \\
&\quad = a\frac{\lambda_1 - \lambda_2}{2}\Big[\frac{\xi_1 - \lambda_1\xi_2}{\xi_2 + \lambda_1\xi_1} - \frac{\xi_1 - \lambda_2\xi_2}{\xi_2 + \lambda_2\xi_1}\Big].
\end{aligned} \tag{1.4.257}$$

This along with Lemma 1.4.26 give

$$\begin{aligned}
\int_{S^1} \partial_\xi^A[\log \varphi_A](\xi)\, \mathrm{d}\mathcal{H}^1(\xi) &= a\frac{\lambda_1 - \lambda_2}{2}(I_{\lambda_1} - I_{\lambda_2}) \\
&= -\pi i a(\lambda_1 - \lambda_2)\big(\operatorname{sign}(\operatorname{Im}\lambda_1) - \operatorname{sign}(\operatorname{Im}\lambda_2)\big) \\
&= -2\pi i a(\lambda_1 - \lambda_2)\operatorname{sign}(\operatorname{Im}\lambda_1),
\end{aligned} \tag{1.4.258}$$

where we have used that $\operatorname{Im}\lambda_1$ and $\operatorname{Im}\lambda_2$ have opposite signs, since $A \in M_2^0$. □

We are now in a position to state the main result concerning an explicit formula for the fundamental solution constructed in Theorem 1.4.2 in relation to differential operators of the form $L_A := \nabla \cdot A\nabla$ for a weakly elliptic matrix A whose symmetric part belongs to M_2^0.

Proposition 1.4.29 *Let $A \in \mathbb{C}^{2\times 2}$ be a weakly elliptic matrix with the property that its symmetric part,* $\operatorname{sym} A := \frac{1}{2}(A + A^\top)$, *belongs to M_2^0. Express the matrix* $\operatorname{sym} A$ *as in* (1.4.246) *and for each point $x = (x_1, x_2) \in \mathbb{R}^2 \setminus \{0\}$ consider*

$$E_A(x) := -\frac{\operatorname{sign}(\operatorname{Im}\lambda_1)}{2\pi i a(\lambda_1 - \lambda_2)} \log\big[a(x_2 + \lambda_1 x_1)(x_2 + \lambda_2 x_1)\big], \tag{1.4.259}$$

where log *is the logarithm branch from Lemma 1.4.27.*

Then $E_A \in L^2_{\mathrm{loc}}(\mathbb{R}^2, \mathcal{L}^2)$ and, when regarded as a distribution in $\mathbb{R}^2$, this is a fundamental solution of $L_A := \nabla \cdot A\nabla$ (the second-order differential operator associated with A as in (1.4.176)*). Finally, there exists some $C_A \in \mathbb{C}$ such that*

$E = E_A + C_A$ where E is the canonical fundamental solution of $L = L_A$ constructed in Theorem 1.4.2.

Proof From (1.4.259), Lemma 1.4.27, and (1.4.256) we conclude that

$$E_A \in \mathscr{C}^\infty(\mathbb{R}^2 \setminus \{0\}) \cap L^2_{\rm loc}(\mathbb{R}^2, \mathcal{L}^2) \cap \mathcal{S}'(\mathbb{R}^2). \tag{1.4.260}$$

From (1.4.256) we also see that the first-order derivatives of E_A are positive homogeneous of degree -1 in $\mathbb{R}^2 \setminus \{0\}$. In addition, it is straightforward to check that $L_A E_A(x) = 0$ at every point $x \in \mathbb{R}^2 \setminus \{0\}$. These properties and Proposition 1.4.1 prove that E_A is indeed a fundamental solution of L_A. Granted this, if E is the canonical fundamental solution associated in relation to the current operator L as in Theorem 1.4.2, then item *(8)* in Theorem 1.4.2 gives that the difference $Q := E_A - E$ is a polynomial. Since both E and E_A have at most logarithmic growth, it follows that Q is a polynomial with at most logarithmic growth. Hence, Q must be constant, call it $C_A \in \mathbb{C}$. Ultimately, $E = E_A + C_A$ as desired. □

Pressing on, we now consider:

Case II: *Assume $A \in M_2^\pm$ with $\lambda_1 \neq \lambda_2$.*

Fix $A \in M_2^\pm$. From the structural result proved in Lemma 1.4.18 and (1.4.189)-(1.4.189) we know that this may be expressed as

$$A = a \begin{pmatrix} 1 & -\frac{\lambda_1+\lambda_2}{2} \\ -\frac{\lambda_1+\lambda_2}{2} & \lambda_1\lambda_2 \end{pmatrix} \quad \text{with } a \in \mathbb{C} \setminus \{0\} \text{ and } \operatorname{Im}\lambda_1 \cdot \operatorname{Im}\lambda_2 > 0. \tag{1.4.261}$$

Here we work under the additional assumption that $\lambda_1 \neq \lambda_2$.

Much as before we consider the differential operator $L_A := \nabla \cdot A\nabla$ associated with A (cf. also (1.4.247)), and recall the definition of ∂^A_ξ given in (1.4.249). In this case, we define the auxiliary function

$$\varphi_A(x) := a\frac{x_2 + \lambda_1 x_1}{x_2 + \lambda_2 x_1}, \qquad x \in \mathbb{R}^2 \setminus \{0\}. \tag{1.4.262}$$

Observe that the denominator never vanishes for $x \in \mathbb{R}^2 \setminus \{0\}$ since $\lambda_2 \in \mathbb{C} \setminus \mathbb{R}$, hence φ_A is well defined and $\mathscr{C}^\infty$ smooth in $\mathbb{R}^2 \setminus \{0\}$. Moreover, φ_A is homogeneous of degree zero, and from (1.4.252) we also conclude that φ_A stays away from zero and infinity.

Lemma 1.4.30 *The auxiliary function φ_A associated with A as in* (1.4.262) *satisfies $\varphi_A(x) \neq at$ whenever $t \leq 0$ and $x \in \mathbb{R}^2 \setminus \{0\}$. Hence, with* log *denoting a branch of the complex logarithm defined as a holomorphic function the simply connected domain $\mathbb{C} \setminus \{at : t \in (-\infty, 0]\}$ (cf.* [112, Remark 5.8.2]), *it follows that* $\log \varphi_A$ *is a well-defined function, belonging to $\mathscr{C}^\infty(\mathbb{R}^2 \setminus \{0\})$.*

Proof Consider the case when $A \in M_2^+$ (the case when $A \in M_2^-$ is completely analogous). Thus, we have $\operatorname{Im}\lambda_1 > 0$ and $\operatorname{Im}\lambda_2 > 0$. Fix $x = (x_1, x_2) \in \mathbb{R}^2 \setminus \{0\}$

arbitrary. If $x_1 > 0$ (respectively, if $x_1 < 0$), then $(x_2 + \lambda_1 x_1)$, $(x_2 + \lambda_1 x_1)$ belong to $\mathbb{C}_+$ (respectively, to $\mathbb{C}_-$). As a consequence, whenever $x_1 \neq 0$ it follows that the argument of $(x_2 + \lambda_1 x_1)/(x_2 + \lambda_2 x_1)$ lies in the open interval $(-\pi, \pi)$. Hence, in this scenario we have

$$\varphi_A(x) = a\frac{x_2 + \lambda_1 x_1}{x_2 + \lambda_2 x_1} \neq at \text{ whenever } t \leq 0. \tag{1.4.263}$$

Finally, if $x_1 = 0$ then $x_2 \neq 0$ from which we conclude that $\varphi_A(x) = a \neq at$ whenever $t \leq 0$. □

As in the past, we next consider the conormal derivative of $\log \varphi_A$ on the unit sphere.

Lemma 1.4.31 *Employing previous notation introduced above in relation to the matrix $A \in M_2^{\pm}$ (cf. (1.4.261) and (1.4.262)), one has*

$$\int_{S^1} \partial_\xi^A[\log \varphi_A](\xi)\, d\mathcal{H}^1(\xi) = -2\pi i a \cdot \operatorname{sign}(\operatorname{Im} \lambda_1)(\lambda_1 - \lambda_2). \tag{1.4.264}$$

Proof Pick some arbitrary $\xi = (\xi_1, \xi_2) \in \mathbb{R}^2 \setminus \{0\}$. Then from (1.4.262) and [112, (5.8.6)] we deduce that

$$\nabla[\log \varphi_A](\xi) = \Big(\frac{\lambda_1}{\xi_2 + \lambda_1\xi_1} - \frac{\lambda_2}{\xi_2 + \lambda_2\xi_1}, \frac{1}{\xi_2 + \lambda_1\xi_1} - \frac{1}{\xi_2 + \lambda_2\xi_1}\Big), \tag{1.4.265}$$

which in concert with (1.4.249) yields

$$\begin{aligned}
&\partial_\xi^A[\log\varphi_A](\xi)\\
&= a\Big[\frac{\lambda_1\xi_1}{\xi_2 + \lambda_1\xi_1} - \frac{\lambda_2\xi_1}{\xi_2 + \lambda_2\xi_1} - \frac{\lambda_1 + \lambda_2}{2}\frac{\xi_1}{\xi_2 + \lambda_1\xi_1} + \frac{\lambda_1 + \lambda_2}{2}\frac{\xi_1}{\xi_2 + \lambda_2\xi_1}\\
&\quad - \frac{\lambda_1 + \lambda_2}{2}\frac{\lambda_1\xi_2}{\xi_2 + \lambda_1\xi_1} + \frac{\lambda_1 + \lambda_2}{2}\frac{\lambda_2\xi_2}{\xi_2 + \lambda_2\xi_1} + \frac{\lambda_1\lambda_2\xi_2}{\xi_2 + \lambda_1\xi_1} - \frac{\lambda_1\lambda_2\xi_2}{\xi_2 + \lambda_2\xi_1}\Big]\\
&= a\frac{\lambda_1 - \lambda_2}{2}\Big[\frac{\xi_1 - \lambda_1\xi_2}{\xi_2 + \lambda_1\xi_1} + \frac{\xi_1 - \lambda_2\xi_2}{\xi_2 + \lambda_2\xi_1}\Big].
\end{aligned} \tag{1.4.266}$$

Along with Lemma 1.4.26, this permits us to compute

$$\begin{aligned}
\int_{S^1} \partial_\xi^A[\log \varphi_A](\xi)\, d\mathcal{H}^1(\xi) &= a\frac{\lambda_1 - \lambda_2}{2}(I_{\lambda_1} + I_{\lambda_2})\\
&= -\pi i a(\lambda_1 - \lambda_2)\big(\operatorname{sign}(\operatorname{Im}\lambda_1) + \operatorname{sign}(\operatorname{Im}\lambda_2)\big)\\
&= -2\pi i a(\lambda_1 - \lambda_2)\operatorname{sign}(\operatorname{Im}\lambda_1),
\end{aligned} \tag{1.4.267}$$

where we have used the fact that $\operatorname{Im}\lambda_1$ and $\operatorname{Im}\lambda_2$ have the same sign, given that $A \in M_2^{\pm}$. □

We can now provide an explicit formula for the fundamental solution constructed in Theorem 1.4.2 for differential operators of the form $L_A := \nabla \cdot A\nabla$ for a weakly elliptic matrix $A \in \mathbb{C}^{2\times 2}$ satisfying $\operatorname{sym} A \in M_2^{\pm}$ with $\lambda_1 \neq \lambda_2$.

Proposition 1.4.32 *Let $A \in \mathbb{C}^{2\times 2}$ be a weakly elliptic matrix with the property that its symmetric part,* $\operatorname{sym} A := \frac{1}{2}(A + A^{\top})$, *belongs to $M_2^{\pm}$. Write* $\operatorname{sym} A$ *as in* (1.4.261) *and assume further that $\lambda_1 \neq \lambda_2$. With* log *denoting the logarithm branch from Lemma 1.4.30, for each $x = (x_1, x_2) \in \mathbb{R}^2 \setminus \{0\}$, define*

$$E_A(x) := -\frac{\operatorname{sign}(\operatorname{Im}\lambda_1)}{2\pi i a(\lambda_1 - \lambda_2)} \log\Big[a\frac{x_2 + \lambda_1 x_1}{x_2 + \lambda_2 x_1}\Big]. \tag{1.4.268}$$

Then $E_A \in L^2_{\mathrm{loc}}(\mathbb{R}^2, \mathcal{L}^2)$ and, when regarded as a distribution in $\mathbb{R}^2$, this is a fundamental solution of $L_A := \nabla \cdot A\nabla$ (the second-order differential operator associated with A as in (1.4.176)*). Moreover, E_A is positive homogeneous of degree zero (in particular, E_A is bounded) in $\mathbb{R}^2 \setminus \{0\}$, and for each multi-index $\alpha \in \mathbb{N}_0^2$ one has*

$$(\partial^\alpha E_A)(x) = O\big(|x|^{-|\alpha|}\big) \text{ as } |x| \to \infty. \tag{1.4.269}$$

Finally, there exists $C_A \in \mathbb{C}$ such that $E = E_A + C_A$ where E is the canonical fundamental solution of $L = L_A$ constructed in Theorem 1.4.2.

Compared to the fundamental solutions for "generic" weakly elliptic systems constructed in Theorem 1.4.2, a significant feature of E_A from (1.4.268) is that this function is positive homogeneous of degree zero (in particular, E_A is bounded in $\mathbb{R}^2 \setminus \{0\}$). In view of (1.4.177) and Lemma 1.4.19, this is in line with the prediction made in (1.4.54).

Also, E_A from (1.4.268) is singular at the origin, but otherwise $\mathscr{C}^\infty$ smooth and bounded in $\mathbb{R}^2 \setminus \{0\}$. This stands in sharp contrast with the nature of the fundamental solution for the Laplacian (in this vein, recall that such singularities are removable for harmonic functions).

Proof of Proposition 1.4.32 As seen from (1.4.268), Lemma 1.4.30, and (1.4.265),

$$E_A \in \mathscr{C}^\infty(\mathbb{R}^2 \setminus \{0\}) \cap L^2_{\mathrm{loc}}(\mathbb{R}^2, \mathcal{L}^2) \cap \mathcal{S}'(\mathbb{R}^2). \tag{1.4.270}$$

That E_A is positive homogeneous of degree zero and satisfies (1.4.269) is clear from (1.4.268) and (1.4.252). In particular, the first-order derivatives of E_A are positive homogeneous of degree -1 in $\mathbb{R}^2 \setminus \{0\}$ (this is also apparent from (1.4.265)). In addition, it is straightforward to check that $L_A E_A(x) = 0$ for every $x \in \mathbb{R}^2 \setminus \{0\}$. Granted these properties, Proposition 1.4.1 guarantees that E_A is a fundamental solution when viewed as a distribution in $\mathbb{R}^2$. Having established this, item *(8)* in Theorem 1.4.2 then shows that $Q := E_A - E$ is a polynomial, where E is the canonical fundamental solution associated with $L = L_A$ as in Theorem 1.4.2. Finally, since E has at most logarithmic growth, while E_A is bounded, Q must be constant, thus $E = E_A + C_A$ for some $C_A \in \mathbb{C}$. □

Finally, we consider:

Case III: *Assume $A \in M_2^{\pm}$ with $\lambda_1 = \lambda_2 = \lambda$.*

Fix $A \in M_2^{\pm}$, such that

$$A = a\begin{pmatrix} 1 & -\lambda \\ -\lambda & \lambda^2 \end{pmatrix} \quad \text{for some } a \in \mathbb{C} \setminus \{0\} \text{ and } \lambda \in \mathbb{C} \setminus \mathbb{R}. \tag{1.4.271}$$

Much as before, associate with A the differential operator $L_A := \nabla \cdot A\nabla$ (see also (1.4.247)), and recall ∂_ξ^A introduced in (1.4.249). In the present case, we consider the auxiliary function φ_A defined as

$$\varphi_A(x) := a\frac{x_1}{x_2 + \lambda x_1}, \qquad \forall x = (x_1, x_2) \in \mathbb{R}^2 \setminus \{0\}. \tag{1.4.272}$$

Since the denominator never vanishes for $x \in \mathbb{R}^2 \setminus \{0\}$, given that we presently assume $\lambda \in \mathbb{C} \setminus \mathbb{R}$, it follows that φ_A is well defined and belongs to $\mathscr{C}^\infty(\mathbb{R}^2 \setminus \{0\})$.

Lemma 1.4.33 *With φ_A defined as in* (1.4.272) *one has*

$$\int_{S^1} (\partial_\xi^A \varphi_A)(\xi)\, d\mathcal{H}^1(\xi) = -2\pi i a^2 \cdot \operatorname{sign}(\operatorname{Im}\lambda). \tag{1.4.273}$$

Proof Fix an arbitrary $\xi = (\xi_1, \xi_2) \in \mathbb{R}^2 \setminus \{0\}$. As seen from (1.4.272) we then have

$$\nabla\varphi_A(\xi) = a\Big(\frac{\xi_2}{(\xi_2 + \lambda\xi_1)^2}, \frac{-\xi_1}{(\xi_2 + \lambda\xi_1)^2}\Big). \tag{1.4.274}$$

In concert with (1.4.249) this implies

$$\begin{aligned}
(\partial_\xi^A \varphi_A)(\xi) &= a^2\Big[\frac{\xi_1\xi_2}{(\xi_2 + \lambda\xi_1)^2} + \frac{\lambda\xi_1^2}{(\xi_2 + \lambda\xi_1)^2} - \frac{\lambda\xi_2^2}{(\xi_2 + \lambda\xi_1)^2} - \frac{\lambda^2\xi_1\xi_2}{(\xi_2 + \lambda\xi_1)^2}\Big] \\
&= a^2\Big[\frac{\xi_1\xi_2(1 - \lambda^2) + (\xi_1^2 - \xi_2^2)\lambda}{(\xi_2 + \lambda\xi_1)^2}\Big] \\
&= a^2\frac{\xi_1 - \lambda\xi_2}{\xi_2 + \lambda\xi_1},
\end{aligned} \tag{1.4.275}$$

where the last inequality holds since $\xi_2 + \lambda\xi_1 \neq 0$ given that $\xi = (\xi_1, \xi_2) \in \mathbb{R}^2 \setminus \{0\}$ and $\lambda \in \mathbb{C} \setminus \mathbb{R}$. Together with Lemma 1.4.26, this permits us to compute

$$\int_{S^1} (\partial_\xi^A \varphi_A)(\xi)\, d\mathcal{H}^1(\xi) = a^2 I_\lambda = -2\pi i a^2 \cdot \operatorname{sign}(\operatorname{Im}\lambda), \tag{1.4.276}$$

completing the proof. □

Here is the result giving an explicit formula for the fundamental solution constructed in Theorem 1.4.2 for differential operators of the form $L_A := \nabla \cdot A\nabla$ for a weakly elliptic matrix $A \in \mathbb{C}^{2\times 2}$ satisfying $\operatorname{sym} A \in M_2^{\pm}$ with $\lambda_1 = \lambda_2$.

Proposition 1.4.34 *Let $A \in \mathbb{C}^{2\times 2}$ be a weakly elliptic matrix with the property that its symmetric part,* $\operatorname{sym} A := \frac{1}{2}(A + A^\top)$*, belongs to $M_2^{\pm}$ and can be written as in* (1.4.271). *In relation to this, define*

$$E_A(x) := -\frac{\operatorname{sign}(\operatorname{Im}\lambda)}{2\pi i a}\,\frac{x_1}{x_2 + \lambda x_1}, \qquad x = (x_1, x_2) \in \mathbb{R}^2 \setminus \{0\}. \tag{1.4.277}$$

Then $E_A \in L^\infty_{\text{loc}}(\mathbb{R}^2, \mathcal{L}^2)$ and, when viewed as a distribution in $\mathbb{R}^2$, this is a fundamental solution of $L_A := \nabla \cdot A\nabla$ (the second-order differential operator associated with A as in (1.4.176)*). Additionally, E_A is positive homogeneous of degree zero (in particular, E_A is bounded) in $\mathbb{R}^2 \setminus \{0\}$, and for each multi-index $\alpha \in \mathbb{N}_0^2$ one has*

$$(\partial^\alpha E_A)(x) = O\big(|x|^{-|\alpha|}\big) \quad \text{as } |x| \to \infty. \tag{1.4.278}$$

Finally, there exists $C_A \in \mathbb{C}$ such that $E = E_A + C_A$, where E is the canonical fundamental solution of $L = L_A$ constructed in Theorem 1.4.2.

In relation to the fundamental solutions for "generic" weakly elliptic systems constructed in Theorem 1.4.2, a significant feature of E_A from (1.4.277) is that this function is positive homogeneous of degree zero (as a consequence, E_A is bounded in $\mathbb{R}^2 \setminus \{0\}$). In light of (1.4.177) and Lemma 1.4.19, this is consistent with the prediction made in (1.4.54). Note that E_A from (1.4.277) is singular at the origin, but otherwise $\mathscr{C}^\infty$ smooth and bounded in $\mathbb{R}^2 \setminus \{0\}$. As with E_A from (1.4.268), this contrasts markedly with the nature of the fundamental solution for the Laplacian (in fact such singularities are removable for harmonic functions).

In relation to Proposition 1.4.34 it is also instructive to consider the special case

$$A := \begin{pmatrix} 1/4 & i/4 \\ i/4 & -1/4 \end{pmatrix}, \tag{1.4.279}$$

in which scenario

$$L_A := \nabla \cdot A\nabla = \tfrac{1}{4}\big(\partial_{x_1}^2 + i\partial_{x_1}\partial_{x_2} + i\partial_{x_2}\partial_{x_1} - \partial_{x_2}^2\big) = \partial_{\bar z}^2 \tag{1.4.280}$$

becomes the square of the Cauchy-Riemann operator $\partial_{\bar z} = \frac{1}{2}\big(\partial_{x_1} + i\partial_{x_2}\big)$. All fundamental solutions of $\partial_{\bar z}^2$ are of the form

$$\frac{\bar z}{\pi z} + c \ \text{ for some } c \in \mathbb{C}. \tag{1.4.281}$$

Under the canonical identification $z = x_1 + i x_2$ and bearing in mind that we presently have $a = 1/4$ and $\lambda = -i$, the fundamental solution E_A from (1.4.277) takes the format predicted in (1.4.281) with $c = -1/\pi$.

Proof of Proposition 1.4.34 First observe from (1.4.277) that

$$E_A \in \mathscr{C}^\infty(\mathbb{R}^2 \setminus \{0\}) \cap L^\infty_{\text{loc}}(\mathbb{R}^2, \mathcal{L}^2) \cap \mathcal{S}'(\mathbb{R}^2). \tag{1.4.282}$$

Also, as seen from (1.4.272) and (1.4.274), the first-order derivatives of E_A are positive homogeneous of degree -1 in $\mathbb{R}^2 \setminus \{0\}$. In addition, elementary calculus shows that $L_A E_A(x) = 0$ for every $x \in \mathbb{R}^2 \setminus \{0\}$. These and Proposition 1.4.1 then imply that E_A is a fundamental solution of L_A. Granted this, item *(8)* in Theorem 1.4.2 gives that $Q := E_A - E$ is a polynomial, where E is the canonical fundamental solution associated with $L = L_A$ as in Theorem 1.4.2. Note that E has at most logarithmic growth, and that E_A is bounded. As such, Q must be constant, hence $E = E_A + C_A$ for some constant $C_A \in \mathbb{C}$. Finally, that the function E_A is positive homogeneous of degree zero and satisfies (1.4.278) is clear from (1.4.277) and (1.4.252). □

This completes the task of providing explicit formulas for the fundamental solution constructed in Theorem 1.4.2 for differential operators of the form $L_A := \nabla \cdot A\nabla$ with $A \in M_2$. Relying on what we have proved so far, we next carry out a similar program in higher dimensions.

Fundamental Solutions for Weakly Elliptic Scalar Operators when $n \geq 3$

For each $n \in \mathbb{N}$ define

$$M_n := \big\{A \in \mathbb{C}^{n\times n} : A^\top = A \text{ and } A \text{ is weakly elliptic}\big\}. \tag{1.4.283}$$

In the proof of the lemma below we employ ideas from the proof of [88, Proposition 1.1, p. 109] (cf. also [142, pp. 159–160]).

Lemma 1.4.35 *Fix an integer $n \geq 3$ and pick an arbitrary matrix $A \in M_n$. Then for each given pair of linearly independent vectors $\xi, \eta \in \mathbb{R}^n$ the 2×2 matrix*

$$B_{\xi,\eta} := \begin{pmatrix} \langle A\xi,\xi\rangle & \langle A\xi,\eta\rangle \\ \langle A\xi,\eta\rangle & \langle A\eta,\eta\rangle \end{pmatrix} \text{ belongs to } M_2^0. \tag{1.4.284}$$

As a corollary of this membership and Lemma 1.4.20 (applied to the matrix $B_{\xi,\eta}$ in M_2^0 and the vector $\mathbf{e}_2 \in \mathbb{R}^2 \setminus \{0\}$), one has

$$\begin{aligned} &\langle A\eta,\eta\rangle \neq t\langle A\xi,\xi\rangle \text{ for every } t \in (-\infty,0] \text{ and} \\ &\text{each pair of linearly independent vectors } \xi,\eta \in \mathbb{R}^n. \end{aligned} \tag{1.4.285}$$

Proof For each $\xi, \eta \in \mathbb{R}^n \setminus \{0\}$ define the quadratic polynomial

$$Q_{\xi,\eta}(\tau) := \langle A(\xi + \tau\eta), \xi + \tau\eta\rangle, \qquad \tau \in \mathbb{C}, \tag{1.4.286}$$

and note that

$$Q_{-\xi,\eta}(\tau) = Q_{\xi,\eta}(-\tau) \text{ for each } \xi,\eta \in \mathbb{R}^n \setminus \{0\} \text{ and } \tau \in \mathbb{C}. \tag{1.4.287}$$

Henceforth fix $\eta \in \mathbb{R}^n \setminus \{0\}$. For every $\xi \in \mathbb{R}^n \setminus \mathbb{R}\eta$ denote by $\tau_1(\xi)$, $\tau_2(\xi)$ the complex roots of the quadratic equation $Q_{\xi,\eta}(\tau) = 0$. From weakly ellipticity and

(1.4.120) we conclude that $\tau_1(\xi), \tau_2(\xi) \in \mathbb{C} \setminus \mathbb{R}$. Thanks to Lemma 1.4.22 and the fact that the set $\mathbb{R}^n \setminus \mathbb{R}\eta$ is connected (since we now assume $n \geq 3$), it follows that the number of roots in, say, $\mathbb{C}_+$ is independent of $\xi \in \mathbb{R}^n \setminus \mathbb{R}\eta$ (as a continuous integer-valued function defined on a connected set). Note (1.4.287) entails that for each $\xi \in \mathbb{R}^n \setminus \mathbb{R}\eta$ we have (with the piece of notation introduced in (1.4.216))

$$\{\tau_1(-\xi), \tau_2(-\xi)\}^* = \{-\tau_1(\xi), -\tau_2(\xi)\}^*. \tag{1.4.288}$$

Collectively, these observations imply that for each $\xi \in \mathbb{R}^n \setminus \mathbb{R}\eta$ we have precisely one root in $\mathbb{C}_+$ and one root in $\mathbb{C}_-$. As a consequence,

$$\operatorname{Im} \tau_1(\xi) \cdot \operatorname{Im} \tau_2(\xi) < 0 \text{ for every } \xi \in \mathbb{R}^n \setminus \mathbb{R}\eta. \tag{1.4.289}$$

After this preamble, we are ready to proceed with the proof of (1.4.284). To set the stage, fix $\eta \in \mathbb{R}^n \setminus \{0\}$ together with $\xi \in \mathbb{R}^n \setminus \mathbb{R}\eta$, and abbreviate $B := B_{\xi,\eta}$. We shall first prove that $B \in M_2$. To this end, note that $B \in \mathbb{C}^{2\times 2}$ and $B^\top = B$, by design. Suppose that there exists $x = (x_1, x_2) \in \mathbb{R}^2 \setminus \{0\}$ such that $\langle Bx, x\rangle = 0$. To proceed, consider the case when $x_2 \neq 0$. Set $\tau := x_1/x_2 \in \mathbb{R}$ and define $\tilde{x} := (1, \tau) \in \mathbb{R}^2$. Then, since $A^\top = A$,

$$\begin{aligned} 0 = \langle B\tilde{x}, \tilde{x}\rangle &= \langle A\xi, \xi\rangle + 2\,\langle A\xi, \eta\rangle\tau + \langle A\eta, \eta\rangle\tau^2 \\ &= \langle A(\xi + \tau\eta), \xi + \tau\eta\rangle. \end{aligned} \tag{1.4.290}$$

Since A is weakly elliptic, this forces $\xi + \tau\eta = 0$ (cf. (1.4.120)) which, in turn, contradicts the assumption that $\xi \in \mathbb{R}^n \setminus \mathbb{R}\eta$.

Consider next the case when $x_1 \neq 0$. In this scenario, let $\tau := x_2/x_1 \in \mathbb{R}$ and define $\tilde{x} := (\tau, 1)$. Much as before, we have

$$0 = \langle B\tilde{x}, \tilde{x}\rangle = \langle A(\eta + \tau\xi), \eta + \tau\xi\rangle, \tag{1.4.291}$$

from which we conclude that $\eta + \tau\xi = 0$, thanks to the weak ellipticity of A and (1.4.120). Again, this contradicts the assumption that $\xi \in \mathbb{R}^n \setminus \mathbb{R}\eta$. Altogether, the above analysis shows that if $\langle Bx, x\rangle \neq 0$, hence B is weakly elliptic (cf. (1.4.120)). This concludes the proof of the fact that $B \in M_2$.

We are left with checking that $B \in M_2$ possesses the property stipulated in (1.4.187) for membership to the class M_2^0. With this goal in mind, we first apply Lemma 1.4.18 to B to conclude that there exist $a_B \in \mathbb{C} \setminus \{0\}$ and $\lambda_1^B, \lambda_2^B \in \mathbb{C} \setminus \mathbb{R}$ such that

$$\langle Bx, x\rangle = a_B(x_1 - \lambda_1^B x_2)(x_1 - \lambda_2^B x_2), \qquad \forall x = (x_1, x_2) \in \mathbb{C}^2. \tag{1.4.292}$$

Applying this to $x := (1, \tau)$ with $\tau \in \mathbb{C}$ arbitrary leads to the conclusion that

$$\begin{aligned} a_B(1 - \lambda_1^B\tau)(1 - \lambda_2^B\tau) = \langle Bx, x\rangle &= \langle A\xi, \xi\rangle + 2\,\langle A\xi, \eta\rangle\tau + \langle A\eta, \eta\rangle\tau^2 \\ &= \langle A(\xi + \tau\eta), \xi + \tau\eta\rangle = Q_{\xi,\eta}(\tau). \end{aligned} \tag{1.4.293}$$

Recall that we have shown that the imaginary parts of the roots of the quadratic polynomial $Q_{\xi,\eta}(\tau)$ have opposite signs. Hence, $\mathrm{Im}\,(1/\lambda_1^B) \cdot \mathrm{Im}\,(1/\lambda_2^B) < 0$ which in turns implies $\mathrm{Im}\,(\lambda_1^B) \cdot \mathrm{Im}\,(\lambda_2^B) < 0$, ultimately proving that $B \in M_2^0$, as desired.□

Given that any $n\times n$ symmetric matrix with complex entries is uniquely determined by its entries lying on or below the main diagonal, we may naturally identify M_n with an open convex subset of $\mathbb{C}^{n(n+1)/2}$. The following result stands in sharp contrast to its two-dimensional counterpart, discussed in Lemma 1.4.23.

Lemma 1.4.36 *Assume $n \geq 3$. Then the set M_n is open (in $\mathbb{C}^{n(n+1)/2}$) and path-connected.*

Proof That M_n may be canonically identified with an open subset of $\mathbb{C}^{n(n+1)/2}$ (under the identification described above) is clear from (1.4.122). To prove the path-connectivity of M_n, pick an arbitrary matrix $A = (a_{jk})_{1\leq j,k\leq n} \in M_n$. Then, by weak ellipticity (cf. (1.4.120)), we have $a_{11} = \langle A\mathbf{e}_1, \mathbf{e}_1\rangle \in \mathbb{C} \setminus \{0\}$. For every $t \in [0, 1]$ define $A(t) := (1 - t)A + ta_{11}I_{n\times n}$. This is a continuous path with $A(0) = A$ and $A(1) = a_{11}I_{n\times n}$. By design, $A(t)^\top = A(t)$ for every $t \in [0, 1]$. Observe that $A(t)$ is weakly elliptic for every $t \in [0, 1]$. Indeed, if this were not to be the case, we would be able to find $t \in [0, 1]$ and $\eta \in \mathbb{R}^n \setminus \{0\}$ such that $\langle A(t)\eta, \eta\rangle = 0$. On the one hand, t cannot be 0 or 1, since A and $a_{11}I_{n\times n}$ are weakly elliptic. On the other hand, for each $t \in (0, 1)$ we may write

$$0 = \langle A(t)\eta, \eta\rangle = (1-t)\langle A\eta, \eta\rangle + ta_{11}|\eta|^2 = (1-t)\langle A\eta, \eta\rangle + t|\eta|^2\langle A\mathbf{e}_1, \mathbf{e}_1\rangle \quad (1.4.294)$$

which implies that $\langle A\eta, \eta\rangle = -(t|\eta|^2/(1-t))\langle A\mathbf{e}_1, \mathbf{e}_1\rangle$. Thanks to Lemma 1.4.35, this further entails $\eta = \lambda\mathbf{e}_1$ for some $\lambda \in \mathbb{R}\setminus\{0\}$. But in that case $\langle A(t)\eta, \eta\rangle = \lambda^2 a_{11} \neq 0$, which is a contradiction. We have therefore shown that $A(t) \in M_n$ for every $t \in [0, 1]$.

To complete the proof, express $a_{11} \in \mathbb{C} \setminus \{0\}$ as $a_{11} = \rho e^{i\theta}$ for some $\rho \in (0, \infty)$ and $\theta \in [0, 2\pi]$. This permits us to define $\widetilde{A}(t) := \rho^{1-t}e^{i\theta(1-t)}I_{n\times n}$ for each $t \in [0, 1]$, which is a continuous path joining $\widetilde{A}(0) = a_{11}I_{n\times n}$ with $\widetilde{A}(1) = I_{n\times n}$. In addition, $\widetilde{A}(t) \in M_n$ for every $t \in [0, 1]$.

Concatenating the first path with the second paths enable us to connect within M_n any $A \in M_n$ with $I_{n\times n}$. This ultimately shows that M_n is indeed path-connected. □

The next lemma elaborates on further algebraic properties of matrices in M_n.

Lemma 1.4.37 *Assume $n \geq 3$. If $A \in M_n$ then A is invertible and $A^{-1} \in M_n$. Moreover, if $A \in M_n$ is Legendre-Hadamard elliptic then A^{-1} is also Legendre-Hadamard elliptic.*

Proof Let $A \in M_n$ and let $\zeta = \xi + i\eta \in \mathbb{C}^n$ with $\xi, \eta \in \mathbb{R}^n$. Notice that since $A^\top = A$ we have

$$\langle A\zeta, \overline{\zeta}\rangle = \langle A\xi, \xi\rangle + i\langle A\eta, \xi\rangle - i\langle A\xi, \eta\rangle + \langle A\eta, \eta\rangle = \langle A\xi, \xi\rangle + \langle A\eta, \eta\rangle. \quad (1.4.295)$$

This and Lemma 1.4.35 imply that

$$\langle A\zeta, \overline{\zeta} \rangle \neq 0 \text{ for every } \zeta \in \mathbb{C}^n \setminus \{0\}. \tag{1.4.296}$$

The invertibility of A becomes a consequence of (1.4.296). Since A is symmetric, $(A^{-1})^\top = A^{-1}$. Let us now show that A^{-1} is weakly elliptic. Seeking a contradiction, suppose the vector $\xi \in \mathbb{R}^n \setminus \{0\}$ satisfies $\langle A^{-1}\xi, \xi \rangle = 0$. Set $\zeta := A^{-1}\xi \in \mathbb{C}^n \setminus \{0\}$ and observe that

$$\langle \overline{A\zeta}, \zeta \rangle = \langle \overline{\xi}, A^{-1}\xi \rangle = \langle A^{-1}\xi, \xi \rangle = 0. \tag{1.4.297}$$

After taking a complex conjugation, this yields $\langle A\zeta, \overline{\zeta} \rangle = 0$, in contradiction with (1.4.296).

The last claim in the statement, to the effect that if $A \in M_n$ is Legendre-Hadamard elliptic then A^{-1} is also Legendre-Hadamard elliptic, is seen from Corollary 1.4.9.□

We next study the numerical range of a matrix $A \in M_n$, when acting on real vectors.

Lemma 1.4.38 *Assume $n \geq 3$. Pick some $A \in M_n$ and set*

$$K := \left\{ \frac{\langle A\xi, \xi \rangle}{|\langle A\xi, \xi \rangle|} : \xi \in S^{n-1} \right\}. \tag{1.4.298}$$

Then there exist $\theta \in [0, 2\pi)$ and $\theta_* \in (0, \pi/2)$ such that

$$e^{i\theta} K \subseteq \left\{ e^{i\omega} : |\omega| \leq \theta_* \right\}. \tag{1.4.299}$$

Proof We begin by observing that, by definition, $K \subseteq S^1$. Since the function

$$S^{n-1} \ni \xi \longmapsto \langle A\xi, \xi \rangle / |\langle A\xi, \xi \rangle| \in \mathbb{C} \tag{1.4.300}$$

is continuous (given that A is weakly elliptic), and S^{n-1} is a compact connected subset of $\mathbb{R}^n$, it follows that K is a connected compact subset of S^1. Hence, K is a closed arc in the unit circle S^1.

Let us now prove that K does not contain diametrically opposite points. Otherwise, there would be $\xi, \eta \in S^{n-1}$ such that

$$\frac{\langle A\xi, \xi \rangle}{|\langle A\xi, \xi \rangle|} = -\frac{\langle A\eta, \eta \rangle}{|\langle A\eta, \eta \rangle|}, \tag{1.4.301}$$

hence $\langle A\eta, \eta \rangle = t \langle A\xi, \xi \rangle$ for some $t < 0$. This and Lemma 1.4.35 then imply that $\xi = \lambda\eta$ for some $\lambda \in \mathbb{R}$. Since $\xi, \eta \in S^{n-1}$ this forces $\xi = \pm\eta$, and in either case (1.4.301) cannot happen.

The above argument proves that K is a closed arc contained in an open half circle, hence there exist $\theta \in [0, 2\pi)$ and $\theta_* \in (0, \pi/2)$ such that (1.4.299) holds. □

As opposed to the two-dimensional case, when $n \geq 3$ any matrix $A \in M_n$ becomes Legendre-Hadamard elliptic after multiplication with a suitable complex number.

Proposition 1.4.39 *Assume $n \geq 3$ and fix some $A \in M_n$. Then there exists $\theta \in [0, 2\pi)$ such that the matrix $A_\theta := e^{i\theta} A$ is Legendre-Hadamard elliptic.*

To illustrate the failure of the above result when $n = 2$, consider the 2×2 matrix

$$A := \frac{1}{4}\begin{pmatrix} 1 & -\mathrm{i} \\ -\mathrm{i} & -1 \end{pmatrix}. \tag{1.4.302}$$

This corresponds to (1.4.179) with $a = 1/4$ and $\lambda_1 = \lambda_2 = \mathrm{i}$, hence $A \in M_2^+ \subseteq M_2$. From (1.4.180) we also see that

$$\langle A\xi, \xi \rangle = 4^{-1}(\xi_1 - \mathrm{i}\xi_2)^2, \qquad \forall \xi = (\xi_1, \xi_2) \in \mathbb{R}^2. \tag{1.4.303}$$

Next, pick an arbitrary $\alpha \in \mathbb{R}$ and define $\xi_\alpha := (\cos\alpha, \sin\alpha) \in S^1$. If $\theta \in [0, 2\pi)$ is such that the matrix $A_\theta := e^{\mathrm{i}\theta} A$ is Legendre-Hadamard elliptic, then (1.4.123) ensures the existence of some $\kappa > 0$ independent of α such that

$$\mathrm{Re}\langle A_\theta \xi_\alpha, \xi_\alpha \rangle \geq \kappa |\xi_\alpha|^2 = \kappa. \tag{1.4.304}$$

Since, as seen from (1.4.303),

$$\langle A_\theta \xi_\alpha, \xi_\alpha \rangle = 4^{-1} e^{\mathrm{i}\theta}(\cos\alpha - \mathrm{i}\sin\alpha)^2 = 4^{-1} e^{\mathrm{i}\theta} e^{-2\mathrm{i}\alpha} = 4^{-1} e^{\mathrm{i}(\theta - 2\alpha)}, \tag{1.4.305}$$

we have

$$\mathrm{Re}\langle A_\theta \xi_\alpha, \xi_\alpha \rangle = 4^{-1}\cos(\theta - 2\alpha). \tag{1.4.306}$$

Choosing $\alpha := \theta/2 - \pi/4$ then contradicts (1.4.304). Ultimately, this shows that

for $A \in M_2$ defined in (1.4.302), there is no $\theta \in [0, 2\pi)$ such that the matrix $A_\theta := e^{\mathrm{i}\theta} A$ is Legendre-Hadamard elliptic. (1.4.307)

Finally, we wish to point out that the second-order operator $L_A := \nabla \cdot A\nabla$ associated with A from (1.4.302) is precisely the Bitsadze operator $\partial_{\bar{z}}^2$.

Here is the proof of Proposition 1.4.39.

Proof of Proposition 1.4.39 Define $C_A := \min_{\xi \in S^{n-1}} |\langle A\xi, \xi \rangle|$. By weakly ellipticity, C_A is a well-defined number in the interval $(0, \infty)$ (cf. (1.4.122)). Lemma 1.4.38 guarantees the existence of some $\theta \in [0, 2\pi)$ and $\theta_* \in (0, \pi/2)$ such that (1.4.299) holds. This implies that for every $\xi \in S^{n-1}$ we have

$$C_A^{-1}\mathrm{Re}\langle A_\theta \xi, \xi \rangle \geq \frac{\mathrm{Re}\langle A_\theta \xi, \xi \rangle}{|\langle A\xi, \xi \rangle|} = \mathrm{Re}\left(e^{\mathrm{i}\theta} \frac{\langle A\xi, \xi \rangle}{|\langle A\xi, \xi \rangle|} \right) \geq \cos\theta_*. \tag{1.4.308}$$

Setting then $\kappa := C_A \cos\theta_* > 0$ we conclude that $\mathrm{Re}\langle A_\theta \xi, \xi \rangle \geq \kappa |\xi|^2$ for all $\xi \in \mathbb{R}^n$, which goes to show that A_θ is Legendre-Hadamard elliptic. □

We shall also need the following result concerning the numerical range of the inverse of the symmetric part of a Legendre-Hadamard elliptic matrix $A \in \mathbb{C}^{n \times n}$ with $n \geq 3$, acting on real vectors.

Lemma 1.4.40 *Assume $n \geq 3$ and pick a Legendre-Hadamard elliptic matrix A in $\mathbb{C}^{n \times n}$. Then* $\mathrm{sym}\, A := \frac{1}{2}(A + A^\top)$ *is invertible and there exists $\vartheta_* \in (0, \pi/2)$ such that* *for every* $x \in \mathbb{R}^n \setminus \{0\}$ *one has*

$$\big\langle (\operatorname{sym} A)^{-1} x, x \big\rangle \in \Sigma_{\vartheta_*} := \big\{ z \in \mathbb{C} \setminus \{0\} : |\arg(z)| < \vartheta_* \big\}. \tag{1.4.309}$$

In particular, since $\Sigma_{\vartheta_*} \ni z \mapsto z^{\frac{2-n}{n}} \in \mathbb{C}$ *may be defined as a holomorphic function (given that the set* $\Sigma_{\vartheta_*} \subseteq \mathbb{C} \setminus \{0\}$ *is simply connected; cf. item (9) in* [112, Proposition 5.8.1]), *one can make sense of the expression* $\big\langle (\operatorname{sym} A)^{-1} x, x \big\rangle^{\frac{2-n}{n}}$ *for every* $x \in \mathbb{R}^n \setminus \{0\}$.

Proof From definitions and (1.4.166) we see that $\operatorname{sym} A$ belongs to M_n. As such, Lemma 1.4.37 guarantees that $\operatorname{sym} A$ is invertible and $(\operatorname{sym} A)^{-1}$ is also Legendre-Hadamard elliptic. Then there exist $\kappa \in (0, \infty)$ and $C \in [\kappa, \infty)$ such that

$$\big|\big\langle (\operatorname{sym} A)^{-1} x, x \big\rangle\big| \leq C \text{ and } \operatorname{Re}\big\langle (\operatorname{sym} A)^{-1} x, x \big\rangle \geq \kappa \text{ for all } x \in S^{n-1}. \tag{1.4.310}$$

Hence, if $\vartheta_* \in (0, \pi/2)$ is such that $\vartheta_* > \arccos(\kappa/C)$, then $\langle (\operatorname{sym} A)^{-1} x, x \rangle \in \Sigma_{\vartheta_*}$ for every $x \in S^{n-1}$. By homogeneity, the same membership then extends to all points $x \in \mathbb{R}^n \setminus \{0\}$. □

Finally, we are prepared to state and prove our main result concerning an explicit formula for the fundamental solution constructed in Theorem 1.4.2 in relation to differential operators of the form $L_A := \operatorname{div} A \nabla = \nabla \cdot A \nabla$ for arbitrary weakly elliptic matrices $A \in \mathbb{C}^{n \times n}$ with $n \geq 3$.

Proposition 1.4.41 *Assume* $n \geq 3$, *and pick an arbitrary weakly elliptic matrix* $A \in \mathbb{C}^{n \times n}$. *Then there exists an angle* $\theta \in [0, 2\pi)$ *such that* $\operatorname{sym} A_\theta$ *is invertible and Legendre-Hadamard elliptic if* $A_\theta := e^{i\theta} A$. *Moreover, the function*

$$E_A(x) := \frac{-e^{i\theta}}{(n-2)\omega_{n-1}\sqrt{\det(\operatorname{sym} A_\theta)}} \big\langle (\operatorname{sym} A_\theta)^{-1} x, x \big\rangle^{\frac{2-n}{2}}, \tag{1.4.311}$$

defined for all $x \in \mathbb{R}^n \setminus \{0\}$, *belongs to* $L^1_{\mathrm{loc}}(\mathbb{R}^n, \mathcal{L}^n)$ *and, when regarded as a distribution in* $\mathbb{R}^n$, *this is a fundamental solution for the second-order operator* $L_A := \nabla \cdot A \nabla$. *In fact,* E_A *agrees with the canonical fundamental solution associated to the weakly elliptic operator* $L := L_A = \operatorname{div} A \nabla$ *as in Theorem 1.4.2.*

In the above context, since $\operatorname{sym} A_\theta$ is Legendre-Hadamard elliptic and symmetric (which amounts to simply saying that $\operatorname{sym} A_\theta \in \mathfrak{M}_n$), Proposition 1.4.12 ensures that $\big(\det(\operatorname{sym} A_\theta)\big)^{-\frac{1}{2}}$ is an unambiguously defined complex number. In fact, (1.4.167) gives the concrete representation

$$\big(\det(\operatorname{sym} A_\theta)\big)^{-\frac{1}{2}} = \pi^{-n/2} \int_{\mathbb{R}^n} e^{-\langle A_\theta \xi, \xi \rangle} \, d\xi. \tag{1.4.312}$$

Proof of Proposition 1.4.41 From (1.4.166), hypotheses, and definitions we see that $\operatorname{sym} A \in M_n$. As such, Proposition 1.4.39 guarantees the existence of some number

$\theta \in [0, 2\pi)$ with the property that $e^{i\theta}\operatorname{sym} A$ is Legendre-Hadamard elliptic. Abbreviate $A_\theta := e^{i\theta}A$, so $\operatorname{sym} A_\theta = e^{i\theta}\operatorname{sym} A$ is symmetric and Legendre-Hadamard elliptic. Proposition 1.4.17 then tells us that

$$E_{\operatorname{sym} A_\theta}(x) = \frac{-1}{(n-2)\omega_{n-1}\sqrt{\det(\operatorname{sym} A_\theta)}}\big\langle(\operatorname{sym} A_\theta)^{-1}x, x\big\rangle^{\frac{2-n}{2}}, \tag{1.4.313}$$

defined for all $x \in \mathbb{R}^n \setminus \{0\}$, is the canonical fundamental solution associated as in Theorem 1.4.2 with the scalar weakly elliptic second-order differential operator $L_{\operatorname{sym} A_\theta} = \nabla \cdot (\operatorname{sym} A_\theta)\nabla$. Comparing (1.4.313) with (1.4.311) reveals that in fact $E_A = e^{i\theta}E_{\operatorname{sym} A_\theta}$. Then the following equalities hold in the sense of tempered distribution:

$$L_A E_A = e^{-i\theta}L_{A_\theta}E_A = L_{\operatorname{sym} A_\theta}E_{\operatorname{sym} A_\theta} = \delta. \tag{1.4.314}$$

Hence, E_A is a fundamental solution of L_A. If E is the canonical fundamental solution associated with $L := L_A = \operatorname{div}A\nabla$ as in Theorem 1.4.2, then item *(8)* in Theorem 1.4.2 implies that $Q := E_A - E$ is a polynomial in $\mathbb{R}^n$. Use (1.4.311) together with item *(4)* in Theorem 1.4.2 to write

$$|Q(x)| \leq |E_A(x)| + |E(x)| \leq C\,|x|^{2-n}, \qquad \forall x \in \mathbb{R}^n \setminus \{0\}, \tag{1.4.315}$$

Since $n \geq 3$, we necessarily have $Q \equiv 0$, from which we conclude that $E_A = E$, as wanted. □

1.5 Boundary Layer Potential Representations for Weakly Elliptic Second-Order Systems

We start by discussing a version of Green's third identity, dealing with an integral representation formula of a function in terms of a double layer, a single layer, and a volume potential, in a very general geometric and analytic setting.

Theorem 1.5.1 *Let $\Omega \subseteq \mathbb{R}^n$, where[31] $n \geq 3$, be an open set with a lower Ahlfors regular boundary, and with the property that $\sigma := \mathcal{H}^{n-1}\lfloor\partial\Omega$ is a doubling measure on $\partial\Omega$. In particular, Ω is a set of locally finite perimeter, and its geometric measure theoretic outward unit normal $\nu = (\nu_1, \dots, \nu_n)$ is defined σ-a.e. on $\partial_*\Omega$. With the summation convention over repeated indices understood throughout, let*

$$L = \big(a_{rs}^{\alpha\beta}\partial_r\partial_s\big)_{1\leq\alpha,\beta\leq M} \tag{1.5.1}$$

be a homogeneous, weakly elliptic, second-order $M \times M$ system in $\mathbb{R}^n$, with complex constant coefficients, and denote by $E = (E_{\gamma\beta})_{1\leq\gamma,\beta\leq M}$ the matrix-valued fundamental solution associated with L as in Theorem 1.4.2. Finally, fix an aperture parameter $\kappa > 0$.

[31] see the last part in the statement for the case $n = 2$

Suppose $u = (u_\beta)_{1\leq\beta\leq M} \in \big[L^1_{\rm loc}(\Omega, \mathcal{L}^n)\big]^M$ *is a vector-valued function satisfying (with all derivatives taken in the sense of distributions in* Ω*)*

$$
\begin{gathered}
Lu \in \Big[L^1\Big(\Omega, \frac{\mathrm{d}y}{1+|y|^{n-2}}\Big)\Big]^M \quad \textit{and} \quad \big(a^{\alpha\beta}_{rs}\partial_s u_\beta\big)_{\substack{1\leq\alpha\leq M\\ 1\leq r\leq n}} \in \big[L^1_{\rm loc}(\Omega, \mathcal{L}^n)\big]^{M\times n},\\
u\big|^{\kappa-\rm n.t.}_{\partial\Omega} \quad \textit{and} \quad \Big((a^{\alpha\beta}_{rs}\partial_s u_\beta)\big|^{\kappa-\rm n.t.}_{\partial\Omega}\Big)_{\substack{1\leq\alpha\leq M\\ 1\leq r\leq n}} \quad \textit{exist at } \sigma\textit{-a.e. point on } \partial_{\rm nta}\Omega.
\end{gathered}
\tag{1.5.2}
$$

In addition, assume the following integrability conditions hold:

$$
\int_{\partial\Omega}\frac{(\mathcal{N}_\kappa u)(y)}{1+|y|^{n-1}}\,\mathrm{d}\sigma(y) < \infty \quad \textit{and} \quad \int_{\partial\Omega}\frac{(\mathcal{N}_\kappa(a^{\alpha\beta}_{rs}\partial_s u_\beta))(y)}{1+|y|^{n-2}}\,\mathrm{d}\sigma(y) < \infty, \tag{1.5.3}
$$

for each $\alpha \in \{1, \dots, M\}$ *and* $r \in \{1, \dots, n\}$.

Then for any $\kappa' > 0$ *the nontangential traces* $u\big|^{\kappa'-\rm n.t.}_{\partial\Omega}$ *and* $\Big((a^{\alpha\beta}_{rs}\partial_s u_\beta)\big|^{\kappa'-\rm n.t.}_{\partial\Omega}\Big)_{\substack{1\leq\alpha\leq M\\ 1\leq r\leq n}}$ *also exist at* σ*-a.e. point on* $\partial_{\rm nta}\Omega$ *and are actually independent of* κ'. *Moreover, for* $\mathcal{L}^n$*-a.e. point* $x \in \Omega$ *one has (with absolutely convergent integrals, and with the dependence on the aperture parameter dropped)*

$$
\begin{aligned}
u(x) = {} & \Big(-\int_{\partial_*\Omega}\nu_s(y)a^{\beta\alpha}_{rs}(\partial_r E_{\gamma\beta})(x-y)\big(u_\alpha\big|^{\rm n.t.}_{\partial\Omega}\big)(y)\,\mathrm{d}\sigma(y)\Big)_{1\leq\gamma\leq M}\\
& - \Big(\int_{\partial_*\Omega}E_{\gamma\alpha}(x-y)\nu_r(y)\Big((a^{\alpha\beta}_{rs}\partial_s u_\beta)\big|^{\rm n.t.}_{\partial\Omega}\Big)(y)\,\mathrm{d}\sigma(y)\Big)_{1\leq\gamma\leq M}\\
& + \Big(\int_{\Omega}E_{\gamma\beta}(x-y)(Lu)_\beta(y)\,\mathrm{d}y\Big)_{1\leq\gamma\leq M}
\end{aligned}
\tag{1.5.4}
$$

if either Ω *is bounded, or* $\partial\Omega$ *is unbounded. In the case when* Ω *is an exterior domain, the same conclusion holds under the additional assumption that there exists* $\lambda \in (1, \infty)$ *such that*

$$
\fint_{B(0,\lambda R)\setminus B(0,R)} |u|\,\mathrm{d}\mathcal{L}^n = o(1) \quad \textit{as } R \to \infty. \tag{1.5.5}
$$

Furthermore, for each $x \in \mathbb{R}^n \setminus \overline{\Omega}$ *one has*

$$
\begin{aligned}
0 = {} & \Big(-\int_{\partial_*\Omega}\nu_s(y)a^{\beta\alpha}_{rs}(\partial_r E_{\gamma\beta})(x-y)\big(u_\alpha\big|^{\rm n.t.}_{\partial\Omega}\big)(y)\,\mathrm{d}\sigma(y)\Big)_{1\leq\gamma\leq M}\\
& - \Big(\int_{\partial_*\Omega}E_{\gamma\alpha}(x-y)\nu_r(y)\Big((a^{\alpha\beta}_{rs}\partial_s u_\beta)\big|^{\rm n.t.}_{\partial\Omega}\Big)(y)\,\mathrm{d}\sigma(y)\Big)_{1\leq\gamma\leq M}\\
& + \Big(\int_{\Omega}E_{\gamma\beta}(x-y)(Lu)_\beta(y)\,\mathrm{d}y\Big)_{1\leq\gamma\leq M}
\end{aligned}
\tag{1.5.6}
$$

with the same caveat as above in the case when Ω *is an exterior domain.*

Finally, similar results are valid in the case when $n = 2$ *provided either*

$$\int_{S^1} \left[L(\xi)\right]^{-1} \mathrm{d}\mathcal{H}^1(\xi) = 0 \in \mathbb{C}^{M\times M}, \tag{1.5.7}$$

or otherwise one assumes that $\partial\Omega$ *is compact, replaces the first membership in* (1.5.2) *by*

$$\int_{\Omega} |(Lu)(y)| \ln(2 + |y|)\, \mathrm{d}y < +\infty, \tag{1.5.8}$$

and, in the case when Ω *is an exterior domain, replaces* (1.5.5) *by (compare with Theorem 1.5.7)*

$$⨍_{B(0,\lambda R)\setminus B(0,R)} |u|\, \mathrm{d}\mathcal{L}^2 = o\big(\tfrac{1}{\ln R}\big) \ \textit{as}\ R \to \infty. \tag{1.5.9}$$

Before presenting the proof of this theorem we make a number of comments. First, we note that the integral decay condition (1.5.5) is satisfied if one assumes the pointwise decay condition

$$u(x) = o(1) \ \text{as}\ |x| \to \infty. \tag{1.5.10}$$

In the particular case when the function u is a null-solution of the system L in an exterior domain Ω, the decay condition (1.5.5) implies (see Lemma 1.5.6) that for each multi-index $\alpha \in \mathbb{N}_0^n$ we actually have

$$(\partial^\alpha u)(x) = o(|x|^{-|\alpha|}) \ \text{as}\ |x| \to \infty. \tag{1.5.11}$$

See also Theorem 1.5.7 for a relaxation of the decay condition (1.5.9) for null-solutions of the system L in exterior domains in the two-dimensional setting.

Second, according to [112, Lemma 3.5.7], the first membership in (1.5.2) is equivalent with having, for $\mathcal{L}^n$-a.e. point $x \in \mathbb{R}^n$,

$$\int_{\Omega} \frac{|(Lu)(y)|}{|x - y|^{n-2}}\, \mathrm{d}y < +\infty. \tag{1.5.12}$$

Moreover, from [112, Lemma 6.2.9] (used with $m := n - 2$) we see that the first membership in (1.5.2) is satisfied provided the components of Lu belong to the Lorentz space $L^{n/2,1}(\Omega, \mathcal{L}^n)$. Here we also wish to note that, according to [112, Lemma 3.5.8], having (1.5.8) is equivalent with the demand that, for $\mathcal{L}^2$-a.e. point $x \in \mathbb{R}^2$,

$$\int_{\Omega} |(Lu)(y)| \big(1 + \big|\ln|x - y|\big|\big)\, \mathrm{d}y < +\infty. \tag{1.5.13}$$

Third, assuming

$$u \in \left[W^{1,1}_{\mathrm{loc}}(\Omega)\right]^M, \tag{1.5.14}$$

if $\partial\Omega$ is upper Ahlfors regular then the integrability conditions recorded in (1.5.3) are satisfies whenever $\mathcal{N}_\kappa u \in L^p(\partial\Omega, \sigma)$ with $p \in [1, \infty)$ and $\mathcal{N}_\kappa(\nabla u) \in L^q(\partial\Omega, \sigma)$ with $q \in [1, n-1)$. Moreover, if (1.5.14) holds and $\partial\Omega$ is actually compact, then the integrability conditions in (1.5.3) are implied by

$$\mathcal{N}_\kappa u,\ \mathcal{N}_\kappa(\nabla u) \in L^1(\partial\Omega, \sigma). \tag{1.5.15}$$

In Remark 1.5.3 we shall show that, even when $\partial\Omega$ is compact and (1.5.14) holds, Theorem 1.5.1 fails if (1.5.3) is relaxed to demanding that $\mathcal{N}_\kappa u$ and $\mathcal{N}_\kappa(\nabla u)$ belong to $L^1(\partial_*\Omega, \sigma)$. This being said, it is possible to further relax (1.5.3) to demanding instead that for each $\alpha \in \{1, \dots, M\}$ and $r \in \{1, \dots, n\}$ we have

$$\begin{aligned} &\mathcal{N}_\kappa u \in L^1_{\text{loc}}(\partial\Omega, \sigma), \quad \mathcal{N}_\kappa(a^{\alpha\beta}_{rs}\partial_s u_\beta) \in L^1_{\text{loc}}(\partial\Omega, \sigma), \\ &\int_{\partial_*\Omega} \frac{\Big|\Big(u\big|^{\kappa-\text{n.t.}}_{\partial\Omega}\Big)(y)\Big|}{1+|y|^{n-1}}\,d\sigma(y) < \infty, \text{ and} \\ &\int_{\partial_*\Omega} \frac{\Big|\Big((a^{\alpha\beta}_{rs}\partial_s u_\beta)\big|^{\kappa-\text{n.t.}}_{\partial\Omega}\Big)(y)\Big|}{1+|y|^{n-2}}\,d\sigma(y) < \infty. \end{aligned} \tag{1.5.16}$$

Such a version would require using [112, Corollary 1.5.2] (in the version recorded in [112, (1.5.23)]) in place of [112, Theorem 1.4.1] (as we do in the proof of Theorem 1.5.1 given below).

Fourth, if [112, Theorem 1.5.1] is employed in lieu of [112, Theorem 1.4.1] in the proof of Theorem 1.5.1, we may relax the doubling assumption on $\sigma := \mathcal{H}^{n-1}\lfloor\partial\Omega$ to simply asking that this is a locally finite measure. In such a scenario, we need to ask that the aperture parameter κ is sufficiently large (depending on Ω), and the flexibility of changing κ when considering nontangential boundary traces may be lost. However, modulo these nuances, the format of the main results (i.e., formulas (1.5.4), (1.5.6)) remains the same. As discussed in Remark 1.5.2, Theorem 1.5.1 fails without the lower Ahlfors regularity assumption on $\partial\Omega$.

Here is the proof of Theorem 1.5.1.

Proof of Theorem 1.5.1 To get started, from (1.5.3) and [112, Lemma 8.3.1] we see that

$$u \in \big[L^\infty_{\text{loc}}(\Omega, \mathcal{L}^n)\big]^M \text{ and } \big(a^{\alpha\beta}_{rs}\partial_s u_\beta\big)_{\substack{1\le\alpha\le M\\ 1\le r\le n}} \in \big[L^\infty_{\text{loc}}(\Omega, \mathcal{L}^n)\big]^{M\times N}. \tag{1.5.17}$$

Also, from (1.5.2), (1.5.3), and [112, Corollary 8.9.9] we deduce that for any $\kappa' > 0$ the nontangential traces

$$\begin{aligned} &u\big|^{\kappa'-\text{n.t.}}_{\partial\Omega} \text{ and } \Big((a^{\alpha\beta}_{rs}\partial_s u_\beta)\big|^{\kappa'-\text{n.t.}}_{\partial\Omega}\Big)_{\substack{1\le\alpha\le M\\ 1\le r\le n}} \text{ exist at } \sigma\text{-a.e. point on } \partial_{\text{nta}}\Omega, \\ &\text{and are actually independent of the chosen aperture parameter } \kappa' > 0. \end{aligned} \tag{1.5.18}$$

Moreover, with the dependence on κ' dropped, based on [112, (8.9.8)], [112, (8.9.44)], and [112, (8.8.52)] we conclude that for each $\alpha \in \{1, \dots, M\}$ and $r \in \{1, \dots, n\}$ we have

$$\int_{\partial_* \Omega} \frac{\left|\left(u\big|_{\partial\Omega}^{\rm n.t.}\right)(y)\right|}{1 + |y|^{n-1}} \, d\sigma(y) < +\infty \text{ and}$$
$$\int_{\partial_* \Omega} \frac{\left|\left((a_{rs}^{\alpha\beta} \partial_s u_\beta)\big|_{\partial\Omega}^{\rm n.t.}\right)(y)\right|}{1 + |y|^{n-2}} \, d\sigma(y) < +\infty. \tag{1.5.19}$$

In particular, from (1.5.19) and (1.4.24) it follows that, for each fixed point $x \in \Omega$, the boundary integrals in the first two lines of (1.5.4) are absolutely convergent.

To proceed, fix an arbitrary index $\gamma \in \{1, \dots, M\}$, pick a Lebesgue point $x \in \Omega$ for the function u with the property that (1.5.12) holds, and consider the vector field $\vec{F}_x = (F_s)_{1 \le s \le n}$ with components (recall that, throughout, the summation convention over repeated indices is in effect)

$$F_s := -a_{rs}^{\beta\alpha} (\partial_r E_{\gamma\beta})(x - \cdot) u_\alpha - E_{\gamma\alpha}(x - \cdot)\big(a_{sr}^{\alpha\beta} \partial_r u_\beta\big)$$
$$\text{at } \mathcal{L}^n\text{-a.e. point in } \Omega. \tag{1.5.20}$$

Then from (1.5.20) and (1.5.17) we see that

$$\vec{F}_x \in \big[L^1_{\rm loc}(\Omega, \mathcal{L}^n)\big]^n. \tag{1.5.21}$$

The next goal is to compute $\operatorname{div}\vec{F}_x$ in the sense of distributions in Ω. To this end, fix an arbitrary scalar-valued function $\varphi \in \mathscr{C}_c^\infty(\Omega)$ and write

$$\begin{aligned}
{}_{\mathcal{D}'(\Omega)}\big\langle \operatorname{div}\vec{F}_x, \varphi\big\rangle_{\mathcal{D}(\Omega)} &= -\int_\Omega F_s(y)(\partial_s\varphi)(y)\, dy \\
&= \int_\Omega a_{rs}^{\beta\alpha} (\partial_r E_{\gamma\beta})(x - y) u_\alpha(y)(\partial_s\varphi)(y)\, dy \\
&\quad + \int_\Omega E_{\gamma\alpha}(x - y)\big(a_{sr}^{\alpha\beta} \partial_r u_\beta\big)(y)(\partial_s\varphi)(y)\, dy \\
&=: I + II.
\end{aligned} \tag{1.5.22}$$

Bring in a scalar-valued function $\theta \in \mathscr{C}^\infty(\mathbb{R}^n)$ with the property that $\theta = 0$ on $B(0, 1)$ and $\theta = 1$ on $\mathbb{R}^n \setminus B(0, 2)$. For each $\varepsilon \in \big(0, \frac{1}{2}\operatorname{dist}(x, \partial\Omega)\big)$ define $\theta_\varepsilon : \mathbb{R}^n \to \mathbb{R}$ by setting

$$\theta_\varepsilon(y) := \theta\Big(\frac{y - x}{\varepsilon}\Big) \text{ for every } y \in \mathbb{R}^n. \tag{1.5.23}$$

Then $\theta_\varepsilon \in \mathscr{C}^\infty(\mathbb{R}^n)$ is a bounded function satisfying

$$\lim_{\varepsilon\to 0^+} \theta_\varepsilon(y) = 1 \text{ for every } y \in \mathbb{R}^n \setminus \{x\}, \tag{1.5.24}$$

and there exists some $C \in (0, \infty)$ such that for each $\varepsilon \in \big(0, \frac{1}{2}\operatorname{dist}(x, \partial\Omega)\big)$ we have

$$\begin{aligned} &1 - \theta_\varepsilon \in \mathscr{C}^\infty_c(\Omega), \quad \theta_\varepsilon \equiv 0 \text{ on } B(x, \varepsilon), \\ &\operatorname{supp}(\nabla\theta_\varepsilon) \subseteq \overline{B(x, 2\varepsilon)} \setminus B(x, \varepsilon), \text{ and} \\ &|(\nabla^j\theta_\varepsilon)(y)| \leq C\varepsilon^{-j} \text{ for every } j \in \mathbb{N}_0 \text{ and } y \in \mathbb{R}^n. \end{aligned} \tag{1.5.25}$$

In particular, for each $\gamma, \beta \in \{1, \dots, M\}$ we have

$$\begin{aligned} &E_{\gamma\beta}(x - \cdot)\theta_\varepsilon \in \mathscr{C}^\infty(\Omega) \text{ and } E_{\gamma\beta}(x - \cdot)\varphi\theta_\varepsilon \in \mathscr{C}^\infty_c(\Omega), \text{ plus} \\ &\text{similar properties for expressions involving partial derivatives} \\ &\text{acting on either of the intervening functions.} \end{aligned} \tag{1.5.26}$$

Let us focus on term I from (1.5.22). Relying on (1.5.17), Lebesgue's Dominated Convergence Theorem, and integrations by parts (keeping in mind (1.5.26) and the fact that $a^{\beta\alpha}_{rs}\partial_s u_\alpha \in L^1_{\rm loc}(\Omega, \mathcal{L}^n)$ for each $\beta \in \{1, \dots, M\}$ and $r \in \{1, \dots, n\}$; cf. (1.5.2)), we may write

$$\begin{aligned} I &= \lim_{\varepsilon\to 0^+} \int_\Omega a^{\beta\alpha}_{rs}(\partial_r E_{\gamma\beta})(x - y)u_\alpha(y)(\partial_s\varphi)(y)\theta_\varepsilon(y)\,\mathrm{d}y \\ &= \lim_{\varepsilon\to 0^+} \Bigg\{ \int_\Omega a^{\beta\alpha}_{rs}(\partial_s\partial_r E_{\gamma\beta})(x - y)u_\alpha(y)\varphi(y)\theta_\varepsilon(y)\,\mathrm{d}y \\ &\qquad - \int_\Omega (\partial_r E_{\gamma\beta})(x - y)\big(a^{\beta\alpha}_{rs}\partial_s u_\alpha\big)(y)\varphi(y)\theta_\varepsilon(y)\,\mathrm{d}y \\ &\qquad - \int_\Omega a^{\beta\alpha}_{rs}(\partial_r E_{\gamma\beta})(x - y)u_\alpha(y)\varphi(y)(\partial_s\theta_\varepsilon)(y)\,\mathrm{d}y \Bigg\}. \end{aligned} \tag{1.5.27}$$

Observe that since (1.4.33) gives that, for each $\alpha \in \{1, \dots, M\}$,

$$a^{\beta\alpha}_{rs}(\partial_s\partial_r E_{\gamma\beta})(x - y) = 0 \text{ for each } y \in \mathbb{R}^n \setminus \{x\}, \tag{1.5.28}$$

the first integral in the curly brackets in (1.5.27) vanishes. As regards the contribution from the very last integral in (1.5.27), we expand

$$\lim_{\varepsilon\to 0^+}\int_\Omega a_{rs}^{\beta\alpha}(\partial_r E_{\gamma\beta})(x-y)u_\alpha(y)\varphi(y)(\partial_s\theta_\varepsilon)(y)\,dy$$

$$= \lim_{\varepsilon\to 0^+}\int_\Omega a_{rs}^{\beta\alpha}(\partial_r E_{\gamma\beta})(x-y)\big((\varphi u_\alpha)(y)-(\varphi u_\alpha)(x)\big)(\partial_s\theta_\varepsilon)(y)\,dy$$

$$+\left\{\lim_{\varepsilon\to 0^+}\int_\Omega a_{rs}^{\beta\alpha}(\partial_r E_{\gamma\beta})(x-y)(\partial_s\theta_\varepsilon)(y)\,dy\right\}(\varphi u_\alpha)(x). \tag{1.5.29}$$

In relation to this, we note that since x is a Lebesgue point for u, we may estimate (using (1.4.24) and (1.5.25))

$$\limsup_{\varepsilon\to 0^+}\left|\int_\Omega a_{rs}^{\beta\alpha}(\partial_r E_{\gamma\beta})(x-y)\big((\varphi u_\alpha)(y)-(\varphi u_\alpha)(x)\big)(\partial_s\theta_\varepsilon)(y)\,dy\right|$$

$$\leq C\limsup_{\varepsilon\to 0^+}⨍_{B(x,2\varepsilon)}\big|(\varphi u_\alpha)(y)-(\varphi u_\alpha)(x)\big|\,dy = 0. \tag{1.5.30}$$

Also, (1.4.33) permits us to conclude that for each index $\alpha\in\{1,\dots,M\}$ we have

$$\lim_{\varepsilon\to 0^+}\int_\Omega a_{rs}^{\beta\alpha}(\partial_r E_{\gamma\beta})(x-y)(\partial_s\theta_\varepsilon)(y)\,dy$$

$$= \lim_{\varepsilon\to 0^+}\int_\Omega a_{rs}^{\beta\alpha}(\partial_r E_{\gamma\beta})(x-y)\partial_s(\theta_\varepsilon-1)(y)\,dy$$

$$= -\lim_{\varepsilon\to 0^+}{}_{\mathcal{D}'(\Omega)}\Big\langle a_{rs}^{\beta\alpha}\partial_r\big[E_{\gamma\beta}(x-\cdot)\big],\partial_s(\theta_\varepsilon-1)\Big\rangle_{\mathcal{D}(\Omega)}$$

$$= \lim_{\varepsilon\to 0^+}{}_{\mathcal{D}'(\Omega)}\Big\langle a_{rs}^{\beta\alpha}\partial_s\partial_r\big[E_{\gamma\beta}(x-\cdot)\big],\theta_\varepsilon-1\Big\rangle_{\mathcal{D}(\Omega)}$$

$$= \delta_{\alpha\gamma}\lim_{\varepsilon\to 0^+}{}_{\mathcal{D}'(\Omega)}\big\langle\delta_x,\theta_\varepsilon-1\big\rangle_{\mathcal{D}(\Omega)}$$

$$= \delta_{\alpha\gamma}\lim_{\varepsilon\to 0^+}(\theta_\varepsilon-1)(x) = -\delta_{\alpha\gamma}. \tag{1.5.31}$$

Altogether,

$$I = (\varphi u_\gamma)(x) - \lim_{\varepsilon\to 0^+}\int_\Omega(\partial_r E_{\gamma\beta})(x-y)\big(a_{rs}^{\beta\alpha}\partial_s u_\alpha\big)(y)\varphi(y)\theta_\varepsilon(y)\,dy. \tag{1.5.32}$$

Next, to treat term II from (1.5.22), we employ Lebesgue's Dominated Convergence Theorem and integrations by parts (bearing in mind (1.5.26)) to express

$$
\begin{aligned}
II &= \lim_{\varepsilon\to 0^+} \int_\Omega E_{\gamma\alpha}(x-y)\big(a_{sr}^{\alpha\beta}\partial_r u_\beta\big)(y)(\partial_s\varphi)(y)\theta_\varepsilon(y)\,dy \\
&= \lim_{\varepsilon\to 0^+} \Big\{ {}_{\mathcal{D}'(\Omega)}\Big\langle a_{sr}^{\alpha\beta}\partial_r u_\beta, \partial_s\big[E_{\gamma\alpha}(x-\cdot)\varphi\theta_\varepsilon\big]\Big\rangle_{\mathcal{D}(\Omega)} \\
&\qquad - \int_\Omega E_{\gamma\alpha}(x-y)\big(a_{sr}^{\alpha\beta}\partial_r u_\beta\big)(y)\varphi(y)(\partial_s\theta_\varepsilon)(y)\,dy \\
&\qquad + \int_\Omega (\partial_s E_{\gamma\alpha})(x-y)\big(a_{sr}^{\alpha\beta}\partial_r u_\beta\big)(y)\varphi(y)\theta_\varepsilon(y)\,dy \Big\}. \qquad (1.5.33)
\end{aligned}
$$

Upon recalling (1.5.26), we compute

$$
\begin{aligned}
&\lim_{\varepsilon\to 0^+} {}_{\mathcal{D}'(\Omega)}\Big\langle a_{sr}^{\alpha\beta}\partial_r u_\beta, \partial_s\big[E_{\gamma\alpha}(x-\cdot)\varphi\theta_\varepsilon\big]\Big\rangle_{\mathcal{D}(\Omega)} \\
&\qquad = -\lim_{\varepsilon\to 0^+} {}_{\mathcal{D}'(\Omega)}\Big\langle a_{sr}^{\alpha\beta}\partial_s\partial_r u_\beta, E_{\gamma\alpha}(x-\cdot)\varphi\theta_\varepsilon\Big\rangle_{\mathcal{D}(\Omega)} \\
&\qquad = -\lim_{\varepsilon\to 0^+} {}_{\mathcal{D}'(\Omega)}\Big\langle (Lu)_\alpha, E_{\gamma\alpha}(x-\cdot)\varphi\theta_\varepsilon\Big\rangle_{\mathcal{D}(\Omega)} \\
&\qquad = -\lim_{\varepsilon\to 0^+} \int_\Omega E_{\gamma\alpha}(x-y)(Lu)_\alpha(y)\varphi(y)\theta_\varepsilon(y)\,dy \\
&\qquad = -\int_\Omega E_{\gamma\alpha}(x-y)(Lu)_\alpha(y)\varphi(y)\,dy, \qquad (1.5.34)
\end{aligned}
$$

where the last two equalities are based on (1.5.12) and Lebesgue's Dominated Convergence Theorem. Since, thanks to (1.4.24) and (1.5.17), we also have

$$
\begin{aligned}
&\limsup_{\varepsilon\to 0^+} \Big| \int_\Omega E_{\gamma\alpha}(x-y)\big(a_{sr}^{\alpha\beta}\partial_r u_\beta\big)(y)\varphi(y)(\partial_s\theta_\varepsilon)(y)\,dy \Big| \qquad (1.5.35) \\
&\qquad \leq C \limsup_{\varepsilon\to 0^+} \Big\{ \varepsilon^{1-n} \int_{B(x,2\varepsilon)} \Big| \big((a_{rs}^{\alpha\beta}\partial_s u_\beta)(y)\big)_{\substack{1\leq\alpha\leq M\\ 1\leq r\leq n}} \Big| |\varphi(y)|\,dy \Big\} = 0,
\end{aligned}
$$

it follows that

$$
\begin{aligned}
II &= -\int_\Omega E_{\gamma\alpha}(x-y)(Lu)_\alpha(y)\varphi(y)\,dy \\
&\qquad + \lim_{\varepsilon\to 0^+} \int_\Omega (\partial_s E_{\gamma\alpha})(x-y)\big(a_{sr}^{\alpha\beta}\partial_r u_\beta\big)(y)\varphi(y)\theta_\varepsilon(y)\,dy. \qquad (1.5.36)
\end{aligned}
$$

Note that by interchanging α with β, and r with s, the limit in (1.5.36) matches the one (1.5.32). Bearing this in mind, from (1.5.22), (1.5.36), and (1.5.32) we ultimately conclude that for each Lebesgue point $x \in \Omega$ for u such that (1.5.12) holds we have

$$\mathrm{div}\vec{F}_x = u_\gamma(x)\delta_x - E_{\gamma\beta}(x-\cdot)(Lu)_\beta \text{ in } \mathcal{D}'(\Omega), \tag{1.5.37}$$

hence

$$\mathrm{div}\vec{F}_x \in \mathscr{E}'(\Omega) + L^1(\Omega, \mathcal{L}^n). \tag{1.5.38}$$

Going further, $K := \overline{B\big(x, \frac{1}{2}\,\mathrm{dist}(x, \partial\Omega)\big)}$ is a compact subset of Ω, and based on (1.5.20), (1.4.24), and [112, Lemma 8.3.7], for each $y \in \partial\Omega$, we may estimate

$$\begin{aligned}\big(\mathcal{N}_\kappa^{\Omega\setminus K}\vec{F}_x\big)(y) &\leq C\big(\mathcal{N}_\kappa u\big)(y)\cdot \sup_{z\in\Gamma_\kappa(y)\setminus K}\big[|x-z|^{1-n}\big]\\ &\quad + C\Big(\mathcal{N}_\kappa\big(a^{\alpha\beta}_{sr}\partial_r u_\beta\big)_{\alpha,s}\Big)(y)\cdot \sup_{z\in\Gamma_\kappa(y)\setminus K}\big[|x-z|^{2-n}\big]\\ &\leq C\left\{\frac{(\mathcal{N}_\kappa u)(y)}{|x-y|^{n-1}} + \frac{\big(\mathcal{N}_\kappa\big(a^{\alpha\beta}_{sr}\partial_r u_\beta\big)_{\alpha,s}\big)(y)}{|x-y|^{n-2}}\right\},\end{aligned} \tag{1.5.39}$$

for some constant $C = C(\Omega, L, n, \kappa) \in (0, \infty)$. From (1.5.39), (1.5.3), and [112, (8.2.26)] it follows that

$$\mathcal{N}_\kappa^{\Omega\setminus K}\vec{F}_x \in L^1(\partial\Omega, \sigma). \tag{1.5.40}$$

Moreover, from (1.5.20), [112, (8.9.10)-(8.9.11)], and (1.5.18) we conclude that

$$\vec{F}_x\Big|^{\text{n.t.}}_{\partial\Omega} \text{ exists at } \sigma\text{-a.e. point on } \partial_{\text{nta}}\Omega \tag{1.5.41}$$

and, in fact, for each $s \in \{1, \dots, n\}$ we have

$$\begin{aligned}\Big(\vec{F}_x\Big|^{\text{n.t.}}_{\partial\Omega}\Big)_s(y) &= -a^{\beta\alpha}_{rs}(\partial_r E_{\gamma\beta})(x-y)\big(u_\alpha\big|^{\text{n.t.}}_{\partial\Omega}\big)(y)\\ &\quad - E_{\gamma\alpha}(x-y)\big((a^{\alpha\beta}_{sr}\partial_r u_\beta)\big|^{\text{n.t.}}_{\partial\Omega}\big)(y)\end{aligned} \tag{1.5.42}$$

at σ-a.e. point $y \in \partial_{\text{nta}}\Omega$.

Collectively, (1.5.21), (1.5.38), and (1.5.41) ensure that, for each Lebesgue point $x \in \Omega$ for u such that (1.5.12) holds, the vector field $\vec{F}_x$ satisfies the hypotheses of [112, Theorem 1.4.1] in the case when either Ω is bounded, or $\partial\Omega$ is unbounded. Assuming this is the case, on account of [112, (4.6.19)], (1.5.37), and (1.5.42), the Divergence Formula [112, (1.4.6)] presently yields

$$
\begin{aligned}
u_\gamma(x) - \int_\Omega E_{\gamma\beta}(x-y)(Lu)_\beta(y)\,\mathrm{d}y
&= {}_{\mathscr{E}'(\Omega)}\big\langle u_\gamma(x)\delta_x, 1\big\rangle_{\mathscr{E}(\Omega)} - \int_\Omega E_{\gamma\beta}(x-y)(Lu)_\beta(y)\,\mathrm{d}y \\
&= {}_{(\mathscr{C}^\infty_b(\Omega))^*}\Big(u_\gamma(x)\delta_x - E_{\gamma\beta}(x-\cdot)(Lu)_\beta, 1\Big)_{\mathscr{C}^\infty_b(\Omega)} \\
&= {}_{(\mathscr{C}^\infty_b(\Omega))^*}\big(\operatorname{div}\vec{F}_x, 1\big)_{\mathscr{C}^\infty_b(\Omega)} = \int_{\partial_*\Omega} \nu\cdot\Big(\vec{F}_x\Big|^{\mathrm{n.t.}}_{\partial\Omega}\Big)\,\mathrm{d}\sigma \\
&= -\int_{\partial_*\Omega} \nu_s(y) a^{\beta\alpha}_{rs}(\partial_r E_{\gamma\beta})(x-y)\Big(u_\alpha\Big|^{\mathrm{n.t.}}_{\partial\Omega}\Big)(y)\,\mathrm{d}\sigma(y) \\
&\quad - \int_{\partial_*\Omega} E_{\gamma\alpha}(x-y)\nu_s(y)\Big((a^{\alpha\beta}_{sr}\partial_r u_\beta)\Big|^{\mathrm{n.t.}}_{\partial\Omega}\Big)(y)\,\mathrm{d}\sigma(y),
\end{aligned}
\tag{1.5.43}
$$

for each Lebesgue point $x \in \Omega$ of the function u with the property that (1.5.12) holds (hence also for $\mathcal{L}^n$-a.e. point $x \in \Omega$), and each $\gamma \in \{1, \dots, M\}$. After a slight adjustment in notation, this finishes the proof of (1.5.4) in the case when either Ω is bounded, or $\partial\Omega$ is unbounded.

Consider next the case when Ω is an exterior domain, under the additional assumption that there exists $\lambda \in (1, \infty)$ such that (1.5.5) holds. In such a scenario, bring in a function $\xi \in \mathscr{C}^\infty_c(\mathbb{R}^n)$ satisfying $\xi \equiv 1$ on $B(0,1)$, $\xi \equiv 0$ on $\mathbb{R}^n \setminus B(0,\lambda)$, and for every $R > 0$ define $\xi_R := \xi(\cdot/R)$ in $\mathbb{R}^n$. For each fixed $R > 0$, sufficiently large so that $x \in B(0,R)$ and $\partial\Omega \subseteq B(0,R)$, run the same argument as before but, this time, in place of (1.5.20) consider now the vector field $\vec{F}^R_x = \big(F^R_s\big)_{1\leq s\leq n}$ with components

$$
F^R_s := a^{\beta\alpha}_{rs}\partial_r\big(\xi_R E_{\gamma\beta}(x-\cdot)\big)u_\alpha - \xi_R E_{\gamma\alpha}(x-\cdot)\big(a^{\alpha\beta}_{sr}\partial_r u_\beta\big) \tag{1.5.44}
$$

defined at $\mathcal{L}^n$-a.e. point in Ω. Upon observing that

$$
\vec{F}^R_x = \xi_R\vec{F}_x + \Big(a^{\beta\alpha}_{rs}(\partial_r\xi_R)E_{\gamma\beta}(x-\cdot)u_\alpha\Big)_{1\leq s\leq n} \tag{1.5.45}
$$

and that

$$
(\partial^\mu\xi_R)E(x-\cdot) \in \big[\mathscr{C}^\infty_c(\Omega)\big]^{M\times M} \quad \text{for any } \mu \in \mathbb{N}^n_0 \text{ with } |\mu| > 0, \tag{1.5.46}
$$

it follows that in place of (1.5.37) we now get (after some algebra that involves canceling those terms which contain first-order derivatives of u)

$$
\begin{aligned}
\operatorname{div}\vec{F}^R_x &= u_\gamma(x)\delta_x - \xi_R E_{\gamma\beta}(x-\cdot)(Lu)_\beta \\
&\quad - a^{\beta\alpha}_{rs}(\partial_r\xi_R)(\partial_s E_{\gamma\beta})(x-\cdot)u_\alpha \\
&\quad - a^{\beta\alpha}_{rs}(\partial_s\xi_R)(\partial_r E_{\gamma\beta})(x-\cdot)u_\alpha \\
&\quad + a^{\beta\alpha}_{rs}(\partial_r\partial_s\xi_R)E_{\gamma\beta}(x-\cdot)u_\alpha \ \text{ in } \mathcal{D}'(\Omega).
\end{aligned}
\tag{1.5.47}
$$

Consequently,

$$
\operatorname{div}\vec{F}^R_x \in \mathscr{E}'(\Omega) + L^1(\Omega,\mathcal{L}^n). \tag{1.5.48}
$$

Also, (1.5.20), [112, (8.9.10)-(8.9.11)], and (1.5.18) ensure that

$$
\vec{F}^R_x\Big|^{\text{n.t.}}_{\partial\Omega} \text{ exists at } \sigma\text{-a.e. point on } \partial_{\text{nta}}\Omega \tag{1.5.49}
$$

Since $\vec{F}^R_x$ vanishes identically outside of a bounded subset of Ω, [112, Theorem 1.4.1] is applicable to this vector field and the Divergence Formula [112, (1.4.6)] gives (bearing in mind (1.5.46))

$$
\begin{aligned}
u_\gamma(x) &- \int_\Omega \xi_R(y)E_{\gamma\beta}(x-y)\big(Lu\big)_\beta(y)\,dy \\
&- \int_\Omega a^{\beta\alpha}_{rs}(\partial_r\xi_R)(y)(\partial_s E_{\gamma\beta})(x-y)u_\alpha(y)\,dy \\
&- \int_\Omega a^{\beta\alpha}_{rs}(\partial_s\xi_R)(y)(\partial_r E_{\gamma\beta})(x-y)u_\alpha(y)\,dy \\
&+ \int_\Omega a^{\beta\alpha}_{rs}(\partial_r\partial_s\xi_R)(y)E_{\gamma\beta}(x-y)u_\alpha(y)\,dy \\
&= {}_{(\mathscr{C}^\infty_b(\Omega))^*}\big(\operatorname{div}\vec{F}^R_x, 1\big)_{\mathscr{C}^\infty_b(\Omega)} = \int_{\partial_*\Omega} \nu\cdot\Big(\vec{F}^R_x\Big|^{\text{n.t.}}_{\partial\Omega}\Big)\,d\sigma \\
&= -\int_{\partial_*\Omega} \nu_s(y)a^{\beta\alpha}_{rs}\xi_R(y)(\partial_r E_{\gamma\beta})(x-y)\Big(u_\alpha\Big|^{\text{n.t.}}_{\partial\Omega}\Big)(y)\,d\sigma(y) \\
&\quad - \int_{\partial_*\Omega} \xi_R(y)E_{\gamma\alpha}(x-y)\nu_s(y)\Big(\big(a^{\alpha\beta}_{sr}\partial_r u_\beta\big)\Big|^{\text{n.t.}}_{\partial\Omega}\Big)(y)\,d\sigma(y).
\end{aligned}
\tag{1.5.50}
$$

There remains to pass to limit $R\to\infty$. In this regard, thanks to (1.4.24), (1.5.5), plus the fact that $\sup_{\mathbb{R}^n}|\nabla\xi_R| \le CR^{-1}$ and $\operatorname{supp}(\nabla\xi_R) \subseteq \overline{B(0,\lambda R)} \setminus B(0,R)$ for each $R>0$, we have

$$\left|\int_{\Omega} a_{rs}^{\beta\alpha}(\partial_s\xi_R)(y)(\partial_r E_{\gamma\beta})(x-y)u_\alpha(y)\,\mathrm{d}y\right|$$
$$+\left|\int_{\Omega} a_{rs}^{\beta\alpha}(\partial_r\xi_R)(y)(\partial_s E_{\gamma\beta})(x-y)u_\alpha(y)\,\mathrm{d}y\right|$$
$$\leq C \fint_{B(0,\lambda R)\setminus B(0,R)} |u|\,\mathrm{d}\mathcal{L}^n = o(1) \text{ as } R \to \infty. \tag{1.5.51}$$

Similarly, since $\sup_{\mathbb{R}^n} |\nabla^2\xi_R| \leq CR^{-2}$ and $\operatorname{supp}(\nabla^2\xi_R) \subseteq \overline{B(0,\lambda R)} \setminus B(0,R)$ for each $R > 0$, the fact that $|E(z)| \leq C|z|^{2-n}$ for each $z \in \mathbb{R}^n \setminus \{0\}$ (cf. (1.4.24)) once again implies (keeping in mind (1.5.5))

$$\left|\int_{\Omega} a_{rs}^{\beta\alpha}(\partial_r\partial_s\xi_R)(y)E_{\gamma\beta}(x-y)u_\alpha(y)\,\mathrm{d}y\right| = o(1) \text{ as } R \to \infty. \tag{1.5.52}$$

Also, (1.5.12), (1.4.24), and Lebesgue's Dominated Convergence Theorem imply

$$\lim_{R\to\infty} \int_{\Omega} \xi_R(y)E_{\gamma\beta}(x-y)\big(Lu\big)_\beta(y)\,\mathrm{d}y = \int_{\Omega} E_{\gamma\beta}(x-y)\big(Lu\big)_\beta(y)\,\mathrm{d}y. \tag{1.5.53}$$

In addition, the fact that $\xi_R \equiv 1$ on $B(0,R)$ and the radius R is sufficiently large so that $\partial\Omega \subseteq B(0,R)$ ensure that

$$\int_{\partial_*\Omega} \nu_s(y)a_{rs}^{\beta\alpha}\xi_R(y)(\partial_r E_{\gamma\beta})(x-y)\big(u_\alpha\big|_{\partial\Omega}^{\text{n.t.}}\big)(y)\,\mathrm{d}\sigma(y)$$
$$= \int_{\partial_*\Omega} \nu_s(y)(\partial_r E_{\gamma\beta})(x-y)\big((a_{rs}^{\beta\alpha}u_\alpha)\big|_{\partial\Omega}^{\text{n.t.}}\big)(y)\,\mathrm{d}\sigma(y), \tag{1.5.54}$$

as well as

$$\int_{\partial_*\Omega} \xi_R(y)E_{\gamma\alpha}(x-y)\nu_s(y)\big((a_{sr}^{\alpha\beta}\partial_r u_\beta)\big|_{\partial\Omega}^{\text{n.t.}}\big)(y)\,\mathrm{d}\sigma(y)$$
$$= \int_{\partial_*\Omega} E_{\gamma\alpha}(x-y)\nu_s(y)\big((a_{sr}^{\alpha\beta}\partial_r u_\beta)\big|_{\partial\Omega}^{\text{n.t.}}\big)(y)\,\mathrm{d}\sigma(y). \tag{1.5.55}$$

On account of (1.5.50)-(1.5.55), formula (1.5.4) follows in the case when Ω is an exterior domain. Parenthetically, we wish to remark that the reason for which we preferred to work with the truncated vector field $\vec{F}_x^R = \big(F_s^R\big)_{1\leq s\leq n}$ (with components as in (1.5.44)) in place of the earlier vector field $\vec{F}_x = \big(F_s\big)_{1\leq s\leq n}$ with components as in (1.5.20), is that the later choice would require (via [112, (1.4.8)]) slightly stronger decay conditions on the original function u than the ones imposed in (1.5.5). This being said, if u is a nullsolution of L then such stronger decay conditions are automatically satisfied (see Lemma 1.5.6), so in this scenario the is no need to resort to employing the truncated vector field $\vec{F}_x^R = \big(F_s^R\big)_{1\leq s\leq n}$ in place of the original $\vec{F}$.

Moving on, formula (1.5.6) is established in a similar (and simpler) fashion, keeping in mind that if $x \in \mathbb{R}^n \setminus \overline{\Omega}$ then $E(x - \cdot)$ is already a $\mathscr{C}^\infty$ matrix-valued function in Ω. As such, there is no need to bring in the cutoff function θ_ε and, this time, the Dirac distribution in (1.5.37) is no longer present.

Finally, the same proof works in the case when $n = 2$ provided either (1.5.7) holds, or otherwise we assume $\partial\Omega$ is compact, replace the first membership in (1.5.2) by (1.5.8), and also replace (1.5.5) by (1.5.9) (such changes now allow us to justify (1.5.52), keeping (1.4.24) in mind). In this scenario, we start by considering a point $x \in \mathbb{R}^2$ with the property (1.5.13) holds and, in the case of (1.5.4), we also demand that $x \in \Omega$ is a Lebesgue point for the function u. □

Remark 1.5.2 *Theorem 1.5.1 fails without the lower Ahlfors regularity assumption on $\partial\Omega$.*

To see that this is the case, consider the (scalar) differential operator $L := \Delta$, the Laplacian in $\mathbb{R}^n$, with fundamental solution

$$E_\Delta(x) := \begin{cases} \dfrac{1}{\omega_{n-1}(2-n)} \dfrac{1}{|x|^{n-2}} & \text{if } n \geq 3, \\[2ex] \dfrac{1}{2\pi} \ln |x| & \text{if } n = 2. \end{cases} \tag{1.5.56}$$

In particular,

$$(\nabla E_\Delta)(x) = \frac{1}{\omega_{n-1}} \cdot \frac{x}{|x|^n} \quad \text{for } x \in \mathbb{R}^n \setminus \{0\}. \tag{1.5.57}$$

Next, fix $n \in \mathbb{N}$ with $n \geq 3$ and consider the open subset of $\mathbb{R}^n$ given by

$$\Omega := B(0,1) \setminus \{0\}. \tag{1.5.58}$$

Note that the set $\partial\Omega = \partial B(0,1) \cup \{0\}$ is upper Ahlfors regular, hence the measure $\sigma := \mathcal{H}^{n-1} \lfloor \partial\Omega$ is locally finite on $\partial\Omega$. Also, $\partial_*\Omega = \partial B(0,1)$ so, in particular, $\mathcal{H}^{n-1}(\partial\Omega \setminus \partial_*\Omega) = 0$. However, $\partial\Omega$ is not lower Ahlfors regular.

Finally, having fixed some $j \in \{1, \dots, n\}$, let us define the function $u : \Omega \to \mathbb{R}$ by setting

$$u(x) := \frac{x_j}{|x|^n} \quad \text{for each } x \in \Omega. \tag{1.5.59}$$

Since u is a constant multiple of $\partial_j E_\Delta$, it follows that u is harmonic in Ω. Also, a direct computation gives

$$(\nabla u)(x) = \left(\frac{\delta_{jk}}{|x|^n} - n \frac{x_j x_k}{|x|^{n+2}} \right)_{1 \leq k \leq n} \quad \text{for } x = (x_1, \dots, x_n) \in \Omega. \tag{1.5.60}$$

Based on these observations, it is easy to check that the function u satisfies the hypotheses of Theorem 1.5.1.

Fix $x \in \Omega$. The goal is to compute the integrals in the right-hand side of (1.5.4) for this choice of Ω, u, and x. To proceed, fix $\varepsilon \in (0, |x|/4)$ and define the auxiliary domain

$$\Omega_\varepsilon := \Omega \setminus \big(\overline{B(0,\varepsilon)} \cup \overline{B(x,\varepsilon)}\ \big). \tag{1.5.61}$$

To simplify notation, introduce

$$I_D^1 := \int_{\partial D} \nu_D(y) \cdot \nabla_y \big[E(x-y)\big] \big(u\big|_{\partial D}^{\rm n.t.}\big)(y)\, d\mathcal{H}^{n-1}(y), \tag{1.5.62}$$

$$I_D^2 := \int_{\partial D} E(x-y)\, \nu_D(y) \cdot \big((\nabla u)\big|_{\partial D}^{\rm n.t.}\big)(y)\, d\mathcal{H}^{n-1}(y), \tag{1.5.63}$$

where D is one of the balls $B(0,1)$, $B(0,\varepsilon)$, or $B(x,\varepsilon)$, ν_D is the outward unit normal to the ball D, and the nontangential boundary traces are taken from within D. Note that Ω_ε is smooth and both u and $E(x-\cdot)$ are harmonic in Ω_ε. Thus, employing the notation in (1.5.62)-(1.5.63), the classical Green's Theorem in Ω_ε implies

$$I^1_{B(0,1)} - I^2_{B(0,1)} - I^1_{B(0,\varepsilon)} + I^2_{B(0,\varepsilon)} - I^1_{B(x,\varepsilon)} + I^2_{B(x,\varepsilon)} = 0. \tag{1.5.64}$$

Consider first the integrals associated with $B(0,\varepsilon)$. Making use of (1.5.57) we may compute

$$\begin{aligned} I^1_{B(0,\varepsilon)} &= -\frac{1}{\omega_{n-1}} \int_{|y|=\varepsilon} \frac{y}{\varepsilon} \cdot \frac{x-y}{|x-y|^n} \frac{y_j}{\varepsilon^n}\, d\mathcal{H}^{n-1}(y) \\ &= -\frac{1}{\omega_{n-1}} \sum_{k=1}^n \int_{|z|=1} \frac{z_k z_j (x_k - \varepsilon z_k)}{|x-\varepsilon z|^n}\, d\mathcal{H}^{n-1}(z), \end{aligned} \tag{1.5.65}$$

where in the second equality we have made the change of variables $y = \varepsilon z$. Relying on Lebesgue's Dominated Convergence Theorem and the fact that

$$\int_{|z|=1} z_k z_j\, d\mathcal{H}^{n-1}(z) = \frac{\delta_{jk}\omega_{n-1}}{n}, \tag{1.5.66}$$

(see [109, Proposition 14.69, p. 581] for more general results of this nature) we further obtain

$$\begin{aligned} \lim_{\varepsilon \to 0^+} I^1_{B(0,\varepsilon)} &= -\frac{1}{\omega_{n-1}} \frac{x_k}{|x|^n} \sum_{k=1}^n \int_{|z|=1} z_k z_j\, d\mathcal{H}^{n-1}(z) \\ &= -\frac{1}{\omega_{n-1}} \sum_{k=1}^n \frac{x_k}{|x|^n} \frac{\delta_{jk}\omega_{n-1}}{n} = -\frac{1}{n} u(x). \end{aligned} \tag{1.5.67}$$

Relying again on (1.5.60) and the change of variables $y = \varepsilon z$ we also have

$$I^2_{B(0,\varepsilon)} = \frac{1}{(2-n)\omega_{n-1}} \int_{|y|=\varepsilon} \frac{1}{|x-y|^{n-2}} \sum_{k=1}^{n} \frac{y_k}{\varepsilon}\Big(\frac{\delta_{jk}}{\varepsilon^n} - n\frac{y_j y_k}{\varepsilon^{n+2}}\Big)\, d\mathcal{H}^{n-1}(y)$$

$$= \frac{n-1}{(n-2)\omega_{n-1}} \int_{|z|=1} \frac{1}{|x-\varepsilon z|^{n-2}} \frac{z_j}{\varepsilon}\, d\mathcal{H}^{n-1}(z). \tag{1.5.68}$$

Since $\int_{|z|=1} z_j \, d\mathcal{H}^{n-1}(z) = 0$, we may further write

$$I^2_{B(0,\varepsilon)} = \frac{n-1}{(n-2)\omega_{n-1}} \int_{|z|=1} \Big[\frac{1}{|x-\varepsilon z|^{n-2}} - \frac{1}{|x|^{n-2}}\Big] \frac{z_j}{\varepsilon}\, d\mathcal{H}^{n-1}(z). \tag{1.5.69}$$

Based on the Fundamental Theorem of Calculus and the Mean Value Formula for integrals, we have

$$\frac{1}{|x-\varepsilon z|^{n-2}} - \frac{1}{|x|^{n-2}} = \int_0^{\varepsilon} \frac{d}{dt}\Big[\frac{1}{|x-tz|^{n-2}}\Big]\, dt$$

$$= \int_0^{\varepsilon} (n-2) z \cdot \frac{x-tz}{|x-tz|^n}\, dt$$

$$= \varepsilon(n-2) z \cdot \frac{x-t_\varepsilon z}{|x-t_\varepsilon z|^n}, \tag{1.5.70}$$

for some $t_\varepsilon \in (0,\varepsilon)$. Combining (1.5.69), (1.5.70), Lebesgue's Dominated Convergence Theorem, and (1.5.66), we arrive at

$$\lim_{\varepsilon\to 0^+} I^2_{B(0,\varepsilon)} = \frac{n-1}{\omega_{n-1}} \sum_{k=1}^{n} \frac{x_k}{|x|^n} \int_{|z|=1} z_j z_k \, d\mathcal{H}^{n-1}(z) = \frac{n-1}{n} u(x). \tag{1.5.71}$$

In the next phase, we treat the integrals over $\partial B(x,\varepsilon)$. First, (1.5.57) implies

$$I^1_{B(x,\varepsilon)} = \frac{1}{\omega_{n-1}} \int_{|y-x|=\varepsilon} \frac{y-x}{\varepsilon} \cdot \frac{y-x}{\varepsilon^n} u(y)\, d\mathcal{H}^{n-1}(y)$$

$$= \frac{1}{\varepsilon^{n-1}\omega_{n-1}} \int_{|y-x|=\varepsilon} \big[u(y)-u(x)\big]\, d\mathcal{H}^{n-1}(y) + u(x). \tag{1.5.72}$$

Since $|u(y)-u(x)| \le C\varepsilon$ for $y \in \partial B(x,\varepsilon)$, from (1.5.72) we obtain

$$\lim_{\varepsilon\to 0^+} I^1_{B(x,\varepsilon)} = u(x). \tag{1.5.73}$$

Second, as regards $I^2_{B(x,\varepsilon)}$, we have

$$\big|I^2_{B(x,\varepsilon)}\big| \le \frac{\|\nabla u\|_{L^\infty(B(x,|x|/4))}}{(n-2)\omega_{n-1}} \int_{|y-x|=\varepsilon} \frac{d\mathcal{H}^{n-1}(y)}{\varepsilon^{n-2}} \xrightarrow[\varepsilon\to 0^+]{} 0. \tag{1.5.74}$$

On account of (1.5.67), (1.5.71), (1.5.73), and (1.5.74), by passing to limit $\varepsilon \to 0^+$ in (1.5.64) we arrive at

$$I^1_{B(0,1)} - I^2_{B(0,1)} = 0. \tag{1.5.75}$$

Since $I^1_{B(0,1)} - I^2_{B(0,1)}$ actually represents the right-hand side of (1.5.4) in the present setting, from (1.5.75) we ultimately conclude that the integral representation formula (1.5.4) does not hold for Ω as in (1.5.58) and u as in (1.5.59). The source of this failure is the lack of lower Ahlfors regularity for $\partial\Omega$.

Remark 1.5.3 *Theorem 1.5.1 fails without* (1.5.3). *In fact, without* (1.5.3), *formula* (1.5.4) *may fail even if $\partial\Omega$ is compact, $u \in \big[W^{1,1}_{\rm loc}(\Omega)\big]^M$ and $\mathcal{N}_\kappa u$, $\mathcal{N}_\kappa(\nabla u)$ belong to $L^1(\partial_*\Omega, \sigma)$ for each $\kappa > 0$.*

To see that this is the case, define $\Omega := B(0,1) \setminus \{x_j = 0\}$. Hence, Ω is an open set with a compact Ahlfors regular boundary, and such that

$$\partial_*\Omega = \partial B(0,1). \tag{1.5.76}$$

For this choice of Ω, take u as in (1.5.59). In particular, u has a smooth extension to $\mathbb{R}^n \setminus \{0\}$ which, in light of (1.5.76), shows that $u \in \big[W^{1,1}_{\rm loc}(\Omega)\big]^M$ and $\mathcal{N}_\kappa u$, $\mathcal{N}_\kappa(\nabla u)$ belong to $L^1(\partial_*\Omega, \sigma)$ for each $\kappa > 0$. However, the equality in (1.5.75) shows (bearing (1.5.76) in mind) that formula (1.5.4) does not hold for Ω and u as specified.

We continue by presenting an integral representation formula in an exterior domain for null-solutions of a weakly elliptic system with control near the boundary in terms of the truncated nontangential maximal function and sub-linear growth at infinity.

Corollary 1.5.4 *Let $\Omega \subseteq \mathbb{R}^n$, where $n \geq 2$, be an unbounded open set with a compact lower Ahlfors regular boundary, and with the property that $\sigma := \mathcal{H}^{n-1}\lfloor\partial\Omega$ is a doubling measure on $\partial\Omega$. In particular, Ω is a set of locally finite perimeter, and its geometric measure theoretic outward unit normal $\nu = (\nu_1, \dots, \nu_n)$ is defined σ-a.e. on $\partial_*\Omega$. With the summation convention over repeated indices understood throughout, let*

$$L = \big(a^{\alpha\beta}_{rs}\partial_r\partial_s\big)_{1\leq\alpha,\beta\leq M} \tag{1.5.77}$$

be a homogeneous, weakly elliptic, second-order $M\times M$ system in $\mathbb{R}^n$, with complex constant coefficients, and denote by $E = (E_{\gamma\beta})_{1\leq\gamma,\beta\leq M}$ the matrix-valued fundamental solution associated with L as in Theorem 1.4.2. Finally, fix an aperture parameter $\kappa > 0$.

Suppose $u = (u_\beta)_{1\leq\beta\leq M}$ is a vector-valued function satisfying

$$\begin{gathered} u \in \big[\mathscr{C}^\infty(\Omega)\big]^M \text{ and } Lu = 0 \text{ in } \Omega, \\ u\big|^{\kappa-\text{n.t.}}_{\partial\Omega} \text{ and } \Big(\big(a^{\alpha\beta}_{rs}\partial_s u_\beta\big)\big|^{\kappa-\text{n.t.}}_{\partial\Omega}\Big)_{\substack{1\leq\alpha\leq M\\1\leq r\leq n}} \text{ exist at } \sigma\text{-a.e. point on } \partial_{\rm nta}\Omega. \end{gathered} \tag{1.5.78}$$

In addition, assume that for some truncation parameter $\varepsilon > 0$ the following integrability conditions hold:

$$\int_{\partial\Omega} \mathcal{N}_\kappa^\varepsilon u \, d\sigma < \infty \text{ and } \int_{\partial\Omega} \mathcal{N}_\kappa^\varepsilon (a_{rs}^{\alpha\beta} \partial_s u_\beta) \, d\sigma < \infty, \tag{1.5.79}$$

for each $\alpha \in \{1, \dots, M\}$ and $r \in \{1, \dots, n\}$. Finally, assume that there exists $\lambda \in (1, \infty)$ such that

$$⨍_{B(0,\lambda R)\setminus B(0,R)} |u| \, d\mathcal{L}^n = o(R) \text{ as } R \to \infty. \tag{1.5.80}$$

Then for any other given aperture parameter $\kappa' \in (0, \infty)$ the nontangential boundary traces $u\big|_{\partial\Omega}^{\kappa'-\text{n.t.}}$ and $\Big((a_{rs}^{\alpha\beta} \partial_s u_\beta)\big|_{\partial\Omega}^{\kappa'-\text{n.t.}}\Big)_{\substack{1\leq\alpha\leq M\\ 1\leq r\leq n}}$ also exist at σ-a.e. point on $\partial_{\text{nta}}\Omega$ and are actually independent of κ'. Furthermore, there exists a constant $c \in \mathbb{C}^M$ with the property that for each point $x \in \Omega$ one has (with absolutely convergent integrals, and with the dependence on the aperture parameter dropped)

$$\begin{aligned} u(x) = c &+ \Big(- \int_{\partial_*\Omega} \nu_s(y) a_{rs}^{\beta\alpha} (\partial_r E_{\gamma\beta})(x-y) \big(u_\alpha\big|_{\partial\Omega}^{\text{n.t.}}\big)(y) \, d\sigma(y) \Big)_{1\leq\gamma\leq M} \\ &- \Big(\int_{\partial_*\Omega} E_{\gamma\alpha}(x-y) \nu_r(y) \Big((a_{rs}^{\alpha\beta} \partial_s u_\beta)\big|_{\partial\Omega}^{\text{n.t.}} \Big)(y) \, d\sigma(y) \Big)_{1\leq\gamma\leq M}. \end{aligned} \tag{1.5.81}$$

Proof We largely reason as in the proof of Theorem 1.5.1. Fix a point $x_0 \in \mathbb{R}^n \setminus \overline{\Omega}$ along with an index $\gamma \in \{1, \dots, M\}$, both arbitrary. This time, in place of (1.5.44), we shall consider the vector field $\vec{F}_{x,x_0}^R := \vec{F}_x^R - \vec{F}_{x_0}^R$, whose s-th component is given, at $\mathcal{L}^n$-a.e. point in Ω, by

$$\begin{aligned} &a_{rs}^{\beta\alpha} \partial_r \big(\xi_R E_{\gamma\beta}(x - \cdot) - \xi_R E_{\gamma\beta}(x_0 - \cdot) \big) u_\alpha \\ &\qquad - \xi_R \big(E_{\gamma\alpha}(x - \cdot) - E_{\gamma\alpha}(x_0 - \cdot) \big) \big(a_{sr}^{\alpha\beta} \partial_r u_\beta \big) \end{aligned} \tag{1.5.82}$$

for each $s \in \{1, \dots, n\}$. Then $\vec{F}_{x,x_0}^R$ is locally integrable in Ω, and in place of (1.5.47) we now get (bearing in mind that both $Lu = 0$ and $L[E(x_0 - \cdot)] = 0$ in Ω):

$$\begin{aligned} \operatorname{div} \vec{F}_{x,x_0}^R &= \operatorname{div} \vec{F}_x^R - \operatorname{div} \vec{F}_{x_0}^R \\ &= u_\gamma(x)\delta_x - a_{rs}^{\beta\alpha} (\partial_r \xi_R) \big((\partial_s E_{\gamma\beta})(x - \cdot) - (\partial_s E_{\gamma\beta})(x_0 - \cdot) \big) u_\alpha \\ &\quad - a_{rs}^{\beta\alpha} (\partial_s \xi_R) \big((\partial_r E_{\gamma\beta})(x - \cdot) - (\partial_r E_{\gamma\beta})(x_0 - \cdot) \big) u_\alpha \\ &\quad + a_{rs}^{\beta\alpha} (\partial_r \partial_s \xi_R) \big(E_{\gamma\beta}(x - \cdot) - E_{\gamma\beta}(x_0 - \cdot) \big) u_\alpha \text{ in } \mathcal{D}'(\Omega). \end{aligned} \tag{1.5.83}$$

As a consequence, for large $R > 0$ we once again have

$$\mathrm{div}\vec{F}^R_{x,x_0} \in \mathscr{E}'(\Omega) + L^1(\Omega, \mathcal{L}^n). \tag{1.5.84}$$

Also, much as before,

$$\vec{F}^R_{x,x_0}\Big|^{\mathrm{n.t.}}_{\partial\Omega} \text{ exists at } \sigma\text{-a.e. point on } \partial_{\mathrm{nta}}\Omega. \tag{1.5.85}$$

Finally, since $\vec{F}^R_{x,x_0}$ vanishes identically outside of a bounded subset of Ω, from (1.5.79), (1.5.82), and [112, (8.2.28)] we see that if $K := \overline{B\big(x, \frac{1}{2}\,\mathrm{dist}(x, \partial\Omega)\big)}$ then

$$\mathcal{N}^{\Omega\setminus K}_{\kappa}\vec{F}_{x,x_0} \in L^1(\partial\Omega, \sigma). \tag{1.5.86}$$

Hence, [112, Theorem 1.4.1] is applicable to this vector field and the Divergence Formula [112, (1.4.6)] presently yields, assuming $R > 0$ is sufficiently large,

$$\begin{aligned}
u_\gamma(x) &- \int_\Omega a^{\beta\alpha}_{rs}(\partial_r\xi_R)(y)\big((\partial_s E_{\gamma\beta})(x-y) - (\partial_s E_{\gamma\beta})(x_0-y)\big)u_\alpha(y)\,\mathrm{d}y \\
&- \int_\Omega a^{\beta\alpha}_{rs}(\partial_s\xi_R)(y)\big((\partial_r E_{\gamma\beta})(x-y) - (\partial_r E_{\gamma\beta})(x_0-y)\big)u_\alpha(y)\,\mathrm{d}y \\
&+ \int_\Omega a^{\beta\alpha}_{rs}(\partial_r\partial_s\xi_R)(y)\big(E_{\gamma\beta}(x-y) - E_{\gamma\beta}(x_0-y)\big)u_\alpha(y)\,\mathrm{d}y \\
&= {}_{(\mathscr{C}^\infty_b(\Omega))^*}\big(\mathrm{div}\vec{F}^R_{x,x_0}, 1\big)_{\mathscr{C}^\infty_b(\Omega)} = \int_{\partial_*\Omega} \nu\cdot\Big(\vec{F}^R_{x,x_0}\Big|^{\mathrm{n.t.}}_{\partial\Omega}\Big)\,\mathrm{d}\sigma \\
&= -\int_{\partial_*\Omega} \nu_s(y)a^{\beta\alpha}_{rs}\xi_R(y)\big((\partial_r E_{\gamma\beta})(x-y) - (\partial_r E_{\gamma\beta})(x_0-y)\big)\times \\
&\qquad\qquad \times\Big(u_\alpha\Big|^{\mathrm{n.t.}}_{\partial\Omega}\Big)(y)\,\mathrm{d}\sigma(y) \\
&\quad - \int_{\partial_*\Omega} \xi_R(y)\big(E_{\gamma\alpha}(x-y) - E_{\gamma\alpha}(x_0-y)\big)\nu_s(y)\times \\
&\qquad\qquad \times\Big((a^{\alpha\beta}_{sr}\partial_r u_\beta)\Big|^{\mathrm{n.t.}}_{\partial\Omega}\Big)(y)\,\mathrm{d}\sigma(y). \\
&= c_\gamma - \int_{\partial_*\Omega} \nu_s(y)a^{\beta\alpha}_{rs}(\partial_r E_{\gamma\beta})(x-y)\Big(u_\alpha\Big|^{\mathrm{n.t.}}_{\partial\Omega}\Big)(y)\,\mathrm{d}\sigma(y) \\
&\quad - \int_{\partial_*\Omega} E_{\gamma\alpha}(x-y)\nu_s(y)\Big((a^{\alpha\beta}_{sr}\partial_r u_\beta)\Big|^{\mathrm{n.t.}}_{\partial\Omega}\Big)(y)\,\mathrm{d}\sigma(y),
\end{aligned} \tag{1.5.87}$$

where

$$
\begin{aligned}
c_\gamma := & \int_{\partial_*\Omega} \nu_s(y) a^{\beta\alpha}_{rs} (\partial_r E_{\gamma\beta})(x_0 - y) \big(u_\alpha\big|^{\rm n.t.}_{\partial\Omega}\big)(y)\, d\sigma(y) \\
& + \int_{\partial_*\Omega} E_{\gamma\alpha}(x_0 - y) \nu_s(y) \big((a^{\alpha\beta}_{sr} \partial_r u_\beta)\big|^{\rm n.t.}_{\partial\Omega}\big)(y)\, d\sigma(y)
\end{aligned}
\tag{1.5.88}
$$

is a constant in $\mathbb{C}$. Let us also note that, as may be seen from Theorem 1.4.2 and the Mean Value Theorem, there exists $C = C(L, x, x_0) \in (0, \infty)$ with the property that

$$
\begin{aligned}
|\nabla \xi_R| \big|(\nabla E)(x - \cdot) - (\nabla E)(x_0 - \cdot)\big| \leq C R^{-n-1} \text{ and} \\
|\nabla^2 \xi_R| \big|E(x - \cdot) - E(x_0 - \cdot)\big| \leq C R^{-n-1}.
\end{aligned}
\tag{1.5.89}
$$

Then (1.5.81) with $c := (c_\gamma)_{1 \leq \gamma \leq M} \in \mathbb{C}^M$ follows from (1.5.87) after passing to limit $R \longrightarrow \infty$ and invoking (1.5.89) and (1.5.80). □

Moving on, we record the following significant result (which will play a role shortly, in the proof of Theorem 1.5.7, as well as later).

Proposition 1.5.5 *Assume that either*

$$
\begin{aligned}
& n \geq 2 \textit{ and } \Omega \subseteq \mathbb{R}^n \textit{ is an open set of locally finite} \\
& \textit{perimeter with a compact toplogical boundary,}
\end{aligned}
\tag{1.5.90}
$$

or

$$
\begin{aligned}
& n \geq 3 \textit{ and } \Omega \subseteq \mathbb{R}^n \textit{ is an open set with a lower Ahlfors regular} \\
& \textit{boundary and such that } \int_{\partial\Omega} [1 + |z|]^{3-2n}\, d\mathcal{H}^{n-1}(z) < +\infty.
\end{aligned}
\tag{1.5.91}
$$

In all cases, abbreviate $\sigma := \mathcal{H}^{n-1} \lfloor \partial\Omega$ and denote by $\nu = (\nu_1, \dots, \nu_n)$ the geometric measure theoretic outward unit normal to Ω. With the summation convention over repeated indices understood throughout, let

$$
L = \big(a^{\alpha\beta}_{rs} \partial_r \partial_s\big)_{1 \leq \alpha,\beta \leq M}
\tag{1.5.92}
$$

be a homogeneous, weakly elliptic, second-order $M \times M$ system in $\mathbb{R}^n$, with complex constant coefficients, and denote by $E = (E_{\gamma\beta})_{1 \leq \gamma,\beta \leq M}$ the matrix-valued fundamental solution associated with L as in Theorem 1.4.2. Finally, select an arbitrary index $\delta \in \{1, \dots, M\}$.

Then for any two points $x, x_o \in \mathbb{R}^n$ one has

$$
\begin{aligned}
& \Big(- \int_{\partial_*\Omega} \nu_s(y) a^{\beta\alpha}_{rs} (\partial_r E_{\gamma\beta})(x - y) E_{\alpha\delta}(y - x_o)\, d\sigma(y) \Big)_{1 \leq \gamma \leq M} \\
& \quad = \Big(\int_{\partial_*\Omega} E_{\gamma\alpha}(x - y) \nu_r(y) a^{\alpha\beta}_{rs} (\partial_s E_{\beta\delta})(y - x_o)\, d\sigma(y) \Big)_{1 \leq \gamma \leq M}
\end{aligned}
\tag{1.5.93}
$$

provided either both x, x_o belong to Ω, or both x, x_o belong to $\mathbb{R}^n \setminus \overline{\Omega}$. Furthermore, if $x \in \Omega$ and $x_o \in \mathbb{R}^n \setminus \overline{\Omega}$ then in place of (1.5.93) *one has*

$$\Big(-\int_{\partial_*\Omega}\nu_s(y)a^{\beta\alpha}_{rs}(\partial_r E_{\gamma\beta})(x-y)E_{\alpha\delta}(y-x_o)\,d\sigma(y)\Big)_{1\leq\gamma\leq M}$$

$$=\Big(\int_{\partial_*\Omega}E_{\gamma\alpha}(x-y)\nu_r(y)a^{\alpha\beta}_{rs}(\partial_s E_{\beta\delta})(y-x_o)\,d\sigma(y)\Big)_{1\leq\gamma\leq M}$$

$$+E_{\bullet\delta}(x-x_o),\tag{1.5.94}$$

where $E_{\bullet\delta}=(E_{\gamma\delta})_{1\leq\gamma\leq M}$.

At least under stronger geometric assumptions than presently assumed, say $n\geq 3$ and $\Omega\subseteq\mathbb{R}^n$ an Ahlfors regular domain, and when $x,x_o\in\Omega$, formula (1.5.93) is a consequence of (1.5.6) in Theorem 1.5.1, written for the open set $\Omega_-:=\mathbb{R}^n\setminus\overline{\Omega}$ in place of Ω, and for the vector-valued function $u:=E_{\bullet\delta}(\cdot-x_o)\in\big[\mathscr{C}^\infty(\overline{\Omega_-})\big]^M$. Specifically, [112, Lemma 5.10.9] ensures that Ω_- is an Ahlfors regular domain, whose topological boundary coincides with that of Ω, whose geometric measure theoretic boundary agrees with that of Ω, and whose geometric measure theoretic outward unit normal is $-\nu$ at σ-a.e. point on $\partial\Omega$. In addition, according to Theorem 1.4.2 and [112, (8.3.47)], for each aperture parameter $\kappa>0$ there exists $C=C(\Omega,\kappa,L)\in(0,\infty)$ such that

$$\big(\mathcal{N}^{\Omega_-}_\kappa u\big)(z)\leq\frac{C}{[1+|z|]^{n-2}}\quad\text{for each } z\in\partial\Omega_-.\tag{1.5.95}$$

Since also $Lu=0$ in Ω_- and u decays at infinity if Ω_- happens to be an exterior domain (cf. Theorem 1.4.2), we conclude that all demands in the statement of Theorem 1.5.1 are presently satisfied. Then (1.5.6) applies and gives (1.5.93). Also, in the same setting, formula (1.5.94) is implied by (1.5.4) written for the vector-valued function $u:=E_{\bullet\delta}(\cdot-x_o)\in\big[\mathscr{C}^\infty(\overline{\Omega})\big]^M$ (which satisfies $Lu=0$ in Ω, plus a nontangential maximal inequality similar to (1.5.95)).

The proof below works in for the more general settings specified in (1.5.90)-(1.5.91).

Proof of Proposition 1.5.5 We distinguish a number of cases. First, assume

$$n\geq 2\text{ and }\Omega\subseteq\mathbb{R}^n\text{ is a bounded open set of locally finite perimeter.}\tag{1.5.96}$$

In the aforementioned setting, fix two distinct points $x,x_o\in\Omega$ and two arbitrary indices $\gamma,\delta\in\{1,\dots,M\}$. Bring in the vector field $\vec{F}=(F_s)_{1\leq s\leq n}$ with components given at each point $y\in\mathbb{R}^n\setminus\{x,x_o\}$ by

$$F_s(y):=a^{\beta\alpha}_{rs}(\partial_r E_{\gamma\beta})(x-y)E_{\alpha\delta}(y-x_o)\mathbf{1}_\Omega(y)$$

$$+E_{\gamma\alpha}(x-y)a^{\alpha\beta}_{sr}(\partial_r E_{\beta\delta})(y-x_o)\mathbf{1}_\Omega(y).\tag{1.5.97}$$

When regarded as a $\mathbb{C}^n$-valued function defined $\mathcal{L}^n$-a.e. in $\mathbb{R}^n$, it follows that

$$\vec{F}\in\big[L^1_{\text{comp}}(\mathbb{R}^n,\mathcal{L}^n)\big]^n\subseteq\big[\mathscr{E}'(\mathbb{R}^n)\big]^n,\tag{1.5.98}$$

since Ω is presently assumed to be bounded, so $\vec{F}$ is compactly supported.

To proceed, consider the compactly supported distribution

$$u := E_{\gamma\delta}(x - x_o)\delta_{x_o} - E_{\gamma\delta}(x - x_o)\delta_x \in \mathscr{E}'(\mathbb{R}^n) \tag{1.5.99}$$

and the complex Borel measure in $\mathbb{R}^n$ given by

$$\begin{aligned} \mu := &-\nu_s a^{\beta\alpha}_{rs}(\partial_r E_{\gamma\beta})(x - \cdot)E_{\alpha\delta}(\cdot - x_o)\sigma_* \\ &- E_{\gamma\alpha}(x - \cdot)\nu_r a^{\alpha\beta}_{sr}(\partial_r E_{\beta\delta})(\cdot - x_o)\,\sigma_* \end{aligned} \tag{1.5.100}$$

where $\sigma_* := \mathcal{H}^{n-1}\lfloor\partial_*\Omega$. In relation to these, we claim that

$$\text{div}\vec{F} = \mu + u \text{ in } \mathcal{D}'(\mathbb{R}^n). \tag{1.5.101}$$

To justify this claim, pick an arbitrary open subset W of Ω which contains x but does not contain x_o. Then, as seen from (1.5.97), the restriction $\vec{F}\big|_W$ (regarded as a locally integrable function in W) is given by

$$\begin{aligned} \vec{F}\big|_W = \Big(&- a^{\beta\alpha}_{rs}\partial_r\big[E_{\gamma\beta}(x - \cdot)\big]E_{\alpha\delta}(\cdot - x_o) \\ &+ E_{\gamma\alpha}(x - \cdot)a^{\alpha\beta}_{sr}\partial_r\big[E_{\beta\delta}(\cdot - x_o)\big]\Big)_{1\leq s\leq n}. \end{aligned} \tag{1.5.102}$$

In the computation below, we treat $E(x - \cdot)$ as a distribution in W and take advantage of the fact that $E(\cdot - x_o)$ is $\mathscr{C}^\infty$-smooth in W. Bearing these in mind, we may then compute

$$\begin{aligned} \text{div}\big(\vec{F}\big|_W\big) &= -a^{\beta\alpha}_{rs}\partial_r\partial_s\big[E_{\gamma\beta}(x - \cdot)\big]E_{\alpha\delta}(\cdot - x_o) \\ &\quad - a^{\beta\alpha}_{rs}\partial_r\big[E_{\gamma\beta}(x - \cdot)\big]\partial_s\big[E_{\alpha\delta}(\cdot - x_o)\big] \\ &\quad + \partial_s\big[E_{\gamma\alpha}(x - \cdot)\big]a^{\alpha\beta}_{sr}\partial_r\big[E_{\beta\delta}(\cdot - x_o)\big] \\ &\quad + E_{\gamma\alpha}(x - \cdot)a^{\alpha\beta}_{sr}\partial_r\partial_s\big[E_{\beta\delta}(\cdot - x_o)\big] \\ &=: \text{I} + \text{II} + \text{III} + \text{IV}. \end{aligned} \tag{1.5.103}$$

Thanks to (1.4.21) and (1.4.33) we have

$$\text{I} = -E_{\gamma\delta}(x - x_o)\delta_x \text{ and IV} = 0, \tag{1.5.104}$$

since x_o does not belong to W. Also, after interchanging α with β and r with s in II, we see that

$$\text{II} + \text{III} = 0. \tag{1.5.105}$$

Gathering terms then leads to the conclusion that

$$\operatorname{div}(\vec{F}\big|_{W}) = -E_{\gamma\delta}(x - x_o)\delta_x = (\mu + u)\big|_{W} \text{ in } \mathcal{D}'(W) \tag{1.5.106}$$

where we have also taken into account that $\mu\big|_W = 0$ (see (1.5.100)). In a very similar fashion, for any open subset W_o of Ω which contains x_o but does to contain x we obtain

$$\operatorname{div}(\vec{F}\big|_{W_o}) = E_{\gamma\delta}(x - x_o)\delta_{x_o} = (\mu + u)\big|_{W_o} \text{ in } \mathcal{D}'(W_o). \tag{1.5.107}$$

Finally, consider an open subset W_∞ of $\mathbb{R}^n$ which does not contain x_o and x, and such that $\partial\Omega \subseteq W_\infty$. Then, keeping in mind that distributional differentiation commutes with restriction to open sets and that σ_* is supported in W_∞, for each $s \in \{1, \dots, n\}$ we write (regarding $\mathbf{1}_\Omega$ as a distribution in $\mathbb{R}^n$)

$$\begin{aligned} \partial_s\big[\mathbf{1}_\Omega\big|_{W_\infty}\big] &= \big[\partial_s \mathbf{1}_\Omega\big]\big|_{W_\infty} = \big[-\nu_s\sigma_*\big]\big|_{W_\infty} \\ &= -\nu_s\sigma_* \text{ regarded as a distribution in } W_\infty \end{aligned} \tag{1.5.108}$$

with the third equality provided by [112, (5.6.10)]. Based on (1.5.108), (1.5.97), and the fact that both functions $E(x - \cdot)$ and $E(\cdot - x_o)$ are $\mathscr{C}^\infty$-smooth in W_∞, we compute

$$\begin{aligned} \operatorname{div}(\vec{F}\big|_{W_\infty}) &= -a^{\beta\alpha}_{rs}(\partial_r\partial_s E_{\gamma\beta})(x - \cdot)E_{\alpha\delta}(\cdot - x_o)(\mathbf{1}_\Omega\big|_{W_\infty}) \\ &\quad + a^{\beta\alpha}_{rs}(\partial_r E_{\gamma\beta})(x - \cdot)(\partial_s E_{\alpha\delta})(\cdot - x_o)(\mathbf{1}_\Omega\big|_{W_\infty}) \\ &\quad - a^{\beta\alpha}_{rs}(\partial_r E_{\gamma\beta})(x - \cdot)E_{\alpha\delta}(\cdot - x_o)\nu_s\sigma_* \\ &\quad - (\partial_s E_{\gamma\alpha})(x - \cdot)a^{\alpha\beta}_{sr}(\partial_r E_{\beta\delta})(\cdot - x_o)(\mathbf{1}_\Omega\big|_{W_\infty}) \\ &\quad + E_{\gamma\alpha}(x - \cdot)a^{\alpha\beta}_{sr}(\partial_r\partial_s E_{\beta\delta})(\cdot - x_o)(\mathbf{1}_\Omega\big|_{W_\infty}) \\ &\quad - E_{\gamma\alpha}(x - \cdot)a^{\alpha\beta}_{sr}(\partial_r E_{\beta\delta})(\cdot - x_o)\nu_s\sigma_* \\ &=: \mathrm{I}_\infty + \mathrm{II}_\infty + \mathrm{III}_\infty + \mathrm{IV}_\infty + \mathrm{V}_\infty + \mathrm{VI}_\infty \end{aligned} \tag{1.5.109}$$

in the sense of distributions in W_∞. Appealing to (1.4.21) and (1.4.33) permits us to conclude that

$$\mathrm{I}_\infty = 0 \text{ and } \mathrm{V}_\infty = 0. \tag{1.5.110}$$

while interchanging α with β and r with s in II_∞, allows us to combine

$$\mathrm{II}_\infty + \mathrm{IV}_\infty = 0. \tag{1.5.111}$$

Lastly, as seen from (1.5.100),

$$\mathrm{III}_\infty + \mathrm{VI}_\infty = \mu\big|_{W_\infty}. \tag{1.5.112}$$

In view of the fact that $u\big|_{W_\infty} = 0$ (since x_o and x are not contained in W_∞), from (1.5.109)-(1.5.112) we see that

$$\mathrm{div}\big(\vec{F}\big|_{W_\infty}\big) = (\mu + u)\big|_{W_\infty} \text{ in } \mathcal{D}'(W_\infty). \tag{1.5.113}$$

Finally, (1.5.106), (1.5.107), (1.5.113), and [109, Proposition 2.52, p. 38] prove the claim made in (1.5.101).

At this stage, observe that on the one hand we have

$$_{\mathscr{E}'(\mathbb{R}^n)}\big\langle \mathrm{div}\vec{F}, 1\big\rangle_{\mathscr{E}(\mathbb{R}^n)} = -_{[\mathscr{E}'(\mathbb{R}^n)]^n}\big\langle \vec{F}, \nabla 1\big\rangle_{[\mathscr{E}(\mathbb{R}^n)]^n} = 0 \tag{1.5.114}$$

while, on the other hand, (1.5.101) gives

$$_{\mathscr{E}'(\mathbb{R}^n)}\big\langle \mathrm{div}\vec{F}, 1\big\rangle_{\mathscr{E}(\mathbb{R}^n)} = {}_{\mathscr{E}'(\mathbb{R}^n)}\big\langle \mu, 1\big\rangle_{\mathscr{E}(\mathbb{R}^n)} + {}_{\mathscr{E}'(\mathbb{R}^n)}\big\langle u, 1\big\rangle_{\mathscr{E}(\mathbb{R}^n)}. \tag{1.5.115}$$

Consequently,

$$_{\mathscr{E}'(\mathbb{R}^n)}\big\langle \mu, 1\big\rangle_{\mathscr{E}(\mathbb{R}^n)} + {}_{\mathscr{E}'(\mathbb{R}^n)}\big\langle u, 1\big\rangle_{\mathscr{E}(\mathbb{R}^n)} = 0. \tag{1.5.116}$$

In this vein, let also observe that (1.5.99) implies

$$_{\mathscr{E}'(\mathbb{R}^n)}\big\langle u, 1\big\rangle_{\mathscr{E}(\mathbb{R}^n)} = E_{\gamma\delta}(x - x_o) - E_{\gamma\delta}(x - x_o) = 0, \tag{1.5.117}$$

while (1.5.100) gives

$$\begin{aligned} _{\mathscr{E}'(\mathbb{R}^n)}\big\langle \mu, 1\big\rangle_{\mathscr{E}(\mathbb{R}^n)} = &- \int_{\partial_*\Omega} \nu_s a^{\beta\alpha}_{rs}(\partial_r E_{\gamma\beta})(x - \cdot)E_{\alpha\delta}(\cdot - x_o)\,\mathrm{d}\sigma \\ &- \int_{\partial_*\Omega} E_{\gamma\alpha}(x - \cdot)\nu_r a^{\alpha\beta}_{sr}(\partial_r E_{\beta\delta})(\cdot - x_o)\,\mathrm{d}\sigma. \end{aligned} \tag{1.5.118}$$

In view of the arbitrariness of the index $\gamma \in \{1, \dots, M\}$, formulas (1.5.117)-(1.5.118) readily imply (1.5.93) (in the case described in (1.5.96)), under the additional assumption that the points x, x_o are distinct. Via continuity, this further implies the version of (1.5.93) when the points $x, x_o \in \Omega$ are identical. The scenario in which $x, x_o \in \mathbb{R}^n \setminus \overline{\Omega}$ is handled very similarly, the task being made a little simpler by the fact that in place of (1.5.99) we now have $u = 0$. Finally, in the case when $x \in \Omega$ and $x_o \in \mathbb{R}^n \setminus \overline{\Omega}$ the same argument leads to (1.5.94), since in place of (1.5.99) we now have $u = -E_{\gamma\delta}(x - x_o)\delta_x \in \mathscr{E}'(\mathbb{R}^n)$.

The next case we consider is the situation when

$$n \geq 2 \text{ and } \Omega \subseteq \mathbb{R}^n \text{ is a unbounded open set of locally finite perimeter with a compact topological boundary.} \tag{1.5.119}$$

In particular, it follows that Ω is an exterior domain. Once again, fix two arbitrary indices $\gamma, \delta \in \{1, \dots, M\}$, along two with points $x, x_o \in \Omega$ (the situations when $x, x_o \in \mathbb{R}^n \setminus \overline{\Omega}$, or when $x \in \Omega$ and $x_o \in \mathbb{R}^n \setminus \overline{\Omega}$, are dealt with analogously).

As before, in view of the conclusions we seek, there is no loss of generality in assuming that said points are distinct. Next, choose a large (open) ball B_o in $\mathbb{R}^n$ with $\partial\Omega \cup \{x, x_o\} \subseteq B_o$, and define the vector field $\vec{F} = (F_s)_{1\leq s\leq n}$ with components given at each point $y \in \mathbb{R}^n \setminus \{x, x_o\}$ by

$$F_s(y) := a^{\beta\alpha}_{rs}(\partial_r E_{\gamma\beta})(x-y)E_{\alpha\delta}(y-x_o)\mathbf{1}_{\Omega\cap B_o}(y) \\ + E_{\gamma\alpha}(x-y)a^{\alpha\beta}_{sr}(\partial_r E_{\beta\delta})(y-x_o)\mathbf{1}_{\Omega\cap B_o}(y). \tag{1.5.120}$$

Such a vector field continues to satisfy (1.5.98). In addition, writing

$$\mathbf{1}_{\Omega\cap B_o} = \mathbf{1}_\Omega - \mathbf{1}_{\mathbb{R}^n\setminus B_o} = \mathbf{1}_\Omega + \mathbf{1}_{B_o} - 1 \tag{1.5.121}$$

permits us to re-run the computation that has led to (1.5.101) (twice: once for the current Ω, and then for $\mathbb{R}^n \setminus B_o$) and conclude that

$$\mathrm{div}\vec{F} = \mu + u + \mu_o \ \text{ in } \mathcal{D}'(\mathbb{R}^n). \tag{1.5.122}$$

Above, u, μ are as in (1.5.99) and (1.5.100), respectively, while μ_o is the complex Borel measure in $\mathbb{R}^n$ given by

$$\mu_o := -N_s a^{\beta\alpha}_{rs}(\partial_r E_{\gamma\beta})(x-\cdot)E_{\alpha\delta}(\cdot - x_o)\sigma_o \\ - E_{\gamma\alpha}(x-\cdot)N_r a^{\alpha\beta}_{sr}(\partial_r E_{\beta\delta})(\cdot - x_o)\,\sigma_o \tag{1.5.123}$$

where $\sigma_o := \mathcal{H}^{n-1}\lfloor\partial B_o$ and $N = (N_s)_{1\leq s\leq n}$ is inward unit normal to B_o. As in (1.5.114), we once again have

$$ {}_{\mathcal{E}'(\mathbb{R}^n)}\big\langle \mathrm{div}\vec{F}, 1\big\rangle_{\mathcal{E}(\mathbb{R}^n)} = 0. \tag{1.5.124}$$

In view of (1.5.122)-(1.5.123), the contribution from μ_o in the context of the left-hand side above is

$$\int_{\partial B_o} N_s(y)a^{\beta\alpha}_{rs}(\partial_r E_{\gamma\beta})(x-y)E_{\alpha\delta}(y-x_o)\,d\sigma_o(y) \tag{1.5.125} \\ + \int_{\partial B_o} E_{\gamma\alpha}(x-y)N_s(y)a^{\alpha\beta}_{sr}(\partial_r E_{\beta\delta})(y-x_o)\,d\sigma_o(y) = 0,$$

with the final equality coming from what we have proved in the first case (with Ω replaced by B_o). Ultimately, this goes to show that the left-hand side of (1.5.124) amounts to

$$ {}_{\mathcal{E}'(\mathbb{R}^n)}\big\langle u+\mu, 1\big\rangle_{\mathcal{E}(\mathbb{R}^n)} = -\int_{\partial_*\Omega} \nu_s a^{\beta\alpha}_{rs}(\partial_r E_{\gamma\beta})(x-\cdot)E_{\alpha\delta}(\cdot - x_o)\,d\sigma \\ - \int_{\partial_*\Omega} E_{\gamma\alpha}(x-\cdot)\nu_r a^{\alpha\beta}_{sr}(\partial_r E_{\beta\delta})(\cdot - x_o)\,d\sigma, \tag{1.5.126}$$

thanks to (1.5.117)-(1.5.118). Together, (1.5.126) and (1.5.114) then establish (1.5.93) in the case considered in (1.5.119). This concludes the treatment of the scenario described in (1.5.90).

Finally, consider the task of proving (1.5.93) in the setting considered in (1.5.91). To set the stage, fix two arbitrary indices $\gamma, \delta \in \{1, \dots, M\}$ and two points $x, x_o \in \Omega$ (the scenario in which $x, x_o \in \mathbb{R}^n \setminus \overline{\Omega}$ goes through virtually verbatim). As before, in view of the conclusion we seek, there is no loss of generality in assuming that said points are distinct. Define the vector field $\vec{F} = (F_s)_{1 \leq s \leq n}$ with components given at each point $y \in \Omega \setminus \{x, x_o\}$ by

$$\begin{aligned} F_s(y) &:= a^{\beta\alpha}_{rs}(\partial_r E_{\gamma\beta})(x - y)E_{\alpha\delta}(y - x_o) \\ &\quad + E_{\gamma\alpha}(x - y)a^{\alpha\beta}_{sr}(\partial_r E_{\beta\delta})(y - x_o). \end{aligned} \tag{1.5.127}$$

As a $\mathbb{C}^n$-valued function defined $\mathcal{L}^n$-a.e. in Ω, this satisfies

$$\vec{F} \in \Big[L^1_{\rm loc}(\mathbb{R}^n, \mathcal{L}^n) \cap \mathscr{C}^\infty\big(\overline{\Omega} \setminus \{x, x_o\}\big)\Big]^n. \tag{1.5.128}$$

Moreover, if K is a compact neighborhood of $\{x, x_o\}$ contained in Ω, then based on (1.5.127), (1.4.24), and [112, (8.3.47)] we conclude (also keeping in mind that we are presently assuming $n \geq 3$) that for each aperture parameter $\kappa > 0$ there exists some constant $C = C(\Omega, \kappa, L) \in (0, \infty)$ such that

$$\big(\mathcal{N}^{\Omega\setminus K}_\kappa \vec{F}\big)(z) \leq \frac{C}{[1 + |z|]^{2n-3}} \quad \text{for each } z \in \partial\Omega. \tag{1.5.129}$$

In particular, $\mathcal{N}^{\Omega\setminus K}_\kappa \vec{F} \in L^1(\partial\Omega, \sigma)$ by the finiteness condition in (1.5.91). Arguing as in the first case, we see that

$$\operatorname{div}\vec{F} = E_{\gamma\delta}(x - x_o)\delta_{x_o} - E_{\gamma\delta}(x - x_o)\delta_x \in \mathscr{E}'(\Omega). \tag{1.5.130}$$

In concert with [112, (4.6.21)], this implies

$${}_{(\mathscr{C}^\infty_b(\Omega))^*}\Big(\operatorname{div}\vec{F}, 1\Big)_{\mathscr{C}^\infty_b(\Omega)} = E_{\gamma\delta}(x - x_o) - E_{\gamma\delta}(x - x_o) = 0. \tag{1.5.131}$$

Lastly, in the case when Ω is an exterior domain, (1.5.127) and (1.4.24) give (again, keeping in mind that $n \geq 3$)

$$\vec{F}(x) = O(|x|^{3-2n}) \text{ as } |x| \to \infty. \tag{1.5.132}$$

Granted these properties, [112, Theorem 1.5.1] applies and the Divergence Formula [112, (1.5.11)] written for the present vector field reduces, in view of (1.5.131) and [112, (8.9.10)], to

$$\int_{\partial_*\Omega} \nu \cdot \big(\vec{F}\big|_{\partial\Omega}\big)\, d\sigma = 0. \tag{1.5.133}$$

In turn, (1.5.133) and (1.5.127) readily imply (1.5.93) in the scenario described in (1.5.91). □

We next include a useful result regarding behavior at infinity for null-solutions of weakly elliptic systems of any order. This has been mentioned, in passing, in relation to Theorem 1.5.1 (cf. the first comment made after its statement), and will play a role shortly, in the proof of Theorem 1.5.7 stated a little further.

Lemma 1.5.6 *Fix $M, m, n \in \mathbb{N}$ with $n \geq 2$. Suppose $\Omega \subseteq \mathbb{R}^n$ is an exterior domain, and assume L is a weakly elliptic homogeneous $M \times M$ system of order $2m$, with constant complex coefficients in $\mathbb{R}^n$. Also, suppose $f : (0, \infty) \to (0, \infty)$ is a function with the property that for each fixed $\lambda \in (1, \infty)$ one has $f(\lambda R) \approx f(R)$, uniformly in $R \in (0, \infty)$ assumed to be sufficiently large (depending on λ). Finally, pick a function $u \in \big[\mathscr{C}^\infty(\Omega)\big]^M$ satisfying $Lu = 0$ in Ω. Then the following conditions are equivalent:*

(1) For some (or any) $\lambda \in (1, \infty)$ one has

$$\fint_{B(0,\lambda R)\setminus B(0,R)} |u|\,\mathrm{d}\mathcal{L}^n = o\big(f(R)\big) \ \text{ as } R \to \infty. \tag{1.5.134}$$

(1') For some (or any) $\lambda \in (1, \infty)$ one has

$$\fint_{B(0,\lambda R)\setminus B(0,R)} |\partial^\alpha u|\,\mathrm{d}\mathcal{L}^n = o\big(R^{-|\alpha|}\cdot f(R)\big) \ \text{ as } R \to \infty, \quad \text{for each multi-index } \alpha \in \mathbb{N}_0^n. \tag{1.5.135}$$

(2) One has

$$u(x) = o\big(f(|x|)\big) \ \text{ as } |x| \to \infty. \tag{1.5.136}$$

(2') One has

$$(\partial^\alpha u)(x) = o\big(|x|^{-|\alpha|}\cdot f(|x|)\big) \ \text{ as } |x| \to \infty, \quad \text{for each multi-index } \alpha \in \mathbb{N}_0^n. \tag{1.5.137}$$

(3) One has

$$\fint_{\partial B(0,R)} |u|\,\mathrm{d}\mathcal{H}^{n-1} = o\big(f(R)\big) \ \text{ as } R \to \infty. \tag{1.5.138}$$

(3') One has

$$\fint_{\partial B(0,R)} |\partial^\alpha u|\,\mathrm{d}\mathcal{H}^{n-1} = o\big(R^{-|\alpha|}\cdot f(R)\big) \ \text{ as } R \to \infty, \quad \text{for each multi-index } \alpha \in \mathbb{N}_0^n. \tag{1.5.139}$$

Furthermore, similar equivalences hold with little "o" replaced by big "O" throughout. Finally, the same results are true if u is a null-solution in Ω of an *injectively elliptic, homogeneous, constant (complex) coefficient, first-order system D in $\mathbb{R}^n$.*

Proof Select some $R_o \in (0, \infty)$ large enough so that $\mathbb{R}^n \setminus \overline{B(0, R_o)} \subseteq \Omega$. Fix some $\lambda \in (1, \infty)$, pick $x \in \Omega$ satisfying $|x| > \lambda R_o$, then define

$$R := 2^{-1}(1 + \lambda^{-1})|x| \text{ and } r := 4^{-1}(1 - \lambda^{-1})|x|. \tag{1.5.140}$$

These choices ensure that $B(x, r) \subseteq B(0, \lambda R) \setminus B(0, R) \subseteq \Omega$. Based on this and interior estimates for higher-order weakly elliptic systems (cf. [112, Theorem 6.5.7]), for each multi-index $\alpha \in \mathbb{N}_0^n$ we may write

$$|(\partial^\alpha u)(x)| \leq \frac{C}{r^{n+|\alpha|}} \int_{B(x,r)} |u| \, d\mathcal{L}^n \leq \frac{C}{r^{|\alpha|}} ⨍_{B(0,\lambda R) \setminus B(0,R)} |u| \, d\mathcal{L}^n \tag{1.5.141}$$

for some constant $C \in (0, \infty)$ which depends only on n, L, λ, and α. Since $R \approx |x|$ and $r \approx |x|$, we conclude from (1.5.141) that *(1)*$\Rightarrow$*(2')*. Via integration, it is clear that *(2')*$\Rightarrow$*(3')* and *(2)*$\Rightarrow$*(1)*. Employing the spherical version of Fubini's Theorem (cf., e.g., [109, Theorem 14.63, p. 577]) permits us to conclude that *(3')*$\Rightarrow$*(1')* and *(3)*$\Rightarrow$*(1)*. Lastly, the implications *(1')*$\Rightarrow$*(1)*, *(2')*$\Rightarrow$*(2)*, and *(3')*$\Rightarrow$*(3')* are trivial.

That similar equivalences hold with little "o" replaced by big "O" throughout, is justified in a very similar fashion. Finally, that the same results are true for null-solutions of injectively elliptic, homogeneous, constant (complex) coefficient first-order systems, is seen from what we have proved so far and Lemma 1.3.5. □

We shall now use Corollary 1.5.4 as a stepping stone to prove integral representations in exterior domains for null-solutions of weakly elliptic systems which, in which in place of sub-linear growth at infinity, are now allowed to have growth of *arbitrary* order. The reader is reminded that for each number $N \in \mathbb{R}$ the symbol $[N]$ denotes the integer part of N (i.e., the largest integer $\leq N$), while

$$\lfloor N \rfloor := \begin{cases} N - 1 \text{ if } N \in \mathbb{Z}, \\ [N] \quad \text{if } N \in \mathbb{R} \setminus \mathbb{Z}, \end{cases} \tag{1.5.142}$$

denotes the largest integer strictly less than N.

Theorem 1.5.7 *Let $\Omega \subseteq \mathbb{R}^n$, where $n \geq 2$, be an unbounded open set with a compact lower Ahlfors regular boundary, and with the property that $\sigma := \mathcal{H}^{n-1} \lfloor \partial\Omega$ is a doubling measure on $\partial\Omega$. In particular, Ω is a set of locally finite perimeter, and its geometric measure theoretic outward unit normal $\nu = (\nu_1, \dots, \nu_n)$ is defined σ-a.e. on $\partial_*\Omega$. With the summation convention over repeated indices understood throughout, let*

$$L = \left(a_{rs}^{\alpha\beta} \partial_r \partial_s\right)_{1 \leq \alpha, \beta \leq M} \tag{1.5.143}$$

be a homogeneous, weakly elliptic, second-order $M \times M$ system in $\mathbb{R}^n$, with complex constant coefficients, and denote by $E = (E_{\gamma\beta})_{1 \leq \gamma, \beta \leq M}$ the matrix-valued fundamental solution associated with L as in Theorem 1.4.2. Finally, fix an aper-

ture parameter $\kappa > 0$, *along with a truncation parameter* $\varepsilon > 0$, *a dilation factor* $\lambda \in (1, \infty)$, *and an exponent* $N \in [0, \infty)$.

Suppose $u = (u_\beta)_{1 \le \beta \le M}$ *is a vector-valued function satisfying*

$$u \in \big[\mathscr{C}^\infty(\Omega)\big]^M \text{ and } Lu = 0 \text{ in } \Omega,$$
$$u\Big|_{\partial\Omega}^{\kappa-\mathrm{n.t.}} \text{ and } \Big((a_{rs}^{\alpha\beta}\partial_s u_\beta)\Big|_{\partial\Omega}^{\kappa-\mathrm{n.t.}}\Big)_{\substack{1\le\alpha\le M\\ 1\le r\le n}} \text{ exist at } \sigma\text{-a.e. point on } \partial_{\mathrm{nta}}\Omega, \tag{1.5.144}$$

plus

$$\int_{\partial\Omega} \mathcal{N}_\kappa^\varepsilon u \, d\sigma < \infty \text{ and } \int_{\partial\Omega} \mathcal{N}_\kappa^\varepsilon (a_{rs}^{\alpha\beta}\partial_s u_\beta)\, d\sigma < \infty, \tag{1.5.145}$$

for each $\alpha \in \{1, \dots, M\}$ *and* $r \in \{1, \dots, n\}$, *as well as*

$$⨍_{B(0,\lambda R)\setminus B(0,R)} |u| \, d\mathcal{L}^n = o(R^N) \text{ as } R \to \infty. \tag{1.5.146}$$

Then for any other given aperture parameter $\kappa' > 0$ *the nontangential boundary traces* $u\Big|_{\partial\Omega}^{\kappa'-\mathrm{n.t.}}$ *and* $\Big((a_{rs}^{\alpha\beta}\partial_s u_\beta)\Big|_{\partial\Omega}^{\kappa'-\mathrm{n.t.}}\Big)_{\substack{1\le\alpha\le M\\ 1\le r\le n}}$ *also exist at* σ*-a.e. point on* $\partial_{\mathrm{nta}}\Omega$ *and are actually independent of the parameter* κ'. *Moreover, there exists some (unique)* $\mathbb{C}^M$*-valued function* P *in* $\mathbb{R}^n$ *satisfying*

$$LP = 0 \in \mathbb{C}^M \text{ in } \mathbb{R}^n, \tag{1.5.147}$$

whose scalar components are polynomials of degree[32] $\le \lfloor N \rfloor$ *(cf. (1.5.142)) and with the property that for each point* $x \in \Omega$ *one has (with absolutely convergent integrals, and with the dependence on the aperture parameter dropped)*

$$\begin{aligned} u(x) = P(x) &+ \Big(-\int_{\partial_*\Omega} \nu_s(y) a_{rs}^{\beta\alpha}(\partial_r E_{\gamma\beta})(x-y)\big(u_\alpha\big|_{\partial\Omega}^{\mathrm{n.t.}}\big)(y)\, d\sigma(y)\Big)_{1\le\gamma\le M} \\ &- \Big(\int_{\partial_*\Omega} E_{\gamma\alpha}(x-y)\nu_r(y)\Big((a_{rs}^{\alpha\beta}\partial_s u_\beta)\Big|_{\partial\Omega}^{\mathrm{n.t.}}\Big)(y)\, d\sigma(y)\Big)_{1\le\gamma\le M}. \end{aligned} \tag{1.5.148}$$

In fact,

$$P(x) = o(|x|^N) \text{ as } |x| \to \infty, \text{ if } n \ge 3, \tag{1.5.149}$$

while in the two-dimensional setting

$$P(x) = \left(\Phi(x) - \frac{\ln|x|}{4\pi^2}\int_{S^1} \big[L(\xi)\big]^{-1} d\mathcal{H}^1(\xi)\right)c + o(|x|^N) \text{ as } |x| \to \infty, \tag{1.5.150}$$

[32] we adopt the convention that a polynomial has a strictly negative degree if and only if said polynomial is identically zero

where $\Phi : \mathbb{R}^2 \setminus \{0\} \to \mathbb{C}^{M\times M}$ *is given in* (1.4.23) *and where the constant* $c \in \mathbb{C}^M$ *is given by*

$$c := \Big(\int_{\partial_*\Omega} \nu_r(y)\Big((a_{rs}^{\alpha\beta}\partial_s u_\beta)\big|_{\partial\Omega}^{\text{n.t.}}\Big)(y)\,d\sigma(y)\Big)_{1\leq\alpha\leq M}. \tag{1.5.151}$$

In addition, for each point $x \in \mathbb{R}^n \setminus \overline{\Omega}$ *one has*

$$\begin{aligned} P(x) = {} & \Big(\int_{\partial_*\Omega} \nu_s(y) a_{rs}^{\beta\alpha}(\partial_r E_{\gamma\beta})(x-y)\big(u_\alpha\big|_{\partial\Omega}^{\text{n.t.}}\big)(y)\,d\sigma(y)\Big)_{1\leq\gamma\leq M} \\ & + \Big(\int_{\partial_*\Omega} E_{\gamma\alpha}(x-y)\nu_r(y)\Big((a_{rs}^{\alpha\beta}\partial_s u_\beta)\big|_{\partial\Omega}^{\text{n.t.}}\Big)(y)\,d\sigma(y)\Big)_{1\leq\gamma\leq M}. \end{aligned} \tag{1.5.152}$$

Finally, if little "o" is replaced by big "O" in (1.5.146) *then all conclusions remain valid with the understanding that* (1.5.149)-(1.5.150) *now hold with little "o" replaced by big "O," and one now has* $\deg P \leq [N]$.

Proof The existence and independence of the boundary traces on the aperture parameter is a consequence of (1.5.144), (1.5.145), and [112, Corollary 8.9.9]. This takes care of the first claim in the statement.

There remains to establish the integral representation formula claimed in (1.5.148). Pick an arbitrary point $x \in \Omega$ and define $K := \overline{B\big(x, \tfrac{1}{2}\text{dist}(x, \partial\Omega)\big)}$. Also, select an arbitrary index $\gamma \in \{1, \dots, M\}$. For each fixed multi-index $\eta \in \mathbb{N}_0^n$, we next consider the vector field $\vec{F}_{x,\eta} = \big(F_s\big)_{1\leq s\leq n}$ with components (recall that, throughout, the summation convention over repeated indices is in effect)

$$F_s := -a_{rs}^{\beta\alpha}\partial^\eta\big[(\partial_r E_{\gamma\beta})(x-\cdot)\big]u_\alpha - \partial^\eta\big[E_{\gamma\alpha}(x-\cdot)\big]\big(a_{sr}^{\alpha\beta}\partial_r u_\beta\big) \in \mathcal{D}'(\Omega). \tag{1.5.153}$$

Above, the package of derivatives ∂^η is applied in the sense of distributions in Ω to $(\partial_r E_{\gamma\beta})(x-\cdot)$ and $E_{\gamma\alpha}(x-\cdot)$ which are locally integrable functions in Ω (hence themselves distributions in Ω). Since $u \in \big[\mathscr{C}^\infty(\Omega)\big]^M$, it then makes sense to further multiply the resulting distributions by smooth functions. The end-result is that $\vec{F}_{x,\eta}$ is a well-defined vector field with components distributions in Ω, i.e.,

$$\vec{F}_{x,\eta} \in \big[\mathcal{D}'(\Omega)\big]^n. \tag{1.5.154}$$

In addition, from (1.5.153) we see that

$$\vec{F}_{x,\eta}\big|_{\Omega\setminus K} \in \big[\mathscr{C}^\infty(\Omega\setminus K)\big]^n \subseteq \big[L^1_{\text{loc}}(\Omega\setminus K, \mathcal{L}^n)\big]^n. \tag{1.5.155}$$

Let us compute the divergence of the vector field $\vec{F}_{x,\eta}$ in the sense of distributions in Ω. Specifically, $\text{div}\vec{F}_{x,\eta} \in \mathcal{D}'(\Omega)$ is given by

$$\begin{aligned}
\operatorname{div}\vec{F}_{x,\eta} &= \partial_s F_s \\
&= a^{\beta\alpha}_{rs}\partial^\eta\partial_r\partial_s\big[E_{\gamma\beta}(x-\cdot)\big]u_\alpha + a^{\beta\alpha}_{rs}\partial^\eta\partial_r\big[E_{\gamma\beta}(x-\cdot)\big](\partial_s u_\alpha) \\
&\quad - \partial^\eta\partial_s\big[E_{\gamma\alpha}(x-\cdot)\big]\big(a^{\alpha\beta}_{sr}\partial_r u_\beta\big) \\
&\quad - \partial^\eta\big[E_{\gamma\alpha}(x-\cdot)\big]\big(a^{\alpha\beta}_{sr}\partial_r\partial_s u_\beta\big).
\end{aligned} \tag{1.5.156}$$

In regard to this, note that

$$\begin{aligned}
a^{\beta\alpha}_{rs}\partial^\eta\partial_r\partial_s\big[E_{\gamma\beta}(x-\cdot)\big]u_\alpha &= u_\alpha\partial^\eta\Big(a^{\beta\alpha}_{rs}\partial_r\partial_s\big[E_{\gamma\beta}(x-\cdot)\big]\Big) \\
&= u_\alpha\partial^\eta\big(\delta_{\alpha\gamma}\delta_x\big) = u_\gamma\partial^\eta\delta_x \in \mathscr{E}'(\Omega),
\end{aligned} \tag{1.5.157}$$

and for each $\alpha \in \{1, \dots, M\}$ we have

$$a^{\alpha\beta}_{sr}\partial_r\partial_s u_\beta = (Lu)_\alpha = 0 \text{ in } \Omega. \tag{1.5.158}$$

Also,

$$a^{\beta\alpha}_{rs}\partial^\eta\partial_r\big[E_{\gamma\beta}(x-\cdot)\big](\partial_s u_\alpha) - \partial^\eta\partial_s\big[E_{\gamma\alpha}(x-\cdot)\big]\big(a^{\alpha\beta}_{sr}\partial_r u_\beta\big) = 0, \tag{1.5.159}$$

as may be seen by interchanging α with β and r with s in the first term. Together, (1.5.156)-(1.5.159) show that

$$\operatorname{div}\vec{F}_{x,\eta} = u_\gamma\partial^\eta\delta_x \in \mathscr{E}'(\Omega). \tag{1.5.160}$$

It is also apparent from (1.5.153) and assumptions that

$$\vec{F}_{x,\eta}\Big|^{\text{n.t.}}_{\partial\Omega} \text{ exists at } \sigma\text{-a.e. point on } \partial_{\text{nta}}\Omega \tag{1.5.161}$$

and for each $s \in \{1, \dots, n\}$ and σ-a.e. point $y \in \partial_{\text{nta}}\Omega$ we actually have

$$\begin{aligned}
\Big(\vec{F}_{x,\eta}\Big|^{\text{n.t.}}_{\partial\Omega}\Big)_s(y) &= -(-1)^{|\eta|}a^{\beta\alpha}_{rs}(\partial^\eta\partial_r E_{\gamma\beta})(x-y)\big(u_\alpha\big|^{\text{n.t.}}_{\partial\Omega}\big)(y) \\
&\quad - (-1)^{|\eta|}(\partial^\eta E_{\gamma\alpha})(x-y)\big((a^{\alpha\beta}_{sr}\partial_r u_\beta)\big|^{\text{n.t.}}_{\partial\Omega}\big)(y).
\end{aligned} \tag{1.5.162}$$

Next, from (1.5.146) and Lemma 1.5.6 we conclude that

$$u(x) = o(|x|^N) \text{ and } (\nabla u)(x) = o(|x|^{N-1}) \text{ as } |x| \to \infty. \tag{1.5.163}$$

In view of these properties, the estimates in (1.4.24), and the definition in (1.5.153) we see that

$$\vec{F}_{x,\eta}(x) = o\big(|x|^{N-n+1-|\eta|}\big) \text{ as } |x| \to \infty. \tag{1.5.164}$$

In particular,

$$\vec{F}_{x,\eta}(x) = o(1) \text{ as } |x| \to \infty, \text{ if } |\eta| > 0 \text{ and } |\eta| \geq N - n + 1. \tag{1.5.165}$$

From this, (1.5.155), (1.5.153), and (1.5.145) we then conclude (bearing in mind that $\partial\Omega$ is compact) that

$$\mathcal{N}_\kappa^{\Omega\setminus K} \vec{F}_{x,\eta} \in L^1(\partial\Omega, \sigma) \text{ if } |\eta| > 0 \text{ and } |\eta| \geq N - n + 1. \tag{1.5.166}$$

Let us now fix an arbitrary multi-index $\eta \in \mathbb{N}_0^n$ with $|\eta| > 0$ and $|\eta| \geq N - n + 1$. Collectively, (1.5.154), (1.5.155), (1.5.160), (1.5.161), and (1.5.166) then permit us to invoke [112, Theorem 1.4.1], and the Divergence Formula recorded in [112, (1.4.6)] presently gives

$$\begin{aligned}
(\partial^\eta u_\gamma)(x) &= {}_{\mathscr{E}'(\Omega)}\big\langle \delta_x, \partial^\eta u_\gamma \big\rangle_{\mathscr{E}(\Omega)} = (-1)^{|\eta|}\, {}_{\mathscr{E}'(\Omega)}\big\langle \partial^\eta \delta_x, u_\gamma \big\rangle_{\mathscr{E}(\Omega)} \\
&= (-1)^{|\eta|}\, {}_{\mathscr{E}'(\Omega)}\big\langle u_\gamma \partial^\eta \delta_x, 1 \big\rangle_{\mathscr{E}(\Omega)} = (-1)^{|\eta|}\, {}_{\mathscr{E}'(\Omega)}\big\langle \operatorname{div}\vec{F}_{x,\eta}, 1 \big\rangle_{\mathscr{E}(\Omega)} \\
&= (-1)^{|\eta|}\, {}_{(\mathscr{C}_b^\infty(\Omega))^*}\big(\operatorname{div}\vec{F}_{x,\eta}, 1\big)_{\mathscr{C}_b^\infty(\Omega)} \\
&= (-1)^{|\eta|} \int_{\partial_*\Omega} \nu \cdot \Big(\vec{F}_{x,\eta}\Big|_{\partial\Omega}^{\text{n.t.}}\Big)\, d\sigma \\
&= -\int_{\partial_*\Omega} \nu_s(y) a_{rs}^{\beta\alpha} (\partial^\eta \partial_r E_{\gamma\beta})(x - y) \big(u_\alpha\big|_{\partial\Omega}^{\text{n.t.}}\big)(y)\, d\sigma(y) \\
&\quad - \int_{\partial_*\Omega} (\partial^\eta E_{\gamma\alpha})(x - y) \nu_s(y) \big((a_{sr}^{\alpha\beta} \partial_r u_\beta)\big|_{\partial\Omega}^{\text{n.t.}}\big)(y)\, d\sigma(y),
\end{aligned} \tag{1.5.167}$$

on account of (1.5.160), [112, (4.6.21)], and (1.5.162). In turn, from (1.5.167) and assumptions we see that if we define the function $w : \Omega \to \mathbb{C}^M$ by setting, at each point $x \in \Omega$,

$$\begin{aligned}
w(x) := u(x) &+ \Big(\int_{\partial_*\Omega} \nu_s(y) a_{rs}^{\beta\alpha} (\partial_r E_{\gamma\beta})(x - y) \big(u_\alpha\big|_{\partial\Omega}^{\text{n.t.}}\big)(y)\, d\sigma(y)\Big)_{1 \leq \gamma \leq M} \\
&+ \Big(\int_{\partial_*\Omega} E_{\gamma\alpha}(x - y) \nu_r(y) \Big((a_{rs}^{\alpha\beta} \partial_s u_\beta)\big|_{\partial\Omega}^{\text{n.t.}}\Big)(y)\, d\sigma(y)\Big)_{1 \leq \gamma \leq M},
\end{aligned} \tag{1.5.168}$$

then

$$\begin{aligned}
&w \text{ belongs to } \big[\mathscr{C}^\infty(\Omega)\big]^M \text{ and satisfies } Lw = 0 \text{ in } \Omega, \text{ as well as} \\
&\partial^\eta w = 0 \text{ in } \Omega \text{ for each } \eta \in \mathbb{N}_0^n \text{ with } |\eta| > 0 \text{ and } |\eta| \geq N - n + 1.
\end{aligned} \tag{1.5.169}$$

Expanding w locally into a Taylor series (with remainder), the last line in (1.5.169) implies that the scalar components of w are locally polynomials in Ω. This and a standard connectivity argument then ultimately show that the scalar components of w are polynomials in each of the connected component of Ω. Denote by Ω_∞ the unbounded connected component of Ω and define $\widetilde{P}$ to be the (unique) extension of $w\big|_{\Omega_\infty}$ to a $\mathbb{C}^M$-valued function in $\mathbb{R}^n$ whose scalar components $(\widetilde{P}_\alpha)_{1\leq\alpha\leq M}$ are polynomials with complex coefficients (of degree $\leq |\eta| - 1$ in $\mathbb{R}^n$). (1.5.170)

Then $L\widetilde{P}$ is a $\mathbb{C}^M$-valued function in $\mathbb{R}^n$ whose scalar components are polynomials with complex coefficients in $\mathbb{R}^n$, and the first line in (1.5.169) gives that $L\widetilde{P}$ vanishes in the (open, nonempty) set Ω_∞. This implies that

$$L\widetilde{P} = 0 \in \mathbb{C}^M \text{ in } \mathbb{R}^n. \tag{1.5.171}$$

If we now define

$$\widetilde{u} := u - \widetilde{P}\big|_{\Omega} \text{ in } \Omega, \tag{1.5.172}$$

if follows that

$$\widetilde{u}\big|_{\Omega_\infty} = u\big|_{\Omega_\infty} - \widetilde{P}\big|_{\Omega_\infty} = (u - w)\big|_{\Omega_\infty} \text{ in } \Omega_\infty \tag{1.5.173}$$

which, together with (1.5.168) and (1.4.24), proves that

$$\widetilde{u}(x) = \begin{cases} O(|x|^{2-n}) & \text{if } n \geq 3, \\ O(\ln|x|) & \text{if } n = 2, \end{cases} \quad \text{as } |x| \to \infty. \tag{1.5.174}$$

In light of (1.5.171), (1.5.172), and the original assumptions on u, we then conclude that $\widetilde{u}$ satisfies all hypotheses in Corollary 1.5.4. The latter guarantees the existence of a constant $c \in \mathbb{C}^M$ with the property that at each point $x \in \Omega$ we have

$$\begin{aligned} \widetilde{u}(x) = c &+ \Big(-\int_{\partial_*\Omega} \nu_s(y) a^{\beta\alpha}_{rs} (\partial_r E_{\gamma\beta})(x-y) \big(\widetilde{u}_\alpha\big|^{\text{n.t.}}_{\partial\Omega}\big)(y)\, d\sigma(y)\Big)_{1\leq\gamma\leq M} \\ &- \Big(\int_{\partial_*\Omega} E_{\gamma\alpha}(x-y)\nu_r(y)\Big((a^{\alpha\beta}_{rs}\partial_s\widetilde{u}_\beta)\big|^{\text{n.t.}}_{\partial\Omega}\Big)(y)\, d\sigma(y)\Big)_{1\leq\gamma\leq M}. \end{aligned} \tag{1.5.175}$$

Going further, from [112, (5.2.3), (5.6.16)] observe that

$\Omega_{ext} := \mathbb{R}^n \setminus \Omega$ is a compact set of locally finite perimeter with the property that $\partial_*(\Omega_{ext}) = \partial_*\Omega$, and the geometric measure theoretic outward unit normal to Ω_{ext} is $-\nu$ at σ-a.e. point on $\partial_*\Omega$. (1.5.176)

Granted these properties, for each index $\gamma \in \{1, \dots, M\}$ and each fixed point $x \in \Omega$ we may now apply the De Giorgi-Federer version of the Divergence Formula recalled in [112, Theorem 1.1.1] to the set Ω_{ext} and the vector field $\vec{G}_x = (G_s)_{1\leq s\leq n}$ with components given by

$$G_s := -a^{\beta\alpha}_{rs}(\partial_r E_{\gamma\beta})(x-\cdot)\widetilde{P}_\alpha - E_{\gamma\alpha}(x-\cdot)a^{\alpha\beta}_{sr}\partial_r\widetilde{P}_\beta \in \mathscr{C}^\infty(\Omega_{ext}). \tag{1.5.177}$$

In view of (1.5.171) and (1.4.33) this is a divergence-free vector field in a neighborhood of Ω_{ext} and, bearing in mind (1.5.176), we deduce from [112, (1.1.8)] that for each $x \in \Omega$ we have

$$\Big(-\int_{\partial_*\Omega}\nu_s(y)a^{\beta\alpha}_{rs}(\partial_r E_{\gamma\beta})(x-y)\big(\widetilde{P}_\alpha\big|_{\partial_*\Omega}\big)(y)\,d\sigma(y)\Big)_{1\le\gamma\le M}$$
$$-\Big(\int_{\partial_*\Omega}E_{\gamma\alpha}(x-y)\nu_r(y)a^{\alpha\beta}_{rs}(\partial_s\widetilde{P}_\beta)\big|_{\partial_*\Omega}(y)\,d\sigma(y)\Big)_{1\le\gamma\le M}=0. \tag{1.5.178}$$

In concert, (1.5.172), (1.5.175), (1.5.178), [112, (8.9.10)], and [112, Proposition 8.8.6] then imply

$$u(x)=c+\widetilde{P}(x)+\Big(-\int_{\partial_*\Omega}\nu_s(y)a^{\beta\alpha}_{rs}(\partial_r E_{\gamma\beta})(x-y)\big(u_\alpha\big|^{\text{n.t.}}_{\partial\Omega}\big)(y)\,d\sigma(y)\Big)_{1\le\gamma\le M}$$
$$-\Big(\int_{\partial_*\Omega}E_{\gamma\alpha}(x-y)\nu_r(y)\Big((a^{\alpha\beta}_{rs}\partial_s u_\beta)\big|^{\text{n.t.}}_{\partial\Omega}\Big)(y)\,d\sigma(y)\Big)_{1\le\gamma\le M} \tag{1.5.179}$$

at each point $x\in\Omega$. If we now define

$$P:=c+\widetilde{P}, \tag{1.5.180}$$

then P is a $\mathbb{C}^M$-valued function in $\mathbb{R}^n$ whose scalar components are polynomials with complex coefficients in $\mathbb{R}^n$. Also, (1.5.171) implies (1.5.147), while (1.5.179) implies (1.5.148).

Pressing on, observe that in all dimensions $n\ge 2$ the last term in (1.5.148) may be recast as

$$-E(x)\Big(\int_{\partial_*\Omega}\nu_r(y)\Big((a^{\alpha\beta}_{rs}\partial_s u_\beta)\big|^{\text{n.t.}}_{\partial\Omega}\Big)(y)\,d\sigma(y)\Big)_{1\le\gamma\le M}$$
$$-\Big(\int_{\partial_*\Omega}\{E_{\gamma\alpha}(x-y)-E_{\gamma\alpha}(x)\}\nu_r(y)\Big((a^{\alpha\beta}_{rs}\partial_s u_\beta)\big|^{\text{n.t.}}_{\partial\Omega}\Big)(y)\,d\sigma(y)\Big)_{1\le\gamma\le M}$$
$$=-E(x)c+O(|x|^{1-n})\ \text{ as }\ |x|\to\infty, \tag{1.5.181}$$

where c is as in (1.5.151), thanks to (1.4.22) and the Mean Value Theorem. In turn, from (1.5.148), (1.5.181), (1.5.146), (1.4.24), and (1.4.22) we see (keeping in mind that $N\ge 0$) that the asymptotic expansions (1.5.149)-(1.5.150) are valid (with c as in (1.5.151)). Since P is a polynomial, from (1.5.149)-(1.5.150) (while also being mindful that the function Φ is positive homogeneous of degree 0 in $\mathbb{R}^2\setminus\{0\}$) we then conclude that the degree of P is $\le\lfloor N\rfloor$.

Consider next the task of proving the validity of formula (1.5.152) at points in $\mathbb{R}^n\setminus\overline{\Omega}$. In the case when $n\ge 3$, define the function $\omega=(\omega_\alpha)_{1\le\alpha\le M}\in\big[\mathscr{C}^\infty(\Omega)\big]^M$ by setting

$$\begin{aligned}\omega(x) &:= u(x) - P(x)\\ &= \Big(-\int_{\partial_*\Omega} \nu_s(y)a_{rs}^{\beta\alpha}(\partial_r E_{\gamma\beta})(x-y)\big(u_\alpha\big|_{\partial\Omega}^{\text{n.t.}}\big)(y)\,d\sigma(y)\Big)_{1\le\gamma\le M}\\ &\quad -\Big(\int_{\partial_*\Omega} E_{\gamma\alpha}(x-y)\nu_r(y)\Big((a_{rs}^{\alpha\beta}\partial_s u_\beta)\big|_{\partial\Omega}^{\text{n.t.}}\Big)(y)\,d\sigma(y)\Big)_{1\le\gamma\le M} \end{aligned}\tag{1.5.182}$$

for each $x \in \Omega$. Then (1.4.24) ensures that

$$\omega(x) = O(|x|^{2-n}) \text{ as } |x| \to \infty. \tag{1.5.183}$$

Collectively, (1.5.183), (1.5.147), plus the original assumptions on u ensure that the hypotheses of Theorem 1.5.1 are satisfied by the function ω, and (1.5.6) gives (keeping in mind that $L\omega = 0$) that for each $x \in \mathbb{R}^n \setminus \overline{\Omega}$ we have

$$\begin{aligned} 0 &= \Big(-\int_{\partial_*\Omega} \nu_s(y)a_{rs}^{\beta\alpha}(\partial_r E_{\gamma\beta})(x-y)\big(\omega_\alpha\big|_{\partial\Omega}^{\text{n.t.}}\big)(y)\,d\sigma(y)\Big)_{1\le\gamma\le M}\\ &\quad -\Big(\int_{\partial_*\Omega} E_{\gamma\alpha}(x-y)\nu_r(y)\Big((a_{rs}^{\alpha\beta}\partial_s \omega_\beta)\big|_{\partial\Omega}^{\text{n.t.}}\Big)(y)\,d\sigma(y)\Big)_{1\le\gamma\le M}\\ &= \Big(-\int_{\partial_*\Omega} \nu_s(y)a_{rs}^{\beta\alpha}(\partial_r E_{\gamma\beta})(x-y)\big(u_\alpha\big|_{\partial\Omega}^{\text{n.t.}}\big)(y)\,d\sigma(y)\Big)_{1\le\gamma\le M}\\ &\quad -\Big(\int_{\partial_*\Omega} E_{\gamma\alpha}(x-y)\nu_r(y)\Big((a_{rs}^{\alpha\beta}\partial_s u_\beta)\big|_{\partial\Omega}^{\text{n.t.}}\Big)(y)\,d\sigma(y)\Big)_{1\le\gamma\le M}\\ &\quad -\Big(-\int_{\partial_*\Omega} \nu_s(y)a_{rs}^{\beta\alpha}(\partial_r E_{\gamma\beta})(x-y)\big(P_\alpha\big|_{\partial_*\Omega}\big)(y)\,d\sigma(y)\Big)_{1\le\gamma\le M}\\ &\quad +\Big(\int_{\partial_*\Omega} E_{\gamma\alpha}(x-y)\nu_r(y)a_{rs}^{\alpha\beta}(\partial_s P_\beta)\big|_{\partial_*\Omega}(y)\,d\sigma(y)\Big)_{1\le\gamma\le M}. \end{aligned}\tag{1.5.184}$$

In relation to the last two lines above, we make the claim that

$$\begin{aligned} &\Big(-\int_{\partial_*\Omega} \nu_s(y)a_{rs}^{\beta\alpha}(\partial_r E_{\gamma\beta})(x-y)\big(P_\alpha\big|_{\partial_*\Omega}\big)(y)\,d\sigma(y)\Big)_{1\le\gamma\le M}\\ &\qquad -\Big(\int_{\partial_*\Omega} E_{\gamma\alpha}(x-y)\nu_r(y)a_{rs}^{\alpha\beta}(\partial_s P_\beta)\big|_{\partial_*\Omega}(y)\,d\sigma(y)\Big)_{1\le\gamma\le M}\\ &\qquad = -P(x) \end{aligned}\tag{1.5.185}$$

for each $x \in \mathbb{R}^n \setminus \overline{\Omega}$. To justify (1.5.185), fix an arbitrary point $x \in \mathbb{R}^n \setminus \overline{\Omega}$ along with an arbitrary index $\gamma \in \{1, \dots, M\}$. Also, bring back the compact set $\Omega_{ext} \subseteq \mathbb{R}^n$ from (1.5.176) and consider a large (open) ball $B \subseteq \mathbb{R}^n$ containing Ω_{ext}. Next, introduce the vector field $\vec{H}_x = (H_s)_{1\le s\le n}$ with scalar components

$$H_s := -a_{rs}^{\beta\alpha}(\partial_r E_{\gamma\beta})(x-\cdot)P_\alpha \mathbf{1}_{\Omega_{ext}} - E_{\gamma\alpha}(x-\cdot)a_{sr}^{\alpha\beta}(\partial_r P_\beta)\mathbf{1}_{\Omega_{ext}} \tag{1.5.186}$$

and note that

$$\vec{H}_x \in \left[L^1_{\rm comp}(\mathbb{R}^n, \mathcal{L}^n)\right]^n \subseteq \left[\mathscr{E}'(\mathbb{R}^n)\right]^n. \tag{1.5.187}$$

Moreover, based on (1.5.171), (1.4.33), [112, (5.6.10)], and (1.5.176), the same type of argument as in the proof of (1.5.99) currently gives

$$\operatorname{div}\vec{H}_x = \eta + \mu \text{ in } \mathcal{D}'(\mathbb{R}^n) \tag{1.5.188}$$

where η is the compactly supported distribution given by

$$\eta := P_\gamma(x)\delta_x \in \mathscr{E}'(\mathbb{R}^n) \tag{1.5.189}$$

and, if $\sigma_* := \mathcal{H}^{n-1}\lfloor\partial_*\Omega$, then μ is the complex Borel measure in $\mathbb{R}^n$ given by

$$\mu := -\nu_s a^{\beta\alpha}_{rs}(\partial_r E_{\gamma\beta})(x-\cdot)P_\alpha\sigma_* - E_{\gamma\alpha}(x-\cdot)\nu_s a^{\alpha\beta}_{sr}(\partial_r P_\beta)\sigma_*. \tag{1.5.190}$$

Granted these properties, we may apply [112, Proposition 2.8.6] (with $\mathcal{O} := \mathbb{R}^n$ and $\Omega := B$), and the Divergence Formula recorded in [112, (2.8.35)] presently gives

$${}_{(\mathscr{C}^\infty_b(B))^*}\Big(\big(\operatorname{div}\vec{H}_x\big)\big|_B, 1\Big)_{\mathscr{C}^\infty_b(B)} = 0. \tag{1.5.191}$$

At this stage, the claim made in (1.5.185) becomes a consequence of (1.5.191) and (1.5.188)-(1.5.190). Lastly, (1.5.184) and (1.5.185) establish (1.5.152) in the case when $n \geq 3$.

There remains to justify (1.5.152) in the two-dimensional case. The novelty in this scenario is that, in contrast to (1.5.183) when $n \geq 3$, now ω may not have enough decay at infinity to justify using Theorem 1.5.1. To remedy this, we fix some point $x_o \in \mathbb{R}^n \setminus \overline{\Omega}$ and, with the constant vector $c \in \mathbb{C}^M$ as in (1.5.151), in place of (1.5.182) now define $\widetilde{\omega} = (\widetilde{\omega}_\alpha)_{1\leq\alpha\leq M} \in \left[\mathscr{C}^\infty(\Omega)\right]^M$ according to

$$\begin{aligned}
\widetilde{\omega}(x) &:= u(x) - P(x) + E(x-x_o)c \qquad (1.5.192)\\
&= \Big(-\int_{\partial_*\Omega}\nu_s(y)a^{\beta\alpha}_{rs}(\partial_r E_{\gamma\beta})(x-y)\big(u_\alpha\big|^{\rm n.t.}_{\partial\Omega}\big)(y)\,d\sigma(y)\Big)_{1\leq\gamma\leq M}\\
&\quad -\Big(\int_{\partial_*\Omega}\{E_{\gamma\alpha}(x-y)-E_{\gamma\alpha}(x-x_o)\}\nu_r(y)\Big(\big(a^{\alpha\beta}_{rs}\partial_s u_\beta\big)\big|^{\rm n.t.}_{\partial\Omega}\Big)(y)\,d\sigma(y)\Big)_{1\leq\gamma\leq M}
\end{aligned}$$

for each $x \in \Omega$. Thanks to (1.4.24) and the Mean Value Theorem we have (recall that we presently assume $n = 2$)

$$\widetilde{\omega}(x) = O(|x|^{-1}) \text{ as } |x| \to \infty, \tag{1.5.193}$$

so Theorem 1.5.1 is applicable to the function $\widetilde{\omega}$. Specifically, formula (1.5.6) written for the function $\widetilde{\omega}$ presently implies (in view of (1.5.192) and the fact that $L\widetilde{\omega} = 0$) that for each $x \in \mathbb{R}^n \setminus \overline{\Omega}$ we have

$$
\begin{aligned}
0 = &\Big(- \int_{\partial_* \Omega} \nu_s(y) a^{\beta\alpha}_{rs} (\partial_r E_{\gamma\beta})(x-y) \big(u_\alpha\big|^{\text{n.t.}}_{\partial\Omega}\big)(y)\, d\sigma(y)\Big)_{1\leq\gamma\leq M} \\
&- \Big(\int_{\partial_* \Omega} E_{\gamma\alpha}(x-y) \nu_r(y) \Big((a^{\alpha\beta}_{rs} \partial_s u_\beta)\big|^{\text{n.t.}}_{\partial\Omega}\Big)(y)\, d\sigma(y)\Big)_{1\leq\gamma\leq M} \\
&- \Big(- \int_{\partial_* \Omega} \nu_s(y) a^{\beta\alpha}_{rs} (\partial_r E_{\gamma\beta})(x-y) \big(P_\alpha\big|_{\partial_*\Omega}\big)(y)\, d\sigma(y)\Big)_{1\leq\gamma\leq M} \\
&+ \Big(\int_{\partial_* \Omega} E_{\gamma\alpha}(x-y) \nu_r(y) a^{\alpha\beta}_{rs} (\partial_s P_\beta)\big|_{\partial_*\Omega}(y)\, d\sigma(y)\Big)_{1\leq\gamma\leq M} \\
&+ \Big(- \int_{\partial_* \Omega} \nu_s(y) a^{\beta\alpha}_{rs} (\partial_r E_{\gamma\beta})(x-y) E_{\alpha\delta}(y-x_o) c_\delta \, d\sigma(y)\Big)_{1\leq\gamma\leq M} \\
&- \Big(\int_{\partial_* \Omega} E_{\gamma\alpha}(x-y) \nu_r(y) a^{\alpha\beta}_{rs} (\partial_s E_{\beta\delta})(y-x_o)\, d\sigma(y)\Big)_{1\leq\gamma\leq M}.
\end{aligned}
\tag{1.5.194}
$$

Much as in (1.5.185) (whose proof continues to be valid in the two-dimensional setting), the combination of lines three and four above amounts to $P(x)$. Also, Proposition 1.5.5 applies (since in the current setting fits into (1.5.90)) and (1.5.93) shows that lines five and six in (1.5.194) amount to zero. All things considered, (1.5.194) becomes (1.5.152), so the latter formula has now been justified in all dimensions $n \geq 2$. Finally, the very last claim in the statement is implicit in what we have proved already. □

We next take up the task of elucidating the behavior at infinity of a null-solution of a weakly elliptic second-order system in an exterior domain. Here is our first result in this regard.

Corollary 1.5.8 *Fix $n \in \mathbb{N}$ with $n \geq 2$, and suppose $\Omega \subseteq \mathbb{R}^n$ is an exterior domain. Let L be a weakly elliptic homogeneous second-order $M \times M$ system (for some $M \in \mathbb{N}$) with constant complex coefficients in $\mathbb{R}^n$ and denote by E the matrix-valued fundamental solution associated with L as in Theorem 1.4.2. Finally, consider a function $u \in \big[\mathscr{C}^\infty(\Omega)\big]^M$ satisfying $Lu = 0$ in Ω as well as*

$$
\fint_{B(0,\lambda R)\setminus B(0,R)} |u| \, d\mathcal{L}^n = o(R^N) \ \text{as} \ R \to \infty, \tag{1.5.195}
$$

for some $\lambda \in (1,\infty)$ and $N \in [0,\infty)$. Then there exist a $\mathbb{C}^M$-valued function P in $\mathbb{R}^n$ satisfying

$$
LP = 0 \in \mathbb{C}^M \ \text{in} \ \mathbb{R}^n, \tag{1.5.196}
$$

whose scalar components are polynomials of degree[33] $\leq \lfloor N \rfloor$ (cf. (1.5.142)), along with a constant vector $c \in \mathbb{C}^M$, such that the following asymptotic expansion holds:

[33] we adopt the convention that a polynomial has a strictly negative degree if and only if said polynomial is identically zero

$$(\partial^\alpha u)(x) = (\partial^\alpha P)(x) + (\partial^\alpha E)(x)c + O(|x|^{1-n-|\alpha|}) \quad as \ |x| \to \infty, \qquad \text{for each multi-index } \alpha \in \mathbb{N}_0^n. \tag{1.5.197}$$

As a consequence, given a multi-index $\alpha \in \mathbb{N}_0^n$, under the assumption that either $|\alpha| > 0$, or $n \geq 3$, or $n = 2$ and $\int_{S^1} \big[L(\xi)\big]^{-1} \, d\mathcal{H}^1(\xi) = 0$, one has

$$(\partial^\alpha u)(x) = \begin{cases} O(|x|^{2-n-|\alpha|}) \ \text{if } |\alpha| \geq \lfloor N \rfloor + 1 \text{ or } \lfloor N \rfloor + n \leq 2, \\ O(|x|^{\lfloor N \rfloor - |\alpha|}) \ \text{if } \lfloor N \rfloor + n \geq 2, \end{cases} \tag{1.5.198}$$

as $|x| \to \infty$.

Proof In light of the conclusions we seek, there is no loss of generality in assuming that Ω is the complement of the closure of a large ball centered at origin in $\mathbb{R}^n$, and that actually $u \in \big[\mathscr{C}^\infty(\overline{\Omega})\big]^M$. Assume this to be the case. Let $\nu = (\nu_1, \dots, \nu_n)$ be the outward unit normal to Ω and abbreviate $\sigma := \mathcal{H}^{n-1} \lfloor \partial\Omega$. Also, let P be associated with the function u as in Theorem 1.5.7, and take $c \in \mathbb{C}^M$ to be the opposite of the vector defined in (1.5.151). As a consequence of (1.5.148) and (1.5.181) we have

$$\begin{aligned} u(x) &= P(x) + E(x)c \\ &\quad + \Big(- \int_{\partial\Omega} \nu_s(y) a_{rs}^{\beta\alpha} (\partial_r E_{\gamma\beta})(x-y) u(y) \, d\sigma(y) \Big)_{1 \leq \gamma \leq M} \\ &\quad - \Big(\int_{\partial\Omega} \big\{ E_{\gamma\alpha}(x-y) - E_{\gamma\alpha}(x) \big\} \nu_r(y) a_{rs}^{\alpha\beta} (\partial_s u_\beta)(y) \, d\sigma(y) \Big)_{1 \leq \gamma \leq M} \end{aligned} \tag{1.5.199}$$

at each point $x \in \Omega$. Then the asymptotic expansion claimed in (1.5.197) is seen by differentiating both sides of formula (1.5.199), and taking into account (1.4.22) and the Mean Value Theorem.

To justify the claim made in (1.5.198), fix an arbitrary multi-index $\alpha \in \mathbb{N}_0^n$. Since $\deg P \leq \lfloor N \rfloor$, we see that $(\partial^\alpha P)(x) = O(|x|^{\lfloor N \rfloor - |\alpha|})$ as $|x| \to \infty$, and actually $\partial^\alpha P = 0$ if $|\alpha| \geq \lfloor N \rfloor + 1$. Furthermore, under the additional assumption that either $|\alpha| > 0$, or $n \geq 3$, we know from (1.4.24) that $(\partial^\alpha E)(x) = O(|x|^{2-n-|\alpha|})$ as $|x| \to \infty$. Keeping these asymptotic formulas in mind, (1.5.198) follows from (1.5.197). □

The version of Corollary 1.5.8 corresponding to functions with at most linear growth at infinity (i.e., with $N \in [0, 1]$ in (1.5.195)) is looked at in greater detail in the theorem below.

Theorem 1.5.9 *Fix $n \in \mathbb{N}$ with $n \geq 2$, and suppose $\Omega \subseteq \mathbb{R}^n$ is an exterior domain. Also, for some $M \in \mathbb{N}$, let L be a weakly elliptic homogeneous second-order $M \times M$ system with constant complex coefficients in $\mathbb{R}^n$. Given a function $u \in \big[\mathscr{C}^\infty(\Omega)\big]^M$ satisfying $Lu = 0$ in Ω, consider the following conditions:*

(1) For some, or any, $\lambda \in (1, \infty)$ one has

$$\fint_{B(0,\lambda R) \setminus B(0,R)} |u| \, d\mathcal{L}^n = o(1) \quad as \ R \to \infty. \tag{1.5.200}$$

(2) One has

$$u(x) = o(1) \ \text{as} \ |x| \to \infty. \tag{1.5.201}$$

(3) One has

$$⨍_{\partial B(0,R)} |u| \, \mathrm{d}\mathcal{H}^{n-1} = o(1) \ \text{as} \ R \to \infty. \tag{1.5.202}$$

(4) One has

$$u(x) = O(|x|^{2-n}) \ \text{as} \ |x| \to \infty. \tag{1.5.203}$$

(5) For some $\lambda \in (1, \infty)$ and some $N \in (0, 1]$ one has

$$⨍_{B(0,\lambda R)\setminus B(0,R)} |u| \, \mathrm{d}\mathcal{L}^n = o(R^N) \ \text{as} \ R \to \infty. \tag{1.5.204}$$

Then

conditions (1), (2), (3) are equivalent and imply $(\partial^\alpha u)(x) = o\big(|x|^{-|\alpha|}\big)$ as $|x| \to \infty$ for each multi-index $\alpha \in \mathbb{N}_0^n$, any of conditions (1)-(4) implies (5) and, in turn, condition (5) implies that, for each multi-index $\alpha \in \mathbb{N}_0^n$ with $|\alpha| > 0$, we have $(\partial^\alpha u)(x) = O\big(|x|^{2-n-|\alpha|}\big)$ as $|x| \to \infty$. (1.5.205)

Also,

if $n \geq 3$ then conditions (1), (2), (3), (4) are equivalent, and any implies that $(\partial^\alpha u)(x) = O\big(|x|^{2-n-|\alpha|}\big)$ as $|x| \to \infty$ for each multi-index $\alpha \in \mathbb{N}_0^n$. (1.5.206)

Furthermore, for some constant $c \in \mathbb{C}^M$,

$$\text{condition (5) implies } u(x) = \begin{cases} O(\ln|x|) & \text{if } n = 2, \\ c + O(|x|^{2-n}) & \text{if } n \geq 3, \end{cases} \ \text{as} \ |x| \to \infty. \tag{1.5.207}$$

Moreover, in the case when $n = 2$, the following two conditions are equivalent:

$$u(x) = o(\ln|x|) \ \text{as} \ |x| \to \infty, \tag{1.5.208}$$

$$u(x) = O(1) \ \text{as} \ |x| \to \infty. \tag{1.5.209}$$

Also, the following two conditions

$$\text{the limit} \ \lim_{x\to\infty} u(x) \ \text{exists (in } \mathbb{C}^M), \tag{1.5.210}$$

$$\text{there exists some } c \in \mathbb{C}^M \text{ such that } u(x) = c + O(|x|^{-1}) \ \text{as} \ |x| \to \infty \tag{1.5.211}$$

are equivalent and either implies

$$\begin{aligned}&(\partial^\alpha u)(x) = O\big(|x|^{-1-|\alpha|}\big) \text{ as } |x| \to \infty\\ &\text{for each multi-index } \alpha \in \mathbb{N}_0^2 \text{ with } |\alpha| > 0.\end{aligned} \tag{1.5.212}$$

In particular,

$$\begin{aligned}&\text{if } n = 2 \text{ then } u(x) = o(1) \text{ as } |x| \to \infty \text{ if and only if } u(x) = O(|x|^{-1})\\ &\text{as } |x| \to \infty, \text{ and either of these two conditions implies that, for each}\\ &\text{multi-index } \alpha \in \mathbb{N}_0^2, \text{ we have, } (\partial^\alpha u)(x) = O\big(|x|^{-1-|\alpha|}\big) \text{ as } |x| \to \infty.\end{aligned} \tag{1.5.213}$$

Finally, if $n = 2$ then (1.5.208)-(1.5.209) *are equivalent to* (1.5.210)-(1.5.211) *for any null-solution u of L in an exterior domain in $\mathbb{R}^2$ if and only if*

$$\text{the } M \times M \text{ matrix } \int_{S^1} \big[L(\xi)\big]^{-1}\, \mathrm{d}\mathcal{H}^1(\xi) \text{ is invertible} \tag{1.5.214}$$

(with the latter condition always true if the system L is assumed to actually satisfy the Legendre-Hadamard strong ellipticity condition, and if $M = 1$ said condition holds if and only if $L = \nabla \cdot A\nabla$ for some $A \in M_2^0$; cf. Lemma 1.4.19 and (1.4.186)).

Before presenting the proof of this theorem we wish to comment on a couple of concrete examples. Consider first the case of the two-dimensional complex Lamé system $L_{\lambda,\mu}$, defined as in (1.3.6) with $\mu \neq 0$ and $\lambda + 2\mu \neq 0$. According to (1.3.9), these conditions ensure that $L_{\lambda,\mu}$ is weakly elliptic. Then a direct computation based on (1.3.10) and (1.5.66) shows that

$$\int_{S^1} \big[L_{\lambda,\mu}(\xi)\big]^{-1}\, \mathrm{d}\mathcal{H}^1(\xi) = -\pi \frac{\lambda + 3\mu}{\mu(\lambda + 2\mu)} I_{2\times 2}. \tag{1.5.215}$$

Consequently, the 2×2 matrix

$$\int_{S^1} \big[L_{\lambda,\mu}(\xi)\big]^{-1}\, \mathrm{d}\mathcal{H}^1(\xi) \text{ is invertible if and only if } \lambda + 3\mu \neq 0. \tag{1.5.216}$$

In particular, this shows that in general (1.5.214) is not expected to hold for generic weakly elliptic systems.

Let us also note that, in the two-dimensional setting, having $\lambda + 3\mu = 0$ makes the logarithmic term in Kelvin's fundamental solution (1.4.72) disappear. As such, Proposition 1.4.4 implies that whenever $\mu \in \mathbb{C} \setminus \{0\}$ and $\lambda = -3\mu$ the matrix-valued function $E = (E_{jk})_{1\leq j,k\leq 2}$ whose (j,k) entry is defined at each point $x = (x_1, x_2)$ in $\mathbb{R}^2 \setminus \{0\}$ according to

$$E_{jk}(x) = -\frac{\mu + \lambda}{4\pi\mu(2\mu + \lambda)} \frac{x_j x_k}{|x|^2} = \frac{1}{2\pi\mu} \frac{x_j x_k}{|x|^2} \tag{1.5.217}$$

is a fundamental solution for the complex Lamé system $L_{\lambda,\mu}$ associated with the Lamé moduli μ, λ as in (1.3.6). In particular, any column of $E \in \big[\mathscr{C}^\infty(\mathbb{R}^2 \setminus \{0\})\big]^{2\times 2}$ is a bounded null-solution of the complex Lamé system $L_{\lambda,\mu}$ in $\mathbb{R}^2 \setminus \{0\}$ which does not possess a limit at infinity. This should be contrasted with the fact that (1.5.209) implies (1.5.210), if (1.5.214) holds (which is not presently the case; cf. (1.5.215)).

Similar phenomena occur for *scalar* differential operators in the two-dimensional setting. To illustrate this, consider the Bitsadze's operator in the complex plane

$$L := \partial_{\bar{z}}^2. \tag{1.5.218}$$

This is a homogeneous, constant (complex) coefficient, weakly elliptic, scalar operator, for which (1.5.214) fails. Indeed, if L is as in (1.5.218) then

$$L(\xi) = -\tfrac{1}{4}(\xi_1 + \mathrm{i}\xi_2)^2 \text{ for each } \xi = (\xi_1, \xi_2) \in \mathbb{R}^2, \tag{1.5.219}$$

which further implies

$$\int_{S^1} \big[L(z)\big]^{-1}\, d\mathcal{H}^1(z) = -4\int_{S^1} \frac{1}{z^2}\, d\mathcal{H}^1(z) = 4\mathrm{i}\int_{S^1} \frac{dz}{z^3} = 0, \tag{1.5.220}$$

in violation of (1.5.214). At the same time, the function

$$u(z) := \frac{\bar{z}}{z} \text{ for each } z \in \mathbb{C} \setminus \{0\} \tag{1.5.221}$$

(which is, up to normalization, a fundamental solution of $\partial_{\bar{z}}^2$ in $\mathbb{C}$) is a bounded null-solution of the Bitsadze's operator $L = \partial_{\bar{z}}^2$ in $\mathbb{C} \setminus \{0\}$ which does not have a limit at infinity. Again, this should be contrasted with the fact that (1.5.209) implies (1.5.210), if (1.5.214) holds (which is not the case here; see (1.5.220)).

Here is the actual proof of Theorem 1.5.9.

Proof of Theorem 1.5.9 First off, the fact that conditions *(1)*, *(2)*, *(3)* are equivalent and imply $(\partial^\alpha u)(x) = o\big(|x|^{-|\alpha|}\big)$ as $|x| \to \infty$ for each multi-index $\alpha \in \mathbb{N}_0^n$ is clear from Lemma 1.5.6. Next, it is clear that any of conditions *(4)* or *(1)* (therefore also *(2)* or *(3)*) implies *(5)*. In addition, since $\lfloor N \rfloor = 0$ (cf. (1.5.142)), Corollary 1.5.8 guarantees that if condition *(5)* holds then $(\partial^\alpha u)(x) = O\big(|x|^{2-n-|\alpha|}\big)$ as $|x| \to \infty$ for any multi-index $\alpha \in \mathbb{N}_0^n$ with $|\alpha| > 0$. This completes the proof of (1.5.205).

Next, the fact that if $n \geq 3$ condition *(1)* implies *(4)* is seen from Corollary 1.5.8 (cf. (1.5.198) written for $\alpha = (0, \dots, 0) \in \mathbb{N}_0^n$, bearing in mind that $\lfloor 0 \rfloor = -1$). The opposite implication is trivial (since $n \geq 3$), so ultimately conditions *(1)*-*(4)* are equivalent if $n \geq 3$. That *(4)* implies $(\partial^\alpha u)(x) = O\big(|x|^{2-n-|\alpha|}\big)$ as $|x| \to \infty$ for any multi-index $\alpha \in \mathbb{N}_0^n$ follows from Lemma 1.5.6. This completes the proof of (1.5.206).

Going further, the implication in (1.5.207) is a consequence of the asymptotic formula (1.5.197) (written for $\alpha = (0, \dots, 0) \in \mathbb{N}_0^n$), the fact that the polynomial P appearing there is presently a constant (since $\deg P \leq \lfloor N \rfloor = 0$ given that $N \in (0, 1]$), and the estimates in (1.4.24).

Clearly, (1.5.209) implies (1.5.208). To prove the opposite implication, first note that (1.5.208) entails

$$u(x) = o(|x|^N) \text{ as } |x| \to \infty, \text{ for each fixed } N \in (0,1]. \tag{1.5.222}$$

This permits us to once again invoke Corollary 1.5.8. Bearing in mind that the polynomial P appearing there is presently a constant (given that $\deg P \leq \lfloor N \rfloor = 0$), the asymptotic formula (1.5.197) written for $\alpha = (0, \dots, 0) \in \mathbb{N}_0^n$ becomes

$$u(x) = P + E(x)c + O(|x|^{-1}) \text{ as } |x| \to \infty. \tag{1.5.223}$$

From this and (1.5.208) we then conclude that $E(x)c = o(\ln|x|)$ as $|x| \to \infty$. In view of (1.4.48) this forces $E(x)c = O(1)$ as $|x| \to \infty$ which, when used back in formula (1.5.197), proves (1.5.209). Thus, conditions (1.5.208), (1.5.209) are indeed equivalent.

The fact that conditions (1.5.210), (1.5.211) are also equivalent is established in a very similar fashion as above, via (1.5.223), now relying on (1.4.47) in place of (1.4.48). Moreover, (1.5.211) implies that $u - c$ is a null-solution of L which is $O(|x|^{-1})$ as $|x| \to \infty$. From this and Lemma 1.5.6 we then see that (1.5.212) holds.

Pressing on, the first claim in (1.5.213) is seen from the equivalence of (1.5.210) with (1.5.211). The second claim in (1.5.213) follows from (1.5.212) if $|\alpha| > 0$ and is trivial if $|\alpha| = 0$. Working in the two-dimensional setting, consider now the issue of whether (1.5.208)-(1.5.209) are equivalent to (1.5.210)-(1.5.211). In one direction, (1.5.210) obviously implies (1.5.209). In the opposite direction, assume $u(x) = O(1)$ as $|x| \to \infty$. Then (1.5.222) holds which, as in the case of (1.5.223), leads to the conclusion that u has the asymptotic expansion

$$u(x) = P + E(x)c + O(|x|^{-1}) \text{ as } |x| \to \infty, \tag{1.5.224}$$

with P a constant in $\mathbb{C}^M$. From this we then conclude that $E(x)c = O(1)$ as $|x| \to \infty$ which, in light of (1.4.48),

$$\Big(\int_{S^1} [L(\xi)]^{-1} \, d\mathcal{H}^1(\xi) \Big) c = 0. \tag{1.5.225}$$

If the invertibility condition (1.5.214) is assumed, (1.5.225) forces $c = 0 \in \mathbb{C}^M$, a scenario in which (1.5.224) reduces to (1.5.210).

Finally, while working in the two-dimensional setting, suppose (1.5.208)-(1.5.209) are equivalent to (1.5.210)-(1.5.211) for any null-solution u of L in an exterior domain in $\mathbb{R}^2$, the goal now being to show that the invertibility property (1.5.214) holds.

Reasoning by contradiction, assume (1.5.209) implies (1.5.210) for any null-solution u of L in an exterior domain in $\mathbb{R}^2$, yet (1.5.214) fails. The latter property guarantees the existence of a non-zero vector $c \in \mathbb{C}^M$ for which (1.5.225) holds. If we define

$$u(x) := E(x)c \text{ for each } x \in \mathbb{R}^2 \setminus \{0\}, \tag{1.5.226}$$

then $u \in \big[\mathscr{C}^\infty(\mathbb{R}^2 \setminus \{0\})\big]^M$ and (1.4.20) gives $Lu = 0$ in $\mathbb{R}^2 \setminus \{0\}$. Moreover, from (1.5.225) and (1.4.22)-(1.4.23) we see that $u(x) = \Phi(x)c$ for each $x \in \mathbb{R}^2 \setminus \{0\}$, so u is bounded in $\mathbb{R}^2 \setminus \{0\}$. In particular, (1.5.209) holds for this function which, according to the current working hypotheses, implies that (1.5.210) holds for the function u defined in (1.5.226). Together with (1.4.47), this ultimately forces $c = 0 \in \mathbb{C}^M$, in contradiction with the fact that the vector c was non-zero to begin with. This finishes the proof of the very last claim in the statement. □

Remarkably, in addition to (1.5.4) there exists an integral representation formula for the gradient of a function involving only at most first-order partial derivatives of the fundamental solution. This is presented in our next theorem in a general geometric setting (but the result is new even in the smooth case). A further generalization is presented later, in Theorem 1.6.6.

Theorem 1.5.10 *Let $\Omega \subseteq \mathbb{R}^n$, where $n \geq 2$, be an open set with a lower Ahlfors regular boundary, and with the property that $\sigma := \mathcal{H}^{n-1}\lfloor\partial\Omega$ is a doubling measure on $\partial\Omega$. In particular, Ω is a set of locally finite perimeter, and its geometric measure theoretic outward unit normal $\nu = (\nu_1, \ldots, \nu_n)$ is defined σ-a.e. on $\partial_*\Omega$. With the summation convention over repeated indices understood throughout, let*

$$L = \big(a_{rs}^{\alpha\beta}\,\partial_r\partial_s\big)_{1\leq\alpha,\beta\leq M} \tag{1.5.227}$$

be a homogeneous, weakly elliptic, second-order $M \times M$ system in $\mathbb{R}^n$, with complex constant coefficients, and denote by $E = (E_{\gamma\beta})_{1\leq\gamma,\beta\leq M}$ the matrix-valued fundamental solution associated with L as in Theorem 1.4.2. In this setting, assume $u = (u_\beta)_{1\leq\beta\leq M} \in \big[W^{1,1}_{\rm loc}(\Omega)\big]^M$ is a vector-valued function which, for some $\kappa > 0$, satisfies

$$\begin{gathered}(\nabla u)\Big|^{\kappa-{\rm n.t.}}_{\partial\Omega} \text{ exists at } \sigma\text{-a.e. point on } \partial_{\rm nta}\Omega,\\ \text{and } \int_{\partial\Omega}\frac{\big(\mathcal{N}_\kappa(\nabla u)\big)(y)}{1+|y|^{n-1}}\,d\sigma(y) < +\infty.\end{gathered} \tag{1.5.228}$$

With Lu considered in the sense of distributions in Ω, suppose also that

$$Lu \in \Big[L^1\Big(\Omega, \frac{dy}{1+|y|^{n-1}}\Big)\Big]^M. \tag{1.5.229}$$

Then the function u is locally Lipschitz in Ω and for any other aperture parameter $\kappa' > 0$ the nontangential trace $(\nabla u)\Big|^{\kappa'-{\rm n.t.}}_{\partial\Omega}$ also exists σ-a.e. on $\partial_{\rm nta}\Omega$ and is actually independent of κ'. Moreover, with the dependence on the parameter κ' dropped and having fixed $\ell \in \{1, \ldots, n\}$ along with $\gamma \in \{1, \ldots, M\}$ arbitrary, it follows that for $\mathcal{L}^n$-a.e. point $x \in \Omega$ one has (with absolutely convergent integrals)

$$
\begin{aligned}
(\partial_\ell u_\gamma)(x) = \int_{\partial_*\Omega} & a_{rs}^{\beta\alpha}(\partial_r E_{\gamma\beta})(x-y) \times \\
& \times \Big\{ \nu_\ell(y)\big((\partial_s u_\alpha)\big|_{\partial\Omega}^{\text{n.t.}}\big)(y) - \nu_s(y)\big((\partial_\ell u_\alpha)\big|_{\partial\Omega}^{\text{n.t.}}\big)(y)\Big\} \, d\sigma(y) \\
& - \int_{\partial_*\Omega} (\partial_\ell E_{\gamma\alpha})(x-y)\nu_r(y) a_{rs}^{\alpha\beta}\big((\partial_s u_\beta)\big|_{\partial\Omega}^{\text{n.t.}}\big)(y)\, d\sigma(y) \\
& + \int_\Omega (\partial_\ell E_{\gamma\beta})(x-y)\big(Lu\big)_\beta(y)\, dy
\end{aligned}
\tag{1.5.230}
$$

if either Ω is bounded, or $\partial\Omega$ is unbounded. In the case when Ω is an exterior domain, the same conclusion holds true under the additional assumption that there exists $\lambda \in (1, \infty)$ such that

$$
⨍_{B(0,\lambda R)\setminus B(0,R)} |\nabla u| \, d\mathcal{L}^n = o(1) \ \textit{as} \ R \to \infty. \tag{1.5.231}
$$

In addition, for each $x \in \mathbb{R}^n \setminus \overline{\Omega}$ one has

$$
\begin{aligned}
0 = \int_{\partial_*\Omega} & a_{rs}^{\beta\alpha}(\partial_r E_{\gamma\beta})(x-y) \times \\
& \times \Big\{ \nu_\ell(y)\big((\partial_s u_\alpha)\big|_{\partial\Omega}^{\text{n.t.}}\big)(y) - \nu_s(y)\big((\partial_\ell u_\alpha)\big|_{\partial\Omega}^{\text{n.t.}}\big)(y)\Big\} \, d\sigma(y) \\
& - \int_{\partial_*\Omega} (\partial_\ell E_{\gamma\alpha})(x-y)\nu_r(y) a_{rs}^{\alpha\beta}\big((\partial_s u_\beta)\big|_{\partial\Omega}^{\text{n.t.}}\big)(y)\, d\sigma(y) \\
& + \int_\Omega (\partial_\ell E_{\gamma\beta})(x-y)\big(Lu\big)_\beta(y)\, dy
\end{aligned}
\tag{1.5.232}
$$

with the same convention as before in the case when Ω is an exterior domain.

Proof Collectively, (1.5.228) and [112, Lemma 8.3.1] imply

$$
u \in \big[W^{1,\infty}_{\text{loc}}(\Omega)\big]^M = \big[\text{Lip}_{\text{loc}}(\Omega)\big]^M. \tag{1.5.233}
$$

Also, from (1.5.228) and [112, Corollary 8.9.9] it follows that for any $\kappa' > 0$ the nontangential traces

$$
\begin{aligned}
&(\nabla u)\big|_{\partial\Omega}^{\kappa'-\text{n.t.}} \ \text{exists at } \sigma\text{-a.e. point on } \partial_{\text{nta}}\Omega \text{ and} \\
&\text{is actually independent of the parameter } \kappa' > 0.
\end{aligned}
\tag{1.5.234}
$$

To proceed, recall from [112, Lemma 3.5.7] that the membership in (1.5.229) is equivalent with having for $\mathcal{L}^n$-a.e. point $x \in \mathbb{R}^n$ the finiteness property

$$
\int_\Omega \frac{|(Lu)(y)|}{|x-y|^{n-1}}\, dy < +\infty. \tag{1.5.235}
$$

Pick some

$$x \in \Omega \text{ for which (1.5.235) holds and which is a Lebesgue point for the locally integrable } \mathbb{C}^{M\times M}\text{-valued function } \nabla u. \tag{1.5.236}$$

Also, fix two arbitrary indexes, $\ell \in \{1, \dots, n\}$ and $\gamma \in \{1, \dots, M\}$. Consider then the vector field defined at $\mathcal{L}^n$-a.e. point in Ω as

$$\begin{aligned} \vec{F}_x := {} & a_{rs}^{\beta\alpha}(\partial_r E_{\gamma\beta})(x-\cdot)\big\{(\partial_s u_\alpha)\mathbf{e}_\ell - (\partial_\ell u_\alpha)\mathbf{e}_s\big\} \\ & - (\partial_\ell E_{\gamma\alpha})(x-\cdot)a_{rs}^{\alpha\beta}(\partial_s u_\beta)\mathbf{e}_r. \end{aligned} \tag{1.5.237}$$

In particular, from (1.5.237), (1.5.233), and (1.4.24) we see that

$$\vec{F}_x \in \big[L^1_{\mathrm{loc}}(\Omega, \mathcal{L}^n)\big]^n. \tag{1.5.238}$$

With the goal of computing $\mathrm{div}\vec{F}_x$ in the sense of distributions in Ω, fix an arbitrary scalar-valued function $\varphi \in \mathscr{C}^\infty_c(\Omega)$ and write

$$\begin{aligned} {}_{\mathcal{D}'(\Omega)}\big\langle \mathrm{div}\vec{F}_x, \varphi\big\rangle_{\mathcal{D}(\Omega)} = {} & -a_{rs}^{\beta\alpha}\int_\Omega (\partial_r E_{\gamma\beta})(x-y)(\partial_s u_\alpha)(y)(\partial_\ell\varphi)(y)\,dy \\ & + a_{rs}^{\beta\alpha}\int_\Omega (\partial_r E_{\gamma\beta})(x-y)(\partial_\ell u_\alpha)(y)(\partial_s\varphi)(y)\,dy \\ & + a_{rs}^{\alpha\beta}\int_\Omega (\partial_\ell E_{\gamma\alpha})(x-\cdot)(\partial_s u_\beta)(\partial_r\varphi)(y)\,dy \\ =: {} & \mathrm{I} + \mathrm{II} + \mathrm{III}. \end{aligned} \tag{1.5.239}$$

Having chosen a scalar-valued function $\theta \in \mathscr{C}^\infty(\mathbb{R}^n)$ with the property that $\theta = 0$ on $B(0,1)$ and $\theta = 1$ on $\mathbb{R}^n \setminus B(0,2)$, for each $\varepsilon \in \big(0, \frac{1}{2}\,\mathrm{dist}(x, \partial\Omega)\big)$ define $\theta_\varepsilon : \mathbb{R}^n \to \mathbb{R}$ by setting

$$\theta_\varepsilon(y) := \theta\Big(\frac{y-x}{\varepsilon}\Big) \quad \text{for every } y \in \mathbb{R}^n. \tag{1.5.240}$$

Then

$$\begin{gathered} \theta_\varepsilon \in \mathscr{C}^\infty(\mathbb{R}^n) \text{ is a bounded function,} \\ \lim_{\varepsilon\to 0^+} \theta_\varepsilon(y) = 1 \text{ for every } y \in \mathbb{R}^n \setminus \{x\}, \\ 1-\theta_\varepsilon \in \mathscr{C}^\infty_c(\Omega), \quad \theta_\varepsilon \equiv 0 \text{ on } B(x,\varepsilon), \\ \mathrm{supp}\,(\nabla\theta_\varepsilon) \subseteq \overline{B(x,2\varepsilon)} \setminus B(x,\varepsilon), \text{ and also} \\ \sup\nolimits_{y\in\mathbb{R}^n} |(\nabla^j\theta_\varepsilon)(y)| \leq C\varepsilon^{-j} \text{ for all } j \in \mathbb{N}_0, \end{gathered} \tag{1.5.241}$$

for some constant $C \in (0, \infty)$ independent of ε. In particular, for each pair of indices $\alpha, \beta \in \{1, \dots, M\}$ we have

$E_{\alpha\beta}(x-\cdot)\theta_\varepsilon \in \mathscr{C}^\infty(\Omega)$ and $E_{\alpha\beta}(x-\cdot)\varphi\theta_\varepsilon \in \mathscr{C}^\infty_c(\Omega)$, plus similar properties for expressions involving partial derivatives acting on either of the intervening functions. (1.5.242)

To further transform the term I from (1.5.239), we employ (1.5.233), Lebesgue's Dominated Convergence Theorem, and integrations by parts (keeping in mind (1.5.242)), on account of which we may write

$$\begin{aligned}
\mathrm{I} &= -\lim_{\varepsilon\to 0^+} a^{\beta\alpha}_{rs}\int_\Omega (\partial_r E_{\gamma\beta})(x-y)(\partial_s u_\alpha)(y)(\partial_\ell\varphi)(y)\theta_\varepsilon(y)\,\mathrm{d}y\\
&= \lim_{\varepsilon\to 0^+}\Big\{-a^{\beta\alpha}_{rs}\int_\Omega (\partial_\ell\partial_r E_{\gamma\beta})(x-y)(\partial_s u_\alpha)(y)\varphi(y)\theta_\varepsilon(y)\,\mathrm{d}y\\
&\qquad + a^{\beta\alpha}_{rs}\,{}_{\mathcal{D}'(\Omega)}\Big\langle \partial_\ell\partial_s u_\alpha, (\partial_r E_{\gamma\beta})(x-\cdot)\varphi\theta_\varepsilon\Big\rangle_{\mathcal{D}(\Omega)}\\
&\qquad + a^{\beta\alpha}_{rs}\int_\Omega (\partial_r E_{\gamma\beta})(x-y)(\partial_s u_\alpha)(y)\varphi(y)(\partial_\ell\theta_\varepsilon)(y)\,\mathrm{d}y\Big\}\\
&= \lim_{\varepsilon\to 0^+}\Big\{\mathrm{I}^{(1)}_\varepsilon + \mathrm{I}^{(2)}_\varepsilon + \mathrm{I}^{(3)}_\varepsilon\Big\}.
\end{aligned} \tag{1.5.243}$$

Likewise,

$$\begin{aligned}
\mathrm{II} &= \lim_{\varepsilon\to 0^+}\Big\{a^{\beta\alpha}_{rs}\int_\Omega (\partial_s\partial_r E_{\gamma\beta})(x-y)(\partial_\ell u_\alpha)(y)\varphi(y)\theta_\varepsilon(y)\,\mathrm{d}y\\
&\qquad - a^{\beta\alpha}_{rs}\,{}_{\mathcal{D}'(\Omega)}\Big\langle \partial_\ell\partial_s u_\alpha, (\partial_r E_{\gamma\beta})(x-\cdot)\varphi\theta_\varepsilon\Big\rangle_{\mathcal{D}(\Omega)}\\
&\qquad - a^{\beta\alpha}_{rs}\int_\Omega (\partial_r E_{\gamma\beta})(x-y)(\partial_\ell u_\alpha)(y)\varphi(y)(\partial_s\theta_\varepsilon)(y)\,\mathrm{d}y\Big\}\\
&= \lim_{\varepsilon\to 0^+}\Big\{\mathrm{II}^{(1)}_\varepsilon + \mathrm{II}^{(2)}_\varepsilon + \mathrm{II}^{(3)}_\varepsilon\Big\}
\end{aligned} \tag{1.5.244}$$

and

$$\begin{aligned}
\mathrm{III} &= \lim_{\varepsilon\to 0^+}\Big\{a^{\alpha\beta}_{rs}\int_\Omega (\partial_r\partial_\ell E_{\gamma\alpha})(x-y)(\partial_s u_\beta)\varphi(y)\theta_\varepsilon(y)\,\mathrm{d}y\\
&\qquad - {}_{\mathcal{D}'(\Omega)}\Big\langle a^{\alpha\beta}_{rs}\partial_r\partial_s u_\beta, (\partial_\ell E_{\gamma\alpha})(x-\cdot)\varphi\theta_\varepsilon\Big\rangle_{\mathcal{D}(\Omega)}\\
&\qquad - a^{\alpha\beta}_{rs}\int_\Omega (\partial_\ell E_{\gamma\alpha})(x-y)(\partial_s u_\beta)\varphi(y)(\partial_r\theta_\varepsilon)(y)\,\mathrm{d}y\Big\}\\
&= \lim_{\varepsilon\to 0^+}\Big\{\mathrm{III}^{(1)}_\varepsilon + \mathrm{III}^{(2)}_\varepsilon + \mathrm{III}^{(3)}_\varepsilon\Big\}.
\end{aligned} \tag{1.5.245}$$

Observe that, after interchanging α and β in $\mathrm{III}_\varepsilon^{(1)}$ the resulting expression matches $-\mathrm{I}_\varepsilon^{(1)}$. Hence,

$$\mathrm{I}_\varepsilon^{(1)} + \mathrm{III}_\varepsilon^{(1)} = 0 \text{ for each } \varepsilon. \tag{1.5.246}$$

It is also clear from definitions that

$$\mathrm{I}_\varepsilon^{(2)} + \mathrm{II}_\varepsilon^{(2)} = 0 \text{ for each } \varepsilon. \tag{1.5.247}$$

Furthermore, (1.4.33) gives

$$\mathrm{II}_\varepsilon^{(1)} = 0 \text{ for each } \varepsilon. \tag{1.5.248}$$

Let us also remark that based on (1.5.229), (1.5.235), (1.4.24), and Lebesgue's Dominated Convergence Theorem, we obtain

$$\begin{aligned}
\lim_{\varepsilon\to 0^+} \mathrm{III}_\varepsilon^{(2)} &= -\lim_{\varepsilon\to 0^+} {}_{\mathcal{D}'(\Omega)}\Big\langle a_{rs}^{\alpha\beta}\partial_r\partial_s u_\beta, (\partial_\ell E_{\gamma\alpha})(x-\cdot)\varphi\theta_\varepsilon\Big\rangle_{\mathcal{D}(\Omega)} \\
&= -\lim_{\varepsilon\to 0^+} {}_{\mathcal{D}'(\Omega)}\Big\langle (Lu)_\alpha, (\partial_\ell E_{\gamma\alpha})(x-\cdot)\varphi\theta_\varepsilon\Big\rangle_{\mathcal{D}(\Omega)} \\
&= -\lim_{\varepsilon\to 0^+} \int_\Omega (\partial_\ell E_{\gamma\alpha})(x-y)\varphi(y)\theta_\varepsilon(y)(Lu)_\alpha(y)\,dy \\
&= -\int_\Omega (\partial_\ell E_{\gamma\alpha})(x-y)\varphi(y)(Lu)_\alpha(y)\,dy.
\end{aligned} \tag{1.5.249}$$

Turning our attention to $\mathrm{I}_\varepsilon^{(3)}$ from (1.5.243), we expand

$$\begin{aligned}
\mathrm{I}_\varepsilon^{(3)} = a_{rs}^{\beta\alpha} &\int_\Omega (\partial_r E_{\gamma\beta})(x-y)\{(\varphi\,\partial_s u_\alpha)(y) - (\varphi\,\partial_s u_\alpha)(x)\}(\partial_\ell\theta_\varepsilon)(y)\,dy \\
&+ \Big\{a_{rs}^{\beta\alpha}\int_\Omega (\partial_r E_{\gamma\beta})(x-y)(\partial_\ell\theta_\varepsilon)(y)\,dy\Big\}(\varphi\,\partial_s u_\alpha)(x).
\end{aligned} \tag{1.5.250}$$

In relation to the first integral above, we note that since (1.5.236) implies that x is a Lebesgue point for $\varphi\nabla u$ we may estimate, using (1.4.24) and (1.5.241),

$$\begin{aligned}
\limsup_{\varepsilon\to 0^+} &\Big| a_{rs}^{\beta\alpha}\int_\Omega (\partial_r E_{\gamma\beta})(x-y)\{(\varphi\,\partial_s u_\alpha)(y) - (\varphi\,\partial_s u_\alpha)(x)\}(\partial_\ell\theta_\varepsilon)(y)\,dy\Big| \\
&\leq C\limsup_{\varepsilon\to 0^+} \fint_{B(x,2\varepsilon)} \big|(\varphi\,\nabla u)(y) - (\varphi\,\nabla u)(x)\big|\,dy = 0.
\end{aligned} \tag{1.5.251}$$

As regards the second term in (1.5.250), observe that for any two fixed indexes $\alpha \in \{1,\dots,M\}$ and $s \in \{1,\dots,n\}$ we have

$$
\begin{aligned}
a^{\beta\alpha}_{rs} &\int_\Omega (\partial_r E_{\gamma\beta})(x-y)(\partial_\ell \theta_\varepsilon)(y)\,\mathrm{d}y \\
&= a^{\beta\alpha}_{rs} \int_\Omega (\partial_r E_{\gamma\beta})(x-y)\partial_\ell(\theta_\varepsilon - 1)(y)\,\mathrm{d}y \\
&= -{}_{\mathcal{D}'(\Omega)}\Big\langle a^{\beta\alpha}_{rs}\partial_r\big[E_{\gamma\beta}(x-\cdot)\big], \partial_\ell(\theta_\varepsilon - 1)\Big\rangle_{\mathcal{D}(\Omega)} \\
&= {}_{\mathcal{D}'(\Omega)}\Big\langle a^{\beta\alpha}_{rs}\partial_\ell\partial_r\big[E_{\gamma\beta}(x-\cdot)\big], \theta_\varepsilon - 1\Big\rangle_{\mathcal{D}(\Omega)}. \qquad (1.5.252)
\end{aligned}
$$

As a result, from (1.5.250), (1.5.251), and (1.5.252) we conclude that

$$
\mathrm{I}^{(3)}_\varepsilon = {}_{\mathcal{D}'(\Omega)}\Big\langle a^{\beta\alpha}_{rs}\partial_\ell\partial_r\big[E_{\gamma\beta}(x-\cdot)\big], \theta_\varepsilon - 1\Big\rangle_{\mathcal{D}(\Omega)}(\varphi\,\partial_s u_\alpha)(x) + o(1) \text{ as } \varepsilon \to 0^+. \tag{1.5.253}
$$

In fact, in a very similar fashion we have

$$
\mathrm{III}^{(3)}_\varepsilon = -{}_{\mathcal{D}'(\Omega)}\Big\langle a^{\alpha\beta}_{rs}\partial_r\partial_\ell\big[E_{\gamma\alpha}(x-\cdot)\big], \theta_\varepsilon - 1\Big\rangle_{\mathcal{D}(\Omega)}(\varphi\,\partial_s u_\beta)(x) + o(1) \text{ as } \varepsilon \to 0^+ \tag{1.5.254}
$$

which, in concert with (1.5.253) (in which we interchange α and β), implies

$$
\lim_{\varepsilon\to 0^+}\Big\{\mathrm{I}^{(3)}_\varepsilon + \mathrm{III}^{(3)}_\varepsilon\Big\} = 0. \tag{1.5.255}
$$

Implementing once more the same type of argument that has produced (1.5.253) yields, on account of (1.4.33) and (1.5.241),

$$
\begin{aligned}
\lim_{\varepsilon\to 0^+}\mathrm{II}^{(3)}_\varepsilon &= -\lim_{\varepsilon\to 0^+} {}_{\mathcal{D}'(\Omega)}\Big\langle a^{\beta\alpha}_{rs}\partial_r\partial_s\big[E_{\gamma\beta}(x-\cdot)\big], \theta_\varepsilon - 1\Big\rangle_{\mathcal{D}(\Omega)}(\varphi\,\partial_\ell u_\alpha)(x) \\
&= -\lim_{\varepsilon\to 0^+}\delta_{\alpha\gamma}(\theta_\varepsilon - 1)(x)(\varphi\,\partial_\ell u_\alpha)(x) = \varphi(x)(\partial_\ell u_\gamma)(x) \\
&= {}_{\mathcal{D}'(\Omega)}\big\langle(\partial_\ell u_\gamma)(x)\delta_x, \varphi\big\rangle_{\mathcal{D}(\Omega)}. \qquad (1.5.256)
\end{aligned}
$$

At this stage, combining (1.5.239), (1.5.243)-(1.5.249), and (1.5.255)-(1.5.256) we ultimately conclude that for each point $x \in \Omega$ as in (1.5.236) we have

$$
\operatorname{div}\vec{F}_x = (\partial_\ell u_\gamma)(x)\delta_x - (\partial_\ell E_{\gamma\alpha})(x-\cdot)(Lu)_\alpha \text{ in } \mathcal{D}'(\Omega). \tag{1.5.257}
$$

In particular,

$$
\operatorname{div}\vec{F}_x \in \mathscr{E}'(\Omega) + L^1(\Omega, \mathcal{L}^n). \tag{1.5.258}
$$

If we now consider the compact subset of Ω given by $K := \overline{B\big(x, \tfrac{1}{2}\operatorname{dist}(x, \partial\Omega)\big)}$, then based on (1.5.237), (1.4.24), and [112, Lemma 8.3.7] we may estimate

$$\begin{aligned}(\mathcal{N}_\kappa^{\Omega\setminus K}\vec{F}_x)(y) &\leq C\big(\mathcal{N}_\kappa(\nabla u)\big)(y)\cdot \sup_{z\in\Gamma_\kappa(y)\setminus K}\big[|x-z|^{1-n}\big]\\ &\leq C\frac{\big(\mathcal{N}_\kappa(\nabla u)\big)(y)}{|x-y|^{n-1}}, \qquad \forall\, y\in\partial\Omega, \end{aligned}\tag{1.5.259}$$

for some constant $C = C(\Omega, L, n, \kappa) \in (0, \infty)$. From (1.5.259), (1.5.228), and [112, (8.2.26)] it follows that

$$\mathcal{N}_\kappa^{\Omega\setminus K}\vec{F}_x \in L^1(\partial\Omega, \sigma), \tag{1.5.260}$$

while from (1.5.237), [112, (8.9.10)-(8.9.11)], and (1.5.234) we conclude that

$$\vec{F}_x\Big|_{\partial\Omega}^{\text{n.t.}} \text{ exists at } \sigma\text{-a.e. point on } \partial_{\text{nta}}\Omega \tag{1.5.261}$$

and, thanks to [112, (8.8.52)], at σ-a.e. point $y \in \partial_*\Omega$ we have

$$\begin{aligned}&\nu(y)\cdot\Big(\vec{F}_x\Big|_{\partial\Omega}^{\text{n.t.}}\Big)(y)\\ &\quad = a_{rs}^{\beta\alpha}(\partial_r E_{\gamma\beta})(x-y)\Big\{\nu_\ell(y)\big((\partial_s u_\alpha)\big|_{\partial\Omega}^{\text{n.t.}}\big)(y) - \nu_s(y)\big((\partial_\ell u_\alpha)\big|_{\partial\Omega}^{\text{n.t.}}\big)(y)\Big\}\\ &\qquad - (\partial_\ell E_{\gamma\alpha})(x-y)\nu_r(y)a_{rs}^{\alpha\beta}\big((\partial_s u_\beta)\big|_{\partial\Omega}^{\text{n.t.}}\big)(y). \end{aligned}\tag{1.5.262}$$

Finally, when Ω is an exterior domain, it follows from (1.5.237) and (1.4.24) that

$$\vec{F}_x(y) = O\big(|y|^{1-n}|(\nabla u)(y)|\big) \text{ as } |y| \to \infty. \tag{1.5.263}$$

Together, (1.5.238), (1.5.258), (1.5.261) (as well as (1.5.231) and (1.5.263) in the case when Ω is an exterior domain) ensure that, for each point $x \in \Omega$ as in (1.5.236), the vector field $\vec{F}_x$ satisfies the hypotheses of [112, Theorem 1.4.1]. As such, on account of [112, (4.6.19)], (1.5.257), and (1.5.262), the Divergence Formula [112, (1.4.6)] currently gives

$$
\begin{aligned}
(\partial_\ell u_\gamma)(x) &- \int_\Omega E_{\gamma\alpha}(x-y)\big(Lu\big)_\alpha(y)\,\mathrm{d}y \\
&= {}_{\mathcal{E}'(\Omega)}\big\langle(\partial_\ell u_\gamma)(x)\delta_x, 1\big\rangle_{\mathcal{E}(\Omega)} - \int_\Omega E_{\gamma\alpha}(x-y)\big(Lu\big)_\alpha(y)\,\mathrm{d}y \\
&= {}_{(\mathscr{C}^\infty_b(\Omega))^*}\Big((\partial_\ell u_\gamma)(x)\delta_x - E_{\gamma\alpha}(x-\cdot)(Lu)_\alpha, 1\Big)_{\mathscr{C}^\infty_b(\Omega)} \\
&= {}_{(\mathscr{C}^\infty_b(\Omega))^*}\big(\mathrm{div}\vec{F}_x, 1\big)_{\mathscr{C}^\infty_b(\Omega)} = \int_{\partial_*\Omega} \nu\cdot\big(\vec{F}_x\big|^{\text{n.t.}}_{\partial\Omega}\big)\,\mathrm{d}\sigma \\
&= \int_{\partial_*\Omega} a^{\beta\alpha}_{rs}(\partial_r E_{\gamma\beta})(x-y)\times \\
&\qquad\times\Big\{\nu_\ell(y)\big((\partial_s u_\alpha)\big|^{\text{n.t.}}_{\partial\Omega}\big)(y) - \nu_s(y)\big((\partial_\ell u_\alpha)\big|^{\text{n.t.}}_{\partial\Omega}\big)(y)\Big\}\,\mathrm{d}\sigma(y) \\
&\quad - \int_{\partial_*\Omega}(\partial_\ell E_{\gamma\alpha})(x-y)\nu_r(y)a^{\alpha\beta}_{rs}\big((\partial_s u_\beta)\big|^{\text{n.t.}}_{\partial\Omega}\big)(y)\,\mathrm{d}\sigma(y) \qquad (1.5.264)
\end{aligned}
$$

for each point $x \in \Omega$ as in (1.5.236), in particular for $\mathcal{L}^n$-a.e. point $x \in \Omega$. This finishes the proof of (1.5.230).

Finally, the validity of (1.5.232) for each $x \in \mathbb{R}^n \setminus \overline{\Omega}$ is established in a similar fashion (this time, the Dirac delta function no longer appears in (1.5.257)). □

1.6 Integral Representation Formulas for Injectively Elliptic First-Order Systems

Our first theorem in this section contains a basic integral representation formula for functions which behave reasonably under the action of a given injectively elliptic first-order system, have a nontangential boundary trace, and with their nontangential maximal function belonging to a suitable weighted Lebesgue space. A more general version is discussed later, in Theorem 1.6.5.

Theorem 1.6.1 *Let $\Omega \subseteq \mathbb{R}^n$, where $n \geq 2$, be an open set with a lower Ahlfors regular boundary, and such that $\sigma := \mathcal{H}^{n-1}\lfloor\partial\Omega$ is a doubling measure on $\partial\Omega$. In particular, Ω is a set of locally finite perimeter, and its geometric measure theoretic outward unit normal ν is defined σ-a.e. on $\partial_*\Omega$.*

Next, consider a homogeneous, first-order $N\times M$ system D with constant complex coefficients in $\mathbb{R}^n$ (where $N, M \in \mathbb{N}$) which is injectively elliptic (cf. (1.3.18)), and suppose $\widetilde{D}$ is a homogeneous first-order $M \times N$ system with constant complex coefficients in $\mathbb{R}^n$ which complements D (i.e., (1.3.21) holds). In particular, since $L := \widetilde{D}D$ is a weakly elliptic second-order $M\times M$ system in $\mathbb{R}^n$ it is meaningful to consider the matrix-valued fundamental solution $E_{L^\top}$ associated with $L^\top$ as in Theorem 1.4.2. Also, with $\widetilde{D}^\top$ acting on the columns of $E_{L^\top}$ define

$$\widetilde{E} := \widetilde{D}^{\top} E_{L^{\top}}. \tag{1.6.1}$$

Finally, fix an aperture parameter $\kappa \in (0, \infty)$ *and suppose* $u : \Omega \to \mathbb{C}^M$ *is a vector-valued function with Lebesgue measurable components satisfying*[34]

$$\begin{aligned} &\mathcal{N}_\kappa u \in L^1\Big(\partial\Omega, \tfrac{\sigma(x)}{1+|x|^{n-1}}\Big), \\ &\Big(u\big|_{\partial\Omega}^{\kappa-\mathrm{n.t.}}\Big)(x) \text{ exists for } \sigma\text{-a.e. } x \in \partial_{\mathrm{nta}}\Omega, \\ &\text{and } Du \in \Big[L^1\Big(\Omega, \frac{dy}{1+|y|^{n-1}}\Big)\Big]^N. \end{aligned} \tag{1.6.2}$$

Then for any $\kappa' > 0$ *the nontangential trace* $u\big|_{\partial\Omega}^{\kappa'-\mathrm{n.t.}}$ *also exists at* σ*-a.e. point on* $\partial_{\mathrm{nta}}\Omega$ *and is actually independent of* κ'. *Moreover, with the dependence on the parameter* κ' *dropped, for* $\mathcal{L}^n$*-a.e.* $x \in \mathbb{R}^n \setminus \partial\Omega$ *one has (with absolutely convergent integrals)*

$$\begin{aligned} &\int_{\partial_*\Omega} \widetilde{E}^{\top}(x-y)(-i)\mathrm{Sym}\big(D; \nu(y)\big)\big(u\big|_{\partial\Omega}^{\mathrm{n.t.}}\big)(y)\,d\sigma(y) \\ &\qquad - \int_{\Omega} \widetilde{E}^{\top}(x-y)(Du)(y)\,dy = \begin{cases} u(x) & \text{if } x \in \Omega, \\ 0 & \text{if } x \in \mathbb{R}^n \setminus \overline{\Omega}, \end{cases} \end{aligned} \tag{1.6.3}$$

if either Ω *is bounded, or* $\partial\Omega$ *is unbounded. In the case when* Ω *is an exterior domain, the same conclusion holds true under the additional assumption that there exists* $\lambda \in (1, \infty)$ *such that*

$$\fint_{B(0,\lambda R)\setminus B(0,R)} |u|\,d\mathcal{L}^n = o(1) \text{ as } R \to \infty. \tag{1.6.4}$$

A couple of comments pertaining to the nature of $\widetilde{E}$ from (1.6.1) are in order. First, Theorem 1.4.2 implies that

$$\begin{aligned} &\widetilde{E} \in \big[\mathcal{S}'(\mathbb{R}^n)\big]^{N\times M}, \quad \widetilde{E}\big|_{\mathbb{R}^n\setminus\{0\}} \in \big[\mathscr{C}^\infty(\mathbb{R}^n \setminus \{0\})\big]^{N\times M}, \\ &\widetilde{E} \text{ is odd and positive homogeneous of degree } 1-n \\ &(\text{in particular, } |(\partial^\alpha \widetilde{E})(x)| \leq C_\alpha |x|^{1-n-|\alpha|} \text{ for each } \alpha \in \mathbb{N}_0^n). \end{aligned} \tag{1.6.5}$$

Second,

$$\widetilde{E} \text{ is a fundamental solution for } D^{\top}. \tag{1.6.6}$$

More specifically, for each fixed point $x \in \mathbb{R}^n$ we have (with all differential operators acting on columns in the "dot" variable)

---

[34] with Du considered in the sense of distributions in Ω

$$D^\top\big[\widetilde{E}(\cdot - x)\big] = D^\top\big[(\widetilde{D}^\top E_{L^\top})(\cdot - x)\big] = (D^\top \widetilde{D}^\top)\big[E_{L^\top}(\cdot - x)\big]$$
$$= L^\top\big[E_{L^\top}(\cdot - x)\big] = \delta_x\, I_{M\times M}, \tag{1.6.7}$$

where δ_x is Dirac's distribution with mass at x, $I_{M\times M}$ is the $M \times M$ identity matrix, and the last equality is comes from (1.4.21).

Here is the proof of Theorem 1.6.1.

Proof of Theorem 1.6.1 With the summation convention over repeated indices assumed throughout, suppose

$$\widetilde{D} = \big(\widetilde{a}_j^{\beta\alpha}\partial_j\big)_{\substack{1\le\beta\le M\\ 1\le\alpha\le N}} \text{ and } D = \big(a_k^{\alpha\gamma}\partial_k\big)_{\substack{1\le\alpha\le N\\ 1\le\gamma\le M}}. \tag{1.6.8}$$

Then the homogeneous second-order $M \times M$ system $L := \widetilde{D}D$ may be written explicitly as

$$L = \widetilde{D}D = \big(A_{jk}^{\beta\gamma}\partial_j\partial_k\big)_{1\le\beta,\gamma\le M} \text{ where each } A_{jk}^{\beta\gamma} := \widetilde{a}_j^{\beta\alpha}a_k^{\alpha\gamma}. \tag{1.6.9}$$

Hence, $L = L_A$, the second-order homogeneous $M \times M$ system associated with the coefficient tensor

$$A := \big(A_{jk}^{\beta\gamma}\big)_{\substack{1\le j,k\le n\\ 1\le\beta,\gamma\le M}} \text{ with } A_{jk}^{\beta\gamma} := \widetilde{a}_j^{\beta\alpha}a_k^{\alpha\gamma}. \tag{1.6.10}$$

Let us also observe that

$$\widetilde{E} = \widetilde{D}^\top E_{L^\top} = -\Big(\widetilde{a}_k^{\gamma\alpha}\partial_k\big(E_{L^\top}\big)_{\gamma\delta}\Big)_{\substack{1\le\alpha\le N\\ 1\le\delta\le M}} = -\Big(\widetilde{a}_k^{\gamma\alpha}\partial_k\big(E_L\big)_{\delta\gamma}\Big)_{\substack{1\le\alpha\le N\\ 1\le\delta\le M}} \tag{1.6.11}$$

where the last equality uses (1.4.32).

Going further, observe that the first property in (1.6.2) and [112, Lemma 8.3.1] guarantee that

$$u \in \big[L^\infty_{\rm loc}(\Omega, \mathcal{L}^n)\big]^N, \tag{1.6.12}$$

while the first two properties in (1.6.2) together with [112, Corollary 8.9.9] imply that

$$\begin{array}{l}\text{for any } \kappa' > 0 \text{ the nontangential trace } u\big|_{\partial\Omega}^{\kappa'-\rm n.t.} \text{ exists at } \sigma\text{-a.e. point}\\ \text{on } \partial_{\rm nta}\Omega \text{ and is actually independent of the parameter } \kappa' > 0.\end{array} \tag{1.6.13}$$

Let us turn our attention to (1.6.3) in earnest. Recall from [112, Lemma 3.5.7] that the last membership in (1.6.2) is equivalent with having for $\mathcal{L}^n$-a.e. point $x \in \mathbb{R}^n$ the finiteness property

$$\int_\Omega \frac{|(Du)(y)|}{|x-y|^{n-1}}\,dy < +\infty. \tag{1.6.14}$$

Fix an arbitrary index $\alpha \in \{1, \dots, M\}$, pick a Lebesgue point $x \in \Omega$ for the function u with the property that (1.6.14) holds, and consider the vector field $\vec{F}_x = \big(F_j\big)_{1 \le j \le n}$ with components (recall that, throughout, the summation convention over repeated indices is in effect)

$$F_j := -A^{\gamma\beta}_{kj}\big(\partial_k(E_L)_{\alpha\gamma}\big)(x - \cdot)u_\beta \text{ at } \mathcal{L}^n\text{-a.e. point in } \Omega. \tag{1.6.15}$$

Collectively, (1.6.15), (1.6.12), and Theorem 1.4.2 ensure that

$$\vec{F}_x \in \big[L^1_{\rm loc}(\Omega, \mathcal{L}^n)\big]^n. \tag{1.6.16}$$

The next goal is to compute $\operatorname{div}\vec{F}_x$ in the sense of distributions in Ω. To this end, fix an arbitrary scalar-valued function $\varphi \in \mathscr{C}^\infty_c(\Omega)$ and write

$$\begin{aligned} {}_{\mathcal{D}'(\Omega)}\big\langle \operatorname{div}\vec{F}_x, \varphi\big\rangle_{\mathcal{D}(\Omega)} &= -\int_\Omega F_j(y)(\partial_j\varphi)(y)\,dy \\ &= \int_\Omega A^{\gamma\beta}_{kj}\big(\partial_k(E_L)_{\alpha\gamma}\big)(x - y)u_\beta(y)(\partial_j\varphi)(y)\,dy. \end{aligned} \tag{1.6.17}$$

Bring in a scalar-valued function $\theta \in \mathscr{C}^\infty(\mathbb{R}^n)$ with the property that $\theta = 0$ on $B(0, 1)$ and $\theta = 1$ on $\mathbb{R}^n \setminus B(0, 2)$. For each $\varepsilon \in \big(0, \frac{1}{2}\operatorname{dist}(x, \partial\Omega)\big)$ define $\theta_\varepsilon : \mathbb{R}^n \to \mathbb{R}$ by setting

$$\theta_\varepsilon(y) := \theta\Big(\frac{y - x}{\varepsilon}\Big) \text{ for every } y \in \mathbb{R}^n. \tag{1.6.18}$$

Then

$$\begin{aligned} &\theta_\varepsilon \in \mathscr{C}^\infty(\mathbb{R}^n) \text{ is a bounded function satisfying} \\ &\lim_{\varepsilon \to 0^+} \theta_\varepsilon(y) = 1 \text{ for every point } y \in \mathbb{R}^n \setminus \{x\}, \end{aligned} \tag{1.6.19}$$

and there exists a constant $C \in (0, \infty)$ such that for each $\varepsilon \in \big(0, \frac{1}{2}\operatorname{dist}(x, \partial\Omega)\big)$ we have

$$\begin{aligned} &1 - \theta_\varepsilon \in \mathscr{C}^\infty_c(\Omega), \quad \theta_\varepsilon \equiv 0 \text{ on } B(x, \varepsilon), \\ &\operatorname{supp}(\nabla\theta_\varepsilon) \subseteq \overline{B(x, 2\varepsilon)} \setminus B(x, \varepsilon), \text{ and} \\ &|(\nabla^\ell\theta_\varepsilon)(y)| \le C\varepsilon^{-\ell} \text{ for every } \ell \in \mathbb{N}_0 \text{ and every } y \in \mathbb{R}^n. \end{aligned} \tag{1.6.20}$$

In particular,

the entries of $E_L(x - \cdot)\theta_\varepsilon$ belong to $\mathscr{C}^\infty(\Omega)$ and the entries of $E_L(x - \cdot)\varphi\theta_\varepsilon$ belong to $\mathscr{C}^\infty_c(\Omega)$; moreover, similar properties are valid for expressions involving partial derivatives acting on either of the intervening functions. (1.6.21)

Relying on (1.6.12), Lebesgue's Dominated Convergence Theorem, and integrations by parts (keeping in mind (1.5.26)) we may write

$$\begin{aligned}
{}_{\mathcal{D}'(\Omega)}\big\langle \mathrm{div}\vec{F}_x, \varphi\big\rangle_{\mathcal{D}(\Omega)} &= \lim_{\varepsilon\to 0^+}\int_\Omega A^{\gamma\beta}_{kj}\big(\partial_k(E_L)_{\alpha\gamma}\big)(x-y)u_\beta(y)(\partial_j\varphi)(y)\theta_\varepsilon(y)\,dy \\
&= \lim_{\varepsilon\to 0^+}\ {}_{\mathcal{D}'(\Omega)}\Big\langle A^{\gamma\beta}_{kj}u_\beta, \big(\partial_k(E_L)_{\alpha\gamma}\big)(x-\cdot)(\partial_j\varphi)\theta_\varepsilon\Big\rangle_{\mathcal{D}(\Omega)} \\
&= \mathrm{I}+\mathrm{II}+\mathrm{III}, \qquad (1.6.22)
\end{aligned}$$

where

$$\mathrm{I} := -\lim_{\varepsilon\to 0^+}\ {}_{\mathcal{D}'(\Omega)}\Big\langle A^{\gamma\beta}_{kj}\partial_j u_\beta, \big(\partial_k(E_L)_{\alpha\gamma}\big)(x-\cdot)\varphi\theta_\varepsilon\Big\rangle_{\mathcal{D}(\Omega)}, \qquad (1.6.23)$$

$$\mathrm{II} := \lim_{\varepsilon\to 0^+}\ {}_{\mathcal{D}'(\Omega)}\Big\langle A^{\gamma\beta}_{kj}u_\beta, \big(\partial_j\partial_k(E_L)_{\alpha\gamma}\big)(x-\cdot)\varphi\theta_\varepsilon\Big\rangle_{\mathcal{D}(\Omega)}, \qquad (1.6.24)$$

$$\mathrm{III} := -\lim_{\varepsilon\to 0^+}\ {}_{\mathcal{D}'(\Omega)}\Big\langle A^{\gamma\beta}_{kj}u_\beta, \big(\partial_k(E_L)_{\alpha\gamma}\big)(x-y)\varphi(\partial_j\theta_\varepsilon)\Big\rangle_{\mathcal{D}(\Omega)}. \qquad (1.6.25)$$

Note that

$$\begin{aligned}
\mathrm{I} &= -\lim_{\varepsilon\to 0^+}\ {}_{\mathcal{D}'(\Omega)}\Big\langle a^{\mu\beta}_j\partial_j u_\beta, \widetilde{a}^{\gamma\mu}_k\big(\partial_k(E_L)_{\alpha\gamma}\big)(x-\cdot)\varphi\theta_\varepsilon\Big\rangle_{\mathcal{D}(\Omega)} \\
&= \lim_{\varepsilon\to 0^+}\ {}_{\mathcal{D}'(\Omega)}\Big\langle (Du)_\mu, \big(\widetilde{D}^\top E_{L^\top}\big)_{\mu\alpha}(x-\cdot)\varphi\theta_\varepsilon\Big\rangle_{\mathcal{D}(\Omega)} \\
&= \lim_{\varepsilon\to 0^+}\ {}_{\mathcal{D}'(\Omega)}\Big\langle (Du)_\mu, \widetilde{E}_{\mu\alpha}(x-\cdot)\varphi\theta_\varepsilon\Big\rangle_{\mathcal{D}(\Omega)} \\
&= \lim_{\varepsilon\to 0^+}\int_\Omega \big(\widetilde{E}^\top(x-y)(Du)(y)\big)_\alpha\varphi(y)\theta_\varepsilon(y)\,dy \\
&= \int_\Omega \big(\widetilde{E}^\top(x-y)(Du)(y)\big)_\alpha\varphi(y)\,dy \\
&= {}_{\mathcal{D}'(\Omega)}\Big\langle \big(\widetilde{E}^\top(x-\cdot)(Du)\big)_\alpha, \varphi\Big\rangle_{\mathcal{D}(\Omega)}, \qquad (1.6.26)
\end{aligned}$$

thanks to (1.6.10), (1.6.8), (1.6.14), and Lebesgue's Dominated Convergence Theorem (which takes into account (1.6.19)). Next,

$$\mathrm{II} = \lim_{\varepsilon\to 0^+}\ {}_{\mathcal{D}'(\Omega)}\Big\langle u_\beta, A^{\gamma\beta}_{kj}\big(\partial_j\partial_k(E_L)_{\alpha\gamma}\big)(x-\cdot)\varphi\theta_\varepsilon\Big\rangle_{\mathcal{D}(\Omega)} = 0, \qquad (1.6.27)$$

since (1.4.33) gives that for each $\beta \in \{1,\dots,M\}$ we have

$$A^{\gamma\beta}_{kj}\big(\partial_j\partial_k(E_L)_{\alpha\gamma}\big)(x-y) = 0 \ \text{ for each } \ y\in\mathbb{R}^n\setminus\{x\}. \qquad (1.6.28)$$

As regards the contribution from the very last term in (1.6.22), we write

$$
\begin{aligned}
\mathrm{III} &= -\lim_{\varepsilon\to 0^+}\int_\Omega A_{kj}^{\gamma\beta}\big(\partial_k(E_L)_{\alpha\gamma}\big)(x-y)u_\beta(y)\varphi(y)(\partial_j\theta_\varepsilon)(y)\,\mathrm{d}y \\
&= \lim_{\varepsilon\to 0^+}\int_\Omega A_{kj}^{\gamma\beta}\big(\partial_k(E_L)_{\alpha\gamma}\big)(x-y)\big((\varphi u_\beta)(x)-(\varphi u_\beta)(y)\big)(\partial_j\theta_\varepsilon)(y)\,\mathrm{d}y \\
&\quad -\Big\{\lim_{\varepsilon\to 0^+}\int_\Omega A_{kj}^{\gamma\beta}\big(\partial_k(E_L)_{\alpha\gamma}\big)(x-y)(\partial_j\theta_\varepsilon)(y)\,\mathrm{d}y\Big\}(\varphi u_\beta)(x). \qquad (1.6.29)
\end{aligned}
$$

In relation to this, we note that since x is a Lebesgue point for u, we may employ (1.6.66) and (1.6.20) to estimate

$$
\begin{aligned}
&\limsup_{\varepsilon\to 0^+}\Big|\int_\Omega A_{kj}^{\gamma\beta}\big(\partial_k(E_L)_{\alpha\gamma}\big)(x-y)\big((\varphi u_\beta)(x)-(\varphi u_\beta)(y)\big)(\partial_j\theta_\varepsilon)(y)\,\mathrm{d}y\Big| \\
&\qquad\qquad \le C\limsup_{\varepsilon\to 0^+}\fint_{B(x,2\varepsilon)}\big|(\varphi u_\beta)(y)-(\varphi u_\beta)(x)\big|\,\mathrm{d}y = 0. \qquad (1.6.30)
\end{aligned}
$$

In addition, (1.4.33) eventually permits us to conclude that for each fixed index $\beta\in\{1,\dots,M\}$ we have

$$
\begin{aligned}
&\lim_{\varepsilon\to 0^+}\int_\Omega A_{kj}^{\gamma\beta}\big(\partial_k(E_L)_{\alpha\gamma}\big)(x-y)(\partial_j\theta_\varepsilon)(y)\,\mathrm{d}y \\
&\qquad = \lim_{\varepsilon\to 0^+}\int_\Omega A_{kj}^{\gamma\beta}\big(\partial_k(E_L)_{\alpha\gamma}\big)(x-y)\partial_j(\theta_\varepsilon-1)(y)\,\mathrm{d}y \\
&\qquad = -\lim_{\varepsilon\to 0^+}\,{}_{\mathcal{D}'(\Omega)}\Big\langle A_{kj}^{\gamma\beta}\partial_k\big[(E_L)_{\alpha\gamma}(x-\cdot)\big],\partial_j(\theta_\varepsilon-1)\Big\rangle_{\mathcal{D}(\Omega)} \\
&\qquad = \lim_{\varepsilon\to 0^+}\,{}_{\mathcal{D}'(\Omega)}\Big\langle A_{kj}^{\gamma\beta}\partial_j\partial_k\big[(E_L)_{\alpha\gamma}(x-\cdot)\big],\theta_\varepsilon-1\Big\rangle_{\mathcal{D}(\Omega)} \\
&\qquad = \delta_{\beta\alpha}\lim_{\varepsilon\to 0^+}\,{}_{\mathcal{D}'(\Omega)}\big\langle\delta_x,\theta_\varepsilon-1\big\rangle_{\mathcal{D}(\Omega)} \\
&\qquad = \delta_{\beta\alpha}\lim_{\varepsilon\to 0^+}(\theta_\varepsilon-1)(x) = -\delta_{\beta\alpha}. \qquad (1.6.31)
\end{aligned}
$$

Altogether,

$$
\mathrm{III} = \delta_{\beta\alpha}(\varphi u_\beta)(x) = \varphi(x)u_\alpha(x) = {}_{\mathcal{D}'(\Omega)}\big\langle u_\alpha(x)\delta_x,\varphi\big\rangle_{\mathcal{D}(\Omega)}. \qquad (1.6.32)
$$

From (1.6.22), (1.6.26), (1.6.27), and (1.6.32) we ultimately conclude that for each Lebesgue point $x\in\Omega$ for u such that (1.6.14) holds we have

$$
\operatorname{div}\vec{F}_x = u_\alpha(x)\delta_x + \big(\widetilde{E}^\top(x-\cdot)(Du)\big)_\alpha \text{ in } \mathcal{D}'(\Omega). \qquad (1.6.33)
$$

In particular,

$$
\operatorname{div}\vec{F}_x \in \mathscr{E}'(\Omega) + L^1(\Omega,\mathcal{L}^n). \qquad (1.6.34)
$$

Going further, $K := \overline{B\big(x, \frac{1}{2}\operatorname{dist}(x, \partial\Omega)\big)}$ is a compact subset of Ω, and based on (1.6.15), (1.4.24), and [112, Lemma 8.3.7] we may estimate

$$
\begin{aligned}
\big(\mathcal{N}_\kappa^{\Omega\setminus K}\vec{F}_x\big)(y) &\leq C\big(\mathcal{N}_\kappa u\big)(y) \cdot \sup_{z\in\Gamma_\kappa(y)\setminus K}\big[|x-z|^{1-n}\big] \\
&\leq C\frac{\big(\mathcal{N}_\kappa u\big)(y)}{|x-y|^{n-1}}, \qquad \forall y \in \partial\Omega,
\end{aligned}
\tag{1.6.35}
$$

for some constant $C = C(\Omega, D, \widetilde{D}, n, \kappa) \in (0, \infty)$. From (1.6.35), (1.6.2), and [112, (8.2.26)] it follows that

$$
\mathcal{N}_\kappa^{\Omega\setminus K}\vec{F}_x \in L^1(\partial\Omega, \sigma). \tag{1.6.36}
$$

Moreover, from (1.6.15), [112, (8.9.10)-(8.9.11)], and (1.6.13) we conclude that

$$
\vec{F}_x\Big|_{\partial\Omega}^{{}^{\kappa-\mathrm{n.t.}}} \text{ exists at } \sigma\text{-a.e. point on } \partial_{\mathrm{nta}}\Omega \tag{1.6.37}
$$

and, in fact, at σ-a.e. point $y \in \partial_*\Omega$ we have

$$
\begin{aligned}
\nu(y)\cdot\Big(\vec{F}_x\Big|_{\partial\Omega}^{{}^{\kappa-\mathrm{n.t.}}}\Big)(y) &= -\nu_j(y)A_{kj}^{\gamma\beta}\big(\partial_k(E_L)_{\alpha\gamma}\big)(x-y)\big(u_\beta\big|_{\partial\Omega}^{{}^{\kappa-\mathrm{n.t.}}}\big)(y) \\
&= -\widetilde{a}_k^{\gamma\mu}\big(\partial_k(E_L)_{\alpha\gamma}\big)(x-y)\nu_j(y)a_j^{\mu\beta}\big(u_\beta\big|_{\partial\Omega}^{{}^{\kappa-\mathrm{n.t.}}}\big)(y) \\
&= \big(\widetilde{D}^\top E_{L^\top}\big)_{\mu\alpha}(x-y)(-\mathrm{i})\Big(\mathrm{Sym}(D;\nu(y))\big(u\big|_{\partial\Omega}^{{}^{\kappa-\mathrm{n.t.}}}\big)(y)\Big)_\mu \\
&= \widetilde{E}_{\mu\alpha}(x-y)(-\mathrm{i})\Big(\mathrm{Sym}(D;\nu(y))\big(u\big|_{\partial\Omega}^{{}^{\kappa-\mathrm{n.t.}}}\big)(y)\Big)_\mu \\
&= \Big(\widetilde{E}^\top(x-y)(-\mathrm{i})\mathrm{Sym}(D;\nu(y))\big(u\big|_{\partial\Omega}^{{}^{\kappa-\mathrm{n.t.}}}\big)(y)\Big)_\alpha,
\end{aligned}
\tag{1.6.38}
$$

by (1.6.15), (1.6.10), (1.3.17).

Finally, when Ω is an exterior domain, it follows from (1.6.15) and (1.4.24) that

$$
\vec{F}_x(y) = O\big(|y|^{1-n}|u(y)|\big) \text{ as } |y| \to \infty \tag{1.6.39}
$$

which, in light of (1.6.4), implies that there exists $\lambda \in (1, \infty)$ such that

$$
\int_{B(0,\lambda R)\setminus B(0,R)} |y\cdot\vec{F}(y)|\,\mathrm{d}\mathcal{L}^n(y) = o(R^2) \text{ as } R \to \infty. \tag{1.6.40}
$$

Collectively, (1.6.16), (1.6.34), (1.6.37), and (1.6.40) ensure that, for each Lebesgue point $x \in \Omega$ for u such that (1.6.14) holds, the vector field $\vec{F}_x$ satisfies the hypotheses of [112, Theorem 1.4.1]. On account of [112, (4.6.19)], (1.6.33), and (1.6.38), the Divergence Formula [112, (1.4.6)] presently yields

$$
\begin{aligned}
&u_\alpha(x) + \int_\Omega \Big(\widetilde{E}^\top(x-y)(Du)(y)\Big)_\alpha \,\mathrm{d}y \\
&\quad = {}_{\mathscr{E}'(\Omega)}\big\langle u_\alpha(x)\delta_x, 1\big\rangle_{\mathscr{E}(\Omega)} + \int_\Omega \Big(\widetilde{E}^\top(x-y)(Du)(y)\Big)_\alpha \,\mathrm{d}y \\
&\quad = {}_{(\mathscr{C}_b^\infty(\Omega))^*}\Big(u_\alpha(x)\delta_x + \big(\widetilde{E}^\top(x-\cdot)(Du)\big)_\alpha, 1\Big)_{\mathscr{C}_b^\infty(\Omega)} \\
&\quad = {}_{(\mathscr{C}_b^\infty(\Omega))^*}\big(\mathrm{div}\vec{F}_x, 1\big)_{\mathscr{C}_b^\infty(\Omega)} = \int_{\partial_*\Omega} \nu\cdot\Big(\vec{F}_x\Big|_{\partial\Omega}^{{}^{\rm n.t.}}\Big)\,\mathrm{d}\sigma \\
&\quad = \int_{\partial_*\Omega} \Big(\widetilde{E}^\top(x-y)(-\mathrm{i})\mathrm{Sym}(D;\nu(y))\big(u\big|_{\partial\Omega}^{{}^{\rm n.t.}}\big)(y)\Big)_\alpha \,\mathrm{d}\sigma(y), \qquad (1.6.41)
\end{aligned}
$$

for each Lebesgue point $x \in \Omega$ of the function u with the property that (1.6.14) holds (in particular, for $\mathcal{L}^n$-a.e. point $x \in \Omega$), and each $\alpha \in \{1, \dots, M\}$. In view of the arbitrariness of $\alpha \in \{1, \dots, M\}$, this establishes (1.6.3) in the case when $x \in \Omega$.The version of (1.6.3) corresponding to $x \in \mathbb{R}^n \setminus \overline{\Omega}$ is proved similarly, the most notable difference being the fact that now III $= 0$, so in place of (1.6.33) we now have

$$
\mathrm{div}\vec{F}_x = \big(\widetilde{E}^\top(x-\cdot)(Du)\big)_\alpha \ \text{ in } \mathcal{D}'(\Omega). \qquad (1.6.42)
$$

The proof of Theorem 1.6.1 is therefore complete. □

As a consequence of Theorem 1.6.1, in our next corollary we elucidate the behavior at infinity for a null-solution of an injectively elliptic first-order system.

Corollary 1.6.2 *Fix $n \in \mathbb{N}$ with $n \geq 2$, and suppose $\Omega \subseteq \mathbb{R}^n$ is an exterior domain. Also, consider an injectively elliptic*[35] *homogeneous first-order $N\times M$ system D with constant complex coefficients in $\mathbb{R}^n$ (where $N, M \in \mathbb{N}$). Then for any given function $u \in \big[\mathscr{C}^\infty(\Omega)\big]^M$ satisfying $Du = 0$ in Ω the following conditions are equivalent:*

(1) For some, or any, $\lambda \in (1, \infty)$ one has

$$
⨍_{B(0,\lambda R)\setminus B(0,R)} |u|\,\mathrm{d}\mathcal{L}^n = o(1) \ \text{ as } R \to \infty. \qquad (1.6.43)
$$

(2) One has

$$
u(x) = o(1) \ \text{ as } |x| \to \infty. \qquad (1.6.44)
$$

(3) One has

$$
⨍_{\partial B(0,R)} |u|\,\mathrm{d}\mathcal{H}^{n-1} = o(1) \ \text{ as } R \to \infty. \qquad (1.6.45)
$$

(4) One has

$$
u(x) = O(|x|^{1-n}) \ \text{ as } |x| \to \infty. \qquad (1.6.46)
$$

[35] i.e., its principal symbol satisfies (1.3.18)

Moreover, if any of the conditions (1)-(4) are satisfied then

$$(\partial^\alpha u)(x) = O\big(|x|^{1-n-|\alpha|}\big) \text{ as } |x| \to \infty. \qquad \text{for each multi-index } \alpha \in \mathbb{N}_0^n. \tag{1.6.47}$$

In light of this result, it is natural to make the following definition.

Definition 1.6.3 *Given an exterior domain $\Omega \subseteq \mathbb{R}^n$, where $n \in \mathbb{N}$ with $n \geq 2$, along with an injectively elliptic $N \times M$ homogeneous first-order system D with constant complex coefficients in $\mathbb{R}^n$, a function $u \in \big[\mathscr{C}^\infty(\Omega)\big]^M$ satisfying $Du = 0$ in Ω is said to* `vanish at infinity` *provided any of the* (equivalent) *conditions (1)-(4) in Corollary 1.6.2 holds.*

Returning to the matter at hand, we now present the proof of Corollary 1.6.2.

Proof of Corollary 1.6.2 Lemma 1.5.6 shows that conditions *(1)-(3)* are equivalent, and that condition *(4)* implies (1.6.47). Trivially, *(4)* implies any of the conditions in *(1)-(3)*, so we are left with proving that *(1)-(3)* imply *(4)*. With this goal in mind, with $R_o \in (0, \infty)$ sufficiently large, introduce $\Omega_o := \mathbb{R}^n \setminus \overline{B(0, R_o)}$. Then $u \in \big[\mathscr{C}^\infty\big(\overline{\Omega_o}\big)\big]^M$ satisfies $Du = 0$ in Ω_o, and (1.6.44) ensures that u is bounded in Ω_o. Granted these qualities, we may write the integral representation formula (1.6.3) for u in Ω_o, which in turn guarantees, on account of (1.4.24), that (1.6.46) holds. This finishes the proof of the fact that *(1)-(4)* are equivalent. □

Recall from (1.6.6) that $\widetilde{E}$ from (1.6.1) is a fundamental solution for $D^\top$. While in general this does not guarantee that $\widetilde{E}^\top$ is a fundamental solution for $-D$, in the proposition below (which refines [92, Proposition 3]) we show that is the case whenever D and $\widetilde{D}$ commute.

Proposition 1.6.4 *Consider a homogeneous, first-order $N \times M$ system D with constant complex coefficients in $\mathbb{R}^n$ (where $N, M \in \mathbb{N}$) which is injectively elliptic (cf.* (1.3.18)), *and suppose $\widetilde{D}$ is a homogeneous first-order $M \times N$ system with constant complex coefficients in $\mathbb{R}^n$ which complements D (in the sense of Definition 1.3.6). In particular, $L := \widetilde{D}D$ is a weakly elliptic second-order $M \times M$ system in $\mathbb{R}^n$. In relation to this, make the assumption that*

$$\mathrm{Sym}(D;\xi)[L(\xi)]^{-1}\mathrm{Sym}(\widetilde{D};\xi) = I_{M\times M} \qquad \text{for each } \xi \in \mathbb{R}^n \setminus \{0\}. \tag{1.6.48}$$

Next, since $L^\top$ is also a weakly elliptic second-order $M \times M$ system in $\mathbb{R}^n$, it is meaningful to consider the matrix-valued fundamental solution $E_{L^\top}$ associated with $L^\top$ as in Theorem 1.4.2. Finally, with $\widetilde{D}^\top$ acting on the columns of $E_{L^\top}$ define[36]

$$\widetilde{E} := \widetilde{D}^\top E_{L^\top}. \tag{1.6.49}$$

[36] from (1.6.6) we know that $\widetilde{E}$ is a fundamental solution for $D^\top$

Then $\widetilde{E}^{\top}$ is a fundamental solution for $-D$, i.e.,

$$D(\widetilde{E}^{\top}) = -\delta I_{M\times M} \text{ as tempered distributions in } \mathbb{R}^n, \tag{1.6.50}$$

where δ is Dirac's distribution with mass at the origin in $\mathbb{R}^n$ and $I_{M\times M}$ is the $M\times M$ identity matrix.

As a corollary, (1.6.50) *holds whenever D commutes with $\widetilde{D}$, i.e., when*[37]

$$D\widetilde{D} = \widetilde{D}D \tag{1.6.51}$$

(*since this turns out to imply* (1.6.48)).

Proof The fact that $L = \widetilde{D}D$ is a weakly elliptic second-order $M \times M$ system in $\mathbb{R}^n$ implies (cf. (1.3.3)) that

$$L(\xi) \text{ is an invertible matrix for each } \xi \in \mathbb{R}^n \setminus \{0\}. \tag{1.6.52}$$

From (1.6.49) and the first formula in (1.4.32) we see that $\widetilde{E} = \widetilde{D}^{\top}(E_L)^{\top}$, so if "hat" denotes the Fourier transform then (1.4.31) gives

$$\widehat{\widetilde{E}} = \widehat{\widetilde{D}^{\top}(E_L)^{\top}} = \mathrm{Sym}(\widetilde{D}^{\top};\xi)\big[L(\xi)^{\top}\big]^{-1} \quad \text{as tempered distributions in } \mathbb{R}^n. \tag{1.6.53}$$

Thanks to (1.6.53) and [112, (1.7.17)], on the one hand, in the sense of tempered distributions in $\mathbb{R}^n$, we have

$$\begin{aligned} \widehat{D(\widetilde{E})^{\top}} &= \mathrm{Sym}(D;\xi)\widehat{(\widetilde{E})^{\top}} = \mathrm{Sym}(D;\xi)\big(\,\widehat{\widetilde{E}}\,\big)^{\top} \\ &= -\mathrm{Sym}(D;\xi)[L(\xi)]^{-1}\mathrm{Sym}(\widetilde{D};\xi). \end{aligned} \tag{1.6.54}$$

On the other hand, since $\mathrm{Sym}(D;\xi)[L(\xi)]^{-1}\mathrm{Sym}(\widetilde{D};\xi)$, regarded as a function defined for $\mathcal{L}^n$-a.e. $\xi \in \mathbb{R}^n$, is locally integrable and bounded, formula (1.6.48) may further be interpreted as saying that

$$\mathrm{Sym}(D;\xi)[L(\xi)]^{-1}\mathrm{Sym}(\widetilde{D};\xi) = I_{M\times M} \quad \text{as tempered distributions in } \mathbb{R}^n. \tag{1.6.55}$$

By combining (1.6.54) with (1.6.55) we then arrive at

$$\widehat{D(\widetilde{E})^{\top}} = -I_{M\times M} \text{ as tempered distributions in } \mathbb{R}^n, \tag{1.6.56}$$

from which (1.6.50) now follows.

Consider next the special case when D and $\widetilde{D}$ commute. Taking the Fourier transform in (1.6.51) then yields

[37] note that this forces $N = M$, i.e., D and $\widetilde{D}$ are square systems

$$\mathrm{Sym}(D;\xi)\,\mathrm{Sym}(\widetilde{D};\xi) = \mathrm{Sym}(\widetilde{D};\xi)\mathrm{Sym}(D;\xi), \qquad \forall \xi \in \mathbb{R}^n. \tag{1.6.57}$$

Also, in view of (1.3.2) and (A.0.118), the identity $L = \widetilde{D}D$ on the Fourier transform side reads

$$L(\xi) = \mathrm{Sym}(\widetilde{D};\xi)\mathrm{Sym}(D;\xi), \qquad \forall \xi \in \mathbb{R}^n. \tag{1.6.58}$$

On account of (1.6.52), (1.6.58), and (1.6.57) we then conclude that

$$\begin{gathered}\text{for each given vector } \xi \in \mathbb{R}^n \setminus \{0\}, \text{ the matrix } [L(\xi)]^{-1} \\ \text{commutes both with } \mathrm{Sym}(\widetilde{D};\xi) \text{ and with } \mathrm{Sym}(D;\xi).\end{gathered} \tag{1.6.59}$$

For each $\xi \in \mathbb{R}^n \setminus \{0\}$ we may then write

$$\begin{aligned}\mathrm{Sym}(D;\xi)[L(\xi)]^{-1}\mathrm{Sym}(\widetilde{D};\xi) &= \mathrm{Sym}(D;\xi)\mathrm{Sym}(\widetilde{D};\xi)[L(\xi)]^{-1} \\ &= \mathrm{Sym}(\widetilde{D};\xi)\mathrm{Sym}(D;\xi)[L(\xi)]^{-1} \\ &= L(\xi)[L(\xi)]^{-1} = I_{M\times M}.\end{aligned} \tag{1.6.60}$$

Hence (1.6.48) holds, so the desired conclusion follows. □

We next present a version of Theorem 1.6.1 phrased exclusively in terms of a given injectively elliptic first-order system D, i.e., without bringing into play an auxiliary system $\widetilde{D}$ which complements D, and without any reference to second-order weakly elliptic systems. Another significant feature is the fact that the present statement involves a rather generic fundamental solution $\widetilde{E}$ of $D^\top$, and not necessarily the specific fundamental solution constructed in (1.6.1) (we shall put this feature to good use shortly, in the very last portion of this section).

Theorem 1.6.5 *Let $\Omega \subseteq \mathbb{R}^n$, where $n \geq 2$, be an open set with a lower Ahlfors regular boundary, and such that $\sigma := \mathcal{H}^{n-1}\lfloor\partial\Omega$ is a doubling measure on $\partial\Omega$; in particular, Ω is a set of locally finite perimeter. Denote by ν the geometric measure theoretic outward unit normal to Ω, and fix an aperture parameter $\kappa \in (0,\infty)$. Finally, consider a homogeneous, first-order $N\times M$ system D with constant complex coefficients in $\mathbb{R}^n$ (where $N, M \in \mathbb{N}$) which is injectively elliptic (cf. (1.3.18)).*

Then the following statements are true. First, there exists a matrix-valued tempered distribution $\widetilde{E}$ satisfying

$$\begin{gathered}\widetilde{E} \in \big[\mathcal{S}'(\mathbb{R}^n)\big]^{N\times M}, \quad \widetilde{E}\big|_{\mathbb{R}^n\setminus\{0\}} \in \big[\mathscr{C}^\infty(\mathbb{R}^n\setminus\{0\})\big]^{N\times M}, \\ \widetilde{E} \textit{ is odd and positive homogeneous of degree } 1-n, \textit{ and} \\ D^\top\widetilde{E} = \delta I_{M\times M} \ (\textit{i.e., } \widetilde{E} \textit{ is a fundamental solution for } D^\top),\end{gathered} \tag{1.6.61}$$

where δ is Dirac's distribution with mass at origin in $\mathbb{R}^n$, and $I_{M\times M}$ is the $M\times M$ identity matrix.

Second, fix a matrix-valued tempered distribution $\widetilde{E}$ as in (1.6.61), and suppose $u : \Omega \to \mathbb{C}^M$ is a vector-valued function with Lebesgue measurable components satisfying

$$\mathcal{N}_\kappa u \in L^1\Big(\partial\Omega, \tfrac{\sigma(x)}{1+|x|^{n-1}}\Big),$$
$$\Big(u\big|_{\partial\Omega}^{\kappa-\mathrm{n.t.}}\Big)(x) \textit{ exists for } \sigma\textit{-a.e. } x \in \partial_{\mathrm{nta}}\Omega, \tag{1.6.62}$$
$$\textit{and } Du \in \Big[L^1\Big(\Omega, \frac{\mathrm{d}y}{1+|y|^{n-1}}\Big)\Big]^N,$$

(with Du considered in the sense of distributions in Ω). Then for any $\kappa' > 0$ the nontangential trace $u\big|_{\partial\Omega}^{\kappa'-\mathrm{n.t.}}$ also exists at σ-a.e. point on $\partial_{\mathrm{nta}}\Omega$ and is actually independent of κ'. Also, with the dependence on the parameter κ' dropped, for $\mathcal{L}^n$-a.e. $x \in \mathbb{R}^n \setminus \partial\Omega$ one has (with absolutely convergent integrals)

$$\int_{\partial_*\Omega} \widetilde{E}^\top(x-y)(-\mathrm{i})\mathrm{Sym}\big(D;\nu(y)\big)\big(u\big|_{\partial\Omega}^{\mathrm{n.t.}}\big)(y)\,\mathrm{d}\sigma(y)$$
$$-\int_{\Omega} \widetilde{E}^\top(x-y)(Du)(y)\,\mathrm{d}y = \begin{cases} u(x) & \textit{if } x \in \Omega, \\ 0 & \textit{if } x \in \mathbb{R}^n \setminus \overline{\Omega}, \end{cases} \tag{1.6.63}$$

if either Ω is bounded, or $\partial\Omega$ is unbounded. In the case when Ω is an exterior domain, the same conclusion holds true under the additional assumption that there exists $\lambda \in (1,\infty)$ such that

$$⨍_{B(0,\lambda R)\setminus B(0,R)} |u|\,\mathrm{d}\mathcal{L}^n = o(1) \textit{ as } R \to \infty. \tag{1.6.64}$$

Observe that if $Du = 0$ then (1.6.63) implies (under the same background assumptions as in the statement of Theorem 1.6.5) that for each $x \in \Omega$ we have

$$u(x) = \int_{\partial_*\Omega} \widetilde{E}^\top(x-y)(-\mathrm{i})\mathrm{Sym}\big(D;\nu(y)\big)\big(u\big|_{\partial\Omega}^{\mathrm{n.t.}}\big)(y)\,\mathrm{d}\sigma(y), \tag{1.6.65}$$

which may be interpreted as a "generalized Cauchy reproducing formula."

Theorem 1.6.5 may be regarded as a higher-dimensional generalization of the version of the Cauchy-Pompeiu's formula presented earlier in Theorem 1.1.1. Theorem 1.6.5 also contains as a special case the integral representation formulas from Theorem 1.2.2 in the context of Clifford algebras.

Proof of Theorem 1.6.5 The existence of a fundamental solution $\widetilde{E}$ for $D^\top$ as in (1.6.61) is seen from (1.6.1), (1.6.5), and (1.6.6). This takes care of the first claim in the statement. To deal with the second claim in the statement, fix such a matrix-valued tempered distribution $\widetilde{E}$. In particular,

$$|(\partial^\alpha \widetilde{E})(x)| \leq \frac{C_\alpha}{|x|^{n-1+|\alpha|}} \text{ for each } \alpha \in \mathbb{N}_0^n. \tag{1.6.66}$$

The first property in (1.6.62) and [112, Lemma 8.3.1] guarantee that

$$u \in \big[L^\infty_{\mathrm{loc}}(\Omega,\mathcal{L}^n)\big]^N, \tag{1.6.67}$$

while the first two properties in (1.6.62) together with [112, Corollary 8.9.9] imply that

$$\text{for any } \kappa' > 0 \text{ the nontangential trace } u\Big|_{\partial\Omega}^{\kappa'-\text{n.t.}} \text{ exists } \sigma\text{-a.e. on } \partial_{\text{nta}}\Omega \text{ and is actually independent of the parameter } \kappa' \in (0,\infty). \tag{1.6.68}$$

Let us establish (1.6.63). Recall from [112, Lemma 3.5.7] that the last membership in (1.6.62) is equivalent with having for $\mathcal{L}^n$-a.e. point $x \in \mathbb{R}^n$ the finiteness property

$$\int_\Omega \frac{|(Du)(y)|}{|x-y|^{n-1}}\,dy < +\infty. \tag{1.6.69}$$

To fix notation, we agree to write

$$\widetilde{E} = \big(\widetilde{E}_{\alpha\beta}\big)_{\substack{1\leq\alpha\leq N\\ 1\leq\beta\leq M}} \text{ and } D = \big(a_j^{\alpha\gamma}\partial_j\big)_{\substack{1\leq\alpha\leq N\\ 1\leq\gamma\leq M}} \tag{1.6.70}$$

with the summation convention over repeated indices assumed throughout. In particular,

$$D^\top = -\big(a_j^{\alpha\gamma}\partial_j\big)_{\substack{1\leq\gamma\leq M\\ 1\leq\alpha\leq N}}. \tag{1.6.71}$$

Next, select an arbitrary index $\alpha \in \{1,\dots,M\}$, pick a Lebesgue point $x \in \Omega$ for the function u with the property that (1.6.69) holds, and consider the vector field $\vec{F}_x = \big(F_j\big)_{1\leq j\leq n}$ with components

$$F_j := a_j^{\gamma\beta}\widetilde{E}_{\gamma\alpha}(x-\cdot)u_\beta \text{ at } \mathcal{L}^n\text{-a.e. point in } \Omega. \tag{1.6.72}$$

Together, (1.6.72), (1.6.67), and Theorem 1.4.2 ensure that

$$\vec{F}_x \in \big[L^1_{\text{loc}}(\Omega,\mathcal{L}^n)\big]^n. \tag{1.6.73}$$

Let us now compute $\text{div}\vec{F}_x$ in the sense of distributions in Ω. With this goal in mind, fix an arbitrary scalar-valued function $\varphi \in \mathscr{C}_c^\infty(\Omega)$ and write

$$\begin{aligned} {}_{\mathcal{D}'(\Omega)}\big\langle \text{div}\vec{F}_x, \varphi\big\rangle_{\mathcal{D}(\Omega)} &= -\int_\Omega F_j(y)(\partial_j\varphi)(y)\,dy \\ &= -\int_\Omega a_j^{\gamma\beta}\widetilde{E}_{\gamma\alpha}(x-y)u_\beta(y)(\partial_j\varphi)(y)\,dy. \end{aligned} \tag{1.6.74}$$

Consider an auxiliary scalar-valued function $\theta \in \mathscr{C}^\infty(\mathbb{R}^n)$ with the property that $\theta = 0$ on $B(0,1)$ and $\theta = 1$ on $\mathbb{R}^n \setminus B(0,2)$. For each $\varepsilon \in \big(0, \frac{1}{2}\,\text{dist}(x,\partial\Omega)\big)$ define $\theta_\varepsilon : \mathbb{R}^n \to \mathbb{R}$ as follows:

$$\theta_\varepsilon(y) := \theta\Big(\frac{y-x}{\varepsilon}\Big) \text{ for every } y \in \mathbb{R}^n. \tag{1.6.75}$$

It follows that

$$\begin{aligned}&\theta_\varepsilon \in \mathscr{C}^\infty(\mathbb{R}^n) \text{ is a bounded function satisfying}\\&\lim_{\varepsilon\to 0^+} \theta_\varepsilon(y) = 1 \text{ for every point } y \in \mathbb{R}^n \setminus \{x\},\end{aligned} \tag{1.6.76}$$

and there exists a constant $C \in (0, \infty)$ such that for each $\varepsilon \in \big(0, \frac{1}{2}\operatorname{dist}(x, \partial\Omega)\big)$ we have

$$\begin{aligned}&1 - \theta_\varepsilon \in \mathscr{C}_c^\infty(\Omega), \quad \theta_\varepsilon \equiv 0 \text{ on } B(x, \varepsilon),\\&\operatorname{supp}(\nabla\theta_\varepsilon) \subseteq \overline{B(x, 2\varepsilon)} \setminus B(x, \varepsilon), \text{ and}\\&|(\nabla^\ell\theta_\varepsilon)(y)| \leq C\varepsilon^{-\ell} \text{ for every } \ell \in \mathbb{N}_0 \text{ and } y \in \mathbb{R}^n.\end{aligned} \tag{1.6.77}$$

As a consequence,

the entries of $E_L(x - \cdot)\theta_\varepsilon$ belong to $\mathscr{C}^\infty(\Omega)$ and the entries of $E_L(x - \cdot)\varphi\theta_\varepsilon$ belong to $\mathscr{C}_c^\infty(\Omega)$; moreover, similar properties are valid for expressions involving partial derivatives acting on either of the intervening functions. (1.6.78)

Making use of (1.6.67), Lebesgue's Dominated Convergence Theorem, and integrations by parts (keeping in mind (1.5.26)) we may write

$$\begin{aligned}{}_{\mathcal{D}'(\Omega)}\big\langle \operatorname{div}\vec{F}_x, \varphi\big\rangle_{\mathcal{D}(\Omega)} &= -\lim_{\varepsilon\to 0^+}\int_\Omega a_j^{\gamma\beta}\widetilde{E}_{\gamma\alpha}(x-y)u_\beta(y)(\partial_j\varphi)(y)\theta_\varepsilon(y)\,dy\\&= -\lim_{\varepsilon\to 0^+}{}_{\mathcal{D}'(\Omega)}\Big\langle a_j^{\gamma\beta}u_\beta, \widetilde{E}_{\gamma\alpha}(x-\cdot)(\partial_j\varphi)\theta_\varepsilon\Big\rangle_{\mathcal{D}(\Omega)}\\&= \mathrm{I} + \mathrm{II} + \mathrm{III},\end{aligned} \tag{1.6.79}$$

where

$$\mathrm{I} := \lim_{\varepsilon\to 0^+}{}_{\mathcal{D}'(\Omega)}\Big\langle a_j^{\gamma\beta}\partial_j u_\beta, \widetilde{E}_{\gamma\alpha}(x-\cdot)(x-\cdot)\varphi\theta_\varepsilon\Big\rangle_{\mathcal{D}(\Omega)}, \tag{1.6.80}$$

$$\mathrm{II} := -\lim_{\varepsilon\to 0^+}{}_{\mathcal{D}'(\Omega)}\Big\langle a_j^{\gamma\beta}u_\beta, \big(\partial_j\widetilde{E}_{\gamma\alpha}\big)(x-\cdot)\varphi\theta_\varepsilon\Big\rangle_{\mathcal{D}(\Omega)}, \tag{1.6.81}$$

$$\mathrm{III} := \lim_{\varepsilon\to 0^+}{}_{\mathcal{D}'(\Omega)}\Big\langle a_j^{\gamma\beta}u_\beta, \widetilde{E}_{\gamma\alpha}(x-y)\varphi(\partial_j\theta_\varepsilon)\Big\rangle_{\mathcal{D}(\Omega)}. \tag{1.6.82}$$

Observe that

$$
\begin{aligned}
\mathrm{I} &= \lim_{\varepsilon\to 0^+} {}_{\mathcal{D}'(\Omega)}\Big\langle (Du)_\gamma, \widetilde{E}_{\gamma\alpha}(x-\cdot)\varphi\theta_\varepsilon\Big\rangle_{\mathcal{D}(\Omega)} \\
&= \lim_{\varepsilon\to 0^+} \int_\Omega \big(\widetilde{E}^\top(x-y)(Du)(y)\big)_\alpha \varphi(y)\theta_\varepsilon(y)\,\mathrm{d}y \\
&= \int_\Omega \big(\widetilde{E}^\top(x-y)(Du)(y)\big)_\alpha \varphi(y)\,\mathrm{d}y \\
&= {}_{\mathcal{D}'(\Omega)}\Big\langle \big(\widetilde{E}^\top(x-\cdot)(Du)\big)_\alpha, \varphi\Big\rangle_{\mathcal{D}(\Omega)},
\end{aligned}
\tag{1.6.83}
$$

thanks to (1.6.70), (1.6.69), and Lebesgue's Dominated Convergence Theorem (which takes into account (1.6.76)). Next,

$$
\mathrm{II} = -\lim_{\varepsilon\to 0^+} {}_{\mathcal{D}'(\Omega)}\Big\langle u_\beta, a_j^{\gamma\beta}\big(\partial_j\widetilde{E}_{\gamma\alpha}\big)(x-\cdot)\varphi\theta_\varepsilon\Big\rangle_{\mathcal{D}(\Omega)} = 0 \tag{1.6.84}
$$

since, according to (1.6.70) and the last line in (1.6.61), for each $\beta\in\{1,\dots,M\}$ we have

$$
a_j^{\gamma\beta}\big(\partial_j\widetilde{E}_{\gamma\alpha}\big)(x-y) = -(D^\top\widetilde{E})_{\beta\alpha}(x-y) = 0 \text{ for each } y\in\mathbb{R}^n\setminus\{x\}. \tag{1.6.85}
$$

Concerning the contribution from the very last term in (1.6.79), we write

$$
\begin{aligned}
\mathrm{III} &= \lim_{\varepsilon\to 0^+}\int_\Omega a_j^{\gamma\beta}\widetilde{E}_{\gamma\alpha}(x-y)u_\beta(y)\varphi(y)(\partial_j\theta_\varepsilon)(y)\,\mathrm{d}y \\
&= \lim_{\varepsilon\to 0^+}\int_\Omega a_j^{\gamma\beta}\widetilde{E}_{\gamma\alpha}(x-y)\big((\varphi u_\beta)(y)-(\varphi u_\beta)(x)\big)(\partial_j\theta_\varepsilon)(y)\,\mathrm{d}y \\
&\quad + \Big\{\lim_{\varepsilon\to 0^+}\int_\Omega a_j^{\gamma\beta}\widetilde{E}_{\gamma\alpha}(x-y)(\partial_j\theta_\varepsilon)(y)\,\mathrm{d}y\Big\}(\varphi u_\beta)(x).
\end{aligned}
\tag{1.6.86}
$$

In relation to this, we note that since x is a Lebesgue point for u, we may estimate (using (1.4.24) and (1.6.77))

$$
\begin{aligned}
&\limsup_{\varepsilon\to 0^+}\Big|\int_\Omega a_j^{\gamma\beta}\widetilde{E}_{\gamma\alpha}(x-y)\big((\varphi u_\beta)(x)-(\varphi u_\beta)(y)\big)(\partial_j\theta_\varepsilon)(y)\,\mathrm{d}y\Big| \\
&\qquad \le C\limsup_{\varepsilon\to 0^+} ⨍_{B(x,2\varepsilon)}\big|(\varphi u_\beta)(y)-(\varphi u_\beta)(x)\big|\,\mathrm{d}y = 0.
\end{aligned}
\tag{1.6.87}
$$

Also, (1.6.71) together with the property recorded in last line in (1.6.61) allow us to conclude that for each fixed index $\beta\in\{1,\dots,M\}$ we have

$$
\begin{aligned}
\lim_{\varepsilon\to 0^+}\int_{\Omega} a_j^{\gamma\beta}\widetilde{E}_{\gamma\alpha}(x-y)(\partial_j\theta_\varepsilon)(y)\,dy
&= \lim_{\varepsilon\to 0^+}\int_{\Omega} a_j^{\gamma\beta}\widetilde{E}_{\gamma\alpha}(x-y)\partial_j(\theta_\varepsilon-1)(y)\,dy \\
&= \lim_{\varepsilon\to 0^+} {}_{\mathcal{D}'(\Omega)}\Big\langle a_j^{\gamma\beta}\partial_j\big[\widetilde{E}_{\gamma\alpha}(x-\cdot)\big],\theta_\varepsilon-1\Big\rangle_{\mathcal{D}(\Omega)} \\
&= -\delta_{\beta\alpha}\lim_{\varepsilon\to 0^+} {}_{\mathcal{D}'(\Omega)}\big\langle \delta_x,\theta_\varepsilon-1\big\rangle_{\mathcal{D}(\Omega)} \\
&= -\delta_{\beta\alpha}\lim_{\varepsilon\to 0^+}(\theta_\varepsilon-1)(x)=\delta_{\beta\alpha}.
\end{aligned}
\tag{1.6.88}
$$

We may therefore conclude that

$$
\mathrm{III}=\delta_{\beta\alpha}(\varphi u_\beta)(x)=\varphi(x)u_\alpha(x)={}_{\mathcal{D}'(\Omega)}\big\langle u_\alpha(x)\delta_x,\varphi\big\rangle_{\mathcal{D}(\Omega)}. \tag{1.6.89}
$$

Based on (1.6.79), (1.6.83), (1.6.84), and (1.6.89) we ultimately see that for each Lebesgue point $x\in\Omega$ for u such that (1.6.69) holds we have

$$
\mathrm{div}\vec{F}_x=u_\alpha(x)\delta_x+\big(\widetilde{E}^{\top}(x-\cdot)(Du)\big)_\alpha \ \text{ in } \mathcal{D}'(\Omega). \tag{1.6.90}
$$

In particular,

$$
\mathrm{div}\vec{F}_x\in\mathscr{E}'(\Omega)+L^1(\Omega,\mathcal{L}^n). \tag{1.6.91}
$$

Pressing on, bring in $K:=\overline{B\big(x,\tfrac12\,\mathrm{dist}(x,\partial\Omega)\big)}$, which is a compact subset of Ω. Based on (1.6.72), (1.6.61), (1.6.66), and [112, Lemma 8.3.7] we may then estimate

$$
\begin{aligned}
\big(\mathcal{N}_\kappa^{\Omega\setminus K}\vec{F}_x\big)(y) &\le C\big(\mathcal{N}_\kappa u\big)(y)\cdot\sup_{z\in\Gamma_\kappa(y)\setminus K}\big[|x-z|^{1-n}\big] \\
&\le C\frac{\big(\mathcal{N}_\kappa u\big)(y)}{|x-y|^{n-1}},\qquad \forall\,y\in\partial\Omega,
\end{aligned}
\tag{1.6.92}
$$

for some constant $C=C(\Omega,D,n,\kappa)\in(0,\infty)$. From (1.6.92), (1.6.62), and [112, (8.2.26)] it follows that

$$
\mathcal{N}_\kappa^{\Omega\setminus K}\vec{F}_x\in L^1(\partial\Omega,\sigma). \tag{1.6.93}
$$

In addition, from (1.6.72), [112, (8.9.10)-(8.9.11)], and (1.6.68) we conclude that

$$
\vec{F}_x\Big|_{\partial\Omega}^{\mathrm{n.t.}} \text{ exists at } \sigma\text{-a.e. point on } \partial_{\mathrm{nta}}\Omega \tag{1.6.94}
$$

and, in fact, at σ-a.e. point $y\in\partial_*\Omega$ we have

$$
\begin{aligned}
\nu(y)\cdot\Big(\vec{F}_x\Big|^{\text{n.t.}}_{\partial\Omega}\Big)(y) &= \nu_j(y)a_j^{\gamma\beta}\widetilde{E}_{\gamma\alpha}(x-y)\big(u_\beta\big|^{\text{n.t.}}_{\partial\Omega}\big)(y)\\
&= \widetilde{E}_{\gamma\alpha}(x-y)(-\mathrm{i})\Big(\mathrm{Sym}(D;\nu(y))\big(u\big|^{\text{n.t.}}_{\partial\Omega}\big)(y)\Big)_\gamma\\
&= \Big(\widetilde{E}^\top(x-y)(-\mathrm{i})\mathrm{Sym}(D;\nu(y))\big(u\big|^{\text{n.t.}}_{\partial\Omega}\big)(y)\Big)_\alpha, \qquad (1.6.95)
\end{aligned}
$$

by (1.6.72) and (1.3.17).

Lastly, when Ω is an exterior domain, it follows from (1.6.72) and (1.6.66) that

$$
\vec{F}_x(y) = O\big(|y|^{1-n}|u(y)|\big) \text{ as } |y|\to\infty \qquad (1.6.96)
$$

which, in view of (1.6.64), implies that there exists $\lambda\in(1,\infty)$ such that

$$
\int_{B(0,\lambda R)\setminus B(0,R)} |y\cdot\vec{F}(y)|\,\mathrm{d}\mathcal{L}^n(y) = o(R^2) \text{ as } R\to\infty. \qquad (1.6.97)
$$

Together, (1.6.73), (1.6.91), (1.6.94), and (1.6.97) ensure that, for each Lebesgue point $x\in\Omega$ for u such that (1.6.69) holds, the vector field $\vec{F}_x$ satisfies the hypotheses of [112, Theorem 1.4.1]. On account of [112, (4.6.19)], (1.6.90), and (1.6.95), the Divergence Formula [112, (1.4.6)] presently yields

$$
\begin{aligned}
u_\alpha(x)+&\int_\Omega \Big(\widetilde{E}^\top(x-y)(Du)(y)\Big)_\alpha\,\mathrm{d}y\\
&= {}_{\mathscr{E}'(\Omega)}\big\langle u_\alpha(x)\delta_x, 1\big\rangle_{\mathscr{E}(\Omega)} + \int_\Omega \Big(\widetilde{E}^\top(x-y)(Du)(y)\Big)_\alpha\,\mathrm{d}y\\
&= {}_{(\mathscr{C}_b^\infty(\Omega))^*}\Big(u_\alpha(x)\delta_x + \big(\widetilde{E}^\top(x-\cdot)(Du)\big)_\alpha, 1\Big)_{\mathscr{C}_b^\infty(\Omega)}\\
&= {}_{(\mathscr{C}_b^\infty(\Omega))^*}\big(\mathrm{div}\vec{F}_x, 1\big)_{\mathscr{C}_b^\infty(\Omega)} = \int_{\partial_*\Omega} \nu\cdot\big(\vec{F}_x\big|^{\text{n.t.}}_{\partial\Omega}\big)\,\mathrm{d}\sigma\\
&= \int_{\partial_*\Omega}\Big(\widetilde{E}^\top(x-y)(-\mathrm{i})\mathrm{Sym}(D;\nu(y))\big(u\big|^{\text{n.t.}}_{\partial\Omega}\big)(y)\Big)_\alpha\,\mathrm{d}\sigma(y), \qquad (1.6.98)
\end{aligned}
$$

for each Lebesgue point $x\in\Omega$ of the function u with the property that (1.6.69) holds (in particular, for $\mathcal{L}^n$-a.e. point $x\in\Omega$), and each $\alpha\in\{1,\dots,M\}$. Bearing in mind that $\alpha\in\{1,\dots,M\}$ is arbitrary, this establishes (1.6.63) in the case when $x\in\Omega$. The version of (1.6.63) corresponding to $x\in\mathbb{R}^n\setminus\overline{\Omega}$ is proved similarly, the most notable difference being the fact that now III $=0$, so in place of (1.6.90) we now have

$$
\mathrm{div}\vec{F}_x = \big(\widetilde{E}^\top(x-\cdot)(Du)\big)_\alpha \text{ in } \mathcal{D}'(\Omega). \qquad (1.6.99)
$$

This concludes the proof of Theorem 1.6.5. □

Let us take another look at the first-order differential operator $\mathbb{D}_A$ from Example 1.3.12. The broader context involved a weakly elliptic, second-order, homogeneous, constant complex coefficient $M \times M$ system L, associated with some coefficient tensor $A = \big(a_{rs}^{\alpha\beta}\big)_{\substack{1\leq r,s\leq n\\ 1\leq\alpha,\beta\leq M}}$ as in (1.3.15). Introduce

$$\widetilde{M} := n \times M \text{ and } \widetilde{N} := (n^2 \times M) + M. \tag{1.6.100}$$

Recall from the definition given in (1.3.39) that $\mathbb{D}_A$ is a homogeneous constant coefficient first-order $\widetilde{N} \times \widetilde{M}$ system, which maps distributions in $\mathbb{R}^n$ with $n \times M$ components $w = \big(w_s^\beta\big)_{\substack{1\leq s\leq n\\ 1\leq\beta\leq M}} \in \big[\mathcal{D}'(\mathbb{R}^n)\big]^{n\times M}$ into

$$\mathbb{D}_A w := \begin{pmatrix} \big(\partial_r w_s^\beta - \partial_s w_r^\beta\big)_{\substack{1\leq r,s\leq n\\ 1\leq\beta\leq M}} \\ \big(a_{rs}^{\alpha\beta}\partial_r w_s^\beta\big)_{1\leq\alpha\leq M} \end{pmatrix}. \tag{1.6.101}$$

It has been noted in (1.3.40) that the first-order system $\mathbb{D}_A$ is injectively elliptic. Moreover, a direct computation based on (1.6.101) shows that the transpose of $\mathbb{D}_A$ is the $\widetilde{M} \times \widetilde{N}$ first-order system acting on distributions in $\mathbb{R}^n$ according to the formula

$$\mathbb{D}_A^\top \begin{pmatrix} \big(U_{rs}^\beta\big)_{\substack{1\leq r,s\leq n\\ 1\leq\beta\leq M}} \\ \big(u_\alpha\big)_{1\leq\alpha\leq M} \end{pmatrix} = \Big(\partial_r U_{sr}^\beta - \partial_r U_{rs}^\beta - a_{rs}^{\alpha\beta}\partial_r u_\alpha\Big)_{\substack{1\leq s\leq n\\ 1\leq\beta\leq M}}. \tag{1.6.102}$$

If we denote by $E = (E_{\gamma\beta})_{1\leq\gamma,\beta\leq M}$ the matrix-valued fundamental solution associated with the second-order weakly elliptic system L as in Theorem 1.4.2, then for each $\ell \in \{1,\dots,n\}$ and $\gamma \in \{1,\dots,M\}$ we may rely on (1.6.102) and (1.4.33) to compute

$$\mathbb{D}_A^\top \begin{pmatrix} \big(a_{js}^{\alpha\beta}\partial_j E_{\gamma\alpha}\big)_{\substack{1\leq r,s\leq n\\ 1\leq\beta\leq M}} \\ \big(-\partial_\ell E_{\gamma\alpha}\big)_{1\leq\alpha\leq M} \end{pmatrix} = \big(\delta\delta_{\beta\gamma}\delta_{\ell s}\big)_{\substack{1\leq s\leq n\\ 1\leq\beta\leq M}} \tag{1.6.103}$$

where δ is Dirac's distribution with mass at origin in $\mathbb{R}^n$. This shows that if we denote by $\widetilde{E}$ the $\widetilde{N} \times \widetilde{M}$ matrix-valued tempered distribution in $\mathbb{R}^n$ whose (ℓ,γ)-th column is given by

$$\begin{pmatrix} \big(a_{js}^{\alpha\beta}\partial_j E_{\gamma\alpha}\big)_{\substack{1\leq r,s\leq n\\ 1\leq\beta\leq M}} \\ \big(-\partial_\ell E_{\gamma\alpha}\big)_{1\leq\alpha\leq M} \end{pmatrix} \tag{1.6.104}$$

for each $\ell \in \{1,\dots,n\}$ and $\gamma \in \{1,\dots,M\}$ then $\widetilde{E}$ is a fundamental solution for $\mathbb{D}_A^\top$, in the sense that

$$\mathbb{D}_A^\top \widetilde{E} = \delta I_{\widetilde{M}\times\widetilde{M}} \tag{1.6.105}$$

in the sense of distributions in $\mathbb{R}^n$, with $\mathbb{D}_A^\top$ acting on the columns of $\widetilde{E}$. Based on this and Theorem 1.4.2 we then see that $\widetilde{E}$ satisfies all properties in (1.6.61) relative to the injectively elliptic first-order $\widetilde{N} \times \widetilde{M}$ system $D := \mathbb{D}_A$.

Granted this observation, Theorem 1.6.5 applies and gives the following result (with Ω as in the statement of Theorem 1.6.5): for each vector-valued function

$$w = \left(w_s^\beta\right)_{\substack{1\le s\le n\\ 1\le\beta\le M}} : \Omega \longrightarrow \mathbb{C}^{\widetilde{M}} \tag{1.6.106}$$

with Lebesgue measurable components satisfying

$$\begin{aligned} &\mathcal{N}_\kappa w \in L^1\Big(\partial\Omega, \tfrac{\sigma(x)}{1+|x|^{n-1}}\Big),\\ &\Big(w\big|_{\partial\Omega}^{\kappa-\mathrm{n.t.}}\Big)(x) \text{ exists for } \sigma\text{-a.e. } x \in \partial_{\mathrm{nta}}\Omega,\\ &\text{and } \mathbb{D}_A w \in \Big[L^1\Big(\Omega, \frac{dy}{1+|y|^{n-1}}\Big)\Big]^{\widetilde{N}}, \end{aligned} \tag{1.6.107}$$

at $\mathcal{L}^n$-a.e. point $x \in \mathbb{R}^n \setminus \partial\Omega$ one has

$$\begin{aligned} &\int_{\partial_*\Omega} \widetilde{E}^\top(x-y)(-\mathrm{i})\mathrm{Sym}\big(\mathbb{D}_A;\nu(y)\big)\big(w\big|_{\partial\Omega}^{\kappa-\mathrm{n.t.}}\big)(y)\,d\sigma(y)\\ &\quad - \int_\Omega \widetilde{E}^\top(x-y)(\mathbb{D}_A w)(y)\,dy = \begin{cases} w(x) & \text{if } x\in\Omega,\\ 0 & \text{if } x\in\mathbb{R}^n\setminus\overline{\Omega}, \end{cases} \end{aligned} \tag{1.6.108}$$

if either Ω is bounded, or $\partial\Omega$ is unbounded. In the case when Ω is an exterior domain, the same conclusion holds true under the additional assumption that there exists $\lambda \in (1,\infty)$ such that

$$⨏_{B(0,\lambda R)\setminus B(0,R)} |w|\,d\mathcal{L}^n = o(1) \text{ as } R \to \infty. \tag{1.6.109}$$

We can formally state the result just established above as follows:

Theorem 1.6.6 *Let $\Omega \subseteq \mathbb{R}^n$, where $n \ge 2$, be an open set with a lower Ahlfors regular boundary, and with the property that $\sigma := \mathcal{H}^{n-1}\lfloor\partial\Omega$ is a doubling measure on $\partial\Omega$; in particular, Ω is a set of locally finite perimeter. Denote by $\nu = (\nu_1,\dots,\nu_n)$ the geometric measure theoretic outward unit normal to Ω. With the summation convention over repeated indices understood throughout, let*

$$L = \left(a_{rs}^{\alpha\beta}\partial_r\partial_s\right)_{1\le\alpha,\beta\le M} \tag{1.6.110}$$

be a homogeneous, weakly elliptic, second-order $M \times M$ system in $\mathbb{R}^n$, with complex constant coefficients, and denote by $E = (E_{\gamma\beta})_{1\le\gamma,\beta\le M}$ the matrix-valued fundamental solution associated with L as in Theorem 1.4.2. In this setting, assume $w = (w_s^\beta)_{\substack{1\le s\le n\\ 1\le\beta\le M}} \in \big[L^1_{\mathrm{loc}}(\Omega,\mathcal{L}^1)\big]^{n\cdot M}$ is a vector-valued function which, for some $\kappa > 0$, satisfies

$$
\begin{aligned}
&w\Big|_{\partial\Omega}^{\kappa-\mathrm{n.t.}} \text{ exists at } \sigma\text{-a.e. point on } \partial_{\mathrm{nta}}\Omega, \\
&\text{and } \int_{\partial\Omega} \frac{(\mathcal{N}_\kappa w)(y)}{1+|y|^{n-1}}\, d\sigma(y) < +\infty.
\end{aligned}
\tag{1.6.111}
$$

With all partial derivatives considered in the sense of distributions in Ω, also suppose that

$$
\begin{aligned}
&(\partial_r w_s^\beta - \partial_s w_r^\beta) \in L^1\Big(\Omega, \tfrac{dy}{1+|y|^{n-1}}\Big) \text{ for all} \\
&\beta \in \{1,\dots,M\},\ r,s \in \{1,\dots,n\}, \text{ and} \\
&a_{rs}^{\alpha\beta}\partial_r w_s^\beta \in L^1\Big(\Omega, \tfrac{dy}{1+|y|^{n-1}}\Big) \text{ for each } \alpha \in \{1,\dots,M\}.
\end{aligned}
\tag{1.6.112}
$$

Then for any other aperture parameter $\kappa' > 0$ the nontangential trace $w\Big|_{\partial\Omega}^{\kappa'-\mathrm{n.t.}}$ also exists σ-a.e. on $\partial_{\mathrm{nta}}\Omega$ and is actually independent of κ'. Moreover, with the dependence on the parameter κ' dropped and having fixed $\ell \in \{1,\dots,n\}$ along with $\gamma \in \{1,\dots,M\}$ arbitrary, it follows that for $\mathcal{L}^n$-a.e. point $x \in \Omega$ one has (with absolutely convergent integrals)

$$
\begin{aligned}
w_\ell^\gamma(x) = &\int_{\partial_*\Omega} a_{js}^{\alpha\beta}(\partial_j E_{\gamma\alpha})(x-y)\Big\{\nu_\ell(y)\big(w_s^\beta\big|_{\partial\Omega}^{\mathrm{n.t.}}\big)(y) - \nu_s(y)\big(w_\ell^\beta\big|_{\partial\Omega}^{\mathrm{n.t.}}\big)(y)\Big\}\, d\sigma(y) \\
&- \int_{\partial_*\Omega} (\partial_\ell E_{\gamma\alpha})(x-y)\nu_r(y) a_{rs}^{\alpha\beta}\big(w_s^\beta\big|_{\partial\Omega}^{\mathrm{n.t.}}\big)(y)\, d\sigma(y) \\
&- \int_\Omega a_{js}^{\alpha\beta}(\partial_j E_{\gamma\alpha})(x-y)\Big\{(\partial_\ell w_s^\beta)(y) - (\partial_s w_\ell^\beta)(y)\Big\}\, dy \\
&+ \int_\Omega (\partial_\ell E_{\gamma\alpha})(x-y) a_{rs}^{\alpha\beta}(\partial_r w_s^\beta)(y)\, dy
\end{aligned}
\tag{1.6.113}
$$

if either Ω is bounded, or $\partial\Omega$ is unbounded. In the case when Ω is an exterior domain, the same conclusion holds true under the additional assumption that there exists $\lambda \in (1,\infty)$ such that

$$
⨍_{B(0,\lambda R)\setminus B(0,R)} |w|\, d\mathcal{L}^n = o(1) \text{ as } R \to \infty. \tag{1.6.114}
$$

In addition, for each $x \in \mathbb{R}^n \setminus \overline{\Omega}$ one has

$$
\begin{aligned}
0 = &\int_{\partial_*\Omega} a_{js}^{\alpha\beta}(\partial_j E_{\gamma\alpha})(x-y)\Big\{\nu_\ell(y)\big(w_s^\beta\big|_{\partial\Omega}^{\text{n.t.}}\big)(y) - \nu_s(y)\big(w_\ell^\beta\big|_{\partial\Omega}^{\text{n.t.}}\big)(y)\Big\}\,\mathrm{d}\sigma(y)\\
&- \int_{\partial_*\Omega} (\partial_\ell E_{\gamma\alpha})(x-y)\nu_r(y)a_{rs}^{\alpha\beta}\big(w_s^\beta\big|_{\partial\Omega}^{\text{n.t.}}\big)(y)\,\mathrm{d}\sigma(y)\\
&- \int_{\Omega} a_{js}^{\alpha\beta}(\partial_j E_{\gamma\alpha})(x-y)\Big\{(\partial_\ell w_s^\beta)(y) - (\partial_s w_\ell^\beta)(y)\Big\}\,\mathrm{d}y\\
&+ \int_{\Omega} (\partial_\ell E_{\gamma\alpha})(x-y)a_{rs}^{\alpha\beta}(\partial_r w_s^\beta)(y)\,\mathrm{d}y
\end{aligned}
\tag{1.6.115}
$$

with the same decay condition as before in the case when Ω is an exterior domain.

Proof This is a direct consequence of (1.6.108), bearing in mind (1.6.101) and the fact that the (ℓ,γ)-th column of $\widetilde{E}$ is described in (1.6.104). □

We make three remarks pertaining to the above theorem. First, Theorem 1.6.6 subsumes our earlier result from Theorem 1.5.10. More specifically, formulas (1.6.113) (1.6.115) from Theorem 1.6.6 applied to

$$
w := \nabla u = \big(\partial_s u_\beta\big)_{\substack{1\le s\le n,\\ 1\le\beta\le M}} \tag{1.6.116}
$$

the Jacobian of a function u as in the statement of Theorem 1.5.10 reduce precisely to (1.5.230) and (1.5.232), respectively (in view of the fact that in this special case we have $\partial_r w_s^\beta - \partial_s w_r^\beta = 0$ for each $\beta \in \{1,\dots,M\}$ and each $r,s \in \{1,\dots,n\}$).

Second, we wish to note that a direct proof of Theorem 1.6.6 may be given by applying the Divergence Formula [112, (1.4.6)] to the vector field $\vec{F}_x = \big(F_k\big)_{1\le k\le n}$ with components given at $\mathcal{L}^n$-a.e. point in Ω by

$$
\begin{aligned}
F_k := {}& a_{js}^{\alpha\beta}(\partial_j E_{\gamma\alpha})(x-\cdot)\delta_{k\ell}w_s^\beta - a_{js}^{\alpha\beta}(\partial_j E_{\gamma\alpha})(x-\cdot)\delta_{ks}w_\ell^\beta\\
&- a_{rs}^{\alpha\beta}(\partial_\ell E_{\gamma\alpha})(x-\cdot)\delta_{kr}w_s^\beta
\end{aligned}
\tag{1.6.117}
$$

and reasoning as in the proof of Theorem 1.6.5.

Lastly, we observe that in the particular case of the scalar Laplacian, i.e., for $M := 1$, $L := \Delta$, and $a_{rs}^{\alpha\beta} = \delta_{\alpha\beta}\delta_{rs}$, the integral representation formula (1.6.113) reduces to the identities established earlier in (1.2.56)-(1.2.57) for vector fields satisfying the generalized Cauchy-Riemann equations.

In the theorem below we re-visit the "generalized Cauchy reproducing formula" (1.6.65), the goal now being to derive a similar integral representation in exterior domains for functions which exhibit power growth at infinity.

Theorem 1.6.7 *Let $\Omega \subseteq \mathbb{R}^n$, where $n \ge 2$, be an unbounded connected open set, with a compact lower Ahlfors regular boundary, and such that $\sigma := \mathcal{H}^{n-1}\lfloor\partial\Omega$ is a doubling measure on $\partial\Omega$; in particular, Ω is a set of locally finite perimeter, and its geometric measure theoretic outward unit normal ν is defined σ-a.e. on $\partial_*\Omega$. Also, assume D is a homogeneous, first-order $N \times M$ system with constant complex* *coefficients in* $\mathbb{R}^n$ *(where* $N, M \in \mathbb{N}$*) which is injectively elliptic (cf.* (1.3.18)*), and consider a matrix-valued tempered distribution* $\widetilde{E}$ *satisfying*

$$\begin{gathered}\widetilde{E} \in \big[\mathcal{S}'(\mathbb{R}^n)\big]^{N\times M}, \quad \widetilde{E}\big|_{\mathbb{R}^n\setminus\{0\}} \in \big[\mathscr{C}^\infty(\mathbb{R}^n\setminus\{0\})\big]^{N\times M}, \\ \widetilde{E} \text{ is odd and positive homogeneous of degree } 1-n, \text{ and} \\ D^\top \widetilde{E} = \delta I_{M\times M} \ \text{(i.e., } \widetilde{E} \text{ is a fundamental solution for } D^\top), \end{gathered} \tag{1.6.118}$$

where δ *is Dirac's distribution with mass at origin in* $\mathbb{R}^n$*, and* $I_{M\times M}$ *is the* $M \times M$ *identity matrix (the existence of* $\widetilde{E}$ *is guaranteed by Theorem 1.6.5). Finally, fix an aperture parameter* $\kappa > 0$*, along with a truncation parameter* $\varepsilon > 0$*, a dilation factor* $\lambda \in (1, \infty)$*, and an exponent* $m \in [0, \infty)$*.*

Next, suppose u is a vector-valued function satisfying

$$\begin{gathered} u \in \big[\mathscr{C}^\infty(\Omega)\big]^M, \quad Du = 0 \text{ in } \Omega, \quad \int_{\partial\Omega} \mathcal{N}_\kappa^\varepsilon u \, \mathrm{d}\sigma < \infty, \\ \text{the nontangential trace } u\big|_{\partial\Omega}^{\kappa-\mathrm{n.t.}} \text{ exists at } \sigma\text{-a.e. point on } \partial_{\mathrm{nta}}\Omega, \end{gathered} \tag{1.6.119}$$

and

$$⨍_{B(0,\lambda R)\setminus B(0,R)} |u| \, \mathrm{d}\mathcal{L}^n = O(R^m) \text{ as } R \to \infty. \tag{1.6.120}$$

Then for any $\kappa' > 0$ *the nontangential trace* $u\big|_{\partial\Omega}^{\kappa'-\mathrm{n.t.}}$ *also exists at* σ*-a.e. point on* $\partial_{\mathrm{nta}}\Omega$ *and is actually independent of the parameter* κ'*. Moreover, there exists some (unique)* $\mathbb{C}^M$*-valued function P in* $\mathbb{R}^n$ *whose scalar components are polynomials and with the property that for each point* $x \in \Omega$ *one has (with absolutely convergent integrals, and with the dependence on the aperture parameter dropped)*

$$u(x) = P(x) + \int_{\partial_*\Omega} \widetilde{E}^\top(x-y)(-\mathrm{i})\mathrm{Sym}\big(D; \nu(y)\big)\big(u\big|_{\partial\Omega}^{\mathrm{n.t.}}\big)(y)\, \mathrm{d}\sigma(y). \tag{1.6.121}$$

Suppose $\widetilde{D}$ *is a homogeneous first-order* $M \times N$ *system with constant complex coefficients in* $\mathbb{R}^n$ *which complements D (in the sense of Definition 1.3.6). In particular,* $L := \widetilde{D}D$ *is a weakly elliptic second-order* $M \times M$ *system in* $\mathbb{R}^n$*. If the fundamental solution* $\widetilde{E}$ *for* $D^\top$ *from* (1.6.118) *is actually taken to be*

$$\widetilde{E} := \widetilde{D}^\top E_{L^\top}, \tag{1.6.122}$$

where $E_{L^\top}$ *is the matrix-valued fundamental solution associated with the weakly elliptic second-order* $M \times M$ *system* $L^\top$ *in* $\mathbb{R}^n$ *as in Theorem 1.4.2, then P also satisfies*

$$LP = 0 \in \mathbb{C}^N \text{ in } \mathbb{R}^n. \tag{1.6.123}$$

Finally, if $\widetilde{D}$ *is such that (with* $L := \widetilde{D}D$ *as before)*

$$\begin{aligned}&\mathrm{Sym}(D;\xi)[L(\xi)]^{-1}\mathrm{Sym}(\widetilde{D};\xi) = I_{M\times M}\\ &\qquad \textit{for each } \xi \in \mathbb{R}^n \setminus \{0\},\end{aligned} \tag{1.6.124}$$

and the fundamental solution $\widetilde{E}$ for $D^\top$ from (1.6.118) *is taken to be as in* (1.6.122), *it follows that P actually satisfies*

$$DP = 0 \in \mathbb{C}^N \ \textit{in}\ \mathbb{R}^n. \tag{1.6.125}$$

Proof The first claim in the statement, regarding the independence of the boundary trace on the aperture parameter, follows from assumptions and [112, Corollary 8.9.9]. There remains to prove the integral representation formula claimed in (1.6.121). This requires some preparations. For specificity, we agree to write

$$\widetilde{E} = \big(\widetilde{E}_{\alpha\beta}\big)_{\substack{1\le\alpha\le N\\ 1\le\beta\le M}} \text{ and } D = \big(a_j^{\alpha\gamma}\partial_j\big)_{\substack{1\le\alpha<N\\ 1\le\gamma\le M}} \tag{1.6.126}$$

with the summation convention over repeated indices assumed throughout. Consequently,

$$D^\top = -\big(a_j^{\alpha\gamma}\partial_j\big)_{\substack{1\le\gamma\le M\\ 1\le\alpha\le N}}. \tag{1.6.127}$$

Introduce the notation $(u_\beta)_{1\le\beta\le M}$ for the scalar components of the vector-valued function $u \in \big[\mathscr{C}^\infty(\Omega)\big]^M$. Next, fix an arbitrary multi-index $\eta \in \mathbb{N}_0^n$ and select an arbitrary index $\alpha \in \{1,\dots,M\}$. Also, pick an arbitrary point $x \in \Omega$ and define $K := \overline{B\big(x, \tfrac{1}{2}\mathrm{dist}(x,\partial\Omega)\big)}$, compact subset of Ω. Use these to define the vector field, whose scalar components are the distributions,

$$\vec{F}_{x,\eta} = \big(F_j\big)_{1\le j\le n} \text{ with } F_j := a_j^{\gamma\beta}\partial^\eta\big[\widetilde{E}_{\gamma\alpha}(x-\cdot)\big]u_\beta \in \mathcal{D}'(\Omega). \tag{1.6.128}$$

Here the package of derivatives ∂^η is applied (in the 'dot' variable) in the sense of distributions in Ω to $\widetilde{E}_{\gamma\alpha}(x-\cdot)$ which is a locally integrable function in Ω (thus, itself a distribution in Ω). It is therefore meaningful to further multiply the resulting distribution by the scalar components of u (which are smooth functions in Ω). All things considered, $\vec{F}_{x,\eta}$ is a well-defined vector field with components distributions in Ω, i.e.,

$$\vec{F}_{x,\eta} \in \big[\mathcal{D}'(\Omega)\big]^n. \tag{1.6.129}$$

From (1.6.128), (1.6.118), and the first property in (1.6.119) we see that

$$\vec{F}_{x,\eta}\big|_{\Omega\setminus K} \in \big[\mathscr{C}^\infty(\Omega\setminus K)\big]^n \subseteq \big[L^1_{\mathrm{loc}}(\Omega\setminus K,\mathcal{L}^n)\big]^n. \tag{1.6.130}$$

In addition, $\mathrm{div}\vec{F}_{x,\eta} \in \mathcal{D}'(\Omega)$, the divergence of the vector field $\vec{F}_{x,\eta}$ in the sense of distributions in Ω is given by

$$\begin{aligned}
\operatorname{div}\vec{F}_{x,\eta} = \partial_j F_j &= a_j^{\gamma\beta}\partial^\eta\partial_j\big[\widetilde{E}_{\gamma\alpha}(x-\cdot)\big]u_\beta + a_j^{\gamma\beta}\partial^\eta\big[\widetilde{E}_{\gamma\alpha}(x-\cdot)\big](\partial_j u_\beta) \\
&= u_\beta\partial^\eta\Big(a_j^{\gamma\beta}\partial_j\big[\widetilde{E}_{\gamma\alpha}(x-\cdot)\big]\Big) + \partial^\eta\big[\widetilde{E}_{\gamma\alpha}(x-\cdot)\big]\big(a_j^{\gamma\beta}\partial_j u_\beta\big) \\
&= -u_\beta\partial^\eta\Big(D^\top\big[\widetilde{E}(x-\cdot)\big]\Big)_{\beta\alpha} + \partial^\eta\big[\widetilde{E}_{\gamma\alpha}(x-\cdot)\big](Du)_\gamma \\
&= u_\beta\partial^\eta\big(\delta_{\beta\alpha}\delta_x\big) = u_\alpha\partial^\eta\delta_x. \qquad (1.6.131)
\end{aligned}$$

Thus,

$$\operatorname{div}\vec{F}_{x,\eta} = u_\alpha\partial^\eta\delta_x \in \mathscr{E}'(\Omega). \qquad (1.6.132)$$

Based on assumptions and (1.6.128) we also see that

$$\vec{F}_{x,\eta}\Big|_{\partial\Omega}^{\text{n.t.}} \text{ exists at } \sigma\text{-a.e. point on } \partial_{\text{nta}}\Omega \qquad (1.6.133)$$

and for each $j \in \{1,\dots,n\}$ and σ-a.e. point $y \in \partial_{\text{nta}}\Omega$ we actually have

$$\Big(\vec{F}_{x,\eta}\Big|_{\partial\Omega}^{\text{n.t.}}\Big)_j(y) = -(-1)^{|\eta|}a_j^{\gamma\beta}(\partial^\eta\widetilde{E}_{\gamma\alpha})(x-y)\big(u_\beta\big|_{\partial\Omega}^{\text{n.t.}}\big)(y). \qquad (1.6.134)$$

Next, from (1.6.120) and Lemma 1.5.6 we conclude that

$$u(x) = O(|x|^m) \text{ as } |x| \to \infty, \qquad (1.6.135)$$

while (1.6.118) implies that

$$(\partial^\eta\widetilde{E})(x) = O\big(|x|^{1-n-|\eta|}\big) \text{ as } |x| \to \infty. \qquad (1.6.136)$$

In view of these properties and the definition in (1.6.128) we see that

$$\vec{F}_{x,\eta}(x) = O\big(|x|^{m+1-n-|\eta|}\big) \text{ as } |x| \to \infty. \qquad (1.6.137)$$

In particular,

$$\vec{F}_{x,\eta}(x) = o(1) \text{ as } |x| \to \infty, \text{ if } |\eta| > m-n+1. \qquad (1.6.138)$$

Collectively, (1.6.138), (1.6.130), (1.6.128), (1.6.118), and (1.6.119) imply (mindful of the fact that $\partial\Omega$ is compact) that

$$\mathcal{N}_\kappa^{\Omega\setminus K}\vec{F}_{x,\eta} \in L^1(\partial\Omega,\sigma) \text{ if } |\eta| > m-n+1. \qquad (1.6.139)$$

Fix now an arbitrary multi-index $\eta \in \mathbb{N}_0^n$ with $|\eta| > m-n+1$. Granted, (1.6.129), (1.6.130), (1.6.132), (1.6.133), and (1.6.139), we may invoke [112, Theorem 1.4.1], and the Divergence Formula recorded in [112, (1.4.6)] presently gives

$$
\begin{aligned}
(\partial^\eta u_\alpha)(x) &= {}_{\mathscr{E}'(\Omega)}\big\langle \delta_x, \partial^\eta u_\alpha \big\rangle_{\mathscr{E}(\Omega)} = (-1)^{|\eta|}\,{}_{\mathscr{E}'(\Omega)}\big\langle \partial^\eta \delta_x, u_\alpha \big\rangle_{\mathscr{E}(\Omega)} \\
&= (-1)^{|\eta|}\,{}_{\mathscr{E}'(\Omega)}\big\langle u_\alpha \partial^\eta \delta_x, 1 \big\rangle_{\mathscr{E}(\Omega)} = (-1)^{|\eta|}\,{}_{\mathscr{E}'(\Omega)}\big\langle \mathrm{div}\vec{F}_{x,\eta}, 1 \big\rangle_{\mathscr{E}(\Omega)} \\
&= (-1)^{|\eta|}\,{}_{(\mathscr{C}^\infty_b(\Omega))^*}\big(\mathrm{div}\vec{F}_{x,\eta}, 1\big)_{\mathscr{C}^\infty_b(\Omega)} = (-1)^{|\eta|}\int_{\partial_*\Omega} \nu \cdot \Big(\vec{F}_{x,\eta}\Big|^{\text{n.t.}}_{\partial\Omega}\Big)\,\mathrm{d}\sigma \\
&= -\int_{\partial_*\Omega} \nu_j(y) a^{\gamma\beta}_j (\partial^\eta \widetilde{E}_{\gamma\alpha})(x-y)\big(u_\beta\big|^{\text{n.t.}}_{\partial\Omega}\big)(y)\,\mathrm{d}\sigma(y),
\end{aligned}
\tag{1.6.140}
$$

on account of (1.6.132), [112, (4.6.21)], (1.6.134), and [112, Proposition 8.8.6]. In turn, from (1.6.140) and assumptions it follows that if we define the function $P : \Omega \longrightarrow \mathbb{C}^M$ by setting, at each point $x \in \Omega$,

$$
P(x) := u(x) + \Big(\int_{\partial_*\Omega} \nu_j(y) a^{\gamma\beta}_j \widetilde{E}_{\gamma\alpha}(x-y)\big(u_\beta\big|^{\text{n.t.}}_{\partial\Omega}\big)(y)\,\mathrm{d}\sigma(y)\Big)_{1\le\alpha\le M}
\tag{1.6.141}
$$

then

$$
\begin{gathered}
P \text{ belongs to } \big[\mathscr{C}^\infty(\Omega)\big]^M \text{ and satisfies } \partial^\eta P = 0 \text{ in } \Omega \\
\text{for each } \eta \in \mathbb{N}_0^n \text{ with } |\eta| > m - n + 1.
\end{gathered}
\tag{1.6.142}
$$

Expanding P locally into a Taylor series (with remainder), (1.6.142) implies that the scalar components of P are locally polynomials in Ω. This and the fact that we are currently assuming Ω to be connected then ultimately show that the scalar components of P are polynomials in Ω. At this stage, (1.6.121) follows from (1.6.141) and (A.0.118).

Next, if $\widetilde{E}$ is as in (1.6.122), then since $Lu = \widetilde{D}(Du) = 0$ in Ω and $L\widetilde{E}^\top = 0$ in $\mathbb{R}^n \setminus \{0\}$, from (1.6.121) we readily conclude that (1.6.123) holds. Finally, when the additional hypothesis (1.6.124) is assumed, (1.6.125) follows from (1.6.121) and (1.6.50). □

In turn, Theorem 1.6.7 allows us to augment the analysis initiated in Corollary 1.6.2, by now including null-solutions of injectively elliptic first-order systems with at most linear growth at infinity.

Corollary 1.6.8 *Pick $n \in \mathbb{N}$ satisfying $n \ge 2$, along with some $N, M \in \mathbb{N}$. Suppose $\Omega \subseteq \mathbb{R}^n$ is an exterior domain. Also, consider an injectively elliptic homogeneous first-order $N \times M$ system D with constant complex coefficients in $\mathbb{R}^n$. Then for any given function $u \in \big[\mathscr{C}^\infty(\Omega)\big]^M$ satisfying $Du = 0$ in Ω the following conditions are equivalent:*

(1) One has $u(x) = o(|x|)$ as $|x| \to \infty$.
(2) One has $u(x) = O(1)$ as $|x| \to \infty$.
(3) The limit $\lim\limits_{|x|\to\infty} u(x)$ exists in $\mathbb{C}^M$.
(4) There exists some $c \in \mathbb{C}^M$ such that $u(x) = c + O(|x|^{1-n})$ as $|x| \to \infty$.

Moreover, if any of the conditions (1)-(4) are satisfied then

$$(\partial^\alpha u)(x) = O\big(|x|^{1-n-|\alpha|}\big) \text{ as } |x| \to \infty. \qquad \text{for each multi-index } \alpha \in \mathbb{N}_0^n \text{ with } |\alpha| > 0. \tag{1.6.143}$$

Proof For the purposes we have in mind, there is no loss of generality in assuming that Ω is the complement of the closure of a large ball centered at the origin in $\mathbb{R}^n$ (in particular, Ω is connected), and that actually $u \in \big[\mathscr{C}^\infty(\overline{\Omega})\big]^M$. Assume this the case, and suppose $u(x) = o(|x|)$ as $|x| \to \infty$. We may then invoke Theorem 1.6.7 (with $m := 1$) and conclude from (1.6.121) and (1.6.118) that there exists a polynomial P in $\mathbb{R}^n$ with the property that $u(x) = P(x) + O(|x|^{1-n})$ as $|x| \to \infty$. The latter property forces $P(x) = o(|x|)$ as $|x| \to \infty$ hence, ultimately, P is a constant, say $P \equiv c \in \mathbb{C}^M$. This argument proves the implication *(1)*$\Rightarrow$*(4)*. Since the implications *(4)*$\Rightarrow$*(3)*$\Rightarrow$*(2)*$\Rightarrow$*(1)* are clear, it follows that conditions *(1)-(4)* are equivalent. Finally, that any of conditions *(1)-(4)* implies (1.6.143) is seen from what we have proved so far, (1.6.121), and (1.6.118). □

1.7 Green-Type Formulas for Second-Order Systems

Given $n, M, N \in \mathbb{N}$, by a `complex coefficient tensor` (of type $(n \times n, M \times N)$) we shall understand any block of the form

$$A = \big(a_{rs}^{\alpha\beta}\big)_{\substack{1\le r,s\le n\\ 1\le\alpha\le M\\ 1\le\beta\le N}} \tag{1.7.1}$$

where each entry $a_{rs}^{\alpha\beta}$ is a complex number. There are several things we shall typically associate with such a complex coefficient tensor A. First, define the transpose of A as being the complex coefficient tensor (of type $(n \times n, N \times M)$) given by

$$A^\top := \big(a_{rs}^{\alpha\beta}\big)_{\substack{1\le s,r\le n,\\ 1\le\beta\le N\\ 1\le\alpha\le M}} \text{ i.e., } \big(A^\top\big)_{sr}^{\beta\alpha} := a_{rs}^{\alpha\beta} \text{ for all } \alpha, \beta, r, s. \tag{1.7.2}$$

Hence, in the case of square coefficient tensors (i.e., when $N = M$),

$$\text{if } A = \big(a_{rs}^{\alpha\beta}\big)_{\substack{1\le r,s\le n\\ 1\le\alpha,\beta\le N}} \text{ then } A^\top := \big(a_{sr}^{\beta\alpha}\big)_{\substack{1\le r,s\le n\\ 1\le\alpha,\beta\le N}}, \text{ in other words } \big(A^\top\big)_{rs}^{\alpha\beta} := a_{sr}^{\beta\alpha} \text{ for all } \alpha, \beta, r, s. \tag{1.7.3}$$

Second, associate with each complex coefficient tensor A as in (1.7.1) the linear mapping

$$A : \mathbb{C}^{N\times n} \longrightarrow \mathbb{C}^{M\times n} \text{ given by}$$
$$A\zeta := \big(a_{rs}^{\alpha\beta}\zeta_s^\beta\big)_{\alpha,r} \in \mathbb{C}^{M\times n} \text{ for each } \zeta := (\zeta_s^\beta)_{\beta,s} \in \mathbb{C}^{N\times n}, \tag{1.7.4}$$

using the summation convention over repeated indices understood here and elsewhere. In particular, given any open set $\Omega \subseteq \mathbb{R}^n$ and any $u \in \left[\mathcal{D}'(\Omega)\right]^N$, with $\nabla u \in \left[\mathcal{D}'(\Omega)\right]^{n\times N}$ denoting the Jacobian matrix of first-order distributional partial derivatives of u in Ω (cf. (A.0.48)), we have

$$A\nabla u = \left(a^{\alpha\beta}_{rs}\,\partial_s u_\beta\right)_{\alpha,r} \in \left[\mathcal{D}'(\Omega)\right]^{M\times n}. \tag{1.7.5}$$

Third, with a complex coefficient tensor A as in (1.7.1), we associated the (real) bilinear form $\langle A\cdot,\cdot\rangle$ defined by

$$\begin{gathered}\langle A\zeta,\eta\rangle := a^{\alpha\beta}_{rs}\,\zeta^\beta_s\,\eta^\alpha_r \ \text{ for any given}\\ \zeta = (\zeta^\beta_s)_{\beta,s} \in \mathbb{C}^{N\times n} \ \text{ and } \ \eta = (\eta^\alpha_r)_{\alpha,r} \in \mathbb{C}^{M\times n}.\end{gathered} \tag{1.7.6}$$

Fourth, with each complex coefficient tensor A as in (1.7.1) we may canonically associate a homogeneous constant (complex) coefficient second-order $M\times N$ system L_A in $\mathbb{R}^n$ which, with the summation convention over repeated indices in place, is expressed as

$$L_A := \left(a^{\alpha\beta}_{rs}\,\partial_r\partial_s\right)_{\substack{1\le\alpha\le M\\ 1\le\beta\le N}}. \tag{1.7.7}$$

Fifth, with each complex coefficient tensor A as in (1.7.1) we also associate a conormal derivative on the boundary of a given domain of the following sort.

Definition 1.7.1 *Suppose $\Omega \subseteq \mathbb{R}^n$ is an open set with the property that $\partial\Omega$ is lower Ahlfors regular and $\sigma := \mathcal{H}^{n-1}\lfloor\partial\Omega$ is a doubling measure on $\partial\Omega$. In particular, Ω is a set of locally finite perimeter, and its geometric measure theoretic outward unit normal $\nu = (\nu_1,\dots,\nu_n)$ is defined σ-a.e. on $\partial_*\Omega$. Given a vector-valued distribution $u = (u_\beta)_{1\le\beta\le N} \in \left[\mathcal{D}'(\Omega)\right]^N$, denote by $\nabla u \in \left[\mathcal{D}'(\Omega)\right]^{n\times N}$ the Jacobian matrix of first-order distributional partial derivatives of u in Ω. In relation to this suppose that for some complex coefficient tensor A of type $(n\times n, M\times N)$, expressed as in (1.7.1), and some aperture parameter $\kappa > 0$,*

$$\begin{gathered}A\nabla u := \left(a^{\alpha\beta}_{rs}\,\partial_s u_\beta\right)_{\alpha,r} \in \left[L^1_{\rm loc}(\Omega,\mathcal{L}^n)\right]^{M\times n} \ \textit{and}\\ \left(A\nabla u\right)\Big|^{\kappa-\rm n.t.}_{\partial\Omega} \ \textit{exists (in } \mathbb{C}^{M\times n}\textit{) at } \sigma\textit{-a.e. point on } \partial_*\Omega.\end{gathered} \tag{1.7.8}$$

In such a setting, define the (pointwise) `conormal derivative` *of u (with respect to the coefficient tensor A and the set Ω) as the $\mathbb{C}^M$-valued function*

$$\partial^A_\nu u := \left(\nu_r\left(a^{\alpha\beta}_{rs}\,\partial_s u_\beta\right)\Big|^{\kappa-\rm n.t.}_{\partial\Omega}\right)_{1\le\alpha\le M} \ \textit{at } \sigma\textit{-a.e. point on } \partial_*\Omega. \tag{1.7.9}$$

A few clarifications are in order. First, the largest subset of $\partial\Omega$ on which one may even attempt to consider the nontangential pointwise limit in (1.7.8) is $A_\kappa(\partial\Omega)$. The current hypotheses on Ω guarantee the applicability of [112, Proposition 8.8.4] which, in turn, ensures that $\partial_*\Omega$ is covered by $A_\kappa(\partial\Omega)$, up to a σ-nullset. Ultimately, this makes the assumption in (1.7.8) meaningfully formulated.

Second, if for some other aperture parameter $\kappa' > 0$ we also have that

$$(A\nabla u)\Big|_{\partial\Omega}^{\kappa'-\mathrm{n.t.}} \text{ exists (in } \mathbb{C}^{M\times n}\text{) at } \sigma\text{-a.e. point on } \partial_*\Omega, \tag{1.7.10}$$

then given that for each $x \in \partial\Omega$ we have

$$\Gamma_{\kappa'}(x) \subseteq \Gamma_\kappa(x) \text{ if } \kappa \geq \kappa', \text{ and } \Gamma_\kappa(x) \subseteq \Gamma_{\kappa'}(x) \text{ if } \kappa \leq \kappa', \tag{1.7.11}$$

it follows that we necessarily have

$$(A\nabla u)\Big|_{\partial\Omega}^{\kappa-\mathrm{n.t.}} = (A\nabla u)\Big|_{\partial\Omega}^{\kappa'-\mathrm{n.t.}} \text{ at } \sigma\text{-a.e. point on } \partial_*\Omega, \tag{1.7.12}$$

hence

$$\Big(\nu_r\big(a_{rs}^{\alpha\beta}\partial_s u_\beta\big)\Big|_{\partial\Omega}^{\kappa-\mathrm{n.t.}}\Big)_{1\leq\alpha\leq M} = \Big(\nu_r\big(a_{rs}^{\alpha\beta}\partial_s u_\beta\big)\Big|_{\partial\Omega}^{\kappa'-\mathrm{n.t.}}\Big)_{1\leq\alpha\leq M} \quad \text{at } \sigma\text{-a.e. point on } \partial_*\Omega. \tag{1.7.13}$$

This ultimately shows that, as the notation already reflects it,

$$\begin{array}{l}\text{the definition of the conormal derivative } \partial_\nu^A u \text{ in (1.7.9)} \\ \text{is actually independent of the aperture parameter } \kappa > 0.\end{array} \tag{1.7.14}$$

Third, in the context of Definition 1.7.1, [112, Corollary 8.9.6] implies that

$$\begin{array}{l}\text{the conormal derivative } \partial_\nu^A u, \text{ defined as in (1.7.9),} \\ \text{is a } \sigma\text{-measurable } (\mathbb{C}^M\text{-valued) function on } \partial_*\Omega.\end{array} \tag{1.7.15}$$

Fourth, retaining the assumptions on the domain Ω and the coefficient tensor A from Definition 1.7.1, as a special case we have

$$\left.\begin{array}{r} u \in \big[W^{1,1}_{\mathrm{loc}}(\Omega)\big]^N \text{ and} \\ (\nabla u)\Big|_{\partial\Omega}^{\kappa-\mathrm{n.t.}} \text{ exists} \\ \sigma\text{-a.e. on } \partial_*\Omega \end{array}\right\} \Longrightarrow \left\{\begin{array}{l} \partial_\nu^A u = \Big(\nu_r a_{rs}^{\alpha\beta}(\partial_s u_\beta)\Big|_{\partial\Omega}^{\kappa-\mathrm{n.t.}}\Big)_{1\leq\alpha\leq M} \\ \text{at } \sigma\text{-a.e. point on } \partial_*\Omega. \end{array}\right. \tag{1.7.16}$$

Occasionally, it is convenient to extend the scope of Definition 1.7.1 by allowing more general domains, though we now restrict to a smaller class of functions.

Convention 1.7.2 *Let $\Omega \subseteq \mathbb{R}^n$ be a Lebesgue measurable set of locally finite perimeter. Abbreviate $\sigma := \mathcal{H}^{n-1}\lfloor\partial\Omega$ and denote by $\nu = (\nu_1, \dots, \nu_n)$ the geometric measure theoretic outward unit normal to Ω. Given a $\mathbb{C}^N$-valued function $u = (u_\beta)_{1\leq\beta\leq M}$ of class $\mathscr{C}^1$ in a neighborhood of $\partial_*\Omega$, we agree to define the conormal derivative $\partial_\nu^A u$ with respect to any given complex coefficient tensor A of type $(n \times n, M \times N)$, expressed as in* (1.7.1)*, according to*

$$\partial_\nu^A u := \Big(\nu_r a_{rs}^{\alpha\beta}(\partial_s u_\beta)\big|_{\partial_*\Omega}\Big)_{1\leq\alpha\leq M} \tag{1.7.17}$$

at σ-a.e. point on $\partial_*\Omega$.

Whenever Ω is open, $\partial\Omega$ is lower Ahlfors regular, and σ is a doubling measure on $\partial\Omega$, there is agreement between the brands of conormal derivatives considered in Definition 1.7.1 and Convention 1.7.2, in the class of function $\big[\mathscr{C}^1(\overline{\Omega})\big]^N$ (here, [112, Proposition 8.8.4] is relevant).

Going further, we note the following useful estimate.

Lemma 1.7.3 *Let $\Omega \subseteq \mathbb{R}^n$ be an open set with the property that $\partial\Omega$ is lower Ahlfors regular and $\sigma := \mathcal{H}^{n-1}\lfloor\partial\Omega$ is a doubling measure on $\partial\Omega$. Also, fix some complex coefficient tensor A of type $(n\times n, M\times N)$ along with an aperture parameter $\kappa > 0$. Then there exists some constant $C \in (0,\infty)$ such that for every distribution $u \in \big[\mathcal{D}'(\Omega)\big]^N$ satisfying (1.7.8), every truncation parameter $\varepsilon > 0$, and every exponents $p, q \in (0,\infty]$ one has*

$$\big\|\partial^A_\nu u\big\|_{[L^{p,q}(\partial_*\Omega,\sigma)]^M} \le C\big\|\mathcal{N}^\varepsilon_\kappa(A\nabla u)\big\|_{L^{p,q}(\partial\Omega,\sigma)}. \tag{1.7.18}$$

Proof This is seen from (1.7.15) and [112, Proposition 8.2.3, (8.9.8), (6.2.16), (6.2.17)]. □

In relation to the conormal derivative operator introduced in Definition 1.7.1 we have the following companion to [113, Proposition 10.2.22], proving that the conormal derivatives associated with any two coefficient tensors producing the same second-order system actually differ by a linear combination of tangential derivatives.

Proposition 1.7.4 *Let Ω be an open nonempty proper subset of $\mathbb{R}^n$ with a lower Ahlfors regular boundary and such that $\sigma := \mathcal{H}^{n-1}\lfloor\partial\Omega$ is a doubling measure on $\partial\Omega$; in particular, Ω is a set of locally finite perimeter. Denote by ν the geometric measure theoretic outward unit normal to Ω, and abbreviate $\sigma_* := \mathcal{H}^{n-1}\lfloor\partial_*\Omega$. Also, fix an aperture parameter $\kappa \in (0,\infty)$, a truncation parameter $\varepsilon > 0$, and two integers $M, N \in \mathbb{N}$. In addition, consider a $\mathbb{C}^N$-valued function $u = (u_\beta)_{1\le\beta\le N}$ in Ω satisfying*

$$\begin{gathered} u \in \big[W^{1,1}_{\rm loc}(\Omega)\big]^N, \quad \mathcal{N}^\varepsilon_\kappa u,\ \mathcal{N}^\varepsilon_\kappa(\nabla u) \in L^1_{\rm loc}(\partial\Omega,\sigma), \\ \text{and the nontangential traces } u\big|^{\kappa-\rm n.t.}_{\partial\Omega},\ (\nabla u)\big|^{\kappa-\rm n.t.}_{\partial\Omega} \\ \text{exist at } \sigma\text{-a.e. point on } \partial_{\rm nta}\Omega. \end{gathered} \tag{1.7.19}$$

Then, when considered on $\partial_*\Omega$, the function $u\big|^{\kappa-\rm n.t.}_{\partial\Omega}$ belongs to $\big[L^1_{1,\rm loc}(\partial_*\Omega,\sigma_*)\big]^N$ and for any two complex coefficient tensors of type $(n\times n, M\times N)$,

$$A = \big(a^{\alpha\beta}_{rs}\big)_{\substack{1\le r,s\le n\\ 1\le\alpha\le M\\ 1\le\beta\le N}} \text{ and } B = \big(b^{\alpha\beta}_{rs}\big)_{\substack{1\le r,s\le n\\ 1\le\alpha\le M\\ 1\le\beta\le N}} \tag{1.7.20}$$

with the property that

$$L_A = L_B \text{ where}$$

$$L_A := \left(a_{rs}^{\alpha\beta}\partial_r\partial_s\right)_{\substack{1\leq\alpha\leq M\\1\leq\beta\leq N}} \text{ and } L_B := \left(b_{rs}^{\alpha\beta}\partial_r\partial_s\right)_{\substack{1\leq\alpha\leq M,\\1\leq\beta\leq N}} \tag{1.7.21}$$

one has

$$\partial_\nu^A u - \partial_\nu^B u = \left(\tfrac{1}{2}\left(a_{rs}^{\alpha\beta} - b_{rs}^{\alpha\beta}\right)\partial_{\tau_{rs}}\left(u_\beta\big|_{\partial\Omega}^{\kappa-\text{n.t.}}\right)\right)_{1\leq\alpha\leq M} \text{ at } \sigma\text{-a.e. point on } \partial_*\Omega. \tag{1.7.22}$$

In particular,

$$\partial_\nu^A u = \partial_\nu^B u \text{ at } \sigma\text{-a.e. point on } \partial_*\Omega \setminus \operatorname{supp}\left(u\big|_{\partial\Omega}^{\kappa-\text{n.t.}}\right). \tag{1.7.23}$$

As a corollary,

with Ω as before, whenever A, B are two coefficient tensors for which $L_A = L_B$ and the function u satisfies the properties listed in (1.7.19) *as well as $u\big|_{\partial\Omega}^{\kappa-\text{n.t.}} = 0$ at σ-a.e. point on $\partial_*\Omega$, it follows that $\partial_\nu^A u = \partial_\nu^B u$ at σ-a.e. point on $\partial_*\Omega$.* (1.7.24)

Proof That $u\big|_{\partial\Omega}^{\kappa-\text{n.t.}} \in \left[L^1_{1,\text{loc}}(\partial_*\Omega, \sigma_*)\right]^N$ follows from [113, Proposition 11.3.2] (used with $p = q = 1$). As regards the claim made in (1.7.22), with the coefficient tensors A, B as in (1.7.20)-(1.7.21), define

$$c_{rs}^{\alpha\beta} := a_{rs}^{\alpha\beta} - b_{rs}^{\alpha\beta} \text{ for each } \alpha, \beta, r, s. \tag{1.7.25}$$

When tested on functions of the form $\left(\delta_{\beta\beta'}e^{\xi\cdot x}\right)_{1\leq\beta'\leq N}$, where $\beta \in \{1, \dots, N\}$ and $\xi \in \mathbb{R}^n$ are arbitrary and fixed, the coincidence of L_A with L_B forces that for each $\alpha \in \{1, \dots, M\}$ and $\beta \in \{1, \dots, N\}$ we have

$$c_{rs}^{\alpha\beta}\xi_r\xi_s = 0 \text{ for each } \xi = (\xi_j)_{1\leq j\leq n} \in \mathbb{R}^n. \tag{1.7.26}$$

After differentiating with respect to ξ_r and ξ_s, this ultimately implies

$$c_{rs}^{\alpha\beta} = -c_{sr}^{\alpha\beta} \text{ for each } r, s \in \{1, \dots, n\}, \quad \alpha \in \{1, \dots, M\}, \text{ and } \beta \in \{1, \dots, N\}. \tag{1.7.27}$$

Granted this, at σ-a.e. point on $\partial_*\Omega$ we may compute

$$
\begin{aligned}
\partial_\nu^A u - \partial_\nu^B u &= \Big(\nu_r a_{rs}^{\alpha\beta}\big(\partial_s u_\beta\big)\Big|_{\partial\Omega}^{\kappa-\mathrm{n.t.}}\Big)_{1\le\alpha\le M} - \Big(\nu_r a_{rs}^{\alpha\beta}\big(\partial_s u_\beta\big)\Big|_{\partial\Omega}^{\kappa-\mathrm{n.t.}}\Big)_{1\le\alpha\le M} \\
&= \Big(\nu_r c_{rs}^{\alpha\beta}\big(\partial_s u_\beta\big)\Big|_{\partial\Omega}^{\kappa-\mathrm{n.t.}}\Big)_{1\le\alpha\le M} \\
&= \Big(\tfrac{1}{2}\nu_r c_{rs}^{\alpha\beta}\big(\partial_s u_\beta\big)\Big|_{\partial\Omega}^{\kappa-\mathrm{n.t.}}\Big)_{1\le\alpha\le M} + \Big(\tfrac{1}{2}\nu_r c_{rs}^{\alpha\beta}\big(\partial_s u_\beta\big)\Big|_{\partial\Omega}^{\kappa-\mathrm{n.t.}}\Big)_{1\le\alpha\le M} \\
&= \Big(\tfrac{1}{2}\nu_r c_{rs}^{\alpha\beta}\big(\partial_s u_\beta\big)\Big|_{\partial\Omega}^{\kappa-\mathrm{n.t.}}\Big)_{1\le\alpha\le M} - \Big(\tfrac{1}{2}\nu_s c_{rs}^{\alpha\beta}\big(\partial_r u_\beta\big)\Big|_{\partial\Omega}^{\kappa-\mathrm{n.t.}}\Big)_{1\le\alpha\le M} \\
&= \Big(\tfrac{1}{2}c_{rs}^{\alpha\beta}\Big\{\nu_r\big(\partial_s u_\beta\big)\Big|_{\partial\Omega}^{\kappa-\mathrm{n.t.}} - \nu_s\big(\partial_r u_\beta\big)\Big|_{\partial\Omega}^{\kappa-\mathrm{n.t.}}\Big\}\Big)_{1\le\alpha\le M} \\
&= \Big(\tfrac{1}{2}c_{rs}^{\alpha\beta}\,\partial_{\tau_{rs}}\big(u_\beta\big|_{\partial\Omega}^{\kappa-\mathrm{n.t.}}\big)\Big)_{1\le\alpha\le M},
\end{aligned}
\tag{1.7.28}
$$

where the first equality originates in (1.7.9), the second one comes from (1.7.25), the third and firth ones involve trivial algebra, the fourth one may be justified by interchanging r with s (in the second vector) then invoking (1.7.27), and the last one is implied by [113, Proposition 11.3.2, (11.3.26)]. From (1.7.28), the claim made in (1.7.22) follows in view of (1.7.25).

Finally, (1.7.23) is a consequence of (1.7.22) and [113, (11.1.38)], while (1.7.24) is a particular case of (1.7.23). □

Remark 1.7.5 *Assume*

$$
\widetilde{D} = \Big(\sum_{r=1}^{n}\widetilde{b}_r^{\alpha\gamma}\,\partial_r\Big)_{\substack{1\le\alpha\le M\\ 1\le\gamma\le\widetilde{M}}} \quad\text{and}\quad D = \Big(\sum_{s=1}^{n} b_s^{\gamma\beta}\,\partial_s\Big)_{\substack{1\le\gamma\le\widetilde{M}\\ 1\le\beta\le N}}
\tag{1.7.29}
$$

are two homogeneous (constant) coefficient first-order systems in $\mathbb{R}^n$ with the property that

$$
\widetilde{D}D = 0. \tag{1.7.30}
$$

Also, suppose the set Ω and the function u are as in Proposition 1.7.4. Then the conormal derivative $\partial_\nu^{A_{\widetilde{D},D}}u$ associated with the coefficient tensor (with the summation convention over repeated indices in effect)

$$
A_{\widetilde{D},D} := \big(a_{rs}^{\alpha\beta}\big)_{\substack{1\le r,s\le n\\ 1\le\alpha\le M\\ 1\le\beta\le N}} \quad\text{where each } a_{rs}^{\alpha\beta} := \widetilde{b}_r^{\alpha\gamma}\, b_s^{\gamma\beta}
\tag{1.7.31}
$$

may be expressed as a linear combination of tangential derivatives of $u\big|_{\partial\Omega}^{\kappa-\mathrm{n.t.}}$. Specifically, at σ-a.e. point on $\partial_*\Omega$ one has

$$
\begin{aligned}
\partial_\nu^{A_{\widetilde{D},D}}u &= \Big(\tfrac{1}{2}a_{rs}^{\alpha\beta}\,\partial_{\tau_{rs}}\big(u_\beta\big|_{\partial\Omega}^{\kappa-\mathrm{n.t.}}\big)\Big)_{1\le\alpha\le M} \\
&= \Big(\tfrac{1}{2}\widetilde{b}_r^{\alpha\gamma}\, b_s^{\gamma\beta}\,\partial_{\tau_{rs}}\big(u_\beta\big|_{\partial\Omega}^{\kappa-\mathrm{n.t.}}\big)\Big)_{1\le\alpha\le M}.
\end{aligned}
\tag{1.7.32}
$$

Indeed, this is seen from (1.7.22) *written for* $A := A_{\widetilde{D},D}$ *and* $B := 0$ (*so that* $L_A = 0 = L_B$).

There is yet a more flexible approach to defining conormal derivatives which we wish to elaborate on. To set the stage, fix

$$\Omega \subseteq \mathbb{R}^n \text{ open set with the property that } \partial\Omega \text{ is lower Ahlfors regular and such that } \sigma := \mathcal{H}^{n-1}\lfloor\partial\Omega \text{ is a doubling measure on the set } \partial\Omega. \tag{1.7.33}$$

In particular, Ω is a set of locally finite perimeter, and its geometric measure theoretic outward unit normal ν is defined σ-a.e. on $\partial_*\Omega$. Next, let L be a given homogeneous, constant (complex) coefficient, second-order $M \times N$ system in $\mathbb{R}^n$. Suppose said system factors as

$$L = \widetilde{D}D \tag{1.7.34}$$

where $\widetilde{D}$ and D are two homogeneous, constant (complex) coefficient, first-order systems in $\mathbb{R}^n$. Then for any given vector-valued distribution $u \in \big[\mathcal{D}'(\Omega)\big]^N$ satisfying (for some aperture parameter $\kappa > 0$)

$$Du \text{ has locally integrable components in } \Omega \text{ and } (Du)\Big|_{\partial\Omega}^{\kappa-\mathrm{n.t.}} \text{ exists at } \sigma\text{-a.e. point on } \partial_*\Omega, \tag{1.7.35}$$

we agree to define the conormal derivative of u associated with the factorization (1.7.34) as

$$\partial_\nu^{\widetilde{D},D} u := (-\mathrm{i})\mathrm{Sym}(\widetilde{D};\nu)(Du)\Big|_{\partial\Omega}^{\kappa-\mathrm{n.t.}}. \tag{1.7.36}$$

To shed light on how this point of view relates to Definition 1.7.1, write $\widetilde{D}$, D explicitly as

$$\widetilde{D} = \Big(\sum_{r=1}^{n} \widetilde{b}_r^{\alpha\gamma}\partial_r\Big)_{\substack{1\leq\alpha\leq M\\ 1\leq\gamma\leq N'}} \tag{1.7.37}$$

and

$$D = \Big(\sum_{s=1}^{n} b_s^{\gamma\beta}\partial_s\Big)_{\substack{1\leq\gamma\leq N'\\ 1\leq\beta\leq M}} \tag{1.7.38}$$

then define the coefficient tensor (with the summation convention over repeated indices in effect)

$$A_{\widetilde{D},D} := \big(a_{rs}^{\alpha\beta}\big)_{\substack{1\leq\alpha\leq M\\ 1\leq\beta\leq N\\ 1\leq r,s\leq n}} \text{ where each } a_{rs}^{\alpha\beta} := \widetilde{b}_r^{\alpha\gamma} b_s^{\gamma\beta}. \tag{1.7.39}$$

Based on these considerations and (1.7.34), we then see that L may be explicitly represented as

$$L = L_{A_{\widetilde{D},D}}, \tag{1.7.40}$$

i.e., (cf. (1.7.7))

$$L = \Big(a_{rs}^{\alpha\beta}\partial_r\partial_s\Big)_{\substack{1\leq\alpha\leq M\\ 1\leq\beta\leq N}}. \tag{1.7.41}$$

Moreover, for any vector-valued distribution $u = (u_\beta)_{1\leq\beta\leq N} \in \big[\mathcal{D}'(\Omega)\big]^N$ the demands in (1.7.8) written for the coefficient tensor $A := A_{\widetilde{D},D}$ are satisfied if the properties stipulated in (1.7.35) hold, and the conormal derivative of u with respect to the coefficient tensor $A := A_{\widetilde{D},D}$ as originally defined in (1.7.9) presently becomes

$$\partial_\nu^{A_{\widetilde{D},D}} u = \Big(\nu_r \big(a_{rs}^{\alpha\beta}\partial_s u_\beta\big)\Big|_{\partial\Omega}^{\kappa-\mathrm{n.t.}}\Big)_{1\leq\alpha\leq M} = \Big(\nu_r \widetilde{b}_r^{\alpha\gamma}\big(b_s^{\gamma\beta}\partial_s u_\beta\big)\Big|_{\partial\Omega}^{\kappa-\mathrm{n.t.}}\Big)_{1\leq\alpha\leq M}$$

$$= (-i)\mathrm{Sym}(\widetilde{D};\nu)(Du)\Big|_{\partial\Omega}^{\kappa-\mathrm{n.t.}} = \partial_\nu^{\widetilde{D},D} u \text{ at } \sigma\text{-a.e. point on } \partial_*\Omega. \qquad (1.7.42)$$

This shows that the "new" conormal derivative (from (1.7.36)) is consistent with the "old" point of view (from Definition 1.7.1).

Let us illustrate how the above philosophy is implemented in the case of the complex Lamé system $L_{\lambda,\mu} = \mu\Delta + (\lambda+\mu)\nabla\mathrm{div}$, for some given Lamé moduli $\mu, \lambda \in \mathbb{C}$. From (1.3.62) we see that for each $\zeta \in \mathbb{C}$ we may express

$$L_{\lambda,\mu} = -2\zeta\,\mathrm{Def}^\top\mathrm{Def} + (\lambda+\mu-\zeta)\nabla\mathrm{div} - (\mu-\zeta)\nabla^\top\nabla. \qquad (1.7.43)$$

For the goals we have in mind, it is convenient to recast this as

$$L_{\lambda,\mu} = \widetilde{D}_\zeta D \qquad (1.7.44)$$

where

$$\widetilde{D}_\zeta := \Big(-2\zeta\,\mathrm{Def}^\top, (\lambda+\mu-\zeta)\nabla, -(\mu-\zeta)\nabla^\top\Big) \text{ and } D := \begin{pmatrix}\mathrm{Def}\\ \mathrm{div}\\ \nabla\end{pmatrix} \qquad (1.7.45)$$

are homogeneous, constant coefficient, first-order systems in $\mathbb{R}^n$. For each $\zeta \in \mathbb{C}$ let us agree to abbreviate

$$A_\zeta := A_{\widetilde{D}_\zeta,D} \qquad (1.7.46)$$

where $A_{\widetilde{D}_\zeta,D}$ is the coefficient tensor associated with the systems $\widetilde{D}_\zeta$, D as in (1.7.39). Assume next that Ω is as in (1.7.33) and consider a vector-valued function $u = (u_j)_{1\leq j\leq n} \in \big[W^{1,1}_{\mathrm{loc}}(\Omega,\mathcal{L}^n)\big]^n$ such that

$$(\nabla u)\Big|_{\partial\Omega}^{\kappa-\mathrm{n.t.}} \text{ exists (in } \mathbb{C}^{n^2}\text{) at } \sigma\text{-a.e. point on } \partial_*\Omega \qquad (1.7.47)$$

(in particular, u satisfies all conditions in (1.7.35), written for D as in (1.7.45)). Then thanks to (1.7.46), (1.7.42), (1.7.36), and (1.7.45) we may write

$$\begin{aligned}\partial_\nu^{A_\zeta} u = \partial_\nu^{A_{\widetilde{D}_\zeta, D}} u = \partial_\nu^{\widetilde{D}_\zeta, D} u &= (-i)\mathrm{Sym}(\widetilde{D}_\zeta; \nu)(Du)\Big|_{\partial\Omega}^{\kappa-\mathrm{n.t.}} \\ &= 2\zeta\big(\mathrm{Def}\, u\big)\Big|_{\partial\Omega}^{\kappa-\mathrm{n.t.}} \nu + (\lambda + \mu - \zeta)(\mathrm{div} u)\Big|_{\partial\Omega}^{\kappa-\mathrm{n.t.}} \nu \\ &\quad + (\mu - \zeta)(\nabla u)\Big|_{\partial\Omega}^{\kappa-\mathrm{n.t.}} \nu, \end{aligned} \tag{1.7.48}$$

where the last step also makes use of the formulas for transpose of the deformation tensor and the Jacobian operator from (1.3.58) and (1.3.57). Classically,

$$\begin{aligned}\partial_\nu^{\mathrm{Trac}} u &:= 2\mu\big(\mathrm{Def}\, u\big)\Big|_{\partial\Omega}^{\kappa-\mathrm{n.t.}} \nu + \lambda(\mathrm{div} u)\Big|_{\partial\Omega}^{\kappa-\mathrm{n.t.}} \nu \\ &= \mu\Big[(\nabla u)\Big|_{\partial\Omega}^{\kappa-\mathrm{n.t.}} + (\nabla u)^\top\Big|_{\partial\Omega}^{\kappa-\mathrm{n.t.}}\Big]\nu + \lambda(\mathrm{div} u)\Big|_{\partial\Omega}^{\kappa-\mathrm{n.t.}} \nu \end{aligned} \tag{1.7.49}$$

is referred to as the *traction* conormal derivative (for the Lamé system). Using this, we may re-frame (1.7.48) as

$$\partial_\nu^{A_\zeta} u = \partial_\nu^{\mathrm{Trac}} u + (\mu - \zeta)\Big[-2\big(\mathrm{Def}\, u\big)\Big|_{\partial\Omega}^{\kappa-\mathrm{n.t.}} \nu + (\mathrm{div} u)\Big|_{\partial\Omega}^{\kappa-\mathrm{n.t.}} \nu + (\nabla u)\Big|_{\partial\Omega}^{\kappa-\mathrm{n.t.}} \nu\Big] \tag{1.7.50}$$

so, in particular, the conormal derivative $\partial_\nu^{A_\zeta} u$ reduces precisely to the traction conormal derivative $\partial_\nu^{\mathrm{Trac}} u$ when $\zeta = \mu$. It is of interest to note (compare with Proposition 1.7.4) that, for any $\zeta \in \mathbb{C}$, the difference between the conormal derivative associated with the coefficient tensor A_ζ and the classical traction conormal derivative is given by

$$\partial_\nu^{A_\zeta} u - \partial_\nu^{\mathrm{Trac}} u = (\mu - \zeta)\Big(\nu_j(\partial_k u_k)\Big|_{\partial\Omega}^{\kappa-\mathrm{n.t.}} - \nu_k(\partial_j u_k)\Big|_{\partial\Omega}^{\kappa-\mathrm{n.t.}}\Big)_{1 \le j \le n} \tag{1.7.51}$$

which, at least if u is reasonably behaved near the boundary (cf. [113, Proposition 11.3.2]), may be further written in terms of tangential derivatives as $(\mu - \zeta)\Big(\partial_{\tau_{jk}}\big(u_k\big|_{\partial\Omega}^{\kappa-\mathrm{n.t.}}\big)\Big)_{1 \le j \le n}$. Finally, we wish to note that the *pseudo-stress conormal derivative operator* is defined when $3\mu + \lambda \neq 0$ as

$$\begin{aligned}\partial_\nu^\psi u &:= \frac{2\mu(\mu + \lambda)}{3\mu + \lambda}(\mathrm{Def}\, u)\Big|_{\partial\Omega}^{\kappa-\mathrm{n.t.}} \nu + \frac{(\mu + \lambda)(2\mu + \lambda)}{3\mu + \lambda}(\mathrm{div} u)\Big|_{\partial\Omega}^{\kappa-\mathrm{n.t.}} \nu \\ &\quad + \frac{2\mu^2}{3\mu + \lambda}(\nabla u)\Big|_{\partial\Omega}^{\kappa-\mathrm{n.t.}} \nu, \end{aligned} \tag{1.7.52}$$

to which (1.7.48) reduces precisely when

$$\zeta := \frac{\mu(\mu + \lambda)}{3\mu + \lambda}. \tag{1.7.53}$$

Moving on, the goal in the next lemma is to decompose the nontangential trace of the full gradient of a function as an algebraically "twisted" sum involving the conormal derivative and "tangential" derivatives of said function (see also Proposition 1.7.7 in this regard).

Lemma 1.7.6 *Suppose $\Omega \subseteq \mathbb{R}^n$ is an open set with the property that $\partial\Omega$ is lower Ahlfors regular and $\sigma := \mathcal{H}^{n-1}\lfloor\partial\Omega$ is a doubling measure on $\partial\Omega$. In particular, Ω is a set of locally finite perimeter, and its geometric measure theoretic outward unit normal $\nu = (\nu_1, \dots, \nu_n)$ is defined σ-a.e. on $\partial_*\Omega$. Having fixed some $M \in \mathbb{N}$, consider a coefficient tensor $A = \big(a_{rs}^{\alpha\beta}\big)_{\substack{1\le r,s\le n\\ 1\le\alpha,\beta\le M}}$ with complex entries, with the property that the $M \times M$ homogeneous second-order system $L = L_A$ associated with A in $\mathbb{R}^n$ as in* (1.3.15) *is weakly elliptic (in the sense of* (1.3.3)*). Finally, fix an aperture parameter $\kappa > 0$ and pick a function $u = (u_\gamma)_{1\le\gamma\le M} \in \big[W^{1,1}_{\rm loc}(\Omega)\big]^M$ for which the nontangential trace $(\nabla u)\big|^{\kappa-{\rm n.t.}}_{\partial\Omega}$ exists (in $\mathbb{C}^{M\times n}$) at σ-a.e. point on $\partial_*\Omega$.*

Then (with notation introduced in (1.3.2), (1.3.7), *and* (1.7.9)*) one has*

$$
\begin{aligned}
(\nabla u)\big|^{\kappa-{\rm n.t.}}_{\partial\Omega} &= -\Big(\big[L(\nu)\big]^{-1}\partial^A_\nu u\Big)\otimes\nu \\
&\quad - \left(\big[L(\nu)\big]^{-1}_{\gamma\alpha}\nu_r a^{\alpha\beta}_{rs}\Big(\nu_s(\partial_j u_\beta)\big|^{\kappa-{\rm n.t.}}_{\partial\Omega} - \nu_j(\partial_s u_\beta)\big|^{\kappa-{\rm n.t.}}_{\partial\Omega}\Big)\right)_{\substack{1\le\gamma\le M\\ 1\le j\le n}}
\end{aligned}
\tag{1.7.54}
$$

at σ-a.e. point on $\partial_*\Omega$.

Proof Based on (1.3.2), (1.3.7), (1.7.9), and (1.7.16), at σ-a.e. point on $\partial_*\Omega$ we may write

$$
\begin{aligned}
-(\nabla u)\big|^{\kappa-{\rm n.t.}}_{\partial\Omega} &- \Big(\big[L(\nu)\big]^{-1}\partial^A_\nu u\Big)\otimes\nu \\
&= \left(-(\partial_j u_\gamma)\big|^{\kappa-{\rm n.t.}}_{\partial\Omega} - \Big(\big[L(\nu)\big]^{-1}\partial^A_\nu u\Big)_\gamma \nu_j\right)_{\substack{1\le\gamma\le M\\ 1\le j\le n}} \\
&= \left(\Big(-\big[L(\nu)\big]^{-1}L(\nu)(\partial_j u)\big|^{\kappa-{\rm n.t.}}_{\partial\Omega} - \nu_j\big[L(\nu)\big]^{-1}\partial^A_\nu u\Big)_\gamma\right)_{\substack{1\le\gamma\le M\\ 1\le j\le n}} \\
&= \left(\Big(\big[L(\nu)\big]^{-1}\big(-L(\nu)(\partial_j u)\big|^{\kappa-{\rm n.t.}}_{\partial\Omega} - \nu_j\partial^A_\nu u\big)\Big)_\gamma\right)_{\substack{1\le\gamma\le M\\ 1\le j\le n}} \\
&= \left(\big[L(\nu)\big]^{-1}_{\gamma\alpha}\Big(\nu_r\nu_s a^{\alpha\beta}_{rs}(\partial_j u_\beta)\big|^{\kappa-{\rm n.t.}}_{\partial\Omega} - \nu_j\nu_r a^{\alpha\beta}_{rs}(\partial_s u_\beta)\big|^{\kappa-{\rm n.t.}}_{\partial\Omega}\right)_{\substack{1\le\gamma\le M\\ 1\le j\le n}} \\
&= \left(\big[L(\nu)\big]^{-1}_{\gamma\alpha}\nu_r a^{\alpha\beta}_{rs}\Big(\nu_s(\partial_j u_\beta)\big|^{\kappa-{\rm n.t.}}_{\partial\Omega} - \nu_j(\partial_s u_\beta)\big|^{\kappa-{\rm n.t.}}_{\partial\Omega}\Big)\right)_{\substack{1\le\gamma\le M\\ 1\le j\le n}}
\end{aligned}
\tag{1.7.55}
$$

from which the desired conclusion readily follows. □

Here is a version of [113, Proposition 11.4.4] adapted to vector-valued functions and weakly elliptic systems.

Proposition 1.7.7 *Suppose $\Omega \subseteq \mathbb{R}^n$ is an open set with the property that $\partial\Omega$ is lower Ahlfors regular and $\sigma := \mathcal{H}^{n-1}\lfloor\partial\Omega$ is a doubling measure on $\partial\Omega$. In particular, Ω is a set of locally finite perimeter, and its geometric measure theoretic outward unit normal $\nu = (\nu_1, \dots, \nu_n)$ is defined σ-a.e. on $\partial_*\Omega$. Having fixed some $M \in \mathbb{N}$, consider a coefficient tensor $A = \big(a_{rs}^{\alpha\beta}\big)_{\substack{1\leq r,s\leq n\\ 1\leq\alpha,\beta\leq M}}$ with complex entries, with the property that the $M \times M$ homogeneous second-order system $L = L_A$ associated with A in $\mathbb{R}^n$ as in* (1.3.15) *is weakly elliptic (in the sense of* (1.3.3)*). Finally, fix an integrability exponent $p \in [1, \infty]$, an aperture parameter $\kappa \in (0, \infty)$, a truncation parameter $\varepsilon > 0$, and consider a function $u = (u_\beta)_{1\leq\beta\leq M} \in \big[W^{1,1}_{\rm loc}(\Omega)\big]^M$ with*

$$\mathcal{N}_\kappa^\varepsilon u,\ \mathcal{N}_\kappa^\varepsilon(\nabla u) \in L^p_{\rm loc}(\partial\Omega, \sigma) \tag{1.7.56}$$

and such that the nontangential traces

$$\begin{aligned} u_\beta\big|^{\kappa-\rm n.t.}_{\partial\Omega} \ \text{ and } \ (\partial_j u_\beta)\big|^{\kappa-\rm n.t.}_{\partial\Omega} \ \text{ exist at } \sigma\text{-a.e. point on } \partial_*\Omega \\ \text{for each } \beta \in \{1, \dots, M\} \text{ and } j \in \{1, \dots, n\}. \end{aligned} \tag{1.7.57}$$

Then $u\big|^{\kappa-\rm n.t.}_{\partial\Omega} \in \big[L^p_{1,\rm loc}(\partial_*\Omega, \sigma)\big]^M$ and, with notation introduced in (1.3.2), (1.3.7), (1.7.9)*, at σ-a.e. point on $\partial_*\Omega$ one has*

$$\begin{aligned} (\nabla u)\big|^{\kappa-\rm n.t.}_{\partial\Omega} &= -\Big(\big[L(\nu)\big]^{-1}\partial^A_\nu u\Big) \otimes \nu \\ &\quad - \Big(\big[L(\nu)\big]^{-1}_{\gamma\alpha}\nu_r a^{\alpha\beta}_{rs}\,\partial_{\tau_{sj}}\big(u_\beta\big|^{\kappa-\rm n.t.}_{\partial\Omega}\big)\Big)_{\substack{1\leq\gamma\leq M\\ 1\leq j\leq n}}. \end{aligned} \tag{1.7.58}$$

Moreover, if one abbreviates

$$\nabla^A_{\tan} u := \Big(\nu_r a^{\alpha\beta}_{rs}\Big(\nabla_{\tan}\big(u_\beta\big|^{\kappa-\rm n.t.}_{\partial\Omega}\big)\Big)_s\Big)_{1\leq\alpha\leq M} \tag{1.7.59}$$

and

$$\nabla_{\tan} u := \Big(\Big(\nabla_{\tan}\big(u_\gamma\big|^{\kappa-\rm n.t.}_{\partial\Omega}\big)\Big)_j\Big)_{\substack{1\leq\gamma\leq M\\ 1\leq j\leq n}} \tag{1.7.60}$$

then under the stronger assumption that Ω is a UR *domain and that $p \in (1, \infty)$ one has*

$$(\nabla u)\big|^{\kappa-\rm n.t.}_{\partial\Omega} = -\Big(\big[L(\nu)\big]^{-1}\big(\partial^A_\nu u - \nabla^A_{\tan} u\big)\Big) \otimes \nu + \nabla_{\tan} u \tag{1.7.61}$$

at σ-a.e. point on $\partial\Omega$.

Proof Formula (1.7.58) is a consequence of Lemma 1.7.6 and [113, Proposition 11.3.2]. Assuming that Ω is actually a UR domain and $p \in (1, \infty)$, formula (1.7.61) follows from (1.7.58), [113, (11.4.8)], (1.7.59), and (1.7.60). □

We continue by deriving a formula relating the boundary integral of the conormal derivative of a given function to the solid integral of the action of the system (associated with the coefficient tensor which has produced the conormal derivative in the first place) on said function.

Proposition 1.7.8 *Assume $\Omega \subseteq \mathbb{R}^n$ is an open set with a lower Ahlfors regular boundary and with the property that $\sigma := \mathcal{H}^{n-1} \lfloor \partial\Omega$ is a doubling measure. In particular, Ω is a set of locally finite perimeter, and its geometric measure theoretic outward unit normal ν is defined σ-a.e. on $\partial_*\Omega$.*

Having fixed some $M, N \in \mathbb{N}$, suppose A is a complex coefficient tensor of type $(n \times n, M \times N)$ and denote by L_A the homogeneous constant (complex) coefficient second-order $M \times N$ system in $\mathbb{R}^n$ associated with A as in (1.7.7)*. Finally, fix an aperture parameter $\kappa > 0$ and consider a complex vector-valued function*

$$u \in \left[L^1_{\text{loc}}(\Omega, \mathcal{L}^n)\right]^N \tag{1.7.62}$$

satisfying (with all derivatives taken in the sense of distributions in Ω):

$$\begin{aligned} &A\nabla u \in \left[L^1_{\text{loc}}(\Omega, \mathcal{L}^n)\right]^{M\cdot n}, \quad \mathcal{N}_\kappa(A\nabla u) \in L^1(\partial\Omega, \sigma), \\ &L_A u \in \left[L^1(\Omega, \mathcal{L}^n)\right]^M, \text{ and the nontangential} \\ &\text{boundary trace } (A\nabla u)\big|^{\kappa-\text{n.t.}}_{\partial\Omega} \text{ exists } \sigma\text{-a.e. on } \partial_{\text{nta}}\Omega. \end{aligned} \tag{1.7.63}$$

Then for any other specified aperture parameter $\kappa' \in (0, \infty)$ the nontangential boundary trace $(A\nabla u)\big|^{\kappa'-\text{n.t.}}_{\partial\Omega}$ exists at σ-a.e. point on $\partial_{\text{nta}}\Omega$ and is actually independent of κ'. Moreover, with the dependence on the aperture parameter dropped and with the conormal derivative considered in the sense of (1.7.8)-(1.7.9)*, one has*

$$\int_{\partial_*\Omega} \partial^A_\nu u \, d\sigma = \int_\Omega L_A u \, d\mathcal{L}^n \tag{1.7.64}$$

when either Ω is bounded, or $\partial\Omega$ is unbounded. In the remaining case, i.e., when Ω is unbounded and $\partial\Omega$ is bounded, formula (1.7.64) *continues to hold under the additional assumption that there exists $\lambda \in (1, \infty)$ such that*[38]

$$\int_{[B(0,\lambda R)\setminus B(0,R)]\cap\Omega} |A\nabla u| \, d\mathcal{L}^n = o(R) \text{ as } R \to \infty. \tag{1.7.65}$$

As a corollary, if the last property in the first line of (1.7.63) *is replaced by*

$$L_A u = 0 \text{ in the sense of distributions in } \Omega, \tag{1.7.66}$$

[38] for example, (1.7.65) holds if $|A(\nabla u)(x)| = o(|x|^{1-n})$ as $|x| \to \infty$

then (1.7.64) *becomes*

$$\int_{\partial_*\Omega} \partial_\nu^A u \, d\sigma = 0. \tag{1.7.67}$$

Proof Assume the coefficient tensor A is explicitly expressed as in (1.7.1). Fix an arbitrary index $\alpha \in \{1, \dots, M\}$ and define the vector field

$$\vec{F}_\alpha := \left(a_{rs}^{\alpha\beta} \partial_s u_\beta\right)_{1 \le r \le n} \quad \text{in } \Omega, \tag{1.7.68}$$

where $(u_\beta)_{1 \le \beta \le N}$ are the scalar components of the vector-valued function u. Then [112, Theorem 1.2.1] applies to $\vec{F}_\alpha$ and gives, under the conditions stipulated in the current statement, that

$$\int_{\partial_*\Omega} \left(\partial_\nu^A u\right)_\alpha d\sigma = \int_\Omega (L_A u)_\alpha \, d\mathcal{L}^n. \tag{1.7.69}$$

In view of the arbitrariness of α, all desired conclusions follow. □

Theorem 1.7.10, stated a little below, establishes a version of Green's second identity for second-order systems in a very general geometric and analytic setting. As a preamble, this requires that we make one quick definition.

Definition 1.7.9 *Suppose*

$$\begin{array}{l} \Sigma \subseteq \mathbb{R}^n \text{ is a } \mathcal{H}^{n-1}\text{-measurable set satisfying} \\ \mathcal{H}^{n-1}(\Sigma \cap K) < \infty \text{ for each compact } K \subset \mathbb{R}^n. \end{array} \tag{1.7.70}$$

As in the past, abbreviate $\sigma := \mathcal{H}^{n-1} \lfloor \Sigma$ *and denote by* $L^0(\Sigma, \sigma)$ *the space of all* μ*-measurable functions* f *on* Σ *with the property that* $|f| < +\infty$ *at* σ*-a.e. on* Σ*. In this setting, call* $(\mathbb{X}, \mathbb{Y})$ *a* `dual system` *on* (Σ, σ) *provided*

$$\begin{array}{l} \mathbb{X}, \mathbb{Y} \text{ are subsets of } L^0(\Sigma, \sigma) \text{ with the property that} \\ \int_\Sigma |f||g| \, d\sigma < \infty \text{ for any given } f \in \mathbb{X} \text{ and } g \in \mathbb{Y}. \end{array} \tag{1.7.71}$$

Finally, call a dual system $(\mathbb{X}, \mathbb{Y})$ *on* (Σ, σ) `regular` *if, in fact,* $\mathbb{X}, \mathbb{Y}$ *are subsets of* $\bigcup_{0<p\le\infty} L^p_{\mathrm{loc}}(\Sigma, \sigma)$.

Basic examples of regular dual systems are offered by: pairs of dual Lebesgue spaces, i.e.,

$$\begin{array}{l} \text{let } \Sigma \text{ be as in (1.7.70) and, with } \sigma := \mathcal{H}^{n-1} \lfloor \Sigma, \text{ take } \mathbb{X} := L^p(\Sigma, \sigma) \text{ and} \\ \mathbb{Y} := L^{p'}(\Sigma, \sigma) \text{ with } p, p' \in [1, \infty] \text{ satisfying } 1/p + 1/p' = 1, \end{array} \tag{1.7.72}$$

pairs of dual Lorentz spaces (cf. O'Neil's inequality recorded in [112, (6.2.61)]), i.e.,

$$\begin{aligned}&\text{let } \Sigma \text{ be as in (1.7.70) and, with } \sigma := \mathcal{H}^{n-1}\lfloor\Sigma, \text{ take } \mathbb{X} := L^{p,q}(\Sigma,\sigma)\\&\text{together with } \mathbb{Y} := L^{p',q'}(\Sigma,\sigma) \text{ where the indices } p,q,p',q' \in [1,\infty]\\&\text{are assumed to satisfy } 1/p + 1/p' = 1 \text{ as well as } 1/q + 1/q' = 1,\end{aligned} \tag{1.7.73}$$

pairs of dual Muckenhoupt weighted Lebesgue spaces (cf. item *(2)* in [112, Lemma 7.7.1]), i.e.,

$$\begin{aligned}&\text{let } \Sigma \text{ be a closed Ahlfors regular set in } \mathbb{R}^n \text{ and, with } \sigma := \mathcal{H}^{n-1}\lfloor\Sigma,\\&\text{take } \mathbb{X} := L^p(\Sigma, w\sigma) \text{ and } \mathbb{Y} := L^{p'}(\Sigma, w^{-p'/p}\sigma) \text{ with } p,p' \in (1,\infty)\\&\text{satisfying } 1/p + 1/p' = 1, \text{ and the weight } w \text{ in Muckenhoupt's class}\\&A_p(\Sigma,\sigma),\end{aligned} \tag{1.7.74}$$

the pair consisting of a Morrey space and its pre-dual (cf. [113, Proposition 6.2.8, (6.2.78)]), i.e.,

$$\begin{aligned}&\text{let } \Sigma \text{ be a closed Ahlfors regular set in } \mathbb{R}^n \text{ and, with } \sigma := \mathcal{H}^{n-1}\lfloor\Sigma, \text{ take}\\&\mathbb{X} := M^{p,\lambda}(\Sigma,\sigma) \text{ and } \mathbb{Y} := \mathcal{B}^{q,\lambda}(\Sigma,\sigma) \text{ with exponents } p,q \in (1,\infty)\\&\text{satisfying } 1/p + 1/q = 1, \text{ and the parameter } \lambda \text{ belonging to } (0, n-1).\end{aligned} \tag{1.7.75}$$

While in all these examples it happens that $\mathbb{X}$ and $\mathbb{Y}$ are linear spaces, we emphasize that this is not a requirement in the original definition of a dual system. In fact, a dual system $(\mathbb{X}, \mathbb{Y})$ may simply be a pair of singletons (which is actually an important particular case).

With this preamble taken care of, here is the general version of Green's second identity promised earlier.

Theorem 1.7.10 *Assume $\Omega \subseteq \mathbb{R}^n$ is an open set with a lower Ahlfors regular boundary and with the property that $\sigma := \mathcal{H}^{n-1}\lfloor\partial\Omega$ is a doubling measure. In particular, Ω is a set of locally finite perimeter, and its geometric measure theoretic outward unit normal ν is defined σ-a.e. on $\partial_*\Omega$.*

Having fixed some $M, N \in \mathbb{N}$, suppose A is a complex coefficient tensor of type $(n \times n, M \times N)$ and denote by L_A the homogeneous constant (complex) coefficient second-order $M \times N$ system in $\mathbb{R}^n$ associated with A as in (1.7.7)*. Finally, fix $\kappa, \kappa' > 0$, assume $(\mathbb{X}_0, \mathbb{Y}_0)$ and $(\mathbb{X}_1, \mathbb{Y}_1)$ are two regular dual systems on $(\partial\Omega, \sigma)$ (in the sense of Definition 1.7.9), and consider two complex vector-valued functions*

$$u \in \big[L^1_{\mathrm{loc}}(\Omega, \mathcal{L}^n)\big]^N, \quad w \in \big[L^1_{\mathrm{loc}}(\Omega, \mathcal{L}^n)\big]^M, \tag{1.7.76}$$

satisfying (with all derivatives taken in the sense of distributions in Ω):

$$\mathcal{N}_\kappa u \in \mathbb{X}_0, \quad A\nabla u \in \big[L^1_{\rm loc}(\Omega, \mathcal{L}^n)\big]^{M\cdot n}, \quad \mathcal{N}_\kappa(A\nabla u) \in \mathbb{X}_1,$$
$$\mathcal{N}_{\kappa'} w \in \mathbb{Y}_1, \quad A^\top\nabla w \in \big[L^1_{\rm loc}(\Omega, \mathcal{L}^n)\big]^{N\cdot n}, \quad \mathcal{N}_{\kappa'}(A^\top\nabla w) \in \mathbb{Y}_0,$$

$$u\big|^{\kappa-{\rm n.t.}}_{\partial\Omega}, \ (A\nabla u)\big|^{\kappa-{\rm n.t.}}_{\partial\Omega}, \ w\big|^{\kappa'-{\rm n.t.}}_{\partial\Omega}, \ (A^\top\nabla w)\big|^{\kappa'-{\rm n.t.}}_{\partial\Omega} \ \text{exist at } \sigma\text{-a.e. point on } \partial_{\rm nta}\Omega, \tag{1.7.77}$$

$$L_A u \in \big[L^1_{\rm loc}(\Omega, \mathcal{L}^n)\big]^M, \quad L_A^\top w \in \big[L^1_{\rm loc}(\Omega, \mathcal{L}^n)\big]^N,$$
$$\langle L_A u, w\rangle \in L^1(\Omega, \mathcal{L}^n), \quad \langle u, L_A^\top w\rangle \in L^1(\Omega, \mathcal{L}^n).$$

Then for any other specified aperture parameter $\kappa'' \in (0,\infty)$ *the nontangential boundary traces* $u\big|^{\kappa''-{\rm n.t.}}_{\partial\Omega}$, $(A\nabla u)\big|^{\kappa''-{\rm n.t.}}_{\partial\Omega}$, $w\big|^{\kappa''-{\rm n.t.}}_{\partial\Omega}$, $(A^\top\nabla w)\big|^{\kappa''-{\rm n.t.}}_{\partial\Omega}$ *exist at* σ*-a.e. point on* $\partial_{\rm nta}\Omega$ *and are independent of* κ''*. Moreover, with the dependence on the aperture parameter dropped and with the conormal derivatives interpreted in the sense of* (1.7.8)-(1.7.9)*, the following Green type formula (involving absolutely convergent integrals) holds*

$$\int_\Omega \langle L_A u, w\rangle \, d\mathcal{L}^n - \int_\Omega \langle u, L_A^\top w\rangle \, d\mathcal{L}^n$$
$$= \int_{\partial_*\Omega} \big\langle \partial^A_\nu u, w\big|^{\rm n.t.}_{\partial\Omega}\big\rangle \, d\sigma - \int_{\partial_*\Omega} \big\langle u\big|^{\rm n.t.}_{\partial\Omega}, \partial^{A^\top}_\nu w\big\rangle \, d\sigma, \tag{1.7.78}$$

when either Ω *is bounded, or* $\partial\Omega$ *is unbounded. In the remaining case, i.e., when* Ω *is unbounded and* $\partial\Omega$ *is bounded, formula* (1.7.78) *continues to hold under the additional assumption that there exists* $\lambda \in (1,\infty)$ *such that*[39]

$$\int_{[B(0,\lambda R)\setminus B(0,R)]\cap\Omega} \big\{|A\nabla u||w| + |u||A^\top\nabla w|\big\} \, d\mathcal{L}^n = o(R) \ \text{ as } R \to \infty. \tag{1.7.79}$$

As a corollary, if the last two lines in (1.7.77) *are replaced by*

$$L_A u = 0 \text{ and } L_A^\top w = 0 \ \text{ in the sense of distributions in } \Omega, \tag{1.7.80}$$

then (1.7.78) *becomes*

$$\int_{\partial_*\Omega} \big\langle \partial^A_\nu u, w\big|^{\rm n.t.}_{\partial\Omega}\big\rangle \, d\sigma = \int_{\partial_*\Omega} \big\langle u\big|^{\rm n.t.}_{\partial\Omega}, \partial^{A^\top}_\nu w\big\rangle \, d\sigma. \tag{1.7.81}$$

Finally, similar results are valid if $(\mathbb{X}_0, \mathbb{Y}_0)$ *and* $(\mathbb{X}_1, \mathbb{Y}_1)$ *are two dual systems on* $(\partial\Omega, \sigma)$ *which are not necessarily regular, with the understanding that, in such a case, all nontangential boundary traces are taken with respect to an aperture parameter which is* $< \min\{\kappa, \kappa'\}$*.*

Before presenting the proof of this theorem we wish to note that a related version holds with the doubling assumption on $\sigma := \mathcal{H}^{n-1}\lfloor\partial\Omega$ relaxed to simply asking

[39] e.g., (1.7.79) holds if $|A(\nabla u)(x)||w(x)| = o(|x|^{1-n})$ and $|u(x)||A^\top(\nabla w)(x)| = o(|x|^{1-n})$

that this is a locally finite measure. In such a scenario, the aperture parameter κ is assumed to be sufficiently large (depending on Ω), and the flexibility of changing κ when considering nontangential boundary traces may be lost. Other than these nuances, the format of the main result (i.e., formula (1.7.78)) remains unchanged. The proof of this version of Theorem 1.7.10 requires that [112, Theorem 1.5.1] is employed in place of [112, Theorem 1.4.1] (which is used below).

In the important particular case when each of the duals systems $(\mathbb{X}_0, \mathbb{Y}_0)$ and $(\mathbb{X}_1, \mathbb{Y}_1)$ is a pair of singletons, the nontangential maximal function memberships in the first two lines of (1.7.77) simply become the integrability conditions

$$\int_{\partial\Omega} \mathcal{N}_\kappa u \cdot \mathcal{N}_{\kappa'}(A^\top \nabla w)\,\mathrm{d}\sigma < +\infty \text{ and } \int_{\partial\Omega} \mathcal{N}_\kappa(A\nabla u) \cdot \mathcal{N}_{\kappa'} w\,\mathrm{d}\sigma < +\infty. \tag{1.7.82}$$

Another basic particular choice is having $(\mathbb{X}_0, \mathbb{Y}_0)$ and $(\mathbb{X}_1, \mathbb{Y}_1)$ pairs of dual Lebesgue spaces on $(\partial\Omega, \sigma)$. Specifically, with the exponents $p, q, p', q' \in [1, \infty]$ such that $1/p + 1/p' = 1 = 1/q + 1/q'$, this amounts to replacing the nontangential maximal function memberships in the first two lines of (1.7.77) by

$$\begin{aligned} &\mathcal{N}_\kappa u \in L^p(\partial\Omega, \sigma), \quad \mathcal{N}_\kappa(A\nabla u) \in L^q(\partial\Omega, \sigma), \\ &\mathcal{N}_{\kappa'} w \in L^{q'}(\partial\Omega, \sigma), \quad \mathcal{N}_{\kappa'}(A^\top \nabla w) \in L^{p'}(\partial\Omega, \sigma). \end{aligned} \tag{1.7.83}$$

Assuming this is the case, we wish to further single out a natural scenario when the integrability conditions in the very last line in (1.7.77) are naturally satisfied. Concretely, if the demands that $L_A u \in \big[L^1_{\mathrm{loc}}(\Omega, \mathcal{L}^n)\big]^M$ and $L_A^\top w \in \big[L^1_{\mathrm{loc}}(\Omega, \mathcal{L}^n)\big]^N$, made in the penultimate line in (1.7.77), are strengthened to

$$\begin{aligned} &L_A u \in \big[L^{nq'/(nq'-n+1)}(\Omega, \mathcal{L}^n)\big]^M \text{ and} \\ &L_A^\top w \in \big[L^{np/(np-n+1)}(\Omega, \mathcal{L}^n)\big]^N, \end{aligned} \tag{1.7.84}$$

then the two integrability conditions in the last line of (1.7.77) are automatically satisfied when Ω is bounded or $\partial\Omega$ is unbounded, since in such a scenario [112, Proposition 8.6.3] and [112, Lemma 8.3.2] permit us to conclude that

$$\begin{aligned} &\mathcal{N}_\kappa u \in L^p(\partial\Omega, \sigma) \Rightarrow u \in \Big[L^{\frac{np}{n-1}}(\Omega, \mathcal{L}^n)\Big]^N = \Big(\big[L^{\frac{np}{np-n+1}}(\Omega, \mathcal{L}^n)\big]^N\Big)^*, \\ &\mathcal{N}_{\kappa'} w \in L^{q'}(\partial\Omega, \sigma) \Rightarrow w \in \Big[L^{\frac{nq'}{n-1}}(\Omega, \mathcal{L}^n)\Big]^M = \Big(\big[L^{\frac{nq'}{nq'-n+1}}(\Omega, \mathcal{L}^n)\big]^M\Big)^*. \end{aligned} \tag{1.7.85}$$

Let us also take a look at the case when Ω is an exterior domain, under the assumption that the functions u, w satisfy

$$\begin{aligned} &u(x) = O(|x|^{1-n}) \text{ and } w(x) = O(|x|^{1-n}) \text{ as } |x| \to \infty, \\ &(\nabla u)(x) = o(1) \text{ and } (\nabla w)(x) = o(1) \text{ as } |x| \to \infty, \end{aligned} \tag{1.7.86}$$

(which, incidentally, implies that condition (1.7.79) holds). Granted this, and further assuming that $p > 1$, $q < \infty$, and that (1.7.84) replaces the penultimate line in (1.7.77), it follows that the two integrability conditions in the last line of (1.7.77) are also satisfied. Indeed, while in this case in place of (1.7.85) we only have (recall that $L^p_{\rm bdd}$ has been introduced in (A.0.65))

$$\begin{aligned} &\mathcal{N}_\kappa u \in L^p(\partial\Omega,\sigma) \Longrightarrow u \in \big[L^{np/(n-1)}_{\rm bdd}(\Omega,\mathcal{L}^n)\big]^N, \\ &\mathcal{N}_{\kappa'} w \in L^{q'}(\partial\Omega,\sigma) \Longrightarrow w \in \big[L^{nq'/(n-1)}_{\rm bdd}(\Omega,\mathcal{L}^n)\big]^M, \end{aligned} \tag{1.7.87}$$

the decay conditions in the first line of (1.7.86) do ensure that both $|u|^{np/(n-1)}$ and $|w|^{nq'/(n-1)}$ are integrable in a neighborhood of infinity so that, ultimately, we still have

$$\begin{aligned} u &\in \big[L^{np/(n-1)}(\Omega,\mathcal{L}^n)\big]^N = \Big(\big[L^{np/(np-n+1)}(\Omega,\mathcal{L}^n)\big]^N\Big)^*, \\ w &\in \big[L^{nq'/(n-1)}(\Omega,\mathcal{L}^n)\big]^M = \Big(\big[L^{nq'/(nq'-n+1)}(\Omega,\mathcal{L}^n)\big]^M\Big)^*, \end{aligned} \tag{1.7.88}$$

in this case as well.

Here is the actual proof of Theorem 1.7.10.

Proof of Theorem 1.7.10 From assumptions and [112, Corollary 8.9.9] it follows that there exists an exponent $p \in (0,\infty]$ with the property that for any given aperture parameter $\kappa'' > 0$ we have

$$\mathcal{N}_{\kappa''} u,\ \mathcal{N}_{\kappa''}(A\nabla u),\ \mathcal{N}_{\kappa''} w,\ \mathcal{N}_{\kappa''}(A^\top\nabla w) \in L^p_{\rm loc}(\partial\Omega,\sigma), \tag{1.7.89}$$

and the nontangential boundary traces

$$\begin{aligned} &u\big|^{\kappa''-{\rm n.t.}}_{\partial\Omega},\ (A\nabla u)\big|^{\kappa''-{\rm n.t.}}_{\partial\Omega},\ w\big|^{\kappa''-{\rm n.t.}}_{\partial\Omega},\ (A^\top\nabla w)\big|^{\kappa''-{\rm n.t.}}_{\partial\Omega} \text{ exist } \sigma\text{-a.e. on } \partial_{\rm nta}\Omega \\ &\text{and agree with the respective boundary traces in (1.7.77).} \end{aligned} \tag{1.7.90}$$

To proceed, note that (1.7.76), the first two lines in (1.7.77) (or (1.7.89)), and [112, Lemma 8.3.1] imply

$$\begin{aligned} &u \in \big[L^\infty_{\rm loc}(\Omega,\mathcal{L}^n)\big]^N, \quad A\nabla u \in \big[L^\infty_{\rm loc}(\Omega,\mathcal{L}^n)\big]^{M\cdot n}, \\ &w \in \big[L^\infty_{\rm loc}(\Omega,\mathcal{L}^n)\big]^M, \quad A^\top\nabla w \in \big[L^\infty_{\rm loc}(\Omega,\mathcal{L}^n)\big]^{N\cdot n}. \end{aligned} \tag{1.7.91}$$

Consider now the vector field $\vec{F} = (F_r)_{1\le r\le n} : \Omega \to \mathbb{C}^n$ with components given (with the summation convention over repeated indices enforced throughout the proof) by

$$
\begin{aligned}
F_r &:= (A\nabla u)_{\alpha,r} w_\alpha - u_\beta (A^\top \nabla w)_{\beta,r} \\
&= (a^{\alpha\beta}_{rs}\partial_s u_\beta) w_\alpha - u_\beta (a^{\alpha\beta}_{sr}\partial_s w_\alpha) \in L^\infty_{\rm loc}(\Omega, \mathcal{L}^n),
\end{aligned}
\tag{1.7.92}
$$

where the final membership is ensured by (1.7.91). In particular,

$$
\vec{F} \in \big[L^1_{\rm loc}(\Omega, \mathcal{L}^n)\big]^n. \tag{1.7.93}
$$

Let us compute $\operatorname{div}\vec{F}$ in the sense of distributions in Ω. With this goal in mind, fix an arbitrary scalar-valued function $\varphi \in \mathscr{C}^\infty_c(\Omega)$ and write

$$
\begin{aligned}
{}_{\mathcal{D}'(\Omega)}\big\langle \operatorname{div}\vec{F}, \varphi\big\rangle_{\mathcal{D}(\Omega)} &= -\int_\Omega F_r \partial_r \varphi \,\mathrm{d}\mathcal{L}^n \\
&= -\int_\Omega (a^{\alpha\beta}_{rs}\partial_s u_\beta) w_\alpha \partial_r \varphi \,\mathrm{d}\mathcal{L}^n \\
&\quad + \int_\Omega u_\beta (a^{\alpha\beta}_{sr}\partial_s w_\alpha)\partial_r \varphi \,\mathrm{d}\mathcal{L}^n \\
&=: I + II.
\end{aligned}
\tag{1.7.94}
$$

For each sufficiently small $\varepsilon > 0$ consider $\Omega_\varepsilon := \big\{x \in \Omega : \operatorname{dist}(x, \partial\Omega) > \varepsilon\big\}$. Using a Friedrichs mollifier, we may construct a sequence $u^\varepsilon = (u^\varepsilon_\beta)_{1\leq\beta\leq N} \in \big[\mathscr{C}^\infty(\Omega_\varepsilon)\big]^N$ such that

$$
\left\{
\begin{array}{l}
u^\varepsilon \xrightarrow[\varepsilon\to 0^+]{} u \text{ at } \mathcal{L}^n\text{-a.e. point in } \Omega, \text{ and} \\
A\nabla u^\varepsilon \xrightarrow[\varepsilon\to 0^+]{} A\nabla u \text{ at } \mathcal{L}^n\text{-a.e. point in } \Omega; \\
\text{moreover, for each fixed compact set } K \subset \Omega \\
\text{we have } L_A u^\varepsilon \xrightarrow[\varepsilon\to 0^+]{} L_A u \text{ in } \big[L^1(K, \mathcal{L}^n)\big]^M \\
\text{and there exists some small } \varepsilon_K > 0 \text{ such that} \\
\sup\limits_{0<\varepsilon<\varepsilon_K} \Big\{\|u^\varepsilon\|_{[L^\infty(K,\mathcal{L}^n)]^N} + \big\|A\nabla u^\varepsilon\big\|_{[L^\infty(K,\mathcal{L}^n)]^{M\cdot n}}\Big\} < \infty,
\end{array}
\right.
\tag{1.7.95}
$$

and also a sequence $w^\varepsilon = (w^\varepsilon_\alpha)_{1\leq\alpha\leq M} \in \big[\mathscr{C}^\infty(\Omega_\varepsilon)\big]^M$ such that

$$\begin{aligned}&w^\varepsilon \xrightarrow[\varepsilon\to 0^+]{} w \text{ at } \mathcal{L}^n\text{-a.e. point in } \Omega, \text{ and}\\ &A^\top\nabla w^\varepsilon \xrightarrow[\varepsilon\to 0^+]{} A^\top\nabla w \text{ at } \mathcal{L}^n\text{-a.e. point in } \Omega;\\ &\text{in addition, for each given compact set } K\subset\Omega\\ &\text{we have } L_A^\top w^\varepsilon \xrightarrow[\varepsilon\to 0^+]{} L_A^\top w \text{ in } \big[L^1(K,\mathcal{L}^n)\big]^N\\ &\text{and there exists some small } \varepsilon_K>0 \text{ such that}\\ &\sup_{0<\varepsilon<\varepsilon_K}\Big\{\|w^\varepsilon\|_{[L^\infty(K,\mathcal{L}^n)]^M}+\big\|A^\top\nabla w^\varepsilon\big\|_{[L^\infty(K,\mathcal{L}^n)]^{N\cdot n}}\Big\}<\infty.\end{aligned} \tag{1.7.96}$$

In relation to these, let us make a useful elementary observation of general nature, to the that effect that

> if (X,μ) is a measure space, $\{f_j\}_{j\in\mathbb{N}}\subseteq L^1(X,\mu)$ is a convergent sequence in $L^1(X,\mu)$ to some $f\in L^1(X,\mu)$, and $\{g_j\}_{j\in\mathbb{N}}$ is a bounded sequence in $L^\infty(X,\mu)$ such that the pointwise limit $g:=\lim_{j\to\infty} g_j$ exists μ-a.e. in X, then $fg\in L^1(X,\mu)$ and $\int_X f_j g_j\,d\mu \longrightarrow \int_X fg\,d\mu$ as $j\to\infty$. (1.7.97)

Then, based on (1.7.91), (1.7.95), (1.7.96), the last two lines in (1.7.77), Lebesgue's Dominated Convergence Theorem, and (1.7.97) we may compute

$$\begin{aligned} I &= -\lim_{\varepsilon\to 0^+}\int_\Omega (a_{rs}^{\alpha\beta}\partial_s u_\beta^\varepsilon) w_\alpha^\varepsilon \partial_r\varphi\,d\mathcal{L}^n\\ &= -\lim_{\varepsilon\to 0^+}\int_\Omega (a_{rs}^{\alpha\beta}\partial_s u_\beta^\varepsilon)\partial_r(w_\alpha^\varepsilon\varphi)\,d\mathcal{L}^n + \lim_{\varepsilon\to 0^+}\int_\Omega (a_{rs}^{\alpha\beta}\partial_s u_\beta^\varepsilon)(\partial_r w_\alpha^\varepsilon)\varphi\,d\mathcal{L}^n\\ &= \lim_{\varepsilon\to 0^+}\int_\Omega (L_A u^\varepsilon)_\alpha w_\alpha^\varepsilon\varphi\,d\mathcal{L}^n + \lim_{\varepsilon\to 0^+}\int_\Omega \big\langle A\nabla u^\varepsilon,\nabla w^\varepsilon\big\rangle\varphi\,d\mathcal{L}^n\\ &= \int_\Omega \big\langle L_A u, w\big\rangle\varphi\,d\mathcal{L}^n + \lim_{\varepsilon\to 0^+}\int_\Omega \big\langle A\nabla u^\varepsilon,\nabla w^\varepsilon\big\rangle\varphi\,d\mathcal{L}^n. \end{aligned} \tag{1.7.98}$$

In a similar fashion,

$$\begin{aligned} II &= -\int_\Omega u_\beta (L_A^\top w)_\beta\varphi\,d\mathcal{L}^n - \lim_{\varepsilon\to 0^+}\int_\Omega (\partial_r u_\beta^\varepsilon)(a_{sr}^{\alpha\beta}\partial_s w_\alpha^\varepsilon)\varphi\,d\mathcal{L}^n\\ &= -\int_\Omega \big\langle u, L_A^\top w\big\rangle\varphi\,d\mathcal{L}^n - \lim_{\varepsilon\to 0^+}\int_\Omega \big\langle A\nabla u^\varepsilon,\nabla w^\varepsilon\big\rangle\varphi\,d\mathcal{L}^n. \end{aligned} \tag{1.7.99}$$

At this stage, from (1.7.94) and (1.7.98)-(1.7.99) we conclude that

$$\operatorname{div}\vec{F} = \big\langle L_A u,\, w\big\rangle - \big\langle u,\, L_A^\top w\big\rangle \text{ in } \mathcal{D}'(\Omega). \tag{1.7.100}$$

In particular, the last line in (1.7.77) ensures that

$$\mathrm{div}\vec{F} \in L^1(\Omega, \mathcal{L}^n). \tag{1.7.101}$$

Also, if we choose an aperture parameter $\kappa_* \in \big(0, \min\{\kappa, \kappa'\}\big)$, from (1.7.92) and [112, (8.2.25)] we see that

$$\mathcal{N}_{\kappa_*}\vec{F} \leq C\Big\{\mathcal{N}_\kappa(A\nabla u)\mathcal{N}_{\kappa'}w + \mathcal{N}_\kappa u\mathcal{N}_{\kappa'}(A^\top\nabla w)\Big\} \text{ on } \partial\Omega, \tag{1.7.102}$$

for some constant $C \in (0, \infty)$ which depends only on the coefficient tensor A. In turn, from (1.7.102), [112, (8.2.26)], the nontangential maximal function memberships in the first two lines of (1.7.77), and (1.7.71) we conclude that

$$\mathcal{N}_{\kappa_*}\vec{F} \in L^1(\partial\Omega, \sigma). \tag{1.7.103}$$

Going further, observe that (1.7.92) and (1.7.90) imply that the nontangential boundary trace

$$\vec{F}\Big|_{\partial\Omega}^{\kappa''-\mathrm{n.t.}} \text{ exists at } \sigma\text{-a.e. point on } \partial_{\mathrm{nta}}\Omega. \tag{1.7.104}$$

In fact, with the dependence on κ'' dropped, for each $r \in \{1, \dots, n\}$ we have

$$\Big(\vec{F}\Big|_{\partial\Omega}^{\mathrm{n.t.}}\Big)_r = \big((a_{rs}^{\alpha\beta}\partial_s u_\beta)\big|_{\partial\Omega}^{\mathrm{n.t.}}\big)\big(w_\alpha\big|_{\partial\Omega}^{\mathrm{n.t.}}\big) - \big(u_\beta\big|_{\partial\Omega}^{\mathrm{n.t.}}\big)\big((a_{sr}^{\alpha\beta}\partial_s w_\alpha)\big|_{\partial\Omega}^{\mathrm{n.t.}}\big) \tag{1.7.105}$$

at σ-a.e. point on $\partial_{\mathrm{nta}}\Omega$. As a consequence of this, (1.7.9), and [112, Proposition 8.8.6], we obtain

$$\nu \cdot \big(\vec{F}\big|_{\partial\Omega}^{\mathrm{n.t.}}\big) = \big\langle \partial_\nu^A u, w\big|_{\partial\Omega}^{\mathrm{n.t.}}\big\rangle - \big\langle u\big|_{\partial\Omega}^{\mathrm{n.t.}}, \partial_\nu^{A^\top} w\big\rangle \text{ at } \sigma\text{-a.e. point on } \partial_*\Omega. \tag{1.7.106}$$

Finally, (1.7.79) ensures that the vector field $\vec{F}$ satisfies the growth condition [112, (1.2.3)]. Granted these properties of $\vec{F}$, [112, Theorem 1.2.1] applies and the Divergence Formula [112, (1.2.2)] presently yields (1.7.78) on account of (1.7.100) and (1.7.106). □

Other versions of Theorem 1.7.10 of practical interest are valid. To state such a result, given an open set $\Omega \subseteq \mathbb{R}^n$, the reader is reminded that the pairing ${}_{\mathscr{E}'(\Omega)}\langle\!\langle \cdot, \cdot \rangle\!\rangle_{\mathscr{E}(\Omega)}$ has been defined in (A.0.104).

Theorem 1.7.11 *Suppose $\Omega \subseteq \mathbb{R}^n$ is an open set with a lower Ahlfors regular boundary and with the property that $\sigma := \mathcal{H}^{n-1}\lfloor\partial\Omega$ is a doubling measure. Hence, Ω is a set of locally finite perimeter, and its geometric measure theoretic outward unit normal ν is defined σ-a.e. on $\partial_*\Omega$.*

Given some arbitrary $M, N \in \mathbb{N}$, suppose A is a complex coefficient tensor of type $(n \times n, M \times N)$ and denote by L_A the homogeneous constant (complex) coefficient second-order $M \times N$ system in $\mathbb{R}^n$ associated with A as in (1.7.7). *Also, pick some aperture parameters $\kappa, \kappa' > 0$, and assume $(\mathbb{X}_0, \mathbb{Y}_0)$ and $(\mathbb{X}_1, \mathbb{Y}_1)$ are two regular dual systems on $(\partial\Omega, \sigma)$ (in the sense of Definition 1.7.9). Finally, consider two complex vector-valued functions*

$$u \in \left[W^{1,1}_{\rm loc}(\Omega)\right]^N, \quad w \in \left[W^{1,1}_{\rm loc}(\Omega)\right]^M, \tag{1.7.107}$$

satisfying (with all derivatives taken in the sense of distributions in Ω*):*

$$\begin{aligned} &\mathcal{N}_\kappa u \in \mathbb{X}_0 \text{ and } \mathcal{N}_\kappa(\nabla u) \in \mathbb{X}_1, \\ &\mathcal{N}_{\kappa'} w \in \mathbb{Y}_1 \text{ and } \mathcal{N}_\kappa(\nabla w) \in \mathbb{Y}_0, \\ &u\big|^{\kappa-\rm n.t.}_{\partial\Omega}, (\nabla u)\big|^{\kappa-\rm n.t.}_{\partial\Omega}, w\big|^{\kappa'-\rm n.t.}_{\partial\Omega}, (\nabla w)\big|^{\kappa'-\rm n.t.}_{\partial\Omega} \text{ exist } \sigma\text{-a.e. on } \partial_{\rm nta}\Omega, \\ &L_A u \in \left[\mathscr{E}'(\Omega)\right]^M \text{ and } w \text{ is of class } \mathscr{C}^\infty \text{ near } \operatorname{supp}(L_A u), \\ &L_A^\top w \in \left[\mathscr{E}'(\Omega)\right]^N \text{ and } u \text{ is of class } \mathscr{C}^\infty \text{ near } \operatorname{supp}(L_{A^\top} w). \end{aligned} \tag{1.7.108}$$

Then for any other specified aperture parameter $\kappa'' \in (0, \infty)$ *the nontangential boundary traces* $u\big|^{\kappa''-\rm n.t.}_{\partial\Omega}$, $(\nabla u)\big|^{\kappa''-\rm n.t.}_{\partial\Omega}$, $w\big|^{\kappa''-\rm n.t.}_{\partial\Omega}$, $(\nabla w)\big|^{\kappa''-\rm n.t.}_{\partial\Omega}$ *exist at* σ*-a.e. point on* $\partial_{\rm nta}\Omega$ *and are independent of* κ''*. In addition, with the dependence on the aperture parameter dropped and with the conormal derivatives interpreted in the sense of* (1.7.8)-(1.7.9)*, the following Green-type formula holds*

$$\begin{aligned} &{}_{[\mathscr{E}'(\Omega)]^M}\big\langle\!\big\langle L_A u, w\big\rangle\!\big\rangle_{[\mathscr{E}(\Omega)]^M} - {}_{[\mathscr{E}'(\Omega)]^N}\big\langle\!\big\langle L_{A^\top} w, u\big\rangle\!\big\rangle_{[\mathscr{E}(\Omega)]^N} \\ &\quad = \int_{\partial_*\Omega} \big\langle \partial^A_\nu u, w\big|^{\rm n.t.}_{\partial\Omega}\big\rangle\, d\sigma - \int_{\partial_*\Omega} \big\langle u\big|^{\rm n.t.}_{\partial\Omega}, \partial^{A^\top}_\nu w\big\rangle\, d\sigma, \end{aligned} \tag{1.7.109}$$

when either Ω *is bounded, or* $\partial\Omega$ *is unbounded. In the remaining case, i.e., when* Ω *is unbounded and* $\partial\Omega$ *is bounded, formula* (1.7.109) *continues to be true under the additional assumption that there exists* $\lambda \in (1, \infty)$ *such that* (1.7.79) *holds.*

Finally, similar results hold when $(\mathbb{X}_0, \mathbb{Y}_0)$ *and* $(\mathbb{X}_1, \mathbb{Y}_1)$ *are two dual systems on* $(\partial\Omega, \sigma)$ *which are not necessarily regular, with the understanding that, in such a case, all nontangential boundary traces are taken with respect to an aperture parameter which is* $< \min\{\kappa, \kappa'\}$.

Proof We proceed along the lines of the argument in the proof of Theorem 1.7.10, so we will only indicate the main changes. Recall the $\mathscr{C}^\infty$-singular support of a distribution from [112, Definition 2.8.5]. Pick $\psi \in \mathscr{C}^\infty_c(\Omega)$ with $\psi \equiv 1$ near $\operatorname{supp}(L_A u)$ and whose support is contained in $\mathscr{C}^\infty$-singsup u, along with $\xi \in \mathscr{C}^\infty_c(\Omega)$ with $\xi \equiv 1$ near $\operatorname{supp}(L_{A^\top} w)$ and whose support is contained in $\mathscr{C}^\infty$-singsup w. Using a Friedrichs mollifier, construct a sequence $u^\varepsilon = (u^\varepsilon_\beta)_{1\leq\beta\leq N} \in \left[\mathscr{C}^\infty(\Omega_\varepsilon)\right]^N$ satisfying

$$\begin{aligned} &u^\varepsilon \xrightarrow[\varepsilon\to0^+]{} u \text{ and } \nabla u^\varepsilon \xrightarrow[\varepsilon\to0^+]{} \nabla u \text{ at } \mathcal{L}^n\text{-a.e. point in } \Omega, \\ &\psi u^\varepsilon \xrightarrow[\varepsilon\to0^+]{} \psi u \text{ in the topology of test functions in } \Omega, \\ &\text{for each compact } K \subset \Omega \text{ there exists } \varepsilon_K > 0 \text{ such that} \\ &\sup_{0<\varepsilon<\varepsilon_K} \Big\{\|u^\varepsilon\|_{[L^\infty(K,\mathcal{L}^n)]^N} + \big\|\nabla u^\varepsilon\big\|_{[L^\infty(K,\mathcal{L}^n)]^{N\cdot n}}\Big\} < +\infty, \end{aligned} \tag{1.7.110}$$

along with a sequence $w^\varepsilon = (w^\varepsilon_\alpha)_{1\leq\alpha\leq M} \in \big[\mathscr{C}^\infty(\Omega_\varepsilon)\big]^M$ such that

$$\begin{aligned}
&w^\varepsilon \xrightarrow[\varepsilon\to 0^+]{} w \ \text{ and } \nabla w^\varepsilon \xrightarrow[\varepsilon\to 0^+]{} \nabla w \ \text{ at } \mathcal{L}^n\text{-a.e. point in } \Omega,\\
&\xi w^\varepsilon \xrightarrow[\varepsilon\to 0^+]{} \xi w \ \text{ in the topology of test functions in } \Omega,\\
&\text{for each compact } K \subset \Omega \text{ there exists } \varepsilon_K > 0 \text{ such that}\\
&\sup_{0<\varepsilon<\varepsilon_K} \Big\{ \|w^\varepsilon\|_{[L^\infty(K,\mathcal{L}^n)]^M} + \big\|\nabla w^\varepsilon\big\|_{[L^\infty(K,\mathcal{L}^n)]^{M\cdot n}} \Big\} < +\infty.
\end{aligned} \tag{1.7.111}$$

In particular,

$$\begin{aligned}
&u^\varepsilon \xrightarrow[\varepsilon\to 0^+]{} u \ \text{ in } \big[L^1_{\rm loc}(\Omega,\mathcal{L}^n)\big]^N\\
&\text{and } w^\varepsilon \xrightarrow[\varepsilon\to 0^+]{} w \ \text{ in } \big[L^1_{\rm loc}(\Omega,\mathcal{L}^n)\big]^M,
\end{aligned} \tag{1.7.112}$$

hence

$$\begin{aligned}
&L_A u^\varepsilon \xrightarrow[\varepsilon\to 0^+]{} L_A u \ \text{ in } \big[\mathcal{D}'(\Omega)\big]^M\\
&\text{and } L_{A^\top} w^\varepsilon \xrightarrow[\varepsilon\to 0^+]{} L_{A^\top} w \ \text{ in } \big[\mathcal{D}'(\Omega)\big]^N.
\end{aligned} \tag{1.7.113}$$

Let us also recall (cf. [66, Theorems 2.1.8, pp. 38-39]) that

$$\begin{aligned}
&\text{if } \{\omega_j\}_{j\in\mathbb{N}} \text{ is a sequence of distributions in } \Omega \text{ converging to some}\\
&\omega \in \mathcal{D}'(\Omega) \text{ in } \mathcal{D}'(\Omega), \text{ then for each sequence } \{\phi_j\}_{j\in\mathbb{N}} \subseteq \mathscr{C}^\infty_c(\Omega)\\
&\text{converging to some } \phi \in \mathscr{C}^\infty_c(\Omega) \text{ in the topology of test functions in } \Omega\\
&\text{we have } \lim_{j\to\infty} \langle \omega_j, \phi_j\rangle = \langle \omega, \phi\rangle.
\end{aligned} \tag{1.7.114}$$

Then, in place of (1.7.98) we now have

$$\begin{aligned}
I &= \lim_{\varepsilon\to 0^+} \int_\Omega (L_A u^\varepsilon)_\alpha w^\varepsilon_\alpha \varphi \, d\mathcal{L}^n + \lim_{\varepsilon\to 0^+} \int_\Omega \big\langle A\nabla u^\varepsilon, \nabla w^\varepsilon \big\rangle \varphi \, d\mathcal{L}^n\\
&= \lim_{\varepsilon\to 0^+} {}_{[\mathcal{D}'(\Omega)]^M}\big\langle (L_A u^\varepsilon)_\alpha, \varphi\,(\psi\, w^\varepsilon)_\alpha \big\rangle_{[\mathcal{D}(\Omega)]^M} + \lim_{\varepsilon\to 0^+} \int_\Omega \big\langle A\nabla u^\varepsilon, \nabla w^\varepsilon \big\rangle \varphi \, d\mathcal{L}^n\\
&= {}_{[\mathcal{D}'(\Omega)]^M}\big\langle (L_A u)_\alpha, \varphi\,(\psi\, w)_\alpha \big\rangle_{[\mathcal{D}(\Omega)]^M} + \lim_{\varepsilon\to 0^+} \int_\Omega \big\langle A\nabla u^\varepsilon, \nabla w^\varepsilon \big\rangle \varphi \, d\mathcal{L}^n\\
&= {}_{\mathcal{D}'(\Omega)}\big\langle (\psi w_\alpha)(L_A u)_\alpha, \varphi \big\rangle_{\mathcal{D}(\Omega)} + \lim_{\varepsilon\to 0^+} \int_\Omega \big\langle A\nabla u^\varepsilon, \nabla w^\varepsilon \big\rangle \varphi \, d\mathcal{L}^n,
\end{aligned} \tag{1.7.115}$$

thanks to (1.7.110)-(1.7.113) and (1.7.114). Similarly, in lieu of (1.7.99) we presently obtain

$$II = -{}_{\mathcal{D}'(\Omega)}\big\langle (\xi u_\beta)(L_{A^\top} w)_\beta, \varphi \big\rangle_{\mathcal{D}(\Omega)} - \lim_{\varepsilon\to 0^+} \int_\Omega \big\langle A\nabla u^\varepsilon, \nabla w^\varepsilon \big\rangle \varphi \, d\mathcal{L}^n. \tag{1.7.116}$$

From (1.7.115)-(1.7.116), in place of (1.7.100) we now arrive at the conclusion that

$$\operatorname{div}\vec{F} = (\psi w_\alpha)(L_A u)_\alpha - (\xi u_\beta)(L_{A^\top} w)_\beta \in \mathscr{E}'(\Omega). \tag{1.7.117}$$

Consequently,

$$\begin{aligned}
&{}_{(\mathscr{C}_b^\infty(\Omega))^*}\big(\operatorname{div}\vec{F}, 1\big)_{\mathscr{C}_b^\infty(\Omega)} \qquad (1.7.118)\\
&= {}_{\mathscr{E}'(\Omega)}\big\langle(\psi w_\alpha)(L_A u)_\alpha, 1\big\rangle_{\mathscr{E}(\Omega)} - {}_{\mathscr{E}'(\Omega)}\big\langle(\xi u_\beta)(L_{A^\top} w)_\beta, 1\big\rangle_{\mathscr{E}(\Omega)}\\
&= {}_{\mathscr{E}'(\Omega)}\big\langle(L_A u)_\alpha, \psi w_\alpha\big\rangle_{\mathscr{E}(\Omega)} - {}_{\mathscr{E}'(\Omega)}\big\langle(L_{A^\top} w)_\beta, \xi u_\beta\big\rangle_{\mathscr{E}(\Omega)}\\
&= {}_{[\mathscr{E}'(\Omega)]^M}\big\langle\!\big\langle L_A u, w\big\rangle\!\big\rangle_{[\mathscr{E}(\Omega)]^M} - {}_{[\mathscr{E}'(\Omega)]^N}\big\langle\!\big\langle L_{A^\top} w, u\big\rangle\!\big\rangle_{[\mathscr{E}(\Omega)]^N}.
\end{aligned}$$

At this stage, [112, Theorem 1.4.1] applies, and the Divergence Formula [112, (1.4.6)] establishes (1.7.109), on account of (1.7.118). □

Moving on, we augment Theorem 1.7.10 by dealing with Green's first identity for second-order systems in the theorem below.

Theorem 1.7.12 *Suppose $\Omega \subseteq \mathbb{R}^n$ is an open set with a lower Ahlfors regular boundary and with the property that $\sigma := \mathcal{H}^{n-1}\lfloor\partial\Omega$ is a doubling measure. Hence, Ω is a set of locally finite perimeter, and its geometric measure theoretic outward unit normal ν is defined σ-a.e. on $\partial_*\Omega$.*

Next, assume A is a complex coefficient tensor of type $(n \times n, M \times N)$, for some $M, N \in \mathbb{N}$, and denote by L_A the homogeneous constant (complex) coefficient second-order $M \times N$ system in $\mathbb{R}^n$ associated with A as in (1.7.7). Finally, pick some $\kappa, \kappa' > 0$, assume $(\mathbb{X}, \mathbb{Y})$ is a regular dual system on $(\partial\Omega, \sigma)$ (in the sense of Definition 1.7.9), and consider two complex vector-valued functions

$$u \in \big[L^1_{\mathrm{loc}}(\Omega, \mathcal{L}^n)\big]^N, \quad w \in \big[W^{1,1}_{\mathrm{loc}}(\Omega)\big]^M, \tag{1.7.119}$$

satisfying (with all derivatives taken in the sense of distributions in Ω):

$$\begin{gathered}
A\nabla u \in \big[L^1_{\mathrm{loc}}(\Omega, \mathcal{L}^n)\big]^{M\cdot n},\ \mathcal{N}_\kappa(A\nabla u) \in \mathbb{X},\ \mathcal{N}_{\kappa'} w \in \mathbb{Y},\\
\text{the nontangential traces } (A\nabla u)\big|^{\kappa-\mathrm{n.t.}}_{\partial\Omega} \text{ and } w\big|^{\kappa'-\mathrm{n.t.}}_{\partial\Omega} \text{ exist } \sigma\text{-a.e. on } \partial_{\mathrm{nta}}\Omega,\\
\big\langle A\nabla u, \nabla w\big\rangle \text{ belongs to } L^1(\Omega, \mathcal{L}^n), \text{ and}\\
L_A u \in \big[L^1_{\mathrm{loc}}(\Omega, \mathcal{L}^n)\big]^M,\ \ \big\langle L_A u, w\big\rangle \in L^1(\Omega, \mathcal{L}^n).
\end{gathered} \tag{1.7.120}$$

Then for any other specified aperture parameter $\kappa'' \in (0, \infty)$ the nontangential boundary traces $(A\nabla u)\big|^{\kappa''-\mathrm{n.t.}}_{\partial\Omega}$ and $w\big|^{\kappa''-\mathrm{n.t.}}_{\partial\Omega}$ exist at σ-a.e. point on $\partial_{\mathrm{nta}}\Omega$ and are independent of κ''. Moreover, with the dependence on the parameter κ'' dropped, the following Green type formula (involving absolutely convergent integrals) holds

$$\int_\Omega \big\langle L_A u, w\big\rangle\,d\mathcal{L}^n + \int_\Omega \big\langle A\nabla u, \nabla w\big\rangle\,d\mathcal{L}^n = \int_{\partial_*\Omega} \big\langle \partial^A_\nu u, w\big|^{\mathrm{n.t.}}_{\partial\Omega}\big\rangle\,d\sigma, \tag{1.7.121}$$

when either Ω is bounded, or $\partial\Omega$ is unbounded. In the remaining case, i.e., when Ω is unbounded and $\partial\Omega$ is bounded, formula (1.7.121) *continues to be true under the additional assumption that there exists $\lambda \in (1, \infty)$ such that*

$$\int_{[B(0,\lambda R)\setminus B(0,R)]\cap\Omega} |A\nabla u||w| \, d\mathcal{L}^n = o(R) \ \ as \ \ R \to \infty. \tag{1.7.122}$$

In particular, corresponding to the case when the last line in (1.7.120) *is strengthened to*

$$L_A u = 0 \ \ \text{in the sense of distributions in } \Omega, \tag{1.7.123}$$

formula (1.7.121) *becomes*

$$\int_\Omega \langle A\nabla u, \nabla w \rangle \, d\mathcal{L}^n = \int_{\partial_*\Omega} \langle \partial_\nu^A u, w\big|_{\partial\Omega}^{\text{n.t.}} \rangle \, d\sigma. \tag{1.7.124}$$

Finally, similar results are valid if $(\mathbb{X}, \mathbb{Y})$ is a dual system on $(\partial\Omega, \sigma)$ which is not necessarily regular, with the understanding that, in such a case, all nontangential boundary traces are taken with respect to an aperture parameter which is $< \min\{\kappa, \kappa'\}$.

At least formally, subtracting from (1.7.121) its own version written with the roles of u and w reversed and with $A^\top$ in place of A would produce (1.7.78). However, such a proof of Green's identity recorded in (1.7.78) is far from economical since this would require imposing conditions on u, w (in the spirit of (1.7.119)-(1.7.120)) that ensure the validity of both (1.7.121) as well as its version with u and w switched and A replaced by $A^\top$. In the proof of Theorem 1.7.12 given below, we avoid this issue by choosing to derive (1.7.121) directly, by applying our version of Divergence Theorem to a suitably chosen vector field.

Proof of Theorem 1.7.12 In concert with [112, Corollary 8.9.9], the present assumptions imply that there exists some $p \in (0, \infty]$ with the property for any given aperture parameter $\kappa'' > 0$ we have

$$\mathcal{N}_{\kappa''}(A\nabla u), \ \mathcal{N}_{\kappa''} w \in L^p_{\text{loc}}(\partial\Omega, \sigma), \tag{1.7.125}$$

and the nontangential boundary traces

$$\begin{gathered} (A\nabla u)\big|_{\partial\Omega}^{\kappa''-\text{n.t.}} \ \text{and} \ w\big|_{\partial\Omega}^{\kappa''-\text{n.t.}}, \ \text{both exist at } \sigma\text{-a.e. point on } \partial_{\text{nta}}\Omega, \\ \text{and agree with the nontangential boundary traces in (1.7.120).} \end{gathered} \tag{1.7.126}$$

Observe that (1.7.119), the first line in (1.7.120) (or (1.7.125)), and [112, Lemma 8.3.1], imply

$$A\nabla u \in \big[L^\infty_{\text{loc}}(\Omega, \mathcal{L}^n)\big]^{M\cdot n} \ \text{and} \ w \in \big[L^\infty_{\text{loc}}(\Omega, \mathcal{L}^n)\big]^M. \tag{1.7.127}$$

With the summation convention over repeated indices in effect throughout the proof, consider now the vector field

$$\vec{F} := \Big((A\nabla u)_{\alpha,r} w_\alpha\Big)_{1\leq r\leq n} = \Big((a^{\alpha\beta}_{rs}\partial_s u_\beta) w_\alpha\Big)_{1\leq r\leq n} \in \big[L^\infty_{\rm loc}(\Omega,\mathcal{L}^n)\big]^n \tag{1.7.128}$$

where the membership is a consequence (1.7.127). In particular,

$$\vec{F} \in \big[L^1_{\rm loc}(\Omega,\mathcal{L}^n)\big]^n. \tag{1.7.129}$$

To compute the divergence of $\vec{F}$, in the sense of distributions in Ω, fix an arbitrary scalar-valued function $\varphi \in \mathscr{C}^\infty_c(\Omega)$ and write

$$_{\mathcal{D}'(\Omega)}\big\langle {\rm div}\vec{F}, \varphi\big\rangle_{\mathcal{D}(\Omega)} = -\int_\Omega F_r\partial_r\varphi\,{\rm d}\mathcal{L}^n = -\int_\Omega (a^{\alpha\beta}_{rs}\partial_s u_\beta) w_\alpha \partial_r\varphi\,{\rm d}\mathcal{L}^n. \tag{1.7.130}$$

Next, for each sufficiently small $\varepsilon > 0$ let $\Omega_\varepsilon := \big\{x \in \Omega : {\rm dist}(x,\partial\Omega) > \varepsilon\big\}$. Using a Friedrichs mollifier, it is possible to construct a sequence of functions $w^\varepsilon = (w^\varepsilon_\alpha)_{1\leq\alpha\leq M} \in \big[\mathscr{C}^\infty(\Omega_\varepsilon)\big]^M$ such that

$$\begin{aligned}
&\text{for each compact set } K \subset \Omega \text{ we have}\\
&w^\varepsilon \xrightarrow[\varepsilon\to 0^+]{} w \text{ in } \big[L^1(K,\mathcal{L}^n)\big]^M \text{ and}\\
&\nabla w^\varepsilon \xrightarrow[\varepsilon\to 0^+]{} \nabla w \text{ in } \big[L^1(K,\mathcal{L}^n)\big]^{n\times M},\\
&\text{and there exists some number } \varepsilon_K > 0\\
&\text{so that } \sup\nolimits_{0<\varepsilon<\varepsilon_K} \|w^\varepsilon\|_{[L^\infty(K,\mathcal{L}^n)]^M} < \infty.
\end{aligned} \tag{1.7.131}$$

By eventually passing to a subsequence there is no loss of generality in assuming that we also have

$$w^\varepsilon \xrightarrow[\varepsilon\to 0^+]{} w \text{ at } \mathcal{L}^n\text{-a.e. point in } \Omega. \tag{1.7.132}$$

On account of (1.7.130), (1.7.127), (1.7.131)-(1.7.132), the last two lines in (1.7.120) and, finally, Lebesgue's Dominated Convergence Theorem, we may then compute

$$
\begin{aligned}
&{}_{\mathcal{D}'(\Omega)}\big\langle \mathrm{div}\vec{F}, \varphi\big\rangle_{\mathcal{D}(\Omega)} \\
&= - \lim_{\varepsilon\to 0^+} \int_\Omega (a^{\alpha\beta}_{rs}\partial_s u_\beta) w^\varepsilon_\alpha \partial_r \varphi \, \mathrm{d}\mathcal{L}^n \\
&= - \lim_{\varepsilon\to 0^+} \int_\Omega (a^{\alpha\beta}_{rs}\partial_s u_\beta) \partial_r (w^\varepsilon_\alpha \varphi) \, \mathrm{d}\mathcal{L}^n + \lim_{\varepsilon\to 0^+} \int_\Omega (a^{\alpha\beta}_{rs}\partial_s u_\beta)(\partial_r w^\varepsilon_\alpha)\varphi \, \mathrm{d}\mathcal{L}^n \\
&= - \lim_{\varepsilon\to 0^+} {}_{\mathcal{D}'(\Omega)}\Big\langle a^{\alpha\beta}_{rs}\partial_s u_\beta, \partial_r (w^\varepsilon_\alpha \varphi)\Big\rangle_{\mathcal{D}(\Omega)} + \int_\Omega (a^{\alpha\beta}_{rs}\partial_s u_\beta)(\partial_r w_\alpha)\varphi \, \mathrm{d}\mathcal{L}^n \\
&= \lim_{\varepsilon\to 0^+} {}_{\mathcal{D}'(\Omega)}\Big\langle a^{\alpha\beta}_{rs}\partial_r\partial_s u_\beta, w^\varepsilon_\alpha \varphi\Big\rangle_{\mathcal{D}(\Omega)} + \int_\Omega \langle A\nabla u, \nabla w\rangle \varphi \, \mathrm{d}\mathcal{L}^n \\
&= \lim_{\varepsilon\to 0^+} {}_{\mathcal{D}'(\Omega)}\Big\langle (L_A u)_\alpha, w^\varepsilon_\alpha \varphi\Big\rangle_{\mathcal{D}(\Omega)} + \int_\Omega \langle A\nabla u, \nabla w\rangle \varphi \, \mathrm{d}\mathcal{L}^n \\
&= \lim_{\varepsilon\to 0^+} \int_\Omega (L_A u)_\alpha w^\varepsilon_\alpha \varphi \, \mathrm{d}\mathcal{L}^n + \int_\Omega \langle A\nabla u, \nabla w\rangle \varphi \, \mathrm{d}\mathcal{L}^n \\
&= \int_\Omega (L_A u)_\alpha w_\alpha \varphi \, \mathrm{d}\mathcal{L}^n + \int_\Omega \langle A\nabla u, \nabla w\rangle \varphi \, \mathrm{d}\mathcal{L}^n \\
&= \int_\Omega \langle L_A u, w\rangle \varphi \, \mathrm{d}\mathcal{L}^n + \int_\Omega \langle A\nabla u, \nabla w\rangle \varphi \, \mathrm{d}\mathcal{L}^n. \qquad (1.7.133)
\end{aligned}
$$

Having established this, we conclude that

$$\mathrm{div}\vec{F} = \big\langle L_A u, w\big\rangle + \big\langle A\nabla u, \nabla w\big\rangle \text{ in } \mathcal{D}'(\Omega), \qquad (1.7.134)$$

so the last line in (1.7.120) implies

$$\mathrm{div}\vec{F} \in L^1(\Omega, \mathcal{L}^n). \qquad (1.7.135)$$

In addition, if we choose an aperture parameter $\kappa_* \in \big(0, \min\{\kappa, \kappa'\}\big)$, from (1.7.128) and [112, (8.2.25)] we see that there exists a constant $C_A \in (0, \infty)$ such that

$$\mathcal{N}_{\kappa_*}\vec{F} \leq C_A\, \mathcal{N}_\kappa(A\nabla u)\mathcal{N}_{\kappa'} w \text{ on } \partial\Omega. \qquad (1.7.136)$$

From (1.7.136), [112, (8.2.26)], the nontangential maximal function memberships in the first line of (1.7.120), and (1.7.71) we conclude that

$$\mathcal{N}_{\kappa_*}\vec{F} \in L^1(\partial\Omega, \sigma). \qquad (1.7.137)$$

In addition, (1.7.126) and (1.7.128) imply that the nontangential boundary trace

$$\vec{F}\Big|^{\kappa''-\text{n.t.}}_{\partial\Omega} \text{ exists at } \sigma\text{-a.e. point on } \partial_{\text{nta}}\Omega. \qquad (1.7.138)$$

Specifically, with the dependence on κ'' dropped, we have

$$\vec{F}\Big|_{\partial\Omega}^{\rm n.t.} = \Big\{\big((a_{rs}^{\alpha\beta}\partial_s u_\beta)\big|_{\partial\Omega}^{\rm n.t.}\big)\big(w_\alpha\big|_{\partial\Omega}^{\rm n.t.}\big)\Big\}_{1\leq r\leq n} \text{ at } \sigma\text{-a.e. point on } \partial_{\rm nta}\Omega. \tag{1.7.139}$$

In concert with (1.7.9) and [112, Proposition 8.8.6] this implies

$$\nu\cdot\big(\vec{F}\big|_{\partial\Omega}^{\rm n.t.}\big) = \big\langle\partial_\nu^A u, w\big|_{\partial\Omega}^{\rm n.t.}\big\rangle \text{ at } \sigma\text{-a.e. point on } \partial_*\Omega. \tag{1.7.140}$$

Let us also observe that (1.7.122) ensures the validity of the integral growth condition [112, (1.2.3)] for the present vector field $\vec{F}$.

Having established the aforementioned properties of $\vec{F}$, [112, Theorem 1.2.1] applies and the Divergence Formula [112, (1.2.2)] currently yields (1.7.121) thanks to (1.7.134) and (1.7.140). □

It turns out that there yet another integral identity for second-order systems relating the bulk with the boundary of a region.

Theorem 1.7.13 *Let $\Omega\subseteq\mathbb{R}^n$ be an open set with a lower Ahlfors regular boundary and with the property that $\sigma := \mathcal{H}^{n-1}\lfloor\partial\Omega$ is a doubling measure. Hence, Ω is a set of locally finite perimeter, and its geometric measure theoretic outward unit normal $\nu = (\nu_1,\dots,\nu_n)$ is defined σ-a.e. on $\partial_*\Omega$.*

Given some arbitrary $M, N\in\mathbb{N}$, suppose A is a complex coefficient tensor of type $(n\times n, M\times N)$ and denote by L_A the homogeneous constant (complex) coefficient second-order $M\times N$ system in $\mathbb{R}^n$ associated with A as in (1.7.7). Also, fix two aperture parameters $\kappa,\kappa'>0$, assume $(\mathbb{X},\mathbb{Y})$ is a regular dual systems on $(\partial\Omega,\sigma)$ (in the sense of Definition 1.7.9), and select some $\ell\in\{1,\dots,n\}$. Finally, consider two complex vector-valued functions

$$u = (u_\alpha)_{1\leq\alpha\leq N}\in\big[W^{2,1}_{\rm loc}(\Omega)\big]^N,\quad w = (w_\beta)_{1\leq\beta\leq M}\in\big[W^{2,1}_{\rm loc}(\Omega)\big]^M, \tag{1.7.141}$$

satisfying (with all derivatives taken in the sense of distributions in Ω):

$$\begin{gathered}\mathcal{N}_\kappa(\nabla u) \text{ belongs to } \mathbb{X},\ \mathcal{N}_{\kappa'}(\nabla w) \text{ belongs to } \mathbb{Y},\\ (\nabla u)\big|_{\partial\Omega}^{\kappa-\rm n.t.} \text{ and } (\nabla w)\big|_{\partial\Omega}^{\kappa'-\rm n.t.} \text{ exist } \sigma\text{-a.e. on } \partial_{\rm nta}\Omega,\\ L_A u\in\big[L^1_{\rm loc}(\Omega,\mathcal{L}^n)\big]^M,\quad L_A^\top w\in\big[L^1_{\rm loc}(\Omega,\mathcal{L}^n)\big]^N,\\ \big\langle L_A u, \partial_\ell w\big\rangle\in L^1(\Omega,\mathcal{L}^n),\quad \big\langle\partial_\ell u, L_A^\top w\big\rangle\in L^1(\Omega,\mathcal{L}^n).\end{gathered} \tag{1.7.142}$$

Then for any other specified aperture parameter $\kappa''\in(0,\infty)$ the nontangential boundary traces $(\nabla u)\big|_{\partial\Omega}^{\kappa''-\rm n.t.}$, $(\nabla w)\big|_{\partial\Omega}^{\kappa''-\rm n.t.}$ exist at σ-a.e. point on $\partial_{\rm nta}\Omega$ and are independent of κ''. Moreover, with the dependence on the parameter κ'' dropped, the following identity (involving absolutely convergent integrals) holds

$$
\begin{aligned}
&\int_{\Omega} \langle L_A u, \partial_\ell w \rangle \, \mathrm{d}\mathcal{L}^n + \int_{\Omega} \langle \partial_\ell u, L_A^\top w \rangle \, \mathrm{d}\mathcal{L}^n \\
&\qquad = \int_{\partial_*\Omega} \langle \partial_\nu^A u, (\partial_\ell w)\big|_{\partial\Omega}^{\text{n.t.}} \rangle \, \mathrm{d}\sigma + \int_{\partial_*\Omega} \langle (\partial_\ell u)\big|_{\partial\Omega}^{\text{n.t.}}, \partial_\nu^{A^\top} w \rangle \, \mathrm{d}\sigma \\
&\qquad\quad - \int_{\partial_*\Omega} \nu_\ell \, a_{rs}^{\beta\alpha} (\partial_r w_\beta)\big|_{\partial\Omega}^{\text{n.t.}} (\partial_s u_\alpha)\big|_{\partial\Omega}^{\text{n.t.}} \, \mathrm{d}\sigma, \qquad (1.7.143)
\end{aligned}
$$

when either Ω is bounded, or $\partial\Omega$ is unbounded. In the remaining case, i.e., when Ω is unbounded and $\partial\Omega$ is bounded, formula (1.7.143) *continues to be true under the additional assumption that there exists $\lambda \in (1, \infty)$ such that*

$$
\int_{[B(0,\lambda R)\setminus B(0,R)]\cap\Omega} |\nabla u||\nabla w| \, \mathrm{d}\mathcal{L}^n = o(R) \ \ \textit{as} \ \ R \to \infty. \qquad (1.7.144)
$$

As a corollary, if the last two lines in (1.7.142) *are replaced by*

$$
L_A u = 0 \ \textit{and} \ L_A^\top w = 0 \ \textit{ in the sense of distributions in } \Omega, \qquad (1.7.145)
$$

then (1.7.143) *becomes*

$$
\begin{aligned}
&\int_{\partial_*\Omega} \langle \partial_\nu^A u, (\partial_\ell w)\big|_{\partial\Omega}^{\text{n.t.}} \rangle \, \mathrm{d}\sigma + \int_{\partial_*\Omega} \langle (\partial_\ell u)\big|_{\partial\Omega}^{\text{n.t.}}, \partial_\nu^{A^\top} w \rangle \, \mathrm{d}\sigma \\
&\qquad = \int_{\partial_*\Omega} \nu_\ell \, a_{rs}^{\beta\alpha} (\partial_r w_\beta)\big|_{\partial\Omega}^{\text{n.t.}} (\partial_s u_\alpha)\big|_{\partial\Omega}^{\text{n.t.}} \, \mathrm{d}\sigma. \qquad (1.7.146)
\end{aligned}
$$

Finally, similar results are valid if $(\mathbb{X}, \mathbb{Y})$ is a dual system on $(\partial\Omega, \sigma)$ which is not necessarily regular, with the understanding that, in such a case, all nontangential boundary traces are taken with respect to an aperture parameter which is $< \min\{\kappa, \kappa'\}$.

Proof Based on assumptions and [112, Corollary 8.9.9] we conclude that there exists $p \in (0, \infty]$ with the property that for any aperture parameter $\kappa'' > 0$ we have

$$
\mathcal{N}_{\kappa''}(\nabla u), \ \mathcal{N}_{\kappa''}(\nabla w) \in L^p_{\mathrm{loc}}(\partial\Omega, \sigma), \qquad (1.7.147)
$$

and the nontangential boundary traces

$$
\begin{aligned}
&(\nabla u)\big|_{\partial\Omega}^{\kappa''-\text{n.t.}}, (\nabla w)\big|_{\partial\Omega}^{\kappa''-\text{n.t.}} \ \text{exist } \sigma\text{-a.e. on } \partial_{\text{nta}}\Omega \text{ and agree with} \\
&\text{the respective nontangential boundary traces in (1.7.142).}
\end{aligned} \qquad (1.7.148)
$$

Also, (1.7.141), the first line in (1.7.142) (or (1.7.147)), and [112, Lemma 8.3.1] imply

$$
\begin{aligned}
u &\in \big[W^{2,1}_{\mathrm{loc}}(\Omega)\big]^N \cap \big[\mathrm{Lip}_{\mathrm{loc}}(\Omega)\big]^N,\\
w &\in \big[W^{2,1}_{\mathrm{loc}}(\Omega)\big]^M \cap \big[\mathrm{Lip}_{\mathrm{loc}}(\Omega)\big]^M.
\end{aligned}
\tag{1.7.149}
$$

To proceed, consider the vector field which, with the summation convention over repeated indices in effect, is given by

$$
\vec{F} := a^{\beta\alpha}_{rs}(\partial_r w_\beta)(\partial_s u_\alpha)\mathbf{e}_\ell - a^{\beta\alpha}_{rs}(\partial_r w_\beta)(\partial_\ell u_\alpha)\mathbf{e}_s - a^{\alpha\beta}_{rs}(\partial_\ell w_\alpha)(\partial_s u_\beta)\mathbf{e}_r. \tag{1.7.150}
$$

Hence, (1.7.150) and (1.7.149) give

$$
\vec{F} \in \big[L^\infty_{\mathrm{loc}}(\Omega, \mathcal{L}^n)\big]^n \subset \big[L^1_{\mathrm{loc}}(\Omega, \mathcal{L}^n)\big]^n. \tag{1.7.151}
$$

Next, based on (1.7.149), (1.7.150), and Leibniz's product formula for weak derivatives from [112, Proposition 4.3.1], we may compute $\operatorname{div}\vec{F}$ in the sense of distributions in Ω as

$$
\begin{aligned}
\operatorname{div}\vec{F} &= a^{\beta\alpha}_{rs}(\partial_\ell\partial_r w_\beta)(\partial_s u_\alpha) + a^{\beta\alpha}_{rs}(\partial_r w_\beta)(\partial_\ell\partial_s u_\alpha)\\
&\quad - a^{\beta\alpha}_{rs}(\partial_s\partial_r w_\beta)(\partial_\ell u_\alpha) - a^{\beta\alpha}_{rs}(\partial_r w_\beta)(\partial_s\partial_\ell u_\alpha)\\
&\quad - a^{\alpha\beta}_{rs}(\partial_r\partial_\ell w_\alpha)(\partial_s u_\beta) - a^{\alpha\beta}_{rs}(\partial_\ell w_\alpha)(\partial_r\partial_s u_\beta).
\end{aligned}
\tag{1.7.152}
$$

After observing that, in the right-hand side above, the first term cancels the fifth and the second term cancels the fourth, we conclude that

$$
\operatorname{div}\vec{F} = -\big\langle \partial_\ell u, L_A^\top w\big\rangle - \big\langle L_A u, \partial_\ell w\big\rangle \text{ in } \mathcal{D}'(\Omega). \tag{1.7.153}
$$

Together with the last line in (1.7.142) this gives

$$
\operatorname{div}\vec{F} \in L^1(\Omega, \mathcal{L}^n). \tag{1.7.154}
$$

Also, if we choose an aperture parameter $\kappa_* \in \big(0, \min\{\kappa, \kappa'\}\big)$, from (1.7.150) and [112, (8.2.25)] we see that

$$
\mathcal{N}_{\kappa_*}\vec{F} \le C \cdot \mathcal{N}_\kappa(\nabla u)\mathcal{N}_{\kappa'}(\nabla w) \text{ on } \partial\Omega, \tag{1.7.155}
$$

for some constant $C \in (0, \infty)$ which depends only on the coefficient tensor A. In turn, (1.7.155), (1.7.147), [112, (8.2.26)], the nontangential maximal function memberships in the first line of (1.7.142), and (1.7.71) ensure that

$$
\mathcal{N}_{\kappa_*}\vec{F} \in L^1(\partial\Omega, \sigma). \tag{1.7.156}
$$

Moving on, observe that (1.7.150) and (1.7.90) imply that the nontangential boundary trace

$$
\vec{F}\Big|^{\kappa''-\mathrm{n.t.}}_{\partial\Omega} \text{ exists at } \sigma\text{-a.e. point on } \partial_{\mathrm{nta}}\Omega. \tag{1.7.157}
$$

In fact, with the dependence on κ'' dropped, we have

$$\vec{F}\Big|_{\partial\Omega}^{\text{n.t.}} = a_{rs}^{\beta\alpha}\big((\partial_r w_\beta)\big|_{\partial\Omega}^{\text{n.t.}}\big)\big((\partial_s u_\alpha)\big|_{\partial\Omega}^{\text{n.t.}}\big)\mathbf{e}_\ell - a_{rs}^{\beta\alpha}\big((\partial_r w_\beta)\big|_{\partial\Omega}^{\text{n.t.}}\big)\big((\partial_\ell u_\alpha)\big|_{\partial\Omega}^{\text{n.t.}}\big)\mathbf{e}_s$$

$$- a_{rs}^{\alpha\beta}\big((\partial_\ell w_\alpha)\big|_{\partial\Omega}^{\text{n.t.}}\big)\big((\partial_s u_\beta)\big|_{\partial\Omega}^{\text{n.t.}}\big)\mathbf{e}_r \text{ at } \sigma\text{-a.e. point on } \partial_{\text{nta}}\Omega. \tag{1.7.158}$$

From (1.7.158), (1.7.9), and [112, Proposition 8.8.6], we therefore see that at σ-a.e. point on $\partial_*\Omega$ we have

$$\nu\cdot\big(\vec{F}\big|_{\partial\Omega}^{\text{n.t.}}\big) = \nu_\ell\, a_{rs}^{\beta\alpha}\big((\partial_r w_\beta)\big|_{\partial\Omega}^{\text{n.t.}}\big)\big((\partial_s u_\alpha)\big|_{\partial\Omega}^{\text{n.t.}}\big)$$

$$- \big\langle(\partial_\ell u)\big|_{\partial\Omega}^{\text{n.t.}}, \partial_\nu^{A^\top} w\big\rangle - \big\langle\partial_\nu^A u, (\partial_\ell w)\big|_{\partial\Omega}^{\text{n.t.}}\big\rangle. \tag{1.7.159}$$

Finally, (1.7.144) guarantees that the vector field $\vec{F}$ satisfies the growth condition [112, (1.2.3)]. Granted these properties of $\vec{F}$, we may invoke [112, Theorem 1.2.1] and the Divergence Formula [112, (1.2.2)] currently produces (1.7.143), thanks to (1.7.153) and (1.7.159). □

A special case of Theorem 1.7.12 which is particularly useful in applications is singled out in our next corollary. In particular, this is the key ingredient in the proof of the uniqueness result for the mixed boundary value problem stated in [115, §1.7].

Corollary 1.7.14 *Assume $\Omega \subseteq \mathbb{R}^n$, where $n \geq 2$, is an open set with a lower Ahlfors regular boundary and with the property that $\sigma := \mathcal{H}^{n-1}\lfloor\partial\Omega$ is a doubling measure. In particular, Ω is a set of locally finite perimeter, and its geometric measure theoretic outward unit normal ν is defined σ-a.e. on $\partial_*\Omega$. Having fixed some $M \in \mathbb{N}$, let A be a complex coefficient tensor of type $(n\times n, M\times M)$ and denote by L_A the homogeneous constant (complex) coefficient second-order $M\times M$ system in $\mathbb{R}^n$ associated with A as in* (1.7.7). *In this context, consider a complex vector-valued function*

$$u \in \big[W^{1,1}_{\text{loc}}(\Omega)\big]^M \tag{1.7.160}$$

with the property that, for some $\kappa > 0$ and $p, q, r \in [1,\infty]$, one has:

$$\mathcal{N}_\kappa(\nabla u) \in L^p(\partial\Omega,\sigma),\ \ \mathcal{N}_\kappa u \in L^q(\partial\Omega,\sigma),$$

$$u\big|_{\partial\Omega}^{\kappa-\text{n.t.}} \text{ and } (\nabla u)\big|_{\partial\Omega}^{\kappa-\text{n.t.}} \text{ exist } \sigma\text{-a.e. on } \partial_{\text{nta}}\Omega, \tag{1.7.161}$$

$$\text{and } L_A u \text{ belongs to the space } \big[L^r(\Omega,\mathcal{L}^n)\big]^M.$$

Then for any other specified aperture parameter $\kappa' \in (0,\infty)$ the nontangential boundary traces $u\big|_{\partial\Omega}^{\kappa'-\text{n.t.}}$ and $(\nabla u)\big|_{\partial\Omega}^{\kappa'-\text{n.t.}}$ exist at σ-a.e. point on $\partial_{\text{nta}}\Omega$ and are independent of κ'. Moreover, with the dependence on the parameter κ' dropped, the formula

$$\int_\Omega \langle L_A u, \overline{u}\rangle\,\mathrm{d}\mathcal{L}^n + \int_\Omega \langle A\nabla u, \overline{\nabla u}\rangle\,\mathrm{d}\mathcal{L}^n = \int_{\partial_*\Omega} \big\langle \partial_\nu^A u, \overline{u}\big|_{\partial\Omega}^{\text{n.t.}}\big\rangle\,\mathrm{d}\sigma \tag{1.7.162}$$

holds (with all integrals involved absolutely convergent) in any of the following scenarios:

(i) $\partial\Omega$ is unbounded, $p = 2(n-1)/n$, $q = 2(n-1)/(n-2)$, and $r = 2n/(n+2)$;
(ii) Ω is bounded, $p \geq 2(n-1)/n$, $q \geq p/(p-1)$, and $r \geq np/(n+p-1)$;
(iii) Ω is an exterior domain, $p \geq 2(n-1)/n$, $q \geq p/(p-1)$, and there exist two real numbers a, b such that

$$(\nabla u)(x) = O(|x|^{-a}) \text{ and } u(x) = O(|x|^{-b}) \text{ as } |x| \to \infty, \tag{1.7.163}$$

$$a > n/2, \quad b > \max\{n-1-a, (n-1)/q\}, \tag{1.7.164}$$

and

$$nq/(nq-n+1) \leq r < \begin{cases} n/(n-b) & \text{if } b < n, \\ \infty & \text{if } b \geq n. \end{cases} \tag{1.7.165}$$

Moreover, if actually $L_A u = 0$ in the sense of distributions in Ω, then (1.7.165) *may be ignored and, in place of* (1.7.164), *one may simply ask that*

$$a > n/2 \text{ and } b > n-1-a. \tag{1.7.166}$$

Proof Consider first the case when $\partial\Omega$ is unbounded, and p, q, r are as in item *(i)*. Then [112, Proposition 8.6.3] and [112, Lemma 8.3.2] permit us to conclude that

$$\begin{aligned} u \in \big[L^{2n/(n-2)}(\Omega, \mathcal{L}^n)\big]^M &= \Big(\big[L^{2n/(n+2)}(\Omega, \mathcal{L}^n)\big]^M\Big)^*, \\ \text{and } \nabla u \in \big[L^2(\Omega, \mathcal{L}^n)\big]^{n\times M}&. \end{aligned} \tag{1.7.167}$$

Consequently,

$$\text{both } \langle A\nabla u, \overline{\nabla u}\rangle \text{ and } \langle L_A u, \overline{u}\rangle \text{ belong to } L^1(\Omega, \mathcal{L}^n). \tag{1.7.168}$$

Given that $p = 2(n-1)/n$ and $q = 2(n-1)/(n-2)$ are Hölder conjugate exponents, all desired conclusions, including formula (1.7.162), are in this case consequences of Theorem 1.7.12 (presently used with $w := \overline{u}$ and $p := 2(n-1)/n$).

Let us now consider the case when Ω is bounded, and p, q, r are as in item *(ii)*. Then [112, Proposition 8.6.3] and [112, Lemma 8.3.2] permit us to conclude that

$$\begin{aligned} u \in \big[L^{nq/(n-1)}(\Omega, \mathcal{L}^n)\big]^M &\subseteq \big[L^{np/[(n-1)(p-1)]}(\Omega, \mathcal{L}^n)\big]^M \\ \text{and } \nabla u \in \big[L^{np/(n-1)}(\Omega, \mathcal{L}^n)\big]^{n\times M} &\subseteq \big[L^2(\Omega, \mathcal{L}^n)\big]^{n\times M}. \end{aligned} \tag{1.7.169}$$

Also,

$$L_A u \in \big[L^r(\Omega, \mathcal{L}^n)\big]^M \subseteq \Big[L^{\frac{np}{n+p-1}}(\Omega, \mathcal{L}^n)\Big]^M = \Big(\big[L^{\frac{np}{(n-1)(p-1)}}(\Omega, \mathcal{L}^n)\big]^M\Big)^*. \tag{1.7.170}$$

It is then apparent from (1.7.169)-(1.7.170) that (1.7.168) holds in this case as well. Given that, with $p' := p/(p-1)$ denoting the Hölder conjugate exponent of p, we also have

$$\mathcal{N}_\kappa(\nabla u) \in L^p(\partial\Omega, \sigma) \text{ and } \mathcal{N}_\kappa u \in L^q(\partial\Omega, \sigma) \subseteq L^{p'}(\partial\Omega, \sigma), \tag{1.7.171}$$

since $\sigma(\partial\Omega) < +\infty$, all desired conclusions are once again seen from Theorem 1.7.12 (employed with $w := \overline{u}$, and the current exponent p).

To deal with case *(iii)*, suppose Ω is an exterior domain, $p \geq 2(n-1)/n$, $q \geq p/(p-1)$, and there exist $a, b \in \mathbb{R}$ such that (1.7.163)-(1.7.165) hold. In particular, $a, b > 0$ and, since $a + b > n - 1$, we also have

$$|u(x)||(\nabla u)(x)| = o(|x|^{1-n}) \text{ as } |x| \to \infty. \tag{1.7.172}$$

In such a scenario, starting with $\mathcal{N}_\kappa(\nabla u) \in L^p(\partial\Omega, \sigma)$ [112, Proposition 8.6.3] now only guarantees that

$$\nabla u \in \big[L^{np/(n-1)}_{\text{bdd}}(\Omega, \mathcal{L}^n)\big]^{n\times M} \subseteq \big[L^{np/(n-1)}_{\text{bdd}}(\Omega, \mathcal{L}^n)\big]^{n\times M}. \tag{1.7.173}$$

However, this may be augmented with the pointwise decay property of ∇u at infinity. Since $a > n/2$, the latter ensures that

$$|\nabla u|^2 \text{ is Lebesgue integrable in a neighborhood of infinity.} \tag{1.7.174}$$

Together, (1.7.173)-(1.7.174) ultimately prove that we still have

$$\nabla u \in \big[L^2(\Omega, \mathcal{L}^n)\big]^{n\times M}. \tag{1.7.175}$$

In particular,

$$\big\langle A\nabla u, \overline{\nabla u}\big\rangle \text{ belongs to } L^1(\Omega, \mathcal{L}^n). \tag{1.7.176}$$

Likewise, starting with $\mathcal{N}_\kappa u \in L^q(\partial\Omega, \sigma)$, [112, Proposition 8.6.3] now gives

$$\begin{gathered} u \in \big[L^{nq/(n-1)}_{\text{bdd}}(\Omega, \mathcal{L}^n)\big]^M \subseteq \big[L^{ns/(n-1)}_{\text{bdd}}(\Omega, \mathcal{L}^n)\big]^M \\ \text{for each } s \in (0, q], \end{gathered} \tag{1.7.177}$$

while the pointwise decay property of u at infinity permits us to deduce that

$$\begin{gathered} \text{the function } |u|^{ns/(n-1)} \text{ is Lebesgue integrable in a} \\ \text{neighborhood of infinity whenever } s > (n-1)/b. \end{gathered} \tag{1.7.178}$$

From (1.7.177)-(1.7.178) we then conclude that

$$u \in \big[L^{ns/(s-1)}(\Omega, \mathcal{L}^n)\big]^M \text{ whenever } (n-1)/b < s \leq q. \tag{1.7.179}$$

Bearing in mind that $L_A u$ belongs to $\big[L^r(\Omega, \mathcal{L}^n)\big]^M$ with r as in (1.7.165), we finally see from (1.7.179) that

$$\left\langle L_A u, \overline{u} \right\rangle \text{ belongs to } L^1(\Omega, \mathcal{L}^n). \tag{1.7.180}$$

With (1.7.176), (1.7.180), and (1.7.172) in hand, and keeping in mind that (1.7.171) continues to be valid, all desired conclusions are presently implied by Theorem 1.7.12 (used with $w := \overline{u}$ and the original exponent p).

Finally, in the case when $L_A u = 0$ the same type of analysis as above applies, this time only making use of $a > n/2$ and $a + b > n - 1$. □

While by now we have established a number of general Green type formulas, when faced with the task of proving a specific identity of this nature, depending on circumstances, one may still find it useful to go back to our original Divergence Theorems in order to ensure that the information available is used in an optimal manner. A concrete scenario is as follows.

Proposition 1.7.15 *Fix $n \in \mathbb{N}$ with $n \geq 2$ and assume u is a harmonic function in $\mathbb{R}^n_+$ which, for some $\kappa > 0$, satisfies*

$$\mathcal{N}_\kappa u,\ \mathcal{N}_\kappa(\nabla u) \in L^2(\mathbb{R}^{n-1}, \mathcal{L}^{n-1}), \tag{1.7.181}$$

where $\mathcal{N}_\kappa$ denotes the nontangential maximal operator (*with aperture parameter κ*) *relative to the open set $\mathbb{R}^n_+$. Then*

$$u\Big|^{\kappa-\mathrm{n.t.}}_{\partial\mathbb{R}^n_+},\ (\nabla u)\Big|^{\kappa-\mathrm{n.t.}}_{\partial\mathbb{R}^n_+} \ \text{ exist } \mathcal{L}^{n-1}\text{-a.e. on } \mathbb{R}^{n-1} \equiv \partial\mathbb{R}^n_+ \tag{1.7.182}$$

and for each upper-graph Lipschitz domain $\mathcal{R} \subseteq \mathbb{R}^n_+$ one has the Green formula

$$\int_{\mathcal{R}} t|(\nabla u)(x', t)|^2\, dx'dt = \int_{\partial\mathcal{R}} \Big\{ t\frac{\partial |u|^2}{\partial \nu} - |u|^2 \frac{\partial t}{\partial \nu} \Big\}\, d\sigma, \tag{1.7.183}$$

where $\nu = (\nu_1, \dots, \nu_n)$ is the outward unit normal to $\mathcal{R}$, and $\sigma := \mathcal{H}^{n-1}\lfloor\partial\mathcal{R}$. Above, at $\mathcal{H}^{n-1}$-a.e. point $(x', t) \in \partial\mathcal{R} \cap \partial\mathbb{R}^n_+$, the expression in the curly brackets is interpreted as

$$2t\big(u\Big|^{\kappa-\mathrm{n.t.}}_{\partial\mathcal{R}}\big)(x', t)\nu(x', t) \cdot \big((\nabla u)\Big|^{\kappa-\mathrm{n.t.}}_{\partial\mathcal{R}}\big)(x', t) - \Big|\big(u\Big|^{\kappa-\mathrm{n.t.}}_{\partial\mathcal{R}}\big)(x', t)\Big|^2 \nu_n(x', t). \tag{1.7.184}$$

This result justifies the claim made by C. Fefferman and E. Stein at the bottom of page 162 in [45] where they take $\mathcal{R}$ to be the sawtooth region in the upper half-space associated with a given closed set $E \subseteq \mathbb{R}^{n-1} \equiv \partial\mathbb{R}^n_+$ and some $\kappa > 0$. Specifically,

$$\mathcal{R} := \bigcup_{x \in E} \Gamma_\kappa(x), \tag{1.7.185}$$

where, for each $x \in \partial\mathbb{R}^n_+$, we have denoted by $\Gamma_\kappa(x)$ the nontangential approach region with aperture parameter κ and vertex at x, relative to the domain $\mathbb{R}^n_+$. The uniform cone property satisfied by $\mathcal{R}$ ensures that this set is the upper-graph of a non-negative Lipschitz function (see [62]).

Harmonic functions in the upper half-space satisfying (1.7.181) arise naturally. To offer a concrete example, recall that the Poisson kernel for the Laplacian in $\mathbb{R}^n_+$ is given by

$$P^{\Delta}(x') = \frac{2}{\omega_{n-1}} \frac{1}{\left(1+|x'|^2\right)^{\frac{n}{2}}}, \qquad \forall x' \in \mathbb{R}^{n-1}. \tag{1.7.186}$$

If for each $x' \in \mathbb{R}^{n-1}$ and $t > 0$ we set $P^{\Delta}_t(x') := t^{1-n} P^{\Delta}(x'/t)$, then given any $f \in W^{1,2}(\mathbb{R}^{n-1})$ the function $u : \mathbb{R}^n_+ \to \mathbb{R}$ defined as $u(x',t) := (P^{\Delta}_t * f)(x')$ for each $(x',t) \in \mathbb{R}^n_+$ satisfies (1.7.181) (cf. [94]).

To give another example (and this is the case considered in [45]), start with a harmonic function u in $\mathbb{R}^n_+$ with the property that $\mathcal{N}_\kappa u \in L^p(\mathbb{R}^{n-1}, \mathcal{L}^{n-1})$ for some $p \in (0,2]$ and $\kappa > 0$ and, for each fixed $\varepsilon > 0$, define $u_\varepsilon(x',t) := u(x', t+\varepsilon)$ for every $(x',t) \in \mathbb{R}^n_+$. Pick an arbitrary point $(x',t) \in \mathbb{R}^n_+$. Using interior estimate in the ball $B\big((x',t), t/2\big) \subset \mathbb{R}^n_+$, for each $k \in \mathbb{N}_0$ we may estimate

$$\begin{aligned}
\big|(\nabla^k u)(x',t)\big| &\leq \frac{C}{t^k} \fint_{B((x',t),t/2)} |u(y',s)|\, \mathrm{d}y' ds \\
&\leq \frac{C}{t^{k+n}} \int_{t/2}^{3t/2} \int_{B_{n-1}(x',t/2)} |u(y',s)|\, \mathrm{d}y' ds \\
&\leq \frac{C}{t^k} \fint_{B_{n-1}(x',t/2)} (\mathcal{N}_\kappa u)(y')\, \mathrm{d}y' \\
&\leq \frac{C}{t^k} \mathcal{M}\big(\mathcal{N}_\kappa u\big)(x'),
\end{aligned} \tag{1.7.187}$$

where $B_{n-1}(a', r)$ denotes the $(n-1)$-dimensional ball centered at $a' \in \mathbb{R}^{n-1}$ of radius r, and $\mathcal{M}$ is the Hardy-Littlewood maximal operator in $\mathbb{R}^{n-1}$. Also, using reverse Hölder estimates for harmonic functions yields

$$\begin{aligned}
\big|u(x',t)\big| &\leq C \left(\fint_{B((x',t),t/2)} |u(y',s)|^p\, \mathrm{d}y' ds \right)^{1/p} \\
&\leq \frac{C}{t^{n/p}} \left(\int_{t/2}^{3t/2} \int_{B_{n-1}(x',\, t/2)} |u(y',s)|^p\, \mathrm{d}y' ds \right)^{1/p} \\
&\leq \frac{C}{t^{(n-1)/p}} \left(\int_{B_{n-1}(x',\, t/2)} (\mathcal{N}_\kappa u)(y')^p\, \mathrm{d}y' \right)^{1/p} \\
&\leq C\, t^{-(n-1)/p} \|\mathcal{N}_\kappa u\|_{L^p(\mathbb{R}^{n-1}, \mathcal{L}^{n-1})}.
\end{aligned} \tag{1.7.188}$$

When used with $t+\varepsilon$ in place of t, estimate (1.7.188) gives

$$|u_\varepsilon(x',t)| \leq C\varepsilon^{-(n-1)/p}\, \|\mathcal{N}_\kappa u\|_{L^p(\mathbb{R}^{n-1}, \mathcal{L}^{n-1})}, \qquad \forall (x',t) \in \mathbb{R}^n_+, \tag{1.7.189}$$

hence $u_\varepsilon \in L^\infty(\mathbb{R}^n_+, \mathcal{L}^n)$ which further implies

$$\mathcal{N}_\kappa u_\varepsilon \in \bigcap_{p \le q \le \infty} L^q(\mathbb{R}^{n-1}, \mathcal{L}^{n-1}) \text{ for each } \varepsilon > 0. \tag{1.7.190}$$

Writing (1.7.187) for $u_{\varepsilon/2}$ in place of u while simultaneously replacing t by $t + \varepsilon/2$ yields

$$\left|(\nabla^k u_\varepsilon)(x', t)\right| \le \frac{C}{\varepsilon^k} \mathcal{M}\big(\mathcal{N}_\kappa u_{\varepsilon/2}\big)(x'), \qquad \forall (x', t) \in \mathbb{R}^n_+. \tag{1.7.191}$$

Bearing in mind that $p \le 2$, in view of (1.7.190) and the boundedness of $\mathcal{M}$, from this we obtain

$$\sup_{t>0} \left\|(\nabla^k u_\varepsilon)(\cdot, t)\right\|_{L^2(\mathbb{R}^{n-1}, \mathcal{L}^{n-1})} < +\infty, \text{ for each } k \in \mathbb{N}_0. \tag{1.7.192}$$

Finally, it is well known (see, e.g., [166, Corollary 1.2.1, p. 200]) that (1.7.192) implies the conditions in (1.7.181) for the function u_ε. The above argument may be summarized as follows:

given a harmonic function u in $\mathbb{R}^n_+$ with $\mathcal{N}_\kappa u \in L^p(\mathbb{R}^{n-1}, \mathcal{L}^{n-1})$ for some exponent $p \in (0, 2]$ and apperture parameter $\kappa > 0$, if for $\varepsilon > 0$ arbitrary we define $u_\varepsilon(x', t) := u(x', t + \varepsilon)$ for each $(x', t) \in \mathbb{R}^n_+$, then u_ε is a harmonic function in $\mathbb{R}^n_+$ which satisfies the conditions listed in (1.7.181). (1.7.193)

After this detour, we are ready to present the proof of Proposition 1.7.15.

Proof of Proposition 1.7.15 That any harmonic function u in $\mathbb{R}^n_+$ which satisfies (1.7.181) also satisfies (1.7.182) is a classical result (see, e.g., [166, Chapter VII]). Next, by assumption, the set $\partial\mathcal{R}$ is the graph of some non-negative Lipschitz function $\phi : \mathbb{R}^{n-1} \to \mathbb{R}$ and we find it useful to abbreviate $M := \|\nabla\phi\|_{L^\infty(\mathbb{R}^{n-1}, \mathcal{L}^{n-1})}$. In particular, if we fix $0 < \kappa < \big(-M + \sqrt{1 + M^2}\big)/M$ then

$$x + \Gamma_\kappa(0) \subseteq \mathcal{R} \text{ for each } x \in \partial\mathcal{R}. \tag{1.7.194}$$

Let us introduce the vector field

$$\vec{G}(x', t) := t u(x', t)(\nabla u)(x', t) - \tfrac{1}{2}|u(x', t)|^2 \mathbf{e}_n, \qquad \forall (x', t) \in \mathbb{R}^n_+. \tag{1.7.195}$$

Hence,

$$\vec{G} \in \big[\mathscr{C}^\infty(\mathbb{R}^n_+)\big]^n. \tag{1.7.196}$$

Simple geometry shows that we may choose $\kappa' > \kappa$ with the property that $B((y', t), t/2) \subseteq \Gamma_{\kappa'}(x)$ for each $x \in \partial\mathbb{R}^n_+$ and each $(y', t) \in \Gamma_\kappa(x)$. Based on this observation and interior estimates, for each $x = (x', 0) \in \partial\mathbb{R}^n_+$ and each $(y', t) \in \Gamma_\kappa(x)$ we may write

$$t|(\nabla u)(y', t)| \le C \fint_{B((y', t), t/2)} |u| \, d\mathcal{L}^n \le C\big(\mathcal{N}_{\kappa'} u\big)(x'), \tag{1.7.197}$$

hence

$$\sup_{(y',t)\in\Gamma_\kappa(x)}\Big[t|(\nabla u)(y',t)|\Big]\le C\big(\mathcal{N}_{\kappa'}u\big)(x'). \tag{1.7.198}$$

From this we then conclude that there exists some dimensional constant $C\in(0,\infty)$ with the property that

$$\mathcal{N}_\kappa\vec{G}\le C\big(\mathcal{N}_{\kappa'}u\big)^2 \text{ everywhere on } \mathbb{R}^{n-1}\equiv\partial\mathbb{R}^n_+. \tag{1.7.199}$$

To proceed, consider the vector field

$$\vec{F}:=\vec{G}\big|_{\mathcal{R}}\in\big[\mathscr{C}^\infty(\mathcal{R})\big]^n \tag{1.7.200}$$

and (keeping (1.7.194) in mind) define

$$\big(\mathring{\mathcal{N}}_\kappa\vec{F}\big)(x):=\sup_{y\in x+\Gamma_\kappa(0)}|\vec{F}(y)| \text{ for each } x\in\partial\mathcal{R}. \tag{1.7.201}$$

Then (1.7.199) implies that for each $x=(x',t)\in\partial\mathcal{R}$ we have

$$\big(\mathring{\mathcal{N}}_\kappa\vec{F}\big)(x)\le\sup_{y\in\Gamma_\kappa(x',0)}|\vec{G}(y)|=\big(\mathcal{N}_\kappa\vec{G}\big)(x')\le C\big(\mathcal{N}_{\kappa'}u\big)^2(x'). \tag{1.7.202}$$

Bearing in mind that $\partial\mathcal{R}$ is the graph of the Lipschitz function $\phi:\mathbb{R}^{n-1}\to\mathbb{R}$, from (1.7.202) we conclude that

$$\big\|\mathring{\mathcal{N}}_\kappa\vec{F}\big\|_{L^1(\partial\mathcal{R},\sigma)}\le C\sqrt{1+M^2}\,\big\|\mathcal{N}_{\kappa'}u\big\|^2_{L^2(\mathbb{R}^{n-1},\mathcal{L}^{n-1})}<+\infty. \tag{1.7.203}$$

In particular, this forces $\mathring{\mathcal{N}}_\kappa\vec{F}\in L^1(\partial\mathcal{R},\sigma)$. Having established this, [122, Proposition 2.2, p. 24] applies and gives that

$$\mathcal{N}_{\mathcal{R},\kappa}\vec{F}\in L^1(\partial\mathcal{R},\sigma) \tag{1.7.204}$$

where $\mathcal{N}_{\mathcal{R},\kappa}$ denotes the nontangential maximal operator (with aperture parameter κ) relative to the domain $\mathcal{R}$. Moreover, if for each $x\in\partial\mathcal{R}$ we let $\Gamma_{\mathcal{R},\kappa}(x)$ stand for the nontangential approach region (with aperture parameter κ) relative to the domain $\mathcal{R}$, the fact that

$$\Gamma_{\mathcal{R},\kappa}(x)\subseteq\Gamma_\kappa(x) \text{ for each point } x\in\partial\mathcal{R}\cap\partial\mathbb{R}^n_+, \tag{1.7.205}$$

together with (1.7.200), (1.7.195)-(1.7.196), and (1.7.182), implies that

$$\vec{F}\Big|^{\kappa-\text{n.t.}}_{\partial\mathcal{R}} \text{ exists } \sigma\text{-a.e. on } \partial\mathcal{R}. \tag{1.7.206}$$

In addition, a direct computation which takes into account that the function u is harmonic (hence, $\Delta(|u|^2)=2|\nabla u|^2$) gives

$$(\text{div}\vec{F})(x',t)=t|(\nabla u)(x',t)|^2 \text{ for each } (x',t)\in\mathcal{R}. \tag{1.7.207}$$

On the other hand, from [166, Corollary 1.2.1, p. 200] and standard L^2-square function estimates for derivatives of the Poisson kernel (see, e.g., [166, Theorem 1, p. 82] when $p = 2$) we see that

$$\int_{\mathbb{R}^n_+} t|(\nabla u)(x',t)|^2 \, dx'dt < +\infty. \tag{1.7.208}$$

Collectively, (1.7.207)-(1.7.208) prove that

$$\operatorname{div}\vec{F} \in L^1(\mathcal{R}, \mathcal{L}^n). \tag{1.7.209}$$

Granted (1.7.200), (1.7.204), (1.7.206), and (1.7.209), [112, Theorem 1.2.1] applies and the Divergence Formula recorded in [112, (1.2.2)] yields precisely the identity in (1.7.183) (interpreting the integrand in the right-hand side as in (1.7.184) for any given parameter $\kappa > 0$). □

In the last part of this section we revisit the Green type formulas for second-order systems from Theorem 1.7.10 and Theorem 1.7.12 and adopt a more general point of view in which the system in question is written as a composition of two first-order differential operators D and $\widetilde{D}$. This adds more flexibility in applications and also leads to an invariant formulation of Green formulas which carries over to the setting of manifolds. We take up the latter task in Theorem 1.7.19 and Theorem 1.7.20. For now, we begin with the following generalization of Theorem 1.7.10.

Theorem 1.7.16 *Suppose $\Omega \subseteq \mathbb{R}^n$ is an open set with a lower Ahlfors regular boundary and with the property that $\sigma := \mathcal{H}^{n-1}\lfloor\partial\Omega$ is a doubling measure. In particular, Ω is a set of locally finite perimeter, and its geometric measure theoretic outward unit normal ν is defined σ-a.e. on $\partial_*\Omega$.*

Having fixed $M, N, N' \in \mathbb{N}$, consider a first-order $M \times N$ system D with constant complex coefficients in $\mathbb{R}^n$, along with a first-order $N' \times M$ system $\widetilde{D}$ with constant complex coefficients in $\mathbb{R}^n$. Then define the constant (complex) coefficient second-order $N' \times N$ system in $\mathbb{R}^n$ given by

$$L := \widetilde{D}D. \tag{1.7.210}$$

Finally, fix some $\kappa, \kappa' > 0$, assume $(\mathbb{X}_0, \mathbb{Y}_0)$ and $(\mathbb{X}_1, \mathbb{Y}_1)$ are two regular dual systems on $(\partial\Omega, \sigma)$ (in the sense of Definition 1.7.9), and consider two complex vector-valued functions

$$u \in \big[L^1_{\mathrm{loc}}(\Omega, \mathcal{L}^n)\big]^N, \quad w \in \big[L^1_{\mathrm{loc}}(\Omega, \mathcal{L}^n)\big]^{N'}, \tag{1.7.211}$$

satisfying (with all derivatives taken in the sense of distributions in Ω):

$$
\begin{aligned}
& Du \in \big[L^1_{\rm loc}(\Omega,\mathcal{L}^n)\big]^M, \quad \widetilde{D}^\top w \in \big[L^1_{\rm loc}(\Omega,\mathcal{L}^n)\big]^M, \\
& \mathcal{N}_\kappa u \in \mathbb{X}_0,\ \mathcal{N}_\kappa(Du) \in \mathbb{X}_1,\ \mathcal{N}_{\kappa'} w \in \mathbb{Y}_1,\ \mathcal{N}_{\kappa'}(\widetilde{D}^\top w) \in \mathbb{Y}_0, \\
& u\big|^{{}^{\kappa-{\rm n.t.}}}_{\partial\Omega},\ (Du)\big|^{{}^{\kappa-{\rm n.t.}}}_{\partial\Omega},\ w\big|^{{}^{\kappa'-{\rm n.t.}}}_{\partial\Omega},\ (\widetilde{D}^\top w)\big|^{{}^{\kappa'-{\rm n.t.}}}_{\partial\Omega} \ \textit{exist } \sigma\textit{-a.e. on } \partial_{\rm nta}\Omega, \\
& Lu \in \big[L^1_{\rm loc}(\Omega,\mathcal{L}^n)\big]^{N'}, \quad L^\top w \in \big[L^1_{\rm loc}(\Omega,\mathcal{L}^n)\big]^N, \\
& \big\langle Lu, w\big\rangle - \big\langle u, L^\top w\big\rangle \ \textit{belongs to } L^1(\Omega,\mathcal{L}^n).
\end{aligned}
\tag{1.7.212}
$$

Then for any other specified aperture parameter $\kappa'' \in (0,\infty)$ *the nontangential boundary traces* $u\big|^{{}^{\kappa''-{\rm n.t.}}}_{\partial\Omega}$, $(Du)\big|^{{}^{\kappa''-{\rm n.t.}}}_{\partial\Omega}$, $w\big|^{{}^{\kappa''-{\rm n.t.}}}_{\partial\Omega}$, $(\widetilde{D}^\top w)\big|^{{}^{\kappa''-{\rm n.t.}}}_{\partial\Omega}$ *exist at* σ*-a.e. point on* $\partial_{\rm nta}\Omega$ *and are independent of* κ''. *Moreover, with the dependence on the parameter* κ'' *dropped, the following Green type formula (involving absolutely convergent integrals) holds*

$$
\begin{aligned}
\int_\Omega & \Big\{\big\langle Lu, w\big\rangle - \big\langle u, L^\top w\big\rangle\Big\}\,{\rm d}\mathcal{L}^n \\
&= \int_{\partial_*\Omega} \Big\langle (-{\rm i})\,{\rm Sym}(\widetilde{D};\nu)\big((Du)\big|^{{}^{\rm n.t.}}_{\partial\Omega}\big), w\big|^{{}^{\rm n.t.}}_{\partial\Omega}\Big\rangle\,{\rm d}\sigma \\
&\quad - \int_{\partial_*\Omega} \Big\langle u\big|^{{}^{\rm n.t.}}_{\partial\Omega}, (-{\rm i})\,{\rm Sym}(D^\top;\nu)\big((\widetilde{D}^\top w)\big|^{{}^{\rm n.t.}}_{\partial\Omega}\big)\Big\rangle\,{\rm d}\sigma,
\end{aligned}
\tag{1.7.213}
$$

when either Ω *is bounded, or* $\partial\Omega$ *is unbounded. In the remaining case, i.e., when* Ω *is unbounded and* $\partial\Omega$ *is bounded, formula* (1.7.213) *continues to be true under the additional assumption that there exists* $\lambda \in (1,\infty)$ *such that*

$$
\int_{[B(0,\lambda R)\setminus B(0,R)]\cap\Omega} \Big\{|u||\widetilde{D}^\top w| + |w||Du|\Big\}\,{\rm d}\mathcal{L}^n = o(R) \ \textit{as } R \to \infty. \tag{1.7.214}
$$

As a consequence, if the last condition in (1.7.212) *is replaced by*

$$
Lu = 0 \ \textit{and } L^\top w = 0 \ \textit{in the sense of distributions in } \Omega, \tag{1.7.215}
$$

then (1.7.213) *reduces to*

$$
\begin{aligned}
\int_{\partial_*\Omega} & \Big\langle (-{\rm i})\,{\rm Sym}(\widetilde{D};\nu)\big((Du)\big|^{{}^{\rm n.t.}}_{\partial\Omega}\big), w\big|^{{}^{\rm n.t.}}_{\partial\Omega}\Big\rangle\,{\rm d}\sigma \\
&= \int_{\partial_*\Omega} \Big\langle u\big|^{{}^{\rm n.t.}}_{\partial\Omega}, (-{\rm i})\,{\rm Sym}(D^\top;\nu)\big((\widetilde{D}^\top w)\big|^{{}^{\rm n.t.}}_{\partial\Omega}\big)\Big\rangle\,{\rm d}\sigma.
\end{aligned}
\tag{1.7.216}
$$

Finally, similar results are valid if $(\mathbb{X}_0,\mathbb{Y}_0)$ *and* $(\mathbb{X}_1,\mathbb{Y}_1)$ *are two dual systems on* $(\partial\Omega,\sigma)$ *which are not necessarily regular, with the understanding that, in such a case, all nontangential boundary traces are taken with respect to an aperture parameter which is* $< \min\{\kappa,\kappa'\}$.

Proof From assumptions and [112, Corollary 8.9.9] it follows that there exists $p \in (0, \infty]$ with the property that for any given aperture parameter $\kappa'' > 0$ we have

$$\mathcal{N}_{\kappa''}u,\ \mathcal{N}_{\kappa''}(Du),\ \mathcal{N}_{\kappa''}w,\ \mathcal{N}_{\kappa''}(\widetilde{D}^{\top}w) \in L^p_{\rm loc}(\partial\Omega, \sigma), \tag{1.7.217}$$

and the nontangential boundary traces

$$\begin{gathered} u\Big|^{\kappa''-\text{n.t.}}_{\partial\Omega},\ (Du)\Big|^{\kappa''-\text{n.t.}}_{\partial\Omega},\ w\Big|^{\kappa''-\text{n.t.}}_{\partial\Omega},\ (\widetilde{D}^{\top}w)\Big|^{\kappa''-\text{n.t.}}_{\partial\Omega} \text{ exist} \\ \text{at } \sigma\text{-a.e. point on } \partial_{\rm nta}\Omega \text{ and agree with the respective} \\ \text{nontangential boundary traces in (1.7.212).} \end{gathered} \tag{1.7.218}$$

To proceed, note that (1.7.211), the first two lines in (1.7.212), and [112, Lemma 8.3.1] imply

$$\begin{aligned} & u \in \big[L^\infty_{\rm loc}(\Omega, \mathcal{L}^n)\big]^N, \quad Du \in \big[L^\infty_{\rm loc}(\Omega, \mathcal{L}^n)\big]^M, \\ & w \in \big[L^\infty_{\rm loc}(\Omega, \mathcal{L}^n)\big]^{N'}, \quad \widetilde{D}^{\top}w \in \big[L^\infty_{\rm loc}(\Omega, \mathcal{L}^n)\big]^M. \end{aligned} \tag{1.7.219}$$

Consider now the vector field $\vec{F} : \Omega \to \mathbb{C}^n$ implicitly defined by the requirement that

$$\begin{aligned} \xi \cdot \vec{F}(x) = & \Big\langle (-\mathrm{i})\,\mathrm{Sym}(\widetilde{D};\xi)(Du)(x), w(x) \Big\rangle \\ & - \Big\langle u(x), (-\mathrm{i})\,\mathrm{Sym}(D^{\top};\xi)(\widetilde{D}^{\top}w)(x) \Big\rangle \\ & \text{for each } \xi \in \mathbb{R}^n \text{ and } \mathcal{L}^n\text{-a.e. } x \in \Omega. \end{aligned} \tag{1.7.220}$$

Then (1.7.219) ensures that

$$\vec{F} \in \big[L^\infty_{\rm loc}(\Omega, \mathcal{L}^n)\big]^N \subseteq \big[L^1_{\rm loc}(\Omega, \mathcal{L}^n)\big]^n. \tag{1.7.221}$$

Let us compute $\mathrm{div}\vec{F}$ in the sense of distributions in Ω. With this goal in mind, fix an arbitrary real-valued function $\varphi \in \mathscr{C}^\infty_c(\Omega)$ and write

$$\begin{aligned} {}_{\mathcal{D}'(\Omega)}\big\langle \mathrm{div}\vec{F}, \varphi \big\rangle_{\mathcal{D}(\Omega)} & = -\int_\Omega \nabla\varphi \cdot \vec{F}\,\mathrm{d}\mathcal{L}^n \\ & = -\int_\Omega \Big\langle (-\mathrm{i})\,\mathrm{Sym}(\widetilde{D};\nabla\varphi)Du, w \Big\rangle\,\mathrm{d}\mathcal{L}^n \\ & \quad + \int_\Omega \Big\langle u, (-\mathrm{i})\,\mathrm{Sym}(D^{\top};\nabla\varphi)\widetilde{D}^{\top}w \Big\rangle\,\mathrm{d}\mathcal{L}^n \\ & =: I + II, \end{aligned} \tag{1.7.222}$$

where the second equality is based on (1.7.220) with $\xi := \nabla\varphi$. For each sufficiently small $\varepsilon > 0$ consider now $\Omega_\varepsilon := \{x \in \Omega : \operatorname{dist}(x, \partial\Omega) > \varepsilon\}$. Using a Friedrichs mollifier, we may construct a sequence of functions $w^\varepsilon = (w^\varepsilon_\alpha)_{1\leq\alpha\leq N'} \in \big[\mathscr{C}^\infty(\Omega_\varepsilon)\big]^{N'}$ such that

$$\begin{aligned}
& w^\varepsilon \xrightarrow[\varepsilon\to 0^+]{} w \text{ uniformly on compact subsets of } \Omega, \\
& \widetilde{D}^\top w^\varepsilon \xrightarrow[\varepsilon\to 0^+]{} \widetilde{D}^\top w \text{ at } \mathcal{L}^n\text{-a.e. point in } \Omega \text{ and, also,} \\
& \text{for each compact set } K \subset \Omega \text{ there exists some } \varepsilon_K > 0 \\
& \text{with the property that } \sup_{0<\varepsilon<\varepsilon_K} \big\|\widetilde{D}^\top w^\varepsilon\big\|_{[L^\infty(K,\mathcal{L}^n)]^M} < \infty.
\end{aligned} \tag{1.7.223}$$

Then, based on [112, (1.7.17), (1.7.20)], (1.7.219), (1.7.223), the penultimate line in (1.7.212), and Lebesgue's Dominated Convergence Theorem, we may compute

$$\begin{aligned}
I &= -\lim_{\varepsilon\to 0^+} \int_\Omega \Big\langle (-\mathrm{i})\operatorname{Sym}(\widetilde{D};\nabla\varphi)Du, w^\varepsilon \Big\rangle \,\mathrm{d}\mathcal{L}^n \\
&= \lim_{\varepsilon\to 0^+} \int_\Omega \Big\langle Du, (-\mathrm{i})\operatorname{Sym}(\widetilde{D}^\top;\nabla\varphi)w^\varepsilon \Big\rangle \,\mathrm{d}\mathcal{L}^n \\
&= \lim_{\varepsilon\to 0^+} \int_\Omega \big\langle Du, \widetilde{D}^\top(\varphi w^\varepsilon)\big\rangle \,\mathrm{d}\mathcal{L}^n - \lim_{\varepsilon\to 0^+} \int_\Omega \big\langle Du, \varphi\widetilde{D}^\top w^\varepsilon\big\rangle \,\mathrm{d}\mathcal{L}^n \\
&= \lim_{\varepsilon\to 0^+} {}_{[\mathcal{D}'(\Omega)]^M}\Big\langle Du, \widetilde{D}^\top(\varphi w^\varepsilon)\Big\rangle_{[\mathcal{D}(\Omega)]^M} - \int_\Omega \big\langle Du, \widetilde{D}^\top w\big\rangle\varphi \,\mathrm{d}\mathcal{L}^n \\
&= \lim_{\varepsilon\to 0^+} {}_{[\mathcal{D}'(\Omega)]^{N'}}\Big\langle \widetilde{D}Du, \varphi w^\varepsilon\Big\rangle_{[\mathcal{D}(\Omega)]^{N'}} - \int_\Omega \big\langle Du, \widetilde{D}^\top w\big\rangle\varphi \,\mathrm{d}\mathcal{L}^n \\
&= \lim_{\varepsilon\to 0^+} {}_{[\mathcal{D}'(\Omega)]^{N'}}\big\langle Lu, \varphi w^\varepsilon\big\rangle_{[\mathcal{D}(\Omega)]^{N'}} - \int_\Omega \big\langle Du, \widetilde{D}^\top w\big\rangle\varphi \,\mathrm{d}\mathcal{L}^n \\
&= \lim_{\varepsilon\to 0^+} \int_\Omega \big\langle Lu, \varphi w^\varepsilon\big\rangle \,\mathrm{d}\mathcal{L}^n - \int_\Omega \big\langle Du, \widetilde{D}^\top w\big\rangle\varphi \,\mathrm{d}\mathcal{L}^n \\
&= \int_\Omega \big\langle Lu, \varphi w\big\rangle \,\mathrm{d}\mathcal{L}^n - \int_\Omega \big\langle Du, \widetilde{D}^\top w\big\rangle\varphi \,\mathrm{d}\mathcal{L}^n \\
&= \int_\Omega \big\langle Lu, w\big\rangle\varphi \,\mathrm{d}\mathcal{L}^n - \int_\Omega \big\langle Du, \widetilde{D}^\top w\big\rangle\varphi \,\mathrm{d}\mathcal{L}^n.
\end{aligned} \tag{1.7.224}$$

In a similar fashion,

$$II = -\int_\Omega \big\langle u, L^\top w\big\rangle\varphi \,\mathrm{d}\mathcal{L}^n + \int_\Omega \big\langle Du, \widetilde{D}^\top w\big\rangle\varphi \,\mathrm{d}\mathcal{L}^n. \tag{1.7.225}$$

At this stage, from (1.7.222) and (1.7.224)-(1.7.225) we conclude that

$$\mathrm{div}\vec{F} = \langle Lu, w\rangle - \langle u, L^\top w\rangle \text{ in } \mathcal{D}'(\Omega). \tag{1.7.226}$$

In particular, the last line in (1.7.212) ensures that

$$\mathrm{div}\vec{F} \in L^1(\Omega, \mathcal{L}^n). \tag{1.7.227}$$

Also, if we choose an aperture parameter $\kappa_* \in (0, \min\{\kappa, \kappa'\})$, from (1.7.220) and [112, (8.2.25)] we obtain

$$\mathcal{N}_{\kappa_*}\vec{F} \leq C\Big\{\mathcal{N}_\kappa(Du)\mathcal{N}_{\kappa'}w + \mathcal{N}_\kappa u\ \mathcal{N}_{\kappa'}(\widetilde{D}^\top w)\Big\} \text{ on } \partial\Omega, \tag{1.7.228}$$

for some constant $C \in (0, \infty)$ which depends only on the coefficient tensor A. In turn, from (1.7.102), (1.7.217), [112, (8.2.26)], the nontangential maximal function memberships in the second line of (1.7.212), and (1.7.71) we conclude that

$$\mathcal{N}_{\kappa_*}\vec{F} \in L^1(\partial\Omega, \sigma). \tag{1.7.229}$$

Going further, observe that (1.7.220) and (1.7.218) imply that the nontangential boundary trace

$$\vec{F}\Big|_{\partial\Omega}^{\kappa''-\mathrm{n.t.}} \text{ exists at } \sigma\text{-a.e. point on } \partial_{\mathrm{nta}}\Omega. \tag{1.7.230}$$

Moreover, with the dependence on κ'' dropped, from (1.7.220) and [112, Proposition 8.8.6] we obtain

$$\begin{aligned}\nu\cdot\Big(\vec{F}\Big|_{\partial\Omega}^{\mathrm{n.t.}}\Big) &= \Big\langle(-\mathrm{i})\,\mathrm{Sym}(\widetilde{D};\nu)\Big((Du)\Big|_{\partial\Omega}^{\mathrm{n.t.}}\Big), w\Big|_{\partial\Omega}^{\mathrm{n.t.}}\Big\rangle \\ &\quad - \Big\langle u\Big|_{\partial\Omega}^{\mathrm{n.t.}}, (-\mathrm{i})\,\mathrm{Sym}(D^\top;\nu)\Big((\widetilde{D}^\top w)\Big|_{\partial\Omega}^{\mathrm{n.t.}}\Big)\Big\rangle\end{aligned} \tag{1.7.231}$$

at σ-a.e. point on $\partial_*\Omega$. Finally, (1.7.214) guarantees that the vector field $\vec{F}$ satisfies the growth condition [112, (1.2.3)]. Granted these properties of $\vec{F}$, [112, Theorem 1.2.1] applies and the Divergence Formula [112, (1.2.2)] presently yields (1.7.213) on account of (1.7.226) and (1.7.231). □

It is of interest to specialize Theorem 1.7.16 to the case when the factorization in (1.7.210) pertains to the zero operator, in which case we obtain an integration by parts formula on the boundary without demanding that the functions involved satisfy a certain PDE (as was the case in (1.7.215)-(1.7.216)). More general formulas of this flavor are established in the context of Riemannian manifolds in [112, §1.12].

Corollary 1.7.17 *Let $\Omega \subseteq \mathbb{R}^n$ be an open set with a lower Ahlfors regular boundary and such that $\sigma := \mathcal{H}^{n-1}\lfloor\partial\Omega$ is a doubling measure. Denote by ν the geometric measure theoretic outward unit normal to Ω Fix $M, N, N' \in \mathbb{N}$ and consider a first-order $M \times N$ system D with constant complex coefficients in $\mathbb{R}^n$, together with a first-order $N' \times M$ system $\widetilde{D}$ with constant complex coefficients in $\mathbb{R}^n$, with the property that*

$$\widetilde{D}D = 0. \tag{1.7.232}$$

Finally, fix some $\kappa, \kappa' > 0$, *assume* $(\mathbb{X}_0, \mathbb{Y}_0)$ *and* $(\mathbb{X}_1, \mathbb{Y}_1)$ *are two regular dual systems on* $(\partial\Omega, \sigma)$ *(in the sense of Definition 1.7.9), and consider two complex vector-valued functions*

$$u \in \big[L^1_{\rm loc}(\Omega, \mathcal{L}^n)\big]^N, \quad w \in \big[L^1_{\rm loc}(\Omega, \mathcal{L}^n)\big]^{N'}, \tag{1.7.233}$$

satisfying (with all derivatives taken in the sense of distributions in Ω*):*

$$\begin{aligned} & Du \in \big[L^1_{\rm loc}(\Omega, \mathcal{L}^n)\big]^M, \quad \widetilde{D}^\top w \in \big[L^1_{\rm loc}(\Omega, \mathcal{L}^n)\big]^M, \\ & \mathcal{N}_\kappa u \in \mathbb{X}_0, \ \mathcal{N}_\kappa(Du) \in \mathbb{X}_1, \ \mathcal{N}_{\kappa'} w \in \mathbb{Y}_1, \ \mathcal{N}_{\kappa'}(\widetilde{D}^\top w) \in \mathbb{Y}_0, \\ & u\big|^{\kappa-{\rm n.t.}}_{\partial\Omega}, \ (Du)\big|^{\kappa-{\rm n.t.}}_{\partial\Omega}, \ w\big|^{\kappa'-{\rm n.t.}}_{\partial\Omega}, \ (\widetilde{D}^\top w)\big|^{\kappa'-{\rm n.t.}}_{\partial\Omega} \ \text{exist } \sigma\text{-a.e. on } \partial_{\rm nta}\Omega. \end{aligned} \tag{1.7.234}$$

Then for any other specified aperture parameter $\kappa'' \in (0, \infty)$ *the nontangential boundary traces* $u\big|^{\kappa''-{\rm n.t.}}_{\partial\Omega}$, $(Du)\big|^{\kappa''-{\rm n.t.}}_{\partial\Omega}$, $w\big|^{\kappa''-{\rm n.t.}}_{\partial\Omega}$, $(\widetilde{D}^\top w)\big|^{\kappa''-{\rm n.t.}}_{\partial\Omega}$ *exist at* σ*-a.e. point on* $\partial_{\rm nta}\Omega$ *and are independent of* κ''. *In addition, with the dependence on the parameter* κ'' *dropped, the following formula holds*

$$\begin{aligned} & \int_{\partial_*\Omega} \Big\langle (-{\rm i})\,{\rm Sym}(\widetilde{D}; \nu)\big((Du)\big|^{\rm n.t.}_{\partial\Omega}\big), w\big|^{\rm n.t.}_{\partial\Omega} \Big\rangle \, {\rm d}\sigma \\ & \qquad = \int_{\partial_*\Omega} \Big\langle u\big|^{\rm n.t.}_{\partial\Omega}, (-{\rm i})\,{\rm Sym}(D^\top; \nu)\big((\widetilde{D}^\top w)\big|^{\rm n.t.}_{\partial\Omega}\big) \Big\rangle \, {\rm d}\sigma \end{aligned} \tag{1.7.235}$$

when either Ω *is bounded, or* $\partial\Omega$ *is unbounded. In the remaining case, i.e., when* Ω *is unbounded and* $\partial\Omega$ *is bounded, formula* (1.7.213) *continues to be true under the additional assumption that there exists* $\lambda \in (1, \infty)$ *such that*

$$\int_{[B(0,\lambda R)\setminus B(0,R)]\cap\Omega} \Big\{ |u||\widetilde{D}^\top w| + |w||Du| \Big\} \, {\rm d}\mathcal{L}^n = o(R) \ \text{as } R \to \infty. \tag{1.7.236}$$

In addition, a similar result is valid if $(\mathbb{X}_0, \mathbb{Y}_0)$ *and* $(\mathbb{X}_1, \mathbb{Y}_1)$ *are two dual systems on* $(\partial\Omega, \sigma)$ *which are not necessarily regular, with the understanding that, in such a case, all nontangential boundary traces are taken with respect to an aperture parameter which is* $< \min\{\kappa, \kappa'\}$.

We wish to remark that the cancelation condition demanded in (1.7.232) naturally occurs in applications. For instance, this is the case for $\widetilde{D} = {\rm curl}$ and $D = \nabla$ or, more generally, taking $\widetilde{D}$ and D to be the exterior derivative operator d. Another basic example is obtained by fixing two indices $j, k \in \{1, \ldots, n\}$ and then taking

$$Du := (\partial_j u)\mathbf{e}_j + (\partial_k u)\mathbf{e}_k, \qquad \widetilde{D}\vec{w} := \partial_j w_k - \partial_k w_j, \tag{1.7.237}$$

for all scalar functions u and vector-valued functions $\vec{w} = (w_i)_{1 \leq i \leq n}$. Formula (1.7.235) corresponding to such a scenario is then related to the result proved in [113, Proposition 11.3.12].

Proof of Corollary 1.7.17 This is an immediate consequence of Theorem 1.7.16. □

Our next result offers a more general perspective on the Green formula established in Theorem 1.7.12.

Theorem 1.7.18 *Let $\Omega \subseteq \mathbb{R}^n$ be an open set with a lower Ahlfors regular boundary and with the property that $\sigma := \mathcal{H}^{n-1} \lfloor \partial\Omega$ is a doubling measure. Hence, Ω is a set of locally finite perimeter, and its geometric measure theoretic outward unit normal ν is defined σ-a.e. on $\partial_*\Omega$.*

Fix $M, N, N' \in \mathbb{N}$ and consider a first-order $M \times N$ system D with constant complex coefficients in $\mathbb{R}^n$, along with a first-order $N' \times M$ system $\widetilde{D}$ with constant complex coefficients in $\mathbb{R}^n$. Then define the constant (complex) coefficient second-order $N' \times N$ system in $\mathbb{R}^n$ given by

$$L := \widetilde{D}D. \tag{1.7.238}$$

Finally, some $\kappa, \kappa' > 0$, assume $(\mathbb{X}, \mathbb{Y})$ is a regular dual system on $(\partial\Omega, \sigma)$ (in the sense of Definition 1.7.9), and consider two complex vector-valued functions

$$u \in \big[L^1_{\rm loc}(\Omega, \mathcal{L}^n)\big]^N, \quad w \in \big[L^1_{\rm loc}(\Omega, \mathcal{L}^n)\big]^{N'}, \tag{1.7.239}$$

satisfying

$$\begin{aligned} & Du \in \big[L^1_{\rm loc}(\Omega, \mathcal{L}^n)\big]^M, \quad \widetilde{D}^\top w \in \big[L^1_{\rm loc}(\Omega, \mathcal{L}^n)\big]^M, \\ & \mathcal{N}_\kappa(Du) \text{ belongs to } \mathbb{X}, \ \mathcal{N}_{\kappa'} w \text{ belongs to } \mathbb{Y}, \\ & (Du)\big|^{\kappa-{\rm n.t.}}_{\partial\Omega} \text{ and } w\big|^{\kappa'-{\rm n.t.}}_{\partial\Omega} \text{ exist } \sigma\text{-a.e. on } \partial_{\rm nta}\Omega, \\ & Lu \text{ belongs to } \big[L^1_{\rm loc}(\Omega, \mathcal{L}^n)\big]^{N'}, \text{ and} \\ & \big\langle Du, \widetilde{D}^\top w\big\rangle + \big\langle Lu, w\big\rangle \text{ belongs to } L^1(\Omega, \mathcal{L}^n). \end{aligned} \tag{1.7.240}$$

Then for any other specified aperture parameter $\kappa'' \in (0, \infty)$ the nontangential boundary traces $(Du)\big|^{\kappa''-{\rm n.t.}}_{\partial\Omega}$ and $w\big|^{\kappa''-{\rm n.t.}}_{\partial\Omega}$ exist at σ-a.e. point on $\partial_{\rm nta}\Omega$ and are independent of κ''. Moreover, with the dependence on the parameter κ'' dropped, the following Green type formula (involving absolutely convergent integrals) holds

$$\begin{aligned} & \int_\Omega \Big\{\big\langle Lu, w\big\rangle + \big\langle Du, \widetilde{D}^\top w\big\rangle\Big\}\, d\mathcal{L}^n \\ & \quad = \int_{\partial_*\Omega} \Big\langle (-i){\rm Sym}(\widetilde{D}; \nu)\big((Du)\big|^{\rm n.t.}_{\partial\Omega}\big), w\big|^{\rm n.t.}_{\partial\Omega}\Big\rangle\, d\sigma. \end{aligned} \tag{1.7.241}$$

when either Ω is bounded, or $\partial\Omega$ is unbounded. In the remaining case, i.e., when Ω is unbounded and $\partial\Omega$ is bounded, formula (1.7.241) *continues to be true under the additional assumption that there exists $\lambda \in (1, \infty)$ such that*

$$\int_{[B(0,\lambda R)\setminus B(0,R)]\cap\Omega} |Du||w| \, d\mathcal{L}^n = o(R) \ \text{as} \ R \to \infty. \tag{1.7.242}$$

In particular, corresponding to the case when the penultimate condition in (1.7.240) *is strengthened to*

$$Lu = 0 \ \text{in the sense of distributions in} \ \Omega, \tag{1.7.243}$$

formula (1.7.241) *becomes*

$$\int_\Omega \langle Du, \widetilde{D}^\top w \rangle \, d\mathcal{L}^n = \int_{\partial_*\Omega} \Big\langle (-i)\mathrm{Sym}(\widetilde{D}; \nu)\big((Du)\big|^{\text{n.t.}}_{\partial\Omega}\big), w\big|^{\text{n.t.}}_{\partial\Omega} \Big\rangle \, d\sigma. \tag{1.7.244}$$

Finally, similar results are valid if $(\mathbb{X}, \mathbb{Y})$ *is a dual system on* $(\partial\Omega, \sigma)$ *which is not necessarily regular, with the understanding that, in such a case, all nontangential boundary traces are taken with respect to an aperture parameter which is* $< \min\{\kappa, \kappa'\}$.

Proof In concert with [112, Corollary 8.9.9], the present assumptions guarantee the existence of some $p \in (0, \infty]$ with the property that for any given aperture parameter $\kappa'' > 0$ we have

$$\mathcal{N}_{\kappa''}(Du), \ \mathcal{N}_{\kappa''} w \in L^p_{\text{loc}}(\partial\Omega, \sigma), \tag{1.7.245}$$

and the nontangential boundary traces

$$\begin{gathered}(Du)\big|^{\kappa''-\text{n.t.}}_{\partial\Omega} \ \text{and} \ w\big|^{\kappa''-\text{n.t.}}_{\partial\Omega}, \ \text{both exist at } \sigma\text{-a.e. point on } \partial_{\text{nta}}\Omega, \\ \text{and agree with the nontangential boundary traces in (1.7.240).}\end{gathered} \tag{1.7.246}$$

Next, observe that (1.7.239), the second line in (1.7.240) (or (1.7.245)), and [112, Lemma 8.3.1] imply

$$Du \in \big[L^\infty_{\text{loc}}(\Omega, \mathcal{L}^n)\big]^M \ \text{and} \ w \in \big[L^\infty_{\text{loc}}(\Omega, \mathcal{L}^n)\big]^{N'}. \tag{1.7.247}$$

Consider now the vector field defined implicitly by the demand that

$$\begin{gathered}\xi \cdot \vec{F}(x) = \Big\langle (-i)\mathrm{Sym}(\widetilde{D}; \xi)(Du)(x), w(x) \Big\rangle \\ \text{for each } \xi \in \mathbb{R}^n \text{ and } \mathcal{L}^n\text{-a.e. } x \in \Omega.\end{gathered} \tag{1.7.248}$$

In particular, from this and (1.7.247) we conclude that

$$\vec{F} \in \big[L^\infty_{\text{loc}}(\Omega, \mathcal{L}^n)\big]^n \subseteq \big[L^1_{\text{loc}}(\Omega, \mathcal{L}^n)\big]^n. \tag{1.7.249}$$

To compute the divergence of $\vec{F}$, in the sense of distributions in Ω, fix an arbitrary real-valued function $\varphi \in \mathscr{C}^\infty_c(\Omega)$ and write

$$
\begin{aligned}
{}_{\mathcal{D}'(\Omega)}\big\langle \operatorname{div}\vec{F}, \varphi \big\rangle_{\mathcal{D}(\Omega)} &= -\int_\Omega \nabla\varphi \cdot \vec{F}\, d\mathcal{L}^n \\
&= -\int_\Omega \Big\langle (-i)\mathrm{Sym}(\widetilde{D}; \nabla\varphi)(Du), w \Big\rangle\, d\mathcal{L}^n, \qquad (1.7.250)
\end{aligned}
$$

where the second equality comes from (1.7.248), used with $\xi := \nabla\varphi$. To proceed, for each sufficiently small $\varepsilon > 0$ consider $\Omega_\varepsilon := \{x \in \Omega : \operatorname{dist}(x, \partial\Omega) > \varepsilon\}$. Using a Friedrichs mollifier, it is possible to construct a sequence of functions $w^\varepsilon = (w^\varepsilon_\alpha)_{1\le\alpha\le M} \in \big[\mathscr{C}^\infty(\Omega_\varepsilon)\big]^M$ such that

$$
\begin{aligned}
& w^\varepsilon \underset{\varepsilon\to 0^+}{\longrightarrow} w \text{ at } \mathcal{L}^n\text{-a.e. point in } \Omega, \\
& \text{for each compact set } K \subset \Omega \text{ we have} \\
& w^\varepsilon \underset{\varepsilon\to 0^+}{\longrightarrow} w \text{ in } \big[L^1(K, \mathcal{L}^n)\big]^{N'} \text{ and} \\
& \widetilde{D}^\top w^\varepsilon \underset{\varepsilon\to 0^+}{\longrightarrow} \widetilde{D}^\top w \text{ in } \big[L^1(K, \mathcal{L}^n)\big]^M, \\
& \text{and there exists some number } \varepsilon_K > 0 \\
& \text{so that } \sup\nolimits_{0<\varepsilon<\varepsilon_K} \|w^\varepsilon\|_{[L^\infty(K,\mathcal{L}^n)]^{N'}} < \infty.
\end{aligned}
\qquad (1.7.251)
$$

On account of (1.7.250), (1.7.247), (1.7.251), (1.7.240), and Lebesgue's Dominated Convergence Theorem, we may then compute

$$
\begin{aligned}
&{}_{\mathcal{D}'(\Omega)}\big\langle \mathrm{div}\vec{F}, \varphi\big\rangle_{\mathcal{D}(\Omega)}\\
&= -\lim_{\varepsilon\to 0^+}\int_\Omega \Big\langle (-\mathrm{i})\mathrm{Sym}(\widetilde{D};\nabla\varphi)Du, w^\varepsilon\Big\rangle\,\mathrm{d}\mathcal{L}^n\\
&= \lim_{\varepsilon\to 0^+}\int_\Omega \Big\langle Du, (-\mathrm{i})\mathrm{Sym}(\widetilde{D}^\top;\nabla\varphi)w^\varepsilon\Big\rangle\,\mathrm{d}\mathcal{L}^n\\
&= \lim_{\varepsilon\to 0^+}\int_\Omega \big\langle Du, \widetilde{D}^\top(\varphi w^\varepsilon)\big\rangle\,\mathrm{d}\mathcal{L}^n - \lim_{\varepsilon\to 0^+}\int_\Omega \big\langle Du, \varphi\widetilde{D}^\top w^\varepsilon\big\rangle\,\mathrm{d}\mathcal{L}^n\\
&= \lim_{\varepsilon\to 0^+}{}_{[\mathcal{D}'(\Omega)]^M}\big\langle Du, \widetilde{D}^\top(\varphi w^\varepsilon)\big\rangle_{[\mathcal{D}(\Omega)]^M} - \int_\Omega \big\langle Du, \widetilde{D}^\top w\big\rangle\varphi\,\mathrm{d}\mathcal{L}^n\\
&= \lim_{\varepsilon\to 0^+}{}_{[\mathcal{D}'(\Omega)]^{N'}}\big\langle \widetilde{D}Du, \varphi w^\varepsilon\big\rangle_{[\mathcal{D}(\Omega)]^{N'}} - \int_\Omega \big\langle Du, \widetilde{D}^\top w\big\rangle\varphi\,\mathrm{d}\mathcal{L}^n\\
&= \lim_{\varepsilon\to 0^+}{}_{[\mathcal{D}'(\Omega)]^{N'}}\big\langle Lu, \varphi w^\varepsilon\big\rangle_{[\mathcal{D}(\Omega)]^{N'}} - \int_\Omega \big\langle Du, \widetilde{D}^\top w\big\rangle\varphi\,\mathrm{d}\mathcal{L}^n\\
&= \lim_{\varepsilon\to 0^+}\int_\Omega \big\langle Lu, \varphi w^\varepsilon\big\rangle\,\mathrm{d}\mathcal{L}^n - \int_\Omega \big\langle Du, \widetilde{D}^\top w\big\rangle\varphi\,\mathrm{d}\mathcal{L}^n\\
&= \int_\Omega \big\langle Lu, w\big\rangle\varphi\,\mathrm{d}\mathcal{L}^n - \int_\Omega \big\langle Du, \widetilde{D}^\top w\big\rangle\varphi\,\mathrm{d}\mathcal{L}^n.
\end{aligned}
\tag{1.7.252}
$$

Having established this, we conclude that

$$
\mathrm{div}\vec{F} = \big\langle Lu, w\big\rangle + \big\langle Du, \widetilde{D}^\top w\big\rangle \text{ in } \mathcal{D}'(\Omega), \tag{1.7.253}
$$

so the last line in (1.7.240) imply

$$
\mathrm{div}\vec{F} \in L^1(\Omega, \mathcal{L}^n). \tag{1.7.254}
$$

In addition, if we choose an aperture parameter $\kappa_* \in \big(0, \min\{\kappa, \kappa'\}\big)$, from (1.7.248) and [112, (8.2.25)] we see that there exists a constant $C \in (0, \infty)$ such that

$$
\mathcal{N}_{\kappa_*}\vec{F} \le C\,\mathcal{N}_\kappa(Du)\;\mathcal{N}_{\kappa'}w \text{ on } \partial\Omega. \tag{1.7.255}
$$

From (1.7.255), (1.7.245), [112, (8.2.26)], the nontangential maximal function memberships in the second line of (1.7.240), and (1.7.71) we conclude that

$$
\mathcal{N}_{\kappa_*}\vec{F} \in L^1(\partial\Omega, \sigma). \tag{1.7.256}
$$

In addition, (1.7.246) and (1.7.248) imply that the nontangential boundary trace

$$
\vec{F}\Big|_{\partial\Omega}^{\kappa''-\mathrm{n.t.}} \text{ exists at } \sigma\text{-a.e. point on } \partial_{\mathrm{nta}}\Omega. \tag{1.7.257}
$$

Moreover, with the dependence on κ'' dropped, from (1.7.248) and [112, Proposition 8.8.6] we conclude that

$$\nu \cdot \Big(\vec{F}\Big|_{\partial\Omega}^{\rm n.t.}\Big) = \Big\langle (-\mathrm{i})\mathrm{Sym}(\widetilde{D};\nu)\big((Du)\big|_{\partial\Omega}^{\rm n.t.}\big), w\big|_{\partial\Omega}^{\rm n.t.}\Big\rangle \text{ at } \sigma\text{-a.e. point on } \partial_*\Omega. \quad (1.7.258)$$

Let us also note that (1.7.242) implies that the present vector field $\vec{F}$ satisfies the growth condition [112, (1.2.3)]. Having established the aforementioned properties of $\vec{F}$, [112, Theorem 1.2.1] applies and the Divergence Formula [112, (1.2.2)] currently yields (1.7.241), thanks to (1.7.253) and (1.7.258). □

One of the virtues of the manner in which Theorem 1.7.16 has been formulated and proved is that virtually the same considerations continue to be valid in the context of manifolds. Specifically, here is the counterpart of Theorem 1.7.16 for differential operators acting on Hermitian vector bundles over a given Riemannian manifold.

Theorem 1.7.19 *Suppose*

$$\begin{array}{l} M \textit{ is a connected, compact, boundaryless, oriented manifold of class} \\ \mathscr{C}^1, \textit{ of real dimension } n, \textit{ equipped with a continuous Riemannian} \\ \textit{metric tensor } g = \sum_{1\leq j,k\leq n} g_{jk}\mathrm{d}x_j \otimes \mathrm{d}x_k. \end{array} \quad (1.7.259)$$

Let Ω be a nonempty, open, proper subset of M with the property that $\partial\Omega$ is lower Ahlfors regular boundary, and that $\sigma_g := \mathcal{H}_g^{n-1}\lfloor\partial\Omega$ is a doubling measure on $\partial\Omega$. In particular, Ω is a set of finite perimeter and its (geometric measure theoretic) outward unit conormal $\nu_g : \partial_*\Omega \to T^*M$ is defined σ_g-a.e. on $\partial_*\Omega$. Denote by $\mathcal{L}_g^n$ the measure induced by the volume element $\mathrm{d}V_g$ on M.

Consider next three Hermitian vector bundles $\mathscr{E}, \mathscr{F}, \mathscr{G} \to M$ and, having fixed two first-order differential operators $D : \mathscr{E} \to \mathscr{F}$, and $\widetilde{D} : \mathscr{F} \to \mathscr{G}$, introduce $L := \widetilde{D}D$. Also, fix $p, q \in [1, \infty]$, denote by p', q' their Hölder conjugate exponents, pick some $\kappa, \kappa' > 0$, and consider two sections

$$u \in L^1_{\rm loc}(\Omega, \mathcal{L}_g^n) \otimes \mathscr{E}, \quad w \in L^1_{\rm loc}(\Omega, \mathcal{L}_g^n) \otimes \mathscr{G}, \quad (1.7.260)$$

satisfying (with all operators applied in the sense of distributions in Ω):

$$\begin{array}{l} Du \in L^1_{\rm loc}(\Omega, \mathcal{L}_g^n) \otimes \mathscr{F}, \quad \widetilde{D}^\top w \in L^1_{\rm loc}(\Omega, \mathcal{L}_g^n) \otimes \mathscr{F}, \\ \mathcal{N}_\kappa u \in L^p(\partial\Omega, \sigma_g), \quad \mathcal{N}_\kappa(Du) \in L^q(\partial\Omega, \sigma_g), \\ \mathcal{N}_{\kappa'} w \in L^{q'}(\partial\Omega, \sigma_g), \quad \mathcal{N}_{\kappa'}(\widetilde{D}^\top w) \in L^{p'}(\partial\Omega, \sigma_g), \\ u\big|_{\partial\Omega}^{\kappa-\rm n.t.}, (Du)\big|_{\partial\Omega}^{\kappa-\rm n.t.}, w\big|_{\partial\Omega}^{\kappa'-\rm n.t.}, (\widetilde{D}^\top w)\big|_{\partial\Omega}^{\kappa'-\rm n.t.} \textit{ exist } \sigma_g\textit{-a.e. on } \partial_{\rm nta}\Omega, \\ Lu \in L^1_{\rm loc}(\Omega, \mathcal{L}_g^n) \otimes \mathscr{G}, \quad L^\top w \in L^1_{\rm loc}(\Omega, \mathcal{L}_g^n) \otimes \mathscr{E}, \\ \langle Lu, w\rangle_{\mathscr{G}} - \langle u, L^\top w\rangle_{\mathscr{E}} \textit{ belongs to } L^1(\Omega, \mathcal{L}_g^n). \end{array} \quad (1.7.261)$$

Then for any other specified aperture parameter $\kappa'' \in (0,\infty)$ *the nontangential boundary traces* $u\big|_{\partial\Omega}^{\kappa''-\text{n.t.}}$, $(Du)\big|_{\partial\Omega}^{\kappa''-\text{n.t.}}$, $w\big|_{\partial\Omega}^{\kappa''-\text{n.t.}}$, $(\widetilde{D}^\top w)\big|_{\partial\Omega}^{\kappa''-\text{n.t.}}$ *exist at* σ_g*-a.e. point on* $\partial_{\text{nta}}\Omega$ *and are independent of* κ''*. Moreover, with the dependence on the parameter* κ'' *dropped, the following Green type formula (involving absolutely convergent integrals) holds:*

$$
\begin{aligned}
&\int_\Omega \Big\{\big\langle Lu, w\big\rangle_{\mathscr{G}} - \big\langle u, L^\top w\big\rangle_{\mathscr{E}}\Big\}\,\mathrm{d}\mathcal{L}^n_g \\
&\quad = \int_{\partial_*\Omega} \Big\langle (-\mathrm{i})\,\mathrm{Sym}(\widetilde{D};\nu)\big((Du)\big|_{\partial\Omega}^{\text{n.t.}}\big), w\big|_{\partial\Omega}^{\text{n.t.}}\Big\rangle_{\mathscr{G}}\,\mathrm{d}\sigma_g \\
&\qquad - \int_{\partial_*\Omega} \Big\langle u\big|_{\partial\Omega}^{\text{n.t.}}, (-\mathrm{i})\,\mathrm{Sym}(D^\top;\nu)\big((\widetilde{D}^\top w)\big|_{\partial\Omega}^{\text{n.t.}}\big)\Big\rangle_{\mathscr{E}}\,\mathrm{d}\sigma_g.
\end{aligned}
\tag{1.7.262}
$$

Moreover, this formula continues to be valid when the nontangential maximal function conditions stipulated in (the second and third lines of) (1.7.261) are replaced by more general memberships to the components of any two regular dual systems on $(\partial\Omega,\sigma_g)$ *(defined in a fashion similar to Definition 1.7.9).*

Proof A cursory inspection reveals that real-variable argument used in the proof of Theorem 1.7.16 may be naturally adapted to the context of manifolds. More specifically, the definition of the vector field $\vec{F}$ in (1.7.220) continues to make sense in the latter context if ξ is now selected from the co-tangent bundle T^*M and the dot product now interpreted as the duality pairing between vectors in TM and co-vectors in T^*M. Since [112, (1.7.17)-(1.7.20)] remain presently valid (see the discussion pertaining to [112, (1.11.21)-(1.11.25)]), virtually the same argument as in the Euclidean case carries over and yields the versions of (1.7.226) and (1.7.231) in the manifold setting. Granted these, [112, Corollary 1.11.5] applies and yields all desired conclusions. □

We conclude by presenting the version of Theorem 1.7.18 in the setting of differential operators acting on Hermitian vector bundles over a given Riemannian manifold.

Theorem 1.7.20 *Suppose the Riemannian manifold* (M,g) *is as in* (1.7.259)*. Let* Ω *be a nonempty, open, proper subset of* M *with the property that* $\partial\Omega$ *is lower Ahlfors regular boundary, and that* $\sigma_g := \mathcal{H}^{n-1}_g\lfloor\partial\Omega$ *is a doubling measure on* $\partial\Omega$*. In particular,* Ω *is a set of finite perimeter and its (geometric measure theoretic) outward unit conormal* $\nu_g : \partial_*\Omega \to T^*M$ *is defined* σ_g*-a.e. on* $\partial_*\Omega$*. Denote by* $\mathcal{L}^n_g$ *the measure induced by the volume element* $\mathrm{d}V_g$ *on* M*.*

Consider next three Hermitian vector bundles $\mathscr{E},\mathscr{F},\mathscr{G} \to M$ *and, having fixed two first-order differential operators* $D : \mathscr{E} \to \mathscr{F}$*, and* $\widetilde{D} : \mathscr{F} \to \mathscr{G}$*, introduce* $L := \widetilde{D}D$*.*

In this setting, fix $p \in [1,\infty]$*, denote by* p' *its Hölder conjugate exponent, pick some* $\kappa,\kappa' > 0$*, and consider two sections*

$$
u \in L^1_{\text{loc}}(\Omega,\mathcal{L}^n_g)\otimes\mathscr{E}, \quad w \in L^1_{\text{loc}}(\Omega,\mathcal{L}^n_g)\otimes\mathscr{G}, \tag{1.7.263}
$$

satisfying

$$
\begin{aligned}
& Du \in L^1_{\rm loc}(\Omega, \mathcal{L}^n_g) \otimes \mathscr{F}, \quad \widetilde{D}^\top w \in L^1_{\rm loc}(\Omega, \mathcal{L}^n_g) \otimes \mathscr{F}, \\
& \mathcal{N}_\kappa(Du) \in L^p(\partial\Omega, \sigma_g), \quad \mathcal{N}_{\kappa'} w \in L^{p'}(\partial\Omega, \sigma_g), \\
& (Du)\Big|^{\kappa-{\rm n.t.}}_{\partial\Omega} \text{ and } w\Big|^{\kappa'-{\rm n.t.}}_{\partial\Omega} \text{ exist } \sigma_g\text{-a.e. on } \partial_{\rm nta}\Omega, \\
& Lu \text{ belongs to } L^1_{\rm loc}(\Omega, \mathcal{L}^n_g) \otimes \mathscr{G}, \text{ and} \\
& \big\langle Du, \widetilde{D}^\top w\big\rangle_{\mathscr{F}} + \big\langle Lu, w\big\rangle_{\mathscr{G}} \text{ belongs to } L^1(\Omega, \mathcal{L}^n_g).
\end{aligned}
\tag{1.7.264}
$$

Then for any other specified aperture parameter $\kappa'' \in (0, \infty)$ *the nontangential boundary traces* $(Du)\big|^{\kappa''-{\rm n.t.}}_{\partial\Omega}$ *and* $w\big|^{\kappa''-{\rm n.t.}}_{\partial\Omega}$ *exist at* σ_g*-a.e. point on* $\partial_{\rm nta}\Omega$ *and are independent of* κ''. *Moreover, with the dependence on the parameter* κ'' *dropped, the following Green type formula (involving absolutely convergent integrals) holds:*

$$
\begin{aligned}
& \int_\Omega \Big\{ \big\langle Lu, w\big\rangle_{\mathscr{G}} + \big\langle Du, \widetilde{D}^\top w\big\rangle_{\mathscr{F}} \Big\}\, d\mathcal{L}^n_g \\
& \qquad = \int_{\partial_*\Omega} \Big\langle (-i){\rm Sym}(\widetilde{D}; \nu_g)\big((Du)\big|^{\rm n.t.}_{\partial\Omega}\big), w\big|^{\rm n.t.}_{\partial\Omega} \Big\rangle_{\mathscr{G}}\, d\sigma_g.
\end{aligned}
\tag{1.7.265}
$$

Furthermore, formula (1.7.265) *remains true when the nontangential maximal function conditions stipulated in (the second and third lines of)* (1.7.264) *are replaced by more general memberships to the components of any regular dual system on* $(\partial\Omega, \sigma_g)$ *(defined in a fashion similar to Definition 1.7.9).*

Proof All desired conclusions follow by observing that the proof of Theorem 1.7.18 naturally adapts to the context of manifolds. The most significant novel aspect is that, this time, in place of [112, Theorem 1.2.1] or [112, Theorem 1.4.1], we rely on [112, Corollary 1.11.5] (see the discussion in the proof of Theorem 1.7.19). □

1.8 Rellich Identities

We begin by discussing an important integral formula in potential theory, originally proved by D. Jerison and C. Kenig in the category of Lipschitz domains in [69] (cf. also [79, Lemma 2.1.7, p. 46]), which has been further studied in [16, Theorem 5.8, p. 80] and [81, Lemma A.3.1, p. 389] (cf. also [80, §6]) where the identity in question has also been shown to be valid in the class of chord arc domains with a sufficiently small constant, as well as the category of δ-Reifenberg flat chord arc domains with the property that the logarithm of the Poisson kernel has vanishing mean oscillations. In all scenarios, the proof involves a suitable approximating sequence $\{\Omega_j\}_{j\in\mathbb{N}}$ of the given set Ω which, after establishing said identity in each Ω_j, allows one to pass to the limit $j \to \infty$ and obtain the desired formula for the original domain Ω.

The availability of such an approximating scheme in the class of chord-arc domains has been established in [81] under suitable additional flatness assumptions. The aforementioned assumptions are then inherited by any result whose proof makes use of such an approximating scheme. Here, by relying on our earlier results on the nature of the Green function, we are able to work directly with the original domain Ω and, most importantly, avoid imposing any artificial flatness conditions on Ω.

Recall that the reverse Hölder classes $B_q(\sigma)$, $1 < q < \infty$, are defined as in item *(v)* of [112, Proposition 7.7.9] (with $\mu := \sigma$ and $X := \partial\Omega$).

Theorem 1.8.1 *Assume $\Omega \subset \mathbb{R}^n$ is a bounded* NTA *domain whose boundary is an upper Ahlfors regular set. Let ν stand for the geometric measure theoretic outward unit normal to Ω, and abbreviate $\sigma := \mathcal{H}^{n-1}\lfloor\partial\Omega$. With ω^{x_0} denoting the harmonic measure with pole at a fixed point $x_0 \in \Omega$, suppose*

$$\omega^{x_0} \in B_2(\sigma) \tag{1.8.1}$$

and consider the Poisson kernel $k^{x_0} = \mathrm{d}\omega^{x_0}/\mathrm{d}\sigma$ with pole at x_0.

Then the following Rellich type identity holds:

$$\int_{\partial\Omega} \frac{k^{x_0}(y)}{|y - x_0|^{n-2}}\,\mathrm{d}\sigma(y) = \frac{1}{\omega_{n-1}} \int_{\partial\Omega} \big(k^{x_0}(y)\big)^2 \langle y - x_0, \nu(y)\rangle\,\mathrm{d}\sigma(y). \tag{1.8.2}$$

Proof We begin by defining $u : \Omega \setminus \{x_0\} \to \mathbb{R}$ by setting

$$u(x) := G(x, x_0) + E_\Delta(x - x_0), \qquad \forall x \in \Omega \setminus \{x_0\}, \tag{1.8.3}$$

where $G_\Omega(\cdot, \cdot)$ is the Green function for the Laplacian in Ω, and E_Δ is the standard fundamental solution for the Laplacian in $\mathbb{R}^n$ defined in (1.5.56). Keeping (5.3.3) in mind, it follows that u extends to a harmonic function in Ω. Retaining the notation u for this extension we therefore have

$$u \in \mathscr{C}^\infty(\Omega) \text{ and } \Delta u = 0 \text{ in } \Omega. \tag{1.8.4}$$

Moreover, (1.8.1) ensures that

$$k^{x_0} \in L^2(\partial\Omega, \sigma) \tag{1.8.5}$$

which, in concert with (5.4.7), the boundedness of the Hardy-Littlewood maximal operator on $L^2(\partial\Omega, \sigma)$ (cf. [112, Corollary 7.6.3]), and [112, Proposition 8.4.1], implies that for each aperture parameter $\kappa > 0$ we have

$$\mathcal{N}_\kappa(\nabla u) \in L^2(\partial\Omega, \sigma). \tag{1.8.6}$$

In addition, from (5.6.1) we deduce that, for each $\kappa > 0$,

$$\begin{array}{c}\text{the nontangential trace } \big(\nabla u\big)\Big|_{\partial\Omega}^{\kappa-\mathrm{n.t.}} \text{ exists } \sigma\text{-a.e. on } \partial\Omega \\ \text{and is actually independent of the parameter } \kappa > 0.\end{array} \tag{1.8.7}$$

Going further, define the function $w : \Omega \to \mathbb{R}$ by setting

$$w(x) := (x - x_0) \cdot (\nabla u)(x), \qquad \forall x \in \Omega. \tag{1.8.8}$$

Then from (1.8.8), (1.8.4), (1.8.6), and (1.8.7) we see that, for each $\kappa > 0$,

$$\begin{aligned} & w \in \mathscr{C}^\infty(\Omega),\ \Delta w = 0 \text{ in } \Omega,\ \mathcal{N}_\kappa w \in L^2(\partial\Omega, \sigma),\ w(x_0) = 0, \text{ and} \\ & w\big|_{\partial\Omega}^{\kappa-\text{n.t.}} \text{ exists } \sigma\text{-a.e. on } \partial\Omega \text{ and is actually independent of } \kappa > 0. \end{aligned} \tag{1.8.9}$$

In fact, as far as the latter nontangential boundary trace is concerned, relying on (1.8.8), (1.8.3), and (5.6.18), at σ-a.e. point $y \in \partial\Omega$ we may compute

$$\begin{aligned} \Big(w\big|_{\partial\Omega}^{\kappa-\text{n.t.}}\Big)(y) &= (y - x_0) \cdot \Big((\nabla u)\big|_{\partial\Omega}^{\kappa-\text{n.t.}}\Big)(y) \\ &= (y - x_0) \cdot \Big\{\Big(\big[\nabla G_\Omega(\cdot, x_0)\big]\Big|_{\partial\Omega}^{\kappa-\text{n.t.}}\Big)(y) + (\nabla E_\Delta)(y - x_0)\Big\} \\ &= -k^{x_0}(y)\langle y - x_0, \nu(y)\rangle + \frac{1}{\omega_{n-1}} \frac{1}{|y - x_0|^{n-2}}. \end{aligned} \tag{1.8.10}$$

Having established (1.8.9)-(1.8.10), Corollary 5.7.8 applies to w (thanks to (1.8.1)) and this permits us to write

$$\begin{aligned} 0 = w(x_0) &= \int_{\partial\Omega} k^{x_0}(y)\Big(w\big|_{\partial\Omega}^{\kappa-\text{n.t.}}\Big)(y)\,d\sigma(y) \\ &= -\int_{\partial\Omega} \big(k^{x_0}(y)\big)^2 \langle y - x_0, \nu(y)\rangle\,d\sigma(y) \\ &\quad + \frac{1}{\omega_{n-1}} \int_{\partial\Omega} \frac{k^{x_0}(y)}{|y - x_0|^{n-2}}\,d\sigma(y). \end{aligned} \tag{1.8.11}$$

With this in hand, (1.8.2) readily follows. □

Shifting perspectives, we shall establish an integral identity on the boundary of an Ahlfors regular domain $\Omega \subseteq \mathbb{R}^n$, involving functions $u \in \mathscr{C}^\infty(\Omega)$ satisfying $\Delta u + \lambda u = 0$ in Ω, for some $\lambda \in \mathbb{R}$. Subsequently, in Corollaries 1.8.3-1.8.4 we shall specialize this result by imposing, respectively, homogeneous Dirichlet and Neumann boundary conditions on u in order to obtain very general formulas in the spirit of those originally proved by Rellich in [154].

Theorem 1.8.2 *Assume Ω is an open nonempty bounded subset of $\mathbb{R}^n$ with a lower Ahlfors regular boundary and such that $\sigma := \mathcal{H}^{n-1}\lfloor\partial\Omega$ is a doubling measure on $\partial\Omega$. In particular, Ω is a set of locally finite perimeter, so its geometric measure theoretic outward unit normal ν is defined σ-a.e. on $\partial_*\Omega$. Fix $\kappa \in (0, \infty)$, $\lambda \in \mathbb{R}$, and assume that $u \in \mathscr{C}^\infty(\Omega)$ is a real-valued function satisfying*

$$\textit{the nontangential traces } u\big|_{\partial\Omega}^{\kappa-\mathrm{n.t.}},\ (\nabla u)\big|_{\partial\Omega}^{\kappa-\mathrm{n.t.}} \textit{ exist } \sigma\textit{-a.e. on } \partial_{\mathrm{nta}}\Omega, \quad \mathcal{N}_\kappa u,\ \mathcal{N}_\kappa(\nabla u) \in L^2(\partial\Omega,\sigma), \textit{ and } \Delta u + \lambda u = 0 \textit{ in } \Omega. \tag{1.8.12}$$

Then for any $\kappa' > 0$ *the nontangential traces* $u\big|_{\partial\Omega}^{\kappa'-\mathrm{n.t.}}$, $(\nabla u)\big|_{\partial\Omega}^{\kappa'-\mathrm{n.t.}}$ *exist* σ*-a.e. on* $\partial_{\mathrm{nta}}\Omega$ *and are actually independent of* κ'. *Moreover, suppressing the dependence on the parameter* κ' *and introducing*

$$\partial_\nu u := \Big\langle \nu, (\nabla u)\big|_{\partial\Omega}^{\mathrm{n.t.}} \Big\rangle \ \textit{at } \sigma\textit{-a.e. point on } \partial_*\Omega, \tag{1.8.13}$$

one has

$$u\big|_{\partial\Omega}^{\mathrm{n.t.}},\ (\nabla u)\big|_{\partial\Omega}^{\mathrm{n.t.}},\ \partial_\nu u \in L^2(\partial_*\Omega,\sigma), \quad u \in L^2(\Omega,\mathcal{L}^n), \tag{1.8.14}$$

and

$$\begin{aligned} \lambda\Bigg\{ \int_{\partial_*\Omega} \big(u\big|_{\partial\Omega}^{\mathrm{n.t.}}\big)^2 \langle \nu, x\rangle \, d\sigma - 2\int_{\Omega} u^2 \, d\mathcal{L}^n \Bigg\} &= \int_{\partial_*\Omega} \big|(\nabla u)\big|_{\partial\Omega}^{\mathrm{n.t.}}\big|^2 \langle \nu, x\rangle \, d\sigma - 2\int_{\partial_*\Omega} \big\langle (\nabla u)\big|_{\partial\Omega}^{\mathrm{n.t.}}, x\big\rangle \partial_\nu u \, d\sigma \\ &\quad - (n-2)\int_{\partial_*\Omega} \big(u\big|_{\partial\Omega}^{\mathrm{n.t.}}\big)\partial_\nu u \, d\sigma, \end{aligned} \tag{1.8.15}$$

with all integrals absolutely convergent.

Proof Given any $\kappa' > 0$, [112, Proposition 8.9.8] implies that $u\big|_{\partial\Omega}^{\kappa'-\mathrm{n.t.}}$, $(\nabla u)\big|_{\partial\Omega}^{\kappa'-\mathrm{n.t.}}$ exist σ-a.e. on $\partial_{\mathrm{nta}}\Omega$ and are actually independent of κ'. Consequently, suppressing the dependence on the aperture parameter when considering said nontangential traces creates no ambiguity. With this convention in mind, the membership of the functions $u\big|_{\partial\Omega}^{\mathrm{n.t.}}$, $(\nabla u)\big|_{\partial\Omega}^{\mathrm{n.t.}}$, $\partial_\nu u$ to $L^2(\partial_*\Omega,\sigma)$ is seen from assumptions, [112, (8.9.44)], [112, Proposition 8.8.6], [112, (8.9.8)], (1.8.13), and [112, (5.6.23)]. In addition, [112, (8.6.51) in Proposition 8.6.3] (used with $p := 2$ and $E := \Omega$) implies

$$\|u\|_{L^{\frac{2n}{n-1}}(\Omega,\mathcal{L}^n)} \le C_\#\|\mathcal{N}_\kappa u\|_{L^2(\partial\Omega,\sigma)} < +\infty. \tag{1.8.16}$$

Since the Lebesgue scale on Ω is nested (given that Ω has finite measure), this allows us to conclude that

$$u \in L^{\frac{2n}{n-1}}(\Omega,\mathcal{L}^n) \hookrightarrow L^2(\Omega,\mathcal{L}^n), \tag{1.8.17}$$

which finishes the proof of (1.8.14).

To proceed, introduce the vector field $\vec{F} : \Omega \to \mathbb{R}^n$ by setting

$$\vec{F}(x) := |(\nabla u)(x)|^2 x - 2\langle(\nabla u)(x), x\rangle(\nabla u)(x) - (n-2)u(x)(\nabla u)(x) - \lambda\, u(x)^2 x, \tag{1.8.18}$$

for each $x \in \Omega$. Then $\vec{F} \in \big[\mathscr{C}^\infty(\Omega)\big]^n$ and a straightforward computation (which twice uses the fact that $\Delta u = -\lambda\, u$) yields

$$\begin{aligned}\mathrm{div}\vec{F} &= -2\langle\nabla u, x\rangle\Delta u - (n-2)u\Delta u - 2\lambda\langle\nabla u, x\rangle u - n\lambda\, u^2 \\ &= -2\lambda\, u^2 \text{ in } \Omega.\end{aligned} \tag{1.8.19}$$

In particular, thanks to (1.8.17),

$$\mathrm{div}\vec{F} \in L^1(\Omega, \mathcal{L}^n). \tag{1.8.20}$$

Moreover, from (1.8.18) and assumptions, it follows that

$$\begin{aligned}&\text{the nontangential trace } \vec{F}\Big|_{\partial\Omega}^{\kappa-\mathrm{n.t.}} \text{ exists at } \sigma\text{-a.e. point on } \partial_{\mathrm{nta}}\Omega \text{ and} \\ &\text{the nontangential maximal function } \mathcal{N}_\kappa\vec{F} \text{ belongs to } L^1(\partial\Omega, \sigma).\end{aligned} \tag{1.8.21}$$

Granted (1.8.19)-(1.8.21), the Divergence Formula [112, (1.2.2)] applies and readily gives (1.8.15). This finishes the proof of Theorem 1.8.2. □

More than 75 years ago, F. Rellich has discovered a basic identity for the Dirichlet eigenvalue problem for the Laplacian in smooth domains (cf. [154]), expressing the eigenvalue as a weighted boundary integral of the square of the normal derivative of the normalized eigenfunction. As evidenced by work in [12], [13], [47], [101], [152], and the references therein, this formula remains an important tool in problems in partial differential equations and spectral theory. The following consequence of Theorem 1.8.2 is a geometrically sharp version of Rellich's celebrated identity which is valid in a much larger category of domains than previously considered in the literature.

Corollary 1.8.3 *Retain the assumptions on Ω and u made in Theorem 1.8.2 and, in addition, suppose u does not vanish identically in Ω and satisfies*

$$u\Big|_{\partial\Omega}^{\mathrm{n.t.}} = 0 \text{ at } \sigma\text{-a.e. point on } \partial_*\Omega. \tag{1.8.22}$$

Then

$$\lambda = \frac{\displaystyle\int_{\partial_*\Omega}(\partial_\nu u)^2\langle\nu, x\rangle\, d\sigma}{\displaystyle 2\int_\Omega u^2\, d\mathcal{L}^n} \tag{1.8.23}$$

with all integrals absolutely convergent.

Proof Identity (1.8.23) is a direct consequence of (1.8.15) and [113, (11.3.45)] (also bearing in mind (1.8.22)). Finally, the fact that the intervening integrals are absolutely convergent is a consequence of (1.8.14). □

Theorem 1.8.2 also yields a sharp version of Rellich's identity for the Neumann eigenvalue problem for the Laplacian in very general domains, of the sort described in the corollary below.

Corollary 1.8.4 *Retain the assumptions on* Ω *and* u *made in Theorem 1.8.2 and, in addition, suppose*

$$\partial_\nu u = 0 \text{ at } \sigma\text{-a.e. point on } \partial_*\Omega. \tag{1.8.24}$$

Then, with all integrals absolutely convergent,

$$\lambda\left\{\int_{\partial_*\Omega}\Big(u\big|_{\partial\Omega}^{{}^{\rm n.t.}}\Big)^2\langle\nu,x\rangle\,d\sigma - 2\int_\Omega u^2\,d\mathcal{L}^n\right\} = \int_{\partial_*\Omega}\Big|(\nabla u)\big|_{\partial\Omega}^{{}^{\rm n.t.}}\Big|^2\langle\nu,x\rangle\,d\sigma \tag{1.8.25}$$

and

$$\lambda\int_{\partial_*\Omega}\Big(u\big|_{\partial\Omega}^{{}^{\rm n.t.}}\Big)^2\langle\nu,x\rangle\,d\sigma = \int_{\partial_*\Omega}\Big|(\nabla u)\big|_{\partial\Omega}^{{}^{\rm n.t.}}\Big|^2\langle\nu,x\rangle\,d\sigma + 2\int_\Omega|\nabla u|^2\,d\mathcal{L}^n. \tag{1.8.26}$$

Moreover, if Ω *is actually a* UR *domain, then* $(\nabla u)\big|_{\partial\Omega}^{{}^{\rm n.t.}}$ *may be replaced by* $\nabla_{\tan}\big(u\big|_{\partial\Omega}^{{}^{\rm n.t.}}\big)$ *in* (1.8.25) *and* (1.8.26).

Proof Formula (1.8.25) is readily implied by (1.8.15) and (1.8.24). In turn, formula (1.8.26) is obtained from (1.8.25) upon observing that

$$\lambda\int_\Omega u^2\,d\mathcal{L}^n = \int_\Omega|\nabla u|^2\,d\mathcal{L}^n. \tag{1.8.27}$$

The latter integral identity is justified by applying the Divergence Formula [112, (1.2.2)] to the vector field $\vec{F} := u\nabla u \in \big[\mathscr{C}^\infty(\Omega)\big]^n$ which, thanks to the present hypotheses, satisfies $\mathcal{N}_\kappa\vec{F} \in L^1(\partial\Omega,\sigma)$,

$$\mathrm{div}\vec{F} = |\nabla u|^2 + u\Delta u = |\nabla u|^2 - \lambda u^2 \in L^1(\Omega,\mathcal{L}^n) \tag{1.8.28}$$

as well as $\vec{F}\big|_{\partial\Omega}^{{}^{\kappa-\rm n.t.}}$ exists at σ-a.e. point on $\partial_{\rm nta}\Omega$, and

$$\nu\cdot\Big(\vec{F}\big|_{\partial\Omega}^{{}^{\kappa-\rm n.t.}}\Big) = \Big(u\big|_{\partial\Omega}^{{}^{\kappa-\rm n.t.}}\Big)\partial_\nu u = 0 \text{ at } \sigma\text{-a.e. point on } \partial_*\Omega. \tag{1.8.29}$$

Finally, the last claim in the statement of the corollary is a consequence of [113, (11.4.41)]. □

The principle behind the proof of Theorem 1.8.2, i.e., applying our Divergence Theorem to a vector field suitably tailored from the original function(s), is robust and may be implemented in a variety of situations of interest. Here is a Rellich type identity involving holomorphic functions.

Theorem 1.8.5 *Suppose Ω is an open nonempty proper subset of $\mathbb{R}^2 \equiv \mathbb{C}$ with a lower Ahlfors regular boundary and such that $\sigma := \mathcal{H}^1 \lfloor \partial\Omega$ is a doubling measure on $\partial\Omega$. Consequently, Ω is a set of locally finite perimeter, so its geometric measure theoretic outward unit normal*

$$\nu = (\nu_1, \nu_2) \equiv \nu_1 + i\nu_2 \tag{1.8.30}$$

is defined σ-a.e. on $\partial_*\Omega$. Fix $\kappa \in (0,\infty)$ and assume $u, w : \Omega \to \mathbb{C}$ are two holomorphic functions satisfying

$$\begin{aligned} &\text{the nontangential traces } u\big|_{\partial\Omega}^{\kappa-\text{n.t.}},\ w\big|_{\partial\Omega}^{\kappa-\text{n.t.}} \text{ exist } \sigma\text{-a.e. on } \partial_{\text{nta}}\Omega, \\ &\text{and the product } \big(\mathcal{N}_\kappa u\big)\cdot\big(\mathcal{N}_\kappa w\big)^2 \text{ belongs to } L^1(\partial\Omega,\sigma). \end{aligned} \tag{1.8.31}$$

Then, with bar denoting complex conjugation, one has

$$\begin{aligned} \operatorname{Re}\int_{\partial_*\Omega} \overline{\nu}\big(u\big|_{\partial\Omega}^{\kappa-\text{n.t.}}\big)\big|w\big|_{\partial\Omega}^{\kappa-\text{n.t.}}\big|^2\,d\sigma \\ = 2\operatorname{Re}\int_{\partial_*\Omega}\big(u\big|_{\partial\Omega}^{\kappa-\text{n.t.}}\big)\big(w\big|_{\partial\Omega}^{\kappa-\text{n.t.}}\big)\operatorname{Re}\Big[\nu\big(w\big|_{\partial\Omega}^{\kappa-\text{n.t.}}\big)\Big]\,d\sigma \end{aligned} \tag{1.8.32}$$

when either Ω is bounded, or $\partial\Omega$ is unbounded. In the remaining cases, i.e., when Ω is unbounded and $\partial\Omega$ is bounded, formula (1.8.32) *continues to hold under the additional assumption that there exists $\lambda \in (1,\infty)$ such that*

$$\int_{[B(0,\lambda R)\setminus B(0,R)]\cap\Omega} |u||w|^2\,d\mathcal{L}^2 = o(R) \text{ as } R \to \infty. \tag{1.8.33}$$

Proof Define two real vector fields $\vec{a}, \vec{b} : \Omega \to \mathbb{R}^2$ by setting

$$\vec{a} := \big(\operatorname{Re} u, \operatorname{Im} u\big) \text{ and } \vec{b} := \big(\operatorname{Re} w, -\operatorname{Im} w\big), \tag{1.8.34}$$

then consider the vector field

$$\vec{F} := |\vec{b}|^2\vec{a} - 2\langle\vec{a},\vec{b}\rangle\vec{b} \in \big[\mathscr{C}^\infty(\Omega)\big]^2. \tag{1.8.35}$$

A direct computation based on the Cauchy-Riemann equations shows that

$$\operatorname{div}\vec{F} = 0 \text{ in } \Omega. \tag{1.8.36}$$

Also, from (1.8.34)-(1.8.35) and (1.8.31) we conclude that

$$\vec{F}\big|_{\partial\Omega}^{\kappa-\text{n.t.}} \text{ exists } \sigma\text{-a.e. on } \partial_{\text{nta}}\Omega, \text{ and } \mathcal{N}_\kappa\vec{F} \in L^1(\partial\Omega,\sigma). \tag{1.8.37}$$

Moreover, at σ-a.e. point on $\partial_*\Omega$ we have

$$\Big\langle \nu, \vec{F}\Big|_{\partial\Omega}^{\kappa-\mathrm{n.t.}}\Big\rangle = \mathrm{Re}\Big[\overline{\nu}\big(u\big|_{\partial\Omega}^{\kappa-\mathrm{n.t.}}\big)\Big]\big|w\big|_{\partial\Omega}^{\kappa-\mathrm{n.t.}}\big|^2 - 2\,\mathrm{Re}\Big[\big(u\big|_{\partial\Omega}^{\kappa-\mathrm{n.t.}}\big)\big(w\big|_{\partial\Omega}^{\kappa-\mathrm{n.t.}}\big)\Big]\mathrm{Re}\Big[\nu\big(w\big|_{\partial\Omega}^{\kappa-\mathrm{n.t.}}\big)\Big]. \tag{1.8.38}$$

Lastly, when Ω is an exterior domain, (1.8.33) implies that

$$\int_{B(0,\lambda R)\setminus B(0,R)} |\vec{F}|\,\mathrm{d}\mathcal{L}^2 = o(R) \text{ as } R \to \infty. \tag{1.8.39}$$

Having established (1.8.35)-(1.8.39), the Divergence Formula [112, (1.2.2)] may be invoked to conclude that (1.8.32) holds. □

Another embodiment of the principle used in the proof of Theorem 1.8.2 is as follows. Assume $\Omega \subseteq \mathbb{R}^n$ is an open set with an Ahlfors regular boundary and abbreviate $\sigma := \mathcal{H}^{n-1}\lfloor\partial\Omega$. Also, let ν stand for the geometric measure theoretic outward unit normal to Ω and fix an aperture parameter $\kappa \in (0,\infty)$. In this setting, we have a Rellich-type identity of the following sort:

$$\int_{\partial_*\Omega} \langle \nu, h\rangle \big|(\nabla u)\big|_{\partial\Omega}^{\kappa-\mathrm{n.t.}}\big|^2\,\mathrm{d}\sigma = 2\int_{\partial_*\Omega} \big\langle h, (\nabla u)\big|_{\partial\Omega}^{\kappa-\mathrm{n.t.}}\big\rangle(\partial_\nu u)\,\mathrm{d}\sigma + \int_\Omega (\mathrm{div} h)|\nabla u|^2\,\mathrm{d}\mathcal{L}^n - 2\int_\Omega \langle Dh, \nabla u \otimes \nabla u\rangle\,\mathrm{d}\mathcal{L}^n. \tag{1.8.40}$$

Above, h is a vector field in $\big[\mathscr{C}_c^1(\mathbb{R}^n)\big]^n$, the symbol Dh denotes the Jacobian matrix of h, and u is a harmonic function in Ω which, for some fixed $\kappa > 0$, satisfies $\mathcal{N}_\kappa(\nabla u) \in L^2_{\mathrm{loc}}(\partial\Omega, \sigma)$ and the nontangential trace $(\nabla u)\big|_{\partial\Omega}^{\kappa-\mathrm{n.t.}}$ exists at σ-a.e. point on $\partial_{\mathrm{nta}}\Omega$. Also, $\partial_\nu u := \nu \cdot \big((\nabla u)\big|_{\partial\Omega}^{\kappa-\mathrm{n.t.}}\big)$ is the normal derivative of u on $\partial_*\Omega$, and $\nabla u \otimes \nabla u$ is (cf. (1.3.7)) the $n \times n$ matrix $\big(\partial_j u \partial_k u\big)_{1\leq j,k\leq n}$.

Identity (1.8.40) is obtained by applying the Divergence Formula [112, (1.2.2)] to the vector field

$$\vec{F} := |\nabla u|^2 h - 2\langle h, \nabla u\rangle \nabla u \in \big[\mathscr{C}^1(\Omega)\big]^n. \tag{1.8.41}$$

The role of the hypothesis $\mathcal{N}_\kappa(\nabla u) \in L^2_{\mathrm{loc}}(\partial\Omega, \sigma)$ is to ensure that $\mathcal{N}_\kappa \vec{F} \in L^1(\partial\Omega, \sigma)$ which is a key assumption in [112, Theorem 1.2.1]. In the particular case when Ω is bounded and $h(x) = x$ for each x near $\overline{\Omega}$ (in which scenario $\mathrm{div} h = n$ and $Dh = I$ in Ω), formula (1.8.40) becomes

$$\int_{\partial_*\Omega} \langle \nu, x\rangle \big|(\nabla u)\big|_{\partial\Omega}^{\kappa-\mathrm{n.t.}}\big|^2\,\mathrm{d}\sigma \tag{1.8.42}$$

$$= 2\int_{\partial_*\Omega} \big\langle x, (\nabla u)\big|_{\partial\Omega}^{\kappa-\mathrm{n.t.}}\big\rangle(\partial_\nu u)\,\mathrm{d}\sigma + (n-2)\int_\Omega |\nabla u|^2\,\mathrm{d}\mathcal{L}^n$$

which refines results of a similar nature derived under more stringent regularity assumptions in [70, (2), p. 204], [79, Lemma 2.1.13, p. 48], [144, §3.7-§3.8], [179].

According to [112, Theorem 1.2.1], formula (1.8.40) continues to be valid when Ω has an unbounded boundary and $h \in \left[\mathscr{C}^1(\mathbb{R}^n)\right]^n$ is bounded, with bounded first-order partial derivatives. For example, if we take Ω to be the domain in $\mathbb{R}^n$ above the graph of a continuous function $\varphi : \mathbb{R}^{n-1} \to \mathbb{R}$ whose first-order partial derivatives belong to $\mathrm{BMO}(\mathbb{R}^{n-1})$, and consider $h \equiv \mathbf{e}_j$ with $j \in \{1, \dots, n\}$, eventually allows us to recast (1.8.40) as

$$\int_{\partial_*\Omega} \nu \left|(\nabla u)\Big|_{\partial\Omega}^{\kappa-\mathrm{n.t.}}\right|^2 \, d\sigma = 2 \int_{\partial_*\Omega} \Big((\nabla u)\Big|_{\partial\Omega}^{\kappa-\mathrm{n.t.}}\Big)(\partial_\nu u) \, d\sigma. \tag{1.8.43}$$

This refines earlier work, including [79, Lemma 2.2.15, p. 55], [105, Théorème 6, p. 496], and [141].

Due to the influence of the book [141] by Nečas on this topic, integral formulas of this nature are occasionally referred to as Nečas-Rellich identities. Here is a Nečas-Rellich identity for systems of differential equations, refining work similar in spirit from [29], [40, Lemma 2.1], [42], [38], [68], [126], [135], [141, (5.47), p. 264].

Proposition 1.8.6 *Let Ω be an open nonempty proper subset of $\mathbb{R}^n$ ($n \in \mathbb{N}$, $n \geq 2$) with a lower Ahlfors regular boundary and such that $\sigma := \mathcal{H}^{n-1} \lfloor \partial\Omega$ is a doubling measure on $\partial\Omega$. Denote by ν the geometric measure theoretic outward unit normal to Ω, and fix some aperture parameter $\kappa \in (0, \infty)$. Next, for some $M \in \mathbb{N}$, consider a coefficient tensor $A = \big(a_{rs}^{\alpha\beta}\big)_{\substack{1 \leq r,s \leq n \\ 1 \leq \alpha,\beta \leq M}}$ with complex entries satisfying the symmetry condition*

$$a_{rs}^{\alpha\beta} = a_{sr}^{\beta\alpha} \text{ for each } \alpha, \beta \in \{1, \dots, M\} \text{ and each } r, s \in \{1, \dots, n\}, \tag{1.8.44}$$

then associate with A the $M \times M$ second-order system $L = \big(a_{rs}^{\alpha\beta} \partial_r \partial_s\big)_{1 \leq \alpha,\beta \leq M}$. In addition, consider a $\mathbb{C}^M$-valued function $u = (u_\beta)_{1 \leq \beta \leq M} \in [\mathscr{C}^2(\Omega)]^M$ with the property that $\mathcal{N}_\kappa(\nabla u)$ belongs to $L^2_{\mathrm{loc}}(\partial\Omega, \sigma)$, the nontangential trace $(\nabla u)\Big|_{\partial\Omega}^{\kappa-\mathrm{n.t.}}$ exists at σ-a.e. point on $\partial_{\mathrm{nta}}\Omega$, and Lu belongs to $\big[L^{2n/(n+1)}_{\mathrm{bdd}}(\Omega, \mathcal{L}^n)\big]^M$. Finally, fix an arbitrary vector field $h = (h_i)_{1 \leq i \leq n} \in [\mathscr{C}^1_c(\mathbb{R}^n)]^n$.

Then for any other aperture parameter parameter $\kappa' > 0$ the nontangential trace $(\nabla u)\Big|_{\partial\Omega}^{\kappa'-\mathrm{n.t.}}$ exists at σ-a.e. point on $\partial_{\mathrm{nta}}\Omega$ and is actually independent of κ'. Moreover, with the dependence on the parameter κ' dropped, one has (with the conormal derivative $\partial_\nu^A u$ defined as in (1.7.9)*)*

$$\begin{aligned}
&\int_{\partial_*\Omega} \Big[\langle \nu, h\rangle a_{rs}^{\alpha\beta} \Big((\partial_r u_\alpha)\Big|_{\partial\Omega}^{\mathrm{n.t.}}\Big)\Big((\partial_s u_\beta)\Big|_{\partial\Omega}^{\mathrm{n.t.}}\Big) - 2h_j \Big((\partial_j u_\alpha)\Big|_{\partial\Omega}^{\mathrm{n.t.}}\Big)(\partial_\nu^A u)_\alpha\Big] \, d\sigma \\
&\quad = \int_\Omega \Big[(\partial_i h_i) a_{rs}^{\alpha\beta} (\partial_r u_\alpha)(\partial_s u_\beta) - 2(\partial_i h_j) a_{is}^{\alpha\beta} (\partial_j u_\alpha)(\partial_s u_\beta) \\
&\qquad\qquad - 2h_j (\partial_j u_\alpha)(Lu)_\alpha\Big] \, d\mathcal{L}^n,
\end{aligned} \tag{1.8.45}$$

where all integrals are absolutely convergent. As a corollary,

$$
\begin{aligned}
&-\int_{\partial_*\Omega}\Big[\langle \nu, h\rangle a^{\alpha\beta}_{rs}\big((\partial_r u_\alpha)\big|^{\text{n.t.}}_{\partial\Omega}\big)\big((\partial_s u_\beta)\big|^{\text{n.t.}}_{\partial\Omega}\big)\\
&\qquad + 2h_j\Big\{\nu_k\big((\partial_j u_\alpha)\big|^{\text{n.t.}}_{\partial\Omega}\big) - \nu_j\big((\partial_k u_\alpha)\big|^{\text{n.t.}}_{\partial\Omega}\big)\Big\}a^{\alpha\beta}_{k\ell}\big((\partial_\ell u_\beta)\big|^{\text{n.t.}}_{\partial\Omega}\big)\Big]\,\mathrm{d}\sigma\\
&= \int_\Omega\Big[(\partial_i h_i)a^{\alpha\beta}_{rs}(\partial_r u_\alpha)(\partial_s u_\beta) - 2(\partial_i h_j)a^{\alpha\beta}_{is}(\partial_j u_\alpha)(\partial_s u_\beta)\\
&\qquad - 2h_j(\partial_j u_\alpha)(Lu)_\alpha\Big]\,\mathrm{d}\mathcal{L}^n.
\end{aligned}
\tag{1.8.46}
$$

Proof For starters, the fact that for each $\kappa' > 0$ the nontangential trace $(\nabla u)\big|^{\kappa'-\text{n.t.}}_{\partial\Omega}$ exists at σ-a.e. point on $\partial_{\text{nta}}\Omega$, belongs to $\big[L^2(\partial_*\Omega, \sigma)\big]^{n\cdot M}$ and is actually independent of κ' follows from assumptions and [112, Propositions 8.9.5, 8.9.8, 8.8.6]. As such, there is no ambiguity in dropping the dependence on the parameter κ' and simply denoting this nontangential trace by $(\nabla u)\big|^{\text{n.t.}}_{\partial\Omega}$. Next, the idea is to apply [112, Theorem 1.2.1] to the vector field $\vec{F} = \big(F_i\big)_{1\le i\le n} : \Omega \to \mathbb{C}^n$ whose scalar components are given by

$$
F_i := h_i a^{\alpha\beta}_{rs}(\partial_r u_\alpha)(\partial_s u_\beta) - 2h_j a^{\alpha\beta}_{is}(\partial_j u_\alpha)(\partial_s u_\beta), \qquad \forall\, i \in \{1,\dots,n\}. \tag{1.8.47}
$$

Thus, $\vec{F} \in \big[\mathscr{C}^1(\Omega)\big]^n$ and a direct computation gives

$$
\begin{aligned}
\partial_i F_i &= (\partial_i h_i)a^{\alpha\beta}_{rs}(\partial_r u_\alpha)(\partial_s u_\beta) + h_i a^{\alpha\beta}_{rs}(\partial_i\partial_r u_\alpha)(\partial_s u_\beta)\\
&\quad + h_i a^{\alpha\beta}_{rs}(\partial_r u_\alpha)(\partial_i\partial_s u_\beta) - 2(\partial_i h_j)a^{\alpha\beta}_{is}(\partial_j u_\alpha)(\partial_s u_\beta)\\
&\quad - 2h_j a^{\alpha\beta}_{is}(\partial_i\partial_j u_\alpha)(\partial_s u_\beta) - 2h_j a^{\alpha\beta}_{is}(\partial_j u_\alpha)(\partial_i\partial_s u_\beta)\\
&=: \mathrm{I} + \mathrm{II} + \mathrm{III} + \mathrm{IV} + \mathrm{V} + \mathrm{VI}.
\end{aligned}
\tag{1.8.48}
$$

Since we are assuming the symmetry condition (1.8.44), in term II we may switch α with β and r with s to conclude that II = III. Hence,

$$
\mathrm{II} + \mathrm{III} = 2h_i a^{\alpha\beta}_{rs}(\partial_i\partial_r u_\alpha)(\partial_s u_\beta) = 2h_j a^{\alpha\beta}_{is}(\partial_j\partial_i u_\alpha)(\partial_s u_\beta) = -\mathrm{V}, \tag{1.8.49}
$$

where the second equality in (1.8.49) is obtained by re-denoting i by j, and then re-denoting r by i. Also, using the definition of L we see that

$$
\mathrm{VI} = -2h_j(\partial_j u_\alpha)(Lu)_\alpha. \tag{1.8.50}
$$

Combining (1.8.48)-(1.8.50) leads to

$$\operatorname{div}\vec{F} = (\partial_i h_i) a^{\alpha\beta}_{rs} (\partial_r u_\alpha)(\partial_s u_\beta) - 2(\partial_i h_j) a^{\alpha\beta}_{is} (\partial_j u_\alpha)(\partial_s u_\beta) \\ - 2h_j (\partial_j u_\alpha)(Lu)_\alpha. \tag{1.8.51}$$

Since $\mathcal{N}_\kappa(\nabla u) \in L^2_{\rm loc}(\partial\Omega, \sigma)$, [112, Proposition 8.6.3] implies that ∇u belongs to $\big[L^{2n/(n-1)}_{\rm bdd}(\Omega, \mathcal{L}^n)\big]^{n\cdot M}$. It is also assumed that $Lu \in \big[L^{2n/(n+1)}_{\rm bdd}(\Omega, \mathcal{L}^n)\big]^M$, so we may ultimately conclude from (1.8.51) that $\operatorname{div}\vec{F} \in L^1(\Omega, \mathcal{L}^n)$. In view of (1.8.47), the hypothesis $\mathcal{N}_\kappa(\nabla u) \in L^2_{\rm loc}(\partial\Omega, \sigma)$ also entails $\mathcal{N}_\kappa \vec{F} \in L^1(\partial\Omega, \sigma)$. By design, $\vec{F}$ vanishes outside of a bounded subset of Ω. Finally, $\vec{F}\big|^{\kappa-{\rm n.t.}}_{\partial\Omega}$ exists at σ-a.e. point on $\partial_{\rm nta}\Omega$ and

$$\nu \cdot \Big(\vec{F}\big|^{\kappa-{\rm n.t.}}_{\partial\Omega}\Big) = \langle \nu, h\rangle a^{\alpha\beta}_{rs} \Big((\partial_r u_\alpha)\big|^{\rm n.t.}_{\partial\Omega}\Big) \Big((\partial_s u_\beta)\big|^{\rm n.t.}_{\partial\Omega}\Big) \\ - 2h_j \Big((\partial_j u_\alpha)\big|^{\rm n.t.}_{\partial\Omega}\Big) (\partial^A_\nu u)_\alpha \tag{1.8.52}$$

at σ-a.e. point on $\partial_*\Omega$. At this stage, (1.8.45) follows from [112, (1.2.2)] and (1.8.51)-(1.8.52). Finally, (1.8.46) follows from (1.8.45) by expanding

$$-2h_j \Big((\partial_j u_\alpha)\big|^{\rm n.t.}_{\partial\Omega}\Big) (\partial^A_\nu u)_\alpha$$
$$= -2h_j \Big((\partial_j u_\alpha)\big|^{\rm n.t.}_{\partial\Omega}\Big) \nu_k a^{\alpha\beta}_{k\ell} \Big((\partial_\ell u_\beta)\big|^{\rm n.t.}_{\partial\Omega}\Big)$$
$$= -2h_j \Big\{\nu_k \Big((\partial_j u_\alpha)\big|^{\rm n.t.}_{\partial\Omega}\Big) - \nu_j \Big((\partial_k u_\alpha)\big|^{\rm n.t.}_{\partial\Omega}\Big)\Big\} a^{\alpha\beta}_{k\ell} \Big((\partial_\ell u_\beta)\big|^{\rm n.t.}_{\partial\Omega}\Big) \\ - \langle \nu, h\rangle a^{\alpha\beta}_{k\ell} \Big((\partial_k u_\alpha)\big|^{\rm n.t.}_{\partial\Omega}\Big) \Big((\partial_\ell u_\beta)\big|^{\rm n.t.}_{\partial\Omega}\Big), \tag{1.8.53}$$

and combining like-terms in the context of (1.8.45). □

Rellich type identities have been extended from the Euclidean space to the setting of Lipschitz subdomains of Riemannian manifolds in [119], [129], [130], [132], [133], [134]. Given the availability of a suitable version of the Divergence Theorem on manifolds (cf. the discussion in [112, §1.11]), such results may be established in considerably larger classes of domains on manifolds. Here we further contribute to this line of work by establishing a very general Rellich-type identity[40] on Riemannian manifolds, of the sort described in the theorem below.

Theorem 1.8.7 *Let (M, g) compact, connected, boundaryless, oriented Riemannian manifold of class $\mathscr{C}^2$, of real dimension n, and denote by $\mathcal{L}^n_g$ the measure induced by the volume element on M. Suppose $\mathcal{E}, \mathcal{F} \to M$ are two Hermitian vector bundles of class $\mathscr{C}^1$, and denote by $\langle\cdot,\cdot\rangle_{\mathcal{E}}$ and $\langle\cdot,\cdot\rangle_{\mathcal{F}}$ the real (i.e., complex bilinear) pointwise pairings in the fibers of $\mathcal{E}$ and $\mathcal{F}$, respectively. Also, assume $\nabla^{\mathcal{E}}$ and $\nabla^{\mathcal{F}}$ are connections on $\mathcal{E}$ and $\mathcal{F}$, respectively. Next, let $D : \mathcal{E} \to \mathcal{F}$ be a first-order differential operator whose coefficients are of class $\mathscr{C}^1$. Also, pick some $A \in \operatorname{Hom}(\mathcal{F}, \mathcal{F})$ with*

[40] which may well be labeled "the mother of all Rellich identities"

coefficients of class $\mathscr{C}^1$ and satisfying

$$A^* = A, \tag{1.8.54}$$

then use it to define the second-order differential operator

$$L := -D^* A D. \tag{1.8.55}$$

Going further, let Ω be a nonempty, open, proper subset of M such that $\partial\Omega$ is a lower Ahlfors regular set, and $\sigma_g := \mathcal{H}_g^{n-1} \lfloor \partial\Omega$ is a doubling measure on $\partial\Omega$. In particular, Ω is a set of finite perimeter, hence its geometric measure theoretic outward unit conormal $\nu_g : \partial_*\Omega \to T^*M$ is defined σ_g-a.e. on $\partial_*\Omega$. Finally, consider a section

$$u \in W^{2,1}_{\rm loc}(\Omega) \otimes \mathcal{E} \tag{1.8.56}$$

which, for some aperture parameter $\kappa > 0$, satisfies

$$\begin{gathered} \mathcal{N}_\kappa u,\ \mathcal{N}_\kappa(\nabla^{\mathcal{E}} u) \text{ belong to } L^2(\partial\Omega, \sigma_g), \\ u\big|_{\partial\Omega}^{\kappa-\text{n.t.}} \text{ and } (\nabla^{\mathcal{E}} u)\big|_{\partial\Omega}^{\kappa-\text{n.t.}} \text{ exist } \sigma_g\text{-a.e. on } \partial_{\rm nta}\Omega, \\ Lu \text{ belongs to the space } L^{\frac{2n}{n+1}}(\Omega, \mathcal{L}_g^n) \otimes \mathcal{E}, \end{gathered} \tag{1.8.57}$$

and define the conormal derivative

$$\partial_\nu^L u := \mathrm{iSym}(D^*; \nu_g) A(Du)\big|_{\partial\Omega}^{\kappa-\text{n.t.}} \text{ at } \sigma_g\text{-a.e. point on } \partial_*\Omega. \tag{1.8.58}$$

Then for each vector field $h \in \mathscr{C}^1(M, TM)$ with real-valued components one has

$$\begin{aligned} & 2\,\mathrm{Re} \int_{\partial_*\Omega} \Big\langle \partial_\nu^L u, \big((\nabla_h^{\mathcal{E}} u)\big|_{\partial\Omega}^{\kappa-\text{n.t.}}\big)^c \Big\rangle_{\mathcal{E}} \, d\sigma_g \\ & \quad - \int_{\partial_*\Omega} {}_{T^*M}(\nu_g, h)_{TM} \Big\langle A(Du)\big|_{\partial\Omega}^{\kappa-\text{n.t.}}, \big((Du)\big|_{\partial\Omega}^{\kappa-\text{n.t.}}\big)^c \Big\rangle_{\mathcal{F}} \, d\sigma_g \\ & \quad = 2\,\mathrm{Re} \int_\Omega \langle Lu, (\nabla_h^{\mathcal{E}} u)^c \rangle_{\mathcal{E}} \, d\mathcal{L}_g^n - \int_\Omega (\mathrm{div}_g h) \langle ADu, (Du)^c \rangle_{\mathcal{F}} \, d\mathcal{L}_g^n \\ & \quad\quad + \int_\Omega O\big(|Du|^2 |[\nabla_h^{\mathcal{F}}, A]|\big) \, d\mathcal{L}_g^n + \int_\Omega O\big(|A||Du||[D, \nabla_h]u|\big) \, d\mathcal{L}_g^n \\ & \quad\quad + \int_\Omega O\big(|A||h||Du|^2\big) \, d\mathcal{L}_g^n, \end{aligned} \tag{1.8.59}$$

where the superscript 'c' indicates complex conjugation, and where the constants implicit in all big "O" terms depend at most on the metrics.

Before we present the proof of this theorem we make a number of comments aimed to shed further light on its scope and nature.

Comment 1. It is of significance to note that the commutator

$$[D, \nabla_h] := D\nabla_h^{\mathcal{E}} - \nabla_h^{\mathcal{F}} D \text{ is a } \textit{first-order} \text{ differential operator,} \tag{1.8.60}$$

since its principal symbol (as a second-order operator) vanishes. Indeed, for each covector $\xi \in T^*M$ we have

$$\begin{aligned} \mathrm{Sym}\big([D, \nabla_h]; \xi\big) &= \mathrm{Sym}(D; \xi) \mathrm{i}_{T^*M}\langle \xi, h\rangle_{TM} I_{\mathcal{E}} \\ &\quad - \mathrm{i}_{T^*M}\langle \xi, h\rangle_{TM} I_{\mathcal{F}} \mathrm{Sym}(D; \xi) = 0. \end{aligned} \tag{1.8.61}$$

As such, all residual terms (appearing in the last three integrals in (1.8.59)) involve at most first-order derivatives on the section u. This is a key feature of the identity recorded in (1.8.59).

Let us also observe that since

$$(-\mathrm{i})\mathrm{Sym}\big(\nabla_h^{\mathcal{F}}; \xi\big) = {}_{T^*M}\langle \xi, h\rangle_{TM} I_{\mathcal{F}} \text{ for each } \xi \in T^*M, \tag{1.8.62}$$

(cf. [118, (9.1.45), p. 376]), it follows that for each covector $\xi \in T^*M$ we have

$$(-\mathrm{i})\mathrm{Sym}\big([\nabla_h^{\mathcal{F}}, A]; \xi\big) = {}_{T^*M}\langle \xi, h\rangle_{TM} I_{\mathcal{F}} A - A_{T^*M}\langle \xi, h\rangle_{TM} I_{\mathcal{F}} = 0. \tag{1.8.63}$$

From this we conclude that

$$[\nabla_h^{\mathcal{F}}, A] \text{ is a } \textit{zero-order} \text{ operator.} \tag{1.8.64}$$

Comment 2. Recall that the connection $\nabla^{\mathcal{F}}$ satisfies a product rule. Specifically, for any two given sections s, t of class $\mathscr{C}^1$ in $\mathcal{F}$ the following Leibniz formula holds:

$$\nabla_h \langle s, t\rangle_{\mathcal{F}} = \langle \nabla_h^{\mathcal{F}} s, t\rangle_{\mathcal{F}} + \langle s, \nabla_h^{\mathcal{F}} t\rangle_{\mathcal{F}} + C_h(s, t) \tag{1.8.65}$$

in a pointwise sense, where C_h is a bilinear form on $\mathcal{F}$. In other words, the connection $\nabla^{\mathcal{F}}$ is "almost" metric (see [118, (9.1.50), p. 377]); if $\nabla^{\mathcal{F}}$ is a metric connection (aka Hermitian connection), then $C_h = 0$ and the very last integral in (1.8.59) (for which C_h is responsible) may be omitted.

Comment 3. The proof of Theorem 1.8.7 given below also works in the case when the manifold M is the entire Euclidean space $\mathbb{R}^n$ (equipped with the standard metric), and $\mathcal{E}$, $\mathcal{F}$ are trivial vector bundles, say $\mathcal{E} \equiv \mathbb{R}^{N'}$ and $\mathcal{F} \equiv \mathbb{R}^{N''}$, provided either the (real, $\mathscr{C}^1$) vector field h is compactly supported, or h is a constant vector field and D, A have constant coefficients. In such a scenario, the set Ω is allowed to be unbounded, with the understanding that it Ω is an exterior domain then ∇u is required to vanish at infinity.

We now take up the task of presenting the proof of Theorem 1.8.7.

Proof of Theorem 1.8.7 To get started, fix a vector field $h \in \mathscr{C}^1(M, TM)$ with real-valued components. Thanks to (1.8.65) and the fact that h has real-valued components, we then have the following sequence of equalities:

$$
\begin{aligned}
&\langle ADu, (\nabla_h^{\mathcal{F}} Du)^c\rangle_{\mathcal{F}} \\
&\quad = \langle ADu, \nabla_h^{\mathcal{F}} (Du)^c\rangle_{\mathcal{F}} \\
&\quad = \nabla_h \langle ADu, (Du)^c\rangle_{\mathcal{F}} - \langle \nabla_h^{\mathcal{F}} ADu, (Du)^c\rangle_{\mathcal{F}} - C_h\big(ADu, (Du)^c\big) \\
&\quad = \nabla_h \langle ADu, (Du)^c\rangle_{\mathcal{F}} - \langle [\nabla_h^{\mathcal{F}}, A]Du, (Du)^c\rangle_{\mathcal{F}} - C_h\big(ADu, (Du)^c\big) \\
&\quad\quad - \langle A\nabla_h^{\mathcal{F}} Du, (Du)^c\rangle_{\mathcal{F}} - C_h\big(ADu, (Du)^c\big),
\end{aligned}
\tag{1.8.66}
$$

pointwise $\mathcal{L}_g^n$-a.e. in Ω. The hypothesis adopted in (1.8.54) allows us to re-write the penultimate term in (1.8.66) as

$$
\langle A\nabla_h^{\mathcal{F}} Du, (Du)^c\rangle_{\mathcal{F}} = \langle \nabla_h^{\mathcal{F}} Du, (ADu)^c\rangle_{\mathcal{F}} = \Big(\langle ADu, (\nabla_h^{\mathcal{F}} Du)^c\rangle_{\mathcal{F}}\Big)^c. \tag{1.8.67}
$$

In view of (1.8.66)-(1.8.67) and bearing in mind the observation made in (1.8.64) (which is relevant in relation to the third-from-last term in (1.8.66)), we may ultimately recast (1.8.66) as

$$
\begin{aligned}
2\,\mathrm{Re}\,\langle ADu, (\nabla_h^{\mathcal{F}} Du)^c\rangle_{\mathcal{F}} &= \nabla_h \langle ADu, (Du)^c\rangle_{\mathcal{F}} \\
&\quad + O\big(|Du|^2 |[\nabla_h^{\mathcal{F}}, A]|\big) + O\big(|A||h||Du|^2\big).
\end{aligned}
\tag{1.8.68}
$$

Going further, the current hypotheses on Ω, the first line in (1.8.57), and [112, (8.6.51)] imply

$$
\nabla^{\mathcal{E}} u \in L^{\frac{2n}{n-1}}(\Omega, \mathcal{L}_g^n) \otimes \mathcal{E}. \tag{1.8.69}
$$

In particular,

$$
Du \in L^{\frac{2n}{n-1}}(\Omega, \mathcal{L}_g^n) \otimes \mathcal{F}. \tag{1.8.70}
$$

Next, define the vector field $\vec{F} : \Omega \to TM$ implicitly, via the requirement that at each point $x \in \Omega$ we have

$$
\begin{aligned}
{}_{T_x^*M}\big(\xi, \vec{F}(x)\big)_{T_xM} &= -\,{}_{T_x^*M}\big(\xi, h(x)\big)_{T_xM} \Big\langle A(x)(Du)(x), (Du)(x)^c \Big\rangle_{\mathcal{F}_x} \\
&\quad - 2\mathrm{Re}\Big\langle (-\mathrm{i})\mathrm{Sym}(D^*;\xi)A(x)(Du)(x), (\nabla_h^{\mathcal{E}} u)(x)^c \Big\rangle_{\mathcal{E}_x}
\end{aligned}
\tag{1.8.71}
$$

for each covector $\xi \in T_x^*M$. Since the right-hand side of (1.8.71) is linear in ξ, this definition is meaningful. Moreover, it is apparent from (1.8.71) and (1.8.56) that

$$
\vec{F} \in L^1_{\mathrm{loc}}(\Omega, \mathcal{L}_g^n) \otimes TM. \tag{1.8.72}
$$

Also,

$$
\mathcal{N}_\kappa \vec{F} \le C\, \mathcal{N}_\kappa(\nabla^{\mathcal{E}} u)^2 \ \text{ pointwise on } \partial\Omega. \tag{1.8.73}
$$

As a consequence of (1.8.73), the first line in (1.8.57), and [112, (8.2.26)] we conclude that

$$\mathcal{N}_\kappa \vec{F} \in L^1(\partial\Omega, \sigma_g). \tag{1.8.74}$$

In addition, the second line in (1.8.57) ensures that

$$\begin{array}{c}\text{the pointwise nontangential boundary}\\ \text{trace } \vec{F}\big|_{\partial\Omega}^{\kappa-\text{n.t.}} \text{ exists } \sigma_g\text{-a.e. on } \partial_{\text{nta}}\Omega,\end{array} \tag{1.8.75}$$

and at σ_g-a.e. point $x \in \partial_*\Omega$ we have

$$\begin{aligned}
&{}_{T_x^*M}\Big(\nu_g(x), \big(\vec{F}\big|_{\partial\Omega}^{\kappa-\text{n.t.}}\big)(x)\Big)_{T_xM}\\
&= -{}_{T_x^*M}\big(\nu_g(x), h(x)\big)_{T_xM}\Big\langle A(x)\big((Du)\big|_{\partial\Omega}^{\kappa-\text{n.t.}}\big)(x), \big((Du)\big|_{\partial\Omega}^{\kappa-\text{n.t.}}\big)(x)^c\Big\rangle_{\mathcal{F}_x}\\
&\quad - 2\mathrm{Re}\Big\langle(-\mathrm{i})\mathrm{Sym}(D^*; \nu_g(x))A(x)\big((Du)\big|_{\partial\Omega}^{\kappa-\text{n.t.}}\big)(x), \big((\nabla_h^{\mathcal{E}} u)\big|_{\partial\Omega}^{\kappa-\text{n.t.}}\big)(x)^c\Big\rangle_{\mathcal{E}_x}\\
&= -{}_{T_x^*M}\big(\nu_g(x), h(x)\big)_{T_xM}\Big\langle A(x)\big((Du)\big|_{\partial\Omega}^{\kappa-\text{n.t.}}\big), \big((Du)\big|_{\partial\Omega}^{\kappa-\text{n.t.}}\big)(x)^c\Big\rangle_{\mathcal{F}_x}\\
&\quad + 2\mathrm{Re}\Big\langle(\partial_\nu^L u)(x), \big((\nabla_h^{\mathcal{E}} u)\big|_{\partial\Omega}^{\kappa-\text{n.t.}}\big)(x)^c\Big\rangle_{\mathcal{E}_x}.
\end{aligned} \tag{1.8.76}$$

To compute the (differential geometric) divergence of $\vec{F}$, we find it convenient to work in the sense of distributions. To this end, select an arbitrary real-valued function $\varphi \in \mathscr{C}_c^1(\Omega)$ and recall (cf. [118, (9.1.4) on p. 372, (9.2.1) on p. 381]) that

$$\nabla_h \varphi = h(\varphi) = d\varphi(h) = {}_{T^*M}(\mathrm{d}\varphi, h)_{TM}, \tag{1.8.77}$$

and that

$$(\nabla_h)^\top = -\nabla_h - \mathrm{div}_g h \tag{1.8.78}$$

(cf. [118, (9.2.16), p. 383]). Use these to compute

$$
\begin{aligned}
\langle \mathrm{div}_g \vec{F}, \varphi \rangle &= -\int_\Omega {}_{T^*M}\big(\mathrm{d}\varphi, \vec{F}\big)_{TM}\, \mathrm{d}\mathcal{L}_g^n \\
&= \int_\Omega {}_{T^*M}\big(\mathrm{d}\varphi, h\big)_{TM} \big\langle ADu, (Du)^c \big\rangle_{\mathcal{F}} \mathrm{d}\mathcal{L}_g^n \\
&\quad + 2\mathrm{Re} \int_\Omega \Big\langle (-\mathrm{i})\mathrm{Sym}(D^*; \mathrm{d}\varphi) ADu, (\nabla_h^{\mathcal{E}} u)^c \Big\rangle_{\mathcal{E}} \mathrm{d}\mathcal{L}_g^n \\
&= \int_\Omega (\nabla_h \varphi) \big\langle ADu, (Du)^c \big\rangle_{\mathcal{F}} \mathrm{d}\mathcal{L}_g^n \\
&\quad + 2\mathrm{Re} \int_\Omega \Big\langle (-\mathrm{i})\mathrm{Sym}(D^*; \mathrm{d}\varphi) ADu, (\nabla_h^{\mathcal{E}} u)^c \Big\rangle_{\mathcal{E}} \mathrm{d}\mathcal{L}_g^n \\
&=: \mathrm{I} + \mathrm{II},
\end{aligned} \tag{1.8.79}
$$

where the second equality is implied by (1.8.77) and by (1.8.71) written for $\xi := d\varphi$. As regards the first term above, use (1.8.78) and integration by parts to write (bearing in mind that φ is compactly supported in Ω)

$$
\mathrm{I} = -\int_\Omega \varphi \nabla_h \Big(\big\langle ADu, (Du)^c \big\rangle_{\mathcal{F}} \Big) \mathrm{d}\mathcal{L}_g^n - \int_\Omega \varphi (\mathrm{div}_g h) \big\langle ADu, (Du)^c \big\rangle_{\mathcal{F}} \mathrm{d}\mathcal{L}_g^n. \tag{1.8.80}
$$

Concerning the second term in the last line of (1.8.79), in view of the commutator identity (cf. [112, (1.11.25)])

$$
\mathrm{i}\mathrm{Sym}(D^*; \mathrm{d}\varphi) = \varphi D^* - D^*(\varphi\, \cdot), \tag{1.8.81}
$$

we obtain

$$
\begin{aligned}
\mathrm{II} &= 2\mathrm{Re}\int_\Omega \Big\langle D^*(\varphi ADu), (\nabla_h^{\mathcal{E}}u)^c\Big\rangle_{\mathcal{E}}\, \mathrm{d}\mathcal{L}_g^n - 2\mathrm{Re}\int_\Omega \varphi\Big\langle D^*ADu, (\nabla_h^{\mathcal{E}}u)^c\Big\rangle_{\mathcal{E}}\, \mathrm{d}\mathcal{L}_g^n \\
&= 2\mathrm{Re}\int_\Omega \varphi\Big\langle ADu, (D\nabla_h^{\mathcal{E}}u)^c\Big\rangle_{\mathcal{E}}\, \mathrm{d}\mathcal{L}_g^n - 2\mathrm{Re}\int_\Omega \varphi\Big\langle D^*ADu, (\nabla_h^{\mathcal{E}}u)^c\Big\rangle_{\mathcal{E}}\, \mathrm{d}\mathcal{L}_g^n \\
&= 2\mathrm{Re}\int_\Omega \varphi\Big\langle ADu, (\nabla_h^{\mathcal{E}}Du)^c\Big\rangle_{\mathcal{E}}\, \mathrm{d}\mathcal{L}_g^n + 2\mathrm{Re}\int_\Omega \varphi\Big\langle ADu, \big([D,\nabla_h^{\mathcal{E}}]u\big)^c\Big\rangle_{\mathcal{E}}\, \mathrm{d}\mathcal{L}_g^n \\
&\quad - 2\mathrm{Re}\int_\Omega \varphi\Big\langle D^*ADu, (\nabla_h^{\mathcal{E}}u)^c\Big\rangle_{\mathcal{E}}\, \mathrm{d}\mathcal{L}_g^n \\
&= \int_\Omega \varphi\nabla_h\Big(\big\langle ADu, (Du)^c\big\rangle_{\mathcal{F}}\Big)\, \mathrm{d}\mathcal{L}_g^n + 2\mathrm{Re}\int_\Omega \varphi\big\langle Lu, (\nabla_h^{\mathcal{E}}u)^c\big\rangle_{\mathcal{E}}\, \mathrm{d}\mathcal{L}_g^n \\
&\quad + \int_\Omega \varphi\cdot O\big(|Du|^2\big|[\nabla_h^{\mathcal{F}}, A]\big|\big)\, \mathrm{d}\mathcal{L}_g^n + \int_\Omega \varphi\cdot O\big(|A||Du|\big|[D,\nabla_h]u\big|\big)\, \mathrm{d}\mathcal{L}_g^n \\
&\quad + \int_\Omega \varphi\cdot O\big(|A||h||Du|^2\big)\, \mathrm{d}\mathcal{L}_g^n,
\end{aligned}
\tag{1.8.82}
$$

where the last equality is based on (1.8.68), the fact that φ is real-valued, and (1.8.55).

In concert, (1.8.79), (1.8.80), and (1.8.82) prove that

$$
\begin{aligned}
\mathrm{div}_g\vec{F} &= 2\mathrm{Re}\,\big\langle Lu, (\nabla_h^{\mathcal{E}}u)^c\big\rangle_{\mathcal{E}} - (\mathrm{div}_g h)\big\langle ADu, (Du)^c\big\rangle_{\mathcal{F}} + O\big(|Du|^2\big|[\nabla_h^{\mathcal{F}}, A]\big|\big) \\
&\quad + O\big(|A||Du|\big|[D,\nabla_h]u\big|\big) + O\big(|A||h||Du|^2\big).
\end{aligned}
\tag{1.8.83}
$$

In view of the last line in (1.8.57) and (1.8.69)-(1.8.70), this implies (keeping in mind that $(2n)/(n+1)$ and $(2n)/(n-1)$ are conjugate exponents) that

$$
\mathrm{div}_g\vec{F} \in L^1(\Omega, \mathcal{L}_g^n). \tag{1.8.84}
$$

At this stage, having established (1.8.72), (1.8.74), (1.8.75), and (1.8.84), [112, Corollary 1.11.5] applies and, on account of (1.8.76) and (1.8.83), the Divergence Formula [112, (1.11.20)] presently yields (1.8.59). □

We would like to further refine the Rellich identity recorded in (1.8.59) under the additional assumption that the second-order differential operator L is weakly elliptic. As in the past, this amounts to having $\mathrm{Sym}(L;\xi)$ an invertible linear map for each $\xi \in T^*M \setminus \{0\}$. Loosely speaking, this refinement is carried out by decomposing D into its tangential and normal component on $\partial\Omega$, analogously to the standard decomposition of the full gradient operator in $\mathbb{R}^n$ into its tangential and normal components on $\partial\Omega$.

Let us describe a procedure which, given an arbitrary first-order differential operator $P : \mathcal{E} \to \mathcal{G}$, where $\mathcal{G} \to M$ is a Hermitian vector bundle of class $\mathscr{C}^1$, allows one to decompose P as the sum of a tangential differential operator on $\partial\Omega$ and a suitable "multiple" of the conormal derivative operator ∂_ν^L. The key observation is that the operator

$$\tau_P := P - \mathrm{i}\mathrm{Sym}(P;\nu_g)\mathrm{Sym}(L;\nu_g)^{-1}\partial_\nu^L \tag{1.8.85}$$

is tangential on $\partial\Omega$, in the sense that this may be expressed as a linear combination of tangential differential operators expressed in local coordinates as (see [113, (11.6.23)])

$$\partial_{\tau_{jk}} := (\nu_g)_j \nabla_{\partial_k} - (\nu_g)_k \nabla_{\partial_j}, \qquad 1 \le j,k \le n, \tag{1.8.86}$$

eventually plus a zero-th order term. This is ensured by the fact that $\mathrm{Sym}(\tau_P;\nu_g) = 0$ at σ_g-a.e. point on $\partial_*\Omega$, which follows readily from definitions:

$$\begin{aligned}
&\mathrm{Sym}(\tau_P;\nu_g)\\
&\quad = \mathrm{Sym}(P;\nu_g) - \mathrm{i}\mathrm{Sym}(P;\nu_g)\mathrm{Sym}(L;\nu_g)^{-1}\mathrm{i}\mathrm{Sym}(D^*;\nu_g)A\,\mathrm{Sym}(D;\nu_g)\\
&\quad = \mathrm{Sym}(P;\nu_g) + \mathrm{Sym}(P;\nu_g)\mathrm{Sym}(L;\nu_g)^{-1}\mathrm{Sym}(-L;\nu_g)\\
&\quad = \mathrm{Sym}(P;\nu_g) - \mathrm{Sym}(P;\nu_g) = 0.
\end{aligned} \tag{1.8.87}$$

Indeed, if a first-order differential operator Q locally written locally as

$$Q = \sum_j A_j \nabla_{\partial_j} + B \tag{1.8.88}$$

satisfies $\mathrm{Sym}(Q;\nu_g) = 0$ at σ_g-a.e. point on $\partial_*\Omega$, then $\sum_j (\nu_g)_j A_j = 0$, and since

$$\sum_k (\nu_g)_k (\nu_g)_k^{\#} = \sum_{j,k} g^{jk}(\nu_g)_k(\nu_g)_j = \langle \nu_g, \nu_g\rangle = |\nu_g|^2 = 1, \tag{1.8.89}$$

we may write

$$\begin{aligned}
Q &= \sum_j A_j \nabla_{\partial_j} + B - \sum_{j,k} A_j (\nu_g)_j (\nu_g)_k^{\#} \nabla_{\partial_k}\\
&= \sum_{j,k} A_j (\nu_g)_k (\nu_g)_k^{\#} \nabla_{\partial_j} + B - \sum_{j,k} A_j (\nu_g)_j (\nu_g)_k^{\#} \nabla_{\partial_k}\\
&= \sum_{j,k} A_j (\nu_g)_k^{\#} \Big((\nu_g)_k \nabla_{\partial_j} - (\nu_g)_j \nabla_{\partial_k} \Big) + B\\
&= -\sum_{j,k} A_j\, (\nu_g)_k^{\#}\, \partial_{\tau_{jk}} + B,
\end{aligned} \tag{1.8.90}$$

which goes to show that the operator Q is, as claimed, tangential.

Having fixed some background aperture parameter $\kappa > 0$, the manner in which the operator τ_P from (1.8.85) acts on a section

$$\begin{aligned}
&u \in W^{1,1}_{\mathrm{loc}}(\Omega) \otimes \mathcal{E} \ \text{ with } \mathcal{N}_\kappa u,\ \mathcal{N}_\kappa(\nabla^{\mathcal{E}} u) \in L^2(\partial\Omega, \sigma_g),\\
&\text{and so that } u\big|^{\kappa-\mathrm{n.t.}}_{\partial\Omega},\ (\nabla^{\mathcal{E}} u)\big|^{\kappa-\mathrm{n.t.}}_{\partial\Omega} \text{ exist } \sigma_g\text{-a.e. on } \partial_{\mathrm{nta}}\Omega,
\end{aligned} \tag{1.8.91}$$

is

$$\tau_P u := (Pu)\Big|_{\partial\Omega}^{\kappa-\mathrm{n.t.}} - \mathrm{i}\,\mathrm{Sym}(P;\nu_g)\mathrm{Sym}(L;\nu_g)^{-1}\partial_\nu^L u. \tag{1.8.92}$$

In the case when this procedure is applied to D (and $\mathcal{G} := \mathcal{F}$), the resulting tangential operator

$$\tau_D := D - \mathrm{i}\,\mathrm{Sym}(D;\nu_g)\mathrm{Sym}(L;\nu_g)^{-1}\partial_\nu^L \tag{1.8.93}$$

has the extra property that

$$\begin{aligned}\mathrm{Sym}(D^*;\nu_g)A\tau_D &= -\mathrm{i}\partial_\nu^L - \mathrm{i}\,\mathrm{Sym}(D^*;\nu_g)A\mathrm{Sym}(D;\nu_g)\mathrm{Sym}(L;\nu_g)^{-1}\partial_\nu^L \\ &= -\mathrm{i}\partial_\nu^L + \mathrm{i}\partial_\nu^L = 0.\end{aligned} \tag{1.8.94}$$

Let us now fix a section u satisfying the properties listed in (1.8.91). Then, as seen from (1.8.93) and (1.8.58),

$$|\partial_\nu^L u| + |\tau_D u| \approx \Big|(Du)\Big|_{\partial\Omega}^{\kappa-\mathrm{n.t.}}\Big| \text{ uniformly, pointwise on } \partial_*\Omega. \tag{1.8.95}$$

Indeed, the design of the conormal derivative (cf. (1.8.58)) implies

$$|\partial_\nu^L u| = \Big|\mathrm{Sym}(D^*;\nu_g)A(Du)\Big|_{\partial\Omega}^{\kappa-\mathrm{n.t.}}\Big| \leq C\Big|(Du)\Big|_{\partial\Omega}^{\kappa-\mathrm{n.t.}}\Big| \tag{1.8.96}$$

at σ_g-a.e. point on $\partial_*\Omega$, while from (1.8.92)-(1.8.93) we see that

$$|\tau_D u| \leq C\Big(\Big|(Du)\Big|_{\partial\Omega}^{\kappa-\mathrm{n.t.}}\Big| + |\partial_\nu^L u|\Big) \text{ at } \sigma_g\text{-a.e. point on } \partial_*\Omega. \tag{1.8.97}$$

Together, (1.8.96) and (1.8.97) establish the left-pointing inequality in (1.8.95). The right-pointing inequality is justified by decomposing

$$D = \mathrm{i}\,\mathrm{Sym}(D;\nu_g)\mathrm{Sym}(L;\nu_g)^{-1}\partial_\nu^L + \tau_D. \tag{1.8.98}$$

Henceforth, work on $\partial_*\Omega$. Using (1.8.98) we may expand

$$\begin{aligned}&\Big\langle A(Du)\Big|_{\partial\Omega}^{\kappa-\mathrm{n.t.}}, \big((Du)\Big|_{\partial\Omega}^{\kappa-\mathrm{n.t.}}\big)^c\Big\rangle_{\mathcal{F}} \\ &\quad = \Big\langle \mathrm{i}A\mathrm{Sym}(D;\nu_g)\mathrm{Sym}(L;\nu_g)^{-1}\partial_\nu^L u, \big((Du)\Big|_{\partial\Omega}^{\kappa-\mathrm{n.t.}}\big)^c\Big\rangle_{\mathcal{F}} \\ &\quad\quad + \Big\langle A\tau_D u, \big(\mathrm{i}\,\mathrm{Sym}(D;\nu_g)\mathrm{Sym}(L;\nu_g)^{-1}\partial_\nu^L u\big)^c\Big\rangle_{\mathcal{F}} \\ &\quad\quad + \big\langle A\tau_D u, (\tau_D u)^c\big\rangle_{\mathcal{F}} \\ &\quad =: \mathrm{I} + \mathrm{II} + \mathrm{III}.\end{aligned} \tag{1.8.99}$$

Observe that, thanks to [112, (1.7.19)] and (1.8.58),

$$\mathrm{I} = \Big\langle \mathrm{Sym}(L;\nu_g)^{-1}\partial_\nu^L u, \Big(- i\mathrm{Sym}(D^*;\nu_g)A\big((Du)\big|_{\partial\Omega}^{\kappa-\mathrm{n.t.}}\big)\Big)^c\Big\rangle_{\mathcal{E}}$$

$$= \Big\langle \mathrm{Sym}(L;\nu)^{-1}\partial_\nu^L u, \big(- \partial_\nu^L u\big)^c\Big\rangle_{\mathcal{E}} \tag{1.8.100}$$

and that, by [112, (1.7.19)] and (1.8.94),

$$\mathrm{II} = \Big\langle \mathrm{Sym}(D^*;\nu_g)A\tau_D u, \big(i\mathrm{Sym}(L;\nu_g)^{-1}\partial_\nu^L u\big)^c\Big\rangle_{\mathcal{E}} = 0. \tag{1.8.101}$$

All in all, we have the following identity[41]:

$$\Big\langle A(Du)\big|_{\partial\Omega}^{\kappa-\mathrm{n.t.}}, \big((Du)\big|_{\partial\Omega}^{\kappa-\mathrm{n.t.}}\big)^c\Big\rangle_{\mathcal{F}} = \Big\langle \mathrm{Sym}(-L;\nu_g)^{-1}\partial_\nu^L u, (\partial_\nu^L u)^c\Big\rangle_{\mathcal{E}}$$
$$+ \big\langle A\tau_D u, (\tau_D u)^c\big\rangle_{\mathcal{F}}. \tag{1.8.102}$$

Similarly, we decompose

$$\nabla_h^{\mathcal{E}} = {}_{T^*M}(\nu_g, h)_{TM}\,\mathrm{Sym}(-L;\nu_g)^{-1}\partial_\nu^L + \tau_h, \tag{1.8.103}$$

where τ_h abbreviates $\tau_{\nabla_h^{\mathcal{E}}}$, the tangential differential operator associated as in (1.8.85) with the first-order differential operator $P := \nabla_h^{\mathcal{E}} : \mathcal{E} \to \mathcal{E}$. Specifically, according to (1.8.85) and (1.8.62),

$$\tau_h := \nabla_h^{\mathcal{E}} - {}_{T^*M}(\nu_g, h)_{TM}\,\mathrm{Sym}(-L;\nu_g)^{-1}\partial_\nu^L. \tag{1.8.104}$$

Thus, bearing in mind that $L^* = L$, we obtain

$$\mathrm{Re}\Big\langle \partial_\nu^L u, \big((\nabla_h^{\mathcal{E}} u)\big|_{\partial\Omega}^{\kappa-\mathrm{n.t.}}\big)^c\Big\rangle_{\mathcal{E}}$$
$$= {}_{T^*M}(\nu_g, h)_{TM}\Big\langle \partial_\nu^L u, \big(\mathrm{Sym}(-L;\nu_g)^{-1}\partial_\nu^L u\big)^c\Big\rangle_{\mathcal{E}} + \mathrm{Re}\,\big\langle \partial_\nu^L u, (\tau_h u)^c\big\rangle_{\mathcal{E}}$$
$$= {}_{T^*M}(\nu_g, h)_{TM}\Big\langle \mathrm{Sym}(-L;\nu_g)^{-1}\partial_\nu^L u, (\partial_\nu^L u)^c\Big\rangle_{\mathcal{E}}$$
$$+ \mathrm{Re}\,\big\langle \partial_\nu^L u, (\tau_h u)^c\big\rangle_{\mathcal{E}}. \tag{1.8.105}$$

Plugging (1.8.102) and (1.8.105) back in (1.8.59) and canceling like-terms (one being the first term in the right side of (1.8.102) and one associated with the first term in the right side of (1.8.105)), finally proves the following general Rellich-type identity:

Theorem 1.8.8 *Retain the assumptions made in Theorem 1.8.7 on the n-dimensional Riemannian manifold (M, g), the two Hermitian vector bundles $\mathcal{E}, \mathcal{F} \to M$, the first order differential operator $D : \mathcal{E} \to \mathcal{F}$, the linear map $A \in \mathrm{Hom}(\mathcal{E}, \mathcal{F})$, and*

[41] reminiscent of the Pythagorean Theorem, according to which the square of the norm of a vector is the sum of the squares of the norms of its normal and tangential parts

assume that

$$L := -D^*AD \text{ is weakly elliptic,} \tag{1.8.106}$$

in the sense that its principal symbol $\mathrm{Sym}(L;\xi)$ *is an invertible linear map for each* $\xi \in T^*M \setminus \{0\}$.

Next, consider a nonempty, open, proper subset Ω *of* M *such that* $\partial\Omega$ *is a lower Ahlfors regular set, and* $\sigma_g := \mathcal{H}_g^{n-1}\lfloor\partial\Omega$ *is a doubling measure on* $\partial\Omega$*. Denote by* $\nu_g : \partial_*\Omega \to T^*M$ *geometric measure theoretic outward unit conormal of the set of locally finite perimeter* Ω*, and denote by* $\mathcal{L}_g^n$ *the measure induced by the volume element on* M*. Finally, consider a section*

$$u \in W^{2,1}_{\mathrm{loc}}(\Omega) \otimes \mathcal{E} \tag{1.8.107}$$

which, for some aperture parameter $\kappa > 0$, *satisfies the properties listed in* (1.8.57), *and pick a vector field* $h \in \mathscr{C}^1(M, TM)$ *with real-valued components.*

Then, with the tangential first-order differential operators τ_D *and* τ_h *defined as in* (1.8.93) *and* (1.8.104), *respectively, and with the conormal derivative* ∂_ν^L *defined as in* (1.8.58), *the following Rellich-type identity holds:*

$$\begin{aligned}
&\int_{\partial_*\Omega} {}_{T^*M}(\nu_g, h)_{TM} \left\langle \mathrm{Sym}(-L;\nu_g)^{-1}\partial_\nu^L u, (\partial_\nu^L u)^c\right\rangle_{\mathcal{E}} \, d\sigma_g \\
&\quad = \int_{\partial_*\Omega} {}_{T^*M}(\nu_g, h)_{TM} \left\langle A\tau_D u, (\tau_D u)^c\right\rangle_{\mathcal{F}} d\sigma_g \\
&\quad\quad - 2\,\mathrm{Re} \int_{\partial_*\Omega} \left\langle \partial_\nu^L u, (\tau_h u)^c\right\rangle_{\mathcal{E}} \, d\sigma_g \\
&\quad\quad + 2\,\mathrm{Re} \int_{\Omega} \left\langle Lu, (\nabla_h^{\mathcal{E}} u)^c\right\rangle_{\mathcal{E}} \, d\mathcal{L}_g^n + \int_{\Omega} O\big(|Du|^2 \big|[\nabla_h^{\mathcal{F}}, A]\big|\big) \, d\mathcal{L}_g^n \\
&\quad\quad + \int_{\Omega} O\big(|Du|^2 |\mathrm{div}_g h|\big) \, d\mathcal{L}_g^n + \int_{\Omega} O\big(|A||Du|\big|[D, \nabla_h]u\big|\big) \, d\mathcal{L}_g^n \\
&\quad\quad + \int_{\Omega} O\big(|A||h||Du|^2\big) \, d\mathcal{L}_g^n,
\end{aligned} \tag{1.8.108}$$

where the superscript 'c' indicates complex conjugation, $\nabla^{\mathcal{F}}$ *is a connection on the vector bundle* $\mathcal{F}$*, and all the big* O*'s involve only constants depending on the metrics.*

It should be noted that the same comments as for the previous theorem apply to Theorem 1.8.8 as well.

Next we discuss a number of consequences of the Rellich-type identities (1.8.59), (1.8.108). These are particularly useful when the vector field h is transverse to $\partial\Omega$, i.e., when

$$\operatorname{essinf}_{\partial_*\Omega} {}_{T^*M}(\nu_g, h)_{TM} > 0. \tag{1.8.109}$$

From [62] we know that the existence of a continuous transverse vector field to $\partial\Omega$ is equivalent to Ω being a Lipschitz domain. Other crucial hypotheses have to do with the (semi or strict) positive definiteness of the homomorphism A, and the strong ellipticity of the second-order operator L.

Corollary 1.8.9 *Retain the hypotheses made in Theorem 1.8.7. Strengthen the assumptions on the set Ω by now assuming that this is a Lipschitz subdomain of the manifold M, and make the assumption that the self-adjoint homomorphism $A \in \mathrm{Hom}(\mathcal{F},\mathcal{F})$ is also uniformly strictly positive definite on $\partial\Omega$ in the sense that there exists some $c > 0$ such that*

$$\langle A(x)\zeta, \zeta^c\rangle_{\mathcal{F}_x} \geq c|\zeta|^2_{\mathcal{F}_x} \text{ for all } x \in \partial\Omega \text{ and } \zeta \in \mathcal{F}_x. \tag{1.8.110}$$

Fix some scalar function $V \in L^\infty(M, \mathcal{L}^n_g)$ along with an aperture parameter $\kappa > 0$, and consider a section $u : \Omega \to \mathcal{E}$ which satisfies

$$\begin{gathered} u \in W^{2,1}_{\mathrm{loc}}(\Omega) \otimes \mathcal{E}, \quad (L - V)u = 0 \text{ in } \Omega, \\ \mathcal{N}_\kappa u \in L^2(\partial\Omega, \sigma_g), \quad \mathcal{N}_\kappa(\nabla^{\mathcal{E}} u) \in L^2(\partial\Omega, \sigma_g), \\ u\big|^{\kappa-\mathrm{n.t.}}_{\partial\Omega} \text{ and } (\nabla^{\mathcal{E}} u)\big|^{\kappa-\mathrm{n.t.}}_{\partial\Omega} \text{ exist } \sigma_g\text{-a.e. on } \partial\Omega. \end{gathered} \tag{1.8.111}$$

Then there exists a constant $C \in (0, \infty)$, independent of u and, for each $\varepsilon > 0$, there exists a constant $C_\varepsilon \in (0, \infty)$ which depends on ε but is independent of u, with the property that

$$\begin{aligned} \int_{\partial\Omega} \Big|(Du)\big|^{\kappa-\mathrm{n.t.}}_{\partial\Omega}\Big|^2 \,\mathrm{d}\sigma_g &\leq C_\varepsilon \int_{\partial\Omega} |\partial^L_\nu u|^2 \,\mathrm{d}\sigma_g + \varepsilon \int_{\partial\Omega} \Big|(\nabla^{\mathcal{E}} u)\big|^{\kappa-\mathrm{n.t.}}_{\partial\Omega}\Big|^2 \,\mathrm{d}\sigma_g \\ &\quad + C \int_\Omega \big(|u|^2 + |\nabla^{\mathcal{E}} u|^2\big) \,\mathrm{d}\mathcal{L}^n_g. \end{aligned} \tag{1.8.112}$$

In the opposite direction, (1.8.96) implies that we also have

$$\int_{\partial\Omega} |\partial^L_\nu u|^2 \,\mathrm{d}\sigma_g \leq C \int_{\partial\Omega} \Big|(Du)\big|^{\kappa-\mathrm{n.t.}}_{\partial\Omega}\Big|^2 \,\mathrm{d}\sigma_g. \tag{1.8.113}$$

Together with (1.8.112) this shows that, in a uniformly fashion for sections u satisfying (1.8.111), there holds

$$\int_{\partial\Omega} |\partial^L_\nu u|^2 \,\mathrm{d}\sigma_g \approx \int_{\partial\Omega} \Big|(Du)\big|^{\kappa-\mathrm{n.t.}}_{\partial\Omega}\Big|^2 \,\mathrm{d}\sigma_g \quad \text{modulo small and lower-order terms.} \tag{1.8.114}$$

Let us now give the proof of Corollary 1.8.9.

Proof of Corollary 1.8.9 The inequality claimed in (1.8.112) is a direct consequence of the Rellich-type identity (1.8.59) written for a real vector field $h \in \mathscr{C}^1(M, TM)$

which is transverse to $\partial\Omega$ (hence (1.8.109) holds), and Cauchy's inequality with epsilon[42]. □

To state another application of a similar flavor, we first make a definition. Concretely, given any function $f \in L^p_1(\partial\Omega, \sigma_g) \otimes \mathcal{E}$ with $1 \leq p \leq \infty$, we define its tangential gradient as the section

$$\nabla_{\tan} f : \partial\Omega \longrightarrow TM \otimes \mathcal{E} \tag{1.8.115}$$

locally expressed as

$$\nabla_{\tan} f := (\nu^\#)_j g^{k\ell} \partial_\ell \otimes (\partial_{\tau_{jk}} f) = \nu_s g^{sj} g^{k\ell} \partial_\ell \otimes (\partial_{\tau_{jk}} f), \tag{1.8.116}$$

where $\partial_{\tau_{jk}}$ are as in (1.8.86) (when acting on smooth functions). In this regard, see also [113, §11.4] and [118, pp. 441-442]. Under the metric identification of TM with T^*M, in place of (1.8.115)-(1.8.116) we may alternatively consider

$$\nabla_{\tan} f : \partial\Omega \longrightarrow T^*M \otimes \mathcal{E} \tag{1.8.117}$$

expressed in local coordinates as

$$\nabla_{\tan} f = (\nu^\#)_j dx_k \otimes (\partial_{\tau_{jk}} f) = \nu_s g^{sj} dx_k \otimes (\partial_{\tau_{jk}} f). \tag{1.8.118}$$

Thus, the elementary tangential differential operators $\partial_{\tau_{jk}}$ serve as building blocks for $\nabla_{\tan}$. Remarkably, it is also possible to extract the former from the latter. Indeed, with the interpretation given in (1.8.118), for each $a, b \in \{1, \ldots, n\}$ we have the following pointwise equality of sections in $\mathcal{E}$:

$$_{T^*M}\big(\nabla_{\tan} f, \nu_a \partial_b\big)_{TM} - {}_{T^*M}\big(\nabla_{\tan} f, \nu_b \partial_a\big)_{TM} = \partial_{\tau_{ab}} f. \tag{1.8.119}$$

To justify this, via density, it suffices to consider the case when $f = \varphi|_{\partial\Omega}$ with φ a smooth section of the bundle $\mathcal{E}$. In such a scenario, we may compute (mindful of (1.8.86) and the fact that $(\nu^\#)_j \nu_j = 1$)

$$\begin{aligned}
&_{T^*M}\big(\nabla_{\tan} f, \nu_a \partial_b\big)_{TM} - {}_{T^*M}\big(\nabla_{\tan} f, \nu_b \partial_a\big)_{TM} \\
&\qquad = \nu_a (\nu^\#)_j (\partial_{\tau_{jb}} f) - \nu_b (\nu^\#)_j (\partial_{\tau_{ja}} f) \\
&\qquad = \nu_a (\nu^\#)_j \big(\nu_j \nabla_{\partial_b} \varphi - \nu_b \nabla_{\partial_j} \varphi\big) - \nu_b (\nu^\#)_j \big(\nu_j \nabla_{\partial_a} \varphi - \nu_a \nabla_{\partial_j} \varphi\big) \\
&\qquad = \nu_a \nabla_{\partial_b} \varphi - \nu_b \nabla_{\partial_a} \varphi = \partial_{\tau_{ab}} f,
\end{aligned} \tag{1.8.120}$$

as wanted.

We are now ready to state the result alluded to earlier.

Corollary 1.8.10 *Preserve the notation and background hypotheses employed in Theorem 1.8.8. This time, strengthen the assumptions by now assuming that* Ω *is a*

[42] to the effect that $ab \leq \frac{\varepsilon}{2} a^2 + \frac{1}{2\varepsilon} b^2$ for each $a, b \in \mathbb{R}$ and each $\varepsilon > 0$

Lipschitz subdomain of the manifold M, and impose the condition that the second-order operator L is strongly elliptic at points on $\partial\Omega$ in a uniform fashion, i.e., there exists some $c > 0$ such that for each $x \in \partial\Omega$ there holds

$$\mathrm{Re}\Big\langle \mathrm{Sym}(-L;\xi)\eta, \eta^c\Big\rangle_{\mathcal{E}_x} \geq c|\xi|^2|\eta|^2, \text{ for all } \xi \in T_x^*M \text{ and } \eta \in \mathcal{E}_x. \tag{1.8.121}$$

Finally, some scalar function $V \in L^\infty(M, \mathcal{L}_g^n)$ along with an aperture parameter $\kappa > 0$, and consider a section $u : \Omega \to \mathcal{E}$ satisfying

$$\begin{gathered} u \in W^{2,1}_{\mathrm{loc}}(\Omega) \otimes \mathcal{E}, \quad (L - V)u = 0 \text{ in } \Omega, \\ \mathcal{N}_\kappa u \in L^2(\partial\Omega, \sigma_g), \quad \mathcal{N}_\kappa(\nabla^{\mathcal{E}} u) \in L^2(\partial\Omega, \sigma_g), \\ u\big|_{\partial\Omega}^{\kappa-\mathrm{n.t.}} \text{ and } (\nabla^{\mathcal{E}} u)\big|_{\partial\Omega}^{\kappa-\mathrm{n.t.}} \text{ exist } \sigma_g\text{-a.e. on } \partial\Omega. \end{gathered} \tag{1.8.122}$$

Then for each real vector field $h \in \mathscr{C}^1(M, TM)$ which is transverse to $\partial\Omega$ there exists a constant $C \in (0, \infty)$, independent of u, with the property that

$$\begin{aligned} \int_{\partial\Omega} |\partial_\nu^L u|^2 \, d\sigma_g \leq C \int_{\partial\Omega} \big\{|\tau_D u|^2 + |\tau_h u|^2\big\} \, d\sigma_g \\ + C \int_\Omega \big(|u|^2 + |\nabla^{\mathcal{E}} u|^2\big) \, d\mathcal{L}_g^n. \end{aligned} \tag{1.8.123}$$

Moreover, there exists a constant $C \in (0, \infty)$, independent of u, with the property that

$$\begin{aligned} \int_{\partial\Omega} \Big|(\nabla^{\mathcal{E}} u)\big|_{\partial\Omega}^{\kappa-\mathrm{n.t.}}\Big|^2 \, d\sigma_g \leq C \int_{\partial\Omega} \Big|\nabla_{\tan}\Big(u\big|_{\partial\Omega}^{\kappa-\mathrm{n.t.}}\Big)\Big|^2 \, d\sigma_g \\ + C \int_\Omega \big(|u|^2 + |\nabla^{\mathcal{E}} u|^2\big) \, d\mathcal{L}_g^n, \end{aligned} \tag{1.8.124}$$

where $\nabla_{\tan}$ is the tangential gradient on $\partial\Omega$.

Proof In the current setting, (1.8.123) is an immediate consequence of the Rellich-type identity (1.8.108), Cauchy's inequality with epsilon, and (1.8.121). As regards (1.8.124), we first note that, through the freedom of choosing h arbitrary, (1.8.103) implies

$$\Big|(\nabla^{\mathcal{E}} u)\big|_{\partial\Omega}^{\kappa-\mathrm{n.t.}}\Big| \leq C\Big(|\partial_\nu^L u| + \Big|\nabla_{\tan}\Big(u\big|_{\partial\Omega}^{\kappa-\mathrm{n.t.}}\Big)\Big|\Big) \tag{1.8.125}$$

at σ_g-a.e. point on $\partial_*\Omega$ (cf. also (1.7.54) and [113, (11.3.26)], in this regard). Then (1.8.124) is implied by (1.8.123) and (1.8.125). □

We want to briefly elaborate on the significance of Corollaries 1.8.9-1.8.10 from the perspective of singular integral operators and boundary value problems. Work in the setting adopted in Corollary 1.8.10. Make the additional assumption that

$$L - V : H^1(M) \otimes \mathcal{E} \longrightarrow H^{-1}(M) \otimes \mathcal{E} \text{ is an isomorphism,} \tag{1.8.126}$$

and denote by $E_L(x, y)$ the Schwartz kernel of $(L - V)^{-1}$. Hence,

$$(L - V)^{-1}w(x) = \int_M \langle E_L(x, y), w(y)\rangle_{\mathcal{E}_y}\, d\mathcal{L}^n_g(y), \qquad x \in M. \tag{1.8.127}$$

Then $E_L(x, y)$ is a fundamental solution for the differential operator $L - V$ whose main singularity (along the diagonal) makes its first-order derivatives amenable to the treatment of singular integral operators from §2.8. As done in Theorem 2.8.4, use this to define the boundary-to-domain single layer potential operator $\mathscr{S}$ acting on each $f \in L^2(\partial\Omega, \sigma_g) \otimes \mathcal{E}$ according to

$$\mathscr{S}f(x) := \int_{\partial\Omega} \langle E_L(x, y), f(y)\rangle_{\mathcal{E}_y}\, d\sigma_g(y), \quad x \in M \setminus \partial\Omega, \tag{1.8.128}$$

as well as its boundary-to-boundary version S, acting on each $f \in L^2(\partial\Omega, \sigma_g) \otimes \mathcal{E}$ according to

$$Sf(x) := \int_{\partial\Omega} \langle E_L(x, y), f(y)\rangle_{\mathcal{E}_y}\, d\sigma_g(y), \quad x \in \partial\Omega. \tag{1.8.129}$$

To proceed, abbreviate

$$\Omega_+ := \Omega, \quad \Omega_- := M \setminus \overline{\Omega}, \tag{1.8.130}$$

and, having fixed an arbitrary $f \in L^2(\partial\Omega, \sigma_g) \otimes \mathcal{E}$, introduce

$$u_\pm := \mathscr{S}f \text{ in } \Omega_\pm. \tag{1.8.131}$$

Then, thanks to Theorem 2.8.4, the sections $u_\pm$ satisfy all conditions listed in (1.8.122) in $\Omega_\pm$ and, in addition,

$$\begin{aligned} &\text{at } \sigma_g\text{-a.e. point on } \partial\Omega \text{ we have } u_+\Big|^{\kappa-\text{n.t.}}_{\partial\Omega} = Sf = u_-\Big|^{\kappa-\text{n.t.}}_{\partial\Omega} \\ &\text{and } \tau_D u_+ = \tau_D u_-, \quad \tau_h u_+ = \tau_h u_-, \quad \partial^L_\nu u_\pm = \big(\mp \tfrac{1}{2}I + K^\#_L\big)f, \end{aligned} \tag{1.8.132}$$

where the conormal derivatives are defined as in (1.8.58) relative to $\Omega_\pm$ (with ν_g retaining its original meaning, i.e., the outward unit conormal to $\Omega = \Omega_+$). The last property in (1.8.132) comes from (2.8.235) and involves the principal-value singular integral operator $K^\#_L$ on $\partial\Omega$ defined as in (2.8.227) corresponding to the factorization of L given by (compare with (1.8.106))

$$L = \widetilde{D}D \text{ where } \widetilde{D} := -D^*A. \tag{1.8.133}$$

In particular, the last property in (1.8.132) implies

$$\partial^L_\nu u_+ - \partial^L_\nu u_- = -f. \tag{1.8.134}$$

Based on these properties and (1.8.123) we then conclude that

$$\|f\|_{L^2(\partial\Omega,\sigma_g)\otimes\mathcal{E}} \leq C\|Sf\|_{L^2_1(\partial\Omega,\sigma_g)\otimes\mathcal{E}} + \|\mathrm{Comp}\, f\| \tag{1.8.135}$$

where Comp is a compact operator mapping $L^2(\partial\Omega, \sigma_g) \otimes \mathcal{E}$ into a Banach space. A word of clarification regarding the latter aspect is in order. In the present scenario, the solid integral in (1.8.123) corresponding to $\Omega_\pm$ in place of just Ω are simply $\|\mathscr{S} f\|^2_{W^{1,2}(\Omega_\pm)\otimes\mathcal{E}}$, so the presence of $\|\mathrm{Comp}\, f\|$ in (1.8.135) is attributable to the fact that

$$\mathscr{S} : L^2(\partial\Omega, \sigma_g) \otimes \mathcal{E} \longrightarrow W^{1,2}(\Omega_\pm) \otimes \mathcal{E} \text{ are compact operators.} \tag{1.8.136}$$

In turn, (1.8.136) is a consequence of the fact that, as seen from (2.8.125),

$$\mathscr{S} : L^2(\partial\Omega, \sigma_g) \otimes \mathcal{E} \longrightarrow B^{2,2}_{3/2}(\Omega_\pm) \otimes \mathcal{E} \text{ are bounded operators} \tag{1.8.137}$$

and that, thanks to the Rellich-Kondrachov theorem, the embeddings

$$B^{2,2}_{3/2}(\Omega_\pm) \otimes \mathcal{E} \hookrightarrow W^{1,2}(\Omega_\pm) \otimes \mathcal{E} \text{ are compact.} \tag{1.8.138}$$

Pressing on, (1.8.135) implies that the linear and bounded operator (cf. (2.8.128))

$$S : L^2(\partial\Omega, \sigma_g) \otimes \mathcal{E} \longrightarrow L^2_1(\partial\Omega, \sigma_g) \otimes \mathcal{E} \tag{1.8.139}$$

has closed range and a finite dimensional kernel. In particular, the boundary-to-boundary single layer S is a semi-Fredholm operator in the context of (1.8.139). To determine its index, we take a continuous family L_t of second order, strongly elliptic operators on M, indexed by $t \in [0, 1]$, such that $L_1 = L$ and L_0 is *scalar*. This gives a norm continuous family

$$S_t : L^2(\partial\Omega, \sigma_g) \otimes \mathcal{E} \longrightarrow L^2_1(\partial\Omega, \sigma_g) \otimes \mathcal{E}, \text{ all semi-Fredholm.} \tag{1.8.140}$$

Since S_0 is Fredholm of index 0 (cf. [129]), so are all the operators S_t in (1.8.140) and, in particular, $S = S_1$. In conclusion, we have the following basic result:

$$S : L^2(\partial\Omega, \sigma_g) \otimes \mathcal{E} \to L^2_1(\partial\Omega, \sigma_g) \otimes \mathcal{E} \text{ is Fredholm with index zero.} \tag{1.8.141}$$

Via duality, we also see that

$$S : L^2_{-1}(\partial\Omega, \sigma_g) \otimes \mathcal{E} \to L^2(\partial\Omega, \sigma_g) \otimes \mathcal{E} \text{ is Fredholm with index zero.} \tag{1.8.142}$$

Ultimately, these results imply the Fredholm solvability of the Dirichlet Problem and Regularity Problem for the operator L in Lipschitz subdomains of the Riemannian manifold M.

Moving on, work in the setting described in Theorem 1.8.8 strengthened by now assuming that Ω is a UR domain, and make the additional assumption formulated in (1.8.126). Having fixed some $f \in L^2(\partial\Omega, \sigma_g) \otimes \mathcal{E}$, define $u_\pm$ as in (1.8.131) and write the Rellich-type identity (1.8.108) for $u_\pm$ in $\Omega_\pm$. Pick some number $\lambda \in \mathbb{R}$, then multiply the formula written for Ω_+ by $\lambda - \frac{1}{2}$, multiply the formula written for Ω_- by $\lambda + \frac{1}{2}$ and, finally, add the resulting identities. Keeping in mind that $u_\pm$ satisfy both the conditions listed in (1.8.122) (written relative to $\Omega_\pm$) and the properties described in (1.8.132), after some algebra (involving canceling like-terms) we arrive at the identity

$$
\begin{aligned}
(\lambda^2 - \tfrac{1}{4}) \int_{\partial\Omega} & {}_{T^*M}(\nu_g, h)_{TM} \left\langle \mathrm{Sym}(-L;\nu_g)^{-1} f, f^c \right\rangle_{\mathcal{E}} \, d\sigma_g \\
&= \int_{\partial\Omega} {}_{T^*M}(\nu_g, h)_{TM} \left\langle \mathrm{Sym}(-L;\nu_g)^{-1} (\lambda I + K_L^{\#}) f, \left[(\lambda I + K_L^{\#}) f\right]^c \right\rangle_{\mathcal{E}} \, d\sigma_g \\
&\quad - \int_{\partial\Omega} {}_{T^*M}(\nu_g, h)_{TM} \left\langle A\tau_D(\mathcal{S}f), \left(\tau_D(\mathcal{S}f)\right)^c \right\rangle_{\mathcal{F}} \, d\sigma_g \\
&\quad + 2\,\mathrm{Re} \int_{\partial_*\Omega} \left\langle (\lambda I + K_L^{\#}) f, \left(\tau_h(\mathcal{S}f)\right)^c \right\rangle_{\mathcal{E}} \, d\sigma_g \\
&\quad + \int_{\Omega_\pm} O\left(|\mathcal{S}f|^2 + \left|\nabla^{\mathcal{E}}(\mathcal{S}f)\right|^2\right) d\mathcal{L}_g^n.
\end{aligned}
\tag{1.8.143}
$$

Suppose next that Ω is actually a Lipschitz domain and that the real vector field h is transverse to $\partial\Omega$ (in the sense of (1.8.109)). Also, suppose L is in fact a strongly elliptic operator (i.e., (1.8.121) holds) and assume the self-adjoint homomorphism $A \in \mathrm{Hom}\,(\mathcal{F}, \mathcal{F})$ is semi-positive definite on $\partial\Omega$, in the sense that

$$
\langle A(x)\zeta, \zeta^c \rangle_{\mathcal{F}_x} \geq 0 \text{ for all } x \in \partial\Omega \text{ and } \zeta \in \mathcal{F}_x. \tag{1.8.144}
$$

Granted these hypotheses, from (1.8.143), Cauchy's inequality with epsilon, and (1.8.136) we obtain that for each $\lambda \in \mathbb{R}$ with $|\lambda| > \frac{1}{2}$ there exists a constant $C_\lambda \in (0, \infty)$, independent of f, with the property that

$$
\|f\|_{L^2(\partial\Omega,\sigma_g)\otimes\mathcal{E}} \leq C_\lambda \left\|(\lambda I + K_L^{\#}) f\right\|_{L^2(\partial\Omega,\sigma_g)\otimes\mathcal{E}} + \|\mathrm{Comp}\, f\| \tag{1.8.145}
$$

where Comp is a compact operator mapping $L^2(\partial\Omega, \sigma_g) \otimes \mathcal{E}$ into a Banach space (equipped with the norm $\|\cdot\|$). As a consequence, the operator $\lambda I + K_L^{\#}$ has closed range and finite dimensional kernel, in particular, $\lambda I + K_L^{\#}$ is semi-Fredholm on $L^2(\partial\Omega, \sigma_g) \otimes \mathcal{E}$. Since $\lambda I + K_L^{\#}$ is also invertible when $|\lambda|$ is large enough, from the homotopic invariance of the index (see item *(8)* of [113, Theorem 2.1.2]) we ultimately conclude that the operator

$$
\begin{gathered}
\lambda I + K_L^{\#} : L^2(\partial\Omega, \sigma_g) \otimes \mathcal{E} \longrightarrow L^2(\partial\Omega, \sigma_g) \otimes \mathcal{E} \\
\text{is Fredholm with index zero for each } \lambda \in \mathbb{R} \text{ with } |\lambda| > \tfrac{1}{2}.
\end{gathered}
\tag{1.8.146}
$$

From (2.8.238) and the fact that we presently have $L = L^*$, conclude that

the (Hermitian) adjoint of $K_L^{\#}$ acting on $L^2(\partial\Omega, \sigma_g) \otimes \mathcal{E}$ is the operator K_L, defined as in (2.8.226) relative to the factorization of L given in (1.8.133), also acting on the space $L^2(\partial\Omega, \sigma_g) \otimes \mathcal{E}$. (1.8.147)

Bearing this in mind, via duality we also obtain that the operator

$$\begin{aligned} &\lambda I + K_L : L^2(\partial\Omega, \sigma_g) \otimes \mathcal{E} \longrightarrow L^2(\partial\Omega, \sigma_g) \otimes \mathcal{E} \\ &\text{is Fredholm with index zero for each } \lambda \in \mathbb{R} \text{ with } |\lambda| > \tfrac{1}{2}. \end{aligned} \tag{1.8.148}$$

To be able to treat the important end-point cases $\lambda = \pm\frac{1}{2}$, assume that the self-adjoint homomorphism $A \in \mathrm{Hom}\,(\mathcal{F}, \mathcal{F})$ is uniformly strictly positive definite on $\partial\Omega$ (in the sense that (1.8.110) holds), and suppose there exists some $C \in (0, \infty)$ such that

$$\begin{aligned} &|\nabla^{\mathcal{E}} w| \leq C\big(|Dw| + |w|\big) \text{ in a pointwise sense,} \\ &\text{for each section } w \in W^{1,1}_{\mathrm{loc}}(M) \otimes \mathcal{E}. \end{aligned} \tag{1.8.149}$$

Once again, fix an arbitrary $f \in L^2(\partial\Omega, \sigma_g) \otimes \mathcal{E}$ and define $u_\pm$ as in (1.8.131). Then (1.8.149) implies

$$\begin{aligned} \int_{\partial\Omega} \Big|\nabla_{\tan}\big(u_\pm\big|^{\kappa-\mathrm{n.t.}}_{\partial\Omega}\big)\Big|^2 \, d\sigma_g &\leq C \int_{\partial\Omega} \Big|(Du_\pm)\big|^{\kappa-\mathrm{n.t.}}_{\partial\Omega}\Big|^2 \, d\sigma_g \\ &\quad + C \int_{\partial\Omega} \Big|u_\pm\big|^{\kappa-\mathrm{n.t.}}_{\partial\Omega}\Big|^2 \, d\sigma_g, \end{aligned} \tag{1.8.150}$$

while from (1.8.112) and (1.8.149) we deduce that

$$\begin{aligned} \int_{\partial\Omega} \Big|(Du_\pm)\big|^{\kappa-\mathrm{n.t.}}_{\partial\Omega}\Big|^2 \, d\sigma_g &\leq C \int_{\partial\Omega} |\partial^L_\nu u_\pm|^2 \, d\sigma_g \\ &\quad + C \int_{\Omega_\pm} \big(|u_\pm|^2 + |\nabla^{\mathcal{E}} u_\pm|^2\big) \, d\mathcal{L}^n_g. \end{aligned} \tag{1.8.151}$$

Together with the trace formulas in (1.8.132) and the mapping properties from (1.8.136) and (1.8.139), these show that

$$\int_{\partial\Omega} \big|\nabla_{\tan}(Sf)\big|^2 \, d\sigma_g \leq C \int_{\partial\Omega} \big|\big(\pm\tfrac{1}{2}I + K^\#_L\big)f\big|^2 \, d\sigma_g + \|\mathrm{Comp}\, f\|, \tag{1.8.152}$$

where Comp is a compact operator mapping $L^2(\partial\Omega, \sigma_g) \otimes \mathcal{E}$ into a Banach space (equipped with the norm $\|\cdot\|$). In the opposite direction, from (1.8.123) written for $u_\pm$ and (1.8.136) we see that

$$\int_{\partial\Omega} \big|\big(\pm\tfrac{1}{2}I + K^\#_L\big)f\big|^2 \, d\sigma_g \leq C \int_{\partial\Omega} \big|\nabla_{\tan}(Sf)\big|^2 \, d\sigma_g + \|\mathrm{Comp}\, f\|. \tag{1.8.153}$$

Combining (1.8.152) with (1.8.153) and using the triangle inequality then proves that, in the current scenario, (1.8.145) also holds for $\lambda = \pm\frac{1}{2}$, i.e., we have

$$\|f\|_{L^2(\partial\Omega,\sigma_g)\otimes\mathcal{E}} \leq C\big\|\big(\pm\tfrac{1}{2}I + K^\#_L\big)f\big\|_{L^2(\partial\Omega,\sigma_g)\otimes\mathcal{E}} + \|\mathrm{Comp}\, f\| \tag{1.8.154}$$

for each $f \in L^2(\partial\Omega, \sigma_g) \otimes \mathcal{E}$. As such, in place of (1.8.146)-(1.8.148) we now obtain

$$\begin{gathered} \lambda I + K_L^{\#},\ \lambda I + K_L : L^2(\partial\Omega, \sigma_g) \otimes \mathcal{E} \longrightarrow L^2(\partial\Omega, \sigma_g) \otimes \mathcal{E} \\ \text{are Fredholm operators with index zero for all } \lambda \in \mathbb{R} \text{ with } |\lambda| \geq \tfrac{1}{2}. \end{gathered} \tag{1.8.155}$$

More can be said when (1.8.145), (1.8.154), and (1.8.141) can be used in tandem. To be specific, (1.8.141) implies (see item *(7)* in [113, Theorem 2.1.2]) that there exists a linear and bounded operator

$$T : L_1^2(\partial\Omega, \sigma_g) \otimes \mathcal{E} \longrightarrow L^2(\partial\Omega, \sigma_g) \otimes \mathcal{E} \tag{1.8.156}$$

such that, with I denoting the identity operator,

$$\begin{gathered} ST - I \text{ is compact from } L_1^2(\partial\Omega, \sigma_g) \otimes \mathcal{E} \text{ into itself,} \\ TS - I \text{ is compact from } L^2(\partial\Omega, \sigma_g) \otimes \mathcal{E} \text{ into itself.} \end{gathered} \tag{1.8.157}$$

Based on these, the fact that S intertwines K_L and $K_L^{\#}$ (cf. Corollary 2.8.13), and (1.8.145) together with (1.8.154), for each $\lambda \in \mathbb{R}$ with $|\lambda| \geq \frac{1}{2}$ we may then estimate (with Comp denoting generic compact operators)

$$\begin{aligned} \|f\|_{L_1^2(\partial\Omega,\sigma_g)\otimes\mathcal{E}} &= \big\|(ST - \text{Comp})f\big\|_{L_1^2(\partial\Omega,\sigma_g)\otimes\mathcal{E}} \\ &\leq \|S(Tf)\|_{L_1^2(\partial\Omega,\sigma_g)\otimes\mathcal{E}} + \|\text{Comp}\, f\| \\ &\leq C\|Tf\|_{L^2(\partial\Omega,\sigma_g)\otimes\mathcal{E}} + \|\text{Comp}\, f\| \\ &\leq C\big\|(\lambda I + K_L^{\#})(Tf)\|_{L^2(\partial\Omega,\sigma_g)\otimes\mathcal{E}} + \|\text{Comp}\, f\| \\ &\leq C\big\|S(\lambda I + K_L^{\#})(Tf)\|_{L_1^2(\partial\Omega,\sigma_g)\otimes\mathcal{E}} + \|\text{Comp}\, f\| \\ &= C\big\|(\lambda I + K_L)(STf)\|_{L_1^2(\partial\Omega,\sigma_g)\otimes\mathcal{E}} + \|\text{Comp}\, f\| \\ &\leq C\big\|(\lambda I + K_L)f\|_{L_1^2(\partial\Omega,\sigma_g)\otimes\mathcal{E}} + \|\text{Comp}\, f\|, \end{aligned} \tag{1.8.158}$$

for each $f \in L_1^2(\partial\Omega, \sigma_g) \otimes \mathcal{E}$. Also, bearing in mind (2.8.239), this ultimately proves that

$$\begin{gathered} \lambda I + K_L : L_1^2(\partial\Omega, \sigma_g) \otimes \mathcal{E} \longrightarrow L_1^2(\partial\Omega, \sigma_g) \otimes \mathcal{E} \\ \text{is Fredholm with index zero for each } \lambda \in \mathbb{R} \text{ with } |\lambda| \geq \tfrac{1}{2}. \end{gathered} \tag{1.8.159}$$

Via duality (cf. (1.8.147)) we therefore obtain

$$\begin{gathered} \lambda I + K_L^{\#} : L_{-1}^2(\partial\Omega, \sigma_g) \otimes \mathcal{E} \longrightarrow L_{-1}^2(\partial\Omega, \sigma_g) \otimes \mathcal{E} \\ \text{is Fredholm with index zero for each } \lambda \in \mathbb{R} \text{ with } |\lambda| \geq \tfrac{1}{2}. \end{gathered} \tag{1.8.160}$$

In this vein, it is also useful to note that from (1.8.141), (1.8.142), (1.8.155), (1.8.159), (1.8.160) and item *(f)* of [113, Theorem 2.2.4] we have the following regularity results:

$$\operatorname{Ker}\Big(S : L^2_{-1}(\partial\Omega, \sigma_g) \otimes \mathcal{E} \to L^2(\partial\Omega, \sigma_g) \otimes \mathcal{E}\Big)$$
$$= \operatorname{Ker}\Big(S : L^2(\partial\Omega, \sigma_g) \otimes \mathcal{E} \to L^2_1(\partial\Omega, \sigma_g) \otimes \mathcal{E}\Big) \tag{1.8.161}$$

and, for each $\lambda \in \mathbb{R}$ with $|\lambda| \geq \frac{1}{2}$,

$$\operatorname{Ker}\Big(\lambda I + K_L^\#; L^2_{-1}(\partial\Omega, \sigma_g) \otimes \mathcal{E}\Big) = \operatorname{Ker}\Big(\lambda I + K_L^\#; L^2(\partial\Omega, \sigma_g) \otimes \mathcal{E}\Big), \tag{1.8.162}$$

$$\operatorname{Ker}\Big(\lambda I + K_L; L^2(\partial\Omega, \sigma_g) \otimes \mathcal{E}\Big) = \operatorname{Ker}\Big(\lambda I + K_L; L^2_1(\partial\Omega, \sigma_g) \otimes \mathcal{E}\Big). \tag{1.8.163}$$

Also, since the quality of being Fredholm and the value of the index are stable on complex interpolation scales (cf. [77, Theorem 2.9]), all results recorded in (1.8.141), (1.8.142), (1.8.155), (1.8.159), (1.8.160) have versions in which the integrability exponent 2 has been replaced by $p \in (2 - \varepsilon, 2 + \varepsilon)$ for some small $\varepsilon = \varepsilon(\Omega) \in (0, 1)$.

In turn, the aforementioned L^p-versions of (1.8.155), (1.8.159), (1.8.160) lead to Fredholm solvability results for the Dirichlet Regularity, Neumann, and Transmission Problems for the differential operator L in Lipschitz subdomains of the Riemannian manifold M, wit data in Lebesgue and Sobolev spaces involving an integrability exponent p near 2. Results allowing arbitrary $p \in (1, \infty)$ are found in [64], in the setting of Lipschitz domains with unit normals in VMO. This topic is re-visited in [116, Chapter 8], where we further refine this work by allowing Ahlfors regular domains which are sufficiently flat either globally, or at infinitesimal scales.

Chapter 2
Calderón-Zygmund Theory on Uniformly Rectifiable Sets

Here we take on the task of methodically developing a Calderón-Zygmund theory for singular integral operators on uniformly rectifiable sets which is applicable to boundary layer potentials associated with elliptic systems both in the Euclidean setting and in the setting of Riemannian manifolds. This body of results continues work of many predecessors, such as Calderón, Zygmund, Mikhlin, Coifman, McIntosh, Meyer, David, Jerison, Kenig, Semmes, Stein, among many others, and refines those in [63] by placing more economical demands on the ambient geometry. An informative historical account of the formative years in the development of the classical Calderón-Zygmund theory of singular integrals, which also touches on the personal perspective of the author, has been written by E.M. Stein in [168]. Among many aspects, it is particularly interesting to learn how this area of mathematics eventually became main-stream, though this accepted was hard-earned. In this regard, Stein writes on [168, p. 1134]: "*I do recall that we, in the small group who were interested in singular integrals then, felt a certain separateness from the larger community of analysts - not that this isolation was self-imposed, but more because our subject matter was seen by our colleagues as somewhat arcane, rarefied, and possibly not very relevant.*" In the early stages, this attitude was not limited to analysts in the US. On page 143 of his autobiographical book [98], V. Maz'ya writes: "*S.G. Mikhlin's highest accomplishment was creation of the theory of multidimensional singular integral equations which was present in a large article, published already in the 1936 edition of Matematicheskiy Sbornik. [...] However, some influential Leningrad experts*[1] *on partial differential equations did not admit that Mikhlin's subject area was part of the 'main trend' and this attitude nagged the life out of him.*"

To set the stage for subsequent work in this chapter, in §2.1 we consider integral operators on Hölder spaces on upper Ahlfors regular sets, and in §2.2 we treat singular integrals on Lebesgue spaces on Ahlfors regular quasi-metric spaces. Principal-value singular integral operators on uniformly rectifiable sets are dealt with in §2.3, while in §2.4 we consider boundary-to-domain integral operators on open sets with uniformly rectifiable boundaries. A powerful and versatile jump-formula is then proved §2.5.

[1] including O.A. Ladyzhenskaya (our note)

D. Mitrea et al., *Geometric Harmonic Analysis III*, Developments in Mathematics 74,
https://doi.org/10.1007/978-3-031-22735-6_2

This describes the nontangential trace of a boundary-to-domain integral operator in an open set with a uniformly rectifiable boundary in terms of a jump-term (involving the Fourier transform of the integral kernel function) and a principal-value singular integral operator on the boundary (of the sort alluded to earlier). A number of applications to integral operators whose kernels are derived from the fundamental solutions of elliptic differential operators are presented in §2.9.

In §2.8, the last section of this chapter, we extend the scope of the Calderón-Zygmund theory developed so far by considering singular integral operators, on uniformly rectifiable sets, with variable coefficient kernels that are Schwartz kernels of classical pseudodifferential operators of order -1 with odd principal symbol, acting between two vector bundles over a given Riemannian manifold.

2.1 Integral Operators Acting on Hölder Spaces on Upper Ahlfors Regular Sets

The main result pertaining to the action of integral operators on Hölder functions in this section is the estimate presented in Proposition 2.1.3. Its proof is divided into two parts, dealt with separately in Lemmas 2.1.1-2.1.2 below.

Lemma 2.1.1 *Suppose Ω is an open subset of $\mathbb{R}^n$ such that $\partial_*\Omega$ is bounded and satisfies an upper Ahlfors regularity condition with constant $c \in (0, \infty)$. In this setting, define $\sigma_* := \mathcal{H}^{n-1}\lfloor\partial_*\Omega$ and consider an integral operator*

$$\mathscr{T}f(x) := \int_{\partial_*\Omega} k(x, y)f(y)\,\mathrm{d}\sigma_*(y), \qquad x \in \Omega, \tag{2.1.1}$$

whose kernel $k : \Omega \times \partial_*\Omega \to \mathbb{R}$ is such that $k(x, \cdot)$ is a σ_*-measurable function for each $x \in \Omega$, and has the property that there exists some finite positive constant C_0 such that

$$|k(x, y)| \le \frac{C_0}{|x - y|^{n-1}} \tag{2.1.2}$$

for each $x \in \Omega$ and σ_*-a.e. $y \in \partial_*\Omega$. Also, suppose the action of $\mathscr{T}$ on the constant function 1 is bounded in Ω, i.e.,

$$M := \sup_{x\in\Omega} \big|\mathscr{T}1(x)\big| < +\infty. \tag{2.1.3}$$

Then for every function $f \in \mathscr{C}^\alpha(\partial_*\Omega)$ with $\alpha \in (0, 1)$ one has

$$\sup_{x\in\Omega} |\mathscr{T}f(x)| \le c\,C_0\Big\{2^{n-1+\alpha} + \frac{2^{2(n-1+2\alpha)}}{\alpha \ln 2}[\mathrm{diam}(\partial_*\Omega)]^\alpha\Big\}\|f\|_{\dot{\mathscr{C}}^\alpha(\partial_*\Omega)}$$
$$+ \max\big\{M, c\,C_0\big\} \cdot \sup_{\partial_*\Omega} |f|. \tag{2.1.4}$$

Proof Pick a function $f \in \mathscr{C}^\alpha(\partial_*\Omega)$ and fix an arbitrary point $x \in \Omega$. Consider first the case when $\operatorname{dist}(x, \partial_*\Omega) \geq \operatorname{diam}(\partial_*\Omega)$, in which scenario we may directly estimate

$$|\mathscr{T} f(x)| \leq C_0 \operatorname{diam}(\partial_*\Omega)^{1-n} \sigma_*(\partial_*\Omega) \sup_{\partial_*\Omega} |f| \leq c\, C_0 \sup_{\partial_*\Omega} |f|. \tag{2.1.5}$$

In the remaining case, i.e., when $\operatorname{dist}(x, \partial_*\Omega) < \operatorname{diam}(\partial_*\Omega) < \infty$, fix $\varepsilon \in (0, 1)$ arbitrary and select a point $x_\varepsilon \in \partial_*\Omega$ such that

$$(1 + \varepsilon) \operatorname{dist}(x, \partial_*\Omega) > |x - x_\varepsilon| =: r \in \big(0, \operatorname{diam}(\partial_*\Omega)\big). \tag{2.1.6}$$

Next, split $\mathscr{T} f(x) = I + II + III$, where

$$I := \int_{\partial_*\Omega \cap B(x_\varepsilon, 2r)} k(x, y)\big[f(y) - f(x_\varepsilon)\big] \, d\sigma_*(y), \tag{2.1.7}$$

$$II := \int_{\partial_*\Omega \setminus B(x_\varepsilon, 2r)} k(x, y)\big[f(y) - f(x_\varepsilon)\big] \, d\sigma_*(y), \tag{2.1.8}$$

and

$$III := (\mathscr{T} 1)(x) f(x_*). \tag{2.1.9}$$

Note that

$$\begin{aligned} |I| &\leq \int_{\partial_*\Omega \cap B(x_\varepsilon, 2r)} |k(x, y)| \, |f(y) - f(x_\varepsilon)| \, d\sigma_*(y) \\ &\leq C_0 \|f\|_{\dot{\mathscr{C}}^\alpha(\partial_*\Omega)} \int_{\partial_*\Omega \cap B(x_\varepsilon, 2r)} \frac{|y - x_\varepsilon|^\alpha}{|x - y|^{n-1}} \, d\sigma_*(y) \\ &\leq C_0 \|f\|_{\dot{\mathscr{C}}^\alpha(\partial_*\Omega)} \frac{(1 + \varepsilon)^{n-1} (2r)^\alpha}{r^{n-1}} \sigma_*\big(\partial_*\Omega \cap B(x_\varepsilon, 2r)\big), \end{aligned} \tag{2.1.10}$$

where the third inequality comes from the fact that $|y - x_\varepsilon|^\alpha < (2r)^\alpha$ on the domain of integration, and the fact that $1/|x - y| \leq (1 + \varepsilon)/|x - x_\varepsilon| = (1 + \varepsilon)/r$, for all $y \in \partial_*\Omega$. Hence,

$$|I| \leq 2^{n-1+\alpha} c\, C_0 (1 + \varepsilon)^{n-1} \|f\|_{\dot{\mathscr{C}}^\alpha(\partial_*\Omega)}, \tag{2.1.11}$$

bearing in mind (2.1.6) and the upper Ahlfors regularity of $\partial_*\Omega$. Also,

$$|II| \leq C_0 \|f\|_{\dot{\mathscr{C}}^\alpha(\partial_*\Omega)} \int_{\partial_*\Omega \setminus B(x_\varepsilon, 2r)} \frac{|y - x_\varepsilon|^\alpha}{|x - y|^{n-1}} \, d\sigma_*(y). \tag{2.1.12}$$

Note that if $y \in \partial_*\Omega \setminus B(x_\varepsilon, 2r)$ then

$$|y - x_\varepsilon| \leq |y - x| + r \quad \text{and} \quad r \leq \frac{|y - x_\varepsilon|}{2} \text{ implies } |y - x_\varepsilon| \leq 2|y - x|. \tag{2.1.13}$$

From this and [112, Lemma 7.2.1] we therefore obtain

$$|II| \leq 2^{n-1} C_0 \|f\|_{\dot{\mathscr{C}}^\alpha(\partial_*\Omega)} \int_{\partial_*\Omega\setminus B(x_\varepsilon,2r)} \frac{1}{|y-x_\varepsilon|^{n-1-\alpha}}\, d\sigma_*(y) \tag{2.1.14}$$

$$\leq 2^{2(n-1+2\alpha)} c\, C_0 (\alpha \ln 2)^{-1} \big[\operatorname{diam}(\partial_*\Omega)\big]^\alpha \|f\|_{\dot{\mathscr{C}}^\alpha(\partial_*\Omega)}. \tag{2.1.15}$$

Clearly, $|III| \leq M \cdot \sup_{\partial_*\Omega} |f|$, so the desired conclusion follows by combining the individual estimates for I, II, III and then letting $\varepsilon \to 0^+$. □

The following is a companion result to Lemma 2.1.1.

Lemma 2.1.2 *Let Ω be an open subset of $\mathbb{R}^n$ such that $\partial_*\Omega$ satisfies an upper Ahlfors regularity condition with constant $c \in (0,\infty)$. Define $\sigma_* := \mathcal{H}^{n-1}\lfloor\partial_*\Omega$ and consider the integral operator*

$$\mathscr{Q}f(x) := \int_{\partial_*\Omega} q(x,y) f(y)\, d\sigma_*(y), \qquad x \in \Omega, \tag{2.1.16}$$

whose kernel $q : \Omega \times \partial_*\Omega \to \mathbb{R}$ is such that $q(x,\cdot)$ is a σ_*-measurable function for each $x \in \Omega$, and has the property that for each $x \in \Omega$ and all σ_*-a.e. $y \in \partial_*\Omega$ one has

$$|q(x,y)| \leq \frac{C_1}{|x-y|^n} \tag{2.1.17}$$

for some finite positive constant C_1. Also, suppose there exists $\alpha \in (0,1)$ with the property that the action of $\mathscr{Q}$ on the constant function 1 satisfies

$$C_2 := \sup_{x\in\Omega} \Big\{ \operatorname{dist}(x,\partial_*\Omega)^{1-\alpha} |(\mathscr{Q}1)(x)| \Big\} < +\infty. \tag{2.1.18}$$

Then for every function $f \in \mathscr{C}^\alpha(\partial_*\Omega)$ one has

$$\sup_{x\in\Omega} \Big\{ \operatorname{dist}(x,\partial_*\Omega)^{1-\alpha} |\mathscr{Q}f(x)| \Big\} \leq c\, C_1 \Big\{ 2^{n-1+\alpha} + \frac{2^{3n-2}}{(1-\alpha)\ln 2} \Big\} \|f\|_{\dot{\mathscr{C}}^\alpha(\partial_*\Omega)} + C_2 \cdot \sup_{\partial_*\Omega} |f|. \tag{2.1.19}$$

Proof Select an arbitrary function $f \in \mathscr{C}^\alpha(\partial_*\Omega)$. Fix some $\varepsilon > 0$ arbitrary, pick a point $x \in \Omega$, and choose $x_\varepsilon \in \partial_*\Omega$ such that $r := |x - x_\varepsilon| < (1+\varepsilon)\operatorname{dist}(x,\partial_*\Omega)$. Split $\mathscr{Q}f(x) = I + II + III$, where

$$I := \int_{\partial_*\Omega \cap B(x_\varepsilon,2r)} q(x,y)\big[f(y) - f(x_\varepsilon)\big]\, d\sigma_*(y), \tag{2.1.20}$$

$$II := \int_{\partial_*\Omega \setminus B(x_\varepsilon,2r)} q(x,y)\big[f(y) - f(x_\varepsilon)\big]\, d\sigma_*(y), \tag{2.1.21}$$

and

$$III := (\mathscr{Q}1)(x)f(x_\varepsilon). \tag{2.1.22}$$

Then since $r/(1+\varepsilon) < \mathrm{dist}(x, \partial_*\Omega) \leq |x-y|$ for each $y \in \partial_*\Omega$, and $r \geq \mathrm{dist}(x, \partial_*\Omega)$, we may estimate

$$\begin{aligned}
|I| &\leq \int_{\partial_*\Omega \cap B(x_\varepsilon, 2r)} |q(x,y)|\,|f(y) - f(x_\varepsilon)|\,\mathrm{d}\sigma_*(y) \\
&\leq C_1 \|f\|_{\dot{\mathscr{C}}^\alpha(\partial_*\Omega)} \int_{\partial_*\Omega \cap B(x_\varepsilon, 2r)} \frac{|y - x_\varepsilon|^\alpha}{|x-y|^n}\,\mathrm{d}\sigma_*(y) \\
&\leq C_1 \|f\|_{\dot{\mathscr{C}}^\alpha(\partial_*\Omega)} \frac{(1+\varepsilon)^n (2r)^\alpha}{r^n}\,\sigma\big(\partial_*\Omega \cap B(x_\varepsilon, 2r)\big) \\
&\leq 2^{n-1+\alpha} c\, C_1 (1+\varepsilon)^n \mathrm{dist}(x, \partial_*\Omega)^{\alpha-1} \|f\|_{\dot{\mathscr{C}}^\alpha(\partial_*\Omega)}.
\end{aligned} \tag{2.1.23}$$

Next, keeping in mind that $1/|x-y|^n \leq 2^n/|y-x_\varepsilon|^n$ on $\partial_*\Omega \setminus B(x_\varepsilon, 2r)$ (cf. (2.1.13)), with the help of [112, Lemma 7.2.1] we may estimate

$$\begin{aligned}
|II| &\leq C_1 \|f\|_{\dot{\mathscr{C}}^\alpha(\partial_*\Omega)} \int_{\partial_*\Omega \setminus B(x_\varepsilon, 2r)} \frac{|y - x_\varepsilon|^\alpha}{|x-y|^n}\,\mathrm{d}\sigma_*(y). \\
&\leq 2^n C_1 \|f\|_{\dot{\mathscr{C}}^\alpha(\partial_*\Omega)} \int_{\partial_*\Omega \setminus B(x_\varepsilon, 2r)} \frac{1}{|y - x_\varepsilon|^{n-\alpha}}\,\mathrm{d}\sigma_*(y) \\
&\leq c\, C_1 \frac{2^{3n-2}}{(1-\alpha)\ln 2}\,\mathrm{dist}(x, \partial_*\Omega)^{\alpha-1} \|f\|_{\dot{\mathscr{C}}^\alpha(\partial_*\Omega)}.
\end{aligned} \tag{2.1.24}$$

Given that $\mathrm{dist}(x, \partial_*\Omega)^{1-\alpha}|III| \leq C_2 \cdot \sup_{\partial_*\Omega} |f|$, estimate (2.1.19) is established after combining the bounds for I, II, III and passing to limit as $\varepsilon \to 0^+$. □

Combining Lemma 2.1.1 and Lemma 2.1.2 then yields the following result which is useful in the treatment of the harmonic double layer potential operator, considered later on.

Proposition 2.1.3 *Let Ω be an open subset of $\mathbb{R}^n$ such that $\partial_*\Omega$ is bounded and satisfies an upper Ahlfors regularity condition with constant $c \in (0, \infty)$. In this setting, define $\sigma_* := \mathcal{H}^{n-1} \lfloor \partial_*\Omega$ and consider the integral operator*

$$\mathcal{T}f(x) := \int_{\partial_*\Omega} K(x,y) f(y)\,\mathrm{d}\sigma_*(y), \qquad x \in \Omega, \tag{2.1.25}$$

whose kernel $K : \Omega \times \partial_*\Omega \to \mathbb{R}$ has the property that there exists a finite constant $B > 0$ such that

$$|K(x,y)| + |x-y||(\nabla_x K)(x,y)| \leq \frac{B}{|x-y|^{n-1}} \tag{2.1.26}$$

for each $x \in \Omega$ and σ_*-a.e. $y \in \partial_*\Omega$. Fix some $\alpha \in (0,1)$ and suppose that

$$A := \sup_{x\in\Omega} \big|(\mathcal{T}1)(x)\big| + \sup_{x\in\Omega} \Big\{\mathrm{dist}(x, \partial_*\Omega)^{1-\alpha}\big|\nabla(\mathcal{T}1)(x)\big|\Big\} < +\infty. \tag{2.1.27}$$

Then there exists a finite constant $C_{n,\alpha,\Omega} > 0$, which depends only on n, α, c, and $\mathrm{diam}(\partial_*\Omega)$, *with the property that for every function $f \in \mathscr{C}^\alpha(\partial_*\Omega)$ one has*

$$\sup_{x\in\Omega} |\mathcal{T}f(x)| + \sup_{x\in\Omega} \Big\{\mathrm{dist}(x, \partial_*\Omega)^{1-\alpha}\big|\nabla(\mathcal{T}f)(x)\big|\Big\}$$
$$\leq C_{n,\alpha,\Omega}(A + B)\|f\|_{\mathscr{C}^\alpha(\partial_*\Omega)}. \tag{2.1.28}$$

Proof This is an immediate consequence of Lemma 2.1.1 and Lemma 2.1.2. □

We conclude by recording a compactness criterion for singular integral operators with weakly singular kernels which is a particular case of a slightly more general result proved on space of homogeneous type in [63, Lemma 2.20, p. 2608].

Lemma 2.1.4 *Suppose $\Sigma \subset \mathbb{R}^n$ is compact and Ahlfors regular, and let $k(\cdot,\cdot)$ be a complex-valued, Borel-measurable function on $\{(x, y) \in \Sigma\times\Sigma : x \neq y\}$ satisfying*

$$|k(x, y)| \leq \frac{\omega(|x-y|)}{|x-y|^{n-1}} \quad \textit{for each } x, y \in \Sigma \textit{ with } x \neq y, \tag{2.1.29}$$

where $\omega : (0,\infty) \longrightarrow (0,\infty)$ is non-decreasing, satisfies Dini's integrability condition

$$\int_0^1 \frac{\omega(t)}{t}\,\mathrm{d}t < \infty, \tag{2.1.30}$$

and is slowly varying, i.e., there exists a finite constant $c > 0$ such that

$$\omega(2t) \leq c\,\omega(t), \qquad \forall t \in (0,\infty). \tag{2.1.31}$$

In this setting, define $\sigma := \mathcal{H}^{n-1}\lfloor\Sigma$ and consider the integral operator given by

$$Rf(x) := \int_\Sigma k(x, y)f(y)\,\mathrm{d}\sigma(y), \qquad x \in \Sigma. \tag{2.1.32}$$

Then for each $p \in (1,\infty)$, the operator

$$R : L^p(\Sigma, \sigma) \longrightarrow L^p(\Sigma, \sigma) \textit{ is compact (hence also bounded)}. \tag{2.1.33}$$

2.2 Singular Integrals on Ahlfors Regular Quasi-Metric Spaces

The most general environment in which a Calderón-Zygmund theory may be developed, in a manner that retains much of its original force from the Euclidean setting, is that of spaces of homogeneous type. In this section we review a number of results pertaining to the nature of singular integral operators of Calderón-Zygmund type on Ahlfors regular quasi-metric spaces. We begin by making the following definition.

Definition 2.2.1 *Given $d \in (0, \infty)$, call a triplet (X, ρ, μ) a d-*`Ahlfors regular (quasi-metric) space` *(or simply, a d-AR space) if the pair (X, ρ) is a quasi-metric space, μ is a non-negative measure defined on a sigma-algebra of subsets of X which contains all ρ-balls and such that there exists $C_{\text{AR}} \in [1, \infty)$ with the property that*

$$(C_{\text{AR}})^{-1} r^d \leq \mu\big(B_\rho(x, r)\big) \leq C_{\text{AR}} r^d, \text{ for all } x \in X \text{ and all finite } r \in \big(0, \operatorname{diam}_\rho(X)\big]. \tag{2.2.1}$$

Clearly, if (X, ρ, μ) is a d-Ahlfors regular quasi-metric space, then (X, ρ, μ) is also a space of homogeneous type. In what follows, given a nonempty set X, we let

$$\operatorname{diag}(X) := \{(x, y) \in X \times X : x = y\} \tag{2.2.2}$$

denote the diagonal in the Cartesian product $X \times X$. If (X, ρ, μ) is a d-Ahlfors regular quasi-metric space, it follows that the diagonal $\operatorname{diag}(X)$ is a $\mu \otimes \mu$-measurable subset of $X \times X$, of zero measure.

Lemma 2.2.2 *Let (X, ρ, μ) be a d-Ahlfors regular quasi-metric space. The diagonal* $\operatorname{diag}(X)$ *is a $\mu \otimes \mu$-measurable subset of $X \times X$ and*

$$(\mu \otimes \mu)\big(\operatorname{diag}(X)\big) = 0. \tag{2.2.3}$$

Proof For each $\varepsilon > 0$ introduce

$$\Delta_\varepsilon := \big\{(x, y) \in X \times X : \rho_\#(x, y) < \varepsilon\big\}. \tag{2.2.4}$$

From [112, (7.1.18)] it follows that the set Δ_ε is open in $\tau_\rho \times \tau_\rho$ for each $\varepsilon > 0$. Since

$$\bigcap_{j \in \mathbb{N}} \Delta_{1/j} = \operatorname{diag}(X), \tag{2.2.5}$$

we conclude that $\operatorname{diag}(X)$ is a $\mu \otimes \mu$-measurable subset of $X \times X$.

To show that the diagonal in question has measure zero, we argue as follows. Fix a small number $\varepsilon \in (0, 1)$ and consider a subset $\mathcal{U}_\varepsilon$ of X such that

$$\rho(x, y) \geq \varepsilon \text{ for every } x, y \in \mathcal{U}_\varepsilon, \tag{2.2.6}$$

and

$$\mathcal{U}_\varepsilon \text{ is maximal (with respect to inclusion) among all subsets of } X \text{ satisfying property (2.2.6).} \tag{2.2.7}$$

The maximality of $\mathcal{U}_\varepsilon$ implies that

$$X = \bigcup_{x \in \mathcal{U}_\varepsilon} B_\rho(x, \varepsilon). \tag{2.2.8}$$

Also, from (2.2.6) we deduce that

$$\text{the family } \left\{B_\rho\big(x,\varepsilon/(2C_\rho)\big)\right\}_{x\in\mathcal{U}_\varepsilon} \text{ consists of mutually disjoint sets.} \tag{2.2.9}$$

To proceed, fix a point $z \in X$. We claim that[2]

$$\#\Big(B_\rho(z,r)\cap\mathcal{U}_\varepsilon\Big) \le (C_{\text{AR}})^2\left(\frac{2(C_\rho)^2 r + C_\rho}{\varepsilon}\right)^d, \qquad \forall r>0. \tag{2.2.10}$$

Indeed, for each $r > 0$ we have

$$x \in B_\rho(z,r)\cap\mathcal{U}_\varepsilon \Longrightarrow B_\rho\big(x,\varepsilon/(2C_\rho)\big) \subseteq B_\rho\big(z, C_\rho r + 1/2\big) \tag{2.2.11}$$

which, in concert with (2.2.1) and (2.2.9), permits us to estimate

$$\begin{aligned} C_{\text{AR}}\big(C_\rho r + 1/2\big)^d &\ge \mu\Big(B_\rho\big(z, C_\rho r + 1/2\big)\Big) \ge \sum_{x\in B_\rho(z,r)\cap\mathcal{U}_\varepsilon} \mu\Big(B_\rho\big(x,\varepsilon/(2C_\rho)\big)\Big) \\ &\ge \#\Big(B_\rho(z,r)\cap\mathcal{U}_\varepsilon\Big)(C_{\text{AR}})^{-1}\big(\varepsilon/(2C_\rho)\big)^d. \end{aligned} \tag{2.2.12}$$

Then (2.2.10) readily follows from this. Note that, as a consequence of (2.2.10), the set $\mathcal{U}_\varepsilon$ is at most countable. Moving on, for z as above define

$$\operatorname{diag}_j(X) := \operatorname{diag}(X) \cap \Big(B_\rho(z,j)\times B_\rho(z,j)\Big), \qquad \forall j \in \mathbb{N}. \tag{2.2.13}$$

From what we have proved so far, it follows that

$$\text{each } \operatorname{diag}_j(X) \text{ is a } \mu\otimes\mu\text{-measurable subset of } X\times X. \tag{2.2.14}$$

For each fixed $j \in \mathbb{N}$ we now claim that

$$\operatorname{diag}_j(X) \subseteq \bigcup_{x\in B_\rho(z,C_\rho j)\cap\mathcal{U}_\varepsilon} B_\rho(x,\varepsilon)\times B_\rho(x,\varepsilon). \tag{2.2.15}$$

This is a consequence of (2.2.8) and the observation that $B_\rho(z,j)\cap B_\rho(x,\varepsilon) \neq \varnothing$ forces $x \in B_\rho(z, C_\rho j)$. Having established (2.2.15), we now invoke (2.2.14), (2.2.1), and (2.2.10), in order to estimate

[2] with $\#(E)$ denoting the cardinality of the set E

$$\begin{aligned}
(\mu\otimes\mu)\big(\mathrm{diag}_j(X)\big) &\leq (\mu\otimes\mu)\Big(\bigcup_{x\in B_\rho(z,C_\rho j)\cap\mathcal{U}_\varepsilon} B_\rho(x,\varepsilon)\times B_\rho(x,\varepsilon)\Big)\\
&\leq \sum_{x\in B_\rho(z,C_\rho j)\cap\mathcal{U}_\varepsilon} \mu\big(B_\rho(x,\varepsilon)\big)^2 \leq \sum_{x\in B_\rho(z,C_\rho j)\cap\mathcal{U}_\varepsilon} (C_{AR})^2\varepsilon^{2d}\\
&= \#\Big(B_\rho(z,C_\rho j)\cap\mathcal{U}_\varepsilon\Big)(C_{AR})^2\varepsilon^{2d}\\
&\leq (C_{AR})^4\big(2(C_\rho)^3 j + C_\rho\big)^d\varepsilon^d
\end{aligned}\tag{2.2.16}$$

for each fixed $j\in\mathbb{N}$. Since $\varepsilon\in(0,1)$ may be taken to be arbitrarily small, this forces

$$(\mu\otimes\mu)\big(\mathrm{diag}_j(X)\big) = 0 \ \text{ for each } \ j\in\mathbb{N}.\tag{2.2.17}$$

Given that, as seen from (2.2.13),

$$\mathrm{diag}(X) = \bigcup_{j\in\mathbb{N}} \mathrm{diag}_j(X),\tag{2.2.18}$$

we ultimately conclude from (2.2.17)-(2.2.18) that (2.2.3) holds. □

The point of the proposition below is that any Ahlfors regular quasi-metric space is a reasonable topological and measure theoretic environment.

Proposition 2.2.3 *Let (X,ρ,μ) be a d-Ahlfors regular quasi-metric space. Then*[3]

$$(X,\tau_\rho) \text{ is a separable topological space,}\tag{2.2.19}$$

and

$$\mu \text{ is a Borel measure on } (X,\tau_\rho).\tag{2.2.20}$$

Moreover[4]*,*

$$(X,\tau_\rho) \text{ is a second countable topological space.}\tag{2.2.21}$$

In particular[5]

$$(X,\tau_\rho) \text{ is a strongly Lindelöf space.}\tag{2.2.22}$$

Proof We begin by observing that, as a consequence of [6, Theorem 6.2] (cf. also [117, Theorem 4.21, p. 184]), the sigma-algebra generated by all ρ-balls contains all open sets in τ_ρ. This renders μ a Borel measure. As a consequence of (2.2.1), we have that μ is a doubling measure. In turn, this implies that (X,ρ) is geometrically

[3] recall that a topological space is called separable if contains a countable dense subset

[4] a topological space is said to be second countable if there exists some countable collection $\mathcal{U}$ of open sets with the property that any open set may be expressed as a union of elements of some subfamily of $\mathcal{U}$

[5] a strongly Lindelöf space is a topological space such that every open cover of any of its open subsets has a countable sub-cover

doubling (cf. [112, Definition 7.5.1]), hence (2.2.19) holds by virtue of [117, (4.49), p. 164]. From (2.2.19) and the fact that the topology induced by any quasi-metric is metrizable (cf. [112, Theorem 7.1.2]), we see that (2.2.21) holds. Finally, (2.2.22) follows from this, given that any second countable space is a strongly Lindelöf space. □

We continue by making a basic definition.

Definition 2.2.4 *Let* (X, ρ, μ) *be a d-Ahlfors regular quasi-metric space for some* $d \in (0, \infty)$, *and consider a function*

$$K : X \times X \setminus \mathrm{diag}(X) \longrightarrow \mathbb{R}. \tag{2.2.23}$$

(i) Define the `transpose kernel` *of K to be the function*

$$\begin{aligned} &K^{\top} : X \times X \setminus \mathrm{diag}(X) \longrightarrow \mathbb{R} \text{ given by} \\ &K^{\top}(x, y) := K(y, x) \quad \forall x, y \in X, \text{ with } x \neq y. \end{aligned} \tag{2.2.24}$$

(ii) Call K a `standard kernel in the first variable` *if K is* $\mu \otimes \mu$*-measurable and there exist finite constants* $\gamma > 0$, $C_0 \geq 0$, $C_1 > 0$ *such that*

$$|K(x, y)| \leq \frac{C_0}{\rho(x, y)^d}, \quad \forall x, y \in X, \text{ with } x \neq y, \tag{2.2.25}$$

and

$$\begin{aligned} &|K(x, y) - K(x', y)| \leq C_0 \frac{\rho(x', x)^{\gamma}}{\rho(x', y)^{d+\gamma}}, \qquad \forall x, x', y \in X, \\ &\text{satisfying } y \notin \{x, x'\} \text{ as well as } \rho(x', y) \geq C_1 \rho(x', x). \end{aligned} \tag{2.2.26}$$

(iii) Call K a `standard kernel in the second variable` *if* $K^{\top}$ *is a standard kernel in the first variable.*

Our next proposition contains a refined version of Cotlar's inequality.

Proposition 2.2.5 *Assume* (X, ρ, μ) *is a d-Ahlfors regular quasi-metric space for some* $d \in (0, \infty)$ *with the property that the quasi-distance* $\rho : X \times X \to [0, +\infty)$ *is continuous in the product topology* $\tau_\rho \times \tau_\rho$. *Suppose*

$$K : X \times X \setminus \mathrm{diag}(X) \longrightarrow \mathbb{R} \tag{2.2.27}$$

is a standard kernel in the first variable (i.e., K is $\mu \otimes \mu$*-measurable and* (2.2.25)-(2.2.26) *are satisfied). Fix* $p \in [1, \infty)$ *and assume that*

$$T : L^p(X, \mu) \longrightarrow L^{p,\infty}(X, \mu) \tag{2.2.28}$$

is an operator[6] *which satisfies the following properties:*

[6] not necessarily linear

(T1) there exists $C_T \in [0, \infty)$ such that for every $f, g \in L^p(X, \mu)$ one has

$$|T(f+g)| \leq C_T\big(|Tf| + |Tg|\big) \ \ \mu\text{-a.e. in } X; \tag{2.2.29}$$

(T2) with the constant C_T as above, for every $f \in L^p(X, \mu)$ there holds

$$\|Tf\|_{L^{p,\infty}(X,\mu)} \leq C_T \|f\|_{L^p(X,\mu)}; \tag{2.2.30}$$

(T3) for each $f \in L^p(X, \mu)$ one has[7]

$$(Tf)(x) = \int_X K(x, y) f(y)\, \mathrm{d}\mu(y) \ \text{ for } \mu\text{-a.e. } x \in X \setminus \operatorname{supp} f. \tag{2.2.31}$$

Next, having fixed some reference point $x_0 \in X$, for each $\varepsilon > 0$ and each function

$$f \in L^1\Big(X, \tfrac{\mu(x)}{1+\rho(x, x_0)^d}\Big) \tag{2.2.32}$$

define

$$(T_\varepsilon f)(x) := \int_{y \in X,\, \rho(x,y) > \varepsilon} K(x, y) f(y)\, \mathrm{d}\mu(y), \qquad \forall x \in X. \tag{2.2.33}$$

Finally, for each f as in (2.2.32)*, set*

$$(T_{\max} f)(x) := \sup_{\varepsilon > 0} \big|(T_\varepsilon f)(x)\big| \ \text{ for every } x \in X. \tag{2.2.34}$$

Then the following statements are true.

(i) If $\mathscr{M}_+(X, \mu)$ denotes the cone of μ-measurable functions defined on X with values in $[0, +\infty]$, then the operator

$$T_{\max} : L^1\Big(X, \tfrac{\mu(x)}{1+\rho(x, x_0)^d}\Big) \longrightarrow \mathscr{M}_+(X, \mu) \tag{2.2.35}$$

is well defined and sub-linear, in the sense that

$$\begin{aligned} &T_{\max}(f+g) \leq T_{\max} f + T_{\max} g, \qquad \forall f, g \in L^1\Big(X, \tfrac{\mu(x)}{1+\rho(x, x_0)^d}\Big), \\ &T_{\max}(\lambda f) = |\lambda|\, T_{\max} f, \quad \forall f \in L^1\Big(X, \tfrac{\mu(x)}{1+\rho(x, x_0)^d}\Big), \quad \forall \lambda \in \mathbb{R}. \end{aligned} \tag{2.2.36}$$

As a consequence, for any two given functions $f, g \in L^1\Big(X, \tfrac{\mu(x)}{1+\rho(x, x_0)^d}\Big)$ one has

$$\big|(T_{\max} f)(x) - (T_{\max} g)(x)\big| \leq \big(T_{\max}(f - g)\big)(x) \tag{2.2.37}$$

at each point $x \in X$ where $(T_{\max} f)(x) < +\infty$ and $(T_{\max} g)(x) < +\infty$.

[7] with $\operatorname{supp} f$ understood in the sense of [112, Definition 3.8.3]

(ii) For each $r \in (0, \infty)$ there exists a constant $C \in (0, \infty)$ which depends only on C_0, C_1, C_ρ, $\widetilde{C}_\rho$, C_{AR}, C_T, d, γ, p, and r, with the property that the following Cotlar-type inequality holds for every function $f \in L^p(X, \mu)$ and every point $x \in X$:

$$(T_{\max} f)(x) \leq C\Big[\mathcal{M}_X\big(|Tf|^r\big)(x)\Big]^{1/r} + C\Big[\mathcal{M}_X\big(|f|^p\big)(x)\Big]^{1/p}$$
$$= C\mathcal{M}_{X,r}(Tf)(x) + C\mathcal{M}_{X,p}(f)(x), \tag{2.2.38}$$

where $\mathcal{M}_X$ is the ordinary Hardy-Littlewood maximal operator on the space of homogeneous type (X, ρ, μ) and, for each $s \in (0, \infty)$, the symbol $\mathcal{M}_{X,s}$ denotes the L^s-based Hardy-Littlewood maximal operator on (X, ρ, μ) (cf. (A.0.83)).

(iii) The operator

$$T_{\max} : L^p(X, \mu) \longrightarrow L^{p,\infty}(X, \mu) \tag{2.2.39}$$

is well defined, sub-linear[8], bounded[9], and continuous[10]. In fact, there exists some constant $C \in (0, \infty)$ with the property that

$$\big\|T_{\max} f - T_{\max} g\big\|_{L^{p,\infty}(X,\mu)} \leq C\|f - g\|_{L^p(X,\mu)} \text{ for all } f, g \in L^p(X, \mu). \tag{2.2.40}$$

As a consequence,

$$\begin{aligned} &\text{the operator } T_\varepsilon : L^p(X, \mu) \longrightarrow L^{p,\infty}(X, \mu) \text{ is} \\ &\text{well-defined linear bounded for every } \varepsilon > 0, \\ &\text{and } \sup_{\varepsilon>0} \|T_\varepsilon\|_{L^p(X,\mu)\to L^{p,\infty}(X,\mu)} < +\infty. \end{aligned} \tag{2.2.41}$$

(iv) The operator

$$T_{\max} : L^1\Big(X, \tfrac{\mu(x)}{1+\rho(x, x_0)^d}\Big) \cap L^p_{\mathrm{loc}}(X, \mu) \longrightarrow L^{p,\infty}_{\mathrm{loc}}(X, \mu) \tag{2.2.42}$$

is well defined, sub-linear, bounded, and continuous.

Proof We proceed in a series of steps, starting with the following claim.

`Step I:` *Fix $x \in X$ along with $\varepsilon > 0$ and $f \in L^1\Big(X, \frac{\mu}{1+\rho(\cdot, x_0)^d}\Big)$. Then*

[8] in the sense that (2.2.36) holds

[9] meaning that bounded sets are mapped into bounded sets

[10] with respect to the topology induced by the respective (quasi-)norms

$$\int_{y\in X,\,\rho(x,y)\geq\varepsilon}|K(x,y)||f(y)|\,\mathrm{d}\mu(y)\leq C_0\cdot M_{\rho,d,x_0,\varepsilon,x}\cdot\|f\|_{L^1\left(X,\frac{\mu}{1+\rho(\cdot,x_0)^d}\right)},$$

$$\text{where } M_{\rho,d,x_0,\varepsilon,x}:=2\Big(\max\big\{\varepsilon^{-1},C_\rho\max\big\{\widetilde{C}_\rho,\rho(x,x_0)/\varepsilon\big\}\big\}\Big)^d. \tag{2.2.43}$$

To justify this, first note that for each $y\in X$ with $\rho(x,y)\geq\varepsilon$ we may estimate (recall (A.0.25) and (A.0.26))

$$\begin{aligned}\rho(y,x_0)&\leq C_\rho\max\big\{\rho(y,x),\rho(x,x_0)\big\}\\&\leq C_\rho\max\big\{\widetilde{C}_\rho,\rho(x,x_0)/\varepsilon\big\}\,\rho(x,y)\end{aligned} \tag{2.2.44}$$

hence

$$\begin{aligned}1+\rho(y,x_0)^d&\leq 2\max\big\{1,\rho(y,x_0)^d\big\}=2\big(\max\big\{1,\rho(y,x_0)\big\}\big)^d\\&\leq M_{\rho,d,x_0,\varepsilon,x}\cdot\rho(x,y)^d.\end{aligned} \tag{2.2.45}$$

On account of (2.2.45) and (2.2.25) we may then write

$$\begin{aligned}&\int_{y\in X,\,\rho(x,y)\geq\varepsilon}|K(x,y)||f(y)|\,\mathrm{d}\mu(y)\\&\qquad\leq C_0\int_{y\in X,\,\rho(x,y)\geq\varepsilon}\frac{|f(y)|}{\rho(x,y)^d}\,\mathrm{d}\mu(y)\\&\qquad\leq C_0\cdot M_{\rho,d,x_0,\varepsilon,x}\int_X\frac{|f(y)|}{1+\rho(y,x_0)^d}\,\mathrm{d}\mu(y),\end{aligned} \tag{2.2.46}$$

as wanted.

`Step II:` *Fix an arbitrary* $f\in L^1\Big(X,\frac{\mu(x)}{1+\rho(x,x_0)^d}\Big)$. *Then* $O_f:=X\setminus\operatorname{supp}f$ *is an open set in the topological space* (X,τ_ρ), *and the following function is well defined and continuous:*

$$\Phi_f:O_f\to\mathbb{R},\quad \Phi_f(x):=\int_X K(x,y)f(y)\,\mathrm{d}\mu(y)\ \textit{for each}\ x\in O_f. \tag{2.2.47}$$

That O_f is open in (X,τ_ρ) is known from [112, Lemma 3.8.4]. We shall show that Φ_f is a well-defined function which actually belongs to the local Hölder space $\dot{\mathscr{C}}^{\gamma}_{\mathrm{loc}}(O_f,\rho)$ (cf. (A.0.30)). In particular, this justifies the claim that Φ_f is continuous on O_f. To this end, pick an arbitrary point $x\in O_f$ and note that (A.0.124) guarantees the existence of some $r_x\in(0,\infty)$ with the property that $B_\rho(x,r_x)\subseteq O_f$, hence

$$\operatorname{supp}f\subseteq X\setminus B_\rho(x,r_x). \tag{2.2.48}$$

Since from [112, Lemma 3.8.4] we know that $f = 0$ at μ-a.e. point in $X \setminus \operatorname{supp} f$, it follows that

$$\int_X |K(x,y)||f(y)|\,\mathrm{d}\mu(y) = \int_{X\setminus B_\rho(x,r_x)} |K(x,y)||f(y)|\,\mathrm{d}\mu(y) < +\infty, \tag{2.2.49}$$

thanks to (2.2.48), (2.2.43), and the fact that $f \in L^1\Big(X, \frac{\mu(x)}{1+\rho(x,x_0)^d}\Big)$. In turn, (2.2.49) proves that for each $x \in O_f$ the expression $\Phi_f(x)$ in (2.2.47) is defined via an absolutely convergent integral. Thus, $\Phi_f : O_f \to \mathbb{R}$ is a well-defined function.

To show that $\Phi_f \in \dot{\mathscr{C}}^\gamma_{\rm loc}(O_f, \rho)$, fix an arbitrary point $x_* \in O_f$. Recall that, by design, $r_{x_*} \in (0,\infty)$ has the property that $\operatorname{supp} f \subseteq X \setminus B_\rho(x_*, r_{x_*})$. Observe that

$$\forall x' \in B_\rho(x_*, r_{x_*}/C_\rho) \Longrightarrow B_\rho(x', r_{x_*}/C_\rho) \subseteq B_\rho(x_*, r_{x_*}), \tag{2.2.50}$$

hence

$$\forall x' \in B_\rho(x_*, r_{x_*}/C_\rho) \Longrightarrow \operatorname{supp} f \subseteq X \setminus B_\rho(x', r_{x_*}/C_\rho). \tag{2.2.51}$$

Define

$$\varepsilon_* := \frac{r_{x_*}}{C_\rho \cdot C_1} \in (0,\infty). \tag{2.2.52}$$

Such a choice ensures that

$$\left.\begin{array}{r}\forall x, x', y \in X \text{ with } \rho(x',x) < \varepsilon_* \\ \text{and } \rho(x',y) \geq r_{x_*}/C_\rho\end{array}\right\} \Longrightarrow \rho(x',y) \geq C_1\rho(x',x). \tag{2.2.53}$$

Granted this, for every $x' \in B_\rho(x_*, r_{x_*}/C_\rho)$ and each $x \in B_\rho(x', \varepsilon_*)$ we may estimate, by relying on (2.2.26),

$$\begin{aligned}
|\Phi_f(x) - \Phi_f(x')| &= \left|\int_{\operatorname{supp} f} \big(K(x,y) - K(x',y)\big) f(y)\,\mathrm{d}\mu(y)\right| \\
&\leq \int_{X\setminus B_\rho(x', r_{x_*}/C_\rho)} \big|K(x,y) - K(x',y)\big||f(y)|\,\mathrm{d}\mu(y) \\
&\leq C_0\,\rho(x',x)^\gamma \int_{X\setminus B_\rho(x', r_{x_*}/C_\rho)} \frac{|f(y)|}{\rho(x',y)^{d+\gamma}}\,\mathrm{d}\mu(y) \\
&\leq Crho(x',x)^\gamma \int_X \frac{|f(y)|}{1+\rho(x',y)^d}\,\mathrm{d}\mu(y),
\end{aligned} \tag{2.2.54}$$

for some constant $C \in (0,\infty)$ which is independent of x and x'. Note that taking

$$R_* := \min\left\{\frac{\varepsilon_*}{C_\rho \widetilde{C}_\rho}, \frac{r_{x_*}}{C_\rho}\right\} \in (0,\infty) \tag{2.2.55}$$

implies

$$
\begin{aligned}
&B_\rho(x_{*}, R_*) \subseteq B_\rho\big(x_{*}, r_{x_*}/C_\rho\big) \subseteq O_f, \ \text{ and} \\
&\rho(x', x) < \varepsilon_* \ \text{ for every } \ x, x' \in B_\rho(x_{*}, R_*).
\end{aligned}
\tag{2.2.56}
$$

Granted these, we deduce from (2.2.54) that

$$
\sup_{\substack{x,x' \in B_\rho(x_{*},R_*)\\ x\neq x'}} \frac{\big|\Phi_f(x) - \Phi_f(x')\big|}{\rho(x',x)^\gamma} < +\infty,
\tag{2.2.57}
$$

hence $\Phi_f \in \dot{\mathscr{C}}^{\gamma}_{\rm loc}\big(O_f, \rho\big)$, as wanted.

`Step III:` *If for each $f \in L^p(X,\mu)$ we define*

$$
\widetilde{T} f := \begin{cases} Tf \ \textit{in} \ \operatorname{supp} f, \\ \Phi_f \ \textit{in} \ O_f, \end{cases}
\tag{2.2.58}
$$

then the operator $\widetilde{T}$ maps $L^p(X,\mu)$ into $L^{p,\infty}(X,\mu)$. In fact, for every given function $f \in L^p(X,\mu)$ one has

$$
\widetilde{T} f = Tf \ \textit{at } \mu\textit{-a.e. point in } X,
\tag{2.2.59}
$$

and (*compare with* (2.2.31))

$$
(\widetilde{T} f)(x) = \int_X K(x,y) f(y)\, d\mu(y) \ \textit{for every} \ x \in X \setminus \operatorname{supp} f.
\tag{2.2.60}
$$

All these claims are clear from definitions, Steps I-II, and (2.2.31).

`Step IV:` *For each fixed $\varepsilon > 0$, the function*

$$
\begin{aligned}
&X \times X \ni (x,y) \longmapsto \big(\mathbf{1}_{\{y \in X,\, \rho(x,y) > \varepsilon\}}\big)(y) \in \mathbb{R} \\
&\textit{is lower-semicontinuous, hence } \mu \otimes \mu\textit{-measurable.}
\end{aligned}
\tag{2.2.61}
$$

To justify this claim, observe that for every number $\lambda \in \mathbb{R}$ the set of points in $X \times X$ where the given function is $> \lambda$ may be described as

$$
\begin{cases}
\varnothing \ \text{ if } \lambda \geq 1, \\
\big\{(x,y) \in X \times X : \rho(x,y) > \varepsilon\big\} \ \text{ if } \lambda \in [0,1), \\
X \times X \ \text{ if } \lambda < 0.
\end{cases}
\tag{2.2.62}
$$

Thanks to [112, (7.1.18)], all sets appearing in (2.2.62) are open in $\tau_\rho \times \tau_\rho$. This proves that the function (2.2.61) is lower-semicontinuous.

`Step V:` *If* $\mathbb{Q}_+$ *denotes the collection of all positive rational numbers, then for each function* f *belonging to the space* $L^1\big(X, \frac{\mu}{1+\rho(\cdot,x_0)^d}\big)$ *we have*

$$(T_{\max} f)(x) = \sup_{\varepsilon\in\mathbb{Q}_+} \big|(T_\varepsilon f)(x)\big| \text{ for every } x \in X. \tag{2.2.63}$$

To justify this, pick some $f \in L^1\big(X, \frac{\mu}{1+\rho(\cdot,x_0)^d}\big)$. We claim that if $x \in X$ is arbitrary and fixed then for each $\varepsilon \in (0,\infty)$ and each sequence $\{\varepsilon_j\}_{j\in\mathbb{N}} \subseteq (0,\infty)$ such that $\varepsilon_j \searrow \varepsilon$ as $j \to \infty$ we have

$$\lim_{j\to\infty} \int_{y\in X,\,\rho(x,y)>\varepsilon_j} K(x,y)f(y)\,d\mu(y) = \int_{y\in X,\,\rho(x,y)>\varepsilon} K(x,y)f(y)\,d\mu(y). \tag{2.2.64}$$

To justify (2.2.64) note that

$$\{y \in X : \rho(x,y) > \varepsilon_j\} \nearrow \{y \in X : \rho(x,y) > \varepsilon\} \text{ as } j \to \infty, \tag{2.2.65}$$

in the sense that

$$\begin{aligned} &\{y \in X : \rho(x,y) > \varepsilon\} = \bigcup_{j\in\mathbb{N}} \{y \in X : \rho(x,y) > \varepsilon_j\} \text{ and} \\ &\{y \in X : \rho(x,y) > \varepsilon_j\} \subseteq \{y \in X : \rho(x,y) > \varepsilon_{j+1}\} \text{ for every } j \in \mathbb{N}. \end{aligned} \tag{2.2.66}$$

Then (2.2.64) follows from (2.2.65), (2.2.43), and Lebesgue's Dominated Convergence Theorem. What we have just proved amounts to saying that for every function $f \in L^1\big(X, \frac{\mu}{1+\rho(\cdot,x_0)^d}\big)$ we have

$$\lim_{j\to\infty} (T_{\varepsilon_j} f)(x) = (T_\varepsilon f)(x) \text{ for every } x \in X, \tag{2.2.67}$$

whenever $\varepsilon \in (0,\infty)$ and $\{\varepsilon_j\}_{j\in\mathbb{N}} \subseteq (0,\infty)$ are such that $\varepsilon_j \searrow \varepsilon$ as $j \to \infty$. Having established this, (2.2.63) readily follows on account of the density of $\mathbb{Q}_+$ in $(0,\infty)$.

`Step VI:` *Given any* $f \in L^1\big(X, \frac{\mu}{1+\rho(\cdot,x_0)^d}\big)$, *the function* $T_{\max} f$ *is* μ*-measurable.* Indeed, granted (2.2.63) and since the supremum of some countable family of μ-measurable functions is itself a μ-measurable function, it suffices to show that

$$\begin{aligned} &T_\varepsilon f \text{ is a } \mu\text{-measurable function, for each fixed} \\ &\varepsilon \in (0,\infty) \text{ and each fixed } f \in L^1\big(X, \tfrac{\mu}{1+\rho(\cdot,x_0)^d}\big). \end{aligned} \tag{2.2.68}$$

With this goal in mind, fix $\varepsilon \in (0,\infty)$ along with $f \in L^1\big(X, \frac{\mu}{1+\rho(\cdot,x_0)^d}\big)$, and for each $j \in \mathbb{N}$ define

$$\begin{aligned} &G_j : X \times X \longrightarrow \mathbb{R} \text{ given at every point } (x,y) \in X \times X \text{ by} \\ &G_j(x,y) := \big(\mathbf{1}_{B_\rho(x_0,j)}\big)(x)K(x,y)f(y)\big(\mathbf{1}_{\{y\in X,\,\rho(x,y)>\varepsilon\}}\big)(y). \end{aligned} \tag{2.2.69}$$

Then, thanks to assumptions, (2.2.61), and bearing in mind that ρ-balls are open, it follows that G_j is a $\mu \otimes \mu$-measurable function for each $j \in \mathbb{N}$. In addition, from (2.2.69), (2.2.43), and since ρ-balls have finite measure, we see that

$$\int_{X\times X} |G_j(x, y)| \,\mathrm{d}\mu(x)\,\mathrm{d}\mu(y) < \infty. \tag{2.2.70}$$

Granted these properties, Fubini's Theorem (whose applicability is ensured by the fact that (X, μ) is a sigma-finite measure space) then guarantees that

$$\begin{gathered} g_j : X \to \mathbb{R}, \quad g_j(x) := \int_X G_j(x, y)\,\mathrm{d}\mu(y), \quad \forall\, x \in X, \\ \text{is a } \mu\text{-measurable function, for each integer } j \in \mathbb{N}. \end{gathered} \tag{2.2.71}$$

On the other hand, from (2.2.69), (2.2.71), and (2.2.33) it is apparent that for each $j \in \mathbb{N}$ we have

$$g_j = \mathbf{1}_{B_\rho(x_0, j)}\, T_\varepsilon f \text{ everywhere in } X. \tag{2.2.72}$$

In particular, this implies

$$\lim_{j\to\infty} g_j = T_\varepsilon f \text{ pointwise everywhere in } X. \tag{2.2.73}$$

At this stage, the fact that $T_\varepsilon f$ is a μ-measurable function follows from (2.2.73) and (2.2.71).

`Step VII:` *Fix a point $x_* \in X$ and select a constant* (*whose relevance will become clear shortly*)

$$A := \big(1 + C_1 C_\rho\big) C_\rho (\widetilde{C}_\rho)^2. \tag{2.2.74}$$

Also, pick some $\varepsilon > 0$, along with some $f \in L^1_{\mathrm{loc}}(X, \mu)$. For these choices, define a pair of functions, $f_1, f_2 : X \to \mathbb{R}$, by setting at each point $x \in X$

$$f_1(x) := \begin{cases} -f(x) & \text{if } \rho(x_*, x) \le A\,\varepsilon, \\ 0 & \text{if } \rho(x_*, x) > A\,\varepsilon, \end{cases} \tag{2.2.75}$$

and

$$f_2(x) := \begin{cases} 0 & \text{if } \rho(x_*, x) \le A\,\varepsilon, \\ f(x) & \text{if } \rho(x_*, x) > A\,\varepsilon. \end{cases} \tag{2.2.76}$$

Then

$$\begin{gathered} f_1, f_2 \in L^1_{\mathrm{loc}}(X, \mu), \quad |f_1|, |f_2| \le |f| \text{ and } f_2 = f + f_1 \text{ on } X, \\ \text{and also } B_\rho(x_*, \varepsilon) \cap \operatorname{supp} f_2 = \varnothing. \end{gathered} \tag{2.2.77}$$

The properties in the first line of (2.2.77) are clear from definitions (upon recalling that we are presently assuming that $\rho : X \times X \to [0, +\infty)$ is continuous in the product topology $\tau_\rho \times \tau_\rho$). To justify the last property in (2.2.77), pick an arbitrary point $z \in B_\rho(x_*, \varepsilon)$ and consider some point $w \in B_\rho(z, \varepsilon)$. Then (A.0.25)-(A.0.26) and the choice of A imply

$$\rho(w,x_*) \le \widetilde{C}_\rho \rho(x_{**},w) \le C_\rho \widetilde{C}_\rho \max\big\{\rho(x_{**},z),\rho(z,w)\big\}$$
$$\le C_\rho(\widetilde{C}_\rho)^2 \max\big\{\rho(z,x_*),\rho(w,z)\big\} < A\,\varepsilon, \tag{2.2.78}$$

proving that $w \in B_\rho(x_{**},A\,\varepsilon)$. Thus, ultimately,

$$B_\rho(z,\varepsilon) \subseteq B_\rho(x_{**},A\,\varepsilon) \tag{2.2.79}$$

which goes to show that $f_2 = 0$ everywhere on $B_\rho(z,\varepsilon)$. Having established this, the last property in (2.2.77) follows from [112, Lemma 3.8.4].

`Step VIII:` *Fix a point* $x_* \in X$ *and pick some* $\varepsilon > 0$. *With* $C_1, A \in (0,\infty)$ *as in* (2.2.26) *and* (2.2.74), *respectively, we have*

$$\left.\begin{array}{l} x,x' \in B_\rho(x_{**},\varepsilon) \text{ and} \\ y \in X,\ \rho(x_{**},y) > A\,\varepsilon \end{array}\right\} \Longrightarrow \left\{\begin{array}{l} \rho(x',y) \ge C_1 \rho(x',x), \\ \rho(x_{**},y) \le C_\rho\widetilde{C}_\rho\,\rho(x',y). \end{array}\right. \tag{2.2.80}$$

Indeed, if x, x', y are as in the left side of (2.2.80) then

$$\rho(x',x) \le C_\rho \max\big\{\rho(x',x_*),\rho(x_{**},x)\big\}$$
$$\le C_\rho\widetilde{C}_\rho \max\big\{\rho(x_{**},x'),\rho(x_{**},x)\big\} < C_\rho\widetilde{C}_\rho\,\varepsilon. \tag{2.2.81}$$

Since $\rho(x_{**},y) > A\,\varepsilon$ by assumption, whereas $\rho(x_{**},x') \le \widetilde{C}_\rho\,\rho(x',x_*) < \widetilde{C}_\rho\,\varepsilon$ and $A\,\varepsilon > C_\rho\widetilde{C}_\rho\,\varepsilon$, the quasi-subadditive estimate

$$\rho(x_{**},y) \le C_\rho \max\big\{\rho(x_{**},x'),\rho(x',y)\big\} \tag{2.2.82}$$

reduces simply to

$$\rho(x_{**},y) \le C_\rho\,\rho(x',y). \tag{2.2.83}$$

Given that

$$\rho(x',y) \le \widetilde{C}_\rho\,\rho(x',y), \tag{2.2.84}$$

we may then conclude from (2.2.83) and assumptions that

$$A\,\varepsilon < C_\rho\widetilde{C}_\rho\,\rho(x',y). \tag{2.2.85}$$

At this stage, the first estimate in the right side of (2.2.80) follows from (2.2.81), (2.2.85), and (2.2.74), while the second estimate in the right side of (2.2.80) follows from (2.2.83) and (2.2.84).

`Step IX:` *Fix a point* $x_* \in X$ *and pick some* $\varepsilon > 0$. *Also, select* $f \in L^p(X,\mu)$ *and define* f_1, f_2 *as in* (2.2.75)-(2.2.76). *Then there exists a constant* $C \in (0,\infty)$ *which* *depends only on* C_0, C_1, C_ρ, $\widetilde{C}_\rho$, C_{AR}, d, *and* γ, *with the property that*

$$\big|(\widetilde{T} f_2)(x) - (\widetilde{T} f_2)(x')\big| \leq C\big(\mathcal{M}_X f\big)(x_*) \text{ for all } x, x' \in B_\rho(x_*, \varepsilon). \tag{2.2.86}$$

To prove this, we begin by noting that, in light of (2.2.60), the properties listed in (2.2.77) ensure that we may express

$$\begin{aligned}(\widetilde{T} f_2)(x) &= \int_X K(x, y) f_2(y) \, d\mu(y) \\ &= \int_{y \in X,\, \rho(x_*, y) > A\varepsilon} K(x, y) f(y) \, d\mu(y) \text{ for all } x \in B_\rho(x_*, \varepsilon).\end{aligned} \tag{2.2.87}$$

Consequently, for every $x, x' \in B_\rho(x_*, \varepsilon)$ we may estimate based on (2.2.80), (2.2.87) (used both for x and x'), (2.2.26), and (2.2.81),

$$\begin{aligned}&\big|(\widetilde{T} f_2)(x) - (\widetilde{T} f_2)(x')\big| \\ &= \left| \int_{y \in X,\, \rho(x_*, y) > A\varepsilon} \big(K(x, y) - K(x', y)\big) f(y) \, d\mu(y) \right| \\ &\leq \int_{y \in X,\, \rho(x_*, y) > A\varepsilon} \big|K(x, y) - K(x', y)\big| |f(y)| \, d\mu(y) \\ &\leq C_0 (C_\rho \widetilde{C}_\rho)^{d+2\gamma} \varepsilon^\gamma \int_{y \in X,\, \rho(x_*, y) > A\varepsilon} \frac{|f(y)|}{\rho(x_*, y)^{d+\gamma}} \, d\mu(y) \\ &= C_0 (C_\rho \widetilde{C}_\rho)^{d+2\gamma} \varepsilon^\gamma \sum_{j=0}^{\infty} \int_{2^{j+1} A\varepsilon \geq \rho(x_*, y) > 2^j A\varepsilon} \frac{|f(y)|}{\rho(x_*, y)^{d+\gamma}} \, d\mu(y) \\ &\leq C_0 (C_\rho \widetilde{C}_\rho)^{d+2\gamma} \varepsilon^\gamma \sum_{j=0}^{\infty} \frac{1}{\big(2^j A\varepsilon\big)^{d+\gamma}} \int_{2^{j+1} A\varepsilon \geq \rho(x_*, y)} |f(y)| \, d\mu(y) \\ &\leq C_0 (C_\rho \widetilde{C}_\rho)^{d+2\gamma} \varepsilon^\gamma \sum_{j=0}^{\infty} \frac{\mu\Big(B_\rho\big(x_*, 2^{j+2} A\varepsilon\big)\Big)}{\big(2^j A\varepsilon\big)^{d+\gamma}} ⨍_{B_\rho(x_*, 2^{j+2} A\varepsilon)} |f(y)| \, d\mu(y) \\ &\leq C_0 (C_\rho \widetilde{C}_\rho)^{d+2\gamma} \varepsilon^\gamma \sum_{j=0}^{\infty} \frac{\mu\Big(B_\rho\big(x_*, 2^{j+2} (C_\rho)^2 A\varepsilon\big)\Big)}{\big(2^j A\varepsilon\big)^{d+\gamma}} \big(\mathcal{M}_X f\big)(x_*) \\ &\leq 4^d A^{-\gamma} C_{\text{AR}} C_0 (C_\rho)^{3d+2\gamma} (\widetilde{C}_\rho)^{d+2\gamma} \big(\mathcal{M}_X f\big)(x_*) \sum_{j=0}^{\infty} 2^{-j\gamma} \\ &= C\big(\mathcal{M}_X f\big)(x_*),\end{aligned} \tag{2.2.88}$$

where the last equality defines the constant $C \in (0, \infty)$.

`Step X:` *There exists a constant $C \in (0, \infty)$ which depends only on C_0, C_1, C_ρ, $\widetilde{C}_\rho$, C_{AR}, and d, with the property that, with all objects involved retaining their earlier significance, there holds*

$$\big|(\widetilde{T} f_2)(x) - (T_\varepsilon f)(x)\big| \leq C\big(\mathcal{M}_X f\big)(x_*) \text{ for every } x \in B_\rho(x_*, \varepsilon). \tag{2.2.89}$$

To see that this is the case, observe first that for every $x \in B_\rho(x_*, \varepsilon)$ we have

$$\begin{aligned}(\widetilde{T} f_2)(x) - (T_\varepsilon f)(x) &= \int_{y \in X,\, \rho(x_*, y) > A\varepsilon} K(x, y) f(y)\, d\mu(y) \\ &\quad - \int_{y \in X,\, \rho(x, y) > \varepsilon} K(x, y) f(y)\, d\mu(y) \\ &= - \int_{\rho(x, y) > \varepsilon,\, \rho(x_*, y) \leq A\varepsilon} K(x, y) f(y)\, d\mu(y). \end{aligned} \tag{2.2.90}$$

Above, the first equality follows from (2.2.60) and (2.2.33). The second equality is a consequence of the fact that

$$\left.\begin{array}{r} x \in B_\rho(x_*, \varepsilon) \text{ and } y \in X \\ \text{such that } \rho(x_*, y) > A\varepsilon \end{array}\right\} \Longrightarrow \rho(x, y) > \varepsilon. \tag{2.2.91}$$

Indeed, if x, y are as in the left side of (2.2.91), then the estimate

$$A\varepsilon < \rho(x_*, y) \leq C_\rho \max\big\{\rho(x_*, x), \rho(x, y)\big\} \tag{2.2.92}$$

reduces simply to

$$A\varepsilon < \rho(x_*, y) \leq C_\rho\, \rho(x, y), \tag{2.2.93}$$

given that $\rho(x_*, x) \leq \widetilde{C}_\rho\, \rho(x, x_*) < \widetilde{C}_\rho\, \varepsilon$ and $A > C_\rho \widetilde{C}_\rho$. Bearing in mind that $\widetilde{C}_\rho \geq 1$ (cf. (A.0.26)) forces $A > C_\rho$, we then deduce from (2.2.93) that $\rho(x, y) > \varepsilon$. This justifies (2.2.91) which, in turn, concludes the proof of (2.2.90).

Combining (2.2.90) with (2.2.25) then permits us to estimate, for every point $x \in B_\rho(x_*, \varepsilon)$,

$$
\begin{aligned}
\big|(\widetilde{T} f_2)(x) - (T_\varepsilon f)(x)\big| &\leq C_0 \int_{\rho(x,y)>\varepsilon,\, \rho(x_*,y)\leq A\,\varepsilon} \frac{|f(y)|}{\rho(x,y)^d}\, \mathrm{d}\mu(y) \\
&\leq C_0 \cdot \varepsilon^{-d} \int_{\rho(x_*,y)\leq A\,\varepsilon} |f(y)|\, \mathrm{d}\mu(y) \\
&\leq C_0 \cdot \varepsilon^{-d} \int_{\rho(y,x_*)<2\widetilde{C}_\rho\, A\,\varepsilon} |f(y)|\, \mathrm{d}\mu(y) \\
&\leq C_0 \cdot \varepsilon^{-d} \mu\Big(B_\rho(x_*, 2\widetilde{C}_\rho\, A\,\varepsilon)\Big) ⨍_{B_\rho(x_*,2\widetilde{C}_\rho\, A\,\varepsilon)} |f(y)|\, \mathrm{d}\mu(y) \\
&\leq C_0 \cdot C_{\mathrm{AR}} \big(2\widetilde{C}_\rho\, A\big)^d (\mathcal{M}_X f)(x_*),
\end{aligned}
\tag{2.2.94}
$$

hence choosing $C := C_0 C_{\mathrm{AR}} \big(2\widetilde{C}_\rho\, A\big)^d$ finishes the proof of (2.2.89).

`Step XI:` *Given any $r \in (0,p)$, there exists a constant $C \in (0,\infty)$ which depends only on C_1, C_ρ, $\widetilde{C}_\rho$, C_{AR}, C_T, d, p, and r, with the property that, if all objects involved retaining their earlier significance,*

$$
\varepsilon^{-d} \int_{B_\rho(x_*,\varepsilon)} \big|(\widetilde{T} f_1)(x)\big|^r \, \mathrm{d}\mu(x) \leq C\Big(\mathcal{M}_X(|f|^p)(x_*)\Big)^{r/p}. \tag{2.2.95}
$$

To justify this estimate, based on (2.2.59), [112, (6.2.37)], and (2.2.30), we write

$$
\begin{aligned}
\int_{B_\rho(x_*,\varepsilon)} \big|(\widetilde{T} f_1)(x)\big|^r \, \mathrm{d}\mu(x) &= \int_{B_\rho(x_*,\varepsilon)} \big|(T f_1)(x)\big|^r \, \mathrm{d}\mu(x) \\
&\leq \Big(\frac{p}{p-r}\Big) \mu\Big(B_\rho(x_*,\varepsilon)\Big)^{1-\frac{r}{p}} \|T f_1\|^r_{L^{p,\infty}(X,\mu)} \\
&\leq (C_T)^r \Big(\frac{p}{p-r}\Big) (C_{\mathrm{AR}})^{1-\frac{r}{p}} \varepsilon^{d(1-\frac{r}{p})} \|f_1\|^r_{L^p(X,\mu)} \\
&= (C_T)^r \Big(\frac{p}{p-r}\Big) (C_{\mathrm{AR}})^{1-\frac{r}{p}} \varepsilon^{d(1-\frac{r}{p})} \bigg(\int_{y\in X,\, \rho(x_*,y)\leq A\,\varepsilon} |f(y)|^p \, \mathrm{d}\mu(y)\bigg)^{\frac{r}{p}} \\
&\leq (C_T)^r \Big(\frac{p}{p-r}\Big) C_{\mathrm{AR}} \big(2\widetilde{C}_\rho\, A\big)^{\frac{dr}{p}} \varepsilon^d \Big(\mathcal{M}_X(|f|^p)(x_*)\Big)^{\frac{r}{p}},
\end{aligned}
\tag{2.2.96}
$$

from which the desired conclusion follows.

`Step XII:` *The end-game in the proof of Proposition 2.2.5.*

Proof of claims in part (i): For starters, observe that given any number $\varepsilon > 0$, any point $x \in X$, and any function $f \in L^1\big(X, \frac{\mu}{1+\rho(\cdot, x_0)^d}\big)$, the integral defining $(T_\varepsilon f)(x)$ in (2.2.33) is absolutely convergent thanks to (2.2.43). In concert with (2.2.68), this shows that each truncated operator $T_\varepsilon : L^q(X,\mu) \to \mathscr{M}_+(X,\mu)$ is well defined and linear. From this and (2.2.34) we then conclude that the maximal operator $T_{\max}$ is well defined and sub-linear in the context of (2.2.35).

Proof of claims in part (ii): First assume some integrability exponent $r \in (0, p)$ has been given. Fix a function $f \in L^p(X, \mu)$, some point $x_* \in X$, and some number $\varepsilon > 0$. With f_1, f_2 defined as in (2.2.75)-(2.2.76) we then have

$$\begin{aligned} \big|(T_\varepsilon f)(x')\big| &\leq \big|(T_\varepsilon f)(x') - (\widetilde{T} f_2)(x')\big| + \big|(\widetilde{T} f_2)(x') - (\widetilde{T} f_2)(x)\big| + \big|(\widetilde{T} f_2)(x)\big| \\ &\leq C\big(\mathcal{M}_X f\big)(x_*) + \big|(\widetilde{T} f_2)(x)\big| \text{ for every } x, x' \in B_\rho(x_*, \varepsilon), \end{aligned} \tag{2.2.97}$$

where the last inequality is based on (2.2.89) (written for x' in place of x) and (2.2.86). Based on (2.2.59), (2.2.77), and (2.2.29), we may further write

$$\begin{aligned} |\widetilde{T} f_2| &= |T f_2| = |T(f + f_1)| \leq C_T\big(|Tf| + |T f_1|\big) \\ &= C_T\Big(|\widetilde{T} f| + |\widetilde{T} f_1|\Big) \text{ at } \mu\text{-a.e. point in } X. \end{aligned} \tag{2.2.98}$$

Combining (2.2.97) with (2.2.98) then proves that

$$\begin{aligned} &\text{for every point } x' \text{ belonging to } B_\rho(x_*, \varepsilon), \text{ the estimate} \\ &\big|(T_\varepsilon f)(x')\big| \leq C\big(\mathcal{M}_X f\big)(x_*) + C_T\big|(\widetilde{T} f)(x)\big| + C_T\big|(\widetilde{T} f_1)(x)\big| \\ &\text{holds for } \mu\text{-a.e. } x \in B_\rho(x_*, \varepsilon). \end{aligned} \tag{2.2.99}$$

Raise both sides of the estimate in (2.2.99) to the power r, then average in the variable x over the ball $B_\rho(x_*, \varepsilon)$. In concert with (2.2.1), [112, (7.1.21)], and (A.0.84), this leads to the conclusion that for every point $x' \in B_\rho(x_*, \varepsilon)$ we have

$$\begin{aligned} \big|(T_\varepsilon f)(x')\big|^r &\leq C\big(\mathcal{M}_X f\big)(x_*)^r + C\mathcal{M}_X\big(|\widetilde{T} f|^r\big)(x_*) \\ &\quad + C\varepsilon^{-d} \int_{B_\rho(x_*, \varepsilon)} \big|(\widetilde{T} f_1)(x)\big|^r \, d\mu(x), \end{aligned} \tag{2.2.100}$$

where $C \in (0, \infty)$ depends only on C_0, C_1, C_ρ, $\widetilde{C}_\rho$, C_{AR}, C_T, d, γ, p, and r. By combining (2.2.100) with (2.2.95) we therefore arrive at the conclusion that

$$\begin{aligned} \big|(T_\varepsilon f)(x')\big|^r &\leq C\big(\mathcal{M}_X f\big)(x_*)^r + C\mathcal{M}_X\big(|\widetilde{T} f|^r\big)(x_*) \\ &\quad + C\Big(\mathcal{M}_X\big(|f|^p\big)(x_*)\Big)^{r/p} \text{ for every } x' \in B_\rho(x_*, \varepsilon), \end{aligned} \tag{2.2.101}$$

with the constant $C \in (0, \infty)$ having the same nature as above. Since Hölder's inequality (recall that $p \in [1, \infty)$) gives

$$\big(\mathcal{M}_X f\big)(x_*) \leq \Big(\mathcal{M}_X\big(|f|^p\big)(x_*)\Big)^{1/p}, \tag{2.2.102}$$

by keeping (2.2.59) in mind we may re-write (2.2.101) simply as

$$\big|(T_\varepsilon f)(x')\big|^r \leq C\mathcal{M}_X\big(|Tf|^r\big)(x_*) + C\Big(\mathcal{M}_X\big(|f|^p\big)(x_*)\Big)^{r/p}$$
$$\text{for every point } x' \in B_\rho(x_*, \varepsilon), \tag{2.2.103}$$

where $C \in (0, \infty)$ has the same nature as before. Specializing (2.2.103) to the case when $x' := x_*$, then taking the supremum over $\varepsilon \in (0, \infty)$ establishes (2.2.38) (written with x_* in place of x) in the case when $r \in (0, p)$. Finally, this version of (2.2.38) self-improves to also allow $r \in [p, \infty)$. Indeed, having fixed $r \in [p, \infty)$, with $r_0 \in (0, p)$ Hölder's inequality gives

$$\Big[\mathcal{M}_X\big(|Tf|^{r_0}\big)(x)\Big]^{1/r_0} \leq \Big[\mathcal{M}_X\big(|Tf|^{r}\big)(x)\Big]^{1/r}, \qquad \forall x \in X. \tag{2.2.104}$$

Proof of claims in part (iii): Given an arbitrary $f \in L^p(X, \mu)$, the function $T_{\max} f$ is μ-measurable and non-negative, thanks to (2.2.35). Granted this, we may then invoke (2.2.38) for some $r \in (0, p)$, [112, (6.2.16)], [112, (6.2.19)], [112, Corollary 7.6.3], and (2.2.30), in order to estimate (bearing in mind that our present choice of r entails $p/r \in (1, \infty)$, and that $|f|^p$ belongs to $L^1(X, \mu)$)

$$\begin{aligned}
\|T_{\max} f\|_{L^{p,\infty}(X,\mu)} &\leq C\Big\|\Big[\mathcal{M}_X\big(|Tf|^r\big)\Big]^{1/r}\Big\|_{L^{p,\infty}(X,\mu)} + C\Big\|\Big[\mathcal{M}_X\big(|f|^p\big)\Big]^{1/p}\Big\|_{L^{p,\infty}(X,\mu)} \\
&= C\Big\|\mathcal{M}_X\big(|Tf|^r\big)\Big\|_{L^{p/r,\infty}(X,\mu)}^{1/r} + C\Big\|\mathcal{M}_X\big(|f|^p\big)\Big\|_{L^{1,\infty}(X,\mu)}^{1/p} \\
&\leq C\big\||Tf|^r\big\|_{L^{p/r,\infty}(X,\mu)}^{1/r} + C\big\||f|^p\big\|_{L^1(X,\mu)}^{1/p} \\
&= C\|Tf\|_{L^{p,\infty}(X,\mu)} + C\|f\|_{L^p(X,\mu)} \\
&\leq C\|f\|_{L^p(X,\mu)},
\end{aligned} \tag{2.2.105}$$

for some constant $C \in (0, \infty)$ which depends only on C_0, C_1, C_ρ, $\widetilde{C}_\rho$, $C_{\rm AR}$, C_T, d, γ, p, and r. This concludes the proof of the fact that the operator (2.2.39) is well defined, sub-linear, and bounded. Granted this, the claims in (2.2.41) readily follow in view of (2.2.34). Next, (2.2.40) is proved observing that (2.2.37), [112, (6.2.16)], and (2.2.105) imply the existence of a constant $C \in (0, \infty)$ such that, as desired,

$$\begin{aligned}
\big\|T_{\max} f - T_{\max} g\big\|_{L^{p,\infty}(X,\mu)} &\leq \big\|T_{\max}(f - g)\big\|_{L^{p,\infty}(X,\mu)} \\
&\leq C\|f - g\|_{L^p(X,\mu)}
\end{aligned} \tag{2.2.106}$$

for every $f, g \in L^p(X, \mu)$. In particular, this shows that $T_{\max}$ is also continuous in the context of (2.2.39).

Proof of claims in part (iv): Fix a sufficiently large constant $C_* \in (0, \infty)$, depending only on ρ, and pick some radius $R \in (0, \infty)$. For an arbitrarily chosen function $f \in L^1\big(X, \frac{\mu(x)}{1+\rho(x,x_o)^d}\big)$, define

$$\begin{aligned} g_R &:= \Big[T_{\max}\big(f\mathbf{1}_{X\setminus B_\rho(x_o,C_*R)}\big)\Big]\Big|_{X\cap B_\rho(x_o,R)} \quad \text{and} \\ h_R &:= \Big[T_{\max}\big(f\mathbf{1}_{B_\rho(x_o,C_*R)}\big)\Big]\Big|_{X\cap B_\rho(x_o,R)}. \end{aligned} \tag{2.2.107}$$

Note that there exists $C_R \in (0, \infty)$ with the property that for each $x \in B_\rho(x_o, R)$ we have

$$\begin{aligned} \int_X |K(x,y)|\big|\big(f\mathbf{1}_{X\setminus B_\rho(x_o,C_*R)}\big)(y)\big|\,d\mu(y) \\ \leq C_R \int_{X\setminus B_\rho(x_o,C_*R)} \frac{|f(y)|}{\rho(x,y)^d}\,d\mu(y) \\ \leq C_R \int_X \frac{|f(y)|}{1+\rho(x,x_o)^d}\,d\mu(y) < +\infty. \end{aligned} \tag{2.2.108}$$

Granted this, it follows that

$$\begin{aligned} g_R \in L^\infty\big(B_\rho(x_o,R),\mu\big) \quad \text{and} \\ \|g_R\|_{L^\infty(B_\rho(x_o,R),\mu)} \leq C_R\|f\|_{L^1\left(X,\frac{\mu(x)}{1+\rho(x,x_o)^d}\right)}. \end{aligned} \tag{2.2.109}$$

From [112, (6.2.27)] we know that the following continuous embeddings hold:

$$L^\infty\big(B_\rho(x_o,R),\mu\big) \hookrightarrow L^p\big(B_\rho(x_o,R),\mu\big) \hookrightarrow L^{p,\infty}\big(B_\rho(x_o,R),\mu\big). \tag{2.2.110}$$

In concert with (2.2.109), this shows that

$$\begin{aligned} g_R \in L^{p,\infty}\big(B_\rho(x_o,R),\mu\big) \quad \text{and} \\ \|g_R\|_{L^{p,\infty}(B_\rho(x_o,R),\mu)} \leq C_R\|f\|_{L^1\left(X,\frac{\mu(x)}{1+\rho(x,x_o)^d}\right)}. \end{aligned} \tag{2.2.111}$$

Next, since

$$\begin{aligned} f\mathbf{1}_{B_\rho(x_o,C_*R)} \in L^p(X,\mu) \quad \text{with} \\ \big\|f\mathbf{1}_{B_\rho(x_o,C_*R)}\big\|_{L^p(X,\mu)} = \big\|f\big|_{B_\rho(x_o,C_*R)}\big\|_{L^p(B_\rho(x_o,C_*R),\mu)}, \end{aligned} \tag{2.2.112}$$

we conclude from (2.2.41) that

$$\begin{aligned} T\big(f\mathbf{1}_{B_\rho(x_o,C_*R)}\big) \in L^{p,\infty}(X,\mu) \quad \text{and we have} \\ \big\|T\big(f\mathbf{1}_{B_\rho(x_o,C_*R)}\big)\big\|_{L^{p,\infty}(X,\mu)} \leq C\big\|f\big|_{B_\rho(x_o,C_*R)}\big\|_{L^p(B_\rho(x_o,C_*R),\mu)}. \end{aligned} \tag{2.2.113}$$

Based on (2.2.113) and the second line in [112, (6.2.17)] we deduce that

$$h_R \in L^{p,\infty}\big(B_\rho(x_o,R),\mu\big) \text{ and } \|h_R\|_{L^{p,\infty}(B_\rho(x_o,R),\mu)} \leq C\big\|f\big|_{B_\rho(x_o,C_*R)}\big\|_{L^p(B_\rho(x_o,C_*R),\mu)}. \tag{2.2.114}$$

In view of the fact that, thanks to what we have already proved in item *(i)*,

$$T_{\max}f \text{ is a } \mu\text{-measurable function on } X \text{ satisfying } 0 \leq (T_{\max}f)\big|_{B_\rho(x_o,R)} \leq g_R + h_R \text{ on } B_\rho(x_o,R), \tag{2.2.115}$$

we ultimately conclude from (2.2.111), (2.2.114), and (2.2.115) that the operator $T_{\max}$ in (2.2.42) is indeed well defined, sub-linear, bounded, and continuous. □

There are a number of other related versions of Cotlar's inequality presented in Proposition 2.2.5, which are of interest. One such version, valid for a larger class of functions, is discussed in the corollary below.

Corollary 2.2.6 *Assume (X,ρ,μ) is a d-Ahlfors regular quasi-metric space for some $d \in (0,\infty)$ with the property that the quasi-distance $\rho : X \times X \to [0,+\infty)$ is continuous in the product topology $\tau_\rho \times \tau_\rho$. Demote by $\mathscr{M}(X,\mu)$ the space of all μ-measurable functions defined on X, and by $\mathscr{M}_+(X,\mu)$ the collection of non-negative functions in $\mathscr{M}(X,\mu)$. Suppose $K : X \times X \setminus \mathrm{diag}(X) \to \mathbb{R}$ is a standard kernel in the first variable (i.e., K is $\mu \otimes \mu$-measurable and* (2.2.25)-(2.2.26) *are satisfied). Fix a reference point $x_0 \in X$ and assume that*

$$T : L^1\Big(X, \tfrac{\mu(x)}{1+\rho(x,x_0)^d}\Big) \longrightarrow \mathscr{M}(X,\mu) \tag{2.2.116}$$

is an operator[11] *which satisfies the following properties:*

(T1) there exists $C_T \in [0,\infty)$ such that for every $f, g \in L^1\Big(X, \tfrac{\mu(x)}{1+\rho(x,x_0)^d}\Big)$ one has

$$|T(f+g)| \leq C_T\big(|Tf| + |Tg|\big) \ \ \mu\text{-a.e. in } X; \tag{2.2.117}$$

(T2) with the constant C_T as above, for every $f \in L^1(X,\mu) \subseteq L^1\Big(X, \tfrac{\mu(x)}{1+\rho(x,x_0)^d}\Big)$ there holds

$$\|Tf\|_{L^{1,\infty}(X,\mu)} \leq C_T\|f\|_{L^1(X,\mu)}; \tag{2.2.118}$$

(T3) for each $f \in L^1\Big(X, \tfrac{\mu(x)}{1+\rho(x,x_0)^d}\Big)$ one has[12]

$$(Tf)(x) = \int_X K(x,y)f(y)\,d\mu(y) \ \ \text{for } \mu\text{-a.e. } x \in X \setminus \operatorname{supp} f. \tag{2.2.119}$$

---

[11] again, not necessarily linear

[12] with $\operatorname{supp} f$ understood in the sense of [112, Definition 3.8.3]

Finally, for each $f \in L^1\Big(X, \frac{\mu(x)}{1+\rho(x,x_0)^d}\Big)$ *set*

$$(T_{\max} f)(x) := \sup_{\varepsilon>0}\Big|\int_{y\in X,\,\rho(x,y)>\varepsilon} K(x,y)f(y)\,\mathrm{d}\mu(y)\Big| \text{ for every } x\in X. \tag{2.2.120}$$

Then for each $r \in (0,\infty)$ *there exists a constant* $C \in (0,\infty)$ *which depends only on* C_0, C_1, C_ρ, $\widetilde{C}_\rho$, C_{AR}, C_T, d, γ, *and* r, *with the property that the following Cotlar-type inequality holds:*

$$\begin{aligned}&(T_{\max} f)(x) \le C\Big[\big(\mathcal{M}_X(|Tf|^r)\big)(x)\Big]^{1/r} + C\big(\mathcal{M}_X f\big)(x)\\ &\text{for every } f\in L^1\Big(X, \tfrac{\mu(x)}{1+\rho(x,x_0)^d}\Big) \text{ and every } x\in X,\end{aligned} \tag{2.2.121}$$

where $\mathcal{M}_X$ *is the Hardy-Littlewood maximal operator on the space of homogeneous type* (X,ρ,μ).

Proof We retrace the steps taken in the proof of Proposition 2.2.5. Specifically, Steps I-II and Steps IV-VIII go through unchanged, and in place of Step III we now have

`New Step III:` *If for each* $f \in L^1\Big(X, \frac{\mu(x)}{1+\rho(x,x_0)^d}\Big)$ *we define*

$$\widetilde{T} f := \begin{cases} Tf \text{ in } \operatorname{supp} f,\\ \Phi_f \text{ in } O_f,\end{cases} \tag{2.2.122}$$

then $\widetilde{T}$ *maps* $L^1\Big(X, \frac{\mu(x)}{1+\rho(x,x_0)^d}\Big)$ *into* $\mathscr{M}(X,\mu)$. *In fact, for every given function* $f \in L^1\Big(X, \frac{\mu(x)}{1+\rho(x,x_0)^d}\Big)$ *one has*

$$\widetilde{T} f = Tf \text{ at } \mu\text{-a.e. point in } X, \tag{2.2.123}$$

and (compare with (2.2.119)*)*

$$(\widetilde{T} f)(x) = \int_X K(x,y)f(y)\,\mathrm{d}\mu(y) \text{ for every } x\in X\setminus \operatorname{supp} f. \tag{2.2.124}$$

Indeed, all these claims are clear from definitions, Steps I-II, and (2.2.119). With this alteration, the old Steps IX-X now go through unchanged. Here are the new formulations of the remaining steps, namely Steps XI-XII:

`New Step XI:` *Fix a point* $x_* \in X$ *and pick some* $\varepsilon > 0$. *Also, select an arbitrary function* $f \in L^1\Big(X, \frac{\mu(x)}{1+\rho(x,x_0)^d}\Big)$ *and define* f_1, f_2 *as in* (2.2.75)-(2.2.76). *Then for each* $r \in (0,1)$ *there exists a constant* $C \in (0,\infty)$ *which depends only on* C_1, C_ρ, $\widetilde{C}_\rho$, C_{AR}, C_T, d, *and* r, *with the property that*

$$\varepsilon^{-d}\int_{B_\rho(x_*,\varepsilon)}\big|(\widetilde{T}f_1)(x)\big|^r\,d\mu(x)\leq C\Big[\big(\mathcal{M}_X f\big)(x_*)\Big]^r. \tag{2.2.125}$$

Indeed, thanks to (2.2.123), [112, (6.2.37)], the fact that (X,ρ,μ) is a d-Ahlfors regular quasi-metric space, and (2.2.118), we may estimate

$$\begin{aligned}
\varepsilon^{-d}\int_{B_\rho(x_*,\varepsilon)}\big|(\widetilde{T}f_1)(x)\big|^r\,d\mu(x) &= \varepsilon^{-d}\int_{B_\rho(x_*,\varepsilon)}\big|(Tf_1)(x)\big|^r\,d\mu(x)\\
&\leq \varepsilon^{-d}\Big(\frac{1}{1-r}\Big)\mu\Big(B_\rho(x_*,\varepsilon)\Big)^{1-r}\|Tf_1\|^r_{L^{1,\infty}(X,\mu)}\\
&\leq (C_T)^r\Big(\frac{1}{1-r}\Big)(C_{\rm AR})^{1-r}\varepsilon^{-dr}\|f_1\|^r_{L^1(X,\mu)}\\
&= (C_T)^r\Big(\frac{1}{1-r}\Big)(C_{\rm AR})^{1-r}\varepsilon^{-dr}\Bigg(\int_{y\in X,\,\rho(x_*,y)\leq A\,\varepsilon}|f(y)|\,d\mu(y)\Bigg)^r\\
&\leq (C_T)^r\Big(\frac{1}{1-r}\Big)C_{\rm AR}\big(2\widetilde{C}_\rho A\big)^{dr}\Big[\big(\mathcal{M}_X f\big)(x_*)\Big]^r,
\end{aligned} \tag{2.2.126}$$

as wanted.

`New Step XII:` *The end-game in the proof of* (2.2.121). Pick an arbitrary function $f\in L^1\Big(X,\frac{\mu(x)}{1+\rho(x,x_0)^d}\Big)$, some point $x_*\in X$, some number $\varepsilon>0$, and define f_1,f_2 as in (2.2.75)-(2.2.76). Also, recall the truncated singular integral operator T_ε from (2.2.33). First, fix some integrability exponent $r\in(0,1)$. Then the same argument as in (2.2.97)-(2.2.100) leads to the conclusion that for every point $x'\in B_\rho(x_*,\varepsilon)$ we have

$$\begin{aligned}
\big|(T_\varepsilon f)(x')\big|^r &\leq C\Big[\big(\mathcal{M}_X f\big)(x_*)\Big]^r+C\mathcal{M}_X\big(|\widetilde{T}f|^r\big)(x_*)\\
&\quad +C\varepsilon^{-d}\int_{B_\rho(x_*,\varepsilon)}\big|(\widetilde{T}f_1)(x)\big|^r\,d\mu(x),
\end{aligned} \tag{2.2.127}$$

where $C\in(0,\infty)$ depends only on C_0, C_1, C_ρ, $\widetilde{C}_\rho$, $C_{\rm AR}$, C_T, d, γ, and r. By combining (2.2.127) with (2.2.125) and keeping (2.2.123) in mind we therefore obtain

$$\begin{aligned}
\big|(T_\varepsilon f)(x')\big|^r \leq C\Big[\big(\mathcal{M}_X f\big)(x_*)\Big]^r+C\mathcal{M}_X\big(|Tf|^r\big)(x_*)&\\
\text{for all }\ x'\in B_\rho(x_*,\varepsilon),&
\end{aligned} \tag{2.2.128}$$

with the constant $C\in(0,\infty)$ having the same nature as above. Specializing (2.2.128) to the case when $x':=x_*$, then taking the supremum over $\varepsilon\in(0,\infty)$ establishes (2.2.121) (written with x_* in place of x) in the case when $r\in(0,1)$. Finally, this version of (2.2.121) self-improves as to also allow $r\in[1,\infty)$. Indeed, having fixed $r\in[1,\infty)$, choose $r_0\in(0,1)$ and apply Hölder's inequality to write

$$\Big[\mathcal{M}_X\big(|Tf|^{r_0}\big)(x)\Big]^{1/r_0} \leq \Big[\mathcal{M}_X\big(|Tf|^{r}\big)(x)\Big]^{1/r}, \qquad \forall x \in X, \tag{2.2.129}$$

from which the desired conclusion follows. □

For operators which are actually linear, the version of Cotlar's inequality contained in Proposition 2.2.5 takes a more streamlined form. It is also of interest to assume boundedness at two different levels. Here is a result of this flavor.

Proposition 2.2.7 *Suppose (X, ρ, μ) is a d-Ahlfors regular quasi-metric space for some $d \in (0, \infty)$ with the property that the quasi-distance $\rho : X \times X \to [0, +\infty)$ is continuous in the product topology $\tau_\rho \times \tau_\rho$. Assume*

$$K : X \times X \setminus \operatorname{diag}(X) \longrightarrow \mathbb{R} \tag{2.2.130}$$

is a standard kernel in the first variable (that is, K is $\mu \otimes \mu$-measurable and (2.2.25)-(2.2.26) are satisfied). Fix $p \in [1, \infty)$ and assume that an operator

$$T : L^p(X, \mu) \longrightarrow L^{p,\infty}(X, \mu), \text{ linear and bounded,} \tag{2.2.131}$$

has been given, with the property that (2.2.31) holds for each $f \in L^p(X, \mu)$. In addition, assume that

there exists some $q \in (p, \infty)$ with the property that the restriction of T to $L^q(X, \mu) \cap L^p(X, \mu)$ extends to a linear and bounded operator from the space $L^q(X, \mu)$ into itself. (2.2.132)

For each $\varepsilon > 0$ define T_ε as in (2.2.32)-(2.2.33).

Then the maximal operator $T_{\max}$, defined as in (2.2.34), induces a sub-linear, bounded, and continuous mapping in each of the following contexts:

$$T_{\max} : L^p(X, \mu) \longrightarrow L^{p,\infty}(X, \mu), \tag{2.2.133}$$

$$T_{\max} : L^r(X, \mu) \longrightarrow L^r(X, \mu) \text{ for each } r \in (p, q], \tag{2.2.134}$$

and, for each fixed reference point $x_0 \in X$,

$$T_{\max} : L^1\Big(X, \tfrac{\mu(x)}{1+\rho(x, x_0)^d}\Big) \cap L^r_{\mathrm{loc}}(X, \mu) \longrightarrow L^r_{\mathrm{loc}}(X, \mu) \quad \text{for each integrability exponent } r \in (p, q]. \tag{2.2.135}$$

As a consequence of (2.2.133)-(2.2.134), for each $\varepsilon > 0$ the linear operator T_ε is bounded in the following contexts:

$$T_\varepsilon : L^p(X, \mu) \longrightarrow L^{p,\infty}(X, \mu), \text{ and} \tag{2.2.136}$$

$$T_\varepsilon : L^r(X, \mu) \longrightarrow L^r(X, \mu) \text{ for each } r \in (p, q]. \tag{2.2.137}$$

In addition,

$$\sup_{\varepsilon>0} \|T_\varepsilon\|_{L^p(X,\mu)\to L^{p,\infty}(X,\mu)} < \infty, \tag{2.2.138}$$

and

$$\sup_{\varepsilon>0} \|T_\varepsilon\|_{L^r(X,\mu)\to L^r(X,\mu)} < \infty \text{ for each } r \in (p,q]. \tag{2.2.139}$$

As is apparent from the proof of Proposition 2.2.7 given below, if in place of (2.2.132) one assumes a weak-(q,q) bound, i.e.,

> there exists some $q \in (p,\infty)$ with the property that the restriction of T to $L^q(X,\mu) \cap L^p(X,\mu)$ extends to a linear and bounded operator from $L^q(X,\mu)$ into $L^{q,\infty}(X,\mu)$, (2.2.140)

then in place of (2.2.133)-(2.2.134) one concludes that

$$T_{\max} : L^r(X,\mu) \longrightarrow L^r(X,\mu) \text{ for each } r \in (p,q), \text{ as well as} \tag{2.2.141}$$

$$T_{\max} : L^p(X,\mu) \to L^{p,\infty}(X,\mu) \text{ and } T_{\max} : L^q(X,\mu) \to L^{q,\infty}(X,\mu) \tag{2.2.142}$$

are well-defined, bounded, continuous, sub-linear operators, and so is $T_{\max}$ in the context of (2.2.135) with $r \in (p,q)$, i.e.,

$$T_{\max} : L^1\Big(X, \tfrac{\mu(x)}{1+\rho(x,x_0)^d}\Big) \cap L^r_{\rm loc}(X,\mu) \longrightarrow L^r_{\rm loc}(X,\mu)$$
is well-defined, bounded, continuous, for each $r \in (p,r)$. (2.2.143)

Here is the actual proof of Proposition 2.2.7.

Proof of Proposition 2.2.7 That $T_{\max}$ induces a sub-linear, bounded, and continuous mapping in the context of (2.2.133) follows directly from Proposition 2.2.5. Assume now that $q \in (p,\infty)$ is an exponent such that $T\big|_{L^q(X,\mu)\cap L^p(X,\mu)}$ extends as a linear and bounded operator from $L^q(X,\mu)$ into itself. Such an extension is unique (cf. [112, (3.1.13)]) and, without ambiguity, we may continue to denote it by T. Hence, in addition to (2.2.131), we also have

$$T : L^q(X,\mu) \longrightarrow L^q(X,\mu), \text{ linear and bounded.} \tag{2.2.144}$$

We claim that the analogue of hypothesis *(T3)* (appearing in the statement of Proposition 2.2.5) presently holds for functions in $L^q(X,\mu)$. Specifically, we claim that

for each function $f \in L^q(X,\mu)$ one has
$$(Tf)(x) = \int_X K(x,y)f(y)\,d\mu(y) \text{ for } \mu\text{-a.e. } x \in X \setminus \operatorname{supp} f. \tag{2.2.145}$$

To justify this claim, fix some function $f \in L^q(X,\mu)$ along with an arbitrary point x' belonging to $X \setminus \operatorname{supp} f$. Since the latter set is open, there exists $r_0 > 0$ such that

$$B_\rho(x', r_0) \subseteq X \setminus \operatorname{supp} f. \tag{2.2.146}$$

From [112, (3.1.13)] we know that there exists a sequence of functions $\{\widetilde{f_j}\}_{j\in\mathbb{N}}$ in $L^q(X,\mu)\cap L^p(X,\mu)$ with the property that $f=\lim_{j\to\infty}\widetilde{f_j}$ in $L^q(X,\mu)$. Introducing $f_j:=\widetilde{f_j}\mathbf{1}_{X\setminus B_\rho(x',r_0)}$ for each $j\in\mathbb{N}$, then produces a sequence

$$\{f_j\}_{j\in\mathbb{N}}\subset L^q(X,\mu)\cap L^p(X,\mu)\ \text{ such that} \tag{2.2.147}$$

$$f=\lim_{j\to\infty}f_j\ \text{ in } L^q(X,\mu),\ \text{ and} \tag{2.2.148}$$

$$B_\rho(x',r_0)\subseteq X\setminus\operatorname{supp}f_j\ \text{ for each } j\in\mathbb{N}. \tag{2.2.149}$$

From (2.2.144) and (2.2.147)-(2.2.148) it follows that $Tf=\lim_{j\to\infty}Tf_j$ in $L^q(X,\mu)$. By eventually passing to a subsequence, we may then write

$$(Tf)(x)=\lim_{j\to\infty}(Tf_j)(x)\ \text{ for } \mu\text{-a.e. } x\in X. \tag{2.2.150}$$

For each $j\in\mathbb{N}$ fixed, the fact that T satisfies (2.2.31) for every function in $L^p(X,\mu)$ implies, in view of (2.2.147) and (2.2.149), that there exists a nullset N_j of μ with the property that

$$(Tf_j)(x)=\int_{X\setminus B_\rho(x',r_0)}K(x,y)f_j(y)\,d\mu(y)\ \text{ for all } x\in B_\rho(x',r_0)\setminus N_j. \tag{2.2.151}$$

Then $N:=\bigcup_{j\in\mathbb{N}}N_j$ is a nullset of μ, and for each $x\in B_\rho\big(x',r_0/(2C_\rho)\big)\setminus N$ we may write

$$\begin{aligned}\lim_{j\to\infty}(Tf_j)(x)&=\lim_{j\to\infty}\int_{X\setminus B_\rho(x',r_0)}K(x,y)f_j(y)\,d\mu(y)\\&=\int_{X\setminus B_\rho(x',r_0)}K(x,y)f(y)\,d\mu(y)\\&=\int_X K(x,y)f(y)\,d\mu(y).\end{aligned} \tag{2.2.152}$$

Above, the first equality is a consequence of (2.2.151) and the definition of N. To justify the second equality, we start by observing that

$$\left.\begin{array}{r}\forall x,y\in X\ \text{ with } \rho(x',y)\geq r_0\\ \text{and } \rho(x',x)\leq r_0/(2C_\rho)\end{array}\right\}\Longrightarrow\rho(x,y)\geq r_0/C_\rho. \tag{2.2.153}$$

Indeed,

$$\left.\begin{array}{l}\rho(x',y)\le C_\rho\max\{\rho(x',x),\rho(x,y)\}\\ \rho(x',y)\ge r_0\ \text{ and }\ \rho(x',x)\le r_0/(2C_\rho)\end{array}\right\}\Longrightarrow \rho(x',y)\le C_\rho\,\rho(x,y) \tag{2.2.154}$$

from which (2.2.153) follows. Having established (2.2.153), if $q'\in(1,\infty)$ denotes the Hölder conjugate exponent of q then for each $x\in B_\rho\big(x',r_0/(2C_\rho\widetilde{C}_\rho)\big)$ we may use (2.2.25), Hölder's inequality, and [112, (7.2.5)] to estimate, for some $C\in(0,\infty)$ independent of j,

$$\begin{aligned}\int_{X\setminus B_\rho(x',r_0)} & |K(x,y)||f_j(y)-f(y)|\,\mathrm{d}\mu(y)\\ &\le C_0\int_{y\in X,\,\rho(x,y)\ge r_0/C_\rho}\frac{|f_j(y)-f(y)|}{\rho(x,y)^d}\,\mathrm{d}\mu(y)\\ &\le C_0\|f_j-f\|_{L^q(X,\mu)}\left(\int_{y\in X,\,\rho(x,y)\ge r_0/C_\rho}\frac{\mathrm{d}\mu(y)}{\rho(x,y)^{dq'}}\right)^{1/q'}\\ &\le C(r_0/C_\rho)^{-d/q}\|f-f_j\|_{L^q(X,\mu)}.\end{aligned} \tag{2.2.155}$$

In concert with (2.2.148), the above estimate implies

$$\lim_{j\to\infty}\int_{X\setminus B_\rho(x',r_0)}|K(x,y)||f_j(y)-f(y)|\,\mathrm{d}\mu(y)=0 \tag{2.2.156}$$

which, in turn, justifies the second equality in (2.2.152). Finally, the third equality in (2.2.152) is a consequence of (2.2.146) and item *(9)* in [112, Lemma 3.8.4] (while also mindful of (2.2.22)). At this stage, from (2.2.150) and (2.2.152) we conclude (bearing in mind that N is a nullset of μ) that

$$\begin{array}{c}\text{for each point } x'\in X\setminus\operatorname{supp}f \text{ there exist some } r_{x'}\in(0,\infty),\\ \text{along with some nullset } N(x') \text{ of } \mu, \text{ with the property that}\\ (Tf)(x)=\displaystyle\int_X K(x,y)f(y)\,\mathrm{d}\mu(y)\ \text{ for each}\\ x\in B_\rho\big(x',r_{x'}/(2C_\rho\widetilde{C}_\rho)\big)\setminus N(x').\end{array} \tag{2.2.157}$$

Invoking (2.2.22), we may now refine the open cover

$$\Big\{B_\rho\big(x',r_{x'}/(2C_\rho\widetilde{C}_\rho)\big)\Big\}_{x'\in X\setminus\operatorname{supp}f} \tag{2.2.158}$$

of $X\setminus\operatorname{supp}f$ to a countable cover of this set, say

$$\Big\{B_\rho\big(x_i,r_i/(2C_\rho\widetilde{C}_\rho)\big)\Big\}_{i\in\mathbb{N}}. \tag{2.2.159}$$

Then $N_0:=\bigcup_{i\in\mathbb{N}}N(x_i)$ is a nullset of μ, and

$$(Tf)(x) = \int_X K(x,y)f(y)\,d\mu(y) \text{ for each } x \in \big(X \setminus \operatorname{supp} f\big) \setminus N_0. \tag{2.2.160}$$

This finishes the proof of (2.2.145).

At this stage, we may conclude that T satisfies hypotheses *(T1)-(T3)* from the statement of Proposition 2.2.5 with q in place of p. Consequently, Proposition 2.2.5 applies (with q in place of p) and gives that $T_{\max}$ induces a sub-linear and bounded mapping in the context

$$T_{\max} : L^q(X,\mu) \longrightarrow L^{q,\infty}(X,\mu). \tag{2.2.161}$$

This is weaker than the conclusion we ultimately seek but, as we shall see momentarily, is nonetheless useful. For now, we note that since the properties listed in (2.2.36) imply

$$\begin{gathered}\big|(T_{\max}f)(x) - (T_{\max}g)(x)\big| \leq \big(T_{\max}(f-g)\big)(x)\\ \text{for every } f,g \in L^q(X,\mu) \text{ and every } x \in X,\end{gathered} \tag{2.2.162}$$

from the boundedness of (2.2.161) as well as [112, (6.2.16)] we conclude that

$$\begin{aligned}\big\|T_{\max}f - T_{\max}g\big\|_{L^{q,\infty}(X,\mu)} &\leq \big\|T_{\max}(f-g)\big\|_{L^{q,\infty}(X,\mu)}\\ &\leq C\|f-g\|_{L^q(X,\mu)}\end{aligned} \tag{2.2.163}$$

for every $f,g \in L^q(X,\mu)$. In particular, this shows that $T_{\max}$ is also continuous in the context of (2.2.161).

To proceed, fix an arbitrary function $f \in L^q(X,\mu) \cap L^p(X,\mu)$. Since the original assumptions on the operator T imply that the hypotheses *(T1)-(T3)* from Proposition 2.2.5 are satisfied for the given exponent p, and since f belongs to the space $L^p(X,\mu)$, we may write the pointwise inequality (2.2.38). Taking L^q norms, much as in (2.2.105) we may then estimate, for some $r \in (0,q)$ and some constant $C \in (0,\infty)$ independent of f,

$$\begin{aligned}\|T_{\max}f\|_{L^q(X,\mu)} &\leq C\Big\|\big[\mathcal{M}_X\big(|Tf|^r\big)\big]^{1/r}\Big\|_{L^q(X,\mu)} + C\Big\|\big[\mathcal{M}_X\big(|f|^p\big)\big]^{1/p}\Big\|_{L^q(X,\mu)}\\ &= C\Big\|\mathcal{M}_X\big(|Tf|^r\big)\Big\|_{L^{q/r}(X,\mu)}^{1/r} + C\Big\|\mathcal{M}_X\big(|f|^p\big)\Big\|_{L^{q/p}(X,\mu)}^{1/p}\\ &\leq C\big\||Tf|^r\big\|_{L^{q/r}(X,\mu)}^{1/r} + C\big\||f|^p\big\|_{L^{q/p}(X,\mu)}^{1/p}\\ &= C\|Tf\|_{L^q(X,\mu)} + C\|f\|_{L^q(X,\mu)}\\ &\leq C\|f\|_{L^q(X,\mu)},\end{aligned} \tag{2.2.164}$$

where the last step uses (2.2.144). Pick now an arbitrary $f \in L^q(X,\mu)$. From [112, (3.1.13)] we know that there exists a sequence $\{f_j\}_{j\in\mathbb{N}} \subset L^q(X,\mu) \cap L^p(X,\mu)$ such that

$$\lim_{j\to\infty} f_j = f \ \text{ in } \ L^q(X,\mu). \tag{2.2.165}$$

Now (2.2.162) implies that $|T_{\max}f_j - T_{\max}f_k| \leq T_{\max}(f_j - f_k)$ pointwise on X, for every $j,k \in \mathbb{N}$. In concert with the estimate obtained in (2.2.164) (written for the difference $f_j - f_k \in L^q(X,\mu) \cap L^p(X,\mu)$ in place of f) this implies

$$\begin{aligned}\big\|T_{\max}f_j - T_{\max}f_k\big\|_{L^q(X,\mu)} &\leq \big\|T_{\max}(f_j - f_k)\big\|_{L^q(X,\mu)} \\ &\leq C\|f_j - f_k\|_{L^q(X,\mu)}, \qquad \forall\, j,k \in \mathbb{N}. \end{aligned} \tag{2.2.166}$$

This proves that $\{T_{\max}f_j\}_{j\in\mathbb{N}}$ is a Cauchy sequence in the Banach space $L^q(X,\mu)$. As such,

$$g := \lim_{j\to\infty} T_{\max}f_j \ \text{ exists in } \ L^q(X,\mu). \tag{2.2.167}$$

Having established this, pass to limit $k \longrightarrow \infty$ in (2.2.166) to obtain

$$\|T_{\max}f_j - g\|_{L^q(X,\mu)} \leq C\|f_j - f\|_{L^q(X,\mu)}, \qquad \forall\, j \in \mathbb{N}. \tag{2.2.168}$$

Based on (2.2.168) and (2.2.164) (written for f_j in place of f), for each $j \in \mathbb{N}$ we may estimate

$$\begin{aligned}\|g\|_{L^q(X,\mu)} &\leq \|T_{\max}f_j - g\|_{L^q(X,\mu)} + \|T_{\max}f_j\|_{L^q(X,\mu)} \\ &\leq C\|f_j - f\|_{L^q(X,\mu)} + C\|f_j\|_{L^q(X,\mu)}. \end{aligned} \tag{2.2.169}$$

By passing to limit $j \longrightarrow \infty$ we therefore arrive at the conclusion that, for some constant $C \in (0,\infty)$ independent of f,

$$\|g\|_{L^q(X,\mu)} \leq C\|f\|_{L^q(X,\mu)}. \tag{2.2.170}$$

We now claim that $g = T_{\max}f$ at μ-a.e. point in X. To see this, observe that, thanks to [112, (6.2.27)] and (2.2.167), we have $g = \lim_{j\to\infty} T_{\max}f_j$ in $L^{q,\infty}(X,\mu)$. Since from (2.2.165) and the continuity of $T_{\max}$ in (2.2.161) we also have

$$\lim_{j\to\infty} T_{\max}f_j = T_{\max}f \ \text{ in } \ L^{q,\infty}(X,\mu), \tag{2.2.171}$$

we ultimately conclude that $g = T_{\max}f$ at μ-a.e. point in X. Utilizing this back in (2.2.170) proves that for every $f \in L^q(X,\mu)$ we have

$$\|T_{\max}f\|_{L^q(X,\mu)} \leq C\|f\|_{L^q(X,\mu)}, \tag{2.2.172}$$

for some constant $C \in (0,\infty)$ independent of f. This proves that $T_{\max}$ induces a sub-linear and bounded mapping in the context

$$T_{\max} : L^q(X,\mu) \longrightarrow L^q(X,\mu). \tag{2.2.173}$$

From this and (2.2.162), we then conclude that there exists a constant $C \in (0, \infty)$ with the property that

$$\begin{aligned} \big\|T_{\max} f - T_{\max} g\big\|_{L^q(X,\mu)} &\leq \big\|T_{\max}(f-g)\big\|_{L^q(X,\mu)} \\ &\leq C\|f-g\|_{L^q(X,\mu)} \end{aligned} \tag{2.2.174}$$

for every $f, g \in L^q(X, \mu)$. In particular, $T_{\max}$ in (2.2.173) is also continuous.

Having proved the boundedness of $T_{\max}$ both in the context of (2.2.133) as well as (2.2.173), the real interpolation result for quasi-subadditive operators recorded in [113, Proposition 1.3.7] applies and, in concert with [112, (6.2.47)], leads to the conclusion that $T_{\max}$ further induces a sub-linear and bounded mapping in the context

$$T_{\max} : L^r(X, \mu) \longrightarrow L^r(X, \mu) \text{ for each } r \in (p, q]. \tag{2.2.175}$$

Parenthetically, we note that we could have arrived at this conclusion by first interpolating between (2.2.131) and (2.2.144) to obtain that T maps $L^r(X, \mu)$ boundedly into itself for each $r \in (p, q]$, then run the same program as above starting from this premise in lieu of (2.2.144).

Returning to the mainstream discussion, the boundedness of (2.2.175) then yields, much as before via (2.2.162) (and the fact that $L^r(X, \mu)$ is a quasi-normed lattice of functions), that for each $r \in (p, q]$ there exists $C \in (0, \infty)$ with the property that

$$\big\|T_{\max} f - T_{\max} g\big\|_{L^{r,\infty}(X,\mu)} \leq C\|f - g\|_{L^r(X,\mu)} \text{ for all } f, g \in L^r(X, \mu). \tag{2.2.176}$$

In particular, the mapping (2.2.175) is also continuous.

The claims pertaining to $T_{\max}$ in the context of (2.2.135) are justified by reasoning as in the proof of (2.2.42), with two important changes. First, in place of (2.2.111) we now conclude from (2.2.109) that

$$\begin{aligned} &g_R \in L^r\big(B_\rho(x_o, R), \mu\big) \text{ and} \\ &\|g_R\|_{L^r(B_\rho(x_o,R),\mu)} \leq C_R \|f\|_{L^1\left(X, \frac{\mu(x)}{1+\rho(x,x_o)^d}\right)}. \end{aligned} \tag{2.2.177}$$

Second, since

$$\begin{aligned} &f\mathbf{1}_{B_\rho(x_o,C_*R)} \in L^r(X, \mu) \text{ with} \\ &\big\|f\mathbf{1}_{B_\rho(x_o,C_*R)}\big\|_{L^r(X,\mu)} = \big\|f\big|_{B_\rho(x_o,C_*R)}\big\|_{L^r(B_\rho(x_o,C_*R),\mu)}, \end{aligned} \tag{2.2.178}$$

in place of (2.2.113)-(2.2.114) we deduce from (2.2.134) that

$$\begin{aligned} &h_R \in L^r\big(B_\rho(x_o, R), \mu\big) \text{ and} \\ &\|h_R\|_{L^r(B_\rho(x_o,R),\mu)} \leq C\big\|f\big|_{B_\rho(x_o,C_*R)}\big\|_{L^r(B_\rho(x_o,C_*R),\mu)}. \end{aligned} \tag{2.2.179}$$

With (2.2.177) and (2.2.179) in hand, we then conclude as before that $T_{\max}$ is a well-defined, sub-linear, bounded, continuous operator in the context of (2.2.135).

Finally, the claims pertaining to (2.2.136)-(2.2.139) are direct consequences of what we have proved so far, (2.2.34), and (2.2.68). □

We proceed to discuss a number of consequences of the Calderón-Zygmund lemma. First, as Proposition 2.2.8 below shows, for a linear operator T being weak-(q,q) bounded for some integrability exponent $q \in (1,\infty)$ and being associated with a standard kernel in the *second* variable extrapolates to strong-(p,p) bounds for T for any $p \in (1,q)$.

Proposition 2.2.8 *Assume (X,ρ,μ) is a d-Ahlfors regular quasi-metric space for some $d \in (0,\infty)$ with the property that the quasi-distance $\rho : X \times X \to [0,+\infty)$ is continuous in the product topology $\tau_\rho \times \tau_\rho$. Suppose*

$$K : X \times X \setminus \operatorname{diag}(X) \longrightarrow \mathbb{R} \tag{2.2.180}$$

is a standard kernel in the second variable (cf. Definition 2.2.4). Fix $q \in (1,\infty)$ and assume that an operator

$$T : L^q(X,\mu) \longrightarrow L^{q,\infty}(X,\mu), \text{ linear and bounded,} \tag{2.2.181}$$

has been given, with the property that T is associated in $L^q(X,\mu)$ with the kernel K, in the sense that for each $f \in L^q(X,\mu)$ one has

$$(Tf)(x) = \int_X K(x,y)f(y)\,d\mu(y) \text{ for } \mu\text{-a.e. } x \in X \setminus \operatorname{supp} f. \tag{2.2.182}$$

Then the following statements are true:

(i) There exists $C \in (0,\infty)$ with the property that

$$\|Tf\|_{L^{1,\infty}(X,\mu)} \le C\|f\|_{L^1(X,\mu)}, \qquad \forall f \in L^q(X,\mu) \cap L^1(X,\mu). \tag{2.2.183}$$

As a consequence, the restriction of T to $L^q(X,\mu) \cap L^1(X,\mu)$ extends to a linear and bounded operator from $L^1(X,\mu)$ into $L^{1,\infty}(X,\mu)$, henceforth, still denoted by T. Thus,

$$T : L^1(X,\mu) \longrightarrow L^{1,\infty}(X,\mu) \text{ linearly and boundedly.} \tag{2.2.184}$$

(ii) For each $p \in (1,q)$, the operator T extends to a mapping

$$T : L^p(X,\mu) \longrightarrow L^p(X,\mu) \text{ linear and bounded.} \tag{2.2.185}$$

To set the stage for presenting the proof of Proposition 2.2.8 we state a lemma containing an estimate for a Marcinkiewicz-type integral.

Lemma 2.2.9 *Assume that (X,ρ,μ) is a d-Ahlfors regular space for some $d > 0$ with the property that the quasi-distance $\rho : X \times X \to [0,+\infty)$ is continuous in the* *product topology $\tau_\rho \times \tau_\rho$. Then for each $\gamma > 0$ there exists $C \in (0, \infty)$ such that whenever $\mathcal{F}$ is a nonempty closed subset of (X, τ_ρ) one has*

$$\int_{\mathcal{F}} \int_X \frac{\operatorname{dist}_\rho(y, \mathcal{F})^\gamma}{\rho(x, y)^{d+\gamma}} \, d\mu(y) \, d\mu(x) \leq C\mu(X \setminus \mathcal{F}). \tag{2.2.186}$$

Proof Since $\operatorname{dist}_\rho(\cdot, \mathcal{F})$ vanishes on $\mathcal{F}$ and $\gamma > 0$, we may write based on Fubini's Theorem

$$\begin{aligned}
&\int_{\mathcal{F}} \Big(\int_X \frac{\operatorname{dist}_\rho(y, \mathcal{F})^\gamma}{\rho(x, y)^{d+\gamma}} \, d\mu(y) \Big) \, d\mu(x) \\
&\quad = \int_{\mathcal{F}} \Big(\int_{X \setminus \mathcal{F}} \frac{\operatorname{dist}_\rho(y, \mathcal{F})^\gamma}{\rho(x, y)^{d+\gamma}} \, d\mu(y) \Big) \, d\mu(x) \\
&\quad = \int_{X \setminus \mathcal{F}} \operatorname{dist}_\rho(y, \mathcal{F})^\gamma \Big(\int_{\mathcal{F}} \frac{d\mu(x)}{\rho(x, y)^{d+\gamma}} \Big) \, d\mu(y) \\
&\quad \leq \int_{X \setminus \mathcal{F}} \operatorname{dist}_\rho(y, \mathcal{F})^\gamma \Big(\int_{X \setminus B_\rho(y, \operatorname{dist}_\rho(y, \mathcal{F}))} \frac{d\mu(x)}{\rho(x, y)^{d+\gamma}} \Big) \, d\mu(y) \\
&\quad \leq C \int_{X \setminus \mathcal{F}} d\mu = C\mu(X \setminus \mathcal{F}),
\end{aligned} \tag{2.2.187}$$

where the last inequality is based on [112, (7.2.5)]. □

We are now ready to present the proof of Proposition 2.2.8.

Proof of Proposition 2.2.8 We shall show the existence of some $C \in (0, \infty)$ with the property that for each function $f \in L^1(X, \mu) \cap L^q(X, \mu)$ we have

$$\mu\Big(\big\{x \in X : |(Tf)(x)| > \lambda\big\}\Big) \leq C \frac{\|f\|_{L^1(X,\mu)}}{\lambda}, \qquad \forall \lambda > 0. \tag{2.2.188}$$

To get started, assume some arbitrary function $f \in L^1(X, \mu) \cap L^q(X, \mu)$ has been fixed. In the case when $0 < \lambda \leq \|f\|_{L^1(X,\mu)}/\mu(X)$ (which may only happen when X is bounded), we have

$$\mu\Big(\big\{x \in X : |(Tf)(x)| > \lambda\big\}\Big) \leq \mu(X) \leq \frac{\|f\|_{L^1(X,\mu)}}{\lambda}, \tag{2.2.189}$$

so (2.2.188) holds in this case if we choose $C \geq 1$.

Consider now the case when $\lambda > \|f\|_{L^1(X,\mu)}/\mu(X)$. There is no loss of generality in assuming that f has bounded support, and we shall perform a Calderón-Zygmund decomposition of f at level λ. More precisely, there exist two finite constants $C > 0$, $N \in \mathbb{N}$ (depending only on geometry), along with an at most countable family of balls $(Q_j)_{j \in J}$, say $Q_j := B_\rho(x_j, r_j)$ for each index $j \in J$, and two real-valued functions $g, b : X \to \mathbb{R}$ satisfying the following properties (cf., e.g., [22]):

$$f = g + b \text{ at } \mu\text{-a.e. point in } X, \tag{2.2.190}$$

$$g \in L^1(X,\mu) \cap L^\infty(X,\mu) \hookrightarrow L^q(X,\mu) \text{ and } b \in L^1(X,\mu) \cap L^q(X,\mu), \tag{2.2.191}$$

$$\|g\|_{L^1(X,\mu)} \leq C\|f\|_{L^1(X,\mu)}, \quad |g(x)| \leq C\lambda \text{ for } \mu\text{-a.e. } x \in X, \tag{2.2.192}$$

$$\text{if } b_j := \Big(f - ⨍_{Q_j} f\Big)\mathbf{1}_{Q_j} \text{ for each } j \in J, \text{ then} \tag{2.2.193}$$

$$b = \sum_{j\in J} b_j \text{ with convergence both in } L^1(X,\mu) \text{ and } L^q(X,\mu), \tag{2.2.194}$$

$$\sum_{j\in J} \|b_j\|_{L^1(X,\mu)} \leq C\|f\|_{L^1(X,\mu)}, \tag{2.2.195}$$

$$b_j = 0 \text{ in } X \setminus Q_j, \quad \int_X b_j \, d\mu = 0, \text{ and } ⨍_{Q_j} |b_j| \, d\mu \leq C\lambda, \quad \forall j \in J, \tag{2.2.196}$$

$$\begin{aligned} &\text{if } O := \textstyle\bigcup_{j\in J} Q_j \text{ and } \mathcal{F} := X \setminus O, \text{ then } \mathbf{1}_O \leq \textstyle\sum_{j\in J} \mathbf{1}_{Q_j} \leq N\mathbf{1}_O, \\ &\mu(O) \leq C\lambda^{-1}\|f\|_{L^1(X,\mu)}, \text{ and } \operatorname{dist}_\rho(Q_j,\mathcal{F}) \approx r_j \text{ uniformly in } j \in J. \end{aligned} \tag{2.2.197}$$

To prove that the series $\sum_{j\in J} b_j$ converges in $L^q(X,\mu)$ to b, there is no loss of generality in assuming that $J = \mathbb{N}$. Suppose this is the case and pick some $\varepsilon > 0$. Given that $|f|^q \, d\mu$ is a finite measure which is absolutely continuous with respect to μ, it follows that there exists $\delta > 0$ with the property that

$$\int_E |f|^q \, d\mu < \varepsilon \text{ whenever } E \subseteq X \text{ is } \mu\text{-measurable and } \mu(E) < \delta. \tag{2.2.198}$$

Next, observe that $\mathbf{1}_O \in L^q(X,\mu)$ since O is open and of finite measure. Mindful of this and the fact that $\sum_{j\in\mathbb{N}} \mathbf{1}_{Q_j} \leq N\mathbf{1}_O$, Lebesgue's Dominated Convergence Theorem implies that

$$\text{the sequence } \Big\{\sum_{j=1}^m \mathbf{1}_{Q_j}\Big\}_m \text{ is convergent in } L^q(X,\mu). \tag{2.2.199}$$

Consequently, the aforementioned sequence is Cauchy in the space $L^q(X,\mu)$. Hence, there exists some integer $M = M(\delta) \in \mathbb{N}$ such that

$$\int_X \Big|\sum_{j=m}^{m+k} \mathbf{1}_{Q_j}\Big|^q \, d\mu < \delta \text{ for all } m,k \in \mathbb{N} \text{ with } m \geq M. \tag{2.2.200}$$

Upon noting that, pointwise on X we have

$$\sum_{j=m}^{m+k} \mathbf{1}_{Q_j} \geq \mathbf{1}_{\bigcup_{j=m}^{m+k} Q_j} \text{ for all } m,k \in \mathbb{N}, \tag{2.2.201}$$

we conclude from (2.2.200) that

$$\mu\Big(\bigcup_{j=m}^{m+k} Q_j\Big) < \delta \text{ for all } m, k \in \mathbb{N} \text{ with } m \geq M. \tag{2.2.202}$$

To proceed, for each $j \in \mathbb{N}$ use the triangle inequality and Hölder's inequality to estimate

$$\begin{aligned}
\int_X |b_j|^q \, d\mu &= \Big\| f\mathbf{1}_{Q_j} - \Big(\fint_{Q_j} f\Big)\mathbf{1}_{Q_j}\Big\|^q_{L^q(X,\mu)} \\
&\leq \Big(\|f\mathbf{1}_{Q_j}\|_{L^q(X,\mu)} + \Big\|\Big(\fint_{Q_j} f\Big)\mathbf{1}_{Q_j}\Big\|_{L^q(X,\mu)}\Big)^q \\
&\leq 2^q \int_{Q_j} |f|^q \, d\mu + 2^q \mu(Q_j)\Big(\fint_{Q_j} f\Big)^q \\
&\leq (2^q + 1) \int_{Q_j} |f|^q \, d\mu.
\end{aligned} \tag{2.2.203}$$

Also, since for each subset I of $\mathbb{N}$ and for each point $x \in X$ the cardinality of the set $\{j \in I : x \in Q_j\}$ is at most N, for every $m, k \in \mathbb{N}$ we may write

$$\begin{aligned}
\Big(\sum_{j=m}^{m+k} |b_j(x)|\Big)^q &= \Big(\sum_{\{m \leq j \leq m+k:\, x \in Q_j\}} |b_j(x)|\Big)^q \leq N^q \sum_{\{m \leq j \leq m+k:\, x \in Q_j\}} |b_j(x)|^q \\
&\leq N^q \sum_{j=m}^{m+k} |b_j(x)|^q \text{ for } \mu\text{-a.e. } x \in X.
\end{aligned} \tag{2.2.204}$$

Together, (2.2.203), (2.2.204), (2.2.202), and (2.2.198) then permit us to estimate

$$
\begin{aligned}
\Big\| \sum_{j=m}^{m+k} b_j \Big\|_{L^q(X,\mu)}^q &\leq \int_X \Big(\sum_{j=m}^{m+k} |b_j| \Big)^q \, d\mu \leq N^q \sum_{j=m}^{m+k} \int_X |b_j|^q \, d\mu \\
&\leq N^q (2^q + 1) \sum_{j=m}^{m+k} \int_{Q_j} |f|^q \, d\mu \\
&\leq N^q (2^q + 1) \int_{\bigcup_{j=m}^{m+k} Q_j} |f|^q \Big(\sum_{j=m}^{m+k} \mathbf{1}_{Q_j} \Big) \, d\mu \\
&\leq N^{q+1} (2^q + 1) \int_{\bigcup_{j=m}^{m+k} Q_j} |f|^q \, d\mu \\
&\leq N^{q+1} (2^q + 1) \varepsilon, \text{ for all } m, k \in \mathbb{N} \text{ with } m \geq M.
\end{aligned} \tag{2.2.205}
$$

Since the number $\varepsilon > 0$ has been arbitrarily chosen, the above estimate proves that the series $\sum_{j \in \mathbb{N}} b_j$ converges in $L^q(X, \mu)$. Since the same series is known to converge in $L^1(X, \mu)$ to the function b, we ultimately conclude that $\sum_{j \in \mathbb{N}} b_j$ converges in $L^q(X, \mu)$ to b.

Let us now return to the main stream discussion. Since

$$
|(Tf)(x)| \leq |(Tg)(x)| + |(Tb)(x)| \text{ for } \mu\text{-a.e. } x \in X, \tag{2.2.206}
$$

as far as (2.2.188) is concerned it suffices to prove that

$$
\mu\Big(\big\{x \in X : |(Tg)(x)| > \lambda/2\big\}\Big) \leq C \frac{\|f\|_{L^1(X,\mu)}}{\lambda}, \tag{2.2.207}
$$

and

$$
\mu\Big(\big\{x \in X : |(Tb)(x)| > \lambda/2\big\}\Big) \leq C \frac{\|f\|_{L^1(X,\mu)}}{\lambda}. \tag{2.2.208}
$$

The boundedness of T in the context of (2.2.181) permits us to estimate

$$
\begin{aligned}
\mu\Big(\big\{x \in X : |(Tg)(x)| > \lambda/2\big\}\Big) &\leq C \left(\frac{\|g\|_{L^q(X,\mu)}}{\lambda} \right)^q \\
&\leq C \frac{\|g\|_{L^\infty(X,\mu)}^{q-1} \|g\|_{L^1(X,\mu)}}{\lambda^q} \\
&\leq C \frac{\|f\|_{L^1(X,\mu)}}{\lambda},
\end{aligned} \tag{2.2.209}
$$

thus (2.2.207) is established. We are therefore left with proving (2.2.208). To justify this, first note that (2.2.197) directly implies

$$\mu\Big(\big\{x \in \mathcal{O} : |(Tb)(x)| > \lambda/2\big\}\Big) \leq \mu(\mathcal{O}) \leq C\frac{\|f\|_{L^1(X,\mu)}}{\lambda}. \tag{2.2.210}$$

Second, Chebytcheff's inequality gives

$$\mu\Big(\big\{x \in \mathcal{F} : |(Tb)(x)| > \lambda/2\big\}\Big) \leq \frac{1}{\lambda}\int_{\mathcal{F}} |Tb|\, d\mu. \tag{2.2.211}$$

Therefore, since $X = \mathcal{O} \cup \mathcal{F}$, in view of (2.2.210) and (2.2.211), estimate (2.2.208) will follow as soon as we prove that

$$\int_{\mathcal{F}} |Tb|\, d\mu \leq C\|f\|_{L^1(X,\mu)} \tag{2.2.212}$$

for some $C \in (0, \infty)$ independent of f. With this goal in mind, we fix $j \in J$ and $x \in \mathcal{F}$ arbitrary and look for a pointwise estimate for

$$(Tb_j)(x) = \int_X K(x, y)b_j(y)\, d\mu(y). \tag{2.2.213}$$

With x_j and r_j denoting, respectively, the center and radius of Q_j, based on the third condition in (2.2.195), we may write

$$\begin{aligned} |(Tb_j)(x)| &= \Big|\int_X K(x, y)b_j(y)\, d\mu(y)\Big| = \Big|\int_X \big[K(x, y) - K(x, x_j)\big]b_j(y)\, d\mu(y)\Big| \\ &= \Big|\int_{Q_j} \big[K(x, y) - K(x, x_j)\big]b_j(y)\, d\mu(y)\Big| \leq I_1 + I_2, \end{aligned} \tag{2.2.214}$$

where we have set

$$I_1 := \int\limits_{\substack{y\in Q_j \\ \rho(y,x_j)<C_1\rho(x,x_j)}} \big|K(x, y) - K(x, x_j)\big||b_j(y)|\, d\mu(y), \tag{2.2.215}$$

$$I_2 := \int\limits_{\substack{y\in Q_j \\ \rho(y,x_j)\geq C_1\rho(x,x_j)}} \big|K(x, y) - K(x, x_j)\big||b_j(y)|\, d\mu(y). \tag{2.2.216}$$

Note that $\rho(y, x_j) < C_1\,\rho(x, x_j)$ on the domain of integration in I_1. Based on this, the fact that the kernel K is standard in the second variable, and (2.2.196), we may then estimate this term as follows

$$
\begin{aligned}
I_1 &\le C_0 \int_{Q_j} \frac{\rho(y,x_j)^\gamma}{\rho(x,x_j)^{d+\gamma}} |b_j(y)|\,\mathrm{d}\mu(y) \\
&\le C \frac{r_j^\gamma}{\rho(x,x_j)^{d+\gamma}} \int_{Q_j} |b_j(y)|\,\mathrm{d}\mu(y) \le C\lambda \frac{r_j^{d+\gamma}}{\rho(x,x_j)^{d+\gamma}} \\
&\le C\lambda \int_{Q_j} \frac{\operatorname{dist}_\rho(y,\mathcal{F})^\gamma}{\rho(x,y)^{d+\gamma}}\,\mathrm{d}\mu(y),
\end{aligned}
\tag{2.2.217}
$$

since $\operatorname{dist}_\rho(y,\mathcal{F}) \approx r_j$ and $\rho(x,x_j) \approx \rho(x,y)$ uniformly for $y \in Q_j$ and $x \in \mathcal{F}$.

Estimating I_2 requires a few geometrical preliminaries. Recall that $x \in \mathcal{F}$ and consider an arbitrary point $y \in Q_j$ such that $\rho(y,x_j) \ge C_1\,\rho(x,x_j)$. Observe that, on the one hand,

$$
r_j \approx \operatorname{dist}_\rho(Q_j,\mathcal{F}) \le \rho(x_j,x) \le \widetilde{C}_\rho\,\rho(x,x_j), \tag{2.2.218}
$$

$$
r_j \approx \operatorname{dist}_\rho(Q_j,\mathcal{F}) \le \rho(y,x) \le \widetilde{C}_\rho\,\rho(x,y). \tag{2.2.219}
$$

On the other hand, the fact that $y \in Q_j$ forces $\rho(x_j,y) < r_j$ which in turn allows us to estimate $\rho(x,x_j) \le C_1^{-1}\rho(y,x_j) \le C_1^{-1}\widetilde{C}_\rho\,\rho(x_j,y) < C_1^{-1}\widetilde{C}_\rho r_j$. Together with (2.2.218), this ultimately implies

$$
r_j \approx \rho(x,x_j), \quad \text{uniformly in } j \in J \text{ and } x \in \mathcal{F}. \tag{2.2.220}
$$

Also, $\rho(x,y) \le C_\rho \max\{\rho(x,x_j),\rho(x_j,y)\} \le C_\rho \max\{C_1^{-1}\widetilde{C}_\rho, 1\} r_j$ which, in light of (2.2.219), entails

$$
r_j \approx \rho(x,y), \quad \text{uniformly in } y \in Q_j \text{ and } x \in \mathcal{F}. \tag{2.2.221}
$$

Consequently, on the domain of integration in I_2 we have thanks to (2.2.220) and (2.2.221)

$$
\begin{aligned}
\big|K(x,y) - K(x,x_j)\big| &\le \big|K(x,y)\big| + \big|K(x,x_j)\big| \\
&\le \frac{C_0}{\rho(x,y)^d} + \frac{C_0}{\rho(x,x_j)^d} \le \frac{C r_j^\gamma}{\rho(x,x_j)^{d+\gamma}}.
\end{aligned}
\tag{2.2.222}
$$

Together with (2.2.216), this allows us to estimate

$$
\begin{aligned}
I_2 &\le C \frac{r_j^\gamma}{\rho(x,x_j)^{d+\gamma}} \int_{Q_j} |b_j(y)|\,\mathrm{d}\mu(y) \le C\lambda \frac{r_j^{d+\gamma}}{\rho(x,x_j)^{d+\gamma}} \\
&\le C\lambda \int_{Q_j} \frac{\operatorname{dist}_\rho(y,\mathcal{F})^\gamma}{\rho(x,y)^{d+\gamma}}\,\mathrm{d}\mu(y).
\end{aligned}
\tag{2.2.223}
$$

From (2.2.214), (2.2.217), and (2.2.223) we conclude that there exists $C \in (0, \infty)$ with the property that for every $j \in J$ we have

$$\big|(Tb_j)(x)\big| \le C\lambda \int_{Q_j} \frac{\operatorname{dist}_\rho(y, \mathcal{F})^\gamma}{\rho(x, y)^{d+\gamma}}\, d\mu(y), \qquad \forall x \in \mathcal{F}. \tag{2.2.224}$$

Summing up inequalities of the form (2.2.224) over $j \in \mathbb{N}$ and using the linearity of the operator T as well as the finite-overlap property in (2.2.197), we obtain

$$\big|(Tb)(x)\big| \le C\lambda \int_{O} \frac{\operatorname{dist}_\rho(y, \mathcal{F})^\gamma}{\rho(x, y)^{d+\gamma}}\, d\mu(y), \qquad \forall x \in \mathcal{F}. \tag{2.2.225}$$

Consequently, from (2.2.225), Lemma 2.2.9, and (2.2.197), we deduce that

$$\begin{aligned} \int_{\mathcal{F}} |Tb|\, d\mu &\le C\lambda \int_{\mathcal{F}} \int_{O} \frac{\operatorname{dist}_\rho(y, \mathcal{F})^\gamma}{\rho(x, y)^{d+\gamma}}\, d\mu(y)\, d\mu(x) \\ &\le C\lambda \int_{\mathcal{F}} \int_{X} \frac{\operatorname{dist}_\rho(y, \mathcal{F})^\gamma}{\rho(x, y)^{d+\gamma}}\, d\mu(y)\, d\mu(x) \\ &\le C\lambda\, \mu\big(X \setminus \mathcal{F}\big) = C\,\lambda\, \mu(O) \le C\|f\|_{L^1(X,\mu)}. \end{aligned} \tag{2.2.226}$$

This proves (2.2.212), thus completing the proof of (2.2.188). From this, the claims in item *(i)* follow.

Interpolating between (2.2.184) and (2.2.181) gives (in light of [112, (6.2.47)])

$$\begin{gathered} T : L^r(X, \mu) \longrightarrow L^r(X, \mu) \\ \text{linearly and boundedly for each } r \in (1, q). \end{gathered} \tag{2.2.227}$$

This takes care of item *(ii)*, and finishes the proof of the proposition. □

The reader is reminded that the space of Hölder functions with bounded support, in the quasi-metric setting, has been defined in (A.0.33). The following result can be thought of as a refinement of [112, Proposition 3.7.2]. Before stating it, the reader is reminded that $\operatorname{dist}_\rho(\cdot, \cdot)$ is defined in (A.0.42).

Lemma 2.2.10 *Suppose (X, ρ) is a quasi-metric space, and fix a finite exponent $\beta \in \big(0, (\log_2 C_\rho)^{-1}\big]$. Assume μ is a locally finite Borel-semiregular measure on X (cf.* [112, Definition 3.4.3]). *Also, let $F \subseteq X$ be a closed set and consider a function*

$$\begin{gathered} h : X \setminus F \longrightarrow \mathbb{R} \text{ which is absolutely integrable with respect to } \mu \text{ on} \\ \text{any bounded closed subset } E \text{ of } X \setminus F \text{ satisfying } \operatorname{dist}_\rho(F, E) > 0, \end{gathered} \tag{2.2.228}$$

and with the property that

$$\int_X h\,\phi\, d\mu = 0 \text{ for each } \phi \in \mathscr{C}^\beta_c(X, \rho) \text{ with } \operatorname{dist}_\rho\big(F, \operatorname{supp}\phi\big) > 0. \tag{2.2.229}$$

Then $h = 0$ at μ-a.e. point in $X \setminus F$.

A particular case of interest is $F = \varnothing$, a scenario in which we set $\mathrm{dist}_\rho(\varnothing, \cdot) := +\infty$.

Proof of Lemma 2.2.10 Fix an arbitrary point x in the open set $X \setminus F$. Then there exists some small radius $r > 0$ such that $B_{\rho_\#}(x, r) \subseteq X \setminus F$. Consider a closed set $E \subseteq B_{\rho_\#}(x, r/4)$. From [117, Lemma 4.14, p. 166] we know that there exists a sequence $\{\phi_j\}_{j\in\mathbb{N}} \subseteq \mathscr{C}_c^\beta(X, \rho)$ with the following properties:

$$\begin{aligned} &\mathrm{supp}\,\phi_j \subseteq B_{\rho_\#}(x, r/2) \subseteq X \setminus F, \quad 0 \leq \phi_j \leq 1 \text{ for all } j \in \mathbb{N}, \\ &\text{and } \phi_j \searrow \mathbf{1}_E \text{ pointwise everywhere on } X, \text{ as } j \to \infty. \end{aligned} \tag{2.2.230}$$

Note that

$$\mathrm{dist}_\rho\big(X \setminus B_{\rho_\#}(x, r), B_{\rho_\#}(x, r/2)\big) > 0. \tag{2.2.231}$$

Indeed, having $y \in X \setminus B_{\rho_\#}(x, r)$ and $z \in B_{\rho_\#}(x, r/2)$ forces $\rho_\#(y, x) \geq r$ as well as $\rho_\#(z, x) \leq r/2$. Granted these properties, we may then use [112, (7.1.19)-(7.1.20)] to estimate

$$r^\beta \leq \rho_\#^\beta(y, x) \leq \rho_\#^\beta(y, z) + \rho_\#^\beta(z, x) \leq \rho_\#^\beta(y, z) + (r/2)^\beta \tag{2.2.232}$$

which, in light of [112, (7.1.21)], further implies

$$r(1 - 2^{-\beta})^{1/\beta} \leq \rho_\#(y, z) \leq \widetilde{C}_\rho \cdot \rho(y, z). \tag{2.2.233}$$

As such,

$$\begin{aligned} &\mathrm{dist}_\rho\big(X \setminus B_{\rho_\#}(x, r), B_{\rho_\#}(x, r/2)\big) \\ &\qquad = \inf\big\{\rho(y, z) : \ y \in X \setminus B_{\rho_\#}(x, r) \text{ and } z \in B_{\rho_\#}(x, r/2)\big\} \\ &\qquad \geq r(1 - 2^{-\beta})^{1/\beta}(\widetilde{C}_\rho)^{-1} > 0, \end{aligned} \tag{2.2.234}$$

justifying (2.2.231). In turn, this implies that

$$\mathrm{dist}_\rho\big(F, \mathrm{supp}\,\phi_j\big) > 0 \text{ for each } j \in \mathbb{N}. \tag{2.2.235}$$

Based on (2.2.235), (2.2.228), and (2.2.229) we conclude that $\int_X h\,\phi_j\,d\mu = 0$ for each $j \in \mathbb{N}$. Since for each $j \in \mathbb{N}$ we have $|h\,\phi_j| \leq \mathbf{1}_{B_{\rho_\#}(x,r/2)} \in L^1(X, \mu)$ we may invoke Lebesgue's Dominated Convergence Theorem to write

$$\int_E h\,d\mu = \lim_{j\to\infty} \int_X h\,\phi_j\,d\mu = 0. \tag{2.2.236}$$

The argument so far proves that

$$\int_E h\,d\mu = 0 \text{ for each closed set } E \subseteq B_{\rho_\#}(x, r/4). \tag{2.2.237}$$

To proceed, introduce

$$A_\pm := \{x \in B_{\rho_\#}(x, r/4) : \pm h(x) \geq 0\}. \tag{2.2.238}$$

Then $A_\pm$ are μ-measurable subsets of $B_{\rho_\#}(x, r/4)$ and $\mu(A_\pm) \leq \mu(B_{\rho_\#}(x, r/4)) < \infty$. Since the measure μ is Borel-semiregular it follows that there exist two Borel sets $B_\pm \subseteq X$ such that

$$\mu(A_\pm \triangle B_\pm) = 0. \tag{2.2.239}$$

Replacing $B_\pm$ by $B_\pm \cap B_{\rho_\#}(x, r/4)$ (and keeping in mind that all $\rho_\#$-balls are open) there is no loss of generality in assuming that we actually have $B_\pm \subseteq B_{\rho_\#}(x, r/4)$. Observe that, having fixed any reference point $x_0 \in X$,

$$\begin{aligned} &X = \bigcup_{j \in \mathbb{N}} B_{\rho_\#}(x_0, j) \text{ and for each } j \in \mathbb{N} \text{ we have } \mu(B_{\rho_\#}(x_0, j)) < \infty \\ &\text{and } B_{\rho_\#}(x_0, j) \text{ is open hence, in particular, it belongs to } Borel_\tau(X). \end{aligned} \tag{2.2.240}$$

Granted (2.2.240) and [112, Lemma 3.4.13], we may then invoke item *(1)* in [112, Proposition 3.4.15] to conclude that

$$\mu(B_\pm) = \sup\{\mu(E_\pm) : E_\pm \text{ closed subset of } B_\pm\}. \tag{2.2.241}$$

Fix $\varepsilon > 0$ arbitrary. Since the $\mathbf{1}_{B_{\rho_\#}(x,r/4)}|h|\,d\mu$ is a finite measure which is absolutely continuous with respect to μ, it follows that there exists $\delta > 0$ with the property that

$$\int_{A \cap B_{\rho_\#}(x,r/4)} |h|\,d\mu < \varepsilon \text{ for each } \mu\text{-measurable set } A \subseteq X \text{ with } \mu(A) < \delta. \tag{2.2.242}$$

For this δ, we may rely on (2.2.241) to pick two closed sets $E_\pm \subseteq B_\pm$ with

$$\mu(B_\pm \setminus E_\pm) < \delta. \tag{2.2.243}$$

Then, thanks to (2.2.239) and (2.2.237), we may write

$$\begin{aligned} \int_{B_{\rho_\#}(x,r/4)} |h|\,d\mu &= \int_{A_+} h\,d\mu - \int_{A_-} h\,d\mu = \int_{B_+} h\,d\mu - \int_{B_-} h\,d\mu \\ &= \left(\int_{B_+ \setminus E_+} h\,d\mu + \int_{E_+} h\,d\mu\right) - \left(\int_{B_- \setminus E_-} h\,d\mu + \int_{E_-} h\,d\mu\right) \\ &= \int_{B_+ \setminus E_+} h\,d\mu - \int_{B_- \setminus E_-} h\,d\mu. \end{aligned} \tag{2.2.244}$$

We may also rely on (2.2.242) and (2.2.243) to estimate

$$\left|\int_{B_\pm \setminus E_\pm} h\,d\mu\right| \leq \int_{B_\pm \setminus E_\pm} |h|\,d\mu < \varepsilon. \tag{2.2.245}$$

Together, (2.2.244) and (2.2.245) prove that

$$\int_{B_{\rho_\#}(x,r/4)} |h|\,\mathrm{d}\mu < 2\varepsilon. \tag{2.2.246}$$

Since $\varepsilon > 0$ is arbitrary, this forces $\int_{B_{\rho_\#}(x,r/4)} |h|\,\mathrm{d}\mu = 0$ hence, further, $h = 0$ at μ-a.e. point in $B_{\rho_\#}(x, r/4)$. Given that the topological space (X, τ_ρ) is Lindelöf (cf. Proposition 2.2.3), we ultimately conclude that $h = 0$ at μ-a.e. point in $X \setminus F$. □

A version of Proposition 2.2.8, for kernels which are standard in *both* variables, is presented next.

Corollary 2.2.11 *Assume (X, ρ, μ) is a d-Ahlfors regular quasi-metric space for some $d \in (0, \infty)$ with the property that the quasi-distance $\rho : X \times X \to [0, +\infty)$ is continuous in the product topology $\tau_\rho \times \tau_\rho$. Also, assume that the measure μ is Borel-semiregular. Next, suppose*

$$K : X \times X \setminus \mathrm{diag}(X) \longrightarrow \mathbb{R} \tag{2.2.247}$$

is a kernel which is standard in both variables, i.e., K is $\mu \otimes \mu$-measurable and both $K(x, y)$ and $K^\top(x, y) := K(y, x)$ satisfy (2.2.25)-(2.2.26). *Fix $q \in (1, \infty)$ and assume that an operator*

$$T : L^q(X, \mu) \longrightarrow L^{q,\infty}(X, \mu), \ \textit{linear and bounded}, \tag{2.2.248}$$

has been given, with the property that T is associated in $L^q(X, \mu)$ with the kernel K, in the sense that for each $f \in L^q(X, \mu)$ one has

$$(Tf)(x) = \int_X K(x, y) f(y)\,\mathrm{d}\mu(y) \ \textit{for } \mu\textit{-a.e. } x \in X \setminus \mathrm{supp}\, f. \tag{2.2.249}$$

For each $\varepsilon > 0$ define T_ε as in (2.2.32)-(2.2.33), *and introduce the maximal operator $T_{\max}$ as in* (2.2.34). *Then*

$$T : L^p(X, \mu) \longrightarrow L^p(X, \mu) \ \textit{linearly and boundedly for each } p \in (1, \infty), \tag{2.2.250}$$

and the maximal operator $T_{\max}$ induces a sub-linear, bounded, and continuous mapping

$$T_{\max} : L^p(X, \mu) \longrightarrow L^p(X, \mu) \ \textit{for each } p \in (1, \infty). \tag{2.2.251}$$

Also, for each fixed reference point $x_0 \in X$,

$$\begin{aligned} &T_{\max} : L^1\Big(X, \tfrac{\mu(x)}{1+\rho(x, x_0)^d}\Big) \cap L^p_{\mathrm{loc}}(X, \mu) \longrightarrow L^p_{\mathrm{loc}}(X, \mu) \\ &\quad \textit{is well-defined, bounded, continuous, for each } p \in (1, \infty). \end{aligned} \tag{2.2.252}$$

Consequently, for each $\varepsilon > 0$ the linear operator T_ε is bounded in the context

$$T_\varepsilon : L^p(X,\mu) \longrightarrow L^p(X,\mu) \text{ for each } p \in (1,\infty), \tag{2.2.253}$$

and

$$\sup_{\varepsilon>0} \|T_\varepsilon\|_{L^p(X,\mu)\to L^p(X,\mu)} < \infty \text{ for each } p \in (1,\infty). \tag{2.2.254}$$

Proof From (2.2.227) we know that

$$\begin{gathered} T : L^r(X,\mu) \longrightarrow L^r(X,\mu), \\ \text{linearly and boundedly for each } r \in (1,q). \end{gathered} \tag{2.2.255}$$

Fix some arbitrary $r_o \in (1,q)$ and denote by r_o' its Hölder conjugate exponent. Then (2.2.255) ensures that T maps $L^{r_o}(X,\mu)$ linearly and boundedly into itself. Via duality, we then conclude that

$$T^\top \text{ maps } L^{r_o'}(X,\mu) \text{ linearly and boundedly into itself.} \tag{2.2.256}$$

We also claim that

$$\begin{gathered} \text{for each function } f \in L^{r_o'}(X,\mu) \text{ we have} \\ (T^\top f)(x) = \int_X K^\top(x,y) f(y)\, d\mu(y) \text{ for } \mu\text{-a.e. } x \in X \setminus \operatorname{supp} f. \end{gathered} \tag{2.2.257}$$

To justify the claim in (2.2.257), pick some $f \in L^{r_o'}(X,\mu)$ and consider an arbitrary test function $\phi \in \mathscr{C}_c^\beta(X,\mu)$, for some finite exponent $\beta \in \big(0, (\log_2 C_\rho)^{-1}\big]$, satisfying

$$\operatorname{dist}_\rho\big(\operatorname{supp} f, \operatorname{supp} \phi\big) > 0. \tag{2.2.258}$$

The latter property then permits us to use Fubini's Theorem which, in concert with (2.2.249), gives

$$\begin{aligned} \int_X (T^\top f)(x)\phi(x)\, d\mu(x) &= \int_X f(y)(T\phi)(y)\, d\mu(y) \\ &= \int_X f(y)\Big(\int_X K(y,x)\phi(x)\, d\mu(x)\Big)\, d\mu(y) \\ &= \int_X \Big(\int_X K^\top(x,y) f(y)\, d\mu(y)\Big)\phi(x)\, d\mu(x). \end{aligned} \tag{2.2.259}$$

Note that $T^\top f \in L^{r_o'}(X,\mu) \subseteq L^1_{\rm loc}(X,\mu)$, thanks to (2.2.256). In addition, (2.2.25) implies that the function defined at μ-a.e. $x \in X \setminus \operatorname{supp} f$ according to $g(x) := \int_X K^\top(x,y) f(y)\, d\mu(y)$ is absolutely integrable with respect to μ on any closed subset E of $X \setminus \operatorname{supp} f$ satisfying $\operatorname{dist}_\rho\big(E, \operatorname{supp} f\big) > 0$. As a consequence, $h := g - \big(T^\top f\big)\big|_{X\setminus \operatorname{supp} f}$ satisfies (2.2.228)-(2.2.229) (with $F := \operatorname{supp} f$). As such, Lemma 2.2.10 applies and proves that (2.2.257) holds. Granted this, and given that $K^\top$ has the same qualities as K, we may then run the program that has produced (2.2.227) in the proof of Proposition 2.2.8 for the transpose operator $T^\top$ and arrive at the conclusion that

$$\begin{aligned}&T^{\top} : L^r(X,\mu) \longrightarrow L^r(X,\mu),\\ &\text{linearly and boundedly for each } r \in (1, r_o').\end{aligned} \tag{2.2.260}$$

Via duality we therefore obtain

$$\begin{aligned}&T : L^r(X,\mu) \longrightarrow L^r(X,\mu),\\ &\text{linearly and boundedly for each } r \in (r_o, \infty).\end{aligned} \tag{2.2.261}$$

Since $r_o \in (1, q)$ has been arbitrarily chosen, this implies

$$\begin{aligned}&T : L^r(X,\mu) \longrightarrow L^r(X,\mu),\\ &\text{linearly and boundedly for each } r \in (1, \infty),\end{aligned} \tag{2.2.262}$$

proving (2.2.250). All other claims now follow from this and Proposition 2.2.7. □

The point of our next result is that smoothly truncating a given kernel retains some of its essential properties, in a quantitatively uniform fashion with respect to the truncation parameter.

Lemma 2.2.12 *Suppose (X, ρ, μ) is a d-Ahlfors regular quasi-metric space for some $d \in (0, \infty)$ and assume $K : X \times X \setminus \operatorname{diag}(X) \to \mathbb{R}$ is a standard kernel in the first variable (i.e., the function K is $\mu\otimes\mu$-measurable and* (2.2.25)*-*(2.2.26) *are satisfied). Pick a real-valued function $\psi \in \mathscr{C}^\infty(\mathbb{R})$ with the property that $\psi \equiv 1$ on $\mathbb{R} \setminus [-1, 1]$ and $\psi \equiv 0$ on $[-1/2, 1/2]$. Having chosen a finite exponent $\beta \in \big(0, (\log_2 C_\rho)^{-1}\big]$, for each $\varepsilon > 0$ define*

$$K_{(\varepsilon)}(x, y) := K(x, y)\psi\Big(\frac{\rho_\#^\beta(x, y)}{\varepsilon^\beta}\Big), \qquad \forall (x, y) \in X \times X \setminus \operatorname{diag}(X). \tag{2.2.263}$$

Then each $K_{(\varepsilon)}$ is also a standard kernel in the first variable, with constants independent of ε. Moreover, if actually $K : X \times X \setminus \operatorname{diag}(X) \to \mathbb{R}$ is a standard kernel in both variables, then each $K_{(\varepsilon)}$ is also a standard kernel in both variables, with constants independent of ε.

Proof Fix an arbitrary $\varepsilon > 0$. First, from (2.2.263) and (2.2.25) we see that for each $x, y \in X$ with $x \neq y$ we have

$$\big|K_{(\varepsilon)}(x, y)\big| \leq \Big(\sup_{\mathbb{R}} |\psi|\Big)\frac{C_0}{\rho(x, y)^d}. \tag{2.2.264}$$

Going further, pick $x_1, x_2, y \in X$ with $y \notin \{x_1, x_2\}$ and $\rho(x_2, y) \geq C_1\rho(x_2, x_1)$. The goal is to estimate the difference $K_{(\varepsilon)}(x_1, y) - K_{(\varepsilon)}(x_2, y)$, much as in (2.2.26). To this end, select a real number

$$C > (C_\rho)^3. \tag{2.2.265}$$

We shall analyze several cases. First, if $\rho(x_2, y) > C\varepsilon$ then

$$\begin{aligned}\rho(x_2, y) &\leq C_\rho \cdot \max\{\rho(x_2, x_1), \rho(x_1, y)\} \\ &\leq C_\rho \cdot \max\{\rho(x_2, y)/C_1, \rho(x_1, y)\}\end{aligned} \tag{2.2.266}$$

which, in view of the fact that $C > C_\rho$ (cf. (2.2.265)), forces $\rho(x_2, y) \leq C_\rho \cdot \rho(x_1, y)$. Hence,

$$C\varepsilon < \rho(x_2, y) \leq C_\rho \cdot \rho(x_1, y). \tag{2.2.267}$$

From this and [112, (7.1.21)] we then conclude that

$$\varepsilon < (C_\rho)^{-3} C\varepsilon \leq (C_\rho)^{-2} \cdot \rho(x_1, y) \leq \rho_\#(x_1, y). \tag{2.2.268}$$

Since we also have

$$\varepsilon < (C_\rho)^{-2} C\varepsilon < (C_\rho)^{-2} \rho(x_2, y) \leq \rho_\#(x_2, y), \tag{2.2.269}$$

it follows that

$$\psi\Big(\frac{\rho_\#^\beta(x_1, y)}{\varepsilon^\beta}\Big) = 0 \text{ and } \psi\Big(\frac{\rho_\#^\beta(x_2, y)}{\varepsilon^\beta}\Big) = 0, \tag{2.2.270}$$

hence $K_{(\varepsilon)}(x_1, y) = 0$ and $K_{(\varepsilon)}(x_2, y) = 0$. Hence, in this case we have

$$K_{(\varepsilon)}(x_1, y) - K_{(\varepsilon)}(x_2, y) = 0. \tag{2.2.271}$$

Assume next that $\rho(x_2, y) \leq C\varepsilon$. In this case, write

$$\begin{aligned}\big|K_{(\varepsilon)}(x_1, y) - K_{(\varepsilon)}(x_2, y)\big| &\leq \big|K_{(\varepsilon)}(x_1, y) - K_{(\varepsilon)}(x_2, y)\big|\Big|\psi\Big(\frac{\rho_\#^\beta(x_1, y)}{\varepsilon^\beta}\Big)\Big| \\ &\quad + \big|K_{(\varepsilon)}(x_2, y)\big|\Big|\psi\Big(\frac{\rho_\#^\beta(x_1, y)}{\varepsilon^\beta}\Big) - \psi\Big(\frac{\rho_\#^\beta(x_2, y)}{\varepsilon^\beta}\Big)\Big| \\ &=: \mathrm{I} + \mathrm{II}.\end{aligned} \tag{2.2.272}$$

Then (2.2.26) gives

$$\mathrm{I} \leq C_0 \Big(\sup_{\mathbb{R}} |\psi|\Big) \frac{\rho(x_2, x_1)^\gamma}{\rho(x_2, y)^{d+\gamma}}. \tag{2.2.273}$$

Also, based on (2.2.25), the Mean Value Theorem, [112, (7.1.19)-(7.1.20)], and the fact that we are presently assuming $\rho(x_2, y) \leq C\varepsilon$ we may estimate

$$\begin{aligned}
\mathrm{II} &\leq \Big(\sup_{\mathbb{R}}|\psi'|\Big)\frac{C_0}{\rho(x_2,y)^d}\frac{\big|\rho_\#^\beta(x_1,y)-\rho_\#^\beta(x_2,y)\big|}{\varepsilon^\beta} \\
&\leq \Big(\sup_{\mathbb{R}}|\psi'|\Big)\frac{C_0}{\rho(x_2,y)^d}\frac{\rho_\#^\beta(x_2,x_1)}{\varepsilon^\beta} \\
&\leq CC_0(\widetilde{C}_\rho)^\beta\Big(\sup_{\mathbb{R}}|\psi'|\Big)\frac{\rho(x_2,x_1)^\beta}{\rho(x_2,y)^{d+\beta}}.
\end{aligned} \tag{2.2.274}$$

If $\gamma \leq (\log_2 C_\rho)^{-1}$ we may choose $\beta := \gamma$ to begin with, then deduce from (2.2.264), (2.2.273), and (2.2.274) that $K_{(\varepsilon)}$ is indeed a standard kernel. In the case when $\gamma > (\log_2 C_\rho)^{-1}$, the fact that we have $\rho(x_2,y) \geq C_1\rho(x_2,x_1)$ entails

$$\frac{\rho(x_2,x_1)^\gamma}{\rho(x_2,y)^{d+\gamma}} \leq (C_1)^{\beta-\gamma}\frac{\rho(x_2,x_1)^\beta}{\rho(x_2,y)^{d+\beta}}. \tag{2.2.275}$$

In concert with (2.2.264), (2.2.273), and (2.2.274) this once again permits us to conclude that $K_{(\varepsilon)}$ is a standard kernel in this case. Ultimately, we conclude that $K_{(\varepsilon)}$ is a standard kernel in the first variable, with constants independent of ε.

Finally, since

$$\begin{aligned}
K_{(\varepsilon)}^\top(x,y) &= K_{(\varepsilon)}(y,x) = K(y,x)\psi\Big(\frac{\rho_\#^\beta(y,x)}{\varepsilon^\beta}\Big) \\
&= K(y,x)\psi\Big(\frac{\rho_\#^\beta(x,y)}{\varepsilon^\beta}\Big) \quad \text{for all } (x,y) \in X\times X\setminus \mathrm{diag}(X),
\end{aligned} \tag{2.2.276}$$

the same type of reasoning as above shows that if K is a standard kernel in both variables then each $K_{(\varepsilon)}^\top$ is also a standard kernel in both variables with constants independent of ε. □

Our next proposition concludes boundedness results for the maximal operator starting from the assumption that the truncated operators are uniformly bounded.

Proposition 2.2.13 *Suppose (X,ρ,μ) is a d-Ahlfors regular quasi-metric space for some $d \in (0,\infty)$ with the property that the quasi-distance $\rho : X\times X \to [0,+\infty)$ is continuous in the product topology $\tau_\rho \times \tau_\rho$. Assume*

$$K : X\times X\setminus \mathrm{diag}(X) \longrightarrow \mathbb{R} \tag{2.2.277}$$

is a standard kernel in both variables (cf. Definition 2.2.4). For each $\varepsilon > 0$ introduce the truncated operator T_ε as in (2.2.32)-(2.2.33), and assume that there exists an exponent $q \in (1,\infty)$ with the property that

$$\sup_{\varepsilon>0}\|T_\varepsilon\|_{L^q(X,\mu)\to L^{q,\infty}(X,\mu)} < \infty. \tag{2.2.278}$$

The maximal operator $T_{\max}$, defined as in (2.2.34)*, induces a sub-linear, bounded, and continuous mapping in the following contexts:*

$$T_{\max} : L^1(X, \mu) \longrightarrow L^{1,\infty}(X, \mu), \text{ and} \tag{2.2.279}$$

$$T_{\max} : L^p(X, \mu) \longrightarrow L^p(X, \mu) \text{ for each } p \in (1, \infty). \tag{2.2.280}$$

Proof Recall that we are assuming that the kernel K is standard in both variables, i.e., both K and $K^\top$ satisfy (2.2.25)-(2.2.26) for some finite constants $\gamma > 0$ and $C_0 \geq 0$, $C_1 > 0$. For each $\varepsilon > 0$ define $K_{(\varepsilon)}$ as in (2.2.263). From Lemma 2.2.12 we know that each $K_{(\varepsilon)}$ is also a standard kernel in both variables, with constants independent of ε. To proceed, fix some reference point $x_0 \in X$. For each $\varepsilon > 0$ and each function

$$f \in L^1\Big(X, \frac{\mu(x)}{1 + \rho(x, x_0)^d}\Big) \tag{2.2.281}$$

define

$$(T_{(\varepsilon)} f)(x) := \int_X K_{(\varepsilon)}(x, y) f(y)\, \mathrm{d}\mu(y), \qquad \forall x \in X. \tag{2.2.282}$$

Let $\mathcal{M}_X$ be the Hardy-Littlewood maximal function on X (cf. (A.0.84)). We claim that there exists some $C \in (0, \infty)$ with the property that for each function f as in (2.2.281) we have

$$\big|T_\varepsilon f - T_{(\varepsilon)} f\big| \leq C \mathcal{M}_X f \text{ on } X, \text{ for each } \varepsilon > 0. \tag{2.2.283}$$

To justify this, fix $\varepsilon > 0$ and pick a function f as in (2.2.281). For each $x \in X$ write

$$\begin{aligned} &\big|(T_\varepsilon f)(x) - (T_{(\varepsilon)} f)(x)\big| \\ &\quad \leq \int_X |K(x, y)| \Big|\psi\Big(\frac{\rho_\#^\beta(x, y)}{\varepsilon^\beta}\Big) - \mathbf{1}_{\{y \in X : \rho(x,y) > \varepsilon\}}(y)\Big| |f(y)|\, \mathrm{d}\mu(y). \end{aligned} \tag{2.2.284}$$

Recall (2.2.25) and observe that $\psi\Big(\frac{\rho_\#^\beta(x,y)}{\varepsilon^\beta}\Big) \neq \mathbf{1}_{\{y \in X : \rho(x,y) > \varepsilon\}}(y)$ forces $\rho(x, y)$ to be comparable to ε, i.e., there exists $\theta \in (0, 1)$ independent of x, y such that $\theta < \rho(x, y)/\varepsilon < \theta^{-1}$. Consequently, there exists a constant $C \in (0, \infty)$ with the property that

$$\big|(T_\varepsilon f)(x) - (T_{(\varepsilon)} f)(x)\big| \leq \frac{C}{\varepsilon^d} \int_{B_\rho(x, C\varepsilon)} |f(y)|\, \mathrm{d}\mu(y) \text{ for each } x \in X. \tag{2.2.285}$$

Since X is d-Ahlfors regular, this readily implies (2.2.283).

Next, from (2.2.278), (2.2.283), and [112, (7.6.18)] we see that

$$\sup_{\varepsilon > 0} \big\|T_{(\varepsilon)}\big\|_{L^q(X,\mu) \to L^{q,\infty}(X,\mu)} < \infty. \tag{2.2.286}$$

Granted this, (2.2.282), and the fact that each $K_{(\varepsilon)}$ is a standard kernel in both variables with constants independent of ε, we may invoke Corollary 2.2.11 to conclude that

$$(T_{(\varepsilon)})_{\max} : L^1(X,\mu) \longrightarrow L^{1,\infty}(X,\mu), \text{ and} \tag{2.2.287}$$

$$(T_{(\varepsilon)})_{\max} : L^p(X,\mu) \longrightarrow L^p(X,\mu) \text{ for } p \in (1,\infty), \tag{2.2.288}$$

are all bounded operators uniformly in $\varepsilon > 0$.

Recall next from (2.2.33)-(2.2.34) and (2.2.282) that for each $\varepsilon > 0$ and each function f as in (2.2.281) we have

$$\big((T_{(\varepsilon)})_{\max} f\big)(x) = \sup_{\varepsilon'>0} \left| \int_{y\in X,\, \rho(x,y)>\varepsilon'} K_{(\varepsilon)}(x,y) f(y)\, \mathrm{d}\mu(y) \right| \tag{2.2.289}$$

for all points $x \in X$. Observe that whenever $\varepsilon' > (C_\rho)^2\varepsilon$ and $\rho(x,y) > \varepsilon'$ then we have $\rho(x,y) > (C_\rho)^2\varepsilon$ which, in view of [112, (7.1.21)], further implies $\rho_\#(x,y) > \varepsilon$. Consequently, $\psi\Big(\frac{\rho_\#^\beta(x,y)}{\varepsilon^\beta}\Big) = 1$ so $K_{(\varepsilon)}(x,y) = K(x,y)$ in this case. Consequently,

$$\begin{gathered} \int_{y\in X,\, \rho(x,y)>\varepsilon'} K_{(\varepsilon)}(x,y) f(y)\, \mathrm{d}\mu(y) = (T_{\varepsilon'} f)(x), \\ \text{for } f \text{ as in (2.2.281)}, \ x \in X, \text{ and } \varepsilon' > (C_\rho)^2\varepsilon. \end{gathered} \tag{2.2.290}$$

From (2.2.289) and (2.2.290) we then conclude that for each f as in (2.2.281) and each $\varepsilon > 0$ we have

$$\big((T_{(\varepsilon)})_{\max} f\big)(x) \geq \sup_{\varepsilon'>(C_\rho)^2\varepsilon} \big|(T_{\varepsilon'} f)(x)\big|, \qquad \forall x \in X. \tag{2.2.291}$$

Since for each $x \in X$ we have

$$\sup_{\varepsilon'>(C_\rho)^2\varepsilon} \big|(T_{\varepsilon'} f)(x)\big| \nearrow \sup_{\varepsilon'>0} \big|(T_{\varepsilon'} f)(x)\big| = \big(T_{\max} f\big)(x) \text{ as } \varepsilon \searrow 0, \tag{2.2.292}$$

Lebesgue's Monotone Convergence Theorem applies and gives that for each f as in (2.2.281) and each $p \in (1,\infty)$ we have

$$\int_X \sup_{\varepsilon'>(C_\rho)^2\varepsilon} (T_{\varepsilon'} f)^p \, \mathrm{d}\mu \nearrow \int_X \big(T_{\max} f\big)^p \, \mathrm{d}\mu \text{ as } \varepsilon \searrow 0. \tag{2.2.293}$$

Collectively, (2.2.291)-(2.2.293) and (2.2.288) then permit us to conclude that for each exponent $p \in (1,\infty)$ there exists some constant $C = C(X,K,p) \in (0,\infty)$ such that

$$\|T_{\max}f\|_{L^p(X,\mu)} = \lim_{\varepsilon \searrow 0} \left(\int_X \sup_{\varepsilon' > (C_\rho)^2 \varepsilon} (T_{\varepsilon'} f)^p \, d\mu \right)^{1/p}$$

$$\leq \limsup_{\varepsilon \searrow 0} \left(\int_X ((T_{(\varepsilon)})_{\max} f)^p \, d\mu \right)^{1/p}$$

$$\leq \sup_{\varepsilon > 0} \|(T_{(\varepsilon)})_{\max} f\|_{L^p(X,\mu)} \leq C\|f\|_{L^p(X,\mu)}, \tag{2.2.294}$$

for each function $f \in L^p(X, \mu)$. From this, the claims about (2.2.280) readily follow. Finally, the claims about (2.2.279) are dealt with similarly, now using (2.2.287). □

Next we elaborate on how the functional analytic properties of the maximal operator and the principal-value singular integral operator associated with a given kernel are related to one another. Essentially, Proposition 2.2.14 shows that "T_{pv} is bounded if and only if $T_{\max}$ is bounded."

Proposition 2.2.14 *Assume (X, ρ, μ) is a d-Ahlfors regular quasi-metric space for some $d \in (0, \infty)$ with the property that the quasi-distance $\rho : X \times X \to [0, +\infty)$ is continuous in the product topology $\tau_\rho \times \tau_\rho$, and such that μ is a complete Borel-semiregular measure. Fix a finite exponent $\beta \in \big(0, (\log_2 C_\rho)^{-1}\big]$ along with some $p \in (1, \infty)$. Also, suppose*

$$K : X \times X \setminus \text{diag}(X) \longrightarrow \mathbb{R} \tag{2.2.295}$$

is a kernel which is standard in the first variable (cf. Definition 2.2.4). Finally, for each $\varepsilon > 0$ associate T_ε with the kernel K as in (2.2.32)-(2.2.33), and introduce the maximal operator $T_{\max}$ as in (2.2.34). Then the following statements are true.

(1) For each function $f \in \mathscr{C}^\beta_c(X, \rho)$ the limit

$$\lim_{\varepsilon \to 0^+} \int_{y \in X,\, \rho(x,y) > \varepsilon} K(x, y) f(y) \, d\mu(y) \ \textit{exists for } \mu\textit{-a.e. } x \in X \tag{2.2.296}$$

if and only if the limit

$$\mathscr{K}(x) := \lim_{\varepsilon \to 0^+} \int_{\substack{y \in X \\ 1 > \rho(x,y) > \varepsilon}} K(x, y) \, d\mu(y) \ \textit{exists for } \mu\textit{-a.e. } x \in X. \tag{2.2.297}$$

(2) If (2.2.297) holds then the mapping assigning to each $f \in \mathscr{C}^\beta_c(X, \rho)$ the function

$$(T_{\text{pv}} f)(x) := \lim_{\varepsilon \to 0^+} \int_{y \in X,\, \rho(x,y) > \varepsilon} K(x, y) f(y) \, d\mu(y) \ \textit{for } \mu\textit{-a.e. } x \in X \tag{2.2.298}$$

extends to a linear and bounded operator from $L^p(X,\mu)$ *into* $L^{p,\infty}(X,\mu)$ *if and only if*

$$T_{\max} : L^p(X,\mu) \longrightarrow L^{p,\infty}(X,\mu) \text{ is a bounded mapping.} \tag{2.2.299}$$

Moreover, if either (hence both) of these conditions happen, then (2.2.296) *holds for each* $f \in L^p(X,\mu)$*, the extension of* $\mathscr{C}_c^\beta(X,\rho) \ni f \mapsto T_{\rm PV} f$ *to a linear and bounded operator from* $L^p(X,\mu)$ *into* $L^{p,\infty}(X,\mu)$ *is*

$$\begin{gathered} T_{\rm PV} : L^p(X,\mu) \longrightarrow L^{p,\infty}(X,\mu), \text{ acting on each } f \in L^p(X,\mu) \text{ by} \\ (T_{\rm PV} f)(x) = \lim_{\varepsilon\to 0^+} \int_{y\in X,\,\rho(x,y)>\varepsilon} K(x,y)f(y)\,d\mu(y) \text{ for } \mu\text{-a.e. } x \in X. \end{gathered} \tag{2.2.300}$$

(3) Strengthen the original hypotheses by now assuming that K is actually a standard kernel in both variables. Once again, suppose (2.2.297) *holds. Then*

$$\begin{gathered} \text{the mapping assigning to each } f \in \mathscr{C}_c^\beta(X,\rho) \text{ the function } T_{\rm PV} f \\ \text{defined as in (2.2.298) extends to a linear and bounded operator} \\ \text{from } L^p(X,\mu) \text{ into } L^{p,\infty}(X,\mu) \end{gathered} \tag{2.2.301}$$

if and only if

$$\begin{gathered} T_{\max} : L^q(X,\mu) \longrightarrow L^q(X,\mu) \text{ is a bounded mapping} \\ \text{for some (or each) index } q \in (1,\infty). \end{gathered} \tag{2.2.302}$$

Moreover, the validity of either (hence both) of these conditions implies that for each exponent $q \in (1,\infty)$ *the operator*

$$\begin{gathered} T_{\rm PV} : L^q(X,\mu) \longrightarrow L^q(X,\mu), \text{ acting on each } f \in L^q(X,\mu) \text{ by} \\ (T_{\rm PV} f)(x) = \lim_{\varepsilon\to 0^+} \int_{y\in X,\,\rho(x,y)>\varepsilon} K(x,y)f(y)\,d\mu(y) \text{ for } \mu\text{-a.e. } x \in X, \end{gathered} \tag{2.2.303}$$

is well defined, linear, bounded,

$$\begin{gathered} \text{for each function } f \in L^q(X,\mu) \text{ one has} \\ T_\varepsilon f \longrightarrow T_{\rm PV} f \text{ in } L^q(X,\mu) \text{ as } \varepsilon \to 0^+, \end{gathered} \tag{2.2.304}$$

and, corresponding to $q = 1$*, the operator*

$$\begin{gathered} T_{\rm PV} : L^1(X,\mu) \longrightarrow L^{1,\infty}(X,\mu), \text{ acting on each } f \in L^1(X,\mu) \text{ by} \\ (T_{\rm PV} f)(x) = \lim_{\varepsilon\to 0^+} \int_{y\in X,\,\rho(x,y)>\varepsilon} K(x,y)f(y)\,d\mu(y) \text{ for } \mu\text{-a.e. } x \in X, \end{gathered} \tag{2.2.305}$$

is also well defined, linear, bounded.

(4) *As above, strengthen the original hypotheses by assuming that K is in fact a standard kernel in both variables and, once more, suppose* (2.2.297) *holds. In addition, assume the limit*

$$\lim_{\varepsilon\to 0^+} \int\limits_{\substack{y\in X\\ 1>\rho(y,x)>\varepsilon}} K^{\top}(x,y)\,\mathrm{d}\mu(y) \text{ exists for } \mu\text{-a.e. } x\in X, \tag{2.2.306}$$

and suppose there exists some integrability exponent $q\in(1,\infty)$ such that

$$T_{\max} : L^q(X,\mu) \longrightarrow L^q(X,\mu) \text{ is a bounded mapping.} \tag{2.2.307}$$

Then, if $q'\in(1,\infty)$ is the Hölder conjugate exponent of q, the (real) transpose of (2.2.303) *is the operator*

$$T^{\top}_{\mathrm{PV}} : L^{q'}(X,\mu) \longrightarrow L^{q'}(X,\mu), \text{ acting on each } f\in L^{q'}(X,\mu) \text{ as}$$

$$(T^{\top}_{\mathrm{PV}} f)(x) = \lim_{\varepsilon\to 0^+} \int_{y\in X,\,\rho(y,x)>\varepsilon} K^{\top}(x,y) f(y)\,\mathrm{d}\mu(y) \text{ for } \mu\text{-a.e. } x\in X. \tag{2.2.308}$$

Proof Given an arbitrary function $f\in\mathscr{C}^{\beta}_{c}(X,\rho)$ along with some $\varepsilon\in(0,1)$, for each $x\in X$ decompose

$$\begin{aligned}\int_{y\in X,\,\rho(x,y)>\varepsilon} K(x,y)f(y)\,\mathrm{d}\mu(y) &= \int_{y\in X,\,\rho(x,y)\geq 1} K(x,y)f(y)\,\mathrm{d}\mu(y)\\ &\quad+ \int\limits_{\substack{y\in X\\ 1>\rho(x,y)>\varepsilon}} K(x,y)[f(y)-f(x)]\,\mathrm{d}\mu(y)\\ &\quad+ \Bigg(\int\limits_{\substack{y\in X\\ 1>\rho(x,y)>\varepsilon}} K(x,y)\,\mathrm{d}\mu(y)\Bigg) f(x).\end{aligned} \tag{2.2.309}$$

The fact that f is Hölder continuous and K satisfies (2.2.25) then ensures (cf. [112, (7.2.5)]) that

$$\begin{aligned}\lim_{\varepsilon\to 0^+} &\int\limits_{\substack{y\in X\\ 1>\rho(x,y)>\varepsilon}} K(x,y)[f(y)-f(x)]\,\mathrm{d}\mu(y)\\ &= \int\limits_{y\in X,\,1>\rho(x,y)} K(x,y)[f(y)-f(x)]\,\mathrm{d}\mu(y).\end{aligned} \tag{2.2.310}$$

Bearing this in mind it follows that (2.2.296) holds if and only if (2.2.297) holds, in which scenario we actually have

$$\begin{aligned}\lim_{\varepsilon\to 0^+}\int_{y\in X,\,\rho(x,y)>\varepsilon} & K(x,y)f(y)\,\mathrm{d}\mu(y) \\ &= \int_{y\in X,\,\rho(x,y)\geq 1} K(x,y)f(y)\,\mathrm{d}\mu(y) \\ &+ \int_{y\in X,\,1>\rho(x,y)} K(x,y)[f(y)-f(x)]\,\mathrm{d}\mu(y) + \mathscr{K}(x)f(x)\end{aligned} \tag{2.2.311}$$

for μ-a.e. $x\in X$. This proves the equivalence claimed in item *(1)*.

Turning attention to item *(2)*, work under the assumption that (2.2.297) holds. Recall from [112, (7.4.13)] that

$$\mathscr{C}_c^{\beta}(X,\rho)\hookrightarrow L^p(X,\mu) \text{ densely.} \tag{2.2.312}$$

Suppose first that the mapping (2.2.298) extends to a linear and bounded operator on $L^p(X,\mu)$ (which is a meaningful assumption in view of (2.2.297) and (2.2.312)). Denote by $\widetilde{T_{\rm PV}}$ the (unique) extension by density of the operator $T_{\rm PV}$ acting on $\mathscr{C}_c^{\beta}(X,\rho)$ as in (2.2.298) as a linear and bounded mapping

$$\widetilde{T_{\rm PV}}: L^p(X,\mu)\longrightarrow L^{p,\infty}(X,\mu). \tag{2.2.313}$$

We claim that

$$\begin{gathered}\text{for each function } f\in L^p(X,\mu) \text{ we have} \\ (\widetilde{T_{\rm PV}}f)(x)=\int_X K(x,y)f(y)\,\mathrm{d}\mu(y) \text{ for } \mu\text{-a.e. } x\in X\setminus \operatorname{supp} f.\end{gathered} \tag{2.2.314}$$

To prove this, fix an arbitrary $f\in L^p(X,\mu)$ and pick some $x_o\in X\setminus\operatorname{supp} f$. From (2.2.312) we know that there exists a sequence $\{f_j\}_{j\in\mathbb{N}}\subseteq\mathscr{C}_c^{\beta}(X,\rho)$ convergent to f in $L^p(X,\mu)$. After eventually multiplying by a suitable cutoff function, matters may be arranged so that, for some $r>0$, we have

$$B_\rho(x_o,r)\cap\operatorname{supp} f_j=\varnothing \text{ for each } j\in\mathbb{N}. \tag{2.2.315}$$

Then, on the one hand, the continuity of (2.2.313) implies

$$\widetilde{T_{\rm PV}}f=\lim_{j\to\infty}\widetilde{T_{\rm PV}}f_j=\lim_{j\to\infty}T_{\rm PV}f_j \text{ in } L^{p,\infty}(X,\mu). \tag{2.2.316}$$

By eventually passing to a subsequence we may therefore assume that

$$\widetilde{T_{\rm PV}}f=\lim_{j\to\infty}T_{\rm PV}f_j \text{ at } \mu\text{-a.e. point in } X. \tag{2.2.317}$$

Consequently, for μ-a.e. $x \in B_\rho(x_o, r)$ we have

$$(\widetilde{T_{\rm PV}} f)(x) = \lim_{j\to\infty} (T_{\rm PV} f_j)(x) = \lim_{j\to\infty} \int_X K(x,y) f_j(y)\,d\mu(y)$$
$$= \int_X K(x,y) f(y)\,d\mu(y), \tag{2.2.318}$$

where the last equality is justified as in (2.2.156). Reasoning as in (2.2.157)-(2.2.160) then finishes the proof of (2.2.314). Granted (2.2.313)-(2.2.314), we may now invoke (2.2.39) to conclude that (2.2.299) is true.

Conversely, assume (2.2.299) is true. Given that for each $f \in \mathscr{C}^\beta_c(X, \rho)$ we have

$$\left|T_{\rm PV} f\right| \leq T_{\max} f \text{ for } \mu\text{-a.e. point on } X, \tag{2.2.319}$$

and $T_{\rm PV} f$ happens to be a μ-measurable function, as the pointwise limit of a sequence of μ-measurable functions (cf. (2.2.298) and [112, Remark 3.1.2], where the latter uses the fact that the measure μ is assumed to be complete). In view of these properties, it follows from [112, (6.2.16)] that the operator (2.2.298) extends to a linear and bounded operator from $L^p(X, \mu)$ into $L^{p,\infty}(X, \mu)$.

Assume next that (2.2.299) holds, with the goal of showing that (2.2.296) holds for each function $f \in L^p(X, \mu)$, and that the extension of $\mathscr{C}^\beta_c(X, \rho) \ni f \mapsto T_{\rm PV} f$ to a linear and bounded operator on $L^p(X, \mu)$ is as described in (2.2.300). To this end, we shall use [112, Proposition 6.2.11] for the ambient $\mathscr{X} := X \times [0, \infty)$ equipped with the canonical product topology, $L^p(X, \mu)$ playing the role of the normed space Y, and $X \times \{0\}$ playing the role of X in the statement of [112, Proposition 6.2.11]. Also, take $\Gamma\big((x,0)\big) := \{x\} \times (0, \infty)$ for each $(x, 0) \in X \times \{0\}$, a choice ensuring that $(x, 0) \in \overline{\Gamma\big((x,0)\big)}$ for every $(x, 0) \in X \times \{0\}$.

Next, consider the linear operator $\mathcal{T}$, mapping functions defined on X into functions defined on $\mathscr{X} \setminus (X \times \{0\}) = X \times (0, \infty)$, of the following sort. For every given $f \in Y = L^p(X, \mu)$ define

$$(\mathcal{T} f)(x, \varepsilon) := \int_{\substack{y\in X\\ \rho(x,y)>\varepsilon}} K(x,y) f(y)\,d\mu(y) \text{ for all } (x, \varepsilon) \in X \times (0, \infty), \tag{2.2.320}$$

and note that this implies

$$(T_\star f)(x, 0) := \sup_{(z,\varepsilon)\in\Gamma((x,0))} |(\mathcal{T} f)(z, \varepsilon)| = \sup_{\varepsilon>0} \left| \int_{\substack{y\in X\\ \rho(x,y)>\varepsilon}} K(x,y) f(y)\,d\mu(y) \right|$$
$$= (T_{\max} f)(x) \text{ for every } x \in X. \tag{2.2.321}$$

In turn, from (2.2.321) and (2.2.299) we conclude that condition [112, (6.2.74)] is verified in the present setting. To finish the implementation of [112, Proposition 6.2.11] we finally choose $\mathcal{V} := \mathscr{C}^{\beta}_{c}(X,\rho)$, which is a dense subset of $L^p(X,\mu)$ (cf. (2.2.312)). Moreover, condition [112, (6.2.75)] is also satisfied since for every $g \in \mathcal{V}$ the limit

$$\lim_{\Gamma((x,0))\ni(x,\varepsilon)\to(x,0)} (\mathcal{T}g)(x,\varepsilon) = \lim_{\varepsilon\to 0^+} \int_{\substack{y\in Y \\ \rho(x,y)>\varepsilon}} K(x,y)g(y)\,\mathrm{d}\mu(y) = \lim_{\varepsilon\to 0^+} (T_\varepsilon g)(x) \tag{2.2.322}$$

exists for μ-a.e. $x \in X$, thanks to the assumption that (2.2.297) holds and item *(1)*. Hence, all hypotheses of [112, Proposition 6.2.11] are satisfied for the present choices, so we may conclude from it that for every function $f \in L^p(X,\mu)$ the limit

$$\big(T_{\mathrm{PV}} f\big)(x) = \lim_{\varepsilon\to 0^+} (T_\varepsilon f)(x) = \lim_{\Gamma((x,0))\ni(x,\varepsilon)\to(x,0)} (\mathcal{T}f)(x,\varepsilon) \tag{2.2.323}$$

exists for μ-a.e. $x \in X$. From this, the desired conclusions follow. This concludes the treatment of item *(2)*.

As far as item *(3)* is concerned, Corollary 2.2.11 applied to $T := T_{\mathrm{PV}}$ proves the equivalence of (2.2.301) with (2.2.302). If either of these conditions holds then items *(1)*-*(2)* imply that for each exponent $q \in (1,\infty)$ the operator (2.2.303) is well defined, linear, and bounded. In concert with (the strongest version of) (2.2.301) and Lebesgue's Dominated Convergence Theorem, this also ensures that (2.2.304) is valid. Consider the very last claim, pertaining to (2.2.305). Assuming (2.2.302), item *(i)* in Proposition 2.2.8 gives that T_{PV} extends to a bounded mapping of $L^1(X,\mu)$ into $L^{1,\infty}(X,\mu)$. With this in hand, what we have proved in the current item *(2)* applies (with $p := 1$) and yields all desired conclusions about (2.2.305).

Finally, consider the duality result claimed in item *(4)*. For each $\varepsilon > 0$ define

$$\begin{gathered} Q_\varepsilon : L^{q'}(X,\mu) \longrightarrow L^{q'}(X,\mu) \text{ given by} \\ (Q_\varepsilon f)(x) := \int_{y\in X,\,\rho(y,x)>\varepsilon} K^\top(x,y) f(y)\,\mathrm{d}\mu(y) \text{ for all } x \in X, \end{gathered} \tag{2.2.324}$$

for each function $f \in L^{q'}(X,\mu)$. Observe that, for each $\varepsilon > 0$, the assumption made in (2.2.307) implies that $T_\varepsilon^\top : L^q(X,\mu) \to L^q(X,\mu)$ is a well-defined, linear, and bounded operator, whose (real) transpose is

$$T_\varepsilon^\top = Q_\varepsilon \text{ on } L^{q'}(X,\mu). \tag{2.2.325}$$

Consequently, for each $\varepsilon > 0$ and each $f \in L^{q'}(X,\mu)$ we have

$$\|Q_\varepsilon f\|_{L^{q'}(X,\mu)} = \sup_{\|g\|_{L^q(X,\mu)}=1} \left|\int_X (Q_\varepsilon f) g \, \mathrm{d}\mu\right| = \sup_{\|g\|_{L^q(X,\mu)}=1} \left|\int_X f(T_\varepsilon g) \, \mathrm{d}\mu\right|$$

$$\leq \|f\|_{L^{q'}(X,\mu)} \|T_\varepsilon g\|_{L^q(X,\mu)} \leq \|f\|_{L^{q'}(X,\mu)} \|T_{\max} g\|_{L^q(X,\mu)}$$

$$\leq \|f\|_{L^{q'}(X,\mu)} \|T_{\max}\|_{L^q(X,\mu)\to L^q(X,\mu)} \tag{2.2.326}$$

hence

$$\sup_{\varepsilon>0} \|Q_\varepsilon\|_{L^{q'}(X,\mu)\to L^{q'}(X,\mu)} \leq \|T_{\max}\|_{L^q(X,\mu)\to L^q(X,\mu)}. \tag{2.2.327}$$

From (2.2.327), the fact that K is a standard kernel in both variables, and Proposition 2.2.13 we conclude that

$$\begin{aligned} &Q_{\max} : L^p(X,\mu) \longrightarrow L^p(X,\mu) \text{ is a bounded mapping} \\ &\quad \text{for each integrability exponent } p \in (1,\infty), \end{aligned} \tag{2.2.328}$$

where

$$(Q_{\max} f)(x) := \sup_{\varepsilon>0} |(Q_\varepsilon f)(x)| \text{ for each } f \in L^p(X,\mu). \tag{2.2.329}$$

Granted (2.2.328), and bearing in mind (2.2.306), what we have proved in the current item *(3)* guarantees that the operator

$$\begin{aligned} &Q : L^{q'}(X,\mu) \longrightarrow L^{q'}(X,\mu), \text{ acting on each } f \in L^{q'}(X,\mu) \text{ by} \\ &(Qf)(x) := \lim_{\varepsilon\to 0^+} \int_{y\in X,\, \rho(y,x)>\varepsilon} K^\top(x,y) f(y) \, \mathrm{d}\mu(y) \text{ for } \mu\text{-a.e. } x \in X, \end{aligned} \tag{2.2.330}$$

is well defined, linear, bounded, and

$$\begin{aligned} &Q_\varepsilon f \longrightarrow Qf \text{ in } L^{q'}(X,\mu) \text{ as } \varepsilon \to 0^+ \\ &\quad \text{for each function } f \in L^{q'}(X,\mu). \end{aligned} \tag{2.2.331}$$

Then the claim in relation to (2.2.308) is seen from this, (2.2.325), (2.2.330), and (2.2.304). □

It is also useful to note that the principal-value singular integral operator associated with a given kernel is bounded whenever the corresponding truncated singular integral operators are bounded in a uniform fashion with respect to the truncation parameter.

Corollary 2.2.15 *Suppose (X, ρ, μ) is a d-Ahlfors regular quasi-metric space for some $d \in (0, \infty)$ with the property that the quasi-distance $\rho : X \times X \to [0, +\infty)$ is continuous in the product topology $\tau_\rho \times \tau_\rho$, and such that μ is a complete Borel-semiregular measure. Assume $K : X \times X \setminus \mathrm{diag}(X) \to \mathbb{R}$ is a standard kernel in both variables (cf. Definition 2.2.4) such that*

$$\textit{the limit } \lim_{\varepsilon\to 0^+} \int_{\substack{y\in X\\ 1>\rho(x,y)>\varepsilon}} K(x,y)\,\mathrm{d}\mu(y) \textit{ exists for } \mu\textit{-a.e. } x\in X. \tag{2.2.332}$$

For each $\varepsilon > 0$ introduce the truncated operator T_ε as in (2.2.32)-(2.2.33), *and assume that there exists some $q \in (1,\infty)$ with the property that*

$$\sup_{\varepsilon>0} \|T_\varepsilon\|_{L^q(X,\mu)\to L^{q,\infty}(X,\mu)} < \infty. \tag{2.2.333}$$

Finally, fix a finite exponent $\beta \in \big(0,(\log_2 C_\rho)^{-1}\big]$.

Then for each $p \in [1,\infty)$ the mapping assigning to each $f \in \mathscr{C}_c^\beta(X,\rho)$ the function

$$\big(T_{\rm PV} f\big)(x) := \lim_{\varepsilon\to 0^+} \int_{y\in X,\,\rho(x,y)>\varepsilon} K(x,y)f(y)\,\mathrm{d}\mu(y) \textit{ for } \mu\textit{-a.e. } x\in X \tag{2.2.334}$$

extends to a linear and bounded operator

$$\begin{aligned} T_{\rm PV} &: L^p(X,\mu) \longrightarrow L^p(X,\mu) \textit{ if } p\in(1,\infty),\\ T_{\rm PV} &: L^1(X,\mu) \longrightarrow L^{1,\infty}(X,\mu) \textit{ if } p=1, \end{aligned} \tag{2.2.335}$$

acting on each $f \in L^p(X,\mu)$ according to

$$\big(T_{\rm PV} f\big)(x) = \lim_{\varepsilon\to 0^+} \int_{y\in X,\,\rho(x,y)>\varepsilon} K(x,y)f(y)\,\mathrm{d}\mu(y) \textit{ for } \mu\textit{-a.e. } x\in X. \tag{2.2.336}$$

Of course, (2.2.333) is satisfied if

$$T_{\max} : L^q(X,\mu) \longrightarrow L^q(X,\mu) \text{ is bounded.} \tag{2.2.337}$$

See Corollary 2.2.11 and Proposition 2.2.7 for more on this.

In the converse direction, if for some fixed $p \in (1,\infty)$ the operator $T_{\rm PV}$ is meaningfully defined as in (2.2.336) for each $f \in L^p(X,\mu)$ and induces a bounded mapping $T_{\rm PV} : L^p(X,\mu) \to L^{p,\infty}(X,\mu)$, then Corollary 2.2.11 and Proposition 2.2.8 applied (to $T := T_{\rm PV}$) plus Proposition 2.2.14 give that $T_{\rm PV}$ continues to be meaningfully defined by the formula in (2.2.336) for any function f as in (2.2.281), and

$$\begin{aligned} T_{\rm PV} &: L^1(X,\mu) \longrightarrow L^{1,\infty}(X,\mu) \text{ boundedly, and}\\ T_{\rm PV} &: L^q(X,\mu) \longrightarrow L^q(X,\mu) \text{ boundedly for each } q\in(1,\infty). \end{aligned} \tag{2.2.338}$$

Here is the proof of Corollary 2.2.15.

Proof of Corollary 2.2.15 All desired properties are obtained by combining Proposition 2.2.13 with Proposition 2.2.14. □

Moving on, the goal is to prove the basic weak convergence result stated a little later, in Theorem 2.2.20. This requires some preparations, and a number of auxiliary tools are developed in Lemmas 2.2.16, 2.2.17, 2.2.19.

Lemma 2.2.16 *Let X, Y be two normed spaces and suppose $a : X \times Y \to \mathbb{C}$ is a bilinear map that is also continuous, i.e., there exists $C \in [0,\infty)$ such that*

$$|a(x,y)| \leq C\|x\|_X\|y\|_Y \text{ for all } (x,y) \in X \times Y. \tag{2.2.339}$$

Then there exists a unique map $R_a \in \mathcal{L}(X \to Y^*)$ such that

$$a(x,y) = {}_{Y^*}\langle R_a x, y\rangle_Y \text{ for all } (x,y) \in X \times Y. \tag{2.2.340}$$

Proof Define the map $R_a : X \to Y^*$ by setting $R_a(x) := a(x,\cdot)$ for each $x \in X$. Then $R_a \in \mathcal{L}(X \to Y^*)$ and for each $(x,y) \in X \times Y$ we have ${}_{Y^*}\langle R_a x, y\rangle_Y = a(x,y)$ as wanted. Also, uniqueness follows from identity (2.2.340). □

Let X be a Banach space and recall the notation $\mathcal{L}(X) := \mathcal{L}(X \to X)$ and the operator norm

$$\|T\|_{\mathcal{L}(X)} := \sup\big\{\|Tx\| : x \in X,\ \|x\|_X = 1\big\}, \quad \text{for each } T \in \mathcal{L}(X). \tag{2.2.341}$$

Then the normed space $\big(\mathcal{L}(X), \|\cdot\|_{\mathcal{L}(X)}\big)$ is a Banach space. In this a setting, fix $(x,\xi) \in X \times X^*$ and define the linear map

$$\omega_{x,\xi} : \mathcal{L}(X) \longrightarrow \mathbb{C}, \quad \omega_{x,\xi}(T) := {}_X\langle Tx, \xi\rangle_{X^*} \text{ for each } T \in \mathcal{L}(X). \tag{2.2.342}$$

Since for each $T \in \mathcal{L}(X)$ we may estimate

$$\big|\omega_{x,\xi}(T)\big| \leq \big|{}_X\langle Tx,\xi\rangle_{X^*}\big| \leq \|\xi\|_{X^*}\|Tx\|_X \leq \|\xi\|_{X^*}\|x\|_X\|T\|_{\mathcal{L}(X)}, \tag{2.2.343}$$

it follows that $\omega_{x,\xi} \in \mathcal{L}(X)^*$ and $\big\|\omega_{x,\xi}\big\|_{\mathcal{L}(X)^*} \leq \|x\|_X\|\xi\|_{X^*}$.

Based on this discussion, we now define the linear space generated in $\mathcal{L}(X)^*$ by the collection $\big\{\omega_{x,\xi} : x \in X,\ \xi \in X^*\big\}$, i.e.,

$$\mathcal{L}_\omega(X) := \Big\{\sum_{j=1}^{N} \omega_{x_j,\xi_j} : N \in \mathbb{N},\ x_j \in X,\ \xi_j \in X^*\Big\}, \tag{2.2.344}$$

and set

$$\mathcal{L}_*(X) := \overline{\mathcal{L}_\omega(X)}^{\mathcal{L}(X)^*}. \tag{2.2.345}$$

Then $\mathcal{L}_*(X)$ is a close subspace of $\mathcal{L}(X)^*$ and is a Banach space when endowed with the norm inherited from $\mathcal{L}(X)^*$.

Lemma 2.2.17 *Let X be a reflexive Banach space. Consider the bilinear form*

$$\begin{aligned} a : \mathcal{L}(X) \times \mathcal{L}_*(X) &\longrightarrow \mathbb{C} \\ a(T,\Lambda) := \Lambda(T) \text{ for all } (T,\Lambda) &\in \mathcal{L}(X) \times \mathcal{L}_*(X), \end{aligned} \tag{2.2.346}$$

and let $R_a \in \mathcal{L}\big(\mathcal{L}(X) \to (\mathcal{L}_*(X))^*\big)$ *be the operator associated with a as in Lemma 2.2.16. Then* $R_a : \mathcal{L}(X) \longrightarrow (\mathcal{L}_*(X))^*$ *is an isometric isomorphism.*

Proof Pick $T \in \mathcal{L}(X)$ arbitrary. Then (2.2.342), (2.2.346), and (2.2.340) allow us to write

$$
\begin{aligned}
\|T\|_{\mathcal{L}(X)} &= \sup\Big\{\|Tx\|_X : x \in X,\ \|x\|_X = 1\Big\}\\
&= \sup\Big\{\big|{}_X\langle Tx, \xi\rangle_{X^*}\big| : x \in X,\ \|x\|_X = 1,\ \xi \in X^*,\ \|\xi\|_{X^*} = 1\Big\}\\
&= \sup\Big\{\big|\omega_{x,\xi}(T)\big| : x \in X,\ \|x\|_X = 1,\ \xi \in X^*,\ \|\xi\|_{X^*} = 1\Big\}\\
&= \sup\Big\{\big|a(T, \omega_{x,\xi})\big| : x \in X,\ \|x\|_X = 1,\ \xi \in X^*,\ \|\xi\|_{X^*} = 1\Big\}\\
&\leq \sup\Big\{\big|a(T, \Lambda)\big| : \Lambda \in \mathcal{L}_*(X),\ \|\Lambda\|_{\mathcal{L}(X)^*} \leq 1\Big\}\\
&= \sup\Big\{\big|{}_{(\mathcal{L}_*(X))^*}\langle R_a(T), \Lambda\rangle_{\mathcal{L}_*(X)}\big| : \Lambda \in \mathcal{L}_*(X),\ \|\Lambda\|_{\mathcal{L}(X)^*} \leq 1\Big\}\\
&\leq \big\|R_a(T)\big\|_{(\mathcal{L}_*(X))^*}
\end{aligned}
\tag{2.2.347}
$$

Also,

$$
\begin{aligned}
\big\|R_a(T)\big\|_{(\mathcal{L}_*(X))^*} &= \sup\Big\{\big|{}_{(\mathcal{L}_*(X))^*}\langle R_a(T), \Lambda\rangle_{\mathcal{L}_*(X)}\big| : \Lambda \in \mathcal{L}_*(X),\ \|\Lambda\|_{\mathcal{L}(X)^*} \leq 1\Big\}\\
&= \sup\Big\{\big|a(T, \Lambda)\big| : \Lambda \in \mathcal{L}_*(X),\ \|\Lambda\|_{\mathcal{L}(X)^*} \leq 1\Big\}\\
&= \sup\Big\{\big|\Lambda(T)\big| : \Lambda \in \mathcal{L}_*(X),\ \|\Lambda\|_{\mathcal{L}(X)^*} \leq 1\Big\}\\
&\leq \|T\|_{\mathcal{L}(X)}.
\end{aligned}
\tag{2.2.348}
$$

Thus,

$$
\big\|R_a(T)\big\|_{(\mathcal{L}_*(X))^*} = \|T\|_{\mathcal{L}(X)} \text{ for all } T \in \mathcal{L}(X). \tag{2.2.349}
$$

In particular, R_a is injective. To finish the proof of the lemma it remains to show that R_a is surjective. To this end, pick $\Theta \in (\mathcal{L}_*(X))^*$ and define $b : X \times X^* \longrightarrow \mathbb{C}$ by

$$
b(x, \xi) := {}_{(\mathcal{L}_*(X))^*}\langle \Theta, \omega_{x,\xi}\rangle_{\mathcal{L}_*(X)} \text{ for all } (x, \xi) \in X \times X^*. \tag{2.2.350}
$$

This is a bilinear and continuous map and we may apply Lemma 2.2.16 to conclude that there exists a unique map $T := R_b$ with $T \in \mathcal{L}\big(X \to (X^*)^*\big) = \mathcal{L}(X)$ (recall that X is reflexive) such that

$$
{}_{(\mathcal{L}_*(X))^*}\langle \Theta, \omega_{x,\xi}\rangle_{\mathcal{L}_*(X)} = {}_X\langle Tx, \xi\rangle_{X^*} \text{ for all } (x, \xi) \in X \times X^*. \tag{2.2.351}
$$

This, in concert with (2.2.342) and (2.2.346) further implies

$$ {}_{(\mathcal{L}_*(X))^*}\langle\Theta, \omega_{x,\xi}\rangle_{\mathcal{L}_*(X)} = \omega_{x,\xi}(T) = a\big(T, \omega_{x,\xi}\big) \tag{2.2.352} $$

for all $(x, \xi) \in X \times X^*$. Since $\{\omega_{x,\xi} : x \in X, \xi \in X^*\}$ spans a dense subspace of $\mathcal{L}_*(X)$ and since Θ and $a(T, \cdot)$ are continuous, it follows that

$$ {}_{(\mathcal{L}_*(X))^*}\langle\Theta, \Lambda\rangle_{\mathcal{L}_*(X)} = a\big(T, \Lambda\big) = {}_{(\mathcal{L}_*(X))^*}\langle R_a(T), \Lambda\rangle_{\mathcal{L}_*(X)} \tag{2.2.353} $$

for all $\Lambda \in \mathcal{L}_*(X)$. Hence, $R_a(T) = \Theta$ as functionals in $(\mathcal{L}_*(X))^*$. This completes the proof of the fact that R_a is onto. □

Next we consider two topologies defined on $\mathcal{L}(X)$. One is the weak operator topology (WOT) which is the topology induced by the family of seminorms

$$ \mathcal{L}(X) \ni T \mapsto \big|{}_X\langle Tx, \xi\rangle_{X^*}\big| \text{ for all } x \in X,\ \xi \in X^*. \tag{2.2.354} $$

The other is the ultraweak operator topology (UWOT) which is defined as the topology induced by the family of seminorms

$$ \mathcal{L}(X) \ni T \mapsto |\Lambda(T)| \text{ for all } \Lambda \in \mathcal{L}_*(X). \tag{2.2.355} $$

The relationship between these topologies on the closed unit ball in $\mathcal{L}(X)$ in the case when X is Banach is discussed in Lemma 2.2.19 below. As a preamble, we record the following result which appears in [173, Lemma, §1.2, p. 14].

Proposition 2.2.18 *Let $(\mathcal{E}, \|\cdot\|_{\mathcal{E}})$ be a Banach space and let $(\mathcal{E}^*, \|\cdot\|_{\mathcal{E}^*})$ be its dual. Consider a linear space $\mathcal{F} \subseteq \mathcal{E}^*$ and denote by $\sigma(\mathcal{E}, \mathcal{F})$ the topology induced on $\mathcal{E}$ by the family of seminorms $\{p_f : f \in \mathcal{F}\}$ defined by $p_f(e) := |f(e)|$ for every $e \in \mathcal{E}$ and each $f \in \mathcal{F}$. Similarly, consider the topology $\sigma(\mathcal{E}, \overline{\mathcal{F}})$, where $\overline{\mathcal{F}}$ denotes the closure of $\mathcal{F}$ in $(\mathcal{E}^*, \|\cdot\|_{\mathcal{E}^*})$.*

Then the topologies $\sigma(\mathcal{E}, \mathcal{F})$ and $\sigma(\mathcal{E}, \overline{\mathcal{F}})$ coincide on the closed unit ball in $\mathcal{E}$.

Here is the lemma alluded to earlier.

Lemma 2.2.19 *Let X be a Banach space and consider the closed unit ball in $(\mathcal{L}(X), \|\cdot\|_{\mathcal{L}(X)})$ denoted by $B_{\mathcal{L}(X)} := \{T \in \mathcal{L}(X) : \|T\|_{\mathcal{L}(X)} \leq 1\}$. Then* WOT *and* UWOT *coincide on $B_{\mathcal{L}(X)}$.*

Proof We apply Proposition 2.2.18 with $\mathcal{E} := \mathcal{L}(X)$ and $\mathcal{F} := \mathcal{L}_\omega(X)$. For this selection, we have (recall (2.2.345))

$$ \overline{\mathcal{F}} = \overline{\mathcal{L}_\omega(X)}^{(\mathcal{L}(X))^*} = \mathcal{L}_*(X) \tag{2.2.356} $$

and since (recall the notation in Proposition 2.2.18) $\sigma\big(\mathcal{L}(X), \mathcal{L}_*(X)\big)$ is the UWOT on $\mathcal{L}(X)$ and $\sigma\big(\mathcal{L}(X), \mathcal{L}_\omega(X)\big)$ is the WOT on $\mathcal{L}(X)$, the desired conclusion follows. □

We are now prepared to prove the following basic weak convergence result.

Theorem 2.2.20 *Let X be a reflexive Banach space. Then the closed unit ball $B_{\mathcal{L}(X)}$ in $\mathcal{L}(X)$ is compact in* WOT. *Consequently, for every sequence $\{T_j\}_{j\in\mathbb{N}}$ of operators in $\mathcal{L}(X)$ with the property that $\sup_{j\in\mathbb{N}} \|T_j\|_{\mathcal{L}(X)} < \infty$, there exists a subsequence $\{T_{j_k}\}_{k\in\mathbb{N}}$ and an operator $T \in \mathcal{L}(X)$ such that, for all $x \in X$ and all $\xi \in X^*$,*

$$ {}_X\langle T_{j_k}x, \xi\rangle_{X^*} \longrightarrow {}_X\langle Tx, \xi\rangle_{X^*} \quad \textit{as } k \to \infty. \tag{2.2.357}$$

Proof Using the isometric isomorphism from Lemma 2.2.17, we may identify $\mathcal{L}(X)$ with $\big(\mathcal{L}_*(X)\big)^*$. From Alaoglu's theorem we know that $B_{\mathcal{L}(X)}$ is compact in UWOT. Invoking Lemma 2.2.19, it follows that $B_{\mathcal{L}(X)}$ is also compact in WOT, as wanted.□

One particularly useful consequence of the above theorem is singled out in the corollary below.

Corollary 2.2.21 *Let (X, μ) be a measure space, fix an exponent $p \in (1, \infty)$, and suppose $\{T_j\}_{j\in\mathbb{N}}$ is a sequence of linear and bounded operators from $L^p(X, \mu)$ into itself with the property that $\sup_{j\in\mathbb{N}} \|T_j\|_{\mathcal{L}(L^p(X,\mu))} < \infty$.*

Then there exists a subsequence $\{T_{j_k}\}_{k\in\mathbb{N}}$ along with some $T \in \mathcal{L}\big(L^p(X, \mu)\big)$ such that

$$\begin{gathered} \lim_{k\to\infty} \int_X T_{j_k} f \cdot g \, d\mu = \int_X T f \cdot g \, d\mu \\ \textit{for all } f \in L^p(X, \mu),\ g \in L^{p'}(X, \mu), \end{gathered} \tag{2.2.358}$$

where $p' := (1 - 1/p)^{-1}$.

Proof This follows from Theorem 2.2.20 applied with $L^p(X, \mu)$ in place of X and Riesz's duality theorem to the effect that $(L^p(X, \mu))^*$ may be identified with $L^{p'}(X, \mu)$. □

Pressing on, in the next proposition we shall show that the only linear, bounded, and local operators are those of pointwise multiplication by an essentially bounded function.

Proposition 2.2.22 *Let $d \in (0, \infty)$ and suppose (X, ρ, μ) is a d-Ahlfors regular quasi-metric space with the property that μ is a Borel-semiregular measure. Fix $p \in (0, \infty)$. Then for each given operator*

$$T : L^p(X, \mu) \longrightarrow L^p(X, \mu) \tag{2.2.359}$$

the following are equivalent:

(1) There exists $b \in L^\infty(X, \mu)$ such that $Tf = b \cdot f$ for all $f \in L^p(X, \mu)$, i.e., T is the operator of pointwise multiplication by an essentially bounded function.
(2) The operator T is linear, bounded, and local in the sense that

$$\operatorname{supp}(Tf) \subseteq \operatorname{supp} f \quad \textit{for each } f \in L^p(X, \mu). \tag{2.2.360}$$

Moreover, if either (1) *or* (2) *holds, then b as in* (1) *is uniquely determined up to a set of μ-measure zero and*

$$\|b\|_{L^\infty(X,\mu)} = \|T\|_{L^p(X,\mu)\to L^p(X,\mu)}. \tag{2.2.361}$$

Before presenting the proof of this result, we make two comments. First, the property described in (2.2.360) may be rephrased, equivalently, as saying that

$$\begin{aligned} &\text{for each function } f \in L^p(X,\mu) \text{ we have} \\ &(Tf)(x) = 0 \text{ for } \mu\text{-a.e. } x \in X \setminus \operatorname{supp} f. \end{aligned} \tag{2.2.362}$$

Second, as a consequence of Proposition 2.2.22, whenever two operators are associated in $L^p(X,\mu)$ with the same kernel (as in item *(T3)* of Proposition 2.2.5) it follows that their difference is the operator of pointwise multiplication by an essentially bounded function.

Here is the proof of Proposition 2.2.22.

Proof of Proposition 2.2.22 The implication *(1)*$\Rightarrow$*(2)* is immediate so we shall focus on proving *(2)*$\Rightarrow$*(1)*. Suppose T as in (2.2.359) is linear, bounded, and local. The first claim we make is that whenever O_1, O_2 are given open subsets of X with $\mu(O_1) < \infty$ and $\mu(O_2) < \infty$, it follows that

$$\big(T\mathbf{1}_{O_1}\big)(x) = \big(T\mathbf{1}_{O_2}\big)(x) \text{ for } \mu\text{-a.e. } x \in O_1 \cap O_2. \tag{2.2.363}$$

To see why this is true, note that $\mathbf{1}_{O_1}, \mathbf{1}_{O_2} \in L^p(X,\mu)$, the set $O_1 \cap O_2$ is open, and $\mathbf{1}_{O_1} - \mathbf{1}_{O_2} \equiv 0$ on $O_1 \cap O_2$. Hence, $O_1 \cap O_2 \subseteq X \setminus \operatorname{supp}\big(\mathbf{1}_{O_1} - \mathbf{1}_{O_2}\big)$ and since T is local, we have $T\big(\mathbf{1}_{O_1} - \mathbf{1}_{O_2}\big)(x) = 0$ for μ-a.e. $x \in X \setminus \operatorname{supp}\big(\mathbf{1}_{O_1} - \mathbf{1}_{O_2}\big)$. Thus, $T\big(\mathbf{1}_{O_1} - \mathbf{1}_{O_2}\big)(x) = 0$ for μ-a.e. $x \in O_1 \cap O_2$. This and the linearity of T now imply (2.2.363).

To proceed, recall the dyadic grid $\mathbb{D}(X) = \big\{Q_\alpha^k : k \in \mathbb{Z}, k \geq \kappa_X, \alpha \in I_k\big\}$ from [112, Proposition 7.5.4]. In particular, for each $k \in \mathbb{Z}$ with $k \geq \kappa_X$ there exists a nullset N_k of μ such that

$$X = \big(\sqcup_{\alpha\in I_k} Q_\alpha^k\big) \sqcup N_k. \tag{2.2.364}$$

Next, fix $k_0 \in \mathbb{Z}$ with $k_0 \geq \kappa_X$ and define the function $b : X \to \mathbb{R}$ by setting

$$b := \begin{cases} T\mathbf{1}_{Q_\alpha^{k_0}} & \text{on } Q_\alpha^{k_0} \text{ for each } \alpha \in I_{k_0}, \\ 0 & \text{on } N_{k_0}. \end{cases} \tag{2.2.365}$$

Then b is μ-measurable and, due to (2.2.363) and the structure of the dyadic grid, we actually have

$$\begin{aligned} &b = T\mathbf{1}_{Q_\alpha^k} \text{ at } \mu\text{-a.e. point in } Q_\alpha^k, \\ &\text{for each } k \in \mathbb{Z} \text{ with } k \geq \kappa_X \text{ and each } \alpha \in I_k. \end{aligned} \tag{2.2.366}$$

In turn, this implies

$$\begin{aligned} &T\mathbf{1}_{Q_\alpha^k} = b \cdot \mathbf{1}_{Q_\alpha^k} \text{ at } \mu\text{-a.e. point in } X, \\ &\text{for each } k \in \mathbb{Z} \text{ with } k \geq \kappa_X \text{ and each } \alpha \in I_k. \end{aligned} \tag{2.2.367}$$

To justify this, observe that on $X \setminus \operatorname{supp}(\mathbf{1}_{Q_\alpha^k})$ both $b \cdot \mathbf{1}_{Q_\alpha^k}$ and $T\mathbf{1}_{Q_\alpha^k}$ are zero μ-a.e. (for the latter we use the fact that T is a local operator). Having proved (2.2.366), and since

$$\operatorname{supp}(\mathbf{1}_{Q_\alpha^k}) = \overline{Q_\alpha^k} \text{ and } 0 = \mu(\partial Q_\alpha^k) = \mu(\overline{Q_\alpha^k} \setminus Q_\alpha^k), \tag{2.2.368}$$

(cf. [112, (3.8.9)] and [112, (7.5.20)]) it follows that (2.2.367) is indeed valid. As a consequence of (2.2.367) and the mapping properties of T we have

$$b \in L^p_{\mathrm{loc}}(X, \mu). \tag{2.2.369}$$

Consequently, $|b|^p \in L^1_{\mathrm{loc}}(X, \mu)$, and we may invoke [112, Proposition 7.4.4] (recall that μ is Borel-semiregular) to conclude

$$|b(x)|^p = \lim_{r \to 0^+} ⨍_{B_{\rho_\#}(x,r)} |b|^p \, d\mu \text{ at } \mu\text{-a.e. } x \in X. \tag{2.2.370}$$

Fix now an arbitrary open set $O \subseteq X$ with $\mu(O) < \infty$. From item *(11)* in [112, Proposition 7.5.4] we know that there exist some μ-nullset $\widetilde{N}$ and some countable set $I_O \subseteq \big\{(\alpha, k) : k \in \mathbb{Z}, k \geq \kappa_X, \alpha \in I_k\big\}$, such that

$$O = \Big(\sqcup_{j \in I_O} Q_j \Big) \sqcup \widetilde{N} \tag{2.2.371}$$

where $Q_j := Q_\alpha^k$ if $j \in I_O$ is such that $j = (\alpha, k)$. By invoking Lebesgue's Dominated Convergence Theorem (with pointwise convergence guaranteed by (2.2.371) and domination by $\mathbf{1}_O$) we further obtain that

$$\sum_{j \in I_O} \mathbf{1}_{O_j} = \mathbf{1}_O \text{ in } L^p(X, \mu). \tag{2.2.372}$$

Since T is linear and continuous as an operator from $L^p(X, \mu)$ into $L^p(X, \mu)$, from (2.2.372) and (2.2.367) we may conclude that, pointwise μ-a.e. on X we have

$$T\mathbf{1}_O = \sum_{j \in I_O} T\mathbf{1}_{O_j} = \sum_{j \in I_O} b \cdot \mathbf{1}_{O_j} = b \cdot \sum_{j \in I_O} T\mathbf{1}_{O_j} = b \cdot \mathbf{1}_O. \tag{2.2.373}$$

In summary, the argument so far proves that

$$\begin{aligned} &T\mathbf{1}_O = b \cdot \mathbf{1}_O \text{ at } \mu\text{-a.e. point in } X, \\ &\text{for every open set } O \subseteq X \text{ with } \mu(O) < \infty. \end{aligned} \tag{2.2.374}$$

Next, for each $x \in X$ and each $r \in (0, \infty)$, with $\rho_\#$ denoting the regularized version of the quasi-distance ρ (cf. [112, (7.1.18)]), we may write

$$\begin{aligned}\int_{B_{\rho_\#}(x,r)} |b|^p \,\mathrm{d}\mu &= \int_{B_{\rho_\#}(x,r)} \big|T\mathbf{1}_{B_{\rho_\#}(x,r)}\big|^p \,\mathrm{d}\mu \leq \int_X \big|T\mathbf{1}_{B_{\rho_\#}(x,r)}\big|^p \,\mathrm{d}\mu \\ &\leq \|T\|^p_{L^p(X,\mu)\to L^p(X,\mu)} \int_X \big|\mathbf{1}_{B_{\rho_\#}(x,r)}\big|^p \,\mathrm{d}\mu \\ &= \|T\|^p_{L^p(X,\mu)\to L^p(X,\mu)} \cdot \mu\big(B_{\rho_\#}(x,r)\big), \end{aligned} \tag{2.2.375}$$

hence

$$⨍_{B_{\rho_\#}(x,r)} |b|^p \,\mathrm{d}\mu \leq \|T\|^p_{L^p(X,\mu)\to L^p(X,\mu)}. \tag{2.2.376}$$

Together, (2.2.376) and (2.2.370) imply

$$b \in L^\infty(X,\mu) \text{ and } \|b\|_{L^\infty(X,\mu)} \leq \|T\|_{L^p(X,\mu)\to L^p(X,\mu)}. \tag{2.2.377}$$

Going further, the goal is to show that (2.2.374) continues to hold with O replaced by any μ-measurable set $E \subseteq X$ satisfying $\mu(E) < \infty$. Fix such a set E. Since μ is Borel-semiregular there exists some Borel set $B \subseteq X$ such that $\mu(E \triangle B) = 0$. In turn, this implies

$$\mu(B) < \infty \text{ and } \mathbf{1}_E = \mathbf{1}_B \text{ at } \mu\text{-a.e. point in } X. \tag{2.2.378}$$

Consequently,

$$T\mathbf{1}_E = T\mathbf{1}_B \text{ at } \mu\text{-a.e. point in } X. \tag{2.2.379}$$

In addition, by relying on [112, (3.4.44)] (whose applicability in the present setting is ensured by [112, Lemma 3.4.13] and the fact that quasi-metric space (X, ρ, μ) is d-Ahlfors regular), it is possible to find a sequence $\{O_j\}_{j\in\mathbb{N}}$ of open sets in X with

$$B \subseteq O_j \text{ for all } j \in \mathbb{N} \text{ and } \mu(O_j \setminus B) \longrightarrow 0 \text{ as } j \to \infty. \tag{2.2.380}$$

Fix some reference point $x_* \in X$ and introduce the sets

$$B_k := B \cap B_{\rho_\#}(x_*, k) \text{ for each } k \in \mathbb{N}, \tag{2.2.381}$$

and

$$O_j^k := O_j \cap B_{\rho_\#}(x_*, k) \text{ for each } j, k \in \mathbb{N}. \tag{2.2.382}$$

Based on these definition and (2.2.380), for each $j, k \in \mathbb{N}$ we see that

$$O_j^k \text{ is open,} \quad B_k \subseteq O_j^k, \text{ and } \mu(O_j^k) \leq \mu\big(B_{\rho_\#}(x_*, k)\big) < \infty. \tag{2.2.383}$$

In addition, since $O_j^k \setminus B_k = (O_j \setminus B) \cap B_{\rho_\#}(x_*, k)$ for each $j, k \in \mathbb{N}$, it follows that

$$\mu(O_j^k \setminus B_k) \leq \mu(O_j \setminus B) \longrightarrow 0 \text{ as } j \to \infty, \text{ for each fixed } k \in \mathbb{N}. \tag{2.2.384}$$

As a consequence of (2.2.384) and the inclusion in (2.2.383), we conclude that

$$\|\mathbf{1}_{O_j^k} - \mathbf{1}_{B_k}\|_{L^p(X,\mu)}^p = \mu(O_j^k \setminus B_k) \longrightarrow 0 \text{ as } j \to \infty, \tag{2.2.385}$$

for each fixed $k \in \mathbb{N}$. Since also $\mathbf{1}_{B_k} \to \mathbf{1}_B$ in $L^p(X,\mu)$ as $k \to \infty$ (by Lebesgue's Dominated Convergence Theorem) and T is assumed to be bounded in the context of (2.2.359), we may further infer that

$$\begin{aligned} T\mathbf{1}_B &= \lim_{k\to\infty} T\mathbf{1}_{B_k} = \lim_{k\to\infty}\Big(\lim_{j\to\infty} T\mathbf{1}_{O_j^k}\Big) = \lim_{k\to\infty}\Big(\lim_{j\to\infty} b\cdot\mathbf{1}_{O_j^k}\Big) \\ &= \lim_{k\to\infty} b\cdot\mathbf{1}_{B_k} = b\cdot\mathbf{1}_B \text{ in } L^p(X,\mu), \end{aligned} \tag{2.2.386}$$

where the third equality in (2.2.386) relies on (2.2.374). The leftmost and rightmost terms in (2.2.386) being equal in $L^p(X,\mu)$ are equal μ-a.e. on X which, in combination with (2.2.378) and (2.2.379), implies

$$T\mathbf{1}_E = T\mathbf{1}_B = b\cdot\mathbf{1}_B = b\cdot\mathbf{1}_E \text{ at } \mu\text{-a.e. point in } X. \tag{2.2.387}$$

Let us record our progress: the argument so far proves that

$$T\mathbf{1}_E = b\cdot\mathbf{1}_B \text{ for every } \mu\text{-measurable set } E \subseteq X \text{ with } \mu(E) < \infty. \tag{2.2.388}$$

Since T is linear, (2.2.388) further extends to simple functions, i.e.,

$$Ts = b\cdot s \text{ for every simple function } s = \sum_{j=1}^{N} \lambda_j \mathbf{1}_{E_j}, \tag{2.2.389}$$

where $N \in \mathbb{N}$, the set $E_j \subseteq X$ is μ-measurable with $\mu(E_j) < \infty$, and $\lambda_j \in \mathbb{C}$, for all $j \in \{1,\dots,N\}$. By the density of simple functions in $L^p(X,\mu)$ (cf. [112, (3.1.11)]) and the boundedness of T from $L^p(X,\mu)$ to $L^p(X,\mu)$, it follows that $Tf = b\cdot f$ for every $f \in L^p(X,\mu)$. This completes the proof of the implication *(2)* $\Rightarrow$ *(1)*.

There remains to prove the final claim in the statement. If (1) holds, it is immediate that a function b as in *(1)* is uniquely determined up to a set of μ-measure zero. In addition,

$$\begin{aligned} \|T\|_{L^p(X,\mu)\to L^p(X,\mu)} &= \sup_{\|f\|_{L^p(X,\mu)}=1} \|Tf\|_{L^p(X,\mu)} \\ &= \sup_{\|f\|_{L^p(X,\mu)}=1} \|b\cdot f\|_{L^p(X,\mu)} \leq \|b\|_{L^\infty(X,\mu)}, \end{aligned} \tag{2.2.390}$$

which when combined with (2.2.377) and the fact that b is unique up to a set of μ-measure zero, yields the equality in (2.2.361). □

Our next proposition amounts to structure result to the effect that if T is a linear operator satisfying both a weak-(p,p) bound as well as a strong (q,q) bound for $1 \leq p < q < \infty$ and which is associated with a kernel K in L^p which is standard kernel in the first variable and satisfies (2.2.332) then necessarily T is of the form $T_{\rm PV} + M_b$ for some essentially bounded function b.

Proposition 2.2.23 *Let $d \in (0, \infty)$ and suppose (X, ρ, μ) is a d-Ahlfors regular quasi-metric space with the property the quasi-distance $\rho : X \times X \to [0, +\infty)$ is continuous in the product topology $\tau_\rho \times \tau_\rho$, and such that μ is a complete Borel-semiregular measure. Fix $p \in [1, \infty)$ and let*

$$T : L^p(X, \mu) \longrightarrow L^{p,\infty}(X, \mu) \tag{2.2.391}$$

be a linear and bounded operator satisfying

$$Tf(x) = \int_X K(x, y) f(y)\, d\mu(y) \text{ for } \mu\text{-a.e. } x \in X \setminus \operatorname{supp} f, \tag{2.2.392}$$

where K is a standard kernel in the first variable (cf. Definition 2.2.4). Also, assume that there exists $q \in (p, \infty)$ such that $T\big|_{L^p(X,\mu)\cap L^q(X,\mu)}$ extends to a linear and bounded operator from $L^q(X, \mu)$ into $L^q(X, \mu)$.

Then for each integrability exponent $r \in (p, q]$ there exist a sequence of numbers $\{\varepsilon_j\}_{j\in\mathbb{N}} \subseteq (0, \infty)$ convergent to zero, an operator $T_{(r)} \in \mathcal{L}\big(L^r(X, \mu)\big)$ and a function $b \in L^\infty(X, \mu)$ such that

$$\begin{gathered} \lim_{j\to\infty} \int_X T_{\varepsilon_j} f \cdot g\, d\mu = \int_X T_{(r)} f \cdot g\, d\mu \\ \text{for all } f \in L^r(X, \mu) \text{ and } g \in L^{r'}(X, \mu), \end{gathered} \tag{2.2.393}$$

where $r' \in (1, \infty)$ is such that $1/r + 1/r' = 1$, and

$$T = T_{(r)} + M_b \text{ in } L^r(X, \mu), \tag{2.2.394}$$

where M_b is the operator of pointwise multiplication by the function b.

Moreover, if in addition to the original hypotheses one also assumes that

$$\text{the limit } \lim_{\varepsilon\to 0^+} \int_{1>\rho(x,y)>\varepsilon} K(x, y)\, d\mu(y) \text{ exists for } \mu\text{-a.e. } x \in X, \tag{2.2.395}$$

then for each $r \in (p, q]$ it follows that

$$\begin{gathered} T_{\text{pv}} : L^r(X, \mu) \longrightarrow L^r(X, \mu), \text{ acting on each } f \in L^r(X, \mu) \text{ by} \\ (T_{\text{pv}} f)(x) = \lim_{\varepsilon\to 0^+} \int_{y\in X,\, \rho(x,y)>\varepsilon} K(x, y) f(y)\, d\mu(y) \text{ for } \mu\text{-a.e. } x \in X, \end{gathered} \tag{2.2.396}$$

is a well-defined, linear, bounded operator, and $T_{(r)} = T_{\text{pv}}$ on $L^r(X, \mu)$. In particular, in this case, (2.2.394) *becomes*

$$T = T_{\text{pv}} + M_b \text{ in } L^r(X, \mu) \text{ for each } r \in (p, q]. \tag{2.2.397}$$

We remark that formula (2.2.397) yields the pointwise inequality

$$|(Tf)(x)| \le (T_{\max} f)(x) + C|f(x)| \text{ for } \mu\text{-a.e. } x \in X, \quad \text{whenever } f \in L^r(X, \mu) \text{ with } r \in (p, q], \tag{2.2.398}$$

where $C := \|b\|_{L^\infty(X,\mu)}$; compare with (2.2.38). We next present the proof of Proposition 2.2.23.

Proof of Proposition 2.2.23 Proposition 2.2.7 ensures that the maximal operator $T_{\max}$ is bounded in the following contexts:

$$T_{\max} : L^p(X, \mu) \longrightarrow L^{p,\infty}(X, \mu), \text{ and} \tag{2.2.399}$$

$$T_{\max} : L^r(X, \mu) \longrightarrow L^r(X, \mu) \text{ for each } r \in (p, q]. \tag{2.2.400}$$

In particular, $\{T_\varepsilon\}_{\varepsilon>0}$ is a bounded sequence in $\mathcal{L}\big(L^r(X, \mu)\big)$ for each fixed exponent $r \in (p, q]$. To proceed, pick a sequence $\{\varepsilon_j\}_{j\in\mathbb{N}}$ such that $\varepsilon_j \searrow 0$ as $j \to \infty$ and apply Corollary 2.2.21 to the sequence of operators $\{T_{\varepsilon_j}\}_{j\in\mathbb{N}}$ acting on $L^r(X, \mu)$, where $r \in (p, q]$ is arbitrary, fixed. This guarantees the existence of an operator $T_{(r)} \in \mathcal{L}\big(L^r(X, \mu)\big)$ and a subsequence $\{\varepsilon_{j_k}\}_{k\in\mathbb{N}}$ of $\{\varepsilon_j\}_{j\in\mathbb{N}}$ such that

$$\lim_{k\to\infty} \int_X T_{\varepsilon_{j_k}} f \cdot g \, d\mu = \int_X T_{(r)} f \cdot g \, d\mu \quad \text{for all } f \in L^r(X, \mu) \text{ and } g \in L^{r'}(X, \mu), \tag{2.2.401}$$

where $r' \in (1, \infty)$ is such that $1/r + 1/r' = 1$. After relabeling, this proves (2.2.393).

Now pick $f \in L^r(X, \mu)$ and $g \in L^{r'}(X, \mu)$ with $\delta := \mathrm{dist}_\rho(\mathrm{supp}\, f, \mathrm{supp}\, g) > 0$. Then, for each $\varepsilon \in (0, \delta/2)$, we have

$$\begin{aligned}
\int_X (T_\varepsilon f)(x) g(x) \, d\mu(x) &= \int_{x\in\mathrm{supp}\, g} \Big(\int_{\substack{\rho(x,y)>\varepsilon \\ y\in\mathrm{supp} f}} K(x, y) f(y) \, d\mu(y) \Big) g(x) \, d\mu(x) \\
&= \int_{x\in\mathrm{supp}\, g} \Big(\int_{y\in\mathrm{supp} f} K(x, y) f(y) \, d\mu(y) \Big) g(x) \, d\mu(x) \\
&= \int_{\mathrm{supp}\, g} (Tf)(x) g(x) \, d\mu(x) = \int_X Tf \cdot g \, d\mu.
\end{aligned} \tag{2.2.402}$$

In concert, (2.2.402) and (2.2.401) imply

$$\int_X Tf \cdot g \, d\mu = \int_X T_{(r)} f \cdot g \, d\mu \text{ for all } f \in L^r(X, \mu) \text{ and } g \in L^{r'}(X, \mu) \text{ with } \mathrm{dist}_\rho(\mathrm{supp} f, \mathrm{supp}\, g) > 0. \tag{2.2.403}$$

From (2.2.403) we further deduce (based on Lemma 2.2.10) that, for every function $f \in L^r(X, \mu)$,

$$Tf = T_{(r)} f \text{ pointwise } \mu\text{-a.e. on } X \setminus \mathrm{supp}\, f. \tag{2.2.404}$$

Thus, $T - T_{(r)} \in \mathcal{L}\big(L^r(X,\mu)\big)$ is a local operator. Proposition 2.2.22 then guarantees the existence of some $b \in L^\infty(X,\mu)$ such that $T - T_{(r)} = M_b$ on $L^r(X,\mu)$. This proves (2.2.394).

As regards the claims in the last part of the statement, assume now that (2.2.395) is also satisfied. Then, thanks to (2.2.400), we may invoke the last claim in part *(2)* of Proposition 2.2.14 to conclude that for each $r \in (p,q]$ the principal-value operator

$$T_{_{\rm PV}} : L^r(X,\mu) \longrightarrow L^{r,\infty}(X,\mu), \text{ acting on each } f \in L^r(X,\mu) \text{ by}$$
$$(T_{_{\rm PV}} f)(x) = \lim_{\varepsilon\to 0^+} \int_{y\in X,\, \rho(x,y)>\varepsilon} K(x,y) f(y)\, d\mu(y) \text{ for } \mu\text{-a.e. } x \in X, \tag{2.2.405}$$

is well defined, linear, and bounded. In view of the fact that $|T_{_{\rm PV}} f| \le T_{\max} f$ for each $f \in L^r(X,\mu)$, another appeal to (2.2.400) shows that (2.2.396) is indeed a well-defined, linear, and bounded operator for each $r \in (p,q]$.

In addition, Lebesgue's Dominated Convergence Theorem plus the fact that $|T_\varepsilon f| \le T_{\max} f$ for each $f \in L^r(X,\mu)$ and each $\varepsilon > 0$ show that for every $f \in L^r(X,\mu)$ with $r \in (p,q]$ we have

$$\lim_{\varepsilon\to 0^+} T_\varepsilon f = T_{_{\rm PV}} f \text{ in } L^r(X,\mu). \tag{2.2.406}$$

From (2.2.406) and (2.2.401) we then obtain that necessarily $T_{_{\rm PV}} = T_{(r)}$ as operators on $L^r(X,\mu)$ for each $r \in (p,q]$. □

The weak boundedness property is introduced next.

Definition 2.2.24 *Assume (X,ρ,μ) is a d-Ahlfors regular quasi-metric space for some $d \in (0,\infty)$, and fix a finite exponent $\beta \in \big(0, (\log_2 C_\rho)^{-1}\big]$. A linear operator*

$$T : \mathscr{C}^\beta_c(X,\rho) \longrightarrow \big(\mathscr{C}^\beta_c(X,\rho)\big)^* \tag{2.2.407}$$

where $\big(\mathscr{C}^\beta_c(X,\rho)\big)^*$ denotes the algebraic dual of $\mathscr{C}^\beta_c(X,\rho)$, is said to satisfies the `weak boundedness property` *(WBP for short) provided there exists some constant $C \in [0,\infty)$ such that for each point $x_o \in X$, each radius $r > 0$, and each functions $f, g \in \mathscr{C}^\beta_c(X,\rho)$ with*

$$\operatorname{supp} f \subseteq B_\rho(x_o,r), \qquad \sup|f| + r^\beta \|f\|_{\dot{\mathscr{C}}^\beta_c(X,\rho)} \le 1 \tag{2.2.408}$$

and

$$\operatorname{supp} g \subseteq B_\rho(x_o,r), \qquad \sup|g| + r^\beta \|g\|_{\dot{\mathscr{C}}^\beta_c(X,\rho)} \le 1, \tag{2.2.409}$$

it follows that

$$\Big|{}_{(\mathscr{C}^\beta_c(X,\rho))^*}\big\langle Tf, g\big\rangle_{\mathscr{C}^\beta_c(X,\rho)}\Big| \le C\mu\big(B_\rho(x_o,r)\big). \tag{2.2.410}$$

With each given antisymmetric kernel is possible to associate an operator mapping "test functions" into "distributions" of the sort described below.

Definition 2.2.25 *Let (X, ρ, μ) be a d-Ahlfors regular quasi-metric space for some $d \in (0, \infty)$, and assume $K : X \times X \setminus \operatorname{diag}(X) \to \mathbb{R}$ is a $\mu \otimes \mu$-measurable function which is antisymmetric, in the sense that*

$$K(y, x) = -K(x, y) \text{ for all } x, y \in X \text{ with } x \neq y, \tag{2.2.411}$$

and with the property that there exists some constant $C \in [0, \infty)$ such that

$$|K(x, y)| \leq C\rho(x, y)^{-d} \text{ for all } x, y \in X \text{ with } x \neq y. \tag{2.2.412}$$

Finally, fix a finite exponent $\beta \in \big(0, (\log_2 C_\rho)^{-1}\big]$.

In this setting, consider the operator T_K canonically induced by (or, associated with) the kernel K in the distributional sense as the mapping

$$T_K : \mathscr{C}^\beta_c(X, \rho) \longrightarrow \big(\mathscr{C}^\beta_c(X, \rho)\big)^* \tag{2.2.413}$$

defined as

$$\begin{aligned} &{}_{(\mathscr{C}^\beta_c(X,\rho))^*}\Big\langle T_K f, g \Big\rangle_{\mathscr{C}^\beta_c(X,\rho)} \\ &\quad := \frac{1}{2}\int_X \int_X K(x, y)\big[f(y)g(x) - f(x)g(y)\big]\, d\mu(y)\, d\mu(x), \end{aligned} \tag{2.2.414}$$

for all $f, g \in \mathscr{C}^\beta_c(X, \rho)$.

That the above definition is indeed meaningful is clarified by the next proposition.

Proposition 2.2.26 *Let (X, ρ, μ) be a d-Ahlfors regular quasi-metric space for some $d \in (0, \infty)$, and assume $K : X \times X \setminus \operatorname{diag}(X) \to \mathbb{R}$ is a $\mu \otimes \mu$-measurable function which is antisymmetric, in the sense of* (2.2.411), *and satisfies* (2.2.412). *Also, fix a finite exponent $\beta \in \big(0, (\log_2 C_\rho)^{-1}\big]$.*

Then the operator T_K canonically induced by (or, associated with) the kernel K in the distributional sense as in (2.2.413)-(2.2.414) *is a well-defined linear mapping*

$$T_K : \mathscr{C}^\beta_c(X, \rho) \longrightarrow \big(\mathscr{C}^\beta_c(X, \rho)\big)^* \tag{2.2.415}$$

which satisfies the weak boundedness property (cf. Definition 2.2.24).

Proof Fix a point $x_o \in X$, a radius $r > 0$, and two functions $f, g \in \mathscr{C}^\beta_c(X, \rho)$ satisfying (2.2.408)-(2.2.409). Then we may rely on [112, (7.2.5)] to estimate

$$\begin{aligned}
&\int_X\int_X |K(x,y)||f(y)g(x)-f(x)g(y)|\,\mathrm{d}\mu(y)\,\mathrm{d}\mu(x)\\
&\qquad\leq \int_{B_{\rho_\#}(x_o,r)}\int_{B_{\rho_\#}(x_o,r)} \frac{C}{\rho_\#(x,y)^d}\Big(\frac{\rho_\#(x,y)}{r}\Big)^\beta\,\mathrm{d}\mu(y)\,\mathrm{d}\mu(x)\\
&\qquad\leq C\mu\big(B_{\rho_\#}(x_o,r)\big)\leq C\mu\big(B_\rho(x_o,r)\big), \qquad (2.2.416)
\end{aligned}$$

from which all desired conclusions follow. □

It turns out that, in an appropriate setting, the operator T_K canonically induced by an antisymmetric kernel K in the distributional sense is associated with this kernel in L^p in the sense described below.

Lemma 2.2.27 *Let (X,ρ,μ) be a d-Ahlfors regular quasi-metric space for some $d\in(0,\infty)$, with the property that μ is a locally finite Borel-semiregular measure on X (cf.* [112, Definition 3.4.3]*). Also, fix a finite exponent $\beta\in\big(0,(\log_2 C_\rho)^{-1}\big]$. Assume $K: X\times X\setminus\mathrm{diag}(X)\to\mathbb{R}$ is a $\mu\otimes\mu$-measurable function which is antisymmetric, in the sense of* (2.2.411)*, and satisfies* (2.2.412)*. Suppose that, for some $p\in(1,\infty)$, the operator T_K canonically induced by the kernel K in the distributional sense (as in* (2.2.413)-(2.2.414)*) extends to a linear operator mapping $L^p(X,\mu)$ into $L^1_{\mathrm{loc}}(X,\mu)$.*
Then for each $f\in L^p(X,\mu)$ one has[13]

$$(T_K f)(x)=\int_X K(x,y)f(y)\,\mathrm{d}\mu(y)\ \textit{for } \mu\textit{-a.e. } x\in X\setminus\operatorname{supp} f. \qquad (2.2.417)$$

Proof Thanks to Lemma 2.2.10, the assignment

$$\begin{gathered}
L^1_{\mathrm{loc}}(X,\mu)\ni\phi\longmapsto\Lambda_\phi\in\big(\mathscr{C}^\beta_c(X,\rho)\big)^*\ \text{ defined by}\\
{}_{(\mathscr{C}^\beta_c(X,\rho))^*}\big\langle\Lambda_\phi,\psi\big\rangle_{\mathscr{C}^\beta_c(X,\rho)}:=\int_X\phi\psi\,\mathrm{d}\mu \qquad (2.2.418)\\
\text{for all } \phi\in L^1_{\mathrm{loc}}(X,\mu) \text{ and } \psi\in\mathscr{C}^\beta_c(X,\rho),
\end{gathered}$$

is injective. This amounts to embedding $L^1_{\mathrm{loc}}(X,\mu)$ into $\big(\mathscr{C}^\beta_c(X,\rho)\big)^*$.

Fix now an arbitrary $f\in\mathscr{C}^\beta_c(X,\rho)\subseteq L^p(X,\mu)$. Then for each

$$g\in\mathscr{C}^\beta_c(X,\rho)\ \text{ with } \operatorname{dist}_\rho\big(\operatorname{supp} f,\operatorname{supp} g\big)>0, \qquad (2.2.419)$$

in view of the embedding (2.2.418) and (2.2.414) we may write

[13] with supp f understood in the sense of [112, Definition 3.8.3]

$$\begin{aligned}\int_X (T_K f) g \, d\mu &= {}_{(\mathscr{C}^\beta_c(X,\rho))^*}\Big\langle T_K f, g \Big\rangle_{\mathscr{C}^\beta_c(X,\rho)} \\ &= \frac{1}{2}\int_X\int_X K(x,y)\big[f(y)g(x) - f(x)g(y)\big]\, d\mu(y)\, d\mu(x) \\ &= \int_X\int_X K(x,y) f(y) g(x)\, d\mu(y)\, d\mu(x), \end{aligned} \tag{2.2.420}$$

where the last equality is obtained by naturally breaking up the previous integral, interchanging the roles of x, y, and using the fact that K is antisymmetric. In turn, (2.2.420) implies

$$\int_X \Big((T_K f)(x) - \int_X K(x,y) f(y)\, d\mu(y) \Big) g(x)\, d\mu(x) = 0 \tag{2.2.421}$$

for each function g as in (2.2.419). Granted this, Lemma 2.2.10 applies and gives that (2.2.417) holds whenever $f \in \mathscr{C}^\beta_c(X,\rho)$. With this in hand, the same type of argument which has established (2.2.145) (see (2.2.146)-(2.2.160)), now using the fact that $\mathscr{C}^\beta_c(X,\rho)$ is dense in $L^p(X,\mu)$ (cf. [112, Proposition 7.4.4] with ρ replaced by $\rho_\#$ from [112, Theorem 7.1.2]) then proves that (2.2.417) holds for each $f \in L^p(X,\mu)$. □

One way to summarize the essence of our next result is to say that "T_K is nice if and only if $T_{\rm PV}$ is nice."

Proposition 2.2.28 *Assume (X, ρ, μ) is a d-Ahlfors regular quasi-metric space for some $d \in (0,\infty)$ such that the quasi-distance $\rho : X \times X \to [0,+\infty)$ is continuous in the product topology $\tau_\rho \times \tau_\rho$, and with the property that μ is a complete Borel-semiregular measure. Fix a finite exponent $\beta \in \big(0, (\log_2 C_\rho)^{-1}\big]$ along with some $p \in (1,\infty)$. Also, suppose*

$$K : X \times X \setminus \text{diag}(X) \longrightarrow \mathbb{R} \tag{2.2.422}$$

is a kernel which is standard in the first variable (cf. Definition 2.2.4), is antisymmetric (in the sense of (2.2.411)), and such that

$$\text{the limit } \lim_{\varepsilon\to 0^+} \int_{\substack{y\in X \\ 1>\rho(x,y)>\varepsilon}} K(x,y)\, d\mu(y) \ \text{ exists for } \mu\text{-a.e. } x \in X. \tag{2.2.423}$$

Finally, recall the operator T_K canonically induced by (or, associated with) the kernel K in the distributional sense as in (2.2.413)-(2.2.414).

Then the following statements are equivalent:

(1) The operator T_K extends to a linear and bounded operator mapping $L^p(X,\mu)$ into $L^{p,\infty}(X,\mu)$ for some $p \in (1,\infty)$.

(2) The operator T_K induces well-defined, linear, and bounded mappings in the following contexts:

$$\begin{aligned} T_K &: L^q(X,\mu) \longrightarrow L^q(X,\mu) \text{ for each } q \in (1,\infty), \\ T_K &: L^1(X,\mu) \longrightarrow L^{1,\infty}(X,\mu). \end{aligned} \tag{2.2.424}$$

(3) There exists $p \in (1,\infty)$ such that principal-value operator assigning to each $f \in \mathscr{C}_c^\beta(X,\rho)$ the function

$$(T_{\rm PV} f)(x) := \lim_{\varepsilon\to 0^+} \int_{y\in X,\, \rho(x,y)>\varepsilon} K(x,y) f(y)\, d\mu(y) \text{ for } \mu\text{-a.e. } x \in X \tag{2.2.425}$$

extends to a linear and bounded mapping from $L^p(X,\mu)$ into $L^{p,\infty}(X,\mu)$.

(4) There exists $p \in (1,\infty)$ such the that principal-value operator

$$T_{\rm PV} : L^p(X,\mu) \longrightarrow L^{p,\infty}(X,\mu), \text{ acting on each } f \in L^p(X,\mu) \text{ by}$$

$$(T_{\rm PV} f)(x) = \lim_{\varepsilon\to 0^+} \int_{y\in X,\, \rho(x,y)>\varepsilon} K(x,y) f(y)\, d\mu(y) \text{ for } \mu\text{-a.e. } x \in X, \tag{2.2.426}$$

is well defined, linear, and bounded.

(5) The operator $T_{\rm PV}$ induces well-defined, linear, and bounded mappings in the following contexts:

$$\begin{aligned} T_{\rm PV} &: L^q(X,\mu) \longrightarrow L^q(X,\mu) \text{ for each } q \in (1,\infty), \\ T_{\rm PV} &: L^1(X,\mu) \longrightarrow L^{1,\infty}(X,\mu). \end{aligned} \tag{2.2.427}$$

Moreover, the veracity of either (hence all) of the above conditions implies that

$$T_K = T_{\rm PV} \text{ on } L^q(X,\mu) \text{ for each } q \in [1,\infty). \tag{2.2.428}$$

Proof Assume *(1)*. Then Lemma 2.2.27 implies that (2.2.417) holds for each function $f \in L^p(X,\mu)$. In turn, this permits us to invoke item *(i)* of Proposition 2.2.8 and Corollary 2.2.11. The former gives that

$$T_K : L^1(X,\mu) \longrightarrow L^{1,\infty}(X,\mu) \text{ linearly and boundedly,} \tag{2.2.429}$$

while the latter presently gives two things. First,

$$\begin{aligned} &T_K : L^q(X,\mu) \longrightarrow L^q(X,\mu) \\ &\text{linearly and boundedly for each } q \in (1,\infty), \end{aligned} \tag{2.2.430}$$

and, second, the maximal operator $T_{\max}$ induces a sub-linear, bounded, and continuous mapping in the context

$$T_{\max} : L^q(X,\mu) \longrightarrow L^q(X,\mu) \text{ for each } q \in (1,\infty). \tag{2.2.431}$$

Collectively, (2.2.429)-(2.2.430) prove the claim made in the current item *(2)*. Since, obviously, *(2)* $\Rightarrow$ *(1)*, we ultimately conclude that *(1)* $\Leftrightarrow$ *(2)*. Granted (2.2.431), from Proposition 2.2.14 we conclude that *(1)* implies either of the items *(3)*, *(4)*, *(5)*. Proposition 2.2.14 also shows that the items *(3)*, *(4)*, *(5)* are equivalent to one another.

Assume next that *(3)-(5)* hold. For each $f, g \in \mathscr{C}^\beta_c(X,\rho)$ write

$$\begin{aligned}
&{}_{(\mathscr{C}^\beta_c(X,\rho))^*}\Big\langle T_K f, g\Big\rangle_{\mathscr{C}^\beta_c(X,\rho)} \\
&\quad= \frac{1}{2}\int_X\int_X K(x,y)\big[f(y)g(x) - f(x)g(y)\big]\,d\mu(y)\,d\mu(x) \\
&\quad= \lim_{\varepsilon\to 0^+}\frac{1}{2}\int_X \int_{y\in X,\,\rho(x,y)>\varepsilon} K(x,y)\big[f(y)g(x) - f(x)g(y)\big]\,d\mu(y)\,d\mu(x) \\
&\quad= \lim_{\varepsilon\to 0^+}\int_X \int_{y\in X,\,\rho(x,y)>\varepsilon} K(x,y)f(y)g(x)\,d\mu(y)\,d\mu(x) \\
&\quad= \lim_{\varepsilon\to 0^+}\int_X (T_\varepsilon f)(x)g(x)\,d\mu \\
&\quad= \int_X (T_{\rm PV} f)(x)g(x)\,d\mu \\
&\quad= {}_{(\mathscr{C}^\beta_c(X,\rho))^*}\Big\langle T_{\rm PV} f, g\Big\rangle_{\mathscr{C}^\beta_c(X,\rho)}.
\end{aligned} \tag{2.2.432}$$

Above, the first equality comes from (2.2.414), the second equality is a consequence of Lebesgue's Dominated Convergence Theorem and (2.2.416), the third equality is obtained by naturally breaking up the previous integral, interchanging the roles of x, y, and using the fact that K is antisymmetric, the fourth equality uses (2.2.33), the fifth equality is implied by (2.2.304), and the last equality is seen from (2.2.418).

In turn, (2.2.432) shows that for each $f \in \mathscr{C}^\beta_c(X,\rho)$ we have $T_K f = T_{\rm PV} f$ as functionals in $\big(\mathscr{C}^\beta_c(X,\rho)\big)^*$. In view of this, the density result in item (5) of [112, Proposition 7.4.4], and the fact that $T_{\rm PV}$ induces linear, bounded, and continuous mappings in the contexts described in (2.2.427), we deduce that the operator T_K induces a well-defined, linear, and bounded mapping from $L^q(X,\mu)$ into itself for each $q \in (1,\infty)$, and that said extension agrees with $T_{\rm PV}$. This proves that *(3)-(5)* imply *(1)* and that $T_K = T_{\rm PV}$ on $L^q(X,\mu)$ for each $q \in (1,\infty)$. Finally, that $T_K f = T_{\rm PV} f$ for each $f \in L^1(X,\mu)$ follows by density and the continuity of the operators involved from the space $L^1(X,\mu)$ into $L^{1,\infty}(X,\mu)$. □

As a preamble to the statement of the $T(1)$ theorem, given a little further below, we make some definitions.

Definition 2.2.29 *Let (X, ρ, μ) be a d-Ahlfors regular quasi-metric space for some $d \in (0, \infty)$, and fix a $\beta \in \big(0, (\log_2 C_\rho)^{-1}\big]$. Assume $K : X \times X \setminus \mathrm{diag}(X) \to \mathbb{R}$ is a standard kernel in the second variable (cf. Definition 2.2.4). In this context, a linear operator*

$$T : \mathscr{C}_c^\beta(X, \rho) \longrightarrow \big(\mathscr{C}_c^\beta(X, \rho)\big)^* \tag{2.2.433}$$

is said to be associated with the kernel K in $\mathscr{C}_c^\beta(X, \rho)$ (or, simply, in a weak sense) provided

$$_{(\mathscr{C}_c^\beta(X,\rho))^*}\big\langle Tf, g\big\rangle_{\mathscr{C}_c^\beta(X,\rho)} = \int_X \int_X K(x, y) f(y) g(x)\, \mathrm{d}\mu(y)\, \mathrm{d}\mu(x) \tag{2.2.434}$$

for all

$$f, g \in \mathscr{C}_c^\beta(X, \rho) \text{ with } \mathrm{dist}_\rho\big(\mathrm{supp}\, f, \mathrm{supp}\, g\big) > 0. \tag{2.2.435}$$

Also, define the weak transpose $T^\top$ of T in (2.2.433) as the mapping

$$\begin{gathered} T^\top : \mathscr{C}_c^\beta(X, \rho) \longrightarrow \big(\mathscr{C}_c^\beta(X, \rho)\big)^* \text{ defined as} \\ _{(\mathscr{C}_c^\beta(X,\rho))^*}\big\langle T^\top f, g\big\rangle_{\mathscr{C}_c^\beta(X,\rho)} = {}_{(\mathscr{C}_c^\beta(X,\rho))^*}\big\langle Tg, f\big\rangle_{\mathscr{C}_c^\beta(X,\rho)} \\ \text{for all functions } f, g \in \mathscr{C}_c^\beta(X, \rho). \end{gathered} \tag{2.2.436}$$

It follows from the above considerations that:

in the context of Definition 2.2.29, if T is weakly associated with K then $T^\top$ is weakly associated with the kernel $K^\top$, defined as $K^\top(x, y) := K(y, x)$ for all points $x, y \in X$ with $x \neq y$. (2.2.437)

Let us also observe that, in the context of Definition 2.2.29, for each

$$f \in \mathscr{C}_c^\beta(X, \rho) \text{ with } \int_X f\, \mathrm{d}\mu = 0 \tag{2.2.438}$$

it is meaningful to extend the action of the functional $Tf \in \big(\mathscr{C}_c^\beta(X, \rho)\big)^*$ to any function

$$\begin{gathered} g \in \mathscr{C}^\beta(X, \rho) \text{ such that there exists } \varepsilon > 0 \text{ with} \\ g \equiv 0 \text{ on } \{x \in X : \mathrm{dist}_\rho\big(x, \mathrm{supp}\, f\big) < \varepsilon\} \end{gathered} \tag{2.2.439}$$

by setting, for some fixed reference point $y_0 \in X$,

$$\begin{gathered} \big\langle Tf, g\big\rangle := \int_X \int_X \big[K(x, y) - K(x, y_0)\big] f(y) g(x)\, \mathrm{d}\mu(y)\, \mathrm{d}\mu(x) \\ \text{for each } f \text{ as in (2.2.438) and each } g \text{ as in (2.2.439).} \end{gathered} \tag{2.2.440}$$

Indeed, the fact that the kernel K is standard in the second variable plus [112, (7.2.5)] ensure that the integral above is absolutely convergent, while the vanishing moment condition of f guarantees that the definition in (2.2.440) is independent of the choice of the reference point $y_0 \in X$.

Definition 2.2.30 *Let (X, ρ, μ) be a d-Ahlfors regular quasi-metric space for some $d \in (0, \infty)$, and fix a finite exponent $\beta \in \big(0, (\log_2 C_\rho)^{-1}\big]$. Consider a linear operator*

$$T : \mathscr{C}^\beta_c(X, \rho) \longrightarrow \big(\mathscr{C}^\beta_c(X, \rho)\big)^* \tag{2.2.441}$$

which is weakly associated with a kernel $K : X \times X \setminus \operatorname{diag}(X) \to \mathbb{R}$ which is standard in the first variable (cf. Definition 2.2.29). In this context, define $T(1)$ as the functional acting on each function

$$f \in \mathscr{C}^\beta_c(X, \rho) \text{ with } \int_X f \, d\mu = 0 \tag{2.2.442}$$

according to[14]

$$\big(T(1), f\big) := {}_{(\mathscr{C}^\beta_c(X,\rho))^*}\Big\langle T\phi, f\Big\rangle_{\mathscr{C}^\beta_c(X,\rho)} + \big\langle T^\top f, (1-\phi)\big\rangle, \tag{2.2.443}$$

where

$$\phi \in \mathscr{C}^\beta_c(X, \rho) \text{ is such that } \phi \equiv 1 \text{ near } \operatorname{supp} f. \tag{2.2.444}$$

In definition (2.2.443), the last pairing is understood in the sense of (2.2.440), with $T^\top$ now playing the role of the operator T (note that $T^\top$ is weakly associated with $K^\top(x, y) := K(y, x)$, which is a standard kernel in the second variable; cf. (2.2.437)), and $1 - \phi$ playing the role of g. It turns out that the above definition of is independent of the auxiliary function ϕ.

Continue to work in the setting of Definition 2.2.30, and let us temporarily assume that X is ρ-bounded. Then $\mathscr{C}^\beta_c(X, \rho) = \mathscr{C}^\beta(X, \rho)$. In particular, $1 \in \mathscr{C}^\beta_c(X, \rho)$, so $T1 \in \big(\mathscr{C}^\beta_c(X, \rho)\big)^*$. Also, for each f as in (2.2.442) and each ϕ as in (2.2.444) we may write

$$\begin{aligned}
\big(T(1), f\big) &= {}_{(\mathscr{C}^\beta(X,\rho))^*}\Big\langle T\phi, f\Big\rangle_{\mathscr{C}^\beta(X,\rho)} + {}_{(\mathscr{C}^\beta(X,\rho))^*}\Big\langle T^\top f, (1-\phi)\Big\rangle_{\mathscr{C}^\beta(X,\rho)} \\
&= {}_{(\mathscr{C}^\beta(X,\rho))^*}\Big\langle T\phi, f\Big\rangle_{\mathscr{C}^\beta(X,\rho)} + {}_{(\mathscr{C}^\beta(X,\rho))^*}\Big\langle T(1-\phi), f\Big\rangle_{\mathscr{C}^\beta(X,\rho)} \\
&= {}_{(\mathscr{C}^\beta(X,\rho))^*}\Big\langle T1, f\Big\rangle_{\mathscr{C}^\beta(X,\rho)}
\end{aligned} \tag{2.2.445}$$

thanks to (2.2.443), the consistency of the pairing in (2.2.440) with the pairing ${}_{(\mathscr{C}^\beta(X,\rho))^*}\langle \cdot, \cdot \rangle_{\mathscr{C}^\beta(X,\rho)}$, and the definition of the transpose given in (2.2.436). We can summarize this observation as follows:

$$\begin{aligned}
&\text{if } X \text{ is } \rho\text{-bounded, then } T(1) = {}_{(\mathscr{C}^\beta(X,\rho))^*}\langle T1, \cdot \rangle_{\mathscr{C}^\beta(X,\rho)} \text{ as} \\
&\text{functionals on } \Big\{f \in \mathscr{C}^\beta_c(X, \rho) : \textstyle\int_X f \, d\mu = 0\Big\}.
\end{aligned} \tag{2.2.446}$$

[14] here, $\big(T(1), f\big)$ simply means the action of the functional $T(1)$ on the function f

Let us assign a precise meaning to the symbol "$T(1) \in \mathrm{BMO}(X, \mu)$."

Definition 2.2.31 *In the context of Definition 2.2.30, say that "$T(1) \in \mathrm{BMO}(X, \mu)$" provided there exists a function $b \in \mathrm{BMO}(X, \mu)$ with the property that*

$$\int_X bf \, d\mu = {}_{(\mathscr{C}^\beta_c(X,\rho))^*}\Big\langle T\phi, f\Big\rangle_{\mathscr{C}^\beta_c(X,\rho)} + \big\langle T^\top f, (1-\phi)\big\rangle \tag{2.2.447}$$

whenever

$$\begin{aligned} &f \in \mathscr{C}^\beta_c(X, \rho) \text{ satisfies } \int_X f \, d\mu = 0 \text{ and} \\ &\phi \in \mathscr{C}^\beta_c(X, \rho) \text{ is such that } \phi \equiv 1 \text{ near } \operatorname{supp} f. \end{aligned} \tag{2.2.448}$$

Owing to the existence of a function $f_o \in \mathscr{C}^\beta_c(X, \rho)$ with $\int_X f_o \, d\mu \neq 0$ (cf., e.g., [117]), it follows that

$$\begin{aligned} &\text{if } \Lambda \in \big(\mathscr{C}^\beta_c(X, \rho)\big)^* \text{ is such that } {}_{(\mathscr{C}^\beta_c(X,\rho))^*}\langle \Lambda, f\rangle_{\mathscr{C}^\beta_c(X,\rho)} = 0 \\ &\text{for each function } f \in \mathscr{C}^\beta_c(X, \rho) \text{ with } \int_X f \, d\mu = 0 \text{ then} \\ &\text{there exists } c \in \mathbb{R} \text{ so that } \Lambda f = c \int_X f \text{ for all } f \in \mathscr{C}^\beta_c(X, \rho). \end{aligned} \tag{2.2.449}$$

In view of this property, it follows that

$$\begin{aligned} &\text{the function } b \in \mathrm{BMO}(X, \mu) \text{ satisfying (2.2.447)-(2.2.448)} \\ &\text{is in fact uniquely determined up to additive constants.} \end{aligned} \tag{2.2.450}$$

In particular, from this, (2.2.446), and (2.2.418) we see that

$$\begin{aligned} &\text{if } X \text{ is } \rho\text{-bounded and "} T(1) \in \mathrm{BMO}(X, \mu)\text{" then the functional} \\ &T1 \text{ agrees (up to additive constants, in the sense of (2.2.418)) with} \\ &\text{the function } b \in L^1_{\mathrm{loc}}(X, \mu) \text{ doing the job in (2.2.447).} \end{aligned} \tag{2.2.451}$$

Conversely, from (2.2.446) and (2.2.447) with $\phi \equiv 1$ we see that

$$\begin{aligned} &\text{whenever } X \text{ is } \rho\text{-bounded and } T1 \in \mathrm{BMO}(X, \mu), \\ &\text{the function } b := T1 \text{ does the job in (2.2.447).} \end{aligned} \tag{2.2.452}$$

Collectively, (2.2.451) and (2.2.452) prove that

$$\begin{aligned} &\text{if } X \text{ is } \rho\text{-bounded then "} T(1) \in \mathrm{BMO}(X, \mu)\text{" (cf. the sense given in} \\ &\text{Definition 2.2.31) if and only if } T1 \in \mathrm{BMO}(X, \mu) \text{ (as in (2.2.418)).} \end{aligned} \tag{2.2.453}$$

The stage is set to state the following version of the $T(1)$ theorem of G. David and J.-L. Journé (cf. the original [33], or [104] for the Euclidean setting, and [8, Theorem 12.2, p. 291], [18], [34] for the setting of spaces of homogeneous type).

Theorem 2.2.32 *Let (X,ρ,μ) be a d-Ahlfors regular quasi-metric space for some $d\in(0,\infty)$ with the property that μ is a Borel-semiregular measure. Fix a finite exponent $\beta\in\big(0,(\log_2 C_\rho)^{-1}\big]$ and suppose*

$$T:\mathscr{C}^\beta_c(X,\rho)\longrightarrow\big(\mathscr{C}^\beta_c(X,\rho)\big)^* \tag{2.2.454}$$

is a linear operator associated in a weak sense with a kernel

$$K:X\times X\setminus\mathrm{diag}(X)\longrightarrow\mathbb{R} \tag{2.2.455}$$

which is standard in both variables (cf. Definition 2.2.29 and Definition 2.2.4).

Then T extends to a linear and bounded operator from $L^2(X,\mu)$ into itself if and only if the following three conditions are satisfied:

(i) T satisfies the weak boundedness property (aka WBP; cf. Definition 2.2.24);
(ii) $T(1)\in\mathrm{BMO}(X,\mu)$ (in the sense of Definition 2.2.31);
(iii) $T^\top(1)\in\mathrm{BMO}(X,\mu)$ (cf. (2.2.436) and Definition 2.2.31).

Remark 2.2.33 *According to Proposition 2.2.26 and (2.2.437), in the case when K is antisymmetric (cf. (2.2.411)), conditions (i)-(iii) above simply reduce to the demand that $T(1)\in\mathrm{BMO}(X,\mu)$ (in the sense of Definition 2.2.31).*

The final result elaborates on the functional analytic properties of T_K, the operator canonically induced by an antisymmetric kernel K in the distributional sense, in light of the $T(1)$ theorem and our earlier results in this section.

Proposition 2.2.34 *Assume (X,ρ,μ) is a d-Ahlfors regular quasi-metric space for some value $d\in(0,\infty)$ with the property that X is ρ-bounded, the quasi-distance $\rho:X\times X\to[0,+\infty)$ is continuous in the product topology $\tau_\rho\times\tau_\rho$, and μ is a complete Borel-semiregular measure. Fix an exponent $\beta\in\big(0,(\log_2 C_\rho)^{-1}\big]$ along with some $p\in(1,\infty)$. Also, suppose*

$$K:X\times X\setminus\mathrm{diag}(X)\longrightarrow\mathbb{R} \tag{2.2.456}$$

is a kernel which is standard in the first variable (cf. Definition 2.2.4) as well as antisymmetric (in the sense of (2.2.411)). Bring in the operator T_K canonically induced by the kernel K in the distributional sense, i.e.,

$$T_K:\mathscr{C}^\beta(X,\rho)\longrightarrow\big(\mathscr{C}^\beta(X,\rho)\big)^* \tag{2.2.457}$$

defined for all $f,g\in\mathscr{C}^\beta(X,\rho)$ as

$$\begin{aligned}&{}_{(\mathscr{C}^\beta(X,\rho))^*}\Big\langle T_K f,g\Big\rangle_{\mathscr{C}^\beta(X,\rho)}\\&:=\frac{1}{2}\int_X\int_X K(x,y)\big[f(y)g(x)-f(x)g(y)\big]\,\mathrm{d}\mu(y)\mathrm{d}\mu(x).\end{aligned} \tag{2.2.458}$$

Then T_K extends to a linear and bounded operator from the space $L^2(X,\mu)$ into itself if and only if $T1 \in \mathrm{BMO}(X,\mu)$ (in the sense of Definition 2.2.31), i.e., if and only if there exists $b \in \mathrm{BMO}(X,\mu)$ such that

$$\int_X bg\,d\mu = \frac{1}{2}\int_X\int_X K(x,y)\big[g(x)-g(y)\big]\,d\mu(y)\,d\mu(x) \tag{2.2.459}$$

for all $g \in \mathscr{C}^\beta(X,\rho)$.

Moreover, if either of the aforementioned conditions holds then:

(a) The operator T_K induces well-defined, linear, and bounded mappings in the following contexts:

$$T_K : L^p(X,\mu) \longrightarrow L^p(X,\mu) \text{ for each } p \in (1,\infty), \tag{2.2.460}$$

$$T_K : L^1(X,\mu) \longrightarrow L^{1,\infty}(X,\mu). \tag{2.2.461}$$

(b) The maximal operator $T_{\max}$ is sub-linear and bounded in the following contexts:

$$T_{\max} : L^p(X,\mu) \longrightarrow L^p(X,\mu) \text{ for each } p \in (1,\infty), \tag{2.2.462}$$

$$T_{\max} : L^1(X,\mu) \longrightarrow L^{1,\infty}(X,\mu). \tag{2.2.463}$$

(c) For each $\varepsilon > 0$ the truncated operator T_ε is linear and bounded in the following contexts:

$$T_\varepsilon : L^p(X,\mu) \longrightarrow L^p(X,\mu) \text{ for each } p \in (1,\infty), \tag{2.2.464}$$

$$T_\varepsilon : L^1(X,\mu) \longrightarrow L^{1,\infty}(X,\mu), \tag{2.2.465}$$

with operator norm bounds uniform with respect to $\varepsilon \in (0,\infty)$.

(d) Under the additional assumption that

$$\text{the limit } \lim_{\varepsilon\to 0^+} \int_{\substack{y\in X \\ 1>\rho(x,y)>\varepsilon}} K(x,y)\,d\mu(y) \text{ exists for } \mu\text{-a.e. } x \in X, \tag{2.2.466}$$

it follows that the principal-value operator

$$(T_{\mathrm{PV}} f)(x) := \lim_{\varepsilon\to 0^+} \int_{y\in X,\,\rho(x,y)>\varepsilon} K(x,y)f(y)\,d\mu(y) \text{ for } \mu\text{-a.e. } x \in X, \tag{2.2.467}$$

is well defined, linear, and bounded in the following contexts:

$$T_{\mathrm{PV}} : L^p(X,\mu) \longrightarrow L^p(X,\mu) \text{ for each } p \in (1,\infty), \tag{2.2.468}$$

$$T_{\mathrm{PV}} : L^1(X,\mu) \longrightarrow L^{1,\infty}(X,\mu). \tag{2.2.469}$$

In addition, T_K *from* (2.2.460)-(2.2.461) *coincides with* T_{PV} *from* (2.2.468)-(2.2.469).

Proof The first claim in the statement is a consequence Theorem 2.2.32, Remark 2.2.33, (2.2.453), and (2.2.452). Assume next that T_K extends to a linear and bounded operator from $L^2(X, \mu)$ into itself. Much as in the proof of Proposition 2.2.28 this implies that the operators (2.2.460)-(2.2.462) are well defined and bounded. In addition from (2.2.462) and Proposition 2.2.13 we also see that (2.2.463) is well defined and bounded. This takes care of items *(a)-(b)*. In turn, (2.2.462)-(2.2.463) readily imply the claims in item *(c)*. Finally, the claims in item *(d)* are consequences of what we have proved so far and Proposition 2.2.28. □

2.3 Principal Value Singular Integral Operators on Uniformly Rectifiable Sets

The results in this section build on the work of many predecessors, including Calderón, Zygmund, Mikhlin, Coifman, McIntosh, Meyer, David, Jerison, Kenig, Semmes, Stein, and a multitude of others. Among other things, here we expand, refine, and sharpen work in [63] by presently placing more economical demands on the ambient geometry.

We begin by recording a useful result, pertaining to the manner in which Euclidean singular integrals are truncated, which appears in [64, Proposition B.2, p. 163].

Proposition 2.3.1 *Let* $A : \mathbb{R}^n \to \mathbb{R}^m$ *be some vector-valued Lipschitz function, and assume that* $F : \mathbb{R}^m \to \mathbb{R}$ *is an odd function of class* $\mathscr{C}^N$*, for some sufficiently large integer* $N = N(m)$*. Also, suppose* $B : \mathbb{R}^n \to \mathbb{R}^{m'}$ *is a bi-Lipschitz function*[15]*, and pick* $p \in (1, \infty)$*. Then for each given function* $f \in L^p(\mathbb{R}^n, \mathcal{L}^n)$ *the limit*

$$\lim_{\varepsilon\to0^+}\int_{\{y\in\mathbb{R}^n:\,|B(x)-B(y)|>\varepsilon\}}\frac{1}{|x-y|^n}F\Big(\frac{A(x)-A(y)}{|x-y|}\Big)f(y)\,dy \tag{2.3.1}$$

exists at $\mathcal{L}^n$*-a.e. point* $x \in \mathbb{R}^n$*. Moreover, this limit is independent of the choice of the function* B *which, in particular, implies that for each given* $f \in L^p(\mathbb{R}^n, \mathcal{L}^n)$ *the limit* (2.3.1) *is equal to*

$$\lim_{\varepsilon\to0^+}\int_{\{y\in\mathbb{R}^n:\,|x-y|>\varepsilon\}}\frac{1}{|x-y|^n}F\Big(\frac{A(x)-A(y)}{|x-y|}\Big)f(y)\,dy \tag{2.3.2}$$

at $\mathcal{L}^n$*-a.e.* $x \in \mathbb{R}^n$.

Our first main result in this section is Theorem 2.3.2, which builds on the work of G. David in [31] and refines [63, Propositions 3.18-3.19, pp. 2639-3240] as well as

[15] in the sense that $|B(x) - B(y)| \approx |x - y|$ uniformly in $x, y \in \mathbb{R}^n$ (which, in the present context, entails $m' \geq n$)

the portion of Theorem 3.33 on p. 2669 in [63] dealing with the pointwise existence of principal-value singular integral operators (by allowing the underlying UR set $\Sigma \subseteq \mathbb{R}^n$ to be more general than the topological boundary of an open set $\Omega \subseteq \mathbb{R}^n$ whose boundary is a UR set satisfying $\mathcal{H}^{n-1}(\partial\Omega \setminus \partial_*\Omega) = 0$, as considered in the aforementioned results in [63], and by allowing a larger class of functions).

Theorem 2.3.2 *For each $n \in \mathbb{N}$ with $n \geq 2$ there exists $N = N(n) \in \mathbb{N}$ with the following significance. Suppose $\Sigma \subseteq \mathbb{R}^n$ is a* UR *set, abbreviate $\sigma := \mathcal{H}^{n-1}\lfloor\Sigma$, and consider a kernel*

$$\begin{aligned} & k \in \mathscr{C}^N(\mathbb{R}^n \setminus \{0\}) \text{ satisfying } k(-x) = -k(x) \text{ as well as} \\ & k(\lambda\, x) = \lambda^{1-n} k(x) \text{ for all } x \in \mathbb{R}^n \setminus \{0\} \text{ and } \lambda \in (0, \infty). \end{aligned} \tag{2.3.3}$$

In this setting, define the maximal singular integral operator $T_{\max}$ acting on functions

$$f \in L^1\Big(\Sigma, \frac{\sigma(x)}{1 + |x|^{n-1}}\Big) \tag{2.3.4}$$

according to

$$(T_{\max} f)(x) := \sup_{\varepsilon>0} \big|(T_\varepsilon f)(x)\big|, \qquad \forall\, x \in \Sigma, \tag{2.3.5}$$

where, for each $\varepsilon > 0$, the truncated singular integral operator T_ε is given by[16]

$$(T_\varepsilon f)(x) := \int\limits_{\substack{y\in\Sigma \\ |x-y|>\varepsilon}} k(x-y) f(y)\, d\sigma(y), \qquad \forall\, x \in \Sigma. \tag{2.3.6}$$

Then the following statements are true.

(1) The maximal operator is a well-defined, bounded[17] *and continuous mapping*

$$T_{\max} : L^1\Big(\Sigma, \frac{\sigma(x)}{1 + |x|^{n-1}}\Big) \longrightarrow L^{1,\infty}_{\rm loc}(\Sigma, \sigma), \tag{2.3.7}$$

which for each $p \in (1, \infty)$ induces a well-defined, bounded and continuous mapping

$$T_{\max} : L^p_{\rm loc}(\Sigma, \sigma) \cap L^1\Big(\Sigma, \frac{\sigma(x)}{1 + |x|^{n-1}}\Big) \longrightarrow L^p_{\rm loc}(\Sigma, \sigma). \tag{2.3.8}$$

Moreover, $T_{\max}$ is sub-linear, in the sense that

$$\begin{aligned} & T_{\max}(f + g) \leq T_{\max} f + T_{\max} g, \qquad \forall\, f, g \in L^1\Big(\Sigma, \tfrac{\sigma(x)}{1+|x|^{n-1}}\Big), \\ & T_{\max}(\lambda f) = |\lambda|\, T_{\max} f, \quad \forall\, f \in L^1\Big(\Sigma, \tfrac{\sigma(x)}{1+|x|^{n-1}}\Big), \quad \forall\, \lambda \in \mathbb{R}. \end{aligned} \tag{2.3.9}$$

---

[16] the fact that the integral in the right-hand side of (2.3.6) is absolutely convergent is a consequence of (2.3.4)

[17] i.e., it maps bounded subsets of $L^1(\Sigma, \frac{\sigma(x)}{1+|x|^{n-1}})$ into bounded subsets of $L^{1,\infty}_{\rm loc}(\Sigma, \sigma)$

As a consequence, given any $f, g \in L^1\Big(\Sigma, \frac{\sigma(x)}{1+|x|^{n-1}}\Big)$ *one has*

$$\begin{aligned} &\big|(T_{\max}f)(x) - (T_{\max}g)(x)\big| \leq \big(T_{\max}(f-g)\big)(x) \text{ at each point} \\ &x \in \Sigma \text{ where } (T_{\max}f)(x) < +\infty \text{ and } (T_{\max}g)(x) < +\infty. \end{aligned} \tag{2.3.10}$$

(2) *For each* $p \in [1, \infty)$ *there exists a constant* $C \in (0, \infty)$ *depending only on* n, p, *and the* UR *character of* Σ, *such that for every function* $f \in L^p(\Sigma, \sigma)$ *one has*

$$\|T_{\max}f\|_{L^p(\Sigma,\sigma)} \leq C\big\|k\big|_{S^{n-1}}\big\|_{\mathscr{C}^N(S^{n-1})}\|f\|_{L^p(\Sigma,\sigma)} \text{ if } 1 < p < \infty, \tag{2.3.11}$$

$$\|T_{\max}f\|_{L^{1,\infty}(\Sigma,\sigma)} \leq C\big\|k\big|_{S^{n-1}}\big\|_{\mathscr{C}^N(S^{n-1})}\|f\|_{L^1(\Sigma,\sigma)} \text{ if } p = 1. \tag{2.3.12}$$

As a consequence of this and (2.3.9)-(2.3.10), *it follows that*

$$T_{\max} : L^p(\Sigma, \sigma) \longrightarrow L^p(\Sigma, \sigma) \text{ if } 1 < p < \infty, \tag{2.3.13}$$

$$T_{\max} : L^{1,\infty}(\Sigma, \sigma) \longrightarrow L^1(\Sigma, \sigma) \text{ if } p = 1, \tag{2.3.14}$$

are Lipschitz (hence, continuous) sub-linear mappings.

(3) *One has that*

$$\begin{aligned} &\text{for each function } f \in L^1\Big(\Sigma, \tfrac{\sigma(x)}{1+|x|^{n-1}}\Big) \text{ the limit} \\ &(Tf)(x) := \lim_{\varepsilon \to 0^+}(T_\varepsilon f)(x) \text{ exists for } \sigma\text{-a.e. } x \in \Sigma, \\ &\text{and the function } Tf \text{ thus defined is } \sigma\text{-measurable.} \end{aligned} \tag{2.3.15}$$

In fact, T *is a well-defined, linear, and continuous*[18] *operator in the context*

$$T : L^1\Big(\Sigma, \frac{\sigma(x)}{1 + |x|^{n-1}}\Big) \longrightarrow L^{1,\infty}_{\text{loc}}(\Sigma, \sigma), \tag{2.3.16}$$

which for each $p \in (1, \infty)$ *induces a well-defined, linear and continuous mapping*

$$T : L^p_{\text{loc}}(\Sigma, \sigma) \cap L^1\Big(\Sigma, \frac{\sigma(x)}{1 + |x|^{n-1}}\Big) \longrightarrow L^p_{\text{loc}}(\Sigma, \sigma). \tag{2.3.17}$$

Moreover, the operators induced by the action of T *from* (2.3.16) *on ordinary Lebesgue spaces (cf.* [112, (7.7.106)]), *i.e.,*

---

[18] This means that for each point $x_o \in \Sigma$ and each radius $r \in (0, \infty)$ there exists $C_{x_o,r} \in (0, \infty)$ with the property that $\Big\|(Tf)\big|_{\Sigma\cap B(x_o,r)}\Big\|_{L^{1,\infty}(\Sigma\cap B(x_o,r),\sigma)} \leq C_{x_o,r}\|f\|_{L^1\big(\Sigma, \frac{\sigma(x)}{1+|x|^{n-1}}\big)}$ for every function $f \in L^1\big(\Sigma, \frac{\sigma(x)}{1+|x|^{n-1}}\big)$.

$$T : L^p(\Sigma,\sigma) \longrightarrow L^p(\Sigma,\sigma), \quad p \in (1,\infty), \tag{2.3.18}$$

$$T : L^1(\Sigma,\sigma) \longrightarrow L^{1,\infty}(\Sigma,\sigma), \tag{2.3.19}$$

are well-defined, linear, bounded, and compatible with one another, with operator norms[19]

$$\|T\|_{L^p(\Sigma,\sigma)\to L^p(\Sigma,\sigma)} \leq C_p \big\|k\big|_{S^{n-1}}\big\|_{\mathscr{C}^N(S^{n-1})} \text{ if } 1 < p < \infty, \tag{2.3.20}$$

$$\|T\|_{L^1(\Sigma,\sigma)\to L^{1,\infty}(\Sigma,\sigma)} \leq C\big\|k\big|_{S^{n-1}}\big\|_{\mathscr{C}^N(S^{n-1})}. \tag{2.3.21}$$

As a consequence of (2.3.21) *and* [112, Lemma 6.2.3], *for each* $\theta \in (0,1)$ *there exists a constant* $C = C(\Sigma,\theta,n,k) \in (0,\infty)$ *with the property that for any* σ*-measurable set* $E \subseteq \Sigma$ *and any function* $f \in L^1(\Sigma,\sigma)$ *one has*

$$\int_E |Tf|^\theta \, d\sigma \leq C\sigma(E)^{1-\theta}\|f\|^\theta_{L^1(\Sigma,\sigma)}. \tag{2.3.22}$$

Finally, for each $r \in (0,\infty)$ *there exists* $C \in (0,\infty)$, *which depends only on n, the* UR *constants of* Σ, *and r, with the property that the following Cotlar-type inequality holds:*

$$\begin{gathered}(T_{\max} f)(x) \leq C\Big[\big(\mathcal{M}_\Sigma(|Tf|^r)\big)(x)\Big]^{1/r} + C\big\|k\big|_{S^{n-1}}\big\|_{\mathscr{C}^N(S^{n-1})}\big(\mathcal{M}_\Sigma f\big)(x)\\ \text{for every function } f \in L^1\Big(\Sigma, \tfrac{\sigma(x)}{1+|x|^{n-1}}\Big) \text{ and every point } x \in \Sigma,\end{gathered} \tag{2.3.23}$$

where $\mathcal{M}_\Sigma$ *is the Hardy-Littlewood maximal operator on* Σ.

(4) *For each fixed function* $f \in L^p(\Sigma,\sigma)$ *with* $1 < p < \infty$ *one has*

$$T_\varepsilon f \longrightarrow Tf \text{ in } L^p(\Sigma,\sigma) \text{ as } \varepsilon \to 0^+. \tag{2.3.24}$$

(5) *One has that*

the (real) transpose of the operator (2.3.18) *is* $-T : L^{p'}(\Sigma,\sigma) \to L^{p'}(\Sigma,\sigma)$, *where the exponent* $p' \in (1,\infty)$ *is the Hölder conjugate of the exponent* $p \in (1,\infty)$. (2.3.25)

In addition, given any two functions

[19] The dependence of the constant C_p appearing in (2.3.20) on the exponent p follows from interpolating between $p = 1$ (cf. (2.3.21)), $p = 2$, and duality: ultimately, $C_p = O(p/(p-1))$ as $p \to \infty$ and as $p \to 1$.

$$
\begin{aligned}
&f \in L^p_{\mathrm{loc}}(\Sigma,\sigma) \cap L^1\Big(\Sigma, \tfrac{\sigma(x)}{1+|x|^{n-1}}\Big) \textit{ and } g \in L^{p'}_{\mathrm{comp}}(\Sigma,\sigma) \\
&\quad\textit{with } p, p' \in (1,\infty) \textit{ such that } 1/p + 1/p' = 1, \\
&\textit{one has } \int_\Sigma |Tf||g|\,\mathrm{d}\sigma < +\infty, \quad \int_\Sigma |f||Tg|\,\mathrm{d}\sigma < +\infty \\
&\quad\textit{and } \int_\Sigma (Tf)g\,\mathrm{d}\sigma = -\int_\Sigma f(Tg)\,\mathrm{d}\sigma.
\end{aligned}
\tag{2.3.26}
$$

(6) *If $H^p(\Sigma,\sigma)$ with $(n-1)/n < p < \infty$ denotes the scale of Hardy spaces on Σ, introduced in* [113, Definition 4.2.1], *then the operator* (2.3.18) *extends uniquely to a linear and bounded mapping*

$$
T : H^p(\Sigma,\sigma) \longrightarrow L^p(\Sigma,\sigma), \quad \tfrac{n-1}{n} < p \le 1, \tag{2.3.27}
$$

according to

$$
\begin{aligned}
&H^p(\Sigma,\sigma) \ni f \mapsto \lim_{j\to\infty} Tf_j \textit{ in } L^p(\Sigma,\sigma) \textit{ where} \\
&\{f_j\}_{j\in\mathbb{N}} \subseteq H^{p,q}_{\mathrm{fin}}(\Sigma,\sigma) \textit{ with } q \in (1,\infty) \textit{ is any} \\
&\textit{sequence satisfying } f = \lim_{j\to\infty} f_j \textit{ in } H^p(\Sigma,\sigma),
\end{aligned}
\tag{2.3.28}
$$

with operator quasi-norm

$$
\|T\|_{H^p(\Sigma,\sigma)\to L^p(\Sigma,\sigma)} \le C_{\Sigma,p} \big\| k\big|_{S^{n-1}} \big\|_{\mathscr{C}^N(S^{n-1})} \textit{ for } \tfrac{n-1}{n} < p \le 1. \tag{2.3.29}
$$

In particular, with P.V. *used to indicate that an integral is taken in the principal-value sense,*

$$
\begin{aligned}
&Tf = \sum_{j=1}^\infty \lambda_j T a_j = \sum_{j=1}^\infty \lambda_j \,\mathrm{P.V.}\int_\Sigma k(\cdot - y) a_j(y)\,\mathrm{d}\sigma(y) \textit{ in } L^p(\Sigma,\sigma), \\
&\textit{when } \tfrac{n-1}{n} < p \le 1 \textit{ and when the distribution } f \in H^p(\Sigma,\sigma) \\
&\textit{is expressed as } f = \textstyle\sum_{j=1}^\infty \lambda_j a_j \textit{ in } H^p(\Sigma,\sigma) \textit{ for some sequence} \\
&\{\lambda_j\}_{j\in\mathbb{N}} \in \ell^p(\mathbb{N}) \textit{ and some sequence } \{a_j\}_{j\in\mathbb{N}} \textit{ of } (p,q)\textit{-atoms on } \Sigma, \\
&\textit{where } q \in [1,\infty] \textit{ with } q > p \textit{ is some fixed background index.}
\end{aligned}
\tag{2.3.30}
$$

Moreover, the operators in (2.3.27) *corresponding to various values of the exponent* $p \in \big(\frac{n-1}{n}, 1\big]$ *are compatible with one another. In addition, the operator* T *defined on Hardy spaces as in* (2.3.27) *acts in a coherent fashion with the operator* T *defined on Lebesgue spaces as in* (2.3.18), *and the operator* T *defined as in* (2.3.27) *with* $p = 1$ *acts in a coherent fashion with the operator* T *defined in* (2.3.19).

(7) *For the end-point* $p = \infty$, *consider the following modified version of the principal-value operator in* (2.3.15)*:*

$$(T_{\text{mod}} f)(x) := \lim_{\varepsilon \to 0^+} \int_{\Sigma} \{k_\varepsilon(x-y) - k_1(-y)\} f(y)\, d\sigma(y) \text{ for } \sigma\text{-a.e. } x \in \Sigma,$$
$$\text{where } k_\varepsilon := k \cdot \mathbf{1}_{\mathbb{R}^n \setminus \overline{B(0,\varepsilon)}} \text{ for each } \varepsilon > 0, \text{ and } f \in L^1\Big(\Sigma, \frac{\sigma(x)}{1+|x|^n}\Big). \tag{2.3.31}$$

Then T_{mod} is meaningfully defined in the above context. In particular, since Hölder's inequality *and* [112, Lemma 7.2.1] *guarantee a continuous embedding*

$$L^p(\Sigma, \sigma) \hookrightarrow L^1\Big(\Sigma, \frac{\sigma(x)}{1+|x|^n}\Big) \text{ for each } p \in [1, \infty], \tag{2.3.32}$$

it follows that T_{mod} acts in a well-defined manner on $L^p(\Sigma, \sigma)$ for each exponent $p \in [1, \infty]$.

Most generally, T_{mod} from (2.3.31) *induces a well-defined, linear and continuous operator*[20]

$$T_{\text{mod}} : L^1\Big(\Sigma, \frac{\sigma(x)}{1+|x|^n}\Big) \longrightarrow L^{1,\infty}_{\text{loc}}(\Sigma, \sigma). \tag{2.3.33}$$

Defined as such, T_{mod} is compatible with T from (2.3.16) (*hence also with T from* (2.3.18)-(2.3.19)), *in the sense that*

for each function f belonging to the space $L^1\big(\Sigma, \frac{\sigma(x)}{1+|x|^{n-1}}\big)$ (hence, in particular, for each $f \in L^p(\Sigma, \sigma)$ with $p \in [1, \infty)$) the difference $C_f := T_{\text{mod}} f - Tf$ is a constant function on Σ with $|C_f| \leq C_{\Sigma,k} \cdot \|f\|_{L^1(\Sigma, \frac{\sigma(x)}{1+|x|^{n-1}})}$ for some finite $C_{\Sigma,k} > 0$. (2.3.34)

Also, in a quantitative fashion,

$$T_{\text{mod}} f \text{ belongs to the space } L^p_{\text{loc}}(\Sigma, \sigma) \text{ whenever}$$
$$f \in L^1\Big(\Sigma, \frac{\sigma(x)}{1+|x|^n}\Big) \cap L^p_{\text{loc}}(\Sigma, \sigma) \text{ for some } p \in (1, \infty), \tag{2.3.35}$$

while (compare with (2.3.26))

$$\text{if } p, p' \in (1, \infty) \text{ are such that } 1/p + 1/p' = 1 \text{ then given any functions}$$
$$f \in L^p_{\text{loc}}(\Sigma, \sigma) \cap L^1\Big(\Sigma, \tfrac{\sigma(x)}{1+|x|^n}\Big) \text{ and } g \in L^{p'}_{\text{comp}}(\Sigma, \sigma) \text{ with } \int_\Sigma g\, d\sigma = 0,$$
$$\text{it follows that } \int_\Sigma |T_{\text{mod}} f||g|\, d\sigma < +\infty, \quad \int_\Sigma |f||Tg|\, d\sigma < +\infty,$$
$$\text{and } \int_\Sigma (T_{\text{mod}} f) g\, d\sigma = -\int_\Sigma f(Tg)\, d\sigma. \tag{2.3.36}$$

[20] with a similar interpretation as in the case of (2.3.16)

Furthermore, there exists a finite constant $C = C(\Sigma, k) > 0$ *with the property that*

$$\|T_{\text{mod}} f\|_{\text{BMO}(\Sigma,\sigma)} \leq C\|f\|_{L^\infty(\Sigma,\sigma)} \text{ for every } f \in L^\infty(\Sigma, \sigma), \tag{2.3.37}$$

In particular,

$$T_{\text{mod}} : L^\infty(\Sigma, \sigma) \longrightarrow \text{BMO}(\Sigma, \sigma) \text{ is well defined, linear and bounded.} \tag{2.3.38}$$

Also,

$$\text{if the set } \Sigma \text{ is bounded then } T \text{ from (2.3.18) induces a linear and bounded mapping } T : L^\infty(\Sigma, \sigma) \longrightarrow \text{BMO}(\Sigma, \sigma). \tag{2.3.39}$$

Finally, with the angled brackets $\langle\cdot,\cdot\rangle$ *denoting the duality pairing between* $H^1(\Sigma, \sigma)$ *and* $\big(H^1(\Sigma, \sigma)\big)^*$ *(cf.* [113, Theorem 4.6.1]), *for any two given functions* $f \in L^\infty(\Sigma, \sigma)$ *and* $g \in H^1(\Sigma, \sigma)$ *one has*

$$\int_\Sigma f\,Tg\,d\sigma = \begin{cases} {}_{H^1(\Sigma,\sigma)}\big\langle g, [T_{\text{mod}} f]\big\rangle_{\text{BMO}(\Sigma,\sigma)/\sim} & \text{if } \Sigma \text{ is unbounded,} \\ {}_{H^1(\Sigma,\sigma)}\big\langle g, Tf\big\rangle_{\text{BMO}(\Sigma,\sigma)} & \text{if } \Sigma \text{ is bounded.} \end{cases} \tag{2.3.40}$$

(8) If $p \in (1, \infty)$ *and* $q \in (0, \infty]$ *then (cf.* [112, (7.7.107)])

$$\text{for each given function } f \in L^{p,q}(\Sigma, \sigma) \text{ the limit } (Tf)(x) = \lim_{\varepsilon\to 0^+}(T_\varepsilon f)(x) \text{ exists for } \sigma\text{-a.e. } x \in \Sigma \text{ and the function } Tf \text{ thus defined is } \sigma\text{-measurable.} \tag{2.3.41}$$

Also,

$$T_{\max} : L^{p,q}(\Sigma, \sigma) \longrightarrow L^{p,q}(\Sigma, \sigma), \tag{2.3.42}$$

$$T : L^{p,q}(\Sigma, \sigma) \longrightarrow L^{p,q}(\Sigma, \sigma), \tag{2.3.43}$$

are well-defined, bounded (sub-linear and linear, respectively) operators. Finally,

$$\text{if } p, q \in (1, \infty) \text{ then the (real) transpose of the operator (2.3.43) is the operator } -T : L^{p',q'}(\Sigma, \sigma) \to L^{p',q'}(\Sigma, \sigma), \text{ where the exponents } p', q' \in (1, \infty) \text{ are such that } 1/p + 1/p' = 1 \text{ and } 1/q + 1/q' = 1. \tag{2.3.44}$$

(9) For each exponent $p \in \big(\tfrac{n-1}{n}, \infty\big)$ *denote by* T_p *the operator (2.3.27)-(2.3.28) if* $\tfrac{n-1}{n} < p \leq 1$ *and, respectively, the operator (2.3.18) if* $1 < p < \infty$. *Then the family of operators* $\{T_p\}_{\frac{n-1}{n}<p<\infty}$ *are mutually compatible with one another, in the sense that*

if $p_0, p_1 \in \big(\frac{n-1}{n}, \infty\big)$ *and* $f \in H^{p_0}(\Sigma, \sigma) \cap H^{p_1}(\Sigma, \sigma)$*, it follows that* $T_{p_0} f = T_{p_1} f$ *at* σ*-a.e. point on* Σ. (2.3.45)

As such, given any $p_0, p_1 \in \big(\frac{n-1}{n}, \infty\big)$*, one may define, in a coherent fashion, a mapping*

$$\begin{aligned}&\widetilde{T}_{p_0,p_1} : H^{p_0}(\Sigma, \sigma) + H^{p_1}(\Sigma, \sigma) \longrightarrow L^{p_0}(\Sigma, \sigma) + L^{p_1}(\Sigma, \sigma),\\ &\widetilde{T}_{p_0,p_1} f := T_{p_0} f_0 + T_{p_1} f_1 \ \text{ whenever } \ f \in H^{p_0}(\Sigma, \sigma) + H^{p_1}(\Sigma, \sigma)\\ &\text{splits as } \ f = f_0 + f_1 \ \text{ with } \ f_0 \in H^{p_0}(\Sigma, \sigma) \ \text{ and } \ f_1 \in H^{p_1}(\Sigma, \sigma).\end{aligned} \tag{2.3.46}$$

Moreover, whenever $\frac{n-1}{n} < p_0 < p_1 < \infty$,

$$\begin{aligned}&\widetilde{T}_{p_0,p_1} \text{ is a linear and bounded operator, which satisfies}\\ &\qquad \widetilde{T}_{p_0,p_1}\Big|_{H^p(\Sigma,\sigma)} = T_p \ \text{ for each } \ p \in [p_0, p_1].\end{aligned} \tag{2.3.47}$$

Also,

$$\begin{aligned}&\text{the operators in the family } \big\{\widetilde{T}_{p_0,p_1}\big\}_{\frac{n-1}{n} < p_0, p_1 < \infty}\\ &\text{are mutually compatible with one another,}\end{aligned} \tag{2.3.48}$$

thus giving rise to a common linear mapping. Specifically,

$$H_{\mathrm{Sum}}(\Sigma, \sigma) := \bigcup_{\frac{n-1}{n} < p_0, p_1 < \infty} \big(H^{p_0}(\Sigma, \sigma) + H^{p_1}(\Sigma, \sigma)\big) \tag{2.3.49}$$

and

$$L_{\mathrm{Sum}}(\Sigma, \sigma) := \bigcup_{\frac{n-1}{n} < p_0, p_1 < \infty} \big(L^{p_0}(\Sigma, \sigma) + L^{p_1}(\Sigma, \sigma)\big) \tag{2.3.50}$$

are linear spaces and there exists a unique linear operator

$$\begin{aligned}&\widetilde{T} : H_{\mathrm{Sum}}(\Sigma, \sigma) \longrightarrow L_{\mathrm{Sum}}(\Sigma, \sigma) \ \text{ with the property that}\\ &\widetilde{T}\Big|_{H^{p_0}(\Sigma,\sigma)+H^{p_1}(\Sigma,\sigma)} = \widetilde{T}_{p_0,p_1} \ \text{ for each } \ p_0, p_1 \in \big(\tfrac{n-1}{n}, \infty\big).\end{aligned} \tag{2.3.51}$$

In addition, the restriction of the operator $\widetilde{T}$ *from* (2.3.51) *to any Lorentz-based Hardy space* $H^{p,q}(\Sigma, \sigma)$ *with* $p \in \big(\frac{n-1}{n}, \infty\big)$ *and* $q \in (0, \infty]$ *maps into Lorentz space* $L^{p,q}(\Sigma, \sigma)$ *in a linear and bounded fashion, and the operator thus induced is compatible with* T *from* (2.3.18) *and* (2.3.27)*. As a consequence, one may employ the same notation for said operator, namely*

$$T : H^{p,q}(\Sigma, \sigma) \longrightarrow L^{p,q}(\Sigma, \sigma)$$
$$\text{for } p \in \left(\tfrac{n-1}{n}, \infty\right) \text{ and } q \in (0, \infty]. \tag{2.3.52}$$

This satisfies the operator quasi-norm estimate

$$\|T\|_{H^{p,q}(\Sigma,\sigma)\to L^{p,q}(\Sigma,\sigma)} \leq C_{\Sigma,p,q} \left\|k\big|_{S^{n-1}}\right\|_{\mathscr{C}^N(S^{n-1})} \tag{2.3.53}$$

and enjoys the following property:

$Tf = Tf_0 + Tf_1$ *at* σ*-a.e. point on* Σ, *whenever* $f \in H^{p,q}(\Sigma, \sigma)$, $f_0 \in H^{p_0,q_0}(\Sigma, \sigma)$, $f_1 \in H^{p_1,q_1}(\Sigma, \sigma)$, *with* $p, p_0, p_1 \in \left(\frac{n-1}{n}, \infty\right)$ *and* $q, q_0, q_1 \in (0, \infty]$, *are such that* $f = f_0 + f_1$ *in* $\left(\mathrm{Lip}_c(\Sigma)\right)'$. (2.3.54)

Finally,

given $p \in \left(\frac{n-1}{n}, \infty\right)$, *if* $q \in (0, \infty)$ *then the operator* (2.3.52) *acts according to* $H^{p,q}(\Sigma, \sigma) \ni f \longmapsto \lim_{j\to\infty} Tf_j$ *in* $L^{p,q}(\Sigma, \sigma)$, *provided* $\{f_j\}_{j\in\mathbb{N}} \subseteq H^{p_0}(\Sigma, \sigma) \cap H^{p_1}(\Sigma, \sigma)$ *with* $\frac{n-1}{n} < p_0 < p < p_1 < \infty$ *is any given sequence satisfying* $f = \lim_{j\to\infty} f_j$ *in the space* $H^{p,q}(\Sigma, \sigma)$. (2.3.55)

(10) Denote by $A_p(\Sigma, \sigma)$, $1 \leq p < \infty$, *the classes of Muckenhoupt weights associated as in* (A.0.3) *with the space of homogeneous type* $\left(\Sigma, |\cdot - \cdot|, \sigma\right)$ (*cf.* [112, Example 7.4.1]). *Then the principal-value singular integral operator T from* (2.3.15) *induces, via restriction* (*cf.* [112, (7.7.104)]), *well-defined linear and bounded mappings on weighted versions of the Lebesgue spaces in* (2.3.18), *for weights in the corresponding Muckenhoupt class* $A_p(\Sigma, \sigma)$, $1 < p < \infty$. *Specifically, the mapping*

$$T : L^p(\Sigma, w\sigma) \longrightarrow L^p(\Sigma, w\sigma), \quad p \in (1, \infty), \ w \in A_p(\Sigma, \sigma), \tag{2.3.56}$$

induced (via restriction; cf. [112, (7.7.104)]) *by the operator T from* (2.3.15) *is well defined, linear, and bounded. Moreover, for each exponent* $p \in (1, \infty)$ *and* $w \in A_p(\Sigma, \sigma)$, *one has*

$$\|T\|_{L^p(\Sigma,w\sigma)\to L^p(\Sigma,w\sigma)} \leq C(\Sigma, p, [w]_{A_p})\left\|k\big|_{S^{n-1}}\right\|_{\mathscr{C}^N(S^{n-1})}. \tag{2.3.57}$$

In addition,

$$T_{\max} : L^p(\Sigma, w\sigma) \longrightarrow L^p(\Sigma, w\sigma), \quad p \in (1, \infty), \ w \in A_p(\Sigma, \sigma), \tag{2.3.58}$$

is a well-defined, sub-linear, bounded, and continuous operator, whose operator norm may be estimated as in (2.3.57), *i.e.,*

$$\|T_{\max}\|_{L^p(\Sigma,w\sigma)\to L^p(\Sigma,w\sigma)} \leq C(\Sigma, p, [w]_{A_p})\left\|k\big|_{S^{n-1}}\right\|_{\mathscr{C}^N(S^{n-1})}. \tag{2.3.59}$$

Also, bearing in mind the natural identification

$$\begin{gathered}\big(L^p(\Sigma, w\sigma)\big)^* = L^{p'}(\Sigma, w^{1-p'}\sigma)\\ \text{for } p, p' \in (1,\infty) \text{ with } 1/p + 1/p' = 1,\end{gathered} \tag{2.3.60}$$

under the canonical integral pairing $(f,g) \mapsto \int_\Sigma fg\,\mathrm{d}\sigma$ *it follows that*[21]

$$\begin{gathered}\text{the (real) transpose of the operator (2.3.56) is}\\ -T : L^{p'}(\Sigma, w^{1-p'}\sigma) \to L^{p'}(\Sigma, w^{1-p'}\sigma), \text{ where}\\ p' \in (1,\infty) \text{ is chosen so that } 1/p + 1/p' = 1.\end{gathered} \tag{2.3.61}$$

Furthermore, for each $p \in (1,\infty)$ *and* $w \in A_p(\Sigma,\sigma)$*, the following mapping is well defined, linear, and continuous:*

$$T : L^p_{\mathrm{loc}}(\Sigma, w\sigma) \cap L^1\Big(\Sigma, \frac{\sigma(x)}{1+|x|^{n-1}}\Big) \longrightarrow L^p_{\mathrm{loc}}(\Sigma, w\sigma). \tag{2.3.62}$$

Finally, for each $p, p' \in (1,\infty)$ *with* $1/p + 1/p' = 1$ *and each* $w \in A_p(\Sigma,\sigma)$ *there exists a constant* $C \in (0,\infty)$ *with the property that*

$$\begin{gathered}\text{if } f \in L^p(\Sigma, w\sigma) \text{ and } g \in L^{p'}(\Sigma, w^{1-p'}\sigma) \text{ then } (Tf)g + f(Tg)\\ \text{belongs to the Hardy space } \dot{H}^1(\Sigma,\sigma) \text{ (cf. [113, (4.2.12)]) and one}\\ \text{has } \big\|(Tf)g + f(Tg)\big\|_{H^1(\Sigma,\sigma)} \leq C\|f\|_{L^p(\Sigma,w\sigma)} \cdot \|g\|_{L^{p'}(\Sigma,w^{1-p'}\sigma)}.\end{gathered} \tag{2.3.63}$$

(11) As before, denote by $\mathcal{M}_\Sigma$ *the Hardy-Littlewood maximal operator associated with the space of homogeneous type* $\big(\Sigma, |\cdot - \cdot|, \sigma\big)$*. Consider a Generalized Banach Function Space* $\mathbb{X}$ *on* (Σ,σ) *and denote by* $\mathbb{X}'$ *its associated space (cf.* [113, Definitions 5.1.4, 5.1.11]*). Then if*

$$\begin{gathered}\mathcal{M}_\Sigma : \mathbb{X} \to \mathbb{X} \text{ and } \mathcal{M}_\Sigma : \mathbb{X}' \to \mathbb{X}'\\ \text{are well-defined bounded mappings}\end{gathered} \tag{2.3.64}$$

it follows that the maximal operator $T_{\max}$*, originally considered as in* (2.3.4)-(2.3.5)*, induces well-defined, bounded and continuous mappings*

$$\begin{gathered}T_{\max} : \mathbb{X} \longrightarrow \mathbb{X} \text{ and } T_{\max} : \mathbb{X}' \longrightarrow \mathbb{X}'\\ \text{with } \max\Big\{\|T_{\max}\|_{\mathbb{X}\to\mathbb{X}}, \|T_{\max}\|_{\mathbb{X}'\to\mathbb{X}'}\Big\} \leq C\big\|k\big|_{S^{n-1}}\big\|_{\mathscr{C}^N(S^{n-1})},\end{gathered} \tag{2.3.65}$$

where $C \in (0,\infty)$ *depends only on* Σ *and the operator norms of* $\mathcal{M}_\Sigma$ *on* $\mathbb{X}$, $\mathbb{X}'$*. Likewise, under the assumptions made in* (2.3.64)*, the principal-value singular integral operator* T*, originally defined as in* (2.3.15)*, induces well-defined, bounded, mappings*

[21] In relation to (2.3.61), it is reassuring to note that if $w \in A_p(\Sigma,\sigma)$ then $w^{1-p'} \in A_{p'}(\Sigma,\sigma)$ (cf. item *(2)* in [112, Lemma 7.7.1])

$$
\begin{gathered}
T : \mathbb{X} \longrightarrow \mathbb{X} \text{ and } T : \mathbb{X}' \longrightarrow \mathbb{X}' \\
\text{with } \max\Big\{\|T\|_{\mathbb{X}\to\mathbb{X}}, \|T\|_{\mathbb{X}'\to\mathbb{X}'}\Big\} \leq C\big\|k\big|_{S^{n-1}}\big\|_{\mathscr{C}^N(S^{n-1})},
\end{gathered}
\tag{2.3.66}
$$

where $C \in (0, \infty)$ depends only on Σ and the operator norms of $\mathcal{M}_\Sigma$ on $\mathbb{X}$ and $\mathbb{X}'$. Next, let $\mathring{\mathbb{X}}$ be defined in relation to $\mathbb{X}$ as in [113, Definition 5.2.6], *i.e., set*

$$
\mathring{\mathbb{X}} := \overline{L^\infty_{\text{comp}}(\Sigma, \sigma)}^{\|\cdot\|_{\mathbb{X}}}. \tag{2.3.67}
$$

Then, under the assumptions made in (2.3.64), *one also has*

$$
\begin{gathered}
T_{\max} : \mathring{\mathbb{X}} \longrightarrow \mathring{\mathbb{X}} \text{ and } T : \mathring{\mathbb{X}} \longrightarrow \mathring{\mathbb{X}} \\
\text{with } \max\Big\{\|T_{\max}\|_{\mathring{\mathbb{X}}\to\mathring{\mathbb{X}}}, \|T\|_{\mathring{\mathbb{X}}\to\mathring{\mathbb{X}}}\Big\} \leq C\big\|k\big|_{S^{n-1}}\big\|_{\mathscr{C}^N(S^{n-1})},
\end{gathered}
\tag{2.3.68}
$$

where, once again, $C \in (0, \infty)$ depends only on Σ and the operator norms of $\mathcal{M}_\Sigma$ on $\mathbb{X}$ and $\mathbb{X}'$. Finally, granted the assumptions in (2.3.64), *the following transposition formula holds:*

$$
\int_\Sigma (Tf)g\,d\sigma = -\int_\Sigma f(Tg)\,d\sigma \text{ for any } f \in \mathring{\mathbb{X}} \text{ and } g \in \mathbb{X}'. \tag{2.3.69}
$$

Proof We begin by noting that the fundamental operator norm estimate recorded in (2.3.11) has been established in [31, Proposition 4 bis]. The claims in (2.3.15) follow directly from [112, Corollary 5.3.6] (bearing in mind [112, Proposition 6.10.5] and the fact that UR sets are closed and upper Ahlfors regular). In concert with (2.3.11), this further implies the fact that T is well defined, linear, and bounded in the context of (2.3.18). Granted this, all claims in item *(1)* are directly implied by Proposition 2.2.5 and (2.2.252) in Corollary 2.2.11.

Before going any further, we wish to present an alternative proof of the claims made in (2.3.15) in a series of steps, presented below.

`Step I:` *The proof of the first claim in* (2.3.15) *when Σ is a Lipschitz graph in $\mathbb{R}^n$ and $f \in L^p(\Sigma, \sigma)$ with $p \in (1, \infty)$.* Concretely, assume that for some Lipschitz function $\phi : \mathbb{R}^{n-1} \to \mathbb{R}$ we have

$$
\Sigma = \big\{(x', \phi(x')) : x' \in \mathbb{R}^{n-1}\big\}, \tag{2.3.70}
$$

and fix some function $f \in L^p(\Sigma, \sigma)$ with $p \in (1, \infty)$. In this scenario, [112, (2.8.69)] gives

$$
\sigma(x', \phi(x')) = \sqrt{1 + |(\nabla'\phi)(x')|^2}\,\mathcal{L}^{n-1}(x'), \qquad x' \in \mathbb{R}^{n-1}. \tag{2.3.71}
$$

Given that the above square-root is bounded away from zero and infinity, if we now define the function $\widetilde{f} : \mathbb{R}^{n-1} \to \mathbb{R}$ by setting

$$
\widetilde{f}(x') := \sqrt{1 + |(\nabla'\phi)(x')|^2}\, f\big(x', \phi(x')\big) \text{ for each } x' \in \mathbb{R}^{n-1}, \tag{2.3.72}
$$

it follows that $\widetilde{f} \in L^p(\mathbb{R}^{n-1}, \mathcal{L}^{n-1})$. Moreover, changing variables in (2.3.6) gives that for each $\varepsilon > 0$ and each $x' \in \mathbb{R}^{n-1}$ we have

$$(T_\varepsilon f)\big(x', \phi(x')\big) = \int\limits_{\substack{y' \in \mathbb{R}^{n-1} \\ |x'-y'|^2 + |\phi(x')-\phi(y')|^2 > \varepsilon^2}} k\big(x'-y', \phi(x')-\phi(y')\big) \widetilde{f}(y')\,dy'. \tag{2.3.73}$$

To study the behavior of the integral in the right side of (2.3.73), the idea is to rely on Proposition 2.3.1 used for a suitable choice of the functions A, B, F. Concretely, we consider $A : \mathbb{R}^{n-1} \to \mathbb{R}^n$ by $A(z') := \big(z', \phi(z')\big)$ for each $z' \in \mathbb{R}^{n-1}$, and take $B := A$. Clearly, A is Lipschitz and B is bi-Lipschitz. To define the function F, first pick a finite number $R > \sqrt{1 + \|\nabla'\phi\|^2_{L^\infty(\mathbb{R}^{n-1}, \mathcal{L}^{n-1})}}$ and consider a cutoff function

$$\begin{aligned} &\psi \in \mathscr{C}^\infty_c(\mathbb{R}^n) \text{ even, such that } \psi \equiv 0 \text{ on } B(0, 1/4), \\ &\quad \text{and } \psi \equiv 1 \text{ on } B(0, R) \setminus B(0, 1/2). \end{aligned} \tag{2.3.74}$$

Having done that, define the function $F := \psi k$ in $\mathbb{R}^n$ and note that this choice entails $F \in \mathscr{C}^N(\mathbb{R}^n)$ and F is odd. In addition, for each $x', y' \in \mathbb{R}^{n-1}$ with $x' \neq y'$, the positive homogeneity of k permits us to express

$$\begin{aligned} k\big(x'-y', \phi(x')-\phi(y')\big) &= \frac{1}{|x'-y'|^{n-1}} k\Big(\frac{x'-y'}{|x'-y'|}, \frac{\phi(x')-\phi(y')}{|x'-y'|}\Big) \\ &= \frac{1}{|x'-y'|^{n-1}} (\psi k)\Big(\frac{x'-y'}{|x'-y'|}, \frac{\phi(x')-\phi(y')}{|x'-y'|}\Big) \\ &= \frac{1}{|x'-y'|^{n-1}} F\Big(\frac{A(x')-A(y')}{|x'-y'|}\Big). \end{aligned} \tag{2.3.75}$$

Granted this, it follows from (2.3.73) and Proposition 2.3.1 that the first claim in (2.3.15) holds in this case.

`Step II:` *The proof of the first claim in* (2.3.15) *when* Σ *is a rotated Lipschitz graph in* $\mathbb{R}^n$ *and* $f \in L^p(\Sigma, \sigma)$ *with* $p \in (1, \infty)$. Given that the original assumptions on the kernel k are rotation invariant, and that the $(n-1)$-dimensional Hausdorff measure is also rotation invariant, this is a consequence of Step I and a change of variables reducing matters to the case when Σ is a genuine Lipschitz graph in $\mathbb{R}^n$.

`First Intermission:` Having dealt at this point with the first claim in (2.3.15) when Σ is a rotated Lipschitz graph in $\mathbb{R}^n$ and $f \in L^p(\Sigma, \sigma)$ with $p \in (1, \infty)$, we now consider the case when $\Sigma \subseteq \mathbb{R}^n$ is a UR set and $f \in L^p(\Sigma, \sigma)$ with $p \in (1, \infty)$. In particular, from [112, Proposition 6.10.5] it follows that Σ is a countably rectifiable set. Using [112, Proposition 5.3.3] (while keeping in mind that Σ is a closed set) we may therefore decompose

$$\Sigma = \Big(\bigcup_{j\in\mathbb{N}} S_j\Big) \cup A, \text{ where } \mathcal{H}^{n-1}(A) = 0 \text{ and, for each } j \in \mathbb{N}, \\ S_j \text{ is a compact subset of a rotated Lipschitz graph } \Sigma_j \text{ in } \mathbb{R}^n. \tag{2.3.76}$$

For further use, in the above context let us also define

$$\widetilde{S_1} := S_1 \text{ then, inductively, set } \widetilde{S_j} := S_j \setminus \bigcup_{i=1}^{j-1} \widetilde{S_i} \text{ for } j \geq 2. \tag{2.3.77}$$

`Step III:` *The proof of the fact that the limit in* (2.3.15) *exists at σ-a.e. point on Σ in the case when Σ is as in* (2.3.76)*, and $f \in L^p(\Sigma, \sigma)$ with $p \in (1, \infty)$ has the additional property that* $\operatorname{supp} f \subseteq S_{j_0}$ *for some* $j_0 \in \mathbb{N}$. In this situation, Step II applies and gives that, on the one hand, $\lim_{\varepsilon\to 0^+} (T_\varepsilon f)(x)$ exists for σ-a.e. $x \in S_{j_0}$. On the other hand, for each $x \in \Sigma \setminus \operatorname{supp} f$ the aforementioned limit trivially exists, as in this latter case we have

$$\begin{aligned}
\lim_{\varepsilon\to 0^+} (T_\varepsilon f)(x) &= \lim_{\varepsilon\to 0^+} \int_{\Sigma\setminus\overline{B(x,\varepsilon)}} k(x-y) f(y)\, d\sigma(y) \\
&= \lim_{\varepsilon\to 0^+} \int_{(\operatorname{supp} f)\setminus\overline{B(x,\varepsilon)}} k(x-y) f(y)\, d\sigma(y) \\
&= \int_{\operatorname{supp} f} k(x-y) f(y)\, d\sigma(y),
\end{aligned} \tag{2.3.78}$$

thanks to item *(1)* in [112, Lemma 3.8.4].

`Step IV:` *The proof of the existence of the limit in* (2.3.15) *at σ-a.e. point on Σ in the case when Σ is as in* (2.3.76)*, and $f \in L^p(\Sigma, \sigma)$ with $p \in (1, \infty)$ has the additional property that*

$$\operatorname{supp} f \subseteq \bigcup_{j=1}^{m} S_j \textit{ for some } m \in \mathbb{N}. \tag{2.3.79}$$

In this scenario, recall the notation introduced in (2.3.77) and define $f_j := f\,\mathbf{1}_{\widetilde{S_j}}$ for $1 \leq j \leq m$. Then, given that each f_j belongs to $L^p(\Sigma, \sigma)$ and is supported on S_j, Step III applies and gives that, for each $j \in \{1, \dots, m\}$, the limit $\lim_{\varepsilon\to 0^+} (T_\varepsilon f_j)(x)$ exists for σ-a.e. $x \in \Sigma$. Since, by design, $f = \sum_{j=1}^{m} f_j$ on Σ, and since T_ε is linear, we ultimately conclude that, indeed, $\lim_{\varepsilon\to 0^+} (T_\varepsilon f)(x)$ exists for σ-a.e. $x \in \Sigma$.

`Step V:` *The proof of the fact that the limit in* (2.3.15) *exists at σ-a.e. point on Σ in the case when Σ is as in* (2.3.76)*, and $f \in L^p(\Sigma, \sigma)$ with $p \in (1, \infty)$ is arbitrary.*

The idea now is to use [112, Proposition 6.2.11] for the ambient $\mathscr{X} := \Sigma \times [0, \infty)$ equipped with the canonical product topology, the normed space $Y := L^p(\Sigma, \sigma)$, $X := \Sigma \times \{0\}$, $\mu := \sigma$ which is a complete measure on X by [112, Lemma 3.6.4], and with $\Gamma\big((x, 0)\big) := \{x\} \times (0, \infty)$ for each $(x, 0) \in X = \Sigma \times \{0\}$. The latter choice ensures that $(x, 0) \in \overline{\Gamma\big((x, 0)\big)}$ for every $(x, 0) \in X = \Sigma \times \{0\}$. Also, consider a linear operator $\mathcal{T}$ mapping functions on X into functions on $\mathscr{X} \setminus X = \Sigma \times (0, \infty)$ of the following sort. For every $f \in Y = L^p(\Sigma, \sigma)$ define

$$
\begin{aligned}
(\mathcal{T} f)(x, \varepsilon) &:= \int\limits_{\substack{y\in\Sigma \\ |x-y|>\varepsilon}} k(x - y) f(y)\, d\sigma(y) \\
&\text{for all } (x, \varepsilon) \in \mathscr{X} \setminus X = \Sigma \times (0, \infty),
\end{aligned}
\tag{2.3.80}
$$

and note that this entails

$$
\begin{aligned}
(T_\star f)(x, 0) &:= \sup_{(z,\varepsilon)\in\Gamma((x,0))} |(\mathcal{T} f)(z, \varepsilon)| \\
&= \sup_{\varepsilon>0} \Bigg| \int\limits_{\substack{y\in\Sigma \\ |x-y|>\varepsilon}} k(x - y) f(y)\, d\sigma(y) \Bigg| \\
&= (T_{\max} f)(x) \text{ for every } x \in \Sigma.
\end{aligned}
\tag{2.3.81}
$$

In turn, from (2.3.81) and (2.3.11) we conclude that condition [112, (6.2.74)] is verified in the present setting. To finish the implementation of [112, Proposition 6.2.11] we finally choose

$$
\mathcal{V} := \Big\{ g \in L^p(\Sigma, \sigma) : \operatorname{supp} g \subseteq \bigcup_{j=1}^{m} S_j \text{ for some } m \in \mathbb{N} \Big\}. \tag{2.3.82}
$$

Then $\mathcal{V}$ is a dense subset of $L^p(\Sigma, \sigma)$ since for every $f \in L^p(\Sigma, \sigma)$ Lebesgue's Dominated Convergence Theorem gives

$$
\mathcal{V} \ni g_m := \sum_{j=1}^{m} f\, \mathbf{1}_{\widetilde{S}_j} \longrightarrow f \text{ in } L^p(\Sigma, \sigma) \text{ as } m \to \infty. \tag{2.3.83}
$$

Moreover, condition [112, (6.2.75)] is also satisfied since for every $g \in \mathcal{V}$ the limit

$$
\begin{aligned}
\lim_{\Gamma((x,0))\ni(x,\varepsilon)\to(x,0)} (\mathcal{T} g)(x, \varepsilon) &= \lim_{\varepsilon\to0^+} \int\limits_{\substack{y\in\Sigma \\ |x-y|>\varepsilon}} k(x - y) g(y)\, d\sigma(y) \\
&= \lim_{\varepsilon\to0^+} (T_\varepsilon g)(x)
\end{aligned}
\tag{2.3.84}
$$

exists for μ-a.e. $(x, 0) \in X = \Sigma \times \{0\}$, thanks to Step IV. At this stage, all hypotheses of [112, Proposition 6.2.11] have been verified for the present choices, so we may conclude from it that for every function $f \in L^p(\Sigma, \sigma) \equiv L^p(X, \mu)$ the limit

$$\lim_{\varepsilon \to 0^+} (T_\varepsilon f)(x) = \lim_{\Gamma((x,0)) \ni (x,\varepsilon) \to (x,0)} (\mathcal{T} f)(x, \varepsilon) \tag{2.3.85}$$

exists for σ-a.e. $x \in \Sigma$.

`Second Intermission:` Let us record our progress. At this point we have proved the first claim in (2.3.15) in the case when $f \in L^p(\Sigma, \sigma)$ with $p \in (1, \infty)$. In particular, we may consider the operator T acting on an arbitrary $f \in L^p(\Sigma, \sigma)$ with $p \in (1, \infty)$ according to

$$(Tf)(x) := \lim_{\varepsilon \to 0^+} (T_\varepsilon f)(x) \text{ for } \sigma\text{-a.e. } x \in \Sigma. \tag{2.3.86}$$

Thanks to this, (2.2.68), the equivalence *(i)* $\Leftrightarrow$ *(iii)* in [112, Remark 3.1.2], and [112, (3.6.26)] it follows that

$$\begin{aligned} &Tf \text{ is a } \sigma\text{-measurable function on } \Sigma \text{ for} \\ &\text{each } f \in L^p(\Sigma, \sigma) \text{ with } p \in (1, \infty). \end{aligned} \tag{2.3.87}$$

Combining (2.3.5), (2.3.11), (2.3.86), and (2.3.87) we arrive at the conclusion that

$$\begin{gathered} \text{the operator } T : L^p(\Sigma, \sigma) \longrightarrow L^p(\Sigma, \sigma) \\ \text{is well defined, linear, and bounded, if } 1 < p < \infty, \end{gathered} \tag{2.3.88}$$

with operator norm

$$\|T\|_{L^p(\Sigma,\sigma) \to L^p(\Sigma,\sigma)} \leq C_{\Sigma,p} \big\|k\big|_{S^{n-1}}\big\|_{\mathscr{C}^N(S^{n-1})}. \tag{2.3.89}$$

These justify (2.3.18) and (2.3.20) in the statement of the theorem. Next, item *(iii)* of Proposition 2.2.8 applied to the operator T from (2.3.86) yields, on account of (2.3.88)-(2.3.89), that

$$\begin{gathered} T_{\max} : L^1(\Sigma, \sigma) \longrightarrow L^{1,\infty}(\Sigma, \sigma) \text{ and there exists } C \in (0, \infty) \text{ such that} \\ \|T_{\max} f\|_{L^{1,\infty}(\Sigma,\sigma)} \leq C \big\|k\big|_{S^{n-1}}\big\|_{\mathscr{C}^N(S^{n-1})} \|f\|_{L^1(\Sigma,\sigma)} \text{ for each } f \in L^1(\Sigma, \sigma). \end{gathered} \tag{2.3.90}$$

This establishes (2.3.12) in the statement of the theorem.

`Step VI:` *The proof of the first claim in* (2.3.15) *when* $\Sigma \subseteq \mathbb{R}^n$ *is an arbitrary* UR *set and when* $f \in L^1(\Sigma, \sigma)$. The idea is to once again implement [112, Proposition 6.2.11], this time with $p = 1$. We make the same choices for all the relevant entities as in treatment of Step V, with the exception of (2.3.82), in place of which we now take

$$Y := L^1(\Sigma, \sigma) \text{ and } \mathcal{V} := L^1(\Sigma, \sigma) \cap L^p(\Sigma, \sigma) \text{ for a fixed } p \in (1, \infty). \tag{2.3.91}$$

Then $\mathcal{V}$ is a dense subset of Y (cf. [112, (3.1.13)]) and, as seen from Step V (cf. (2.3.86)), condition [112, (6.2.75)] is presently verified. Moreover, [112, (6.2.74)] currently holds thanks to (2.3.90). Granted these, [112, Proposition 6.2.11] applies (with $p = 1$) and proves that the limit in (2.3.15) exists, for each fixed function $f \in L^1(\Sigma, \sigma)$, at σ-a.e. point on Σ.

`Step VII:` *The proof of the claims in* (2.3.15) *when* $\Sigma \subseteq \mathbb{R}^n$ *is an arbitrary* UR *set and* f *is an arbitrary function belonging to the space* $L^1\Big(\Sigma, \frac{\sigma(x)}{1+|x|^{n-1}}\Big)$. In such a scenario, fix $x_0 \in \Sigma$ and, for each $j \in \mathbb{N}$, abbreviate $\Delta_r := B(x_0, j) \cap \Sigma$. Having fixed a function $f \in L^1\Big(\Sigma, \frac{\sigma(x)}{1+|x|^{n-1}}\Big)$, and given $\varepsilon \in (0,1)$ arbitrary, for each $j \in \mathbb{N}$ and each $x \in \Delta_j$ split

$$(T_\varepsilon f)(x) = \mathrm{I}_\varepsilon(x) + \int\limits_{\substack{y\in\Sigma \\ |x-y|\geq 1}} k(x-y) f(y)\, d\sigma(y) \tag{2.3.92}$$

where, with $f_j := f \cdot \mathbf{1}_{\Delta_{j+1}}$,

$$\begin{aligned} \mathrm{I}_\varepsilon(x) &:= \int\limits_{\substack{y\in\Sigma \\ 1>|x-y|>\varepsilon}} k(x-y) f(y)\, d\sigma(y) = \int\limits_{\substack{y\in\Sigma \\ 1>|x-y|>\varepsilon}} k(x-y) f_j(y)\, d\sigma(y) \\ &= (T_\varepsilon f_j)(x) - \int\limits_{\substack{y\in\Sigma \\ |x-y|\geq 1}} k(x-y) f_j(y)\, d\sigma(y). \end{aligned} \tag{2.3.93}$$

Since the function $f_j = f \cdot \mathbf{1}_{\Delta_{j+1}}$ belongs to $L^1(\Sigma, \sigma)$, the result proved in Step VI ensures the existence of some σ-measurable set $N_j \subseteq \Sigma$ such that $\sigma(N_j) = 0$ and $\lim\limits_{\varepsilon\to 0^+} (T_\varepsilon f_j)(x)$ exists for each $x \in \Sigma \setminus N_j$. Granted this, it follows from (2.3.92)-(2.3.93) that $\lim\limits_{\varepsilon\to 0^+} (T_\varepsilon f)(x)$ exists for each $x \in \Delta_j \setminus N_j$. Given that $\Sigma = \bigcup\limits_{j\in\mathbb{N}} \Delta_j$, this ultimately shows that for each function $f \in L^1\Big(\Sigma, \frac{\sigma(x)}{1+|x|^{n-1}}\Big)$ the limit $(Tf)(x) := \lim\limits_{\varepsilon\to 0^+} (T_\varepsilon f)(x)$ exists for σ-a.e. $x \in \Sigma$. With this in hand, from (2.2.68), the equivalence *(i)* ⇔ *(iii)* in [112, Remark 3.1.2], and [112, (3.6.26)] we then conclude that the function Tf thus defined is σ-measurable. Hence, the claims in (2.3.15) hold as stated.

At this stage, the claims in (2.3.15) have been fully justified. Having accomplished this task, arguing as in (2.3.86)-(2.3.87) and availing ourselves of (2.3.90) then proves (2.3.19) and (2.3.21). As regards the claim made in (2.3.16), fix a point $x_o \in \Sigma$ and pick some radius $r \in (0, \infty)$. For an arbitrarily chosen function $f \in L^1\big(\Sigma, \frac{\sigma(x)}{1+|x|^{n-1}}\big)$, define

$$g_r := \Big[T\big(f\mathbf{1}_{\Sigma\setminus B(x_o,2r)}\big)\Big]\Big|_{\Sigma\cap B(x_o,r)} \quad\text{and}$$
$$h_r := \Big[T\big(f\mathbf{1}_{\Sigma\cap B(x_o,2r)}\big)\Big]\Big|_{\Sigma\cap B(x_o,r)}. \tag{2.3.94}$$

Since there exists $C_{k,r} \in (0,\infty)$ with the property that for each $x \in \Sigma \cap B(x_o, r)$ we have

$$\begin{aligned}\int_\Sigma |k(x-y)|\big|\big(f\mathbf{1}_{\Sigma\setminus B(x_o,2r)}\big)(y)\big|\,d\sigma(y)\\ &\leq C_{k,r}\int_{\Sigma\setminus B(x_o,2r)} \frac{|f(y)|}{|x-y|^{n-1}}\,d\sigma(y)\\ &\leq C_{k,r}\int_\Sigma \frac{|f(y)|}{1+|y|^{n-1}}\,d\sigma(y) < +\infty,\end{aligned} \tag{2.3.95}$$

it follows that

$$g_r \in L^\infty\big(\Sigma\cap B(x_o,r),\sigma\big) \text{ and}$$
$$\|g_r\|_{L^\infty(\Sigma\cap B(x_o,r),\sigma)} \leq C_{k,r}\|f\|_{L^1\left(\Sigma,\frac{\sigma(x)}{1+|x|^{n-1}}\right)}. \tag{2.3.96}$$

Given that, according to [112, (6.2.27)], we have the continuous embeddings

$$L^\infty\big(\Sigma\cap B(x_o,r),\sigma\big) \hookrightarrow L^1\big(\Sigma\cap B(x_o,r),\sigma\big) \hookrightarrow L^{1,\infty}\big(\Sigma\cap B(x_o,r),\sigma\big), \tag{2.3.97}$$

we deduce from (2.3.96) that

$$g_r \in L^{1,\infty}\big(\Sigma\cap B(x_o,r),\sigma\big) \text{ and}$$
$$\|g_r\|_{L^{1,\infty}(\Sigma\cap B(x_o,r),\sigma)} \leq C_{k,x_o,r}\|f\|_{L^1\left(\Sigma,\frac{\sigma(x)}{1+|x|^{n-1}}\right)}. \tag{2.3.98}$$

Also, since there exists $C_{x_o,r} \in (0,\infty)$ such that

$$f\mathbf{1}_{\Sigma\cap B(x_o,2r)} \in L^1(\Sigma,\sigma) \text{ with the property that}$$
$$\big\|f\mathbf{1}_{\Sigma\cap B(x_o,2r)}\big\|_{L^1(\Sigma,\sigma)} \leq C_{x_o,r}\|f\|_{L^1\left(\Sigma,\frac{\sigma(x)}{1+|x|^{n-1}}\right)}, \tag{2.3.99}$$

we conclude from (2.3.19) (which has already been established at this point) that

$$T\big(f\mathbf{1}_{\Sigma\cap B(x_o,2r)}\big) \in L^{1,\infty}(\Sigma,\sigma) \text{ and we have}$$
$$\big\|T\big(f\mathbf{1}_{\Sigma\cap B(x_o,2r)}\big)\big\|_{L^{1,\infty}(\Sigma,\sigma)} \leq C_{k,x_o,r}\|f\|_{L^1\left(\Sigma,\frac{\sigma(x)}{1+|x|^{n-1}}\right)}. \tag{2.3.100}$$

In turn, from (2.3.100) and the second line in [112, (6.2.17)] we see that

$$
\begin{aligned}
& h_r \in L^{1,\infty}\big(\Sigma \cap B(x_o, r), \sigma\big) \text{ and} \\
& \|h_r\|_{L^{1,\infty}(\Sigma\cap B(x_o,r),\sigma)} \leq C_{k,x_o,r}\|f\|_{L^1\left(\Sigma,\frac{\sigma(x)}{1+|x|^{n-1}}\right)}.
\end{aligned}
\tag{2.3.101}
$$

In view of the fact that, by design, $(Tf)\big|_{B(x_o,r)\cap\Sigma} = g_r + h_r$ on $\Sigma \cap B(x_o, r)$, we ultimately conclude from (2.3.98) and (2.3.101) that the operator T in (2.3.16) is indeed well defined, linear, and continuous.

Note that if the above argument is carried out for a function f which is now assumed to belong to the intersection $L^p_{\rm loc}(\Sigma, \sigma) \cap L^1\left(\Sigma, \frac{\sigma(x)}{1+|x|^{n-1}}\right)$ for some exponent $p \in (1, \infty)$, then in place of (2.3.99) we now have

$$
\begin{aligned}
& f\mathbf{1}_{\Sigma\cap B(x_o,2r)} \in L^p(\Sigma, \sigma) \text{ with} \\
& \big\|f\mathbf{1}_{\Sigma\cap B(x_o,2r)}\big\|_{L^p(\Sigma,\sigma)} = \|f\|_{L^p(\Sigma\cap B(x_o,2r),\sigma)}.
\end{aligned}
\tag{2.3.102}
$$

Thanks to this and (2.3.18), in place of (2.3.100) we now obtain

$$
\begin{aligned}
& T\big(f\mathbf{1}_{\Sigma\cap B(x_o,2r)}\big) \in L^p(\Sigma, \sigma) \text{ with} \\
& \big\|T\big(f\mathbf{1}_{\Sigma\cap B(x_o,2r)}\big)\big\|_{L^p(\Sigma,\sigma)} \leq C_{k,x_o,r}\|f\|_{L^p(\Sigma\cap B(x_o,2r),\sigma)}.
\end{aligned}
\tag{2.3.103}
$$

Hence, (2.3.101) presently improves to

$$
\begin{aligned}
& h_r \in L^p\big(\Sigma \cap B(x_o, r), \sigma\big) \text{ with} \\
& \|h_r\|_{L^p(\Sigma\cap B(x_o,r),\sigma)} \leq C_{k,x_o,r}\|f\|_{L^p(\Sigma\cap B(x_o,2r),\sigma)}.
\end{aligned}
\tag{2.3.104}
$$

Together with (2.3.96), this ultimately shows that the mapping in (2.3.17) is, as claimed, well defined, linear, and continuous.

Next, the Cotlar-type inequality claimed in (2.3.23) is a direct consequence of Corollary 2.2.6 and (2.3.21). Going further, (2.3.24) is a consequence of (2.3.11), the fact that the limit in (2.3.15) exists at σ-a.e. point, and Lebesgue's Dominated Convergence Theorem. As regards the claim in (2.3.25), pick two arbitrary functions, $f \in L^p(\Sigma, \sigma)$ and $g \in L^{p'}(\Sigma, \sigma)$, with $p, p' \in (1, \infty)$ Hölder conjugate exponents. Then for every $\varepsilon > 0$ Fubini's Theorem gives

$$
\begin{aligned}
\int_\Sigma (T_\varepsilon f)(x)g(x)\,d\sigma(x) &= \int_\Sigma\int_\Sigma k(x-y)f(y)g(x)\mathbf{1}_{|x-y|>\varepsilon}\,d\sigma(y)\,d\sigma(x) \\
&= -\int_\Sigma f(x)(T_\varepsilon g)(x)\,d\sigma(x).
\end{aligned}
\tag{2.3.105}
$$

Keeping in mind that $T_\varepsilon f \to Tf$ in $L^p(\Sigma, \sigma)$ and $T_\varepsilon g \to Tg$ in $L^{p'}(\Sigma, \sigma)$ as $\varepsilon \to 0^+$ (cf. (2.3.24)), we deduce from (2.3.105) that

$$
\int_\Sigma (Tf)g\,d\sigma = -\int_\Sigma f(Tg)\,d\sigma,
\tag{2.3.106}
$$

proving (2.3.25). As regards the claims in (2.3.26), pick some arbitrary functions $f \in L^p_{\rm loc}(\Sigma, \sigma) \cap L^1\big(\Sigma, \frac{\sigma(x)}{1+|x|^{n-1}}\big)$ and $g \in L^{p'}_{\rm comp}(\Sigma, \sigma)$ with $p, p' \in (1, \infty)$ such that $1/p + 1/p' = 1$, and select $x_o \in \Sigma$ along with $R > 0$ with the property that $\operatorname{supp} g \subseteq \Sigma \cap B(x_o, R)$. Then since $f\mathbf{1}_{B(x_o,2R)\cap\Sigma} \in L^p(\Sigma, \sigma)$ we conclude from (2.3.18) that $T\big(f\mathbf{1}_{\Sigma\cap B(x_o,2R)}\big) \in L^p(\Sigma, \sigma)$, hence

$$\int_\Sigma \big|T\big(f\mathbf{1}_{\Sigma\cap B(x_o,2R)}\big)\big||g|\,d\sigma < +\infty, \tag{2.3.107}$$

by Hölder's inequality. Also, there exists a constant $C = C(k, x_o, R) \in (0, \infty)$ for which

$$\begin{aligned}
&\int_\Sigma \big|T\big(f\mathbf{1}_{\Sigma\setminus B(x_o,2R)}\big)(x)\big||g(x)|\,d\sigma(x) \qquad\qquad (2.3.108)\\
&\quad\le \int_{x\in\Sigma\cap B(x_o,R)}\int_{y\in\Sigma\setminus B(x_o,2R)} |k(x-y)||f(y)||g(x)|\,d\sigma(y)\,d\sigma(x)\\
&\quad\le C\Big(\int_{\Sigma\setminus B(x_o,2R)}\frac{|f(y)|}{1+|y|^{n-1}}\,d\sigma(y)\Big)\Big(\int_{\Sigma\cap B(x_o,R)}|g(x)|\,d\sigma(x)\Big) < +\infty,
\end{aligned}$$

given the integrability properties assumed of f and g. Combining (2.3.107) and (2.3.108) then proves $\int_\Sigma |Tf||g|\,d\sigma < +\infty$. That we also have $\int_\Sigma |f||Tg|\,d\sigma < +\infty$ is then established in a very similar fashion. In turn, the finiteness properties just proved permit us to write

$$\begin{aligned}
\int_\Sigma (Tf)g\,d\sigma &= \int_\Sigma T\big(f\mathbf{1}_{\Sigma\cap B(x_o,2R)}\big)g\,d\sigma + \int_\Sigma T\big(f\mathbf{1}_{\Sigma\setminus B(x_o,2R)}\big)g\,d\sigma\\
&= -\int_\Sigma f\mathbf{1}_{\Sigma\cap B(x_o,2R)}(Tg)\,d\sigma\\
&\quad + \int_{\Sigma\cap B(x_o,R)}\Big(\int_{\Sigma\setminus B(x_o,2R)} k(x-y)f(y)\,d\sigma(y)\Big)g(x)\,d\sigma(x)\\
&= -\int_\Sigma f\mathbf{1}_{\Sigma\cap B(x_o,2R)}(Tg)\,d\sigma\\
&\quad - \int_{\Sigma\setminus B(x_o,2R)} f(y)\Big(\int_{\Sigma\cap B(x_o,R)} k(y-x)g(x)\,d\sigma(x)\Big)\,d\sigma(y)\\
&= -\int_\Sigma f\mathbf{1}_{\Sigma\cap B(x_o,2R)}(Tg)\,d\sigma - \int_\Sigma f\mathbf{1}_{\Sigma\setminus B(x_o,2R)}(Tg)\,d\sigma\\
&= -\int_\Sigma f(Tg)\,d\sigma,
\end{aligned} \tag{2.3.109}$$

where we have also used (2.3.25) in the second equality above (bearing in mind that the function $f\mathbf{1}_{B(x_o,2R)\cap\Sigma}$ belongs to $L^p(\Sigma,\sigma)$ and that $g \in L^{p'}(\Sigma,\sigma)$), as well as Fubini's Theorem and the fact that k is odd in the third equality. This establishes the final formula in (2.3.25). Hence, at this point in the proof we have justified all claims in items *(1)*-*(5)*.

To deal with the claims in item *(6)*, fix some $q \in (1,\infty)$ and let a be an arbitrary (p,q)-atom on Σ. Hence, there exist $x_o \in \Sigma$ and $r \in \big(0, \operatorname{diam}\Sigma\big]$ such that

$$\begin{aligned}&\operatorname{supp} a \subseteq B(x_o,r)\cap\Sigma, \qquad \int_\Sigma a\,\mathrm{d}\sigma = 0,\\ &\text{and } \|a\|_{L^q(\Sigma,\sigma)} \leq \sigma\big(B(x_o,r)\cap\Sigma\big)^{1/q-1/p}.\end{aligned} \tag{2.3.110}$$

If $\sigma(\Sigma) < +\infty$ then, by definition, the constant function $a \equiv \sigma(\Sigma)^{-1/p}$ is also a (p,q)-atom. We may then estimate

$$\begin{aligned}\int_{B(x_o,2r)\cap\Sigma} |Ta|^p\,\mathrm{d}\sigma &\leq \|Ta\|^p_{L^q(\Sigma,\sigma)}\cdot\sigma\big(B(x_o,2r)\cap\Sigma\big)^{1-p/q}\\ &\leq C\big\|k\big|_{S^{n-1}}\big\|^p_{\mathscr{C}^N(S^{n-1})}\|a\|^p_{L^q(\Sigma,\sigma)}\cdot\sigma\big(B(x_o,2r)\cap\Sigma\big)^{1-p/q}\\ &\leq C(\Sigma,p,q)\big\|k\big|_{S^{n-1}}\big\|^p_{\mathscr{C}^N(S^{n-1})}.\end{aligned} \tag{2.3.111}$$

Above, the first estimate is based on Hölder's inequality, the second estimate is a consequence of (2.3.18) (with $p := q$), and the third estimate relies on (2.3.110) and the Ahlfors regularity of Σ. Next, if $x \in \Sigma$ is such that $|x - x_o| \geq 2r$ then based on (2.3.110), the Mean Value Theorem, and Hölder's inequality we may estimate

$$\begin{aligned}|(Ta)(x)| &= \Big|\int_\Sigma k(x-y)a(y)\,\mathrm{d}\sigma(y)\Big|\\ &\leq \int_{B(x_o,r)\cap\Sigma} |k(x-y)-k(x-x_o)||a(y)|\,\mathrm{d}\sigma(y)\\ &\leq C\big\|k\big|_{S^{n-1}}\big\|_{\mathscr{C}^1(S^{n-1})}\frac{r}{|x-x_o|^n}\int_{B(x_o,r)\cap\Sigma}|a(y)|\,\mathrm{d}\sigma(y)\\ &\leq C\big\|k\big|_{S^{n-1}}\big\|_{\mathscr{C}^1(S^{n-1})}\frac{r}{|x-x_o|^n}\sigma\big(B(x_o,r)\cap\Sigma\big)^{1-1/p}.\end{aligned} \tag{2.3.112}$$

With this in hand, [112, Lemma 7.2.1] applies and, since $pn > n-1$, gives

$$\int_{\Sigma\setminus B(x_o,2r)} |(Ta)(x)|^p\,\mathrm{d}\sigma(x) \leq C(\Sigma,p)\big\|k\big|_{S^{n-1}}\big\|^p_{\mathscr{C}^1(S^{n-1})}. \tag{2.3.113}$$

Collectively, (2.3.111) and (2.3.113) imply that, whenever the atom a is as in (2.3.110), then

$$\|Ta\|_{L^p(\Sigma,\sigma)} \leq C\big\|k\big|_{S^{n-1}}\big\|_{\mathscr{C}^1(S^{n-1})}, \tag{2.3.114}$$

for some finite constant $C = C(\Sigma, p, q) > 0$ independent of the atom a. In the scenario when $\sigma(\Sigma) < +\infty$, it follows from (2.3.18) (with $p = 2$) and Hölder's inequality that the same estimate holds for the constant function $a \equiv \sigma(\Sigma)^{-1/p}$. Having established this, we may now invoke the general boundedness criterion described in [113, Theorem 4.4.7] (presently used with $q := 2$, $X := L^2(\Sigma, \sigma)$, and $Y := L^p(\Sigma, \sigma)$) in order to conclude that the operator (2.3.18) extends uniquely (as indicated in [113, (4.4.144)] with $Y := L^p(\Sigma, \sigma)$) to a linear and bounded mapping as in (2.3.27), with operator quasi-norm satisfying (2.3.29). That the operators in (2.3.27) corresponding to various values of $p \in \left(\frac{n-1}{n}, 1\right]$ are compatible with one another is a consequence of the simultaneous approximation result established in [113, Theorem 4.4.3]. In fact, [113, Theorem 4.4.3] also implies that the operator T defined on Hardy spaces as in (2.3.27) acts in a coherent fashion with the operator T defined on Lebesgue spaces as in (2.3.18).

To also see that the operator T defined as in (2.3.27) with $p = 1$ acts in a coherent fashion with the operator T defined in (2.3.19), pick an arbitrary function $f \in H^1(\Sigma, \sigma)$ and consider a sequence $\{f_j\}_{j\in\mathbb{N}} \subseteq H^{1,q}_{\rm fin}(\Sigma, \sigma)$ for some fixed $q \in (1, \infty)$ such that $f = \lim\limits_{j\to\infty} f_j$ in $H^1(\Sigma, \sigma)$. Then, on the one hand, (2.3.28) implies that Tf_j converges as $j \to \infty$ in $L^1(\Sigma, \sigma)$ to Tf defined as in (2.3.27). On the other hand, since $H^1(\Sigma, \sigma) \hookrightarrow L^1(\Sigma, \sigma)$ continuously, the continuity of the operator in (2.3.19) ensures that Tf_j converges as $j \to \infty$ in $L^{1,\infty}(\Sigma, \sigma)$ to Tf defined as in (2.3.15). The desired compatibility property now readily follows, finishing the treatment of item *(6)*.

Let us now deal with the claims in item *(7)* in the statement of the theorem. For starters, recall that

$$k_\varepsilon := k \cdot \mathbf{1}_{\mathbb{R}^n\setminus\overline{B(0,\varepsilon)}} \text{ for each } \varepsilon > 0, \tag{2.3.115}$$

and observe that, thanks to the smoothness and the homogeneity of the kernel k, for each multi-index $\alpha \in \mathbb{N}_0^n$ we have

$$\big|(\partial^\alpha k)(z)\big| \leq \frac{\sup_{S^{n-1}} |(\partial^\alpha k)|}{|z|^{n-1+|\alpha|}}, \qquad \forall z \in \mathbb{R}^n \setminus \{0\}. \tag{2.3.116}$$

In relation to (2.3.115) we claim that for each fixed $x \in \mathbb{R}^n$ and $\varepsilon > 0$ there exists a constant $C_{x,\varepsilon} \in (0, \infty)$ such that

$$\big|k_\varepsilon(x - y) - k_1(-y)\big| \leq \frac{C_{x,\varepsilon}}{1 + |y|^n}, \qquad \forall y \in \mathbb{R}^n. \tag{2.3.117}$$

This is seen by analyzing two cases. First, if either $|y| \leq \max\{2|x|, 1\}$ or $|x - y| \leq \varepsilon$ then (2.3.117) readily follows from (2.3.115) and (2.3.116) (with $|\alpha| = 0$), bearing in mind that y is bounded. Second, consider the case when both $|y| > \max\{2|x|, 1\}$ and $|x - y| > \varepsilon$. In particular, the former inequality implies that for each $t \in [0, 1]$

we have $|tx - y| \geq |y| - |x| \geq \frac{1}{2}|y|$. As such, $F(t) := k(tx - y)$ for each $t \in [0, 1]$ is a well-defined function which is continuously differentiable on $[0, 1]$. Granted this, the Mean Value Theorem together with (2.3.116) permit us to estimate

$$\begin{aligned} \big|k_\varepsilon(x-y) - k_1(-y)\big| &= |F(1) - F(0)| \leq \sup_{0<t<1} |F'(t)| \\ &\leq |x| \sup_{0<t<1} |(\nabla k)(tx-y)| \leq \frac{2^n|x|}{|y|^n} \\ &\leq \frac{2^{2n}|x|}{(1+|y|)^n}, \qquad \forall\, y \in \mathbb{R}^n. \end{aligned} \tag{2.3.118}$$

This completes the proof of (2.3.117). In concert with [112, Lemma 7.2.1], estimate (2.3.117) implies that

$$\begin{gathered} \int_\Sigma \big|k_\varepsilon(x-y) - k_1(-y)\big||f(y)|\,d\sigma(y) < +\infty \\ \text{for each } \varepsilon > 0, \text{ each } x \in \Sigma, \text{ and each } f \in L^1\Big(\Sigma, \frac{\sigma}{1+|\cdot|^n}\Big). \end{gathered} \tag{2.3.119}$$

Hence, the integrals appearing in the definition of $(T_{\rm mod} f)(x)$ in (2.3.31) are all absolutely convergent for each $x \in \Sigma$ and each $f \in L^1\big(\Sigma, \frac{\sigma}{1+|\cdot|^n}\big)$.

To proceed, fix an arbitrary number $R \in (1, \infty)$ and select an arbitrary point $x \in \Sigma \cap B(0, R)$. Also, pick some $\varepsilon \in (0, 1)$. Having selected some $f \in L^1\big(\Sigma, \frac{\sigma}{1+|\cdot|^n}\big)$, we then proceed to split

$$\int_\Sigma \big\{k_\varepsilon(x-y) - k_1(-y)\big\} f(y)\,d\sigma(y) = \mathrm{I}_\varepsilon + \mathrm{II}_\varepsilon \tag{2.3.120}$$

where the term

$$\begin{aligned} \mathrm{I}_\varepsilon &:= \int_{\substack{y\in\Sigma \\ |x-y|>1}} \big\{k_\varepsilon(x-y) - k_1(-y)\big\} f(y)\,d\sigma(y) \\ &= \int_{\substack{y\in\Sigma \\ |x-y|\leq 1}} \big\{k(x-y) - k_1(-y)\big\} f(y)\,d\sigma(y) \end{aligned} \tag{2.3.121}$$

turns out to be independent of $\varepsilon > 0$, and

$$\mathrm{II}_\varepsilon := \int_{\substack{y\in\Sigma \\ |x-y|\leq 1}} \big\{k_\varepsilon(x-y) - k_1(-y)\big\} f(y)\,d\sigma(y). \tag{2.3.122}$$

Given that $|y| \leq |x-y| + |x| \leq 1 + R < 2R$ on the domain of integration in (2.3.122), we may replace f by the function

$$f_R := \mathbf{1}_{B(0,2R)} \cdot f \text{ which belongs to } L^1(\Sigma, \sigma) \text{ and has bounded support,} \tag{2.3.123}$$

and write

$$\mathrm{II}_\varepsilon = \int\limits_{\substack{y\in\Sigma \\ |x-y|\leq 1}} \big\{k_\varepsilon(x-y) - k_1(-y)\big\} f_R(y)\, d\sigma(y). \tag{2.3.124}$$

Pressing on, observe that since k_1 is bounded in $\mathbb{R}^n$ (cf. (2.3.116)), we have $k_1(-\cdot) f_R \in L^1(\Sigma, \sigma)$. In light of this and (2.3.119), we may further decompose (bearing (2.3.6) in mind)

$$\begin{aligned}
\mathrm{II}_\varepsilon &= \int\limits_{\substack{y\in\Sigma \\ |x-y|\leq 1}} k_\varepsilon(x-y) f_R(y)\, d\sigma(y) - \int\limits_{\substack{y\in\Sigma \\ |x-y|\leq 1}} k_1(-y) f_R(y)\, d\sigma(y) \\
&= \int\limits_{\substack{y\in\Sigma \\ |x-y|>\varepsilon}} k(x-y) f_R(y)\, d\sigma(y) - \int\limits_{\substack{y\in\Sigma \\ |x-y|>1}} k(x-y) f_R(y)\, d\sigma(y) \\
&\quad - \int\limits_{\substack{y\in\Sigma \\ |x-y|\leq 1}} k_1(-y) f_R(y)\, d\sigma(y) \\
&= (T_\varepsilon f_R)(x) - \int\limits_{\substack{y\in\Sigma \\ |x-y|>1}} k(x-y) f_R(y)\, d\sigma(y) \\
&\quad - \int\limits_{\substack{y\in\Sigma \\ |x-y|\leq 1}} k_1(-y) f_R(y)\, d\sigma(y).
\end{aligned} \tag{2.3.125}$$

Note that the last two integrals above are independent of ε (as well as absolutely convergent). Also, thanks to (2.3.15) and (2.3.123), the limit $\lim\limits_{\varepsilon\to 0^+} (T_\varepsilon f_R)(x)$ exists for σ-a.e. $x \in \Sigma \cap B(0, R)$. Let us record our progress. In light of the arbitrariness of $R \in (1, \infty)$, the argument so far shows that

$$\begin{gathered}
\text{for each given function } f \text{ in the space } L^1\big(\Sigma, \tfrac{\sigma}{1+|\cdot|^n}\big), \text{ the limit} \\
\lim_{\varepsilon\to 0^+} \int_\Sigma \big\{k_\varepsilon(x-y) - k_1(-y)\big\} f(y)\, d\sigma(y) \text{ exists for } \sigma\text{-a.e. } x \in \Sigma.
\end{gathered} \tag{2.3.126}$$

Upon also invoking (2.2.68), the equivalence *(i)* ⇔ *(iii)* in [112, Remark 3.1.2], and [112, (3.6.26)], ultimately this proves that the operator T_{mod} introduced in (2.3.31) is well defined on the weighted Lebesgue space $L^1\Big(\Sigma, \frac{\sigma(x)}{1+|x|^n}\Big)$ which it maps linearly into the space of σ-measurable functions on Σ. (2.3.127)

Next, the claims made in relation to the operator (2.3.33) may be proved in an analogous manner to the earlier justification of similar claims for the operator in (2.3.16).

Observe now that if $f \in L^1\Big(\Sigma, \frac{\sigma(x)}{1+|x|^{n-1}}\Big)$ then based on (2.3.115) and (2.3.116) (with $|\alpha| = 0$) we may estimate

$$\int_\Sigma |k_1(-y)f(y)|\,d\sigma(y) \leq C\int_\Sigma \frac{|f(y)|}{1+|y|^{n-1}}\,d\sigma(y) < +\infty. \tag{2.3.128}$$

Having proved that $\int_\Sigma k_1(-y)f(y)\,d\sigma(y)$ is absolutely convergent, from (2.3.126), (2.3.31), (2.3.6), and (2.3.15) we conclude that

$$\begin{aligned}(T_{\text{mod}}f)(x) &= \lim_{\varepsilon\to 0^+}\int_\Sigma k_\varepsilon(x-y)f(y)\,d\sigma(y) - \int_\Sigma k_1(-y)f(y)\,d\sigma(y)\\ &= (Tf)(x) - C_f \quad \text{at } \sigma\text{-a.e. } x\in\Sigma,\end{aligned} \tag{2.3.129}$$

where the constant C_f is given by

$$C_f := \int_\Sigma k_1(-y)f(y)\,d\sigma(y) \in \mathbb{C}. \tag{2.3.130}$$

In particular, from this and (2.3.128) we have $|C_f| \leq C_{\Sigma,k}\cdot \|f\|_{L^1(\Sigma,\frac{\sigma(x)}{1+|x|^{n-1}})}$ for some finite constant $C_{\Sigma,k} > 0$. This finishes the proof of (2.3.34).

Suppose next that

$$f \in L^1\Big(\Sigma, \frac{\sigma(x)}{1+|x|^n}\Big) \cap L^p_{\text{loc}}(\Sigma,\sigma) \text{ for some } p\in(1,\infty), \tag{2.3.131}$$

and pick an arbitrary surface ball $\Delta \subseteq \Sigma$. For each $x\in\Delta$ decompose

$$(T_{\text{mod}}f)(x) = \big(T_{\text{mod}}(\mathbf{1}_{2\Delta}f)\big)(x) + \big(T_{\text{mod}}(\mathbf{1}_{\Sigma\setminus 2\Delta}f)\big)(x). \tag{2.3.132}$$

Note that since $\mathbf{1}_{2\Delta}f \in L^p(\Sigma,\sigma)$ it follows from (2.3.34) that $T_{\text{mod}}(\mathbf{1}_{2\Delta}f)$ and $T(\mathbf{1}_{2\Delta}f)$ differ by a constant. On account of this and (2.3.18) we then conclude that

$$T_{\text{mod}}(\mathbf{1}_{2\Delta}f) \in L^p_{\text{loc}}(\Sigma,\sigma). \tag{2.3.133}$$

There remains to observe that, thanks to the nature of the constant in (2.3.117), for each $x\in\Delta$ we may estimate

$$
\begin{aligned}
\big|\big(T_{\rm mod}\big(\mathbf{1}_{\Sigma\setminus 2\Delta}f\big)\big)(x)\big| &\leq \int_{\Sigma\setminus 2\Delta} \big|k(x-y)-k_1(-y)\big||f(y)|\,d\sigma(y) \\
&\leq c_\Delta \int_\Sigma \frac{|f(y)|}{1+|y|^n}\,d\sigma(y) < +\infty, \qquad (2.3.134)
\end{aligned}
$$

for some constant $c_\Delta \in (0,\infty)$ which depends on k and Δ but not on x. Since the last quantity above is a finite constant independent of x, we conclude that

$$
T_{\rm mod}\big(\mathbf{1}_{\Sigma\setminus 2\Delta}f\big) \text{ is bounded on } \Delta. \qquad (2.3.135)
$$

At this point, (2.3.35) is clear from (2.3.133)-(2.3.135).

To deal with the claim in (2.3.36) we refine the argument which has produced (2.3.26). To get started, suppose $p, p' \in (1,\infty)$ satisfy $1/p + 1/p' = 1$ and pick two functions,

$$
f \in L^p_{\rm loc}(\Sigma,\sigma) \cap L^1\Big(\Sigma, \tfrac{\sigma(x)}{1+|x|^n}\Big) \qquad (2.3.136)
$$

and

$$
g \in L^{p'}_{\rm comp}(\Sigma,\sigma) \text{ with } \int_\Sigma g\,d\sigma = 0. \qquad (2.3.137)
$$

To fix ideas, choose $x_o \in \Sigma$ and $r \in \big(0, 2\mathrm{diam}\Sigma\big)$ such that

$$
\mathrm{supp}\, g \subseteq \Delta(x_o,r) := B(x_o,r)\cap\Sigma. \qquad (2.3.138)
$$

Granted (2.3.136), from (2.3.35) we conclude that

$$
T_{\rm mod} f \in L^p_{\rm loc}(\Sigma,\sigma). \qquad (2.3.139)
$$

Together with the membership stipulated (2.3.137), this property shows that we do have $\int_\Sigma |T_{\rm mod} f||g|\,d\sigma < +\infty$, which is one of the claims in (2.3.36). Next, much as in (2.3.112), for each $x \in \Sigma$ such that $|x - x_o| \geq 2r$ we obtain (based on (2.3.137)-(2.3.138), (2.3.3), the Mean Value Theorem, and Hölder's inequality)

$$
\begin{aligned}
|(Tg)(x)| &= \Big|\int_\Sigma k(x-y)g(y)\,d\sigma(y)\Big| \\
&\leq \int_{B(x_o,r)\cap\Sigma} |k(x-y)-k(x-x_o)||g(y)|\,d\sigma(y) \\
&\leq C\big\|k\big|_{S^{n-1}}\big\|_{\mathscr{C}^1(S^{n-1})} \frac{r\cdot\sigma(\Delta(x_o,r))}{|x-x_o|^n} ⨍_{\Delta(x_o,r)} |g(y)|\,d\sigma(y) \\
&\leq C\big\|k\big|_{S^{n-1}}\big\|_{\mathscr{C}^1(S^{n-1})} \frac{r\cdot\sigma(\Delta(x_o,r))}{|x-x_o|^n} \Big(⨍_{\Delta(x_o,r)} |g|^{p'}\,d\sigma\Big)^{1/p'} \\
&\leq \frac{C(\Sigma,n,p,k,x_o,r,g)}{1+|x|^n}, \qquad (2.3.140)
\end{aligned}
$$

for some finite constant $C(\Sigma, n, p, k, x_o, r, g) > 0$. In concert with (2.3.136), this pointwise inequality proves that $\int_{\Sigma\setminus B(x_o,2r)} |f||Tg|\,d\sigma < +\infty$. Since we also have $\int_{\Sigma\cap B(x_o,2r)} |f||Tg|\,d\sigma < +\infty$, thanks to (2.3.137), (2.3.18), and the fact that $f \cdot \mathbf{1}_{\Sigma\cap B(x_o,2r)}$ belongs to $L^p(\Sigma,\sigma)$ (cf. (2.3.136)), we ultimately deduce that $\int_\Sigma |f||Tg|\,d\sigma < +\infty$, which is another claim in (2.3.36). Bearing this in mind, we may then write

$$\int_\Sigma f(Tg)\,d\sigma = \int_{\Sigma\cap B(x_o,2r)} f(Tg)\,d\sigma + \int_{\Sigma\setminus B(x_o,2r)} f(Tg)\,d\sigma =: \mathrm{I} + \mathrm{II}, \tag{2.3.141}$$

with the last equality defining I and II. In relation to these, note that on account of (2.3.25) we may recast I as

$$\begin{aligned}\mathrm{I} &= \int_\Sigma (f\cdot \mathbf{1}_{\Sigma\cap B(x_o,2r)})(Tg)\,d\sigma = -\int_\Sigma \Big(T\big(f\cdot \mathbf{1}_{\Sigma\cap B(x_o,2r)}\big)\Big)\,g\,d\sigma\\ &= -\int_\Sigma \Big(T_{\mathrm{mod}}\big(f\cdot \mathbf{1}_{\Sigma\cap B(x_o,2r)}\big)\Big)g\,d\sigma\end{aligned} \tag{2.3.142}$$

since $T\big(f\cdot \mathbf{1}_{\Sigma\cap B(x_o,2r)}\big)$ differs from $T_{\mathrm{mod}}\big(f\cdot \mathbf{1}_{\Sigma\cap B(x_o,2r)}\big)$ by a constant on Σ (thanks to (2.3.34), bearing in mind that $f\cdot \mathbf{1}_{\Sigma\cap B(x_o,2r)} \in L^p(\Sigma,\sigma)$) and the fact that g has integral zero (cf. the last property in (2.3.137)). Also,

$$\begin{aligned}\mathrm{II} &= \int_{\Sigma\setminus B(x_o,2r)} f(x)\bigg(\int_\Sigma k(x-y)g(y)\,d\sigma(y)\bigg)\,d\sigma(x)\\
&= \int_{\Sigma\setminus B(x_o,2r)} f(x)\bigg(\int_\Sigma k_\varepsilon(x-y)g(y)\,d\sigma(y)\bigg)\,d\sigma(x)\\
&= \int_{\Sigma\setminus B(x_o,2r)} f(x)\bigg(\int_\Sigma \{k_\varepsilon(x-y)+k_1(-x)\}g(y)\,d\sigma(y)\bigg)\,d\sigma(x)\\
&= \int_{\Sigma\setminus B(x_o,2r)} f(x)\bigg(\int_{\Sigma\cap B(x_o,r)} \{k_\varepsilon(x-y)+k_1(-x)\}g(y)\,d\sigma(y)\bigg)\,d\sigma(x)\\
&= \int_{\Sigma\cap B(x_o,r)} \bigg(\int_{\Sigma\setminus B(x_o,2r)} \{k_\varepsilon(x-y)+k_1(-x)\}f(x)\,d\sigma(x)\bigg)g(y)\,d\sigma(y)\\
&= \int_\Sigma \lim_{\varepsilon\to 0^+}\bigg(\int_{\Sigma\setminus B(x_o,2r)} \{k_\varepsilon(x-y)+k_1(-x)\}f(x)\,d\sigma(x)\bigg)g(y)\,d\sigma(y)\\
&= -\int_\Sigma \Big(T_{\mathrm{mod}}\big(f\cdot \mathbf{1}_{\Sigma\setminus B(x_o,2r)}\big)\Big)(y)\,g(y)\,d\sigma(y).\end{aligned} \tag{2.3.143}$$

The first equality in (2.3.143) is implied by (2.3.15) bearing in mind that, thanks to (2.3.138), the variables x, y are uniformly separated. The second equality in (2.3.143) uses the definition of k_ε in the second line of (2.3.31) and is valid for each choice $\varepsilon \in (0, r)$. The third equality in (2.3.143) is a consequence of the cancelation property of the function g (cf. the last property in (2.3.137)), while the fourth equality in (2.3.143) is seen from (2.3.138). The fifth equality in (2.3.143) follows from Fubini's Theorem whose applicability is presently ensured by the fact that the double integral is absolutely convergent, thanks to the properties listed in (2.3.137)-(2.3.138), the estimate in (2.3.117) (with the roles of x, y reversed, and a constant which stays bounded for y in a compact subset of Σ; applicability here uses the fact that k is odd), and (2.3.136). The sixth equality in (2.3.143) uses the fact that the inner integral is actually independent of $\varepsilon \in (0, r)$, and also the support property of g (cf. (2.3.138)). Finally, the last equality in (2.3.143) is seen from (2.3.31) (bearing in mind that k is odd). Collectively, from (2.3.141)-(2.3.143) we conclude that

$$\int_\Sigma f(Tg)\,d\sigma = -\int_\Sigma (T_{\text{mod}} f)g\,d\sigma, \tag{2.3.144}$$

finishing the proof of (2.3.36).

Moving on, fix a function $f \in L^\infty(\Sigma, \sigma)$ and pick an arbitrary surface ball $\Delta \subseteq \Sigma$, of center $x_\Delta \in \Sigma$ and radius $r_\Delta \in (0, \infty)$. Introduce

$$C_{f,\Delta} := \int_{2\Delta} k_1(-y)f(y)\,d\sigma(y) \in \mathbb{C}, \tag{2.3.145}$$

and for each $x \in \Delta$ define

$$\mathrm{I}(x) := \big(T_{\text{mod}}(\mathbf{1}_{2\Delta}f)\big)(x) = \big(T(\mathbf{1}_{2\Delta}f)\big)(x) + C_{f,\Delta} \tag{2.3.146}$$

where the second equality is a consequence of (2.3.129)-(2.3.145), bearing in mind that $\mathbf{1}_{2\Delta}f$ belongs to $L^1(\Sigma, \sigma)$. Then, on account of (2.3.146), (2.3.20) (with $p := 2$), and the fact that σ is doubling, we may estimate

$$\begin{aligned} ⨍_\Delta |\mathrm{I}(x) - C_{f,\Delta}|\,d\sigma(x) &\leq \Big(⨍_\Delta \big|\big(T(\mathbf{1}_{2\Delta}f)\big)(x)\big|^2\,d\sigma(x)\Big)^{1/2} \\ &\leq \sigma(\Delta)^{-1/2}\big\|T(\mathbf{1}_{2\Delta}f)\big\|_{L^2(\Sigma,\sigma)} \\ &\leq C\sigma(\Delta)^{-1/2}\big\|\mathbf{1}_{2\Delta}f\big\|_{L^2(\Sigma,\sigma)} \leq C\|f\|_{L^\infty(\Sigma,\sigma)}. \end{aligned} \tag{2.3.147}$$

Next, introduce

$$\widetilde{C}_{f,\Delta} := \int_{\Sigma\setminus 2\Delta} \{k(x_\Delta - y) - k_1(-y)\}f(y) \in \mathbb{C}, \tag{2.3.148}$$

(with the membership guaranteed by (2.3.119) and [112, (7.4.118)]) and for each $x \in \Delta$ define

$$\begin{aligned}\mathrm{II}(x) &:= \big(T_{\rm mod}(\mathbf{1}_{\Sigma\setminus 2\Delta}f)\big)(x) = \int_{\Sigma\setminus 2\Delta}\{k(x-y)-k_1(-y)\}f(y)\,d\sigma(y)\\ &= \int_{\Sigma\setminus 2\Delta}\{k(x-y)-k(x_\Delta-y)\}f(y)\,d\sigma(y)+\widetilde{C}_{f,\Delta} \qquad (2.3.149)\end{aligned}$$

where the second equality is a consequence of the fact that the variables $x\in\Delta$ and $y\in\Sigma\setminus 2\Delta$ are separated. Based on (2.3.149), the Mean Value Theorem, (2.3.116), and [112, Lemma 7.2.1], for each $x\in\Delta$ we may then estimate

$$\begin{aligned}\big|\mathrm{II}(x)-\widetilde{C}_{f,\Delta}\big| &\le \int_{\Sigma\setminus 2\Delta}\big|k(x-y)-k(x_\Delta-y)\big||f(y)|\,d\sigma(y)\\ &\le C\|f\|_{L^\infty(\Sigma,\sigma)}\int_{\Sigma\setminus 2\Delta}\frac{r_\Delta}{|y-x_\Delta|^n}\,d\sigma(y)\\ &\le C\|f\|_{L^\infty(\Sigma,\sigma)}, \qquad (2.3.150)\end{aligned}$$

for some $C\in(0,\infty)$ which depends only on Σ. Consequently,

$$⨍_\Delta \big|\mathrm{II}(x)-\widetilde{C}_{f,\Delta}\big|\,d\sigma(x)\le C\|f\|_{L^\infty(\Sigma,\sigma)}. \qquad (2.3.151)$$

If we now set

$$c_{f,\Delta}:=C_{f,\Delta}+\widetilde{C}_{f,\Delta}\in\mathbb{C} \qquad (2.3.152)$$

then from (2.3.146), (2.3.149), (2.3.147), and (2.3.151) we conclude that

$$\begin{aligned}⨍_\Delta \big|T_{\rm mod}f-c_{f,\Delta}\big|\,d\sigma &\le ⨍_\Delta\big|\mathrm{I}(x)-C_{f,\Delta}\big|\,d\sigma + ⨍_\Delta\big|\mathrm{II}(x)-\widetilde{C}_{f,\Delta}\big|\,d\sigma\\ &\le C\|f\|_{L^\infty(\Sigma,\sigma)}. \qquad (2.3.153)\end{aligned}$$

In view of [112, (7.4.57)] this ultimately establishes (2.3.37).

We shall next consider the claim formulated in (2.3.39). Work under the assumption that Σ is bounded, a scenario guaranteeing that the space $L^\infty(\Sigma,\sigma)$ is contained in $\bigcap_{1\le p\le\infty} L^p(\Sigma,\sigma)$. Granted this, we see from (2.3.128) that if C_f is associated with f as in (2.3.130) then there exists a constant $c\in(0,\infty)$, which depends only on Σ and k, such

$$|C_f|\le c\|f\|_{L^\infty(\Sigma,\sigma)} \text{ for every } f\in L^\infty(\Sigma,\sigma). \qquad (2.3.154)$$

With this in hand, (2.3.39) follows with the help of (2.3.34) and (2.3.37).

To deal with the claim in (2.3.40), recall from (2.3.32) that

$$L^\infty(\Sigma,\sigma)\subseteq L^1\Big(\Sigma,\frac{\sigma(x)}{1+|x|^n}\Big)\cap\Big(\bigcap_{1<p<\infty}L^p_{\rm loc}(\Sigma,\sigma)\Big). \qquad (2.3.155)$$

Pick an arbitrary function $f \in L^\infty(\Sigma, \sigma)$. Let us also fix a background exponent $q \in (1, \infty)$ and consider an arbitrary $(1, q)$-atom a on Σ with vanishing moment[22]. Specifically, this a σ-measurable function defined on Σ, with the property that there exist $x_o \in \Sigma$ and $r \in \big(0, 2\mathrm{diam}\Sigma\big)$ such that

$$\operatorname{supp} a \subseteq B(x_o, r) \cap \Sigma, \quad \|a\|_{L^q(\Sigma,\sigma)} \leq \sigma\big(B(x_o, r) \cap \Sigma\big)^{1/q-1}, \\ \text{and} \int_\Sigma a \, d\sigma = 0. \tag{2.3.156}$$

From (2.3.36), (2.3.155), (2.3.156) we then conclude that

$$\int_\Sigma |T_{\rm mod} f||a| \, d\sigma < +\infty, \qquad \int_\Sigma |f||Ta| \, d\sigma < +\infty, \tag{2.3.157}$$

and

$$\int_\Sigma f(Ta) d\sigma = -\int_\Sigma (T_{\rm mod} f) a \, d\sigma. \tag{2.3.158}$$

Having proved (2.3.157)-(2.3.158), the claim in (2.3.40) now follows with the help of [113, Theorem 4.4.1], the duality result from [113, Theorem 4.6.1, (4.6.8)] (cf. [113, (4.6.8)], plus the very last claim in its statement), (2.3.27) with $p = 1$, and the current item *(5)* (in the case when Σ is bounded). This completes the treatment of item *(7)*.

Consider next the claims made in item *(8)*. For starters, (2.3.41) is a direct corollary of [112, (7.7.107)] and (2.3.15). Next, thanks to [112, (7.7.106)] and (2.3.18), it follows that

$$T : L^{p_0}(\Sigma, \sigma) + L^{p_1}(\Sigma, \sigma) \longrightarrow L^{p_0}(\Sigma, \sigma) + L^{p_1}(\Sigma, \sigma) \\ \text{is well defined, linear, and bounded for all } p_0, p_1 \in (1, \infty). \tag{2.3.159}$$

Based on (2.3.159), [112, (6.2.48)], and [113, Proposition 1.3.7] we then conclude that the claims about the operator (2.3.43) are valid. With this in hand, the Cotlar-type inequality recorded in (2.2.38) (presently used with $p := 1$ and $r := 1$) together with the boundedness of the Hardy-Littlewood maximal operator on Lorentz spaces (cf. [112, (6.2.20)]) justify the claims made about the operator (2.3.42). Finally, from (2.3.25) and [112, (6.2.51), (6.2.62)-(6.2.63)] we see that (2.3.44) holds.

Regarding item *(9)*, since the operators from (2.3.27) and (2.3.18) are compatible with one another (cf. the compatibility claim in item *(3)* of Theorem 2.3.2 and the very last part of item *(6)* of Theorem 2.3.2), the claim in (2.3.45) follows. In turn, for each $p_0, p_1 \in \big(\frac{n-1}{n}, \infty\big)$, this permits us to define, in a coherent fashion, the mapping $\widetilde{T}_{p_0,p_1}$ as in (2.3.46).

Going further, fix two exponents p_0, p_1 with $\frac{n-1}{n} < p_0 < p_1 < \infty$. As a byproduct of the lack of ambiguity in the definition of $\widetilde{T}_{p_0,p_1}$ above, plus the fact that T_{p_0}, T_{p_1} are linear and bounded, it follows that

[22] this is always the case if Σ is unbounded

$\widetilde{T}_{p_0,p_1}$ is a linear and bounded operator, which satisfies

$$\widetilde{T}_{p_0,p_1}\Big|_{H^{p_0}(\Sigma,\sigma)} = T_{p_0} \text{ as well as } \widetilde{T}_{p_0,p_1}\Big|_{H^{p_1}(\Sigma,\sigma)} = T_{p_1}. \tag{2.3.160}$$

Granted (2.3.46) and (2.3.160), we may use real interpolation (cf. [113, Proposition 1.3.7, (1.3.66)]), together with [113, (4.3.3)] and [112, (6.2.48)], to conclude that the operator $\widetilde{T}_{p_0,p_1}$ from (2.3.46) induces a linear and bounded mapping

$$\widetilde{T}_{p_0,p_1} : H^p(\Sigma,\sigma) \longrightarrow L^p(\Sigma,\sigma) \text{ for each } p \in (p_0,p_1). \tag{2.3.161}$$

Moreover, given any $p \in (p_0,p_1)$, since $H^{p_0}(\Sigma,\sigma) \cap H^{p_1}(\Sigma,\sigma)$ is a dense subspace of $H^p(\Sigma,\sigma)$ (cf. [113, (4.3.145)]) and since

$$\begin{gathered}\widetilde{T}_{p_0,p_1} f = T_{p_0} f = T_p f \\ \text{for each } f \in H^{p_0}(\Sigma,\sigma) \cap H^{p_1}(\Sigma,\sigma),\end{gathered} \tag{2.3.162}$$

where the first equality comes from (2.3.160) and the second equality is implied by (2.3.45), we ultimately conclude that the properties claimed in (2.3.47) are all valid.

Consider now the case when we have $\frac{n-1}{n} < q_0 < p_0 < p_1 < q_1 < \infty$ and pick some arbitrary $f \in H^{p_0}(\Sigma,\sigma) + H^{p_1}(\Sigma,\sigma)$. In particular, there exist $f_0 \in H^{p_0}(\Sigma,\sigma)$ and $f_1 \in H^{p_1}(\Sigma,\sigma)$ such that $f = f_0 + f_1$. Then repeated applications of property recorded in the second line of (2.3.47) allows us to compute

$$\widetilde{T}_{q_0,q_1} f = \widetilde{T}_{q_0,q_1} f_0 + \widetilde{T}_{q_0,q_1} f_1 = T_{p_0} f_0 + T_{p_1} f_1 = \widetilde{T}_{p_0,p_1} f. \tag{2.3.163}$$

This goes to show that $\widetilde{T}_{q_0,q_1}$ is an extension of $\widetilde{T}_{p_0,p_1}$, from which the claim in (2.3.48) follows. Next, since [113, (1.3.41)] and the real interpolation result established in [113, (4.3.4)] ensure that

$$\begin{gathered}H^p(\Sigma,\sigma) \subseteq H^{p_0}(\Sigma,\sigma) + H^{p_1}(\Sigma,\sigma) \\ \text{whenever } \tfrac{n-1}{n} < p_0 < p < p_1 < \infty,\end{gathered} \tag{2.3.164}$$

we conclude that $H_{\text{Sum}}(\Sigma,\sigma)$ defined in (2.3.49) is indeed a linear space. Likewise, $L_{\text{Sum}}(\Sigma,\sigma)$ defined in (2.3.50) is a linear space, thanks to [112, (6.2.25), (6.2.52)]. Thus, introducing

$$\begin{gathered}\widetilde{T} : H_{\text{Sum}}(\Sigma,\sigma) \longrightarrow L_{\text{Sum}}(\Sigma,\sigma) \text{ by setting } \widetilde{T} f := \widetilde{T}_{p_0,p_1} f \\ \text{whenever } f \in H^{p_0}(\Sigma,\sigma) + H^{p_1}(\Sigma,\sigma) \text{ for some } p_0,p_1 \in \big(\tfrac{n-1}{n},\infty\big),\end{gathered} \tag{2.3.165}$$

yields an unambiguously defined linear mapping which satisfies

$$\widetilde{T}\Big|_{H^{p_0}(\Sigma,\sigma)+H^{p_1}(\Sigma,\sigma)} = \widetilde{T}_{p_0,p_1} \text{ for each } p_0,p_1 \in \big(\tfrac{n-1}{n},\infty\big), \tag{2.3.166}$$

and which is unique with this property. If, much as in (2.3.161), we now use real interpolation (again, relying on [113, Proposition 1.3.7, (1.3.66), (4.3.3)] and [112,

(6.2.48)]), we conclude that the operator $\widetilde{T}$ from (2.3.51) induces a linear and bounded mapping

$$\widetilde{T} : H^{p,q}(\Sigma, \sigma) \longrightarrow L^{p,q}(\Sigma, \sigma)$$
$$\text{for each } p \in \left(\tfrac{n-1}{n}, \infty\right) \text{ and } q \in (0, \infty], \tag{2.3.167}$$

obeying (cf. [113, (1.3.64)]) the operator quasi-norm estimate

$$\|\widetilde{T}\|_{H^{p,q}(\Sigma,\sigma)\to L^{p,q}(\Sigma,\sigma)} \leq C_{\Sigma,p,q}\big\|k\big|_{S^{n-1}}\big\|_{\mathscr{C}^N(S^{n-1})}. \tag{2.3.168}$$

Given any $p, p_* \in \left(\frac{n-1}{n}, \infty\right)$ and $q \in (0, \infty]$, choose p_0, p_1 such that

$$\tfrac{n-1}{n} < p_0 < \min\{p, p_*\} \leq \max\{p, p_*\} < p_1 < \infty, \tag{2.3.169}$$

then for each $f \in H^{p_*}(\Sigma, \sigma) \cap H^{p,q}(\Sigma, \sigma)$ write

$$\widetilde{T}f = \widetilde{T}_{p_0,p_1}f = T_{p^*}f, \tag{2.3.170}$$

thanks to the definition of $\widetilde{T}$ and the second line in (2.3.47). This shows that $\widetilde{T}$ from (2.3.167) is compatible with T from (2.3.18) and (2.3.27). As such, there is no ambiguity if we may retain the same notation, i.e., T, for the operator (2.3.167). In such a scenario, (2.3.52)-(2.3.53) become consequences of (2.3.167)-(2.3.168). Also, the property claimed in (2.3.54) is a manifestation of the fact that T acting on any $H^{p,q}(\Sigma, \sigma)$ with $p \in \left(\frac{n-1}{n}, \infty\right)$ and $q \in (0, \infty]$ is merely the restriction of the linear operator $\widetilde{T} : H_{\rm Sum}(\Sigma, \sigma) \to L_{\rm Sum}(\Sigma, \sigma)$ to $H^{p,q}(\Sigma, \sigma)$. Finally, from [113, (4.3.145)] we see that (2.3.55) holds.

Let us now deal with item *(10)*. As regards (2.3.56)-(2.3.57), see [19], [104, Corollaire 2, p. 254] for the Euclidean setting, and [93] for the setting of spaces of homogeneous type. In turn, from (2.3.56), Cotlar's inequality from item *(ii)* in Proposition 2.2.5 (which, thanks to (2.3.19), is presently applicable with $p := 1$ and $r := 1$), and item *(1)* in [112, Lemma 7.7.1] we also see that (2.3.58) holds. Parenthetically, we note that we could have also obtained this conclusion based on good-λ inequalities for $T_{\max}$ and $\mathcal{M}_\Sigma$, in the spirit of [104, Théorème 9, pp. 252-253]; cf. also [104, Corollaire 1, Corollaire 2, p. 254]. Next, (2.3.61) is a consequence of (2.3.25) and a standard density argument. Also, the claim pertaining to (2.3.62) may be justified by reasoning much as in the proof of (2.3.17) (now keeping in mind (2.3.56)). Lastly, (2.3.63) is a consequence of (2.7.115), (2.3.57), and (2.3.61).

As regards item *(11)*, the claims made in (2.3.65)-(2.3.66) are consequences of what we have proved in the current item *(10)* and [113, Corollary 5.2.3]. Let us turn our attention to (2.3.68). To set the stage, bring in an exponent $q' \in (1, \infty)$ associated with $\mathbb{X}$ as in [113, Proposition 5.2.7, (5.2.90)], and recall from the current item *(3)* that T is a well-defined linear and bounded operator from the space $L^{q'}(\Sigma, \sigma)$ into itself. In particular,

$$T\big(L^{q'}_{\rm comp}(\Sigma, \sigma)\big) \subseteq L^{q'}(\Sigma, \sigma). \tag{2.3.171}$$

To proceed, fix a point $x_0 \in \Sigma$, and select an arbitrary function $f \in L^{q'}_{\rm comp}(\Sigma, \sigma)$. From (2.3.41) and the nature of the kernel k (cf. (2.3.3)), we see that there exists a constant $C_f \in (0, \infty)$, which is allowed to depend on f, with the property that for each sufficiently large $j \in \mathbb{N}$ we have

$$\mathbf{1}_{\Sigma\setminus\Delta(x_0,j)} \cdot |Tf| \leq C_f \cdot \frac{\mathbf{1}_{\Sigma\setminus\Delta(x_0,j)}}{1 + |\cdot - x_0|^{n-1}} \text{ at } \sigma\text{-a.e. point in } \Sigma. \tag{2.3.172}$$

In addition, (2.3.171) implies that

$$\phi_j := \mathbf{1}_{\Delta(x_0,j)} \cdot Tf \in L^{q'}_{\rm comp}(\Sigma, \sigma) \text{ for each } j \in \mathbb{N}. \tag{2.3.173}$$

For each large $j \in \mathbb{N}$ now estimate

$$\begin{aligned}
\|Tf - \phi_j\|_{\mathbb{X}} &= \big\|\mathbf{1}_{\Sigma\setminus\Delta(x_0,j)} \cdot |Tf|\big\|_{\mathbb{X}} \leq C_f \Big\| \frac{\mathbf{1}_{\Sigma\setminus\Delta(x_0,j)}}{1 + |\cdot - x_0|^{n-1}} \Big\|_{\mathbb{X}} \\
&= C_f \Big\| \frac{\mathbf{1}_{\Sigma\setminus\Delta(x_0,j)}}{1 + \log_+|\cdot - x_0|} \cdot \frac{1 + \log_+|\cdot - x_0|}{1 + |\cdot - x_0|^{n-1}} \Big\|_{\mathbb{X}} \\
&\leq \frac{C_f}{\ln j} \Big\| \frac{1 + \log_+|\cdot - x_0|}{1 + |\cdot - x_0|^{n-1}} \Big\|_{\mathbb{X}} \\
&\leq \frac{C_f}{\ln j} \big\|(\mathcal{M}_\Sigma \circ \mathcal{M}_\Sigma)(\mathbf{1}_{\Delta(x_0,1)})\big\|_{\mathbb{X}} \\
&\leq \frac{C_f}{\ln j} \|\mathcal{M}_\Sigma\|^2_{\mathbb{X}\to\mathbb{X}} \big\|\mathbf{1}_{\Delta(x_0,1)}\big\|_{\mathbb{X}}
\end{aligned} \tag{2.3.174}$$

where the first equality is seen from (2.3.173), the subsequent inequality is a consequence of (2.3.172) (keeping in mind property *(P2)* in [113, Definition 5.1.1]), the second equality is clear, the penultimate inequality is based on [112, (7.6.69)] and once again the monotonicity of the norm in $\mathbb{X}$, while the final inequality follows from (2.3.64). Given that (2.3.64) entails $\|\mathcal{M}_\Sigma\|_{\mathbb{X}\to\mathbb{X}} < \infty$, and that [113, (5.2.75)] implies $\big\|\mathbf{1}_{\Delta(x_0,1)}\big\|_{\mathbb{X}} < \infty$, this proves that

$$\phi_j \to Tf \text{ in } \mathbb{X} \text{ as } j \to \infty. \tag{2.3.175}$$

Together with (2.3.173) and (2.3.67), this ultimately proves that $Tf \in \mathring{\mathbb{X}}$. Thus,

$$T\big(L^{q'}_{\rm comp}(\Sigma, \sigma)\big) \subseteq \mathring{\mathbb{X}}. \tag{2.3.176}$$

From (2.3.176), (2.3.67), and item *(1)* in [113, Lemma 1.2.20] we then deduce that T maps $\mathring{\mathbb{X}}$ linearly and boundedly into itself. Thanks to the fact that $\mathring{\mathbb{X}}$ is a closed linear subspace of $\mathbb{X}$, it follows that $\|T\|_{\mathring{\mathbb{X}}\to\mathring{\mathbb{X}}}$ is dominated by $\|T\|_{\mathbb{X}\to\mathbb{X}}$ (cf. [113, (1.2.20)]), and the latter has been shown to be $\leq C\big\|k\big|_{S^{n-1}}\big\|_{\mathscr{C}^N(S^{n-1})}$ where the constant $C \in (0, \infty)$ depends only on Σ and the operator norms of $\mathcal{M}_\Sigma$ on $\mathbb{X}$

and $\mathbb{X}'$. Once this has been established, the corresponding properties for $T_{\max}$ listed in (2.3.68) become consequences of the Cotlar-type inequality recorded in (2.3.23) (presently used with $r := 1$), whose applicability is ensured by the embedding $\mathbb{X} \hookrightarrow L^1\Big(\Sigma, \frac{\sigma(x)}{1+|x|^{n-1}}\Big)$ (see [113, (5.2.88)]), the assumption made in (2.3.64), and the monotonicity of the norm in $\mathbb{X}$ (see *(P2)* in [113, Definition 5.1.1]).

Finally, to justify the transposition formula claimed in (2.3.69), fix $f \in \mathring{\mathbb{X}}$ and $g \in \mathbb{X}'$, both arbitrary. From [113, (5.2.90)] we know that there exists some $p \in (1, \infty)$ such that

$$g \in \mathbb{X}' \subseteq L^p_{\rm loc}(\Sigma, \sigma) \cap L^1\Big(\Sigma, \frac{\sigma(x)}{1+|x|^{n-1}}\Big). \tag{2.3.177}$$

In addition, since $L^\infty_{\rm comp}(\Sigma, \sigma)$ is dense in $\mathring{\mathbb{X}}$ (cf. (2.3.67)), there exists a sequence $\{f_j\}_{j\in\mathbb{N}} \subseteq L^\infty_{\rm comp}(\Sigma, \sigma)$ which converges to f in $\mathbb{X}$. In particular, if $p' \in (1, \infty)$ is such that $1/p + 1/p' = 1$ then each f_j belongs to $L^{p'}_{\rm comp}(\Sigma, \sigma)$. Granted these properties we may invoke (2.3.26) and conclude that

$$\int_\Sigma (Tf_j)g \, d\sigma = -\int_\Sigma f_j(Tg) \, d\sigma \quad \text{for each } j \in \mathbb{N}. \tag{2.3.178}$$

Passing to limit $j \to \infty$ in (2.3.178) then yields (2.3.69), on account of the fact that T is continuous both on $\mathbb{X}$ and on $\mathbb{X}'$ (cf. (2.3.66) and (2.3.66)) plus the generalized Hölder inequality recorded in [113, Proposition 5.1.12]. All claims in item *(11)* are therefore justified. This finishes the proof of Theorem 2.3.2. □

In the last part of this section we wish to further elaborate on $T_{\rm mod}$, the modified version of the principal-value singular integral operator T, originally introduced in item *(7)* of Theorem 2.3.2.

Proposition 2.3.3 *Suppose $\Sigma \subseteq \mathbb{R}^n$, $n \in \mathbb{N}$ with $n \geq 2$, is a* UR *set, and abbreviate $\sigma := \mathcal{H}^{n-1}\lfloor\Sigma$. Having picked a sufficiently large integer $N = N(n) \in \mathbb{N}$, consider a kernel $k \in \mathscr{C}^N(\mathbb{R}^n \setminus \{0\})$ which is odd and positive homogeneous of degree $1 - n$. In this setting, associate with Σ and k the principal-value singular integral operator T as in* (2.3.15)*, together with its modified version $T_{\rm mod}$, defined as in* (2.3.31). *Then the following statements are true:*

(i) For each function $f \in L^1\Big(\Sigma, \frac{\sigma(x)}{1+|x|^n}\Big)$ and each surface ball $\Delta \subseteq \Sigma$ there exists a constant $C_{f,\Delta} \in \mathbb{R}$ with the property that for σ-a.e. point $x \in \Delta$ one has

$$(T_{\rm mod} f)(x) = T(f\mathbf{1}_\Delta)(x) + \int_{\Sigma\setminus\Delta} \big\{k(x-y) - k(x_\Delta - y)\big\} f(y) \, d\sigma(y) + C_{f,\Delta}, \tag{2.3.179}$$

where $x_\Delta \in \Sigma$ is the center of the surface ball Δ.

(ii) Fix a reference point $x_0 \in \Sigma$ and for each $j \in \mathbb{N}$ set $\Delta_j := B(x_0, j) \cap \Sigma$. Given an arbitrary function $f \in L^1\Big(\Sigma, \frac{\sigma(x)}{1+|x|^n}\Big)$, for each $j \in \mathbb{N}$ define

$$g_j(x) := T(f\mathbf{1}_{\Delta_j})(x) - \int_{\Delta_j\setminus\Delta_1} k(x_0 - y) f(y) \, d\sigma(y) \quad \textit{for } \sigma\textit{-a.e. } x \in \Sigma. \tag{2.3.180}$$

Then the sequence $\{g_j\}_{j\in\mathbb{N}}$ *converges pointwise* σ*-a.e. on* Σ*. In particular,*

$$(\widetilde{T}_{\rm mod}f)(x) := \lim_{j\to\infty}\Big\{T(f\mathbf{1}_{\Delta_j})(x) - \int_{\Delta_j\setminus\Delta_1} k(x_0-y)f(y)\,d\sigma(y)\Big\} \tag{2.3.181}$$

is well defined at σ*-a.e. point* $x\in\Sigma$*. Moreover, there exists a constant* $\widetilde{C}_{f,x_0}\in\mathbb{R}$ *with the property that*

$$(\widetilde{T}_{\rm mod}f)(x) = (T_{\rm mod}f)(x) + \widetilde{C}_{f,x_0} \text{ for } \sigma\text{-a.e. } x\in\Sigma. \tag{2.3.182}$$

Proof Pick a function $f\in L^1\big(\Sigma, \frac{\sigma(x)}{1+|x|^n}\big)$ and a surface ball $\Delta\subseteq\Sigma$, with center $x_\Delta\in\Sigma$ and radius R. Define the constant

$$C_{f,\Delta} := \int_\Sigma \big\{k_R(x_\Delta-y) - k_1(-y)\big\}f(y)\,d\sigma(y)\in\mathbb{R}. \tag{2.3.183}$$

At σ-a.e. point $x\in\Delta$ we may then use (2.3.31), (2.3.183), and (2.3.15) to write

$$\begin{aligned}
(T_{\rm mod}f)(x) &= \lim_{\varepsilon\to0^+}\int_\Sigma\big\{k_\varepsilon(x-y)-k_1(-y)\big\}f(y)\,d\sigma(y)\\
&= \lim_{\varepsilon\to0^+}\int_\Sigma\big\{k_\varepsilon(x-y)-k_R(x_\Delta-y)\big\}f(y)\,d\sigma(y)+C_{f,\Delta}\\
&= \lim_{\varepsilon\to0^+}\int_\Delta\big\{k_\varepsilon(x-y)-k_R(x_\Delta-y)\big\}f(y)\,d\sigma(y)\\
&\quad+\lim_{\varepsilon\to0^+}\int_{\Sigma\setminus\Delta}\big\{k_\varepsilon(x-y)-k_R(x_\Delta-y)\big\}f(y)\,d\sigma(y)+C_{f,\Delta}\\
&= T\big(f\mathbf{1}_\Delta\big)(x)+\int_{\Sigma\setminus\Delta}\big\{k(x-y)-k(x_\Delta-y)\big\}f(y)\,d\sigma(y)\\
&\quad+C_{f,\Delta}.
\end{aligned}\tag{2.3.184}$$

The last equality is a consequence of two observations. First, $k_R(x_\Delta-y)=0$ for each $y\in\Delta$. Second, since $x\in\Delta$ it follows that if $\varepsilon>0$ is small enough then $k_\varepsilon(x-y)=k(x-y)$ for each $y\in\Sigma\setminus\Delta$. This finishes the proof of (2.3.179).

Turning attention to item *(ii)*, fix an arbitrary compact set $F\subseteq\Sigma$ and choose $j_F\in\mathbb{N}$ such that $F\subseteq\Delta_{j_F}$. Then for each $j\in\mathbb{N}$ with $j\geq j_F$ and σ-a.e. point $x\in F$ we have

$$
\begin{aligned}
g_j(x) - g_{j_F}(x) &= T\big(f\mathbf{1}_{\Delta_j\setminus\Delta_{j_F}}\big)(x) - \int_{\Delta_j\setminus\Delta_{j_F}} k(x_0-y)f(y)\,\mathrm{d}\sigma(y) \\
&= \int_{\Delta_j\setminus\Delta_{j_F}} k(x-y)f(y)\,\mathrm{d}\sigma(y) - \int_{\Delta_j\setminus\Delta_{j_F}} k(x_0-y)f(y)\,\mathrm{d}\sigma(y) \\
&= \int_{\Delta_j\setminus\Delta_{j_F}} \big\{k(x-y)-k(x_0-y)\big\}f(y)\,\mathrm{d}\sigma(y). \qquad (2.3.185)
\end{aligned}
$$

Granted this, Lebesgue's Dominated Convergence Theorem applies and shows that at σ-a.e. point $x \in F$ we have

$$
\lim_{j\to\infty} g_j(x) = g_{j_F}(x) + \int_{\Sigma\setminus\Delta_{j_F}} \big\{k(x-y)-k(x_0-y)\big\}f(y)\,\mathrm{d}\sigma(y). \qquad (2.3.186)
$$

Hence, at σ-a.e. point $x \in F$ we may write

$$
\begin{aligned}
(\widetilde{T}_{\text{mod}}f)(x) &= \lim_{j\to\infty} g_j(x) = g_{j_F}(x) + \int_{\Sigma\setminus\Delta_{j_F}} \big\{k(x-y)-k(x_0-y)\big\}f(y)\,\mathrm{d}\sigma(y) \\
&= T\big(f\mathbf{1}_{\Delta_{j_F}}\big)(x) - \int_{\Delta_{j_F}\setminus\Delta_1} k(x_0-y)f(y)\,\mathrm{d}\sigma(y) \\
&\quad + \int_{\Sigma\setminus\Delta_{j_F}} \big\{k(x-y)-k(x_0-y)\big\}f(y)\,\mathrm{d}\sigma(y) \\
&= (T_{\text{mod}}f)(x) + C_{F,f,x_0}, \qquad (2.3.187)
\end{aligned}
$$

where $C_{F,f,x_0} \in \mathbb{R}$ is a constant, defined as

$$
C_{F,f,x_0} := -\int_{\Delta_{j_F}\setminus\Delta_1} k(x_0-y)f(y)\,\mathrm{d}\sigma(y) - C_{f,\Delta_{j_F}}, \qquad (2.3.188)
$$

with $C_{f,\Delta_{j_F}}$ as in (2.3.183) (with j_F in place of R, and x_0 in place of x_Δ). In (2.3.187), the first equality comes from (2.3.180)-(2.3.181), the second equality is (2.3.186), the third equality uses (2.3.180) (with j_F in place of j), and the final equality is implied by (2.3.179) (presently employed with $\Delta := B(x_0, j_F) \cap \Sigma$) plus the definition made in (2.3.188).

In summary, (2.3.187) proves that $\widetilde{T}_{\text{mod}}f - T_{\text{mod}}f$ is a constant function on each compact subset of Σ. In turn, this shows that the constant C_{F,f,x_0} is actually independent of F, which then establishes (2.3.182). □

2.4 Boundary-to-Domain Integral Operators on Open sets with Uniformly Rectifiable Boundaries

Theorem 2.3.2 is one of the main ingredients in the proof of our main result in this section, contained in Theorem 2.4.1 below. Items *(1)-(5)* deal with nontangential maximal function estimates for integral operators mapping functions defined on a UR set $\Sigma \subseteq \mathbb{R}^n$ into functions defined in $\mathbb{R}^n \setminus \Sigma$ (refining [63, Propositions 3.20-3.21, pp. 2641-3243], originally proved in a more restrictive geometric setting; see the preamble to the statement of Theorem 2.3.2). Items *(6)-(8)* in this theorem, containing mixed-norm, area-function, square-function, and Carleson estimates, make essential use of work in [61]. Items *(9)-(10)* in Theorem 2.4.1 deal with off-diagonal Carleson measure estimates for singular integral operators on UR sets. Recall that given an open set $\Omega \subseteq \mathbb{R}^n$ with an Ahlfors regular boundary, a Borel measure μ in Ω is said to be a `Carleson measure` provided its Carleson constant is finite, i.e.,

$$\sup_{\substack{r\in(0,2\mathrm{diam}(\partial\Omega))\\ \text{and } x\in\partial\Omega}} \frac{\mu\big(B(x,r)\cap\Omega\big)}{\mathcal{H}^{n-1}\big(B(x,r)\cap\partial\Omega\big)} < +\infty. \tag{2.4.1}$$

For example, a simple application of Hölder's inequality shows that

$$\begin{array}{l}\text{if } \Omega \subseteq \mathbb{R}^n \text{ is an open set with an Ahlfors regular boundary}\\ \text{and if } F \in L^n(\Omega,\mathcal{L}^n) \text{ is some non-negative function, then}\\ \mu := F\mathcal{L}^n \text{ is a Carleson measure in } \Omega.\end{array} \tag{2.4.2}$$

Finally, item *(11)* contains regularity results, with smoothness measured on the scale of Triebel-Lizorkin spaces.

Theorem 2.4.1 *Fix $n \in \mathbb{N}$ with $n \geq 2$ and suppose $\Sigma \subseteq \mathbb{R}^n$ is a* UR *set. Abbreviate $\Sigma^c := \mathbb{R}^n \setminus \Sigma$, which entails $\Sigma = \partial(\Sigma^c)$, and pick an arbitrary aperture parameter $\kappa \in (0,\infty)$. Let $\Gamma_\kappa^{\Sigma^c}(x)$, with $x \in \Sigma$, stand for the nontangential approach regions with aperture parameter κ in the open set Σ^c. Also, denote by $\mathcal{N}_\kappa^{\Sigma^c}$ the corresponding nontangential maximal operator with aperture parameter κ in the open set Σ^c. Finally, define $\sigma := \mathcal{H}^{n-1}\lfloor\Sigma$, and let $H^p(\Sigma,\sigma)$ with $\frac{n-1}{n} < p < \infty$ denote the scale of Hardy spaces on Σ, as introduced in* [113, Definition 4.2.1].

Consider next a kernel $k \in \mathscr{C}^N(\mathbb{R}^n \setminus \{0\})$, with $N \geq 2$, which is odd and positive homogeneous of degree $1-n$ (cf. (2.3.3)*), then define the integral operator $\mathcal{T}$ acting on $L^1\big(\Sigma, \frac{\sigma(y)}{1+|y|^{n-1}}\big)$, i.e., on σ-measurable functions $f : \Sigma \to \mathbb{C}$ satisfying*

$$\int_\Sigma \frac{|f(y)|}{1+|y|^{n-1}}\,\mathrm{d}\sigma(y) < +\infty, \tag{2.4.3}$$

according to

$$(\mathcal{T}f)(x) := \int_\Sigma k(x-y)f(y)\,\mathrm{d}\sigma(y), \qquad \forall x \in \mathbb{R}^n \setminus \Sigma. \tag{2.4.4}$$

In relation to this, the following statements are true.

(1) There exists some constant $C = C(\Sigma, \kappa, n) \in (0, \infty)$ *such that*

$$\mathcal{N}_\kappa^{\Sigma^c}(\mathcal{T} f)(x) \leq (T_{\max} f)(x) + C\big\|k\big|_{S^{n-1}}\big\|_{\mathscr{C}^1(S^{n-1})}(\mathcal{M}_\Sigma f)(x),$$
$$\text{for each function } f \in L^1\Big(\Sigma, \tfrac{\sigma(y)}{1+|y|^{n-1}}\Big) \text{ and each point } x \in \Sigma, \tag{2.4.5}$$

where $T_{\max}$ *is the maximal singular integral operator defined as in* (2.3.4)-(2.3.6), *and* $\mathcal{M}_\Sigma$ *is the Hardy-Littlewood maximal operator on the set*[23] Σ.
As a consequence of (2.4.5) *and the Cotlar-type inequality from* (2.3.23), *in the case when the smoothness exponent N of the kernel k is sufficiently large (depending only on the dimension* $n \in \mathbb{N}$ *of the ambient Euclidean space) it follows that for each* $r \in (0, \infty)$ *one has*

$$\mathcal{N}_\kappa^{\Sigma^c}(\mathcal{T} f)(x) \leq C\Big[\big(\mathcal{M}_\Sigma(|Tf|^r)\big)(x)\Big]^{1/r} + C\big\|k\big|_{S^{n-1}}\big\|_{\mathscr{C}^N(S^{n-1})}(\mathcal{M}_\Sigma f)(x),$$
$$\text{for each function } f \in L^1\Big(\Sigma, \tfrac{\sigma(y)}{1+|y|^{n-1}}\Big) \text{ and each point } x \in \Sigma, \tag{2.4.6}$$

where T is the principal-value singular integral operator associated with the kernel k and the set Σ *as in* (2.3.15), *and* $C \in (0, \infty)$ *depends only on n, the* UR *constants of* Σ, *and r.*

(2) Assume the smoothness exponent N of the kernel k is sufficiently large, depending only on the dimension $n \in \mathbb{N}$ *of the ambient Euclidean space. Then the assignment*

$$L^1\Big(\Sigma, \frac{\sigma(x)}{1+|x|^{n-1}}\Big) \ni f \longmapsto \mathcal{N}_\kappa^{\Sigma^c}(\mathcal{T} f) \in L^{1,\infty}_{\mathrm{loc}}(\Sigma, \sigma) \tag{2.4.7}$$

is well defined, sub-linear, bounded[24] *and continuous, and for each* $p \in (1, \infty)$ *it induces a well-defined, sub-linear, bounded, and continuous mapping*

$$L^p_{\mathrm{loc}}(\Sigma, \sigma) \cap L^1\Big(\Sigma, \frac{\sigma(x)}{1+|x|^{n-1}}\Big) \ni f \longmapsto \mathcal{N}_\kappa^{\Sigma^c}(\mathcal{T} f) \in L^p_{\mathrm{loc}}(\Sigma, \sigma). \tag{2.4.8}$$

(3) Assume the smoothness exponent N of the kernel k is sufficiently large, depending only on the dimension $n \in \mathbb{N}$ *of the ambient Euclidean space. Then for each* $p \in [1, \infty)$ *there exists a constant* $C = C(\Sigma, p, \kappa, n) \in [0, \infty)$ *such that for every* $f \in L^p(\Sigma, \sigma)$ *one has*

---

[23] regarded as a space of homogeneous type when equipped with the Euclidean distance and the measure σ; cf. the discussion in [112, Chapter 7]

[24] i.e., it maps bounded subsets of $L^1\big(\Sigma, \frac{\sigma(x)}{1+|x|^{n-1}}\big)$ into bounded subsets of $L^{1,\infty}_{\mathrm{loc}}(\Sigma, \sigma)$

$$\|\mathcal{N}_\kappa^{\Sigma^c}(\mathcal{T}f)\|_{L^p(\Sigma,\sigma)} \leq C\|k|_{S^{n-1}}\|_{\mathscr{C}^N(S^{n-1})}\|f\|_{L^p(\Sigma,\sigma)} \text{ if } 1 < p < \infty, \quad (2.4.9)$$

$$\|\mathcal{N}_\kappa^{\Sigma^c}(\mathcal{T}f)\|_{L^{1,\infty}(\Sigma,\sigma)} \leq C\|k|_{S^{n-1}}\|_{\mathscr{C}^N(S^{n-1})}\|f\|_{L^1(\Sigma,\sigma)} \text{ if } p = 1. \quad (2.4.10)$$

In addition, in view of work in [112, §8.5]*, similar estimates to* (2.4.9)-(2.4.10) *are valid with the nontangential maximal operator replaced by the tangential maximal operator* (*associated as in* [112, Definition 8.5.1] *with* $\Omega := \Sigma^c$ *and a sufficiently large power* M*; cf.* (A.0.85)). *Also, if for each* $p \in (0, \infty]$ *one considers* (*in agreement with* (A.0.98))

$$N_\kappa^p(\Sigma^c;\sigma) := \Big\{u \in L^\infty_{\mathrm{loc}}(\Sigma^c, \mathcal{L}^n) : \|u\|_{N_\kappa^p(\Sigma^c;\sigma)} := \|\mathcal{N}_\kappa^{\Sigma^c} u\|_{L^p(\Sigma,\sigma)} < +\infty\Big\}, \quad (2.4.11)$$

then

$$\mathcal{T} : L^p(\Sigma,\sigma) \longrightarrow N_\kappa^p(\Sigma^c;\sigma) \text{ is a well-defined, linear and bounded operator for each exponent } p \in (1,\infty). \quad (2.4.12)$$

Moreover, the operator $\mathcal{T}$ *as originally defined in* (2.4.3)-(2.4.4) *further extends in a unique and coherent fashion to a linear mapping*

$$\mathcal{T} : H^p(\Sigma,\sigma) \longrightarrow N_\kappa^p(\Sigma^c;\sigma) \text{ for each } p \in \big(\tfrac{n-1}{n}, 1\big], \quad (2.4.13)$$

which, for each $f \in H^p(\Sigma,\sigma)$ *with* $\frac{n-1}{n} < p \leq 1$*, satisfies* (*for some* $C_p \in (0,\infty)$ *independent of* f)

$$\|\mathcal{N}_\kappa^{\Sigma^c}(\mathcal{T}f)\|_{L^p(\Sigma,\sigma)} \leq C_p\|k|_{S^{n-1}}\|_{\mathscr{C}^N(S^{n-1})}\|f\|_{H^p(\Sigma,\sigma)}, \quad (2.4.14)$$

plus a similar estimate in which the nontangential maximal operator is replaced by the tangential maximal operator (*associated as in* [112, Definition 8.5.1] *with* $\Omega := \Sigma^c$ *and a sufficiently large power* M*; cf.* (A.0.85)).
An actual formula for said extension is given by

$$(\mathcal{T}f)(x) = \sum_{j=1}^{\infty} \lambda_j \int_\Sigma k(x-y)a_j(y)\,\mathrm{d}\sigma(y), \qquad \forall x \in \mathbb{R}^n \setminus \Sigma, \quad (2.4.15)$$

with uniform convergence of the series on compact subsets of $\mathbb{R}^n \setminus \Sigma$*, whenever* $f = \sum_{j=1}^{\infty} \lambda_j a_j$ *in* $H^p(\Sigma,\sigma)$ *for some sequence* $\{\lambda_j\}_{j\in\mathbb{N}} \in \ell^p(\mathbb{N})$ *and family* $\{a_j\}_{j\in\mathbb{N}}$ *of* H^p*-atoms on* Σ*. In fact, the same formula remains valid if the* a_j*'s happen to be* H^p*-molecules on* Σ (*in the sense of* [113, Definition 4.5.1]). *Alternatively, the aforementioned extension acts on each given* $f \in H^p(\Sigma,\sigma)$ *with* $p \in \big(\frac{n-1}{n}, 1\big]$ *according to*

$$(\mathcal{T}f)(x) = \begin{cases} \langle k(x-\cdot), f\rangle & \text{if } \Sigma \text{ is bounded,} \\ \langle [k(x-\cdot)], f\rangle & \text{if } \Sigma \text{ is unbounded,} \end{cases} \qquad \forall x \in \mathbb{R}^n \setminus \Sigma. \quad (2.4.16)$$

Above, $\langle\cdot,\cdot\rangle$ *stands for the duality bracket described in* [113, Theorem 4.6.1]. *In addition, if* Σ *is bounded,* $k(x-\cdot)$ *is regarded as a function in* $\mathscr{C}^{\alpha}(\Sigma)$ *with* $\alpha:=(n-1)\big(\frac{1}{p}-1\big)$ *if* $p\in\big(\frac{n-1}{n},1\big)$, *and as a function in* $L^{\infty}(\Sigma,\sigma)\subseteq \mathrm{BMO}(\Sigma,\sigma)$ *if* $p=1$. *Also, if* Σ *is unbounded,* $[k(x-\cdot)]$ *is the equivalence class (modulo constants) of the function* $k(x-\cdot)\big|_{\Sigma}$, *considered in the quotient space* $\dot{\mathscr{C}}^{\alpha}(\Sigma)/\sim$ *if* $p\in\big(\frac{n-1}{n},1\big)$, *and in the quotient space* $\widetilde{\mathrm{BMO}}(\Sigma,\sigma)$ *in the case when* $p=1$. *Finally, whenever* $f\in H^{p}(\Sigma,\sigma)$ *has compact support (as a distribution, which is automatically the case if* Σ *is bounded) and* $\psi\in \mathrm{Lip}_c(\Sigma)$ *is identically one near* $\operatorname{supp} f$ *one has*

$$\begin{gathered}(\mathcal{T}f)(x)={}_{\mathrm{Lip}_c(\Sigma)}\Big\langle \psi k(x-\cdot)\big|_{\partial\Omega},f\Big\rangle_{(\mathrm{Lip}_c(\Sigma))'}\ \text{ for each }\ x\in\mathbb{R}^n\setminus\Sigma,\\ \text{where }\psi\in \mathrm{Lip}_c(\Sigma)\text{ is identically one near }\operatorname{supp} f.\end{gathered}\tag{2.4.17}$$

(4) Assume the smoothness exponent N *of the kernel* k *is large enough (depending on* $n\in\mathbb{N}$*), and recall that* $A_p(\Sigma,\sigma)$, $1\le p<\infty$, *are the classes of Muckenhoupt weights associated as in* (A.0.3) *with the space of homogeneous type* $\big(\Sigma,|\cdot-\cdot|,\sigma\big)$ *(cf.* [112, Example 7.4.1]). *Then*

$$\begin{gathered}\text{for each } p\in(1,\infty)\text{ and } w\in A_p(\Sigma,\sigma)\text{ there exists}\\ C=C(\Sigma,p,[w]_{A_p})\in(0,\infty)\text{ with the property that}\\ \big\|\mathcal{N}_{\kappa}^{\Sigma^c}(\mathcal{T}f)\big\|_{L^p(\Sigma,w\sigma)}\le C\big\|k\big|_{S^{n-1}}\big\|_{\mathscr{C}^N(S^{n-1})}\|f\|_{L^p(\Sigma,w\sigma)}\\ \text{for each function } f\in L^p(\Sigma,w\sigma).\end{gathered}\tag{2.4.18}$$

In fact, a more general result is valid. To state it, bring back the Hardy-Littlewood maximal operator $\mathcal{M}_{\Sigma}$ *on* Σ, *and assume* $\mathbb{X}$ *is a Generalized Banach Function Space on* (Σ,σ) *such that*

$$\begin{gathered}\mathcal{M}_{\Sigma}:\mathbb{X}\to\mathbb{X}\text{ and }\mathcal{M}_{\Sigma}:\mathbb{X}'\to\mathbb{X}'\\ \text{are well-defined bounded mappings,}\end{gathered}\tag{2.4.19}$$

where $\mathbb{X}'$ *is the associated space of* $\mathbb{X}$ *(cf.* [113, Definitions 5.1.4, 5.1.11]). *Then there exists a constant* $C\in(0,\infty)$, *which depends only on* Σ *and the operator norms of* $\mathcal{M}_{\Sigma}$ *on* $\mathbb{X}$ *and* $\mathbb{X}'$, *with the property that*

$$\begin{gathered}\big\|\mathcal{N}_{\kappa}^{\Sigma^c}(\mathcal{T}f)\big\|_{\mathbb{X}}\le C\big\|k\big|_{S^{n-1}}\big\|_{\mathscr{C}^N(S^{n-1})}\|f\|_{\mathbb{X}}\text{ for each } f\in\mathbb{X},\\ \big\|\mathcal{N}_{\kappa}^{\Sigma^c}(\mathcal{T}f)\big\|_{\mathbb{X}'}\le C\big\|k\big|_{S^{n-1}}\big\|_{\mathscr{C}^N(S^{n-1})}\|f\|_{\mathbb{X}'}\text{ for each } f\in\mathbb{X}'.\end{gathered}\tag{2.4.20}$$

Also, if $\mathring{\mathbb{X}}$ *is defined in relation to* $\mathbb{X}$ *as in* [113, Definition 5.2.6], *i.e.,*

$$\mathring{\mathbb{X}}:=\overline{L^{\infty}_{\mathrm{comp}}(\Sigma,\sigma)}^{\|\cdot\|_{\mathbb{X}}},\tag{2.4.21}$$

then, under the assumptions made in (2.4.19), *one also has*

$$\mathcal{N}_\kappa^{\Sigma^c}(\mathcal{T} f) \in \mathring{\mathbb{X}} \text{ for each } f \in \mathring{\mathbb{X}}. \tag{2.4.22}$$

Finally, in light of work in [112, §8.5], *estimates similar to* (2.4.18) *and* (2.4.20) *are valid with the nontangential maximal operator replaced by the tangential maximal operator* (*associated as in* [112, Definition 8.5.1] *with* $\Omega := \Sigma^c$ *and a sufficiently large power* M; *cf.* (A.0.85)).

(5) Assume the smoothness exponent N *of the kernel* k *is sufficiently large, depending on the dimension* n *of the ambient Euclidean space. Then for each* $p \in (1, \infty)$ *and* $q \in (0, \infty]$ *there exists a constant* $C \in (0, \infty)$ *such that*

$$\big\| \mathcal{N}_\kappa^{\Sigma^c}(\mathcal{T} f) \big\|_{L^{p,q}(\Sigma,\sigma)} \leq C \| f \|_{L^{p,q}(\Sigma,\sigma)} \text{ for all } f \in L^{p,q}(\Sigma, \sigma). \tag{2.4.23}$$

More generally, the operator $\mathcal{T}$ *originally defined as in* (2.4.3)-(2.4.4) *further extends, in a coherent fashion, to a well-defined mapping acting on each arbitrary* $f \in H^{p,q}(\Sigma, \sigma)$ *with* $p \in \big(\frac{n-1}{n}, \infty\big)$ *and* $q \in (0, \infty]$ *according to*

$$(\mathcal{T} f)(x) = {}_{(H^{p,q}(\Sigma,\sigma))^*}\big\langle k(x - \cdot)\big|_\Sigma, f \big\rangle_{H^{p,q}(\Sigma,\sigma)} \text{ for all } x \in \mathbb{R}^n \setminus \Sigma. \tag{2.4.24}$$

Then

the identity recorded in (2.4.17) *continues to be valid in the case when* $f \in H^{p,q}(\Sigma, \sigma)$ *has compact support (as a distribution).* (2.4.25)

In particular, (2.4.25) *and* [113, (4.2.34)] *imply that*

$$\big(\mathcal{T} \delta_{x_o}\big)(x) = k(x - x_o) \text{ for each } x_o \in \Sigma \text{ and each } x \in \mathbb{R}^n \setminus \Sigma. \tag{2.4.26}$$

Also, interpreting $\mathcal{T}$ *as in* (2.4.24), *it follows that for each* $p \in \big(\frac{n-1}{n}, \infty\big)$ *and* $q \in (0, \infty]$ *there exists a constant* $C \in (0, \infty)$ *such that*

$$\big\| \mathcal{N}_\kappa^{\Sigma^c}(\mathcal{T} f) \big\|_{L^{p,q}(\Sigma,\sigma)} \leq C \| f \|_{H^{p,q}(\Sigma,\sigma)} \quad \text{for each } f \in H^{p,q}(\Sigma, \sigma). \tag{2.4.27}$$

As a consequence of (2.4.27) *and* [112, (8.6.50)], *for each Lebesgue-measurable set* $E \subseteq \mathbb{R}^n \setminus \Sigma$ *which is also assumed to satisfy* $\mathcal{L}^n(E) < +\infty$ *in the case when* (2.4.43) *holds, and for each* $p \in \big(\frac{n-1}{n}, \infty\big)$ *and* $q \in (0, \infty]$ *there exists a constant* $C \in (0, \infty)$, *which is actually independent of* E *in the case when* $\mathcal{H}^{n-1}(\Sigma) = +\infty$, *with the property that the operator*

$$\mathcal{T} : H^{p,q}(\Sigma, \sigma) \longrightarrow L^{np/(n-1),q}(E, \mathcal{L}^n) \quad \text{is well defined, linear, and bounded.} \tag{2.4.28}$$

Finally, thanks to work in [112, §8.5], *similar estimates to* (2.4.23) *and* (2.4.27) *hold with the nontangential maximal operator replaced by the tangential maximal operator* (*associated as in* [112, Definition 8.5.1] *with* $\Omega := \Sigma^c$ *and a sufficiently large power* M; *cf.* (A.0.85)).

(6) *Given any integrability expoents* $p \in \left(\frac{n-1}{n}, \infty\right)$ *and* $q \in (1, \infty)$, *there exists some constant* $C = C(\Sigma, k, p, q, \kappa, n) \in [0, \infty)$ *such that for every* $f \in H^p(\Sigma, \sigma)$ *the following mixed-norm estimate holds:*

$$\left(\int_\Sigma \left(\int_{\Gamma_\kappa^{\Sigma^c}(x)} |(\nabla \mathcal{T} f)(z)|^q \operatorname{dist}(z, \Sigma)^{q-n} \, dz\right)^{p/q} d\sigma(x)\right)^{1/p}$$

$$\leq C\|f\|_{H^p(\Sigma,\sigma)}, \tag{2.4.29}$$

where the function $\mathcal{T} f$ *is defined as in* (2.4.4) *if* $p \in (1, \infty)$, *and as in* (2.4.15) *or* (2.4.16) *if* $p \in \left(\frac{n-1}{n}, 1\right]$.

In addition, corresponding to the case when $p = 1$, *for every fixed exponent* $q \in (1, \infty)$ *there exists a constant* $C = C(\Sigma, k, q, \kappa, n) \in [0, \infty)$ *with the property that for each* $f \in L^1(\Sigma, \sigma)$ *the following weak estimate holds:*

$$\sup_{\lambda>0} \left[\lambda \cdot \sigma\left(\left\{x \in \Sigma : \int_{\Gamma_\kappa^{\Sigma^c}(x)} |\nabla(\mathcal{T} f)|^q \operatorname{dist}(\cdot, \Sigma)^{q-n} \, d\mathcal{L}^n > \lambda^q\right\}\right)\right]$$

$$\leq C\|f\|_{L^1(\Sigma,\sigma)}. \tag{2.4.30}$$

In the language of the L^q*-based area-function* $(0 < q < \infty)$, *whose action on a generic function* $u \in W^{1,1}_{\text{loc}}(\mathbb{R}^n \setminus \Sigma)$ *is defined as*

$$(\mathcal{A}_{q,\kappa} u)(x) := \left(\int_{\Gamma_\kappa^{\Sigma^c}(x)} |(\nabla u)(y)|^q |x - y|^{q-n} \, dy\right)^{1/q}, \qquad \forall x \in \Sigma, \tag{2.4.31}$$

(*compare with* [113, (9.2.206)]), *the estimate in* (2.4.29) *becomes the statement that for each integrability exponents* $p \in \left(\frac{n-1}{n}, \infty\right)$ *and* $q \in (1, \infty)$ *there exists a constant* $C \in [0, \infty)$ *such that for every* $f \in H^p(\Sigma, \sigma)$ *one has*

$$\left\|\mathcal{A}_{q,\kappa}(\mathcal{T} f)\right\|_{L^p(\Sigma,\sigma)} \leq C\|f\|_{H^p(\Sigma,\sigma)}, \tag{2.4.32}$$

while the estimate in (2.4.30) *translates into saying that for each* $q \in (1, \infty)$ *there exists some constant* $C \in [0, \infty)$ *such that for each* $f \in L^1(\Sigma, \sigma)$ *one has*

$$\left\|\mathcal{A}_{q,\kappa}(\mathcal{T} f)\right\|_{L^{1,\infty}(\Sigma,\sigma)} \leq C\|f\|_{L^1(\Sigma,\sigma)}. \tag{2.4.33}$$

(7) *For each* $p \in (1, \infty)$ *there exists a constant* $C = C(\Sigma, p, n) \in [0, \infty)$ *with the property that for each function* $f \in L^p(\Sigma, \sigma)$ *one has the* L^p*-square function estimate*

$$\left(\int_{\mathbb{R}^n \setminus \Sigma} |(\nabla \mathcal{T} f)(x)|^p \operatorname{dist}(x, \Sigma)^{p-1} \, dx\right)^{1/p}$$

$$\leq C\left\|k\big|_{S^{n-1}}\right\|_{\mathscr{C}^2(S^{n-1})}\|f\|_{L^p(\Sigma,\sigma)}. \tag{2.4.34}$$

In particular, corresponding to $p = 2$, the following L^2-square function estimate holds: there exists a constant $C = C(\Sigma, k, n) \in [0, \infty)$ with the property that for each $f \in L^2(\Sigma, \sigma)$ one has

$$\int_{\mathbb{R}^n \setminus \Sigma} |(\nabla \mathcal{T} f)(x)|^2 \operatorname{dist}(x, \Sigma)\,\mathrm{d}x \le C \int_\Sigma |f(x)|^2\,\mathrm{d}\sigma(x). \tag{2.4.35}$$

Also, for each $p \in (1, \infty)$ there exists some $C = C(\Sigma, k, p, n) \in [0, \infty)$ such that for each function $f \in L^p(\Sigma, \sigma)$ there holds

$$\sup_{x \in \mathbb{R}^n \setminus \Sigma} \Big\{ \big|(\mathcal{T} f)(x)\big| \operatorname{dist}(x, \Sigma)^{\frac{n-1}{p}} + \big|\nabla(\mathcal{T} f)(x)\big| \operatorname{dist}(x, \Sigma)^{1+\frac{n-1}{p}} \Big\} \le C\|f\|_{L^p(\Sigma,\sigma)}. \tag{2.4.36}$$

Finally, whenever $1 < q < \infty$ and $q < p < \infty$ there exists a finite constant $C = C(\Sigma, k, p, q) > 0$ such that the following type of square function estimate

$$\Big\| \sup_{r>0} \Big(\frac{1}{\sigma\big(\Sigma \cap B(z,r)\big)} \int_{B(z,r)\setminus\Sigma} |\nabla \mathcal{T} f|^q \operatorname{dist}(\cdot, \Sigma)^{q-1}\,\mathrm{d}\mathcal{L}^n \Big)^{1/q} \Big\|_{L^p_z(\Sigma,\sigma)} \le C\|f\|_{L^p(\Sigma,\sigma)} \tag{2.4.37}$$

holds for every $f \in L^p(\Sigma, \sigma)$. In particular, corresponding to $q = 2$ and the end-point $p = \infty$, for each function $f \in L^\infty(\Sigma, \sigma)$ one has the following type of Carleson measure estimate:

$$\sup_{z \in \Sigma,\, r>0} \Big(\frac{1}{\sigma\big(\Sigma \cap B(z,r)\big)} \int_{B(z,r)\setminus\Sigma} \big|\nabla \mathcal{T} f\big|^2 \operatorname{dist}(\cdot, \Sigma)\,\mathrm{d}\mathcal{L}^n \Big) \le C\|f\|^2_{L^\infty(\Sigma,\sigma)}. \tag{2.4.38}$$

(8) Define

$$(\nabla \mathcal{T}) f(x) := \int_\Sigma (\nabla k)(x-y) f(y)\,\mathrm{d}\sigma(y) \ \text{ for all } \ x \in \mathbb{R}^n \setminus \Sigma, \quad \text{for each given function } f \in L^1\Big(\Sigma, \frac{\sigma(y)}{1+|y|^n}\Big). \tag{2.4.39}$$

Then for each $q \in (1, \infty)$ there exists some constant $C = C(\Sigma, k, n, q) \in [0, \infty)$ with the property that for each function $f \in L^\infty(\Sigma, \sigma)$ one has the Carleson measure type estimate

$$\sup_{x \in \Sigma,\, r>0} \Big(\frac{1}{\sigma\big(\Sigma \cap B(x,r)\big)} \int_{B(x,r)\setminus\Sigma} |(\nabla \mathcal{T}) f|^q \operatorname{dist}(\cdot, \Sigma)\,\mathrm{d}\mathcal{L}^n \Big)^{1/q} \le C\|f\|_{L^\infty(\Sigma,\sigma)}, \tag{2.4.40}$$

as well as the uniform estimate

$$\sup_{x\in\mathbb{R}^n\setminus\Sigma}\Big\{\big|(\nabla\mathcal{T})f(x)\big|\,\mathrm{dist}(x,\Sigma)\Big\}\leq C\|f\|_{L^\infty(\Sigma,\sigma)}. \tag{2.4.41}$$

(9) Assume the smoothness exponent N of the kernel k is sufficiently large, depending only on the dimension $n\in\mathbb{N}$ of the ambient Euclidean space, and suppose μ is a Borel measure on $\mathbb{R}^n\setminus\Sigma$ with the property that there exists a number $\alpha\geq 1$ such that

$$C_\mu := \sup_{\substack{r\in(0,2\mathrm{diam}(\Sigma))\\ \text{and } x\in\Sigma}}\frac{\mu\big(B(x,r)\setminus\Sigma\big)}{r^{(n-1)\alpha}}<+\infty. \tag{2.4.42}$$

Also, let $E\subseteq\mathbb{R}^n\setminus\Sigma$ be a μ-measurable set which is also assumed to satisfy

$$\mu(E)<+\infty \text{ in the case when } \mathcal{H}^{n-1}(\Sigma)<+\infty. \tag{2.4.43}$$

Then for each $p\in(1,\infty)$ there exists $C=C(\Sigma,n,p,\alpha,C_\mu,\mu(E))>0$ which is actually independent of E in the case when $\mathcal{H}^{n-1}(\Sigma)=+\infty$, with the property that every $f\in L^p(\Sigma,\sigma)$ one has

$$\Big(\int_E|\mathcal{T}f|^{\alpha p}\,\mathrm{d}\mu\Big)^{\frac{1}{\alpha p}}\leq C\big\|k\big|_{S^{n-1}}\big\|_{\mathscr{C}^N(S^{n-1})}\|f\|_{L^p(\Sigma,\sigma)}. \tag{2.4.44}$$

(10) Assume the smoothness exponent N of the kernel k is sufficiently large, depending only on the dimension $n\in\mathbb{N}$ of the ambient Euclidean space. Then whenever

$$p\in(1,\infty),\quad \theta\in[0,\infty),\quad \alpha:=\tfrac{n+\theta}{n-1},\quad 0<q\leq\alpha p, \tag{2.4.45}$$

there exists some constant $C=C(\Sigma,n,p,q,\theta,k)\in(0,\infty)$ such that for each point $x\in\Sigma$ and each radius $r\in\big(0,2\,\mathrm{diam}(\Sigma)\big)$ one has

$$\frac{1}{\sigma\big(B(x,r)\cap\Sigma\big)^{\alpha-(q/p)}}\int_{B(x,r)\setminus\Sigma}|\mathcal{T}f|^q\,\mathrm{dist}(\cdot,\Sigma)^\theta\,\mathrm{d}\mathcal{L}^n\leq C\|f\|^q_{L^p(\Sigma,\sigma)},$$
$$\text{for every function } f \text{ belonging to the space } L^p(\Sigma,\sigma). \tag{2.4.46}$$

Corresponding to the special case when $q:=\big(\frac{1+\theta}{n-1}\big)p$, there exists some constant $C=C(\Sigma,n,p,\theta,k)\in(0,\infty)$ with the property that for every $f\in L^p(\Sigma,\sigma)$ the expression

$$|\mathcal{T}f|^{\left(\frac{1+\theta}{n-1}\right)p}\,\mathrm{dist}(\cdot,\Sigma)^\theta\mathcal{L}^n \text{ is a Carleson measure in the open } \mathbb{R}^n\setminus\Sigma, \text{ with constant bounded by } C\|f\|^{\left(\frac{1+\theta}{n-1}\right)p}_{L^p(\Sigma,\sigma)}. \tag{2.4.47}$$

In particular, if $n\geq 3$, there exists $C=C(\Sigma,n,k)\in(0,\infty)$ such that *given any* $f \in L^{n-1}(\Sigma, \sigma)$, *it follows that* $|\mathcal{T} f|^2 \operatorname{dist}(\cdot, \Sigma)\,\mathcal{L}^n$ *is a Carleson measure in* $\mathbb{R}^n \setminus \Sigma$ *with constant* $\leq C\|f\|^2_{L^{n-1}(\Sigma,\sigma)}$. (2.4.48)

(11) For each $p \in (1, \infty)$ *and each bounded Lipschitz domain* $\Omega \subseteq \mathbb{R}^n$ *satisfying* $\Omega \cap \Sigma = \varnothing$ *there exists a constant* $C \in (0, \infty)$ *with the property that*

$$\|\mathcal{T} f\|_{B^{p,p}_{1/p}(\Omega)} \leq C\|f\|_{L^p(\Sigma,\sigma)} \text{ for every } f \in L^p(\Sigma, \sigma). \tag{2.4.49}$$

Finally, suppose the kernel k is $\mathbb{C}^{M\times M}$*-valued and is a null-solution in* $\mathbb{R}^n \setminus \{0\}$ *of a homogeneous, constant (complex) coefficient,* $M \times M$ *system L, which is weakly elliptic (in the sense that its* $M \times M$ *symbol matrix is invertible; cf.* (1.3.2)). *Also, suppose* $\Omega \subseteq \mathbb{R}^n$ *is a bounded one-sided* NTA *domain satisfying* $\Omega \cap \Sigma = \varnothing$, *and fix*

$$1 < p < \infty, \quad \tfrac{n}{n+1/p} < q_1 \leq \infty, \quad p \leq q_2 \leq \infty. \tag{2.4.50}$$

Then there exists a constant $C = C(\Sigma, \Omega, k, L, p, q_1, q_2) \in [0, \infty)$ *with the property that*

$$\max\Big\{\|\mathcal{T} f\|_{[F^{p,q_1}_{1/p}(\Omega)]^M}, \|\mathcal{T} f\|_{[B^{p,q_2}_{1/p}(\Omega)]^M}\Big\} \leq C\|f\|_{[L^p(\Sigma,\sigma)]^M} \text{ for each function } f \in \big[L^p(\Sigma, \sigma)\big]^M. \tag{2.4.51}$$

In particular, corresponding to $p = q_1 = q_2 = 2$, *one has (recall* (A.0.57))

$$\|\mathcal{T} f\|_{[H^{1/2}(\Omega)]^M} \leq C\|f\|_{[L^2(\Sigma,\sigma)]^M} \text{ for each } f \in \big[L^2(\Sigma, \sigma)\big]^M. \tag{2.4.52}$$

Before presenting the proof of this theorem we make a few comments.

Condition (2.4.3) ensures that the integral defining $(\mathcal{T} f)(x)$ in (2.4.4) is absolutely convergent for each fixed point $x \in \mathbb{R}^n \setminus \Sigma$. Regarding (2.4.9)-(2.4.10), it is reassuring to recall [112, (7.7.106)-(7.7.107)]. We also want to point out that a natural version of Theorem 2.4.1 holds when the kernel k is matrix-valued and the functions/distributions this is paired with are vector-valued.

Two comments pertaining to item *(9)* of Theorem 2.4.1 are as follows. First assume that μ is the Lebesgue measure on $\mathbb{R}^n \setminus \Sigma$. In such a scenario, (2.4.42) holds with $\alpha = n/(n-1)$. Granted this, (2.4.44) then gives that for each $p \in (1, \infty)$ and each $\mathcal{L}^n$-measurable set $E \subseteq \mathbb{R}^n \setminus \Sigma$ which, in the case when Σ is bounded and $\pi_{\Sigma^c,\kappa}(E) = \Sigma$ for each $\kappa > 0$, is also assumed to satisfy $\mathcal{L}^n(E) < +\infty$, the operator

$$\mathcal{T} : L^p(\Sigma, \sigma) \longrightarrow L^{np/(n-1)}(E, \mathcal{L}^n) \text{ is well defined, linear, and bounded.} \tag{2.4.53}$$

Of course, this also may be seen by specializing (2.4.28) to the case $p = q \in (1, \infty)$ (bearing in mind [112, (6.2.27)].

Second, if Σ is unbounded and μ is a Carleson measure in $\mathbb{R}^n \setminus \Sigma$ then for each $p \in (1, \infty)$ the operator

$$\begin{gathered}\mathcal{T} : L^p(\Sigma, \sigma) \longrightarrow L^p(\mathbb{R}^n \setminus \Sigma, \mu)\\ \text{is well defined, linear, and bounded.}\end{gathered} \tag{2.4.54}$$

We next turn to the proof of Theorem 2.4.1.

Proof of Theorem 2.4.1 A few preliminary observations are as follows. First, since Σ is closed and has an empty interior (due to the upper Ahlfors regularity condition), it follows that $\Sigma^c = \mathbb{R}^n \setminus \Sigma$ is an open set whose topological boundary is precisely Σ. Second, hypotheses (2.3.3) imposed on the kernel k imply

$$|k(x)| \leq \Big(\sup_{S^{n-1}} |k|\Big)|x|^{1-n} \text{ for each } x \in \mathbb{R}^n \setminus \{0\}, \tag{2.4.55}$$

and

$$|(\nabla k)(x)| \leq \Big(\sup_{S^{n-1}} |\nabla k|\Big)|x|^{-n} \text{ for each } x \in \mathbb{R}^n \setminus \{0\}. \tag{2.4.56}$$

Third, for each function f as in (2.4.3), the integral in (2.4.4) is meaningfully defined and

$$\mathcal{T} f \in \mathscr{C}^N\big(\mathbb{R}^n \setminus \Sigma\big). \tag{2.4.57}$$

To deal with item *(1)*, pick some function $f \in L^1\big(\Sigma, \frac{\sigma}{1+|\cdot|^{n-1}}\big)$. Also, assume that $z \in \Sigma$ and $x \in \Sigma^c$ are two fixed points such that $x \in \Gamma_\kappa^{\Sigma^c}(z)$, i.e.,

$$|x - z| < (1 + \kappa)\,\mathrm{dist}(x, \Sigma). \tag{2.4.58}$$

Recall the truncated singular integral operator from (2.3.6). We presently consider the threshold

$$\varepsilon := |x - z| > 0 \tag{2.4.59}$$

and estimate the difference

$$\begin{aligned}
\big|(\mathcal{T} f)(x)&-(T_{2\varepsilon} f)(z)\big|\\
&= \Bigg|\int_\Sigma k(x-y) f(y)\,\mathrm{d}\sigma(y) - \int_{\substack{y\in\Sigma\\ |z-y|>2\varepsilon}} k(z-y) f(y)\,\mathrm{d}\sigma(y)\Bigg|\\
&\leq \Bigg|\int_{\substack{y\in\Sigma\\ |z-y|<2\varepsilon}} k(x-y) f(y)\,\mathrm{d}\sigma(y)\Bigg|\\
&\quad + \Bigg|\int_{\substack{y\in\Sigma\\ |z-y|>2\varepsilon}} \Big(k(x-y) - k(z-y)\Big) f(y)\,\mathrm{d}\sigma(y)\Bigg|\\
&=: I + II.
\end{aligned} \tag{2.4.60}$$

Since

$$|x - y| \geq \operatorname{dist}(x, \Sigma) > \frac{|x - z|}{1 + \kappa} = \frac{\varepsilon}{1 + \kappa}, \tag{2.4.61}$$

making use of (2.4.55), (2.4.61), and the Ahlfors regularity of Σ, we obtain

$$\begin{aligned} |I| &\leq \Big(\sup_{S^{n-1}} |k|\Big)\frac{C(\kappa, n)}{\varepsilon^{n-1}} \int\limits_{\substack{y\in\Sigma \\ |z-y|<2\varepsilon}} |f(y)|\, d\sigma(y) \\ &\leq C(\Sigma, \kappa, n)\Big(\sup_{S^{n-1}} |k|\Big)(\mathcal{M}_\Sigma f)(z), \end{aligned} \tag{2.4.62}$$

where $\mathcal{M}_\Sigma$ denotes the Hardy-Littlewood maximal operator on Σ. On the other hand, if $y \in \Sigma$ is such that $|y - z| > 2\varepsilon$, then for each $t \in [0, 1]$ we have

$$\begin{aligned} \varepsilon + \tfrac{1}{2}|z - y| &\leq \tfrac{1}{2}|z - y| + \tfrac{1}{2}|z - y| = |z - y| \\ &\leq |z - (tx + (1 - t)z)| + |tx + (1 - t)z - y| \\ &= t|x - z| + |tx + (1 - t)z - y| \\ &\leq \varepsilon + |tx + (1 - t)z - y|, \end{aligned} \tag{2.4.63}$$

hence

$$|tx + (1 - t)z - y| \geq \tfrac{1}{2}|z - y| \ \text{ for all } \ t \in [0, 1]. \tag{2.4.64}$$

Granted this, the Mean Value Theorem applies and, on account of (2.4.59) and (2.4.56), gives

$$\begin{aligned} |k(x - y) - k(z - y)| &\leq |x - z| \sup_{0\leq t\leq 1} \big|(\nabla k)\big(tx + (1 - t)z - y\big)\big| \\ &\leq 2^n\Big(\sup_{S^{n-1}} |\nabla k|\Big)\frac{\varepsilon}{|z - y|^n}. \end{aligned} \tag{2.4.65}$$

Consequently, a familiar argument gives

$$
\begin{aligned}
|II| &\le 2^n\Big(\sup_{S^{n-1}}|\nabla k|\Big)\int\limits_{\substack{y\in\Sigma\\ |z-y|>2\varepsilon}}\frac{\varepsilon}{|z-y|^n}|f(y)|\,d\sigma(y)\\
&= 2^n\Big(\sup_{S^{n-1}}|\nabla k|\Big)\sum_{j=1}^{\infty}\int\limits_{\substack{y\in\Sigma\\ 2^j\varepsilon<|z-y|\le 2^{j+1}\varepsilon}}\frac{\varepsilon}{|z-y|^n}|f(y)|\,d\sigma(y)\\
&\le 2^n\Big(\sup_{S^{n-1}}|\nabla k|\Big)\sum_{j=1}^{\infty}2^{-j}(2^{j+1}\varepsilon)^{1-n}\int\limits_{\substack{y\in\Sigma\\ |z-y|\le 2^{j+1}\varepsilon}}|f(y)|\,d\sigma(y)\\
&\le C(\Sigma,n)\Big(\sup_{S^{n-1}}|\nabla k|\Big)\sum_{j=1}^{\infty}2^{-j}\big(\mathcal{M}_\Sigma f\big)(z)\\
&= C(\Sigma,n)\Big(\sup_{S^{n-1}}|\nabla k|\Big)\big(\mathcal{M}_\Sigma f\big)(z),
\end{aligned}
\tag{2.4.66}
$$

where the next-to-last step utilizes the upper Ahlfors regularity of Σ.

In summary, the above analysis (consisting of (2.4.60), (2.4.62), and (2.4.66)) proves that

$$
\begin{gathered}
\big|(\mathcal{T}f)(x)-(T_{2\varepsilon}f)(z)\big|\le C(\Sigma,\kappa,n)\big\|k\big|_{S^{n-1}}\big\|_{\mathscr{C}^1(S^{n-1})}\big(\mathcal{M}_\Sigma f\big)(z)\\
\text{for any two points } x\in\Sigma^c \text{ and } z\in\Sigma \text{ as in (2.4.58)-(2.4.59).}
\end{gathered}
\tag{2.4.67}
$$

In view of (2.3.5), this further entails

$$
\begin{gathered}
\big|(\mathcal{T}f)(x)\big|\le\big|(T_{\max}f)(z)\big|+C(\Sigma,\kappa,n)\big\|k\big|_{S^{n-1}}\big\|_{\mathscr{C}^1(S^{n-1})}\big(\mathcal{M}_\Sigma f\big)(z)\\
\text{whenever } z\in\Sigma \text{ and } x\in\Gamma_\kappa^{\Sigma^c}(z).
\end{gathered}
\tag{2.4.68}
$$

For each fixed point $z\in\Sigma$, with the property that $\Gamma_\kappa^{\Sigma^c}(z)\neq\varnothing$, take the supremum in $x\in\Gamma_\kappa^{\Sigma^c}(z)$ in (2.4.68). Bearing (2.4.57) in mind, we arrive at the conclusion that there exists some constant $C=C(\Sigma,\kappa,n)\in(0,\infty)$ such that

$$
\mathcal{N}_\kappa^{\Sigma^c}(\mathcal{T}f)(z)\le(T_{\max}f)(z)+C\big\|k\big|_{S^{n-1}}\big\|_{\mathscr{C}^1(S^{n-1})}\big(\mathcal{M}_\Sigma f\big)(z).
\tag{2.4.69}
$$

Since the above estimate is trivially valid when $\Gamma_\kappa^{\Sigma^c}(z)=\varnothing$, it follows that (2.4.5) holds for all points in Σ.

Next, we shall deal with items *(2)-(3)* simultaneously (in an intertwined fashion). Make the assumption that the smoothness exponent N of the kernel k is large enough. Then (2.4.9)-(2.4.10) in item *(3)* become consequences of (2.4.5), (2.3.11)-(2.3.12), and [112, Corollary 7.6.3].

Returning now to item *(2)*, consider first the claim made in (2.4.7). To set the stage, fix a reference point $x_o\in\Sigma$ and choose some radius $r>0$. For an arbitrary function $f \in L^1(\Sigma, \frac{\sigma}{1+|\cdot|^{n-1}})$, define

$$
\begin{aligned}
g_r &:= \mathcal{N}_\kappa^{\Sigma^c}\Big(\mathcal{T}\big(f\mathbf{1}_{\Sigma\setminus B(x_o,4r)}\big)\Big) \text{ and} \\
h_r &:= \mathcal{N}_\kappa^{\Sigma^c}\Big(\mathcal{T}\big(f\mathbf{1}_{\Sigma\cap B(x_o,4r)}\big)\Big) \text{ on } \Sigma\cap B(x_o,r).
\end{aligned}
\tag{2.4.70}
$$

Fix an arbitrary point $x \in \Sigma \cap B(x_o, r)$. Note that if $\Gamma_\kappa(x) \neq \varnothing$ then [112, (8.1.12)] permits us to estimate

$$
\begin{aligned}
&|y - x_o| \approx |y - x| \leq (2+\kappa)\,\mathrm{dist}\big(y, \Gamma_\kappa(x)\big) \leq (2+\kappa)|y - z|, \\
&\qquad \text{uniformly for } y \in \Sigma\setminus B(x_o,4r) \text{ and } z \in \Gamma_\kappa(x).
\end{aligned}
\tag{2.4.71}
$$

Consequently, there exists some $C = C(k, r, x_o) \in (0, \infty)$ with the property that if $x \in \Sigma \cap B(x_o, r)$ and $z \in \Gamma_\kappa(x)$ then

$$
\begin{aligned}
\mathcal{T}\big(f\mathbf{1}_{\Sigma\setminus B(x_o,4r)}\big)(z) &\leq \int_\Sigma |k(z-y)|\big|\big(f\mathbf{1}_{\Sigma\setminus B(x_o,4r)}\big)(y)\big|\,d\sigma(y) \\
&\leq C\int_{\Sigma\setminus B(x_o,4r)} \frac{|f(y)|}{|z-y|^{n-1}}\,d\sigma(y) \\
&\leq C\int_{\Sigma\setminus B(x_o,4r)} \frac{|f(y)|}{|x_o-y|^{n-1}}\,d\sigma(y) \\
&\leq C\int_\Sigma \frac{|f(y)|}{1+|y|^{n-1}}\,d\sigma(y) < +\infty.
\end{aligned}
\tag{2.4.72}
$$

This proves that there exists $C = C(k, r, x_o) \in (0, \infty)$ such that

$$
\begin{aligned}
&g_r \in L^\infty\big(\Sigma\cap B(x_o,r), \sigma\big) \text{ and} \\
&\|g_r\|_{L^\infty(\Sigma\cap B(x_o,r),\sigma)} \leq C\|f\|_{L^1\big(\Sigma, \frac{\sigma(x)}{1+|x|^{n-1}}\big)}.
\end{aligned}
\tag{2.4.73}
$$

In turn, from this and (2.3.97) we deduce that

$$
\begin{aligned}
&g_r \in L^{1,\infty}\big(\Sigma\cap B(x_o,r), \sigma\big) \text{ and} \\
&\|g_r\|_{L^{1,\infty}(\Sigma\cap B(x_o,r),\sigma)} \leq C\|f\|_{L^1\big(\Sigma, \frac{\sigma(x)}{1+|x|^{n-1}}\big)}.
\end{aligned}
\tag{2.4.74}
$$

Also, since there exists $C = C(r, x_o) \in (0, \infty)$ such that

$$
\begin{aligned}
&f\mathbf{1}_{\Sigma\cap B(x_o,4r)} \in L^1(\Sigma, \sigma) \text{ with} \\
&\big\|f\mathbf{1}_{\Sigma\cap B(x_o,4r)}\big\|_{L^1(\Sigma,\sigma)} \leq C\|f\|_{L^1\big(\Sigma, \frac{\sigma(x)}{1+|x|^{n-1}}\big)},
\end{aligned}
\tag{2.4.75}
$$

we conclude from (2.4.10) (which has already been established at this point) and the second line in [112, (6.2.17)] that there exists $C = C(k, r, x_o) \in (0, \infty)$ with the property that

$$
\begin{aligned}
&h_r \in L^{1,\infty}\big(\Sigma \cap B(x_o,r),\sigma\big) \ \text{and} \\
&\|h_r\|_{L^{1,\infty}(\Sigma\cap B(x_o,r),\sigma)} \leq C\|f\|_{L^1\left(\Sigma,\frac{\sigma(x)}{1+|x|^{n-1}}\right)}.
\end{aligned}
\tag{2.4.76}
$$

Altogether, from (2.4.74), (2.4.76), [112, (8.2.28)], and the fact that, at each point on $\Sigma \cap B(x_o,r)$, we have $0 \leq \mathcal{N}_\kappa^{\Sigma^c}\big(\mathcal{T}f\big) \leq g_r + h_r$, we ultimately conclude that the assignment in (2.4.7) is indeed well defined, sub-linear, bounded, and continuous.

Next, the claims about the mapping (2.4.8) are dealt with in a similar fashion, taking into account that if now $f \in L^p_{\rm loc}(\Sigma,\sigma) \cap L^1\big(\Sigma, \frac{\sigma(x)}{1+|x|^{n-1}}\big)$ then, thanks to (2.4.9), in place of (2.4.76) we now have

$$
\begin{aligned}
&h_r \in L^p\big(\Sigma \cap B(x_o,r),\sigma\big) \ \text{and} \\
&\|h_r\|_{L^p(\Sigma\cap B(x_o,r),\sigma)} \leq C\|f\|_{L^p(\Sigma\cap B(x_o,4r),\sigma)},
\end{aligned}
\tag{2.4.77}
$$

and that (2.4.73) implies

$$
\begin{aligned}
&g_r \in L^p\big(\Sigma \cap B(x_o,r),\sigma\big) \ \text{and} \\
&\|g_r\|_{L^p(\Sigma\cap B(x_o,r),\sigma)} \leq C\|f\|_{L^1\left(\Sigma,\frac{\sigma(x)}{1+|x|^{n-1}}\right)}.
\end{aligned}
\tag{2.4.78}
$$

This finishes the justification of the claims in item *(2)*.

Going back to item *(3)*, consider the Hardy space estimate in (2.4.14). To justify this, fix some $q \in (1,\infty)$ with the goal of estimating $\mathcal{N}_\kappa^{\Sigma^c}(\mathcal{T}a)$ when the function a is an arbitrary (p,q)-atom on $\big(\Sigma, |\cdot - \cdot|, \sigma\big)$, viewed as a space of homogeneous type (cf. [112, Example 7.4.1]). Recall that this means that there exist $x_o \in \Sigma$ along with some finite $r \in \big(0, \operatorname{diam}\Sigma\big]$ with the property that

$$
\begin{aligned}
&\operatorname{supp} a \subseteq B(x_o,r) \cap \Sigma, \qquad \int_\Sigma a\,\mathrm{d}\sigma = 0, \\
&\text{and } \|a\|_{L^q(\Sigma,\sigma)} \leq \sigma\big(B(x_o,r)\cap\Sigma\big)^{1/q-1/p}.
\end{aligned}
\tag{2.4.79}
$$

When $\sigma(\Sigma) < +\infty$, the constant function $a \equiv \sigma(\Sigma)^{-1/p}$ is also considered to be a (p,q)-atom.

First, estimate (2.4.9) used with $p := q$ and $f := a$ as in (2.4.79) gives that, on the one hand,

$$
\begin{aligned}
&\int_{B(x_o,2(2+\kappa)r)\cap\Sigma} |\mathcal{N}_\kappa^{\Sigma^c}(\mathcal{T}a)|^p\,\mathrm{d}\sigma \\
&\quad \leq \|\mathcal{N}_\kappa^{\Sigma^c}(\mathcal{T}a)\|^p_{L^q(\Sigma,\sigma)} \cdot \sigma\big(B(x_o,2(2+\kappa)r)\cap\Sigma\big)^{1-p/q} \\
&\quad \leq C\big\|k\big|_{S^{n-1}}\big\|^p_{\mathscr{C}^N(S^{n-1})}\|a\|^p_{L^q(\Sigma,\sigma)} \cdot \sigma\big(B(x_o,2(2+\kappa)r)\cap\Sigma\big)^{1-p/q} \\
&\quad \leq C(\Sigma,\kappa,p,q)\big\|k\big|_{S^{n-1}}\big\|^p_{\mathscr{C}^N(S^{n-1})},
\end{aligned}
\tag{2.4.80}
$$

where we have also used Hölder's inequality, (2.4.79), and the Ahlfors regularity of Σ. On the other hand, whenever $z \in \Sigma^c$ is such that $|z - x_o| \geq 2r$, and whenever a is as in (2.4.79), based on the vanishing moment condition for the atom, the Mean Value Theorem, and Hölder's inequality we may estimate

$$
\begin{aligned}
|(\mathcal{T}a)(z)| &\leq \int_{B(x_o,r)\cap\Sigma} |k(z-y) - k(z-x_o)||a(y)|\,d\sigma(y) \\
&\leq C\big\|k\big|_{S^{n-1}}\big\|_{\mathscr{C}^1(S^{n-1})}\frac{r}{|z-x_o|^n}\int_{B(x_o,r)\cap\Sigma}|a(y)|\,d\sigma(y) \\
&\leq C\big\|k\big|_{S^{n-1}}\big\|_{\mathscr{C}^1(S^{n-1})}\frac{r}{|z-x_o|^n}\sigma\big(B(x_o,r)\cap\Sigma\big)^{1-1/p}. \qquad (2.4.81)
\end{aligned}
$$

Now, if $x \in \Sigma$ and $z \in \Sigma^c$ are such that

$$
|x - x_o| \geq 2(2+\kappa)r \text{ and } |z - x| < (1+\kappa)\,\text{dist}(z,\Sigma) \qquad (2.4.82)
$$

then

$$
2(2+\kappa)r \leq |x - x_o| \leq |x - z| + |z - x_o| \leq (2+\kappa)|z - x_o|. \qquad (2.4.83)
$$

In particular,

$$
|z - x_o| \geq 2r \text{ and } |z - x_o| \geq \frac{1}{2+\kappa}|x - x_o|. \qquad (2.4.84)
$$

In light of (2.4.84), after taking the supremum in (2.4.81) over $z \in \Sigma^c$ satisfying $|z - x| < (1+\kappa)\,\text{dist}(z,\Sigma)$, we obtain

$$
\begin{aligned}
\big|\mathcal{N}_\kappa^{\Sigma^c}(\mathcal{T}a)(x)\big| \leq C\big\|k\big|_{S^{n-1}}\big\|_{\mathscr{C}^1(S^{n-1})}\frac{r}{|x-x_o|^n}\sigma\big(B(x_o,r)\cap\Sigma\big)^{1-1/p}, \\
\text{for every point } x \in \Sigma \setminus B\big(x_o, 2(2+\kappa)r\big). \qquad (2.4.85)
\end{aligned}
$$

Keeping in mind that $pn > n - 1$, on account of (2.4.85) and [112, Lemma 7.2.1] we may then estimate

$$
\int_{\Sigma\setminus B(x_o,2(2+\kappa)r)} \big|\mathcal{N}_\kappa^{\Sigma^c}(\mathcal{T}a)(x)\big|^p\,d\sigma(x) \leq C(\Sigma,\kappa,p)\big\|k\big|_{S^{n-1}}\big\|^p_{\mathscr{C}^1(S^{n-1})}, \qquad (2.4.86)
$$

for some finite constant. From (2.4.80) and (2.4.86) it follows that

$$
\big\|\mathcal{N}_\kappa^{\Sigma^c}(\mathcal{T}a)\big\|_{L^p(\Sigma,\sigma)} \leq C\big\|k\big|_{S^{n-1}}\big\|_{\mathscr{C}^N(S^{n-1})}, \qquad (2.4.87)
$$

for some finite constant $C = C(\Sigma,\kappa,n,p,q) > 0$, independent of the atom a as in (2.4.79). When $\sigma(\Sigma) < +\infty$, it follows from (2.4.9) (with $p = 2$) and Hölder's inequality that the same estimate holds for the constant function $a \equiv \sigma(\Sigma)^{-1/p}$. Having established this, we may now invoke the general boundedness criterion from [113, Theorem 4.4.7] (see also [7, Proposition 8.32, pp. 421-422] for a general result of this flavor), used here with

$$q := 2,\ X := N_\kappa^2(\Sigma^c;\sigma),\ \text{and}\ Y := N_\kappa^p(\Sigma^c;\sigma), \tag{2.4.88}$$

(bearing in mind [112, Proposition 8.3.5]), in order to conclude that (2.4.14) holds.

Next, formula (2.4.15) is implied by [113, (4.4.144)] (keeping in mind [112, (8.3.33)]). Formula (2.4.16) follows from (2.4.15) and the duality result from [113, Theorem 4.6.1], upon noting that since for each fixed point $x \in \mathbb{R}^n \setminus \Sigma$ the function $\Sigma \ni y \mapsto k(x-y) \in \mathbb{C}$ is both bounded and Lipschitz (cf. (2.4.55)-(2.4.56)), we infer from [112, (7.3.25)] that

$$k(x-\cdot)\big|_\Sigma \in \bigcap_{0<\alpha<1} \mathscr{C}^\alpha(\Sigma), \qquad \forall x \in \mathbb{R}^n \setminus \Sigma. \tag{2.4.89}$$

That formula (2.4.15) remains valid if the a_j's happen to be H^p-molecules on Σ is seen from (2.4.16), (2.4.89), and [113, (4.8.68), (4.8.70)]. Also, that the operator $\mathcal{T}$ in (2.4.16) acts in a coherent fashion with $\mathcal{T}$ in (2.4.4) considered on functions from $L^1\big(\Sigma, \frac{\sigma(y)}{1+|y|^{n-1}}\big)$ is implied when $p \in \big(\frac{n-1}{n}, 1\big)$ by the compatibility result established in [113, Proposition 4.8.8] (bearing in mind [113, (4.8.57)] and (2.4.55)-(2.4.56)), and when $p = 1$ by the compatibility result proved in [113, Proposition 4.8.6].

To deal with the claim made in (2.4.17), assume $f \in H^p(\Sigma,\sigma)$ is a compactly supported distribution, and pick some function $\psi \in \mathrm{Lip}_c(\Sigma)$ which is identically one near $\operatorname{supp} f$. Also, choose a function $\theta \in \mathscr{C}_c^\infty(\mathbb{R}^n)$ satisfying $\theta \equiv 1$ near the origin in $\mathbb{R}^n$, and for each $R > 0$ define $\theta_R(x) := \theta(x/R)$ for every $x \in \mathbb{R}^n$. Finally, fix a point $x \in \mathbb{R}^n \setminus \Sigma$. We may then rely on [112, (7.3.17)] (with $\alpha := 1$) to conclude that

$$\lim_{R\to\infty} \theta_R k(x-\cdot)\big|_\Sigma = k(x-\cdot)\big|_\Sigma \ \text{ in } \mathscr{C}^\alpha(\Sigma) \ \text{ for each } \alpha \in (0,1). \tag{2.4.90}$$

In concert with (2.4.16) and [113, Lemma 4.6.4], this permits us to write

$$\begin{aligned}(\mathcal{T} f)(x) &= \lim_{R\to\infty} {}_{\mathrm{Lip}_c(\Sigma)}\Big\langle \theta_R k(x-\cdot)\big|_\Sigma, f\Big\rangle_{(\mathrm{Lip}_c(\Sigma))'}\\ &= \lim_{R\to\infty} {}_{\mathrm{Lip}_c(\Sigma)}\Big\langle \theta_R \psi k(x-\cdot)\big|_\Sigma, f\Big\rangle_{(\mathrm{Lip}_c(\Sigma))'}\\ &= {}_{\mathrm{Lip}_c(\Sigma)}\Big\langle \psi k(x-\cdot)\big|_\Sigma, f\Big\rangle_{(\mathrm{Lip}_c(\Sigma))'}\end{aligned} \tag{2.4.91}$$

which proves (2.4.17). At this point, the justification of all claims made in item *(3)* of the statement of the theorem is complete.

As regards item *(4)*, that $\mathcal{T} f$ in (2.4.4) is well defined to being with, for each $f \in L^p(\Sigma, w\sigma)$, is seen from [112, (7.7.104)], while the claim in (2.4.18) is imply by (2.3.58)-(2.3.57), (2.4.5), and item *(1)* in [112, Lemma 7.7.1]. Then (2.4.20) becomes a consequence of (2.4.18) and [113, Corollary 5.2.3]. Moreover, the claim made in (2.4.22) is seen from (2.4.5), (2.3.68), and [113, (5.2.86), (5.2.102)].

Let us now turn to item *(5)*. That, to being with, $\mathcal{T} f$ in (2.4.4) is well defined for each given function $f \in L^{p,q}(\Sigma,\sigma)$ is a consequence of [112, (7.7.107)]. The estimate claimed in (2.4.23) follows from [113, Proposition 1.3.7], [112, (6.2.48)], and (2.4.9). Since for each fixed point $x \in \Sigma^c = \mathbb{R}^n \setminus \Sigma$ we have (cf. [112, (7.2.5),

(7.3.17)])

$$k(x-\cdot)\big|_{\Sigma} \in \operatorname{Lip}(\Sigma) \cap \Big(\bigcap_{1<p\leq\infty} L^p(\Sigma,\sigma)\Big) \cap \Big(\bigcap_{0<\alpha<1} \mathscr{C}^{\alpha}(\Sigma)\Big), \tag{2.4.92}$$

From [113, Proposition 4.8.9] we see that the operator $\mathcal{T}$, initially defined as in (2.4.3)-(2.4.4) may be further extended, in a coherent fashion, to a well-defined mapping acting on each $f \in H^{p,q}(\Sigma,\sigma)$ with $p \in \big(\frac{n-1}{n},\infty\big)$ and $q \in (0,\infty]$ as in (2.4.24). To justify (2.4.25), consider a compactly supported distribution f in $H^{p,q}(\Sigma,\sigma)$ along with $\psi \in \operatorname{Lip}_c(\Sigma)$ satisfying $\psi \equiv 1$ near $\operatorname{supp} f$. As before, pick some $\theta \in \mathscr{C}^{\infty}_c(\mathbb{R}^n)$ with $\theta \equiv 1$ near the origin in $\mathbb{R}^n$, and for each $R > 0$ set $\theta_R(x) := \theta(x/R)$ for every $x \in \mathbb{R}^n$. Let us also choose $p_0, p_1 \in \big(\frac{n-1}{n},\infty\big)$ such that $p_0 < p < p_1$ and decompose $f = f^{(0)} + f^{(1)}$ where $f^{(i)} \in H^{p_i}(\Sigma,\sigma)$ for $i \in \{0,1\}$. Fix an arbitrary point $x \in \mathbb{R}^n \setminus \Sigma$. Combining (2.4.90) with [113, Lemma 4.6.4] and results from [115, §2.2] allows us to write

$$(\mathcal{T} f^{(i)})(x) = \lim_{R\to\infty} {}_{\operatorname{Lip}_c(\Sigma)}\Big\langle \theta_R k(x-\cdot)\big|_{\Sigma}, f^{(i)}\Big\rangle_{(\operatorname{Lip}_c(\Sigma))'} \text{ for } i \in \{0,1\}. \tag{2.4.93}$$

Summing up the two formulas in (2.4.93) then yields

$$\begin{aligned}(\mathcal{T} f)(x) &= \lim_{R\to\infty} {}_{\operatorname{Lip}_c(\Sigma)}\Big\langle \theta_R k(x-\cdot)\big|_{\Sigma}, f\Big\rangle_{(\operatorname{Lip}_c(\Sigma))'} \\ &= \lim_{R\to\infty} {}_{\operatorname{Lip}_c(\Sigma)}\Big\langle \theta_R \psi k(x-\cdot)\big|_{\Sigma}, f\Big\rangle_{(\operatorname{Lip}_c(\Sigma))'} \\ &= {}_{\operatorname{Lip}_c(\Sigma)}\Big\langle \psi k(x-\cdot)\big|_{\Sigma}, f\Big\rangle_{(\operatorname{Lip}_c(\Sigma))'} \end{aligned} \tag{2.4.94}$$

which proves (2.4.25). Finally, the estimate in (2.4.27) is a consequence of (2.4.9), (2.4.14), [113, (4.3.3)], and the real interpolation result from [113, Proposition 1.3.7].

Consider next the claim made in item *(6)*. According to [61, Theorem 1.3(1), pp. 9-10] the L^2-Square Function Estimate stated in [61, Theorem 1.1(1), p. 6] holds for the choices

$$\begin{aligned} &\mathscr{X} := \mathbb{R}^n,\ E := \Sigma,\ m := n,\ d := n-1,\ \upsilon := 1,\ \alpha := 1, \\ &\text{and } \theta(x,y) := (\nabla k)(x-y) \text{ (which entails } \Theta = \nabla\mathcal{T}). \end{aligned} \tag{2.4.95}$$

Granted this, [61, Theorem 1.1(13), pp. 6-8] then applies and readily gives the estimate recorded in (2.4.29) (after unraveling notation, and keeping (2.4.95) in mind). In turn, once (2.4.29) has been established, the weak estimate in (2.4.30) follows by invoking [61, Theorem 6.17, p. 94].

To deal with the first claim in item *(7)*, fix some $f \in L^p(\Sigma,\sigma)$ with $p \in (1,\infty)$. Then, on the one hand, (2.4.29) used with $q := p$ reads

$$\left(\int_{\Sigma}\Big(\int_{\Gamma^{\Sigma^c}_{\kappa}(x)} |\nabla\mathcal{T} f|^p \operatorname{dist}(\cdot,\Sigma)^{p-n}\, d\mathcal{L}^n\Big)\, d\sigma(x)\right)^{\frac{1}{p}} \leq C\|f\|_{L^p(\Sigma,\sigma)}. \tag{2.4.96}$$

On the other hand, using Fubini's Theorem as in [112, (8.1.43)], we may recast the left-hand side above as

$$\Big(\int_{\Sigma}\Big(\int_{\Gamma_{\kappa}^{\Sigma^c}(x)}|(\nabla\mathcal{T}f)(z)|^p\operatorname{dist}(z,\Sigma)^{p-n}\,dz\Big)\,d\sigma(x)\Big)^{1/p}$$
$$=\Big(\int_{\mathbb{R}^n\setminus\Sigma}|(\nabla\mathcal{T}f)(z)|^p\operatorname{dist}(z,\Sigma)^{p-n}\sigma\big(\pi_\kappa(\{z\})\big)\,dz\Big)^{1/p} \tag{2.4.97}$$

where for each point $z\in\mathbb{R}^n\setminus\Sigma$, the "(reverse) conical projection" $\pi_\kappa(\{z\})$ of z onto Σ is defined as

$$\pi_\kappa(\{z\}):=\big\{x\in\Sigma:\ z\in\Gamma_{\kappa}^{\Sigma^c}(x)\big\}. \tag{2.4.98}$$

In view of the Ahlfors regularity of Σ, from [112, (8.1.19)] it follows that

$$\sigma\big(\pi_\kappa(\{z\})\big)\approx\operatorname{dist}(z,\Sigma)^{n-1},\ \text{ uniformly in } z\in\mathbb{R}^n\setminus\Sigma. \tag{2.4.99}$$

Combining (2.4.96), (2.4.97), and (2.4.99) then yields (2.4.34). In turn, (2.4.35) follows by specializing (2.4.34) to the case when $p=2$. Next, the estimate claimed in (2.4.36) is a consequence of [112, Proposition 8.7.13] used twice (with $\Omega:=\mathbb{R}^n\setminus\Sigma$).

Finally, consider (2.4.37). To this end, pick $q\in(1,\infty)$ and $p\in(q,\infty)$, and select an arbitrary function $f\in L^p(\Sigma,\sigma)$. Also, let $z\in\Sigma$ and $r>0$ be arbitrary. The goal is to establish the pointwise estimate

$$r^{1-n}\int_{B(z,r)\setminus\Sigma}\big|\nabla\mathcal{T}f(x)\big|^q\operatorname{dist}(x,\Sigma)^{q-1}\,dx\leq C\big(\mathcal{M}_\Sigma|f|^q\big)(z), \tag{2.4.100}$$

where $\mathcal{M}_\Sigma$ stands for the Hardy-Littlewood maximal operator on Σ, i.e.,

$$\big(\mathcal{M}_\Sigma\,g\big)(x):=\sup_{\rho>0}⨍_{\Delta(x,\rho)}|g|\,d\sigma,\qquad\forall\,x\in\Sigma. \tag{2.4.101}$$

Here and elsewhere, $\Delta(x,\rho):=B(x,\rho)\cap\Sigma$ is the surface ball centered at $x\in\Sigma$ and having radius $\rho>0$. Once (2.4.100) has been justified, (2.4.37) is readily obtained from this and the fact that $\mathcal{M}_\Sigma$ is bounded on $L^{p/q}(\Sigma,\sigma)$ since $p/q\in(1,\infty)$ (cf. [112, (7.6.18)]).

As regards (2.4.100), decompose

$$f=\mathbf{1}_{\Delta_{2r}}f+\mathbf{1}_{\Sigma\setminus\Delta_{2r}}f \tag{2.4.102}$$

where, for each $R>0$, we let Δ_R abbreviate $\Delta(z,R)$. Making use of the square function estimate (2.4.34) (written for q in place of p) and the doubling property of the surface measure σ, we may then estimate

$$
\begin{aligned}
r^{1-n}\int_{B(z,r)\setminus\Sigma} &\left|\nabla\mathcal{T}\big(\mathbf{1}_{\Delta_{2r}}f\big)(x)\right|^q \operatorname{dist}(x,\Sigma)^{q-1}\,\mathrm{d}x \\
&\leq r^{1-n}\int_{\mathbb{R}^n\setminus\Sigma}\left|\nabla\mathcal{T}\big(\mathbf{1}_{\Delta_{2r}}f\big)(x)\right|^q \operatorname{dist}(x,\Sigma)^{q-1}\,\mathrm{d}x \\
&\leq Cr^{1-n}\int_\Sigma |\mathbf{1}_{\Delta_{2r}}f|^q\,\mathrm{d}\sigma \leq C\fint_{\Delta(z,2r)}|f|^q\,\mathrm{d}\sigma \\
&\leq C\big(\mathcal{M}_\Sigma|f|^q\big)(z),
\end{aligned}
\tag{2.4.103}
$$

which suits our purposes. Next, observe that if $x \in B(z,r)\setminus\Sigma$ is fixed then

$$
|x-y| \approx |z-y| \text{ uniformly for } y \in \Sigma\setminus\Delta_{2r}. \tag{2.4.104}
$$

Also, by homogeneity,

$$
|(\nabla k)(x-y)| \leq \frac{C}{|x-y|^n}. \tag{2.4.105}
$$

In turn, (2.4.104), (2.4.105), the Ahlfors regularity of $\partial\Omega$, Hölder's inequality, and (2.4.101) can be used to justify the following sequence of inequalities:

$$
\begin{aligned}
\left|\nabla\mathcal{T}\big(\mathbf{1}_{\Sigma\setminus\Delta_{2r}}f\big)(x)\right| &\leq C\int_{\Sigma\setminus\Delta_{2r}}\frac{1}{|z-y|^n}|f(y)|\,\mathrm{d}\sigma(y) \\
&\leq C\sum_{j=1}^\infty\frac{1}{(2^jr)^n}\int_{\Delta_{2^{j+1}r}\setminus\Delta_{2^jr}}|f(y)|\,\mathrm{d}\sigma(y) \\
&\leq C\sum_{j=1}^\infty\frac{1}{2^jr}\fint_{\Delta_{2^{j+1}r}}|f|\,\mathrm{d}\sigma \leq C\sum_{j=1}^\infty\frac{1}{2^jr}\Big(\fint_{\Delta_{2^{j+1}r}}|f|^q\,\mathrm{d}\sigma\Big)^{1/q} \\
&\leq Cr^{-1}\Big[\big(\mathcal{M}_\Sigma|f|^q\big)(z)\Big]^{1/q}.
\end{aligned}
\tag{2.4.106}
$$

Hence, after raising the most extreme sides to the power q, we obtain the pointwise inequality

$$
\left|\nabla\mathcal{T}\big(\mathbf{1}_{\Sigma\setminus\Delta_{2r}}f\big)(x)\right|^q \leq Cr^{-q}\big(\mathcal{M}_\Sigma|f|^q\big)(z) \tag{2.4.107}
$$

for each $x \in B(z,r)\setminus\Sigma$. In turn, this implies (keeping in mind that $\operatorname{dist}(x,\Sigma) \leq r$ on the domain of integration) that

$$
r^{1-n}\int_{B(z,r)\setminus\Sigma}\left|\nabla\mathcal{T}\big(\mathbf{1}_{\Sigma\setminus\Delta_{2r}}f\big)(x)\right|^q \operatorname{dist}(x,\Sigma)^{q-1}\,\mathrm{d}x \leq C\big(\mathcal{M}_\Sigma|f|^q\big)(z), \tag{2.4.108}
$$

which is of the right order. Now, (2.4.37) becomes an immediate consequence of (2.4.103) and (2.4.108).

Moving on, at least when $q = 2$, the Carleson measure estimate (2.4.40) in item *(8)* is contained in [61, Theorem 1.3(3), pp. 9-10]. We may directly justify (2.4.40) for any $q \in (1, \infty)$ by re-running the proof of (2.4.37) with $p := \infty$. The uniform estimate (2.4.41) is implied by (the very last claim in) [112, Proposition 8.7.13] used with $p := \infty$, $\alpha := 1$, and $\Omega := \mathbb{R}^n \setminus \Sigma$.

Turning to the claims in part *(9)*, assume the smoothness exponent N of the kernel k is large enough, and pick a set E as specified in the statement of the theorem. Also, select an integrability exponent $p \in (1, \infty)$ and fix some arbitrary aperture parameter $\kappa > 0$. In the current setting, condition (2.4.42) amounts to saying that (μ, σ) is an $(\alpha, 0)$-Carleson pair for the ambient (Σ^c, Σ) (cf. [112, Definition 8.6.1])). Having observed this, we may invoke [112, Theorem 8.6.2, (8.6.13)] and (2.4.9) in order to estimate

$$\Big(\int_E |\mathcal{T} f|^{\alpha p}\,\mathrm{d}\mu\Big)^{\frac{1}{\alpha p}} \leq \big\|\mathcal{N}_\kappa^{\Sigma^c}(\mathcal{T} f)\big\|_{L^p(\Sigma,\sigma)}$$
$$\leq C\big\|k\big|_{S^{n-1}}\big\|_{\mathscr{C}^N(S^{n-1})}\|f\|_{L^p(\Sigma,\sigma)}, \tag{2.4.109}$$

for each $f \in L^p(\Sigma, \sigma)$, where the constant $C = C(\Sigma, n, p, \alpha, C_\mu, \kappa)$ is a finite positive number. This establishes (2.4.44).

Consider next the claims made in item *(10)*. First we observe that whenever $r \in \big(0, \operatorname{diam}(\Sigma)\big)$ the estimate in (2.4.46) holds under the assumptions made on p, θ, α, q in (2.4.45). Indeed, this is a consequence of (2.4.44) (applied for the set $E := B(x, r) \setminus \Sigma$ and the measure $\mu := \operatorname{dist}(\cdot, \Sigma)^\theta \mathcal{L}^n$ in $\mathbb{R}^n \setminus \Sigma$ which satisfies (2.4.42) with $\alpha := \frac{n+\theta}{n-1} \geq 1$, and which also satisfies $\mu(E) < +\infty$ in the case when Σ is compact, thus rendering (2.4.43) valid), followed by an application of Hölder's inequality (used to increase the integrability exponent from q to αp). If $\operatorname{diam}(\Sigma) = +\infty$ this proves (2.4.46) as stated. In turn, (2.4.47) is obtained by specializing (2.4.46), and (2.4.48) is a particular case of (2.4.47).

As far as item *(11)* is concerned, the first claim is a direct consequence of [113, Lemma 9.2.40], the L^p-square function estimate (2.4.34), and the nontangential maximal function estimate (2.4.9) together with [112, (8.6.51) in Proposition 8.6.3].

To treat the second claim in item *(11)*, we shall work under the additional assumptions on the kernel specified there. In particular, the fact that k is a null-solution of a weakly elliptic $M \times M$ system L implies that $k \in \big[\mathscr{C}^\infty(\mathbb{R}^n \setminus \{0\})\big]^M$ (cf. [112, Theorem 6.5.7]). Next, fix an arbitrary function $f \in \big[L^p(\Sigma, \sigma)\big]^M$ with $p \in (1, \infty)$ and consider

$$u := \mathcal{T} f\big|_\Omega \in \big[\mathscr{C}^\infty(\Omega)\big]^M. \tag{2.4.110}$$

Since $Lu = 0$ in Ω, [112, Theorem 6.5.7] implies that u is subaveraging in Ω. To proceed, observe that

$$\operatorname{dist}(x, \partial\Omega) \leq \operatorname{dist}(x, \Sigma) \text{ for each } x \in \Omega. \tag{2.4.111}$$

Indeed, given $x \in \Omega$ arbitrary, if $r := \operatorname{dist}(x, \partial\Omega)$ then a simple argument (using the fact that Ω is open and $B(x, r)$ is connected) implies that $B(x, r) \subseteq \Omega$. Hence, $B(x, r) \subseteq \mathbb{R}^n \setminus \Sigma$ which goes to show that $\operatorname{dist}(x, \Sigma) \geq r = \operatorname{dist}(x, \partial\Omega)$. Having justified (2.4.111), we may then estimate

$$\Big(\int_{\Omega}|\nabla u|^p\,\mathrm{dist}(\cdot,\partial\Omega)^{p-1}\,\mathrm{d}\mathcal{L}^n\Big)^{1/p}\leq\Big(\int_{\mathbb{R}^n\setminus\Sigma}|\nabla(\mathcal{T}f)|^p\,\mathrm{dist}(\cdot,\Sigma)^{p-1}\,\mathrm{d}\mathcal{L}^n\Big)^{1/p}$$
$$\leq C\|f\|_{[L^p(\Sigma,\sigma)]^M}, \tag{2.4.112}$$

where the last step comes from (2.4.35). Also, [112, (8.6.51) in Proposition 8.6.3] presently implies

$$\|u\|_{\big[L^{\frac{pn}{n-1}}(\Omega,\mathcal{L}^n)\big]^M}\leq C\big\|\mathcal{N}_\kappa^\Omega u\big\|_{L^p(\Sigma,\sigma)} \tag{2.4.113}$$

for some constant $C\in[0,\infty)$ independent of u. In concert with Hölder's inequality, (2.4.110), and (2.4.9), this implies

$$\|u\|_{[L^p(\Omega,\mathcal{L}^n)]^M}\leq C\|u\|_{\big[L^{\frac{pn}{n-1}}(\Omega,\mathcal{L}^n)\big]^M}\leq C\big\|\mathcal{N}_\kappa^\Omega u\big\|_{L^p(\Sigma,\sigma)}$$
$$\leq C\big\|\mathcal{N}_\kappa^{\Sigma^c}(\mathcal{T}f)\big\|_{L^p(\Sigma,\sigma)}\leq C\|f\|_{[L^p(\Sigma,\sigma)]^M}. \tag{2.4.114}$$

Hence, if $n/(n+1/p)<q_1\leq\infty$, $p\leq q_2\leq\infty$, and $k:=0$, $\alpha:=1/p$, we may invoke [113, Theorem 9.2.30] (with $q\in\{q_1,q_2\}$) which, in concert with (2.4.112) and (2.4.114), yields (2.4.51). Finally, (2.4.52) is a consequence of (2.4.51) and [113, (9.2.22)]. The proof of Theorem 2.4.1 is therefore complete. □

In the next corollary we establish a Carleson measure estimate which brings into play the role of the Divergence Theorems from [112, §1.2].

Corollary 2.4.2 *Assume $\Omega\subseteq\mathbb{R}^n$ (where $n\in\mathbb{N}$ with $n\geq 2$) is an open set with the property that $\partial\Omega$ is a* UR *set; in particular, Ω is a set of locally finite perimeter. Denote by $\nu=(\nu_1,\dots,\nu_n)$ the geometric measure theoretic outward unit normal to Ω and abbreviate $\sigma:=\mathcal{H}^{n-1}\lfloor\partial\Omega$. Assume $k\in\mathscr{C}^2(\mathbb{R}^n\setminus\{0\})$ is an odd function which is positive homogeneous of degree $1-n$ and, for each $i,j\in\{1,\dots,n\}$, consider the integral operator acting on functions $f\in$* BMO$(\partial\Omega,\sigma)$ *according to*

$$Q_{ij}f(x):=\int_{\partial_*\Omega}\Big\{\nu_i(y)\partial_{y_j}\big[k(x-y)\big]-\nu_j(y)\partial_{y_i}\big[k(x-y)\big]\Big\}f(y)\,\mathrm{d}\sigma(y), \tag{2.4.115}$$

for each $x\in\Omega$.

Then for each $f\in$ BMO$(\partial\Omega,\sigma)$ *the function $Q_{ij}f$ is well defined in Ω and, given any exponent $p\in(1,\infty)$, there exists some $C=C(\Omega,k,p)\in(0,\infty)$, independent of f, with the property that*

$$\sum_{1\leq i,j\leq n}\big|Q_{ij}f\big|^p\,\mathrm{dist}(\cdot,\partial\Omega)^{p-1}\,\mathrm{d}\mathcal{L}^n \tag{2.4.116}$$

is a Carleson measure in Ω, in the quantitative sense that

$$\sup_{x\in\partial\Omega,\, r>0} \frac{1}{\sigma\big(B(x,r)\cap\partial\Omega\big)} \int_{B(x,r)\cap\Omega} \sum_{i,j=1}^{n} |Q_{ij}f|^p \operatorname{dist}(\cdot,\partial\Omega)^{p-1}\, d\mathcal{L}^n$$

$$\leq C\|f\|^p_{\dot{\mathrm{BMO}}(\partial\Omega,\sigma)}, \tag{2.4.117}$$

with the piece of notation introduced in (A.0.14). *In particular, corresponding to* $p = 2$ *it follows that, in a quantitative fashion,*

$$\sum_{1\leq i,j\leq n} \big|Q_{ij}f\big|^2 \operatorname{dist}(\cdot,\partial\Omega)\, d\mathcal{L}^n \text{ is a Carleson measure in } \Omega, \quad \text{for each function } f \in \mathrm{BMO}(\partial\Omega,\sigma). \tag{2.4.118}$$

Proof Pick $i, j \in \{1,\dots,n\}$ and fix an arbitrary function $f \in \mathrm{BMO}(\partial\Omega,\sigma)$. Then for each fixed $x \in \Omega$ we may estimate

$$\int_{\partial_*\Omega} \Big|\nu_i(y)\partial_{y_j}\big[k(x-y)\big] - \nu_j(y)\partial_{y_i}\big[k(x-y)\big]\Big| |f(y)|\, d\sigma(y)$$

$$\leq 2 \sup_{S^{n-1}} |\nabla k| \cdot \int_{\partial\Omega} \frac{|f(y)|}{|x-y|^n}\, d\sigma(y) < +\infty, \tag{2.4.119}$$

thanks to the homogeneity of the kernel k, the fact that $\partial\Omega$ is an Ahlfors regular set, and [112, (7.4.116)]. This goes to show that the function $Q_{ij}f$ is well defined in Ω, via an absolutely convergent integral.

Next, we claim that

$$Q_{ij}1 \equiv 0 \text{ in } \Omega. \tag{2.4.120}$$

To prove this, fix an arbitrary point $x \in \Omega$ and note that $k(x-\cdot) \in L^1_{\mathrm{loc}}(\Omega,\mathcal{L}^n)$. As such, it is meaningful to consider the vector field

$$\vec{F} := \partial_j\big[k(x-\cdot)\big]\mathbf{e}_i - \partial_i\big[k(x-\cdot)\big]\mathbf{e}_j \tag{2.4.121}$$

where the partial derivatives are taken in the sense of distributions in Ω (in the "dot" variable). This places $\vec{F}$ in $\big[\mathcal{D}'(\Omega)\big]^n$. Since its singular support is the singleton $\{x\}$, we see that actually

$$\vec{F} \in \big[L^1_{\mathrm{loc}}(\Omega,\mathcal{L}^n) + \mathscr{E}'(\Omega)\big]^n. \tag{2.4.122}$$

Clearly,

$$\operatorname{div}\vec{F} = \partial_i\partial_j\big[k(x-\cdot)\big] - \partial_j\partial_i\big[k(x-\cdot)\big] = 0 \text{ in } \mathcal{D}'(\Omega). \tag{2.4.123}$$

Moreover, with $d := \operatorname{dist}(x,\partial\Omega) > 0$ and $\kappa > 0$ arbitrary, [112, Lemma 8.3.7] ensures that there exists a constant $C \in (0,\infty)$ such that

$$\Big(\mathcal{N}_\kappa^{\Omega\setminus\overline{B(x,d/2)}}\vec{F}\Big)(y) \leq C|y-x|^{-n} \text{ for each } y \in \partial\Omega. \tag{2.4.124}$$

From this, [112, (8.2.26)], and [112, Lemma 7.2.1], we may therefore conclude that

$$\mathcal{N}_\kappa^{\Omega\setminus\overline{B(x,d/2)}}\vec{F} \in L^1(\partial\Omega, \sigma). \tag{2.4.125}$$

Furthermore, since $\vec{F}$ is continuous in a neighborhood of $\partial\Omega$ it follows that the pointwise nontangential boundary trace $\vec{F}\big|_{\partial\Omega}^{\kappa-\mathrm{n.t.}}$ exists at every point in $\partial_{\mathrm{nta}}\Omega$ and, in fact,

$$\Big(\vec{F}\big|_{\partial\Omega}^{\kappa-\mathrm{n.t.}}\Big)(y) = \partial_{y_j}\big[k(x-y)\big]\mathbf{e}_i - \partial_{y_i}\big[k(x-y)\big]\mathbf{e}_j \text{ at every } y \in \partial_{\mathrm{nta}}\Omega. \tag{2.4.126}$$

In concert with [112, Proposition 8.8.6] this also shows that

$$\Big(\vec{F}\big|_{\partial\Omega}^{\kappa-\mathrm{n.t.}}\Big)(y) = \partial_{y_j}\big[k(x-y)\big]\mathbf{e}_i - \partial_{y_i}\big[k(x-y)\big]\mathbf{e}_j \text{ for } \sigma\text{-a.e. } y \in \partial_*\Omega. \tag{2.4.127}$$

Our final observation is that in the case when the set Ω is unbounded we have $|\vec{F}(y)| = O(|y|^{-n})$ for $y \in \Omega$ with $|y| \to \infty$. Thus, in such a scenario,

$$\int_{[B(0,2R)\setminus B(0,R)]\cap\Omega} |y\cdot\vec{F}(y)|\,\mathrm{d}\mathcal{L}^n(y) = O(R) \text{ as } R\to\infty. \tag{2.4.128}$$

Granted (2.4.122), (2.4.123), (2.4.125), (2.4.127), and (2.4.128), [112, Theorem 1.4.1] applies and formula [112, (1.4.6)] presently yields

$$\int_{\partial_*\Omega} \nu(y)\cdot\Big(\vec{F}\big|_{\partial\Omega}^{\kappa-\mathrm{n.t.}}\Big)(y)\,\mathrm{d}\sigma(y) = 0. \tag{2.4.129}$$

In view of (2.4.127), the identity claimed in (2.4.120) now readily follows from (2.4.129). For further reference we wish to note that, as an inspection of the above argument shows,

> the cancelation property recorded in (2.4.120) holds even when we merely assume that $\partial\Omega$ is an Ahlfors regular set and only requiring that the kernel $k \in \mathscr{C}^2(\mathbb{R}^n\setminus\{0\})$ is positive homogeneous of degree $1-N$ for some $N \geq n$. (2.4.130)

Going forward, select an integrability exponent $p \in (1,\infty)$. Also, fix a point $x_o \in \partial\Omega$ and for each $R > 0$ abbreviate

$$S_R := B(x_o,R)\cap\partial\Omega,\ T(S_R) := B(x_o,R)\cap\Omega, \text{ and } f_{S_R} := \fint_{S_R} f\,\mathrm{d}\sigma. \tag{2.4.131}$$

In addition, pick a scale $r \in \big(0, 2\,\mathrm{diam}(\partial\Omega)\big)$ and consider a cutoff function η in $\mathbb{R}^n$ satisfying

$$\begin{gathered} \eta \in \mathscr{C}_c^\infty(\mathbb{R}^n), \quad 0 \leq \eta \leq 1, \\ \eta \equiv 1 \text{ on } B(x_o,2r), \quad \eta \equiv 0 \text{ outside } B(x_o,4r), \\ |\partial^\alpha\eta(x)| \leq C_\alpha r^{-|\alpha|}, \quad \forall x \in \mathbb{R}^n \text{ and } \forall \alpha \in \mathbb{N}_0^n. \end{gathered} \tag{2.4.132}$$

We then proceed to split

$$f = \eta(f - f_{S_{4r}}) + (1-\eta)(f - f_{S_{4r}}) + f_{S_{4r}}. \tag{2.4.133}$$

From (2.4.133) and the cancelation property (2.4.120) we see that

$$Q_{ij}f = Q_{ij}\big(\eta(f - f_{S_{4r}})\big) + Q_{ij}\big((1-\eta)(f - f_{S_{4r}})\big). \tag{2.4.134}$$

Thus,

$$\begin{aligned}
&\int_{T(S_r)} \big|Q_{ij}f(x)\big|^p \operatorname{dist}(x, \partial\Omega)^{p-1}\,\mathrm{d}x \\
&\qquad \le C\int_{T(S_r)} \big|Q_{ij}\big(\eta(f - f_{S_{4r}})\big)(x)\big|^p \operatorname{dist}(x, \partial\Omega)^{p-1}\,\mathrm{d}x \\
&\qquad\quad + C\int_{T(S_r)} \big|Q_{ij}\big((1-\eta)(f - f_{S_{4r}})\big)(x)\big|^p \operatorname{dist}(x, \partial\Omega)^{p-1}\,\mathrm{d}x \\
&\qquad =: I + II.
\end{aligned} \tag{2.4.135}$$

We shall show that

$$I \le C\sigma(S_r) f_p^{\#}(x_o)^p \ \text{ and } \ II \le C\sigma(S_r) f_1^{\#}(x_o)^p \tag{2.4.136}$$

where the L^q-based Fefferman-Stein sharp maximal function $f_q^{\#}$, with $q \in [1,\infty)$, has been defined in (A.0.123). To justify the first inequality in (2.4.136), we write

$$\begin{aligned}
I &\le C\int_{\Omega} \big|Q_{ij}\big(\eta(f - f_{S_{4r}})\big)(x)\big|^p \operatorname{dist}(x, \partial\Omega)^{p-1}\,\mathrm{d}x \\
&\le C\int_{\partial\Omega} |\eta(f - f_{S_{4r}})|^p\,\mathrm{d}\sigma \le C\int_{S_{4r}} |f - f_{S_{4r}}|^p\,\mathrm{d}\sigma \\
&= C\sigma(S_{4r}) \fint_{S_{4r}} |f - f_{S_{4r}}|^p\,\mathrm{d}\sigma \le C\sigma(S_r) f_p^{\#}(x_o)^p.
\end{aligned} \tag{2.4.137}$$

Above, the first inequality follows from the definition of I in (2.4.135), the second inequality follows from (2.4.34), the third inequality is clear from the support properties of the function η introduced in (2.4.132), the subsequent equality is obvious, and the fourth inequality is due to the fact that σ is doubling (itself, a consequence of the Ahlfors regularity of $\partial\Omega$), and the definition of $f_p^{\#}$ in (A.0.123).

To justify the second inequality in (2.4.136), we first observe that for each point $x \in T(S_r)$ we have

$$\begin{aligned}
\big|Q_{ij}\big((1-\eta)(f - f_{S_{4r}})\big)(x)\big| &\le C\int_{\partial\Omega\setminus S_{2r}} \frac{|f(y) - f_{S_{4r}}|}{|x_o - y|^n}\,\mathrm{d}\sigma(y) \\
&\le Cr^{-1} f_1^{\#}(x_o).
\end{aligned} \tag{2.4.138}$$

Indeed, the first inequality above follows from the definition of $Q_{ij}\big((1-\eta)(f-f_{S_{4r}})\big)$, the properties of the function η from (2.4.132), and the fact that

$$|x-y| \approx |x_o - y| \text{ uniformly for } x \in T(S_r) \text{ and } y \in \partial\Omega \setminus S_{2r}. \tag{2.4.139}$$

The second inequality in (2.4.138) is a direct consequence of [112, (7.4.115)]. In turn, from (2.4.135) and (2.4.138) we obtain

$$\begin{aligned} II &\leq C r^{-p} f_1^{\#}(x_o)^p \int_{T(S_r)} \operatorname{dist}(x, \partial\Omega)^{p-1}\, dx \\ &\leq C r^{n-1} f_1^{\#}(x_o)^p \leq C\sigma(S_r) f_1^{\#}(x_o)^p, \end{aligned} \tag{2.4.140}$$

where the last equality is a consequence of the lower Ahlfors regularity of $\partial\Omega$ and the fact that $r \in \big(0, 2\operatorname{diam}(\partial\Omega)\big)$. This establishes the second inequality in (2.4.136).

At this stage, from (2.4.135), (2.4.136), and [112, (7.4.111)] we obtain

$$\begin{aligned} \sup_{\substack{x\in\partial\Omega \text{ and} \\ 0<r<2\operatorname{diam}(\partial\Omega)}} \frac{1}{\sigma\big(B(x,r)\cap\partial\Omega\big)} \int_{B(x,r)\cap\Omega} \sum_{i,j=1}^{n} \big|Q_{ij} f\big|^p \operatorname{dist}(\cdot, \partial\Omega)^{p-1}\, d\mathcal{L}^n \\ \leq C\|f\|^p_{\widetilde{\mathrm{BMO}}(\partial\Omega,\sigma)}. \end{aligned} \tag{2.4.141}$$

From this, (2.4.117) immediately follows in the case when $\partial\Omega$ is unbounded, or Ω is bounded. To complete the proof of the corollary there remains to establish a similar estimate in the case when Ω is an exterior domain and when the supremum is taken in the regime $r \in \big[2\operatorname{diam}(\partial\Omega), \infty\big)$.

In such a scenario, the inequality we seek to establish reads

$$\sup_{x\in\partial\Omega} \frac{1}{\sigma(\partial\Omega)} \int_{B(x,r)\cap\Omega} \sum_{i,j=1}^{n} \big|Q_{ij} f\big|^p \operatorname{dist}(\cdot, \partial\Omega)^{p-1}\, d\mathcal{L}^n \leq C\|f\|^p_{\widetilde{\mathrm{BMO}}(\partial\Omega,\sigma)}. \tag{2.4.142}$$

Using the abbreviation $f_{\partial\Omega} := ⨍_{\partial\Omega} f\, d\sigma$ we can write

$$\begin{aligned} \sup_{x\in\partial\Omega} &\frac{1}{\sigma(\partial\Omega)} \int_{B(x,r)\cap\Omega} \sum_{i,j=1}^{n} \big|Q_{ij} f\big|^p \operatorname{dist}(\cdot, \partial\Omega)^{p-1}\, d\mathcal{L}^n \\ &\leq \frac{1}{\sigma(\partial\Omega)} \int_{\Omega} \sum_{i,j=1}^{n} \big|Q_{ij}(f - f_{\partial\Omega})\big|^p \operatorname{dist}(\cdot, \partial\Omega)^{p-1}\, d\mathcal{L}^n \\ &\leq C ⨍_{\partial\Omega} |f - f_{\partial\Omega}|^p\, d\sigma \\ &\leq C\|f\|^p_{\widetilde{\mathrm{BMO}}(\partial\Omega,\sigma)}, \end{aligned} \tag{2.4.143}$$

where the first inequality is based on (2.4.120), and the second one comes from (2.4.34). This justifies (2.4.142) and finishes the proof of (2.4.117). In turn, (2.4.118) follows by specializing (2.4.117) to the case $p = 2$. □

2.5 The Jump-Formula for Boundary-to-Domain Integral Operators in Open Sets with Uniformly Rectifiable Boundaries

Having proved operator bounds on UR sets in Theorem 2.3.2 and Theorem 2.4.1, we shall now use these results in order to establish jump-formulas for integral operators of the sort described in Theorem 2.5.1 below[25]. Once again, this refines work in the first part of [63, Theorem 3.33, p. 2669] carried out for a smaller class of functions (we presently identify the optimal class of functions for which said jump-formulas hold), and in the more restrictive setting of UR domains $\Omega \subseteq \mathbb{R}^n$ (in the terminology of [63, Definition 3.7, p. 2631]; in particular, the condition that the geometric measure theoretic boundary of Ω has full $\mathcal{H}^{n-1}$-measure in its topological boundary – as stipulated in [63, (3.1.22), p. 2631] – is no longer required here).

Theorem 2.5.1 *For each $n \in \mathbb{N}$ with $n \geq 2$ there exists an integer $N = N(n) \geq 4$ with the following significance. Suppose $\Omega \subseteq \mathbb{R}^n$ is a nonempty open set with the property that $\partial\Omega$ is a* UR *set; in particular, Ω is a set of locally finite perimeter. Abbreviate $\sigma := \mathcal{H}^{n-1} \lfloor \partial\Omega$ and denote by ν the geometric measure theoretic outward unit normal to Ω.*

Next, consider a kernel $k \in \mathscr{C}^N(\mathbb{R}^n \setminus \{0\})$ which is odd and positive homogeneous of degree $1 - n$ (cf. (2.3.3)) and define the singular integral operator acting on arbitrary functions

$$f \in L^1\Big(\partial\Omega, \frac{\sigma(x)}{1 + |x|^{n-1}}\Big) \tag{2.5.1}$$

according to

$$(\mathcal{T} f)(x) := \int_{\partial\Omega} k(x - y) f(y) \, d\sigma(y) \ \textit{for each} \ x \in \Omega, \tag{2.5.2}$$

along with its principal-value version

$$(T f)(x) := \lim_{\varepsilon \to 0^+} \int_{\substack{y \in \partial\Omega \\ |x-y| > \varepsilon}} k(x - y) f(y) \, d\sigma(y) \ \textit{for } \sigma\textit{-a.e.} \ x \in \partial\Omega. \tag{2.5.3}$$

[25] the roots of this results go back in time to the work of Julian Karol Sokhotski who first proved a special version in 1868, and Josip Plemelj who rediscovered it and used it as the main ingredient in his solution of a Riemann-Hilbert Problem in 1908

Then for each aperture parameter $\kappa > 0$ *and each function* $f \in L^1\Big(\partial\Omega, \frac{\sigma(x)}{1+|x|^{n-1}}\Big)$ *the following jump-formula*[26] *holds:*

$$\big(\mathcal{T} f\big)\Big|_{\partial\Omega}^{\kappa-\mathrm{n.t.}}(x) = \frac{1}{2\sqrt{-1}}\widehat{k}\big(\nu(x)\big) f(x) + (Tf)(x) \ \textit{for } \sigma\textit{-a.e. } x \in \partial_*\Omega, \tag{2.5.4}$$

where $\sqrt{-1} = \mathrm{i} \in \mathbb{C}$, *and 'hat' denotes the Fourier transform in* $\mathbb{R}^n$ (*cf.* (1.4.14)).

A few comments to help clarify the nature of the statement of Theorem 2.5.1 are in order.

Comment 1. For each function f as in (2.5.1), the integral in (2.5.2) is absolutely convergent and

$$\mathcal{T} f \in \mathscr{C}^\infty(\Omega). \tag{2.5.5}$$

As such, for each given $\kappa > 0$, the nontangential boundary trace in (2.5.4) is defined as

$$\big(\mathcal{T} f\big)\Big|_{\partial\Omega}^{\kappa-\mathrm{n.t.}}(x) := \lim_{\substack{z \in \Gamma_\kappa(x)\\ z \to x}} (\mathcal{T} f)(z) \ \text{ for } \sigma\text{-a.e. } x \in \partial_*\Omega. \tag{2.5.6}$$

That for σ-a.e. $x \in \partial_*\Omega$ it is actually meaningful to consider the limit in the right-hand side of (2.5.6) follows from [112, (8.8.52)] in [112, Proposition 8.8.6] and the current assumptions on Ω.

Comment 2. The fact that the principal-value singular integral operator in (2.5.3) is well defined when acting on functions as in (2.5.1) is guaranteed by Theorem 2.3.2 (presently used with $\Sigma := \partial\Omega$).

Comment 3. It is natural to restrict attention to points on the geometric measure theoretic boundary $\partial_*\Omega$ (as opposed to, say, points on the topological boundary $\partial\Omega$) when considering the jump-formula (2.5.4) since this involves the geometric measure theoretic outward unit normal $\nu(x)$ which is well defined for σ-a.e. $x \in \partial_*\Omega$ (but not necessarily in a larger context).

Comment 4. The Fourier transform in $\mathbb{R}^n$ employed here is the one defined earlier in (1.4.14). The reader is alerted to the fact that this is different from the normalizations used in [175] and [166].

Comment 5. Estimate (2.4.55) guarantees that k defines a tempered distribution in $\mathbb{R}^n$ (via integration against Schwartz functions). Moreover, $\widehat{k}$ originally considered in the class of tempered distributions in $\mathbb{R}^n$ satisfies

$$\begin{aligned} &\widehat{k} \in \mathscr{C}^m(\mathbb{R}^n \setminus \{0\}) \text{ if } N \in \mathbb{N} \text{ is even} \\ &\text{and } m \in \mathbb{N}_0 \text{ is such that } m < N - 1. \end{aligned} \tag{2.5.7}$$

Indeed, according to [109, Exercise 4.62, p. 144],

[26] or *saltus-condition*, using an older piece of terminology; see, e.g., [4, p. 275]

if u is a tempered distribution in $\mathbb{R}^n$ with $u\big|_{\mathbb{R}^n\setminus\{0\}} \in \mathscr{C}^N(\mathbb{R}^n \setminus \{0\})$ for some even $N \in \mathbb{N}$, and $u\big|_{\mathbb{R}^n\setminus\{0\}}$ is positive homogeneous of degree $\theta < N - n$, then $\widehat{u}\big|_{\mathbb{R}^n\setminus\{0\}} \in \mathscr{C}^m(\mathbb{R}^n \setminus \{0\})$ whenever $m \in \mathbb{N}_0$ is such that $m < N - n - \theta$. (2.5.8)

In particular, (2.5.7) ensures that

if $N \geq 2$ then the function $\widehat{k}(\nu(\cdot))$ is meaningfully defined σ-a.e. on $\partial_*\Omega$ and, in fact, said function belongs to $L^\infty(\partial_*\Omega, \sigma)$. (2.5.9)

Moreover, given that k is odd, it follows that

$$\widehat{k} : \mathbb{R}^n \setminus \{0\} \longrightarrow \mathbb{C} \text{ is an odd function.} \tag{2.5.10}$$

<u>Comment 6.</u> If for each given $\omega \in S^{n-1}$ we consider the hyperplane in $\mathbb{R}^n$ consisting of vectors which are orthogonal to ω, i.e., $\langle\omega\rangle^\perp := \{\xi \in \mathbb{R}^n : \langle \xi, \omega\rangle = 0\}$, then according to [109, Proposition 4.80, p. 172] we have

$$\widehat{k}(\omega) = -2\mathrm{i} \lim_{r\to\infty} \int\limits_{\xi\in\langle\omega\rangle^\perp,\, |\xi|<r} k(\xi + \omega)\, \mathrm{d}\mathcal{H}^{n-1}(\xi), \qquad \forall\, \omega \in S^{n-1}. \tag{2.5.11}$$

Since for each $r \in (0, \infty)$ and each $\omega \in S^{n-1}$ the parity of k implies that

$$\int\limits_{\xi\in\langle\omega\rangle^\perp,\, |\xi|<r} k(\xi + \omega)\, \mathrm{d}\mathcal{H}^{n-1}(\xi) \tag{2.5.12}$$

$$= \int\limits_{\xi\in\langle\omega\rangle^\perp,\, |\xi|<r} \tfrac{1}{2}\big\{k(\xi + \omega) - k(\xi - \omega)\big\}\, \mathrm{d}\mathcal{H}^{n-1}(\xi),$$

and since the integrand in (2.5.12) exhibits better decay than the integrand in (2.5.11) (specifically, we now have $k(\xi + \omega) - k(\xi - \omega) = O(|\xi|^{-n})$ as $|\xi| \to \infty$), we may recast (2.5.11) as

$$\widehat{k}(\omega) = -\mathrm{i} \int\limits_{\xi\in\langle\omega\rangle^\perp} \big\{k(\xi + \omega) - k(\xi - \omega)\big\}\, \mathrm{d}\mathcal{H}^{n-1}(\xi), \qquad \forall\, \omega \in S^{n-1}. \tag{2.5.13}$$

In addition, if we introduce

$$\Theta(x) := k\Big(\frac{x}{|x|}\Big) \text{ for each } x \in \mathbb{R}^n \setminus \{0\}, \tag{2.5.14}$$

then

$$\begin{aligned}&\Theta \in \mathscr{C}^N(\mathbb{R}^n \setminus \{0\}) \text{ is odd, positive homogeneous of degree zero,}\\ &\text{and } k(x) = \frac{\Theta(x)}{|x|^{n-1}} \text{ for each } x \in \mathbb{R}^n \setminus \{0\}.\end{aligned} \tag{2.5.15}$$

In turn, the latter formula implies that

$$\begin{aligned}k(\xi + \omega) - k(\xi - \omega) = \frac{\Theta(\xi + \omega) - \Theta(\xi - \omega)}{(1 + |\xi|^2)^{(n-1)/2}}\\ \text{whenever } \omega \in S^{n-1} \text{ and } \xi \in \langle\omega\rangle^{\perp}.\end{aligned} \tag{2.5.16}$$

As a consequence, for σ-a.e. $x \in \partial_*\Omega$ the quantity $\widehat{k}\big(\nu(x)\big)$ appearing in (2.5.4) may be alternatively expressed as

$$\begin{aligned}\widehat{k}\big(\nu(x)\big) &= -2\mathrm{i} \lim_{r\to\infty} \int_{\xi\in\langle\nu(x)\rangle^{\perp},\ |\xi|<r} k\big(\xi + \nu(x)\big)\, d\mathcal{H}^{n-1}(\xi)\\ &= -\mathrm{i} \int_{\xi\in\langle\nu(x)\rangle^{\perp}} \big\{k(\xi + \nu(x)) - k(\xi - \nu(x))\big\}\, d\mathcal{H}^{n-1}(\xi)\\ &= -\mathrm{i} \int_{\xi\in\langle\nu(x)\rangle^{\perp}} \frac{\Theta(\xi + \nu(x)) - \Theta(\xi - \nu(x))}{(1 + |\xi|^2)^{(n-1)/2}}\, d\mathcal{H}^{n-1}(\xi).\end{aligned} \tag{2.5.17}$$

Comment 7. In relation to (2.5.8), for further reference it is also useful to make two elementary observations. First, based on (2.5.8) and [109, Exercise 4.63] we have that

> if $N \geq 4$ is even and $b \in \mathscr{C}^N(\mathbb{R}^n \setminus \{0\})$ is positive homogeneous of degree $\theta \in (1 - n, N - 1 - n)$, then for each $j \in \{1, \ldots, n\}$ the Fourier transforms of the tempered distributions induced by $\partial_j b$ and b in $\mathbb{R}^n$ (via integration against Schwartz functions) are continuous when restricted to $\mathbb{R}^n \setminus \{0\}$ and satisfy $(\widehat{\partial_j b})(\xi) = \mathrm{i}\xi_j \widehat{b}(\xi)$ for all $\xi \in \mathbb{R}^n \setminus \{0\}$. (2.5.18)

Second, from (2.5.8) and [109, Theorem 4.26(b), p. 132] we may also conclude that

> if u is a tempered distribution in $\mathbb{R}^n$ such that $u\big|_{\mathbb{R}^n\setminus\{0\}} \in \mathscr{C}^N(\mathbb{R}^n \setminus \{0\})$ for some $N > 1$ odd, and $(\nabla u)\big|_{\mathbb{R}^n\setminus\{0\}}$ is positive homogeneous of degree $\theta < N - 1 - n$, then $\widehat{\nabla u}\big|_{\mathbb{R}^n\setminus\{0\}}$ is continuous in $\mathbb{R}^n \setminus \{0\}$ and $\xi_k(\widehat{\partial_j u})(\xi) = \xi_j(\widehat{\partial_k u})(\xi)$ for each $j, k \in \{1, \ldots, n\}$ and $\xi \in \mathbb{R}^n \setminus \{0\}$. (2.5.19)

Comment 8. Recall (cf., e.g., [109, Corollary 4.65, p. 147]) that whenever $n \in \mathbb{N}$ with $n \geq 2$ and $j \in \{1, \ldots, n\}$ then

$$\begin{aligned} &\widehat{k_j}(\xi) = -\mathrm{i}\omega_{n-1}\tfrac{\xi_j}{|\xi|^2} \text{ in } \mathcal{S}'(\mathbb{R}^n) \\ &\text{if } k_j(x) := \tfrac{x_j}{|x|^n} \text{ for all } x \in \mathbb{R}^n \setminus \{0\}. \end{aligned} \tag{2.5.20}$$

More generally, we recall from [166, p. 73] that if P_j is a homogeneous harmonic polynomial of degree $j \in \mathbb{N}$ in $\mathbb{R}^n$ (where $n \in \mathbb{N}$ satisfies $n \geq 2$), then

for each $\alpha \in (0, n)$ it follows that the Fourier transform (considered with the normalization given in (1.4.14)) of the tempered distribution induced by the locally integrable function $P_j(x)/|x|^{n+j-\alpha}$ which decays at infinity in $\mathbb{R}^n$ (via integration against functions from the Schwartz class) is the tempered distribution induced by the locally integrable function which decays at infinity in $\mathbb{R}^n$ given by (with Γ the Gamma function) (2.5.21)

$$2^\alpha \pi^{n/2}(-\mathrm{i})^j \frac{\Gamma(j/2+\alpha/2)}{\Gamma(j/2+n/2-\alpha/2)} \frac{P_j(x)}{|x|^{j+\alpha}}, \qquad \forall x \in \mathbb{R}^n \setminus \{0\}.$$

In the case when $n \in \mathbb{N}$ satisfies $n \geq 2$, the exponent $j \in \mathbb{N}$ is odd, and

$$\begin{aligned} &k(x) := P_j(x)/|x|^{n-1+j} \text{ for each } x \in \mathbb{R}^n \setminus \{0\} \text{ where} \\ &P_j \text{ is a homogeneous harmonic polynomial of degree } j \text{ in } \mathbb{R}^n \end{aligned} \tag{2.5.22}$$

we conclude from (2.5.21) used with $\alpha := 1$ that the Fourier transform (considered with the normalization given in (1.4.14)) of k is the tempered distribution induced, via integration against Schwartz functions, by the locally integrable function

$$\begin{aligned} \widehat{k}(\xi) &= 2\pi^{n/2}(-\mathrm{i})^j \frac{\Gamma(j/2+1/2)}{\Gamma(j/2+n/2-1/2)} \frac{P_j(\xi)}{|\xi|^{j+1}} \qquad (2.5.23) \\ &= (-\mathrm{i})^j \omega_{n-1} \frac{\Gamma(j/2+1/2)\Gamma(n/2)}{\Gamma(j/2+n/2-1/2)} \frac{P_j(\xi)}{|\xi|^{j+1}}, \qquad \forall \xi \in \mathbb{R}^n \setminus \{0\}. \end{aligned}$$

Hence, in such a scenario, for σ-a.e. $x \in \partial_*\Omega$ the quantity $\widehat{k}(\nu(x))$ appearing in (2.5.4) may be computed as

$$\widehat{k}\big(\nu(x)\big) = (-\mathrm{i})^j \omega_{n-1} \frac{\Gamma(j/2+1/2)\Gamma(n/2)}{\Gamma(j/2+n/2-1/2)} P_j\big(\nu(x)\big). \tag{2.5.24}$$

For example, in the special situation when $j = 1$ and, for some fixed vector $a \in \mathbb{R}^n$, we now take $P_1(x) := \langle x, a\rangle$ for each $x \in \mathbb{R}^n$ then (2.5.24) becomes

$$\widehat{k}\big(\nu(x)\big) = -\mathrm{i}\omega_{n-1}\big\langle \nu(x), a\big\rangle \text{ for } \sigma\text{-a.e. } x \in \partial_*\Omega. \tag{2.5.25}$$

Also, in the particular case when $n = 2$ (so that $\mathbb{R}^2 \equiv \mathbb{C}$) and k is as in (2.5.22) with $j \in \mathbb{N}$ odd number and where, for some $a, b \in \mathbb{C}$, we now take $P_j(z) := az^j + b\overline{z}^j$

for each $z \in \mathbb{C}$, then (2.5.24) reduces to

$$\widehat{k}\big(\nu(x)\big) = 2\pi(-\mathrm{i})^j\big(a\nu(x)^j + b\overline{\nu(x)}^j\big) \text{ for } \sigma\text{-a.e. } x \in \partial_*\Omega. \tag{2.5.26}$$

Finally, we wish to note that whenever $n \in \mathbb{N}$ with $n \geq 3$ and $a, b, c \in \{1, \dots, n\}$ then the Fourier transform of the tempered distribution induced by the locally integrable function in $\mathbb{R}^n$ defined as

$$k_{abc}(x) := \frac{x_a x_b x_c}{|x|^{n+2}} \text{ for each } x \in \mathbb{R}^n \setminus \{0\} \tag{2.5.27}$$

is the tempered distribution $\widehat{k_{abc}} \in \mathcal{S}'(\mathbb{R}^n)$ which is of function type outside the origin, where it is explicitly given by

$$\widehat{k_{abc}}(\xi) = -\mathrm{i}\frac{\omega_{n-1}}{n}\left\{\delta_{ab}\frac{\xi_c}{|\xi|^2} + \delta_{ac}\frac{\xi_b}{|\xi|^2} + \delta_{bc}\frac{\xi_a}{|\xi|^2} - 2\frac{\xi_a\xi_b\xi_c}{|\xi|^4}\right\}. \tag{2.5.28}$$

(Indeed, this may be seen from (2.5.20), [109, (7.14.9), p. 320], and a straightforward computation based of properties of the Fourier transform.) Note that the kernel (2.5.27) falls under the scope of Theorem 2.5.1, so (2.5.28) gives an explicit formula for the jump-term in (2.5.4) in this significant speacial case.

Comment 9. The jump-formula (2.5.4) readily extends to the case when the kernel k is a matrix-valued function and the function f is vector-valued. In such a scenario, $\widehat{k}$ is computed entry-by-entry and $\widehat{k}\big(\nu(x)\big)f(x)$ is considered in the sense of the natural action of a matrix on a vector.

Comment 10. In the case when $\partial\Omega$ is bounded we have $\frac{1}{1+|x|^{n-1}} \approx 1$ uniformly for $x \in \partial\Omega$, hence the space $L^1\Big(\partial\Omega, \frac{\sigma(x)}{1+|x|^{n-1}}\Big)$ simply reduces to $L^1(\partial\Omega, \sigma)$ in such a situation.

Comment 11. It is relevant to contrast the hypotheses guaranteeing the validity of our jump-formula (2.5.4) to the most accommodating geometric setting in which its very formulation is meaningful. Two key aspects in this regard are as follows. First, the very presence of $\nu(x)$ in the right-hand side limits x to $\partial_*\Omega$ since, up to a σ-nullset, this is the largest ambient where the geometric measure theoretic outward unit normal may be considered. Second, on the analytic side, assuming as we presently do that f belongs to the weighted Lebesgue space $L^1\big(\partial\Omega, \frac{\sigma(x)}{1+|x|^{n-1}}\big)$ is optimal since this is the largest class of functions on which the integral operator $\mathcal{T}$ is meaningfully defined as in (2.5.2) via an absolutely convergent integral.

Before turning to the proof Theorem 2.5.1, in the next several results (Corollaries 2.5.2-2.5.7) we shall discuss a variety of its special cases and natural adaptations, which are of particular interest in applications. We begin by presenting a version of Theorem 2.5.1 pertaining to the scales of ordinary Lebesgue and Lorentz spaces.

Corollary 2.5.2 *In the context of Theorem 2.5.1 the jump-formula* (2.5.4) *holds for each function f in the Lebesgue space $L^p(\partial\Omega, \sigma)$ with $1 \leq p < \infty$, as well as each function f belonging to the Lorentz space $L^{p,q}(\partial\Omega, \sigma)$ with $1 < p < \infty$ and $0 < q \leq \infty$.*

Proof The embedding in [112, (7.7.106)] (used with $\Sigma := \partial\Omega$) ensures that the jump-formula (2.5.4) holds, in particular, for every function $f \in L^p(\partial\Omega, \sigma)$ with $1 \leq p < \infty$, while [112, (7.7.107)] (again, used with $\Sigma := \partial\Omega$) guarantees that the jump-formula (2.5.4) is also valid for every function $f \in L^{p,q}(\partial\Omega, \sigma)$ with $p \in (1, \infty)$ and $q \in (0, \infty]$. □

Next we present a result which, in particular, gives a version of the jump-formula (2.5.4) for bounded functions.

Corollary 2.5.3 *Let $\Omega \subseteq \mathbb{R}^n$ (where $n \in \mathbb{N}$ with $n \geq 2$) be a nonempty open set with the property that $\partial\Omega$ is an Ahlfors regular set, and abbreviate $\sigma := \mathcal{H}^{n-1}\lfloor\partial\Omega$. Given a kernel $k \in \mathscr{C}^1(\mathbb{R}^n \setminus \{0\})$ which is odd and positive homogeneous of degree $1-n$, consider the following modified version of the operator in* (2.5.1)-(2.5.2)*:*

$$\begin{aligned}&(\mathcal{T}_{\text{mod}} f)(x) := \int_{\partial\Omega} \big\{k(x-y) - k_1(-y)\big\} f(y)\, d\sigma(y) \text{ for each } x \in \Omega,\\&\text{where } k_1 := k \cdot \mathbf{1}_{\mathbb{R}^n \setminus \overline{B(1,0)}} \text{ and } f \in L^1\Big(\partial\Omega, \frac{\sigma(x)}{1+|x|^n}\Big) \text{ is arbitrary.}\end{aligned} \tag{2.5.29}$$

Then $\mathcal{T}_{\text{mod}}$ is meaningfully defined in the above context and is compatible with $\mathcal{T}$ from (2.5.1)-(2.5.2)*, in the sense that for each function f belonging to the smaller space $L^1\big(\partial\Omega, \frac{\sigma(x)}{1+|x|^{n-1}}\big)$ (hence, in particular, for each function $f \in L^p(\partial\Omega, \sigma)$ with $p \in [1, \infty)$) the difference*

$$C_f := \mathcal{T}_{\text{mod}} f - \mathcal{T} f \text{ is a constant in } \Omega. \tag{2.5.30}$$

As a consequence, $\nabla \mathcal{T}_{\text{mod}} f = \nabla \mathcal{T} f$ in Ω for each such function f.

Next, strengthen the original hypotheses by assuming now that $k \in \mathscr{C}^N(\mathbb{R}^n \setminus \{0\})$ for some integer $N = N(n) \geq 4$ large enough and that $\partial\Omega$ is a UR *set. In this setting, denote by ν the geometric measure theoretic outward unit normal to Ω, and bring in the operator T_{mod} from* (2.3.31) *with $\Sigma := \partial\Omega$. Then, with the same notational conventions as in* (2.5.4)*, the following jump-formula holds:*

$$\begin{aligned}&\big(\mathcal{T}_{\text{mod}} f\big)\Big|_{\partial\Omega}^{\kappa-\text{n.t.}}(x) = \frac{1}{2\sqrt{-1}}\widehat{k}\big(\nu(x)\big) f(x) + (T_{\text{mod}} f)(x) \text{ for } \sigma\text{-a.e. } x \in \partial_*\Omega,\\&\text{for each given } \kappa > 0 \text{ and each given function } f \in L^1\Big(\partial\Omega, \frac{\sigma(x)}{1+|x|^n}\Big).\end{aligned} \tag{2.5.31}$$

Also, for each aperture parameter $\kappa \in (0, \infty)$ and each truncation parameter $\varepsilon \in (0, \infty)$, in a quantitative way one has

$$\begin{aligned}&\mathcal{N}_\kappa^\varepsilon(\mathcal{T}_{\text{mod}} f) \in L^p_{\text{loc}}(\partial\Omega, \sigma) \text{ for each function}\\&f \in L^1\big(\partial\Omega, \tfrac{\sigma(x)}{1+|x|^n}\big) \cap L^p_{\text{loc}}(\partial\Omega, \sigma) \text{ with } p \in (1, \infty).\end{aligned} \tag{2.5.32}$$

Proof Note that if $x \in \Omega$, $y \in \partial\Omega$, and $0 < \varepsilon < |x - y|$, then $k(x - y) = k_\varepsilon(x - y)$, where k_ε is as in (2.3.115). Keeping this in mind, the same argument used to prove (2.3.119) presently gives that

$$\begin{gathered}\int_{\partial\Omega} \big|k(x - y) - k_1(-y)\big||f(y)|\, d\sigma(y) < +\infty \\ \text{for each } x \in \Omega \text{ and each } f \in L^1\Big(\partial\Omega, \frac{\sigma(x)}{1 + |x|^n}\Big).\end{gathered} \tag{2.5.33}$$

Thus, the integral in the definition of $(\mathcal{T}_{\text{mod}} f)(x)$ in (2.5.29) is absolutely convergent for each $x \in \Omega$ and each $f \in L^1\big(\partial\Omega, \frac{\sigma(x)}{1+|x|^n}\big)$. From (2.3.116) (used with $|\alpha| = 0$) we also see that

$$\begin{gathered}\text{if actually } f \text{ belongs to the space } L^1\Big(\partial\Omega, \frac{\sigma(x)}{1 + |x|^{n-1}}\Big) \text{ then} \\ \int_{\partial\Omega} |k_1(-y) f(y)|\, d\sigma(y) \leq \int_{\partial\Omega} \frac{|f(y)|}{1 + |y|^{n-1}}\, d\sigma(y) < +\infty.\end{gathered} \tag{2.5.34}$$

In concert, (2.5.33)-(2.5.33) prove that

$$\begin{gathered}C_f := \mathcal{T}_{\text{mod}} f - \mathcal{T} f = \int_{\partial\Omega} k_1(-y) f(y)\, d\sigma(y) \in \mathbb{C}, \\ \text{is a constant in } \Omega, \text{ whenever } f \in L^1\Big(\partial\Omega, \frac{\sigma(x)}{1 + |x|^{n-1}}\Big).\end{gathered} \tag{2.5.35}$$

Assuming the stronger hypotheses made in the second part of the statement, there remains to prove the jump-formula recorded in (2.5.31). To this end, fix a point $x_0 \in \partial\Omega$ and pick an arbitrary number $r \in (0, \infty)$. Next, having selected some function $f \in L^1\big(\partial\Omega, \frac{\sigma(x)}{1+|x|^n}\big)$, split

$$\begin{gathered}f = f_1 + f_2 \text{ on } \partial\Omega, \text{ where} \\ f_1 := \mathbf{1}_{\partial\Omega\cap B(x_0,4r)} \cdot f \text{ and } f_2 := \mathbf{1}_{\partial\Omega\setminus B(x_0,4r)} \cdot f,\end{gathered} \tag{2.5.36}$$

then decompose

$$\mathcal{T}_{\text{mod}} f = \mathcal{T}_{\text{mod}} f_1 + \mathcal{T}_{\text{mod}} f_2 \text{ in } \Omega. \tag{2.5.37}$$

Note that since $\mathcal{T}_{\text{mod}} f_2$ has a continuous extension to $B(x_0, r)$ we trivially have that the nontangential trace $\mathcal{T}_{\text{mod}} f_2\Big|_{\partial\Omega}^{\kappa-\text{n.t.}}$ exists at every point in $\partial_{\text{nta}}\Omega \cap B(x_0, r)$. More precisely,

$$\begin{aligned}\mathcal{T}_{\text{mod}} f_2\Big|_{\partial\Omega}^{\kappa-\text{n.t.}}(x) &= \int_{\partial\Omega} \big\{k(x - y) - k_1(-y)\big\} f_2(y)\, d\sigma(y) \\ &= \big(T_{\text{mod}} f_2\big)(x) \text{ for each } x \in \partial_{\text{nta}}\Omega \cap B(x_0, r).\end{aligned} \tag{2.5.38}$$

In particular, from (2.5.38) and item *(iii)* in [112, Proposition 8.8.6] we conclude that, on the one hand,

$$\mathcal{T}_{\rm mod} f_2\Big|_{\partial\Omega}^{\kappa-{\rm n.t.}}(x) = \big(T_{\rm mod} f_2\big)(x) \text{ at } \sigma\text{-a.e. } x \in \partial_*\Omega \cap B(x_0, r). \tag{2.5.39}$$

On the other hand, since $f_1 \in L^1(\partial\Omega, \sigma)$, at σ-a.e. point $x \in \partial_*\Omega \cap B(x_0, r)$ we may write

$$\begin{aligned}\mathcal{T}_{\rm mod} f_1\Big|_{\partial\Omega}^{\kappa-{\rm n.t.}}(x) &= \mathcal{T} f_1\Big|_{\partial\Omega}^{\kappa-{\rm n.t.}}(x) + C_{f_1}\\ &= \frac{1}{2\sqrt{-1}}\widehat{k}\big(\nu(x)\big) f_1(x) + (T f_1)(x) + C_{f_1}\\ &= \frac{1}{2\sqrt{-1}}\widehat{k}\big(\nu(x)\big) f(x) + \big(T_{\rm mod} f_1\big)(x).\end{aligned} \tag{2.5.40}$$

Above, the first equality uses (2.5.35) (with f replaced by f_1), the second equality is provided by Theorem 2.5.1, and the final equality is seen from (2.3.34) (with $\Sigma := \partial\Omega$, a scenario in which C_f from (2.3.34) coincides with C_f from (2.5.35)).

In turn, from (2.5.39)-(2.5.40) and (2.5.36) we conclude that

$$\begin{aligned}\big(\mathcal{T}_{\rm mod} f\big)\Big|_{\partial\Omega}^{\kappa-{\rm n.t.}}(x) &= \tfrac{1}{2\sqrt{-1}}\widehat{k}\big(\nu(x)\big) f(x) + (T_{\rm mod} f)(x)\\ &\text{for } \sigma\text{-a.e. } x \in \partial_*\Omega \cap B(x_0, r).\end{aligned} \tag{2.5.41}$$

Given that $r > 0$ has been arbitrarily chosen, this ultimately proves (2.5.31).

Finally, to justify the claim made in (2.5.32), fix some $\kappa, \varepsilon \in (0, \infty)$ and pick an arbitrary function $f \in L^1\big(\partial\Omega, \frac{\sigma(x)}{1+|x|^n}\big) \cap L^p_{\rm loc}(\partial\Omega, \sigma)$ with $p \in (1, \infty)$. Having picked a reference point $x_0 \in \partial\Omega$ along with a radius $r > \max\Big\{(1+\kappa)\varepsilon, \frac{1}{4}(1+|x_0|)\Big\}$, split f as in (2.5.36) then decompose $\mathcal{T}_{\rm mod} f$ as in (2.5.37). For each $x \in \partial\Omega \cap B(x_0, r)$, each $z \in \Gamma_\kappa(x)$ with $\operatorname{dist}(z, \partial\Omega) < \varepsilon$, and each $\xi \in [z, x]$ we have

$$|z - x| < (1+\kappa)\operatorname{dist}(z, \partial\Omega) \le (1+\kappa)\varepsilon < r, \tag{2.5.42}$$

and

$$\begin{aligned}|x - y| &\le |x - \xi| + |\xi - y| \le |x - z| + |\xi - y|\\ &\le r + |\xi - y| \le \tfrac{1}{3}|x - y| + |\xi - y|,\end{aligned} \tag{2.5.43}$$

hence, $|x - y| \le \frac{3}{2}|\xi - y|$. Also, $|z| \le |z - x| + |x - x_0| + |x_0| \le 2r + |x_0|$, and the fact that we have $4r \le |x_0 - y| \le |x_0| + |y|$ forces $|y| \ge 1$. On account of these estimates and the Mean Value Theorem we may then write

$$\begin{aligned}(\mathcal{T}_{\rm mod} f_2)(z) &\leq \int_{\partial\Omega} |k(z-y) - k_1(-y)| \big|\big(f\mathbf{1}_{\partial\Omega\setminus B(x_0,4r)}\big)(y)\big| \, d\sigma(y) \\ &= \int_{\partial\Omega} |k(z-y) - k(-y)| \big|\big(f\mathbf{1}_{\partial\Omega\setminus B(x_0,4r)}\big)(y)\big| \, d\sigma(y) \\ &\leq C(k,n) \int_{\partial\Omega\setminus B(x_0,4r)} \Big(\sup_{\xi\in[z,x]} \frac{|z|}{|\xi - y|^n}\Big) |f(y)| \, d\sigma(y) \\ &\leq C(k,n,r,x_0) \int_{\partial\Omega\setminus B(x_0,4r)} \frac{|f(y)|}{|x-y|^n} \, d\sigma(y) \\ &\leq C(k,n,r,x_0) \int_{\partial\Omega} \frac{|f(y)|}{1+|y|^n} \, d\sigma(y) < +\infty. \end{aligned} \tag{2.5.44}$$

In concert with [112, (8.2.28)] this proves that, in a quantitative fashion,

$$\mathcal{N}_\kappa^\varepsilon\big(\mathcal{T}_{\rm mod} f_2\big) \in L^\infty\big(\partial\Omega \cap B(x_0,r), \sigma\big). \tag{2.5.45}$$

Since $f_1 \in L^p(\partial\Omega, \sigma)$ with $p \in (1,\infty)$, we conclude from (2.5.30) and (2.4.9) that (once again in a quantitative manner)

$$\mathcal{N}_\kappa^\varepsilon\big(\mathcal{T}_{\rm mod} f_1\big) \in L^p(\partial\Omega,\sigma) + L^\infty_{\rm loc}(\partial\Omega,\sigma). \tag{2.5.46}$$

At this stage, the claim made in (2.5.32) follows from (2.5.45) and (2.5.46) (keeping in mind (2.5.37) and [112, (8.2.9), (8.2.28)]). □

There is also a suitable version of Theorem 2.5.1 on the scale of Hardy spaces, as described next.

Corollary 2.5.4 *Let $\Omega \subseteq \mathbb{R}^n$ (where $n \in \mathbb{N}$ with $n \geq 2$) be a nonempty open set with the property that $\partial\Omega$ is a* UR *set. Abbreviate $\sigma := \mathcal{H}^{n-1}\lfloor\partial\Omega$ and denote by ν the geometric measure theoretic outward unit normal to Ω.*

Next, fix a kernel $k \in \mathscr{C}^N(\mathbb{R}^n\setminus\{0\})$ (for some integer $N = N(n) \geq 4$ large enough) which is odd and positive homogeneous of degree $1-n$, and pick $p \in \big(\frac{n-1}{n}, 1\big]$ along with $\kappa \in (0,\infty)$. Recall from item (3) in Theorem 2.4.1 (with $\Sigma := \partial\Omega$) that if for each $f \in H^p(\partial\Omega,\sigma)$ one defines

$$(\mathcal{T} f)(x) := \begin{cases} \langle k(x-\cdot), f\rangle & \text{if } \partial\Omega \text{ is bounded,} \\ \langle [k(x-\cdot)], f\rangle & \text{if } \partial\Omega \text{ is unbounded,} \end{cases} \qquad \forall x \in \Omega, \tag{2.5.47}$$

where $\langle\cdot,\cdot\rangle$ stands for the duality bracket described in [113, Theorem 4.6.1] (with $\Sigma := \partial\Omega$), then $\mathcal{T}$ is a well-defined linear and bounded operator

$$\mathcal{T} : H^p(\partial\Omega,\sigma) \longrightarrow N_\kappa^p(\Omega;\sigma). \tag{2.5.48}$$

(recall that $N_\kappa^p(\Omega;\sigma)$ has been introduced in (A.0.98)*) which acts in a coherent fashion with $\mathcal{T}$ as originally defined in* (2.4.3)-(2.4.4) *(with $\Sigma := \partial\Omega$).*

Also, recall from item (6) of Theorem 2.3.2 that the principal-value singular integral of formal convolution type with the kernel k on $\partial\Omega$ (cf. (2.5.3)) induces a well-defined linear and bounded operator

$$T : H^p(\partial\Omega, \sigma) \longrightarrow L^p(\partial\Omega, \sigma). \tag{2.5.49}$$

Finally, recall the L^p-filtering operator $\mathfrak{H} : H^p(\partial\Omega, \sigma) \to L^p(\partial\Omega, \sigma)$ from [113, Theorem 4.9.1] *(used here with $\Sigma := \partial\Omega$).*

Then for each $f \in H^p(\partial\Omega, \sigma)$ the following jump-formula holds:

$$\big(\mathcal{T}f\big)\Big|^{\kappa-\mathrm{n.t.}}_{\partial\Omega}(x) = \frac{1}{2\sqrt{-1}}\widehat{k}\big(\nu(x)\big)(\mathfrak{H}f)(x) + (Tf)(x) \text{ for } \sigma\text{-a.e. } x \in \partial_*\Omega. \tag{2.5.50}$$

One significant consequence of the above result is that, in the context of Corollary 2.5.4, the operator T from (2.5.49) acts in a compatible fashion with its principal-value version (defined as in (2.3.15) with $\Sigma := \partial\Omega$), when the outputs are compared on $\partial_*\Omega$, i.e.,

$$\begin{aligned} &\text{if } f \in L^1\Big(\partial\Omega, \tfrac{\sigma(x)}{1+|x|^{n-1}}\Big) \cap H^p(\partial\Omega, \sigma) \text{ for some } p \in \big(\tfrac{n-1}{n}, 1\big] \\ &\text{then } Tf \in L^p(\partial\Omega, \sigma) \text{ defined in the sense of (2.5.49) satisfies} \\ &(Tf)(x) = \lim_{\varepsilon\to 0^+} \int_{\substack{y\in\partial\Omega \\ |x-y|>\varepsilon}} k(x-y)f(y)\,d\sigma(y) \text{ at } \sigma\text{-a.e. } x \in \partial_*\Omega. \end{aligned} \tag{2.5.51}$$

Indeed, this is seen from (2.5.50), [113, (4.9.3)], and the fact that, as noted earlier, the operator $\mathcal{T}$ from (2.5.48) acts in a coherent fashion with its version originally defined as in (2.4.3)-(2.4.4) with $\Sigma := \partial\Omega$.

In the case when Ω is actually a UR domain, the identity recorded in (2.5.50) also provides an alternative formula for the action of the operator (2.5.49), namely

$$\begin{aligned} &\text{if } f \in H^p(\partial\Omega, \sigma) \text{ with } p \in \big(\tfrac{n-1}{n}, 1\big], \text{ then } Tf \in L^p(\partial\Omega, \sigma) \text{ is given by} \\ &(Tf)(x) = -\frac{1}{2\sqrt{-1}}\widehat{k}\big(\nu(x)\big)(\mathfrak{H}f)(x) + \big(\mathcal{T}f\big)\Big|^{\kappa-\mathrm{n.t.}}_{\partial\Omega}(x) \text{ for } \sigma\text{-a.e. } x \in \partial\Omega. \end{aligned} \tag{2.5.52}$$

For example, upon recalling from [113, (4.2.17)] that for each given pair of points $x_0, x_1 \in \partial\Omega$ the distribution $\delta_{x_0} - \delta_{x_1} \in \big(\mathrm{Lip}_c(\partial\Omega)\big)'$ belongs to the Hardy space $H^p(\partial\Omega, \sigma)$ for every $p \in \big(\frac{n-1}{n}, 1\big)$, from formula (2.5.52) and [113, (4.9.9)] we conclude that

$$\big(T(\delta_{x_0} - \delta_{x_1})\big)(x) = k(x - x_0) - k(x - x_1) \text{ for } \sigma\text{-a.e. } x \in \partial\Omega. \tag{2.5.53}$$

Parenthetically, given the properties of the kernel k (cf. (2.4.55)-(2.4.56)), [112, Lemma 7.2.1] (cf. [112, (7.2.5)]) provides independent confirmation of the fact that $\big(k(\cdot - x_0) - k(\cdot - x_1)\big)\Big|_{\partial\Omega}$ belongs to $L^p(\partial\Omega, \sigma)$ for each $p \in \big(\frac{n-1}{n}, 1\big)$.

Proof of Corollary 2.5.4 Fix some integrability exponent $q \in (1, \infty)$. Then for each function $f \in H^p(\partial\Omega, \sigma) \cap L^q(\partial\Omega, \sigma)$ the function in left-hand side of (2.5.50) becomes (thanks to [113, (4.9.3)] and the very last claim in item *(6)* of Theorem 2.3.2)

$$\frac{1}{2\sqrt{-1}}\widehat{k}\big(\nu(x)\big)f(x) + \lim_{\varepsilon\to 0^+} \int\limits_{\substack{y\in\partial\Omega \\ |x-y|>\varepsilon}} k(x-y)f(y)\,d\sigma(y) \text{ for } \sigma\text{-a.e. } x \in \partial_*\Omega. \tag{2.5.54}$$

On the other hand, on account of (2.5.47) and the compatibility results from [113, Lemmas 4.6.5-4.6.6] we have

$$(\mathcal{T}f)(x) = \int_{\partial\Omega} k(x-y)f(y)\,d\sigma(y) \text{ for each } x \in \Omega. \tag{2.5.55}$$

In turn, from (2.5.55) and (2.5.54) we see that the jump-formula (2.5.50) holds for each function $f \in H^p(\partial\Omega, \sigma) \cap L^q(\partial\Omega, \sigma)$. Consequently (cf. [113, (4.4.114)]),

$$\begin{array}{l}\text{the jump-formula recorded in (2.5.50) holds for any} \\ \text{function } f \text{ belonging to the space } H^{p,q}_{\text{fin}}(\partial\Omega, \sigma).\end{array} \tag{2.5.56}$$

Granted this, the idea now is to apply [112, Proposition 6.2.11] with $\mathscr{X} := \Omega \cup \partial_*\Omega$ equipped with the topology inherited from the ambient Euclidean space, $X := \partial_*\Omega$, $\mu := \mathcal{H}^{n-1}\lfloor\partial_*\Omega$ which is a locally finite complete Borel-regular measure on X by virtue of [112, Lemma 3.6.4] used with $s := n-1$ (that the hypotheses in [112, (3.6.25)] are presently satisfied is seen from [112, (5.6.33), (5.6.35)]), the sets $\Gamma(x) := \Gamma_\kappa(x)$ for each $x \in \partial_*\Omega$ which satisfy [112, (6.2.71)] thanks to [112, (8.8.45)], the quasi-Banach space $Y := H^p(\partial\Omega, \sigma)$, the linear operator $\mathcal{T}$ as in (2.5.48), and $\mathcal{V} := H^{p,q}_{\text{fin}}(\partial\Omega, \sigma)$, which is a dense subset of Y, as indicated in [113, (4.4.114)]. In view of these choices, (2.5.56) implies that [112, (6.2.75)] holds. Also, the maximal operator associated with the present choices as in [112, (6.2.73)] is (bearing in mind that $\mathcal{T}$ maps distributions from $H^p(\partial\Omega, \sigma)$ into continuous functions on Ω)

$$\begin{array}{c}T_\star f(x) := \sup_{y\in\Gamma_\kappa(x)} |(\mathcal{T}f)(y)| = \big(\mathcal{N}_\kappa(\mathcal{T}f)\big)(x) \\ \text{at } \sigma\text{-a.e. } x \in \partial_*\Omega, \text{ for each } f \in H^p(\partial\Omega, \sigma).\end{array} \tag{2.5.57}$$

Since (2.4.14) implies that

$$\mathcal{N}_\kappa(\mathcal{T}\cdot) : H^p(\partial\Omega, \sigma) \longrightarrow L^p(\partial\Omega, \sigma) \tag{2.5.58}$$

is a well-defined, sub-linear and bounded operator, and since the restriction operator

$$L^p(\partial\Omega, \sigma) \ni g \longmapsto g\big|_{\partial_*\Omega} \in L^{p,\infty}(\partial_*\Omega, \sigma) \tag{2.5.59}$$

is well defined, linear, and bounded (cf. [112, (6.2.17)] and [112, (6.2.27)]), it follows that [112, (6.2.74)] holds. As such, all hypotheses of [112, Proposition 6.2.11] are verified, which then permits us to conclude that

$$\begin{aligned}&\text{for each } f \in H^p(\partial\Omega, \sigma), \text{ the nontangential limit}\\&\big(\mathcal{T} f\big)\Big|_{\partial\Omega}^{\kappa-\mathrm{n.t.}}(x) \text{ exists at } \sigma\text{-a.e. point } x \in \partial_*\Omega.\end{aligned} \tag{2.5.60}$$

Next, thanks to (2.5.60), (2.5.58), [112, Corollary 8.9.6], and [112, (8.9.8)], we see that the operator

$$H^p(\partial\Omega, \sigma) \ni f \longmapsto \big(\mathcal{T} f\big)\Big|_{\partial\Omega}^{\kappa-\mathrm{n.t.}} \in L^p(\partial_*\Omega, \sigma) \tag{2.5.61}$$

is well defined, linear and bounded. Moreover, from [113, (4.9.2)], (2.5.9), and (2.5.49) we also have that

$$H^p(\partial\Omega, \sigma) \ni f \longmapsto \frac{1}{2\sqrt{-1}}\widehat{k}\big(\nu(\cdot)\big)\mathfrak{H} f + T f \in L^p(\partial_*\Omega, \sigma) \tag{2.5.62}$$

is a well-defined, linear and bounded operator. Combining (2.5.61), (2.5.62), (2.5.56), and [113, (4.4.114)] then establishes (2.5.50) for each $f \in H^p(\partial\Omega, \sigma)$. □

Corollary 2.5.4 also extends naturally to the more inclusive scale of Lorentz-based Hardy spaces, as indicated below.

Corollary 2.5.5 *Suppose* $\Omega \subseteq \mathbb{R}^n$ *(where* $n \in \mathbb{N}$ *with* $n \geq 2$*) is a nonempty open set with the property that* $\partial\Omega$ *is a* UR *set. Abbreviate* $\sigma := \mathcal{H}^{n-1}\lfloor\partial\Omega$ *and denote by* ν *the geometric measure theoretic outward unit normal to* Ω*. Next, fix a kernel* $k \in \mathscr{C}^N(\mathbb{R}^n \setminus \{0\})$ *(for some integer* $N = N(n) \geq 4$ *large enough) which is odd and positive homogeneous of degree* $1 - n$*, and pick integrability exponents* $p \in \big(\frac{n-1}{n}, 1\big]$ *and* $q \in (0, \infty]$ *along with an aperture parameter* $\kappa \in (0, \infty)$*. Recall from* (2.3.52)-(2.3.55) *that the principal-value singular integral of formal convolution type with the kernel* k *on* $\partial\Omega$ *(cf.* (2.5.3)*) induces a well-defined linear and bounded operator*

$$T : H^{p,q}(\partial\Omega, \sigma) \longrightarrow L^{p,q}(\partial\Omega, \sigma). \tag{2.5.63}$$

Also, as in (2.4.24) *(with* $\Sigma := \partial\Omega$*), for each* $f \in H^{p,q}(\partial\Omega, \sigma)$ *define*

$$(\mathcal{T} f)(x) := {}_{(H^{p,q}(\partial\Omega,\sigma))^*}\big\langle k(x - \cdot)\big|_{\partial\Omega}, f\, \big\rangle_{H^{p,q}(\partial\Omega,\sigma)} \quad \textit{for all } x \in \Omega \tag{2.5.64}$$

which, according to [113, Proposition 4.8.9]*, acts in a coherent fashion with the original version of* $\mathcal{T}$ *defined as in* (2.4.3)-(2.4.4) *with* $\Sigma := \partial\Omega$*. Finally, recall from* [113, (4.9.5) in Theorem 4.9.1] *(used here with* $\Sigma := \partial\Omega$*) the* $L^{p,q}$*-filtering operator* $\mathfrak{H} : H^{p,q}(\partial\Omega, \sigma) \to L^{p,q}(\partial\Omega, \sigma)$.

Then for each $f \in H^{p,q}(\partial\Omega, \sigma)$ *the following jump-formula holds:*

$$\big(\mathcal{T} f\big)\Big|_{\partial\Omega}^{\kappa-\mathrm{n.t.}}(x) = \frac{1}{2\sqrt{-1}}\widehat{k}\big(\nu(x)\big)(\mathfrak{H} f)(x) + (T f)(x) \ \textit{for } \sigma\textit{-a.e. } x \in \partial_*\Omega. \tag{2.5.65}$$

A byproduct of Corollary 2.5.5 is that, in the above context, the operator T from (2.5.63) acts in a coherent manner with its principal-value counterpart (defined as in (2.3.15) with $\Sigma := \partial\Omega$), when the outputs are compared on $\partial_*\Omega$, i.e.,

$$\begin{gathered}\text{if } f \in L^1\Big(\partial\Omega, \tfrac{\sigma(x)}{1+|x|^{n-1}}\Big) \cap H^{p,q}(\partial\Omega,\sigma) \text{ for } p \in \big(\tfrac{n-1}{n}, 1\big] \text{ and } q \in (0,\infty],\\ \text{then with } Tf \in L^{p,q}(\partial\Omega,\sigma) \text{ considered in the sense of (2.5.63) we have}\\ (Tf)(x) = \lim_{\varepsilon\to 0^+} \int_{\substack{y\in\partial\Omega\\ |x-y|>\varepsilon}} k(x-y)f(y)\,d\sigma(y) \text{ at } \sigma\text{-a.e. point } x \in \partial_*\Omega.\end{gathered} \tag{2.5.66}$$

This follows from (2.5.65), [113, (4.9.7)], and the fact that, as remarked above, the operator $\mathcal{T}$ from (2.5.64) acts in a coherent fashion with its version originally defined as in (2.4.3)-(2.4.4) with $\Sigma := \partial\Omega$.

When Ω is in fact a UR domain, identity (2.5.65) may also be construed as an alternative formula for the action of the operator (2.5.63), namely

$$\begin{gathered}\text{if } f \in H^{p,q}(\partial\Omega,\sigma) \text{ with } p \in \big(\tfrac{n-1}{n}, 1\big] \text{ and } q \in (0,\infty]\\ \text{then } Tf \in L^{p,q}(\partial\Omega,\sigma) \text{ is given at } \sigma\text{-a.e. } x \in \partial\Omega \text{ by}\\ (Tf)(x) = -\frac{1}{2\sqrt{-1}}\widehat{k}\big(\nu(x)\big)(\mathfrak{H}f)(x) + \big(\mathcal{T}f\big)\Big|^{\kappa-\mathrm{n.t.}}_{\partial\Omega}(x).\end{gathered} \tag{2.5.67}$$

To provide a concrete example, recall from [113, (4.2.34)] that for each fixed point $x_0 \in \partial\Omega$ the distribution $\delta_{x_0} \in \big(\mathrm{Lip}_c(\partial\Omega)\big)'$ belongs to the weak Hardy space $H^{1,\infty}(\partial\Omega,\sigma)$. Then from (2.5.67), [113, (4.9.8)], and (2.4.26) we see that

$$\begin{gathered}\text{the operator } T\text{, considered as in (2.5.63) (with } p := 1 \text{ and } q := \infty),\\ \text{satisfies } (T\delta_{x_0})(x) = k(x-x_0) \text{ for } \sigma\text{-a.e. point } x \in \partial\Omega.\end{gathered} \tag{2.5.68}$$

Incidentally, this yields independent confirmation of the fact that $k(\cdot - x_0)\big|_{\partial\Omega}$ belongs to the weak Lebesgue space $L^{1,\infty}(\partial\Omega,\sigma)$ (see [112, Example 6.2.2] in this regard).

Proof of Corollary 2.5.5 When $q \in (0,\infty)$, we may follow in broad outline the same argument used earlier in the proof of Corollary 2.5.4. Once again, the idea is to apply [112, Proposition 6.2.11] for the same choices of the parameters $\mathscr{X}$, X, and μ as before, except that we now take $Y := H^{p,q}(\partial\Omega,\sigma)$, consider the linear operator $\mathcal{T}$ as in (2.5.64) and, this time, define $\mathcal{V} := H^{q_0}(\partial\Omega,\sigma) \cap L^{q_1}(\partial\Omega,\sigma)$ for some $q_0 \in \big(\tfrac{n-1}{n}, p\big)$ and $q_1 \in (1,\infty)$. According to [113, (4.3.145)], $\mathcal{V}$ is a dense subset of the present choice of Y, since $q < \infty$. Thanks to (2.4.27) and Corollary 2.5.4, all hypotheses of [112, Proposition 6.2.11] are then satisfied. This permits us to conclude that

$$\begin{gathered}\text{for each } f \in H^{p,q}(\partial\Omega,\sigma)\text{, the nontangential limit}\\ \big(\mathcal{T}f\big)\Big|^{\kappa-\mathrm{n.t.}}_{\partial\Omega}(x) \text{ exists at } \sigma\text{-a.e. point } x \in \partial_*\Omega.\end{gathered} \tag{2.5.69}$$

Granted this, (2.4.27), [112, Corollary 8.9.6], [112, (8.9.8)], [112, (6.2.16)], and [112, (6.2.17)] then imply that we have a well-defined, linear, and bounded operator

$$H^{p,q}(\partial\Omega,\sigma) \ni f \longmapsto \left(\mathcal{T}f\right)\Big|_{\partial\Omega}^{\kappa-\mathrm{n.t.}} \in L^{p,q}(\partial_*\Omega,\sigma). \tag{2.5.70}$$

In addition, from [113, (4.9.5) in Theorem 4.9.1], (2.5.9), (2.5.63), and [112, (6.2.17)] we also know that

$$H^{p,q}(\partial\Omega,\sigma) \ni f \longmapsto \frac{1}{2\sqrt{-1}}\widehat{k}\big(\nu(\cdot)\big)\mathfrak{H}f + Tf \in L^{p,q}(\partial_*\Omega,\sigma) \tag{2.5.71}$$

is a well-defined, linear and bounded operator. From (2.5.70), (2.5.71), (2.5.56), (2.5.50), and [113, (4.3.145)] we may now conclude that the jump-formula (2.5.65) is indeed valid for each $f \in H^{p,q}(\partial\Omega,\sigma)$ when $q < \infty$.

An argument which works for the full range $q \in (0,\infty]$ is based on the observation that once we have fixed $p_0 \in \left(\frac{n-1}{n}, p\right)$ and $p_1 \in (1,\infty)$ then [113, (1.3.41)], [113, (4.3.3)], and [112, (3.6.27)] ensure that the space $H^{p,q}(\partial\Omega,\sigma)$ embeds (continuously) into $H^{p_0}(\partial\Omega,\sigma) + L^{p_1}(\partial\Omega,\sigma)$. Hence, any given distribution $f \in H^{p,q}(\partial\Omega,\sigma)$ may be decomposed as $f = f_0 + f_1$ (in the sense of distributions on $\partial\Omega$) with $f_0 \in H^{p_0}(\partial\Omega,\sigma)$ and $f_1 \in L^{p_1}(\partial\Omega,\sigma)$. Then [113, Lemma 4.7.1] implies that

$$\left(\mathcal{T}f\right)(x) = \left(\mathcal{T}f_0\right)(x) + \left(\mathcal{T}f_1\right)(x) \text{ for each } x \in \Omega, \tag{2.5.72}$$

where $\mathcal{T}f_0$ is considered as in (2.5.47) (with p replaced by p_0) and $\mathcal{T}f_1$ is considered as in (2.5.1)-(2.5.2) (cf. also [112, (7.7.106)]). In turn, from (2.5.72), (2.5.50), and (2.5.4) we conclude that at σ-a.e. point $x \in \partial_*\Omega$ we have

$$\begin{aligned}\left(\mathcal{T}f\right)\Big|_{\partial\Omega}^{\kappa-\mathrm{n.t.}}(x) &= \frac{1}{2\sqrt{-1}}\widehat{k}\big(\nu(x)\big)(\mathfrak{H}f_0)(x) + (Tf_0)(x)\\ &\quad + \frac{1}{2\sqrt{-1}}\widehat{k}\big(\nu(x)\big)f_1(x) + (Tf_1)(x).\end{aligned} \tag{2.5.73}$$

From [113, (4.9.3), (4.9.6)] we also know that

$$(\mathfrak{H}f_0) + f_1 = \mathfrak{H}f_0 + \mathfrak{H}f_1 = \mathfrak{H}(f_0 + f_1) = \mathfrak{H}f \text{ at } \sigma\text{-a.e. point on } \partial\Omega, \tag{2.5.74}$$

where as (2.3.54) implies

$$Tf = Tf_0 + Tf_1 \text{ at } \sigma\text{-a.e. point on } \partial\Omega. \tag{2.5.75}$$

At this stage, (2.5.65) follows from (2.5.73), (2.5.74), and (2.5.75). □

We continue by presenting a version of Theorem 2.5.1 adapted to the scale of Muckenhoupt weighted Lebesgue spaces, as well as to the scale of Morrey spaces.

Corollary 2.5.6 *Suppose $\Omega \subseteq \mathbb{R}^n$ (where $n \in \mathbb{N}$ with $n \geq 2$) is a nonempty open set with the property that $\partial\Omega$ is a* UR *set. Abbreviate $\sigma := \mathcal{H}^{n-1}\lfloor\partial\Omega$ and denote by ν the geometric measure theoretic outward unit normal to Ω. Also, fix a sufficiently large integer $N = N(n)$ and consider a kernel $k \in \mathscr{C}^N(\mathbb{R}^n \setminus \{0\})$ as in* (2.3.3)*. Finally, define the integral operator $\mathcal{T}$, along with its principal version T, in relation to the domain Ω and the kernel k as in* (2.5.2)-(2.5.3)*, and fix an aperture parameter $\kappa \in (0, \infty)$.*

Then for each integrability exponent $p \in (1, \infty)$, each Muckenhoupt weight $w \in A_p(\partial\Omega, \sigma)$, and each function $f \in L^p(\partial\Omega, w\sigma)$, the following jump-formula holds:

$$\left(\mathcal{T}f\right)\Big|_{\partial\Omega}^{\kappa-\text{n.t.}}(x) = \frac{1}{2\sqrt{-1}}\widehat{k}(\nu(x))f(x) + (Tf)(x) \text{ for } \sigma\text{-a.e. } x \in \partial_*\Omega. \tag{2.5.76}$$

Moreover, the jump-formula (2.5.76) *also holds for each function f belonging to the Morrey space $M^{p,\lambda}(\partial\Omega, \sigma)$ with $p \in (1, \infty)$ and $\lambda \in (0, n-1)$.*

Proof Thanks to [112, (7.7.104)] and [113, (6.2.25)], we may invoke Theorem 2.5.1 to conclude that the jump-formula (2.5.76) holds both when $f \in L^p(\partial\Omega, w\sigma)$ with $p \in (1, \infty)$ and $w \in A_p(\partial\Omega, \sigma)$, as well as when $f \in M^{p,\lambda}(\partial\Omega, \sigma)$ with $p \in (1, \infty)$ and $\lambda \in (0, n-1)$.

An alternative proof of (2.5.76) in the former scenario, which avoids [112, (7.7.104)] and, instead, involves approximating functions in weighted Lebesgue spaces by functions in ordinary Lebesgue spaces, goes as follows. The fact that simple functions are dense in Lebesgue space associated with any given generic measure space (cf. [112, (3.1.11)]) implies that $L^p(\partial\Omega, \sigma) \cap L^p(\partial\Omega, w\sigma)$ is dense in $L^p(\partial\Omega, w\sigma)$. As such, having fixed an arbitrary function $f \in L^p(\partial\Omega, w\sigma)$ we can find a sequence

$$\begin{aligned} &\{f_j\}_{j\in\mathbb{N}} \subseteq L^p(\partial\Omega, \sigma) \cap L^p(\partial\Omega, w\sigma) \text{ such that} \\ &f_j \longrightarrow f \text{ in } L^p(\partial\Omega, w\sigma) \text{ as } j \to \infty. \end{aligned} \tag{2.5.77}$$

In concert with item *(10)* in Theorem 2.3.2 and item *(5)* in Theorem 2.4.1, the convergence result in (2.5.77) implies that

$$\begin{aligned} &\mathcal{N}_\kappa\big(\mathcal{T}(f - f_j)\big) \to 0 \text{ in } L^p(\partial\Omega, w\sigma) \text{ and} \\ &Tf_j \to Tf \text{ in } L^p(\partial\Omega, w\sigma) \text{ as } j \to \infty. \end{aligned} \tag{2.5.78}$$

As such, by passing to a subsequence if necessary, matters may be arranged so that, as $j \to \infty$,

$$\begin{aligned} &f_j(x) \to f(x), \quad (Tf_j)(x) \to (Tf)(x), \text{ and} \\ &\mathcal{N}_\kappa\big(\mathcal{T}(f - f_j)\big)(x) \to 0, \text{ for } \sigma\text{-a.e. } x \in \partial\Omega. \end{aligned} \tag{2.5.79}$$

Fix now $j \in \mathbb{N}$ arbitrary. Then for σ-a.e. $x \in \partial_*\Omega$ we may rely on Corollary 2.5.2 to write

$$\begin{aligned}
&\limsup_{\substack{z\in\Gamma_\kappa(x)\\ z\to x}}\Big|(\mathcal{T}f)(z)-\Big(\tfrac{1}{2\sqrt{-1}}\widehat{k}\big(\nu(x)\big)f(x)+(Tf)(x)\Big)\Big| \\
&\qquad \le \limsup_{\substack{z\in\Gamma_\kappa(x)\\ z\to x}}\Big|\big(\mathcal{T}(f-f_j)\big)(z)\Big| \\
&\qquad\quad + \limsup_{\substack{z\in\Gamma_\kappa(x)\\ z\to x}}\Big|(\mathcal{T}f_j)(z)-\Big(\tfrac{1}{2\sqrt{-1}}\widehat{k}\big(\nu(x)\big)f_j(x)+(Tf_j)(x)\Big)\Big| \\
&\qquad\quad + \Big|\Big(\tfrac{1}{2\sqrt{-1}}\widehat{k}\big(\nu(x)\big)f_j(x)+(Tf_j)(x)\Big)-\Big(\tfrac{1}{2\sqrt{-1}}\widehat{k}\big(\nu(x)\big)f(x)+(Tf)(x)\Big)\Big| \\
&\qquad \le \Big(\mathcal{N}_\kappa\big(\mathcal{T}(f-f_j)\big)\Big)(x)+C_k|f_j(x)-f(x)|+\big|(Tf_j)(x)-(Tf)(x)\big|,
\end{aligned} \tag{2.5.80}$$

where $C_k \in (0,\infty)$ appearing above is a constant which depends only on the kernel k. In light of (2.5.79) we may now conclude that for each $f \in L^p(\partial\Omega, w\sigma)$ the jump-formula (2.5.76) is valid at σ-a.e. point $x \in \partial_*\Omega$. □

Here is a two-sided jump-formula, across the boundary of a UR domain. Results similar in spirit to Corollary 2.5.7 below are also valid in relation to the jump-formulas from Corollary 2.5.4, Corollary 2.5.5, and Corollary 2.5.6.

Corollary 2.5.7 *For each $n \in \mathbb{N}$ with $n \ge 2$ there exists an integer $N = N(n) \ge 4$ with the following significance. Let $k \in \mathscr{C}^N(\mathbb{R}^n \setminus \{0\})$ be an odd, positive homogeneous kernel of degree $1-n$ (cf. (2.3.3)), and let $\Omega \subsetneq \mathbb{R}^n$ be a* UR *domain. Abbreviate $\sigma := \mathcal{H}^{n-1}\lfloor\partial\Omega$ and denote by ν the geometric measure theoretic outward unit normal to Ω. Next, define*

$$\Omega_+ := \Omega \text{ and } \Omega_- := \mathbb{R}^n \setminus \overline{\Omega}, \tag{2.5.81}$$

then consider the integral operators acting on functions $f \in L^1\Big(\partial\Omega, \frac{\sigma(x)}{1+|x|^{n-1}}\Big)$ according to

$$(\mathcal{T}_\pm f)(x) := \int_{\partial\Omega} k(x-y)f(y)\,\mathrm{d}\sigma(y) \text{ for all } x \in \Omega_\pm \text{ and} \tag{2.5.82}$$

$$(Tf)(x) := \lim_{\varepsilon\to 0^+} \int_{\substack{y\in\partial\Omega\\ |x-y|>\varepsilon}} k(x-y)f(y)\,\mathrm{d}\sigma(y) \text{ for } \sigma\text{-a.e. } x \in \partial\Omega. \tag{2.5.83}$$

Then Ω_- is also a UR *domain, whose topological boundary coincides with that of Ω, and whose geometric measure theoretic boundary agrees with that of Ω (i.e., $\partial(\Omega_-) = \partial\Omega$ and $\partial_*(\Omega_-) = \partial_*\Omega$). In addition, the geometric measure theoretic outward unit normal to Ω_- is $-\nu$ at σ-a.e. point on $\partial\Omega$. Furthermore, for each $\kappa > 0$ and each $f \in L^1\Big(\partial\Omega, \frac{\sigma(x)}{1+|x|^{n-1}}\Big)$ the following jump-formulas hold:*

$$\left(\mathcal{T}_{\pm} f\right)\Big|_{\partial\Omega_{\pm}}^{\kappa-\mathrm{n.t.}}(x) = \pm\frac{1}{2\sqrt{-1}}\widehat{k}\big(\nu(x)\big)f(x) + (Tf)(x) \textit{ for } \sigma\textit{-a.e. } x \in \partial\Omega. \tag{2.5.84}$$

Proof All claims about the nature of the set Ω_- have been already established in [112, Lemma 5.10.9]. Granted these, Theorem 2.5.1 applies both to Ω_+ and Ω_-, a scenario in which the jump-formula (2.5.4) yields (2.5.84), thanks to (2.5.10) (keeping in mind that the geometric measure theoretic outward unit normal to Ω_- is $-\nu$ at σ-a.e. point on $\partial\Omega$). □

The proof of Theorem 2.5.1 will be presented later on, as a corollary of the more general result recorded in Theorem 2.5.38 where we deal with variable coefficient kernels. In turn, Theorem 2.5.38 is proved by first treating in Proposition 2.5.37 kernels of a special algebraic structure, i.e., functions of the form

$$\begin{gathered} k(x) = \frac{P(x)}{|x|^{n-1+\deg P}}, \qquad \text{for each } x \in \mathbb{R}^n \setminus \{0\}, \\ \text{with } P \text{ homogeneous, odd, harmonic polynomial in } \mathbb{R}^n, \end{gathered} \tag{2.5.85}$$

which then permits us to deal with arbitrary kernels as in (2.3.3) via a spherical harmonic expansion. Next, treating the case of kernels as in (2.5.85) is done via induction on $\deg P$, the degree of the homogeneous, odd, harmonic polynomial P in $\mathbb{R}^n$. The initial step in this induction involves polynomials of the form $P(x) = x_j$, with $j \in \{1, \dots, n\}$. Up to normalization, these produce Riesz transforms in the context of (2.5.85) which are treated in Proposition 2.5.35. As a preamble to the proof of this latter result, in Propositions 2.5.21-2.5.27 we deal with the harmonic double layer and other related operators (whose integral kernels are obtained by taking certain "tangential" derivatives on the fundamental solution for the Laplacian). Finally, the proofs of Propositions 2.5.21-2.5.27 require certain auxiliary results, presented in Lemmas 2.5.8-2.5.20.

The result in our lemma next refines and strengthens [63, Proposition 3.3, p. 2628] and [120, Lemma 2.5]. The main conclusion (see (2.5.88) below) may be interpreted as saying that, for a given set of locally finite perimeter E, the surface measure of the portion contained in E of any sphere with center on $\partial^* E$ is roughly the area of a half-sphere, and the degree of accuracy improves indefinitely as the radius shrinks to zero trough a suitable set of admissible values. For this reason we shall refer to a result of this type as a "half-sphere lemma."

Lemma 2.5.8 *Let $E \subseteq \mathbb{R}^n$ be a set of locally finite perimeter. Then for each $x \in \partial^* E$ there exists a Lebesgue measurable set $O_x \subset (0, 1)$ of density 1 at 0, i.e., satisfying*

$$\lim_{\varepsilon \to 0^+} \frac{\mathcal{L}^1\big(O_x \cap (0, \varepsilon)\big)}{\varepsilon} = 1, \tag{2.5.86}$$

with the property that

$$\begin{aligned}&\textit{the sets } E\cap\partial B(x,r) \textit{ and } \partial B(x,r)\setminus E\\&\textit{are } \mathcal{H}^{n-1}\textit{-measurable for } \mathcal{L}^1\textit{-a.e. } r\in O_x\end{aligned} \tag{2.5.87}$$

and, with ω_{n-1} *denoting the surface area of the unit sphere in* $\mathbb{R}^n$,

$$\lim_{O_x\ni r\to 0}\frac{\mathcal{H}^{n-1}\big(E\cap\partial B(x,r)\big)}{\omega_{n-1}r^{n-1}}=\frac{1}{2}, \tag{2.5.88}$$

$$\lim_{O_x\ni r\to 0}\frac{\mathcal{H}^{n-1}\big(\partial B(x,r)\setminus E\big)}{\omega_{n-1}r^{n-1}}=\frac{1}{2}. \tag{2.5.89}$$

Proof Having fixed some $x\in\partial^*E$ and $R\in(0,\infty)$ arbitrary, invoke [112, (5.3.7) in Proposition 5.3.2] with $m:=1$, $A:=B(x,R)$, $g:=\mathbf{1}_{B(x,R)\cap E}$, and the Lipschitz function

$$f:\mathbb{R}^n\longrightarrow\mathbb{R} \text{ given by } f(z):=|z-x| \text{ for each } z\in\mathbb{R}^n. \tag{2.5.90}$$

Bearing in mind that $R>0$ is arbitrary, this eventually proves that

$$E\cap\partial B(x,r) \text{ is } \mathcal{H}^{n-1}\text{-measurable for } \mathcal{L}^1\text{-a.e. } r\in(0,\infty). \tag{2.5.91}$$

Moving on, fix $x\in\partial^*E$ and define the half-spaces

$$H_x^{\pm}:=\{y\in\mathbb{R}^n:\ \pm\nu(x)\cdot(y-x)>0\}. \tag{2.5.92}$$

For each number $r>0$, set $\partial^{\pm}B(x,r):=\partial B(x,r)\cap H_x^{\pm}$ and introduce

$$\begin{aligned}W(x,r)&:=\partial^-B(x,r)\triangle\big[E\cap\partial B(x,r)\big]\\&=\partial B(x,r)\cap\big[E\triangle H_x^-\big]\end{aligned} \tag{2.5.93}$$

where, generally speaking, $U\triangle V:=(U\setminus V)\cup(V\setminus U)$ denotes the symmetric difference of any two given sets U,V. For an arbitrary number $R\in(0,\infty)$, the idea is now to invoke formula [112, (5.3.7) in Proposition 5.3.2] with $m:=1$, $A:=B(x,R)$, the Lipschitz function f from (2.5.90), and $g:=\mathbf{1}_{B(x,R)\cap(E\triangle H_x^-)}$. In particular, these choices entail $A\cap f^{-1}\big(\{r\}\big)=\varnothing$ whenever the number $r\in\mathbb{R}\setminus[0,R]$, and $(Df)(z)=(z-x)/|z-x|$ which implies $Jf=1$ at every point in $\mathbb{R}^n\setminus\{x\}$. From [112, Proposition 5.3.2] we then conclude that

$$\text{the set } W(x,r) \text{ is } \mathcal{H}^{n-1}\text{-measurable for } \mathcal{L}^1\text{-a.e. } r\in(0,R) \tag{2.5.94}$$

and

$$\begin{aligned}\int_0^R\mathcal{H}^{n-1}\big(W(x,r)\big)\,dr&=\mathcal{L}^n\big(B(x,R)\cap(E\triangle H_x^-)\big)\\&=\mathcal{L}^n\Big(\big(B(x,R)\cap E\big)\cap H_x^+\Big)+\mathcal{L}^n\Big(\big(B(x,R)\setminus E\big)\cap H_x^-\Big).\end{aligned} \tag{2.5.95}$$

In turn, from (2.5.95) and [112, (5.6.57)-(5.6.58)] we conclude that

$$\int_0^R \mathcal{H}^{n-1}\big(W(x,r)\big)\,dr = o(R^n) \text{ as } R \to 0^+. \tag{2.5.96}$$

To proceed, consider the function

$$\begin{gathered}\phi : (0,1) \longrightarrow [0,\infty) \text{ given by} \\ \phi(r) := r^{1-n}\mathcal{H}^{n-1}\big(W(x,r)\big) \text{ for } \mathcal{L}^1\text{-a.e. } r \in (0,1).\end{gathered} \tag{2.5.97}$$

It follows from (2.5.96) that

$$\int_{R/2}^R \phi(r)\,dr \leq \Big(\frac{R}{2}\Big)^{1-n} \int_0^R \mathcal{H}^{n-1}\big(W(x,r)\big)\,dr = o(R) \text{ as } R \to 0^+. \tag{2.5.98}$$

Bring in the dyadic intervals $I_k := \big[2^{-(k+1)}, 2^{-k}\big]$ for $k \in \mathbb{N}_0$ and note that (2.5.98) entails

$$\delta_k := \fint_{I_k} \phi(r)\,dr \longrightarrow 0^+ \text{ as } k \to \infty. \tag{2.5.99}$$

For each $k \in \mathbb{N}_0$ split

$$I_k = A_k \cup B_k, \text{ with } B_k := \big\{r \in I_k : \phi(r) > \sqrt{\delta_k}\big\} \text{ and } A_k := I_k \setminus B_k. \tag{2.5.100}$$

Then Chebytcheff's inequality permits us to estimate

$$\frac{\mathcal{L}^1(B_k)}{\mathcal{L}^1(I_k)} \leq \frac{1}{\sqrt{\delta_k}} \fint_{I_k} \phi(r)\,dr = \sqrt{\delta_k}, \qquad \forall k \in \mathbb{N}_0. \tag{2.5.101}$$

In light of (2.5.99), this implies that if we now define

$$O_x := \bigcup_{k\in\mathbb{N}_0} A_k \subset (0,1), \tag{2.5.102}$$

then

$$\lim_{O_x \ni r \to 0} \phi(r) = 0, \tag{2.5.103}$$

hence

$$\lim_{O_x \ni r \to 0} \frac{\mathcal{H}^{n-1}\big(W(x,r)\big)}{r^{n-1}} = 0. \tag{2.5.104}$$

We claim that (2.5.86) also holds for this choice of O_x. To see that this is the case, assume some arbitrary $\theta > 0$ has been fixed. For each $\varepsilon \in (0,1)$, let $N_\varepsilon \in \mathbb{N}_0$ be such that $2^{-N_\varepsilon - 1} < \varepsilon \leq 2^{-N_\varepsilon}$. Since $N_\varepsilon \to \infty$ as $\varepsilon \to 0^+$, it follows from (2.5.99) that there exists $\varepsilon_\theta > 0$ with the property that

$$\delta_k \leq \theta^2 \text{ whenever } 0 < \varepsilon < \varepsilon_\theta \text{ and } k \geq N_\varepsilon. \tag{2.5.105}$$

Assuming that $0 < \varepsilon < \varepsilon_\theta$ we may then estimate

$$\begin{aligned} 0 &\le \frac{\varepsilon - \mathcal{L}^1\big(O_x \cap (0,\varepsilon)\big)}{\varepsilon} = \varepsilon^{-1}\mathcal{L}^1\big((0,\varepsilon)\setminus O_x\big) \\ &\le \varepsilon^{-1}\mathcal{L}^1\big((0,2^{-N_\varepsilon})\setminus O_x\big) = \varepsilon^{-1}\sum_{k=N_\varepsilon}^{\infty}\mathcal{L}^1(B_k) \\ &\le \varepsilon^{-1}\sum_{k=N_\varepsilon}^{\infty}\mathcal{L}^1(I_k)\sqrt{\delta_k} \le \varepsilon^{-1}\theta\, 2^{-N_\varepsilon} \le \theta/2. \end{aligned} \tag{2.5.106}$$

This finishes the proof of (2.5.86).

At this stage there remains to observe that since, generally speaking,

$$\big|\mathcal{H}^{n-1}(U) - \mathcal{H}^{n-1}(V)\big| \le \mathcal{H}^{n-1}(U\Delta V) \tag{2.5.107}$$

for any $\mathcal{H}^{n-1}$-measurable sets $U, V \subseteq \mathbb{R}^n$ of finite $\mathcal{H}^{n-1}$-measure, from (2.5.91), (2.5.107), and (2.5.93) we obtain

$$\begin{aligned} \left|\frac{\mathcal{H}^{n-1}\big(E\cap\partial B(x,r)\big)}{\omega_{n-1}r^{n-1}} - \frac{1}{2}\right| &= \frac{\big|\mathcal{H}^{n-1}\big(E\cap\partial B(x,r)\big) - \mathcal{H}^{n-1}\big(\partial^- B(x,r)\big)\big|}{\omega_{n-1}r^{n-1}} \\ &\le \frac{\mathcal{H}^{n-1}\big(W(x,r)\big)}{\omega_{n-1}r^{n-1}} \quad \text{for } \mathcal{L}^1\text{-a.e. } r\in(0,1). \end{aligned} \tag{2.5.108}$$

Then formula (2.5.88) is a consequence of this and (2.5.104). Moreover, (2.5.89) is readily implied by (2.5.88) and (2.5.91). Finally, taking (2.5.91) into account and eventually adjusting O_x by an $\mathcal{L}^1$-nullset, we may also ensure that (2.5.87) holds. □

It turns out that the "half-sphere lemma" proved earlier for arbitrary sets of locally finite perimeter (cf. Lemma 2.5.8) has a version in which the radius approaches zero unrestrictedly, except for an $\mathcal{L}^1$-nullset. En route to Lemma 2.5.12, where such a version is formulated, we first establish several pointwise convergence results.

Lemma 2.5.9 *Let $\Omega \subseteq \mathbb{R}^n$ be a set of locally finite perimeter whose topological boundary, $\partial\Omega$, is compact. Abbreviate $\sigma_* := \mathcal{H}^{n-1}\lfloor\partial_*\Omega$ and denote by ν the geometric measure theoretic outward unit normal to Ω. Make the assumption that either Ω is bounded or $n \ge 2$. Also, consider a vector-valued function*

$$\begin{gathered} \vec{k} = (k_j)_{1\le j\le n} \in \big[\mathscr{C}^2(\mathbb{R}^n\setminus\{0\})\big]^n \ \textit{which is} \\ \textit{odd, positive homogeneous of degree } 1-n, \\ \textit{and which satisfies } \operatorname{div}\vec{k} = 0 \ \textit{in } \mathbb{R}^n\setminus\{0\}. \end{gathered} \tag{2.5.109}$$

Finally, set

$$\vartheta := \int_{S^{n-1}} \langle \omega, \vec{k}(\omega) \rangle \, d\mathcal{H}^{n-1}(\omega) \in \mathbb{C}. \tag{2.5.110}$$

Then for σ_*-a.e. point $x \in \partial_*\Omega$ there holds

$$\lim_{\varepsilon \to 0^+} \int_{\substack{y \in \partial_*\Omega \\ |x-y|>\varepsilon}} \langle \nu(y), \vec{k}(x-y) \rangle \, d\sigma_*(y) = \begin{cases} -\dfrac{\vartheta}{2} & \text{if } \Omega \text{ is bounded,} \\[2ex] +\dfrac{\vartheta}{2} & \text{if } \Omega \text{ is unbounded.} \end{cases} \tag{2.5.111}$$

In particular, for σ_*-a.e. point $x \in \partial_*\Omega$ one has

$$\lim_{\varepsilon \to 0^+} \frac{1}{\omega_{n-1}} \int_{\substack{y \in \partial_*\Omega \\ |x-y|>\varepsilon}} \frac{\langle \nu(y), y-x \rangle}{|x-y|^n} \, d\sigma_*(y) = \begin{cases} +\frac{1}{2} & \text{if } \Omega \text{ is bounded,} \\ -\frac{1}{2} & \text{if } \Omega \text{ is unbounded.} \end{cases} \tag{2.5.112}$$

Proof Let us first work under the assumption that $\Omega \subseteq \mathbb{R}^n$ is a bounded set of locally finite perimeter. Pick some $x \in \mathbb{R}^n$ and, for each $\ell \in \mathbb{N}$, select a scalar function $\psi_\ell \in \mathscr{C}^\infty(\mathbb{R}^n)$ satisfying

$$\psi_\ell \equiv 0 \text{ in } B\big(x, 1/(4\ell)\big) \text{ and } \psi_\ell \equiv 1 \text{ in } \mathbb{R}^n \setminus B\big(x, 1/(2\ell)\big). \tag{2.5.113}$$

The idea is to use [112, Lemma 5.7.2] for the bounded set $E := \Omega$ and the vector field

$$\vec{F}_\ell \in \big[\mathscr{C}^2(\mathbb{R}^n)\big]^n, \quad \vec{F}_\ell(y) := \psi_\ell(y)\vec{k}(x-y) \text{ for all } y \in \mathbb{R}^n. \tag{2.5.114}$$

Given that the vector field $\vec{F}_\ell$ is divergence-free near $\Omega \setminus B(x, 1/\ell)$ (cf. the last line in (2.5.109)), this lemma implies that there exists an $\mathcal{L}^1$-nullset $N_{x,\ell} \subset (0,\infty)$ such that for each $\varepsilon \in (1/\ell, \infty) \setminus N_{x,\ell}$ the set $\Omega \cap \partial B(x,\varepsilon)$ is $\mathcal{H}^{n-1}$-measurable and we have

$$\begin{aligned} \int_{\substack{|x-y|>\varepsilon \\ y \in \partial_*\Omega}} \langle \nu(y), \vec{k}(x-y) \rangle \, d\sigma_*(y) &= \int_{(\partial_*\Omega)\setminus\overline{B(x,\varepsilon)}} \nu(y) \cdot \vec{F}_\ell(y) \, d\sigma_*(y) \\ &= \int_{\Omega \cap \partial B(x,\varepsilon)} \Big(\frac{y-x}{\varepsilon}\Big) \cdot \vec{F}_\ell(y) \, d\mathcal{H}^{n-1}(y) \\ &= \int_{\substack{|x-y|=\varepsilon \\ y \in \Omega}} \big((y-x)_j/\varepsilon\big) \, k_j(x-y) \, d\mathcal{H}^{n-1}(y). \end{aligned} \tag{2.5.115}$$

For further use, introduce

$$N_x := \bigcup_{\ell\in\mathbb{N}} N_{x,\ell} \tag{2.5.116}$$

and note that $N_x \subset (0,\infty)$ is an $\mathcal{L}^1$-nullset. In particular, [112, Lemma 5.7.2] also implies (via the same type of argument as in (2.5.115)) that for each $\varepsilon \in (0,\infty)\setminus N_x$ we have

$$\int_{\substack{|x-y|\geq\varepsilon\\ y\in\partial_*\Omega}} \langle \nu(y), \vec{k}(x-y)\rangle \, d\sigma_*(y) = \int_{\substack{|x-y|=\varepsilon\\ y\in\Omega}} \big((y-x)_j/\varepsilon\big)\, k_j(x-y)\, d\mathcal{H}^{n-1}(y). \tag{2.5.117}$$

Assume next that in fact $x \in \partial^*\Omega$ and, for each $r \in (0,\infty)$, abbreviate

$$\begin{aligned} &\partial^- B(x,r) := \partial B(x,r) \cap H_x^- \ \text{ where} \\ &H_x^- := \{y \in \mathbb{R}^n : \nu(x)\cdot(y-x) < 0\}. \end{aligned} \tag{2.5.118}$$

For each $r \in (0,\infty)$ let us also introduce

$$W(x,r) := \partial^- B(x,r) \triangle \big[\Omega \cap \partial B(x,r)\big] = \partial B(x,r) \cap \big[\Omega \triangle H_x^-\big] \tag{2.5.119}$$

where, as in the past, $\triangle$ denotes the set-theoretic symmetric difference operator. Since, generally speaking,

$$\Big|\int_A f\, d\mathcal{H}^{n-1} - \int_B f\, d\mathcal{H}^{n-1}\Big| \leq \int_{A\triangle B} |f|\, d\mathcal{H}^{n-1} \tag{2.5.120}$$

for any two $\mathcal{H}^{n-1}$-measurable sets $A, B \subseteq \mathbb{R}^n$ and any function $f \in L^1(A\cup B, \mathcal{H}^{n-1})$, on the one hand we obtain, for each $\varepsilon \in (0,\infty)\setminus N_x$,

$$\begin{aligned} \Bigg| &\int_{\Omega\cap\partial B(x,\varepsilon)} \big((y-x)_j/\varepsilon\big)\, k_j(x-y)\, d\mathcal{H}^{n-1}(y) \\ &\qquad - \int_{\partial^- B(x,\varepsilon)} \big((y-x)_j/\varepsilon\big)\, k_j(x-y)\, d\mathcal{H}^{n-1}(y)\Bigg| \\ &\leq \frac{\mathcal{H}^{n-1}\big(W(x,\varepsilon)\big)}{\varepsilon^{n-1}} \cdot \sup_{S^{n-1}} |\vec{k}|. \end{aligned} \tag{2.5.121}$$

Recall that $\vec{k}$ is odd and positive homogeneous of degree $1-n$ (cf. (2.5.109)). Fix $\varepsilon \in (0,\infty)\setminus N_x$. Making the changes of variables $\omega := \pm(y-x)/\varepsilon \in S^{n-1}$ yields, respectively,

$$\int_{\partial^- B(x,\varepsilon)} \big((y-x)_j/\varepsilon\big)\, k_j(x-y)\, d\mathcal{H}^{n-1}(y) = -\int_{\substack{\omega\in S^{n-1}\\ \nu(x)\cdot\omega<0}} \big\langle \omega, \vec{k}(\omega)\big\rangle\, d\mathcal{H}^{n-1}(y) \tag{2.5.122}$$

and

$$\int_{\partial^- B(x,\varepsilon)} \big((y-x)_j/\varepsilon\big)\, k_j(x-y)\, d\mathcal{H}^{n-1}(y) = -\int_{\substack{\omega\in S^{n-1}\\ \nu(x)\cdot\omega>0}} \big\langle \omega, \vec{k}(\omega)\big\rangle\, d\mathcal{H}^{n-1}(y). \tag{2.5.123}$$

From (2.5.122)-(2.5.123) and (2.5.110) we then conclude that

$$\int_{\partial^- B(x,\varepsilon)} \big((y-x)_j/\varepsilon\big)\, k_j(x-y)\, d\mathcal{H}^{n-1}(y) = -\frac{\vartheta}{2} \tag{2.5.124}$$

for each $\varepsilon \in (0,\infty)\setminus N_x$.

In addition, from (2.5.104) we know that there exists a Lebesgue measurable set $O_x \subset (0,1)$ of density 1 at 0, i.e., satisfying (2.5.86), with the property that the set $\Omega \cap \partial B(x,\varepsilon)$ happens to be $\mathcal{H}^{n-1}$-measurable for $\mathcal{L}^1$-a.e. $\varepsilon \in O_x$ and

$$\lim_{O_x \ni \varepsilon \to 0} \frac{\mathcal{H}^{n-1}\big(W(x,\varepsilon)\big)}{\varepsilon^{n-1}} = 0. \tag{2.5.125}$$

From (2.5.121), (2.5.124), and (2.5.125) we then obtain

$$\lim_{O_x\setminus N_x \ni \varepsilon \to 0} \int_{\substack{|x-y|=\varepsilon\\ y\in\Omega}} \big((y-x)_j/\varepsilon\big)\, k_j(x-y)\, d\mathcal{H}^{n-1}(y) = -\frac{\vartheta}{2}. \tag{2.5.126}$$

In turn, from (2.5.115) and (2.5.126) we conclude that, on the one hand,

$$\lim_{O_x\setminus N_x \ni \varepsilon \to 0} \int_{\substack{y\in\partial_*\Omega\\ |x-y|>\varepsilon}} \big\langle \nu(y), \vec{k}(x-y)\big\rangle\, d\sigma_*(y) = -\frac{\vartheta}{2} \quad \text{for each } x \in \partial^*\Omega. \tag{2.5.127}$$

On the other hand, [112, Proposition 5.6.7] is presently applicable (with f taken to be the scalar components of the geometric measure theoretic outward unit normal to Ω; indeed, given that the set $\partial\Omega$ is bounded, the finite perimeter assumption ensures that $\sigma_*(\partial_*\Omega) = \mathcal{H}^{n-1}(\partial_*\Omega) < +\infty$). As such, [112, (5.6.45)] may be used to conclude that the limit

$$\lim_{\varepsilon\to 0^+} \int_{\substack{y\in\partial_*\Omega\\ |x-y|>\varepsilon}} \big\langle \nu(y), \vec{k}(x-y)\big\rangle\, d\sigma_*(y) \ \text{ exists for } \mathcal{H}^{n-1}\text{-a.e. } x \in \partial_*\Omega. \tag{2.5.128}$$

Having established (2.5.127) and (2.5.128), the version of (2.5.111) corresponding to the case when Ω is bounded follows, on account of [112, (5.6.21)].

Next, the version of (2.5.111) corresponding to the case when the set Ω is unbounded and $n \geq 2$ becomes a consequence of what we have proved so far, applied to $\Omega^c := \mathbb{R}^n \setminus \Omega$. Indeed, since $n \geq 2$ and $\partial\Omega$ is compact, from [112, Lemma 5.10.10], [112, (5.2.3)], [112, (5.6.16)], and assumptions, we conclude that $\Omega^c := \mathbb{R}^n \setminus \Omega$ is a bounded set of locally finite perimeter whose geometric measure theoretic outward unit normal is $-\nu$ at each point on $\partial^*(\Omega^c) = \partial^*\Omega$. This completes the proof of (2.5.111).

Finally, specializing (2.5.111) to the case when

$$\vec{k}(x) := -\frac{1}{\omega_{n-1}} \frac{x}{|x|^n} \text{ for all } x \in \mathbb{R}^n \setminus \{0\}, \tag{2.5.129}$$

yields (2.5.112). □

For further use, we isolate a technical result in the following lemma.

Lemma 2.5.10 *Let $\Omega \subseteq \mathbb{R}^n$ be a set of locally finite perimeter and consider a vector-valued function $\vec{k} \in \left[\mathscr{C}^2(\mathbb{R}^n \setminus \{0\})\right]^n$ which is odd, positive homogeneous of degree $1 - n$, and satisfies* $\operatorname{div}\vec{k} = 0$ *in $\mathbb{R}^n \setminus \{0\}$. Also, abbreviate*

$$\vartheta := \int_{S^{n-1}} \langle \omega, \vec{k}(\omega) \rangle \, d\mathcal{H}^{n-1}(\omega) \in \mathbb{C}. \tag{2.5.130}$$

Then for $\mathcal{H}^{n-1}$-a.e. point $x \in \partial_*\Omega$ there exists some $\mathcal{L}^1$-nullset $N_x \subset (0, \infty)$ with the property that for each $\varepsilon \in (0, \infty) \setminus N_x$ the set $\Omega \cap \partial B(x, \varepsilon)$ is $\mathcal{H}^{n-1}$-measurable and

$$\lim_{\substack{\varepsilon \in (0,\infty)\setminus N_x \\ \varepsilon \to 0}} \int_{\Omega \cap \partial B(x,\varepsilon)} \Big\langle \frac{y - x}{\varepsilon}, \vec{k}(x - y) \Big\rangle \, d\mathcal{H}^{n-1}(y) = -\frac{\vartheta}{2}. \tag{2.5.131}$$

Proof Assume first that Ω is bounded. As in the past, abbreviate $\sigma_* := \mathcal{H}^{n-1} \lfloor \partial_*\Omega$. From (2.5.115) and (2.5.117) we know that if for each $x \in \mathbb{R}^n$ we define N_x as in (2.5.116) then $N_x \subset (0, \infty)$ is an $\mathcal{L}^1$-nullset with the property that for each $\varepsilon \in (0, \infty) \setminus N_x$ the set $\Omega \cap \partial B(x, \varepsilon)$ is $\mathcal{H}^{n-1}$-measurable and

$$\begin{aligned} \int_{\Omega \cap \partial B(x,\varepsilon)} \Big\langle \frac{y - x}{\varepsilon}, \vec{k}(x - y) \Big\rangle \, d\mathcal{H}^{n-1}(y) &= \int_{\substack{|x-y|>\varepsilon \\ y \in \partial_*\Omega}} \langle \nu(y), \vec{k}(x - y) \rangle \, d\sigma_*(y) \\ &= \int_{\substack{|x-y|\geq\varepsilon \\ y \in \partial_*\Omega}} \langle \nu(y), \vec{k}(x - y) \rangle \, d\sigma_*(y). \end{aligned} \tag{2.5.132}$$

Then (2.5.131) in the case when Ω is bounded is seen from (2.5.132) and (2.5.111).

To treat the case when Ω is not necessarily bounded we reason as follows. From [112, Lemma 5.7.3] we know that there exists some $\mathcal{L}^1$-nullset $N_0 \subset (0, \infty)$ such that for each $r \in (0, \infty) \setminus N_0$ the set $\Omega_r := \Omega \cap B(0, r)$ is bounded, has finite perimeter, and

$$\partial_*\Omega_r \equiv \big(\Omega \cap \partial B(0, r)\big) \cup \big(\partial_*\Omega \cap B(0, r)\big) \text{ modulo } \mathcal{H}^{n-1}. \tag{2.5.133}$$

Select a sequence $\{r_j\}_{j\in\mathbb{N}} \subseteq (2, \infty) \setminus N_0$ with $r_j \nearrow \infty$ as $j \to \infty$. From what we have established in the first part of the current proof and (2.5.133) we conclude that, given any integer $j \in \mathbb{N}$, there exists an $\mathcal{H}^{n-1}$-nullset A_j such that for each point $x \in \big(\partial_*\Omega \cap B(0, r_j)\big) \setminus A_j$ we may find some $\mathcal{L}^1$-nullset $N_{x,j} \subset (0, \infty)$ with the property that for each $\varepsilon \in (0, \infty) \setminus N_{x,j}$ the set $\Omega_{r_j} \cap \partial B(x, \varepsilon)$ is $\mathcal{H}^{n-1}$-measurable and

$$\lim_{\substack{\varepsilon \in (0,\infty)\setminus N_{x,j} \\ \varepsilon \to 0}} \int_{\Omega_{r_j} \cap \partial B(x,\varepsilon)} \Big\langle \frac{y - x}{\varepsilon}, \vec{k}(x - y) \Big\rangle \, d\mathcal{H}^{n-1}(y) = -\frac{\vartheta}{2}. \tag{2.5.134}$$

Moving on, introduce

$$A := \bigcup_{j\in\mathbb{N}} A_j \tag{2.5.135}$$

which continues to be an $\mathcal{H}^{n-1}$-nullset. In addition, for each $x \in \partial_*\Omega \setminus A$ define

$$N_x := \bigcup_{j\in\mathbb{N}:\, |x|<r_j} N_{x,j} \tag{2.5.136}$$

so N_x is a meaningfully define $\mathcal{L}^1$-nullset subset of $(0, \infty)$. Let us also observe that

$$\begin{gathered} \Omega_{r_j} \cap \partial B(x, \varepsilon) = \Omega \cap \partial B(x, \varepsilon) \\ \text{for each } x \in B(0, r_j/2) \text{ and } \varepsilon \in (0, r_j/2). \end{gathered} \tag{2.5.137}$$

Combining (2.5.134)-(2.5.137) then proves the following result: Given any $j \in \mathbb{N}$ and any $x \in \big(\partial_*\Omega \cap B(0, r_j/2)\big) \setminus A$, whenever $\varepsilon \in (0, 1) \setminus N_x$ the set $\Omega \cap \partial B(x, \varepsilon)$ is $\mathcal{H}^{n-1}$-measurable and

$$\lim_{\substack{\varepsilon \in (0,1)\setminus N_x \\ \varepsilon \to 0}} \int_{\Omega \cap \partial B(x,\varepsilon)} \Big\langle \frac{y - x}{\varepsilon}, \vec{k}(x - y) \Big\rangle \, d\mathcal{H}^{n-1}(y) = -\frac{\vartheta}{2}. \tag{2.5.138}$$

Since $r_j \nearrow \infty$ as $j \to \infty$, this ultimately shows that for any $x \in \partial_*\Omega \setminus A$ and any $\varepsilon \in (0, 1) \setminus N_x$ the set $\Omega \cap \partial B(x, \varepsilon)$ is $\mathcal{H}^{n-1}$-measurable and (2.5.138) holds. From this, the claim made in (2.5.131) readily follows. □

Particularizing (2.5.132) yields a number of significant results.

Remark 2.5.11 *Throughout, assume $\Omega \subseteq \mathbb{R}^n$ is a bounded set of locally finite perimeter, abbreviate $\sigma_* := \mathcal{H}^{n-1}\lfloor\partial_*\Omega$, and denote by $\nu = (\nu_1, \ldots, \nu_n)$ the geometric measure theoretic outward unit normal to Ω.*

(a) *When $\vec{k}$ is as in* (2.5.129) *we obtain the following result: For each $x \in \mathbb{R}^n$ there exists an $\mathcal{L}^1$-nullset $N_x \subset (0, \infty)$ such that, for each $r \in (0, \infty) \setminus N_x$,*

$$\int_{\substack{y\in\partial_*\Omega\\|x-y|>r}} \frac{\langle\nu(y), y - x\rangle}{|x - y|^n}\,d\sigma_*(y) = \int_{\substack{y\in\partial_*\Omega\\|x-y|\geq r}} \frac{\langle\nu(y), y - x\rangle}{|x - y|^n}\,d\sigma_*(y)$$
$$= \frac{\mathcal{H}^{n-1}\big(\Omega \cap \partial B(x, r)\big)}{r^{n-1}}. \tag{2.5.139}$$

(b) *Fix $j, k \in \{1, \ldots, n\}$ and specialize* (2.5.132) *to the case when*

$$\vec{k} := (\partial_k E_\Delta)\mathbf{e}_j - (\partial_j E_\Delta)\mathbf{e}_k, \tag{2.5.140}$$

where E_Δ is the standard fundamental solution for the Laplacian in $\mathbb{R}^n$ defined in (1.5.56). *This allows us to conclude that for each $x \in \mathbb{R}^n$ there exists an $\mathcal{L}^1$-nullset $N_x \subset (0, \infty)$ such that, for each $r \in (0, \infty) \setminus N_x$,*

$$\begin{aligned}&\int_{\substack{|x-y|>r\\y\in\partial_*\Omega}} \Big[\nu_j(y)(\partial_k E_\Delta)(y - x) - \nu_k(y)(\partial_j E_\Delta)(y - x)\Big]\,d\sigma_*(y) = 0 \text{ and}\\&\int_{\substack{|x-y|\geq r\\y\in\partial_*\Omega}} \Big[\nu_j(y)(\partial_k E_\Delta)(y - x) - \nu_k(y)(\partial_j E_\Delta)(y - x)\Big]\,d\sigma_*(y) = 0.\end{aligned} \tag{2.5.141}$$

Here is the version of the "half-sphere lemma" mentioned earlier, valid in the class of sets of locally finite perimeter, refining Lemma 2.5.8.

Lemma 2.5.12 *Assume $\Omega \subseteq \mathbb{R}^n$ is a set of locally finite perimeter and abbreviate $\sigma := \mathcal{H}^{n-1}\lfloor\partial\Omega$. Then for σ-a.e. $x \in \partial_*\Omega$ there exists some $\mathcal{L}^1$-nullset $N_x \subset (0, \infty)$ with the property that*

$$\begin{aligned}&\text{the sets } \Omega \cap \partial B(x, \varepsilon) \text{ and } \partial B(x, \varepsilon) \setminus \Omega\\&\text{are } \mathcal{H}^{n-1}\text{-measurable for each } \varepsilon \in (0, \infty) \setminus N_x\end{aligned} \tag{2.5.142}$$

and such that (with ω_{n-1} denoting the surface area of the unit sphere in $\mathbb{R}^n$)

$$\lim_{\substack{\varepsilon\in(0,\infty)\setminus N_x\\\varepsilon\to0}} \frac{\mathcal{H}^{n-1}\big(\Omega \cap \partial B(x, \varepsilon)\big)}{\omega_{n-1}\varepsilon^{n-1}} = \frac{1}{2}, \tag{2.5.143}$$

$$\lim_{\substack{\varepsilon\in(0,\infty)\setminus N_x\\\varepsilon\to0}} \frac{\mathcal{H}^{n-1}\big(\partial B(x, \varepsilon) \setminus \Omega\big)}{\omega_{n-1}\varepsilon^{n-1}} = \frac{1}{2}. \tag{2.5.144}$$

Proof Much as in the proof of Lemma 2.5.8, the claim culminating in (2.5.142) is a consequence of the co-area theorem. Next, specializing (2.5.131) to the case when $\vec{k}$ is as in (2.5.129) yields (2.5.143). Finally, (2.5.144) is an immediate consequence of (2.5.143). □

In the lemma below we prove that a certain maximal operator on the boundary of a Lebesgue measurable set Ω maps Hölder functions into bounded functions provided $\partial\Omega$ is compact and upper Ahlfors regular.

Lemma 2.5.13 *Let $\Omega \subseteq \mathbb{R}^n$ (where $n \in \mathbb{N}$, $n \geq 2$) be a Lebesgue measurable set such that $\partial\Omega$ is bounded and $\partial_*\Omega$ is upper Ahlfors regular. Abbreviate $\sigma_* := \mathcal{H}^{n-1}\lfloor\partial_*\Omega$ and denote by ν the geometric measure theoretic outward unit normal to Ω. Also, consider a divergence-free vector-valued function $\vec{k} \in \big[\mathscr{C}^1(\mathbb{R}^n \setminus \{0\})\big]^n$ with the property that there exists some $C \in (0,\infty)$ such that*

$$|\vec{k}(x)| \leq C|x|^{1-n} \text{ for all } x \in \mathbb{R}^n \setminus \{0\}. \tag{2.5.145}$$

In this setting, define the action of the maximal operator $Z_{\max}$ on an arbitrary function $f \in L^1(\partial_*\Omega, \sigma_*)$ according to

$$(Z_{\max} f)(x) := \sup_{\varepsilon>0}\left|\int_{\substack{y\in\partial_*\Omega\\ |x-y|>\varepsilon}} \big\langle \nu(y), \vec{k}(x-y)\big\rangle f(y)\,\mathrm{d}\sigma_*(y)\right| \tag{2.5.146}$$

for all $x \in \partial\Omega$.

Then for each given exponent $\alpha > 0$ there exists some number $N \in (0,\infty)$, depending only on $\vec{k}$, n, α, and the upper Ahlfors regularity constant of $\partial_*\Omega$, with the property that

$$\sup_{x\in\partial_*\Omega} (Z_{\max} f)(x) \leq N\big[\mathrm{diam}(\partial_*\Omega)\big]^{\alpha} \cdot \|f\|_{\dot{\mathscr{C}}^{\alpha}(\partial_*\Omega)} + N \cdot \sup_{\partial_*\Omega} |f|, \tag{2.5.147}$$

for each function $f \in \mathscr{C}^{\alpha}(\partial_*\Omega)$.

Proof The present assumptions imply that Ω is a set of locally finite perimeter (cf. [112, (5.9.16)]), so the geometric measure theoretic outward unit normal ν to Ω is well defined at σ_*-a.e. point on $\partial_*\Omega$. In a first stage, make the additional assumption that Ω is bounded. Fix an arbitrary point $x \in \mathbb{R}^n$ and observe that (2.5.132) implies that for $\mathcal{L}^1$-a.e. $\varepsilon > 0$ we have

$$
\begin{aligned}
&\Big|\int_{\partial_*\Omega\setminus\overline{B(x,\varepsilon)}} \big\langle \nu(y), \vec{k}(x-y)\big\rangle \, d\sigma_*(y)\Big| \\
&\qquad = \left| \int_{\Omega\cap\partial B(x,\varepsilon)} \Big\langle \frac{y-x}{\varepsilon}, \vec{k}(x-y)\Big\rangle \, d\mathcal{H}^{n-1}(y)\right| \\
&\qquad \leq C \cdot \frac{\mathcal{H}^{n-1}\big(\partial B(x,\varepsilon)\big)}{\varepsilon^{n-1}} \\
&\qquad = C \cdot \omega_{n-1}.
\end{aligned}
\tag{2.5.148}
$$

Let us now pick an arbitrary $r \in (0,\infty)$. Based on (2.5.148) we conclude that there exists $\varepsilon \in (r/2, r)$ such that

$$
\Big|\int_{\partial_*\Omega\setminus\overline{B(x,\varepsilon)}} \big\langle \nu(y), \vec{k}(x-y)\big\rangle \, d\sigma_*(y)\Big| \leq C \cdot \omega_{n-1}.
\tag{2.5.149}
$$

For this choice of ε we then proceed to estimate

$$
\begin{aligned}
&\Big|\int_{\partial_*\Omega\setminus\overline{B(x,r)}} \big\langle \nu(y), \vec{k}(x-y)\big\rangle \, d\sigma_*(y)\Big| \\
&\qquad \leq \Big|\int_{\partial_*\Omega\setminus\overline{B(x,\varepsilon)}} \big\langle \nu(y), \vec{k}(x-y)\big\rangle \, d\sigma_*(y)\Big| \\
&\qquad\quad + \Big|\int_{[\overline{B(x,r)}\setminus\overline{B(x,\varepsilon)}]\cap\partial_*\Omega} \big\langle \nu(y), \vec{k}(x-y)\big\rangle \, d\sigma_*(y)\Big| \\
&\qquad \leq C\Big\{\omega_{n-1} + \int_{[\overline{B(x,r)}\setminus\overline{B(x,\varepsilon)}]\cap\partial_*\Omega} \frac{1}{|x-y|^{n-1}} \, d\mathcal{H}^{n-1}(y)\Big\} \\
&\qquad \leq C\Big\{\omega_{n-1} + \int_{[B(x,2\varepsilon)\setminus\overline{B(x,\varepsilon)}]\cap\partial_*\Omega} \frac{1}{|x-y|^{n-1}} \, d\mathcal{H}^{n-1}(y)\Big\} \\
&\qquad \leq C\Big\{\omega_{n-1} + \varepsilon^{-(n-1)}\mathcal{H}^{n-1}\big(B(x,2\varepsilon)\cap\partial_*\Omega\big)\Big\} \\
&\qquad \leq C = C(\partial_*\Omega, \vec{k}, n) < +\infty.
\end{aligned}
\tag{2.5.150}
$$

The argument so far shows that

$$
\sup_{x\in\mathbb{R}^n}\ \sup_{r\in(0,\infty)} \Big|\int_{\partial_*\Omega\setminus\overline{B(x,r)}} \big\langle \nu(y), \vec{k}(x-y)\big\rangle \, d\sigma_*(y)\Big| \leq C,
\tag{2.5.151}
$$

where $C \in (0,\infty)$ depends only on $\vec{k}$, the dimension n, and the upper Ahlfors regularity constant of $\partial_*\Omega$.

To proceed, select $f \in \mathscr{C}^\alpha(\partial_*\Omega)$ with $\alpha > 0$, and fix $x \in \partial_*\Omega$ along with $r > 0$ arbitrary. Split

$$\int_{\partial_*\Omega\setminus\overline{B(x,r)}} \langle \nu(y), \vec{k}(x-y)\rangle f(y)\,d\sigma_*(y) = \mathrm{I} + \mathrm{II}, \tag{2.5.152}$$

where

$$\mathrm{I} := \int_{\partial_*\Omega\setminus\overline{B(x,r)}} \langle \nu(y), \vec{k}(x-y)\rangle [f(y) - f(x)]\,d\sigma_*(y) \tag{2.5.153}$$

and

$$\mathrm{II} := f(x)\int_{\partial_*\Omega\setminus\overline{B(x,r)}} \langle \nu(y), \vec{k}(x-y)\rangle\,d\sigma_*(y). \tag{2.5.154}$$

To estimate I, we use (2.5.145), (2.5.153), the fact that $\sigma_* = \mathcal{H}^{n-1}\lfloor\partial_*\Omega$, and [112, Lemma 7.2.1] to write

$$\begin{aligned}|\mathrm{I}| &\le C\cdot\int_{\partial_*\Omega}\frac{1}{|x-y|^{n-1}}|f(y)-f(x)|\,d\sigma_*(y)\\ &\le C\cdot\|f\|_{\dot{\mathscr{C}}^\alpha(\partial_*\Omega)}\int_{\partial_*\Omega}\frac{1}{|x-y|^{n-1-\alpha}}\,d\mathcal{H}^{n-1}(y)\\ &\le C\cdot\big[\mathrm{diam}(\partial_*\Omega)\big]^\alpha\cdot\|f\|_{\dot{\mathscr{C}}^\alpha(\partial_*\Omega)},\end{aligned} \tag{2.5.155}$$

for some $C = C(\partial_*\Omega, \vec{k}, n, \alpha) \in (0,\infty)$ independent of f. Also, (2.5.154) and (2.5.151) imply

$$|\mathrm{II}| \le C\cdot\sup_{\partial_*\Omega}|f|. \tag{2.5.156}$$

Together, (2.5.146), (2.5.152), (2.5.155), and (2.5.156) establish (2.5.147) in the case when Ω is bounded.

Finally, the case when the set Ω unbounded is a consequence of what we have proved so far, applied to $\Omega^c := \mathbb{R}^n\setminus\Omega$. Indeed, since $n \ge 2$ and $\partial\Omega$ is compact, from [112, Lemma 5.10.10], [112, (5.2.3)], [112, (5.6.16)], and assumptions, we conclude that Ω^c is a bounded set of locally finite perimeter whose geometric measure theoretic outward unit normal is $-\nu$ at $\mathcal{H}^{n-1}$-a.e. point on $\partial_*(\Omega^c) = \partial_*\Omega$. □

As a preamble to Proposition 2.5.16, in the next couple of lemmas we study the boundary-to-domain version of the integral operator with a kernel as in (2.5.146).

Lemma 2.5.14 *Let $\Omega \subseteq \mathbb{R}^n$ (where $n \in \mathbb{N}$, $n \ge 2$) be an open set of locally finite perimeter with the property that $\partial\Omega$ is bounded. Denote by ν the geometric measure theoretic outward unit normal to Ω and abbreviate $\sigma_* := \mathcal{H}^{n-1}\lfloor\partial_*\Omega$. Next, bring in a divergence-free vector-valued function $\vec{k} \in \big[\mathscr{C}^1(\mathbb{R}^n\setminus\{0\})\big]^n$ which is positive* *homogeneous of degree* $1-n$. *Use this to define the integral operator acting on each function* $f \in L^1(\partial_*\Omega, \sigma_*)$ *according to*

$$\mathcal{Z}f(x) := \int_{\partial_*\Omega} \big\langle \nu(y), \vec{k}(x-y)\big\rangle f(y)\,\mathrm{d}\sigma_*(y) \ \textit{for all} \ x \in \Omega. \tag{2.5.157}$$

Then, with $\vartheta := \int_{S^{n-1}} \big\langle \omega, \vec{k}(\omega)\big\rangle\,\mathrm{d}\mathcal{H}^{n-1}(\omega) \in \mathbb{C}$, *one has*

$$\mathcal{Z}1 \equiv \begin{cases} -\vartheta & \textit{in } \Omega, \textit{ if } \Omega \textit{ is bounded,} \\ 0 & \textit{in } \Omega, \textit{ if } \Omega \textit{ is unbounded.} \end{cases} \tag{2.5.158}$$

Proof In a first stage, assume that Ω is bounded and fix an arbitrary point $x \in \Omega$. The idea is to apply [112, Corollary 2.8.8] to a vector field $\vec{F} \in \big[\mathscr{E}'(\mathbb{R}^n)\big]^n$ whose $\mathscr{C}^1$-singular support is a compact subset of Ω and which coincides near the compact set $\overline{\Omega}$ with

$$\vec{k}(x-y) \ \text{for } \mathcal{L}^n\text{-a.e. } y \in \Omega. \tag{2.5.159}$$

Since from [112, (4.5.46)] we know that $\big(\mathrm{div}\vec{F}\big)\big|_{\Omega} = -\vartheta\delta_x$ where δ_x is the Dirac distribution with mass at x in Ω, the Divergence Formula [112, (2.8.57)] reduces precisely to $(\mathcal{Z}1)(x) = 1$. As such, the version of (2.5.158) when Ω is bounded follows on account of the arbitrariness of $x \in \Omega$.

In the case when the set Ω is unbounded and $\partial\Omega$ is bounded, [112, Lemma 5.10.10] and [112, (5.2.3), (5.6.16)] imply that $\Omega^c := \mathbb{R}^n \setminus \Omega$ is a compact set of locally finite perimeter, satisfying $\partial_*(\Omega^c) = \partial_*\Omega$, $\partial^*(\Omega^c) = \partial^*\Omega$, and whose geometric measure theoretic outward unit normal is $-\nu$. Having fixed an arbitrary $x \in \Omega$, apply the De Giorgi-Federer version of the Divergence Theorem (see [112, Theorem 1.1.1]) to the set Ω^c and a vector field $\vec{F} \in \big[\mathscr{C}^1_c(\mathbb{R}^n)\big]^n$ which coincides in an open neighborhood $O \subseteq \mathbb{R}^n \setminus \{x\}$ of $\mathbb{R}^n \setminus \Omega$ with

$$O \ni y \longmapsto \vec{k}(x-y) \in \mathbb{C}^n. \tag{2.5.160}$$

In view of the fact that $\big(\mathrm{div}\vec{F}\big)\big|_{\Omega^c} = 0$, the Divergence Formula [112, (1.1.8)] now simply reads $(\mathcal{Z}1)(x) = 0$, finishing the proof of (2.5.158). □

Here is the second auxiliary lemma alluded to earlier.

Lemma 2.5.15 *Let* $\Omega \subseteq \mathbb{R}^n$ *(where* $n \in \mathbb{N}$, $n \geq 2$*) be an open set such that* $\partial\Omega$ *is bounded and* $\partial_*\Omega$ *is upper Ahlfors regular (in particular* Ω *has locally finite perimeter; cf.* [112, (5.9.16)]*). Consider a divergence-free vector-valued function* $\vec{k} \in \big[\mathscr{C}^1(\mathbb{R}^n \setminus \{0\})\big]^n$ *which is positive homogeneous of degree* $1-n$ *and use it to define the integral operator* $\mathcal{Z}$ *as in* (2.5.157). *Finally, fix an arbitrary exponent* $\alpha \in (0,1)$.

Then there exists some $C \in (0,\infty)$, *depending only on* $\vec{k}$, n, α, $\mathrm{diam}(\partial_*\Omega)$, *and the upper Ahlfors regularity constant of* $\partial_*\Omega$, *such that for every function* $f \in \mathscr{C}^\alpha(\partial_*\Omega)$ *one has*

$$\sup_{x\in\Omega}\big|(\mathcal{Z}f)(x)\big| + \sup_{x\in\Omega}\Big\{\mathrm{dist}(x,\partial_*\Omega)^{1-\alpha}\big|\nabla(\mathcal{Z}f)(x)\big|\Big\} \leq C\|f\|_{\mathscr{C}^\alpha(\partial_*\Omega)}. \tag{2.5.161}$$

Moreover, under the additional assumption that Ω *is a uniform domain, the operator*

$$\mathcal{Z} : \mathscr{C}^\alpha(\partial_*\Omega) \longrightarrow \mathscr{C}^\alpha\big(\overline{\Omega}\big) \tag{2.5.162}$$

is well defined, linear, and bounded; hence so is

$$\mathcal{Z} : \mathscr{C}^\alpha(\partial\Omega) \longrightarrow \mathscr{C}^\alpha\big(\overline{\Omega}\big). \tag{2.5.163}$$

Proof The estimate in (2.5.161) is a consequence of (2.5.158) and Proposition 2.1.3. In particular, since $\partial_*\Omega \subseteq \partial\Omega$, this further entails

$$\sup_{x\in\Omega}\big|(\mathcal{Z}f)(x)\big| + \sup_{x\in\Omega}\Big\{\mathrm{dist}(x,\partial\Omega)^{1-\alpha}\big|\nabla(\mathcal{Z}f)(x)\big|\Big\} \leq C\|f\|_{\mathscr{C}^\alpha(\partial_*\Omega)}, \tag{2.5.164}$$

for each $f \in \mathscr{C}^\alpha(\partial_*\Omega)$. In the case when the set $\Omega \subseteq \mathbb{R}^n$ is also assumed to be a uniform domain, the fact that the operator $\mathcal{Z}$ in the context of (2.5.162) is well defined, linear, and bounded, becomes a consequence of (2.5.164) and [112, (5.11.78)] (keeping in mind that $\mathcal{Z}f \in \mathscr{C}^1(\Omega)$ for each function $f \in \mathscr{C}^\alpha(\partial_*\Omega)$). □

We are now in a position to prove the following basic result for singular integral operators whose kernel is the "dot product" between the geometric measure theoretic outward unit normal to a given Lebesgue measurable set with a compact upper Ahlfors boundary boundary and a smooth, divergence-free, vector-valued function which is odd and positive homogeneous of degree $1-n$ in $\mathbb{R}^n \setminus \{0\}$. One way to think of this class of singular integral operators is as a natural generalization of the standard harmonic double layer operator.

Proposition 2.5.16 *Suppose* $\Omega \subseteq \mathbb{R}^n$ *(where* $n \in \mathbb{N}$ *with* $n \geq 2$*) is a Lebesgue measurable set such that* $\partial\Omega$ *is compact and* $\partial_*\Omega$ *is upper Ahlfors boundary. Denote by* ν *the geometric measure theoretic outward unit normal to* Ω *and abbreviate* $\sigma_* := \mathcal{H}^{n-1}\lfloor\partial_*\Omega$. *Also, consider* $\vec{k} \in \big[\mathscr{C}^2(\mathbb{R}^n \setminus \{0\})\big]^n$ *a divergence-free vector-valued function which is odd and positive homogeneous of degree* $1-n$. *Finally, fix an exponent* $\alpha \in (0,1)$ *and define*

$$\vartheta := \int_{S^{n-1}} \big\langle \omega, \vec{k}(\omega)\big\rangle \, d\mathcal{H}^{n-1}(\omega) \in \mathbb{C}. \tag{2.5.165}$$

Then the following statements are true.

(i) For each function $f \in \mathscr{C}^\alpha(\partial_*\Omega)$ *the limit*

$$Zf(x) := \lim_{\varepsilon\to 0^+} \int_{\substack{y\in\partial_*\Omega \\ |x-y|>\varepsilon}} \big\langle \nu(y), \vec{k}(x-y)\big\rangle f(y)\, d\sigma_*(y) \tag{2.5.166}$$

exists for σ_**-a.e. point* $x \in \partial_*\Omega$.

(ii) After possibly redefining Zf on a σ_*-nullset contained in $\partial_*\Omega$, the assignment $f \mapsto Zf$ thus defined induces a well-defined, linear, and bounded operator

$$Z : \mathscr{C}^\alpha(\partial_*\Omega) \longrightarrow \mathscr{C}^\alpha(\partial_*\Omega) \tag{2.5.167}$$

with the property that

$$Z1 \equiv \begin{cases} -\frac{\vartheta}{2} \text{ on } \partial_*\Omega, \text{ if } \Omega \text{ is bounded,} \\ +\frac{\vartheta}{2} \text{ on } \partial_*\Omega, \text{ if } \Omega \text{ is unbounded.} \end{cases} \tag{2.5.168}$$

(iii) Under the additional assumption that the set Ω is open and $\partial\Omega$ is Ahlfors regular, the following jump-formula (where $\mathcal{Z}$ is the integral operator introduced in (2.5.157)*, I denotes the identity operator, and $\kappa > 0$ is an arbitrary fixed aperture parameter) is valid:*

$$\text{for every given function } f \in \mathscr{C}^\alpha(\partial_*\Omega) \text{ one has} \\ \big(\mathcal{Z}f\big)\Big|_{\partial\Omega}^{\kappa-\text{n.t.}} = (-\tfrac{\vartheta}{2}I + Z)f \text{ at } \sigma\text{-a.e. point on } \partial_*\Omega. \tag{2.5.169}$$

(iv) In the case when it is also assumed that $\mathcal{H}^{n-1}(\partial\Omega \setminus \partial_*\Omega) = 0$, one may replace $\partial_*\Omega$ by $\partial\Omega$ in the formulation of (2.5.167), (2.5.168), *and* (2.5.169).

Proof To deal with the claim in part *(i)*, pick an arbitrary function $f \in \mathscr{C}^\alpha(\partial_*\Omega)$. Then (2.5.111) gives that for σ_*-a.e. point $x \in \partial_*\Omega$ we have

$$\begin{aligned} &\lim_{\varepsilon\to 0^+} \int_{\substack{y\in\partial_*\Omega \\ |x-y|>\varepsilon}} \langle \nu(y), \vec{k}(x-y)\rangle f(y)\,d\sigma_*(y) \\ &= \lim_{\varepsilon\to 0^+} \int_{\substack{y\in\partial_*\Omega \\ |x-y|>\varepsilon}} \langle \nu(y), \vec{k}(x-y)\rangle \big(f(y) - f(x)\big)\,d\sigma_*(y) \mp \frac{\vartheta}{2} f(x) \\ &= \int_{\partial_*\Omega} \langle \nu(y), \vec{k}(x-y)\rangle \big(f(y) - f(x)\big)\,d\sigma_*(y) \mp \frac{\vartheta}{2} f(x), \end{aligned} \tag{2.5.170}$$

where the sign of $\frac{\vartheta}{2}f(x)$ is minus if Ω is bounded and plus if Ω is unbounded. For the last equality in (2.5.170) we have used Lebesgue's Dominated Convergence Theorem. More concretely, given that $f(y) - f(x) = O(|x-y|^\alpha)$, [112, Lemma 7.2.1] shows that the last integrand above is absolutely integrable for each fixed $x \in \partial_*\Omega$. In turn, (2.5.170) allows us to conclude that the limit defining $(Zf)(x)$ in (2.5.166) exists for σ_*-a.e. $x \in \partial_*\Omega$. This takes care of the claim made in part *(i)*.

As regards the claims in part *(ii)*, we first observe from (2.5.166) and (2.5.170) that for every given $f \in \mathscr{C}^\alpha(\partial_*\Omega)$ the function Zf may be redefined on a σ_*-nullset so that

$$(Zf)(x) = \mp\frac{\vartheta}{2}f(x) + \int_{\partial_*\Omega} \big\langle \nu(y), \vec{k}(x-y)\big\rangle \big(f(y) - f(x)\big)\, d\sigma_*(y) \tag{2.5.171}$$

for all $x \in \partial_*\Omega$, with the sign dictated by whether Ω is bounded (minus), or Ω is unbounded (plus). Then (2.5.171) readily gives (2.5.168). We next proceed to show that, in the context of (2.5.167), the operator $f \mapsto Zf$ defined as in (2.5.171) is well defined and bounded. To this end, fix two arbitrary distinct points $x_1, x_2 \in \partial_*\Omega$ and, bearing (2.5.171) in mind, write

$$(Zf)(x_1) - (Zf)(x_2) = \mathrm{I} + \mathrm{II} \tag{2.5.172}$$

where

$$\mathrm{I} := \mp\frac{\vartheta}{2}\big(f(x_1) - f(x_2)\big) \tag{2.5.173}$$

and

$$\begin{aligned}\mathrm{II} := \int_{\partial_*\Omega} \Big\{ &\big\langle \nu(y), \vec{k}(x_1-y)\big\rangle \big(f(y) - f(x_1)\big) \\ &- \big\langle \nu(y), \vec{k}(x_2-y)\big\rangle \big(f(y) - f(x_2)\big)\Big\}\, d\sigma_*(y).\end{aligned} \tag{2.5.174}$$

Next, introduce $r := |x_1 - x_2| > 0$ and estimate

$$\big|\mathrm{II}\big| \leq \mathrm{II}_1 + \mathrm{II}_2 + \mathrm{II}_3, \tag{2.5.175}$$

where

$$\begin{aligned}\mathrm{II}_1 := \Bigg| \int_{\substack{y\in\partial_*\Omega \\ |x_1-y|>2r}} &\big\langle \nu(y), \vec{k}(x_1-y)\big\rangle \big(f(y) - f(x_1)\big) \\ &- \big\langle \nu(y), \vec{k}(x_2-y)\big\rangle \big(f(y) - f(x_2)\big)\, d\sigma_*(y)\Bigg|,\end{aligned} \tag{2.5.176}$$

while

$$\mathrm{II}_2 := \int_{\substack{y\in\partial_*\Omega \\ |x_1-y|\leq 2r}} \Big|\big\langle \nu(y), \vec{k}(x_1-y)\big\rangle \big(f(y) - f(x_1)\big)\Big|\, d\sigma_*(y), \tag{2.5.177}$$

$$\mathrm{II}_3 := \int_{\substack{y\in\partial_*\Omega \\ |x_1-y|\leq 2r}} \Big|\big\langle \nu(y), \vec{k}(x_2-y)\big\rangle \big(f(y) - f(x_2)\big)\Big|\, d\sigma_*(y). \tag{2.5.178}$$

Note that

$$\mathrm{II}_2 \leq \Big(\sup_{S^{n-1}} |\vec{k}|\Big)\|f\|_{\dot{\mathscr{C}}^\alpha(\partial_*\Omega)} \int\limits_{\substack{y\in\partial_*\Omega\\ |x_1-y|\leq 2r}} \frac{d\mathcal{H}^{n-1}(y)}{|x_1-y|^{n-1-\alpha}}, \tag{2.5.179}$$

and, given that $|x_1 - y| \leq 2r$ forces $|x_2 - y| \leq |x_1 - x_2| + |x_1 - y| \leq 3r$,

$$\begin{aligned} \mathrm{II}_3 &\leq \int\limits_{\substack{y\in\partial_*\Omega\\ |x_2-y|\leq 3r}} \Big|\big\langle \nu(y), \vec{k}(x_2-y)\big\rangle\big(f(y)-f(x_2)\big)\Big|\,d\sigma_*(y) \\ &\leq \Big(\sup_{S^{n-1}} |\vec{k}|\Big)\|f\|_{\dot{\mathscr{C}}^\alpha(\partial_*\Omega)} \int\limits_{\substack{y\in\partial_*\Omega\\ |x_2-y|\leq 3r}} \frac{d\mathcal{H}^{n-1}(y)}{|x_2-y|^{n-1-\alpha}}. \end{aligned} \tag{2.5.180}$$

Invoking [112, Lemma 7.2.1] (cf. [112, (7.2.5)]), we obtain from (2.5.179) and (2.5.180) (keeping in mind the significance of the number r) that there exists some constant $M = M(\partial\Omega, \vec{k}, n, \alpha) \in (0, \infty)$ with the property that

$$\mathrm{II}_2 + \mathrm{II}_3 \leq M\|f\|_{\dot{\mathscr{C}}^\alpha(\partial_*\Omega)}|x_1 - x_2|^\alpha. \tag{2.5.181}$$

Going further, bound

$$\mathrm{II}_1 \leq \mathrm{II}_1^a + \mathrm{II}_1^b, \tag{2.5.182}$$

with

$$\begin{aligned} \mathrm{II}_1^a &:= \Bigg|\int\limits_{\substack{y\in\partial_*\Omega\\ |x_1-y|>2r}} \big\langle \nu(y), \vec{k}(x_1-y)\big\rangle\big(f(x_2)-f(x_1)\big)\,d\sigma_*(y)\Bigg| \\ &= \Bigg|\int\limits_{\substack{y\in\partial_*\Omega\\ |x_1-y|>2r}} \big\langle \nu(y), \vec{k}(x_1-y)\big\rangle\,d\sigma_*(y)\Bigg|\big|f(x_2)-f(x_1)\big| \\ &\leq (Z_{\max}1)(x_1)\big|f(x_2)-f(x_1)\big| \leq M\,r^\alpha\,\|f\|_{\dot{\mathscr{C}}^\alpha(\partial_*\Omega)}, \end{aligned} \tag{2.5.183}$$

where the last inequality is based on (2.5.147) (with $f := 1$), and

$$
\begin{aligned}
\mathrm{II}_1^b &:= \Bigg| \int_{\substack{y\in\partial_*\Omega\\ |x_1-y|>2r}} \big\langle \nu(y), \vec{k}(x_1-y) - \vec{k}(x_2-y)\big\rangle \big(f(y)-f(x_2)\big)\, d\sigma_*(y) \Bigg| \\
&\leq \int_{\substack{y\in\partial_*\Omega\\ |x_1-y|>2r}} \big|\vec{k}(x_1-y) - \vec{k}(x_2-y)\big| |f(y)-f(x_2)|\, d\sigma_*(y) \\
&\leq C r \|f\|_{\dot{\mathscr{C}}^\alpha(\partial_*\Omega)} \int_{\substack{y\in\partial_*\Omega\\ |x_1-y|>2r}} \frac{d\mathcal{H}^{n-1}(y)}{|x_1-y|^{n-\alpha}},
\end{aligned}
\tag{2.5.184}
$$

using the Mean Value Theorem and the fact that the function f is Hölder of order α on $\partial_*\Omega$. Here it helps to note that if $|x_1 - y| > 2r$ then $|\xi - y| \approx |x_1 - y|$ for all $\xi \in [x_1, x_2]$, and $|y - x_2| < 3|y - x_1|/2$. Combining (2.5.182), (2.5.183), (2.5.184), and [112, (7.2.5)] we may then conclude that there exists some constant $M = M(\partial\Omega, \vec{k}, n, \alpha) \in (0, \infty)$ with the property that

$$
\mathrm{II}_1 \leq M \|f\|_{\dot{\mathscr{C}}^\alpha(\partial_*\Omega)} |x_1 - x_2|^\alpha. \tag{2.5.185}
$$

From (2.5.172), (2.5.173), (2.5.175), (2.5.181), and (2.5.185) we may then conclude that

$$
\big|(Zf)(x_1) - (Zf)(x_2)\big| \leq M \|f\|_{\dot{\mathscr{C}}^\alpha(\partial_*\Omega)} |x_1 - x_2|^\alpha, \quad \forall x_1, x_2 \in \partial_*\Omega, \tag{2.5.186}
$$

for some constant $M = M(\partial\Omega, \vec{k}, n, \alpha) \in (0, \infty)$. The argument so far gives that the principal-value singular integral operator Z is well defined and bounded in the context

$$
Z : \dot{\mathscr{C}}^\alpha(\partial_*\Omega) \longrightarrow \dot{\mathscr{C}}^\alpha(\partial_*\Omega). \tag{2.5.187}
$$

Let us also note that (2.5.171) and [112, Lemma 7.2.1] imply that for every function $f \in \mathscr{C}^\alpha(\partial_*\Omega)$ we have

$$
\begin{aligned}
\sup_{x\in\partial_*\Omega} |(Zf)(x)| &\leq C \sup_{x\in\partial_*\Omega} |f(x)| + C \|f\|_{\dot{\mathscr{C}}^\alpha(\partial_*\Omega)} \sup_{x\in\partial_*\Omega} \int_{\partial_*\Omega} \frac{d\sigma_*(y)}{|x-y|^{n-1-\alpha}} \\
&\leq C \sup_{\partial_*\Omega} |f| + C \|f\|_{\dot{\mathscr{C}}^\alpha(\partial_*\Omega)} \cdot \big[\mathrm{diam}(\partial_*\Omega)\big]^\alpha,
\end{aligned}
\tag{2.5.188}
$$

for some $C \in (0, \infty)$ which depends only on $\vec{k}$, n, α, and the upper Ahlfors regularity constant of $\partial_*\Omega$. Together, (2.5.187) and (2.5.188) complete the proof of the claim made about (2.5.167).

Turning our attention to the claim in part *(iii)*, make the additional assumption that the set Ω is open and that $\partial\Omega$ is Ahlfors regular. As far as the jump-formula (2.5.169) is concerned, it has been already noted in Lemma 2.5.15 that the action of the integral operator $\mathcal{Z}$ (cf. (2.5.157)) is presently meaningful on Hölder functions. Also, given that $\partial\Omega$ is presently assumed to be Ahlfors regular, item *(iii)* in [112, Proposition 8.8.6] ensures that it is meaningful to consider the nontangential boundary trace in the left-hand side of the equality in (2.5.169) in the present setting. Assume now that some $f \in \mathscr{C}^\alpha(\partial_*\Omega)$ with $\alpha \in (0,1)$ has been given and observe from (2.5.157) that $\mathcal{Z}f$ is continuous in Ω.

Also, recall from the current item *(i)* that there exists a σ_*-measurable set $A \subseteq \partial_*\Omega$ satisfying $\sigma_*(A) = 0$ and with the property that the limit

$$\lim_{\varepsilon\to 0^+} \int_{\substack{|x-y|>\varepsilon \\ y\in\partial_*\Omega}} \big\langle \nu(y), \vec{k}(x-y)\big\rangle f(y)\,d\sigma_*(y) \text{ exists for each } x \in \partial_*\Omega \setminus A. \tag{2.5.189}$$

Fix an arbitrary point

$$x \in \big(\partial_*\Omega \setminus A\big) \cap \partial_{\text{nta}}\Omega \tag{2.5.190}$$

and let $N_x \subset (0,\infty)$ be the union between the $\mathcal{L}^1$-nullset associated with x as in Lemma 2.5.10 (used with Ω replaced by $\mathbb{R}^n \setminus \Omega$) and the $\mathcal{L}^1$-nullset associated with x as in [112, Lemma 5.7.2, (5.7.25)] (used with $E := \mathbb{R}^n \setminus \Omega$). In particular, for each $\varepsilon \in (0,\infty) \setminus N_x$ the set $\partial B(x,\varepsilon) \setminus \Omega$ is $\mathcal{H}^{n-1}$-measurable and formula (2.5.131) holds with Ω replaced by $\mathbb{R}^n \setminus \Omega$. For some arbitrary $\kappa > 0$ fixed, write

$$\begin{aligned}
&\lim_{\substack{z\in\Gamma_\kappa(x)\\ z\to x}} \mathcal{Z}f(z)\\
&\quad= \lim_{\substack{\varepsilon\in(0,\infty)\setminus N_x\\ \varepsilon\to 0}} \lim_{\substack{z\in\Gamma_\kappa(x)\\ z\to x}} \int_{\substack{|x-y|>\varepsilon \\ y\in\partial_*\Omega}} \big\langle \nu(y), \vec{k}(z-y)\big\rangle f(y)\,d\sigma_*(y)\\
&\qquad+ \lim_{\substack{\varepsilon\in(0,\infty)\setminus N_x\\ \varepsilon\to 0}} \lim_{\substack{z\in\Gamma_\kappa(x)\\ z\to x}} \int_{\substack{|x-y|\le\varepsilon \\ y\in\partial_*\Omega}} \big\langle \nu(y), \vec{k}(z-y)\big\rangle (f(y)-f(x))\,d\sigma_*(y)\\
&\qquad+ \left(\lim_{\substack{\varepsilon\in(0,\infty)\setminus N_x\\ \varepsilon\to 0}} \lim_{\substack{z\in\Gamma_\kappa(x)\\ z\to x}} \int_{\substack{|x-y|\le\varepsilon \\ y\in\partial_*\Omega}} \big\langle \nu(y), \vec{k}(z-y)\big\rangle\,d\sigma_*(y) \right) f(x)\\
&\quad=: \mathrm{I}_1 + \mathrm{I}_2 + \mathrm{I}_3.
\end{aligned} \tag{2.5.191}$$

For each fixed threshold $\varepsilon > 0$, Lebesgue's Dominated Convergence Theorem applies to the limit as $\Gamma_\kappa(x) \ni z \to x$ in I_1 and yields

$$\mathrm{I}_1 = \lim_{\substack{\varepsilon\in(0,\infty)\setminus N_x \\ \varepsilon\to 0}} \int_{\substack{|x-y|>\varepsilon \\ y\in\partial_*\Omega}} \langle\nu(y),\vec{k}(x-y)\rangle f(y)\,d\sigma_*(y)$$

$$= \lim_{\varepsilon\to 0^+} \int_{\substack{|x-y|>\varepsilon \\ y\in\partial_*\Omega}} \langle\nu(y),\vec{k}(x-y)\rangle f(y)\,d\sigma_*(y) = Zf(x), \tag{2.5.192}$$

where the second equality is a consequence of (2.5.189), and the last equality comes from (2.5.166).

To handle I_2, we first observe that for every $y \in \partial_*\Omega \subseteq \partial\Omega$ and $z \in \Gamma_\kappa(x)$ we may estimate

$$|x-y| \le |z-y| + |z-x| \le |z-y| + (1+\kappa)\,\mathrm{dist}(z,\partial\Omega)$$

$$\le |z-y| + (1+\kappa)|z-y| = (2+\kappa)|z-y|. \tag{2.5.193}$$

Hence, since f is Hölder of order α on $\partial_*\Omega$,

$$\Big|\langle\nu(y),\vec{k}(z-y)\rangle\Big|\,|f(y)-f(x)| \le \Big(\sup_{S^{n-1}}|\vec{k}|\Big)\|f\|_{\dot{\mathscr{C}}^\alpha(\partial_*\Omega)}\frac{(2+\kappa)^{n-1}}{|x-y|^{n-1-\alpha}}. \tag{2.5.194}$$

Based on this, [112, (7.2.5)], the upper Ahlfors regularity of $\partial_*\Omega$, and once again Lebesgue's Dominated Convergence Theorem, we obtain that

$$|\mathrm{I}_2| = \left|\lim_{\substack{\varepsilon\in(0,\infty)\setminus N_x \\ \varepsilon\to 0}} \int_{\substack{|x-y|\le\varepsilon \\ y\in\partial_*\Omega}} \langle\nu(y),\vec{k}(x-y)\rangle(f(y)-f(x))\,d\sigma_*(y)\right|$$

$$\le \Big(\sup_{S^{n-1}}|\vec{k}|\Big)\cdot\|f\|_{\dot{\mathscr{C}}^\alpha(\partial_*\Omega)}\limsup_{\varepsilon\to 0^+}\int_{\substack{|x-y|\le\varepsilon \\ y\in\partial_*\Omega}}\frac{d\sigma_*(y)}{|x-y|^{n-1-\alpha}}$$

$$\le C\cdot\|f\|_{\dot{\mathscr{C}}^\alpha(\partial_*\Omega)}\limsup_{\varepsilon\to 0^+}\varepsilon^\alpha = 0. \tag{2.5.195}$$

Thus,

$$\mathrm{I}_2 = 0. \tag{2.5.196}$$

As for I_3 in (2.5.191), for each fixed $z \in \Gamma_\kappa(x)$ pick $r \in \big(0,\mathrm{dist}(z,\partial\Omega)\big)$ and consider a scalar function $\psi \in \mathscr{C}^\infty(\mathbb{R}^n)$ satisfying

$$\psi \equiv 0 \text{ in } B(z,r/2) \text{ and } \psi \equiv 1 \text{ in } \mathbb{R}^n\setminus B(z,r). \tag{2.5.197}$$

The idea is now to use [112, Lemma 5.7.2] with $E := \mathbb{R}^n\setminus\Omega$ and the vector field

$$\vec{F} \in \big[\mathscr{C}^2(\mathbb{R}^n)\big]^n,\quad \vec{F}(y) := \psi(y)\vec{k}(z-y) \text{ for all } y\in\mathbb{R}^n. \tag{2.5.198}$$

Specifically, bearing in mind that $\vec{F}$ is divergence-free near E (given that $\psi \equiv 1$ near E), [112, Lemma 5.7.2] applied to this vector field implies (upon recalling that, by design, N_x contains the $\mathcal{L}^1$-nullset associated as in [112, Lemma 5.7.2, (5.7.25)] with the point x from (2.5.190) and the set $E := \mathbb{R}^n \setminus \Omega$) that for each $\varepsilon \in (0, \infty) \setminus N_x$ we have

$$\int_{\substack{|x-y|\leq\varepsilon \\ y\in\partial_*\Omega}} \big\langle \nu(y), \vec{k}(z-y)\big\rangle \, d\sigma_*(y) = \int_{\substack{|x-y|=\varepsilon \\ y\in\mathbb{R}^n\setminus\Omega}} \big\langle (y-x)/\varepsilon, \vec{k}(z-y)\big\rangle \, d\mathcal{H}^{n-1}(y), \tag{2.5.199}$$

bearing in mind that the geometric measure theoretic outward unit normal to the set $E := \mathbb{R}^n \setminus \Omega$ is $-\nu$. From (2.5.199) and (2.5.131) (used with Ω replaced by $\mathbb{R}^n \setminus \Omega$) we therefore conclude that

$$\lim_{\substack{\varepsilon\in(0,\infty)\setminus N_x \\ \varepsilon\to 0}} \lim_{\substack{z\in\Gamma_\kappa(x) \\ z\to x}} \int_{\substack{|x-y|\leq\varepsilon \\ y\in\partial_*\Omega}} \big\langle \nu(y), \vec{k}(z-y)\big\rangle \, d\sigma_*(y) \tag{2.5.200}$$
$$= \lim_{\substack{\varepsilon\in(0,\infty)\setminus N_x \\ \varepsilon\to 0}} \int_{\substack{|x-y|=\varepsilon \\ y\in\mathbb{R}^n\setminus\Omega}} \big\langle (y-x)/\varepsilon, \vec{k}(x-y)\big\rangle \, d\mathcal{H}^{n-1}(y) = -\frac{\vartheta}{2}.$$

A combination of (2.5.191), (2.5.192), (2.5.196), and (2.5.200) shows that the limit in the left-hand side of (2.5.191) exists and matches $(-\frac{\vartheta}{2}I + Z)f(x)$. In view of (2.5.190) and [112, (8.8.52)], this proves that the formula in the second line of (2.5.169) holds for each function $f \in \mathscr{C}^\alpha(\partial_*\Omega)$ at σ_*-a.e. point $x \in \partial_*\Omega$.

Finally, under the additional assumption that $\sigma(\partial\Omega \setminus \partial_*\Omega) = 0$, the claim in part *(iv)* is seen from what we have proved so far together with [112, (7.3.24)] and [112, (5.9.18)] which, in concert, show that $\mathscr{C}^\alpha(\partial_*\Omega)$ may be canonically identified with $\mathscr{C}^\alpha(\partial\Omega)$ in the present setting. □

Next, we turn our attention to the study the harmonic double layer. In the most general setting, we introduce the latter operator as follows.

Definition 2.5.17 *Given a set $\Omega \subseteq \mathbb{R}^n$ (where $n \in \mathbb{N}$, $n \geq 2$) of locally finite perimeter, denote by ν its geometric measure theoretic outward unit normal, and abbreviate $\sigma_* := \mathcal{H}^{n-1}\lfloor\partial_*\Omega$. In this setting, define the action of the harmonic double layer $\mathcal{D}$ associated with Ω on an arbitrary σ_*-measurable function $f : \partial_*\Omega \to \mathbb{C}$ satisfying*

$$\int_{\partial_*\Omega} \frac{|f(y)|}{1+|y|^{n-1}} \, d\sigma_*(y) < +\infty \tag{2.5.201}$$

according to

$$\mathcal{D}f(x) := \frac{1}{\omega_{n-1}} \int_{\partial_*\Omega} \frac{\langle \nu(y), y - x \rangle}{|x - y|^n} f(y)\, d\sigma_*(y), \qquad \forall x \in \mathring{\Omega}, \tag{2.5.202}$$

where ω_{n-1} is the surface area of the unit sphere in $\mathbb{R}^n$.

In the same geometric context, [112, Proposition 5.6.7] *guarantees that the principal-value limit*

$$Kf(x) := \lim_{\varepsilon \to 0^+} \frac{1}{\omega_{n-1}} \int_{\substack{y \in \partial_*\Omega \\ |x-y|>\varepsilon}} \frac{\langle \nu(y), y - x \rangle}{|x - y|^n} f(y)\, d\sigma_*(y) \tag{2.5.203}$$

exists for any given function $f \in L^1\big(\partial_*\Omega, \frac{\sigma_*}{1+|\cdot|^{n-1}}\big)$ at σ_*-a.e. point $x \in \partial_*\Omega$. The assignment $f \mapsto Kf$, mapping functions from the space $L^1\big(\partial_*\Omega, \frac{\sigma_*}{1+|\cdot|^{n-1}}\big)$ into σ_*-measurable functions on $\partial_*\Omega$ (again, see [112, Proposition 5.6.7]*), is then referred to as the* **principal-value**, *or* **boundary-to-boundary, harmonic double layer potential operator**.

Finally, if one strengthens the original hypotheses on the underlying domain by assuming that Ω is a Lebesgue measurable set whose topological boundary $\partial\Omega$ is countably rectifiable (of dimension $n-1$) and has locally finite $\mathcal{H}^{n-1}$ measure, then the last claim in [112, Proposition 5.6.7] *guarantees that for each function $f \in L^1\big(\partial_*\Omega, \frac{\sigma_*(x)}{1+|x|^{n-1}}\big)$ the limit in* (2.5.203) *actually exists for $\mathcal{H}^{n-1}$-a.e. $x \in \partial\Omega$ and gives rise to a $\mathcal{H}^{n-1}$-measurable function on $\partial\Omega$. This is how Kf, the action of the principal-value (or boundary-to-boundary) harmonic double layer potential operator on the function f, is to be understood in this context.*

A few comments pertaining to Definition 2.5.17 are as follows. First, it is worth noting that the conditions imposed on f ensure that the integral in the right-hand side of (2.5.202) is absolutely convergent for each $x \in \mathring{\Omega}$. Second, if $\partial\Omega$ is upper Ahlfors regular then any function $f \in L^p(\partial_*\Omega, \sigma_*)$ with $p \in [1, \infty)$ is σ_*-measurable and satisfies (2.5.201) (as maybe seen using [112, Lemma 7.2.1]). Third, recall from (1.5.56) that E_Δ denotes the standard fundamental solution for the Laplacian in $\mathbb{R}^n$, where $n \in \mathbb{N}$ with $n \geq 2$. If Ω is as in Definition 2.5.17, then the integral kernel of the harmonic double layer $\mathcal{D}$ associated with Ω may be expressed in terms of (1.5.56) as

$$\frac{1}{\omega_{n-1}} \frac{\langle \nu(y), y - x \rangle}{|x - y|^n} = \nu(y) \cdot (\nabla E_\Delta)(y - x) = \partial_{\nu(y)}\big[E_\Delta(x - y)\big] \quad \text{for all } x \in \Omega \text{ and } \sigma_*\text{-a.e. } y \in \partial_*\Omega, \tag{2.5.204}$$

where $\partial_\nu := \nu \cdot \nabla$ denotes the normal derivative on $\partial_*\Omega$ (i.e., the directional derivative along the unit vector ν). Lastly, we wish to remark that whenever Ω is a Lebesgue measurable set whose topological boundary is a UR set then automatically $\partial\Omega$ is countably rectifiable (of dimension $n-1$) and has locally finite $\mathcal{H}^{n-1}$ measure (cf. [112, Definition 5.10.1] and [112, Proposition 6.10.5]).

Some trade-mark properties of the integral operator (2.5.202) are discussed below. Specifically, the harmonic double layer is a mechanism for producing lots of harmonic functions in Ω, using σ_*-measurable functions satisfying (2.5.201) (called "densities") as inputs, and the harmonic double layer reproduces constants if Ω is bounded.

Lemma 2.5.18 *Let $\Omega \subseteq \mathbb{R}^n$ (where $n \in \mathbb{N}$, $n \geq 2$) be a set of locally finite perimeter. Denote by ν its geometric measure theoretic outward unit normal, and abbreviate $\sigma_* := \mathcal{H}^{n-1} \lfloor \partial_* \Omega$. Consider the harmonic double layer associated with Ω as in* (2.5.202).

Then for each σ_*-measurable function $f : \partial_* \Omega \to \mathbb{C}$ satisfying the integrability condition in (2.5.201) *one has*

$$\mathcal{D}f \in \mathscr{C}^\infty\big(\mathring{\Omega}\big) \text{ and } \Delta(\mathcal{D}f) = 0 \text{ in } \mathring{\Omega}, \tag{2.5.205}$$

where $\Delta := \partial_1^2 + \cdots + \partial_n^2$ is the Laplacian in $\mathbb{R}^n$. Moreover,

$$\text{whenever } \Omega \text{ is open and } \partial\Omega \text{ is bounded then}$$
$$\mathcal{D}1 \equiv \begin{cases} 1 \text{ in } \Omega, \text{ if } \Omega \text{ is bounded,} \\ 0 \text{ in } \Omega, \text{ if } \Omega \text{ is unbounded.} \end{cases} \tag{2.5.206}$$

Proof The claims in (2.5.205) are seen straight from definitions. The claim in (2.5.206) is a direct consequence of Lemma 2.5.14 used with $\vec{k}$ as in (2.5.129). □

As seen from our next lemma, the harmonic double layer is reasonably behaved on Hölder spaces, especially when considered on uniform domains (cf. [112, Definition 5.11.10]) with compact Ahlfors regular boundaries.

Lemma 2.5.19 *Let $\Omega \subseteq \mathbb{R}^n$ (where $n \in \mathbb{N}$, $n \geq 2$) be an open set such that $\partial\Omega$ is bounded and $\partial_*\Omega$ is upper Ahlfors regular. In particular* (*cf.* [112, (5.9.16)]), *Ω has locally finite perimeter, so one may associate the harmonic double layer $\mathcal{D}$ with Ω as in* (2.5.202).

Then for each $\alpha \in (0, 1)$ there exists some $C \in (0, \infty)$, depending only on n, α, diam$(\partial_*\Omega)$, *and the upper Ahlfors regularity constant of $\partial_*\Omega$, such that for every function $f \in \mathscr{C}^\alpha(\partial_*\Omega)$ one has*

$$\sup_{x\in\Omega} \big|(\mathcal{D}f)(x)\big| + \sup_{x\in\Omega} \Big\{\text{dist}(x, \partial_*\Omega)^{1-\alpha}\big|\nabla(\mathcal{D}f)(x)\big|\Big\} \leq C\|f\|_{\mathscr{C}^\alpha(\partial_*\Omega)}. \tag{2.5.207}$$

In particular, if $\Omega \subseteq \mathbb{R}^n$ is a uniform domain whose boundary is compact and upper Ahlfors regular then, for each $\alpha \in (0, 1)$,

$$\mathcal{D} : \mathscr{C}^\alpha(\partial_*\Omega) \longrightarrow \mathscr{C}^\alpha\big(\overline{\Omega}\big) \tag{2.5.208}$$

is a well-defined, linear, and bounded operator; hence so is

$$\mathcal{D} : \mathscr{C}^\alpha(\partial\Omega) \longrightarrow \mathscr{C}^\alpha\big(\overline{\Omega}\big). \tag{2.5.209}$$

Proof All claims are direct consequence of Lemma 2.5.15 specialized to the case when the vector-valued function $\vec{k}$ is defined as in (2.5.129). □

Given a set $\Omega \subseteq \mathbb{R}^n$ (where $n \in \mathbb{N}$, $n \geq 2$) of locally finite perimeter, abbreviate $\sigma_* := \mathcal{H}^{n-1} \lfloor \partial_* \Omega$ and denote by ν the geometric measure theoretic outward unit normal to Ω. In this setting, we let the maximal harmonic double layer $K_{\max}$ act on functions $f \in L^1\big(\partial_*\Omega, \frac{\sigma_*}{1+|\cdot|^{n-1}}\big)$ according to

$$(K_{\max} f)(x) := \sup_{\varepsilon>0} \left| \frac{1}{\omega_{n-1}} \int\limits_{\substack{y\in\partial_*\Omega \\ |x-y|>\varepsilon}} \frac{\langle \nu(y), y-x\rangle}{|x-y|^n} f(y)\, d\sigma_*(y) \right|, \tag{2.5.210}$$

for all $x \in \partial\Omega$.

In the lemma below we prove that $K_{\max}$ maps Hölder functions into bounded functions provided Ω is a Lebesgue measurable set whose boundary is compact and upper Ahlfors regular.

Lemma 2.5.20 *Let $\Omega \subseteq \mathbb{R}^n$ (where $n \in \mathbb{N}$, $n \geq 2$) be a Lebesgue measurable set such that $\partial\Omega$ is bounded and $\partial_*\Omega$ is upper Ahlfors regular. Then for each given exponent $\alpha > 0$ there exists some number $N \in (0, \infty)$, depending only on n, α, and the upper Ahlfors regularity constant of $\partial_*\Omega$, with the property that*

$$\sup_{x\in\partial_*\Omega} (K_{\max} f)(x) \leq N\big[\mathrm{diam}(\partial_*\Omega)\big]^{\alpha} \cdot \|f\|_{\dot{\mathscr{C}}^{\alpha}(\partial_*\Omega)} + N \cdot \sup_{\partial_*\Omega} |f|, \tag{2.5.211}$$

for each function $f \in \mathscr{C}^{\alpha}(\partial_*\Omega)$.

Proof This is a direct consequence of Lemma 2.5.13 used with $\vec{k}$ as in (2.5.129). □

The next order of business is to study the boundary-to-boundary harmonic double layer potential operator from (2.5.203). In this vein, we shall first prove the following result, dealing with its action on Hölder functions.

Proposition 2.5.21 *Let $\Omega \subseteq \mathbb{R}^n$ (where $n \in \mathbb{N}$ with $n \geq 2$) be a Lebesgue measurable set with a compact upper Ahlfors regular boundary and abbreviate $\sigma := \mathcal{H}^{n-1} \lfloor \partial\Omega$. Then for each $\alpha \in (0, 1)$ the following statements are true.*

(i) For each $f \in \mathscr{C}^{\alpha}(\partial_*\Omega)$ the limit defining the principal-value harmonic double layer potential $(Kf)(x)$ in (2.5.203) exists for σ-a.e. point $x \in \partial_*\Omega$.
(ii) After possibly redefining Kf on a σ-nullset contained in $\partial_*\Omega$, the assignment $f \mapsto Kf$ thus defined induces a well-defined, linear, and bounded operator

$$K : \mathscr{C}^{\alpha}(\partial_*\Omega) \longrightarrow \mathscr{C}^{\alpha}(\partial_*\Omega) \tag{2.5.212}$$

with the property that

$$K1 \equiv \begin{cases} +\frac{1}{2} \text{ on } \partial_*\Omega, \text{ if } \Omega \text{ is bounded}, \\ -\frac{1}{2} \text{ on } \partial_*\Omega, \text{ if } \Omega \text{ is unbounded}. \end{cases} \tag{2.5.213}$$

(iii) Under the additional assumption that the set Ω is open and $\partial\Omega$ is also lower Ahlfors regular, the following jump-formula (where I denotes the identity operator and $\kappa > 0$ is an arbitrary fixed aperture parameter) is valid:

$$\begin{gathered}\text{for every given function } f \in \mathscr{C}^{\alpha}(\partial_*\Omega) \text{ one has} \\ \big(\mathcal{D}f\big)\Big|_{\partial\Omega}^{\kappa-\text{n.t.}} = \big(\tfrac{1}{2}I + K\big)f \text{ at } \sigma\text{-a.e. point on } \partial_*\Omega.\end{gathered} \tag{2.5.214}$$

(iv) In the case when it is also assumed that $\mathcal{H}^{n-1}(\partial\Omega \setminus \partial_*\Omega) = 0$, one may replace $\partial_*\Omega$ by $\partial\Omega$ in the formulation of (2.5.212), (2.5.213), *and* (2.5.214).

Proof All claims are direct consequences of Proposition 2.5.16 used for the vector-valued function $\vec{k}$ defined as in (2.5.129). □

As the proposition below illustrates, the most natural environment in which the principal-value harmonic double layer (2.5.203) is reasonably behaved on Lebesgue spaces is that of uniformly rectifiable sets.

Proposition 2.5.22 *Consider a Lebesgue measurable set $\Omega \subseteq \mathbb{R}^n$ (where $n \in \mathbb{N}$, $n \geq 2$) whose boundary is a* UR *set. Abbreviate $\sigma := \mathcal{H}^{n-1}\lfloor\partial\Omega$, $\sigma_* := \mathcal{H}^{n-1}\lfloor\partial_*\Omega$, and denote by ν the geometric measure theoretic outward unit normal to Ω. In this setting, consider the harmonic double layer potential operator $\mathcal{D}$ and the maximal operator $K_{\max}$ associated with Ω as in* (2.5.202) *and* (2.5.210), *respectively.*

Then for each $p \in [1, \infty)$ one can select a constant $C \in (0, \infty)$ depending only on n, p, and the UR *character of $\partial\Omega$, such that for every $f \in L^p(\partial_*\Omega, \sigma_*)$ one has*

$$\|K_{\max}f\|_{L^p(\partial\Omega,\sigma)} \leq C\|f\|_{L^p(\partial_*\Omega,\sigma_*)} \text{ if } 1 < p < \infty, \tag{2.5.215}$$

$$\|K_{\max}f\|_{L^{1,\infty}(\partial\Omega,\sigma)} \leq C\|f\|_{L^1(\partial_*\Omega,\sigma_*)} \text{ if } p = 1, \tag{2.5.216}$$

Furthermore, for each $f \in L^p(\partial_*\Omega, \sigma_*)$ with $p \in [1, \infty)$ the limit

$$(Kf)(x) = \lim_{\varepsilon\to 0^+} \frac{1}{\omega_{n-1}} \int\limits_{\substack{y\in\partial_*\Omega \\ |x-y|>\varepsilon}} \frac{\langle\nu(y), y - x\rangle}{|x - y|^n} f(y)\,\mathrm{d}\sigma_*(y) \text{ exists for } \sigma\text{-a.e. } x \in \partial\Omega, \tag{2.5.217}$$

and the operator K is well defined, linear, and bounded in the following settings:

$$K : L^p(\partial_*\Omega, \sigma_*) \longrightarrow L^p(\partial\Omega, \sigma), \quad p \in (1, \infty), \tag{2.5.218}$$

$$K : L^1(\partial_*\Omega, \sigma_*) \longrightarrow L^{1,\infty}(\partial\Omega, \sigma). \tag{2.5.219}$$

Finally, under the additional assumption that Ω is open, for each aperture parameter $\kappa \in (0, \infty)$ and integrability exponent $p \in [1, \infty)$, there exists a constant $C \in (0, \infty)$ depending only on n, p, κ, and the UR *character of $\partial\Omega$, such that for every $f \in L^p(\partial_*\Omega, \sigma_*)$ one has*

$$\|\mathcal{N}_\kappa(\mathcal{D}f)\|_{L^p(\partial\Omega,\sigma)} \leq C\|f\|_{L^p(\partial_*\Omega,\sigma_*)} \text{ if } 1 < p < \infty, \tag{2.5.220}$$

$$\|\mathcal{N}_\kappa(\mathcal{D}f)\|_{L^{1,\infty}(\partial\Omega,\sigma)} \leq C\|f\|_{L^1(\partial_*\Omega,\sigma_*)} \text{ if } p = 1. \tag{2.5.221}$$

Proof The claims about $K_{\max}$ and K are seen from Theorem 2.3.2 applied with $\Sigma := \partial\Omega$ and a natural choice of the kernel, keeping in mind that any given $f \in L^p(\partial_*\Omega, \sigma_*)$ may be canonically viewed as a function in $L^p(\partial\Omega, \sigma)$ simply extending it by zero on $\partial\Omega \setminus \partial_*\Omega$. In a similar fashion, the nontangential maximal function estimates from (2.5.220)-(2.5.221) are consequences of Theorem 2.4.1. □

Remark 2.5.23 *In the setting of Proposition 2.5.22, it is worth noting that we may also view K as a well-defined, linear, and bounded operator*

$$K : L^p(\partial\Omega, \sigma) \longrightarrow L^p(\partial\Omega, \sigma), \quad p \in (1, \infty), \tag{2.5.222}$$

$$K : L^1(\partial\Omega, \sigma) \longrightarrow L^{1,\infty}(\partial\Omega, \sigma), \tag{2.5.223}$$

simply by composing (2.5.218)-(2.5.219) *to the left with the canonical restriction*

$$L^p(\partial\Omega, \sigma) \ni f \longmapsto f\big|_{\partial_*\Omega} \in L^p(\partial_*\Omega, \sigma_*), \quad p \in [1, \infty). \tag{2.5.224}$$

In other words, for $f \in L^p(\partial\Omega, \sigma)$ with $p \in [1, \infty)$ we let Kf retain the definition (2.5.217), *simply by now regarding f as being restricted to $\partial_*\Omega$ in the integrand in the right-hand side of* (2.5.217).

To further shed some light on the nature of the principal-value limit in (2.5.217), consider the simple case of the harmonic double layer K_{poly} associated with an open planar polygon region $\Omega_{\text{poly}} \subset \mathbb{R}^2$, with vertexes x_j, $1 \leq j \leq N$, and corresponding angles $\alpha_j \in (0, 2\pi)$, $1 \leq j \leq N$. In particular, Ω_{poly} is a bounded open subset of $\mathbb{R}^2$, whose topological boundary is a compact UR set, satisfying $\partial_*\Omega_{\text{poly}} = \partial\Omega_{\text{poly}}$ and $\partial^*\Omega_{\text{poly}} = \partial\Omega_{\text{poly}} \setminus \{x_1, \dots, x_N\}$. As such, the general result from (2.5.217) presently ensures that

$$(K_{\text{poly}}1)(x) = \lim_{\varepsilon \to 0^+} \frac{1}{2\pi} \int\limits_{\substack{y \in \partial\Omega_{\text{poly}} \\ |x-y|>\varepsilon}} \frac{\langle \nu(y), y - x\rangle}{|x-y|^2} \, d\mathcal{H}^1(y) \text{ exists for } \mathcal{H}^1\text{-a.e. } x \in \partial\Omega. \tag{2.5.225}$$

This being said, from (2.5.139) we see that actually for every point $x \in \partial\Omega_{\text{poly}}$ the limit defining $(K_{\text{poly}}1)(x)$ exists and, in fact,

$$\begin{aligned}(K_{\text{poly}}1)(x) &= \lim_{\varepsilon \to 0^+} \frac{\mathcal{H}^1\big(\Omega_{\text{poly}} \cap \partial B(x,r)\big)}{2\pi r} \\ &= \begin{cases} 1/2 & \text{if } x \in \partial\Omega_{\text{poly}} \setminus \{x_1, \dots, x_N\}, \\ \alpha_j/(2\pi) & \text{if } x = x_j \text{ for some } 1 \leq j \leq N. \end{cases}\end{aligned} \tag{2.5.226}$$

Moreover, writing $(K_{\rm poly}f)(x) = \big(K_{\rm poly}(f - f(x))\big)(x) + f(x)(K_{\rm poly}1)(x)$ we see that, given any function $f \in \mathscr{C}^\alpha(\partial\Omega_{\rm poly})$ with $\alpha \in (0,1)$, the limit defining $(K_{\rm poly}1)(x)$ exists for every $x \in \partial\Omega_{\rm poly}$. In fact, with an absolutely convergent integral we have

$$(K_{\rm poly}f)(x) = \frac{1}{2\pi}\int_{\partial\Omega_{\rm poly}} \frac{\langle \nu(y), y - x\rangle}{|x-y|^2}\big(f(y) - f(x)\big)\,d\mathcal{H}^1(y)$$

$$+\begin{cases} f(x)/2 & \text{if } x \in \partial\Omega_{\rm poly}\setminus\{x_1,\dots,x_N\} \\ f(x)\alpha_j/(2\pi) & \text{if } x = x_j \text{ for some } 1 \le j \le N. \end{cases} \tag{2.5.227}$$

In our next proposition we discuss Carleson measure estimates for the harmonic double layer acting on functions in L^{n-1} as well as for its gradient acting on functions of bounded mean oscillations.

Proposition 2.5.24 *Let $\Omega \subseteq \mathbb{R}^n$ (where $n \in \mathbb{N}$, $n \ge 2$) be an open set with the property that $\partial\Omega$ is a UR set; in particular, Ω is a set of locally finite perimeter. Denote by ν the geometric measure theoretic outward unit normal to Ω and abbreviate $\sigma := \mathcal{H}^{n-1}\lfloor\partial\Omega$.*

Then there exists a constant $C = C(\partial\Omega) \in (0,\infty)$ with the property that

$$\text{given any } f \in L^{n-1}(\partial\Omega,\sigma), \text{ it follows that } |\mathcal{D}f|^2\,\text{dist}(\cdot,\partial\Omega)\,d\mathcal{L}^n \text{ is a Carleson measure in } \Omega, \text{ with constant} \le C\|f\|^2_{L^{n-1}(\partial\Omega,\sigma)}. \tag{2.5.228}$$

Furthermore, if in the same setting one defines the action of the gradient of the harmonic double layer $\nabla\mathcal{D}$ on functions $f \in \text{BMO}(\partial\Omega,\sigma)$ according to

$$(\nabla\mathcal{D})f(x) := \Big\{\frac{1}{\omega_{n-1}}\int_{\partial_*\Omega}\Big\langle \nu(y), \partial_{x_j}\Big(\frac{y-x}{|x-y|^n}\Big)\Big\rangle f(y)\,d\sigma(y)\Big\}_{1\le j\le n} \tag{2.5.229}$$

for all $x \in \Omega$, then for each $f \in \text{BMO}(\partial\Omega,\sigma)$ the function $(\nabla\mathcal{D})f$ is well defined in Ω and, given any $p \in (1,\infty)$, it follows that

$$\big|(\nabla\mathcal{D})f\big|^p\,\text{dist}(\cdot,\partial\Omega)^{p-1}\,d\mathcal{L}^n \tag{2.5.230}$$

is a Carleson measure in Ω, in the quantitative sense that there exists a constant $C \in (0,\infty)$ independent of f with the property that

$$\sup_{x\in\partial\Omega,\, r>0}\frac{1}{\sigma\big(B(x,r)\cap\partial\Omega\big)}\int_{B(x,r)\cap\Omega}\big|(\nabla\mathcal{D})f\big|^p\text{dist}(\cdot,\partial\Omega)^{p-1}\,d\mathcal{L}^n \le C\|f\|^p_{\text{BMO}(\partial\Omega,\sigma)}. \tag{2.5.231}$$

In particular, corresponding to $p = 2$,

$$\big|(\nabla\mathcal{D})f\big|^2\,\text{dist}(\cdot,\partial\Omega)\,d\mathcal{L}^n \text{ is a Carleson measure in } \Omega, \text{ for each function } f \in \text{BMO}(\partial\Omega,\sigma). \tag{2.5.232}$$

Proof The claim in (2.5.228) is clear from (2.4.48) and Definition 2.5.17. To proceed, fix an arbitrary function $f \in \mathrm{BMO}(\partial\Omega, \sigma)$. Since for each $x \in \Omega$ we have

$$\int_{\partial_*\Omega} \Big| \Big\langle \nu(y), \partial_{x_j}\Big(\frac{y-x}{|x-y|^n}\Big)\Big\rangle \Big| |f(y)|\, d\sigma(y) \leq c_n \int_{\partial\Omega} \frac{|f(y)|}{|x-y|^n}\, d\sigma(y) < +\infty, \tag{2.5.233}$$

thanks to the fact that $\partial\Omega$ is an Ahlfors regular set, and [112, (7.4.116)], we conclude that the function $(\nabla\mathcal{D})f$ is well defined in Ω, via absolutely convergent integrals.

We next claim that

$$(\nabla\mathcal{D})1 \equiv 0 \text{ in } \Omega. \tag{2.5.234}$$

To justify this claim, fix some index $j \in \{1, \dots, n\}$ and pick an arbitrary point $x \in \Omega$. Then the function $(\partial_j E)(x - \cdot)$ is locally absolutely integrable in Ω and this permits us to meaningfully define the vector field

$$\vec{F}_j := \nabla\big[(\partial_j E_\Delta)(x - \cdot)\big] \tag{2.5.235}$$

by taking the gradient in the sense of distributions in Ω (in the "dot" variable). Hence, $\vec{F}_j$ belongs to $\big[\mathcal{D}'(\Omega)\big]^n$. In fact, since the singular support of $\vec{F}_j$ is the singleton $\{x\}$, we see that actually

$$\vec{F}_j \in \big[L^1_{\mathrm{loc}}(\Omega, \mathcal{L}^n) + \mathscr{E}'(\Omega)\big]^n. \tag{2.5.236}$$

With the divergence taken in the sense of distributions in Ω we may compute (bearing in mind that E_Δ is a fundamental solution for Δ in $\mathbb{R}^n$)

$$\mathrm{div}\vec{F}_j = \Delta\big[(\partial_j E_\Delta)(x - \cdot)\big] = -\partial_j \Delta\big[E_\Delta(x - \cdot)\big] = -\partial_j \delta_x \text{ in } \mathcal{D}'(\Omega), \tag{2.5.237}$$

where δ_x is the Dirac distribution in Ω with mass at $x \in \Omega$. Also, if we introduce $d := \mathrm{dist}(x, \partial\Omega) > 0$ and select some arbitrary $\kappa > 0$, [112, Lemma 8.3.7] shows that there exists a constant $C \in (0, \infty)$ such that

$$\Big(\mathcal{N}_\kappa^{\Omega\setminus\overline{B(x,d/2)}} \vec{F}_j\Big)(y) \leq C|y - x|^{-n} \text{ for each } y \in \partial\Omega. \tag{2.5.238}$$

In turn, from (2.5.238), [112, (8.2.26)], and [112, Lemma 7.2.1], we conclude that

$$\mathcal{N}_\kappa^{\Omega\setminus\overline{B(x,d/2)}} \vec{F}_j \in L^1(\partial\Omega, \sigma). \tag{2.5.239}$$

In addition, since $\vec{F}_j$ is continuous in a neighborhood of $\partial\Omega$ it follows that the pointwise nontangential boundary trace $\vec{F}_j\big|^{\kappa-\mathrm{n.t.}}_{\partial\Omega}$ exists at every point in $\partial_{\mathrm{nta}}\Omega$ and, in fact,

$$\Big(\vec{F}_j\big|^{\kappa-\mathrm{n.t.}}_{\partial\Omega}\Big)(y) = \nabla_y\big[(\partial_j E_\Delta)(x - y)\big] \text{ at every } y \in \partial_{\mathrm{nta}}\Omega. \tag{2.5.240}$$

In concert with (1.5.56) and [112, Proposition 8.8.6] this ultimately proves that

$$\Big(\vec{F}_j\Big|_{\partial\Omega}^{\kappa-\mathrm{n.t.}}\Big)(y)=\frac{1}{\omega_{n-1}}\partial_{x_j}\Big(\frac{y-x}{|x-y|^n}\Big)\ \text{ for }\sigma\text{-a.e. } y\in\partial_*\Omega. \tag{2.5.241}$$

Finally, we note that in the case when Ω is unbounded we have $|\vec{F}_j(y)|=O(|y|^{-n})$ for $y\in\Omega$ with $|y|\to\infty$. Hence, in such a scenario,

$$\int_{[B(0,2R)\setminus B(0,R)]\cap\Omega}|y\cdot\vec{F}_j(y)|\,\mathrm{d}\mathcal{L}^n(y)=O(R)\ \text{ as } R\to\infty. \tag{2.5.242}$$

From (2.5.236), (2.5.237), (2.5.239), (2.5.241), and (2.4.128), we see that the hypotheses of [112, Theorem 1.4.1] are satisfied by $\vec{F}_j$. As such formula [112, (1.4.6)] applies and presently gives

$$_{(\mathscr{C}_b^\infty(\Omega))^*}\big(\mathrm{div}\vec{F}_j,1\big)_{\mathscr{C}_b^\infty(\Omega)}=\int_{\partial_*\Omega}\nu(y)\cdot\Big(\vec{F}_j\Big|_{\partial\Omega}^{\kappa-\mathrm{n.t.}}\Big)(y)\,\mathrm{d}\sigma(y). \tag{2.5.243}$$

Now, on the one hand, thanks to (2.5.237) and the compatibility property recorded in [112, (4.6.21)], we have

$$\begin{aligned}_{(\mathscr{C}_b^\infty(\Omega))^*}\big(\mathrm{div}\vec{F}_j,1\big)_{\mathscr{C}_b^\infty(\Omega)}&=-\,_{\mathscr{E}'(\Omega)}\langle\partial_j\delta_x,1\rangle_{\mathscr{E}(\Omega)}\\&=\,_{\mathscr{E}'(\Omega)}\langle\delta_x,\partial_j 1\rangle_{\mathscr{E}(\Omega)}=0.\end{aligned} \tag{2.5.244}$$

On the other hand, (2.5.241) implies that

$$\begin{aligned}\int_{\partial_*\Omega}&\nu(y)\cdot\Big(\vec{F}_j\Big|_{\partial\Omega}^{\kappa-\mathrm{n.t.}}\Big)(y)\,\mathrm{d}\sigma(y)\\&=\frac{1}{\omega_{n-1}}\int_{\partial_*\Omega}\Big\langle\nu(y),\partial_{x_j}\Big(\frac{y-x}{|x-y|^n}\Big)\Big\rangle\,\mathrm{d}\sigma(y).\end{aligned} \tag{2.5.245}$$

Collectively, (2.5.243), (2.5.244), (2.5.245), and (2.5.229) imply the cancelation property claimed in (2.5.234).

With (2.5.234) in hand, the remaining argument proceeds very much as in the case of the proof of Corollary 2.4.2. Specifically, having fixed a point $x_o\in\partial\Omega$ and a scale $r>0$, the same arguments which have produced (2.4.136) (based on the estimate recorded in (2.4.34) and the decay of the integral kernel of $\nabla\mathcal{D}$) now give

$$\begin{aligned}\int_{B(x_o,r)\cap\Omega}&\big|(\nabla\mathcal{D}f)(x)\big|^p\,\mathrm{dist}(x,\partial\Omega)^{p-1}\,\mathrm{d}x\\&\le Cr^{n-1}f_p^{\#}(x_o)^p+Cr^{n-1}f_1^{\#}(x_o)^p,\end{aligned} \tag{2.5.246}$$

where the L^q-based Fefferman-Stein maximal function $f_q^{\#}$ has been defined in (A.0.123). From this, (2.5.231) now follows on account of [112, (7.4.112)]. The final claim is obtained by specializing the earlier result to the case when $p = 2$. □

We desire to augment Proposition 2.5.22 and Proposition 2.5.24 with a jump-formula relating the harmonic double layers $\mathcal{D}$ and K much as in (2.5.214) but when the operators in question now act on functions from Lebesgue spaces. This is done in the proposition below where we consider the nontangential trace of $\mathcal{D}$ acting on L^p functions at almost all points in the geometric measure theoretic boundary of the underlying domain.

Proposition 2.5.25 *Suppose $\Omega \subseteq \mathbb{R}^n$ (where $n \in \mathbb{N}$ with $n \geq 2$) is a nonempty open set with the property that $\partial\Omega$ is a* UR *set. In particular, Ω is a set of locally finite perimeter. Abbreviate $\sigma_* := \mathcal{H}^{n-1}\lfloor\partial_*\Omega$ and denote by ν the geometric measure theoretic outward unit normal to Ω. In this context, define the (harmonic) double layer potential operator associated with Ω as in* (2.5.202), *and also consider its principal-value version acting on functions $f \in L^p(\partial_*\Omega, \sigma_*)$ with $p \in [1, \infty)$ as in* (2.5.217) *(for σ_*-a.e. $x \in \partial_*\Omega$). Finally, fix an arbitrary $\kappa > 0$.*

Then the double layer potential (2.5.202) *has the property that for every function $f \in L^p(\partial_*\Omega, \sigma_*)$ with $1 \leq p < \infty$ the following jump-formula holds (where I denotes the identity operator):*

$$\lim_{\substack{z\in\Gamma_\kappa(x)\\ z\to x}} (\mathcal{D}f)(z) = \left(\tfrac{1}{2}I + K\right)f(x) \text{ for } \sigma_*\text{-a.e. } x \in \partial_*\Omega. \tag{2.5.247}$$

In the setting of Proposition 2.5.25, it has been noted in Remark 2.5.23 that the principal-value harmonic double layer K may be naturally considered as an operator acting on functions belonging to $L^p(\partial\Omega, \sigma)$ with $p \in [1, \infty)$ and where $\sigma := \mathcal{H}^{n-1}\lfloor\partial\Omega$. Since this is also the case for the boundary-to-domain harmonic double layer $\mathcal{D}$ (originally defined in (2.5.202)), one may wonder whether the jump-formula (2.5.247) written for generic functions $f \in L^p(\partial\Omega, \sigma)$ with $p \in [1, \infty)$ is actually valid σ-a.e. on E for a possibly larger set $E \subseteq \partial\Omega$ than $\partial_*\Omega$. However, this is *not* the case since such an eventuality would imply that

$$f = 2(\mathcal{D}f)\Big|_{\partial\Omega}^{\kappa-\text{n.t.}} - 2Kf \text{ at } \sigma\text{-a.e. point on } E. \tag{2.5.248}$$

Given that both $\mathcal{D}f$ and Kf depend only on $f\big|_{\partial_*\Omega}$, this would ultimately imply that $f\big|_{\partial_*\Omega}$ always determines f on the bigger set E, which is obviously not the case. Hence, from this point of view, the jump-formula (2.5.247) is optimally formulated.

Proof of Proposition 2.5.25 First note that [112, Proposition 8.8.6] ensures that it is meaningful to consider the limit in (2.5.247). Second, as seen from (2.5.220)-(2.5.221), the nontangential maximal operator associated with the type of limit implicit in (2.5.247), i.e.,

$$f \longmapsto \mathcal{N}_\kappa(\mathcal{D}f)(x) := \sup_{z\in\Gamma_\kappa(x)} \big|(\mathcal{D}f)(z)\big|, \qquad x \in \partial_*\Omega, \tag{2.5.249}$$

is bounded from $L^p(\partial_*\Omega, \sigma_*)$ into itself if $1 < p < \infty$, and from $L^1(\partial_*\Omega, \sigma_*)$ into $L^{1,\infty}(\partial_*\Omega, \sigma_*)$.

Third, [112, Proposition 6.2.11] may be presently applied with $\mathscr{X} := \Omega \cup \partial_*\Omega$ endowed with the topology inherited from the ambient Euclidean space, $X := \partial_*\Omega$, $\mu := \sigma_*$ which is a locally finite complete Borel-regular measure on X by virtue of [112, Lemma 3.6.4] (the fact that the hypotheses in [112, (3.6.25)] are presently satisfied is seen from [112, (5.6.35)] and [112, (5.6.33)]), the sets $\Gamma(x) := \Gamma_\kappa(x)$ for each $x \in \partial_*\Omega$ which satisfy [112, (6.2.71)] thanks to item *(iii)* in [112, Proposition 8.8.6], $Y := L^p(\partial_*\Omega, \sigma_*)$, the linear operator $\mathcal{T} := \mathcal{D}$, and

$$\mathcal{V} := \{f|_{\partial_*\Omega} : f \in \operatorname{Lip}_c(\partial\Omega)\} \tag{2.5.250}$$

which is a dense subset of Y by [112, (3.7.3)] and [112, Proposition 3.7.1]. These choices imply that the associated maximal operator (cf. [112, (6.2.73)]) is (2.5.249) which, according to earlier remarks, satisfies [112, (6.2.74)]. As such, the only hypothesis in [112, Proposition 6.2.11] left to check at this point is [112, (6.2.75)]. This follows once we show that

$$\begin{gathered}\text{for each function } f \text{ belonging to } \operatorname{Lip}_c(\partial\Omega),\\ \lim_{\substack{z\in\Gamma_\kappa(x)\\ z\to x}} (\mathcal{D}f)(z) = (\tfrac{1}{2}I + K)f(x) \text{ for } \sigma_*\text{-a.e. } x \in \partial_*\Omega.\end{gathered} \tag{2.5.251}$$

However, this may be seen from a cursory inspection of the argument used in the proof of (2.5.169) in Proposition 2.5.16 (with $\vec{k}$ as in (2.5.129)) which continues to work in the present setting since the compactness of $\partial\Omega$ stipulated in Proposition 2.5.16 is not currently needed. The only significant difference is that now, having fixed an arbitrary function $f \in \operatorname{Lip}_c(\partial\Omega)$, the existence of a σ_*-measurable set $A \subseteq \partial_*\Omega$ with the property that (2.5.189) holds is implied by (2.5.217) (in lieu of item *(i)* of Proposition 2.5.16, as was the case in the past).

The application of [112, Proposition 6.2.11] in the manner just described then guarantees that

$$\begin{gathered}\text{for each given function } f \in L^p(\partial_*\Omega, \sigma_*) \text{ with } 1 \leq p < \infty \text{ the}\\ \text{nontangential limit } \lim_{\substack{z\in\Gamma_\kappa(x)\\ z\to x}} (\mathcal{D}f)(z) \text{ exists for } \sigma_*\text{-a.e. } x \in \partial_*\Omega, \text{ and}\\ \Big|\lim_{\substack{z\in\Gamma_\kappa(x)\\ z\to x}} (\mathcal{D}f)(z)\Big| \leq \mathcal{N}_\kappa(\mathcal{D}f)(x) \text{ at } \sigma_*\text{-a.e. point } x \in \partial_*\Omega.\end{gathered} \tag{2.5.252}$$

With this in hand, [112, Corollary 8.9.6] additionally gives that

$$\begin{gathered}\text{for each } f \in L^p(\partial_*\Omega, \sigma_*) \text{ with } 1 \leq p < \infty \text{ the function defined}\\ \sigma_*\text{-a.e. as } \partial_*\Omega \ni x \mapsto \lim_{\substack{z\in\Gamma_\kappa(x)\\ z\to x}} (\mathcal{D}f)(z) \text{ is } \sigma_*\text{-measurable on } \partial_*\Omega.\end{gathered} \tag{2.5.253}$$

In concert, (2.5.252), (2.5.253), and the boundedness of the mapping in (2.5.249) ensure that the assignment

$$f \longmapsto \lim_{\substack{z\in\Gamma_\kappa(x)\\ z\to x}} (\mathcal{D}f)(z), \qquad x \in \partial_*\Omega, \tag{2.5.254}$$

is linear and bounded from $L^p(\partial_*\Omega, \sigma_*)$ into itself if $1 < p < \infty$, and from $L^1(\partial_*\Omega, \sigma_*)$ into the Lorentz space $L^{1,\infty}(\partial_*\Omega, \sigma_*)$. Thanks to (2.3.18)-(2.3.19), the operator $\frac{1}{2}I+K$ is also bounded in the very same context. Granted these boundedness properties, the identity recorded in (2.5.251), and the density of $\mathcal{V}$ in $L^p(\partial_*\Omega, \sigma_*)$ for each $p \in [1, \infty)$, we ultimately conclude that (2.5.247) holds. □

Fix an integer $n \in \mathbb{N}$ with $n \geq 2$ and recall the standard fundamental solution E_Δ for the Laplace operator in $\mathbb{R}^n$ introduced in (1.5.56). Given an open set of locally finite perimeter $\Omega \subseteq \mathbb{R}^n$, abbreviate $\sigma_* := \mathcal{H}^{n-1}\lfloor\partial_*\Omega$ and denote by $\nu = (\nu_1, \dots, \nu_n)$ the geometric measure theoretic outward unit normal to Ω. In this setting, we are next interested in considering integral operators of a similar nature to the harmonic double layer $\mathcal{D}$ but whose integral kernels are obtained by applying certain "tangential" derivatives to $E_\Delta(x - y)$ (instead of taking the normal derivative as done in the case of $\mathcal{D}$ in (2.5.204)). Specifically, with Ω as above, for each $j, k \in \{1, \dots, n\}$ we presently consider the operator $\mathcal{R}_{jk}$ acting on σ_*-measurable functions $f : \partial_*\Omega \to \mathbb{C}$ satisfying

$$\int_{\partial_*\Omega} \frac{|f(y)|}{1 + |y|^{n-1}}\, d\sigma_*(y) < +\infty \tag{2.5.255}$$

according to

$$\begin{aligned}\mathcal{R}_{jk}f(x) &:= \int_{\partial_*\Omega} \Big\{\nu_j(y)\partial_{y_k}\big[E_\Delta(x - y)\big] - \nu_k(y)\partial_{y_j}\big[E_\Delta(x - y)\big]\Big\} f(y)\, d\sigma_*(y)\\ &= \frac{-1}{\omega_{n-1}} \int_{\partial_*\Omega} \frac{\nu_j(y)(x_k - y_k) - \nu_k(y)(x_j - y_j)}{|x - y|^n} f(y)\, d\sigma_*(y)\end{aligned} \tag{2.5.256}$$

for each $x \in \Omega$. In particular,

$$\mathcal{R}_{jj} = 0 \text{ and } \mathcal{R}_{kj} = -\mathcal{R}_{jk} \text{ for every } j, k \in \{1, \dots, n\}. \tag{2.5.257}$$

Proposition 2.5.26 *Suppose $\Omega \subseteq \mathbb{R}^n$ (where $n \in \mathbb{N}$, $n \geq 2$) is an open set of locally finite perimeter. Denote by $\nu = (\nu_1, \dots, \nu_n)$ the geometric measure theoretic outward unit normal to Ω, and abbreviate $\sigma_* := \mathcal{H}^{n-1}\lfloor\partial_*\Omega$, $\sigma := \mathcal{H}^{n-1}\lfloor\partial\Omega$. Then for each $j, k \in \{1, \dots, n\}$ the following properties hold.*

(i) For every σ_*-measurable function $f : \partial_*\Omega \to \mathbb{C}$ satisfying the integrability condition demanded in (2.5.255) one has

$$\mathcal{R}_{jk}f \in \mathscr{C}^\infty(\Omega) \text{ and } \Delta(\mathcal{R}_{jk}f) = 0 \text{ in } \Omega. \tag{2.5.258}$$

(ii) Under the additional assumption that $\partial\Omega$ is bounded one has

$$\mathcal{R}_{jk}1 \equiv \begin{cases} 1 \ \textit{in} \ \Omega, \ \textit{if} \ \Omega \ \textit{is bounded}, \\ 0 \ \textit{in} \ \Omega, \ \textit{if} \ \Omega \ \textit{is unbounded}. \end{cases} \tag{2.5.259}$$

(iii) Under the additional assumption that $\partial\Omega$ is bounded and $\partial_*\Omega$ is upper Ahlfors regular, it follows that for each $\alpha \in (0,1)$ there exists some $C \in (0,\infty)$, depending only on n, α, $\operatorname{diam}(\partial_*\Omega)$, and the upper Ahlfors regularity constant of $\partial_*\Omega$, such that for every function $f \in \mathscr{C}^\alpha(\partial_*\Omega)$ one has

$$\sup_{x\in\Omega} \big|(\mathcal{R}_{jk}f)(x)\big| + \sup_{x\in\Omega} \Big\{\operatorname{dist}(x,\partial_*\Omega)^{1-\alpha}\big|\nabla(\mathcal{R}_{jk}f)(x)\big|\Big\} \leq C\|f\|_{\mathscr{C}^\alpha(\partial_*\Omega)}. \tag{2.5.260}$$

In particular, if $\Omega \subseteq \mathbb{R}^n$ is a uniform domain whose boundary is compact and upper Ahlfors regular then $\mathcal{R}_{jk}$ induces a well-defined, linear, and bounded operator in the context

$$\mathcal{R}_{jk} : \mathscr{C}^\alpha(\partial_*\Omega) \longrightarrow \mathscr{C}^\alpha\big(\overline{\Omega}\big), \qquad \forall\alpha \in (0,1). \tag{2.5.261}$$

As a corollary, if $\Omega \subseteq \mathbb{R}^n$ is simultaneously a uniform domain, and an Ahlfors regular domain with compact boundary, then for each $\alpha \in (0,1)$ the operator $\mathcal{R}_{jk}$ is well defined, linear, and bounded in the context

$$\mathcal{R}_{jk} : \mathscr{C}^\alpha(\partial\Omega) \longrightarrow \mathscr{C}^\alpha\big(\overline{\Omega}\big). \tag{2.5.262}$$

(iv) If $\partial\Omega$ is a UR *set then for each aperture parameter $\kappa \in (0,\infty)$ and integrability exponent $p \in [1,\infty)$ there exists a constant $C = C(\Omega, n, \kappa, p) \in (0,\infty)$ with the property that for each function $f \in L^p(\partial_*\Omega, \sigma_*)$ one has*

$$\|\mathcal{N}_\kappa(\mathcal{R}_{jk}f)\|_{L^p(\partial\Omega,\sigma)} \leq C\|f\|_{L^p(\partial_*\Omega,\sigma_*)} \ \textit{if} \ 1 < p < \infty, \tag{2.5.263}$$

$$\|\mathcal{N}_\kappa(\mathcal{R}_{jk}f)\|_{L^{1,\infty}(\partial\Omega,\sigma)} \leq C\|f\|_{L^1(\partial_*\Omega,\sigma_*)} \ \textit{if} \ p = 1. \tag{2.5.264}$$

(v) Suppose $\partial\Omega$ is a UR *set. Then there exists $C = C(\partial\Omega) \in (0,\infty)$ such that*

$$\begin{array}{l} \textit{given any } f \in L^{n-1}(\partial\Omega,\sigma), \textit{ it follows that } \big|\mathcal{R}_{jk}f\big|^2 \operatorname{dist}(\cdot,\partial\Omega)\,\mathrm{d}\mathcal{L}^n \textit{ is a} \\ \textit{Carleson measure in } \Omega, \textit{ with constant} \leq C\|f\|^2_{L^{n-1}(\partial\Omega,\sigma)}. \end{array} \tag{2.5.265}$$

In addition, if one defines the action of the operator $\nabla\mathcal{R}_{jk}$ on arbitrary functions f belonging to the space $\mathrm{BMO}(\partial\Omega,\sigma)$ *according to*

$$(\nabla\mathcal{R}_{jk})f(x) := \int_{\partial_*\Omega} \Big[\nu_k(y)(\nabla\partial_j E_\Delta)(x-y) - \nu_j(y)(\nabla\partial_k E_\Delta)(x-y)\Big] f(y)\,\mathrm{d}\sigma_*(y) \tag{2.5.266}$$

at each $x \in \Omega$, then for every $f \in \mathrm{BMO}(\partial\Omega,\sigma)$ the function $(\nabla\mathcal{R}_{jk})f$ is well defined and $\big|(\nabla\mathcal{R}_{jk})f\big|^p \operatorname{dist}(\cdot,\partial\Omega)^{p-1}\,\mathrm{d}\mathcal{L}^n$ is a Carleson measure in Ω for each $p \in (1,\infty)$, in the quantitative sense that there exists some constant $C \in (0,\infty)$, *independent of f, with the property that*

$$\sup_{x\in\partial\Omega,\, r>0} \frac{1}{\sigma\big(B(x,r)\cap\partial\Omega\big)} \int_{B(x,r)\cap\Omega} \big|(\nabla\mathcal{R}_{jk})f\big|^p \operatorname{dist}(\cdot,\partial\Omega)^{p-1}\,d\mathcal{L}^n$$
$$\leq C\|f\|^p_{\mathrm{BMO}(\partial\Omega,\sigma)}. \tag{2.5.267}$$

In particular, for $p = 2$, one has that $\big|(\nabla\mathcal{R}_{jk})f\big|^2 \operatorname{dist}(\cdot,\partial\Omega)\,d\mathcal{L}^n$ is a Carleson measure in Ω, for each function $f \in \mathrm{BMO}(\partial\Omega,\sigma)$.

Proof The claims in items *(i)-(iii)* are consequences of (2.5.256) and Proposition 2.5.16 used with $\vec{k}$ as in (2.5.140). Next, item *(iv)* is a consequence of Theorem 2.4.1. Finally, the first claim in item *(v)* is a direct consequence of (2.4.48) and (2.5.256), while the second claim is implied by Corollary 2.4.2. □

Let us now bring in the principal-value versions of the integral operators considered in (2.5.256). Specifically, fix $n \in \mathbb{N}$ with $n \geq 2$ and suppose $\Omega \subseteq \mathbb{R}^n$ is a set of locally finite perimeter. Abbreviate $\sigma_* := \mathcal{H}^{n-1}\lfloor\partial_*\Omega$ and denote by $\nu = (\nu_1,\dots,\nu_n)$ the geometric measure theoretic outward unit normal to Ω. Pick $j,k \in \{1,\dots,n\}$ arbitrary. Then [112, Proposition 5.6.7] ensures that, given any function $f \in L^1\big(\partial_*\Omega, \frac{\sigma_*}{1+|\cdot|^{n-1}}\big)$, the principal-value limit

$$R_{jk}f(x) := \lim_{\varepsilon\to0^+} \int_{\substack{y\in\partial_*\Omega\\ |x-y|>\varepsilon}} \Big[\nu_j(y)(\partial_k E_\Delta)(y-x) - \nu_k(y)(\partial_j E_\Delta)(y-x)\Big] f(y)\,d\sigma_*(y) \tag{2.5.268}$$

exists at σ_*-a.e. point $x \in \partial_*\Omega$ and gives rise to a σ_*-measurable function on $\partial_*\Omega$. Note that, by design,

$$R_{jj} = 0 \text{ and } R_{kj} = -R_{jk} \text{ for every } j,k \in \{1,\dots,n\}. \tag{2.5.269}$$

From [112, Proposition 5.6.7] we also know that, if Ω is a Lebesgue measurable set whose topological boundary $\partial\Omega$ is countably rectifiable (of dimension $n-1$) and has locally finite $\mathcal{H}^{n-1}$ measure, then for each function $f \in L^1\big(\partial_*\Omega, \frac{\sigma_*(x)}{1+|x|^{n-1}}\big)$ the limit in (2.5.268) actually exists for $\mathcal{H}^{n-1}$-a.e. $x \in \partial\Omega$ and gives rise to a $\mathcal{H}^{n-1}$-measurable function on $\partial\Omega$.

Finally, we define the corresponding maximal operator acting on each function $f \in L^1\big(\partial_*\Omega, \frac{\sigma_*}{1+|\cdot|^{n-1}}\big)$ according to

$$(R_{jk})_{\max}f(x) \tag{2.5.270}$$
$$:= \sup_{\varepsilon>0} \left| \int_{\substack{y\in\partial_*\Omega\\ |x-y|>\varepsilon}} \Big[\nu_j(y)(\partial_k E_\Delta)(y-x) - \nu_k(y)(\partial_j E_\Delta)(y-x)\Big] f(y)\,d\sigma_*(y) \right|$$

at each $x \in \partial\Omega$. In relation to the operators (2.5.268), (2.5.270), we have the following result.

Proposition 2.5.27 *Assume $\Omega \subseteq \mathbb{R}^n$ (where $n \in \mathbb{N}$ with $n \geq 2$) is a set of locally finite perimeter, and abbreviate $\sigma := \mathcal{H}^{n-1}\lfloor\partial\Omega$, $\sigma_* := \mathcal{H}^{n-1}\lfloor\partial_*\Omega$. Then for each $j, k \in \{1, \dots, n\}$ the following statements are true.*

(i) If $\partial\Omega$ is bounded and $\partial_*\Omega$ is upper Ahlfors regular then for each $\alpha > 0$ there exists a number $N \in (0, \infty)$, depending only on the dimension n, exponent α, and the upper Ahlfors regularity constant of $\partial_*\Omega$, with the property that for every function $f \in \mathscr{C}^\alpha(\partial_*\Omega)$ one has

$$\sup_{x \in \partial_*\Omega} \big((R_{jk})_{\max} f\big)(x) \leq N\big[\mathrm{diam}(\partial_*\Omega)\big]^\alpha \cdot \|f\|_{\dot{\mathscr{C}}^\alpha(\partial_*\Omega)} + N \cdot \sup_{\partial_*\Omega} |f|. \tag{2.5.271}$$

(ii) If $\partial\Omega$ is a bounded upper Ahlfors regular set then for each $f \in \mathscr{C}^\alpha(\partial_*\Omega)$ with $\alpha \in (0, 1)$ the limit defining $R_{jk} f(x)$ in (2.5.268) exists for σ-a.e. point $x \in \partial_*\Omega$. Moreover, after possibly redefining $R_{jk} f$ on a σ-nullset contained in $\partial_*\Omega$, the assignment $f \mapsto R_{jk} f$ thus defined induces a well-defined, linear, and bounded operator

$$R_{jk} : \mathscr{C}^\alpha(\partial_*\Omega) \longrightarrow \mathscr{C}^\alpha(\partial_*\Omega), \qquad \forall \alpha \in (0, 1), \tag{2.5.272}$$

with the property that

$$R_{jk} 1 \equiv 0 \ \text{ on } \partial_*\Omega. \tag{2.5.273}$$

(iii) Suppose Ω is actually open and $\partial\Omega$ is a compact Ahlfors regular set. Then for each arbitrary fixed number $\kappa > 0$, the following nontangential boundary trace formula hods:

$$\begin{gathered} \text{for each } f \in \mathscr{C}^\alpha(\partial_*\Omega) \text{ with } \alpha \in (0,1) \text{ one has} \\ \big(\mathcal{R}_{jk} f\big)\Big|_{\partial\Omega}^{\kappa-\text{n.t.}} = R_{jk} f \ \text{ at } \sigma\text{-a.e. point on } \partial_*\Omega. \end{gathered} \tag{2.5.274}$$

(iv) In the case when it is also assumed that $\sigma(\partial\Omega \setminus \partial_*\Omega) = 0$, one may replace $\partial_*\Omega$ by $\partial\Omega$ in the formulation of (2.5.272), (2.5.273), and (2.5.274).

(v) Strengthen the assumptions by requiring that $\partial\Omega$ is a UR set. Then for each $p \in [1, \infty)$ there exists a constant $C = C(\Omega, n, p) \in (0, \infty)$ with the property that for each $f \in L^p(\partial_*\Omega, \sigma_*)$ one has

$$\|(R_{jk})_{\max} f\|_{L^p(\partial\Omega,\sigma)} \leq C\|f\|_{L^p(\partial_*\Omega,\sigma_*)} \ \text{ if } 1 < p < \infty, \tag{2.5.275}$$

$$\|(R_{jk})_{\max} f\|_{L^{1,\infty}(\partial\Omega,\sigma)} \leq C\|f\|_{L^1(\partial_*\Omega,\sigma_*)} \ \text{ if } p = 1. \tag{2.5.276}$$

Furthermore, given any function $f \in L^p(\partial_*\Omega, \sigma_*)$ with $p \in [1, \infty)$ the limit defining $R_{jk} f(x)$ in (2.5.268) exists for σ-a.e. $x \in \partial\Omega$ and the operator R_{jk} is well defined, linear, and bounded in the following settings:

$$R_{jk} : L^p(\partial_*\Omega, \sigma_*) \longrightarrow L^p(\partial\Omega, \sigma), \quad p \in (1, \infty), \tag{2.5.277}$$

$$R_{jk} : L^1(\partial_*\Omega, \sigma_*) \longrightarrow L^{1,\infty}(\partial\Omega, \sigma). \tag{2.5.278}$$

(vi) Assume that actually Ω is open and that $\partial\Omega$ is a UR *set. Then for each fixed arbitrary number $\kappa > 0$, then following nontangential boundary trace formula holds:*

$$\begin{gathered} \text{for each } f \in L^p(\partial_*\Omega, \sigma_*) \text{ with } p \in [1, \infty) \text{ one has} \\ \left(\mathcal{R}_{jk} f\right)\Big|_{\partial\Omega}^{\kappa-\text{n.t.}} = R_{jk} f \text{ at } \sigma_*\text{-a.e. point on } \partial_*\Omega. \end{gathered} \tag{2.5.279}$$

Proof The claim made in item *(i)* is a direct consequence of Lemma 2.5.13 specialized to the case when $\vec{k}$ is as in (2.5.140). The claims in items *(ii)-(iv)* follows from Proposition 2.5.16 once again with $\vec{k}$ as in (2.5.140) (bearing in mind that $\vartheta = 0$ in this case). Next, all claims in item *(v)* are direct consequences of Theorem 2.3.2. Finally, the nontangential boundary trace formula in item *(vi)* may be proved by arguing as in Proposition 2.5.25 (making use of (2.5.274) the same way (2.5.214) was used in the end-game of the proof of (2.5.247)). □

Let $\Omega \subset \mathbb{R}^n$ be a set of locally finite perimeter and denote by $\nu = (\nu_1, \dots, \nu_n)$ its geometric measure theoretic outward unit normal and abbreviate $\sigma_* := \mathcal{H}^{n-1} \lfloor \partial_*\Omega$. Embed $\mathbb{R}^n$ into the Clifford algebra $\mathcal{C}\ell_n$ as in [112, (6.4.3)]. In particular, we agree to identify ν with the $\mathcal{C}\ell_n$-valued function defined σ_*-a.e. on $\partial_*\Omega$ as

$$\nu = \nu_1 \mathbf{e}_1 + \cdots + \nu_n \mathbf{e}_n. \tag{2.5.280}$$

The action of the (boundary-to-domain) `Cauchy-Clifford integral operator` on any given σ_*-measurable Clifford algebra-valued function $f : \partial_*\Omega \to \mathcal{C}\ell_n$ satisfying

$$\int_{\partial_*\Omega} \frac{|f(y)|}{1 + |y|^{n-1}} \, d\sigma_*(y) < +\infty \tag{2.5.281}$$

is defined as

$$\mathcal{C}f(x) := \frac{1}{\omega_{n-1}} \int_{\partial_*\Omega} \frac{x - y}{|x - y|^n} \odot \nu(y) \odot f(y) \, d\sigma_*(y), \qquad \forall x \in \mathring{\Omega}, \tag{2.5.282}$$

while the action of the (boundary-to-boundary) maximal operator on f is defined as

$$\mathfrak{C}_{\max} f(x) := \sup_{\varepsilon > 0} \left| \frac{1}{\omega_{n-1}} \int_{\substack{y \in \partial_*\Omega \\ |x-y| > \varepsilon}} \frac{x - y}{|x - y|^n} \odot \nu(y) \odot f(y) \, d\sigma_*(y) \right|, \tag{2.5.283}$$

for all $x \in \partial\Omega$. From [112, Proposition 5.6.7] we also see that for any function $f \in L^1\big(\partial_*\Omega, \frac{\sigma_*}{1+|\cdot|^{n-1}}\big) \otimes \mathcal{C}\ell_n$ the principal-value limit

$$\mathfrak{C} f(x) := \lim_{\varepsilon \to 0^+} \frac{1}{\omega_{n-1}} \int\limits_{\substack{y \in \partial_* \Omega \\ |x-y|>\varepsilon}} \frac{x-y}{|x-y|^n} \odot \nu(y) \odot f(y)\, d\sigma_*(y) \tag{2.5.284}$$

exists at σ_*-a.e. point $x \in \partial_*\Omega$ and gives rise to a σ_*-measurable Clifford algebra-valued function on $\partial_*\Omega$. Then $\mathfrak{C}$ is a well-defined operator, sending functions from $L^1\big(\partial_*\Omega, \frac{\sigma_*}{1+|\cdot|^{n-1}}\big) \otimes \mathcal{C}\ell_n$ into σ_*-measurable Clifford algebra-valued function on $\partial_*\Omega$.

In fact, the last claim in [112, Proposition 5.6.7] implies that if Ω is a Lebesgue measurable set whose topological boundary $\partial\Omega$ is countably rectifiable (of dimension $n-1$) and has locally finite $\mathcal{H}^{n-1}$ measure, then for each function $f \in L^1\big(\partial_*\Omega, \frac{\sigma_*(x)}{1+|x|^{n-1}}\big) \otimes \mathcal{C}\ell_n$ the limit in (2.5.284) actually exists for $\mathcal{H}^{n-1}$-a.e. $x \in \partial\Omega$ and gives rise to a $\mathcal{H}^{n-1}$-measurable Clifford algebra-valued function on $\partial\Omega$.

Remark 2.5.28 *Note that in* (2.5.282), (2.5.284), *the factor* $(x-y)/|x-y|^n$ *is located to the left. For this reason, we shall occasionally, use notation designed to emphasize this specific feature, i.e., write* C_L *and* $\mathfrak{C}_L$ *in place of* C *and* $\mathfrak{C}$, *respectively. This becomes particularly relevant upon noting that there are also "right-handed" versions of the operators* (2.5.282), (2.5.284). *Specifically, in the same geometric context as before, for each* $f \in L^1\big(\partial_*\Omega, \frac{\sigma_*}{1+|\cdot|^{n-1}}\big) \otimes \mathcal{C}\ell_n$ *we define*

$$C_R f(x) := \frac{1}{\omega_{n-1}} \int_{\partial_*\Omega} f(y) \odot \nu(y) \odot \frac{x-y}{|x-y|^n}\, d\sigma_*(y), \qquad \forall x \in \mathring{\Omega}, \tag{2.5.285}$$

and, for σ_**-a.e. point* $x \in \partial_*\Omega$,

$$\mathfrak{C}_R f(x) := \lim_{\varepsilon \to 0^+} \frac{1}{\omega_{n-1}} \int\limits_{\substack{y \in \partial_* \Omega \\ |x-y|>\varepsilon}} f(y) \odot \nu(y) \odot \frac{x-y}{|x-y|^n}\, d\sigma_*(y). \tag{2.5.286}$$

It is then apparent from the above definitions and [112, (6.4.29), (6.4.32)] *that for each given function* $f \in L^1\big(\partial_*\Omega, \frac{\sigma_*}{1+|\cdot|^{n-1}}\big) \otimes \mathcal{C}\ell_n$ *we have (with 'bar' denoting Clifford conjugation)*

$$C_R f = \overline{C_L(\overline{f})} \ \text{ in } \mathring{\Omega}, \tag{2.5.287}$$

and

$$\mathfrak{C}_R f = \overline{\mathfrak{C}_L(\overline{f})} \ \text{ at } \sigma_*\text{-a.e. point on } \partial_*\Omega. \tag{2.5.288}$$

Consequently, the operators C_R, $\mathfrak{C}_R$ *enjoy very similar properties of their standard, left-handed counterparts* C, $\mathfrak{C}$. *for this reason, we shall primarily just focus on the latter. Here we only wish to note that for each function* $f \in L^1\big(\partial_*\Omega, \frac{\sigma_*}{1+|\cdot|^{n-1}}\big) \otimes \mathcal{C}\ell_n$ *we have (cf.* (2.5.290) *below)*

$$D_R C_R f = 0 \ \text{ and } \ D_L C_L f = 0 \ \text{ in } \mathring{\Omega}, \tag{2.5.289}$$

where D_L and D_R denote the Dirac operator acting from the right and from the left, respectively (*cf.* (A.0.40)-(A.0.41)).

The operators (2.5.282)-(2.5.284) are studied in the proposition below.

Proposition 2.5.29 *Assume $\Omega \subseteq \mathbb{R}^n$ (where $n \in \mathbb{N}$, $n \geq 2$) is a given set of locally finite perimeter. Denote by ν its geometric measure theoretic outward unit normal, and abbreviate $\sigma_* := \mathcal{H}^{n-1}\lfloor\partial_*\Omega$, $\sigma := \mathcal{H}^{n-1}\lfloor\partial\Omega$. Then, for the Cauchy-Clifford integral operators associated with Ω as in* (2.5.282)-(2.5.284), *the following statements are true.*

(i) For each σ_*-measurable function $f : \partial_*\Omega \to \mathcal{C}\ell_n$ satisfying the integrability condition in (2.5.281) *one has*

$$\mathcal{C}f \in \mathscr{C}^\infty(\mathring{\Omega}) \text{ and } D(\mathcal{C}f) = 0 \text{ in } \mathring{\Omega}, \tag{2.5.290}$$

where $D := \mathbf{e}_1\partial_1 + \cdots + \mathbf{e}_n\partial_n$ is the Dirac operator in $\mathbb{R}^n$.

(ii) Whenever Ω is open and $\partial\Omega$ is bounded, one has

$$\mathcal{C}1 \equiv \begin{cases} 1 \text{ in } \Omega, \text{ if } \Omega \text{ is bounded,} \\ 0 \text{ in } \Omega, \text{ if } \Omega \text{ is unbounded.} \end{cases} \tag{2.5.291}$$

(iii) Under the additional assumption that $\partial\Omega$ is bounded and $\partial_*\Omega$ is upper Ahlfors regular, it follows that for each $\alpha \in (0,1)$ there exists some $C \in (0,\infty)$, depending only on n, α, diam$(\partial_*\Omega)$, *and the upper Ahlfors regularity constant of $\partial_*\Omega$, such that for every function $f \in \mathscr{C}^\alpha(\partial_*\Omega)$ one has*

$$\sup_{x\in\Omega} \big|(\mathcal{C}f)(x)\big| + \sup_{x\in\Omega} \Big\{\text{dist}(x,\partial_*\Omega)^{1-\alpha}\big|\nabla(\mathcal{C}f)(x)\big|\Big\} \leq C\|f\|_{\mathscr{C}^\alpha(\partial_*\Omega)}. \tag{2.5.292}$$

In particular, if $\Omega \subseteq \mathbb{R}^n$ is a uniform domain whose boundary is compact and upper Ahlfors regular then $\mathcal{C}$ induces a well-defined, linear, and bounded operator in the context

$$\mathcal{C} : \mathscr{C}^\alpha(\partial_*\Omega) \otimes \mathcal{C}\ell_n \longrightarrow \mathscr{C}^\alpha\big(\overline{\Omega}\big) \otimes \mathcal{C}\ell_n, \qquad \forall\,\alpha \in (0,1). \tag{2.5.293}$$

As a corollary, if $\Omega \subseteq \mathbb{R}^n$ is simultaneously a uniform domain, and an Ahlfors regular domain with compact boundary, then for each $\alpha \in (0,1)$ the operator $\mathcal{C}$ is well defined, linear, and bounded in the context

$$\mathcal{C} : \mathscr{C}^\alpha(\partial\Omega) \otimes \mathcal{C}\ell_n \longrightarrow \mathscr{C}^\alpha\big(\overline{\Omega}\big) \otimes \mathcal{C}\ell_n. \tag{2.5.294}$$

(iv) If $\partial\Omega$ is a UR set then for each aperture parameter $\kappa \in (0,\infty)$ and integrability exponent $p \in [1,\infty)$ there exists a constant $C = C(\Omega, n, \kappa, p) \in (0,\infty)$ with the property that for each function $f \in L^p(\partial_*\Omega, \sigma_*) \otimes \mathcal{C}\ell_n$ one has

$$\|\mathcal{N}_\kappa(\mathcal{C}f)\|_{L^p(\partial\Omega,\sigma)} \leq C\|f\|_{L^p(\partial_*\Omega,\sigma_*)\otimes\mathcal{C}\ell_n} \text{ if } 1 < p < \infty, \tag{2.5.295}$$

$$\|\mathcal{N}_\kappa(\mathcal{C}f)\|_{L^{1,\infty}(\partial\Omega,\sigma)} \leq C\|f\|_{L^1(\partial_*\Omega,\sigma_*)\otimes\mathcal{C}\ell_n} \text{ if } p = 1. \tag{2.5.296}$$

(v) *Suppose $\partial\Omega$ is a* UR *set. Then there exists $C = C(\partial\Omega) \in (0, \infty)$ such that*

$$\begin{array}{l} \textit{given any } f \in L^{n-1}(\partial\Omega,\sigma)\otimes \mathcal{C}\ell_n \textit{ it follows that} \\ |\mathcal{C}f|^2 \operatorname{dist}(\cdot,\partial\Omega)\,\mathrm{d}\mathcal{L}^n \textit{ is a Carleson measure in } \Omega, \\ \textit{with constant} \leq C\|f\|^2_{L^{n-1}(\partial\Omega,\sigma)\otimes\mathcal{C}\ell_n}. \end{array} \tag{2.5.297}$$

Moreover, if one defines the action of the operator $\nabla\mathcal{C}$ on arbitrary functions f belonging to the space $\mathrm{BMO}(\partial\Omega,\sigma)\otimes\mathcal{C}\ell_n$ *according to*

$$(\nabla\mathcal{C})f(x) := \left\{\frac{1}{\omega_{n-1}}\int_{\partial_*\Omega} \partial_{x_j}\Big(\frac{x-y}{|x-y|^n}\Big)\odot\nu(y)\odot f(y)\,\mathrm{d}\sigma_*(y)\right\}_{1\leq j\leq n} \tag{2.5.298}$$

at each point $x \in \Omega$, then for each $f \in \mathrm{BMO}(\partial\Omega,\sigma)\otimes\mathcal{C}\ell_n$ the function $(\nabla\mathcal{C})f$ is well defined and for each $p \in (1,\infty)$ it follows that

$$\big|(\nabla\mathcal{C})f\big|^p \operatorname{dist}(\cdot,\partial\Omega)^{p-1}\,\mathrm{d}\mathcal{L}^n \tag{2.5.299}$$

is a Carleson measure in Ω, in the quantitative sense that there exists some constant $C \in (0,\infty)$, independent of f, such that

$$\begin{aligned} \sup_{x\in\partial\Omega,\, r>0} \frac{1}{\sigma\big(B(x,r)\cap\partial\Omega\big)}\int_{B(x,r)\cap\Omega} &\big|(\nabla\mathcal{C})f\big|^p \operatorname{dist}(\cdot,\partial\Omega)^{p-1}\,\mathrm{d}\mathcal{L}^n \\ &\leq C\|f\|^p_{\mathrm{BMO}(\partial\Omega,\sigma)\otimes\mathcal{C}\ell_n}. \end{aligned} \tag{2.5.300}$$

Thus, corresponding to $p = 2$, it follows that $\big|(\nabla\mathcal{C})f\big|^2 \operatorname{dist}(\cdot,\partial\Omega)\,\mathrm{d}\mathcal{L}^n$ is a Carleson measure in Ω, for each function $f \in \mathrm{BMO}(\partial\Omega,\sigma)\otimes\mathcal{C}\ell_n$.

(vi) *If $\partial\Omega$ is bounded and $\partial_*\Omega$ is upper Ahlfors regular then for each $\alpha > 0$ there exists a number $N \in (0,\infty)$, depending only on the dimension n, exponent α, and the upper Ahlfors regularity constant of $\partial_*\Omega$, with the property that for every function $f \in \mathscr{C}^\alpha(\partial_*\Omega)\otimes\mathcal{C}\ell_n$ one has*

$$\sup_{x\in\partial_*\Omega}\big(\mathfrak{C}_{\max}f\big)(x) \leq N\big[\operatorname{diam}(\partial_*\Omega)\big]^\alpha\cdot\|f\|_{\dot{\mathscr{C}}^\alpha(\partial_*\Omega)} + N\cdot\sup_{\partial_*\Omega}|f|. \tag{2.5.301}$$

(vii) *Assume that $\partial\Omega$ is a bounded upper Ahlfors regular set. Then for each function $f \in \mathscr{C}^\alpha(\partial_*\Omega)\otimes\mathcal{C}\ell_n$ with $\alpha \in (0,1)$ the limit defining $\mathfrak{C}f(x)$ in (2.5.284) exists for σ-a.e. point $x \in \partial_*\Omega$. Moreover, after possibly redefining $\mathfrak{C}f$ on a σ_*-nullset, the assignment $f \mapsto \mathfrak{C}f$ thus defined induces a well-defined, linear, and bounded operator*

$$\mathfrak{C} : \mathscr{C}^\alpha(\partial_*\Omega)\otimes\mathcal{C}\ell_n \longrightarrow \mathscr{C}^\alpha(\partial_*\Omega)\otimes\mathcal{C}\ell_n, \qquad \forall\alpha\in(0,1), \tag{2.5.302}$$

with the property that

$$\mathfrak{C}1 \equiv \begin{cases} +\frac{1}{2} \text{ on } \partial_*\Omega, \text{ if } \Omega \text{ is bounded}, \\ -\frac{1}{2} \text{ on } \partial_*\Omega, \text{ if } \Omega \text{ is unbounded}. \end{cases} \tag{2.5.303}$$

In particular, all these properties hold whenever Ω is Lebesgue measurable with a compact Ahlfors regular boundary.

(viii) Suppose Ω is actually open and $\partial\Omega$ is a compact Ahlfors regular set. Then for each arbitrary fixed number $\kappa > 0$, the following nontangential boundary trace formula hods:

$$\begin{aligned} &\text{if } f \in \mathscr{C}^\alpha(\partial_*\Omega) \otimes C\ell_n \text{ with } \alpha \in (0,1) \text{ then} \\ &(\mathcal{C}f)\Big|_{\partial\Omega}^{\kappa-\text{n.t.}} = \big(\tfrac{1}{2}I + \mathfrak{C}\big)f \text{ at } \sigma_*\text{-a.e. point in } \partial_*\Omega. \end{aligned} \tag{2.5.304}$$

(ix) In the case when it is also assumed that $\sigma(\partial\Omega \setminus \partial_*\Omega) = 0$, one may replace $\partial_*\Omega$ by $\partial\Omega$ in the formulation of (2.5.302), (2.5.303), *and* (2.5.304).

(x) Strengthen the assumptions by requiring that $\partial\Omega$ is a UR *set. Then for each $p \in [1,\infty)$ there exists a constant $C = C(\Omega, n, p) \in (0,\infty)$ with the property that for each $f \in L^p(\partial_*\Omega, \sigma_*) \otimes C\ell_n$ one has*

$$\|\mathfrak{C}_{\max} f\|_{L^p(\partial\Omega,\sigma)} \leq C\|f\|_{L^p(\partial_*\Omega,\sigma_*)\otimes C\ell_n} \text{ if } 1 < p < \infty, \tag{2.5.305}$$

$$\|\mathfrak{C}_{\max} f\|_{L^{1,\infty}(\partial\Omega,\sigma)} \leq C\|f\|_{L^1(\partial_*\Omega,\sigma_*)\otimes C\ell_n} \text{ if } p = 1. \tag{2.5.306}$$

Furthermore, given any function $f \in L^p(\partial_*\Omega, \sigma_*) \otimes C\ell_n$ with $p \in [1,\infty)$ the limit defining $\mathfrak{C}f(x)$ in (2.5.284) *exists for σ-a.e. $x \in \partial\Omega$ and the operator $\mathfrak{C}$ is well defined, linear, and bounded in the following settings:*

$$\mathfrak{C} : L^p(\partial_*\Omega, \sigma_*) \otimes C\ell_n \longrightarrow L^p(\partial\Omega, \sigma) \otimes C\ell_n, \quad p \in (1,\infty), \tag{2.5.307}$$

$$\mathfrak{C} : L^1(\partial_*\Omega, \sigma_*) \otimes C\ell_n \longrightarrow L^{1,\infty}(\partial\Omega, \sigma) \otimes C\ell_n. \tag{2.5.308}$$

(xi) Assume that actually Ω is open and that $\partial\Omega$ is a UR *set. Then for each fixed arbitrary number $\kappa > 0$, then following nontangential boundary trace formula holds:*

$$\begin{aligned} &\text{for each } f \in L^p(\partial_*\Omega, \sigma_*) \otimes C\ell_n \text{ with } p \in [1,\infty) \text{ one has} \\ &(\mathcal{C}f)\Big|_{\partial\Omega}^{\kappa-\text{n.t.}} = \big(\tfrac{1}{2}I + \mathfrak{C}\big)f \text{ at } \sigma_*\text{-a.e. point on } \partial_*\Omega. \end{aligned} \tag{2.5.309}$$

Proof The claims in item *(i)* are seen directly from definitions. To proceed, denote by $(\nu_1, \dots, \nu_n)$ the components of ν. Then from [112, (6.4.2)-(6.4.3)] it follows that for each $y = (y_1, \dots, y_n) \in \partial^*\Omega$ and $x = (x_1, \dots x_n) \in \mathbb{R}^n \setminus \{y\}$ we have

$$\frac{x-y}{|x-y|^n} \odot \nu(y) = \frac{\langle \nu(y), y-x\rangle}{|x-y|^n} \tag{2.5.310}$$

$$- \frac{1}{2} \cdot \sum_{j,k=1}^{n} \frac{\nu_j(y)(x_k - y_k) - \nu_k(y)(x_j - y_j)}{|x-y|^n} \mathbf{e}_j \odot \mathbf{e}_k.$$

From (2.5.310), (2.5.282), (2.5.202), and (2.5.256) we then conclude that, for each σ_*-measurable function $f : \partial_*\Omega \to \mathcal{C}\ell_n$ satisfying the integrability condition in (2.5.281),

$$\mathcal{C}f(x) = \mathcal{D}f(x) + \frac{1}{2} \cdot \sum_{j,k=1}^{n} \mathbf{e}_j \odot \mathbf{e}_k \odot \mathcal{R}_{jk} f(x), \qquad \forall x \in \mathring{\Omega}. \tag{2.5.311}$$

Also, whenever the individual entities are meaningful,

$$\mathfrak{C}f(x) = Kf(x) + \frac{1}{2} \cdot \sum_{j,k=1}^{n} \mathbf{e}_j \odot \mathbf{e}_k \odot R_{jk} f(x), \qquad x \in \partial\Omega. \tag{2.5.312}$$

Granted these observations, all other claims become consequences of Lemma 2.5.20, Proposition 2.5.21, Proposition 2.5.22, Proposition 2.5.24, Proposition 2.5.25, Proposition 2.5.26, and Proposition 2.5.27. □

Incidentally, it is possible to first prove Proposition 2.5.29 (by suitably specializing Proposition 2.5.16) and then read off the corresponding properties for $\mathcal{D}$, $\mathcal{R}_{jk}$, K, R_{jk} directly from those of the Cauchy-Clifford operators via (2.5.311)-(2.5.312).

The thrust of the next result is that, up to normalization, the principal-value Cauchy-Clifford operator is its own inverse when acting on Hölder space.

Proposition 2.5.30 *Suppose $\Omega \subseteq \mathbb{R}^n$ (where $n \in \mathbb{N}$, $n \geq 2$) is an open set, with the property that $\partial\Omega$ is a compact Ahlfors regular set, and such that*

$$\mathcal{H}^{n-1}\big(\partial_{\text{nta}}\Omega \setminus \partial_*\Omega\big) = 0. \tag{2.5.313}$$

Then for each $\alpha \in (0, 1)$ the principal-value Cauchy-Clifford integral operator $\mathfrak{C}$ associated with Ω as in (2.5.284) satisfies

$$\mathfrak{C}^2 = \tfrac{1}{4} I \text{ on } \mathscr{C}^\alpha(\partial_*\Omega) \otimes \mathcal{C}\ell_n. \tag{2.5.314}$$

As a corollary, whenever $\Omega \subseteq \mathbb{R}^n$, with $n \geq 2$, is an Ahlfors regular domain with compact boundary for each $\alpha \in (0, 1)$ one has

$$\mathfrak{C}^2 = \tfrac{1}{4} I \text{ on } \mathscr{C}^\alpha(\partial\Omega) \otimes \mathcal{C}\ell_n. \tag{2.5.315}$$

Proof Introduce $\sigma := \mathcal{H}^{n-1} \lfloor \partial\Omega$ and select some apperture parameter $\kappa > 0$. Also, fix some $\alpha \in (0, 1)$ and pick an arbitrary function $f \in \mathscr{C}^\alpha(\partial_*\Omega) \otimes \mathcal{C}\ell_n$. In concert with (2.5.313), Proposition 2.5.29 ensures that if we now define $u := \mathcal{C}f$ in Ω then

$$\begin{aligned}
&u \in \mathscr{C}^\infty(\Omega) \otimes C\ell_n, \quad Du = 0 \text{ in } \Omega,\\
&u\Big|_{\partial\Omega}^{\kappa-\text{n.t.}} = (\tfrac{1}{2}I + \mathfrak{C})f \text{ at } \sigma\text{-a.e. point on } \partial_{\text{nta}}\Omega, \\
&\text{and } \sup_{x\in\Omega} \big|u(x)\big| \leq C\|f\|_{\mathscr{C}^\alpha(\partial_*\Omega)} < +\infty.
\end{aligned} \tag{2.5.316}$$

Thanks to [112, (8.2.26)], the last property above implies (bearing in mind that $\partial\Omega$ is compact)

$$\mathcal{N}_\kappa u \in L^\infty(\partial\Omega, \sigma), \text{ hence } \int_{\partial\Omega} \frac{(\mathcal{N}_\kappa u)(x)}{1+|x|^{n-1}}\, d\sigma(x) < +\infty. \tag{2.5.317}$$

Granted (2.5.325)-(2.5.326) and the current assumptions on Ω, Theorem 1.2.2 applies and gives that

$$\begin{aligned}
u(x) &= \frac{1}{\omega_{n-1}} \int_{\partial_*\Omega} \frac{x-y}{|x-y|^n} \odot \nu(y) \odot \Big(u\Big|_{\partial\Omega}^{\kappa-\text{n.t.}}\Big)(y)\, d\sigma(y)\\
&= \big(\mathcal{C}(\tfrac{1}{2}I + \mathfrak{C})f\big)(x), \qquad \forall x \in \Omega.
\end{aligned} \tag{2.5.318}$$

Taking the nontangential boundary traces of the most extreme sides in (2.5.318) then yields (on account of (2.5.316) and (2.5.304))

$$(\tfrac{1}{2}I + \mathfrak{C})f = (\tfrac{1}{2}I + \mathfrak{C})f(\tfrac{1}{2}I + \mathfrak{C})f \text{ at } \sigma\text{-a.e. point on } \partial_*\Omega. \tag{2.5.319}$$

From this, (2.5.314) readily follows. Finally, for the very last claim in the statement, pertaining to (2.5.315), is justified similarly, with the help of item *(ix)* in Proposition 2.5.29. □

To state our next result, the reader is reminded the notation employed for the commutator $[A; B] := AB - BA$ and anti-commutator $\{A; B\} := AB + BA$ of two operators A, B (cf. [112, (6.4.60)-(6.4.61)]).

Proposition 2.5.31 *Suppose $\Omega \subseteq \mathbb{R}^n$ (where $n \in \mathbb{N}$, $n \geq 2$) is an open set, with the property that its topological boundary $\partial\Omega$ is a compact Ahlfors regular set, and such that $\mathcal{H}^{n-1}\big(\partial_{\text{nta}}\Omega \setminus \partial_*\Omega\big) = 0$. Then for each $\alpha \in (0, 1)$ the following operator identities are valid on $\mathscr{C}^\alpha(\partial_*\Omega)$:*

$$K^2 - \frac{1}{2}\sum_{j,k=1}^n (R_{jk})^2 = \tfrac{1}{4}I, \tag{2.5.320}$$

$$\sum_{\ell=1}^n [R_{j\ell}; R_{k\ell}] = \{K; R_{jk}\} \quad \textit{for each } j, k \in \{1, \dots, n\}, \tag{2.5.321}$$

and

$$\{R_{ij}; R_{k\ell}\} + \{R_{i\ell}; R_{jk}\} = \{R_{ik}; R_{j\ell}\} \quad \textit{for each } i, j, k, \ell \in \{1, \dots, n\}. \tag{2.5.322}$$

As a corollary, whenever $\Omega \subseteq \mathbb{R}^n$, with $n \geq 2$, is an Ahlfors regular domain with compact boundary the operator identities (2.5.320)-(2.5.322) *hold on $\mathscr{C}^\alpha(\partial\Omega)$ for each $\alpha \in (0, 1)$.*

Proof This is a direct consequence of Proposition 2.5.30 in the sense that (2.5.320)-(2.5.322) are seen by untangling the Clifford algebra identity $\mathfrak{C}^2 = \frac{1}{4}I$ component-wise, as in [112, Lemma 6.4.2]. □

We next present a version of Proposition 2.5.30 on Lebesgue space.

Proposition 2.5.32 *Suppose $\Omega \subseteq \mathbb{R}^n$ (where $n \in \mathbb{N}$, $n \geq 2$) is an open set such that $\partial\Omega$ is a* UR *set and*

$$\mathcal{H}^{n-1}\big(\partial_{\text{nta}}\Omega \setminus \partial_*\Omega\big) = 0. \tag{2.5.323}$$

Abbreviate $\sigma_* := \mathcal{H}^{n-1}\lfloor\partial_*\Omega$ and fix $p \in (1, \infty)$ arbitrary. Then the principal-value Cauchy-Clifford integral operator $\mathfrak{C}$ associated with Ω as in (2.5.284) *satisfies*

$$\mathfrak{C}^2 = \tfrac{1}{4}I \ \textit{ on } \ L^p(\partial_*\Omega, \sigma_*) \otimes \mathcal{C}\ell_n. \tag{2.5.324}$$

Equivalently, the operator identities (2.5.320)-(2.5.322) *are valid on the space $L^p(\partial_*\Omega, \sigma_*) \otimes \mathcal{C}\ell_n$.*

Proof Introduce $\sigma := \mathcal{H}^{n-1}\lfloor\partial\Omega$ and pick an arbitrary $\kappa > 0$. Also, fix $p \in (1, \infty)$ and select an arbitrary function $f \in L^p(\partial_*\Omega, \sigma_*) \otimes \mathcal{C}\ell_n$. If we now define $u := \mathcal{C}f$ in Ω then, in light of (2.5.323), Proposition 2.5.29 guarantees that

$$\begin{aligned}
& u \in \mathscr{C}^\infty(\Omega) \otimes \mathcal{C}\ell_n, \quad Du = 0 \text{ in } \Omega, \\
& u\big|_{\partial\Omega}^{\kappa-\text{n.t.}} = (\tfrac{1}{2}I + \mathfrak{C})f \text{ at } \sigma\text{-a.e. point on } \partial_{\text{nta}}\Omega, \\
& \mathcal{N}_\kappa u \text{ belongs to the space } L^p(\partial\Omega, \sigma), \text{ and} \\
& u(x) = O(|x|^{1-n}) \text{ if } \Omega \text{ is an exterior domain.}
\end{aligned} \tag{2.5.325}$$

In particular, the next-to-last property recorded above implies (thanks to Hölder's inequality and [112, Lemma 7.2.1])

$$\int_{\partial\Omega} \frac{(\mathcal{N}_\kappa u)(x)}{1 + |x|^{n-1}}\, d\sigma(x) < +\infty, \tag{2.5.326}$$

while the last property in (2.5.325) implies that, in the case when Ω is an exterior domain,

$$\int_{[B(0,2R)\setminus B(0,R)]\cap\Omega} |u(x)|\, d\mathcal{L}^n = o(R^n) \text{ as } R \to \infty. \tag{2.5.327}$$

Having established (2.5.325)-(2.5.327), we may now invoke Theorem 1.2.2 in order to conclude that

$$u(x) = \frac{1}{\omega_{n-1}} \int_{\partial_* \Omega} \frac{x-y}{|x-y|^n} \odot \nu(y) \odot \left(u\big|_{\partial\Omega}^{\kappa-\mathrm{n.t.}}\right)(y)\,d\sigma(y)$$
$$= \left(\mathcal{C}(\tfrac{1}{2}I + \mathfrak{C})f\right)(x), \qquad \forall x \in \Omega. \tag{2.5.328}$$

Taking the nontangential boundary traces of the most extreme sides in (2.5.328) then yields (on account of (2.5.325) and (2.5.309))

$$(\tfrac{1}{2}I + \mathfrak{C})f = (\tfrac{1}{2}I + \mathfrak{C})f(\tfrac{1}{2}I + \mathfrak{C})f \text{ at } \sigma_*\text{-a.e. point on } \partial_*\Omega. \tag{2.5.329}$$

This establishes (2.5.324). Finally, the operator identities (2.5.320)-(2.5.322) are seen by considering the components of the Clifford algebra expressions involved in the formula (2.5.324), as in [112, Lemma 6.4.2]. □

In the class of UR domains, we have the following version of Proposition 2.5.32 on Muckenhoupt weighted Lebesgue spaces.

Corollary 2.5.33 *Let* $\Omega \subseteq \mathbb{R}^n$, *where* $n \geq 2$, *be a* UR *domain. Set* $\sigma := \mathcal{H}^{n-1}\lfloor\partial\Omega$ *and fix* $p \in (1,\infty)$ *along with* $w \in A_p(\partial\Omega,\sigma)$. *Then*

$$\mathfrak{C} : L^p(\partial\Omega, w) \otimes \mathcal{C}\ell_n \longrightarrow L^p(\partial\Omega, w) \otimes \mathcal{C}\ell_n \tag{2.5.330}$$

is a well-defined, linear, and bounded operator, which satisfies

$$\mathfrak{C}^2 = \tfrac{1}{4}I \text{ on } L^p(\partial\Omega, w) \otimes \mathcal{C}\ell_n. \tag{2.5.331}$$

Equivalently, the operator identities (2.5.320)-(2.5.322) *are valid on the Muckenhoupt weighted Lebesgue space* $L^p(\partial\Omega, w)$ *for each* $p \in (1,\infty)$ *and* $w \in A_p(\partial\Omega,\sigma)$.

Proof The claims regarding (2.5.330) are seen from (2.5.284) and item *(10)* in Theorem 2.3.2. Also, since we are currently assuming that Ω is a UR domain, (2.5.324) becomes

$$\mathfrak{C}^2 = \tfrac{1}{4}I \text{ on } L^p(\partial\Omega, \sigma) \otimes \mathcal{C}\ell_n. \tag{2.5.332}$$

In turn, formula (2.5.331) follows from this, the density of $L^p(\partial\Omega,\sigma) \cap L^p(\partial\Omega, w)$ in the space $L^p(\partial\Omega, w)$, and (2.5.330). □

In the two-dimensional setting, we may recast results on the Cauchy-Clifford operator in the manner described below.

Corollary 2.5.34 *Given a* UR *domain* $\Omega \subseteq \mathbb{R}^2$, *set* $\sigma := \mathcal{H}^1\lfloor\partial\Omega$ *and denote by* $\nu = (\nu_1, \nu_2)$ *its geometric measure theoretic outward unit normal. Also, for each function* $f \in L^1\big(\partial\Omega, \frac{\sigma(x)}{1+|x|}\big)$ *define*

$$Rf(x) := \lim_{\varepsilon \to 0^+} \int_{\substack{y \in \partial\Omega \\ |x-y|>\varepsilon}} \Big[\nu_1(y)(\partial_2 E_\Delta)(y-x) - \nu_2(y)(\partial_1 E_\Delta)(y-x)\Big] f(y)\,d\sigma(y), \tag{2.5.333}$$

for σ-a.e. $x \in \partial\Omega$, where E_Δ is the standard fundamental solution for the two-dimensional Laplacian (cf. (1.5.56)). *Then for each $p \in (1, \infty)$ one has*

$$KR = -RK \text{ and } \big(\tfrac{1}{2}I + K\big)\big(-\tfrac{1}{2}I + K\big) = R^2 \text{ on } L^p(\partial\Omega, \sigma). \tag{2.5.334}$$

Proof These identities are readily implied by the versions of (2.5.320)-(2.5.321) on Lebesgue spaces (cf. the last claim in Proposition 2.5.32) bearing in mind that, as seen from (2.5.269), (2.5.268), and (2.5.333), we have $R_{11} = R_{22} = 0$ and $R_{12} = -R_{21} = R$. □

Let $\Sigma \subseteq \mathbb{R}^n$ (where $n \in \mathbb{N}$ with $n \geq 2$) be a closed set which is countably rectifiable (of dimension $n-1$) and satisfies $\mathcal{H}^{n-1}(K \cap \Sigma) < +\infty$ for each compact set $K \subseteq \mathbb{R}^n$. Abbreviate $\sigma := \mathcal{H}^{n-1}\lfloor\Sigma$. In this setting, for $j \in \{1, \dots, n\}$ introduce the boundary-to-domain `Riesz transforms` acting on functions $f \in L^1\big(\Sigma, \frac{\sigma}{1+|\cdot|^{n-1}}\big)$ according to

$$\mathcal{R}_j f(x) := \frac{2}{\omega_{n-1}} \int_\Sigma \frac{x_j - y_j}{|x-y|^n} f(y)\, d\sigma(y), \qquad \forall x \in \mathbb{R}^n \setminus \Sigma, \tag{2.5.335}$$

along with their boundary-to-boundary versions

$$R_j f(x) := \lim_{\varepsilon \to 0^+} \frac{2}{\omega_{n-1}} \int_{\substack{|x-y|>\varepsilon \\ y \in \Sigma}} \frac{x_j - y_j}{|x-y|^n} f(y)\, d\sigma(y) \text{ for } \sigma\text{-a.e. } x \in \Sigma. \tag{2.5.336}$$

Granted the current assumptions, [112, Proposition 5.6.7] guarantees that the latter principal-value limit exists for σ-a.e. point $x \in \Sigma$ and gives rise to a σ-measurable function on Σ.

Suppose next that $\Omega \subseteq \mathbb{R}^n$ (where $n \in \mathbb{N}$ with $n \geq 2$) is an open set such that $\partial\Omega$ is a UR set. All previous considerations work with $\Sigma := \partial\Omega$, and for each integrability exponent $p \in [1, \infty)$ we now have $L^p(\partial\Omega, \sigma) \subseteq L^1\big(\partial\Omega, \frac{\sigma}{1+|\cdot|^{n-1}}\big)$. In such a setting, the Riesz transforms (2.5.336) (with $\Sigma := \partial\Omega$) fall directly under the scope of Theorem 2.3.2 which gives that the operators

$$R_j : L^p(\partial\Omega, \sigma) \longrightarrow L^p(\partial\Omega, \sigma), \quad p \in (1, \infty), \tag{2.5.337}$$

$$R_j : L^1(\partial\Omega, \sigma) \longrightarrow L^{1,\infty}(\partial\Omega, \sigma). \tag{2.5.338}$$

are well defined, linear, and bounded. Moreover, Theorem 2.4.1 implies that for each $\kappa > 0$ we have the estimates

$$\sum_{j=1}^n \|\mathcal{N}_\kappa(\mathcal{R}_j f)\|_{L^p(\partial\Omega,\sigma)} \leq C\|f\|_{L^p(\partial\Omega,\sigma)}, \quad 1 < p < \infty, \tag{2.5.339}$$

$$\sum_{j=1}^n \|\mathcal{N}_\kappa(\mathcal{R}_j f)\|_{L^{1,\infty}(\partial\Omega,\sigma)} \leq C\|f\|_{L^1(\partial\Omega,\sigma)}, \tag{2.5.340}$$

for some finite constant $C > 0$ depending only on p, κ as well as the Ahlfors regularity and UR constants of $\partial\Omega$.

In addition, for these Riesz transforms we have the jump-formulas stated in the proposition below.

Proposition 2.5.35 *Assume $\Omega \subseteq \mathbb{R}^n$ (where $n \in \mathbb{N}$ with $n \geq 2$) is an open set with the property that $\partial\Omega$ is a* UR *set (in particular, Ω is a set of locally finite perimeter). Abbreviate $\sigma := \mathcal{H}^{n-1}\lfloor\partial\Omega$, $\sigma_* := \mathcal{H}^{n-1}\lfloor\partial_*\Omega$ and denote by $\nu = (\nu_1, \dots, \nu_n)$ the geometric measure theoretic outward unit normal to Ω.*

Then, for every $f \in L^p(\partial_*\Omega, \sigma_*)$ with $1 \leq p < \infty$ (canonically regarded as a function in the space $L^p(\partial\Omega, \sigma)$ by extending it by zero outside $\partial_*\Omega$ to the entire $\partial\Omega$), and every $j \in \{1, \dots, n\}$, the limit in (2.5.336) *exists at σ_*-almost every $x \in \partial_*\Omega$, and*

$$\lim_{\substack{z\in\Gamma_\kappa(x)\\ z\to x}} (\mathcal{R}_j f)(z) = -\nu_j(x)f(x) + R_j f(x) \ \textit{at}\ \sigma_*\textit{-a.e.}\ x \in \partial_*\Omega. \tag{2.5.341}$$

Proof Fix some arbitrary function $f \in L^p(\partial_*\Omega, \sigma_*)$ where $1 \leq p < \infty$, along with $j \in \{1, \dots, n\}$. That the limit in (2.5.336) exists σ_*-a.e. on $\partial_*\Omega$ is a direct consequence of Theorem 2.3.2. Turning our attention to the jump-formula (2.5.341), let $\partial_\nu := \nu \cdot \nabla$ denote the normal derivative. Then for each $x \in \Omega$ and σ_*-a.e. $y \in \partial_*\Omega$ we may write

$$\begin{aligned} \frac{1}{\omega_{n-1}} \frac{x_j - y_j}{|x-y|^n} &= (\partial_j E_\Delta)(x-y) \\ &= -\nu_j(y)\partial_{\nu(y)}[E_\Delta(x-y)] \\ &\quad + \sum_{k=1}^n \nu_k(y)\Big[\nu_k(y)(\partial_j E_\Delta)(x-y) - \nu_j(y)(\partial_k E_\Delta)(x-y)\Big]. \end{aligned} \tag{2.5.342}$$

In view of the fact that (cf. (2.5.204))

$$\partial_{\nu(y)}[E_\Delta(x-y)] = \frac{1}{\omega_{n-1}} \frac{\langle \nu(y), y-x\rangle}{|x-y|^n}, \tag{2.5.343}$$

the decomposition in (2.5.342) entails

$$\mathcal{R}_j f = -2\mathcal{D}(\nu_j f) + 2\sum_{k=1}^n \mathcal{R}_{jk}(\nu_k f) \ \text{in}\ \Omega. \tag{2.5.344}$$

Granted (2.5.344), Proposition 2.5.25 and Proposition 2.5.27 may be used (with f replaced by $-\nu_j f$ and $\nu_k f$, respectively) in order to compute

$$\lim_{\substack{z\in\Gamma_\kappa(x)\\ z\to x}} (\mathcal{R}_j f)(z) = -\nu_j(x)f(x) - 2K(\nu_j f)(x) + 2\sum_{k=1}^n R_{jk}(\nu_k f)(x) \tag{2.5.345}$$

at σ_*-a.e. $x \in \partial_*\Omega$. Since the decomposition (2.5.342) also holds for σ_*-a.e. points $x, y \in \partial_*\Omega$, from (2.5.342), (2.5.343), (2.5.217), (2.5.268), and (2.5.336) we conclude that

$$-2K(\nu_j f) + 2\sum_{k=1}^{n} R_{jk}(\nu_k f) = R_j f \quad \text{at } \sigma_*\text{-a.e. point on } \partial_*\Omega. \tag{2.5.346}$$

At this stage, the jump-formula in (2.5.341) follows from (2.5.345)-(2.5.346). □

Changing course, recall the Clifford algebra ambient $(\mathcal{C}\ell_n, +, \odot)$ from [112, §6.4], which serves as an enlargement of the Euclidean space $\mathbb{R}^n$ to a non-commutative unitary real algebra. In this regard, we shall require one additional piece of notation. Specifically, for each $s \in \{1, \dots, n\}$ we agree to let $[\cdot]_s$ denote the projection onto the s-th Euclidean coordinate, i.e.,

$$[x]_s := x_s \quad \text{if } x = (x_1, \dots, x_n) \in \mathbb{R}^n. \tag{2.5.347}$$

The lemma below, originating in [159], will play an important role for us shortly. Here we shall follow the exposition in [120] where the following more refined version has been established.

Lemma 2.5.36 *For any odd, harmonic, homogeneous polynomial $P(x)$, $x \in \mathbb{R}^n$ (with $n \geq 2$), of degree $\ell \geq 3$, there exist a family $P_{rs}(x)$, $1 \leq r, s \leq n$, of harmonic, homogeneous polynomials, which are either zero or have degree $\ell - 2$, as well as a family of odd, $\mathscr{C}^\infty$ functions*

$$k_{rs} : \mathbb{R}^n \setminus \{0\} \longrightarrow \mathbb{R}^n \hookrightarrow \mathcal{C}\ell_n, \qquad \text{with } r, s \in \{1, \dots, n\}, \tag{2.5.348}$$

which are homogeneous of degree $1 - n$, and for each $x \in \mathbb{R}^n \setminus \{0\}$ satisfy (with the piece of notation introduced in (2.5.347))

$$\frac{P(x)}{|x|^{n-1+\ell}} = \sum_{r,s=1}^{n} [k_{rs}(x)]_s \tag{2.5.349}$$

as well as (with D_R denoting the Dirac operator acting from the right, as in (A.0.41))

$$(D_R\, k_{rs})(x) = \frac{\ell - 1}{n + \ell - 3} \frac{\partial}{\partial x_r} \left(\frac{P_{rs}(x)}{|x|^{n+\ell-3}} \right) \quad \text{for all } r, s \in \{1, \dots, n\}. \tag{2.5.350}$$

Moreover, there exists a finite purely dimensional constant $c_n > 0$ such that

$$\max_{1 \leq r,s \leq n} \Big\{ \sup_{S^{n-1}} |k_{rs}| \Big\} + \max_{1 \leq r,s \leq n} \Big\{ \sup_{S^{n-1}} |\nabla k_{rs}| \Big\} \leq c_n\, 2^\ell \|P\|_{L^1(S^{n-1}, \mathcal{H}^{n-1})}. \tag{2.5.351}$$

Proof Given now an odd, harmonic, homogeneous polynomial $P(x)$ of degree $\ell \geq 3$ in $\mathbb{R}^n$, for each $r, s \in \{1, \dots, n\}$ introduce

$$P_{rs}(x) := \frac{1}{\ell(\ell-1)}(\partial_r \partial_s P)(x), \qquad \forall\, x \in \mathbb{R}^n. \tag{2.5.352}$$

Then each P_{rs} is an odd, harmonic, homogeneous polynomial of degree $\ell - 2$ in $\mathbb{R}^n$, and Euler's formula for homogeneous functions gives

$$P(x) = \sum_{r,s=1}^{n} x_r x_s P_{rs}(x), \qquad \forall\, x \in \mathbb{R}^n, \tag{2.5.353}$$

and, for each $r, s \in \{1, \dots, n\}$,

$$\langle (\nabla P_{rs})(x), x \rangle = (\ell - 2) P_{rs}(x), \qquad \forall\, x \in \mathbb{R}^n. \tag{2.5.354}$$

To proceed, assume first that $n \geq 3$ and, for each pair of indices $r, s \in \{1, \dots, n\}$, define the function $k_{rs} : \mathbb{R}^n \setminus \{0\} \longrightarrow \mathbb{R}^n \hookrightarrow \mathcal{C}\ell_n$ by setting

$$k_{rs}(x) := \frac{1}{(n+\ell-3)(n+\ell-5)} \sum_{j=1}^{n} \partial_r \partial_j \Big(\frac{P_{rs}(x)}{|x|^{n+\ell-5}} \Big) \mathbf{e}_j \quad \forall\, x \in \mathbb{R}^n \setminus \{0\}. \tag{2.5.355}$$

The fact that $n, \ell \geq 3$ ensure that both $n + \ell - 3 \neq 0$ and $n + \ell - 5 \neq 0$ so each k_{rs} is well defined, odd, $\mathscr{C}^\infty$ and homogeneous of degree $-(n-1)$ in $\mathbb{R}^n \setminus \{0\}$. In addition,

$$k_{rs}(x) = \frac{1}{(n+\ell-3)(n+\ell-5)} D_R \Big[\partial_r \Big(\frac{P_{rs}(x)}{|x|^{n+\ell-5}} \Big) \Big], \quad \forall\, x \in \mathbb{R}^n \setminus \{0\}, \tag{2.5.356}$$

hence for all $x \in \mathbb{R}^n \setminus \{0\}$ we may write

$$\begin{aligned} (D_R k_{rs})(x) &= \frac{1}{(n+\ell-3)(n+\ell-5)} D_R^2 \Big[\partial_r \Big(\frac{P_{rs}(x)}{|x|^{n+\ell-5}} \Big) \Big] \\ &= \frac{-1}{(n+\ell-3)(n+\ell-5)} \Delta \Big[\partial_r \Big(\frac{P_{rs}(x)}{|x|^{n+\ell-5}} \Big) \Big] \\ &=: I + II + III, \end{aligned} \tag{2.5.357}$$

where

$$
\begin{aligned}
I &:= \frac{-1}{(n+\ell-3)(n+\ell-5)}\partial_r\Big[\frac{(\Delta P_{rs})(x)}{|x|^{n+\ell-5}}\Big] = 0,\\
II &:= \frac{-1}{(n+\ell-3)(n+\ell-5)}\partial_r\Big[2\big\langle(\nabla P_{rs})(x), \nabla\big[|x|^{-(n+\ell-5)}\big]\big\rangle\Big]\\
&= \frac{2}{n+\ell-3}\partial_r\Big[\frac{\langle(\nabla P_{rs})(x), x\rangle}{|x|^{n+\ell-3}}\Big]\\
&= \frac{2(\ell-2)}{n+\ell-3}\partial_r\Big[\frac{P_{rs}(x)}{|x|^{n+\ell-3}}\Big],\\
III &:= \frac{-1}{(n+\ell-3)(n+\ell-5)}\partial_r\Big[P_{rs}(x)\Delta\big[|x|^{-(n+\ell-5)}\big]\Big]\\
&= \frac{-\ell+3}{n+\ell-3}\partial_r\Big[\frac{P_{rs}(x)}{|x|^{n+\ell-3}}\Big],
\end{aligned}
\tag{2.5.358}
$$

by the harmonicity of P, (2.5.354), and straightforward algebra. This proves that (2.5.349) holds when $n \geq 3$. Going further, from (2.5.355) and the fact that

$$
\sum_{r=1}^{n}(\partial_r P_{rs})(x) = \sum_{s=1}^{n}(\partial_s P_{rs})(x) = 0 \text{ and } \sum_{r=1}^{n} P_{rr}(x) = 0 \tag{2.5.359}
$$

(as seen from formula (2.5.352) and the harmonicity of P), we deduce that for each point $x \in \mathbb{R}^n \setminus \{0\}$

$$
\begin{aligned}
\sum_{r,s=1}^{n}[k_{rs}(x)]_s &= \frac{1}{(n+\ell-3)(n+\ell-5)}\sum_{r,s=1}^{n}\partial_r\partial_s\Big(\frac{P_{rs}(x)}{|x|^{n+\ell-5}}\Big)\\
&= \frac{1}{(n+\ell-3)(n+\ell-5)}\sum_{r,s=1}^{n}P_{rs}(x)\partial_r\partial_s\big[|x|^{-(n+\ell-5)}\big]\\
&= \frac{-1}{n+\ell-3}\sum_{r,s=1}^{n}P_{rs}(x)\Big\{\frac{\delta_{rs}}{|x|^{n+\ell-3}} - (n+\ell-3)\frac{x_r x_s}{|x|^{n+\ell-1}}\Big\}\\
&= \frac{P(x)}{|x|^{n-1+\ell}}.
\end{aligned}
\tag{2.5.360}
$$

This establishes (2.5.350) for $n \geq 3$. Moving on, for each $\gamma \in \mathbb{N}_0^n$, interior estimates for the harmonic function P give

$$
\begin{aligned}
\|\partial^\gamma P\|_{L^\infty(S^{n-1},\mathcal{H}^{n-1})} &\le c_{n,\gamma}\int_{B(0,2)} |P(x)|\,\mathrm{d}x r^{n-1+\ell}\,\mathrm{d}r\Big)\,\mathrm{d}\mathcal{H}^{n-1}(\omega)\\
&= c_{n,\gamma}\int_{S^{n-1}} |P(\omega)|\Big(\int_0^2 r^{n-1+\ell}\,\mathrm{d}r\Big)\,\mathrm{d}\mathcal{H}^{n-1}(\omega)\\
&= c_{n,\gamma}\frac{2^\ell}{n+\ell}\|P\|_{L^1(S^{n-1},\mathcal{H}^{n-1})},
\end{aligned} \tag{2.5.361}
$$

where we have also used the fact that P is homogeneous of degree ℓ. The estimates in (2.5.351) now readily follow on account of (2.5.355), (2.5.352), and (2.5.361).

To treat the two-dimensional case, first we observe that if $Q_m(x)$ is an arbitrary homogeneous polynomial of degree $m \in \mathbb{N}_0$ in $\mathbb{R}^n$ with $n \ge 2$ and $\lambda > 0$ then

$$
\frac{Q_m(x)}{|x|^{n+m-\lambda}} \text{ is a tempered distribution in } \mathbb{R}^n. \tag{2.5.362}
$$

If, in addition, $Q_m(x)$ is harmonic and $\lambda < n$ then (cf. [166, p. 73]) also

$$
\mathcal{F}_{x\to\xi}\Big(\frac{Q_m(x)}{|x|^{n+m-\lambda}}\Big) = \gamma_{n,m,\lambda}\frac{Q_m(\xi)}{|\xi|^{m+\lambda}} \text{ as tempered distributions in } \mathbb{R}^n, \tag{2.5.363}
$$

where $\mathcal{F}_{x\to\xi}$ is an alternative notation for the Fourier transform in $\mathbb{R}^n$ from (1.4.14) and

$$
\gamma_{n,m,\lambda} := (-1)^{3m/2}\pi^{n/2}2^\lambda\frac{\Gamma(m/2+\lambda/2)}{\Gamma(m/2+n/2-\lambda/2)}. \tag{2.5.364}
$$

Pick now an odd, harmonic, homogeneous polynomial $P(x)$ of degree $\ell \ge 3$ in $\mathbb{R}^2$ and define P_{rs} for $r, s \in \{1, \dots, n\}$ as in (2.5.352). Hence, once again each P_{rs} is an odd, harmonic, homogeneous polynomial of degree $\ell - 2$ in $\mathbb{R}^2$, and (2.5.353) holds. Moreover, (2.5.363) used for $n = 2$, $m = \ell - 2$, $\lambda = 1$, and $Q_m = P_{rs}$ yields

$$
\frac{P_{rs}(x)}{|x|^{\ell-1}} = -(-1)^{3\ell/2}2\pi\,\mathcal{F}^{-1}_{\xi\to x}\Big(\frac{P_{rs}(\xi)}{|\xi|^{\ell-1}}\Big). \tag{2.5.365}
$$

Now, for each $r, s \in \{1, 2\}$ define the function $k_{rs} : \mathbb{R}^2 \setminus \{0\} \longrightarrow \mathbb{R}^2 \hookrightarrow \mathcal{C}\ell_2$ by setting

$$
k_{rs}(x) := (-1)^{3\ell/2}2\pi\sum_{j=1}^{2}\mathcal{F}^{-1}_{\xi\to x}\Big(\xi_r\xi_j\frac{P_{rs}(\xi)}{|\xi|^{\ell+1}}\Big)\mathbf{e}_j, \quad \forall\, x \in \mathbb{R}^2 \setminus \{0\}. \tag{2.5.366}
$$

By (2.5.362) used with $n = 2$, $m = \ell$, $\lambda = 1$, and $Q_m(\xi) = \xi_r\xi_j P_{rs}(\xi)$, it follows that $\xi_r\xi_j\frac{P_{rs}(\xi)}{|\xi|^{\ell+1}}$ is a tempered distribution in $\mathbb{R}^2$. Consequently, k_{rs} in (2.5.366) is meaningfully defined and, from [109, Proposition 4.60, p. 143], we deduce that $k_{rs} \in \mathscr{C}^\infty(\mathbb{R}^2 \setminus \{0\})$. Also, based on standard properties of the Fourier transform (cf., e.g., [109, Chapter 4, p. 117]) it follows that k_{rs} is odd and homogeneous of degree -1 in $\mathbb{R}^2 \setminus \{0\}$. In addition,

$$
\begin{aligned}
(D_R k_{rs})(x) &= (-1)^{3\ell/2} 2\pi \sum_{i,j=1}^{2} \partial_{x_i} \mathcal{F}^{-1}_{\xi\to x}\Big(\xi_r \xi_j \frac{P_{rs}(\xi)}{|\xi|^{\ell+1}}\Big) \mathbf{e}_j \odot \mathbf{e}_i \\
&= \sqrt{-1}(-1)^{3\ell/2} 2\pi \sum_{i,j=1}^{2} \mathcal{F}^{-1}_{\xi\to x}\Big(\xi_r \xi_j \xi_i \frac{P_{rs}(\xi)}{|\xi|^{\ell+1}}\Big) \mathbf{e}_j \odot \mathbf{e}_i \\
&=: I + II, \qquad (2.5.367)
\end{aligned}
$$

where I and II are the pieces produced by summing up over $j = i$ and $j \neq i$, respectively. Since in the latter scenario $\xi_i \xi_j = \xi_j \xi_i$ while $\mathbf{e}_j \odot \mathbf{e}_i = -\mathbf{e}_i \odot \mathbf{e}_j$ it follows that $II = 0$. Given that $\mathbf{e}_j \odot \mathbf{e}_j = -1$ for each $j \in \{1, 2\}$, we conclude that

$$
\begin{aligned}
(D_R k_{rs})(x) &= -\sqrt{-1}(-1)^{3\ell/2} 2\pi \sum_{j=1}^{2} \mathcal{F}^{-1}_{\xi\to x}\Big(\xi_r \xi_j^2 \frac{P_{rs}(\xi)}{|\xi|^{\ell+1}}\Big) \\
&= -\sqrt{-1}(-1)^{3\ell/2} 2\pi \mathcal{F}^{-1}_{\xi\to x}\Big(\xi_r \frac{P_{rs}(\xi)}{|\xi|^{\ell-1}}\Big) \\
&= -(-1)^{3\ell/2} 2\pi\, \partial_{x_r} \Big[\mathcal{F}^{-1}_{\xi\to x}\Big(\frac{P_{rs}(\xi)}{|\xi|^{\ell-1}}\Big)\Big] = \partial_{x_r}\Big[\frac{P_{rs}(x)}{|x|^{\ell-1}}\Big], \qquad (2.5.368)
\end{aligned}
$$

where the last step uses (2.5.365). Hence, (2.5.349) holds when $n = 2$. Finally, from (2.5.355), (2.5.353), and (2.5.363) (used for P) we deduce that for each $x \in \mathbb{R}^2 \setminus \{0\}$ we have

$$
\begin{aligned}
\sum_{r,s=1}^{2} [k_{rs}(x)]_s &= (-1)^{3\ell/2} 2\pi \sum_{r,s=1}^{2} \mathcal{F}^{-1}_{\xi\to x}\Big(\xi_r \xi_s \frac{P_{rs}(\xi)}{|\xi|^{\ell+1}}\Big) \\
&= (-1)^{3\ell/2} 2\pi \mathcal{F}^{-1}_{\xi\to x}\Big(\frac{P(\xi)}{|\xi|^{\ell+1}}\Big) = \frac{P(x)}{|x|^{\ell+1}}. \qquad (2.5.369)
\end{aligned}
$$

This establishes (2.5.350) when $n = 2$.

At this stage, there remains to justify (2.5.351) in the case $n = 2$. To this end, pick $\psi \in \mathscr{C}^\infty_c(\mathbb{R}^2)$ with $0 \leq \psi \leq 1$, $\psi = 1$ on $B(0, 1)$ and $\psi = 0$ on $\mathbb{R}^2 \setminus \overline{B(0, 2)}$. Fix $r, s, j \in \{1, 2\}$ and abbreviate $u(\xi) := \xi_r \xi_j P_{rs}(\xi)/|\xi|^{\ell+1}$ for $\xi \in \mathbb{R}^2 \setminus \{0\}$. Then u is locally integrable and defines a tempered distribution in $\mathbb{R}^2$. Hence, for each $\alpha \in \mathbb{N}_0^2$ with $|\alpha| = 2$ and $\xi \in \overline{B(0, 1)}$ we may write

$$
\begin{aligned}
\big|\mathcal{F}_{x\to\xi}\big(\psi(x)\partial^\alpha u(x)\big)\big| &= \big|\big\langle \psi\partial^\alpha u, \mathrm{e}^{-\sqrt{-1}\langle\xi,\cdot\rangle}\big\rangle\big| = \big|\big\langle u, \partial^\alpha\big(\psi \mathrm{e}^{-\sqrt{-1}\langle\xi,\cdot\rangle}\big)\big\rangle\big| \\
&\leq C \int_{B(0,2)} |u(x)|\, \mathrm{d}x \leq C \int_{S^1} |P_{rs}(\omega)|\, \mathrm{d}\mathcal{H}^1(\omega) \\
&\leq C 2^\ell \|P\|_{L^1(S^1,\mathcal{H}^1)}, \qquad (2.5.370)
\end{aligned}
$$

and

$$
\begin{aligned}
\big|\mathcal{F}_{x\to\xi}\big((1-\psi(x))\partial^\alpha u(x)\big)\big| &\le \|(1-\psi)\partial^\alpha u\|_{L^1(\mathbb{R}^2,\mathcal{L}^2)} \le \int_{\mathbb{R}^2\setminus B(0,1)} |\partial^\alpha u(x)|\,\mathrm{d}x \\
&\le C\int_{S^1} |\partial^\alpha u(\omega)|\,\mathrm{d}\mathcal{H}^1(\omega) \\
&\le C2^\ell \|P\|_{L^1(S^1,\mathcal{H}^1)}. \qquad (2.5.371)
\end{aligned}
$$

Collectively, (2.5.370) and (2.5.371) give that, for each $\alpha \in \mathbb{N}_0^2$ with $|\alpha| = 2$ and $\xi \in \overline{B(0,1)}$,

$$
\begin{aligned}
\big|\mathcal{F}_{x\to\xi}\big(\partial^\alpha u(x)\big)\big| &\le \big|\mathcal{F}_{x\to\xi}\big(\psi(x)\partial^\alpha u(x)\big)\big| + \big|\mathcal{F}_{x\to\xi}\big((1-\psi(x))\partial^\alpha u(x)\big)\big| \\
&\le C2^\ell \|P\|_{L^1(S^1,\mathcal{H}^1)}, \qquad (2.5.372)
\end{aligned}
$$

hence for each $\xi \in \overline{B(0,1)}$ we have

$$
|\xi|^2\big|\widehat{u}(\xi)\big| = \sum_{i=1}^{2}\big|\xi_i^2\widehat{u}(\xi)\big| = \sum_{i=1}^{2}\big|\mathcal{F}_{x\to\xi}\big(\partial_i^2 u(x)\big)\big| \le C2^\ell \|P\|_{L^1(S^1,\mathcal{H}^1)}. \qquad (2.5.373)
$$

In particular, $\|k_{rs}\|_{L^\infty(S^1,\mathcal{H}^1)} \le C\sup_{|\xi|=1}\big|\widehat{u}(\xi)\big| \le C2^\ell\|P\|_{L^1(S^1,\mathcal{H}^1)}$. A similar circle of ideas also yields $\|\nabla k_{rs}\|_{L^\infty(S^1,\mathcal{H}^1)} \le C2^\ell\|P\|_{L^1(S^1,\mathcal{H}^1)}$. This proves (2.5.351) in the case $n = 2$ and completes the proof of Lemma 2.5.36. □

We are now ready to establish jump-formulas for integral operators associated with kernels which have the special algebraic structure stipulated in (2.5.85).

Proposition 2.5.37 *Suppose $\Omega \subseteq \mathbb{R}^n$ (where $n \in \mathbb{N}$ with $n \ge 2$) is a nonempty open set with the property that $\partial\Omega$ is a UR set. In particular, Ω is a set of locally finite perimeter. Abbreviate $\sigma := \mathcal{H}^{n-1}\lfloor\partial\Omega$ and denote by ν the geometric measure theoretic outward unit normal to Ω. Also, consider an odd, harmonic, homogeneous polynomial $P(x)$ of degree $\ell \in \mathbb{N}$ in $\mathbb{R}^n$ and define*

$$
k(x) := \frac{P(x)}{|x|^{n-1+\ell}} \quad \text{for each } x \in \mathbb{R}^n \setminus \{0\}. \qquad (2.5.374)
$$

In relation to the kernel k and the domain Ω, associate the integral operators

$$
(\mathcal{T}f)(x) := \int_{\partial\Omega} k(x-y)f(y)\,\mathrm{d}\sigma(y), \quad \forall\, x \in \Omega, \qquad (2.5.375)
$$

and

$$
Tf(x) := \lim_{\varepsilon\to 0^+} \int_{\substack{y\in\partial\Omega \\ |x-y|>\varepsilon}} k(x-y)f(y)\,\mathrm{d}\sigma(y), \quad \forall\, x \in \partial\Omega. \qquad (2.5.376)
$$

Then for each fixed $\kappa > 0$, *and each* $f \in L^p(\partial\Omega, \sigma)$ *with* $1 \leq p < \infty$, *the jump-formula*

$$\lim_{\substack{z\in\Gamma_\kappa(x)\\ z\to x}} (\mathcal{T} f)(z) = \frac{1}{2\sqrt{-1}}\widehat{k}\big(\nu(x)\big) f(x) + T f(x) \tag{2.5.377}$$

is valid at σ*-a.e.* $x \in \partial_*\Omega$.

Proof That for each $f \in L^p(\partial\Omega, \sigma)$ the limit in (2.5.376) exists for σ-a.e. $x \in \partial\Omega$, and the induced operators

$$T : L^p(\partial\Omega, \sigma) \longrightarrow L^p(\partial\Omega, \sigma), \quad p \in (1, \infty), \tag{2.5.378}$$

$$T : L^1(\partial\Omega, \sigma) \longrightarrow L^{1,\infty}(\partial\Omega, \sigma), \tag{2.5.379}$$

are well-defined, linear, and bounded, follows from Theorem 2.3.2. Moreover, for each $\kappa > 0$ and $f \in L^p(\partial\Omega, \sigma)$ with $p \in [1, \infty)$, Theorem 2.4.1 also ensures that

$$\|\mathcal{N}_\kappa(\mathcal{T} f)\|_{L^p(\partial\Omega,\sigma)} \leq C\|f\|_{L^p(\partial\Omega,\sigma)}, \quad 1 < p < \infty, \tag{2.5.380}$$

$$\|\mathcal{N}_\kappa(\mathcal{T} f)\|_{L^{1,\infty}(\partial\Omega,\sigma)} \leq C\|f\|_{L^1(\partial\Omega,\sigma)}. \tag{2.5.381}$$

Pressing on, fix some $\kappa > 0$ and an exponent $p \in [1, \infty)$ arbitrary. Also, abbreviate $\sigma_* := \mathcal{H}^{n-1}\lfloor\partial_*\Omega$. In a first stage, we propose to prove by induction on $\ell \in \mathbb{N}$ that

given an integral kernel k as in (2.5.374) for an odd, harmonic, homogeneous polynomial $P(x)$ of degree ℓ in $\mathbb{R}^n$, and given any $f \in L^p(\partial_*\Omega, \sigma_*)$, canonically regarded as a function in $L^p(\partial\Omega, \sigma)$ by extending it by zero outside $\partial_*\Omega$, the jump-formula (2.5.377) holds at σ_*-a.e. point $x \in \partial_*\Omega$. (2.5.382)

The case $\ell = 1$ has been already dealt with in Proposition 2.5.35, so we may assume that $\ell \geq 3$ and that the corresponding jump-formula is true for any kernel k associated with an odd harmonic homogeneous polynomial P of degree $\ell - 2$ as in (2.5.374). Now, let P, k be as in the statement of the current proposition and recall the kernels $\{k_{rs}\}_{1\leq r,s\leq n}$ introduced in relation to these as in Lemma 2.5.36. For future purposes it will be convenient to set, for each $r, s \in \{1, \dots, n\}$,

$$k^{rs}(x) := \frac{\ell - 1}{n + \ell - 3}\frac{P_{rs}(x)}{|x|^{n+\ell-3}} \quad \text{for each } x \in \mathbb{R}^n \setminus \{0\}, \tag{2.5.383}$$

where P_{rs} are the harmonic homogeneous polynomials in $\mathbb{R}^n$ which are either zero or have degree $\ell - 2$, associated with the given function k as in Lemma 2.5.36.

For each $r, s \in \{1, \dots, n\}$ then define

$$(\mathcal{T}_{rs} f)(x) := \int_{\partial_*\Omega} k_{rs}(x - y) f(y)\, d\sigma_*(y), \quad \forall\, x \in \Omega. \tag{2.5.384}$$

On the one hand, identity (2.5.349) gives

$$\mathcal{T} f = \sum_{r,s=1}^{n} [\mathcal{T}_{rs} f]_s \text{ for every } f \in L^p(\partial_*\Omega, \sigma_*). \tag{2.5.385}$$

On the other hand, for every $f \in L^p(\partial_*\Omega, \sigma_*) \otimes \mathcal{C}\ell_n$ with $1 \leq p < \infty$ from (2.3.15) in Theorem 2.3.2 we know that there exists a set $A_f \subseteq \partial\Omega$ satisfying $\mathcal{H}^{n-1}(A_f) = 0$ and such that for each $r, s \in \{1, \dots, n\}$ the limits

$$\lim_{\varepsilon\to 0^+} \int\limits_{\substack{y\in\partial_*\Omega \\ |x-y|>\varepsilon}} k_{rs}(x-y) f(y)\, d\sigma_*(y) \text{ and } \lim_{\varepsilon\to 0^+} \int\limits_{\substack{y\in\partial_*\Omega \\ |x-y|>\varepsilon}} k^{rs}(x-y) f(y)\, d\sigma_*(y) \tag{2.5.386}$$

exist for each point $x \in \partial_*\Omega \setminus A_f$.

From (2.5.385)-(2.5.386) it is then clear that it suffices to show that, having fixed $r, s \in \{1, \dots, n\}$ and $f \in L^p(\partial_*\Omega, \sigma_*)$ with $1 \leq p < \infty$, at σ_*-a.e. $x \in \partial_*\Omega$ we have

$$\begin{aligned} \lim_{\substack{z\in\Gamma_\kappa(x) \\ z\to x}} (\mathcal{T}_{rs} f)(z) &= \frac{1}{2\sqrt{-1}} \widehat{k_{rs}}\big(\nu(x)\big) f(x) \\ &\quad + \lim_{\varepsilon\to 0^+} \int\limits_{\substack{y\in\partial_*\Omega \\ |x-y|>\varepsilon}} k_{rs}(x-y) f(y)\, d\sigma_*(y). \end{aligned} \tag{2.5.387}$$

In fact, by arguing as in the first part of the proof of Proposition 2.5.25 (a process in which the properties recorded in (2.5.378)-(2.5.381) are relevant) we see that it suffices to prove (2.5.387) in the case when

$$f = \nu \odot g \text{ with } g \in \mathrm{Lip}_c(\partial_*\Omega) \otimes \mathcal{C}\ell_n \text{ arbitrary.} \tag{2.5.388}$$

To this end, fix an arbitrary point

$$x \in \big(\partial^*\Omega \setminus A_f\big) \cap \partial_{\text{nta}}\Omega \tag{2.5.389}$$

and let $O_x \subset (0, 1)$ be associated with the point x and the set $E := \Omega$ (a choice which places the fixed point x in $\partial^* E$) as in Lemma 2.5.8. Also, let $N_x \subset (0, \infty)$ be an $\mathcal{L}^1$-measurable set such that $\mathcal{L}^1(N_x) = 0$ associated with the point x and the set $E := \mathbb{R}^n \setminus \Omega$ as in [112, Lemma 5.7.2]. In this scenario, paralleling the treatment in (2.5.191) we decompose

$$\lim_{\substack{z\in\Gamma_\kappa(x)\\ z\to x}} (\mathcal{T}_{rs}f)(z)$$

$$= \lim_{\substack{\varepsilon\in(O_x\setminus N_x)\\ \varepsilon\to 0}} \lim_{\substack{z\in\Gamma_\kappa(x)\\ z\to x}} \int_{\substack{|x-y|>\varepsilon\\ y\in\partial_*\Omega}} k_{rs}(z-y)f(y)\,d\sigma_*(y)$$

$$+ \lim_{\substack{\varepsilon\in(O_x\setminus N_x)\\ \varepsilon\to 0}} \lim_{\substack{z\in\Gamma_\kappa(x)\\ z\to x}} \frac{1}{\omega_{n-1}} \int_{\substack{|x-y|\leq\varepsilon\\ y\in\partial_*\Omega}} k_{rs}(z-y)\odot\nu(y)\odot[g(y)-g(x)]\,d\sigma_*(y)$$

$$+ \left(\lim_{\substack{\varepsilon\in(O_x\setminus N_x)\\ \varepsilon\to 0}} \lim_{\substack{z\in\Gamma_\kappa(x)\\ z\to x}} \int_{\substack{|x-y|\leq\varepsilon\\ y\in\partial_*\Omega}} k_{rs}(z-y)\odot\nu(y)\,d\sigma_*(y) \right)\odot g(x)$$

$$=: I_1 + I_2 + I_3. \tag{2.5.390}$$

Bearing in mind (2.5.386), as in the case of (2.5.191) (with $\vec{k}$ given by (2.5.129)) we obtain

$$I_1 = \lim_{\varepsilon\to 0^+} \int_{\substack{y\in\partial_*\Omega\\ |x-y|>\varepsilon}} k_{rs}(x-y)f(y)\,d\sigma_*(y) \text{ and } I_2 = 0. \tag{2.5.391}$$

There remains to treat I_3 and we begin by rewriting the last integral in (2.5.390) in a more convenient fashion. To accomplish this, To accomplish this, fix an arbitrary point $z \in \Gamma_\kappa(x)$ then consider the vector field

$$\begin{gathered} \vec{F} = (F_j)_{1\leq j\leq n} \text{ with } \mathcal{C}\ell_n\text{-valued components given by} \\ F_j(y) := k_{rs}(z-y)\odot \mathbf{e}_j - k^{rs}(z-y)\delta_{rj} \text{ for each } y\in\mathbb{R}^n\setminus\{z\}, \end{gathered} \tag{2.5.392}$$

(no sum over r) where the kernels k^{rs} are as in (2.5.383). Then, thanks to (2.5.392), (2.5.350), and (2.5.383), for each $y \in \mathbb{R}^n \setminus \{z\}$ we have

$$\begin{aligned} \mathrm{div}\vec{F}(y) &= \sum_{j=1}^{n} (\partial_j F_j)(y) \\ &= -(D_R\, k_{rs})(z-y) + (\partial_r k^{rs})(z-y) = 0. \end{aligned} \tag{2.5.393}$$

On account of this, [112, Lemma 5.7.2] used with the set $E := \mathbb{R}^n \setminus \Omega$ (which ensures that the vector field $\vec{F}$ is divergence-free near E, and that the geometric measure theoretic outward unit normal to E is $-\nu$), and the point x as in (2.5.389), guarantees, in view of the current choice of the $\mathcal{L}^1$-nullset $N_x \subset (0,\infty)$, that for each $\varepsilon \in (0,\infty) \setminus N_x$ we have

$$
\begin{aligned}
\int\limits_{\substack{|x-y|\leq\varepsilon\\ y\in\partial_*\Omega}} & k_{rs}(z_i-y)\odot\nu(y)\,\mathrm{d}\sigma_*(y)\\
&= \int\limits_{\substack{|x-y|\leq\varepsilon\\ y\in\partial_*\Omega}} k^{rs}(z_i-y)\nu_r(y)\,\mathrm{d}\sigma_*(y)\\
&\quad+\int\limits_{(\partial B(x,\varepsilon))\setminus\Omega} k_{rs}(z_i-y)\odot\Big(\frac{y-x}{\varepsilon}\Big)\,\mathrm{d}\mathcal{H}^{n-1}(y)\\
&\quad-\int\limits_{(\partial B(x,\varepsilon))\setminus\Omega} k^{rs}(z_i-y)\Big(\frac{y_r-x_r}{\varepsilon}\Big)\,\mathrm{d}\mathcal{H}^{n-1}(y).
\end{aligned}
\tag{2.5.394}
$$

To proceed, for each $\varepsilon > 0$ define

$$
\partial^- B(x,\varepsilon) := \{y\in\partial B(x,\varepsilon):\ \nu(x)\cdot(y-x)>0\} \tag{2.5.395}
$$

and

$$
W(x,\varepsilon) := \partial^- B(x,\varepsilon)\triangle\big[(\partial B(x,\varepsilon))\setminus\Omega\big]. \tag{2.5.396}
$$

Then (2.5.104) in the proof of Lemma 2.5.8 presently gives

$$
\lim_{\mathcal{O}_x\ni\varepsilon\to 0}\frac{\mathcal{H}^{n-1}\big(W(x,\varepsilon)\big)}{\varepsilon^{n-1}} = 0. \tag{2.5.397}
$$

Also, for each $\varepsilon > 0$ fixed, [112, (6.4.35)] and the positive homogeneity of the kernel k_{rs} permit us to estimate

$$
\begin{aligned}
\Bigg|\int\limits_{(\partial B(x,\varepsilon))\setminus\Omega} & k_{rs}(x-y)\odot\Big(\frac{y-x}{\varepsilon}\Big)\,\mathrm{d}\mathcal{H}^{n-1}(y)\\
&-\int\limits_{\partial^- B(x,\varepsilon)} k_{rs}(x-y)\odot\Big(\frac{y-x}{\varepsilon}\Big)\,\mathrm{d}\mathcal{H}^{n-1}(y)\Bigg|\\
&\leq \int\limits_{W(x,\varepsilon)} |k_{rs}(x-y)|\,\mathrm{d}\mathcal{H}^{n-1}(y)\\
&\leq \Big(\sup_{S^{n-1}}|k_{rs}|\Big)\frac{\mathcal{H}^{n-1}\big(W(x,\varepsilon)\big)}{\varepsilon^{n-1}}.
\end{aligned}
\tag{2.5.398}
$$

Based on (2.5.398) and (2.5.397) we may therefore write

$$\lim_{\substack{\varepsilon\in(O_x\setminus N_x)\\ \varepsilon\to 0}}\ \lim_{\substack{z\in\Gamma_\kappa(x)\\ z\to x}} \int_{(\partial B(x,\varepsilon))\setminus\Omega} k_{rs}(z-y)\odot\Big(\frac{y-x}{\varepsilon}\Big)\,d\mathcal{H}^{n-1}(y)$$

$$= \lim_{\substack{\varepsilon\in(O_x\setminus N_x)\\ \varepsilon\to 0}} \int_{(\partial B(x,\varepsilon))\setminus\Omega} k_{rs}(x-y)\odot\Big(\frac{y-x}{\varepsilon}\Big)\,d\mathcal{H}^{n-1}(y)$$

$$= \lim_{\substack{\varepsilon\in(O_x\setminus N_x)\\ \varepsilon\to 0}} \int_{\partial^- B(x,\varepsilon)} k_{rs}(x-y)\odot\Big(\frac{y-x}{\varepsilon}\Big)\,d\mathcal{H}^{n-1}(y)$$

$$= \int_{\partial^- B(x,1)} k_{rs}(x-\widetilde{y}\,)\odot(\widetilde{y}-x)\,d\mathcal{H}^{n-1}(\widetilde{y}\,) \tag{2.5.399}$$

where, the last equality, is seen by making the change of variables

$$\partial^- B(x,\varepsilon)\ni y\longmapsto \widetilde{y}:=x+\Big(\frac{y-x}{\varepsilon}\Big)\in\partial^- B(x,1), \tag{2.5.400}$$

while bearing in mind the positive homogeneity of the kernel k_{rs}. Moreover, in a very similar fashion, we have

$$\lim_{\substack{\varepsilon\in(O_x\setminus N_x)\\ \varepsilon\to 0}}\ \lim_{\substack{z\in\Gamma_\kappa(x)\\ z\to x}} \int_{(\partial B(x,\varepsilon))\setminus\Omega} k^{rs}(z-y)\Big(\frac{y_r-x_r}{\varepsilon}\Big)\,d\mathcal{H}^{n-1}(y)$$

$$= \int_{\partial^- B(x,1)} k^{rs}(x-\widetilde{y}\,)(\widetilde{y}_r-x_r)\,d\mathcal{H}^{n-1}(\widetilde{y}\,). \tag{2.5.401}$$

In summary, the above argument (which, starting with the decomposition in (2.5.390), has produced (2.5.391), (2.5.394), (2.5.399), and (2.5.401)) gives (keeping in mind (2.5.389) and [112, (8.8.52)]) that for σ_*-a.e. $x\in\partial_*\Omega$ we have (after a slight adjustment in notation)

$$\lim_{\substack{z\in\Gamma_\kappa(x)\\ z\to x}} (\mathcal{T}_{rs}f)(z)$$

$$= \lim_{\varepsilon\to 0^+} \int\limits_{\substack{y\in\partial_*\Omega\\ |x-y|>\varepsilon}} k_{rs}(x-y)f(y)\,d\sigma_*(y)$$

$$+ \left(\lim_{\substack{\varepsilon\in(O_x\setminus N_x)\\ \varepsilon\to 0}} \lim_{\substack{z\in\Gamma_\kappa(x)\\ z\to x}} \int\limits_{\substack{|x-y|\leq\varepsilon\\ y\in\partial_*\Omega}} k^{rs}(z-y)\nu_r(y)\,d\sigma_*(y) \right) \odot g(x)$$

$$+ \left(\int\limits_{\partial^- B(x,1)} k_{rs}(x-y)\odot(y-x)\,d\mathcal{H}^{n-1}(y) \right) \odot g(x)$$

$$+ \left(\int\limits_{\partial^- B(x,1)} k^{rs}(x-y)(y_r-x_r)\,d\mathcal{H}^{n-1}(y) \right) \odot g(x). \tag{2.5.402}$$

To proceed, for each $h \in L^p(\partial_*\Omega, \sigma_*)$ introduce

$$(\mathcal{T}^{rs}h)(x) := \int_{\partial_*\Omega} k^{rs}(x-y)h(y)\,d\sigma_*(y), \quad \forall x \in \Omega. \tag{2.5.403}$$

Two notable things about this family of integral operators are as follows. On the one hand, given the nature of the kernels k^{rs} from (2.5.383) (in particular, recall that P_{rs} is a harmonic homogeneous polynomial in $\mathbb{R}^n$ which is either zero or has degree $\ell-2$), the current induction hypothesis pertaining to the validity of the jump-formula (2.5.377) allows us to write that, for each function $h \in L^p(\partial_*\Omega, \sigma_*)$,

$$\lim_{\substack{z\in\Gamma_\kappa(x)\\ z\to x}} \mathcal{T}^{rs}(\nu_r h)(z) = \frac{1}{2\sqrt{-1}}\widehat{k^{rs}}(\nu(x))\nu_r(x)h(x)$$

$$+ \lim_{\varepsilon\to 0^+} \int\limits_{\substack{y\in\partial_*\Omega\\ |x-y|>\varepsilon}} k^{rs}(x-y)\nu_r(y)h(y)\,d\sigma_*(y) \tag{2.5.404}$$

at σ_*-almost every $x \in \partial_*\Omega$. On the other hand, by breaking up the domain of integration in (2.5.403) into $\partial_*\Omega \setminus \overline{B(x,\varepsilon)}$ and $(\partial_*\Omega) \cap B(x,\varepsilon)$, then passing to the limit $\Gamma_\kappa(x) \ni z \to x$ and $\varepsilon \to 0^+$, we obtain that for every function $h \in L^p(\partial_*\Omega, \sigma_*)$ we have

$$\lim_{\substack{z\in\Gamma_\kappa(x)\\ z\to x}} \mathcal{T}^{rs}(\nu_r h)(z) = \lim_{\varepsilon\to0^+} \int\limits_{\substack{y\in\partial_*\Omega\\ |x-y|>\varepsilon}} k^{rs}(x-y)\nu_r(y)h(y)\,d\sigma_*(y)$$

$$+ \lim_{\varepsilon\to0^+} \lim_{\substack{z\in\Gamma_\kappa(x)\\ z\to x}} \int\limits_{\substack{|x-y|\le\varepsilon\\ y\in\partial_*\Omega}} k^{rs}(z-y)\nu_r(y)h(y)\,d\sigma_*(y) \quad (2.5.405)$$

for σ_*-a.e. $x \in \partial_*\Omega$. By comparing (2.5.404) with (2.5.405) (and choosing h to be identically 1 near the point x) we eventually arrive at the conclusion that

$$\lim_{\varepsilon\to0^+} \lim_{\substack{z\in\Gamma_\kappa(x)\\ z\to x}} \int\limits_{\substack{|x-y|\le\varepsilon\\ y\in\partial_*\Omega}} k^{rs}(z-y)\nu_r(y)\,d\sigma_*(y) = \frac{1}{2\sqrt{-1}}\widehat{k^{rs}}\big(\nu(x)\big)\nu_r(x) \qquad (2.5.406)$$

for σ_*-a.e. $x \in \partial_*\Omega$. In turn, from (2.5.406), (2.5.402), and the fact that $g = -\nu \odot f$ at σ_*-a.e. point on $\partial_*\Omega$ (as seen from (2.5.388) and [112, (6.4.59)]), we finally conclude that for σ_*-a.e. $x \in \partial_*\Omega$ we have

$$\lim_{\substack{z\in\Gamma_\kappa(x)\\ z\to x}} (\mathcal{T}_{rs} f)(z) = \lim_{\varepsilon\to0^+} \int\limits_{\substack{y\in\partial_*\Omega\\ |x-y|>\varepsilon}} k_{rs}(x-y)f(y)\,d\sigma_*(y)$$

$$- \frac{1}{2\sqrt{-1}}\widehat{k^{rs}}\big(\nu(x)\big)\nu_r(x) \odot \nu(x) \odot f(x)$$

$$- \left(\int\limits_{\partial^- B(x,1)} k_{rs}(x-y) \odot (y-x)\,d\mathcal{H}^{n-1}(y) \right) \odot \nu(x) \odot f(x)$$

$$- \left(\int\limits_{\partial^- B(x,1)} k^{rs}(x-y)(y_r - x_r)\,d\mathcal{H}^{n-1}(y) \right) \odot \nu(x) \odot f(x). \qquad (2.5.407)$$

In short, for every $f \in L^p(\partial_*\Omega, \sigma_*)$ with $1 \le p < \infty$, at σ_*-a.e. $x \in \partial_*\Omega$ we have

$$\lim_{\substack{z\in\Gamma_\kappa(x)\\ z\to x}} (\mathcal{T}_{rs} f)(z) = \alpha_{rs}(x) \odot f(x) + \lim_{\varepsilon\to0^+} \int\limits_{\substack{y\in\partial_*\Omega\\ |x-y|>\varepsilon}} k_{rs}(x-y)f(y)\,d\sigma_*(y) \quad (2.5.408)$$

where, at σ_*-a.e. $x \in \partial_*\Omega$, the coefficient $\alpha_{rs}(x)$ is given by

$$\alpha_{rs}(x) = -\frac{1}{2\sqrt{-1}}\widehat{k^{rs}}\big(\nu(x)\big)\nu_r(x) \odot \nu(x)$$
$$- \left(\int_{\partial^- B(x,1)} k_{rs}(x-y) \odot (y-x)\, \mathrm{d}\mathcal{H}^{n-1}(y) \right) \odot \nu(x)$$
$$- \left(\int_{\partial^- B(x,1)} k^{rs}(x-y)(y_r - x_r)\, \mathrm{d}\mathcal{H}^{n-1}(y) \right) \odot \nu(x). \tag{2.5.409}$$

Hence, in order to fully justify (2.5.387), it remains to show that

$$\alpha_{rs}(x) = \frac{1}{2\sqrt{-1}}\widehat{k}_{rs}\big(\nu(x)\big) \text{ at } \sigma_*\text{-a.e. } x \in \partial_*\Omega. \tag{2.5.410}$$

A key characteristic of $\alpha_{rs}(x)$, visible from (2.5.409), is that, as far as the geometry is concerned, this quantity depends only on the vector $\nu(x)$ and not on the set Ω itself. Consequently, in the process of computing the actual value of $\alpha_{rs}(x_0)$ for a given point $x_0 \in \partial^*\Omega$, using (2.5.408), we may replace Ω by any other open set in $\mathbb{R}^n$ with a UR boundary and which has the same geometric measure theoretic outward unit normal $\nu(x_0)$ at x_0 as Ω does. Of course, the simplest such replacement is a suitably rotated and translated half-space, namely,

$$\widetilde{\Omega} := \{x \in \mathbb{R}^n : \nu(x_0) \cdot (x - x_0) < 0\}. \tag{2.5.411}$$

In this latter scenario, we agree to decorate with tilde objects associated with $\widetilde{\Omega}$ in the same manner as their plain counterparts have been introduced in relation to the original set Ω. With this convention in mind, for every $f \in L^p(\partial_*\widetilde{\Omega}, \widetilde{\sigma}_*)$ where $1 \le p < \infty$, the jump-formula

$$\lim_{\substack{z\in\widetilde{\Gamma}_\kappa(x)\\ z\to x}} (\widetilde{\mathcal{T}}_{rs} f)(z) \tag{2.5.412}$$
$$= \frac{1}{2\sqrt{-1}}\widehat{k}_{rs}\big(\nu(x_0)\big) f(x) + \lim_{\varepsilon\to 0^+} \int_{\substack{y\in\partial_*\widetilde{\Omega}\\ |x-y|>\varepsilon}} k_{rs}(x-y) f(y)\, \mathrm{d}\widetilde{\sigma}_*(y),$$

at $\widetilde{\sigma}_*$-a.e. $x \in \partial_*\widetilde{\Omega}$, is well known (see, e.g., [109, Corollary 4.81, p. 174] and the discussion in [129]). Altogether, from (2.5.412) and (2.5.408) we then conclude that (2.5.410) holds. This justifies (2.5.387) and finishes the proof of (2.5.382).

Having established (2.5.382), the goal now is to prove the validity of the jump-formula (2.5.377) at σ_*-a.e. point in $\partial_*\Omega$ in the case when f is an arbitrary function in $L^p(\partial\Omega, \sigma)$. With this goal in mind, pick some arbitrary $f \in L^p(\partial\Omega, \sigma)$, fix a point $x_0 \in \partial\Omega$ and, for each $m \in \mathbb{N}$, define

$$f_m := f \cdot \mathbf{1}_{B(x_0,m)\cap\partial\Omega} \in L^p(\partial\Omega, \sigma). \tag{2.5.413}$$

Lebesgue's Dominated Convergence Theorem then ensures that

$$f_m \longrightarrow f \text{ in } L^p(\partial\Omega, \sigma) \text{ as } m \to \infty. \tag{2.5.414}$$

In turn, from (2.5.414), Theorem 2.3.2, and Theorem 2.4.1 we conclude that, as $m \longrightarrow \infty$,

$$Tf_m \longrightarrow Tf \text{ in } \begin{cases} L^p(\partial\Omega, \sigma) & \text{if } 1 < p < \infty, \\ L^{1,\infty}(\partial\Omega, \sigma) & \text{if } p = 1, \end{cases} \tag{2.5.415}$$

and

$$\mathcal{N}_\kappa\big(\mathcal{T}(f - f_m)\big) \to 0 \text{ in } \begin{cases} L^p(\partial\Omega, \sigma) & \text{if } 1 < p < \infty, \\ L^{1,\infty}(\partial\Omega, \sigma) & \text{if } p = 1. \end{cases} \tag{2.5.416}$$

In view of [112, Proposition 6.2.7], by eventually passing to a subsequence there is no loss of generality in assuming that there exists $A \subseteq \partial\Omega$, which is σ-measurable and satisfies $\sigma(A) = 0$, with the property that, as $m \longrightarrow \infty$, we also have

$$\begin{aligned} &f_m(x) \to f(x), \quad (Tf_m)(x) \to (Tf)(x), \text{ and} \\ &\mathcal{N}_\kappa\big(\mathcal{T}(f - f_m)\big)(x) \to 0, \text{ for each } x \in \partial\Omega \setminus A. \end{aligned} \tag{2.5.417}$$

To proceed, make the working assumption that

$$\begin{aligned} &\text{for each } m \in \mathbb{N} \text{ there exists some } \sigma_*\text{-nullset } A_m \subseteq \partial_*\Omega \\ &\text{with the property that for each } x \in \partial_*\Omega \setminus A_m \text{ we have} \\ &\lim_{\substack{z \in \Gamma_\kappa(x) \\ z \to x}} (\mathcal{T} f_m)(z) = \tfrac{1}{2\sqrt{-1}}\widehat{k}\big(\nu(x)\big) f_m(x) + (Tf_m)(x). \end{aligned} \tag{2.5.418}$$

Introduce

$$\widetilde{A} := (\partial_*\Omega \cap A) \cup \Big(\bigcup_{m \in \mathbb{N}} A_m\Big) \tag{2.5.419}$$

so that

$$\widetilde{A} \subseteq \partial_*\Omega \text{ is } \sigma_*\text{-measurable and satisfies } \sigma_*(\widetilde{A}) = 0. \tag{2.5.420}$$

Then, for each fixed $x \in \partial_*\Omega \setminus \widetilde{A}$ and each $m \in \mathbb{N}$ we may write

$$
\begin{aligned}
&\limsup_{\substack{z\in\Gamma_\kappa(x)\\ z\to x}} \Big|(\mathcal{T}f)(z) - \Big(\tfrac{1}{2\sqrt{-1}}\widehat{k}(\nu(x))f(x) + (Tf)(x)\Big)\Big| \\
&\qquad \leq \limsup_{\substack{z\in\Gamma_\kappa(x)\\ z\to x}} \Big|(\mathcal{T}(f-f_m))(z)\Big| \\
&\qquad\quad + \limsup_{\substack{z\in\Gamma_\kappa(x)\\ z\to x}} \Big|(\mathcal{T}f_m)(z) - \Big(\tfrac{1}{2\sqrt{-1}}\widehat{k}(\nu(x))f_m(x) + (Tf_m)(x)\Big)\Big| \\
&\qquad\quad + \Big|\Big(\tfrac{1}{2\sqrt{-1}}\widehat{k}(\nu(x))f_m(x) + (Tf_m)(x)\Big) - \Big(\tfrac{1}{2\sqrt{-1}}\widehat{k}(\nu(x))f(x) + (Tf)(x)\Big)\Big| \\
&\qquad =: \mathrm{I}_m(x) + \mathrm{II}_m(x) + \mathrm{III}_m(x).
\end{aligned}
\tag{2.5.421}
$$

Note that

$$
0 \leq \mathrm{I}_m(x) \leq \Big(\mathcal{N}_\kappa\big(\mathcal{T}(f-f_m)\big)\Big)(x) \tag{2.5.422}
$$

which, in concert with the last property in (2.5.417), implies

$$
\lim_{m\to\infty} \mathrm{I}_m(x) = 0. \tag{2.5.423}
$$

Moreover, thanks to the working assumption (2.5.418), we have

$$
\mathrm{II}_m(x) = 0 \ \text{ for each } m \in \mathbb{N}, \tag{2.5.424}
$$

whereas the first line in (2.5.417) entails

$$
\lim_{m\to\infty} \mathrm{III}_m(x) = 0. \tag{2.5.425}
$$

From (2.5.420)-(2.5.425) we then conclude that, for any given $f \in L^p(\partial\Omega, \sigma)$ with $1 \leq p < \infty$, the jump-formula (2.5.377) is valid at σ_*-a.e. point $x \in \partial_*\Omega$.

At this stage, there remains to dispense with the extra working assumption made in (2.5.418) where, as before, f is an arbitrary function in $L^p(\partial\Omega, \sigma)$ with $1 \leq p < \infty$. To this end, fix $m \in \mathbb{N}$ arbitrary and introduce

$$
g_m := f_m\big|_{\partial_*\Omega} \in L^p(\partial_*\Omega, \sigma_*). \tag{2.5.426}
$$

Let us also denote by $\widetilde{g}_m$ the extension of g_m by zero outside $\partial_*\Omega$ to the entire $\partial\Omega$, and define

$$
h_m := f_m - \widetilde{g}_m \in L^p(\partial\Omega, \sigma). \tag{2.5.427}
$$

From (2.5.382) we know that for σ_*-a.e. $x \in \partial_*\Omega$ we have

$$\begin{aligned}\lim_{\substack{z\in\Gamma_\kappa(x)\\ z\to x}}\big(\mathcal{T}\widetilde{g}_m\big)(z) &= \tfrac{1}{2\sqrt{-1}}\widehat{k}\big(\nu(x)\big)\widetilde{g}_m(x)+\big(T\widetilde{g}_m\big)(x)\\ &= \tfrac{1}{2\sqrt{-1}}\widehat{k}\big(\nu(x)\big)f_m(x)+\big(T\widetilde{g}_m\big)(x).\end{aligned} \tag{2.5.428}$$

As regards the function defined in (2.5.427), note that

$$h_m = 0 \text{ on } \partial_*\Omega\cup\big(\partial\Omega\setminus B(x_0,m)\big) = \partial\Omega\setminus E \tag{2.5.429}$$

where

$$E := \big(\partial\Omega\setminus\partial_*\Omega\big)\cap B(x_0,m). \tag{2.5.430}$$

Observe that the set E is σ-measurable and, thanks to the fact that $\partial\Omega$ is upper Ahlfors regular, satisfies

$$\sigma(E)\le\sigma\big(\partial\Omega\cap B(x_0,m)\big)<+\infty. \tag{2.5.431}$$

In view of [112, Lemma 3.4.13] and the upper Ahlfors regularity of $\partial\Omega$, the hypotheses in item *(3)* of [112, Proposition 3.4.15] are satisfied by the ambient $X := \partial\Omega$ (a closed set regarded as a topological space in its own right when equipped with the topology inherited from $\mathbb{R}^n$), and the measure σ which is Borel-regular on X (thanks to [112, Lemma 3.6.4]). Since $E\subseteq\partial\Omega$ is σ-measurable, the inner-regularity property [112, (3.4.47)] presently implies

$$\sigma(E)=\sup\big\{\sigma(C):\ C \text{ closed subset of } \partial\Omega \text{ contained in } E\big\}. \tag{2.5.432}$$

Bearing in mind that $\sigma(E)<+\infty$, we conclude from (2.5.432) that there exists a nested sequence $\{C_j\}_{j\in\mathbb{N}}$ of closed subsets of $\partial\Omega$ with the property that

$$\begin{aligned} C_j\subseteq E \ \text{ for each } j\in\mathbb{N}, \text{ and}\\ \mathbf{1}_{C_j}\nearrow\mathbf{1}_E \ \text{ as } j\to\infty, \ \text{ for } \sigma\text{-a.e. point in } \partial\Omega.\end{aligned} \tag{2.5.433}$$

Hence, if we define

$$h_{m,j} := h_m\cdot\mathbf{1}_{C_j}\in L^p(\partial\Omega,\sigma) \text{ for each } j\in\mathbb{N}, \tag{2.5.434}$$

then from (2.5.433), Lebesgue's Dominated Convergence Theorem, and (2.5.429) we conclude that

$$h_{m,j}\longrightarrow h_m\cdot\mathbf{1}_E = h_m \text{ in } L^p(\partial\Omega,\sigma) \text{ as } j\to\infty. \tag{2.5.435}$$

In turn, with the help of Theorem 2.3.2 and Theorem 2.4.1, from this we conclude that, as $j\to\infty$,

$$Th_{m,j}\to Th_m \text{ in } \begin{cases} L^p(\partial\Omega,\sigma) & \text{if } 1<p<\infty,\\ L^{1,\infty}(\partial\Omega,\sigma) & \text{if } p=1,\end{cases} \tag{2.5.436}$$

and

$$\mathcal{N}_\kappa\big(\mathcal{T}(h_m - h_{m,j})\big) \to 0 \text{ in } \begin{cases} L^p(\partial\Omega, \sigma) & \text{if } 1 < p < \infty, \\ L^{1,\infty}(\partial\Omega, \sigma) & \text{if } p = 1. \end{cases} \tag{2.5.437}$$

As in the past, by eventually passing to a subsequence there is no loss of generality in assuming that

$$\begin{gathered}(Th_{m,j})(x) \to (Th_m)(x) \text{ and } \mathcal{N}_\kappa\big(\mathcal{T}(h_m - h_{m,j})\big)(x) \to 0 \\ \text{as } j \to \infty, \text{ for } \sigma\text{-a.e. point } x \in \partial\Omega.\end{gathered} \tag{2.5.438}$$

Going further, recall the set $A_\kappa(\partial\Omega)$ defined in (A.0.2). We claim that

$$\begin{gathered}\lim_{\substack{z\in\Gamma_\kappa(x)\\ z\to x}} (\mathcal{T}h_{m,j})(z) = (Th_{m,j})(x) \\ \text{for each } j \in \mathbb{N} \text{ and each } x \in \partial_*\Omega \cap A_\kappa(\partial\Omega).\end{gathered} \tag{2.5.439}$$

To prove this claim, fix $j \in \mathbb{N}$ and $x \in \partial_*\Omega \cap A_\kappa(\partial\Omega)$ arbitrary. In particular, $x \notin C_j$. Since C_j is a closed set, this implies that there exists some $\varepsilon_j(x) > 0$ with the property that

$$B\big(x, \varepsilon_j(x)\big) \cap C_j = \varnothing. \tag{2.5.440}$$

Let us also note that for any $z \in \Gamma_\kappa(x)$ and any $y \in \partial\Omega$ we have

$$\begin{aligned}|x - y| &\le |x - z| + |z - y| < (1 + \kappa)\,\mathrm{dist}(z, \partial\Omega) + |z - y| \\ &\le (1 + \kappa)|z - y| + |z - y| = (2 + \kappa)|z - y|,\end{aligned} \tag{2.5.441}$$

hence (bearing in mind the homogeneity property of the kernel k),

$$|k(z - y)| \le \Big(\sup_{S^{n-1}} |k|\Big) \cdot \frac{(2 + \kappa)^{n-1}}{|x - y|^{n-1}} =: F_x(y). \tag{2.5.442}$$

On the one hand, from (2.5.440) and the definition of $F_x(y)$ just given above we see that

$$\sup_{y\in C_j} |F_x(y)| \le \Big(\sup_{S^{n-1}} |k|\Big) \cdot \frac{(2 + \kappa)^{n-1}}{\varepsilon_j(x)^{n-1}} < +\infty, \tag{2.5.443}$$

while on the other hand for each integrability exponent $q \in (1, \infty)$ we may rely on the definition of F_x from (2.5.442) (which among other things, shows that F_x is continuous), (2.5.440), and [112, Lemma 7.2.1] to estimate

$$\int_{C_j} |F_x(y)|^q \, d\sigma(y) \tag{2.5.444}$$
$$\leq \Big(\sup_{S^{n-1}} |k| \Big)^q (2+\kappa)^{q(n-1)} \int_{\partial\Omega \setminus B(x,\varepsilon_j(x))} \frac{1}{|x-y|^{q(n-1)}} \, d\sigma(y) < +\infty.$$

Collectively (2.5.443)-(2.5.444) imply

$$F_x \in \bigcap_{1<q\leq\infty} L^q(C_j, \sigma). \tag{2.5.445}$$

From (2.5.442), (2.5.445), and (2.5.434) (keep in mind that $1 \leq p < \infty$), we conclude that for each point $z \in \Gamma_\kappa(x)$ we have

$$\begin{aligned} &\big|k(z-y)h_{m,j}(y)\big| \leq \big|F_x(y)\big|\big|h_{m,j}(y)\big| \ \text{ for every } \ y \in C_j, \\ &\text{and the function } \big|F_x\big|\big|h_{m,j}\big| \text{ belongs to } L^1(C_j, \sigma). \end{aligned} \tag{2.5.446}$$

For each point $x \in \partial_*\Omega \cap A_\kappa(\partial\Omega)$ we are now prepared to compute

$$\begin{aligned} \lim_{\substack{z\in\Gamma_\kappa(x)\\ z\to x}} \big(\mathcal{T} h_{m,j}\big)(z) &= \lim_{\substack{z\in\Gamma_\kappa(x)\\ z\to x}} \int_{\partial\Omega} k(z-y)h_{m,j}(y)\, d\sigma(y) \\ &= \lim_{\substack{z\in\Gamma_\kappa(x)\\ z\to x}} \int_{C_j} k(z-y)h_{m,j}(y)\, d\sigma(y) \\ &= \int_{C_j} k(x-y)h_{m,j}(y)\, d\sigma(y) \\ &= \lim_{\varepsilon\to 0^+} \int_{\partial\Omega\setminus B(x,\varepsilon)} k(x-y)h_{m,j}(y)\, d\sigma(y) \\ &= (T h_{m,j})(x). \end{aligned} \tag{2.5.447}$$

Above, the first equality uses the definition of the operator $\mathcal{T}$ (cf. (2.5.375)), while the second equality is clear from (2.5.434). The third equality in (2.5.447) is a consequence of Lebesgue's Dominated Convergence Theorem, whose applicability is ensured by the continuity of k on $\mathbb{R}^n \setminus \{0\}$, the fact that $x \in \overline{\Gamma_\kappa(x)}$, and (2.5.446). The fourth equality in (2.5.447) is implied by the fact that the set C_j (outside of which $h_{m,j}$ vanishes identically) is contained in $\partial\Omega \setminus B(x,\varepsilon)$ whenever $0 < \varepsilon < \varepsilon_j(x)$ (as seen from (2.5.440)). Finally, the fifth equality in (2.5.447) relies on the definition of the operator T (cf. (2.5.376)). In summary, (2.5.447) finishes the proof of (2.5.439).

With (2.5.439) in hand, for each $x \in \partial_*\Omega \cap A_\kappa(\partial\Omega)$ and each $j \in \mathbb{N}$ we may write

$$\begin{aligned}\limsup_{\substack{z\in\Gamma_\kappa(x)\\ z\to x}}\Big|(\mathcal{T}h_m)(z)-(Th_m)(x)\Big| &\le \limsup_{\substack{z\in\Gamma_\kappa(x)\\ z\to x}}\Big|\big(\mathcal{T}(h_m-h_{m,j})\big)(z)\Big|\\ &\quad+\limsup_{\substack{z\in\Gamma_\kappa(x)\\ z\to x}}\Big|\big(\mathcal{T}h_{m,j}\big)(z)-(Th_{m,j})(x)\Big|\\ &\quad+\Big|(Th_{m,j})(x)-(Th_m)(x)\Big|\\ &\le \Big(\mathcal{N}_\kappa\big(\mathcal{T}(h_m-h_{m,j})\big)\Big)(x)\\ &\quad+\Big|(Th_{m,j})(x)-(Th_m)(x)\Big|. \end{aligned}\tag{2.5.448}$$

In concert with (2.5.438) and [112, (8.8.45)], this ultimately implies that

$$\lim_{\substack{z\in\Gamma_\kappa(x)\\ z\to x}}(\mathcal{T}h_m)(z)=(Th_m)(x)\ \text{ for }\sigma_*\text{-a.e. } x\in\partial_*\Omega.\tag{2.5.449}$$

The end-game in the proof is as follows. First, (2.5.427) implies that

$$f_m=\widetilde{g}_m+h_m \text{ on } \partial\Omega.\tag{2.5.450}$$

Second, based on (2.5.450), (2.5.428), and (2.5.449), for σ_*-a.e. $x\in\partial_*\Omega$ we may write

$$\begin{aligned}\lim_{\substack{z\in\Gamma_\kappa(x)\\ z\to x}}\big(\mathcal{T}f_m\big)(z) &= \lim_{\substack{z\in\Gamma_\kappa(x)\\ z\to x}}\big(\mathcal{T}\widetilde{g}_m\big)(z)+\lim_{\substack{z\in\Gamma_\kappa(x)\\ z\to x}}\big(\mathcal{T}h_m\big)(z)\\ &=\tfrac{1}{2\sqrt{-1}}\widehat{k}\big(\nu(x)\big)f_m(x)+\big(T\widetilde{g}_m\big)(x)+\big(Th_m\big)(x)\\ &=\tfrac{1}{2\sqrt{-1}}\widehat{k}\big(\nu(x)\big)f_m(x)+\big(Tf_m\big)(x).\end{aligned}\tag{2.5.451}$$

This proves that (2.5.418) holds, thus the proof of Proposition 2.5.37 is complete. □

Having established Proposition 2.5.37, the general case described in Theorem 2.5.1 may be dealt via a spherical harmonics expansion. We execute this in the more general context described in Theorem 2.5.38 below, which may be regarded as a variable coefficient version of Theorems 2.3.2, 2.4.1, 2.5.1.

Theorem 2.5.38 *For each $n\in\mathbb{N}$ with $n\geq 2$ there exists a positive integer $M=M(n)$ with the following significance. Let $b(x,z)$ be a function which is odd and positive homogeneous of degree $1-n$ in the variable $z\in\mathbb{R}^n\setminus\{0\}$, and such that $\partial_z^\alpha b(x,z)$ is continuous and bounded on $\mathbb{R}^n\times S^{n-1}$ for each multi-index $\alpha\in\mathbb{N}_0^n$ satisfying $|\alpha|\leq M$. Also, let $\Omega\subseteq\mathbb{R}^n$ be a nonempty open set with the property that its topological boundary $\partial\Omega$ is a UR set; in particular, Ω is a set of locally finite perimeter. Abbreviate $\sigma:=\mathcal{H}^{n-1}\lfloor\partial\Omega$ and denote by ν the geometric measure theoretic outward unit normal to Ω.*

Then the following statements are true.

(a) For every $f \in L^1\Big(\partial\Omega, \frac{\sigma(x)}{1+|x|^{n-1}}\Big)$ *the limit*

$$(Bf)(x) := \lim_{\varepsilon\to 0^+} \big(B_\varepsilon f\big)(x) \text{ where, for each } \varepsilon > 0, \tag{2.5.452}$$

$$\big(B_\varepsilon f\big)(x) := \int\limits_{\substack{y\in\partial\Omega \\ |x-y|>\varepsilon}} b(x, x-y)f(y)\,\mathrm{d}\sigma(y), \quad \forall\, x \in \partial\Omega, \tag{2.5.453}$$

exists at σ*-a.e.* $x \in \partial\Omega$*, and the principal-value singular integral operator* B *thus defined is bounded from* $L^p(\partial\Omega, \sigma)$ *into itself for every* $p \in (1, \infty)$*, and from* $L^1(\partial\Omega, \sigma)$ *into* $L^{1,\infty}(\partial\Omega, \sigma)$*. In fact, if for each* $f \in L^1\Big(\partial\Omega, \frac{\sigma(x)}{1+|x|^{n-1}}\Big)$ *one defines*

$$\big(B_{\max} f\big)(x) := \sup_{\varepsilon>0} \big|(B_\varepsilon f)(x)\big|, \qquad \forall\, x \in \partial\Omega, \tag{2.5.454}$$

then, for every $p \in (1, \infty)$ *there exists* $C_{\Omega,p} \in (0, \infty)$ *such that*

$$\|B_{\max} f\|_{L^p(\partial\Omega,\sigma)} \leq C_{\Omega,p} \cdot \sup_{\substack{(x,z)\in\mathbb{R}^n\times S^{n-1} \\ |\alpha|\leq M}} \big|(\partial_z^\alpha b)(x, z)\big| \|f\|_{L^p(\partial\Omega,\sigma)}. \tag{2.5.455}$$

Corresponding to $p = 1$*, estimate* (2.5.455) *holds if the weak-*L^1 *norm is used in the left-hand side.*

(b) Given any $f \in L^1\Big(\partial\Omega, \frac{\sigma(x)}{1+|x|^{n-1}}\Big)$ *define*

$$(\mathcal{B}f)(x) := \int_{\partial\Omega} b(x, x-y)f(y)\,\mathrm{d}\sigma(y) \text{ for each } x \in \Omega. \tag{2.5.456}$$

Further extend the action of the above boundary-to-domain integral operator to distributions $f \in H^p(\partial\Omega, \sigma)$ *with* $p \in \big(\frac{n-1}{n}, 1\big)$ *by setting, at each* $x \in \Omega$*,*

$$(\mathcal{B}f)(x) := \begin{cases} \big\langle b(x, x-\cdot),\, f\big\rangle & \text{if } \partial\Omega \text{ is bounded,} \\ \big\langle [b(x, x-\cdot)],\, f\big\rangle & \text{if } \partial\Omega \text{ is unbounded,} \end{cases} \quad \text{if } \tfrac{n-1}{n} < p \leq 1, \tag{2.5.457}$$

where $\langle\cdot,\cdot\rangle$ *stands for the duality bracket described in* [113, Theorem 4.6.1]*. Then for each aperture parameter* $\kappa \in (0, \infty)$ *and each integrability exponent* $p \in \big(\frac{n-1}{n}, \infty\big)$ *there exists some constant* $C = C(\Omega, \kappa, p) \in (0, \infty)$ *such that*

$$\|\mathcal{N}_\kappa(\mathcal{B}f)\|_{L^p(\partial\Omega,\sigma)} \leq C \cdot \sup_{\substack{(x,z)\in\mathbb{R}^n\times S^{n-1} \\ |\alpha|\leq M}} \big|(\partial_z^\alpha b)(x, z)\big| \|f\|_{H^p(\partial\Omega,\sigma)} \tag{2.5.458}$$

for each $f \in H^p(\partial\Omega, \sigma)$*. Also, there exists a constant* $C = C(\Omega, \kappa) \in (0, \infty)$ *for which*

$$\|\mathcal{N}_\kappa(\mathcal{B}f)\|_{L^{1,\infty}(\partial\Omega,\sigma)} \leq C \cdot \sup_{\substack{(x,z)\in\mathbb{R}^n\times S^{n-1}\\ |\alpha|\leq M}} \big|(\partial_z^\alpha b)(x,z)\big| \|f\|_{L^1(\partial\Omega,\sigma)} \tag{2.5.459}$$

whenever $f \in L^1(\partial\Omega, \sigma)$.

(c) *For every* $f \in L^1\Big(\partial\Omega, \frac{\sigma(x)}{1+|x|^{n-1}}\Big)$ *and every* $\kappa \in (0, \infty)$, *the following jump-formula holds:*

$$\lim_{\substack{z\in\Gamma_\kappa(x)\\ z\to x}} (\mathcal{B}f)(z) = \frac{1}{2\sqrt{-1}}\widehat{b}\big(x, \nu(x)\big) f(x) + (Bf)(x) \tag{2.5.460}$$

at σ*-a.e.* $x \in \partial_*\Omega$, *where 'hat' stands for the Fourier transform in the second variable.*

Moreover, in the range $p \in \big(\frac{n-1}{n}, 1\big]$ *the operator* B *induces a linear and bounded mapping*

$$B : H^p(\partial\Omega, \sigma) \longrightarrow L^p(\partial\Omega, \sigma) \tag{2.5.461}$$

according to

$$Bf := \sum_{j=1}^{\infty} \lambda_j B a_j \ \text{ in } L^p(\partial\Omega, \sigma), \tag{2.5.462}$$

whenever the given distribution $f \in H^p(\partial\Omega, \sigma)$ *is expressed as* $f = \sum_{j=1}^{\infty} \lambda_j a_j$ *in* $H^p(\partial\Omega, \sigma)$ *for some sequence* $\{\lambda_j\}_{j\in\mathbb{N}} \in \ell^p(\mathbb{N})$ *and some sequence* $\{a_j\}_{j\in\mathbb{N}}$ *of* (p, q)*-atoms on* $\partial\Omega$ *(where* $q \in [1, \infty]$ *with* $q > p$ *is some fixed background integrability exponent). Also, having fixed an aperture parameter* $\kappa \in (0, \infty)$, *if* $\mathfrak{H} : H^p(\partial\Omega, \sigma) \to L^p(\partial\Omega, \sigma)$ *stands for the* L^p*-filtering operator from* [113, Theorem 4.9.1] *(currently used with* $\Sigma := \partial\Omega$*) then for each* $f \in H^p(\partial\Omega, \sigma)$ *with* $p \in \big(\frac{n-1}{n}, 1\big]$ *the following jump-formula holds:*

$$(\mathcal{B}f)\Big|_{\partial\Omega}^{\kappa-\text{n.t.}}(x) = \frac{1}{2\sqrt{-1}}\widehat{b}\big(x, \nu(x)\big)(\mathfrak{H}f)(x) + (Bf)(x) \tag{2.5.463}$$

for σ*-a.e.* $x \in \partial_*\Omega$.

(d) *Similar results to those stated in items (a)-(c) are valid for the operators defined for each function* $f \in L^1\Big(\partial\Omega, \frac{\sigma(x)}{1+|x|^{n-1}}\Big)$ *as*

$$(B^\#_{\max} f)(x) := \sup_{\varepsilon>0} \big|(B^\#_\varepsilon f)(x)\big| \ \text{ at each } x \in \partial\Omega, \text{ and} \tag{2.5.464}$$

$$(B^\# f)(x) := \lim_{\varepsilon\to 0^+} (B^\#_\varepsilon f)(x) \ \text{ at } \sigma\text{-a.e. } x \in \partial\Omega, \text{ where} \tag{2.5.465}$$

$$(B^\#_\varepsilon f)(x) := \int_{\substack{y\in\partial\Omega\\ |x-y|>\varepsilon}} b(y, x - y) f(y)\, d\sigma(y), \ \forall\, x \in \partial\Omega, \ \forall\, \varepsilon > 0, \tag{2.5.466}$$

as well as

$$(\mathcal{B}^{\#} f)(x) := \int_{\partial\Omega} b(y, x-y) f(y)\, \mathrm{d}\sigma(y) \text{ for each } x \in \Omega. \tag{2.5.467}$$

In particular, the limit in (2.5.465) exists at σ-a.e. point in $\partial\Omega$ and, given integrability exponent $p \in [1, \infty)$ along with any aperture parameter $\kappa \in (0, \infty)$, there exists a constant $C = C(\Omega, \kappa, p) \in (0, \infty)$ such that

$$\|B^{\#}_{\max} f\|_{L^p(\partial\Omega,\sigma)} \leq C \cdot \sup_{\substack{(y,z)\in\mathbb{R}^n\times S^{n-1} \\ |\alpha|\leq M}} \big|(\partial_z^\alpha b)(y,z)\big| \|f\|_{L^p(\partial\Omega,\sigma)}, \tag{2.5.468}$$

$$\big\|\mathcal{N}_\kappa(\mathcal{B}^{\#} f)\big\|_{L^p(\partial\Omega,\sigma)} \leq C \cdot \sup_{\substack{(y,z)\in\mathbb{R}^n\times S^{n-1} \\ |\alpha|\leq M}} \big|(\partial_z^\alpha b)(y,z)\big| \|f\|_{L^p(\partial\Omega,\sigma)}, \tag{2.5.469}$$

if $p \in (1, \infty)$ plus similar estimates (involving the space $L^{1,\infty}(\partial\Omega, \sigma)$ in the left-hand side) when $p = 1$. Also,

the (real) transpose of B acting on $L^p(\partial\Omega, \sigma)$ with $p \in (1, \infty)$ is $-B^{\#}$ acting on $L^{p'}(\partial\Omega, \sigma)$, provided $p, p' \in (1, \infty)$ satisfy $1/p + 1/p' = 1$. (2.5.470)

Moreover, for each $f \in L^1\Big(\partial\Omega, \frac{\sigma(x)}{1+|x|^{n-1}}\Big)$ and each $\kappa > 0$, the following jump-formula holds:

$$\lim_{\substack{z\in\Gamma_\kappa(x) \\ z\to x}} (\mathcal{B}^{\#} f)(z) = \frac{1}{2\sqrt{-1}} \widehat{b}\big(x, \nu(x)\big) f(x) + (B^{\#} f)(x) \tag{2.5.471}$$

at σ-a.e. $x \in \partial_*\Omega$, again, with 'hat' denoting the Fourier transform in the second variable.

(e) In all results stated in items (a)-(d) above which involve Lebesgue spaces $L^p(\partial\Omega, \sigma)$ with $p \in (1, \infty)$ one may use instead the more inclusive scale of Muckenhoupt weighted Lebesgue spaces $L^p(\partial\Omega, w\sigma)$ with $p \in (1, \infty)$ and $w \in A_p(\partial\Omega, \sigma)$ arbitrary.

(f) In a slightly more restrictive setting than originally assumed, there are also natural Hardy space estimates for the operators $\mathcal{B}^{\#}$, $B^{\#}$. Specifically, suppose $\partial\Omega$ is compact, $p \in \big(\frac{n-1}{n}, 1\big]$, $r > (n-1)\big(\frac{1}{p} - 1\big)$, and whenever $\alpha \in \mathbb{N}_0^n$ has $|\alpha| \leq M$ the function $\partial_z^\alpha b(\cdot, z)$ belongs to $\mathscr{C}^r(\mathbb{R}^n)$, uniformly for $z \in S^{n-1}$. Then the operator $\mathcal{B}^{\#}$, now interpreted as

$$(\mathcal{B}^{\#} f)(x) := \big\langle b(\cdot, x - \cdot),\ f\big\rangle \text{ for each } f \in H^p(\partial\Omega, \sigma) \text{ and } x \in \Omega, \tag{2.5.472}$$

where $\langle\cdot,\cdot\rangle$ *stands for the duality bracket described in* [113, Theorem 4.6.1], *satisfies*

$$\|\mathcal{N}_\kappa(\mathcal{B}^\# f)\|_{L^p(\partial\Omega,\sigma)} \tag{2.5.473}$$
$$\leq C\cdot \max_{|\alpha|\leq M}\ \sup_{z\in S^{n-1}} \big\|(\partial_z^\alpha b)(\cdot,z)\big\|_{\mathscr{C}^r(\mathbb{R}^n)}\|f\|_{H^p(\partial\Omega,\sigma)},$$

for each $f\in H^p(\partial\Omega,\sigma)$, *where* $C\in(0,\infty)$ *is independent of* f. *Also,* $B^\#$ *may be extended to a linear and bounded operator*

$$B^\# : H^p(\partial\Omega,\sigma)\longrightarrow L^p(\partial\Omega,\sigma) \tag{2.5.474}$$

by setting

$$B^\# f := \sum_{j=1}^{\infty}\lambda_j B^\# a_j \ \text{ in } L^p(\partial\Omega,\sigma) \tag{2.5.475}$$

whenever the given distribution $f\in H^p(\partial\Omega,\sigma)$ *is expressed as* $f=\sum_{j=1}^{\infty}\lambda_j a_j$ *in* $H^p(\partial\Omega,\sigma)$ *for some sequence* $\{\lambda_j\}_{j\in\mathbb{N}}\in\ell^p(\mathbb{N})$ *and some sequence* $\{a_j\}_{j\in\mathbb{N}}$ *of* (p,q)*-atoms on* $\partial\Omega$ *(where* $q\in[1,\infty]$ *with* $q>p$ *is some fixed background integrability exponent).*
Moreover, with $\mathfrak{H} : H^p(\partial\Omega,\sigma)\to L^p(\partial\Omega,\sigma)$ *denoting the* L^p*-filtering operator from* [113, Theorem 4.9.1] *(presently used with* $\Sigma:=\partial\Omega$*) and having fixed an aperture parameter* $\kappa\in(0,\infty)$, *it follows that for each* $f\in H^p(\partial\Omega,\sigma)$ *the following jump-formula holds:*

$$\big(\mathcal{B}^\# f\big)\Big|_{\partial\Omega}^{\kappa-\text{n.t.}}(x) = \frac{1}{2\sqrt{-1}}\,\widehat{b}\big(x,\nu(x)\big)(\mathfrak{H}f)(x) + (B^\# f)(x) \tag{2.5.476}$$

for σ*-a.e.* $x\in\partial_*\Omega$.

(g) There is a natural L^p*-square function estimate for the operator* $\mathcal{B}^\#$. *Concretely, for each* $p\in(1,\infty)$ *there exists a constant* $C\in(0,\infty)$ *depending only on* n, p, *and the* UR *character of* $\partial\Omega$ *with the property that for each* $f\in L^p(\partial\Omega,\sigma)$ *one has*

$$\int_\Omega |\nabla\mathcal{B}^\# f(x)|^p \operatorname{dist}(x,\partial\Omega)^{p-1}\,dx \tag{2.5.477}$$
$$\leq C\cdot\Bigg(\sup_{\substack{(y,z)\in\mathbb{R}^n\times S^{n-1}\\ |\alpha|\leq M}} \big|(\partial_z^\alpha b)(y,z)\big|\Bigg)^p \int_{\partial\Omega}|f|^p\,d\sigma.$$

Also, under the stronger assumption that Ω *is actually a bounded Lipschitz domain,*

$$\mathcal{B}^\# : L^p(\partial\Omega,\sigma)\longrightarrow B^{p,p}_{1/p}(\Omega) \tag{2.5.478}$$

is a well-defined, linear, and bounded operator, for each $p\in(1,\infty)$.

(h) *A similar L^p-square function estimate also holds for the variable coefficient operator $\mathcal{B}$ defined in* (2.5.456)*, with some caveats. Specifically, under the additional assumptions that Ω is bounded and whenever $\alpha, \beta \in \mathbb{N}_0^n$ satisfy $|\alpha| \leq M$ and $|\beta| \leq 1$ the function $\partial_x^\beta \partial_z^\alpha b(x,z)$ is continuous and bounded on $\mathbb{R}^n \times S^{n-1}$, it follows that for each $p \in (1,\infty)$ there exists a constant $C \in (0,\infty)$ depending only on n, p, the* UR *character of $\partial\Omega$, and the diameter of Ω, with the property that for each $f \in L^p(\partial\Omega,\sigma)$ one has the following L^p-square function estimate:*

$$\int_\Omega |\nabla \mathcal{B} f(x)|^p \operatorname{dist}\big(x,\partial\Omega\big)^{p-1}\,dx$$

$$\leq C \cdot \max_{\substack{|\alpha|\leq M\\ |\beta|\leq 1}} \sup_{\substack{x\in\mathbb{R}^n\\ z\in S^{n-1}}} \big|(\partial_x^\beta \partial_z^\alpha b)(x,z)\big| \int_{\partial\Omega} |f|^p\,d\sigma. \tag{2.5.479}$$

Furthermore, if Ω is actually a bounded Lipschitz domain then for each exponent $p \in (1,\infty)$ the operator

$$\mathcal{B} : L^p(\partial\Omega,\sigma) \longrightarrow B^{p,p}_{1/p}(\Omega) \tag{2.5.480}$$

is well defined, linear, and bounded.

Proof Define

$$H_0 := 1, \quad H_1 := n, \text{ and } H_\ell := \binom{n-1+\ell}{\ell} - \binom{n+\ell-3}{\ell-2} \text{ if } \ell \geq 2, \tag{2.5.481}$$

and, for each $\ell \in \mathbb{N}_0$, let $\{\Psi_{i\ell}\}_{1\leq i\leq H_\ell}$ be an orthonormal basis for the space of spherical harmonics of degree ℓ on the $(n-1)$-dimensional sphere S^{n-1} equipped with the measure $\mathcal{H}^{n-1}$. In particular,

$$H_\ell \leq (\ell+1)\cdot(\ell+2)\cdots(n+\ell-2)\cdot(n+\ell-1) \leq C_n\,\ell^{n-1} \text{ for } \ell \geq 1 \tag{2.5.482}$$

and, if $\Delta_{S^{n-1}}$ denotes the Laplace-Beltrami operator on S^{n-1}, then for each $\ell \in \mathbb{N}_0$ and $1 \leq i \leq H_\ell$,

$$\begin{aligned} &\Delta_{S^{n-1}}\Psi_{i\ell} = -\ell(n+\ell-2)\Psi_{i\ell} \text{ on } S^{n-1}, \text{ and}\\ &\Psi_{i\ell}\Big(\frac{x}{|x|}\Big) = \frac{P_{i\ell}(x)}{|x|^\ell} \text{ for every } x \in \mathbb{R}^n \setminus \{0\}, \end{aligned} \tag{2.5.483}$$

for some homogeneous harmonic polynomial $P_{i\ell}$ of degree ℓ in $\mathbb{R}^n$, thus

$$\Psi_{i\ell} : S^{n-1} \to \mathbb{R} \text{ is an even function whenever } \ell \text{ is even.} \tag{2.5.484}$$

Also,

$$\{\Psi_{i\ell}\}_{\ell\in\mathbb{N}_0,\,1\leq i\leq H_\ell} \text{ is an orthonormal basis for } L^2(S^{n-1},\mathcal{H}^{n-1}), \tag{2.5.485}$$

hence,

$$\|\Psi_{i\ell}\|_{L^2(S^{n-1},\mathcal{H}^{n-1})} = 1 \text{ for each } \ell \in \mathbb{N}_0 \text{ and } 1 \le i \le H_\ell. \tag{2.5.486}$$

More details on these matters may be found in, e.g., [171, pp. 137–152] and [167, pp. 68–75].

Next, recall the integer $N = N(n) \in \mathbb{N}$ from Theorems 2.3.2-2.5.1 and fix

$$\text{an even integer } d \in \mathbb{N} \text{ with } d > [(n-1)/2] + N. \tag{2.5.487}$$

Sobolev's embedding theorem then gives that for each $\ell \in \mathbb{N}_0$ and $1 \le i \le H_\ell$ we have (with I standing for the identity operator on S^{n-1})

$$\|\Psi_{i\ell}\|_{\mathscr{C}^N(S^{n-1})} \le C_n \big\|(I - \Delta_{S^{n-1}})^{d/2}\Psi_{i\ell}\big\|_{L^2(S^{n-1},\mathcal{H}^{n-1})} \le C_n \max\{1,\ell\}^d, \tag{2.5.488}$$

where the last inequality is a consequence of (2.5.483)-(2.5.486).

Going further, pick a number

$$M = M(n) \in \mathbb{N} \text{ even, such that } M \ge 2(n+1) + d, \tag{2.5.489}$$

and consider a function $b(\cdot,\cdot) : \mathbb{R}^n \times S^{n-1} \to \mathbb{R}$ which is odd and positive homogeneous of degree $1-n$ in the second variable and such that $\partial_z^\alpha b(x,z)$ is continuous and bounded on $\mathbb{R}^n \times S^{n-1}$ for each $\alpha \in \mathbb{N}_0^n$ with $|\alpha| \le M$. Consider now $\ell \in \mathbb{N}_0$ and $1 \le i \le H_\ell$ arbitrary. If we define

$$a_{i\ell}(x) := \int_{S^{n-1}} b(x,\omega)\Psi_{i\ell}(\omega)\, d\mathcal{H}^{n-1}(\omega), \quad \text{for each } x \in \mathbb{R}^n, \tag{2.5.490}$$

it follows from (2.5.484) and the assumptions on $b(x,z)$ that

$$\begin{aligned}&\text{the function } a_{i\ell} \text{ is continuous in } \mathbb{R}^n \text{ and}\\ &\text{is identically zero whenever } \ell \text{ is even.}\end{aligned} \tag{2.5.491}$$

In addition, for each number $m \in \mathbb{N}$ with $2m \le M$ we have

$$\begin{aligned}[-\ell(n+\ell-2)]^m a_{i\ell}(x) &= \int_{S^{n-1}} b(x,\omega)\big(\Delta^m_{S^{n-1}}\Psi_{i\ell}\big)(\omega)\, d\mathcal{H}^{n-1}(\omega)\\ &= \int_{S^{n-1}} \big(\Delta^m_{S^{n-1}} b\big)(x,\omega)\Psi_{i\ell}(\omega)\, d\mathcal{H}^{n-1}(\omega),\end{aligned} \tag{2.5.492}$$

from which we conclude that

$$\begin{aligned}\sup_{x\in\mathbb{R}^n}\big|[-\ell(n+\ell-2)]^m a_{i\ell}(x)\big| &\leq \sup_{x\in\mathbb{R}^n}\big\|\big(\Delta^m_{S^{n-1}}b\big)(x,\cdot)\big\|_{L^2(S^{n-1},\mathcal{H}^{n-1})}\\ &\leq C_n \sup_{\substack{(x,z)\in\mathbb{R}^n\times S^{n-1}\\ |\alpha|\leq M}}\big|\big(\partial_z^\alpha b\big)(x,z)\big|\\ &=: C_b\in(0,\infty).\end{aligned} \tag{2.5.493}$$

Hence, the coefficients $a_{i\ell}$ are rapidly decreasing, in a uniform manner, in the sense that for each number $m\in\mathbb{N}$ with $2m\leq M$ there exists a constant $C_{n,m}\in(0,\infty)$ such that

$$\sup_{x\in\mathbb{R}^n}\big|a_{i\ell}(x)\big|\leq C_{n,m}\cdot C_b\cdot\max\{1,\ell\}^{-2m},\qquad \ell\in\mathbb{N}_0,\quad 1\leq i\leq H_\ell. \tag{2.5.494}$$

For each fixed $x\in\mathbb{R}^n$, expand the function $b(x,\cdot)\in L^2(S^{n-1},\mathcal{H}^{n-1})$ with respect to the orthonormal basis (2.5.485) to obtain that (in the sense of $L^2(S^{n-1},\mathcal{H}^{n-1})$ in the variable $z/|z|\in S^{n-1}$)

$$\begin{aligned}b(x,z)=b\Big(x,\frac{z}{|z|}\Big)|z|^{1-n}&=\sum_{\ell\in\mathbb{N}_0}\sum_{i=1}^{H_\ell}a_{i\ell}(x)\Psi_{i\ell}\Big(\frac{z}{|z|}\Big)|z|^{1-n}\\ &=\sum_{\ell\in2\mathbb{N}_0+1}\sum_{i=1}^{H_\ell}a_{i\ell}(x)\Psi_{i\ell}\Big(\frac{z}{|z|}\Big)|z|^{1-n},\end{aligned} \tag{2.5.495}$$

where the last equality is a consequence of (2.5.491). For each $\ell\in\mathbb{N}_0$ and $1\leq i\leq H_\ell$ let us now set

$$k_{i\ell}(z):=\Psi_{i\ell}\Big(\frac{z}{|z|}\Big)|z|^{1-n}=\frac{P_{i\ell}(z)}{|z|^{n-1+\ell}}\ \text{ for } z\in\mathbb{R}^n\setminus\{0\}. \tag{2.5.496}$$

In this notation, (2.5.495) streamlines to

$$b(x,z)=\sum_{\ell\in2\mathbb{N}_0+1}\sum_{i=1}^{H_\ell}a_{i\ell}(x)k_{i\ell}(z),\qquad \forall\,(x,z)\in\mathbb{R}^n\times\big(\mathbb{R}^n\setminus\{0\}\big). \tag{2.5.497}$$

Moreover, from definition (2.5.496) and estimate (2.5.488) we conclude that if d is as in (2.5.487) then there exists some constant $C_{n,N}\in(0,\infty)$ with the property that

$$\big\|k_{i\ell}\big|_{S^{n-1}}\big\|_{\mathscr{C}^N(S^{n-1})}\leq C_{n,N}\cdot\max\{1,\ell\}^d\ \text{ if } \ell\in\mathbb{N}_0 \text{ and } 1\leq i\leq H_\ell. \tag{2.5.498}$$

For future use, let us also note here that, as seen from the expansion (2.5.497) and the estimates in (2.5.494) and (2.5.498), we have

$$\widehat{b}(x,\xi) = \sum_{\ell\in 2\mathbb{N}_0+1} \sum_{i=1}^{H_\ell} a_{i\ell}(x)\widehat{k}_{i\ell}(\xi) \text{ with uniform convergence}$$
$$\text{for } (x,\xi) \text{ in compact subsets of } \mathbb{R}^n \times \big(\mathbb{R}^n \setminus \{0\}\big). \tag{2.5.499}$$

Moving on, given any $f \in L^1\Big(\partial\Omega, \frac{\sigma(x)}{1+|x|^{n-1}}\Big)$, for each $\ell \in \mathbb{N}_0$ and $1 \leq i \leq H_\ell$ define

$$\begin{aligned} B_\varepsilon^{i\ell} f(x) &:= \int_{\substack{y\in\partial\Omega \\ |x-y|>\varepsilon}} k_{i\ell}(x-y)f(y)\,d\sigma(y), \quad x \in \partial\Omega, \\ B_{\max}^{i\ell} f(x) &:= \sup_{\varepsilon>0} \big|B_\varepsilon^{i\ell} f(x)\big|, \quad x \in \partial\Omega, \\ B^{i\ell} f(x) &:= \lim_{\varepsilon\to 0^+} B_\varepsilon^{i\ell} f(x), \quad x \in \partial\Omega. \end{aligned} \tag{2.5.500}$$

Then for each $p \in [1,\infty)$ Theorem 2.3.2 guarantees the existence of a constant $C_{\Omega,p} \in (0,\infty)$ such that whenever $\ell \in \mathbb{N}_0$ and $1 \leq i \leq H_\ell$ it follows that for each $f \in L^p(\partial\Omega,\sigma)$ we have

$$\begin{aligned} \|B_{\max}^{i\ell} f\|_{L^p(\partial\Omega,\sigma)} &\leq C_{\Omega,p} \big\|k_{i\ell}\big|_{S^{n-1}}\big\|_{\mathscr{C}^N(S^{n-1})} \|f\|_{L^p(\partial\Omega,\sigma)} \\ &\leq C_{\Omega,p}\cdot C_{n,N}\cdot \max\{1,\ell\}^d \|f\|_{L^p(\partial\Omega,\sigma)} \end{aligned} \tag{2.5.501}$$

if $1 < p < \infty$ and, corresponding to $p = 1$, a constant $C_\Omega \in (0,\infty)$ such that for every $f \in L^1(\partial\Omega,\sigma)$ we have

$$\begin{aligned} \|B_{\max}^{i\ell} f\|_{L^{1,\infty}(\partial\Omega,\sigma)} &\leq C_{\Omega} \big\|k_{i\ell}\big|_{S^{n-1}}\big\|_{\mathscr{C}^N(S^{n-1})} \|f\|_{L^1(\partial\Omega,\sigma)} \\ &\leq C_{\Omega}\cdot C_{n,N}\cdot \max\{1,\ell\}^d \|f\|_{L^1(\partial\Omega,\sigma)}. \end{aligned} \tag{2.5.502}$$

Next, fix $f \in L^p(\partial\Omega,\sigma)$ with $1 \leq p < \infty$ arbitrary along with $\varepsilon > 0$ and, for each threshold $\mu \in \mathbb{N}$, split

$$(B_\varepsilon f)(x) = \sum_{\substack{\ell\in 2\mathbb{N}_0+1 \\ \ell\leq\mu}} \sum_{1\leq i\leq H_\ell} a_{i\ell}(x) B_\varepsilon^{i\ell} f(x) + \sum_{\substack{\ell\in 2\mathbb{N}_0+1 \\ \ell>\mu}} \sum_{1\leq i\leq H_\ell} a_{i\ell}(x) B_\varepsilon^{i\ell} f(x), \tag{2.5.503}$$

for each $x \in \partial\Omega$. In relation to this splitting, observe that if $1 < p < \infty$, thanks to (2.5.500), Minkowski's inequality, (2.5.494), (2.5.501), and (2.5.482), for each $m \in \mathbb{N}$ such that

$$n + 1 + d \leq 2m \leq M \tag{2.5.504}$$

(a viable choice given (2.5.489)) we have, with all multiplicative constants positive and finite,

$$
\begin{aligned}
&\Big\| \sup_{\varepsilon>0} \Big| \sum_{\substack{\ell\in 2\mathbb{N}_0+1\\ \ell>\mu}} \sum_{1\leq i\leq H_\ell} a_{i\ell} B^{i\ell}_\varepsilon f \Big| \Big\|_{L^p(\partial\Omega,\sigma)} \\
&\qquad \leq \sum_{\substack{\ell\in 2\mathbb{N}_0+1\\ \ell>\mu}} \sum_{1\leq i\leq H_\ell} \sup_{x\in\mathbb{R}^n} |a_{i\ell}(x)| \, \big\| B^{i\ell}_{\max} f \big\|_{L^p(\partial\Omega,\sigma)} \\
&\qquad \leq C(\Omega,b,p,n) \sum_{\ell>\mu} \ell^{n-1+d-2m} \|f\|_{L^p(\partial\Omega,\sigma)} \\
&\qquad \leq C(\Omega,b,p,n) \Big(\sum_{\ell>\mu} \ell^{-2} \Big) \|f\|_{L^p(\partial\Omega,\sigma)} \\
&\qquad \leq \frac{C(\Omega,b,p,n)}{\mu} \|f\|_{L^p(\partial\Omega,\sigma)}.
\end{aligned} \tag{2.5.505}
$$

Corresponding to $p = 1$, there is a similar estimate, involving the weak-L^1 space $L^{1,\infty}(\partial\Omega,\sigma)$. Since the latter is only quasi-Banach, more care is required in the step when Minkowski's inequality has been used in the past. Specifically, from the equivalence in [112, (6.2.21)] and [112, Proposition 6.2.10] (used with $p = 1$, which forces $\alpha_1 = 1/2$) we conclude that if we now take $m \in \mathbb{N}$ such that

$$
2(n+1) + d \leq 2m \leq M \tag{2.5.506}
$$

(once again, a viable choice thanks to (2.5.489)) then, with all multiplicative constants positive and finite,

$$
\begin{aligned}
&\Big\| \sup_{\varepsilon>0} \Big| \sum_{\substack{\ell\in 2\mathbb{N}_0+1\\ \ell>\mu}} \sum_{1\leq i\leq H_\ell} a_{i\ell} B^{i\ell}_\varepsilon f \Big| \Big\|_{L^{1,\infty}(\partial\Omega,\sigma)} \\
&\qquad \leq C \Big(\sum_{\substack{\ell\in 2\mathbb{N}_0+1\\ \ell>\mu}} \sum_{1\leq i\leq H_\ell} \sup_{x\in\mathbb{R}^n} |a_{i\ell}(x)|^{1/2} \, \big\| B^{i\ell}_{\max} f \big\|^{1/2}_{L^{1,\infty}(\partial\Omega,\sigma)} \Big)^2 \\
&\qquad \leq C(\Omega,b,n) \Big(\sum_{\ell>\mu} \ell^{n-1+(d/2)-m} \Big)^2 \|f\|_{L^1(\partial\Omega,\sigma)} \\
&\qquad \leq C(\Omega,b,n) \Big(\sum_{\ell>\mu} \ell^{-2} \Big)^2 \|f\|_{L^1(\partial\Omega,\sigma)} \\
&\qquad \leq \frac{C(\Omega,b,n)}{\mu^2} \|f\|_{L^1(\partial\Omega,\sigma)}.
\end{aligned} \tag{2.5.507}
$$

On the other hand, Theorem 2.3.2 gives that, for each fixed $\mu \in \mathbb{N}$, the limit as $\varepsilon \to 0^+$ of the first sum in (2.5.503), i.e.,

$$\lim_{\varepsilon \to 0^+} \sum_{\substack{\ell \in 2\mathbb{N}_0+1 \\ \ell \leq \mu}} \sum_{1 \leq i \leq H_\ell} a_{i\ell}(x) B_\varepsilon^{i\ell} f(x) = \sum_{\substack{\ell \in 2\mathbb{N}_0+1 \\ \ell \leq \mu}} \sum_{1 \leq i \leq H_\ell} a_{i\ell}(x) \lim_{\varepsilon \to 0^+} B_\varepsilon^{i\ell} f(x)$$
$$= \sum_{\substack{\ell \in 2\mathbb{N}_0+1 \\ \ell \leq \mu}} \sum_{1 \leq i \leq H_\ell} a_{i\ell}(x) B^{i\ell} f(x), \qquad (2.5.508)$$

exists at σ-a.e. $x \in \partial\Omega$. Thus, if we consider the disagreement function

$$D_f(x) := \limsup_{\varepsilon \to 0^+} B_\varepsilon f(x) - \liminf_{\varepsilon \to 0^+} B_\varepsilon f(x), \qquad x \in \partial\Omega, \qquad (2.5.509)$$

then for every $\mu \in \mathbb{N}$ we have

$$D_f(x) = \limsup_{\varepsilon \to 0^+} \Big(\sum_{\substack{\ell \in 2\mathbb{N}_0+1 \\ \ell > \mu}} \sum_{1 \leq i \leq H_\ell} a_{i\ell}(x) B_\varepsilon^{i\ell} f(x) \Big)$$
$$- \liminf_{\varepsilon \to 0^+} \Big(\sum_{\substack{\ell \in 2\mathbb{N}_0+1 \\ \ell > \mu}} \sum_{1 \leq i \leq H_\ell} a_{i\ell}(x) B_\varepsilon^{i\ell} f(x) \Big) \qquad (2.5.510)$$

at σ-a.e. $x \in \partial\Omega$. Granted this, in the case when $1 < p < \infty$, from Chebytcheff's inequality and (2.5.505) we conclude that for every $\mu \in \mathbb{N}$ we have

$$\sigma\Big(\{x \in \partial\Omega : D_f(x) > \lambda\}\Big) \leq \Big(\frac{C(\Omega, b, p, n)}{\mu \cdot \lambda}\Big)^p \|f\|^p_{L^p(\partial\Omega,\sigma)}, \quad \forall \lambda > 0. \quad (2.5.511)$$

Similarly, corresponding to $p = 1$, from (2.5.510) and (2.5.507) we deduce that

$$\sigma\Big(\{x \in \partial\Omega : D_f(x) > \lambda\}\Big) \leq \frac{C(\Omega, b, n)}{\mu^2 \cdot \lambda} \|f\|_{L^1(\partial\Omega,\sigma)}, \qquad \forall \lambda > 0. \qquad (2.5.512)$$

Passing to limit $\mu \to \infty$ in (2.5.511)-(2.5.512) then shows that $D_f(x) = 0$ for σ-a.e. $x \in \partial\Omega$ and this shows that

the limit in (2.5.452) exists at σ-a.e. point on $\partial\Omega$
for any function $f \in L^p(\partial\Omega, \sigma)$ with $1 \leq p < \infty$. (2.5.513)

Let us now turn our attention to the jump-formula (2.5.460) in the case when $f \in L^p(\partial\Omega, \sigma)$ with $1 \leq p < \infty$. To set the stage we note that, as seen from (2.5.497), for each $\mu \in \mathbb{N}$ the difference

$$B - \sum_{\substack{\ell \in 2\mathbb{N}_0+1 \\ \ell \leq \mu}} \sum_{1 \leq i \leq H_\ell} a_{i\ell} B^{i\ell} \qquad (2.5.514)$$

is a principal-value integral operator on $\partial\Omega$ with kernel

$$\sum_{\substack{\ell\in 2\mathbb{N}_0+1\\ \ell>\mu}}\sum_{1\leq i\leq H_\ell} a_{i\ell}(x)k_{i\ell}(x-y), \qquad x,y\in\partial\Omega. \tag{2.5.515}$$

Arguing as in (2.5.505) and (2.5.507), based on (2.3.20)-(2.3.21), this permits us to conclude that

$$\lim_{\mu\to\infty}\Big\|B-\sum_{\substack{\ell\in 2\mathbb{N}_0+1\\ \ell\leq\mu}}\sum_{1\leq i\leq H_\ell} a_{i\ell}B^{i\ell}\Big\|_{L^p(\partial\Omega,\sigma)\to L^p(\partial\Omega,\sigma)}=0 \text{ if } 1<p<\infty \tag{2.5.516}$$

and, corresponding to $p=1$,

$$\lim_{\mu\to\infty}\Big\|B-\sum_{\substack{\ell\in 2\mathbb{N}_0+1\\ \ell\leq\mu}}\sum_{1\leq i\leq H_\ell} a_{i\ell}B^{i\ell}\Big\|_{L^1(\partial\Omega,\sigma)\to L^{1,\infty}(\partial\Omega,\sigma)}=0. \tag{2.5.517}$$

Hence,

$$B=\sum_{\ell\in 2\mathbb{N}_0+1}\sum_{1\leq i\leq H_\ell} a_{i\ell}B^{i\ell} \tag{2.5.518}$$

with convergence in $\mathcal{B}\Big(L^p(\partial\Omega,\sigma)\to L^p(\partial\Omega,\sigma)\Big)$ if $1<p<\infty$ and, corresponding to $p=1$, in $\mathcal{B}\Big(L^1(\partial\Omega,\sigma)\to L^{1,\infty}(\partial\Omega,\sigma)\Big)$.

To proceed, given any function $f\in L^1\Big(\partial\Omega,\frac{\sigma(x)}{1+|x|^{n-1}}\Big)$, for each $\ell\in\mathbb{N}_0$ and $1\leq i\leq H_\ell$ we now introduce

$$(\mathcal{B}_{i\ell}f)(x):=\int_{\partial\Omega}k_{i\ell}(x-y)f(y)\,d\sigma(y), \qquad \forall\,x\in\Omega. \tag{2.5.519}$$

Having fixed an arbitrary threshold $\mu\in\mathbb{N}$ then split

$$\begin{aligned}\mathcal{B}f(z)&=\sum_{\substack{\ell\in 2\mathbb{N}_0+1\\ \ell\leq\mu}}\sum_{1\leq i\leq H_\ell} a_{i\ell}(z)\mathcal{B}_{i\ell}f(z)+\sum_{\substack{\ell\in 2\mathbb{N}_0+1\\ \ell>\mu}}\sum_{1\leq i\leq H_\ell} a_{i\ell}(z)\mathcal{B}_{i\ell}f(z)\\ &=:I_\mu(z)+II_\mu(z), \qquad \forall\,z\in\Omega.\end{aligned} \tag{2.5.520}$$

Then whenever $f\in L^p(\partial\Omega,\sigma)$ with $1\leq p<\infty$, the jump-formula (2.5.377) applies (given (2.5.496)) and yields

$$\lim_{\substack{z\in\Gamma_\kappa(x)\\ z\to x}} I_\mu(z)=\sum_{\substack{\ell\in 2\mathbb{N}_0+1\\ \ell\leq\mu}}\sum_{1\leq i\leq H_\ell} a_{i\ell}(z)\Big\{\tfrac{1}{2\sqrt{-1}}\widehat{k_{i\ell}}\big(\nu(x)\big)f(x)+B^{i\ell}f(x)\Big\} \tag{2.5.521}$$

at σ-a.e. $x\in\partial_*\Omega$. From this, (2.5.499), and (2.5.516)-(2.5.517) we then conclude that there exists a strictly increasing sequence $\{\mu_j\}_{j\in\mathbb{N}}\subseteq\mathbb{N}$ with the property that

$$\lim_{\substack{z\in\Gamma_\kappa(x)\\ z\to x}} I_{\mu_j}(z) \text{ converges at } \sigma\text{-a.e. } x\in\partial_*\Omega \text{ to} \\ \frac{1}{2\sqrt{-1}}\widehat{b}\big(x,\nu(x)\big)f(x)+Bf(x) \text{ as } j\to\infty. \tag{2.5.522}$$

On the other hand, for each $x\in\partial\Omega$ we have

$$\sup_{z\in\Gamma_\kappa(x)}\big|II_\mu(z)\big| \leq \sum_{\substack{\ell\in 2\mathbb{N}_0+1\\ \ell>\mu}}\sum_{1\leq i\leq H_\ell}\sup_{x\in\mathbb{R}^n}|a_{i\ell}(x)|\,\mathcal{N}_\kappa\big(\mathcal{B}_{i\ell}f\big)(x) \tag{2.5.523}$$

which, in concert with Theorem 2.4.1, (2.5.494), and (2.5.498), proves that

$$\lim_{\mu\to\infty}\Big\|\sup_{z\in\Gamma_\kappa(x)}\big|II_\mu(z)\big|\Big\|_{L^p(\partial\Omega,\sigma)}=0. \tag{2.5.524}$$

Hence, after possibly refining our earlier strictly increasing sequence $\{\mu_j\}_{j\in\mathbb{N}}\subseteq\mathbb{N}$, matters may be arranged so that

$$\lim_{\substack{z\in\Gamma_\kappa(x)\\ z\to x}} II_{\mu_j}(z) \text{ converges to } 0 \text{ at } \sigma\text{-a.e. } x\in\partial\Omega \text{ as } j\to\infty. \tag{2.5.525}$$

Ultimately, from (2.5.520), (2.5.522), and (2.5.525), we conclude that

$$\begin{array}{l}\text{the jump-formula (2.5.460) holds whenever the func-}\\ \text{tion } f \text{ belongs to } L^p(\partial\Omega,\sigma) \text{ for some } 1\leq p<\infty.\end{array} \tag{2.5.526}$$

We shall now show that both the σ-a.e. existence of the limit in (2.5.452) and the jump-formula (2.5.460) are actually valid for functions f belonging to the larger space $L^1\big(\partial\Omega,\frac{\sigma(x)}{1+|x|^{n-1}}\big)$ (which contains any $L^p(\partial\Omega,\sigma)$ with $1\leq p<\infty$). In preparation to dealing with the former claim, given any surface ball $\Delta\subseteq\partial\Omega$, for each truncation parameter $\varepsilon>0$ and each function $f\in L^1\big(\partial\Omega,\frac{\sigma(x)}{1+|x|^{n-1}}\big)$ define

$$\big(B_{\varepsilon,\Delta}f\big)(x):=\int_{\substack{y\in\partial\Omega\\ |x-y|>\varepsilon}} b(x,x-y)f(y)\,d\sigma(y),\quad \forall\,x\in\Delta. \tag{2.5.527}$$

As far as the σ-a.e. existence on $\partial\Omega$ of the limit in (2.5.452) is concerned, it suffices to show that whenever $f\in L^1\big(\partial\Omega,\frac{\sigma(x)}{1+|x|^{n-1}}\big)$ the limit

$$\lim_{\varepsilon\to 0^+}\big(B_{\varepsilon,\Delta}f\big)(x) \text{ exists for } \sigma\text{-a.e. } x\in\Delta. \tag{2.5.528}$$

With this goal in mind, pick an arbitrary $f\in L^1\big(\partial\Omega,\frac{\sigma(x)}{1+|x|^{n-1}}\big)$ and for each fixed $\varepsilon>0$ decompose

$$B_{\varepsilon,\Delta}f=B_{\varepsilon,\Delta}\big(\mathbf{1}_{2\Delta}\cdot f\big)+B_{\varepsilon,\Delta}\big(\mathbf{1}_{\partial\Omega\setminus 2\Delta}\cdot f\big). \tag{2.5.529}$$

Note that the first term in the right side of (2.5.529) is simply $B_\varepsilon\big(\mathbf{1}_{2\Delta}\cdot f\big)\Big|_{\Delta}$ and $\mathbf{1}_{2\Delta}\cdot f\in L^1(\partial\Omega,\sigma)$. In view of (2.5.513) we may then conclude that, on the one hand,

$$\lim_{\varepsilon\to 0^+} B_{\varepsilon,\Delta}\big(\mathbf{1}_{2\Delta}\cdot f\big)(x) \text{ exists for } \sigma\text{-a.e. } x\in\Delta. \tag{2.5.530}$$

On the other hand, for each $x\in\Delta$ the number $B_{\varepsilon,\Delta}\big(\mathbf{1}_{\partial\Omega\setminus 2\Delta}\cdot f\big)(x)$ is actually independent of $\varepsilon>0$ whenever this truncation parameter is smaller than the radius of Δ. In concert with (2.5.530) and (2.5.529), this establishes (2.5.528). Hence, the proof of the σ-a.e. existence of the limit in (2.5.452) for any $f\in L^1\big(\partial\Omega,\frac{\sigma(x)}{1+|x|^{n-1}}\big)$ is complete.

Let us now show that the jump-formula stated in (2.5.460) is in fact true for any given function $f\in L^1\big(\partial\Omega,\frac{\sigma(x)}{1+|x|^{n-1}}\big)$. To get started, fix a point $x_0\in\partial\Omega$ and pick an arbitrary number $r\in(0,\infty)$. Next, having selected some arbitrary function $f\in L^1\big(\partial\Omega,\frac{\sigma(x)}{1+|x|^{n-1}}\big)$, decompose

$$\begin{gathered} f=f_1+f_2 \text{ on } \partial\Omega, \text{ where} \\ f_1:=\mathbf{1}_{\partial\Omega\cap B(x_0,2r)}\cdot f \text{ and } f_2:=\mathbf{1}_{\partial\Omega\setminus B(x_0,2r)}\cdot f, \end{gathered} \tag{2.5.531}$$

then split

$$\mathcal{B}f=\mathcal{B}f_1+\mathcal{B}f_2 \text{ in } \Omega. \tag{2.5.532}$$

Fix an aperture parameter $\kappa>0$. Since the function $\mathcal{B}f_2$ has a continuous extension to $B(x_0,r)$, we trivially have that the nontangential trace $\mathcal{B}f_2\Big|^{\kappa-\mathrm{n.t.}}_{\partial\Omega}$ exists at every point in $\partial_{\mathrm{nta}}\Omega\cap B(x_0,r)$. More precisely,

$$\begin{aligned} \big(\mathcal{B}f_2\big)\Big|^{\kappa-\mathrm{n.t.}}_{\partial\Omega}(x) &= \int_{\partial\Omega} b(x,x-y)f_2(y)\,d\sigma(y) \\ &= \big(Bf_2\big)(x) \text{ for each } x\in\partial_{\mathrm{nta}}\Omega\cap B(x_0,r). \end{aligned} \tag{2.5.533}$$

In particular, from (2.5.533) and item *(iii)* in [112, Proposition 8.8.6] we conclude that, on the one hand,

$$\big(\mathcal{B}f_2\big)\Big|^{\kappa-\mathrm{n.t.}}_{\partial\Omega}(x) = \big(Bf_2\big)(x) \text{ at } \sigma\text{-a.e. } x\in\partial_*\Omega\cap B(x_0,r). \tag{2.5.534}$$

On the other hand, since $f_1\in L^1(\partial\Omega,\sigma)$, we may invoke (2.5.526) to obtain that, at σ-a.e. point $x\in\partial_*\Omega\cap B(x_0,r)$, we have

$$\big(\mathcal{B}f_1\big)\Big|^{\kappa-\mathrm{n.t.}}_{\partial\Omega}(x) = \frac{1}{2\sqrt{-1}}\,\widehat{b}\big(x,\nu(x)\big)f_1(x)+(Bf_1)(x). \tag{2.5.535}$$

In turn, from (2.5.534), (2.5.535), and (2.5.531) we conclude that

$$(\mathcal{B}f)\Big|_{\partial\Omega}^{\kappa-\text{n.t.}}(x) = \frac{1}{2\sqrt{-1}}\widehat{b}\big(x,\nu(x)\big)f(x) + (Bf)(x) \tag{2.5.536}$$

for σ-a.e. $x \in \partial_*\Omega \cap B(x_0, r)$. In view of the fact that $r > 0$ has been arbitrarily chosen, this ultimately proves that the jump-formula (2.5.460) is in fact true for any function $f \in L^1\big(\partial\Omega, \frac{\sigma(x)}{1+|x|^{n-1}}\big)$.

Save for the L^p-square function estimate (2.5.479), all remaining claims in the statement of the theorem pertaining to the operators (2.5.452), (2.5.456), (2.5.457), are either implicit in what we have done so far, or may be proved in a similar fashion, using Theorems 2.3.2, 2.4.1, 2.5.1, Corollary 2.5.4, and the expansion in spherical harmonics given in (2.5.497). Also, in the case of the operators (2.5.464), (2.5.465), (2.5.467), the same sort of analysis works with x replaced by y in the spherical harmonic expansion (2.5.495).

The claims from item *(f)*, when the operators in question act from the Hardy space $H^p(\partial\Omega,\sigma)$ with $p \in \big(\frac{n-1}{n}, 1\big]$, also make use of [113, Proposition 4.4.8]. In this regard, it is of relevance to note that starting from (2.5.492), in place of (2.5.493) we now conclude that for each $m \in \mathbb{N}$ with $2m \leq M$ we have

$$[\ell(n+\ell-2)]^m \|a_{i\ell}\|_{\mathscr{C}^r(\mathbb{R}^n)} \leq C_n \cdot \max_{|\alpha|\leq M} \sup_{z\in S^{n-1}} \big\|(\partial_z^\alpha b)(\cdot, z)\big\|_{\mathscr{C}^r(\mathbb{R}^n)}. \tag{2.5.537}$$

Thus, the size of the coefficients $a_{i\ell}$ measured on the Hölder space $\mathscr{C}^r(\mathbb{R}^n)$ in the variable x, in a uniform manner in the variable $z \in S^{n-1}$, is rapidly decreasing. Specifically, (2.5.537) implies that for each number $m \in \mathbb{N}$ with $2m \leq M$ there exists a constant $C_{n,m} \in (0,\infty)$ such that

$$\|a_{i\ell}\|_{\mathscr{C}^r(\mathbb{R}^n)} \leq C_{n,m} \cdot \max\{1,\ell\}^{-2m} \cdot \max_{|\alpha|\leq M} \sup_{z\in S^{n-1}} \big\|(\partial_z^\alpha b)(\cdot, z)\big\|_{\mathscr{C}^r(\mathbb{R}^n)} \quad \text{whenever } \ell \in \mathbb{N}_0 \text{ and } 1 \leq i \leq H_\ell. \tag{2.5.538}$$

In view of [113, (4.4.159)], this tells that each $M_{a_{i\ell}}$, the operator of multiplication by $a_{i\ell}$, has the property that for each number $m \in \mathbb{N}$ with $2m \leq M$ there exists a constant $C_{n,m} \in (0,\infty)$ such that

$$\begin{aligned}\big\|M_{a_{i\ell}}\big\|_{H^p(\partial\Omega,\sigma)\to H^p(\partial\Omega,\sigma)} &\leq C_{n,m} \cdot \max\{1,\ell\}^{-2m} \cdot \max_{|\alpha|\leq M} \sup_{z\in S^{n-1}} \big\|(\partial_z^\alpha b)(\cdot, z)\big\|_{\mathscr{C}^r(\mathbb{R}^n)} \\ &\quad \text{whenever } \ell \in \mathbb{N}_0 \text{ and } 1 \leq i \leq H_\ell.\end{aligned} \tag{2.5.539}$$

Granted this, the same type of argument based on the spherical harmonic expansion (2.5.497) with x replaced by y goes through and, on account of Theorems 2.3.2, 2.4.1, and Corollary 2.5.4, all claims in item *(f)* follow.

Moreover, the L^p-square function estimate (2.5.477) becomes a direct consequence of (2.4.34), keeping in mind that the coefficients $a_{i\ell}$ are rapidly decreasing (in the sense described in (2.5.494)) and the "plain" convolution kernels $k_{i\ell}$ satisfy the growth condition (2.5.498). Having proved (2.5.477), the claim made in (2.5.478) follows, when Ω is actually a bounded Lipschitz domain, from [113, (9.2.215)] (used with $\theta := 1/p$) plus (2.5.469) and [112, (8.6.51)].

In contrast to (2.5.477), when dealing with the L^p-square function estimate (2.5.479) via a similar approach, based on the expansion in spherical harmonics given in (2.5.497), there is one additional series to consider, in which the gradient ∇_x now falls on the coefficients $a_{i\ell}(x)$ rather than the kernels $k_{i\ell}(x-y)$. This leads to the issue of estimating

$$\sum_{\ell\in 2\mathbb{N}_0+1}\sum_{i=1}^{H_\ell}\int_\Omega\Big|\int_{\partial\Omega}(\nabla a_{i\ell})(x)k_{i\ell}(x-y)f(y)\,\mathrm{d}\sigma(y)\Big|^p\,\mathrm{dist}(x,\partial\Omega)^{p-1}\,\mathrm{d}x \tag{2.5.540}$$

$$=\sum_{\ell\in 2\mathbb{N}_0+1}\sum_{i=1}^{H_\ell}\int_\Omega|(\nabla a_{i\ell})(x)|^p\Big|\int_{\partial\Omega}k_{i\ell}(x-y)f(y)\,\mathrm{d}\sigma(y)\Big|^p\,\mathrm{dist}(x,\partial\Omega)^{p-1}\,\mathrm{d}x$$

by a finite multiple of $\int_{\partial\Omega}|f|^p\,\mathrm{d}\sigma$. In this regard, under the assumption that $\partial_x^\beta\partial_z^\alpha b(x,z)$ is continuous and bounded on $\mathbb{R}^n\times S^{n-1}$ whenever $\alpha,\beta\in\mathbb{N}_0^n$ satisfy $|\alpha|\leq M$ and $|\beta|\leq 1$, we may differentiate (2.5.490) to obtain

$$(\nabla a_{i\ell})(x)=\int_{S^{n-1}}(\nabla_x b)(x,\omega)\Psi_{i\ell}(\omega)\,\mathrm{d}\mathcal{H}^{n-1}(\omega)\ \text{ for all } x\in\mathbb{R}^n, \tag{2.5.541}$$

and then reason as in (2.5.493) to conclude that for each number $m\in\mathbb{N}$ with $2m\leq M$ we may find a constant $C_{n,m}\in(0,\infty)$ for which

$$\sup_{x\in\mathbb{R}^n}\big|(\nabla a_{i\ell})(x)\big|\leq C_{n,m}\cdot\Big(\sup_{\substack{|\alpha|\leq M\\|\beta|\leq 1}}\ \sup_{\substack{x\in\mathbb{R}^n\\ z\in S^{n-1}}}\big|(\partial_x^\beta\partial_z^\alpha b)(x,z)\big|\Big)\cdot\max\{1,\ell\}^{-2m}$$
$$\text{whenever } \ell\in\mathbb{N}_0 \text{ and } 1\leq i\leq H_\ell. \tag{2.5.542}$$

In view of the format of (2.5.540), this serves our current purpose well. To deal with the last boundary integral appearing in (2.5.540), let $p'\in(1,\infty)$ be such that $1/p+1/p'=1$. For any two distinct points $x,y\in\mathbb{R}^n$ if $r:=|x-y|/2$ then the homogeneity of $k_{i\ell}$ together with (2.5.498) permit us to estimate

$$\|k_{i\ell}(\cdot-y)\|_{L^p(B(x,r),\mathcal{L}^n)}\leq C_{n,p}\Big(\sup_{S^{n-1}}|k_{i\ell}|\Big)|x-y|^{1-n/p'}$$
$$\leq C_{n,p}\,\ell^d\cdot|x-y|^{1-n/p'}, \tag{2.5.543}$$

for all $\ell\in 2\mathbb{N}_0+1$ and $1\leq i\leq H_\ell$. In particular, if we make the additional assumption that Ω is bounded, it follows that [112, (8.7.67)] is satisfied, for any choice $\varepsilon\in(0,\frac{1}{p})$, by the function $b(x,y):=k_{i\ell}(x-y)$. As such, we may rely on [112, Lemma 8.7.11] which, in concert with (2.5.542), ultimately shows that in the current scenario the series (2.5.540) is dominated by $C\int_{\partial\Omega}|f|^p\,d\sigma$, with $C\in(0,\infty)$ independent of f. This finishes the proof of the L^p-square function estimate (2.5.479).

Subsequently, when Ω is actually a bounded Lipschitz domain, the claims made in relation to (2.5.480) follow from (2.5.479), [113, (9.2.215)] (used with $\theta := 1/p$), (2.5.458) (bearing in mind [112, (3.6.27)]), and [112, (8.6.51)]. This completes the proof of Theorem 2.5.38. □

At last, we are ready to present the proof of Theorem 2.5.1.

Proof of Theorem 2.5.1 By invoking the jump-formula in item *(c)* of Theorem 2.5.38 with

$$\begin{gathered} b(\cdot,\cdot):\mathbb{R}^n\times\big(\mathbb{R}^n\setminus\{0\}\big)\longrightarrow\mathbb{R} \text{ given by} \\ b(x,z):=k(z) \text{ for all } x\in\mathbb{R}^n \text{ and } z\in\mathbb{R}^n\setminus\{0\}, \end{gathered} \tag{2.5.544}$$

the desired result follows. □

We augment the results in this section with a proposition pertaining to the non-tangential behavior of integral operators with weakly singular kernels.

Proposition 2.5.39 *Assume $\Omega\subseteq\mathbb{R}^n$ (where $n\in\mathbb{N}$ with $n\geq 2$) is an open set with an upper Ahlfors regular boundary and abbreviate $\sigma:=\mathcal{H}^{n-1}\lfloor\partial\Omega$. Also, suppose $b(\cdot,\cdot)$ is a Borel measurable function on $\Omega\times\partial\Omega$ with the property that there exist some exponent $\alpha\in(0,n-1)$ and some constant $C\in(0,\infty)$ such that*

$$\begin{gathered} \text{for } \mathcal{L}^n\text{-a.e. point } x\in\Omega \text{ one has } |b(x,y)|\leq C\,|x-y|^{-(n-1-\alpha)} \\ \text{at } \sigma\text{-a.e. point } y \text{ belonging to } \partial\Omega. \end{gathered} \tag{2.5.545}$$

In this context, define the integral operator

$$\begin{gathered} \mathscr{B}f(x):=\int_{\partial\Omega}b(x,y)f(y)\,d\sigma(y) \text{ at } \mathcal{L}^n\text{-a.e. } x\in\Omega, \\ \text{for each function } f\in L^1\Big(\partial\Omega,\frac{\sigma(x)}{1+|x|^{n-1-\alpha}}\Big). \end{gathered} \tag{2.5.546}$$

Finally, denote by I_α the fractional integral operator of order α on $\partial\Omega$, acting on each non-negative σ-measurable function g on $\partial\Omega$ according to

$$I_\alpha g(z):=\int_{\partial\Omega}\frac{g(y)}{|z-y|^{n-1-\alpha}}\,d\sigma(y),\qquad \forall\,z\in\partial\Omega. \tag{2.5.547}$$

Then for each function $f\in L^1\Big(\partial\Omega,\frac{\sigma(x)}{1+|x|^{n-1-\alpha}}\Big)$ the integral in (2.5.546) is absolutely convergent and for each aperture parameter $\kappa>0$ there exists some constant $C=C(\Omega,n,\alpha,\kappa)\in(0,\infty)$ with the property that

$$\big(\mathcal{N}_\kappa(\mathscr{B}f)\big)(z)\leq C\,I_\alpha(|f|)(z),\qquad \forall\,z\in\partial\Omega. \tag{2.5.548}$$

Also, one has the continuous embedding

$$L^p(\partial\Omega,\sigma) \hookrightarrow L^1\Big(\partial\Omega, \frac{\sigma(x)}{1+|x|^{n-1-\alpha}}\Big) \text{ for each } p \in \big[1,(n-1)/\alpha\big), \tag{2.5.549}$$

and for each integrability exponent $p \in \big[1,(n-1)/\alpha\big)$ *and each aperture parameter* $\kappa > 0$ *there exists* $C \in (0,\infty)$ *such that for every function* $f \in L^p(\partial\Omega,\sigma)$ *one has*

$$\begin{gathered} \big\|\mathcal{N}_\kappa(\mathscr{B}f)\big\|_{L^{p_*}(\partial\Omega,\sigma)} \leq C\|f\|_{L^p(\partial\Omega,\sigma)} \\ \text{with } 1/p_* = 1/p - \alpha/(n-1) \text{ if } p > 1, \end{gathered} \tag{2.5.550}$$

and, corresponding to the end-point $p = 1$,

$$\big\|\mathcal{N}_\kappa(\mathscr{B}f)\big\|_{L^{(n-1)/(n-1-\alpha),\infty}(\partial\Omega,\sigma)} \leq C\|f\|_{L^1(\partial\Omega,\sigma)}. \tag{2.5.551}$$

In addition, for each $p \in \big[1,(n-1)/\alpha\big)$ *there exists* $C \in (0,\infty)$ *with the property that*

$$\Big\|\delta_{\partial\Omega}^{\frac{n-1}{p}-\alpha}\mathscr{B}f\Big\|_{L^\infty(\Omega,\mathcal{L}^n)} \leq C\|f\|_{L^p(\partial\Omega,\sigma)} \text{ for each } f \in L^p(\partial\Omega,\sigma). \tag{2.5.552}$$

Moreover, under the additional assumption that

$$\begin{gathered} \text{for } \sigma\text{-a.e. point } y \in \partial\Omega \text{ the function} \\ b(\cdot,y) \text{ extends continuously to } \overline{\Omega}\setminus\{y\} \end{gathered} \tag{2.5.553}$$

it follows that for each function $f \in L^1\big(\partial\Omega, \frac{\sigma(x)}{1+|x|^{n-1-\alpha}}\big)$ *one has*

$$\int_{\partial\Omega} |b(x,y)||f(y)|\,d\sigma(y) < +\infty \text{ at } \sigma\text{-a.e. } x \in \partial\Omega, \tag{2.5.554}$$

and for each aperture parameter $\kappa > 0$ *the nontangential boundary limit* $\mathscr{B}f\big|_{\partial\Omega}^{\kappa-\text{n.t.}}$ *exists at* σ*-a.e. point on* $A_\kappa(\partial\Omega)$*; in fact, for each aperture parameter* $\kappa > 0$,

$$\Big(\mathscr{B}f\big|_{\partial\Omega}^{\kappa-\text{n.t.}}\Big)(x) = \int_{\partial\Omega} b(x,y)f(y)\,d\sigma(y) \text{ at } \sigma\text{-a.e. point } x \in A_\kappa(\partial\Omega). \tag{2.5.555}$$

We wish to note that if $\partial\Omega$ is *bounded* then estimates in the spirit of (2.5.550)-(2.5.551) continue to hold for the range $p \in \big[(n-1)/\alpha,\infty\big)$. Specifically, for each $p \in \big((n-1)/\alpha,\infty\big)$ Hölder's inequality and the first estimate in [112, (7.2.5)] imply that there exists $C_p \in (0,\infty)$, which is now also allowed to depend on $\operatorname{diam}(\partial\Omega)$, such that

$$\big\|\mathscr{B}f\big\|_{L^\infty(\Omega,\mathcal{L}^n)} \leq C_p\|f\|_{L^p(\partial\Omega,\sigma)}, \qquad \forall f \in L^p(\partial\Omega,\sigma). \tag{2.5.556}$$

Also, corresponding to the critical value $p = (n-1)/\alpha$, from (2.5.551) and Hölder's inequality we obtain

$$\|\mathcal{N}_\kappa(\mathscr{B}f)\|_{L^q(\partial\Omega,\sigma)} \leq C_q\|f\|_{L^{(n-1)/\alpha}(\partial\Omega,\sigma)}, \quad \forall f \in L^{(n-1)/\alpha}(\partial\Omega,\sigma), \quad \forall q \in (1,\infty). \tag{2.5.557}$$

Proof of Proposition 2.5.39 That the integral in (2.5.546) is absolutely convergent for each $x \in \Omega$ is clear from (2.5.545). Also, given any $p \in \big[1,(n-1)/\alpha\big)$, if p' denotes its Hölder conjugate exponent then $\frac{n-1}{n-1-\alpha} < p' \leq \infty$ so for each function $f \in L^p(\partial\Omega,\sigma)$ we may estimate

$$\int_{\partial\Omega} \frac{|f(x)|}{1+|x|^{n-1-\alpha}}\,d\sigma(x) \leq C\|f\|_{L^p(\partial\Omega,\sigma)}\Big(\int_{\partial\Omega}\frac{d\sigma(x)}{1+|x|^{(n-1-\alpha)p'}}\Big)^{1/p'} < +\infty, \tag{2.5.558}$$

thanks to Hölder's inequality and [112, Lemma 7.2.1]. This takes care of (2.5.549).

To proceed, recall from [112, (8.1.8)] that

$$|z-y| < (2+\kappa)|x-y|, \quad \forall z \in \partial\Omega, \quad \forall x \in \Gamma_\kappa(z), \quad \forall y \in \partial\Omega. \tag{2.5.559}$$

Consequently, for each $f \in L^1\Big(\partial\Omega, \frac{\sigma(x)}{1+|x|^{n-1-\alpha}}\Big)$ we have

$$\begin{aligned}\big(\mathcal{N}_\kappa(\mathscr{B}f)\big)(z) = \|\mathscr{B}f\|_{L^\infty(\Gamma_\kappa(z),\mathcal{L}^n)} &\leq C\int_{\partial\Omega}\frac{|f(y)|}{|z-y|^{n-1-\alpha}}\,d\sigma(y)\\ &= C\,I_\alpha(|f|)(z), \qquad \forall z \in \partial\Omega,\end{aligned} \tag{2.5.560}$$

proving (2.5.548). With this in hand, the estimates claimed in (2.5.550)-(2.5.551) now follow with the help of the Fractional Integration Theorem (cf. [112, (7.8.7)-(7.8.9)]).

As regards (2.5.552), given $f \in L^p(\partial\Omega,\sigma)$ with $p \in \big[1,(n-1)/\alpha\big)$, for $\mathcal{L}^n$-a.e. point $x \in \Omega$ we may estimate

$$\begin{aligned}\big|(\mathscr{B}f)(x)\big| &\leq C\|f\|_{L^p(\partial\Omega,\sigma)}\Big(\int_{\partial\Omega}\frac{d\sigma(x)}{|x-y|^{(n-1-\alpha)p'}}\Big)^{1/p'}\\ &\leq C\|f\|_{L^p(\partial\Omega,\sigma)}\delta_{\partial\Omega}(x)^{\frac{n-1}{p'}-(n-1-\alpha)},\end{aligned} \tag{2.5.561}$$

thanks to (2.5.546), Hölder's inequality, (2.5.545), and [112, Corollary 8.7.12]. The estimate claimed in (2.5.552) now follows from (2.5.561) upon observing that $\frac{n-1}{p'} - (n-1-\alpha) = -\frac{n-1}{p} + \alpha$. Parenthetically, we wish to note that, under the additional assumptions that $\partial\Omega$ is lower Ahlfors regular and $p > 1$, the estimate in (2.5.552) may be alternatively deduced from (2.5.550) and [112, (8.6.62)].

Henceforth, work under the additional assumption made in (2.5.553). We claim that

$$\begin{aligned}&\text{for each } x \in \partial\Omega \text{ we have } |b(x,y)| \leq C\,|x-y|^{-(n-1-\alpha)}\\ &\text{at } \sigma\text{-a.e. point } y \text{ belonging to } \partial\Omega.\end{aligned} \tag{2.5.562}$$

To see that this is the case, denote by N the exceptional subset of Ω satisfying $\mathcal{L}^n(N) = 0$ and such that for each $x \in \Omega \setminus N$ we have $|b(x,y)| \leq C\,|x-y|^{-(n-1-\alpha)}$

at σ-a.e. point $y \in \partial\Omega$. Next, fix an arbitrary point $x \in \partial\Omega$, and observe that there exists a sequence $\{x_j\}_{j\in\mathbb{N}} \subseteq \Omega \setminus N$ which converges to x. For each $j \in \mathbb{N}$ use (2.5.562) to find a σ-measurable set $N_j \subseteq \partial\Omega$ such that $\sigma(N_j) = 0$ and $|b(x_j, y)| \leq C\,|x_j - y|^{-(n-1-\alpha)}$ for each $y \in \partial\Omega \setminus N_j$. The hypothesis listed in (2.5.553) also guarantees that there exists a σ-measurable set $N_0 \subseteq \partial\Omega$ such that $\sigma(N_0) = 0$ and for each $y \in \partial\Omega \setminus N_0$ the function $b(\cdot, y)$ has a continuous extension to $\overline{\Omega} \setminus \{y\}$. Consider now an arbitrary point

$$y \in \partial\Omega \setminus \Big\{\big(\cup_{j\in\mathbb{N}} N_j\big) \cup N_0 \cup \{x\}\Big\}. \tag{2.5.563}$$

Then $b(\cdot, y)$ has a continuous extension to $\overline{\Omega} \setminus \{y\}$ and $|b(x_j, y)| \leq C\,|x_j - y|^{-(n-1-\alpha)}$ for each $j \in \mathbb{N}$. Passing to limit $j \to \infty$ yields $|b(x, y)| \leq C\,|x - y|^{-(n-1-\alpha)}$, from which the conclusion in (2.5.562) follows given that that $\sigma\Big(\big(\cup_{j\in\mathbb{N}} N_j\big) \cup N_0\Big) = 0$.

Having proved (2.5.562), that (2.5.554) holds for any $f \in L^1\big(\partial\Omega, \frac{\sigma(x)}{1+|x|^{n-1-\alpha}}\big)$ becomes a consequence of [112, (7.8.5)]. Finally, consider the claim made in (2.5.555). To this end, fix some arbitrary $f \in L^1\big(\partial\Omega, \frac{\sigma(x)}{1+|x|^{n-1-\alpha}}\big)$. Then [112, (7.8.5)] implies that there exists some σ-measurable set $N \subseteq \partial\Omega$ satisfying

$$I_\alpha(|f|)(z) < \infty \text{ for every } z \in \partial\Omega \setminus N, \text{ and } \sigma(N) = 0. \tag{2.5.564}$$

For each $z \in \partial\Omega$ bring in the function

$$G_z : \partial\Omega \to [0, \infty], \quad G_z(y) := \frac{|f(y)|}{|z - y|^{n-1-\alpha}} \text{ for every } y \in \partial\Omega. \tag{2.5.565}$$

In particular, since $\int_{\partial\Omega} G_z(y)\,d\sigma(y) = I_\alpha(|f|)(z)$ for each $z \in \partial\Omega$, it follows from (2.5.564) that

$$G_z \in L^1(\partial\Omega, \sigma) \text{ for every } z \in \partial\Omega \setminus N. \tag{2.5.566}$$

Fix now $z \in A_\kappa(\partial\Omega) \setminus N$ and consider an arbitrary sequence $\{x_j\}_{j\in\mathbb{N}} \subset \Gamma_\kappa(z)$ which converges to z. For each $j \in \mathbb{N}$ introduce the function $F_j : \partial\Omega \to \mathbb{R}$ by setting $F_j(y) := b(x_j, y)f(y)$ for σ-a.e. $y \in \partial\Omega$. Finally, define $F : \partial\Omega \to \mathbb{R}$ by setting $F(y) := b(z, y)f(y)$ for σ-a.e. $y \in \partial\Omega$. The fact that for σ-a.e. $y \in \partial\Omega$ the function $b(\cdot, y)$ extends continuously to $\overline{\Omega} \setminus \{y\}$ then implies

$$\lim_{j\to\infty} F_j(y) = F(y) \text{ for } \sigma\text{-a.e. } y \in \partial\Omega. \tag{2.5.567}$$

In addition, from (2.5.545), (2.5.559) and definitions we deduce that for each $j \in \mathbb{N}$ we have

$$|F_j(y)| \leq C \cdot G_z(y) \text{ for } \sigma\text{-a.e. } y \in \partial\Omega. \tag{2.5.568}$$

In concert with Lebesgue's Dominated Convergence Theorem, (2.5.566), (2.5.567), (2.5.568) then prove that for every $z \in \partial\Omega \setminus N$ we have

$$F_j \in L^1(\partial\Omega, \sigma) \text{ for every } j \in \mathbb{N}, \quad F \in L^1(\partial\Omega, \sigma),$$
$$\text{and } \lim_{j\to\infty} \int_{\partial\Omega} F_j(y)\, d\sigma(y) = \int_{\partial\Omega} F(y)\, d\sigma(y). \tag{2.5.569}$$

Unraveling definitions this ultimately yields

$$\Big(\mathscr{B} f\Big|_{\partial\Omega}^{\kappa-\text{n.t.}}\Big)(z) = \int_{\partial\Omega} k(z, y) f(y)\, d\sigma(y) \text{ for every } z \in A_\kappa(\partial\Omega) \setminus N. \tag{2.5.570}$$

Upon recalling from (2.5.564) that $\sigma(N) = 0$, we conclude that (2.5.555) holds. □

A significant consequence of Proposition 2.5.39, pertaining to global integrability properties of functions expressed as integrals involving weakly singular kernels, is presented below.

Proposition 2.5.40 *Assume $\Omega \subseteq \mathbb{R}^n$, where $n \in \mathbb{N}$ with $n \geq 2$, is an arbitrary* UR *domain and abbreviate $\sigma := \mathcal{H}^{n-1}\lfloor\partial\Omega$. Also, for a sufficiently large number $N = N(n) \in \mathbb{N}$, consider a complex-valued function $b \in \mathscr{C}^N(\mathbb{R}^n \setminus \{0\})$ with the property that there exist $C \in (0, \infty)$ and $\alpha \in (0, n-1)$ such that*

$$|b(x)| \leq C|x|^{-(n-1-\alpha)} \textit{ for each } x \in \mathbb{R}^n \setminus \{0\} \tag{2.5.571}$$

and such that ∇b is odd and positive homogeneous of degree $1 - n$ in $\mathbb{R}^n \setminus \{0\}$. Fix a function

$$f \in L^p(\partial\Omega, \sigma) \textit{ with } p \in \big(1, (n-1)/\alpha\big). \tag{2.5.572}$$

Finally, define

$$u(x) := \int_{\partial\Omega} b(x - y) f(y)\, d\sigma(y) \textit{ for each } x \in \mathbb{R}^n \setminus \partial\Omega, \tag{2.5.573}$$

and regard u as a function defined $\mathcal{L}^n$-a.e. in $\mathbb{R}^n$.

Then there exists some constant $C \in (0, \infty)$ independent of f such that with

$$p_\alpha := \Big(\frac{1}{p} - \frac{\alpha}{n-1}\Big)^{-1} \in (1, \infty) \tag{2.5.574}$$

one has

$$u \in L^{np_\alpha/(n-1)}(\mathbb{R}^n, \mathcal{L}^n) \textit{ and } \|u\|_{L^{np_\alpha/(n-1)}(\mathbb{R}^n, \mathcal{L}^n)} \leq C\|f\|_{L^p(\partial\Omega,\sigma)}. \tag{2.5.575}$$

Also, for each $j \in \{1, \dots, n\}$, the distribution $\partial_j u \in \mathcal{D}'(\mathbb{R}^n)$ actually belongs the Lebesgue space $L^{np/(n-1)}(\mathbb{R}^n, \mathcal{L}^n)$ and

$$\sum_{j=1}^{n} \|\partial_j u\|_{L^{np/(n-1)}(\mathbb{R}^n, \mathcal{L}^n)} \leq C\|f\|_{L^p(\partial\Omega,\sigma)} \tag{2.5.576}$$

for some constant $C \in (0, \infty)$ independent of f. In particular,

$$u \in W^{1,\,np/(n-1)}_{\mathrm{loc}}(\mathbb{R}^n). \tag{2.5.577}$$

Proof Denote by $\nu = (\nu_1, \dots, \nu_n)$ the geometric measure theoretic outward unit normal to Ω. If we introduce $\Omega_+ := \Omega$ and $\Omega_- := \mathbb{R}^n \setminus \overline{\Omega}$ then [112, Lemma 5.10.9] ensures that Ω_- is also a UR domain, whose topological boundary coincides with that of Ω, whose geometric measure theoretic boundary agrees with that of Ω, and whose geometric measure theoretic outward unit normal is $-\nu$ at σ-a.e. point on $\partial\Omega$. To proceed, define $u_\pm := u\big|_{\Omega_\pm}$. Also, fix an aperture parameter $\kappa > 0$. From (2.5.550) we know that

$$\big\|\mathcal{N}_\kappa u_\pm\big\|_{L^{p_\alpha}(\partial\Omega,\sigma)} \leq C\|f\|_{L^p(\partial\Omega,\sigma)}. \tag{2.5.578}$$

In turn, from (2.5.578) and [112, (8.6.51) in Proposition 8.6.3] we conclude that, in a quantitative fashion,

$$u_\pm \in \begin{cases} L^{np_\alpha/(n-1)}(\Omega_\pm, \mathcal{L}^n) \text{ if } \Omega_\pm \text{ is bounded, or } \partial\Omega \text{ is unbounded,} \\ L^{np_\alpha/(n-1)}_{\mathrm{bdd}}(\Omega_\pm, \mathcal{L}^n) \text{ if } \Omega_\pm \text{ is an exterior domain.} \end{cases} \tag{2.5.579}$$

Also, it is apparent from (2.5.573) that

$$u_\pm(x) = O(|x|^{-(n-1-\alpha)}) \text{ as } |x| \to \infty \text{ if } \Omega_\pm \text{ is an exterior domain.} \tag{2.5.580}$$

At this stage, the claims in (2.5.575) become consequences of (2.5.578), (2.5.579), and (2.5.580).

Moving on, fix an arbitrary $j \in \{1, \dots, n\}$. Then Theorem 2.4.1 applies and gives that

$$\big\|\mathcal{N}_\kappa(\partial_j u_\pm)\big\|_{L^p(\partial\Omega,\sigma)} \leq C\|f\|_{L^p(\partial\Omega,\sigma)} \tag{2.5.581}$$

for some constant $C \in (0, \infty)$ independent of f. Much as before, as a consequence of (2.5.581), the decay property $(\partial_j u_\pm)(x) = O(|x|^{1-n})$ as $|x| \to \infty$ if $\Omega_\pm$ is an exterior domain, and [112, Proposition 8.6.3], we may conclude that the function defined as

$$w_j := \begin{cases} \partial_j u_+ \text{ in } \Omega_+, \\ \partial_j u_- \text{ in } \Omega_-, \end{cases} \tag{2.5.582}$$

belongs to the Lebesgue space $L^{np/(n-1)}(\mathbb{R}^n, \mathcal{L}^n)$ (again, in a quantitative fashion). We now claim that $\partial_j u \in \mathcal{D}'(\mathbb{R}^n)$ coincides with the distribution induced by the locally integrable function w_j in $\mathbb{R}^n$ (via integration against test functions). To see that this is the case, pick an arbitrary $\varphi \in \mathscr{C}^\infty_c(\mathbb{R}^n)$ and write

$$
\begin{aligned}
{}_{\mathcal{D}'(\mathbb{R}^n)}\big\langle \partial_j u, \varphi \big\rangle_{\mathcal{D}(\mathbb{R}^n)} &= -{}_{\mathcal{D}'(\mathbb{R}^n)}\big\langle u, \partial_j \varphi \big\rangle_{\mathcal{D}(\mathbb{R}^n)} = -\int_{\mathbb{R}^n} u\, \partial_j \varphi \, \mathrm{d}\mathcal{L}^n \\
&= -\int_{\Omega_+} u_+ \partial_j \varphi \, \mathrm{d}\mathcal{L}^n - \int_{\Omega_-} u_- \partial_j \varphi \, \mathrm{d}\mathcal{L}^n \\
&= \int_{\Omega_+} (\partial_j u_+) \varphi \, \mathrm{d}\mathcal{L}^n + \int_{\partial\Omega} \nu_j \big(u_+\big|^{\kappa-\mathrm{n.t.}}_{\partial\Omega} \big) \varphi \, \mathrm{d}\sigma \\
&\quad + \int_{\Omega_-} (\partial_j u_-) \varphi \, \mathrm{d}\mathcal{L}^n - \int_{\partial\Omega} \nu_j \big(u_-\big|^{\kappa-\mathrm{n.t.}}_{\partial\Omega} \big) \varphi \, \mathrm{d}\sigma \\
&= \int_{\mathbb{R}^n} w_j \varphi \, \mathrm{d}\mathcal{L}^n. \qquad (2.5.583)
\end{aligned}
$$

The fourth equality above is implied by the integration by parts formula recorded in [112, Theorem 1.7.1] (whose applicability in the present circumstances is guaranteed by Proposition 2.5.39 and [112, Proposition 8.8.4]), while the fifth equality above makes use of (2.5.582) and (2.5.555). From (2.5.583) we then conclude that $\partial_j u = w_j$ in $\mathcal{D}'(\mathbb{R}^n)$, from which all desired conclusions follow. □

Here is a version of Proposition 2.5.40 for UR domains with compact boundaries. This makes less demanding assumptions on the function b, albeit the given function f now has vanishing moment.

Proposition 2.5.41 *Assume $\Omega \subseteq \mathbb{R}^n$, where $n \in \mathbb{N}$ with $n \geq 2$, is a* UR *domain with compact boundary and abbreviate $\sigma := \mathcal{H}^{n-1}\lfloor\partial\Omega$. Also, for a sufficiently large number $N = N(n) \in \mathbb{N}$, consider a complex-valued function $b \in \mathscr{C}^N(\mathbb{R}^n \setminus \{0\})$ with the property that there exist $C \in (0, \infty)$ and $\alpha \in (0, n-1)$ such that*

$$
|b(x)| \leq C|x|^{-(n-1-\alpha)} \text{ for each } x \in B(0,1) \setminus \{0\} \qquad (2.5.584)
$$

and such that ∇b is odd and positive homogeneous of degree $1-n$ in $\mathbb{R}^n \setminus \{0\}$. Fix a function

$$
f \in L^p(\partial\Omega, \sigma) \text{ with } p \in (1, \infty) \text{ and } \int_{\partial\Omega} f \, \mathrm{d}\sigma = 0. \qquad (2.5.585)
$$

Finally, define

$$
u(x) := \int_{\partial\Omega} b(x-y) f(y) \, \mathrm{d}\sigma(y) \text{ for each } x \in \mathbb{R}^n \setminus \partial\Omega, \qquad (2.5.586)
$$

and regard u as a function defined $\mathcal{L}^n$-a.e. in $\mathbb{R}^n$.

Then there exists some constant $C \in (0, \infty)$ independent of f such that with

$$
p_\alpha := \Big(\frac{1}{p} - \frac{\alpha}{n-1} \Big)^{-1} \in (1, \infty) \qquad (2.5.587)
$$

one has, in a quantitative sense,

$$u \in L^{np_\alpha/(n-1)}(\mathbb{R}^n, \mathcal{L}^n) \text{ if } p \in \big(1, (n-1)/\alpha\big), \tag{2.5.588}$$

and

$$u \in \bigcap_{\frac{n}{n-1}<q<\infty} L^q(\mathbb{R}^n, \mathcal{L}^n) \text{ if } p \in \big[(n-1)/\alpha, \infty\big). \tag{2.5.589}$$

Moreover, for each $j \in \{1,\dots,n\}$, the distribution $\partial_j u \in \mathcal{D}'(\mathbb{R}^n)$ actually belongs any Lebesgue space $L^q(\mathbb{R}^n, \mathcal{L}^n)$ with $1 < q \leq np/(n-1)$.

Proof The argument proceeds along the lines of the proof of Proposition 2.5.40, so we will elaborate only on the main differences. To get started, there is no loss of generality in assuming that Ω_+ is bounded (and Ω_- is an exterior domain). Fix a cutoff function $\psi \in \mathscr{C}_c^\infty(\mathbb{R}^n)$ with $\psi \equiv 1$ near $\overline{\Omega_+}$ and define $b(x,y) := \psi(x)b(x-y)$ for each distinct points $x, y \in \mathbb{R}^n$. Next, for f as in (2.5.585), define

$$\mathscr{B}_\pm f(x) := \int_{\partial\Omega} b(x,y) f(y)\,\mathrm{d}\sigma(y) \text{ for each } x \in \Omega_\pm. \tag{2.5.590}$$

As in the past, set $u_\pm := u\big|_{\Omega_\pm}$ where u is as in (2.5.586). From definitions, we have $\mathscr{B}_\pm f = \psi u_\pm$ in $\Omega_\pm$. Since $\partial\Omega$ is compact and ψ is compactly supported, $b(x,y)$ satisfies (2.5.545) both relative to Ω_+ and Ω_-. As a consequence of (2.5.550) we then have

$$\big\|\mathcal{N}_\kappa(\psi u_\pm)\big\|_{L^{p_\alpha}(\partial\Omega,\sigma)} \leq C\|f\|_{L^p(\partial\Omega,\sigma)} \text{ if } 1 < p < (n-1)/\alpha. \tag{2.5.591}$$

In concert with [112, (8.6.51)], this eventually leads to the conclusion that

$$u_\pm \in L^{np_\alpha/(n-1)}_{\text{bdd}}(\Omega_\pm, \mathcal{L}^n) \text{ if } 1 < p < (n-1)/\alpha. \tag{2.5.592}$$

Upon taking into account the cancelation property of f and the fact that ∇b is positive homogeneous of degree $1-n$ in $\mathbb{R}^n \setminus \{0\}$, we conclude that

$$(\nabla^k u_-)(x) = O(|x|^{-(n-1+k)}) \text{ as } |x| \to \infty, \text{ for all } k \in \{0, 1, \cdots, N\}. \tag{2.5.593}$$

At this stage, (2.5.589) follows from (2.5.592) and (2.5.593) (used with $k := 0$). In the case when $p \in \big[(n-1)/\alpha, \infty\big)$, in place of (2.5.591) we now obtain

$$\big\|\mathcal{N}_\kappa(\psi u_\pm)\big\|_{L^q(\partial\Omega,\sigma)} \leq C_q\|f\|_{L^p(\partial\Omega,\sigma)} \text{ for each } q \in (1,\infty), \tag{2.5.594}$$

thanks to (2.5.556)-(2.5.557). By once again using [112, (8.6.51)] and (2.5.593), we thus arrive at the conclusion that (2.5.589) holds.

Moving on, the same argument as in the proof of Proposition 2.5.40 gives that the distributional derivatives $\partial_j u \in \mathcal{D}'(\mathbb{R}^n)$ with $j \in \{1,\dots,n\}$ belong to $L^{np/(n-1)}(\mathbb{R}^n, \mathcal{L}^n)$. Then the last claim in the statement follows from this and (2.5.593) used with $k := 1$. □

2.6 Singular Integrals on Morrey Spaces and Their Pre-Duals

This section contains our main results regarding singular integral operators on Morrey spaces, as well as their pre-duals, considered on uniformly rectifiable sets. We begin treating singular integral operators of a general nature in the theorem below, which we subsequently apply to the study of boundary layer potentials in [115, §3.3].

Theorem 2.6.1 *Suppose* $\Sigma \subseteq \mathbb{R}^n$ *(where* $n \in \mathbb{N}$ *with* $n \geq 2$*) is a* UR *set and abbreviate* $\sigma := \mathcal{H}^{n-1}\lfloor\Sigma$. *Consider a kernel* $k \in \mathscr{C}^N(\mathbb{R}^n \setminus \{0\})$, *with* $N = N(n) \in \mathbb{N}$ *sufficiently large, which is odd and positive homogeneous of degree* $1-n$ *(cf.* (2.3.3)*). Associated with* Σ *and* k, *define the principal-value singular integral operator* T *as in* (2.3.15) *as well as its boundary-to-domain version,* $\mathcal{T}$ *defined as in* (2.4.3)-(2.4.4)*. Finally, fix two integrability exponents* $p, q \in (1, \infty)$ *such that* $1/p + 1/q = 1$ *along with some parameter* $\lambda \in (0, n-1)$. *Then the following claims are true:*

(1) The restriction (cf. [113, (6.2.25)]*) of the principal-value singular integral operator* T *to the Morrey space* $M^{p,\lambda}(\Sigma, \sigma)$ *induces a well-defined, linear, and bounded mapping*

$$T : M^{p,\lambda}(\Sigma, \sigma) \longrightarrow M^{p,\lambda}(\Sigma, \sigma), \tag{2.6.1}$$

whose operator norm satisfies, for some $C = C(\Sigma, n, p, \lambda) \in (0, \infty)$,

$$\|T\|_{M^{p,\lambda}(\Sigma,\sigma)\to M^{p,\lambda}(\Sigma,\sigma)} \leq C\big\|k\big|_{S^{n-1}}\big\|_{\mathscr{C}^N(S^{n-1})}. \tag{2.6.2}$$

Also, the maximal operator $T_{\max}$ *(cf.* (2.3.5), (2.3.7)*) induces a well-defined, sub-linear, bounded, and continuous mapping in the context*

$$T_{\max} : M^{p,\lambda}(\Sigma, \sigma) \longrightarrow M^{p,\lambda}(\Sigma, \sigma), \tag{2.6.3}$$

and its norm may be estimated as in (2.6.2)*. Furthermore, for each given aperture parameter* $\kappa \in (0, \infty)$ *there exists some constant* $C = C(\Sigma, n, q, \lambda, \kappa) \in (0, \infty)$ *with the property that boundary-to-domain integral operator* $\mathcal{T}$ *satisfies*

$$\begin{gathered}\big\|\mathcal{N}_\kappa^{\Sigma^c}(\mathcal{T} f)\big\|_{M^{p,\lambda}(\Sigma,\sigma)} \leq C\big\|k\big|_{S^{n-1}}\big\|_{\mathscr{C}^N(S^{n-1})}\|f\|_{M^{p,\lambda}(\Sigma,\sigma)} \\ \text{for each function } f \text{ belonging to the Morrey space } M^{p,\lambda}(\Sigma, \sigma).\end{gathered} \tag{2.6.4}$$

Also, in view of work in [112, §8.5]*, an estimate analogous to* (2.6.4) *holds if the nontangential maximal operator is replaced by the tangential maximal operator (associated as in* [112, Definition 8.5.1] *with* $\Omega := \Sigma^c$ *and a sufficiently large power* M*; cf.* (A.0.85)*).*

In addition, there exists some $C \in (0, \infty)$, *depending only on* n, p, λ, k, *and the Ahlfors regularity constants of* Σ, *with the property that for each multi-index* $\alpha \in \mathbb{N}_0^n$ *with* $|\alpha| \leq N$ *one has*

$$\sup_{x\in\mathbb{R}^n\setminus\Sigma}\Big\{\operatorname{dist}(x,\Sigma)^{|\alpha|+\frac{n-1-\lambda}{p}}\big|\partial^\alpha(\mathcal{T}f)(x)\big|\Big\}\le C\big\|k|_{S^{n-1}}\big\|_{\mathscr{C}^N(S^{n-1})}\|f\|_{M^{p,\lambda}(\Sigma,\sigma)}$$
$$\textit{for each function } f\in M^{p,\lambda}(\Sigma,\sigma). \tag{2.6.5}$$

Finally, similar results are valid with the Morrey space $M^{p,\lambda}(\Sigma,\sigma)$ *replaced by its version* $\mathring{M}^{p,\lambda}(\Sigma,\sigma)$ *introduced in* (A.0.88).

(2) *More generally, let* $b(x,z)$ *be a function which is odd and positive homogeneous of degree* $1-n$ *in the variable* $z\in\mathbb{R}^n\setminus\{0\}$, *and such that* $\partial_z^\alpha b(x,z)$ *is continuous and bounded on* $\mathbb{R}^n\times S^{n-1}$ *for each multi-index* $\alpha\in\mathbb{N}_0^n$ *satisfying* $|\alpha|\le N$. *If one then defines the variable-coefficient kernel integral operators acting on functions* $f\in L^1\Big(\Sigma,\frac{\sigma(x)}{1+|x|^{n-1}}\Big)$ *according to*

$$(Bf)(x):=\lim_{\varepsilon\to0^+}\int_{\substack{y\in\Sigma\\|x-y|>\varepsilon}}b(x,x-y)f(y)\,\mathrm{d}\sigma(y)\ \textit{ for } \sigma\textit{-a.e. } x\in\Sigma, \tag{2.6.6}$$

$$\big(B_{\max}f\big)(x):=\sup_{\varepsilon>0}\Bigg|\int_{\substack{y\in\Sigma\\|x-y|>\varepsilon}}b(x,x-y)f(y)\,\mathrm{d}\sigma(y)\Bigg|\ \textit{ for each } x\in\Sigma, \tag{2.6.7}$$

and

$$(\mathcal{B}f)(x):=\int_\Sigma b(x,x-y)f(y)\,\mathrm{d}\sigma(y)\ \textit{ for each } x\in\mathbb{R}^n\setminus\Sigma, \tag{2.6.8}$$

it follows that there exists a constant $C=C(\Sigma,p,\lambda)\in(0,\infty)$ *such that for each function* $f\in M^{p,\lambda}(\Sigma,\sigma)$ *one has*

$$\max\Big\{\|Bf\|_{M^{p,\lambda}(\Sigma,\sigma)},\|B_{\max}f\|_{M^{p,\lambda}(\Sigma,\sigma)}\Big\} \tag{2.6.9}$$
$$\le C\cdot\sup_{\substack{(x,z)\in\mathbb{R}^n\times S^{n-1}\\|\alpha|\le N}}\big|(\partial_z^\alpha b)(x,z)\big|\|f\|_{M^{p,\lambda}(\Sigma,\sigma)}.$$

Furthermore, for any given aperture parameter κ *belonging to* $(0,\infty)$ *one can find some constant* $C=C(\Sigma,n,p,\lambda,\kappa)\in(0,\infty)$ *with the property that for each function* $f\in M^{p,\lambda}(\Sigma,\sigma)$ *one has*

$$\big\|\mathcal{N}_\kappa^{\Sigma^c}(\mathcal{B}f)\big\|_{M^{p,\lambda}(\Sigma,\sigma)}\le C\cdot\sup_{\substack{(x,z)\in\mathbb{R}^n\times S^{n-1}\\|\alpha|\le N}}\big|(\partial_z^\alpha b)(x,z)\big|\|f\|_{M^{p,\lambda}(\Sigma,\sigma)}. \tag{2.6.10}$$

Once again, thanks to work in [112, §8.5], *a similar estimate to* (2.6.10) *holds with the nontangential maximal operator replaced by the tangential maximal operator (associated as in* [112, Definition 8.5.1] *with* $\Omega:=\Sigma^c$ *and a sufficiently large power* M*; cf.* (A.0.85)).

Lastly, analogous results are true with the Morrey space $M^{p,\lambda}(\Sigma,\sigma)$ *replaced by its version* $\mathring{M}^{p,\lambda}(\Sigma,\sigma)$ *introduced in* (A.0.88).

(3) *With the function b as in item (2), similar results to those described item (2) are valid for the variable-coefficient kernel integral operators acting on functions* $f \in L^1\Big(\Sigma, \frac{\sigma(x)}{1+|x|^{n-1}}\Big)$ *according to*

$$(B^{\#} f)(x) := \lim_{\varepsilon\to 0^+} \int_{\substack{y\in\Sigma\\ |x-y|>\varepsilon}} b(y, x-y) f(y)\, d\sigma(y) \ \textit{for } \sigma\textit{-a.e. } x \in \Sigma, \tag{2.6.11}$$

$$\big(B^{\#}_{\max} f\big)(x) := \sup_{\varepsilon>0} \left| \int_{\substack{y\in\Sigma\\ |x-y|>\varepsilon}} b(y, x-y) f(y)\, d\sigma(y) \right| \ \textit{for each } x \in \Sigma, \tag{2.6.12}$$

and

$$(\mathscr{B}^{\#} f)(x) := \int_\Sigma b(y, x-y) f(y)\, d\sigma(y) \ \textit{for each } x \in \mathbb{R}^n \setminus \Sigma. \tag{2.6.13}$$

(4) *The principal-value singular integral operator T from* (2.3.18) *induces (in view of the embedding in* [113, (6.2.71)]) *a well-defined linear and bounded mapping*

$$T : \mathcal{B}^{q,\lambda}(\Sigma,\sigma) \longrightarrow \mathcal{B}^{q,\lambda}(\Sigma,\sigma), \tag{2.6.14}$$

and

$$\begin{array}{l}\textit{the real transpose of (2.6.14) is the opposite of the operator}\\ T \textit{ acting on Morrey spaces as in (2.6.1) (that is, } T^\top = -T\textit{),}\end{array} \tag{2.6.15}$$

while

$$\begin{array}{l}\textit{the real transpose of the operator } T \textit{ acting on } \mathring{M}^{p,\lambda}(\Sigma,\sigma) \textit{ is}\\ \textit{the opposite of the operator } T \textit{ from (2.6.14) (i.e., } T^\top = -T\textit{).}\end{array} \tag{2.6.16}$$

Also,

$$\begin{array}{l}\textit{if } f \in \mathring{M}^{p,\lambda}(\Sigma,\sigma) \textit{ and } g \in \mathcal{B}^{q,\lambda}(\Sigma,\sigma) \textit{ then } (Tf)g + f(Tg) \textit{ belongs to the}\\ \textit{homogeneous Hardy space } \dot{H}^1(\Sigma,\sigma) \textit{ (cf. [113, (4.2.12)]) and one has}\\ \big\|(Tf)g + f(Tg)\big\|_{H^1(\Sigma,\sigma)} \leq C\|f\|_{M^{p,\lambda}(\Sigma,\sigma)} \cdot \|g\|_{\mathcal{B}^{q,\lambda}(\Sigma,\sigma)}.\end{array} \tag{2.6.17}$$

Moreover, the operator norm of T in (2.6.14) *is* $\leq C\big\|k\big|_{S^{n-1}}\big\|_{\mathscr{C}^N(S^{n-1})}$ *where the constant* $C \in (0,\infty)$ *depends only on* Σ, n, q, λ. *Also, the maximal operator* $T_{\max}$ *(cf.* (2.3.5), (2.3.7)) *induces a well-defined, sub-linear, bounded, and continuous mapping in the context*

$$T_{\max} : \mathcal{B}^{q,\lambda}(\Sigma,\sigma) \longrightarrow \mathcal{B}^{q,\lambda}(\Sigma,\sigma) \tag{2.6.18}$$

and its norm may be estimated in a similar fashion as before.
In addition, if one also considers the boundary-to-domain integral operator $\mathcal{T}$ *with kernel k, as in* (2.4.3)-(2.4.4), *then for each given aperture parameter* $\kappa > 0$ *there exists some constant* $C = C(\Sigma, n, q, \lambda, \kappa) \in (0, \infty)$ *with the property that*

$$\left\|\mathcal{N}_\kappa^{\Sigma^c}(\mathcal{T} f)\right\|_{\mathcal{B}^{q,\lambda}(\Sigma,\sigma)} \leq C\left\|k\right|_{S^{n-1}}\left\|_{\mathscr{C}^N(S^{n-1})}\right\|f\|_{\mathcal{B}^{q,\lambda}(\Sigma,\sigma)} \quad \text{for each function } f \text{ belonging to the space } \mathcal{B}^{q,\lambda}(\Sigma,\sigma). \tag{2.6.19}$$

In light of work in [112, §8.5], *a similar estimate to* (2.6.19) *holds with the nontangential maximal operator replaced by the tangential maximal operator (associated as in* [112, Definition 8.5.1] *with* $\Omega := \Sigma^c$ *and a sufficiently large power M; cf.* (A.0.85)).
More generally, if the function b(x, z) is as described in item (2) above, then for each function $f \in \mathcal{B}^{q,\lambda}(\Sigma, \sigma)$ *the operators* (2.6.6)-(2.6.8) *presently satisfy*

$$\max\left\{\|Bf\|_{\mathcal{B}^{q,\lambda}(\Sigma,\sigma)}, \|B_{\max} f\|_{\mathcal{B}^{q,\lambda}(\Sigma,\sigma)}, \left\|\mathcal{N}_\kappa^{\Sigma^c}(\mathcal{B}f)\right\|_{\mathcal{B}^{q,\lambda}(\Sigma,\sigma)}\right\} \leq C \cdot \sup_{\substack{(x,z)\in\mathbb{R}^n\times S^{n-1}\\ |\alpha|\leq N}} \left|(\partial_z^\alpha b)(x,z)\right| \|f\|_{\mathcal{B}^{q,\lambda}(\Sigma,\sigma)}, \tag{2.6.20}$$

where the constant $C \in (0, \infty)$ *depends only on the ambient. Finally, an estimate similar to* (2.6.20) *is valid for the variable-coefficient kernel integral operators* (2.6.11)-(2.6.13).

(5) Fix $q \in (1, \infty)$ *along with* $\lambda \in (0, n-1)$ *and recall the* (q, λ)*-midway space* $\mathscr{M}^{q,\lambda}(\Sigma, \sigma)$ *introduced in* (A.0.89)-(A.0.90). *Also, recall that* $\mathscr{H}^{q,\lambda}(\Sigma, \sigma)$, *the pre-dual of a Morrey-Campanato space, has been introduced in* (A.0.55)-(A.0.56). *Finally, fix an aperture parameter* $\kappa > 0$. *Then there exists some constant* $C = C(\Sigma, k, p, \lambda, \kappa, n) \in [0, \infty)$ *with the property that*

$$\left\|\mathcal{N}_\kappa^{\Sigma^c}(\mathcal{T} f)\right\|_{\mathscr{M}^{q,\lambda}(\Sigma,\sigma)} \leq C\left\|k\right|_{S^{n-1}}\left\|_{\mathscr{C}^N(S^{n-1})}\right\|f\|_{\mathscr{H}^{q,\lambda}(\Sigma,\sigma)} \quad \text{for each function } f \text{ belonging to the space } \mathscr{H}^{q,\lambda}(\Sigma,\sigma). \tag{2.6.21}$$

Finally, a similar estimate holds for the variable-coefficient kernel integral operator (2.6.13).

The results pertaining to the integral operators with variable coefficient kernels considered in Theorem 2.6.1 apply to the Schwartz kernels of certain pseudo-differential operators. In particular, this yields results in the spirit of Theorem 2.8.2 on Morrey (and related) spaces on manifolds.

We also wish to note that, thanks to [112, (6.2.23)], given any open set $\Omega \subseteq \mathbb{R}^n$ whose topological boundary $\partial\Omega$ is a UR set the jump-formula (2.5.4) continues to hold for each function f belonging to the Morrey space $M^{p,\lambda}(\partial\Omega, \sigma)$ with $p \in (1, \infty)$ and $\lambda \in (0, n-1)$. More generally, the jump-formulas (2.5.460), (2.5.471) remain true for each $f \in M^{p,\lambda}(\partial\Omega, \sigma)$.

Proof of Theorem 2.6.1 The fact that the operator T is well defined, linear, and bounded in the context of (2.6.1) is a consequence of [113, Proposition 6.2.12] and (2.3.56) (also bearing in mind the monotonicity of the Muckenhoupt classes; cf. item *(5)* in [112, Lemma 7.7.1]). Alternatively, we could have used (2.3.56) in combination with the extrapolation result from [113, Corollary 5.2.3], whose applicability in the present setting with $\mathbb{X} := M^{p,\lambda}(\partial\Omega, \sigma)$ is ensured by [113, Proposition 6.2.17] together with [113, Corollaries 6.2.11, 6.2.13].

Next, the estimate in (2.6.2) is then implied by [113, (6.2.121)-(6.2.122)] and (2.3.57). In fact, thanks to (2.3.58), the same argument works in the case of $T_{\max}$ in (2.6.3). As another justification for (2.6.3) we may rely on Cotlar's inequality from (2.3.23) (written for $r := 1$), (2.6.1), [113, (6.2.3), (6.2.25)], [113, Corollary 6.2.13], and (2.3.7).

Likewise, the estimate in (2.6.4) is a consequence of [113, Proposition 6.2.12] and (2.4.18). An alternative proof of (2.6.4) is obtained by relying on (2.4.5), [113, (6.2.25)], (2.6.3), [113, Corollary 6.2.13], [113, (6.2.3)], and [112, (8.2.28)]. Yet another justification is seen from (2.4.18), [113, Corollary 5.2.3], [113, Proposition 6.2.17], and [113, Corollaries 6.2.11, 6.2.13].

To justify (2.6.5), fix some $\alpha \in \mathbb{N}_0^n$ with $|\alpha| \leq N$ and note that

$$\begin{gathered}|(\partial^\alpha k)(x-y)| \leq \big\|k\big|_{S^{n-1}}\big\|_{\mathscr{C}^N(S^{n-1})} \cdot |x-y|^{-(n-1+|\alpha|)}\\ \text{for all points } x \in \mathbb{R}^n \setminus \Sigma \text{ and } y \in \Sigma.\end{gathered} \tag{2.6.22}$$

Consequently, given a function $f \in M^{p,\lambda}(\Sigma, \sigma)$, for any $x \in \mathbb{R}^n \setminus \Sigma$ we may estimate

$$\begin{aligned}\big|\partial^\alpha(\mathcal{T}f)(x)\big| &\leq \int_\Sigma |(\partial^\alpha k)(x-y)||f(y)|\,d\sigma(y)\\ &\leq \big\|k\big|_{S^{n-1}}\big\|_{\mathscr{C}^N(S^{n-1})} \int_\Sigma \frac{|f(y)|}{|x-y|^{n-1+|\alpha|}}\,d\sigma(y).\end{aligned} \tag{2.6.23}$$

To proceed, pick $x_* \in \Sigma$ such that $|x - x_*| = \operatorname{dist}(x, \Sigma) := \delta_\Sigma(x)$, and write

$$
\begin{aligned}
\int_{\Sigma} \frac{|f(y)|}{|x-y|^{n-1+|\alpha|}}\,\mathrm{d}\sigma(y) &= \int_{\Delta(x_*,2\delta_\Sigma(x))} \frac{|f(y)|}{|x-y|^{n-1+|\alpha|}}\,\mathrm{d}\sigma(y) \\
&\quad + \sum_{j=1}^{\infty} \int_{\Delta(x_*,2^{j+1}\delta_\Sigma(x))\setminus\Delta(x_*,2^j\delta_\Sigma(x))} \frac{|f(y)|}{|x-y|^{n-1+|\alpha|}}\,\mathrm{d}\sigma(y) \\
&\leq C\delta_\Sigma(x)^{-|\alpha|} \fint_{\Delta(x_*,2\delta_\Sigma(x))} |f(y)|\,\mathrm{d}\sigma(y) \\
&\quad + C\sum_{j=1}^{\infty} (2^j\delta_\Sigma(x))^{-|\alpha|} \fint_{\Delta(x_*,2^{j+1}\delta_\Sigma(x))} |f(y)|\,\mathrm{d}\sigma(y) \\
&\leq C\delta_\Sigma(x)^{-|\alpha|-\frac{n-1-\lambda}{p}} \cdot \delta_\Sigma(x)^{\frac{n-1-\lambda}{p}} \Big(\fint_{\Delta(x_*,2\delta_\Sigma(x))} |f|^p\,\mathrm{d}\sigma \Big)^{1/p} \\
&\quad + C\sum_{j=1}^{\infty} (2^j\delta_\Sigma(x))^{-|\alpha|-\frac{n-1-\lambda}{p}} \times \\
&\qquad \times (2^j\delta_\Sigma(x))^{\frac{n-1-\lambda}{p}} \Big(\fint_{\Delta(x_*,2^{j+1}\delta_\Sigma(x))} |f|^p\,\mathrm{d}\sigma \Big)^{1/p} \\
&\leq C\delta_\Sigma(x)^{-|\alpha|-\frac{n-1-\lambda}{p}} \Big\{ \sum_{j=0}^{\infty} (2^j)^{-|\alpha|-\frac{n-1-\lambda}{p}} \Big\} \|f\|_{M^{p,\lambda}(\Sigma,\sigma)} \\
&= C\delta_\Sigma(x)^{-|\alpha|-\frac{n-1-\lambda}{p}} \|f\|_{M^{p,\lambda}(\Sigma,\sigma)} \qquad (2.6.24)
\end{aligned}
$$

for some constant $C \in (0, \infty)$ independent of x and f, based on (2.6.22), the Ahlfors regularity of Σ, Hölder's inequality, the definition of the Morrey norm from [113, (6.2.2)], and bearing in mind that $|\alpha| + (n-1-\lambda)/p > 0$. With (2.6.24) in hand, the claim made in (2.6.5) readily follows.

The very last claim in item *(1)* is clear from what we have proved so far, the definition of $\mathring{M}^{p,\lambda}(\Sigma, \sigma)$ in (A.0.88), and the fact that the operators involved are continuous on $L^s(\Sigma, \sigma)$ with $s := \frac{p(n-1)}{n-1-\lambda} \in (1, \infty)$ (cf. Theorem 2.3.2 and Theorem 2.4.1).

The claims in item *(2)* about the integral operators (2.6.6)-(2.6.8), recorded in (2.6.9)-(2.6.10), may be justified based on what we have proved so far, making use of the same spherical harmonics expansion technique used in the proof of Theorem 2.5.38 and taking into account [113, (6.2.5)]. Alternatively, we may rely on the very last property in the statement of Theorem 2.5.38 together with [113, Proposition 6.2.12], as before. The very last claim in item *(2)* is handled as before. Next, the claims in item *(3)* are dealt with in a very analogous fashion.

On to item *(4)*, let us deal with the claims made in relation to the operator in (2.6.14). These follow by combining (2.3.56) with the extrapolation result from [113, Corollary 5.2.3] whose applicability in the present setting with $\mathbb{X} := \mathcal{B}^{q,\lambda}(\partial\Omega, \sigma)$

is guaranteed by [113, Proposition 6.2.17] together with [113, Corollaries 6.2.11, 6.2.13].

Another, more direct, proof of the claims made in relation to the operator in (2.6.14) is as follows. Fix an arbitrary $\mathcal{B}^{q,\lambda}$-block b on Σ. Then there exist some point $x_o \in \Sigma$ and some radius $R \in \big(0, 2\,\mathrm{diam}(\Sigma)\big)$ such that

$$\operatorname{supp} b \subseteq B(x_o, R) \cap \Sigma \text{ and } \|b\|_{L^q(\Sigma,\sigma)} \leq R^{\lambda\left(\frac{1}{q}-1\right)}. \tag{2.6.25}$$

We may then use (2.3.18) (with $p := q$) to estimate

$$\begin{aligned}\Big(\int_\Sigma |(Tb)(x)|^q \, d\sigma(x)\Big)^{1/q} &\leq C\big\|k\big|_{S^{n-1}}\big\|_{\mathscr{C}^N(S^{n-1})}\|b\|_{L^q(\Sigma,\sigma)} \\ &\leq C\big\|k\big|_{S^{n-1}}\big\|_{\mathscr{C}^N(S^{n-1})} R^{\lambda\left(\frac{1}{q}-1\right)}.\end{aligned} \tag{2.6.26}$$

Next, if $x \in \Sigma$ is such that $|x - x_o| \geq 2R$ then based on (2.6.25), the homogeneity of the kernel, and the Ahlfors regularity of Σ we may estimate

$$\begin{aligned}|(Tb)(x)| &= \Big|\int_{B(x_o,R)\cap\Sigma} k(x-y)b(y)\, d\sigma(y)\Big| \\ &\leq C\big(\sup_{S^{n-1}}|k|\big)|x-x_o|^{-(n-1)}\int_{B(x_o,R)\cap\Sigma}|b(y)|\, d\sigma(y) \\ &\leq C\big(\sup_{S^{n-1}}|k|\big)|x-x_o|^{-(n-1)}R^{n-1}\fint_{B(x_o,R)\cap\Sigma}|b|\, d\sigma \\ &\leq C\big(\sup_{S^{n-1}}|k|\big)|x-x_o|^{-(n-1)}R^{n-1}\Big(\fint_{B(x_o,R)\cap\Sigma}|b|^q\, d\sigma\Big)^{1/q} \\ &\leq C\big(\sup_{S^{n-1}}|k|\big)|x-x_o|^{-(n-1)}R^{n-1}R^{-\frac{n-1}{q}}\|b\|_{L^q(\Sigma,\sigma)} \\ &\leq C\big(\sup_{S^{n-1}}|k|\big)|x-x_o|^{-(n-1)}R^{n-1}R^{-\frac{n-1}{q}}R^{\lambda\left(\frac{1}{q}-1\right)} \\ &\leq C\big(\sup_{S^{n-1}}|k|\big)\frac{R^{-(n-1-\lambda)\left(\frac{1}{q}-1\right)}}{|x-x_o|^{n-1}}.\end{aligned} \tag{2.6.27}$$

Collectively, (2.6.26), (2.6.27), and [113, Lemma 6.2.10] (used with $\theta := 0$, which is a permissible value in [113, (6.2.108)]) imply that there exists some finite constant $C = C(\Sigma, n, q, \lambda) > 0$ with the property that

$$\begin{aligned}&Tb \in \mathcal{B}^{q,\lambda}(\Sigma,\sigma) \text{ and } \|Tb\|_{\mathcal{B}^{q,\lambda}(\Sigma,\sigma)} \leq C\big\|k\big|_{S^{n-1}}\big\|_{\mathscr{C}^N(S^{n-1})} \\ &\qquad \text{for each given } \mathcal{B}^{q,\lambda}\text{-block } b \text{ on } \Sigma.\end{aligned} \tag{2.6.28}$$

Suppose now that some function $f \in \mathcal{B}^{q,\lambda}(\Sigma, \sigma)$ has been given. Then there exist some numerical sequence $\{\lambda_j\}_{j\in\mathbb{N}} \in \ell^1(\mathbb{N})$ along with a sequence $\{b_j\}_{j\in\mathbb{N}}$ of $\mathcal{B}^{q,\lambda}$-blocks on Σ such that $f = \sum_{j=1}^{\infty} \lambda_j b_j$ in the sense of distributions on Σ. As noted in [113, (6.2.73)], the series $\sum_{j=1}^{\infty} \lambda_j b_j$ actually converges to f in $L^r(\Sigma, \sigma)$ with $r := \frac{q(n-1)}{n-1+\lambda(q-1)} \in (1, q)$. Also, (2.6.28) readily implies that the sequence $\{T\big(\sum_{j=1}^{J} \lambda_j b_j\big)\}_{J\in\mathbb{N}}$ is Cauchy, hence convergent, in the Banach space $\mathcal{B}^{q,\lambda}(\Sigma, \sigma)$. In light of [113, (6.2.71)], this latter convergence takes place in $L^r(\Sigma, \sigma)$ as well. However, since T is continuous on $L^r(\Sigma, \sigma)$, it follows that $\{T\big(\sum_{j=1}^{J} \lambda_j b_j\big)\}_{J\in\mathbb{N}}$ is, in fact, convergent to Tf in $L^r(\Sigma, \sigma)$. Bearing (2.6.28) and (A.0.22) in mind, this argument ultimately proves the following: given any $f \in \mathcal{B}^{q,\lambda}(\Sigma, \sigma)$, regarding f as a function in $L^r(\Sigma, \sigma)$ with $r := \frac{q(n-1)}{n-1+\lambda(q-1)} \in (1, q)$ and considering Tf in the principal-value sense (cf. (2.3.18)) yields a function which satisfies

$$\begin{aligned} &\text{the function } Tf \text{ belongs to } \mathcal{B}^{q,\lambda}(\Sigma, \sigma) \text{ and} \\ &\|Tf\|_{\mathcal{B}^{q,\lambda}(\Sigma,\sigma)} \leq C\big\|k\big|_{S^{n-1}}\big\|_{\mathscr{C}^N(S^{n-1})}\|f\|_{\mathcal{B}^{q,\lambda}(\Sigma,\sigma)}, \end{aligned} \tag{2.6.29}$$

for some constant $C = C(\Sigma, n, q, \lambda) \in (0, \infty)$ independent of f. Hence, the operator T in (2.6.14) is well defined, linear, continuous, and satisfied the desired operator norm bound.

To proceed, observe that for each function $f \in M^{p,\lambda}(\Sigma, \sigma)$ and each $\mathcal{B}^{q,\lambda}$-block b on Σ we may write

$$\int_{\Sigma} (Tf) b \, d\sigma = - \int_{\Sigma} f(Tb) \, d\sigma, \tag{2.6.30}$$

thanks to (2.3.26) and [113, (6.2.25)]. On account of [113, Proposition 6.2.8], (A.0.21), [113, (6.2.67)], the fact that the operator in (2.6.1) is well defined and the operator in (2.6.14) is continuous, this further implies that

$$\begin{aligned} &\int_{\Sigma} (Tf) g \, d\sigma = - \int_{\Sigma} f(Tg) \, d\sigma \text{ for each} \\ &\text{functions } f \in M^{p,\lambda}(\Sigma, \sigma) \text{ and } g \in \mathcal{B}^{q,\lambda}(\Sigma, \sigma). \end{aligned} \tag{2.6.31}$$

We then conclude from (2.6.31) and [113, Proposition 6.2.8] that the real transpose of the operator T in (2.6.14) is the opposite of the operator T in (2.6.1). The claim in (2.6.15) is therefore established. In light of [113, (6.2.15)], we also deduce from (2.6.31) that

$$\begin{aligned} &\int_{\Sigma} (Tf) g \, d\sigma = - \int_{\Sigma} f(Tg) \, d\sigma \text{ for each} \\ &\text{functions } f \in \mathring{M}^{p,\lambda}(\Sigma, \sigma) \text{ and } g \in \mathcal{B}^{q,\lambda}(\Sigma, \sigma). \end{aligned} \tag{2.6.32}$$

In concert with [113, Proposition 6.2.16], this proves (2.6.16). Having established this, (2.6.17) follows from (2.6.16) and the abstract (commutator type) estimate (2.7.72), established independently of the current considerations.

There is an alternative approach to the transposition formulas just proved, which is more resourceful. Specifically, recall from [113, (6.2.203)] that

$$\Big(M^{p,\lambda}(\Sigma,\sigma)\Big)' = \mathcal{B}^{q,\lambda}(\Sigma,\sigma) \text{ and } \Big(\mathcal{B}^{q,\lambda}(\Sigma,\sigma)\Big)' = M^{p,\lambda}(\Sigma,\sigma), \tag{2.6.33}$$

and note from [113, (5.2.100), (6.2.20), (6.2.70)] that

$$\begin{aligned} \Big(M^{p,\lambda}(\Sigma,\sigma)\Big)^{\circ} &:= \overline{L^{\infty}_{\rm comp}(\Sigma,\sigma)}^{\|\cdot\|_{M^{p,\lambda}(\Sigma,\sigma)}} \text{ coincides with } \mathring{M}^{p,\lambda}(\Sigma,\sigma), \\ \Big(\mathcal{B}^{q,\lambda}(\Sigma,\sigma)\Big)^{\circ} &:= \overline{L^{\infty}_{\rm comp}(\Sigma,\sigma)}^{\|\cdot\|_{\mathcal{B}^{q,\lambda}(\Sigma,\sigma)}} \text{ coincides with } \mathcal{B}^{q,\lambda}(\Sigma,\sigma). \end{aligned} \tag{2.6.34}$$

Then (2.6.15) and (2.6.16) follow from (2.6.33)-(2.6.34) and the general transposition formula obtained in (2.3.69).

Next, Cotlar's inequality recorded in (2.3.23) (presently employed with $r := 1$) shows that there exists a constant $C \in (0,\infty)$ with the property that for each function $g \in L^1\Big(\Sigma, \frac{\sigma(x)}{1+|x|^{n-1}}\Big)$ we have

$$(T_{\max} g)(x) \leq C\mathcal{M}_{\Sigma}(Tg)(x) + C\big(\mathcal{M}_{\Sigma} g\big)(x), \qquad \forall\, x \in \Sigma, \tag{2.6.35}$$

where $\mathcal{M}_{\Sigma}$ is the Hardy-Littlewood maximal operator on Σ. From this, [112, (6.2.72)], (2.6.14), [113, Corollary 6.2.11], [113, (6.2.75)], and (2.3.7) we conclude that $T_{\max}$ in (2.6.18) is a well-defined, sub-linear, bounded, and continuous mapping.

As regards the nontangential maximal function estimate in (2.6.19), one route is to rely on (2.4.5), [112, (6.2.72)], (2.6.18), [113, Corollary 6.2.11], [113, (6.2.75)], and [112, (8.2.28)]. Another route is to combine (2.4.18), [113, Corollary 5.2.3], [113, Proposition 6.2.17], and [113, Corollaries 6.2.11, 6.2.13].

Yest another argument for (2.6.19) is as follows. Fix some aperture parameter $\kappa > 0$ and pick an arbitrary $\mathcal{B}^{q,\lambda}$-block b on Σ. If $x_o \in \Sigma$ and $R \in \big(0, 2\,{\rm diam}(\Sigma)\big)$ are such that the properties in (2.6.25) hold, we may employ (2.4.9) (with $f := a$ and $p := q$) to write

$$\begin{aligned} \Big(\int_{\Sigma} \big|\mathcal{N}_{\kappa}^{\Sigma^c}(\mathcal{T}b)(x)\big|^q\, d\sigma(x)\Big)^{1/q} &\leq C\big\|k\big|_{S^{n-1}}\big\|_{\mathscr{C}^N(S^{n-1})}\|b\|_{L^q(\Sigma,\sigma)} \\ &\leq C\big\|k\big|_{S^{n-1}}\big\|_{\mathscr{C}^N(S^{n-1})} R^{\lambda\left(\frac{1}{q}-1\right)}. \end{aligned} \tag{2.6.36}$$

Also, whenever $z \in \Sigma^c$ is such that $|z-x_o| \geq 2R$, based on support and normalization properties of the atom, the homogeneity of the kernel, and the Ahlfors regularity of Σ we may estimate

$$
\begin{aligned}
|(\mathcal{T}b)(z)| &\le \int_{\Sigma\cap B(x_o,R)} |k(z-y)||b(y)|\,d\sigma(y)\\
&\le C\Big(\sup_{S^{n-1}}|k|\Big)|z-x_o|^{-(n-1)}\int_{\Sigma\cap B(x_o,R)}|b(y)|\,d\sigma(y)\\
&\le C\Big(\sup_{S^{n-1}}|k|\Big)|z-x_o|^{-(n-1)}R^{n-1}\fint_{\Sigma\cap B(x_o,R)}|b|\,d\sigma\\
&\le C\Big(\sup_{S^{n-1}}|k|\Big)|z-x_o|^{-(n-1)}R^{n-1}\Big(\fint_{\Sigma\cap B(x_o,R)}|b|^q\,d\sigma\Big)^{1/q}\\
&\le C\Big(\sup_{S^{n-1}}|k|\Big)|z-x_o|^{-(n-1)}R^{n-1}R^{-\frac{n-1}{q}}\|b\|_{L^q(\Sigma,\sigma)}\\
&\le C\Big(\sup_{S^{n-1}}|k|\Big)|z-x_o|^{-(n-1)}R^{n-1}R^{-\frac{n-1}{q}}R^{\lambda\left(\frac{1}{q}-1\right)}\\
&\le C\Big(\sup_{S^{n-1}}|k|\Big)\frac{R^{-(n-1-\lambda)\left(\frac{1}{q}-1\right)}}{|z-x_o|^{n-1}}.
\end{aligned}
\tag{2.6.37}
$$

Now, if $x\in\Sigma\setminus B\big(x_o,2(2+\kappa)R\big)$ and $z\in\Gamma_\kappa^{\Sigma^c}(x)$, then

$$
\begin{aligned}
2(2+\kappa)R &\le |x-x_o|\le|x-z|+|z-x_o|\le(1+\kappa)\operatorname{dist}(z,\Sigma)+|z-x_o|\\
&\le(1+\kappa)|z-x_o|+|z-x_o|=(2+\kappa)|z-x_o|.
\end{aligned}
\tag{2.6.38}
$$

In particular,

$$
|z-x_o|\ge 2R \text{ and } |z-x_o|\ge\frac{1}{2+\kappa}|x-x_o|,
\tag{2.6.39}
$$

so (2.6.37) gives, after taking the supremum over $z\in\Gamma_\kappa^{\Sigma^c}(x)$,

$$
\big|\mathcal{N}_\kappa^{\Sigma^c}(\mathcal{T}b)(x)\big|\le C\Big(\sup_{S^{n-1}}|k|\Big)\frac{R^{-(n-1-\lambda)\left(\frac{1}{q}-1\right)}}{|x-x_o|^{n-1}}
\quad\text{at each point } x\in\Sigma\setminus B\big(x_o,2(2+\kappa)R\big).
\tag{2.6.40}
$$

From (2.6.36), (2.6.40), and [113, Lemma 6.2.10] used with $\theta:=0$ (which is a permissible value in [113, (6.2.108)]) we obtain that there exists some $C=C(\Sigma,n,\kappa,q,\lambda)\in(0,\infty)$ for which

$$
\mathcal{N}_\kappa^{\Sigma^c}(\mathcal{T}b)\in\mathcal{B}^{q,\lambda}(\Sigma,\sigma) \text{ and } \|\mathcal{N}_\kappa^{\Sigma^c}(\mathcal{T}b)\|_{\mathcal{B}^{q,\lambda}(\Sigma,\sigma)}\le C\big\|k\big|_{S^{n-1}}\big\|_{\mathscr{C}^N(S^{n-1})}.
\quad\text{for each given } \mathcal{B}^{q,\lambda}\text{-block } b \text{ on } \Sigma.
\tag{2.6.41}
$$

The end-game in the (second) proof of (2.6.19) is as follows. Having fixed a function $f\in\mathcal{B}^{q,\lambda}(\Sigma,\sigma)$, there exist a numerical sequence $\{\lambda_j\}_{j\in\mathbb{N}}\in\ell^1(\mathbb{N})$ along with a sequence $\{b_j\}_{j\in\mathbb{N}}$ of $\mathcal{B}^{q,\lambda}$-blocks on Σ such that $f = \sum_{j=1}^{\infty}\lambda_j b_j$ in the sense of distributions on Σ. Thanks to [113, (6.2.73)], the series $\sum_{j=1}^{\infty}\lambda_j b_j$ actually converges to f in $L^r(\Sigma,\sigma)$ with $r := \frac{q(n-1)}{n-1+\lambda(q-1)} \in (1,q)$. Collectively, [112, (8.2.11)], [112, (8.2.9)], [113, (6.2.3)], and (2.6.41) also imply that the sequence $\Big\{\mathcal{N}_\kappa^{\Sigma^c}\Big(\mathcal{T}\big(\sum_{j=1}^{J}\lambda_j b_j\big)\Big)\Big\}_{J\in\mathbb{N}}$ is Cauchy, hence convergent, in the Banach space $\mathcal{B}^{q,\lambda}(\Sigma,\sigma)$. In light of [113, (6.2.71)], this latter convergence takes place in $L^r(\Sigma,\sigma)$ as well. However, thanks to (2.4.9) (with $p := r$), [112, (8.2.11)], and [113, (6.2.3)], it follows that $\Big\{\mathcal{N}_\kappa^{\Sigma^c}\Big(\mathcal{T}\big(\sum_{j=1}^{J}\lambda_j b_j\big)\Big)\Big\}_{J\in\mathbb{N}}$ actually converges to $\mathcal{N}_\kappa^{\Sigma^c}(\mathcal{T}f)$ in $L^r(\Sigma,\sigma)$. By also taking (2.6.41) into account, this argument ultimately gives that

$$
\begin{aligned}
&\mathcal{N}_\kappa^{\Sigma^c}(\mathcal{T}f) \in \mathcal{B}^{q,\lambda}(\Sigma,\sigma) \text{ and} \\
&\big\|\mathcal{N}_\kappa^{\Sigma^c}(\mathcal{T}f)\big\|_{\mathcal{B}^{q,\lambda}(\Sigma,\sigma)} \le C\big\|k|_{S^{n-1}}\big\|_{\mathscr{C}^N(S^{n-1})}\|f\|_{\mathcal{B}^{q,\lambda}(\Sigma,\sigma)}
\end{aligned}
\tag{2.6.42}
$$

for some constant $C = C(\Sigma, n, \kappa, q, \lambda) \in (0,\infty)$. This finishes the proof of (2.6.19).

Going further, the estimate recorded in (2.6.20) may be justified based on what we have proved so far, making use of the same spherical harmonics expansion technique used in the proof of Theorem 2.5.38 (as well as the boundedness results established in Theorem 2.5.38). Finally, on account of [113, (6.2.74)], the same type of argument may be used to justify the very last claim made in item *(4)*.

Let us now deal with item *(5)*. To justify the estimate claimed in (2.6.21), consider first the case when Σ is unbounded. Pick an arbitrary $\mathscr{H}^{q,\lambda}$-atom a on Σ. Then there exist some point $x_o \in \Sigma$ and some radius $R \in (0,\infty)$ such that

$$
\operatorname{supp} a \subseteq B(x_o,R)\cap\Sigma, \quad \|a\|_{L^q(\Sigma,\sigma)} \le R^{\lambda\left(\frac{1}{q}-1\right)}, \quad \int_\Sigma a\,\mathrm{d}\sigma = 0. \tag{2.6.43}
$$

In particular, $b := a$ satisfies the properties in (2.6.25), so according to (2.6.36) we have

$$
\Big(\int_\Sigma \big|\mathcal{N}_\kappa^{\Sigma^c}(\mathcal{T}a)\big|^q\,\mathrm{d}\sigma\Big)^{1/q} \le C\big\|k|_{S^{n-1}}\big\|_{\mathscr{C}^N(S^{n-1})}R^{\lambda\left(\frac{1}{q}-1\right)}. \tag{2.6.44}
$$

In addition, whenever $z \in \Sigma^c$ is such that $|z-x_o| \ge 2R$, based on support, cancelation, and normalization properties of the atom, the homogeneity of the kernel, the Mean Value Theorem, and the Ahlfors regularity of Σ we may estimate

$$
\begin{aligned}
|(\mathcal{T}a)(z)| &= \left|\int_{\Sigma\cap B(x_o,R)} \{k(z-y)-k(z-x_o)\}a(y)\,d\sigma(y)\right| \\
&\le \int_{\Sigma\cap B(x_o,R)} |k(z-y)-k(z-x_o)||a(y)|\,d\sigma(y) \\
&\le CR\big(\sup_{S^{n-1}}|\nabla k|\big)|z-x_o|^{-n}\int_{\Sigma\cap B(x_o,R)}|a(y)|\,d\sigma(y) \\
&\le CR^n\big(\sup_{S^{n-1}}|\nabla k|\big)|z-x_o|^{-n}⨍_{\Sigma\cap B(x_o,R)}|a|\,d\sigma \\
&\le CR^n\big(\sup_{S^{n-1}}|\nabla k|\big)|z-x_o|^{-n}\Big(⨍_{\Sigma\cap B(x_o,R)}|a|^q\,d\sigma(y)\Big)^{1/q} \\
&\le CR^n\big(\sup_{S^{n-1}}|\nabla k|\big)|z-x_o|^{-n}R^{-\frac{n-1}{q}}\|a\|_{L^q(\Sigma,\sigma)} \\
&\le CR^n\big(\sup_{S^{n-1}}|\nabla k|\big)|z-x_o|^{-n}R^{-\frac{n-1}{q}}R^{\lambda\left(\frac{1}{q}-1\right)} \\
&\le CR\big(\sup_{S^{n-1}}|\nabla k|\big)\frac{R^{-(n-1-\lambda)\left(\frac{1}{q}-1\right)}}{|z-x_o|^n}.
\end{aligned}
\tag{2.6.45}
$$

Next, if $x\in\Sigma\setminus B\big(x_o,2(2+\kappa)R\big)$ and $z\in\Gamma_\kappa^{\Sigma^c}(x)$ then (2.6.39) holds. Combining this with (2.6.45) and taking the supremum over $z\in\Gamma_\kappa^{\Sigma^c}(x)$ we arrive at the conclusion that

$$
\big|\mathcal{N}_\kappa^{\Sigma^c}(\mathcal{T}a)(x)\big|\le C\big(\sup_{S^{n-1}}|\nabla k|\big)\frac{R^{1-(n-1-\lambda)\left(\frac{1}{q}-1\right)}}{|x-x_o|^n}
\quad\text{at each point } x\in\Sigma\setminus B\big(x_o,2(2+\kappa)R\big).
\tag{2.6.46}
$$

In turn, from (2.6.44), (2.6.46), and [113, Definition 6.2.23] we conclude that $\mathcal{N}_\kappa^{\Sigma^c}(\mathcal{T}a)$ is a fixed multiple of what we call a (q,λ)-dull molecule concentrated near $\Delta(x_o,R):=\Sigma\cap B(x_o,R)$. In view of this and the nature of the constants involved, from [113, (6.2.251)] we see that there exists some constant $C=C(\Sigma,n,\kappa,q,\lambda,\lambda)\in(0,\infty)$ for which

$$
\begin{gathered}
\mathcal{N}_\kappa^{\Sigma^c}(\mathcal{T}a)\in\mathcal{M}^{q,\lambda}(\Sigma,\sigma)\ \text{ and} \\
\|\mathcal{N}_\kappa^{\Sigma^c}(\mathcal{T}a)\|_{\mathcal{M}^{q,\lambda}(\Sigma,\sigma)}\le C\big\|k\big|_{S^{n-1}}\big\|_{\mathscr{C}^N(S^{n-1})}, \\
\text{for each } \mathscr{H}^{q,\lambda}\text{-atom } a \text{ on } \Sigma.
\end{gathered}
\tag{2.6.47}
$$

When Σ is compact, the constant 1 is also considered to be an $\mathscr{H}^{q,\lambda}$-atom on Σ. That, in such a scenario, the properties listed in (2.6.47) are also valid for the function $a=1$ may be seen directly from (2.4.9), [113, Definition 6.2.23], and [113, (6.2.251)].

Having established (2.6.47) in all cases, the end-game in the proof of (2.6.21) is very similar to the argument that, starting with (2.6.41), has produced (2.6.42). Finally, the variable coefficient version of (2.6.21) may be established much as before.

We close with some comments on the particular case when Σ is compact. In such a scenario, we know from [113, (6.2.77)] that $\mathcal{B}^{q,\lambda}(\Sigma,\sigma) = \mathcal{H}^{q,\lambda}(\Sigma,\sigma)$. As such, the results in item *(4)* may be formulated in terms of the latter space. In fact, it is possible to give an alternative proof for the boundedness of T on $\mathcal{H}^{q,\lambda}(\Sigma,\sigma)$ when Σ is compact, based on the atomic description of this space given in (A.0.55). Specifically, having fixed a parameter $\theta \in \big[0, qn - n + 1\big)$ and with $a \in L^q(\Sigma,\sigma)$ an $\mathcal{H}^{q,\lambda}$-atom on Σ as in [113, (6.1.15)], we may use (2.3.18) (with $p := q$) to estimate

$$\begin{aligned}
\int_{B(x_o,2R)\cap\Sigma} |(Ta)(x)|^q |x-x_o|^\theta \, d\sigma(x) \\
&\leq CR^\theta \cdot \int_{B(x_o,2R)\cap\Sigma} |Ta|^q \, d\sigma \leq CR^\theta \|Ta\|^q_{L^q(\Sigma,\sigma)} \\
&\leq C\big\|k\big|_{S^{n-1}}\big\|^q_{\mathscr{C}^N(S^{n-1})} R^\theta \|a\|^q_{L^q(\Sigma,\sigma)} \\
&\leq C\big\|k\big|_{S^{n-1}}\big\|^q_{\mathscr{C}^N(S^{n-1})} R^{\lambda(1-q)+\theta}. && (2.6.48)
\end{aligned}$$

Next, if $x \in \Sigma$ is such that $|x - x_o| \geq 2R$ then based on [113, (6.1.15)], the Mean Value Theorem, and the Ahlfors regularity of Σ we may estimate

$$\begin{aligned}
|(Ta)(x)| &= \Big|\int_\Sigma k(x-y)a(y)\, d\sigma(y)\Big| \\
&\leq \int_{B(x_o,R)\cap\Sigma} |k(x-y) - k(x-x_o)||a(y)| \, d\sigma(y) \\
&\leq C\big\|k\big|_{S^{n-1}}\big\|_{\mathscr{C}^1(S^{n-1})} \frac{R}{|x-x_o|^n} \int_{B(x_o,R)\cap\Sigma} |a(y)| \, d\sigma(y) \\
&\leq C\big\|k\big|_{S^{n-1}}\big\|_{\mathscr{C}^1(S^{n-1})} \frac{R}{|x-x_o|^n} \|a\|_{L^q(\Sigma,\sigma)} \sigma\big(B(x_o,r)\cap\Sigma\big)^{1-1/q} \\
&\leq C\big\|k\big|_{S^{n-1}}\big\|_{\mathscr{C}^1(S^{n-1})} \frac{R^{1+(n-1-\lambda)(1-1/q)}}{|x-x_o|^n}. && (2.6.49)
\end{aligned}$$

With this in hand, [112, Lemma 7.2.1] applies since $qn - \theta > n - 1$ and gives

$$
\begin{aligned}
\int_{\Sigma\setminus B(x_o,2R)} & |(Ta)(x)|^q |x-x_o|^\theta \, d\sigma(x) \\
&\leq C\big\|k\big|_{S^{n-1}}\big\|^q_{\mathscr{C}^1(S^{n-1})} R^{1+(n-1-\lambda)(q-1)} \int_{\Sigma\setminus B(x_o,2R)} \frac{d\sigma(x)}{|x-x_o|^{qn-\theta}} \\
&\leq C\big\|k\big|_{S^{n-1}}\big\|^q_{\mathscr{C}^1(S^{n-1})} R^{1+(n-1-\lambda)(q-1)} \cdot R^{n-1-nq+\theta} \\
&= C\big\|k\big|_{S^{n-1}}\big\|^q_{\mathscr{C}^1(S^{n-1})} R^{\lambda(1-q)+\theta}.
\end{aligned} \tag{2.6.50}
$$

Collectively, (2.6.48) and (2.6.50) imply that whenever a is an in [113, (6.1.15)] we have

$$
\int_\Sigma |(Ta)(x)|^q |x-x_o|^\theta \, d\sigma(x) \leq C\big\|k\big|_{S^{n-1}}\big\|^q_{\mathscr{C}^N(S^{n-1})} R^{\lambda(1-q)+\theta}, \tag{2.6.51}
$$

for some finite constant $C = C(\Sigma, n, q, \lambda, \theta) > 0$ independent of a. Specializing (2.6.51) first to the case when $\theta := 0$ and then fixing $\theta \in \big((n-1)(q-1), qn-n+1\big)$ proves that the function Ta is a fixed multiple of some $\mathscr{H}^{q,\lambda,\theta}$-dome on Σ (cf. [113, (6.1.33)]). Since (2.3.18) (with $p := q$) implies that $T1 \in L^q(\Sigma, \sigma)$, we conclude from this and [113, (6.2.54)] that the same conclusion is valid when $a \equiv$ constant. With $\theta \in \big((n-1)(q-1), qn-n+1\big)$ fixed, the argument so far shows that

$$
\begin{gathered}
\text{for each } \mathscr{H}^{q,\lambda}\text{-atom } a \text{ on } \Sigma, \text{ the function } Ta \text{ is} \\
\text{a fixed multiple of a } \mathscr{H}^{q,\lambda,\theta}\text{-dome on } \Sigma.
\end{gathered} \tag{2.6.52}
$$

Ultimately, from (2.6.52) and [113, (6.2.59)] we conclude that there exists some $C = C(\Sigma, n, q, \lambda) \in (0, \infty)$ with the property that

$$
\begin{gathered}
Ta \in \mathscr{H}^{q,\lambda}(\Sigma,\sigma) \text{ and } \|Ta\|_{\mathscr{H}^{q,\lambda}(\Sigma,\sigma)} \leq C\big\|k\big|_{S^{n-1}}\big\|_{\mathscr{C}^N(S^{n-1})}, \\
\text{for each given } \mathscr{H}^{q,\lambda}\text{-atom } a \text{ on } \Sigma.
\end{gathered} \tag{2.6.53}
$$

Granted this, the same type of reasoning which starting with (2.6.28) has produced (2.6.29) now leads to the conclusion that T is a bounded endomorphism of $\mathscr{H}^{q,\lambda}(\Sigma,\sigma)$ when Σ is compact.

In fact, considerations of a similar nature (cf. also (2.4.79)-(2.4.87) for relevant estimates in fairly analogous circumstances) may be used to show that if Σ is compact then, having picked a parameter $\theta \in \big((n-1)(q-1), qn-n+1\big)$, we have that

$$
\begin{gathered}
\text{for each } \mathscr{H}^{q,\lambda}\text{-atom } a \text{ on } \Sigma, \text{ the function } \mathcal{N}_\kappa^{\Sigma^c}(\mathcal{T}a) \text{ is} \\
\text{a fixed multiple of some } \mathscr{H}^{q,\lambda,\theta}\text{-dome on } \Sigma.
\end{gathered} \tag{2.6.54}
$$

In turn, from (2.6.54) and [113, (6.2.59)] we deduce that there exists some constant $C = C(\Sigma, n, \kappa, q, \lambda) \in (0, \infty)$ for which

$$
\begin{gathered}
\mathcal{N}_\kappa^{\Sigma^c}(\mathcal{T}a) \in \mathscr{H}^{q,\lambda}(\Sigma,\sigma) \text{ and } \|\mathcal{N}_\kappa^{\Sigma^c}(\mathcal{T}a)\|_{\mathscr{H}^{q,\lambda}(\Sigma,\sigma)} \leq C\big\|k\big|_{S^{n-1}}\big\|_{\mathscr{C}^N(S^{n-1})}, \\
\text{for each given } \mathscr{H}^{q,\lambda}\text{-atom } a \text{ on } \Sigma.
\end{gathered} \tag{2.6.55}
$$

Ultimately, this proves that for each function $f \in \mathscr{H}^{q,\lambda}(\Sigma, \sigma)$ we have

$$\begin{aligned} &\mathcal{N}_\kappa^{\Sigma^c}(\mathcal{T} f) \in \mathscr{H}^{q,\lambda}(\Sigma, \sigma) \text{ and} \\ &\big\|\mathcal{N}_\kappa^{\Sigma^c}(\mathcal{T} f)\big\|_{\mathscr{H}^{q,\lambda}(\Sigma,\sigma)} \leq C\big\|k\big|_{S^{n-1}}\big\|_{\mathscr{C}^N(S^{n-1})}\|f\|_{\mathscr{H}^{q,\lambda}(\Sigma,\sigma)} \end{aligned} \tag{2.6.56}$$

for some constant $C = C(\Sigma, n, \kappa, q, \lambda) \in (0, \infty)$. This provides an alternative proof of (2.6.19) in the case when Σ is compact. □

For a multitude of reasons it is important to augment Theorem 2.6.1 with similar results for weakly singular integral operators. We do so in the next two propositions. First, we deal with boundary-to-boundary weakly singular integral operators, as indicated below.

Proposition 2.6.2 *Suppose $\Sigma \subseteq \mathbb{R}^n$ (where $n \in \mathbb{N}$ with $n \geq 2$) is a closed Ahlfors regular set and abbreviate $\sigma := \mathcal{H}^{n-1}\lfloor\Sigma$. Fix a parameter $\lambda \in (0, n-1)$ along with an exponent $\alpha \in (0, n-1-\lambda)$, and consider some $\sigma \otimes \sigma$-measurable function $\theta : \Sigma \times \Sigma \to \mathbb{C}$ with the property that there exists a constant $C \in (0, \infty)$ such that*

$$|\theta(x, y)| \leq C|x - y|^{-(n-1-\alpha)} \text{ for a.e. } x, y \in \Sigma. \tag{2.6.57}$$

Use this to define the integral operator

$$\begin{aligned} &\Theta : L^1\big(\Sigma, \tfrac{\sigma(x)}{1+|x|^{n-1-\alpha}}\big) \longrightarrow L^1_{\text{loc}}(\Sigma, \sigma) \\ &\text{acting on each } f \in L^1\big(\Sigma, \tfrac{\sigma(x)}{1+|x|^{n-1-\alpha}}\big) \text{ according to} \\ &(\Theta f)(x) := \int_\Sigma \theta(x, y) f(y)\, d\sigma(y) \text{ for } \sigma\text{-a.e. } x \in \Sigma. \end{aligned} \tag{2.6.58}$$

Finally, assume that

$$1 < p < \frac{n-1-\lambda}{\alpha} \text{ and } p_* := \Big(\frac{1}{p} - \frac{\alpha}{n-1-\lambda}\Big)^{-1} \tag{2.6.59}$$

and recall from [113, (6.2.140), (6.2.230)] *that there are well-defined, linear, and continuous embeddings*

$$M^{p,\lambda}(\Sigma, \sigma) \hookrightarrow L^1\Big(\Sigma, \frac{\sigma(x)}{1 + |x|^{n-1-\alpha}}\Big), \tag{2.6.60}$$

$$\mathcal{B}^{p,\lambda}(\Sigma, \sigma) \hookrightarrow L^1\Big(\Sigma, \frac{\sigma(x)}{1 + |x|^{n-1-\alpha}}\Big). \tag{2.6.61}$$

Then the operators induced via (2.6.60)-(2.6.61) *by* Θ *from* (2.6.58) *in the contexts*

$$\Theta : M^{p,\lambda}(\Sigma, \sigma) \longrightarrow M^{p_*,\lambda}(\Sigma, \sigma), \tag{2.6.62}$$

$$\Theta : \mathcal{B}^{p,\lambda}(\Sigma, \sigma) \longrightarrow \mathcal{B}^{p_*,\lambda}(\Sigma, \sigma), \tag{2.6.63}$$

$$\Theta : \mathring{M}^{p,\lambda}(\Sigma, \sigma) \longrightarrow \mathring{M}^{p_*,\lambda}(\Sigma, \sigma), \tag{2.6.64}$$

are well-defined, linear, and bounded.

Proof For starters, under the assumptions made in (2.6.59), the embedding recorded in [112, (6.2.23)] guarantees that (2.6.60) holds. Going further, let $I_{\Sigma,\alpha}$ denote the fractional integral operator of order α on Σ. Since there exists a constant $C \in (0, \infty)$ with the property that

$$\begin{gathered} \text{for each function } f \in L^1\big(\Sigma, \tfrac{\sigma(x)}{1+|x|^{n-1-\alpha}}\big) \text{ we have} \\ |(\Theta f)(x)| \leq C\, I_{\Sigma,\alpha}(|f|)(x) \text{ at } \sigma\text{-a.e. point } x \in \Sigma, \end{gathered} \tag{2.6.65}$$

the fact that the operator Θ in (2.6.62) is well defined, linear, and bounded is implied by [113, Proposition 6.2.14] and [113, (6.2.3)], while the fact that the operator Θ in (2.6.63) is a consequence of [113, Proposition 6.2.20] and [113, (6.2.75)].

As regards the claims made in relation to the operator Θ in (2.6.64), recall from (A.0.88) that $\mathring{M}^{p,\lambda}(\Sigma, \sigma)$ is the closure of $L^s(\Sigma, \sigma)$ with $s := \frac{p(n-1)}{n-1-\lambda}$ in $M^{p,\lambda}(\Sigma, \sigma)$, and that $\mathring{M}^{p_*,\lambda}(\Sigma, \sigma)$ is the closure of $L^{s_*}(\Sigma, \sigma)$ with $s_* := \frac{p_*(n-1)}{n-1-\lambda}$ in $M^{p_*,\lambda}(\Sigma, \sigma)$. In particular, from definitions and (2.6.59) we see that

$$s \in (1, n-1) \text{ and } \frac{1}{s_*} = \frac{1}{s} - \frac{1}{n-1}. \tag{2.6.66}$$

Granted this, we conclude from (2.6.65) and [112, (7.8.7)] that Θ maps $L^s(\partial\Omega, \sigma)$ linearly and boundedly into $L^{s_*}(\partial\Omega, \sigma)$. In concert with (2.6.62) and the definitions of the spaces $\mathring{M}^{p,\lambda}(\Sigma, \sigma)$, $\mathring{M}^{p_*,\lambda}(\Sigma, \sigma)$ just recalled above, this further implies that Θ maps the space $\mathring{M}^{s,\lambda}(\partial\Omega, \sigma)$ in a linear and bounded fashion into the space $\mathring{M}^{s_*,\lambda}(\partial\Omega, \sigma)$. □

The second proposition alluded to above, further complementing the body of results from Theorem 2.6.1 and Proposition 2.6.2, deals with boundary-to-domain weakly singular integral operators on Morrey spaces and their pre-duals.

Proposition 2.6.3 *Let $\Omega \subseteq \mathbb{R}^n$ (where $n \in \mathbb{N}$ with $n \geq 2$) be an open set with an Ahlfors regular boundary and abbreviate $\sigma := \mathcal{H}^{n-1}\lfloor\partial\Omega$. Also, pick $\lambda \in (0, n-1)$ and $\alpha \in (0, n-1-\lambda)$. In this context, suppose $b(\cdot,\cdot)$ is a Borel-measurable function on $\Omega \times \partial\Omega$ with the property that there exists some constant $C \in (0, \infty)$ such that*

$$\begin{gathered} \text{for } \mathcal{L}^n\text{-a.e. point } x \in \Omega \text{ one has } |b(x,y)| \leq C\,|x-y|^{-(n-1-\alpha)} \\ \text{at } \sigma\text{-a.e. point } y \text{ belonging to } \partial\Omega. \end{gathered} \tag{2.6.67}$$

Use this function as a kernel to define the integral operator

$$\mathscr{B}f(x) := \int_{\partial\Omega} b(x,y)f(y)\,\mathrm{d}\sigma(y) \ \textit{at } \mathcal{L}^n\textit{-a.e. } x \in \Omega, \quad \textit{for each function } f \in L^1\Big(\partial\Omega, \frac{\sigma(x)}{1+|x|^{n-1-\alpha}}\Big). \tag{2.6.68}$$

Finally, fix an aperture parameter $\kappa > 0$*, assume that*

$$1 < p < \frac{n-1-\lambda}{\alpha} \ \textit{and } p_* := \Big(\frac{1}{p} - \frac{\alpha}{n-1-\lambda}\Big)^{-1}. \tag{2.6.69}$$

Then the integral operator $\mathscr{B}$ *acts in a meaningful manner on any of the spaces* $M^{p,\lambda}(\partial\Omega,\sigma)$, $\mathcal{B}^{p,\lambda}(\partial\Omega,\sigma)$, $\mathring{M}^{p,\lambda}(\partial\Omega,\sigma)$, *and there exists a constant* $C \in (0,\infty)$ *such that*

$$\big\|\mathcal{N}_\kappa(\mathscr{B}f)\big\|_{M^{p_*,\lambda}(\partial\Omega,\sigma)} \le C\|f\|_{M^{p,\lambda}(\partial\Omega,\sigma)} \ \textit{for each } f \in M^{p,\lambda}(\partial\Omega,\sigma), \quad \textit{hence, in particular, for each function } f \in \mathring{M}^{p,\lambda}(\partial\Omega,\sigma), \tag{2.6.70}$$

and

$$\big\|\mathcal{N}_\kappa(\mathscr{B}f)\big\|_{\mathcal{B}^{p_*,\lambda}(\partial\Omega,\sigma)} \le C\|f\|_{\mathcal{B}^{p,\lambda}(\partial\Omega,\sigma)} \ \textit{for each } f \in \mathcal{B}^{p,\lambda}(\partial\Omega,\sigma). \tag{2.6.71}$$

Moreover, under the additional assumption that

$$\textit{for } \sigma\textit{-a.e. point } y \in \partial\Omega \textit{ the function } b(\cdot,y) \textit{ extends continuously to } \overline{\Omega}\setminus\{y\} \tag{2.6.72}$$

it follows that the nontangential boundary limit $\mathscr{B}f\big|_{\partial\Omega}^{\kappa-\mathrm{n.t.}}$ *exists at* σ*-a.e. point on* $A_\kappa(\partial\Omega)$ *and, in fact,*

$$\Big(\mathscr{B}f\big|_{\partial\Omega}^{\kappa-\mathrm{n.t.}}\Big)(x) = \int_{\partial\Omega} b(x,y)f(y)\,\mathrm{d}\sigma(y) \ \textit{at } \sigma\textit{-a.e. point } x \in A_\kappa(\partial\Omega), \tag{2.6.73}$$

for each function f *belonging to any of the spaces* $M^{p,\lambda}(\partial\Omega,\sigma)$, $\mathcal{B}^{p,\lambda}(\partial\Omega,\sigma)$, $\mathring{M}^{p,\lambda}(\partial\Omega,\sigma)$.

Proof For starters, the fact that the integral operator $\mathscr{B}$ acts in a meaningful manner on any of the spaces $M^{p,\lambda}(\partial\Omega,\sigma)$, $\mathcal{B}^{p,\lambda}(\partial\Omega,\sigma)$, $\mathring{M}^{p,\lambda}(\partial\Omega,\sigma)$ follows from observing that the embeddings from [113, (6.2.140), (6.2.230)] hold with $\Sigma := \partial\Omega$. In concert with (2.5.555), this observation also guarantees the validity of the nontangential boundary trace formula stated in (2.6.73), under the additional assumption made in (2.6.72). As such, there remains to justify the estimates claimed in (2.6.70)-(2.6.71). To this end, let $I_{\partial\Omega,\alpha}$ denote the fractional integral operator of order α on $\partial\Omega$ and recall from (2.5.560) that there exists $C \in (0,\infty)$ with the property that for each $f \in L^1\big(\partial\Omega, \frac{\sigma(x)}{1+|x|^{n-1-\alpha}}\big)$ we have

$$\big(\mathcal{N}_\kappa(\mathscr{B}f)\big)(x) \le C\, I_{\partial\Omega,\alpha}(|f|)(x), \qquad \forall\, x \in \partial\Omega. \tag{2.6.74}$$

Granted this, (2.6.70) follows on account of [113, Proposition 6.2.14], [113, (6.2.3)], and [112, (8.2.26)]. Likewise, (2.6.71) is implied by (2.6.74), [113, Proposition 6.2.20], [113, (6.2.75)], and [112, (8.2.26)]. □

2.7 Commutator Estimates

Our first result, which refines [63, Theorem 2.16, p. 2603], is a combination of the extrapolation theorem of Rubio de Francia with the commutator theorem of Coifman et al., [21], suitably adapted to setting of spaces of homogeneous type. The reader is reminded that the semi-norm $\|\cdot\|_{\dot{\mathrm{BMO}}}$ has been introduced in (A.0.14).

Theorem 2.7.1 *Let $\Sigma \subseteq \mathbb{R}^n$ be a closed Ahlfors regular set, and set $\sigma := \mathcal{H}^{n-1}\lfloor\Sigma$. Fix $p_0 \in (1,\infty)$ along with some non-decreasing function $\Phi : (0,\infty) \longrightarrow (0,\infty)$ and let T be a linear operator which is bounded on $L^{p_0}(\Sigma,\omega)$ for every $\omega \in A_{p_0}(\Sigma,\sigma)$, with operator norm $\leq \Phi([\omega]_{A_{p_0}})$.*

Then for each integrability exponent $p \in (1,\infty)$ there exist $C_1, C_2 \in (0,\infty)$ which depend exclusively on the dimension n, the exponents p_0, p, and the Ahlfors regularity constants of Σ, with the property that for any Muckenhoupt weight $w \in A_p(\Sigma,\sigma)$ the operator

$$T : L^p(\Sigma,w) \longrightarrow L^p(\Sigma,w) \tag{2.7.1}$$

is well defined, linear, and bounded, with operator norm[27]

$$\|T\|_{L^p(\Sigma,w)\to L^p(\Sigma,w)} \leq C_1 \cdot \Phi\Big(C_2 \cdot [w]_{A_p}^{1+(p_0-1)/(p-1)}\Big). \tag{2.7.2}$$

In addition, given any exponent $p \in (1,\infty)$ along with some weight $w \in A_p(\Sigma,\sigma)$, there exists a constant $C = C(\Sigma, n, p_0, p, [w]_{A_p}) \in (0,\infty)$, which stays bounded if $[w]_{A_p}$ stays bounded, and with the property that for every complex-valued function $b \in L^\infty(\Sigma,\sigma)$ one has (with C_1 as before)

$$\big\|[M_b,T]\big\|_{L^p(\Sigma,w)\to L^p(\Sigma,w)} \leq C_1 \cdot \Phi(C)\|b\|_{\dot{\mathrm{BMO}}(\Sigma,\sigma)}, \tag{2.7.3}$$

where $[M_b,T]$ is the commutator of M_b, the operator of pointwise multiplication on $L^p(\Sigma,w)$ by the function b, with T considered as in (2.7.1), i.e.,

$$[M_b,T]f := b(Tf) - T(bf) \text{ for each } f \in L^p(\Sigma,w). \tag{2.7.4}$$

Of course, ordinary Lebesgue spaces are included in (2.7.3). Indeed, specializing (2.7.3) to the case when $w \equiv 1$ yields

[27] by adapting to the setting of measure metric spaces the Euclidean argument employed in [36, Theorem 3.2], [24, Theorem 3.22, p.40], and also making use of the estimates recorded in [112, (7.7.14)], it is possible to improve (2.7.2) to $\|T\|_{L^p(\Sigma,w)\to L^p(\Sigma,w)} \leq C_1 \cdot \Phi\Big(C_2 \cdot [w]_{A_p}^{\max\{1,(p_0-1)/(p-1)\}}\Big)$

$$\|[M_b,T]\|_{L^p(\Sigma,\sigma)\to L^p(\Sigma,\sigma)} \leq C_1 \cdot \Phi(C)\|b\|_{\dot{\mathrm{BMO}}(\Sigma,\sigma)}. \tag{2.7.5}$$

In particular, from (2.7.5) and real interpolation we obtain the following commutator estimate on Lorentz spaces: For each $p \in (1,\infty)$ and $q \in (0,\infty]$ there exists $C = C(\Sigma, n, p, q) \in (0,\infty)$ with the property that for every complex-valued function $b \in L^\infty(\Sigma,\sigma)$ one has

$$\|[M_b,T]\|_{L^{p,q}(\Sigma,\sigma)\to L^{p,q}(\Sigma,\sigma)} \leq C_1 \cdot \Phi(C)\|b\|_{\dot{\mathrm{BMO}}(\Sigma,\sigma)}. \tag{2.7.6}$$

We also wish to note that the formulation of Theorem 2.7.1 is such that this result may be iterated. Specifically, in view of (2.7.3), the commutator $[M_b, T]$ has the same nature as the original operator T, so another application of Theorem 2.7.1 gives that there exists a constant $C \in (0,\infty)$ with the property that for any two functions $b_1, b_2 \in L^\infty(\Sigma,\sigma)$ we have

$$\Big\|\big[M_{b_1},[M_{b_2},T]\big]\Big\|_{L^p(\Sigma,w)\to L^p(\Sigma,w)} \leq C\|b_1\|_{\dot{\mathrm{BMO}}(\Sigma,\sigma)}\|b_2\|_{\dot{\mathrm{BMO}}(\Sigma,\sigma)}. \tag{2.7.7}$$

Inductively, it follows that for each $m \in \mathbb{N}$ there exists a constant $C \in (0,\infty)$ with the property that for any family of functions $b_1,\dots,b_m \in L^\infty(\Sigma,\sigma)$ we have

$$\Big\|\Big[M_{b_1},\cdots\big[M_{b_{m-1}},[M_{b_m},T]\big]\cdots\Big]\Big\|_{L^p(\Sigma,w)\to L^p(\Sigma,w)} \leq C\prod_{j=1}^{m}\|b_j\|_{\dot{\mathrm{BMO}}(\Sigma,\sigma)}. \tag{2.7.8}$$

Parenthetically, we wish to note that if T is an integral operator on Σ with kernel $k(x,y)$, then the iterated commutator $\Big[M_{b_1},\cdots\big[M_{b_{m-1}},[M_{b_m},T]\big]\cdots\Big]$ is also an integral operator on Σ, whose kernel is $k(x,y)\prod_{j=1}^m\big(b_j(x)-b_j(y)\big)$.

The proof of Theorem 2.7.1 is given next.

Proof of Theorem 2.7.1 From [112, (7.7.82)] we know that

$$\bigcup_{\omega\in A_{p_0}(\Sigma,\sigma)} L^{p_0}(\Sigma,\omega) = \bigcup_{\substack{1<p<\infty\\ w\in A_p(\Sigma,\sigma)}} L^p(\Sigma,w), \tag{2.7.9}$$

and the current assumptions ensure that the operator T is a well-defined mapping in the context

$$T: \bigcup_{\omega\in A_{p_0}(\Sigma,\sigma)} L^{p_0}(\Sigma,\omega) \longrightarrow \bigcup_{\omega\in A_{p_0}(\Sigma,\sigma)} L^{p_0}(\Sigma,\omega). \tag{2.7.10}$$

For an arbitrary

$$f \in \bigcup_{\omega\in A_{p_0}(\Sigma,\sigma)} L^{p_0}(\Sigma,\omega), \tag{2.7.11}$$

it follows that Tf is a well-defined σ-measurable function on Σ and, as hypothesized in the statement of the theorem, we have

$$\Big(\int_\Sigma |Tf|^{p_0}\,d\omega\Big)^{1/p_0} \leq \Phi\big([\omega]_{A_{p_0}}\big)\Big(\int_\Sigma |f|^{p_0}\,d\omega\Big)^{1/p_0} \text{ for each } \omega\in A_{p_0}(\Sigma,\sigma). \tag{2.7.12}$$

Granted this, Rubio de Francia's extrapolation theorem (recalled in [112, Proposition 7.7.6]) applies and gives that for each $p\in(1,\infty)$ there exists $C\in(0,\infty)$, which depends only on p, p_0, and the Ahlfors regularity constants of Σ, with the property that for each $w\in A_p(\Sigma,\sigma)$ we have

$$\Big(\int_\Sigma |Tf|^{p}\,dw\Big)^{1/p} \leq 4\cdot\Phi\Big(C\cdot[w]_{A_p}^{1+(p_0-1)/(p-1)}\Big)\Big(\int_\Sigma |f|^{p}\,dw\Big)^{1/p}. \tag{2.7.13}$$

From (2.7.9), (2.7.11), and (2.7.13) we then conclude that the linear operator T induces a bounded linear mapping on $L^p(\Sigma,w)$, whose operator norm is controlled as in (2.7.2). This takes care of first claim in the statement of the theorem.

To proceed, fix $p\in(1,\infty)$ and $w\in A_p(\Sigma,\sigma)$. We propose to show that

$$\big\|[M_b,T]\big\|_{L^p(\Sigma,w)\to L^p(\Sigma,w)} \leq C_1\cdot\Phi(C)\|b\|_{\mathrm{BMO}(\Sigma,\sigma)}. \tag{2.7.14}$$

To justify this estimate, we shall adapt arguments from [21], [75], [63]. First, from simple linearity and homogeneity considerations, there is no loss of generality in assuming that $b\in L^\infty(\Sigma,\sigma)$ is actually real-valued and satisfies $\|b\|_{\mathrm{BMO}(\Sigma,\sigma)}=1$ (the case when b is constant[28] is trivial). From item *(12)* of [112, Lemma 7.7.1] we know that there exists some small $\varepsilon=\varepsilon(\Sigma,p,[w]_{A_p})>0$ with the property that for each complex number z with $|z|\leq\varepsilon$ we have

$$w\cdot e^{(\mathrm{Re}\,z)b}\in A_p(\Sigma,\sigma) \text{ with } \big[w\cdot e^{(\mathrm{Re}\,z)b}\big]_{A_p}\leq C, \tag{2.7.15}$$

where the constant $C=C(\Sigma,p,[w]_{A_p})\in(0,\infty)$ is independent of z. The idea is now to observe that

$$\begin{aligned} &\Phi:\big\{z\in\mathbb{C}:\ |z|<\varepsilon/2\big\}\longrightarrow \mathrm{Bd}\big(L^p(\Sigma,w)\big) \text{ defined as}\\ &\Phi(z):=M_{e^{zb}}TM_{e^{-zb}} \text{ for each } z\in\mathbb{C} \text{ with } |z|<\varepsilon/2, \end{aligned} \tag{2.7.16}$$

is an analytic map which, for each $z\in\mathbb{C}$ with $|z|<\varepsilon/2$ and each $f\in L^p(\Sigma,w)$, satisfies

[28] the only scenario which would prevent dividing by $\|b\|_{\mathrm{BMO}(\Sigma,\sigma)}=0$

$$
\begin{aligned}
\int_\Sigma & |\Phi(z)f(x)|^p\, w(x)\, \mathrm{d}\sigma(x) \\
&= \int_\Sigma |T(\mathrm{e}^{-zb} f)(x)|^p\, w(x)\cdot \mathrm{e}^{(\mathrm{Re}\, z)b(x)}\, \mathrm{d}\sigma(x) \\
&\le \|T\|^p_{L^p(\Sigma, w\cdot \mathrm{e}^{(\mathrm{Re}\, z)b})\to L^p(\Sigma, w\cdot \mathrm{e}^{(\mathrm{Re}\, z)b})} \times \\
&\qquad \times \int_\Sigma |\mathrm{e}^{-zb(x)} f(x)|^p\, w(x)\cdot \mathrm{e}^{(\mathrm{Re}\, z)b(x)}\, \mathrm{d}\sigma(x) \\
&\le C_1^p \cdot \Phi\Big(C_2\cdot C^{1+(p_0-1)/(p-1)}\Big)^p \|f\|^p_{L^p(\Sigma,w)}, \qquad (2.7.17)
\end{aligned}
$$

thanks to (2.7.16), (2.7.15), and (2.7.2). In addition, from (2.7.16) and Cauchy's reproducing formula for analytic functions we see that

$$
[M_b, T] = \Phi_j'(0) = \frac{1}{2\pi \mathrm{i}} \int_{|z|=\varepsilon/4} \frac{\Phi(z)}{z^2}\, \mathrm{d}z. \qquad (2.7.18)
$$

Consequently, for each $f \in L^p(\Sigma, w)$ and $x \in \Sigma$, we have

$$
\big|[M_b, T]f(x)\big| \le \frac{8}{\pi\varepsilon^2} \int_{|z|=\varepsilon/4} \big|\Phi(z)f(x)\big|\, \mathrm{d}\mathcal{H}^1(z), \qquad (2.7.19)
$$

hence

$$
\big|[M_b, T]f(x)\big|^p \le \Big(\frac{8}{\pi\varepsilon^2}\Big)^p \int_{|z|=\varepsilon/4} \big|\Phi(z)f(x)\big|^p\, \mathrm{d}\mathcal{H}^1(z). \qquad (2.7.20)
$$

From (2.7.1) and (2.7.4) we see that for each $f \in L^p(\Sigma, w)$ the function $[M_b, T]f$ is σ-measurable. In concert with (2.7.20) and (2.7.17), this permits us to estimate

$$
\begin{aligned}
\int_\Sigma & \big|[M_b, T]f(x)\big|^p\, \mathrm{d}w(x) \\
&\le \Big(\frac{8}{\pi\varepsilon^2}\Big)^p \int_\Sigma \Bigg(\int_{|z|=\varepsilon/4} \big|\Phi(z)f(x)\big|^p\, \mathrm{d}\mathcal{H}^1(z)\Bigg)\, \mathrm{d}w(x) \\
&= \Big(\frac{8}{\pi\varepsilon^2}\Big)^p \int_{|z|=\varepsilon/4} \Bigg(\int_\Sigma \big|\Phi(z)f(x)\big|^p\, \mathrm{d}w(x)\Bigg)\, \mathrm{d}\mathcal{H}^1(z) \\
&\le \Big(\frac{2^{3p-1}}{\pi^{p-1}\varepsilon^{2p-1}}\Big) C_1^p \cdot \Phi\Big(C_2\cdot C^{1+(p_0-1)/(p-1)}\Big)^p \|f\|^p_{L^p(\Sigma,w)}. \qquad (2.7.21)
\end{aligned}
$$

In turn, (2.7.21) implies (2.7.14).

The end-game in the proof is as follows. If Σ is unbounded, then (2.7.3) is a direct consequence of (2.7.14) and (A.0.15). If Σ is bounded, then (2.7.3) is obtained by applying (2.7.14) to the function $\widetilde{b} := b - ⨍_\Sigma b\,d\sigma$. Indeed, $\widetilde{b} \in L^\infty(\Sigma, \sigma)$ and, as seen from (A.0.15) and (2.7.4), we have $\|\widetilde{b}\|_{\mathrm{BMO}(\Sigma,\sigma)} = \|b\|_{\dot{\mathrm{BMO}}(\Sigma,\sigma)}$ and $[M_{\widetilde{b}}, T] = [M_b, T]$. Then (2.7.3) readily follows from these observations. □

In turn, Theorem 2.7.1 is a basic ingredient in the proof of the following commutator estimate on Muckenhoupt weighted Lebesgue spaces for convolution type singular integral operators.

Theorem 2.7.2 *Suppose $\Sigma \subseteq \mathbb{R}^n$ is a* UR *set, and abbreviate $\sigma := \mathcal{H}^{n-1}\lfloor\Sigma$. Given a sufficiently large integer $N = N(n) \in \mathbb{N}$, fix a function $k \in \mathscr{C}^N(\mathbb{R}^n \setminus \{0\})$ which is complex-valued, odd, and positive homogeneous of degree $1 - n$. Also, pick an exponent $p \in (1, \infty)$ along with a Muckenhoupt weight $w \in A_p(\Sigma, \sigma)$ and recall from item (10) in Theorem 2.3.2 (cf.* (2.3.56)) *that the singular integral operator defined originally on functions $f \in L^1\big(\Sigma, \frac{\sigma(x)}{1+|x|^{n-1}}\big)$ as*

$$Tf(x) := \lim_{\varepsilon\to 0^+} \int_{\substack{y\in\Sigma\\ |x-y|>\varepsilon}} k(x-y)f(y)\,d\sigma(y) \ \textit{for } \sigma\textit{-a.e. } x \in \Sigma, \tag{2.7.22}$$

induces a well-defined, linear, and bounded mapping

$$T : L^p(\Sigma, w) \longrightarrow L^p(\Sigma, w). \tag{2.7.23}$$

Then there exists a constant $C \in (0, \infty)$, which depends only on n, k, p, $[w]_{A_p}$, and the UR *character of Σ, such that for each function $b \in L^\infty(\Sigma, \sigma)$ one has*

$$\big\|[T, M_b]\big\|_{L^p(\Sigma,w)\to L^p(\Sigma,w)} \leq C\|b\|_{\dot{\mathrm{BMO}}(\Sigma,\sigma)}. \tag{2.7.24}$$

More generally, consider a function $b(x, z)$ which is odd and positive homogeneous of degree $1 - n$ in the variable $z \in \mathbb{R}^n \setminus \{0\}$, and such that $\partial_z^\alpha b(x, z)$ is continuous and bounded on $\mathbb{R}^n \times S^{n-1}$ for each multi-index $\alpha \in \mathbb{N}_0^n$ satisfying $|\alpha| \leq M$ where $M = M(n)$ is a sufficiently large positive integer. Then the singular integral operators acting on functions $f \in L^1\big(\Sigma, \frac{\sigma(x)}{1+|x|^{n-1}}\big)$ via

$$Bf(x) := \lim_{\varepsilon\to 0^+} \int_{\substack{y\in\Sigma\\ |x-y|>\varepsilon}} b(x, x-y)f(y)\,d\sigma(y) \ \textit{for } \sigma\textit{-a.e. } x \in \Sigma, \tag{2.7.25}$$

and, respectively,

$$B^\# f(x) := \lim_{\varepsilon\to 0^+} \int_{\substack{y\in\Sigma\\ |x-y|>\varepsilon}} b(y, x-y)f(y)\,d\sigma(y) \ \textit{for } \sigma\textit{-a.e. } x \in \Sigma, \tag{2.7.26}$$

induce well-defined, linear, and bounded mappings

$$B, B^{\#} : L^p(\Sigma, w) \longrightarrow L^p(\Sigma, w) \tag{2.7.27}$$

and there exists a constant $C \in (0, \infty)$, which depends only on n, b, p, $[w]_{A_p}$, and the UR *character of Σ, such that for each function $\psi \in L^\infty(\Sigma, \sigma)$ one has*

$$\max\Big\{\big\|[B, M_\psi]\big\|_{L^p(\Sigma,w)\to L^p(\Sigma,w)}, \big\|[B^{\#}, M_\psi]\big\|_{L^p(\Sigma,w)\to L^p(\Sigma,w)}\Big\}$$
$$\leq C\|\psi\|_{\dot{\mathrm{BMO}}(\Sigma,\sigma)}. \tag{2.7.28}$$

Proof The claims pertaining to (2.7.23)-(2.7.24) are consequences of Theorem 2.7.1 and item *(10)* in Theorem 2.3.2, while the claims made in (2.7.27)-(2.7.28) are seen from Theorem 2.7.1 and item *(e)* in Theorem 2.5.38. □

In the following companion result to Theorem 2.7.2 we estimate the *essential norm* of the commutator (2.7.4) in the case when the underlying set is compact.

Theorem 2.7.3 *Let $\Sigma \subseteq \mathbb{R}^n$ be a compact* UR *set, and abbreviate $\sigma := \mathcal{H}^{n-1}\lfloor\Sigma$. Having fixed a sufficiently large integer $N = N(n) \in \mathbb{N}$, consider a complex-valued function $k \in \mathscr{C}^N(\mathbb{R}^n \setminus \{0\})$ is which is odd and positive homogeneous of degree $1 - n$. Pick an exponent $p \in (1, \infty)$ along with a Muckenhoupt weight $w \in A_p(\Sigma, \sigma)$ and recall from item (10) in Theorem 2.3.2 (cf. (2.3.56)) that the singular integral operator defined originally on functions $f \in L^1(\Sigma, \sigma)$ as*

$$Tf(x) := \lim_{\varepsilon\to 0^+} \int_{\substack{y\in\Sigma \\ |x-y|>\varepsilon}} k(x-y)f(y)\,d\sigma(y) \ \textit{for } \sigma\textit{-a.e. } x \in \Sigma, \tag{2.7.29}$$

induces a well-defined, linear, and bounded mapping

$$T : L^p(\Sigma, w) \longrightarrow L^p(\Sigma, w). \tag{2.7.30}$$

Then there exists a constant $C \in (0, \infty)$, which depends only on n, p, $[w]_{A_p}$, and the UR *character of Σ, such that for each function $b \in L^\infty(\Sigma, \sigma)$ one has*

$$\mathrm{dist}\Big([T, M_b], \mathrm{Cp}\big(L^p(\Sigma, w)\big)\Big) \leq C\Big(\sum_{|\alpha|\leq N} \sup_{S^{n-1}} |\partial^\alpha k|\Big) \cdot \mathrm{dist}\big(b, \mathrm{VMO}(\Sigma, \sigma)\big) \tag{2.7.31}$$

where the distance in the left-hand side is measured in $\mathrm{Bd}\big(L^p(\Sigma, w)\big)$ *and the distance in the right-hand side is measured in* $\mathrm{BMO}(\Sigma, \sigma)$.

As a consequence of (2.7.31) *with $w \equiv 1$, real interpolation, and* [113, Proposition 1.4.24] (*whose applicability for Lebesgue spaces is ensured by item (3) in* [113, Proposition 7.3.6] *together with the identification in* [113, (7.1.55)]), *for each $q \in (0, \infty]$ there exists a constant $C \in (0, \infty)$, which depends only on n, p, q, and the* UR *character of Σ, such that for each function $b \in L^\infty(\Sigma, \sigma)$ one has*

$$\operatorname{dist}\Big([T, M_b], \operatorname{Cp}\big(L^{p,q}(\Sigma,\sigma)\big)\Big) \leq C\Big(\sum_{|\alpha|\leq N}\sup_{S^{n-1}}|\partial^\alpha k|\Big)\operatorname{dist}\big(b, \operatorname{VMO}(\Sigma,\sigma)\big) \tag{2.7.32}$$

where the distance in the left-hand side is measured in $\operatorname{Bd}\big(L^{p,q}(\Sigma,\sigma)\big)$ *and the distance in the right-hand side is measured in* $\operatorname{BMO}(\Sigma,\sigma)$.

In particular,

$$[T, M_b] \in \operatorname{Cp}\big(L^p(\Sigma, w)\big) \text{ whenever } b \in L^\infty(\Sigma,\sigma)\cap \operatorname{VMO}(\Sigma,\sigma), \tag{2.7.33}$$

and, for each $q \in (0,\infty]$,

$$[T, M_b] \in \operatorname{Cp}\big(L^{p,q}(\Sigma, \sigma)\big) \text{ whenever } b \in L^\infty(\Sigma,\sigma)\cap \operatorname{VMO}(\Sigma,\sigma). \tag{2.7.34}$$

Utilizing the piece of notation introduced in (A.0.102), we may rephrase (2.7.31) and (2.7.32) as

$$\big\|[T, M_b]\big\|^{\text{ess}}_{L^p(\Sigma,w)\to L^p(\Sigma,w)} \leq C\Big(\sum_{|\alpha|\leq N}\sup_{S^{n-1}}|\partial^\alpha k|\Big)\cdot\operatorname{dist}\big(b, \operatorname{VMO}(\Sigma,\sigma)\big) \tag{2.7.35}$$

and, respectively,

$$\big\|[T, M_b]\big\|^{\text{ess}}_{L^{p,q}(\Sigma,\sigma)\to L^{p,q}(\Sigma,\sigma)} \leq C\Big(\sum_{|\alpha|\leq N}\sup_{S^{n-1}}|\partial^\alpha k|\Big)\cdot\operatorname{dist}\big(b, \operatorname{VMO}(\Sigma,\sigma)\big). \tag{2.7.36}$$

We can also prove a version of Theorem 2.7.3 for commutators involving singular integral operators with variable coefficient kernels. To elaborate, assume that

$b(x,z)$ is a function which is odd and positive homogeneous of degree $1-n$ in the variable $z \in \mathbb{R}^n\setminus\{0\}$, and such that $\partial_z^\alpha b(x,z)$ is continuous and bounded on $\mathbb{R}^n\times S^{n-1}$ for each multi-index $\alpha \in \mathbb{N}_0^n$ satisfying $|\alpha| \leq M$, where $M = M(n)$ is a sufficiently large positive integer. (2.7.37)

Given a compact UR set $\Sigma \subseteq \mathbb{R}^n$, abbreviate $\sigma := \mathcal{H}^{n-1}\lfloor\Sigma$. For each function $f \in L^1\big(\Sigma, \frac{\sigma(x)}{1+|x|^{n-1}}\big)$ define

$$Bf(x) := \lim_{\varepsilon\to 0^+}\int_{\substack{y\in\Sigma\\ |x-y|>\varepsilon}} b(x, x-y)f(y)\,d\sigma(y) \text{ for } \sigma\text{-a.e. } x \in \Sigma. \tag{2.7.38}$$

Then for each $p \in (1,\infty)$ and $w \in A_p(\Sigma,\sigma)$ there exists some $C \in (0,\infty)$, depending only on n, p, $[w]_{A_p}$, and Σ, with the property that for any function $\psi \in L^\infty(\Sigma,\sigma)$ we have

$$\begin{aligned}\big\|[B, M_\psi]\big\|^{\text{ess}}_{L^p(\Sigma,w)\to L^p(\partial\Omega,w)} &= \operatorname{dist}\big([B, M_\psi], \operatorname{Cp}(L^p(\Sigma,w))\big)\\ &\leq C\cdot C_b\cdot\operatorname{dist}\big(\psi, \operatorname{VMO}(\Sigma,\sigma)\big),\end{aligned} \tag{2.7.39}$$

where

$$C_b := \sup_{\substack{(x,z)\in\mathbb{R}^n\times S^{n-1}\\ |\alpha|\leq M}} \big|(\partial_z^\alpha b)(x,z)\big| \in (0,\infty). \tag{2.7.40}$$

To justify (2.7.39), we shall avail ourselves of notation and results from the proof of Theorem 2.5.38. Much as with the decomposition (2.5.518), we presently obtain

$$B = \sum_{\ell\in 2\mathbb{N}_0+1}\ \sum_{1\leq i\leq H_\ell} a_{i\ell} B^{i\ell} \tag{2.7.41}$$

with convergence in $\mathscr{B}\Big(L^p(\Sigma,w) \to L^p(\Sigma,w)\Big)$. For an arbitrary fixed threshold $\mu\in\mathbb{N}$, split

$$B = \sum_{\substack{\ell\in 2\mathbb{N}_0+1\\ \ell\leq\mu}}\ \sum_{1\leq i\leq H_\ell} a_{i\ell} B^{i\ell} + \sum_{\substack{\ell\in 2\mathbb{N}_0+1\\ \ell>\mu}}\ \sum_{1\leq i\leq H_\ell} a_{i\ell} B^{i\ell}. \tag{2.7.42}$$

Similarly to (2.5.505), we may estimate

$$\Bigg\|\sum_{\substack{\ell\in 2+1\mathbb{N}_0\\ \ell>\mu}}\ \sum_{1\leq i\leq H_\ell} a_{i\ell} B^{i\ell}\Bigg\|_{L^p(\Sigma,w)\to L^p(\Sigma,w)} \leq \frac{C(\Omega,b,p,w,n)}{\mu}. \tag{2.7.43}$$

Fix an arbitrary function $\psi\in L^\infty(\Sigma,\sigma)$. For each $\ell\in 2\mathbb{N}_0+1$ and $1\leq i\leq H_\ell$, let $C^{i\ell}\in \mathrm{Cp}\big(L^p(\Sigma,w)\big)$ be such that

$$\big\|[B^{i\ell},M_\psi]-C^{i\ell}\big\|_{L^p(\Sigma,w)\to L^p(\Sigma,w)} \leq 2\cdot \mathrm{dist}\,\big([B^{i\ell},M_\psi],\,\mathrm{Cp}(L^p(\Sigma,w))\big). \tag{2.7.44}$$

By (2.7.31), for each $\ell\in 2\mathbb{N}_0+1$ and $1\leq i\leq H_\ell$ we have

$$\begin{aligned}&\mathrm{dist}\,\big([B^{i\ell},M_\psi],\,\mathrm{Cp}(L^p(\Sigma,w))\big)\\ &\qquad\qquad \leq C\Big(\sum_{|\alpha|\leq N}\sup_{S^{n-1}}|\partial^\alpha k_{i\ell}|\Big)\,\mathrm{dist}\,\big(\psi,\,\mathrm{VMO}(\Sigma,\sigma)\big).\end{aligned} \tag{2.7.45}$$

Also, (2.5.494), (2.5.498), and (2.5.482) ensure that

$$\sum_{\ell\in\mathbb{N}_0}\ \sum_{1\leq i\leq H_\ell}\Big(\sup_{\mathbb{R}^n}|a_{i\ell}|\Big)\Big(\sum_{|\alpha|\leq N}\sup_{S^{n-1}}|\partial^\alpha k_{i\ell}|\Big) \leq C_n\cdot C_b. \tag{2.7.46}$$

If we now define

$$C_\mu := \sum_{\substack{\ell\in 2\mathbb{N}_0+1\\ \ell\leq\mu}}\ \sum_{1\leq i\leq H_\ell} a_{i\ell} C^{i\ell} \in \mathrm{Cp}\big(L^p(\Sigma,w)\big), \tag{2.7.47}$$

then (2.7.42)-(2.7.47) imply

$$\operatorname{dist}\big([B, M_\psi],\ \mathrm{Cp}(L^p(\Sigma, w))\big) \leq \big\|[B, M_\psi] - C_\mu\big\|_{L^p(\Sigma,w)\to L^p(\Sigma,w)} \tag{2.7.48}$$

$$\leq C \cdot C_b \cdot \operatorname{dist}\big(\psi,\ \mathrm{VMO}(\Sigma, \sigma)\big) + \frac{C(\Omega, b, p, w, n)}{\mu}.$$

Passing to limit $\mu \to \infty$ now yields (2.7.39).

Lastly, we note that an estimate similar to (2.7.39) is also valid for commutators involving the variable coefficient kernel singular integral operator acting on functions $f \in L^1\big(\Sigma, \frac{\sigma(x)}{1+|x|^{n-1}}\big)$ according to

$$B^{\#} f(x) := \lim_{\varepsilon\to 0^+} \int_{\substack{y\in\Sigma \\ |x-y|>\varepsilon}} b(y, y-x) f(y)\, d\sigma(y) \ \text{ for } \sigma\text{-a.e. } x \in \Sigma. \tag{2.7.49}$$

We now turn to the proof of Theorem 2.7.3.

Proof of Theorem 2.7.3 Fix some $\alpha \in (0, 1)$ and consider an arbitrary function $\phi \in \mathrm{VMO}(\Sigma, \sigma)$. Based on [113, (3.1.50)] we may then find a sequence of functions $\{b_j\}_{j\in\mathbb{N}} \subseteq \mathscr{C}^\alpha(\Sigma)$ with the property that

$$\lim_{j\to\infty} \|\phi - b_j\|_{\mathrm{BMO}(\Sigma,\sigma)} = 0. \tag{2.7.50}$$

From (2.7.3) we conclude that there exists constant $C = C(\Sigma, n, p, [w]_{A_p}) \in (0, \infty)$ with the property that for each $j \in \mathbb{N}$ we have

$$\big\|[M_b, T] - [M_{b_j}, T]\big\|_{L^p(\Sigma,w)\to L^p(\Sigma,w)} \tag{2.7.51}$$

$$= \big\|[M_{b-b_j}, T]\big\|_{L^p(\Sigma,w)\to L^p(\Sigma,w)}$$

$$\leq C\Big(\sum_{|\alpha|\leq N} \sup_{S^{n-1}} |\partial^\alpha k|\Big) \|b - b_j\|_{\mathrm{BMO}(\Sigma,\sigma)}$$

$$\leq C\Big(\sum_{|\alpha|\leq N} \sup_{S^{n-1}} |\partial^\alpha k|\Big) \Big\{\|b - \phi\|_{\mathrm{BMO}(\Sigma,\sigma)} + \|\phi - b_j\|_{\mathrm{BMO}(\Sigma,\sigma)}\Big\}.$$

To proceed, we make the claim

$$[M_{b_j}, T] \in \mathrm{Cp}\big(L^p(\Sigma, w)\big) \ \text{ for each } j \in \mathbb{N}. \tag{2.7.52}$$

Postponing the proof of (2.7.52), we conclude from (2.7.51) that

$$\operatorname{dist}\Big([T, M_b], \mathrm{Cp}\big(L^p(\Sigma, w)\big)\Big) \tag{2.7.53}$$

$$\leq C\Big(\sum_{|\alpha|\leq N} \sup_{S^{n-1}} |\partial^\alpha k|\Big) \Big\{\|b - \phi\|_{\mathrm{BMO}(\Sigma,\sigma)} + \|\phi - b_j\|_{\mathrm{BMO}(\Sigma,\sigma)}\Big\}$$

for each $j \in \mathbb{N}$. Passing to the limit $j \to \infty$ in (2.7.53) then yields

$$\operatorname{dist}\Big([T, M_b], \operatorname{Cp}\big(L^p(\Sigma, w)\big)\Big) \leq C\Big(\sum_{|\alpha|\leq N} \sup_{S^{n-1}} |\partial^\alpha k|\Big)\|b - \phi\|_{\mathrm{BMO}(\Sigma,\sigma)} \tag{2.7.54}$$

on account of (2.7.50). Finally, taking the infimum over all $\phi \in \mathrm{VMO}(\Sigma, \sigma)$ establishes (2.7.31).

There remains to justify (2.7.52). Fix $j \in \mathbb{N}$ and observe that for each $f \in L^p(\Sigma, w)$ we have

$$\begin{aligned}[M_{b_j}, T]f(x) &= \lim_{\varepsilon\to 0^+} \int_{\substack{y\in\Sigma \\ |x-y|>\varepsilon}} \big(b_j(x) - b_j(y)\big)k(x-y)f(y)\,\mathrm{d}\sigma(y) \\ &= \int_\Sigma \big(b_j(x) - b_j(y)\big)k(x-y)f(y)\,\mathrm{d}\sigma(y) \ \text{ for } \sigma\text{-a.e. } x \in \Sigma.\end{aligned} \tag{2.7.55}$$

Above, the last equality is a consequence of Lebesgue's Dominated Convergence Theorem, whose applicability in the presence setting is ensured by the membership of b_j to $\mathscr{C}^\alpha(\Sigma)$, the properties of the kernel k, [112, (7.8.5)], and the fact that (cf. [112, (7.7.103)])

$$L^p(\Sigma, w) \hookrightarrow L^1(\Sigma, \sigma). \tag{2.7.56}$$

To proceed, fix some $\psi \in \mathscr{C}^\infty(\mathbb{R})$ with $\psi \equiv 0$ on $(-1, 1)$ and $\psi \equiv 1$ on $\mathbb{R} \setminus (-2, 2)$. For every $\varepsilon > 0$, consider the truncated singular integral operator acting on each function $f \in L^p(\Sigma, w)$ according to

$$Q_\varepsilon f(x) := \int_\Sigma q_\varepsilon(x, y)f(y)\,\mathrm{d}\sigma(y) \ \text{ for } \sigma\text{-a.e. } x \in \Sigma, \tag{2.7.57}$$

where

$$q_\varepsilon(x, y) := \psi\Big(\frac{|x-y|}{\varepsilon}\Big)\big(b_j(x) - b_j(y)\big)k(x-y) \ \text{ for } x, y \in \Sigma. \tag{2.7.58}$$

Then [112, Lemma 7.7.16] ensures that there exists some $C \in (0, \infty)$ with the property that for each $\varepsilon > 0$ and each $f \in L^p(\Sigma, w)$ we have

$$\big\|[M_{b_j}, T]f - Q_\varepsilon f\big\|_{L^p(\Sigma,w)} \leq C\varepsilon^\alpha \|f\|_{L^p(\Sigma,w)}. \tag{2.7.59}$$

In particular,

$$\big\|[M_{b_j}, T] - Q_\varepsilon\big\|_{L^p(\Sigma,w)\to L^p(\Sigma,w)} \leq C\varepsilon^\alpha \ \text{ for each } \varepsilon > 0, \tag{2.7.60}$$

hence, on the one hand,

$$[M_{b_j}, T] = \lim_{\varepsilon\to 0^+} Q_\varepsilon \ \text{ in } \operatorname{Bd}\big(L^p(\Sigma, w) \to L^p(\Sigma, w)\big). \tag{2.7.61}$$

On the other hand, for each fixed $\varepsilon > 0$ the kernel $q_\varepsilon(x, y)$ of the integral operator Q_ε is bounded and there exists some $C \in (0, \infty)$ such that

$$\big|q_\varepsilon(x_0, y) - q_\varepsilon(x_1, y)\big| \leq C|x_0 - x_1|^\alpha \text{ for all } x_0, x_1, y \in \Sigma. \tag{2.7.62}$$

Indeed, (2.7.62) is a simple consequence of the fact that $F(z) := \psi\big(\frac{|z|}{\varepsilon}\big)k(z)$ for $z \in \mathbb{R}^n$ is bounded and Lipschitz (hence, Hölder continuous of order α), and that $b_j \in \mathscr{C}^\alpha(\Sigma)$. In turn, from (2.7.62) and the Arzelà-Ascoli Theorem we conclude that Q_ε maps $L^1(\Sigma, \sigma)$ compactly into the space of continuous functions on Σ, equipped with the supremum norm. As a consequence,

$$Q_\varepsilon \in \mathrm{Cp}\big(L^1(\Sigma, \sigma) \to L^\infty(\Sigma, \sigma)\big) \text{ for each } \varepsilon > 0. \tag{2.7.63}$$

From this, (2.7.56), and the fact that $L^\infty(\Sigma, \sigma)$ embeds continuously into $L^p(\Sigma, w)$, we ultimately see that

$$Q_\varepsilon \in \mathrm{Cp}\big(L^p(\Sigma, w) \to L^p(\Sigma, w)\big) \text{ for each } \varepsilon > 0. \tag{2.7.64}$$

At this stage, the claim made in (2.7.52) becomes a consequence of (2.7.61), (2.7.64), and [113, (1.2.53)]. □

We next turn to the task of establishing commutator estimates in Morrey and block spaces. Here is a version of the result from Theorem 2.7.1.

Theorem 2.7.4 *Suppose $\Sigma \subseteq \mathbb{R}^n$ is a closed Ahlfors regular set and abbreviate $\sigma := \mathcal{H}^{n-1}\lfloor\Sigma$. Fix an exponent $p_0 \in (1, \infty)$ along with some non-decreasing function $\Phi : (0, \infty) \to (0, \infty)$ and let T be a linear operator which is bounded on $L^{p_0}(\Sigma, w)$ for every $w \in A_{p_0}(\Sigma, \sigma)$, with operator norm $\leq \Phi\big([w]_{A_{p_0}}\big)$.*

Then for each integrability exponent $p \in (1, \infty)$ and each parameter $\lambda \in (0, n-1)$ the operator T induces well-defined, linear, and bounded mappings in the contexts

$$T : M^{p,\lambda}(\Sigma, \sigma) \longrightarrow M^{p,\lambda}(\Sigma, \sigma), \tag{2.7.65}$$

$$T : \mathring{M}^{p,\lambda}(\Sigma, \sigma) \longrightarrow \mathring{M}^{p,\lambda}(\Sigma, \sigma). \tag{2.7.66}$$

Also, given any exponent $p \in (1, \infty)$ along with some $\lambda \in (0, n-1)$, there exist two constants, $C_1 = C_1(\Sigma, n, p_0, p, \lambda) \in (0, \infty)$ and $C_2 = C_2(\Sigma, n, p_0, p) \in (0, \infty)$, with the property that for every complex-valued function $b \in L^\infty(\Sigma, \sigma)$ one has

$$\begin{aligned}\big\|[M_b, T]\big\|_{\mathring{M}^{p,\lambda}(\Sigma,\sigma)\to\mathring{M}^{p,\lambda}(\Sigma,\sigma)} &\leq \big\|[M_b, T]\big\|_{M^{p,\lambda}(\Sigma,\sigma)\to M^{p,\lambda}(\Sigma,\sigma)} \\ &\leq C_1\Phi(C_2)\|b\|_{\mathrm{B\dot{M}O}(\Sigma,\sigma)},\end{aligned} \tag{2.7.67}$$

where $[M_b, T] := bT(\cdot) - T(b\,\cdot)$ is the commutator of the operator M_b of pointwise multiplication (either on $M^{p,\lambda}(\Sigma, \sigma)$, or on $\mathring{M}^{p,\lambda}(\Sigma, \sigma)$) by the function b and the operator T (considered either as in (2.7.65), or as in (2.7.66)).

Moreover, as a consequence of the duality result from [113, Proposition 6.2.16], *the (real) transpose of* (2.7.66), *i.e.,*

$$T^{\top} : \mathcal{B}^{q,\lambda}(\Sigma, \sigma) \longrightarrow \mathcal{B}^{q,\lambda}(\Sigma, \sigma), \tag{2.7.68}$$

is also well defined, linear, and bounded if $q := (1 - 1/p)^{-1} \in (1, \infty)$ *is the Hölder conjugate exponent of* p. *In addition, there exist two positive finite constants, say* $C_1 = C_1(\Sigma, n, p_0, q, \lambda)$ *and* $C_2 = C_2(\Sigma, n, p_0, q)$, *with the property that for every complex-valued function* $b \in L^\infty(\Sigma, \sigma)$ *one has*

$$\big\|[M_b, T^{\top}]\big\|_{\mathcal{B}^{q,\lambda}(\Sigma,\sigma)\to\mathcal{B}^{q,\lambda}(\Sigma,\sigma)} \leq C_1 \Phi(C_2) \|b\|_{\dot{\mathrm{BMO}}(\Sigma,\sigma)}. \tag{2.7.69}$$

Finally, the (*real*) *transpose of* (2.7.68), *i.e.*,

$$\widetilde{T} := \big(T^{\top}\big)^{\top} : M^{p,\lambda}(\Sigma, \sigma) \longrightarrow M^{p,\lambda}(\Sigma, \sigma), \tag{2.7.70}$$

is an extension of (2.7.66), *and there exists a constant* $C \in (0, \infty)$ *with the property that*

if $f \in M^{p,\lambda}(\Sigma, \sigma)$ *and* $g \in \mathcal{B}^{q,\lambda}(\Sigma, \sigma)$ *then* $(\widetilde{T} f)g - f(T^{\top} g)$ *belongs to the homogeneous Hardy space* $\dot{H}^1(\Sigma, \sigma)$ (*cf.* [113, (4.2.12)]) *and one has* $\big\|(\widetilde{T} f)g - f(T^{\top} g)\big\|_{H^1(\Sigma,\sigma)} \leq C\|f\|_{M^{p,\lambda}(\Sigma,\sigma)} \cdot \|g\|_{\mathcal{B}^{q,\lambda}(\Sigma,\sigma)}$. (2.7.71)

In particular,

if $f \in \mathring{M}^{p,\lambda}(\Sigma, \sigma)$ *and* $g \in \mathcal{B}^{q,\lambda}(\Sigma, \sigma)$ *then* $(T f)g - f(T^{\top} g)$ *belongs to the homogeneous Hardy space* $\dot{H}^1(\Sigma, \sigma)$ (*cf.* [113, (4.2.12)]) *and one has* $\big\|(T f)g - f(T^{\top} g)\big\|_{H^1(\Sigma,\sigma)} \leq C\|f\|_{M^{p,\lambda}(\Sigma,\sigma)} \cdot \|g\|_{\mathcal{B}^{q,\lambda}(\Sigma,\sigma)}$. (2.7.72)

When Σ is the boundary of a UR domain $\Omega \subseteq \mathbb{R}^n$ and T is a Riesz transform on Σ (which makes T skew-symmetric), the results in (2.7.71)-(2.7.72) are already contained in the version of [113, (10.1.137)] corresponding to the scenario described in item *(2)* of [113, Corollary 10.1.12].

Proof of Theorem 2.7.4 The present hypotheses ensure that it makes sense to define the following family of pairs of σ-measurable functions on Σ:

$$\mathcal{F} := \left\{ (T f, f) : f \in \bigcup_{w \in A_{p_0}(\Sigma,\sigma)} L^{p_0}(\Sigma, \sigma) \right\}. \tag{2.7.73}$$

In view of the monotonicity property of the classes of Muckenhoupt weights on Σ (cf. item *(5)* in [112, Lemma 7.7.1]), the current assumptions also guarantee that

$$\|T f\|_{L^p(\Sigma,w)} \leq \Phi\big([w]_{A_1}\big)\|f\|_{L^p(\Sigma,w)} \\ \text{for each pair } (T f, f) \in \mathcal{F} \text{ and weight } w \in A_1(\Sigma, \sigma). \tag{2.7.74}$$

Granted this, once some $p \in (1, \infty)$ and $\lambda \in (0, n - 1)$ have been fixed, the extrapolation result established in [113, Proposition 6.2.12] applies and gives a constant $C \in (0, \infty)$ with the property that

$$\|Tf\|_{M^{p,\lambda}(\Sigma,\sigma)} \leq C\|f\|_{M^{p,\lambda}(\Sigma,\sigma)} \text{ for each pair } (Tf, f) \in \mathcal{F}. \tag{2.7.75}$$

In view of [113, (6.2.246) in Proposition 6.2.22], we conclude from (2.7.75) that we actually have $\|Tf\|_{M^{p,\lambda}(\Sigma,\sigma)} \leq C\|f\|_{M^{p,\lambda}(\Sigma,\sigma)}$ for each function $f \in M^{p,\lambda}(\Sigma,\sigma)$. Thus, T induces a well-defined, linear, and bounded operator in the context of (2.7.65). Since from (2.7.1) we also know that

$$T : L^s(\Sigma,\sigma) \longrightarrow L^s(\Sigma,\sigma) \text{ with } s := \tfrac{p(n-1)}{n-1-\lambda} \tag{2.7.76}$$

is a well-defined, linear, and bounded operator, and since $\mathring{M}^{p,\lambda}(\Sigma,\sigma)$ is the closure of $L^s(\Sigma,\sigma)$ in $M^{p,\lambda}(\Sigma,\sigma)$ (cf. (A.0.88)), we additionally conclude that T induces a well-defined, linear, and bounded operator in the context of (2.7.66).

In a similar fashion, the second estimate in (2.7.67) is implied by Theorem 2.7.1, and [113, Propositions 6.2.12, 6.2.22]. The first estimate in (2.7.67) is a simple consequence of [113, (6.2.15)].

At this stage, the claims regarding (2.7.68) and (2.7.69) follow from what we have proved already and the duality result from [113, Proposition 6.2.8]. Alternatively, we could have used the extrapolation result from [113, Corollary 5.2.3], whose applicability in the present setting with $\mathbb{X} := \mathcal{B}^{q,\lambda}(\partial\Omega,\sigma)$ is ensured by [113, Proposition 6.2.17] together with [113, Corollaries 6.2.11, 6.2.13]. Next, the fact that $\widetilde{T} := \left(T^\top\right)^\top$ from (2.7.70) is an extension of (2.7.66) is seen from definitions, (A.0.88), [113, (6.2.80)] and [113, (6.2.155)].

There remains to justify the claims made in (2.7.71)-(2.7.72). To this end, fix arbitrary functions, $f \in M^{p,\lambda}(\Sigma,\sigma)$ and $g \in \mathcal{B}^{q,\lambda}(\Sigma,\sigma)$. Then from [113, Proposition 6.2.8, (6.2.78)] we conclude that the function $h := (\widetilde{T}f)g - f(T^\top g)$ belongs to $L^1(\Sigma,\sigma)$ and satisfies

$$\begin{aligned}\int_\Sigma h\,d\sigma &= \int_\Sigma \Big(\big(T^\top\big)^\top f\Big) g\,d\sigma - \int_\Sigma f(T^\top g)\,d\sigma \\ &= \int_\Sigma f(T^\top g)\,d\sigma - \int_\Sigma f(T^\top g)\,d\sigma = 0. \end{aligned}\tag{2.7.77}$$

To proceed, pick an arbitrary $\phi \in \mathrm{Lip}_c(\Sigma)$ and let M_ϕ denote the operator of pointwise multiplication by the function ϕ. Then [113, (6.2.74)] ensures that ϕg belongs to $\mathcal{B}^{q,\lambda}(\Sigma,\sigma)$. Keeping this in mind and using [113, Proposition 6.2.8, (6.2.78)] we may then compute

$$\begin{aligned}\Big|\int_\Sigma \phi\, h\,d\sigma\Big| &= \Big|\int_\Sigma \phi((\widetilde{T}f)g - f(T^\top g))\,d\sigma\Big| \\ &= \Big|\int_\Sigma f\big[T^\top, M_\phi\big]g\,d\sigma\Big| \leq \|f\|_{M^{p,\lambda}(\Sigma,\sigma)} \cdot \big\|\big[T^\top, M_\phi\big]g\big\|_{\mathcal{B}^{q,\lambda}(\Sigma,\sigma)} \\ &\leq C\|f\|_{M^{p,\lambda}(\Sigma,\sigma)} \cdot \|g\|_{\mathcal{B}^{q,\lambda}(\Sigma,\sigma)} \cdot \|\phi\|_{\dot{\mathrm{BMO}}(\Sigma,\sigma)},\end{aligned}\tag{2.7.78}$$

with the last inequality provided by (2.7.69). Granted this, we conclude from [113, (4.4.10), (4.6.26)] and (2.7.77) that all conclusions in (2.7.71) are valid. Finally, (2.7.72) follows from this and the fact that $\widetilde{T}$ is an extension of T in (2.7.66). □

In turn, Theorem 2.7.4 is one of the key ingredients in the proof of the following commutator estimates on Morrey and block spaces for convolution type singular integral operators.

Theorem 2.7.5 *Assume $\Sigma \subseteq \mathbb{R}^n$ is a* UR *set, and abbreviate $\sigma := \mathcal{H}^{n-1}\lfloor\Sigma$. Having fixed a sufficiently large integer $N = N(n) \in \mathbb{N}$, consider a complex-valued function $k \in \mathscr{C}^N(\mathbb{R}^n \setminus \{0\})$ which is odd and positive homogeneous of degree $1-n$. Also, pick two integrability exponents $p, q \in (1, \infty)$ along with some parameter $\lambda \in (0, n-1)$. Recall from Theorem 2.6.1 that, in this setting, the singular integral operator defined originally on functions $f \in L^1\big(\Sigma, \frac{\sigma(x)}{1+|x|^{n-1}}\big)$ as*

$$Tf(x) := \lim_{\varepsilon \to 0^+} \int_{\substack{y\in\Sigma\\ |x-y|>\varepsilon}} k(x-y)f(y)\,\mathrm{d}\sigma(y) \text{ for } \sigma\text{-a.e. } x \in \Sigma, \tag{2.7.79}$$

induces well-defined, linear, and bounded mappings

$$T : M^{p,\lambda}(\Sigma,\sigma) \longrightarrow M^{p,\lambda}(\Sigma,\sigma), \tag{2.7.80}$$

$$T : \mathring{M}^{p,\lambda}(\Sigma,\sigma) \longrightarrow \mathring{M}^{p,\lambda}(\Sigma,\sigma), \tag{2.7.81}$$

$$T : \mathcal{B}^{q,\lambda}(\Sigma,\sigma) \longrightarrow \mathcal{B}^{q,\lambda}(\Sigma,\sigma). \tag{2.7.82}$$

Then there exists a constant $C \in (0, \infty)$, which depends only on n, p, q, λ, and the UR *character of Σ, such that for each function $b \in L^\infty(\Sigma, \sigma)$ one has*

$$\big\|[T, M_b]\big\|_{M^{p,\lambda}(\Sigma,\sigma)\to M^{p,\lambda}(\Sigma,\sigma)} \leq C\|b\|_{\dot{\mathrm{BMO}}(\Sigma,\sigma)}, \tag{2.7.83}$$

$$\big\|[T, M_b]\big\|_{\mathring{M}^{p,\lambda}(\Sigma,\sigma)\to \mathring{M}^{p,\lambda}(\Sigma,\sigma)} \leq C\|b\|_{\dot{\mathrm{BMO}}(\Sigma,\sigma)}, \tag{2.7.84}$$

$$\big\|[T, M_b]\big\|_{\mathcal{B}^{q,\lambda}(\Sigma,\sigma)\to \mathcal{B}^{q,\lambda}(\Sigma,\sigma)} \leq C\|b\|_{\dot{\mathrm{BMO}}(\Sigma,\sigma)}. \tag{2.7.85}$$

More generally, similar norm estimates are valid for commutators involving the variable coefficient kernel singular integral operators B from (2.7.25) *and $B^\#$ from* (2.7.26).

Proof All estimates in (2.7.83)-(2.7.85) are consequence of Theorem 2.7.4 and Theorem 2.6.1. The very last claim is justified in a similar fashion, using the last part in Theorem 2.7.2. □

In the case when the underlying set is compact, we may use Theorem 2.7.5 to estimate the essential norm of commutators on Morrey and block spaces.

Theorem 2.7.6 *Let $\Sigma \subseteq \mathbb{R}^n$ be a compact* UR *set, and abbreviate $\sigma := \mathcal{H}^{n-1}\lfloor\Sigma$. Fix a sufficiently large integer $N = N(n) \in \mathbb{N}$ and consider a complex-valued function $k \in \mathscr{C}^N(\mathbb{R}^n \setminus \{0\})$ which is odd and positive homogeneous of degree $1-n$. Pick two integrability exponents $p, q \in (1, \infty)$ along with some parameter $\lambda \in (0, n-1)$, and recall from Theorem 2.6.1 that the singular integral operator defined originally on functions $f \in L^1(\Sigma, \sigma)$ as*

$$Tf(x) := \lim_{\varepsilon \to 0^+} \int_{\substack{y \in \Sigma \\ |x-y|>\varepsilon}} k(x-y)f(y)\,\mathrm{d}\sigma(y) \text{ for } \sigma\text{-a.e. } x \in \Sigma, \tag{2.7.86}$$

induces well-defined, linear, and bounded mappings

$$T : M^{p,\lambda}(\Sigma, \sigma) \longrightarrow M^{p,\lambda}(\Sigma, \sigma), \tag{2.7.87}$$

$$T : \mathring{M}^{p,\lambda}(\Sigma, \sigma) \longrightarrow \mathring{M}^{p,\lambda}(\Sigma, \sigma), \tag{2.7.88}$$

$$T : \mathcal{B}^{q,\lambda}(\Sigma, \sigma) \longrightarrow \mathcal{B}^{q,\lambda}(\Sigma, \sigma). \tag{2.7.89}$$

Then there exists a constant $C \in (0, \infty)$, which depends only on n, p, λ, and the UR *character of Σ, such that for each function $b \in L^\infty(\Sigma, \sigma)$ one has*

$$\operatorname{dist}\Big([T, M_b], \operatorname{Cp}\big(M^{p,\lambda}(\Sigma, \sigma)\big)\Big) \leq C\Big(\sum_{|\alpha|\leq N} \sup_{S^{n-1}} |\partial^\alpha k|\Big) \cdot \operatorname{dist}\big(b, \operatorname{VMO}(\Sigma, \sigma)\big) \tag{2.7.90}$$

and

$$\operatorname{dist}\Big([T, M_b], \operatorname{Cp}\big(\mathring{M}^{p,\lambda}(\Sigma, \sigma)\big)\Big) \leq C\Big(\sum_{|\alpha|\leq N} \sup_{S^{n-1}} |\partial^\alpha k|\Big) \cdot \operatorname{dist}\big(b, \operatorname{VMO}(\Sigma, \sigma)\big), \tag{2.7.91}$$

where the distances in the left-hand side are measured in $\operatorname{Bd}\big(M^{p,\lambda}(\Sigma, \sigma)\big)$ *and* $\operatorname{Bd}\big(\mathring{M}^{p,\lambda}(\Sigma, \sigma)\big)$*, respectively, and the distances in the right-hand side are measured in* $\operatorname{BMO}(\Sigma, \sigma)$.

Moreover, there exists a constant $C \in (0, \infty)$, which depends only on n, q, λ, and the UR *character of Σ, such that for each function $b \in L^\infty(\Sigma, \sigma)$ one has*

$$\operatorname{dist}\Big([T, M_b], \operatorname{Cp}\big(\mathcal{B}^{q,\lambda}(\Sigma, \sigma)\big)\Big) \leq C\Big(\sum_{|\alpha|\leq N} \sup_{S^{n-1}} |\partial^\alpha k|\Big) \cdot \operatorname{dist}\big(b, \operatorname{VMO}(\Sigma, \sigma)\big), \tag{2.7.92}$$

where the distance in the left-hand side is measured in $\operatorname{Bd}\big(\mathcal{B}^{q,\lambda}(\Sigma, \sigma)\big)$ *and the distance in the right-hand side is measured in* $\operatorname{BMO}(\Sigma, \sigma)$.

In particular,

$$\begin{gathered} [T, M_b] \text{ is compact on } M^{p,\lambda}(\Sigma, \sigma),\ \mathring{M}^{p,\lambda}(\Sigma, \sigma),\ \mathcal{B}^{q,\lambda}(\Sigma, \sigma) \\ \text{whenever } b \in L^\infty(\Sigma, \sigma) \cap \operatorname{VMO}(\Sigma, \sigma). \end{gathered} \tag{2.7.93}$$

Finally, similar results are valid for commutators involving the variable coefficient kernel singular integral operators B *from* (2.7.38) *and* $B^{\#}$ *from* (2.7.49), *in relation to a kernel function as in* (2.7.37).

In view of the piece of notation introduced in (A.0.102), we may recast (2.7.90)-(2.7.92) as

$$\big\|[T,M_b]\big\|^{\mathrm{ess}}_{M^{p,\lambda}(\Sigma,\sigma)\to M^{p,\lambda}(\Sigma,\sigma)} \leq C\Big(\sum_{|\alpha|\leq N}\sup_{S^{n-1}}|\partial^\alpha k|\Big)\cdot\mathrm{dist}\big(b,\mathrm{VMO}(\Sigma,\sigma)\big), \tag{2.7.94}$$

$$\big\|[T,M_b]\big\|^{\mathrm{ess}}_{\mathring{M}^{p,\lambda}(\Sigma,\sigma)\to \mathring{M}^{p,\lambda}(\Sigma,\sigma)} \leq C\Big(\sum_{|\alpha|\leq N}\sup_{S^{n-1}}|\partial^\alpha k|\Big)\cdot\mathrm{dist}\big(b,\mathrm{VMO}(\Sigma,\sigma)\big), \tag{2.7.95}$$

and

$$\big\|[T,M_b]\big\|^{\mathrm{ess}}_{\mathcal{B}^{q,\lambda}(\Sigma,\sigma)\to \mathcal{B}^{q,\lambda}(\Sigma,\sigma)} \leq C\Big(\sum_{|\alpha|\leq N}\sup_{S^{n-1}}|\partial^\alpha k|\Big)\cdot\mathrm{dist}\big(b,\mathrm{VMO}(\Sigma,\sigma)\big), \tag{2.7.96}$$

respectively.

Proof of Theorem 2.7.6 Fix an arbitrary function $\phi \in \mathrm{VMO}(\Sigma,\sigma)$. As in the proof of Theorem 2.7.3, fix some exponent $\alpha \in (0,1)$ and consider a sequence $\{b_j\}_{j\in\mathbb{N}} \subseteq \mathscr{C}^\alpha(\Sigma)$ such that $\lim_{j\to\infty}\|\phi - b_j\|_{\mathrm{BMO}(\Sigma,\sigma)} = 0$. From (2.7.51) and [113, Proposition 6.2.12] (cf. Comment 2 following its statement) we conclude that there exists some $C = C(\Sigma,p,\lambda) \in (0,\infty)$ with the property that for each $j \in \mathbb{N}$ we have

$$\begin{aligned}\big\|[M_b,T]-[M_{b_j},T]\big\|_{M^{p,\lambda}(\Sigma,\sigma)\to M^{p,\lambda}(\Sigma,\sigma)} \\ \leq C\Big(\sum_{|\alpha|\leq N}\sup_{S^{n-1}}|\partial^\alpha k|\Big)\Big\{\|b-\phi\|_{\mathrm{BMO}(\Sigma,\sigma)} + \|\phi - b_j\|_{\mathrm{BMO}(\Sigma,\sigma)}\Big\}.\end{aligned} \tag{2.7.97}$$

Likewise, from (2.7.60) and [113, Proposition 6.2.12] we see that, if Q_ε is as in (2.7.57)-(2.7.58), there exists some $C \in (0,\infty)$ with the property that for each $j \in \mathbb{N}$ we have

$$\big\|[M_{b_j},T] - Q_\varepsilon\big\|_{M^{p,\lambda}(\Sigma,\sigma)\to M^{p,\lambda}(\Sigma,\sigma)} \leq C\varepsilon^\alpha \ \text{ for each } \varepsilon > 0. \tag{2.7.98}$$

Consequently, for each $j \in \mathbb{N}$ we obtain

$$[M_{b_j},T] = \lim_{\varepsilon\to 0^+} Q_\varepsilon \ \text{ in } \mathrm{Bd}\big(M^{p,\lambda}(\Sigma,\sigma)\to M^{p,\lambda}(\Sigma,\sigma)\big). \tag{2.7.99}$$

Since (2.7.63) and [113, (6.2.7)] imply that

$$Q_\varepsilon \in \mathrm{Cp}\big(M^{p,\lambda}(\Sigma,\sigma)\to M^{p,\lambda}(\Sigma,\sigma)\big) \ \text{ for each } \varepsilon > 0, \tag{2.7.100}$$

we conclude from (2.7.99) and [113, (1.2.53)] that

$$[M_{b_j}, T] \in \mathrm{Cp}\big(M^{p,\lambda}(\Sigma,\sigma)\big) \ \text{ for each } \ j \in \mathbb{N}. \tag{2.7.101}$$

Together with (2.7.97), this proves that for each $j \in \mathbb{N}$ we have

$$\begin{aligned} &\mathrm{dist}\Big([T, M_b], \mathrm{Cp}\big(M^{p,\lambda}(\Sigma,\sigma)\big)\Big) \\ &\qquad \leq C\Big(\sum_{|\alpha|\leq N} \sup_{S^{n-1}} |\partial^\alpha k|\Big)\Big\{\|b-\phi\|_{\mathrm{BMO}(\Sigma,\sigma)} + \|\phi - b_j\|_{\mathrm{BMO}(\Sigma,\sigma)}\Big\}. \end{aligned} \tag{2.7.102}$$

Passing to the limit $j \to \infty$ in (2.7.102) then yields (keeping in mind (2.7.50))

$$\mathrm{dist}\Big([T, M_b], \mathrm{Cp}\big(M^{p,\lambda}(\Sigma,\sigma)\big)\Big) \leq C\Big(\sum_{|\alpha|\leq N} \sup_{S^{n-1}} |\partial^\alpha k|\Big)\|b-\phi\|_{\mathrm{BMO}(\Sigma,\sigma)}. \tag{2.7.103}$$

At this stage, taking the infimum over all $\phi \in \mathrm{VMO}(\Sigma,\sigma)$ establishes (2.7.90).

In view of (A.0.88)-[113, (6.2.15)] and [113, (1.2.20)], the same proof as above works, and gives (2.7.91), as soon as we show that

$$Q_\varepsilon \in \mathrm{Cp}\big(\mathring{M}^{p,\lambda}(\Sigma,\sigma) \to \mathring{M}^{p,\lambda}(\Sigma,\sigma)\big) \ \text{ for each } \ \varepsilon > 0. \tag{2.7.104}$$

To justify this, observe first that since with $s := \frac{p(n-1)}{n-1-\lambda} \in (p,\infty)$ we have

$$Q_\varepsilon \in \mathrm{Bd}\big(M^{p,\lambda}(\Sigma,\sigma) \to M^{p,\lambda}(\Sigma,\sigma)\big) \cap \mathrm{Bd}\big(L^s(\Sigma,\sigma) \to L^s(\Sigma,\sigma)\big) \tag{2.7.105}$$

(cf. (2.7.100), (2.7.63), and [113, (1.2.52)]), it follows from (A.0.88) and [113, (6.2.15)] that

$$Q_\varepsilon \in \mathrm{Bd}\big(\mathring{M}^{p,\lambda}(\Sigma,\sigma) \to \mathring{M}^{p,\lambda}(\Sigma,\sigma)\big). \tag{2.7.106}$$

From (2.7.100), (2.7.106), and [113, (1.1.17), (6.2.15)] we then conclude that the claim made in (2.7.104) is true. This completes the proof of (2.7.91).

Let us now show (2.7.92). To this end, assume that $1/p + 1/q = 1$. Via duality, from (2.7.104), Schauder's theorem (cf. [113, (1.2.57)]), and [113, (6.2.155)] we conclude that the transpose of the operator Q_ε satisfies

$$Q_\varepsilon^\top \in \mathrm{Cp}\big(\mathcal{B}^{q,\lambda}(\Sigma,\sigma) \to \mathcal{B}^{q,\lambda}(\Sigma,\sigma)\big) \ \text{ for each } \ \varepsilon > 0. \tag{2.7.107}$$

Next, based on duality, (2.7.99), (2.7.107), [113, (6.2.155)], [113, (1.2.53)], and (2.6.15) we see that

$$[M_{b_j}, T] \in \mathrm{Cp}\big(\mathcal{B}^{q,\lambda}(\Sigma,\sigma)\big) \ \text{ for each } \ j \in \mathbb{N}. \tag{2.7.108}$$

Note that (2.7.97), [113, (6.2.15)], and [113, (1.2.20)] imply that

$$\begin{aligned}\big\|[M_b,T]-[M_{b_j},T]\big\|_{\mathring{M}^{p,\lambda}(\Sigma,\sigma)\to\mathring{M}^{p,\lambda}(\Sigma,\sigma)} \quad (2.7.109)\\ \leq C\Big(\sum_{|\alpha|\leq N}\sup_{S^{n-1}}|\partial^\alpha k|\Big)\Big\{\|b-\phi\|_{\mathrm{BMO}(\Sigma,\sigma)}+\|\phi-b_j\|_{\mathrm{BMO}(\Sigma,\sigma)}\Big\}.\end{aligned}$$

Using duality, (2.7.109), [113, (1.2.68), (6.2.155)], and (2.6.15) we next deduce that

$$\begin{aligned}\big\|[M_b,T]-[M_{b_j},T]\big\|_{\mathcal{B}^{q,\lambda}(\Sigma,\sigma)\to\mathcal{B}^{q,\lambda}(\Sigma,\sigma)} \quad (2.7.110)\\ \leq C\Big(\sum_{|\alpha|\leq N}\sup_{S^{n-1}}|\partial^\alpha k|\Big)\Big\{\|b-\phi\|_{\mathrm{BMO}(\Sigma,\sigma)}+\|\phi-b_j\|_{\mathrm{BMO}(\Sigma,\sigma)}\Big\}.\end{aligned}$$

In particular, (2.7.110) and (2.7.108) prove that for each $j\in\mathbb{N}$ we have

$$\begin{aligned}\mathrm{dist}\Big([T,M_b],\mathrm{Cp}\big(\mathcal{B}^{q,\lambda}(\Sigma,\sigma)\big)\Big) \quad (2.7.111)\\ \leq C\Big(\sum_{|\alpha|\leq N}\sup_{S^{n-1}}|\partial^\alpha k|\Big)\Big\{\|b-\phi\|_{\mathrm{BMO}(\Sigma,\sigma)}+\|\phi-b_j\|_{\mathrm{BMO}(\Sigma,\sigma)}\Big\}.\end{aligned}$$

Upon sending $j\to\infty$ in (2.7.111) and taking into account (2.7.50) we obtain

$$\begin{aligned}\mathrm{dist}\Big([T,M_b],\mathrm{Cp}\big(\mathcal{B}^{q,\lambda}(\Sigma,\sigma)\big)\Big) \quad (2.7.112)\\ \leq C\Big(\sum_{|\alpha|\leq N}\sup_{S^{n-1}}|\partial^\alpha k|\Big)\|b-\phi\|_{\mathrm{BMO}(\Sigma,\sigma)}.\end{aligned}$$

At this stage, taking the infimum over all $\phi\in\mathrm{VMO}(\Sigma,\sigma)$ yields (2.7.92).

Finally, the very last claim in the statement of the theorem is justified in a similar fashion, relying on the discussion just prior to the proof of Theorem 2.7.3. □

The following theorem refines some of the work in [21].

Theorem 2.7.7 *Suppose $\Sigma\subseteq\mathbb{R}^n$ is a closed Ahlfors regular set, and abbreviate $\sigma:=\mathcal{H}^{n-1}\lfloor\Sigma$. Fix an exponent $p_0\in(1,\infty)$ along with some non-decreasing function $\Phi:(0,\infty)\to(0,\infty)$ and let T be a linear operator which is bounded on $L^{p_0}(\Sigma,\omega)$ for every $\omega\in A_{p_0}(\Sigma,\sigma)$, with operator norm $\leq\Phi([\omega]_{A_{p_0}})$.*

Then for each exponent $p\in(1,\infty)$ and Muckenhoupt weight $w\in A_p(\Sigma,\sigma)$ the operator

$$T:L^p(\Sigma,w)\longrightarrow L^p(\Sigma,w) \quad (2.7.113)$$

is well defined, linear, and bounded. In particular, its transpose

$$T^\top:L^{p'}(\Sigma,w')\longrightarrow L^{p'}(\Sigma,w') \quad (2.7.114)$$

is also well defined, linear, and bounded, where $p':=(1-1/p)^{-1}\in(1,\infty)$ is the conjugate exponent of p and $w':=w^{1-p'}\in A_{p'}(\Sigma,\sigma)$ is the conjugate weight of w

(cf. item (2) in [112, Lemma 7.7.1]). *Finally, there exists a constant* $C \in (0, \infty)$ *with the property that*

$$\begin{aligned}&\text{if } f \in L^p(\Sigma, w) \text{ and } g \in L^{p'}(\Sigma, w') \text{ then } (Tf)g - f(T^\top g) \text{ belongs}\\&\text{to the homogeneous Hardy space } \dot{H}^1(\Sigma, \sigma) \text{ (cf. [113, (4.2.12)]) and}\\&\text{one has } \big\|(Tf)g - f(T^\top g)\big\|_{H^1(\Sigma,\sigma)} \leq C\|f\|_{L^p(\Sigma,w)} \cdot \|g\|_{L^{p'}(\Sigma,w')}.\end{aligned} \tag{2.7.115}$$

In the case when Σ is the boundary of a UR domain $\Omega \subseteq \mathbb{R}^n$, (2.7.115) contains as a particular case [113, (10.1.137)] in the scenario described in item *(1)* of [113, Corollary 10.1.12] (since each Riesz transform R_j is skew-symmetric). See also item *(10)* in Theorem 2.3.2.

Proof of Theorem 2.7.7 That T is well defined, linear, and bounded in the context of (2.7.113) has been seen in Theorem 2.7.1. Via duality (cf. (2.3.60)), this further implies that $T^\top$ is also well defined, linear, and bounded in the context of (2.7.114). To justify the claims made in (2.7.115), fix two arbitrary functions $f \in L^p(\Sigma, w)$ and $g \in L^{p'}(\Sigma, w')$. In particular, the difference $h := (Tf)g - f(T^\top g)$ belongs to $L^1(\Sigma, \sigma)$ and satisfies

$$\int_\Sigma h \, d\sigma = 0. \tag{2.7.116}$$

Next, pick an arbitrary $\phi \in \mathrm{Lip}_c(\Sigma)$ and denote by M_ϕ the operator of pointwise multiplication by the function ϕ. Then $\phi f \in L^p(\Sigma, w)$ and we may then write

$$\begin{aligned}\Big|\int_\Sigma \phi\, h \, d\sigma\Big| &= \Big|\int_\Sigma \phi\big((Tf)g - f(T^\top g)\big) \, d\sigma\Big|\\&= \Big|\int_\Sigma g[M_\phi, T]f \, d\sigma\Big| \leq \|g\|_{L^{p'}(\Sigma,w')} \cdot \big\|[M_\phi, T]f\big\|_{L^p(\Sigma,w)}\\&\leq C\|g\|_{L^{p'}(\Sigma,w')} \cdot \|f\|_{L^p(\Sigma,w)} \cdot \|\phi\|_{\dot{\mathrm{BMO}}(\Sigma,\sigma)},\end{aligned} \tag{2.7.117}$$

using Hölder's inequality and (2.7.3). With this in hand, we conclude from [113, (4.4.10)], (2.7.116), and [113, (4.6.26)] that all conclusions in (2.7.115) hold. □

In the last part of this section we shall extend our earlier commutator estimates to abstract Generalized Banach Function Spaces. Here is the first result of this flavor, generalizing Theorem 2.7.4 (here [113, Proposition 6.2.17] is relevant) and Theorem 2.7.7.

Theorem 2.7.8 *Suppose* $\Sigma \subseteq \mathbb{R}^n$ *is a closed Ahlfors regular set, and abbreviate* $\sigma := \mathcal{H}^{n-1}\lfloor\Sigma$. *With* $\mathcal{M}_\Sigma$ *denoting the Hardy-Littlewood maximal operator on* Σ, *assume* $\mathbb{X}$ *is a Generalized Banach Function Space on* (Σ, σ) *such that*

$$\begin{aligned}\mathcal{M}_\Sigma : \mathbb{X} \to \mathbb{X} \text{ and } \mathcal{M}_\Sigma : \mathbb{X}' \to \mathbb{X}'\\\text{are well-defined bounded mappings,}\end{aligned} \tag{2.7.118}$$

where $\mathbb{X}'$ *is the associated space of* $\mathbb{X}$ *(cf.* [113, Definitions 5.1.4, 5.1.11]). *Finally, fix an integrability exponent* $p_0 \in (1, \infty)$ *along with some non-decreasing function*

$\Phi : (0, \infty) \to (0, \infty)$ and let T be a linear operator which is bounded on $L^{p_0}(\Sigma, w)$ for every $w \in A_{p_0}(\Sigma, \sigma)$, with operator norm $\leq \Phi([w]_{A_{p_0}})$.

Then the operator T induces well-defined, linear, and bounded mappings in the contexts

$$T : \mathbb{X} \longrightarrow \mathbb{X}, \tag{2.7.119}$$

$$T : \mathbb{X}' \longrightarrow \mathbb{X}'. \tag{2.7.120}$$

Also, there exist two constants $C_1, C_2 \in (0, \infty)$, depending only on Σ, p_0, and the operator norms of $\mathcal{M}_\Sigma$ on $\mathbb{X}$ and $\mathbb{X}'$, with the property that for every complex-valued function $b \in L^\infty(\Sigma, \sigma)$ one has

$$\big\|[M_b, T]\big\|_{\mathbb{X}\to\mathbb{X}} \leq C_1\Phi(C_2)\|b\|_{\dot{\mathrm{BMO}}(\Sigma,\sigma)}, \tag{2.7.121}$$

$$\big\|[M_b, T]\big\|_{\mathbb{X}'\to\mathbb{X}'} \leq C_1\Phi(C_2)\|b\|_{\dot{\mathrm{BMO}}(\Sigma,\sigma)}, \tag{2.7.122}$$

where $[M_b, T] := bT(\cdot) - T(b\,\cdot)$ is the commutator of the operator M_b of pointwise multiplication by the function b and the operator T (considered in either of the scenarios described in (2.7.119)-(2.7.120)*). Moreover,*

similar results to (2.7.119), (2.7.120) *and* (2.7.121), (2.7.122) *are also valid for $T^\top$, the (real) transpose of the original operator T.* (2.7.123)

Finally, there exists a constant $C \in (0, \infty)$ with the property that

whenever $f \in \mathring{\mathbb{X}}$ and $g \in \mathbb{X}'$ it follows that $(Tf)g - f(T^\top g)$ belongs to the homogeneous Hardy space $\dot{H}^1(\Sigma, \sigma)$ (cf. [113, (4.2.12)]*) and one has the estimate $\big\|(Tf)g - f(T^\top g)\big\|_{H^1(\Sigma,\sigma)} \leq C\|f\|_{\mathbb{X}} \cdot \|g\|_{\mathbb{X}'}$,* (2.7.124)

and

whenever $f \in \mathbb{X}$ and $g \in (\mathbb{X}')^\circ$ it follows that $(Tf)g - f(T^\top g)$ belongs to the homogeneous Hardy space $\dot{H}^1(\Sigma, \sigma)$ (cf. [113, (4.2.12)]*) and one has the estimate $\big\|(Tf)g - f(T^\top g)\big\|_{H^1(\Sigma,\sigma)} \leq C\|f\|_{\mathbb{X}} \cdot \|g\|_{\mathbb{X}'}$.* (2.7.125)

In relation to Theorem 2.7.8, we make two comments. First, thanks to item *(10)* in Theorem 2.3.2, the conclusions in Theorem 2.7.8 apply to all principal-value singular integral operators of the sort considered in (2.3.15). Given the significance of this special case, we shall state it as a stand-alone result a little later, subsumed in Corollary 2.7.9.

Our second comment regarding Theorem 2.7.8 is that, as noted in [113, Proposition 5.3.14], the conditions demanded in (2.7.118) automatically hold in the case when $\mathbb{X}$ is a rearrangement invariant Banach function space on the (non-atomic, sigma-finite measure space (Σ, σ)) whose Boyd indices satisfy $1 < p_{\mathbb{X}} \le q_{\mathbb{X}} < \infty$. (2.7.126)

Consequently, *all conclusions in Theorem 2.7.8 are valid in the class of rearrangement invariant Banach function spaces* $\mathbb{X}$ *whose Boyd indices satisfy* $1 < p_{\mathbb{X}} \le q_{\mathbb{X}} < \infty$. In particular, this is the case when the Generalized Banach Function Space $\mathbb{X}$ is actually an Orlicz space, $\mathbb{X} := L^{\Phi}(\Sigma, \sigma)$, associated with the (non-atomic, sigma-finite) measure space (Σ, σ) and a Young function Φ satisfying $1 < i(\Phi) \le I(\Phi) < \infty$; see [113, Proposition 5.3.15] and [113, (5.3.62)]. As a concrete example, one may take $\Phi(t) \approx t^p[\ln(\mathrm{e} + t)]^{\alpha}$ for some fixed $p \in (1, \infty)$ and $\alpha \in \mathbb{R}$, a choice for which the Orlicz space L^{Φ} actually becomes Zygmund's space $L^p(\log L)^{\alpha}$ (see [113, Example 5.3.10]).

Here is the proof of Theorem 2.7.8.

Proof of Theorem 2.7.8 All claims in (2.7.119)-(2.7.120) and (2.7.121)-(2.7.122) are consequences of Theorem 2.7.1 and [113, Corollary 5.2.3]. Also, (2.7.123) is justified in a similar manner, keeping in mind that $T^{\top}$ satisfies the same properties as the original operator T, with p_0 replaced by $q_0 := (1 - 1/p_0)^{-1} \in (1, \infty)$, its Hölder conjugate exponent.

As a preamble to the proof of (2.7.124), we first establish that

$$\int_{\Sigma} (Tf)g \, d\sigma = \int_{\Sigma} f(T^{\top}g) \, d\sigma \quad \text{for each } f \in \mathring{\mathbb{X}} \text{ and } g \in \mathbb{X}'. \tag{2.7.127}$$

With this aim in mind, fix $f \in \mathring{\mathbb{X}}$ and $g \in \mathbb{X}'$, arbitrary. Since $L^{\infty}_{\text{comp}}(\Sigma, \sigma)$ is dense in $\mathring{\mathbb{X}}$ (cf. [113, (5.2.82)]), there exists a sequence $\{f_j\}_{j \in \mathbb{N}} \subseteq L^{\infty}_{\text{comp}}(\Sigma, \sigma)$ which converges to f in $\mathbb{X}$. Also, [113, Corollary 5.2.3] implies that there exists a weight $\omega \in A_{q_0}(\Sigma, \sigma)$ such that $g \in L^{q_0}(\Sigma, \omega\sigma)$. Then item *(2)* in [112, Lemma 7.7.1] ensures that $w := \omega^{1-p_0} \in A_{p_0}(\Sigma, \sigma)$ so, bearing in mind (2.7.119) as well as [113, Proposition 5.1.12], we may write

$$\begin{aligned} \int_{\Sigma} (Tf)g \, d\sigma &= \lim_{j \to \infty} \int_{\Sigma} (Tf_j)g \, d\sigma = \lim_{j \to \infty} \int_{\Sigma} f_j(T^{\top}g) \, d\sigma \\ &= \int_{\Sigma} f(T^{\top}g) \, d\sigma, \end{aligned} \tag{2.7.128}$$

where the second equality in (2.7.128) uses the fact that the transposed of the operator T is considered at the level of Muckenhoupt weighted Lebesgue spaces and that $L^{\infty}_{\text{comp}}(\Sigma, \sigma) \subseteq L^{p_0}(\Sigma, w\sigma) = \big(L^{q_0}(\Sigma, \omega\sigma)\big)^*$. This finishes the proof of (2.7.127).

To proceed, select two arbitrary functions, $f \in \mathring{\mathbb{X}}$ and $g \in \mathbb{X}'$. Since $\mathring{\mathbb{X}} \subseteq \mathbb{X}$, from (2.7.119)-(2.7.120) and [113, Proposition 5.1.12] we conclude that the function $h := (Tf)g - f(T^{\top}g)$ is absolutely integrable on Σ (with respect to the measure σ). In addition, (2.7.127) permits us to write

$$\begin{aligned}\int_\Sigma h\,d\sigma &= \int_\Sigma (Tf)g\,d\sigma - \int_\Sigma f(T^\top g)\,d\sigma \\ &= \int_\Sigma (Tf)g\,d\sigma - \int_\Sigma (Tf)g\,d\sigma = 0.\end{aligned} \tag{2.7.129}$$

Fix now an arbitrary function $\phi \in \mathrm{Lip}_c(\Sigma)$ and denote by M_ϕ the operator of pointwise multiplication by ϕ. From [113, (5.2.85)] we know that $\phi f \in \mathring{\mathbb{X}}$. Granted this and using (2.7.127) we may then compute

$$\begin{aligned}\Big|\int_\Sigma \phi\, h\,d\sigma\Big| &= \Big|\int_\Sigma \phi\big((Tf)g - f(T^\top g)\big)\,d\sigma\Big| = \Big|\int_\Sigma \phi(Tf)g - T(\phi f)g\big)\,d\sigma\Big| \\ &= \Big|\int_\Sigma \Big(\big[M_\phi, T\big]f\Big)g\,d\sigma\Big| \leq \big\|[T, M_\phi]f\big\|_{\mathbb{X}} \cdot \|g\|_{\mathbb{X}'} \\ &\leq C\|f\|_{\mathbb{X}} \cdot \|g\|_{\mathbb{X}'} \cdot \|\phi\|_{\mathrm{B\dot{M}O}(\Sigma,\sigma)},\end{aligned} \tag{2.7.130}$$

with the penultimate inequality coming from [113, (5.1.37)], and last inequality provided by (2.7.121). At this stage, we conclude from [113, (4.4.10)], (2.7.129), (2.7.130), [113, (4.6.26)] that all conclusions in (2.7.124) are valid. With this in hand, (2.7.125) follows from (2.7.124) by replacing $\mathbb{X}$ with $\mathbb{X}'$ (a permissible choice in light of [113, Proposition 5.1.14]), interchanging the roles of f and g, and replacing T by $T^\top$ (which enjoys the same properties as the original operator). □

It is of interest to specialize Theorem 2.7.8 to convolution-type singular integral operators on UR sets, a scenario in which we have the result described in the following corollary, which also encompasses $\mathring{\mathbb{X}}$ is defined in relation to a given Generalized Banach Function space $\mathbb{X}$ as in [113, Definition 5.2.6].

Corollary 2.7.9 *Let $\Sigma \subseteq \mathbb{R}^n$ be a* UR *set, and abbreviate $\sigma := \mathcal{H}^{n-1}\lfloor\Sigma$. With $\mathcal{M}_\Sigma$ denoting the Hardy-Littlewood maximal operator on Σ, assume $\mathbb{X}$ is a Generalized Banach Function Space on (Σ, σ) such that*

$$\begin{gathered}\mathcal{M}_\Sigma : \mathbb{X} \to \mathbb{X} \text{ and } \mathcal{M}_\Sigma : \mathbb{X}' \to \mathbb{X}' \\ \text{are well-defined bounded mappings,}\end{gathered} \tag{2.7.131}$$

where $\mathbb{X}'$ is the associated space of $\mathbb{X}$ (cf. [113, Definitions 5.1.4, 5.1.11]). *Finally, fix a sufficiently large integer $N = N(n) \in \mathbb{N}$ and consider a complex-valued function $k \in \mathscr{C}^N(\mathbb{R}^n \setminus \{0\})$ which is odd and positive homogeneous of degree $1 - n$, and consider the singular integral operator acting on functions $f \in L^1\big(\Sigma, \frac{\sigma(x)}{1+|x|^{n-1}}\big)$ according to*

$$Tf(x) := \lim_{\varepsilon \to 0^+} \int_{\substack{y\in\Sigma \\ |x-y|>\varepsilon}} k(x-y)f(y)\,d\sigma(y) \ \textit{for } \sigma\textit{-a.e. } x \in \Sigma. \tag{2.7.132}$$

Then the operator T induces well-defined, linear, and bounded mappings in the contexts

$$T : \mathbb{X} \longrightarrow \mathbb{X}, \quad T : \mathring{\mathbb{X}} \longrightarrow \mathring{\mathbb{X}}, \tag{2.7.133}$$

$$T : \mathbb{X}' \longrightarrow \mathbb{X}', \quad T : (\mathbb{X}')^\circ \longrightarrow (\mathbb{X}')^\circ, \tag{2.7.134}$$

where

$$\mathring{\mathbb{X}} := \overline{L^\infty_{\rm comp}(\Sigma,\sigma)}^{\|\cdot\|_{\mathbb{X}}} \text{ and } (\mathbb{X}')^\circ := \overline{L^\infty_{\rm comp}(\Sigma,\sigma)}^{\|\cdot\|_{\mathbb{X}'}}. \tag{2.7.135}$$

In addition, there exists a constant $C \in (0,\infty)$, depending only on Σ, k, plus the operator norms of $\mathcal{M}_\Sigma$ on $\mathbb{X}$ and $\mathbb{X}'$, with the property that for every complex-valued function $b \in L^\infty(\Sigma,\sigma)$ one has

$$\big\|[M_b, T]\big\|_{\mathring{\mathbb{X}}\to\mathring{\mathbb{X}}} \le \big\|[M_b, T]\big\|_{\mathbb{X}\to\mathbb{X}} \le C\|b\|_{\dot{\rm BMO}(\Sigma,\sigma)}, \tag{2.7.136}$$

$$\big\|[M_b, T]\big\|_{(\mathbb{X}')^\circ\to(\mathbb{X}')^\circ} \le \big\|[M_b, T]\big\|_{\mathbb{X}'\to\mathbb{X}'} \le C\|b\|_{\dot{\rm BMO}(\Sigma,\sigma)}, \tag{2.7.137}$$

where $[M_b, T] := bT(\cdot) - T(b\,\cdot)$ is the commutator of the operator M_b of pointwise multiplication by the function b and the operator T (considered in either of the scenarios described in (2.7.133)-(2.7.134)).

Furthermore, there exists a constant $C \in (0,\infty)$ with the property that

whenever either $f \in \mathring{\mathbb{X}}$ and $g \in \mathbb{X}'$, or $f \in \mathbb{X}$ and $g \in (\mathbb{X}')^\circ$, it follows that the function $(Tf)g + f(Tg)$ belongs to the homogeneous Hardy space $\dot{H}^1(\Sigma,\sigma)$ (defined in [113, (4.2.12)]) *and one has the estimate $\big\|(Tf)g + f(Tg)\big\|_{H^1(\Sigma,\sigma)} \le C\|f\|_{\mathbb{X}} \cdot \|g\|_{\mathbb{X}'}$.* (2.7.138)

More generally, similar mapping properties and norm estimates are valid for commutators involving the variable coefficient kernel singular integral operators B from (2.7.25) *and $B^\#$ from* (2.7.26).

We wish to remark that, in light of [113, Proposition 5.3.14], *all conclusions in Corollary 2.7.9 are valid in the class of rearrangement invariant Banach function spaces $\mathbb{X}$ whose Boyd indices satisfy* $1 < p_{\mathbb{X}} \le q_{\mathbb{X}} < \infty$. For example, this is the case when the Generalized Banach Function Space $\mathbb{X}$ is actually an Orlicz space, $\mathbb{X} := L^\Phi(\Sigma,\sigma)$, associated with the (non-atomic, sigma-finite) measure space (Σ,σ) and a Young function Φ satisfying $1 < i(\Phi) \le I(\Phi) < \infty$; see [113, Proposition 5.3.15] and [113, (5.3.62)].

Proof of Corollary 2.7.9 All claims pertaining to constant coefficient kernel singular integral operators are consequences of item *(11)* in Theorem 2.3.2, and Theorem 2.7.8, bearing in mind [113, (1.2.20)] and (2.3.25) The more general case, corresponding to singular integral operators with variable coefficient kernels, are dealt with virtually verbatim, while keeping in mind item *(e)* in Theorem 2.5.38 and (2.5.470). □

Finally, in the case when the underlying set is compact, we have the following version of Theorem 2.7.6 with abstract Generalized Banach Function Spaces playing the role of Morrey and block spaces.

Theorem 2.7.10 *Let* $\Sigma \subseteq \mathbb{R}^n$ *be a compact* UR *set, and abbreviate* $\sigma := \mathcal{H}^{n-1}\lfloor\Sigma$. *With* $\mathcal{M}_\Sigma$ *denoting the Hardy-Littlewood maximal operator on* Σ, *assume* $\mathbb{X}$ *is a Generalized Banach Function Space on* (Σ, σ) *such that*

$$\begin{gathered}\mathcal{M}_\Sigma : \mathbb{X} \to \mathbb{X} \text{ and } \mathcal{M}_\Sigma : \mathbb{X}' \to \mathbb{X}' \\ \text{are well-defined bounded mappings,}\end{gathered} \tag{2.7.139}$$

where $\mathbb{X}'$ *is the associated space of* $\mathbb{X}$ (*cf.* [113, Definitions 5.1.4, 5.1.11]). *Lastly, fix a sufficiently large integer* $N = N(n) \in \mathbb{N}$ *and consider a complex-valued function* $k \in \mathscr{C}^N(\mathbb{R}^n \setminus \{0\})$ *which is odd and positive homogeneous of degree* $1-n$, *and recall from Corollary 2.7.9 that the singular integral operator defined originally on functions* $f \in L^1(\Sigma, \sigma)$ *as*

$$Tf(x) := \lim_{\varepsilon \to 0^+} \int_{\substack{y \in \Sigma \\ |x-y|>\varepsilon}} k(x-y)f(y)\,\mathrm{d}\sigma(y) \text{ for } \sigma\text{-a.e. } x \in \Sigma, \tag{2.7.140}$$

has $\mathbb{X}$, $\mathbb{X}'$, *as well as*

$$\mathring{\mathbb{X}} := \overline{L^\infty(\Sigma, \sigma)}^{\|\cdot\|_{\mathbb{X}}} \text{ and } (\mathbb{X}')^\circ := \overline{L^\infty(\Sigma, \sigma)}^{\|\cdot\|_{\mathbb{X}'}} \tag{2.7.141}$$

as invariant subspaces (*cf.* [113, (5.2.76)]), *hence*

$$T : \mathbb{X} \longrightarrow \mathbb{X}, \quad T : \mathring{\mathbb{X}} \longrightarrow \mathring{\mathbb{X}}, \tag{2.7.142}$$

$$T : \mathbb{X}' \longrightarrow \mathbb{X}', \quad T : (\mathbb{X}')^\circ \longrightarrow (\mathbb{X}')^\circ, \tag{2.7.143}$$

are well-defined, linear, bounded operators.

Then there exists a constant $C \in (0, \infty)$, *which depends only on the operator norms of* $\mathcal{M}_\Sigma$ *on* $\mathbb{X}$ *and* $\mathbb{X}'$ *and the* UR *character of* Σ, *with the property that for each function* $b \in L^\infty(\Sigma, \sigma)$ *one has*

$$\begin{aligned}&\operatorname{dist}\Big([T, M_b], \operatorname{Cp}\big(\mathring{\mathbb{X}}\big)\Big) \\ &\qquad \le C\Big(\sum_{|\alpha|\le N} \sup_{S^{n-1}} |\partial^\alpha k|\Big) \cdot \operatorname{dist}\big(b, \mathrm{VMO}(\Sigma, \sigma)\big),\end{aligned} \tag{2.7.144}$$

and

$$\operatorname{dist}\Big([T, M_b], \operatorname{Cp}(\mathbb{X})\Big)$$
$$\leq C\Big(\sum_{|\alpha|\leq N}\sup_{S^{n-1}}|\partial^\alpha k|\Big)\cdot\operatorname{dist}\big(b, \mathrm{VMO}(\Sigma,\sigma)\big), \tag{2.7.145}$$

where the distances in the left are measured in $\operatorname{Bd}(\mathring{\mathbb{X}})$ *and* $\operatorname{Bd}(\mathbb{X})$*, respectively, while the distances in the right are measured in* $\mathrm{BMO}(\Sigma,\sigma)$*, as well as*

$$\operatorname{dist}\Big([T, M_b], \operatorname{Cp}\big((\mathbb{X}')^\circ\big)\Big)$$
$$\leq C\Big(\sum_{|\alpha|\leq N}\sup_{S^{n-1}}|\partial^\alpha k|\Big)\cdot\operatorname{dist}\big(b, \mathrm{VMO}(\Sigma,\sigma)\big), \tag{2.7.146}$$

and

$$\operatorname{dist}\Big([T, M_b], \operatorname{Cp}(\mathbb{X}')\Big)$$
$$\leq C\Big(\sum_{|\alpha|\leq N}\sup_{S^{n-1}}|\partial^\alpha k|\Big)\cdot\operatorname{dist}\big(b, \mathrm{VMO}(\Sigma,\sigma)\big), \tag{2.7.147}$$

where the distances in the left are measured in $\operatorname{Bd}\big((\mathbb{X}')^\circ\big)$ *and* $\operatorname{Bd}(\mathbb{X}')$*, respectively, and the distances in the right are measured in* $\mathrm{BMO}(\Sigma,\sigma)$.

In particular,

$$\begin{aligned}&[T, M_b] \text{ is compact on } \mathbb{X},\ \mathbb{X}',\ \mathring{\mathbb{X}},\ (\mathbb{X}')^\circ\\ &\text{whenever } b\in L^\infty(\Sigma,\sigma)\cap\mathrm{VMO}(\Sigma,\sigma).\end{aligned} \tag{2.7.148}$$

Finally, analogous estimates are true for commutators involving the more general variable coefficient kernel singular integral operators B *from* (2.7.38) *and* $B^\#$ *from* (2.7.49).

Upon recalling the piece of notation introduced in (A.0.102), we may reformulate (2.7.144)-(2.7.145) and (2.7.146)-(2.7.147) simply as

$$\big\|[T, M_b]\big\|^{\mathrm{ess}}_{\mathring{\mathbb{X}}\to\mathring{\mathbb{X}}}\leq C\Big(\sum_{|\alpha|\leq N}\sup_{S^{n-1}}|\partial^\alpha k|\Big)\cdot\operatorname{dist}\big(b, \mathrm{VMO}(\Sigma,\sigma)\big), \tag{2.7.149}$$

$$\big\|[T, M_b]\big\|^{\mathrm{ess}}_{\mathbb{X}\to\mathbb{X}}\leq C\Big(\sum_{|\alpha|\leq N}\sup_{S^{n-1}}|\partial^\alpha k|\Big)\cdot\operatorname{dist}\big(b, \mathrm{VMO}(\Sigma,\sigma)\big), \tag{2.7.150}$$

and

$$\big\|[T, M_b]\big\|^{\text{ess}}_{(\mathbb{X}')^\circ \to (\mathbb{X}')^\circ} \leq C\Big(\sum_{|\alpha|\leq N} \sup_{S^{n-1}} |\partial^\alpha k|\Big) \cdot \operatorname{dist}\big(b, \mathrm{VMO}(\Sigma, \sigma)\big), \tag{2.7.151}$$

$$\big\|[T, M_b]\big\|^{\text{ess}}_{\mathbb{X}' \to \mathbb{X}'} \leq C\Big(\sum_{|\alpha|\leq N} \sup_{S^{n-1}} |\partial^\alpha k|\Big) \cdot \operatorname{dist}\big(b, \mathrm{VMO}(\Sigma, \sigma)\big), \tag{2.7.152}$$

respectively.

It is also of significance to point out that, thanks to [113, Proposition 5.3.14], *all conclusions in Theorem 2.7.10 remain true for rearrangement invariant Banach function spaces* $\mathbb{X}$ *whose Boyd indices satisfy* $1 < p_{\mathbb{X}} \leq q_{\mathbb{X}} < \infty$. In particular, said conclusions hold for $\mathbb{X} := L^\Phi(\Sigma, \sigma)$, the Orlicz space associated with the (non-atomic, sigma-finite) measure space (Σ, σ) and a Young function Φ satisfying $1 < i(\Phi) \leq I(\Phi) < \infty$ (cf. [113, Proposition 5.3.15] and [113, (5.3.62)]).

Proof of Theorem 2.7.10 Fix an arbitrary function $\phi \in \mathrm{VMO}(\Sigma, \sigma)$. Having picked an exponent $\alpha \in (0, 1)$, the characterization of the Sarason space given in [113, (3.1.50)] guarantees the existence of a sequence $\{b_j\}_{j\in\mathbb{N}} \subseteq \mathscr{C}^\alpha(\Sigma)$ such that $\lim_{j\to\infty} \|\phi - b_j\|_{\mathrm{BMO}(\Sigma,\sigma)} = 0$. Also, from (2.7.133), [113, (5.2.85)], [113, (1.2.20)], (2.7.51), and [113, Corollary 5.2.3] we see that there exists a constant $C = C(\Sigma, \mathbb{X}) \in (0, \infty)$ with the property that for each $j \in \mathbb{N}$ we have

$$\begin{aligned} \big\|[M_b, T]-[M_{b_j}, T]\big\|_{\dot{\mathbb{X}}\to\dot{\mathbb{X}}} &\leq \big\|[M_b, T] - [M_{b_j}, T]\big\|_{\mathbb{X}\to\mathbb{X}} \\ &\leq C\Big(\sum_{|\alpha|\leq N} \sup_{S^{n-1}} |\partial^\alpha k|\Big)\Big\{\|b - \phi\|_{\mathrm{BMO}(\Sigma,\sigma)} + \|\phi - b_j\|_{\mathrm{BMO}(\Sigma,\sigma)}\Big\}, \end{aligned} \tag{2.7.153}$$

and

$$\begin{aligned} \big\|[M_b, T]-[M_{b_j}, T]\big\|_{(\mathbb{X}')^\circ\to(\mathbb{X}')^\circ} &\leq \big\|[M_b, T] - [M_{b_j}, T]\big\|_{\mathbb{X}'\to\mathbb{X}'} \\ &\leq C\Big(\sum_{|\alpha|\leq N} \sup_{S^{n-1}} |\partial^\alpha k|\Big)\Big\{\|b - \phi\|_{\mathrm{BMO}(\Sigma,\sigma)} + \|\phi - b_j\|_{\mathrm{BMO}(\Sigma,\sigma)}\Big\}. \end{aligned} \tag{2.7.154}$$

Next, if Q_ε is as in (2.7.57)-(2.7.58), then from [113, (1.2.20)], (2.7.60), and [113, Corollary 5.2.3] we conclude (also bearing in mind (2.7.63) and [113, (5.2.76)]) that there exists some constant $C \in (0, \infty)$ with the property that for each $j \in \mathbb{N}$ we have

$$\big\|[M_{b_j}, T] - Q_\varepsilon\big\|_{\dot{\mathbb{X}}\to\dot{\mathbb{X}}} \leq \big\|[M_{b_j}, T] - Q_\varepsilon\big\|_{\mathbb{X}\to\mathbb{X}} \leq C\varepsilon^\alpha \text{ for each } \varepsilon > 0, \tag{2.7.155}$$

and

$$\big\|[M_{b_j}, T] - Q_\varepsilon\big\|_{(\mathbb{X}')^\circ\to(\mathbb{X}')^\circ} \leq \big\|[M_{b_j}, T] - Q_\varepsilon\big\|_{\mathbb{X}'\to\mathbb{X}'} \leq C\varepsilon^\alpha \tag{2.7.156}$$

for each $\varepsilon > 0$.

Consequently, for each $j \in \mathbb{N}$ we obtain

$$[M_{b_j}, T] = \lim_{\varepsilon \to 0^+} Q_\varepsilon \text{ in } \mathrm{Bd}(\mathbb{X}),\ \mathrm{Bd}(\mathbb{X}'),\ \mathrm{Bd}(\mathring{\mathbb{X}}),\ \text{and in } \mathrm{Bd}((\mathbb{X}')^\circ). \tag{2.7.157}$$

In view of the fact that (2.7.63) and [113, (5.2.76), (5.2.84)] imply

$$Q_\varepsilon \in \mathrm{Cp}(\mathbb{X}) \cap \mathrm{Cp}(\mathbb{X}') \cap \mathrm{Cp}(\mathring{\mathbb{X}}) \cap \mathrm{Cp}((\mathbb{X}')^\circ) \text{ for each } \varepsilon > 0, \tag{2.7.158}$$

we conclude from (2.7.157) and [113, (1.2.53)] (bearing in mind the final conclusion in [113, Proposition 5.1.8]) that for each $j \in \mathbb{N}$ we have

$$[M_{b_j}, T] \in \mathrm{Cp}(\mathbb{X}) \cap \mathrm{Cp}(\mathbb{X}') \cap \mathrm{Cp}(\mathring{\mathbb{X}}) \cap \mathrm{Cp}((\mathbb{X}')^\circ). \tag{2.7.159}$$

Together with (2.7.153), this proves that for each $j \in \mathbb{N}$ we have

$$\begin{aligned} &\mathrm{dist}\Big([T, M_b], \mathrm{Cp}(\mathring{\mathbb{X}})\Big) \\ &\quad\le C\Big(\sum_{|\alpha|\le N} \sup_{S^{n-1}} |\partial^\alpha k|\Big)\Big\{\|b-\phi\|_{\mathrm{BMO}(\Sigma,\sigma)} + \|\phi - b_j\|_{\mathrm{BMO}(\Sigma,\sigma)}\Big\}, \end{aligned} \tag{2.7.160}$$

and

$$\begin{aligned} &\mathrm{dist}\Big([T, M_b], \mathrm{Cp}(\mathbb{X})\Big) \\ &\quad\le C\Big(\sum_{|\alpha|\le N} \sup_{S^{n-1}} |\partial^\alpha k|\Big)\Big\{\|b-\phi\|_{\mathrm{BMO}(\Sigma,\sigma)} + \|\phi - b_j\|_{\mathrm{BMO}(\Sigma,\sigma)}\Big\}. \end{aligned} \tag{2.7.161}$$

Also, from (2.7.154) and (2.7.159) we conclude that for each $j \in \mathbb{N}$ we have

$$\begin{aligned} &\mathrm{dist}\Big([T, M_b], \mathrm{Cp}((\mathbb{X}')^\circ)\Big) \\ &\quad\le C\Big(\sum_{|\alpha|\le N} \sup_{S^{n-1}} |\partial^\alpha k|\Big)\Big\{\|b-\phi\|_{\mathrm{BMO}(\Sigma,\sigma)} + \|\phi - b_j\|_{\mathrm{BMO}(\Sigma,\sigma)}\Big\}, \end{aligned} \tag{2.7.162}$$

and

$$\begin{aligned} &\mathrm{dist}\Big([T, M_b], \mathrm{Cp}(\mathbb{X}')\Big) \\ &\quad\le C\Big(\sum_{|\alpha|\le N} \sup_{S^{n-1}} |\partial^\alpha k|\Big)\Big\{\|b-\phi\|_{\mathrm{BMO}(\Sigma,\sigma)} + \|\phi - b_j\|_{\mathrm{BMO}(\Sigma,\sigma)}\Big\}. \end{aligned} \tag{2.7.163}$$

Keeping in mind (2.7.50) and sending $j \to \infty$ in (2.7.160)-(2.7.163) then yields

$$
\begin{aligned}
&\operatorname{dist}\Big([T, M_b], \operatorname{Cp}(\mathring{\mathbb{X}})\Big) \\
&\qquad \le C\Big(\sum_{|\alpha|\le N} \sup_{S^{n-1}} |\partial^\alpha k|\Big)\|b-\phi\|_{\mathrm{BMO}(\Sigma,\sigma)},
\end{aligned}
\tag{2.7.164}
$$

and

$$
\begin{aligned}
&\operatorname{dist}\Big([T, M_b], \operatorname{Cp}(\mathbb{X})\Big) \\
&\qquad \le C\Big(\sum_{|\alpha|\le N} \sup_{S^{n-1}} |\partial^\alpha k|\Big)\|b-\phi\|_{\mathrm{BMO}(\Sigma,\sigma)},
\end{aligned}
\tag{2.7.165}
$$

as well as

$$
\begin{aligned}
&\operatorname{dist}\Big([T, M_b], \operatorname{Cp}((\mathbb{X}')^\circ)\Big) \\
&\qquad \le C\Big(\sum_{|\alpha|\le N} \sup_{S^{n-1}} |\partial^\alpha k|\Big)\|b-\phi\|_{\mathrm{BMO}(\Sigma,\sigma)},
\end{aligned}
\tag{2.7.166}
$$

and

$$
\begin{aligned}
&\operatorname{dist}\Big([T, M_b], \operatorname{Cp}(\mathbb{X}')\Big) \\
&\qquad \le C\Big(\sum_{|\alpha|\le N} \sup_{S^{n-1}} |\partial^\alpha k|\Big)\|b-\phi\|_{\mathrm{BMO}(\Sigma,\sigma)}.
\end{aligned}
\tag{2.7.167}
$$

After taking the infimum over all $\phi \in \mathrm{VMO}(\Sigma, \sigma)$, we now arrive at (2.7.144)-(2.7.147). Let us also note that (2.7.148) is a consequence of [113, (1.2.53)] (again, bearing in mind the final conclusion in [113, Proposition 5.1.8]) and (2.7.144)-(2.7.147).

Lastly, the final claim in the statement of the theorem is seen from what we have proved so far and the spherical harmonics expansion employed in the proof of Theorem 2.5.38, by reasoning much as in (2.7.37)-(2.7.49) (with $L^p(\Sigma, w)$ replaced by $\mathbb{X}$, $\mathbb{X}'$, $\mathring{\mathbb{X}}$, $(\mathbb{X}')^\circ$), while bearing in mind [113, (5.1.13)]. □

2.8 Calderón-Zygmund Theory for Singular Integrals on Riemannian Manifolds

The Calderón-Zygmund machinery developed so far may be naturally adapted to integral operators, on uniformly rectifiable sets, with variable coefficient kernels that are actually the Schwartz kernels of classical pseudodifferential operators of order -1 with odd principal symbol, acting between vector bundles over a manifold. Introducing this very general class of integral operators requires some preparations.

To set the stage, recall that there are several symbol classes of importance. One class, denoted $S^m_{1,0}$, is defined by

$$q(x,\xi) \in S^m_{1,0} \iff \left|D^\beta_x D^\alpha_\xi q(x,\xi)\right| \leq C_{\alpha,\beta}(1+|\xi|)^{m-|\alpha|}. \tag{2.8.1}$$

With each symbol $q(x,\xi)$ in Hörmander's class $S^m_{1,0}$ associate a pseudodifferential operator

$$q(x,D) := Q(x,D) \tag{2.8.2}$$

acting[29] on each Schwartz function[30] u as

$$\begin{aligned} Q(x,D)u(x) &:= (2\pi)^{-n}\int q(x,\xi)\widehat{u}(\xi)e^{i\langle x,\xi\rangle}\,d\xi \\ &= (2\pi)^{-n}\iint q(x,\xi)e^{i\langle x-y,\xi\rangle}u(y)\,dy\,d\xi. \end{aligned} \tag{2.8.3}$$

The latter is an oscillatory integral, in the sense, e.g., of [65]. It is also of interest to note that (see [176, (2.1)-(2.2), p. 5])

$$\begin{aligned} &\text{the Schwartz kernel of } q(x,D) \text{ is } b(x,x-y) \text{ where} \\ &b(x,z) := (2\pi)^{-n}\int q(x,\xi)e^{i\langle z,\xi\rangle}\,d\xi, \end{aligned} \tag{2.8.4}$$

with the integral once again interpreted in an oscillatory sense. It turns out that b is of class $\mathscr{C}^\infty$ in $\mathbb{R}^n \times (\mathbb{R}^n \setminus \{0\})$ (cf. [176, Proposition 2.1, p. 5]), and satisfies (see [176, Proposition 2.2, p. 6]) the estimate

$$\left|\partial^\alpha_x \partial^\beta_z b(x,z)\right| \leq C_{\alpha,\beta}|z|^{-m-n-|\alpha|-|\beta|} \text{ if } |\alpha|+|\beta| > -m-n. \tag{2.8.5}$$

After taking a partial Fourier transform in the second variable, (2.8.4) also gives

$$\widehat{b}(x,\xi) = q(x,\xi). \tag{2.8.6}$$

Here, we are concerned with a smaller class of symbols, $S^m_{\rm cl}$ with $m \in \mathbb{Z}$, defined by requiring that the (matrix-valued) function $q(x,\xi)$ has an asymptotic expansion of the form

$$q(x,\xi) \sim q_m(x,\xi) + q_{m-1}(x,\xi) + \cdots, \tag{2.8.7}$$

with q_j smooth in x and ξ and positive homogeneous of degree j in ξ for $|\xi| \geq 1$ (in the sense that $q_j(x,\lambda\xi) = \lambda^j q_j(x,\xi)$ if $\lambda \geq 1$ and $|\xi| \geq 1$). The meaning of (2.8.7) is that, for each $k \in \mathbb{N}$, the difference between the left side and the sum of the first k terms on the right belong to $S^{m-k}_{1,0}$. In particular, the "error"

[29] with "hat" denoting the Fourier transform in $\mathbb{R}^n$, with the normalization given in (1.4.14)

[30] this subsequently extends to tempered distributions in a natural fashion; see [176, Lemma 1.1, p. 3]

$$r_{m-1}(x,\xi) := q(x,\xi) - q_m(x,\xi) \text{ belongs to } S^{m-1}_{1,0}. \tag{2.8.8}$$

Define

$$\mathrm{OP}S^m_{\mathrm{cl}} := \big\{q(x,D) : q(x,\xi) \in S^m_{\mathrm{cl}}\big\}. \tag{2.8.9}$$

Also, given any $q(x,D) \in \mathrm{OP}S^m_{\mathrm{cl}}$, call $q_m(x,\xi)$, i.e., the leading term in (2.8.7), the **principal symbol** of $q(x,D) = Q(x,D)$. In the sequel, we shall use the notation

$$\mathrm{Sym}(Q;\xi) := q_m(x,\xi). \tag{2.8.10}$$

Let us illustrate how this information ties up with earlier work. Start with a classical pseudodifferential operator $q(x,D)$ of order m, and consider its Schwartz kernel $b(x, x-y)$ as well as its principal symbol $q_m(x,\xi)$. The latter is known to be homogeneous of degree m in the ξ variable only for $|\xi| \geq 1$ (in the sense that $q_m(x,\lambda\xi) = \lambda^m q_m(x,\xi)$ if $\lambda \geq 1$ and $|\xi| \geq 1$). To correct this, consider $\widetilde{q}_m(x,\xi) := |\xi|^m q_m\big(x, \frac{\xi}{|\xi|}\big)$ for $\xi \in \mathbb{R}^n \setminus \{0\}$. Then $\widetilde{q}_m(x,\xi)$ is smooth and (genuinely) positive homogeneous of degree m in $\mathbb{R}^n \setminus \{0\}$, and also matches $q_m(x,\xi)$ when $|\xi| \geq 1$. Moreover, if $q_m(x,\xi)$ is odd in ξ then so is $\widetilde{q}_m(x,\xi)$. Hence, $\widetilde{q}_m(x,\xi)$ induces a unique tempered distribution in $\mathbb{R}^n$, which is positive homogeneous of degree m. If we now define

$$b^{\mathrm{top}}(x,z) := (2\pi)^{-n} \int \widetilde{q}_m(x,\xi) e^{\mathrm{i}\langle z,\xi\rangle}\, d\xi, \tag{2.8.11}$$

then (see [175, Proposition 8.1, p. 239 in Chapter 3])

$$\begin{gathered} b^{\mathrm{top}}(x,z) \text{ is positive homogeneous of degree } -m-n \\ \text{as well as smooth in the variable } z \in \mathbb{R}^n \setminus \{0\}, \\ \text{and it is also odd in } z \text{ if } q_m(x,\xi) \text{ is odd in } \xi. \end{gathered} \tag{2.8.12}$$

In addition, if

$$\begin{aligned} b^{\mathrm{err}}(x,z) &:= (2\pi)^{-n} \int \big[q_m(x,\xi) - \widetilde{q}_m(x,\xi)\big] e^{\mathrm{i}\langle z,\xi\rangle}\, d\xi \\ &\quad + (2\pi)^{-n} \int r_{m-1}(x,\xi) e^{\mathrm{i}\langle z,\xi\rangle}\, d\xi, \end{aligned} \tag{2.8.13}$$

then the first term in the right-hand side belongs to $\mathscr{C}^\infty(\mathbb{R}^n)$ (as the Fourier transform of a compactly supported distribution; cf., e.g., [109, Theorem 4.35, p. 135], [175, Proposition 4.3, p. 208]), while the second term in the right-hand side of (2.8.13) satisfies estimates of the sort described in (2.8.5) with m replaced by $m-1$. Altogether we obtain

$$\big|\partial_x^\alpha \partial_z^\beta b^{\mathrm{err}}(x,z)\big| \leq C_{\alpha,\beta} |z|^{1-m-n-|\alpha|-|\beta|} \text{ for } |\alpha| + |\beta| > 1-m-n. \tag{2.8.14}$$

This argument shows that we have the decomposition

$$b(x,z) = b^{\rm top}(x,z) + b^{\rm err}(x,z) \tag{2.8.15}$$

of the Schwartz kernel of $q(x, D)$ into a top part $b^{\rm top}$ as in (2.8.12) and the error term $b^{\rm err}$ exhibiting a milder singularity at the origin, manifested by the estimate recorded in (2.8.14). We also wish to make the important observation that

if $m = -1$ and the principal symbol $q_{-1}(x, \xi)$ of $q(x, D) \in {\rm OP}S^{-1}_{\rm cl}$ is odd in ξ then Theorem 2.5.38 applies to $b^{\rm top}(x, z)$ since this is odd and positive homogeneous of degree $1 - n$ in z. (2.8.16)

From (2.8.6) and (2.8.10) we also see that

$$\widehat{b^{\rm top}}(x,\xi) = {\rm Sym}(Q;\xi) \ \text{ for } \ |\xi| \geq 1, \tag{2.8.17}$$

where once again 'hat' stands for the partial Fourier transform in the second variable.

In fact, we shall find it convenient to work with classes of symbols that only exhibit a limited amount of regularity in the spatial variable (while still $\mathscr{C}^\infty$ in the Fourier variable). Specifically, for each $r \geq 0$ define

$$C^r S^m_{1,0} := \big\{q(x,\xi) : \|D^\alpha_\xi q(\cdot,\xi)\|_{\mathscr{C}^r} \leq C_\alpha (1+|\xi|)^{m-|\alpha|}, \quad \forall \alpha \in \mathbb{N}_0^n\big\}. \tag{2.8.18}$$

Denote by ${\rm OP}C^r S^m_{1,0}$ the class of pseudodifferential operators associated with such symbols. As before, we write ${\rm OP}C^r S^m_{\rm cl}$ for the subclass of `classical pseudodifferential operators` in ${\rm OP}C^r S^m_{1,0}$ whose symbols can be expanded as in (2.8.7), where $q_j(x, \xi) \in C^r S^{m-j}_{1,0}$ is homogeneous of degree j in ξ for $|\xi| \geq 1$, $j = m, m-1, \dots$. Basic material on these symbol classes, as well as other symbol classes with limited regularity in x, can be found in [174].

Another type of pseudodifferential operator is defined by

$$Q(D,x)u(x) := (2\pi)^{-n} \iint q(\xi,y) e^{i\langle x-y,\xi\rangle} u(y)\,dy\,d\xi \tag{2.8.19}$$

(often, we simply write $q(D, x)$ in place of $Q(D, x)$). The Schwartz kernel of $Q(D, x)$ is $\widetilde{b}(y, x - y)$ where

$$\widetilde{b}(y,z) := (2\pi)^{-n} \int q(\xi,y) e^{i\langle z,\xi\rangle}\,d\xi, \tag{2.8.20}$$

also interpreted as an oscillatory integral. Any operator of the form (2.8.19) can be rewritten in the form (2.8.3) if $q(\xi, x)$ belongs to $S^m_{\rm cl}$, but for symbols in $C^r S^m_{\rm cl}$ with $r < \infty$, these operator classes do not coincide. If $q(\xi, y)$ belongs to such a symbol class, we write $q(D, x) \in \emptyset{\rm P}C^r S^m_{\rm cl}$. Hence, $\emptyset{\rm P}C^r S^m_{\rm cl}$ consists of formal adjoints of elements of ${\rm OP}C^r S^m_{\rm cl}$. There is also a natural formula for composition of operators in these two classes, namely

$$p(x,D)q(D,x)u = (2\pi)^{-n} \iint p(x,\xi) q(\xi,y) e^{i\langle x-y,\xi\rangle} u(y)\,dy\,d\xi. \tag{2.8.21}$$

Let us also remark that similar considerations are valid for operators in $\text{ØPC}^0 S^{-1}_{\text{cl}}$ with an odd principal symbol in the cotangent variable. We therefore have the following basic result:

$$\begin{array}{l}\text{operators in } \mathrm{OPC}^0 S^{-1}_{\mathrm{cl}} \text{ and } \text{ØPC}^0 S^{-1}_{\mathrm{cl}} \text{ with odd principal symbols in the} \\ \text{cotangent variable have Schwartz kernels that differ from the variable} \\ \text{coefficient kernels treated in Theorem 2.5.38 by error terms with weak} \\ \text{singularities along the diagonal.}\end{array} \tag{2.8.22}$$

This is going to be a guiding principle in extending our Euclidean results to the setting of manifolds in this section.

Consider next an n-dimensional smooth oriented manifold $\mathcal{M}$ equipped with a Riemannian metric tensor g. Let dV_g be the volume element on $\mathcal{M}$ induced by the metric g, and denote by $\mathcal{L}^n_g$ the measure canonically associated with dV_g. Also, let $\mathcal{H}^{n-1}_g$ stand for the $(n-1)$-dimensional Hausdorff measure associated with the metric g on $\mathcal{M}$. By $T\mathcal{M}$ and $T^*\mathcal{M}$ we denote, respectively, the tangent and cotangent bundles on $\mathcal{M}$.

Given two Hermitian vector bundles $\mathcal{E}, \mathcal{F} \to \mathcal{M}$, denote by $\mathrm{OPC}^r S^m_{1,0}(\mathcal{E}, \mathcal{F})$ the collection of pseudodifferential operators mapping sections of the vector bundle $\mathcal{E}$ into sections of the vector bundle $\mathcal{F}$ which, in local coordinates of $\mathcal{M}$ and over local trivializations of $\mathcal{E}$, $\mathcal{F}$, can be represented as matrices with entries from $\mathrm{OPC}^r S^m_{1,0}$. Finally, let $\text{ØPC}^r S^m_{\mathrm{cl}}(\mathcal{E}, \mathcal{F})$ consist of all formal adjoints of operators in $\mathrm{OPC}^r S^m_{\mathrm{cl}}(\mathcal{F}, \mathcal{E})$.

Next, given a classical pseudodifferential operator $Q(x, D) \in \mathrm{OPC}^0 S^{-1}_{\mathrm{cl}}(\mathcal{E}, \mathcal{F})$, denote its principal symbol by

$$\mathrm{Sym}(Q; \cdot) : T^*\mathcal{M} \setminus 0 \to \mathrm{Hom}\,(\mathcal{E}, \mathcal{F}), \quad \mathrm{Sym}(Q; \cdot) \in C^0 S^{-1}_{\mathrm{cl}}(\mathcal{E}, \mathcal{F}). \tag{2.8.23}$$

Also, let

$$k_Q \in \mathcal{D}'(\mathcal{M} \times \mathcal{M}, \mathcal{F} \otimes \mathcal{E}) \cap \mathscr{C}^0(\mathcal{M} \times \mathcal{M} \setminus \mathrm{diag}, \mathcal{F} \otimes \mathcal{E}) \tag{2.8.24}$$

stand for the Schwartz kernel of $Q(x, D)$. Hence, for all test functions ϕ, ψ on $\mathcal{M}$,

$$_{\mathcal{D}'(\mathcal{M})}\langle Q\phi, \psi \rangle_{\mathcal{D}(\mathcal{M})} = {}_{\mathcal{D}'(\mathcal{M}\times\mathcal{M})}\langle k_Q, \psi \otimes \phi \rangle_{\mathcal{D}(\mathcal{M}\times\mathcal{M})}, \tag{2.8.25}$$

and

$$Q(x, D)\phi(x) = \int_{\mathcal{M}} \langle k_Q(x, y), \phi(y) \rangle_{\mathcal{E}_y} \, d\mathcal{L}^n_g(y), \quad \forall\, x \in \mathcal{M} \setminus \mathrm{supp}\, \phi. \tag{2.8.26}$$

In view of (2.8.26) and the fact that $d\mathcal{L}^n_g = \sqrt{g}\, d\mathcal{L}^n$ locally, where $\mathcal{L}^n$ is the ordinary n-dimensional Lebesgue measure in $\mathbb{R}^n$, we conclude that

$$\begin{array}{l}\text{in local coordinates, the Schwartz kernel of } Q(x, D) \text{ with respect to} \\ \text{the ordinary Euclidean Lebesgue measure } \mathcal{L}^n \text{ is } k_Q(x, y)\sqrt{g(y)}.\end{array} \tag{2.8.27}$$

Going further, fix an open set $\Omega \subseteq \mathcal{M}$ satisfying $\mathcal{H}_g^{n-1}(\partial\Omega) < \infty$ abbreviate $\sigma_g := \mathcal{H}_g^{n-1} \lfloor \partial\Omega$. In this context, introduce the boundary-to-domain integral operator acting on $f \in L^1(\partial\Omega, \sigma_g) \otimes \mathcal{E}$ according to

$$\mathscr{K}_Q f(x) := \int_{\partial\Omega} \langle k_Q(x,y), f(y) \rangle_{\mathcal{E}_y} \, d\sigma_g(y), \quad \forall\, x \in \mathcal{M} \setminus \partial\Omega. \tag{2.8.28}$$

With $d_g(x,y)$ standing for the geodesic distance between x and y in $\mathcal{M}$, with respect to the Riemannian metric g (cf. [112, (1.11.5)]), let us also introduce the maximal operator

$$K_{Q,\max} f(x) := \sup_{\varepsilon>0} \left| \int_{\substack{y\in\partial\Omega \\ d_g(x,y)>\varepsilon}} \langle k_Q(x,y), f(y) \rangle_{\mathcal{E}_y} \, d\sigma_g(y) \right|, \quad \forall\, x \in \partial\Omega, \tag{2.8.29}$$

for each $f \in L^1(\partial\Omega, \sigma_g) \otimes \mathcal{E}$. Similar considerations apply to pseudodifferential operators in the class $\mathit{Ø}\mathrm{P}C^0 S^{-1}_{\mathrm{cl}}(\mathcal{E}, \mathcal{F})$.

Going further, we make the following definition which basically stipulates that for singular integral operators on a Riemannian manifold the principal-value limit is considered in local Euclidean coordinates.

Definition 2.8.1 *In the same geometric setting as above, given a kernel*

$$k \in \mathscr{C}^0(\mathcal{M} \times \mathcal{M} \setminus \mathrm{diag}, \mathcal{F} \otimes \mathcal{E}), \tag{2.8.30}$$

for each $f \in L^1(\partial\Omega, \sigma_g) \otimes \mathcal{E}$ the meaning assigned to the formal expression

$$\mathrm{P.V.} \int_{\partial\Omega} \langle k(x,y), f(y) \rangle_{\mathcal{E}_y} \, d\sigma_g(y), \quad \forall\, x \in \partial\Omega, \tag{2.8.31}$$

is as follows. First, for each $x \in \partial\Omega \setminus \mathrm{supp}\, f$ define (2.8.31) *as being the absolutely convergent integral*

$$\int_{\partial\Omega} \langle k(x,y), f(y) \rangle_{\mathcal{E}_y} \, d\sigma_g(y). \tag{2.8.32}$$

Second, suppose $\mathrm{supp}\, f$ *is contained in a local coordinate patch over which the vector-bundles $\mathcal{E}, \mathcal{F}$ trivialize. Using local orthonormal frames, identify the kernel $k(x,y)$ with a matrix $\big(k_{\alpha\beta}(x,y)\big)_{\alpha,\beta}$ and f with a vector-valued function with components $\{f_\beta\}_\beta$. Work in local Euclidean coordinates. Hence, in addition to the original surface measure σ_g induced by the Riemannian metric, one can also speak of the Euclidean surface measure σ^{E} on $\partial\Omega$. Locally, these measures are related to one another as in* [112, (1.11.8)], *(so $\rho := d\sigma_g / d\sigma^{\mathrm{E}}$ is meaningfully defined). Then, in local Euclidean coordinates,* (2.8.31) *is identified with the vector whose α-component is given by*

$$\lim_{\varepsilon\to 0^+} \int_{\substack{y\in\partial\Omega \\ |x-y|>\varepsilon}} k_{\alpha\beta}(x,y) f_\beta(y)\rho(y)\, d\sigma^{\mathrm{E}}(y), \tag{2.8.33}$$

at points $x \in \operatorname{supp} f$ for which the limit exists. Finally, in the remaining case, (2.8.31) *is defined using a partition of unity which reduces matters to the previous situations.*

In this vein, we want to point out that, as seen from [112, (1.11.10)], the density ρ appearing in (2.8.33) satisfies

$$\rho = \sqrt{g}\Big(\sum_{r,s} g^{rs}\nu_r^{\mathrm{E}}\nu_s^{\mathrm{E}}\Big)^{1/2} \quad \text{on } \partial_*\Omega, \tag{2.8.34}$$

where ν^{E} is the Euclidean geometric measure theoretic outward unit normal to Ω.

Lastly, we observe that the same blueprint for fashioning smoothness spaces in open sets out of their global versions in the entire Euclidean space used in [113, §9.2] works in setting of a given manifold[31] $\mathcal{M}$ as well. For example, first the global Bessel potential space $L^p_s(\mathcal{M})$ is transported to the manifold $\mathcal{M}$ from the Euclidean setting by means of local coordinate charts after localization via a sufficiently smooth partition of unity. Then the Bessel potential spaces in an open set $\Omega \subseteq \mathcal{M}$ is defined via restriction, as

$$L^p_s(\Omega) := \big\{u|_\Omega : u \in L^p_s(\mathcal{M})\big\}. \tag{2.8.35}$$

Similar considerations apply to other scales of interest, including Besov spaces and Triebel-Lizorkin spaces.

Theorem 2.8.2 *Suppose $\mathcal{M}$ is an n-dimensional compact boundaryless oriented manifold of class $\mathscr{C}^2$ equipped with a continuous Riemannian metric tensor g, and let $\mathcal{L}^n_g$, $\mathcal{H}^{n-1}_g$ denote, respectively, the n-dimensional Lebesgue measure and the $(n-1)$-dimensional Hausdorff measure associated with the metric g on $\mathcal{M}$.*

Next, consider two vector bundles $\mathcal{E}, \mathcal{F} \to \mathcal{M}$ endowed with continuous Hermitian metrics, and let $Q(x, D) \in \mathrm{OPC}^0 S^{-1}_{\mathrm{cl}}(\mathcal{E}, \mathcal{F})$ be a classical pseudodifferential operator with the property that its principal symbol $\mathrm{Sym}(Q;\xi)$ *is odd in the cotangent variable $\xi \in T^*\mathcal{M}$.*

Having fixed an open set $\Omega \subset \mathcal{M}$ such that $\partial\Omega$ is a UR *set, denote the geometric measure theoretic outward unit conormal to Ω by $\nu_g \in T^*\mathcal{M}$ and let $\sigma_g := \mathcal{H}^{n-1}_g \lfloor \partial\Omega$. Finally, recall the integral operators from* (2.8.28)-(2.8.29).

Then the following statements are true:

(1) For each function $f \in L^1(\partial\Omega, \sigma_g) \otimes \mathcal{E}$ the principal-value integral

$$K_Q f(x) := \mathrm{P.V.}\int_{\partial\Omega} \big\langle k_Q(x,y), f(y)\big\rangle_{\mathcal{E}_y}\, d\sigma_g(y), \tag{2.8.36}$$

considered in the sense of Definition 2.8.1, is well defined at σ-a.e. point $x \in \partial\Omega$, and

[31] assumed to be sufficiently regular, relative to the smoothness indices one has in mind

$$K_Q : L^p(\partial\Omega, \sigma_g) \otimes \mathcal{E} \longrightarrow L^p(\partial\Omega, \sigma_g) \otimes \mathcal{F} \tag{2.8.37}$$

is a well defined, linear, and bounded operator whenever $p \in (1, \infty)$. *Also, corresponding to* $p = 1$,

$$K_Q : L^1(\partial\Omega, \sigma_g) \otimes \mathcal{E} \longrightarrow L^{1,\infty}(\partial\Omega, \sigma_g) \otimes \mathcal{F} \text{ is bounded.} \tag{2.8.38}$$

Finally, for each $p \in \left(\frac{n-1}{n}, 1\right]$ *the operator* K_Q *induces a well-defined, linear, and bounded mapping*

$$K_Q : H^p(\partial\Omega, \sigma_g) \otimes \mathcal{E} \longrightarrow L^p(\partial\Omega, \sigma_g) \otimes \mathcal{F}. \tag{2.8.39}$$

(2) The maximal operators

$$\begin{aligned} &K_{Q,\max} : L^p(\partial\Omega, \sigma_g) \otimes \mathcal{E} \longrightarrow L^p(\partial\Omega, \sigma_g) \otimes \mathcal{F} \text{ if } 1 < p < \infty, \\ &K_{Q,\max} : L^1(\partial\Omega, \sigma_g) \otimes \mathcal{E} \longrightarrow L^{1,\infty}(\partial\Omega, \sigma_g) \otimes \mathcal{F}, \end{aligned} \tag{2.8.40}$$

are well defined, sub-linear, and bounded.

(3) With $K_{Q^\top}$ *denoting the singular integral operator associated with* $Q^\top$ *in the same manner* K_Q *has been associated with* Q *in* (2.8.36), *for each* $p \in (1, \infty)$ *one has*

$$\begin{aligned} &(K_Q)^\top = K_{Q^\top}, \text{ as operators from} \\ &L^p(\partial\Omega, \sigma_g) \otimes \mathcal{F} \text{ into } L^p(\partial\Omega, \sigma_g) \otimes \mathcal{E}. \end{aligned} \tag{2.8.41}$$

(4) For every $\kappa > 0$ *and* $p \in (1, \infty)$ *there exists a constant* $C \in (0, \infty)$ *such that*

$$\|\mathcal{N}_\kappa(\mathscr{K}_Q f)\|_{L^p(\partial\Omega, \sigma_g)} \leq C\|f\|_{L^p(\partial\Omega, \sigma_g)\otimes\mathcal{E}}, \tag{2.8.42}$$

for all $f \in L^p(\partial\Omega, \sigma_g) \otimes \mathcal{E}$, *and, corresponding to* $p = 1$, *there exists* $C \in (0, \infty)$ *such that*

$$\|\mathcal{N}_\kappa(\mathscr{K}_Q f)\|_{L^{1,\infty}(\partial\Omega, \sigma_g)} \leq C\|f\|_{L^1(\partial\Omega, \sigma_g)\otimes\mathcal{E}} \tag{2.8.43}$$

for all $f \in L^1(\partial\Omega, \sigma_g) \otimes \mathcal{E}$. *Moreover, if* $p \in \left(\frac{n-1}{n}, 1\right]$ *then*

$$\|\mathcal{N}_\kappa(\mathscr{K}_Q f)\|_{L^p(\partial\Omega, \sigma_g)} \leq C\|f\|_{H^p(\partial\Omega, \sigma_g)\otimes\mathcal{E}}, \tag{2.8.44}$$

for all $f \in H^p(\partial\Omega, \sigma_g) \otimes \mathcal{E}$, *where the boundary-to-domain integral operator* $\mathscr{K}_Q$ *from* (2.8.28) *now acts on any given* $f \in H^p(\partial\Omega, \sigma_g) \otimes \mathcal{E}$ *according to (with* $\langle \cdot, \cdot \rangle$ *denoting the natural duality bracket)*

$$\mathscr{K}_Q f(x) := \langle k_Q(x, \cdot), f \rangle, \quad \forall\, x \in M \setminus \partial\Omega. \tag{2.8.45}$$

(5) Given any $f \in L^1(\partial\Omega, \sigma_g) \otimes \mathcal{E}$, *the function* $\mathscr{K}_Q f$ *possesses a nontangential boundary trace at* σ_g*-a.e. point on* $\partial_*\Omega$, *in the precise sense that for any aperture parameter* $\kappa > 0$ *one has*

$$\mathscr{K}_Q f\Big|_{\partial\Omega}^{\kappa-\mathrm{n.t.}} = \frac{1}{2\sqrt{-1}}\,\mathrm{Sym}(Q;\nu_g)f + K_Q f \text{ at } \sigma_g\text{-a.e. point on } \partial_*\Omega, \tag{2.8.46}$$

with the singular integral operator K_Q considered as in (2.8.38). *Furthermore, for any given $f \in H^p(\partial\Omega,\sigma_g)\otimes\mathcal{E}$ with $p \in \big(\frac{n-1}{n}, 1\big]$, the function $\mathscr{K}_Q f$, now defined as in* (2.8.45)*, also possesses a nontangential boundary trace at σ_g-a.e. point on $\partial_*\Omega$, in the precise sense that for any aperture parameter $\kappa > 0$ one has*

$$\mathscr{K}_Q f\Big|_{\partial\Omega}^{\kappa-\mathrm{n.t.}} = \frac{1}{2\sqrt{-1}}\,\mathrm{Sym}(Q;\nu_g)\mathfrak{H}f + K_Q f \text{ at } \sigma_g\text{-a.e. point on } \partial_*\Omega, \tag{2.8.47}$$

where $\mathfrak{H} : H^p(\partial\Omega,\sigma_g)\otimes\mathcal{E} \to L^p(\partial\Omega,\sigma_g)\otimes\mathcal{E}$ is the L^p-filtering operator on $\partial\Omega$ (cf. the discussion in [113, §4.9]*), and the operator K_Q is now considered as in* (2.8.39).

(6) On the scale of Lebesgue spaces with integrability exponents $p \in [1,\infty)$, similar results to those in items (1)-(6) above are valid for integral operators associated with pseudodifferential operators $Q(D,x)$ from the class $\varnothing\mathrm{PC}^0 S^{-1}_{\mathrm{cl}}(\mathcal{E},\mathcal{F})$, while for Hardy spaces corresponding to $p \in \big(\frac{n-1}{n}, 1\big]$ similar results to those in items (1)-(6) above are valid for integral operators associated with pseudodifferential operators $Q(D,x) \in \varnothing\mathrm{PC}^r S^{-1}_{\mathrm{cl}}(\mathcal{E},\mathcal{F})$ where it is now assumed that the smoothness index satisfies $r > (n-1)\big(\frac{1}{p}-1\big)$.
In addition, given any $Q(D,x) \in \varnothing\mathrm{PC}^0 S^{-1}_{\mathrm{cl}}(\mathcal{E},\mathcal{F})$ it follows that for each $p \in (1,\infty)$ there exists a constant $C \in (0,\infty)$ with the property that for each $f \in L^p(\partial\Omega,\sigma_g)\otimes\mathcal{E}$ one has the L^p-square function estimate

$$\int_\Omega \big|\nabla(\mathscr{K}_Q f)(x)\big|^p \,\mathrm{dist}_g(x,\partial\Omega)^{p-1}\,d\mathcal{L}^n_g(x) \le C\int_{\partial\Omega}|f|^p\,d\sigma_g, \tag{2.8.48}$$

where ∇ is a connection on the vector bundle $\mathcal{F}$, and $\mathrm{dist}_g(\cdot,\partial\Omega)$ stands for the distance to $\partial\Omega$ naturally associated with the metric g (or any other sufficiently smooth background Riemannian metric on $\mathcal{M}$).
Finally, if Ω is a Lipschitz domain, then for each $Q(D,x) \in \varnothing\mathrm{PC}^0 S^{-1}_{\mathrm{cl}}(\mathcal{E},\mathcal{F})$ the operator

$$\mathscr{K}_Q : L^p(\partial\Omega,\sigma_g)\otimes\mathcal{E} \longrightarrow B^{p,p}_{1/p}(\Omega)\otimes\mathcal{F} \tag{2.8.49}$$

is well defined, linear, and bounded; in particular, corresponding to the case when $p = 2$, the following is a well defined, linear, and bounded operator:

$$\mathscr{K}_Q : L^2(\partial\Omega,\sigma_g)\otimes\mathcal{E} \longrightarrow H^{1/2}(\Omega)\otimes\mathcal{F}. \tag{2.8.50}$$

(7) The L^p-square function estimate (2.8.48) *is also valid if we actually have $Q(x,D) \in \mathrm{OPC}^1 S^{-1}_{\mathrm{cl}}(\mathcal{E},\mathcal{F})$. Also, if Ω is a Lipschitz domain then the mapping properties recorded in* (2.8.49)-(2.8.50) *remain true when $\mathscr{K}_Q$ is now associated with $Q(x,D) \in \mathrm{OPC}^1 S^{-1}_{\mathrm{cl}}(\mathcal{E},\mathcal{F})$.*

The jump-formula (2.8.46) is particularly elegant and beautiful, as it brings together in an invariant, coordinate-free formalism, the algebraic and analytic structures underpinning the current setting, through the involvement of the principal symbol and, respectively, the principal-value singular integral operator associated with the given pseudodifferential operator (we wish to emphasize that it is only for pseudodifferential operators of order -1 that a genuine jump-formula like the one recorded in (2.8.46) is to be expected). The fact that the pseudodifferential operator in question acts between generic Hermitian vector bundles[32] makes this result especially general and versatile.

We now turn to the task of providing the proof of Theorem 2.8.2.

Proof of Theorem 2.8.2 The problem localizes and, given the invariance of the class of pseudodifferential operators and symbols under discussion, it can be transported in to the Euclidean setting via coordinate mappings. By fixing local frames in $\mathcal{E}$ and $\mathcal{F}$, the operator $Q(x, D)$ can be locally identified, in a canonical manner, with a matrix $\big(Q_{jk}(x, D)\big)_{j,k}$ whose entries $Q_{jk}(x, D)$ are classical pseudodifferential operators having principal symbols odd in the cotangent variable. Also, $k_Q(x, y)$, the Schwartz kernel of $Q(x, D)$, can be identified with a matrix-valued distribution $\big(\theta_{jk}(x, y)\big)_{j,k}$. Thanks to (2.8.27), the latter is such that

$$\text{the Schwartz kernel } b_{jk}(x, x-y) \text{ of } Q_{jk}(x, D) \text{ with respect to the ordinary Euclidean Lebesgue measure is } \theta_{jk}(x, y)\sqrt{g(y)}. \tag{2.8.51}$$

As in (2.8.15), decompose each $b_{jk}(x, z)$ as

$$b_{jk}(x, z) = b^{\rm top}_{jk}(x, z) + b^{\rm err}_{jk}(x, z) \tag{2.8.52}$$

with

$$\text{the top term } b^{\rm top}_{jk}(x, z) \text{ odd and positive homogeneous of degree } 1-n \text{ in } z, \text{ and the error term } b^{\rm err}_{jk}(x, z) \text{ only weakly singular (cf. (2.8.14)).} \tag{2.8.53}$$

Also, with 'hat' denoting the partial Fourier transform in the second variable, from (2.8.17) we see that each top term satisfies

$$\widehat{b^{\rm top}_{jk}}(x, \xi) = \mathrm{Sym}(Q_{jk}; \xi) \ \text{ for } \ |\xi| \geq 1. \tag{2.8.54}$$

Going further, Ω may be locally identified with an open set having a UR boundary in $\mathbb{R}^n$. While we retain the same notation, Ω, for the latter set, there are now two measures operating on $\partial\Omega$, namely the original σ_g induced by the Riemannian metric, and the Euclidean surface measure which we denote by $\sigma^{\rm E}$. They are related to one another as in [112, (1.11.10)]. Moreover, the original differential geometric outward unit conormal ν_g and the Euclidean outward unit normal, denoted by $\nu^{\rm E}$,

[32] as the old adage goes, working on a vector bundle is like doing algebra vertically (along fibers, which are vector spaces), and doing analysis horizontally (relative to the base point, in the underlying manifold)

are related to one another as in [112, (1.11.10)]. Consequently, $\mathscr{K}_Q$ can be identified with the operator-valued matrix $(\mathscr{B}_{jk})_{j,k}$ where, for $x \in \Omega$,

$$\mathscr{B}_{jk} f(x) := \int_{\partial\Omega} \theta_{jk}(x,y) f(y) \Bigg[\sqrt{g(y)} \Big(\sum_{r,s} g^{rs}(y) \nu_r^{\mathrm{E}}(y) \nu_s^{\mathrm{E}}(y) \Big)^{1/2} \Bigg] d\sigma^{\mathrm{E}}(y)$$

$$= \int_{\partial\Omega} b_{jk}(x, x-y) \Big(\sum_{r,s} g^{rs}(y) \nu_r^{\mathrm{E}}(y) \nu_s^{\mathrm{E}}(y) \Big)^{1/2} f(y)\, d\sigma^{\mathrm{E}}(y), \tag{2.8.55}$$

with the last equality coming from (2.8.51). To justify the jump-formula claimed in (2.8.46), at σ^{E}-a.e. point $x \in \partial\Omega$ write

$$\begin{aligned}
&\Big(\mathscr{B}_{jk} f\Big|_{\partial\Omega}^{\text{n.t.}}\Big)(x) \\
&= \frac{1}{2\mathrm{i}} \widehat{b_{jk}^{\,\text{top}}}(x, \nu^{\mathrm{E}}(x)) \Big(\sum_{r,s} g^{rs}(x) \nu_r^{\mathrm{E}}(x) \nu_s^{\mathrm{E}}(x) \Big)^{1/2} f(x) \\
&\quad + \lim_{\varepsilon\to 0^+} \int_{\substack{y\in\partial\Omega \\ |x-y|>\varepsilon}} b_{jk}(x, x-y) \Big(\sum_{r,s} g^{rs}(y) \nu_r^{\mathrm{E}}(y) \nu_s^{\mathrm{E}}(y) \Big)^{1/2} f(y)\, d\sigma^{\mathrm{E}}(y) \\
&= -\tfrac{1}{2}\mathrm{i}\,\mathrm{Sym}\big(Q_{jk}; \nu^{\mathrm{E}}(x)\big) \Big(\sum_{r,s} g^{rs}(x) \nu_r^{\mathrm{E}}(x) \nu_s^{\mathrm{E}}(x) \Big)^{1/2} f(x) \\
&\quad + \lim_{\varepsilon\to 0^+} \int_{\substack{y\in\partial\Omega \\ |x-y|>\varepsilon}} \theta_{jk}(x,y) \sqrt{g(y)} \Big(\sum_{r,s} g^{rs}(y) \nu_r^{\mathrm{E}}(y) \nu_s^{\mathrm{E}}(y) \Big)^{1/2} f(y)\, d\sigma^{\mathrm{E}}(y) \\
&= -\tfrac{1}{2}\mathrm{i}\,\mathrm{Sym}\big(Q_{jk}; \nu_g(x)\big) f(x) + \lim_{\varepsilon\to 0^+} \int_{\substack{y\in\partial\Omega \\ |x-y|>\varepsilon}} \theta_{jk}(x,y) f(y) \rho(y)\, d\sigma^{\mathrm{E}}(y) \\
&= -\tfrac{1}{2}\mathrm{i}\,\mathrm{Sym}\big(Q_{jk}; \nu_g(x)\big) f(x) + \text{P.V.} \int_{\partial\Omega} \theta_{jk}(x,y) f(y)\, d\sigma_g(y) \\
&= -\tfrac{1}{2}\mathrm{i}\,\mathrm{Sym}\big(Q_{jk}; \nu_g(x)\big) f(x) + K_{jk} f(x).
\end{aligned} \tag{2.8.56}$$

Above, the first equality is a consequence of (2.8.55), (2.8.52)-(2.8.53), and the jump-formulas proved in Theorem 2.5.38 in the class of variable coefficient Calderón-Zygmund operators in the Euclidean setting; here, 'hat' stands for partial Fourier transform in the second variable. The second equality in (2.8.56) is implied by (2.8.54) and (2.8.51). In the third equality in (2.8.56) we have used the fact that $\mathrm{Sym}(Q_{jk}; \xi)$ is homogeneous of degree -1 in ξ, plus the formula relating ν_g to ν^{E} in [112, (1.11.10)], and (2.8.34) (keeping in mind that currently $\partial_*\Omega$ has full measure in

$\partial\Omega$). The fourth equality in (2.8.56) is seen from Definition 2.8.1. The final equality in (2.8.56) actually defines the operator K_{jk} as the principal-value integral in the penultimate line of (2.8.56). Under the identification $K_Q \equiv (K_{jk})_{j,k}$, the analysis above amounts to

$$\Big(\mathscr{K}_Q f\Big|_{\partial\Omega}^{\rm n.t.}\Big)(x) = -\tfrac{1}{2}\mathrm{iSym}\big(Q;\nu_g(x)\big)f(x) \qquad (2.8.57)$$
$$+\,\mathrm{P.V.}\int_{\partial\Omega}\langle k_Q(x,y), f(y)\rangle_{\mathcal{E}_y}\,d\sigma_g(y) \text{ at } \sigma_g\text{-a.e. } x\in\partial\Omega,$$

which is precisely (2.8.46).

Next, the claims pertaining to (2.8.36)-(2.8.40), (2.8.42)-(2.8.44) and (2.8.47) follow in a similar fashion from the corresponding theory for singular integrals with variable kernel in UR domains in the Euclidean setting from Theorem 2.5.38. Also, the L^p-square function estimates claimed in items *(6)-(7)* may be established in an analogous manner by relying on the Euclidean result recorded in Theorem 2.5.38 for the top terms and using [112, Lemma 8.7.11] to control less singular terms. Once these have been established, then we may invoke [113, Lemma 9.2.40] together with the nontangential maximal function estimates from the current item *(4)* and [112, (8.6.51)] to conclude that, if Ω is actually a Lipschitz domain, then the global Besov regularity result claimed in (2.8.49) holds. With this in hand, (2.8.50) is implied by [113, (9.2.22)].

Finally, consider the claim made in (2.8.41). With this in mind, first note that

$$(k_Q)^\top(y,x) = k_{Q^\top}(x,y) \text{ in } \mathcal{D}'(\mathcal{M}\times\mathcal{M},\,\mathcal{E}\otimes\mathcal{F}). \qquad (2.8.58)$$

Indeed, for two arbitrary test functions $\phi\in\mathscr{C}^2(\mathcal{M},\mathcal{E})$ and $\psi\in\mathscr{C}^2(\mathcal{M},\mathcal{F})$ we have

$$\begin{aligned}\Big\langle (k_Q)^\top(y,x), \phi(x)\otimes\psi(y)\Big\rangle_{\mathcal{E}\otimes\mathcal{F}} &= \Big\langle k_Q(x,y), \psi(x)\otimes\phi(y)\Big\rangle_{\mathcal{F}\otimes\mathcal{E}}\\ &= \langle Q\phi,\psi\rangle_{\mathcal{F}} = \langle Q^\top\psi,\phi\rangle_{\mathcal{E}}\\ &= \big\langle k_{Q^\top}(x,y), \phi(x)\otimes\psi(y)\big\rangle_{\mathcal{E}\otimes\mathcal{F}} \end{aligned}\qquad (2.8.59)$$

from which (2.8.58) follows. To start the proof of (2.8.41) in earnest, observe that using a partition of unity and the linearity of the operators involved, matters are reduced to showing that

$$(K_Q)^\top h = K_{Q^\top} h \qquad (2.8.60)$$

for each $h\in L^p(\partial\Omega,\sigma_g)\otimes\mathcal{F}$ such that supp h is contained in a local coordinate patch O (which we will freely identify with an open set in $\mathbb{R}^n$, still denoted by O) over which the vector-bundles $\mathcal{E},\mathcal{F}$ trivialize. Using local orthonormal frames, identify h with a vector-valued function with components $\{h_\alpha\}_\alpha$, and identify the kernel $k_Q(x,y)$ with a matrix $\big(k_Q^{\alpha\beta}(x,y)\big)_{\alpha,\beta}$. If the kernel $k_{Q^\top}(x,y)$ is also identified in O with the matrix $\big(k_{Q^\top}^{\alpha\beta}(x,y)\big)_{\alpha,\beta}$, then (2.8.58) implies

$$k_{Q^{\top}}^{\alpha\beta}(x,y) = k_{Q}^{\beta\alpha}(y,x). \tag{2.8.61}$$

Pick an exponent $p' \in (1,\infty)$ satisfying $1/p + 1/p' = 1$, and bring in another function, $f \in L^{p'}(\partial\Omega, \sigma_g) \otimes \mathcal{E}$ supported in the local coordinate patch $\mathcal{O}$. Use the local orthonormal frame in $\mathcal{E}$ to identify f with a vector-valued function with components $\{f_\beta\}_\beta$. Working in local Euclidean coordinates, for each $\varepsilon > 0$ define

$$K_{Q,\varepsilon} f(x) := \left(\int_{\substack{y\in\partial\Omega \\ |x-y|>\varepsilon}} k_Q^{\alpha\beta}(x,y) f_\beta(y)\rho(y)\, d\sigma^{\mathrm{E}}(y) \right)_\alpha \tag{2.8.62}$$

for each $x \in \partial\Omega \cap \mathcal{O}$, and also

$$K_{Q,*} f(x) := \sup_{\varepsilon>0} \left| \left(\int_{\substack{y\in\partial\Omega \\ |x-y|>\varepsilon}} k_Q^{\alpha\beta}(x,y) f_\beta(y)\rho(y)\, d\sigma^{\mathrm{E}}(y) \right)_\alpha \right| \tag{2.8.63}$$

for each $x \in \partial\Omega \cap \mathcal{O}$. Then (2.8.36) and Definition 2.8.1 ensure that

$$\lim_{\varepsilon\to 0^+} K_{Q,\varepsilon} f(x) = K_Q f(x) \text{ at } \sigma^{\mathrm{E}}\text{-a.e. point } x \in \partial\Omega \cap \mathcal{O}, \tag{2.8.64}$$

while the Euclidean Calderón-Zygmund theory for variable coefficient kernels (cf. Theorem 2.5.38) guarantees that

$$K_{Q,*} f \in L^p(\partial\Omega \cap \mathcal{O}, \sigma^{\mathrm{E}}). \tag{2.8.65}$$

Granted (2.8.64)-(2.8.65), Lebesgue's Dominated Convergence Theorem applies and gives

$$\lim_{\varepsilon\to 0^+} K_{Q,\varepsilon} f = K_Q f \text{ in } L^p(\partial\Omega \cap \mathcal{O}, \sigma^{\mathrm{E}}). \tag{2.8.66}$$

If, much as in (2.8.62)-(2.8.63), we also define, for each $y \in \partial\Omega \cap \mathcal{O}$,

$$K_{Q^{\top},\varepsilon} h(y) := \left(\int_{\substack{x\in\partial\Omega \\ |y-x|>\varepsilon}} k_{Q^{\top}}^{\beta\alpha}(y,x) h_\alpha(x)\rho(x)\, d\sigma^{\mathrm{E}}(x) \right)_\beta \tag{2.8.67}$$

then the same type of argument based on Lebesgue's Dominated Convergence Theorem implies that

$$\lim_{\varepsilon\to 0^+} K_{Q^{\top},\varepsilon} h = K_{Q^{\top}} h \text{ in } L^{p'}(\partial\Omega \cap \mathcal{O}, \sigma^{\mathrm{E}}). \tag{2.8.68}$$

Thanks to the aforementioned identifications, for each $\varepsilon > 0$ we may now write

$$
\begin{aligned}
&\int_{\partial\Omega\cap O} h_\alpha(x)\big(K_{Q,\varepsilon}f\big)_\alpha(x)\rho(x)\,d\sigma^{\mathrm{E}}(x)\\
&\quad=\int_{x\in\partial\Omega\cap O} h_\alpha(x)\Bigg(\int_{\substack{y\in\partial\Omega\\|x-y|>\varepsilon}} k_Q^{\alpha\beta}(x,y)f_\beta(y)\rho(y)\,d\sigma^{\mathrm{E}}(y)\Bigg)\rho(x)\,d\sigma^{\mathrm{E}}(x)\\
&\quad=\int_{y\in\partial\Omega\cap O} f_\beta(y)\Bigg(\int_{\substack{x\in\partial\Omega\\|x-y|>\varepsilon}} k_Q^{\alpha\beta}(x,y)h_\alpha(x)\rho(x)\,d\sigma^{\mathrm{E}}(x)\Bigg)\rho(y)\,d\sigma^{\mathrm{E}}(y)\\
&\quad=\int_{y\in\partial\Omega\cap O} f_\beta(y)\Bigg(\int_{\substack{x\in\partial\Omega\\|y-x|>\varepsilon}} k_{Q^{\top}}^{\beta\alpha}(y,x)h_\alpha(x)\rho(x)\,d\sigma^{\mathrm{E}}(x)\Bigg)\rho(y)\,d\sigma^{\mathrm{E}}(y)\\
&\quad=\int_{\partial\Omega\cap O} f_\beta(y)\big(K_{Q^{\top},\varepsilon}h\big)_\beta(y)\rho(y)\,d\sigma^{\mathrm{E}}(y),
\end{aligned}
\tag{2.8.69}
$$

where we have used Fubini's Theorem together with (2.8.61). Use (2.8.66) and (2.8.68) to pass to limit $\varepsilon \to 0^+$ in (2.8.69), then move the integrals back to the manifold $\mathcal{M}$. This gives

$$
\int_{\partial\Omega}\langle f,(K_Q)^{\top}h\rangle_{\mathcal{E}}\,d\sigma_g=\int_{\partial\Omega}\langle h,K_Q f\rangle_{\mathcal{F}}\,d\sigma_g=\int_{\partial\Omega}\langle f,K_{Q^{\top}}h\rangle_{\mathcal{E}}\,d\sigma_g,
\tag{2.8.70}
$$

with the first equality based on the actual definition of $(K_Q)^{\top}$. In view of the arbitrariness of f, this proves that

$$
(K_Q)^{\top}h=K_{Q^{\top}}h\ \text{ on }\ \partial\Omega\cap O.
\tag{2.8.71}
$$

Recall that the support of h is contained in $\partial\Omega\cap O$. Bearing this in mind, it follows from (2.8.32) in Definition 2.8.1, (2.8.58), and an argument based on Fubini's Theorem in the spirit of (2.8.69) (in which we now take the function f to be supported in $\partial\Omega\setminus O$) that we also have

$$
(K_Q)^{\top}h=K_{Q^{\top}}h\ \text{ on }\ \partial\Omega\setminus O.
\tag{2.8.72}
$$

Altogether, this shows that $(K_Q)^{\top}h=K_{Q^{\top}}h$ on $\partial\Omega$, finishing the proof of (2.8.60).□

It is interesting to contrast the results in items *(6)-(7)* of Theorem 2.8.2 with the following global regularity result in NTA domains on manifolds for null-solutions of weakly elliptic systems. Recall that a differential operator $L:\mathcal{E}\to\mathcal{F}$ acting between two vector bundles $\mathcal{E},\mathcal{F}\to\mathcal{M}$ is called weakly elliptic if the symbol map $\mathrm{Sym}(L;\xi):\mathcal{E}_x\to\mathcal{F}_x$ is invertible for each $x\in\mathcal{M}$ and $\xi\in T_x^*\mathcal{M}\setminus\{0\}$. This should be contrasted with the Legendre-Hadamard strong ellipticity condition, demanding the existence of a constant $c\in(0,\infty)$ such that

$$\begin{aligned}&\operatorname{Re}\big\langle \operatorname{Sym}(-L;\xi)\eta,\eta^c\big\rangle_{\mathcal{F}_x} \geq c|\eta|^2_{\mathcal{F}_x}\\ &\text{for all } x\in\mathcal{M},\ \eta\in\mathcal{F}_x \text{ and } \xi\in T^*_x\mathcal{M} \text{ with } |\xi|=1.\end{aligned} \tag{2.8.73}$$

Proposition 2.8.3 *Let $\mathcal{M}$ be an n-dimensional compact boundaryless oriented manifold of class $\mathscr{C}^2$ equipped with a Lipschitz Riemannian metric tensor g. Denote by $\mathcal{L}^n_g$ the n-dimensional Lebesgue measure associated with the metric g on $\mathcal{M}$, and let $d_g(\cdot,\cdot)$ stand for the geodesic distance between points in $\mathcal{M}$. Also, let $\mathcal{E},\mathcal{F}\to\mathcal{M}$ be two Hermitian vector bundles with Lipschitz metrics, and denote by ∇ a connection on $\mathcal{F}$. Consider a weakly elliptic, second-order, differential operator L mapping sections of $\mathcal{E}$ into sections of $\mathcal{F}$ such that, when written in local coordinates as*

$$Lu=\Big(\sum_{j,k}\sum_{\beta}\partial_j a^{\alpha\beta}_{jk}\partial_k u_\beta+\sum_j\sum_\beta b^{\alpha\beta}_j\partial_j u_\beta+\sum_\beta d^{\alpha\beta}u_\beta\Big)_\alpha \tag{2.8.74}$$

where $u=(u_\beta)_\beta$, its coefficients exhibit the following type of regularity:

$$a^{\alpha\beta}_{jk}\in\mathrm{Lip},\quad b^{\alpha\beta}_j\in L^r_1,\quad d^{\alpha\beta}\in L^r \text{ for some } r>n. \tag{2.8.75}$$

Then, given any one-sided NTA *domain $\Omega\subset\mathcal{M}$ along with $p\in(r/(r-1),r)$ and $s\in(0,1)$, there exists some constant $C=C(\Omega,L,p,s)\in(0,\infty)$ such that for each section $u\in L^p(\Omega,\mathcal{L}^n_g)\otimes\mathcal{E}$ satisfying $Lu=0$ in Ω one has*

$$\begin{aligned}\|u\|_{B^{p,p}_s(\Omega)\otimes\mathcal{E}}&\leq C\|u\|_{L^p(\Omega,\mathcal{L}^n_g)\otimes\mathcal{E}}\\ &\quad+C\Big(\int_\Omega|(\nabla u)(x)|^p\operatorname{dist}_g(x,\partial\Omega)^{p(1-s)}\,d\mathcal{L}^n_g(x)\Big)^{1/p}.\end{aligned} \tag{2.8.76}$$

As a corollary, corresponding to the particular choice $p:=2$ and $s:=1/2$, one has

$$\begin{aligned}\|u\|_{H^{1/2}(\Omega)\otimes\mathcal{E}}&\leq C\|u\|_{L^2(\Omega,\mathcal{L}^n_g)\otimes\mathcal{E}}\\ &\quad+C\Big(\int_\Omega|(\nabla u)(x)|^2\operatorname{dist}_g(x,\partial\Omega)\,d\mathcal{L}^n_g(x)\Big)^{1/2}.\end{aligned} \tag{2.8.77}$$

Proof Work in local coordinates in a small neighborhood of a boundary point $x_*\in\partial\Omega$, over which the vector bundles trivialize, and drop the dependence of various pieces of notations on $\mathcal{E}$, $\mathcal{F}$ in what follows. We shall need an interior elliptic estimate that appears in [131, Proposition 3.4] to the effect that, under the regularity assumptions on the coefficients imposed in (2.8.75), if $p\in(r/(r-1),r)$ then whenever $x\in\Omega$ and $0<t<2\operatorname{dist}(x,\partial\Omega)/3$ one has

$$\sup_{B(x,t)}|\nabla u|\leq Ct^{-1}\Big(⨍_{B(x,3t/2)}|u(y)-u(x)|^p\,dy\Big)^{1/p}+Ct^{1-n/r}|u(x)|, \tag{2.8.78}$$

where the constant $C\in(0,\infty)$ is independent of u, x and t. For each point x, let us abbreviate

$$\rho(x) := \operatorname{dist}(x, \partial\Omega) \tag{2.8.79}$$

and, whenever $x \in \Omega$ and $0 < t < \rho(x)/2$, define the oscillation

$$\operatorname{osc}^0(u, x, t) := \fint_{B(x,t)} |u(y) - u(x)| \, dy. \tag{2.8.80}$$

Then [158, Corollary 1, p. 398] implies that whenever $1 < p < \infty$ and $s \in (0, 1)$ we have

$$\begin{aligned} \|u\|_{B_s^{p,p}(\Omega)} &\approx \|u\|_{L^p(\Omega, \mathcal{L}^n)} \\ &\quad + \left\| \left(\int_0^{\frac{1}{2}\rho(\cdot)} \left[\operatorname{osc}^0(u, \cdot, t)\right]^p \frac{dt}{t^{1+sp}} \right)^{1/p} \right\|_{L^p(\Omega, \mathcal{L}^n)} . \end{aligned} \tag{2.8.81}$$

Next, for every $x \in \Omega$ and $t \in \big(0, \rho(x)/2\big)$ we estimate

$$\begin{aligned} \operatorname{osc}^0(u, x, t) &= \fint_{B(x,t)} |u(y) - u(x)| \, dy \leq Ct \fint_{B(x,t)} |(\nabla u)(y)| \, dy \\ &\leq Ct \cdot \sup_{B(x,t)} |\nabla u| \leq Ct \cdot \sup_{B(x,\rho(x)/2)} |\nabla u| \\ &\leq Ct\rho(x)^{-1} \Big(\fint_{B(x,3\rho(x)/4)} |u(y) - u(x)|^p \, dy \Big)^{1/p} \\ &\quad + Ct\rho(x)^{1-n/r} |u(x)| \\ &\leq Ct \Big(\fint_{B(x,3\rho(x)/4)} |(\nabla u)(y)|^p \, dy \Big)^{1/p} + Ct\rho(x)^{1-n/r} |u(x)|, \end{aligned} \tag{2.8.82}$$

by Poincaré's inequality (used twice) and the interior estimate recorded in (2.8.78). This implies that whenever $x \in \Omega$ and $t \in \big(0, \rho(x)/2\big)$ we have

$$\begin{aligned} \left[\operatorname{osc}^0(u, x, t)\right]^p \frac{1}{t^p} &\leq C \fint_{B(x,3\rho(x)/4)} |(\nabla u)(y)|^p \, dy \\ &\quad + C\rho(x)^{p-(np)/r} |u(x)|^p . \end{aligned} \tag{2.8.83}$$

Consequently,

$$\left\| \left(\int_0^{\rho(\cdot)/2} \left[\operatorname{osc}^0(u, \cdot, t)\right]^p \frac{dt}{t^{1+sp}} \right)^{1/p} \right\|_{L^p(\Omega, \mathcal{L}^n)} \leq \mathrm{I} + \mathrm{II}, \tag{2.8.84}$$

where

$$\mathrm{I} := C\left(\int_\Omega\left(\int_0^{\rho(x)/2} t^{p-1-sp}\left(\fint_{B(x,3\rho(x)/4)}|(\nabla u)(y)|^p\,\mathrm{d}y\right)\mathrm{d}t\right)\mathrm{d}x\right)^{1/p} \tag{2.8.85}$$

and

$$\mathrm{II} := C\left(\int_\Omega\left(\int_0^{\rho(x)/2} t^{p-1-sp}\left(\rho(x)^{p-(np)/r}|u(x)|^p\right)\mathrm{d}t\right)\mathrm{d}x\right)^{1/p}. \tag{2.8.86}$$

Note that

$$\begin{aligned}\mathrm{I} &\le C\left(\int_\Omega \rho(x)^{p(1-s)}\left(\fint_{B(x,3\rho(x)/4)}|(\nabla u)(y)|^p\,\mathrm{d}y\right)\mathrm{d}x\right)^{1/p}\\ &\le C\left(\int_\Omega\int_\Omega \rho(x)^{p(1-s)-n}|(\nabla u)(y)|^p\,\mathbf{1}_{\mathrm{dist}(x,y)\le 3\rho(x)/4}\,\mathrm{d}y\,\mathrm{d}x\right)^{1/p}.\end{aligned} \tag{2.8.87}$$

Since whenever $\mathrm{dist}(x,y) \le 3\rho(x)/4$ we necessarily have $\rho(x) \approx \rho(y)$, we may further estimate

$$\begin{aligned}\mathrm{I} &\le C\left(\int_\Omega\int_\Omega \rho(y)^{p(1-s)-n}|(\nabla u)(y)|^p\,\mathbf{1}_{\mathrm{dist}(x,y)\le c\rho(y)}\,\mathrm{d}y\,\mathrm{d}x\right)^{1/p}\\ &\le C\left(\int_\Omega \rho(y)^{p(1-s)-n}|(\nabla u)(y)|^p\left(\int_\Omega \mathbf{1}_{\mathrm{dist}(x,y)\le c\rho(y)}\,\mathrm{d}x\right)\mathrm{d}y\right)^{1/p}\\ &\le C\left(\int_\Omega \rho(y)^{p(1-s)}|(\nabla u)(y)|^p\,\mathrm{d}y\right)^{1/p},\end{aligned} \tag{2.8.88}$$

which suits our purposes. There remains to observe that, since $r > n$ and $s < 1$,

$$\mathrm{II} \le C\left(\int_\Omega \rho(x)^{p(2-s-n/r)}|u(x)|^p\,\mathrm{d}x\right)^{1/p} \le C\|u\|_{L^p(\Omega,\mathcal{L}^n)}. \tag{2.8.89}$$

At this stage, (2.8.76) follows from [113, (9.2.149)], (2.8.84), (2.8.88), and (2.8.89).□

The stage is set to fashion out of Theorem 2.8.2 a similar result for boundary layer potentials associated with a given elliptic operator with coefficients of limited smoothness. Before stating this result we make the following convention. Suppose $\mathcal{E}, \mathcal{F}, \mathcal{G}, \mathcal{H} \to \mathcal{M}$ are four vector bundles. Given some arbitrary double distribution $E \in \mathcal{D}'(\mathcal{M}\times\mathcal{M}, \mathcal{E}\otimes\mathcal{F})$ along with two differential operators $\widetilde{P} : \mathcal{E} \to \mathcal{G}$ and $P : \mathcal{F} \to \mathcal{H}$, writing

$$(\widetilde{P}_x \otimes P_y)E(x,y) \text{ or, simply } (\widetilde{P}\otimes P)E(x,y), \tag{2.8.90}$$

indicates that $\widetilde{P}$ acts in the variable x on the columns of $E(x,y)$ while P acts in the variable y on the rows of $E(x,y)$. More specifically, work in local coordinates over which $\mathcal{E}, \mathcal{F}$ trivialize, and let $\{e_\alpha\}_\alpha$ and $\{f_\beta\}_\beta$ be arbitrary local orthonormal frames for $\mathcal{E}$ and $\mathcal{F}$, respectively. Use these to identify the distribution $E(x, y)$ with a matrix $E_{\alpha\beta}(x, y)e_\alpha \otimes f_\beta$, in the sense that for each $\varphi \in \mathscr{D}(\mathcal{M}, \mathcal{F})$ locally expressed as $\varphi = \varphi_\beta f_\beta$ we have

$$\langle E(x, y), \varphi(y)\rangle = \langle E_{\alpha\beta}(x, y), \varphi_\beta(y)\rangle\, e_\alpha. \tag{2.8.91}$$

With I denoting the identity operator then define

$$\widetilde{P}_x E(x, y) := (\widetilde{P}_x \otimes I_y)E(x, y) := \Big(\widetilde{P}_x\big[E_{\alpha\beta}(x, y)\, e_\alpha\big]\Big) \otimes f_\beta, \tag{2.8.92}$$

$$P_y E(x, y) := (I_x \otimes P_y)E(x, y) := e_\alpha \otimes \Big(P_y\big[E_{\alpha\beta}(x, y)\, f_\beta\big]\Big). \tag{2.8.93}$$

Otherwise said, we are declaring that

$$\begin{aligned} \widetilde{P}_x = (\widetilde{P}_x \otimes I_y) &\text{ acts on the columns of } E(x, y), \text{ and} \\ P_y = (I_x \otimes P_y) &\text{ acts on the rows of } E(x, y). \end{aligned} \tag{2.8.94}$$

For further reference let us also point out here that

$$\begin{aligned} &\big\langle (\widetilde{P}_x \otimes I_y)E(x, y), \varphi(x) \otimes \psi(y)\big\rangle = \big\langle E(x, y), (\widetilde{P}^\top \varphi)(x) \otimes \psi(y)\big\rangle \\ &\qquad \text{for any } \varphi \in \mathcal{D}(\mathcal{M}, \mathcal{G}) \text{ and } \psi \in \mathcal{D}(\mathcal{M}, \mathcal{F}), \end{aligned} \tag{2.8.95}$$

and

$$\begin{aligned} &\big\langle (I_x \otimes P_y)E(x, y), \varphi(x) \otimes \psi(y)\big\rangle = \big\langle E(x, y), \varphi(x) \otimes (P^\top \psi)(y)\big\rangle \\ &\qquad \text{for any } \varphi \in \mathcal{D}(\mathcal{M}, \mathcal{E}) \text{ and } \psi \in \mathcal{D}(\mathcal{M}, \mathcal{H}). \end{aligned} \tag{2.8.96}$$

Theorem 2.8.4 *Let $\mathcal{M}$ be an n-dimensional compact boundaryless oriented manifold of class $\mathscr{C}^2$ equipped with a $\mathscr{C}^1$ Riemannian metric tensor g. Let $\mathcal{L}_g^n$ and $\mathcal{H}_g^{n-1}$ be, respectively, the n-dimensional Lebesgue measure and the $(n-1)$-dimensional Hausdorff measure associated with the metric g on $\mathcal{M}$. Also, let $\mathcal{E}, \mathcal{F} \to \mathcal{M}$ be two Hermitian vector bundles equipped with $\mathscr{C}^1$ metrics, and use ∇ to denote connections on $\mathcal{E}$ and $\mathcal{F}$.*

In this context, consider a weakly elliptic, second-order, differential operator L mapping sections of $\mathcal{E}$ into sections of $\mathcal{F}$ such that, when written in local coordinates as

$$Lu = \Bigg(\sum_{j,k}\sum_{\beta} \partial_j a_{jk}^{\alpha\beta} \partial_k u_\beta + \sum_{j}\sum_{\beta} b_j^{\alpha\beta} \partial_j u_\beta + \sum_{\beta} d^{\alpha\beta} u_\beta\Bigg)_\alpha, \tag{2.8.97}$$

its coefficients satisfy

$$a_{jk}^{\alpha\beta} \in \mathscr{C}^1, \quad b_j^{\alpha\beta} \in L^\infty, \quad d^{\alpha\beta} \in L^r \textit{ for some } r > n. \tag{2.8.98}$$

In addition, assume that in local coordinates the coefficients of L^*, the Hermitian adjoint of L, also satisfy[33] (2.8.98). *Also, suppose that*

$$L : H^1(\mathcal{M}) \otimes \mathcal{E} \longrightarrow H^{-1}(\mathcal{M}) \otimes \mathcal{F} \text{ is invertible} \tag{2.8.99}$$

and let

$$E \in \mathcal{D}'(\mathcal{M} \times \mathcal{M}, \mathcal{E} \otimes \mathcal{F}) \text{ denote the Schwartz kernel of } L^{-1}. \tag{2.8.100}$$

Next, let Ω be an arbitrary UR *domain in $\mathcal{M}$. Denote its geometric measure theoretic outward unit conormal by $\nu_g \in T^*\mathcal{M}$ and abbreviate $\sigma_g := \mathcal{H}_g^{n-1}\lfloor\partial\Omega$. Also, let $d_g(\cdot,\cdot)$ stand for the geodesic distance between points in $\mathcal{M}$. In this setting, define action of the single layer operator $\mathscr{S}$ and of its boundary version S on an arbitrary function $f \in L^1(\partial\Omega, \sigma_g) \otimes \mathcal{F}$ as*

$$\mathscr{S}f(x) := \int_{\partial\Omega} \big\langle E(x,y), f(y)\big\rangle_{\mathcal{F}_y}\, \mathrm{d}\sigma_g(y), \qquad x \in \Omega, \tag{2.8.101}$$

and, respectively,

$$Sf(x) := \int_{\partial\Omega} \big\langle E(x,y), f(y)\big\rangle_{\mathcal{F}_y}\, \mathrm{d}\sigma_g(y), \qquad x \in \partial\Omega. \tag{2.8.102}$$

In addition, for a first-order differential operator $P : \mathcal{F} \to \mathcal{F}$ with continuous coefficients, consider the integral operator with kernel $(I_x \otimes P_y)E(x,y)$, i.e., for each function $f \in L^1(\partial\Omega, \sigma_g) \otimes \mathcal{F}$ set

$$\mathscr{A}f(x) := \int_{\partial\Omega} \big\langle (I_x \otimes P_y)E(x,y), f(y)\big\rangle_{\mathcal{F}_y}\, \mathrm{d}\sigma_g(y), \qquad x \in \Omega. \tag{2.8.103}$$

Define the maximal integral operator $A_{\max}$ acting on each $f \in L^1(\partial\Omega, \sigma_g) \otimes \mathcal{F}$ according to

$$A_{\max}f(x) := \sup_{\varepsilon>0}\left| \int_{\substack{y\in\partial\Omega \\ d_g(x,y)>\varepsilon}} \big\langle (I_x \otimes P_y)E(x,y), f(y)\big\rangle_{\mathcal{F}_y}\, \mathrm{d}\sigma_g(y)\right|, \tag{2.8.104}$$

for all $x \in \partial\Omega$, and consider

$$Af(x) := \text{P.V.}\int_{\partial\Omega} \big\langle (I_x \otimes P_y)E(x,y), f(y)\big\rangle_{\mathcal{F}_y}\, \mathrm{d}\sigma_g(y) \tag{2.8.105}$$

for σ_g-a.e. $x \in \partial\Omega$, with the principal-value integral understood in the sense of Definition 2.8.1.

Then the following claims are true.

[33] this is the case if in place of (2.8.98) we start by assuming that $a_{jk}^{\alpha\beta} \in \mathscr{C}^1$, $b_j^{\alpha\beta} \in L_1^r$, and $d^{\alpha\beta} \in L^r$ for some $r > n$

(1) The boundary-to-domain single layer, originally acting on $L^1(\partial\Omega, \sigma_g) \otimes \mathcal{F}$ *as in* (2.8.101) *extends to a linear mapping acting on each* $f \in L^p_{-1}(\partial\Omega, \sigma_g) \otimes \mathcal{F}$ *with* $p \in (1, \infty)$ *as*

$$\mathscr{S} f(x) := {}_{L^{p'}_1(\partial\Omega,\sigma_g)\otimes\mathcal{F}}\Big\langle E(x,\cdot)\big|_{\partial\Omega}, f\Big\rangle_{L^p_{-1}(\partial\Omega,\sigma_g)\otimes\mathcal{F}}, \qquad \forall x \in \Omega, \tag{2.8.106}$$

where $p' \in (1, \infty)$ *is such that* $1/p + 1/p' = 1$. *Defined as such, for each* $p \in (1, \infty)$ *one then has*

$$L(\mathscr{S} f) = 0 \ \text{in} \ \Omega, \qquad \forall f \in L^p_{-1}(\partial\Omega, \sigma_g) \otimes \mathcal{F}, \tag{2.8.107}$$

and

$$\mathscr{S} f \in \mathscr{C}^0_{\mathrm{loc}}(\Omega) \otimes \mathcal{E}, \qquad \forall f \in L^p_{-1}(\partial\Omega, \sigma_g) \otimes \mathcal{F}, \tag{2.8.108}$$

while

$$\mathscr{S} f \in \mathscr{C}^1_{\mathrm{loc}}(\Omega) \otimes \mathcal{E}, \qquad \forall f \in L^1(\partial\Omega, \sigma_g) \otimes \mathcal{F}. \tag{2.8.109}$$

Also, for each aperture parameter $\kappa > 0$ *and any* $p \in [1, \infty)$ *there exists a constant* $C \in (0, \infty)$ *with the property that*

$$\begin{gathered} \|\mathcal{N}_\kappa(\mathscr{S} f)\|_{L^p(\partial\Omega,\sigma_g)} \leq C\|f\|_{L^p(\partial\Omega,\sigma_g)\otimes\mathcal{F}} \\ \text{for each } f \in L^p(\partial\Omega, \sigma_g) \otimes \mathcal{F}, \ \text{if } p \in [1, \infty], \end{gathered} \tag{2.8.110}$$

$$\begin{gathered} \|\mathcal{N}_\kappa(\mathscr{S} f)\|_{L^p(\partial\Omega,\sigma_g)} \leq C\|f\|_{L^p_{-1}(\partial\Omega,\sigma_g)\otimes\mathcal{F}} \\ \text{for each } f \in L^p_{-1}(\partial\Omega, \sigma_g) \otimes \mathcal{F}, \ \text{if } p \in (1, \infty), \end{gathered} \tag{2.8.111}$$

$$\begin{gathered} \|\mathcal{N}_\kappa(\nabla\mathscr{S} f)\|_{L^p(\partial\Omega,\sigma_g)} \leq C\|f\|_{L^p(\partial\Omega,\sigma_g)\otimes\mathcal{F}} \\ \text{for each } f \in L^p(\partial\Omega, \sigma_g) \otimes \mathcal{F}, \ \text{if } p \in (1, \infty), \end{gathered} \tag{2.8.112}$$

and, corresponding to $p = 1$,

$$\begin{gathered} \|\mathcal{N}_\kappa(\nabla\mathscr{S} f)\|_{L^{1,\infty}(\partial\Omega,\sigma_g)} \leq C\|f\|_{L^1(\partial\Omega,\sigma_g)\otimes\mathcal{F}} \\ \text{for each function } f \in L^1(\partial\Omega, \sigma_g) \otimes \mathcal{F}. \end{gathered} \tag{2.8.113}$$

In fact, a more nuanced version of (2.8.110) *holds. Namely, if* $n = \dim \mathcal{M} \geq 3$ *one has:*

if $p \in (1, n-1)$ *and* $1/q = 1/p - 1/(n-1)$ *there exists* $C \in (0, \infty)$ *so that* $\|\mathcal{N}_\kappa(\mathcal{S}f)\|_{L^q(\partial\Omega,\sigma_g)} \leq C\|f\|_{L^p(\partial\Omega,\sigma_g)\otimes\mathcal{F}}$, $\forall f \in L^p(\partial\Omega, \sigma_g) \otimes \mathcal{F}$,

while for each $p \in (n-1, \infty)$ *there exists* $C \in (0, \infty)$ *such that*

$$\big\|\mathcal{N}_\kappa(\mathcal{S}f)\big\|_{L^\infty(\partial\Omega,\sigma_g)} \leq C\|f\|_{L^p(\partial\Omega,\sigma_g)\otimes\mathcal{F}}, \qquad \forall f \in L^p(\partial\Omega, \sigma_g) \otimes \mathcal{F},$$

and, corresponding to $p = n-1$, *for every* $q \in (1, \infty)$ *there holds*

$$\big\|\mathcal{N}_\kappa(\mathcal{S}f)\big\|_{L^q(\partial\Omega,\sigma_g)} \leq C_q\|f\|_{L^{n-1}(\partial\Omega,\sigma_g)\otimes\mathcal{F}}, \quad \forall f \in L^{n-1}(\partial\Omega, \sigma_g) \otimes \mathcal{F},$$

while corresponding to the critical value $p = 1$ *one has the estimate*

$$\big\|\mathcal{N}_\kappa(\mathcal{S}f)\big\|_{L^{(n-1)/(n-2),\infty}(\partial\Omega,\sigma_g)} \leq C\|f\|_{L^1(\partial\Omega,\sigma_g)\otimes\mathcal{F}}, \quad \forall f \in L^1(\partial\Omega, \sigma_g) \otimes \mathcal{F}. \tag{2.8.114}$$

In addition, in the case when $n = 2$ *one has*

$$\begin{aligned} &\big\|\mathcal{N}_\kappa(\mathcal{S}f)\big\|_{L^\infty(\partial\Omega,\sigma_g)} \leq C\|f\|_{L^p(\partial\Omega,\sigma_g)\otimes\mathcal{F}} \\ &\text{for each } f \in L^p(\partial\Omega, \sigma_g) \otimes \mathcal{F} \text{ with } p \in (1, \infty), \end{aligned} \tag{2.8.115}$$

and

$$\begin{aligned} &\big\|\mathcal{N}_\kappa(\mathcal{S}f)\big\|_{L^q(\partial\Omega,\sigma_g)} \leq C_q\|f\|_{L^1(\partial\Omega,\sigma_g)\otimes\mathcal{F}} \\ &\text{for each } f \in L^1(\partial\Omega, \sigma_g) \otimes \mathcal{F} \text{ and each } q \in (1, \infty). \end{aligned} \tag{2.8.116}$$

(2) *The single layer operator satisfies a number of square function styled estimates. First, if hypothesis* (2.8.98) *regarding the regularity of the coefficients of the differential operator* L *is strengthened to*

$$\begin{aligned} &a_{jk}^{\alpha\beta} \in \mathscr{C}^{1+\gamma}, \quad b_j^{\alpha\beta} \in L_1^r, \quad d^{\alpha\beta} \in L^r, \\ &\text{for some } \gamma > 0 \text{ and } r > n, \end{aligned} \tag{2.8.117}$$

then for each integrability exponent $p \in (1, \infty)$ *there exists some constant* $C \in (0, \infty)$ *with the property that for each* $f \in L^p_{-1}(\partial\Omega, \sigma_g) \otimes \mathcal{F}$ *one has*

$$\Big(\int_\Omega |\nabla(\mathcal{S}f)(x)|^p \operatorname{dist}_g(x, \partial\Omega)^{p-1}\, d\mathcal{L}_g^n(x)\Big)^{1/p} \leq C\|f\|_{L^p_{-1}(\partial\Omega,\sigma_g)\otimes\mathcal{F}}, \tag{2.8.118}$$

while for each $q \in (1, \infty)$ *and* $p \in (q, \infty]$ *there exists a finite constant* $C > 0$ *such that for every* $f \in L^p_{-1}(\partial\Omega, \sigma_g) \otimes \mathcal{F}$ *one has*

$$\begin{aligned} &\Big\|\sup_{r>0}\Big(r^{1-n}\int_{B(z,r)\cap\Omega} |\nabla\mathcal{S}f|^q \operatorname{dist}_g(\cdot, \partial\Omega)^{q-1}\, d\mathcal{L}_g^n\Big)^{1/q}\Big\|_{L^p_z(\partial\Omega,\sigma_g)} \\ &\qquad \leq C\|f\|_{L^p_{-1}(\partial\Omega,\sigma_g)\otimes\mathcal{F}}. \end{aligned} \tag{2.8.119}$$

Second, if hypothesis (2.8.98) *is strengthened to*

$$a_{jk}^{\alpha\beta} \in \mathscr{C}^{1+\gamma}, \quad b_j^{\alpha\beta} \in L^\infty, \quad d^{\alpha\beta} \in L^r, \\ \textit{for some } \gamma > 0 \textit{ and } r > n, \tag{2.8.120}$$

then for each integrability exponent $p \in (1, r)$ there exists some $C \in (0, \infty)$ with the property that for each $f \in L^p_{-1}(\partial\Omega, \sigma_g) \otimes \mathcal{F}$ one has

$$\Big(\int_\Omega |\nabla^2(\mathscr{S} f)(x)|^p \operatorname{dist}_g(x, \partial\Omega)^{p-1} \, d\mathcal{L}^n_g(x)\Big)^{1/p} \leq C\|f\|_{L^p(\partial\Omega,\sigma_g)\otimes\mathcal{F}}. \tag{2.8.121}$$

Third, if in place of (2.8.98) *one now assumes*

$$a_{jk}^{\alpha\beta} \in \mathscr{C}^{1+\gamma}, \quad b_j^{\alpha\beta} \in L^r_1, \quad d^{\alpha\beta} \in \mathscr{C}^\gamma, \\ \textit{for some } \gamma > 0 \textit{ and } r > n, \tag{2.8.122}$$

then for each $q \in (1, r)$ and $p \in (q, \infty]$ there exists a finite constant $C > 0$ such that for every function $f \in L^p(\partial\Omega, \sigma_g) \otimes \mathcal{F}$ one has

$$\Big\|\sup_{r>0}\Big(r^{1-n}\int_{B(z,r)\cap\Omega} |\nabla^2\mathscr{S} f|^q \operatorname{dist}_g(\cdot, \partial\Omega)^{q-1} \, d\mathcal{L}^n_g\Big)^{1/q}\Big\|_{L^p_z(\partial\Omega,\sigma_g)} \tag{2.8.123}$$
$$\leq C\|f\|_{L^p(\partial\Omega,\sigma_g)\otimes\mathcal{F}}.$$

Finally, if Ω is actually a Lipschitz domain, then for each $p \in (1, \infty)$ the single layer operator induces well-defined, linear, and bounded mappings

$$\mathscr{S} : L^p_{-1}(\partial\Omega, \sigma_g) \otimes \mathcal{F} \longrightarrow B^{p,p}_{1/p}(\Omega) \otimes \mathcal{E} \textit{ for each } p \in (1, \infty) \\ \textit{provided the regularity conditions in (2.8.117) hold,} \tag{2.8.124}$$

and

$$\mathscr{S} : L^p(\partial\Omega, \sigma_g) \otimes \mathcal{F} \longrightarrow B^{p,p}_{1+1/p}(\Omega) \otimes \mathcal{E} \textit{ for each } p \in (1, r) \\ \textit{provided the regularity conditions in (2.8.120) hold.} \tag{2.8.125}$$

(3) For each function $f \in L^1(\partial\Omega, \sigma_g) \otimes \mathcal{F}$ and aperture parameter $\kappa > 0$, one has

$$\mathscr{S} f\Big|_{\partial\Omega}^{\kappa-\mathrm{n.t.}} = Sf \textit{ at } \sigma_g\textit{-a.e. point on } \partial\Omega, \tag{2.8.126}$$

$$\textit{and } \big(\nabla\mathscr{S} f\big)\Big|_{\partial\Omega}^{\kappa-\mathrm{n.t.}} \textit{ exists at } \sigma_g\textit{-a.e. point on } \partial\Omega. \tag{2.8.127}$$

Moreover, for each $p \in (1, \infty)$, the operators

$$S : L^p(\partial\Omega, \sigma_g) \otimes \mathcal{F} \longrightarrow L^p_1(\partial\Omega, \sigma_g) \otimes \mathcal{E}, \tag{2.8.128}$$

$$S : L^p_{-1}(\partial\Omega, \sigma_g) \otimes \mathcal{F} \longrightarrow L^p(\partial\Omega, \sigma_g) \otimes \mathcal{E}, \tag{2.8.129}$$

are well defined, linear, and bounded. In fact, the nontangential trace formula in (2.8.126) *continues to hold for each* $f \in L^p_{-1}(\partial\Omega, \sigma_g) \otimes \mathcal{F}$ *with* $p \in (1, \infty)$. *Also, for all continuous vector fields* X, Y, *and every* $f \in L^p(\partial\Omega, \sigma_g) \otimes \mathcal{F}$ *with* $p \in (1, \infty)$, *one has*

$$\begin{aligned}\partial_{\tau_{XY}}(Sf)(x) &= \langle \nu_g(x), X^\flat(x)\rangle \Big(\text{P.V.} \int_{\partial\Omega} \langle (\nabla_{Y(x)} \otimes I_y) E(x,y), f(y)\rangle \, d\sigma_g(y)\Big) \\ &\quad - \langle \nu_g(x), Y^\flat(x)\rangle \Big(\text{P.V.} \int_{\partial\Omega} \langle (\nabla_{X(x)} \otimes I_y) E(x,y), f(y)\rangle \, d\sigma_g(y)\Big) \\ &\qquad \text{at } \sigma_g\text{-almost every point } x \text{ belonging to } \partial\Omega, \end{aligned} \tag{2.8.130}$$

where $\partial_{\tau_{XY}}$ *is understood in the sense of* [112, (1.12.89)] *(while keeping* (2.8.128) *in mind).*

In addition, for each $p \in (1, \infty)$ *the mapping*

$$S : L^p(\partial\Omega, \sigma_g) \otimes \mathcal{F} \longrightarrow L^p(\partial\Omega, \sigma_g) \otimes \mathcal{E} \tag{2.8.131}$$

is compact, and its (real) transpose is the operator

$$S^\top : L^{p'}(\partial\Omega, \sigma_g) \otimes \mathcal{E} \longrightarrow L^{p'}(\partial\Omega, \sigma_g) \otimes \mathcal{F}, \quad 1/p + 1/p' = 1, \tag{2.8.132}$$

acting on each $g \in L^{p'}(\partial\Omega, \sigma_g) \otimes \mathcal{E}$ *according to*

$$S^\top g(x) = \int_{\partial\Omega} \langle E_{L^\top}(x,y), g(y)\rangle_{\mathcal{E}_y} \, d\sigma_g(y), \qquad x \in \partial\Omega, \tag{2.8.133}$$

where $E_{L^\top} \in \mathcal{D}'(\mathcal{M} \times \mathcal{M}, \mathcal{F} \otimes \mathcal{E})$ *denotes the Schwartz kernel of*

$$(L^\top)^{-1} : H^1(\mathcal{M}) \otimes \mathcal{E} \longrightarrow H^{-1}(\mathcal{M}) \otimes \mathcal{F}. \tag{2.8.134}$$

(4) For each aperture parameter $\kappa > 0$ *and each* $p \in [1, \infty)$ *there exists a constant* $C \in (0, \infty)$ *with the property that*

$$\begin{aligned} &\|\mathcal{N}_\kappa(\mathcal{A}f)\|_{L^p(\partial\Omega,\sigma_g)} \leq C\|f\|_{L^p(\partial\Omega,\sigma_g)\otimes\mathcal{F}} \\ &\text{for each } f \in L^p(\partial\Omega, \sigma_g) \otimes \mathcal{F}, \text{ if } p \in (1, \infty), \end{aligned} \tag{2.8.135}$$

and, corresponding to $p = 1$,

$$\begin{aligned} &\|\mathcal{N}_\kappa(\mathcal{A}f)\|_{L^{1,\infty}(\partial\Omega,\sigma_g)} \leq C\|f\|_{L^1(\partial\Omega,\sigma_g)\otimes\mathcal{F}} \\ &\text{for each } f \in L^1(\partial\Omega, \sigma_g) \otimes \mathcal{F}. \end{aligned} \tag{2.8.136}$$

Moreover,

$$L(\mathcal{A}f) = 0 \text{ in } \Omega, \qquad \forall f \in L^1(\partial\Omega, \sigma_g) \otimes \mathcal{F}, \tag{2.8.137}$$

and

$$\mathcal{A}f \in \mathscr{C}^0_{\text{loc}}(\Omega) \otimes \mathcal{E}, \qquad \forall f \in L^1(\partial\Omega, \sigma_g) \otimes \mathcal{F}. \tag{2.8.138}$$

(5) For each $f \in L^1(\partial\Omega, \sigma_g) \otimes \mathcal{F}$, the principal-value integral operator

$$Af(x) := \text{P.V.} \int_{\partial\Omega} \big\langle (I_x \otimes P_y) E(x, y), f(y) \big\rangle_{\mathcal{E}_y} \, d\sigma_g(y), \tag{2.8.139}$$

considered in the sense of Definition 2.8.1, is well defined at σ_g-a.e. $x \in \partial\Omega$. Moreover,

$$A : L^p(\partial\Omega, \sigma_g) \otimes \mathcal{F} \longrightarrow L^p(\partial\Omega, \sigma_g) \otimes \mathcal{E} \text{ if } 1 < p < \infty, \tag{2.8.140}$$

$$A : L^1(\partial\Omega, \sigma_g) \otimes \mathcal{F} \longrightarrow L^{1,\infty}(\partial\Omega, \sigma_g) \otimes \mathcal{E}, \tag{2.8.141}$$

are linear and bounded mappings. In fact, the maximal operator (defined much as in (2.8.104)) *induces sub-linear bounded mappings*

$$A_{\max} : L^p(\partial\Omega, \sigma_g) \otimes \mathcal{F} \longrightarrow L^p(\partial\Omega, \sigma_g) \otimes \mathcal{E} \text{ if } 1 < p < \infty, \tag{2.8.142}$$

$$A_{\max} : L^1(\partial\Omega, \sigma_g) \otimes \mathcal{F} \longrightarrow L^{1,\infty}(\partial\Omega, \sigma_g) \otimes \mathcal{E}. \tag{2.8.143}$$

(6) For every $f \in L^1(\partial\Omega, \sigma_g) \otimes \mathcal{F}$ and every aperture parameter $\kappa > 0$ the following jump-formula holds:

$$\mathcal{A}f\Big|^{\kappa-\text{n.t.}}_{\partial\Omega} = \tfrac{1}{2} \mathrm{i} \,\text{Sym}(L; \nu_g)^{-1} \,\text{Sym}(P; \nu_g)^{\top} f + Af \tag{2.8.144}$$

at σ_g-a.e. point on $\partial\Omega$, where the superscript $^{\top}$ denotes (real) transposition.

(7) If hypothesis (2.8.98) *regarding the regularity of the coefficients of the differential operator L is strengthened to*

$$\begin{gathered} a^{\alpha\beta}_{jk} \in \mathscr{C}^{1+\gamma}, \quad b^{\alpha\beta}_j \in L^r_1, \quad d^{\alpha\beta} \in L^r, \\ \text{for some } \gamma > 0 \text{ and } r > n, \end{gathered} \tag{2.8.145}$$

then

$$\mathcal{A}f \in \mathscr{C}^1_{\text{loc}}(\Omega) \otimes \mathcal{E}, \qquad \forall f \in L^1(\partial\Omega, \sigma_g) \otimes \mathcal{F}, \tag{2.8.146}$$

and for each $p \in (1, \infty)$ there exists $C \in (0, \infty)$ such that for every function $f \in L^p(\partial\Omega, \sigma_g) \otimes \mathcal{F}$ one has

$$\int_{\Omega} |\nabla(\mathcal{A}f)(x)|^p \,\text{dist}_g(x, \partial\Omega)^{p-1} \, d\mathcal{L}^n_g(x) \le C \int_{\partial\Omega} |f|^p \, d\sigma_g. \tag{2.8.147}$$

In addition, under the stronger regularity assumptions imposed in (2.8.145), *whenever one has $1 < q < \infty$ and $q < p \le \infty$ it follows that there exists some finite constant $C > 0$ for which the following type of square function estimate*

$$\left\| \sup_{r>0} \left(r^{1-n} \int_{B(z,r)\cap\Omega} |\nabla \mathcal{A} f|^q \operatorname{dist}_g(\cdot, \partial\Omega)^{q-1} \, \mathrm{d}\mathcal{L}_g^n \right)^{1/q} \right\|_{L_z^p(\partial\Omega,\sigma_g)} \leq C\|f\|_{L^p(\partial\Omega,\sigma_g)\otimes\mathcal{F}} \tag{2.8.148}$$

holds for every $f \in L^p(\partial\Omega, \sigma_g) \otimes \mathcal{F}$; in particular, corresponding to $q := 2$ and the end-point $p = \infty$, for each function $f \in L^\infty(\partial\Omega, \sigma_g) \otimes \mathcal{F}$ one has the following Carleson measure estimate:

$$\sup_{z\in\partial\Omega,\, r>0} \left(r^{1-n} \int_{B(z,r)\cap\Omega} |\nabla \mathcal{A} f|^2 \operatorname{dist}_g(\cdot, \partial\Omega) \, \mathrm{d}\mathcal{L}_g^n \right) \leq C\|f\|^2_{L^\infty(\partial\Omega,\sigma_g)\otimes\mathcal{F}}. \tag{2.8.149}$$

Finally, if in addition to (2.8.145) *one also assumes that Ω is a Lipschitz domain, then for each $p \in (1, \infty)$ the operator*

$$\mathcal{A} : L^p(\partial\Omega, \sigma_g) \otimes \mathcal{F} \longrightarrow B^{p,p}_{1/p}(\Omega) \otimes \mathcal{E} \tag{2.8.150}$$

is well defined, linear, and bounded.

(8) *Similar results to those described in items (4)-(6) above (with the exception of* (2.8.137)) *are valid for the integral operators*

$$\widetilde{\mathcal{A}} f(x) := \int_{\partial\Omega} \langle (\widetilde{P}_x \otimes I_y) E(x, y), f(y) \rangle_{\mathcal{F}_y} \, \mathrm{d}\sigma_g(y), \quad x \in \Omega, \tag{2.8.151}$$

and

$$\widetilde{A} f(x) := \text{P.V.} \int_{\partial\Omega} \langle (\widetilde{P}_x \otimes I_y) E(x, y), f(y) \rangle_{\mathcal{F}_y} \, \mathrm{d}\sigma_g(y), \quad x \in \partial\Omega, \tag{2.8.152}$$

where $\widetilde{P} : \mathcal{E} \to \mathcal{E}$ is a first-order differential operator with continuous coefficients. In this scenario, in place of the jump-formula (2.8.144), *for each function $f \in L^1(\partial\Omega, \sigma_g) \otimes \mathcal{F}$ and at σ_g-a.e. point on $\partial\Omega$ one now has*

$$\widetilde{\mathcal{A}} f \Big|_{\partial\Omega}^{\kappa-\text{n.t.}} = -\tfrac{1}{2} \mathrm{i} \operatorname{Sym}(\widetilde{P}; \nu_g) \operatorname{Sym}(L; \nu_g)^{-1} f + \widetilde{A} f. \tag{2.8.153}$$

Moreover, if the top terms in $\widetilde{P}$ have $\mathscr{C}^1$ coefficients while the lower-order ones are continuous, and if hypothesis (2.8.98) *is strengthened to*

$$a_{jk}^{\alpha\beta} \in \mathscr{C}^{1+\gamma}, \quad b_j^{\alpha\beta} \in L^\infty, \quad d^{\alpha\beta} \in L^r, \quad \text{for some } \gamma > 0 \text{ and } r > n, \tag{2.8.154}$$

then the analogues of (2.8.147) *and* (2.8.148), *albeit in a restricted range, are also valid in this case: first, for each $p \in (1, r)$ there exists some $C \in (0, \infty)$ such that for every function $f \in L^p(\partial\Omega, \sigma_g) \otimes \mathcal{F}$ one has*

$$\int_\Omega |\nabla(\widetilde{\mathcal{A}} f)(x)|^p \operatorname{dist}_g(x, \partial\Omega)^{p-1} \, \mathrm{d}\mathcal{L}_g^n(x) \leq C \int_{\partial\Omega} |f|^p \, \mathrm{d}\sigma_g \tag{2.8.155}$$

and, second, for each $q \in (1, r)$ and $p \in (q, \infty]$ there exists $C \in (0, \infty)$ such that for every $f \in L^p(\partial\Omega, \sigma_g) \otimes \mathcal{F}$ one has

$$\left\| \sup_{r>0} \left(r^{1-n} \int_{B(z,r)\cap\Omega} |\nabla \widetilde{\mathcal{A}} f|^q \operatorname{dist}_g(\cdot, \partial\Omega)^{q-1} \, d\mathcal{L}^n_g \right)^{1/q} \right\|_{L^p_z(\partial\Omega,\sigma_g)} \leq C\|f\|_{L^p(\partial\Omega,\sigma_g)\otimes\mathcal{F}}. \tag{2.8.156}$$

Finally, if in addition to (2.8.154) *and the latest requirements on the coefficients of $\widetilde{P}$ one also assumes that Ω is a Lipschitz domain then for each $p \in (1, r)$ the operator*

$$\widetilde{\mathcal{A}} : L^p(\partial\Omega, \sigma_g) \otimes \mathcal{F} \longrightarrow B^{p,p}_{1/p}(\Omega) \otimes \mathcal{E} \tag{2.8.157}$$

is well defined, linear, and bounded.

(9) The (*real*) *transpose $A^\top$ of A in* (2.8.140) *is the operator*

$$A^\top : L^p(\partial\Omega, \sigma_g) \otimes \mathcal{E} \longrightarrow L^p(\partial\Omega, \sigma_g) \otimes \mathcal{F}, \quad 1 < p < \infty, \tag{2.8.158}$$

acting on each $h \in L^p(\partial\Omega, \sigma_g) \otimes \mathcal{E}$ at σ_g-a.e. point $x \in \partial\Omega$ according to

$$A^\top h(x) = \text{P.V.} \int_{\partial\Omega} \big\langle (P_x \otimes I_y) E_{L^\top}(x, y), h(y) \big\rangle_{\mathcal{E}_y} \, d\sigma_g(y) \tag{2.8.159}$$

where, as before, $E_{L^\top} \in \mathcal{D}'(\mathcal{M} \times \mathcal{M}, \mathcal{F} \otimes \mathcal{E})$ denotes the Schwartz kernel of $(L^\top)^{-1}$.

Similarly, the (*real*) *transpose $(\widetilde{A})^\top$ of $\widetilde{A}$ from* (2.8.152) *is the operator sending each function $h \in L^p(\partial\Omega, \sigma_g) \otimes \mathcal{E}$ with $1 < p < \infty$ into*

$$(\widetilde{A})^\top h(x) = \text{P.V.} \int_{\partial\Omega} \big\langle (I_x \otimes \widetilde{P}_y) E_{L^\top}(x, y), h(y) \big\rangle_{\mathcal{E}_y} \, d\sigma_g(y), \tag{2.8.160}$$

for σ_g-a.e. $x \in \partial\Omega$.

(10) If Ω is also a one-sided NTA *domain and hypothesis* (2.8.98) *for L is strengthened to*

$$a^{\alpha\beta}_{jk} \in \mathscr{C}^1, \quad b^{\alpha\beta}_j \in L^r_1, \quad d^{\alpha\beta} \in L^r, \quad \textit{for some } r > n, \tag{2.8.161}$$

then for each $p \in \big(r/(r-1), r\big)$ the following operator is well defined, linear, and bounded:

$$\mathcal{S} : L^p_{-1}(\partial\Omega, \sigma_g) \otimes \mathcal{F} \longrightarrow B^{p,p}_{1/p}(\Omega) \otimes \mathcal{E}. \tag{2.8.162}$$

Moreover, if hypothesis (2.8.98) *for L is further strengthened to*

$$\begin{gathered} a^{\alpha\beta}_{jk} \in \mathscr{C}^{1+\gamma}, \quad b^{\alpha\beta}_{j} \in L^r_1, \quad d^{\alpha\beta} \in L^r, \\ \textit{for some } \gamma > 0 \textit{ and } r > n, \end{gathered} \tag{2.8.163}$$

while Ω *is once again assumed to also be a one-sided* NTA *domain, then the operator*

$$\mathcal{A} : L^p(\partial\Omega, \sigma_g) \otimes \mathcal{F} \longrightarrow B^{p,p}_{1/p}(\Omega) \otimes \mathcal{E} \tag{2.8.164}$$

is well defined, linear, and bounded for each $p \in \big(r/(r-1), r\big)$.

If the all the underlying differentiability assumptions for the Riemannian manifold $\mathcal{M}$, the Hermitian vector bundles $\mathcal{E}$, $\mathcal{F}$, and the coefficients of the differential operators L, P, $\widetilde{P}$ are upgraded to $\mathscr{C}^\infty$, then $Q := L^{-1}P^\top$ and $\widetilde{Q} := \widetilde{P}L^{-1}$ are both classical pseudodifferential operators in $\mathrm{OP}S^{-1}_{\mathrm{cl}}$, with Schwartz kernels and principal symbols given by

$$\begin{gathered} k_Q(x,y) = (I_x \otimes P_y)E(x,y), \\ \mathrm{Sym}(Q;\xi) = \mathrm{Sym}(L;\xi)^{-1}\mathrm{Sym}(P^\top;\xi) \end{gathered} \tag{2.8.165}$$

and, respectively,

$$\begin{gathered} k_{\widetilde{Q}}(x,y) = (\widetilde{P}_x \otimes I_y)E(x,y), \\ \mathrm{Sym}(\widetilde{Q};\xi) = \mathrm{Sym}(\widetilde{P};\xi)\mathrm{Sym}(L;\xi)^{-1}. \end{gathered} \tag{2.8.166}$$

As such, in the current scenario, the "mother" jump-formula for integral operators associated with Schwartz kernels of classical pseudodifferential operators in $\mathrm{OP}S^{-1}_{\mathrm{cl}}$ possessing odd principal symbols in the cotangent variable, recorded in (2.8.46), directly gives the jump-formulas in (2.8.153) and (2.8.144) (also keeping in mind [112, (1.7.17)] in the case of the latter).

In this regard, we also wish to mention that, in principle, P, $\widetilde{P}$ are allowed to be zero-th order differential operators (hence, plain homomorphisms). In such a case, their principal symbols vanish identically. As a result, the jump-terms in (2.8.144), (2.8.153) disappear, rendering these jump-formulas in line with the boundary trace result for the single layer, recorded in (2.8.126). This is to be expected, and logically consistent, since now the integral kernels of $\mathcal{A}$, $\widetilde{\mathcal{A}}$ exhibit the same type of main singularity as for the single layer operator.

We conclude this preamble with several useful observations regarding the algebraic nature of the double distribution E_L. For one thing, the fact that E_L is the Schwartz kernel of L^{-1} means that

$$E_L \in \mathcal{D}'\big(\mathcal{M}\times\mathcal{M}, \mathcal{E}\otimes\mathcal{F}\big) \tag{2.8.167}$$

has the property that for arbitrary test sections $\varphi \in \mathcal{D}(\mathcal{M},\mathcal{E})$ and $\psi \in \mathcal{D}(\mathcal{M},\mathcal{F})$ we have

$$\big\langle E_L(x,y), \varphi(x)\otimes\psi(y)\big\rangle = \langle L^{-1}\psi, \varphi\rangle. \tag{2.8.168}$$

Based on this it may be checked that

$$E_{\overline{L}}(x,y) = \overline{E_L(x,y)}, \tag{2.8.169}$$

$$E_{L^\top}(x,y) = \big[E_L(y,x)\big]^\top, \tag{2.8.170}$$

$$E_{L^*}(x,y) = \Big[\overline{E_L(y,x)}\Big]^\top. \tag{2.8.171}$$

We also wish to bring in the diagonal Dirac distribution

$$\delta_\Delta = \delta_\Delta^{\mathcal{E}} \in \mathcal{D}'\big(\mathcal{M}\times\mathcal{M}, \mathcal{E}\otimes\mathcal{E}\big), \tag{2.8.172}$$

defined by setting

$$\langle\delta_\Delta, \Theta\rangle := \int_{\mathcal{M}} \operatorname{Trace}\Theta(x,x)\,\mathrm{d}\mathcal{L}_g^n(x) \text{ for each } \Theta \in \mathcal{D}\big(\mathcal{M}\times\mathcal{M}, \mathcal{E}\otimes\mathcal{E}\big), \tag{2.8.173}$$

where $\operatorname{Trace}\Theta(x,x) := \sum_\alpha \theta_{\alpha\alpha}(x,x)$ if the test section Θ is expressed in an arbitrary local orthonormal frame $\{e_\alpha\}_\alpha$ of $\mathcal{E}$ as $\Theta(x,y) = \theta_{\alpha\beta}(x,y)\, e_\alpha \otimes e_\beta$. In particular,

$$\big\langle\delta_\Delta(x,y), \varphi(x)\otimes\psi(y)\big\rangle = \int_{\mathcal{M}} \langle\varphi(x), \psi(x)\rangle\,\mathrm{d}\mathcal{L}_g^n(x) \tag{2.8.174}$$

for any $\varphi, \psi \in \mathcal{D}(\mathcal{M}, \mathcal{E})$. Otherwise said,

$$\big\langle\delta_\Delta(x,y), \psi(y)\big\rangle = \psi(x) \text{ for each } \psi \in \mathcal{D}(\mathcal{M}, \mathcal{E}). \tag{2.8.175}$$

Let us also note that δ_Δ is supported on the diagonal, hence

$$\delta_\Delta(x,y) = 0 \text{ for } x \neq y. \tag{2.8.176}$$

Lastly, with the help of (2.8.95)-(2.8.96), (2.8.168), (2.8.174) it may be readily verified that

$$(L_x \otimes I_y)E_L(x,y) = \delta_\Delta^{\mathcal{F}}(x,y) \text{ in } \mathcal{D}'\big(\mathcal{M}\times\mathcal{M}, \mathcal{F}\otimes\mathcal{F}\big), \tag{2.8.177}$$

$$(I_x \otimes L_y^\top)E_L(x,y) = \delta_\Delta^{\mathcal{E}}(x,y) \text{ in } \mathcal{D}'\big(\mathcal{M}\times\mathcal{M}, \mathcal{E}\otimes\mathcal{E}\big), \tag{2.8.178}$$

and, more generally, for any differential operators $P : \mathcal{E} \to \mathcal{G}$ and $Q \in \mathcal{F} \to \mathcal{H}$ we have

$$(L_x \otimes Q_y)E_L(x,y) = (I_x \otimes Q_y)\delta_\Delta^{\mathcal{F}}(x,y) \text{ in } \mathcal{D}'\big(\mathcal{M}\times\mathcal{M}, \mathcal{F}\otimes\mathcal{H}\big), \tag{2.8.179}$$

$$(P_x \otimes L_y^\top)E_L(x,y) = (P_x \otimes I_y)\delta_\Delta^{\mathcal{E}}(x,y) \text{ in } \mathcal{D}'\big(\mathcal{M}\times\mathcal{M}, \mathcal{G}\otimes\mathcal{E}\big). \tag{2.8.180}$$

Here is the proof of Theorem 2.8.4.

Proof of Theorem 2.8.4 We begin with a review of the nature of the singularity the Schwartz kernel of L^{-1} along the diagonal, which is of independent interest. Denote the Schwartz kernel of L^{-1} by $E(x,y)$. Then, in local coordinates,

$$L_x\big[E(x,y)\sqrt{g(y)}\,\big] = \delta_y(x), \tag{2.8.181}$$

where the operator L acts in the variable x on columns, and

$$L_y^\top\big[E(x,y)\sqrt{g(x)}\,\big] = \delta_x(y), \tag{2.8.182}$$

where the transposed operator $L^\top$ acts in the variable y on rows. Granted the regularity conditions from (2.8.98), the analysis[34] of the Schwartz kernel $E(x,y)$ of L^{-1} provided in [119, Propositions 2.2-2.6] (cf. also the discussion in [124, Appendix]) applies and yields the following Hölder and Sobolev regularity properties:

$$E(\cdot,y) \in \mathscr{C}^\alpha_{\rm loc}\big(\mathcal{M}\setminus\{y\}, \mathcal{E}\otimes\mathcal{F}_y\big), \quad \forall\, y\in\mathcal{M}, \quad \forall\,\alpha < 2 - n/r, \tag{2.8.183}$$

$$E(\cdot,\cdot) \in \mathscr{C}^{1+\theta}_{\rm loc}\big(\mathcal{M}\times\mathcal{M}\setminus{\rm diag},\, \mathcal{E}\otimes\mathcal{F}\big) \text{ for some } \theta = \theta(n,r) > 0, \tag{2.8.184}$$

$$E(\cdot,y) \in L^p_{s,{\rm loc}}\big(\mathcal{M}\setminus\{y\}, \mathcal{E}\otimes\mathcal{F}_y\big), \quad \forall\, y\in\mathcal{M},\ \forall\, s<2,\ \forall\, p<r, \tag{2.8.185}$$

$$E(\cdot,y) \in L^p_1\big(\mathcal{M}, \mathcal{E}\otimes\mathcal{F}_y\big),\ \forall\, p < n/(n-1), \text{ uniformly in } y\in\mathcal{M}. \tag{2.8.186}$$

The same analysis also yields the decomposition

$$E(x,y) = \frac{1}{\sqrt{g(y)}}\,E_0(y, x-y) + E_1(x,y) \tag{2.8.187}$$

where the residual piece $E_1(x,y)$ obeys, for each $\varepsilon > 0$,

$$|E_1(x,y)| \le C_\varepsilon |x-y|^{-(n-3+\varepsilon)}, \tag{2.8.188}$$

$$|\nabla_x E_1(x,y)| \le C_\varepsilon |x-y|^{-(n-2+\varepsilon)}, \tag{2.8.189}$$

$$|\nabla_y E_1(x,y)| \le C_\varepsilon |x-y|^{-(n-2+\varepsilon)}, \tag{2.8.190}$$

while the top singularity depends exclusively on the top coefficients of L and is explicitly given by

$$E_0(y,z) = -(2\pi)^{-n}\int\Big(\sum_{j,k}\xi_j\xi_k A_{jk}(y)\Big)^{-1} e^{i\langle z,\xi\rangle}\,d\xi, \tag{2.8.191}$$

where we have set $A_{jk} := \big(a^{\alpha\beta}_{jk}\big)_{\alpha,\beta}$. As such, if $n \ge 3$ it follows that

$$\begin{array}{c} E_0(y,z) \text{ is even and smooth in } z\in\mathbb{R}^n\setminus\{0\}, \text{ and is} \\ \text{homogeneous of degree } 2-n \text{ in the variable } z, \\ \text{with } \mathscr{C}^1 \text{ dependence in the variable } y. \end{array} \tag{2.8.192}$$

[34] the argument in [119] is based on microlocal analysis: symbol decomposition techniques, local elliptic regularity, pseudodifferential operators

In particular, this implies that for $n \geq 3$ we have

$$E_0(y, \rho z) = |\rho|^{-(n-2)} E_0(z, y) \text{ for any } \rho \in \mathbb{R} \text{ and } z \in \mathbb{R}^n \setminus \{0\}, \tag{2.8.193}$$

and

$$\begin{gathered}(\nabla_z^m \nabla_y^\ell E_0)(y, z) = O\big(|z|^{-(n-2+m)}\big) \text{ as } z \longrightarrow 0, \\ \text{uniformly in } y, \text{ for every } m \in \mathbb{N}_0 \text{ and } \ell \in \{0, 1\}.\end{gathered} \tag{2.8.194}$$

The singularity in the top term $E_0(y, z)$ when $n = 2$ is at most of logarithmic type, i.e., $|E_0(y, z)| \leq C|\ln|z||$, with $\mathscr{C}^1$ dependence in the variable y.

Let us also remark here that if $\mathscr{F}_{z\to\xi}$ denotes the Fourier transform turning a function in the variable z into a function in the variable ξ, then (2.8.191) gives

$$\mathscr{F}_{z\to\xi}\big[E_0(y, \cdot)\big](\xi) = -\Big(\sum_{j,k} \xi_j \xi_k A_{jk}(y)\Big)^{-1} \tag{2.8.195}$$

which, in turn, shows (cf. item *(6)* of Theorem 1.4.2) that

$$\begin{gathered}\text{for each } y \text{ fixed, the matrix-valued function } E_0(y, \cdot) \text{ is a} \\ \text{fundamental solution for the system } \sum_{j,k=1}^{n} A_{jk}(y)\partial_j\partial_k.\end{gathered} \tag{2.8.196}$$

Based on (2.8.195) we may also check that each row $E_0^{\alpha\bullet}(y, \cdot)$ of $E_0(y, \cdot)$ satisfies the PDE

$$\sum_{j,k=1}^{n} A_{jk}^{\top}(y)\partial_{z_j}\partial_{z_k}\big[E_0^{\alpha\bullet}(y, z)\big] = 0 \text{ for } z \in \mathbb{R}^n \setminus \{0\}. \tag{2.8.197}$$

In this connection, it has been noted in [119] that (with the operator L acting on columns)

$$\begin{gathered}L_x\big[E_0(y, x-y)\big] = \delta_y(x) + R(x, y) \text{ where, if } V := (d^{\alpha\beta})_{\alpha,\beta}, \\ |R(x, y)| \leq C|x-y|^{-(n-1)} + C|V(x)||x-y|^{-(n-2)},\end{gathered} \tag{2.8.198}$$

assuming $n \geq 3$. Also, from (2.8.187), (2.8.188), and the above discussion pertaining to the nature of the singularity in $E_0(y, z)$ it follows that for each $x, y \in \mathcal{M}$ with $x \neq y$ we have

$$|E(x, y)| \leq \begin{cases} C\,\mathrm{dist}(x, y)^{-(n-2)} & \text{if } n \geq 3, \\ C_\alpha \mathrm{dist}(x, y)^{-(1-\alpha)} & \forall \alpha \in (0, 1) \text{ if } n = 2. \end{cases} \tag{2.8.199}$$

More can be said if the leading coefficients of L exhibit more smoothness. Specifically, [119, Proposition 2.6, p. 19] gives that for any two distinct points $x, y \in \mathcal{M}$ we have

$$\left(\fint_{B(x,\,|x-y|/2)} |\nabla_x \nabla_x E_1(w, y)|^p \, dw\right)^{1/p} \leq C_{p,\,\varepsilon}|x-y|^{-(n-1+\varepsilon)}$$
$$\text{for each exponent } p \in (1, r) \text{ and each } \varepsilon > 0, \text{ provided}$$
$$a_{jk}^{\alpha\beta} \in \mathscr{C}^{1+\gamma},\ b_j^{\alpha\beta} \in L^\infty,\ \mathrm{d}^{\alpha\beta} \in L^r \text{ with } \gamma > 0 \text{ and } r > n. \tag{2.8.200}$$

Let us also note here that under the stronger regularity assumptions we can further augment the list of properties in (2.8.183)-(2.8.188). Specifically,

$$|\nabla_x \nabla_y E_1(x, y)| \leq C_\varepsilon |x-y|^{-(n-1+\varepsilon)} \text{ for each } \varepsilon > 0, \text{ and}$$
$$E(\cdot, y) \in L^p_{2,\mathrm{loc}}(\mathcal{M} \setminus \{y\}, \mathcal{E} \otimes \mathcal{F}_y), \quad \forall y \in \mathcal{M}, \quad \forall p \in (1, r), \tag{2.8.201}$$
$$\text{if } a_{jk}^{\alpha\beta} \in \mathscr{C}^{1+\gamma}, \quad b_j^{\alpha\beta} \in L^r_1, \quad d^{\alpha\beta} \in L^r \text{ with } \gamma > 0 \text{ and } r > n.$$

These follow from [119, Proposition 2.8, p. 21] and, respectively, (2.8.181) together with [119, Proposition 2.7, p. 20]. Finally, it has been remarked in [119, (2.53)-(2.54), p. 19] that the following pointwise estimate (which should be compared with (2.8.200)) holds:

$$|\nabla_x \nabla_x E_1(x, y)| \leq C_\varepsilon |x-y|^{-(n-1+\varepsilon)} \text{ for each } \varepsilon > 0 \text{ whenever}$$
$$a_{jk}^{\alpha\beta} \in \mathscr{C}^{1+\gamma}, \quad b_j^{\alpha\beta} \in L^r_1, \quad d^{\alpha\beta} \in \mathscr{C}^\gamma \text{ with } \gamma > 0 \text{ and } r > n. \tag{2.8.202}$$

After this preamble, we are ready to turn to the proof of Theorem 2.8.4 in earnest. First, in view of (2.8.184), the extension of the single layer operator proposed in (2.8.106) is meaningful, and the operator so defined is linear and bounded in the context

$$\mathscr{S} : L^p_{-1}(\partial\Omega, \sigma_g) \otimes \mathcal{F} \longrightarrow \mathscr{C}^0_{\mathrm{loc}}(\Omega) \otimes \mathcal{E}. \tag{2.8.203}$$

In particular, (2.8.108) holds. The claim in (2.8.109) also follows from (2.8.184). Since the embedding of $L^1(\partial\Omega, \sigma_g) \otimes \mathcal{F}$ into $L^p_{-1}(\partial\Omega, \sigma_g) \otimes \mathcal{F}$ is dense (cf. [113, (11.8.4)]), and since (2.8.101) and the fact that $E(x, y)$ is the Schwartz kernel of L^{-1} imply $L(\mathscr{S} f) = 0$ in Ω for each $f \in L^1(\partial\Omega, \sigma_g) \otimes \mathcal{F}$, we then conclude from (2.8.203) that (2.8.107) holds as well.

Next, (2.8.199) shows that the single layer (2.8.102) behaves like a fractional integral operator of order 1 if $n \geq 3$, and of any order $\alpha \in (0, 1)$ if $n = 2$. Bearing this in mind, an argument based on the Fractional Integration Theorem (see Proposition 2.5.39) then establishes (2.8.114)-(2.8.116). As a byproduct, we also have (2.8.110). The trace formula in (2.8.126) is also a consequence of the above observation and Proposition 2.5.39.

Moving on, by working in local coordinates and then reassembling the pieces by using a smooth partition of unity, we can construct

$$Q_0 \in \mathit{Ø}\mathrm{PC}^1 S^{-2}_{\mathrm{cl}}(\mathcal{F}, \mathcal{E}) \tag{2.8.204}$$

such that $\mathrm{Sym}(Q_0;\xi) = \mathrm{Sym}(L;\xi)^{-1}$ and the remainder $R := Q_0 - L^{-1}$ is relatively tame, i.e., its Schwartz kernel $E_1(x,y)$ satisfies (2.8.188)-(2.8.190). This happens under the assumptions made in (2.8.98). Under stronger regularity assumptions, $E_1(x,y)$ satisfies additional properties, as described in (2.8.200)-(2.8.202).

The idea is now that if we employ the notation k_{Q_0} for the Schwartz kernel of Q_0, then $(I_x \otimes P_y)k_{Q_0}(x,y)$ becomes the Schwartz kernel of $Q := Q_0 P^\top$. Since this satisfies

$$Q \in \mathit{\emptyset}\mathrm{PC}^0 S^{-1}_{\mathrm{cl}}(\mathcal{F},\mathcal{E}) \text{ and } Q \text{ has an odd principal symbol in the cotangent variable } \xi \in T^*\mathcal{M}, \tag{2.8.205}$$

the desired conclusions for the contribution of $(I_x \otimes P_y)k_{Q_0}(x,y)$ follow directly from Theorem 2.8.2, while the contribution due to $(I_x \otimes P_y)E_1(x,y)$ is, because of the aforementioned estimates, amenable to a much more straightforward analysis. Essentially, a similar analysis applies to the kernel $(\widetilde{P}_x \otimes I_y)E(x,y)$.

This principle applies to a sizable portion of the statement of the theorem, including all claims in item *(5)-(6)*, the claims in item *(8)* up to, and including, the jump-formula (2.8.153), and (2.8.135)-(2.8.136). In turn, these also cover the claims made in (2.8.127), (2.8.112)-(2.8.113), and also (2.8.111) (bearing in mind the structure of the Sobolev spaces of negative order on $\partial\Omega$; cf. [113, (11.8.7)-(11.8.8)]).

Let us next exemplify the manner in which the aforementioned principle is implemented in relation to the various L^p-square function estimates recorded in the statement of the theorem. First, having fixed some $p \in (1,\infty)$, consider the task of proving (2.8.147), working under the stronger regularity assumptions made in (2.8.145). These imply that the error term $E_1(x,y)$ satisfies the estimate recorded in the first line of (2.8.201). Denote by $p' \in (1,\infty)$ the Hölder conjugate exponent of p and choose some $\varepsilon \in (0, \frac{1}{p})$. Also, pick a pair of arbitrary distinct points x, y, and abbreviate $\rho := |x-y|/2$. Since $|w-y| \approx |x-y|$, uniformly for $w \in B(x,\rho)$, based on (2.8.201) we may estimate

$$\begin{aligned}\Big(\int_{B(x,\rho)} |\nabla_x\nabla_y E_1(w,y)|^p\,dw\Big)^{1/p} &\le C\Big(\int_{B(x,\rho)} |x-y|^{-p(n-1+\varepsilon)}\,dw\Big)^{1/p}\\ &= C|x-y|^{-(n-1+\varepsilon)}\rho^{n/p}\\ &= C|x-y|^{1-\varepsilon-n/p'}.\end{aligned} \tag{2.8.206}$$

Thanks to (2.8.206), [112, Lemma 8.7.11] applies (with $a := p-1$) to the contribution to the left-hand side of (2.8.147) coming from $\nabla_x\nabla_y E_1(x,y)$, whereas the contribution to the left-hand side of (2.8.147) coming from $\nabla_y\big[E_0(y,x-y)\big]g(y)^{-1/2}$ is, thanks to (2.8.192), directly amenable to the L^p-square function estimate from Theorem 2.5.38 (alternatively, we may invoke Theorem 2.8.2). Granted these and keeping in mind (2.8.187), the L^p-square function estimate (2.8.147) follows. A virtually identical argument works in the case of (2.8.118). The L^p-square function estimate in (2.8.121) is proved using the same decomposition of the fundamental solution, upon observing that Theorem 2.5.38 applies to the expression $\nabla_x\big[E_0(y,x-y)\big]g(y)^{-1/2}$

while [112, Lemma 8.7.11] can now be used directly, thanks to (2.8.200), to control the contribution from $\nabla_x \nabla_x E_1(x, y)$.

Once (2.8.147) has been established, the L^p-square function estimate in (2.8.118) follows as a special case, keeping in mind the structure of the Sobolev spaces of negative order on $\partial\Omega$ (cf. [113, (11.8.7)-(11.8.8)]). In turn, (2.8.119) may be deduced from (2.8.118) by the same type of argument used in the proof of (2.4.37). This time, in place of (2.4.105), we use the fact that

$$|\nabla_x \nabla_y E(x, y)| \leq C \operatorname{dist}(x, y)^{-n}, \tag{2.8.207}$$

which follows from (2.8.187), (2.8.192), and (2.8.201).

Likewise, (2.8.121) directly implies the L^p-square function estimate claimed in (2.8.121) which, in turn, further implies (2.8.123) by reasoning as in the proof of (2.4.37) and keeping in mind that under the assumptions in (2.8.122) we have

$$|\nabla_x \nabla_x E(x, y)| \leq C \operatorname{dist}(x, y)^{-n}, \tag{2.8.208}$$

as may be seen from (2.8.187), (2.8.192), and (2.8.202).

Moving on, if Ω is actually a Lipschitz domain then the mapping properties claimed in (2.8.125) become consequences of the L^p-square function estimates in (2.8.118)-(2.8.121), [113, Lemma 9.2.40], and known lifting results on the Besov scale in Lipschitz domains (cf. [73, Proposition 2.18(b), p. 173]). Similarly, if Ω is Lipschitz, then the claims pertaining to the operator in (2.8.124) are direct consequences of the L^p-square function estimate (2.8.147) and [113, Lemma 9.2.40].

Next, the first claims in item *(8)* may be established by reasoning as in the proof of item *(3)* in Theorem 2.8.2, bearing in mind that

$$\big[(I \otimes P)E\big]^{\top}(y, x) = (P \otimes I)E_{L^{\top}}(x, y), \tag{2.8.209}$$

which may be checked directly from definitions. A similar argument also allows us to identify the transpose of the single layer as claimed in (2.8.131).

Regarding the single layer on Sobolev spaces, from (2.8.112), (2.8.110), (2.8.127), the existence of the pointwise nontangential trace in (2.8.126), and [112, Proposition 1.12.9] it follows that the operator (2.8.128) is well-defined, linear, and bounded. Via duality, the same conclusions are valid for (2.8.129).

Going further, the claims pertaining to the operator (2.8.162) (under the assumption (2.8.161)) follow from Proposition 2.8.3, (2.8.107), (2.8.111), and [112, (8.6.51)], while the claims pertaining to the operator (2.8.164) (under the assumption (2.8.163)) follow from Proposition 2.8.3, (2.8.137), (2.8.135), and [112, (8.6.51)].

Consider now (2.8.148). Thanks to the L^p-square function estimate recorded in (2.8.147), the same type of argument as in the proof of (2.4.37) works virtually verbatim in the current context; the most notable novelty is that in place of (2.4.105) we presently use (2.8.207).

The Carleson measure estimate (2.8.149) is, of course, a particular case of (2.8.148). Also, the L^p-square function estimate (2.8.119) for the single layer operator is a consequence of (2.8.148), bearing in mind the structure of the Sobolev spaces of negative order on $\partial\Omega$ (cf. [113, (11.8.7)-(11.8.8)]).

As regards formula (2.8.130), fix a pair of continuous vector fields X, Y, along with a function $f \in L^p(\partial\Omega, \sigma_g) \otimes \mathcal{F}$ for some $p \in (1, \infty)$. From the jump-formula (2.8.153) and [112, (1.12.52)] we know that for every continuous vector field Z we have

$$\left(\nabla_Z \mathscr{S} f\right)\Big|_{\partial\Omega}^{\kappa-\mathrm{n.t.}} = \tfrac{1}{2}\nu_g(Z)\,\mathrm{Sym}(L;\nu_g)^{-1} f + \mathrm{P.V.}\int_{\partial\Omega} \big\langle(\nabla_Z \otimes I)E, f\big\rangle_{\mathcal{F}}\, d\sigma_g \qquad (2.8.210)$$

at σ-a.e. point on $\partial\Omega$. Then (2.8.130) follows from this (used with $Z = X$ and $Z = Y$), and [112, (1.12.101)], keeping in mind [112, (1.12.140)].

The first claim in item *(10)*, pertaining to the operator (2.8.162), is a consequence of Proposition 2.8.3 whose applicability in the current setting is ensured by (2.8.107) and the L^p-square function estimate (2.8.118). Similarly, the second claim in item *(10)*, regarding the operator (2.8.164), is once again implied by Proposition 2.8.3, now bearing in mind (2.8.137) and the L^p-square function estimate (2.8.147). □

Finally, we elaborate on the definition and nature of *double layer potential operators* associated with various (quasi-)factorizations of weakly elliptic second order operators acting between sections of a vector bundle over a given Riemannian manifold. The reader is reminded that L^p_s stands for the L^p-based Sobolev space of (smoothness) order $s \in \mathbb{R}$, and that $L^p_{s,\mathrm{loc}}$ denotes the local version of this scale. As in the past, we shall also abbreviate $H^s := L^2_s$.

Hypothesis 2.8.5 (Analytic Assumptions and Notation) *Let $\mathcal{M}$ denote a compact, oriented, boundaryless manifold of class $\mathscr{C}^2$ and real dimension $n \in \mathbb{N}$ (where $n \geq 2$), equipped with a Riemannian metric tensor of class $\mathscr{C}^1$. Consider a second order differential operator $L : \mathcal{E} \to \mathcal{E}$ acting between sections of a given Hermitian vector bundle $\mathcal{E} \to \mathcal{M}$, satisfying the following properties:*

(i) One has the quasi-factorization

$$L = \widetilde{D}D + W, \qquad (2.8.211)$$

where $\widetilde{D}$, D are first-order differential operators

$$D : \mathcal{E} \longrightarrow \mathcal{G}, \qquad \widetilde{D} : \mathcal{G} \longrightarrow \mathcal{E} \qquad (2.8.212)$$

for some Hermitian vector bundle $\mathcal{G} \to \mathcal{M}$ which, in any local coordinate chart U on $\mathcal{M}$ and with respect to local trivializations of $\mathcal{E}, \mathcal{G}$, may be represented as

$$\begin{gathered} Du(x) = \sum_j A_j(x)\partial_j u(x) + B(x)u(x) \ \text{ where, for some } r > n, \\ A_j \in L^r_{2,\mathrm{loc}}\big(U, \mathbb{C}^{\mathrm{rank}\mathcal{G}\times\mathrm{rank}\mathcal{E}}\big), \quad B \in L^r_{1,\mathrm{loc}}\big(U, \mathbb{C}^{\mathrm{rank}\mathcal{G}\times\mathrm{rank}\mathcal{E}}\big), \end{gathered} \qquad (2.8.213)$$

and

$$\widetilde{D}w(x) = \sum_j \widetilde{A}_j(x)\partial_j w(x) + \widetilde{B}(x)w(x) \text{ where, for some } r > n,$$
$$\widetilde{A}_j \in L^r_{2,\mathrm{loc}}\big(U, \mathbb{C}^{\mathrm{rank}\mathcal{E}\times\mathrm{rank}\mathcal{G}}\big), \quad \widetilde{B} \in L^r_{1,\mathrm{loc}}\big(U, \mathbb{C}^{\mathrm{rank}\mathcal{E}\times\mathrm{rank}\mathcal{G}}\big), \tag{2.8.214}$$

while

$$W \in \mathrm{Hom}\big(\mathcal{E}, \mathcal{E}\big) \text{ has coefficients in } L^r \text{ for some } r > n. \tag{2.8.215}$$

(ii) The operator L is weakly elliptic, in the sense that its principal symbol has the property that

$$\mathrm{Sym}(L;\xi) : \mathcal{E} \to \mathcal{E} \text{ is invertible for each } \xi \in T^*\mathcal{M} \setminus 0. \tag{2.8.216}$$

(iii) The operator L is invertible as a mapping

$$L : H^1(\mathcal{M}) \otimes \mathcal{E} \longrightarrow H^{-1}(\mathcal{M}) \otimes \mathcal{E}. \tag{2.8.217}$$

In view of the quasi-factorization (2.8.211) and the subsequent assumptions on the operators $\widetilde{D}, D$, it follows that L may be locally written as

$$Lu(x) = \sum_{j,k} \partial_j\big(\mathbb{A}_{jk}(x)\partial_k u(x)\big) + \sum_j \mathbb{B}_j(x)\partial_j u(x) + \mathbb{V}(x)u(x), \tag{2.8.218}$$

with coefficients

$$\begin{aligned} &\mathbb{A}_{jk} := \widetilde{A}_j A_k \in \mathscr{C}^{1+\gamma}_{\mathrm{loc}}, \text{ for some } \gamma > 0, \\ &\mathbb{B}_j := -\sum_k (\partial_k \widetilde{A}_k) A_j + \widetilde{A}_j B + \widetilde{B} A_j \in L^r_{1,\mathrm{loc}}, \\ &\mathbb{V} := \sum_j \widetilde{A}_j \partial_j B + \widetilde{B} B + W \in L^r_{\mathrm{loc}}. \end{aligned} \tag{2.8.219}$$

If E_L denotes the Schwartz kernel of the inverse L^{-1} of L in (2.8.217) then

$$E_L \in \mathcal{D}'(\mathcal{M} \times \mathcal{M}, \mathcal{E} \otimes \mathcal{E}) \cap \mathscr{C}^{1+\theta}_{\mathrm{loc}}\big(\mathcal{M} \times \mathcal{M} \setminus \mathrm{diag}, \mathcal{E} \otimes \mathcal{E}\big) \tag{2.8.220}$$

for some $\theta > 0$; see (2.8.184) in this regard.

Associated with the quasi-factorization (2.8.211), introduce the family of first order differential operators indexed by sections in the cotangent bundle

$$\partial^L_\xi := (-\mathrm{i})\,\mathrm{Sym}\big(\widetilde{D};\xi\big)D, \qquad \xi \in T^*\mathcal{M}. \tag{2.8.221}$$

Whenever L is as in Hypothesis 2.8.5, it follows that $L^\top$, the real transposed of L, also satisfies all conditions in Hypothesis 2.8.5. In particular, we now have the quasi-factorization

$$L^\top = D^\top \widetilde{D}^\top + W^\top. \tag{2.8.222}$$

We shall denote by $E_{L^\top}$ the Schwartz kernel of the inverse $\big(L^\top\big)^{-1}$ of the operator

$$L^\top : H^1(\mathcal{M}) \otimes \mathcal{E} \longrightarrow H^{-1}(\mathcal{M}) \otimes \mathcal{E} \tag{2.8.223}$$

which continues to enjoy a regularity property analogous to (2.8.220). Also, associated with the quasi-factorization (2.8.222), we introduce the family of first order differential operators indexed by sections in the cotangent bundle

$$\partial_\xi^{L^\top} := (-\mathrm{i})\,\mathrm{Sym}\big(D^\top;\xi\big)\widetilde{D}^\top, \qquad \xi \in T^*\mathcal{M}. \tag{2.8.224}$$

We now turn to assumptions of a geometric nature.

Hypothesis 2.8.6 (Geometric Assumptions and Notation) *Let $\mathcal{M}$ denote a compact, oriented, boundaryless manifold of class $\mathscr{C}^2$ and real dimension $n \in \mathbb{N}$ (where $n \geq 2$), equipped with a Riemannian metric tensor g of class $\mathscr{C}^1$. Assume $\Omega \subset \mathcal{M}$ is an Ahlfors regular domain. With $\mathcal{H}_g^{n-1}$ denoting the $(n-1)$-dimensional Hausdorff measure associated with the metric g on $\mathcal{M}$, abbreviate $\sigma_g := \mathcal{H}_g^{n-1}\lfloor\partial\Omega$. Also, denote by $\nu_g \in T^*\mathcal{M}$ the geometric measure theoretic outward unit conormal to Ω (which is well defined at σ_g-a.e. point on $\partial\Omega$). Finally, given any Hermitian vector bundle $\mathcal{E} \to \mathcal{M}$ and any exponent $p \in (0,\infty]$, the space of all σ_g-measurable sections $f : \partial\Omega \to \mathcal{E}$ which are p-th power integrable with respect to the surface measure σ_g is denoted by $L^p(\partial\Omega,\sigma_g) \otimes \mathcal{E}$.*

We are now prepared to introduce the boundary-to-domain double layer potential operator and its principal value version, as indicated below.

Definition 2.8.7 (Double Layers) *Assume Hypotheses 2.8.5-2.8.6. In this context, define the double layer associated with the quasi-factorization of L in* (2.8.211) *as the integral operator sending any $f \in L^1(\partial\Omega,\sigma_g) \otimes \mathcal{E}$ into the function defined at each $x \in \mathcal{M} \setminus \partial\Omega$ by*

$$\begin{aligned}\mathcal{D}_L f(x) &:= \int_{\partial\Omega} \Big\langle \big(I_x \otimes (-\mathrm{i})\,\mathrm{Sym}\big(D^\top;\nu_g(y)\big)\widetilde{D}_y^\top\big)E_L(x,y), f(y)\Big\rangle_{\mathcal{E}_y}\,\mathrm{d}\sigma_g(y)\\ &= \int_{\partial\Omega} \Big\langle \big(I_x \otimes \partial_{\nu_g(y)}^{L^\top}\big)E_L(x,y), f(y)\Big\rangle_{\mathcal{E}_y}\,\mathrm{d}\sigma_g(y),\end{aligned} \tag{2.8.225}$$

(where I denotes the identity, here acting in the variable x, etc.). In addition, consider its principal value version on $\partial\Omega$ (in the sense of Definition 2.8.1) acting on each $f \in L^1(\partial\Omega,\sigma_g) \otimes \mathcal{E}$ according to

$$\begin{aligned}K_L f(x) &:= \mathrm{P.V.}\int_{\partial\Omega} \Big\langle \big(I_x \otimes (-\mathrm{i})\,\mathrm{Sym}\big(D^\top;\nu_g(y)\big)\widetilde{D}_y^\top\big)E_L(x,y), f(y)\Big\rangle_{\mathcal{E}_y}\,\mathrm{d}\sigma_g(y)\\ &= \mathrm{P.V.}\int_{\partial\Omega} \Big\langle \big(I_x \otimes \partial_{\nu_g(y)}^{L^\top}\big)E_L(x,y), f(y)\Big\rangle_{\mathcal{E}_y}\,\mathrm{d}\sigma_g(y),\end{aligned} \tag{2.8.226}$$

at σ_g-a.e. point $x \in \partial\Omega$. Finally, define the singular integral operator acting on each function $f \in L^1(\partial\Omega,\sigma_g) \otimes \mathcal{E}$ according to

$$K_L^{\#} f(x) := \text{P.V.} \int_{\partial\Omega} \Big\langle \big((-\mathrm{i})\mathrm{Sym}\big(\widetilde{D}; \nu_g(x)\big) D_x \otimes I_y\big) E_L(x, y), f(y) \Big\rangle_{\mathcal{E}_y} \, d\sigma_g(y)$$

$$= \text{P.V.} \int_{\partial\Omega} \Big\langle \big(\partial^L_{\nu_g(x)} \otimes I_x\big) E_L(x, y), f(y) \Big\rangle_{\mathcal{E}_y} \, d\sigma_g(y), \tag{2.8.227}$$

at σ_g-a.e. point $x \in \partial\Omega$.

The remark below further clarifies some nuances in the above definition.

Remark 2.8.8 *In the context of Hypothesis 2.8.5, the given second-order operator L does not determine uniquely the first-order operators $\widetilde{D}$, D. Indeed, L can have multiple (infinitely many, actually) quasi-factorizations as in* (2.8.211). *This being said, the formulas of the double layer operators (both boundary-to-boundary and boundary-to-domain) in Definition 2.8.7 strongly depend on the choice of $\widetilde{D}$ and D. As such, in place of $\mathcal{D}_L$, K_L, $K_L^{\#}$ (which seems to suggest that L alone determines said operators) it would be perhaps more appropriate to write $\mathcal{D}_{\widetilde{D},D}$, $K_{\widetilde{D},D}$, $K^{\#}_{\widetilde{D},D}$, respectively. Same comment applies in relation to the conormal derivative from* (2.8.221). *The reader may want to keep in mind this aspect throughout this section.*

In the subsequent discussion, adopt the stronger hypothesis that Ω is actually a UR domain. Then the boundary-to-domain double layer $\mathcal{D}_L$ falls under the general template of the operator $\mathscr{A}$ discussed in (2.8.103), corresponding to the special case when the first-order differential operator P is given by[35]

$$P := (-\mathrm{i})\,\mathrm{Sym}\big(D^{\top}; \nu_g\big)\widetilde{D}^{\top}. \tag{2.8.228}$$

Also, the principal-value double layer K_L is a particular embodiment of (2.8.105), corresponding to the choice of P given in (2.8.228), while the singular integral operator $K_L^{\#}$ is a particular case of (2.8.152) corresponding to the choice

$$\widetilde{P} := (-\mathrm{i})\mathrm{Sym}(\widetilde{D}; \nu_g)D. \tag{2.8.229}$$

Granted this interpretation, Theorem 2.8.4 then yields a wealth of properties of our brand of double layer operators. For example, under Hypotheses 2.8.5-2.8.6 and assuming that Ω is actually a UR domain, it follows that

$$L(\mathcal{D}_L f) = 0 \text{ in } \Omega, \text{ for each } f \in L^1(\partial\Omega, \sigma_g) \otimes \mathcal{E}. \tag{2.8.230}$$

and for each integrability exponent $p \in (1, \infty)$ and each aperture parameter $\kappa > 0$ there exists a constant $C \in (0, \infty)$ with the property that

---

[35] Strictly speaking, P given in (2.8.228) does not have continuous coefficients due to the presence of ν_g but, working in local coordinates, we may expand $(-\mathrm{i})\,\mathrm{Sym}\big(D^{\top}; \nu_g\big) = \sum_j (\nu_g)_j (-\mathrm{i})\,\mathrm{Sym}\big(D^{\top}; dx_j\big)$ and focus on the first-order differential operators $\mathrm{Sym}\big(D^{\top}; dx_j\big)$ as playing the role of P, while the components $(\nu_g)_j$ are regarded as pointwise multipliers on the density f.

$$\big\|\mathcal{N}_\kappa(\mathcal{D}_L f)\big\|_{L^p(\partial\Omega,\sigma_g)} \leq C\|f\|_{L^p(\partial\Omega,\sigma_g)\otimes\mathcal{E}}, \qquad \forall f \in L^p(\partial\Omega,\sigma_g)\otimes\mathcal{E}, \tag{2.8.231}$$

Also, for each $p \in (1,\infty)$, the operators

$$\begin{aligned} K_L &: L^p(\partial\Omega,\sigma_g)\otimes\mathcal{E} \longrightarrow L^p(\partial\Omega,\sigma_g)\otimes\mathcal{E}, \\ K_L^\# &: L^p(\partial\Omega,\sigma_g)\otimes\mathcal{E} \longrightarrow L^p(\partial\Omega,\sigma_g)\otimes\mathcal{E}, \end{aligned} \tag{2.8.232}$$

are well defined, linear, and bounded. In addition, for each $f \in L^1(\partial\Omega,\sigma_g)\otimes\mathcal{E}$ the following jump-formula holds:

$$(\mathcal{D}_L f)\Big|_{\partial\Omega}^{\kappa-\mathrm{n.t.}} = \big(\tfrac{1}{2}I + K_L\big)f \ \text{ at } \sigma_g\text{-a.e. point on } \partial\Omega. \tag{2.8.233}$$

Indeed, (2.8.233) follows from the jump-formula (2.8.144) written for P as in (2.8.228), upon noting that

$$\begin{aligned} &\tfrac{1}{2}\mathrm{i}\mathrm{Sym}(L;\nu_g)^{-1}\,\mathrm{Sym}(P;\nu_g)^\top \\ &\quad= \tfrac{1}{2}\mathrm{i}\mathrm{Sym}(L;\nu_g)^{-1}\Big[(-\mathrm{i})\,\mathrm{Sym}\big(D^\top;\nu_g\big)\mathrm{Sym}\big(\widetilde{D}^\top;\nu_g\big)\Big]^\top \\ &\quad= \tfrac{1}{2}\mathrm{i}\mathrm{Sym}(L;\nu_g)^{-1}\,(-\mathrm{i})\,\mathrm{Sym}\big(\widetilde{D};\nu_g\big)\mathrm{Sym}\big(D;\nu_g\big) \\ &\quad= \tfrac{1}{2}\,\mathrm{Sym}(L;\nu_g)^{-1}\mathrm{Sym}\big(\widetilde{D}D;\nu_g\big) \\ &\quad= \tfrac{1}{2}\,\mathrm{Sym}(L;\nu_g)^{-1}\mathrm{Sym}\big(L;\nu_g\big) = \tfrac{1}{2}I, \end{aligned} \tag{2.8.234}$$

thanks to [112, (1.7.17)] and (2.8.211). In summary, the boundary-to-domain double layer is a mechanism for producing lots of null-solutions of the operator L, all of which have quantitative control of their nontangential maximal functions, and have nontangential pointwise boundary traces given explicitly (in terms of the principal value double layer and a constant jump term).

It is also of significance to remark that, in the same setting as above, for each $f \in L^1(\partial\Omega,\sigma_g)\otimes\mathcal{E}$ the jump-formula (2.8.153) used with $\widetilde{P}$ as in (2.8.229) presently gives

$$\begin{aligned} \partial^L_{\nu_g}(\mathscr{S}_L f) &:= (-\mathrm{i})\mathrm{Sym}(\widetilde{D};\nu_g)\Big(D\mathscr{S}_L f\Big)\Big|_{\partial\Omega}^{\kappa-\mathrm{n.t.}} \\ &= -\tfrac{1}{2}i(-\mathrm{i})\mathrm{Sym}(\widetilde{D};\nu_g)\,\mathrm{Sym}(D;\nu_g)\,\mathrm{Sym}(L;\nu_g)^{-1}f + K_L^\# f \\ &= -\tfrac{1}{2}\mathrm{Sym}(\widetilde{D}D;\nu_g)\,\mathrm{Sym}(L;\nu_g)^{-1}f + K_L^\# f \\ &= -\tfrac{1}{2}\mathrm{Sym}(L;\nu_g)\,\mathrm{Sym}(L;\nu_g)^{-1}f + K_L^\# f \\ &= \big(-\tfrac{1}{2}I + K_L^\#\big)f \ \text{ at } \sigma_g\text{-a.e. point on } \partial\Omega. \end{aligned} \tag{2.8.235}$$

Our next comment has to do with $K^{\#}_{L^\top}$, the principal value singular integral operator associated with $L^\top$ much as $K^{\#}_L$ has been associated with L in Definition 2.8.7, this time making use of the quasi-factorization (2.8.222) for $L^\top$. Specifically, for each $f \in L^1(\partial\Omega, \sigma_g) \otimes \mathcal{E}$ we have

$$\begin{aligned} K^{\#}_{L^\top} f(x) &= \text{P.V.} \int_{\partial\Omega} \Big\langle \big((-\mathrm{i})\mathrm{Sym}\big(D^\top; \nu_g(x)\big)\widetilde{D}^\top_x \otimes I_y\big) E_{L^\top}(x,y), f(y)\Big\rangle_{\mathcal{E}_y} \, d\sigma_g(y) \\ &= \text{P.V.} \int_{\partial\Omega} \Big\langle \big(\partial^{L^\top}_{\nu_g(x)} \otimes I_x\big) E_{L^\top}(x,y), f(y)\Big\rangle_{\mathcal{E}_y} \, d\sigma_g(y), \end{aligned} \tag{2.8.236}$$

at σ_g-a.e. point $x \in \partial\Omega$. Then, granted the same analytic and geometric conditions as before, item *(9)* of Theorem 2.8.4 presently implies

$$\big(K_L\big)^\top = K^{\#}_{L^\top} \tag{2.8.237}$$

in the precise sense that, for each $p, p' \in (1, \infty)$ satisfying $1/p + 1/p' = 1$,

> the (real) transpose of K_L acting on $L^p(\partial\Omega, \sigma_g) \otimes \mathcal{E}$ is the operator $K^{\#}_{L^\top}$ acting on $L^{p'}(\partial\Omega, \sigma_g) \otimes \mathcal{E}$ (see (2.8.232)). (2.8.238)

We also want to comment that, in the same setting as above, for any $p \in (1, \infty)$ the operator K_L originally acting from $L^p(\partial\Omega, \sigma_g) \otimes \mathcal{E}$ into itself leaves the Sobolev space $L^p_1(\partial\Omega, \sigma_g) \otimes \mathcal{E}$ invariant, and the following map is well defined, linear, and bounded:

$$K_L : L^p_1(\partial\Omega, \sigma_g) \otimes \mathcal{E} \longrightarrow L^p_1(\partial\Omega, \sigma_g) \otimes \mathcal{E}. \tag{2.8.239}$$

See [118, Theorem 4.9, p. 155], as well as [118, Theorem 4.10, p. 169] and [128], for related results.

Here we give a proof of the aforementioned mapping property of the principal value double layer on boundary Sobolev spaces in the case when all differentiability structures related to the Riemannian manifold, the Hermitian vector bundles, and the coefficients of the differential operator L are $\mathscr{C}^\infty$. Specifically, (2.8.239) is a consequence of the jump-formula (2.8.233), [112, Proposition 1.12.9], and Theorem 2.8.10 stated a little further below. In the latter theorem we consider the double layer $\mathcal{D}_L$ acting from L^p-based Sobolev spaces of order one, i.e., $L^p_1(\partial\Omega, \sigma_g) \otimes \mathcal{E}$ with $1 < p < \infty$, where Ω is a UR domain of a smooth Riemannian manifold, and show that the covariant derivative of the double layer is well-behaved near the boundary. We begin with a few preliminary results, setting the stage for the proof of Theorem 2.8.10.

First, we record a useful fact regarding the nature of the principal symbol of a commutator. Work on a smooth manifold $\mathcal{M}$, of real dimension n. Consider a classical pseudodifferential operator $\Psi \in OPS^m_{\mathrm{cl}}$, with m arbitrary, acting between two vector bundles over $\mathcal{M}$, and denote the principal symbol of Ψ by

$$q(x, \xi) := \mathrm{Sym}(\Psi; \xi), \quad \forall \xi \in T^*_x\mathcal{M}. \tag{2.8.240}$$

Also, having fixed a smooth vector field $X = \sum_j X_j \partial_j \in T\mathcal{M}$, denote the principal symbol of the commutator $[\nabla_X, \Psi] := \nabla_X \Psi - \Psi \nabla_X$ by

$$p(x, \xi) := \mathrm{Sym}\big([\nabla_X, \Psi]; \xi\big), \quad \forall \xi \in T^*_x\mathcal{M}. \tag{2.8.241}$$

Then the discussion in [176, p. 13] shows that

$$p(x, \xi) = \mathrm{i} \sum_j X_j \{\xi_j, q(x, \xi)\}, \quad \forall \xi \in T^*_x\mathcal{M}, \tag{2.8.242}$$

where, in general, the Poisson bracket of $p_1(x, \xi)$ and $p_2(x, \xi)$ is defined (cf., e.g., [176, (3.26), p. 13]) by

$$\{p_1(x, \xi), p_2(x, \xi)\} := \sum_{\ell=1}^{n} \big(\partial_{\xi_\ell} p_1 \partial_{x_\ell} p_2 - \partial_{x_\ell} p_1 \partial_{\xi_\ell} p_2\big), \quad \forall \xi \in T^*_x\mathcal{M}. \tag{2.8.243}$$

Next, let $\mathcal{E}$, $\mathcal{F}$, and $\mathcal{G}$ be Hermitian vector bundles of class $\mathscr{C}^\infty$ over a given Riemannian manifold $\mathcal{M}$ of class $\mathscr{C}^\infty$. We make the following convention:

$$\begin{aligned} &\text{given } T : \mathcal{D}(\mathcal{M}, \mathcal{E}) \to \mathcal{D}'(\mathcal{M}, \mathcal{F}) \text{ linear and continuous,} \\ &\text{we agree to let } k_T(x, y) \text{ denote the Schwartz kernel of } T. \end{aligned} \tag{2.8.244}$$

Then, if T is as in (2.8.244), for any differential operators $P : \mathcal{F} \to \mathcal{G}$ and $Q : \mathcal{G} \to \mathcal{E}$ we have (bearing in mind the conventions made in (2.8.90)-(2.8.94))

$$k_{PT}(x, y) = (P_x \otimes I_y) k_T(x, y) \text{ in } \mathcal{D}'\big(\mathcal{M} \times \mathcal{M}, \mathcal{G} \otimes \mathcal{E}\big), \tag{2.8.245}$$

$$k_{TQ}(x, y) = (I_x \otimes Q_y^\top) k_T(x, y) \text{ in } \mathcal{D}'\big(\mathcal{M} \times \mathcal{M}, \mathcal{F} \otimes \mathcal{G}\big), \tag{2.8.246}$$

$$k_{PTQ}(x, y) = (P_x \otimes Q_y^\top) k_T(x, y) \text{ in } \mathcal{D}'\big(\mathcal{M} \times \mathcal{M}, \mathcal{G} \otimes \mathcal{G}\big). \tag{2.8.247}$$

The following lemma plays a key role in the proof of Theorem 2.8.10.

Lemma 2.8.9 *Let $(\mathcal{M}, g)$ be a compact Riemannian manifold of class $\mathscr{C}^\infty$, having (real) dimension n. Suppose $\mathcal{E}, \mathcal{G} \to \mathcal{M}$ are two Hermitian vector bundles of class $\mathscr{C}^\infty$ equipped with connections*

$$\nabla^{\mathcal{E}} : T\mathcal{M} \times \mathcal{E} \longrightarrow \mathcal{E}, \quad \nabla^{\mathcal{G}} : T\mathcal{M} \times \mathcal{G} \longrightarrow \mathcal{G}. \tag{2.8.248}$$

Assume $L : \mathcal{E} \to \mathcal{E}$ is a weakly elliptic second-order partial differential operator with smooth coefficients such that

$$L : H^1(\mathcal{M}) \otimes \mathcal{E} \longrightarrow H^{-1}(\mathcal{M}) \otimes \mathcal{E} \textit{ is an isomorphism.} \tag{2.8.249}$$

In addition, suppose L has the quasi-factorization

$$L = \widetilde{D} D + W \tag{2.8.250}$$

for some first-order differential operators with smooth coefficients

$$D : \mathcal{E} \longrightarrow \mathcal{G}, \quad \widetilde{D} : \mathcal{G} \longrightarrow \mathcal{E}, \tag{2.8.251}$$

and a homomorphism $W \in \mathrm{Hom}(\mathcal{E}, \mathcal{E})$. *Let* $E_L(x, y)$ *denote the Schwartz kernel of* L^{-1}.

Next, let $\Omega \subset \mathcal{M}$ *be* UR *domain with outward unit conormal* $\nu_g : \partial\Omega \to T^*\mathcal{M}$ *and surface measure* $\sigma_g := \mathcal{H}_g^{n-1}\lfloor\partial\Omega$. *Choose an arbitrary vector field* $X \in T\mathcal{M}$, *and consider the kernel function*

$$\begin{aligned}\omega_X(x, y) := &\Big(\nabla^{\mathcal{E}}_{X_x} \otimes (-\mathrm{i})\mathrm{Sym}\big(D^\top; \nu_g(y)\big)\widetilde{D}^\top_y\Big)E_L(x, y)\\ &- \Big(I_x \otimes (-\mathrm{i})\mathrm{Sym}\big(D^\top; \nu_g(y)\big)(\nabla^{\mathcal{G}}_X)^\top_y\widetilde{D}^\top_y\Big)E_L(x, y).\end{aligned} \tag{2.8.252}$$

Use this to define an integral operator $\mathscr{K}_X$ *mapping any integrable section* f *in* $L^1(\partial\Omega, \sigma_g) \otimes \mathcal{E}$ *into the section* $\mathscr{K}_X f : \Omega \to \mathcal{E}$ *defined by*

$$(\mathscr{K}_X f)(x) := \int_{\partial\Omega} \big\langle \omega_X(x, y), f(y)\big\rangle_{\mathcal{E}}\, d\sigma_g(y), \quad \forall x \in \Omega. \tag{2.8.253}$$

Finally, fix an aperture parameter $\kappa > 0$.

Then for each $f \in L^1(\partial\Omega, \sigma_g) \otimes \mathcal{E}$ *the nontangential boundary trace*

$$(\mathscr{K}_X f)\Big|^{\kappa-\mathrm{n.t.}}_{\partial\Omega} \quad \text{exists } \sigma_g\text{-a.e. on } \partial\Omega \tag{2.8.254}$$

and, in fact, using the abbreviation $[\nabla_X, D]$ *for the first-order differential operator* $\nabla^{\mathcal{G}}_X D - D\nabla^{\mathcal{E}}_X$, *at* σ_g*-a.e. point* $x \in \partial\Omega$ *one has*

$$\begin{aligned}\Big((\mathscr{K}_X f)\big|^{\kappa-\mathrm{n.t.}}_{\partial\Omega}\Big)(x) = &-\frac{1}{2}\mathrm{Sym}(L; \nu_g)^{-1}\mathrm{Sym}(\widetilde{D}; \nu_g)\mathrm{Sym}\big([\nabla_X, D]; \nu_g\big)f(x)\\ &+ \mathrm{P.V.}\int_{\partial\Omega} \big\langle \omega_X(x, y), f(y)\big\rangle_{\mathcal{E}}\, d\sigma_g(y).\end{aligned} \tag{2.8.255}$$

Also, for each $p \in (1, \infty)$ *there exists* $C = C(\Omega, L, X, \kappa, p) \in (0, \infty)$ *such that*

$$\|\mathcal{N}_\kappa(\mathscr{K}_X f)\|_{L^p(\partial\Omega,\sigma_g)} \leq C\|f\|_{L^p(\partial\Omega,\sigma_g)\otimes\mathcal{E}}, \quad \forall f \in L^p(\partial\Omega, \sigma_g) \otimes \mathcal{E}. \tag{2.8.256}$$

Proof First we introduce some notation by rewriting $\omega_X(x, y)$ as

$$\omega_X(x, y) = \Big(I_x \otimes (-\mathrm{i})\mathrm{Sym}\big(D^\top; \nu_g(y)\big)\Big)\widetilde{\omega}_X(x, y), \tag{2.8.257}$$

where

$$\widetilde{\omega}_X(x, y) := (\nabla^{\mathcal{E}}_{X_x} \otimes \widetilde{D}^\top_y)E_L(x, y) - (I_x \otimes (\nabla^{\mathcal{G}}_X)^\top_y\widetilde{D}^\top_y)E_L(x, y). \tag{2.8.258}$$

Further defining

$$\Theta(x,y) := \big(I_x \otimes \widetilde{D}_y^\top\big) E_L(x,y), \tag{2.8.259}$$

allows us to recast $\widetilde{\omega}_X(x,y)$ as

$$\widetilde{\omega}_X(x,y) = \big(\nabla^{\mathcal{E}}_{X_x} \otimes I_y\big)\Theta(x,y) - \big(I_x \otimes (\nabla^{\mathcal{G}}_X)_y^\top\big)\Theta(x,y). \tag{2.8.260}$$

The idea is to now interpret the expressions in (2.8.259) and (2.8.260) as Schwartz kernels. First recall that

$$E_L(x,y) = k_{L^{-1}}(x,y), \tag{2.8.261}$$

where $k_{L^{-1}}(x,y)$ denotes the Schwartz kernel of L^{-1}. Taking $Q := \widetilde{D}$ in (2.8.246) then yields

$$\Theta(x,y) = k_{L^{-1}\widetilde{D}}(x,y). \tag{2.8.262}$$

This suggests considering the composition

$$\Psi := L^{-1}\widetilde{D}. \tag{2.8.263}$$

Since the differential geometric structures of $\mathcal{M}, \mathcal{E}, \mathcal{G}$ and the coefficients of L are all $\mathscr{C}^\infty$, we have

$$L^{-1} \in \mathrm{OP}S^{-2}_{\mathrm{cl}} \text{ and } \widetilde{D} \in \mathrm{OP}S^{1}_{\mathrm{cl}}, \tag{2.8.264}$$

which implies

$$\Psi = L^{-1}\widetilde{D} \in \mathrm{OP}S^{-1}_{\mathrm{cl}} \text{ with } \Theta(x,y) = k_\Psi(x,y). \tag{2.8.265}$$

Moreover,

$$\mathrm{Sym}(\Psi;\xi) = \mathrm{Sym}(L^{-1};\xi)\mathrm{Sym}(\widetilde{D};\xi) = \mathrm{Sym}(L;\xi)^{-1}\mathrm{Sym}(\widetilde{D};\xi). \tag{2.8.266}$$

Since L is a second-order operator and $\widetilde{D}$ is first-order, their principal symbols are even and odd in ξ, respectively. Hence,

$$\mathrm{Sym}(\Psi;\xi) \text{ is odd in } \xi. \tag{2.8.267}$$

Moving on, by (2.8.245) it follows that

$$\big(\nabla^{\mathcal{E}}_{X_x} \otimes I_y\big)\Theta(x,y) = k_{\nabla^{\mathcal{E}}_X \Psi}(x,y). \tag{2.8.268}$$

Also,

$$\big(I_x \otimes (\nabla^{\mathcal{G}}_X)_y^\top\big)\Theta(x,y) = k_{\Psi\nabla^{\mathcal{G}}_X}(x,y), \tag{2.8.269}$$

which follows from (2.8.246) with $Q := (\nabla^{\mathcal{G}}_X)^\top$. Then by (2.8.260), (2.8.268), and (2.8.269) it follows that

$$\begin{aligned}\widetilde{\omega}_X(x,y) &= k_{\nabla^{\mathcal{E}}_X\Psi}(x,y) - k_{\Psi\nabla^{\mathcal{G}}_X}(x,y) = k_{\nabla^{\mathcal{E}}_X\Psi-\Psi\nabla^{\mathcal{G}}_X}(x,y)\\ &= k_{[\nabla_X,\Psi]}(x,y). \end{aligned}\tag{2.8.270}$$

Going further, if $q(x,\xi) := \mathrm{Sym}(\Psi;\xi)$, then based on (2.8.242) and (2.8.243) we have

$$\begin{aligned}\mathrm{Sym}\big([\nabla_X,\Psi];\xi\big) &= \mathrm{i}\sum_j X_j\{\xi_j, q(x,\xi)\}\\ &= \mathrm{i}\sum_j X_j\sum_{\ell=1}^{n}\big((\partial_{\xi_\ell}\xi_j)(\partial_{x_\ell}q)(x,\xi) - (\partial_{x_\ell}\xi_j)(\partial_{\xi_\ell}q)(x,\xi)\big)\\ &= \mathrm{i}\sum_j X_j\sum_{\ell=1}^{n}\delta_{j\ell}(\partial_{x_\ell}q)(x,\xi) = \mathrm{i}\sum_j X_j(\partial_{x_j}q)(x,\xi)\\ &= \mathrm{i}\nabla_{X_x}\big(q(x,\xi)\big), \quad \forall\,\xi\in T^*_x\mathcal{M}. \end{aligned}\tag{2.8.271}$$

Hence, comparing (2.8.240), (2.8.241), and (2.8.271) yields

$$\mathrm{Sym}\big([\nabla_X;\Psi],\xi\big) = \mathrm{i}\nabla_X\big(\mathrm{Sym}(\Psi;\xi)\big), \quad \forall\,\xi\in T^*\mathcal{M}. \tag{2.8.272}$$

Recalling from (2.8.267) that $\mathrm{Sym}(\Psi;\xi)$ is odd in ξ, and also observing that the covariant derivative along X which appears in (2.8.272) does not affect the parity of $\mathrm{Sym}(\Psi;\xi)$ in the variable ξ, we then conclude that

$$\mathrm{Sym}\big([\nabla_X,\Psi];\xi\big) \text{ is odd in } \xi. \tag{2.8.273}$$

To proceed, recall (see, e.g., the discussion in [176, pp. 12-13]) that, in general,

given two classical pseudodifferential operators, P_1, P_2 of order r, such that $\mathrm{Sym}(P_1;\xi) = \mathrm{Sym}(P_2;\xi)$ for every cotangent variable ξ, it follows that $P_1 - P_2$ is a classical pseudodifferential operator of order $\leq r-1$. (2.8.274)

We also note that since $\mathrm{Sym}(\nabla^{\mathcal{E}}_X;\xi) = \mathrm{i}\xi(X)\,I_{\mathcal{E}}$ and $\mathrm{Sym}(\nabla^{\mathcal{G}}_X;\xi) = \mathrm{i}\xi(X)\,I_{\mathcal{G}}$ by [112, (1.12.52)], where $I_{\mathcal{E}}$ is the identity on $\mathcal{E}$ and $I_{\mathcal{G}}$ is the identity on $\mathcal{G}$, we have

$$\begin{aligned}\mathrm{Sym}(\nabla^{\mathcal{E}}_X;\xi)\mathrm{Sym}(\Psi;\xi) &= \big(\mathrm{i}\xi(X)I_{\mathcal{E}}\big)\mathrm{Sym}(\Psi;\xi) = \mathrm{Sym}(\Psi;\xi)\big(\mathrm{i}\xi(X)\,I_{\mathcal{G}}\big)\\ &= \mathrm{Sym}(\Psi;\xi)\mathrm{Sym}(\nabla^{\mathcal{G}}_X;\xi), \end{aligned}\tag{2.8.275}$$

i.e., the symbols of Ψ intertwines the symbols of $\nabla^{\mathcal{E}}_X$ and $\nabla^{\mathcal{G}}_X$. It then follows that

$$[\nabla_X,\Psi] := \nabla^{\mathcal{E}}_X\Psi - \Psi\nabla^{\mathcal{G}}_X \in \mathrm{OP}S^{-1}_{\mathrm{cl}}, \tag{2.8.276}$$

from (2.8.274), (2.8.275), plus the fact that both $\nabla^{\mathcal{E}}_X\Psi$ and $\Psi\nabla^{\mathcal{G}}_X$ belong to $\mathrm{OP}S^0_{\mathrm{cl}}$ (as seen from (2.8.265)).

Next, fix $p \in (1, \infty)$ and let $f \in L^p(\partial\Omega, \sigma_g) \otimes \mathcal{E}$ be arbitrary. By (2.8.273) and (2.8.276), we may apply (2.8.42) with $Q := [\nabla_X, \Psi]$ to conclude that the singular integral operator defined by

$$(\mathscr{K}_{[\nabla_X,\Psi]}f)(x) := \int_{\partial\Omega} \langle k_{[\nabla_X,\Psi]}(x,y), f(y)\rangle_{\mathcal{E}_y}\, d\sigma_g(y), \quad \forall x \in \Omega, \tag{2.8.277}$$

satisfies the nontangential maximal function estimate

$$\|\mathcal{N}_\kappa(\mathscr{K}_{[\nabla_X,\Psi]}f)\|_{L^p(\partial\Omega,\sigma_g)} \leq C\|f\|_{L^p(\partial\Omega,\sigma_g)\otimes\mathcal{E}}, \tag{2.8.278}$$

for all $f \in L^p(\partial\Omega, \sigma_g) \otimes \mathcal{E}$, where the constant $C \in (0, \infty)$ is independent of f. Recalling (2.8.270), it follows from (2.8.278) that for each $p \in (1, \infty)$ there exists a constant $C \in (0, \infty)$ such that

$$\big\|\mathcal{N}_\kappa\big(\widetilde{\mathscr{K}}_X f\big)\big\|_{L^p(\partial\Omega,\sigma_g)} \leq C\|f\|_{L^p(\partial\Omega,\sigma_g)\otimes\mathcal{E}}, \quad \forall f \in L^p(\partial\Omega, \sigma_g) \otimes \mathcal{E}, \tag{2.8.279}$$

where

$$\big(\widetilde{\mathscr{K}}_X f\big)(x) := \int_{\partial\Omega} \langle \widetilde{\omega}_X(x,y), f(y)\rangle_{\mathcal{E}_y}\, d\sigma_g(y), \quad \forall x \in \Omega. \tag{2.8.280}$$

In turn, one may deduce that the same type of nontangential maximal function estimate, namely (2.8.256), for the singular integral operator $\mathscr{K}_X$ defined in (2.8.253). Indeed, since $\omega_X(x, y)$ is of the form (2.8.257), we see that

$$\begin{aligned}(\mathscr{K}_X f)(x) &= \int_{\partial\Omega} \Big\langle \Big(I_x \otimes (-\mathrm{i})\mathrm{Sym}\big(D^\top; \nu_g(y)\big)\Big)\widetilde{\omega}_X(x,y), f(y)\Big\rangle_{\mathcal{E}_y} d\sigma_g(y) \\ &= \int_{\partial\Omega} \Big\langle \widetilde{\omega}_X(x,y), \mathrm{i}\mathrm{Sym}\big(D; \nu_g(y)\big) f(y)\Big\rangle_{\mathcal{E}_y} d\sigma_g(y) \\ &= \Big(\widetilde{\mathscr{K}}_X\big(\mathrm{i}\mathrm{Sym}(D; \nu_g) f\big)\Big)(x), \quad \forall x \in \Omega.\end{aligned} \tag{2.8.281}$$

From this and (2.8.279) it follows that there exists $C = C(\Omega, X, L, \kappa, p) \in (0, \infty)$ with the property that

$$\begin{aligned}\|\mathcal{N}_\kappa(\mathscr{K}_X f)\|_{L^p(\partial\Omega,\sigma_g)} &= \Big\|\mathcal{N}_\kappa\Big(\widetilde{\mathscr{K}}_X\big(\mathrm{i}\mathrm{Sym}(D; \nu_g) f\big)\Big)\Big\|_{L^p(\partial\Omega,\sigma_g)} \\ &\leq C\big\|\mathrm{i}\mathrm{Sym}(D; \nu_g) f\big\|_{L^p(\partial\Omega,\sigma_g)\otimes\mathcal{E}} \\ &\leq C\|f\|_{L^p(\partial\Omega,\sigma_g)\otimes\mathcal{E}},\end{aligned} \tag{2.8.282}$$

for all $f \in L^p(\partial\Omega, \sigma_g) \otimes \mathcal{E}$, as desired. Finally, the analysis above combined with (2.8.46) also implies that for each section $f \in L^1(\partial\Omega, \sigma_g) \otimes \mathcal{E}$ the nontangential trace $(\mathscr{K}_X f)\big|_{\partial\Omega}^{\kappa-\mathrm{n.t.}}$ exists, and at σ_g-a.e. point $x \in \partial\Omega$ we have

$$
\begin{aligned}
\Big((\mathscr{K}_X f)\Big|_{\partial\Omega}^{\kappa-\text{n.t.}}\Big)(x) &= \frac{1}{2\mathrm{i}}\mathrm{Sym}\big([\nabla_X,\Psi];\nu_g\big)\mathrm{i}\,\mathrm{Sym}(D;\nu_g)f(x) \\
&\quad + \mathrm{P.V.}\int_{\partial\Omega}\big\langle\omega_X(x,y),f(y)\big\rangle_{\mathcal{E}}\,d\sigma_g(y) \\
&= -\frac{1}{2}\mathrm{Sym}(L;\nu_g)^{-1}\mathrm{Sym}(\widetilde{D};\nu_g)\mathrm{Sym}\big([\nabla_X,D];\nu_g\big)f(x) \\
&\quad + \mathrm{P.V.}\int_{\partial\Omega}\big\langle\omega_X(x,y),f(y)\big\rangle_{\mathcal{E}}\,d\sigma_g(y), \qquad (2.8.283)
\end{aligned}
$$

since

$$
\begin{aligned}
\mathrm{Sym}\big([\nabla_X,\Psi];\nu_g\big)\,\mathrm{Sym}(D;\nu_g) &= \mathrm{Sym}\big([\nabla_X,\Psi]D;\nu_g\big) \qquad (2.8.284)\\
&= \mathrm{Sym}\Big(\nabla_X\Psi D - \Psi\nabla_X D;\nu_g\Big) \\
&= \mathrm{Sym}\Big(\nabla_X L^{-1}\widetilde{D}D - L^{-1}\widetilde{D}\nabla_X D;\nu_g\Big) \\
&= \mathrm{Sym}\Big(\nabla_X - L^{-1}\widetilde{D}[\nabla_X,D] + L^{-1}\widetilde{D}D\nabla_X;\nu_g\Big) \\
&= \mathrm{Sym}\Big(-L^{-1}\widetilde{D}[\nabla_X,D];\nu_g\Big) \\
&= -\mathrm{Sym}(L;\nu_g)^{-1}\mathrm{Sym}(\widetilde{D};\nu_g)\mathrm{Sym}\big([\nabla_X,D];\nu_g\big),
\end{aligned}
$$

thanks to the fact that the operators L and $\widetilde{D}D$ differ by a lower-order term. This concludes the proof of Lemma 2.8.9. □

We are now ready to state and prove the result alluded to earlier, regarding size estimates and boundary behavior for the boundary-to-domain double layer potential operator acting from Sobolev spaces defined on the boundary of a UR domain in a smooth Riemannian manifold.

Theorem 2.8.10 *Let $(\mathcal{M},g)$ be a compact Riemannian manifold of class $\mathscr{C}^\infty$. Suppose $\mathcal{E},\mathcal{G}\to\mathcal{M}$ are Hermitian vector bundles of class $\mathscr{C}^\infty$ equipped with connections*

$$
\nabla^{\mathcal{E}}:T\mathcal{M}\times\mathcal{E}\longrightarrow\mathcal{E},\quad \nabla^{\mathcal{G}}:T\mathcal{M}\times\mathcal{G}\longrightarrow\mathcal{G}. \tag{2.8.285}
$$

Assume $L:\mathcal{E}\to\mathcal{E}$ is a weakly elliptic second-order partial differential operator with smooth coefficients such that

$$
L:H^1(\mathcal{M})\otimes\mathcal{E}\longrightarrow H^{-1}(\mathcal{M})\otimes\mathcal{E}\ \text{ is an isomorphism.} \tag{2.8.286}
$$

In addition, suppose L has the quasi-factorization

$$
L=\widetilde{D}D+W \tag{2.8.287}
$$

for some first-order differential operators with smooth coefficients

$$D : \mathcal{E} \longrightarrow \mathcal{G}, \quad \widetilde{D} : \mathcal{G} \longrightarrow \mathcal{E}, \tag{2.8.288}$$

and some homomorphism $W \in \mathrm{Hom}(\mathcal{E}, \mathcal{E})$. Let $E_L(x, y)$ denote the Schwartz kernel of L^{-1}.

Next, let $\Omega \subset \mathcal{M}$ be UR *domain with outward unit conormal $\nu_g : \partial\Omega \to T^*\mathcal{M}$ and surface measure $\sigma_g := \mathcal{H}_g^{n-1}\lfloor\partial\Omega$. Finally, fix an integrability exponent $p \in (1, \infty)$ and an aperture parameter $\kappa > 0$.*

Then for each $f \in L_1^p(\partial\Omega, \sigma_g) \otimes \mathcal{E}$, the nontangential pointwise trace

$$(\nabla^{\mathcal{E}}\mathcal{D}_L f)\Big|_{\partial\Omega}^{\kappa-\mathrm{n.t.}} \text{ exist at } \sigma_g\text{-a.e. point on } \partial\Omega. \tag{2.8.289}$$

Moreover, there exists a constant $C = C(\Omega, L, p, \kappa) \in (0, \infty)$ such that

$$\|\mathcal{N}_\kappa(\nabla^{\mathcal{E}}\mathcal{D}_L f)\|_{L^p(\partial\Omega,\sigma_g)} \leq C\|f\|_{L_1^p(\partial\Omega,\sigma_g)\otimes\mathcal{E}}, \tag{2.8.290}$$

for all $f \in L_1^p(\partial\Omega, \sigma_g) \otimes \mathcal{E}$.

Proof Let $f \in L_1^p(\partial\Omega, \sigma_g) \otimes \mathcal{E}$ be arbitrary. Also fix an arbitrary vector field $X \in T\mathcal{M}$. For any point $x \in \Omega$, we have

$$\begin{aligned}(\nabla_X^{\mathcal{E}}\mathcal{D}_L f)(x) &= \int_{\partial\Omega}\Big\langle\Big(\nabla_{X_x}^{\mathcal{E}} \otimes (-\mathrm{i})\mathrm{Sym}\big(D^\top; \nu_g(y)\big)\widetilde{D}_y^\top\Big)E_L(x, y), f(y)\Big\rangle_{\mathcal{E}_y} \, d\sigma_g(y) \\ &= (\mathscr{K}_X f)(x) + (\mathscr{K}_X^\sharp f)(x),\end{aligned} \tag{2.8.291}$$

where $\mathscr{K}_X$ is the singular integral operator defined by (2.8.253) in Lemma 2.8.9 and $\mathscr{K}_X^\sharp$ is the singular integral operator defined as

$$(\mathscr{K}_X^\sharp f)(x) := \int_{\partial\Omega}\Big\langle\Big(I_x \otimes (-\mathrm{i})\mathrm{Sym}\big(D^\top; \nu_g(y)\big)(\nabla_X^{\mathcal{G}})_y^\top\widetilde{D}_y^\top\Big)E_L(x, y), f(y)\Big\rangle_{\mathcal{E}_y} \, d\sigma_g(y) \tag{2.8.292}$$

at each point $x \in \Omega$. Observing that

$$0 \leq \mathcal{N}_\kappa(\nabla_X^{\mathcal{E}}\mathcal{D}_L f) \leq \mathcal{N}_\kappa(\mathscr{K}_X f) + \mathcal{N}_\kappa(\mathscr{K}_X^\sharp f), \tag{2.8.293}$$

one may infer

$$\begin{aligned}\|\mathcal{N}_\kappa(\nabla_X^{\mathcal{E}}\mathcal{D}_L f)\|_{L^p(\partial\Omega,\sigma_g)} &\leq \|\mathcal{N}_\kappa(\mathscr{K}_X f)\|_{L^p(\partial\Omega,\sigma_g)} \\ &\quad + \|\mathcal{N}_\kappa(\mathscr{K}_X^\sharp f)\|_{L^p(\partial\Omega,\sigma_g)}.\end{aligned} \tag{2.8.294}$$

From Lemma 2.8.9, there exists a constant $C = C(\Omega, L, X, p, \kappa) \in (0, \infty)$ such that

$$\|\mathcal{N}_\kappa(\mathscr{K}_X f)\|_{L^p(\partial\Omega,\sigma_g)} \leq C\|f\|_{L^p(\partial\Omega,\sigma_g)\otimes\mathcal{E}} \leq C\|f\|_{L_1^p(\partial\Omega,\sigma_g)\otimes\mathcal{E}}, \tag{2.8.295}$$

where the second inequality is due to $L^p_1(\partial\Omega,\sigma_g)\otimes\mathcal{E}\hookrightarrow L^p(\partial\Omega,\sigma_g)\otimes\mathcal{E}$. Furthermore, (2.8.254) shows that at σ_g-a.e. point $x\in\partial\Omega$ we have

$$\Big((\mathscr{K}_X f)\big|^{\kappa-\mathrm{n.t.}}_{\partial\Omega}\Big)(x)=-\frac{1}{2}\mathrm{Sym}(L;\nu_g)^{-1}\mathrm{Sym}(\widetilde{D};\nu_g)\mathrm{Sym}\big([\nabla_X,D];\nu_g\big)f(x)$$
$$+\,\mathrm{P.V.}\int_{\partial\Omega}\big\langle\omega_X(x,y),f(y)\big\rangle_{\mathcal{E}}\,d\sigma_g(y),\tag{2.8.296}$$

where the kernel $\omega_X(x,y)$ is as in (2.8.252). Thus, it remains to prove

$$\|\mathcal{N}_\kappa(\mathscr{K}^\sharp_X f)\|_{L^p(\partial\Omega,\sigma_g)}\le C\|f\|_{L^p_1(\partial\Omega,\sigma_g)\otimes\mathcal{E}},\tag{2.8.297}$$

for some constant $C=C(\Omega,L,X,p,\kappa)\in(0,\infty)$, and find a suitable formula for $(\mathscr{K}^\sharp_X f)\big|^{\kappa-\mathrm{n.t.}}_{\partial\Omega}$ at σ_g-a.e. point on $\partial\Omega$.

To achieve these goals, the integral kernel of $\mathscr{K}^\sharp_X$ requires a closer examination. The delicate issue with the integral kernel of $\mathscr{K}^\sharp_X$ is that there are two derivatives on $E_L(x,y)$, increasing its singularity from order $n-2$ to order n, and thus making the integral kernel of $\mathscr{K}^\sharp_X$ hypersingular (due to the dimension of $\partial\Omega$ being $n-1$). Our strategy involves using the quasi-factorization $L^\top=D^\top\widetilde{D}^\top+W^\top$ and then applying integration by parts on the boundary to transfer the burden of one of the derivatives on $E_L(x,y)$ to the function f in the form of a tangential derivative. This strategy is only effective because $f\in L^p_1(\partial\Omega,\sigma_g)\otimes\mathcal{E}$, so it can take a tangential derivative within the context of the nontangential maximal function estimate we are trying to prove in (2.8.297) (as well as the subsequent boundary trace). With only one remaining derivative on $E_L(x,y)$, the Calderón-Zygmund theory in Theorem 2.8.4 can be invoked to finish the job.

To handle $\mathscr{K}^\sharp_X f$, for ease of writing abbreviate

$$\Upsilon(x,y):=\Big(I_x\otimes\Big((-\mathrm{i})\mathrm{Sym}\big(D^\top;\nu_g(y)\big)(\nabla^{\mathcal{G}}_X)^\top_y+\nu_g(X)(y)D^\top_y\Big)\widetilde{D}^\top_y\Big)E_L(x,y)\tag{2.8.298}$$

for each $x\in\Omega$ and $y\in\partial\Omega$. To proceed in earnest, fix $x\in\Omega$, add and subtract

$$\int_{\partial\Omega}\Big\langle\Big(I_x\otimes\nu_g(X)(y)D^\top_y\widetilde{D}^\top_y\Big)E_L(x,y),f(y)\Big\rangle_{\mathcal{E}_y}\,d\sigma_g(y)\tag{2.8.299}$$

from $(\mathscr{K}^\sharp_X f)(x)$, and then use the fact that $D^\top\widetilde{D}^\top=L^\top-W^\top$ to obtain

$$
\begin{aligned}
(\mathscr{K}^{\#}_{X} f)(x) &= \int_{\partial\Omega} \Big\langle \Big(I_x \otimes (-\mathrm{i})\mathrm{Sym}\big(D^{\top}; \nu_g(y)\big)(\nabla^{\mathcal{G}}_{X})^{\top}_{y} \widetilde{D}^{\top}_{y}\Big) E_L(x,y), f(y) \Big\rangle_{\mathcal{E}_y} \, d\sigma_g(y) \\
&= \int_{\partial\Omega} \Big\langle \Upsilon(x,y), f(y) \Big\rangle_{\mathcal{E}_y} \, d\sigma_g(y) \\
&\quad - \int_{\partial\Omega} \Big\langle \Big(I_x \otimes \nu_g(X)(y) D^{\top}_{y} \widetilde{D}^{\top}_{y}\Big) E_L(x,y), f(y) \Big\rangle_{\mathcal{E}_y} \, d\sigma_g(y) \\
&= \int_{\partial\Omega} \Big\langle \Upsilon(x,y), f(y) \Big\rangle_{\mathcal{E}_y} \, d\sigma_g(y) \\
&\quad - \int_{\partial\Omega} \Big\langle \Big(I_x \otimes \nu_g(X)(y) L^{\top}_{y}\Big) E_L(x,y), f(y) \Big\rangle_{\mathcal{E}_y} \, d\sigma_g(y) \\
&\quad + \int_{\partial\Omega} \Big\langle \Big(I_x \otimes \nu_g(X)(y) W^{\top}_{y}\Big) E_L(x,y), f(y) \Big\rangle_{\mathcal{E}_y} \, d\sigma_g(y) \\
&=: \mathrm{I}_1(x) + \mathrm{I}_2(x) + \mathrm{I}_3(x). \qquad (2.8.300)
\end{aligned}
$$

Next, observe that

$$
\begin{aligned}
\mathrm{I}_2(x) &= - \int_{\partial\Omega} \Big\langle \Big(I_x \otimes \nu_g(X)(y) L^{\top}_{y}\Big) E_L(x,y), f(y) \Big\rangle_{\mathcal{E}_y} \, d\sigma_g(y) \\
&= - \int_{\partial\Omega} \big\langle (I_x \otimes L^{\top}_{y}) E_L(x,y), \nu_g(X)(y) f(y) \big\rangle_{\mathcal{E}_y} \, d\sigma_g(y) \\
&= 0, \qquad (2.8.301)
\end{aligned}
$$

where the third equality is a consequence of the fact that $\big(I_x \otimes L^{\top}_{y}\big) E_L(x,y) = 0$ for $x \in \Omega$ and $y \in \partial\Omega$. Moving on, the term $\mathrm{I}_3(x)$ can be handled by rewriting

$$
\begin{aligned}
\mathrm{I}_3(x) &= \int_{\partial\Omega} \Big\langle \Big(I_x \otimes \nu_g(X)(y) W^{\top}_{y}\Big) E_L(x,y), f(y) \Big\rangle_{\mathcal{E}_y} \, d\sigma_g(y) \\
&= \int_{\partial\Omega} \big\langle E_L(x,y), \nu_g(X)(y) W_y f(y) \big\rangle_{\mathcal{E}_y} \, d\sigma_g(y). \qquad (2.8.302)
\end{aligned}
$$

Since the integral kernel of $\mathrm{I}_3(x)$ contains no derivatives on $E_L(x,y)$, it is weakly singular and the term $\mathrm{I}_3(x)$ falls under the scope of Proposition 2.5.39 with $(\mathscr{B} f)(x) := \mathrm{I}_3(x)$. As such,

$$
\Big(\mathrm{I}_3\Big|^{\kappa-\mathrm{n.t.}}_{\partial\Omega}\Big)(x) = \int_{\partial\Omega} \big\langle E_L(x,y), \, \nu_g(X)(y) W_y f(y) \big\rangle_{\mathcal{E}_y} \, d\sigma_g(y) \qquad (2.8.303)
$$

at σ_g-a.e. point $x \in \partial\Omega$, and

$$\|\mathcal{N}_\kappa(\mathrm{I}_3)\|_{L^p(\partial\Omega,\sigma_g)} \le C\|f\|_{L^p(\partial\Omega,\sigma_g)\otimes\mathcal{E}} \le C\|f\|_{L_1^p(\partial\Omega,\sigma_g)\otimes\mathcal{E}}, \tag{2.8.304}$$

where the fact that $L_1^p(\partial\Omega,\sigma_g)\otimes\mathcal{E} \hookrightarrow L^p(\partial\Omega,\sigma_g)\otimes\mathcal{E}$ is a continuous embedding is used the last inequality.

To treat the term $\mathrm{I}_1(x)$, in a first stage we shall use integration by parts on the boundary, as described in [112, Corollary 1.12.12], in which we apply [112, (1.12.132)] with $P := D^\top$, the vector bundle $\mathcal{G}$ in place of $\mathcal{F}$, and with the mapping $y \mapsto (I_x \otimes \widetilde{D}_y^\top)E_L(x,\cdot)$ playing the role of the test function[36] φ. In this regard, bring in the tangential derivative operator $\partial_\tau^{\mathbf{Q},\mathbf{P}}$ defined as in [112, (1.12.131)] for $P := D^\top$ (and with the vector bundle $\mathcal{G}$ in place of $\mathcal{F}$). The point of performing the aforementioned integration by parts on the boundary is that the Calderón-Zygmund theory from Theorem 2.8.4 may then be brought to bear, keeping in mind that, much as in [112, (1.12.131)], the mapping

$$\partial_\tau^{\mathbf{Q},\mathbf{P}} : L_1^p(\partial\Omega,\sigma_g)\otimes\mathcal{E} \longrightarrow L^p(\partial\Omega,\sigma_g)\otimes\mathcal{E} \text{ is continuous.} \tag{2.8.305}$$

Turning to specifics, we use [112, (1.12.132)] as explained earlier (recall the notation in (2.8.298)) to write

$$\begin{aligned}
\mathrm{I}_1(x) &= \int_{\partial\Omega} \Big\langle \Upsilon(x,y), f(y) \Big\rangle_{\mathcal{E}_y} \, d\sigma_g(y) \\
&= -\int_{\partial\Omega} \Big\langle (I_x \otimes \widetilde{D}_y^\top)E_L(x,y), (\partial_\tau^{\mathbf{Q},\mathbf{P}} f)(y) \Big\rangle_{\mathcal{E}_y} \, d\sigma_g(y) \\
&\quad + \int_{\partial\Omega} \Big\langle \Big(I_x \otimes \mathrm{i}\,\mathrm{Sym}(\nabla_X^{\mathcal{G}} D - D\nabla_X^{\mathcal{E}}; \nu_g(y))^\top \widetilde{D}_y^\top\Big)E_L(x,y), f(y) \Big\rangle_{\mathcal{E}_y} \, d\sigma_g(y) \\
&=: \mathrm{I}_1^{(a)}(x) + \mathrm{I}_1^{(b)}(x).
\end{aligned} \tag{2.8.306}$$

Having decomposed $\mathrm{I}_1(x)$ as in (2.8.306), the Calderón-Zygmund theory (pertaining to the operator $\mathscr{A}$) from Theorem 2.8.4 can be invoked for the piece $\mathrm{I}_1^{(a)}(x)$, since $\partial_\tau^{\mathbf{Q},\mathbf{P}} f \in L^p(\partial\Omega,\sigma_g)\otimes\mathcal{E}$. This guarantees that there exists a constant $C = C(\Omega, L, X, p, \kappa) \in (0,\infty)$ with the property that

$$\big\|\mathcal{N}_\kappa(\mathrm{I}_1^{(a)})\big\|_{L^p(\partial\Omega,\sigma_g)} \le C\big\|\partial_\tau^{\mathbf{Q},\mathbf{P}} f\big\|_{L^p(\partial\Omega,\sigma_g)\otimes\mathcal{E}} \le C\|f\|_{L_1^p(\partial\Omega,\sigma_g)\otimes\mathcal{E}}, \tag{2.8.307}$$

where the last inequality in (2.8.307) follows from (2.8.305), and (2.8.144) gives

$$\begin{aligned}
\Big(\mathrm{I}_1^{(a)}\Big|_{\partial\Omega}^{\kappa-\mathrm{n.t.}}\Big)(x) &= \frac{\mathrm{i}}{2}\mathrm{Sym}(L;\nu_g)^{-1}\mathrm{Sym}(\widetilde{D};\nu_g)(\partial_\tau^{\mathbf{Q},\mathbf{P}} f)(x) \\
&\quad - \mathrm{P.V.}\int_{\partial\Omega} \Big\langle (I_x \otimes \widetilde{D}_y^\top)E_L(x,y), (\partial_\tau^{\mathbf{Q},\mathbf{P}} f)(y) \Big\rangle_{\mathcal{E}_y} \, d\sigma_g(y)
\end{aligned} \tag{2.8.308}$$

---

[36] the singularity at x is away from $\partial\Omega$, so it may be safely ignored

at σ_g-a.e. point $x \in \partial\Omega$.

Next, we observe that from [112, (1.12.52)] that $\mathrm{Sym}(\nabla_X^{\mathcal{G}};\xi) = \mathrm{i}\xi(X)\, I_{\mathcal{G}}$ and $\mathrm{Sym}(\nabla_X^{\mathcal{E}};\xi) = \mathrm{i}\xi(X)\, I_{\mathcal{E}}$, where $I_{\mathcal{G}}$ is the identity on $\mathcal{G}$ and $I_{\mathcal{E}}$ is the identity on $\mathcal{E}$, for each $\xi \in T^*\mathcal{M}$. Consequently, for each $\xi \in T^*\mathcal{M}$ we have

$$\begin{aligned}\mathrm{Sym}(\nabla_X^{\mathcal{G}};\xi)\mathrm{Sym}(D;\xi) &= \big(\mathrm{i}\xi(X)I_{\mathcal{G}}\big)\mathrm{Sym}(D;\xi) = \mathrm{Sym}(D;\xi)\big(\mathrm{i}\xi(X)\, I_{\mathcal{E}}\big)\\ &= \mathrm{Sym}(D;\xi)\mathrm{Sym}(\nabla_X^{\mathcal{E}};\xi)\end{aligned} \tag{2.8.309}$$

which further implies that

$$[\nabla_X, D] := \nabla_X^{\mathcal{G}} D - D\nabla_X^{\mathcal{E}} \text{ is a first-order differential operator.} \tag{2.8.310}$$

Upon expressing

$$\begin{aligned}&\mathrm{I}_1^{(b)}(x) \\ &= \int_{\partial\Omega}\Big\langle\Big(I_x\otimes \mathrm{i}\mathrm{Sym}\big(\nabla_X^{\mathcal{G}}D - D\nabla_X^{\mathcal{E}};\nu_g(y)\big)^{\top}\widetilde{D}_y^{\top}\Big)E_L(x,y), f(y)\Big\rangle_{\mathcal{E}_y}\, d\sigma_g(y)\\ &= \int_{\partial\Omega}\Big\langle\big(I_x\otimes \widetilde{D}_y^{\top}\big)E_L(x,y), \mathrm{i}\mathrm{Sym}\big(\nabla_X^{\mathcal{G}}D - D\nabla_X^{\mathcal{E}};\nu_g(y)\big)f(y)\Big\rangle_{\mathcal{E}_y}\, d\sigma_g(y)\end{aligned} \tag{2.8.311}$$

and bearing in mind (2.8.310), we may once again call upon Theorem 2.8.4 which guarantees the existence of a constant $C = C(\Omega, L, X, p, \kappa) \in (0,\infty)$ with

$$\|\mathcal{N}_\kappa(\mathrm{I}_1^{(b)})\|_{L^p(\partial\Omega,\sigma_g)} \le C\|f\|_{L^p(\partial\Omega,\sigma_g)\otimes\mathcal{E}} \le C\|f\|_{L_1^p(\partial\Omega,\sigma_g)\otimes\mathcal{E}}, \tag{2.8.312}$$

where the fact that $L_1^p(\partial\Omega,\sigma_g)\otimes\mathcal{E} \hookrightarrow L^p(\partial\Omega,\sigma_g)\otimes\mathcal{E}$ is a continuous embedding is used in the second inequality. Furthermore, granted the format of $\mathrm{I}_1^{(b)}(x)$ in (2.8.312), the jump-formula (2.8.144) from Theorem 2.8.4 also ensures that

$$\begin{aligned}&\Big(\mathrm{I}_1^{(b)}\Big|_{\partial\Omega}^{\kappa-\mathrm{n.t.}}\Big)(x)\\ &= \frac{1}{2}\mathrm{Sym}(L;\nu_g)^{-1}\mathrm{Sym}(\widetilde{D};\nu_g)\mathrm{Sym}\big([\nabla_X, D];\nu_g\big)f(x)\\ &\quad + \mathrm{P.V.}\int_{\partial\Omega}\Big\langle\big(I_x\otimes \widetilde{D}_y^{\top}\big)E_L(x,y),\ \mathrm{i}\,\mathrm{Sym}\big([\nabla_X, D];\nu_g(y)\big)f(y)\Big\rangle_{\mathcal{E}_y}\, d\sigma_g(y)\end{aligned} \tag{2.8.313}$$

at σ_g-a.e. point $x \in \partial\Omega$. Note that the jump term in (2.8.313) cancels the one in (2.8.255). Based on (2.8.301), (2.8.303), (2.8.308), (2.8.300), (2.8.255), (2.8.306), and (2.8.291) we therefore conclude that at σ_g-a.e. point $x \in \partial\Omega$ we have

$$\Big((\nabla^{\mathcal{E}}_X \mathcal{D}_L f)\Big|^{\kappa-\mathrm{n.t.}}_{\partial\Omega}\Big)(x) \tag{2.8.314}$$

$$\begin{aligned}
&= \frac{i}{2}\mathrm{Sym}(L;\nu_g)^{-1}\mathrm{Sym}(\widetilde{D};\nu_g)(\partial_\tau^{\mathbf{Q},\mathbf{P}} f)(x)\\
&\quad - \mathrm{P.V.}\int_{\partial\Omega}\Big\langle (I_x\otimes \widetilde{D}_y^\top)E_L(x,y), (\partial_\tau^{\mathbf{Q},\mathbf{P}} f)(y)\Big\rangle_{\mathcal{G}_y}\,d\sigma_g(y)\\
&\quad + \mathrm{P.V.}\int_{\partial\Omega}\Big\langle (I_x\otimes \widetilde{D}_y^\top)E_L(x,y),\, i\,\mathrm{Sym}\big([\nabla_X, D];\nu_g(y)\big)f(y)\Big\rangle_{\mathcal{G}_y}\,d\sigma_g(y)\\
&\quad + \mathrm{P.V.}\int_{\partial\Omega}\big\langle E_L(x,y),\, \nu_g(X)(y)W_y f(y)\big\rangle_{\mathcal{E}_y}\,d\sigma_g(y)\\
&\quad + \mathrm{P.V.}\int_{\partial\Omega}\big\langle \omega_X(x,y), f(y)\big\rangle_{\mathcal{E}_y}\,d\sigma_g(y),
\end{aligned} \tag{2.8.315}$$

where the tangential derivative operator $\partial_\tau^{\mathbf{Q},\mathbf{P}}$ is defined as in [112, (1.12.131)] for $P := D^\top$ (and with the vector bundle $\mathcal{G}$ in place of $\mathcal{F}$), and where the kernel $\omega_X(x, y)$ is as in (2.8.252). This establishes (2.8.289). From (2.8.301), (2.8.304), (2.8.307), (2.8.312), (2.8.300), and (2.8.306) we also see that (2.8.297) does in fact hold. Hence, the result claimed in (2.8.290) follows from (2.8.294), (2.8.295), and (2.8.297). The proof of Theorem 2.8.10 is therefore complete. □

Continue to work in the context of Theorem 2.8.10. To set the stage, let us agree that

$\partial_\tau^{(\nabla_X;P)}$ stands for tangential derivative operator $\partial_\tau^{\mathbf{Q},\mathbf{P}}$ defined as in [112, (1.12.131)] for a given vector field X on the manifold $\mathcal{M}$ and a given first-order differential operator $P : \mathcal{G} \to \mathcal{E}$. (2.8.316)

Introduce

$$\Omega_+ := \Omega \text{ and } \Omega_- := \mathcal{M}\setminus\overline{\Omega}. \tag{2.8.317}$$

Given $f \in L^p_1(\partial\Omega, \sigma_g)\otimes\mathcal{E}$ with $1 < p < \infty$, define

$$\mathcal{D}^\pm_L f := (\mathcal{D}_L f)\big|_{\Omega_\pm}. \tag{2.8.318}$$

Then Theorem 2.8.10 together with (2.8.231) and (2.8.233) imply that if $u_\pm := \mathcal{D}^\pm_L f$ we have

$$\begin{aligned}
&\mathcal{N}_\kappa u_\pm \in L^p(\partial\Omega,\sigma_g),\quad \mathcal{N}_\kappa(\nabla u_\pm)\in L^p(\partial\Omega,\sigma_g),\\
&\text{both } u_\pm\Big|^{\kappa-\mathrm{n.t.}}_{\partial\Omega},\ (\nabla u_\pm)\Big|^{\kappa-\mathrm{n.t.}}_{\partial\Omega} \text{ exist } \sigma_g\text{-a.e. on } \partial\Omega,
\end{aligned} \tag{2.8.319}$$

and, in fact,

$$u_\pm\Big|^{\kappa-\mathrm{n.t.}}_{\partial\Omega} = \big(\pm\tfrac{1}{2}I + K_L\big)f \text{ at } \sigma_g\text{-a.e. point on } \partial\Omega. \tag{2.8.320}$$

To proceed, observe from (2.8.314) that for each vector field X we have

$$(\nabla_X^{\mathcal{E}}\mathcal{D}_L^+ f)\Big|_{\partial\Omega}^{\kappa-\mathrm{n.t.}} - (\nabla_X^{\mathcal{E}}\mathcal{D}_L^- f)\Big|_{\partial\Omega}^{\kappa-\mathrm{n.t.}} = \mathrm{i}\,\mathrm{Sym}(L;\nu_g)^{-1}\mathrm{Sym}(\widetilde{D};\nu_g)\partial_\tau^{(\nabla_X;D^\top)} f. \tag{2.8.321}$$

Use (2.8.320) to decompose the section f as

$$f = u_+\Big|_{\partial\Omega}^{\kappa-\mathrm{n.t.}} - u_-\Big|_{\partial\Omega}^{\kappa-\mathrm{n.t.}}. \tag{2.8.322}$$

In this regard, we observe from (2.8.316) and [112, (1.12.137) in Theorem 1.12.13] (whose applicability is ensured by (2.8.319)) employed with $\mathbf{P}$, $\mathbf{Q}$ in place of P, Q, where

$$\begin{aligned} &\mathbf{P} : \mathcal{E} \to \mathcal{G} \oplus \mathcal{E}, && \mathbf{P}u := \big(Du, -\nabla_X^{\mathcal{E}} u\big), \\ &\mathbf{Q} : \mathcal{G} \oplus \mathcal{E} \to \mathcal{G}, && \mathbf{Q}(v,w) := \nabla_X^{\mathcal{G}} v + Dw, \end{aligned} \tag{2.8.323}$$

that for any vector field X we have

$$\begin{aligned} \partial_\tau^{(\nabla_X;D^\top)}\Big(u_+\Big|_{\partial\Omega}^{\kappa-\mathrm{n.t.}}\Big) &= \mathrm{i}\,\mathrm{Sym}(\mathbf{Q};\nu_g)(\mathbf{P}u_+)\Big|_{\partial\Omega}^{\kappa-\mathrm{n.t.}} \\ &= -(-\mathrm{i})\mathrm{Sym}(\nabla_X;\nu_g)(Du_+)\Big|_{\partial\Omega}^{\kappa-\mathrm{n.t.}} \\ &\quad + (-\mathrm{i})\mathrm{Sym}(D;\nu_g)(\nabla_X u_+)\Big|_{\partial\Omega}^{\kappa-\mathrm{n.t.}}. \end{aligned} \tag{2.8.324}$$

Thus, if we locally express

$$D = A_j \nabla_{\partial_j} + B \tag{2.8.325}$$

where the A_j's and B are in $\mathrm{Hom}(\mathcal{E},\mathcal{G})$, then (2.8.324) implies that for each j we have

$$\begin{aligned} \partial_\tau^{(\nabla_{\partial_j};D^\top)}\Big(u_+\Big|_{\partial\Omega}^{\kappa-\mathrm{n.t.}}\Big) &= -(-\mathrm{i})\mathrm{Sym}(\nabla_{\partial_j};\nu_g)(Du_+)\Big|_{\partial\Omega}^{\kappa-\mathrm{n.t.}} \\ &\quad + (-\mathrm{i})\mathrm{Sym}(D;\nu_g)(\nabla_{\partial_j} u_+)\Big|_{\partial\Omega}^{\kappa-\mathrm{n.t.}} \end{aligned} \tag{2.8.326}$$

hence at σ_g-a.e. point on $\partial\Omega$ we may write

$$
\begin{aligned}
&\mathrm{Sym}(\widetilde{D};\nu_g)A_j\,\mathrm{i}\,\mathrm{Sym}(L;\nu_g)^{-1}\mathrm{Sym}(\widetilde{D};\nu_g)\partial_\tau^{(\nabla_{\partial_j};D^\top)}\Big(u_+\big|_{\partial\Omega}^{\kappa-\mathrm{n.t.}}\Big)\\
&\quad= -\mathrm{Sym}(\widetilde{D};\nu_g)\,A_j\,\mathrm{Sym}(L;\nu_g)^{-1}\mathrm{Sym}(\widetilde{D};\nu_g)\mathrm{Sym}(\nabla_{\partial_j};\nu_g)(Du_+)\big|_{\partial\Omega}^{\kappa-\mathrm{n.t.}}\\
&\qquad+\mathrm{Sym}(\widetilde{D};\nu_g)\,A_j\mathrm{Sym}(L;\nu_g)^{-1}\mathrm{Sym}(\widetilde{D};\nu_g)\mathrm{Sym}(D;\nu_g)(\nabla_{\partial_j}u_+)\big|_{\partial\Omega}^{\kappa-\mathrm{n.t.}}\\
&\quad= -\mathrm{Sym}(\widetilde{D};\nu_g)\,A_j\mathrm{Sym}(\nabla_{\partial_j};\nu_g)\mathrm{Sym}(L;\nu_g)^{-1}\mathrm{Sym}(\widetilde{D};\nu_g)(Du_+)\big|_{\partial\Omega}^{\kappa-\mathrm{n.t.}}\\
&\qquad+\mathrm{Sym}(\widetilde{D};\nu_g)\,A_j\mathrm{Sym}(L;\nu_g)^{-1}\mathrm{Sym}(\widetilde{D}D;\nu_g)(\nabla_{\partial_j}u_+)\big|_{\partial\Omega}^{\kappa-\mathrm{n.t.}}\\
&\quad= -\mathrm{Sym}(\widetilde{D};\nu_g)\,\mathrm{Sym}(D;\nu_g)\mathrm{Sym}(L;\nu_g)^{-1}\mathrm{Sym}(\widetilde{D};\nu_g)(Du_+)\big|_{\partial\Omega}^{\kappa-\mathrm{n.t.}}\\
&\qquad+\mathrm{Sym}(\widetilde{D};\nu_g)\,A_j(\nabla_{\partial_j}u_+)\big|_{\partial\Omega}^{\kappa-\mathrm{n.t.}}\\
&\quad= -\mathrm{Sym}(\widetilde{D}D;\nu_g)\mathrm{Sym}(L;\nu_g)^{-1}\mathrm{Sym}(\widetilde{D};\nu_g)(Du_+)\big|_{\partial\Omega}^{\kappa-\mathrm{n.t.}}\\
&\qquad+\mathrm{Sym}(\widetilde{D};\nu_g)(Du_+)\big|_{\partial\Omega}^{\kappa-\mathrm{n.t.}}-\mathrm{Sym}(\widetilde{D};\nu_g)\,B\big(u_+\big|_{\partial\Omega}^{\kappa-\mathrm{n.t.}}\big)\\
&\quad= -\mathrm{Sym}(\widetilde{D};\nu_g)(Du_+)\big|_{\partial\Omega}^{\kappa-\mathrm{n.t.}}\\
&\qquad+\mathrm{Sym}(\widetilde{D};\nu_g)(Du_+)\big|_{\partial\Omega}^{\kappa-\mathrm{n.t.}}-\mathrm{Sym}(\widetilde{D};\nu_g)\,B\big(u_+\big|_{\partial\Omega}^{\kappa-\mathrm{n.t.}}\big)\\
&\quad= -\mathrm{Sym}(\widetilde{D};\nu_g)\,B\big(u_+\big|_{\partial\Omega}^{\kappa-\mathrm{n.t.}}\big)
\end{aligned}
\tag{2.8.327}
$$

plus a similar formula when u_- is used in place of u_+. In concert, these two formulas imply

$$
\begin{aligned}
&\mathrm{Sym}(\widetilde{D};\nu_g)A_j\,\mathrm{i}\,\mathrm{Sym}(L;\nu_g)^{-1}\mathrm{Sym}(\widetilde{D};\nu_g)\partial_\tau^{(\nabla_{\partial_j};D^\top)}\Big(u_+\big|_{\partial\Omega}^{\kappa-\mathrm{n.t.}}\Big)\\
&\qquad-\mathrm{Sym}(\widetilde{D};\nu_g)\,A_j\,\mathrm{i}\,\mathrm{Sym}(L;\nu_g)^{-1}\mathrm{Sym}(\widetilde{D};\nu_g)\partial_\tau^{(\nabla_{\partial_j};D^\top)}\Big(u_-\big|_{\partial\Omega}^{\kappa-\mathrm{n.t.}}\Big)\\
&\qquad= -\mathrm{Sym}(\widetilde{D};\nu_g)\,B\Big(u_+\big|_{\partial\Omega}^{\kappa-\mathrm{n.t.}}-u_-\big|_{\partial\Omega}^{\kappa-\mathrm{n.t.}}\Big)\\
&\qquad= -\mathrm{Sym}(\widetilde{D};\nu_g)\,Bf,
\end{aligned}
\tag{2.8.328}
$$

where the last equality comes from (2.8.322). Finally, from (2.8.325), (2.8.321), (2.8.328), and (2.8.322) we conclude that

$$\begin{aligned}
&\mathrm{Sym}(\widetilde{D};\nu_g)(D\mathcal{D}_L^+f)\Big|_{\partial\Omega}^{\kappa-\mathrm{n.t.}} - \mathrm{Sym}(\widetilde{D};\nu_g)(D\mathcal{D}_L^-f)\Big|_{\partial\Omega}^{\kappa-\mathrm{n.t.}}\\
&\quad= \mathrm{Sym}(\widetilde{D};\nu_g)\,A_j\,(\nabla_{\partial_j}\mathcal{D}_L^+f)\Big|_{\partial\Omega}^{\kappa-\mathrm{n.t.}} - \mathrm{Sym}(\widetilde{D};\nu_g)\,A_j\,(\nabla_{\partial_j}\mathcal{D}_L^-f)\Big|_{\partial\Omega}^{\kappa-\mathrm{n.t.}}\\
&\qquad+ B(\mathcal{D}_L^+f)\Big|_{\partial\Omega}^{\kappa-\mathrm{n.t.}} - B(\mathcal{D}_L^-f)\Big|_{\partial\Omega}^{\kappa-\mathrm{n.t.}}\\
&\quad= \mathrm{Sym}(\widetilde{D};\nu_g)\,A_j\,\mathrm{i}\,\mathrm{Sym}(L;\nu_g)^{-1}\mathrm{Sym}(\widetilde{D};\nu_g)\partial_\tau^{(\nabla_{\partial_j};D^\top)}\Big(u_+\Big|_{\partial\Omega}^{\kappa-\mathrm{n.t.}}\Big)\\
&\qquad- \mathrm{Sym}(\widetilde{D};\nu_g)\,A_j\,\mathrm{i}\,\mathrm{Sym}(L;\nu_g)^{-1}\mathrm{Sym}(\widetilde{D};\nu_g)\partial_\tau^{(\nabla_{\partial_j};D^\top)}\Big(u_-\Big|_{\partial\Omega}^{\kappa-\mathrm{n.t.}}\Big)\\
&\qquad+ B\Big(u_+\Big|_{\partial\Omega}^{\kappa-\mathrm{n.t.}}\Big) - B\Big(u_-\Big|_{\partial\Omega}^{\kappa-\mathrm{n.t.}}\Big)\\
&\quad= -\mathrm{Sym}(\widetilde{D};\nu_g)\,Bf + \mathrm{Sym}(\widetilde{D};\nu_g)\,Bf = 0.
\end{aligned} \tag{2.8.329}$$

Then Theorem 2.8.10 together with the above argument prove the following result, which may be regarded as a far-reaching generalization of the classical Lyapunov-Tauber Theorem regarding the lack of jump for the normal derivative of the standard harmonic double layer potential operator in the flat Euclidean setting (cf. [56, Theorem 1,pp. 72-73], [90, p. 58]).

Theorem 2.8.11 *Work in the context of Theorem 2.8.10 and retain notation introduced above in* (2.8.317)-(2.8.318). *Then for each* $f \in L_1^p(\partial\Omega,\sigma_g)\otimes\mathcal{E}$ *with* $1 < p < \infty$ *one has*

$$\mathrm{Sym}(\widetilde{D};\nu_g)(D\mathcal{D}_L^+f)\Big|_{\partial\Omega}^{\kappa-\mathrm{n.t.}} = \mathrm{Sym}(\widetilde{D};\nu_g)(D\mathcal{D}_L^-f)\Big|_{\partial\Omega}^{\kappa-\mathrm{n.t.}} \tag{2.8.330}$$

at σ_g*-a.e. point on* $\partial\Omega$*. In particular, the conormal derivatives*

$$\partial_{\nu_g}^L\mathcal{D}_L^\pm f := (-\mathrm{i})\mathrm{Sym}(\widetilde{D};\nu_g)(D\mathcal{D}_L^\pm f)\Big|_{\partial\Omega}^{\kappa-\mathrm{n.t.}} \tag{2.8.331}$$

agree with one another, so it makes sense to simply write

$$\partial_{\nu_g}^L\mathcal{D}_L f := (-\mathrm{i})\mathrm{Sym}(\widetilde{D};\nu_g)(D\mathcal{D}_L^\pm f)\Big|_{\partial\Omega}^{\kappa-\mathrm{n.t.}}. \tag{2.8.332}$$

Consider as such, the conormal derivative operator

$$\partial_{\nu_g}^L\mathcal{D}_L : L_1^p(\partial\Omega,\sigma_g)\otimes\mathcal{E} \longrightarrow L^p(\partial\Omega,\sigma_g)\otimes\mathcal{E} \tag{2.8.333}$$

is then well-defined, linear, and bounded.

The single and double layer potential operators play a basic role in the context of the following Green type integral representation formula for null-solutions of a weakly elliptic second-order differential operator L acting between vector bundles over a Riemannian manifold.

Theorem 2.8.12 *Assume Hypotheses 2.8.5-2.8.6. Bring in the double layer operator $\mathcal{D}_L$ associated as in Definition 2.8.7 with the quasi-factorization of L in* (2.8.211), *and recall the single layer operator $\mathcal{S}$ associated with L as in* (2.8.101). *Fix an aperture parameter $\kappa > 0$ and suppose*

$$\begin{aligned} & u \in W^{1,1}_{\mathrm{loc}}(\Omega) \otimes \mathcal{E}, \quad Lu = 0 \text{ in } \Omega, \\ & \mathcal{N}_\kappa u \text{ and } \mathcal{N}_\kappa(Du) \text{ belong to } L^1(\partial\Omega, \sigma_g), \\ & u\big|^{\kappa-\mathrm{n.t.}}_{\partial\Omega} \text{ and } (Du)\big|^{\kappa-\mathrm{n.t.}}_{\partial\Omega} \text{ exist at } \sigma_g\text{-a.e. point on } \partial\Omega. \end{aligned} \tag{2.8.334}$$

Then, with the conormal derivative defined as

$$\partial^L_{\nu_g} u := (-\mathrm{i})\mathrm{Sym}(\widetilde{D}; \nu_g)(Du)\big|^{\kappa-\mathrm{n.t.}}_{\partial\Omega}, \tag{2.8.335}$$

the following integral representation formula holds:

$$u = \mathcal{D}_L\Big(u\big|^{\kappa-\mathrm{n.t.}}_{\partial\Omega}\Big) - \mathcal{S}\big(\partial^L_{\nu_g} u\big) \text{ in } \Omega. \tag{2.8.336}$$

Proof We begin by fixing an arbitrary point $x \in \Omega$, and considering the finite dimensional vector space $\mathscr{V} := \mathcal{E}_x$. Next, define the mapping $\vec{F} : \Omega \to \mathscr{V} \otimes T\mathcal{M}$ via the requirement that for each $y \in \Omega$ and each $\xi \in T^*_y\mathcal{M}$ we have

$$\begin{aligned} {}_{T^*_y\mathcal{M}}\big(\xi, \vec{F}(y)\big)_{T_y\mathcal{M}} &= \Big\langle \big(I_x \otimes (-\mathrm{i})\,\mathrm{Sym}\big(D^\top; \xi\big)\widetilde{D}^\top_y\big) E_L(x,y), u(y)\Big\rangle_{\mathcal{E}_y} \\ &\quad - \Big\langle E_L(x,y), (-\mathrm{i})\mathrm{Sym}(\widetilde{D}; \xi)(Du)(y)\Big\rangle_{\mathcal{E}_y} \end{aligned} \tag{2.8.337}$$

as vectors in $\mathscr{V}$. Thanks to the first property in (2.8.334) and (2.8.186), this implies that $\vec{F}$ belongs to $L^1_{\mathrm{loc}}(\Omega, \mathcal{L}^n_g) \otimes (\mathscr{V} \otimes T\mathcal{M})$. In fact, we may locally express $\vec{F} = F_j\partial_j$ with

$$\begin{aligned} F_j(y) &= \Big\langle \big(I_x \otimes (-\mathrm{i})\,\mathrm{Sym}\big(D^\top; dx_j\big)\widetilde{D}^\top_y\big) E_L(x,y), u(y)\Big\rangle_{\mathcal{E}_y} \\ &\quad - \Big\langle E_L(x,y), (-\mathrm{i})\mathrm{Sym}(\widetilde{D}; dx_j)(Du)(y)\Big\rangle_{\mathcal{E}_y}. \end{aligned} \tag{2.8.338}$$

Moreover, for any scalar-valued function $\psi \in \mathscr{C}^1_c(\Omega)$ we may write

$$
\begin{aligned}
&{}_{\mathcal{D}'(\Omega)}\big(\mathrm{div}_g \vec{F}, \psi\big)_{\mathcal{D}(\Omega)} \\
&= -{}_{\mathcal{D}'(\Omega)}\big(\vec{F}, \mathrm{grad}\,\psi\big)_{\mathcal{D}(\Omega)} \\
&= -\int_\Omega {}_{T_y\mathcal{M}}\big\langle(\mathrm{grad}\,\psi)(y), \vec{F}(y)\big\rangle_{T_y\mathcal{M}}\,\mathrm{d}\mathcal{L}^n_g(y) \\
&= -\int_\Omega {}_{T_y^*\mathcal{M}}\big((\mathrm{grad}\,\psi)^\flat(y), \vec{F}(y)\big)_{T_y\mathcal{M}}\,\mathrm{d}\mathcal{L}^n_g(y) \\
&= -\int_\Omega {}_{T_y^*\mathcal{M}}\big(\mathrm{d}\psi(y), \vec{F}(y)\big)_{T_y\mathcal{M}}\,\mathrm{d}\mathcal{L}^n_g(y) \\
&= -\int_\Omega \Big\langle\big(I_x \otimes (-\mathrm{i})\,\mathrm{Sym}\big(D^\top; \mathrm{d}\psi(y)\big)\widetilde{D}_y^\top\big)E_L(x,y), u(y)\Big\rangle_{\mathcal{E}_y}\,\mathrm{d}\mathcal{L}^n_g(y) \\
&\quad + \int_\Omega \Big\langle E_L(x,y), (-\mathrm{i})\mathrm{Sym}\big(\widetilde{D}; \mathrm{d}\psi(y)\big)(Du)(y)\Big\rangle_{\mathcal{E}_y}\,\mathrm{d}\mathcal{L}^n_g(y) \\
&=: \mathrm{I} + \mathrm{II}.
\end{aligned}
\tag{2.8.339}
$$

Based on [112, (1.7.17), (1.7.20)], and integration by parts we may compute

$$
\begin{aligned}
\mathrm{I} &= -\int_\Omega \Big\langle\big(I \otimes (-\mathrm{i})\,\mathrm{Sym}\big(D^\top; \mathrm{d}\psi\big)\widetilde{D}^\top\big)E_L(x,\cdot), u\Big\rangle_{\mathcal{E}}\,\mathrm{d}\mathcal{L}^n_g \\
&= \int_\Omega \Big\langle\big(I \otimes (-\mathrm{i})\,\mathrm{Sym}\big(D; \mathrm{d}\psi\big)^\top\widetilde{D}^\top\big)E_L(x,\cdot), u\Big\rangle_{\mathcal{E}}\,\mathrm{d}\mathcal{L}^n_g \\
&= \int_\Omega \Big\langle\big(I \otimes \widetilde{D}^\top\big)E_L(x,\cdot), (-\mathrm{i})\,\mathrm{Sym}\big(D; \mathrm{d}\psi\big)u\Big\rangle_{\mathcal{E}}\,\mathrm{d}\mathcal{L}^n_g \\
&= \int_\Omega \big\langle\big(I \otimes \widetilde{D}^\top\big)E_L(x,\cdot), D(\psi u)\big\rangle_{\mathcal{E}}\,\mathrm{d}\mathcal{L}^n_g \\
&\quad - \int_\Omega \big\langle\big(I \otimes \widetilde{D}^\top\big)E_L(x,\cdot), \psi(Du)\big\rangle_{\mathcal{E}}\,\mathrm{d}\mathcal{L}^n_g \\
&= (\psi u)(x) - \int_\Omega \psi\big\langle\big(I \otimes W^\top\big)E_L(x,\cdot), u\big\rangle_{\mathcal{E}}\,\mathrm{d}\mathcal{L}^n_g \\
&\quad - \int_\Omega \psi\big\langle\big(I \otimes \widetilde{D}^\top\big)E_L(x,\cdot), Du\big\rangle_{\mathcal{E}}\,\mathrm{d}\mathcal{L}^n_g,
\end{aligned}
\tag{2.8.340}
$$

where we have also used (2.8.222) and the fact that $E_L(x, y)$ is the Schwartz kernel of L^{-1}. Likewise, from [112, (1.7.17), (1.7.20)], integration by parts, (2.8.211), and the fact that $Lu = 0$ in the sense of distributions in Ω, we obtain

$$
\begin{aligned}
\mathrm{II} &= \int_\Omega \Big\langle E_L(x,\cdot), (-\mathrm{i})\mathrm{Sym}\big(\widetilde{D}; \mathrm{d}\psi\big) Du \Big\rangle_{\mathcal{E}} \,\mathrm{d}\mathcal{L}^n_g \\
&= -\int_\Omega \Big\langle \big(I \otimes (-\mathrm{i})\mathrm{Sym}(\widetilde{D}^\top; \mathrm{d}\psi)\big) E_L(x,\cdot), Du \Big\rangle_{\mathcal{E}} \,\mathrm{d}\mathcal{L}^n_g \\
&= -\int_\Omega \Big\langle \big(I \otimes \widetilde{D}^\top\big)\big(\psi E_L(x,\cdot)\big), Du \Big\rangle_{\mathcal{E}} \,\mathrm{d}\mathcal{L}^n_g \\
&\quad + \int_\Omega \psi \Big\langle \big(I \otimes \widetilde{D}^\top\big) E_L(x,\cdot), Du \Big\rangle_{\mathcal{E}} \,\mathrm{d}\mathcal{L}^n_g \\
&= \int_\Omega \psi \langle E_L(x,\cdot), Wu \rangle_{\mathcal{E}} \,\mathrm{d}\mathcal{L}^n_g + \int_\Omega \psi \langle (I \otimes \widetilde{D}^\top) E_L(x,\cdot), Du \rangle_{\mathcal{E}} \,\mathrm{d}\mathcal{L}^n_g \\
&= \int_\Omega \psi \langle (I \otimes W^\top) E_L(x,\cdot), u \rangle_{\mathcal{E}} \,\mathrm{d}\mathcal{L}^n_g \\
&\quad + \int_\Omega \psi \langle (I \otimes \widetilde{D}^\top) E_L(x,\cdot), Du \rangle_{\mathcal{E}} \,\mathrm{d}\mathcal{L}^n_g .
\end{aligned}
\tag{2.8.341}
$$

Altogether, (2.8.339)-(2.8.341) imply

$$
{}_{\mathcal{D}'(\Omega)}\big(\mathrm{div}_g \vec{F}, \psi\big)_{\mathcal{D}(\Omega)} = (\psi u)(x) = {}_{\mathcal{D}'(\Omega)}\big(u(x)\delta_x, \psi\big)_{\mathcal{D}(\Omega)},
\tag{2.8.342}
$$

where δ_x denotes the Dirac distribution with mass at x. Hence,

$$
\mathrm{div}_g \vec{F} = u(x)\delta_x \in \mathcal{V} \otimes \mathscr{E}'(\Omega).
\tag{2.8.343}
$$

In addition, it follows from (2.8.337) and the third line in (2.8.334) that the nontangential trace $\big(\vec{F}\big|^{\kappa-\mathrm{n.t.}}_{\partial\Omega}\big)(y)$ exists at σ_g-a.e. point $y \in \partial\Omega$ and

$$
\begin{aligned}
&{}_{T_y^*M}\Big(\nu_g(y), \big(\vec{F}\big|^{\kappa-\mathrm{n.t.}}_{\partial\Omega}\big)(y)\Big)_{T_yM} \\
&\quad = \Big\langle \big(I_x \otimes (-\mathrm{i})\,\mathrm{Sym}\big(D^\top; \nu_g(y)\big)\widetilde{D}_y^\top\big) E_L(x,y), \big(u\big|^{\kappa-\mathrm{n.t.}}_{\partial\Omega}\big)(y) \Big\rangle_{\mathcal{E}_y} \\
&\qquad - \Big\langle E_L(x,y), (-\mathrm{i})\mathrm{Sym}\big(\widetilde{D}; \nu_g(y)\big)\Big((Du)\big|^{\kappa-\mathrm{n.t.}}_{\partial\Omega}\Big)(y) \Big\rangle_{\mathcal{E}_y} .
\end{aligned}
\tag{2.8.344}
$$

Also, if $\mathcal{N}^\rho_\kappa$ denotes the nontangential maximal operator truncated at some height $\rho \in \big(0, \mathrm{dist}(x, \partial\Omega)\big)$, then from (2.8.337), (2.8.334), and (2.8.220) we see that

$$
\mathcal{N}^\rho_\kappa \vec{F} \leq C\big(\mathcal{N}_\kappa u + \mathcal{N}_\kappa(Du)\big) \text{ pointwise on } \partial\Omega.
\tag{2.8.345}
$$

In light of [112, (8.2.28)] and the second line in (2.8.334), this gives the membership $\mathcal{N}^\rho_\kappa \vec{F} \in L^1(\partial\Omega, \sigma_g)$. At this point the version of [112, Theorem 1.11.3] mentioned in [112, Remark 1.11.4] applies and, account of (2.8.343)-(2.8.344),

(2.8.225), (2.8.335), (2.8.101), ultimately yields the integral representation formula (2.8.336). □

In turn, the Green type integral representation formula established in Theorem 2.8.12 has a wealth of useful consequences. Here is one such relevant application, which plays a role in the discussion in the last part of §1.8.

Corollary 2.8.13 *Work under Hypotheses 2.8.5-2.8.6 and strengthen the demand on Ω by assuming that this is actually a* UR *domain. Associate with the differential operator L the principal value singular integral operator $K_L^\#$ as in* (2.8.236)*, and the principal value singular integral operator K_L as in* (2.8.226)*. Also, let S be the boundary-to-boundary single layer potential operator, associated with L much as in* (2.8.102)*. Then the following intertwining formula is true:*

$$K_L S = S K_L^\# \tag{2.8.346}$$

as operators from $L^p(\partial\Omega, \sigma_g) \otimes \mathcal{E}$ into $L_1^p(\partial\Omega, \sigma_g) \otimes \mathcal{E}$, for each $p \in (1, \infty)$.

Proof Given any $f \in L^p(\partial\Omega, \sigma_g) \otimes \mathcal{E}$ with $p \in (1, \infty)$, Theorem 2.8.4 ensures that the function $u := \mathcal{S} f$ satisfies all properties listed in (2.8.334). As such, Theorem 2.8.12 guarantees that integral representation formula (2.8.336) holds for this u. Going nontangentially to the boundary in (2.8.336) then ultimately yields $K_L(Sf) = S(K_L^\# f)$, thanks to (2.8.126), (2.8.233), and (2.8.235). That both $K_L S$ and $S K_L^\#$ map $L^p(\partial\Omega, \sigma_g) \otimes \mathcal{E}$ into $L_1^p(\partial\Omega, \sigma_g) \otimes \mathcal{E}$ is seen from (2.8.128), (2.8.232), and (2.8.239). □

Work in the setting described in Hypotheses 2.8.5-2.8.6 and consider an additional Hermitian vector bundle $\mathcal{F} \longrightarrow \mathcal{M}$. We shall describe a procedure which allows us to decompose a generic first-order differential operator $P : \mathcal{E} \longrightarrow \mathcal{F}$ as the sum of a tangential differential operator on $\partial\Omega$ and a suitable "multiple" of the conormal derivative operator $\partial_{\nu_g}^L$ defined in (2.8.335). The key observation is that the operator

$$\tau_P := P - \mathrm{i}\,\mathrm{Sym}(P; \nu_g)\mathrm{Sym}(L; \nu_g)^{-1}\partial_{\nu_g}^L \tag{2.8.347}$$

is tangential on $\partial\Omega$, in the sense that this may be expressed as a linear combination of tangential differential operators expressed in local coordinates as (see [113, (11.6.23)])

$$\partial_{\tau_{jk}} := (\nu_g)_j \nabla_{\partial_k} - (\nu_g)_k \nabla_{\partial_j}, \qquad 1 \le j, k \le n, \tag{2.8.348}$$

eventually plus a zero-th order term. This is ensured by the fact that $\mathrm{Sym}(\tau_P; \nu_g) = 0$ at σ_g-a.e. point on $\partial\Omega$, which follows readily from definitions:

$$
\begin{aligned}
&\mathrm{Sym}(\tau_P;\nu_g)\\
&\quad= \mathrm{Sym}(P;\nu_g) - i\,\mathrm{Sym}(P;\nu_g)\mathrm{Sym}(L;\nu_g)^{-1}(-\mathrm{i})\mathrm{Sym}(\widetilde{D};\nu_g)\mathrm{Sym}(D;\nu_g)\\
&\quad= \mathrm{Sym}(P;\nu_g) - \mathrm{Sym}(P;\nu_g)\mathrm{Sym}(L;\nu_g)^{-1}\mathrm{Sym}(L;\nu_g)\\
&\quad= \mathrm{Sym}(P;\nu_g) - \mathrm{Sym}(P;\nu_g) = 0. \qquad (2.8.349)
\end{aligned}
$$

In turn, this fits into a more general narrative to the ffect that if a first-order differential operator Q written locally as

$$
Q = \sum_j A_j \nabla_{\partial_j} + B \qquad (2.8.350)
$$

satisfies

$$
\mathrm{Sym}(Q;\nu_g) = 0 \text{ at } \sigma_g\text{-a.e. point on } \partial\Omega, \qquad (2.8.351)
$$

then

$$
\sum_j (\nu_g)_j A_j = 0, \qquad (2.8.352)
$$

and since

$$
\sum_k (\nu_g)_k (\nu_g)^\sharp_k = \sum_{j,k} g^{jk}(\nu_g)_k(\nu_g)_j = \langle \nu_g, \nu_g\rangle = |\nu_g|^2 = 1, \qquad (2.8.353)
$$

we may write

$$
\begin{aligned}
Q &= \sum_j A_j \nabla_{\partial_j} + B - \sum_{j,k} A_j (\nu_g)_j (\nu_g)^\sharp_k \nabla_{\partial_k}\\
&= \sum_{j,k} A_j (\nu_g)_k (\nu_g)^\sharp_k \nabla_{\partial_j} + B - \sum_{j,k} A_j (\nu_g)_j (\nu_g)^\sharp_k \nabla_{\partial_k}\\
&= \sum_{j,k} A_j (\nu_g)^\sharp_k \Big((\nu_g)_k \nabla_{\partial_j} - (\nu_g)_j \nabla_{\partial_k} \Big) + B\\
&= -\sum_{j,k} A_j\,(\nu_g)^\sharp_k\,\partial_{\tau_{jk}} + B, \qquad (2.8.354)
\end{aligned}
$$

which goes to show that the operator Q is tangential. This observation then permits to think of τ_P as a linear and continuous operator

$$
\tau_P : L^p_1(\partial\Omega, \sigma_g) \otimes \mathcal{E} \longrightarrow L^p(\partial\Omega, \sigma_g) \otimes \mathcal{E} \text{ for } p \in [1, \infty] \qquad (2.8.355)
$$

defined as

$$
\tau_P f := -\sum_{j,k} A_j\,(\nu_g)^\sharp_k (\partial_{\tau_{jk}} f) + Bf \text{ for each } f \in L^p_1(\partial\Omega, \sigma_g) \otimes \mathcal{E}, \qquad (2.8.356)
$$

where the homomorphisms A_j's and B are as in (2.8.351) when $Q := \tau_P$ (and where each $\partial_{\tau_{jk}} f$ is considered as in [112, Definition 1.12.7] with $X := \partial_j$ and $Y := \partial_k$). On the other hand, we can also consider the action of τ_P from (1.8.85) on any section

$$\begin{aligned} & u \in W^{1,1}_{\rm loc}(\Omega) \otimes \mathcal{E} \ \text{ with } \ \mathcal{N}_\kappa u,\ \mathcal{N}_\kappa(\nabla^{\mathcal{E}} u) \in L^1(\partial\Omega, \sigma_g), \\ & \text{and such that } u\Big|^{\kappa-\text{n.t.}}_{\partial\Omega},\ (\nabla^{\mathcal{E}} u)\Big|^{\kappa-\text{n.t.}}_{\partial\Omega} \ \text{ exist } \sigma_g\text{-a.e. on } \partial\Omega, \end{aligned} \tag{2.8.357}$$

(where $\kappa > 0$ is a some fixed background aperture parameter) according to

$$\tau_P u := (Pu)\Big|^{\kappa-\text{n.t.}}_{\partial\Omega} - \mathrm{i}\,\mathrm{Sym}(P; \nu_g)\mathrm{Sym}(L; \nu_g)^{-1}\partial^L_{\nu_g} u, \tag{2.8.358}$$

with the conormal derivative understood in the sense of (2.8.335). It is significant to note that, as seen from (2.8.358) and (2.8.335), we have

$$\mathrm{Sym}(\widetilde{D}; \nu_g)\tau_D u = 0 \ \text{ for each } u \text{ as in } (2.8.357). \tag{2.8.359}$$

In regard to the two definitions of τ_P given above we wish to note that:

given any section u as in (2.8.357), it follows that the nontangential boundary trace $u\big|^{\kappa-\text{n.t.}}_{\partial\Omega}$ belongs to the Sobolev space $L^1_1(\partial\Omega, \sigma_g) \otimes \mathcal{E}$ and $\tau_P\big(u\big|^{\kappa-\text{n.t.}}_{\partial\Omega}\big)$ (considered as in (2.8.355)-(2.8.356)) is equal to $\tau_P u$ (considered as in (2.8.358)). (2.8.360)

To justify the claims made above, suppose u is as in (2.8.357). That $u\big|^{\kappa-\text{n.t.}}_{\partial\Omega}$ belongs to the Sobolev space $L^1_1(\partial\Omega, \sigma_g)\otimes\mathcal{E}$ is then a consequence of [112, Proposition 1.12.9]. Furthermore, based on (2.8.355), (2.8.356), and [112, (1.12.102)] we may write

$$\begin{aligned} \tau_P\Big(u\Big|^{\kappa-\text{n.t.}}_{\partial\Omega}\Big) &= -\sum_{j,k} A_j\,(\nu_g)^\#_k \partial_{\tau_{jk}}\Big(u\Big|^{\kappa-\text{n.t.}}_{\partial\Omega}\Big) + B\Big(u\Big|^{\kappa-\text{n.t.}}_{\partial\Omega}\Big) \\ &= -\sum_{j,k} A_j\,(\nu_g)^\#_k \Big\{(\nu_g)_j(\nabla_{\partial_k} u)\Big|^{\kappa-\text{n.t.}}_{\partial\Omega} - (\nu_g)_k(\nabla_{\partial_j} u)\Big|^{\kappa-\text{n.t.}}_{\partial\Omega}\Big\} + B\Big(u\Big|^{\kappa-\text{n.t.}}_{\partial\Omega}\Big) \\ &= \sum_j A_j\,(\nabla_{\partial_j} u)\Big|^{\kappa-\text{n.t.}}_{\partial\Omega} + B\Big(u\Big|^{\kappa-\text{n.t.}}_{\partial\Omega}\Big) = \tau_P u, \end{aligned} \tag{2.8.361}$$

thanks to (2.8.352), (2.8.353), and (2.8.358). This establishes (2.8.360).

Additional basic properties of τ_P and other related tangential differential operators are contained in the lemma below.

Lemma 2.8.14 *Assume Hypotheses 2.8.5-2.8.6. Then, having fixed an aperture parameter $\kappa > 0$, given any section*

$$\begin{aligned} & u \in W^{1,1}_{\rm loc}(\Omega) \otimes \mathcal{E} \ \textit{ with } \ \mathcal{N}_\kappa u,\ \mathcal{N}_\kappa(\nabla^{\mathcal{E}} u) \in L^1(\partial\Omega, \sigma_g), \\ & \textit{and such that } u\Big|^{\kappa-\text{n.t.}}_{\partial\Omega},\ (\nabla^{\mathcal{E}} u)\Big|^{\kappa-\text{n.t.}}_{\partial\Omega} \ \textit{ exist } \sigma_g\textit{-a.e. on } \partial\Omega, \end{aligned} \tag{2.8.362}$$

it follows that the nontangential boundary trace $u\big|_{\partial\Omega}^{\kappa-\mathrm{n.t.}}$ *belongs to the Sobolev space* $L^1_1(\partial\Omega,\sigma_g)\otimes\mathcal{E}$ *and for any two vector fields* $X, Y \in TM$ *one has*

$$\nu_g(X)\tau_{\nabla_Y}u - \nu_g(Y)\tau_{\nabla_X}u = \partial_{\tau_{XY}}\Big(u\big|_{\partial\Omega}^{\kappa-\mathrm{n.t.}}\Big) \tag{2.8.363}$$

where τ_{∇_X}, τ_{∇_Y} *are considered as in* (2.8.358), *and where* $\partial_{\tau_{XY}}$ *acts on the space* $L^1_1(\partial\Omega,\sigma_g)\otimes\mathcal{E}$ *as in* [112, Definition 1.12.7]. *In particular, corresponding to* $X := \partial_j$ *and* $Y := \partial_k$ *for some* $j, k \in \{1,\dots,n\}$, *one obtains*

$$(\nu_g)_j\tau_{\nabla_{\partial_k}}u - (\nu_g)_k\tau_{\nabla_{\partial_j}}u = \partial_{\tau_{jk}}\Big(u\big|_{\partial\Omega}^{\kappa-\mathrm{n.t.}}\Big) \tag{2.8.364}$$

for any section u as in (2.8.362).

Henceforth strengthen the demand on Ω *by assuming that this is actually a* UR *domain. Then any two vector fields* $X, Y \in TM$ *one has*

$$\begin{gathered}\nu_g(X)\tau_{\nabla_Y}f - \nu_g(Y)\tau_{\nabla_X}f = \partial_{\tau_{XY}}f\\ \textit{for each } f \in L^p_1(\partial\Omega,\sigma_g)\otimes\mathcal{E} \textit{ with } p\in(1,\infty),\end{gathered} \tag{2.8.365}$$

where the actions of τ_{∇_X}, τ_{∇_Y} *on* f *are considered as in* (2.8.355)-(2.8.356), *and where* $\partial_{\tau_{XY}}f$ *is considered as in* [112, Definition 1.12.7]. *As a consequence, corresponding to* $X := \partial_j$ *and* $Y := \partial_k$ *for some* $j, k \in \{1,\dots,n\}$, *one has*

$$\begin{gathered}(\nu_g)_j\tau_{\nabla_{\partial_k}}f - (\nu_g)_k\tau_{\nabla_{\partial_j}}f = \partial_{\tau_{jk}}f\\ \textit{for each } f \in L^p_1(\partial\Omega,\sigma_g)\otimes\mathcal{E} \textit{ with } p\in(1,\infty).\end{gathered} \tag{2.8.366}$$

Furthermore, the tangential differential operator τ_D *defined as in* (2.8.355)-(2.8.356) *with* $P := D$ *satisfies*

$$\mathrm{Sym}(\widetilde{D};\nu_g)\tau_D f = 0 \textit{ for each } f \in L^p_1(\partial\Omega,\sigma_g)\otimes\mathcal{E} \textit{ with } p\in(1,\infty). \tag{2.8.367}$$

Also, for any vector field $X \in TM$ *one has*

$$\begin{gathered}\mathrm{i}\,\mathrm{Sym}(L;\nu_g)^{-1}\mathrm{Sym}(\widetilde{D};\nu_g)\partial_\tau^{(\nabla_X;D^\top)}f = \tau_{\nabla_X}f\\ \textit{for each } f \in L^p_1(\partial\Omega,\sigma_g)\otimes\mathcal{E} \textit{ with } p\in(1,\infty),\end{gathered} \tag{2.8.368}$$

where $\partial_\tau^{(\nabla_X;D^\top)}$ *is defined as in* (2.8.316) *with* $P := D^\top$. *Finally, for any two vector fields* $X, Y \in TM$ *one has*

$$\begin{gathered}\nu_g(X)\partial_\tau^{(\nabla_Y;D^\top)}f - \nu_g(Y)\partial_\tau^{(\nabla_X;D^\top)}f = (-\mathrm{i})\mathrm{Sym}(D;\nu_g)\partial_{\tau_{XY}}f\\ \textit{for each } f \in L^p_1(\partial\Omega,\sigma_g)\otimes\mathcal{E} \textit{ with } p\in(1,\infty)\end{gathered} \tag{2.8.369}$$

so, in particular,

$$(\nu_g)_j \partial_\tau^{(\nabla_{\partial_k};D^\top)} f - (\nu_g)_k \partial_\tau^{(\nabla_{\partial_j};D^\top)} f = (-\mathrm{i})\mathrm{Sym}(D;\nu_g)\partial_{\tau_{jk}} f$$
$$\textit{for each } f \in L_1^p(\partial\Omega,\sigma_g)\otimes\mathcal{E} \textit{ with } p\in(1,\infty). \tag{2.8.370}$$

Proof Select an arbitrary section u satisfying the properties listed in (2.8.362). Then [112, (1.12.100)] ensures that $u\big|_{\partial\Omega}^{\kappa-\mathrm{n.t.}}$ belongs to $L_1^1(\partial\Omega,\sigma_g)\otimes\mathcal{E}$, while (2.8.358) and [112, (1.12.101)] permit us to compute

$$\begin{aligned}
&\nu_g(X)\tau_{\nabla_Y}u - \nu_g(Y)\tau_{\nabla_X}u \\
&\quad= \nu_g(X)(\nabla_Y u)\big|_{\partial\Omega}^{\kappa-\mathrm{n.t.}} - \nu_g(X)\,\mathrm{i}\,\mathrm{Sym}(\nabla_Y;\nu_g)\mathrm{Sym}(L;\nu_g)^{-1}\partial_{\nu_g}^L u \\
&\qquad - \nu_g(Y)(\nabla_X u)\big|_{\partial\Omega}^{\kappa-\mathrm{n.t.}} + \nu_g(Y)\,\mathrm{i}\,\mathrm{Sym}(\nabla_X;\nu_g)\mathrm{Sym}(L;\nu_g)^{-1}\partial_{\nu_g}^L u \\
&\quad= \nu_g(X)(\nabla_Y u)\big|_{\partial\Omega}^{\kappa-\mathrm{n.t.}} + \nu_g(X)\nu_g(Y)\mathrm{Sym}(L;\nu_g)^{-1}\partial_{\nu_g}^L u \\
&\qquad - \nu_g(Y)(\nabla_X u)\big|_{\partial\Omega}^{\kappa-\mathrm{n.t.}} - \nu_g(Y)\nu_g(X)\mathrm{Sym}(L;\nu_g)^{-1}\partial_{\nu_g}^L u \\
&\quad= \nu_g(X)(\nabla_Y u)\big|_{\partial\Omega}^{\kappa-\mathrm{n.t.}} - \nu_g(Y)(\nabla_X u)\big|_{\partial\Omega}^{\kappa-\mathrm{n.t.}} \\
&\quad= \partial_{\tau_{XY}}\Big(u\big|_{\partial\Omega}^{\kappa-\mathrm{n.t.}}\Big).
\end{aligned} \tag{2.8.371}$$

This proves (2.8.363).

For the remainder of the proof work under the stronger assumption that Ω is actually a UR domain, and pick an integrability exponent $p\in(1,\infty)$. In such a scenario, (2.8.365) becomes a consequence of the decomposition in (2.8.322), (2.8.363), and (2.8.360). We next observe that whenever $f = u\big|_{\partial\Omega}^{\kappa-\mathrm{n.t.}}$ for some

$$\begin{aligned}
&u\in W^{1,1}_{\mathrm{loc}}(\Omega)\otimes\mathcal{E} \text{ with } \mathcal{N}_\kappa u,\ \mathcal{N}_\kappa(\nabla^{\mathcal{E}}u)\in L^p(\partial\Omega,\sigma_g), \\
&\text{and such that } u\big|_{\partial\Omega}^{\kappa-\mathrm{n.t.}},\ (\nabla^{\mathcal{E}}u)\big|_{\partial\Omega}^{\kappa-\mathrm{n.t.}} \text{ exist } \sigma_g\text{-a.e. on } \partial\Omega,
\end{aligned} \tag{2.8.372}$$

it follows that

$$\mathrm{Sym}(\widetilde{D};\nu_g)\tau_D f = \mathrm{Sym}(\widetilde{D};\nu_g)\tau_D\Big(u\big|_{\partial\Omega}^{\kappa-\mathrm{n.t.}}\Big) = \mathrm{Sym}(\widetilde{D};\nu_g)\tau_D u = 0, \tag{2.8.373}$$

thanks to (2.8.360) and (2.8.359). The claim made in (2.8.367) then becomes a consequence of this special case and the decomposition (2.8.322).

As regards the claim made in (2.8.368), we once again first treat the case when the section $f\in L_1^p(\partial\Omega,\sigma_g)\otimes\mathcal{E}$ is of the form $f = u\big|_{\partial\Omega}^{\kappa-\mathrm{n.t.}}$ for some u as in (2.8.372). In such a scenario we may employ (2.8.324), the fact that ∇_X has a scalar principal symbol, and (2.8.358) with $P := \nabla_X$ to write

$$
\begin{aligned}
&\mathrm{i}\,\mathrm{Sym}(L;\nu_g)^{-1}\mathrm{Sym}(\widetilde{D};\nu_g)\partial_\tau^{(\nabla_X;D^\top)}f \\
&\qquad= -\mathrm{Sym}(L;\nu_g)^{-1}\mathrm{Sym}(\widetilde{D};\nu_g)\mathrm{Sym}(\nabla_X;\nu_g)(Du)\Big|_{\partial\Omega}^{\kappa-\mathrm{n.t.}} \\
&\qquad\quad+ \mathrm{Sym}(L;\nu_g)^{-1}\mathrm{Sym}(\widetilde{D};\nu_g)\mathrm{Sym}(D;\nu_g)(\nabla_X u)\Big|_{\partial\Omega}^{\kappa-\mathrm{n.t.}} \\
&\qquad= -\mathrm{Sym}(\nabla_X;\nu_g)\mathrm{Sym}(L;\nu_g)^{-1}\mathrm{Sym}(\widetilde{D};\nu_g)(Du)\Big|_{\partial\Omega}^{\kappa-\mathrm{n.t.}} \\
&\qquad\quad+ \mathrm{Sym}(L;\nu_g)^{-1}\mathrm{Sym}(\widetilde{D}D;\nu_g)(\nabla_X u)\Big|_{\partial\Omega}^{\kappa-\mathrm{n.t.}} \\
&\qquad= -i\,\mathrm{Sym}(\nabla_X;\nu_g)\mathrm{Sym}(L;\nu_g)^{-1}\partial_{\nu_g}^L u + (\nabla_X u)\Big|_{\partial\Omega}^{\kappa-\mathrm{n.t.}} \\
&\qquad= \tau_{\nabla_X} f,
\end{aligned}
\tag{2.8.374}
$$

as wanted. The general case of (2.8.368) then follows from this and the decomposition (2.8.322).

In the proof (2.8.369) we shall adopt the same strategy of first checking the case when the section $f \in L_1^p(\partial\Omega,\sigma_g)\otimes\mathcal{E}$ has the particular structure $f = u\big|_{\partial\Omega}^{\kappa-\mathrm{n.t.}}$ where u as in (2.8.372). In this special situation we make use of (2.8.324), (2.8.358) with $P := \nabla_X$, the formulas for the principal symbols of ∇_X and ∇_Y, as well as [112, (1.12.101)] to compute

$$
\begin{aligned}
&\nu_g(X)\partial_\tau^{(\nabla_Y;D^\top)}f - \nu_g(Y)\partial_\tau^{(\nabla_X;D^\top)}f \\
&\qquad= \nu_g(X)\partial_\tau^{(\nabla_Y;D^\top)}\Big(u\Big|_{\partial\Omega}^{\kappa-\mathrm{n.t.}}\Big) - \nu_g(Y)\partial_\tau^{(\nabla_X;D^\top)}\Big(u\Big|_{\partial\Omega}^{\kappa-\mathrm{n.t.}}\Big) \\
&\qquad= -\nu_g(X)(-\mathrm{i})\mathrm{Sym}(\nabla_Y;\nu_g)(Du)\Big|_{\partial\Omega}^{\kappa-\mathrm{n.t.}} \\
&\qquad\quad+ \nu_g(X)(-\mathrm{i})\mathrm{Sym}(D;\nu_g)(\nabla_Y u)\Big|_{\partial\Omega}^{\kappa-\mathrm{n.t.}} \\
&\qquad\quad+ \nu_g(Y)(-\mathrm{i})\mathrm{Sym}(\nabla_X;\nu_g)(Du)\Big|_{\partial\Omega}^{\kappa-\mathrm{n.t.}} \\
&\qquad\quad- \nu_g(Y)(-\mathrm{i})\mathrm{Sym}(D;\nu_g)(\nabla_X u)\Big|_{\partial\Omega}^{\kappa-\mathrm{n.t.}} \\
&\qquad= (-\mathrm{i})\mathrm{Sym}(D;\nu_g)\Big\{\nu_g(X)(\nabla_Y u)\Big|_{\partial\Omega}^{\kappa-\mathrm{n.t.}} - \nu_g(Y)(\nabla_X u)\Big|_{\partial\Omega}^{\kappa-\mathrm{n.t.}}\Big\} \\
&\qquad= (-\mathrm{i})\mathrm{Sym}(D;\nu_g)\partial_{\tau_{XY}}\Big(u\Big|_{\partial\Omega}^{\kappa-\mathrm{n.t.}}\Big) \\
&\qquad= (-\mathrm{i})\mathrm{Sym}(D;\nu_g)\partial_{\tau_{XY}} f,
\end{aligned}
\tag{2.8.375}
$$

as desired. Having established this, then (2.8.369) follows in full generality by availing ourselves of the decomposition in (2.8.322). □

We are now in a position to give global, coordinate-free formulas for nontangential traces of generic first-order derivatives of the boundary-to-domain double layer introduced in Definition 2.8.7.

Theorem 2.8.15 *Work in the context of Theorem 2.8.10. Consider an additional Hermitian vector bundle $\mathcal{F} \to \mathcal{M}$ and suppose $P : \mathcal{E} \to \mathcal{F}$ is a generic first-order differential operator. Then for each $f \in L^p_1(\partial\Omega, \sigma_g) \otimes \mathcal{E}$ with $1 < p < \infty$ one has*

$$(P\mathcal{D}_L f)\Big|_{\partial\Omega}^{\kappa-\mathrm{n.t.}} = \tau_P\big[(\tfrac{1}{2}I + K_L)f\big] + \mathrm{i}\,\mathrm{Sym}(P;\nu_g)\mathrm{Sym}(L;\nu_g)^{-1}\partial^L_{\nu_g}(\mathcal{D}_L f) \quad (2.8.376)$$

at σ_g-a.e. point on $\partial\Omega$, where the tangential derivative operator τ_P is as in (2.8.355)-(2.8.356). *In particular, using notation introduced in* (2.8.317)-(2.8.318), *from* (2.8.376) *and Theorem 2.8.11 one obtains*

$$(P\mathcal{D}^+_L f)\Big|_{\partial\Omega}^{\kappa-\mathrm{n.t.}} - (P\mathcal{D}^-_L f)\Big|_{\partial\Omega}^{\kappa-\mathrm{n.t.}} = \tau_P f. \quad (2.8.377)$$

Furthermore, given any vector field $X \in T\mathcal{M}$ and any $f \in L^p_1(\partial\Omega, \sigma_g) \otimes \mathcal{E}$ with $1 < p < \infty$, at σ_g-a.e. point $x \in \partial\Omega$ one has

$$\begin{aligned}
&\Big((\nabla^{\mathcal{E}}_X \mathcal{D}_L f)\Big|_{\partial\Omega}^{\kappa-\mathrm{n.t.}}\Big)(x)\\
&= \frac{1}{2}(\tau_{\nabla_X} f)(x)\\
&\quad - \mathrm{P.V.}\int_{\partial\Omega} \Big\langle (I_x \otimes \widetilde{D}^\top_y)E_L(x,y), (\partial_\tau^{(\nabla_X;D^\top)} f)(y)\Big\rangle_{\mathcal{G}_y} \, d\sigma_g(y)\\
&\quad + \mathrm{P.V.}\int_{\partial\Omega} \Big\langle (I_x \otimes \widetilde{D}^\top_y)E_L(x,y), \mathrm{i}\,\mathrm{Sym}\big([\nabla_X, D];\nu_g(y)\big)f(y)\Big\rangle_{\mathcal{G}_y} \, d\sigma_g(y)\\
&\quad + \mathrm{P.V.}\int_{\partial\Omega} \big\langle E_L(x,y), \nu_g(X)(y)W_y f(y)\big\rangle_{\mathcal{E}_y} \, d\sigma_g(y)\\
&\quad + \mathrm{P.V.}\int_{\partial\Omega} \big\langle \omega_X(x,y), f(y)\big\rangle_{\mathcal{E}_y} \, d\sigma_g(y),
\end{aligned} \quad (2.8.378)$$

where τ_{∇_X} is considered as in (2.8.355)-(2.8.356), *the tangential derivative operator $\partial_\tau^{(\nabla_X;D^\top)}$ is defined as in* (2.8.316), *and where the kernel $\omega_X(x,y)$ has been introduced in* (2.8.252).

Proof Pick $f \in L^p_1(\partial\Omega, \sigma_g) \otimes \mathcal{E}$ with $1 < p < \infty$, arbitrary. From Theorem 2.8.10, (2.8.231), (2.8.233) we see that we may employ (2.8.358) with $u := \mathcal{D}_L f$. In concert with (2.8.360) this permits us to write

$$\begin{aligned}(P\mathcal{D}_L f)\Big|_{\partial\Omega}^{\kappa-\mathrm{n.t.}} &= \tau_P(\mathcal{D}_L f) + (-\mathrm{i})\mathrm{Sym}(P;\nu_g)\mathrm{Sym}(-L;\nu_g)^{-1}\partial^L_{\nu_g}(\mathcal{D}_L f)\\ &= \tau_P\Big[(\mathcal{D}_L f)\Big|_{\partial\Omega}^{\kappa-\mathrm{n.t.}}\Big] + (-\mathrm{i})\mathrm{Sym}(P;\nu_g)\mathrm{Sym}(-L;\nu_g)^{-1}\partial^L_{\nu_g}(\mathcal{D}_L f)\\ &= \tau_P\big[(\tfrac{1}{2}I + K_L)f\big] + (-\mathrm{i})\mathrm{Sym}(P;\nu_g)\mathrm{Sym}(-L;\nu_g)^{-1}\partial^L_{\nu_g}(\mathcal{D}_L f)\end{aligned} \tag{2.8.379}$$

as wanted. Also, (2.8.378) is a direct consequence of (2.8.314), (2.8.368), and (2.8.316). □

Parenthetically, we wish to note that, thanks to (2.8.367), the jump-formula recorded in (2.8.377) written for $P := D$ is consistent with the lack of jump for the conormal derivative of the double layer (see (2.8.330)).

We continue by presenting an important commutator identity to the effect that one can pass tangential derivatives through the principal value double layer all the way to the section on which this acts (itself assumed to be in a boundary Sobolev space), at the expense of operators which may be regarded (at least in a favorable geometric setting) as being residual.

Theorem 2.8.16 *Retain the setting of Theorem 2.8.10 and consider two vector fields $X, Y \in T\mathcal{M}$. Then, for each $p \in (1,\infty)$,*

$$\begin{gathered}\textit{the commutator } \big[\partial_{\tau_{XY}}, K_L\big] : L^p_1(\partial\Omega,\sigma_g)\otimes\mathcal{E} \longrightarrow L^p(\partial\Omega,\sigma_g)\otimes\mathcal{E}\\ \textit{may be written as a linear combination of operators of the form}\\ [M_{\nu_g}, T]\circ\partial_\tau \textit{ and a bounded map } B : L^p(\partial\Omega,\sigma_g)\otimes\mathcal{E}\to L^p(\partial\Omega,\sigma_g)\otimes\mathcal{E},\end{gathered} \tag{2.8.380}$$

where $[M_{\nu_g}, T]$ is the commutator between the pointwise multiplication with either $\nu_g(X)$ or $\nu_g(Y)$ and a Calderón-Zygmund singular integral operator T on $\partial\Omega$ (whose kernel involves at most first-order derivatives of $E_L(x,y)$), and where the map $\partial_\tau : L^p_1(\partial\Omega,\sigma_g)\otimes\mathcal{E}\to L^p(\partial\Omega,\sigma_g)\otimes\mathcal{G}$ is bounded.

Proof Fix an arbitrary $f \in L^p_1(\partial\Omega,\sigma_g)\otimes\mathcal{E}$. Based on [112, (1.12.101) in Proposition 1.12.9] (whose applicability in the present setting is guaranteed by Theorem 2.8.10, (2.8.231), and (2.8.233)) and the jump-formula (2.8.233) we may write

$$\begin{aligned}\partial_{\tau_{XY}}(K_L f) &= \partial_{\tau_{XY}}\big[(\tfrac{1}{2}I + K_L)f\big] - \tfrac{1}{2}\partial_{\tau_{XY}} f = \partial_{\tau_{XY}}\Big(\mathcal{D}_L f\Big|_{\partial\Omega}^{\kappa-\mathrm{n.t.}}\Big) - \tfrac{1}{2}\partial_{\tau_{XY}} f\\ &= \nu_g(X)\big(\nabla_Y\mathcal{D}_L f\big)\Big|_{\partial\Omega}^{\kappa-\mathrm{n.t.}} - \nu_g(Y)\big(\nabla_X\mathcal{D}_L f\big)\Big|_{\partial\Omega}^{\kappa-\mathrm{n.t.}} - \tfrac{1}{2}\partial_{\tau_{XY}} f.\end{aligned} \tag{2.8.381}$$

At this stage, we shall employ (2.8.378) to compute, at σ_g-a.e. point $x \in \partial\Omega$,

$$
\begin{aligned}
&\nu_g(X)(x)\big(\nabla_Y \mathcal{D}_L f\big)\Big|_{\partial\Omega}^{\kappa-\text{n.t.}}(x) - \nu_g(Y)(x)\big(\nabla_X \mathcal{D}_L f\big)\Big|_{\partial\Omega}^{\kappa-\text{n.t.}}(x) \\
&\quad = \frac{1}{2}\Big\{\nu_g(X)(x)\big(\tau_{\nabla_Y} f\big)(x) - \nu_g(Y)(x)\big(\tau_{\nabla_X} f\big)(x)\Big\} \\
&\qquad - \nu_g(X)(x)\,\text{P.V.}\int_{\partial\Omega}\Big\langle\big(I_x\otimes \widetilde{D}_y^{\top}\big)E_L(x,y), \big(\partial_\tau^{(\nabla_Y;D^{\top})} f\big)(y)\Big\rangle_{\mathcal{G}_y}\, d\sigma_g(y) \\
&\qquad + \nu_g(Y)(x)\,\text{P.V.}\int_{\partial\Omega}\Big\langle\big(I_x\otimes \widetilde{D}_y^{\top}\big)E_L(x,y), \big(\partial_\tau^{(\nabla_X;D^{\top})} f\big)(y)\Big\rangle_{\mathcal{G}_y}\, d\sigma_g(y) \\
&\qquad + (Bf)(x),
\end{aligned}
\tag{2.8.382}
$$

where B is a linear and bounded operator from $L^p(\partial\Omega, \sigma_g)\otimes\mathcal{E}$ into itself, originating from the last three lines in (2.8.378). In relation to this formula, we make two key observations. First, thanks to (2.8.365), the expression in the first line in the right-hand side of (2.8.382) is simply $\frac{1}{2}\big(\partial_{\tau_{XY}} f\big)(x)$ (which ultimately is going to cancel the last term in (2.8.381)). Second, up to operators of the form $[M_{\nu_g}, T]\circ\partial_\tau$ where $[M_{\nu_g}, T]$ is the commutator between the pointwise multiplication with $\nu_g(X)$ or $\nu_g(Y)$ and the Calderón-Zygmund singular integral operator T on $\partial\Omega$, acting on each $\phi\in L^p(\partial\Omega,\sigma_g)\otimes\mathcal{G}$ according to

$$
(T\phi)(x) := -\text{P.V.}\int_{\partial\Omega}\big\langle\big(I_x\otimes\widetilde{D}_y^{\top}\big)E_L(x,y), \phi(y)\big\rangle_{\mathcal{G}_y}\, d\sigma_g(y), \tag{2.8.383}
$$

(note that the integral kernel of T involves at most first-order derivatives on $E_L(x,y)$) and ∂_τ is either $\partial_\tau^{(\nabla_X;D^{\top})}$ or $\partial_\tau^{(\nabla_Y;D^{\top})}$, the expressions in the second and third lines in the right-hand side of (2.8.382) combine to:

$$
\begin{aligned}
&\text{P.V.}\int_{\partial\Omega}\Big\langle\big(I_x\otimes\widetilde{D}_y^{\top}\big)E_L(x,y), \nu_g(Y)(y)\big(\partial_\tau^{(\nabla_X;D^{\top})} f\big)(y) \\
&\qquad\qquad - \nu_g(X)(y)\big(\partial_\tau^{(\nabla_Y;D^{\top})} f\big)(y)\Big\rangle_{\mathcal{G}_y}\, d\sigma_g(y) \\
&\quad = \text{P.V.}\int_{\partial\Omega}\Big\langle\big(I_x\otimes\widetilde{D}_y^{\top}\big)E_L(x,y), \mathrm{i}\,\mathrm{Sym}(D;\nu_g(y))\big(\partial_{\tau_{XY}} f\big)(y)\Big\rangle_{\mathcal{G}_y}\, d\sigma_g(y) \\
&\quad = \text{P.V.}\int_{\partial\Omega}\Big\langle\big(I_x\otimes(-\mathrm{i})\,\mathrm{Sym}(D^{\top};\nu_g(y))\widetilde{D}_y^{\top}\big)E_L(x,y),\ f(y)\Big\rangle_{\mathcal{E}_y}\, d\sigma_g(y) \\
&\quad = \Big(K_L\big(\partial_{\tau_{XY}} f\big)\Big)(x).
\end{aligned}
\tag{2.8.384}
$$

Above, the second equality comes from (2.8.369) and the last equality is just the definition of K_L from (2.8.226).

Altogether, this argument establishes the claim made in (2.8.380). □

We conclude with the following remark, pointing to further extensions.

Remark 2.8.17 *In view of the fact that integral operators with variable coefficient kernels are allowed in the formulation of Theorem 2.6.1, the same type localization argument as in the proofs of Theorem 2.8.2 and Theorem 2.8.4 shows that similar results to those described in Theorem 2.8.2, Theorem 2.8.4, Theorem 2.8.10, Theorem 2.8.11, Theorem 2.8.15, and Theorem 2.8.16 are also valid in the context of Morrey and block spaces, naturally defined on the boundaries of* UR *domains in Riemannian manifolds.*

2.9 Some Applications to Singular Integrals Associated with Elliptic Differential Operators

The powerful Calderón-Zygmund theory developed as the collection of results centered around Theorem 2.3.2, Theorem 2.4.1, and Theorem 2.5.1 (as well as their proofs and various extensions) has a multitude of applications of special interest. In this section we shall elaborate on some of these applications, with the understanding that much more will follow in Volumes III-V.

To set the stage for our first application, presented in Theorem 2.9.1 below, consider next a second-order, homogeneous, constant (complex) coefficient, elliptic differential operator

$$L = \sum_{r,s=1}^{n} a_{rs}\partial_r\partial_s \tag{2.9.1}$$

in $\mathbb{R}^n$, where $n \in \mathbb{N}$ with $n \geq 2$. The aforementioned ellipticity condition means that the coefficient matrix $A := (a_{rs})_{1\leq r,s\leq n} \in \mathbb{C}^{n\times n}$ has the property that there exists some constant $c > 0$ such that

$$\mathrm{Re}\Big\{\sum_{r,s=1}^{n} a_{rs}\xi_r\xi_s\Big\} \geq c|\xi|^2, \qquad \forall\,\xi = (\xi_1,\dots,\xi_n) \in \mathbb{R}^n. \tag{2.9.2}$$

Then the locally integrable function E defined in (1.4.171) is a fundamental solution for the differential operator L in $\mathbb{R}^n$.

Theorem 2.9.1 below provides natural examples of vector fields satisfying the hypotheses of [112, Theorem 1.2.1]. Theorem 2.9.1 is also very much relevant in the treatment of the Neumann Problem for elliptic differential operators in rough domains, via boundary layer potentials, as done in [28] in the case of the Laplace operator in Lipschitz domains.

Theorem 2.9.1 *Suppose $\Omega \subseteq \mathbb{R}^n$ (where $n \in \mathbb{N}$ with $n \geq 2$) is a nonempty open set with the property that $\partial\Omega$ is a* UR *set; in particular, Ω is a set of locally finite perimeter. Abbreviate $\sigma := \mathcal{H}^{n-1}\lfloor\partial\Omega$, and denote by $\nu = (\nu_1,\dots,\nu_n)$ the geometric measure theoretic outward unit normal to Ω. Also, consider a second-order, homogeneous, constant* (*complex*) *coefficient, elliptic differential operator L as in* (2.9.1), *and recall its fundamental solution E defined in* (1.4.171).

In relation to the differential operator L and the domain Ω, introduce the vector field

$$\vec{F}(x) := \left(\int_{\partial\Omega} \sum_{s=1}^{n} a_{rs}(\partial_s E)(x-y) f(y)\, \mathrm{d}\sigma(y) \right)_{1\leq r\leq n} \quad \textit{for } x \in \Omega, \tag{2.9.3}$$

where $(a_{rs})_{1\leq r,s\leq n} \in \mathbb{C}^{n\times n}$ is the coefficient matrix of L, and $f \in H^1(\partial\Omega, \sigma)$ is an arbitrary function.

Then $\vec{F} \in \left[\mathscr{C}^\infty(\Omega)\right]^n$ is a divergence-free vector field in Ω, with the property that for each fixed parameter $\kappa \in (0,\infty)$ there exists a constant $C_{\Omega,\kappa} \in (0,\infty)$ such that

$$\mathcal{N}_\kappa \vec{F} \in L^1(\partial\Omega, \sigma) \textit{ and } \|\mathcal{N}_\kappa \vec{F}\|_{L^1(\partial\Omega,\sigma)} \leq C_{\Omega,\kappa} \|f\|_{H^1(\partial\Omega,\sigma)}, \tag{2.9.4}$$

and such that the nontangential boundary trace $\vec{F}\big|_{\partial\Omega}^{\kappa-\mathrm{n.t.}}$ exists σ-a.e. on $\partial_*\Omega$ and is independent of κ. In fact, as regards the latter property, at σ-a.e. point $x \in \partial_*\Omega$ one actually has (with the dependence on κ dropped)

$$\left(\vec{F}\big|_{\partial\Omega}^{\mathrm{n.t.}}\right)(x) = \left(-\frac{1}{2} \frac{\sum_{s=1}^{n} a_{rs}\nu_s(x)}{\sum_{s,\ell=1}^{n} a_{s\ell}\nu_s(x)\nu_\ell(x)} f(x) + \sum_{s=1}^{n} a_{rs}(T_s f)(x) \right)_{1\leq r\leq n} \tag{2.9.5}$$

where, for each $s \in \{1,\dots,n\}$,

$$T_s f(x) := \lim_{\varepsilon\to 0^+} \int_{\substack{y\in\partial\Omega \\ |x-y|>\varepsilon}} (\partial_s E)(x-y) f(y)\, \mathrm{d}\sigma(y) \quad \textit{at } \sigma\textit{-a.e. } x \in \partial\Omega. \tag{2.9.6}$$

Moreover, if the set Ω is actually a UR *domain (i.e., if the additional condition that $\mathcal{H}^{n-1}(\partial\Omega \setminus \partial_*\Omega) = 0$ also holds), then there exists a constant $C_\Omega \in (0,\infty)$ such that*

$$\nu \cdot \left(\vec{F}\big|_{\partial\Omega}^{\mathrm{n.t.}}\right) \in H^1(\partial\Omega,\sigma) \textit{ and } \left\|\nu \cdot \left(\vec{F}\big|_{\partial\Omega}^{\mathrm{n.t.}}\right)\right\|_{H^1(\partial\Omega,\sigma)} \leq C_\Omega \|f\|_{H^1(\partial\Omega,\sigma)}. \tag{2.9.7}$$

Proof By design, $\vec{F}$ is a $\mathscr{C}^\infty$ smooth, divergence-free vector field in Ω. The properties in (2.9.4) are direct consequences of (2.4.14) (used with $p = 1$, $\Sigma := \partial\Omega$, and $k := \partial_s E$) in Theorem 2.4.1, whereas the nontangential boundary trace formula (2.9.5) is implied by (2.5.4) (with $k := \partial_s E$), bearing in mind that the tempered distribution $\widehat{E}$ is of function type in $\mathbb{R}^n \setminus \{0\}$ and, in fact (see, e.g., [109, §7.12, p. 305]),

$$\widehat{E}(\xi) = -\left(\sum_{r,s=1}^{n} a_{rs}\xi_r\xi_s \right)^{-1}, \qquad \forall\, \xi = (\xi_1,\dots,\xi_n) \in \mathbb{R}^n \setminus \{0\}. \tag{2.9.8}$$

There remains to deal with the claims made in (2.9.7) under the assumption that Ω is actually a UR domain. To this end, consider the singular integral operator defined for each $f \in L^p(\partial\Omega, \sigma)$ with $1 \le p < \infty$ according to

$$Tf(x) := \lim_{\varepsilon \to 0^+} \int_{\substack{y \in \partial\Omega \\ |x-y|>\varepsilon}} \sum_{r,s=1}^{n} a_{rs} \nu_r(x) (\partial_s E)(x-y) f(y) \, d\sigma(y) \tag{2.9.9}$$

at σ-a.e. $x \in \partial\Omega$. In particular, Theorem 2.3.2 guarantees that

$$\begin{aligned}&T : L^p(\partial\Omega, \sigma) \longrightarrow L^p(\partial\Omega, \sigma) \text{ with } p \in (1, \infty), \text{ as well as}\\ &T : L^1(\partial\Omega, \sigma) \longrightarrow L^{1,\infty}(\partial\Omega, \sigma), \text{ are well-defined, linear, and}\\ &\text{bounded operators.}\end{aligned} \tag{2.9.10}$$

To proceed, observe that if $\vec{F}$ is the vector field associated with a given function $f \in L^p(\partial\Omega, \sigma)$ with $1 \le p < \infty$ as in (2.9.3), then much as in the case of (2.9.5) we have

$$\begin{aligned}\nu(x) \cdot \Big(\vec{F}\Big|_{\partial\Omega}^{\rm n.t.}\Big)(x) &= \sum_{r=1}^{n} \nu_r(x) \left(-\frac{1}{2} \frac{\sum_{s=1}^{n} a_{rs} \nu_s(x)}{\sum_{s,\ell=1}^{n} a_{s\ell} \nu_s(x) \nu_\ell(x)} f(x) + \sum_{s=1}^{n} a_{rs} (T_s f)(x) \right)\\ &= -\tfrac{1}{2} f(x) + (Tf)(x) \text{ for } \sigma\text{-a.e. } x \in \partial\Omega.\end{aligned} \tag{2.9.11}$$

Suppose now that we fix some function

$$f \in L^p_{\rm comp}(\partial\Omega, \sigma), \text{ with } p \in (1, \infty), \text{ satisfying } \int_{\partial\Omega} f \, d\sigma = 0 \tag{2.9.12}$$

and, once again, associate the vector field $\vec{F}$ with this f as in (2.9.3). Then the function $f \in H^1(\partial\Omega, \sigma)$ (cf. [7, Lemma 5.8, p. 183]) hence, according to what we have established in the first part of the present proof, $\vec{F} \in \big[\mathscr{C}^\infty(\Omega)\big]^n$ is a divergence-free vector field in Ω, satisfying (2.9.4), and such that (2.9.5) holds for each $\kappa > 0$. We may therefore compute

$$\int_{\partial\Omega} Tf \, d\sigma = \int_{\partial\Omega} (-\tfrac{1}{2} f + Tf) \, d\sigma = \int_{\partial\Omega} \nu \cdot \Big(\vec{F}\Big|_{\partial\Omega}^{\rm n.t.}\Big) \, d\sigma = 0, \tag{2.9.13}$$

on account of (2.9.11), the cancelation condition in (2.9.12), and the Divergence Formula [112, (1.2.2)] from [112, Theorem 1.2.1] (whose current applicability in the present setting is ensured by the hypotheses on Ω and the aforementioned properties of $\vec{F}$). In turn, (2.9.9), (2.9.10), and (2.9.13) allow us to invoke the Hardy space boundedness criterion from [7, Theorem 8.40, p. 438] according to which

$$\begin{aligned}&\text{the operator } T \text{ from (2.9.10) induces a unique linear}\\ &\text{bounded mapping } T : H^1(\partial\Omega, \sigma) \longrightarrow H^1(\partial\Omega, \sigma).\end{aligned} \tag{2.9.14}$$

Then the claims in (2.9.7) are seen from (2.9.11), the fact that we have the inclusion $H^1(\partial\Omega, \sigma) \subset L^1(\partial\Omega, \sigma)$, and (2.9.14). □

It turns out that if Ω is a UR domain in $\mathbb{R}^n$ then the entire Hardy space $H^1(\partial\Omega, \sigma)$ is actually spanned by functions of the form $\nu \cdot \big(\vec{F}\big|_{\partial\Omega}^{^{\rm n.t.}}\big)$ where $\vec{F}$ is a divergence-free vector field whose nontangential maximal operator is integrable on the boundary, along with their counterparts defined relative to $\mathbb{R}^n \setminus \overline{\Omega}$. Specifically, we have the following result.

Corollary 2.9.2 *Assume $\Omega \subseteq \mathbb{R}^n$ (where $n \in \mathbb{N}$ with $n \geq 2$) is* UR *domain, with geometric measure theoretic outward unit normal ν, and abbreviate $\sigma := \mathcal{H}^{n-1}\lfloor\partial\Omega$. Also, fix $\kappa \in (0, \infty)$ and define*

$$\Omega_+ := \Omega \text{ and } \Omega_- := \mathbb{R}^n \setminus \overline{\Omega}. \tag{2.9.15}$$

Then there exists a constant $C_{\Omega,\kappa} \in (0, \infty)$ with the property that for every $f \in H^1(\partial\Omega, \sigma)$ one can find two divergence-free vector fields $\vec{F}_\pm \in \big[\mathscr{C}^\infty(\Omega_\pm)\big]^n$ satisfying:

$$\|\mathcal{N}_\kappa \vec{F}_\pm\|_{L^1(\partial\Omega,\sigma)} \leq C_{\Omega,\kappa}\|f\|_{H^1(\partial\Omega,\sigma)}, \tag{2.9.16}$$

$$\vec{F}_\pm\Big|_{\partial\Omega_\pm}^{^{\kappa-\rm n.t.}} \text{ exist } \sigma\text{-a.e. on } \partial\Omega, \tag{2.9.17}$$

$$\nu \cdot \Big(\vec{F}_\pm\Big|_{\partial\Omega_\pm}^{^{\kappa-\rm n.t.}}\Big) \in H^1(\partial\Omega, \sigma), \tag{2.9.18}$$

$$\Big\|\nu \cdot \Big(\vec{F}_\pm\Big|_{\partial\Omega_\pm}^{^{\kappa-\rm n.t.}}\Big)\Big\|_{H^1(\partial\Omega,\sigma)} \leq C_{\Omega,\kappa}\|f\|_{H^1(\partial\Omega,\sigma)}, \tag{2.9.19}$$

$$\nu \cdot \Big(\vec{F}_+\Big|_{\partial\Omega_+}^{^{\kappa-\rm n.t.}}\Big) - \nu \cdot \Big(\vec{F}_-\Big|_{\partial\Omega_-}^{^{\kappa-\rm n.t.}}\Big) = f \ \ \sigma\text{-a.e. on } \partial\Omega. \tag{2.9.20}$$

Proof Select some second-order, homogeneous, constant coefficient, elliptic differential operator L as in (2.9.1), and let E be its fundamental solution defined in (1.4.171). Also, fix an arbitrary $f \in H^1(\partial\Omega, \sigma)$. If we now define $\vec{F}_\pm$ as in (2.9.3) relative to $\Omega_\pm$ and the function $-f$, Theorem 2.9.1 and [112, Lemma 5.10.9] give that the vector fields $\vec{F}_\pm$ are $\mathscr{C}^\infty$ smooth and divergence-free in $\Omega_\pm$, and satisfy (2.9.16)-(2.9.19). Next, observe that (2.9.5) presently gives (bearing in mind that f is currently replaced by $-f$, and taking [112, Lemma 5.10.9] once again into account) that at σ-a.e. $x \in \partial\Omega$ we have

$$\Big(\vec{F}_\pm\Big|_{\partial\Omega_\pm}^{^{\kappa-\rm n.t.}}\Big)(x) = \left(\pm\frac{1}{2}\frac{\sum_{s=1}^n a_{rs}\nu_s(x)}{\sum_{s,\ell=1}^n a_{s\ell}\nu_s(x)\nu_\ell(x)}f(x) - \sum_{s=1}^n a_{rs}(T_s f)(x)\right)_{1\leq r\leq n}. \tag{2.9.21}$$

Then (2.9.20) readily follows from (2.9.21) and simple algebra. □

We proceed by recalling the fundamental solution associated with higher-order homogeneous constant (complex) coefficient elliptic systems from [109, Theorem 11.1, pp. 393-395]. This is going to be relevant in the formulation and proof of Corollary 2.9.4, discussed shortly.

Theorem 2.9.3 *Fix $n, m, M \in \mathbb{N}$ with $n \geq 2$, and assume that L is an $M \times M$ system in $\mathbb{R}^n$ of order $2m$ of the form*

$$L = \sum_{|\alpha|=|\beta|=m} \partial^\alpha A_{\alpha\beta} \partial^\beta \tag{2.9.22}$$

where each $A_{\alpha\beta} \in \mathbb{C}^{M\times M}$. Assume that the characteristic matrix of L, i.e., the $M \times M$ matrix-valued function[37]

$$L(\xi) := (-1)^m \sum_{|\alpha|=|\beta|=m} \xi^{\alpha+\beta} A_{\alpha\beta}, \qquad \forall \xi \in \mathbb{R}^n, \tag{2.9.23}$$

satisfies

$$\det\,[L(\xi)] \neq 0, \qquad \forall \xi \in \mathbb{R}^n \setminus \{0\}. \tag{2.9.24}$$

Then the $M \times M$ matrix $\mathbb{E}$ defined at each $x \in \mathbb{R}^n \setminus \{0\}$ by

$$\mathbb{E}(x) := \frac{(-1)^m}{4(2\pi \mathrm{i})^{n-1}(2m-1)!} \Delta_x^{(n-1)/2} \int_{S^{n-1}} (x\cdot\xi)^{2m-1} \mathrm{sgn}\,(x\cdot\xi) \big[L(\xi)\big]^{-1} \, d\mathcal{H}^{n-1}(\xi) \tag{2.9.25}$$

if n is odd, and

$$\mathbb{E}(x) := \frac{(-1)^{m+1}}{(2\pi \mathrm{i})^n (2m)!} \Delta_x^{n/2} \int_{S^{n-1}} (x\cdot\xi)^{2m} \ln|x\cdot\xi| \big[L(\xi)\big]^{-1} \, d\mathcal{H}^{n-1}(\xi) \tag{2.9.26}$$

if n is even, and satisfies the following properties.

(1) The entries in $\mathbb{E}$ are real-analytic functions in $\mathbb{R}^n \setminus \{0\}$ (hence, in particular, functions in $\mathscr{C}^\infty(\mathbb{R}^n \setminus \{0\})$) and

$$\mathbb{E}(-x) = \mathbb{E}(x) \text{ for each } x \in \mathbb{R}^n \setminus \{0\}. \tag{2.9.27}$$

(2) Define the $M \times M$ matrix-valued function

$$\mathbb{P}(x) := \frac{(-1)^{m+1}}{(2\pi \mathrm{i})^n (2m-n)!} \int_{S^{n-1}} (x\cdot\xi)^{2m-n} \big[L(\xi)\big]^{-1} \, d\mathcal{H}^{n-1}(\xi) \tag{2.9.28}$$

at each $x \in \mathbb{R}^n$. Then the entries of $\mathbb{P}$ are identically zero when either n is odd or $n > 2m$, and are homogeneous polynomials of degree $2m - n$ when $n \leq 2m$. Moreover, there exists a matrix-valued function $\Phi : \mathbb{R}^n \setminus \{0\} \to \mathbb{C}^{M\times M}$, whose entries belong to $\mathscr{C}^\infty(\mathbb{R}^n \setminus \{0\})$ and are positive homogeneous of degree $2m - n$, such that

$$\mathbb{E}(x) = \Phi(x) + \big(\ln|x|\big)\mathbb{P}(x), \qquad \forall x \in \mathbb{R}^n \setminus \{0\}. \tag{2.9.29}$$

[37] The reader is alerted to the fact that our present definition of $L(\xi)$ differs by a factor of $(-1)^m$ from the one in [109, (11.3.2), p. 391]

(3) *For each multi-index* $\gamma \in \mathbb{N}_0^n$ *there exists* $C_\gamma \in (0, \infty)$ *such that for every* $x \in \mathbb{R}^n \setminus \{0\}$ *one has*

$$\big|(\partial^\gamma \mathbb{E})(x)\big| \leq \begin{cases} \dfrac{C_\gamma}{|x|^{n-2m+|\gamma|}} & \text{if either } n \text{ is odd, or } n > 2m, \text{ or if } |\gamma| > 2m-n, \\[2ex] \dfrac{C_\gamma(1+|\ln|x||)}{|x|^{n-2m+|\gamma|}} & \text{if } 0 \leq |\gamma| \leq 2m-n. \end{cases} \tag{2.9.30}$$

In particular,

the entries of $\mathbb{E}$ *belong to* $L^1_{\rm loc}(\mathbb{R}^n, \mathcal{L}^n)$, *hence they may be canonically regarded as distributions in* $\mathbb{R}^n$; *in the latter scenario, the entries of* $\mathbb{E}$ *are actually tempered distributions in* $\mathbb{R}^n$. (2.9.31)

Also,

for each multi-index $\gamma \in \mathbb{N}_0^n$ *with* $|\gamma| \geq 2m-1$, *the function* $\partial^\gamma \mathbb{E}$ *is of class* $\mathscr{C}^\infty$, *as well as positive homogeneous of degree* $2m-n-|\gamma|$, *in* $\mathbb{R}^n \setminus \{0\}$. (2.9.32)

(4) *For each fixed* $y \in \mathbb{R}^n$ *one has (in the sense of tempered distributions in* $\mathbb{R}^n$*)*

$$L_x\big[\mathbb{E}(x-y)\big] = \delta_y(x)\, I_{M\times M}. \tag{2.9.33}$$

Above, the subscript x *denotes the fact that the operator* L *in* (2.9.33) *is applied to each column of* $\mathbb{E}$ *in the variable* x. *Also,* δ_y *stands for the Dirac distribution with mass at* y *in* $\mathbb{R}^n$, *and* $I_{M\times M}$ *is the* $M \times M$ *identity matrix.*

(5) *When restricted to* $\mathbb{R}^n \setminus \{0\}$, *the entries of* $\widehat{\mathbb{E}}$ *(with "hat" denoting the Fourier transform in* $\mathbb{R}^n$; *cf.* (1.4.14)) *are* $\mathscr{C}^\infty$ *functions and, in fact,*

$$\widehat{\mathbb{E}}(\xi) = \big[L(\xi)\big]^{-1} \text{ for each } \xi \in \mathbb{R}^n \setminus \{0\}. \tag{2.9.34}$$

(6) *Writing* $\mathbb{E}_L$ *in place of* $\mathbb{E}$ *to emphasize the dependence on* L, *the fundamental solution* $\mathbb{E}_L$ *with entries as in* (2.9.25)-(2.9.26) *satisfies*

$$\begin{gathered} (\mathbb{E}_L)^\top = \mathbb{E}_{L^\top}, \quad \overline{\mathbb{E}_L} = \mathbb{E}_{\overline{L}}, \quad (\mathbb{E}_L)^* = \mathbb{E}_{L^*}, \\ \text{and } \mathbb{E}_{\lambda L} = \lambda^{-1}\mathbb{E}_L \text{ for each } \lambda \in \mathbb{C} \setminus \{0\}. \end{gathered} \tag{2.9.35}$$

(7) *Any* $M \times M$ *matrix* $\mathbb{U}$ *whose entries are tempered distributions in* $\mathbb{R}^n$ *and which is a fundamental solution of the system* L *in* $\mathbb{R}^n$ *is of the form* $\mathbb{U} = \mathbb{E} + \mathbb{Q}$ *where* $\mathbb{E}$ *is as in* (2.9.25)-(2.9.26) *and* $\mathbb{Q}$ *is an* $M \times M$ *matrix whose entries are polynomials in* $\mathbb{R}^n$ *and whose columns,* $\mathbb{Q}_k$, $k \in \{1, \dots, M\}$, *satisfy the pointwise equations* $L\mathbb{Q}_k = 0 \in \mathbb{C}^M$ *in* $\mathbb{R}^n$ *for each* $k \in \{1, \dots, M\}$.

Collectively, Theorem 2.3.2, Theorem 2.4.1, and Theorem 2.5.1 permit us to deal with integral operators on uniformly rectifiable sets whose kernels consists of derivatives of the fundamental solution of a higher-order system of differential operators, of the sort described in the corollary below.

Corollary 2.9.4 *Fix $n, m, M \in \mathbb{N}$ with $n \geq 2$, and suppose L is an $M \times M$ system in $\mathbb{R}^n$ of order $2m$ of the form*

$$L = \sum_{|\alpha|=|\beta|=m} \partial^\alpha A_{\alpha\beta} \partial^\beta, \qquad \text{where } A_{\alpha\beta} \in \mathbb{C}^{M\times M}, \tag{2.9.36}$$

satisfying the weak ellipticity condition

$$\det\Bigg[\sum_{|\alpha|=|\beta|=m} \xi^{\alpha+\beta} A_{\alpha\beta}\Bigg] \neq 0, \qquad \forall \xi \in \mathbb{R}^n \setminus \{0\}. \tag{2.9.37}$$

Denote by $\mathbb{E}_L$ the fundamental solution of L defined as in (2.9.25)-(2.9.26). Next, assume $\Omega \subseteq \mathbb{R}^n$ is a nonempty open set with the property that $\partial\Omega$ is a UR set; in particular, Ω is a set of locally finite perimeter. Abbreviate $\sigma := \mathcal{H}^{n-1}\lfloor\partial\Omega$ and denote by ν the geometric measure theoretic outward unit normal to Ω.

Finally, fix a multi-index $\gamma \in \mathbb{N}_0^n$ with $|\gamma| = 2m-1$ and define the singular integral operator acting on $\mathbb{C}^M$-valued functions

$$f \in \Big[L^1\Big(\partial\Omega, \frac{\sigma(x)}{1+|x|^{n-1}}\Big)\Big]^M \tag{2.9.38}$$

according to

$$(\mathcal{T}_{L,\gamma} f)(x) := \int_{\partial\Omega} (\partial^\gamma \mathbb{E}_L)(x-y) f(y)\, d\sigma(y), \qquad \forall x \in \Omega, \tag{2.9.39}$$

along with its principal-value version

$$(T_{L,\gamma} f)(x) := \lim_{\varepsilon\to 0^+} \int_{\substack{y\in\partial\Omega \\ |x-y|>\varepsilon}} (\partial^\gamma \mathbb{E}_L)(x-y) f(y)\, d\sigma(y) \text{ for } \sigma\text{-a.e. } x \in \partial\Omega. \tag{2.9.40}$$

Then

$$\begin{array}{c}\text{for each function } f \text{ as in (2.9.38), the limit defining} \\ (T_{L,\gamma} f)(x) \text{ in (2.9.40) exists at } \sigma\text{-a.e. point } x \in \partial\Omega\end{array} \tag{2.9.41}$$

and the induced operators

$$T_{L,\gamma} : \big[L^p(\partial\Omega,\sigma)\big]^M \longrightarrow \big[L^p(\partial\Omega,\sigma)\big]^M, \quad p \in (1,\infty), \tag{2.9.42}$$

$$T_{L,\gamma} : \big[L^1(\partial\Omega,\sigma)\big]^M \longrightarrow \big[L^{1,\infty}(\partial\Omega,\sigma)\big]^M, \tag{2.9.43}$$

are well-defined, linear, and bounded. The operator (2.9.42) further extends to a linear and bounded mapping

$$T_{L,\gamma} : \big[H^p(\partial\Omega,\sigma)\big]^M \longrightarrow \big[L^p(\partial\Omega,\sigma)\big]^M \quad \textit{for } \tfrac{n-1}{n} < p \leq 1, \tag{2.9.44}$$

and, corresponding to the end-point $p = \infty$,

$$T_{L,\gamma} : \big[L^\infty(\partial\Omega,\sigma)\big]^M \longrightarrow \big[\mathrm{BMO}(\partial\Omega,\sigma)\big]^M. \tag{2.9.45}$$

Moreover, for each $p \in (1,\infty)$ there exists $C = C(\Omega, L, p, n) \in [0,\infty)$ with the property that for each function $f \in \big[L^p(\partial\Omega,\sigma)\big]^M$ one has

$$\Big(\int_\Omega |(\nabla \mathcal{T}_{L,\gamma} f)(x)|^p \operatorname{dist}(x,\partial\Omega)^{p-1}\,dx\Big)^{1/p} \leq C\|f\|_{[L^p(\partial\Omega,\sigma)]^M}. \tag{2.9.46}$$

Also, for each $\kappa > 0$ and $p \in \big(\frac{n-1}{n},\infty\big)$ there exists $C = C(\Omega, L, p, \kappa, n) \in [0,\infty)$ such that

$$\big\|\mathcal{N}_\kappa(\mathcal{T}_{L,\gamma} f)\big\|_{L^p(\partial\Omega,\sigma)} \leq C\|f\|_{[L^p(\partial\Omega,\sigma)]^M} \ \textit{ if } 1 < p < \infty, \tag{2.9.47}$$

$$\big\|\mathcal{N}_\kappa(\mathcal{T}_{L,\gamma} f)\big\|_{L^{1,\infty}(\partial\Omega,\sigma)} \leq C\|f\|_{[L^1(\partial\Omega,\sigma)]^M}, \tag{2.9.48}$$

$$\big\|\mathcal{N}_\kappa(\mathcal{T}_{L,\gamma} f)\big\|_{L^p(\partial\Omega,\sigma)} \leq C\|f\|_{[H^p(\partial\Omega,\sigma)]^M} \ \textit{ if } \tfrac{n-1}{n} < p \leq 1. \tag{2.9.49}$$

Finally, for each $\kappa > 0$ and each $f \in \big[L^1\big(\partial\Omega, \frac{\sigma(x)}{1+|x|^{n-1}}\big)\big]^M$ the following jump-formula holds at σ-a.e. point $x \in \partial_*\Omega$:

$$\big(\mathcal{T}_{L,\gamma} f\big)\Big|_{\partial\Omega}^{\kappa-\mathrm{n.t.}}(x) = -\frac{\nu(x)^\gamma}{2}\Big(\sum_{|\alpha|=|\beta|=m} \nu(x)^{\alpha+\beta} A_{\alpha\beta}\Big)^{-1} f(x) + (T_{L,\gamma} f)(x). \tag{2.9.50}$$

Proof Let k be a generic entry in the matrix $\partial^\gamma \mathbb{E}_L$. From Theorem 2.9.3 we know that each such function belongs to $\mathscr{C}^\infty(\mathbb{R}^n \setminus \{0\})$, and is odd and positive homogeneous of degree $1-n$ in $\mathbb{R}^n \setminus \{0\}$. In turn, this permits us to invoke Theorem 2.3.2 and Theorem 2.4.1 which yield all conclusions in (2.9.41)-(2.9.49). Formula (2.9.50) is going to be a consequence of Theorem 2.5.1. Concretely, thanks to (2.9.34) and standard Fourier analysis, for each $x \in \partial^*\Omega$ we have

$$\begin{aligned} \widehat{\partial^\gamma \mathbb{E}_L}\big(\nu(x)\big) &= (\sqrt{-1})^{|\gamma|}\nu(x)^\gamma \widehat{\mathbb{E}_L}\big(\nu(x)\big) \\ &= -\sqrt{-1}\,\nu(x)^\gamma \Big(\sum_{|\alpha|=|\beta|=m} \nu(x)^{\alpha+\beta} A_{\alpha\beta}\Big)^{-1}. \end{aligned} \tag{2.9.51}$$

Granted this, the jump-formula (2.9.50) now follows on account of (2.5.4). □

It is worth singling out the formula giving the nontangential traces of generic derivatives of the single layer potential operator. We do so in the corollary below.

Corollary 2.9.5 *Assume $\Omega \subseteq \mathbb{R}^n$ is a nonempty open set with the property that $\partial\Omega$ is a UR set; in particular, Ω is a set of locally finite perimeter. Let $\sigma := \mathcal{H}^{n-1}\lfloor\partial\Omega$ and* *denote by* $\nu = (\nu_1, \ldots, \nu_n)$ *the geometric measure theoretic outward unit normal to* Ω. *Also, let* L *be a homogeneous, second-order, constant complex coefficient, weakly elliptic* $M \times M$ *system in* $\mathbb{R}^n$ *and consider the matrix-valued fundamental solution* $E = (E_{\alpha\beta})_{1\leq\alpha,\beta\leq M}$ *associated with* L *as in Theorem 1.4.2. Finally, pick a coefficient tensor* $A = \left(a_{rs}^{\alpha\beta}\right)_{\substack{1\leq\alpha,\beta\leq M\\1\leq r,s\leq n}}$ *for which* $L_A = L$, *and consider the matrix-valued function*

$$\begin{aligned} B = (b_{\alpha\beta})_{1\leq\alpha,\beta\leq M} : \partial_*\Omega \to \mathbb{C}^{M\times M} \quad \textit{defined } \sigma\textit{-a.e. on } \partial_*\Omega \textit{ as} \\ B := [L(\nu)]^{-1} = \big[-(a_{rs}^{\alpha\beta}\nu_r\nu_s)_{1\leq\alpha,\beta\leq M}\big]^{-1}. \end{aligned} \tag{2.9.52}$$

For each $\alpha, \beta \in \{1, \ldots, M\}$ *and* $r \in \{1, \ldots, n\}$ *introduce an operator, denoted by* $(\partial_r \mathcal{S}_{\alpha\beta})$, *which acts on each function* $f \in L^1\big(\partial\Omega, \frac{\sigma(x)}{1+|x|^{n-1}}\big)$ *according to*[38]

$$(\partial_r \mathcal{S}_{\alpha\beta})f(x) := \int_{\partial\Omega} (\partial_r E_{\alpha\beta})(x-y)f(y)\,d\sigma(y) \quad \textit{for all } x \in \Omega. \tag{2.9.53}$$

Then for each indexes $\alpha, \beta \in \{1, \ldots, M\}$ *and* $r \in \{1, \ldots, n\}$, *each function* $f \in L^1\big(\partial\Omega, \frac{\sigma(x)}{1+|x|^{n-1}}\big)$, *and each aperture parameter* $\kappa > 0$, *at* σ*-a.e. point* $x \in \partial_*\Omega$ *one has*

$$(\partial_r \mathcal{S}_{\alpha\beta})f\Big|_{\partial\Omega}^{\kappa-\text{n.t.}}(x) = \frac{1}{2}\nu_r(x)b_{\alpha\beta}(x)f(x) + \lim_{\varepsilon\to 0^+} \int_{\substack{y\in\partial\Omega\\|x-y|>\varepsilon}} (\partial_r E_{\alpha\beta})(x-y)f(y)\,d\sigma(y). \tag{2.9.54}$$

Proof This is an immediate consequence of (2.9.50). □

---

[38] we emphasize that $(\partial_r \mathcal{S}_{\alpha\beta})f$ is simply a convenient piece of notation, and should *not* be interpreted as $\partial_r(\mathcal{S}_{\alpha\beta}f)$

Chapter 3
Quantitative Fatou-Type Theorems in Arbitrary UR Domains

The trademark blueprint of a Fatou-type theorem is that size/integrability properties of the nontangential maximal operator for a null-solution of an elliptic equation in a certain domain implies the a.e. existence of the pointwise nontangential boundary trace of said function. It is natural to call such a theorem quantitative if the boundary trace does not just simply exist, but also encodes significant information regarding the size and regularity of the original function.

One of the main results in this chapter is Theorem 3.1.6, where we establish a quantitative Fatou-type theorem for any null-solution u of an injectively elliptic first-order (homogeneous, constant complex coefficient) system of differential operators in an arbitrary UR domain $\Omega \subseteq \mathbb{R}^n$, assuming that the nontangential maximal operator of u is p-th power integrable on $\partial\Omega$ (with respect to the Hausdorff measure $\mathcal{H}^{n-1}$) for some integrability exponent $p \in \left(\frac{n-1}{n}, \infty\right)$. Such a result has a wide range of applications, including the theory of Hardy spaces associated with injectively elliptic first-order systems in UR domains. We also prove here a quantitative Fatou-type theorem for the gradient of null-solutions of second-order systems in UR domains (cf. Theorem 3.3.4).

Quantitative Fatou-type result for null-solutions u of weakly elliptic second-order system in arbitrary UR domains $\Omega \subseteq \mathbb{R}^n$, in which integrability properties of the nontangential maximal functions of u and ∇u together ensure the existence of non-tangential pointwise traces of both u and ∇u on $\partial\Omega$, are presented in Theorem 3.3.9. In addition we show that, for a null-solution of a weakly elliptic system, having sufficient regularity on the Sobolev/Besov/Triebel-Lizorkin scales in a given bounded two-sided NTA domain with an Ahlfors regular boundary guarantees the existence of its nontangential pointwise trace on the boundary of said domain. In the final section in this chapter we establish Fatou-type theorems for first and second order differential operators on Riemannian manifolds.

D. Mitrea et al., *Geometric Harmonic Analysis III*, Developments in Mathematics 74,
https://doi.org/10.1007/978-3-031-22735-6_3

3.1 Quantitative Fatou-Type Theorems in UR Domains for First-Order Systems

Recall the class of injectively elliptic first-order systems from Definition 1.3.4. The integral representation formula in the theorem below is central to the considerations in this section.

Theorem 3.1.1 *Fix $n \in \mathbb{N}$ with $n \geq 2$, and suppose $\Omega \subseteq \mathbb{R}^n$ is an Ahlfors regular domain. Denote by ν the geometric measure theoretic outward unit normal to Ω and abbreviate $\sigma := \mathcal{H}^{n-1} \lfloor \partial\Omega$. Next, consider a homogeneous, first-order $N \times M$ system D with constant complex coefficients in $\mathbb{R}^n$ (where $N, M \in \mathbb{N}$) which is injectively elliptic (cf. (1.3.18)), and suppose $\widetilde{D}$ is a homogeneous first-order $M \times N$ system with constant complex coefficients in $\mathbb{R}^n$ which complements D (i.e., (1.3.21) holds). In particular, since $L := \widetilde{D}D$ is a weakly elliptic second-order $M \times M$ system in $\mathbb{R}^n$ it is meaningful to consider the matrix-valued fundamental solution $E_{L^\top}$ associated with $L^\top$ as in Theorem 1.4.2. Also, with $\widetilde{D}^\top$ acting on the columns of $E_{L^\top}$ define*

$$\widetilde{E} := \widetilde{D}^\top E_{L^\top}. \tag{3.1.1}$$

Next, suppose $u : \Omega \to \mathbb{C}^M$ is a vector-valued function with Lebesgue measurable components satisfying, for an aperture parameter $\kappa \in (0, \infty)$,

$$\mathcal{N}_\kappa u \in L^p(\partial\Omega, \sigma) \text{ for some } p \in \left(\tfrac{n-1}{n}, \infty\right). \tag{3.1.2}$$

In addition, with Du considered in the sense of distributions in Ω, assume that

$$Du \in \left[L^1_{\mathrm{loc}}(\Omega, \mathcal{L}^n)\right]^N \text{ and } \mathfrak{P}(Du) \in L^p(\partial\Omega, \sigma). \tag{3.1.3}$$

Then $\mathrm{Sym}(D; \nu) \bullet u$ belongs to the Hardy space $\left[H^p(\partial\Omega, \sigma)\right]^N$ and there exists some constant $C = C(\Omega, D, p, \kappa) \in (0, \infty)$, independent of u, such that

$$\left\|\mathrm{Sym}(D; \nu) \bullet u\right\|_{[H^p(\partial\Omega,\sigma)]^N} \leq C\left\|\mathcal{N}_\kappa u\right\|_{L^p(\partial\Omega,\sigma)} + C\left\|\mathfrak{P}(Du)\right\|_{L^p(\partial\Omega,\sigma)}. \tag{3.1.4}$$

Moreover, if $p \in \left(\frac{n-1}{n}, 1\right]$ and $\partial\Omega$ is unbounded then for $\mathcal{L}^n$-a.e. $x \in \mathbb{R}^n \setminus \partial\Omega$ with the property that

$$\int_\Omega \frac{|(Du)(y)|}{|x - y|^{n-1}}\, dy < +\infty \tag{3.1.5}$$

which is also assumed to be a Lebesgue point for the function u in the case when $x \in \Omega$, one has

$$\Big\langle \big[\widetilde{E}^{\top}(x-\cdot)\big|_{\partial\Omega}\big], (-\mathrm{i})\mathrm{Sym}\big(D;\nu\big)\bullet u\Big\rangle$$

$$-\int_{\Omega}\big\langle \widetilde{E}^{\top}(x-y), (Du)(y)\big\rangle\,\mathrm{d}y = \begin{cases} u(x) & \text{if } x\in\Omega,\\ 0 & \text{if } x\in\mathbb{R}^n\setminus\overline{\Omega}.\end{cases} \tag{3.1.6}$$

The bracket $\langle\cdot,\cdot\rangle$ in the first line of (3.1.6) *is viewed as the duality pairing (in the sense of* [113, Theorem 4.6.1], *presently used with* $\Sigma := \partial\Omega$*) between the rows of the* $M\times N$ *matrix* $\big[\widetilde{E}^{\top}(x-\cdot)\big|_{\partial\Omega}\big]$, *each of which belonging to*[1]

$$\Big(\big[H^p(\partial\Omega,\sigma)\big]^N\Big)^* = \begin{cases} \big[\dot{\mathscr{C}}^{(n-1)(\frac{1}{p}-1)}(\partial\Omega)/\sim\big]^N & \text{if } p<1,\\ \big[\widetilde{\mathrm{BMO}}(\partial\Omega,\sigma)\big]^N \big(\supset \big[L^\infty(\partial\Omega,\sigma)/\sim\big]^N\big) & \text{if } p=1, \end{cases} \tag{3.1.7}$$

and $(-\mathrm{i})\mathrm{Sym}\big(D;\nu\big)\bullet u \in \big[H^p(\partial\Omega,\sigma)\big]^N$. *The bracket inside the solid integral in the second line of* (3.1.6) *is simply the pointwise (real inner product) pairing between the columns of the* $M\times N$ *matrix* $\widetilde{E}^{\top}(x-y)$ *and the vector* $(Du)(y)\in\mathbb{C}^N$.

A similar integral representation formula holds if $p\in\big(\frac{n-1}{n},1\big]$ *and* $\partial\Omega$ *is bounded, this time without having to take the equivalence class of* $\widetilde{E}^{\top}(x-\cdot)\big|_{\partial\Omega}$ *modulo constants in the first line of* (3.1.6), *provided one also assumes that there exists* $\lambda\in(1,\infty)$ *such that*

$$⨍_{B(0,\lambda R)\setminus B(0,R)} |u|\,\mathrm{d}\mathcal{L}^n = o(1) \ \text{ as } R\to\infty \tag{3.1.8}$$

in the case when Ω *happens to be an exterior domain.*

Finally, an analogous integral representation formula holds when $p\in(1,\infty)$, *this time interpreting the duality pairings as integration on* $\partial\Omega$ *with respect to* σ, *i.e.,*

$$\int_{\partial\Omega}\big\langle \widetilde{E}^{\top}(x-y), (-\mathrm{i})\big(\mathrm{Sym}\big(D;\nu\big)\bullet u\big)(y)\big\rangle\,\mathrm{d}\sigma(y)$$

$$-\int_{\Omega}\big\langle \widetilde{E}^{\top}(x-y), (Du)(y)\big\rangle\,\mathrm{d}y = \begin{cases} u(x) & \text{if } x\in\Omega,\\ 0 & \text{if } x\in\mathbb{R}^n\setminus\overline{\Omega},\end{cases} \tag{3.1.9}$$

with the same caveat as before when Ω *is an exterior domain.*

Proof The fact that we have the membership

$$\mathrm{Sym}\big(D;\nu\big)\bullet u \in \big[H^p(\partial\Omega,\sigma)\big]^N \tag{3.1.10}$$

in the quantitative manner, indicated in (3.1.4), is a direct consequence of [113, Corollary 10.2.13] (used with $q := p$), [112, (6.2.25)], and [113, (4.2.25)]. Let us also remark here that (3.1.2) and [112, Lemma 8.3.1] ensure that

[1] membership implied by the estimates in Theorem 1.4.2

$$u \in \big[L^{\infty}_{\rm loc}(\Omega, \mathcal{L}^n)\big]^M. \tag{3.1.11}$$

To proceed, we distinguish two cases.

`Case I:` *Suppose* $p \in \big(\frac{n-1}{n}, 1\big]$. To fix ideas, assume $\partial\Omega$ is unbounded, and pick a Lebesgue point $x \in \Omega$ for the function u, with the additional property that (3.1.5) is satisfied. Also, consider a real-valued function $\eta \in \mathscr{C}^\infty(\mathbb{R}^n)$ satisfying $\eta \equiv 0$ on $B(0,1)$ and $\eta \equiv 1$ on $\mathbb{R}^n \setminus B(0,2)$. For each number $\varepsilon \in \big(0, \frac{1}{2}\,\mathrm{dist}(x,\partial\Omega)\big)$ define $\eta_\varepsilon : \mathbb{R}^n \to \mathbb{R}$ by setting

$$\eta_\varepsilon(y) := \eta\Big(\frac{y-x}{\varepsilon}\Big) \text{ for every } y \in \mathbb{R}^n. \tag{3.1.12}$$

Then

$$\begin{aligned}
&\eta_\varepsilon \in \mathscr{C}^\infty(\mathbb{R}^n) \text{ is a bounded function,}\\
&\lim_{\varepsilon\to 0^+} \eta_\varepsilon(y) = 1 \text{ for every } y \in \mathbb{R}^n \setminus \{x\},\\
&1-\eta_\varepsilon \in \mathscr{C}^\infty_c(\Omega), \quad \eta_\varepsilon \equiv 0 \text{ on } B(x,\varepsilon),\\
&\mathrm{supp}\,(\nabla\eta_\varepsilon) \subseteq \overline{B(x,2\varepsilon)} \setminus B(x,\varepsilon), \text{ and also}\\
&\sup\nolimits_{y\in\mathbb{R}^n} |(\nabla^j\eta_\varepsilon)(y)| \leq C\varepsilon^{-j} \text{ for all } j \in \mathbb{N}_0,
\end{aligned} \tag{3.1.13}$$

for some constant $C \in (0,\infty)$ independent of the parameter ε. In particular, the function $\widetilde{E}(x-\cdot)\eta_\varepsilon$ belongs to $\big[\mathscr{C}^\infty(\mathbb{R}^n)\big]^{N\times M}$ and coincides with $\widetilde{E}(x-\cdot)$ near $\partial\Omega$. As such,

$$\begin{aligned}
&\Big\langle \big[\widetilde{E}^\top(x-\cdot)\big|_{\partial\Omega}\big], (-\mathrm{i})\mathrm{Sym}(D;\nu)\bullet u\Big\rangle\\
&\qquad = \Big\langle \big[\eta_\varepsilon\widetilde{E}^\top(x-\cdot)\big|_{\partial\Omega}\big], (-\mathrm{i})\mathrm{Sym}(D;\nu)\bullet u\Big\rangle.
\end{aligned} \tag{3.1.14}$$

To proceed, fix a dilation factor $\lambda \in (1,\infty)$ and choose a real-valued function $\theta \in \mathscr{C}^\infty_c(\mathbb{R}^n)$ satisfying $0 \leq \theta \leq 1$ as well as $\theta \equiv 1$ on $B(0,1)$, and $\theta \equiv 0$ on $\mathbb{R}^n \setminus B(0,\lambda)$. For each given $R > 0$, define $\theta_R(y) := \theta(y/R)$ at every point $y \in \mathbb{R}^n$. Fix now some number $\varepsilon \in \big(0, \frac{1}{2}\,\mathrm{dist}(x,\partial\Omega)\big)$ and observe that

$$\begin{aligned}
&\lim_{R\to\infty} \Big[\big(\theta_R\eta_\varepsilon\widetilde{E}^\top(x-\cdot)\big)\big|_{\partial\Omega}\Big] = \Big[\big(\eta_\varepsilon\widetilde{E}^\top(x-\cdot)\big)\big|_{\partial\Omega}\Big]\\
&\text{weak-}*\text{ in } \begin{cases} \Big[\dot{\mathscr{C}}^{(n-1)(1/p-1)}(\partial\Omega)\big/\sim\Big]^{M\times N} & \text{if } p<1,\\[4pt] \big[\widetilde{\mathrm{BMO}}(\partial\Omega,\sigma)\big]^{M\times N} & \text{if } p=1. \end{cases}
\end{aligned} \tag{3.1.15}$$

This is implied by the general weak-$*$ convergence results established in [113, Lemma 4.8.4] and, respectively, [113, Lemma 4.8.1] (in the latter case also bearing in mind the trivial bounded embedding $L^\infty(\partial\Omega,\sigma) \hookrightarrow \mathrm{BMO}(\partial\Omega,\sigma)$). At this stage, from (3.1.14), (3.1.15), (3.1.10), the duality result from [113, Theorem 4.6.1], and

[113, Lemma 4.6.4] we conclude that for each index $\alpha \in \{1, \dots, M\}$ we have

$$\Big\langle \big[\widetilde{E}^{\top}(x-\cdot)\big|_{\partial\Omega}\big], (-\mathrm{i})\mathrm{Sym}\big(D;\nu\big) \bullet u\Big\rangle_{\alpha} \tag{3.1.16}$$

$$= \Big\langle \big[(\widetilde{E}^{\top})_{\alpha\bullet}(x-\cdot)\big|_{\partial\Omega}\big], (-\mathrm{i})\mathrm{Sym}\big(D;\nu\big) \bullet u\Big\rangle$$

$$= \lim_{R\to\infty} {}_{[(\mathrm{Lip}_c(\partial\Omega))']^N}\Big\langle (-\mathrm{i})\mathrm{Sym}\big(D;\nu\big) \bullet u, \big(\theta_R\eta_\varepsilon\widetilde{E}_{\bullet\alpha}(x-\cdot)\big)\big|_{\partial\Omega}\Big\rangle_{[\mathrm{Lip}_c(\partial\Omega)]^N}.$$

Based on (3.1.16) and [113, Proposition 10.2.11] (presently used with $F := u$) we obtain (cf. [113, (10.2.94)-(10.2.95)])

$$\Big\langle \big[\widetilde{E}^{\top}(x-\cdot)\big|_{\partial\Omega}\big], (-\mathrm{i})\mathrm{Sym}\big(D;\nu\big) \bullet u\Big\rangle$$

$$= \lim_{R\to\infty}\int_{\Omega}\big\langle Du, \theta_R\eta_\varepsilon\widetilde{E}(x-\cdot)\big\rangle\,\mathrm{d}\mathcal{L}^n$$

$$- \lim_{R\to\infty}\int_{\Omega}\big\langle u, D^{\top}\big(\theta_R\eta_\varepsilon\widetilde{E}(x-\cdot)\big)\big\rangle\,\mathrm{d}\mathcal{L}^n, \tag{3.1.17}$$

where $D^{\top}$ acts on the columns of the matrix $\theta_R\eta_\varepsilon\widetilde{E}(x-\cdot)$. Thanks to (3.1.5), (1.4.24), and Lebesgue's Dominated Convergence Theorem,

$$\lim_{\varepsilon\to 0^+}\lim_{R\to\infty}\int_{\Omega}\big\langle Du, \theta_R\eta_\varepsilon\widetilde{E}(x-\cdot)\big\rangle\,\mathrm{d}\mathcal{L}^n = \int_{\Omega}\big\langle Du, \widetilde{E}(x-\cdot)\big\rangle\,\mathrm{d}\mathcal{L}^n$$

$$= \int_{\Omega}\big\langle \widetilde{E}^{\top}(x-\cdot), Du\big\rangle\,\mathrm{d}\mathcal{L}^n. \tag{3.1.18}$$

In order to be specific, suppose D is expressed as in (1.3.19). Upon recalling (1.3.22), from (3.1.17), (3.1.18), and (1.3.22) we then deduce that

$$\Big\langle \big[\widetilde{E}(x-\cdot)\big|_{\partial\Omega}\big], (-\mathrm{i})\mathrm{Sym}\big(D;\nu\big) \bullet u\Big\rangle - \int_{\Omega}\big\langle \widetilde{E}^{\top}(x-\cdot), Du\big\rangle\,\mathrm{d}\mathcal{L}^n$$

$$= \Big(\lim_{\varepsilon\to 0^+}\lim_{R\to\infty}\int_{\Omega} u_\beta a_j^{\alpha\beta}\partial_j\big(\theta_R\eta_\varepsilon\widetilde{E}_{\alpha\gamma}(x-\cdot)\big)\,\mathrm{d}\mathcal{L}^n\Big)_{1\leq\gamma\leq M}. \tag{3.1.19}$$

Going further, observe that (1.6.7) entails

$$a_j^{\alpha\beta}(\partial_j\widetilde{E})_{\alpha\gamma}(x-\cdot) = 0 \text{ in } \mathbb{R}^n\setminus\{x\}, \tag{3.1.20}$$

for each $\beta,\gamma \in \{1,\dots,M\}$. Make use of (3.1.20) to split the last expression in (3.1.19) as

$$\Big(\lim_{\varepsilon\to 0^+}\lim_{R\to\infty}\int_{\Omega} u_\beta a_j^{\alpha\beta}\partial_j\big(\theta_R\eta_\varepsilon\widetilde{E}_{\alpha\gamma}(x-\cdot)\big)\,\mathrm{d}\mathcal{L}^n\Big)_{1\leq\gamma\leq M} = \mathrm{I} + \mathrm{II}, \tag{3.1.21}$$

where

$$\begin{aligned}\mathrm{I} &:= \Big(\lim_{\varepsilon\to 0^+} \lim_{R\to\infty} \int_\Omega u_\beta a_j^{\alpha\beta}(\partial_j\theta_R)\eta_\varepsilon \widetilde{E}_{\alpha\gamma}(x-\cdot)\,\mathrm{d}\mathcal{L}^n \Big)_{1\le\gamma\le M} \\ &= \Big(\lim_{R\to\infty} \int_\Omega u_\beta a_j^{\alpha\beta}(\partial_j\theta_R) \widetilde{E}_{\alpha\gamma}(x-\cdot)\,\mathrm{d}\mathcal{L}^n \Big)_{1\le\gamma\le M}, \end{aligned} \tag{3.1.22}$$

and

$$\begin{aligned}\mathrm{II} &:= \Big(\lim_{\varepsilon\to 0^+} \lim_{R\to\infty} \int_\Omega u_\beta a_j^{\alpha\beta}\theta_R(\partial_j\eta_\varepsilon) \widetilde{E}_{\alpha\gamma}(x-\cdot)\,\mathrm{d}\mathcal{L}^n \Big)_{1\le\gamma\le M} \\ &= \Big(\lim_{\varepsilon\to 0^+} \int_\Omega u_\beta a_j^{\alpha\beta}(\partial_j\eta_\varepsilon) \widetilde{E}_{\alpha\gamma}(x-\cdot)\,\mathrm{d}\mathcal{L}^n \Big)_{1\le\gamma\le M}. \end{aligned} \tag{3.1.23}$$

Note that since we are presently assuming $\partial\Omega$ to be unbounded, estimate [112, (8.6.51)] is applicable with $E := \Omega \cap \big[B(0,\lambda R) \setminus B(0,R)\big]$ and $p := (n-1)/n$. Together with Hölder's inequality and (1.4.24), this gives

$$\begin{aligned}\big|\mathrm{I}\big| &\le C \limsup_{R\to\infty} \Big\{ R^{-n} \int_{\Omega\cap[B(0,\lambda R)\setminus B(0,R)]} |u|\,\mathrm{d}\mathcal{L}^n \Big\} \\ &\le C \limsup_{R\to\infty} \Big\{ R^{-n} \Big(\int_{\Omega\cap[B(0,\lambda R)\setminus B(0,R)]} |u|^{\frac{np}{n-1}}\,\mathrm{d}\mathcal{L}^n \Big)^{\frac{n-1}{np}} \Big(\int_{B(0,\lambda R)} \mathrm{d}\mathcal{L}^n \Big)^{1-\frac{n-1}{np}} \Big\} \\ &\le C \limsup_{R\to\infty} \Big\{ R^{-(n-1)/p} \|\mathcal{N}_\kappa u\|_{L^p(\partial\Omega,\sigma)} \Big\} = 0. \end{aligned} \tag{3.1.24}$$

Hence,

$$\mathrm{I} = 0. \tag{3.1.25}$$

Next, we decompose

$$\mathrm{II} = \mathrm{II}^{(a)} + \mathrm{II}^{(b)} \tag{3.1.26}$$

where

$$\begin{aligned}\mathrm{II}^{(a)} &:= \Big(\lim_{\varepsilon\to 0^+} \int_\Omega \big(u_\beta(y) - u_\beta(x)\big) a_j^{\alpha\beta}(\partial_j\eta_\varepsilon)(y) \widetilde{E}_{\alpha\gamma}(x-y)\,\mathrm{d}y \Big)_{1\le\gamma\le M}, \\ \mathrm{II}^{(b)} &:= u_\beta(x) \Big(\lim_{\varepsilon\to 0^+} \int_\Omega a_j^{\alpha\beta}(\partial_j\eta_\varepsilon)(y) \widetilde{E}_{\alpha\gamma}(x-y)\,\mathrm{d}y \Big)_{1\le\gamma\le M}. \end{aligned} \tag{3.1.27}$$

In relation to $\mathrm{II}^{(a)}$ observe that since $x \in \Omega$ is a Lebesgue point for the function u we may estimate (using (3.1.13) and (1.4.24))

$$\left|\mathrm{II}^{(a)}\right| \leq C \cdot \limsup_{\varepsilon \to 0^+} \fint_{B(x,2\varepsilon)} \left|u(y) - u(x)\right| \, dy = 0. \tag{3.1.28}$$

Also, for each $\beta, \gamma \in \{1, \ldots, M\}$ fixed, keeping (3.1.13) in mind permits us to compute

$$\begin{aligned}
&\lim_{\varepsilon \to 0^+} \int_\Omega a_j^{\alpha\beta} \widetilde{E}_{\alpha\gamma}(x-y)(\partial_j \eta_\varepsilon)(y) \, dy \\
&\quad = \lim_{\varepsilon \to 0^+} \int_\Omega a_j^{\alpha\beta} \widetilde{E}_{\alpha\gamma}(x-y)\partial_j(\eta_\varepsilon - 1)(y) \, dy \\
&\quad = \lim_{\varepsilon \to 0^+} {}_{\mathcal{D}'(\Omega)}\Big\langle a_j^{\alpha\beta} \widetilde{E}_{\alpha\gamma}(x - \cdot), \partial_j(\eta_\varepsilon - 1)\Big\rangle_{\mathcal{D}(\Omega)} \\
&\quad = -\lim_{\varepsilon \to 0^+} {}_{\mathcal{D}'(\Omega)}\Big\langle a_j^{\alpha\beta} \partial_j \big[\widetilde{E}_{\alpha\gamma}(x - \cdot)\big], \eta_\varepsilon - 1\Big\rangle_{\mathcal{D}(\Omega)} \\
&\quad = \lim_{\varepsilon \to 0^+} {}_{\mathcal{D}'(\Omega)}\Big\langle \big(D^\top \big[\widetilde{E}(x - \cdot)\big]\big)_{\beta\gamma}, \eta_\varepsilon - 1\Big\rangle_{\mathcal{D}(\Omega)} \\
&\quad = -\lim_{\varepsilon \to 0^+} \delta_{\beta\gamma}(\eta_\varepsilon - 1)(x) = \delta_{\beta\gamma}.
\end{aligned} \tag{3.1.29}$$

where the penultimate step makes use of (1.6.7). From (3.1.26)-(3.1.29) we then conclude that

$$\mathrm{II} = u(x). \tag{3.1.30}$$

Combining (3.1.19)-(3.1.21) with (3.1.25) and (3.1.30) proves (3.1.6) in the case when $\partial\Omega$ is unbounded. The case when Ω is bounded is dealt with in an absolutely similar fashion. Finally, in the case when Ω is an exterior domain, the same type of reasoning applies, this time with (3.1.25) justified based on the additional assumption made in (3.1.8) and the first line of (3.1.24).

`Case II:` *Suppose* $p \in (1, \infty)$. By and large, we reason as before, bearing in mind that (3.1.10) now becomes (cf. [112, (3.6.27)])

$$\mathrm{Sym}(D; \nu) \bullet u \in \big[L^p(\partial\Omega, \sigma)\big]^N, \tag{3.1.31}$$

that in the current case all duality pairings used in the past are now ordinary integration on $\partial\Omega$ with respect to σ, and that if p' is the Hölder conjugate exponent of p then (1.4.24) and [112, (7.2.5)] imply $\int_{\partial\Omega} |\widetilde{E}(x - \cdot)|^{p'} \, d\sigma < +\infty$. The integral representation formula (3.1.9) eventually follows. □

There is also a natural version of Theorem 3.1.1 which employs the more inclusive scale of Lorentz spaces in place of Lebesgue spaces in the hypotheses made on the given function u, of the sort described below.

Theorem 3.1.2 *Retain the assumptions from Theorem 3.1.1 on the set* Ω*, as well as the first-order systems* D*,* $\widetilde{D}$*, and continue to define* $\widetilde{E}$ *as in* (3.1.1)*. This time, assume*

$u : \Omega \to \mathbb{C}^M$ is a vector-valued function with Lebesgue measurable components satisfying, for an aperture parameter $\kappa \in (0, \infty)$,

$$\begin{aligned} &\mathcal{N}_\kappa u \in L^{p,q}(\partial\Omega, \sigma) \text{ for some } p \in \left(\tfrac{n-1}{n}, \infty\right) \text{ and } q \in (0, \infty), \\ &Du \in \left[L^1_{\text{loc}}(\Omega, \mathcal{L}^n)\right]^N \text{ and } \mathfrak{P}(Du) \in L^{p,q}(\partial\Omega, \sigma). \end{aligned} \tag{3.1.32}$$

Then $\operatorname{Sym}(D; \nu) \bullet u$ *belongs to the Lorentz-based Hardy space* $\left[H^{p,q}(\partial\Omega, \sigma)\right]^N$ *and there exists some constant* $C = C(\Omega, D, p, q, \kappa) \in (0, \infty)$, *independent of* u, *such that*

$$\begin{aligned} &\left\|\operatorname{Sym}(D; \nu) \bullet u\right\|_{[H^{p,q}(\partial\Omega,\sigma)]^N} \\ &\qquad \leq C\left\|\mathcal{N}_\kappa u\right\|_{L^{p,q}(\partial\Omega,\sigma)} + C\left\|\mathfrak{P}(Du)\right\|_{L^{p,q}(\partial\Omega,\sigma)}. \end{aligned} \tag{3.1.33}$$

In addition,

$$\widetilde{E}(x - \cdot)\big|_{\partial\Omega} \in \left[H^{p,q}(\partial\Omega, \sigma)^*\right]^{N\times M} \text{ for each } x \in \mathbb{R}^n \setminus \partial\Omega, \tag{3.1.34}$$

and for $\mathcal{L}^n$*-a.e.* $x \in \mathbb{R}^n \setminus \partial\Omega$ *with the property that*

$$\int_\Omega \frac{|(Du)(y)|}{|x - y|^{n-1}}\,dy < +\infty \tag{3.1.35}$$

which is also assumed to be a Lebesgue point for the function u *in the case when* $x \in \Omega$ *one has*

$$\begin{aligned} &{}_{([H^{p,q}(\partial\Omega,\sigma)]^N)^*}\Big\langle \widetilde{E}^\top(x - \cdot)\big|_{\partial\Omega}, (-i)\operatorname{Sym}(D; \nu) \bullet u\Big\rangle_{[H^{p,q}(\partial\Omega,\sigma)]^N} \\ &\qquad - \int_\Omega \big\langle \widetilde{E}^\top(x - y), (Du)(y)\big\rangle\,dy = \begin{cases} u(x) & \text{if } x \in \Omega, \\ 0 & \text{if } x \in \mathbb{R}^n \setminus \overline{\Omega} \end{cases} \end{aligned} \tag{3.1.36}$$

provided u *also vanishes at infinity (in the sense of Definition 1.6.3) in the case when* Ω *happens to be an exterior domain.*

Proof We follow in broad outline the argument used in the proof of Theorem 3.1.1, so we shall only discuss the main differences. First, that actually $\operatorname{Sym}(D; \nu) \bullet u$ belongs to the Lorentz-based Hardy space $\left[H^{p,q}(\partial\Omega, \sigma)\right]^N$ and the estimate in (3.1.33) holds is seen from [113, Corollary 10.2.13]. Second, the membership in (3.1.34) is a consequence of [113, (4.7.2)], whose applicability in the present setting is ensured by (3.1.1), Theorem 1.4.2, and (2.4.92). Third, the integral representation formula (3.1.36) is established much as (3.1.6). In this regard, there are two important distinctions. Specifically, in place of (3.1.15) we now have

$$\begin{aligned} &\lim_{R\to\infty} \big(\theta_R \eta_\varepsilon \widetilde{E}^\top(x - \cdot)\big)\big|_{\partial\Omega} = \big(\eta_\varepsilon \widetilde{E}^\top(x - \cdot)\big)\big|_{\partial\Omega} \\ &\qquad \text{weak-}* \text{ in } \left[H^{p,q}(\partial\Omega, \sigma)^*\right]^{M\times N}. \end{aligned} \tag{3.1.37}$$

This is implied by the general weak-∗ convergence result established in [113, Lemma 4.8.13]. Also, in place of (3.1.16) we now have

$$\begin{aligned}
&\Big({}_{([H^{p,q}(\partial\Omega,\sigma)]^N)^*}\Big\langle \widetilde{E}^{\top}(x-\cdot)\big|_{\partial\Omega}, (-\mathrm{i})\mathrm{Sym}(D;\nu)\bullet u\Big\rangle_{[H^{p,q}(\partial\Omega,\sigma)]^N}\Big)_{\alpha} \qquad (3.1.38)\\
&= {}_{([H^{p,q}(\partial\Omega,\sigma)]^N)^*}\Big\langle (\widetilde{E}^{\top})_{\alpha\,\cdot}(x-\cdot)\big|_{\partial\Omega}, (-\mathrm{i})\mathrm{Sym}(D;\nu)\bullet u\Big\rangle_{[H^{p,q}(\partial\Omega,\sigma)]^N}\\
&= \lim_{R\to\infty} {}_{[(\mathrm{Lip}_c(\partial\Omega))']^N}\Big\langle (-\mathrm{i})\mathrm{Sym}(D;\nu)\bullet u, \big(\theta_R\eta_\varepsilon\widetilde{E}_{\cdot\,\alpha}(x-\cdot)\big)\big|_{\partial\Omega}\Big\rangle_{[\mathrm{Lip}_c(\partial\Omega)]^N}
\end{aligned}$$

for each $\alpha\in\{1,\dots,M\}$. This is justified as before, with [113, Lemma 4.8.14] presently employed in lieu of [113, Lemma 4.6.4]. Finally, to see that (3.1.25) holds in the case when $\partial\Omega$ is unbounded, in place of (3.1.24) we now estimate, with Ω_R abbreviating $\Omega\cap[B(0,\lambda R)\setminus B(0,R)]$ for each radius $R>0$, some fixed exponent $r\in\big(\max\{1,q\},\infty\big)$, and with prime indicating Hölder conjugation,

$$\begin{aligned}
|\mathrm{I}| &\le C\limsup_{R\to\infty}\Big\{R^{-n}\int_{\Omega\cap[B(0,\lambda R)\setminus B(0,R)]}|u|\,\mathrm{d}\mathcal{L}^n\Big\}\\
&= C\limsup_{R\to\infty}\Big\{R^{-n}\|u\|_{L^1(\Omega_R,\mathcal{L}^n)}\Big\}\\
&\le C\limsup_{R\to\infty}\Big\{R^{-n}\|u\|_{L^{\frac{np}{n-1},r}(\Omega,\mathcal{L}^n)}\|\mathbf{1}_{\Omega_R}\|_{L^{(\frac{np}{n-1})',r'}(\Omega,\mathcal{L}^n)}\Big\}\\
&\le C\limsup_{R\to\infty}\Big\{R^{-n}\|u\|_{L^{\frac{np}{n-1},q}(\Omega,\mathcal{L}^n)}\big[\mathcal{L}^n(\Omega_R)\big]^{1-\frac{n-1}{np}}\Big\}\\
&\le C\limsup_{R\to\infty}\Big\{R^{-(n-1)/p}\|\mathcal{N}_\kappa u\|_{L^{p,q}(\partial\Omega,\sigma)}\Big\}=0, \qquad (3.1.39)
\end{aligned}$$

thanks to [112, (6.2.26), (6.2.40), (6.2.61), (8.6.50)] and (3.1.32). Modulo these alterations, the argument in the proof of Theorem 3.1.1 goes through, and presently yields (3.1.36). □

There is actually a more general[2] phenomenon at play here, valid for arbitrary (not necessarily injectively elliptic) first-order systems, which we would like to single out in the next two propositions.

Proposition 3.1.3 *Fix $n\in\mathbb{N}$ with $n\ge 2$, and suppose $\Omega\subseteq\mathbb{R}^n$ is an Ahlfors regular domain. Denote by ν the geometric measure theoretic outward unit normal to Ω and abbreviate $\sigma:=\mathcal{H}^{n-1}\lfloor\partial\Omega$. Next, consider a homogeneous, first-order $N\times M$ system D with constant complex coefficients in $\mathbb{R}^n$ (where $N,M\in\mathbb{N}$) and suppose u is a vector-valued satisfying, for an aperture parameter $\kappa\in(0,\infty)$,*

[2] at least in the class of smooth functions

$$
\begin{aligned}
& u \in \big[\mathscr{C}^\infty(\Omega)\big]^M, \quad Du = 0 \ \text{in } \Omega, \text{ and} \\
& \mathcal{N}_\kappa u \in L^p(\partial\Omega, \sigma) \text{ for some } p \in \big(\tfrac{n-1}{n}, \infty\big),
\end{aligned}
\tag{3.1.40}
$$

as well as

$$
\fint_{B(0,\lambda R)\setminus B(0,R)} |u| \, d\mathcal{L}^n = o(1) \ \text{as } R \to \infty \tag{3.1.41}
$$

for some $\lambda \in (1, \infty)$ in the case when Ω happens to be an exterior domain. In addition, assume that

$$
\begin{aligned}
& w \in \big[\mathcal{D}'(\Omega)\big]^N \text{ has } D^\top w \in \big[\mathscr{E}'(\Omega)\big]^M \text{ as well as} \\
& w\big|_{\Omega\setminus K} \in \big[\mathrm{Lip}(\Omega \setminus K)\big]^N \text{ for a compact set } K \subseteq \Omega,
\end{aligned}
\tag{3.1.42}
$$

and

$$
\begin{aligned}
& w(x) = O(|x|^{1-n}) \ \text{as } |x| \to \infty \\
& \text{in the case when } \Omega \text{ is unbounded.}
\end{aligned}
\tag{3.1.43}
$$

Then $(-\mathrm{i})\mathrm{Sym}(D;\nu) \bullet u$ belongs to the Hardy space $\big[H^p(\partial\Omega, \sigma)\big]^N$. Also, the nontangential trace $w\big|_{\partial\Omega}^{\kappa-\mathrm{n.t.}}$ exists at σ-a.e. point on $\partial\Omega$, as a function this trace belongs to the space

$$
\Big(\big[H^p(\partial\Omega,\sigma)\big]^N\Big)^* = \begin{cases}
\big[\mathscr{C}^{(n-1)(\frac{1}{p}-1)}(\partial\Omega)\big]^N & \text{if } \partial\Omega \text{ compact and } p < 1, \\
\big[\mathrm{BMO}(\partial\Omega,\sigma)\big]^N & \text{if } \partial\Omega \text{ compact and } p = 1, \\
\big[L^{p'}(\partial\Omega,\sigma)\big]^N & \text{if } p > 1 \text{ and } p' := \big(1 - \tfrac{1}{p}\big)^{-1},
\end{cases}
\tag{3.1.44}
$$

and, in all cases described above,

$$
\Big\langle (-\mathrm{i})\mathrm{Sym}(D;\nu) \bullet u, w\big|_{\partial\Omega}^{\kappa-\mathrm{n.t.}} \Big\rangle = -{}_{[\mathscr{E}(\Omega)]^M}\Big\langle u, D^\top w \Big\rangle_{[\mathscr{E}'(\Omega)]^M} \tag{3.1.45}
$$

where the bracket $\langle\cdot,\cdot\rangle$ in the left-hand side of (3.1.45) *is viewed as the duality pairing (in the sense of* [113, Theorem 4.6.1], *presently used with $\Sigma := \partial\Omega$) between $(-\mathrm{i})\mathrm{Sym}(D;\nu) \bullet u \in \big[H^p(\partial\Omega,\sigma)\big]^N$ and $w\big|_{\partial\Omega}^{\kappa-\mathrm{n.t.}} \in \Big(\big[H^p(\partial\Omega,\sigma)\big]^N\Big)^*$.*

Finally, if $\partial\Omega$ is unbounded then $\big[w\big|_{\partial\Omega}^{\kappa-\mathrm{n.t.}}\big]$, the equivalence class of $w\big|_{\partial\Omega}^{\kappa-\mathrm{n.t.}}$ modulo constant vectors (in $\mathbb{C}^N$), belongs to

$$
\Big(\big[H^p(\partial\Omega,\sigma)\big]^N\Big)^* = \begin{cases}
\big[\dot{\mathscr{C}}^{(n-1)(\frac{1}{p}-1)}(\partial\Omega)/\sim\big]^N & \text{if } p < 1, \\
\big[\widetilde{\mathrm{BMO}}(\partial\Omega,\sigma)\big]^N & \text{if } p = 1,
\end{cases}
\tag{3.1.46}
$$

and in all these cases one has

$$
\Big\langle (-\mathrm{i})\mathrm{Sym}(D;\nu) \bullet u, \big[w\big|_{\partial\Omega}^{\kappa-\mathrm{n.t.}}\big] \Big\rangle = -{}_{[\mathscr{E}(\Omega)]^M}\Big\langle u, D^\top w \Big\rangle_{[\mathscr{E}'(\Omega)]^M} \tag{3.1.47}
$$

with a similar interpretation of the duality bracket in the left-hand side of (3.1.47).

The version of Theorem 3.1.1 in which the function u is a smooth null-solution of the system D becomes a consequence of Proposition 3.1.3 applied with w taken to be an arbitrary column of the matrix $\widetilde{E}(x - \cdot)$.

There is also a natural variant of Proposition 3.1.3 which employs the more inclusive scale of Lorentz spaces in place of Lebesgue spaces in the hypotheses made on the given function u. As indicated earlier, this variant yields a direct proof of the special case of Theorem 3.1.2 in which u is a smooth null-solution of the system D.

We now present the proof of Proposition 3.1.3.

Proof of Proposition 3.1.3 That $(-\mathrm{i})\mathrm{Sym}(D;\nu) \bullet u$ belongs to $\big[H^p(\partial\Omega,\sigma)\big]^N$ is seen from [113, Corollary 10.2.13] (presently employed with $q := p$), [112, (6.2.25)], and [113, (4.2.25)]. From (3.1.42)-(3.1.43) and a rudimentary extension theorem[3] we see that there exists

$$\begin{aligned} &\text{a bounded function } W \in \big[\mathrm{Lip}(\overline{\Omega})\big]^N \\ &\text{such that } W\big|_{\Omega\setminus K} = w\big|_{\Omega\setminus K}. \end{aligned} \tag{3.1.48}$$

Based on this and [112, Proposition 8.8.4] we then conclude that the nontangential trace $w\big|_{\partial\Omega}^{\kappa-\mathrm{n.t.}}$ exists at σ-a.e. point on $\partial\Omega$. In fact,

$$w\big|_{\partial\Omega}^{\kappa-\mathrm{n.t.}} = W\big|_{\partial\Omega} \in \big[L^\infty(\partial\Omega,\sigma) \cap \mathrm{Lip}(\partial\Omega)\big]^N \tag{3.1.49}$$

and (here (3.1.43) is also relevant) there exists $C \in (0,\infty)$ with the property that for each $x \in \partial\Omega$ we have

$$\Big|\Big(w\big|_{\partial\Omega}^{\kappa-\mathrm{n.t.}}\Big)(x)\Big| \le \frac{C}{1+|x|^{n-1}}. \tag{3.1.50}$$

Since the function in the right side of (3.1.50) belongs to any $L^q(\partial\Omega,\sigma)$ with $q \in (1,\infty)$, and since

$$L^\infty(\partial\Omega,\sigma) \cap \mathrm{Lip}(\partial\Omega) \subseteq \mathrm{BMO}(\partial\Omega,\sigma) \cap \Big(\bigcap_{0<\alpha<1} \mathscr{C}^\alpha(\partial\Omega)\Big), \tag{3.1.51}$$

all claims made about the memberships of $w\big|_{\partial\Omega}^{\kappa-\mathrm{n.t.}}$ and $\Big[w\big|_{\partial\Omega}^{\kappa-\mathrm{n.t.}}\Big]$ in the statement of the proposition follow.

To proceed, pick a scalar function

$$\eta \in \mathscr{C}_c^\infty(\Omega) \text{ such that } \eta \equiv 1 \text{ near } K \cup \mathrm{supp}\,(D^\top w). \tag{3.1.52}$$

Then

$$(1-\eta)w = (1-\eta)W \text{ as distributions in } \big[\mathcal{D}'(\Omega)\big]^N, \tag{3.1.53}$$

[3] for, say, bounded uniformly continuous functions

since the distributions in questions agree on $\Omega \setminus K$ (by (3.1.48)), and they both vanish near K (by (3.1.52)). Moreover,

$$(1-\eta)W \in \big[\mathrm{Lip}(\overline{\Omega})\big]^N \text{ and } \big((1-\eta)W\big)\big|_{\partial\Omega} = w\Big|_{\partial\Omega}^{\kappa-\text{n.t.}} \text{ at } \sigma\text{-a.e. point on } \partial\Omega, \tag{3.1.54}$$

given that $1-\eta \equiv 1$ near $\partial\Omega$. Let us choose a scalar function $\theta \in \mathscr{C}_c^\infty(\mathbb{R}^n)$ satisfying $0 \leq \theta \leq 1$, $\theta \equiv 1$ on $B(0,1)$, $\theta \equiv 0$ on $\mathbb{R}^n \setminus B(0,\lambda)$ and, for each $R > 0$, define $\theta_R(x) := \theta(x/R)$ for every $x \in \mathbb{R}^n$. Then for each $R > 0$ it follows that

$$\begin{aligned} &\text{the function } \theta_R(1-\eta)W \text{ belongs to } \big[\mathrm{Lip}_c(\overline{\Omega})\big]^N, \\ &\text{vanishes identically outside a bounded subset of } \overline{\Omega}, \\ &\text{and } \theta_R\Big(w\Big|_{\partial\Omega}^{\kappa-\text{n.t.}}\Big) = \big(\theta_R(1-\eta)W\big)\big|_{\partial\Omega} \in \big[\mathrm{Lip}_c(\partial\Omega)\big]^N. \end{aligned} \tag{3.1.55}$$

Also, for each $\alpha \in (0,1)$ we may write

$$\sup_{R>0}\Big\|\big(\theta_R(1-\eta)W\big)\big|_{\partial\Omega}\Big\|_{[\mathscr{C}^\alpha(\partial\Omega)]^N} = \sup_{R>0}\Big\|\theta_R\Big(w\Big|_{\partial\Omega}^{\kappa-\text{n.t.}}\Big)\Big\|_{[\mathscr{C}^\alpha(\partial\Omega)]^N} < \infty, \tag{3.1.56}$$

thanks to (3.1.49) and (3.1.51). If $\partial\Omega$ is unbounded we therefore have

$$\lim_{R\to\infty}\Big[\big(\theta_R(1-\eta)W\big)\big|_{\partial\Omega}\Big] = \Big[w\Big|_{\partial\Omega}^{\kappa-\text{n.t.}}\Big] \quad \text{weak-}* \text{ in} \begin{cases} \Big[\dot{\mathscr{C}}^{(n-1)(1/p-1)}(\partial\Omega)\big/\sim\Big]^N & \text{if } p<1, \\ \big[\widetilde{\mathrm{BMO}}(\partial\Omega,\sigma)\big]^N & \text{if } p=1, \end{cases} \tag{3.1.57}$$

while if $\partial\Omega$ is bounded we have

$$\lim_{R\to\infty}\big(\theta_R(1-\eta)W\big)\big|_{\partial\Omega} = w\Big|_{\partial\Omega}^{\kappa-\text{n.t.}} \quad \text{weak-}* \text{ in} \begin{cases} \big[\mathscr{C}^{(n-1)(1/p-1)}(\partial\Omega)\big]^N & \text{if } p<1, \\ \big[\mathrm{BMO}(\partial\Omega,\sigma)\big]^N & \text{if } p=1. \end{cases} \tag{3.1.58}$$

Indeed, all these weak convergence properties are implied by (3.1.57) plus the general weak-$*$ convergence results established in [113, Lemma 4.8.4] and, respectively, [113, Lemma 4.8.1] (in the latter case also bearing in mind the trivial bounded embedding $L^\infty(\partial\Omega,\sigma) \hookrightarrow \mathrm{BMO}(\partial\Omega,\sigma)$). Furthermore, in the case when $p > 1$, since the function in the right side of (3.1.50) belongs to the space $L^{p'}(\partial\Omega,\sigma)$ for $p' := \big(1-\frac{1}{p}\big)^{-1} \in (1,\infty)$, Lebesgue's Dominated Convergence Theorem ensures that

$$\lim_{R\to\infty}\big(\theta_R(1-\eta)W\big)\big|_{\partial\Omega} = w\Big|_{\partial\Omega}^{\kappa-\text{n.t.}} \text{ in } \big[L^{p'}(\partial\Omega,\sigma)\big]^N. \tag{3.1.59}$$

For clarity of exposition consider first the case when $\partial\Omega$ is unbounded (which forces Ω to be unbounded). In this scenario, we may write

$$
\begin{aligned}
&{}_{[H^p(\partial\Omega,\sigma)]^N}\Big\langle (-\mathrm{i})\mathrm{Sym}(D;\nu)\bullet u,\, \Big[w\big|^{\kappa-\mathrm{n.t.}}_{\partial\Omega}\Big]\Big\rangle_{([H^p(\partial\Omega,\sigma)]^N)^*}\\
&= \lim_{R\to\infty} {}_{[H^p(\partial\Omega,\sigma)]^N}\Big\langle (-\mathrm{i})\mathrm{Sym}(D;\nu)\bullet u,\, \Big[\big(\theta_R(1-\eta)W\big)\big|_{\partial\Omega}\Big]\Big\rangle_{([H^p(\partial\Omega,\sigma)]^N)^*}\\
&= \lim_{R\to\infty} {}_{[(\mathrm{Lip}_c(\partial\Omega))']^N}\Big\langle (-\mathrm{i})\mathrm{Sym}(D;\nu)\bullet u,\, \Big[\big(\theta_R(1-\eta)W\big)\big|_{\partial\Omega}\Big]\Big\rangle_{[\mathrm{Lip}_c(\partial\Omega)]^N}\\
&= -\lim_{R\to\infty}\int_\Omega \Big\langle u, D^\top\big(\theta_R(1-\eta)W\big)\Big\rangle\, d\mathcal{L}^n,
\end{aligned}
\tag{3.1.60}
$$

where the first equality is a consequence of (3.1.57) and (3.1.59) (depending on how p compares with 1), the second equality in (3.1.60) is implied by [113, Lemma 4.6.4], and the third equality in (3.1.60) follows from (3.1.55) and the definition in [113, Proposition 10.2.11] (cf. [113, (10.2.94)]).

To continue, for each fixed $R > 0$ we compute

$$
\begin{aligned}
&D^\top\big(\theta_R(1-\eta)W\big)\\
&\quad= [D^\top,\theta_R]\big((1-\eta)W\big) + \theta_R D^\top\big((1-\eta)W\big)\\
&\quad= (1-\eta)(-\mathrm{i})\,\mathrm{Sym}\big(D^\top;\nabla\theta_R\big)W + \theta_R D^\top\big((1-\eta)w\big)\\
&\quad= (1-\eta)(-\mathrm{i})\,\mathrm{Sym}\big(D^\top;\nabla\theta_R\big)W + \theta_R[D^\top,1-\eta]w + \theta_R(1-\eta)D^\top w\\
&\quad= (1-\eta)(-\mathrm{i})\,\mathrm{Sym}\big(D^\top;\nabla\theta_R\big)W + \theta_R(-\mathrm{i})\,\mathrm{Sym}\big(D^\top;\nabla(1-\eta)\big)w\\
&\quad= (1-\eta)(-\mathrm{i})\,\mathrm{Sym}\big(D^\top;\nabla\theta_R\big)W - \theta_R(-\mathrm{i})\,\mathrm{Sym}\big(D^\top;\nabla\eta\big)w\\
&\quad= (1-\eta)(-\mathrm{i})\,\mathrm{Sym}\big(D^\top;\nabla\theta_R\big)W - \theta_R(-\mathrm{i})\,\mathrm{Sym}\big(D^\top;\nabla\eta\big)W,
\end{aligned}
\tag{3.1.61}
$$

making repeated applications of formula [112, (1.7.20)], the observation that we currently have $(1-\eta)D^\top w = 0$ since $1-\eta \equiv 0$ on the support of the distribution $D^\top w$, and the fact that $\nabla\eta$ is supported in $\Omega\setminus K$, an open set where W and w coincide (cf. (3.1.48)). In turn, we may use (3.1.61) to decompose

$$
\int_\Omega \Big\langle u, D^\top\big(\theta_R(1-\eta)W\big)\Big\rangle\, d\mathcal{L}^n = \mathrm{I}_R + \mathrm{II}_R \tag{3.1.62}
$$

where

$$
\mathrm{I}_R := (-\mathrm{i})\int_\Omega \Big\langle u, (1-\eta)\,\mathrm{Sym}\big(D^\top;\nabla\theta_R\big)W\Big\rangle\, d\mathcal{L}^n, \tag{3.1.63}
$$

and

$$
\mathrm{II}_R := \mathrm{i}\int_\Omega \Big\langle u, \theta_R\,\mathrm{Sym}\big(D^\top;\nabla\eta\big)W\Big\rangle\, d\mathcal{L}^n. \tag{3.1.64}
$$

In view of the fact that we are presently assuming the set $\partial\Omega$ to be unbounded, the estimate recorded in [112, (8.6.51) in Proposition 8.6.3] is applicable for the choice $p := (n-1)/n$ and $E := \Omega \cap \big[B(0,\lambda R) \setminus B(0,R)\big]$. In concert with Hölder's inequality, (3.1.48), and (3.1.43), this gives

$$\begin{aligned}
\big|\mathrm{I}_R\big| &\le CR^{-n} \int_{\Omega\cap[B(0,\lambda R)\setminus B(0,R)]} |u|\,\mathrm{d}\mathcal{L}^n \\
&\le CR^{-n} \Big(\int_{\Omega\cap[B(0,\lambda R)\setminus B(0,R)]} |u|^{\frac{np}{n-1}}\,\mathrm{d}\mathcal{L}^n \Big)^{\frac{n-1}{np}} \Big(\int_{B(0,\lambda R)} \mathrm{d}\mathcal{L}^n \Big)^{1-\frac{n-1}{np}} \\
&\le CR^{-(n-1)/p} \|\mathcal{N}_\kappa u\|_{L^p(\partial\Omega,\sigma)} \longrightarrow 0 \text{ as } R \to \infty.
\end{aligned} \tag{3.1.65}$$

Thus, on the one hand,

$$\lim_{R\to\infty} \mathrm{I}_R = 0. \tag{3.1.66}$$

On the other hand, another appeal to [112, (1.7.20)] permits us to compute

$$\begin{aligned}
D^\top w &= D^\top\big((1-\eta)w\big) + D^\top(\eta w) \\
&= [D^\top, 1-\eta]w + (1-\eta)D^\top w + D^\top(\eta w) \\
&= \mathrm{i}\,\mathrm{Sym}\big(D^\top; \nabla\eta\big)w + D^\top(\eta w)
\end{aligned} \tag{3.1.67}$$

since, as noted earlier, $(1-\eta)D^\top w = 0$ in Ω. Next, we make use of (3.1.67) to write

$$\begin{aligned}
{}_{[\mathscr{E}(\Omega)]^M}\Big\langle u, D^\top w\Big\rangle_{[\mathscr{E}'(\Omega)]^M} &= {}_{[\mathscr{E}(\Omega)]^M}\Big\langle u, \mathrm{i}\,\mathrm{Sym}\big(D^\top; \nabla\eta\big)w\Big\rangle_{[\mathscr{E}'(\Omega)]^M} \\
&\quad + {}_{[\mathscr{E}(\Omega)]^M}\Big\langle u, D^\top(\eta w)\Big\rangle_{[\mathscr{E}'(\Omega)]^M} \\
&= {}_{[\mathscr{E}(\Omega)]^M}\Big\langle u, \mathrm{i}\,\mathrm{Sym}\big(D^\top; \nabla\eta\big)W\Big\rangle_{[\mathscr{E}'(\Omega)]^M} \\
&= \int_\Omega \Big\langle u, \mathrm{i}\,\mathrm{Sym}\big(D^\top; \nabla\eta\big)W\Big\rangle\,\mathrm{d}\mathcal{L}^n \\
&= \lim_{R\to\infty} \mathrm{II}_R,
\end{aligned} \tag{3.1.68}$$

where we have used the fact that

$${}_{[\mathscr{E}(\Omega)]^M}\Big\langle u, D^\top(\eta w)\Big\rangle_{[\mathscr{E}'(\Omega)]^M} = {}_{[\mathscr{E}(\Omega)]^M}\Big\langle Du, \eta w\Big\rangle_{[\mathscr{E}'(\Omega)]^M} = 0 \tag{3.1.69}$$

since $\eta w \in \big[\mathscr{E}'(\Omega)\big]^M$ and $Du = 0$ (cf. (3.1.40)), that $w = W$ on the support of $\nabla\eta$, the definition of II_R from (3.1.64), and the observation that θ_R becomes identically 1 on the support of $\nabla\eta$ when R is large.

Hence, in the case when $\partial\Omega$ is unbounded, gathering (3.1.60), (3.1.62)-(3.1.64), (3.1.66), and (3.1.68) proves (3.1.46). Finally, the case when $\partial\Omega$ is bounded is dealt with in a similar fashion, with one important difference. Specifically, when Ω is an exterior domain then (3.1.66) is proved by using (3.1.63), (3.1.43), and (3.1.41). The proof of Proposition 3.1.3 is therefore complete. □

Here is the second proposition alluded to earlier, a close companion to Proposition 3.1.3 which we shall use later in Corollary 3.1.5 to further expand on Theorem 3.1.1.

Proposition 3.1.4 *Pick $n \in \mathbb{N}$ with $n \geq 2$, and assume $\Omega \subseteq \mathbb{R}^n$ is an Ahlfors regular domain. Denote by ν the geometric measure theoretic outward unit normal to Ω and abbreviate $\sigma := \mathcal{H}^{n-1}\lfloor\partial\Omega$. Next, suppose $\mathbb{X}$ is a Generalized Banach Function Space on $(\partial\Omega, \sigma)$ with the property that*

$$\begin{gathered} \mathcal{M}_{\partial\Omega} : \mathbb{X} \to \mathbb{X} \text{ and } \mathcal{M}_{\partial\Omega} : \mathbb{X}' \to \mathbb{X}' \\ \text{are well-defined bounded mappings,} \end{gathered} \tag{3.1.70}$$

where $\mathcal{M}_{\partial\Omega}$ is the Hardy-Littlewood maximal operator on $\partial\Omega$, and $\mathbb{X}'$ is the associated space of $\mathbb{X}$ (cf. [113, Definitions 5.1.4, 5.1.11]). *In addition, consider a homogeneous first-order $N \times M$ system D with constant complex coefficients in $\mathbb{R}^n$ (where $N, M \in \mathbb{N}$) and assume u is a vector-valued satisfying, for some aperture parameter $\kappa \in (0, \infty)$,*

$$u \in \big[\mathscr{C}^\infty(\Omega)\big]^M, \quad Du = 0 \text{ in } \Omega, \text{ and } \mathcal{N}_\kappa u \in \mathbb{X}, \tag{3.1.71}$$

as well as

$$⨍_{B(0,\lambda R)\setminus B(0,R)} |u| \, d\mathcal{L}^n = o(1) \text{ as } R \to \infty \tag{3.1.72}$$

for some $\lambda \in (1, \infty)$ in the case when Ω happens to be an exterior domain. Finally, assume that

$$\begin{gathered} w \in \big[\mathcal{D}'(\Omega)\big]^N \text{ has } D^\top w \in \big[\mathscr{E}'(\Omega)\big]^M \text{ as well as} \\ w\big|_{\Omega\setminus K} \in \big[\mathrm{Lip}(\Omega \setminus K)\big]^N \text{ for a compact set } K \subseteq \Omega, \end{gathered} \tag{3.1.73}$$

and also

$$\begin{gathered} w(x) = O(|x|^{1-n}) \text{ as } |x| \to \infty \\ \text{in the case when } \Omega \text{ is unbounded.} \end{gathered} \tag{3.1.74}$$

Then the (vector) distribution $(-\mathrm{i})\mathrm{Sym}(D;\nu) \bullet u$ actually belongs to the space $\big[\mathbb{X}\big]^N$ and satisfies the estimate

$$\big\|\mathrm{Sym}(D;\nu) \bullet u\big\|_{[\mathbb{X}]^N} \leq C\big\|\mathcal{N}_\kappa u\big\|_{\mathbb{X}}, \tag{3.1.75}$$

the nontangential trace $w\big|_{\partial\Omega}^{\kappa-\mathrm{n.t.}}$ exists at σ-a.e. point on $\partial\Omega$, as a function this trace belongs to the space $\big[\mathbb{X}'\big]^N$, and one has (with an absolutely convergent integral in the left side)

$$\int_{\partial\Omega}\Big\langle(-\mathrm{i})\mathrm{Sym}(D;\nu)\bullet u,\, w\Big|_{\partial\Omega}^{\kappa-\mathrm{n.t.}}\Big\rangle\,d\sigma = -{}_{[\mathscr{E}(\Omega)]^M}\Big\langle u,\, D^{\top}w\Big\rangle_{[\mathscr{E}'(\Omega)]^M}. \tag{3.1.76}$$

Proof Combining (3.1.70) with [113, Proposition 5.2.7] shows that

$$\begin{aligned}&\text{there exist } q\in(1,\infty) \text{ and } \varepsilon\in(0,1) \text{ such that}\\ &\mathbb{X}\hookrightarrow L^q\Big(\partial\Omega,\,\frac{\sigma(x)}{1+|x|^{n-1-\varepsilon}}\Big) \text{ continuously.}\end{aligned} \tag{3.1.77}$$

In light of the current assumptions, [113, Proposition 5.2.7] also guarantees that if $q' := (1-1/q)^{-1}\in(1,\infty)$ denotes the Hölder conjugate exponent of q, then

$$L^{q'}\Big(\partial\Omega,\,(1+|x|^{n-1-\varepsilon})^{q'-1}\sigma(x)\Big)\hookrightarrow\mathbb{X}'. \tag{3.1.78}$$

As notes in (3.1.48), there exists

$$\begin{aligned}&\text{a bounded function } W\in\big[\mathrm{Lip}(\overline{\Omega})\big]^N\\ &\text{such that } W\big|_{\Omega\setminus K} = w\big|_{\Omega\setminus K}.\end{aligned} \tag{3.1.79}$$

From this, [112, Proposition 8.8.4], and (3.1.74) we then conclude that the nontangential trace $w\Big|_{\partial\Omega}^{\kappa-\mathrm{n.t.}}$ exists at σ-a.e. point on $\partial\Omega$, satisfies

$$w\Big|_{\partial\Omega}^{\kappa-\mathrm{n.t.}} = W\big|_{\partial\Omega}\in\big[\mathrm{Lip}(\partial\Omega)\big]^N, \tag{3.1.80}$$

and there exists $C\in(0,\infty)$ such that for each $x\in\partial\Omega$ we have

$$\Big|\Big(w\Big|_{\partial\Omega}^{\kappa-\mathrm{n.t.}}\Big)(x)\Big| \leq \frac{C}{1+|x|^{n-1}}. \tag{3.1.81}$$

In particular, (3.1.80) implies that $w\Big|_{\partial\Omega}^{\kappa-\mathrm{n.t.}}$ is σ-measurable which, in concert with (3.1.81), may be used to conclude that

$$\begin{aligned}\int_{\partial\Omega}&\Big|\Big(w\Big|_{\partial\Omega}^{\kappa-\mathrm{n.t.}}\Big)(x)\Big|^{q'}(1+|x|^{n-1-\varepsilon})^{q'-1}\,d\sigma(x)\\ &\leq C\int_{\partial\Omega}\frac{d\sigma(x)}{1+|x|^{n-1+\varepsilon(q'-1)}}<\infty\end{aligned} \tag{3.1.82}$$

since $\partial\Omega$ is Ahlfors regular. Together with (3.1.78), this ultimately shows that

$$w\Big|_{\partial\Omega}^{\kappa-\mathrm{n.t.}} \text{ belongs to } \mathbb{X}'. \tag{3.1.83}$$

To proceed, suppose the homogeneous first-order $N\times M$ system D with constant complex coefficients is expressed as

$$D = \Big(\sum_{j=1}^{n} a_j^{\alpha\beta}\partial_j\Big)_{\substack{1\leq\alpha\leq N\\ 1\leq\beta\leq M}} \tag{3.1.84}$$

and that the vector-valued function u has scalar components $(u_\beta)_{1\leq\beta\leq M}$ in Ω. For each fixed index $\alpha \in \{1,\dots,N\}$, consider the vector field

$$\vec{F}_\alpha = (F_j^\alpha)_{1\leq j\leq n} : \Omega \longrightarrow \mathbb{C}^n \tag{3.1.85}$$

with components given by

$$F_j^\alpha := \sum_{\beta=1}^{M} a_j^{\alpha\beta} u_\beta \text{ in } \Omega, \quad \text{for } j \in \{1,\dots,n\}. \tag{3.1.86}$$

Then

$$\mathcal{N}_\kappa \vec{F}_\alpha \in \mathbb{X} \subseteq L^1_{\rm loc}(\partial\Omega,\sigma), \tag{3.1.87}$$

and

$$\operatorname{div}\vec{F}_\alpha = \sum_{j=1}^{n}\partial_j F_j^\alpha = \sum_{\beta=1}^{M}\sum_{j=1}^{n} a_j^{\alpha\beta}\partial_j u_\beta = (Du)_\alpha = 0 \text{ in } \mathcal{D}'(\Omega). \tag{3.1.88}$$

Given any $\psi = (\psi_\alpha)_{1\leq\alpha\leq N} \in \big[\mathrm{Lip}_c(\partial\Omega)\big]^N$, if we let $\Psi = (\Psi_\alpha)_{1\leq\alpha\leq N}$ be any $\mathbb{C}^N$-valued function satisfying

$$\begin{gathered}\Psi \in \big[\mathrm{Lip}(\overline{\Omega})\big]^N,\ \Psi\big|_{\partial\Omega} = \psi, \text{ and } \Psi \equiv 0\\ \text{outside of some compact subset of } \overline{\Omega},\end{gathered} \tag{3.1.89}$$

then according to the definitions given in [113, (10.2.94)] and (A.0.95) we have (bearing in mind that we are currently assuming $Du = 0$ and $\operatorname{div}\vec{F}_\alpha = 0$ for each α)

$$\begin{aligned}\big\langle(-\mathrm{i})\mathrm{Sym}(D;\nu)\bullet u,\psi\big\rangle &= -\int_\Omega \langle u, D^\top\Psi\rangle\,\mathrm{d}\mathcal{L}^n = \int_\Omega a_j^{\alpha\beta}u_\beta(\partial_j\Psi_\alpha)\,\mathrm{d}\mathcal{L}^n\\ &= \int_\Omega \vec{F}_\alpha\cdot\nabla\Psi_\alpha\,\mathrm{d}\mathcal{L}^n = \big\langle \nu\bullet\vec{F}_\alpha,\psi_\alpha\big\rangle,\end{aligned} \tag{3.1.90}$$

where the brackets stand for (vector) distributional pairing on $\partial\Omega$. In view of the arbitrariness of ψ, this proves that

$$(-\mathrm{i})\mathrm{Sym}(D;\nu)\bullet u = \big(\nu\bullet\vec{F}_\alpha\big)_{1\leq\alpha\leq N} \in \big[\mathrm{Lip}_c(\partial\Omega)'\big]^N. \tag{3.1.91}$$

Having established this, from [113, Proposition 10.2.6] and (3.1.97) we then conclude that

$$\mathrm{Sym}\big(D;\nu\big)\bullet u \text{ belongs to } \big[\mathbb{X}\big]^N \tag{3.1.92}$$

and the estimate claimed in (3.1.75) holds. As in the proof of Proposition 3.1.3, pick a scalar function

$$\eta \in \mathscr{C}_c^\infty(\Omega) \text{ such that } \eta \equiv 1 \text{ near } K \cup \operatorname{supp}(D^\top w). \tag{3.1.93}$$

and also choose a scalar function $\theta \in \mathscr{C}_c^\infty(\mathbb{R}^n)$ satisfying $0 \le \theta \le 1$, as well as $\theta \equiv 1$ on $B(0,1)$, and $\theta \equiv 0$ on $\mathbb{R}^n \setminus B(0,\lambda)$. For each given dilation factor $R > 0$, define $\theta_R(x) := \theta(x/R)$ at every point $x \in \mathbb{R}^n$. Then

$$\begin{aligned}
&\int_{\partial\Omega} \Big\langle (-\mathrm{i})\mathrm{Sym}(D;\nu) \bullet u,\, \Big[w\big|^{\kappa-\mathrm{n.t.}}_{\partial\Omega}\Big]\Big\rangle \, \mathrm{d}\sigma \\
&= \lim_{R\to\infty} \int_{\partial\Omega} \Big\langle (-\mathrm{i})\mathrm{Sym}(D;\nu) \bullet u,\, \Big[\big(\theta_R(1-\eta)W\big)\big|_{\partial\Omega}\Big]\Big\rangle \, \mathrm{d}\sigma \\
&= \lim_{R\to\infty} {}_{[(\mathrm{Lip}_c(\partial\Omega))']^N}\Big\langle (-\mathrm{i})\mathrm{Sym}(D;\nu) \bullet u,\, \Big[\big(\theta_R(1-\eta)W\big)\big|_{\partial\Omega}\Big]\Big\rangle_{[\mathrm{Lip}_c(\partial\Omega)]^N} \\
&= -\lim_{R\to\infty} \int_{\Omega} \Big\langle u,\, D^\top\big(\theta_R(1-\eta)W\big)\Big\rangle \, \mathrm{d}\mathcal{L}^n.
\end{aligned} \tag{3.1.94}$$

Above, the first equality is a consequence of (3.1.83), (3.1.92), the generalized Hölder inequality from [113, Proposition 5.1.12], and Lebesgue's Dominated Convergence Theorem. The second equality in (3.1.94) is implied by (3.1.55) and [112, Proposition 4.1.4]. The third equality in (3.1.94) follows from (3.1.55) and the definition in [113, Proposition 10.2.11, (10.2.94)].

Formula (3.1.94) should be compared with (3.1.60). From this point on we follow the proof of Proposition 3.1.3 with one key alteration. Specifically, in the case when $\partial\Omega$ is unbounded, in place of (3.1.65) we now make use of Hölder's inequality, [112, Proposition 8.6.3], [112, (8.1.17)], the last membership in (3.1.71), and (3.1.77) to estimate (with q and ε as in (3.1.77)):

$$\begin{aligned}
\big|\mathrm{I}_R\big| &\le CR^{-n} \int_{\Omega\cap[B(0,\lambda R)\setminus B(0,R)]} |u| \, \mathrm{d}\mathcal{L}^n \\
&\le CR^{-n}\Big(\int_{\Omega\cap[B(0,\lambda R)\setminus B(0,R)]} |u|^{\frac{nq}{n-1}} \, \mathrm{d}\mathcal{L}^n\Big)^{\frac{n-1}{nq}} \Big(\int_{B(0,\lambda R)} \mathrm{d}\mathcal{L}^n\Big)^{1-\frac{n-1}{nq}} \\
&\le CR^{-\frac{n-1}{q}}\Big(\int_{\partial\Omega\cap B(0,\lambda(2+\kappa)R)} |\mathcal{N}_\kappa u|^q \, \mathrm{d}\sigma\Big)^{1/q} \\
&\le CR^{-\frac{n-1}{q}}\Big(\int_{\partial\Omega\cap B(0,\lambda(2+\kappa)R)} R^{n-1-\varepsilon}\frac{|(\mathcal{N}_\kappa u)(y)|^q}{1+|y|^{n-1-\varepsilon}} \, \mathrm{d}\sigma(y)\Big)^{1/q} \\
&\le CR^{-\frac{\varepsilon}{q}} \big\|\mathcal{N}_\kappa u\big\|_{L^q\left(\partial\Omega,\, \frac{\sigma(y)}{1+|y|^{n-1-\varepsilon}}\right)} \le CR^{-\frac{\varepsilon}{q}} \|\mathcal{N}_\kappa u\|_{\mathbb{X}} \\
&= o(1) \text{ as } R \to \infty.
\end{aligned} \tag{3.1.95}$$

Consequently, (3.1.66) continues to be true in the current setting. Having established this, the rest of the argument in the proof of Proposition 3.1.3 goes through and gives (3.1.76). □

As promised earlier, we shall now employ Proposition 3.1.4 to produce the following version of Theorem 3.1.1 with the role of the Lebesgue spaces played by Generalized Banach Function Spaces.

Corollary 3.1.5 *Pick $n \in \mathbb{N}$ with $n \geq 2$, and suppose $\Omega \subseteq \mathbb{R}^n$ is an Ahlfors regular domain. Denote by ν the geometric measure theoretic outward unit normal to Ω and abbreviate $\sigma := \mathcal{H}^{n-1}\lfloor\partial\Omega$. Consider a homogeneous, first-order $N \times M$ system D with constant complex coefficients in $\mathbb{R}^n$ (where $N, M \in \mathbb{N}$) which is injectively elliptic (cf. (1.3.18)), and suppose $\widetilde{D}$ is a homogeneous first-order $M \times N$ system with constant complex coefficients in $\mathbb{R}^n$ which complements D (i.e., (1.3.21) holds). In particular, since $L := \widetilde{D}D$ is a weakly elliptic second-order $M \times M$ system in $\mathbb{R}^n$ it is meaningful to consider the matrix-valued fundamental solution $E_{L^\top}$ associated with $L^\top$ as in Theorem 1.4.2. With $\widetilde{D}^\top$ acting on the columns of $E_{L^\top}$, define*

$$\widetilde{E} := \widetilde{D}^\top E_{L^\top}. \tag{3.1.96}$$

Next, suppose $\mathbb{X}$ is a Generalized Banach Function Space on $(\partial\Omega, \sigma)$ with the property that

$$\begin{gathered} \mathcal{M}_{\partial\Omega} : \mathbb{X} \to \mathbb{X} \text{ and } \mathcal{M}_{\partial\Omega} : \mathbb{X}' \to \mathbb{X}' \\ \text{are well-defined bounded mappings,} \end{gathered} \tag{3.1.97}$$

where $\mathcal{M}_{\partial\Omega}$ is the Hardy-Littlewood maximal operator on $\partial\Omega$, and $\mathbb{X}'$ is the associated space of $\mathbb{X}$ (cf. [113, Definitions 5.1.4, 5.1.11]*). Finally, let $u : \Omega \to \mathbb{C}^M$ be a vector-valued function with Lebesgue measurable components satisfying, for an aperture parameter $\kappa \in (0, \infty)$,*

$$\mathcal{N}_\kappa u \in \mathbb{X} \text{ and } Du = 0 \text{ in } [\mathcal{D}'(\Omega)]^N. \tag{3.1.98}$$

Then the distribution $\mathrm{Sym}(D; \nu) \bullet u$ *actually belongs to the space* $[\mathbb{X}]^N$ *and there exists some constant $C \in (0, \infty)$, which depends only on Ω, D, $\mathbb{X}$, and κ, such that*

$$\|\mathrm{Sym}(D; \nu) \bullet u\|_{[\mathbb{X}]^N} \leq C\|\mathcal{N}_\kappa u\|_{\mathbb{X}}. \tag{3.1.99}$$

Moreover, with an absolutely convergent integral in the right side one has

$$u(x) = \int_{\partial\Omega} \langle \widetilde{E}^\top(x - y), (-\mathrm{i})(\mathrm{Sym}(D; \nu) \bullet u)(y)\rangle \, d\sigma(y) \quad \forall\, x \in \Omega, \tag{3.1.100}$$

if one also assumes that there exists $\lambda \in (1, \infty)$ such that

$$\fint_{B(0,\lambda R)\setminus B(0,R)} |u| \, d\mathcal{L}^n = o(1) \text{ as } R \to \infty \tag{3.1.101}$$

in the case when Ω happens to be an exterior domain.

Proof All claims follow from Proposition 3.1.4 and Theorem 1.4.2 (also bearing in mind the elliptic regularity result recorded in [112, (6.5.40)]) taking the vector distribution w to be an arbitrary column of the matrix $\widetilde{E}(x - \cdot)$. □

We are now in a position to discuss a rather powerful and elegant quantitative Fatou-type theorem, for null-solutions of injectively elliptic first-order systems in arbitrary UR domains in $\mathbb{R}^n$. At this level of generality, such a result is the first of its kind. In particular, we wish to stress that the underlying domain may fail to satisfy a corkscrew condition (which has typically been "de rigueur" even in the scalar case), and may even fail to be n-thick (in the sense of [112, Definition 5.1.1]). The "bullet product" technology, initiated in [112, §4.2] (cf. [112, Proposition 4.2.3]) and subsequently developed more fully in [113, §10.2] (starting with [113, Theorem 10.2.1]), is key to the proof of our result, even though the qualitative aspect of the theorem (i.e., the a.e. existence of the nontangential pointwise boundary trace; see item *(i)* below) does not involve this notion in any way. Finally, we note that for the membership of the nontangential maximal operator to Lebesgue spaces arbitrary exponents $p \in \big(\frac{n-1}{n}, \infty\big)$ are allowed, and that said membership may also be prescribed in large classes of Generalized Banach Function Spaces.

Theorem 3.1.6 *Fix $n \in \mathbb{N}$ with $n \geq 2$, and suppose $\Omega \subseteq \mathbb{R}^n$ is an arbitrary* UR *domain. Abbreviate $\sigma := \mathcal{H}^{n-1}\lfloor\partial\Omega$ and denote by ν the geometric measure theoretic outward unit normal to Ω. Next, consider an injectively elliptic homogeneous first-order $N \times M$ system D with constant complex coefficients in $\mathbb{R}^n$ (where $N, M \in \mathbb{N}$), and suppose $u : \Omega \to \mathbb{C}^M$ is a vector-valued function with Lebesgue measurable components satisfying, for some aperture parameter $\kappa \in (0, \infty)$,*

$$\begin{aligned} &\mathcal{N}_\kappa u \in L^p(\partial\Omega, \sigma) \text{ with } p \in \big(\tfrac{n-1}{n}, \infty\big), \\ &\text{and } Du = 0 \text{ in } [\mathcal{D}'(\Omega)]^N. \end{aligned} \tag{3.1.102}$$

Then the following statements are true.

(i) The nontangential boundary trace

$$\begin{aligned} &u\big|_{\partial\Omega}^{\kappa-\text{n.t.}} \text{ exists (in } \mathbb{C}^M\text{) at } \sigma\text{-a.e. point on } \partial\Omega \text{ and} \\ &\text{the function } u\big|_{\partial\Omega}^{\kappa-\text{n.t.}} \text{ is } \sigma\text{-measurable on } \partial\Omega; \\ &\text{also, } u\big|_{\partial\Omega}^{\kappa-\text{n.t.}} \text{ belongs to the space } \big[L^p(\partial\Omega, \sigma)\big]^M \\ &\text{and one has } \Big\| u\big|_{\partial\Omega}^{\kappa-\text{n.t.}} \Big\|_{[L^p(\partial\Omega,\sigma)]^M} \leq \|\mathcal{N}_\kappa u\|_{L^p(\partial\Omega,\sigma)}. \end{aligned} \tag{3.1.103}$$

Moreover, for any other given aperture parameter $\kappa' \in (0, \infty)$ the nontangential boundary trace $u\big|_{\partial\Omega}^{\kappa'-\text{n.t.}}$ also exists, and agrees with $u\big|_{\partial\Omega}^{\kappa-\text{n.t.}}$, at σ-a.e. point on $\partial\Omega$. Henceforth, it is therefore meaningful to drop the dependence of $u\big|_{\partial\Omega}^{\kappa-\text{n.t.}}$ on the aperture parameter κ, and simply denote this function by $u\big|_{\partial\Omega}^{\text{n.t.}}$.

(ii) Suppose $p \in (1,\infty)$ and, in the case when Ω is an exterior domain, make the additional assumption that u vanishes at infinity (in the sense of Definition 1.6.3). Then one has

$$\|\mathcal{N}_\kappa u\|_{L^p(\partial\Omega,\sigma)} \approx \big\|\mathrm{Sym}(D;\nu)\big(u\big|^{{}^{\mathrm{n.t.}}}_{\partial\Omega}\big)\big\|_{[L^p(\partial\Omega,\sigma)]^N} \approx \big\|u\big|^{{}^{\mathrm{n.t.}}}_{\partial\Omega}\big\|_{[L^p(\partial\Omega,\sigma)]^M}. \tag{3.1.104}$$

Moreover, if $\widetilde{D}$ is a homogeneous first-order $M\times N$ system with constant complex coefficients in $\mathbb{R}^n$ which complements D (i.e., (1.3.21) holds) and $\mathcal{D}$, K are the boundary double layer potential operators associated with the weakly elliptic second-order $M\times M$ system $L := \widetilde{D}D$ and the set Ω as in [115, §1.4] then

$$u = \mathcal{D}\big(u\big|^{{}^{\mathrm{n.t.}}}_{\partial\Omega}\big) \ \textit{in } \Omega \ \textit{and} \ \big(-\tfrac{1}{2}I + K\big)\big(u\big|^{{}^{\mathrm{n.t.}}}_{\partial\Omega}\big) = 0 \ \textit{on } \partial\Omega. \tag{3.1.105}$$

In addition,

if $\mathrm{Sym}(D;\xi)[L(\xi)]^{-1}\mathrm{Sym}(\widetilde{D};\xi) = I_{M\times M}$ for each $\xi \in \mathbb{R}^n\setminus\{0\}$ (in particular, if D and $\widetilde{D}$ commute), then the double layer $\mathcal{D}$ has the property that $D(\mathcal{D}f) = 0$ in Ω for each $f \in \Big[L^1\big(\partial\Omega, \frac{\sigma(x)}{1+|x|^{n-1}}\big)\Big]^M$. (3.1.106)

Furthermore, for each $q \in (1,\infty)$ one has

$$u\big|^{{}^{\mathrm{n.t.}}}_{\partial\Omega} \in \big[L^q(\partial\Omega,\sigma)\big]^M \Longleftrightarrow \mathcal{N}_\kappa u \in L^q(\partial\Omega,\sigma) \tag{3.1.107}$$

and

$$u\big|^{{}^{\mathrm{n.t.}}}_{\partial\Omega} \in \big[L^{p,q}_1(\partial\Omega,\sigma)\big]^M \Longleftrightarrow \mathcal{N}_\kappa(\nabla u) \in L^q(\partial\Omega,\sigma), \tag{3.1.108}$$

in a quantitative fashion. In particular, the nontangential pointwise trace $u\big|^{{}^{\mathrm{n.t.}}}_{\partial\Omega}$ belongs to the boundary Sobolev space $\big[L^p_1(\partial\Omega,\sigma)\big]^M$ if and only if $\mathcal{N}_\kappa(\nabla u)$ belongs to $L^p(\partial\Omega,\sigma)$, once again in a quantitative fashion, namely

$$\big\|u\big|^{{}^{\mathrm{n.t.}}}_{\partial\Omega}\big\|_{[L^p_1(\partial\Omega,\sigma)]^M} \approx \|\mathcal{N}_\kappa u\|_{L^p(\partial\Omega,\sigma)} + \|\mathcal{N}_\kappa(\nabla u)\|_{L^p(\partial\Omega,\sigma)}, \tag{3.1.109}$$

where the proportionality constants are independent of u.

(iii) In the case $p = 1$, one has

$$\mathrm{Sym}(D;\nu)\big(u\big|^{{}^{\mathrm{n.t.}}}_{\partial\Omega}\big) \in \big[H^1(\partial\Omega,\sigma)\big]^N \tag{3.1.110}$$

and, for some constant $C = C(\Omega, D, \kappa) \in (0,\infty)$ independent of u,

$$\big\|\mathrm{Sym}(D;\nu)\big(u\big|^{{}^{\mathrm{n.t.}}}_{\partial\Omega}\big)\big\|_{[H^1(\partial\Omega,\sigma)]^N} \leq C\|\mathcal{N}_\kappa u\|_{L^1(\partial\Omega,\sigma)}. \tag{3.1.111}$$

In fact, one also has

$$\begin{aligned}\Big\|u\Big|^{\rm n.t.}_{\partial\Omega}\Big\|_{[L^1(\partial\Omega,\sigma)]^M} &\leq \big\|\mathcal{N}_\kappa u\big\|_{L^1(\partial\Omega,\sigma)} \leq C\Big\|\mathrm{Sym}(D;\nu)\big(u\big|^{\rm n.t.}_{\partial\Omega}\big)\Big\|_{[H^1(\partial\Omega,\sigma)]^N}\\ &\leq C\Big\|u\Big|^{\rm n.t.}_{\partial\Omega}\Big\|_{[L^1(\partial\Omega,\sigma)]^M}\end{aligned} \tag{3.1.112}$$

provided, in the case when Ω is an exterior domain, it is assumed that u vanishes at infinity (in the sense of Definition 1.6.3).

(iv) In the case $p \in \big(\frac{n-1}{n}, 1\big)$, one has

$$\begin{aligned}&\mathrm{Sym}(D;\nu)\bullet u \in \big[H^p(\partial\Omega,\sigma)\big]^N \text{ and}\\ &\big\|\mathrm{Sym}(D;\nu)\bullet u\big\|_{[H^p(\partial\Omega,\sigma)]^N} \leq C\big\|\mathcal{N}_\kappa u\big\|_{L^p(\partial\Omega,\sigma)}\end{aligned} \tag{3.1.113}$$

for some constant $C = C(\Omega, D, \kappa, p) \in (0,\infty)$ independent of u. Henceforth, in the case when Ω is an exterior domain, make the additional assumption that u vanishes at infinity (in the sense of Definition 1.6.3). Then one also has

$$\big\|\mathcal{N}_\kappa u\big\|_{L^p(\partial\Omega,\sigma)} \leq C\big\|\mathrm{Sym}(D;\nu)\bullet u\big\|_{[H^p(\partial\Omega,\sigma)]^N}. \tag{3.1.114}$$

Furthermore, if $\widetilde{D}$ is a homogeneous first-order $M\times N$ system with constant complex coefficients in $\mathbb{R}^n$ which complements D (i.e., (1.3.21) holds) then[4]

$$\begin{aligned}u\Big|^{\rm n.t.}_{\partial\Omega} = \big[\mathrm{Sym}(\widetilde{D}D;\nu)\big]^{-1}\mathrm{Sym}(\widetilde{D};\nu)\mathfrak{H}\Big(\mathrm{Sym}(D;\nu)\bullet u\Big)\\ \text{at } \sigma\text{-a.e. point on } \partial\Omega,\end{aligned} \tag{3.1.115}$$

where $\mathfrak{H}$ is the L^p-filtering operator from [113, (4.9.2) in Theorem 4.9.1] *(presently used with $\Sigma := \partial\Omega$). Consequently, with D^* denoting the Hermitian adjoint of D, one has*[5]

$$\begin{aligned}u\Big|^{\rm n.t.}_{\partial\Omega} = \big[\mathrm{Sym}(D^*D;\nu)\big]^{-1}\mathrm{Sym}(D^*;\nu)\mathfrak{H}\Big(\mathrm{Sym}(D;\nu)\bullet u\Big)\\ \text{at } \sigma\text{-a.e. point on } \partial\Omega.\end{aligned} \tag{3.1.116}$$

In particular, if the operator D is of Dirac type (i.e., if D^*D has scalar principal symbol) then

$$\begin{aligned}\mathfrak{H}\Big(\mathrm{Sym}(D;\nu)\bullet u\Big) = \mathrm{Sym}(D;\nu)\big(u\big|^{\rm n.t.}_{\partial\Omega}\big)\\ \text{at } \sigma\text{-a.e. point on } \partial\Omega.\end{aligned} \tag{3.1.117}$$

[4] with $\mathrm{Sym}(\widetilde{D}D;\xi)$ denoting the principal symbol of $\widetilde{D}D$, viewed as a *second-order* differential operator

[5] again, with $\mathrm{Sym}(D^*D;\xi)$ denoting the principal symbol of D^*D, regarded as a *second-order* differential operator

(v) *For each* $q \in (1, \infty)$ *there exists some* $C = C(\Omega, D, p, q, \kappa) \in (0, \infty)$, *independent of* u, *such that the following mixed-norm estimate holds:*

$$\left(\int_{\partial\Omega}\left(\int_{\Gamma_\kappa(x)} |\nabla u|^q \operatorname{dist}(\cdot, \partial\Omega)^{q-n}\, d\mathcal{L}^n\right)^{\frac{p}{q}} d\sigma(x)\right)^{\frac{1}{p}} \leq C\big\|\mathcal{N}_\kappa u\big\|_{L^p(\partial\Omega,\sigma)}. \tag{3.1.118}$$

Alternatively, with the L^q*-based area-function defined as in* [113, (9.2.206)], *i.e.,*

$$(\mathcal{A}_{q,\kappa}u)(x) := \left(\int_{\Gamma_\kappa(x)} |(\nabla u)(y)|^q |x - y|^{q-n}\, dy\right)^{1/q}, \qquad \forall x \in \partial\Omega, \tag{3.1.119}$$

the estimate in (3.1.118) *may be recast as*

$$\big\|\mathcal{A}_{q,\kappa}u\big\|_{L^p(\partial\Omega,\sigma)} \leq C\big\|\mathcal{N}_\kappa u\big\|_{L^p(\partial\Omega,\sigma)}. \tag{3.1.120}$$

In addition, when $p \in (1, \infty)$ *there exists a constant* $C = C(\Omega, D, p, \kappa) \in (0, \infty)$ *with the property that*

$$\left(\int_{\Omega} |(\nabla u)(x)|^p \operatorname{dist}(x, \partial\Omega)^{p-1}\, dx\right)^{1/p} \leq C\big\|\mathcal{N}_\kappa u\big\|_{L^p(\partial\Omega,\sigma)}. \tag{3.1.121}$$

(vi) *In the case when* Ω *is actually a bounded* NTA *domain with Ahlfors regular boundary in* $\mathbb{R}^n$ *and* $p \in (1, \infty)$, *for each* $q \in \left(\frac{n}{n+1/p}, \infty\right]$ *there exists a constant* $C = C(\Omega, D, p, q) \in (0, \infty)$ *with the property that*

$$\max\left\{\|u\|_{[F^{p,q}_{1/p}(\Omega)]^M}, \|u\|_{[B^{p,p}_{1/p}(\Omega)]^M}\right\} \leq C\Big\|u\Big|^{\text{n.t.}}_{\partial\Omega}\Big\|_{[L^p(\partial\Omega,\sigma)]^M}. \tag{3.1.122}$$

In particular, corresponding to the case when $p = q = 2$, *one has*

$$\|u\|_{[H^{1/2}(\Omega)]^M} \leq C\Big\|u\Big|^{\kappa-\text{n.t.}}_{\partial\Omega}\Big\|_{[L^2(\partial\Omega,\sigma)]^M}. \tag{3.1.123}$$

(vii) *Suppose in place of the first line in* (3.1.102) *one now assumes*

$$\mathcal{N}_\kappa u \in L^{p,q}(\partial\Omega, \sigma) \text{ for some } p \in \left(\tfrac{n-1}{n}, \infty\right) \text{ and } q \in (0, \infty). \tag{3.1.124}$$

Then the nontangential boundary trace $u\big|^{\kappa-\text{n.t.}}_{\partial\Omega}$ *exists at* σ*-a.e. point on* $\partial\Omega$, *is independent of the given aperture parameter* κ, *belongs to the Lorentz space* $\left[L^{p,q}(\partial\Omega, \sigma)\right]^M$, *and satisfies the estimate*

$$\Big\|u\Big|^{\kappa-\text{n.t.}}_{\partial\Omega}\Big\|_{[L^{p,q}(\partial\Omega,\sigma)]^M} \leq \|\mathcal{N}_\kappa u\|_{L^{p,q}(\partial\Omega,\sigma)}. \tag{3.1.125}$$

Henceforth, make the additional assumption that u *vanishes at infinity* (*in the sense of Definition 1.6.3*) *in the case when* Ω *is an exterior domain. Then one has*

$$\begin{aligned}&\mathrm{Sym}(D;\nu)\bullet u\in\big[H^{p,q}(\partial\Omega,\sigma)\big]^N \text{ and}\\ &\big\|\mathrm{Sym}(D;\nu)\bullet u\big\|_{[H^{p,q}(\partial\Omega,\sigma)]^N}\approx\|\mathcal{N}_\kappa u\|_{L^{p,q}(\partial\Omega,\sigma)},\end{aligned}\tag{3.1.126}$$

and formulas (3.1.115)-(3.1.117) *remain valid, where* $\mathfrak{H}$ *is now interpreted as the* $L^{p,q}$*-filtering operator from* [113, (4.9.5) in Theorem 4.9.1] *(again, used with* $\Sigma := \partial\Omega$*). Also, in the particular case when* $p \in (1, \infty)$*, the formulas in* (3.1.105) *continue to hold and one has*

$$\begin{aligned}\|\mathcal{N}_\kappa u\|_{L^{p,q}(\partial\Omega,\sigma)}&\approx\big\|\mathrm{Sym}(D;\nu)\big(u\big|^{\kappa-\mathrm{n.t.}}_{\partial\Omega}\big)\big\|_{[L^{p,q}(\partial\Omega,\sigma)]^N}\\ &\approx\big\|u\big|^{\kappa-\mathrm{n.t.}}_{\partial\Omega}\big\|_{[L^{p,q}(\partial\Omega,\sigma)]^M}.\end{aligned}\tag{3.1.127}$$

Finally, with the L^r*-based area-function defined similarly to* (3.1.119) *for some* $r \in (1, \infty)$*, one has*

$$\|\mathcal{A}_{r,\kappa}u\|_{L^{p,q}(\partial\Omega,\sigma)}\leq C\|\mathcal{N}_\kappa u\|_{L^{p,q}(\partial\Omega,\sigma)},\tag{3.1.128}$$

for some constant $C \in (0, \infty)$ *independent of* u.

(viii) *Suppose in place of the first line in* (3.1.102) *one now assumes*

$$\mathcal{N}_\kappa u\in\mathbb{X},\tag{3.1.129}$$

where $\mathbb{X}$ *is a Generalized Banach Function Space on* $(\partial\Omega, \sigma)$ *with the property that*

$$\begin{aligned}&\mathcal{M}_{\partial\Omega}:\mathbb{X}\to\mathbb{X}\text{ and }\mathcal{M}_{\partial\Omega}:\mathbb{X}'\to\mathbb{X}'\\ &\text{are well-defined bounded mappings}\end{aligned}\tag{3.1.130}$$

(with $\mathcal{M}_{\partial\Omega}$ *denoting the Hardy-Littlewood maximal operator on* $\partial\Omega$*, and with* $\mathbb{X}'$ *denoting the associated space of* $\mathbb{X}$*; cf.* [113, Definitions 5.1.4, 5.1.11]*). Then the nontangential boundary trace* $u\big|^{\kappa-\mathrm{n.t.}}_{\partial\Omega}$ *exists at* σ*-a.e. point on* $\partial\Omega$ *and is independent of the given aperture parameter* κ*. Also, as a function,* $u\big|^{\kappa-\mathrm{n.t.}}_{\partial\Omega}$ *belongs to the space* $[\mathbb{X}]^M$*, and satisfies* $\big\|u\big|^{\kappa-\mathrm{n.t.}}_{\partial\Omega}\big\|_{[\mathbb{X}]^M} \leq \|\mathcal{N}_\kappa u\|_{\mathbb{X}}$*. Furthermore,*

$$\mathrm{Sym}(D;\nu)\bullet u=\mathrm{Sym}(D;\nu)\big(u\big|^{\kappa-\mathrm{n.t.}}_{\partial\Omega}\big)\in[\mathbb{X}]^N\tag{3.1.131}$$

and, under the additional assumption that u *vanishes at infinity (in the sense of Definition 1.6.3) in the case when* Ω *is an exterior domain, one has*

$$\|\mathcal{N}_\kappa u\|_{\mathbb{X}}\approx\big\|\mathrm{Sym}(D;\nu)\big(u\big|^{\kappa-\mathrm{n.t.}}_{\partial\Omega}\big)\big\|_{[\mathbb{X}]^N}\approx\big\|u\big|^{\kappa-\mathrm{n.t.}}_{\partial\Omega}\big\|_{[\mathbb{X}]^M}.\tag{3.1.132}$$

We continue by including several comments which further clarify the context and scope of the above theorem.

`Comment 1.` In light of [112, (8.3.6) in Lemma 8.3.2], for each given integrability exponent $p \in (1, \infty)$, the equivalence (cf. (3.1.104))

$$\|\mathcal{N}_\kappa u\|_{L^p(\partial\Omega,\sigma)} \approx \|u\big|^{\text{n.t.}}_{\partial\Omega}\|_{[L^p(\partial\Omega,\sigma)]^M} \tag{3.1.133}$$

may be naturally interpreted as an L^p-**Maximum Principle** for null-solutions u of D in Ω (which also vanish at infinity if Ω is an exterior domain). Likewise, (3.1.109) may be thought of as an L^p_1-Maximum Principle, while (3.1.127) may be construed as an $L^{p,q}$-Maximum Principle. Finally, (3.1.132) contains what may be called the $\mathbb{X}$-**Maximum Principle** for null-solutions u of D in Ω, namely

$$\|\mathcal{N}_\kappa u\|_{\mathbb{X}} \approx \|u\big|^{\kappa-\text{n.t.}}_{\partial\Omega}\|_{[\mathbb{X}]^M}. \tag{3.1.134}$$

Comment 2. The first identity in (3.1.105) may be regarded as Cauchy's reproducing formula. To add substance to this perspective, let $D := \partial_x + i\partial_y$ be the Cauchy-Riemann operator in the plane, and pick $\widetilde{D} := D^*$. Since $D^* = -\partial_x + i\partial_y$, we have $L := D^*D = -\Delta$ in $\mathbb{R}^2$. The fundamental solution associated with L as in Theorem 1.4.2 then becomes $E_L(x,y) := -(2\pi)^{-1}\ln\sqrt{x^2+y^2}$. In particular,

$$(\partial_x E_L)(x,y) := -\frac{1}{2\pi}\frac{x}{x^2+y^2} \quad\text{and}\quad (\partial_y E_L)(x,y) := -\frac{1}{2\pi}\frac{y}{x^2+y^2}. \tag{3.1.135}$$

Expressing $D = \bar{\partial}$ as $a_j\partial_j$ with complex coefficients $a_1 := 1$ and $a_2 := i$, it follows that $L = D^*D = A_{jk}\partial_j\partial_k$, where $A_{jk} = -a_j\overline{a_k}$, for $1 \le j,k \le 2$, i.e.,

$$A_{11} = -1, \quad A_{12} = i, \quad A_{21} = -i, \quad A_{22} = -1. \tag{3.1.136}$$

Then the boundary-to-domain double layer $\mathcal{D}$ associated with the coefficient tensor $A = (A_{jk})_{1\le j,k\le 2}$ acts on functions $f \in L^p(\partial\Omega,\sigma)$ with $1 < p < \infty$ according to

$$\begin{aligned}
(\mathcal{D}f)(z) &= \frac{1}{2\pi}\int_{\partial\Omega}\Big\{\frac{-\nu_1(\zeta)(z_1-\zeta_1) + i\nu_1(\zeta)(z_2-\zeta_2)}{|z-\zeta|^2} \\
&\qquad + \frac{-i\nu_2(\zeta)(z_1-\zeta_1) - \nu_2(\zeta)(z_2-\zeta_2)}{|z-\zeta|^2}\Big\} f(\zeta)\,d\sigma(\zeta) \\
&= \frac{1}{2\pi}\int_{\partial\Omega}\frac{-\big(\nu_1(\zeta)+i\nu_2(\zeta)\big)\big((z_1-\zeta_1) - i(z_2-\zeta_2)\big)}{|z-\zeta|^2} f(\zeta)\,d\sigma(\zeta) \\
&= \frac{-1}{2\pi}\int_{\partial\Omega}\frac{\nu(\zeta)\overline{(z-\zeta)}}{(z-\zeta)\overline{(z-\zeta)}} f(\zeta)\,d\sigma(\zeta) = \frac{-1}{2\pi i}\int_{\partial\Omega}\frac{f(\zeta)}{z-\zeta} i\nu(\zeta)\,d\sigma(\zeta) \\
&= \frac{1}{2\pi i}\int_{\partial\Omega}\frac{f(\zeta)}{\zeta - z}\,d\zeta \quad\text{for each } z \in \Omega,
\end{aligned} \tag{3.1.137}$$

which is the familiar format of the Cauchy operator on $\partial\Omega$. In this setting, the first identity in (3.1.105) simply asserts that the holomorphic function u may be recovered from its (nontangential) boundary trace via the action of Cauchy operator on the latter function.

However, as opposed to the case of the Cauchy operator which maps L^p-functions on $\partial\Omega$ into holomorphic functions (hence, null-solutions of the Cauchy-Riemann operator) in Ω, in general it is *not* true that $D(\mathcal{D}f) = 0$ in Ω for arbitrary functions $f \in \Big[L^1\big(\partial\Omega, \frac{\sigma(x)}{1+|x|^{n-1}}\big)\Big]^M$. This being said, we see from (3.1.106) that

$$\text{if } D \text{ is an injectively elliptic constant complex coefficient first-order system in } \mathbb{R}^n \text{ which commutes with its Hermitian adjoint (i.e., one has } D^*D = DD^*) \text{ then the double layer } \mathcal{D} \text{ associated with the } M \times M \text{ weakly elliptic system } L := D^*D \text{ satisfies } D(\mathcal{D}f) = 0 \text{ in } \Omega \text{ for each function } f \in \Big[L^1\big(\partial\Omega, \tfrac{\sigma(x)}{1+|x|^{n-1}}\big)\Big]^M. \tag{3.1.138}$$

See [67, §4.5] for a multitude of examples of first-order elliptic systems which commute with their Hermitian adjoints.

Comment 3. Formula (3.1.116), relating the nontangential pointwise trace of u on $\partial\Omega$ to the bullet product $\mathrm{Sym}(D;\nu) \bullet u$ (itself a distribution in the Hardy space H^p on the boundary, whose definition is inspired by a variational approach) via the L^p-filtering operator $\mathfrak{H}$, is remarkable in view of the nature of the objects involved. In this vein, it is worth pointing out that $\big[\mathrm{Sym}(D^*D;\nu)\big]^{-1}\mathrm{Sym}(D^*;\nu)$ is actually a left-inverse for the (injective) mapping $\mathrm{Sym}(D;\nu)$.

Comment 4. Specializing Theorem 3.1.6 to the case when $n = 2$ and $D := \overline{\partial}$, the Cauchy-Riemann operator (hence, $M = N = 1$), yields the following corollary:

$$\text{any holomorphic function } u \text{ in a UR domain } \Omega \subseteq \mathbb{C} \text{ satisfying } \int_{\partial\Omega} (\mathcal{N}_\kappa u)^p \, d\mathcal{H}^1 < +\infty, \text{ for some } \kappa \in (0,\infty) \text{ and } p \in \big(\tfrac{1}{2},\infty\big), \text{ has the property that } u\big|^{\kappa-\mathrm{n.t.}}_{\partial\Omega} \text{ exists in } \mathbb{C} \text{ at } \mathcal{H}^1\text{-a.e. point on } \partial\Omega. \tag{3.1.139}$$

A similar property is valid if the integrability condition imposed above on the nontangential maximal function is replaced by membership to a Lorentz space (cf. (3.1.124)).

A large category of UR domains in the complex plane has been singled out in [112, Proposition 5.10.7]. While this covers many mundane cases, which arise routinely in applications, the reader is reminded that the topology of a UR domain can be quite wild. See [123, §A.4, p. 755] where a class of UR domains in $\mathbb{R}^n$, with $n \geq 2$ arbitrary, of infinite topological type has been produced.

Combining Theorem 1.1.1 with (3.1.139) then proves that if u is a holomorphic function in a UR domain $\Omega \subseteq \mathbb{C}$ satisfying $\int_{\partial\Omega} (\mathcal{N}_\kappa u)^p \, d\mathcal{H}^1 < +\infty$ for some $\kappa \in (0,\infty)$ and $p \in (1,\infty)$, and which also vanishes at infinity in the case when Ω is an exterior domain, then $u\big|^{\kappa-\mathrm{n.t.}}_{\partial\Omega}$ exists at $\mathcal{H}^1$-a.e. point on $\partial\Omega$ and the following Cauchy integral representation formula holds:

$$u(z) = \frac{1}{2\pi i} \int_{\partial\Omega} \frac{\big(u\big|_{\partial\Omega}^{\kappa-\text{n.t.}}\big)(\zeta)}{\zeta - z}\, d\zeta, \qquad \forall z \in \Omega. \tag{3.1.140}$$

Let us also point out that a version of the Fatou-type result for holomorphic functions recorded in (3.1.139) continues to hold even in the presence of singularities. Concretely, we have the following result (which, in particular, allows the consideration of meromorphic functions with singularities contained in a compact subset of their domain):

if $\Omega \subseteq \mathbb{C}$ is a UR domain and $u \in \mathcal{D}'(\Omega)$ verifies $\overline{\partial} u = 0$ in $\Omega \backslash K$, where K is a compact subset of Ω, as well as $\int_{\partial\Omega} \big(\mathcal{N}_\kappa^{\Omega\backslash K} u\big)^p \, d\mathcal{H}^1 < +\infty$, for some $\kappa > 0$ and $p \in \big(\frac{1}{2}, \infty\big)$, then $u\big|_{\partial\Omega}^{\kappa-\text{n.t.}}$ exists at $\mathcal{H}^1$-a.e. point on $\partial\Omega$. (3.1.141)

This is seen by applying (3.1.139) to the restriction of u to $\Omega \setminus \overline{O}$, where O is a bounded Lipschitz domain satisfying $K \subseteq O$ and $\overline{O} \subseteq \Omega$. When used together with Theorem 1.1.3, the Fatou-type result recorded in (3.1.141) yields a version of the residue formula (1.1.66) in which the existence of nontangential pointwise boundary traces need *not* be assumed a priori, but rather becomes a consequence of the hypotheses.

Finally, we wish to note that a result in the spirit of (3.1.139), in which we now allow *any* integrability exponent in the interval $(0, \infty)$, holds as described below:

given any simply connected UR domain $\Omega \subseteq \mathbb{C}$ along with any non-vanishing holomorphic function u in Ω with $\int_{\partial\Omega} \big(\mathcal{N}_\kappa u\big)^p \, d\mathcal{H}^1 < +\infty$, for some $\kappa \in (0, \infty)$ and $p \in (0, \infty)$, it follows that $u\big|_{\partial\Omega}^{\kappa-\text{n.t.}}$ exists in $\mathbb{C}$ at $\mathcal{H}^1$-a.e. point on $\partial\Omega$. (3.1.142)

Indeed, this follows by applying (3.1.139) to $u^{1/m}$, which is a well-defined holomorphic function in Ω (cf., e.g., [156, Theorem 13.11, p.274]), with $m \in \mathbb{N}$ sufficiently large so that $mp > 1/2$.

Comment 5. Another natural candidate for the (homogeneous, constant coefficient) first-order operator with injective principal symbol intervening in the statement of Theorem 3.1.6 is the Dirac operator $D := \sum_{j=1}^{n} \mathbf{e}_j \odot \partial_j$ (cf. the discussion in [112, §6.4]). Upon recalling that $\mathcal{C}\ell_n$-valued functions which are null-solutions of D are called monogenic, Theorem 3.1.6 implies the following Fatou-type result:

any monogenic function u in a UR domain $\Omega \subseteq \mathbb{R}^n$ satisfying $\int_{\partial\Omega} \big(\mathcal{N}_\kappa u\big)^p \, d\mathcal{H}^{n-1} < +\infty$, for some $\kappa \in (0, \infty)$ and $p \in \big(\frac{n-1}{n}, \infty\big)$, has the property that $u\big|_{\partial\Omega}^{\kappa-\text{n.t.}}$ exists in $\mathcal{C}\ell_n$ at $\mathcal{H}^{n-1}$-a.e. point on $\partial\Omega$. (3.1.143)

A similar result is valid when the above integrability condition imposed on the nontangential maximal function is replaced by membership to a Lorentz space (cf. (3.1.124)).

`Comment 6.` Holomorphic functions of several complex variables also fit into the framework of Theorem 3.1.6 by considering the first-order operator $D := (\partial_{\bar{z}_j})_{1\leq j\leq n}$. Since from Example 1.3.10 we know that this is injectively elliptic, Theorem 3.1.6 implies the following Fatou-type result:

$$\begin{array}{l}\text{any holomorphic function } u \text{ in a UR domain } \Omega \subseteq \mathbb{C}^n \text{ satisfying}\\ \int_{\partial\Omega} (\mathcal{N}_\kappa u)^p \, d\mathcal{H}^{2n-1} < +\infty, \text{ for some } \kappa \in (0,\infty) \text{ and } p \in \big(\tfrac{2n-1}{2n}, \infty\big),\\ \text{has the property that } u\big|_{\partial\Omega}^{\kappa-\text{n.t.}} \text{ exists in } \mathbb{C} \text{ at } \mathcal{H}^{2n-1}\text{-a.e. point on } \partial\Omega.\end{array} \tag{3.1.144}$$

A similar Fatou-type result is valid if the integrability condition imposed above on the nontangential maximal function is replaced by membership to a Lorentz space (cf. (3.1.124)), and also for separately monogenic functions of m Clifford variables in UR domains in $(\mathbb{R}^n)^m$, where $m \in \mathbb{N}$ is arbitrary (see [92, §6.8] for a brief introduction to Clifford analysis for functions of several variables).

`Comment 7.` Specializing Theorem 3.1.6 to the case when D is the Hodge-Dirac operator $d + \delta$ (where d, δ are, respectively, the exterior derivative operator and its formal adjoint; cf. [112, (6.6.141)]) acting on differential forms yields the following Fatou-type result:

$$\begin{array}{l}\text{any (mixed degree) differential form } u \text{ defined in a UR domain } \Omega \subseteq \mathbb{R}^n\\ \text{and satisfying } (d+\delta)u = 0 \text{ in } \Omega, \text{ as well as } \int_{\partial\Omega} (\mathcal{N}_\kappa u)^p \, d\mathcal{H}^{n-1} < \infty\\ \text{for some } \kappa \in (0,\infty) \text{ and some } p \in \big(\tfrac{n-1}{n}, \infty\big), \text{ has the property that the}\\ \text{pointwise nontangential trace } u\big|_{\partial\Omega}^{\kappa-\text{n.t.}} \text{ exists at } \mathcal{H}^{n-1}\text{-a.e. point on } \partial\Omega.\end{array} \tag{3.1.145}$$

In turn, this further implies that

$$\begin{array}{l}\text{any (mixed degree) differential form } u \text{ defined in a UR domain } \Omega \subseteq \mathbb{R}^n\\ \text{and satisfying } \Delta u = 0 \text{ in } \Omega, \text{ as well as } \int_{\partial\Omega} (\mathcal{N}_\kappa(du))^p \, d\mathcal{H}^{n-1} < \infty \text{ and}\\ \int_{\partial\Omega} (\mathcal{N}_\kappa(\delta u))^p \, d\mathcal{H}^{n-1} < \infty \text{ for some } \kappa \in (0,\infty) \text{ and } p \in \big(\tfrac{n-1}{n}, \infty\big),\\ \text{has the property that } (du)\big|_{\partial\Omega}^{\kappa-\text{n.t.}} \text{ and } (\delta u)\big|_{\partial\Omega}^{\kappa-\text{n.t.}} \text{ exist } \mathcal{H}^{n-1}\text{-a.e. on } \partial\Omega.\end{array} \tag{3.1.146}$$

Indeed, thanks to the above hypotheses and the conclusion we now seek, there is no loss of generality in assuming that u is a homogeneous differential form, having a (well-defined) degree $\ell \in \{0, 1, \dots, n\}$. Since $(d + \delta)^2 = -\Delta$, we may invoke (3.1.145) for the (mixed degree) differential form $w := du + \delta u \in \Lambda^{\ell+1} \oplus \Lambda^{\ell-1}$ and conclude that $w\big|_{\partial\Omega}^{\kappa-\text{n.t.}}$ exists at $\mathcal{H}^{n-1}$-a.e. point on $\partial\Omega$. In terms of the individual components of w of degree $\ell + 1$ and $\ell - 1$, this translates precisely as saying that $(du)\big|_{\partial\Omega}^{\kappa-\text{n.t.}}$ and $(\delta u)\big|_{\partial\Omega}^{\kappa-\text{n.t.}}$ exist at $\mathcal{H}^{n-1}$-a.e. point on $\partial\Omega$, proving (3.1.146).

Again, similar properties are valid if the integrability conditions imposed above on the nontangential maximal function are replaced by memberships to Lorentz spaces (cf. (3.1.124)).

`Comment 8.` Recall the d-bar operator $\bar\partial$ in the several complex variable theory, and its Hermitian adjoint ϑ (cf. [115, Chapter 7]). Since the so-called complex Laplacian in $\mathbb{C}^n$, defined as $\Box := \bar\partial\vartheta + \vartheta\bar\partial = (\bar\partial + \vartheta)^2$ turns out a scalar multiple of the real Laplacian Δ, the same type of argument as in Comment 7 above produces the following Fatou-type results:

$$\begin{aligned}&\text{any complex differential form } u \text{ defined in a UR domain } \Omega \subseteq \mathbb{C}^n \text{ and}\\ &\text{satisfying } (\bar\partial + \vartheta)u = 0 \text{ in } \Omega, \text{ as well as } \int_{\partial\Omega} \big(\mathcal{N}_\kappa u\big)^p \, d\mathcal{H}^{2n-1} < +\infty\\ &\text{for some } \kappa \in (0,\infty) \text{ and some } p \in \big(\tfrac{2n-1}{2n}, \infty\big), \text{ has the property that the}\\ &\text{pointwise nontangential trace } u\big|_{\partial\Omega}^{\kappa-\text{n.t.}} \text{ exists at } \mathcal{H}^{2n-1}\text{-a.e. point on } \partial\Omega,\end{aligned} \tag{3.1.147}$$

which further implies that

$$\begin{aligned}&\text{any complex differential form } u \text{ defined in a UR domain } \Omega \subseteq \mathbb{C}^n \text{ and}\\ &\text{satisfying } \Delta u = 0 \text{ in } \Omega, \text{ as well as } \int_{\partial\Omega} \big(\mathcal{N}_\kappa(\bar\partial u)\big)^p \, d\mathcal{H}^{2n-1} < \infty \text{ and}\\ &\int_{\partial\Omega} \big(\mathcal{N}_\kappa(\vartheta u)\big)^p \, d\mathcal{H}^{2n-1} < \infty \text{ for some } \kappa \in (0,\infty) \text{ and } p \in \big(\tfrac{2n-1}{2n}, \infty\big),\\ &\text{has the property that } (\bar\partial u)\big|_{\partial\Omega}^{\kappa-\text{n.t.}} \text{ and } (\vartheta u)\big|_{\partial\Omega}^{\kappa-\text{n.t.}} \text{ exist } \mathcal{H}^{2n-1}\text{-a.e. on } \partial\Omega.\end{aligned} \tag{3.1.148}$$

Once more, similar Fatou-type results hold when the integrability condition imposed above on the nontangential maximal function is replaced by membership to a Lorentz space (cf. (3.1.124)).

<u>Comment 9.</u> It turns out that family of functions satisfying the Moisil-Teodorescu system (aka the generalized Cauchy-Riemann equations), namely $\partial_j u_k = \partial_k u_j$ for $1 \le j,k \le n$ and $\sum_{j=1}^n \partial_j u_j = 0$ in an arbitrary UR domain $\Omega \subseteq \mathbb{R}^n$ with $n \ge 2$, and satisfying $\int_{\partial\Omega} \sum_{j=1}^n \big(\mathcal{N}_\kappa u_j\big)^p \, d\mathcal{H}^{n-1} < \infty$ for some $\kappa \in (0,\infty)$ and some $p \in \big(\frac{n-1}{n}, \infty\big)$, have nontangential pointwise traces $u_j\big|_{\partial\Omega}^{\kappa-\text{n.t.}}$ at $\mathcal{H}^{n-1}$-a.e. point on $\partial\Omega$, for $1 \le j \le n$. Indeed, this follows by applying Theorem 3.1.6 to the Moisil-Teodorescu system D from Example 1.3.8.

A similar Fatou-type result holds in connection with the modified Moisil-Teodorescu system associated with a matrix $A = (a_{jk})_{1 \le j,k \le n} \in \mathbb{C}^{n \times n}$ which is weakly elliptic, namely for vector-valued functions $u = (u_j)_{1 \le j \le n}$ defined in an arbitrary UR domain $\Omega \subseteq \mathbb{R}^n$ with $n \ge 2$ and satisfying in Ω:

$$\partial_j u_k = \partial_k u_j \ \text{ for } 1 \le j,k \le n, \ \text{ and } \sum_{j,k=1}^n a_{jk}\partial_j u_k = 0. \tag{3.1.149}$$

Then, based on Example 1.3.9 and Theorem 3.1.6, we conclude that whenever $\int_{\partial\Omega} \sum_{j=1}^n \big(\mathcal{N}_\kappa u_j\big)^p \, d\mathcal{H}^{n-1} < \infty$ for some $\kappa \in (0,\infty)$ and some $p \in \big(\frac{n-1}{n}, \infty\big)$ it follows that the nontangential pointwise traces $u_j\big|_{\partial\Omega}^{\kappa-\text{n.t.}}$, with $1 \le j \le n$, exist at $\mathcal{H}^{n-1}$-a.e. point on $\partial\Omega$.

<u>Comment 10.</u> Recall from [112, Definition 5.10.6] that for a nonempty open subset Ω of $\mathbb{R}^n$ to be a UR domain means that $\partial\Omega$ is a UR set (cf. [112, Definition 5.10.1]) and that $\partial_*\Omega$ has full $\mathcal{H}^{n-1}$ measure in $\partial\Omega$, i.e., $\mathcal{H}^{n-1}\big(\partial\Omega \setminus \partial_*\Omega\big) = 0$. It turns out that the failure of the latter condition may lead to the failure of the conclusion recorded in the first line of (3.1.103). Indeed, this is seen by taking $n = 2$ and $D := \bar\partial$ the Cauchy-Riemann operator in the plane, take Ω to be the slit disk defined in (1.1.24)

(which a bounded open set and $\partial\Omega$ is a UR set, but Ω is not a UR domain), and consider the function u as in (1.1.25). Then (1.1.29) implies the property in the first line of (3.1.103) fails in such a setting.

It is easy to see that having D injectively elliptic is also a necessary condition for the validity of the Fatou-type property in the first line of (3.1.103). For example, if $n := 2$, $M := 1$, $N := 1$, $D := \partial_x$, $\Omega := (0,1) \times (0,1)$, and

$$u(x,y) := \sin(1/y) \text{ for each } (x,y) \in \Omega, \tag{3.1.150}$$

then Ω is a bounded UR domain in $\mathbb{R}^2$, D is a homogeneous, constant coefficient, first-order operator, and u is a bounded function satisfying $Du = 0$ in Ω. Hence, for each aperture parameter $\kappa > 0$ we have

$$\mathcal{N}_\kappa u \in \bigcap_{0<p\leq\infty} L^p(\partial\Omega, \mathcal{H}^1), \tag{3.1.151}$$

but $u\big|_{\partial\Omega}^{\kappa-\text{n.t.}}$ does not exist at any point on $A := (0,1) \times \{0\} \subset \partial\Omega$. Thus, the Fatou-type property in the first line of (3.1.103) presently fails, since $\mathcal{H}^1(A) > 0$. The source of this failure is precisely the lack of injectivity for the principal symbol of $D = \partial_x$. Indeed, for each $\xi = (\xi_1, \xi_2) \in \mathbb{R}^2$ we have $\text{Sym}(\partial_x; \xi) = i\xi_1$ which vanishes identically on $\{0\} \times \mathbb{R}$.

Comment 11. We wish to stress that in the statement of Theorem 3.1.6 no (interior, or exterior) corkscrew condition is imposed on the underlying set Ω. In particular, Ω may fail to be an NTA domain, which stands in sharp contrast to the Fatou-type result from Proposition 5.5.2. Examples of UR domains which are not NTA include heart-shaped regions, since the presence of cusps invalidates the corkscrew condition.

Fig. 3.1 A UR domain which is not NTA for which Theorem 3.1.6 holds

Comment 12. Suppose $L = \left(a_{rs}^{\alpha\beta}\partial_r\partial_s\right)_{\substack{1\leq\alpha\leq M\\1\leq\beta\leq M}}$ (with the summation convention over repeated indices in effect) is a weakly elliptic, constant (complex) coefficient, second-order, $M \times M$ system in $\mathbb{R}^n$. To this, associate the first-order, homogeneous, constant coefficient, $(n \cdot M) \times M$ system

$$Du := A\nabla u := \big(a^{\alpha\beta}_{rs}\partial_s u_\beta\big)_{\substack{1\leq\alpha\leq M\\ 1\leq r\leq n}} \tag{3.1.152}$$

which, from Example 1.3.11, is known to be injectively elliptic. Specializing Theorem 3.1.6 to this system then yields the following result: *Let $\Omega \subseteq \mathbb{R}^n$ (where $n \geq 2$) be a* UR *domain and set $\sigma := \mathcal{H}^{n-1}\lfloor\partial\Omega$. Also, consider a coefficient tensor A with the property that L_A is a weakly elliptic $M \times M$ system. Then each function $u \in \big[\mathscr{C}^\infty(\Omega)\big]^M$ satisfying $A\nabla u = 0$ in Ω and $\mathcal{N}_\kappa u \in L^{p,q}(\partial\Omega, \sigma)$ for some $\kappa > 0$, $p \in \big(\frac{n-1}{n}, \infty\big)$, and $q \in (0, \infty]$, has a nontangential boundary trace $u\big|^{\kappa-\mathrm{n.t.}}_{\partial\Omega}$ at σ-a.e. point on $\partial\Omega$. Moreover, if actually $p \in (1, \infty)$, then $u = \mathcal{D}\big(u\big|^{\kappa-\mathrm{n.t.}}_{\partial\Omega}\big)$ in Ω, where $\mathcal{D}$ is the double layer potential operator associated with the coefficient tensor A as in* [115, §1.3].

`Comment 13.` It turns out that the proof of Theorem 3.1.6 may be extended, by reasoning largely along the lines of the flat case discussed below, to the setting of UR subdomains of compact boundaryless manifolds of class $\mathscr{C}^2$, and for injectively elliptic first-order differential operators acting between vector bundles over said manifold. On example of such an operator is the so-called deformation tensor, a choice which leads to Fatou-type theorems for Killing fields on UR domains in Riemannian manifolds. Such body of results complements work in [118, §6] which contains Fatou theorems for the Hodge-Laplacian on regular SKT domains on Riemannian manifolds.

We now turn to the proof of Theorem 3.1.6.

Proof of Theorem 3.1.6 For starters, from [113, Corollary 10.2.13] (presently used with $q := p$), [112, (6.2.25)], [113, (4.2.25)], and (3.1.102) it follows that

$$\begin{gathered}\mathrm{Sym}(D;\nu)\bullet u \text{ belongs to the space } \big[H^p(\partial\Omega,\sigma)\big]^N\\ \text{and } \big\|\mathrm{Sym}(D;\nu)\bullet u\big\|_{[H^p(\partial\Omega,\sigma)]^N} \leq C\big\|\mathcal{N}_\kappa u\big\|_{L^p(\partial\Omega,\sigma)}\\ \text{for some } C = C(\Omega, D, \kappa, p) \in (0,\infty) \text{ independent of } u.\end{gathered} \tag{3.1.153}$$

To deal with the claims in item *(i)*, assume first that either $\partial\Omega$ is unbounded, or Ω is bounded. Also, assume that $p \in \big(\frac{n-1}{n}, 1\big]$. Granted these assumptions, and bearing in mind that $Du = 0$ in Ω and that u is continuous in Ω, formula (3.1.6) currently implies that for each $x \in \Omega$ we have

$$u(x) = \begin{cases} \Big\langle \big[\widetilde{E}^{\top}(x-\cdot)\big|_{\partial\Omega}\big], (-\mathrm{i})\mathrm{Sym}(D;\nu)\bullet u\Big\rangle & \text{if } \partial\Omega \text{ is unbounded,}\\[2ex] \Big\langle \widetilde{E}^{\top}(x-\cdot)\big|_{\partial\Omega}, (-\mathrm{i})\mathrm{Sym}(D;\nu)\bullet u\Big\rangle & \text{if } \partial\Omega \text{ is bounded.}\end{cases} \tag{3.1.154}$$

In turn, from (3.1.153), (3.1.154), (3.1.1), Theorem 1.4.2, Corollary 2.5.4, and [112, Corollary 8.9.6] we conclude that $u\big|^{\kappa-\mathrm{n.t.}}_{\partial\Omega}$ exists (in $\mathbb{C}^M$) at σ-a.e. point on $\partial\Omega$ and the function $u\big|^{\kappa-\mathrm{n.t.}}_{\partial\Omega}$ is σ-measurable on $\partial\Omega$. Moreover, on account of [112, (8.9.8)] we also have $\big\|u\big|^{\kappa-\mathrm{n.t.}}_{\partial\Omega}\big\|_{[L^p(\partial\Omega,\sigma)]^M} \leq \|\mathcal{N}_\kappa u\|_{L^p(\partial\Omega,\sigma)} < +\infty$, hence $u\big|^{\kappa-\mathrm{n.t.}}_{\partial\Omega}$ belongs to $\big[L^p(\partial\Omega,\sigma)\big]^M$. This establishes (3.1.103) in the case when either Ω is bounded, or $\partial\Omega$ is unbounded, under the assumption that $p \in \big(\frac{n-1}{n}, 1\big]$.

When either Ω is bounded, or $\partial\Omega$ is unbounded, but $p \in (1,\infty)$, we reason as above with natural alterations. First, (3.1.153) and [112, (3.6.27)] we have $\mathrm{Sym}(D;\nu) \bullet u \in \big[L^p(\partial\Omega,\sigma)\big]^N$. Thus, if we now employ the integral representation formula (3.1.9), in place of (3.1.154), which presently reads

$$u(x) = \int_{\partial\Omega} \big\langle \widetilde{E}^\top(x-y), (-\mathrm{i})\big(\mathrm{Sym}(D;\nu) \bullet u\big)(y) \big\rangle \, d\sigma(y) \ \text{ for all } x \in \Omega, \tag{3.1.155}$$

from Theorem 2.5.1, Theorem 1.4.2, and [112, Corollary 8.9.6] we again conclude that $u\big|^{\kappa-\mathrm{n.t.}}_{\partial\Omega}$ exists σ-a.e. on $\partial\Omega$ and the function $u\big|^{\kappa-\mathrm{n.t.}}_{\partial\Omega}$ is σ-measurable on $\partial\Omega$. Together with [112, (8.9.8)], these also imply that $u\big|^{\kappa-\mathrm{n.t.}}_{\partial\Omega}$ belongs to $\big[L^p(\partial\Omega,\sigma)\big]^M$ and satisfies $\big\|u\big|^{\kappa-\mathrm{n.t.}}_{\partial\Omega}\big\|_{[L^p(\partial\Omega,\sigma)]^M} \le \|\mathcal{N}_\kappa u\|_{L^p(\partial\Omega,\sigma)}$.

Thus, as far as (3.1.103) is concerned, there remains to consider the case when Ω is unbounded and $\partial\Omega$ is bounded. Assuming this is the case, invoke [112, Lemma 5.10.10] to find $R \in (0,\infty)$ such that $\mathbb{R}^n \setminus \Omega \subset B(0, R/2)$, then run the first part of the current proof with Ω replaced by $\Omega_R := \Omega \cap B(0,R)$ which is a bounded UR domain with $\partial\Omega_R = (\partial\Omega) \cup \partial B(0,R)$. Specifically, observe first that since $Lu = 0$ in $\big[\mathcal{D}'(\Omega)\big]^M$, with $L := D^*D$ weakly elliptic $M \times M$ system, from [112, (6.5.40) in Theorem 6.5.7] we see that actually $u \in [\mathscr{C}^\infty(\Omega)]^M$. As a consequence,

$$u \text{ is continuous and bounded near } \partial B(0,R). \tag{3.1.156}$$

Next, if for each $x \in \partial\Omega_R$ we denote by

$$\Gamma^R_\kappa(x) := \big\{y \in \Omega_R : |x-y| < (1+\kappa)\,\mathrm{dist}(y, \partial\Omega_R)\big\} \tag{3.1.157}$$

the nontangential approach region with vertex at x in the set Ω_R, then the fact that $\partial\Omega \subset \partial\Omega_R$ forces $\Gamma^R_\kappa(x) \subseteq \Gamma_\kappa(x)$ for each $x \in \partial\Omega$, where $\Gamma_\kappa(x)$ is the nontangential approach region with vertex at x in the original set Ω. As such, if $\mathcal{N}^R_\kappa$ denotes the nontangential maximal operator associated with the domain Ω_R, we have $\mathcal{N}^R_\kappa u \le \mathcal{N}_\kappa u$ pointwise on $\partial\Omega$. In view of [112, (8.2.26)], (3.1.102), and (3.1.156), this implies

$$\mathcal{N}^R_\kappa u \in L^p(\partial\Omega_R, \sigma_R) \text{ where } \sigma_R := \mathcal{H}^{n-1}\lfloor\partial\Omega_R. \tag{3.1.158}$$

Since clearly $Du = 0$ in Ω_R, the first part of the proof gives that the nontangential boundary trace of u, considered from within nontangential approach regions $\Gamma^R_\kappa(x)$ defined as in (3.1.157), exists at σ-a.e. point $x \in \partial\Omega$ and the resulting function is σ-measurable. In concert with [112, (8.9.8)], these properties further imply that said function belongs to $\big[L^p(\partial\Omega,\sigma)\big]^M$ and its quasi-norm in this space is dominated by $\|\mathcal{N}_\kappa u\|_{L^p(\partial\Omega,\sigma)}$. At this stage, there remains to observe that there exists $\varepsilon > 0$ with the property that

$$\Gamma_\kappa^R(x) \cap B(x,\varepsilon) = \Gamma_\kappa(x) \cap B(x,\varepsilon) \text{ for each } x \in \partial\Omega, \tag{3.1.159}$$

so the nontangential boundary trace of u at points on $\partial\Omega$, taken from within Ω_R, agrees, whenever meaningful, with $u\big|_{\partial\Omega}^{\kappa-\text{n.t.}}$. This completes the proof of (3.1.103).

The last claim in item *(i)*, i.e., that $u\big|_{\partial\Omega}^{\kappa-\text{n.t.}}$ is actually independent of the aperture parameter κ, then becomes a consequence of what we have just proved, the first property in (3.1.102), [112, Proposition 8.9.8], and [112, Proposition 8.8.6].

Turning to item *(ii)*, assume that $p \in (1,\infty)$. Then from [113, (10.2.96)-(10.2.97)], (3.1.102)-(3.1.103), (3.1.153), [112, (3.6.27)], and [112, Corollary 3.7.3] we conclude that

$$\begin{gathered} \mathrm{Sym}(D;\nu) \bullet u = \mathrm{Sym}(D;\nu)\big(u\big|_{\partial\Omega}^{\kappa-\text{n.t.}}\big) \\ \text{at } \sigma\text{-a.e. point on } \partial\Omega. \end{gathered} \tag{3.1.160}$$

Since we are presently assuming that u also vanishes at infinity in the case when Ω is an exterior domain, the integral representation formula (3.1.9) together with (3.1.160) then give (bearing in mind that u is a null-solution of D)

$$u(x) = \int_{\partial\Omega} \Big\langle \widetilde{E}^\top(x-y), (-\mathrm{i})\Big(\mathrm{Sym}(D;\nu)\big(u\big|_{\partial\Omega}^{\kappa-\text{n.t.}}\big)\Big)(y)\Big\rangle \, d\sigma(y) \text{ for all } x \in \Omega. \tag{3.1.161}$$

In concert with item *(3)* in Theorem 2.4.1 (whose applicability is ensured by (3.1.1) and Theorem 1.4.2), this further implies

$$\big\|\mathcal{N}_\kappa u\big\|_{L^p(\partial\Omega,\sigma)} \leq C\big\|\mathrm{Sym}(D;\nu)\big(u\big|_{\partial\Omega}^{\kappa-\text{n.t.}}\big)\big\|_{[L^p(\partial\Omega,\sigma)]^N} \tag{3.1.162}$$

for some constant $C = C(\Omega, D, \kappa, p) \in (0,\infty)$ independent of u. Together with the estimate in (3.1.103), this proves the first equivalence in (3.1.104). The second equivalence in (3.1.104) is a direct consequence of the fact that, since

$$\mathrm{Sym}(D;\xi) : \mathbb{C}^M \longrightarrow \mathbb{C}^N \text{ is injective for each } \xi \in \mathbb{R}^n \setminus \{0\}, \tag{3.1.163}$$

there exists $\lambda > 0$ such that

$$\big|\mathrm{Sym}(D;\xi)w\big| \geq \lambda|\xi||u| \text{ for all } \xi \in \mathbb{R}^n \text{ and } w \in \mathbb{C}^M. \tag{3.1.164}$$

Next, the first formula in (3.1.105) is seen from the Cauchy-type reproducing formula for null-solutions of first-order injectively elliptic systems from [115, §1.4]. The second identity in (3.1.105) then becomes a consequence of the first, and the jump-formula for the boundary-to-domain double layer proved in [115, §1.5]. The claim in (3.1.106) is implied by the definition of the boundary-to-domain double layer operator (cf. [115, §1.4]) and Proposition 1.6.4.

Going further, if $\mathcal{N}_\kappa u \in L^q(\partial\Omega,\sigma)$ for some $q \in (1,\infty)$ then item *(i)* implies that $u\big|_{\partial\Omega}^{\kappa-\text{n.t.}}$ belongs to $\big[L^q(\partial\Omega,\sigma)\big]^M$ with control of the norm. Conversely, assume that

$u\big|_{\partial\Omega}^{\kappa-\mathrm{n.t.}} \in \big[L^q(\partial\Omega,\sigma)\big]^M$ for some $q \in (1,\infty)$. Then from the integral representation formula (3.1.161) and item *(3)* in Theorem 2.4.1 we conclude that there exists some constant $C = C(\Omega, D, \kappa, q) \in (0,\infty)$ independent of u such that

$$\begin{aligned}\|\mathcal{N}_\kappa u\|_{L^q(\partial\Omega,\sigma)} &\leq C\Big\|\mathrm{Sym}(D;\nu)\big(u\big|_{\partial\Omega}^{\kappa-\mathrm{n.t.}}\big)\Big\|_{[L^q(\partial\Omega,\sigma)]^N} \\ &\leq C\Big\|u\big|_{\partial\Omega}^{\kappa-\mathrm{n.t.}}\Big\|_{[L^q(\partial\Omega,\sigma)]^M}. \end{aligned} \tag{3.1.165}$$

This finishes the proof of the equivalence in (3.1.107). Next, in the case when we actually have $u\big|_{\partial\Omega}^{\kappa-\mathrm{n.t.}} \in \big[L^{p,q}_1(\partial\Omega,\sigma)\big]^M$ for some $q \in (1,\infty)$ the nontangential maximal function estimates for the boundary-to-domain double layer potential operator established in [115, §1.5] imply $\mathcal{N}_\kappa(\nabla u) \in L^q(\partial\Omega,\sigma)$ in a quantitative fashion. This justifies the right-pointing implication in (3.1.108). The left-pointing implication in (3.1.108) is a consequence of [113, Proposition 11.3.4].

Regarding the claims in item *(iii)*, assume that $p = 1$. First, from [113, (10.2.96)-(10.2.97)], (3.1.102)-(3.1.103), and (3.1.153), we see that

$$\begin{aligned}&\mathrm{Sym}(D;\nu)\big(u\big|_{\partial\Omega}^{\kappa-\mathrm{n.t.}}\big) = \mathrm{Sym}(D;\nu)\bullet u \in \big[H^1(\partial\Omega,\sigma)\big]^N \\ &\text{and } \Big\|\mathrm{Sym}(D;\nu)\big(u\big|_{\partial\Omega}^{\kappa-\mathrm{n.t.}}\big)\Big\|_{[H^1(\partial\Omega,\sigma)]^N} \leq C\|\mathcal{N}_\kappa u\|_{L^1(\partial\Omega,\sigma)}\end{aligned} \tag{3.1.166}$$

for some $C = C(\Omega, D, \kappa) \in (0,\infty)$ independent of u. This takes care of (3.1.110)-(3.1.111). Assuming that u also vanishes at infinity in the case when Ω is an exterior domain, we may invoke the integral representation formula (3.1.154), together with the nontangential maximal function estimates from (2.4.14) (with $p := 1$, keeping in mind (3.1.1) and Theorem 1.4.2), as well as the first line in (3.1.166), the estimate in (3.1.103) (with $p = 1$), and the first embedding in [113, (4.2.28)] to conclude that (3.1.112) holds.

Moving on, to deal with item *(iv)* suppose $p \in \big(\frac{n-1}{n}, 1\big)$. The claims in (3.1.113) have been already justified in (3.1.153). For the remainder of the proof make the additional assumption that u vanishes at infinity in the case when Ω is an exterior domain. Granted this, and taking into account that u is a null-solution of D in Ω, formula (3.1.6) presently becomes

$$\Big\langle \big[\widetilde{E}^{\top}(x-\cdot)\big|_{\partial\Omega}\big], (-\mathrm{i})\mathrm{Sym}(D;\nu)\bullet u\Big\rangle = \begin{cases} u(x) & \text{if } x \in \Omega, \\ 0 & \text{if } x \in \mathbb{R}^n \setminus \overline{\Omega}. \end{cases} \tag{3.1.167}$$

In concert with the jump-formula (2.5.50), at σ-a.e. point $x \in \partial\Omega$ this allows us to write

$$\begin{aligned}&\mp\frac{1}{2}\widehat{k}(\nu(x))\mathfrak{H}\big(\mathrm{Sym}(D;\nu)\bullet u\big)(x) \\ &\qquad + T\big((-\mathrm{i})\mathrm{Sym}(D;\nu)\bullet u\big)(x) = \begin{cases}\big(u\big|_{\partial\Omega}^{\kappa-\mathrm{n.t.}}\big)(x) \\ 0\end{cases}\end{aligned} \tag{3.1.168}$$

where

$$k := \widetilde{E}^{\top} \in \left[\mathscr{C}^{\infty}(\mathbb{R}^n \setminus \{0\})\right]^{M\times N} \tag{3.1.169}$$

and

$$T : \left[H^p(\partial\Omega, \sigma)\right]^N \longrightarrow \left[L^p(\partial\Omega, \sigma)\right]^M \tag{3.1.170}$$

is the mapping induced as in item *(6)* of Theorem 2.3.2 by the principal-value singular integral of formal convolution type with the matrix kernel k on $\partial\Omega$ (cf. (2.5.3)). Subtracting the two versions of (3.1.168) then yields

$$-\widehat{k}\big(\nu(x)\big)\mathfrak{H}\Big(\mathrm{Sym}(D;\nu)\bullet u\Big)(x) = \big(u\big|_{\partial\Omega}^{\kappa-\mathrm{n.t.}}\big)(x) \tag{3.1.171}$$

at σ-a.e. point $x \in \partial\Omega$. Observe that since $\widetilde{E} = \widetilde{D}^{\top} E_{L^{\top}}$, we may express the Fourier transform of $\widetilde{E}^{\top}$ at each $\xi \in \mathbb{R}^n \setminus \{0\}$ as

$$\begin{aligned}
\Big[\mathrm{Sym}\big(\widetilde{D}^{\top};\xi\big)\widehat{E_{L^{\top}}}(\xi)\Big]^{\top} &= -\Big[\mathrm{Sym}\big(\widetilde{D};\xi\big)^{\top}\big(\widehat{E_L}(\xi)\big)^{\top}\Big]^{\top} \\
&= -\Big[\mathrm{Sym}\big(\widetilde{D};\xi\big)^{\top}\Big(\big[\mathrm{Sym}\big(\widetilde{D}D;\xi\big)\big]^{-1}\Big)^{\top}\Big]^{\top} \\
&= -\big[\mathrm{Sym}\big(\widetilde{D}D;\xi\big)\big]^{-1}\mathrm{Sym}\big(\widetilde{D};\xi\big),
\end{aligned} \tag{3.1.172}$$

where we have also used the properties of E_L recorded in Theorem 1.4.2. Hence,

$$\widehat{k}(\xi) = -\big[\mathrm{Sym}\big(\widetilde{D}D;\xi\big)\big]^{-1}\mathrm{Sym}\big(\widetilde{D};\xi\big) \text{ for each } \xi \in \mathbb{R}^n \setminus \{0\}. \tag{3.1.173}$$

Together with (3.1.171), this proves (3.1.115). Specializing (3.1.115) to the case when $\widetilde{D} := D^*$ (which, according to (1.3.28), is an admissible choice) yields (3.1.116). In the case when D^*D has scalar principal symbol it follows that $\big[\mathrm{Sym}\big(D^*D;\nu\big)\big]^{-1}$ commutes with $\mathrm{Sym}\big(D;\nu\big)$. As such, applying $\mathrm{Sym}\big(D;\nu\big)$ to both sides in (3.1.171) produces (3.1.117), bearing in mind that

$$\big[\mathrm{Sym}\big(D^*D;\nu\big)\big]^{-1}\mathrm{Sym}\big(D;\nu\big)\mathrm{Sym}\big(D^*;\nu\big) = I, \tag{3.1.174}$$

the identity operator. The claims in the current items *(v)-(vi)* are consequences of the integral representation formulas (3.1.154), (3.1.161), items *(4)*, *(6)*, *(7)* in Theorem 2.4.1, as well as (3.1.104) and (3.1.113).

Going further, the claims in item *(vii)* may be justified much as their counterparts in items *(i)-(vi)*, making use of Theorem 3.1.2 in place of Theorem 3.1.1.

Turning our attention to item *(viii)*, work under the assumptions made in (3.1.129)-(3.1.130). All claims with the exception of (3.1.131)-(3.1.132) then follow from Corollary 3.1.5, [113, Proposition 5.2.7], Theorem 2.5.1, Theorem 1.4.2, [112, Corollary 8.9.9], [113, (5.1.12)], [112, (8.9.8)], and [112, Corollary 8.9.6]. To justify (3.1.131), we use what we have proved up to this point, [113, Proposition 10.2.9], and (3.1.91) to write, with the family of vector fields

$$\vec{F}_\alpha = (F_j^\alpha)_{1\leq j\leq n} : \Omega \longrightarrow \mathbb{C}^n \tag{3.1.175}$$

indexed by $\alpha \in \{1, \dots, N\}$ as in (3.1.86),

$$\begin{aligned}(-\mathrm{i})\mathrm{Sym}(D;\nu)\big(u\big|_{\partial\Omega}^{\kappa-\mathrm{n.t.}}\big) &= \Big(a_j^{\alpha\beta}\nu_j\big(u_\beta\big|_{\partial\Omega}^{\kappa-\mathrm{n.t.}}\big)\Big)_{1\leq\alpha\leq N} \\ &= \Big(\nu_j\big(F_j^\alpha\big|_{\partial\Omega}^{\kappa-\mathrm{n.t.}}\big)\Big)_{1\leq\alpha\leq N} = \Big(\nu\cdot\big(\vec{F}^\alpha\big|_{\partial\Omega}^{\kappa-\mathrm{n.t.}}\big)\Big)_{1\leq\alpha\leq N} \\ &= \Big(\nu\bullet\big(\vec{F}^\alpha\big|_{\partial\Omega}^{\kappa-\mathrm{n.t.}}\big)\Big)_{1\leq\alpha\leq N} = (-\mathrm{i})\mathrm{Sym}(D;\nu)\bullet u,\end{aligned} \tag{3.1.176}$$

as wanted. Finally, (3.1.132) is a consequence of what we have proved so far, the integral representation formula from Corollary 3.1.5, Theorem 1.4.2, the estimate recorded in (2.4.20), and [113, (5.1.13)]. The proof of Theorem 3.1.6 is therefore complete. □

Theorem 3.1.6 is surprisingly robust, and a version of its item *(i)*, in which in place of u being a null-solution for the system D in Ω we now ask that Du has components in $L^q(\Omega, \mathcal{L}^n)$ with $q > n$, is presented below, in the class of bounded UR domains $\Omega \subseteq \mathbb{R}^n$.

Theorem 3.1.7 *Let $\Omega \subseteq \mathbb{R}^n$, where $n \in \mathbb{N}$ with $n \geq 2$, be a bounded* UR *domain, and abbreviate $\sigma := \mathcal{H}^{n-1}\lfloor\partial\Omega$. Also, consider an injectively elliptic homogeneous first-order $N \times M$ system D with constant complex coefficients in $\mathbb{R}^n$ (where $N, M \in \mathbb{N}$), and suppose $u : \Omega \to \mathbb{C}^M$ is a vector-valued function with Lebesgue measurable components satisfying, for some aperture parameter $\kappa \in (0, \infty)$,*

$$\begin{aligned}&\mathcal{N}_\kappa u \in L^p(\partial\Omega, \sigma) \text{ for some } p \in \big(\tfrac{n-1}{n}, \infty\big], \\ &\text{and } Du \in \big[L^q(\Omega, \mathcal{L}^n)\big]^N \text{ for some } q \in (n, \infty].\end{aligned} \tag{3.1.177}$$

Then the nontangential boundary trace $u\big|_{\partial\Omega}^{\kappa-\mathrm{n.t.}}$ exists (in $\mathbb{C}^M$) at σ-a.e. point on $\partial\Omega$, and is in fact independent of the aperture parameter κ. Moreover, the function $u\big|_{\partial\Omega}^{\kappa-\mathrm{n.t.}}$ is σ-measurable on $\partial\Omega$ and, in fact, $u\big|_{\partial\Omega}^{\kappa-\mathrm{n.t.}} \in \big[L^p(\partial\Omega, \sigma)\big]^M$.

In light of [112, (8.3.6) in Lemma 8.3.2], the case $p = \infty$ of Theorem 3.1.7 asserts that:

if D is an injectively elliptic homogeneous first-order $N \times M$ system D with constant complex coefficients in $\mathbb{R}^n$, and u is an essentially bounded function in a bounded UR *domain $\Omega \subseteq \mathbb{R}^n$ such that Du has components in $L^q(\Omega, \mathcal{L}^n)$ for some $q > n$ then the nontangential boundary trace $u\big|_{\partial\Omega}^{\mathrm{n.t.}}$ exists at $\mathcal{H}^{n-1}$-a.e. point on $\partial\Omega$ and, in fact, $u\big|_{\partial\Omega}^{\mathrm{n.t.}} \in \big[L^\infty(\partial\Omega, \mathcal{H}^{n-1})\big]^M$.* (3.1.178)

It is interesting to compare this Fatou-type theorem with an earlier result of A. Nagel and W. Rudin (cf. [155, Theorem 11.1.2, p. 235]) to the effect that

$$\begin{aligned}&\textit{if } a,b,c\in\mathbb{R} \textit{ satisfy } a<b \textit{ and } c>0, \textit{ and if } u\in\mathscr{C}^1\big((a,b)\times(0,c)\big)\\ &\textit{is a bounded function satisfying } \partial u/\partial\bar{z}\in L^p\big((a,b)\times(0,c),\mathcal{L}^2\big) \textit{ for}\\ &\textit{some } p>1, \textit{ then the normal limit } \lim_{y\to0^+}u(x+iy) \textit{ exists for } \mathcal{H}^1\textit{-a.e.}\\ &\textit{number } x\in(a,b).\end{aligned} \tag{3.1.179}$$

By further specializing (3.1.178) to scalar functions and $D:=\nabla$, the gradient operator, we obtain the following result, about nontangential pointwise boundary traces of bounded functions belonging to Sobolev spaces in bounded UR domains:

Corollary 3.1.8 *Let $\Omega\subseteq\mathbb{R}^n$, where $n\in\mathbb{N}$ with $n\geq2$, be an arbitrary bounded* UR *domain. Abbreviate $\sigma:=\mathcal{H}^{n-1}\lfloor\partial\Omega$ and fix an aperture parameter $\kappa\in(0,\infty)$. Then any function*

$$u\in W^{1,q}(\Omega)\cap L^\infty(\Omega,\mathcal{L}^n) \textit{ for some } q>n \tag{3.1.180}$$

has a nontangential boundary trace $u\big|_{\partial\Omega}^{\kappa-\mathrm{n.t.}}$ at σ-a.e. point on $\partial\Omega$. Moreover, this trace is in fact independent of the aperture parameter κ and belongs to $L^\infty(\partial\Omega,\sigma)$.

In connection to Corollary 3.1.8 we wish to note if $\Omega\subseteq\mathbb{R}^n$ is a Sobolev extension domain[6] then any function $u\in W^{1,q}(\Omega)$ with $q>n$ necessarily belongs to $\mathscr{C}^0(\overline{\Omega})$. Hence, in the case when Ω is also bounded, any such function u is globally bounded. However, in the class of bounded UR domains, while any function $u\in W^{1,q}(\Omega)$ with $q>n$ is locally continuous (hence, locally bounded), it is not true in general that u belongs to $L^\infty(\Omega,\mathcal{L}^n)$. For example, given $n\in\mathbb{N}$ with $n\geq2$ along with any $q\in(1,\infty)$, if $a>q-1$ and if $b>0$ is small enough so that $a>q(1+b)-1$, then taking

$$\begin{aligned}&\Omega:=\big\{(x_1,\dots,x_n)\in\mathbb{R}^n:\ 0<x_1,\dots,x_{n-1}<1 \text{ and } 0<x_n<x_{n-1}^a\big\},\\ &\qquad\text{and } u(x):=x_{n-1}^{-b} \text{ for each } x=(x_1,\dots,x_n)\in\Omega,\end{aligned} \tag{3.1.181}$$

yields a bounded UR domain $\Omega\subseteq\mathbb{R}^n$ and an unbounded function $u\in W^{1,q}(\Omega)$. In summary, in the class of bounded UR domains $\Omega\subseteq\mathbb{R}^n$, membership to $W^{1,q}(\Omega)$ does not automatically imply global essential boundedness in Ω, even when $q\in(0,\infty)$ is large. Hence, the hypothesis in (3.1.180) is not formulated in a redundant manner.

Another point worth making is that without the hypothesis that u is essentially bounded, one cannot expect the conclusion in Corollary 3.1.8 to retain its full strength. For example, if Ω and u are as in (3.1.181) then for each fixed $\kappa>0$ the nontangential boundary trace $u\big|_{\partial\Omega}^{\kappa-\mathrm{n.t.}}$ exists at σ-a.e. point on $\partial\Omega$ but since

$$\begin{aligned}&\Big(u\big|_{\partial\Omega}^{\kappa-\mathrm{n.t.}}\Big)(x_1,\dots,x_{n-1},0)=x_{n-1}^{-b}\\ &\text{for each } (x_1,\dots,x_{n-1},0)\in(0,1)^{n-1}\times\{0\}\subseteq\partial\Omega,\end{aligned} \tag{3.1.182}$$

---

[6] i.e., there exists an operator $E: W^{1,q}(\Omega)\to W^{1,q}(\mathbb{R}^n)$ with the property that $(Eu)\big|_\Omega=u$ for each $u\in W^{1,q}(\Omega)$

this nontangential trace no longer belongs to the space $L^\infty(\partial\Omega, \sigma)$, even in the case when the integrability exponent q is large.

We now turn to the task of presenting the proof of Theorem 3.1.7.

Proof of Theorem 3.1.7 From the second property in (3.1.177), [113, Lemma 10.1.1], and the fact that $\sigma(\partial\Omega) < +\infty$, we see that

$$\mathfrak{P}(Du) \in L^\infty(\partial\Omega, \sigma) \subseteq L^p(\partial\Omega, \sigma). \tag{3.1.183}$$

Denote by $q' \in \big[1, \frac{n}{n-1}\big)$ the conjugate exponent of $q \in (n, \infty]$. Then the second property in (3.1.177) also implies, thanks to Hölder's inequality and the boundedness of Ω, that for each $x \in \mathbb{R}^n$ we have

$$\int_\Omega \frac{|(Du)(y)|}{|x-y|^{n-1}}\,dy \leq \|Du\|_{[L^q(\Omega,\mathcal{L}^n)]^N}\bigg(\int_\Omega \frac{dy}{|x-y|^{(n-1)q'}}\bigg)^{1/q'} < +\infty. \tag{3.1.184}$$

To proceed, recall $\widetilde{E}$ from (3.1.1) and observe that Theorem 1.4.2 implies

$$\big|\widetilde{E}^\top(x-y)\big| \leq \frac{C}{|x-y|^{n-1}} \quad \text{for each } x, y \in \mathbb{R}^n \text{ with } x \neq y. \tag{3.1.185}$$

Granted (3.1.184)-(3.1.185), the second property in (3.1.177) then permits us to conclude (also bearing in mind that composition with translations is continuous on the Lebesgue scale) that

$$\begin{gathered}\text{the mapping } \mathbb{R}^n \ni x \longmapsto \int_\Omega \big\langle \widetilde{E}^\top(x-y), (Du)(y)\big\rangle\,dy \in \mathbb{C}^M \\ \text{is a well-defined } \mathbb{C}^M\text{-valued function which is continuous in } \mathbb{R}^n.\end{gathered} \tag{3.1.186}$$

With (3.1.183) and (3.1.184) in hand, and granted the first property in (3.1.177), Theorem 3.1.1 applies and gives that $\mathrm{Sym}(D;\nu) \bullet u$ belongs to $\big[H^p(\partial\Omega,\sigma)\big]^N$ and that the representation formula (3.1.6) holds at each Lebesgue point (hence, $\mathcal{L}^n$-a.e., since u is locally integrable) in Ω. In turn, from (3.1.6), (3.1.186), Theorem 2.5.1, Theorem 1.4.2, Corollary 2.5.4, the first property in (3.1.177), and [112, Corollary 8.9.6] we conclude that $u\big|_{\partial\Omega}^{\kappa-\mathrm{n.t.}}$ exists at σ-a.e. point on $\partial\Omega$ and the function $u\big|_{\partial\Omega}^{\kappa-\mathrm{n.t.}}$ is σ-measurable on $\partial\Omega$. Finally, that $u\big|_{\partial\Omega}^{\kappa-\mathrm{n.t.}}$ is actually independent of the aperture parameter is a consequence of what we have just proved, the first property in (3.1.177), [112, Proposition 8.9.8], and [112, Proposition 8.8.6]. □

Here is a local version of Theorem 3.1.6 which, among other things, allows the end-point $p = \infty$ to be considered in (3.1.102). In particular, this implies that any bounded null-solution of an injectively elliptic first-order (homogeneous, constant coefficient) system in an arbitrary UR domain has a nontangential trace at $\mathcal{H}^{n-1}$-a.e. point on the boundary.

Theorem 3.1.9 *Suppose $\Omega \subseteq \mathbb{R}^n$, where $n \in \mathbb{N}$ with $n \geq 2$, is an arbitrary* UR *domain and abbreviate $\sigma := \mathcal{H}^{n-1} \lfloor \partial\Omega$. Also, consider an injectively elliptic homogeneous first-order $N \times M$ system D with constant complex coefficients in $\mathbb{R}^n$ (where $N, M \in \mathbb{N}$), and suppose $u \in \left[\mathscr{C}^\infty(\Omega)\right]^M$ is a vector-valued function satisfying, for some aperture parameter $\kappa \in (0, \infty)$ and some truncation parameter $\varepsilon > 0$,*

$$\mathcal{N}_\kappa^\varepsilon u \in L^p_{\mathrm{loc}}(\partial\Omega, \sigma) \text{ for some } p \in (1, \infty], \text{ and } Du = 0 \text{ in } \Omega. \tag{3.1.187}$$

Then the nontangential boundary trace $u\big|_{\partial\Omega}^{\kappa-\mathrm{n.t.}}$ exists (in $\mathbb{C}^M$) at σ-a.e. point on $\partial\Omega$, is actually independent of the aperture parameter κ, the function $u\big|_{\partial\Omega}^{\kappa-\mathrm{n.t.}}$ is σ-measurable on $\partial\Omega$ and belongs to the space $\left[L^p_{\mathrm{loc}}(\partial\Omega, \sigma)\right]^M$.

As a corollary, the same conclusions stated above remain valid for each given function $u \in \left[L^\infty_{\mathrm{bdd}}(\Omega, \mathcal{L}^n)\right]^M$ satisfying $Du = 0$ in $[\mathcal{D}'(\Omega)]^N$ hence, in particular, for any bounded null-solution $u \in \left[\mathscr{C}^\infty(\Omega)\right]^M$ of the system D in Ω.

If $\partial\Omega$ is bounded then, of course, $L^p_{\mathrm{loc}}(\partial\Omega, \sigma) = L^p(\partial\Omega, \sigma)$, so all conclusions are readily implied Theorem 3.1.6. Hence, the crux of the matter here is allowing $\partial\Omega$ to be an unbounded set.

To offer an example, let Ω be a UR domain in the complex plane, and consider a holomorphic mapping $\varphi : \Omega \to \mathbb{D}$, where $\mathbb{D} := \{z \in \mathbb{C} : |z| < 1\}$ is the unit disk in the plane. In particular, φ is bounded and satisfies $\partial_{\bar{z}}\varphi = 0$ in Ω. Then the last part in the statement of Theorem 3.1.9 (used with $n = 2$ and D the Cauchy-Riemann operator) guarantees that, for each aperture parameter $\varkappa \in (0, \infty)$,

> the nontangential boundary trace $\varphi\big|_{\partial\Omega}^{\kappa-\mathrm{n.t.}}$ exists (in $\mathbb{C}$) at $\mathcal{H}^1$-a.e. point on $\partial\Omega$, is actually independent of the aperture parameter κ, and the function $\varphi\big|_{\partial\Omega}^{\kappa-\mathrm{n.t.}}$ is $\mathcal{H}^1$-measurable and (essentially) bounded on $\partial\Omega$. (3.1.188)

In particular, this applies to the conformal mapping of Ω onto $\mathbb{D}$, from the classical Riemann Mapping Theorem. The latter result should be compared with Carathéodory's theorem (cf., e.g., [49, Theorem 3.1, p 13]), which states that if $\psi : \mathbb{D} \to \Omega$ is a conformal mapping of the unit disk onto a Jordan domain $\Omega \subseteq \mathbb{R}^2 \equiv \mathbb{C}$ then ψ may be extended, as a continuous one-to-one mapping, from $\overline{\mathbb{D}}$ onto $\overline{\Omega}$.

We now present the proof of Theorem 3.1.9.

Proof of Theorem 3.1.9 Fix a point $x_0 \in \partial\Omega$ along with an arbitrary radius $R > 0$, and consider a scalar function $\varphi \in \mathscr{C}^\infty_c(\mathbb{R}^n)$ such that $\operatorname{supp}\varphi \subseteq B(x_0, 2R)$ and $\varphi \equiv 1$ on $B(x_0, R)$. Since from (A.0.92) and [112, (8.1.17)] (also keeping in mind the second line in [112, (8.2.26)]) we have

$$\mathcal{N}_\kappa(\varphi u) \leq \Big(\sup_{\mathbb{R}^n} |\varphi|\Big) \cdot \max\Big\{\mathcal{N}_\kappa^\varepsilon u, \sup_{B(x_0,2R)\setminus O_\varepsilon} |u|\Big\} \cdot \mathbf{1}_{\partial\Omega \cap B(x_0,(4+2\kappa)R)} \text{ on } \partial\Omega, \tag{3.1.189}$$

we conclude from [112, Proposition 8.2.3] and the first property in (3.1.187) that

$$\mathcal{N}_\kappa(\varphi u) \in L^p_{\rm comp}(\partial\Omega, \sigma) \subseteq L^r(\partial\Omega, \sigma) \text{ for each } r \in (0, p]. \tag{3.1.190}$$

In concert with [112, (8.6.51) from Proposition 8.6.3], the first property in (3.1.187) also implies (bearing in mind that $u \in \big[\mathscr{C}^\infty(\Omega)\big]^M \subseteq \big[L^\infty_{\rm loc}(\Omega, \mathcal{L}^n)\big]^M$) that

$$u \in \big[L^{np/(n-1)}_{\rm bdd}(\Omega, \mathcal{L}^n)\big]^M. \tag{3.1.191}$$

In addition, thanks to [112, (1.7.20)] and the second property in (3.1.187), we have

$$D(\varphi u) = (-{\rm i})\,{\rm Sym}(D; \nabla\varphi)u \text{ in } \Omega. \tag{3.1.192}$$

By combining (3.1.191) and (3.1.192) we arrive at the conclusion that

$$\begin{aligned} &D(\varphi u) \text{ belongs to } \big[L^1(\Omega, \mathcal{L}^n)\big]^N \text{ and} \\ &D(\varphi u) = 0 \text{ on } \Omega \setminus \big[B(x_0, 2R) \setminus B(x_0, R)\big]. \end{aligned} \tag{3.1.193}$$

In particular, for each point $x \in \Omega \cap B(x_0, R/2)$ we have

$$\begin{aligned} \int_\Omega \frac{\big|(D(\varphi u))(y)\big|}{|x-y|^{n-1}}\,{\rm d}y &= \int_{\Omega\cap[B(x_0,2R)\setminus B(x_0,R)]} \frac{\big|(D(\varphi u))(y)\big|}{|x-y|^{n-1}}\,{\rm d}y \\ &\leq (R/2)^{1-n} \int_\Omega \big|D(\varphi u)\big|\,{\rm d}\mathcal{L}^n < \infty. \end{aligned} \tag{3.1.194}$$

Going further, keeping (3.1.192) in mind and reasoning as in (3.1.189)-(3.1.190) gives

$$\mathcal{N}_\kappa\big(D(\varphi u)\big) \in L^p_{\rm comp}(\partial\Omega, \sigma) \subseteq L^q(\partial\Omega, \sigma) \text{ for each } q \in (0, p]. \tag{3.1.195}$$

Suppose now that $q \in \big(\frac{n-1}{n}, 1\big)$ is fixed and invoke [113, (10.1.8)-(10.1.9)] to conclude from (3.1.195) that

$$\mathfrak{P}\big(D(\varphi u)\big) \in L^{q^*}(\partial\Omega, \sigma) \text{ where } q^* := \big(\tfrac{1}{q} - \tfrac{1}{n-1}\big)^{-1}. \tag{3.1.196}$$

Given that the assignment $\big(\frac{n-1}{n}, 1\big) \ni q \mapsto q^* \in \big(1, \frac{n-1}{n-2}\big)$ is bijective, we deduce from (3.1.190) and (3.1.196) (since $p > 1$) that

$$\begin{aligned} &\text{there exists some } r \in (1, \infty) \text{ with the property that} \\ &\mathcal{N}_\kappa(\varphi u) \in L^r(\partial\Omega, \sigma) \text{ and } \mathfrak{P}\big(D(\varphi u)\big) \in L^r(\partial\Omega, \sigma). \end{aligned} \tag{3.1.197}$$

Collectively, (3.1.197), (3.1.193), and (3.1.194) ensure that Theorem 3.1.1 is applicable to the function φu, with the old integrability exponent p currently replaced by some $r \in (1, \infty)$. With ν denoting the geometric measure theoretic outward unit normal to Ω, this theorem implies that

$${\rm Sym}\big(D; \nu\big) \bullet (\varphi u) \text{ belongs to the Lebesgue space } \big[L^r(\partial\Omega, \sigma)\big]^N. \tag{3.1.198}$$

Also, keeping in mind that the decay condition (3.1.8) is trivially satisfied by the function φu, and that each point $x \in \Omega \cap B(x_0, R/2)$ is, trivially, a Lebesgue point for the (continuous) function φu, the integral representation formula (3.1.9) gives

$$u(x) = \int_{\partial\Omega} \big\langle \widetilde{E}^{\top}(x-y), (-\mathrm{i})\big(\mathrm{Sym}(D;\nu) \bullet (\varphi u)\big)(y)\big\rangle \, d\sigma(y) - \int_{\Omega \cap [B(x_0,2R) \setminus B(x_0,R)]} \big\langle \widetilde{E}^{\top}(x-y), \big(D(\varphi u)\big)(y)\big\rangle \, dy \tag{3.1.199}$$

for each point $x \in \Omega \cap B(x_0, R/2)$. In addition, from (3.1.193), Theorem 1.4.2, and Lebesgue's Dominated Convergence Theorem it follows that the assignment

$$B(x_0, R/2) \ni x \longmapsto \int_{\Omega \cap [B(x_0,2R) \setminus B(x_0,R)]} \big\langle \widetilde{E}^{\top}(x-y), \big(D(\varphi u)\big)(y)\big\rangle \, dy \tag{3.1.200}$$

is a well-defined, $\mathbb{C}^M$-valued, continuous function.

At this stage, from (3.1.198), (3.1.199), (3.1.200), and Corollary 2.5.2 we deduce that $u\big|_{\partial\Omega}^{\kappa-\mathrm{n.t.}}$ exists at σ-a.e. point on $\partial\Omega \cap B(x_0, R/2)$. Given that $R > 0$ has been taken to be an arbitrary number this, in turn, implies that said nontangential boundary trace actually exists at σ-a.e. point on $\partial\Omega$. Having proved this, we may invoke [112, Corollary 8.9.6] to conclude (also mindful of the fact that we currently have $\sigma(\partial\Omega \setminus \partial_*\Omega) = 0$) that the function $u\big|_{\partial\Omega}^{\kappa-\mathrm{n.t.}}$ is σ-measurable on $\partial\Omega$, and we may rely on [112, Corollary 8.9.9] (also bearing [112, (8.8.52)] in mind) to justify that $u\big|_{\partial\Omega}^{\kappa-\mathrm{n.t.}}$ is actually independent of the aperture parameter κ. Finally, [112, (8.9.8)] and the first property in (3.1.187) guarantee that, in fact, $u\big|_{\partial\Omega}^{\kappa-\mathrm{n.t.}} \in \big[L^p_{\mathrm{loc}}(\partial\Omega, \sigma)\big]^M$. □

We continue by presenting a version of Theorem 3.1.6 which amounts to a quantitative Fatou theorem emphasizing the scale of Morrey spaces, in place of Lebesgue spaces as done in the past.

Theorem 3.1.10 *Suppose $\Omega \subseteq \mathbb{R}^n$, where $n \in \mathbb{N}$ with $n \geq 2$, is an arbitrary* UR *domain and abbreviate $\sigma := \mathcal{H}^{n-1}\lfloor\partial\Omega$. Also, consider an injectively elliptic homogeneous first-order $N \times M$ system D with constant complex coefficients in $\mathbb{R}^n$ (where $N, M \in \mathbb{N}$), and suppose $u \in \big[\mathscr{C}^\infty(\Omega)\big]^M$ is a vector-valued function satisfying, for some $\kappa \in (0,\infty)$, $p \in (1,\infty)$, and $\lambda \in (0, n-1)$,*

$$\mathcal{N}_\kappa u \in M^{p,\lambda}(\partial\Omega, \sigma) \text{ and } Du = 0 \text{ in } \Omega. \tag{3.1.201}$$

Then the nontangential boundary trace $u\big|_{\partial\Omega}^{\kappa-\mathrm{n.t.}}$ exists (in $\mathbb{C}^M$) at σ-a.e. point on $\partial\Omega$, is actually independent of the aperture parameter κ, the function $u\big|_{\partial\Omega}^{\kappa-\mathrm{n.t.}}$ happens to be σ-measurable on $\partial\Omega$, belongs to the Morrey space $\big[M^{p,\lambda}(\partial\Omega, \sigma)\big]^M$, and satisfies

$$\Big\| u\big|_{\partial\Omega}^{\kappa-\mathrm{n.t.}} \Big\|_{[M^{p,\lambda}(\partial\Omega,\sigma)]^M} \leq \big\|\mathcal{N}_\kappa u\big\|_{M^{p,\lambda}(\partial\Omega,\sigma)}. \tag{3.1.202}$$

In addition, under the additional assumption that u vanishes at infinity (in the sense of Definition 1.6.3) in the case when Ω is an exterior domain, one actually has

$$\Big\| u\big|_{\partial\Omega}^{\kappa-\mathrm{n.t.}} \Big\|_{[M^{p,\lambda}(\partial\Omega,\sigma)]^M} \approx \big\|\mathcal{N}_\kappa u\big\|_{M^{p,\lambda}(\partial\Omega,\sigma)}. \tag{3.1.203}$$

Proof One way to justify these claims is to invoke item *(viii)* of Theorem 3.1.6 with $\mathbb{X} := M^{p,\lambda}(\partial\Omega,\sigma)$, a permissible choice in light of [113, Proposition 6.2.17], and [113, Corollaries 6.2.11, 6.2.13].

An alternative, more self-contained, proof goes as follows. Since we have the inclusion $M^{p,\lambda}(\partial\Omega,\sigma) \subseteq L^p_{\mathrm{loc}}(\partial\Omega,\sigma)$, Theorem 3.1.9 applies and gives that the nontangential boundary trace $u\big|_{\partial\Omega}^{\kappa-\mathrm{n.t.}}$ exists at σ-a.e. point on $\partial\Omega$, is independent of the aperture parameter κ, and the function $u\big|_{\partial\Omega}^{\kappa-\mathrm{n.t.}}$ is σ-measurable on $\partial\Omega$. Granted these properties, on account of [113, (6.2.3)], [112, (8.9.8)], and the first property in (3.1.201) we further conclude that $u\big|_{\partial\Omega}^{\kappa-\mathrm{n.t.}}$ belongs to $\big[M^{p,\lambda}(\partial\Omega,\sigma)\big]^M$ and (3.1.202) holds. From what we have proved up to this stage, [113, (10.2.96)-(10.2.97)] (used with $F := u$), [113, (6.2.5)], and [112, (6.2.23)] we also see that

$$\begin{gathered}\mathrm{Sym}(D;\nu)\bullet u = \mathrm{Sym}(D;\nu)\Big(u\big|_{\partial\Omega}^{\kappa-\mathrm{n.t.}}\Big) \in \big[M^{p,\lambda}(\partial\Omega,\sigma)\big]^N,\\ \text{hence } \mathrm{Sym}(D;\nu)\bullet u \in \Big[L^1\Big(\partial\Omega,\frac{\sigma(x)}{1+|x|^{n-1}}\Big)\Big]^N.\end{gathered} \tag{3.1.204}$$

At this stage, there remains to prove (3.1.203) (in fact, only the right-pointing inequality requires justification), working under the additional assumption that u vanishes at infinity in the case when Ω is an exterior domain. We make the observation that there is a natural version of Theorem 3.1.1 which employs Morrey spaces in place of Lebesgue spaces. Specifically, with $\widetilde{E}$ defined as in (3.1.1), we claim that in the current setting we have

$$u(x) = \int_{\partial\Omega} \big\langle \widetilde{E}^{\top}(x-y), (-\mathrm{i})\big(\mathrm{Sym}(D;\nu)\bullet u\big)(y)\big\rangle\, d\sigma(y) \quad \text{for each } x\in\Omega. \tag{3.1.205}$$

To justify (3.1.205), we reason as in the proof of Theorem 3.1.1 (and retain notation introduced on that occasion). For starters, having fixed an arbitrary point $x \in \Omega$, in place of (3.1.16) we now have, for each $\alpha \in \{1,\dots,M\}$,

$$\begin{aligned}&\int_{\partial\Omega} \Big\langle \widetilde{E}^{\top}(x-y), (-\mathrm{i})\big(\mathrm{Sym}(D;\nu)\bullet u\big)(y)\Big\rangle_{\alpha}\, d\sigma(y)\\ &= \lim_{R\to\infty} {}_{[(\mathrm{Lip}_c(\partial\Omega))']^N}\Big\langle (-\mathrm{i})\mathrm{Sym}(D;\nu)\bullet u, \big(\theta_R\eta_\varepsilon \widetilde{E}_{\cdot\alpha}(x-\cdot)\big)\big|_{\partial\Omega}\Big\rangle_{[\mathrm{Lip}_c(\partial\Omega)]^N}\end{aligned} \tag{3.1.206}$$

thanks to the second line in (3.1.204), the fact that there exists a constant $C \in (0,\infty)$ with the property that $\big|\widetilde{E}(x-y)\big| \le C(1+|y|)^{-(n-1)}$ for each $y \in \partial\Omega$ (cf. (3.1.1) and (1.4.24)), Lebesgue's Dominated Convergence Theorem, and [112, (4.1.47)]. We then proceed along the lines of (3.1.17)-(3.1.30), bearing in mind that we are now assuming $Du = 0$. The only aspect we need to reconsider is the fact that the term

$$\mathrm{I} = \Big(\lim_{R\to\infty}\int_\Omega u_\beta a_j^{\alpha\beta}(\partial_j\theta_R)\widetilde{E}_{\alpha\gamma}(x-\cdot)\,\mathrm{d}\mathcal{L}^n\Big)_{1\leq\gamma\leq M} \tag{3.1.207}$$

continues to vanish (much as it did in (3.1.25)) in the present case as well. To see that this is indeed the case, assume first that $\partial\Omega$ is unbounded and, without loss of generality suppose $0 \in \partial\Omega$. Then we may invoke [112, (8.6.51)] for the choice $E := \Omega \cap \big[B(0, CR) \setminus B(0, R)\big]$ (with $C \in (1, \infty)$ standing for what used to be λ in the proof of Theorem 3.1.1) and $p := (n-1)/n$. In concert with (1.4.24), Hölder's inequality, [112, (8.1.17)], and (A.0.87), this permits us to estimate

$$\begin{aligned}
|\mathrm{I}| &\leq C\limsup_{R\to\infty}\Big\{R^{-n}\int_{\Omega\cap[B(0,CR)\setminus B(0,R)]}|u|\,\mathrm{d}\mathcal{L}^n\Big\}\\
&\leq C\limsup_{R\to\infty}\Big\{R^{-n}\Big(\int_{\Omega\cap[B(0,CR)\setminus B(0,R)]}|u|^{\frac{np}{n-1}}\,\mathrm{d}\mathcal{L}^n\Big)^{\frac{n-1}{np}}\Big(\int_{B(0,CR)}\mathrm{d}\mathcal{L}^n\Big)^{1-\frac{n-1}{np}}\Big\}\\
&\leq C\limsup_{R\to\infty}\Big\{R^{-\frac{n-1}{p}}\Big(\int_{\partial\Omega\cap B(0,C(2+\kappa)R)}|\mathcal{N}_\kappa u|^p\,\mathrm{d}\sigma\Big)^{\frac{1}{p}}\Big\}\\
&\leq C\limsup_{R\to\infty}R^{-\frac{n-1-\lambda}{p}}\Big\{R^{\frac{n-1-\lambda}{p}}\Big(⨍_{\partial\Omega\cap B(0,C(2+\kappa)R)}|\mathcal{N}_\kappa u|^p\,\mathrm{d}\sigma\Big)^{\frac{1}{p}}\Big\}\\
&\leq C\limsup_{R\to\infty}R^{-\frac{n-1-\lambda}{p}}\|\mathcal{N}_\kappa u\|_{M^{p,\lambda}(\partial\Omega,\sigma)} = 0.
\end{aligned} \tag{3.1.208}$$

This proves that $\mathrm{I} = 0$, as wanted. That the same conclusion remains valid when Ω is bounded is clear since $\Omega \cap [B(0, CR) \setminus B(0, R)] = \varnothing$ if R is sufficiently large. Finally, in the case when Ω is an exterior domain, we once again have $\mathrm{I} = 0$ thanks to the assumption that u vanishes at infinity and the first line of (3.1.208). This finishes the proof of (3.1.205).

At this stage, from (3.1.205), (3.1.1), Theorem 1.4.2, (2.6.4), (3.1.204), and [113, (6.2.5)] we see that there exists $C = C(\Omega, D, n, \kappa, p, \lambda) \in (0, \infty)$ with the property that

$$\begin{aligned}
\|\mathcal{N}_\kappa u\|_{M^{p,\lambda}(\partial\Omega,\sigma)} &\leq C\big\|\mathrm{Sym}(D;\nu)\bullet u\big\|_{[M^{p,\lambda}(\partial\Omega,\sigma)]^N}\\
&\leq C\Big\|u\Big|_{\partial\Omega}^{\kappa-\mathrm{n.t.}}\Big\|_{[M^{p,\lambda}(\partial\Omega,\sigma)]^M}.
\end{aligned} \tag{3.1.209}$$

Together with (3.1.202), this completes the alternative proof of Theorem 3.1.10. □

We next study the issue whether the L^p-Maximum Principle for null-solutions u of an injectively elliptic first-order system D in a UR domain Ω singled out in (3.1.133) has a suitable pointwise version. In Proposition 3.1.11 below we show that this is indeed the case provided the underlying domain also satisfies an exterior corkscrew condition. Specifically, in such a geometric setting, we prove that the nontangential maximal function of u lies in between (the absolute value of) the nontangential boundary trace of u and (a fixed multiple of) the Hardy-Littlewood maximal operator acting on the nontangential boundary trace of u; see (3.1.212) below (which, in concert with [112, Proposition 8.2.3] and [112, (7.6.18)], actually implies (3.1.133)).

Proposition 3.1.11 *Having fixed $n \in \mathbb{N}$ with $n \geq 2$, assume $\Omega \subseteq \mathbb{R}^n$ is a* UR *domain satisfying an exterior corkscrew condition* (*cf.* [112, Definition 5.1.3]). *Abbreviate $\sigma := \mathcal{H}^{n-1}\lfloor\partial\Omega$ and pick an arbitrary aperture parameter $\kappa \in (0,\infty)$. Also, suppose D is an injectively elliptic homogeneous first-order $N \times M$ system with constant complex coefficients in $\mathbb{R}^n$ (where $N, M \in \mathbb{N}$), and consider a vector-valued function $u : \Omega \to \mathbb{C}^M$ with Lebesgue measurable components satisfying*

$$\begin{gathered}\mathcal{N}_\kappa u \in L^1\Big(\partial\Omega, \frac{\sigma(x)}{1+|x|^{n-1}}\Big) \cap L^p_{\mathrm{loc}}(\partial\Omega,\sigma) \text{ for some } p \in (1,\infty),\\ Du = 0 \text{ in } [\mathcal{D}'(\Omega)]^N, \text{ and } u \text{ vanishes at infinity}\\ (\text{cf. Definition 1.6.3}) \text{ if } \Omega \text{ is an exterior domain.}\end{gathered} \tag{3.1.210}$$

Then the nontangential boundary trace $u\big|^{\kappa-\mathrm{n.t.}}_{\partial\Omega}$ exists (in $\mathbb{C}^M$) at σ-a.e. point on $\partial\Omega$ and is independent of the aperture parameter, the function

$$u\big|^{\kappa-\mathrm{n.t.}}_{\partial\Omega} \text{ belongs to } \Big[L^1\Big(\partial\Omega, \frac{\sigma(x)}{1+|x|^{n-1}}\Big) \cap L^p_{\mathrm{loc}}(\partial\Omega,\sigma)\Big]^M \tag{3.1.211}$$

and, with $\mathcal{M}_{\partial\Omega}$ denoting the Hardy-Littlewood maximal operator on $\partial\Omega$, one has

$$\Big|u\big|^{\kappa-\mathrm{n.t.}}_{\partial\Omega}\Big| \leq \mathcal{N}_\kappa u \leq C \cdot \mathcal{M}_{\partial\Omega}\Big(u\big|^{\kappa-\mathrm{n.t.}}_{\partial\Omega}\Big) \text{ at } \sigma\text{-a.e. point on } \partial\Omega, \tag{3.1.212}$$

for some constant $C = C(\Omega, D, \kappa) \in (0,\infty)$ independent of u.

Proof From Theorem 3.1.9 and [112, (8.9.8)] we see that $u\big|^{\kappa-\mathrm{n.t.}}_{\partial\Omega}$ is a well-defined function, independent of the aperture parameter, and (3.1.211) holds. Granted this property and the current assumptions, the integral representation formula from Theorem 1.6.1 gives (with ν denoting the geometric measure theoretic outward unit normal to Ω)

$$\int_{\partial\Omega}\Big\langle \widetilde{E}^{\top}(x-y), (-\mathrm{i})\Big(\mathrm{Sym}(D;\nu)\Big(u\big|^{\kappa-\mathrm{n.t.}}_{\partial\Omega}\Big)\Big)(y)\Big\rangle\, d\sigma(y) = \begin{cases} u(x) & \text{if } x \in \Omega, \\ 0 & \text{if } x \in \mathbb{R}^n \setminus \overline{\Omega}. \end{cases} \tag{3.1.213}$$

To proceed, fix an arbitrary point $x \in \Omega$ and consider some $z \in \partial\Omega$ such that $r := \mathrm{dist}(x, \partial\Omega) = |x - z|$. Denote by $x_* \in \mathbb{R}^n \setminus \Omega$ a corkscrew point relative to z at scale r. That is, for some $\theta \in (0, 1)$ independent of z and r, we have

$$B(x_*, \theta r) \subseteq B(z, r) \setminus \overline{\Omega}. \tag{3.1.214}$$

Subtracting the two versions of (3.1.213), written for the current point x and for x_* then permits us to estimate

$$|u(x)| \leq C \int_{\partial\Omega} \big|\widetilde{E}(x - y) - \widetilde{E}(x_* - y)\big| \Big|\big(u\big|_{\partial\Omega}^{\kappa-\mathrm{n.t.}}\big)\Big|(y)\, d\sigma(y), \tag{3.1.215}$$

for some $C = C(D) \in (0, \infty)$ independent of u, z, x, x_*. We claim that

$$\big|\widetilde{E}(x - y) - \widetilde{E}(x_* - y)\big| \leq \frac{Cr}{\big(r + |z - y|\big)^n} \text{ for each } y \in \partial\Omega, \tag{3.1.216}$$

where $C = C(\Omega, D, \kappa) \in (0, \infty)$ is independent of z, x, x_*, y, r. To justify this, we study two cases.

Case I: *Assume* $y \in \partial\Omega$ *is such that* $|z - y| \geq 4r$. Then, for each $\xi \in [x, x_*]$, we have

$$\begin{aligned} |\xi - z| &\leq |x - \xi| + |x - z| \leq |x - x_*| + |x - z| \leq |z - x_*| + 2|x - z| \\ &< r + 2r = 3r < \tfrac{3}{4}|z - y|, \end{aligned} \tag{3.1.217}$$

hence

$$|\xi - y| \geq |z - y| - |\xi - z| \geq \big(1 - \tfrac{3}{4}\big)|z - y| = \tfrac{1}{4}|z - y|. \tag{3.1.218}$$

In addition, $|x - x_*| \leq |x - z| + |z - x_*| < r + r = 2r$. Granted these, we may invoke the Mean Value Theorem to estimate (bearing in mind (3.1.1), (1.4.24), and the working hypotheses)

$$\begin{aligned} \big|\widetilde{E}(x - y) - \widetilde{E}(x_* - y)\big| &\leq |x - x_*| \cdot \sup_{\xi \in [x, x_*]} \big|(\nabla\widetilde{E})(\xi - y)\big| \\ &\leq 2r \cdot \sup_{\xi \in [x, x_*]} \Big(\frac{C}{|\xi - y|^n}\Big) \leq \frac{Cr}{|z - y|^n} \\ &\leq \frac{Cr}{\big(r + |z - y|\big)^n}, \end{aligned} \tag{3.1.219}$$

as wanted.

Case II: *Assume* $y \in \partial\Omega$ *is such that* $|z - y| < 4r$. In this scenario, the right-hand side in (3.1.216) is $\geq C/r^{n-1}$, and thanks to (3.1.1), (1.4.24), and assumptions we have

$$
\begin{aligned}
\big|\widetilde{E}(x-y)-\widetilde{E}(x_*-y)\big| &\leq \big|\widetilde{E}(x-y)\big|+\big|\widetilde{E}(x_*-y)\big| \\
&\leq \frac{C}{|x-y|^{n-1}}+\frac{C}{|x_*-y|^{n-1}} \\
&\leq \frac{C}{\operatorname{dist}(x,\partial\Omega)^{n-1}}+\frac{C}{\operatorname{dist}(x_*,\partial\Omega)^{n-1}} \\
&\leq \frac{C}{r^{n-1}},
\end{aligned}
\tag{3.1.220}
$$

from which (3.1.216) follows.

Having established (3.1.216), we make the observation that since for each point $y \in \partial\Omega$ we have $|x-y| \leq |x-z|+|z-y| \leq r+|z-y|$, it follows that

$$
\big|\widetilde{E}(x-y)-\widetilde{E}(x_*-y)\big| \leq \frac{C\cdot\operatorname{dist}(x,\partial\Omega)}{|x-y|^{n}} \quad \text{for each } y \in \partial\Omega. \tag{3.1.221}
$$

When used in concert with (3.1.215) and [112, (8.4.152)] (with $\alpha := 1$) from [112, Proposition 8.4.12], this permits us to conclude that the second estimate in (3.1.212) holds. Since the first estimate in (3.1.212) is also true (by virtue of [112, (8.9.8)] and [112, Proposition 8.8.4]), the proof of the proposition is complete. □

We conclude this section by including the purely real variable Fatou type result from Proposition 3.1.13, As a preamble, in the lemma below we isolate a useful geometric result employed in its proof.

Lemma 3.1.12 *Let $\Omega \subseteq \mathbb{R}^n$, where $n \in \mathbb{N}$ with $n \geq 2$, be an open set of locally finite perimeter with a lower Ahlfors regular boundary $\partial\Omega$, and such that*[7]

$$
\mathcal{H}^{n-1}(\partial\Omega \setminus \partial_*\Omega) = 0. \tag{3.1.222}
$$

In particular, the geometric measure theoretic outward unit normal ν is well defined $\mathcal{H}^{n-1}$-a.e. on $\partial\Omega$.

Then for any $\kappa \in (0,\infty)$ there exist an angle $\theta \in (0,\pi)$ and an aperture parameter $\widetilde{\kappa} \in (0,\infty)$, both depending only on κ and the lower Ahlfors regularity constant of $\partial\Omega$, with the property that for $\mathcal{H}^{n-1}$-a.e. point $x \in \partial\Omega$ one can find a radius $r(x) \in (0,\infty)$, depending only on x, κ, and the lower Ahlfors regularity constant of $\partial\Omega$, such that

$$
\Gamma_\kappa(x) \cap B\big(x,r(x)\big) \subseteq \mathrm{C}_\theta\big(x,\nu(x),r(x)\big) \subseteq \Gamma_{\widetilde{\kappa}}(x) \cap B\big(x,r(x)\big) \tag{3.1.223}
$$

where, generally speaking,

$$
\mathrm{C}_\theta(x,h,r) := \big\{y \in \mathbb{R}^n \cap B(x,r) : (y-x)\cdot h > \cos(\theta/2)\,|y-x|\big\} \tag{3.1.224}
$$

[7] hence, any Ahlfors regular domain will do

denotes the one-component circular (roundly) truncated cone with vertex at $x \in \mathbb{R}^n$, *symmetry axis along the vector* $h \in S^{n-1}$, *full aperture angle* $\theta \in (0, \pi)$, *and truncation radius* $r \in (0, \infty)$.

Proof From (3.1.222) and [112, Proposition 5.6.11] we see that at each point $x \in \partial^*\Omega$ there exists an approximate tangent $(n-1)$-plane to the set $\partial\Omega$ (cf. [112, (5.4.1), (5.4.2)]). Moreover, the vector $\nu(x)$ is perpendicular to the approximate tangent plane in question, call it $\pi(x)$.

To proceed, denote by $c \in (0, \infty)$ the lower Ahlfors regularity constant of $\partial\Omega$, and fix an arbitrary number $s \in (0, 1)$. Also, choose some $s' \in (0, s)$. Finally, pick an arbitrary point $x \in \partial^*\Omega$. Then [112, (5.4.2)] implies that there exists a radius $r(x) \in (0, \infty)$, depending only on x, s', and c, such that

$$\mathcal{H}^{n-1}\big(\Delta(x,r) \cap \mathrm{C}_{x,s'}\big) < \big(c/2^{n-1}\big)r^{n-1} \qquad \forall\, r \in \big(0, r(x)\big), \tag{3.1.225}$$

where

$$\begin{aligned} \mathrm{C}_{x,s'} &:= \big\{y \in \mathbb{R}^n : \operatorname{dist}\big(y, \pi(x)\big) > s'|y-x|\big\} \\ &= \big\{y \in \mathbb{R}^n : \big|(y-x)\cdot \nu(x)\big| > s'|y-x|\big\} \end{aligned} \tag{3.1.226}$$

is the double cone with vertex at x, axis along $\nu(x)$, and full aperture angle $2\arccos s' \in (0, \pi)$. Suppose there exists a point

$$z \in \partial\Omega \cap \mathrm{C}_{x,s} \cap B\big(x, r(x)/2\big), \tag{3.1.227}$$

where the double cone $\mathrm{C}_{x,s}$ is defined as in (3.1.226) with s' replaced by s. Then $r := |z-x| > 0$ (since x belongs to $\partial\Omega$ and the set Ω is open) hence, for one thing, $r \in \big(0, r(x)/2\big)$. For another thing, there exists some small number $\lambda' \in (0,1)$, depending only on n, s, and s', with the property that $B(z, \lambda' r) \subseteq \mathrm{C}_{x,s'}$. In particular,

$$\Delta(z, \lambda' r) \subseteq B(z, \lambda' r) \subseteq B\big(x, (1+\lambda')r\big) \cap \mathrm{C}_{x,s'}, \tag{3.1.228}$$

Given that $(1+\lambda')r \in \big(0, r(x)\big)$, from (3.1.228) and (3.1.225) we obtain

$$\begin{aligned} cr^{n-1} &\leq \mathcal{H}^{n-1}\big(\Delta(z, \lambda r)\big) \leq \mathcal{H}^{n-1}\Big(B\big(x, (1+\lambda')r\big) \cap \mathrm{C}_{x,s'}\Big) \\ &< \big(c/2^{n-1}\big)\big((1+\lambda')r\big)^{n-1} \leq \big(c/2^{n-1}\big)(2r)^{n-1} = cr^{n-1}. \end{aligned} \tag{3.1.229}$$

This is a contradiction which invalidates the happenstance assumed in (3.1.227). As a result, the above argument shows that

$$\partial\Omega \cap \mathrm{C}_{x,s} \cap B\big(x, r(x)/2\big) = \varnothing. \tag{3.1.230}$$

Going further, denote by $\mathrm{C}^{\pm}_{x,s}$ the two connected components of the open double cone $\mathrm{C}_{x,s}$. Then (3.1.230) implies that $\Omega_+ := \Omega$ together with $\Omega_- := \mathbb{R}^n \setminus \overline{\Omega}$ constitute an open cover of $\mathrm{C}^{+}_{x,s} \cap B\big(x, r(x)/2\big)$ as well as of $\mathrm{C}^{-}_{x,s} \cap B\big(x, r(x)/2\big)$. Since the latter sets are connected, it follows that each of them is solely contained in either Ω_+ or Ω_-. We claim that they cannot be both simultaneously contained in Ω_+, or in Ω_- for that matter. To justify this, we shall analyze the case when both intersections $C^{\pm}_{x,s} \cap B\big(x, r(x)/2\big)$ are contained in Ω_+ (the other case, when said intersections are both in Ω_-, is similar[8]). Then $C_{x,s} \cap B\big(x, r(x)/2\big) \subseteq \Omega$, a scenario in which [112, Lemma 5.6.12] together with the first equality in [112, (5.6.69)] (both used with $E := \Omega$) lead to the conclusion that

$$\begin{aligned} 0 &= \lim_{r\to 0^+} \frac{\mathcal{L}^n\big(\{y \in B(x,r) \cap \Omega : (y-x)\cdot\nu(x) > 0\}\big)}{\mathcal{L}^n\big(B(x,r)\big)} \\ &\geq \lim_{r\to 0^+} \frac{\mathcal{L}^n\big(\{y \in B(x,r) \cap C_{x,s} : (y-x)\cdot\nu(x) > 0\}\big)}{\mathcal{L}^n\big(B(x,r)\big)} > 0, \end{aligned} \tag{3.1.231}$$

which is obviously a contradiction. Hence, $C^+_{x,s} \cap B\big(x, r(x)/2\big)$ is contained in one of the sets $\Omega_\pm$, while $C^-_{x,s} \cap B\big(x, r(x)/2\big)$ is contained in the other one. To make a choice, suppose the components of $C_{x,s}$ are labeled so that

$$C^+_{x,s} \cap B\big(x, r(x)/2\big) \subseteq \Omega \text{ and } C^-_{x,s} \cap B\big(x, r(x)/2\big) \subseteq \mathbb{R}^n \setminus \overline{\Omega}. \tag{3.1.232}$$

Form (3.1.226), the first inclusion in (3.1.232), and [112, Lemma 5.6.15] we also conclude that

$$\begin{aligned} C^+_{x,s'} &= \big\{y \in \mathbb{R}^n : (x-y)\cdot\nu(x) > s'|y-x|\big\}, \\ C^-_{x,s'} &= \big\{y \in \mathbb{R}^n : (y-x)\cdot\nu(x) > s'|y-x|\big\}, \end{aligned} \tag{3.1.233}$$

hence the simple cones $C^+_{x,s'}$, $C^-_{x,s'}$ open up in the direction of $-\nu(x)$ and $\nu(x)$, respectively.

Next, pick some number $s'' \in (s,1)$. Then $C_{x,s''} \subseteq C_{x,s}$ and, in fact, there exists some small number $\lambda'' \in (0,1)$, depending only on n, s, and s'', with the property that $B\big(y, \lambda''|y-x|\big) \subseteq C^+_{x,s}$ for each $y \in C^+_{x,s''}$. Thus, given any point $y \in C^+_{x,s''} \cap B\big(x, r(x)/4\big)$ we have $\rho := |x-y| \in \big(0, r(x)/4\big)$ and

$$\begin{aligned} B(y, \lambda''\rho) &\subseteq B\big(x, (1+\lambda'')\rho\big) \cap C^+_{x,s} \subseteq B(x, 2\rho) \cap C^+_{x,s} \\ &\subseteq B\big(x, r(x)/2\big) \cap C^+_{x,s} \subseteq \Omega. \end{aligned} \tag{3.1.234}$$

Hence, $y \in \Omega$ and $\operatorname{dist}(y, \partial\Omega) \geq \lambda''|x-y|$, two properties placing y in $\Gamma_{\widetilde{\kappa}}(x)$ if $\widetilde{\kappa} > 0$ is such that $\lambda'' = (1+\widetilde{\kappa})^{-1}$. This proves that

$$C^+_{x,s} \cap B\big(x, r(x)/2\big) \subseteq \Gamma_{\widetilde{\kappa}}(x) \text{ if } \widetilde{\kappa} := (1/\lambda'') - 1 > 0. \tag{3.1.235}$$

In the opposite direction, start now by fixing an aperture parameter $\kappa > 0$. For any point $y \in \Gamma_\kappa(x) \cap B\big(x, r(x)/4\big)$ we therefore have $\rho := |x-y| \in \big(0, r(x)/4\big)$ and $\operatorname{dist}(y, \partial\Omega) \geq \rho(1+\kappa)^{-1}$. As such, $B\big(y, \rho(1+\kappa)^{-1}\big) \subseteq \Omega$ and also

[8] now using the second equality in [112, (5.6.69)] and keeping in mind that $\mathcal{L}^n(\partial\Omega) = 0$

$$B\big(y, \rho(1+\kappa)^{-1}\big) \subseteq B(x, 2\rho) \subseteq B\big(x, r(x)/2\big) \tag{3.1.236}$$

so that, ultimately,

$$B\big(y, \rho(1+\kappa)^{-1}\big) \subseteq \Omega \cap B\big(x, r(x)/2\big). \tag{3.1.237}$$

In view of (3.1.232), this goes to show that

$$B\big(y, \rho(1+\kappa)^{-1}\big) \cap \Big(\mathrm{C}^-_{x,s} \cap B\big(x, r(x)/2\big)\Big) = \varnothing \tag{3.1.238}$$

which, thanks to (3.1.236) and the definition of ρ, self-improves to

$$B\big(y, |x-y|(1+\kappa)^{-1}\big) \cap \mathrm{C}^-_{x,s} = \varnothing. \tag{3.1.239}$$

Based on elementary Euclidean geometry, it transpires that if $s \in (0,1)$ is small enough to beging with such that

$$\arccos s > \tfrac{\pi}{2} - \arcsin\big((1+\kappa)^{-1}\big) \tag{3.1.240}$$

then the collection of all points $y \in \mathbb{R}^n$ satisfying (3.1.239) is

$$\mathrm{C}_\theta\big(x, -\nu(x)\big) := \big\{y \in \mathbb{R}^n : (x-y)\cdot\nu(x) > \cos(\theta/2)\,|x-y|\big\} \tag{3.1.241}$$

i.e., the infinite circular one-component cone with vertex at $x \in \mathbb{R}^n$, symmetry axis along the vector $-\nu(x)$, and full aperture

$$\theta := 2\pi - 2\arcsin\big((1+\kappa)^{-1}\big) - 2\arccos s \in (0,\pi). \tag{3.1.242}$$

Keeping in mind the fact that the point $y \in \Gamma_\kappa(x) \cap B\big(x, r(x)/4\big)$ has been arbitrarily chosen, we arrive at the conclusion that

$$\begin{gathered}\text{if } \theta \in (0,\pi) \text{ is as in (3.1.242) (with } s \in (0,1) \text{ as in (3.1.240))}\\ \text{then } \Gamma_\kappa(x) \cap B\big(x, r(x)/4\big) \subseteq \mathrm{C}_\theta\big(x, -\nu(x)\big), \text{ the cone from (3.1.241).}\end{gathered} \tag{3.1.243}$$

Keeping in mind the fact that we presently have $\mathcal{H}^{n-1}(\partial\Omega \setminus \partial^*\Omega) = 0$, the first inclusion in (3.1.223) is seen from (3.1.243), while the second inclusion in (3.1.223) is a consequence of (the more general result established in) (3.1.235). □

Here is the Fatou type result of a purely real variable nature, alluded to earlier.

Proposition 3.1.13 *Let $\Omega \subseteq \mathbb{R}^n$, where $n \in \mathbb{N}$ with $n \geq 2$, be an Ahlfors regular domain, and abbreviate $\sigma := \mathcal{H}^{n-1}\lfloor\partial\Omega$. Also, assume $u \in \mathscr{C}^1(\Omega)$ is a function with the property that, for some aperture parameter $\kappa \in (0,\infty)$, some integrability exponent $p \in (0,\infty]$, and some truncation parameter $\varepsilon > 0$ satisfies*

$$\mathcal{N}^\varepsilon_\kappa(\nabla u) \in L^p_{\mathrm{loc}}(\partial\Omega, \sigma). \tag{3.1.244}$$

Then

$$\text{the nontangential boundary trace } u\Big|_{\partial\Omega}^{\kappa-\text{n.t.}} \text{ exists at } \sigma\text{-a.e. point on } \partial\Omega. \tag{3.1.245}$$

Moreover, for any other given aperture parameter $\kappa' \in (0, \infty)$ *the nontangential boundary trace* $u\Big|_{\partial\Omega}^{\kappa'-\text{n.t.}}$ *also exists, and agrees with* $u\Big|_{\partial\Omega}^{\kappa-\text{n.t.}}$, *at* σ*-a.e. point on* $\partial\Omega$. *Finally, with the dependence on the aperture parameter dropped, the function* $u\Big|_{\partial\Omega}^{\text{n.t.}}$ *is* σ*-measurable on* $\partial\Omega$.

Proof From Lemma 3.1.12 we know that there exist an angle $\theta \in (0, \pi)$ and an aperture parameter $\widetilde{\kappa} \in (0, \infty)$ with the property that for σ-a.e. point $x \in \partial\Omega$ one can find a radius $r(x) \in (0, \infty)$ such that the inclusions in (3.1.223) hold. Observe that this implies that for any given radius $r \in \big(0, \min\{\varepsilon, r(x)\}\big)$ and any given pair of points $y_1, y_2 \in \Gamma_\kappa(x) \cap B(x, r)$ the line segment $[y_1, y_2]$ is contained in the intersection $\Gamma_{\widetilde{\kappa}}(x) \cap B(x, r)$, and the Man Value Theorem allows us to estimate

$$|u(y_1) - u(y_2)| \leq |y_1 - y_2| \cdot \sup_{\xi \in (y_1, y_2)} |(\nabla u)(\xi)| \leq 2r \cdot \Big(\mathcal{N}^{\varepsilon}_{\widetilde{\kappa}}(\nabla u)\Big)(x). \tag{3.1.246}$$

Based on this, Cauchy's criterion for convergence, and Definition A.0.99, whenever we also have $x \in A_\kappa(\partial\Omega)$ and $\big(\mathcal{N}^{\varepsilon}_{\widetilde{\kappa}}(\nabla u)\big)(x) < \infty$ we see that the nontangential boundary trace $\Big(u\Big|_{\partial\Omega}^{\kappa-\text{n.t.}}\Big)(x)$. Since [112, Proposition 8.8.4] guarantees that we presently have $\sigma\big(\partial\Omega \setminus A_\kappa(\partial\Omega)\big) = 0$, while [112, Corollary 8.4.2] together with (3.1.244) ensure that $\big(\mathcal{N}^{\varepsilon}_{\widetilde{\kappa}}(\nabla u)\big)(x) < \infty$ at σ-a.e. point $x \in \partial\Omega$, we ultimately conclude that the claim made in (3.1.245) is true. Granted this, the fact that Ω is an Ahlfors regular domain together with [112, Corollary 8.9.6] and [112, Lemma 3.6.4] then imply that $u\Big|_{\partial\Omega}^{\kappa-\text{n.t.}}$ is a σ-measurable function on $\partial\Omega$.

Finally, the remaining claims in the statement follow from what we have proved so far and the fact that, thanks to [112, Corollary 8.4.2], the assumption made in (3.1.244) is unaffected by changing the aperture parameter. □

3.2 Brief Look at Hardy Spaces Associated with Injectively Elliptic First-Order Systems

Let D be first-order homogeneous $N \times M$ system in $\mathbb{R}^n$ (where $n \in \mathbb{N}$ satisfies $n \geq 2$) with constant (complex) coefficients, which is injectively elliptic. Also, consider an open set $\Omega \subseteq \mathbb{R}^n$ such that $\sigma := \mathcal{H}^{n-1} \lfloor \partial\Omega$ has the property that

$$\sigma\big(B(x, r) \cap \partial\Omega\big) > 0 \text{ for each } x \in \partial\Omega \text{ and } r > 0. \tag{3.2.1}$$

In this setting, having fixed a background parameter $\kappa \in (0, \infty)$, for each integrability exponent $p \in (0, \infty]$ define the `Hardy space associated with` the operator D in Ω as follows:

$$\mathcal{H}^p(\Omega; D) \text{ is the collection of all functions } u \in \big[\mathscr{C}^\infty(\Omega)\big]^M \text{ satisfying } \mathcal{N}_\kappa u \in L^p(\partial\Omega, \sigma) \text{ and } Du = 0 \text{ in } \Omega, \text{ and which also vanish at infinity (in the sense described in Definition 1.6.3) when } \Omega \text{ is an exterior domain.} \tag{3.2.2}$$

Equip the space $\mathcal{H}^p(\Omega; D)$ introduced in (3.2.2) with the quasi-norm

$$\|u\|_{\mathcal{H}^p(\Omega;D)} := \big\|\mathcal{N}_\kappa u\big\|_{L^p(\partial\Omega,\sigma)}, \qquad \forall\, u \in \mathcal{H}^p(\Omega; D). \tag{3.2.3}$$

Hence, $\mathcal{H}^p(\Omega; D)$ is a subspace of $N^p_\kappa(\Omega; \sigma)$ (defined as in (A.0.98)). Thanks to (3.2.1), we may then rely on [112, Proposition 8.3.5] to conclude that

$$\mathcal{H}^p(\Omega; D) \text{ is a quasi-Banach space and } \|\cdot\|_{\mathcal{H}^p(\Omega;D)} \text{ is a } p\text{-norm for } p \in (0,1); \text{ also, } \mathcal{H}^p(\Omega; D) \text{ is an actual Banach space if } p \in [1,\infty]. \tag{3.2.4}$$

It is comforting to note that, whenever σ is a doubling measure on $\partial\Omega$, [112, Proposition 8.4.1] guarantees that the definition of the Hardy space $\mathcal{H}^p(\Omega; D)$ in (3.2.2) is actually independent of the aperture parameter $\kappa \in (0, \infty)$.

Under the stronger hypothesis that the underlying set Ω is actually a UR domain, assumed henceforth, Theorem 3.1.6 ensures that the Hardy spaces defined in (3.2.2) enjoy a wealth of remarkable properties. Among other things, in such a scenario the nontangential pointwise trace

$$\mathcal{H}^p(\Omega; D) \ni u \longmapsto u\big|^{\text{n.t.}}_{\partial\Omega} \in \big[L^p(\partial\Omega, \sigma)\big]^M \tag{3.2.5}$$

defines a well-defined, linear, and bounded operator. In analogy with more classical settings, we define the boundary Hardy space associated with the operator D as the image of this mapping, i.e.,

$$\mathcal{H}^p(\partial\Omega; D) := \Big\{u\big|^{\text{n.t.}}_{\partial\Omega} : u \in \mathcal{H}^p(\Omega; D)\Big\}, \tag{3.2.6}$$

regarded as a subspace of $\big[L^p(\partial\Omega, \sigma)\big]^M$ and equipped with the quasi-norm

$$\|u\|_{\mathcal{H}^p(\partial\Omega;D)} := \big\|u\big|^{\text{n.t.}}_{\partial\Omega}\big\|_{[L^p(\partial\Omega,\sigma)]^M}, \qquad \forall\, u \in \mathcal{H}^p(\partial\Omega; D). \tag{3.2.7}$$

In particular,

$$\text{the mapping } \mathcal{H}^p(\Omega; D) \ni u \longmapsto u\big|^{\text{n.t.}}_{\partial\Omega} \in \mathcal{H}^p(\partial\Omega; D) \text{ is well defined, linear, bounded, surjective, and actually an isomorphism whenever } p \in (1,\infty), \tag{3.2.8}$$

with the very last claim implied by item *(ii)* of Theorem 3.1.6. Let us additionally note here that if $\mathcal{D}$ and K are the boundary double layer potential operators associated as in [115, §1.4] with the set Ω and the weakly elliptic second-order $M \times M$ system

$$L := \widetilde{D}D, \text{ where } \widetilde{D} \text{ is a } M \times N \text{ homogeneous first-order system with constant complex coefficients in } \mathbb{R}^n \text{ which complements } D \text{ (in the sense that (1.3.21) holds)} \tag{3.2.9}$$

then (3.1.105) implies that, for each $p \in (1, \infty)$,

$$\mathcal{D} : \mathrm{Ker}\Big(-\tfrac{1}{2}I + K; \big[L^p(\partial\Omega, \sigma)\big]^M\Big) \longrightarrow \mathcal{H}^p(\Omega; D) \tag{3.2.10}$$

is a well-defined, linear, bounded, and surjective operator, with the nontangential pointwise trace

$$\mathcal{H}^p(\Omega; D) \ni u \longmapsto u\big|^{\text{n.t.}}_{\partial\Omega} \in \mathrm{Ker}\Big(-\tfrac{1}{2}I + K; \big[L^p(\partial\Omega, \sigma)\big]^M\Big) \tag{3.2.11}$$

serving as a right-inverse. Since the jump-formula for the double layer (cf. [115, §1.5]) shows that the operator (3.2.10) is also injective, we ultimately conclude that, for each $p \in (1, \infty)$, the operator

$$\begin{gathered} \mathcal{D} : \mathrm{Ker}\Big(-\tfrac{1}{2}I + K; \big[L^p(\partial\Omega, \sigma)\big]^M\Big) \to \mathcal{H}^p(\Omega; D) \\ \text{is an isomorphism with (3.2.11) as inverse, and} \\ \mathcal{H}^p(\partial\Omega; D) = \mathrm{Ker}\Big(-\tfrac{1}{2}I + K; \big[L^p(\partial\Omega, \sigma)\big]^M\Big). \end{gathered} \tag{3.2.12}$$

We also wish to remark that [112, Corollary 1.7.3] used with $p = 1$ and w arbitrary constant in $\mathbb{C}^N$ establishes, keeping in mind the decay property (1.6.46), the cancelation property to the effect that

$$\int_{\partial\Omega} (-\mathrm{i})\mathrm{Sym}(D; \nu)\big(u\big|^{\text{n.t.}}_{\partial\Omega}\big)\, d\sigma = 0 \ \text{ for each } u \in \mathcal{H}^1(\Omega; D). \tag{3.2.13}$$

In fact, if D is actually elliptic, in the sense that its principal symbol $\mathrm{Sym}(D; \xi)$ is invertible for each $\xi \in \mathbb{R}^n \setminus \{0\}$, then formula [112, (1.7.35) in Corollary 1.7.3] ensures (in view of (3.2.2) and Definition 1.6.3) that

$$\begin{gathered} \int_{\partial\Omega} \Big\langle (-\mathrm{i})\mathrm{Sym}(D; \nu)\big(u\big|^{\text{n.t.}}_{\partial\Omega}\big), w\big|^{\text{n.t.}}_{\partial\Omega} \Big\rangle\, d\sigma = 0 \ \text{ for each} \\ u \in \mathcal{H}^p(\Omega; D) \text{ and } w \in \mathcal{H}^{p'}(\Omega; D^\top) \text{ with } 1/p + 1/p' = 1. \end{gathered} \tag{3.2.14}$$

In light of (3.2.6), we may simply rephrase this cancelation property as

$$\begin{gathered} \int_{\partial\Omega} \big\langle (-\mathrm{i})\mathrm{Sym}(D; \nu) f, g \big\rangle\, d\sigma = 0 \ \text{ for any two functions} \\ f \in \mathcal{H}^p(\partial\Omega; D) \text{ and } g \in \mathcal{H}^{p'}(\partial\Omega; D^\top) \text{ with } 1/p + 1/p' = 1. \end{gathered} \tag{3.2.15}$$

Here is one last comment pertaining to our brand of Hardy spaces, when considered in the category of UR domains satisfying an exterior corkscrew condition (which, in particular, includes all two-sided NTA domains with Ahlfors regular boundaries). Specifically, Proposition 3.1.11 implies that

$$\begin{array}{l}\text{if } \Omega \subseteq \mathbb{R}^n \text{ (with } n \geq 2\text{) is a UR domain satisfying an exterior corkscrew}\\ \text{condition, } D \text{ is an injectively elliptic homogeneous first-order system}\\ \text{with constant complex coefficients in } \mathbb{R}^n, \text{ and } \kappa \in (0,\infty), \text{ then there}\\ \text{exists some } C = C(\Omega, D, \kappa) \in (0,\infty) \text{ such that for each } u \in \mathcal{H}^p(\Omega; D)\\ \text{with } p \in (1,\infty) \text{ we have } \Big|u\big|^{\kappa-\mathrm{n.t.}}_{\partial\Omega}\Big| \leq \mathcal{N}_\kappa u \leq C \cdot \mathcal{M}_{\partial\Omega}\big(u\big|^{\kappa-\mathrm{n.t.}}_{\partial\Omega}\big) \text{ at almost}\\ \text{every point on } \partial\Omega \text{ (with respect to } \mathcal{H}^{n-1}).\end{array} \tag{3.2.16}$$

The history of Hardy spaces is particularly rich and influential, going back to its origins involving holomorphic functions of one complex variable in the planar unit disk (and the upper half-plane; cf. [48], [49], [59], [84], [156] and the references therein) to the modern theory of Hardy spaces associated with the Moisil-Teodorescu system (aka generalized Cauchy-Riemann equations) (as originally developed in [45], [166], [167], [169]-[171]) and, further, to more general domains and differential operators (cf. [78], [72], [125]), including extensions to null-solutions for Dirac-type operators on manifolds (see [118], [124], [123], [127]). Here we only briefly elaborate on the impact of our current results on the theory of Hardy spaces for the standard Dirac operator in arbitrary UR domains. To state a significant result in this regard, the reader is advised to recall the Clifford bullet product defined in (A.0.97). The reader is also alerted to the fact that we shall occasionally use notation and results discussed later (independently of the current considerations), such as those pertaining to the Clifford-Riesz transform from [115, §2.1].

Theorem 3.2.1 *Fix $n \in \mathbb{N}$ with $n \geq 2$, and suppose $\Omega \subseteq \mathbb{R}^n$ is an arbitrary* UR *domain. Abbreviate $\sigma := \mathcal{H}^{n-1}\lfloor\partial\Omega$ and denote by ν the geometric measure theoretic outward unit normal to Ω. Pick a background aperture parameter $\kappa \in (0,\infty)$ and, for each $p \in \big(\frac{n-1}{n},\infty\big)$, let $\mathcal{H}^p(\Omega; D)$ be the Hardy space associated as in* (3.2.2) *with the Dirac operator D from* (A.0.37).

(1) For each $u \in \mathcal{H}^p(\Omega; D)$, the nontangential boundary trace $u\big|^{\kappa-\mathrm{n.t.}}_{\partial\Omega}$ exists (in $\mathcal{C}\ell_n$) at σ-a.e. point on $\partial\Omega$, and is actually independent of the aperture parameter κ. Simplifying notation by dropping the dependence on κ, the function $u\big|^{\mathrm{n.t.}}_{\partial\Omega}$ belongs to the space $L^p(\partial\Omega,\sigma)\otimes\mathcal{C}\ell_n$ and $\big\|u\big|^{\mathrm{n.t.}}_{\partial\Omega}\big\|_{L^p(\partial\Omega,\sigma)\otimes\mathcal{C}\ell_n} \leq \|\mathcal{N}_\kappa u\|_{L^p(\partial\Omega,\sigma)}$. In addition, if $p \in (1,\infty)$ one actually has

$$\|\mathcal{N}_\kappa u\|_{L^p(\partial\Omega,\sigma)} \approx \big\|u\big|^{\mathrm{n.t.}}_{\partial\Omega}\big\|_{L^p(\partial\Omega,\sigma)\otimes\mathcal{C}\ell_n} \tag{3.2.17}$$

as well as

$$\begin{aligned} &u\big|^{\mathrm{n.t.}}_{\partial\Omega} \in L^p_1(\partial\Omega,\sigma)\otimes\mathcal{C}\ell_n \Longleftrightarrow \mathcal{N}_\kappa(\nabla u) \in L^p(\partial\Omega,\sigma), \text{ and}\\ &\big\|u\big|^{\mathrm{n.t.}}_{\partial\Omega}\big\|_{L^p_1(\partial\Omega,\sigma)\otimes\mathcal{C}\ell_n} \approx \|\mathcal{N}_\kappa u\|_{L^p(\partial\Omega,\sigma)} + \|\mathcal{N}_\kappa(\nabla u)\|_{L^p(\partial\Omega,\sigma)} \end{aligned} \tag{3.2.18}$$

with proportionality constants independent of the function u in both cases.

(2) *With the Clifford-Riesz transform operator* $\mathcal{R}_{C\ell}$ *originally defined as in* (A.0.116), *and subsequently extended to Hardy spaces, for each function* $u \in \mathcal{H}^p(\Omega; D)$ *the following version of Cauchy's Reproducing Formula holds*

$$u = \tfrac{1}{2}\mathcal{R}_{C\ell}(\nu \bullet u) \ \text{in}\ \Omega, \tag{3.2.19}$$

and one has

$$\begin{aligned} &\nu \bullet u \in H^p(\partial\Omega, \sigma) \otimes C\ell_n \ \text{and} \\ &\|\nu \bullet u\|_{H^p(\partial\Omega,\sigma)\otimes C\ell_n} \approx \|\mathcal{N}_\kappa u\|_{L^p(\partial\Omega,\sigma)}, \end{aligned} \tag{3.2.20}$$

with proportionality constants independent of the function u.

(3) *If one defines the "bullet" boundary Hardy space*

$$\mathscr{H}^p_\bullet(\partial\Omega; D) := \{\nu \bullet u : u \in \mathcal{H}^p(\Omega; D)\} \tag{3.2.21}$$

and equip it with the quasi-norm inherited from $H^p(\partial\Omega, \sigma) \otimes C\ell_n$, *then* $\mathscr{H}^p_\bullet(\partial\Omega; D)$ *is a closed linear subspace of* $H^p(\partial\Omega, \sigma) \otimes C\ell_n$. *Also, if* $\mathfrak{C}^\#$ *is the "transpose" Cauchy-Clifford operator (originally defined as in* (A.0.36) *and then extended to Hardy spaces) then* $\frac{1}{2}I + \mathfrak{C}^\#$ *is a projection*[9] *of* $H^p(\partial\Omega, \sigma) \otimes C\ell_n$ *onto* $\mathscr{H}^p_\bullet(\partial\Omega; D)$. *In addition,*

$$\begin{aligned} \mathscr{H}^p_\bullet(\partial\Omega; D) &= \mathrm{Im}\Big[\tfrac{1}{2}I - \mathfrak{C}^\# : H^p(\partial\Omega, \sigma) \otimes C\ell_n \to H^p(\partial\Omega, \sigma) \otimes C\ell_n\Big] \\ &= \mathrm{Ker}\Big[\tfrac{1}{2}I + \mathfrak{C}^\# : H^p(\partial\Omega, \sigma) \otimes C\ell_n \to H^p(\partial\Omega, \sigma) \otimes C\ell_n\Big] \end{aligned} \tag{3.2.22}$$

and

$$\begin{aligned} &\mathcal{R}_{C\ell} : \mathscr{H}^p_\bullet(\partial\Omega; D) \longrightarrow \mathcal{H}^p(\Omega; D) \ \text{isomorphically,} \\ &\text{with inverse}\ \mathcal{H}^p(\Omega; D) \ni u \mapsto \tfrac{1}{2}\nu \bullet u \in \mathscr{H}^p_\bullet(\partial\Omega; D). \end{aligned} \tag{3.2.23}$$

In particular, a monogenic[10] *function* u *belongs to the Hardy space* $\mathcal{H}^p(\Omega; D)$ *if and only if* $u = \mathcal{R}_{C\ell} f$ *for some* $f \in H^p(\partial\Omega, \sigma) \otimes C\ell_n$, *hence*

$$\begin{aligned} &\mathcal{R}_{C\ell} : H^p(\partial\Omega, \sigma) \otimes C\ell_n \longrightarrow \mathcal{H}^p(\Omega; D) \\ &\text{is a well-defined, linear, bounded, surjective operator.} \end{aligned} \tag{3.2.24}$$

(4) *With* $\mathfrak{H}$ *denoting the* L^p*-filtering operator from* [113, Theorem 4.9.1] *(used here with* $\Sigma := \partial\Omega$*), for each* $u \in \mathcal{H}^p(\Omega; D)$ *one has*

$$\mathfrak{H}\big(\nu \bullet u\big) = \nu \odot \big(u\big|^{\text{n.t.}}_{\partial\Omega}\big) \ \text{at}\ \sigma\text{-a.e. point on}\ \partial\Omega. \tag{3.2.25}$$

---

[9] i.e., a linear, bounded, and idempotent operator

[10] i.e., a null-solution of the Dirac operator (A.0.37)

As a consequence, with the boundary Hardy space $\mathcal{H}^p(\partial\Omega; D)$ defined as in (3.2.6) (*where now D stands for the Dirac operator from* (A.0.37)), *the mapping*

$$\mathfrak{H} : \mathscr{H}^p_\bullet(\partial\Omega; D) \longrightarrow \nu \odot \mathcal{H}^p(\partial\Omega; D) \tag{3.2.26}$$

is well defined, linear, bounded, and surjective.

(5) *Introduce $\Omega_+ := \Omega$ and $\Omega_- := \mathbb{R}^n \setminus \overline{\Omega}$. Then for each $f \in H^p(\partial\Omega, \sigma) \otimes \mathcal{C}\ell_n$ there exist unique functions $u_\pm \in \mathcal{H}^p(\Omega_\pm; D)$ with the property that*

$$f = \nu \circledbullet u_+ + \nu \circledbullet u_- \text{ in } H^p(\partial\Omega, \sigma) \otimes \mathcal{C}\ell_n. \tag{3.2.27}$$

Consequently, with the direct sum topological, the following Calderón-type decomposition formula holds:

$$H^p(\partial\Omega, \sigma) \otimes \mathcal{C}\ell_n = \mathscr{H}^p_\bullet(\partial\Omega_+; D) \oplus \mathscr{H}^p_\bullet(\partial\Omega_-; D). \tag{3.2.28}$$

(6) *Suppose $\partial\Omega$ is bounded and recall the boundary-to-boundary "transpose" Cauchy-Clifford operator $\mathfrak{C}^\#$ (originally defined as in* (A.0.36) *and then extended to Hardy spaces). Then, whenever $\frac{n-1}{n} < p < p^* \le 1$, the p^*-envelope of the "bullet" boundary Hardy space $\mathscr{H}^p_\bullet(\partial\Omega; D)$ may be described as*

$$\begin{aligned}\mathcal{E}_{p^*}\big(\mathscr{H}^p_\bullet(\partial\Omega; D)\big) &= \operatorname{Im}\Big[\tfrac{1}{2}I - \mathfrak{C}^\# : B^{p^*,p^*}_{-(n-1)(\frac{1}{p}-\frac{1}{p^*})}(\partial\Omega, \sigma) \otimes \mathcal{C}\ell_n \\ &\qquad\qquad \to B^{p^*,p^*}_{-(n-1)(\frac{1}{p}-\frac{1}{p^*})}(\partial\Omega, \sigma) \otimes \mathcal{C}\ell_n\Big] \\ &= \operatorname{Ker}\Big[\tfrac{1}{2}I + \mathfrak{C}^\# : B^{p^*,p^*}_{-(n-1)(\frac{1}{p}-\frac{1}{p^*})}(\partial\Omega, \sigma) \otimes \mathcal{C}\ell_n \\ &\qquad\qquad \to B^{p^*,p^*}_{-(n-1)(\frac{1}{p}-\frac{1}{p^*})}(\partial\Omega, \sigma) \otimes \mathcal{C}\ell_n\Big].\end{aligned} \tag{3.2.29}$$

Finally, strengthen the original hypotheses by assuming that Ω is a bounded NTA *domain with an Ahlfors regular boundary, and recall the Clifford-Riesz transform $\mathcal{R}_{\mathcal{C}\ell}$ from* (A.0.116). *Then, whenever $\frac{n-1}{n} < p < p^* \le 1$, it follows that p^*-envelope of the Hardy space $\mathcal{H}^p(\Omega; D)$ may be described as*

$$\begin{aligned}&\mathcal{E}_{p^*}\big(\mathcal{H}^p(\Omega; D)\big) \\ &= \operatorname{Im}\Big[\mathcal{R}_{\mathcal{C}\ell} : B^{p^*,p^*}_{-(n-1)(\frac{1}{p}-\frac{1}{p^*})}(\partial\Omega, \sigma) \otimes \mathcal{C}\ell_n \to B^{p^*,p^*}_{\frac{1}{p^*}-(n-1)(\frac{1}{p}-\frac{1}{p^*})}(\Omega)\Big] \\ &= \Big\{\mathcal{R}_{\mathcal{C}\ell} f : f \in B^{p^*,p^*}_{-(n-1)(\frac{1}{p}-\frac{1}{p^*})}(\partial\Omega, \sigma) \otimes \mathcal{C}\ell_n \text{ and } \big(\tfrac{1}{2}I + \mathfrak{C}^\#\big)f = 0\Big\}.\end{aligned} \tag{3.2.30}$$

Before presenting the proof of this theorem, we make some comments pointing to further research on this topic. Specifically, it is natural to consider Hardy spaces in UR domains $\Omega \subseteq \mathbb{R}^n$ associated with a given injectively elliptic first-order homogeneous constant (complex) coefficient system D in $\mathbb{R}^n$, by demanding the membership of the nontangential maximal operator to a variety of other function spaces on $\partial\Omega$. For example, as apparent from the proof of Theorem 3.2.1 given below, similar results to those described in items *(1)-(5)* above are valid for Muckenhoupt weighted Lebesgue-styled Hardy spaces, and Morrey-styled Hardy spaces (involving super-unital integrability exponents). The same is true for Lorentz-based Hardy spaces associated with the Dirac operator D in Ω, i.e.,

$\mathcal{H}^{p,q}(\Omega; D)$ defined for each $p \in \left(\frac{n-1}{n}, \infty\right)$ and $q \in (0, \infty)$ as the space of functions $u \in \left[\mathscr{C}^\infty(\Omega)\right]^M$ satisfying $\mathcal{N}_\kappa u \in L^{p,q}(\partial\Omega, \sigma)$ and $Du = 0$ in Ω, and which also vanish at infinity (in the sense of Definition 1.6.3) in the case when Ω is an exterior domain. (3.2.31)

It is also of interest to consider more general sets and larger ranges of indices. To offer a concrete example, work in the two-dimensional setting under the canonical identification $\mathbb{R}^2 \equiv \mathbb{C}$, and take D to be $\partial_{\bar{z}} := \frac{1}{2}\left(\partial_x + i\partial_y\right)$, the Cauchy-Riemann operator. Then, if $\Omega \subseteq \mathbb{C}$ is an open set with an Ahlfors regular boundary and $z_o \in \partial\Omega$ is a fixed point, then for each $m \in \mathbb{N}$ the function

$$u(z) := \frac{1}{(z - z_o)^m}, \qquad \forall z \in \Omega, \tag{3.2.32}$$

belongs to the Lorentz-based Hardy space $\mathcal{H}^{1/m,\infty}(\Omega; \partial_{\bar{z}})$. This is seen from [112, Lemma 8.3.7] and [112, Example 6.2.2].

We now turn to the task of providing the proof of Theorem 3.2.1.

Proof of Theorem 3.2.1 Since the Dirac operator D is injectively elliptic, the claims in item *(1)* are direct consequences of items *(i)-(ii)* in Theorem 3.1.6. Next, the discussion in [113, Example 10.2.14] implies that

$$\begin{aligned} &\nu \odot u \in H^p(\partial\Omega, \sigma) \otimes \mathcal{C}\ell_n \ \text{ and} \\ &\|\nu \odot u\|_{H^p(\partial\Omega,\sigma)\otimes\mathcal{C}\ell_n} \leq C\|\mathcal{N}_\kappa u\|_{L^p(\partial\Omega,\sigma)}, \end{aligned} \tag{3.2.33}$$

for some finite constant $C > 0$ independent of u. Also, in view of the properties of the Clifford-Riesz transform from [115, §2.1] and (A.0.97), formula (3.1.154) reduces precisely to the integral representation formula (3.2.19). Having established this, the nontangential maximal function estimates for the Clifford-Riesz transform established in [115, §2.1] then permit us to write (for some $C \in (0, \infty)$ independent of u)

$$\begin{aligned} \left\|\mathcal{N}_\kappa u\right\|_{L^p(\partial\Omega,\sigma)} &= \tfrac{1}{2}\left\|\mathcal{N}_\kappa\left(\mathcal{R}_{\mathcal{C}\ell}(\nu \odot u)\right)\right\|_{L^p(\partial\Omega,\sigma)} \\ &\leq C\|\nu \odot u\|_{H^p(\partial\Omega,\sigma)\otimes\mathcal{C}\ell_n}. \end{aligned} \tag{3.2.34}$$

Together with (3.2.33), this finishes the proof of the claims made in (3.2.20). Then the fact that $\mathscr{H}^p_{\bullet}(\partial\Omega; D)$ is a closed subspace of $H^p(\partial\Omega, \sigma) \otimes \mathcal{C}\ell_n$ becomes a consequence of (3.2.20) and the completeness of $\mathcal{H}^p(\Omega; D)$ (cf. (3.2.4)). Next, by once again relying on the properties of the Clifford-Riesz transform from [115, §2.1] we conclude that

$$\begin{gathered} \mathcal{R}_{\mathcal{C}\ell} : H^p(\partial\Omega, \sigma) \otimes \mathcal{C}\ell_n \longrightarrow \mathcal{H}^p(\Omega; D) \\ \text{is well defined, linear, bounded, and surjective.} \end{gathered} \tag{3.2.35}$$

In fact, it is shown in [115, §2.1] that

$$\nu \odot (\mathcal{R}_{\mathcal{C}\ell} f) = (I - 2\mathfrak{C}^{\#}) f \quad \text{for each } f \in H^p(\partial\Omega, \sigma) \otimes \mathcal{C}\ell_n. \tag{3.2.36}$$

In concert, (3.2.35), (3.2.36), (3.2.21), and (3.2.19) imply

$$\mathscr{H}^p_{\bullet}(\partial\Omega; D) = \Big\{ \big(\tfrac{1}{2} I - \mathfrak{C}^{\#}\big) f : f \in H^p(\partial\Omega, \sigma) \otimes \mathcal{C}\ell_n \Big\}, \tag{3.2.37}$$

which proves the first equality in (3.2.22). The second equality in (3.2.22) is a consequence of the this and the fact that, as readily seen from the fact that $2\mathfrak{C}^{\#}$ is idempotent when acting on Hardy spaces, $\frac{1}{2} I - \mathfrak{C}^{\#}$ is a projection on $H^p(\partial\Omega, \sigma) \otimes \mathcal{C}\ell_n$.

Moving on, (3.2.23) is implied by (3.2.35), (3.2.19), and (3.2.21). Also, (3.2.25) is a consequence of (3.1.117) when $p \in \big(\frac{n-1}{n}, 1\big)$, and of [113, (4.9.3)] together with [113, (10.2.104)] in the case when $1 \leq p < \infty$.

Consider next the claim made in item *(5)*. Given a function $f \in H^p(\partial\Omega, \sigma) \otimes \mathcal{C}\ell_n$, the jump-formula (3.2.36) written both in Ω_+ and Ω_- implies that (3.2.27) holds for $u_\pm := \frac{1}{2} \mathcal{R}_{\mathcal{C}\ell} f$ in $\Omega_\pm$. To prove the uniqueness of the functions $u_\pm \in \mathcal{H}^p(\Omega_\pm; D)$ doing the job in (3.2.27), assume $f = 0$. Then the function

$$u = \begin{cases} u_+ & \text{in } \Omega_+, \\ u_- & \text{in } \Omega_-, \end{cases} \tag{3.2.38}$$

is locally integrable in $\mathbb{R}^n$ and, thanks to [113, (10.2.101)] and the present working hypotheses, satisfies $Du = 0$ in the sense of distributions in $\mathbb{R}^n$. If $\partial\Omega$ is bounded, then u also vanishes at infinity, so Liouville's theorem ultimately implies that u vanishes in $\mathbb{R}^n$. Hence, in this case we have $u_\pm = 0$ in $\Omega_\pm$, as wanted. An alternative approach, which works regardless of whether $\partial\Omega$ is bounded or not, is to observe that if

$$u_\pm \in \mathcal{H}^p(\Omega_\pm; D) \text{ are such that } \nu \odot u_+ + \nu \odot u_- = 0, \tag{3.2.39}$$

then representation formula (3.1.167) presently implies that for each $x \in \Omega_-$ we have

$$u_-(x) = \tfrac{1}{2} \big(\mathcal{R}_{\mathcal{C}\ell}(\nu \odot u_-)\big)(x) = -\tfrac{1}{2} \big(\mathcal{R}_{\mathcal{C}\ell}(\nu \odot u_+)\big)(x) = 0. \tag{3.2.40}$$

Thus, $u_- = 0$ in Ω_- and, analogously, $u_+ = 0$ in Ω_+.

Finally, consider the claims in item *(6)*. From (3.2.22) we know that the "transpose" Cauchy-Clifford operator $\mathfrak{C}^{\#}$ (originally defined as in (A.0.36) and then extended to Hardy spaces) induces is a well-defined, linear, bounded, and surjective mapping

$$\tfrac{1}{2}I - \mathfrak{C}^{\#} : H^p(\partial\Omega, \sigma) \otimes C\ell_n \longrightarrow \mathscr{H}^p_{\bullet}(\partial\Omega; D). \tag{3.2.41}$$

Assume that $\partial\Omega$ is bounded and $\frac{n-1}{n} < p < p^* \le 1$. By taking p^*-envelopes in (3.2.41) we then conclude, based on item *(2)* in [113, Proposition 7.8.15], [113, (7.8.153)], [113, Proposition 7.8.13], and the mapping properties of $\mathfrak{C}^{\#}$ on the Besov scale (established in [115, §4.1]), that the mapping

$$\tfrac{1}{2}I - \mathfrak{C}^{\#} : B^{p^*,p^*}_{-(n-1)(\frac{1}{p}-\frac{1}{p^*})} \otimes C\ell_n \longrightarrow \mathcal{E}_{p^*}\big(\mathscr{H}^p_{\bullet}(\partial\Omega; D)\big) \tag{3.2.42}$$

continues to be surjective, where $\mathfrak{C}^{\#}$ is now the "transpose" Cauchy-Clifford operator in the context of boundary Besov spaces (with negative smoothness). From this, the first equality in (3.2.29) readily follows. Also, the second equality in (3.2.29) is a consequence of the first and the fact that $2\mathfrak{C}^{\#}$ is idempotent when acting on Besov spaces.

Finally, work under the assumption that Ω is a bounded NTA domain with an Ahlfors regular boundary. Also, suppose $\frac{n-1}{n} < p < p^* \le 1$. Taking p^*-envelopes in (3.2.24) yields, on account of based on item *(2)* in [113, Proposition 7.8.15], [113, (7.8.153)], and (A.0.116) that

$$\begin{gathered} \mathcal{R}_{C\ell} : B^{p^*,p^*}_{-(n-1)(\frac{1}{p}-\frac{1}{p^*})} \otimes C\ell_n \longrightarrow \mathcal{E}_{p^*}\big(\mathcal{H}^p(\Omega; D)\big) \\ \text{is a well-defined, linear, bounded, surjective mapping,} \end{gathered} \tag{3.2.43}$$

where the Clifford-Riesz transform $\mathcal{R}_{C\ell}$ now acts from boundary Besov spaces with a negative amount of smoothness. Granted this, the first equality in (3.2.30) follows. One way to see that $\mathcal{E}_{p^*}\big(\mathcal{H}^p(\Omega; D)\big)$ may also be described as in the second line of (3.2.30) is to take p^*-envelopes in the first line of (3.2.23), then invoke (3.2.29) and item *(1)* in [113, Proposition 7.8.15]. □

3.3 Quantitative Fatou-Type Theorems in UR Domains for Second-Order Systems

Here we establish certain Fatou-type theorems in arbitrary UR domains, involving weakly elliptic second-order systems. To put matters in the proper perspective we begin by recalling that T. Wolff has proved (cf. [181, Theorem 3, p. 321] and subsequent comments) that, given $n \in \mathbb{N}$ with $n \ge 3$, there exists some harmonic function u in $\mathbb{R}^n_+$ with the property that

$$\sup_{t>0}\Big(\int_{\mathbb{R}^{n-1}}|(\nabla u)(x',t)|^p\,dx'\Big)^{1/p}<+\infty \text{ for some } p>0, \text{ and yet} \tag{3.3.1}$$
$$\nabla u \text{ does not have nontangential limits } \mathcal{H}^{n-1}\text{-a.e. on } \mathbb{R}^{n-1}\equiv\partial\mathbb{R}^n_+.$$

The actual value of critical exponent p in (3.3.1) is unknown, but E. Stein and G. Weiss proved (cf. [169, Theorem A, p. 28]) that if u is harmonic in some open subset of $\mathbb{R}^n$ then $|\nabla u|^p$ is subharmonic if $p\geq\frac{n-2}{n-1}$. This, in turn, implies that harmonic gradients have limits almost everywhere if the uniform L^p-integrability condition in (3.3.1) is satisfied with $p\geq\frac{n-2}{n-1}$ (cf. [169, Theorem B, p. 46]). Hence, one necessarily has $0<p<\frac{n-2}{n-1}$ in (3.3.1). In this vein, let us also note that C. Fefferman and E. Stein have shown in [45, Theorem 9, p. 168] that, for a harmonic gradient ∇u, the uniform L^p-integrability condition in (3.3.1) is equivalent, in the regime $p>\frac{n-2}{n-1}$, to actually having $\mathcal{N}_\kappa(\nabla u)\in L^p(\mathbb{R}^{n-1},\mathcal{H}^{n-1})$ (for some, or any, aperture parameter $\kappa>0$). Our goal is to generalize the positive Fatou-type result of Stein-Weiss, in the Fefferman-Stein formulation, by showing that

> if u is a (vector-valued) null-solution in a UR domain $\Omega\subseteq\mathbb{R}^n$ (with $n\geq 2$) of a weakly elliptic homogeneous constant coefficient second-order system in $\mathbb{R}^n$, having the property that the nontangential maximal function of its Jacobian belongs to the space $L^p(\partial\Omega,\mathcal{H}^{n-1})$ for some $p\in\big(\frac{n-1}{n},\infty\big)$, then ∇u has nontangential limits $\mathcal{H}^{n-1}$-a.e. on $\partial\Omega$. (3.3.2)

In light of Wolff's counterexample in (3.3.1), this goal seems reasonably sharp, and in Theorem 3.3.4 we shall accomplish just that. This theorem is stated later, and is given two proofs. A key ingredient in one of these proofs is the representation formula contained in the theorem below.

Theorem 3.3.1 *Let $\Omega\subseteq\mathbb{R}^n$, where $n\in\mathbb{N}$ with $n\geq 2$, be an Ahlfors regular domain, and abbreviate $\sigma:=\mathcal{H}^{n-1}\lfloor\partial\Omega$. For some $M\in\mathbb{N}$, consider a coefficient tensor $A=\big(a^{\alpha\beta}_{rs}\big)_{\substack{1\leq r,s\leq n\\ 1\leq\alpha,\beta\leq M}}$ with complex entries, with the property that the $M\times M$ homogeneous second-order system $L=L_A$ associated with A in $\mathbb{R}^n$ as in* (1.3.15) *is weakly elliptic (in the sense of* (1.3.3)*). Also, denote by $E=(E_{\gamma\beta})_{1\leq\gamma,\beta\leq M}$ the matrix-valued fundamental solution associated with L as in Theorem 1.4.2.*

In this setting, assume $u=(u_\beta)_{1\leq\beta\leq M}\in\big[W^{1,1}_{\rm loc}(\Omega)\big]^M$ is a vector-valued function which, for some aperture parameter $\kappa>0$, satisfies

$$\mathcal{N}_\kappa(\nabla u)\in L^p(\partial\Omega,\sigma) \text{ for some } p\in\big(\tfrac{n-1}{n},\infty\big). \tag{3.3.3}$$

With Lu considered in the sense of distributions in Ω, suppose also that

$$Lu\in\big[L^1_{\rm loc}(\Omega,\mathcal{L}^n)\big]^M \text{ and } \mathfrak{P}(Lu)\in L^p(\partial\Omega,\sigma). \tag{3.3.4}$$

Then for any $\ell,s\in\{1,\dots,n\}$ one has

$$\mathring{\partial}_{\tau_{\ell s}} u \in \big[H^p(\partial\Omega,\sigma)\big]^M \ \text{ and } \ \mathring{\partial}^A_\nu u \in \big[H^p(\partial\Omega,\sigma)\big]^M. \tag{3.3.5}$$

Moreover, given any $\ell \in \{1,\dots,n\}$ *and* $\gamma \in \{1,\dots,M\}$, *it follows that for* $\mathcal{L}^n$*-a.e. Lebesgue point* $x \in \Omega$ *of* ∇u *with the property that*

$$\int_\Omega \frac{|(Lu)(y)|}{|x-y|^{n-1}}\,dy < +\infty \tag{3.3.6}$$

one has (with the duality pairings understood in the sense of [113, Theorem 4.6.1] *with* $\Sigma := \partial\Omega$, *and the summation convention over repeated indices in effect)*

$$\begin{aligned}(\partial_\ell u_\gamma)(x) &= a^{\beta\alpha}_{rs}\Big\langle \big[(\partial_r E_{\gamma\beta})(x-\cdot)\big|_{\partial\Omega}\big], \mathring{\partial}_{\tau_{\ell s}} u_\alpha\Big\rangle \\ &\quad - \Big\langle \big[(\partial_\ell E_{\gamma\beta})(x-\cdot)\big|_{\partial\Omega}\big], (\mathring{\partial}^A_\nu u)_\beta\Big\rangle \\ &\quad + \int_\Omega (\partial_\ell E_{\gamma\beta})(x-y)\big(Lu\big)_\beta(y)\,dy \end{aligned} \tag{3.3.7}$$

assuming that $p \in \big(\frac{n-1}{n},1\big]$ *and* $\partial\Omega$ *is unbounded. A similar formula holds when* $p \in \big(\frac{n-1}{n},1\big]$ *and* Ω *is bounded, this time omitting taking equivalence classes of functions modulo constants in the duality pairings in* (3.3.7). *In the case when* $p \in \big(\frac{n-1}{n},1\big]$ *and* Ω *is an exterior domain, the latter formula remains true under the additional assumption that there exists* $\lambda \in (1,\infty)$ *such that*

$$\fint_{B(0,\lambda R)\setminus B(0,R)} |\nabla u|\,d\mathcal{L}^n = o(1) \ \text{ as } \ R \to \infty. \tag{3.3.8}$$

In fact, a similar integral representation formula holds when $p \in (1,\infty)$ *(with the same caveat as before when* Ω *is an exterior domain), this time interpreting the duality pairings as integration on* $\partial\Omega$ *with respect to* σ, *i.e.,*

$$\begin{aligned}(\partial_\ell u_\gamma)(x) &= \int_{\partial\Omega} a^{\beta\alpha}_{rs}(\partial_r E_{\gamma\beta})(x-y)\big(\mathring{\partial}_{\tau_{\ell s}} u_\alpha\big)(y)\,d\sigma(y) \\ &\quad - \int_{\partial\Omega} (\partial_\ell E_{\gamma\beta})(x-y)\big(\mathring{\partial}^A_\nu u\big)_\beta(y)\,d\sigma(y) \\ &\quad + \int_\Omega (\partial_\ell E_{\gamma\beta})(x-y)\big(Lu\big)_\beta(y)\,dy. \end{aligned} \tag{3.3.9}$$

Finally, if in place of (3.3.3)-(3.3.4) *one now assumes*

$$\begin{gathered}\mathcal{N}_\kappa(\nabla u) \in L^{p,q}(\partial\Omega,\sigma) \text{ for some } p \in \big(\tfrac{n-1}{n},\infty\big) \text{ and } q \in (0,\infty), \\ \text{as well as } Lu \in \big[L^1_{\mathrm{loc}}(\Omega,\mathcal{L}^n)\big]^M \text{ and } \mathfrak{P}(Lu) \in L^{p,q}(\partial\Omega,\sigma), \end{gathered} \tag{3.3.10}$$

then

$$\begin{aligned} \overset{\bullet}{\partial}_{\tau_{\ell s}} u \in \big[H^{p,q}(\partial\Omega,\sigma)\big]^M \textit{ for all } \ell, s \in \{1,\dots,n\} \\ \textit{and } \overset{\bullet}{\partial}{}^A_\nu u \in \big[H^{p,q}(\partial\Omega,\sigma)\big]^M, \end{aligned} \tag{3.3.11}$$

and given any $\ell \in \{1,\dots,n\}$, $\gamma \in \{1,\dots,M\}$, it follows that for $\mathcal{L}^n$-a.e. Lebesgue point $x \in \Omega$ of ∇u with the property that (3.3.6) *holds one has*

$$\begin{aligned} (\partial_\ell u_\gamma)(x) &= {}_{(H^{p,q}(\partial\Omega,\sigma))^*}\Big\langle a^{\beta\alpha}_{rs}(\partial_r E_{\gamma\beta})(x-\cdot)\big|_{\partial\Omega}, \overset{\bullet}{\partial}_{\tau_{\ell s}} u_\alpha \Big\rangle_{H^{p,q}(\partial\Omega,\sigma)} \\ &\quad - {}_{(H^{p,q}(\partial\Omega,\sigma))^*}\Big\langle (\partial_\ell E_{\gamma\beta})(x-\cdot)\big|_{\partial\Omega}, \big(\overset{\bullet}{\partial}{}^A_\nu u\big)_\beta \Big\rangle_{H^{p,q}(\partial\Omega,\sigma)} \\ &\quad + \int_\Omega (\partial_\ell E_{\gamma\beta})(x-y)\big(Lu\big)_\beta(y)\,dy \end{aligned} \tag{3.3.12}$$

with the duality pairings understood in the sense of [113, Lemma 4.7.1, (4.7.2)] *(with $\Sigma := \partial\Omega$), the summation convention over repeated indices in effect, and also assuming the decay condition* (3.3.8) *in the case when Ω is an exterior domain.*

A significant consequence of Theorem 3.3.1, which may be justified along the lines of the proof of Corollary 1.6.2, is that if u is a null-solution of a weakly elliptic homogeneous constant coefficient second-order system in an exterior domain, then having ∇u vanish at infinity in the sense of (3.3.8) is equivalent to having

$$\begin{aligned} (\partial^\alpha u)(x) = O\big(|x|^{2-n-|\alpha|}\big) \text{ as } |x| \to \infty, \\ \text{for each multi-index } \alpha \in \mathbb{N}_0^n \text{ with } |\alpha| \geq 1. \end{aligned} \tag{3.3.13}$$

Proof of Theorem 3.3.1 First we shall work under the assumptions made in (3.3.3)-(3.3.4). From [113, (10.2.14) in Example 10.2.2], [113, (10.2.182)], and (3.3.3)-(3.3.4) we see that

$$\begin{aligned} \overset{\bullet}{\partial}_{\tau_{\ell s}} u \in \big[H^p(\partial\Omega,\sigma)\big]^M \text{ for each } \ell, s \in \{1,\dots,n\}, \\ \text{and } \overset{\bullet}{\partial}{}^A_\nu u \in \big[H^p(\partial\Omega,\sigma)\big]^M. \end{aligned} \tag{3.3.14}$$

The remainder of the proof of (3.3.7) and (3.3.9) is divided into two cases, depending on the size of p.

`Case I:` *Suppose $p \in \big(\frac{n-1}{n}, 1\big]$*. To fix ideas, suppose for now that $\partial\Omega$ is unbounded, and choose a Lebesgue point $x \in \Omega$ for ∇u with the property that (3.3.6) holds. Also, select two arbitrary indexes, $\gamma \in \{1,\dots,M\}$ and $\ell \in \{1,\dots,n\}$. Bring in a scalar-valued function $\eta \in \mathscr{C}^\infty(\mathbb{R}^n)$ satisfying $\eta = 0$ on $B(0,1)$, $\eta = 1$ on $\mathbb{R}^n \setminus B(0,2)$, then for each $\varepsilon \in \big(0, \frac{1}{2}\operatorname{dist}(x,\partial\Omega)\big)$ define $\eta_\varepsilon(y) := \eta\big((y-x)/\varepsilon\big)$ for every $y \in \mathbb{R}^n$. Then there exists a constant $C \in (0,\infty)$ such that for each $\varepsilon \in \big(0, \frac{1}{2}\operatorname{dist}(x,\partial\Omega)\big)$ we have

$$
\begin{aligned}
&\eta_\varepsilon \in \mathscr{C}^\infty(\mathbb{R}^n), \ \text{bounded function,}\\
&\lim_{\varepsilon\to 0^+} \eta_\varepsilon(y) = 1 \ \text{ for every } \ y \in \mathbb{R}^n \setminus \{x\},\\
&1 - \eta_\varepsilon \in \mathscr{C}^\infty_c(\Omega), \quad \eta_\varepsilon \equiv 0 \ \text{ on } B(x,\varepsilon),\\
&\operatorname{supp}(\nabla\eta_\varepsilon) \subseteq \overline{B(x,2\varepsilon)} \setminus B(x,\varepsilon), \ \text{ and}\\
&|(\nabla^j \eta_\varepsilon)(y)| \le C\varepsilon^{-j} \ \text{ for all } j \in \mathbb{N}_0 \ \text{ and } \ y \in \mathbb{R}^n.
\end{aligned}
\tag{3.3.15}
$$

In particular, for each $\beta \in \{1,\dots,M\}$ the function $E_{\gamma\beta}(x-\cdot)\eta_\varepsilon$ belongs to $\mathscr{C}^\infty(\mathbb{R}^n)$ and coincides with $E_{\gamma\beta}(x-\cdot)$ near $\partial\Omega$.

Let us also fix some number $\lambda \in (1,\infty)$ and select a function $\theta \in \mathscr{C}^\infty_c(\mathbb{R}^n)$ satisfying $0 \le \theta \le 1$, $\theta \equiv 1$ on $B(0,1)$, $\theta \equiv 0$ on $\mathbb{R}^n \setminus B(0,\lambda)$ and, for each $R > 0$, define $\theta_R(y) := \theta(y/R)$ for every $y \in \mathbb{R}^n$. Then the general weak-$*$ convergence results established in [113, Lemma 4.8.4] and [113, Lemma 4.8.1] (also bearing in mind the trivial bounded embedding $L^\infty(\partial\Omega,\sigma) \hookrightarrow \mathrm{BMO}(\partial\Omega,\sigma)$) imply that for each fixed $\varepsilon \in \big(0, \frac12 \operatorname{dist}(x,\partial\Omega)\big)$, and any two indexes $\beta \in \{1,\dots,M\}$ and $j \in \{1,\dots,n\}$, we have

$$
\begin{aligned}
\big[(\partial_j E_{\gamma\beta})(x-\cdot)\big|_{\partial\Omega}\big] &= \big[\big((\partial_j E_{\gamma\beta})(x-\cdot)\eta_\varepsilon\big)\big|_{\partial\Omega}\big]\\
&= \lim_{R\to\infty} \Big[\big(\theta_R(\partial_j E_{\gamma\beta})(x-\cdot)\eta_\varepsilon\big)\big|_{\partial\Omega}\Big]
\end{aligned}
\tag{3.3.16}
$$

$$
\text{weak-}*\text{ in } \big(H^p(\partial\Omega,\sigma)\big)^* = \begin{cases} \dot{\mathscr{C}}^{(n-1)(1/p-1)}(\partial\Omega)/\sim & \text{if } p < 1,\\[4pt] \widetilde{\mathrm{BMO}}(\partial\Omega,\sigma) & \text{if } p = 1. \end{cases}
\tag{3.3.17}
$$

For future use, let us also observe that, thanks to (1.4.24), Hölder's inequality, and [112, Proposition 8.6.3] we have

$$
\begin{aligned}
&\limsup_{R\to\infty} \int_\Omega |\nabla\theta_R||(\nabla E)(x-\cdot)||\eta_\varepsilon||\nabla u| \, d\mathcal{L}^n\\
&\quad \le \limsup_{R\to\infty} \frac{1}{R^n} \int_{\Omega\cap[B(0,\lambda R)\setminus B(0,R)]} |\nabla u| \, d\mathcal{L}^n\\
&\quad \le \limsup_{R\to\infty} \frac{C_{\lambda,n}}{R^n} \Big(\int_{\Omega\cap[B(0,\lambda R)\setminus B(0,R)]} |\nabla u|^{\frac{np}{n-1}} \, d\mathcal{L}^n \Big)^{\frac{n-1}{np}} \cdot R^{n(1-(n-1)/(np))}\\
&\quad \le C_\# C_{\lambda,n} \big\|\mathcal{N}_\kappa(\nabla u)\big\|_{L^p(\partial\Omega,\sigma)} \cdot \limsup_{R\to\infty} R^{-(n-1)/p} = 0,
\end{aligned}
\tag{3.3.18}
$$

with $C_{\lambda,n} \in (0,\infty)$ depending only on λ, n, and with $C_\#$ as in [112, (8.6.49)] (hence independent of R, since $\sigma(\partial\Omega) = +\infty$ in the current case).

To proceed, fix $\varepsilon \in \big(0, \frac12 \operatorname{dist}(x,\partial\Omega)\big)$ and use (3.3.16) with $j := r$, along with the first membership in (3.3.5), the compatibility condition in [113, Lemma 4.6.4], formula [113, (10.2.12], and (3.3.18) to express the first term in the right-hand side of (3.3.7) as

$$
\begin{aligned}
&d_{rs}^{\beta\alpha}\Big\langle\big[(\partial_r E_{\gamma\beta})(x-\cdot)\big|_{\partial\Omega}\big], \overset{\bullet}{\partial}_{\tau_{\ell s}} u_\alpha\Big\rangle \\
&= \lim_{R\to\infty}\Big\langle d_{rs}^{\beta\alpha}\Big[\big(\theta_R(\partial_r E_{\gamma\beta})(x-\cdot)\eta_\varepsilon\big)\Big|_{\partial\Omega}\Big], \overset{\bullet}{\partial}_{\tau_{\ell s}} u_\alpha\Big\rangle \\
&= \lim_{R\to\infty} {}_{(\mathrm{Lip}_c(\partial\Omega))'}\Big\langle \overset{\bullet}{\partial}_{\tau_{\ell s}} u_\alpha, d_{rs}^{\beta\alpha}\theta_R(\partial_r E_{\gamma\beta})(x-\cdot)\eta_\varepsilon\Big\rangle_{\mathrm{Lip}_c(\partial\Omega)} \\
&= \lim_{R\to\infty}\int_\Omega d_{rs}^{\beta\alpha}\Big\{\partial_\ell\big(\theta_R(\partial_r E_{\gamma\beta})(x-\cdot)\eta_\varepsilon\big)(\partial_s u_\alpha) \\
&\qquad\qquad - \partial_s\big(\theta_R(\partial_r E_{\gamma\beta})(x-\cdot)\eta_\varepsilon\big)(\partial_\ell u_\alpha)\Big\}\,\mathrm{d}\mathcal{L}^n \\
&= \lim_{R\to\infty}\int_\Omega d_{rs}^{\beta\alpha}\Big\{-\theta_R(\partial_\ell\partial_r E_{\gamma\beta})(x-\cdot)\eta_\varepsilon(\partial_s u_\alpha) \\
&\qquad\qquad + (\partial_r E_{\gamma\beta})(x-\cdot)(\partial_\ell\eta_\varepsilon)(\partial_s u_\alpha) \\
&\qquad\qquad + \theta_R(\partial_s\partial_r E_{\gamma\beta})(x-\cdot)\eta_\varepsilon(\partial_\ell u_\alpha) \\
&\qquad\qquad - (\partial_r E_{\gamma\beta})(x-\cdot)(\partial_s\eta_\varepsilon)(\partial_\ell u_\alpha)\Big\}\,\mathrm{d}\mathcal{L}^n.
\end{aligned}
\tag{3.3.19}
$$

In relation to the last integrand above observe that, thanks to (1.4.33), we have

$$
d_{rs}^{\beta\alpha}\theta_R(\partial_s\partial_r E_{\gamma\beta})(x-\cdot)\eta_\varepsilon(\partial_\ell u_\alpha) = 0. \tag{3.3.20}
$$

Moving on, we wish to treat the second term appearing in the right-hand side of (3.3.7). To do so, recall from (A.0.113)-(A.0.114) that the weak conormal derivative of u associated with the coefficient tensor A is defined as the distribution

$$
\begin{aligned}
&\overset{\bullet}{\partial}{}_\nu^A u := \big(\nu\bullet\vec{F}^\alpha\big)_{1\le\alpha\le M} \text{ in } \Big[\big(\mathrm{Lip}_c(\partial\Omega)\big)'\Big]^M \text{ where} \\
&\vec{F}^\alpha := (A\nabla u)_\alpha = \big(a_{rs}^{\alpha\beta}\partial_s u_\beta\big)_{1\le r\le n} \text{ for each } \alpha\in\{1,\dots,M\}.
\end{aligned}
\tag{3.3.21}
$$

Combining (3.3.16) (presently used with $j := \ell$), [113, Lemma 4.6.4], and [112, Proposition 4.2.3], for each fixed number $\varepsilon\in\big(0,\tfrac12\,\mathrm{dist}(x,\partial\Omega)\big)$ we obtain

$$
\begin{aligned}
&-\Big\langle \big[(\partial_\ell E_{\gamma\beta})(x-\cdot)\big|_{\partial\Omega}\big], \big(\overset{\bullet}{\partial}{}^A_\nu u\big)_\beta\Big\rangle \\
&\quad = -\lim_{R\to\infty} \Big\langle \Big[\big(\theta_R(\partial_\ell E_{\gamma\alpha})(x-\cdot)\eta_\varepsilon\big)\Big|_{\partial\Omega}\Big], \big(\overset{\bullet}{\partial}{}^A_\nu u\big)_\alpha\Big\rangle \\
&\quad = -\lim_{R\to\infty} {}_{(\mathrm{Lip}_c(\partial\Omega))'}\Big\langle \big(\overset{\bullet}{\partial}{}^A_\nu u\big)_\alpha, \theta_R(\partial_\ell E_{\gamma\alpha})(x-\cdot)\eta_\varepsilon\Big\rangle_{\mathrm{Lip}_c(\partial\Omega)} \\
&\quad = -\lim_{R\to\infty} \int_\Omega (\mathrm{div}\vec{F}^\alpha)\theta_R(\partial_\ell E_{\gamma\alpha})(x-\cdot)\eta_\varepsilon \, d\mathcal{L}^n \\
&\qquad -\lim_{R\to\infty} \int_\Omega \big\langle \vec{F}^\alpha, \nabla\big(\theta_R(\partial_\ell E_{\gamma\alpha})(x-\cdot)\eta_\varepsilon\big)\big\rangle \, d\mathcal{L}^n \\
&\quad = -\int_\Omega (Lu)_\alpha(\partial_\ell E_{\gamma\alpha})(x-\cdot)\, d\mathcal{L}^n \\
&\qquad -\lim_{R\to\infty} \int_\Omega a^{\alpha\beta}_{rs}(\partial_s u_\beta)\partial_r\big(\theta_R(\partial_\ell E_{\gamma\alpha})(x-\cdot)\eta_\varepsilon\big)\, d\mathcal{L}^n,
\end{aligned}
\tag{3.3.22}
$$

where the last equality is based on Lebesgue's Dominated Convergence Theorem (whose applicability is ensured by (3.3.6) and (1.4.24)). By combining (3.3.22) with (3.3.18) we therefore arrive at

$$
\begin{aligned}
&-\Big\langle \big[(\partial_\ell E_{\gamma\beta})(x-\cdot)\big|_{\partial\Omega}\big], \big(\overset{\bullet}{\partial}{}^A_\nu u\big)_\beta\Big\rangle \\
&\quad = -\int_\Omega (\partial_\ell E_{\gamma\beta})(x-\cdot)(Lu)_\beta \, d\mathcal{L}^n \\
&\qquad -\lim_{R\to\infty} \int_\Omega a^{\alpha\beta}_{rs}\Big\{-\theta_R(\partial_r\partial_\ell E_{\gamma\alpha})(x-\cdot)\eta_\varepsilon \\
&\qquad\qquad + (\partial_\ell E_{\gamma\alpha})(x-\cdot)(\partial_r\eta_\varepsilon)\Big\}(\partial_s u_\beta)\, d\mathcal{L}^n \\
&\quad = -\int_\Omega (\partial_\ell E_{\gamma\beta})(x-\cdot)(Lu)_\beta \, d\mathcal{L}^n \\
&\qquad -\lim_{R\to\infty} \int_\Omega a^{\beta\alpha}_{rs}\Big\{-\theta_R(\partial_r\partial_\ell E_{\gamma\beta})(x-\cdot)\eta_\varepsilon \\
&\qquad\qquad + (\partial_\ell E_{\gamma\beta})(x-\cdot)(\partial_r\eta_\varepsilon)\Big\}(\partial_s u_\alpha)\, d\mathcal{L}^n,
\end{aligned}
\tag{3.3.23}
$$

where in the last step we have simply interchanged the roles of α and β in the very last integral.

Adding (3.3.19) with (3.3.23) then yields, after taking (3.3.20) into account and canceling the terms involving $\partial_r\partial_\ell E_{\gamma\beta}$,

$$d_{rs}^{\beta\alpha}\Big\langle \big[(\partial_r E_{\gamma\beta})(x-\cdot)\big|_{\partial\Omega}\big], \partial_{\tau_{\ell s}} u_\alpha\Big\rangle - \Big\langle \big[(\partial_\ell E_{\gamma\beta})(x-\cdot)\big|_{\partial\Omega}\big], (\partial_\nu^A u)_\beta\Big\rangle$$

$$= \int_\Omega d_{rs}^{\beta\alpha}\Big\{(\partial_r E_{\gamma\beta})(x-\cdot)(\partial_\ell\eta_\varepsilon)(\partial_s u_\alpha) - (\partial_r E_{\gamma\beta})(x-\cdot)(\partial_s\eta_\varepsilon)(\partial_\ell u_\alpha)$$

$$- (\partial_\ell E_{\gamma\beta})(x-\cdot)(\partial_r\eta_\varepsilon)(\partial_s u_\alpha)\Big\}\, d\mathcal{L}^n$$

$$- \int_\Omega (\partial_\ell E_{\gamma\beta})(x-\cdot)(Lu)_\beta\, d\mathcal{L}^n. \tag{3.3.24}$$

Note that

$$\int_\Omega d_{rs}^{\beta\alpha}\Big\{(\partial_r E_{\gamma\beta})(x-\cdot)(\partial_\ell\eta_\varepsilon)(\partial_s u_\alpha) - (\partial_\ell E_{\gamma\beta})(x-\cdot)(\partial_r\eta_\varepsilon)(\partial_s u_\alpha)\Big\}\, d\mathcal{L}^n$$

$$= \int_\Omega d_{rs}^{\beta\alpha}\Big\{(\partial_r E_{\gamma\beta})(x-y)(\partial_\ell\eta_\varepsilon)(y) - (\partial_\ell E_{\gamma\beta})(x-y)(\partial_r\eta_\varepsilon)\Big\}(\partial_s u_\alpha)(y)\, dy$$

$$= \mathrm{I}_\varepsilon + \mathrm{II}_\varepsilon \tag{3.3.25}$$

where

$$\mathrm{I}_\varepsilon := \int_\Omega d_{rs}^{\beta\alpha}\Big\{(\partial_r E_{\gamma\beta})(x-y)(\partial_\ell\eta_\varepsilon)(y) - (\partial_\ell E_{\gamma\beta})(x-y)(\partial_r\eta_\varepsilon)\Big\}\times$$

$$\times\Big\{(\partial_s u_\alpha)(y) - (\partial_s u_\alpha)(x)\Big\}\, dy \tag{3.3.26}$$

and

$$\mathrm{II}_\varepsilon := d_{rs}^{\beta\alpha}\Bigg\{\int_\Omega \Big((\partial_r E_{\gamma\beta})(x-y)(\partial_\ell\eta_\varepsilon)(y)$$

$$- (\partial_\ell E_{\gamma\beta})(x-y)(\partial_r\eta_\varepsilon)\Big)\, dy\Bigg\}(\partial_s u_\alpha)(x). \tag{3.3.27}$$

Since

$$\int_\Omega \Big((\partial_r E_{\gamma\beta})(x-y)(\partial_\ell\eta_\varepsilon)(y) - (\partial_\ell E_{\gamma\beta})(x-y)(\partial_r\eta_\varepsilon)\Big)\, dy$$

$$= {}_{\mathcal{D}'(\Omega)}\Big\langle (\partial_r E_{\gamma\beta})(x-\cdot), \partial_\ell(\eta_\varepsilon - 1)\Big\rangle_{\mathcal{D}(\Omega)}$$

$$- {}_{\mathcal{D}'(\Omega)}\Big\langle (\partial_\ell E_{\gamma\beta})(x-\cdot), \partial_r(\eta_\varepsilon - 1)\Big\rangle_{\mathcal{D}(\Omega)}$$

$$= {}_{\mathcal{D}'(\Omega)}\Big\langle E_{\gamma\beta}(x-\cdot), \partial_r\partial_\ell(\eta_\varepsilon - 1)\Big\rangle_{\mathcal{D}(\Omega)}$$

$$- {}_{\mathcal{D}'(\Omega)}\Big\langle E_{\gamma\beta}(x-\cdot), \partial_\ell\partial_r(\eta_\varepsilon - 1)\Big\rangle_{\mathcal{D}(\Omega)} = 0, \tag{3.3.28}$$

it follows that for each $\varepsilon \in \big(0, \frac{1}{2}\operatorname{dist}(x, \partial\Omega)\big)$ we have

$$\mathrm{II}_\varepsilon = 0. \tag{3.3.29}$$

Also, making use of (3.3.15) and the fact that x is a Lebesgue point for ∇u, we may write

$$\limsup_{\varepsilon\to 0^+} \big|\mathrm{I}_\varepsilon\big| \leq C \cdot \limsup_{\varepsilon\to 0^+} \fint_{B(x,2\varepsilon)} \big|(\nabla u)(y) - (\nabla u)(x)\big|\,\mathrm{d}y = 0. \tag{3.3.30}$$

Finally, decompose

$$-\int_\Omega a^{\beta\alpha}_{rs}(\partial_r E_{\gamma\beta})(x-\cdot)(\partial_s\eta_\varepsilon)(\partial_\ell u_\alpha)\,\mathrm{d}\mathcal{L}^n = \mathrm{III}_\varepsilon + \mathrm{IV}_\varepsilon \tag{3.3.31}$$

where

$$\mathrm{III}_\varepsilon := -\int_\Omega a^{\beta\alpha}_{rs}(\partial_r E_{\gamma\beta})(x-y)(\partial_s\eta_\varepsilon)(y)\big\{(\partial_\ell u_\alpha)(y) - (\partial_\ell u_\alpha)(x)\big\}\,\mathrm{d}y \tag{3.3.32}$$

and

$$\mathrm{IV}_\varepsilon := -\left\{\int_\Omega a^{\beta\alpha}_{rs}(\partial_r E_{\gamma\beta})(x-y)(\partial_s\eta_\varepsilon)(y)\,\mathrm{d}y\right\}(\partial_\ell u_\alpha)(x). \tag{3.3.33}$$

Since x is a Lebesgue point for ∇u,

$$\limsup_{\varepsilon\to 0^+} \big|\mathrm{III}_\varepsilon\big| \leq C \cdot \limsup_{\varepsilon\to 0^+} \fint_{B(x,2\varepsilon)} \big|(\nabla u)(y) - (\nabla u)(x)\big|\,\mathrm{d}y = 0. \tag{3.3.34}$$

In addition, (1.4.33) eventually permits us to conclude that for each $\beta \in \{1, \dots, M\}$ we have

$$\begin{aligned}
&\lim_{\varepsilon\to 0^+} \int_\Omega a^{\beta\alpha}_{rs}(\partial_r E_{\gamma\beta})(x-y)(\partial_s\eta_\varepsilon)(y)\,\mathrm{d}y \\
&\quad = \lim_{\varepsilon\to 0^+} \int_\Omega a^{\beta\alpha}_{rs}(\partial_r E_{\gamma\beta})(x-y)\partial_s(\eta_\varepsilon - 1)(y)\,\mathrm{d}y \\
&\quad = -\lim_{\varepsilon\to 0^+} {}_{\mathcal{D}'(\Omega)}\Big\langle a^{\beta\alpha}_{rs}\partial_r\big[E_{\gamma\beta}(x-\cdot)\big], \partial_s(\eta_\varepsilon - 1)\Big\rangle_{\mathcal{D}(\Omega)} \\
&\quad = \lim_{\varepsilon\to 0^+} {}_{\mathcal{D}'(\Omega)}\Big\langle a^{\beta\alpha}_{rs}\partial_s\partial_r\big[E_{\gamma\beta}(x-\cdot)\big], \eta_\varepsilon - 1\Big\rangle_{\mathcal{D}(\Omega)} \\
&\quad = \lim_{\varepsilon\to 0^+} \delta_{\alpha\gamma}(\eta_\varepsilon - 1)(x) = -\delta_{\alpha\gamma}.
\end{aligned} \tag{3.3.35}$$

Thus,

$$\lim_{\varepsilon\to 0^+} \mathrm{IV}_\varepsilon = (\partial_\ell u_\gamma)(x). \tag{3.3.36}$$

Collectively, (3.3.24), (3.3.25), (3.3.29), (3.3.30), (3.3.31), (3.3.34), and (3.3.36) prove (3.3.7) in the case when $\partial\Omega$ is unbounded.

The case when Ω is bounded is treated very similarly, since (3.3.18) continues to be valid in such a scenario. Finally, when Ω is an exterior domain, estimate [112, (8.6.51)] (on which (3.3.18) is based) may fail. Nonetheless, the decay condition (3.3.8) turns out to be an appropriate substitute, as this implies

$$\begin{aligned}\limsup_{R\to\infty} \int_\Omega |\nabla\theta_R||(\nabla E)(x-\cdot)||\eta_\varepsilon||\nabla u|\,\mathrm{d}\mathcal{L}^n \\ \leq \limsup_{R\to\infty} \fint_{B(0,\lambda R)\setminus B(0,R)} |\nabla u|\,\mathrm{d}\mathcal{L}^n = 0.\end{aligned} \tag{3.3.37}$$

`Case II:` *Suppose $p \in (1,\infty)$.* We reason largely as before, with natural alterations. Two key aspects to bear in mind are as follows. First, [112, (3.6.27)] implies that the memberships in (3.3.14) currently read

$$\begin{aligned}\overset{\bullet}{\partial}_{\tau_{\ell s}} u \in \big[L^p(\partial\Omega,\sigma)\big]^M \ \text{ for each } \ell, s \in \{1,\dots,n\}, \\ \text{and } \overset{\bullet}{\partial}{}^A_\nu u \in \big[L^p(\partial\Omega,\sigma)\big]^M.\end{aligned} \tag{3.3.38}$$

Second, all duality pairings used in the past now become ordinary integral pairings on $\partial\Omega$, with respect to the measure σ. Upon also observing that if p' is the Hölder conjugate exponent of p, we have $\int_{\partial\Omega} |(\nabla E)(x-\cdot)|^{p'}\,\mathrm{d}\sigma < +\infty$ thanks to (1.4.24) and [112, (7.2.5)], the proof proceeds as before, and the integral representation formula (3.3.9) follows.

Finally, consider the case when in place of (3.3.3)-(3.3.4) we now assume the conditions imposed in (3.3.10). In such a scenario, the memberships claimed in (3.3.11) make use of the full force of [113, (10.2.13), (10.2.182)]. Granted these, formula (3.3.12) may be justified much as (3.3.7), making similar adjustments as in the proof of Theorem 3.1.2. □

We momentarily digress for the purpose of making the integral representation formulas from Theorem 3.3.1 corresponding to null-solutions of the given weakly elliptic system subject to the technique developed earlier in Proposition 3.1.3.

Remark 3.3.2 *It is of interest to point out that the version of Theorem 3.3.1 corresponding to the case when the given function u is a smooth null-solution of the system L may be directly deduced from Proposition 3.1.3 for a suitable choice of u, w, and D. To be specific, work in the context of Theorem 3.3.1 in which we now assume that $u \in \big[\mathscr{C}^\infty(\Omega)\big]^M$ and $Lu = 0$ in Ω. Fix $x \in \Omega$, along with $\gamma \in \{1,\dots,M\}$ and $\ell \in \{1,\dots,n\}$, all arbitrary. With the summation convention over repeated indices in effect, define the $\mathbb{C}^{n\cdot M+M}$-valued function*

$$w := \begin{pmatrix} \big(a^{\beta\alpha}_{rs}(\partial_r E_{\gamma\beta})(x-\cdot)\big)_{\substack{1\le s\le n\\ 1\le\alpha\le M}} \\ -\big((\partial_\ell E_{\gamma\beta})(x-\cdot)\big)_{1\le\beta\le M} \end{pmatrix}, \tag{3.3.39}$$

the $\mathbb{C}^{n\cdot M}$*-valued function*

$$U := \big(\partial_j u_\alpha\big)_{\substack{1\le j\le n\\ 1\le\alpha\le M}}, \tag{3.3.40}$$

and the first-order $(n\cdot M+M)\times(n\cdot M)$ *system*

$$D := \begin{pmatrix} D_1 \\ D_2 \end{pmatrix} \tag{3.3.41}$$

where, with $\pi_{s,\alpha}$ *denoting the canonical coordinate projection selecting the* (s,α)*-entry of a matrix-valued function,*

$$D_1 := \big(\partial_\ell\circ\pi_{s,\alpha} - \partial_s\circ\pi_{\ell,\alpha}\big)_{\substack{1\le s\le n\\ 1\le\alpha\le M}} \text{ and } D_2 := \big(a^{\alpha\beta}_{rs}\partial_r\circ\pi_{s,\beta}\big)_{1\le\alpha\le M}. \tag{3.3.42}$$

We then have

$$DU = \begin{pmatrix} \big(\partial_\ell\partial_s u_\alpha - \partial_s\partial_\ell u_\alpha\big)_{\substack{1\le s\le n\\ 1\le\alpha\le M}} \\ \big(a^{\alpha\beta}_{rs}\partial_r\partial_s u_\beta\big)_{1\le\alpha\le M} \end{pmatrix} = 0 \tag{3.3.43}$$

since $\big(a^{\alpha\beta}_{rs}\partial_r\partial_s u_\beta\big)_{1\le\alpha\le M} = Lu = 0$*, thanks to the current assumptions. Next, a direct computation based on* (3.3.41)-(3.3.42) *shows that* $D^\top$*, the real transpose of* D*, is given by*

$$D^\top\begin{pmatrix} (\phi_{s\alpha})_{\substack{1\le s\le n\\ 1\le\alpha\le M}} \\ (\psi_\beta)_{1\le\beta\le M} \end{pmatrix} = \big(-\partial_\ell\phi_{s\alpha} + \delta_{\ell s}\partial_j\phi_{j\alpha} - a^{\mu\alpha}_{rs}\partial_r\psi_\mu\big)_{\substack{1\le s\le n\\ 1\le\alpha\le M}} \tag{3.3.44}$$

so if w *is as in* (3.3.39) *we obtain*

$$D^\top w = -\big(\delta_{\ell s}\delta_{\alpha\gamma}\delta_x\big)_{\substack{1\le s\le n\\ 1\le\alpha\le M}}. \tag{3.3.45}$$

In particular, $D^\top w\in\big[\mathscr{E}'(\Omega)\big]^{n\cdot M}$ *and, for* U *as in* (3.3.40),

$${}_{[\mathscr{E}(\Omega)]^{n\cdot M}}\Big\langle U, D^\top w\Big\rangle_{[\mathscr{E}'(\Omega)]^{n\cdot M}} = -(\partial_\ell u_\gamma)(x). \tag{3.3.46}$$

In addition,

$$(-i)\mathrm{Sym}(D;\nu)\bullet U = \begin{pmatrix} (-i)\mathrm{Sym}(D_1;\nu) \\ (-i)\mathrm{Sym}(D_2;\nu) \end{pmatrix}\bullet U = \begin{pmatrix} \big(\partial_{\tau_{\ell s}} u_\alpha\big)_{\substack{1\le s\le n\\ 1\le\alpha\le M}} \\ \big(\partial^A_\nu u\big)_{1\le\beta\le M} \end{pmatrix}. \tag{3.3.47}$$

Hence, in the case when $\partial\Omega$ *is unbounded and* $p\le 1$*, we have*

$$\Big\langle (-\mathrm{i})\mathrm{Sym}(D;\nu)\bullet U, \big[w\big|^{\kappa-\mathrm{n.t.}}_{\partial\Omega}\big]\Big\rangle$$
$$= d^{\beta\alpha}_{rs}\Big\langle \big[(\partial_r E_{\gamma\beta})(x-\cdot)\big|_{\partial\Omega}\big], \partial_{\tau_{\ell s}} u_\alpha\Big\rangle$$
$$- \Big\langle \big[(\partial_\ell E_{\gamma\beta})(x-\cdot)\big|_{\partial\Omega}\big], \big(\partial^A_\nu u\big)_\beta\Big\rangle \tag{3.3.48}$$

which in concert with (3.3.46) *proves, thanks to* (3.1.46) *in Proposition 3.1.3* (*written with U in place of u*), *the versions of* (3.3.7) *and* (3.3.9) *corresponding to situation when $Lu = 0$ in Ω. The case when $\partial\Omega$ is bounded, or $p > 1$, is deal with similarly, now employing* (3.1.45).

It is also possible to prove a version of Theorem 3.3.1 in which the membership of the nontangential maximal function to Lebesgue spaces is now replaced by a membership to a suitable Generalized Banach Function Space.

Proposition 3.3.3 *Let $\Omega \subseteq \mathbb{R}^n$, where $n \in \mathbb{N}$ with $n \geq 2$, be an Ahlfors regular domain, and abbreviate $\sigma := \mathcal{H}^{n-1}\lfloor\partial\Omega$. For some $M \in \mathbb{N}$, consider a coefficient tensor $A = \big(a^{\alpha\beta}_{rs}\big)_{\substack{1\leq r,s\leq n\\ 1\leq\alpha,\beta\leq M}}$ with complex entries, with the property that the $M \times M$ homogeneous second-order system $L = L_A$ associated with A in $\mathbb{R}^n$ as in* (1.3.15) *is weakly elliptic (in the sense of* (1.3.3)*). Also, denote by $E = (E_{\gamma\beta})_{1\leq\gamma,\beta\leq M}$ the matrix-valued fundamental solution associated with L as in Theorem 1.4.2.*

In addition, suppose $\mathbb{X}$ is a Generalized Banach Function Space on $(\partial\Omega, \sigma)$ such that

$$\mathcal{M}_{\partial\Omega} : \mathbb{X} \to \mathbb{X} \text{ and } \mathcal{M}_{\partial\Omega} : \mathbb{X}' \to \mathbb{X}' \quad \text{are well-defined bounded mappings,} \tag{3.3.49}$$

where $\mathcal{M}_{\partial\Omega}$ is the Hardy-Littlewood maximal operator on $\partial\Omega$, and $\mathbb{X}'$ denotes the associated space of $\mathbb{X}$ (cf. [113, Definitions 5.1.4, 5.1.11]).

In this setting, assume $u = (u_\beta)_{1\leq\beta\leq M} \in \big[\mathscr{C}^\infty(\Omega)\big]^M$ is a vector-valued function which, for some aperture parameter $\kappa > 0$, satisfies

$$\mathcal{N}_\kappa(\nabla u) \in \mathbb{X} \text{ and } Lu = 0 \text{ in } \Omega. \tag{3.3.50}$$

In the case when Ω is an exterior domain, make the additional assumption that there exists $\lambda \in (1, \infty)$ such that

$$⨍_{B(0,\lambda R)\setminus B(0,R)} |\nabla u| \, d\mathcal{L}^n = o(1) \text{ as } R \to \infty. \tag{3.3.51}$$

Then

$$\partial^A_\nu u \in \big[\mathbb{X}\big]^M \text{ and } \partial_{\tau_{\ell s}} u \in \big[\mathbb{X}\big]^M \text{ for any } \ell, s \in \{1, \ldots, n\}, \tag{3.3.52}$$

and, given any $\ell \in \{1, \ldots, n\}$ and $\gamma \in \{1, \ldots, M\}$, at each point $x \in \Omega$ one has

$$(\partial_\ell u_\gamma)(x) = \int_{\partial\Omega} a_{rs}^{\beta\alpha}(\partial_r E_{\gamma\beta})(x-y)\big(\partial_{\tau_{\ell s}}^{\circ} u_\alpha\big)(y)\,d\sigma(y)$$

$$- \int_{\partial\Omega} (\partial_\ell E_{\gamma\beta})(x-y)\big(\partial_\nu^{A} u\big)_\beta(y)\,d\sigma(y). \tag{3.3.53}$$

Proof The argument presented in Remark 3.3.2 (for the same choice of D, U, and w) allows us to reduce matters to the case of first-order systems, treated earlier in Proposition 3.1.4. □

We now turn to the task of presenting the Fatou-type theorem elaborating on the result advertised earlier, in (3.3.2).

Theorem 3.3.4 *Let $\Omega \subseteq \mathbb{R}^n$, where $n \in \mathbb{N}$ with $n \geq 2$, be an arbitrary* UR *domain, and abbreviate $\sigma := \mathcal{H}^{n-1}\lfloor\partial\Omega$. Consider a homogeneous constant (complex) coefficient second-order $M \times M$ system L in $\mathbb{R}^n$ (for some $M \in \mathbb{N}$) which is weakly elliptic (in the sense of (1.3.3)), and assume $u \in \big[W^{1,1}_{\mathrm{loc}}(\Omega)\big]^M$ is a vector-valued function which, for some aperture parameter $\kappa > 0$, satisfies*

$$\mathcal{N}_\kappa(\nabla u) \in L^p(\partial\Omega, \sigma) \text{ with } p \in \big(\tfrac{n-1}{n}, \infty\big) \\ \text{and } Lu = 0 \text{ in } \big[\mathcal{D}'(\Omega)\big]^M. \tag{3.3.54}$$

Then the following statements are true.

(a) The nontangential boundary trace

$$(\nabla u)\Big|_{\partial\Omega}^{\kappa-\mathrm{n.t.}} \text{ exists (in } \mathbb{C}^{n\cdot M}) \text{ at } \sigma\text{-a.e. point on } \partial\Omega \text{ and} \\ \text{the function } (\nabla u)\Big|_{\partial\Omega}^{\kappa-\mathrm{n.t.}} \text{ is } \sigma\text{-measurable on } \partial\Omega; \\ \text{also, } (\nabla u)\Big|_{\partial\Omega}^{\kappa-\mathrm{n.t.}} \text{ belongs to the space } \big[L^p(\partial\Omega,\sigma)\big]^{n\cdot M} \\ \text{and one has } \Big\|(\nabla u)\Big|_{\partial\Omega}^{\kappa-\mathrm{n.t.}}\Big\|_{[L^p(\partial\Omega,\sigma)]^{n\cdot M}} \leq \|\mathcal{N}_\kappa(\nabla u)\|_{L^p(\partial\Omega,\sigma)}. \tag{3.3.55}$$

In addition, for any other given aperture parameter $\kappa' \in (0,\infty)$ the nontangential boundary trace $(\nabla u)\Big|_{\partial\Omega}^{\kappa'-\mathrm{n.t.}}$ also exists, and agrees with $(\nabla u)\Big|_{\partial\Omega}^{\kappa-\mathrm{n.t.}}$, at σ-a.e. point on $\partial\Omega$.
Furthermore,

$$\text{the nontangential boundary trace } u\Big|_{\partial\Omega}^{\kappa-\mathrm{n.t.}} \text{ exists} \\ \text{at } \sigma\text{-a.e. point on } \partial\Omega \text{ and is a } \sigma\text{-measurable function,} \tag{3.3.56}$$

and for any other given aperture parameter $\kappa' \in (0,\infty)$ the nontangential boundary trace $u\Big|_{\partial\Omega}^{\kappa'-\mathrm{n.t.}}$ exists and agrees with $u\Big|_{\partial\Omega}^{\kappa-\mathrm{n.t.}}$ at σ-a.e. point on $\partial\Omega$.

(b) Given any coefficient tensor A which allows representing $L = L_A$, one has:

$$\text{if } p \in \left(\tfrac{n-1}{n}, 1\right] \text{ then } \mathring{\partial}^A_\nu u \in \left[H^p(\partial\Omega, \sigma)\right]^M \text{ and}$$
$$\mathring{\partial}_{\tau_{jk}} u \in \left[H^p(\partial\Omega, \sigma)\right]^M \text{ for each } j, k \in \{1, \dots, n\}. \tag{3.3.57}$$

(c) The following H^p/L^p-styled boundary maximum principle holds:

$$\|\mathcal{N}_\kappa(\nabla u)\|_{L^p(\partial\Omega,\sigma)} \approx \|\mathring{\partial}^A_\nu u\|_{[H^p(\partial\Omega,\sigma)]^M} + \sum_{j,k=1}^{n} \|\mathring{\partial}_{\tau_{jk}} u\|_{[H^p(\partial\Omega,\sigma)]^M} \text{ if } p \in \left(\tfrac{n-1}{n}, 1\right], \tag{3.3.58}$$

and

$$\begin{aligned}\|\mathcal{N}_\kappa(\nabla u)\|_{L^p(\partial\Omega,\sigma)} &\approx \left\|(\nabla u)\Big|^{\kappa-\text{n.t.}}_{\partial\Omega}\right\|_{[L^p(\partial\Omega,\sigma)]^{n\cdot M}} \\ &\approx \sum_{j,k=1}^{n} \left\|\nu_j(\partial_k u)\Big|^{\kappa-\text{n.t.}}_{\partial\Omega} - \nu_k(\partial_j u)\Big|^{\kappa-\text{n.t.}}_{\partial\Omega}\right\|_{[L^p(\partial\Omega,\sigma)]^M} \\ &\quad + \|\partial^A_\nu u\|_{[L^p(\partial\Omega,\sigma)]^M} \text{ if } p \in (1, \infty),\end{aligned} \tag{3.3.59}$$

provided, in the case when Ω is an exterior domain, it is also assumed (for both equivalences) that ∇u vanishes at infinity (in the sense of (3.3.8)*).*
Moreover, if $\Omega \subseteq \mathbb{R}^n$ is actually an NTA *domain with an Ahlfors regular boundary, then if $p \in \left(\frac{n-1}{n}, 1\right]$ it follows that*

the nontangential boundary trace $u\big|^{\kappa-\text{n.t.}}_{\partial\Omega}$ exists (in $\mathbb{C}^M$) at σ-a.e. point on $\partial\Omega$, as a vector-valued function said trace belongs to the homogeneous Hardy-based Sobolev space $\left[\dot{H}^p_1(\partial\Omega, \sigma)\right]^M$, (3.3.60)

and, if ∇u also vanishes at infinity (in the sense of (3.3.8)*) when Ω is an exterior domain, one has (compare with* (3.3.58)*)*

$$\|\mathcal{N}_\kappa(\nabla u)\|_{L^p(\partial\Omega,\sigma)} \approx \|\mathring{\partial}^A_\nu u\|_{[H^p(\partial\Omega,\sigma)]^M} + \left\|u\Big|^{\kappa-\text{n.t.}}_{\partial\Omega}\right\|_{[\dot{H}^p_1(\partial\Omega,\sigma)]^M}, \tag{3.3.61}$$

while if $p \in (1, \infty)$ it follows that

the nontangential boundary trace $u\big|^{\kappa-\text{n.t.}}_{\partial\Omega}$ exists (in $\mathbb{C}^M$) at σ-a.e. point on $\partial\Omega$, as a vector-valued function said trace belongs to the homogeneous Sobolev space $\left[\dot{L}^p_1(\partial\Omega, \sigma)\right]^M$, (3.3.62)

and if ∇u also vanishes at infinity (in the sense of (3.3.8)*) when Ω is an exterior domain, one has (compare with* (3.3.59)*)*

$$\|\mathcal{N}_\kappa(\nabla u)\|_{L^p(\partial\Omega,\sigma)} \approx \|\partial^A_\nu u\|_{[L^p(\partial\Omega,\sigma)]^M} + \left\|u\Big|^{\kappa-\text{n.t.}}_{\partial\Omega}\right\|_{[\dot{L}^p_1(\partial\Omega,\sigma)]^M}. \tag{3.3.63}$$

(d) Given any $q \in (1, \infty)$, there exists some $C = C(\Omega, L, p, q, \kappa) \in (0, \infty)$, independent of u, such that the following mixed-norm estimate holds:

$$\left(\int_{\partial\Omega}\left(\int_{\Gamma_\kappa(x)} |\nabla^2 u|^q \operatorname{dist}(\cdot, \partial\Omega)^{q-n}\, d\mathcal{L}^n\right)^{\frac{p}{q}} d\sigma(x)\right)^{\frac{1}{p}} \leq C\|\mathcal{N}_\kappa(\nabla u)\|_{L^p(\partial\Omega,\sigma)}. \tag{3.3.64}$$

In particular, with the L^q-based area-function defined as in (3.1.119)*, the estimate in* (3.3.64) *may be recast as*

$$\|\mathcal{A}_{q,\kappa}(\nabla u)\|_{L^p(\partial\Omega,\sigma)} \leq C\|\mathcal{N}_\kappa(\nabla u)\|_{L^p(\partial\Omega,\sigma)}. \tag{3.3.65}$$

Furthermore, when the integrability exponent p belongs to $(1, \infty)$ there exists some constant $C = C(\Omega, L, p, \kappa) \in (0, \infty)$ with the property that

$$\left(\int_\Omega |(\nabla^2 u)(x)|^p \operatorname{dist}(x, \partial\Omega)^{p-1}\, dx\right)^{1/p} \leq C\|\mathcal{N}_\kappa(\nabla u)\|_{L^p(\partial\Omega,\sigma)}. \tag{3.3.66}$$

(e) In the scenario when the set Ω is actually a bounded NTA *domain with an Ahlfors regular boundary in $\mathbb{R}^n$ and when $p \in (1, \infty)$, it follows that for each exponent $q \in \left(\frac{n}{n+1/p}, \infty\right]$ there exists some constant $C = C(\Omega, L, \kappa, p, q) \in (0, \infty)$ with the property that*

$$\max\left\{\|\nabla u\|_{[F^{p,q}_{1/p}(\Omega)]^{n\cdot M}}, \|\nabla u\|_{[B^{p,p}_{1/p}(\Omega)]^{n\cdot M}}\right\} \leq C\left\|(\nabla u)\big|^{\kappa-\text{n.t.}}_{\partial\Omega}\right\|_{[L^p(\partial\Omega,\sigma)]^{n\cdot M}}. \tag{3.3.67}$$

In particular, corresponding to the case when $p = q = 2$, one has

$$\|\nabla u\|_{[H^{1/2}(\Omega)]^{n\cdot M}} \leq C\left\|(\nabla u)\big|^{\kappa-\text{n.t.}}_{\partial\Omega}\right\|_{[L^2(\partial\Omega,\sigma)]^{n\cdot M}}. \tag{3.3.68}$$

(f) Suppose the first line in (3.3.54) *is replaced by*

$$\mathcal{N}_\kappa(\nabla u) \in L^{p,q}(\partial\Omega, \sigma) \text{ for some } p \in \left(\tfrac{n-1}{n}, \infty\right) \text{ and } q \in (0, \infty). \tag{3.3.69}$$

Then nontangential boundary trace $(\nabla u)\big|^{\kappa-\text{n.t.}}_{\partial\Omega}$ exists at σ-a.e. point on $\partial\Omega$, belongs to the Lorentz space $\left[L^{p,q}(\partial\Omega, \sigma)\right]^{n\cdot M}$, is independent of the aperture parameter, and satisfies the estimate

$$\left\|(\nabla u)\big|^{\kappa-\text{n.t.}}_{\partial\Omega}\right\|_{[L^{p,q}(\partial\Omega,\sigma)]^{n\cdot M}} \leq \|\mathcal{N}_\kappa(\nabla u)\|_{L^{p,q}(\partial\Omega,\sigma)}. \tag{3.3.70}$$

Moreover, under the additional assumption that ∇u vanishes at infinity (in the sense of (3.3.8)*) in the case when Ω is an exterior domain, one has*

$$\|\mathcal{N}_\kappa(\nabla u)\|_{L^{p,q}(\partial\Omega,\sigma)} \approx \|\mathring{\partial}^A_\nu u\|_{[H^{p,q}(\partial\Omega,\sigma)]^M} + \sum_{j,k=1}^n \|\mathring{\partial}_{\tau_{jk}} u\|_{[H^{p,q}(\partial\Omega,\sigma)]^M}, \tag{3.3.71}$$

$$\|\mathcal{N}_\kappa(\nabla u)\|_{L^{p,q}(\partial\Omega,\sigma)} \approx \|(\nabla u)\big|_{\partial\Omega}^{\kappa-\text{n.t.}}\|_{[L^{p,q}(\partial\Omega,\sigma)]^{n\cdot M}} \text{ if } p \in (1,\infty), \tag{3.3.72}$$

and

$$\|\mathcal{A}_\kappa(\nabla u)\|_{L^{p,q}(\partial\Omega,\sigma)} \leq C\|\mathcal{N}_\kappa(\nabla u)\|_{L^{p,q}(\partial\Omega,\sigma)}. \tag{3.3.73}$$

(g) Suppose the membership in the first line in (3.3.54) *is now replaced by*

$$\mathcal{N}_\kappa(\nabla u) \in \mathbb{X}, \tag{3.3.74}$$

where $\mathbb{X}$ is a Generalized Banach Function Space on $(\partial\Omega,\sigma)$ with the property that

$$\begin{gathered} \mathcal{M}_{\partial\Omega} : \mathbb{X} \to \mathbb{X} \text{ and } \mathcal{M}_{\partial\Omega} : \mathbb{X}' \to \mathbb{X}' \\ \text{are well-defined bounded mappings} \end{gathered} \tag{3.3.75}$$

(with $\mathcal{M}_{\partial\Omega}$ denoting the Hardy-Littlewood maximal operator on $\partial\Omega$, and with $\mathbb{X}'$ denoting the associated space of $\mathbb{X}$; cf. [113, Definitions 5.1.4, 5.1.11]). *Then, in place of* (3.3.55), *one has*

$$\begin{gathered} (\nabla u)\big|_{\partial\Omega}^{\kappa-\text{n.t.}} \text{ exists (in } \mathbb{C}^{n\cdot M}) \text{ at } \sigma\text{-a.e. point on } \partial\Omega \text{ and} \\ \text{the function } (\nabla u)\big|_{\partial\Omega}^{\kappa-\text{n.t.}} \text{ is } \sigma\text{-measurable on } \partial\Omega; \\ \text{also, } (\nabla u)\big|_{\partial\Omega}^{\kappa-\text{n.t.}} \text{ belongs to the space } [\mathbb{X}]^{n\cdot M} \\ \text{and one has } \|(\nabla u)\big|_{\partial\Omega}^{\kappa-\text{n.t.}}\|_{[\mathbb{X}]^{n\cdot M}} \leq \|\mathcal{N}_\kappa(\nabla u)\|_{\mathbb{X}}. \end{gathered} \tag{3.3.76}$$

Moreover, for any other given aperture parameter $\kappa' \in (0,\infty)$ the nontangential boundary trace $(\nabla u)\big|_{\partial\Omega}^{\kappa'-\text{n.t.}}$ also exists, and agrees with $(\nabla u)\big|_{\partial\Omega}^{\kappa-\text{n.t.}}$, at σ-a.e. point on $\partial\Omega$. Finally, in place of (3.3.59) *one presentaly has*

$$\begin{aligned} \|\mathcal{N}_\kappa(\nabla u)\|_{\mathbb{X}} &\approx \|(\nabla u)\big|_{\partial\Omega}^{\kappa-\text{n.t.}}\|_{[\mathbb{X}]^{n\cdot M}} \\ &\approx \sum_{j,k=1}^{n} \Big\|\nu_j(\partial_k u)\big|_{\partial\Omega}^{\kappa-\text{n.t.}} - \nu_k(\partial_j u)\big|_{\partial\Omega}^{\kappa-\text{n.t.}}\Big\|_{[\mathbb{X}]^M} + \|\partial_\nu^A u\|_{[\mathbb{X}]^M} \end{aligned} \tag{3.3.77}$$

if, in the case when Ω is an exterior domain, one also assumes (for both equivalences) that ∇u vanishes at infinity (in the sense of (3.3.8)).

In the regime $p \in [1,\infty)$, part *(a)* of the above theorem together with [113, Theorem 10.2.24] imply that if the $M \times M$ system L is associated with the coefficient tensor $A = \big(a_{rs}^{\alpha\beta}\big)_{\substack{1\leq r,s\leq n\\ 1\leq\alpha,\beta\leq M}}$ as in (1.3.15), then for Ω and u as in the statement of Theorem 3.3.4 we have (with $\nu = (\nu_r)_{1\leq r\leq n}$ denoting the geometric measure theoretic outward unit normal to Ω)

$$\begin{gathered} \mathring{\partial}_\nu^A u \text{ belongs to the Lebesgue space } \big[L^p(\partial\Omega,\sigma)\big]^M \text{ and} \\ \mathring{\partial}_\nu^A u = \Big(\nu_r a_{rs}^{\alpha\beta}(\partial_s u_\beta)\big|_{\partial\Omega}^{\kappa-\text{n.t.}}\Big)_{1\leq\alpha\leq M} \text{ at } \mathcal{H}^{n-1}\text{-a.e. point on } \partial\Omega. \end{gathered} \tag{3.3.78}$$

As mentioned earlier, we shall give Theorem 3.3.4 two proofs.

First proof of Theorem 3.3.4 The claims in (3.3.55) are proved in the same manner as Theorem 3.1.6, making use of the integral representation formula from Theorem 3.3.1 (in lieu of Theorem 3.1.1). The integrability properties and independence of the nontangential boundary trace on the aperture parameter follow from assumptions, [112, Proposition 8.9.8], [112, Corollary 8.9.6], [112, (8.9.8)], and [112, Proposition 8.8.6]. The claims made in (3.3.56) (and just below) are direct consequences of Proposition 3.1.13. Also, the claims in (3.3.57) are implied by item *(1)* in [113, Theorem 10.2.24] and [113, (10.2.14) in Example 10.2.2].

Next, (3.3.58) is a consequence of (3.3.7), (2.4.14), and the estimates that accompany the memberships in (3.3.57) (cf. [113, (10.2.14), (10.2.182)]). Moving on, (3.3.59) is seen from (3.3.9), (2.4.9), and (1.7.54), keeping in mind (3.3.55) and [113, (10.2.92), (10.2.186)]. Also, the claims made in (3.3.60)-(3.3.61) are implied by the trace result established in the last theorem of [115, §2.2] and (3.3.58), while the claims made in (3.3.62)-(3.3.63) follow from (3.3.59) and [113, Propositions 11.3.2, 11.5.12] (bearing in mind (3.3.54), (3.3.55), and [113, (11.5.129)]).

Going further, the claims in items *(d)*-*(e)* are implied by the integral representation formula from Theorem 3.3.1, items *(4)*, *(6)*, *(7)* in Theorem 2.4.1, together with (3.3.58) and (3.3.59). The claims in item *(f)* may be justified along the lines of the proofs of their earlier counterparts, making use of (3.3.11)-(3.3.12) in Theorem 3.3.1.

Finally, the claims in item *(g)* are established reasoning as before, now making use of the integral representation formula from Proposition 3.3.3, and bearing in mind [113, (5.1.12), (5.1.13)], [113, Proposition 5.2.7], and item *(4)* in Theorem 2.4.1. □

In the second proof of Theorem 3.3.4, given next, the strategy is to reduce matters to a suitable application of the Fatou-type theorem for first-order injectively elliptic systems from Theorem 3.1.6.

Second proof of Theorem 3.3.4 To get things started, suppose the $M \times M$ system L is associated with some coefficient tensor $A = \big(a_{rs}^{\alpha\beta}\big)_{\substack{1\leq r,s\leq n\\ 1\leq\alpha,\beta\leq M}}$ as in (1.3.15). Bring in the first-order system $\mathbb{D}_A$ defined as in (1.3.39). From Example 1.3.12 we know that $\mathbb{D}_A$ is injectively elliptic. Going further, assume $u = (u_\beta)_{1\leq\beta\leq M}$ is as in the statement of Theorem 3.3.4 and set $w := \nabla u$, i.e.,

$$\begin{aligned} w = \big(w_s^\beta\big)_{\substack{1\leq s\leq n\\ 1\leq\beta\leq M}} \ \text{ with } \ w_s^\beta := \partial_s u_\beta \ \text{ for each} \\ s \in \{1,\dots,n\} \ \text{ and } \ \beta \in \{1,\dots,M\}. \end{aligned} \tag{3.3.79}$$

Then $w \in \big[L^1_{\mathrm{loc}}(\Omega, \mathcal{L}^n)\big]^{n\times M}$ is a vector-valued function which, thanks to (1.3.41), satisfies

$$\begin{aligned} \mathcal{N}_\kappa w \approx \mathcal{N}_\kappa(\nabla u) \ \text{ and} \\ \mathbb{D}_A w = 0 \ \text{ in the sense of distributions in } \ \Omega. \end{aligned} \tag{3.3.80}$$

Moreover, from (1.3.39), (A.0.107), and [113, Proposition 10.2.20] we see that

$$(-\mathrm{i})\mathrm{Sym}(\mathbb{D}_A;\nu)\bullet w=\begin{pmatrix}\big(\partial_{\tau_{sr}}u_\beta\big)_{\substack{1\le r,s\le n\\ 1\le\beta\le M}}\\ \partial^A_\nu u\end{pmatrix}. \tag{3.3.81}$$

Granted (1.3.40), (3.3.80), and (3.3.81), Theorem 3.1.6 applies (with u replaced by w) and all desired conclusions follow directly from it (with the exception of (3.3.56), (3.3.60)-(3.3.61) and (3.3.62)-(3.3.63) which are established as before, based on Proposition 3.1.13, (3.3.58), (3.3.59), the trace result established in the last theorem of [115, §2.2], and [113, Propositions 11.3.2, 11.5.12]). □

There is also a local version of Theorem 3.3.4, of the following sort.

Theorem 3.3.5 *Suppose $\Omega\subseteq\mathbb{R}^n$, where $n\in\mathbb{N}$ with $n\ge 2$, is an arbitrary* UR *domain and abbreviate $\sigma:=\mathcal{H}^{n-1}\lfloor\partial\Omega$. Also, consider a homogeneous constant (complex) coefficient second-order $M\times M$ system L in $\mathbb{R}^n$ (for some $M\in\mathbb{N}$) which is weakly elliptic (in the sense of* (1.3.3)*), and assume $u\in\big[W^{1,1}_{\mathrm{loc}}(\Omega)\big]^M$ is a vector-valued function which, for some aperture parameter $\kappa>0$ and some truncation parameter $\varepsilon>0$, satisfies*

$$\mathcal{N}^{\varepsilon}_{\kappa}(\nabla u)\in L^p_{\mathrm{loc}}(\partial\Omega,\sigma)\ \text{for some}\ p\in(1,\infty]\quad\text{and}\ Lu=0\ \text{in}\ \big[\mathcal{D}'(\Omega)\big]^M. \tag{3.3.82}$$

Then the nontangential boundary trace $(\nabla u)\big|^{\kappa-\mathrm{n.t.}}_{\partial\Omega}$ exists (in $\mathbb{C}^{M\times n}$) at σ-a.e. point on $\partial\Omega$, is actually independent of the aperture parameter κ, the function $(\nabla u)\big|^{\kappa-\mathrm{n.t.}}_{\partial\Omega}$ is σ-measurable on $\partial\Omega$ and belongs to the space $\big[L^p_{\mathrm{loc}}(\partial\Omega,\sigma)\big]^{M\times n}$.

Furthermore, the nontangential boundary trace $u\big|^{\kappa-\mathrm{n.t.}}_{\partial\Omega}$ also exists at σ-a.e. point on $\partial\Omega$ and is a σ-measurable function which is actually independent of the aperture parameter.

It is worth noting that the membership condition in the first line of (3.3.82) is satisfied if, for example, $\mathcal{N}^{\varepsilon}_{\kappa}(\nabla u)\in L^p(\partial\Omega,w\sigma)$ for some $p\in(1,\infty)$ and $w\in A_p(\partial\Omega,\sigma)$ (see [112, (7.7.105)] in this regard), or if $\mathcal{N}^{\varepsilon}_{\kappa}(\nabla u)\in M^{p,\lambda}(\partial\Omega,\sigma)$ for some $p\in(1,\infty)$ and $\lambda\in(0,n-1)$ (cf. [113, (6.2.25)]).

Proof of Theorem 3.3.5 All claims pertaining to the nontangential boundary trace $(\nabla u)\big|^{\kappa-\mathrm{n.t.}}_{\partial\Omega}$ may be deduced directly from Theorem 3.1.9 by reasoning as in the second proof of Theorem 3.3.4 presented above. The claims regarding $u\big|^{\kappa-\mathrm{n.t.}}_{\partial\Omega}$ are seen from Proposition 3.1.13. □

We continue by making yet another remark pertaining to Theorem 3.3.4. Specifically, in the case when $\Omega\subseteq\mathbb{R}^n$ is actually a *bounded* UR domain, we may relax the hypotheses in (3.3.54) to

$$\mathcal{N}_{\kappa}(\nabla u)\in L^p(\partial\Omega,\sigma)\ \text{with}\ p\in\big(\tfrac{n-1}{n},\infty\big)\quad\text{and}\ Lu\in\big[L^q(\Omega,\mathcal{L}^n)\big]^M\ \text{for some}\ q\in(n,\infty] \tag{3.3.83}$$

and still conclude[11] that the nontangential boundary trace $(\nabla u)\big|_{\partial\Omega}^{\kappa-\text{n.t.}}$ exists at σ-a.e. point on $\partial\Omega$ and is independent of the aperture parameter κ.

Moving on, by combining the Fatou-type result from Theorem 3.3.5 with the integral representation formulas from Theorem 1.5.10 we obtain the following result:

Theorem 3.3.6 *Suppose $\Omega \subseteq \mathbb{R}^n$, where $n \in \mathbb{N}$ with $n \geq 2$, is an arbitrary* UR *domain. Abbreviate $\sigma := \mathcal{H}^{n-1}\lfloor\partial\Omega$ and denote by $\nu = (\nu_1, \dots, \nu_n)$ the geometric measure theoretic outward unit normal to Ω. With the summation convention over repeated indices understood throughout, let*

$$L = \big(a^{\alpha\beta}_{rs}\partial_r\partial_s\big)_{1\leq\alpha,\beta\leq M} \tag{3.3.84}$$

be a homogeneous, weakly elliptic, second-order $M \times M$ system in $\mathbb{R}^n$, with complex constant coefficients, and denote by $E = (E_{\gamma\beta})_{1\leq\gamma,\beta\leq M}$ the matrix-valued fundamental solution associated with L as in Theorem 1.4.2. In this setting, assume $u = (u_\beta)_{1\leq\beta\leq M} \in \big[W^{1,1}_{\text{loc}}(\Omega)\big]^M$ is a vector-valued function which, for some aperture parameter $\kappa > 0$ and some truncation parameter $\varepsilon > 0$, satisfies

$$\mathcal{N}^{\varepsilon}_{\kappa}(\nabla u) \in L^p_{\text{loc}}(\partial\Omega, \sigma) \text{ for some } p \in (1, \infty],$$
$$\text{and } \int_{\partial\Omega} \frac{\big(\mathcal{N}_\kappa(\nabla u)\big)(y)}{1 + |y|^{n-1}}\, d\sigma(y) < +\infty. \tag{3.3.85}$$

With Lu considered in the sense of distributions in Ω, suppose also that

$$Lu = 0 \text{ in } \Omega. \tag{3.3.86}$$

Then the function u is of class $\mathscr{C}^\infty$ in Ω and for any aperture parameter $\kappa' > 0$ the nontangential trace $(\nabla u)\big|_{\partial\Omega}^{\kappa'-\text{n.t.}}$ exists σ-a.e. on $\partial\Omega$ and is actually independent of κ'. Moreover, with the dependence on the parameter κ' dropped and having fixed $\ell \in \{1, \dots, n\}$ along with $\gamma \in \{1, \dots, M\}$ arbitrary, it follows that for each point $x \in \Omega$ one has (with absolutely convergent integrals)

$$\begin{aligned}(\partial_\ell u_\gamma)(x) = &\int_{\partial\Omega} a^{\beta\alpha}_{rs}(\partial_r E_{\gamma\beta})(x - y)\times \\ &\times \Big\{\nu_\ell(y)\big((\partial_s u_\alpha)\big|_{\partial\Omega}^{\text{n.t.}}\big)(y) - \nu_s(y)\big((\partial_\ell u_\alpha)\big|_{\partial\Omega}^{\text{n.t.}}\big)(y)\Big\}\, d\sigma(y) \\ &- \int_{\partial\Omega}(\partial_\ell E_{\gamma\alpha})(x - y)\nu_r(y)a^{\alpha\beta}_{rs}\big((\partial_s u_\beta)\big|_{\partial\Omega}^{\text{n.t.}}\big)(y)\, d\sigma(y)\end{aligned} \tag{3.3.87}$$

if either Ω is bounded, or $\partial\Omega$ is unbounded. In the case when Ω is an exterior domain, the same conclusion holds true under the additional assumption that there exists $\lambda \in (1, \infty)$ such that

[11] based on Theorem 3.3.1, the same way Theorem 3.1.7 has been proved with the help of Theorem 3.1.6

$$\fint_{B(0,\lambda R)\setminus B(0,R)} |\nabla u| \, d\mathcal{L}^n = o(1) \text{ as } R \to \infty. \tag{3.3.88}$$

In addition, for each $x \in \mathbb{R}^n \setminus \overline{\Omega}$ one has

$$\begin{aligned} 0 = & \int_{\partial\Omega} a_{rs}^{\beta\alpha}(\partial_r E_{\gamma\beta})(x-y)\Big\{\nu_\ell(y)\big((\partial_s u_\alpha)\big|_{\partial\Omega}^{\text{n.t.}}\big)(y) - \nu_s(y)\big((\partial_\ell u_\alpha)\big|_{\partial\Omega}^{\text{n.t.}}\big)(y)\Big\}\, d\sigma(y) \\ & - \int_{\partial\Omega} (\partial_\ell E_{\gamma\alpha})(x-y)\nu_r(y) a_{rs}^{\alpha\beta}\big((\partial_s u_\beta)\big|_{\partial\Omega}^{\text{n.t.}}\big)(y)\, d\sigma(y) \end{aligned} \tag{3.3.89}$$

with the same convention as before in the case when Ω is an exterior domain.

Proof All claims are consequence of elliptic regularity (cf. [109]), the Fatou-type result from Theorem 3.3.5, the integral formulas from Theorem 1.5.10, [112, Proposition 8.8.6], and [112, Corollary 8.9.9]. □

We continue by presenting an adaptation of Theorem 3.3.4 to the scale of Morrey spaces.

Theorem 3.3.7 *Let $\Omega \subseteq \mathbb{R}^n$, where $n \in \mathbb{N}$ with $n \geq 2$, be an arbitrary* UR *domain, and abbreviate $\sigma := \mathcal{H}^{n-1}\lfloor\partial\Omega$. Consider a homogeneous constant (complex) coefficient second-order $M \times M$ system L in $\mathbb{R}^n$ (for some $M \in \mathbb{N}$) which is weakly elliptic (in the sense of (1.3.3)), and assume $u \in \big[W^{1,1}_{\text{loc}}(\Omega)\big]^M$ is a vector-valued function which, for some aperture parameter $\kappa > 0$, satisfies*

$$\begin{aligned} \mathcal{N}_\kappa(\nabla u) \in M^{p,\lambda}(\partial\Omega,\sigma) \text{ with } p \in (1,\infty),\ \lambda \in (0,n-1), \\ \text{and } Lu = 0 \text{ in } \big[\mathcal{D}'(\Omega)\big]^M. \end{aligned} \tag{3.3.90}$$

Then the nontangential boundary trace

$$\begin{aligned} (\nabla u)\big|_{\partial\Omega}^{\kappa-\text{n.t.}} \text{ exists (in } \mathbb{C}^{n\cdot M}) \text{ at } \sigma\text{-a.e. point on } \partial\Omega \text{ and} \\ \text{the function } (\nabla u)\big|_{\partial\Omega}^{\kappa-\text{n.t.}} \text{ is } \sigma\text{-measurable on } \partial\Omega. \end{aligned} \tag{3.3.91}$$

In addition, for any other given aperture parameter $\kappa' \in (0,\infty)$ the nontangential boundary trace $(\nabla u)\big|_{\partial\Omega}^{\kappa'-\text{n.t.}}$ also exists, and agrees with $(\nabla u)\big|_{\partial\Omega}^{\kappa-\text{n.t.}}$, at σ-a.e. point on $\partial\Omega$. Also,

$$\begin{aligned} (\nabla u)\big|_{\partial\Omega}^{\kappa-\text{n.t.}} \text{ belongs to the Morrey space } \big[M^{p,\lambda}(\partial\Omega,\sigma)\big]^{n\cdot M} \\ \text{and one has } \Big\|(\nabla u)\big|_{\partial\Omega}^{\kappa-\text{n.t.}}\Big\|_{[M^{p,\lambda}(\partial\Omega,\sigma)]^{n\cdot M}} \leq \|\mathcal{N}_\kappa(\nabla u)\|_{M^{p,\lambda}(\partial\Omega,\sigma)}. \end{aligned} \tag{3.3.92}$$

In fact, the following Morrey-styled boundary maximum principle holds:

$$\big\|\mathcal{N}_\kappa(\nabla u)\big\|_{M^{p,\lambda}(\partial\Omega,\sigma)} \approx \Big\|(\nabla u)\big|_{\partial\Omega}^{\kappa-\text{n.t.}}\Big\|_{[M^{p,\lambda}(\partial\Omega,\sigma)]^{n\cdot M}} \tag{3.3.93}$$

provided, in the case when Ω *is an exterior domain, it is also assumed that* ∇u *vanishes at infinity (in the sense of* (3.3.8)*).*

Finally, if (3.3.90) *is strengthened to*

$$\begin{aligned} &\mathcal{N}_\kappa(\nabla u) \in \mathring{M}^{p,\lambda}(\partial\Omega,\sigma) \text{ with } p \in (1,\infty),\ \lambda \in (0, n-1), \\ &\qquad \text{and } Lu = 0 \text{ in } \big[\mathcal{D}'(\Omega)\big]^M, \end{aligned} \tag{3.3.94}$$

then (3.3.91) *continues to hold and one now has*

$$(\nabla u)\Big|_{\partial\Omega}^{\kappa-\text{n.t.}} \in \big[\mathring{M}^{p,\lambda}(\partial\Omega,\sigma)\big]^{n\cdot M}. \tag{3.3.95}$$

Proof All claims, except (3.3.95), may be justified by reasoning as in the second proof of Theorem 3.3.4 (in order to reduce matters to working with a null-solution of an injectively elliptic first-order system) and then invoking Theorem 3.1.10. Alternatively, we invoke item *(g)* in Theorem 3.3.4 with $\mathbb{X} := M^{p,\lambda}(\partial\Omega,\sigma)$, a permissible choice in light of [113, Proposition 6.2.17], and [113, Corollaries 6.2.11, 6.2.13]. Finally, (3.3.95) follows from (3.3.91), [112, (8.9.8)], and [113, (6.2.18)]. Let us also remark that a natural version of Theorem 3.3.1 holds in the context of Morrey spaces, and such a result may be used to provide an alternative proof of Theorem 3.3.7. □

Finally, here is a version of Theorem 3.3.4 for the scale of block spaces, which states as follows.

Theorem 3.3.8 *Let* $\Omega \subseteq \mathbb{R}^n$*, where* $n \in \mathbb{N}$ *with* $n \geq 2$*, be an arbitrary* UR *domain, and abbreviate* $\sigma := \mathcal{H}^{n-1}\lfloor\partial\Omega$*. Consider a homogeneous constant (complex) coefficient second-order* $M \times M$ *system* L *in* $\mathbb{R}^n$ *(for some* $M \in \mathbb{N}$*) which is weakly elliptic (in the sense of* (1.3.3)*), and assume* $u \in \big[W^{1,1}_{\text{loc}}(\Omega)\big]^M$ *is a vector-valued function which, for some aperture parameter* $\kappa > 0$*, satisfies*

$$\begin{aligned} &\mathcal{N}_\kappa(\nabla u) \in \mathcal{B}^{q,\lambda}(\partial\Omega,\sigma) \text{ with } q \in (1,\infty),\ \lambda \in (0, n-1), \\ &\qquad \text{and } Lu = 0 \text{ in } \big[\mathcal{D}'(\Omega)\big]^M. \end{aligned} \tag{3.3.96}$$

Then the nontangential boundary trace

$$\begin{aligned} &(\nabla u)\Big|_{\partial\Omega}^{\kappa-\text{n.t.}} \text{ exists (in } \mathbb{C}^{n\cdot M}) \text{ at } \sigma\text{-a.e. point on } \partial\Omega \text{ and} \\ &\quad \text{the function } (\nabla u)\Big|_{\partial\Omega}^{\kappa-\text{n.t.}} \text{ is } \sigma\text{-measurable on } \partial\Omega. \end{aligned} \tag{3.3.97}$$

Moreover, for any other given aperture parameter $\kappa' \in (0,\infty)$ *the nontangential boundary trace* $(\nabla u)\Big|_{\partial\Omega}^{\kappa'-\text{n.t.}}$ *also exists, and agrees with* $(\nabla u)\Big|_{\partial\Omega}^{\kappa-\text{n.t.}}$*, at* σ*-a.e. point on* $\partial\Omega$*. In addition,*

$$\begin{aligned} &(\nabla u)\Big|_{\partial\Omega}^{\kappa-\text{n.t.}} \text{ belongs to the block space } \big[\mathcal{B}^{q,\lambda}(\partial\Omega,\sigma)\big]^{n\cdot M} \\ &\text{and one has } \Big\|(\nabla u)\Big|_{\partial\Omega}^{\kappa-\text{n.t.}}\Big\|_{[\mathcal{B}^{q,\lambda}(\partial\Omega,\sigma)]^{n\cdot M}} \leq \|\mathcal{N}_\kappa(\nabla u)\|_{\mathcal{B}^{q,\lambda}(\partial\Omega,\sigma)}. \end{aligned} \tag{3.3.98}$$

In fact, the following block-styled boundary maximum principle holds:

$$\big\|\mathcal{N}_\kappa(\nabla u)\big\|_{\mathcal{B}^{q,\lambda}(\partial\Omega,\sigma)} \approx \big\|(\nabla u)\big|^{\kappa-\mathrm{n.t.}}_{\partial\Omega}\big\|_{[\mathcal{B}^{q,\lambda}(\partial\Omega,\sigma)]^{n\cdot M}} \tag{3.3.99}$$

provided, in the case when Ω is an exterior domain, it is also assumed that ∇u vanishes at infinity (in the sense of (3.3.8)*).*

Proof One possible way to see this is to rely on item *(g)* in Theorem 3.3.4 with $\mathbb{X} := \mathcal{B}^{q,\lambda}(\partial\Omega,\sigma)$, a valid choice thanks to [113, Proposition 6.2.17], and [113, Corollaries 6.2.11, 6.2.13]. Alternatively, we may reason as follows. From (3.3.96) and [113, (6.2.71)] we see that

$$\mathcal{N}_\kappa(\nabla u) \in L^r(\partial\Omega,\sigma) \text{ with } r := \frac{q(n-1)}{n-1+\lambda(q-1)} \in (1,q). \tag{3.3.100}$$

Granted this, Theorem 3.3.4 applies and proves that (3.3.97) holds. The independence of the boundary trace on the aperture parameter is then provided by [112, Proposition 8.9.8]. Next, the claims in (3.3.98) are implied by [113, (6.2.75)] and [112, (8.9.8)]. Finally, the right-pointing inequality in (3.3.99) is contained in (3.3.98), while the left-pointing inequality in (3.3.99) is implied by (3.3.9) and (2.6.19). □

The remaining material in this section concerns quantitative Fatou-type results for null-solutions u of a weakly elliptic second-order system L in arbitrary UR domains $\Omega \subseteq \mathbb{R}^n$, describing scenarios in which integrability properties of the nontangential maximal functions of u and ∇u together ensure the existence of nontangential pointwise traces of both u and ∇u on $\partial\Omega$ with quantitative control. Here is our first theorem of this flavor.

Theorem 3.3.9 *Let $\Omega \subseteq \mathbb{R}^n$, where $n \in \mathbb{N}$ with $n \geq 3$, be an arbitrary* UR *domain, and abbreviate $\sigma := \mathcal{H}^{n-1}\lfloor\partial\Omega$. For some integer $M \in \mathbb{N}$, consider a coefficient tensor $A = \big(a^{\alpha\beta}_{rs}\big)_{\substack{1\leq r,s\leq n\\ 1\leq\alpha,\beta\leq M}}$ with complex entries, with the property that the $M \times M$ homogeneous second-order system $L = L_A$ associated with A in $\mathbb{R}^n$ as in* (1.3.15) *is weakly elliptic (in the sense of* (1.3.3)*). Finally, fix some aperture parameter $\kappa > 0$. In this setting, consider a vector-valued function satisfying*

$$\begin{aligned} &u = (u_\beta)_{1\leq\beta\leq M} \in \big[\mathscr{C}^\infty(\Omega)\big]^M, \quad Lu = 0 \text{ in } \Omega, \\ &\mathcal{N}_\kappa(\nabla u) \in L^p(\partial\Omega,\sigma) \text{ for some } p \in \big(\tfrac{n-1}{n}, n-1\big) \\ &\text{and } \mathcal{N}_\kappa u \in L^{p^*}(\partial\Omega,\sigma) \text{ with } p^* := \big(\tfrac{1}{p} - \tfrac{1}{n-1}\big)^{-1}. \end{aligned} \tag{3.3.101}$$

Then

$$
\begin{aligned}
&\textit{the nontangential traces } u\big|_{\partial\Omega}^{\kappa-\text{n.t.}},\ (\nabla u)\big|_{\partial\Omega}^{\kappa-\text{n.t.}} \textit{ exist } \sigma\textit{-a.e. on } \partial\Omega,\\
&u\big|_{\partial\Omega}^{\kappa-\text{n.t.}} \in \big[L^{p^*}(\partial\Omega,\sigma)\big]^M \textit{ and } (\nabla u)\big|_{\partial\Omega}^{\kappa-\text{n.t.}} \in \big[L^{p}(\partial\Omega,\sigma)\big]^{n\cdot M},\\
&\Big\|u\big|_{\partial\Omega}^{\kappa-\text{n.t.}}\Big\|_{[L^{p^*}(\partial\Omega,\sigma)]^M} \leq \|\mathcal{N}_\kappa u\|_{L^{p^*}(\partial\Omega,\sigma)} \textit{ and}\\
&\Big\|(\nabla u)\big|_{\partial\Omega}^{\kappa-\text{n.t.}}\Big\|_{[L^{p}(\partial\Omega,\sigma)]^{n\cdot M}} \leq \|\mathcal{N}_\kappa (\nabla u)\|_{L^{p}(\partial\Omega,\sigma)}.
\end{aligned}
\tag{3.3.102}
$$

In addition, for any other given aperture parameter $\kappa' \in (0,\infty)$ *the nontangential boundary traces* $u\big|_{\partial\Omega}^{\kappa'-\text{n.t.}}$, $(\nabla u)\big|_{\partial\Omega}^{\kappa'-\text{n.t.}}$ *also exist and agree with* $u\big|_{\partial\Omega}^{\kappa-\text{n.t.}}$ *and* $(\nabla u)\big|_{\partial\Omega}^{\kappa-\text{n.t.}}$, *respectively, at* σ*-a.e. point on* $\partial\Omega$. *Also, for some* $C = C(\Omega,\kappa,p) \in (0,\infty)$, *one has (recall the scale of Hardy-based inhomogeneous Sobolev spaces from* [113, Definition 11.10.6]) *that*

$$
\begin{aligned}
&u\big|_{\partial\Omega}^{\kappa-\text{n.t.}} \textit{ belongs to the Hardy-based Sobolev space } \big[H_1^{p^*,p}(\partial\Omega,\sigma)\big]^M,\\
&\textit{and } \Big\|u\big|_{\partial\Omega}^{\kappa-\text{n.t.}}\Big\|_{[H_1^{p^*,p}(\partial\Omega,\sigma)]^M} \leq C\|\mathcal{N}_\kappa u\|_{L^{p^*}(\partial\Omega,\sigma)} + C\|\mathcal{N}_\kappa(\nabla u)\|_{L^{p}(\partial\Omega,\sigma)},
\end{aligned}
\tag{3.3.103}
$$

while the weak conormal derivative of u associated with the coefficient tensor A as in [113, Definition 10.2.18] *satisfies*

$$
\begin{aligned}
&\dot{\partial}_\nu^A u \in \big[H^p(\partial\Omega,\sigma)\big]^M \textit{ and}\\
&\big\|\dot{\partial}_\nu^A u\big\|_{[H^p(\partial\Omega,\sigma)]^M} \leq C\big\|\mathcal{N}_\kappa(\nabla u)\big\|_{L^p(\partial\Omega,\sigma)}
\end{aligned}
\tag{3.3.104}
$$

for some constant $C \in (0,\infty)$ *which depends only on* Ω, n, A, κ, p.

Moreover, with the single and double layer potential operators $\mathscr{S}$, $\mathcal{D}$, *associated with A and* Ω *as in* [115, §1.3], *one has the integral representation formula*

$$
u = \mathcal{D}\big(u\big|_{\partial\Omega}^{\kappa-\text{n.t.}}\big) - \mathscr{S}(\dot{\partial}_\nu^A u) \textit{ in } \Omega,
\tag{3.3.105}
$$

with the understanding that if Ω *is an exterior domain one also assumes that there exists* $\lambda \in (1,\infty)$ *such that*

$$
⨍_{B(0,\lambda R)\setminus B(0,R)} |u|\, d\mathcal{L}^n = o(1) \textit{ as } R \to \infty.
\tag{3.3.106}
$$

Furthermore,

if actually $\partial\Omega$ *is bounded and in place of* (3.3.101) *one now assumes that* $u \in \big[\mathscr{C}^\infty(\Omega)\big]^M$ *with* $Lu = 0$ *in* Ω, $\mathcal{N}_\kappa(\nabla u) \in L^p(\partial\Omega,\sigma)$ *for some* $p \in \big(\frac{n-1}{n},\infty\big]$, *and* $\mathcal{N}_\kappa u \in L^q(\partial\Omega,\sigma)$ *for some* $q \in \big(\frac{n-1}{n},\infty\big]$, *it follows that the conclusions in* (3.3.102) *hold with q replacing* p^*. (3.3.107)

Finally, the same results are valid in the case $n = 2$ *under the additional assumption that* Ω *is bounded.*

Before presenting the proof of this result we make a couple of comments. First, if $p \geq 1$ then item *(3)* in [113, Theorem 10.2.24] guarantees that the weak conormal derivative $\overset{\bullet}{\partial}{}^A_\nu u$ actually belongs to the Lebesgue space $\big[L^p(\partial\Omega,\sigma)\big]^M$ and has the pointwise formula

$$\overset{\bullet}{\partial}{}^A_\nu u = \Big(\nu_r a^{\alpha\beta}_{rs}(\partial_s u_\beta)\big|^{\kappa-\mathrm{n.t.}}_{\partial\Omega}\Big)_{1\leq\alpha\leq M} \quad \text{at } \sigma\text{-a.e. point on } \partial\Omega. \tag{3.3.108}$$

Second, it is only the range $q \in \big(\frac{n-1}{n}, 1\big]$ that is genuinely new in (3.3.107). Indeed, if one assumes $q \in (1, \infty]$ in (3.3.107) then it is possible to choose

$$r \in \big(\tfrac{n-1}{n}, \min\{p, n-1\}\big) \text{ such that } \tfrac{1}{q} + \tfrac{1}{n-1} < \tfrac{1}{r} < \tfrac{n}{n-1}. \tag{3.3.109}$$

As a consequence, if we define $r^* := \big(\frac{1}{r} - \frac{1}{n-1}\big)^{-1}$ then $r^* < q$. Given that the Lebesgue scale on $\partial\Omega$ is nested (since presently we have $\sigma(\partial\Omega) < +\infty$), this choice implies $\mathcal{N}_\kappa(\nabla u) \in L^r(\partial\Omega,\sigma)$ and $\mathcal{N}_\kappa u \in L^{r^*}(\partial\Omega,\sigma)$. Hence, the hypotheses in (3.3.101) are satisfied with $r \in \big(\frac{n-1}{n}, n-1\big)$ replacing the integrability exponent p. Granted this, (3.3.102) ensures that the nontangential traces $u\big|^{\kappa-\mathrm{n.t.}}_{\partial\Omega}$, $(\nabla u)\big|^{\kappa-\mathrm{n.t.}}_{\partial\Omega}$ exist at σ-a.e. on point on $\partial\Omega$, and the corresponding functions are σ-measurable. Having established this, the conclusions in (3.3.107) become consequences of [112, (8.9.8)] and assumptions.

Let us now turn to the task of giving the proof of Theorem 3.3.9.

Proof of Theorem 3.3.9 The claims in (3.3.104) are implied by item *(1)* in [113, Theorem 10.2.24] and the first two lines in (3.3.101). For the existence of $u\big|^{\kappa-\mathrm{n.t.}}_{\partial\Omega}$ we may invoke Proposition 3.1.13. An alternative approach which also yields the integral representation formula (3.3.105) goes as follows. To set the stage, fix $\beta \in \{1,\dots,M\}$ together with $j \in \{1,\dots,n\}$. From (3.3.101) and [112, Proposition 8.6.3] we conclude that the vector field $u_\beta \mathbf{e}_j$ belongs to the space of "admissible" vector fields,

$$\mathscr{A} := \Big\{\vec{F} \in \big[L^1_{\mathrm{bdd}}(\Omega,\mathcal{L}^n)\big]^n : \operatorname{div}\vec{F} \in L^1_{\mathrm{bdd}}(\Omega,\mathcal{L}^n)\Big\}. \tag{3.3.110}$$

As such, [112, Proposition 4.2.3] guarantees that $\nu \bullet (u_\beta \mathbf{e}_j)$ makes sense as a distribution, in $\big(\mathrm{Lip}_c(\partial\Omega)\big)'$. In relation to this, we next claim that for each two indices, $\beta \in \{1,\dots,M\}$ and $j \in \{1,\dots,n\}$, we have

$$\begin{gathered}\text{the distribution } \nu \bullet (u_\beta \mathbf{e}_j) \text{ belongs to the space } L^{p^*}(\partial\Omega,\sigma) \text{ and} \\ \big\|\nu \bullet (u_\beta \mathbf{e}_j)\big\|_{L^{p^*}(\partial\Omega,\sigma)} \leq C\|\mathcal{N}_\kappa u\|_{L^{p^*}(\partial\Omega,\sigma)} + C\|\mathcal{N}_\kappa(\nabla u)\|_{L^p(\partial\Omega,\sigma)}\end{gathered} \tag{3.3.111}$$

for some constant $C \in (0,\infty)$ depending only on Ω, n, κ, p. Indeed, the vector field $u_\beta \mathbf{e}_j$ satisfies $\mathcal{N}_\kappa(u_\beta \mathbf{e}_j) \leq \mathcal{N}_\kappa u$ which, in view of [112, Proposition 8.2.3], implies

$$\begin{gathered}\mathcal{N}_\kappa(u_\beta \mathbf{e}_j) \in L^{p^*}(\partial\Omega,\sigma) \text{ and} \\ \big\|\mathcal{N}_\kappa(u_\beta \mathbf{e}_j)\big\|_{L^{p^*}(\partial\Omega,\sigma)} \leq \big\|\mathcal{N}_\kappa u\big\|_{L^{p^*}(\partial\Omega,\sigma)}.\end{gathered} \tag{3.3.112}$$

Also, $\operatorname{div}(u_\beta \mathbf{e}_j) = \partial_j u_\beta$, so on the one hand we have

$$\mathfrak{P}\big(\operatorname{div}(u_\beta \mathbf{e}_j)\big) \leq \mathfrak{P}(\nabla u). \tag{3.3.113}$$

On the other hand, from the second line in (3.3.101), [113, (10.1.8)-(10.1.9)] (used with ∇u in place of u), and [112, (6.2.26)] we know that

$$\begin{gathered} \mathfrak{P}(\nabla u) \in L^{p^*,p}(\partial\Omega,\sigma) \hookrightarrow L^{p^*}(\partial\Omega,\sigma) \text{ and} \\ \big\|\mathfrak{P}(\nabla u)\big\|_{L^{p^*}(\partial\Omega,\sigma)} \leq C\big\|\mathcal{N}_\kappa(\nabla u)\big\|_{L^p(\partial\Omega,\sigma)}. \end{gathered} \tag{3.3.114}$$

Combining (3.3.113), (3.3.114), and [113, (10.1.3)] proves that

$$\begin{gathered} \mathfrak{P}\big(\operatorname{div}(u_\beta \mathbf{e}_j)\big) \in L^{p^*}(\partial\Omega,\sigma) \text{ and} \\ \big\|\mathfrak{P}\big(\operatorname{div}(u_\beta \mathbf{e}_j)\big)\big\|_{L^{p^*}(\partial\Omega,\sigma)} \leq C\big\|\mathcal{N}_\kappa(\nabla u)\big\|_{L^p(\partial\Omega,\sigma)}. \end{gathered} \tag{3.3.115}$$

From (3.3.112), (3.3.115), and [113, Theorem 10.2.1] we conclude that the claims in (3.3.111) are true.

Moving on, let us temporarily work under the additional assumption that if Ω is an exterior domain then there exists $\lambda \in (1,\infty)$ such that (3.3.106) holds. Denote by $E = (E_{\gamma\beta})_{1\leq\gamma,\beta\leq M}$ the matrix-valued fundamental solution associated with L as in Theorem 1.4.2. Fix an arbitrary index $\gamma \in \{1,\dots,M\}$, pick an arbitrary point $x \in \Omega$, and consider the vector field $\vec{F}_x = \big(F_s\big)_{1\leq s\leq n}$ with components (throughout, the summation convention over repeated indices is assumed)

$$\begin{gathered} F_s := -a^{\beta\alpha}_{rs}(\partial_r E_{\gamma\beta})(x-\cdot)u_\alpha - E_{\gamma\alpha}(x-\cdot)a^{\alpha\beta}_{sr}(\partial_r u_\beta) \\ \text{at } \mathcal{L}^n\text{-a.e. point in } \Omega. \end{gathered} \tag{3.3.116}$$

The goal is to apply [112, Theorem 1.9.4] to the vector field $\vec{F}_x$. Checking that this vector field satisfies the hypotheses of [112, Theorem 1.9.4] takes some work. For starters, observe that (3.3.116) and Theorem 1.4.2 imply

$$\vec{F}_x \in \big[L^1_{\text{bdd}}(\Omega,\mathcal{L}^n)\big]^n \subseteq \big[L^1_{\text{bdd}}(\Omega,\mathcal{L}^n) + \mathscr{E}'(\Omega)\big]^n. \tag{3.3.117}$$

Moreover, from (3.3.116) and (1.4.21) we see that the divergence of $\vec{F}_x$, computed in the sense of distributions in Ω, is given by

$$\begin{aligned} \operatorname{div}\vec{F}_x = \partial_s F_s &= a^{\beta\alpha}_{rs}\partial_s\partial_r[E_{\gamma\beta}(\cdot - x)]u_\alpha \\ &\quad - a^{\beta\alpha}_{rs}(\partial_r E_{\gamma\beta})(x-\cdot)(\partial_s u_\alpha) + (\partial_s E_{\gamma\alpha})(x-\cdot)a^{\alpha\beta}_{sr}(\partial_r u_\beta) \\ &\quad - E_{\gamma\alpha}(x-\cdot)a^{\alpha\beta}_{sr}(\partial_s\partial_r u_\beta) \\ &= u_\alpha\delta_{\gamma\alpha}\delta_x = u_\gamma(x)\delta_x. \end{aligned} \tag{3.3.118}$$

Hence,

$$\operatorname{div} \vec{F}_x \in \mathscr{E}'(\Omega) \subseteq L^1(\Omega, \mathcal{L}^n) + \mathscr{E}'(\Omega). \tag{3.3.119}$$

Note that (3.3.117) and (3.3.119) are in agreement with [112, (1.9.29)].

Pressing on, with λ as in (3.3.106) if Ω is an exterior domain and $\lambda := 2$ otherwise, pick a function $\phi \in \mathscr{C}_c^\infty\big(B(0,\lambda)\big)$ with the property that $\phi \equiv 1$ on $B(0,1)$. Then the family

$$\mathscr{F} := \{\phi_R\}_{R>0}, \text{ where } \phi_R := \phi(\cdot/R) \text{ for each } R \in (0,\infty), \tag{3.3.120}$$

becomes a system of auxiliary functions (in the sense of [112, (1.3.3)]). Recall from [112, (1.9.30)] that

$$[\vec{F}_x]_{\mathscr{F}} = -\lim_{R\to\infty} R^{-1} \int_\Omega (\nabla\phi)(y/R) \cdot \vec{F}_x(y)\,\mathrm{d}\mathcal{L}^n(y) \tag{3.3.121}$$

Under the assumption that Ω is bounded, or $\partial\Omega$ is unbounded, we may rely on [112, Proposition 8.6.3] (while also mindful of (3.3.101)), and Hölder's inequality to estimate, for R large,

$$\begin{aligned}
R^{-1} \int_\Omega &|(\nabla\phi)(y/R)||(\nabla E)(x-y)||u(y)|\,\mathrm{d}\mathcal{L}^n(y)\\
&\leq CR^{-n} \int_{\Omega\cap[B(0,\lambda R)\setminus B(0,R)]} |u|\,\mathrm{d}\mathcal{L}^n\\
&\leq CR^{-n}\Big(\int_\Omega |u|^{\frac{np^*}{n-1}}\,\mathrm{d}\mathcal{L}^n\Big)^{\frac{n-1}{np^*}} (R^n)^{1-\frac{n-1}{np^*}}\\
&\leq CR^{-\frac{n-1}{p^*}} \|\mathcal{N}_\kappa u\|_{L^{p^*}(\partial\Omega,\sigma)}\\
&= o(1) \text{ as } R\to\infty,
\end{aligned} \tag{3.3.122}$$

and

$$\begin{aligned}
R^{-1} \int_\Omega &|(\nabla\phi)(y/R)||E(x-y)||(\nabla u)(y)|\,\mathrm{d}\mathcal{L}^n(y)\\
&\leq CR^{-(n-1)} \int_{\Omega\cap[B(0,\lambda R)\setminus B(0,R)]} |\nabla u|\,\mathrm{d}\mathcal{L}^n\\
&\leq CR^{-(n-1)}\Big(\int_\Omega |\nabla u|^{\frac{np}{n-1}}\,\mathrm{d}\mathcal{L}^n\Big)^{\frac{n-1}{np}} (R^n)^{1-\frac{n-1}{np}}\\
&\leq CR^{1-\frac{n-1}{p}} \|\mathcal{N}_\kappa(\nabla u)\|_{L^{p}(\partial\Omega,\sigma)}\\
&= o(1) \text{ as } R\to\infty,
\end{aligned} \tag{3.3.123}$$

since $1 - \frac{n-1}{p} < 0$ given that $p < n-1$. Collectively, (3.3.121), (3.3.116), (3.3.122), (3.3.123) prove that

$$\text{if either } \Omega \text{ is bounded, or } \partial\Omega \text{ is unbounded, then } [\vec{F}_x]_{\mathscr{F}} = 0. \tag{3.3.124}$$

In the case when Ω is an exterior domain, Lemma 1.5.6 and (3.3.106) imply

$$(\nabla u)(x) = o\big(|x|^{-1}\big) \text{ as } |x| \to \infty. \tag{3.3.125}$$

Hence,

$$\begin{aligned} R^{-1}\int_{\Omega} |(\nabla\phi)(y/R)||\vec{F}(y)|\,d\mathcal{L}^n(y) &\leq \fint_{B(0,\lambda R)\setminus B(0,R)} \big\{|u| + R|\nabla u|\big\}\,d\mathcal{L}^n \\ &= o(1) \text{ as } R \to \infty, \end{aligned} \tag{3.3.126}$$

by (3.3.106) and (3.3.125). Thus,

$$\text{if } \Omega \text{ is an exterior domain then } [\vec{F}_x]_{\mathscr{F}} = 0. \tag{3.3.127}$$

Together, (3.3.124) and (3.3.127) show that in all cases the limit in [112, (1.9.30)] exists, and is actually zero.

To proceed, choose a function $\eta \in \mathscr{C}_c^\infty(\Omega)$ satisfying $\eta \equiv 1$ near x. Granted (3.3.118), (3.3.119), [112, Definition 4.2.6] permits us to define $\nu \bullet \vec{F}_x$ as

$$\nu \bullet \vec{F}_x = \nu \bullet \big((1-\eta)\vec{F}_x\big) \text{ in } \big(\mathrm{Lip}_c(\partial\Omega)\big)' \tag{3.3.128}$$

where $\nu \bullet \big((1-\eta)\vec{F}_x\big)$ is now interpreted in the sense of [112, Proposition 4.2.3] (bearing in mind that $(1-\eta)\vec{F}_x$ belongs to $\mathscr{A}$, the space of vector fields defined in (3.3.110)). Hence, with the piece of notation introduced in [112, (1.9.31)], we have

$$\begin{aligned} \big(\nu \bullet \vec{F}_x, 1\big)_{\mathscr{F}} &= \Big(\nu \bullet \big((1-\eta)\vec{F}_x\big), 1\Big)_{\mathscr{F}} \\ &= \lim_{R\to\infty} {}_{(\mathrm{Lip}_c(\partial\Omega))'}\Big\langle \nu \bullet \big((1-\eta)\vec{F}_x\big), \phi_R\big|_{\partial\Omega}\Big\rangle_{\mathrm{Lip}_c(\partial\Omega)} \\ &= -\lim_{R\to\infty} {}_{(\mathrm{Lip}_c(\partial\Omega))'}\Big\langle \nu \bullet \big((1-\eta)a_{rs}^{\beta\alpha}(\partial_r E_{\gamma\beta})(x-\cdot)u_\alpha \mathbf{e}_s\big), \phi_R\big|_{\partial\Omega}\Big\rangle_{\mathrm{Lip}_c(\partial\Omega)} \\ &\quad -\lim_{R\to\infty} {}_{(\mathrm{Lip}_c(\partial\Omega))'}\Big\langle \nu \bullet \big((1-\eta)E_{\gamma\alpha}(x-\cdot)a_{sr}^{\alpha\beta}(\partial_r u_\beta)\mathbf{e}_s\big), \phi_R\big|_{\partial\Omega}\Big\rangle_{\mathrm{Lip}_c(\partial\Omega)} \\ &=: \mathrm{I} + \mathrm{II}. \end{aligned} \tag{3.3.129}$$

We may further compute

$$
\begin{aligned}
\mathrm{I} &= -\lim_{R\to\infty} {}_{(\mathrm{Lip}_c(\partial\Omega))'}\Big\langle d^{\beta\alpha}_{rs}\Big((1-\eta)(\partial_r E_{\gamma\beta})(x-\cdot)\Big)\Big|_{\partial\Omega} \nu\bullet\big(u_\alpha \mathbf{e}_s\big),\, \phi_R\big|_{\partial\Omega}\Big\rangle_{\mathrm{Lip}_c(\partial\Omega)} \\
&= -\lim_{R\to\infty} {}_{(\mathrm{Lip}_c(\partial\Omega))'}\Big\langle d^{\beta\alpha}_{rs}(\partial_r E_{\gamma\beta})(x-\cdot)\big|_{\partial\Omega} \nu\bullet\big(u_\alpha \mathbf{e}_s\big),\, \phi_R\big|_{\partial\Omega}\Big\rangle_{\mathrm{Lip}_c(\partial\Omega)} \\
&= -\lim_{R\to\infty} {}_{(\mathrm{Lip}_c(\partial\Omega))'}\Big\langle \nu\bullet\big(u_\alpha \mathbf{e}_s\big),\, d^{\beta\alpha}_{rs}\big((\partial_r E_{\gamma\beta})(x-\cdot)\phi_R\big)\big|_{\partial\Omega}\Big\rangle_{\mathrm{Lip}_c(\partial\Omega)} \\
&= -\lim_{R\to\infty} \int_{\partial\Omega} \nu\bullet\big(u_\alpha \mathbf{e}_s\big) d^{\beta\alpha}_{rs}(\partial_r E_{\gamma\beta})(x-\cdot)\phi_R \,\mathrm{d}\sigma \\
&= -\int_{\partial\Omega} \nu\bullet\big(u_\alpha \mathbf{e}_s\big) d^{\beta\alpha}_{rs}(\partial_r E_{\gamma\beta})(x-\cdot)\,\mathrm{d}\sigma.
\end{aligned}
\tag{3.3.130}
$$

Above, the first equality comes from [112, (4.2.14)], whose applicability in the present setting is ensured by the fact that $u_\beta \mathbf{e}_j$ belongs to the space $\mathscr{A}$ of "admissible" vector fields, defined in (3.3.110). The second equality is simply the observation that $1-\eta = 1$ on $\partial\Omega$, while the third equality is implied by [112, (4.1.43)]. The fourth equality is a consequence of (3.3.111) and [112, Proposition 4.1.4]. Finally, the fifth equality is justified by relying on Lebesgue's Dominated Convergence Theorem (keeping in mind the estimate for ∇E from Theorem 1.4.2, and [112, Lemma 7.2.1]).

To treat the last term in (3.3.129), let us first consider the case when $\partial\Omega$ is bounded. We then have

$$
\begin{aligned}
\mathrm{II} &= -\lim_{R\to\infty} {}_{(\mathrm{Lip}_c(\partial\Omega))'}\Big\langle \big((1-\eta)E_{\gamma\alpha}(x-\cdot)\big)\big|_{\partial\Omega} \nu\bullet\big(a^{\alpha\beta}_{sr}(\partial_r u_\beta)\mathbf{e}_s\big),\, \phi_R\big|_{\partial\Omega}\Big\rangle_{\mathrm{Lip}_c(\partial\Omega)} \\
&= -\lim_{R\to\infty} {}_{(\mathrm{Lip}_c(\partial\Omega))'}\Big\langle \big(E_{\gamma\alpha}(x-\cdot)\big)\big|_{\partial\Omega} \nu\bullet\big(a^{\alpha\beta}_{sr}(\partial_r u_\beta)\mathbf{e}_s\big),\, \phi_R\big|_{\partial\Omega}\Big\rangle_{\mathrm{Lip}_c(\partial\Omega)} \\
&= -\lim_{R\to\infty} {}_{(\mathrm{Lip}_c(\partial\Omega))'}\Big\langle \nu\bullet\big(a^{\alpha\beta}_{sr}(\partial_r u_\beta)\mathbf{e}_s\big),\, \big(E_{\gamma\alpha}(x-\cdot)\phi_R\big)\big|_{\partial\Omega}\Big\rangle_{\mathrm{Lip}_c(\partial\Omega)} \\
&= -\lim_{R\to\infty} {}_{(\mathrm{Lip}_c(\partial\Omega))'}\Big\langle \big(\dot{\partial}^A_\nu u\big)_\alpha,\, \big(E_{\gamma\alpha}(x-\cdot)\phi_R\big)\big|_{\partial\Omega}\Big\rangle_{\mathrm{Lip}_c(\partial\Omega)} \\
&= -\lim_{R\to\infty} {}_{H^p(\partial\Omega,\sigma)}\Big\langle \big(\dot{\partial}^A_\nu u\big)_\alpha,\, \Big[\big(E_{\gamma\alpha}(x-\cdot)\phi_R\big)\big|_{\partial\Omega}\Big]\Big\rangle_{(H^p(\partial\Omega,\sigma))^*}.
\end{aligned}
\tag{3.3.131}
$$

Here, the first equality uses [112, (4.2.14)], whose present applicability is guaranteed by the fact that, for each fixed $\alpha\in\{1,\dots,M\}$, the vector field $a^{\alpha\beta}_{sr}(\partial_r u_\beta)\mathbf{e}_s$ belongs to the space $\mathscr{A}$ from (3.3.110). Indeed, the first two lines in (3.3.101) together with [112, Proposition 8.6.3] ensure that this is a divergence-free vector field whose components are absolutely integrable on bounded measurable subsets of Ω. The second equality relies on the observation that $1-\eta = 1$ on $\partial\Omega$, and the third equality is a consequence of [112, (4.1.43)]. The fourth equality is provided by [113, Definition 10.2.18], while the fifth equality follows on account of (3.3.104) and [113, Lemma 4.6.4]. Going further, we also claim that for each $\alpha\in\{1,\dots,M\}$ we have

$$\lim_{R\to\infty}\Big[\big(E_{\gamma\alpha}(x-\cdot)\phi_R\big)\big|_{\partial\Omega}\Big]=\big[E_{\gamma\alpha}(x-\cdot)\big|_{\partial\Omega}\big]\ \text{weak-}*\text{ in}$$

$$\big(H^p(\partial\Omega,\sigma)\big)^*=\begin{cases}\dot{\mathscr{C}}^{(n-1)(1/p-1)}(\partial\Omega)/\sim & \text{if } \frac{n-1}{n}<p<1,\\ \widetilde{\mathrm{BMO}}(\partial\Omega,\sigma) & \text{if } p=1,\\ L^{p'}(\partial\Omega,\sigma) & \text{if } p>1 \text{ and } p':=(1-1/p)^{-1}.\end{cases} \tag{3.3.132}$$

This claim is implied by the general weak-$*$ convergence results established in [113, Lemma 4.8.4] if $p<1$, [113, Lemma 4.8.1] (in the latter case also bearing in mind the trivial bounded embedding $L^\infty(\partial\Omega,\sigma)\hookrightarrow \mathrm{BMO}(\partial\Omega,\sigma)$) if $p=1$, and Lebesgue's Dominated Convergence Theorem (plus Theorem 1.4.2) if $1<p<n-1$. In turn, from (3.3.131), (3.3.132), the manner in which the boundary-to-domain single layer acts from Hardy spaces (cf. [115, §2.2]) if $\frac{n-1}{n}<p\le 1$, plus the definition of the boundary-to-domain single layer (cf. [115, §1.3]), and (2.5.549) (with $\alpha:=1$) if $1<p<n-1$, we conclude that

$$\begin{aligned}\mathrm{II}&=-{}_{H^p(\partial\Omega,\sigma)}\Big\langle\big(\dot{\partial}^A_\nu u\big)_\alpha,\ \Big[\big(E_{\gamma\alpha}(x-\cdot)\big)\big|_{\partial\Omega}\Big]\Big\rangle_{(H^p(\partial\Omega,\sigma))^*}\\ &=-\Big(\mathscr{S}(\dot{\partial}^A_\nu u)(x)\Big)_\gamma.\end{aligned} \tag{3.3.133}$$

The same type of argument proves (3.3.133) in the case when $\partial\Omega$ is bounded (a scenario in which the dual of H^p and the corresponding duality bracket no longer involve classes of functions modulo constants). Thus, in all cases, from (3.3.129), (3.3.130), and (3.3.133) we obtain

$$\big(\nu\bullet\vec{F}_x,1\big)_{\mathscr{F}}=-\int_{\partial\Omega}\nu\bullet\big(u_\alpha\mathbf{e}_s\big)a^{\beta\alpha}_{rs}(\partial_rE_{\gamma\beta})(x-\cdot)\,\mathrm{d}\sigma-\Big(\mathscr{S}(\dot{\partial}^A_\nu u)(x)\Big)_\gamma. \tag{3.3.134}$$

Granted this, [112, Theorem 1.9.4] applies and, on account of (3.3.118), (3.3.124), (3.3.127), the Divergence Formula [112, (1.9.32)] written for $\vec{F}_x$ presently gives

$$u_\gamma(x)=-\int_{\partial\Omega}\nu\bullet\big(u_\alpha\mathbf{e}_s\big)a^{\beta\alpha}_{rs}(\partial_rE_{\gamma\beta})(x-\cdot)\,\mathrm{d}\sigma-\Big(\mathscr{S}(\dot{\partial}^A_\nu u)(x)\Big)_\gamma. \tag{3.3.135}$$

In turn, from (3.3.135), (3.3.111), Theorem 2.5.1, the trace formula (lack of jump) for the single layer (cf. [115, §2.2]), and (3.3.104), we conclude that

the nontangential trace $u\big|^{\kappa-\mathrm{n.t.}}_{\partial\Omega}$ exists at σ-a.e. point on $\partial\Omega$ if either Ω is bounded, or $\partial\Omega$ is unbounded, or Ω is an exterior domain and the decay condition (3.3.106) holds. (3.3.136)

In fact, this Fatou-type property further self-improves. Specifically, in the case when Ω is an exterior domain, working with $\Omega_R:=B(0,R)\cap\Omega$ in place of Ω (where

$R \in (0, \infty)$ is a sufficiently large number), much as in the proof of Theorem 3.1.6, eliminates the need of asking that the decay condition (3.3.106) holds. Simply put, at this stage we have shown that, in all cases,

$$\text{the nontangential trace } u\big|_{\partial\Omega}^{\kappa-\text{n.t.}} \text{ exists } \sigma\text{-a.e. on } \partial\Omega. \tag{3.3.137}$$

With this in hand, we may now apply [113, Proposition 10.2.9] to each vector field $u_\alpha \mathbf{e}_s$ to conclude that if $\nu = (\nu_1, \dots, \nu_n)$ is the geometric measure theoretic outward unit normal to Ω then

$$\nu \bullet \big(u_\alpha \mathbf{e}_s\big) = \nu_s \big(u_\alpha\big|_{\partial\Omega}^{\kappa-\text{n.t.}}\big) \text{ for each } s \in \{1, \dots, n\} \text{ and } \alpha \in \{1, \dots, M\}. \tag{3.3.138}$$

Returning with (3.3.138) in (3.3.135) ultimately produces (3.3.105). Having proved (3.3.137), we may rely on [113, Definition 11.10.6] and [113, Corollary 11.10.11] to conclude that the claims in (3.3.103) hold as well. The claims about the existence and integrability properties of the nontangential trace $(\nabla u)\big|_{\partial\Omega}^{\kappa-\text{n.t.}}$ made in (3.3.102) are consequences of Theorem 3.3.4. The independence of the nontangential traces $u\big|_{\partial\Omega}^{\kappa-\text{n.t.}}$, $(\nabla u)\big|_{\partial\Omega}^{\kappa-\text{n.t.}}$ on the aperture parameter is a consequence of [112, Propositions 8.8.6, 8.9.8].

The next goal is to justify the claim made in (3.3.107). The fact that $\partial\Omega$ is currently assumed to be compact makes the Lebesgue scale on $\partial\Omega$ nested. Given the aims we have in mind, there is no loss of generality in assuming that actually $p \in \big(\frac{n-1}{n}, n-1\big)$. Define p^* as before. The idea is now to run the same proof as above with a few key adjustments. First, in place of (3.3.111) we now get that, for each $\beta \in \{1, \dots, M\}$ and $j \in \{1, \dots, n\}$,

$$\begin{gathered}\text{the distribution } \nu \bullet (u_\beta \mathbf{e}_j) \text{ belongs} \\ \text{to the Hardy space } H^{\min\{q,p^*\}}(\partial\Omega, \sigma),\end{gathered} \tag{3.3.139}$$

thanks to [113, Theorem 10.2.4], (3.3.114), and the definition of the scale of Hardy spaces given in [113, Definition 4.2.1]. Second, the same conclusion in (3.3.122) holds (with an identical proof, now using q in place of p^*). Third, in place of (3.3.130) we now have

$$\begin{aligned} \mathrm{I} &= -\lim_{R\to\infty} {}_{(\mathrm{Lip}_c(\partial\Omega))'}\Big\langle \nu \bullet \big(u_\alpha \mathbf{e}_s\big), a_{rs}^{\beta\alpha}\big((\partial_r E_{\gamma\beta})(x-\cdot)\phi_R\big)\big|_{\partial\Omega}\Big\rangle_{\mathrm{Lip}_c(\partial\Omega)} \\ &= -{}_{H^{\min\{q,p^*\}}(\partial\Omega,\sigma)}\Big\langle \nu \bullet \big(u_\alpha \mathbf{e}_s\big), a_{rs}^{\beta\alpha}(\partial_r E_{\gamma\beta})(x-\cdot)\big|_{\partial\Omega}\Big\rangle_{(H^{\min\{q,p^*\}}(\partial\Omega,\sigma))^*}. \end{aligned} \tag{3.3.140}$$

The last equality in (3.3.140) is a consequence of the membership in (3.3.139) and the fact that for each indexes $\beta \in \{1, \dots, M\}$ and $r \in \{1, \dots, n\}$, we have (reasoning as in (3.3.16), and also employing Lebesgue's Dominated Convergence Theorem together with Theorem 1.4.2)

$$(\partial_r E_{\gamma\beta})(x-\cdot)\big|_{\partial\Omega} = \lim_{R\to\infty} \big((\partial_r E_{\gamma\beta})(x-\cdot)\phi_R\big)\big|_{\partial\Omega}$$
$$\text{weak-}*\text{ in } \big(H^{\min\{q,p^*\}}(\partial\Omega,\sigma)\big)^*. \tag{3.3.141}$$

Hence, in place of (3.3.135) we now obtain

$$u_\gamma(x) = -_{H^{\min\{q,p^*\}}(\partial\Omega,\sigma)}\Big\langle \nu \bullet \big(u_\alpha \mathbf{e}_s\big),\, a^{\beta\alpha}_{rs}(\partial_r E_{\gamma\beta})(x-\cdot)\big|_{\partial\Omega}\Big\rangle_{(H^{\min\{q,p^*\}}(\partial\Omega,\sigma))^*}$$
$$-\Big(\mathscr{S}(\overset{\bullet}{\partial}{}^A_\nu u)(x)\Big)_\gamma. \tag{3.3.142}$$

Granted this representation formula, the conclusion in (3.3.136) follows on account of (3.3.139), Corollary 2.5.4 (if $q \le 1$) and Theorem 2.5.1 (if $q > 1$), the trace formula (lack of jump) for the single layer (cf. [115, §2.2]), and (3.3.104). Everything else is as before.

Finally, a cursory inspection reveals that the proof given above also works in the case $n = 2$ under the additional assumption that Ω is bounded. □

In the corollary below we present a Fatou-type result for a null-solution u of a weakly elliptic system L in which no a priori assumptions are made on the nontangential maximal function of u. Compared with Theorem 3.3.9, we shall strengthen the geometric assumptions by demanding that the given UR domain is bounded and also happens to be a one-sided NTA domain. A related, more nuanced, result is proved later, in Theorem 3.4.4, in the class of bounded two-sided NTA domains with an Ahlfors regular boundary.

Corollary 3.3.10 *Let $\Omega \subseteq \mathbb{R}^n$, where $n \in \mathbb{N}$ with $n \ge 2$, be a* UR *domain with compact boundary which is also a one-sided* NTA *domain (for example, any* NTA *domain with a compact Ahlfors regular boundary will do). Abbreviate $\sigma := \mathcal{H}^{n-1}\lfloor\partial\Omega$ and fix some aperture parameter $\kappa > 0$. Also, for some $M \in \mathbb{N}$, consider a weakly elliptic, homogeneous, constant complex coefficient, second-order $M \times M$ system L in $\mathbb{R}^n$. Finally, fix a truncation parameter $\varepsilon > 0$ and assume u is a vector-valued function satisfying*

$$u \in \big[\mathscr{C}^\infty(\Omega)\big]^M, \quad Lu = 0 \text{ in } \Omega, \text{ and}$$
$$\mathcal{N}^\varepsilon_\kappa(\nabla u) \in L^p(\partial\Omega,\sigma) \text{ for some } p \in \big(\tfrac{n-1}{n}, \infty\big]. \tag{3.3.143}$$

Then $\mathcal{N}^\varepsilon_\kappa u$ also belongs to $L^p(\partial\Omega,\sigma)$ and the nontangential traces $u\big|^{\kappa-\text{n.t.}}_{\partial\Omega}$ and $(\nabla u)\big|^{\kappa-\text{n.t.}}_{\partial\Omega}$ exist at σ-a.e. point on $\partial\Omega$.

Proof Suppose first that Ω is bounded. Since u and ∇u are locally bounded, from assumptions and [112, (8.4.109)] we conclude that $\mathcal{N}_\kappa u$ and $\mathcal{N}_\kappa(\nabla u)$ belong to $L^p(\partial\Omega,\sigma)$. This permits us to invoke (3.3.107) in Theorem 3.3.9 (with $q := p$) to conclude that the nontangential traces $u\big|^{\kappa-\text{n.t.}}_{\partial\Omega}$ and $(\nabla u)\big|^{\kappa-\text{n.t.}}_{\partial\Omega}$ exist at σ-a.e. point on $\partial\Omega$. Finally, in the situation when Ω is unbounded, apply what we have just proved to the bounded set $\Omega \cap B(0,R)$ with $R > 2\max\{|x| : x \in \partial\Omega\}$. □

We close this section by discussing a version of Theorem 3.3.9 adapted to the scale of Morrey spaces.

Theorem 3.3.11 *Suppose $\Omega \subseteq \mathbb{R}^n$, where $n \in \mathbb{N}$ with $n \geq 3$, is an arbitrary* UR *domain, and abbreviate $\sigma := \mathcal{H}^{n-1}\lfloor\partial\Omega$. For some $M \in \mathbb{N}$, consider an $M \times M$ weakly elliptic, homogeneous, second-order system $L = L_A$ in $\mathbb{R}^n$, associated with some coefficient tensor A as in* (1.3.15). *Also, fix some aperture parameter $\kappa > 0$ and consider a vector-valued function $u = (u_\beta)_{1\leq\beta\leq M}$ satisfying*

$$\begin{aligned} &u \in \big[\mathscr{C}^\infty(\Omega)\big]^M, \quad Lu = 0 \text{ in } \Omega, \quad \mathcal{N}_\kappa(\nabla u) \in M^{p,\lambda}(\partial\Omega,\sigma) \\ &\qquad \text{for some } \lambda \in \big(0, \tfrac{(n-1)^2}{n}\big) \text{ and } p \in \big(\tfrac{n-1}{n}, n-1-\lambda\big), \\ &\text{and } \mathcal{N}_\kappa u \in M^{p_\star,\lambda}(\partial\Omega,\sigma) \text{ with } p_\star := \big(\tfrac{1}{p} - \tfrac{1}{n-1-\lambda}\big)^{-1}. \end{aligned} \tag{3.3.144}$$

Then

$$\begin{aligned} &\text{the nontangential traces } u\big|_{\partial\Omega}^{\kappa-\text{n.t.}}, \ (\nabla u)\big|_{\partial\Omega}^{\kappa-\text{n.t.}} \text{ exist } \sigma\text{-a.e. on } \partial\Omega, \\ &u\big|_{\partial\Omega}^{\kappa-\text{n.t.}} \in \big[M^{p_\star,\lambda}(\partial\Omega,\sigma)\big]^M \text{ and } (\nabla u)\big|_{\partial\Omega}^{\kappa-\text{n.t.}} \in \big[M^{p,\lambda}(\partial\Omega,\sigma)\big]^{n\cdot M}, \\ &\Big\|u\big|_{\partial\Omega}^{\kappa-\text{n.t.}}\Big\|_{[M^{p_\star,\lambda}(\partial\Omega,\sigma)]^M} \leq \|\mathcal{N}_\kappa u\|_{M^{p_\star,\lambda}(\partial\Omega,\sigma)} \text{ and} \\ &\Big\|(\nabla u)\big|_{\partial\Omega}^{\kappa-\text{n.t.}}\Big\|_{[M^{p,\lambda}(\partial\Omega,\sigma)]^{n\cdot M}} \leq \|\mathcal{N}_\kappa(\nabla u)\|_{M^{p,\lambda}(\partial\Omega,\sigma)}. \end{aligned} \tag{3.3.145}$$

Moreover, for any other aperture parameter $\kappa' \in (0,\infty)$ the nontangential boundary traces $u\big|_{\partial\Omega}^{\kappa'-\text{n.t.}}$, $(\nabla u)\big|_{\partial\Omega}^{\kappa'-\text{n.t.}}$ also exist and agree with $u\big|_{\partial\Omega}^{\kappa-\text{n.t.}}$ and $(\nabla u)\big|_{\partial\Omega}^{\kappa-\text{n.t.}}$, respectively, at σ-a.e. point on $\partial\Omega$. Also, for some $C = C(\Omega,\kappa,p,\lambda) \in (0,\infty)$, it follows that

$$\begin{aligned} &u\big|_{\partial\Omega}^{\kappa-\text{n.t.}} \text{ lies in the off-diagonal Morrey-based Sobolev space } \big[M_1^{p_\star,p,\lambda}(\partial\Omega,\sigma)\big]^M \\ &\text{and } \Big\|u\big|_{\partial\Omega}^{\kappa-\text{n.t.}}\Big\|_{[M_1^{p_\star,p,\lambda}(\partial\Omega,\sigma)]^M} \leq C\|\mathcal{N}_\kappa u\|_{M^{p_\star,\lambda}(\partial\Omega,\sigma)} + C\|\mathcal{N}_\kappa(\nabla u)\|_{M^{p,\lambda}(\partial\Omega,\sigma)} \end{aligned} \tag{3.3.146}$$

(*cf.* [113, (11.7.14)-(11.7.15)]), *while the weak conormal derivative of u associated with the coefficient tensor A as in* [113, Definition 10.2.18] *satisfies*

$$\begin{aligned} &\mathring{\partial}^A_\nu u \text{ belongs to } \big[M^{p,\lambda}(\partial\Omega,\sigma)\big]^M \text{ and} \\ &\big\|\mathring{\partial}^A_\nu u\big\|_{[M^{p,\lambda}(\partial\Omega,\sigma)]^M} \leq C\big\|\mathcal{N}_\kappa(\nabla u)\big\|_{M^{p,\lambda}(\partial\Omega,\sigma)}, \end{aligned} \tag{3.3.147}$$

for some constant $C \in (0,\infty)$ which depends only on $\Omega, n, A, \kappa, p, \lambda$.

Finally, with the single and double layer potential operators $\mathscr{S}$, $\mathcal{D}$, associated with A and Ω as in [115, §1.3], *one has the integral representation formula*

$$u = \mathcal{D}\big(u\big|_{\partial\Omega}^{\kappa-\text{n.t.}}\big) - \mathscr{S}(\mathring{\partial}^A_\nu u) \text{ in } \Omega, \tag{3.3.148}$$

with the agreement that if Ω is an exterior domain one also assumes that also assumes that there exists $C \in (1, \infty)$ such that

$$\fint_{B(0,CR)\setminus B(0,R)} |u| \, d\mathcal{L}^n = o(1) \text{ as } R \to \infty. \tag{3.3.149}$$

Proof We largely proceed as in the proof of Theorem 3.3.9, so we will only indicate the main changes needed to carry out the same game-plan. First, Theorem 3.3.7 applies and gives that the nontangential boundary trace $(\nabla u)\big|_{\partial\Omega}^{\kappa-\text{n.t.}}$ exists σ-a.e. on $\partial\Omega$, and the function $(\nabla u)\big|_{\partial\Omega}^{\kappa-\text{n.t.}}$ is independent of the aperture parameter, belongs to the Morrey space $\big[M^{p,\lambda}(\partial\Omega,\sigma)\big]^{n\cdot M}$, and satisfy the estimate $\big\|(\nabla u)\big|_{\partial\Omega}^{\kappa-\text{n.t.}}\big\|_{[M^{p,\lambda}(\partial\Omega,\sigma)]^{n\cdot M}} \leq \|\mathcal{N}_\kappa(\nabla u)\|_{M^{p,\lambda}(\partial\Omega,\sigma)}$. Second, the claims in (3.3.147) are consequences of item *(5)* in [113, Theorem 10.2.24] and the first two lines in (3.3.144). For further reference, from (3.3.147) and [112, (6.2.23)] it is useful to note that

$$\overset{\bullet}{\partial}{}^A_\nu u \in M^{p_\star,\lambda}(\partial\Omega,\sigma) \hookrightarrow L^1\Big(\partial\Omega, \frac{\sigma(y)}{1+|y|^{n-2}}\Big). \tag{3.3.150}$$

To proceed, fix $\beta \in \{1, \dots, M\}$ and $j \in \{1, \dots, n\}$. As before, $\nu \bullet (u_\beta \mathbf{e}_j)$ makes sense as a distribution in $\big(\mathrm{Lip}_c(\partial\Omega)\big)'$ and we claim that

$$\begin{gathered} \nu \bullet (u_\beta \mathbf{e}_j) \text{ belongs to the Morrey space } M^{p_\star,\lambda}(\partial\Omega,\sigma) \text{ and satisfies} \\ \big\|\nu \bullet (u_\beta \mathbf{e}_j)\big\|_{M^{p_\star,\lambda}(\partial\Omega,\sigma)} \leq C\|\mathcal{N}_\kappa u\|_{M^{p_\star,\lambda}(\partial\Omega,\sigma)} + C\|\mathcal{N}_\kappa(\nabla u)\|_{M^{p,\lambda}(\partial\Omega,\sigma)} \end{gathered} \tag{3.3.151}$$

for some constant $C \in (0, \infty)$ depending only on $\Omega, n, \kappa, p, \lambda$. To justify this, observe that in a pointwise sense we have $\mathcal{N}_\kappa(u_\beta \mathbf{e}_j) \leq \mathcal{N}_\kappa u$ which, in view of (3.3.144), [113, (6.2.3)], and [112, Proposition 8.2.3], implies

$$\begin{gathered} \mathcal{N}_\kappa(u_\beta \mathbf{e}_j) \in M^{p_\star,\lambda}(\partial\Omega,\sigma) \text{ and} \\ \big\|\mathcal{N}_\kappa(u_\beta \mathbf{e}_j)\big\|_{M^{p_\star,\lambda}(\partial\Omega,\sigma)} \leq \big\|\mathcal{N}_\kappa u\big\|_{M^{p_\star,\lambda}(\partial\Omega,\sigma)}. \end{gathered} \tag{3.3.152}$$

Next, observe that $\mathrm{div}(u_\beta \mathbf{e}_j) = \partial_j u_\beta$ hence

$$\mathfrak{P}\big(\mathrm{div}(u_\beta \mathbf{e}_j)\big) \leq \mathfrak{P}(\nabla u). \tag{3.3.153}$$

In addition, from the last property in the top line of (3.3.144) and [113, (10.1.11)-(10.1.12)] (used with ∇u in place of u) we see that

$$\begin{gathered} \mathfrak{P}(\nabla u) \text{ belongs to } M^{p_\star,\lambda}(\partial\Omega,\sigma) \text{ and} \\ \big\|\mathfrak{P}(\nabla u)\big\|_{M^{p_\star,\lambda}(\partial\Omega,\sigma)} \leq C\big\|\mathcal{N}_\kappa(\nabla u)\big\|_{M^{p,\lambda}(\partial\Omega,\sigma)}. \end{gathered} \tag{3.3.154}$$

Collectively, (3.3.153), (3.3.154), and [113, (6.2.3), (10.1.3)] establish that

$$\mathfrak{P}\big(\mathrm{div}(u_\beta \mathbf{e}_j)\big) \text{ belongs to } M^{p_\star,\lambda}(\partial\Omega,\sigma) \text{ and}$$
$$\big\|\mathfrak{P}\big(\mathrm{div}(u_\beta \mathbf{e}_j)\big)\big\|_{M^{p_\star,\lambda}(\partial\Omega,\sigma)} \leq C\big\|\mathcal{N}_\kappa(\nabla u)\big\|_{M^{p,\lambda}(\partial\Omega,\sigma)}. \tag{3.3.155}$$

At this stage, from (3.3.152), (3.3.155), and [113, Corollary 10.2.7] we conclude that the claims in (3.3.151) hold.

Pressing on, fix an arbitrary index $\gamma \in \{1,\dots,M\}$, pick some arbitrary point $x \in \Omega$, and define the vector field $\vec{F}_x = \big(F_s\big)_{1\leq s\leq n}$ with components as in (3.3.116). This continues to satisfy (3.3.117)-(3.3.119). With the family $\mathscr{F} := \{\phi_R\}_{R>0}$ as in (3.3.120) and with $[\vec{F}_x]_{\mathscr{F}}$ defined as in (3.3.121), we claim that

$$[\vec{F}_x]_{\mathscr{F}} = 0. \tag{3.3.156}$$

To see that this is the case, assume first that either Ω is bounded, or $\partial\Omega$ is unbounded. Also, without any loss of generality assume that $0 \in \partial\Omega$. In such a scenario, we may rely on [112, Proposition 8.6.3], (3.3.144), Hölder's inequality, [112, (8.2.26)], [112, (8.1.17)], and (A.0.87) to estimate (for $R \in (0,\infty)$ large and with $C \in (0,\infty)$ presently denoting what used to be λ in the proof of Theorem 3.3.9)

$$\begin{aligned}
R^{-1}&\int_\Omega |(\nabla\phi)(y/R)||(\nabla E)(x-y)||u(y)|\,\mathrm{d}\mathcal{L}^n(y)\\
&\leq CR^{-n}\int_{\Omega\cap[B(0,CR)\setminus B(0,R)]}|u|\,\mathrm{d}\mathcal{L}^n\\
&\leq CR^{-n}\Big(\int_{\Omega\cap B(0,CR)}|u|^{\frac{np_\star}{n-1}}\,\mathrm{d}\mathcal{L}^n\Big)^{\frac{n-1}{np_\star}}\big(R^n\big)^{1-\frac{n-1}{np_\star}}\\
&\leq CR^{-\frac{n-1}{p_\star}}\Big(\int_{\partial\Omega\cap B(0,C(2+\kappa)R)}|\mathcal{N}_\kappa u|^{p_\star}\,\mathrm{d}\sigma\Big)^{1/p_\star}\\
&\leq C\Big(⨍_{\partial\Omega\cap B(0,C(2+\kappa)R)}|\mathcal{N}_\kappa u|^{p_\star}\,\mathrm{d}\sigma\Big)^{1/p_\star}\\
&\leq CR^{-\frac{n-1-\lambda}{p_\star}}\big\|\mathcal{N}_\kappa u\big\|_{M^{p_\star,\lambda}(\partial\Omega,\sigma)}\\
&= o(1) \text{ as } R\to\infty,
\end{aligned} \tag{3.3.157}$$

since $-\frac{n-1-\lambda}{p_\star} < 0$ given that $\lambda < n-1$, and

$$
\begin{aligned}
R^{-1}\int_{\Omega}|(\nabla\phi)(y/R)||E(x-y)||(\nabla u)(y)|\,\mathrm{d}\mathcal{L}^n(y)& \\
&\leq CR^{-(n-1)}\int_{\Omega\cap[B(0,CR)\setminus B(0,R)]}|\nabla u|\,\mathrm{d}\mathcal{L}^n \\
&\leq CR^{-(n-1)}\Big(\int_{\Omega\cap B(0,CR)}|\nabla u|^{\frac{np}{n-1}}\,\mathrm{d}\mathcal{L}^n\Big)^{\frac{n-1}{np}}(R^n)^{1-\frac{n-1}{np}} \\
&\leq CR^{1-\frac{n-1}{p}}\Big(\int_{\partial\Omega\cap B(0,C(2+\kappa)R)}|\mathcal{N}_\kappa(\nabla u)|^p\,\mathrm{d}\sigma\Big)^{1/p} \\
&\leq CR\Big(\fint_{\partial\Omega\cap B(0,C(2+\kappa)R)}|\mathcal{N}_\kappa(\nabla u)|^p\,\mathrm{d}\sigma\Big)^{1/p} \\
&\leq CR^{1-\frac{n-1-\lambda}{p}}\|\mathcal{N}_\kappa(\nabla u)\|_{M^{p,\lambda}(\partial\Omega,\sigma)} \\
&= o(1) \text{ as } R\to\infty,
\end{aligned}
\tag{3.3.158}
$$

since $1-\frac{n-1-\lambda}{p}<0$ given that $p<n-1-\lambda$. Together, (3.3.157)-(3.3.158) establish (3.3.156) when either Ω is bounded, or $\partial\Omega$ is unbounded. The justification of (3.3.156) in the case when Ω is an exterior domain is similar to (3.3.125)-(3.3.126), based on interior estimates and (3.3.149). This finishes the proof of (3.3.156).

We then follow the computations in (3.3.128)-(3.3.131) almost verbatim, the only two significant differences being as follows. First, the last equality in (3.3.130) now uses the fact that, as seen from (3.3.151), [112, (6.2.23)] and the current assumptions on $p, \lambda, p_\star$,

$$
\nu\bullet(u_\beta\mathbf{e}_j)\in M^{p_\star,\lambda}(\partial\Omega,\sigma)\hookrightarrow L^1\Big(\partial\Omega,\frac{\sigma(y)}{1+|y|^{n-1}}\Big), \tag{3.3.159}
$$

that there exists $C=C_x\in(0,\infty)$ such that (cf. Theorem 1.4.2)

$$
|(\nabla E)(x-y)|\leq\frac{C}{1+|y|^{n-1}}\quad\text{for each } y\in\partial\Omega, \tag{3.3.160}
$$

and Lebesgue's Dominated Convergence Theorem. Hence, we once again have

$$
\mathrm{I}=-\int_{\partial\Omega}\nu\bullet\big(u_\alpha\mathbf{e}_s\big)a^{\beta\alpha}_{rs}(\partial_r E_{\gamma\beta})(x-\cdot)\,\mathrm{d}\sigma. \tag{3.3.161}
$$

Second, in place of the last equality in (3.3.131) we now have

$$\begin{aligned}
\mathrm{II} &= -\lim_{R\to\infty} {}_{(\mathrm{Lip}_c(\partial\Omega))'}\Big\langle \big(\partial^A_\nu u\big)_\alpha, \big(E_{\gamma\alpha}(x-\cdot)\phi_R\big)\big|_{\partial\Omega}\Big\rangle_{\mathrm{Lip}_c(\partial\Omega)} \\
&= -\lim_{R\to\infty}\int_{\partial\Omega}\big\langle \big(\partial^A_\nu u\big)_\alpha, E_{\gamma\alpha}(x-\cdot)\phi_R\big\rangle\,\mathrm{d}\sigma \\
&= -\int_{\partial\Omega}\big\langle \big(\partial^A_\nu u\big)_\alpha, E_{\gamma\alpha}(x-\cdot)\big\rangle\,\mathrm{d}\sigma = \mathscr{S}(\partial^A_\nu u)(x),
\end{aligned} \tag{3.3.162}$$

thanks to (3.3.150), [112, (4.1.47)], the fact that there exists $C = C_x \in (0,\infty)$ such that (cf. Theorem 1.4.2)

$$|E(x-y)| \leq \frac{C}{1+|y|^{n-2}} \quad \text{for each } y \in \partial\Omega, \tag{3.3.163}$$

Lebesgue's Dominated Convergence Theorem, and the definition of the boundary-to-domain single layer (cf. [115, §1.3]).

Granted (3.3.161)-(3.3.162), we may rely on [112, Theorem 1.9.4] to conclude, as in (3.3.135), that

$$u_\gamma(x) = -\int_{\partial\Omega}\nu\bullet\big(u_\alpha\mathbf{e}_s\big)a^{\beta\alpha}_{rs}(\partial_r E_{\gamma\beta})(x-\cdot)\,\mathrm{d}\sigma - \Big(\mathscr{S}(\partial^A_\nu u)(x)\Big)_\gamma. \tag{3.3.164}$$

In turn, from (3.3.164), (3.3.159), Theorem 2.5.1, (3.3.150), and the existence of the nontangential boundary trace for the single layer (cf. [115, §2.2]) we conclude that the nontangential trace $u\big|^{\kappa-\mathrm{n.t.}}_{\partial\Omega}$ exists at σ-a.e. point on $\partial\Omega$ if either Ω is bounded, or $\partial\Omega$ is unbounded, or Ω is an exterior domain and the decay condition (3.3.149) holds. With this in hand, the same argument that has led to (3.3.137) once again shows that, in all cases,

$$\text{the nontangential trace } u\big|^{\kappa-\mathrm{n.t.}}_{\partial\Omega} \text{ exists } \sigma\text{-a.e. on } \partial\Omega. \tag{3.3.165}$$

In view of this property, we may invoke [113, Proposition 10.2.9] to conclude that if $\nu = (\nu_1, \dots, \nu_n)$ is the geometric measure theoretic outward unit normal to Ω then

$$\begin{gathered}
\nu\bullet\big(u_\alpha\mathbf{e}_s\big) = \nu_s\big(u_\alpha\big|^{\kappa-\mathrm{n.t.}}_{\partial\Omega}\big) \text{ for each} \\
s \in \{1,\dots,n\} \text{ and } \alpha \in \{1,\dots,M\}.
\end{gathered} \tag{3.3.166}$$

With (3.3.166) in hand, the integral representation formula (3.3.164) reduces precisely to (3.3.148). Let us also note that, on account of (3.3.165), the last line in (3.3.144), [113, (6.2.3)], and [112, Corollary 8.9.6], we have

$$\begin{gathered}
u\big|^{\kappa-\mathrm{n.t.}}_{\partial\Omega} \in \big[M^{p_\star,\lambda}(\partial\Omega,\sigma)\big]^M \text{ and} \\
\big\|u\big|^{\kappa-\mathrm{n.t.}}_{\partial\Omega}\big\|_{[M^{p_\star,\lambda}(\partial\Omega,\sigma)]^M} \leq \|\mathcal{N}_\kappa u\|_{M^{p_\star,\lambda}(\partial\Omega,\sigma)}.
\end{gathered} \tag{3.3.167}$$

Let us now fix $j, k \in \{1, \dots, n\}$, arbitrary. On the one hand, from the first part of the current proof we know that the functions

$$\begin{aligned} &(\partial_j u)\Big|_{\partial\Omega}^{\kappa-\text{n.t.}}, (\partial_k u)\Big|_{\partial\Omega}^{\kappa-\text{n.t.}} \text{ belong to } \big[M^{p,\lambda}(\partial\Omega, \sigma)\big]^M \\ &\text{and each has norm } \leq \|\mathcal{N}_\kappa(\nabla u)\|_{M^{p,\lambda}(\partial\Omega,\sigma)}. \end{aligned} \tag{3.3.168}$$

On the other hand, from (3.3.144), the second inclusion in [113, (6.2.7)], and [113, Proposition 11.3.2] we see that $u\Big|_{\partial\Omega}^{\kappa-\text{n.t.}}$ belongs to $\big[L^{p_\star,p}_{1,\text{loc}}(\partial\Omega, \sigma)\big]^M$ and

$$\partial_{\tau_{jk}}\Big(u\Big|_{\partial\Omega}^{\kappa-\text{n.t.}}\Big) = \nu_j\Big((\partial_k u)\Big|_{\partial\Omega}^{\kappa-\text{n.t.}}\Big) - \nu_k\Big((\partial_j u)\Big|_{\partial\Omega}^{\kappa-\text{n.t.}}\Big) \text{ at } \sigma\text{-a.e. point on } \partial\Omega. \tag{3.3.169}$$

Combining (3.3.168), (3.3.169), and [113, (6.2.3)] proves that

$$\begin{aligned} &\partial_{\tau_{jk}}\Big(u\Big|_{\partial\Omega}^{\kappa-\text{n.t.}}\Big) \in \big[M^{p,\lambda}(\partial\Omega, \sigma)\big]^M \text{ and} \\ &\Big\|\partial_{\tau_{jk}}\Big(u\Big|_{\partial\Omega}^{\kappa-\text{n.t.}}\Big)\Big\|_{[M^{p,\lambda}(\partial\Omega,\sigma)]^M} \leq C\|\mathcal{N}_\kappa(\nabla u)\|_{M^{p,\lambda}(\partial\Omega,\sigma)}. \end{aligned} \tag{3.3.170}$$

Together with (3.3.167), this establishes the claims in (3.3.146). The proof of Theorem 3.3.11 is therefore complete. □

3.4 Fatou-Type Theorems in Two-Sided NTA Domains for Second-Order Systems

Our first result in this section reveals that, for a null-solution of a weakly elliptic system, having sufficient regularity on the Sobolev/Besov/Triebel-Lizorkin scales in a given two-sided NTA domain with a compact Ahlfors regular boundary guarantees the existence of its nontangential pointwise trace on the boundary of said domain.

Theorem 3.4.1 *Let $\Omega \subseteq \mathbb{R}^n$ be a two-sided* NTA *domain with a compact Ahlfors regular boundary. Abbreviate $\sigma := \mathcal{H}^{n-1}\lfloor\partial\Omega$ and fix an aperture parameter $\kappa > 0$. Also, let L be a homogeneous, constant (complex) second-order $M \times M$ weakly elliptic system in $\mathbb{R}^n$, and consider a null-solution $u \in \big[\mathscr{C}^\infty(\Omega)\big]^M$ of the system L in Ω. Then*

$$\begin{aligned} &\textit{the nontangential boundary trace } u\Big|_{\partial\Omega}^{\kappa-\text{n.t.}} \textit{ exists (in } \mathbb{C}^M\textit{) at } \sigma\textit{-a.e. point} \\ &\textit{on } \partial\Omega, \textit{ and the function } u\Big|_{\partial\Omega}^{\kappa-\text{n.t.}} \textit{ is } \sigma\textit{-measurable on } \partial\Omega, \end{aligned} \tag{3.4.1}$$

provided either

$$\begin{aligned} &u \textit{ belongs to } \big[A^{p,q}_\alpha(\Omega)_{\text{bdd}}\big]^M \textit{ with } A \in \{B, F\} \textit{ and} \\ &0 < p < \infty, \quad 0 < q \leq \infty, \quad (n-1)\big(\tfrac{1}{p} - 1\big)_+ + \tfrac{1}{p} < \alpha, \end{aligned} \tag{3.4.2}$$

or

$$u \in \big[W^{1,p}_a(\Omega)_{\rm bdd}\big]^M \ \textit{with}\ p \in (1,\infty)\ \textit{and}\ a \in \big(-\tfrac{1}{p}, 1-\tfrac{1}{p}\big). \tag{3.4.3}$$

Moreover, in either of the scenarios described in (3.4.2), (3.4.3) *the nontangential boundary trace* $u\big|_{\partial\Omega}^{\kappa-\mathrm{n.t.}}$ *actually agrees with the trace considered in the sense of Sobolev/Besov/Triebel-Lizorkin spaces, i.e.,*

$$\big(u\big|_{\partial\Omega}^{\kappa-\mathrm{n.t.}}\big)(x) = \lim_{r\to 0^+} \fint_{B(x,r)\cap\Omega} u\,d\mathcal{L}^n \ \textit{at}\ \sigma\textit{-a.e.}\ x \in \partial\Omega, \tag{3.4.4}$$

and for each truncation parameter $\varepsilon > 0$ *one has*

$$\mathcal{N}^{\varepsilon}_{\kappa} u \in L^{p_o}(\partial\Omega,\sigma)\ \textit{for some}\ p_o \in (1,\infty). \tag{3.4.5}$$

Before presenting the proof of Theorem 3.4.1 we make a few remarks further elaborating on the nature and scope of this result.

Remark 3.4.2 A careful inspection of the argument that yields (3.4.5) gives a more precise version of this result, namely

$$\begin{gathered}\textit{in the scenario described in (3.4.2) we have } \mathcal{N}^{\varepsilon}_{\kappa} u \in L^{p_o}(\partial\Omega,\sigma) \textit{ for any}\\ p_o < \tfrac{n-1}{\left(\frac{n}{p}-\alpha\right)_+} \textit{ if } p \in (0,1], \textit{ and any } p_o < \tfrac{n-1}{\left(\frac{n-1}{p}-\min\left\{1,\alpha-\frac{1}{p}\right\}\right)_+} \textit{ if } p \in (1,\infty),\end{gathered} \tag{3.4.6}$$

and

$$\begin{gathered}\textit{in the scenario described in (3.4.3) we have}\\ \mathcal{N}^{\varepsilon}_{\kappa} u \in L^{p_o}(\partial\Omega,\sigma) \textit{ for any } p_o < \tfrac{n-1}{\left(\frac{n}{p}-1+a\right)_+}.\end{gathered} \tag{3.4.7}$$

Remark 3.4.3 It suffices to assume Ω is an NTA domain satisfying an exterior local John condition and whose boundary is a compact Ahlfors regular set.

The proof of Theorem 3.4.1, given below, makes essential use of Green's integral representation formula from [115, §4.4] and the boundary behavior of layer potential operators.

Proof of Theorem 3.4.1 By eventually replacing Ω by $\Omega \cap B(0,R)$ for some large $R \in (0,\infty)$, there is no loss of generality in assuming that Ω is actually a bounded two-sided NTA domain with an Ahlfors regular boundary. Thanks to [112, (5.11.66)], we then see that $\Omega \subseteq \mathbb{R}^n$ is a bounded (ε,δ)-domain. Consider the scenario described in (3.4.2) with $A := B$. Also, from [113, Corollary 9.2.1], [113, Theorems 9.1.1, 9.1.3, 9.1.4], and the present assumptions on p, q, α we conclude that there exist $p_o, q_o \in (1,\infty)$ and $s_o \in (0,1)$ with the property that

$$B^{p,q}_{\alpha}(\Omega) \hookrightarrow B^{p_o,q_o}_{s_o+\frac{1}{p_o}}(\Omega). \tag{3.4.8}$$

As such, work in [115, §4.4] guarantees that we have the integral representation formula

$$u = \mathcal{D}\big(\mathrm{Tr}_{\Omega\to\partial\Omega}\, u\big) - \mathcal{S}\big(\partial_\nu^A(u,0)\big) \ \text{ in } \Omega, \tag{3.4.9}$$

where A is a coefficient tensor used to write the given system as $L = L_A$, where $\mathcal{D}$, $\mathcal{S}$ are the boundary-to-domain double and single layer potential operators associated with A and Ω, where

$$\mathrm{Tr}_{\Omega\to\partial\Omega}\, u \in \big[B^{p_o,q_o}_{s_o}(\partial\Omega,\sigma)\big]^M \hookrightarrow \big[L^{p_o}(\partial\Omega,\sigma)\big]^M, \tag{3.4.10}$$

with the last inclusion provided by [113, (7.9.10)], and where

$$\partial_\nu^A(u,0) \in \big[B^{p_o,q_o}_{s_o-1}(\partial\Omega,\sigma)\big]^M \hookrightarrow \big[L^{p_o}_{-1}(\partial\Omega,\sigma)\big]^M, \tag{3.4.11}$$

with the last inclusion provided by [113, Proposition 11.11.3] (bearing in mind that the two-sided NTA domain Ω satisfies a two-sided local John condition; cf. [112, (5.11.28)]).

In turn, from (3.4.9), (3.4.10), (3.4.11), as well as the Calderón-Zygmund theory developed in relation to the boundary-to-domain double and single layer potential operators established in [115, §1.5] we conclude that the nontangential boundary trace $u\big|^{\kappa-\mathrm{n.t.}}_{\partial\Omega}$ exists at σ-a.e. point on $\partial\Omega$ and, in fact,

$$u\big|^{\kappa-\mathrm{n.t.}}_{\partial\Omega} = \big(\tfrac{1}{2}I + K\big)\big(\mathrm{Tr}_{\Omega\to\partial\Omega}\, u\big) - S\big(\partial_\nu^A(u,0)\big) \in \big[L^{p_o}(\partial\Omega,\sigma)\big]^M. \tag{3.4.12}$$

In particular, the function $u\big|^{\kappa-\mathrm{n.t.}}_{\partial\Omega}$ is σ-measurable on $\partial\Omega$. This concludes the proof of (3.4.1) under the hypotheses stipulated in (3.4.2) with $A := B$. All other scenarios described in (3.4.2), (3.4.3) are dealt with similarly, once again relying on Green's integral representation formulas derived in [115, §4.4].

Next, based on [113, Theorems 8.3.6, 9.4.5] and [113, (8.3.64), (9.4.99)] we conclude that (3.4.4) holds. Finally, the claim made in (3.4.5) is seen from (3.4.9)-(3.4.11), and the nontangential maximal function estimates for the double and single layers proved in [115, §1.5]. □

In turn, we may use Theorem 3.4.1 to establish the following companion result for Theorem 3.3.9.

Theorem 3.4.4 *Suppose $\Omega \subseteq \mathbb{R}^n$ (where $n \in \mathbb{N}$, $n \geq 3$) is a bounded two-sided* NTA *domain with an Ahlfors regular boundary. Abbreviate $\sigma := \mathcal{H}^{n-1}\lfloor\partial\Omega$ and fix an aperture parameter $\kappa > 0$. In addition, let L be a homogeneous, constant (complex) second-order $M \times M$ weakly elliptic system in $\mathbb{R}^n$, and consider a null-solution $u \in \big[\mathscr{C}^\infty(\Omega)\big]^M$ of the system L in Ω with the property that $\mathcal{N}_\kappa(\nabla u) \in L^p(\partial\Omega,\sigma)$ for some $p \in \big(\frac{n-1}{n}, \infty\big]$. Then the nontangential boundary traces*

$$u\big|^{\kappa-\mathrm{n.t.}}_{\partial\Omega} \ \textit{and}\ (\nabla u)\big|^{\kappa-\mathrm{n.t.}}_{\partial\Omega} \ \textit{exist at } \sigma\textit{-a.e. point on } \partial\Omega, \tag{3.4.13}$$

and

$$\mathcal{N}_\kappa u \in \begin{cases} L^{p^*}(\partial\Omega,\sigma) \text{ if } p \in \left(\frac{n-1}{n}, n-1\right) \text{ and } p^* := \left(\frac{1}{p} - \frac{1}{n-1}\right)^{-1}, \\ L^q(\partial\Omega,\sigma) \text{ if } p = n-1 \text{ and } q \in (0,\infty), \\ L^\infty(\partial\Omega,\sigma) \text{ if } p \in (n-1,\infty]. \end{cases} \tag{3.4.14}$$

An inspection of proof given below reveals that it suffices to assume Ω is a bounded NTA domain satisfying an exterior local John condition and whose boundary is Ahlfors regular.

Proof of Theorem 3.4.4 Since the current hypotheses imply that Ω is a UR domain (cf. [112, (5.2.4), (5.10.24), Definition 5.10.6]), we may invoke Theorem 3.3.4 to conclude that the nontangential boundary trace $(\nabla u)\big|_{\partial\Omega}^{\kappa-\text{n.t.}}$ exists at σ-a.e. point on $\partial\Omega$. To proceed, from assumptions, [112, Proposition 8.4.9], and [112, (8.2.28)] we see that $\mathcal{N}_\kappa u \in L^p(\partial\Omega,\sigma)$. On account of [112, (8.6.51)] we therefore have $u \in \left[L^{np/(n-1)}(\Omega,\mathcal{L}^n)\right]^M$ and $\nabla u \in \left[L^{np/(n-1)}(\Omega,\mathcal{L}^n)\right]^{M\cdot n}$, hence

$$u \in \left[W^{1,np/(n-1)}(\Omega)\right]^M. \tag{3.4.15}$$

As such, the hypotheses in (3.4.3) are satisfied (with p replaced by $np/(n-1) > 1$ and $a := 0$) so we may rely on Theorem 3.4.1 to conclude that the nontangential boundary trace $u\big|_{\partial\Omega}^{\kappa-\text{n.t.}}$ also exists at σ-a.e. point on $\partial\Omega$. This completes the proof of (3.4.13).

Turning our attention to (3.4.14), first work under the assumption that the integrability exponent p belongs to $(1, n-1)$. Pick a coefficient tensor which permits to express the given system as $L = L_A$, and bring in $\mathcal{D}$ and $\mathcal{S}$, the boundary-to-domain double and single layer potential operators associated with A and Ω. Granted what we have proved up to this point, we may invoke the Green-type integral representation formula for null-solutions of L in terms of double and single layers from [115, §1.3] in order to express the given function u as

$$u = \mathcal{D}\big(u\big|_{\partial\Omega}^{\kappa-\text{n.t.}}\big) - \mathcal{S}(\partial_\nu^A u) \text{ in } \Omega, \tag{3.4.16}$$

where the conormal derivative $\partial_\nu^A u$ is considered in a pointwise sense, as in Definition 1.7.1. In particular, Lemma 1.7.3 implies that

$$\partial_\nu^A u \in \left[L^p(\partial\Omega,\sigma)\right]^M. \tag{3.4.17}$$

In addition, if $p^* := \left(\frac{1}{p} - \frac{1}{n-1}\right)^{-1}$ we claim that

$$u\big|_{\partial\Omega}^{\kappa-\text{n.t.}} \in \left[L_1^p(\partial\Omega,\sigma)\right]^M \hookrightarrow \left[L^{p^*}(\partial\Omega,\sigma)\right]^M. \tag{3.4.18}$$

Indeed, the membership above is provided by [113, Proposition 11.3.2]. As regards the embedding in (3.4.18), recall from [112, (5.11.28)] that the present hypotheses ensure that Ω satisfies a two-sided local John condition, so we may rely on item *(ii)* in [113, Theorem 11.5.18] to see that said embedding presently holds. Having established (3.4.16)-(3.4.18), we may use the nontangential maximal function estimates for the double and single layers proved in [115, Chapter 1] to conclude that $\mathcal{N}_\kappa u \in L^{p^*}(\partial\Omega,\sigma)$.

Consider now the task of establishing the membership claimed in (3.4.13) when $p \in \big(\frac{n-1}{n}, 1\big]$. For starters, the argument so far guarantees (3.4.13) and Theorem 3.4.1 further implies (3.4.5). Granted these, we may invoke the Green-type integral representation formula for null-solutions of L in terms of double and single layers from [115, §2.3] to presently write

$$u = \mathcal{D}\big(u\big|_{\partial\Omega}^{\kappa-\mathrm{n.t.}}\big) - \mathcal{S}(\mathring{\partial}_\nu^A u) \text{ in } \Omega, \tag{3.4.19}$$

where the weak conormal derivative $\mathring{\partial}_\nu^A u$ is now considered in the sense of [113, Definition 10.2.18]. In particular, [113, Theorem 10.2.24] ensures that

$$\mathring{\partial}_\nu^A u \in \big[H^p(\partial\Omega,\sigma)\big]^M. \tag{3.4.20}$$

Also, with $q := pn/(n-1) \in \big(1, \frac{n}{n-1}\big]$ and $p^* := \big(\frac{1}{p} - \frac{1}{n-1}\big)^{-1} \in \big(1, \frac{n-1}{n-2}\big] \subseteq (1,2]$, we may write

$$u\big|_{\partial\Omega}^{\kappa-\mathrm{n.t.}} = \mathrm{Tr}_{\Omega\to\partial\Omega}\, u \in \big[B^{q,q}_{1-\frac{1}{q}}(\partial\Omega,\sigma)\big]^M \hookrightarrow \big[L^{p^*}(\partial\Omega,\sigma)\big]^M. \tag{3.4.21}$$

The equality above is a consequence of (3.4.15), (3.4.4), and (A.0.126), while the first membership in (3.4.21) is implied by [113, (8.3.38)]. The second membership in (3.4.21) may be justified by writing

$$B^{q,q}_{1-\frac{1}{q}}(\partial\Omega,\sigma) \hookrightarrow B^{p^*,2}_0(\partial\Omega,\sigma) \hookrightarrow F^{p^*,2}_0(\partial\Omega,\sigma) = L^{p^*}(\partial\Omega,\sigma), \tag{3.4.22}$$

where the first embedding is provided by [113, Theorem 7.7.4] upon observing that

$$\tfrac{1}{q} - \tfrac{(1-1/q)}{n-1} = \tfrac{n-1}{np} - \tfrac{1}{n-1}\big(1 - \tfrac{n-1}{np}\big) = \tfrac{1}{p} - \tfrac{1}{n-1} = \tfrac{1}{p^*}, \tag{3.4.23}$$

the second embedding is a consequence of [113, (7.7.5)] (keeping in mind that $p^* \leq 2$), and the final equality comes from [113, (7.1.55)]. This concludes the proof of (3.4.21). In turn, from (3.4.19), (3.4.20), and (3.4.21) we conclude, on account of the nontangential maximal function estimates for the double and single layers proved in [115, Chapters 1 and 2], that $\mathcal{N}_\kappa u \in L^{p^*}(\partial\Omega,\sigma)$.

Next, consider the membership in (3.4.13) corresponding to $p = n-1$. Since $\partial\Omega$ is bounded, having $\mathcal{N}_\kappa(\nabla u) \in L^{n-1}(\partial\Omega,\sigma)$ implies that $\mathcal{N}_\kappa(\nabla u) \in L^q(\partial\Omega,\sigma)$ for any $q \in \big(\frac{n-1}{n}, n-1\big)$. Based on what we have proved already, this entails

$$\begin{gathered} \mathcal{N}_\kappa u \in L^{q^*}(\partial\Omega,\sigma) \text{ whenever} \\ q^* := \big(\tfrac{1}{q} - \tfrac{1}{n-1}\big)^{-1} \text{ with } q \in \big(\tfrac{n-1}{n}, n-1\big) \text{ arbitrary.} \end{gathered} \tag{3.4.24}$$

Thus $\mathcal{N}_\kappa u \in L^r(\partial\Omega, \sigma)$ for each $r \in (1, \infty)$ hence, since $\partial\Omega$ has finite measure, $\mathcal{N}_\kappa u \in L^r(\partial\Omega, \sigma)$ for each $r \in (0, \infty)$.

To complete the proof of the theorem there remains to consider the membership in (3.4.13) in the case when $p \in (n-1, \infty]$. In such a scenario, from (3.4.15) and [113, Theorem 8.3.2] (also keeping in mind [112, (5.11.66)]) we see that $u = U\big|_\Omega$ where $U \in \big[W^{1,q}(\mathbb{R}^n)\big]^M$ with $q := pn/(n-1) > n$. Since classical embedding results for Sobolev spaces in the Euclidean setting imply that $U \in \big[\mathscr{C}^\alpha(\mathbb{R}^n)\big]^M$ with $\alpha := 1 - n/q \in (0,1)$, we ultimately conclude that

$$u \in \big[\mathscr{C}^\alpha(\overline{\Omega})\big]^M \text{ with } \alpha := 1 - (n-1)/p \in (0,1) \tag{3.4.25}$$

in this case. In particular, u is bounded in Ω so $\mathcal{N}_\kappa u \in L^\infty(\partial\Omega, \sigma)$ (cf. [112, Lemma 8.3.2]). □

3.5 Fatou-Type Theorems on Riemannian Manifolds

The main goal in this section is to prove Fatou-type theorems on Riemannian manifolds in the spirit of our earlier work from §3.1 and §3.3 from the Euclidean setting. As in the past, we shall treat both first-order and second-order operators. We begin by treating the former class, with our main result in this regard being Theorem 3.5.4, stated a little later.

To set the stage, we first extend the notion of "bullet product" from the Euclidean ambient (cf. [112, Proposition 4.2.3]) to the setting of Riemannian manifolds. Specifically, the integration by parts formula [112, (1.11.29)] suggests that we make the following definition:

Definition 3.5.1 *Let $\mathcal{M}$ denote a compact, oriented, boundaryless manifold of class $\mathscr{C}^2$ and real dimension $n \in \mathbb{N}$ (where $n \geq 2$), equipped with a Riemannian metric tensor g of class $\mathscr{C}^1$. As in the past, denote by $d\mathcal{L}^n_g$ the Lebesgue measure induced by the metric tensor g (expressed in local coordinates by $\sqrt{g}\,d\mathcal{L}^n$). Next, suppose $\mathcal{E} \to \mathcal{M}$, $\mathcal{G} \to \mathcal{M}$ are two Hermitian vector bundles of class $\mathscr{C}^1$ and consider a first-order differential operator $D : \mathcal{E} \to \mathcal{G}$ which, in local coordinates, has top coefficients of class $\mathscr{C}^1$ and lower-order coefficients of class $\mathscr{C}^0$. Finally, pick an arbitrary open set $\Omega \subseteq \mathcal{M}$.*

Then for each given section

$$u \in L^1(\Omega, \mathcal{L}^n_g) \otimes \mathcal{E} \text{ with } Du \in L^1(\Omega, \mathcal{L}^n_g) \otimes \mathcal{G} \tag{3.5.1}$$

define the $\mathcal{G}$-valued distribution $(-i)\,\mathrm{Sym}(D; \nu_g) \bullet u$ on $\partial\Omega$ by setting (dropping the dependence on the vector bundle in all pairings)

$$_{\mathrm{Lip}(\partial\Omega)}\big\langle \varphi, (-i)\,\mathrm{Sym}(D; \nu_g) \bullet u \big\rangle_{\mathrm{Lip}'(\partial\Omega)} := \int_\Omega \langle Du, \Phi\rangle \, d\mathcal{L}^n_g - \int_\Omega \langle u, D^\top \Phi\rangle \, d\mathcal{L}^n_g \tag{3.5.2}$$

for each $\varphi \in \mathrm{Lip}(\partial\Omega) \otimes \mathcal{G}$ *and* $\Phi \in \mathrm{Lip}(\mathcal{M}) \otimes \mathcal{G}$ *with* $\Phi\big|_{\partial\Omega} = \varphi$.

Here and elsewhere, $D^\top$ *is the (real) transpose of D, the symbol* $\langle\cdot,\cdot\rangle$ *in the right-hand side denotes the (real) pointwise inner product in the fibers of the vector bundles considered, and* $\mathrm{Lip}'(\partial\Omega)$ *denotes the space of vector bundle-valued distributions on the set* $\partial\Omega$.

Regarding Definition 3.5.1, we wish to emphasize that $\Omega \subseteq \mathcal{M}$ is a completely arbitrary open set. In particular, Ω may not be a set of locally finite perimeter and, as such, it may not have a geometric measure theoretic outward unit normal ν_g. Even though the latter symbol appears in the notation $(-\mathrm{i})\,\mathrm{Sym}(D;\nu_g) \bullet u$, its presence here is purely formal, as the object in question is simply the name of the distribution defined in (3.5.2).

Reasoning as in the proof of [112, Lemma 4.2.1] it may be shown that the definition in (3.5.2) is unambiguous, i.e., the right side of (3.5.2) does not depend of the particular choice of an extension $\Phi \in \mathrm{Lip}(\mathcal{M}) \otimes \mathcal{G}$ of the given "test function" $\varphi \in \mathrm{Lip}(\partial\Omega) \otimes \mathcal{G}$.

Finally, we not that the integrals in the right side of (3.5.2) are absolutely convergent, since u and Du are absolutely integrable while Φ and $D^\top\Phi$ are essentially bounded.

Moving on, let $(\mathcal{M}, g)$ be a Riemannian manifold as above. Fix an arbitrary Ahlfors regular domain $\Omega \subseteq \mathcal{M}$ and abbreviate $\sigma_g := \mathcal{H}_g^{n-1} \lfloor \partial\Omega$, where $\mathcal{H}_g^{n-1}$ is the $(n-1)$-dimensional Hausdorff measure induced by the metric g on $\mathcal{M}$. Also, consider a Hermitian vector bundle $\mathcal{E} \to \mathcal{M}$. In this geometric setting, bring back the maximal operator acting on any given $\mathcal{L}_g^n$-measurable section $u : \Omega \to \mathcal{E}$ according to

$$(\mathfrak{P}_g u)(x) := \sup_{r>0} \left\{ \frac{1}{\sigma_g\big(\partial\Omega \cap B_g(x,r)\big)} \int_{\Omega\cap B_g(x,r)} |u| \, \mathrm{d}\mathcal{L}_g^n \right\}, \qquad x \in \partial\Omega, \quad (3.5.3)$$

where $B_g(x,r)$ denotes the geodesic ball of radius r centered at x. We are now ready to state the following basic weak ("bullet product") trace result, extending to the manifold setting work done in [113, §10.2] in the Euclidean space.

Theorem 3.5.2 *Let* $\mathcal{M}$ *denote a compact, oriented, boundaryless manifold of class* $\mathscr{C}^2$ *and real dimension* $n \in \mathbb{N}$ *(where* $n \geq 2$*), equipped with a Riemannian metric tensor* g *of class* $\mathscr{C}^1$*. As in the past, denote by* $\mathcal{L}_g^n$ *the Lebesgue measure induced by the metric tensor* g *(expressed in local coordinates as* $\sqrt{g}\,\mathrm{d}\mathcal{L}^n$*). Next, suppose* $\mathcal{E} \to \mathcal{M}$, $\mathcal{G} \to \mathcal{M}$ *are two Hermitian vector bundles of class* $\mathscr{C}^1$ *and consider a first-order differential operator* $D : \mathcal{E} \to \mathcal{G}$ *which, in local coordinates, has top coefficients of class* $\mathscr{C}^1$ *and lower-order coefficients of class* $\mathscr{C}^0$.

In addition, let $\Omega \subseteq \mathcal{M}$ *be an Ahlfors regular domain and set* $\sigma_g := \mathcal{H}_g^{n-1} \lfloor \partial\Omega$, *where* $\mathcal{H}_g^{n-1}$ *is the* $(n-1)$*-dimensional Hausdorff measure induced by the metric* g *on* $\mathcal{M}$*. Also, denote by* $\nu_g \in T^*\mathcal{M}$ *the geometric measure theoretic outward unit conormal (with respect to the metric tensor* g*) to the set of locally finite perimeter* Ω*. Finally, assume* $u : \Omega \to \mathcal{E}$ *is a section having* $\mathcal{L}^n$*-measurable components and with the property that for some* $\kappa \in (0,\infty)$ *one has*

$$\mathcal{N}_\kappa u \in L^p(\partial\Omega, \sigma_g) \text{ for some } p \in \left(\tfrac{n-1}{n}, \infty\right). \tag{3.5.4}$$

In particular, $u \in L^1(\Omega, \mathcal{L}_g^n) \otimes \mathcal{E}$ (cf. [112, Proposition 8.6.3]), *and one also assumes that Du, computed in the sense of distributions in Ω, belongs to $L^1(\Omega, \mathcal{L}_g^n) \otimes \mathcal{G}$ and satisfies*

$$\mathfrak{P}_g(Du) \in L^p(\partial\Omega, \sigma_g). \tag{3.5.5}$$

Then with the bullet-product sysmbol interpreted as in Definition 3.5.1

$$\mathrm{Sym}(D; \nu_g) \bullet u \in H^p(\partial\Omega, \sigma_g) \otimes \mathcal{G} \tag{3.5.6}$$

and there exists a constant $C \in (0, \infty)$ independent of u such that

$$\begin{aligned} &\left\|\mathrm{Sym}(D; \nu_g) \bullet u\right\|_{H^p(\partial\Omega,\sigma_g)\otimes\mathcal{G}} \\ &\qquad \le C\left\|\mathcal{N}_\kappa u\right\|_{L^p(\partial\Omega,\sigma_g)} + C\|\mathfrak{P}_g(Du)\|_{L^p(\partial\Omega,\sigma_g)}. \end{aligned} \tag{3.5.7}$$

Finally, under the additional assumptions that $\mathcal{N}_\kappa u \in L^1(\partial\Omega, \sigma_g)$ and $u\big|_{\partial\Omega}^{\kappa-\mathrm{n.t.}}$ exists at σ_g-a.e. point on $\partial\Omega$ it follows that

$$\mathrm{Sym}(D; \nu_g) \bullet u = \mathrm{Sym}(D; \nu_g)\Big(u\big|_{\partial\Omega}^{\kappa-\mathrm{n.t.}}\Big) \tag{3.5.8}$$

as distributions on $\partial\Omega$.

Proof Fix two parameters

$$\gamma \in (0, 1) \text{ and } \alpha \in \left(0, \frac{n-1+\gamma}{n-1}\right). \tag{3.5.9}$$

The crux of the matter is establishing the existence of a finite constant $C > 0$, independent of the section u, with the property that the following pointwise maximal inequality holds:

$$\begin{aligned} \big(\mathrm{Sym}(D; \nu_g) \bullet u\big)^\sharp_\gamma(x) \le C\Big[\mathcal{M}_{\partial\Omega}\big(|\mathcal{N}_\kappa u|^{\frac{1}{\alpha}}\big)(x)\Big]^\alpha + C\,\mathfrak{P}_g\big(Du\big)(x) \\ \text{at every point } x \in \partial\Omega, \end{aligned} \tag{3.5.10}$$

where $(\cdot)^\sharp_\gamma$ is the Fefferman-Stein grand maximal operator (defined much as in [113, (4.1.6)], now in the manifold setting), $\mathcal{M}_{\partial\Omega}$ is the Hardy-Littlewood maximal operator on $\partial\Omega$, and $\mathfrak{P}_g$ is maximal operator defined in (3.5.3). Granted this, the conclusions in (3.5.6)-(3.5.7) are then implied by the definition of the scale of Hardy spaces given in [113, Definition 4.2.1] and the boundedness of the Hardy-Littlewood maximal operator on Lebesgue spaces. To justify (3.5.10), we reason along the lines of the proof of [113, Theorem 10.2.4]. The only significant difference is that in place of [113, (10.2.31)] we now have, thanks to Definition 3.5.1,

$$ {}_{\mathrm{Lip}(\partial\Omega)}\big\langle \psi\,,\,(-\mathrm{i})\,\mathrm{Sym}(D;\nu_g)\bullet u\big\rangle_{\mathrm{Lip}'(\partial\Omega)} = \int_\Omega \langle Du,\widetilde{\psi}\rangle\,\mathrm{d}\mathcal{L}_g^n - \int_\Omega \langle u, D^{\top}\widetilde{\psi}\rangle\,\mathrm{d}\mathcal{L}_g^n \tag{3.5.11} $$

where $\psi \in \mathrm{Lip}(\partial\Omega)\otimes \mathcal{G}$ is arbitrary and $\widetilde{\psi} \in \mathrm{Lip}(\mathcal{M})\otimes\mathcal{G}$ is a suitable quantitative extension of ψ constructed using [113, Lemma 10.1.3], much as ψ and $\widetilde{\psi}$ are related to each other in [113, (10.2.29)-(10.2.30)]. The rest of the argument in the proof of (3.5.10) may then be adapted from that of [113, Theorem 10.2.4] to the manifold setting in a straightforward fashion. Finally, the claim made in relation to (3.5.8) may be justified as in the proof of [113, Proposition 10.2.11]. □

Recall that a differential operator D acting between two vector bundles over a manifold $\mathcal{M}$ is said to have the `unique continuation property` (UCP) if whenever $u \in H^1(\mathcal{M})$ (the L^2-based Sobolev space of order 1 on $\mathcal{M}$ is such that $Du = 0$ on $\mathcal{M}$ and $\mathcal{M}$ and $u|_{\mathcal{O}} = 0$ for some open subset $\mathcal{O}$ of $\mathcal{M}$ then necessarily $u = 0$ in any connected component of $\mathcal{M}$ overlapping with $\mathcal{O}$. Observe that if D is an injectively elliptic first-order homogeneous constant complex coefficient system in $\mathbb{R}^n$ then [109, Theorem 11.7, p. 409] (applied to the weakly elliptic second order operator $L := D^*D$) implies that any nullsolution u of D is real-analytic. In view of [109, Theorem 6.25, pp. 229-230], this implies that

> any injectively elliptic first-order homogeneous constant complex coefficient system in $\mathbb{R}^n$ has UCP (i.e., the unique continuation property). (3.5.12)

Our next proposition contains a basic representation formula for nullsolutions of nullsolutions of an injectively elliptic first-order differential operator in an arbitrary open subset of a given Riemannian manifold. The reader is advised to recall the conventions made in (2.8.90)-(2.8.94).

Proposition 3.5.3 *Let $\mathcal{M}$ denote a compact, oriented, boundaryless manifold of class $\mathscr{C}^3$ and real dimension $n \in \mathbb{N}$ (where $n \geq 2$), equipped with a Riemannian metric tensor g of class $\mathscr{C}^3$. Suppose $\mathcal{E} \to \mathcal{M}$, $\mathcal{G} \to \mathcal{M}$ are two Hermitian vector bundles of class $\mathscr{C}^3$ and let $D : \mathcal{E} \to \mathcal{G}$ be a first-order differential operator which, in any local coordinate chart U on $\mathcal{M}$ and with respect to any local trivializations of $\mathcal{E}, \mathcal{G}$, may be represented as*

$$ \begin{gathered} D = \sum_j A_j(x)\partial_j + B(x) \text{ where the coefficients satisfy} \\ A_j \in \mathscr{C}^3_{\mathrm{loc}}\big(U, \mathbb{C}^{\mathrm{rank}\mathcal{G}\times\mathrm{rank}\mathcal{E}}\big), \quad B \in \mathscr{C}^2_{\mathrm{loc}}\big(U, \mathbb{C}^{\mathrm{rank}\mathcal{G}\times\mathrm{rank}\mathcal{E}}\big). \end{gathered} \tag{3.5.13} $$

In addition suppose D is injectively elliptic and has the unique continuation property. Next, consider

$$ \begin{gathered} 0 \leq V \in \mathrm{Hom}\,(\mathcal{E},\mathcal{E}) \text{ with coefficients in } \mathscr{C}^1, \\ \text{with the property that } \mathrm{supp}\, V \cap \overline{\Omega} = \varnothing \text{ and } V > 0 \\ \text{in a nonempty open subset of each component of } \mathcal{M}, \end{gathered} \tag{3.5.14} $$

and introduce the second-order differential operator

$$L := D^*D + V : \mathcal{E} \longrightarrow \mathcal{E}. \tag{3.5.15}$$

where D^* denotes the Hermitian adjoint (i.e., the complex conjugate transposed) of D. Finally, having fixed an arbitrary open set $\Omega \subseteq \mathcal{M}$, select a section

$$u \in \big[L^{\infty}_{\rm loc}(\Omega, \mathcal{L}^n_g) \cap L^1(\Omega, \mathcal{L}^n_g)\big] \otimes \mathcal{E} \text{ with } Du = 0 \text{ in } \Omega. \tag{3.5.16}$$

Then[12]

$$L : H^1(\mathcal{M}) \otimes \mathcal{E} \longrightarrow H^{-1}(\mathcal{M}) \otimes \mathcal{E} \text{ is an isomorphism} \tag{3.5.17}$$

and, with $E_L(\cdot,\cdot)$ denoting the Schwartz kernel of L^{-1} (the inverse of L in (3.5.17)), at each point $x \in \Omega$ one has (dropping the dependence on the vector bundle in the distributional pairing)

$$u(x) = -_{\mathrm{Lip}(\partial\Omega)}\big\langle \big(I \otimes (D^*)^{\top}\big) E_L(x,\cdot),\, (-\mathrm{i})\,\mathrm{Sym}(D;\nu_g) \bullet u \big\rangle_{\mathrm{Lip}'(\partial\Omega)} \tag{3.5.18}$$

where I is the identity and the distribution involved in the pairing is considered in the sense of Definition 3.5.1.

Proof The fact that D is injectively elliptic makes L in (3.5.15) weakly elliptic. In turn, this ensures that L has a two-sided parametrix $E^{\#} \in \mathrm{OP}S^{-2}_{1,\delta}$ (for some $\delta \in (0,1)$) which inverts L in (3.5.17) up to smoothing operators (see the proof of [119, Proposition 2.1, p. 13]). Since the latter are compact, by the Rellich-Kondrachov theorem, we conclude that L in (3.5.17) is invertible modulo compact operators, hence ultimately Fredholm. The same is true for L^*, and since $L - L^*$ is lower-order we conclude that L has the same index as L^*. However, functional analysis tells that these numbers are opposite to each other, hence the index of L in (3.5.17) is zero. Since L is also injective in this context (an easy consequence of (3.5.15), the fact that the operator D has the unique continuation property, and the potential V is nonnegative as well as strictly positive in some nonempty open subset of each component of $\mathcal{M}$), we finally conclude that L is actually an isomorphism in the context of (3.5.17).

The regularity properties of the Schwartz kernel $E_L(\cdot,\cdot)$ of L^{-1} (including the nature of the singularity along the diagonal) have been discussed in the proof of Theorem 2.8.4. The present stronger smoothness assumptions adopted for the metric tensor g and the coefficients in (3.5.13) ensure similar regularity properties one unit higher. Also, granted the weak ellipticity of L and the conditions imposed on the coefficients in (3.5.13), the (local) elliptic regularity result established in [119, Proposition 2.7, p. 20] may be used to conclude that

$$u \in \mathscr{C}^2(\Omega) \otimes \mathcal{E}. \tag{3.5.19}$$

We continue by making two observations. First, we recall that the principal symbol of D satisfies the properties

[12] throughout this section $H^{\pm 1}(\mathcal{M})$ denote L^2-based Sobolev spaces on $\mathcal{M}$; alternatively, $H^1(\mathcal{M}) = W^{1,2}(\mathcal{M})$ and $H^{-1}(\mathcal{M}) = W^{-1,2}(\mathcal{M})$

$$\mathrm{Sym}(D^\top;\xi) = -\mathrm{Sym}(D;\xi)^\top, \tag{3.5.20}$$

$$(-\mathrm{i})\,\mathrm{Sym}(D;d\varphi) = [D,\varphi], \tag{3.5.21}$$

for every cotangent vector ξ and, respectively, for every scalar-valued function φ of class $\mathscr{C}^1$ (also identified with the operator of pointwise multiplication by φ) where, generally speaking,

$$[A,B] := AB - BA \tag{3.5.22}$$

stands for the commutator of the operators A, B. Second, we note that in the current setting Definition 3.5.1 gives

$${}_{\mathrm{Lip}(\partial\Omega)}\big\langle\varphi,\,(-\mathrm{i})\,\mathrm{Sym}(D;\nu_g)\bullet u\big\rangle_{\mathrm{Lip}'(\partial\Omega)} = -\int_\Omega \langle u, D^\top\Phi\rangle\,\mathrm{d}\mathcal{L}^n_g \tag{3.5.23}$$

for each $\varphi\in\mathrm{Lip}(\partial\Omega)\otimes\mathcal{G}$ and $\Phi\in\mathrm{Lip}(\mathcal{M})\otimes\mathcal{G}$ with $\Phi\big|_{\partial\Omega}=\varphi$.

To justify (3.5.18), fix an arbitrary point $x\in\Omega$ and pick some scalar function $\eta\in\mathscr{C}^2_c(\Omega)$ such that $\eta\equiv 1$ near x. Then we may write

$$\begin{aligned}
&-{}_{\mathrm{Lip}(\partial\Omega)}\big\langle\big(I\otimes(D^*)^\top\big)E_L(x,\cdot),\,(-\mathrm{i})\,\mathrm{Sym}(D;\nu_g)\bullet u\big\rangle_{\mathrm{Lip}'(\partial\Omega)}\\
&\quad= -{}_{\mathrm{Lip}(\partial\Omega)}\big\langle\big(I\otimes(1-\eta)(D^*)^\top\big)E_L(x,\cdot),\,(-\mathrm{i})\,\mathrm{Sym}(D;\nu_g)\bullet u\big\rangle_{\mathrm{Lip}'(\partial\Omega)}\\
&\quad= \int_\Omega\big\langle\big(I\otimes D^\top(1-\eta)(D^*)^\top\big)E_L(x,\cdot),\,u\big\rangle\,\mathrm{d}\mathcal{L}^n_g\\
&\quad= \int_\Omega\big\langle\big(I\otimes[D^\top,1-\eta](D^*)^\top\big)E_L(x,\cdot),\,u\big\rangle\,\mathrm{d}\mathcal{L}^n_g\\
&\qquad+\int_\Omega\big\langle\big(I\otimes(1-\eta)D^\top(D^*)^\top\big)E_L(x,\cdot),\,u\big\rangle\,\mathrm{d}\mathcal{L}^n_g\\
&\quad= \int_\Omega\big\langle\big(I\otimes(D^*)^\top\big)E_L(x,\cdot),\,(-\mathrm{i})\,\mathrm{Sym}(D;d\eta)u\big\rangle\,\mathrm{d}\mathcal{L}^n_g\\
&\quad= \int_\Omega\big\langle\big(I\otimes(D^*)^\top\big)E_L(x,\cdot),\,D(\eta\,u)\big\rangle\,\mathrm{d}\mathcal{L}^n_g\\
&\quad= {}_{\mathcal{D}'(\Omega)}\big\langle\big(I\otimes(D^*)^\top\big)E_L(x,\cdot),\,D(\eta\,u)\big\rangle_{\mathcal{D}(\Omega)}\\
&\quad= {}_{\mathcal{D}'(\Omega)}\big\langle\big(I\otimes D^\top(D^*)^\top\big)E_L(x,\cdot),\,\eta\,u\big\rangle_{\mathcal{D}(\Omega)}\\
&\quad= {}_{\mathcal{D}'(\Omega)}\big\langle\big(I\otimes L^\top\big)E_L(x,\cdot),\,\eta\,u\big\rangle_{\mathcal{D}(\Omega)} = (\eta\,u)(x) = u(x).
\end{aligned} \tag{3.5.24}$$

The first equality in (3.5.24) is based on the fact that $1-\eta\equiv 1$ on $\partial\Omega$, while the second equality in (3.5.24) is a consequence of formula (3.5.23) applied with $\Phi := \big(I\otimes(1-\eta)(D^*)^\top\big)E_L(x,\cdot)$. The third equality in (3.5.24) simply follows from (3.5.22). The fourth equality in (3.5.24) is based on two facts. First, relying on (3.5.21) and (3.5.20) we may write

$$[D^\top,1-\eta] = -(-\mathrm{i})\,\mathrm{Sym}(D^\top;d\eta) = (-\mathrm{i})\,\mathrm{Sym}(D;d\eta)^\top. \tag{3.5.25}$$

Second, since $V = 0$ on Ω, we have

$$\begin{aligned}\big(I \otimes (1-\eta)D^{\top}(D^{*})^{\top}\big)E_L(x,\cdot) &= \big(I \otimes (1-\eta)L^{\top}\big)E_L(x,\cdot) \\ &= I \otimes (1-\eta)\delta_{\Delta}(x,\cdot) = 0 \text{ in } \mathcal{D}'(\Omega), \end{aligned} \tag{3.5.26}$$

by (2.8.178) and (2.8.176). For the fifth equality in (3.5.24) we used again (3.5.21) to the effect that

$$(-\mathrm{i})\,\mathrm{Sym}(D; d\eta)u = [D,\eta]u = D(\eta u) - \eta(Du) = D(\eta u) \text{ in } \Omega, \tag{3.5.27}$$

given that, by assumption, $Du = 0$ in Ω. The sixth equality in (3.5.24) is simply the reinterpretation of the integral as a distributional pairing on Ω (suitably understood, for test functions of limited regularity). The seventh equality in (3.5.24) is the result of "integration by parts" in the sense of distributions as in (2.8.96), also relying on the fact that $\eta u \in \mathscr{C}^2_c(\Omega) \otimes \mathcal{E}$ (cf. (3.5.19)). Finally, the last equality in (3.5.24) is a consequence of $\big(I \otimes L^{\top}\big)E_L(x,\cdot) = \delta_{\Delta}(x,\cdot)$ (cf. (2.8.178)) and the formula recorded in (2.8.175). □

We are now in a position to state and prove the following remarkable quantitative Fatou-type result for nullsolutions of injectively elliptic first-order differential operators in UR subdomains of Riemannian manifolds. In view of (3.5.12), this is a most natural generalization of Theorem 3.1.6 from the Euclidean setting.

Theorem 3.5.4 *Let $\mathcal{M}$ denote a compact, oriented, boundaryless manifold of class $\mathscr{C}^3$ and real dimension $n \in \mathbb{N}$ (where $n \geq 2$), equipped with a Riemannian metric tensor g of class $\mathscr{C}^3$. Suppose $\mathcal{E} \to \mathcal{M}$, $\mathcal{G} \to \mathcal{M}$ are two Hermitian vector bundles of class $\mathscr{C}^3$ and*

$$\begin{gathered} D : \mathcal{E} \to \mathcal{G} \text{ is a first-order differential operator which is} \\ \text{injectively elliptic and has the unique continuation property.} \end{gathered} \tag{3.5.28}$$

Assume that, in any local coordinate chart U on $\mathcal{M}$ and with respect to any local trivializations of $\mathcal{E}, \mathcal{G}$, the operator D may be represented as

$$\begin{gathered} D = \sum_j A_j(x)\partial_j + B(x) \text{ where the coefficients satisfy} \\ A_j \in \mathscr{C}^3_{\mathrm{loc}}\big(U, \mathbb{C}^{\mathrm{rank}\mathcal{G}\times\mathrm{rank}\mathcal{E}}\big), \quad B \in \mathscr{C}^2_{\mathrm{loc}}\big(U, \mathbb{C}^{\mathrm{rank}\mathcal{G}\times\mathrm{rank}\mathcal{E}}\big). \end{gathered} \tag{3.5.29}$$

Next, suppose $\Omega \subseteq \mathcal{M}$ is a UR *domain. Denote the $(n-1)$-dimensional Hausdorff measure associated with the metric g on $\mathcal{M}$ by $\mathcal{H}^{n-1}_g$, abbreviate $\sigma_g := \mathcal{H}^{n-1}_g \lfloor \partial\Omega$, and denote by $\nu_g \in T^*\mathcal{M}$ the geometric measure theoretic outward unit conormal (with respect to the metric tensor g) to Ω. Finally, select a section $u : \Omega \to \mathcal{E}$*

$$\mathcal{N}_\kappa u \in L^p(\partial\Omega, \sigma_g) \text{ and } Du = 0 \text{ in } \Omega, \tag{3.5.30}$$

for an aperture parameter $\kappa > 0$ and an integrability exponent $p \in \big(\frac{n}{n-1}, \infty\big)$.

Then the nontangential boundary trace

$$\begin{aligned} &u\big|_{\partial\Omega}^{\kappa-\mathrm{n.t.}} \ \textit{exists (in } \mathcal{E}) \textit{ at } \sigma_g\textit{-a.e. point on } \partial\Omega \textit{ and} \\ &\textit{the function } u\big|_{\partial\Omega}^{\kappa-\mathrm{n.t.}} \textit{ is } \sigma_g\textit{-measurable on } \partial\Omega; \\ &\textit{also, } u\big|_{\partial\Omega}^{\kappa-\mathrm{n.t.}} \textit{ belongs to the space } L^p(\partial\Omega,\sigma_g)\otimes\mathcal{E} \\ &\textit{and one has } \Big\|u\big|_{\partial\Omega}^{\kappa-\mathrm{n.t.}}\Big\|_{L^p(\partial\Omega,\sigma_g)\otimes\mathcal{E}} \leq \|\mathcal{N}_\kappa u\|_{L^p(\partial\Omega,\sigma_g)}. \end{aligned} \tag{3.5.31}$$

Moreover, for any other given aperture parameter $\kappa' \in (0,\infty)$ *the nontangential boundary trace* $u\big|_{\partial\Omega}^{\kappa'-\mathrm{n.t.}}$ *also exists, and agrees with* $u\big|_{\partial\Omega}^{\kappa-\mathrm{n.t.}}$, *at* σ_g*-a.e. point on* $\partial\Omega$. *Finally, if* $p \in (1,\infty)$ *then*

$$\|\mathcal{N}_\kappa u\|_{L^p(\partial\Omega,\sigma_g)} \approx \Big\|u\big|_{\partial\Omega}^{\kappa-\mathrm{n.t.}}\Big\|_{L^p(\partial\Omega,\sigma_g)\otimes\mathcal{E}}. \tag{3.5.32}$$

In particular, Theorem 3.5.4 is applicable to Dirac operators with sufficiently regular coefficients, such as the Hodge-Dirac operator $\mathrm{d}+\delta$ acting on differential forms on a sufficiently regular Riemannian manifold. For D taken to be Def, the deformation tensor acting on vector fields on a sufficiently regular Riemannian manifold, we obtain a Fatou-type theorem for Killing fields in arbitrary UR domains.

Here is the proof of Theorem 3.5.4.

Proof of Theorem 3.5.4 Having chosen a real-valued scalar function

$$\begin{aligned} &0 \leq V \in \mathscr{C}^1(\mathcal{M}) \text{ such that } \operatorname{supp} V \cap \overline{\Omega} = \varnothing \text{ and } V > 0 \\ &\text{in a nonempty open subset of each component of } \mathcal{M}, \end{aligned} \tag{3.5.33}$$

define the second-order differential operator $L := D^*D + V : \mathcal{E} \to \mathcal{E}$, as in (3.5.15). Then Proposition 3.5.3 ensures that

$$L : H^1(\mathcal{M}) \otimes \mathcal{E} \longrightarrow H^{-1}(\mathcal{M}) \otimes \mathcal{E} \text{ is an isomorphism,} \tag{3.5.34}$$

and we shall denote by E_L the Schwartz kernel of L^{-1}, the inverse of L in (3.5.34).

Next, from the membership in (3.5.30), [112, Lemma 8.3.1], and the last estimate in [112, Proposition 8.6.3] we deduce (bearing in mind that Ω is relatively compact) that

$$u \in \big[L^\infty_{\mathrm{loc}}(\Omega, \mathcal{L}^n_g) \cap L^1(\Omega, \mathcal{L}^n_g)\big] \otimes \mathcal{E}. \tag{3.5.35}$$

In view of the properties in (3.5.30), Theorem 3.5.2 also guarantees that

$$\operatorname{Sym}(D;\nu_g) \bullet u \text{ belongs to } \begin{cases} H^p(\partial\Omega,\sigma_g)\otimes\mathcal{G} \text{ if } p \in \big(\tfrac{n-1}{n}, 1\big], \\ L^p(\partial\Omega,\sigma_g)\otimes\mathcal{G} \text{ if } p \in (1,\infty), \end{cases} \tag{3.5.36}$$

and if $p \in (1,\infty)$ there exists a constant $C \in (0,\infty)$ independent of u such that

$$\Big\|\operatorname{Sym}(D;\nu_g) \bullet u\Big\|_{L^p(\partial\Omega,\sigma_g)\otimes\mathcal{G}} \leq C\big\|\mathcal{N}_\kappa u\big\|_{L^p(\partial\Omega,\sigma_g)}. \tag{3.5.37}$$

Thanks to (3.5.35), (3.5.36), and the compatibility between the duality pairing from [113, Theorem 4.6.1] and the distributional pairing (cf. [113, Lemma 4.6.4]), Proposition 3.5.3 applies and gives that if $p \in \big(\frac{n-1}{n}, 1\big]$ then at each point $x \in \Omega$ we have

$$u(x) = -\big\langle \big(I \otimes (D^*)^\top\big) E_L(x,\cdot),\, (-\mathrm{i})\,\mathrm{Sym}(D;\nu_g) \bullet u \big\rangle \tag{3.5.38}$$

with the brackets denoting the pairing between the Hardy space H^p and its dual $(H^p)^*$ (cf. [113, Theorem 4.6.1]), while for $p \in (1,\infty)$ we have

$$u(x) = -\int_{\partial\Omega} \big\langle \big(I \otimes (D^*)^\top\big) E_L(x,\cdot),\, (-\mathrm{i})\,\mathrm{Sym}(D;\nu_g) \bullet u \big\rangle_{\mathcal{G}}\, \mathrm{d}\sigma_g \tag{3.5.39}$$

at each point $x \in \Omega$. From (3.5.36), (3.5.38), (3.5.39), and item *(5)* in Theorem 2.8.2 we are now able to see that $u\big|_{\partial\Omega}^{\kappa-\mathrm{n.t.}}$ exists at σ_g-a.e. point on $\partial\Omega$. All other properties in (3.5.31) (as well as the subsequent claim in the statement) are then seen from this, the membership in (3.5.30), [112, (8.9.8)], and [112, Proposition 8.9.8].

Furthermore, from (3.5.36), (3.5.39) item *(4)* in Theorem 2.8.2, and the very last property in the statement of Theorem 3.5.2 we conclude that if $p \in (1,\infty)$ there

$$\Big\| \mathrm{Sym}(D;\nu_g)\big(u\big|_{\partial\Omega}^{\kappa-\mathrm{n.t.}}\big) \Big\|_{L^p(\partial\Omega,\sigma_g)\otimes\mathcal{G}} \approx \|\mathcal{N}_\kappa u\|_{L^p(\partial\Omega,\sigma_g)}. \tag{3.5.40}$$

Granted this, the equivalence claimed in (3.5.32) follows with the help of the estimate in the last line of (3.5.31) and the fact that

$$\|\mathrm{Sym}(D;\nu_g)\,\big(u\big|_{\partial\Omega}^{\kappa-\mathrm{n.t.}}\big)\|_{L^p(\partial\Omega,\sigma_g)\otimes\mathcal{G}} \le C \|u\big|_{\partial\Omega}^{\kappa-\mathrm{n.t.}}\|_{L^p(\partial\Omega,\sigma_g)\otimes\mathcal{E}}. \tag{3.5.41}$$

The proof of Theorem 3.5.4 is now complete. □

In the remaining portion of this section we shall concern ourselves with Fatou-type theorems for *second-order* differential operators on UR subdomains of Riemannian manifolds. To set the stage, we make a couple of definitions.

Definition 3.5.5 *Let $\mathcal{M}$ be a compact, boundaryless, oriented $\mathscr{C}^2$ manifold equipped with a Riemannian metric g, and consider two $\mathscr{C}^2$ Hermitian vector bundles $\mathcal{E}$, $\mathcal{G}$ over $\mathcal{M}$. Suppose all metrics involved are of class $\mathscr{C}^1$, and equip $\mathcal{E}$, $\mathcal{G}$ with connections*

$$\nabla^{\mathcal{E}} : T\mathcal{M} \times \mathcal{E} \longrightarrow \mathcal{E}, \quad \nabla^{\mathcal{G}} : T\mathcal{M} \times \mathcal{G} \longrightarrow \mathcal{G}. \tag{3.5.42}$$

Consider a first-order differential operator $D : \mathcal{E} \to \mathcal{G}$ with top coefficients of class $\mathscr{C}^1$ and zero-order coefficients of class $\mathscr{C}^0$. Finally, fix an arbitrary $\mathscr{C}^1$ vector field $X \in T\mathcal{M}$ and introduce the first-order differential operators

$$\mathbf{P} : \mathscr{C}^1(\mathcal{M}) \otimes \mathcal{E} \to \mathscr{C}^0(\mathcal{M}) \otimes \mathcal{G} \times \mathscr{C}^0(\mathcal{M}) \otimes \mathcal{E}, \quad \mathbf{P} := \big(D, -\nabla_X^{\mathcal{E}}\big), \tag{3.5.43}$$

and

$$\begin{aligned}&\mathbf{Q}:\mathscr{C}^1(\mathcal{M})\otimes\mathcal{G}\times\mathscr{C}^1(\mathcal{M})\otimes\mathcal{E}\longrightarrow\mathscr{C}^0(\mathcal{M})\otimes\mathcal{G},\\&\mathbf{Q}(a,b):=\nabla_X^{\mathcal{G}}a+Db,\quad\forall\,a\in\mathscr{C}^1(\mathcal{M})\otimes\mathcal{G},\quad\forall\,b\in\mathscr{C}^1(\mathcal{M})\otimes\mathcal{E}.\end{aligned}\tag{3.5.44}$$

In addition, having picked a set of locally finite perimeter $\Omega\subseteq\mathcal{M}$, introduce $\sigma_g:=\mathcal{H}_g^{n-1}\lfloor\partial\Omega$ and denote by ν_g the geometric measure theoretic outward unit conormal to Ω. In this setting, given any section $u\in W^{1,1}(\Omega)\otimes\mathcal{E}$ define the $\mathcal{G}$-valued distribution $\overset{\bullet}{\partial}{}_\tau^{\mathbf{P},\mathbf{Q}}u$ on $\partial\Omega$ as

$$\overset{\bullet}{\partial}{}_\tau^{\mathbf{P},\mathbf{Q}}u:=\mathrm{i}\,\mathrm{Sym}(\mathbf{Q};\nu_g)\bullet(\mathbf{P}u)\in\mathrm{Lip}'(\partial\Omega).\tag{3.5.45}$$

A few comments are in order here. First, the composition

$$\begin{aligned}&\mathbf{QP}:\mathscr{C}^2(\mathcal{M})\otimes\mathcal{E}\longrightarrow\mathscr{C}^0(\mathcal{M})\otimes\mathcal{G}\\&\text{is a first-order differential operator,}\\&\text{namely }\mathbf{QP}=\nabla_X^{\mathcal{G}}D-D\nabla_X^{\mathcal{E}}=:[\nabla_X,D].\end{aligned}\tag{3.5.46}$$

Granted this and the current assumptions, Definition 3.5.1 then guarantees that the "bullet tangential derivative" $\overset{\bullet}{\partial}{}_\tau^{\mathbf{P},\mathbf{Q}}u$ is indeed a well-defined $\mathcal{G}$-valued distribution on the set $\partial\Omega$. To better understand its action on arbitrary test functions, we need to take a closer look at the operators $\mathbf{P}$, $\mathbf{Q}$ from (3.5.43)-(3.5.44). Observe that for any

$$a\in\mathscr{C}^1(\mathcal{M})\otimes\mathcal{G},\quad b\in\mathscr{C}^1(\mathcal{M})\otimes\mathcal{E},\quad w\in\mathscr{C}^1(\mathcal{M})\otimes\mathcal{G}\tag{3.5.47}$$

we may compute, based on definitions,

$$\begin{aligned}\big\langle\mathbf{Q}^\top w,(a,b)\big\rangle&=\big\langle w,\mathbf{Q}(a,b)\big\rangle=\langle w,\nabla_X a\rangle+\langle w,Db\rangle\\&=\langle\nabla_X^\top w,a\rangle+\langle D^\top w,b\rangle=\big\langle(\nabla_X^\top,D^\top)w,(a,b)\big\rangle,\end{aligned}\tag{3.5.48}$$

thus

$$\mathbf{Q}^\top=\big(\nabla_X^\top,D^\top\big).\tag{3.5.49}$$

As such, for each $u\in W^{1,1}(\Omega)\otimes\mathcal{E}$ and $\Phi\in\mathrm{Lip}(\mathcal{M})\otimes\mathcal{G}$, we see from (3.5.43) and (3.5.49) that

$$\begin{aligned}\big\langle\mathbf{P}u,\mathbf{Q}^\top\Phi\big\rangle&=\big\langle(Du,-\nabla_X u),(\nabla_X^\top,D^\top)\Phi\big\rangle\\&=\big\langle Du,\nabla_X^\top\Phi\big\rangle-\big\langle\nabla_X u,D^\top\Phi\big\rangle.\end{aligned}\tag{3.5.50}$$

At this point, we may explicitly write the action of the $\mathcal{G}$-valued distribution $\overset{\bullet}{\partial}{}_\tau^{\mathbf{P},\mathbf{Q}}u$ on an any given test function $\varphi\in\mathrm{Lip}(\partial\Omega)\otimes\mathcal{G}$ as

$$
\begin{aligned}
{}_{\mathrm{Lip}(\partial\Omega)}\big\langle \varphi\,,\,\overset{\bullet}{\partial}{}^{\mathbf{P},\mathbf{Q}}_{\tau} u\big\rangle_{\mathrm{Lip}'(\partial\Omega)} &= {}_{\mathrm{Lip}(\partial\Omega)}\big\langle \varphi\,,\,\mathrm{i}\,\mathrm{Sym}(\mathbf{Q};\nu_g)\bullet(\mathbf{P}u)\big\rangle_{\mathrm{Lip}'(\partial\Omega)} \\
&= \int_\Omega \big\langle \mathbf{P}u\,,\,\mathbf{Q}^\top\Phi\big\rangle\,\mathrm{d}\mathcal{L}^n_g - \int_\Omega \big\langle \mathbf{Q}\mathbf{P}u\,,\,\Phi\big\rangle\,\mathrm{d}\mathcal{L}^n_g \\
&= \int_\Omega \big\langle Du, \nabla_X^\top\Phi\big\rangle\,\mathrm{d}\mathcal{L}^n_g - \int_\Omega \big\langle \nabla_X u, D^\top\Phi\big\rangle\,\mathrm{d}\mathcal{L}^n_g \\
&\quad - \int_\Omega \big\langle [\nabla_X, D]u\,,\,\Phi\big\rangle\,\mathrm{d}\mathcal{L}^n_g,
\end{aligned}
\tag{3.5.51}
$$

where $\Phi \in \mathrm{Lip}(\mathcal{M}) \otimes \mathcal{G}$ is any global Lipschitz section of $\mathcal{E}$ extending the given φ, i.e., such that $\Phi\big|_{\partial\Omega} = \varphi$. Indeed, the first equality in (3.5.51) comes from (3.5.45), the second equality in (3.5.51) is implied by the definition of the bullet product introduced in Definition 3.5.1, while the final equality in (3.5.51) is a consequence of (3.5.50) and (3.5.46).

Going further, the second definition we wish to make pertains to the notion of weak ("bullet product") conormal derivative associated with a any (quasi-)factorization of a given second-order differentential operator.

Definition 3.5.6 *Retain the above hypotheses on the Riemannian manifold $(\mathcal{M}, g)$ and the Hermitian vector bundles $\mathcal{E}, \mathcal{G} \to \mathcal{M}$. Also, assume $L : \mathcal{E} \to \mathcal{E}$ is a second-order differential operator which has the quasi-factorization*

$$
L = \widetilde{D}D + W \tag{3.5.52}
$$

for some first-order differential operators

$$
D : \mathcal{E} \longrightarrow \mathcal{G}, \quad \widetilde{D} : \mathcal{G} \longrightarrow \mathcal{E}, \tag{3.5.53}
$$

with top coefficients of class $\mathscr{C}^1$ and lower-order coefficients of class $\mathscr{C}^0$, and a zero-order operator $W \in \mathrm{Hom}\,(\mathcal{E},\mathcal{E})$ with bounded coefficients. Finally, let $\Omega \subset \mathcal{M}$ be an Ahlfors regular domain with outward unit conormal $\nu_g : \partial\Omega \to T^*\mathcal{M}$ and surface measure $\sigma_g := \mathcal{H}^{n-1}_g \lfloor \partial\Omega$.

Then, for each section $u \in W^{1,1}(\Omega) \otimes \mathcal{E}$ such that $Lu = 0$ in Ω, define the weak (bullet product) conormal derivative $\overset{\bullet}{\partial}{}^L_{\nu_g} u$ as the $\mathcal{E}$-valued distribution on $\partial\Omega$ given by

$$
\overset{\bullet}{\partial}{}^L_{\nu_g} u := (-\mathrm{i})\mathrm{Sym}(\widetilde{D};\nu_g) \bullet (Du) \in \mathrm{Lip}'(\partial\Omega), \tag{3.5.54}
$$

where $(-\mathrm{i})\mathrm{Sym}(\widetilde{D};\nu_g) \bullet (Du)$ is considered as in Definition 3.5.1.

Hence, in the context of the above definition, from (3.5.54) and (3.5.2) (written with D replaced by $\widetilde{D}$) we see that for every test function $\varphi \in \mathrm{Lip}(\partial\Omega) \otimes \mathcal{E}$ we have

$$
\begin{aligned}
{}_{\mathrm{Lip}(\partial\Omega)}\big\langle \varphi\,,\,\overset{\bullet}{\partial}{}^L_{\nu_g} u\big\rangle_{\mathrm{Lip}'(\partial\Omega)} &= {}_{\mathrm{Lip}(\partial\Omega)}\Big\langle \varphi\,,\,(-\mathrm{i})\mathrm{Sym}(\widetilde{D};\nu_g)\bullet(Du)\Big\rangle_{\mathrm{Lip}'(\partial\Omega)} \\
&= -\int_\Omega \big\langle Du\,,\,\widetilde{D}^\top\Phi\big\rangle\,\mathrm{d}\mathcal{L}^n_g - \int_\Omega \big\langle Wu\,,\,\Phi\big\rangle\,\mathrm{d}\mathcal{L}^n_g
\end{aligned}
\tag{3.5.55}
$$

for every $\Phi \in \mathrm{Lip}(\mathcal{M}) \otimes \mathcal{E}$ such that $\Phi\big|_{\partial\Omega} = \varphi$.

We are now prepared to state and prove the following basic representation formula for nullsolutions of second-order differential operators acting between two vector-bundles over a Riemannian manifold. Once again, reader is advised to recall the conventions made in (2.8.90)-(2.8.94).

Theorem 3.5.7 *Let $\mathcal{M}$ be a compact, boundaryless, oriented $\mathscr{C}^3$ manifold of (real) dimension $n \geq 2$, equipped with a Riemannian metric g of class $\mathscr{C}^3$. Consider two Hermitian vector bundles, $\mathcal{E}, \mathcal{G}$ over $\mathcal{M}$, of class $\mathscr{C}^3$ and equip them with connections*

$$\nabla = \nabla^{\mathcal{E}} : T\mathcal{M} \times \mathcal{E} \to \mathcal{E}, \quad \nabla = \nabla^{\mathcal{G}} : T\mathcal{M} \times \mathcal{G} \to \mathcal{G}. \tag{3.5.56}$$

Next, assume $L : \mathcal{E} \to \mathcal{E}$ is a weakly elliptic second-order partial differential operator which has the quasi-factorization

$$L = \widetilde{D}D + W \tag{3.5.57}$$

for some first-order differential operators

$$D : \mathcal{E} \longrightarrow \mathcal{G}, \quad \widetilde{D} : \mathcal{G} \longrightarrow \mathcal{E}, \tag{3.5.58}$$

with top coefficients of class $\mathscr{C}^3$ and lower-order coefficients of class $\mathscr{C}^2$, and a zero-order operator $W \in \mathrm{Hom}\,(\mathcal{E}, \mathcal{E})$ with coefficients of class $\mathscr{C}^1$. In addition, suppose[13]

$$L : H^1(\mathcal{M}) \otimes \mathcal{E} \longrightarrow H^{-1}(\mathcal{M}) \otimes \mathcal{E} \text{ is an isomorphism,} \tag{3.5.59}$$

and denote by $E_L(x, y)$ the Schwartz kernel of L^{-1}, the inverse of L in (3.5.59).

Suppose $\Omega \subset \mathcal{M}$ is an arbitrary Ahlfors regular domain, with outward unit conormal $\nu_g : \partial\Omega \to T^*\mathcal{M}$ and surface measure $\sigma_g := \mathcal{H}_g^{n-1} \lfloor \partial\Omega$. Finally, fix an aperture parameter $\kappa \in (0, \infty)$ and assume the section $u \in W^{1,1}(\Omega) \otimes \mathcal{E}$ is such that

$$\begin{gathered} \mathcal{N}_\kappa u \text{ belongs to the space } L^1(\partial\Omega, \sigma_g), \\ u\big|_{\partial\Omega}^{\kappa-\mathrm{n.t.}} \text{ exists } \sigma_g\text{-a.e. on } \partial\Omega, \\ \text{and } Lu = 0 \text{ in } \Omega. \end{gathered} \tag{3.5.60}$$

Then for each $\mathscr{C}^1$ vector field $X \in T\mathcal{M}$ and each point $x \in \Omega$ one has (with notation introduced in Definition 3.5.5 and Definition 3.5.6)

[13] recall that $H^{\pm 1}(\mathcal{M})$ denote L^2-based Sobolev spaces on $\mathcal{M}$; alternatively, $H^1(\mathcal{M}) = W^{1,2}(\mathcal{M})$ and $H^{-1}(\mathcal{M}) = W^{-1,2}(\mathcal{M})$

$$
\begin{aligned}
(\nabla_X u)(x) = &-{}_{\mathrm{Lip}(\partial\Omega)}\Big\langle \big(I \otimes \widetilde{D}^{\top}\big) E_L(x,\cdot),\ \mathring{\partial}_{\tau}^{\mathbf{P},\mathbf{Q}} u\Big\rangle_{\mathrm{Lip}'(\partial\Omega)} \\
&- {}_{\mathrm{Lip}(\partial\Omega)}\Big\langle \big(\nabla_X \otimes I\big) E_L(x,\cdot),\ \mathring{\partial}_{\nu_g}^{L} u\Big\rangle_{\mathrm{Lip}'(\partial\Omega)} \\
&+ \int_{\partial\Omega} \big\langle \omega_X(x,y), \big(u\big|_{\partial\Omega}^{\kappa-\mathrm{n.t.}}\big)(y)\big\rangle \, d\sigma_g(y) \\
&+ \int_{\partial\Omega} \big\langle \big(I \otimes \mathrm{i}\,\mathrm{Sym}\big([\nabla_X, D]; \nu_g\big)^{\top} \widetilde{D}^{\top}\big) E_L(x,y), \big(u\big|_{\partial\Omega}^{\kappa-\mathrm{n.t.}}\big)(y)\big\rangle \, d\sigma_g(y) \\
&- \int_{\partial\Omega} \big\langle \big(I \otimes (-\mathrm{i})\mathrm{Sym}(\nabla_X^{\top}; \nu_g) W^{\top}\big) E_L(x,y), \big(u\big|_{\partial\Omega}^{\kappa-\mathrm{n.t.}}\big)(y)\big\rangle \, d\sigma_g(y)
\end{aligned}
\tag{3.5.61}
$$

where ω_X is the kernel function from (2.8.252).

Proof Recall that the regularity properties of the Schwartz kernel $E_L(\cdot,\cdot)$ of L^{-1} (including the nature of the singularity along the diagonal) have been discussed in the proof of Theorem 2.8.4. The current stronger smoothness assumptions adopted for the metric tensor g and the coefficients of L ensure similar regularity properties one unit higher. Also, granted the weak ellipticity of L and the conditions imposed on its coefficients, the (local) elliptic regularity result established in [119, Proposition 2.7, p. 20] may be used to conclude that

$$
u \in \mathscr{C}^2(\Omega) \otimes \mathcal{E}. \tag{3.5.62}
$$

To proceed, fix a point $x \in \Omega$ along with a $\mathscr{C}^1$ vector field $X \in T\mathcal{M}$, and pick a scalar function $\eta \in \mathscr{C}_c^2(\Omega)$ such that $\eta \equiv 1$ near x. Then the properties of η and the definition of $\mathring{\partial}_{\tau}^{\mathbf{P},\mathbf{Q}} u$ from (3.5.51) combined with (3.5.50) written for $\Phi := \big(I \otimes (1-\eta)\widetilde{D}^{\top}\big) E_L(x,\cdot)$ imply

$$
\begin{aligned}
&{}_{\mathrm{Lip}(\partial\Omega)}\Big\langle \big(I\otimes \widetilde{D}^{\top}\big)E_L(x,\cdot),\ \mathring{\partial}_{\tau}^{\mathbf{P},\mathbf{Q}}u\Big\rangle_{\mathrm{Lip}'(\partial\Omega)} \\
&\quad = -{}_{\mathrm{Lip}(\partial\Omega)}\Big\langle \big(I\otimes (1-\eta)\widetilde{D}^{\top}\big)E_L(x,\cdot),\ \mathring{\partial}_{\tau}^{\mathbf{P},\mathbf{Q}}u\Big\rangle_{\mathrm{Lip}'(\partial\Omega)} \\
&\quad = -\int_{\Omega}\Big\langle \big(I\otimes [\nabla_X^{\top},1-\eta]\widetilde{D}^{\top}\big)E_L(x,\cdot),\ Du\Big\rangle\,\mathrm{d}\mathcal{L}_g^n \\
&\qquad -\int_{\Omega}\Big\langle \big(I\otimes (1-\eta)\nabla_X^{\top}\widetilde{D}^{\top}\big)E_L(x,\cdot),\ Du\Big\rangle\,\mathrm{d}\mathcal{L}_g^n \\
&\qquad +\int_{\Omega}\Big\langle \big(I\otimes [D^{\top},1-\eta]\widetilde{D}^{\top}\big)E_L(x,\cdot),\ \nabla_X u\Big\rangle\,\mathrm{d}\mathcal{L}_g^n \\
&\qquad +\int_{\Omega}\Big\langle \big(I\otimes (1-\eta)D^{\top}\widetilde{D}^{\top}\big)E_L(x,\cdot),\ \nabla_X u\Big\rangle\,\mathrm{d}\mathcal{L}_g^n \\
&\qquad +\int_{\Omega}\Big\langle \big(I\otimes \widetilde{D}^{\top}\big)E_L(x,\cdot),\ (1-\eta)[\nabla_X,D]u\Big\rangle\,\mathrm{d}\mathcal{L}_g^n.
\end{aligned}
\tag{3.5.63}
$$

Next, apply (3.5.21) and (3.5.20) to rewrite the commutators involving $1-\eta$ in (3.5.63) in the form

$$
\begin{aligned}
&[\nabla_X^{\top},1-\eta] = (-\mathrm{i})\mathrm{Sym}\big(\nabla_X^{\top}; d(1-\eta)\big) = (-\mathrm{i})\,\mathrm{Sym}\big(\nabla_X; d\eta\big)^{\top} \\
&\quad\text{and } [D^{\top},1-\eta] = \mathrm{i}\,\mathrm{Sym}(D^{\top}; d\eta) = (-\mathrm{i})\,\mathrm{Sym}(D; d\eta)^{\top}.
\end{aligned}
\tag{3.5.64}
$$

Also, since $D^{\top}\widetilde{D}^{\top} = L^{\top} - W^{\top}$ and $\big(I\otimes(1-\eta)L^{\top}\big)E_L(x,\cdot) = 0$ in Ω thanks to (2.8.178) and (2.8.176), we have

$$
\begin{aligned}
&\int_{\Omega}\Big\langle \big(I\otimes (1-\eta)D^{\top}\widetilde{D}^{\top}\big)E_L(x,\cdot),\ \nabla_X u\Big\rangle\,\mathrm{d}\mathcal{L}_g^n \\
&\quad = -\int_{\Omega}\Big\langle \big(I\otimes (1-\eta)W^{\top}\big)E_L(x,\cdot),\ \nabla_X u\Big\rangle\,\mathrm{d}\mathcal{L}_g^n.
\end{aligned}
\tag{3.5.65}
$$

Moving on, in light of (3.5.55) applied with $\Phi := \big(\nabla_X\otimes I\big)E_L(x,\cdot)$ we may write

$$
\begin{aligned}
&{}_{\mathrm{Lip}(\partial\Omega)}\Big\langle \big(\nabla_X\otimes I\big)E_L(x,\cdot),\ \mathring{\partial}_{\nu_g}^{L}u\Big\rangle_{\mathrm{Lip}'(\partial\Omega)} \\
&\quad = -{}_{\mathrm{Lip}(\partial\Omega)}\Big\langle \big(\nabla_X\otimes (1-\eta)I\big)E_L(x,\cdot),\ \mathring{\partial}_{\nu_g}^{L}u\Big\rangle_{\mathrm{Lip}'(\partial\Omega)} \\
&\quad = \int_{\Omega}\Big\langle \big(\nabla_X\otimes [\widetilde{D}^{\top},1-\eta]\big)E_L(x,\cdot), Du\Big\rangle\,\mathrm{d}\mathcal{L}_g^n \\
&\qquad +\int_{\Omega}\Big\langle \big(\nabla_X\otimes (1-\eta)\widetilde{D}^{\top}\big)E_L(x,\cdot), Du\Big\rangle\,\mathrm{d}\mathcal{L}_g^n \\
&\qquad +\int_{\Omega}\Big\langle \big(\nabla_X\otimes I\big)E_L(x,\cdot), (1-\eta)Wu\Big\rangle\,\mathrm{d}\mathcal{L}_g^n.
\end{aligned}
\tag{3.5.66}
$$

Using again (3.5.21) and (3.5.20), we have

$$[\widetilde{D}^{\top}, 1-\eta] = \mathrm{i}\,\mathrm{Sym}(\widetilde{D}^{\top}; d\eta) = (-\mathrm{i})\,\mathrm{Sym}(\widetilde{D}; d\eta)^{\top}. \tag{3.5.67}$$

A combination of (3.5.63)-(3.5.67) and subsequent use to replace the first two terms in the right-hand side of (3.5.61), allow us to conclude that (3.5.61) will follow once we show the following new identity:

$$\begin{aligned}
(\nabla_X u)(x) = &-\int_{\Omega}\Big\langle \big(I\otimes \widetilde{D}^{\top}\big)E_L(x,\cdot),\,(-\mathrm{i})\,\mathrm{Sym}\big(\nabla_X; d\eta\big)Du\Big\rangle\,\mathrm{d}\mathcal{L}^n_g \\
&-\int_{\Omega}\Big\langle \big(I\otimes \nabla_X^{\top}\widetilde{D}^{\top}\big)E_L(x,\cdot),\,(1-\eta)Du\Big\rangle\,\mathrm{d}\mathcal{L}^n_g \\
&+\int_{\Omega}\Big\langle \big(I\otimes \widetilde{D}^{\top}\big)E_L(x,\cdot),\,(-\mathrm{i})\,\mathrm{Sym}(D; d\eta)\nabla_X u\Big\rangle\,\mathrm{d}\mathcal{L}^n_g \\
&-\int_{\Omega}\Big\langle \big(I\otimes (1-\eta)W^{\top}\big)E_L(x,\cdot),\,\nabla_X u\Big\rangle\,\mathrm{d}\mathcal{L}^n_g \\
&+\int_{\Omega}\Big\langle \big(I\otimes \widetilde{D}^{\top}\big)E_L(x,\cdot),\,(1-\eta)[\nabla_X, D]u\Big\rangle\,\mathrm{d}\mathcal{L}^n_g \\
&+\int_{\Omega}\Big\langle \big(\nabla_X\otimes I\big)E_L(x,\cdot),(-\mathrm{i})\,\mathrm{Sym}(\widetilde{D}; d\eta)Du\Big\rangle\,\mathrm{d}\mathcal{L}^n_g \\
&+\int_{\Omega}\Big\langle \big(\nabla_X\otimes \widetilde{D}^{\top}\big)E_L(x,\cdot),(1-\eta)Du\Big\rangle\,\mathrm{d}\mathcal{L}^n_g \\
&+\int_{\Omega}\Big\langle \big(\nabla_X\otimes I\big)E_L(x,\cdot),(1-\eta)Wu\Big\rangle\,\mathrm{d}\mathcal{L}^n_g \\
&+\int_{\partial\Omega}\Big\langle \omega_X(x,\cdot),\,u\big|^{\kappa-\mathrm{n.t.}}_{\partial\Omega}\Big\rangle\,\mathrm{d}\sigma_g \\
&+\int_{\partial\Omega}\Big\langle \big(I\otimes \mathrm{i}\,\mathrm{Sym}\big([\nabla_X, D]; \nu_g\big)^{\top}\widetilde{D}^{\top}\big)E_L(x,\cdot),u\big|^{\kappa-\mathrm{n.t.}}_{\partial\Omega}\Big\rangle\,\mathrm{d}\sigma_g \\
&-\int_{\partial\Omega}\Big\langle \big(I\otimes (-\mathrm{i})\mathrm{Sym}\big(\nabla_X^{\top}; \nu_g\big)W^{\top}\big)E_L(x,\cdot),u\big|^{\kappa-\mathrm{n.t.}}_{\partial\Omega}\Big\rangle\,\mathrm{d}\sigma_g \\
=:\,& I_1^{\Omega}+I_2^{\Omega}+I_3^{\Omega}+I_4^{\Omega}+I_5^{\Omega}+I_6^{\Omega}+I_7^{\Omega}+I_8^{\Omega}+I_1^{\partial\Omega}+I_2^{\partial\Omega}+I_3^{\partial\Omega}.
\end{aligned} \tag{3.5.68}$$

Above, for each $j\in\{1,\dots,8\}$ the symbol I_j^{Ω} denotes the j-th solid integral over Ω, including the sign in front, as listed in the right-hand side of (3.5.68), while for each $k\in\{1,2,3\}$ the symbol $I_k^{\partial\Omega}$ denotes the k-th boundary integral over $\partial\Omega$, including the sign in front, as listed in the right-hand side of (3.5.68).

The crux of the matter now is to invoke the Divergence Formula established in [112, Theorem 1.11.1] in the context of Riemannian manifolds in order to convert the boundary integrals into solid integrals. We shall define the vector field $\vec{F}$ (to which [112, Theorem 1.11.1] applies) implicitly, asking that for each point $y\in\Omega$

the inner product of $\vec{F}(y) \in T_y\mathcal{M}$ with an arbitrary cotangent vector $\xi \in T_y^*\mathcal{M}$ is given by

$$\begin{aligned}(\vec{F}(y), \xi) := &\Big\langle\Big(\nabla_X \otimes (-\mathrm{i})\mathrm{Sym}\big(D^\top;\xi\big)\widetilde{D}^\top\Big)E_L(x,y),\, (1-\eta(y))u(y)\Big\rangle\\ &-\Big\langle\Big(I \otimes (-\mathrm{i})\mathrm{Sym}\big(D^\top;\xi\big)\nabla_X^\top\widetilde{D}^\top\Big)E_L(x,y),\, (1-\eta(y))u(y)\Big\rangle\\ &-\Big\langle\Big(I \otimes (-\mathrm{i})\mathrm{Sym}\big([\nabla_X, D];\xi\big)^\top\widetilde{D}^\top\Big)E_L(x,y),\, (1-\eta(y))u(y)\Big\rangle\\ &-\Big\langle\Big(I \otimes (-\mathrm{i})\mathrm{Sym}\big(\nabla_X^\top;\xi\big)W^\top\Big)E_L(x,y),\, (1-\eta(y))u(y)\Big\rangle. \qquad (3.5.69)\end{aligned}$$

From this writing, assumptions, and (3.5.62) it transpires that $\vec{F} \in L^1_{\mathrm{loc}}(\Omega, \mathcal{L}_g^n)\otimes T\mathcal{M}$, that the nontangential trace that $\vec{F}\big|_{\partial\Omega}^{\kappa-\mathrm{n.t.}}$ exists at σ_g-a.e. point on $\partial\Omega$, and that $\mathcal{N}_\kappa\vec{F} \in L^1(\partial\Omega, \sigma_g)$. that $\vec{F}\big|_{\partial\Omega}^{\kappa-\mathrm{n.t.}}$ exists at σ_g-a.e. point on $\partial\Omega$. In relation to the third term in the right side of (3.5.69) we wish to note that, based on (3.5.20) and the fact that

$$[A, B]^\top = -[A^\top, B^\top] \text{ for generic operators } A, B, \qquad (3.5.70)$$

we have

$$\mathrm{Sym}\big([\nabla_X; D];\xi\big)^\top = \mathrm{Sym}\big([\nabla_X^\top; D^\top];\xi\big) \qquad (3.5.71)$$

for every $\xi \in T_y^*\mathcal{M}$. This is going to be relevant shortly.

The applicability of the Divergence Formula from [112, Theorem 1.11.1] requires that $\mathrm{div}_g\,\vec{F}$, the differential geometric divergence of $\vec{F}$ computed in the sense of distributions in Ω, belongs to $L^1(\Omega, \mathcal{L}_g^n)$. To see why this is indeed the case, we shall identify $\mathrm{div}_g\,\vec{F}$ explicitly. To this end, pick arbitrary scalar-valued function $\varphi \in \mathscr{C}_c^2(\Omega)$ and use (3.5.69) and (3.5.71) to write

$$\begin{aligned}\langle \mathrm{div}_g\,\vec{F}, \varphi\rangle &= -(\vec{F}, d\varphi) \qquad (3.5.72)\\ &= -\int_\Omega \Big\langle\Big(\nabla_X \otimes (-\mathrm{i})\mathrm{Sym}\big(D^\top; d\varphi\big)\widetilde{D}^\top\Big)E_L(x,\cdot),\, (1-\eta)u\Big\rangle\, \mathrm{d}\mathcal{L}_g^n\\ &\quad+\int_\Omega \Big\langle\Big(I \otimes (-\mathrm{i})\mathrm{Sym}\big(D^\top; d\varphi\big)\nabla_X^\top\widetilde{D}^\top\Big)E_L(x,\cdot),\, (1-\eta)u\Big\rangle\, \mathrm{d}\mathcal{L}_g^n\\ &\quad+\int_\Omega \Big\langle\Big(I \otimes (-\mathrm{i})\mathrm{Sym}\big([\nabla_X^\top, D^\top]; d\varphi\big)\widetilde{D}^\top\Big)E_L(x,\cdot),\, (1-\eta)u\Big\rangle\, \mathrm{d}\mathcal{L}_g^n\\ &\quad+\int_\Omega \Big\langle\Big(I \otimes (-\mathrm{i})\mathrm{Sym}\big(\nabla_X^\top; d\varphi\big)W^\top\Big)E_L(x,\cdot),\, (1-\eta)u\Big\rangle\, \mathrm{d}\mathcal{L}_g^n.\end{aligned}$$

Invoking (3.5.21), each one of the principal symbols in (3.5.72) may be written as commutators in the form

$$\begin{aligned}(-\mathrm{i})\mathrm{Sym}\big(D^{\top}; d\varphi\big) &= \big[D^{\top}, \varphi\big],\\ (-\mathrm{i})\mathrm{Sym}\big([\nabla^{\top}_{X}, D^{\top}]; d\varphi\big) &= \big[[\nabla^{\top}_{X}, D^{\top}], \varphi\big],\\ (-\mathrm{i})\mathrm{Sym}\big(\nabla^{\top}_{X}; d\varphi\big) &= \big[\nabla^{\top}_{X}, \varphi\big].\end{aligned} \tag{3.5.73}$$

Using (3.5.73) back in (3.5.72) and then integrating by parts (note that the integration by parts does not yield any boundary terms due to the fact that φ is compactly supported in Ω) further gives

$$\begin{aligned}\langle \mathrm{div}_g \vec{F}, \varphi\rangle = &-\int_{\Omega}\Big\langle\Big(\nabla_X \otimes \varphi\,\widetilde{D}^{\top}\Big)E_L(x,\cdot),\, D\big((1-\eta)u\big)\Big\rangle\,\mathrm{d}\mathcal{L}^n_g\\ &+\int_{\Omega}\Big\langle\Big(\nabla_X \otimes \varphi\, D^{\top}\widetilde{D}^{\top}\Big)E_L(x,\cdot),\, (1-\eta)u\Big\rangle\,\mathrm{d}\mathcal{L}^n_g\\ &+\int_{\Omega}\Big\langle\Big(I \otimes \varphi\nabla^{\top}_X\widetilde{D}^{\top}\Big)E_L(x,\cdot),\, D\big((1-\eta)u\big)\Big\rangle\,\mathrm{d}\mathcal{L}^n_g\\ &-\int_{\Omega}\Big\langle\Big(I \otimes \varphi\, D^{\top}\nabla^{\top}_X\widetilde{D}^{\top}\Big)E_L(x,\cdot),\, (1-\eta)u\Big\rangle\,\mathrm{d}\mathcal{L}^n_g\\ &-\int_{\Omega}\Big\langle\Big(I \otimes \varphi\,\widetilde{D}^{\top}\Big)E_L(x,\cdot),\, [\nabla_X, D]\big((1-\eta)u\big)\Big\rangle\,\mathrm{d}\mathcal{L}^n_g\\ &-\int_{\Omega}\Big\langle\Big(I \otimes \varphi[\nabla^{\top}_X, D^{\top}]\widetilde{D}^{\top}\Big)E_L(x,\cdot),\, (1-\eta)u\Big\rangle\,\mathrm{d}\mathcal{L}^n_g\\ &+\int_{\Omega}\Big\langle\Big(I \otimes \varphi W^{\top}\Big)E_L(x,\cdot),\, \nabla_X\big((1-\eta)u\big)\Big\rangle\,\mathrm{d}\mathcal{L}^n_g\\ &-\int_{\Omega}\Big\langle\Big(I \otimes \varphi\nabla^{\top}_X W^{\top}\Big)E_L(x,\cdot),\, (1-\eta)u\Big\rangle\,\mathrm{d}\mathcal{L}^n_g,\end{aligned} \tag{3.5.74}$$

keeping in mind for the fifth integral in (3.5.74) that $[\nabla^{\top}_X, D^{\top}] = -[\nabla_X, D]^{\top}$ (cf. (3.5.70)). Invoking the quasi-factorization assumed in (3.5.57), the fact that we have $\big(I \otimes L^{\top}\big)E_L(x,\cdot) = \delta_{\Delta}(x,\cdot)$ (cf. (2.8.178)), and that $\eta \equiv 1$ near x, we also obtain

$$\begin{aligned}&\int_{\Omega}\Big\langle\Big(\nabla_X \otimes \varphi\, D^{\top}\widetilde{D}^{\top}\Big)E_L(x,\cdot),\, (1-\eta)u\Big\rangle\,\mathrm{d}\mathcal{L}^n_g\\ &\qquad = -\int_{\Omega}\Big\langle\Big(\nabla_X \otimes \varphi\, W^{\top}\Big)E_L(x,\cdot),\, (1-\eta)u\Big\rangle\,\mathrm{d}\mathcal{L}^n_g.\end{aligned} \tag{3.5.75}$$

Similarly, since

$$\begin{aligned}&D^{\top}\nabla^{\top}_X\widetilde{D}^{\top} + [\nabla^{\top}_X, D^{\top}]\widetilde{D}^{\top} + \nabla^{\top}_X W^{\top}\\ &\qquad = D^{\top}\nabla^{\top}_X\widetilde{D}^{\top} + \nabla^{\top}_X D^{\top}\widetilde{D}^{\top} - D^{\top}\nabla^{\top}_X\widetilde{D}^{\top} + \nabla^{\top}_X W^{\top}\\ &\qquad = \nabla^{\top}_X L^{\top}\end{aligned} \tag{3.5.76}$$

and again making use of the fact that $(I \otimes L^\top)E_L(x,\cdot) = \delta_\Delta(x,\cdot)$ (cf. (2.8.178)) and $\eta \equiv 1$ near x, we obtain

$$
\begin{aligned}
&-\int_\Omega \Big\langle \Big(I \otimes \varphi D^\top \nabla_X^\top \widetilde{D}^\top\Big) E_L(x,\cdot),\, (1-\eta)u \Big\rangle \, d\mathcal{L}_g^n \\
&-\int_\Omega \Big\langle \Big(I \otimes \varphi [\nabla_X^\top, D^\top] \widetilde{D}^\top\Big) E_L(x,\cdot),\, (1-\eta)u \Big\rangle \, d\mathcal{L}_g^n \\
&-\int_\Omega \Big\langle \Big(I \otimes \varphi \nabla_X^\top W^\top\Big) E_L(x,\cdot),\, (1-\eta)u \Big\rangle \, d\mathcal{L}_g^n \\
&= -\int_\Omega \varphi \Big\langle \Big(I \otimes \nabla_X^\top L^\top\Big) E_L(x,\cdot),\, (1-\eta)u \Big\rangle \, d\mathcal{L}_g^n = 0. \qquad (3.5.77)
\end{aligned}
$$

Together, (3.5.74), (3.5.75), and (3.5.77) imply

$$
\begin{aligned}
\langle \mathrm{div}_g \vec{F}, \varphi \rangle = &-\int_\Omega \varphi \Big\langle \Big(\nabla_X \otimes \widetilde{D}^\top\Big) E_L(x,\cdot),\, D\big((1-\eta)u\big) \Big\rangle \, d\mathcal{L}_g^n \\
&-\int_\Omega \varphi \Big\langle \Big(\nabla_X \otimes W^\top\Big) E_L(x,\cdot),\, (1-\eta)u \Big\rangle \, d\mathcal{L}_g^n \\
&+\int_\Omega \varphi \Big\langle \Big(I \otimes \nabla_X^\top \widetilde{D}^\top\Big) E_L(x,\cdot),\, D\big((1-\eta)u\big) \Big\rangle \, d\mathcal{L}_g^n \\
&-\int_\Omega \varphi \Big\langle \Big(I \otimes \widetilde{D}^\top\Big) E_L(x,\cdot),\, [\nabla_X, D]\big((1-\eta)u\big) \Big\rangle \, d\mathcal{L}_g^n \\
&+\int_\Omega \varphi \Big\langle \Big(I \otimes W^\top\Big) E_L(x,\cdot),\, \nabla_X\big((1-\eta)u\big) \Big\rangle \, d\mathcal{L}_g^n. \qquad (3.5.78)
\end{aligned}
$$

Since $\varphi \in \mathscr{C}_c^2(\Omega)$ is arbitrary, from (3.5.78) we ultimately conclude that

$$
\begin{aligned}
\mathrm{div}_g \vec{F} = &-\Big\langle \Big(\nabla_X \otimes \widetilde{D}^\top\Big) E_L(x,\cdot),\, D\big((1-\eta)u\big) \Big\rangle \\
&-\Big\langle \Big(\nabla_X \otimes W^\top\Big) E_L(x,\cdot),\, (1-\eta)u \Big\rangle \\
&+\Big\langle \Big(I \otimes \nabla_X^\top \widetilde{D}^\top\Big) E_L(x,\cdot),\, D\big((1-\eta)u\big) \Big\rangle \\
&-\Big\langle \Big(I \otimes \widetilde{D}^\top\Big) E_L(x,\cdot),\, [\nabla_X, D]\big((1-\eta)u\big) \Big\rangle \\
&+\Big\langle \Big(I \otimes W^\top\Big) E_L(x,\cdot),\, \nabla_X\big((1-\eta)u\big) \Big\rangle. \qquad (3.5.79)
\end{aligned}
$$

An inspection of (3.5.79) reveals that since $\eta \equiv 1$ near x, there is no blow up term in $\mathrm{div}_g \vec{F}$ and, in the final analysis, we obtain

$$
\mathrm{div}\, \vec{F} \in L^1(\Omega, \mathcal{L}_g^n). \qquad (3.5.80)
$$

To continue, write

$$\int_\Omega \operatorname{div}_g \vec{F}\, d\mathcal{L}_g^n = II_1 + II_2 + II_3 + II_4 + II_5, \tag{3.5.81}$$

where

$$\begin{aligned}
II_1 &:= -\int_\Omega \Big\langle \Big(\nabla_X \otimes \widetilde{D}^\top\Big) E_L(x,\cdot),\, D\big((1-\eta)u\big)\Big\rangle \, d\mathcal{L}_g^n \\
II_2 &:= -\int_\Omega \Big\langle \Big(\nabla_X \otimes W^\top\Big) E_L(x,\cdot),\, (1-\eta)u\Big\rangle \, d\mathcal{L}_g^n \\
II_3 &:= \int_\Omega \Big\langle \Big(I \otimes \nabla_X^\top \widetilde{D}^\top\Big) E_L(x,\cdot),\, D\big((1-\eta)u\big)\Big\rangle \, d\mathcal{L}_g^n \\
II_4 &:= -\int_\Omega \Big\langle \Big(I \otimes \widetilde{D}^\top\Big) E_L(x,\cdot),\, [\nabla_X, D]\big((1-\eta)u\big)\Big\rangle \, d\mathcal{L}_g^n \\
II_5 &:= \int_\Omega \Big\langle \Big(I \otimes W^\top\Big) E_L(x,\cdot),\, \nabla_X\big((1-\eta)u\big)\Big\rangle \, d\mathcal{L}_g^n .
\end{aligned} \tag{3.5.82}$$

To proceed further, observe that

$$[D, 1-\eta]u = -[D,\eta]u \text{ is a compactly supported function in } \Omega. \tag{3.5.83}$$

Using (3.5.83) and integrating by parts, we obtain

$$\begin{aligned}
II_1 + I_7^\Omega &= -\int_\Omega \Big\langle \Big(\nabla_X \otimes \widetilde{D}^\top\Big) E_L(x,\cdot),\, [D, 1-\eta]u\Big)\Big\rangle \, d\mathcal{L}_g^n \\
&= \int_\Omega \Big\langle \Big(\nabla_X \otimes \widetilde{D}^\top\Big) E_L(x,\cdot),\, [D,\eta]u\Big)\Big\rangle \, d\mathcal{L}_g^n \\
&= \int_\Omega \Big\langle \Big(\nabla_X \otimes I\Big) E_L(x,\cdot),\, \widetilde{D}[D,\eta]u\Big)\Big\rangle \, d\mathcal{L}_g^n .
\end{aligned} \tag{3.5.84}$$

In addition, since in general $A[B,C] = [AB,C] - [A,C]B$, we write

$$\begin{aligned}
\widetilde{D}[D,\eta]u &= \big[\widetilde{D}D,\eta\big]u - \big[\widetilde{D},\eta\big]Du \\
&= [L,\eta]u - [W,\eta]u - (-\mathrm{i})\mathrm{Sym}\big(\widetilde{D}; d\eta\big)Du \\
&= L(\eta u) - (-\mathrm{i})\mathrm{Sym}\big(\widetilde{D}; d\eta\big)Du,
\end{aligned} \tag{3.5.85}$$

making use of the fact that $Lu = 0$ in Ω, the operator W commutes with multiplication by η, and also relying on (3.5.21). Together, (3.5.84), (3.5.85), and integration by parts (with no boundary terms since ηu is compactly supported in Ω) yield

$$
\begin{aligned}
II_1 + I_7^{\Omega} &= \int_{\Omega} \Big\langle \Big(\nabla_X \otimes I\Big) E_L(x,\cdot),\ L(\eta u)\Big\rangle \,\mathrm{d}\mathcal{L}_g^n \\
&\quad - \int_{\Omega} \Big\langle \Big(\nabla_X \otimes I\Big) E_L(x,\cdot),\ (-\mathrm{i})\mathrm{Sym}\big(\widetilde{D}; d\eta\big) Du\Big\rangle \,\mathrm{d}\mathcal{L}_g^n \\
&= \int_{\Omega} \Big\langle \Big(\nabla_X \otimes L^{\top}\Big) E_L(x,\cdot),\ \eta u\Big\rangle \,\mathrm{d}\mathcal{L}_g^n - I_6^{\Omega} \\
&= (\nabla_X u)(x) - I_6^{\Omega}.
\end{aligned}
\tag{3.5.86}
$$

Next, (3.5.83) and integration by parts allows us to write (in light of the fact that $[D,\eta]u$ is compactly supported in Ω)

$$
\begin{aligned}
II_3 + I_2^{\Omega} &= \int_{\Omega} \Big\langle \Big(I \otimes \nabla_X^{\top} \widetilde{D}^{\top}\Big) E_L(x,\cdot),\ [D, 1-\eta]u\Big\rangle \,\mathrm{d}\mathcal{L}_g^n \\
&= - \int_{\Omega} \Big\langle (I \otimes \nabla_X^{\top} \widetilde{D}^{\top}) E_L(x,\cdot),\ [D,\eta]u\Big\rangle \,\mathrm{d}\mathcal{L}_g^n \\
&= - \int_{\Omega} \Big\langle (I \otimes \widetilde{D}^{\top}) E_L(x,\cdot),\ \nabla_X [D,\eta]u\Big\rangle \,\mathrm{d}\mathcal{L}_g^n.
\end{aligned}
\tag{3.5.87}
$$

We may also combine II_4 and I_5^{Ω} in the form

$$
\begin{aligned}
II_4 + I_5^{\Omega} &= - \int_{\Omega} \Big\langle \Big(I \otimes \widetilde{D}^{\top}\Big) E_L(x,\cdot),\ \big[[\nabla_X, D], 1-\eta\big]u\Big\rangle \,\mathrm{d}\mathcal{L}_g^n \\
&= \int_{\Omega} \Big\langle (I \otimes \widetilde{D}^{\top}) E_L(x,\cdot),\ [\nabla_X, D]\eta u - \eta[\nabla_X, D]u\Big\rangle \,\mathrm{d}\mathcal{L}_g^n.
\end{aligned}
\tag{3.5.88}
$$

On account of (3.5.21) we may trade the principal symbols in I_1^{Ω} and I_3^{Ω} for commutators and obtain

$$
\begin{aligned}
I_1^{\Omega} + I_3^{\Omega} &= - \int_{\Omega} \Big\langle (I \otimes \widetilde{D}^{\top}) E_L(x,\cdot),\ \big[\nabla_X, \eta\big] Du\Big\rangle \,\mathrm{d}\mathcal{L}_g^n \\
&\quad + \int_{\Omega} \Big\langle (I \otimes \widetilde{D}^{\top}) E_L(x,\cdot),\ [D,\eta]\nabla_X u\Big\rangle \,\mathrm{d}\mathcal{L}_g^n.
\end{aligned}
\tag{3.5.89}
$$

A close inspection of II_2 and I_8^{Ω} reveals that they cancel, thus

$$
\begin{aligned}
II_2 + I_8^{\Omega} &= - \int_{\Omega} \Big\langle \Big(\nabla_X \otimes W^{\top}\Big) E_L(x,\cdot),\ (1-\eta)u\Big\rangle \,\mathrm{d}\mathcal{L}_g^n \\
&\quad + \int_{\Omega} \Big\langle (\nabla_X \otimes I) E_L(x,\cdot), (1-\eta) W u \Big) \Big\rangle \,\mathrm{d}\mathcal{L}_g^n = 0,
\end{aligned}
\tag{3.5.90}
$$

while

$$
\begin{aligned}
II_5 + I_4^{\Omega} &= \int_{\Omega} \Big\langle \Big(I \otimes W^{\top} \Big) E_L(x, \cdot), \ \big[\nabla_X, 1-\eta\big] u \Big\rangle \, d\mathcal{L}_g^n \\
&= \int_{\Omega} \Big\langle \big(I \otimes (1-\eta) W^{\top} \big) E_L(x, \cdot), \ \big[\eta, \nabla_X\big] u \Big\rangle \, d\mathcal{L}_g^n.
\end{aligned} \tag{3.5.91}
$$

Thus combined, (3.5.86), (3.5.90), (3.5.87), (3.5.88), (3.5.91), and (3.5.89) imply

$$
\begin{aligned}
&\int_{\Omega} \mathrm{div}_g \vec{F} \, d\mathcal{L}_g^n + I_1^{\Omega} + I_2^{\Omega} + I_3^{\Omega} + I_4^{\Omega} + I_5^{\Omega} + I_6^{\Omega} + I_7^{\Omega} + I_8^{\Omega} \\
&\quad = (II_1 + I_7^{\Omega} + I_6^{\Omega}) + (II_2 + I_8^{\Omega}) + (II_3 + I_2^{\Omega}) \\
&\qquad + (II_4 + I_5^{\Omega}) + (II_5 + I_4^{\Omega}) + (I_1^{\Omega} + I_3^{\Omega}) \\
&\quad = (\nabla_X u)(x) - \int_{\Omega} \Big\langle \big(I \otimes \widetilde{D}^{\top} \big) E_L(x, \cdot), \ \nabla_X [D, \eta] u \Big\rangle \, d\mathcal{L}_g^n \\
&\qquad + \int_{\Omega} \Big\langle \big(I \otimes \widetilde{D}^{\top} \big) E_L(x, \cdot), \ [\nabla_X, D]\eta u - \eta[\nabla_X, D] u \Big\rangle \, d\mathcal{L}_g^n \\
&\qquad + \int_{\Omega} \Big\langle \Big(I \otimes W^{\top} \Big) E_L(x, \cdot), \ \big[\eta, \nabla_X\big] u \Big\rangle \, d\mathcal{L}_g^n \\
&\qquad - \int_{\Omega} \Big\langle \big(I \otimes \widetilde{D}^{\top} \big) E_L(x, \cdot), \ \big[\nabla_X, \eta\big] D u \Big\rangle \, d\mathcal{L}_g^n \\
&\qquad + \int_{\Omega} \Big\langle \big(I \otimes \widetilde{D}^{\top} \big) E_L(x, \cdot), \ [D, \eta] \nabla_X u \Big\rangle \, d\mathcal{L}_g^n \\
&\quad = (\nabla_X u)(x) + \int_{\Omega} \Big\langle \big(I \otimes \widetilde{D}^{\top} \big) E_L(x, \cdot), \ D[\eta, \nabla_X] u \Big\rangle \, d\mathcal{L}_g^n \\
&\qquad + \int_{\Omega} \Big\langle \Big(I \otimes W^{\top} \Big) E_L(x, \cdot), \ \big[\eta, \nabla_X\big] u \Big\rangle \, d\mathcal{L}_g^n,
\end{aligned} \tag{3.5.92}
$$

where the last equality is a consequence of

$$
\begin{aligned}
&-\nabla_X [D, \eta] + [\nabla_X, D]\eta - \eta[\nabla_X, D] - \big[\nabla_X; \eta\big] D + [D; \eta] \nabla_X \\
&\quad = -\nabla_X D \, \eta + \nabla_X \eta \, D + \nabla_X D \, \eta - D \nabla_X \eta - \eta \nabla_X D + \eta \, D \nabla_X \\
&\qquad - \nabla_X \eta D + \eta \nabla_X D + D \, \eta \nabla_X - \eta \, D \nabla_X \\
&\quad = -D \nabla_X \eta + D \, \eta \nabla_X = D[\eta, \nabla_X].
\end{aligned} \tag{3.5.93}
$$

Given that $[\eta, \nabla_X] u$ is of class $\mathscr{C}^1$ and compactly supported in Ω, we may again integrate by parts

$$
\begin{aligned}
&\int_\Omega \Big\langle (I\otimes \widetilde{D}^\top)E_L(x,\cdot),\, D[\eta,\nabla_X]u\Big\rangle\,\mathrm{d}\mathcal{L}^n_g\\
&= {}_{\mathcal{D}'(\Omega)}\Big\langle (I\otimes \widetilde{D}^\top)E_L(x,\cdot),\, D[\eta,\nabla_X]u\Big\rangle_{\mathcal{D}(\Omega)}\\
&= {}_{\mathcal{D}'(\Omega)}\Big\langle (I\otimes D^\top\widetilde{D}^\top)E_L(x,\cdot),\, [\eta,\nabla_X]u\Big\rangle_{\mathcal{D}(\Omega)}\\
&= {}_{\mathcal{D}'(\Omega)}\Big\langle (I\otimes (L^\top - W^\top))E_L(x,\cdot),\, [\eta,\nabla_X]u\Big\rangle_{\mathcal{D}(\Omega)}\\
&= \langle \delta_x, [\eta,\nabla_X]u\rangle - \int_\Omega \Big\langle (I\otimes W^\top)E_L(x,\cdot),\, [\eta,\nabla_X]u\Big\rangle\,\mathrm{d}\mathcal{L}^n_g\\
&= -\int_\Omega \Big\langle (I\otimes W^\top)E_L(x,\cdot),\, [\eta,\nabla_X]u\Big\rangle\,\mathrm{d}\mathcal{L}^n_g \qquad (3.5.94)
\end{aligned}
$$

again recalling that $[\eta,\nabla_X]u$ is identically zero near x. Finally, (3.5.92), (3.5.94), and the Divergence Theorem imply

$$
\begin{aligned}
I_1^{\partial\Omega} + I_2^{\partial\Omega} + I_3^{\partial\Omega} &= \int_{\partial\Omega} \big\langle \nu_g, \vec{F}\big|^{\kappa-\mathrm{n.t.}}_{\partial\Omega}\big\rangle\,\mathrm{d}\sigma_g = \int_\Omega \mathrm{div}_g\,\vec{F}\,\mathrm{d}\mathcal{L}^n_g\\
&= (\nabla_X u)(x) - I_1^\Omega - I_2^\Omega - I_3^\Omega - I_4^\Omega - I_5^\Omega - I_6^\Omega - I_7^\Omega - I_8^\Omega \qquad (3.5.95)
\end{aligned}
$$

which proves (3.5.68), as wanted. □

As a corollary of Theorem 3.5.7, we obtain the following Fatou-type result for nullsolutions of weakly elliptic second-order differential operators, in UR subdomains Riemannian manifolds.

Corollary 3.5.8 *Let $\mathcal{M}$ be a compact, boundaryless, oriented $\mathscr{C}^3$ manifold of (real) dimension n, where $n\in\mathbb{N}$ satisfies $n\geq 2$, equipped with a Riemannian metric g of class $\mathscr{C}^3$. Consider a Hermitian vector bundle $\mathcal{E}\to\mathcal{M}$ of class $\mathscr{C}^2$ and denote by $\nabla=\nabla^{\mathcal{E}}: T\mathcal{M}\times\mathcal{E}\to\mathcal{E}$ a connection on $\mathcal{E}$.*

Next, assume $L:\mathcal{E}\to\mathcal{E}$ is a weakly elliptic second-order partial differential operator with the property that, when written in local coordinates as

$$
L = \Big(\sum_{j,k}\partial_j a^{\alpha\beta}_{jk}\partial_k + \sum_j b^{\alpha\beta}_j\partial_j + d^{\alpha\beta}\Big)_{\alpha,\beta} \tag{3.5.96}
$$

its coefficients satisfy

$$
a^{\alpha\beta}_{jk}\in\mathscr{C}^3,\quad b^{\alpha\beta}_j\in\mathscr{C}^2,\quad d^{\alpha\beta}\in\mathscr{C}^1. \tag{3.5.97}
$$

In addition, suppose

$$
L: H^1(\mathcal{M})\otimes\mathcal{E}\longrightarrow H^{-1}(\mathcal{M})\otimes\mathcal{E} \text{ is an isomorphism.} \tag{3.5.98}
$$

Let $\Omega \subseteq \mathcal{M}$ be an arbitrary UR *domain, and abbreviate $\sigma_g := \mathcal{H}_g^{n-1}\lfloor\partial\Omega$. Finally, fix an aperture parameter $\kappa > 0$ and assume the section $u \in W^{1,1}_{\rm loc}(\Omega) \otimes \mathcal{E}$ is such that the nontangential boundary trace $u\big|^{\kappa-{\rm n.t.}}_{\partial\Omega}$ exists at σ_g-a.e. point on $\partial\Omega$ and*

$$
\begin{aligned}
&Lu = 0 \text{ in } \Omega, \text{ as well as } \mathcal{N}_\kappa u \in L^1(\partial\Omega, \sigma_g)\\
&\text{and } \mathcal{N}_\kappa(\nabla u) \in L^p(\partial\Omega, \sigma_g) \text{ for some } p \in \big(\tfrac{n}{n-1}, \infty\big).
\end{aligned}
\tag{3.5.99}
$$

Then the nontangential trace $(\nabla u)\big|^{\kappa-{\rm n.t.}}_{\partial\Omega}$ exists at σ_g-a.e. point on $\partial\Omega$. Moreover, as a function, $(\nabla u)\big|^{\kappa-{\rm n.t.}}_{\partial\Omega}$ is σ_g-measurable and actually independent of the aperture parameter κ.

Proof We debut by making two observations. First, since the scale of Lebesgue spaces on $\partial\Omega$ is nested, as far as the conclusion we seek is concerned we may assume that $p \in \big(\frac{n}{n-1}, 1\big]$. Second, we can find a $\mathscr{C}^3$ Hermitian vector bundle $\mathcal{G} \to \mathcal{M}$ along with two first-order differential operators

$$
D : \mathcal{E} \longrightarrow \mathcal{G} \text{ and } \widetilde{D} : \mathcal{G} \longrightarrow \mathcal{E}, \tag{3.5.100}
$$

with top coefficients of class $\mathscr{C}^3$ and lower-order coefficients of class $\mathscr{C}^2$, along with a zeroth-order operator $W \in \mathrm{Hom}\,(\mathcal{E}, \mathcal{E})$ with coefficients of class $\mathscr{C}^1$ which allow us to express L as

$$
L = \widetilde{D}D + W. \tag{3.5.101}
$$

To proceed, fix an arbitrary $\mathscr{C}^1$ vector field $X \in T\mathcal{M}$. In relation to D and X introduce the first-order differential operators $\mathbf{P}, \mathbf{Q}$ as in (3.5.43)-(3.5.44). Granted the current assumptions, we may invoke Theorem 3.5.2 which, in light of Definition 3.5.5 and Definition 3.5.6, guarantees that

$$
\mathring{\partial}^{\mathbf{P},\mathbf{Q}}_\tau u \in H^p(\partial\Omega, \sigma_g) \otimes \mathcal{G} \text{ and } \mathring{\partial}^{L}_{\nu_g} u \in H^p(\partial\Omega, \sigma_g) \otimes \mathcal{E}. \tag{3.5.102}
$$

Having established these memberships and keeping in mind the compatibility between the duality pairing from [113, Theorem 4.6.1] and the distributional pairing (cf. [113, Lemma 4.6.4]), we may now write the representation formula (3.5.61) from Theorem 3.5.7 with the brackets in the right-hand side now denoting the pairing between the Hardy space H^p and its dual $(H^p)^*$ (cf. [113, Theorem 4.6.1]). With this interpretation, item *(5)* in Theorem 2.8.2 applies and, together with Lemma 2.8.9, gives that $(\nabla_X u)\big|^{\kappa-{\rm n.t.}}_{\partial\Omega}$ exists σ_g-a.e. on $\partial\Omega$. In view of the arbitrariness of X, this ultimately shows that the nontangential boundary trace $(\nabla u)\big|^{\kappa-{\rm n.t.}}_{\partial\Omega}$ exists at σ_g-a.e. point on $\partial\Omega$. The remaining properties claimed in the statement then follow from this, the last membership in (3.5.99), [112, (8.9.8)], and [112, Proposition 8.9.8]. □

It is of interest to prove Fatou-type results in the spirit of Corollary 3.5.8 with no assumptions on the existence of the nontangential trace of the function in question. We succeed in accomplishing this later, in Theorem 3.5.10 which, in effect, is our main result for second-order operators in this section. As a preamble, we first establish two basic representation formulas with weak traces in the theorem below (the reader is advised to recall the conventions made in (2.8.90)-(2.8.94)).

Theorem 3.5.9 *Let $\mathcal{M}$ be a compact, boundaryless, oriented $\mathscr{C}^3$ manifold of (real) dimension $n \geq 2$, equipped with a Riemannian metric g of class $\mathscr{C}^3$. Consider two Hermitian vector bundles, $\mathcal{E}$ and $\mathcal{G}$ over $\mathcal{M}$, of class $\mathscr{C}^3$ and equip them with connections*

$$\nabla = \nabla^{\mathcal{E}} : T\mathcal{M} \times \mathcal{E} \to \mathcal{E}, \quad \nabla = \nabla^{\mathcal{G}} : T\mathcal{M} \times \mathcal{G} \to \mathcal{G}. \tag{3.5.103}$$

Next, assume $L : \mathcal{E} \to \mathcal{E}$ is a weakly elliptic second-order partial differential operator which has the quasi-factorization

$$L = \widetilde{D}D + W \tag{3.5.104}$$

for some first-order differential operators

$$D : \mathcal{E} \longrightarrow \mathcal{G}, \quad \widetilde{D} : \mathcal{G} \longrightarrow \mathcal{E}, \tag{3.5.105}$$

with top coefficients of class $\mathscr{C}^3$ and lower-order coefficients of class $\mathscr{C}^2$, and a zero-order operator $W \in \mathrm{Hom}(\mathcal{E}, \mathcal{E})$ with coefficients of class $\mathscr{C}^1$. In addition, suppose

$$L : H^1(\mathcal{M}) \otimes \mathcal{E} \longrightarrow H^{-1}(\mathcal{M}) \otimes \mathcal{E} \text{ is an isomorphism,} \tag{3.5.106}$$

and denote by $E_L(x, y)$ the Schwartz kernel of L^{-1}, the inverse of L in (3.5.106).

Finally, having fixed an Ahlfors regular domain $\Omega \subseteq \mathcal{M}$ with geometric measure theoretic outward unit conormal $\nu_g \in T^*\mathcal{M}$, pick a section satisfying

$$u \in W^{1,1}(\Omega) \otimes \mathcal{E} \text{ and } Lu = 0 \text{ in } \Omega. \tag{3.5.107}$$

Then for each $\mathscr{C}^1$ vector field $X \in T\mathcal{M}$ and each point $x \in \Omega$ one has (with notation introduced in Definition 3.5.1, Definition 3.5.5, and Definition 3.5.6)

$$\begin{aligned}
(\nabla_X u)(x) = &-{}_{\mathrm{Lip}(\partial\Omega)}\Big\langle \big(I \otimes \widetilde{D}^{\top}\big) E_L(x, \cdot),\, \mathring{\partial}_{\tau}^{\mathbf{P},\mathbf{Q}} u \Big\rangle_{\mathrm{Lip}'(\partial\Omega)} \\
&- {}_{\mathrm{Lip}(\partial\Omega)}\Big\langle \big(\nabla_X \otimes I\big) E_L(x, \cdot),\, \mathring{\partial}_{\nu_g}^{L} u \Big\rangle_{\mathrm{Lip}'(\partial\Omega)} \\
&- {}_{\mathrm{Lip}(\partial\Omega)}\Big\langle \big(\nabla_X \otimes \widetilde{D}^{\top}\big) E_L(x, \cdot),\, (-\mathrm{i})\mathrm{Sym}(D; \nu_g) \bullet u \Big\rangle_{\mathrm{Lip}'(\partial\Omega)} \\
&+ {}_{\mathrm{Lip}(\partial\Omega)}\Big\langle \big(I \otimes \nabla_X^{\top} \widetilde{D}^{\top}\big) E_L(x, \cdot),\, (-\mathrm{i})\mathrm{Sym}(D; \nu_g) \bullet u \Big\rangle_{\mathrm{Lip}'(\partial\Omega)} \\
&- {}_{\mathrm{Lip}(\partial\Omega)}\Big\langle \big(I \otimes \widetilde{D}^{\top}\big) E_L(x, \cdot),\, (-\mathrm{i})\mathrm{Sym}\big([\nabla_X, D]; \nu_g\big) \bullet u \Big\rangle_{\mathrm{Lip}'(\partial\Omega)} \\
&+ {}_{\mathrm{Lip}(\partial\Omega)}\Big\langle \big(I \otimes W^{\top}\big) E_L(x, \cdot),\, (-\mathrm{i})\mathrm{Sym}(\nabla_X; \nu_g) \bullet u \Big\rangle_{\mathrm{Lip}'(\partial\Omega)}.
\end{aligned} \tag{3.5.108}$$

In addition, under the same assumptions as above, at each point $x \in \Omega$ one has (with notation introduced in Definition 3.5.1 and Definition 3.5.6)

$$u(x) = -{}_{\mathrm{Lip}(\partial\Omega)}\Big\langle \big(I \otimes \widetilde{D}^{\top}\big)E_L(x,\cdot),\ (-\mathrm{i})\,\mathrm{Sym}\big(D;\nu_g\big) \bullet u\Big\rangle_{\mathrm{Lip}'(\partial\Omega)}$$
$$- {}_{\mathrm{Lip}(\partial\Omega)}\Big\langle E_L(x,\cdot),\ \mathring{\partial}^{L}_{\nu_g} u\Big\rangle_{\mathrm{Lip}'(\partial\Omega)}. \tag{3.5.109}$$

Proof The regularity properties of the Schwartz kernel $E_L(\cdot,\cdot)$ of L^{-1} (including the nature of the singularity along the diagonal) have been reviewed during the course of the proof of Theorem 2.8.4. The present stronger smoothness assumptions adopted for the metric tensor g and the coefficients of L guarantee similar regularity properties one unit higher. In addition, granted the weak ellipticity of L and the conditions imposed on its coefficients, the (local) elliptic regularity result established in [119, Proposition 2.7, p. 20] may be employed to conclude that

$$u \in \mathscr{C}^2(\Omega) \otimes \mathcal{E}. \tag{3.5.110}$$

To continue, select an arbitrary point $x \in \Omega$ together with a $\mathscr{C}^1$ vector field $X \in T\mathcal{M}$, and pick a scalar function $\eta \in \mathscr{C}^2_c(\Omega)$ such that $\eta \equiv 1$ near x. Then the properties of η and formula (3.5.51) written for $\Phi := \big(I \otimes (1-\eta)\widetilde{D}^{\top}\big)E_L(x,\cdot)$ imply

$$-{}_{\mathrm{Lip}(\partial\Omega)}\Big\langle \big(I \otimes \widetilde{D}^{\top}\big)E_L(x,\cdot),\ \mathring{\partial}^{\mathbf{P},\mathbf{Q}}_{\tau} u\Big\rangle_{\mathrm{Lip}'(\partial\Omega)}$$
$$= -{}_{\mathrm{Lip}(\partial\Omega)}\Big\langle \big(I \otimes (1-\eta)\widetilde{D}^{\top}\big)E_L(x,\cdot),\ \mathring{\partial}^{\mathbf{P},\mathbf{Q}}_{\tau} u\Big\rangle_{\mathrm{Lip}'(\partial\Omega)}$$
$$= -\int_{\Omega} \Big\langle \big(I \otimes [\nabla^{\top}_{X}, 1-\eta]\widetilde{D}^{\top}\big)E_L(x,\cdot),\ Du\Big\rangle\, d\mathcal{L}^n_g$$
$$- \int_{\Omega} \Big\langle \big(I \otimes (1-\eta)\nabla^{\top}_{X}\widetilde{D}^{\top}\big)E_L(x,\cdot),\ Du\Big\rangle\, d\mathcal{L}^n_g$$
$$+ \int_{\Omega} \Big\langle \big(I \otimes [D^{\top}, 1-\eta]\widetilde{D}^{\top}\big)E_L(x,\cdot),\ \nabla_X u\Big\rangle\, d\mathcal{L}^n_g$$
$$+ \int_{\Omega} \Big\langle \big(I \otimes (1-\eta)D^{\top}\widetilde{D}^{\top}\big)E_L(x,\cdot),\ \nabla_X u\Big\rangle\, d\mathcal{L}^n_g$$
$$+ \int_{\Omega} \Big\langle \big(I \otimes \widetilde{D}^{\top}\big)E_L(x,\cdot),\ (1-\eta)[\nabla_X, D]u\Big\rangle\, d\mathcal{L}^n_g. \tag{3.5.111}$$

Next, we invoke (3.5.21) and (3.5.20) to rewrite the commutators involving $1-\eta$ in (3.5.111) in the form

$$[\nabla^{\top}_{X}, 1-\eta] = (-\mathrm{i})\mathrm{Sym}\big(\nabla^{\top}_{X}; d(1-\eta)\big) = (-\mathrm{i})\,\mathrm{Sym}\big(\nabla_X; d\eta\big)^{\top}$$
$$\text{and } [D^{\top}, 1-\eta] = \mathrm{i}\,\mathrm{Sym}(D^{\top}; d\eta) = (-\mathrm{i})\,\mathrm{Sym}(D; d\eta)^{\top}. \tag{3.5.112}$$

Also, since $D^\top \widetilde{D}^\top = L^\top - W^\top$ and $\big(I \otimes (1-\eta)L^\top\big)E_L(x,\cdot) = 0$ in Ω (as seen from (2.8.178) and (2.8.176)), we have

$$\int_\Omega \Big\langle \big(I \otimes (1-\eta)D^\top \widetilde{D}^\top\big)E_L(x,\cdot),\, \nabla_X u\Big\rangle\, d\mathcal{L}^n_g$$
$$= -\int_\Omega \Big\langle \big(I \otimes (1-\eta)W^\top\big)E_L(x,\cdot),\, \nabla_X u\Big\rangle\, d\mathcal{L}^n_g. \tag{3.5.113}$$

Moving on, in light of (3.5.55) applied with $\Phi := \big(\nabla_X \otimes I\big)E_L(x,\cdot)$, we may write

$$-{}_{\mathrm{Lip}(\partial\Omega)}\Big\langle \big(\nabla_X \otimes I\big)E_L(x,\cdot),\, \dot{\partial}^L_{\nu_g} u\Big\rangle_{\mathrm{Lip}'(\partial\Omega)}$$
$$= -{}_{\mathrm{Lip}(\partial\Omega)}\Big\langle \big(\nabla_X \otimes (1-\eta)I\big)E_L(x,\cdot),\, \dot{\partial}^L_{\nu_g} u\Big\rangle_{\mathrm{Lip}'(\partial\Omega)}$$
$$= \int_\Omega \Big\langle \big(\nabla_X \otimes [\widetilde{D}^\top, 1-\eta]\big)E_L(x,\cdot), Du\Big\rangle\, d\mathcal{L}^n_g$$
$$+ \int_\Omega \Big\langle \big(\nabla_X \otimes (1-\eta)\widetilde{D}^\top\big)E_L(x,\cdot), Du\Big\rangle\, d\mathcal{L}^n_g$$
$$+ \int_\Omega \Big\langle \big(\nabla_X \otimes I\big)E_L(x,\cdot), (1-\eta)Wu\Big\rangle\, d\mathcal{L}^n_g. \tag{3.5.114}$$

Recalling (3.5.21) and (3.5.20), we have

$$[\widetilde{D}^\top, 1-\eta] = \mathrm{i}\,\mathrm{Sym}(\widetilde{D}^\top; d\eta) = (-\mathrm{i})\,\mathrm{Sym}(\widetilde{D}; d\eta)^\top \tag{3.5.115}$$

which when used in (3.5.114) yields

$$-{}_{\mathrm{Lip}(\partial\Omega)}\Big\langle \big(\nabla_X \otimes I\big)E_L(x,\cdot),\, \dot{\partial}^L_{\nu_g} u\Big\rangle_{\mathrm{Lip}'(\partial\Omega)}$$
$$= \int_\Omega \Big\langle \big(\nabla_X \otimes I\big)E_L(x,\cdot), (-\mathrm{i})\,\mathrm{Sym}(\widetilde{D}; d\eta)Du\Big\rangle\, d\mathcal{L}^n_g$$
$$+ \int_\Omega \Big\langle \big(\nabla_X \otimes (1-\eta)\widetilde{D}^\top\big)E_L(x,\cdot), Du\Big\rangle\, d\mathcal{L}^n_g$$
$$+ \int_\Omega \Big\langle \big(\nabla_X \otimes I\big)E_L(x,\cdot), (1-\eta)Wu\Big\rangle\, d\mathcal{L}^n_g. \tag{3.5.116}$$

For the remaining four terms in the right hand-side of (3.5.108) we use the same circle of ideas and the definition in (3.5.2) to rewrite the pairings as combinations of solid integrals over Ω. First, corresponding to $\Phi := \big(\nabla_X \otimes (1-\eta)\widetilde{D}^\top\big)E_L(x,\cdot)$ in (3.5.2), we obtain

$$
\begin{aligned}
&-{}_{\mathrm{Lip}(\partial\Omega)}\Big\langle \big(\nabla_X \otimes \widetilde{D}^\top\big)E_L(x,\cdot),\, (-\mathrm{i})\mathrm{Sym}\big(D;\nu_g\big)\bullet u\Big\rangle_{\mathrm{Lip}'(\partial\Omega)}\\
&\quad = -{}_{\mathrm{Lip}(\partial\Omega)}\Big\langle \big(\nabla_X \otimes (1-\eta)\widetilde{D}^\top\big)E_L(x,\cdot),\, (-\mathrm{i})\mathrm{Sym}\big(D;\nu_g\big)\bullet u\Big\rangle_{\mathrm{Lip}'(\partial\Omega)}\\
&\quad = -\int_\Omega \Big\langle \big(\nabla_X \otimes (1-\eta)\widetilde{D}^\top\big)E_L(x,\cdot),\, Du\Big\rangle\,\mathrm{d}\mathcal{L}_g^n\\
&\qquad + \int_\Omega \Big\langle \big(\nabla_X \otimes [D^\top, 1-\eta]\widetilde{D}^\top\big)E_L(x,\cdot),\, u\Big\rangle\,\mathrm{d}\mathcal{L}_g^n\\
&\qquad + \int_\Omega \Big\langle \big(\nabla_X \otimes (1-\eta)D^\top\widetilde{D}^\top\big)E_L(x,\cdot),\, u\Big\rangle\,\mathrm{d}\mathcal{L}_g^n\\
&\quad = -\int_\Omega \Big\langle \big(\nabla_X \otimes (1-\eta)\widetilde{D}^\top\big)E_L(x,\cdot),\, Du\Big\rangle\,\mathrm{d}\mathcal{L}_g^n\\
&\qquad + \int_\Omega \Big\langle \big(\nabla_X \otimes \widetilde{D}^\top\big)E_L(x,\cdot),\, (-\mathrm{i})\,\mathrm{Sym}(D;d\eta)u\Big\rangle\,\mathrm{d}\mathcal{L}_g^n\\
&\qquad - \int_\Omega \Big\langle \big(\nabla_X \otimes (1-\eta)W^\top\big)E_L(x,\cdot),\, u\Big\rangle\,\mathrm{d}\mathcal{L}_g^n
\end{aligned}
\tag{3.5.117}
$$

where for the last equality we have also relied on (3.5.112). Second, definition (3.5.2) applied with $\Phi := \big(I \otimes (1-\eta)\nabla_X^\top\widetilde{D}^\top\big)E_L(x,\cdot)$ and (3.5.112) allow us to write

$$
\begin{aligned}
&{}_{\mathrm{Lip}(\partial\Omega)}\Big\langle \big(I \otimes \nabla_X^\top\widetilde{D}^\top\big)E_L(x,\cdot),\, (-\mathrm{i})\mathrm{Sym}\big(D;\nu_g\big)\bullet u\Big\rangle_{\mathrm{Lip}'(\partial\Omega)}\\
&\quad = {}_{\mathrm{Lip}(\partial\Omega)}\Big\langle \big(I \otimes (1-\eta)\nabla_X^\top\widetilde{D}^\top\big)E_L(x,\cdot),\, (-\mathrm{i})\mathrm{Sym}\big(D;\nu_g\big)\bullet u\Big\rangle_{\mathrm{Lip}'(\partial\Omega)}\\
&\quad = \int_\Omega \Big\langle \big(I \otimes (1-\eta)\nabla_X^\top\widetilde{D}^\top\big)E_L(x,\cdot),\, Du\Big\rangle\,\mathrm{d}\mathcal{L}_g^n\\
&\qquad - \int_\Omega \Big\langle \big(I \otimes [D^\top, 1-\eta]\nabla_X^\top\widetilde{D}^\top\big)E_L(x,\cdot),\, u\Big\rangle\,\mathrm{d}\mathcal{L}_g^n\\
&\qquad - \int_\Omega \Big\langle \big(I \otimes (1-\eta)D^\top\nabla_X^\top\widetilde{D}^\top\big)E_L(x,\cdot),\, u\Big\rangle\,\mathrm{d}\mathcal{L}_g^n\\
&\quad = \int_\Omega \Big\langle \big(I \otimes (1-\eta)\nabla_X^\top\widetilde{D}^\top\big)E_L(x,\cdot),\, Du\Big\rangle\,\mathrm{d}\mathcal{L}_g^n\\
&\qquad - \int_\Omega \Big\langle \big(I \otimes \nabla_X^\top\widetilde{D}^\top\big)E_L(x,\cdot),\, (-\mathrm{i})\mathrm{Sym}(D;d\eta)u\Big\rangle\,\mathrm{d}\mathcal{L}_g^n\\
&\qquad - \int_\Omega \Big\langle \big(I \otimes (1-\eta)D^\top\nabla_X^\top\widetilde{D}^\top\big)E_L(x,\cdot),\, u\Big\rangle\,\mathrm{d}\mathcal{L}_g^n.
\end{aligned}
\tag{3.5.118}
$$

Third, (3.5.2) applied with $\Phi := \big(I \otimes (1-\eta)\widetilde{D}^\top\big)E_L(x,\cdot)$ and with the operator $\big[\nabla_X, D\big]$ in place of D, implies

$$
\begin{aligned}
&-{}_{\mathrm{Lip}(\partial\Omega)}\Big\langle \big(I\otimes \widetilde{D}^{\top}\big)E_L(x,\cdot),\ (-\mathrm{i})\mathrm{Sym}\big([\nabla_X,D];\nu_g\big)\bullet u\Big\rangle_{\mathrm{Lip}'(\partial\Omega)}\\
&\quad= -{}_{\mathrm{Lip}(\partial\Omega)}\Big\langle \big(I\otimes (1-\eta)\widetilde{D}^{\top}\big)E_L(x,\cdot),\ (-\mathrm{i})\mathrm{Sym}\big([\nabla_X,D];\nu_g\big)\bullet u\Big\rangle_{\mathrm{Lip}'(\partial\Omega)}\\
&\quad= -\int_\Omega \Big\langle \big(I\otimes (1-\eta)\widetilde{D}^{\top}\big)E_L(x,\cdot),\ \big[\nabla_X,D\big]u\Big\rangle\,\mathrm{d}\mathcal{L}^n_g\\
&\qquad+\int_\Omega \Big\langle \big(I\otimes \big[[\nabla_X,D]^{\top},1-\eta\big]\widetilde{D}^{\top}\big)E_L(x,\cdot),\ u\Big\rangle\,\mathrm{d}\mathcal{L}^n_g\\
&\qquad+\int_\Omega \Big\langle \big(I\otimes (1-\eta)[\nabla_X,D]^{\top}\widetilde{D}^{\top}\big)E_L(x,\cdot),\ u\Big\rangle\,\mathrm{d}\mathcal{L}^n_g\\
&\quad= -\int_\Omega \Big\langle \big(I\otimes (1-\eta)\widetilde{D}^{\top}\big)E_L(x,\cdot),\ \big[\nabla_X,D\big]u\Big\rangle\,\mathrm{d}\mathcal{L}^n_g\\
&\qquad+\int_\Omega \Big\langle \big(I\otimes \widetilde{D}^{\top}\big)E_L(x,\cdot),\ (-\mathrm{i})\mathrm{Sym}([\nabla_X,D];d\eta)u\Big\rangle\,\mathrm{d}\mathcal{L}^n_g\\
&\qquad+\int_\Omega \Big\langle \big(I\otimes (1-\eta)[\nabla_X,D]^{\top}\widetilde{D}^{\top}\big)E_L(x,\cdot),\ u\Big\rangle\,\mathrm{d}\mathcal{L}^n_g
\end{aligned}
\tag{3.5.119}
$$

since by (3.5.21) and (3.5.20),

$$
\big[[\nabla_X,D]^{\top},1-\eta\big] = -\big[[\nabla_X,D]^{\top},\eta\big] = (-\mathrm{i})\,\mathrm{Sym}([\nabla_X,D];d\eta)^{\top}. \tag{3.5.120}
$$

Finally, a fourth application of (3.5.2) with $\Phi := \big(I\otimes(1-\eta)W^{\top}\big)E_L(x,\cdot)$ and with the operator ∇_X in place of D, permits us to write

$$
\begin{aligned}
&{}_{\mathrm{Lip}(\partial\Omega)}\Big\langle \big(I\otimes W^{\top}\big)E_L(x,\cdot),\ (-\mathrm{i})\mathrm{Sym}\big(\nabla_X;\nu_g\big)\bullet u\Big\rangle_{\mathrm{Lip}'(\partial\Omega)}\\
&\quad= {}_{\mathrm{Lip}(\partial\Omega)}\Big\langle \big(I\otimes (1-\eta)W^{\top}\big)E_L(x,\cdot),\ (-\mathrm{i})\mathrm{Sym}\big(\nabla_X;\nu_g\big)\bullet u\Big\rangle_{\mathrm{Lip}'(\partial\Omega)}\\
&\quad= \int_\Omega \Big\langle \big(I\otimes (1-\eta)W^{\top}\big)E_L(x,\cdot),\ \nabla_X u\Big\rangle\,\mathrm{d}\mathcal{L}^n_g\\
&\qquad-\int_\Omega \Big\langle \big(I\otimes [\nabla_X^{\top},1-\eta]W^{\top}\big)E_L(x,\cdot),\ u\Big\rangle\,\mathrm{d}\mathcal{L}^n_g\\
&\qquad-\int_\Omega \Big\langle \big(I\otimes (1-\eta)\nabla_X^{\top}W^{\top}\big)E_L(x,\cdot),\ u\Big\rangle\,\mathrm{d}\mathcal{L}^n_g\\
&\quad= \int_\Omega \Big\langle \big(I\otimes (1-\eta)W^{\top}\big)E_L(x,\cdot),\ \nabla_X u\Big\rangle\,\mathrm{d}\mathcal{L}^n_g\\
&\qquad-\int_\Omega \Big\langle \big(I\otimes W^{\top}\big)E_L(x,\cdot),\ (-\mathrm{i})\,\mathrm{Sym}\big(\nabla_X;d\eta\big)u\Big\rangle\,\mathrm{d}\mathcal{L}^n_g\\
&\qquad-\int_\Omega \Big\langle \big(I\otimes (1-\eta)\nabla_X^{\top}W^{\top}\big)E_L(x,\cdot),\ u\Big\rangle\,\mathrm{d}\mathcal{L}^n_g
\end{aligned}
\tag{3.5.121}
$$

by also invoking (3.5.112) for the last equality.

Next, a combination of (3.5.111), (3.5.113), (3.5.116), (3.5.117)-(3.5.119), and (3.5.121) allows us to conclude that (3.5.108) will follow once we show the following new identity:

$$
\begin{aligned}
(\nabla_X u)(x) = &- \int_\Omega \Big\langle (I \otimes \widetilde{D}^\top) E_L(x,\cdot),\, (-\mathrm{i})\,\mathrm{Sym}(\nabla_X; d\eta) Du \Big\rangle \, \mathrm{d}\mathcal{L}^n_g \\
&- \int_\Omega \Big\langle (I \otimes \nabla_X^\top \widetilde{D}^\top) E_L(x,\cdot),\, (1-\eta) Du \Big\rangle \, \mathrm{d}\mathcal{L}^n_g \\
&+ \int_\Omega \Big\langle (I \otimes \widetilde{D}^\top) E_L(x,\cdot),\, (-\mathrm{i})\,\mathrm{Sym}(D; d\eta) \nabla_X u \Big\rangle \, \mathrm{d}\mathcal{L}^n_g \\
&- \int_\Omega \Big\langle (I \otimes (1-\eta) W^\top) E_L(x,\cdot),\, \nabla_X u \Big\rangle \, \mathrm{d}\mathcal{L}^n_g \\
&+ \int_\Omega \Big\langle (I \otimes \widetilde{D}^\top) E_L(x,\cdot),\, (1-\eta)[\nabla_X, D] u \Big\rangle \, \mathrm{d}\mathcal{L}^n_g \\
&+ \int_\Omega \Big\langle (\nabla_X \otimes I) E_L(x,\cdot), (-\mathrm{i})\,\mathrm{Sym}(\widetilde{D}; d\eta) Du \Big\rangle \, \mathrm{d}\mathcal{L}^n_g \\
&+ \int_\Omega \Big\langle (\nabla_X \otimes \widetilde{D}^\top) E_L(x,\cdot), (1-\eta) Du \Big\rangle \, \mathrm{d}\mathcal{L}^n_g \\
&+ \int_\Omega \Big\langle (\nabla_X \otimes I) E_L(x,\cdot), (1-\eta) W u \Big\rangle \, \mathrm{d}\mathcal{L}^n_g \\
&- \int_\Omega \Big\langle (\nabla_X \otimes (1-\eta) \widetilde{D}^\top) E_L(x,\cdot),\, Du \Big\rangle \, \mathrm{d}\mathcal{L}^n_g \\
&+ \int_\Omega \Big\langle (\nabla_X \otimes \widetilde{D}^\top) E_L(x,\cdot),\, (-\mathrm{i})\,\mathrm{Sym}(D; d\eta) u \Big\rangle \, \mathrm{d}\mathcal{L}^n_g \\
&- \int_\Omega \Big\langle (\nabla_X \otimes (1-\eta) W^\top) E_L(x,\cdot),\, u \Big\rangle \, \mathrm{d}\mathcal{L}^n_g \\
&+ \int_\Omega \Big\langle (I \otimes (1-\eta) \nabla_X^\top \widetilde{D}^\top) E_L(x,\cdot),\, Du \Big\rangle \, \mathrm{d}\mathcal{L}^n_g \\
&- \int_\Omega \Big\langle (I \otimes \nabla_X^\top \widetilde{D}^\top) E_L(x,\cdot),\, (-\mathrm{i})\mathrm{Sym}(D; d\eta) u \Big\rangle \, \mathrm{d}\mathcal{L}^n_g \\
&- \int_\Omega \Big\langle (I \otimes (1-\eta) D^\top \nabla_X^\top \widetilde{D}^\top) E_L(x,\cdot),\, u \Big\rangle \, \mathrm{d}\mathcal{L}^n_g \\
&- \int_\Omega \Big\langle (I \otimes (1-\eta) \widetilde{D}^\top) E_L(x,\cdot),\, [\nabla_X, D] u \Big\rangle \, \mathrm{d}\mathcal{L}^n_g \\
&+ \int_\Omega \Big\langle (I \otimes \widetilde{D}^\top) E_L(x,\cdot),\, (-\mathrm{i})\mathrm{Sym}\big([\nabla_X, D]; d\eta\big) u \Big\rangle \, \mathrm{d}\mathcal{L}^n_g \\
&+ \int_\Omega \Big\langle (I \otimes (1-\eta) [\nabla_X, D]^\top \widetilde{D}^\top) E_L(x,\cdot),\, u \Big\rangle \, \mathrm{d}\mathcal{L}^n_g \\
&+ \int_\Omega \Big\langle (I \otimes (1-\eta) W^\top) E_L(x,\cdot),\, \nabla_X u \Big\rangle \, \mathrm{d}\mathcal{L}^n_g \\
&- \int_\Omega \Big\langle (I \otimes W^\top) E_L(x,\cdot),\, (-\mathrm{i})\,\mathrm{Sym}(\nabla_X; d\eta) u \Big\rangle \, \mathrm{d}\mathcal{L}^n_g \\
&- \int_\Omega \Big\langle (I \otimes (1-\eta) \nabla_X^\top W^\top) E_L(x,\cdot),\, u \Big\rangle \, \mathrm{d}\mathcal{L}^n_g \\
=:& \sum_{j=1}^{20} I_j.
\end{aligned}
\tag{3.5.122}
$$

Above, for each $j \in \{1, \dots, 20\}$ the symbol I_j stands for the j-th solid integral over Ω, including the sign in front, in the order listed in the right-hand side of (3.5.122).

An inspection of (3.5.122) reveals some immediate cancellations:

$$I_2 + I_{12} = 0, \quad I_4 + I_{18} = 0, \quad I_5 + I_{15} = 0, \quad I_7 + I_9 = 0, \quad I_8 + I_{11} = 0. \tag{3.5.123}$$

Also, recalling (3.5.21) we may combine

$$\begin{aligned} I_1 &+ I_3 + I_{16} \\ &= \int_\Omega \Big\langle (I \otimes \widetilde{D}^\top) E_L(x, \cdot), \, -\big[\nabla_X, \eta\big] Du + [D, \eta] \nabla_X u + \big[[\nabla_X, D], \eta\big] u \Big\rangle \, d\mathcal{L}_g^n. \end{aligned} \tag{3.5.124}$$

Observe that

$$\begin{aligned} -\big[\nabla_X, \eta\big] Du &+ [D, \eta] \nabla_X u + \big[[\nabla_X, D], \eta\big] u \\ &= -\nabla_X \eta\, Du + \eta\, \nabla_X Du + D\eta\, \nabla_X u - \eta\, D\nabla_X u \\ &\quad + \nabla_X D\, \eta u - D\nabla_X \eta u - \eta \nabla_X Du + \eta\, D\nabla_X u \\ &= \nabla_X [D, \eta] u - D[\nabla_X, \eta] u \ \text{ in } \Omega. \end{aligned} \tag{3.5.125}$$

This identity may be used in (3.5.124) to further obtain

$$\begin{aligned} I_1 + I_3 + I_{16} + I_{19} &= \int_\Omega \Big\langle (I \otimes \widetilde{D}^\top) E_L(x, \cdot), \, \nabla_X [D, \eta] u \Big\rangle \, d\mathcal{L}_g^n \\ &\quad - \int_\Omega \Big\langle (I \otimes \widetilde{D}^\top) E_L(x, \cdot), \, D[\nabla_X, \eta] u \Big\rangle \, d\mathcal{L}_g^n \\ &\quad - \int_\Omega \Big\langle (I \otimes W^\top) E_L(x, \cdot), \, [\nabla_X, \eta] u \Big\rangle \, d\mathcal{L}_g^n \\ &= \int_\Omega \Big\langle (I \otimes \nabla_X^\top \widetilde{D}^\top) E_L(x, \cdot), \, [D, \eta] u \Big\rangle \, d\mathcal{L}_g^n \\ &\quad - \int_\Omega \Big\langle (I \otimes D^\top \widetilde{D}^\top) E_L(x, \cdot), \, [\nabla_X, \eta] u \Big\rangle \, d\mathcal{L}_g^n \\ &\quad - \int_\Omega \Big\langle (I \otimes W^\top) E_L(x, \cdot), \, [\nabla_X, \eta] u \Big\rangle \, d\mathcal{L}_g^n \\ &= -I_{13} - \int_\Omega \Big\langle (I \otimes L^\top) E_L(x, \cdot), \, [\nabla_X, \eta] u \Big\rangle \, d\mathcal{L}_g^n \\ &= -I_{13} - \big\langle \delta_x, [\nabla_X, \eta] u \big\rangle. \end{aligned} \tag{3.5.126}$$

The second equality in (3.5.126) is based on integration by parts (keeping in mind that $[D, \eta] u$ and $[\nabla_X, \eta] u$ are compactly supported), and the last equality uses (2.8.178). Upon observing that $[\nabla_X, \eta] u$ is equal to zero near x, from (3.5.126) and

(2.8.176) we conclude that

$$I_1 + I_3 + I_{16} + I_{19} + I_{13} = 0. \tag{3.5.127}$$

Next, we claim that

$$I_{14} + I_{17} + I_{20} = 0. \tag{3.5.128}$$

To see why this is true, note that

$$-D^\top \nabla_X^\top \widetilde{D}^\top + [\nabla_X, D]^\top \widetilde{D}^\top - \nabla_X^\top W^\top \tag{3.5.129}$$
$$= -D^\top \nabla_X^\top \widetilde{D}^\top + D^\top \nabla_X^\top \widetilde{D}^\top - \nabla_X^\top D^\top \widetilde{D}^\top - \nabla_X^\top W^\top$$
$$= -\nabla_X^\top (D^\top \widetilde{D}^\top + W^\top) = -\nabla_X^\top L^\top \tag{3.5.130}$$

which implies

$$I_{14} + I_{17} + I_{20} = -\int_\Omega \Big\langle \big(I \otimes \nabla_X^\top L^\top\big) E_L(x,\cdot),\, (1-\eta)u \Big\rangle \, d\mathcal{L}_g^n = 0, \tag{3.5.131}$$

since $\big(I \otimes (1-\eta)\nabla_X^\top L^\top\big) E_L(x,\cdot) = \big(I \otimes (1-\eta)\nabla_X^\top\big)\delta_\Delta(x,\cdot) = 0$ thanks to (2.8.178), (2.8.176), and the fact that $1-\eta$ is zero near x. This justifies the claim made in (3.5.128).

We are left with simplifying $I_6 + I_{10}$. In this regard, it is useful to observe since $Lu = 0$ in Ω we have

$$\big[\widetilde{D}, \eta\big] Du = \widetilde{D}\eta\, Du - \eta \widetilde{D} Du = \widetilde{D}\,\eta\, Du - \eta(L - W)u$$
$$= \widetilde{D}\,\eta\, Du + \eta\, W u \ \text{ in } \Omega, \tag{3.5.132}$$

thus

$$\widetilde{D}[D,\eta]u = \widetilde{D}D\,\eta u - \widetilde{D}\,\eta Du = (L - W)\eta u - \widetilde{D}\,\eta Du$$
$$= L(\eta u) - \big[\widetilde{D}, \eta\big] Du \ \text{ in } \Omega. \tag{3.5.133}$$

Consequently, based on (3.5.133), integration by parts, (2.8.180), and (2.8.175) we have

$$
\begin{aligned}
I_6 &= \int_\Omega \Big\langle (\nabla_X \otimes I) E_L(x,\cdot),\ \big[\widetilde{D},\eta\big] Du \Big\rangle \, \mathrm{d}\mathcal{L}_g^n \\
&= -\int_\Omega \Big\langle (\nabla_X \otimes I) E_L(x,\cdot),\ \widetilde{D}[D,\eta]u \Big\rangle \, \mathrm{d}\mathcal{L}_g^n \\
&\quad + {}_{\mathcal{D}'(\Omega)}\Big\langle (\nabla_X \otimes I) E_L(x,\cdot),\ L(\eta u) \Big\rangle_{\mathcal{D}(\Omega)} \\
&= -I_{10} + {}_{\mathcal{D}'(\Omega)}\Big\langle (\nabla_X \otimes L^\top) E_L(x,\cdot),\ \eta u \Big\rangle_{\mathcal{D}(\Omega)} \\
&= -I_{10} + (\nabla_X u)(x).
\end{aligned} \tag{3.5.134}
$$

Together, (3.5.123), (3.5.127), (3.5.128), and (3.5.134) imply

$$
\begin{aligned}
&(I_2 + I_{12}) + (I_4 + I_{18}) + (I_5 + I_{15}) + (I_7 + I_9) + (I_8 + I_{11}) \\
&\quad + (I_1 + I_3 + I_{16} + I_{19} + I_{13}) + (I_{14} + I_{17} + I_{20}) + (I_6 + I_{10}) = (\nabla_X u)(x),
\end{aligned} \tag{3.5.135}
$$

establishing (3.5.122). This completes the proof of (3.5.108).

The justification of (3.5.109) uses the same circle of ideas as those employed in the proof (3.5.108). First, we note that the reasoning that lead to (3.5.117) now leads, after applying (3.5.2) with $\Phi := \big(I \otimes (1-\eta)\widetilde{D}^\top\big) E_L(x,\cdot)$, to the following sequence of identities:

$$
\begin{aligned}
&-{}_{\mathrm{Lip}(\partial\Omega)}\Big\langle \big(I \otimes \widetilde{D}^\top\big) E_L(x,\cdot),\ (-\mathrm{i})\mathrm{Sym}\big(D;\nu_g\big) \bullet u \Big\rangle_{\mathrm{Lip}'(\partial\Omega)} \\
&\quad = -{}_{\mathrm{Lip}(\partial\Omega)}\Big\langle \big(I \otimes (1-\eta)\widetilde{D}^\top\big) E_L(x,\cdot),\ (-\mathrm{i})\mathrm{Sym}\big(D;\nu_g\big) \bullet u \Big\rangle_{\mathrm{Lip}'(\partial\Omega)} \\
&\quad = -\int_\Omega \Big\langle \big(I \otimes (1-\eta)\widetilde{D}^\top\big) E_L(x,\cdot),\ Du \Big\rangle \, \mathrm{d}\mathcal{L}_g^n \\
&\qquad + \int_\Omega \Big\langle \big(I \otimes [D^\top, 1-\eta]\widetilde{D}^\top\big) E_L(x,\cdot),\ u \Big\rangle \, \mathrm{d}\mathcal{L}_g^n \\
&\qquad + \int_\Omega \Big\langle \big(I \otimes (1-\eta)D^\top\widetilde{D}^\top\big) E_L(x,\cdot),\ u \Big\rangle \, \mathrm{d}\mathcal{L}_g^n \\
&\quad = -\int_\Omega \Big\langle \big(I \otimes (1-\eta)\widetilde{D}^\top\big) E_L(x,\cdot),\ Du \Big\rangle \, \mathrm{d}\mathcal{L}_g^n \\
&\qquad + \int_\Omega \Big\langle \big(I \otimes \widetilde{D}^\top\big) E_L(x,\cdot),\ (-\mathrm{i})\,\mathrm{Sym}(D;d\eta)u \Big\rangle \, \mathrm{d}\mathcal{L}_g^n \\
&\qquad - \int_\Omega \Big\langle E_L(x,\cdot),\ (1-\eta)Wu \Big\rangle \, \mathrm{d}\mathcal{L}_g^n.
\end{aligned} \tag{3.5.136}
$$

Second, observe that the argument that resulted in (3.5.116) now implies

$$
\begin{aligned}
-{}_{\mathrm{Lip}(\partial\Omega)}\Big\langle E_L(x,\cdot),\ \mathring{\partial}^{L}_{\nu_g}u\Big\rangle_{\mathrm{Lip}'(\partial\Omega)}
&= \int_\Omega \Big\langle E_L(x,\cdot), (-\mathrm{i})\,\mathrm{Sym}(\widetilde{D}; d\eta)Du\Big\rangle\,\mathrm{d}\mathcal{L}^n_g \\
&\quad + \int_\Omega \Big\langle (1-\eta)\widetilde{D}^{\top}E_L(x,\cdot), Du\Big\rangle\,\mathrm{d}\mathcal{L}^n_g \\
&\quad + \int_\Omega \Big\langle E_L(x,\cdot), (1-\eta)Wu\Big\rangle\,\mathrm{d}\mathcal{L}^n_g.
\end{aligned}
\tag{3.5.137}
$$

Combined, (3.5.136), (3.5.137) and (3.5.133) yield

$$
\begin{aligned}
&-{}_{\mathrm{Lip}(\partial\Omega)}\Big\langle \big(I\otimes \widetilde{D}^{\top}\big)E_L(x,\cdot),\ (-\mathrm{i})\mathrm{Sym}\big(D;\nu_g\big)\bullet u\Big\rangle_{\mathrm{Lip}'(\partial\Omega)} \\
&\quad - {}_{\mathrm{Lip}(\partial\Omega)}\Big\langle E_L(x,\cdot),\ \mathring{\partial}^{L}_{\nu_g}u\Big\rangle_{\mathrm{Lip}'(\partial\Omega)} \\
&= -\int_\Omega \Big\langle \big(I\otimes (1-\eta)\widetilde{D}^{\top}\big)E_L(x,\cdot),\ Du\Big\rangle\,\mathrm{d}\mathcal{L}^n_g \\
&\quad + \int_\Omega \Big\langle \big(I\otimes \widetilde{D}^{\top}\big)E_L(x,\cdot),\ (-\mathrm{i})\,\mathrm{Sym}(D;d\eta)u\Big\rangle\,\mathrm{d}\mathcal{L}^n_g \\
&\quad - \int_\Omega \Big\langle \big(I\otimes (1-\eta)W^{\top}\big)E_L(x,\cdot),\ u\Big\rangle\,\mathrm{d}\mathcal{L}^n_g \\
&\quad + \int_\Omega \Big\langle E_L(x,\cdot), (-\mathrm{i})\,\mathrm{Sym}(\widetilde{D}; d\eta)Du\Big\rangle\,\mathrm{d}\mathcal{L}^n_g \\
&\quad + \int_\Omega \Big\langle \big(I\otimes (1-\eta)\widetilde{D}^{\top}\big)E_L(x,\cdot), Du\Big\rangle\,\mathrm{d}\mathcal{L}^n_g \\
&\quad + \int_\Omega \Big\langle E_L(x,\cdot), (1-\eta)Wu\Big\rangle\,\mathrm{d}\mathcal{L}^n_g \\
&= \int_\Omega \Big\langle \big(I\otimes \widetilde{D}^{\top}\big)E_L(x,\cdot),\ [D,\eta]u\Big\rangle\,\mathrm{d}\mathcal{L}^n_g \\
&\quad + \int_\Omega \Big\langle E_L(x,\cdot), \big[\widetilde{D},\eta\big]Du\Big\rangle\,\mathrm{d}\mathcal{L}^n_g \\
&= \int_\Omega \Big\langle E_L(x,\cdot),\ \widetilde{D}[D,\eta]u + \big[\widetilde{D},\eta\big]Du\Big\rangle\,\mathrm{d}\mathcal{L}^n_g \\
&= {}_{\mathcal{D}'(\Omega)}\Big\langle \big(I\otimes L^{\top}\big)E_L(x,\cdot),\ \eta u\Big\rangle_{\mathcal{D}(\Omega)} = u(x),
\end{aligned}
\tag{3.5.138}
$$

with the last equality provided by (2.8.178).This establishes (3.5.109) and finishes the proof of Theorem 3.5.9. □

In turn, the representation formula from Theorem 3.5.9 is a key ingredient in the proof of our main Fatou-type result for nullsolutions of weakly elliptic second-order differential operators, in arbitrary UR subdomains of a Riemannian manifold. Theorem 3.5.10, stated next, further refines work in Corollary 3.5.8, and extends to the realm of manifolds earlier Euclidean results from Theorem 3.3.9.

Theorem 3.5.10 *Suppose $\mathcal{M}$ is a compact, boundaryless, oriented $\mathscr{C}^3$ manifold of (real) dimension n, where $n \in \mathbb{N}$ satisfies $n \geq 2$, equipped with a Riemannian metric g of class $\mathscr{C}^3$. Consider a $\mathscr{C}^3$ Hermitian vector bundle $\mathcal{E} \to \mathcal{M}$ and denote by $\nabla = \nabla^{\mathcal{E}} : T\mathcal{M} \times \mathcal{E} \to \mathcal{E}$ a connection on $\mathcal{E}$.*

Next, suppose $L : \mathcal{E} \to \mathcal{E}$ is a weakly elliptic second-order differential operator with the property that[14]

$$L : H^1(\mathcal{M}) \otimes \mathcal{E} \longrightarrow H^{-1}(\mathcal{M}) \otimes \mathcal{E} \text{ is an isomorphism,} \tag{3.5.139}$$

and which may be expressed in local coordinates as

$$L = \left(\sum_{j,k} \partial_j a_{jk}^{\alpha\beta} \partial_k + \sum_j b_j^{\alpha\beta} \partial_j + d^{\alpha\beta} \right)_{\alpha,\beta} \tag{3.5.140}$$

for coefficients satisfying

$$a_{jk}^{\alpha\beta} \in \mathscr{C}^3, \quad b_j^{\alpha\beta} \in \mathscr{C}^2, \quad d^{\alpha\beta} \in \mathscr{C}^1. \tag{3.5.141}$$

Also, let $\Omega \subseteq \mathcal{M}$ be a UR *domain, and abbreviate $\sigma_g := \mathcal{H}_g^{n-1} \lfloor \partial\Omega$. Finally, fix an aperture parameter $\kappa \in (0, \infty)$ and assume the section $u \in W^{1,1}_{\mathrm{loc}}(\Omega) \otimes \mathcal{E}$ satisfies*

$$\begin{gathered} \mathcal{N}_\kappa u,\ \mathcal{N}_\kappa(\nabla u) \in L^p(\partial\Omega, \sigma_g) \text{ for some } p \in \left(\tfrac{n}{n-1}, \infty\right) \\ \text{and } Lu = 0 \text{ in } \Omega. \end{gathered} \tag{3.5.142}$$

Then the nontangential traces $u\big|_{\partial\Omega}^{\kappa-\mathrm{n.t.}}$ and $(\nabla u)\big|_{\partial\Omega}^{\kappa-\mathrm{n.t.}}$ exist at σ_g-a.e. point on $\partial\Omega$. Also, as functions, they are σ_g-measurable and are actually independent of the aperture parameter κ.

Proof As noted in the proof of Corollary 3.5.8, there exist a Hermitian vector bundle $\mathcal{G} \to \mathcal{M}$ of class $\mathscr{C}^3$ together with two first-order differential operators

$$D : \mathcal{E} \longrightarrow \mathcal{G} \text{ and } \widetilde{D} : \mathcal{G} \longrightarrow \mathcal{E}, \tag{3.5.143}$$

having top coefficients of class $\mathscr{C}^3$ and lower-order coefficients of class $\mathscr{C}^2$, plus a zeroth-order operator $W \in \mathrm{Hom}\,(\mathcal{E}, \mathcal{E})$ with coefficients of class $\mathscr{C}^1$ allowing us to express L as

$$L = \widetilde{D}D + W. \tag{3.5.144}$$

Given that the scale of Lebesgue spaces on the compact set $\partial\Omega$ is nested, as far as the existence of nontangential boundary traces is concerned we may assume without

---

[14] recall that $H^{\pm 1}(\mathcal{M})$ denote L^2-based Sobolev spaces on $\mathcal{M}$; alternatively, $H^1(\mathcal{M}) = W^{1,2}(\mathcal{M})$ and $H^{-1}(\mathcal{M}) = W^{-1,2}(\mathcal{M})$

loss of generality that actually $p \in \big(\frac{n}{n-1}, 1\big)$. Thanks to the memberships in the first line of (3.5.142) and the last estimate in [112, Proposition 8.6.3] we deduce (bearing in mind that Ω is relatively compact) that

$$u \in W^{\frac{np}{n-1},1}(\Omega) \otimes \mathcal{E} \subseteq W^{1,1}(\Omega) \otimes \mathcal{E}. \tag{3.5.145}$$

In particular,

$$u \in L^1(\Omega, \mathcal{L}^n_g) \otimes \mathcal{E} \text{ and } Du \in L^1(\Omega, \mathcal{L}^n_g) \otimes \mathcal{G}. \tag{3.5.146}$$

Also, from the current assumptions and [113, (10.1.9) in Lemma 10.1.1] we deduce that

$$\begin{aligned} &\mathfrak{P}_g(Du) \in L^{p^*,p}(\partial\Omega, \sigma_g) \subseteq L^p(\partial\Omega, \sigma_g), \\ &\text{where } p^* := \big(\tfrac{1}{p} - \tfrac{1}{n-1}\big)^{-1} \in (1, \infty). \end{aligned} \tag{3.5.147}$$

Granted (3.5.146)-(3.5.147), we then conclude from Theorem 3.5.2 that

$$(-i)\mathrm{Sym}\big(D; \nu_g\big) \bullet u \text{ belongs to } H^p(\partial\Omega, \sigma_g) \otimes \mathcal{G}. \tag{3.5.148}$$

To continue, pick an arbitrary $\mathscr{C}^1$ vector field $X \in TM$ and bring in the first-order differential operators $\mathbf{P}$, $\mathbf{Q}$ constructed as in (3.5.43)-(3.5.44) for the present D and X. In light of the current assumptions, Theorem 3.5.2, Definition 3.5.5, and Definition 3.5.6 imply that

$$\mathring{\partial}^{\mathbf{P},\mathbf{Q}}_{\tau} u \in H^p(\partial\Omega, \sigma_g) \otimes \mathcal{G} \text{ and } \mathring{\partial}^{L}_{\nu_g} u \in H^p(\partial\Omega, \sigma_g) \otimes \mathcal{E}. \tag{3.5.149}$$

Since, much as in (3.5.146)-(3.5.147), we also have

$$\nabla_X u \in L^1(\Omega, \mathcal{L}^n_g) \otimes \mathcal{E} \text{ and } [\nabla_X, D]u \in L^1(\Omega, \mathcal{L}^n_g) \otimes \mathcal{G}, \tag{3.5.150}$$

as well as

$$\mathfrak{P}_g\big(\nabla_X u\big) \in L^p(\partial\Omega, \sigma_g) \text{ and } \mathfrak{P}_g\big([\nabla_X, D]u\big) \in L^p(\partial\Omega, \sigma_g), \tag{3.5.151}$$

we may once again invoke Theorem 3.5.2 to conclude that

$$\begin{aligned} &(-i)\mathrm{Sym}\big([\nabla_X, D]; \nu_g\big) \bullet u \in H^p(\partial\Omega, \sigma_g) \otimes \mathcal{G}, \\ &\text{and } (-i)\mathrm{Sym}\big(\nabla_X; \nu_g\big) \bullet u \in H^p(\partial\Omega, \sigma_g) \otimes \mathcal{E}. \end{aligned} \tag{3.5.152}$$

Once the memberships in (3.5.148), (3.5.149), (3.5.152) have been established we may proceed to write, keeping in mind the compatibility between the duality pairing from [113, Theorem 4.6.1] and the distributional pairing (cf. [113, Lemma 4.6.4]), the representation formulas (3.5.108) and (3.5.109) from Theorem 3.5.9 with all brackets in the right side now denoting the pairing between the Hardy space H^p and its dual $(H^p)^*$ (cf. [113, Theorem 4.6.1]). With this interpretation, item *(5)* in Theorem 2.8.2 applies gives that both $(\nabla_X u)\big|^{\kappa-\mathrm{n.t.}}_{\partial\Omega}$ and $u\big|^{\kappa-\mathrm{n.t.}}_{\partial\Omega}$ exist σ_g-a.e. on $\partial\Omega$. In view of the arbitrariness of X, the former ultimately proves that the nontangential boundary trace $(\nabla u)\big|_{\partial\Omega}^{\kappa-\mathrm{n.t.}}$ exists at σ_g-a.e. point on $\partial\Omega$. The remaining properties claimed in the statement of the theorem then follow from this, the memberships in the first line of (3.5.142), [112, (8.9.8)], and [112, Proposition 8.9.8]. □

Chapter 4
Green Functions and Uniqueness for Boundary Problems for Second-Order Systems

The main topic in this chapter is the study of Green functions (associated, in a rather inclusive sense, with second-order systems) with special emphasis on the role they play in establishing uniqueness for boundary problems. Basically all results in this chapter make essential use of the brand of Divergence Theorem developed earlier, in Volume I ([112]).

Any object G referred to as Green function[1] (in relation to a given differential operator L and an open set $\Omega \subseteq \mathbb{R}^n$) should satisfy two basic properties. First, G should be a fundamental solution for L with pole (or singularity) at some point $x_0 \in \Omega$ and, second, the boundary trace of G (understood in a suitable sense) should vanish on $\partial\Omega$. These two properties alone typically fail to identify such an object uniquely (thus justifying referring to it as "a Green function," as opposed to "the Green function"). The reason for adopting such a level of generality is to allow for additional flexibility in adapting these basic Green functions to the specific demands made in relation to various aims (such as proving uniqueness for boundary value problems, or establishing integral representation formulas for null-solutions of L in Ω involving Green functions). This is the point of view is pervasive in this chapter.

In §4.1 we examine the role of Green functions in establishing uniqueness for the Dirichlet Problem for weakly elliptic homogeneous, constant (complex) coefficient, second-order systems, formulated in a very broad geometric setting. In particular, in Theorem 4.1.1 and Theorem 4.1.6 we present some sharp embodiments of the heuristic principle asserting that uniqueness holds in the Dirichlet Problem for a given system L and domain Ω granted the existence of Green functions which "pair well" with the null-solutions of this BVP. By not having to actually specify any concrete function spaces in the formulation of these results substantially increases their range of applicability; see the discussion in Examples 4.1.2-4.1.5 and Examples 4.1.7-4.1.10.

The topic of §4.2 is the Reciprocity Principle (aka Law of Reciprocity). This essentially asserts that Green functions satisfy an adjoint symmetry property with respect to its pole and its actual variable (regarded as two independent arguments),

[1] in the case of the Laplacian, Green introduced the function which bears his name in [53]

D. Mitrea et al., *Geometric Harmonic Analysis III*, Developments in Mathematics 74,
https://doi.org/10.1007/978-3-031-22735-6_4

which also takes into account the differential operators with respect to which they are considered. This is made precise in Theorem 4.2.1, where a geometrically general version of this Reciprocity Principle for second-order systems, also placing minimal analytical demands on the Green functions involved, is formally stated. In this regard, we have discovered that, in its most general format, the Reciprocity Principle actually involves *two* Green functions, whose properties are somewhat reminiscent of those exhibited by the solutions of the Dirichlet and Regularity boundary value problems for the system L and its transpose $L^\top$.

In §4.3 we elaborate on the actual technology of constructing Green functions in relation to a given domain Ω and second-order system L which are effective tools in proving uniqueness results for boundary value problems for the system L in Ω. Specifically, the main point of Theorem 4.3.1 is that the solvability of the Regularity Problem for $L^\top$ permits us to construct a Green function which then yields uniqueness for the Dirichlet Problem for the original system L, formulated in a dual setting. Likewise, Theorem 4.3.2 essentially asserts that uniqueness for the Regularity Problem for a given system L is implied by the solvability of the Dirichlet Problem for $L^\top$, formulated in a dual setting.

The main result in §4.4 is the Poisson integral representation formula from Theorem 4.4.1, to the effect that the value of any function, which is a null-solution of a given second-order system L in a domain Ω, at a point $x_0 \in \Omega$ is given by the integral pairing between the boundary trace of said function with the conormal derivative of the Green function in Ω with pole at x_0 for the system $L^\top$. This result is formulated under very general analytic and geometric hypotheses. In particular, no concrete function spaces are involved in the statement (which simply requires that the aforementioned integral is absolutely convergent), which makes this theorem widely applicable to specific cases of interest.

Finally, in §4.5-§4.7 we introduce and study the Poisson kernel associated with a given weakly elliptic system, in various geometric settings.

4.1 The Role of Green Functions in Uniqueness Issues

Theorem 4.1.1 is a sharp embodiment of the heuristic principle asserting that uniqueness holds in the Dirichlet Problem for a given system L and domain Ω granted the existence of Green functions which "pair well" with the null-solutions of the respective boundary value problem. The reader is reminded that, given an arbitrary open set $\Omega \subseteq \mathbb{R}$ along with some Lebesgue measurable function u defined in Ω, by $\mathcal{N}_\kappa^\rho u := \mathcal{N}_\kappa(\mathbf{1}_{O_\rho} \cdot u)$ we denote the truncated version of the nontangential maximal operator at height $\rho > 0$ (cf. (A.0.92)), and by $\mathcal{N}_\kappa^E u := \mathcal{N}_\kappa(\mathbf{1}_E \cdot u)$ we denote the restricted version of the nontangential maximal operator to a Lebesgue measurable subset E of Ω (cf. (A.0.93)).

Theorem 4.1.1 *Suppose $\Omega \subset \mathbb{R}^n$, where $n \in \mathbb{N}$ with*[2] *$n \geq 3$, is a locally pathwise nontangentially accessible set*[3] *(in the sense of* [112, Definition 8.9.14]) *with a lower Ahlfors regular boundary and such that $\sigma := \mathcal{H}^{n-1}\lfloor\partial\Omega$ is a doubling measure on $\partial\Omega$. Also, suppose L is a weakly elliptic homogeneous, constant (complex) coefficient, second-order $M \times M$ system in $\mathbb{R}^n$. Finally, fix an aperture parameter $\kappa > 0$, an arbitrary point $x_0 \in \Omega$, and a truncation parameter $\rho \in \big(0, \frac{1}{4}\operatorname{dist}(x_0, \partial\Omega)\big)$.*

Then there exists some $\widetilde{\kappa} > 0$, which depends only on Ω and κ, with the following significance. Assume G is a matrix-valued function satisfying[4]

$$\begin{cases} G = \big(G_{\alpha\beta}\big)_{1\leq\alpha,\beta\leq M} \in \big[L^1_{\mathrm{loc}}(\Omega, \mathcal{L}^n)\big]^{M\times M}, \\ \big(L^\top G_{\bullet\beta}\big)_\alpha = -\delta_{x_0}\delta_{\alpha\beta} \text{ in } \mathcal{D}'(\Omega) \text{ for all } \alpha, \beta \in \{1, \dots, M\}, \\ G\Big|_{\partial\Omega}^{\widetilde{\kappa}-\mathrm{n.t.}} = 0 \text{ at } \sigma\text{-a.e. point on } \partial\Omega, \\ \text{and also } G(x) = o(1) \text{ as } |x| \longrightarrow \infty \\ \text{whenever } \Omega \text{ is an exterior domain,} \end{cases} \tag{4.1.1}$$

and assume u is a vector-valued function satisfying

$$\begin{cases} u \in \big[\mathscr{C}^\infty(\Omega)\big]^M, \quad Lu = 0 \text{ in } \Omega, \\ u\Big|_{\partial\Omega}^{\kappa-\mathrm{n.t.}} = 0 \text{ at } \sigma\text{-a.e. point on } \partial\Omega, \\ \displaystyle\int_{\partial\Omega} \mathcal{N}_\kappa^\rho u \cdot \mathcal{N}_{\widetilde{\kappa}}^\rho(\nabla G)\,d\sigma < +\infty, \\ \text{and also } u(x) = o(1) \text{ as } |x| \longrightarrow \infty \\ \text{whenever } \Omega \text{ is an exterior domain,} \end{cases} \tag{4.1.2}$$

(with the convention that $0 \cdot \infty = \infty \cdot 0 = \infty$ used for the integrand in the third line of (4.1.2)). *In the case when $\partial\Omega$ is unbounded, strengthen the initial geometric hypotheses on Ω and also strengthen the integral condition in the third line of* (4.1.2) *by assuming that (cf.* [112, (8.9.179)])

$$\begin{cases} \Omega \text{ is a globally pathwise nontangentially accessible set,} \\ \text{and } \displaystyle\int_{\partial\Omega} \mathcal{N}_\kappa u \cdot \mathcal{N}_{\widetilde{\kappa}}^{\Omega\setminus K}(\nabla G)\,d\sigma < +\infty \text{ where } K := \overline{B(x_0, \rho)}, \end{cases} \tag{4.1.3}$$

(once again with the convention that $0 \cdot \infty = \infty \cdot 0 = \infty$).

Then necessarily

---

[2] see the last part in the statement for the two-dimensional case

[3] in particular, a one-sided NTA domain will do; cf. [112, (8.9.180)] and the subsequent comment

[4] Here and elsewhere, $G_{\bullet\beta}$ denotes the β-th column of the matrix G. As such, the second condition in (4.1.1) may be recast as $L^\top G = -\delta_{x_0} I_{M\times M}$ where $L^\top$ acts on the columns of G, and $I_{M\times M}$ is the $M \times M$ identity matrix.

$$u(x_0) = 0 \in \mathbb{C}^M. \tag{4.1.4}$$

Finally, the same result is also valid in the case $n = 2$ provided either Ω is bounded, or $\partial\Omega$ is unbounded. Also, if $n = 2$ and Ω is an exterior domain, the same result is true provided one assumes

$$\begin{gathered}\int_{S^1} \big[L(\xi)\big]^{-1}\, d\mathcal{H}^1(\xi) \\ \text{is an invertible } M \times M \text{ matrix}\end{gathered} \tag{4.1.5}$$

(which is always the case if the system L is assumed to actually satisfy the Legendre-Hadamard strong ellipticity condition, and if $M = 1$ this is true if and only if $L = \nabla \cdot A\nabla$ for some $A \in M_2^0$; cf. Lemma 1.4.19 and (1.4.186)*), and with the demand that u and G are $o(1)$ at infinity replaced by*

$$u(x) = O(1) \text{ and } G(x) = O(1) \text{ as } |x| \to \infty. \tag{4.1.6}$$

In relation to the above theorem we wish to make several comments.

Comment 1. It is worth pointing out that, generally speaking,

$$\begin{gathered}\text{if } \Omega \subseteq \mathbb{R}^n \text{ is an open set, } x_0 \in \Omega \text{ is an arbitrary point, } L \text{ is a weakly elliptic homogeneous, constant coefficient, second-order } M \times M \text{ system} \\ \text{in } \mathbb{R}^n, \text{ and } U \in \big[\mathcal{D}'(\Omega)\big]^{M\times M} \text{ satisfies } LU = -\delta_{x_0} I_{M\times M} \text{ in the} \\ \text{sense of distributions in } \Omega \text{ (where } L \text{ acts on the columns of } U\text{) then} \\ U \in \big[\mathscr{C}^\infty(\Omega \setminus \{x_0\}) \cap W^{1,1}_{\rm loc}(\Omega)\big]^{M\times M}.\end{gathered} \tag{4.1.7}$$

Indeed, the elliptic regularity result recorded in [112, (6.5.40) in Theorem 6.5.7] implies that the matrix-distribution defined as $W := U + E(\cdot - x_0)$, where E is the fundamental solution associated with the given weakly elliptic system L as in Theorem 1.4.2, satisfies $W \in \big[\mathscr{C}^\infty(\Omega)\big]^{M\times M}$. Thus, as claimed, the matrix-valued function $U = W - E(\cdot - x_0)$ belongs to $\big[\mathscr{C}^\infty(\Omega \setminus \{x_0\}) \cap W^{1,1}_{\rm loc}(\Omega)\big]^{M\times M}$.

When (4.1.7) is used in the setting of Theorem 4.1.1, for the matrix-distribution G and the system $L^\top$, the first two lines of (4.1.1) imply that

$$G \text{ belongs to } \big[\mathscr{C}^\infty(\Omega \setminus \{x_0\}) \cap W^{1,1}_{\rm loc}(\Omega)\big]^{M\times M}. \tag{4.1.8}$$

In particular, this membership allows us to make sense of the property formulated in the third line of (4.1.1).

Comment 2. We know that $\mathcal{H}^{n-1}\big(\partial\Omega \setminus \partial_{\rm nta}\Omega\big) = 0$ (cf. [112, Proposition 8.9.16]), which makes it meaningful to consider nontangential boundary traces σ-a.e. on $\partial\Omega$. Also, from [112, (8.2.28)] we see (bearing in mind the convention $0 \cdot \infty = \infty \cdot 0 = \infty$) that $\mathcal{N}^\rho_\kappa u \cdot \mathcal{N}^\rho_{\widetilde{\kappa}}(\nabla G)$ is a well-defined non-negative σ-measurable function on $\partial\Omega$. In particular, the integral condition formulated in the third line of (4.1.2) is meaningful and, since any integrable function is finite almost everywhere, it implies (again, thanks to the aforementioned convention)

$$\big(\mathcal{N}_\kappa^\rho u\big)(x) < +\infty \text{ and } \big(\mathcal{N}_{\widetilde{\kappa}}^\rho(\nabla G)\big)(x) < +\infty \text{ at } \sigma\text{-a.e. } x \in \partial\Omega. \tag{4.1.9}$$

Likewise, $\mathcal{N}_\kappa u \cdot \mathcal{N}_{\widetilde{\kappa}}^{\Omega\setminus K}(\nabla G)$ is a well-defined non-negative σ-measurable function on $\partial\Omega$ and, in the scenario in which $\partial\Omega$ is unbounded, the integral condition formulated in the second line of (4.1.3) is meaningful and implies

$$\big(\mathcal{N}_\kappa u\big)(x) < +\infty \text{ and } \big(\mathcal{N}_{\widetilde{\kappa}}^{\Omega\setminus K}(\nabla G)\big)(x) < +\infty \text{ at } \sigma\text{-a.e. } x \in \partial\Omega. \tag{4.1.10}$$

<u>Comment 3.</u> The demands imposed in (4.1.1) are far from ensuring that such a function G is unique. For example, if we work in the two-dimensional setting where we take $\Omega := B(0,1) \subseteq \mathbb{R}^2 \equiv \mathbb{C}$ and $L := \Delta$, then adding[5]

$$w(z) := \mathrm{Re}\Big(\frac{1+z}{1-z}\Big) = \frac{1-|z|^2}{|z-1|^2}, \qquad \forall z \in \Omega, \tag{4.1.11}$$

to a function G as in (4.1.1) does not change any of these properties.

We now turn to the actual proof of Theorem 4.1.1.

Proof of Theorem 4.1.1 For starters, the reader is reminded that [112, Proposition 8.9.16] ensures that we presently have

$$\sigma\big(\partial\Omega \setminus \partial_{\text{nta}}\Omega\big) = 0, \tag{4.1.12}$$

while (4.1.8) guarantees that G belongs to $\big[\mathscr{C}^\infty(\Omega \setminus \{x_0\}) \cap W^{1,1}_{\text{loc}}(\Omega)\big]^{M\times M}$. Also, recall the family of one-sided collar neighborhoods $\{\mathcal{O}_\varepsilon\}_{\varepsilon>0}$ of $\partial\Omega$ defined in (A.0.92).

To proceed in earnest, we claim that there exists some threshold $\varepsilon_o \in (0,\rho)$, some magnification factor $\lambda \in (1,\infty)$, and some aperture parameter $\widetilde{\kappa} \geq \kappa$ with the property that for each $\varepsilon \in (0, \varepsilon_o/\lambda)$ we may write

$$\begin{aligned}
\int_{\mathcal{O}_\varepsilon} |u||G| \, d\mathcal{L}^n &\leq C\varepsilon \int_{\partial\Omega} \mathcal{N}_\kappa^\varepsilon(|u||G|) \, d\sigma \leq C\varepsilon \int_{\partial\Omega} \mathcal{N}_\kappa^\varepsilon u \, \mathcal{N}_\kappa^\varepsilon G \, d\sigma \\
&\leq C\lambda\varepsilon^2 \int_{\partial\Omega} \mathcal{N}_\kappa^\varepsilon u \, \mathcal{N}_{\widetilde{\kappa}}^{\lambda\varepsilon}(\nabla G) \, d\sigma \\
&\leq C\lambda\varepsilon^2 \int_{\partial\Omega} \mathcal{N}_\kappa^\rho u \, \mathcal{N}_{\widetilde{\kappa}}^\rho(\nabla G) \, d\sigma < +\infty,
\end{aligned} \tag{4.1.13}$$

for some constant $C = C(\partial\Omega, \kappa) \in (0,\infty)$. Indeed, the first inequality in (4.1.13) comes from [112, Proposition 8.6.10] (which provides an initial specification of the threshold $\varepsilon_o \in (0,\rho)$), the second one follows from [112, (8.2.10)], the third one is ensured by (4.1.12) and [112, Proposition 8.9.17] (which provides the magnification factor $\lambda \in (1,\infty)$, the aperture parameter $\widetilde{\kappa} \geq \kappa$, and eventually re-sets the threshold

[5] note that the function w is closely related to the Poisson kernel for the unit disk

ε_o to a smaller value), the fourth one is implied by [112, (8.2.25)] since we presently have $\rho > \lambda\varepsilon > \varepsilon$, and the last one is guaranteed by hypotheses.

With $C \in (0, \infty)$ of the same nature as above, for each truncation parameter $\varepsilon \in (0, \varepsilon_o)$ we also have

$$\begin{aligned}\int_{O_\varepsilon} |u||\nabla G| \, d\mathcal{L}^n &\leq C\varepsilon \int_{\partial\Omega} \mathcal{N}_\kappa^\varepsilon(|u||\nabla G|) \, d\sigma \leq C\varepsilon \int_{\partial\Omega} \mathcal{N}_\kappa^\varepsilon u \, \mathcal{N}_\kappa^\varepsilon(\nabla G) \, d\sigma \\ &\leq C\varepsilon \int_{\partial\Omega} \mathcal{N}_\kappa^\rho u \, \mathcal{N}_{\widetilde{\kappa}}^\rho(\nabla G) \, d\sigma < +\infty. \end{aligned} \tag{4.1.14}$$

Above, the first inequality is provided by [112, Proposition 8.6.10], the second inequality is implied by [112, (8.2.10)], the third inequality is a consequence of [112, (8.2.25)] given that $\widetilde{\kappa} \geq \kappa$ and $\rho > \varepsilon$, while the last inequality is ensured by hypotheses. Collectively, (4.1.13) and (4.1.14) prove that

$$\int_{O_\varepsilon} \{|u||G| + |u||\nabla G|\} \, d\mathcal{L}^n < +\infty \text{ for each } \varepsilon \in (0, \varepsilon_o). \tag{4.1.15}$$

With the goal of showing that $u(x_0) = 0$, bring in the family of functions $\{\Phi_\varepsilon\}_{\varepsilon>0}$ constructed in [112, Lemma 6.1.2]. In addition, select a complex coefficient tensor $A = \big(a_{rs}^{\alpha\beta}\big)_{\substack{1\leq r,s\leq n\\ 1\leq\alpha,\beta\leq M}}$ which permits us to express $L = L_A$, where L_A is the homogeneous constant coefficient second-order $M \times M$ system in $\mathbb{R}^n$ associated with A as in (1.7.7). Finally, denote by $(u_\beta)_{1\leq\beta\leq M}$ the scalar components of the vector-valued function u. Then fix $\gamma \in \{1, \dots, M\}$ along with $\varepsilon \in (0, \varepsilon_o)$ and define the vector field (using the summation convention over repeated indices)

$$\vec{F} := \Big(\Phi_\varepsilon \, G_{\alpha\gamma} \, a_{jk}^{\alpha\beta} \, \partial_k u_\beta - \Phi_\varepsilon \, u_\alpha a_{kj}^{\beta\alpha} \, \partial_k G_{\beta\gamma} - u_\beta G_{\alpha\gamma} \, a_{kj}^{\alpha\beta} \, \partial_k \Phi_\varepsilon\Big)_{1\leq j\leq n} \quad \text{in } \Omega. \tag{4.1.16}$$

From (4.1.16), (4.1.8), and assumptions it follows that

$$\vec{F} \in \big[L^1_{\rm loc}(\Omega, \mathcal{L}^n)\big]^n. \tag{4.1.17}$$

Also, since $\Phi_\varepsilon \equiv 0$ in the one-sided collar neighborhood $O_{\varepsilon/N}$ of $\partial\Omega$ (cf. [112, (6.1.5)]), from (4.1.16) we see that

$$\begin{gathered} \vec{F} \text{ vanishes identically in } O_{\varepsilon/N} \text{ thus, in particular,} \\ \mathcal{N}_\kappa^{\varepsilon/N} \vec{F} = 0 \text{ everywhere on } \partial\Omega, \text{ and } \vec{F}\Big|_{\partial\Omega}^{\kappa-\text{n.t.}} = 0 \text{ everywhere on } \partial_{\text{nta}}\Omega. \end{gathered} \tag{4.1.18}$$

Next, a direct calculation shows that $\text{div}\vec{F}$ (considered in the sense of distributions in Ω) is given by

$$\begin{aligned}
\operatorname{div}\vec{F} &= (\partial_j\Phi_\varepsilon)G_{\alpha\gamma}a^{\alpha\beta}_{jk}(\partial_k u_\beta) + \Phi_\varepsilon(\partial_j G_{\alpha\gamma})a^{\alpha\beta}_{jk}(\partial_k u_\beta)\\
&\quad + \Phi_\varepsilon G_{\alpha\gamma}a^{\alpha\beta}_{jk}(\partial_j\partial_k u_\beta) - (\partial_j\Phi_\varepsilon)u_\alpha a^{\beta\alpha}_{kj}(\partial_k G_{\beta\gamma})\\
&\quad - \Phi_\varepsilon(\partial_j u_\alpha)a^{\beta\alpha}_{kj}(\partial_k G_{\beta\gamma}) - \Phi_\varepsilon u_\alpha a^{\beta\alpha}_{kj}(\partial_j\partial_k G_{\beta\gamma})\\
&\quad - (\partial_j u_\beta)G_{\alpha\gamma}a^{\alpha\beta}_{kj}(\partial_k\Phi_\varepsilon) - u_\beta(\partial_j G_{\alpha\gamma})a^{\alpha\beta}_{kj}(\partial_k\Phi_\varepsilon)\\
&\quad - u_\beta G_{\alpha\gamma}a^{\alpha\beta}_{kj}(\partial_j\partial_k\Phi_\varepsilon)\\
&=: I_1 + I_2 + I_3 + I_4 + I_5 + I_6 + I_7 + I_8 + I_9,
\end{aligned} \tag{4.1.19}$$

where the last equality defines the I_i's with $i \in \{1, \dots, 9\}$. Let us analyze some of these terms. Changing variables $j' = k$ and $k' = j$ in I_1 yields

$$I_1 = (\partial_{k'}\Phi_\varepsilon)\, G_{\alpha\gamma}\, a^{\alpha\beta}_{k'j'}\, (\partial_{j'}u_\beta) = -I_7. \tag{4.1.20}$$

For I_2 we change variables $j' = k$, $k' = j$, $\alpha' = \beta$, $\beta' = \alpha$ in order to write

$$I_2 = \Phi_\varepsilon\, (\partial_{k'}G_{\beta'\gamma})\, a^{\beta'\alpha'}_{k'j'}\, (\partial_{j'}u_{\alpha'}) = -I_5. \tag{4.1.21}$$

As regards I_3, we have

$$I_3 = \Phi_\varepsilon\, G_{\alpha\gamma}\, (L_A u)_\alpha = \Phi_\varepsilon\, G_{\alpha\gamma}\, (Lu)_\alpha = 0, \tag{4.1.22}$$

by the assumptions on u. For I_6 we observe that (with $G_{\bullet\gamma} := (G_{\mu\gamma})_{1\leq\mu\leq M}$ denoting the γ-th column of the matrix G)

$$\begin{aligned}
I_6 &= -\Phi_\varepsilon\, u_\alpha(L_{A^\top}G_{\bullet\gamma})_\alpha = -\Phi_\varepsilon\, u_\alpha(L^\top G_{\bullet\gamma})_\alpha\\
&= \Phi_\varepsilon\, u_\alpha\delta_{\alpha\gamma}\delta_{x_0} = \Phi_\varepsilon\, u_\gamma\delta_{x_0} = u_\gamma(x_0)\delta_{x_0},
\end{aligned} \tag{4.1.23}$$

thanks to the first condition in (4.1.1). Collectively, these equalities permit us to conclude that, in the sense of distributions in Ω,

$$\begin{aligned}
\operatorname{div}\vec{F} &= u_\gamma(x_0)\delta_{x_0} - (\partial_j\Phi_\varepsilon)\, u_\alpha a^{\beta\alpha}_{kj}\, (\partial_k G_{\beta\gamma})\\
&\quad - u_\beta(\partial_j G_{\alpha\gamma})a^{\alpha\beta}_{kj}\, (\partial_k\Phi_\varepsilon) - u_\beta G_{\alpha\gamma}\, a^{\alpha\beta}_{kj}\, (\partial_j\partial_k\Phi_\varepsilon).
\end{aligned} \tag{4.1.24}$$

Note that the first term in the right-hand side is a distribution supported at the singleton $\{x_0\}$, while the remaining terms are in $L^1(\Omega, \mathcal{L}^n)$, as seen from (4.1.15). As such,

$$\operatorname{div}\vec{F} \in \mathscr{E}'(\Omega) + L^1(\Omega, \mathcal{L}^n). \tag{4.1.25}$$

Let us now focus on the case when $\partial\Omega$ is bounded and $n \geq 3$. When Ω is an exterior domain, from assumptions and Theorem 1.5.9 (applied both to $L^\top$ and L) we conclude that as $|x| \to \infty$ we have

$$\begin{aligned} G(x) &= O(|x|^{2-n}), \quad (\nabla G)(x) = O(|x|^{1-n}), \\ u(x) &= O(|x|^{2-n}), \quad (\nabla u)(x) = O(|x|^{1-n}). \end{aligned} \tag{4.1.26}$$

In concert with (4.1.16) this implies that $\vec{F}(x) = O(|x|^{3-2n})$ as $|x| \to \infty$. Hence, since $n \geq 3$, for each $\mu \in (1, \infty)$ we have

$$\int_{B(0,\mu R)\setminus B(0,R)} |\vec{F}(x)| \, \mathrm{d}\mathcal{L}^n(x) = o(R) \text{ as } R \to \infty \tag{4.1.27}$$

which, in turn, shows that condition [112, (1.5.22)] presently holds, if Ω is an exterior domain. Having established (4.1.17), (4.1.18), (4.1.25), and (4.1.27), [112, Corollary 1.5.2] in the version recorded in [112, (1.5.23)] applies and, if ν denotes the geometric measure theoretic outward unit normal to Ω, the Divergence Formula [112, (1.5.20)] gives, in view of (4.1.24) and [112, (4.6.19)], that for each $\varepsilon > 0$ small enough we have

$$\begin{aligned} 0 &= \int_{\partial_*\Omega} \nu \cdot \big(\vec{F}\big|_{\partial\Omega}^{\kappa-\mathrm{n.t.}}\big) \, \mathrm{d}\sigma = {}_{(\mathscr{C}_b^\infty(\Omega))^*}\big(\mathrm{div}\vec{F}, 1\big)_{\mathscr{C}_b^\infty(\Omega)} \\ &= u_\gamma(x_0) - \int_\Omega (\partial_j \Phi_\varepsilon) u_\alpha a_{kj}^{\beta\alpha} (\partial_k G_{\beta\gamma}) \, \mathrm{d}\mathcal{L}^n \\ &\quad - \int_\Omega u_\beta (\partial_j G_{\alpha\gamma}) a_{kj}^{\alpha\beta} (\partial_k \Phi_\varepsilon) \, \mathrm{d}\mathcal{L}^n - \int_\Omega u_\beta G_{\alpha\gamma} a_{kj}^{\alpha\beta} (\partial_j \partial_k \Phi_\varepsilon) \, \mathrm{d}\mathcal{L}^n. \end{aligned} \tag{4.1.28}$$

Based on (4.1.28), [112, (6.1.5)-(6.1.6)], and re-using portions of (4.1.14)-(4.1.13), we may then estimate (also bearing in mind [112, (8.2.25)] as well as $\widetilde{\kappa} \geq \kappa$ and $\rho > \lambda\varepsilon > \varepsilon$)

$$\begin{aligned} |u_\gamma(x_0)| &\leq C \int_{O_\varepsilon} \{\varepsilon^{-2}|u||G| + \varepsilon^{-1}|u||\nabla G|\} \, \mathrm{d}\mathcal{L}^n \\ &\leq C \int_{\partial\Omega} \mathcal{N}_\kappa^\varepsilon u \cdot \mathcal{N}_\kappa^\varepsilon (\nabla G) \, \mathrm{d}\sigma + C \int_{\partial\Omega} \mathcal{N}_\kappa^\varepsilon u \cdot \mathcal{N}_{\widetilde{\kappa}}^{\lambda\varepsilon} (\nabla G) \, \mathrm{d}\sigma \\ &\leq C \int_{\partial\Omega} \mathcal{N}_\kappa^\varepsilon u \cdot \mathcal{N}_{\widetilde{\kappa}}^\rho (\nabla G) \, \mathrm{d}\sigma \text{ for each } \varepsilon \in (0, \varepsilon_o). \end{aligned} \tag{4.1.29}$$

There remains to observe that, thanks to Lebesgue's Dominated Convergence Theorem we have

$$\lim_{\varepsilon \to 0^+} \int_{\partial\Omega} \mathcal{N}_\kappa^\varepsilon u \cdot \mathcal{N}_{\widetilde{\kappa}}^\rho (\nabla G) \, \mathrm{d}\sigma = 0. \tag{4.1.30}$$

Indeed, from [112, (8.9.13)], (A.0.111), (4.1.12), and the hypothesis made in the second line of (4.1.2) we see that

$$\lim_{\varepsilon\to 0^+} \mathcal{N}_\kappa^\varepsilon u = 0 \ \text{ at } \sigma\text{-a.e. point on } \partial\Omega. \tag{4.1.31}$$

Together with the finiteness condition in the third line of (4.1.1) this proves that

$$\lim_{\varepsilon\to 0^+} \mathcal{N}_\kappa^\varepsilon u \cdot \mathcal{N}_{\tilde{\kappa}}^\rho(\nabla G) = 0 \ \text{ at } \sigma\text{-a.e. point on } \partial\Omega. \tag{4.1.32}$$

Since for each $\varepsilon \in (0, \varepsilon_o) \subset (0, \rho)$ we also have

$$0 \leq \mathcal{N}_\kappa^\varepsilon u \cdot \mathcal{N}_{\tilde{\kappa}}^\rho(\nabla G) \leq \mathcal{N}_\kappa^\rho u \cdot \mathcal{N}_{\tilde{\kappa}}^\rho(\nabla G) \in L^1(\partial\Omega, \sigma), \tag{4.1.33}$$

by virtue of [112, (8.2.25)] and the integrability condition in the third line of (4.1.2), Lebesgue's Dominated Convergence Theorem applies and gives (4.1.30). With this in hand, upon letting $\varepsilon \to 0^+$ in (4.1.29) we conclude that $u_\gamma(x_0) = 0$. Since $\gamma \in \{1, \dots, M\}$ has been arbitrarily chosen, this establishes (4.1.4) in the case when $\partial\Omega$ is bounded and $n \geq 3$. In fact, the same argument works in the case when Ω is bounded and $n \geq 2$, since the decay at infinity is no longer a concern.

There remains to consider the case when

$$\begin{aligned} &\text{either } \partial\Omega \text{ is unbounded and } n \geq 2, \\ &\text{or } \Omega \text{ is an exterior domain and } n = 2. \end{aligned} \tag{4.1.34}$$

Consider first the case when $\partial\Omega$ is unbounded and $n \geq 2$, a scenario in which the stronger hypotheses made in (4.1.3) take effect. We reason as before, with some important alterations. Now the idea is to use the Divergence Formula [112, (1.5.20) in Corollary 1.5.2], and this requires that we verify the existence of an aperture parameter $\kappa' \in (0, \infty)$ and a compact set $K' \subseteq \Omega$ with the property that (cf. [112, (1.5.21)])

$$\mathcal{N}_{\kappa'}^{\Omega\setminus K'} \vec{F} \in L^1(\partial\Omega, \sigma). \tag{4.1.35}$$

We claim that (4.1.35) holds provided

$$\kappa' \in (0, \kappa) \ \text{ and } \ K' := \overline{B(x_0, 3\rho)}. \tag{4.1.36}$$

To justify this claim, first observe from (4.1.16) that there exists a constant $C \in (0, \infty)$ independent of $\varepsilon > 0$ such that

$$\begin{aligned} |\vec{F}| \leq{}& C\mathbf{1}_{\{\delta_{\partial\Omega} \geq \varepsilon/N\}}|G|\,|\nabla u| + C\mathbf{1}_{\{\delta_{\partial\Omega} \geq \varepsilon/N\}}|u|\,|\nabla G| \\ &+ C\varepsilon^{-1}\mathbf{1}_{\{\varepsilon/N \leq \delta_{\partial\Omega} \leq \varepsilon\}}|u||G| \ \text{ in } \Omega. \end{aligned} \tag{4.1.37}$$

Note that at σ-a.e. point on $\partial\Omega$ we have

$$\begin{aligned}\mathcal{N}_{\kappa'}^{\Omega\setminus K'}(|G|\,|\nabla u|) &= \mathcal{N}_{\kappa'}^{\Omega\setminus K'}\big((|G|/\delta_{\partial\Omega})\cdot(\delta_{\partial\Omega}|\nabla u|)\big)\\ &\leq \Big(\mathcal{N}_{\kappa'}^{\Omega\setminus K'}(G/\delta_{\partial\Omega})\Big)\cdot\Big(\mathcal{N}_{\kappa'}^{\Omega\setminus K'}(\delta_{\partial\Omega}|\nabla u|)\Big)\\ &\leq \Big(\mathcal{N}_{\kappa}^{\Omega\setminus K'}(G/\delta_{\partial\Omega})\Big)\cdot\Big(\mathcal{N}_{\kappa'}(\delta_{\partial\Omega}|\nabla u|)\Big)\\ &\leq C\big(\mathcal{N}_{\kappa}u\big)\big(\mathcal{N}_{\widetilde{\kappa}}^{\Omega\setminus K}(\nabla G)\big),\end{aligned}\tag{4.1.38}$$

where we have used [112, (8.2.10), (8.2.25), (8.9.147)] (with $|\alpha| = 1$ and $\lambda = 0$), [112, (8.9.227)] (with $u := G$ and with the understanding that $\widetilde{\kappa}$ has been taken, to begin with, greater than the aperture parameter $\widetilde{\kappa}_o$ appearing in the statement of [112, Proposition 8.9.21]), and (4.1.12). Also, at σ-a.e. point on $\partial\Omega$ we have

$$\begin{aligned}\mathcal{N}_{\kappa'}^{\Omega\setminus K'}\big(\varepsilon^{-1}\mathbf{1}_{\{\varepsilon/N\leq\delta_{\partial\Omega}\leq\varepsilon\}}|u||G|\big) &\leq \mathcal{N}_{\kappa'}^{\Omega\setminus K'}\big(|u|(|G|/\delta_{\partial\Omega})\big)\\ &\leq \big(\mathcal{N}_{\kappa'}^{\Omega\setminus K'}u\big)\big(\mathcal{N}_{\kappa'}^{\Omega\setminus K'}(G/\delta_{\partial\Omega})\big)\\ &\leq C\big(\mathcal{N}_{\kappa}u\big)\big(\mathcal{N}_{\widetilde{\kappa}}^{\Omega\setminus K}(\nabla G)\big),\end{aligned}\tag{4.1.39}$$

thanks to [112, (8.2.10), (8.2.25)], as well as [112, (8.9.227)] (with $u := G$ and the same convention as above regarding the size of $\widetilde{\kappa}$), and (4.1.12). From (4.1.37), (4.1.38), and (4.1.39) we then deduce that

$$\mathcal{N}_{\kappa'}^{\Omega\setminus K'}\vec{F} \leq C\,\mathcal{N}_{\kappa}u\cdot\mathcal{N}_{\widetilde{\kappa}}^{\Omega\setminus K}(\nabla G) \text{ on } \partial\Omega.\tag{4.1.40}$$

Based on (4.1.40), (4.1.3), and [112, (8.2.28)] we finally conclude that the membership in (4.1.35) is indeed valid.

At last, we are left with dealing with the second case in (4.1.34), i.e., when Ω is an exterior domain and $n = 2$. In such a scenario, the additional assumptions made in (4.1.5) and (4.1.6) come in effect. The argument in the first part of the proof applies and yields the desired conclusion as soon as we check (4.1.27) in the present setting. To this end, note that the last part in Theorem 1.5.9 (applied both to $L^\top$ and L) shows that as $|x| \to \infty$ we have

$$\begin{aligned}G(x) &= O(1), \quad (\nabla G)(x) = O(|x|^{-2}),\\ u(x) &= O(1), \quad (\nabla u)(x) = O(|x|^{-2}).\end{aligned}\tag{4.1.41}$$

Together with (4.1.16) this implies that $\vec{F}(x) = O(|x|^{-2})$ as $|x| \to \infty$, hence for each $\mu \in (1,\infty)$ we have

$$\int_{B(0,\mu R)\setminus B(0,R)} |\vec{F}(x)|\,d\mathcal{L}^2(x) = o(R) \text{ as } R \to \infty,\tag{4.1.42}$$

as wanted. This finishes the proof of Theorem 4.1.1. □

Remarkably, the only assumptions of quantitative nature made[6] in the statement of Theorem 4.1.1 are the pointwise a.e. finiteness condition in the third line of (4.1.1), the integrability condition in the third line of (4.1.2), as well as the pointwise finiteness conditions and integrability property stipulated in (4.1.3). These demands are naturally satisfied in many cases of practical interest, like the ones on which we elaborate in next series of examples.

Example 4.1.2 (Ordinary Lebesgue Spaces) *In Theorem 4.1.1, the uniqueness result remains valid when the third line in* (4.1.2) *is replaced by*

$$\begin{gathered}\mathcal{N}_\kappa u \in L^p(\partial\Omega, \sigma) \ \textit{and}\ \mathcal{N}_\kappa^\rho(\nabla G) \in L^{p'}(\partial\Omega, \sigma),\\ \textit{where } p, p' \in [1, \infty] \textit{ are such that } 1/p + 1/p' = 1,\end{gathered} \tag{4.1.43}$$

and, in the case when $\partial\Omega$ *is unbounded, the integrability condition in the second line of* (4.1.3) *is replaced by*

$$\begin{gathered}\mathcal{N}_\kappa u \in L^p(\partial\Omega, \sigma) \ \textit{and}\ \mathcal{N}_\kappa^{\Omega\setminus K}(\nabla G) \in L^{p'}(\partial\Omega, \sigma),\\ \textit{where } p, p' \in [1, \infty] \textit{ are such that } 1/p + 1/p' = 1.\end{gathered} \tag{4.1.44}$$

Indeed, [112, Proposition 8.4.1] shows that $\mathcal{N}_\kappa^\rho(\nabla G) \in L^{p'}(\partial\Omega, \sigma)$ implies $\mathcal{N}_{\widetilde{\kappa}}^\rho(\nabla G) \in L^{p'}(\partial\Omega, \sigma)$ for any other $\widetilde{\kappa} > 0$, so the integrability condition in the third line of (4.1.2) becomes a consequence of Hölder's inequality. Also, thanks to [112, Corollary 8.9.9] and the last condition hypothesized in (4.1.43), in the third line of (4.1.1) we may simply ask that $G\big|_{\partial\Omega}^{\kappa-\mathrm{n.t.}} = 0$ at σ-a.e. point on $\partial\Omega$. Similar comments apply in relation to (4.1.44).

Example 4.1.3 (Muckenhoupt Weighted Lebesgue Spaces) *The uniqueness result formulated in Theorem 4.1.1 remains valid when the third line in* (4.1.2) *is replaced by*

$$\begin{gathered}\mathcal{N}_\kappa u \in L^p(\partial\Omega, w\sigma) \ \textit{and}\ \mathcal{N}_\kappa^\rho(\nabla G) \in L^{p'}(\partial\Omega, w^{1-p'}\sigma) \textit{ where}\\ p, p' \in (1, \infty) \textit{ are such that } 1/p + 1/p' = 1, \textit{ and } w \in A_p(\partial\Omega, \sigma),\end{gathered} \tag{4.1.45}$$

and, in the case when $\partial\Omega$ *is unbounded, the integrability condition in the second line of* (4.1.3) *is replaced by*

$$\begin{gathered}\mathcal{N}_\kappa u \in L^p(\partial\Omega, w\sigma) \ \textit{and}\ \mathcal{N}_\kappa^{\Omega\setminus K}(\nabla G) \in L^{p'}(\partial\Omega, w^{1-p'}\sigma) \textit{ where}\\ p, p' \in (1, \infty) \textit{ are such that } 1/p + 1/p' = 1, \textit{ and } w \in A_p(\partial\Omega, \sigma).\end{gathered} \tag{4.1.46}$$

This time we rely on [112, Corollary 8.4.4], which ensures that having $\mathcal{N}_\kappa^\rho(\nabla G)$ in $L^{p'}(\partial\Omega, w^{1-p'}\sigma)$ implies $\mathcal{N}_{\widetilde{\kappa}}^\rho(\nabla G) \in L^{p'}(\partial\Omega, w^{1-p'}\sigma)$ for any other $\widetilde{\kappa} > 0$, since $w^{1-p'} \in A_{p'}(\partial\Omega, \sigma)$ (cf. item *(2)* in [112, Lemma 7.7.1]), and the fact that $\big(L^p(\partial\Omega, w\sigma)\big)^*$ is canonically identified with $L^{p'}(\partial\Omega, w^{1-p'}\sigma)$ under the integral

[6] in addition to specifying the behavior at infinity, in the case when Ω is an exterior domain

pairing on $\partial\Omega$. In addition, as seen from [112, Corollary 8.9.9] and the last condition hypothesized in (4.1.45), in the third line of (4.1.1) we may simply ask that $G\big|_{\partial\Omega}^{\kappa-\mathrm{n.t.}} = 0$ at σ-a.e. point on $\partial\Omega$. Similar comments apply in relation to (4.1.46). It is worth noting that the only relevant property of a Muckenhoupt weight for the uniqueness result just described is the fact that the measures $w\,\sigma$ and $w^{1-p'}\sigma$ are doubling on $\partial\Omega$.

Example 4.1.4 (Lorentz Spaces) *The uniqueness result in Theorem 4.1.1 remains valid when the third line in* (4.1.2) *is replaced by*

$$\begin{aligned}&\mathcal{N}_\kappa u \in L^{p,q}(\partial\Omega,\sigma) \text{ and } \mathcal{N}_\kappa^{\rho}(\nabla G) \in L^{p',q'}(\partial\Omega,\sigma) \text{ where}\\&p,q,p',q' \in [1,\infty] \text{ satisfy } 1/p + 1/p' = 1 = 1/q + 1/q',\end{aligned}\tag{4.1.47}$$

and, in the case when $\partial\Omega$ is unbounded, the integrability condition in the second line of (4.1.3) *is replaced by*

$$\begin{aligned}&\mathcal{N}_\kappa u \in L^{p,q}(\partial\Omega,\sigma) \text{ and } \mathcal{N}_\kappa^{\Omega\setminus K}(\nabla G) \in L^{p',q'}(\partial\Omega,\sigma) \text{ where}\\&p,q,p',q' \in [1,\infty] \text{ satisfy } 1/p + 1/p' = 1 = 1/q + 1/q',\end{aligned}\tag{4.1.48}$$

We now bring in [112, Proposition 8.4.1], which shows that having $\mathcal{N}_\kappa^{\rho}(\nabla G)$ in $L^{p',q'}(\partial\Omega,\sigma)$ implies $\mathcal{N}_{\widetilde{\kappa}}^{\rho}(\nabla G) \in L^{p',q'}(\partial\Omega,\sigma)$ for any other $\widetilde{\kappa} > 0$, and O'Neil's inequality (cf. [112, (6.2.61)]). Once again, [112, Corollary 8.9.9] and the last condition demanded in (4.1.47) show that in the third line of (4.1.1) we may simply ask that $G\big|_{\partial\Omega}^{\kappa-\mathrm{n.t.}} = 0$ at σ-a.e. point on $\partial\Omega$. Similar comments are valid in relation to (4.1.48).

Here is a final example, whose justification now involves [113, Proposition 6.2.8, (6.2.78)] and [112, Corollary 8.4.8] (see also [112, Corollary 8.4.5]).

Example 4.1.5 (Morrey Spaces and Their Pre-Duals) *In the case when $\partial\Omega$ is assumed to be an Ahlfors regular set, the uniqueness result in Theorem 4.1.1 remains valid when the third line in* (4.1.1) *is replaced by*

$$G\big|_{\partial\Omega}^{\kappa-\mathrm{n.t.}} = 0 \text{ at } \sigma\text{-a.e. point on } \partial\Omega,\tag{4.1.49}$$

and the third line in (4.1.2) *is replaced by*

$$\begin{aligned}&\mathcal{N}_\kappa u \in \mathcal{B}^{p,\lambda}(\partial\Omega,\sigma) \text{ and } \mathcal{N}_\kappa^{\rho}(\nabla G) \in M^{p',\lambda}(\partial\Omega,\sigma) \text{ where}\\&p,p' \in (1,\infty) \text{ are such that } 1/p + 1/p' = 1 \text{ and } \lambda \in (0,n-1).\end{aligned}\tag{4.1.50}$$

Also, in the case when the set $\partial\Omega$ happens to be unbounded, one may replace (4.1.3) *by*

$$\begin{aligned}&\mathcal{N}_\kappa u \in \mathcal{B}^{p,\lambda}(\partial\Omega,\sigma) \text{ and } \mathcal{N}_\kappa^{\Omega\setminus K}(\nabla G) \in M^{p',\lambda}(\partial\Omega,\sigma) \text{ where}\\&p,p' \in (1,\infty) \text{ are such that } 1/p + 1/p' = 1 \text{ and } \lambda \in (0,n-1).\end{aligned}\tag{4.1.51}$$

Alternatively, retaining the assumption that $\partial\Omega$ is an Ahlfors regular set, the uniqueness result in Theorem 4.1.1 continues to hold when the third line in (4.1.2) *is replaced by*

$$\begin{aligned}&\mathcal{N}_\kappa u \in M^{p,\lambda}(\partial\Omega,\sigma) \text{ and } \mathcal{N}_\kappa^\rho(\nabla G) \in \mathcal{B}^{p',\lambda}(\partial\Omega,\sigma) \text{ where}\\ &p, p' \in (1,\infty) \text{ are such that } 1/p + 1/p' = 1 \text{ and } \lambda \in (0, n-1),\end{aligned} \tag{4.1.52}$$

and, in the case when the set $\partial\Omega$ is unbounded, with (4.1.3) *replaced by*

$$\begin{aligned}&\mathcal{N}_\kappa u \in M^{p,\lambda}(\partial\Omega,\sigma) \text{ and } \mathcal{N}_\kappa^{\Omega\setminus K}(\nabla G) \in \mathcal{B}^{p',\lambda}(\partial\Omega,\sigma) \text{ where}\\ &p, p' \in (1,\infty) \text{ are such that } 1/p + 1/p' = 1 \text{ and } \lambda \in (0, n-1).\end{aligned} \tag{4.1.53}$$

Moving on, we present a companion result to Theorem 4.1.1, pertaining to uniqueness for the Dirichlet Problem for second-order systems in which an integrability condition is imposed on the gradient of the solution, with a weight involving a Green function.

Theorem 4.1.6 *Fix $n \in \mathbb{N}$ with*[7] *$n \geq 3$. Suppose $\Omega \subset \mathbb{R}^n$ is a locally pathwise nontangentially accessible set*[8] *(in the sense of* [112, Definition 8.9.14]) *with a lower Ahlfors regular boundary and such that $\sigma := \mathcal{H}^{n-1}\lfloor\partial\Omega$ is a doubling measure on $\partial\Omega$. Also, suppose L is a weakly elliptic homogeneous, constant (complex) coefficient, second-order $M \times M$ system in $\mathbb{R}^n$. Finally, fix some point $x_0 \in \Omega$, a truncation parameter $\rho \in \big(0, \frac{1}{4}\operatorname{dist}(x_0, \partial\Omega)\big)$, and an aperture parameter $\kappa \in (0,\infty)$.*

Then there exists some $\widetilde{\kappa} > 0$, which depends only on Ω and κ, with the following significance. Suppose G is a matrix-valued function satisfying

$$\begin{cases} G = \big(G_{\alpha\beta}\big)_{1\leq\alpha,\beta\leq M} \in \big[L^1_{\mathrm{loc}}(\Omega, \mathcal{L}^n)\big]^{M\times M},\\ \big(L^\top G_{\cdot\beta}\big)_\alpha = -\delta_{x_0}\delta_{\alpha\beta} \text{ in } \mathcal{D}'(\Omega) \text{ for all } \alpha,\beta \in \{1,\dots,M\},\\ G\Big|_{\partial\Omega}^{\kappa-\mathrm{n.t.}} = 0 \text{ at } \sigma\text{-a.e. point on } \partial\Omega,\\ \text{and also } G(x) = o(1) \text{ as } |x| \longrightarrow \infty\\ \text{whenever } \Omega \text{ is an exterior domain,}\end{cases} \tag{4.1.54}$$

and suppose u is a vector-valued function satisfying

---

[7] see the last part in the statement for the case $n = 2$

[8] in particular, this is the case if Ω is a one-sided NTA domain (see [112, (8.9.180)] and the subsequent comment)

$$
\begin{cases}
u \in \big[\mathscr{C}^\infty(\Omega)\big]^M, \quad Lu = 0 \ \text{in} \ \Omega, \\
u\big|_{\partial\Omega}^{\widetilde{\kappa}-\text{n.t.}} = 0 \ \text{at} \ \sigma\text{-a.e. point on} \ \partial\Omega, \\
\displaystyle\int_{\partial\Omega} \mathcal{N}_{\widetilde{\kappa}}^{\rho}(\nabla u) \cdot \mathcal{N}_{\kappa}^{\rho} G \, d\sigma < +\infty, \\
\text{and also} \ u(x) = o(1) \ \text{as} \ |x| \longrightarrow \infty \\
\text{whenever} \ \Omega \ \text{is an exterior domain,}
\end{cases}
\tag{4.1.55}
$$

(with the convention that $0 \cdot \infty = \infty \cdot 0 = \infty$ *used for the integrand in the third line of* (4.1.55)*). In the case when* $\partial\Omega$ *is unbounded, strengthen the initial geometric hypotheses on* Ω *and also strengthen the integral condition in the third line of* (4.1.55) *by assuming that (cf.* [112, (8.9.179)]*)*

$$
\begin{cases}
\Omega \ \text{is a globally pathwise nontangentially accessible set,} \\
\text{and} \ \displaystyle\int_{\partial\Omega} \mathcal{N}_{\widetilde{\kappa}}(\nabla u) \cdot \mathcal{N}_{\kappa}^{\Omega\setminus K}(\nabla G) \, d\sigma < +\infty \ \text{where} \ K := \overline{B(x_0, \rho)}
\end{cases}
\tag{4.1.56}
$$

(once again with the convention that $0 \cdot \infty = \infty \cdot 0 = \infty$*).*
Then necessarily

$$
u(x_0) = 0 \in \mathbb{C}^M. \tag{4.1.57}
$$

Moreover, the same result is also valid in the case $n = 2$ *provided either* Ω *is bounded, or* $\partial\Omega$ *is unbounded. Also, if* $n = 2$ *and* Ω *is an exterior domain, the same result is true provided one assumes*

$$
\begin{gathered}
\int_{S^1} \big[L(\xi)\big]^{-1} \, d\mathcal{H}^1(\xi) \\
\text{is an invertible } M \times M \text{ matrix}
\end{gathered}
\tag{4.1.58}
$$

(which is always the case if the system L is assumed to actually satisfy the Legendre-Hadamard strong ellipticity condition, and if $M = 1$ *this is true if and only if* $L = \nabla \cdot A\nabla$ *for some* $A \in M_2^0$*; cf. Lemma 1.4.19 and* (1.4.186)*), and with the demand that u and G are* $o(1)$ *at infinity replaced by*

$$
u(x) = O(1) \ \text{and} \ G(x) = O(1) \ \text{as} \ |x| \to \infty. \tag{4.1.59}
$$

Proof The main idea is to re-run the proof of Theorem 4.1.1, with the roles of u and G reversed, and with the coefficient tensor A now replaced by $A^\top$. To elaborate on these adjustments, denote by $(u_\beta)_{1\leq\beta\leq M}$ the scalar components of the vector-valued function u. Then, in place of (4.1.14), for each $\varepsilon > 0$ small we now have

$$
\int_{O_\varepsilon} |G||\nabla u| \, d\mathcal{L}^n \leq C\varepsilon \int_{\partial\Omega} \mathcal{N}_{\kappa}^{\varepsilon} G \cdot \mathcal{N}_{\widetilde{\kappa}}^{\varepsilon}(\nabla u) \, d\sigma < +\infty, \tag{4.1.60}
$$

while in place of (4.1.13) we now have

$$\int_{O_\varepsilon} |u||G|\,\mathrm{d}\mathcal{L}^n \le C\lambda\varepsilon^2 \int_{\partial\Omega} \mathcal{N}_\kappa^\varepsilon G \cdot \mathcal{N}_{\widetilde{\kappa}}^{\lambda\varepsilon}(\nabla u)\,\mathrm{d}\sigma < +\infty. \tag{4.1.61}$$

In particular,

$$\int_{O_\varepsilon} \big\{|u||G| + |G||\nabla u|\big\}\,\mathrm{d}\mathcal{L}^n < +\infty \text{ for each } \varepsilon > 0 \text{ small}. \tag{4.1.62}$$

Let us also observe that (4.1.7) presently implies that actually

$$G \in \big[\mathscr{C}^\infty(\Omega \setminus \{x_0\}) \cap W^{1,1}_{\mathrm{loc}}(\Omega)\big]^{M\times M}. \tag{4.1.63}$$

To proceed, fix $\gamma \in \{1,\dots,M\}$ along with $\varepsilon > 0$ sufficiently small and recall the family of functions $\{\Phi_\varepsilon\}_{\varepsilon>0}$ introduced in [112, Lemma 6.1.2]. Using the summation convention over repeated indices, we now define the vector field

$$\vec{F} := \Big(\Phi_\varepsilon\, u_\alpha\, a^{\beta\alpha}_{kj}\, \partial_k G_{\beta\gamma} - \Phi_\varepsilon\, G_{\alpha\gamma}\, a^{\alpha\beta}_{jk}\, \partial_k u_\beta - G_{\beta\gamma}\, u_\alpha\, a^{\beta\alpha}_{jk}\, \partial_k \Phi_\varepsilon\Big)_{1\le j\le n}. \tag{4.1.64}$$

From (4.1.64), (4.1.63), and assumptions we have

$$\begin{aligned} &\vec{F} \text{ belongs to } \big[L^1_{\mathrm{loc}}(\Omega,\mathcal{L}^n)\big]^n \text{ and } \vec{F} \equiv 0 \text{ in } O_{\varepsilon/N} \text{ thus, in particular,}\\ &\mathcal{N}_\kappa^{\varepsilon/N}\vec{F} = 0 \text{ everywhere on } \partial\Omega \text{ and } \vec{F}\big|^{\kappa-\mathrm{n.t.}}_{\partial\Omega} = 0 \text{ everywhere on } \partial_{\mathrm{nta}}\Omega. \end{aligned} \tag{4.1.65}$$

Moreover, $\mathrm{div}\vec{F}$ considered in the sense of distributions in Ω is now given by

$$\begin{aligned} \mathrm{div}\vec{F} &= -u_\gamma(x_0)\delta_{x_0} - G_{\alpha\gamma}(\partial_k u_\beta) a^{\alpha\beta}_{jk}(\partial_j \Phi_\varepsilon)\\ &\quad - G_{\beta\gamma}(\partial_j u_\alpha) a^{\beta\alpha}_{jk}(\partial_k \Phi_\varepsilon) - G_{\beta\gamma} u_\alpha a^{\beta\alpha}_{jk}(\partial_j\partial_k \Phi_\varepsilon). \end{aligned} \tag{4.1.66}$$

Hence, once again, $\mathrm{div}\vec{F} \in \mathscr{E}'(\Omega) + L^1(\Omega,\mathcal{L}^n)$, thanks to (4.1.62).

In addition, when Ω is an exterior domain the current assumptions and Theorem 1.5.9 (applied both to $L^\top$ and L) imply the decay properties recorded in (4.1.26) (cf. also (4.1.42)) as $|x| \to \infty$. In turn, much as in (4.1.27), we ultimately conclude that condition [112, (1.5.13)] presently holds, if Ω is an exterior domain. Finally, in the case when $\partial\Omega$ is unbounded (a scenario in which (4.1.56) comes into effect), we claim that if

$$\kappa' \in (0,\kappa) \text{ and } K' := \overline{B(x_0, 3\rho)} \tag{4.1.67}$$

then there exist some aperture parameter $\kappa' > 0$ and some compact set $K' \subseteq \Omega$ such that

$$\mathcal{N}_{\kappa'}^{\Omega\setminus K'}\vec{F} \in L^1(\partial\Omega,\sigma). \tag{4.1.68}$$

To prove this claim, first note from (4.1.64) and [112, Lemma 6.1.2] that there exists a constant $C \in (0,\infty)$ independent of $\varepsilon > 0$ such that

$$
\begin{aligned}
|\vec{F}| \le C\mathbf{1}_{\{\delta_{\partial\Omega}\ge\varepsilon/N\}}|u|\,|\nabla G| &+ C\mathbf{1}_{\{\delta_{\partial\Omega}\ge\varepsilon/N\}}|G|\,|\nabla u| \\
&+ C\varepsilon^{-1}\mathbf{1}_{\{\varepsilon/N\le\delta_{\partial\Omega}\le\varepsilon\}}|u||G| \text{ in } \Omega.
\end{aligned}
\tag{4.1.69}
$$

In relation to this, we observe that at σ-a.e. point on $\partial\Omega$ we have

$$
\begin{aligned}
\mathcal{N}_{\kappa'}^{\Omega\setminus K'}(|u|\,|\nabla G|) &= \mathcal{N}_{\kappa'}^{\Omega\setminus K'}((|u|/\delta_{\partial\Omega})\cdot(\delta_{\partial\Omega}|\nabla G|)) \\
&\le \Big(\mathcal{N}_{\kappa'}^{\Omega\setminus K'}(u/\delta_{\partial\Omega})\Big)\cdot\Big(\mathcal{N}_{\kappa'}^{\Omega\setminus K'}(\delta_{\partial\Omega}|\nabla G|)\Big) \\
&\le \Big(\mathcal{N}_{\kappa'}(u/\delta_{\partial\Omega})\Big)\cdot\Big(\mathcal{N}_{\kappa'}^{\Omega\setminus K'}(\delta_{\partial\Omega}|\nabla G|)\Big) \\
&\le C\big(\mathcal{N}_{\widetilde{\kappa}}(\nabla u)\big)\big(\mathcal{N}_{\kappa}^{\Omega\setminus K}G\big),
\end{aligned}
\tag{4.1.70}
$$

thanks to [112, (8.2.10), (8.2.25)], as well as [112, (8.9.167)] (with $u := G$, $|\alpha| = 1$, and $\lambda = 0$), [112, (8.9.210)] (with the understanding that the current $\widetilde{\kappa}$ has been taken, from the beginning, to be larger than the aperture parameter $\widetilde{\kappa}$ appearing in the statement of [112, Proposition 8.9.18]), and the fact that $\sigma\big(\partial\Omega\setminus\partial_{\text{nta}}\Omega\big) = 0$ (cf. [112, Proposition 8.9.16]). In addition, at σ-a.e. point on $\partial\Omega$ we have

$$
\begin{aligned}
\mathcal{N}_{\kappa'}^{\Omega\setminus K'}\big(\varepsilon^{-1}\mathbf{1}_{\{\varepsilon/N\le\delta_{\partial\Omega}\le\varepsilon\}}|u||G|\big) &\le \mathcal{N}_{\kappa}^{\Omega\setminus K}(|G|(|u|/\delta_{\partial\Omega})) \\
&\le \big(\mathcal{N}_{\kappa}^{\Omega\setminus K}G\big)\big(\mathcal{N}_{\kappa}(u/\delta_{\partial\Omega})\big) \\
&\le C\big(\mathcal{N}_{\kappa}^{\Omega\setminus K}G\big)\big(\mathcal{N}_{\widetilde{\kappa}}(\nabla u)\big),
\end{aligned}
\tag{4.1.71}
$$

where we used [112, (8.2.10)], [112, (8.2.25)], [112, (8.9.227)] (with the same convention as above regarding the size of $\widetilde{\kappa}$), and the fact that $\sigma\big(\partial\Omega\setminus\partial_{\text{nta}}\Omega\big) = 0$. From (4.1.37), (4.1.38), and (4.1.39) we then conclude that

$$
\mathcal{N}_{\kappa'}^{\Omega\setminus K'}\vec{F} \le C\,\mathcal{N}_{\kappa}^{\Omega\setminus K}G\cdot\mathcal{N}_{\widetilde{\kappa}}(\nabla u) \text{ on } \partial\Omega.
\tag{4.1.72}
$$

Collectively, (4.1.72), (4.1.56), and [112, (8.2.28)] permit us to conclude that the membership claimed in (4.1.68) is indeed true.

Granted the aforementioned properties of the vector field $\vec{F}$, [112, Corollary 1.5.2] (in the version recorded in [112, (1.5.23)]) applies and, with ν denoting the geometric measure theoretic outward unit normal to Ω, the Divergence Formula [112, (1.5.20)] gives that for each $\varepsilon > 0$ small enough we have

$$
\begin{aligned}
0 &= \int_{\partial_*\Omega} \nu \cdot \big(\vec{F}\big|^{\kappa-\text{n.t.}}_{\partial\Omega}\big)\,\mathrm{d}\sigma = {}_{(\mathscr{C}^\infty_b(\Omega))^*}\big(\mathrm{div}\vec{F}, 1\big)_{\mathscr{C}^\infty_b(\Omega)} \\
&= -u_\gamma(x_0) - \int_\Omega G_{\alpha\gamma}(\partial_k u_\beta) a^{\alpha\beta}_{jk}(\partial_j \Phi_\varepsilon)\,\mathrm{d}\mathcal{L}^n \qquad (4.1.73) \\
&\quad - \int_\Omega G_{\beta\gamma}(\partial_j u_\alpha) a^{\beta\alpha}_{jk}(\partial_k \Phi_\varepsilon)\,\mathrm{d}\mathcal{L}^n - \int_\Omega G_{\beta\gamma} u_\alpha a^{\beta\alpha}_{jk}(\partial_j\partial_k \Phi_\varepsilon)\,\mathrm{d}\mathcal{L}^n.
\end{aligned}
$$

Thus,

$$
\begin{aligned}
u_\gamma(x_0) = &- \int_\Omega G_{\alpha\gamma}(\partial_k u_\beta) a^{\alpha\beta}_{jk}(\partial_j \Phi_\varepsilon)\,\mathrm{d}\mathcal{L}^n - \int_\Omega G_{\beta\gamma}(\partial_j u_\alpha) a^{\beta\alpha}_{jk}(\partial_k \Phi_\varepsilon)\,\mathrm{d}\mathcal{L}^n \\
&- \int_\Omega G_{\beta\gamma} u_\alpha a^{\beta\alpha}_{jk}(\partial_j\partial_k \Phi_\varepsilon)\,\mathrm{d}\mathcal{L}^n \qquad (4.1.74)
\end{aligned}
$$

which, in concert with [112, (6.1.5)-(6.1.6)] and (4.1.60)-(4.1.61), allows us to estimate

$$
\begin{aligned}
|u_\gamma(x_0)| &\leq C \int_{O_\varepsilon} \big\{\varepsilon^{-2}|u||G| + \varepsilon^{-1}|G||\nabla u|\big\}\,\mathrm{d}\mathcal{L}^n \\
&\leq C \int_{\partial\Omega} \mathcal{N}^\varepsilon_\kappa G \cdot \mathcal{N}^\rho_{\bar{\kappa}}(\nabla u)\,\mathrm{d}\sigma \ \text{ for each } \varepsilon > 0 \text{ small enough.} \qquad (4.1.75)
\end{aligned}
$$

On the other hand, from [112, (8.9.13)], (A.0.111), (4.1.12), and the assumption made in the third line of (4.1.54) we see that

$$
\lim_{\varepsilon\to 0^+} \mathcal{N}^\varepsilon_\kappa G = 0 \ \text{ at } \sigma\text{-a.e. point on } \partial\Omega. \qquad (4.1.76)
$$

In concert with the finiteness condition in the third line of (4.1.55) this gives

$$
\lim_{\varepsilon\to 0^+} \mathcal{N}^\varepsilon_\kappa G \cdot \mathcal{N}^\rho_{\bar{\kappa}}(\nabla u) = 0 \ \text{ at } \sigma\text{-a.e. point on } \partial\Omega. \qquad (4.1.77)
$$

Given that for each $\varepsilon > 0$ small we also have

$$
0 \leq \mathcal{N}^\varepsilon_\kappa G \cdot \mathcal{N}^\rho_{\bar{\kappa}}(\nabla u) \leq \mathcal{N}^\rho_\kappa G \cdot \mathcal{N}^\rho_{\bar{\kappa}}(\nabla u) \in L^1(\partial\Omega, \sigma), \qquad (4.1.78)
$$

by virtue of [112, (8.2.25)] and the integrability condition in the fourth line of (4.1.55), Lebesgue's Dominated Convergence Theorem applies and proves that

$$
\lim_{\varepsilon\to 0^+} \int_{\partial\Omega} \mathcal{N}^\varepsilon_\kappa G \cdot \mathcal{N}^\rho_{\bar{\kappa}}(\nabla u)\,\mathrm{d}\sigma = 0. \qquad (4.1.79)
$$

Having established (4.1.79), upon letting $\varepsilon \to 0^+$ in (4.1.75) we conclude that $u_\gamma(x_0) = 0$. Since $\gamma \in \{1, \dots, M\}$ has been arbitrarily chosen, the desired conclusion follows. □

As in the case of Theorem 4.1.1, the only assumptions of quantitative nature made in the formulation of Theorem 4.1.6 are the pointwise a.e. finiteness conditions and the integrability conditions made in the third and fourth lines of (4.1.55), as well as in (4.1.56). Once again, these demands are naturally satisfied in many cases of practical interest. For instance, with the same type of justifications as in Examples 4.1.2-4.1.5 we now have:

Example 4.1.7 (Ordinary Lebesgue Spaces) *In Theorem 4.1.6, the uniqueness result remains valid when the second and third lines in* (4.1.55) *are replaced by*

$$\begin{aligned} &u\Big|_{\partial\Omega}^{\kappa-\mathrm{n.t.}} = 0 \text{ at } \sigma\text{-a.e. point on } \partial\Omega,\\ &\mathcal{N}_\kappa(\nabla u) \in L^p(\partial\Omega,\sigma) \text{ and } \mathcal{N}_\kappa^\rho G \in L^{p'}(\partial\Omega,\sigma),\\ &\text{where } p,p' \in [1,\infty] \text{ are such that } 1/p + 1/p' = 1, \end{aligned} \tag{4.1.80}$$

and, in the case when $\partial\Omega$ *is unbounded, the finiteness condition in* (4.1.56) *is replaced by*

$$\begin{aligned} &\mathcal{N}_\kappa u \in L^p(\partial\Omega,\sigma) \text{ and } \mathcal{N}_\kappa^{\Omega\setminus K} G \in L^{p'}(\partial\Omega,\sigma)\\ &\text{where } p,p' \in [1,\infty] \text{ are such that } 1/p + 1/p' = 1. \end{aligned} \tag{4.1.81}$$

Example 4.1.8 (Muckenhoupt Weighted Lebesgue Spaces) *The uniqueness result formulated in Theorem 4.1.6 remains valid when the second and third lines in* (4.1.55) *are replaced by*

$$\begin{aligned} &u\Big|_{\partial\Omega}^{\kappa-\mathrm{n.t.}} = 0 \text{ at } \sigma\text{-a.e. point on } \partial\Omega,\\ &\mathcal{N}_\kappa(\nabla u) \in L^p(\partial\Omega,w\sigma) \text{ and } \mathcal{N}_\kappa^\rho G \in L^{p'}(\partial\Omega,w^{1-p'}\sigma) \text{ where}\\ &p,p' \in (1,\infty) \text{ are such that } 1/p + 1/p' = 1, \text{ and } w \in A_p(\partial\Omega,\sigma), \end{aligned} \tag{4.1.82}$$

and, in the case when $\partial\Omega$ *is unbounded, the finiteness condition in* (4.1.56) *is replaced by*

$$\begin{aligned} &\mathcal{N}_\kappa(\nabla u) \in L^p(\partial\Omega,w\sigma) \text{ and } \mathcal{N}_\kappa^{\Omega\setminus K} G \in L^{p'}(\partial\Omega,w^{1-p'}\sigma) \text{ where}\\ &p,p' \in (1,\infty) \text{ are such that } 1/p + 1/p' = 1, \text{ and } w \in A_p(\partial\Omega,\sigma). \end{aligned} \tag{4.1.83}$$

Example 4.1.9 (Lorentz Spaces) *The uniqueness result in Theorem 4.1.6 remains valid when the second and third lines in* (4.1.55) *are replaced by*

$$\begin{aligned} &u\Big|_{\partial\Omega}^{\kappa-\mathrm{n.t.}} = 0 \text{ at } \sigma\text{-a.e. point on } \partial\Omega,\\ &\mathcal{N}_\kappa(\nabla u) \in L^{p,q}(\partial\Omega,\sigma) \text{ and } \mathcal{N}_\kappa^\rho G \in L^{p',q'}(\partial\Omega,\sigma) \text{ where}\\ &p,q,p',q' \in [1,\infty] \text{ satisfy } 1/p + 1/p' = 1 = 1/q + 1/q', \end{aligned} \tag{4.1.84}$$

and, in the case when $\partial\Omega$ *is unbounded, the finiteness condition in* (4.1.56) *is replaced by*

$$\begin{aligned}&\mathcal{N}_\kappa(\nabla u) \in L^{p,q}(\partial\Omega, \sigma) \text{ and } \mathcal{N}_\kappa^{\Omega\setminus K} G \in L^{p',q'}(\partial\Omega, \sigma) \text{ where}\\ &p, q, p', q' \in [1, \infty] \text{ satisfy } 1/p + 1/p' = 1 = 1/q + 1/q'.\end{aligned} \tag{4.1.85}$$

Example 4.1.10 (Morrey Spaces and Their Pre-Duals) *In the case when $\partial\Omega$ is assumed to be an Ahlfors regular set, the uniqueness result in Theorem 4.1.6 remains valid when the second and third lines in* (4.1.55) *are replaced by*

$$\begin{aligned}&u\Big|_{\partial\Omega}^{\kappa-\text{n.t.}} = 0 \text{ at } \sigma\text{-a.e. point on } \partial\Omega,\\ &\mathcal{N}_\kappa(\nabla u) \in M^{p,\lambda}(\partial\Omega, \sigma) \text{ and } \mathcal{N}_\kappa^{\rho} G \in \mathcal{B}^{p',\lambda}(\partial\Omega, \sigma) \text{ where}\\ &p, p' \in (1, \infty) \text{ are such that } 1/p + 1/p' = 1, \text{ and } \lambda \in (0, n-1),\end{aligned} \tag{4.1.86}$$

and, in the case when $\partial\Omega$ is unbounded, the finiteness condition in (4.1.56) *is replaced by*

$$\begin{aligned}&\mathcal{N}_\kappa(\nabla u) \in M^{p,\lambda}(\partial\Omega, \sigma) \text{ and } \mathcal{N}_\kappa^{\Omega\setminus K} G \in \mathcal{B}^{p',\lambda}(\partial\Omega, \sigma) \text{ where}\\ &p, p' \in (1, \infty) \text{ are such that } 1/p + 1/p' = 1, \text{ and } \lambda \in (0, n-1).\end{aligned} \tag{4.1.87}$$

Alternatively, one may replace the second and third lines in (4.1.55) *by*

$$\begin{aligned}&u\Big|_{\partial\Omega}^{\kappa-\text{n.t.}} = 0 \text{ at } \sigma\text{-a.e. point on } \partial\Omega,\\ &\mathcal{N}_\kappa(\nabla u) \in \mathcal{B}^{p,\lambda}(\partial\Omega, \sigma) \text{ and } \mathcal{N}_\kappa^{\rho} G \in M^{p',\lambda}(\partial\Omega, \sigma) \text{ where}\\ &p, p' \in (1, \infty) \text{ are such that } 1/p + 1/p' = 1, \text{ and } \lambda \in (0, n-1),\end{aligned} \tag{4.1.88}$$

and, in the case when $\partial\Omega$ is unbounded, one may replace the finiteness condition in (4.1.56) *by*

$$\begin{aligned}&\mathcal{N}_\kappa(\nabla u) \in \mathcal{B}^{p,\lambda}(\partial\Omega, \sigma) \text{ and } \mathcal{N}_\kappa^{\Omega\setminus K} G \in M^{p',\lambda}(\partial\Omega, \sigma) \text{ where}\\ &p, p' \in (1, \infty) \text{ are such that } 1/p + 1/p' = 1, \text{ and } \lambda \in (0, n-1).\end{aligned} \tag{4.1.89}$$

4.2 The Reciprocity Principle

The Reciprocity Principle (aka Law of Reciprocity) roughly states that Green functions satisfy an adjoint symmetry property with respect to the pole and the actual variable (regarded as two independent arguments), which also takes into account the differential operators with respect to which they are considered. Our next theorem contains a geometrically general version of this Reciprocity Principle for second-order systems, which also places minimal analytical demands on the Green functions involved. Regarding the latter aspect, Theorem 4.2.1 brings to the forefront the fact that, in its most general format, the Reciprocity Principle actually involves *two* Green functions, whose properties are somewhat reminiscent of those exhibited by the solutions of the Dirichlet and Regularity boundary value problems for the system L and its transpose $L^{\top}$.

Theorem 4.2.1 *Let $\Omega \subset \mathbb{R}^n$, where $n \in \mathbb{N}$ with*[9]*, be a locally pathwise nontangentially accessible set*[10] *(in the sense of* [112, Definition 8.9.14]) *with a lower Ahlfors regular boundary and such that $\sigma := \mathcal{H}^{n-1}\lfloor\partial\Omega$ is a doubling measure on $\partial\Omega$. Also, suppose L is a weakly elliptic homogeneous, constant (complex) coefficient, second-order $M \times M$ system in $\mathbb{R}^n$. Finally, fix two points $x_0, x_1 \in \Omega$ with $x_0 \neq x_1$, an aperture parameter $\kappa > 0$, and a truncation parameter*

$$0 < \rho < \tfrac{1}{4} \min\Big\{\mathrm{dist}(x_0, \partial\Omega), \mathrm{dist}(x_1, \partial\Omega)\Big\}. \tag{4.2.1}$$

Then there exists some $\widetilde{\kappa} > 0$, which depends only on Ω and κ, with the following significance. Suppose $G \in \big[L^1_{\mathrm{loc}}(\Omega, \mathcal{L}^n)\big]^{M\times M}$ is a matrix-valued function satisfying[11]

$$\begin{cases} L^{\top}G = -\delta_{x_0} I_{M\times M} \text{ in } \big[\mathcal{D}'(\Omega)\big]^{M\times M}, \\ G\Big|_{\partial\Omega}^{\widetilde{\kappa}-\mathrm{n.t.}} = 0 \text{ at } \sigma\text{-a.e. point on } \partial\Omega, \\ \text{and also } G(x) = o(1) \text{ as } |x| \longrightarrow \infty \\ \text{whenever } \Omega \text{ is an exterior domain,} \end{cases} \tag{4.2.2}$$

and suppose $H \in \big[L^1_{\mathrm{loc}}(\Omega, \mathcal{L}^n)\big]^{M\times M}$ is a matrix-valued function satisfying[12]

$$\begin{cases} LH = -\delta_{x_1} I_{M\times M} \text{ in } \big[\mathcal{D}'(\Omega)\big]^{M\times M}, \\ H\Big|_{\partial\Omega}^{\kappa-\mathrm{n.t.}} = 0 \text{ at } \sigma\text{-a.e. point on } \partial\Omega, \\ \displaystyle\int_{\partial\Omega} \mathcal{N}^{\rho}_{\kappa} H \cdot \mathcal{N}^{\rho}_{\widetilde{\kappa}}(\nabla G)\, d\sigma < +\infty, \\ \text{and also } H(x) = o(1) \text{ as } |x| \longrightarrow \infty \\ \text{whenever } \Omega \text{ is an exterior domain,} \end{cases} \tag{4.2.3}$$

(with the convention that $0 \cdot \infty = \infty \cdot 0 = \infty$ used for the integrand in the third line of (4.2.3)*). In the case when $\partial\Omega$ is unbounded, strengthen the initial geometric hypotheses on Ω and also strengthen the integral condition in the third line of* (4.2.3) *by assuming that* (*cf.* [112, (8.9.179)])

---

[9] see the last part in the statement for the case $n = 2$

[10] in particular, this is the case if Ω is a one-sided NTA domain (cf. [112, (8.9.180)] and the subsequent comment)

[11] with $L^{\top}$ acting on the columns of G

[12] with L acting on the columns of H

$$\begin{cases} \Omega \text{ is a globally pathwise nontangentially accessible set, and} \\ \displaystyle\int_{\partial\Omega} \mathcal{N}_{\kappa}^{\Omega\setminus K} H \cdot \mathcal{N}_{\widetilde{\kappa}}^{\Omega\setminus K}(\nabla G)\,\mathrm{d}\sigma < \infty \text{ where } K := \overline{B(x_0,\rho)} \cup \overline{B(x_1,\rho)} \end{cases} \tag{4.2.4}$$

(once again with the convention that $0 \cdot \infty = \infty \cdot 0 = \infty$*).*

Then necessarily

$$G(x_1) = H(x_0)^{\top}. \tag{4.2.5}$$

Moreover, the same result continues to be valid in the case when $n = 2$ *provided either* Ω *is bounded, or* $\partial\Omega$ *is unbounded. Also, if* $n = 2$ *and* Ω *is an exterior domain, the same result is true provided one assumes*

$$\begin{gathered} \int_{S^1} \big[L(\xi)\big]^{-1}\,\mathrm{d}\mathcal{H}^1(\xi) \\ \text{is an invertible } M \times M \text{ matrix} \end{gathered} \tag{4.2.6}$$

(which is always the case if the system L is assumed to actually satisfy the Legendre-Hadamard strong ellipticity condition, and if $M = 1$ *this is true if and only if* $L = \nabla \cdot A\nabla$ *for some* $A \in M_2^0$*; cf. Lemma 1.4.19 and* (1.4.186)*), and with the demand that G and H are* $o(1)$ *at infinity replaced by*

$$G(x) = O(1) \text{ and } H(x) = O(1) \text{ as } |x| \to \infty. \tag{4.2.7}$$

Before presenting the proof of this theorem we make several remarks.

Remark 1. If, in the notation for Green functions G, H, we also include the dependence on the pole as a second variable, as well as the dependence on the system as a superscript, then the Reciprocity Relation recorded in (4.2.5) takes a more precise form, namely the transposition identity

$$G^{L^{\top}}(x_1, x_0) = [H^L(x_0, x_1)]^{\top}. \tag{4.2.8}$$

Remark 2. It is known from [112, Proposition 8.9.16] that we presently have $\mathcal{H}^{n-1}(\partial\Omega \setminus \partial_{\mathrm{nta}}\Omega) = 0$, so it is meaningful to consider nontangential boundary traces σ-a.e. on $\partial\Omega$. In addition, in view of [112, (8.2.28)] and the convention that $0 \cdot \infty = \infty \cdot 0 = \infty$, it follows that $\mathcal{N}_{\kappa}^{\rho} H \cdot \mathcal{N}_{\widetilde{\kappa}}^{\rho}(\nabla G)$ is a well-defined non-negative σ-measurable function on $\partial\Omega$. As such, the integral condition formulated in the third line of (4.2.3) is meaningful and, since any integrable function is finite almost everywhere, it implies (again, thanks to the aforementioned convention)

$$\big(\mathcal{N}_{\kappa}^{\rho} H\big)(x) < +\infty \text{ and } \big(\mathcal{N}_{\widetilde{\kappa}}^{\rho}(\nabla G)\big)(x) < +\infty \text{ at } \sigma\text{-a.e. } x \in \partial\Omega. \tag{4.2.9}$$

Similarly, $\mathcal{N}_{\kappa}^{\Omega\setminus K} H \cdot \mathcal{N}_{\widetilde{\kappa}}^{\Omega\setminus K}(\nabla G)$ is a well-defined non-negative σ-measurable function on $\partial\Omega$ and, in the case when $\partial\Omega$ is unbounded, the finiteness condition in (4.2.4) is meaningful and implies

$$\big(\mathcal{N}_{\kappa}^{\Omega\setminus K} H\big)(x) < +\infty \text{ and } \big(\mathcal{N}_{\widetilde{\kappa}}^{\Omega\setminus K}(\nabla G)\big)(x) < +\infty \text{ at } \sigma\text{-a.e. } x \in \partial\Omega. \tag{4.2.10}$$

Remark 3. The integral condition imposed in the third line of (4.2.3), and the integral condition in (4.2.4) if $\partial\Omega$ is unbounded, are the only demands linking G and H directly. It is remarkable that when G, H are quantitatively compatible in this specific fashion we can conclude that G, H are *algebraically* compatible, in the sense of (4.2.5). Indeed, without imposing said integral conditions there could be infinitely many pairs of (generally completely unrelated) functions G, H satisfying the remaining conditions in (4.2.2)-(4.2.3)[13]. On the other hand, insisting on the full set of requirements listed in (4.2.2), (4.2.3), (4.2.4) guarantees that the matrices $G(x_1)$ and $H(x_0)$ uniquely determine one another (cf. (4.2.5)).

We now turn to the task of presenting the proof of Theorem 4.2.1.

Proof of Theorem 4.2.1 The regularity result from (4.1.8) guarantees that

$$\begin{aligned} G &= (G_{\alpha\beta})_{1\leq\alpha,\beta\leq M} \in \big[\mathscr{C}^\infty(\Omega\setminus\{x_0\})\cap W^{1,1}_{\rm loc}(\Omega)\big]^{M\times M}, \\ H &= (H_{\alpha\beta})_{1\leq\alpha,\beta\leq M} \in \big[\mathscr{C}^\infty(\Omega\setminus\{x_1\})\cap W^{1,1}_{\rm loc}(\Omega)\big]^{M\times M}. \end{aligned} \tag{4.2.11}$$

The idea is to re-do the proof of Theorem 4.1.1 for G as in (4.2.2) and u an arbitrary column of H. To be specific, fix $\gamma, \mu \in \{1, \dots, M\}$ and for each $\varepsilon \in (0, \rho)$ sufficiently small define the vector field (using the summation convention over repeated indices)

$$\begin{aligned} \vec{F} := \Big(\Phi_\varepsilon\, G_{\alpha\gamma}\, a^{\alpha\beta}_{jk}\, \partial_k H_{\beta\mu} - \Phi_\varepsilon\, H_{\alpha\mu}\, a^{\beta\alpha}_{kj}\, \partial_k G_{\beta\gamma} \\ - H_{\beta\mu}\, G_{\alpha\gamma}\, a^{\alpha\beta}_{kj}\, \partial_k \Phi_\varepsilon\Big)_{1\leq j\leq n} \quad \text{in } \Omega. \end{aligned} \tag{4.2.12}$$

This is the analogue of (4.1.16) in which we take $u := H_{\bullet\mu}$, the μ-th column of the matrix-valued function H. Granted (4.2.2), (4.2.3), (4.2.11), the same arguments as in the proof of Theorem 4.1.1 give that

$$\begin{aligned} &\vec{F} \in \big[L^1_{\rm loc}(\Omega, \mathcal{L}^n)\big]^n \text{ and } \vec{F} \equiv 0 \text{ in } O_{\varepsilon/N} \text{ thus, in particular,} \\ &\mathcal{N}^{\varepsilon/N}_\kappa \vec{F} = 0 \text{ everywhere on } \partial\Omega, \text{ and } \vec{F}\Big|^{\kappa-{\rm n.t.}}_{\partial\Omega} = 0 \text{ everywhere on } \partial_{\rm nta}\Omega. \end{aligned} \tag{4.2.13}$$

We may also compute $\operatorname{div}\vec{F}$, in the sense of distributions in Ω, much as in (4.1.19)-(4.1.23) with $u := H_{\bullet\mu}$. Given the latter choice, this time we get

$$\begin{aligned} I_3 &= \Phi_\varepsilon\, G_{\alpha\gamma}\, (L_A H_{\bullet\mu})_\alpha = \Phi_\varepsilon\, G_{\alpha\gamma}\, (L H_{\bullet\mu})_\alpha \\ &= -\Phi_\varepsilon\, G_{\alpha\gamma}\delta_{x_1}\delta_{\alpha\mu} = -\Phi_\varepsilon\, G_{\mu\gamma}\delta_{x_1} = -G_{\mu\gamma}(x_1)\delta_{x_1} \end{aligned} \tag{4.2.14}$$

by the assumptions on H, and

[13] For example, assuming $n := 3$, $M := 1$, $L := \Delta$, and $\Omega := \mathbb{R}^3_+$, then $u(x, y, z) := z/(y^2 + z^2)$ for each $(x, y, z) \in \Omega$ satisfies $Lu = 0$ in Ω, as well as $u\big|^{\kappa-{\rm n.t.}}_{\partial\Omega} = 0$ and $\mathcal{N}_\kappa(\nabla u) < +\infty$ at every point in $\partial\Omega \setminus \{(x, 0, 0) : x \in \mathbb{R}\}$, hence σ-a.e. on $\partial\Omega$, for each $\kappa > 0$. As such, adding multiples of u to either G or H does not change their properties in (4.2.2), (4.2.3), (4.2.4), with the exception of the finiteness condition in the scond line of (4.2.4).

$$
\begin{aligned}
I_6 &= -\Phi_\varepsilon\, H_{\alpha\mu}(L_{A^\top} G_{\bullet\gamma})_\alpha = -\Phi_\varepsilon\, H_{\alpha\mu}(L^\top G_{\bullet\gamma})_\alpha \\
&= \Phi_\varepsilon\, H_{\alpha\mu}\delta_{\alpha\gamma}\delta_{x_0} = \Phi_\varepsilon\, H_{\gamma\mu}\delta_{x_0} = H_{\gamma\mu}(x_0)\delta_{x_0}
\end{aligned} \tag{4.2.15}
$$

by the assumptions on G. Ultimately, in place of (4.1.24) we now have

$$
\begin{aligned}
\operatorname{div}\vec{F} &= -G_{\mu\gamma}(x_1)\delta_{x_1} + H_{\gamma\mu}(x_0)\delta_{x_0} - (\partial_j\Phi_\varepsilon)\, H_{\alpha\mu}\, a^{\beta\alpha}_{kj}\, (\partial_k G_{\beta\gamma}) \\
&\quad - H_{\beta\mu}\, (\partial_j G_{\alpha\gamma}) a^{\alpha\beta}_{kj}\, (\partial_k \Phi_\varepsilon) - H_{\beta\mu}\, G_{\alpha\gamma}\, a^{\alpha\beta}_{kj}\, (\partial_j\partial_k\Phi_\varepsilon).
\end{aligned} \tag{4.2.16}
$$

The first two terms in the right-hand side of (4.2.16) belong to $\mathscr{E}'(\Omega)$. Also, since much as in (4.1.15) we now have

$$
\int_{O_\varepsilon} \big\{|H||G| + |H||\nabla G|\big\}\, \mathrm{d}\mathcal{L}^n < +\infty \text{ for each } \varepsilon > 0 \text{ small,} \tag{4.2.17}
$$

it follows that the last three terms in the right-hand side of (4.2.16) belong to $L^1(\Omega, \mathcal{L}^n)$. Altogether, this proves that

$$
\operatorname{div}\vec{F} \in \mathscr{E}'(\Omega) + L^1(\Omega, \mathcal{L}^n). \tag{4.2.18}
$$

Let us also observe that, much as in (4.1.29)-(4.1.33), we now have

$$
\int_{O_\varepsilon} \big\{\varepsilon^{-2}|H||G| + \varepsilon^{-1}|H||\nabla G|\big\}\, \mathrm{d}\mathcal{L}^n = o(1) \text{ as } \varepsilon \to 0^+. \tag{4.2.19}
$$

Moreover, in the case when Ω is an exterior domain, from assumptions and Theorem 1.5.9 (applied both to $L^\top$ and L) we conclude that

$$
\int_{B(0,\,\mu R)\setminus B(0,R)} |\vec{F}(x)|\, \mathrm{d}\mathcal{L}^n(x) = o(R) \text{ as } R \to \infty. \tag{4.2.20}
$$

This, in turn, shows that condition [112, (1.5.22)] presently holds (assuming Ω is an exterior domain). Finally, if $\partial\Omega$ is unbounded then, much as in (4.1.35), we continue to have the property that

$$
\begin{gathered}
\mathcal{N}^{\Omega\setminus K'}_{\kappa'} \vec{F} \text{ belongs to the space } L^1(\partial\Omega, \sigma), \\
\text{for some aperture parameter } \kappa' > 0 \text{ and some compact set } K' \subseteq \Omega.
\end{gathered} \tag{4.2.21}
$$

Having proved (4.2.13), (4.2.13), (4.2.18) (and also [112, (1.5.22)] when Ω is an exterior domain, and (4.2.21) when $\partial\Omega$ is unbounded), [112, Corollary 1.5.2] applies (in the version recorded in [112, (1.5.23)]) and, if ν denotes the geometric measure theoretic outward unit normal to Ω, the Divergence Formula [112, (1.5.20)] gives, in view of (4.2.16) and (4.2.13), that for each $\varepsilon > 0$ small enough we have

$$
\begin{aligned}
0 &= \int_{\partial_*\Omega} \nu \cdot \big(\vec{F}\big|_{\partial\Omega}^{\kappa-\mathrm{n.t.}}\big)\, d\sigma = {}_{(\mathscr{C}_b^\infty(\Omega))^*}\big(\mathrm{div}\vec{F}, 1\big)_{\mathscr{C}_b^\infty(\Omega)} \\
&= -G_{\mu\gamma}(x_1) + H_{\gamma\mu}(x_0) - \int_\Omega (\partial_j \Phi_\varepsilon)\, H_{\alpha\mu}\, a_{kj}^{\beta\alpha}\, (\partial_k G_{\beta\gamma})\, d\mathcal{L}^n \qquad (4.2.22) \\
&\quad - \int_\Omega H_{\beta\mu}\, (\partial_j G_{\alpha\gamma}) a_{kj}^{\alpha\beta}\, (\partial_k \Phi_\varepsilon)\, d\mathcal{L}^n - \int_\Omega H_{\beta\mu}\, G_{\alpha\gamma}\, a_{kj}^{\alpha\beta}\, (\partial_j \partial_k \Phi_\varepsilon)\, d\mathcal{L}^n.
\end{aligned}
$$

Together with [112, (6.1.5)-(6.1.6)] and (4.2.19), this permits us to estimate

$$
\begin{aligned}
|G_{\mu\gamma}(x_1) - H_{\gamma\mu}(x_0)| &\le C \int_{O_\varepsilon} \big\{\varepsilon^{-2}|H||G| + \varepsilon^{-1}|H||\nabla G|\big\}\, d\mathcal{L}^n \\
&= o(1) \text{ as } \varepsilon \to 0^+. \qquad (4.2.23)
\end{aligned}
$$

Granted this, upon letting $\varepsilon \to 0^+$ we conclude that $G_{\mu\gamma}(x_1) = H_{\gamma\mu}(x_0)$. Since $\gamma, \mu \in \{1, \dots, M\}$ have been arbitrarily chosen, this finishes the proof of (4.2.5). □

4.3 How to Construct Green Functions and Use Them in Uniqueness Issues

The main point of our first result in this section is that uniqueness for the Dirichlet Problem for a given system L is implied by the solvability of the Regularity Problem for $L^\top$, formulated in a dual setting.

Theorem 4.3.1 *Given $n \in \mathbb{N}$ with*[14] *$n \ge 3$, let $\Omega \subset \mathbb{R}^n$ be an Ahlfors regular domain, which is locally pathwise nontangentially accessible*[15] *(cf.* [112, Definition 8.9.14]*), and abbreviate $\sigma := \mathcal{H}^{n-1}\lfloor\partial\Omega$. Also, consider a homogeneous constant (complex) coefficient second-order $M \times M$ system L in $\mathbb{R}^n$, which is weakly elliptic (in the sense of* (1.3.3)*). Finally, fix an aperture parameter $\kappa > 0$ and assume that the spaces $\mathscr{X}$, $\mathscr{Y}$, $\mathscr{Z}$ are as in any of the following cases:*

$$
\begin{aligned}
&\mathscr{X} := L^{p'}(\partial\Omega, \sigma),\ \mathscr{Y} := \big[L_1^{p'}(\partial\Omega, \sigma)\big]^M,\ \mathscr{Z} := \mathscr{X}^* = L^p(\partial\Omega, \sigma), \\
&\quad \textit{where } p, p' \in (1, \infty) \textit{ are such that } 1/p + 1/p' = 1,
\end{aligned} \qquad (4.3.1)
$$

or

$$
\begin{aligned}
&\mathscr{X} := L^{p'}(\partial\Omega, w^{1-p'}\sigma),\ \mathscr{Y} := \big[L_1^{p'}(\partial\Omega, w^{1-p'}\sigma)\big]^M,\ \mathscr{Z} := \mathscr{X}^* = L^p(\partial\Omega, w\sigma), \\
&\quad \textit{where } p, p' \in (1, \infty) \textit{ are such that } 1/p + 1/p' = 1 \textit{ and } w \in A_p(\partial\Omega, \sigma),
\end{aligned} \qquad (4.3.2)
$$

[14] see the last part in the statement for the two-dimensional case

[15] in particular, this is the case if Ω is a one-sided NTA domain (see [112, (8.9.180)] and the subsequent comment)

or

$$\mathscr{X} := L^{p',q'}(\partial\Omega,\sigma),\ \ \mathscr{Z} := \mathscr{X}^* = L^{p,q}(\partial\Omega,\sigma),\ \ \text{and}$$
$$\mathscr{Y} := \big\{f \in \big[L^{p',q'}(\partial\Omega,\sigma)\big]^M : \partial_{\tau_{jk}} f \in \big[L^{p',q'}(\partial\Omega,\sigma)\big]^M,\ 1 \le j,k \le n\big\},$$
$$\text{where } p,p',q,q' \in (1,\infty) \text{ are such that } 1/p + 1/p' = 1 = 1/q + 1/q', \tag{4.3.3}$$

or

$$\mathscr{X} := M^{p',\lambda}(\partial\Omega,\sigma),\ \ \mathscr{Y} := \big[M_1^{p',\lambda}(\partial\Omega,\sigma)\big]^M,\ \ \mathscr{Z} := \mathcal{B}^{p,\lambda}(\partial\Omega,\sigma),$$
$$\text{where } p,p' \in (1,\infty) \text{ are such that } 1/p + 1/p' = 1, \tag{4.3.4}$$

or

$$\mathscr{X} := \mathcal{B}^{p',\lambda}(\partial\Omega,\sigma),\ \ \mathscr{Y} := \big[\mathcal{B}_1^{p',\lambda}(\partial\Omega,\sigma)\big]^M,\ \ \mathscr{Z} := \mathscr{X}^* = M^{p,\lambda}(\partial\Omega,\sigma),$$
$$\text{where } p,p' \in (1,\infty) \text{ are such that } 1/p + 1/p' = 1. \tag{4.3.5}$$

Then if the Regularity-type boundary value problem for $L^\top$

$$\begin{cases} w \in \big[\mathscr{C}^\infty(\Omega)\big]^M, \quad L^\top w = 0 \text{ in } \Omega, \\ \mathcal{N}_\kappa(\nabla w) \text{ belongs to the space } \mathscr{X}, \\ w\big|_{\partial\Omega}^{\kappa-\text{n.t.}} = f \text{ at } \sigma\text{-a.e. point on } \partial\Omega, \\ \text{and also } w(x) = o(1) \text{ as } |x| \to \infty \\ \text{when } \Omega \text{ is an exterior domain,} \end{cases} \tag{4.3.6}$$

is solvable for each boundary datum $f \in \mathscr{Y}$*, the following uniqueness result for the dual Dirichlet-type boundary value problem for L is valid:*

$$\left.\begin{array}{r} u \in \big[\mathscr{C}^\infty(\Omega)\big]^M, \quad Lu = 0 \text{ in } \Omega, \\ u\big|_{\partial\Omega}^{\kappa-\text{n.t.}} = 0 \text{ at } \sigma\text{-a.e. point on } \partial\Omega, \\ \mathcal{N}_\kappa u \text{ belongs to the space } \mathscr{Z}, \\ \text{and also } u(x) = o(1) \text{ as } |x| \to \infty \\ \text{when } \Omega \text{ is an exterior domain,} \end{array}\right\} \Longrightarrow u \equiv 0 \text{ in } \Omega. \tag{4.3.7}$$

Furthermore, similar results are valid in the case $n = 2$ *provided either* Ω *is bounded, or* $\partial\Omega$ *is unbounded. Also, if* $n = 2$ *and* Ω *is an exterior domain, the same result is true provided one assumes*

$$\int_{S^1} \big[L(\xi)\big]^{-1}\, d\mathcal{H}^1(\xi)$$
$$\text{is an invertible } M \times M \text{ matrix} \tag{4.3.8}$$

(which is always the case if the system L is assumed to actually satisfy the Legendre-Hadamard strong ellipticity condition, and if $M = 1$ *this is true if and only if* $L = \nabla \cdot A\nabla$ *for some* $A \in M_2^0$*; cf. Lemma 1.4.19 and* (1.4.186)*), and with the demand that u and w are* $o(1)$ *at infinity replaced by*

$$u(x) = O(1) \text{ and } w(x) = O(1) \text{ as } |x| \to \infty. \tag{4.3.9}$$

Proof Let $E = (E_{\alpha\beta})_{1 \le \alpha,\beta \le M}$ be the fundamental solution associated with the system $L^\top$ as in Theorem 1.4.2. Fix $x_\star \in \mathbb{R}^n \setminus \overline{\Omega}$ along with $x_0 \in \Omega$, arbitrary. Then, for each fixed $\beta \in \{1, \dots, M\}$, consider the $\mathbb{C}^M$-valued function

$$f_\beta(x) := \big(E_{\beta\alpha}(x - x_0) - E_{\beta\alpha}(x - x_\star)\big)_{1 \le \alpha \le M}, \qquad \forall x \in \partial\Omega. \tag{4.3.10}$$

In relation to this, we claim that

$$f_\beta \text{ belongs to the space } \mathscr{Y}\text{, as given in any of the scenarios described in (4.3.1)-(4.3.5).} \tag{4.3.11}$$

To justify this, bring in the set $\mathcal{O} := \mathbb{R}^n \setminus \{x_0, x_\star\}$, which is an open neighborhood of $\partial\Omega$, then define the function $\psi \in \mathscr{C}^\infty(\mathcal{O})$ by setting

$$\psi(x) := E_{\beta\alpha}(x - x_0) - E_{\beta\alpha}(x - x_\star) \text{ for each } x \in \mathcal{O}. \tag{4.3.12}$$

Thanks to the Mean Value Theorem and item *(4)* in Theorem 1.4.2, this satisfies

$$|\psi(x)| \le C_{x_\star, x_0}(1 + |x|)^{1-n} \text{ and } |(\nabla\psi)(x)| \le C_{x_\star, x_0}(1 + |x|)^{-n}, \quad \text{for each point } x \in \partial\Omega, \tag{4.3.13}$$

where $C_{x_\star, x_0}$ is a positive finite constant which depends only on $x_\star$, x_0, and L. Granted these, the claim in (4.3.11) becomes a consequence of [113, Lemma 11.1.5] together with [112, Lemma 7.2.1] if $\mathscr{Y}$ is as in (4.3.1), a consequence of [112, (7.7.101)] together with item *(2)* in [112, Lemma 7.7.1] if $\mathscr{Y}$ is as in (4.3.2), a consequence of [112, (7.7.118)] if $\mathscr{Y}$ is as in (4.3.3), a consequence of [113, (6.2.7)] if $\mathscr{Y}$ is as in (4.3.4), and a consequence of [113, Proposition 6.2.21] (or [113, (6.2.245)]) if $\mathscr{Y}$ is as in (4.3.5).

Having established (4.3.11), our current working hypotheses then make it possible to consider the solution $w_\beta = (w_{\beta\alpha})_{1 \le \alpha \le M}$ of the boundary value problem (4.3.6), formulated for this choice of the boundary datum. Subsequently, for each pair of indices $\alpha, \beta \in \{1, \dots, M\}$ define

$$G_{\alpha\beta}(x) := w_{\beta\alpha}(x) - \big(E_{\beta\alpha}(x - x_0) - E_{\beta\alpha}(x - x_\star)\big), \qquad \forall x \in \Omega \setminus \{x_0\}. \tag{4.3.14}$$

When viewing $G := (G_{\alpha\beta})_{1 \le \alpha,\beta \le M}$ as a $\mathbb{C}^{M \times M}$-valued function defined $\mathcal{L}^n$-a.e. in Ω, from (4.3.14) and Theorem 1.4.2 we see that $G \in \big[L^1_{\text{loc}}(\Omega, \mathcal{L}^n)\big]^{M \times M}$. Also, by design,

$$\begin{aligned} L^{\top}G = -\delta_{x_0} I_{M\times M} \ \text{ in } \ \big[\mathcal{D}'(\Omega)\big]^{M\times M} \\ \text{and } G\Big|_{\partial\Omega}^{\kappa-\text{n.t.}} = 0 \ \text{ at } \sigma\text{-a.e. point on } \partial\Omega, \end{aligned} \tag{4.3.15}$$

while from (4.3.14), (4.3.6), (4.3.9), the Mean Value Theorem, and Theorem 1.4.2 we see that, in the case when Ω is an exterior domain, as $|x| \to \infty$ we have

$$G(x) = \begin{cases} o(1) + O(|x|^{1-n}) = o(1) & \text{if } n \geq 3, \\ O(1) + O(|x|^{-1}) = O(1) & \text{if } n = 2. \end{cases} \tag{4.3.16}$$

Finally, from item *(7)* in Theorem 1.4.2 and [112, Lemma 8.3.7] it follows that for any choice of a radius $\rho \in \big(0, \operatorname{dist}(x_0, \partial\Omega)\big)$ we have

$$\begin{aligned} &\big(\mathcal{N}_{\widetilde{\kappa}}^{\Omega\setminus K} G\big)(x) \leq \max\nolimits_{1\leq\beta\leq M} \big(\mathcal{N}_{\kappa} w_\beta\big)(x) + C_{x_0,\rho}(1+|x|)^{1-n} \ \text{ and} \\ &\big(\mathcal{N}_{\widetilde{\kappa}}^{\Omega\setminus K}(\nabla G)\big)(x) \leq \max\nolimits_{1\leq\beta\leq M} \big(\mathcal{N}_{\kappa}(\nabla w_\beta)\big)(x) + C_{x_0,\rho}(1+|x|)^{-n}, \\ &\qquad \text{at each point } x \in \partial\Omega, \ \text{ where } K := \overline{B(x_0,\rho)}. \end{aligned} \tag{4.3.17}$$

Consequently, the conditions listed in (4.1.43)-(4.1.44) are satisfied, thanks to (4.3.15)-(4.3.17) and [112, Lemma 7.2.1] when $\mathscr{X}$, $\mathscr{Y}$, $\mathscr{Z}$ are as in (4.3.1), the conditions listed in (4.1.45)-(4.1.46) are satisfied thanks to (4.3.15)-(4.3.17) and [112, (7.7.101)] when $\mathscr{X}$, $\mathscr{Y}$, $\mathscr{Z}$ are as in (4.3.2), the conditions listed in (4.1.47)-(4.1.48) are satisfied thanks to (4.3.15)-(4.3.17) and [112, (7.7.118)] when $\mathscr{X}$, $\mathscr{Y}$, $\mathscr{Z}$ are as in (4.3.3), the conditions listed in (4.1.50)-(4.1.51) are satisfied thanks to (4.3.15)-(4.3.17) and [113, (6.2.7)] when $\mathscr{X}$, $\mathscr{Y}$, $\mathscr{Z}$ are as in (4.3.4), and the conditions listed in (4.1.52)-(4.1.53) are satisfied thanks to (4.3.15)-(4.3.17) and [113, Proposition 6.2.21] when $\mathscr{X}$, $\mathscr{Y}$, $\mathscr{Z}$ are as in (4.3.5).

As indicated in Examples 4.1.2-4.1.5, these always force $u(x_0) = 0$, and since the point $x_0 \in \Omega$ has been arbitrarily chosen the implication in (4.3.7) follows. □

Theorem 4.3.1 has a natural counterpart, essentially asserting that uniqueness for the Regularity Problem for a given system L is implied by the solvability of the Dirichlet Problem for $L^{\top}$, formulated in a dual setting.

Theorem 4.3.2 *Fix $n \in \mathbb{N}$ with[16] $n \geq 3$. Assume $\Omega \subset \mathbb{R}^n$ is an Ahlfors regular domain which is locally pathwise nontangentially accessible[17] (cf.* [112, Definition 8.9.14]). *Abbreviate $\sigma := \mathcal{H}^{n-1}\lfloor\partial\Omega$ and fix an aperture parameter $\kappa > 0$. Also, consider a homogeneous constant (complex) coefficient second-order $M\times M$ system L in $\mathbb{R}^n$, which is weakly elliptic (in the sense of* (1.3.3)). *In addition, assume that the spaces $\mathscr{X}$, $\mathscr{Y}$ are as in any of the following cases:*

$$\begin{aligned} &\mathscr{X} := L^{p'}(\partial\Omega,\sigma) \ \textit{and} \ \mathscr{Y} := \mathscr{X}^* = L^p(\partial\Omega,\sigma), \\ &\textit{where } p, p' \in (1,\infty) \textit{ are such that } 1/p + 1/p' = 1, \end{aligned} \tag{4.3.18}$$

[16] see the last part in the statement for the two-dimensional case

[17] in particular, this is the case if Ω is a one-sided NTA domain; see [112, (8.9.180)]

or

$$\mathscr{X} := L^{p'}(\partial\Omega, w^{1-p'}\sigma) \text{ and } \mathscr{Y} := \mathscr{X}^* = L^p(\partial\Omega, w\sigma),$$
$$\text{where } p, p' \in (1, \infty) \text{ satisfy } 1/p + 1/p' = 1 \text{ and } w \in A_p(\partial\Omega, \sigma), \tag{4.3.19}$$

or

$$\mathscr{X} := L^{p',q'}(\partial\Omega, \sigma) \text{ and } \mathscr{Y} := \mathscr{X}^* = L^{p,q}(\partial\Omega, \sigma),$$
$$\text{where } p, p', q, q' \in (1, \infty) \text{ satisfy } 1/p + 1/p' = 1 = 1/q + 1/q', \tag{4.3.20}$$

or

$$\mathscr{X} := \mathcal{B}^{p',\lambda}(\partial\Omega, \sigma) \text{ and } \mathscr{Y} := \mathscr{X}^* = M^{p,\lambda}(\partial\Omega, \sigma),$$
$$\text{where } p, p' \in (1, \infty) \text{ are such that } 1/p + 1/p' = 1, \tag{4.3.21}$$

or

$$\mathscr{X} := M^{p',\lambda}(\partial\Omega, \sigma) \text{ and } \mathscr{Y} := \mathscr{X}^* = \mathcal{B}^{p,\lambda}(\partial\Omega, \sigma),$$
$$\text{where } p, p' \in (1, \infty) \text{ are such that } 1/p + 1/p' = 1. \tag{4.3.22}$$

Then if the Dirichlet-type boundary value problem for $L^\top$

$$\begin{cases} w \in \big[\mathscr{C}^\infty(\Omega)\big]^M, \quad L^\top w = 0 \text{ in } \Omega, \\ \mathcal{N}_\kappa w \text{ belongs to the space } \mathscr{X}, \\ w\big|_{\partial\Omega}^{\kappa-\mathrm{n.t.}} = f \text{ at } \sigma\text{-a.e. point on } \partial\Omega, \\ \text{and also } w(x) = o(1) \text{ as } |x| \to \infty \\ \text{when } \Omega \text{ is an exterior domain,} \end{cases} \tag{4.3.23}$$

is solvable for each boundary datum $f \in \mathscr{X}$, the following uniqueness result for the dual Regularity-type boundary value problem for L is valid:

$$\left.\begin{array}{r} u \in \big[\mathscr{C}^\infty(\Omega)\big]^M, \quad Lu = 0 \text{ in } \Omega, \\ u\big|_{\partial\Omega}^{\kappa-\mathrm{n.t.}} = 0 \text{ at } \sigma\text{-a.e. point on } \partial\Omega, \\ \mathcal{N}_\kappa(\nabla u) \text{ belongs to the space } \mathscr{Y}, \\ \text{and also } u(x) = o(1) \text{ as } |x| \to \infty \\ \text{when } \Omega \text{ is an exterior domain,} \end{array}\right\} \Longrightarrow u \equiv 0 \text{ in } \Omega. \tag{4.3.24}$$

Furthermore, similar results are valid in the case $n = 2$ provided either Ω is bounded or $\partial\Omega$ is unbounded. Also, if $n = 2$ and Ω is an exterior domain, the same result is true provided one assumes

$$\int_{S^1} \big[L(\xi)\big]^{-1}\, d\mathcal{H}^1(\xi)$$
$$\text{is an invertible } M \times M \text{ matrix} \tag{4.3.25}$$

(which is always the case if the system L is assumed to actually satisfy the Legendre-Hadamard strong ellipticity condition, and if $M = 1$ *this is true if and only if* $L = \nabla \cdot A\nabla$ *for some* $A \in M_2^0$*; cf. Lemma 1.4.19 and* (1.4.186)*), and with the demand that u and w are* $o(1)$ *at infinity replaced by*

$$u(x) = O(1) \text{ and } w(x) = O(1) \text{ as } |x| \to \infty. \tag{4.3.26}$$

Proof This is proved similarly to Theorem 4.3.1, this time relying on Theorem 4.1.6 in the version given in Examples 4.1.7-4.1.10. □

We continue by discussing how the joint solvability of the Dirichlet Problem and Regularity Problem in bounded uniformly rectifiable domains, formulated for dual operators and with boundary data in Lebesgue and Sobolev spaces involving Hölder conjugate integrability exponents, entails genuine well-posedness for the Regularity Problem in such a setting.

Proposition 4.3.3 *Let* $\Omega \subset \mathbb{R}^n$, *with* $n \geq 2$, *be a bounded* UR *domain which is locally pathwise nontangentially accessible* (*in the sense of* [112, Definition 8.9.14]), *and abbreviate* $\sigma := \mathcal{H}^{n-1}\lfloor\partial\Omega$. *Also, consider a homogeneous constant* (*complex*) *coefficient second-order* $M \times M$ *system* L *in* $\mathbb{R}^n$, *which is weakly elliptic* (*in the sense of* (1.3.3)). *Finally, pick* $p, p' \in (1, \infty)$ *with* $1/p + 1/p' = 1$ *and fix an aperture parameter* $\kappa > 0$.

In this context, consider the Dirichlet Problem for the system L with boundary data from the Lebesgue space $[L^p(\partial\Omega, \sigma)]^M$, *formulated as*

$$\begin{cases} u \in \left[\mathscr{C}^\infty(\Omega)\right]^M, \quad Lu = 0 \text{ in } \Omega, \\ \mathcal{N}_\kappa u \in L^p(\partial\Omega, \sigma), \\ u\big|_{\partial\Omega}^{\kappa-\mathrm{n.t.}} = f \in \left[L^p(\partial\Omega, \sigma)\right]^M \text{ at } \sigma\text{-a.e. point on } \partial\Omega, \end{cases} \tag{4.3.27}$$

together with the Regularity Problem for the transpose system $L^\top$ *with boundary data from the boundary Sobolev space* $[L_1^{p'}(\partial\Omega, \sigma)]^M$, *formulated as*

$$\begin{cases} u \in \left[\mathscr{C}^\infty(\Omega)\right]^M, \quad L^\top u = 0 \text{ in } \Omega, \\ \mathcal{N}_\kappa u,\ \mathcal{N}_\kappa(\nabla u) \in L^{p'}(\partial\Omega, \sigma), \\ u\big|_{\partial\Omega}^{\kappa-\mathrm{n.t.}} = f \in \left[L_1^{p'}(\partial\Omega, \sigma)\right]^M \text{ at } \sigma\text{-a.e. point on } \partial\Omega. \end{cases} \tag{4.3.28}$$

Then the solvability of (4.3.27) *and* (4.3.28) *implies that the Regularity Problem* (4.3.28) *is actually well-posed* (*i.e., one has existence, uniqueness, and continuous dependence of the solution on the boundary datum*).

Proof Uniqueness for (4.3.28) is a consequence of Theorem 4.3.2. Since existence for (4.3.28) is assumed to begin with, there remains to establish the continuous dependence of the solution on the boundary datum in this problem. To deal with this issue, we begin by considering the space

$$\mathscr{V} := \Big\{ u \in \big[\mathscr{C}^\infty(\Omega)\big]^M : L^\top u = 0 \text{ in } \Omega \text{ and } \mathcal{N}_\kappa u,\ \mathcal{N}_\kappa(\nabla u) \in L^{p'}(\partial\Omega,\sigma) \Big\}, \tag{4.3.29}$$

and set

$$\|u\|_{\mathscr{V}} := \big\|\mathcal{N}_\kappa u\big\|_{L^{p'}(\partial\Omega,\sigma)} + \big\|\mathcal{N}_\kappa(\nabla u)\big\|_{L^{p'}(\partial\Omega,\sigma)} \text{ for each } u \in \mathscr{V}. \tag{4.3.30}$$

When equipped with this norm, we claim that $\mathscr{V}$ becomes complete (hence, a Banach space). To justify this, recall the Banach space $N_\kappa^{p'}(\Omega;\sigma)$ (defined as in (A.0.98) with $\mu := \sigma$). We may then embed $\mathscr{V}$ into the Banach space

$$\mathscr{W} := \underbrace{\big[N_\kappa^{p'}(\Omega;\sigma)\big]^M \oplus \cdots \oplus \big[N_\kappa^{p'}(\Omega;\sigma)\big]^M}_{n+1 \text{ terms}} \tag{4.3.31}$$

via

$$\mathscr{V} \ni u \longmapsto (u, \partial_1 u, \dots, \partial_n u) \in \mathscr{W} \tag{4.3.32}$$

which satisfies

$$\|u\|_{\mathscr{V}} \approx \big\|(u, \partial_1 u, \dots, \partial_n u)\big\|_{\mathscr{W}} \text{ uniformly in } u \in \mathscr{V}. \tag{4.3.33}$$

Assume now that $\{u_\mu\}_{\mu\in\mathbb{N}}$ is a Cauchy sequence in $\mathscr{V}$. Then (4.3.33) and the completeness of $\mathscr{W}$ imply that there exist $w, w_1, \dots, w_n \in \big[N_\kappa^{p'}(\Omega;\sigma)\big]^M$ such that $u_\mu \to w$ in $\big[N_\kappa^{p'}(\Omega;\sigma)\big]^M$ as $\mu \to \infty$ and $\partial_j u_\mu \to w_j$ in $\big[N_\kappa^{p'}(\Omega;\sigma)\big]^M$ as $\mu \to \infty$ for each $j \in \{1,\dots,n\}$. Thanks to [112, (8.3.33)], the above convergences also take place in $\big[L^\infty_{\rm loc}(\Omega,\mathcal{L}^n)\big]^M$. Consequently, given an arbitrary $\varphi \in \big[\mathscr{C}^\infty_c(\Omega)\big]^M$, for each $j \in \{1,\dots,n\}$ we may write

$$\begin{aligned} \int_\Omega \langle w_j, \varphi\rangle \, d\mathcal{L}^n &= \lim_{\mu\to\infty} \int_\Omega \langle \partial_j u_\mu, \varphi\rangle \, d\mathcal{L}^n \\ &= -\lim_{\mu\to\infty} \int_\Omega \langle u_\mu, \partial_j\varphi\rangle \, d\mathcal{L}^n = -\int_\Omega \langle w, \partial_j\varphi\rangle \, d\mathcal{L}^n. \end{aligned} \tag{4.3.34}$$

This proves that for each $j \in \{1,\dots,n\}$ we have $\partial_j w = w_j$ in the sense of distributions in Ω. Let us also observe that since

$$\int_\Omega \langle w, L\varphi\rangle \, d\mathcal{L}^n = \lim_{\mu\to\infty} \int_\Omega \langle u_\mu, L\varphi\rangle \, d\mathcal{L}^n = \lim_{\mu\to\infty} \int_\Omega \langle L^\top u_\mu, \varphi\rangle \, d\mathcal{L}^n = 0, \tag{4.3.35}$$

it follows that $L^\top w = 0$ in the sense of distributions in Ω. On account of this and elliptic regularity (see the first claim in [112, Theorem 6.5.7]), we conclude that $w \in \big[\mathscr{C}^\infty(\Omega)\big]^M$. Altogether, this argument shows that $w \in \mathscr{V}$ and $u_\mu \to w$ in $\mathscr{V}$ as $\mu \to \infty$. Hence, the space $\mathscr{V}$ is complete.

To proceed, define the operator

$$T : \mathscr{V} \longrightarrow \big[L_1^{p'}(\partial\Omega,\sigma)\big]^M, \qquad Tu := u\big|_{\partial\Omega}^{\kappa-\text{n.t.}} \text{ for each } u \in \mathscr{V}. \tag{4.3.36}$$

Thanks to [113, Proposition 11.3.4], the version of the Fatou-type theorem recorded in (3.3.107), and the fact that, as already noted, the Regularity Problem (4.3.28) is uniquely solvable, it follows that the operator T is well defined, linear, bounded, and bijective. Since $\mathscr{V}$ is a Banach space, the Open Mapping Theorem implies that the inverse of T is also bounded. This, in turn, guarantees the existence of a constant $C \in (0, \infty)$ with the property that the (unique) solution u of (4.3.28) satisfies

$$\|\mathcal{N}_\kappa u\|_{L^{p'}(\partial\Omega,\sigma)} + \|\mathcal{N}_\kappa(\nabla u)\|_{L^{p'}(\partial\Omega,\sigma)} \leq C\|f\|_{[L_1^{p'}(\partial\Omega,\sigma)]^M}. \tag{4.3.37}$$

This shows that the Regularity Problem (4.3.28) is indeed well-posed, finishing the proof of the proposition. □

4.4 A Sharp Poisson Integral Representation Formula

In Theorem 4.4.1 below we derive a Poisson integral representation formula for solutions of the Dirichlet Problem for a given weakly elliptic second-order system L, which involves the conormal derivative of the Green function for the transpose system $L^\top$ as integral kernel.

There are several features of this result which make the case for labeling it as being sharp. First, the underlying domain is of a very general geometric nature. Second, the brand of Green function employed here is minimalist, requiring only basic characteristics typically associated with such as object, like the property of being a fundamental solution with a vanishing (nontangential pointwise) trace, in addition to the demand of the existence of the nontangential pointwise trace for its gradient (which is needed in the very definition of the conormal derivative). Third, our type of Poisson integral formula is "spaceless," in the sense that no recognizable function space measuring size is employed in the formulation of its hypotheses. Instead, the function u being represented via this Poisson integral formula is only required to pair well, at the level on nontangential maximal operators, with the gradient of the Green function (in the sense that their product is absolutely integrable on the boundary). This ensures that the integral of the boundary trace of u coupled with the conormal derivative of the Green function is actually integrable, which is needed to simply make sense of our brand of Poisson integral representation formula. In Comment 6, following the statement of Theorem 4.4.1, we give examples indicating that this integrability condition may not be weakened much (and certainly cannot be eliminated).

Theorem 4.4.1 *Let $\Omega \subset \mathbb{R}^n$, where $n \in \mathbb{N}$ with*[18] *$n \geq 3$, be a locally pathwise nontangentially accessible set*[19] *(in the sense of* [112, Definition 8.9.14]*) with a lower Ahlfors regular boundary and with the property that $\sigma := \mathcal{H}^{n-1}\lfloor\partial\Omega$ is a doubling*

[18] see the last claim in the statement for the case $n = 2$

[19] in particular, this is the case if Ω is a one-sided NTA domain; see [112, (8.9.180)] and the subsequent comment

measure on $\partial\Omega$. Suppose L is a weakly elliptic, homogeneous, constant (complex) coefficient, second-order, $M\times M$ system in $\mathbb{R}^n$. Fix an aperture parameter $\kappa \in (0,\infty)$, along with an arbitrary point $x_0 \in \Omega$, and suppose $0 < \rho < \frac{1}{4}\operatorname{dist}(x_0, \partial\Omega)$.

Then there exists some $\widetilde{\kappa} > 0$, which depends only on Ω and κ, with the following significance. Assume G is a matrix-valued function satisfying

$$\begin{cases} G = (G_{\alpha\beta})_{1\leq\alpha,\beta\leq M} \in \big[L^1_{\mathrm{loc}}(\Omega, \mathcal{L}^n)\big]^{M\times M}, \\ \big(L^\top G_{\cdot\beta}\big)_\alpha = -\delta_{x_0}\delta_{\alpha\beta} \text{ in } \mathcal{D}'(\Omega) \text{ for all } \alpha,\beta \in \{1,\dots,M\}, \\ (\nabla G)\Big|^{\widetilde{\kappa}-\mathrm{n.t.}}_{\partial\Omega} \text{ exists (in } \mathbb{C}^{n\cdot M^2}) \text{ at } \sigma\text{-a.e. point on } \partial\Omega, \\ G\Big|^{\widetilde{\kappa}-\mathrm{n.t.}}_{\partial\Omega} = 0 \in \mathbb{C}^{M\times M} \text{ at } \sigma\text{-a.e. point on } \partial\Omega, \\ \text{and also } G(x) = o(1) \text{ as } |x| \longrightarrow \infty \\ \text{when } \Omega \text{ is an exterior domain,} \end{cases} \tag{4.4.1}$$

and assume $u = (u_\beta)_{1\leq\beta\leq M}$ is a $\mathbb{C}^M$-valued function satisfying

$$\begin{cases} u \in \big[\mathscr{C}^\infty(\Omega)\big]^M, \quad Lu = 0 \text{ in } \Omega, \\ u\Big|^{\kappa-\mathrm{n.t.}}_{\partial\Omega} \text{ exists at } \sigma\text{-a.e. point on } \partial\Omega, \\ \displaystyle\int_{\partial\Omega} \mathcal{N}^\rho_\kappa u \cdot \mathcal{N}^\rho_{\widetilde{\kappa}}(\nabla G)\,d\sigma < +\infty, \\ \text{and also } u(x) = o(1) \text{ as } |x| \longrightarrow \infty \\ \text{when } \Omega \text{ is an exterior domain.} \end{cases} \tag{4.4.2}$$

In the case when $\partial\Omega$ is unbounded, strengthen the initial geometric hypotheses on Ω and also strengthen the integral condition in the third line of (4.4.2) *by assuming that (cf.* [112, (8.9.179)])

$$\begin{cases} \Omega \text{ is a globally pathwise nontangentially accessible set,} \\ \text{and } \displaystyle\int_{\partial\Omega} \mathcal{N}_\kappa u \cdot \mathcal{N}^{\Omega\setminus K}_{\widetilde{\kappa}}(\nabla G)\,d\sigma < +\infty \text{ where } K := \overline{B(x_0,\rho)}, \end{cases} \tag{4.4.3}$$

(with the convention that $0\cdot\infty = \infty\cdot 0 = \infty$).

Then for any choice of a coefficient tensor $A = \big(a^{\alpha\beta}_{rs}\big)_{\substack{1\leq r,s\leq n\\ 1\leq\alpha,\beta\leq M}}$ which permits writing L as L_A, the system associated with A as in (1.7.7), *one has the Poisson integral representation formula*

$$u_\beta(x_0) = -\int_{\partial_*\Omega} \Big\langle u\Big|^{\kappa-\mathrm{n.t.}}_{\partial\Omega}, \partial^{A^\top}_\nu G_{\cdot\beta}\Big\rangle\,d\sigma, \qquad \forall\,\beta \in \{1,\dots,M\}, \tag{4.4.4}$$

where ν denotes the geometric measure theoretic outward unit normal to Ω and $\partial_\nu^{A^\top}$ stands for the conormal derivative associated with $A^\top$ as in (1.7.9)*, acting on the columns of the matrix-valued function G.*

Moreover, the same result remains true in the two-dimensional setting if either Ω is bounded, or $\partial\Omega$ is unbounded. Also, if $n = 2$ and Ω is an exterior domain, the same result is true provided one assumes

$$\begin{gathered}\int_{S^1} \big[L(\xi)\big]^{-1}\, d\mathcal{H}^1(\xi) \\ \textit{is an invertible } M \times M \textit{ matrix}\end{gathered} \tag{4.4.5}$$

(which is always the case if the system L is assumed to actually satisfy the Legendre-Hadamard strong ellipticity condition, and if $M = 1$ this is true if and only if $L = \nabla \cdot A\nabla$ for some $A \in M_2^0$; cf. Lemma 1.4.19 and (1.4.186)*), and with the demand that u and G are $o(1)$ at infinity replaced by*

$$u(x) = O(1) \textit{ and } G(x) = O(1) \textit{ as } |x| \to \infty. \tag{4.4.6}$$

Before presenting the proof of this theorem we make a number of comments.

Comment 1. Recall from (4.1.7)-(4.1.8) that the conditions demanded in the first two lines of (4.4.1) imply that actually

$$G \in \big[\mathscr{C}^\infty(\Omega \setminus \{x_0\}) \cap W^{1,1}_{\rm loc}(\Omega)\big]^{M\times M}. \tag{4.4.7}$$

In particular, the condition in the third line of (4.4.1) makes sense. Together with (4.4.7), said condition permits us to define each conormal derivative $\partial_\nu^{A^\top} G_{\bullet\beta}$ as in (1.7.9) (with the aperture parameter $\widetilde{\kappa}$ now playing the role of the old κ). Moreover, [112, Proposition 8.9.16] guarantees that we presently have $\sigma\big(\partial\Omega \setminus \partial_{\rm nta}\Omega\big) = 0$, so it makes sense to consider nontangential boundary traces (associated with any aperture parameter) at σ-a.e. point on $\partial\Omega$.

Bearing this in mind, from the first two lines in (4.4.2), (4.4.7), the third line of (4.4.1), [112, Lemma 8.9.3], and [112, Proposition 8.9.16] we conclude that

$$\big(\mathcal{N}_\kappa^\rho u\big)(x) < +\infty \text{ and } \big(\mathcal{N}_{\widetilde{\kappa}}^\rho(\nabla G)\big)(x) < +\infty \text{ for } \sigma\text{-a.e. } x \in \partial\Omega. \tag{4.4.8}$$

In particular, (4.4.8) ensures that the product $\big(\mathcal{N}_\kappa^\rho u\big)(x) \cdot \big(\mathcal{N}_{\widetilde{\kappa}}^\rho(\nabla G)\big)(x)$ is a well-defined number in $[0, +\infty)$ for σ-a.e. $x \in \partial\Omega$. In concert with [112, (8.2.28)], this proves that $\mathcal{N}_\kappa^\rho u \cdot \mathcal{N}_{\widetilde{\kappa}}^\rho(\nabla G)$ is a well-defined non-negative σ-measurable function on $\partial\Omega$. Consequently, the integral condition formulated in the third line of (4.4.2) is meaningful. In turn, based on [112, (8.9.8)] and the hypothesis in the third line of (4.4.2) we see that the integral in (4.4.4) is absolutely convergent, so the conclusion in the theorem has a clear meaning.

Finally, from the second line in (4.4.3) it follows (bearing in mind the convention $0 \cdot \infty = \infty \cdot 0 = \infty$) that

$$\big(\mathcal{N}_\kappa u\big)(x) < +\infty \text{ and } \big(\mathcal{N}_{\widetilde{\kappa}}^{\Omega\setminus K}(\nabla G)\big)(x) < +\infty \text{ for } \sigma\text{-a.e. } x \in \partial\Omega. \tag{4.4.9}$$

Comment 2. Remarkably, the only assumptions of quantitative nature made[20] in the statement of Theorem 4.4.1 are the finiteness integral condition imposed in the third line of (4.4.2), and its version from (4.4.3) in the case when $\partial\Omega$ is unbounded. These demands are automatically satisfied in any of the instances described in (4.1.43)-(4.1.44), (4.1.45)-(4.1.46), (4.1.47)-(4.1.48), (4.1.50)-(4.1.51), and (4.1.52)-(4.1.53). Also, the aforementioned integral conditions may be recast as memberships to weighted Lebesgue spaces of the sort

$$\mathcal{N}_\kappa^\rho u \in L^1\Big(\partial\Omega, \mathcal{N}_{\overline{\kappa}}^\rho(\nabla G)\,\sigma\Big) \text{ and } \mathcal{N}_\kappa u \in L^1\Big(\partial\Omega, \mathcal{N}_{\overline{\kappa}}^{\Omega\setminus K}(\nabla G)\,\sigma\Big). \tag{4.4.10}$$

Comment 3. The integral representation formula recorded in (4.4.4) readily implies the vanishing property $u(x_0) = 0$ for any null-solution u of the Dirichlet Problem for the system L in Ω, in a formulation which uses (4.4.10) as the demand imposed on the size of the solution. We have already established this in Theorem 4.1.1 asking $\mathcal{N}_{\overline{\kappa}}^\rho(\nabla G) < +\infty$ at σ-a.e. point on $\partial\Omega$ in place of the σ-a.e. existence of $(\nabla G)\big|_{\partial\Omega}^{\overline{\kappa}-\mathrm{n.t.}}$ on $\partial\Omega$ which, according to [112, (8.9.22)], is a less demanding assumption on the corresponding Green function. This being said, the latter (stronger) condition is automatically satisfied if, for example, Ω is a UR domain and we assume that $\mathcal{N}_{\overline{\kappa}}^\rho(\nabla G) \in L^p(\partial\Omega, \sigma)$ for some $p \in \big(\frac{n-1}{n}, \infty\big)$. Indeed, this is seen by applying the Fatou-type result from Theorem 3.3.4 to the columns of the matrix-valued function G restricted to $\Omega \setminus \overline{B\big(x_0, \frac{1}{2}\operatorname{dist}(x_0, \partial\Omega)\big)}$.

Comment 4. Different choices of the coefficient tensor A which permit us to represent the given system L as L_A yield differently-looking integral representations formulas (4.4.4). At least heuristically, this somewhat curious aspect may be explained by recalling that the conormal derivatives associated with any two such coefficient tensors differ by a linear combination of tangential derivatives which, in the case of the Green function, should vanish since the nontangential boundary trace of G is zero (see the condition in the fourth line of (4.4.1)). More precisely, if the condition in the third line of (4.4.1) is strengthened to $\mathcal{N}_{\overline{\kappa}}^\rho(\nabla G) \in L^1_{\mathrm{loc}}(\partial\Omega, \sigma)$ then on account of (1.7.24) (whose applicability is presently ensured by [112, Proposition 8.9.17] and the condition in the fourth line of (4.4.1)) we see that

$$\text{if } A, B \text{ are two coefficient tensors such that the original system may be represented both as } L = L_A \text{ and as } L = L_B, \text{ then for each index } \beta \in \{1, \dots, M\} \text{ we have } \partial_\nu^{A^\top} G_{\cdot\beta} = \partial_\nu^{B^\top} G_{\cdot\beta} \text{ at } \sigma\text{-a.e. point on } \partial_*\Omega. \tag{4.4.11}$$

Comment 5. In the context of Theorem 4.4.1, the integral representation formula (4.4.4) yields the estimate

$$|u(x_0)| \le C \int_{\partial_*\Omega} \Big|u\big|_{\partial\Omega}^{\kappa-\mathrm{n.t.}}\Big| \Big|(\nabla G)\big|_{\partial\Omega}^{\overline{\kappa}-\mathrm{n.t.}}\Big| \, d\sigma. \tag{4.4.12}$$

---

[20] apart from decay at infinity in the case when Ω is an exterior domain

Suppose now that there exist $\alpha > 0$ and $C \in (0,\infty)$ such that a Green function G as in (4.4.1) exists for each pole $x_0 \in \Omega$ and has the additional property that[21]

$$|(\nabla G)(x)| \leq C \cdot \operatorname{dist}(x_0, \partial\Omega)^{\alpha} |x - x_0|^{1-n-\alpha}, \qquad \forall x \in \Omega. \tag{4.4.13}$$

We may then rely on (4.4.12) and [112, Proposition 8.4.12] to conclude that, with $\mathcal{M}_{\partial\Omega}$ denoting the Hardy-Littlewood maximal operator on $\partial\Omega$ (cf. (A.0.84)), the following pointwise estimate holds:

$$\mathcal{N}_{\kappa'} u \leq C \cdot \mathcal{M}_{\partial\Omega}\big(u\big|_{\partial\Omega}^{\kappa-\mathrm{n.t.}}\big) \text{ on } \partial\Omega, \tag{4.4.14}$$

for each given aperture parameter $\kappa' > 0$, where $C \in (0,\infty)$ is independent of u. In turn, estimate (4.4.14) is particularly versatile. For example, this may be used to establish a Weak Maximum Principle for null-solutions u of the system L in Ω asserting that, for some background constant $C \in (0,\infty)$, we have

$$\sup_{\Omega} |u| \leq C \big\| u\big|_{\partial\Omega}^{\kappa-\mathrm{n.t.}} \big\|_{[L^\infty(\partial\Omega,\sigma)]^M}, \tag{4.4.15}$$

and also prove regularity results to the effect that better integrability properties of the nontangential boundary trace $u\big|_{\partial\Omega}^{\kappa-\mathrm{n.t.}}$ entail better integrability properties of the nontangential maximal function of the null-solution u of the system L.

Comment 6. Consider $\Omega := \mathbb{D} := B(0,1)$ the unit disk in the plane $\mathbb{R}^2 \equiv \mathbb{C}$ and introduce

$$\begin{gathered}\text{the function } u : \mathbb{D} \longrightarrow \mathbb{R} \text{ defined as} \\ u(z) := \operatorname{Re}\Big(\tfrac{z+1}{z-1}\Big) = \tfrac{|z|^2-1}{|z-1|^2} \text{ for each } z \in \mathbb{D}.\end{gathered} \tag{4.4.16}$$

With $\mathbb{T} := \partial\mathbb{D}$, this implies

$$u \in \mathscr{C}^\infty\big(\overline{\mathbb{D}} \setminus \{1\}\big), \quad \Delta u = 0 \text{ in } \mathbb{D}, \text{ and } u\big|_{\mathbb{T}\setminus\{1\}} = 0. \tag{4.4.17}$$

Having fixed some aperture parameter $\kappa > 0$, we also claim that

$$\mathcal{N}_\kappa u \in L^{1,\infty}\big(\mathbb{T}, \mathcal{H}^1 \lfloor \mathbb{T}\big) \setminus L^1\big(\mathbb{T}, \mathcal{H}^1 \lfloor \mathbb{T}\big). \tag{4.4.18}$$

To prove this, we start by observing that

$$|u(z)| = \frac{(1-|z|)(1+|z|)}{|z-1|^2} \leq \frac{2\operatorname{dist}(z,\mathbb{T})}{|z-1|^2} \text{ for each } z \in \mathbb{D}. \tag{4.4.19}$$

For each $\zeta \in \mathbb{T}$ and each $z \in \Gamma_\kappa(\zeta)$ we may invoke [112, (8.1.23)] to estimate

[21] such an estimate entails (cf. [112, Lemma 8.3.7]) that for each fixed point $x \in \Omega$ and any given aperture parameter $\kappa' > 0$ we have $\big(\mathcal{N}_{\kappa'}^{\Omega\setminus K_{x_0}}(\nabla G)\big)(x) \leq C_{x_0,\kappa'}(1+|x|)^{-(n-1+\alpha)}$ at every $x \in \partial\Omega$, where $K_{x_0} := B\big(x, \tfrac{1}{2}\operatorname{dist}(x_0,\partial\Omega)\big)$

$$|z-1|^2 \approx \big[|z-\zeta| + |\zeta - 1|\big]^2 \approx \big[\mathrm{dist}(z,\mathbb{T}) + |\zeta - 1|\big]^2$$
$$\approx \big[\mathrm{dist}(z,\mathbb{T})\big]^2 + |\zeta - 1|^2 \geq 2\big[\mathrm{dist}(z,\mathbb{T})\big]\cdot|\zeta - 1|. \tag{4.4.20}$$

Together with (4.4.19) this permits us to conclude that

$$|u(z)| \leq \frac{C}{|\zeta - 1|} \text{ for each } \zeta \in \mathbb{T} \text{ and } z \in \Gamma_\kappa(\zeta), \tag{4.4.21}$$

which ultimately implies that, for some universal constant $C \in (0,\infty)$,

$$(\mathcal{N}_\kappa u)(\zeta) \leq \frac{C}{|\zeta - 1|} \text{ for each } \zeta \in \mathbb{T}. \tag{4.4.22}$$

In fact, we claim that an inequality of similar nature in the opposite direction is also valid. Indeed, if for each $\zeta \in \mathbb{T} \cap B(1, 100^{-1})$ in the first quadrant we let z_ζ denote the intersection point of the line $x + y = 1$ with the line joining the origin with ζ, then $z_\zeta \in \Gamma_\kappa(\zeta)$ hence

$$(\mathcal{N}_\kappa u)(\zeta) \geq |u(z_\zeta)|. \tag{4.4.23}$$

Regarding z_ζ as a point on the line $x + y = 1$, permits us to express $z_\zeta = (1-t, t)$ for some $t \in (0, 1/2)$. Since elementary geometry gives $|z_\zeta - 1| \approx |\zeta - 1|$, it follows that $t \approx |\zeta - 1|$. In view of these properties and the formula for u given in (4.4.16) we may then compute

$$|u(z_\zeta)| = \frac{1 - |z_\zeta|^2}{|z_\zeta - 1|^2} \approx \frac{1 - |(1-t,t)|^2}{t^2} \approx \frac{1}{t} \approx \frac{1}{|\zeta - 1|}. \tag{4.4.24}$$

Together with (4.4.23) this gives

$$(\mathcal{N}_\kappa u)(\zeta) \geq \frac{C}{|\zeta - 1|}. \tag{4.4.25}$$

By symmetry, a similar estimate holds for each $\zeta \in \mathbb{T} \cap B(1, 100^{-1})$ in the fourth quadrant, ergo

$$(\mathcal{N}_\kappa u)(\zeta) \geq \frac{C}{|\zeta - 1|} \text{ for each } \zeta \in \mathbb{T}. \tag{4.4.26}$$

In concert with (4.4.22) this proves

$$(\mathcal{N}_\kappa u)(\zeta) \approx \frac{1}{|\zeta - 1|} \text{ uniformly for } \zeta \in \mathbb{T}, \tag{4.4.27}$$

from which the claim made in (4.4.18) follows.

Since the harmonic function u defined in (4.4.16) is nonzero in $\mathbb{D}$ but its nontangential trace vanishes $\mathcal{H}^1$-a.e. on $\mathbb{T}$, it follows that u fails to be representable via the version of the Poisson integral representation formula (4.4.4) corresponding to $L := \Delta$ and $\Omega := \mathbb{D}$. The only hypothesis in the statement of Theorem 4.4.1 that is currently violated is the absolute integrability condition demanded in the third line of (4.4.2). To briefly elaborate on this aspect, recall that the Green function G for the Laplacian in the unit disk in the plane with pole at a fixed point $x_0 \in \mathbb{D}$ is given at each $x \in \mathbb{D}$ by

$$G(x) = \frac{1}{2\pi} \ln|x - x_0| - \frac{1}{2\pi} \ln\big(|x_0^*||x - x_0^*|\big), \quad \text{where } x_0^* := \frac{x_0}{|x_0|^2} \in \mathbb{C} \setminus \overline{\mathbb{D}}. \tag{4.4.28}$$

As such, G belongs to $\mathscr{C}^\infty\big(\overline{\mathbb{D}} \setminus \{x_0\}\big)$ and its normal derivative is given by

$$x \cdot (\nabla G)(x) = \frac{1}{2\pi} \frac{1 - |x_0|^2}{|x - x_0|^2} \quad \text{for each } x \in \mathbb{T} = \partial\mathbb{D}. \tag{4.4.29}$$

In particular, at each point $x \in \mathbb{T} = \partial\mathbb{D}$ we may estimate

$$|(\nabla G)(x)| \geq |x \cdot (\nabla G)(x)| = \frac{1}{2\pi} \frac{1 - |x_0|}{1 + |x_0|}, \tag{4.4.30}$$

which goes to show that for any given aperture parameter $\widetilde{\kappa} > 0$ and any given truncation parameter $\rho \in \big(0, 1 - |x_0|\big)$ there exists a constant $C_\rho \in (0, \infty)$ such that

$$C_\rho \geq \big(\mathcal{N}_{\widetilde{\kappa}}^{\rho}(\nabla G)\big)(x) \geq \frac{1}{2\pi} \frac{1 - |x_0|}{1 + |x_0|} > 0 \text{ for all } x \in \mathbb{T} = \partial\mathbb{D}. \tag{4.4.31}$$

In view of this and (4.4.18), we conclude that the absolute integrability condition demanded in the third line of (4.4.2) fails, though just barely. This points to the fact that Theorem 4.4.1 is in the nature of best possible.

Comment 7. Similar counterexamples to those discussed in Comment 6 above may be constructed in higher dimensions and for unbounded domains. For instance, corresponding to $L := \Delta$ and $\Omega := \mathbb{R}^n_+$ with $n \geq 2$ arbitrary, we may consider $u : \mathbb{R}^n_+ \to \mathbb{R}$ defined as

$$u(x) := \frac{x_n}{|x|^n} \quad \text{for each } x = (x', x_n) \in \mathbb{R}^{n-1} \times (0, \infty) = \mathbb{R}^n_+. \tag{4.4.32}$$

This is a harmonic function which is never zero in $\mathbb{R}^n_+$, and whose nontangential boundary trace vanishes $\mathcal{L}^{n-1}$-a.e. on $\mathbb{R}^{n-1} \equiv \partial\mathbb{R}^n_+$. As such, the version of the Poisson integral representation formula (4.4.4) for the current choices fails. The source of this failure is the fact that the absolute integrability condition stipulated in the third line of (4.4.2) does not materialize, and this is the only hypothesis in the statement of Theorem 4.4.1 that is currently violated. To see that this is indeed the case, recall that the Green function for the Laplacian in the upper half-space with pole at an arbitrary fixed point $x_0 = (x_0', t_0) \in \mathbb{R}^{n-1} \times (0, \infty) = \mathbb{R}^n_+$ is given by

$$G(x) = E_\Delta(x - x_0) - E_\Delta(x' - x_0', t + t_0) \text{ for each } x = (x', t) \in \mathbb{R}^n_+ \setminus \{x_0\}, \tag{4.4.33}$$

where E_Δ is the standard fundamental solution for the Laplacian in $\mathbb{R}^n$ (cf. (1.5.56)). Hence, G belongs to $\mathscr{C}^\infty\big(\overline{\mathbb{R}^n_+} \setminus \{x_0\}\big)$ and its normal derivative is given by

$$(\partial_{x_n} G)(x', 0) = \frac{2}{\omega_{n-1}} \frac{t_0}{(t_0 + |x' - x_0'|^2)^{n/2}} \text{ for each } x' \in \mathbb{R}^{n-1}. \tag{4.4.34}$$

As a consequence, there exists some constant $C(n, x_0) \in (0, \infty)$ with the property that

$$\big|(\partial_{x_n} G)(x', 0)\big| \geq \frac{C(n, x_0)}{1 + |x'|^n} \text{ for all } x' \in \mathbb{R}^{n-1}. \tag{4.4.35}$$

If we now fix some aperture parameter $\widetilde{\kappa} > 0$ and abbreviate $K := \overline{B(x_0, t_0/2)}$, from (4.4.35) we then deduce that

$$\big(\mathcal{N}_{\widetilde{\kappa}}^{\mathbb{R}^n_+ \setminus K}(\nabla G)\big)(x) \geq \frac{C(n, x_0)}{1 + |x'|^n} \text{ for all } x' \in \mathbb{R}^{n-1}. \tag{4.4.36}$$

In the opposite direction, using the fact that there exists a dimensional constant $C_n \in (0, \infty)$ with the property that

$$\big|(\nabla G)(x)\big| \leq C_n \frac{t_0}{|x - x_0|^n} \text{ for each } x \in \mathbb{R}^n_+ \text{ with } |x - x_0| > 4t_0 \tag{4.4.37}$$

(itself a consequence of (4.4.33) and (1.5.56)), we may employ [112, Lemma 8.3.7] to conclude that there exists some $C(n, \widetilde{\kappa}, x_0) \in (0, \infty)$ for which

$$\big(\mathcal{N}_{\widetilde{\kappa}}^{\mathbb{R}^n_+ \setminus K}(\nabla G)\big)(x') \leq \frac{C(n, \widetilde{\kappa}, x_0)}{1 + |x'|^n} \text{ for all } x' \in \mathbb{R}^{n-1}. \tag{4.4.38}$$

Altogether, (4.4.36) and (4.4.38) prove that

$$\big(\mathcal{N}_{\widetilde{\kappa}}^{\mathbb{R}^n_+ \setminus K}(\nabla G)\big)(x') \approx \frac{1}{1 + |x'|^n} \text{ uniformly in } x' \in \mathbb{R}^{n-1}. \tag{4.4.39}$$

As far as the function u defined in (4.4.32) is concerned, fix an aperture parameter $\kappa > 0$ and observe that for each $x' \in \mathbb{R}^{n-1} \setminus \{0'\}$ and each $y = (y', y_n) \in \Gamma_\kappa(x')$ we have

$$|u(y)| \leq |y|^{1-n} \leq \big[\mathrm{dist}(0, \Gamma_\kappa(x'))\big]^{1-n} \leq C_\kappa |x'|^{1-n} \tag{4.4.40}$$

for some constant $C_\kappa \in (0, \infty)$ which depends exclusively on κ. In turn, this implies that

$$(\mathcal{N}_\kappa u)(x') \leq \frac{C_\kappa}{|x'|^{n-1}} \text{ for all } x' \in \mathbb{R}^{n-1}. \tag{4.4.41}$$

Since for each $x' \in \mathbb{R}^{n-1} \setminus \{0'\}$ we may also estimate

$$(\mathcal{N}_\kappa u)(x') \geq \big|u(x', |x'|)\big| = \frac{|x'|}{\big|(x', |x'|)\big|^n} = \frac{2^{-n/2}}{|x'|^{n-1}}, \tag{4.4.42}$$

we deduce from (4.4.41) and (4.4.42) that the function u defined in (4.4.32) satisfies

$$(\mathcal{N}_\kappa u)(x') \approx \frac{1}{|x'|^{n-1}} \text{ uniformly in } x' \in \mathbb{R}^{n-1}. \tag{4.4.43}$$

Together, (4.4.39) and (4.4.43) prove that

$$(\mathcal{N}_{\widetilde{\kappa}}^{\mathbb{R}^n_+\setminus K}(\nabla G))(x')\cdot(\mathcal{N}_\kappa u)(x') \approx \frac{1}{|x'|^{n-1}(1+|x'|^n)} \quad \text{uniformly in } x'\in\mathbb{R}^{n-1}. \tag{4.4.44}$$

In light of the fact that for each $x'\in\mathbb{R}^{n-1}\setminus\{0'\}$ we have

$$\frac{1}{2|x'|^{n-1}}\cdot\mathbf{1}_{B_{n-1}(0',1)} \le \frac{1}{|x'|^{n-1}(1+|x'|^n)} \le \frac{1}{|x'|^{n-1}}, \tag{4.4.45}$$

we conclude from (4.4.44)-(4.4.45) that

$$\big(\mathcal{N}_{\widetilde{\kappa}}^{\mathbb{R}^n_+\setminus K}(\nabla G)\big)\cdot(\mathcal{N}_\kappa u) \text{ belongs to } L^{1,\infty}(\mathbb{R}^{n-1},\mathcal{L}^{n-1})\setminus L^1(\mathbb{R}^{n-1},\mathcal{L}^{n-1}). \tag{4.4.46}$$

Hence, the absolute integrability condition demanded in the third line of (4.4.2) fails, though very narrowly. Once again, this highlights the optimality of Theorem 4.4.1.

Comment 8. Theorem 4.4.1 naturally yields a uniqueness result for the Dirichlet Problem, as the conclusion in Theorem 4.1.1 is implied by Theorem 4.4.1. The reason for which Theorem 4.1.1 is not a corollary of Theorem 4.4.1 is the fact that, as opposed to the latter, the former theorem involves an additional hypothesis, demanding the existence of the nontangential boundary trace of ∇G.

We now turn to the task of presenting the proof of Theorem 4.4.1.

Proof of Theorem 4.4.1 Recall that (4.4.7) holds. Fix some $\beta\in\{1,\dots,M\}$ and, using the summation convention over repeated indices, define the vector field

$$\vec{F} := \Big(u_\alpha a_{kj}^{\gamma\alpha}\partial_k G_{\gamma\beta} - G_{\alpha\beta}a_{jk}^{\alpha\gamma}\partial_k u_\gamma\Big)_{1\le j\le n} \quad \text{at } \mathcal{L}^n\text{-a.e. point in } \Omega. \tag{4.4.47}$$

From (4.4.47), (4.4.7), and the fact that $u\in\big[\mathscr{C}^\infty(\Omega)\big]^M$ it follows that

$$\vec{F}\in\big[L^1_{\rm loc}(\Omega,\mathcal{L}^n)\big]^n \tag{4.4.48}$$

and, in the sense of distributions in Ω, we have

$$\begin{aligned}\mathrm{div}\vec{F} &= (\partial_j u_\alpha)a_{kj}^{\gamma\alpha}(\partial_k G_{\gamma\beta}) + u_\alpha a_{kj}^{\gamma\alpha}(\partial_j\partial_k G_{\gamma\beta})\\ &\quad - (\partial_j G_{\alpha\beta})a_{jk}^{\alpha\gamma}(\partial_k u_\gamma) - G_{\alpha\beta}a_{jk}^{\alpha\gamma}(\partial_j\partial_k u_\gamma)\\ &=: I_1+I_2+I_3+I_4,\end{aligned} \tag{4.4.49}$$

where the last equality defines the I_i's. Changing variables $j'=k$, $k'=j$, $\alpha'=\gamma$, and $\gamma'=\alpha$ in I_3 yields

$$I_3 = -(\partial_{k'}G_{\gamma'\beta})\,a_{k'j'}^{\gamma'\alpha'}\,(\partial_{j'}u_{\alpha'}) = -I_1. \tag{4.4.50}$$

As regards I_4, we have

$$I_4 = -G_{\alpha\beta}\,(L_A u)_\alpha = -G_{\alpha\beta}\,(Lu)_\alpha = 0, \tag{4.4.51}$$

by the assumptions on u. Finally, thanks to the second line in (4.4.1), we have

$$\begin{aligned} I_2 &= u_\alpha (L_{A^\top} G_{\bullet\beta})_\alpha = u_\alpha (L^\top G_{\bullet\beta})_\alpha \\ &= -u_\alpha \delta_{\alpha\beta} \delta_{x_0} = -u_\beta \delta_{x_0} = -u_\beta(x_0)\delta_{x_0}. \end{aligned} \tag{4.4.52}$$

Collectively, these equalities permit us to conclude that, in the sense of distributions in Ω,

$$\mathrm{div}\vec{F} = -u_\beta(x_0)\delta_{x_0} \in \mathscr{E}'(\Omega). \tag{4.4.53}$$

In particular,

$$\mathrm{div}\vec{F} \in \mathcal{D}'(\Omega) \text{ induces a continuous functional in } \left(\mathscr{C}_b^\infty(\Omega)\right)^*. \tag{4.4.54}$$

Moving on, recall from [112, Proposition 8.9.16] that in the present context we have

$$\sigma\big(\partial\Omega \setminus \partial_{\mathrm{nta}}\Omega\big) = 0. \tag{4.4.55}$$

Starting with the aperture parameter κ given in the statement of the theorem, choose $a \in (0,1)$ small enough so that $\kappa > 2a/(1-a)$, then define

$$\kappa' := \kappa(1-a) - 2a \in (0,\kappa). \tag{4.4.56}$$

Next, let $\widetilde{\kappa} \geq \kappa'$ along with $d > 0$ and $c \in [1,\infty)$ be parameters associated as in [112, (8.9.178)] with κ' (now playing the role of κ in that context). If necessary, further increase $\widetilde{\kappa}$ as to ensure that $\widetilde{\kappa} \geq \kappa$. Also, set

$$\theta := [c(1+\kappa')]^{-1} \in (0,1), \tag{4.4.57}$$

define

$$N_1 := \left\{x \in \partial\Omega : \mathcal{N}_{\widetilde{\kappa}}^{\rho}(\nabla G)(x) = +\infty\right\}, \tag{4.4.58}$$

and introduce

$$N_2 := \left\{x \in \partial_{\mathrm{nta}}\Omega : \big(u\big|_{\partial\Omega}^{\kappa-\mathrm{n.t.}}\big)(x) \text{ fails to exist}\right\}. \tag{4.4.59}$$

Then (4.4.8) implies $\sigma(N_1) = 0$. Also, from the second line in (4.4.2) and [112, (3.6.26)] we see that N_2 is a σ-measurable set and $\sigma(N_2) = 0$.

To proceed, fix $x \in \partial_{\mathrm{nta}}\Omega \setminus (N_1 \cup N_2)$ and pick an arbitrary point

$$y \in \Gamma_{\kappa'}(x) \text{ with } \delta_{\partial\Omega}(y) < \min\left\{\theta \cdot \rho, \mathrm{d}\right\}. \tag{4.4.60}$$

The property in the fourth line of (4.4.1) ensures that the condition formulated in [112, (8.9.196)] is satisfied by (the entries of) the matrix-valued function G. Granted this, [112, (8.9.197)] written with κ' in place of κ and for any

$$\varepsilon \in \Big(0, \min\{\rho, \mathrm{d}/\theta\}\Big) \text{ with } \varepsilon > \theta^{-1}\delta_{\partial\Omega}(y) \tag{4.4.61}$$

guarantees the existence of a σ-measurable set $N_3 \subseteq \partial\Omega$ satisfying $\sigma(N_3) = 0$ such that

$$\begin{aligned}|G(y)| &\leq (\mathcal{N}^{\theta\varepsilon}_{\kappa'} G)(x) \leq \varepsilon \cdot \mathcal{N}^{\varepsilon}_{\widetilde{\kappa}}(\nabla G)(x) \\ &\leq \varepsilon \cdot \mathcal{N}^{\rho}_{\widetilde{\kappa}}(\nabla G)(x) \text{ if } x \notin N_3,\end{aligned} \tag{4.4.62}$$

where the first inequality takes into account [112, (8.2.13)], the second inequality relies on [112, Remark 8.9.19], and the last inequality is a consequence of (4.4.61) together with [112, (8.2.25)]. Sending $\varepsilon \searrow \theta^{-1}\delta_{\partial\Omega}(y)$ and keeping in mind our choice of θ in (4.4.57) then yields

$$|G(y)| \leq c(1+\kappa')\delta_{\partial\Omega}(y) \cdot \mathcal{N}^{\rho}_{\widetilde{\kappa}}(\nabla G)(x) \text{ provided } x \notin N_3. \tag{4.4.63}$$

Next, the fact that x belongs to the set $A_\kappa(\partial\Omega) \setminus N_2$ ensures that the nontangential boundary trace $\big(u\big|^{\kappa-\mathrm{n.t.}}_{\partial\Omega}\big)(x)$ exists in $\mathbb{C}^M$. Using interior estimates (cf. [112, Theorem 6.5.7]) in the ball $B\big(y, a\cdot\delta_{\partial\Omega}(y)\big)$ for the null-solution of the weakly elliptic system L defined as

$$w(z) := u(z) - \big(u\big|^{\kappa-\mathrm{n.t.}}_{\partial\Omega}\big)(x) \text{ for each } z \in \Omega, \tag{4.4.64}$$

we may estimate, for some constant $C = C(L, n, a) \in (0, \infty)$,

$$\begin{aligned}|(\nabla u)(y)| = |(\nabla w)(y)| &\leq \frac{C}{\delta_{\partial\Omega}(y)} ⨍_{B(y,a\cdot\delta_{\partial\Omega}(y))} \Big|u - \big(u\big|^{\kappa-\mathrm{n.t.}}_{\partial\Omega}\big)(x)\Big| \,\mathrm{d}\mathcal{L}^n \\ &\leq C\delta_{\partial\Omega}(y)^{-1} \cdot \sup_{\substack{z\in\Gamma_\kappa(x) \\ |x-z|<(1+\kappa'+a)\delta_{\partial\Omega}(y)}} \Big|u(z) - \big(u\big|^{\kappa-\mathrm{n.t.}}_{\partial\Omega}\big)(x)\Big|,\end{aligned} \tag{4.4.65}$$

since having $z \in B\big(y, a\cdot\delta_{\partial\Omega}(y)\big)$ entails (in view of the fact that the distance function $\delta_{\partial\Omega}$ is Lipschitz with constant ≤ 1),

$$\delta_{\partial\Omega}(y) \leq \delta_{\partial\Omega}(z) + |z-y| < \delta_{\partial\Omega}(z) + a\cdot\delta_{\partial\Omega}(y) \Longrightarrow \delta_{\partial\Omega}(y) < \frac{1}{1-a}\delta_{\partial\Omega}(z), \tag{4.4.66}$$

which, bearing in mind that y is as in (4.4.60) and the formula for κ' from (4.4.56), permits us to conclude that

$$|z-x| \le |z-y| + |y-x| < a \cdot \delta_{\partial\Omega}(y) + (1+\kappa')\delta_{\partial\Omega}(y)$$

$$= (1+\kappa'+a)\delta_{\partial\Omega}(y) < \frac{1+\kappa'+a}{1-a}\delta_{\partial\Omega}(z) = (1+\kappa)\delta_{\partial\Omega}(z),$$

$$\text{hence } z \in \Gamma_\kappa(x) \text{ and } |z-x| < (1+\kappa'+a)\delta_{\partial\Omega}(y). \tag{4.4.67}$$

By combining (4.4.63) and (4.4.65) we arrive at

$$|G(y)||(\nabla u)(y)| \le C\mathcal{N}_{\widetilde{\kappa}}^{\rho}(\nabla G)(x) \cdot \sup_{\substack{z\in\Gamma_\kappa(x)\\ |x-z|<(1+\kappa'+a)\delta_{\partial\Omega}(y)}} \Big|u(z) - \big(u\big|_{\partial\Omega}^{\kappa-\mathrm{n.t.}}\big)(x)\Big| \tag{4.4.68}$$

which, in light of [112, Definition 8.9.1] (cf. (A.0.99)) and (4.4.58), implies

$$\lim_{\Gamma_{\kappa'}(x)\ni y\to x} |G(y)||(\nabla u)(y)| = 0 \text{ for each } x \in \partial_{\mathrm{nta}}\Omega \setminus (N_1 \cup N_2 \cup N_3). \tag{4.4.69}$$

In turn, from (4.4.47), (4.4.1), (4.4.2), (4.4.69), and [112, (8.9.12)] we conclude (upon recalling that $0 < \kappa' < \kappa \le \widetilde{\kappa}$ and $\sigma(N_1 \cup N_2 \cup N_3) = 0$) that

$$\vec{F}\Big|_{\partial\Omega}^{\kappa'-\mathrm{n.t.}} \text{ exists at } \sigma\text{-a.e. point on } \partial_{\mathrm{nta}}\Omega \tag{4.4.70}$$

and, in fact,

$$\begin{aligned}\vec{F}\Big|_{\partial\Omega}^{\kappa'-\mathrm{n.t.}} &= \Big(\big(u_\alpha\big|_{\partial\Omega}^{\kappa'-\mathrm{n.t.}}\big)a_{kj}^{\gamma\alpha}\big(\partial_k G_{\gamma\beta}\big)\Big|_{\partial\Omega}^{\kappa'-\mathrm{n.t.}}\Big)_{1\le j\le n} \\ &= \Big(\big(u_\alpha\big|_{\partial\Omega}^{\kappa-\mathrm{n.t.}}\big)a_{kj}^{\gamma\alpha}\big(\partial_k G_{\gamma\beta}\big)\Big|_{\partial\Omega}^{\widetilde{\kappa}-\mathrm{n.t.}}\Big)_{1\le j\le n}.\end{aligned} \tag{4.4.71}$$

Next, fix an arbitrary point $x \in \partial_{\mathrm{nta}}\Omega$ and, for each $y \in \Gamma_{\kappa'}(x)$ with

$$\delta_{\partial\Omega}(y) < \varepsilon_o := \min\{\theta\cdot\rho, \mathrm{d}, \rho/(1+\kappa'+a)\}, \tag{4.4.72}$$

in place of (4.4.65) now write, using interior estimates for the function u, [112, (8.2.13)], and keeping in mind that the conclusions in (4.4.66) are valid for each point $z \in B(y, a\cdot\delta_{\partial\Omega}(y))$,

$$\begin{aligned}|(\nabla u)(y)| &\le \frac{C}{\delta_{\partial\Omega}(y)} \fint_{B(y,a\cdot\delta_{\partial\Omega}(y))} |u| \, d\mathcal{L}^n \\ &\le C\delta_{\partial\Omega}(y)^{-1} \cdot \sup_{\substack{z\in\Gamma_\kappa(x)\\ |x-z|<(1+\kappa'+a)\delta_{\partial\Omega}(y)}} |u(z)| \le C\delta_{\partial\Omega}(y)^{-1}\cdot\big(\mathcal{N}_\kappa^\rho u\big)(x).\end{aligned} \tag{4.4.73}$$

Then combining (4.4.63) with (4.4.73) gives, on account of (4.4.55),

$$\mathcal{N}_{\kappa'}^{\varepsilon_o}\big(|G||\nabla u|\big) \le C\big(\mathcal{N}_{\widetilde{\kappa}}^{\rho}(\nabla G)\big)\cdot\mathcal{N}_\kappa^\rho u \text{ at } \sigma\text{-a.e. point on } \partial\Omega. \tag{4.4.74}$$

Since thanks to [112, (8.2.10)], [112, (8.2.25)], $\rho > \varepsilon_o$, and $\widetilde{\kappa} > \kappa'$, we also have

$$\begin{aligned} \mathcal{N}_{\kappa'}^{\varepsilon_o}\big(|\nabla G||u|\big) &\leq \big(\mathcal{N}_{\kappa'}^{\varepsilon_o}(\nabla G)\big) \cdot \mathcal{N}_{\kappa'}^{\varepsilon_o} u \\ &\leq \big(\mathcal{N}_{\widetilde{\kappa}}^{\rho}(\nabla G)\big) \cdot \mathcal{N}_{\kappa}^{\rho} u \text{ at each point on } \partial\Omega, \end{aligned} \tag{4.4.75}$$

we conclude from (4.4.47), (4.4.74), (4.4.75), the third line in (4.4.2), and [112, Proposition 8.2.3] that the truncated nontangential maximal function of $\vec{F}$ (at height ε_o and aperture κ') satisfies

$$\mathcal{N}_{\kappa'}^{\varepsilon_o} \vec{F} \in L^1(\partial\Omega, \sigma). \tag{4.4.76}$$

Let us now specialize the discussion to the case when Ω is an exterior domain. When $n \geq 3$, from assumptions and Theorem 1.5.9 (applied both to $L^\top$ and L) it follows that as $|x| \to \infty$ we also have

$$\begin{aligned} G(x) &= O(|x|^{2-n}), \quad (\nabla G)(x) = O(|x|^{1-n}), \\ u(x) &= O(|x|^{2-n}), \quad (\nabla u)(x) = O(|x|^{1-n}). \end{aligned} \tag{4.4.77}$$

Together with (4.4.47), this implies that $\vec{F}(x) = O(|x|^{3-2n})$ as $|x| \to \infty$. Thus, since $n \geq 3$, for each $\lambda \in (1, \infty)$ we have

$$\int_{B(0,\lambda R)\setminus B(0,R)} |\vec{F}(x)| \, d\mathcal{L}^n(x) = o(R) \text{ as } R \to \infty. \tag{4.4.78}$$

When $n = 2$, the additional assumptions made in (4.4.5) and (4.4.6) come in effect. As such, the last part in Theorem 1.5.9 (applied both to $L^\top$ and L) applies and gives that as $|x| \to \infty$ we have

$$\begin{aligned} G(x) &= O(1), \quad (\nabla G)(x) = O(|x|^{-2}), \\ u(x) &= O(1), \quad (\nabla u)(x) = O(|x|^{-2}). \end{aligned} \tag{4.4.79}$$

In concert with (4.4.47) this implies that $\vec{F}(x) = O(|x|^{-2})$ as $|x| \to \infty$, hence once again (4.4.78) holds (with $n = 2$) for each $\mu \in (1, \infty)$. To sum up, this analysis shows that the condition formulated in [112, (1.5.22)] presently holds, when Ω is an exterior domain, in all dimensions $n \geq 2$.

Having established (4.4.48), (4.4.54), (4.4.76), (4.4.70), and [112, (1.5.22)] (when Ω is an exterior domain), [112, Corollary 1.5.2] applies and the Divergence Formula [112, (1.5.20)] gives, in view of (4.4.53), the last line in (4.4.71), (1.7.9), and with $(\nu_j)_{1\leq j\leq n}$ denoting the components of the geometric measure theoretic outward unit normal ν to Ω,

$$
\begin{aligned}
-u_\beta(x_0) &= {}_{(\mathscr{C}^\infty_b(\Omega))^*}\big(\mathrm{div}\vec{F}, 1\big)_{\mathscr{C}^\infty_b(\Omega)} = \int_{\partial_*\Omega} \nu \cdot \Big(\vec{F}\Big|^{\kappa'-\mathrm{n.t.}}_{\partial\Omega}\Big)\, d\sigma \\
&= \int_{\partial_*\Omega} \Big(u_\alpha\Big|^{\kappa-\mathrm{n.t.}}_{\partial\Omega}\Big) \nu_j a^{\gamma\alpha}_{kj} (\partial_k G_{\gamma\beta})\Big|^{\widetilde{\kappa}-\mathrm{n.t.}}_{\partial\Omega}\, d\sigma \\
&= \int_{\partial_*\Omega} \Big\langle u\Big|^{\kappa-\mathrm{n.t.}}_{\partial\Omega}, \partial^{A^\top}_\nu G_{\cdot\beta}\Big\rangle\, d\sigma.
\end{aligned}
\tag{4.4.80}
$$

Since $\beta \in \{1, \dots, M\}$ has been arbitrarily chosen, this establishes (4.4.4) in the case when $\partial\Omega$ is bounded and $n \geq 3$. In fact, the same argument works in the case when Ω is bounded and $n = 2$ (since (4.4.78) is now no longer relevant).

To finish the proof of Theorem 4.4.1 there remains to treat the case when $\partial\Omega$ is unbounded and $n \geq 2$. To this end, work under the stronger assumptions made in (4.4.3) (in particular, Ω is a globally pathwise nontangentially accessible set). We claim that, in this setting,

$$
\mathcal{N}^{\Omega\setminus(2K)}_{\kappa'}\vec{F} \in L^1(\partial\Omega, \sigma). \tag{4.4.81}
$$

To prove this claim, observe from (4.4.47) that

$$
|\vec{F}| \leq C|G|\,|\nabla u| + C|u|\,|\nabla G| \ \text{ in } \Omega, \tag{4.4.82}
$$

and that at σ-a.e. point on $\partial\Omega$ we have

$$
\begin{aligned}
\mathcal{N}^{\Omega\setminus(2K)}_{\kappa'}\big(|G|\,|\nabla u|\big) &= \mathcal{N}^{\Omega\setminus(2K)}_{\kappa'}\big((|G|/\delta_{\partial\Omega})\cdot(\delta_{\partial\Omega}|\nabla u|)\big) \\
&\leq \Big(\mathcal{N}^{\Omega\setminus(2K)}_{\kappa'}(G/\delta_{\partial\Omega})\Big)\cdot\Big(\mathcal{N}^{\Omega\setminus(2K)}_{\kappa'}(\delta_{\partial\Omega}|\nabla u|)\Big) \\
&\leq \Big(\mathcal{N}^{\Omega\setminus(2K)}_{\kappa}(G/\delta_{\partial\Omega})\Big)\cdot\Big(\mathcal{N}_{\kappa'}(\delta_{\partial\Omega}|\nabla u|)\Big) \\
&\leq C\big(\mathcal{N}_\kappa u\big)\big(\mathcal{N}^{\Omega\setminus K}_{\widetilde{\kappa}}(\nabla G)\big),
\end{aligned}
\tag{4.4.83}
$$

thanks to [112, (8.2.10), (8.2.25), (8.9.147)] (with $|\alpha| = 1$ and $\lambda = 0$), [112, (8.9.227)] (with $u := G$ and with the understanding that $\widetilde{\kappa}$ has been chosen, from the beginning, to be greater than the aperture parameter $\widetilde{\kappa}_o$ appearing in the statement of [112, Proposition 8.9.21]), and (4.4.55). From (4.4.82) and (4.4.83) we then see that

$$
\mathcal{N}^{\Omega\setminus(2K)}_{\kappa'}\vec{F} \leq C\,\mathcal{N}_\kappa u \cdot \mathcal{N}^{\Omega\setminus K}_{\widetilde{\kappa}}(\nabla G) \ \text{ at } \sigma\text{-a.e. point on } \partial\Omega. \tag{4.4.84}
$$

On account of (4.4.84), (4.4.3), and [112, (8.2.28)] we then conclude that the membership in (4.4.81) is indeed true.

At this stage, (4.4.48), (4.4.54), (4.4.81), and (4.4.70) ensure that [112, Theorem 1.4.1] (in the version recorded in Note 8 following its statement) applies. Then the Divergence Formula [112, (1.4.6)] gives, in view of (4.4.53), the last line in (4.4.71), and (1.7.9), that

$$
\begin{aligned}
-u_\beta(x_0) &= {}_{(\mathscr{C}_b^\infty(\Omega))^*}\big(\mathrm{div}\vec{F}, 1\big)_{\mathscr{C}_b^\infty(\Omega)} = \int_{\partial_*\Omega} \nu \cdot \Big(\vec{F}\Big|_{\partial\Omega}^{\kappa'-\mathrm{n.t.}}\Big)\,\mathrm{d}\sigma \\
&= \int_{\partial_*\Omega} \Big(u_\alpha\Big|_{\partial\Omega}^{\kappa-\mathrm{n.t.}}\Big)\nu_j a_{kj}^{\gamma\alpha}\big(\partial_k G_{\gamma\beta}\big)\Big|_{\partial\Omega}^{\widetilde{\kappa}-\mathrm{n.t.}}\,\mathrm{d}\sigma \\
&= \int_{\partial_*\Omega} \Big\langle u\Big|_{\partial\Omega}^{\kappa-\mathrm{n.t.}}, \partial_\nu^{A^\top} G_{\bullet\beta}\Big\rangle\,\mathrm{d}\sigma. \qquad (4.4.85)
\end{aligned}
$$

In view of the arbitrariness of $\beta \in \{1, \dots, M\}$, formula (4.4.4) follows in the case when $\partial\Omega$ is unbounded and $n \geq 2$. This concludes the proof of Theorem 4.4.1. □

4.5 The Poisson Kernel Associated with a System: A First Look

The goal here is to introduce and study the Poisson kernel associated with a given weakly elliptic system, in various geometric settings.

Having fixed some $n \in \mathbb{N}$ with $n \geq 2$, let $\Omega \subset \mathbb{R}^n$ be a UR domain which is globally pathwise nontangentially accessible (in the sense of [112, Definition 8.9.14]). Abbreviate $\sigma := \mathcal{H}^{n-1}\lfloor\partial\Omega$ and denote by ν the geometric measure theoretic outward unit normal to Ω. Also, fix a sufficiently large aperture parameter $\kappa \in (0, \infty)$. Finally, suppose L is a homogeneous, constant (complex) coefficient, second-order, $M \times M$ system in $\mathbb{R}^n$, satisfying the Legendre-Hadamard strong ellipticity condition.

In this context, assume that for each given $x \in \Omega$ there exists a matrix-valued function $G^L(\cdot, x)$ which satisfies

$$
\begin{cases}
G^L(\cdot, x) = \big(G^L_{\alpha\beta}(\cdot, x)\big)_{1\leq\alpha,\beta\leq M} \in \big[L^1_{\mathrm{loc}}(\Omega, \mathcal{L}^n)\big]^{M\times M}, \\
L\big[G^L_{\bullet\beta}(\cdot, x)\big] = -\big(\delta_x\delta_{\alpha\beta}\big)_{1\leq\alpha\leq M} \text{ in } \big[\mathcal{D}'(\Omega)\big]^M \text{ for each } \beta \in \{1, \dots, M\}, \\
\big(\nabla[G^L(\cdot, x)]\big)\Big|_{\partial\Omega}^{\kappa-\mathrm{n.t.}} \text{ exists (in } \mathbb{C}^{n\cdot M^2}) \text{ at } \sigma\text{-a.e. point on } \partial\Omega, \\
G^L(\cdot, x)\Big|_{\partial\Omega}^{\kappa-\mathrm{n.t.}} = 0 \in \mathbb{C}^{M\times M} \text{ at } \sigma\text{-a.e. point on } \partial\Omega, \\
\text{and also } G^L(\cdot, x) = O(|x|^{2-n}) \text{ at infinity} \\
\text{when } \Omega \text{ is an exterior domain.}
\end{cases}
\qquad (4.5.1)
$$

For each $\beta \in \{1, \dots, M\}$ define

$$
\mathcal{P}^L_{\bullet\beta}(x, y) := -\partial^A_{\nu(y)}\big[G^L_{\bullet\beta}(y, x)\big] \text{ for each } x \in \Omega \text{ and } \sigma\text{-a.e. } y \in \partial\Omega. \qquad (4.5.2)
$$

Assuming that a function as in (4.5.1) exists for L replaced by $L^\top$, we shall call the $\mathbb{C}^{M\times M}$-valued function

$$
P_L(x, y) := \big(\mathcal{P}^{L^\top}_{\beta\alpha}(x, y)\big)_{1\leq\alpha,\beta\leq M} \text{ for } x \in \Omega \text{ arbitrary and } \sigma\text{-a.e. } y \in \partial\Omega, \qquad (4.5.3)
$$

the `Poisson kernel` associated with the system L in the domain Ω. In terms of this piece of notation, (4.4.4) becomes the integral representation formula

$$u(x) = \int_{\partial\Omega} P_L(x,y)\big(u\big|_{\partial\Omega}^{\kappa-\mathrm{n.t.}}\big)(y)\,d\sigma(y), \qquad \forall\, x \in \Omega, \tag{4.5.4}$$

for each $\mathbb{C}^M$-valued function u as in (4.4.2).

Suppose the Green functions $G^L(\cdot,\cdot)$ and $G^{L^\top}(\cdot,\cdot)$ both exist (as matrix-valued functions satisfying the conditions listed in (4.5.1) for L and $L^\top$, respectively). In addition, assume that for any two distinct points $x_0, x_1 \in \Omega$ we have

$$\int_{\partial\Omega} \Big(\mathcal{N}_\kappa^{\Omega\setminus K}\big(\nabla G^L(\cdot,x_0)\big)\Big)(y)\frac{d\sigma(y)}{1+|y|^{n-2}} < +\infty, \tag{4.5.5}$$

$$\int_{\partial\Omega} \mathcal{N}_\kappa^{\Omega\setminus K} G^{L^\top}(\cdot,x_1)\cdot \mathcal{N}_\kappa^{\Omega\setminus K}\big(\nabla G^L(\cdot,x_0)\big)\,d\sigma < +\infty, \tag{4.5.6}$$

and

$$\int_{\partial\Omega} \mathcal{N}_\kappa^{\Omega\setminus K} G^{L^\top}(\cdot,x_1)\cdot \mathcal{N}_\kappa^{\Omega\setminus K} G^L(\cdot,x_0)\,d\sigma < +\infty, \tag{4.5.7}$$

where $K \subseteq \Omega$ is some compact neighborhood of $\{x_0, x_1\}$. If $\neq 3$ we may apply the integral representation formula (4.5.4) (written for $L^\top$ in place of L) to the columns of $G^{L^\top}(\cdot,x_1) + E_{L^\top}(\cdot - x_1)$, where E is the fundamental solution associated with the system $L^\top$ as in Theorem 1.4.2 which presently reads

$$G^{L^\top}(x_0,x_1) + E_{L^\top}(x_1 - x_0) = \int_{\partial_*\Omega} P_{L^\top}(x_0,y)E_{L^\top}(x_1-y)\,d\sigma(y). \tag{4.5.8}$$

Passing to transpose and invoking the Reciprocity Relation from (4.2.8) which currently reads

$$G^{L^\top}(x_1,x_0) = [G^L(x_0,x_1)]^\top, \tag{4.5.9}$$

we therefore obtain

$$G^L(x_1,x_0) + E(x_1 - x_0) = \int_{\partial_*\Omega} E(x_1-y)\big[P_{L^\top}(x_0,y)\big]^\top d\sigma(y). \tag{4.5.10}$$

To proceed, we shall need notation and results in connection with boundary layer potential operators associated with the system L and the set Ω. The (formal) convolution-type integral operator on $\partial\Omega$ whose kernel is the fundamental solution $E = E_L$ is referred to as single layer potential operator. If we denote by $\mathscr{S}$ the boundary-to-domain single layer potential operator, then (4.5.10) simply reads

$$G^L(\cdot,x_0) + E(\cdot - x_0) = \mathscr{S}\Big(\big[P_{L^\top}(x_0,\cdot)\big]^\top\Big) \text{ in } \Omega, \tag{4.5.11}$$

or, equivalently,

$$G^L_{\bullet\beta}(\cdot, x_0) + E_{\bullet\beta}(\cdot - x_0) = \mathscr{S}\big(\mathcal{P}^L_{\bullet\beta}(x_0, \cdot)\big) \text{ in } \Omega, \text{ for } 1 \leq \beta \leq M. \tag{4.5.12}$$

If $S = S_L$ denotes the boundary-to-boundary single layer potential operator on $\partial\Omega$ associated with the system L, then taking the nontangential boundary traces in (4.5.12) leads to the conclusion that

$$S\big(\mathcal{P}^L_{\bullet\beta}(x_0, \cdot)\big) = E_{\bullet\beta}(\cdot - x_0)\big|_{\partial\Omega} \text{ for } 1 \leq \beta \leq M. \tag{4.5.13}$$

In other words,

$$S\Big(\big[P_{L^\top}(x_0, \cdot)\big]^\top\Big) = E(\cdot - x_0)\big|_{\partial\Omega}. \tag{4.5.14}$$

To proceed, we need the notion of double layer. Such an operator may be associated with any coefficient tensor A which may be used to write the given system L. Once again, associated with any such coefficient tensor A, there is a boundary-to-boundary double layer potential operator, denoted by K_A, and a boundary-to-domain double layer potential operator, denoted by $\mathcal{D}_A$. Finally, we shall denote by $K^\#_A$ the "transpose" double layer associated with the coefficient tensor A. All these are discussed at length in [115, Chapter 1]. What we need here is the fact that taking conormal derivatives in (4.5.11) yields

$$\big(\tfrac{1}{2}I + K^\#_{A^\top}\big)\big(\mathcal{P}^L_{\bullet\beta}(x_0, \cdot)\big) = \partial^A_\nu\big[E_{\bullet\beta}(\cdot - x_0)\big] \text{ for } 1 \leq \beta \leq M, \tag{4.5.15}$$

or, in matrix formalism (with the operators acting on columns),

$$\big(\tfrac{1}{2}I + K^\#_{A^\top}\big)\Big(\big[P_{L^\top}(x_0, \cdot)\big]^\top\Big) = \partial^A_\nu\big[E(\cdot - x_0)\big]. \tag{4.5.16}$$

Going further, suppose we work in an environment where, say, the L^2 Dirichlet boundary value problem for the system L is well-posed. For example, such a setting is described in [116, Chapter 8] where the solution is expressed as

$$u = \mathscr{S}\Big(S^{-1}\big(u\big|^{\kappa-\text{n.t.}}_{\partial\Omega}\big)\Big) \text{ in } \Omega. \tag{4.5.17}$$

It is interesting to note that this is compatible with the Poisson integral representation formula (4.5.4). Indeed, for each $x \in \Omega$ and each $\beta \in \{1, \ldots, M\}$ we may use (4.5.17), the fact that the transpose of S^{-1} is $\big(S_{L^\top}\big)^{-1}$, (4.5.13) (written with $L^\top$ in place of L, (1.4.32), and (4.5.3) to write

$$
\begin{aligned}
u_\beta(x) &= \int_{\partial\Omega} \Big\langle E_{\beta\bullet}(x-\cdot)\big|_{\partial\Omega}, S^{-1}\big(u\big|_{\partial\Omega}^{\kappa-\mathrm{n.t.}}\big)\Big\rangle \, d\sigma \\
&= \int_{\partial\Omega} \Big\langle (S_{L^\top})^{-1} E_{\beta\bullet}(x-\cdot)\big|_{\partial\Omega}, u\big|_{\partial\Omega}^{\kappa-\mathrm{n.t.}}\Big\rangle \, d\sigma \\
&= \int_{\partial\Omega} \Big\langle \mathcal{P}^{L^\top}_{\bullet\beta}(x,\cdot), u\big|_{\partial\Omega}^{\kappa-\mathrm{n.t.}}\Big\rangle \, d\sigma \\
&= \int_{\partial\Omega} \Big(P_L(x,\cdot)\big(u\big|_{\partial\Omega}^{\kappa-\mathrm{n.t.}}\big)\Big)_\beta \, d\sigma, \qquad (4.5.18)
\end{aligned}
$$

from which (4.5.4) follows.

When $\frac{1}{2}I + K_A$ is an invertible operator on, say, $\big[L^2(\partial\Omega,\sigma)\big]^M$, then an alternative formula for the solution u of the L^2 Dirichlet boundary value problem for the system L in Ω is

$$
u = \mathcal{D}_A\Big(\big(\tfrac{1}{2}I + K_A\big)^{-1}\big(u\big|_{\partial\Omega}^{\kappa-\mathrm{n.t.}}\big)\Big) \text{ in } \Omega. \qquad (4.5.19)
$$

Again, we wish to check that this is compatible with the Poisson integral representation formula (4.5.4). To justify this, given any $x \in \Omega$ and $\beta \in \{1,\dots,M\}$ we write, thanks to the formula for the boundary-to-domain double layer (cf. [115, §1.3]), the fact that the transpose of K_A is $K_A^\#$, (4.5.15) (written for $A^\top$ in place of A), and (1.4.32),

$$
\begin{aligned}
u_\beta(x) &= \int_{\partial\Omega} \Big\langle \partial_\nu^{A^\top}\big[(E_{L^\top})_{\bullet\beta}(x-\cdot)\big], \big(\tfrac{1}{2}I + K_A\big)^{-1}\big(u\big|_{\partial\Omega}^{\kappa-\mathrm{n.t.}}\big)\Big\rangle \, d\sigma \\
&= \int_{\partial\Omega} \Big\langle \big(\tfrac{1}{2}I + K_A^\#\big)^{-1}\Big(\partial_\nu^{A^\top}\big[(E_{L^\top})_{\bullet\beta}(x-\cdot)\big]\Big), u\big|_{\partial\Omega}^{\kappa-\mathrm{n.t.}}\Big\rangle \, d\sigma \\
&= \int_{\partial\Omega} \Big\langle \mathcal{P}^{L^\top}_{\bullet\beta}(x,\cdot), u\big|_{\partial\Omega}^{\kappa-\mathrm{n.t.}}\Big\rangle \, d\sigma \\
&= \int_{\partial\Omega} \Big(P_L(x,\cdot)\big(u\big|_{\partial\Omega}^{\kappa-\mathrm{n.t.}}\big)\Big)_\beta \, d\sigma, \qquad (4.5.20)
\end{aligned}
$$

as wanted.

Moving on, we wish to check directly the veracity of formula (4.5.15) for the Laplacian $L := \Delta$ in the unit ball $\Omega := B(0,1) \subseteq \mathbb{R}^n$ with $n \geq 3$. To this end, recall first that the Green function this special setting is given by

$$
G(y,x) := -\frac{1}{(2-n)\omega_{n-1}|x-y|^{n-2}} + \frac{1}{(2-n)\omega_{n-1}|x|^{n-2}\big|y - x/|x|^2\big|^{n-2}} \qquad (4.5.21)
$$

for any two distinct points $x, y \in \overline{B(0,1)}$. Of course, the first term above is just $-E_\Delta(x-y)$, where E_Δ is the standard fundamental solution of the Laplacian in $\mathbb{R}^n$ (defined in (1.5.56)), while the second term is a harmonic correction, designed as to ensure that $G(\cdot,x)\big|_{S^{n-1}} = 0$ for each fixed $x \in B(0,1)$. Indeed, the latter hinges on the fact that, as it may be easily checked, $|x|\big|y-x/|x|^2\big| = |x-y|$ for each $x \in B(0,1)$ and $y \in S^{n-1}$. Also, in the same setting, the Poisson kernel is given by

$$P(x,y) := -\partial_{\nu(y)}\big[G(y,x)\big] = \frac{1}{\omega_{n-1}}\frac{1-|x|^2}{|x-y|^n} \quad \text{for each point } x \in B(0,1) \text{ and each point } y \in S^{n-1}, \tag{4.5.22}$$

where the vector ν is the outward unit normal to $B(0,1)$. The goal is to directly check that, for each $x \in B(0,1)$,

$$\big(\tfrac{1}{2}I + K_\Delta^\#\big)\big(P(x,\cdot)\big) = \partial_\nu\big[E_\Delta(\cdot - x)\big] \text{ on } S^{n-1}, \tag{4.5.23}$$

where I is the identity, and $K_\Delta^\#$ is the transpose of the harmonic double layer potential operator K_Δ (associated with $\Omega := B(0,1)$ as in (2.5.203)). To this end, use the fact that (see the discussion in the preamble of [116, Chapter 7])

$$K_\Delta f = K_\Delta^\# f = \big(\tfrac{2-n}{2}\big)S_\Delta f \text{ if } n \geq 3, \tag{4.5.24}$$

where S_Δ is the boundary-to-boundary harmonic single layer potential operator on $\partial B(0,1) = S^{n-1}$, to write

$$\big(\tfrac{1}{2}I + K_\Delta^\#\big)\big(P(x,\cdot)\big) = \tfrac{1}{2}P(x,\cdot) + \big(\tfrac{2-n}{2}\big)S_\Delta\big(P(x,\cdot)\big). \tag{4.5.25}$$

Next, if $\mathscr{S}_\Delta$ denotes the boundary-to-domain harmonic single layer potential operator associated with the unit ball $B(0,1)$, then for each point $z \in B(0,1)$ we may use the integration by parts formula from Theorem 1.7.11 and the fact that we have $G(\cdot,x)\big|_{S^{n-1}} = 0$ for each fixed $x \in B(0,1)$ to compute

$$\begin{aligned}
\mathscr{S}_\Delta\big(P(x,\cdot)\big)(z) &= \int_{\partial B(0,1)} E_\Delta(z-y)P(x,y)\,d\sigma(y) \\
&= -\int_{\partial B(0,1)} E_\Delta(z-y)\partial_{\nu(y)}\big[G(y,x)\big]\,d\sigma(y) \\
&= {}_{\mathscr{E}'(B(0,1))}\big\langle\!\big\langle \Delta_y\big[E_\Delta(z-y)\big], G(y,x)\big\rangle\!\big\rangle_{\mathscr{E}(B(0,1))} \\
&\quad - {}_{\mathscr{E}'(B(0,1))}\big\langle\!\big\langle \Delta_y\big[G(y,x)\big], E_\Delta(z-y)\big\rangle\!\big\rangle_{\mathscr{E}(B(0,1))} \\
&= G(z,x) + E_\Delta(z-x).
\end{aligned} \tag{4.5.26}$$

Consequently,

$$\begin{aligned}
S_\Delta\big(P(x,\cdot)\big) &= \Big(\mathscr{S}_\Delta\big(P(x,\cdot)\big)\Big)\Big|_{\partial B(0,1)}^{\text{n.t.}} = G(\cdot,x)\big|_{S^{n-1}} + E_\Delta(\cdot - x)\big|_{S^{n-1}} \\
&= E_\Delta(\cdot - x)\big|_{S^{n-1}}.
\end{aligned} \tag{4.5.27}$$

With this in hand, we return to (4.5.25) and conclude that at each point $y \in S^{n-1}$ we have

$$\begin{aligned}\Big(\big(\tfrac{1}{2}I + K^{\#}_{\Delta}\big)\big(P(x,\cdot)\big)\Big)(y) &= \tfrac{1}{2}P(x,y) + \big(\tfrac{2-n}{2}\big)E_{\Delta}(x-y)\\ &= \frac{1}{2\omega_{n-1}}\frac{1-|x|^2}{|x-y|^n} + \frac{1}{2\omega_{n-1}|x-y|^{n-2}}\\ &= \frac{1}{\omega_{n-1}}\frac{\langle y, y-x\rangle}{|x-y|^n} = \partial_{\nu(y)}\big[E_{\Delta}(y-x)\big],\end{aligned} \tag{4.5.28}$$

which completes the verification of (4.5.16).

4.6 The Poisson Kernel Associated with a System in the Upper Half-Space

Theorem 4.4.1 yields nontrivial new results even in the case when the set Ω is simply the upper half-space $\mathbb{R}^n_+$. To describe them in detail, consider a constant (complex) coefficient, second-order, $M \times M$ system L in $\mathbb{R}^n$ satisfying the Legendre-Hadamard strong ellipticity condition. From [35] we then know that there exists a unique $\mathbb{C}^{M\times M}$-valued function $G^L(\cdot,\cdot)$ defined in $\mathbb{R}^n_+$ such that for each $y = (y', y_n) \in \mathbb{R}^n_+$ the following properties hold (for some aperture parameter $\kappa > 0$):

$$G^L(\cdot, y) \in \big[L^1_{\mathrm{loc}}(\mathbb{R}^n_+, \mathcal{L}^n)\big]^{M\times M}, \tag{4.6.1}$$

$$G^L(\cdot, y)\Big|^{\kappa-\mathrm{n.t.}}_{\partial\mathbb{R}^n_+} = 0 \ \ \mathcal{L}^{n-1}\text{-a.e. in } \mathbb{R}^{n-1} \equiv \partial\mathbb{R}^n_+, \tag{4.6.2}$$

$$\int_{\mathbb{R}^{n-1}} \Big(\mathcal{N}_\kappa^{\mathbb{R}^n_+\setminus \overline{B(y, y_n/2)}}\, G^L(\cdot, y)\Big)(x')\frac{dx'}{1+|x'|^{n-1}} < \infty, \tag{4.6.3}$$

$$L\big[G^L(\cdot, y)\big] = -\delta_y\, I_{M\times M} \ \text{ in } \big[\mathcal{D}'(\mathbb{R}^n_+)\big]^{M\times M}, \tag{4.6.4}$$

where L acts in the "dot" variable on the columns of G^L (with the outcomes also arranged as columns). In addition, for each fixed $y \in \mathbb{R}^n_+$ we have

$$G^L(\cdot, y) \in \big[\mathscr{C}^\infty\big(\overline{\mathbb{R}^n_+} \setminus B(y,\varepsilon)\big)\big]^{M\times M} \ \text{ for every } \varepsilon > 0. \tag{4.6.5}$$

Also, given $\kappa > 0$, for each $y \in \mathbb{R}^n_+$ and each compact neighborhood K of y in $\mathbb{R}^n_+$ there exists a finite constant $C = C(n, L, \kappa, K, y) > 0$ such that

$$\Big(\mathcal{N}_\kappa^{\mathbb{R}^n_+\setminus K} G^L(\cdot, y)\Big)(x') \leq C\frac{1+\log_+|x'|}{1+|x'|^{n-1}} \ \text{ for each } x' \in \mathbb{R}^{n-1}. \tag{4.6.6}$$

In fact,

if L may be written using a coefficient tensor $A = \big(a^{\alpha\beta}_{jk}\big)_{\substack{1\le j,k\le n\\ 1\le\alpha,\beta\le M}}$ which has the following block-structure: $a^{\alpha\beta}_{jn} = a^{\alpha\beta}_{nj} = 0$ whenever $j < n$, then the logarithm in the right side of (4.6.6) may actually be omitted and, with $\overline{y} \in \mathbb{R}^n_-$ denoting the reflection of $y \in \mathbb{R}^n_+$ across $\partial\mathbb{R}^n_+$, we have $G^L(x,y) = E(x-y) - E(x-\overline{y})$ for all points $(x,y) \in \mathbb{R}^n_+ \times \mathbb{R}^n_+ \setminus \text{diag}$. (4.6.7)

Moreover, if for any given multi-indices $\alpha, \beta \in \mathbb{N}^n_0$ we write $\partial^\alpha_X \partial^\beta_Y G^L$ to denote the derivative of order α in the first variable together with the derivative of order β in the second variable of the function $G^L(\cdot,\cdot)$, then for any multi-indices $\alpha,\beta \in \mathbb{N}^n_0$ such that $|\alpha| + |\beta| > 0$ there exists some constant $C = C(n, L, \kappa, \alpha, \beta, K, y) \in (0,\infty)$ for which

$$\Big(\mathcal{N}_\kappa^{\mathbb{R}^n_+\setminus K}\big(\partial^\alpha_X\partial^\beta_Y G^L)(\cdot,y)\big)\Big)(x') \le \frac{C}{1+|x'|^{n-2+|\alpha|+|\beta|}}. \tag{4.6.8}$$

For each $\alpha,\beta \in \mathbb{N}^n_0$ we also have that the function $\partial^\alpha_X\partial^\beta_Y G^L$ is translation invariant in the tangential variables, in the sense that

$$\begin{gathered}\big(\partial^\alpha_X\partial^\beta_Y G^L\big)\big(x-(z',0), y-(z',0)\big) = \big(\partial^\alpha_X\partial^\beta_Y G^L\big)(x,y)\\ \text{for each } (x,y) \in \mathbb{R}^n_+ \times \mathbb{R}^n_+ \setminus \text{diag and } z' \in \mathbb{R}^{n-1},\end{gathered} \tag{4.6.9}$$

and is positive homogeneous, in the sense that

$$\begin{gathered}\big(\partial^\alpha_X\partial^\beta_Y G^L\big)(\lambda x, \lambda y) = \lambda^{2-n-|\alpha|-|\beta|}\big(\partial^\alpha_X\partial^\beta_Y G^L\big)(x,y)\\ \text{for each } x, y \in \mathbb{R}^n_+ \text{ with } x \neq y \text{ and } \lambda \in (0,\infty),\\ \text{provided either } n \ge 3, \text{ or } |\alpha| + |\beta| > 0.\end{gathered} \tag{4.6.10}$$

Finally, if $G^{L^\top}(\cdot,\cdot)$ denotes the (unique) Green function for the transpose system $L^\top$ in $\mathbb{R}^n_+$, then

$$G^L(x,y) = \Big[G^{L^\top}(y,x)\Big]^\top, \qquad \forall (x,y) \in \mathbb{R}^n_+ \times \mathbb{R}^n_+ \setminus \text{diag}. \tag{4.6.11}$$

Then Theorem 4.4.1 (in the version when the conditions in (4.4.3) are assumed) specialized to this setting gives that any function u satisfying, for some aperture parameter $\kappa > 0$,

$$\begin{cases} u \in \big[\mathscr{C}^\infty(\mathbb{R}^n_+)\big]^M, \quad Lu = 0 \text{ in } \mathbb{R}^n_+,\\ u\big|^{\kappa-\text{n.t.}}_{\partial\mathbb{R}^n_+} \text{ exists at } \mathcal{L}^{n-1}\text{-a.e. point in } \mathbb{R}^{n-1},\\ \displaystyle\int_{\mathbb{R}^{n-1}} (\mathcal{N}_\kappa u)(x') \frac{dx'}{1+|x'|^{n-1}} < +\infty, \end{cases} \tag{4.6.12}$$

has the Poisson integral representation formula

$$\begin{aligned} u(x) &= \int_{\mathbb{R}^{n-1}} P_t^L(x'-y')\big(u\big|_{\partial\mathbb{R}^n_+}^{\kappa-\text{n.t.}}\big)(y')\,dy' \\ &= \Big(P_t^L * \big(u\big|_{\partial\mathbb{R}^n_+}^{\kappa-\text{n.t.}}\big)\Big)(x'), \qquad \forall x=(x',t)\in\mathbb{R}^n_+, \end{aligned} \tag{4.6.13}$$

where $P^L = \big(P^L_{\beta\alpha}\big)_{1\leq\beta,\alpha\leq M}$ is the Agmon-Douglis-Nirenberg Poisson kernel for the system L in $\mathbb{R}^n_+$ (cf. [2], [3], and the discussion in [95, Theorem 2.4]) and $P_t^L(x') := t^{1-n}P^L(x'/t)$ for each $x'\in\mathbb{R}^{n-1}$ and $t>0$.

In particular, (4.6.13) shows that if $u\big|_{\partial\mathbb{R}^n_+}^{\kappa-\text{n.t.}}$ vanishes at $\mathcal{L}^{n-1}$-a.e. point in $\mathbb{R}^{n-1}$ then actually $u\equiv 0$ in $\mathbb{R}^n_+$, which is a uniqueness result established in [95, Theorem 3.2, p. 935] (via a proof which does not yield an actual integral representation formula). We also wish to remark that the integrability condition in the last line of (4.6.12) cannot be weakened to membership to $L^{1,\infty}(\mathbb{R}^{n-1},\mathcal{L}^{n-1})$. This may be seen by considering the case when $n:=2$, $L:=\Delta$, and the nonzero harmonic function $u(x',t) := t/(t^2+(x')^2)$ for each $(x',t)\in\mathbb{R}^2_+$ whose nontangential trace vanishes $\mathcal{L}^1$-a.e. on $\partial\mathbb{R}^2_+$ (since, in this case, $\big(\mathcal{N}_\kappa u\big)(x')\approx|x'|^{-1}$ uniformly for $x'\in\mathbb{R}\equiv\partial\mathbb{R}^2_+$). This points to the sharpness of the hypotheses made in (4.6.12).

To justify (4.6.13), having fixed a coefficient tensor $A=\big(a^{\alpha\beta}_{jk}\big)_{\substack{1\leq j,k\leq n\\ 1\leq\alpha,\beta\leq M}}$ which permits writing L as L_A, the system associated with A as in (1.3.15), and denoting by $(u_\alpha)_{1\leq\alpha\leq M}$ the scalar components of u, we may write

$$\begin{aligned} u(x) &= \left(-\int_{\mathbb{R}^{n-1}}\big(u_\alpha\big|_{\partial\mathbb{R}^n_+}^{\kappa-\text{n.t.}}\big)(y')\big(\partial_\nu^{A^\top}G^{L^\top}_{\cdot\beta}\big)_\alpha((y',0),x)\,dy'\right)_{1\leq\beta\leq M} \\ &= \left(\int_{\mathbb{R}^{n-1}}\big(u_\alpha\big|_{\partial\mathbb{R}^n_+}^{\kappa-\text{n.t.}}\big)(y')a^{\gamma\alpha}_{nn}\big(\partial_{X_n}G^{L^\top}_{\gamma\beta}\big)((y',0),x)\,dy'\right)_{1\leq\beta\leq M} \\ &= \left(\int_{\mathbb{R}^{n-1}}\big(u_\alpha\big|_{\partial\mathbb{R}^n_+}^{\kappa-\text{n.t.}}\big)(y')a^{\gamma\alpha}_{nn}\big(\partial_{Y_n}G^{L}_{\beta\gamma}\big)(x,(y',0))\,dy'\right)_{1\leq\beta\leq M}. \end{aligned} \tag{4.6.14}$$

The fact that (4.6.12) implies the first equality in (4.6.14) is a consequence of (4.4.4) in Theorem 4.4.1 (bearing in mind the versions of (4.6.1)-(4.6.2) and (4.6.4)-(4.6.5) for $G^{L^\top}$). The second equality in (4.6.14) is seen from the formula for the conormal derivative (cf. Definition 1.7.1), the fact that the outward unit normal to $\mathbb{R}^n_+$ is $\nu:=(0,\dots,0,-1)$, and the observation that $\big(\partial_{X_k}G^{L^\top}\big)((y',0),x)=0$ whenever $k<n$ since $G^{L^\top}(\cdot,x)$ vanishes (in a smooth fashion) on $\mathbb{R}^{n-1}\times\{0\}$ (cf. (4.6.2) and (4.6.5)). Finally, the last equality in (4.6.14) is based on (4.6.11). Once (4.6.14) has been justified, the Poisson integral representation formula (4.6.13) follows on account of (4.6.9)-(4.6.10) and a formula from [95, Theorem A.4] to the effect that

$$\begin{aligned} P^L_{\beta\alpha}(z') = a^{\gamma\alpha}_{nn}\big(\partial_{Y_n}G^L_{\beta\gamma}\big)((z',1),0), \quad \forall z'\in\mathbb{R}^{n-1}, \\ \text{for each } \beta,\alpha\in\{1,\dots,M\}. \end{aligned} \tag{4.6.15}$$

Finally, as regards the nature of the Poisson kernel P^L in the case when the fundamental solution $E = \big(E_{\gamma\beta}\big)_{1\leq\gamma,\beta\leq M}$ associated with the system L as in Theorem 1.4.2 happens to be a radial function, we also wish to mention that for each $\beta, \alpha \in \{1, \dots, M\}$ we have (again, see [95, Theorem A.4])

$$P^L_{\beta\alpha}(z') = 2a^{\gamma\alpha}_{nn}(\partial_n E_{\beta\gamma})(z', 1), \qquad \forall z' \in \mathbb{R}^{n-1}. \tag{4.6.16}$$

Remarkably, the Poisson integral representation formula (4.6.13) continues to hold even when the existence of the pointwise nontangential boundary trace is no longer assumed to begin with, since it turns out that this is actually implied by the other conditions imposed in (4.6.12). Specifically, having

$$\begin{cases} u \in \big[\mathscr{C}^\infty(\mathbb{R}^n_+)\big]^M, \quad Lu = 0 \text{ in } \mathbb{R}^n_+, \\ \displaystyle\int_{\mathbb{R}^{n-1}} (\mathcal{N}_\kappa u)(x') \frac{dx'}{1 + |x'|^{n-1}} < +\infty, \end{cases} \tag{4.6.17}$$

implies that

$$\begin{cases} u\big|^{\kappa-\text{n.t.}}_{\partial\mathbb{R}^n_+} \text{ exists at } \mathcal{L}^{n-1}\text{-a.e. point in } \mathbb{R}^{n-1}, \\ u\big|^{\kappa-\text{n.t.}}_{\partial\mathbb{R}^n_+} \text{ belongs to } \Big[L^1\Big(\mathbb{R}^{n-1}, \dfrac{dx'}{1 + |x'|^{n-1}}\Big)\Big]^M, \\ u(x', t) = \Big(P^L_t * \big(u\big|^{\kappa-\text{n.t.}}_{\partial\mathbb{R}^n_+}\big)\Big)(x') \text{ for each } (x', t) \in \mathbb{R}^n_+. \end{cases} \tag{4.6.18}$$

This refines [95, Theorem 3.2, p. 935] and [95, Theorem 6.1, p. 956]. In fact, the result just mentioned interfaces tightly with the topic of boundary value problems for strongly elliptic systems in the upper half-space considered in [95]. Indeed, in view of the assumptions made in (4.6.17) it is natural to consider the linear space

$$\begin{aligned} \mathscr{Z} &:= \Big\{f \in \big[L^1\big(\mathbb{R}^{n-1}, \tfrac{dx'}{1+|x'|^{n-1}}\big)\big]^M : \mathcal{M}_{\mathbb{R}^{n-1}} f \in \big[L^1\big(\mathbb{R}^{n-1}, \tfrac{dx'}{1+|x'|^{n-1}}\big)\big]^M\Big\} \\ &= \Big\{f : \mathbb{R}^{n-1} \to \mathbb{C}^M : \text{ measurable and } \mathcal{M}_{\mathbb{R}^{n-1}} f \in \big[L^1\big(\mathbb{R}^{n-1}, \tfrac{dx'}{1+|x'|^{n-1}}\big)\big]^M\Big\}, \end{aligned} \tag{4.6.19}$$

where $\mathcal{M}_{\mathbb{R}^{n-1}}$ is the Hardy-Littlewood maximal operator in $\mathbb{R}^{n-1}$, equipped with the norm

$$\begin{aligned} \|f\|_{\mathscr{Z}} &:= \|f\|_{[L^1(\mathbb{R}^{n-1}, \frac{dx'}{1+|x'|^{n-1}})]^M} + \|\mathcal{M}_{\mathbb{R}^{n-1}} f\|_{[L^1(\mathbb{R}^{n-1}, \frac{dx'}{1+|x'|^{n-1}})]^M} \\ &\approx \|\mathcal{M}_{\mathbb{R}^{n-1}} f\|_{[L^1(\mathbb{R}^{n-1}, \frac{dx'}{1+|x'|^{n-1}})]^M}, \qquad \forall f \in \mathscr{Z}. \end{aligned} \tag{4.6.20}$$

Then [95, Theorem 1.1, p. 915], with $\mathbb{X} := \mathscr{Z}$ and $\mathbb{Y} := \big[L^1\big(\mathbb{R}^{n-1}, \frac{dx'}{1+|x'|^{n-1}}\big)\big]^M$, gives that the following boundary-value problem is well posed:

$$\begin{cases} u \in \left[\mathscr{C}^\infty(\mathbb{R}^n_+)\right]^M, \quad Lu = 0 \text{ in } \mathbb{R}^n_+, \\ \displaystyle\int_{\mathbb{R}^{n-1}} (\mathcal{N}_\kappa u)(x') \frac{dx'}{1+|x'|^{n-1}} < +\infty, \\ u\Big|^{\kappa-\text{n.t.}}_{\partial\mathbb{R}^n_+} = f \in \mathscr{X}. \end{cases} \tag{4.6.21}$$

The relevance of the fact that (4.6.17) implies (4.6.18) in the context of (4.6.21) is that the nontangential boundary trace $u\big|^{\kappa-\text{n.t.}}_{\partial\mathbb{R}^n_+}$ is guaranteed to exist by the other conditions imposed on the function u in the formulation of the aforementioned boundary value problem, and that the solution may be recovered from the boundary datum via convolution with the Poisson kernel canonically associated with the system L.

To prove that (4.6.17) implies (4.6.18), for each $\varepsilon > 0$ consider

$$u_\varepsilon \in \left[\mathscr{C}^\infty(\overline{\mathbb{R}^n_+})\right]^M \text{ given by } u_\varepsilon(x) := u(x', x_n + \varepsilon) \\ \text{at each point } x = (x', x_n) \in \overline{\mathbb{R}^n_+}, \tag{4.6.22}$$

and note that this function satisfies the conditions stipulated in (4.6.12). As such, we may write the Poisson integral representation formula (4.6.13) for each u_ε and obtain that

$$u(x', t + \varepsilon) = \int_{\mathbb{R}^{n-1}} P^L_t(x' - y') f_\varepsilon(y')\, dy' \text{ for each } x = (x', t) \in \mathbb{R}^n_+, \tag{4.6.23}$$

with $f_\varepsilon := u(\cdot, \varepsilon) : \mathbb{R}^{n-1} \to \mathbb{C}^M$ for each $\varepsilon > 0$. If we also consider the weight $\upsilon : \mathbb{R}^{n-1} \to (0, \infty)$ defined as $\upsilon(x') := \frac{1}{1+|x'|^{n-1}}$ for each $x' \in \mathbb{R}^{n-1}$, then the last condition in (4.6.17) entails

$$\sup_{\varepsilon>0} |f_\varepsilon| \le \mathcal{N}_\kappa u \in L^1\big(\mathbb{R}^{n-1}, \upsilon\, \mathcal{L}^{n-1}\big). \tag{4.6.24}$$

To proceed, we need a suitable weak compactness result, so we momentarily digress to elaborate on this issue.

Lemma 4.6.1 *Let $\upsilon : \mathbb{R}^{n-1} \to (0, \infty)$ be a Lebesgue measurable function and consider a sequence $\{f_j\}_{j\in\mathbb{N}} \subset L^1(\mathbb{R}^{n-1}, \upsilon\mathcal{L}^{n-1})$ such that*

$$F := \sup_{j\in\mathbb{N}} |f_j| \in L^1(\mathbb{R}^{n-1}, \upsilon\mathcal{L}^{n-1}). \tag{4.6.25}$$

Then there exists a subsequence $\{f_{j_k}\}_{k\in\mathbb{N}}$ of $\{f_j\}_{j\in\mathbb{N}}$ together with some function $f \in L^1(\mathbb{R}^{n-1}, \upsilon\mathcal{L}^{n-1})$ with the property that

$$\int_{\mathbb{R}^{n-1}} f_{j_k} \varphi\, \upsilon\, d\mathcal{L}^{n-1} \longrightarrow \int_{\mathbb{R}^{n-1}} f \varphi\, \upsilon\, d\mathcal{L}^{n-1} \text{ as } k \to \infty, \\ \text{for every function } \varphi \in \mathscr{C}^0(\mathbb{R}^{n-1}) \cap L^\infty(\mathbb{R}^{n-1}, \mathcal{L}^{n-1}). \tag{4.6.26}$$

Proof Begin by fixing an arbitrary function $\varphi \in \mathscr{C}^0(\mathbb{R}^{n-1}) \cap L^\infty(\mathbb{R}^{n-1}, \mathcal{L}^{n-1})$. Also, having picked some function $\theta \in \mathscr{C}_c^\infty(\mathbb{R}^{n-1})$ with the property that $0 \leq \theta \leq 1$ and $\theta \equiv 1$ near the origin, for each $R > 0$ define $\theta_R(x') := \theta(x'/R)$ for every $x' \in \mathbb{R}^{n-1}$. In particular, for each fixed $R > 0$ the function $\theta_R \varphi$ is continuous with compact support. Keeping this in mind, [95, Lemma 6.2, p. 956] implies that there exists a subsequence $\{f_{j_k}\}_{k \in \mathbb{N}}$ of $\{f_j\}_{j \in \mathbb{N}}$ and a function $f \in L^1(\mathbb{R}^{n-1}, \upsilon \mathcal{L}^{n-1})$ such for each fixed $R > 0$ we have

$$\int_{\mathbb{R}^{n-1}} \theta_R f_{j_k} \varphi \upsilon \, d\mathcal{L}^{n-1} \longrightarrow \int_{\mathbb{R}^{n-1}} \theta_R f \varphi \upsilon \, d\mathcal{L}^{n-1} \text{ as } k \to \infty. \tag{4.6.27}$$

Next, fix an arbitrary threshold $\delta > 0$. Since $F, f \in L^1(\mathbb{R}^{n-1}, \upsilon \mathcal{L}^{n-1})$ while φ is continuous and bounded, Lebesgue's Dominated Convergence Theorem guarantees that there exists $R_\delta \in (0, \infty)$ with the property that

$$\int_{\mathbb{R}^{n-1}} (1 - \theta_R)|F||\varphi|\upsilon \, d\mathcal{L}^{n-1} < \delta \text{ whenever } R > R_\delta, \tag{4.6.28}$$

and

$$\int_{\mathbb{R}^{n-1}} (1 - \theta_R)|f||\varphi|\upsilon \, d\mathcal{L}^{n-1} < \delta \text{ whenever } R > R_\delta. \tag{4.6.29}$$

For each $R > R_\delta$ we may then rely on (4.6.27)-(4.6.29) to write

$$\begin{aligned}
\limsup_{k\to\infty} & \left| \int_{\mathbb{R}^{n-1}} f_{j_k} \varphi \, \upsilon \, d\mathcal{L}^{n-1} - \int_{\mathbb{R}^{n-1}} f \varphi \upsilon \, d\mathcal{L}^{n-1} \right| \\
& \leq \limsup_{k\to\infty} \left| \int_{\mathbb{R}^{n-1}} \theta_R f_{j_k} \varphi \upsilon \, d\mathcal{L}^{n-1} - \int_{\mathbb{R}^{n-1}} \theta_R f \varphi \upsilon \, d\mathcal{L}^{n-1} \right| \\
& \quad + \limsup_{k\to\infty} \int_{\mathbb{R}^{n-1}} (1 - \theta_R)|f_{j_k}||\varphi|\upsilon \, d\mathcal{L}^{n-1} \\
& \quad + \int_{\mathbb{R}^{n-1}} (1 - \theta_R)|f||\varphi|\upsilon \, d\mathcal{L}^{n-1} \\
& \leq \int_{\mathbb{R}^{n-1}} (1 - \theta_R)|F||\varphi|\upsilon \, d\mathcal{L}^{n-1} + \int_{\mathbb{R}^{n-1}} (1 - \theta_R)|f||\varphi|\upsilon \, d\mathcal{L}^{n-1} \\
& \leq 2\delta.
\end{aligned} \tag{4.6.30}$$

In view of the arbitrariness of $\delta > 0$, this establishes the claim made in (4.6.26). □

Returning to the mainstream discussion, the weak-$*$ convergence result from Lemma 4.6.1 may be used for the sequence $\{f_\varepsilon\}_{\varepsilon > 0} \subset L^1(\mathbb{R}^{n-1}, \upsilon \, \mathcal{L}^{n-1})$ to conclude that there exist $f \in L^1(\mathbb{R}^{n-1}, \upsilon \, \mathcal{L}^{n-1})$ and a sequence $\{\varepsilon_j\}_{j \in \mathbb{N}} \subset (0, \infty)$ which converges to zero with the property that

$$\lim_{j\to\infty}\int_{\mathbb{R}^{n-1}}\varphi(y')f_{\varepsilon_j}(y')\frac{dy'}{1+|y'|^{n-1}}=\int_{\mathbb{R}^{n-1}}\varphi(y')f(y')\frac{dy'}{1+|y'|^{n-1}} \tag{4.6.31}$$

for every bounded continuous function $\varphi:\mathbb{R}^{n-1}\to\mathbb{C}^{M\times M}$. The fact that there exists a constant $C\in(0,\infty)$ for which

$$|P^L(z')|\le\frac{C}{1+|z'|^n}\ \text{ for each }\ z'\in\mathbb{R}^{n-1} \tag{4.6.32}$$

(cf. [95, Theorem 2.4, p. 934]) ensures that for each fixed point $(x',t)\in\mathbb{R}^n_+$ the assignment

$$\begin{gathered}\mathbb{R}^{n-1}\ni y'\mapsto\varphi(y'):=P^L_t(x'-y')(1+|y'|^{n-1})\in\mathbb{C}^{M\times M}\\ \text{is a bounded continuous matrix-valued function.}\end{gathered} \tag{4.6.33}$$

At this stage, from (4.6.23) and (4.6.31) used for the function φ defined in (4.6.33) we obtain (bearing in mind that u is continuous in $\mathbb{R}^n_+$) that

$$u(x',t)=\int_{\mathbb{R}^{n-1}}P^L_t(x'-y')f(y')\,dy'\ \text{ for each }\ x=(x',t)\in\mathbb{R}^n_+. \tag{4.6.34}$$

With this in hand, and since $L^1(\mathbb{R}^{n-1},\upsilon\,\mathcal{L}^{n-1})\subseteq L^1\Big(\mathbb{R}^{n-1},\frac{dx'}{1+|x'|^n}\Big)$, we may invoke [95, Theorem 3.1, p. 934] to conclude that[22]

$$u\Big|^{\kappa-\text{n.t.}}_{\partial\mathbb{R}^n_+}\ \text{ exists and equals } f \text{ at } \mathcal{L}^{n-1}\text{-a.e. point in } \mathbb{R}^{n-1}. \tag{4.6.35}$$

Once this has been established, all conclusions in (4.6.18) are implied by (4.6.12)-(4.6.13).

Lastly, we wish to remark that even in the classical case when $L:=\Delta$, the Laplacian in $\mathbb{R}^n$, our Poisson integral representation formula (4.6.17)-(4.6.18) is more general (in the sense that it allows for a larger class of functions) than the existing results in the literature. Indeed, the latter typically assume an L^p integrability condition for the harmonic function which, in the range $1<p<\infty$, implies our weighted L^1 integrability condition for the nontangential maximal function demanded in (4.6.17). In this vein see, e.g., [48, Theorems 4.8-4.9, pp. 174-175], [166, Corollary, p. 200], [167, Proposition 1, p. 119].

4.7 More on Uniqueness and Poisson Integral Representations

Here we shall consider the Poisson integral operator for a given second-order weakly elliptic system, in the context described below.

[22] convergence holds, for instance, in the set of Lebesgue points of f

Theorem 4.7.1 *Let $\Omega \subset \mathbb{R}^n$, where $n \in \mathbb{N}$ satisfies*[23] *$n \geq 3$, be a* UR *domain which is locally pathwise nontangentially accessible (in the sense of* [112, Definition 8.9.14]*), and set $\sigma := \mathcal{H}^{n-1} \lfloor \partial\Omega$. Consider a homogeneous constant (complex) coefficient second-order $M \times M$ system L in $\mathbb{R}^n$, which is weakly elliptic (in the sense of* (1.3.3)*). Also, fix an aperture parameter $\kappa > 0$ and assume that the spaces $\mathscr{X}$, $\mathscr{Y}$, $\mathscr{Z}$ are as in any of the following cases:*

$$\begin{gathered}\mathscr{X} := L^{p'}(\partial\Omega, \sigma),\ \ \mathscr{Y} := \big[L_1^{p'}(\partial\Omega, \sigma)\big]^M,\ \ \mathscr{Z} := \mathscr{X}^* = L^p(\partial\Omega, \sigma),\\ \text{where } p, p' \in (1, \infty) \text{ are such that } 1/p + 1/p' = 1,\end{gathered} \tag{4.7.1}$$

or

$$\begin{gathered}\mathscr{X} := L^{p'}(\partial\Omega,\ w^{1-p'}\sigma),\ \ \mathscr{Y} := \big[L_1^{p'}(\partial\Omega,\ w^{1-p'}\sigma)\big]^M,\ \ \mathscr{Z} := \mathscr{X}^* = L^p(\partial\Omega, w\sigma),\\ \text{where } p, p' \in (1, \infty) \text{ are such that } 1/p + 1/p' = 1 \text{ and } w \in A_p(\partial\Omega, \sigma),\end{gathered} \tag{4.7.2}$$

or

$$\begin{gathered}\mathscr{X} := L^{p',q'}(\partial\Omega, \sigma),\ \ \mathscr{Z} := \mathscr{X}^* = L^{p,q}(\partial\Omega, \sigma),\ \text{ and}\\ \mathscr{Y} := \big\{f \in \big[L^{p',q'}(\partial\Omega, \sigma)\big]^M :\ \partial_{\tau_{jk}} f \in \big[L^{p',q'}(\partial\Omega, \sigma)\big]^M,\ 1 \leq j, k \leq n\big\},\\ \text{where } p, p', q, q' \in (1, \infty) \text{ are such that } 1/p + 1/p' = 1 = 1/q + 1/q',\end{gathered} \tag{4.7.3}$$

or

$$\begin{gathered}\mathscr{X} := M^{p',\lambda}(\partial\Omega, \sigma),\ \ \mathscr{Y} := \big[M_1^{p',\lambda}(\partial\Omega, \sigma)\big]^M,\ \ \mathscr{Z} := \mathscr{X}^* = \mathcal{B}^{p,\lambda}(\partial\Omega, \sigma),\\ \text{where } p, p' \in (1, \infty) \text{ are such that } 1/p + 1/p' = 1,\end{gathered} \tag{4.7.4}$$

or

$$\begin{gathered}\mathscr{X} := \mathcal{B}^{p',\lambda}(\partial\Omega, \sigma),\ \ \mathscr{Y} := \big[\mathcal{B}_1^{p',\lambda}(\partial\Omega, \sigma)\big]^M,\ \ \mathscr{Z} := M^{p,\lambda}(\partial\Omega, \sigma),\\ \text{where } p, p' \in (1, \infty) \text{ are such that } 1/p + 1/p' = 1.\end{gathered} \tag{4.7.5}$$

Then if the Regularity Problem for the transpose system $L^\top$, formulated as

$$\begin{cases} w \in \big[\mathscr{C}^\infty(\Omega)\big]^M, \quad L^\top w = 0 \text{ in } \Omega,\\ \mathcal{N}_\kappa(\nabla w) \text{ belong to the space } \mathscr{X},\\ w\big|_{\partial\Omega}^{\kappa-\text{n.t.}} = g \text{ at } \sigma\text{-a.e. point on } \partial\Omega,\\ \text{and also } w(x) = o(1) \text{ as } |x| \to \infty\\ \text{when } \Omega \text{ is an exterior domain,}\end{cases} \tag{4.7.6}$$

[23] see the last part in the statement for the case $n = 2$

has a solution for each boundary datum $g \in \mathscr{Y}$, and if the Dirichlet Problem for the original system L, formulated as

$$\begin{cases} u \in \big[\mathscr{C}^\infty(\Omega)\big]^M, \quad Lu = 0 \ \textit{in} \ \Omega, \\ \mathcal{N}_\kappa u \ \textit{belongs to the space} \ \mathscr{Z}, \\ u\big|_{\partial\Omega}^{\kappa-\mathrm{n.t.}} = f \ \textit{at} \ \sigma\textit{-a.e. point on} \ \partial\Omega, \\ \textit{and also} \ u(x) = o(1) \ \textit{as} \ |x| \to \infty \\ \textit{when} \ \Omega \ \textit{is an exterior domain,} \end{cases} \tag{4.7.7}$$

has a solution for each boundary datum $f \in \big[\mathscr{Z}\big]^M$, then for each $x \in \Omega$ there exists a unique function $G(\cdot, x)$ satisfying (with all derivatives, as well as the nontangential maximal function and the nontangential traces, taken in the "dot" variable)

$$\begin{gathered} G(\cdot, x) \in \big[\mathscr{C}^\infty(\Omega \setminus \{x\}) \cap W^{1,1}_{\mathrm{loc}}(\Omega)\big]^{M\times M}, \\ L^\top G(\cdot, x) = -\delta_x I_{M\times M} \ \textit{in} \ \big[\mathcal{D}'(\Omega)\big]^{M\times M}, \\ \mathcal{N}_\kappa^{\Omega\setminus K_x}\big(\nabla G(\cdot, x)\big) \in \mathscr{X} \ \textit{where} \ K_x := \overline{B\big(x, \tfrac{1}{2}\,\mathrm{dist}(x, \partial\Omega)\big)}, \\ \big(\nabla G(\cdot, x)\big)\big|_{\partial\Omega}^{\kappa-\mathrm{n.t.}} \ \textit{exists at} \ \sigma\textit{-a.e. point on} \ \partial\Omega, \\ G(\cdot, x)\big|_{\partial\Omega}^{\kappa-\mathrm{n.t.}} = 0 \ \textit{at} \ \sigma\textit{-a.e. point on} \ \partial\Omega, \\ \textit{and also} \ G(y, x) = o(1) \ \textit{as} \ |y| \to \infty, \\ \textit{in the case when} \ \Omega \ \textit{is an exterior domain.} \end{gathered} \tag{4.7.8}$$

Furthermore, if in place of the solvability of (4.7.6) *one now assumes that the following type of Regularity Problem*

$$\begin{cases} w \in \big[\mathscr{C}^\infty(\Omega)\big]^M, \quad L^\top w = 0 \ \textit{in} \ \Omega, \\ \mathcal{N}_\kappa w \ \textit{and} \ \mathcal{N}_\kappa(\nabla w) \ \textit{belong to} \ \mathscr{X}, \\ w\big|_{\partial\Omega}^{\kappa-\mathrm{n.t.}} = g \ \textit{at} \ \sigma\textit{-a.e. point on} \ \partial\Omega, \\ \textit{and also} \ w(x) = o(1) \ \textit{as} \ |x| \to \infty \\ \textit{when} \ \Omega \ \textit{is an exterior domain,} \end{cases} \tag{4.7.9}$$

is solvable for each boundary datum $g \in \mathscr{Y}$, then in addition to the properties listed in (4.7.8) *one also has*

$$\mathcal{N}_\kappa^{\Omega\setminus K_x}\big(G(\cdot, x)\big) \in \mathscr{X}. \tag{4.7.10}$$

Moreover, if $P^L(x, y)$ is the Poisson kernel associated as in (4.5.2)-(4.5.3) *with the Green function G from* (4.7.8)*, then the Poisson integral operator for the system L, defined for each function $f \in \big[\mathscr{Z}\big]^M$ as*

$$(\mathrm{PI}_L f)(x) := \int_{\partial\Omega} P^L(x,y) f(y)\,\mathrm{d}\sigma(y), \qquad \forall x \in \Omega, \tag{4.7.11}$$

is precisely the unique solution of the Dirichlet Problem (4.7.7) *with boundary datum* f.

Finally, the same results are also valid in the case $n = 2$ *provided either* Ω *is bounded, or* $\partial\Omega$ *is unbounded. Also, if* $n = 2$ *and* Ω *is an exterior domain, the same results are true provided one assumes*

$$\int_{S^1} \big[L(\xi)\big]^{-1}\,\mathrm{d}\mathcal{H}^1(\xi) \quad \text{is an invertible } M \times M \text{ matrix} \tag{4.7.12}$$

(*which is always the case if the system L is assumed to actually satisfy the Legendre-Hadamard strong ellipticity condition, and if* $M = 1$ *this is true if and only if* $L = \nabla\cdot A\nabla$ *for some* $A \in M_2^0$; *cf. Lemma 1.4.19 and* (1.4.186)), *and with the demand that u and* $G(\cdot, x)$ *are* $o(1)$ *at infinity replaced by*

$$u = O(1) \ \text{and} \ G(\cdot, x) = O(1) \ \text{at infinity.} \tag{4.7.13}$$

Proof Granted the solvability of the Regularity Problem for $L^\top$ formulated in (4.7.6), we may construct a Green function for $L^\top$ as in the proof of Theorem 4.3.1 (cf. (4.3.14)). For each fixed pole $x \in \Omega$, this yields a function $G(\cdot, x)$ satisfying all but the fourth condition in (4.7.8). That said condition also holds in the present setting is then seen by applying the Fatou-type result from Theorem 3.3.5 to the columns of the matrix-valued function $G(\cdot, x)$ restricted to $\Omega \setminus \overline{B\big(x, \frac{1}{2}\,\mathrm{dist}(x, \partial\Omega)\big)}$. This ultimately shows that $\big(\nabla G(\cdot, x)\big)\big|_{\partial\Omega}^{\kappa-\mathrm{n.t.}}$ exists at σ-a.e. point on $\partial\Omega$, completing the proof of the existence of a Green function as in (4.7.8).

To prove its uniqueness, suppose $\widetilde{G}(\cdot, x)$ is another function enjoying all properties listed in (4.7.8). Then $w := G(\cdot, x) - \widetilde{G}(\cdot, x)$ satisfies $L^\top w = 0$ in Ω, hence w is smooth in Ω, thanks to elliptic regularity (cf. [112, (6.5.40) in Theorem 6.5.7]). Also, $w\big|_{\partial\Omega}^{\kappa-\mathrm{n.t.}} = 0$ at σ-a.e. point on $\partial\Omega$ and $\mathcal{N}_\kappa(\nabla w) \in \mathscr{X}$. Given that we are assuming the solvability of the Dirichlet Problem (4.7.7), from Theorem 4.3.2 (used with $u := w$ and $L^\top$ in place of L) we conclude that $w \equiv 0$ in Ω. This finishes the proof of the existence of a unique Green function as in (4.7.8). Next, if in place of the solvability of (4.7.6) one now assumes the solvability of (4.7.9) for each $g \in \mathscr{Y}$, then from (4.3.17) we see that (4.7.10) holds.

The fact that the solution of the Dirichlet Problem (4.7.7) is unique is implied by the solvability of the Regularity Problem (4.7.6) (cf. Theorem 4.3.1). Granted this, Theorem 4.4.1 (cf. (4.5.4)) implies that the unique solution of the Dirichlet Problem (4.7.7) with boundary datum $f \in \big[\mathscr{X}\big]^M$ is given by (4.7.11). □

Chapter 5
Green Functions and Poisson Kernels for the Laplacian

Good references for background material pertaining to the topics in this chapter are [71], [49], [51], [58], [79]. The developments here are motivated by questions in potential theory for the Laplacian in $\mathbb{R}^n$ of the following sort:

• When is the Poisson kernel associated with a domain $\Omega \subseteq \mathbb{R}^n$ (as the Radon-Nikodym derivative of the harmonic measure with respect to the surface measure) well-defined and equal to the (minus) normal derivative of the Green function?

• When is said Poisson kernel bounded from below away from zero?

• When can one represent the solution of the Dirichlet Problem with L^p data as an integral involving the Poisson kernel?

Of course, having Ω sufficiently smooth suffices, but we are interested in minimal regularity conditions. This is the perspective through which such questions have been raised, for example, by J. Garnett and D.E. Marshall in [49].

5.1 Upper/Lower Semicontinuous Functions and Super/Sub Harmonic Functions

This section contains a brief review of some basic properties of upper-semicontinuous and lower-semicontinuous functions, as well as superharmonic and subharmonic functions, meant to facilitate the subsequent discussion.

Let (X, τ) be a topological space. For each $x \in X$ denote by $\mathscr{A}_x$ the collection of all neighborhoods of x, and let $\mathscr{O}_x \subseteq \mathscr{A}_x$ be the collection of all open neighborhoods of x (in the topology τ). Given a function $f : X \longrightarrow [-\infty, +\infty]$, for each $x \in X$ define

$$\begin{aligned} \liminf_{y\to x} f(y) &:= \sup\nolimits_{A\in\mathscr{A}_x} \big(\inf\nolimits_{y\in A} f(y)\big) = \sup\nolimits_{O\in\mathscr{O}_x} \big(\inf\nolimits_{y\in O} f(y)\big) \\ \limsup_{y\to x} f(y) &:= \inf\nolimits_{A\in\mathscr{A}_x} \big(\sup\nolimits_{y\in A} f(y)\big) = \inf\nolimits_{O\in\mathscr{O}_x} \big(\sup\nolimits_{y\in O} f(y)\big). \end{aligned} \tag{5.1.1}$$

D. Mitrea et al., *Geometric Harmonic Analysis III*, Developments in Mathematics 74,
https://doi.org/10.1007/978-3-031-22735-6_5

A function $f : X \to (-\infty, +\infty]$ is said to be lower-semicontinuous at $x \in X$ if $\liminf_{y \to x} f(y) \geq f(x)$. Then

$$\begin{aligned}&f \text{ is lower-semicontinuous at } x \in X \text{ if and only if}\\&\{f > a\} \text{ is a neighborhood of } x \text{ whenever } a < f(x).\end{aligned} \tag{5.1.2}$$

Also,

$$\begin{aligned}&\text{if the function } f \text{ is lower-semicontinuous at the point } x \in X\\&\text{then } f(x) \leq \liminf_{j\to\infty} f(x_j) := \sup_{k\in\mathbb{N}}\big(\inf_{j\geq k} f(x_j)\big) \text{ for any sequence}\\&\{x_j\}_{j\in\mathbb{N}} \subseteq X \text{ such that } \lim_{j\to\infty} x_j = x \text{ in } \tau\text{; moreover, the converse is}\\&\text{also true provided the topology } \tau \text{ is metrizable.}\end{aligned} \tag{5.1.3}$$

Simply call $f : X \to (-\infty, +\infty]$ lower-semicontinuous if it is lower-semicontinuous at each point $x \in X$. Then

$$\begin{aligned}&f \text{ is lower-semicontinuous if and only if } \{f > a\} \text{ is open for any}\\&a \in \mathbb{R} \text{ or, equivalently, } \{f \leq a\} \text{ is closed for any } a \in \mathbb{R}.\end{aligned} \tag{5.1.4}$$

In particular,

$$\begin{aligned}&\text{if } (X,\tau) \text{ is a topological space and } f : X \to (-\infty,+\infty] \text{ is lower-}\\&\text{semicontinuous, then for every set } Y \subseteq X \text{ the restriction } f\big|_Y \text{ is a}\\&\text{lower-semicontinuous function on } \big(Y, \tau\big|_Y\big).\end{aligned} \tag{5.1.5}$$

It is also useful to note that

$$\begin{aligned}&\text{if the topology } \tau \text{ on } X \text{ is metrizable and } f : X \to (-\infty,+\infty] \text{ is}\\&\text{lower-semicontinuous, then on each compact set } f \text{ is bounded from}\\&\text{below and its infimum (allowed to be } +\infty\text{) is attained.}\end{aligned} \tag{5.1.6}$$

To justify this, fix $K \subseteq X$ compact and abbreviate $m := \inf_{x\in K} f(x) \in [-\infty, +\infty]$. Then there exists a sequence $\{x_j\}_{j\in\mathbb{N}} \subseteq K$ such that $f(x_j) \searrow m$ as $j \to \infty$. Since τ is metrizable, it follows that we can select a subsequence $\{x_{j_k}\}_{k\in\mathbb{N}}$ which converges in τ to some point $x_* \in K$. Then the property recorded in (5.1.3) permits us to write $m \leq f(x_*) \leq \liminf_{k\to\infty} f(x_{j_k}) = m$. Consequently, $m = f(x_*) > -\infty$, proving (5.1.6).

Remark 5.1.1 *If (X, τ) is a metrizable topological space, μ is a locally finite Borel measure on X, and $f : X \to (-\infty, +\infty]$ is a lower-semicontinuous function, then for each compact set $K \subseteq X$ the integral*

$$\int_K f \, d\mu \text{ is well defined in } (-\infty, +\infty]. \tag{5.1.7}$$

Indeed, based (5.1.4) we deduce that f is μ-measurable on K, while (5.1.6) ensures that there exists $m \in \mathbb{R}$ such that $m \leq f(x)$ for each $x \in K$. Then we set

$$\int_K f\,\mathrm{d}\mu := \int_K (f - m)\,\mathrm{d}\mu + m \cdot \mu(K) \in (-\infty, +\infty], \tag{5.1.8}$$

where $\int_K (f - m)\,\mathrm{d}\mu \in [0, +\infty]$ is the integral of the non-negative, μ-measurable function $f - m$ on K.

Lemma 5.1.2 *Assume (X, d) is a metric space, and let $f : X \to (-\infty, +\infty]$ be lower-semicontinuous and bounded from below. Then there exists a sequence $\{f_j\}_{j\in\mathbb{N}}$ of Lipschitz functions on X with the property that for each $x \in X$ one has $f_j(x) \nearrow f(x)$ as $j \to \infty$.*

Proof If $f \equiv +\infty$ then taking $f_j(x) := j$ for each $j \in \mathbb{N}$ and each $x \in X$ will do. Consider next the case when $f \not\equiv +\infty$. In such a scenario, define

$$f_j(x) := \inf\big\{f(y) + j \cdot d(x, y) : \ y \in X\big\} \text{ for each } j \in \mathbb{N} \text{ and each } x \in X. \tag{5.1.9}$$

Since f is bounded from below and the distance function is non-negative, it follows that the assignment $X \ni y \mapsto f(y) + j \cdot d(x, y)$ is bounded from below. Hence, each $f_j(x)$ is well defined in $\mathbb{R}$ and

$$f_1(x) \le f_2(x) \le \cdots \le f(x) \text{ for each } x \in X. \tag{5.1.10}$$

Also, we claim that for each $j \in \mathbb{N}$ fixed, the function f_j is Lipschitz. Indeed, from definitions, $f_j(z) \le f(y) + j \cdot d(z, y)$ for each $z, y \in X$, and given any $x \in X$ along with $\varepsilon > 0$ there exists some $y_\varepsilon \in X$ such that

$$f_j(x) + \varepsilon > f(y_\varepsilon) + j \cdot d(x, y_\varepsilon). \tag{5.1.11}$$

Combining these estimates yields

$$\begin{aligned} f_j(x) + \varepsilon &> f(y_\varepsilon) + j \cdot d(x, y_\varepsilon) \\ &\ge \big(f_j(z) - j \cdot d(z, y_\varepsilon)\big) + j\big(d(z, y_\varepsilon) - d(x, z)\big) \\ &= f_j(z) - j \cdot d(x, z). \end{aligned} \tag{5.1.12}$$

By letting $\varepsilon \to 0^+$ we arrive at $f_j(x) \ge f_j(z) - j \cdot d(x, z)$ which, after interchanging z and x, ultimately shows that $|f_j(x) - f_j(z)| \le j \cdot d(x, z)$. This proves that, indeed, each f_j is Lipschitz.

We claim that $\lim\limits_{j\to\infty} f_j(x) = f(x)$ for each $x \in X$. To prove this, fix an arbitrary $x \in X$ along with an arbitrary $\varepsilon > 0$. Then for each $j \in \mathbb{N}$ there exists $x_{j,\varepsilon} \in X$ such that

$$f_j(x) \ge f(x_{j,\varepsilon}) + j \cdot d(x, x_{j,\varepsilon}) - \varepsilon. \tag{5.1.13}$$

To proceed, assume $\lim\limits_{j\to\infty} x_{j,\varepsilon} = x$. In such a scenario, using (5.1.10) and the lower-semicontinuity of f (cf. (5.1.3)) we may write

$$f(x) \geq \limsup_{j\to\infty} f_j(x) \geq \liminf_{j\to\infty} f_j(x) \geq \liminf_{j\to\infty} \big\{f(x_{j,\varepsilon}) + j \cdot d(x, x_{j,\varepsilon})\big\} - \varepsilon$$
$$\geq \liminf_{j\to\infty} f(x_{j,\varepsilon}) - \varepsilon \geq f(x) - \varepsilon. \tag{5.1.14}$$

This ultimately shows that $\lim_{j\to\infty} f_j(x) = f(x)$ in this case. Finally, suppose $\big\{x_{j,\varepsilon}\big\}_{j\in\mathbb{N}}$ does not converge to x. By eventually passing to a subsequence, we may assume that there exist a strictly increasing sequence $\{j_k\}_{k\in\mathbb{N}} \subseteq \mathbb{N}$ and some $\delta > 0$ such that $d(x, x_{j_k,\varepsilon}) \geq \delta$ for each $k \in \mathbb{N}$. Let m be a lower bound for f. On account of (5.1.10) and (5.1.13) we may then write

$$f(x) \geq f_{j_k}(x) \geq f(x_{j_k,\varepsilon}) + j_k \cdot d(x, x_{j_k,\varepsilon}) - \varepsilon \geq m + j_k \cdot \delta - \varepsilon. \tag{5.1.15}$$

Passing to limit as $k \to \infty$ then forces $\lim_{j\to\infty} f_j(x) = f(x) = +\infty$ in this case. □

In turn, Lemma 5.1.2 permits us to establish the following useful characterization of lower-semicontinuity.

Proposition 5.1.3 *Let (X, τ) be a metrizable locally compact topological space, and consider a function $f : X \to (-\infty, +\infty]$. Then the following are equivalent:*

(1) the function f is lower-semicontinuous on X;
(2) for each compact set $K \subseteq X$ there exists a sequence $\{f_j\}_{j\in\mathbb{N}}$ of continuous functions on K such that $f_j(x) \nearrow f(x)$ as $j \to \infty$ for each $x \in K$.

Proof To prove that *(1)*⇒*(2)*, fix a compact set $K \subseteq X$. Then $\big(K, \tau\big|_K\big)$ is a compact, metrizable topological space and $f\big|_K$ is lower-semicontinuous function which is bounded from below on K, thanks to (5.1.5) and (5.1.6). Granted this, Lemma 5.1.2 applies and the desired conclusion follows. Consider the implication *(2)*⇒*(1)*. Fix an arbitrary point $x \in X$ along with a sequence $\{x_i\}_{i\in\mathbb{N}} \subseteq X$ which converges to x in the topology τ. We want to show that

$$f(x) \leq \liminf_{i\to\infty} f(x_i). \tag{5.1.16}$$

Since (X, τ) is locally compact, there exists a compact neighborhood K of x. Given the nature of the conclusion we presently seek, there is no loss of generality in assuming that $\{x_i\}_{i\in\mathbb{N}} \subseteq K$. The working hypotheses guarantee that there exists a sequence $\{f_j\}_{j\in\mathbb{N}}$ of continuous functions on K such that $f_j \nearrow f$ pointwise on K as $j \to \infty$. Then for each $j \in \mathbb{N}$ we may write

$$\liminf_{i\to\infty} f(x_i) \geq \liminf_{i\to\infty} f_j(x_i) = f_j(x), \tag{5.1.17}$$

given that $f \geq f_j$ on K and f_j is continuous at x. Since $f_j(x) \nearrow f(x)$ as $j \to \infty$, passing to limit as $j \to \infty$ in (5.1.17) yields (5.1.16). Having proved (5.1.16), we conclude from (5.1.3) that the function f is indeed lower-semicontinuous on X. □

Given topological space (X, τ), call a function $f : X \to [-\infty, +\infty)$ upper-semicontinuous at a point $x \in X$ provided $\limsup_{y\to x} f(y) \leq f(x)$. Also, f is simply called upper-semicontinuous if it is upper-semicontinuous at each point $x \in X$. Since f is upper-semicontinuous if and only if $-f$ is lower-semicontinuous, similar properties to (5.1.2)-(5.1.6), Remark 5.1.1, Lemma 5.1.2, and Proposition 5.1.3 are valid in the class of upper-semicontinuous functions.

Definition 5.1.4 *Assume $\Omega \subseteq \mathbb{R}^n$ is an arbitrary nonempty open set. A given function $u : \Omega \to (-\infty, +\infty]$ is called* `superharmonic` *if the following conditions are satisfied:*

(i) u is lower-semicontinuous;
(ii) given any $x \in \Omega$ together with $0 < r < \text{dist}(x, \partial\Omega)$ and a real-valued continuous function h on $\overline{B(x,r)}$ that is harmonic in $B(x,r)$ and satisfies $h(y) \leq u(y)$ for all $y \in \partial B(x,r)$, one has $h(y) \leq u(y)$ for all $y \in B(x,r)$;
(iii) $u \not\equiv +\infty$.

Also, call $u : \Omega \to [-\infty, +\infty)$ `subharmonic` *if $-u$ is superharmonic.*

It turns out that there is quite a bit of flexibility in the manner in which condition *(ii)* is formulated. This is made precise in the proposition below.

Proposition 5.1.5 *Let Ω be an open set in $\mathbb{R}^n$ and consider a lower-semicontinuous function $u : \Omega \to (-\infty, +\infty]$. Then condition (ii) in Definition 5.1.4 is further equivalent with each of the following properties (all integrals are considered in the sense of Remark 5.1.1):*

(ii.a) given any compact set $K \subseteq \Omega$ together with a real-valued continuous function h on K that is harmonic in $\mathring{K}$ and satisfying $h(y) \geq u(y)$ for all $y \in \partial K$, one has $h(y) \geq u(y)$ for all $y \in K$;
(ii.b) given any $x \in \Omega$ and $0 < r < \text{dist}(x, \partial\Omega)$ one has $u(x) \geq \fint_{\partial B(x,r)} u \, d\mathcal{H}^{n-1}$;
(ii.c) given any $x \in \Omega$ and $0 < r < \text{dist}(x, \partial\Omega)$ one has $u(x) \geq \fint_{B(x,r)} u \, d\mathcal{L}^n$.

Proof The implication *(ii.a)* $\Rightarrow$ *(ii)* is obvious. To justify the implication *(ii)* $\Rightarrow$ *(ii.b)*, fix $x_0 \in \Omega$ along with $0 < r < \text{dist}(x, \partial\Omega)$. Then Proposition 5.1.3 guarantees the existence of a sequence $\{f_j\}_{j\in\mathbb{N}}$ of continuous functions on $\overline{B(x_0,r)}$ such that $f_j(x) \nearrow u(x)$ as $j \to \infty$ for each $x \in \overline{B(x_0,r)}$. For each $j \in \mathbb{N}$, denote by $P[f_j]$ the Poisson integral of $f_j\big|_{\partial B(x_0,r)}$ in the ball $B(x_0,r)$. Since each $P[f_j]$ is harmonic in $B(x_0,r)$, continuous on $\overline{B(x_0,r)}$, and satisfies $P[f_j](y) = f_j(y) \leq u(y)$ for all $y \in \partial B(x,r)$, item *(ii)* implies that $P[f_j] \leq u$ in $B(x_0,r)$ for each $j \in \mathbb{N}$. Consequently,

$$u(x_0) \geq P[f_j](x_0) = \fint_{\partial B(x_0,r)} f_j \, d\mathcal{H}^{n-1} \nearrow \fint_{\partial B(x_0,r)} u \, d\mathcal{H}^{n-1} \text{ as } j \to \infty, \tag{5.1.18}$$

by Lebesgue's Monotone Convergence Theorem. This establishes *(ii.b)*. Next, the implication *(ii.b)* ⇒ *(ii.c)* is a consequence of the "spherical" Fubini formula (cf. [109, Theorem 14.63, p. 577]).

At this stage, there remains to prove that *(ii.c)* ⇒ *(ii.a)*. To this end, let $K \subseteq \Omega$ be a compact set and consider a real-valued continuous function h on K that is harmonic in $\mathring{K}$ and which satisfies $h(y) \leq u(y)$ for all $y \in \partial K$. Then $w := u - h$ is a lower-semicontinuous function on K. Also, according to (5.1.6), the function w is bounded from below and its infimum (allowed to be $+\infty$) is attained, say $\inf_K w = w(x_0) \in (-\infty, +\infty]$, for some point $x_0 \in K$. If x_0 actually belongs to ∂K, then $\inf_K w = w(x_0) = u(x_0) - h(x_0) \geq 0$. This ultimately implies $w \geq 0$ on K, hence $h(y) \leq u(y)$ for all $y \in K$. On the other hand, if $x_0 \notin \partial K$ then x_0 belongs to $K \setminus \partial K = \mathring{K}$, which forces $r := \operatorname{dist}(x_0, \partial K) > 0$. Since $x_0 \in \Omega$ and $0 < r < \operatorname{dist}(x_0, \partial\Omega)$, the current working hypothesis guarantees that

$$w(x_0) \geq \fint_{B(x_0,r)} w \, \mathrm{d}\mathcal{L}^n \geq \inf_K w = w(x_0). \tag{5.1.19}$$

Thus, there exists a set $A \subseteq B(x_0, r)$ which is Lebesgue measurable and satisfies $\mathcal{L}^n(A) = 0$, such that $w(x) = w(x_0)$ for each $x \in B(x_0, r) \setminus A$. Given the choice of r, we have $\partial K \cap \partial B(x_0, r) \neq \varnothing$, hence there exists $x_* \in \partial K \cap \partial B(x_0, r)$. Choose a sequence $\{x_j\}_{j \in \mathbb{N}} \subseteq B(x_0, r) \setminus A$ such that $\lim_{j \to \infty} x_j = x_*$. Keeping in mind that $x_* \in \partial K$ and (5.1.3), we see that

$$0 \leq w(x_*) \leq \liminf_{j \to \infty} w(x_j) = w(x_0) = \inf_K w. \tag{5.1.20}$$

Hence $w \geq 0$ on K from which we finally conclude that $u(y) - h(y) = w(y) \geq 0$ for all $y \in K$, as wanted. □

In relation to Definition 5.1.4 we also wish to remark that if $\Omega \subseteq \mathbb{R}^n$ is a nonempty open set and $u \in \mathscr{C}^0(\Omega)$, then u is subharmonic if $\Delta u \geq 0$ in the sense of distributions in Ω, i.e.,

$$\int_\Omega u \Delta\varphi \, \mathrm{d}\mathcal{L}^n \geq 0 \text{ for each non-negative function } \varphi \in \mathscr{C}_c^\infty(\Omega). \tag{5.1.21}$$

To justify this classical fact, assume first that actually $u \in \mathscr{C}^\infty(\Omega)$ and $\Delta u \geq 0$ pointwise in Ω. For some arbitrary fixed $x \in \Omega$, consider the function

$$f(r) := \fint_{\partial B(x,r)} u \, \mathrm{d}\mathcal{H}^{n-1} = \fint_{\partial B(0,1)} u(x + r\xi) \, \mathrm{d}\mathcal{H}^{n-1}(\xi)$$
$$\text{for each } r \in \big(0, \operatorname{dist}(x, \partial\Omega)\big). \tag{5.1.22}$$

Using the classical Divergence Theorem we see that this satisfies

$$f'(r) = \fint_{\partial B(0,1)} \xi \cdot (\nabla u)(x + r\xi)\, \mathrm{d}\mathcal{H}^{n-1}(\xi)$$
$$= \frac{1}{\omega_{n-1} r^{n-1}} \int_{B(x,r)} \Delta u \,\mathrm{d}\mathcal{L}^n = \frac{r}{n} \fint_{B(x,r)} \Delta u \,\mathrm{d}\mathcal{L}^n. \tag{5.1.23}$$

Consequently, $f'(r) \geq 0$ which shows that f is non-decreasing. In turn, this allows us to estimate

$$u(x) = \lim_{\rho \to 0^+} f(\rho) \leq f(r) = \fint_{\partial B(x,r)} u \,\mathrm{d}\mathcal{H}^{n-1} \tag{5.1.24}$$

for each $r \in \big(0, \mathrm{dist}(x, \partial\Omega)\big)$, proving that u is subharmonic in this case. In the general case, when u only belongs to $\mathscr{C}^0(\Omega)$ and satisfies (5.1.21), a standard mollifier argument (involving a smooth non-negative compactly supported function) produces a sequence $\{u_\varepsilon\}_{\varepsilon>0} \subseteq \mathscr{C}^\infty(\Omega_\varepsilon)$, where $\Omega_\varepsilon := \{x \in \Omega : \mathrm{dist}(x, \partial\Omega) > \varepsilon\}$, which satisfies $\Delta u_\varepsilon \geq 0$ pointwise in Ω_ε and which converges uniformly to u on compact subsets of Ω. Based on what we have just proved, each u_ε satisfies the sub-mean property. Passing to the limit as $\varepsilon \to 0^+$ yields that u itself satisfies the sub-mean property, hence u is subharmonic.

We conclude this review of preliminary matters by recording the following version of the Maximum Principle.

Theorem 5.1.6 *Let $\Omega \subseteq \mathbb{R}^n$ be bounded open set, and suppose $u : \Omega \to [-\infty, +\infty)$ is a subharmonic function while $w : \Omega \to (-\infty, +\infty]$ is a superharmonic function. Then the following statements are true:*

(i) If Ω is connected and $u - w$ has a maximum in Ω then $u - w$ is constant.
(ii) If $\limsup_{y \to x} u(y) \leq \liminf_{y \to x} w(y)$ for all $x \in \partial\Omega$, then $u \leq w$ in Ω.

Proof Working with $u - w$ in place of u, there is no loss of generality in assuming that $w \equiv 0$ throughout. To prove the claim in part *(i)*, suppose Ω is connected and there exists some point $x_0 \in \Omega$ with the property that $\sup_\Omega u = u(x_0)$. In particular, if we abbreviate $M := \sup_\Omega u$ then $M \in (-\infty, +\infty)$. Define

$$A := \{x \in \Omega : u(x) < M\} \text{ and } B := \{x \in \Omega : u(x) = M\}. \tag{5.1.25}$$

Since the function u is upper-semicontinuous, the set A is open. Also, by design, $x_0 \in B$, hence $B \neq \varnothing$. Fix $0 < r < \mathrm{dist}(x_0, \partial\Omega)$. By the sub-mean value property (as in Proposition 5.1.5 applied to $-u$), we have

$$M = u(x_0) \leq \fint_{B(x_0,r)} u \,\mathrm{d}\mathcal{L}^n \leq M. \tag{5.1.26}$$

In turn, this ensures the existence of a Lebesgue measurable set $A \subseteq B(x_0, r)$ satisfying $\mathcal{L}^n(A) = 0$ and $u(x) = M$ at each point $x \in B(x_0, r) \setminus A$. Given any point $a \in A$, it is possible to select a sequence $\{x_j\}_{j\in\mathbb{N}} \subseteq B(x_0, r) \setminus A$ which converges to a. Since u is upper-semicontinuous, (5.1.3) permits us to estimate $M \geq u(a) \geq \limsup_{j\to\infty} u(x_j) = M$ which forces $u(a) = M$. Thus, $u(x) = M$ for all $x \in B(x_0, r)$ which goes to shows that B is open as well. In view of the fact that Ω is connected, this ultimately implies that $B = \Omega$, as wanted.

As regards the claim in part *(ii)*, by working in each connected component of Ω we may assume that Ω is connected to begin with. Suppose this is the case and set $M := \sup_\Omega u$. Then there exists a sequence $\{x_j\}_{j\in\mathbb{N}} \subseteq \Omega$ with the property that $u(x_j) \to M$ as $j \to \infty$. Since Ω is bounded, so is the sequence $\{x_j\}_{j\in\mathbb{N}}$. By Bolzano-Weierstrass' Theorem, by eventually passing to a subsequence there is no loss of generality in assuming that $\{x_j\}_{j\in\mathbb{N}}$ converges to a point $x_* \in \overline{\Omega}$. If actually $x_* \in \Omega$, then the current part *(i)* gives that u is constant in Ω and since $\limsup_{y\to x} u(y) \leq 0$ for all $x \in \partial\Omega$ we ultimately conclude that $u \leq 0$ in Ω. On the other hand, if $x_* \in \overline{\Omega} \setminus \Omega = \partial\Omega$, then the working hypothesis implies that $M = \limsup_{j\to\infty} u(x_j) \leq 0$. This, in turn, once again forces $u \leq 0$ in Ω. □

5.2 The Harmonic Measure

We start by briefly recalling the construction of harmonic measure associated with some open subset Ω of $\mathbb{R}^n$ which is bounded (the boundedness assumption, although pervasive in this section, is not essential). Given $f : \partial\Omega \to \mathbb{R}$ define U_f, the `upper class` of functions associated with f, as the collection of all superharmonic functions $u : \Omega \to (-\infty, +\infty]$ satisfying

$$\inf_\Omega u > -\infty \text{ and } \liminf_{\Omega\ni x\to y} u(x) \geq f(y) \text{ for each point } y \in \partial\Omega, \tag{5.2.1}$$

with the agreement that the function $u \equiv +\infty$ in Ω is also contained in U_f. The `lower class` of functions associated with f is then defined as

$$L_f := \{-u : u \in U_{-f}\}. \tag{5.2.2}$$

Note that the Maximum Principle from item *(ii)* in Theorem 5.1.6 ensures that for each given f,

$$w \leq u \text{ for each } w \in L_f \text{ and } u \in U_f. \tag{5.2.3}$$

In particular, if at each $x \in \Omega$ we define the upper and lower Poisson integral operators associated with a given function f as

$$(\overline{\mathrm{PI}}\, f)(x) := \inf\{u(x) : u \in U_f\} \text{ and } (\underline{\mathrm{PI}}\, f)(x) := \sup\{w(x) : w \in L_f\}, \tag{5.2.4}$$

i.e., if $\overline{\mathrm{PI}}\, f$ and $\underline{\mathrm{PI}}\, f$ are, respectively, the upper and the lower solution of the generalized Dirichlet Problem for f, then

$$(\underline{\mathrm{PI}}\, f)(x) \leq (\overline{\mathrm{PI}}\, f)(x) \text{ for each } x \in \Omega. \tag{5.2.5}$$

Call f a **resolutive boundary function** if $(\overline{\mathrm{PI}}\, f)(x) = (\underline{\mathrm{PI}}\, f)(x)$ at every point $x \in \Omega$ and if $\overline{\mathrm{PI}}\, f$, $\underline{\mathrm{PI}}\, f$ are harmonic functions in the set[1] Ω. Whenever f is a resolutive boundary function, its **Poisson integral** is defined as the harmonic function

$$(\mathrm{PI}\, f)(x) := (\overline{\mathrm{PI}}\, f)(x) = (\underline{\mathrm{PI}}\, f)(x) \text{ for each } x \in \Omega. \tag{5.2.6}$$

In [180], N. Wiener has proved that

$$\text{every continuous real-valued function on } \partial\Omega \text{ is resolutive.} \tag{5.2.7}$$

Together with the Maximum Principle and the Riesz Representation Theorem, this result allows us to define the harmonic measure on $\partial\Omega$. Specifically, (5.2.7) and the Maximum Principle for harmonic functions (cf. Theorem 5.1.6) ensures that for each fixed $x \in \Omega$ the assignment

$$\mathscr{C}^0(\partial\Omega) \ni f \longmapsto (\mathrm{PI}\, f)(x) \in \mathbb{R} \text{ is a positive linear functional.} \tag{5.2.8}$$

As such, Riesz's Representation Theorem (cf., e.g., [46, Theorem 7.2, p. 212], [156, Theorem 2.14, pp. 40-41]) ensures that there exists a sigma-algebra $\mathfrak{M}^x$ which contains all Borel sets on $\partial\Omega$, along with a unique measure $\omega^x : \mathfrak{M}^x \to [0, +\infty]$ satisfying

$$(\mathrm{PI}\, f)(x) = \int_{\partial\Omega} f \, d\omega^x, \qquad \forall f \in \mathscr{C}^0(\partial\Omega), \tag{5.2.9}$$

and which enjoys the following additional properties:

(1) The measure ω^x is a complete probability Borel measure on $\partial\Omega$ which satisfies

$$\begin{gathered}\omega^x(O) = \sup\Big\{(\mathrm{PI}\, f)(x) : f \in \mathscr{C}^0(\partial\Omega),\ 0 \leq f \leq 1,\ \operatorname{supp} f \subseteq O\Big\} \\ \text{for each relatively open subset } O \text{ of } \partial\Omega,\end{gathered} \tag{5.2.10}$$

as well as

$$\begin{gathered}\omega^x(K) = \inf\Big\{(\mathrm{PI}\, f)(x) : f \in \mathscr{C}^0(\partial\Omega),\ f \geq \mathbf{1}_K\Big\} \\ \text{for each compact subset } K \text{ of } \partial\Omega.\end{gathered} \tag{5.2.11}$$

(2) The measure ω^x is Radon (cf. [112, Definition 3.5.1]). In fact, ω^x satisfies the outer-regularity property

[1] There is a version of this procedure that leads to the upper and the lower solution of the generalized Dirichlet Problem which are harmonic for each given bounded function f on $\partial\Omega$. This requires asking that the superharmonic functions in the upper class U_f (hence also the subharmonic functions in the lower class L_f) are continuous on $\overline{\Omega}$. See, e.g., [51, Theorem 2.12, p. 24] in this regard.

$$\omega^x(E) = \inf_{\substack{E \subseteq O \subseteq \partial\Omega \\ O \text{ relatively open}}} \omega^x(O), \qquad \forall E \in \mathfrak{M}^x, \tag{5.2.12}$$

as well as the inner-regularity property

$$\omega^x(E) = \sup_{\substack{K \subseteq \partial\Omega \text{ compact} \\ K \subseteq E}} \omega^x(K), \qquad \forall E \in \mathfrak{M}^x. \tag{5.2.13}$$

Other significant properties of the family of measures $\{\omega^x\}_{x\in\Omega}$ are presented in our next lemma.

Lemma 5.2.1 *Let Ω be a bounded open set in $\mathbb{R}^n$. Then the following statements are true:*

(i) The sigma-algebra $\mathfrak{M}^x$ is independent of $x \in \Omega$. Henceforth, it is therefore meaningful to abbreviate $\mathfrak{M} := \mathfrak{M}^x$ for some, or any, $x \in \Omega$.

(ii) For each given point $x \in \Omega$,

$$\omega^x : \mathfrak{M} \longrightarrow [0, 1] \text{ is a complete Borel-regular measure.} \tag{5.2.14}$$

(iii) For each $x_0, x_1 \in \Omega$ one has

$$\omega^{x_0}(E) \approx \omega^{x_1}(E), \text{ uniformly for } E \in \mathfrak{M}, \tag{5.2.15}$$

where the intervening proportionality constants stay bounded away from zero and infinity as long as x_0, x_1 remain in a compact subset of Ω.

Proof We proceed in a number of steps.

`Step I:` *For each $x \in \Omega$, the measure $\omega^x : \mathfrak{M}^x \to [0, 1]$ is Borel-regular.* To justify this, fix $x \in \Omega$ and pick some arbitrary $E \in \mathfrak{M}^x$. From the outer-regularity property (5.2.12) we conclude that there exists a sequence $\{O_j\}_{j\in\mathbb{N}}$ of relatively open subsets of $\partial\Omega$ such that $E \subseteq O_j$ for each $j \in \mathbb{N}$ and $\lim_{j\to\infty} \omega^x(O_j) = \omega^x(E)$. If we define $B := \bigcap_{j\in\mathbb{N}} O_j$ then B is a Borel subset of $\partial\Omega$ which contains E and satisfies $\omega^x(E) \le \omega^x(B) \le \omega^x(O_j) \to \omega^x(E)$ as $j \to \infty$. Hence, $\omega^x(E) = \omega^x(B)$, proving the claim in Step I (cf. [112, Definition 3.4.1]).

`Step II:` *For each $x_0, x_1 \in \Omega$, we have $\int_{\partial\Omega} f \, d\omega^{x_0} \approx \int_{\partial\Omega} f \, d\omega^{x_1}$ uniformly for non-negative functions $f \in \mathscr{C}^0(\partial\Omega)$, with constants bounded away from zero and infinity as long as x_0, x_1 remain in a compact subset of Ω.* Having fixed f as above, the desired equivalence is a consequence of Harnack's inequality applied to the non-negative harmonic function PI f in Ω (cf. (5.2.7)-(5.2.8)), bearing in mind (5.2.9).

`Step III:` *For each $x_0, x_1 \in \Omega$, we have $\omega^{x_0}(K) \approx \omega^{x_1}(K)$ uniformly for compacts $K \subseteq \partial\Omega$, with constants bounded away from zero and infinity as long as x_0, x_1 remain in a compact subset of Ω.* To see that this is the case, fix a compact set $K \subseteq \partial\Omega$ and consider a sequence of functions $\{f_j\}_{j\in\mathbb{N}} \subseteq \mathscr{C}^0(\partial\Omega)$ satisfying $0 \le f_j \le 1$ for each

$j \in \mathbb{N}$ and $f_j \searrow \mathbf{1}_K$ pointwise on $\partial\Omega$ as $j \to \infty$ (cf. [117, Lemma 4.14, p. 166]). Then we may write

$$\omega^{x_0}(K) \le (\mathrm{PI}\, f_j)(x_0) = \int_{\partial\Omega} f_j \,\mathrm{d}\omega^{x_0} \approx \int_{\partial\Omega} f_j \,\mathrm{d}\omega^{x_1}$$

$$\xrightarrow[j\to\infty]{} \int_{\partial\Omega} \mathbf{1}_K \,\mathrm{d}\omega^{x_1} = \omega^{x_1}(K), \tag{5.2.16}$$

thanks to (5.2.11), (5.2.9), Step II, and Lebesgue's Dominated Convergence Theorem. Changing the roles of x_0 and x_1 then finishes the proof of the claim made in Step III.

`Step IV:` *For each $x_0, x_1 \in \Omega$, we have $\omega^{x_0}(B) \approx \omega^{x_1}(B)$ uniformly for Borel sets $B \subseteq \partial\Omega$, with constants bounded away from zero and infinity as long as x_0, x_1 remain in a compact subset of Ω.* This is a consequence of Step III and the inner-regularity property (5.2.13).

`Step V:` *Fix $x_0 \in \Omega$ and consider $E \in \mathfrak{M}^{x_0}$ with the property that $\omega^{x_0}(E) = 0$. Then for any other point $x_1 \in \Omega$ we have $E \in \mathfrak{M}^{x_1}$ and $\omega^{x_1}(E) = 0$.* To prove this, recall from Step I that for each $x_0 \in \Omega$ the measure $\omega^{x_0} : \mathfrak{M}^{x_0} \to [0,1]$ is Borel-regular. Thus, given $E \in \mathfrak{M}^{x_0}$ with $\omega^{x_0}(E) = 0$, we can find a Borel set $B \subseteq \partial\Omega$ such that $E \subseteq B$ and $\omega^{x_0}(B) = \omega^{x_0}(E) = 0$. Since $B \in \mathfrak{M}^{x_0}$ and $B = (B \setminus E) \sqcup E$, it follows that $\omega^{x_0}(B) = \omega^{x_0}(B \setminus E) + \omega^{x_0}(E)$, which ultimately forces $\omega^{x_0}(B \setminus E) = 0$. Pick an arbitrary $x_1 \in \Omega$. Given that $\omega^{x_1} : \mathfrak{M}^{x_1} \to [0,1]$ is a complete measure, $B \in \mathfrak{M}^{x_1}$, and $0 = \omega^{x_0}(B) \approx \omega^{x_1}(B)$ (cf. Step IV), we conclude (cf. [112, (3.1.23)]) that $B \setminus E \in \mathfrak{M}^{x_1}$ and $\omega^{x_1}(B \setminus E) = 0$. Thus, $E = B \setminus (B \setminus E) \in \mathfrak{M}^{x_1}$ and

$$0 = \omega^{x_1}(B) = \omega^{x_1}(B \setminus E) + \omega^{x_1}(E) = \omega^{x_1}(E), \tag{5.2.17}$$

hence $\omega^{x_1}(E) = 0$ as wanted.

`Step VI:` *For each $x_0, x_1 \in \Omega$ we have $\mathfrak{M}^{x_0} = \mathfrak{M}^{x_1}$ and, abbreviating $\mathfrak{M} := \mathfrak{M}^x$ for some, or any, $x \in \Omega$, it follows that $\omega^{x_0}(E) \approx \omega^{x_1}(E)$ uniformly for $E \in \mathfrak{M}$, with constants bounded away from zero and infinity as long as x_0, x_1 remain in a compact subset of Ω.* To justify these claims, fix $x_0, x_1 \in \Omega$ along with some arbitrary $E \in \mathfrak{M}^{x_0}$. Since $\omega^{x_0} : \mathfrak{M}^{x_0} \to [0,1]$ is a Borel-regular measure (cf. Step I), we can find a Borel set $B \subseteq \partial\Omega$ such that $E \subseteq B$ and $\omega^{x_0}(B) = \omega^{x_0}(E)$. This implies that $B \setminus E \in \mathfrak{M}^{x_0}$ and $\omega^{x_0}(B \setminus E) = 0$. Granted this, Step V implies that we also have $B \setminus E \in \mathfrak{M}^{x_1}$ and $\omega^{x_1}(B \setminus E) = 0$. As such, $E = B \setminus (B \setminus E) \in \mathfrak{M}^{x_1}$ and

$$\omega^{x_1}(E) = \omega^{x_1}(B) \approx \omega^{x_0}(B) = \omega^{x_0}(E), \tag{5.2.18}$$

thanks to Step IV. At this stage, all desired conclusions follow. □

At this stage, we can properly introduce the notion of harmonic measure.

Definition 5.2.2 *Given a bounded open set $\Omega \subseteq \mathbb{R}^n$ along with an arbitrary point $x \in \Omega$, call the complete, Borel-regular, probability measure $\omega^x : \mathfrak{M} \to [0,1]$*

(where the sigma-algebra $\mathfrak{M}$ is as in item (i) of Lemma 5.2.1) the harmonic measure *on $\partial\Omega$ with pole at x.*

We are interested in the class of bounded open sets $\Omega \subseteq \mathbb{R}^n$ for which the Poisson integral $f \mapsto \mathrm{PI}\, f$ is the solution operator for the classical Dirichlet Problem

$$u \in \mathscr{C}^\infty(\Omega) \cap \mathscr{C}^0\big(\overline{\Omega}\big), \quad \Delta u = 0 \text{ in } \Omega, \quad u\big|_{\partial\Omega} = f \in \mathscr{C}^0(\partial\Omega). \tag{5.2.19}$$

In this vein, we make the following definition.

Definition 5.2.3 *A bounded open set Ω in $\mathbb{R}^n$ is called* regular *(for the Dirichlet Problem for the Laplacian) if, given any $f \in \mathscr{C}^0(\partial\Omega)$ one has $\mathrm{PI}\, f \in \mathscr{C}^0(\overline{\Omega})$ and $(\mathrm{PI}\, f)(y) = f(y)$ for each $y \in \partial\Omega$.*

Regularity in the sense of the above definition turns out to be equivalent to the well-posedness of the classical Dirichlet Problem.

Proposition 5.2.4 *Let Ω be a bounded open set in $\mathbb{R}^n$. Then Ω is regular (for the Dirichlet Problem for the Laplacian) if and only if the classical Dirichlet Problem* (5.2.19) *is solvable.*

Hence, if $\Omega \subseteq \mathbb{R}^n$ is a bounded open set which is regular, then the classical Dirichlet Problem (5.2.19) *is well posed and for each $f \in \mathscr{C}^0(\partial\Omega)$ its unique solution is given by $u := \mathrm{PI}\, f$.*

Proof Assume the classical Dirichlet Problem is solvable. Then for each function $f \in \mathscr{C}^0(\partial\Omega)$, the solution u of (5.2.19) belongs both to the upper class U_f, and the lower class L_f, of functions associated with f. In concert with (5.2.4), this implies

$$\overline{\mathrm{PI}}\, f \leq u \text{ in } \Omega, \text{ and } u \leq \underline{\mathrm{PI}}\, f \text{ in } \Omega. \tag{5.2.20}$$

Combining (5.2.20) with (5.2.5) then proves that

$$\overline{\mathrm{PI}}\, f = \underline{\mathrm{PI}}\, f = u \text{ in } \Omega, \tag{5.2.21}$$

which further goes to show that $\mathrm{PI}\, f = u \in \mathscr{C}^0(\overline{\Omega})$ and $(\mathrm{PI}\, f)(y) = u(y) = f(y)$ for each $y \in \partial\Omega$. Thus, Ω is regular for the Dirichlet Problem for the Laplacian (in the sense of Definition 5.2.3).

Conversely, if Ω is regular then, tautologically, for each $f \in \mathscr{C}^0(\partial\Omega)$, the function $u := \mathrm{PI}\, f$ solves the classical Dirichlet Problem (5.2.19). Finally, uniqueness for this classical Dirichlet Problem is guaranteed by the Maximum Principle (cf. Theorem 5.1.6). □

For example, from Proposition 5.2.4 and [49, Theorem 1.1, pp. 37-38] it follows that

$$\text{any finitely connected Jordan domain in the plane is regular for the Dirichlet Problem for the Laplacian.} \tag{5.2.22}$$

To give a simple example of a domain which is not regular for the Laplacian, consider the bounded, open, connected set $\Omega := \{x \in \mathbb{R}^n : 0 < |x| < 1\} = B(0, 1) \setminus \{0\}$ with

$n \geq 2$. If $f \in \mathscr{C}^0(\partial\Omega)$ is defined at each $x \in \partial\Omega = S^{n-1} \cup \{0\}$ by $f(x) := 0$ if $|x| = 1$ and $f(0) := 1$, then any function $u \in \mathscr{C}^0(\overline{\Omega})$ which is harmonic in Ω and satisfies $u\big|_{S^{n-1}} = f\big|_{S^{n-1}} = 0$ necessarily vanishes everywhere in Ω. Indeed, if for each $\varepsilon > 0$ we consider the harmonic function

$$u_\varepsilon(x) := \begin{cases} \varepsilon\big(|x|^{2-n} - 1\big) & \text{if } n \geq 3, \\ -\varepsilon \ln|x| & \text{if } n = 2, \end{cases} \qquad \forall x \in \Omega, \tag{5.2.23}$$

then Maximum Principle (cf. Theorem 5.1.6) implies that $0 \leq u(x) \leq u_\varepsilon(x)$ for each $x \in \Omega$. Upon letting $\varepsilon \to 0^+$ we conclude that, indeed, $u \equiv 0$ in Ω. Hence, $u(0) = 0 \neq 1 = f(0)$ so the classical Dirichlet Problem in Ω is not solvable for the boundary datum f. A more sophisticated example due to H. Lebesgue of a bounded, connected, open set in $\mathbb{R}^3$ with a sufficiently sharp inward cusp for which the classical Dirichlet Problem (5.2.19) fails to be solvable (hence the domain in question is not regular) is discussed in [23].

Necessary and sufficient conditions for regularity are known. To state one of them, recall the following concept of capacity.

Definition 5.2.5 *Given a compact set $K \subset \mathbb{R}^n$, define*

$$\operatorname{cap}(K) := \inf\Big\{\int_{\mathbb{R}^n} |\nabla\varphi|^2 \, d\mathcal{L}^n : \varphi \in \mathscr{C}_c^\infty(\mathbb{R}^n),\ \varphi \geq 1 \text{ on } K\Big\}. \tag{5.2.24}$$

See, e.g., [39, § 4.7, pp. 146-158] and [51, pp. 27-28]. For example, one may check without difficulty that

$$\begin{aligned} &\operatorname{cap}(K) \leq \operatorname{cap}(\widetilde{K}) \text{ for each compacts } K, \widetilde{K} \subset \mathbb{R}^n \text{ with } K \subseteq \widetilde{K}, \\ &\operatorname{cap}(rK) = r^{n-2}\operatorname{cap}(K) \text{ for each compact } K \subset \mathbb{R}^n \text{ and } r > 0. \end{aligned} \tag{5.2.25}$$

For other basic properties see, e.g., [39, Theorem 2, p. 151]. In relation to this notion of capacity, following Wiener criterion holds.

Theorem 5.2.6 *Assume that Ω is a bounded, open, connected set in $\mathbb{R}^n$. Then Ω is regular for the Dirichlet Problem for the Laplacian if and only if*

$$\int_0^1 \operatorname{cap}\big(\overline{B(x,r)} \setminus \Omega\big) \frac{dr}{r^{n-1}} = +\infty \text{ for every } x \in \partial\Omega. \tag{5.2.26}$$

See also [55, Theorem 2.5, p. 25] and the discussion in [51, § 2.9, pp. 27–28] in this regard. A discrete version of (5.2.26) reads[2]

$$\sum_{j=0}^\infty 2^{(n-2)j} \operatorname{cap}\big(\overline{B(x,2^{-j})} \setminus \Omega\big) = +\infty \text{ for each } x \in \partial\Omega. \tag{5.2.27}$$

See [55, Corollary 2.5, p. 27]. In fact, there is a point-by-point version of Wiener's criterion. Specifically, if for some given point $x \in \partial\Omega$ we have

[2] if $n = 2$ then $2^{(n-2)j}$ is replaced by j in (5.2.27)

$$\sum_{j=0}^{\infty} 2^{(n-2)j} \operatorname{cap}\big(\overline{B(x, 2^{-j})} \setminus \Omega\big) = +\infty \tag{5.2.28}$$

then for arbitrary $f \in \mathscr{C}^0(\partial\Omega)$ the solution to the generalized Dirichlet Problem $\Delta u = 0$ in Ω and "$u = f$ on $\partial\Omega$" constructed by the Perron-Wiener-Brelot method has the property that $u(y) \to f(x)$ whenever $\Omega \ni y \to x$. Moreover, if for some $x \in \partial\Omega$ we have $\sum_{j=0}^{\infty} 2^{(n-2)j} \operatorname{cap}\big(\overline{B(x, 2^{-j})} \setminus \Omega\big) < +\infty$ then there exists $f \in \mathscr{C}^0(\partial\Omega)$ for which $u(y)$ fails to converge to $f(x)$ as $\Omega \ni y \to x$.

In particular, from (5.2.25) and (5.2.27) it follows that

if $\Omega \subseteq \mathbb{R}^n$ is some bounded open set satisfying an exterior corkscrew condition then Ω is regular for the Dirichlet Problem for the Laplacian. (5.2.29)

Another useful criterion for regularity is based on the notion of barrier, recalled next.

Definition 5.2.7 *Let $\Omega \subseteq \mathbb{R}^n$ be a bounded open set and fix $x_0 \in \partial\Omega$. A function $w \in \mathscr{C}^0\big(\overline{\Omega}\big)$ is said to be a* `barrier` *at x_0 (relative to Ω) provided*

$$\begin{aligned} &w \text{ is superharmonic in } \Omega, \\ &w > 0 \text{ in } \overline{\Omega} \setminus \{x_0\}, \text{ and } w(x_0) = 0. \end{aligned} \tag{5.2.30}$$

Then (cf., e.g., [51, Theorem 2.14, p. 26]) we have:

Theorem 5.2.8 *Let Ω be a bounded open set in $\mathbb{R}^n$. Then the classical Dirichlet Problem* (5.2.19) *is solvable (i.e., Ω is regular for the Dirichlet Problem for the Laplacian) if and only if for each $x \in \partial\Omega$ there exists a barrier function at x, relative to Ω.*

5.3 The Green Function for the Laplacian

The `Green function` for the Laplacian in arbitrary bounded open sets is defined in the theorem below, where we also collect its most basic properties in such a setting (for proofs, we refer to [55]). In two subsequent theorems we then indicate how some of these properties improve as the underlying domain becomes more regular. The reader is reminded that $W^{k,p}(\Omega)$ is the L^p-based Sobolev space of order $k \in \mathbb{N}$ in the open set $\Omega \subseteq \mathbb{R}^n$, and $\mathring{W}^{k,p}(\Omega)$ is the closure of $\mathscr{C}_c^\infty(\Omega)$ in $W^{k,p}(\Omega)$.

Theorem 5.3.1 *Assume that Ω is an arbitrary bounded open set in $\mathbb{R}^n$. Then there exists a unique function $G_\Omega : \Omega \times \Omega \to [0, +\infty]$, henceforth referred to as the Green function for the Laplacian in Ω, such that*

$$G_\Omega(\cdot, y) \in W^{1,2}\big(\Omega \setminus \overline{B(y, r)}\big) \cap \mathring{W}^{1,1}(\Omega), \qquad \forall y \in \Omega, \ \forall r > 0, \tag{5.3.1}$$

and

$$\int_\Omega \langle \nabla_x G_\Omega(x,y), \nabla\varphi(x)\rangle \, dx = \varphi(y), \qquad \forall \varphi \in \mathscr{C}_c^\infty(\Omega). \tag{5.3.2}$$

Furthermore, the Green function also satisfies the following additional properties:

(i) $G_\Omega(x,y) = G_\Omega(y,x)$ *for all* $x, y \in \Omega$ *with* $x \neq y$*;*
(ii) $0 < G_\Omega(x,y) \le C_n|x-y|^{2-n}$ *for all* $x, y \in \Omega$ *with* $x \neq y$*;*
(iii) $G_\Omega(x,y) \ge C_n|x-y|^{2-n}$ *for all* $x, y \in \Omega$ *with* $x \neq y$ *and* $|x-y| \le \frac{1}{2}\operatorname{dist}(y, \partial\Omega)$*;*
(iv) $G_\Omega(\cdot, y) \in \mathring{W}^{1,p}(\Omega)$ *for all* $1 \le p < \frac{n}{n-1}$*, uniformly in* $y \in \Omega$*;*
(v) $G_\Omega(\cdot, y) \in L^{\frac{n}{n-2},\infty}(\Omega, \mathcal{L}^n)$*, uniformly in* $y \in \Omega$*;*
(vi) $\nabla G_\Omega(\cdot, y) \in L^{\frac{n}{n-1},\infty}(\Omega, \mathcal{L}^n)$*, uniformly in* $y \in \Omega$*.*

From (5.3.1)-(5.3.2) we see that, in the sense of distributions in Ω,

$$\Delta G_\Omega(\cdot, y) = -\delta_y \quad \text{for each fixed } y \in \Omega, \tag{5.3.3}$$

where δ_y is the Dirac distribution in Ω with mass at y. In particular,

$$G_\Omega(\cdot, y) \in \mathscr{C}^\infty(\Omega \setminus \{y\}) \quad \text{for each fixed } y \in \Omega. \tag{5.3.4}$$

In fact, in [55] the authors established Theorem 5.3.1 for arbitrary scalar second-order elliptic operators L in divergence form, with bounded measurable coefficients. Using the same template for the construction of the Green function as in [55], this result has been further adapted in [60] to a class containing all second-order strongly elliptic systems L in divergence form with bounded measurable coefficients such that weak solutions of both L and $L^\top$ are locally Hölder (a property which, for scalar operators, is automatically guaranteed by the classical De Giorgi-Nash-Moser theory). We are interested in the results of [60] for the case $L := \Delta$. Specialized as such, [60, Theorem 4.1] implies, among other things, that if Ω is a bounded connected open set in $\mathbb{R}^n$ then

$$\begin{aligned}&\text{for any } \eta \in \mathscr{C}_c^\infty(\Omega) \text{ satisfying } \eta \equiv 1 \text{ on } B(x_0, r) \text{ for some} \\ &r \in \big(0, \operatorname{dist}(x_0, \partial\Omega)\big) \text{ we have } (1-\eta)G_\Omega(\cdot, x_0) \in \mathring{W}^{1,2}(\Omega),\end{aligned} \tag{5.3.5}$$

and

$$\begin{aligned}&\text{for each given } f \in \mathscr{C}_c^\infty(\Omega), \text{ the function defined for each } x \in \Omega \\ &\text{as } u(x) := \textstyle\int_\Omega G_\Omega(x,y) f(y)\, dy \text{ belongs to } \mathring{W}^{1,2}(\Omega) \text{ and satisfies} \\ &\Delta u = -f \text{ in the sense of distributions in the open set } \Omega.\end{aligned} \tag{5.3.6}$$

Additional geometric assumptions on the underlying set Ω ensure extra regularity properties for the associated Green function.

Theorem 5.3.2 *Let* $\Omega \subseteq \mathbb{R}^n$ *be a bounded open set which is regular for the Dirichlet Problem for the Laplacian (in the sense of Definition 5.2.3). Then for every* $x_0 \in \Omega$ *fixed,* $G_\Omega(\cdot, x_0)$ *extends uniquely to a function belonging to* $\mathscr{C}^0\big(\overline{\Omega} \setminus \{x_0\}\big)$ *and, retaining the same notation for this extension, one has* $G_\Omega(\cdot, x_0)\big|_{\partial\Omega} = 0$.

Proof Working in each connected component of Ω, there is no loss of generality in assuming that Ω itself is a connected set to begin with. Suppose this is the case and fix a point $x_0 \in \Omega$. With $d := \mathrm{dist}(x_0, \partial\Omega) > 0$, define

$$M := \sup_{|x-x_0|=d/2} G_\Omega(x, x_0), \quad \Omega_0 := \Omega \setminus \overline{B(x_0, d/2)} \\ \text{and introduce } u := M^{-1} \cdot G_\Omega(\cdot, x_0)\big|_{\Omega_0}. \tag{5.3.7}$$

Then from (5.3.1), (5.3.4), (5.3.3), and the positivity of the Green function it follows that u satisfies

$$u \in W^{1,2}(\Omega_0) \cap \mathscr{C}^\infty(\Omega_0), \quad u > 0, \quad \Delta u = 0 \text{ in } \Omega_0, \tag{5.3.8}$$

while (5.3.5) implies that

$$0 \leq u \leq 1 \text{ on } \partial\Omega_0 \text{ and } u = 0 \text{ on } \partial\Omega \subseteq \overline{\Omega_0}, \tag{5.3.9}$$

in the sense of $W^{1,2}(\Omega_0)$ (cf. [79, Definition 1.1.13, p. 3], [89], [60, § 4.2]). Granted these, [55, Lemma 2.3, p. 23] applies and gives that there exist $c \in (0, \infty)$ and $r_0 \in (0, d/2)$ with the property that for each $x \in \partial\Omega$ we have

$$0 \leq \sup_{\Omega \cap B(x,r)} u \leq \exp\Big\{ - c \int_r^{r_0} \frac{\mathrm{cap}\big(\overline{B(x,t)} \setminus \Omega\big)}{t^{n-1}}\, dt \Big\}, \qquad \forall r \in (0, r_0). \tag{5.3.10}$$

From this, Theorem 5.2.6, and assumptions, we see that u vanishes in a continuous fashion at each point $x \in \partial\Omega$. Since u is a fixed multiple of $G_\Omega(\cdot, x_0)$ near $\partial\Omega$, the desired conclusion follows. □

The result below is contained in [79, Theorem 1.2.8, p. 7] (cf. also [60, Theorem 4.8]).

Theorem 5.3.3 *Assume $\Omega \subseteq \mathbb{R}^n$ is a bounded open set with the property that $\mathbb{R}^n \setminus \Omega$ is n-thick. Then one can find some exponent $\alpha = \alpha(\Omega) > 0$ along with some constant $C \in (0, \infty)$ with the property that the Green function associated with Ω as in Theorem 5.3.1 satisfies*

(i) $G_\Omega(x, y) \leq C\, \mathrm{dist}(y, \partial\Omega)^\alpha |x - y|^{2-n-\alpha}$ for all $x, y \in \Omega$;
(ii) $|G_\Omega(x, y) - G_\Omega(z, y)| \leq C|x - z|^\alpha/(|x - y|^{2-n-\alpha} + |z - y|^{2-n-\alpha})$ for all $x, y, z \in \Omega$.

In particular,

if $\Omega \subseteq \mathbb{R}^n$ is a bounded open set whose complement is n-thick then, for every $x_0 \in \Omega$ fixed, $G_\Omega(\cdot, x_0)$ extends to a Hölder continuous function in a neighborhood of the topological boundary and $G_\Omega(\cdot, x_0)\big|_{\partial\Omega} = 0$. (5.3.11)

Lemma 5.3.4 *Suppose $\Omega \subseteq \mathbb{R}^n$ is a bounded open set which is regular for the Dirichlet Problem for the Laplacian. Recall the standard fundamental solution E_Δ for the Laplacian in $\mathbb{R}^n$ defined in* (1.5.56). *Then*

$$G_\Omega(\cdot, x_0) = \mathrm{PI}\Big(E_\Delta(\cdot - x_0)\big|_{\partial\Omega}\Big) - E_\Delta(\cdot - x_0)$$

$$= \int_{\partial\Omega} E_\Delta(\cdot - x_0)\, d\omega^{x_0} - E_\Delta(\cdot - x_0) \textit{ for each } x_0 \in \Omega, \tag{5.3.12}$$

and for each $x_0, x_1 \in \Omega$ one has

$$\int_{\partial\Omega} E_\Delta(y - x_0)\, d\omega^{x_1}(y) = \int_{\partial\Omega} E_\Delta(y - x_1)\, d\omega^{x_0}(y). \tag{5.3.13}$$

Proof Fix $x_0 \in \Omega$ and consider

$$H(x, x_0) := G_\Omega(x, x_0) + E_\Delta(x - x_0) \text{ for } \mathcal{L}^n\text{-a.e. } x \in \Omega. \tag{5.3.14}$$

Then $H(\cdot, x_0) \in L^1_{\mathrm{loc}}(\Omega, \mathcal{L}^n)$ and $\Delta H(\cdot, x_0) = 0$ in $\mathcal{D}'(\Omega)$, hence $H(\cdot, x_0) \in \mathscr{C}^\infty(\Omega)$. From assumptions and Theorem 5.3.2, we also see that $H(\cdot, x_0) \in \mathscr{C}^0(\overline{\Omega})$ and $H(\cdot, x_0)\big|_{\partial\Omega} = E_\Delta(\cdot - x_0)\big|_{\partial\Omega}$. Consequently,

$$H(\cdot, x_0) = \mathrm{PI}\Big(E_\Delta(\cdot - x_0)\big|_{\partial\Omega}\Big) = \int_{\partial\Omega} E_\Delta(\cdot - x_0)\, d\omega^{x_0}. \tag{5.3.15}$$

This establishes (5.3.12). Moreover, (5.3.15) permits us to write

$$H(x, x_0) = \int_{\partial\Omega} E_\Delta(y - x_0)\, d\omega^{x}(y), \qquad \forall x \in \Omega. \tag{5.3.16}$$

On the other hand, from (5.3.14), item *(i)* in Theorem 5.3.1, and (1.5.56) we see that

$$H(x, x_0) = H(x_0, x) \text{ for each } x, x_0 \in \Omega \text{ with } x \neq x_0. \tag{5.3.17}$$

Then (5.3.13) readily follows from (5.3.16) and (5.3.17). □

Our next theorem elaborates on the boundary behavior of the gradient Green function associated with an arbitrary UR domain in $\mathbb{R}^n$. In particular, it refines work in [25, Theorem 3, p. 275], [81, Lemma 3.1, p. 336], [81, Lemmas 3.1-3.2, pp. 336-337] [81, §A.2, pp. 384-389], [82, Lemma 3.2, p. 10], where more restrictive classes of domains have been considered.

Theorem 5.3.5 *Assume $\Omega \subseteq \mathbb{R}^n$ is an arbitrary bounded* UR *domain. In particular, Ω is a set of locally finite perimeter and, if $\sigma := \mathcal{H}^{n-1}\lfloor\partial\Omega$, its geometric measure theoretic outward unit normal $\nu = (\nu_1, \dots, \nu_n)$ is defined σ-a.e. on $\partial\Omega$. Also, fix a point $x_0 \in \Omega$ along with some aperture parameter $\kappa \in (0, \infty)$, and assume that the Green function $G_\Omega(\cdot, \cdot)$ associated with Ω as in Theorem 5.3.1 satisfies*

$$\mathcal{N}_\kappa^\varepsilon\big(\nabla G_\Omega(\cdot, x_0)\big) \in L^1(\partial\Omega, \sigma) \textit{ for some } \varepsilon \in \big(0, \mathrm{dist}(x_0, \partial\Omega)\big). \tag{5.3.18}$$

Then for each $\kappa' > 0$, the nontangential trace

$$\left(\nabla G_\Omega(\cdot, x_0)\right)\Big|_{\partial\Omega}^{\kappa'-\mathrm{n.t.}} \ \textit{exists at } \sigma\textit{-a.e. point on } \partial\Omega \\ \textit{and is actually independent on the parameter } \kappa'. \tag{5.3.19}$$

In particular, the normal derivative of $G_\Omega(\cdot, x_0)$ *may be meaningfully defined as*

$$\partial_\nu G_\Omega(\cdot, x_0) := \left\langle \nu, \left(\nabla G_\Omega(\cdot, x_0)\right)\Big|_{\partial\Omega}^{\kappa'-\mathrm{n.t.}} \right\rangle \ \textit{at } \sigma\textit{-a.e. point on } \partial\Omega, \tag{5.3.20}$$

and this definition is independent of the aperture parameter κ'.

Moreover, with the dependence on the aperture parameter dropped, for each $j, k \in \{1, \dots, n\}$ *one has*

$$\nu_j\Big(\left(\partial_k G_\Omega(\cdot, x_0)\right)\Big|_{\partial\Omega}^{\mathrm{n.t.}}\Big) = \nu_k\Big(\left(\partial_j G_\Omega(\cdot, x_0)\right)\Big|_{\partial\Omega}^{\mathrm{n.t.}}\Big) \ \textit{at } \sigma\textit{-a.e. point on } \partial\Omega. \tag{5.3.21}$$

As a consequence,

$$\left(\nabla G_\Omega(\cdot, x_0)\right)\Big|_{\partial\Omega}^{\mathrm{n.t.}} = \partial_\nu G_\Omega(\cdot, x_0)\nu \ \textit{at } \sigma\textit{-a.e. point on } \partial\Omega, \tag{5.3.22}$$

and for any vector field $\vec{T} : \partial\Omega \to \mathbb{C}^n$ *which is tangential, in the sense that* $\langle \nu, \vec{T}\rangle = 0$ *at* σ*-a.e. point on* $\partial\Omega$, *one has*

$$\left\langle \vec{T}, \left(\nabla G_\Omega(\cdot, x_0)\right)\Big|_{\partial\Omega}^{\mathrm{n.t.}} \right\rangle = 0 \ \textit{at } \sigma\textit{-a.e. point on } \partial\Omega. \tag{5.3.23}$$

Finally, for each $\kappa' > 0$, *the nontangential trace*

$$G_\Omega(\cdot, x_0)\Big|_{\partial\Omega}^{\kappa'-\mathrm{n.t.}} = 0 \ \textit{at } \sigma\textit{-a.e. point on } \partial\Omega. \tag{5.3.24}$$

Proof Observe that $\Omega_0 := \Omega \setminus \overline{B\big(x_0, \frac{1}{2}\operatorname{dist}(x_0, \partial\Omega)\big)}$ is a bounded UR domain, and $G_\Omega(\cdot, x_0)$ is a harmonic in Ω_0. In addition, (5.3.18) implies that the nontangential maximal function of $G_\Omega(\cdot, x_0)$ relative to Ω_0 belongs to $L^1(\partial\Omega_0, \sigma_0)$, where we have set $\sigma_0 := \mathcal{H}^{n-1}\lfloor\partial\Omega_0$. Granted this, the Fatou-type result from Theorem 3.3.4 applies (with $L := \Delta$) and gives that the nontangential boundary trace of $\nabla G_\Omega(\cdot, x_0)$, taken from within Ω_0 with any aperture parameter $\kappa' > 0$, exists at σ_0-a.e. point on $\partial\Omega_0$ and is actually independent of κ'. Upon observing that there exists $\varepsilon > 0$ with the property that

$$\Gamma^0_{\kappa'}(x) \cap B(x, \varepsilon) = \Gamma_{\kappa'}(x) \cap B(x, \varepsilon) \ \text{for each } x \in \partial\Omega \subseteq \partial\Omega_0, \\ \text{where } \Gamma^0_{\kappa'}(x), \Gamma_{\kappa'}(x) \text{ are the nontangential approach regions} \\ \text{with apex at } x, \text{ considered relative to } \Omega_0 \text{ and } \Omega, \text{ respectively,} \tag{5.3.25}$$

it follows that the nontangential boundary trace of $\nabla G_\Omega(\cdot, x_0)$ at points on $\partial\Omega$, taken from within Ω_0, agrees σ-a.e. with $\left(\nabla G_\Omega(\cdot, x_0)\right)\Big|_{\partial\Omega}^{\kappa'-\mathrm{n.t.}}$. This completes the proof of (5.3.19).

Next, observe that item *(ii)* in Theorem 5.3.1 ensures that the harmonic function $G_\Omega(\cdot, x_0)$ is also bounded in the UR domain Ω_0. As such, we may invoke (3.3.107) in Theorem 3.3.9 to conclude (also bearing in mind (5.3.25)) that for each $\kappa' > 0$ the nontangential trace

$$G_\Omega(\cdot, x_0)\Big|_{\partial\Omega}^{\kappa'-\mathrm{n.t.}} \text{ exists at } \sigma\text{-a.e. point on } \partial\Omega \\ \text{and is actually independent on the parameter } \kappa'. \tag{5.3.26}$$

To prove the stronger conclusion claimed in (5.3.24), let us select an arbitrary index $j \in \{1, \dots, n\}$ along with any test function $\varphi \in \mathscr{C}_c^\infty(\mathbb{R}^n)$ having the property that $x_0 \notin \operatorname{supp}\varphi$. Since $G_\Omega(\cdot, x_0) \in \mathring{W}^{1,1}(\Omega)$ (cf. (5.3.1)), it follows that

$$\text{there exists a sequence } \{\psi_\mu\}_{\mu\in\mathbb{N}} \subseteq \mathscr{C}_c^\infty(\Omega) \text{ with the property that } \\ \psi_\mu \to G_\Omega(\cdot, x_0) \text{ and } \partial_j\psi_\mu \to \partial_j G_\Omega(\cdot, x_0) \text{ in } L^1(\Omega, \mathcal{L}^n) \text{ as } \mu \to \infty. \tag{5.3.27}$$

Then

$$\begin{aligned} &G_\Omega(\cdot, x_0)\varphi \in W^{1,1}(\Omega), \text{ and} \\ &\partial_j\big(G_\Omega(\cdot, x_0)\varphi\big) = \big(\partial_j G_\Omega(\cdot, x_0)\big)\varphi + G_\Omega(\cdot, x_0)(\partial_j\varphi). \end{aligned} \tag{5.3.28}$$

Consequently, for each $\kappa' > 0$,

$$\begin{aligned} \int_{\partial\Omega}\Big(G_\Omega(\cdot, x_0)\Big|_{\partial\Omega}^{\kappa'-\mathrm{n.t.}}\Big)\varphi\nu_j\,\mathrm{d}\sigma &= \int_\Omega \partial_j\big(G_\Omega(\cdot, x_0)\varphi\big)\,\mathrm{d}\mathcal{L}^n \\ &= \int_\Omega \Big\{\big(\partial_j G_\Omega(\cdot, x_0)\big)\varphi + G_\Omega(\cdot, x_0)(\partial_j\varphi)\Big\}\,\mathrm{d}\mathcal{L}^n \\ &= \lim_{\mu\to\infty}\int_\Omega \Big\{(\partial_j\psi_\mu)\varphi + \psi_\mu(\partial_j\varphi)\Big\}\,\mathrm{d}\mathcal{L}^n \\ &= \lim_{\mu\to\infty}\int_\Omega \partial_j\big(\psi_\mu\varphi\big)\,\mathrm{d}\mathcal{L}^n \\ &= 0. \end{aligned} \tag{5.3.29}$$

Above, the first equality is a consequence of the Divergence Formula [112, (1.2.2)] applied to the vector field

$$\vec{F} := G_\Omega(\cdot, x_0)\varphi\,\mathbf{e}_j \in \big[\mathscr{C}_c^\infty(\Omega)\big]^n. \tag{5.3.30}$$

In this regard, we note that from item *(ii)* in Theorem 5.3.1 and our choice of φ, it follows that $\vec{F}$ is also bounded in Ω. Based on this and [112, (8.3.6) in Lemma 8.3.2] we conclude that

$$\mathcal{N}_{\kappa'}\vec{F} \in L^\infty(\partial\Omega, \sigma) \subseteq L^1(\partial\Omega, \sigma). \tag{5.3.31}$$

Thanks to assumptions, (5.3.26), (5.3.28), and (5.3.31) the vector field $\vec{F}$ satisfies all hypotheses in [112, Theorem 1.2.1]. This justifies the applicability of Divergence

Formula [112, (1.2.2)] in the present setting. Next, the second equality in (5.3.29) comes from (5.3.28), while the third equality in (5.3.29) is implied by (5.3.27). Finally, the fourth equality in (5.3.29) is simply Chain Rule for smooth functions, and the last equality in (5.3.29) is justified by noting that $\psi_\mu\varphi \in \mathscr{C}_c^\infty(\Omega)$ for each $\mu \in \mathbb{N}$.

Having established (5.3.29), we may invoke [112, Corollary 3.7.3, (3.7.23)] (with $X := \partial\Omega$, $\mu := \sigma$, and $O := \mathbb{R}^n \setminus \{x_0\}$) to conclude that, for each $\kappa' > 0$ and each $j \in \{1, \dots, n\}$ we have

$$\Big(G_\Omega(\cdot, x_0)\Big|_{\partial\Omega}^{\kappa'-\text{n.t.}}\Big)\nu_j = 0 \text{ at } \sigma\text{-a.e. point on } \partial\Omega. \tag{5.3.32}$$

Given that $|\nu| = 1$ at σ-a.e. point in $\partial_*\Omega$ (cf. [112, (5.6.20), (5.6.21)]), and since we are presently assuming (cf. the definition of a UR domain from [112, Definition 5.10.6])

$$\sigma(\partial\Omega \setminus \partial_*\Omega) = 0, \tag{5.3.33}$$

we ultimately arrive at the conclusion that (5.3.24) holds.

Finally, with (5.3.19) and (5.3.24) in hand, the claims in (5.3.21)-(5.3.23) follow from [113, Corollary 11.3.3)], bearing in mind (5.3.33) and the fact that item *(ii)* in Theorem 5.3.1 together with [112, (8.3.6) in Lemma 8.3.2] imply

$$\begin{gathered} \mathcal{N}_{\kappa'}^{\varepsilon} G(\cdot, x_0) \in L^\infty(\partial\Omega, \sigma) \subseteq L^1(\partial\Omega, \sigma) \\ \text{for each } \varepsilon \in \big(0, \text{dist}(x_0, \partial\Omega)\big) \text{ and } \kappa' > 0. \end{gathered} \tag{5.3.34}$$

This finishes the proof of Theorem 5.3.5. □

The next result in this section provides an answer to the first question raised in the preamble to §5 (cf. [49, Question 2, p. 49]). Essentially, if a given domain is regular for the Dirichlet Problem for the Laplacian in the sense of Definition 5.2.3, satisfies a local pathwise nontangential accessibility condition, has an Ahlfors regular boundary, and its Green function (considered as in Theorem 5.3.1) has a reasonable boundary behavior (which basically ensures that its normal derivative is meaningful), then the harmonic measure is absolutely continuous with respect to the surface measure and the Radon-Nikodym derivative is the (minus) normal derivative of the Green function. Our result refines work in [16, Theorem 5.7, p. 79], [81, Proposition A.1.1, p. 382], [81, Proposition A.2.2, p. 385], and [82, Lemma 3.4, p. 16], where more restrictive classes of sets have been considered.

Theorem 5.3.6 *Let $\Omega \subset \mathbb{R}^n$, where $n \in \mathbb{N}$ with $n \geq 2$, be a bounded open set which is regular for the Dirichlet Problem for the Laplacian (in the sense of Definition 5.2.3), is locally pathwise nontangentially accessible (in the sense of [112, Definition 8.9.14]), has a lower Ahlfors regular boundary, and for which $\sigma := \mathcal{H}^{n-1}\lfloor\partial\Omega$ is a doubling measure on $\partial\Omega$. Also, fix a point $x_0 \in \Omega$ along with some aperture parameter $\kappa \in (0, \infty)$, and assume that the Green function $G_\Omega(\cdot, \cdot)$ associated with Ω as in Theorem 5.3.1 satisfies*

$$\mathcal{N}_\kappa^\varepsilon\big(\nabla G_\Omega(\cdot,x_0)\big)\in L^1(\partial\Omega,\sigma) \text{ for some } \varepsilon\in\big(0,\operatorname{dist}(x_0,\partial\Omega)\big),$$
$$\text{and } \big(\nabla G_\Omega(\cdot,x_0)\big)\Big|_{\partial\Omega}^{\kappa-\mathrm{n.t.}} \text{ exists at } \sigma\text{-a.e. point on } \partial\Omega. \tag{5.3.35}$$

Then

ω^{x_0}, the harmonic measure on $\partial\Omega$ with pole at x_0, is absolutely continuous with respect to the surface measure $\sigma = \mathcal{H}^{n-1}\lfloor\partial\Omega$, in the sense that any σ-measurable set belongs to the sigma-algebra $\mathfrak{M}$, and if $E\subseteq\partial\Omega$ is σ-measurable and $\sigma(E)=0$ then $\omega^{x_0}(E)=0$. (5.3.36)

Moreover, if ν denotes the geometric measure theoretic outward unit normal to Ω, then the Radon-Nikodym derivative of the harmonic measure ω^{x_0} with respect to the surface measure σ is given by

$$\frac{d\omega^{x_0}}{d\sigma} = -\mathbf{1}_{\partial_*\Omega}\cdot\partial_\nu G_\Omega(\cdot,x_0) \text{ at } \sigma\text{-a.e. point on } \partial\Omega,$$
$$\text{where } \partial_\nu G_\Omega(\cdot,x_0) := \Big\langle \nu, \big(\nabla G_\Omega(\cdot,x_0)\big)\Big|_{\partial\Omega}^{\kappa-\mathrm{n.t.}}\Big\rangle. \tag{5.3.37}$$

Before giving the proof of this theorem, a few comments are in order.

Comment 1. Under the additional assumption that Ω is a UR domain, Theorem 5.3.5 guarantees that the condition in the second line of (5.3.35) is automatically satisfied. Of course, if Ω is a UR domain then, by design, $\partial\Omega$ is Ahlfors regular, so Ω has a lower Ahlfors regular boundary and $\sigma=\mathcal{H}^{n-1}\lfloor\partial\Omega$ is a doubling measure on $\partial\Omega$.

Comment 2. Whenever $\omega^{x_0} \ll \sigma$, the Radon-Nikodym derivative

$$k^{x_0} := \frac{d\omega^{x_0}}{d\sigma} \tag{5.3.38}$$

is called the `Poisson kernel` for the domain Ω. Hence, whenever $\omega^{x_0} \ll \sigma$ it follows that

the Poisson kernel $k^{x_0} := d\omega^{x_0}/d\sigma$ is a well-defined, non-negative function, which belongs to $L^1(\partial\Omega,\sigma)$, and satisfies $\int_{\partial\Omega} k^{x_0}\,d\sigma = 1$, (5.3.39)

(since ω^{x_0} is a probability measure). Assuming that the Poisson kernel exists, Riesz's duality theorem also implies that given any $p,p'\in(1,\infty)$ satisfying $1/p+1/p'=1$ we have

$$L^p(\partial\Omega,\sigma)\hookrightarrow L^1(\partial\Omega,\omega^{x_0}) \iff k^{x_0}\in L^{p'}(\partial\Omega,\sigma). \tag{5.3.40}$$

Comment 3. In the context of Theorem 5.3.6, the conclusion in (5.3.36) ensures that (5.3.39) holds, while (5.3.37) gives the pointwise representation formula

$$k^{x_0} = -\mathbf{1}_{\partial_*\Omega}\cdot\partial_\nu G_\Omega(\cdot,x_0) \text{ at } \sigma\text{-a.e. point on } \partial\Omega. \tag{5.3.41}$$

Since, as noted in (5.3.39), the very existence of the Poisson kernel entails $k^{x_0} \in L^1(\partial\Omega, \sigma)$, it follows from (5.3.41) that the quantitative assumption made on $\nabla G_\Omega(\cdot, x_0)$ in the first line of (5.3.35) is actually natural.

We now turn to the proof of Theorem 5.3.6.

Proof of Theorem 5.3.6 Since Ω is a bounded open set which is regular for the Dirichlet Problem for the Laplacian, for each $f \in \mathscr{C}^0(\partial\Omega)$ the function $u := \mathrm{PI} f$ is the unique solution of the classical Dirichlet Problem (5.2.19). In concert with [112, Proposition 8.9.16] and [112, Lemma 8.3.2], this implies that u satisfies

$$\begin{cases} u \in \mathscr{C}^\infty(\Omega), \quad \Delta u = 0 \text{ in } \Omega, \\ u\big|_{\partial\Omega}^{\kappa-\text{n.t.}} = f \text{ at } \sigma\text{-a.e. point on } \partial\Omega, \\ \mathcal{N}_\kappa u \text{ belongs to } L^\infty(\partial\Omega, \sigma). \end{cases} \tag{5.3.42}$$

Let us also observe that, by virtue of item *(ii)* in Theorem 5.3.1, whenever we have $0 < \varepsilon < \operatorname{dist}(x_0, \partial\Omega)$ it follows that the function $G_\Omega(\cdot, x_0)$ is bounded in the one-sided collar neighborhood $\mathcal{O}_\varepsilon := \{x \in \Omega : \operatorname{dist}(x, \partial\Omega) < \varepsilon\}$ of $\partial\Omega$. In particular, thanks to [112, (8.2.26)], we have $\mathcal{N}_\kappa^\varepsilon G_\Omega(\cdot, x_0) \in L^\infty(\partial\Omega, \sigma)$. This property may be also justified based on Theorem 5.3.2 which, in view of [112, (8.9.10)] and the fact that we presently have $\mathcal{H}^{n-1}(\partial\Omega \setminus \partial_{\text{nta}}\Omega) = 0$ (cf. [112, (8.9.193)]), additionally gives that

$$G_\Omega(\cdot, x_0)\Big|_{\partial\Omega}^{\kappa-\text{n.t.}} = 0 \text{ at } \sigma\text{-a.e. point on } \partial\Omega. \tag{5.3.43}$$

Thanks to these properties, the original assumptions, and [112, Corollary 8.9.9], we may invoke Theorem 4.4.1, presently used with $M := 1$ and $L := \Delta$. In view of (5.2.9) and the Poisson formula (4.4.4), we therefore obtain

$$\int_{\partial\Omega} f \, d\omega^{x_0} = u(x_0) = -\int_{\partial_*\Omega} f \cdot \partial_\nu G_\Omega(\cdot, x_0) \, d\sigma. \tag{5.3.44}$$

Hence, ultimately

$$\int_{\partial\Omega} f \, d\omega^{x_0} = -\int_{\partial_*\Omega} f \cdot \partial_\nu G_\Omega(\cdot, x_0) \, d\sigma, \qquad \forall f \in \mathscr{C}^0(\partial\Omega). \tag{5.3.45}$$

To proceed, fix an arbitrary compact set $K \subseteq \partial\Omega$. From [117, Lemma 4.14, p. 166] we know that there exists a sequence of functions $\{f_j\}_{j\in\mathbb{N}} \subseteq \mathscr{C}^0(\partial\Omega)$ satisfying $0 \le f_j \le 1$ for each $j \in \mathbb{N}$ and $f_j \searrow \mathbf{1}_K$ pointwise on $\partial\Omega$ as $j \to \infty$. Granted these properties, we may use Lebesgue's Dominated Convergence Theorem (twice) and (5.3.45) to write

$$\begin{aligned} \omega^{x_0}(K) &= \int_{\partial\Omega} \mathbf{1}_K \, d\omega^{x_0} = \lim_{j\to\infty} \int_{\partial\Omega} f_j \, d\omega^{x_0} \\ &= -\lim_{j\to\infty} \int_{\partial_*\Omega} f_j \cdot \partial_\nu G_\Omega(\cdot, x_0) \, d\sigma = -\int_{\partial_*\Omega} \mathbf{1}_K \cdot \partial_\nu G_\Omega(\cdot, x_0) \, d\sigma. \end{aligned} \tag{5.3.46}$$

From this, (5.2.13), and item *(1)* in [112, Proposition 3.4.15] (also keeping in mind [112, Lemma 3.4.13]), we then conclude that

$$\omega^{x_0}(B) = -\int_{\partial_*\Omega} \mathbf{1}_B \cdot \partial_\nu G_\Omega(\cdot, x_0)\, d\sigma \quad \text{for each Borel set } B \subseteq \partial\Omega. \tag{5.3.47}$$

Consider next an arbitrary set $E \subseteq \partial\Omega$ which is σ-measurable and satisfies $\sigma(E) = 0$. Since σ is a Borel-regular measure (cf. [112, Lemma 3.6.4]), it follows that there exists a Borel set $B \subseteq \partial\Omega$ such that $E \subseteq B$ and $\sigma(B) = \sigma(E) = 0$. Together with (5.3.47), this forces $\omega^{x_0}(B) = 0$ thus, ultimately, $E \in \mathfrak{M}$ and $\omega^{x_0}(E) = 0$, since ω^{x_0} is a complete measure. Let us record our progress. The argument so far shows that

$$\begin{array}{l}\text{if } E \subseteq \partial\Omega \text{ is } \sigma\text{-measurable and satisfies } \sigma(E) = 0 \text{ then} \\ E \text{ belongs to the sigma-algebra } \mathfrak{M} \text{ and } \omega^{x_0}(E) = 0.\end{array} \tag{5.3.48}$$

Assume now that $E \subseteq \partial\Omega$ is an arbitrary σ-measurable set. The fact that σ is a Borel-regular measure (cf. [112, Lemma 3.6.4]) guarantees the existence of a Borel set $B \subseteq \partial\Omega$ such that $E \subseteq B$ and $\sigma(B) = \sigma(E)$. In particular,

$$B \setminus E \ \text{ is } \sigma\text{-measurable, and } \sigma(B \setminus E) = \sigma(B) - \sigma(E) = 0, \tag{5.3.49}$$

since σ is a finite Borel measure. Together with (5.3.48), this proves that

$$B \setminus E \in \mathfrak{M} \ \text{ and } \ \omega^{x_0}(B \setminus E) = 0. \tag{5.3.50}$$

Since $B \in \mathfrak{M}$, this further implies that

$$E = B \setminus (B \setminus E) \in \mathfrak{M} \ \text{ and } \ \omega^{x_0}(E) = \omega^{x_0}(B). \tag{5.3.51}$$

From (5.3.48) and the first property in (5.3.51) the claim in (5.3.36) follows. Finally, based on (5.3.49), (5.3.51), and (5.3.47) we may write

$$\begin{aligned}\omega^{x_0}(E) = \omega^{x_0}(B) &= -\int_{\partial_*\Omega} \mathbf{1}_B \cdot \partial_\nu G_\Omega(\cdot, x_0)\, d\sigma \\ &= -\int_{\partial_*\Omega} \mathbf{1}_E \cdot \partial_\nu G_\Omega(\cdot, x_0)\, d\sigma.\end{aligned} \tag{5.3.52}$$

Thus,

$$\omega^{x_0}(E) = -\int_{\partial_*\Omega} \mathbf{1}_E \cdot \partial_\nu G_\Omega(\cdot, x_0)\, d\sigma \quad \text{for each } \sigma\text{-measurable set } E \subseteq \partial\Omega. \tag{5.3.53}$$

In turn, this formula may be interpreted (cf. [156, Theorem 6.9(b), p. 122]) as saying that (5.3.37) holds. □

Moving on, the basic fact that the harmonic measure in an NTA domain is doubling has been established in [71]. The following lemma, borrowed from [71,

Lemma 4.8, p. 86], is a generalization of a similar result proved by B. Dahlberg in the category of Lipschitz domains in [25, Lemma 1, p. 276]. To state it, form here on we agree to denote by

$$\Delta(x,r) := B(x,r) \cap \partial\Omega, \text{ with } x \in \partial\Omega \text{ and } r > 0, \tag{5.3.54}$$

the `surface balls` on $\partial\Omega$.

Lemma 5.3.7 *Let Ω be a bounded* NTA *domain in $\mathbb{R}^n$ and denote by $G_\Omega(\cdot,\cdot)$ the Green function for the Laplacian in Ω. Then there exist constants $R > 0$, $M > 1$ which depend only on Ω such that whenever $0 < r < R/2$, $z \in \partial\Omega$, and $x \in \Omega \setminus B(z, 2r)$, one has*

$$M^{-1} < \frac{\omega^x(\Delta(z,r))}{r^{n-2} G_\Omega(A_r(z), x)} < M, \tag{5.3.55}$$

for each point $A_r(z) \in \Omega$ satisfying

$$|A_r(z) - z| < Mr \text{ and } \operatorname{dist}\big(A_r(z), \partial\Omega\big) > M^{-1} r \tag{5.3.56}$$

(making $A_r(z)$ a corkscrew point relative to z, at scale r; cf. [112, (5.1.5)]).

A useful estimate for the truncated nontangential maximal operator of the gradient of the Green function, in terms of the harmonic measure, is contained in the next lemma.

Lemma 5.3.8 *Suppose Ω is a bounded* NTA *domain in $\mathbb{R}^n$ and let $G_\Omega(\cdot,\cdot)$ be the Green function for the Laplacian in Ω. Then there exist $\kappa > 0$ and a constant $C \in (0,\infty)$, both depending only on Ω, such that for each point $x_0 \in \Omega$ one can find $0 < \varepsilon_0 < \operatorname{dist}(x_0, \partial\Omega)$ with the property that for each $\varepsilon \in (0, \varepsilon_0)$ one has*

$$\mathcal{N}_\kappa^\varepsilon\big(\nabla G_\Omega(\cdot, x_0)\big)(z) \le C \cdot \sup_{0<r<\varepsilon} \left\{ \frac{\omega^{x_0}(\Delta(z,r))}{r^{n-1}} \right\} \text{ for every } z \in \partial\Omega. \tag{5.3.57}$$

Proof Recall the constants R, M from Lemma 5.3.7. Since $M > 1$ we may pick some small $\kappa > 0$ along with some large $\widetilde{C} > 1 + \kappa$ such that

$$\widetilde{C}^{-1} + 1 < M \text{ and } (1+\kappa)^{-1} - \widetilde{C}^{-1} > M^{-1}. \tag{5.3.58}$$

Fix a point $x_0 \in \Omega$, introduce $\varepsilon_0 := \frac{1}{2} \min\{R, \operatorname{dist}(x_0, \partial\Omega)\}$, and take $\varepsilon \in (0, \varepsilon_0)$. Let $z \in \partial\Omega$ be arbitrary, select $y \in B(z, \varepsilon) \cap \Gamma_\kappa(z)$, and set $r := |y - z| \in (0, \varepsilon)$. Observe that the choices just made imply $x_0 \in \Omega \setminus B(z, 2r)$. Also, on the one hand, we have $\operatorname{dist}(y, \partial\Omega) > |y - z|/(1+\kappa) = r/(1+\kappa) > r$, hence $B(y, r) \subseteq \Omega$. On the other hand, since $\operatorname{dist}(y, \partial\Omega) \le |y - z| < \varepsilon$ and $r < \varepsilon$, it follows that $x_0 \notin B(y, r)$. Consequently, $G_\Omega(\cdot, x_0)$ is harmonic in a neighborhood of the closure of $B(y, r/\widetilde{C})$. As such, standard interior estimates give

$$|\nabla G_\Omega(y, x_0)| \le \frac{c}{r} \fint_{B(y, r/\widetilde{C})} G_\Omega(x, x_0)\, \mathrm{d}x. \tag{5.3.59}$$

Note that, thanks to (5.3.58), any $x \in B(y, r/\widetilde{C})$ satisfies

$$|x - z| \le |x - y| + |y - z| < r/\widetilde{C} + r < Mr \tag{5.3.60}$$

as well as

$$\begin{aligned} r/(1+\kappa) = |y - z|/(1+\kappa) &< \mathrm{dist}(y, \partial\Omega) \le \mathrm{dist}(x, \partial\Omega) + |x - y| \\ &\le \mathrm{dist}(x, \partial\Omega) + r/\widetilde{C}, \end{aligned} \tag{5.3.61}$$

hence $\mathrm{dist}(x, \partial\Omega) > \big[(1+\kappa)^{-1} - \widetilde{C}^{-1}\big] r > M^{-1} r$. Granted these, estimate (5.3.55) (used with $x := x_0$ and $A_r(z) := x$) yields

$$G_\Omega(x, x_0) < M r^{2-n} \omega^{x_0}\big(\Delta(z, r)\big) \;\text{ for every }\; x \in B(y, r/\widetilde{C}). \tag{5.3.62}$$

Using this back in (5.3.59) gives

$$|\nabla G_\Omega(y, x_0)| \le c M r^{1-n} \omega^{x_0}\big(\Delta(z, r)\big) \;\text{ for every }\; y \in B(z, \varepsilon) \cap \Gamma_\kappa(z). \tag{5.3.63}$$

Abbreviating $C := cM$ and eventually taking the supremum in y ultimately yields (5.3.57). □

5.4 The Poisson Kernel in NTA Domains

Defining the Poisson kernel for the domain Ω as the Radon-Nikodym derivative (5.3.38), as done earlier, requires having $\omega^{x_0} \ll \sigma$. The result below describes a scenario where this absolute continuity property is valid.

Proposition 5.4.1 *Assume $\Omega \subseteq \mathbb{R}^n$ is a bounded open set, satisfying a two sided corkscrew condition, and possessing an upper Ahlfors regular boundary. Abbreviate $\sigma := \mathcal{H}^{n-1} \lfloor \partial\Omega$ and, having fixed some $x_0 \in \Omega$, denote by ω^{x_0} the harmonic measure on $\partial\Omega$ with pole at x_0.*

Then ω^{x_o} and σ are mutually absolutely continuous. Hence, the Poisson kernel of the domain Ω, defined as the Radon-Nikodym derivative

$$k^{x_0} := \frac{\mathrm{d}\omega^{x_0}}{\mathrm{d}\sigma}, \tag{5.4.1}$$

exists, is non-negative, belongs to $L^1(\partial\Omega, \sigma)$, and satisfies $\int_{\partial\Omega} k^{x_0}\, \mathrm{d}\sigma = 1$.

Proof This is a consequence of [11, Theorem 1] in concert with [112, Proposition 5.9.16]. □

We note that [112, (5.10.24)] implies that if Ω is as in Proposition 5.4.1 then $\partial\Omega$ is actually a UR set (this observation comes in handy frequently in this section). It turns out that both the harmonic measure and the Poisson kernel exhibit finer properties under suitable connectivity assumptions. To elaborate on this aspect, recall that a quantitative, scale invariant, version of the mutual absolute continuity of the harmonic measure with respect to the surface measure is the membership condition[3] $\omega^{x_0} \in A_\infty(\sigma)$ (cf. [112, Definition 7.7.8]). This A_∞ property has been studied extensively and [10, Theorem 1.2] asserts (compare with [112, (5.11.6)]) that:

$$\text{if } \Omega \subseteq \mathbb{R}^n \text{ is an one-sided NTA domain with an Ahlfors regular boundary, then the harmonic measure is in } A_\infty \text{ with respect to the surface measure if and only if } \partial\Omega \text{ is a UR set, if and only if } \Omega \text{ satisfies an exterior corkscrew condition.} \tag{5.4.2}$$

In addition, J. Azzam has proved the following related result in [9]:

$$\text{in the class of open connected sets } \Omega \subseteq \mathbb{R}^n \text{ with an Ahlfors regular boundary, the harmonic measure is in } A_\infty \text{ with respect to the surface measure if and only if } \Omega \text{ is semi-uniform (cf. [112, (8.9.181)]) and } \partial\Omega \text{ is a UR set.} \tag{5.4.3}$$

As a consequence, we have the following result (originating in [160] and [32, Theorem 2, p. 842]).

Corollary 5.4.2 *Let $\Omega \subseteq \mathbb{R}^n$ be a bounded* NTA *domain with an upper Ahlfors regular boundary. Abbreviate $\sigma := \mathcal{H}^{n-1} \lfloor \partial\Omega$, fix some $x_0 \in \Omega$. Then the Poisson kernel for Ω satisfies*

$$k^{x_0} \in L^p(\partial\Omega, \sigma) \text{ for some } p > 1. \tag{5.4.4}$$

Indeed, there is no loss of generality in assuming that Ω is connected. If ω^{x_0} denotes the harmonic measure on $\partial\Omega$ with pole at x_0, then (5.4.3) (whose applicability is now ensured by [112, (5.10.24), (8.9.183)]) implies that

$$\omega^{x_0} \text{ belongs to a reverse Hölder class } B_p(\sigma) \text{ for some } p > 1 \tag{5.4.5}$$

(cf. [112, Proposition 7.7.9]). With this in hand, (5.4.4) follows.

Lemma 5.4.3 *Let $\Omega \subseteq \mathbb{R}^n$ be a bounded* NTA *domain with an upper Ahlfors regular boundary and abbreviate $\sigma := \mathcal{H}^{n-1} \lfloor \partial\Omega$. Also, denote by $G_\Omega(\cdot,\cdot)$ the Green function for the Laplacian in Ω, and fix an arbitrary point $x_0 \in \Omega$. Then there exist an aperture parameter $\kappa > 0$, a constant $C \in (0, \infty)$, and a truncation parameter $0 < \varepsilon_0 < \operatorname{dist}(x_0, \partial\Omega)$ such that for every $\varepsilon \in (0, \varepsilon_0)$ and every $z \in \partial\Omega$ one has*

$$\mathcal{N}_\kappa^\varepsilon\big(\nabla G_\Omega(\cdot, x_0)\big)(z) \le C \cdot \sup_{0<r<\varepsilon} \fint_{\Delta(z,r)} k^{x_0}\, d\sigma. \tag{5.4.6}$$

In particular,

[3] Throughout, the membership of ω^{x_0} to $A_\infty(\sigma)$ should be understood in a scale invariant fashion (see, e.g., the formulation in [11])

$$\mathcal{N}_\kappa^\varepsilon\big(\nabla G_\Omega(\cdot, x_0)\big) \leq C\mathcal{M}_{\partial\Omega} k^{x_0} \ \text{on } \partial\Omega, \tag{5.4.7}$$

where $\mathcal{M}_{\partial\Omega}$ denotes the Hardy-Littlewood maximal operator on $\partial\Omega$ (cf. (A.0.84)), *hence*

$$\mathcal{N}_\kappa^\varepsilon\big(\nabla G_\Omega(\cdot, x_0)\big) \in L^p(\partial\Omega, \sigma) \ \text{for some } p > 1. \tag{5.4.8}$$

Proof Since for each $z \in \partial\Omega$ and $r > 0$ we have

$$\begin{aligned} \frac{\omega^{x_0}\big(\Delta(z,r)\big)}{r^{n-1}} &= \frac{1}{r^{n-1}} \int_{\Delta(z,r)} \mathrm{d}\omega^{x_0} \\ &= \frac{1}{r^{n-1}} \int_{\Delta(z,r)} k^{x_0}\,\mathrm{d}\sigma \leq C \fint_{\Delta(z,r)} k^{x_0}\,\mathrm{d}\sigma, \end{aligned} \tag{5.4.9}$$

on account of the upper Ahlfors regularity of $\partial\Omega$, estimate (5.4.6) follows by combining (5.3.57) with (5.4.9). With (5.4.6) in hand, (5.4.8) then follows from (5.4.4) and the boundedness of the Hardy-Littlewood maximal operator on $\partial\Omega$ (cf. [112, Corollary 7.6.3]). □

Finally, we record an estimate for the truncated nontangential maximal operators of the Green function and its gradient, which is going to be very useful later on.

Lemma 5.4.4 *Consider a bounded* NTA *domain $\Omega \subseteq \mathbb{R}^n$ with an upper Ahlfors regular boundary, and abbreviate $\sigma := \mathcal{H}^{n-1}\lfloor\partial\Omega$. Also, let $G_\Omega(\cdot,\cdot)$ denote the Green function for the Laplacian in Ω, and fix an arbitrary point $x_0 \in \Omega$ along with an aperture parameter $\kappa > 0$. Then there exist a constant $\theta \in (0,1)$ and an aperture parameter $\widetilde{\kappa} > 0$ with the property that for each truncation parameter $\varepsilon \in \big(0, \operatorname{dist}(x_0, \partial\Omega)\big)$ one has*

$$\mathcal{N}_\kappa^{\theta\,\varepsilon}\big(G_\Omega(\cdot, x_0)\big)(z) \leq \varepsilon \cdot \mathcal{N}_{\widetilde{\kappa}}^\varepsilon\big(\nabla G_\Omega(\cdot, x_0)\big)(z) \ \text{for } \sigma\text{-a.e. } z \in \partial\Omega. \tag{5.4.10}$$

Proof This follows from [112, Proposition 8.9.17], bearing in mind the observation made in [112, (8.9.180)] as well as Theorem 5.3.2, or (5.3.11). □

5.5 Hardy Spaces of Harmonic Functions in NTA Domains

We say that a function $u \in \mathscr{C}^0(\Omega)$ is `nontangentially bounded from below at` $x \in \partial\Omega$ if there exists $\kappa > 0$, $\varepsilon > 0$, and $M \geq 0$, such that $u(y) \geq -M$ for every $y \in B(x,\varepsilon) \cap \Gamma_\kappa(x)$. Given $E \subseteq \partial\Omega$, call $u \in \mathscr{C}^0(\Omega)$ `nontangentially bounded from below` on E provided u is nontangentially bounded from below at every point in E. The following appears in [71, Theorem 6.4, p. 112].

Proposition 5.5.1 *Let $\Omega \subset \mathbb{R}^n$ be a bounded* NTA *domain and fix some point $x_0 \in \Omega$ along with a set $E \subseteq \partial\Omega$. If u is a harmonic function in Ω which is nontangentially bounded from below on E then for every $\kappa > 0$ it follows that*

$$u\Big|_{\partial\Omega}^{\kappa-\text{n.t.}} \text{ exists at } \omega^{x_0}\text{-a.e. point on } E. \tag{5.5.1}$$

A version of the Fatou-type result from Proposition 5.5.1, with the harmonic measure replaced by the "surface measure" $\sigma := \mathcal{H}^{n-1}\lfloor\partial\Omega$, is stated below.

Proposition 5.5.2 *Let $\Omega \subset \mathbb{R}^n$ be a bounded* NTA *domain with an upper Ahlfors regular boundary and abbreviate $\sigma := \mathcal{H}^{n-1}\lfloor\partial\Omega$. If u is a harmonic function in Ω which is nontangentially bounded from below on some set $E \subseteq \partial\Omega$ then for every $\kappa > 0$ it follows that*

$$u\Big|_{\partial\Omega}^{\kappa-\text{n.t.}} \text{ exists } \sigma\text{-a.e. on } E, \text{ and is independent of } \kappa. \tag{5.5.2}$$

In particular, (5.5.2) *holds for any $\kappa > 0$, any $E \subseteq \partial\Omega$, and any harmonic function u in Ω which, for some $\varepsilon > 0$, satisfies*

$$\big(\mathcal{N}_\kappa^{\varepsilon} u\big)(x) < +\infty \text{ at } \sigma\text{-a.e. point } x \in E. \tag{5.5.3}$$

Proof The existence part in (5.5.2) is a consequence of Proposition 5.5.1 and Proposition 5.4.1. The independence on the parameter κ then follows from this and [112, (8.1.20)]. □

Given an open set $\Omega \subseteq \mathbb{R}^n$, for each $\kappa > 0$ and $p \in (0, \infty]$ consider the space of harmonic functions

$$H_\kappa^p(\Omega) := \big\{u \in \mathscr{C}^\infty(\Omega) : \Delta u = 0 \text{ in } \Omega, \text{ and } \mathcal{N}_\kappa u \in L^p(\partial\Omega, \sigma)\big\} \tag{5.5.4}$$

where $\sigma := \mathcal{H}^{n-1}\lfloor\partial\Omega$. In terms of this, we have the following global, quantitative Fatou-type result. Before stating it, the reader is alerted to the fact that the space $N_\kappa^p(\Omega; \sigma)$ is defined as in (A.0.98) (corresponding to $\mu := \sigma$).

Proposition 5.5.3 *Let $\Omega \subset \mathbb{R}^n$ be a bounded* NTA *domain with an upper Ahlfors regular boundary and abbreviate $\sigma := \mathcal{H}^{n-1}\lfloor\partial\Omega$. Then for each aperture parameter $\kappa > 0$ and each integrability exponent $p \in (0, \infty]$,*

$$H_\kappa^p(\Omega) \text{ is a closed subspace of } N_\kappa^p(\Omega; \sigma) \tag{5.5.5}$$

and, when considering $H_\kappa^p(\Omega)$ equipped with the quasi-norm from the larger ambient $N_\kappa^p(\Omega; \sigma)$ (cf. (A.0.98)*), the nontangential boundary trace induces a well-defined, linear, and continuous operator in the context*

$$H_\kappa^p(\Omega) \ni u \longmapsto u\Big|_{\partial\Omega}^{\kappa-\text{n.t.}} \in L^p(\partial\Omega, \sigma). \tag{5.5.6}$$

Moreover, both the space $H_\kappa^p(\Omega)$ and the above nontangential trace operator are actually independent of κ.

Proof First, the claim in (5.5.5) is seen by expressing

$$H_\kappa^p(\Omega) = \big\{u \in N_\kappa^p(\Omega; \sigma) : \Delta u = 0 \text{ in } \Omega\big\}, \tag{5.5.7}$$

then invoking [112, Proposition 8.3.5] (bearing in mind that harmonicity is preserved under uniform convergence on compact sets). Next, that the operator (5.5.6) is well defined, linear, and continuous follows from (5.5.4), Proposition 5.5.2, [112, (8.9.44)], [112, (8.9.8)], [112, (8.8.52)], and [112, (5.2.4)]. Lastly, the independence on the parameter κ of the objects involved in (5.5.6) may be justified with the help of [112, Proposition 8.9.8]. □

5.6 Boundary Behavior of the Green Function in NTA Domains

The results on the nontangential boundary trace of the derivatives of the Green function recorded in (5.6.1)-(5.6.2) below extend work in [25, Theorem 3, p. 275] dealing with the case of bounded Lipschitz domains, and also refine [81, Lemma 3.1, p. 336] where δ-Reifenberg flat chord-arc domains with $\delta > 0$ sufficiently small have been considered.

Lemma 5.6.1 *Assume $\Omega \subseteq \mathbb{R}^n$ is a bounded* NTA *domain with an upper Ahlfors regular boundary. Abbreviate $\sigma := \mathcal{H}^{n-1}\lfloor\partial\Omega$ and denote by $G_\Omega(\cdot,\cdot)$ the Green function for the Laplacian in Ω. Also, fix a point $x_0 \in \Omega$ and an aperture parameter $\kappa > 0$. Then the nontangential boundary trace*

$$\begin{aligned}&\big[\nabla G_\Omega(\cdot,x_0)\big]\Big|_{\partial\Omega}^{\kappa-\mathrm{n.t.}} \text{ exists (in } \mathbb{R}^n) \text{ at } \sigma\text{-a.e. point on } \partial\Omega\\ &\text{and is actually independent of the parameter } \kappa > 0.\end{aligned} \tag{5.6.1}$$

In particular, if ν denotes the geometric measure theoretic outward unit normal to Ω, then the normal derivative of $G_\Omega(\cdot,x_0)$ may be meaningfully defined as

$$\partial_\nu G_\Omega(\cdot,x_0) := \Big\langle \nu, \big(\nabla G_\Omega(\cdot,x_0)\big)\Big|_{\partial\Omega}^{\kappa-\mathrm{n.t.}}\Big\rangle \text{ at } \sigma\text{-a.e. point on } \partial\Omega, \tag{5.6.2}$$

and this definition is independent of the aperture parameter κ.

We emphasize that the same result holds for unbounded NTA domains, and also for the Green function with pole at infinity.

Proof of Lemma 5.6.1 This is a consequence of Theorem 5.3.5, bearing in mind (5.4.8) and the fact that

$$\begin{aligned}&\text{any bounded NTA domain with an upper}\\ &\text{Ahlfors regular boundary is a UR domain,}\end{aligned} \tag{5.6.3}$$

as seen from [112, (5.11.4)], [112, (5.10.24)], and the definition of a UR domain from [112, Definition 5.10.6]. □

In [49, Question 2, p. 49] Garnett and Marshall raise the issue of determining minimal regularity conditions on an given open set ensuring that the Poisson kernel may be expressed as the normal derivative of the Green function. Our next theorem shows that merely being an NTA domain with upper Ahlfors regular boundary will do. In particular, this sharpens [16, Theorem 5.7, p. 79], [81, Proposition A.1.1, p. 382], [81, Proposition A.2.2, p. 385], and [82, Lemma 3.4, p. 16], where Kenig and Toro have considered δ-Reifenberg flat chord-arc domains in $\mathbb{R}^n$ and/or imposed the condition that the Poisson kernel is locally $\mathcal{H}^{n-1}$-square-integrable. Our result dispenses with these additional assumptions; for example, Theorem 5.6.2 applies to arbitrary BMO_1 domains in $\mathbb{R}^n$.

Theorem 5.6.2 *Assume $\Omega \subseteq \mathbb{R}^n$ is a bounded* NTA *domain with an upper Ahlfors regular boundary. Let ν denote the geometric measure theoretic outward unit normal to Ω, and abbreviate $\sigma := \mathcal{H}^{n-1}\lfloor\partial\Omega$. Also, denote by $G_\Omega(\cdot,\cdot)$ the Green function for the Laplacian in Ω, and fix an arbitrary point $x_0 \in \Omega$. Then, with the normal derivative of $G_\Omega(\cdot, x_0)$ defined as in (5.6.2), one has*

$$-\partial_\nu G_\Omega(\cdot, x_0) = k^{x_0} := \frac{d\omega^{x_0}}{d\sigma} \quad \textit{at } \sigma\textit{-a.e. point on } \partial\Omega. \tag{5.6.4}$$

We wish to stress that the same result holds for unbounded NTA domains, in which case we may also allow the Green function and the Poisson kernel to have poles at infinity (in the sense of [80, Lemma 3.7, p. 390]). We shall give Theorem 5.6.2 two proofs, the first of which casts this result as a direct corollary of Theorem 5.3.6 and Lemma 5.4.3.

First Proof of Theorem 5.6.2 From (5.2.29), [112, (8.9.180)], [112, (5.11.4)], and [112, (5.2.4)] we see that

any bounded NTA domain Ω with an upper Ahlfors regular boundary is regular for the Dirichlet Problem for the Laplacian (in the sense of Definition 5.2.3), is locally pathwise nontangentially accessible (in the sense of [112, Definition 8.9.14]), has a lower Ahlfors regular boundary (hence σ is a doubling measure on $\partial\Omega$), and satisfies $\partial_*\Omega = \partial\Omega$. (5.6.5)

Granted this, and keeping in mind (5.6.3) as well as (5.4.8), we may invoke Theorem 5.3.6 to conclude that (5.6.4) holds. □

The second proof of Theorem 5.6.2 relies on the version of the Divergence Formula recorded in [112, Theorem 1.2.1] in rather transparent fashion.

Second Proof of Theorem 5.6.2 Bearing in mind that any bounded NTA domain is a bounded open set which is regular for the Dirichlet Problem for the Laplacian and whose complement is n-thick, from (5.3.14) and (5.3.13) we see that

$$\begin{aligned} G_\Omega(x, x_0) &= H(x, x_0) - E_\Delta(x - x_0) = \int_{\partial\Omega} E_\Delta(x_0 - z)\, d\omega^x(z) - E_\Delta(x - x_0) \\ &= \int_{\partial\Omega} E_\Delta(x - z)\, d\omega^{x_0}(z) - E_\Delta(x - x_0), \qquad \forall x \in \Omega \setminus \{x_0\}, \end{aligned} \tag{5.6.6}$$

where E_Δ is the standard fundamental solution for the Laplacian in $\mathbb{R}^n$ defined in (1.5.56). Fix now $\varphi \in \mathscr{C}_c^\infty(\mathbb{R}^n)$ arbitrary and define

$$u(z) := \varphi(z) - \int_\Omega E_\Delta(x - z)(\Delta\varphi)(x)\,\mathrm{d}x, \qquad \forall z \in \overline{\Omega}. \tag{5.6.7}$$

Then, by design, $u \in \mathscr{C}^0(\overline{\Omega})$ and $\Delta u = 0$ in Ω. Consequently,

$$\begin{aligned}\varphi(x_0) - \int_\Omega E_\Delta(x - x_0)(\Delta\varphi)(x)\,\mathrm{d}x &= u(x_0) = \int_{\partial\Omega} u(z)\,\mathrm{d}\omega^{x_0}(z)\\ &= \int_{\partial\Omega} \varphi(z)\,\mathrm{d}\omega^{x_0}(z) - \int_{\partial\Omega}\Big(\int_\Omega E_\Delta(x - z)(\Delta\varphi)(x)\,\mathrm{d}x\Big)\,\mathrm{d}\omega^{x_0}(z)\\ &= \int_{\partial\Omega} \varphi\,\mathrm{d}\omega^{x_0} - \int_\Omega G_\Omega(x, x_0)(\Delta\varphi)(x)\,\mathrm{d}x - \int_\Omega E_\Delta(x - x_0)(\Delta\varphi)(x)\,\mathrm{d}x.\end{aligned} \tag{5.6.8}$$

Ultimately this goes to show that for each fixed point $x_0 \in \Omega$ the Green function satisfies

$$\int_\Omega G_\Omega(x, x_0)(\Delta\varphi)(x)\,\mathrm{d}x = \int_{\partial\Omega} \varphi\,\mathrm{d}\omega^{x_0} - \varphi(x_0), \qquad \forall \varphi \in \mathscr{C}_c^\infty(\mathbb{R}^n). \tag{5.6.9}$$

To proceed, pick now an arbitrary function $\varphi \in \mathscr{C}_c^\infty(\mathbb{R}^n \setminus \{x_0\})$ and consider the vector field

$$\vec{F} := G_\Omega(\cdot, x_0)\nabla\varphi - \varphi\nabla G_\Omega(\cdot, x_0) \text{ in } \Omega. \tag{5.6.10}$$

Then, thanks to (5.3.4) and the fact that φ vanishes near x_0, we have $\vec{F} \in \big[\mathscr{C}^\infty(\Omega)\big]^n$ and

$$\operatorname{div}\vec{F} = G_\Omega(\cdot, x_0)\Delta\varphi \in L^1(\Omega, \mathcal{L}^n), \tag{5.6.11}$$

where the membership is a consequence of the estimates for the Green function recorded in Theorem 5.3.1. Also, from Theorem 5.3.2 (or (5.3.11)) and Lemma 5.6.1 we see that, for each $\kappa > 0$ fixed,

$$\vec{F}\Big|_{\partial\Omega}^{\kappa-\mathrm{n.t.}} \text{ exists (in } \mathbb{R}^n\text{) at } \sigma\text{-a.e. point on } \partial\Omega \tag{5.6.12}$$

and, in fact,

$$\vec{F}\Big|_{\partial\Omega}^{\kappa-\mathrm{n.t.}} = -\varphi\big[\nabla G_\Omega(\cdot, x_0)\big]\Big|_{\partial\Omega}^{\kappa-\mathrm{n.t.}} \text{ at } \sigma\text{-a.e. point on } \partial\Omega. \tag{5.6.13}$$

In particular,

$$\nu \cdot \Big(\vec{F}\Big|_{\partial\Omega}^{\kappa-\mathrm{n.t.}}\Big) = -\varphi\,\partial_\nu G_\Omega(\cdot, x_0) \text{ at } \sigma\text{-a.e. point on } \partial\Omega. \tag{5.6.14}$$

Granted these properties, [112, Theorem 1.2.1] applies and gives that, on the one hand,

$$\int_\Omega G_\Omega(x,x_0)(\Delta\varphi)(x)\,dx = \int_\Omega \operatorname{div}\vec{F}\,d\mathcal{L}^n = \int_{\partial\Omega} \nu\cdot\Big(\vec{F}\Big|_{\partial\Omega}^{\kappa-\mathrm{n.t.}}\Big)\,d\sigma$$
$$= -\int_{\partial\Omega} \varphi\,\partial_\nu G_\Omega(\cdot,x_0)\,d\mathcal{H}^{n-1}. \tag{5.6.15}$$

On the other hand, from (5.6.9), the current choice of φ, and Proposition 5.4.1 we conclude that

$$\int_\Omega G_\Omega(x,x_0)(\Delta\varphi)(x)\,dx = \int_{\partial\Omega} \varphi\,d\omega^{x_0} = \int_{\partial\Omega} k^{x_0}\varphi\,d\mathcal{H}^{n-1}. \tag{5.6.16}$$

Comparing (5.6.16) with (5.6.15) then yields (5.6.4) on account of [112, Corollary 3.7.3]. □

The following theorem generalizes results from [81, Lemmas 3.1-3.2, pp. 336-337] [81, §A.2, pp. 384-389] (where the authors attribute them to G. David) and [82, Lemma 3.2, p. 10] which, compared with Theorem 5.6.3 below, have been established under certain additional flatness assumptions on the underlying domain and/or assuming local $\mathcal{H}^{n-1}$-square-integrability for the associated Poisson kernel. Here we eliminate these assumptions; in particular, Theorem 5.6.3 applies to arbitrary BMO_1 domains in $\mathbb{R}^n$.

Theorem 5.6.3 *Assume $\Omega \subseteq \mathbb{R}^n$ is a bounded* NTA *domain with an upper Ahlfors regular boundary; in particular, Ω is a set of locally finite perimeter. Denote by $\nu = (\nu_1,\dots,\nu_n)$ its geometric measure theoretic outward unit normal and abbreviate $\sigma := \mathcal{H}^{n-1}\lfloor\partial\Omega$. Also, denote by $G_\Omega(\cdot,\cdot)$ the Green function for the Laplacian in Ω, and fix an arbitrary point $x_0 \in \Omega$. Then (with the dependence on the aperture parameter dropped) for each $j,k \in \{1,\dots,n\}$ one has*

$$\nu_j\Big(\big(\partial_k G_\Omega(\cdot,x_0)\big)\Big|_{\partial\Omega}^{\mathrm{n.t.}}\Big) = \nu_k\Big(\big(\partial_j G_\Omega(\cdot,x_0)\big)\Big|_{\partial\Omega}^{\mathrm{n.t.}}\Big) \text{ at } \sigma\text{-a.e. point on } \partial\Omega. \tag{5.6.17}$$

In fact, with the Poisson kernel k^{x_0} defined as in (5.4.1), *one has*

$$\big(\nabla G_\Omega(\cdot,x_0)\big)\Big|_{\partial\Omega}^{\mathrm{n.t.}} = -k^{x_0}\nu = -\Big(\frac{d\omega^{x_0}}{d\sigma}\Big)\nu \text{ at } \sigma\text{-a.e. point on } \partial\Omega. \tag{5.6.18}$$

In particular, for any vector field $\vec{T} : \partial\Omega \to \mathbb{C}^n$ which is tangential, in the sense that $\langle \nu, \vec{T}\rangle = 0$ at σ-a.e. point on $\partial\Omega$, one has

$$\Big\langle \vec{T}, \big(\nabla G_\Omega(\cdot,x_0)\big)\Big|_{\partial\Omega}^{\mathrm{n.t.}}\Big\rangle = 0 \text{ at } \sigma\text{-a.e. point on } \partial\Omega. \tag{5.6.19}$$

We emphasize that the same results are valid for unbounded NTA domains, in which scenario one may also allow the Green function and the Poisson kernel to have poles at infinity (cf. [80, Lemma 3.7, p. 390]).

Proof of Theorem 5.6.3 This follows from Theorem 5.3.5, bearing in mind (5.4.8) and (5.6.3), as well as Theorem 5.6.2. □

5.7 The L^p Dirichlet Problem for the Laplacian and Integral Representations

The following result appears as [71, Lemma 4.11, p. 101].

Lemma 5.7.1 *If Ω is a bounded* NTA *domain in $\mathbb{R}^n$, then there exists $R > 0$ with the following significance. Assume $x_0, x_1 \in \partial\Omega$ and $0 < r_0, r_1 < R$ are such that $\Delta(x_1, r_1) \subset \Delta(x_0, r_0/2)$. Then, with $A_r(x_0)$ denoting a corkscrew point for x_0 at scale r, one has*

$$\omega^{A_r(x_0)}\big(\Delta(x_1, r_1)\big) \approx \frac{\omega^x(\Delta(x_1, r_1))}{\omega^x(\Delta(x_0, r_0))}, \tag{5.7.1}$$

uniformly for $x \in \Omega \setminus B(x_0, 2r_0)$.

From item *(iii)* in Lemma 5.2.1 it follows that if x_1, x_2 are two arbitrary points in Ω, then the measures ω^{x_1} and ω^{x_2} are mutually absolutely continuous. As such, the Radon-Nikodym derivative $d\omega^{x_1}/d\omega^{x_2}$ exists and, hence, the following definition is meaningful.

Definition 5.7.2 *Let Ω be a bounded* NTA *domain in $\mathbb{R}^n$ and fix $x, x_0 \in \Omega$. Then the kernel function associated with Ω is defined as the Radon-Nikodym derivative*

$$K_{x_0}(x, \cdot) := \frac{d\omega^x}{d\omega^{x_0}} \quad \textit{on } \partial\Omega. \tag{5.7.2}$$

In particular, if $x_0 \in \Omega$ is fixed, then for every $x \in \Omega$ one has

$$K_{x_0}(x, y) = \lim_{r \searrow 0} \frac{\omega^x\big(\Delta(y, r)\big)}{\omega^{x_0}\big(\Delta(y, r)\big)} \quad \text{for } \omega^{x_0}\text{-a.e. } y \in \partial\Omega. \tag{5.7.3}$$

Even though, a priori, $K_{x_0}(x, y)$ is only defined for ω^{x_0}-a.e. $y \in \partial\Omega$ according to [71, Theorem 5.5, p. 104] it turns out that

if $\Omega \subseteq \mathbb{R}^n$ is a bounded NTA domain, then $K_{x_0}(x, y)$ exists for all $y \in \partial\Omega$, is a harmonic function in $x \in \Omega$, and a continuous function in the variable $y \in \partial\Omega$. (5.7.4)

In fact, [71, Corollary 7.1, p. 115] implies that

if $\Omega \subseteq \mathbb{R}^n$ is a bounded NTA domain, then there exists $\alpha > 0$ which depends only on the NTA constants of Ω such that $K_{x_0}(x, \cdot) \in \mathscr{C}^\alpha(\partial\Omega)$ for each fixed point $x \in \Omega$. (5.7.5)

In particular, from (5.7.5), or (5.2.15), we see that

if $\Omega \subseteq \mathbb{R}^n$ is a bounded NTA domain, then the space $L^1(\partial\Omega, \omega^{x_0})$ is independent of the choice of $x_0 \in \Omega$. (5.7.6)

We continue by recording several other properties of the kernel function in NTA domains which are due to D. Jerison and C. Kenig ([71], [79]).

Lemma 5.7.3 *Let Ω be a bounded* NTA *domain in $\mathbb{R}^n$, and fix $x_0 \in \Omega$. Then there exists $r_0 > 0$ such that*

$$K_{x_0}(A_r(z), y) \approx \frac{1}{\omega^{x_0}(\Delta(z,r))}, \tag{5.7.7}$$

uniformly for $z \in \partial\Omega \setminus B(x_0, 2r)$, $y \in \Delta(z,r)$, and $0 < r < r_0$, where $A_r(z)$ denotes a corkscrew point for z at scale r.

Indeed, estimate (5.7.7) is a consequence of (5.7.3) and Lemma 5.7.1.

We next recall [71, Theorem 5.8, p. 105] which provides a way of extending boundary functions (which are absolutely integrable with respect to the harmonic measure) harmonically inside of a given an NTA domain. Before stating this, we recall that, given a bounded NTA domain $\Omega \subseteq \mathbb{R}^n$ and some fixed point $x_0 \in \Omega$, the Hardy-Littlewood maximal operator $\mathcal{M}_{\omega^{x_0}}$ associated with the harmonic measure ω^{x_0} on $\partial\Omega$ is (cf. [112, (7.7.87)])

$$\big(\mathcal{M}_{\omega^{x_0}} f\big)(x) := \sup_{r>0}\Big(\fint_{\Delta(x,r)} |f|\,\mathrm{d}\omega^{x_0}\Big), \qquad \forall x \in \partial\Omega, \tag{5.7.8}$$

for each $f \in L^1(\partial\Omega, \omega^{x_0})$.

Theorem 5.7.4 *Suppose Ω is a bounded* NTA *domain in $\mathbb{R}^n$, and select $\kappa > 0$ sufficiently large (depending on the interior corkscrew constants of Ω). Also, fix a point $x_0 \in \Omega$ and recall the kernel function $K_{x_0}(\cdot,\cdot)$ from (5.7.2). In this context, if for $f \in L^1(\partial\Omega, \omega^{x_0})$ one defines*

$$u(x) := \int_{\partial\Omega} f\,\mathrm{d}\omega^x = \int_{\partial\Omega} K_{x_0}(x,y) f(y)\,\mathrm{d}\omega^{x_0}(y), \qquad \forall x \in \Omega, \tag{5.7.9}$$

then u is a well-defined function, which is harmonic in Ω and satisfies

$$u\big|_{\partial\Omega}^{\kappa-\mathrm{n.t.}} = f \text{ at } \omega^{x_0}\text{-a.e. point on } \partial\Omega, \tag{5.7.10}$$

and

$$\begin{gathered}\text{(the truncated version of) } \mathcal{N}_\kappa u \text{ is comparable} \\ \text{with (the local version of) } \mathcal{M}_{\omega^{x_0}} f.\end{gathered} \tag{5.7.11}$$

This is essentially [71, Theorem 5.8, p. 105]. Strictly speaking, the theorem just cited only contains the left-pointing inequality in (5.7.11), but the opposite one is established much as in the proof of [79, Lemma 1.4.2, p. 14]. To give the main step in this argument, let $f \in L^1(\partial\Omega, \omega^{x_0})$, $f \geq 0$, and let r_0 be as in Lemma 5.7.3. Then for each $z \in \partial\Omega$, $0 < r < \min\big\{r_0, \frac{1}{2}\operatorname{dist}(x_0, \partial\Omega)\big\}$, and some point x belonging to the set $[B(z,2r) \setminus B(z,r)] \cap \Gamma_\kappa(z)$ (which, thanks to the interior corkscrew condition, is nonempty if the aperture κ is large enough), we may write

$$\fint_{\Delta(z,r)} f(y)\,\mathrm{d}\omega^{x_0}(y) \le C \int_{\partial\Omega} K_{x_0}(x,y) f(y)\,\mathrm{d}\omega^{x_0}(y)$$

$$= C \int_{\partial\Omega} f(y)\,\mathrm{d}\omega^{x}(y) = Cu(x) \le C(\mathcal{N}_\kappa u)(z), \tag{5.7.12}$$

by (5.7.7). A slight variant of this estimate then yields the desired conclusion.

Our next result deals with the issue of uniqueness for the Dirichlet Problem with the boundary trace considered in the nontangential sense and with nontangential maximal function control.

Theorem 5.7.5 *Let $\Omega \subset \mathbb{R}^n$ be a bounded* NTA *domain with an upper Ahlfors regular boundary and abbreviate $\sigma := \mathcal{H}^{n-1}\lfloor\partial\Omega$. Denote by ω^{x_0} the harmonic measure on $\partial\Omega$ with pole at the point $x_0 \in \Omega$. Also, select two integrability exponents $p, p' \in (1,\infty)$ satisfying $1/p + 1/p' = 1$, and fix an aperture parameter $\kappa > 0$.*

Then, if the Poisson kernel $k^{x_0} = \frac{\mathrm{d}\omega^{x_0}}{\mathrm{d}\sigma}$ belongs to $L^{p'}(\partial\Omega,\sigma)$, it follows that

$$\left.\begin{array}{r} u \in \mathscr{C}^\infty(\Omega), \quad \Delta u = 0 \ \text{in } \Omega \\ u\big|_{\partial\Omega}^{\kappa-\text{n.t.}} = 0 \ \text{at } \sigma\text{-a.e. point on } \partial\Omega \\ \text{and } \mathcal{N}_\kappa u \in L^p(\partial\Omega,\sigma) \end{array}\right\} \Longrightarrow u \equiv 0 \ \text{in } \Omega. \tag{5.7.13}$$

Proof Denote by $G_\Omega(\cdot,\cdot)$ the Green function associated with the Laplacian in Ω, and recall that we are assuming

$$k^{x_0} \in L^{p'}(\partial\Omega,\sigma). \tag{5.7.14}$$

Then Lemma 5.4.3, Lemma 5.4.4, [112, Proposition 8.4.1], and [112, Corollary 7.6.3] guarantee the existence of a constant $C \in (0,\infty)$ with the property that for every $\varepsilon > 0$ sufficiently small we have

$$\Big\|\mathcal{N}_\kappa^\varepsilon\big(\nabla G_\Omega(\cdot,x_0)\big)\Big\|_{L^{p'}(\partial\Omega,\sigma)} \le C\big\|k^{x_0}\big\|_{L^{p'}(\partial\Omega,\sigma)}. \tag{5.7.15}$$

In particular, for $\varepsilon > 0$ small, from (5.7.15), (5.4.10), and [112, (8.2.26)] we conclude that

$$\mathcal{N}_\kappa^\varepsilon G_\Omega(\cdot,x_0),\ \mathcal{N}_\kappa^\varepsilon\big(\nabla G_\Omega(\cdot,x_0)\big) \in L^{p'}(\partial\Omega,\sigma). \tag{5.7.16}$$

Thanks to (5.7.16), [112, Proposition 5.9.16], and [112, (5.2.4), (8.8.52), (8.9.180)], all hypotheses of Theorem 4.1.1 are satisfied for $L := \Delta$. As such, (4.1.2) presently implies that whenever u is as in the left-hand side of (5.7.13) we necessarily have $u(x_0) = 0$. Since for any other point $x_1 \in \Omega$ we have

$$k^{x_1} = K_{x_0}(x_1,\cdot)k^{x_0} \in L^{p'}(\partial\Omega,\sigma), \tag{5.7.17}$$

by virtue of (5.7.2) and (5.7.4) (which implies that $K_{x_0}(x_1,\cdot)$ is continuous and bounded on $\partial\Omega$), the same type of argument as above with x_1 playing the role of

x_0 yields $u(x_1) = 0$. Given that the point $x_1 \in \Omega$ is arbitrary, this establishes the implication in (5.7.13), finishing the proof of the theorem. □

The next theorem in this section refines and expands the scope of [79, Theorem 1.7.3, p. 29]. Its proof relies essentially on results due to G. David, D. Jerison, and C. Kenig (cf. [32], [71]). Among other things, this result builds a bridge between the well-posedness of the L^p-Dirichlet Problem (for the Laplacian) and the availability of L^p-nontangential maximal function estimates for the solution of the classical Dirichlet Problem. Before stating it, the reader is reminded that the space $H^p_\kappa(\Omega)$ has been defined in (5.5.4).

Theorem 5.7.6 *Let $\Omega \subset \mathbb{R}^n$ be a bounded* NTA *domain with the property that $\partial\Omega$ is an upper Ahlfors regular set, and abbreviate $\sigma := \mathcal{H}^{n-1}\lfloor\partial\Omega$. Denote by ω^{x_0} the harmonic measure with pole at the point $x_0 \in \Omega$. Also, fix some $\kappa > 0$ along with $p, p' \in (1, \infty)$ such that $\frac{1}{p} + \frac{1}{p'} = 1$. Then the following statements are equivalent:*

(i) There exists a constant $C \in (0, \infty)$ such that whenever $f \in \mathscr{C}^0(\partial\Omega)$ and u is the unique solution of the classical Dirichlet Problem

$$\begin{cases} u \in \mathscr{C}^0(\overline{\Omega}) \cap \mathscr{C}^\infty(\Omega), \\ \Delta u = 0 \ \text{ in } \Omega, \\ u\big|_{\partial\Omega} = f \ \text{ on } \partial\Omega, \end{cases} \tag{5.7.18}$$

one has

$$\|\mathcal{N}_\kappa u\|_{L^p(\partial\Omega,\sigma)} \leq C\|f\|_{L^p(\partial\Omega,\sigma)}. \tag{5.7.19}$$

In other words, there exists $C \in (0, \infty)$ with the property that

$$\big\|\mathcal{N}_\kappa(\mathrm{PI}\, f)\big\|_{L^p(\partial\Omega,\sigma)} \leq C\|f\|_{L^p(\partial\Omega,\sigma)}, \qquad \forall f \in \mathscr{C}^0(\partial\Omega). \tag{5.7.20}$$

(ii) One has

$$\omega^{x_0} \in B_{p'}(\sigma), \tag{5.7.21}$$

in the sense that for each scale $r \in \big(0, \operatorname{diam}\partial\Omega\big)$, each location $x \in \partial\Omega$, and each pole $x_0 \in B(x, r) \cap \Omega$ with $\operatorname{dist}(x_0, \partial\Omega) \approx r$, it follows that the harmonic measure ω^{x_0} is absolutely continuous with respect to the surface measure σ, the Poisson kernel $k^{x_0} := \frac{d\omega^{x_0}}{d\sigma}$ belongs to $L^{p'}(\partial\Omega, \sigma)$, and there exists some $C \in (0, \infty)$ such that the following reverse Hölder condition is satisfied:

$$\Big(⨍_{\Delta(x,r)} (k^{x_0})^{p'}\, d\sigma\Big)^{\frac{1}{p'}} \leq C ⨍_{\Delta(x,r)} k^{x_0}\, d\sigma. \tag{5.7.22}$$

(iii) For every given $f \in L^p(\partial\Omega, \sigma)$ there exists a unique function $u \in H^p_\kappa(\Omega)$ with the property that $u\big|^{\kappa-\mathrm{n.t.}}_{\partial\Omega} = f$ at σ-a.e. point on $\partial\Omega$. In other words,

$$\text{the nontangential boundary trace operator is an isomorphism in the context of (5.5.6).} \tag{5.7.23}$$

(iv) One has

$$\|\mathcal{N}_\kappa u\|_{L^p(\partial\Omega,\sigma)} \approx \Big\|u\big|_{\partial\Omega}^{\kappa-\mathrm{n.t.}}\Big\|_{L^p(\partial\Omega,\sigma)}, \text{ uniformly for } u \in H^p_\kappa(\Omega). \tag{5.7.24}$$

From Proposition 5.2.4 and (5.2.29) we see that the classical Dirichlet Problem (cf. (5.7.18)) is uniquely solvable in the set Ω, so the statement of Theorem 5.7.6 is meaningful. Before presenting its proof, we wish to comment on terminology. Specifically, at least two labels have been in use for the property described in item *(i)*. Originally, the validity of this property is indicated in [79, Definition 1.7.4] by simply saying that "$(D)_p$ *holds (for the Laplacian)*" but, subsequently, the phrase "(D_p) *is solvable*" has become prevalent in the literature. We find the latter practice unfortunate since it redefines the meaning of the word "solvable" in a rather unconventional (and somewhat misleading) manner.

Here is the proof of Theorem 5.7.6.

Proof of Theorem 5.7.6 To show that *(i)* implies *(ii)*, fix a scale $r \in \big(0, \operatorname{diam}\partial\Omega\big)$ along with a location $x \in \partial\Omega$ and some point $x_0 \in B(x,r) \cap \Omega$ satisfying $\operatorname{dist}(x_0, \partial\Omega) \approx r$. From Proposition 5.4.1 we know that ω^{x_o} and σ are mutually absolutely continuous. We shall estimate

$$\Big(⨍_{\Delta(x,r)} (k^{x_0})^{p'}\, d\sigma\Big)^{\frac{1}{p'}} = \sigma\big(\Delta(x,r)\big)^{-\frac{1}{p'}} \|k^{x_0}\|_{L^{p'}(\Delta(x,r),\sigma)} \tag{5.7.25}$$

via duality. Specifically, pick an arbitrary non-negative function $f \in \mathscr{C}^0(\partial\Omega)$ with $\operatorname{supp} f \subseteq \Delta(x,r)$ and write

$$\int_{\Delta(x,r)} f k^{x_0}\, d\sigma = \int_{\partial\Omega} f\, d\omega^{x_0} = u(x_0), \tag{5.7.26}$$

where u is the unique solution of the classical Dirichlet Problem (5.7.18) for the boundary datum f. Since there exists some large aperture parameter $\widetilde{\kappa} > 0$ and some small $\varepsilon > 0$ such that $x_0 \in \Gamma_{\widetilde{\kappa}}(z)$ for each $z \in \Delta(x, \varepsilon r)$, we conclude from (5.7.26) that

$$\int_{\Delta(x,r)} f k^{x_0}\, d\sigma \leq (\mathcal{N}_{\widetilde{\kappa}} u)(z) \text{ for each } z \in \Delta(x, \varepsilon r). \tag{5.7.27}$$

Taking integral averages, then using Hölder's inequality, [112, Proposition 8.4.1], the fact that σ is doubling, and (5.7.19), we therefore obtain

$$\begin{aligned}\int_{\Delta(x,r)} f k^{x_0}\,d\sigma &\le \fint_{\Delta(x,\varepsilon r)} \mathcal{N}_{\widetilde{\kappa}} u\,d\sigma \le \Big(\fint_{\Delta(x,\varepsilon r)} (\mathcal{N}_{\widetilde{\kappa}} u)^p\,d\sigma\Big)^{\frac{1}{p}}\\ &\le \sigma\big(\Delta(x,\varepsilon r)\big)^{-\frac{1}{p}} \|\mathcal{N}_{\widetilde{\kappa}} u\|_{L^p(\partial\Omega,\sigma)}\\ &\le C\sigma\big(\Delta(x,r)\big)^{-\frac{1}{p}} \|\mathcal{N}_{\kappa} u\|_{L^p(\partial\Omega,\sigma)}\\ &\le C\sigma\big(\Delta(x,r)\big)^{-\frac{1}{p}} \|f\|_{L^p(\partial\Omega,\sigma)}\\ &= C\sigma\big(\Delta(x,r)\big)^{-\frac{1}{p}} \|f\|_{L^p(\Delta(x,r),\sigma)}. \end{aligned} \tag{5.7.28}$$

This ultimately proves that

$$\sigma\big(\Delta(x,r)\big)^{-\frac{1}{p'}} \|k^{x_0}\|_{L^{p'}(\Delta(x,r),\sigma)} \le C\sigma\big(\Delta(x,r)\big)^{-1}. \tag{5.7.29}$$

Since the right-hand side in (5.7.22) is

$$C\fint_{\Delta(x,r)} k^{x_0}\,d\sigma = C\frac{\omega^{x_0}\big(\Delta(x,r)\big)}{\sigma\big(\Delta(x,r)\big)}, \tag{5.7.30}$$

the desired conclusions now follow from (5.7.25), (5.7.29), (5.7.30), and Bourgain's Lemma.

Let us now show that *(ii)* implies *(i)*. Given a function $f \in \mathscr{C}^0(\partial\Omega)$, we know from (5.2.29) and Definitions 5.2.2-5.2.3 that the unique solution of the classical Dirichlet Problem (5.7.18) is given by

$$u(x) = (\mathrm{PI}\, f)(x) = \int_{\partial\Omega} f\,d\omega^x, \qquad \forall x \in \Omega. \tag{5.7.31}$$

Granted this, and bearing in mind (5.7.11), it follows that estimate (5.7.19) is implied by the boundedness on $L^p(\partial\Omega,\sigma)$ of the Hardy-Littlewood maximal operator $\mathcal{M}_{\omega^{x_0}}$ associated with the harmonic measure ω^{x_0} on $\partial\Omega$ (cf. (cf. [112, (7.7.87)]). In turn, according to item *(vii)* in [112, Proposition 7.7.9], the latter boundedness property holds if $\omega^{x_0} \in B_{p'}(\sigma)$. The argument so far shows that we have the equivalence *(i)* $\Leftrightarrow$ *(ii)*.

We next propose to show that *(i)*-*(ii)* imply *(iii)*. With this goal in mind, fix an arbitrary function $f \in L^p(\partial\Omega,\sigma)$. Then [112, Corollary 3.7.3] implies that there exists a sequence

$$\begin{gathered}\{f_j\}_{j\in\mathbb{N}} \subseteq \mathscr{C}^0(\partial\Omega) \text{ such that}\\ f_j \to f \text{ in } L^p(\partial\Omega,\sigma) \text{ as } j \to \infty.\end{gathered} \tag{5.7.32}$$

If for each $j \in \mathbb{N}$ we now define

$$u_j(x) := (\mathrm{PI}\, f_j)(x) = \int_{\partial\Omega} f_j\,d\omega^x, \qquad \forall x \in \Omega, \tag{5.7.33}$$

then

$$u_j \in \mathscr{C}^0(\overline{\Omega}) \cap \mathscr{C}^\infty(\Omega), \quad \Delta u_j = 0 \text{ in } \Omega, \quad \text{and } u_j\big|_{\partial\Omega} = f_j \text{ on } \partial\Omega. \tag{5.7.34}$$

Since for each $j, j' \in \mathbb{N}$ we have $u_j - u_{j'} = \mathrm{PI}(f_j - f_{j'})$, from (5.7.19) and [112, (8.9.8)] we conclude that

$$\|u_j - u_{j'}\|_{N_\kappa^p(\Omega;\sigma)} = \|\mathcal{N}_\kappa(u_j - u_{j'})\|_{L^p(\partial\Omega,\sigma)} \approx \|f_j - f_{j'}\|_{L^p(\partial\Omega,\sigma)}, \text{ uniformly in } j, j' \in \mathbb{N}. \tag{5.7.35}$$

In particular, from (5.7.35) and (5.7.32) we conclude that the sequence $\{u_j\}_{j\in\mathbb{N}}$ is Cauchy in $N_\kappa^p(\Omega;\sigma)$. Granted this, we may invoke [112, Proposition 8.3.5] to conclude that

$$\text{there exists } u \in N_\kappa^p(\Omega;\sigma) \text{ such that } u_j \longrightarrow u \text{ in } N_\kappa^p(\Omega;\sigma) \text{ as } j \longrightarrow \infty. \tag{5.7.36}$$

Given that each u_j belongs to $H_\kappa^p(\Omega)$ which is a closed subspace of $N_\kappa^p(\Omega;\sigma)$ (cf. (5.5.5)), it follows that actually

$$u \in H_\kappa^p(\Omega). \tag{5.7.37}$$

Granted this and recalling (5.5.6), it follows that the nontangential boundary trace

$$g := u\big|_{\partial\Omega}^{\kappa-\mathrm{n.t.}} \text{ is well defined in } L^p(\partial\Omega,\sigma). \tag{5.7.38}$$

Upon passing to the limit $j' \longrightarrow \infty$ in (5.7.35) we obtain

$$\|u_j - u\|_{N_\kappa^p(\Omega;\sigma)} \approx \|f_j - f\|_{L^p(\partial\Omega,\sigma)}, \text{ uniformly in } j \in \mathbb{N}, \tag{5.7.39}$$

on account of (5.7.36) and (5.7.32). In concert with [112, (8.9.8)], (5.7.34), [112, (8.9.10)], and (5.7.38), this permits us to estimate

$$\begin{aligned} \|f_j - f\|_{L^p(\partial\Omega,\sigma)} &\approx \|u_j - u\|_{N_\kappa^p(\Omega;\sigma)} \\ &= \|\mathcal{N}_\kappa(u_j - u)\|_{L^p(\partial\Omega,\sigma)} \geq \Big\| u_j\big|_{\partial\Omega}^{\kappa-\mathrm{n.t.}} - u\big|_{\partial\Omega}^{\kappa-\mathrm{n.t.}} \Big\|_{L^p(\partial\Omega,\sigma)} \\ &= \|f_j - g\|_{L^p(\partial\Omega,\sigma)}, \text{ uniformly in } j \in \mathbb{N}. \end{aligned} \tag{5.7.40}$$

From (5.7.40) and (5.7.32) we then conclude that necessarily

$$f = g \text{ at } \sigma\text{-a.e. point on } \partial\Omega. \tag{5.7.41}$$

At this stage, from (5.7.37) and (5.7.41) we see that u is a function in $H_\kappa^p(\Omega)$ with the property that $u\big|_{\partial\Omega}^{\kappa-\mathrm{n.t.}} = f$ at σ-a.e. point on $\partial\Omega$. That actually u is the only function with these two qualities is, thanks to the current working assumptions, a consequence of the uniqueness result established in Theorem 5.7.5. The proof of the fact that *(i)-(ii)* imply *(iii)* is therefore complete.

Going further, note that (5.7.24) is implied by (5.7.23) and the Open Mapping Theorem (whose applicability is ensured by the completeness result proved in [112, Proposition 8.3.5]). Hence, *(iii)* $\Rightarrow$ *(iv)*.

To conclude, there remains to show that *(iv)* implies *(i)*. To this end, observe that

$$\mathscr{C}^0(\partial\Omega) \subseteq \bigcap_{0<q\leq\infty} L^q(\partial\Omega, \sigma) \tag{5.7.42}$$

and that

$$\mathrm{PI} : \mathscr{C}^0(\partial\Omega) \longrightarrow \big\{u \in \mathscr{C}^0(\overline{\Omega}) \cap \mathscr{C}^\infty(\Omega) : \Delta u = 0 \text{ in } \Omega\big\} \subseteq \bigcap_{0<q\leq\infty} H^q_\kappa(\Omega). \tag{5.7.43}$$

In addition,

$$(\mathrm{PI}\, f)\big|_{\partial\Omega}^{\kappa-\mathrm{n.t.}} = (\mathrm{PI}\, f)\big|_{\partial\Omega} = f, \qquad \forall f \in \mathscr{C}^0(\partial\Omega). \tag{5.7.44}$$

Granted (5.7.42)-(5.7.44), estimate (5.7.20) is then readily implied by (5.7.24). This establishes the implication *(iv)* $\Rightarrow$ *(i)*, and finishes the proof of Theorem 5.7.6. □

Our next result extends [71, Theorem 10.1, p. 132], dealing with the special case of bounded BMO_1 domains in $\mathbb{R}^n$, and also refines [79, Theorem 1.7.3, p. 29]. In particular, the integral representation formula (5.7.49) involving the normal derivative of the Green function is new in the current context (compare with [81] where the more restrictive class of δ-Reifenberg flat chord-arc domains has been considered).

Theorem 5.7.7 *Suppose $\Omega \subset \mathbb{R}^n$ is a bounded* NTA *domain whose boundary is an upper Ahlfors regular set. Abbreviate $\sigma := \mathcal{H}^{n-1}\lfloor\partial\Omega$ and denote by ω^{x_0} the harmonic measure with pole at the point $x_0 \in \Omega$. Consider the critical exponent*

$$q_\Omega := \sup\big\{q \in (1,\infty) : \omega^{x_0} \in B_q(\sigma)\big\} \tag{5.7.45}$$

(which, according to (5.4.5)*, is well defined and belongs to the interval* $(1,\infty]$*), then let $p_\Omega \in [1,\infty)$ be its Hölder conjugate exponent, i.e.,*

$$p_\Omega := \begin{cases} \frac{q_\Omega}{q_\Omega - 1} & \text{if } q_\Omega < \infty, \\ 1 & \text{if } q_\Omega = \infty. \end{cases} \tag{5.7.46}$$

Then, for each fixed $\kappa > 0$ and $p \in (p_\Omega, \infty)$, the Dirichlet Problem with the boundary trace considered in the nontangential sense and with nontangential maximal function control,

$$\begin{cases} u \in \mathscr{C}^\infty(\Omega), \\ \Delta u = 0 \ \text{in } \Omega, \\ \mathcal{N}_\kappa u \in L^p(\partial\Omega, \sigma), \\ u\big|_{\partial\Omega}^{\kappa-\mathrm{n.t.}} = f \ \sigma\text{-a.e. on } \partial\Omega, \quad f \in L^p(\partial\Omega, \sigma), \end{cases} \tag{5.7.47}$$

has a unique solution. Moreover, there exists a constant $C = C(\Omega, \kappa, p) \in (0, \infty)$ *such that said solution satisfies*

$$\|\mathcal{N}_\kappa u\|_{L^p(\partial\Omega,\sigma)} \leq C\|f\|_{L^p(\partial\Omega,\sigma)}. \tag{5.7.48}$$

Also, if ν *stands for the geometric measure theoretic outward unit normal to* Ω *and if* $G_\Omega(\cdot,\cdot)$ *denotes the Green associated with the Laplacian in* Ω *then, for each fixed point* $x_0 \in \Omega$, *the unique solution* u *of* (5.7.47) *may be represented as*

$$\begin{aligned} u(x) &= \int_{\partial\Omega} f\,\mathrm{d}\omega^x = \int_{\partial\Omega} f k^x\,\mathrm{d}\sigma = \int_{\partial\Omega} K_{x_0}(x,y) f(y)\,\mathrm{d}\sigma(y) \\ &= -\int_{\partial\Omega} \partial_\nu G_\Omega(\cdot, x) f\,\mathrm{d}\sigma, \qquad \forall\, x \in \Omega, \end{aligned} \tag{5.7.49}$$

where the normal derivative of the Green function is defined as in (5.6.2).

Conversely, wherever $p \in (1, \infty)$ *is such that the Dirichlet Problem* (5.7.47) *is uniquely solvable and the solution satisfies* (5.7.48) *it follows that* $p > p_\Omega$. *In particular,*

> p_Ω *may be characterized as the infimum of all integrability exponents* $p \in (1, \infty)$ *with the property that the* L^p *Dirichlet Problem in* Ω *(cf.* (5.7.47)*) is well posed.* (5.7.50)

Proof For starters, from (5.7.45) and item *(vi)* in [112, Proposition 7.7.9] we conclude that

$$\{q \in (1, \infty) : \omega^{x_0} \in B_q(\sigma)\} = (1, q_\Omega). \tag{5.7.51}$$

Let $p' \in (1, \infty)$ be such that $\frac{1}{p} + \frac{1}{p'} = 1$. The fact that $p \in (p_\Omega, \infty)$ forces $p' \in (1, q_\Omega)$, hence

$$\omega^{x_0} \in B_{p'}(\sigma). \tag{5.7.52}$$

In particular, (cf. (5.7.17)) we have

$$k^x = K(x, \cdot) k^{x_0} \in L^{p'}(\partial\Omega, \sigma), \ \text{ for each } x \in \Omega. \tag{5.7.53}$$

To proceed, pick an arbitrary $f \in L^p(\partial\Omega, \sigma)$. On account of (5.7.52), Theorem 5.7.6 ensures that there is a unique function $u \in H^p_\kappa(\Omega)$ with the property that $u\big|^{\kappa-\mathrm{n.t.}}_{\partial\Omega} = f$ at σ-a.e. point on $\partial\Omega$, and such that

$$\|\mathcal{N}_\kappa u\|_{L^p(\partial\Omega,\sigma)} \approx \Big\|u\big|^{\kappa-\mathrm{n.t.}}_{\partial\Omega}\Big\|_{L^p(\partial\Omega,\sigma)} \tag{5.7.54}$$

where the proportionality constants depend only on Ω, p, and κ. This proves that the Dirichlet Problem (5.7.47) is uniquely solvable and the solution satisfies (5.7.48).

From the proof of Theorem 5.7.6 (cf. (5.7.36) in particular) we also know that if a sequence $\{f_j\}_{j\in\mathbb{N}} \subseteq \mathscr{C}^0(\partial\Omega)$ has been chosen such that $f_j \to f$ in $L^p(\partial\Omega, \sigma)$ as $j \to \infty$ then at each fixed point $x \in \Omega$ the solution u of (5.7.47) may be compute as

$$u(x) = \lim_{j\to\infty} (\mathrm{PI}\, f_j)(x) = \lim_{j\to\infty} \int_{\partial\Omega} f_j \, d\omega^x. \tag{5.7.55}$$

Together with (5.7.53) and Theorem 5.6.2, for each $x \in \Omega$ this permits us to write

$$\begin{aligned} u(x) &= \lim_{j\to\infty} \int_{\partial\Omega} f_j \, d\omega^x = \lim_{j\to\infty} \int_{\partial\Omega} f_j k^x \, d\sigma \\ &= \int_{\partial\Omega} f k^x \, d\sigma = -\int_{\partial\Omega} \partial_\nu G_\Omega(\cdot, x) f \, d\sigma \end{aligned} \tag{5.7.56}$$

which, bearing in mind (5.4.1) and (5.7.9), proves (5.7.49).

In the converse direction, assume $p \in (1, \infty)$ is such that the Dirichlet Problem (5.7.47) is uniquely solvable and the solution satisfies (5.7.48). Since $\mathscr{C}^0(\partial\Omega)$ is a subspace of $L^p(\partial\Omega, \sigma)$, it follows that (5.7.19) holds whenever $f \in \mathscr{C}^0(\partial\Omega)$ and u is the unique solution of the classical Dirichlet Problem (5.7.18). In concert with *(i)* ⇒ *(ii)* in Theorem 5.7.6, this implies (cf. (5.7.21)) that $\omega^{x_0} \in B_{p'}(\sigma)$. In light of (5.7.45)-(5.7.46), this forces $p > p_\Omega$. □

Here is an integral representation formula augmenting the Fatou-type result from Proposition 5.5.3.

Corollary 5.7.8 *Consider a bounded* NTA *domain $\Omega \subset \mathbb{R}^n$ whose boundary is an upper Ahlfors regular set and let $p_\Omega \in [1, \infty)$ be associated with Ω as in* (5.7.46). *Denote by ν the geometric measure theoretic outward unit normal to Ω and let $G_\Omega(\cdot, \cdot)$ stand for the Green associated with the Laplacian in Ω. Lastly, fix some aperture parameter $\kappa > 0$ along with some integrability exponent $p \in (p_\Omega, \infty)$. Then any function $u \in H^p_\kappa(\Omega)$ may be expressed as*

$$\begin{aligned} u(x) &= \int_{\partial\Omega} u\big|^{\kappa-\mathrm{n.t.}}_{\partial\Omega} \, d\omega^x = \int_{\partial\Omega} k^x \Big(u\big|^{\kappa-\mathrm{n.t.}}_{\partial\Omega}\Big) \, d\sigma = \int_{\partial\Omega} K_{x_0}(x, y) \Big(u\big|^{\kappa-\mathrm{n.t.}}_{\partial\Omega}\Big)(y) \, d\sigma(y) \\ &= -\int_{\partial\Omega} \partial_\nu G_\Omega(\cdot, x) \Big(u\big|^{\kappa-\mathrm{n.t.}}_{\partial\Omega}\Big) \, d\sigma, \qquad \forall\, x \in \Omega, \end{aligned} \tag{5.7.57}$$

where the normal derivative of the Green function is defined as in (5.6.2).

Proof This is a direct consequence of Proposition 5.5.3 (cf. (5.5.6)) and Theorem 5.7.7. □

5.8 The Nature of the Critical Index p_Ω and Further Results on the Green Function

In order to further place Theorem 5.7.7 and Corollary 5.7.8 into perspective, we wish to comment on the nature of the critical index $p_\Omega \in [1, \infty)$ associated with a bounded NTA domain $\Omega \subseteq \mathbb{N}$ possessing an Ahlfors regular boundary as in (5.7.46). First, we note that

$$\text{if } \Omega \subseteq \mathbb{R}^n \text{ is a bounded Lipschitz domain then } 1 \leq p_\Omega < 2. \tag{5.8.1}$$

This result is due to B. Dahlberg (cf. [25]) who also proved that the range $1 \leq p_\Omega < 2$ is sharp in the category of bounded Lipschitz domains $\Omega \subseteq \mathbb{R}^n$. A new approach to (5.8.1), based on a Rellich identity involving the Poisson kernel, has been proposed by D. Jerison and C. Kenig in [69]. For the class a bounded Lipschitz domains with connected boundaries, alternative proofs of (5.8.1), based on boundary layer potentials, may be found in [28] and [179]. Incidentally, the latter approach yields the following representation of the Poisson kernel for a bounded Lipschitz domain $\Omega \subseteq \mathbb{R}^n$ with connected boundary:

$$k^{x_0} = \big(\tfrac{1}{2}I + K_\Delta^\#\big)^{-1}\Big(\partial_\nu\big[E_\Delta(\cdot - x_0)\big]\Big), \qquad x_0 \in \Omega, \tag{5.8.2}$$

where $K_\Delta^\#$ is the "transpose" of the harmonic double layer potential operator associated with Ω as in (2.5.203), the vector ν is the outward unit normal to Ω, and E_Δ is the standard fundamental solution for the Laplacian in $\mathbb{R}^n$ defined in (1.5.56). Formula (5.8.2) should be compared with (4.5.16).

Second, it has been proved by B. Dahlberg in [26] that

$$\text{if } \Omega \subseteq \mathbb{R}^n \text{ is a bounded } \mathscr{C}^1 \text{ domain then } p_\Omega = 1. \tag{5.8.3}$$

Under the additional assumption that the bounded $\mathscr{C}^1$ domain in question has a connected boundary, this has also been established by E.B. Fabes, M. Jodeit, and N.M. Rivière in [41] (at least implicit, via (5.7.50)). Subsequently, S. Hofmann, M. Mitrea, and M. Taylor have succeeded in extending (5.8.3) to the larger class of regular SKT domains (cf. [112, (5.11.93)]) with compact boundaries. Specifically, the well-posedness result established in [63, Theorem 7.2, p. 2831], in concert with (5.7.50), implies that

$$\text{if } \Omega \subseteq \mathbb{R}^n \text{ is a bounded regular SKT domain then } p_\Omega = 1. \tag{5.8.4}$$

In fact, the theorems just mentioned yield a more nuanced result, in the category of ε-`regular` SKT `domains` (introduced in [63]). The reader is reminded that the membership of a set $\Omega \subseteq \mathbb{R}^n$ to this class is characterized by the following geometric measure theoretic conditions:

Ω is an open set with an Ahlfors regular boundary, satisfying a two-sided local John condition, and whose geometric measure theoretic outward unit normal ν satisfies $\operatorname{dist}\big(\nu, [\mathrm{VMO}(\partial\Omega, \sigma)]^n\big) < \varepsilon$, (5.8.5)

where $\mathrm{VMO}(\partial\Omega, \sigma)$ stands for the Sarason class of functions of vanishing mean oscillation on $\partial\Omega$, relative to the surface measure $\sigma := \mathcal{H}^{n-1}\lfloor\partial\Omega$, and the distance is measured in $\big[\mathrm{BMO}(\partial\Omega, \sigma)\big]^n$. It turns out that in the class of bounded ε-regular SKT domains $\Omega \subseteq \mathbb{R}^n$ the critical exponent p_Ω can made as closed to the value 1 as wanted by taking $\varepsilon > 0$ small enough (relative to the desired degree of proximity of p_Ω to 1, as well as the NTA and the Ahlfors regularity character of $\partial\Omega$). For example, this allows the consideration of bounded Lipschitz domains with small Lipschitz constant, as a particular case.

Continuing the discussion aimed at further elucidating the nature of the critical index $p_\Omega \in [1,\infty)$ associated as in (5.7.46) with a bounded NTA domain $\Omega \subseteq \mathbb{N}$ having an Ahlfors regular boundary, we note the following result.

Proposition 5.8.1 *Let $\Omega \subset \mathbb{R}^n$ be a bounded* NTA *domain, possessing an upper Ahlfors regular boundary, and with the property that there exist $\alpha > 0$ and $C \in (0,\infty)$ such that the Green function G_Ω associated with the Laplacian in Ω satisfies the estimate*

$$|\nabla_y G_\Omega(y,x)| \leq C\,\mathrm{dist}(x,\partial\Omega)^\alpha |x-y|^{1-n-\alpha}, \qquad \forall x,y \in \Omega. \tag{5.8.6}$$

Then $p_\Omega = 1$.

Proof Abbreviate $\sigma := \mathcal{H}^{n-1}\lfloor\partial\Omega$ and denote by ν the geometric measure theoretic outward unit normal to Ω. From (5.8.6) and (5.6.1) we conclude that, for each fixed $x \in \Omega$,

$$\Big|\Big(\big[\nabla G_\Omega(\cdot,x)\big]\Big|^{\mathrm{n.t.}}_{\partial\Omega}\Big)(y)\Big| \leq C\,\mathrm{dist}(x,\partial\Omega)^\alpha |x-y|^{1-n-\alpha} \quad \text{for } \sigma\text{-a.e. } y \in \partial\Omega. \tag{5.8.7}$$

This further entails that, for each fixed $x \in \Omega$,

$$\big|\partial_{\nu(y)} G_\Omega(y,x)\big| \leq C\,\mathrm{dist}(x,\partial\Omega)^\alpha |x-y|^{1-n-\alpha} \text{ for } \sigma\text{-a.e. } y \in \partial\Omega. \tag{5.8.8}$$

Suppose now that some $f \in \mathscr{C}^0(\partial\Omega)$ has been given and consider

$$(\mathrm{PI}\, f)(x) = \int_{\partial\Omega} f\, d\omega^x, \qquad \forall x \in \Omega. \tag{5.8.9}$$

Theorem 5.6.2 then allows us to express

$$(\mathrm{PI}\, f)(x) = -\int_{\partial\Omega} \partial_\nu G_\Omega(\cdot,x) f\, d\sigma, \qquad \forall x \in \Omega. \tag{5.8.10}$$

Combining (5.8.8) with (5.8.10) therefore yields

$$|(\mathrm{PI}\, f)(x)| \leq C \int_{\partial\Omega} \frac{\mathrm{dist}(x,\partial\Omega)^\alpha}{|x-y|^{n-1+\alpha}} |f(y)|\, d\sigma(y), \qquad \forall x \in \Omega. \tag{5.8.11}$$

In concert with [112, Proposition 8.4.12] this proves that for each $p \in (1,\infty)$ we have

$$\|\mathcal{N}_\kappa(\mathrm{PI}\, f)\|_{L^p(\partial\Omega,\sigma)} \leq C(\Omega,\kappa,p)\|f\|_{L^p(\partial\Omega,\sigma)}. \tag{5.8.12}$$

On account of items *(i)-(ii)* in Theorem 5.7.6, from the fact that we have an estimate as in (5.8.12) for each $p \in (1,\infty)$ we ultimately conclude that $p_\Omega = 1$. □

To give a concrete example when the Green function estimate hypothesized in Proposition 5.8.1 materializes, we first recall a couple of definitions. First, a set

$\Omega \subseteq \mathbb{R}^n$ is said to satisfy a `uniform exterior ball condition` (UEBC for short), if there exists a number $r > 0$ such that at each boundary point x it is possible to find an open ball $B_{x,r}$ of radius r which is contained in $\mathbb{R}^n \setminus \Omega$ and such that $x \in \partial B_{x,r}$. Second, the concept of `uniform interior ball condition` (UIBC for short) is defined analogously, but now insisting that the ball $B_{x,r}$ is actually contained in Ω.

The uniform interior/exterior ball condition has been originally introduced by Poincaré in [147] and, through the many intervening years, it has proved to be a natural, useful condition in the study of the boundary behavior of solutions of large classes of elliptic differential equations. While it is known (see, e.g., [5] where more general results of this nature may be found) that for an open set Ω in $\mathbb{R}^n$

$$\Omega \text{ satisfies both a UIBC and a UEBC} \Longleftrightarrow \Omega \text{ is a } \mathscr{C}^{1,1} \text{ domain,} \tag{5.8.13}$$

a mere one-sided ball condition permits the boundary of the set in question to be quite rough. For instance, any bounded, open, convex set in $\mathbb{R}^n$ is a bounded NTA domain, with an Ahlfors regular boundary, and which satisfies a UEBC (for every given radius $r > 0$). Of course, bounded, open, convex subsets of $\mathbb{R}^n$ can be quite rough, though they are, nonetheless, Lipschitz domains. In this vein we wish to note that, in the presence of certain background geometric characteristics, additionally imposing UEBC on a given set may enhance its regularity in surprising ways. To illustrate this phenomenon, we first recall a couple of definitions. First, given an open set $\mathcal{O}$ in $\mathbb{R}^n$, call a function $u : \mathcal{O} \to \mathbb{R}$ `semiconvex` provided it is continuous and has the property that there exists $C \in (0, \infty)$ such that

$$\begin{gathered} 2u(x) - u(x+h) - u(x-h) \leq C|h|^2 \\ \text{for all } x, h \in \mathbb{R}^n \text{ with } [x-h, x+h] \subseteq \mathcal{O}. \end{gathered} \tag{5.8.14}$$

Second, call a nonempty, closed set $E \subseteq \mathbb{R}^n$ `proximally regular` provided there exists some $r > 0$ such that for each $x \in \mathbb{R}^n$ with $0 < \delta_E(x) < r$ there exists a unique point $x_* \in E$ such that $\delta_E(x) = |x - x_*|$. This condition turns out to be equivalent to E being a set of positive reach in the sense of H. Federer (cf. [44, Definition 4.1, p. 432]). The following theorem is proved in [110]:

Theorem 5.8.2 *Let $\Omega \subseteq \mathbb{R}^n$ be an open set with compact boundary. Then the following conditions are equivalent:*

(i) Ω is a Lipschitz domain satisfying a uniform exterior ball condition (UEBC)*;*
(ii) Ω is a semiconvex domain (i.e., Ω is an open set satisfying $\partial\Omega = \partial(\overline{\Omega})$ and whose boundary may be locally described, up to a rigid transformation, by graphs of semiconvex functions);
(iii) Ω is an NTA *domain satisfying a* UEBC*;*
(iv) Ω satisfies an interior corkscrew condition as well as a UEBC*;*
(v) Ω is a Lipschitz domain with the property that $\overline{\Omega}$ is proximally regular (the latter condition also being equivalent to $\overline{\Omega}$ being a set of positive reach);
(vi) $\overline{\Omega}$ is proximally regular (or of positive reach), $\partial\Omega = \partial(\overline{\Omega})$ and the following boundary density condition holds

$$\limsup_{r\to 0^+} \frac{\mathcal{L}^n(\Omega \cap B(x,r))}{r^n} > 0, \qquad \forall x \in \partial\Omega; \tag{5.8.15}$$

(vii) Ω *is n-tick,* $\partial\Omega = \partial(\overline{\Omega})$*, and the set* $\overline{\Omega}$ *is proximally regular (or of positive reach);*

(viii) Ω *is an n-thick set satisfying a* UEBC*;*

(ix) Ω *satisfies a* UEBC *and (cf.* (A.0.2))

$$\partial\Omega = \bigcup_{\kappa>0} A_\kappa(\partial\Omega); \tag{5.8.16}$$

(x) Ω *is a set of locally finite perimeter with* $\partial\Omega = \partial(\overline{\Omega})$*, satisfying* (5.8.16) *and such that there exists* $C \in (0,\infty)$ *with the property that if* ν *denotes the geometric measure theoretic outward unit normal to* Ω *then*

$$\begin{gathered}(x-y)\cdot\nu(y) + \mathrm{dist}(x,\partial\Omega) \le C|x-y|^2 \\ \text{for each } x \in \overline{\Omega} \text{ and each } y \in \partial^*\Omega;\end{gathered} \tag{5.8.17}$$

(xi) Ω *is a set of locally finite perimeter with* $\partial\Omega = \partial(\overline{\Omega})$*, satisfying* (5.8.16) *and such that there exists* $C \in (0,\infty)$ *with the property that if* ν *denotes the geometric measure theoretic outward unit normal to* Ω *then*

$$(x-y)\cdot\nu(y) \le C|x-y|^2 \text{ for all } x \in \partial\Omega \text{ and } y \in \partial^*\Omega. \tag{5.8.18}$$

Parenthetically, we wish to remark that condition (5.8.18) amounts to a pointwise *lower* bound for the integral kernel of the harmonic double layer K associated with a bounded semiconvex domain $\Omega \subseteq \mathbb{R}^n$ as in (2.5.203) of the following sort:

$$-\frac{1}{\omega_{n-1}}\frac{C}{|x-y|^{n-2}} \le \frac{1}{\omega_{n-1}}\frac{\langle y-x, \nu(y)\rangle}{|x-y|^n} \text{ for all } x \in \partial\Omega \text{ and } y \in \partial^*\Omega. \tag{5.8.19}$$

As regards the nature of the Green function, the relevance of Ω satisfying a UEBC stems from from the fact that the latter guarantees that $\nabla_x G_\Omega(x,y)$ exhibits a behavior akin to that of the the Poisson kernel for the upper half-space. Specifically, we have the following theorem proved by M. Grüter and K.-O. Widman in [55, Theorem 3.3, p. 31], expanding on the properties of the Green function discussed in Theorem 5.3.1.

Theorem 5.8.3 *If* $\Omega \subset \mathbb{R}^n$ *is a bounded open set which satisfies a* UEBC *then the Green function associated with the Laplacian in* Ω *satisfies the following properties for all distinct points* $x, y \in \Omega$*:*

(i) $G_\Omega(x,y) \le C|x-y|^{2-n}$*;*

(ii) $G_\Omega(x,y) \le C\,\mathrm{dist}(x,\partial\Omega)|x-y|^{1-n}$*;*

(iii) $G_\Omega(x,y) \le C\,\mathrm{dist}(y,\partial\Omega)|x-y|^{1-n}$*;*

(iv) $G_\Omega(x,y) \le C\,\mathrm{dist}(x,\partial\Omega)\,\mathrm{dist}(y,\partial\Omega)|x-y|^{-n}$*;*

(v) $|\nabla_x G_\Omega(x,y)| \le C|x-y|^{1-n}$*;*

(vi) $|\nabla_x G_\Omega(x,y)| \le C\,\mathrm{dist}(y,\partial\Omega)|x-y|^{-n}$;
(vii) $|\nabla_y G_\Omega(x,y)| \le C|x-y|^{1-n}$;
(viii) $|\nabla_y G_\Omega(x,y)| \le C\,\mathrm{dist}(x,\partial\Omega)|x-y|^{-n}$;
(ix) $|\nabla_x \nabla_y G_\Omega(x,y)| \le C|x-y|^{-n}$,

for some constant $C \in (0,\infty)$ *independent of* x, y.

By combining Theorem 5.8.2, Theorem 5.8.3, Proposition 5.8.1 (used here with $\alpha = 1$), and Theorem 5.7.7, we obtain the following well-posedness result for the L^p Dirichlet Problem.

Corollary 5.8.4 *Let* $\Omega \subset \mathbb{R}^n$ *be a bounded* NTA *domain with an upper Ahlfors regular boundary and which satisfies a* UEBC *(equivalently, let* $\Omega \subset \mathbb{R}^n$ *be a bounded semiconvex domain). Then for each fixed* $\kappa > 0$ *and* $p \in (1,\infty)$*, the Dirichlet Problem with the boundary trace considered in the nontangential sense and with nontangential maximal function control, formulated in* (5.7.47)*, is uniquely solvable. Moreover, the solution* u *satisfies* (5.7.48) *and is representable as in* (5.7.49).

Being defined as a Radon-Nikodym derivative of two positive measures the Poisson kernel is a non-negative function, but under suitable additional geometric assumptions this turns out to actually be bounded away from zero in a uniform fashion. To facilitate the statement of such a result, following [5] we define a `pseudo-ball` in $\mathbb{R}^n$ with apex at the point $x \in \mathbb{R}^n$, symmetry axis along the vector vector $h \in S^{n-1}$, truncated at height $b \in (0,\infty)$, and whose profile is modeled by the "shape function" $\delta : [0,R] \to [0,\infty)$ to be the set

$$\mathscr{P}_{\delta,b}(x,h) := \big\{y \in B(x,R) : |y-x|\delta(|y-x|) < h\cdot(y-x) < b\big\} \subseteq \mathbb{R}^n. \tag{5.8.20}$$

Theorem 5.8.5 *Let* $\Omega \subseteq \mathbb{R}^n$ *be a bounded* NTA *domain with an upper Ahlfors regular boundary. satisfying a uniform interior pseudo-ball condition of the following sort: there exist parameters* $b, R \in (0,\infty)$*, a unit vector field* $h : \partial\Omega \to S^{n-1}$*, and a continuous function* $\delta : [0,R] \to [0,\infty)$ *which is positive and non-decreasing on* $(0,R]$ *and satisfies Dini's integrability condition*

$$\int_0^R \frac{\delta(t)}{t}\,dt < +\infty, \tag{5.8.21}$$

with the property that

$$\mathscr{P}_{\delta,b}\big(x,h(x)\big) \subseteq \Omega \text{ for each point } x \in \partial\Omega. \tag{5.8.22}$$

Set $\sigma := \mathcal{H}^{n-1}\lfloor\partial\Omega$ *and let* ν *stand for the geometric measure theoretic outward unit normal to* Ω*. Also, denote by* $G_\Omega(\cdot,\cdot)$ *the Green function for the Laplacian in* Ω*, and fix an arbitrary point* $x_0 \in \Omega$*.*

Then

$$h(x) = -\nu(x) \text{ for each } x \in \partial^*\Omega, \tag{5.8.23}$$

hence

$$h = -\nu \ \textit{at } \sigma\textit{-a.e. point on } \partial\Omega, \tag{5.8.24}$$

and there exists a constant $c > 0$ *such that, with the normal derivative defined as in* (5.6.2), *one has*

$$k^{x_0} = -\partial_\nu G_\Omega(\cdot, x_0) \geq c \ \textit{at } \sigma\textit{-a.e. point on } \partial\Omega. \tag{5.8.25}$$

The above theorem addresses the issue raised in [49, Question 2, p. 49] where Garnett and Marshall ask what regularity conditions on $\partial\Omega$ ensure the validity of (5.8.25). In this regard, we also wish to note that any bounded domain of class $\mathscr{C}^{1,\alpha}$ with $\alpha \in (0, 1]$ satisfies a uniform interior pseudo-ball condition with shape function $\delta(t) := Ct^\alpha$ in which case Dini's integrability condition (5.8.21) holds (in particular, if Ω satisfies a genuine uniform interior ball condition then Ω satisfies a uniform interior pseudo-ball condition with shape function $\delta(t) := Ct$). From this point of view, Theorem 5.8.5 is sharp in the sense that if Ω is merely a bounded $\mathscr{C}^1$ domain then k^{x_0} (hence also $-\partial_\nu G_\Omega(\cdot, x_0)$) may not be bounded away from zero (this follows from [49, Exercise 14, p. 71]).

Proof of Theorem 5.8.5 To begin with, observe that the assumptions made on δ ensure that this function vanishes continuously at 0. This implies that for each angle $\theta \in (0, \pi)$ there exists a small height $b_\theta \in (0, \infty)$ with the property that

$$\mathrm{C}_{\theta,b_\theta}\big(x, h(x)\big) \subseteq \mathscr{P}_{\delta,b}\big(x, h(x)\big) \ \text{ for each } x \in \partial\Omega \tag{5.8.26}$$

where, generally speaking,

$$\mathrm{C}_{\theta,b}(x, h) := \big\{y \in \mathbb{R}^n : \cos(\theta/2)\,|y - x| < (y - x)\cdot h < b\big\} \tag{5.8.27}$$

is the cone with vertex at $x \in \mathbb{R}^n$, symmetry axis along $h \in S^{n-1}$, and full aperture $\theta \in (0, \pi)$. In view of (5.8.26), the uniform interior pseudo-ball condition (5.8.22), and [112, Lemma 5.6.15], we may conclude that (5.8.23) holds. Hence, (5.8.24) also holds (thanks to [112, (5.2.4)] and [112, (5.6.21)]). Keeping this and (5.8.26) in mind, the implication recorded in [112, (8.1.6)] presently gives that for each $\kappa > 0$ there exists a small $\varepsilon > 0$ such that

$$\big\{x - t\nu(x) : 0 < t < \varepsilon\big\} \subseteq \Gamma_\kappa(x) \ \text{ for } \sigma\text{-a.e. } x \in \partial\Omega. \tag{5.8.28}$$

Having established (5.8.24), the Boundary Point Principle proved in [5] then provides some constant $c > 0$ with the property that

$$\liminf_{t\to 0^+} \frac{G_\Omega\big(x - t\nu(x), x_0\big) - G_\Omega(x, x_0)}{t} \geq c \ \text{ at } \sigma\text{-a.e. } x \in \partial\Omega. \tag{5.8.29}$$

Then (5.8.25) follows from (5.8.29) with the help of (5.8.28) and Lemma 5.6.1. □

The notion of ε-regular SKT domain (cf. (5.8.5)) plays a significant role in the formulation of Theorem 5.8.6 below. In light of [113, (9.2.22)], said theorem refines a result of E. Fabes who has shown in [40] that if $\Omega \subseteq \mathbb{R}^n$ is a bounded Lipschitz domain, σ denotes its surface measure and $\kappa > 0$ is a fixed aperture parameter, then $\|\mathcal{N}_\kappa u\|_{L^2(\partial\Omega,\sigma)} \approx \|u\|_{H^{1/2}(\Omega)}$, uniformly for u harmonic function in Ω.

Theorem 5.8.6 *Let Ω be a bounded* NTA *domain in $\mathbb{R}^n$, with $n \geq 2$, whose boundary is upper Ahlfors regular, and abbreviate $\sigma := \mathcal{H}^{n-1}\lfloor\partial\Omega$. Also, fix an arbitrary aperture parameter $\kappa > 0$ along with some integrability exponent $p \in (1,2]$.*

Then any harmonic function $u \in B^{p,p}_{1/p}(\Omega)$ satisfies $\mathcal{N}_\kappa u \in L^p(\partial\Omega,\sigma)$, has the property that the nontangential boundary trace $u\big|^{\kappa-\mathrm{n.t.}}_{\partial\Omega}$ exists at σ-a.e. point on $\partial\Omega$, and there exists a finite constant $C > 0$, independent of u, such that

$$\|\mathcal{N}_\kappa u\|_{L^p(\partial\Omega,\sigma)} \leq C\|u\|_{B^{p,p}_{1/p}(\Omega)}. \tag{5.8.30}$$

Moreover, as regards the opposite inequality in (5.8.30),

> *given any $p \in (1,\infty)$ and $M \in (0,\infty)$ there exists $\varepsilon > 0$ such that if Ω is a bounded ε-regular* SKT *domain whose local John constants and Ahlfors regularity character are controlled by M then one can find some $C \in (0,\infty)$ for which $\|u\|_{B^{p,p}_{1/p}(\Omega)} \leq C\|\mathcal{N}_\kappa u\|_{L^p(\partial\Omega,\sigma)}$ whenever u is harmonic in Ω.* (5.8.31)

In particular,

> *given any $p \in (1,2]$ and $M \in (0,\infty)$ there exists $\varepsilon > 0$ such that if Ω is a bounded ε-regular* SKT *domain whose local John constants and Ahlfors regularity character are controlled by M then actually $\|\mathcal{N}_\kappa u\|_{L^p(\partial\Omega,\sigma)} \approx \|u\|_{B^{p,p}_{1/p}(\Omega)}$ uniformly for u harmonic in Ω,* (5.8.32)

hence, as a corollary,

> *if Ω is a bounded regular* SKT *domain and $p \in (1,2]$ then $\|\mathcal{N}_\kappa u\|_{L^p(\partial\Omega,\sigma)} \approx \|u\|_{B^{p,p}_{1/p}(\Omega)}$ uniformly for u harmonic in Ω.* (5.8.33)

Proof For starters, observe that [112, (5.11.4)] presently ensures that $\partial\Omega$ is an Ahlfors regular set. To proceed, recall the L^2-based area-function defined as in (3.1.119) with $q := 2$. From the version of [27, Theorem, p. 98] given in [27, item (A), p. 106] (as well as Proposition 5.4.1) we see that there exists some finite constant $C > 0$ such that for any harmonic function u in Ω we have

$$\|\mathcal{N}_\kappa u\|_{L^p(\partial\Omega,\sigma)} \leq C\|\mathcal{A}_{2,\kappa} u\|_{L^p(\partial\Omega,\sigma)} + C\|u\|_{L^p(\Omega,\mathcal{L}^n)}. \tag{5.8.34}$$

Granted this, (5.8.30) follows by invoking [113, Theorem 9.2.39] (with $q := 2$). Also, the existence of the nontangential boundary trace $u\big|^{\kappa-\mathrm{n.t.}}_{\partial\Omega}$ at σ-a.e. point on $\partial\Omega$ is a consequence of (5.8.34) and Proposition 5.5.3.

To deal with (5.8.31), fix $p \in (1,\infty)$ and work under the assumption that Ω is actually a bounded ε-regular SKT domain for some sufficiently small $\varepsilon = \varepsilon(p) > 0$. In this setting, suppose u is a harmonic function in Ω satisfying $\mathcal{N}_\kappa u \in L^p(\partial\Omega,\sigma)$.

From the Fatou-type result from Proposition 5.5.2 and the well-posedness result for the L^p Dirichlet Problem in Ω from [63] we then conclude that there exists some $f \in L^p(\partial\Omega, \sigma)$ such that

$$\|f\|_{L^p(\partial\Omega,\sigma)} \leq C\|\mathcal{N}_\kappa u\|_{L^p(\partial\Omega,\sigma)} \text{ and } u = \mathcal{D}f \text{ in } \Omega, \tag{5.8.35}$$

where the constant $C \in (0, \infty)$ is independent of u, and where $\mathcal{D}$ is the boundary-to-domain harmonic double layer potential operator associated with Ω. Combining (5.8.35) with the mapping properties of the boundary-to-domain double layer acting from Lebesgue spaces into Besov spaces from [115, §4.3] then establishes that $u \in B^{p,p}_{1/p}(\Omega)$ and $\|u\|_{B^{p,p}_{1/p}(\Omega)} \leq C\|\mathcal{N}_\kappa u\|_{L^p(\partial\Omega,\sigma)}$. This completes the proof of (5.8.31), from which (5.8.32) follows on account of (5.8.30). □

We conclude this section with an application to the mapping properties of the Poisson integral operator on Besov-Triebel-Lizorkin spaces, which has significant implications for the solvability of the Dirichlet Problem for the Laplacian in this context (for related results the interested reader is referred to [37], [43], [54], [73], [85], [100], [99], [97], [96], [121] and the references therein).

Theorem 5.8.7 *Let Ω be a bounded Lipschitz domain in $\mathbb{R}^n$ satisfying a* UEBC. *Denote its outward unit normal ν, and abbreviate $\sigma := \mathcal{H}^{n-1}\lfloor\partial\Omega$. Also, consider the Green function $G_\Omega(\cdot,\cdot)$ for the Laplacian in Ω and for each $f \in L^1(\partial\Omega, \sigma)$ define*

$$(\mathrm{PI}f)(x) := -\int_{\partial\Omega} \partial_{\nu(y)} G_\Omega(x, y) f(y)\, d\sigma(y) \text{ for each } x \in \Omega. \tag{5.8.36}$$

Then the operators

$$\mathrm{PI} : B^{p,q}_s(\partial\Omega, \sigma) \to B^{p,q}_{s+\frac{1}{p}}(\Omega) \text{ and } \mathrm{PI} : B^{p,p}_s(\partial\Omega, \sigma) \to F^{p,q}_{s+\frac{1}{p}}(\Omega) \tag{5.8.37}$$

are bounded whenever $0 < p, q \leq \infty$ and $(n-1)(\frac{1}{p} - 1)_+ < s < 1$, with the additional condition that $p, q \neq \infty$ in the case of Triebel-Lizorkin spaces.

It is worth recalling from Theorem 5.8.2 within the class of bounded open sets satisfying a UEBC, being an NTA domain, or being a Lipschitz domain, or being a semiconvex domain are all equivalent conditions.

Proof We claim that there exist a constant $C \in (0, \infty)$ and some σ-measurable set $E \subseteq \partial\Omega$ with $\sigma(E) = 0$ such that

$$|\nabla_x \nabla_y G_\Omega(x, y)| \leq \frac{C}{|x - y|^n} \qquad \forall x \in \Omega, \quad \forall y \in \partial\Omega \setminus E. \tag{5.8.38}$$

Indeed, from Theorem 5.8.3 we know that the estimate in (5.8.38) holds for every $x, y \in \Omega$. Then, keeping $x \in \Omega$, by Fatou's Theorem proved in [25] for bounded, harmonic functions in Lipschitz domains, it follows that there exists $\underline{E_x} \subseteq \partial\Omega$ with $\sigma(E_x) = 0$ and such that the estimate in (5.8.38) holds for every $y \in \overline{\Omega} \setminus E_x$. Pick a countable dense subset D of Ω, and set $E := \bigcup_{x\in D} E_x$. Then $\sigma(E) = 0$ and the estimate in (5.8.38) holds for every $x \in D$ and $y \in \overline{\Omega} \setminus E$. Keeping now $y \in \overline{\Omega} \setminus E$ fixed, by density then the estimate in (5.8.38) holds for every $x \in \Omega$.

Next, recall that we have

$$\mathrm{PI}(1) = 1 \text{ in } \Omega \text{ and } \Delta \circ \mathrm{PI} = 0 \text{ in } \Omega. \tag{5.8.39}$$

As a consequence, for each fixed $\theta \in (0, 1)$ we may rely on the mapping properties for certain boundary-to-domain integral operators acting from Besov spaces with a positive amount of smoothness into Lebesgue spaces suitably weighted in terms of the distance to the boundary established in [115, §4.2] to the operator $\nabla\mathrm{PI}$ and conclude that if $\frac{n-1}{n} < p \leq \infty$ and $(n-1)(\frac{1}{p}-1)_+ < s < 1$ then there exists $C \in (0,\infty)$ such that for each $f \in B^{p,p}_s(\partial\Omega,\sigma)$ we have

$$\Big\| \delta_{\partial\Omega}^{1-\frac{1}{p}-s} \big|\nabla\mathrm{PI} f\big|_{\star,\theta} \Big\|_{L^p(\Omega,\mathcal{L}^n)} \leq C \|f\|_{B^{p,p}_s(\partial\Omega,\sigma)}. \tag{5.8.40}$$

We claim that there exists $C \in (0,\infty)$ such that for each $f \in B^{p,p}_s(\partial\Omega,\sigma)$ we also have

$$\big\|\mathrm{PI} f\big\|_{L^p(\Omega,\mathcal{L}^n)} \leq C \|f\|_{B^{p,p}_s(\partial\Omega,\sigma)}. \tag{5.8.41}$$

To justify this, assume first that $p \in (1,\infty)$ and $s \in (0,1)$. Since as is apparent from (5.8.38), the integral kernel $k(x,y) := \nabla_x \big[\partial_{\nu(y)} G_\Omega(x,y)\big]$ of the operator $\nabla\mathrm{PI}$ satisfies [112, (8.7.52)] for each $\varepsilon_* \in (0,1)$, we may invoke [112, (8.7.54) in Proposition 8.7.10] (with $\beta := 0$ and $\varepsilon_* > 1 - \frac{1}{p}$) to conclude (also keeping in mind [113, (7.9.10)]) that (5.8.41) holds. The case when $\frac{n-1}{n} < p \leq 1$ and $(n-1)(\frac{1}{p}-1) < s < 1$ may be reduced to this by using a suitable embedding (cf. [113, §7.7]), and this completes the proof of (5.8.41).

From (5.8.40)-(5.8.41), the second formula in (5.8.39), and [113, (9.2.159)] (presently used with $L := \Delta$) we may then conclude that PI maps $B^{p,p}_s(\partial\Omega,\sigma)$ boundedly into $F^{p,q}_{s+\frac{1}{p}}(\Omega)$ whenever $0 < p < \infty$, $(n-1)(\frac{1}{p}-1)_+ < s < 1$, and $\frac{n}{n+s+1/p} < q \leq \infty$. The extension to $q \in (0,\infty]$ then follows by invoking [76, Theorem 1.6]. Finally, the boundedness of the first operator in (5.8.37) now follows from this and real interpolation (cf. [122, Proposition 2.57, pp. 102-103] and [76, Theorem 9.4]). □

Chapter 6
Scattering by Rough Obstacles

The Gauss-Green formula (aka Divergence Theorem) has proved to be a very useful tool in scattering theory. For scattering by rough obstacles one needs such a formula for rough integrands and rough boundaries, of the sort we have developed in Volume I ([112]). The behavior at infinity is of particular significance, and various decay conditions (aka radiation conditions) have been used in the literature.

In this chapter the goal is to provide a unified approach to scattering theory in rough settings which is also aimed at identifying the broadest possible spectrum of radiation conditions for null-solutions of the vector Helmholtz operator $\Delta + k^2$. The latter contains, as particular cases, the Sommerfeld, Silver-Müller, and McIntosh-Mitrea radiation conditions corresponding to scattering by acoustic waves, electromagnetic waves, and null-solutions of perturbed Dirac operators, respectively. In this regard, we succeeded in producing a family of radiation conditions, (RC_A), indexed by all possible writings of the vector Laplace operator as $\Delta = \mathrm{div} A \nabla$ involving complex coefficient tensors A. In turn, this opens the door for proving well-posedness results and for boundary value problems as well as integral representation formulas for the vector Helmholtz operator in some very general types of exterior domains.

6.1 Integral Representations for Null-Solutions of the Helmholtz Operator

To discuss certain scattering phenomena, throughout we assume that $n \in \mathbb{N}$, $n \geq 2$, and fix a wave number $k \in (0, \infty)$. We also agree to abbreviate

$$\widehat{x} := x/|x| \quad \text{for each} \quad x \in \mathbb{R}^n \setminus \{0\}. \tag{6.1.1}$$

Definition 6.1.1 *Call a complex-valued smooth function u defined in an exterior domain in $\mathbb{R}^n$* `radiating at infinity` *provided u is a null-solution of the Helmholtz operator $\Delta + k^2$ and satisfies Sommerfeld's radiation condition*

D. Mitrea et al., *Geometric Harmonic Analysis III*, Developments in Mathematics 74,
https://doi.org/10.1007/978-3-031-22735-6_6

$$\int_{|x|=R} \left|iku(x) - \widehat{x} \cdot (\nabla u)(x)\right|^2 \, d\mathcal{H}^{n-1}(x) = o(1) \text{ as } R \to \infty. \tag{6.1.2}$$

Furthermore, call a vector-valued function radiating at infinity provided each of its scalar components is a null-solution of the Helmholtz operator $\Delta + k^2$ and satisfies Sommerfeld's radiation condition (6.1.2).

Originally (6.1.2) was augmented by the Sommerfeld's finiteness condition

$$\int_{|x|=R} |u(x)|^2 \, d\mathcal{H}^{n-1}(x) = O(1) \text{ as } R \to \infty, \tag{6.1.3}$$

However, it is now known that (6.1.2) and (6.1.3) are not independent in the sense that, for null-solutions of the Helmholtz operator, the former implies the latter. This is clear from the lemma below, which is a particular case of a more general result proved later in Proposition 6.5.1.

Lemma 6.1.2 *Let Ω be an exterior domain in $\mathbb{R}^n$, where $n \in \mathbb{N}$ with $n \geq 2$, and fix $k \in (0, \infty)$. Suppose $u \in \mathscr{C}^\infty(\Omega)$ satisfies $(\Delta + k^2)u = 0$ in Ω and*

$$\int_{|x|=R} \left|iku(x) - \widehat{x} \cdot (\nabla u)(x)\right|^2 \, d\mathcal{H}^{n-1}(x) = O(1) \text{ as } R \to \infty. \tag{6.1.4}$$

Then

$$\int_{|x|=R} |u(x)|^2 \, d\mathcal{H}^{n-1}(x) = O(1) \text{ as } R \to \infty \tag{6.1.5}$$

and

$$\int_{|x|=R} \left|\widehat{x} \cdot (\nabla u)(x)\right|^2 \, d\mathcal{H}^{n-1}(x) = O(1) \text{ as } R \to \infty. \tag{6.1.6}$$

For example (see the discussion in [91]), the only fundamental solution of the Helmholtz operator $\Delta + k^2$ in $\mathbb{R}^n$ which is radiating at infinity is given at each $x \in \mathbb{R}^n \setminus \{0\}$ by

$$\Phi_k(x) := c_n k^{(n-2)/2} \frac{H^{(1)}_{(n-2)/2}(k|x|)}{|x|^{(n-2)/2}}, \text{ with } c_n := \frac{1}{4i(2\pi)^{(n-2)/2}}, \tag{6.1.7}$$

where, generally speaking, $H^{(1)}_\lambda(\cdot)$ denotes the Hankel function of the first kind with index $\lambda \in \mathbb{R}$ (cf. [1, § 9.1]).

To state our first major result in this section, recall the truncated nontangential maximal operator from (A.0.92).

Theorem 6.1.3 *Let $n \in \mathbb{N}$, $n \geq 2$, and fix a wave number $k \in (0, \infty)$. Assume Ω is an exterior domain in $\mathbb{R}^n$ with a lower Ahlfors regular boundary, such that $\sigma := \mathcal{H}^{n-1} \lfloor \partial\Omega$ is a doubling measure on $\partial\Omega$. This implies that Ω is a set of locally finite perimeter, and its geometric measure theoretic outward unit normal ν is defined σ-a.e. on $\partial_*\Omega$.*

Suppose $u, w \in \mathscr{C}^\infty(\Omega)$ are two null-solutions of the Helmholtz operator $\Delta + k^2$ in Ω which are radiating at infinity and such that, for some parameters $\kappa, \kappa' > 0$, and $\varepsilon > 0$, satisfy

$$\begin{gathered}\mathcal{N}_\kappa^\varepsilon u \cdot \mathcal{N}_{\kappa'}^\varepsilon(\nabla w) \in L^1(\partial\Omega, \sigma), \quad \mathcal{N}_\kappa^\varepsilon(\nabla u) \cdot \mathcal{N}_{\kappa'}^\varepsilon w \in L^1(\partial\Omega, \sigma),\\ \text{and the traces } u\big|_{\partial\Omega}^{\kappa-\text{n.t.}}, \ (\nabla u)\big|_{\partial\Omega}^{\kappa-\text{n.t.}}, \ w\big|_{\partial\Omega}^{\kappa'-\text{n.t.}}, \ (\nabla w)\big|_{\partial\Omega}^{\kappa'-\text{n.t.}}\\ \text{exist at } \sigma\text{-almost every point on } \partial_{\text{nta}}\Omega.\end{gathered} \tag{6.1.8}$$

Then for any other $\kappa'' > 0$ the nontangential traces $u\big|_{\partial\Omega}^{\kappa''-\text{n.t.}}$, $(\nabla u)\big|_{\partial\Omega}^{\kappa''-\text{n.t.}}$, $w\big|_{\partial\Omega}^{\kappa''-\text{n.t.}}$, $(\nabla w)\big|_{\partial\Omega}^{\kappa''-\text{n.t.}}$ also exist at σ-a.e. point on $\partial_{\text{nta}}\Omega$, and are actually independent of κ''. Moreover, with the dependence on the parameter κ'' dropped, the following Green type formula holds:

$$\int_{\partial_*\Omega} \big(u\big|_{\partial\Omega}^{\text{n.t.}}\big)\nu \cdot \big((\nabla w)\big|_{\partial\Omega}^{\text{n.t.}}\big)\, d\sigma = \int_{\partial_*\Omega} \big(w\big|_{\partial\Omega}^{\text{n.t.}}\big)\nu \cdot \big((\nabla u)\big|_{\partial\Omega}^{\text{n.t.}}\big)\, d\sigma. \tag{6.1.9}$$

Parenthetically we wish to note that the conditions imposed in the first line of (6.1.8) are automatically satisfied if

$$\begin{aligned}&\mathcal{N}_\kappa^\varepsilon u \in L^p(\partial\Omega, \sigma), \quad \mathcal{N}_\kappa^\varepsilon(\nabla u) \in L^q(\partial\Omega, \sigma),\\ &\mathcal{N}_{\kappa'}^\varepsilon w \in L^{q'}(\partial\Omega, \sigma), \ \mathcal{N}_{\kappa'}^\varepsilon(\nabla w) \in L^{p'}(\partial\Omega, \sigma),\end{aligned} \tag{6.1.10}$$

for some $p, p', q, q' \in [1, \infty]$ with $1/p + 1/p' = 1/q + 1/q' = 1$.

Proof of Theorem 6.1.3 The first conclusion in the statement of the theorem is a direct consequence of assumptions and the comment following the statement of [112, Proposition 8.9.8], so we focus on establishing the Green type formula formulated in (6.1.9). To this end, consider the vector field

$$\vec{F} := u\nabla w - w\nabla u \in \big[\mathscr{C}^\infty(\Omega)\big]^n \tag{6.1.11}$$

and note that $\vec{F}$ is divergence-free since

$$\begin{aligned}\operatorname{div}\vec{F} &= \nabla u \cdot \nabla w + u\Delta w - \nabla w \cdot \nabla u - w\Delta u\\ &= u(\Delta + k^2)w - w(\Delta + k^2)u = 0 \ \text{ in } \Omega.\end{aligned} \tag{6.1.12}$$

In addition, the properties listed in (6.1.8) imply (with the dependence on the aperture parameter dropped)

$$\mathcal{N}^\varepsilon \vec{F} \in L^1(\partial\Omega, \sigma) \text{ and } \vec{F}\big|_{\partial\Omega}^{\text{n.t.}} \text{ exists } \sigma\text{-a.e. on } \partial_{\text{nta}}\Omega. \tag{6.1.13}$$

In fact, at σ-a.e. point on $\partial_*\Omega$ we have

$$\nu \cdot \big(\vec{F}\big|_{\partial\Omega}^{\text{n.t.}}\big) = \big(u\big|_{\partial\Omega}^{\text{n.t.}}\big)\nu \cdot \big((\nabla w)\big|_{\partial\Omega}^{\text{n.t.}}\big) - \big(w\big|_{\partial\Omega}^{\text{n.t.}}\big)\nu \cdot \big((\nabla u)\big|_{\partial\Omega}^{\text{n.t.}}\big). \tag{6.1.14}$$

On account of (6.1.11)-(6.1.14), the Green type identity (6.1.9) is implied by the Divergence Formula [112, (1.4.5)] as soon as we prove that the contribution on $\vec{F}$ at infinity, i.e., the complex number (cf. [112, (4.7.5)])

$$[\vec{F}]_\infty = \lim_{R\to\infty} \int_{|x|=R} \widehat{x}\cdot \vec{F}(x)\, d\mathcal{H}^{n-1}(x), \tag{6.1.15}$$

presently vanishes. To see that this is the case, we write

$$\begin{aligned}
&\Big|\int_{|x|=R} \widehat{x}\cdot \vec{F}(x)\, d\mathcal{H}^{n-1}(x)\Big| \\
&= \Big|\int_{|x|=R} \big\{u(x)\widehat{x}\cdot(\nabla w)(x) - w(x)\widehat{x}\cdot(\nabla u)(x)\big)\, d\mathcal{H}^{n-1}(x)\Big| \\
&= \Big|\int_{|x|=R} \big\{u(x)\big(\widehat{x}\cdot(\nabla w)(x) - ikw(x)\big) \\
&\qquad - w(x)\big(\widehat{x}\cdot(\nabla u)(x) - iku(x)\big)\big\}\, d\mathcal{H}^{n-1}(x)\Big| \\
&\le \Big(\int_{|x|=R} |u(x)|^2\, d\mathcal{H}^{n-1}(x)\Big)^{1/2} \times \\
&\qquad \times \Big(\int_{|x|=R} \big|\widehat{x}\cdot(\nabla w)(x) - ikw(x)\big|^2\, d\mathcal{H}^{n-1}(x)\Big)^{1/2} \\
&\quad + \Big(\int_{|x|=R} |w(x)|^2\, d\mathcal{H}^{n-1}(x)\Big)^{1/2} \times \\
&\qquad \times \Big(\int_{|x|=R} \big|\widehat{x}\cdot(\nabla u)(x) - iku(x)\big|^2\, d\mathcal{H}^{n-1}(x)\Big)^{1/2} \\
&= O(1)\cdot o(1) + O(1)\cdot o(1) = o(1) \text{ as } R\to\infty,
\end{aligned} \tag{6.1.16}$$

where the last equality is based on the Sommerfeld's finiteness and radiation conditions for u and w (cf. (6.1.2)-(6.1.3)). From (6.1.15) and (6.1.16) we then conclude that $[\vec{F}]_\infty = 0$, which finishes the proof of (6.1.9). □

In the next theorem we present a boundary layer integral representation formula for null-solutions of the Helmholtz operator which are radiating at infinity.

Theorem 6.1.4 *Fix $n \in \mathbb{N}$, $n \ge 2$, along with a wave number $k \in (0,\infty)$. Suppose Ω is an exterior domain in $\mathbb{R}^n$ with a lower Ahlfors regular boundary and such that $\sigma := \mathcal{H}^{n-1}\lfloor\partial\Omega$ is a doubling measure on $\partial\Omega$. In particular, Ω is a set of locally finite perimeter, and its geometric measure theoretic outward unit normal ν is defined σ-a.e. on $\partial_*\Omega$.*

Assume $u \in \mathscr{C}^\infty(\Omega)$ is a null-solution of the Helmholtz operator $\Delta + k^2$ in Ω which is radiating at infinity and such that, for some parameters $\kappa > 0$, $\varepsilon > 0$, satisfies

$$\begin{aligned}&\mathcal{N}_\kappa^\varepsilon u \in L^1(\partial\Omega, \sigma), \quad \mathcal{N}_\kappa^\varepsilon(\nabla u) \in L^1(\partial\Omega, \sigma),\\ &\textit{and } u\big|_{\partial\Omega}^{\kappa-\text{n.t.}}, \ (\nabla u)\big|_{\partial\Omega}^{\kappa-\text{n.t.}} \ \textit{exist } \sigma\textit{-a.e. on } \partial_{\text{nta}}\Omega.\end{aligned} \tag{6.1.17}$$

Then for any other $\kappa' > 0$ the nontangential traces $u\big|_{\partial\Omega}^{\kappa'-\text{n.t.}}$, $(\nabla u)\big|_{\partial\Omega}^{\kappa'-\text{n.t.}}$ also exist at σ-a.e. point on $\partial_{\text{nta}}\Omega$, and are actually independent of κ'. Furthermore, with the dependence on the parameter κ' dropped, the function u admits the following boundary layer integral representation formula:

$$\begin{aligned}u(x) = &\int_{\partial_*\Omega} \nu(y) \cdot \nabla_y \big[\Phi_k(x-y)\big] \big(u\big|_{\partial\Omega}^{\text{n.t.}}\big)(y)\, d\sigma(y)\\ &- \int_{\partial_*\Omega} \Phi_k(x-y)\nu(y) \cdot \big((\nabla u)\big|_{\partial\Omega}^{\text{n.t.}}\big)(y)\, d\sigma(y) \ \textit{for each } x \in \Omega.\end{aligned} \tag{6.1.18}$$

As a preamble to the proof of this theorem, we record the following lemma (cf. [91]), pertaining to the nature of the function Φ_k.

Lemma 6.1.5 *Fix $k \in (0,\infty)$ and $n \in \mathbb{N}$, $n \geq 2$. Then for each fixed $R \in (0,\infty)$ there exists some finite constant $C = C(R,n,k) > 0$ with the property that the function Φ_k from* (6.1.7) *satisfies*

$$|\Phi_k(x)| \leq \begin{cases} C\big[1 + \big|\ln|x|\big|\big], & n = 2,\\ C|x|^{2-n}, & n \geq 3,\end{cases} \qquad \textit{and} \qquad |(\nabla\Phi_k)(x)| \leq C|x|^{1-n}, \tag{6.1.19}$$

whenever $x \in \mathbb{R}^n$ is such that $0 < |x| < R$.

We now turn to the proof of Theorem 6.1.4.

Proof of Theorem 6.1.4 The first conclusion in the statement of the theorem is given by [112, Proposition 8.9.8] (see the comment following its statement), so we focus on proving the integral representation formula (6.1.18). With this goal in mind, fix an arbitrary point $x \in \Omega$. The idea is to adapt the proof of Theorem 6.1.3 to the case when $w(y) := \Phi_k(x-y)$ for each $y \in \Omega \setminus \{x\}$. While this has a locally integrable singularity at x, the function w may be nonetheless canonically viewed as a distribution in $\mathcal{D}'(\Omega)$ with compact singular support, which satisfies

$$(\Delta + k^2)w = \delta_x \ \text{ in } \mathcal{D}'(\Omega), \tag{6.1.20}$$

where δ_x is the Dirac distribution in Ω with mass at x, and which radiates at infinity. If analogously to (6.1.11) we now define the vector field

$$\vec{F}(y) := u(y)\nabla_y\big[\Phi_k(x-y)\big] - \Phi_k(x-y)(\nabla u)(y), \qquad \forall y \in \Omega \setminus \{x\}, \tag{6.1.21}$$

then Lemma 6.1.5 implies that $\vec{F} \in \big[L^1_{\text{loc}}(\Omega, \mathcal{L}^n)\big]^n$ and, in place of (6.1.12), we now have

$$\text{div}\vec{F} = u(x)\delta_x \ \text{ in } \mathcal{D}'(\Omega). \tag{6.1.22}$$

The current assumptions also ensure that $\vec{F}$ has the properties listed in (6.1.13). In fact, at σ-a.e. point $y \in \partial_*\Omega$ we presently have

$$\nu(y) \cdot \big(\vec{F}\big|^{\text{n.t.}}_{\partial\Omega}\big)(y) = \nu(y) \cdot \nabla_y \big[\Phi_k(x-y)\big] \big(u\big|^{\text{n.t.}}_{\partial\Omega}\big)(y) \\ - \Phi_k(x-y)\nu(y) \cdot \big((\nabla u)\big|^{\text{n.t.}}_{\partial\Omega}\big)(y). \tag{6.1.23}$$

Finally, the contribution of $\vec{F}$ at infinity may be estimated as in the past (cf. (6.1.16)) bearing in mind that, as noted earlier, $\Phi_k(x-\cdot)$ radiates at infinity. Hence, as before, we see that $[\vec{F}]_\infty = 0$. Granted this, the Divergence Formula [112, (1.4.5)] then yields (6.1.18) on account of (6.1.22) and (6.1.23). □

6.2 Radiation Conditions: Motivation

Throughout, assume $n \in \mathbb{N}$ satisfies $n \geq 2$, and fix a wave number $k \in (0, \infty)$. Suppose $\Omega \subseteq \mathbb{R}^n$ is an exterior domain (i.e., a set whose complement is compact) which is connected, and consider the Dirichlet Problem for the Helmholtz operator:

$$\begin{cases} u \in \mathscr{C}^\infty(\overline{\Omega}), \\ (\Delta + k^2)u = 0 \text{ in } \Omega, \\ u\big|_{\partial\Omega} = f \text{ on } \partial\Omega. \end{cases} \tag{6.2.1}$$

Typically this is accompanied by a suitable decay condition for u at infinity, designed to ensure uniqueness.

With J_λ denoting the regular Bessel function of the first kind of order $\lambda \in \mathbb{R}$, let $r > 0$ be such that $J_{(n-2)/2}(r) = 0$. If we then define[1]

$$\Omega := \mathbb{R}^n \setminus \overline{B(0, r/k)} \text{ and } u(x) := \frac{J_{(n-2)/2}(k|x|)}{|x|^{(n-2)/2}}, \qquad \forall x \in \Omega, \tag{6.2.2}$$

then $u \in \mathscr{C}^\infty(\overline{\Omega})$, $(\Delta + k^2)u = 0$ in Ω, $u\big|_{\partial\Omega} = 0$, and

$$(\partial^\alpha u)(x) = O\big(|x|^{-(n-1)/2}\big) \text{ as } |x| \to \infty, \\ \text{for each multi-index } \alpha \in \mathbb{N}_0^n. \tag{6.2.3}$$

This goes to show that imposing the decay condition

$$u(x) = O\big(|x|^{-(n-1)/2}\big) \text{ as } |x| \to \infty \tag{6.2.4}$$

in the boundary value problem (6.2.1) does not lead to uniqueness. Since (6.2.4) is too weak, one may be tempted to ask

[1] for $n = 3$ this takes the form $u(x) = \frac{\sin(k|x|)}{|x|}$ for $x \in \Omega := \mathbb{R}^3 \setminus \overline{B(0, \pi/k)}$

$$u(x) = o\big(|x|^{-(n-1)/2}\big) \text{ as } |x| \to \infty. \tag{6.2.5}$$

However, in such a scenario Rellich's uniqueness lemma forces $u \equiv 0$ in Ω. Hence, (6.2.5) amounts to too much decay in order to have a nontrivial solution.

In order to ensure uniqueness of the solutions of the Dirichlet Problem for the Helmholtz operator in exterior domains (cf. (6.2.1)), in 1912 A. Sommerfeld has introduced in [164] the condition

$$iku(x) - \sum_{j=1}^{n} \widehat{x}_j(\partial_j u)(x) = o\big(|x|^{-(n-1)/2}\big) \text{ as } |x| \to \infty, \tag{6.2.6}$$

where $\widehat{x} := \frac{x}{|x|} \in S^{n-1}$ is the normalization of $x \in \mathbb{R}^n \setminus \{0\}$ to a unit vector (cf. (6.1.1)). This is referred to as Sommerfeld's radiation condition.

Let us briefly review some other known radiation conditions in the literature. Radiation conditions for the Maxwell's system have been identified by C. Müler (in 1945; cf. [140]) and S. Silver (in 1949; cf. [162]). Specifically, if $\vec{E} = (E_1, E_2, E_3)$ and $\vec{H} = (H_1, H_2, H_3)$ are two vector fields satisfying the time-harmonic Maxwell system in a three-dimensional exterior domain Ω

$$\operatorname{curl} \vec{E} - ik\,\vec{H} = 0 \text{ and } \operatorname{curl} \vec{H} + ik\,\vec{E} = 0 \text{ in } \Omega, \tag{6.2.7}$$

the classical Silver-Müller radiation conditions read

$$\widehat{x} \times \vec{H}(x) + \vec{E}(x) = o(|x|^{-1}) \text{ as } |x| \to \infty, \tag{6.2.8}$$

$$\widehat{x} \times \vec{E}(x) - \vec{H}(x) = o(|x|^{-1}) \text{ as } |x| \to \infty. \tag{6.2.9}$$

Another known radiation condition in the literature is that for the perturbed Dirac equation, identified by A. McIntosh and M. Mitrea (in 1999; cf. [102]). To elaborate on this topic, let $(\mathcal{C}\ell_{n+1}, +, \odot)$ be the Clifford algebra generated by $n+1$ anti-commuting imaginary units $(\mathbf{e}_i)_{1\leq i\leq n+1}$ (see the discussion in [112, §6.4]). Recall that

$$\mathbf{D} := \sum_{j=1}^{n} \mathbf{e}_j \odot \partial_j \tag{6.2.10}$$

is the homogeneous the Dirac operator in $\mathbb{R}^n$ and let

$$\mathbf{D}_k := \mathbf{D} + k\mathbf{e}_{n+1} \tag{6.2.11}$$

be the perturbed Dirac operator in $\mathbb{R}^n$ (hence, $\mathbf{D}_k u = \mathbf{D}u + k\mathbf{e}_{n+1} \odot u$ for each $u = \sum_I u_I e_I$). Then the McIntosh-Mitrea radiation condition for Clifford algebra-valued null-solutions u of the perturbed Dirac operator $\mathbf{D}_k$ in an exterior domain $\Omega \subseteq \mathbb{R}^n$ reads

$$(i - \widehat{x} \odot \mathbf{e}_{n+1}) \odot u(x) = o\big(|x|^{-(n-1)/2}\big) \text{ as } |x| \to \infty. \tag{6.2.12}$$

Observe that

any null-solution of the Maxwell system or the perturbed Dirac operator is also a null-solution of the (vector) Helmholtz operator $\Delta + k^2$. (6.2.13)

This prompts the question:

given that both the Maxwell system and the perturbed Dirac equation are "sub-problems" of the (vector) Helmholtz equation, why do the radiation conditions identified for each of these PDE's look so different? (6.2.14)

Our answer is that, as we shall see, *Sommerfeld's radiation condition is just one of a more flexible and inclusive family of (equivalent) radiation conditions for null-solutions of the Helmholtz equation*. Our goal is to find them all. To this end, we shall need to introduce some proper algebraic formalism.

For some fixed $M \in \mathbb{N}$, consider the vector Laplacian Δ in $\mathbb{R}^n$, viewed as a second-order $M \times M$ system in $\mathbb{R}^n$, acting componentwise on vector-valued functions $u = \big(u_I\big)_{1 \le I \le M}$ according to

$$\Delta u = \big(\Delta u_I\big)_{\le I \le M}. \tag{6.2.15}$$

In other words, we identify

$$\Delta \equiv \Delta\, I_{M \times M} = \big(\delta_{JI}\Delta\big)_{1 \le J, I \le M}. \tag{6.2.16}$$

Suppose a coefficient tensor $A := \big(a_{rs}^{JI}\big)_{\substack{1 \le J, I \le M\\ 1 \le r,s \le n}}$ with complex constant entries has been chosen so that we may express said vector Laplacian as

$$\Delta = \Big(\sum_{r,s=1}^{n} a_{rs}^{JI}\partial_r\partial_s\Big)_{1 \le J, I \le M}. \tag{6.2.17}$$

Henceforth, we agree to succinctly write

$$\Delta = \mathrm{div} A \nabla \tag{6.2.18}$$

in place of (6.2.17). For further use, observe that

$$\begin{gathered} a_{rs}^{IJ}\xi_s\xi_r = \delta_{IJ}|\xi|^2 \text{ for each } \xi = (\xi_1, \dots, \xi_n) \in \mathbb{R}^n \\ \text{and each } I, J \in \{1, \dots, M\}. \end{gathered} \tag{6.2.19}$$

Indeed, since for each $I, J \in \{1, \dots, M\}$ we know from (6.2.16)-(6.2.17) that $a_{rs}^{IJ}\partial_r\partial_s$ and $\delta_{IJ}\partial_j\partial_j$ are identical partial differential operators, they have identical principal symbols, so (6.2.19) follows. In particular, (6.2.19) implies

$$a_{rs}^{IJ}\widehat{x}_s\widehat{x}_r = \delta_{IJ} \text{ for each } x \in \mathbb{R}^n \setminus \{0\} \text{ and each } I, J \in \{1, \dots, M\}. \tag{6.2.20}$$

Going forward, the conormal derivative ∂_ξ^A associated with each coefficient tensor $A := \big(a_{rs}^{JI}\big)_{\substack{1 \le J, I \le M\\ 1 \le r,s \le n}}$ in the writing of the vector Laplacian as $\Delta = \mathrm{div} A \nabla$ and each vector $\xi = (\xi_r)_r \in \mathbb{C}^n$ is defined as

$$\partial_\xi^A u := \Big(\sum_{I=1}^{M}\sum_{r,s=1}^{n} a_{rs}^{JI}\xi_r\partial_s u_I\Big)_{1\leq J\leq M} \tag{6.2.21}$$

for each given vector-valued function $u = \big(u_I\big)_{1\leq I\leq M}$. Finally, we note that

$$\begin{aligned}&\text{for } A_0 := \big(\delta_{JI}\delta_{rs}\big)_{\substack{1\leq J,I\leq M\\ 1\leq r,s\leq n}} \text{ we may express } \Delta = \operatorname{div}A_0\nabla, \text{ and we have}\\ &\partial_\xi^{A_0}u = \big(\xi_r\partial_r u_I\big)_{1\leq I\leq M} \text{ for each function } u = (u_I)_{1\leq I\leq M} \text{ and each}\\ &\text{vector } \xi = (\xi_r)_r \in \mathbb{C}^n.\end{aligned} \tag{6.2.22}$$

Next, fix an arbitrary integer $M \in \mathbb{N}$ and consider a complex coefficient tensor $A := \big(a_{rs}^{JI}\big)_{\substack{1\leq J,I\leq M\\ 1\leq r,s\leq n}}$ in the writing of the vector Laplacian as $\Delta = \operatorname{div}A\nabla$. For each (doubly indexed) vector $\zeta = (\zeta_{sI})_{\substack{1\leq s\leq n\\ 1\leq I\leq M}}$ with complex components define

$$A\zeta := \Big(\sum_{I=1}^{M}\sum_{r,s=1}^{n} a_{rs}^{JI}\zeta_{sI}\Big)_{\substack{1\leq r\leq n\\ 1\leq J\leq M}}. \tag{6.2.23}$$

Consequently, if $\langle\cdot,\cdot\rangle$ denotes the real-pairing of vectors with complex components, we have

$$\langle A\zeta,\overline{\zeta}\rangle = \sum_{I,J=1}^{M}\sum_{r,s=1}^{n} a_{rs}^{JI}\zeta_{sI}\overline{\zeta_{rJ}}. \tag{6.2.24}$$

This is going to be relevant later on (e.g., when discussing L^2-finiteness results).

6.3 The Family of Radiation Conditions (RC_A)

Fix $M \in \mathbb{N}$ arbitrary. For each complex coefficient tensor $A = \big(a_{rs}^{JI}\big)_{\substack{1\leq J,I\leq M\\ 1\leq r,s\leq n}}$ in the writing of the vector Laplacian as $\Delta = \operatorname{div}A\nabla$ we shall consider the radiation condition

$$\Big(\sum_{1\leq I\leq M}\sum_{r,s=1}^{n} a_{rs}^{JI}\widehat{x}_r\partial_s u_I(x) - iku_J(x)\Big)_{1\leq J\leq M} = o\big(|x|^{-(n-1)/2}\big) \quad \text{as } |x| \to \infty. \tag{6.3.1}$$

Equivalently, this may be expressed in terms of the conormal derivative (6.2.21) as

$$iku(x) - \big(\partial_{\widehat{x}}^A u\big)(x) = o\big(|x|^{-(n-1)/2}\big) \text{ as } |x| \to \infty. \tag{6.3.2}$$

We shall say that u satisfies the radiation condition (RC_A) provided (6.3.2) holds.

As is apparent from (6.2.21), the radiation condition (6.3.2) involves the conormal derivative associated with the factorization of the vector Laplacian $\Delta = \operatorname{div} A \nabla$, evaluated at unit vector S^{n-1} in the direction of x.

In summary: *The vector Laplacian has many writings as* $\Delta = \operatorname{div} A \nabla$ *for various choices of a complex coefficient tensor A and, associated with each of those writings, we have proposed the radiation condition* (RC_A). *We therefore have a family of radiation conditions indexed by these complex coefficient tensors A.*

In relation to this, we seek to address the following two questions:

Question 1: How are the radiation conditions (RC_A), corresponding to various choices of of complex coefficient tensor A (used to represent vector Laplacian as $\Delta = \operatorname{div} A \nabla$) related to one another? In particular, how do they relate to the classical Sommerfeld's (pointwise) radiation condition?

Question 2: Under what kind of assumptions on its behavior at infinity and near the boundary does a given null-solution of the (vector) Helmholtz operator $\Delta + k^2$ in an exterior domain admit a Green type integral representation formula?

We shall provide satisfactory answers to both questions in Theorem 6.6.1, formulated a little later. For now, recall that Φ_k, introduced in (6.1.7), is the unique distribution in $\mathbb{R}^n$ which is a fundamental solution of $\Delta + k^2$ in $\mathbb{R}^n$ and satisfies Sommerfeld's radiation condition. Corresponding to the limiting case $k = 0$ (when the Helmholtz operator $\Delta + k^2$ is simply the Laplacian Δ), let us define

$$\Phi_0(x) := \Phi_0(x; n) := \begin{cases} \dfrac{1}{(2-n)\omega_{n-1}} |x|^{2-n}, & x \in \mathbb{R}^n \setminus \{0\}, \text{ if } n \geq 3, \\ \dfrac{1}{2\pi} \ln |x|, & x \in \mathbb{R}^2 \setminus \{0\}, \text{ if } n = 2. \end{cases} \tag{6.3.3}$$

We shall temporarily adopt a more general point of view and work with the Helmholtz operator $\Delta + z$ involving a generic wave number $z \in \mathbb{C} \setminus (-\infty, 0]$. To be specific, fix $n \in \mathbb{N}$ satisfying $n \geq 2$. For each $z \in \mathbb{C} \setminus (-\infty, 0]$, let $\Phi_z(x; n)$ be the fundamental solution of the Helmholtz operator $\Delta + z$ in $\mathbb{R}^n$ given by (compare with (6.1.7))

$$\Phi_z(x; n) := -\frac{i}{4} \left(\frac{z^{1/2}}{2\pi |x|} \right)^{(n-2)/2} H^{(1)}_{(n-2)/2}\big(z^{1/2}|x|\big) \text{ for all } x \in \mathbb{R}^n \setminus \{0\}, \tag{6.3.4}$$

where $H^{(1)}_\lambda(\cdot)$ denotes the Hankel function of the first kind with index $\lambda \in \mathbb{R}$ (cf. [1, § 9.1]). For each fixed point $x \in \mathbb{R}^n \setminus \{0\}$, the function $\Phi_z(x; n)$ happens to have a limit as $\big(\mathbb{C} \setminus (-\infty, 0]\big) \ni z \to 0$ in higher dimensions, namely

$$\lim_{z \to 0} \Phi_z(x; n) = \Phi_0(x; n) = \frac{1}{(2-n)\omega_{n-1}} |x|^{2-n} \text{ if } n \geq 3, \tag{6.3.5}$$

but this connection between $\Phi_z(x; n)$ and $\Phi_0(x; n)$ fails to materialize for $n = 2$, since

$$\begin{aligned}\Phi_z(x;2) = \tfrac{1}{2\pi}\ln\big(z^{1/2}|x|/2\big)\big[1+O\big(z|x|^2\big)\big] - \tfrac{1}{2\pi}\psi(1) + O\big(z|x|^2\big),\\ \text{as } z \longrightarrow 0, \text{ uniformly for } x \text{ in compact subsets of } \mathbb{R}^2 \setminus \{0\}.\end{aligned} \tag{6.3.6}$$

Here $\psi(w) := \Gamma'(w)/\Gamma(w)$ denotes the digamma function (cf. [1, § 6.3, (6.3.1), p. 258]), where $\Gamma(w)$ with $w \in \mathbb{C} \setminus \{-n : n \in \mathbb{N}\}$ is the classical Gamma function (cf., e.g., [109, § 14.5, p. 566]).

In the lemma below (which expands on [50, Lemma C.1]) we elucidate the asymptotic behavior at zero and infinity of the fundamental solution of the Helmholtz operator recalled in (6.3.4).

Lemma 6.3.1 *Fix $n \in \mathbb{N}$ with $n \geq 2$, along with $z \in \mathbb{C} \setminus (-\infty, 0]$ and $R \in (0, \infty)$. In addition, select some arbitrary multi-index $\alpha \in \mathbb{N}_0^n$. Then there exists some constant $C = C(n, z, R, \alpha) \in (0, \infty)$ with the following significance. If $|\alpha|$ is even then for each $x \in \mathbb{R}^n$ with $0 < |x| < R$ one has*

$$\big|\partial_x^\alpha \Phi_z(x;n) - \partial_x^\alpha \Phi_0(x;n)\big| \leq \begin{cases} C & \textit{if } n = 2, 3 \textit{ and } |\alpha| = 0,\\ C\big[\big|\ln|x|\big| + 1\big] & \textit{if } n + |\alpha| = 4,\\ C|x|^{4-n-|\alpha|} & \textit{if } n + |\alpha| \geq 5, \end{cases} \tag{6.3.7}$$

while if $|\alpha|$ is odd then for each $x \in \mathbb{R}^n$ with $0 < |x| < R$ one has

$$\big|\partial_x^\alpha \Phi_z(x;n) - \partial_x^\alpha \Phi_0(x;n)\big| \leq \begin{cases} C & \textit{if } n + |\alpha| = 3,\\ C|x|^{4-n-|\alpha|} & \textit{if } n + |\alpha| \geq 4. \end{cases} \tag{6.3.8}$$

Finally, for large values of $|x|$, formula (6.3.4) *implies (cf.* [1, § 9.1]) *the following asymptotic behavior at infinity:*

$$\begin{aligned}\Phi_z(x;n) = -\frac{\mathrm{i}}{2z^{1/2}}\left(\frac{z^{1/2}}{2\pi|x|}\right)^{(n-1)/2} e^{\mathrm{i}\left(z^{1/2}|x| - \pi((n-1)/4)\right)}\big[1 + O\big(|x|^{-1}\big)\big]\\ \textit{as } |x| \longrightarrow \infty, \textit{ for all } z \in \mathbb{C} \setminus (-\infty, 0] \textit{ and } n \geq 2.\end{aligned} \tag{6.3.9}$$

In particular, as long as $\operatorname{Im}(z^{1/2}) > 0$ it follows that $\Phi_z(x;n)$ decays exponentially with respect to $x \in \mathbb{R}^n \setminus \{0\}$ as $|x| \longrightarrow \infty$.

Proof We begin by recalling (cf. [1, § 9.1]) that

$$H^{(1)}_{(n-2)/2}(\cdot) = J_{(n-2)/2}(\cdot) + \mathrm{i}Y_{(n-2)/2}(\cdot), \tag{6.3.10}$$

where J_λ and Y_λ denote the regular and irregular Bessel functions, respectively. Let us also recall the following absolutely convergent expansions (cf. [1, § 9.1, (9.1.10)-(9.1.11), p. 360]):

$$J_\lambda(\zeta) = \left(\frac{\zeta}{2}\right)^\lambda \sum_{j=0}^{\infty} \frac{(-1)^j \zeta^{2j}}{4^j j! \Gamma(\lambda + j + 1)}, \quad \zeta \in \mathbb{C} \setminus (-\infty, 0], \ \lambda \in \mathbb{R} \setminus (-\mathbb{N}), \tag{6.3.11}$$

$$J_{-m}(\zeta) = (-1)^m J_m(\zeta), \quad \zeta \in \mathbb{C}, \ m \in \mathbb{N}_0, \tag{6.3.12}$$

$$Y_\lambda(\zeta) = \frac{J_\lambda(\zeta)\cos(\lambda\pi) - J_{-\lambda}(\zeta)}{\sin(\lambda\pi)}, \quad \zeta \in \mathbb{C} \setminus (-\infty, 0], \ \lambda \in (0, \infty) \setminus \mathbb{N}, \tag{6.3.13}$$

$$\begin{aligned} Y_m(\zeta) &= -\frac{\zeta^{-m}}{2^m \pi} \sum_{j=0}^{m-1} \frac{(m-j-1)!}{j!} \cdot \frac{\zeta^{2j}}{4^j} + \frac{2}{\pi} J_m(\zeta) \ln(\zeta/2) \\ &\quad - \frac{\zeta^m}{2^m \pi} \sum_{j=0}^{\infty} [\psi(j+1) + \psi(m+j+1)] \frac{(-1)^j \zeta^{2j}}{4^j j! (m+j)!}, \end{aligned} \tag{6.3.14}$$

for $\zeta \in \mathbb{C} \setminus (-\infty, 0]$, $m \in \mathbb{N}_0$.

We note that all functions in (6.3.11), (6.3.13), and (6.3.14) are analytic in $\mathbb{C}\setminus(-\infty, 0]$ and that $J_m(\cdot)$ is entire for $m \in \mathbb{Z}$. In addition, all functions in (6.3.11)-(6.3.14) have continuous limits as $\zeta \in \mathbb{C} \setminus (-\infty, 0]$ approaches nontangentially points in the cut $(-\infty, 0]$, with generally different values on either side of the cut $(-\infty, 0]$ (due to the presence of the functions ζ^λ and $\ln(\zeta)$).

Due to the presence of the logarithmic term for even dimensions we next distinguish two cases, corresponding to even and odd values of n:

(i) Assume $n = 2m + 2$ for some $m \in \mathbb{N}_0$. Then for each $z \in \mathbb{C} \setminus (-\infty, 0]$ fixed, for each $x \in \mathbb{R}^n \setminus \{0\}$ we have

$$\begin{aligned} \Phi_z(x; 2m+2) &= -\frac{i}{4}\left(\frac{z^{1/2}}{2\pi|x|}\right)^m H_m^{(1)}(z^{1/2}|x|) \\ &= -\frac{i}{4}\left(\frac{z^{1/2}}{2\pi|x|}\right)^m \big[J_m(z^{1/2}|x|) + iY_m(z^{1/2}|x|)\big]. \end{aligned} \tag{6.3.15}$$

This implies that, as $|x| \to 0$,

$$\begin{aligned} \Phi_z(x; 2m+2) = -\frac{i}{4}\left(\frac{z^{1/2}}{2\pi|x|}\right)^m \bigg\{ &O(|x|^m) + \frac{2i}{\pi} \ln\left(\frac{z^{1/2}|x|}{2}\right) O(|x|^m) \\ &- \frac{i}{\pi}\left(\frac{2}{z^{1/2}|x|}\right)^m (1 - \delta_{m,0}) \left[(m-1)! + (1 - \delta_{m,1})(m-2)! \left(\frac{z|x|^2}{4}\right)\right] \\ &- \frac{i}{\pi}\left(\frac{2}{z^{1/2}|x|}\right)^m (1 - \delta_{m,0}) O(|x|^4) \bigg\}, \end{aligned} \tag{6.3.16}$$

where $\delta_{j,k}$ is the Kronecker symbol.

(ii) Assume $n = 2m + 1$ for some $m \in \mathbb{N}$. Then for each $z \in \mathbb{C} \setminus (-\infty, 0]$ fixed, making use of [1, (10.1.1), p. 437], [1, (10.1.16), p. 439], and [1, (10.1.9), p. 437] we may write

$$\begin{aligned}\Phi_z(x; 2m+1) &= -\frac{\mathrm{i}}{4}\left(\frac{z^{1/2}}{2\pi|x|}\right)^{m-(1/2)} H^{(1)}_{m-(1/2)}\big(z^{1/2}|x|\big) \\ &= -\frac{\mathrm{i}}{4} z^{m/2}\pi^{-m}2^{1-m}|x|^{1-m}h^{(1)}_{m-1}\big(z^{1/2}|x|\big) \qquad (6.3.17)\\ &= -\frac{\mathrm{i}}{2z^{1/2}}\left(\frac{z^{1/2}}{2\pi\mathrm{i}|x|}\right)^{m} e^{\mathrm{i}z^{1/2}|x|}\sum_{j=0}^{m-1}\frac{(m+j-1)!}{j!(m-j-1)!}\big(-2\mathrm{i}z^{1/2}|x|\big)^{-j}\end{aligned}$$

for each $x \in \mathbb{R}^n \setminus \{0\}$, where $h^{(1)}_\ell(\cdot)$ is the first Spherical Bessel function of the third kind, defined in [1, § 10.1]. Note that

$$e^{\mathrm{i}z^{1/2}|x|} = 1 + \mathrm{i}z^{1/2}|x| + O\big(|x|^2\big) \ \text{ as } \ |x| \to 0, \qquad (6.3.18)$$

and, as $|x| \to 0$,

$$\begin{aligned}&\sum_{j=0}^{m-1}\frac{(m+j-1)!}{j!(m-j-1)!}\big(-2\mathrm{i}z^{1/2}|x|\big)^{-j} \qquad (6.3.19)\\ &= \big(-2\mathrm{i}z^{1/2}|x|\big)^{1-m}\sum_{j=0}^{m-1}\frac{(m+j-1)!}{j!(m-j-1)!}\big(-2\mathrm{i}z^{1/2}|x|\big)^{m-1-j} \\ &= \big(-2\mathrm{i}z^{1/2}|x|\big)^{1-m}\Big\{\frac{(2m-2)!}{(m-1)!} + \frac{(2m-3)!}{(m-2)!}\big(-2\mathrm{i}z^{1/2}|x|\big) + O(|x|^2)\Big\}.\end{aligned}$$

Consequently, from (6.3.17)-(6.3.19) and algebra we see that, as $|x| \to 0$,

$$\Phi_z(x; 2m+1) = \begin{cases} -(4\pi|x|)^{-1}\big[1 + \mathrm{i}z^{1/2}|x| + O\big(|x|^2\big)\big], & m = 1, \\ -[(2m-1)\omega_{2m}]^{-1}|x|^{1-2m}\big[1 + O\big(|x|^2\big)\big], & m \geq 2. \end{cases} \qquad (6.3.20)$$

We now claim that for each fixed $n \in \mathbb{N}$, each fixed $z \in \mathbb{C} \setminus (-\infty, 0]$, each fixed $R > 0$ there exists a constant $C = C(n, z, R) \in (0, \infty)$ with the property that whenever $0 < |x| < R$ we have

$$\big|\Phi_z(x; n) - \Phi_0(x; n)\big| \leq \begin{cases} C, & n = 2, 3, \\ C\big[|\ln(|x|)| + 1\big], & n = 4, \\ C|x|^{4-n}, & n \geq 5. \end{cases} \qquad (6.3.21)$$

Indeed, when $n = 2$ this follows from (6.3.6) and (6.3.3), while if $n \geq 3$ this may be seen from (6.3.3), (6.3.16), and (6.3.20).

Next, we observe that

$$\begin{aligned} &\partial_{x_j} \Phi_z(x;n) = -2\pi\, x_j \Phi_z(x;n+2), \text{ for all} \\ &z \in \mathbb{C} \setminus (-\infty,0),\ x \in \mathbb{R}^n \setminus \{0\},\ j \in \{1,\dots,n\},\ n \geq 2. \end{aligned} \tag{6.3.22}$$

Indeed, if $z = 0$ this is seen straight from definitions, while for $z \in \mathbb{C} \setminus (-\infty, 0]$ the recursion relation (6.3.22) is a consequence of the well-known identity (cf. [1, § 9.1])

$$\frac{d}{d\zeta}\big(\zeta^{-\lambda} C_\lambda(\zeta)\big) = -\zeta^{-\lambda} C_{\lambda+1}(\zeta), \quad \zeta \in \mathbb{C} \setminus \{0\},\ \lambda \in \mathbb{R}, \tag{6.3.23}$$

where $C_\lambda(\cdot)$ denotes any linear combination of Bessel functions of order λ with coefficients independent of ζ and λ.

Given an arbitrary multi-index $\alpha \in \mathbb{N}_0^n$, based on (6.3.22) and induction we see that there exist homogeneous polynomials $P_\ell^{(\alpha)}$ in $\mathbb{R}^n$ of degree $\ell \in \mathbb{N}_0$ such that if $|\alpha| = 2m$ for $m \in \mathbb{N}$ then

$$\begin{aligned} \partial_x^\alpha \Phi_z(x;n) &= P_0^{(\alpha)}(x) \cdot \Phi_z(x;n+|\alpha|) + P_2^{(\alpha)}(x) \cdot \Phi_z(x;n+|\alpha|+2) \\ &\quad + \cdots + P_{2m-2}^{(\alpha)}(x) \cdot \Phi_z(x;n+|\alpha|+2m-2) \\ &\quad + (-2\pi)^{|\alpha|} x^\alpha \Phi_z(x;n+2|\alpha|), \\ &\quad \text{for all } z \in \mathbb{C} \setminus (-\infty,0),\ x \in \mathbb{R}^n \setminus \{0\},\ n \geq 2, \end{aligned} \tag{6.3.24}$$

and if $|\alpha| = 2m+1$ for $m \in \mathbb{N}_0$ then

$$\begin{aligned} \partial_x^\alpha \Phi_z(x;n) &= P_1^{(\alpha)}(x) \cdot \Phi_z(x;n+|\alpha|+1) + P_3^{(\alpha)}(x) \cdot \Phi_z(x;n+|\alpha|+3) \\ &\quad + \cdots + P_{2m-1}^{(\alpha)}(x) \cdot \Phi_z(x;n+|\alpha|+2m-1) \\ &\quad + (-2\pi)^{|\alpha|} x^\alpha \Phi_z(x;n+2|\alpha|), \\ &\quad \text{for all } z \in \mathbb{C} \setminus (-\infty,0),\ x \in \mathbb{R}^n \setminus \{0\},\ n \geq 2. \end{aligned} \tag{6.3.25}$$

Hence, in the case when $|\alpha| = 2m$ for some $m \in \mathbb{N}$, for each $z \in \mathbb{C} \setminus (-\infty, 0]$, $x \in \mathbb{R}^n \setminus \{0\}$, and $n \geq 2$, we may use (6.3.24) to estimate

$$\begin{aligned}
&\left|\partial_x^\alpha \Phi_z(x;n) - \partial_x^\alpha \Phi_0(x;n)\right| \\
&\qquad \le \left|P_0^{(\alpha)}(x)\right| \cdot \left|\Phi_z(x;n+|\alpha|) - \Phi_0(x;n+|\alpha|)\right| \\
&\qquad\quad + \left|P_2^{(\alpha)}(x)\right| \cdot \left|\Phi_z(x;n+|\alpha|+2) - \Phi_0(x;n+|\alpha|+2)\right| \\
&\qquad\quad + \cdots + \left|P_{2m}^{(\alpha)}(x)\right| \cdot \left|\Phi_z(x;n+2|\alpha|) - \Phi_0(x;n+2|\alpha|)\right|. \qquad (6.3.26)
\end{aligned}$$

Now, (6.3.7) follows from this and (6.3.21). Finally, (6.3.8) is justified in a similar fashion, making use of (6.3.25) and (6.3.21). □

Shortly, we shall need the information regarding the asymptotic behavior of Φ_k and its derivatives at infinity contained in the next lemma (see [91] for a proof).

Lemma 6.3.2 *Pick $k \in (0, \infty)$ and consider $n \in \mathbb{N}$ with $n \ge 2$. Then for each fixed $x \in \mathbb{R}^n$ one has*

$$\Phi_k(x-y) = O\big(|y|^{-(n-1)/2}\big) \ \text{as} \ |y| \to \infty. \qquad (6.3.27)$$

Also, for each multi-index $\alpha \in \mathbb{N}_0^n$ with $|\alpha| > 0$ one has

$$\begin{aligned}
(\partial^\alpha \Phi_k)(x-y) = (-ik\widehat{y})^\alpha \Phi_k(x-y) + O\big(|y|^{-(n+1)/2}\big) \ \text{as} \ |y| \to \infty, \\
\text{uniformly for } x \text{ in compact subsets of } \mathbb{R}^n.
\end{aligned} \qquad (6.3.28)$$

Finally, for any two multi-indices $\alpha, \beta \in \mathbb{N}_0^n$ one has

$$\begin{aligned}
(\partial^{\alpha+\beta} \Phi_k)(x-y) = (ik\widehat{x}\,)^\alpha (\partial^\beta \Phi_k)(x-y) + O\big(|x|^{-(n+1)/2}\big) \ \text{as} \ |x| \to \infty, \\
\text{uniformly for } y \text{ in compact subsets of } \mathbb{R}^n.
\end{aligned} \qquad (6.3.29)$$

We next use the asymptotics of Φ_k to study the behavior at infinity of functions that are representable as formal convolution of derivatives of Φ_k with finite measures.

Lemma 6.3.3 *Fix $k \in (0, \infty)$ and pick $n \in \mathbb{N}$ with $n \ge 2$. Let $\Omega \subseteq \mathbb{R}^n$ be an exterior domain and consider a Borel measure μ on $\partial\Omega$. Given $f \in L^1(\partial\Omega, \mu)$ and $\alpha \in \mathbb{N}_0^n$ arbitrary define*

$$u(x) := \int_{\partial\Omega} (\partial^\alpha \Phi_k)(x-y) f(y)\, d\mu(y), \qquad \forall x \in \Omega. \qquad (6.3.30)$$

Then u belongs to $\mathscr{C}^\infty(\Omega)$, satisfies $(\Delta + k^2)u = 0$ in Ω, and

$$\begin{aligned}
ik\widehat{x}_j\, u(x) - \big(\partial_j u\big)(x) = O\big(|x|^{-(n+1)/2}\big) \ \text{as} \ |x| \to \infty, \\
\text{for each index } j \in \{1, \dots, n\}.
\end{aligned} \qquad (6.3.31)$$

Proof The background assumptions readily imply that $u \in \mathscr{C}^\infty(\Omega)$ and for each multi-index $\beta \in \mathbb{N}_0^n$ we have

$$(\partial^\beta u)(x) = \int_{\partial\Omega} (\partial^{\alpha+\beta}\Phi_k)(x-y)f(y)\,d\mu(y), \qquad \forall x \in \Omega. \tag{6.3.32}$$

In particular, $(\Delta + k^2)u = 0$ in Ω. To show that u also satisfies (6.3.31), fix some $j \in \{1,\dots,n\}$. Then based on (6.3.32) and (6.3.29) we may write (keeping in mind the fact that $f \in L^1(\partial\Omega, \mu)$)

$$\begin{aligned}(\partial_j u)(x) &= \int_{\partial\Omega} (\partial^{\alpha+e_j}\Phi_k)(x-y)f(y)\,d\mu(y)\\ &= \int_{\partial\Omega} \Big\{ik\widehat{x}_j(\partial^\alpha\Phi_k)(x-y) + O\big(|x|^{-(n+1)/2}\big)\Big\}f(y)\,d\mu(y)\\ &= ik\widehat{x}_j\,u(x) + O\big(|x|^{-(n+1)/2}\big) \text{ as } |x| \to \infty,\end{aligned} \tag{6.3.33}$$

from which (6.3.31) follows. □

It turns out that derivatives of radiating functions are radiating themselves. Here is a formal statement:

Proposition 6.3.4 *If u is radiates at infinity (in the sense of Definition 6.1.1), then for each multi-index $\alpha \in \mathbb{R}^n$ the function $\partial^\alpha u$ also radiates at infinity.*

Proof This is a consequence of the integral representation formula from Theorem 6.1.4 (applied with $\Omega := \mathbb{R}^n \setminus \overline{B(0,R)}$ for a sufficiently large $R \in (0,\infty)$), and Lemma 6.3.3. □

For further use, we isolate a technical result in the following lemma.

Lemma 6.3.5 *Suppose $k \in (0,\infty)$ and consider $n \in \mathbb{N}$ with $n \geq 2$ along with $M \in \mathbb{N}$ arbitrary. Let $u = (u_J)_{1\leq J\leq M}$ be vector-valued function defined in an exterior domain $\Omega \subseteq \mathbb{R}^n$ whose components are in $\mathscr{C}^\infty(\Omega)$ and which satisfies $(\Delta + k^2)u = 0$ in Ω. Also, let $A = \big(a^{JI}_{rs}\big)_{\substack{1\leq J,I\leq M\\ 1\leq r,s\leq n}}$ be a complex coefficient tensor which allows expressing the vector Laplacian as $\Delta = \operatorname{div} A\nabla$. Then for each fixed point $x \in \mathbb{R}^n$ the limit*

$$\begin{aligned}[u]^A_\infty(x) := \lim_{R\to\infty}\Bigg(\int_{|y|=R} \widehat{y}_s\Big\{&a^{IJ}_{rs}(\partial_r\Phi_k)(x-y)u_J(y)\\ &+ a^{IJ}_{sr}\Phi_k(x-y)(\partial_r u_J)(y)\Big\}\,d\mathcal{H}^{n-1}(x)\Bigg)_{1\leq I\leq M}\end{aligned} \tag{6.3.34}$$

exists in $\mathbb{C}^M$ (in fact there exists some $R_x \in (0,\infty)$ with the property that the expression in the round parentheses is independent of R if $R \geq R_x$).

Proof Pick an arbitrary point $x \in \mathbb{R}^n$ and consider a radius $R_x > |x|$ with the property that $\partial\Omega \subseteq B(x, R_x)$. Also, fix an index $I \in \{1,\dots,M\}$ and suppose R_1, R_2 are such that $R_x < R_1 < R_2 < \infty$. We may then use the standard version of the

Divergence Theorem to express the difference between the expression in the round parentheses in (6.3.34) written, respectively, for $R = R_2$ and $R = R_1$, as

$$\int_{R_1<|y|<R_2} \partial_{y_s}\Big\{a^{IJ}_{rs}(\partial_r\Phi_k)(x-y)u_J(y) + a^{IJ}_{sr}\Phi_k(x-y)(\partial_r u_J)(y)\Big\}\,\mathrm{d}\mathcal{L}^n(x). \tag{6.3.35}$$

Since $a^{JI}_{rs}\partial_r\partial_s = \delta_{JI}\Delta$ for each index $J \in \{1,\dots,M\}$ (cf. (6.2.17)), on the domain of integration we have

$$\begin{aligned}
&\partial_{y_s}\Big\{a^{IJ}_{rs}(\partial_r\Phi_k)(x-y)u_J(y) + a^{IJ}_{sr}\Phi_k(x-y)(\partial_r u_J)(y)\Big\}\\
&\quad= -a^{IJ}_{rs}(\partial_s\partial_r\Phi_k)(x-y)u_J(y)\\
&\qquad + a^{IJ}_{rs}(\partial_r\Phi_k)(x-y)(\partial_s u_J)(y) - a^{IJ}_{sr}(\partial_s\Phi_k)(x-y)(\partial_r u_J)(y)\\
&\qquad + a^{IJ}_{sr}\Phi_k(x-y)(\partial_s\partial_r u_J)(y)\\
&\quad= -\delta_{JI}(\Delta\Phi_k)(x-y)u_J(y) + \delta_{IJ}\Phi_k(x-y)(\Delta u_J)(y)\\
&\quad= k^2\Phi_k(x-y)u_I(y) - k^2\Phi_k(x-y)u_I(y) = 0,
\end{aligned} \tag{6.3.36}$$

so the desired conclusion follows from (6.3.35). □

6.4 Single and Double Acoustic Layer Potentials

Fix an arbitrary wave number $k \in (0,\infty)$ and consider $n \in \mathbb{N}$ satisfying $n \geq 2$ along with some $M \in \mathbb{N}$ arbitrary. Suppose $\Omega \subseteq \mathbb{R}^n$ is an exterior domain which is a set of locally finite perimeter. Abbreviate $\sigma := \mathcal{H}^{n-1}\lfloor\partial\Omega$ and denote by $\nu = (\nu_1,\dots,\nu_n)$ the geometric measure theoretic outward unit normal to Ω. In this setting, define the action of the (vector) acoustic single layer potential operator on any function $f = (f_I)_{1\leq I\leq M} \in \big[L^1(\partial_*\Omega,\sigma)\big]^M$ as

$$\mathscr{S}f(x) := \Big(\int_{\partial_*\Omega}\Phi_k(x-y)f_I(y)\,\mathrm{d}\sigma(y)\Big)_{1\leq I\leq M} \quad \text{for all } x \in \Omega. \tag{6.4.1}$$

Also, having fixed a complex coefficient tensor $A := \big(a^{JI}_{rs}\big)_{\substack{1\leq J,I\leq M\\ 1\leq r,s\leq n}}$ that allows us to express the vector Laplacian as $\Delta = \mathrm{div}A\nabla$, associated the acoustic double layer potential operator

$$\mathcal{D}_A f(x) := \Big(-\int_{\partial_*\Omega}\nu_s(y)a^{IJ}_{rs}(\partial_r\Phi_k)(x-y)f_J(y)\,\mathrm{d}\sigma(y)\Big)_{1\leq I\leq M} \quad \text{for all } x \in \Omega, \tag{6.4.2}$$

for each vector-valued function $f = (f_J)_{1\le J\le M} \in [L^1(\partial_*\Omega, \sigma)]^M$.

In [115, §1.5] we shall discuss fundamental properties of acoustic boundary layer potentials considered in open sets with uniformly rectifiable boundaries. For now, in the theorem below, the goal is to describe a basic Green-type integral representation formula for null-solutions of the Helmholtz operator in an exterior domain in which no decay demands for the functions involved are imposed (cf. (6.4.5)).

Theorem 6.4.1 *Select $k \in (0,\infty)$ and consider $n \in \mathbb{N}$ with $n \ge 2$ along with some arbitrary integer $M \in \mathbb{N}$. Let $\Omega \subseteq \mathbb{R}^n$ be an exterior domain with a lower Ahlfors regular boundary and such that $\sigma := \mathcal{H}^{n-1}\lfloor\partial\Omega$ is a doubling measure on $\partial\Omega$. Also, write the vector Laplacian as $\Delta = \mathrm{div}A\nabla$ for some complex coefficient tensor $A = \big(a^{JI}_{rs}\big)_{\substack{1\le J,I\le M\\ 1\le r,s\le n}}$. In this setting, consider a vector-valued function*

$$\begin{aligned} & u = (u_I)_{1\le I\le M} \in \big[\mathscr{C}^\infty(\Omega)\big]^M \text{ with } (\Delta + k^2)u = 0 \text{ in } \Omega, \\ & \text{such that } u\Big|_{\partial\Omega}^{\kappa-\mathrm{n.t.}},\ (A\nabla u)\Big|_{\partial\Omega}^{\kappa-\mathrm{n.t.}} \text{ exist } \sigma\text{-a.e. on } \partial_{\mathrm{nta}}\Omega, \\ & \text{and } \mathcal{N}^\rho_\kappa u,\ \mathcal{N}^\rho_\kappa(A\nabla u) \in L^1(\partial\Omega,\sigma) \text{ for some } \kappa,\rho > 0, \end{aligned} \tag{6.4.3}$$

(with $A\nabla u$ understood in the sense of (6.2.23)*). Finally, recall* (6.3.34) *and, with $\nu = (\nu_1, \dots, \nu_n)$ denoting the geometric measure theoretic outward unit normal to Ω, at σ-a.e. point on $\partial_*\Omega$ define (compare with* (6.2.21)*)*

$$\begin{aligned} \partial^A_\nu u &:= \Big(\sum_{r=1}^n \nu_r (A\nabla u)_{r,J}\Big|_{\partial\Omega}^{\kappa-\mathrm{n.t.}}\Big)_{1\le J\le M} \\ &= \Big(\sum_{r=1}^n \nu_r \Big(\sum_{I=1}^M\sum_{s=1}^n a^{JI}_{rs}\partial_s u_I\Big)\Big|_{\partial\Omega}^{\kappa-\mathrm{n.t.}}\Big)_{1\le J\le M}. \end{aligned} \tag{6.4.4}$$

Then for each $x \in \Omega$ one has

$$u(x) = \mathcal{D}_A\big(u\big|_{\partial\Omega}^{\kappa-\mathrm{n.t.}}\big)(x) - \mathscr{S}\big(\partial^A_\nu u\big)(x) - [u]^A_\infty(x). \tag{6.4.5}$$

Furthermore, if one additionally assumes that

$$\int_{|x|=R} \big|iku(x) - (\partial^A_{\frac{x}{|x|}} u)(x)\big|^2 \, d\mathcal{H}^{n-1}(x) = o(1) \ \text{ as } R \to \infty, \tag{6.4.6}$$

and

$$\int_{|x|=R} |u(x)|^2 \, d\mathcal{H}^{n-1}(x) = O(1) \ \text{ as } R \to \infty, \tag{6.4.7}$$

it follows that $[u]^A_\infty(x) = 0$ for each $x \in \Omega$, in which case (6.4.5) *becomes the integral representation formula*

$$u = \mathcal{D}_A\big(u\big|_{\partial\Omega}^{\kappa-\mathrm{n.t.}}\big) - \mathscr{S}\big(\partial^A_\nu u\big) \ \text{ in } \Omega. \tag{6.4.8}$$

Proof Fix a point $x \in \Omega$ and an index $I \in \{1, \dots, M\}$. Consider the vector field $\vec{F} = \big(F_s\big)_{1 \le s \le n}$ whose components defined at $\mathcal{L}^n$-a.e. point $y \in \Omega$ according to

$$F_s(y) := -a^{IJ}_{rs}(\partial_r \Phi_k)(x-y)u_J(y) - a^{IJ}_{sr}\Phi_k(x-y)\partial_r u_J(y), \qquad 1 \le s \le n. \tag{6.4.9}$$

Then this definition readily implies that

$$\vec{F} \text{ belongs to } \big[\mathscr{C}^\infty(\Omega \setminus \{x\}) \cap L^1_{\rm loc}(\Omega, \mathcal{L}^n)\big]^n, \tag{6.4.10}$$

$$\mathcal{N}^\varepsilon_\kappa \vec{F} \in L^1(\partial\Omega, \sigma) \text{ if } \varepsilon \in \big(0, \min\{\rho, \operatorname{dist}(x, \partial\Omega)\}\big), \tag{6.4.11}$$

and the nontangential trace

$$\vec{F}\Big|^{\kappa-\mathrm{n.t.}}_{\partial\Omega} \text{ exists at } \sigma\text{-a.e. point on } \partial_{\rm nta}\Omega. \tag{6.4.12}$$

We shall next compute $\operatorname{div}\vec{F}$ in the sense of distributions in Ω. Concretely,

$$\begin{aligned}\operatorname{div}\vec{F}(y) &= a^{IJ}_{rs}\partial_{y_r}\partial_{y_s}\big[\Phi_k(x-y)\big]u_J(y)\\ &\quad - a^{IJ}_{rs}(\partial_r\Phi_k)(x-y)\partial_s u_J(y) + a^{IJ}_{sr}(\partial_s\Phi_k)(x-y)\partial_r u_J(y)\\ &\quad - a^{IJ}_{sr}\Phi_k(x-y)\partial_s\partial_r u_J(y)\\ &=: \mathbf{I} + \mathbf{II} + \mathbf{III} + \mathbf{IV}.\end{aligned} \tag{6.4.13}$$

Note that $\mathbf{II} + \mathbf{III} = 0$. Also, in view of the fact that $\Delta u = -k^2 u$ we may write

$$\mathbf{IV} = \big(-\Phi_k(x-y)(\Delta u)(y)\big)_I = k^2\Phi_k(x-y)u_I(y). \tag{6.4.14}$$

Finally, based on (6.2.17) and the fact that Φ_k is a fundamental solution of the Helmholtz operator $\Delta + k^2$, for each index $J \in \{1, \dots, M\}$ we may write

$$\begin{aligned}a^{IJ}_{rs}\partial_{y_r}\partial_{y_s}\big[\Phi_k(x-\cdot)\big] &= \delta_{IJ}\Delta_y\big[\Phi_k(x-\cdot)\big]\\ &= -k^2\Phi_k(x-\cdot)\delta_{IJ} + \delta_x\delta_{IJ} \text{ in } \mathcal{D}'(\Omega).\end{aligned} \tag{6.4.15}$$

Thus,

$$\mathbf{I} = -k^2\Phi_k(x-y)u_I(y) + \delta_x(y)u_I(y) \text{ in } \mathcal{D}'(\Omega). \tag{6.4.16}$$

Altogether, this proves that

$$\operatorname{div}\vec{F} = u_I(x)\delta_x \text{ in } \mathcal{D}'(\Omega). \tag{6.4.17}$$

In particular, $\operatorname{div}\vec{F} \in \mathscr{E}'(\Omega)$.

Granted the aforementioned properties, the version of the Divergence Theorem established in [112, Corollary 1.5.2] applies. First, this guarantees that the contribution of $\vec{F}$ at infinity is meaningfully and unambiguously defined. Based on [112, Proposition 4.7.1] (cf. also [112, (4.7.6)]) as well as (6.4.9) and (6.3.34) we may then write

$$
\begin{aligned}
[\vec{F}]_\infty &= \lim_{R\to\infty} \int_{|y|=R} \widehat{y}_s F_s(y)\, d\mathcal{H}^{n-1}(y) \\
&= \lim_{R\to\infty} \int_{|y|=R} \widehat{y}_s \big\{ -a_{rs}^{IJ} (\partial_r \Phi_K)(x-y) u_J(y) \\
&\qquad\qquad -a_{sr}^{IJ} \Phi_K(x-y) \partial_r u_J(y) \big\}\, d\mathcal{H}^{n-1}(y) \\
&= -\big([u]_\infty^A(x)\big)_I.
\end{aligned} \tag{6.4.18}
$$

In view of (6.4.16), the Divergence Formula recorded in [112, (1.5.19)] presently gives

$$
u_I(x) = {}_{(\mathscr{C}_b^\infty(\Omega))^*}\big(\mathrm{div}\vec{F}, 1\big)_{\mathscr{C}_b^\infty(\Omega)} = \int_{\partial_*\Omega} \nu_s \big(F_s\big|_{\partial\Omega}^{\kappa-\mathrm{n.t.}}\big)\, d\sigma + [\vec{F}]_\infty. \tag{6.4.19}
$$

In view of (6.4.9), (6.4.4), (6.4.1), and (6.4.2) we may further write

$$
\begin{aligned}
&\int_{\partial_*\Omega} \nu_s(y) \big(F_s\big|_{\partial\Omega}^{\kappa-\mathrm{n.t.}}\big)(y)\, d\sigma(y) \\
&\quad = -\int_{\partial_*\Omega} \nu_s(y) a_{rs}^{IJ} (\partial_r \Phi_K)(x-y) \big(u_J\big|_{\partial\Omega}^{\kappa-\mathrm{n.t.}}\big)(y)\, d\sigma(y) \\
&\quad\quad - \int_{\partial_*\Omega} \Phi_K(x-y) \nu_s(y) \Big((a_{sr}^{IJ} \partial_r u_J)\big|_{\partial\Omega}^{\kappa-\mathrm{n.t.}} \Big)(y)\, d\sigma(y) \\
&\quad = \Big(\mathcal{D}_A\big(u\big|_{\partial\Omega}^{\kappa-\mathrm{n.t.}}\big)(x) - \mathscr{S}\big(\partial_\nu^A u\big)(x) \Big)_I .
\end{aligned} \tag{6.4.20}
$$

Collectively, (6.4.18), (6.4.19), and (6.4.20) imply that

$$
u_I(x) = \Big(\mathcal{D}_A\big(u\big|_{\partial\Omega}^{\kappa-\mathrm{n.t.}}\big)(x) - \mathscr{S}\big(\partial_\nu^A u\big)(x) - [u]_\infty^A(x) \Big)_I . \tag{6.4.21}
$$

Since the index $I \in \{1, \dots, M\}$ was arbitrary, this completes the proof of the Green-type integral representation formula (with a correction taking into account the behavior at infinity) for u claimed in (6.4.5).

Next we further analyze the contribution at infinity $[u]_\infty^A(x)$ defined in (6.3.34). The idea is to use the asymptotic formulas for Hankel functions and their derivatives from Lemma 6.3.2. Specifically, for each $x \in \Omega$ fixed we may use (6.3.34) and (6.3.28) to write

$$
\begin{aligned}
[u]^A_\infty(x) &= \lim_{R\to\infty}\int_{|y|=R} \widehat{y}_s\Big\{a^{IJ}_{rs}(\partial_r\Phi_k)(x-y)u_J(y) \\
&\qquad\qquad + a^{IJ}_{sr}\Phi_k(x-y)\partial_r u_J(y)\Big\}\,\mathrm{d}\mathcal{H}^{n-1}(y) \\
&= \lim_{R\to\infty}\int_{|y|=R} \Phi_k(x-y)\Big\{-ika^{IJ}_{rs}\widehat{y}_s\widehat{y}_r u_J(y) + a^{IJ}_{sr}\widehat{y}_s\partial_r u_J(y)\Big\}\,\mathrm{d}\mathcal{H}^{n-1}(y) \\
&\quad + \lim_{R\to\infty}\int_{|y|=R} O\big(R^{-(n+1)/2}\big)O\big(|u(y)|\big)\,\mathrm{d}\mathcal{H}^{n-1}(y) =: \mathrm{I}+\mathrm{II}.
\end{aligned} \tag{6.4.22}
$$

Assume in addition that u satisfies (6.4.6). Then the Cauchy-Schwarz inequality and (6.4.7) permit us to conclude that

$$
\mathrm{II} = \lim_{R\to\infty}\left\{O(R^{-1})\Big(\int_{|y|=R}|u(y)|^2\,\mathrm{d}\mathcal{H}^{n-1}(y)\Big)^{1/2}\right\} = 0. \tag{6.4.23}
$$

As for I, keeping (6.2.20) in mind we may express

$$
\mathrm{I} = \lim_{R\to\infty}\int_{|y|=R}\Phi_k(x-y)\left\{-iku_I(y) + \big(\partial^A_{\widehat{y}}u\big)_I(y)\right\}\,\mathrm{d}\mathcal{H}^{n-1}(y). \tag{6.4.24}
$$

Hence, using the decay of Φ_k from the Lemma 6.3.2, and keeping in mind that u satisfies (6.4.6), we obtain

$$
\begin{aligned}
|\mathrm{I}| &\le \lim_{R\to\infty}\left(\int_{|y|=R}|\Phi_k(x-y)|^2\,\mathrm{d}\mathcal{H}^{n-1}(y)\right)^{1/2}\times \\
&\qquad \times\left(\int_{|y|=R}\Big|iku(y) - \big(\partial^A_{\widehat{y}}u\big)(y)\Big|^2\,\mathrm{d}\mathcal{H}^{n-1}(y)\right)^{\frac{1}{2}} \\
&\le \limsup_{R\to\infty}\big\{O(1)\cdot o(1)\big\} = 0.
\end{aligned} \tag{6.4.25}
$$

This proves that $[u]^A_\infty(x) = 0$ whenever u also satisfies (6.4.6). □

6.5 L^2-Finiteness Results

As is apparent from Lemma 6.1.2, sometimes it happens that radiation conditions, such as (6.4.6), imply L^2-finiteness properties, like the one in (6.4.7). Our goal in this section is to understand when this is the case.

With the piece of notation introduced in (6.2.24), we have the following basic L^2-finiteness result.

Proposition 6.5.1 *Pick $k \in (0, \infty)$ and consider $n \in \mathbb{N}$ with $n \geq 2$ along with $M \in \mathbb{N}$ arbitrary. Write the vector Laplacian as $\Delta = \mathrm{div} A \nabla$ for some coefficient tensor A satisfying $\mathrm{Im}\,\langle A\zeta, \overline{\zeta}\rangle \leq 0$ for every complex vector ζ. Also, let u be a $\mathbb{C}^M$-valued null-solution of the vector Helmholtz operator $\Delta + k^2$ in an exterior domain $\Omega \subseteq \mathbb{R}^n$ with the property that*

$$\int_{|x|=R} \left|iku(x) - (\partial^A_{\frac{x}{|x|}} u)(x)\right|^2 \, d\mathcal{H}^{n-1}(x) = O(1) \quad \text{as } R \to \infty. \tag{6.5.1}$$

Then

$$\int_{|x|=R} |u(x)|^2 \, d\mathcal{H}^{n-1}(x) = O(1) \quad \text{as } R \to \infty, \tag{6.5.2}$$

and

$$\int_{|x|=R} \left|(\partial^A_{\frac{x}{|x|}} u)(x)\right|^2 \, d\mathcal{H}^{n-1}(x) = O(1) \quad \text{as } R \to \infty. \tag{6.5.3}$$

A few comments in relation to this result are in order. First, since it is clear from the triangle inequality that (6.5.1)-(6.5.2) imply (6.5.3), the conclusion in Proposition 6.5.1 amounts to a reverse triangle inequality. Second, replacing $O(1)$ by $o(1)$ in the hypothesis formulated in (6.5.1) does *not* imply $o(1)$ in the conclusions formulated in (6.5.2), (6.5.3).

As a preamble to the proof of Proposition 6.5.1, we establish the following auxiliary result.

Lemma 6.5.2 *Let $k \in (0, \infty)$ and consider $n \in \mathbb{N}$ with $n \geq 2$. Suppose $D \subseteq \mathbb{R}^n$ is a bounded $\mathscr{C}^1$ domain with surface measure σ and outward unit normal ν, and $u \in \left[\mathscr{C}^1(\overline{D})\right]^M$, for some $M \in \mathbb{N}$, is a vector-valued function such that $(\Delta + k^2)u = 0$ in D. If $A := \left(a^{JI}_{rs}\right)_{\substack{1 \leq J,I \leq M \\ 1 \leq r,s \leq n}}$ is a complex coefficient tensor in the writing of the vector Laplacian as $\Delta = \mathrm{div} A \nabla$ with the property that*

$$\mathrm{Im}\,\langle A\zeta, \overline{\zeta}\rangle \leq 0 \text{ for every complex vector } \zeta \in \mathbb{C}^{nM}, \tag{6.5.4}$$

then

$$\int_{\partial D} \left|iku - \partial^A_\nu u\right|^2 \, d\sigma \geq \int_{\partial D} \left\{k^2|u|^2 + \left|\partial^A_\nu u\right|^2\right\} d\sigma. \tag{6.5.5}$$

Proof Expand the left-hand side of (6.5.5) as

$$\int_{\partial D} \left|iku - \partial^A_\nu u\right|^2 \, d\sigma = \int_{\partial D} \left\{k^2|u|^2 + \left|\partial^A_\nu u\right|^2\right\} d\sigma - 2k\mathrm{Im} \int_{\partial D} \langle \partial^A_\nu u, \overline{u}\rangle \, d\sigma. \tag{6.5.6}$$

Hence, since $k > 0$, it suffices to show that

$$\mathrm{Im} \int_{\partial D} \langle \partial^A_\nu u, \overline{u}\rangle \, d\sigma \leq 0. \tag{6.5.7}$$

Integration by parts and using the fact that $(\Delta + k^2)u = 0$ in D, we obtain (with $\mathcal{L}^n$ denoting the Lebesgue measure in $\mathbb{R}^n$, and with the convention of summation for repeated indices in effect)

$$\begin{aligned}\int_D \langle A\nabla u, \overline{\nabla u}\rangle \, \mathrm{d}\mathcal{L}^n &= \int_D a_{rs}^{JI}(\partial_s u_I)\overline{\partial_r u_J} \, \mathrm{d}\mathcal{L}^n \\ &= -\int_D a_{rs}^{JI}(\partial_r \partial_s u_I)\overline{u_J} \, \mathrm{d}\mathcal{L}^n + \int_{\partial D} a_{rs}^{JI} \nu_r (\partial_s u_I)\overline{u_J} \, \mathrm{d}\sigma \\ &= \int_D k^2 |u|^2 \, \mathrm{d}\mathcal{L}^n + \int_{\partial D} \langle \partial_\nu^A u, \overline{u}\rangle \, \mathrm{d}\sigma. \end{aligned} \tag{6.5.8}$$

Taking the imaginary parts of the most extreme sides of this identity yields, bearing in mind that $\operatorname{Im}\langle A\zeta, \overline{\zeta}\rangle \leq 0$ for every vector ζ with complex components,

$$\begin{aligned} 0 &\geq \int_D \operatorname{Im}\langle A\nabla u, \overline{\nabla u}\rangle \, \mathrm{d}\mathcal{L}^n \\ &= \operatorname{Im}\int_D k^2|u|^2 \, \mathrm{d}\mathcal{L}^n + \operatorname{Im}\int_{\partial D} \langle \partial_\nu^A u, \overline{u}\rangle \, \mathrm{d}\sigma \\ &= \operatorname{Im}\int_{\partial D} \langle \partial_\nu^A u, \overline{u}\rangle \, \mathrm{d}\sigma, \end{aligned} \tag{6.5.9}$$

since k is real. This establishes (6.5.7) and finishes the proof of the lemma. □

We are now prepared to present the proof of the L^2-finiteness result from Proposition 6.5.1.

Proof of Proposition 6.5.1 Apply Lemma 6.5.2 with $D := B(0, R) \setminus \overline{B(0, r)}$, where $0 < r < R$ are such that $\overline{D} \subset \Omega$. Since $\partial D = \partial B(0, R) \cup \partial B(0, r)$, we have

$$\begin{aligned} &\int_{|x|=R} \left|iku(x) - (\partial_{\frac{x}{x}}^A u)(x)\right|^2 \mathrm{d}\mathcal{H}^{n-1}(x) \\ &\quad + \int_{|x|=r} \left|iku(x) + (\partial_{\frac{x}{x}}^A u)(x)\right|^2 \mathrm{d}\mathcal{H}^{n-1}(x) \\ &\quad \geq \int_{|x|=R} k^2|u(x)|^2 \, \mathrm{d}\mathcal{H}^{n-1}(x) + \int_{|x|=R} \left|(\partial_{\frac{x}{x}}^A u)(x)\right|^2 \mathrm{d}\mathcal{H}^{n-1}(x) \\ &\quad + \int_{|x|=r} \left\{k^2|u(x)|^2 + \left|(\partial_{\frac{x}{x}}^A u)(x)\right|^2\right\} \mathrm{d}\mathcal{H}^{n-1}(x). \end{aligned} \tag{6.5.10}$$

Since for each r fixed the left-hand side of (6.5.10) is $O(1)$ as $R \to \infty$, it follows that for each r fixed the right-hand side of (6.5.10) must be $O(1)$ as $R \to \infty$. From this, the desired conclusion follows. □

6.6 The Principal Result

The main result identifying all radiation conditions for the vector Helmholtz operator is presented next. Among other things, this provides satisfactory answers to Questions 1-2 formulated earlier in this section.

Theorem 6.6.1 *Pick $k \in (0, \infty)$, fix $n \in \mathbb{N}$ with $n \geq 2$, and let $M \in \mathbb{N}$ be arbitrary. Suppose $\Omega \subseteq \mathbb{R}^n$ and consider a function $u \in \big[\mathscr{C}^\infty(\Omega)\big]^M$ satisfying $(\Delta + k^2)u = 0$ in Ω. Then items (1)-(6) below are equivalent:*

(1) The function u satisfies

$$ik\widehat{x}_j\, u(x) - \big(\partial_j u\big)(x) = O\big(|x|^{-(n+1)/2}\big) \text{ as } |x| \to \infty, \quad \text{for each index } j \in \{1, \dots, n\}. \tag{6.6.1}$$

(2) The function u satisfies

$$ik\widehat{x}_j\, u(x) - \big(\partial_j u\big)(x) = o\big(|x|^{-(n-1)/2}\big) \text{ as } |x| \to \infty, \quad \text{for each index } j \in \{1, \dots, n\}. \tag{6.6.2}$$

(3) For some, or any, complex coefficient tensor A such that $\Delta = \mathrm{div} A \nabla$ and with the property that $\mathrm{Im}\,\langle A\zeta, \overline{\zeta}\rangle \leq 0$ for every vector ζ with complex components, one has

$$iku(x) - \big(\partial^A_{\widehat{x}} u\big)(x) = o\big(|x|^{-(n-1)/2}\big) \text{ as } |x| \to \infty. \tag{6.6.3}$$

(4) For some, or any, complex coefficient tensor A such that $\Delta = \mathrm{div} A \nabla$ and with the property that $\mathrm{Im}\,\langle A\zeta, \overline{\zeta}\rangle \leq 0$ for every vector ζ with complex components, one has

$$iku(x) - \big(\partial^A_{\widehat{x}} u\big)(x) = O\big(|x|^{-(n+1)/2}\big) \text{ as } |x| \to \infty. \tag{6.6.4}$$

(5) For some, or any, complex coefficient tensor A such that $\Delta = \mathrm{div} A \nabla$ and with the property that $\mathrm{Im}\,\langle A\zeta, \overline{\zeta}\rangle \leq 0$ for every vector ζ with complex components, one has

$$\int_{|x|=R} \big|iku(x) - \big(\partial^A_{\widehat{x}} u\big)(x)\big|^2 \, d\mathcal{H}^{n-1}(x) = o(1) \text{ as } R \to \infty. \tag{6.6.5}$$

(6) For some, or any, complex coefficient tensor A such that $\Delta = \mathrm{div} A \nabla$ and with the property that $\mathrm{Im}\,\langle A\zeta, \overline{\zeta}\rangle \leq 0$ for every vector ζ with complex components, one has

$$\int_{|x|=R} \big|iku(x) - \big(\partial^A_{\widehat{x}} u\big)(x)\big|^2 \, d\mathcal{H}^{n-1}(x) = O(R^{-2}) \text{ as } R \to \infty. \tag{6.6.6}$$

Strengthen the hypotheses on the underlying set Ω by additionally assuming that $\partial\Omega$ is a lower Ahlfors regular set and $\sigma := \mathcal{H}^{n-1} \lfloor \partial\Omega$ is a doubling measure on $\partial\Omega$. Also, denote by ν the geometric measure theoretic outward unit normal to Ω and fix *some aperture parameter* $\kappa \in (0,\infty)$ *along with a truncation parameter* $\rho \in (0,\infty)$. *Then if u also satisfies*

$$\begin{gathered}\text{the traces } u\Big|_{\partial\Omega}^{\kappa-\mathrm{n.t.}},\ (\nabla u)\Big|_{\partial\Omega}^{\kappa-\mathrm{n.t.}} \text{ exist } \sigma\text{-a.e. on } \partial_{\mathrm{nta}}\Omega,\\ \text{and } \mathcal{N}_\kappa^\rho u,\ \mathcal{N}_\kappa^\rho(\nabla u) \in L^1(\partial\Omega,\sigma),\end{gathered}\tag{6.6.7}$$

it follows that each of the radiation conditions described in items (1)-(6) is further equivalent with having

$$\begin{gathered}u = \mathcal{D}_A\Big(u\Big|_{\partial\Omega}^{\kappa-\mathrm{n.t.}}\Big) - \mathcal{S}(\partial_\nu^A u) \text{ in } \Omega\\ \text{for some, or any, } A \text{ such that } \Delta = \mathrm{div}A\nabla,\end{gathered}\tag{6.6.8}$$

where $\mathcal{D}_A$, $\mathcal{S}$ *are the acoustic double and single layer potential operators associated with the wave number k, the coefficient tensor A, and the exterior domain* Ω *as in* (6.4.1)-(6.4.2).

Finally, if the function u is radiating at infinity (*cf. Definition 6.1.1*) *then for any given complex coefficient tensor A such that* $\Delta = \mathrm{div}A\nabla$ *one has*

$$\begin{gathered}u = \mathcal{D}_A\Big(u\Big|_{\partial\Omega}^{\kappa-\mathrm{n.t.}}\Big) - \mathcal{S}(\partial_\nu^A u) \text{ in } \Omega \text{ if both traces } u\Big|_{\partial\Omega}^{\kappa-\mathrm{n.t.}},\ (A\nabla u)\Big|_{\partial\Omega}^{\kappa-\mathrm{n.t.}}\\ \text{exist } \sigma\text{-a.e. on } \partial_{\mathrm{nta}}\Omega, \text{ and if } \mathcal{N}_\kappa^\rho u,\ \mathcal{N}_\kappa^\rho(A\nabla u) \in L^1(\partial\Omega,\sigma).\end{gathered}\tag{6.6.9}$$

A few comments shedding further light on this theorem are in order. First, note that (6.6.1), (6.6.2) are intrinsic conditions (independent of the choice of a coefficient tensor A used to write the vector Laplacian as $\Delta = \mathrm{div}A\nabla$), which may be equivalently be considered on each of the coefficients of the vector-valued function u.

Second, writing $\Delta = \mathrm{div}A\nabla$ with $A := A_0 = \big(\delta_{JI}\delta_{rs}\big)_{\substack{1\le J,I\le M\\ 1\le r,s\le n}}$ the identity, we see from (6.2.22) that any of the items *(1)-(6)* in Theorem 6.6.1 is further equivalent to any of the conditions

$$iku(x) - \sum_{j=1}^n \widehat{x}_j(\partial_j u)(x) = o\big(|x|^{-(n-1)/2}\big) \text{ as } |x|\to\infty,\tag{6.6.10}$$

$$iku(x) - iku(x) - \sum_{j=1}^n \widehat{x}_j(\partial_j u)(x) = O\big(|x|^{-(n+1)/2}\big) \text{ as } |x|\to\infty,\tag{6.6.11}$$

$$\int_{|x|=R}\Big|iku(x) - \sum_{j=1}^n \widehat{x}_j(\partial_j u)(x)\Big|^2\,d\mathcal{H}^{n-1}(x) = o(1) \text{ as } R\to\infty,\tag{6.6.12}$$

$$\int_{|x|=R}\Big|iku(x) - \sum_{j=1}^n \widehat{x}_j(\partial_j u)(x)\Big|^2\,d\mathcal{H}^{n-1}(x) = O(R^{-2}) \text{ as } R\to\infty.\tag{6.6.13}$$

In particular, any of the items *(1)-(6)* in Theorem 6.6.1 is further equivalent to Sommerfeld's radiation condition for the function u.

Third, as is apparent from (A.0.118), for any function u satisfying the radiation condition (6.6.2) and any homogeneous first-order system D with constant complex coefficients we have

$$k\,\mathrm{Sym}(D;\widehat{x})u(x) - (Du)(x) = O\big(|x|^{-(n+1)/2}\big) \text{ as } |x| \to \infty. \tag{6.6.14}$$

Here is the proof of Theorem 6.6.1.

Proof of Theorem 6.6.1 To set the stage for the subsequent discussion, fix some sufficiently large radius $R \in (0,\infty)$ (with the property that $\partial\Omega \subseteq B(0,R)$) and introduce the open set $\Omega_R := \mathbb{R}^n \setminus \overline{B(0,R)}$. Then Ω_R is an exterior domain in $\mathbb{R}^n$ whose closure is contained in Ω. In particular, all components of u are in $\mathscr{C}^\infty(\overline{\Omega_R})$.

Let us now begin in earnest. The fact that *(1)*$\Rightarrow$*(2)* is obvious. Assume now that the function u satisfies (6.6.2). Then for each index $I \in \{1,\dots,M\}$ we may write

$$\begin{aligned} iku_I(x) - \widehat{x}\cdot(\nabla u_I)(x) &= iku_I(x) - \sum_{j=1}^n \widehat{x}_j(\partial_j u_I)(x) \\ &= iku_I(x) - \sum_{j=1}^n \widehat{x}_j\Big(ik\widehat{x}_j\, u_I(x) + o\big(|x|^{-(n-1)/2}\big)\Big) \\ &= o\big(|x|^{-(n-1)/2}\big) \text{ as } |x|\to\infty, \end{aligned} \tag{6.6.15}$$

i.e.,

$$iku_I(x) - \widehat{x}\cdot(\nabla u_I)(x) = o\big(|x|^{-(n-1)/2}\big) \text{ as } |x|\to\infty. \tag{6.6.16}$$

This guarantees that Sommerfeld's radiation condition (6.1.2) is satisfied by each component of u. As such, we may invoke Theorem 6.1.4 for each component of u restricted to Ω_R and obtain the integral representation formula (6.1.18) written for Ω_R in place of Ω for said function. In turn, from this and Lemma 6.3.3 it follows that (6.6.1) holds, hence *(2)*$\Rightarrow$*(1)*. The argument so far shows that *(1)*$\Leftrightarrow$*(2)*.

Assume now that u is a function as in item *(1)* and consider an arbitrary complex coefficient tensor $A := \big(a_{rs}^{JI}\big)_{\substack{1\le J,I\le M\\ 1\le r,s\le n}}$ which permits us to express the vector Laplacian as $\Delta = \mathrm{div}A\nabla$. Thanks to (6.2.21), (6.6.2), and (6.2.20) we may then write

$$\begin{aligned} iku(x) - \big(\partial^A_{\widehat{x}}u\big)(x) &= iku(x) - \big(a_{rs}^{JI}\widehat{x}_r\partial_s u_I(x)\big)_J \\ &= iku(x) - ik\big(a_{rs}^{JI}\widehat{x}_r\widehat{x}_s u_I(x)\big)_J + O\big(|x|^{-(n+1)/2}\big) \\ &= iku(x) - ik\big(|\widehat{x}|^2\delta_{JI}u_I(x)\big)_J + O\big(|x|^{-(n+1)/2}\big) \\ &= O\big(|x|^{-(n+1)/2}\big) \text{ as } |x|\to\infty. \end{aligned} \tag{6.6.17}$$

This goes to show that (6.6.4) holds, i.e., *(1)* implies the strongest version of *(4)*, which clearly implies the strongest version of *(6)*, which further implies the strongest version of *(5)*.

Going further, suppose A is a complex coefficient tensor such that $\Delta = \mathrm{div}A\nabla$ and with the property that $\mathrm{Im}\,\langle A\zeta, \overline{\zeta}\rangle \leq 0$ for every vector ζ with complex components. Also, assume u satisfies (6.6.5) for said coefficient tensor A. Then from Proposition 6.5.1 and Theorem 6.4.1 applied to u restricted to Ω_R we conclude that the integral representation formula (6.4.8) written for Ω_R in place of Ω is valid. In concert with Lemma 6.3.3, this implies that (6.6.1) holds, hence *(5)*$\Rightarrow$*(1)*. At this stage, we have proved that items *(1)-(6)* are equivalent.

For the remainder of the proof strengthen the hypotheses on the underlying set Ω as indicated in the statement of the theorem. For now, assume that u has the additional qualities listed in (6.6.7). Then, as may be seen from Lemma 6.3.3, the property recorded in (6.6.8) implies the claim made in item *(1)*.

To deal with the opposite implication, assume u satisfies the radiation condition formulated in (6.6.1) and pick a complex coefficient tensor A which permits us to express the vector Laplacian as $\Delta = \mathrm{div}A\nabla$. It suffices to prove the claim in (6.6.9). To this end, assume that the nontangential traces $u\big|_{\partial\Omega}^{\kappa-\mathrm{n.t.}}$, $(A\nabla u)\big|_{\partial\Omega}^{\kappa-\mathrm{n.t.}}$ exist at σ-a.e. point on $\partial_{\mathrm{nta}}\Omega$, and that $\mathcal{N}_\kappa^\rho u$, $\mathcal{N}_\kappa^\rho(A\nabla u) \in L^1(\partial\Omega, \sigma)$. As proved earlier in (6.6.17), from (6.6.1) we deduce that

$$iku(x) - \big(\partial_{\frac{x}{}}^{A} u\big)(x) = O\big(|x|^{-(n+1)/2}\big) \text{ as } |x| \to \infty. \tag{6.6.18}$$

As a consequence,

$$\int_{|x|=R} \big|iku(x) - \big(\partial_{\widehat{x}}^{A} u\big)(x)\big|^2 \, d\mathcal{H}^{n-1}(x) = O(R^{-2}) \text{ as } R \to \infty. \tag{6.6.19}$$

We already know that (6.6.1) implies the version of (6.6.5) written for the identity coefficient tensor=, i.e., $A := \big(\delta_{JI}\delta_{rs}\big)_{\substack{1\leq J,I\leq M\\ 1\leq r,s\leq n}}$. In view of (6.2.22) this yields

$$\int_{|x|=R} \Big|iku(x) - \sum_{j=1}^{n} \widehat{x}_j(\partial_j u)(x)\Big|^2 \, d\mathcal{H}^{n-1}(x) = o(1) \text{ as } R \to \infty. \tag{6.6.20}$$

In concert with Lemma 6.1.2, this also implies

$$\int_{|x|=R} |u(x)|^2 \, d\mathcal{H}^{n-1}(x) = O(1) \text{ as } R \to \infty. \tag{6.6.21}$$

Granted (6.6.19) and (6.6.21), we may invoke Theorem 6.4.1 to conclude that (6.4.8) holds. This establishes the claim in (6.6.9).

To finish the proof of the theorem there remains to observe that if u radiates at infinity in the sense of Definition 6.1.1 (as assumed in the preamble to (6.6.9)) then the weakest version of the claim in item *(3)* (corresponding to the choice $A := \big(\delta_{JI}\delta_{rs}\big)_{\substack{1\leq J,I\leq M\\ 1\leq r,s\leq n}}$) holds and, as already proved above, this implies (6.6.9). $\square$

6.7 Coordinate-Free Formalism and Examples

Select some $k \in (0, \infty)$ and pick some $n \in \mathbb{N}$ with $n \geq 2$. Also, fix $M, N \in \mathbb{N}$ arbitrary. We are interested in the case when the vector Laplacian[2] in $\mathbb{R}^n$ factors as

$$\Delta = \widetilde{D}D \tag{6.7.1}$$

where D is a first-order constant complex constant coefficient $N \times M$ system, acting on vector-valued function $u = (u_I)_{1 \leq I \leq M}$ according to

$$Du = \big(b_s^{LI}\partial_s u_I + c_{LI}u_I\big)_{1 \leq L \leq N} \tag{6.7.2}$$

for some constant coefficients b_s^{LI}, $c_{LI} \in \mathbb{C}$, with $1 \leq L \leq N$, $1 \leq I \leq M$, $1 \leq s \leq n$, and where $\widetilde{D}$ is a first-order constant complex constant coefficient $M \times N$ system acting on vector-valued function $w = (w_J)_{1 \leq J \leq N}$ according to

$$\widetilde{D}w = \big(\widetilde{b}_r^{LJ}\partial_r w_L + \widetilde{c}_{LJ}w_J\big)_{1 \leq J \leq M}. \tag{6.7.3}$$

Introducing the complex coefficient tensor

$$\begin{aligned} &A := \big(a_{rs}^{JI}\big)_{\substack{1 \leq J, I \leq M \\ 1 \leq r, s \leq n}} \text{ with } a_{rs}^{JI} := \widetilde{b}_r^{LJ} b_s^{LI} \in \mathbb{C} \\ &\text{for all } J, I \in \{1, \dots, M\} \text{ and } r, s \in \{1, \dots, n\}, \end{aligned} \tag{6.7.4}$$

then permits us to express the vector Laplacian as $\Delta = \mathrm{div} A \nabla$. In this regard, note that

> the condition $\mathrm{Im}\langle A\zeta, \overline{\zeta}\rangle \leq 0$ for all vectors ζ with complex entries is automatically satisfied if $\widetilde{D} = \lambda \overline{D}^{\top}$ for some $\lambda \in \mathbb{C}$ with $\mathrm{Im}\,\lambda \geq 0$ (where 'bar' denotes complex conjugation, and where $\top$ stands for real transposition). (6.7.5)

Indeed, if all coefficients b_r^{LJ} are real numbers and if there exists $\lambda \in \mathbb{C}$ with $\mathrm{Im}\,\lambda \geq 0$ such that

$$\widetilde{b}_r^{LJ} = -\lambda \overline{b_r^{LJ}} \text{ for all } J, I \in \{1, \dots, M\} \text{ and } r, s \in \{1, \dots, n\}, \tag{6.7.6}$$

then for each (doubly indexed) vector $\zeta = (\zeta_{sI})_{\substack{1 \leq s \leq n \\ 1 \leq I \leq M}}$ with complex components (6.2.24) becomes

$$\langle A\zeta, \overline{\zeta}\rangle = -\lambda \sum_{I,J=1}^{M} \sum_{L=1}^{N} \sum_{r,s=1}^{n} \overline{b_r^{LJ}} b_s^{LI} \zeta_{sI} \overline{\zeta_{rJ}} = -\lambda \sum_{L=1}^{N} \Big| \sum_{I=1}^{M} \sum_{s=1}^{n} b_s^{LI} \zeta_{sI} \Big|^2 \tag{6.7.7}$$

which proves that in this case we indeed have

[2] recall that we tacitly identify Δ with the $M \times M$ system $\Delta I_{M \times M}$, where $I_{M \times M}$ is the $M \times M$ identity matrix

$$\operatorname{Im}\langle A\zeta, \overline{\zeta}\rangle = (-\operatorname{Im}\lambda)\sum_{L=1}^{N}\Big|\sum_{I=1}^{M}\sum_{s=1}^{n} b_s^{LI}\zeta_{sI}\Big|^2 \leq 0. \tag{6.7.8}$$

Next, bearing in mind (6.7.2)-(6.7.4), (6.2.21), and (A.0.118) we may express the conormal derivative associated with the coefficient tensor A from (6.7.4) in an invariant fashion, namely

$$\partial_\xi^A = (-\mathrm{i})\mathrm{Sym}\big(\widetilde{D};\xi\big)D \text{ for each } \xi \in \mathbb{C}^n. \tag{6.7.9}$$

This further gives

$$\partial_{\widehat{x}}^A = (-\mathrm{i})\mathrm{Sym}\big(\widetilde{D};\widehat{x}\big)D \text{ for each } x \in \mathbb{R}^n \setminus \{0\}. \tag{6.7.10}$$

As such, the corresponding radiation condition (RC_A) (associated as in (6.3.2) with thc coefficient tensor A from (6.7.4)) for a vector-valued null-solution u of the vector Helmholtz operator $\Delta + k^2$ in an exterior domain in $\mathbb{R}^n$ reads

$$\mathrm{i}ku(x) - (-\mathrm{i})\mathrm{Sym}\big(\widetilde{D};\widehat{x}\big)(Du)(x) = o\big(|x|^{-(n-1)/2}\big) \text{ as } |x| \to \infty. \tag{6.7.11}$$

For further use let us also remark that, thanks to (6.2.23), (A.0.48), (6.7.4), and (6.7.2), for each function $u = (u_I)_{1\leq I\leq M}$ we have

$$\begin{aligned} A\nabla u &= \big(a_{rs}^{JI}\partial_s u_I\big)_{\substack{1\leq J\leq M\\ 1\leq r\leq n}} = \big(\widetilde{b}_r^{LJ} b_s^{LI}\partial_s u_I\big)_{\substack{1\leq J\leq M\\ 1\leq r\leq n}} \\ &= \big(\widetilde{b}_r^{LJ}(Du)_L\big)_{\substack{1\leq J\leq M\\ 1\leq r\leq n}}. \end{aligned} \tag{6.7.12}$$

In turn, if $\Omega \subseteq \mathbb{R}^n$ is an arbitrary open set, $\sigma := \mathcal{H}^{n-1}\lfloor\partial\Omega$, and $\kappa, \rho \in (0,\infty)$ are arbitrary, then (6.7.12) implies (also bearing in mind [112, (8.2.28)]) that for each function $u \in \big[\mathscr{C}^\infty(\Omega)\big]^M$ we see that

if $\Big((Du)\big|_{\partial\Omega}^{\kappa-\mathrm{n.t.}}\Big)(x)$ exists at some point $x \in \partial_{\mathrm{nta}}\Omega$, then the nontangential trace $\Big((A\nabla u)\big|_{\partial\Omega}^{\kappa-\mathrm{n.t.}}\Big)(x)$ also exists, and if $\mathcal{N}_\kappa^\rho(Du) \in L^1(\partial\Omega,\sigma)$ then the nontangential maximal function $\mathcal{N}_\kappa^\rho(A\nabla u)$ also belongs to $L^1(\partial\Omega,\sigma)$. (6.7.13)

Next we reconsider the acoustic double and single layer potential operators in the formalism associated with the factorization (6.7.1). First, let us take a look at the integrand in (6.4.2). Observe that for each point $x \in \Omega$ and σ-a.e. $y \in \partial_*\Omega$ we may write (keeping (6.7.2)-(6.7.4) and (A.0.118) in mind)

$$
\begin{aligned}
&\Big(a^{IJ}_{rs}(\partial_r\Phi_k)(x-y)\nu_s(y)f_J(y)\Big)_{1\le I\le M}\\
&=\Big(\widetilde{b}^{LI}_s\nu_s(y)b^{LJ}_r(\partial_r\Phi_k)(x-y)\nu_s(y)f_J(y)\Big)_{1\le I\le M}\\
&=(-\mathrm{i})\mathrm{Sym}\big(\widetilde{D};\nu(y)\big)D_x[\Phi_k(x-y)f(y)]\\
&=-\,\mathrm{Sym}\big(\widetilde{D};\nu(y)\big)\mathrm{Sym}\big(D;(\nabla\Phi_k)(x-y)\big)f(y).
\end{aligned}
\tag{6.7.14}
$$

In view of this and (6.4.4), the integral representation formula

$$
u=\mathcal{D}_A\big(u\big|^{\kappa-\mathrm{n.t.}}_{\partial\Omega}\big)-\mathcal{S}\big(\partial^A_\nu u\big)\ \text{ in }\Omega
\tag{6.7.15}
$$

appearing in (6.6.8) may be re-fashioned as

$$
\begin{aligned}
u(x)=&\int_{\partial_*\Omega}\mathrm{Sym}\big(\widetilde{D};\nu(y)\big)\mathrm{Sym}\big(D;(\nabla\Phi_k)(x-y)\big)\big(u\big|^{\kappa-\mathrm{n.t.}}_{\partial\Omega}\big)(y)\,d\sigma(y)\\
&+\mathrm{i}\int_{\partial_*\Omega}\Phi_k(x-y)\mathrm{Sym}\big(\widetilde{D};\nu(y)\big)\big((Du)\big|^{\kappa-\mathrm{n.t.}}_{\partial\Omega}\big)(y)\,d\sigma(y),\qquad\forall\,x\in\Omega.
\end{aligned}
\tag{6.7.16}
$$

Collectively, Theorem 6.6.1 and (6.7.13) then prove the following result:

Theorem 6.7.1 *Pick $k\in(0,\infty)$ and choose $n\in\mathbb{N}$ with $n\ge 2$. Also, fix $M,N\in\mathbb{N}$ arbitrary and consider a first-order constant complex constant coefficient $M\times N$ system $\widetilde{D}$ along with a first-order constant complex constant coefficient $N\times M$ system D with the property that the vector Laplacian in $\mathbb{R}^n$ may be factored as $\Delta=\widetilde{D}D$. Then the integral representation formula* (6.7.16) *holds whenever*

$$
\begin{aligned}
&\Omega\subseteq\mathbb{R}^n\text{ is an exterior domain with a lower Ahlfors regular boundary}\\
&\text{and such that }\sigma:=\mathcal{H}^{n-1}\lfloor\partial\Omega\text{ is a doubling measure on }\partial\Omega,
\end{aligned}
\tag{6.7.17}
$$

and the function u is a $\mathbb{C}^M$-valued null-solution in Ω of the vector Helmholtz operator $\Delta+k^2$ which radiates at infinity (in the sense of Definition 6.1.1), and satisfies

$$
\begin{aligned}
&\text{the traces } u\big|^{\kappa-\mathrm{n.t.}}_{\partial\Omega},\ (Du)\big|^{\kappa-\mathrm{n.t.}}_{\partial\Omega}\ \text{ exist }\sigma\text{-a.e. on }\partial_{\mathrm{nta}}\Omega,\\
&\text{and } \mathcal{N}^{\rho}_{\kappa}u,\ \mathcal{N}^{\rho}_{\kappa}(Du)\in L^1(\partial\Omega,\sigma)\text{ for some }\kappa,\rho>0.
\end{aligned}
\tag{6.7.18}
$$

Example 6.7.2 Start with $\Delta=\partial_r\partial_r$ acting on scalar functions u, which we interpret as the factorization $\Delta=\widetilde{D}D$ with

$$
D:=\begin{pmatrix}\partial_1\\ \vdots\\ \partial_n\end{pmatrix}
\tag{6.7.19}
$$

and

$$
\widetilde{D}:=-D^{\top}=(\partial_1,\dots,\partial_n).
\tag{6.7.20}
$$

Since now we have $(-\mathrm{i})\mathrm{Sym}(\widetilde{D};\widehat{x})(Du)(x) = \widehat{x}\cdot(\nabla u)(x)$ for every scalar function u, the corresponding radiation condition (RC_A) (originally introduced as in (6.3.2) and subsequently equivalently rephrased in (6.7.11)) in this scenario simply becomes $iku(x) - \widehat{x}\cdot(\nabla u)(x) = o\big(|x|^{-(n-1)/2}\big)$ as $|x| \to \infty$, which is precisely the pointwise version of Sommerfeld's radiation condition.

Also, the integral representation formula (6.7.16) becomes precisely (6.1.18).

Example 6.7.3 Start with $\Delta = -\delta\mathrm{d} - \mathrm{d}\delta$ acting on differential forms (see the discussion in [112, §6.4]; cf. [112, (6.4.145)] in particular), where d is the exterior derivative operator, and δ its formal transpose (cf. [112, (6.4.140)]). We may interpret this as the factorization $\Delta = \widetilde{D}D$ with

$$D := \begin{pmatrix}\mathrm{d}\\ \delta\end{pmatrix} \tag{6.7.21}$$

(which is a first-order system with constant real coefficients) and

$$\widetilde{D} := -D^{\top} = (-\delta, -\mathrm{d}). \tag{6.7.22}$$

In this case, $(-\mathrm{i})\mathrm{Sym}(\widetilde{D};\widehat{x})(Du)(x) = \widehat{x} \vee (\mathrm{d}u)(x) - \widehat{x}\wedge(\delta u)(x)$ for each differential form u, with $\wedge$ and $\vee$ denoting the exterior and interior product, respectively, of differential forms (again, see [112, §6.4]). As such, the corresponding radiation condition (RC_A) (originally introduced in (6.3.2) and then equivalently rephrased in (6.7.11)) presently takes the form

$$iku(x) - \widehat{x} \vee (\mathrm{d}u)(x) + \widehat{x}\wedge(\delta u)(x) = o\big(|x|^{-(n-1)/2}\big) \text{ as } |x| \to \infty. \tag{6.7.23}$$

In view of (6.7.21) and (6.7.5), Theorem 6.6.1 ensures that for any differential form u which is a null-solution of the Helmholtz operator (6.7.23) is equivalent to the demand that u radiates at infinity (in the sense of Definition 6.1.1). In fact, according to (6.6.4)-(6.6.6), if u is a differential form which is a null-solution of the Helmholtz operator then any of the conditions

$$iku(x) - \widehat{x} \vee (\mathrm{d}u)(x) + \widehat{x}\wedge(\delta u)(x) = O\big(|x|^{-(n+1)/2}\big) \text{ as } |x| \to \infty, \tag{6.7.24}$$

$$\int_{|x|=R}\Big|iku(x) - \widehat{x} \vee (\mathrm{d}u)(x) + \widehat{x}\wedge(\delta u)(x)\Big|^2\,\mathrm{d}\mathcal{H}^{n-1}(x) = o(1) \text{ as } R \to \infty, \tag{6.7.25}$$

$$\int_{|x|=R}\Big|iku(x) - \widehat{x} \vee (\mathrm{d}u)(x) + \widehat{x}\wedge(\delta u)(x)\Big|^2\,\mathrm{d}\mathcal{H}^{n-1}(x) = O(R^{-2}) \text{ as } R \to \infty, \tag{6.7.26}$$

is also equivalent to the demand that u radiates at infinity in the sense of Definition 6.1.1.

In addition, the integral representation formula (6.7.16) presently becomes

$$u(x) = -\int_{\partial_*\Omega} \nu(y) \vee \Big((\nabla\Phi_k)(x-y) \wedge \big(u\big|_{\partial\Omega}^{\kappa-\text{n.t.}} \big)(y) \Big)\, d\sigma(y)$$
$$- \int_{\partial_*\Omega} \nu(y) \wedge \Big((\nabla\Phi_k)(x-y) \vee \big(u\big|_{\partial\Omega}^{\kappa-\text{n.t.}} \big)(y) \Big)\, d\sigma(y)$$
$$+ \int_{\partial_*\Omega} \Phi_k(x-y) \Big\{ \big(\nu \wedge (\delta u)\big|_{\partial\Omega}^{\kappa-\text{n.t.}} \big)(y) - \big(\nu \vee (du)\big|_{\partial\Omega}^{\kappa-\text{n.t.}} \big)(y) \Big\}\, d\sigma(y) \quad (6.7.27)$$

which is valid at each point $x \in \Omega$ under the assumption that Ω is as in (6.7.17) and u is a differential form in Ω which is a null-solution of the Helmholtz operator $\Delta + k^2$, radiates at infinity (in the sense of Definition 6.1.1 or, equivalently, in the sense of any of the conditions in (6.7.23), (6.7.24)-(6.7.26)), and satisfies

$$\begin{aligned} &\text{the traces } u\big|_{\partial\Omega}^{\kappa-\text{n.t.}},\ (du)\big|_{\partial\Omega}^{\kappa-\text{n.t.}}\ (\delta u)\big|_{\partial\Omega}^{\kappa-\text{n.t.}} \text{ exist } \sigma\text{-a.e. on } \partial_{\text{nta}}\Omega, \\ &\text{and } \mathcal{N}_\kappa^\rho u,\ \mathcal{N}_\kappa^\rho(du),\ \mathcal{N}_\kappa^\rho(\delta u) \in L^1(\partial\Omega, \sigma) \text{ for some } \kappa, \rho > 0. \end{aligned} \quad (6.7.28)$$

Example 6.7.4 Suppose E and H are differential forms satisfying the higher-dimensional version of the Maxwell system for differential forms in an exterior domain $\Omega \subseteq \mathbb{R}^n$. For a given wave number $k \in (0, \infty)$, this reads (cf. [68])

$$\delta E - ikH = 0 \text{ and } dH + ikE = 0 \text{ in } \Omega. \quad (6.7.29)$$

Since $d^2 = 0$ and $\delta^2 = 0$ (cf. [112, (6.4.145)]), this forces

$$dE = 0 \text{ and } \delta H = 0 \text{ in } \Omega, \quad (6.7.30)$$

as well as

$$d\delta E = ikdH = k^2 E \text{ and } \delta dH = -ik\delta E = k^2 H \text{ in } \Omega. \quad (6.7.31)$$

As a consequence of (6.7.30), (6.7.32), and [112, (6.4.145)], we have

$$\begin{aligned} (\Delta + k^2)E &= -\delta dE - d\delta E + k^2 E = 0 \text{ and} \\ (\Delta + k^2)H &= -\delta dH - d\delta H + k^2 H = 0 \text{ in } \Omega. \end{aligned} \quad (6.7.32)$$

Hence, E and H may be regarded as differential forms which, individually, are null-solutions of the Helmholtz operator $\Delta + k^2$ in Ω. From (6.7.29), (6.7.32), and Proposition 6.3.4 we then see that

$$\begin{aligned} &E \text{ radiates at infinity if and only } H \text{ radiates} \\ &\text{at infinity (in the sense of Definition 6.1.1).} \end{aligned} \quad (6.7.33)$$

Granted (6.7.29)-(6.7.30), the radiation condition (6.7.23) written for the choice $u := E$ presently becomes equivalent to

$$E(x) + \widehat{x} \wedge H(x) = o\big(|x|^{-(n-1)/2}\big) \text{ as } |x| \to \infty, \quad (6.7.34)$$

whereas the conditions in (6.7.24)-(6.7.26) presently become equivalent to, respectively,

$$E(x) + \widehat{x} \wedge H(x) = O\big(|x|^{-(n+1)/2}\big) \text{ as } |x| \to \infty, \tag{6.7.35}$$

$$\int_{|x|=R} \big|E(x) + \widehat{x} \wedge H(x)\big|^2 \, d\mathcal{H}^{n-1}(x) = o(1) \text{ as } R \to \infty, \tag{6.7.36}$$

$$\int_{|x|=R} \big|E(x) + \widehat{x} \wedge H(x)\big|^2 \, d\mathcal{H}^{n-1}(x) = O(R^{-2}) \text{ as } R \to \infty. \tag{6.7.37}$$

From the discussion in Example 6.7.3 above we then conclude that

any of the conditions (6.7.34)-(6.7.37) is equivalent to the demand that E radiates at infinity (in the sense of Definition 6.1.1). (6.7.38)

Moreover, the integral representation formula (6.7.27) written for $u := E$ presently becomes

$$\begin{aligned} E(x) = &- \int_{\partial_*\Omega} \nu(y) \vee \Big((\nabla \Phi_k)(x-y) \wedge \big(E\big|_{\partial\Omega}^{\kappa-\mathrm{n.t.}}\big)(y)\Big) \, d\sigma(y) \\ &- \int_{\partial_*\Omega} \nu(y) \wedge \Big((\nabla \Phi_k)(x-y) \vee \big(E\big|_{\partial\Omega}^{\kappa-\mathrm{n.t.}}\big)(y)\Big) \, d\sigma(y) \\ &+ ik \int_{\partial_*\Omega} \Phi_k(x-y) \big(\nu \wedge H\big|_{\partial\Omega}^{\kappa-\mathrm{n.t.}}\big)(y) \, d\sigma(y), \end{aligned} \tag{6.7.39}$$

which is valid at each point $x \in \Omega$ under the assumption that Ω is as in (6.7.17) and the differential forms E, H solve the Maxwell system (6.7.29), radiate at infinity (in the sense of Definition 6.1.1 or, equivalently, in the sense of any of the conditions in (6.7.34)-(6.7.37)), and satisfy

$$\begin{aligned} &\text{the traces } E\big|_{\partial\Omega}^{\kappa-\mathrm{n.t.}}, \ H\big|_{\partial\Omega}^{\kappa-\mathrm{n.t.}} \text{ exist } \sigma\text{-a.e. on } \partial_{\mathrm{nta}}\Omega, \\ &\text{and } \mathcal{N}_\kappa^\rho E, \ \mathcal{N}_\kappa^\rho H \in L^1(\partial\Omega, \sigma) \text{ for some } \kappa, \rho > 0. \end{aligned} \tag{6.7.40}$$

Likewise, thanks to (6.7.29)-(6.7.30) the radiation condition (6.7.23) written for $u := H$ presently becomes equivalent to

$$H(x) + \widehat{x} \vee E(x) = o\big(|x|^{-(n-1)/2}\big) \text{ as } |x| \to \infty, \tag{6.7.41}$$

while the conditions in (6.7.24)-(6.7.26) presently become equivalent to, respectively,

$$H(x) + \widehat{x} \vee E(x) = O\big(|x|^{-(n+1)/2}\big) \text{ as } |x| \to \infty, \tag{6.7.42}$$

$$\int_{|x|=R} \big|H(x) + \widehat{x} \vee E(x)\big|^2 \, d\mathcal{H}^{n-1}(x) = o(1) \text{ as } R \to \infty, \tag{6.7.43}$$

$$\int_{|x|=R} \left|H(x) + \widehat{x} \vee E(x)\right|^2 \, d\mathcal{H}^{n-1}(x) = O(R^{-2}) \text{ as } R \to \infty. \tag{6.7.44}$$

Then the discussion in Example 6.7.3 above permits us to conclude that

any of the conditions (6.7.41)-(6.7.44) is equivalent to the demand that H radiates at infinity (in the sense of Definition 6.1.1). (6.7.45)

Furthermore, the integral representation formula (6.7.27) written for $u := H$ presently becomes

$$\begin{aligned} H(x) = &- \int_{\partial_* \Omega} \nu(y) \vee \Big((\nabla \Phi_k)(x-y) \wedge \big(H\big|_{\partial\Omega}^{\kappa-\text{n.t.}}\big)(y) \Big) \, d\sigma(y) \\ &- \int_{\partial_* \Omega} \nu(y) \wedge \Big((\nabla \Phi_k)(x-y) \vee \big(H\big|_{\partial\Omega}^{\kappa-\text{n.t.}}\big)(y) \Big) \, d\sigma(y) \\ &+ ik \int_{\partial_* \Omega} \Phi_k(x-y) \big(\nu \wedge E\big|_{\partial\Omega}^{\kappa-\text{n.t.}}\big)(y) \, d\sigma(y) \quad \forall x \in \Omega, \end{aligned} \tag{6.7.46}$$

under the same assumptions on Ω, E, H as in the case of (6.7.39).

Finally, we wish to note that (6.7.34) and (6.7.41) considered collectively, i.e.,

$$\begin{cases} E(x) + \widehat{x} \wedge H(x) = o\big(|x|^{-(n-1)/2}\big) \text{ as } |x| \to \infty, \\ H(x) + \widehat{x} \vee E(x) = o\big(|x|^{-(n-1)/2}\big) \text{ as } |x| \to \infty, \end{cases} \tag{6.7.47}$$

constitute a higher-dimensional generalization of the classical three-dimensional Sylver-Müler radiation condition from (6.2.8)-(6.2.9), and that the integral representation formulas in (6.7.39) and (6.7.46) are higher dimensional generalizations of the classical Stratton-Chu integral representation formulas for solutions the standard Maxwell system in the three-dimensional setting (cf. [172]).

Example 6.7.5 Denote by

$$\mathbf{D} := \sum_{j=1}^{n} \mathbf{e}_j \odot \partial_j \tag{6.7.48}$$

the standard homogeneous Dirac operator in $\mathbb{R}^n$, and observe that [112, (6.4.52)] entails

$$\mathbf{D}^\top = \mathbf{D}. \tag{6.7.49}$$

Consider the Laplacian acting componentwise on Clifford algebra-valued functions. Then (6.7.49) and [112, (6.4.56)] imply that $\Delta = -\mathbf{DD}$. We shall interpret this factorization as

$$\Delta = \widetilde{D} D \text{ with } \widetilde{D} := -\mathbf{D} \text{ and } D := \mathbf{D}. \tag{6.7.50}$$

In this case, $(-i)\mathrm{Sym}(\widetilde{D}; \widehat{x})(Du)(x) = -\widehat{x} \odot (\mathbf{D}u)(x)$ hence the corresponding radiation condition (RC_A) (originally introduced in (6.3.2) and then equivalently rephrased in (6.7.11)) presently has the form

$$iku(x) + \widehat{x} \odot (\mathbf{D}u)(x) = o\big(|x|^{-(n-1)/2}\big) \text{ as } |x| \to \infty. \tag{6.7.51}$$

Note that (6.7.49) and the choices made in (6.7.50) imply that we have

$$\widetilde{D} = -D^{\top}, \tag{6.7.52}$$

which is relevant in the context of (6.7.5). With this in mind, Theorem 6.6.1 then guarantees that, for any Clifford algebra-valued function u which is a null-solution of the Helmholtz operator, the radiation condition recorded in (6.7.51) is equivalent to the demand that u radiates at infinity (in the sense of Definition 6.1.1). In fact, according to (6.6.4)-(6.6.6), if u is a Clifford algebra-valued function which is a null-solution of the Helmholtz operator then any of the conditions

$$iku(x) + \widehat{x} \odot (\mathbf{D}u)(x) = O\big(|x|^{-(n+1)/2}\big) \text{ as } |x| \to \infty, \tag{6.7.53}$$

$$\int_{|x|=R} \big|iku(x) + \widehat{x} \odot (\mathbf{D}u)(x)\big|^2 \, d\mathcal{H}^{n-1}(x) = o(1) \text{ as } R \to \infty, \tag{6.7.54}$$

$$\int_{|x|=R} \big|iku(x) + \widehat{x} \odot (\mathbf{D}u)(x)\big|^2 \, d\mathcal{H}^{n-1}(x) = O(R^{-2}) \text{ as } R \to \infty, \tag{6.7.55}$$

is also equivalent to the demand that u radiates at infinity in the sense of Definition 6.1.1.

Finally, the integral representation formula (6.7.16) currently takes the form

$$\begin{aligned} u(x) = &\int_{\partial_*\Omega} (\mathbf{D}\Phi_k)(x-y) \odot \nu(y) \odot \big(u\big|^{\kappa-\mathrm{n.t.}}_{\partial\Omega}\big)(y)\, d\sigma(y) \\ &+ \int_{\partial_*\Omega} \Phi_k(x-y) \odot \nu(y) \odot \big((\mathbf{D}u)\big|^{\kappa-\mathrm{n.t.}}_{\partial\Omega}\big)(y)\, d\sigma(y), \qquad \forall\, x \in \Omega, \end{aligned} \tag{6.7.56}$$

assuming that Ω is as in (6.7.17) and that u is a Clifford algebra-valued function in Ω which is a null-solution of the Helmholtz operator $\Delta + k^2$ in Ω, radiates at infinity (in the sense of Definition 6.1.1 or, equivalently, in the sense of any of the conditions in (6.7.51), (6.7.53)-(6.7.55)), and satisfies

$$\begin{aligned} &\text{the traces } u\big|^{\kappa-\mathrm{n.t.}}_{\partial\Omega},\ (\mathbf{D}u)\big|^{\kappa-\mathrm{n.t.}}_{\partial\Omega} \text{ exist } \sigma\text{-a.e. on } \partial_{\mathrm{nta}}\Omega, \\ &\text{and } \mathcal{N}^{\rho}_{\kappa} u,\ \mathcal{N}^{\rho}_{\kappa}(\mathbf{D}u) \in L^1(\partial\Omega, \sigma) \text{ for some } \kappa, \rho > 0. \end{aligned} \tag{6.7.57}$$

Example 6.7.6 Recall that a k-monogenic function in an open set $\Omega \subseteq \mathbb{R}^n$ is a smooth Clifford algebra-valued function u in Ω satisfying $\mathbf{D}u + k\mathbf{e}_{n+1} \odot u = 0$ in Ω (where $\mathbf{D}$ is the standard homogeneous Dirac operator in $\mathbb{R}^n$; cf. (6.7.48)). Since

$$\big(\mathbf{D} + k\mathbf{e}_{n+1}\big)^2 = -(\Delta + k^2) \tag{6.7.58}$$

it follows that any k-monogenic function u is a null-solution of $\Delta + k^2$ with the additional property that $\mathbf{D}u = -k\mathbf{e}_{n+1} \odot u$. In such a case, (6.7.51) reduces precisely to the McIntosh-Mitrea radiation condition for k-monogenic functions, namely

$$(i - \widehat{x} \odot \mathbf{e}_{n+1}) \odot u(x) = o\big(|x|^{-(n-1)/2}\big) \text{ as } |x| \to \infty, \tag{6.7.59}$$

whereas (6.7.53)-(6.7.55) become

$$(i - \widehat{x} \odot \mathbf{e}_{n+1}) \odot u(x) = O\big(|x|^{-(n+1)/2}\big) \text{ as } |x| \to \infty, \tag{6.7.60}$$

$$\int_{|x|=R} \big|(i - \widehat{x} \odot \mathbf{e}_{n+1}) \odot u(x)\big|^2 \, d\mathcal{H}^{n-1}(x) = o(1) \text{ as } R \to \infty, \tag{6.7.61}$$

$$\int_{|x|=R} \big|(i - \widehat{x} \odot \mathbf{e}_{n+1}) \odot u(x)\big|^2 \, d\mathcal{H}^{n-1}(x) = O(R^{-2}) \text{ as } R \to \infty. \tag{6.7.62}$$

Each of these conditions is equivalent to the demand that the k-monogenic function u radiates at infinity in the sense of Definition 6.1.1.

Moreover, for a k-monogenic function u in Ω which radiates at infinity (in the sense of Definition 6.1.1 or, equivalently, in the sense of any of the conditions in (6.7.59)-(6.7.62)) and satisfies

$$\begin{gathered} \text{the nontangential trace } u\big|_{\partial\Omega}^{\kappa-\text{n.t.}} \text{ exists } \sigma\text{-a.e. on } \partial_{\text{nta}}\Omega \\ \text{and } \mathcal{N}_\kappa^\rho u \in L^1(\partial\Omega, \sigma) \text{ for some } \kappa, \rho > 0, \end{gathered} \tag{6.7.63}$$

the integral representation formula (6.7.56) further reduces to

$$u(x) = \int_{\partial_*\Omega} \big((\mathbf{D}\Phi_k)(x-y) + k\Phi_k(x-y)\mathbf{e}_{n+1}\big) \odot \nu(y) \odot \big(u\big|_{\partial\Omega}^{\kappa-\text{n.t.}}\big)(y) \, d\sigma(y) \tag{6.7.64}$$

at each $x \in \Omega$.

We continue by discussing a general integral representation result for radiating null-solutions of the Helmholtz operator in an exterior domain, whose specific format depends on the manner in which the Helmholtz operator is factored as the product of two first-order systems.

Theorem 6.7.7 *Let $\Omega \subseteq \mathbb{R}^n$, where $n \in \mathbb{N}$ satisfies $n \geq 2$, be an exterior domain with a lower Ahlfors regular boundary, and such that $\sigma := \mathcal{H}^{n-1}\lfloor\partial\Omega$ is a doubling measure on $\partial\Omega$. Denote by ν the geometric measure theoretic outward unit normal to Ω, and fix an aperture parameter $\kappa \in (0,\infty)$ along with a truncation parameter $\rho \in (0,\infty)$.*

Next, fix a wave number $k \in (0,\infty)$ and select $N, M \in \mathbb{N}$ arbitrary. Assume the vector Helmholtz operator $\Delta + k^2$, regarded as a second-order $M \times M$ system in $\mathbb{R}^n$, factors as

$$\Delta + k^2 = \widetilde{\mathbb{D}}\mathbb{D} \tag{6.7.65}$$

for some first-order $M \times N$ system $\widetilde{\mathbb{D}}$ in $\mathbb{R}^n$ and some $N \times M$ system $\mathbb{D}$ in $\mathbb{R}^n$ (not necessarily homogeneous; see [112, (1.7.12)]) *both with constant complex coefficients. Finally, pick a vector-valued function u satisfying*

$$\begin{aligned} & u \in \big[\mathscr{C}^\infty(\Omega)\big]^M, \quad (\Delta + k^2)u = 0 \text{ in } \Omega, \\ & u \text{ radiates at infinity (cf. Definition 6.1.1)}, \\ & \mathcal{N}_\kappa^\rho u \text{ and } \mathcal{N}_\kappa^\rho(\mathbb{D}u) \text{ belong to } L^1(\partial\Omega, \sigma), \\ & u\big|_{\partial\Omega}^{\kappa-\text{n.t.}} \text{ and } (\mathbb{D}u)\big|_{\partial\Omega}^{\kappa-\text{n.t.}} \text{ exist } \sigma\text{-a.e. on } \partial_{\text{nta}}\Omega. \end{aligned} \tag{6.7.66}$$

Then for any $\kappa' > 0$ the nontangential traces $u\big|_{\partial\Omega}^{\kappa'-\text{n.t.}}$ and $(\mathbb{D}u)\big|_{\partial\Omega}^{\kappa'-\text{n.t.}}$ also exist at σ-a.e. point on $\partial_{\text{nta}}\Omega$ and are actually independent of κ'. Moreover, with the dependence on the aperture parameter dropped, the following integral representation formula holds:

$$\begin{aligned} u(x) = & \int_{\partial_*\Omega} \mathrm{Sym}\big(\widetilde{\mathbb{D}}; \nu(y)\big)\mathrm{Sym}\big(\mathbb{D}; (\nabla\Phi_k)(x-y)\big)\big(u\big|_{\partial\Omega}^{\text{n.t.}}\big)(y)\,d\sigma(y) \\ & + \mathrm{i}\int_{\partial_*\Omega} \Phi_k(x-y)\mathrm{Sym}\big(\widetilde{\mathbb{D}}; \nu(y)\big)\big((\mathbb{D}u)\big|_{\partial\Omega}^{\text{n.t.}}\big)(y)\,d\sigma(y), \qquad \forall x \in \Omega. \end{aligned} \tag{6.7.67}$$

Before presenting the proof of this result we wish to note that the integral representation formula (6.7.67) may be though of a generalization of (6.7.16). Indeed, given a first-order $M \times N$ system $\widetilde{D}$ along with a first-order $N \times M$ system D having constant complex coefficients and satisfying (6.7.1), it follows that[3]

$$\widetilde{\mathbb{D}} := \big(\widetilde{D}, kI_{M\times M}\big) \text{ and } \mathbb{D} := \begin{pmatrix} D \\ kI_{M\times M} \end{pmatrix} \tag{6.7.68}$$

are constant complex coefficient first-order ($M \times (N+M)$ and $(N+M) \times M$, respectively) systems satisfying

$$\widetilde{\mathbb{D}}\mathbb{D} = \big(\widetilde{D}, kI_{M\times M}\big)\begin{pmatrix} D \\ kI_{M\times M} \end{pmatrix} = \widetilde{D}D + k^2 = \Delta^2 + k^2. \tag{6.7.69}$$

Hence, the factorization (6.7.65) is valid for $\widetilde{\mathbb{D}}$ and $\mathbb{D}$ as in (6.7.68). Since for this choice in (6.7.68) we also have that (6.7.18) implies the properties in the last two lines of (6.7.66), formula (6.7.67) holds and (6.7.16) readily follows from it.

Let us now turn to the proof of Theorem 6.7.7.

Proof of Theorem 6.7.7 The fact that, for any other aperture parameter $\kappa' > 0$, the nontangential traces $u\big|_{\partial\Omega}^{\kappa'-\text{n.t.}}$ and $(\mathbb{D}u)\big|_{\partial\Omega}^{\kappa'-\text{n.t.}}$ exist at σ-a.e. point on $\partial_{\text{nta}}\Omega$ and are actually independent of κ' is seen from the present hypotheses and [112, Corollary 8.9.9].

[3] recall that $I_{M\times M}$ denotes the $M \times M$ identity matrix

To proceed, with $I_{M\times M}$ denoting the $M\times M$ identity matrix, observe that if we define the first-order ($M \times (N+M)$ and $(N+M)\times M$, respectively) systems

$$\widetilde{D} := \big(\widetilde{\mathbb{D}}, ikI_{M\times M}\big) \text{ and } D := \begin{pmatrix} \mathbb{D} \\ ikI_{M\times M} \end{pmatrix} \tag{6.7.70}$$

then

$$\widetilde{D}D = \big(\mathbb{D}^\top, ikI_{M\times M}\big)\begin{pmatrix} \mathbb{D} \\ ikI_{M\times M} \end{pmatrix} = \mathbb{D}^\top\mathbb{D} - k^2 = (\Delta^2 + k^2) - k^2 = \Delta. \tag{6.7.71}$$

Also, for each $\xi \in \mathbb{C}^n$,

$$\mathrm{Sym}(\widetilde{D};\xi) = \big(\mathrm{Sym}(\widetilde{\mathbb{D}};\xi), 0\big) \text{ and } \mathrm{Sym}(D;\xi) = \begin{pmatrix} \mathrm{Sym}(\mathbb{D};\xi) \\ 0 \end{pmatrix}. \tag{6.7.72}$$

Moreover, since

$$Du = \begin{pmatrix} \mathbb{D}u \\ iku \end{pmatrix}, \tag{6.7.73}$$

the last two lines in (6.7.66) guarantee that

$$\begin{aligned} &\mathcal{N}_\kappa^\rho u \text{ and } \mathcal{N}_\kappa^\rho(Du) \text{ belong to } L^1(\partial\Omega,\sigma), \\ &u\big|_{\partial\Omega}^{\kappa-\mathrm{n.t.}} \text{ and } (Du)\big|_{\partial\Omega}^{\kappa-\mathrm{n.t.}} \text{ exist } \sigma\text{-a.e. on } \partial_{\mathrm{nta}}\Omega. \end{aligned} \tag{6.7.74}$$

Then from (6.7.74), (6.7.72), and (6.7.73) we see that at σ-a.e. point $y \in \partial_*\Omega$ we have

$$\mathrm{Sym}\big(\widetilde{D};\nu(y)\big)\big((Du)\big|_{\partial\Omega}^{\kappa-\mathrm{n.t.}}\big)(y) = \mathrm{Sym}\big(\widetilde{\mathbb{D}};\nu(y)\big)\big((\mathbb{D}u)\big|_{\partial\Omega}^{\kappa-\mathrm{n.t.}}\big)(y), \tag{6.7.75}$$

and

$$\begin{aligned} &\mathrm{Sym}\big(\widetilde{D};\nu(y)\big)\mathrm{Sym}\big(D;(\nabla\Phi_k)(x-y)\big)\big(u\big|_{\partial\Omega}^{\kappa-\mathrm{n.t.}}\big)(y) \\ &\qquad = \mathrm{Sym}\big(\widetilde{\mathbb{D}};\nu(y)\big)\mathrm{Sym}\big(\mathbb{D};(\nabla\Phi_k)(x-y)\big)\big(u\big|_{\partial\Omega}^{\kappa-\mathrm{n.t.}}\big)(y). \end{aligned} \tag{6.7.76}$$

Granted (6.7.71), (6.7.74), (6.7.75), (6.7.76), we may invoke Theorem 6.7.1 and conclude from (6.7.16) that the integral representation formula claimed in (6.7.67) holds. □

It is of interest to single out the following special case of Theorem 6.7.7:

Corollary 6.7.8 *Pick a wave number $k \in (0,\infty)$, a dimension $n \in \mathbb{N}$ with $n \geq 2$, and select $N, M \in \mathbb{N}$ arbitrary. Let $\Omega \subseteq \mathbb{R}^n$ be an exterior domain with a lower Ahlfors regular boundary, and such that $\sigma := \mathcal{H}^{n-1}\lfloor\partial\Omega$ is a doubling measure. Denote by ν the geometric measure theoretic outward unit normal to Ω, and fix an aperture parameter $\kappa \in (0,\infty)$ along with a truncation parameter $\rho \in (0,\infty)$. Next, assume the vector Helmholtz operator $\Delta + k^2$, regarded as a second-order $M \times M$ system in*

$\mathbb{R}^n$, *factors as*

$$\Delta + k^2 = \widetilde{\mathbb{D}}\mathbb{D} \tag{6.7.77}$$

for some first-order $M \times N$ *system* $\widetilde{\mathbb{D}}$ *in* $\mathbb{R}^n$ *and some* $N \times M$ *system* $\mathbb{D}$ *in* $\mathbb{R}^n$ (*not necessarily homogeneous; see* [112, (1.7.12)]) *both with constant complex coefficients. Finally, consider a vector-valued function* u *satisfying*

$$\begin{gathered} u \in \left[\mathscr{C}^\infty(\Omega)\right]^M, \quad \mathbb{D}u = 0 \ \textit{in} \ \Omega, \\ u \textit{ radiates at infinity (cf. Definition 6.1.1)}, \\ \mathcal{N}_\kappa^\rho u \textit{ belongs to the space } L^1(\partial\Omega, \sigma), \\ \textit{and the trace } u\big|_{\partial\Omega}^{\kappa-\mathrm{n.t.}} \textit{ exists } \sigma\textit{-a.e. on } \partial_{\mathrm{nta}}\Omega. \end{gathered} \tag{6.7.78}$$

Then for any other aperture parameter $\kappa' > 0$ *the nontangential trace* $u\big|_{\partial\Omega}^{\kappa'-\mathrm{n.t.}}$ *also exists at* σ*-a.e. point on* $\partial_{\mathrm{nta}}\Omega$ *and, in fact, is independent of* κ'*. Furthermore, with the dependence on the aperture parameter dropped, the following integral representation formula holds at each point* $x \in \Omega$*:*

$$u(x) = \int_{\partial_*\Omega} \mathrm{Sym}\big(\widetilde{\mathbb{D}}; \nu(y)\big)\mathrm{Sym}\big(\mathbb{D}; (\nabla\Phi_k)(x-y)\big)\big(u\big|_{\partial\Omega}^{\mathrm{n.t.}}\big)(y)\,\mathrm{d}\sigma(y). \tag{6.7.79}$$

Proof Since having $\mathbb{D}u = 0$ in Ω entails

$$(\Delta + k^2)u = \widetilde{\mathbb{D}}\mathbb{D}u = 0 \ \text{ in } \Omega, \tag{6.7.80}$$

this is a direct consequence of Theorem 6.7.7. □

To offer an example, with $\mathbf{D}$ as in (6.7.48) abbreviate

$$\mathbf{D}_k := \mathbf{D} + k\mathbf{e}_{n+1}. \tag{6.7.81}$$

In this piece of notation, (6.7.58) becomes

$$\mathbf{D}_k^2 = -(\Delta + k^2). \tag{6.7.82}$$

We may regard this as the factorization (6.7.77) written for

$$\widetilde{\mathbb{D}} := -\mathbf{D}_k \ \text{ and } \ \mathbb{D} := \mathbf{D}_k. \tag{6.7.83}$$

Then Corollary 6.7.8 gives that any k-monogenic function u in Ω (i.e., any Clifford algebra-valued null-solution of $\mathbf{D}_k$ in Ω) which radiates at infinity (in the sense of Definition 6.1.1 or, equivalently, in the sense of any of the conditions in (6.7.59)-(6.7.62)) and with the property that the nontangential trace $u\big|_{\partial\Omega}^{\kappa-\mathrm{n.t.}}$ exists σ-a.e. on $\partial_{\mathrm{nta}}\Omega$ and $\mathcal{N}_\kappa^\rho u \in L^1(\partial\Omega, \sigma)$ (for some $\kappa, \rho > 0$), may be represented as in (6.7.79) for $\widetilde{\mathbb{D}}$, $\mathbb{D}$ as in (6.7.83). A cursory inspection shows that said integral representation formula is precisely the one recorded in (6.7.64).

In closing we wish to remark that the analysis in this section has brought to prominence a number of singular integral operators naturally associated with the Helmholtz equation, such as (6.4.2). We shall take a closer look at some of these in [115, §1.5].

Appendix A
Terms and notation used in Volume III

A

Ahlfors regular domain (cf. [112, Definition 5.9.15]):

$$\text{a nonempty open subset } \Omega \text{ of } \mathbb{R}^n \text{ such that } \partial\Omega \text{ is an Ahlfors regular set and } \mathcal{H}^{n-1}(\partial\Omega \setminus \partial_*\Omega) = 0 \tag{A.0.1}$$

$A_\kappa(\partial\Omega)$, the κ-accessible (from within Ω) subset of $\partial\Omega$ (cf. [112, (8.8.2)]):

$$A_\kappa(\partial\Omega) := \big\{x \in \partial\Omega : \ x \in \overline{\Gamma_\kappa(x)}\big\} \tag{A.0.2}$$

$[w]_{A_p}$, the characteristic of the weight w on a space of homogeneous type (X, ρ, μ) (cf. [112, §7.7]):

$$[w]_{A_p} := \sup_{B\ \rho\text{-ball}} \Big(⨍_B w \, d\mu\Big)\Big(⨍_B w^{-1/(p-1)} \, d\mu\Big)^{p-1} \tag{A.0.3}$$

$A_p(X, \rho, \mu)$, the Muckenhoupt A_p-class on a space of homogeneous type (X, ρ, μ) (cf. [112, §7.7]):

$$A_p(X, \rho, \mu) := \big\{w \ \text{ weight function } : \ [w]_{A_p} < \infty\big\} \tag{A.0.4}$$

$A_\infty(X, \rho, \mu)$, the Muckenhoupt A_∞-class on a space of homogeneous type (X, ρ, μ) (cf. [112, §7.7]):

$$A_\infty(X, \rho, \mu) := \bigcup_{1 \le p < \infty} A_p(X, \mu) \tag{A.0.5}$$

$V^\perp$, the annihilator $V^\perp$ of a subspace V of a Banach space X:

$$V^\perp := \big\{\Lambda \in X^* : \ \Lambda(x) = 0 \ \text{ for all } \ x \in V\big\} \tag{A.0.6}$$

${}^\perp W$, the annihilator of a subspace W of X^* (where X is Banach):

D. Mitrea et al., *Geometric Harmonic Analysis II*, Developments in Mathematics 74,
https://doi.org/10.1007/978-3-031-22735-6

$$ {}^{\perp}W := \{x \in X : \Lambda(x) = 0 \text{ for all } \Lambda \in W\} \tag{A.0.7} $$

T^*, the adjoint of the operator $T \in \mathcal{L}(X \to Y)$:

$$ T^* : Y^* \longrightarrow X^*, \quad T^*(\Lambda) := \Lambda \circ T \text{ for each } \Lambda \in Y^* \tag{A.0.8} $$

$\langle A \rangle_p$, the absolutely p-convex hull of a subset A of a vector space X, with $p \in (0,1]$, defined as (cf. [112, §7.8.3]):

$$ \langle A \rangle_p := \Big\{ \sum_{j=1}^{M} \lambda_j v_j : M \in \mathbb{N},\ \{v_j\}_{1\le j\le M} \subseteq A, \\ \{\lambda_j\}_{1\le j\le M} \subseteq \mathbb{C} \text{ with } \sum_{j=1}^{M} |\lambda_j|^p \le 1 \Big\} \tag{A.0.9} $$

$a \otimes b$, the tensor product of vectors $a = (a_j)_{1\le j\le N} \in \mathbb{C}^N$ and $b = (b_k)_{1\le k\le M} \in \mathbb{C}^M$, defined as the $N \times M$ matrix defined as in (1.3.7):

$$ a \otimes b := \big(a_j b_k\big)_{\substack{1\le j\le N\\ 1\le k\le M}} $$

$A := \big(a_{rs}^{\alpha\beta}\big)_{\substack{1\le r,s\le n\\ 1\le\alpha\le M\\ 1\le\beta\le N}}$, a coefficient tensor (of type $(n\times n, M\times N)$) with complex number entries (1.7.1)

$A^{\top}$, the transpose of a coefficient tensor A defined as $\big(A^{\top}\big)_{sr}^{\beta\alpha} := a_{rs}^{\alpha\beta}$ for all α, β, r, s if $A = \big(a_{rs}^{\alpha\beta}\big)$ (1.7.3)

$A\nabla u := \big(a_{rs}^{\alpha\beta}\partial_s u_\beta\big)_{\substack{1\le\alpha\le M\\ 1\le r\le n}}$, the action of the linear mapping $A = \big(a_{rs}^{\alpha\beta}\big)_{\substack{1\le r,s\le n\\ 1\le\alpha\le M\\ 1\le\beta\le N}}$ on the Jacobian matrix $\nabla u \in \big[\mathcal{D}'(\Omega)\big]^{n\times N}$ of the distribution $u = (u_\beta)_{1\le\beta\le N} \in \big[\mathcal{D}'(\Omega)\big]^N$ (1.7.5)

$\big\langle A\cdot,\cdot\big\rangle$, the bilinear form associated with the coefficient tensor A as in (1.7.6):

$$ \big\langle A\zeta, \eta\big\rangle := a_{rs}^{\alpha\beta}\zeta_s^{\beta}\eta_r^{\alpha} \text{ for all } \zeta := (\zeta_s^{\beta})_{\beta,s} \in \mathbb{C}^{N\times n} \text{ and } \eta := (\eta_r^{\alpha})_{\alpha,r} \in \mathbb{C}^{M\times n} $$

$\mathcal{A}_{q,\kappa}$, the L^q-based area-function, defined as in (2.4.31):

$$ (\mathcal{A}_{q,\kappa}u)(x) := \Big(\int_{\Gamma_\kappa^{\Sigma^c}(x)} |(\nabla u)(y)|^q |x-y|^{q-n}\, dy \Big)^{1/q} $$

for $x \in \Sigma$ if $u \in W^{1,1}_{\rm loc}(\mathbb{R}^n \setminus \Sigma)$

$A_s^{p,q}(\Omega)$, the Besov/Triebel-Lizorkin space in the open set $\Omega \subseteq \mathbb{R}^n$ (with $A := B$ corresponding to Besov spaces, and with $A := F$ corresponding to Triebel-Lizorkin spaces), with $0 < p, q \le \infty$ and $s \in \mathbb{R}$, defined as (cf. [113, §9.2]):

$$A_s^{p,q}(\Omega) := \left\{u \in \mathcal{D}'(\Omega) : \text{there exists } U \in A_s^{p,q}(\mathbb{R}^n) \text{ such that } U\big|_{\Omega} = u\right\} \tag{A.0.10}$$

and equipped with the quasi-norm

$$\|u\|_{A_s^{p,q}(\Omega)} := \inf\left\{\|U\|_{A_s^{p,q}(\mathbb{R}^n)} : U \in A_s^{p,q}(\mathbb{R}^n),\ U\big|_{\Omega} = u\right\} \tag{A.0.11}$$

$A_s^{p,q}(\Omega)_{\mathrm{bdd}}$, the space of all distributions u in Ω with the property that $(\psi\big|_{\Omega})u$ belongs to $A_s^{p,q}(\Omega)$ for each cutoff function $\psi \in \mathscr{C}_c^\infty(\mathbb{R}^n)$ (cf. [113, Convention 8.3.7] and (A.0.132)).

B

$B_\rho(x, r)$, the ρ-ball with center at $x \in X$ and radius $r > 0$ in the quasi-metric space (X, ρ) (cf. [112, §7.1]):

$$B_\rho(x, r) := \{y \in X : \rho(x, y) < r\} \tag{A.0.12}$$

$B_{n-1}(x', r) := \{y' \in \mathbb{R}^{n-1} : |y' - x'| < r\}$, the $(n-1)$-dimensional (open) ball in $\mathbb{R}^{n-1}$ centered at $x' \in \mathbb{R}^{n-1}$ and of radius $r \in (0, \infty)$

BMO_1, the BMO-based Sobolev spaces of order one (locally integrable functions with distributional first-order partial derivatives in BMO)

$\|f\|_*(\Delta)$, the local BMO norm of the function f on the surface ball Δ (cf. [112, §7.4]):

$$\|f\|_*(\Delta) := \sup_{\Delta' \subseteq \Delta} \fint_{\Delta'} |f - f_{\Delta'}|\, d\mu \tag{A.0.13}$$

$\|f\|_{\dot{\mathrm{BMO}}(X,\mu)}$, the homogeneous BMO semi-norm of the function f in the context of a space of homogeneous type (X, ρ, μ) (cf. [112, §7.4]):

$$\|f\|_{\dot{\mathrm{BMO}}(X,\mu)} := \sup_{x\in X,\, r>0} \fint_{B_\rho(x,r)} \left|f - f_{B_\rho(x,r)}\right| d\mu \tag{A.0.14}$$

$\|\cdot\|_{\mathrm{BMO}(X,\mu)}$, the inhomogeneous BMO “norm” in the context of a space of homogeneous type (X, ρ, μ) (cf. [112, §7.4]):

$$\|f\|_{\mathrm{BMO}(X,\mu)} := \begin{cases} \|f\|_{\dot{\mathrm{BMO}}(X,\mu)} & \text{if } X \text{ is unbounded} \\ \|f\|_{\dot{\mathrm{BMO}}(X,\mu)} + \left|\int_X f\, d\mu\right| & \text{if } X \text{ is bounded} \end{cases} \tag{A.0.15}$$

$\mathrm{BMO}(X, \mu)$, the space of functions of bounded mean oscillations for a space of homogeneous type (X, ρ, μ) (cf. [112, §7.4]):

$$\mathrm{BMO}(X, \mu) := \left\{f \in L^1_{\mathrm{loc}}(X, \mu) : \|f\|_{\mathrm{BMO}(X,\mu)} < +\infty\right\} \tag{A.0.16}$$

$\widetilde{\mathrm{BMO}}(X,\mu)$, the space BMO modulo constants for a space of homogeneous type (X,ρ,μ) (cf. [112, (7.4.96)]):

$$\widetilde{\mathrm{BMO}}(X,\mu) := \mathrm{BMO}(X,\mu)/\sim = \big\{[f] : f \in \mathrm{BMO}(X,\mu)\big\} \tag{A.0.17}$$

∂E, the topological boundary of the set E

$Borel_\tau(X)$, the Borelians of the topological space (X,τ)

$[A;B] := [A,B] := AB - BA$, the commutator of A and B

$\{A;B\} := AB + BA$, the anti-commutator of A and B

$\mathrm{Bd}(X \to Y)$, the space of linear and bounded operators from the quasi-normed vector space $(X,\|\cdot\|_X)$ into the quasi-normed vector space $(Y,\|\cdot\|_Y)$ (cf. [113, §1.2]):

$$\mathrm{Bd}(X \to Y) := \big\{T : X \to Y : T \text{ linear mapping with } \|T\|_{X\to Y} < +\infty\big\} \tag{A.0.18}$$

$\mathrm{Bd}(X)$, the space of linear and bounded operators from the quasi-normed vector space X into itself:

$$\mathrm{Bd}(X) := \mathrm{Bd}(X \to X) \tag{A.0.19}$$

$\mathcal{B}(X \to Y)$, the space of linear and bounded operators from X to Y, where X, Y are two linear topological spaces (cf. [113, §1.1]):

$$\mathcal{B}(X \to Y) := \big\{T : X \longrightarrow Y : T \text{ linear and bounded}\big\} \tag{A.0.20}$$

$\mathcal{B}^{q,\lambda}(\Sigma,\sigma)$, the block space on the Ahlfors regular set $\Sigma \subseteq \mathbb{R}^n$, defined for $q \in (1,\infty)$ and $\lambda \in (0,n-1)$ as (cf. [113, §6.2]):

$$\begin{aligned} &\mathcal{B}^{q,\lambda}(\Sigma,\sigma) \\ &\quad := \Big\{f \in \big(\mathrm{Lip}_c(\Sigma)\big)' : \text{there exist a numerical sequence } \{\lambda_j\}_{j\in\mathbb{N}} \in \ell^1(\mathbb{N}) \\ &\qquad\qquad \text{and a family } \{b_j\}_{j\in\mathbb{N}} \text{ of } \mathcal{B}^{q,\lambda}\text{-blocks on } \Sigma \text{ so that} \\ &\qquad\qquad f = \sum_{j=1}^{\infty} \lambda_j b_j \text{ with convergence in } \big(\mathrm{Lip}_c(\Sigma)\big)'\Big\} \end{aligned} \tag{A.0.21}$$

and equipped with the norm

$$\begin{aligned} &\|f\|_{\mathcal{B}^{q,\lambda}(\Sigma,\sigma)} \\ &\quad := \inf\Big\{\sum_{j=1}^{\infty}|\lambda_j| : f = \sum_{j=1}^{\infty}\lambda_j b_j \text{ in } \big(\mathrm{Lip}_c(\Sigma)\big)' \text{ with} \\ &\qquad\qquad \{\lambda_j\}_{j\in\mathbb{N}} \in \ell^1(\mathbb{N}) \text{ and each } b_j \text{ a } \mathcal{B}^{q,\lambda}\text{-block on } \Sigma\Big\} \end{aligned} \tag{A.0.22}$$

$B^{p,q}_s(\Sigma,\sigma)$, the inhomogeneous Besov space on the Ahlfors regular set $\Sigma \subseteq \mathbb{R}^n$, defined (cf. [113, §7.1]) for

$$s \in (-1,1), \qquad \max\left\{\tfrac{n-1}{n}, \tfrac{n-1}{n+s}\right\} < p \le \infty, \qquad 0 < q \le \infty,$$
$$\max\left\{(s)_+, -s+(n-1)\left(\tfrac{1}{p}-1\right)_+\right\} < \beta < 1, \quad (n-1)\left(\tfrac{1}{p}-1\right)_+ < \gamma < 1, \tag{A.0.23}$$

as the collection of all "distributions" f on Σ such that, if $\{E_k\}_{k\in\mathbb{Z},\,k\ge\widetilde{\kappa}_\Sigma}$ is the family of conditional expectation operators on the Ahlfors regular set Σ, then

$$\|f\|_{B^{p,q}_s(\Sigma,\sigma)} := \left\{\sum_{\tau\in I_{\widetilde{\kappa}_\Sigma}} \sum_{\nu=1}^{N(\widetilde{\kappa}_\Sigma,\tau)} \sigma(Q_\tau^{\widetilde{\kappa}_\Sigma,\nu})\Big[m_{Q_\tau^{\widetilde{\kappa}_\Sigma,\nu}}(|E_{\widetilde{\kappa}_\Sigma} f|)\Big]^p\right\}^{1/p}$$
$$+\left\{\sum_{\substack{k\in\mathbb{Z}\\ k\ge\widetilde{\kappa}_\Sigma+1}} \big[2^{ks}\|E_k f\|_{L^p(\Sigma,\sigma)}\big]^q\right\}^{1/q} < \infty \tag{A.0.24}$$

$B^{p,q}_s(\mathbb{R}^n)$, the Besov space in $\mathbb{R}^n$ equipped with the quasi-norm $\|\cdot\|_{B^{p,q}_s(\mathbb{R}^n)}$ for $0<p,q\le\infty$ and $s\in\mathbb{R}$ (cf. [113, §9.1])

$B^{p,q}_s(\Omega)$, the Besov space in the open set $\Omega\subseteq\mathbb{R}^n$ with $0<p,q\le\infty$ and $s\in\mathbb{R}$, defined as the collection of $u\in\mathcal{D}'(\Omega)$ for which there exists $U\in B^{p,q}_s(\mathbb{R}^n)$ such that $U\big|_\Omega = u$, and equipped with the quasi-norm (cf. [113, (9.2.1)])

$$\|u\|_{B^{p,q}_s(\Omega)} := \inf\left\{\|U\|_{B^{p,q}_s(\mathbb{R}^n)} : U\in B^{p,q}_s(\mathbb{R}^n),\ U\big|_\Omega = u\right\}$$

C

$\overline{U}$, the closure of the set $U\subseteq\mathbb{R}^n$

$\mathscr{C}^k(\overline{\Omega})$, the space of functions of class $\mathscr{C}^k$ in an open neighborhood of $\overline{\Omega}$

$\mathscr{C}^k_c(\Omega)$, the space of functions of class $\mathscr{C}^k$ with compact support in the open set Ω

$\mathscr{C}^k_b(\Omega)$, the space of bounded functions of class $\mathscr{C}^k$ in Ω

$\left(\mathscr{C}^\infty_b(\Omega)\right)^*$, the algebraic dual of $\mathscr{C}^\infty_b(\Omega)$

$\mathrm{CBM}(\Omega)$, the space of complex Borel measures in the open set $\Omega\subseteq\mathbb{R}^n$

$\mathrm{CBM}(X,\tau)$, the space of complex Borel measures in the topological space (X,τ)

C_ρ, the triangle inequality "penalty" constant associated with the quasi-distance ρ as follows (cf. [112, (7.1.3)]):

$$C_\rho := \sup_{\substack{x,y,z\in X\\ \text{not all equal}}} \frac{\rho(x,y)}{\max\{\rho(x,z),\rho(z,y)\}} \in [1,\infty) \tag{A.0.25}$$

$\widetilde{C}_\rho$, the symmetry "penalty" constant associated with the quasi-distance ρ as follows (cf. [112, (7.1.4)]):

$$\widetilde{C}_\rho := \sup_{\substack{x,y\in X\\ x\neq y}} \frac{\rho(y,x)}{\rho(x,y)} \in [1,\infty) \tag{A.0.26}$$

$\|\cdot\|_{\dot{\mathscr{C}}^\alpha(U,\rho)}$, the homogeneous Hölder space semi-norm of order $\alpha > 0$ in the set $U \subseteq X$, in the context of a quasi-metric space (X,ρ), defined for each function $f : U \to \mathbb{R}$ as (cf. [112, (7.3.2)]):

$$\|f\|_{\dot{\mathscr{C}}^\alpha(U,\rho)} := \sup_{\substack{x,y\in U\\ x\neq y}} \frac{|f(x)-f(y)|}{\rho(x,y)^\alpha} \tag{A.0.27}$$

$\dot{\mathscr{C}}^\alpha(U,\rho)$, the homogeneous Hölder space of order $\alpha > 0$ in the set $U \subseteq X$, defined in the context of a quasi-metric space (X,ρ) as (cf. [112, (7.3.1)]):

$$\dot{\mathscr{C}}^\alpha(U,\rho) := \Big\{ f : U \to \mathbb{R} : \|f\|_{\dot{\mathscr{C}}^\alpha(U,\rho)} < +\infty \Big\} \tag{A.0.28}$$

$\dot{\mathscr{C}}^\alpha(U,\rho)/\sim$, the homogeneous Hölder space of order $\alpha > 0$ modulo constants, in the set $U \subseteq X$, defined in the context of a quasi-metric space (X,ρ) as (cf. [112, (7.3.6)]):

$$\dot{\mathscr{C}}^\alpha(U,\rho)/\sim := \big\{ [f] : f \in \dot{\mathscr{C}}^\alpha(U,\rho) \big\} \tag{A.0.29}$$

$\dot{\mathscr{C}}^\alpha_{\rm loc}(U,\rho)$, the local homogeneous Hölder space of order $\alpha > 0$ in the set $U \subseteq X$, defined in the context of a quasi-metric space (X,ρ) as (cf. [112, (7.3.7)]):

$$\dot{\mathscr{C}}^\alpha_{\rm loc}(U,\rho) := \Big\{ f : U \to \mathbb{C} : f\big|_{B_\rho(x,r)\cap U} \in \dot{\mathscr{C}}^\alpha\big(B_\rho(x,r)\cap U,\rho\big) \\ \text{for each } x \in U \text{ and } r \in (0,\infty) \Big\} \tag{A.0.30}$$

$\|\cdot\|_{\mathscr{C}^\alpha(U,\rho)}$, the inhomogeneous Hölder space norm of order $\alpha > 0$ in the set $U \subseteq X$, in the context of a quasi-metric space (X,ρ), defined for each function $f : U \to \mathbb{R}$ as (cf. [112, (7.3.20)]):

$$\|f\|_{\mathscr{C}^\alpha(U,\rho)} := \sup_U |f| + \|f\|_{\dot{\mathscr{C}}^\alpha(U,\rho)}, \qquad \forall f \in \mathscr{C}^\alpha(U,\rho) \tag{A.0.31}$$

$\mathscr{C}^\alpha(U,\rho)$, the inhomogeneous Hölder space of order $\alpha > 0$ in the set $U \subseteq X$, defined in the context of a quasi-metric space (X,ρ) as (cf. [112, (7.3.19)]):

$$\mathscr{C}^\alpha(U,\rho) := \big\{ f \in \dot{\mathscr{C}}^\alpha(U,\rho) : f \text{ is bounded in } U \big\} \tag{A.0.32}$$

$\mathscr{C}^\alpha_c(U,\rho)$, the space of Hölder functions of order $\alpha > 0$ with ρ-bounded support in the set $U \subseteq X$, defined in the context of a quasi-metric space (X,ρ) as (cf. [112,

(7.3.26), (7.3.27)]):

$$\mathscr{C}_c^\alpha(U,\rho) := \{f \in \dot{\mathscr{C}}^\alpha(U,\rho) : f \text{ vanishes outside of a } \rho\text{-bounded subset of } U\} \tag{A.0.33}$$

$\mathrm{Cp}(X \to Y)$, the space of compact linear operators from the topological vector space X into the topological vector space Y:

$$\mathrm{Cp}(X \to Y) := \{T : X \to Y : T \text{ linear compact mapping}\} \tag{A.0.34}$$

$\mathrm{Cp}(X)$, the space of compact linear operators from the topological vector space X into itself:

$$\mathrm{Cp}(X) := \mathrm{Cp}(X \to X) \tag{A.0.35}$$

$\operatorname{cap}(K)$, the capacity of a compact set $K \subset \mathbb{R}^n$ defined as the infimum of all values $\int_{\mathbb{R}^n} |\nabla\varphi|^2 \, d\mathcal{L}^n$ corresponding to $\varphi \in \mathscr{C}_c^\infty(\mathbb{R}^n)$ satisfying $\varphi \geq 1$ on K (cf. Definition 5.2.5)

$\mathcal{C}\ell_n$, the Clifford algebra $(\mathcal{C}\ell_n, +, \odot)$ generated by n imaginary units defined as the minimal enlargement of $\mathbb{R}^n$ to a unitary real algebra which is not generated (as an algebra) by any proper subspace of $\mathbb{R}^n$, and such that $x \odot x = -|x|^2$ for each $x \in \mathbb{R}^n \hookrightarrow \mathcal{C}\ell_n$ (cf. [112, §6.4])

$C(= C_L)$, the (left-handed) boundary-to-domain Cauchy-Clifford integral operator whose action on each $f \in L^1\big(\partial_*\Omega, \frac{\sigma_*(x)}{1+|x|^{n-1}}\big) \otimes \mathcal{C}\ell_n$ at each $x \in \mathring{\Omega}$ is given by (2.5.282)

$$Cf(x) := \frac{1}{\omega_{n-1}} \int_{\partial_*\Omega} \frac{x-y}{|x-y|^n} \odot \nu(y) \odot f(y) \, d\sigma_*(y)$$

$\mathfrak{C}(= \mathfrak{C}_L)$, the (left-handed) boundary-to-boundary Cauchy-Clifford integral operator whose action on each $f \in L^1\big(\partial_*\Omega, \frac{\sigma_*(x)}{1+|x|^{n-1}}\big) \otimes \mathcal{C}\ell_n$ at σ_*-a.e. point $x \in \partial_*\Omega$ is given by (2.5.284)

$$\mathfrak{C}f(x) := \lim_{\varepsilon\to 0^+} \frac{1}{\omega_{n-1}} \int_{\substack{y\in\partial_*\Omega \\ |x-y|>\varepsilon}} \frac{x-y}{|x-y|^n} \odot \nu(y) \odot f(y) \, d\sigma_*(y)$$

$\mathfrak{C}_{\max}$, the maximal Cauchy-Clifford integral operator whose action on each function $f \in L^1\big(\partial_*\Omega, \frac{\sigma_*(x)}{1+|x|^{n-1}}\big) \otimes \mathcal{C}\ell_n$ at each $x \in \partial\Omega$ is given by (2.5.283)

$$\mathfrak{C}_{\max}f(x) := \sup_{\varepsilon>0} \left| \frac{1}{\omega_{n-1}} \int_{\substack{y\in\partial_*\Omega \\ |x-y|>\varepsilon}} \frac{x-y}{|x-y|^n} \odot \nu(y) \odot f(y) \, d\sigma_*(y) \right|$$

C_R, the right-handed boundary-to-domain Cauchy-Clifford integral operator whose action on each $f \in L^1\big(\partial_*\Omega, \frac{\sigma_*(x)}{1+|x|^{n-1}}\big) \otimes C\ell_n$ at each $x \in \mathring{\Omega}$ is given by (2.5.285)

$$C_R f(x) := \frac{1}{\omega_{n-1}} \int_{\partial_*\Omega} f(y) \odot \nu(y) \odot \frac{x-y}{|x-y|^n}\, d\sigma_*(y)$$

$\mathfrak{C}_R$, the right-handed boundary-to-boundary Cauchy-Clifford integral operator whose action on $f \in L^1\big(\partial_*\Omega, \frac{\sigma_*(x)}{1+|x|^{n-1}}\big) \otimes C\ell_n$ at σ_*-a.e. point $x \in \partial_*\Omega$ is given by (2.5.286)

$$\mathfrak{C}_R f(x) := \lim_{\varepsilon\to 0^+} \frac{1}{\omega_{n-1}} \int_{\substack{y\in\partial_*\Omega\\ |x-y|>\varepsilon}} f(y) \odot \nu(y) \odot \frac{x-y}{|x-y|^n}\, d\sigma_*(y)$$

$\mathfrak{C}^\#$, the transpose Cauchy-Clifford integral operator whose action on each function $f \in L^1\big(\partial\Omega, \frac{\sigma(y)}{1+|y|^{n-1}}\big) \otimes C\ell_n$ at each point σ-a.e. $x \in \partial_*\Omega$ is given by

$$\mathfrak{C}^\# f(x) := \lim_{\varepsilon\to 0^+} \frac{-1}{\omega_{n-1}} \int_{\substack{y\in\partial\Omega\\ |x-y|>\varepsilon}} \nu(x) \odot \frac{x-y}{|x-y|^n} \odot f(y)\, d\sigma(y) \tag{A.0.36}$$

D

$u \cdot w = \langle u, w\rangle$, the dot product of two vectors $u, w \in \mathbb{R}^n$

$\operatorname{div}\vec{F}$, the divergence of the vector field $\vec{F}$

div_g, the differential geometric divergence (associated with the metric tensor g; cf. [112, §1.11])

$d_g(x, y)$, the geodesic distance (induced by the metric tensor g) between x and y (cf. [112, §1.11])

$\mathcal{D}'(\Omega)$, the space of distributions in the open set Ω

${}_{\mathcal{D}'(\Omega)}\langle\cdot,\cdot\rangle_{\mathcal{D}(\Omega)}$, the distributional pairing in Ω

$\Delta := \partial_1^2 + \cdots + \partial_n^2$, the Laplace operator in $\mathbb{R}^n$

δ_x, the Dirac distribution with mass at x

D, the classical (homogeneous) Dirac operator in $\mathbb{R}^n$ defined as (cf. [112, (6.4.139)]):

$$D = \sum_{j=1}^n \mathbf{e}_j \odot \partial_j \tag{A.0.37}$$

$D = \left(\sum_{j=1}^n a_j^{\alpha\beta} \partial_j + b^{\alpha\beta} \right)_{\substack{1\leq\alpha\leq N \\ 1\leq\beta\leq N'}}$, a (generic) $N \times N'$ first-order system

$D^\top$, the (real) transpose of the first-order system D:

$$D^\top := \left(- \sum_{j=1}^n a_j^{\alpha\beta} \partial_j + b^{\alpha\beta} \right)_{\substack{1\leq\beta\leq N' \\ 1\leq\alpha\leq N}} \tag{A.0.38}$$

$\overline{D}$, the complex conjugate of the first-order system D

D^*, the Hermitian adjoint of the first-order system D

d, the exterior derivative operator acting on the differential form $u = \sum_J u_J \mathrm{d}x^J$ according to (cf. [112, (1.11.32)], and also [112, (6.4.140)-(6.4.141)] for the Clifford algebra context):

$$\mathrm{d}u = \sum_{j=1}^n \sum_J \frac{\partial u_J}{\partial x_j} \mathrm{d}x_j \wedge \mathrm{d}x^J \tag{A.0.39}$$

δ, the formal adjoint of the exterior derivative operator d on differential forms (see also [112, (6.4.142)] for the Clifford algebra context)

δ_{jk}, the Kronecker symbol, i.e., $\delta_{jk} := 1$ if $j = k$ and $\delta_{jk} := 0$ if $j \neq k$

$\delta_F(\cdot)$, the distance function to the set F

$\delta_{\partial\Omega}(\cdot)$, the distance function to the boundary of Ω

$U \triangle V := (U \setminus V) \cup (V \setminus U)$, the symmetric difference of the sets U and V

$\Delta(x, r) := B(x, r) \cap \partial\Omega$, the surface ball on $\partial\Omega$ with center at $x \in \partial\Omega$ and radius $r > 0$

$\Delta(x, r) := B(x, r) \cap \Sigma$, the surface ball on Σ with center at $x \in \Sigma$ and radius $r > 0$

D_L, the Dirac operator acting from the left on each Clifford algebra-valued distribution $u \in \mathcal{D}'(\Omega) \otimes \mathcal{C}\ell_n$ according to (cf. [112, (6.4.48)]):

$$D_L u := \sum_{j=1}^n \mathbf{e}_j \odot (\partial_j u) \tag{A.0.40}$$

D_R, the Dirac operator acting from the right on each Clifford algebra-valued distribution $u \in \mathcal{D}'(\Omega) \otimes \mathcal{C}\ell_n$ according to (cf. [112, (6.4.49)]):

$$D_R u := \sum_{j=1}^n (\partial_j u) \odot \mathbf{e}_j \tag{A.0.41}$$

$\mathrm{diam}_\rho(A)$, the ρ-diameter of the set $A \subseteq X$, in the context of a quasi-metric space (X, ρ), defined as (cf. [112, (7.1.6)]):

$$\mathrm{diam}_\rho(A) := \sup \{\rho(x, y) : x, y \in A\} \tag{A.0.42}$$

$\mathbb{D}_k(X)$, the k-th generation of dyadic cubes in the geometrically doubling quasi-metric space X, defined as in [112, Proposition 7.5.4]:

$$\mathbb{D}_k(X) := \{Q_\alpha^k\}_{\alpha \in I_k} \tag{A.0.43}$$

$\mathbb{D}(X)$, the dyadic grid on the geometrically doubling quasi-metric space X, defined as in [112, Proposition 7.5.4]:

$$\mathbb{D}(X) := \bigcup_{k \in \mathbb{Z},\, k \geq \kappa_X} \mathbb{D}_k(X) \tag{A.0.44}$$

$\dim X$, the dimension of a vector space X

$d\zeta := i\nu(\zeta)\, d\sigma(\zeta)$, the complex arc-length (1.1.3)

$\operatorname{Def} \vec{u} := \frac{1}{2}\big(\partial_j u_k + \partial_k u_j\big)_{1 \leq j,k \leq n}$, the deformation tensor acting on the vector-distribution $\vec{u} = (u_1, \dots, u_n)$ (1.3.49)

$\mathbf{D}_k := \mathbf{D} + k\mathbf{e}_{n+1}$, the perturbed Dirac operator in $\mathbb{R}^n$, where $k \in (0, \infty)$ (6.2.11)

$\Delta + k^2$, the Helmholtz operator with wave number k

$\operatorname{diag}(X) := \{(x, y) \in X \times X : x = y\}$, the diagonal in the Cartesian product $X \times X$ (2.2.2)

E

(ε, δ)-domain (cf. [112, Definition 5.11.8]):

a nonempty, open, proper subset Ω of $\mathbb{R}^n$ such that for any $x, y \in \Omega$ with $|x-y| < \delta$ there exists $\gamma : [0, 1] \to \Omega$ rectifiable curve satisfying $\gamma(0) = x$, $\gamma(1) = y$, as well as the conditions $\operatorname{length}(\gamma) \leq \frac{1}{\varepsilon}|x - y|$ and $\frac{|z-x|\,|z-y|}{|x-y|} \leq \frac{1}{\varepsilon}\operatorname{dist}(z, \partial\Omega)$ for each $z \in \gamma([0, 1])$ (A.0.45)

$\mathbf{e}_j := (\delta_{jk})_{1 \leq k \leq n} \in \mathbb{R}^n$, where δ_{jk} is the Kronecker symbol, standard unit vector in the j-th direction in $\mathbb{R}^n$

$\{\mathbf{e}_j\}_{1 \leq j \leq n}$, the standard orthonormal basis in $\mathbb{R}^n$

E_Δ, the standard fundamental solution for the Laplacian in $\mathbb{R}^n$ (1.5.56)

E_{Δ^2}, the standard fundamental solution for the bi-Laplacian (1.4.93)

$E = E_L = \big(E_{\alpha\beta}\big)_{1 \leq \alpha,\beta \leq M}$, the canonical fundamental solution of the weakly elliptic $M \times M$ system L (1.4.18)

$\mathscr{E}'_K(\Omega)$, the space of distributions in Ω supported in the compact set $K \subset \Omega$

$\mathscr{E}'(\Omega)$, the space of distributions compactly supported in the open set $\Omega \subseteq \mathbb{R}^n$

$[x]_{X/Y} := x + Y$, the equivalence class of the vector $x \in X$ in the quotient space X/Y

$\mathcal{E}_p(X)$, the p-envelope (with $p \in (0,1]$) of a quasi-normed space X whose dual separates points, defined as the completion of X in the quasi-norm $|\!|\!| \cdot |\!|\!|_p$ (cf. [113, Definition 7.8.4])

$\mathrm{Ex}_{\partial\Omega\to\Omega}$, the extension operator from $\partial\Omega$ to Ω (cf. [113, Theorem 8.4.1])

F

$[\vec{F}]_\infty$, the contribution of the vector field $\vec{F}$ at infinity, defined for any family $\{\phi_R\}_{R>0}$ (referred to as a system of auxiliary functions) consisting of smooth compactly supported functions in $\mathbb{R}^n$ which are globally bounded and progressively equal to 1 on compact sets in a uniform fashion, as (cf. [112, (1.3.2)-(1.3.3)]):

$$[\vec{F}]_\infty := -\lim_{R\to\infty}\int_\Omega \nabla\phi_R\cdot\vec{F}\,\mathrm{d}\mathcal{L}^n \tag{A.0.46}$$

f_E^*, the non-increasing rearrangement of $f : E \to \mathbb{R}$ defines as (cf. [112, (6.2.2)]):

$$f_E^*(t) := \inf\Big\{\lambda \geq 0 : \mu(\{x\in E : |f(x)| > \lambda\}) \leq t\Big\},\quad \forall\, t\in[0,\infty) \tag{A.0.47}$$

$\Phi(X \to Y)$, the collection of Fredholm operators from the linear topological space X into the linear topological space Y (cf. [113, Definition 2.2.1])

$f_\gamma^\#(x) := \sup\limits_{\psi\in\mathcal{T}_\gamma(x)} \big|\langle f,\psi\rangle\big|$, the Fefferman-Stein grand maximal function (at the point $x\in\Sigma$) of the "distribution" $f \in \big(\mathrm{Lip}_c(\Sigma)\big)'$ on the Ahlfors regular set Σ (cf. [113, (4.1.6)])

$F_s^{p,q}(\Sigma,\sigma)$, the inhomoageneous Triebel-Lizorkin space on the Ahlfors regular set $\Sigma\subseteq\mathbb{R}^n$, defined as the collection of all "distributions" f on Σ with the property that $\|f\|_{F_s^{p,q}(\Sigma,\sigma)} < +\infty$ (cf. [113, Definition 7.1.2])

$F_s^{p,q}(\mathbb{R}^n)$, the (inhomogeneous) Triebel-Lizorkin space in $\mathbb{R}^n$ for $0<p,q\leq\infty$ and $s\in\mathbb{R}$, equipped with the quasi-norm $\|\cdot\|_{F_s^{p,q}(\mathbb{R}^n)}$ (cf. [113, (9.1.4)-(9.1.5)])

$F_s^{p,q}(\Omega) := \big\{u \in \mathcal{D}'(\Omega) :$ there exists $U \in F_s^{p,q}(\mathbb{R}^n)$ such that $U\big|_\Omega = u\big\}$, the Triebel-Lizorkin space in the (arbitrary) open set $\Omega\subseteq\mathbb{R}^n$ for $0<p,q\leq\infty$ and $s\in\mathbb{R}$, equipped with the quasi-norm (cf. [113, (9.2.1)])

$$\|u\|_{F_s^{p,q}(\Omega)} := \inf\Big\{\|U\|_{F_s^{p,q}(\mathbb{R}^n)} : U\in F_s^{p,q}(\mathbb{R}^n),\ U\big|_\Omega = u\Big\}$$

$\widehat{\phi}(\xi) := (\mathscr{F}\phi)(x) := \int\limits_{\mathbb{R}^n} e^{-i\langle x,\xi\rangle}\phi(x)\,\mathrm{d}x$, the Fourier transform of ϕ in $\mathbb{R}^n$ (1.4.14)

$\Phi_k(\cdot)$, the (unique) radiating fundamental solution of the Helmholtz operator $\Delta+k^2$ in $\mathbb{R}^n$, given by formula (6.1.7)

G

$g = \sum\limits_{1\leq j,k\leq n} g_{jk}\,\mathrm{d}x_j\otimes\mathrm{d}x_k$, the Riemannian metric tensor

∇u, the gradient (Jacobian matrix) of a $\mathbb{C}^M$-valued function $u = (u_\alpha)_{1\leq\alpha\leq M}$ defined in an open subset of $\mathbb{R}^n$, defined as:

$$\nabla u := \big(\partial_j u_\alpha\big)_{\substack{1\leq\alpha\leq M\\ 1\leq j\leq n}} = \begin{bmatrix} \partial_1 u_1 & \cdots & \partial_n u_1 \\ \vdots & \vdots & \vdots \\ \partial_1 u_M & \cdots & \partial_n u_M \end{bmatrix} \tag{A.0.48}$$

∇', the gradient operator in $\mathbb{R}^{n-1}$

$\Gamma_\kappa(x) = \Gamma_{\Omega,\kappa}(x)$, the ($\kappa$-)nontangential approach region with vertex at $x \in \partial\Omega$, defined as (cf. [112, (8.1.2)]):

$$\Gamma_\kappa(x) = \Gamma_{\Omega,\kappa}(x) := \big\{y \in \Omega : |x-y| < (1+\kappa)\delta_{\partial\Omega}(y)\big\}, \quad \forall x \in \partial\Omega \tag{A.0.49}$$

$\mathbb{X}$, a Generalized Banach Function Space on the measure space $(X, \mathfrak{M}, \mu)$ equipped with the norm $\|\cdot\|_{\mathbb{X}}$ (cf. [113, Definition 5.1.4])

$\mathring{\mathbb{X}} := \overline{L^\infty_{\text{comp}}(X,\mu)}^{\|\cdot\|_{\mathbb{X}}}$, the closure of L^∞_{comp} in the Generalized Banach Function Space $(\mathbb{X}, \|\cdot\|_{\mathbb{X}})$ (cf. [113, Definition 5.2.6])

$\nabla_{\tan} f$, the tangential gradient of the function $f \in L^1_{1,\text{loc}}(\partial_*\Omega, \sigma_*)$ (cf. [113, §11.4]):

$$\nabla_{\tan} f := \Big(\sum_{k=1}^n \nu_k \partial_{\tau_{kj}} f\Big)_{1\leq j\leq n} \quad \text{at } \sigma_*\text{-a.e. point on } \partial_*\Omega \tag{A.0.50}$$

$\nabla^A_{\tan}$, the tangential gradient of $u = (u_\beta)_\beta$ associated with the coefficient tensor $A := \big(a^{\alpha\beta}_{rs}\big)_{\substack{1\leq\alpha,\beta\leq M\\ 1\leq r,s\leq n}}$, defined by $\nabla^A_{\tan} u := \Big(\nu_r a^{\alpha\beta}_{rs}\big(\nabla_{\tan}\big(u_\beta\big|^{\kappa-\text{n.t.}}_{\partial\Omega}\big)\big)_s\Big)_{1\leq\alpha\leq M}$ where $\nabla_{\tan}$ is the tangential gradient (acting on scalar functions) along $\partial\Omega$ (1.7.59), (1.7.60)

$G_\Omega(\cdot,\cdot)$, the Green function for the Laplacian, where Ω is a bounded open set in $\mathbb{R}^n$ Theorem 5.3.1

H

$\mathcal{H}^{n-1}$, the $(n-1)$-dimensional Hausdorff measure in $\mathbb{R}^n$

$\mathcal{H}^{n-1}_g$, the $(n-1)$-dimensional Hausdorff measure induced by the metric tensor g

$\mathcal{H}^s$, the s-dimensional Hausdorff measure in $\mathbb{R}^n$

$\mathcal{H}^s_*$, the s-dimensional Hausdorff outer measure in $\mathbb{R}^n$

$H^p(\Sigma,\sigma)$, the Lebesgue-based Hardy space on the Ahlfors regular set $\Sigma \subseteq \mathbb{R}^n$, defined for $p \in \big(\frac{n-1}{n}, \infty\big)$ and $\gamma \in \big((n-1)(\frac{1}{p}-1)_+, 1\big)$ as (cf. [113, Definition 4.2.1])

$$H^p(\Sigma,\sigma) := \Big\{f \in \big(\text{Lip}_c(\Sigma)\big)' : f^\#_\gamma \in L^p(\Sigma,\sigma)\Big\} \tag{A.0.51}$$

and equipped with the quasi-norm

$$\|f\|_{H^p(\Sigma,\sigma)} := \|f_\gamma^\#\|_{L^p(\Sigma,\sigma)}, \qquad \forall f \in H^p(\Sigma,\sigma) \tag{A.0.52}$$

$H^{p,q}(\Sigma,\sigma)$, the Lorentz-based Hardy space on the Ahlfors regular set $\Sigma \subseteq \mathbb{R}^n$, defined for $p \in \left(\frac{n-1}{n}, \infty\right)$, $p \in (0,\infty]$, and $\gamma \in \left((n-1)\left(\frac{1}{p}-1\right)_+, 1\right)$ as (cf. [113, Definition 4.2.3])

$$H^{p,q}(\Sigma,\sigma) := \left\{ f \in \left(\mathrm{Lip}_c(\Sigma)\right)' : f_\gamma^\# \in L^{p,q}(\Sigma,\sigma) \right\} \tag{A.0.53}$$

and equipped with the quasi-norm

$$\|f\|_{H^{p,q}(\Sigma,\sigma)} := \|f_\gamma^\#\|_{L^{p,q}(\Sigma,\sigma)}, \qquad \forall f \in H^{p,q}(\Sigma,\sigma) \tag{A.0.54}$$

$H^{p,q}_{\mathrm{fin}}(\Sigma,\sigma)$, the vector space of all finite linear combinations of (p,q)-atoms on Σ equipped with the quasi-norm $\|f\|_{H^{p,q}_{\mathrm{fin}}(\Sigma,\sigma)}$ defined as the infimum of all $\left(\sum_{j=1}^N |\lambda_j|^p\right)^{1/p}$ such that $f = \sum_{j=1}^N \lambda_j a_j$ for $\{\lambda_j\}_{1\le j\le N} \subseteq \mathbb{C}$ and (p,q)-atoms $\{a_j\}_{1\le j\le N}$ (cf. [113, (4.4.113)])

$\mathfrak{H}$, the L^p-filtering operator, acting on each given distribution $f \in H^p(\Sigma,\sigma)$ with $\frac{n-1}{n} < p < \infty$ according to $(\mathfrak{H} f)(x) := \lim_{t\to 0^+} {}_{(H^p(\Sigma,\sigma))^*}\langle S_t(x,\cdot), f\rangle_{H^p(\Sigma,\sigma)}$ at each $x \in \Sigma$, where $\{S_t(\cdot,\cdot)\}_t$ are the integral kernels of a suitable approximation to the identity (cf. [113, Theorem 4.9.1])

$\mathscr{H}^{q,\lambda}(\Sigma,\sigma)$, the space defined for any given $q \in (1,\infty)$ and $\lambda \in (0, n-1)$ as (cf. [113, §6.1]):

$$\begin{aligned} &\mathscr{H}^{q,\lambda}(\Sigma,\sigma) \\ &:= \Big\{ f \in \left(\mathrm{Lip}_c(\Sigma)\right)' : \text{there exist a numerical sequence } \{\lambda_j\}_{j\in\mathbb{N}} \in \ell^1(\mathbb{N}) \\ &\qquad \text{and a family } \{a_j\}_{j\in\mathbb{N}} \text{ of } \mathscr{H}^{q,\lambda}\text{-atoms on } \Sigma \text{ so that} \\ &\qquad f = \sum_{j=1}^\infty \lambda_j a_j \text{ with convergence in } \left(\mathrm{Lip}_c(\Sigma)\right)' \Big\}, \end{aligned} \tag{A.0.55}$$

and equipped with the norm

$$\begin{aligned} &\|f\|_{\mathscr{H}^{q,\lambda}(\Sigma,\sigma)} \\ &:= \inf\Big\{ \sum_{j=1}^\infty |\lambda_j| : f = \sum_{j=1}^\infty \lambda_j a_j \text{ in } \left(\mathrm{Lip}_c(\Sigma)\right)' \text{ with} \\ &\qquad \{\lambda_j\}_{j\in\mathbb{N}} \in \ell^1(\mathbb{N}) \text{ and each } a_j \text{ a } \mathscr{H}^{q,\lambda}\text{-atom on } \Sigma \Big\} \end{aligned} \tag{A.0.56}$$

$H^s(\Omega)$, the L^2-based fractional Sobolev space of order $s \in \mathbb{R}$ in an open set $\Omega \subseteq \mathbb{R}^n$ (cf. [113, §9.2]):

$$H^s(\Omega) := \left\{U\big|_\Omega : U \in H^s(\mathbb{R}^n)\right\} = \left\{U\big|_\Omega : U \in F^{2,2}_s(\mathbb{R}^n)\right\}, \qquad \forall s \in \mathbb{R}, \quad \text{(A.0.57)}$$

and equipped with the norm

$$\|u\|_{H^s(\Omega)} := \inf\left\{\|U\|_{H^s(\mathbb{R}^n)} : U \in H^s(\mathbb{R}^n) \text{ such that } u = U\big|_\Omega\right\}$$
$$\approx \inf\left\{\|U\|_{F^{2,2}_s(\mathbb{R}^n)} : U \in F^{2,2}_s(\mathbb{R}^n) \text{ such that } u = U\big|_\Omega\right\} \qquad \text{(A.0.58)}$$

$H^p_\kappa(\Omega)$, the Hardy space of harmonic functions u in the open set $\Omega \subseteq \mathbb{R}^n$ satisfying $\mathcal{N}_\kappa u \in L^p(\partial\Omega, \sigma)$ (5.5.4)

$H^{(1)}_\lambda(\cdot)$, the Hankel function of the first kind with index $\lambda \in \mathbb{R}$

$\mathcal{H}^p(\Omega; D)$, the Hardy space in Ω associated with the first-order $N \times M$ system D, defined as the collection of all $u \in \left[\mathscr{C}^\infty(\Omega)\right]^M$ satisfying $\mathcal{N}_\kappa u \in L^p(\partial\Omega, \sigma)$ and $Du = 0$ in Ω (which also vanish at infinity when Ω is an exterior domain), equipped with the quasi-norm $\|u\|_{\mathcal{H}^p(\Omega;D)} := \left\|\mathcal{N}_\kappa u\right\|_{L^p(\partial\Omega,\sigma)}$ (3.2.2), (3.2.3)

$\mathscr{H}^p_\bullet(\partial\Omega; D) := \{\nu \bullet u : u \in \mathcal{H}^p(\Omega; D)\}$, the "bullet" boundary Hardy space associated with the Dirac operator $D := \sum_{j=1}^n \mathbf{e}_j \odot \partial_j$, equipped with the quasi-norm inherited from $H^p(\partial\Omega, \sigma) \otimes \mathcal{C\ell}_n$ (3.2.21)

I

$\langle \cdot, \cdot \rangle$, the (real) inner product in $\mathbb{C}^M$ defined for any vectors $u = (u_k)_{1\le k\le M} \in \mathbb{C}^M$ and $w = (w_k)_{1\le k\le M} \in \mathbb{C}^M$ as:

$$\langle u, w \rangle := \sum_{k=1}^M u_k w_k \qquad \text{(A.0.59)}$$

$\mathrm{i} := \sqrt{-1} \in \mathbb{C}$, the complex imaginary unit

$\mathring{U}$, the interior of the set $U \subseteq \mathbb{R}^n$

$⨍_E f\,d\mu := \frac{1}{\mu(E)} \int_E f\,d\mu$, the integral average of the function f on the set $E \subseteq X$, in a measure space (X, μ)

$f_{B_\rho(x,r)}$, the integral average of f over the ρ-ball $B_\rho(x, r)$, in the context of a space of homogeneous type (X, ρ, μ), defined as (cf. [112, (7.4.9)]):

$$f_{B_\rho(x,r)} := ⨍_{B_\rho(x,r)} f\,d\mu := \frac{1}{\mu\big(B_\rho(x,r)\big)} \int_{B_\rho(x,r)} f(y)\,d\mu(y) \qquad \text{(A.0.60)}$$

$f_\Delta := ⨍_\Delta f\,d\sigma$, the integral average of f in the "surface ball" Δ

$I_{E,\alpha}$, the fractional integral operator of order α on the set E contained in a metric space (X, ρ) equipped with an upper d-dimensional Borel measure μ on (X, τ_ρ), acting on functions $f \in L^1\big(E, \frac{\mu(x)}{1+\rho(x,x_0)^{d-\alpha}}\big)$ according to (cf. [112, (7.8.3)]):

$$I_{E,\alpha}f(x) := \int_E \frac{f(y)}{\rho(x,y)^{d-\alpha}}\,d\mu(y) \quad \text{for } \mu\text{-a.e. } x \in E \tag{A.0.61}$$

$\mathrm{Im}\,(T : X \to Y) := \{Tx : x \in X\}$, the image (or range) of the operator T

$\mathrm{ind}\,T := \dim(\mathrm{Ker}\,T) - \mathrm{codim}\,(\mathrm{Im}\,T)$ the index of a (semi-)Fredholm operator T (cf. [113, Definition 2.1.1])

$i(\Phi) := \sup\limits_{0<t<1} \frac{\ln h_\Phi(t)}{\ln t} = \lim\limits_{t\to0^+} \frac{\ln h_\Phi(t)}{\ln t}$ where $h_\Phi(t) := \sup\limits_{s>0} \frac{\Phi(st)}{\Phi(s)}$ for $t \in (0,\infty)$, the lower dilation index of the Young function Φ (cf. [113, (5.3.14)])

$I(\Phi) := \inf\limits_{1<t<\infty} \frac{\ln h_\Phi(t)}{\ln t} = \lim\limits_{t\to\infty} \frac{\ln h_\Phi(t)}{\ln t}$ where $h_\Phi(t) := \sup\limits_{s>0} \frac{\Phi(st)}{\Phi(s)}$ for $t \in (0,\infty)$, the upper dilation index of the Young function Φ (cf. [113, (5.3.15)])

K

K_Δ, the boundary-to-boundary harmonic double layer potential, defined as (cf. [112, (1.1.32)]):

$$K_\Delta f(x) := \lim_{\varepsilon\to0^+} \frac{1}{\omega_{n-1}} \int_{\partial\Omega\setminus\overline{B(x,\varepsilon)}} \frac{\langle \nu(y), y-x\rangle}{|x-y|^n} f(y)\,d\mathcal{H}^{n-1}(y), \qquad x \in \partial\Omega \tag{A.0.62}$$

$K^\#_\Delta$, the transpose harmonic double layer potential, defined as (cf. [112, (1.1.33)]):

$$K^\#_\Delta f(x) := \lim_{\varepsilon\to0^+} \frac{1}{\omega_{n-1}} \int_{\partial\Omega\setminus\overline{B(x,\varepsilon)}} \frac{\langle \nu(x), x-y\rangle}{|x-y|^n} f(y)\,d\mathcal{H}^{n-1}(y), \qquad x \in \partial\Omega \tag{A.0.63}$$

$\mathrm{Ker}\,(T : X \to Y) := \{x \in X : Tx = 0\}$ the kernel (or null-space) of the operator T

$\mathrm{Ker}\,L := \Big\{u \in \big[\mathscr{C}^\infty(\Omega)\big]^M : Lu = 0 \text{ in } \Omega\Big\}$ the null-space of the $M \times M$ system L, in an open set Ω

k^{x_0} Poisson kernel for the Laplacian for the domain Ω, defined as the Radon-Nikodym derivative of the harmonic measure ω^{x_0} with pole at $x_0 \in \Omega$ with respect to the surface measure σ on $\partial\Omega$ (5.3.38)

L

local John condition, satisfied by an open set $\Omega \subseteq \mathbb{R}^n$ (cf. [112, Definition 5.11.7]):

> there exist $\theta \in (0,1)$, $M_o \in (1,\infty)$, and $R \in \big(0, \operatorname{diam}\partial\Omega\big]$ (the latter required to be ∞ if $\partial\Omega$ is unbounded) such that for every $x \in \partial\Omega$ and $r \in (0, R_*)$ one can find a point $x_r \in B(x,r) \cap \Omega$ with the property that $B(x_r, \theta r) \subseteq \Omega$ and for each $y \in \Delta(x,r)$ it is possible to find a rectifiable path $\gamma_y : [0,1] \to \overline{\Omega}$ whose length is $\leq M_o \cdot r$, which satisfies $\gamma_y(0) = y$, $\gamma_y(1) = x_r$, and such that $\operatorname{dist}\big(\gamma_y(t), \partial\Omega\big) > \theta \cdot |\gamma_y(t) - y|$ for every $t \in (0,1]$ (A.0.64)

$\mathcal{L}^n$, the (n-dimensional) Lebesgue measure in $\mathbb{R}^n$

L^∞_{comp}, the space of essentially bounded functions with compact support

$d\mathcal{L}^n_g := \sqrt{g}\, d\mathcal{L}^n$ Lebesgue measure induced by the metric tensor g

$\Lambda^\ell TM$, the ℓ-th exterior power of the vector bundle on the manifold M

$L^0(X,\mu)$, the space of measurable functions which are pointwise finite μ-a.e. on X

$L^p_{\rm bdd}(\Omega,\mathcal{L}^n)$, the space of functions which are p-th power integrable on bounded subsets of Ω (cf. [112, (4.2.4)]):

$$L^p_{\rm bdd}(\Omega,\mathcal{L}^n) \text{ be the collection of all } \mathcal{L}^n\text{-measurable functions defined in } \Omega \text{ which are } p\text{-th power absolutely integrable with respect to the Lebesgue measure on each bounded } \mathcal{L}^n\text{-measurable subset of the set } \Omega. \tag{A.0.65}$$

$\|\cdot\|_{{\rm Lip}(X)}$, the natural semi-norm on Lip(X), defined in the context of a metric space $(\mathscr{X}, d)$ as (cf. [112, (3.7.1)]):

$$\|f\|_{{\rm Lip}(X)} := \sup_{x,y\in X,\, x\neq y} \frac{|f(x)-f(y)|}{d(x,y)} \tag{A.0.66}$$

${\rm Lip}_c(X)$, the space of Lipschitz functions with bounded support in the (quasi-)metric space X

$\big({\rm Lip}_c(\Sigma)\big)'$, the space distributions on a given set $\Sigma \subseteq \mathbb{R}^n$, defined as (cf. [112, (4.1.34)]):

$$\text{the topological dual of } \big({\rm Lip}_c(\Sigma), \tau_{\mathscr{D}}\big) \tag{A.0.67}$$

${}_{({\rm Lip}_c(\Sigma))'}\langle\cdot,\cdot\rangle_{{\rm Lip}_c(\Sigma)}$, or simply $\langle\cdot,\cdot\rangle$, the distributional pairing on the set Σ

$\|\cdot\|_{L^{p,q}(X,\mu)}$, the Lorentz space quasi-norm, defined as (cf. [112, (6.2.14)]):

$$\|f\|_{L^{p,q}(X,\mu)} := \begin{cases} \left(\displaystyle\int_0^\infty \big[t^{1/p} f^*_X(t)\big|^q \frac{dt}{t}\right)^{1/q} & \text{if } 0<p,q<\infty, \\ \sup_{t>0}\big[t^{1/p} f^*_X(t)\big] & \text{if } 0<p\leq\infty,\ q=\infty, \\ \|f\|_{L^\infty(X,\mu)} & \text{if } p=\infty,\ 0<q\leq\infty \end{cases} \tag{A.0.68}$$

$L^{p,q}(X,\mu)$, the Lorentz space on X with respect to the measure μ defined as (cf. [112, (6.2.13)]):

$$L^{p,q}(X,\mu) := \Big\{ f : X \to \overline{\mathbb{R}}\ \mu\text{-measurable} : \|f\|_{L^{p,q}(X,\mu)} < +\infty \Big\} \tag{A.0.69}$$

$L^{p,q}_\star(\Omega,\mu)$, the maximal Lorentz space with respect to the Borel measure μ in the open set $\Omega \subseteq \mathbb{R}^n$, defined as (cf. [112, (6.6.41)]):

$$L^{p,q}_\star(\Omega,\mu) := \big\{ u : \Omega \to \mathbb{C} : u \text{ is } \mathcal{L}^n\text{-measurable and } u_{\star,\theta} \in L^{p,q}(\Omega,\mu) \big\} \tag{A.0.70}$$

$L^p_\star(\Omega,\mu)$, the maximal Lebesgue space with respect to the Borel measure μ in the open set $\Omega \subseteq \mathbb{R}^n$, defined as (cf. [112, (6.6.43)]):

$$
\begin{aligned}
L^p_\star(\Omega,\mu) &:= L^{p,p}_\star(\Omega,\mu) \qquad \text{(A.0.71)}\\
&= \big\{u:\Omega\to\mathbb{C} : u \text{ is } \mathcal{L}^n\text{-measurable and } u_{\star,\theta}\in L^p(\Omega,\mu)\big\}
\end{aligned}
$$

$\log_+$, the positive part of ln, defined for each $t\in[0,\infty)$ as (cf. [112, (7.6.68)]):

$$
\log_+ t := \begin{cases} 0 & \text{if } t\in[0,1],\\ \ln t & \text{if } t\in[1,\infty)\end{cases} \qquad \text{(A.0.72)}
$$

$\mathcal{L}(X\to Y)$ the space of linear and continuous operators from X to Y

$L^\Phi(X,\mu)$, the Orlicz space on the sigma-finite measure space $(X,\mathfrak{M},\mu)$ associated with a Young function Φ (cf. [113, §5.3]):

$$
L^\Phi(X,\mu) := \big\{f\in\mathcal{M}(X,\mu) : \|f\|_{L^\Phi(X,\mu)}<\infty\big\} \qquad \text{(A.0.73)}
$$

where the Luxemburg norm $\|\cdot\|_{L^\Phi(X,\mu)}$ is defined as

$$
\begin{aligned}
\|f\|_{L^\Phi(X,\mu)} &:= \inf\Big\{\lambda>0 : \int_X \Phi(|f(x)|/\lambda)\,\mathrm{d}\mu(x)\le 1\Big\} \qquad \text{(A.0.74)}\\
&= \inf\Big\{\lambda>0 : \int_0^\infty \mu\Big(\big\{x\in X : |f(x)|>\lambda\Phi^{-1}(t)\big\}\Big)\,\mathrm{d}t\le 1\Big\} \qquad \text{(A.0.75)}
\end{aligned}
$$

$L^p(\log L)^\alpha(X,\mu) = \Big\{f\in\mathcal{M}(X,\mu) : \int_X |f(x)|^p\big[\ln(\mathrm{e}+|f(x)|)\big]^\alpha\,\mathrm{d}\mu(x) < +\infty\Big\}$ for $p\in(1,\infty)$ and $\alpha\in\mathbb{R}$, Zygmund's space (cf. [113, §5.3])

$L^p(\Omega, w\mathcal{L}^n)$, the weighted L^p Lebesgue space on the set $\Omega\subseteq\mathbb{R}^n$, equipped with the natural norm $\|u\|_{L^p(\Omega,w\mathcal{L}^n)} := \Big(\int_\Omega |u|^p\, w\,\mathrm{d}\mathcal{L}^n\Big)^{1/p}$ (cf. [113, §8.3])

$L^p_s(\Omega) := \Big\{U\big|_\Omega : U\in L^p_s(\mathbb{R}^n)\Big\}$, the Bessel potential space in an open set $\Omega\subseteq\mathbb{R}^n$ for $p\in(1,\infty)$ and $s\in\mathbb{R}$, equipped with the norm (cf. [113, §9.2]):

$$
\|u\|_{L^p_s(\Omega)} := \inf\Big\{\|U\|_{L^p_s(\mathbb{R}^n)} : U\in L^p_s(\mathbb{R}^n),\ u=U\big|_\Omega\Big\}
$$

$L^p_{1,\mathrm{loc}}(\partial_*\Omega,\sigma_*) = L^p_{1,\mathrm{loc}}(\partial_*\Omega,\sigma)$, the local Sobolev space of order one on $\partial_*\Omega$, defined for $p\in[1,\infty]$ as (cf. [113, Definition 11.1.2]):

$$
\begin{aligned}
L^p_{1,\mathrm{loc}}(\partial_*\Omega,\sigma_*) := \Big\{f\in L^p_{\mathrm{loc}}(\partial_*\Omega,\sigma_*) : \partial_{\tau_{jk}}f \text{ exists in } L^p_{\mathrm{loc}}(\partial_*\Omega,\sigma_*)&\\
\text{for each } j,k\in\{1,\dots,n\}\Big\}& \qquad \text{(A.0.76)}
\end{aligned}
$$

$\dot{L}^p_1(\partial\Omega,\sigma)$, the L^p-based homogeneous Sobolev space of order one on $\partial\Omega$ as (cf. [113, Definition 11.5.3]):

$$\dot{L}^p_1(\partial\Omega,\sigma) := \Big\{ f \in L^1\Big(\partial\Omega, \frac{\sigma(x)}{1+|x|^n}\Big) : \partial_{\tau_{jk}} f \in L^p(\partial_*\Omega,\sigma) \text{ for each } j,k \in \{1,\dots,n\}\Big\} \tag{A.0.77}$$

and equipped with the semi-norm

$$\dot{L}^p_1(\partial\Omega,\sigma) \ni f \mapsto \|f\|_{\dot{L}^p_1(\partial\Omega,\sigma)} := \sum_{j,k=1}^{n} \|\partial_{\tau_{jk}} f\|_{L^p(\partial_*\Omega,\sigma)} \tag{A.0.78}$$

$L^p_1(\partial_*\Omega,\sigma_*)$, the L^p-based Sobolev space of order one on $\partial_*\Omega$, defined for each $p \in [1,\infty]$ as (cf. [113, Definition 11.1.2]):

$$L^p_1(\partial_*\Omega,\sigma_*) := \Big\{ f \in L^p(\partial_*\Omega,\sigma_*) : \partial_{\tau_{jk}} f \text{ exists in } L^p(\partial_*\Omega,\sigma_*) \text{ for each } j,k \in \{1,\dots,n\}\Big\} \tag{A.0.79}$$

and equipped with the norm (cf. [113, Proposition 11.1.9]):

$$L^p_1(\partial_*\Omega,\sigma_*) \ni f \mapsto \|f\|_{L^p_1(\partial_*\Omega,\sigma_*)} := \|f\|_{L^p(\partial_*\Omega,\sigma_*)} + \sum_{1\le j,k\le n} \|\partial_{\tau_{jk}} f\|_{L^p(\partial_*\Omega,\sigma_*)} \tag{A.0.80}$$

$L^p_{-1}(\partial_*\Omega,\sigma_*)$, the ($L^p$-based) negative Sobolev space of order minus one on $\partial_*\Omega$ defined for $p, p' \in (1,\infty)$ with $\frac{1}{p} + \frac{1}{p'} = 1$ as (cf. [113, Definition 11.8.1]):

$$L^p_{-1}(\partial_*\Omega,\sigma_*) := \Big(L^{p'}_1(\partial_*\Omega,\sigma_*)\Big)^* \tag{A.0.81}$$

$L = (a^{\alpha\beta}_{rs}\partial_r\partial_s)_{1\le\alpha,\beta\le M}$ homogeneous, second-order, constant (complex) coefficient, $M\times M$ system in $\mathbb{R}^n$ (1.3.1)

$L_A := \Big(a^{\alpha\beta}_{rs}\partial_r\partial_s\Big)_{\substack{1\le\alpha\le M\\ 1\le\beta\le N}}$ second-order system in $\mathbb{R}^n$ canonically associated with a given coefficient tensor $A := (a^{\alpha\beta}_{rs})_{\substack{1\le r,s\le n\\ 1\le\alpha\le M\\ 1\le\beta\le N}}$ (1.7.7)

$L(\xi)$, the characteristic matrix of a given homogeneous (constant-coefficient) higher-order system $L = \sum_{|\alpha|=|\beta|=m} \partial^\alpha A_{\alpha\beta}\partial^\beta$, defined as (cf. [112, (6.5.39)]):

$$L(\xi) := (-1)^m \sum_{|\alpha|=|\beta|=m} \xi^{\alpha+\beta} A_{\alpha\beta}, \qquad \forall \xi \in \mathbb{R}^n \tag{A.0.82}$$

$L(\xi) := -\big(a^{\alpha\beta}_{rs}\xi_r\xi_s\big)_{1\leq\alpha,\beta\leq M}$ for all $\xi = (\xi_r)_{1\leq r\leq n} \in \mathbb{R}^n$, the characteristic matrix of the homogeneous, complex constant coefficient, second-order $M \times M$ system $L = \big(a^{\alpha\beta}_{rs}\partial_r\partial_s\big)_{1\leq\alpha,\beta\leq M}$ in $\mathbb{R}^n$ (1.3.2)

$L_{\lambda,\mu} := \mu\Delta + (\lambda+\mu)\nabla\mathrm{div}$ the complex Lamé system, with Lamé moduli $\lambda, \mu \in \mathbb{C}$ (1.3.6)

$L_{\lambda,\mu}(\xi) := -\mu|\xi|^2 I_{n\times n} - (\lambda+\mu)\xi\otimes\xi$ for each $\xi \in \mathbb{R}^n$, the characteristic matrix of the complex Lamé system $L_{\lambda,\mu} := \mu\Delta + (\lambda+\mu)\nabla\mathrm{div}$ (1.3.8)

M

$\mathcal{M}_{X,s}$, the L^s-based Hardy-Littlewood maximal operator in the space of homogeneous type (X, ρ, μ), defined for each μ-measurable function f on X as (cf. [112, (7.6.7)]):

$$\mathcal{M}_{X,s}f(x) := \sup_{r>0}\Big(⨍_{B_\rho(x,r)}|f|^s\,d\mu\Big)^{\frac{1}{s}}, \quad \forall\,x\in X \tag{A.0.83}$$

$\mathcal{M}_X$, the Hardy-Littlewood maximal operator of the function f on the space of homogeneous type (X, ρ, μ), defined for each μ-measurable function f on X as (cf. [112, (7.6.16)]):

$$\mathcal{M}_X f(x) := \sup_{r\in(0,\infty)}\frac{1}{\mu\big(B_\rho(x,r)\big)}\int_{B_\rho(x,r)}|f|\,d\mu, \quad \forall\,x\in X \tag{A.0.84}$$

$u^{\max}_M$, the tangential maximal function of u (with exponent M), defined at each $x \in \partial\Omega$ as (cf. [112, (8.5.2)]):

$$\begin{gathered} u^{\max}_M : \partial\Omega \longrightarrow [0,+\infty] \ \text{ defined by} \\ u^{\max}_M(x) := \Big\| u(y)\Big(\frac{\delta_{\partial\Omega}(y)}{|x-y|}\Big)^M\Big\|_{L^\infty_y(\Omega,\mathcal{L}^n)} \quad \text{for each } x\in\partial\Omega \end{gathered} \tag{A.0.85}$$

$\mathscr{M}(X,\mu)$, the collection of μ-measurable functions defined on an arbitrary measure space $(X, \mathfrak{M}, \mu)$

$\mathscr{M}_+(X,\mu)$ the collection of non-negative μ-measurable functions defined on an arbitrary measure space $(X, \mathfrak{M}, \mu)$

$M^{p,\lambda}(\Sigma,\sigma)$, the Morrey space on the Ahlfors regular set $\Sigma \subseteq \mathbb{R}^n$, defined for each $p \in (0,\infty)$ and $\lambda \in (0, n-1)$ as (cf. [113, §6.2]):

$$M^{p,\lambda}(\Sigma,\sigma) := \Big\{f:\Sigma\to\mathbb{C} :\ f \text{ is } \sigma\text{-measurable and } \|f\|_{M^{p,\lambda}(\Sigma,\sigma)} < +\infty\Big\} \tag{A.0.86}$$

where, for each σ-measurable function f on Σ,

$$\|f\|_{M^{p,\lambda}(\Sigma,\sigma)} := \sup_{\substack{x\in\Sigma \text{ and} \\ 0<R<2\,\mathrm{diam}(\Sigma)}}\Big\{R^{\frac{n-1-\lambda}{p}}\Big(⨍_{\Sigma\cap B(x,R)}|f|^p\,d\sigma\Big)^{\frac{1}{p}}\Big\} \tag{A.0.87}$$

$\mathring{M}^{p,\lambda}(\Sigma,\sigma)$, the vanishing Morrey space on the Ahlfors regular set $\Sigma \subseteq \mathbb{R}^n$ with $p \in (0,\infty)$ and $\lambda \in (0, n-1)$ (cf. [113, §6.2]):

$$\mathring{M}^{p,\lambda}(\Sigma,\sigma) := \text{the closure of } L^s(\Sigma,\sigma) \text{ with } s := \tfrac{p(n-1)}{n-1-\lambda} \text{ in } M^{p,\lambda}(\Sigma,\sigma) \qquad \text{(A.0.88)}$$

$\mathcal{M}^{q,\lambda}(\Sigma,\sigma)$, the (q,λ)-mid space on the Ahlfors regular set $\Sigma \subseteq \mathbb{R}^n$, defined for $q \in (1,\infty)$ and $\lambda \in (0, n-1)$ as (cf. [113, §6.2]):

$$\begin{aligned}\mathcal{M}^{q,\lambda}(\Sigma,\sigma) & \qquad\qquad\qquad\qquad\qquad\qquad\qquad\qquad\qquad \text{(A.0.89)}\\ := \Big\{ f \in \big(\mathrm{Lip}_c(\Sigma)\big)' : & \text{ there exist a sequence } \{\lambda_j\}_{j\in\mathbb{N}} \in \ell^1(\mathbb{N}) \text{ and a family}\\ & \{m_j\}_{j\in\mathbb{N}} \text{ of } (q,\lambda)\text{-dull molecules on } \Sigma \text{ such that}\\ & f = \sum_{j=1}^{\infty} \lambda_j m_j \text{ with convergence in } \big(\mathrm{Lip}_c(\Sigma)\big)' \Big\}\end{aligned}$$

and equipped with the norm

$$\begin{aligned}\|f\|_{\mathcal{M}^{q,\lambda}(\Sigma,\sigma)} & \qquad\qquad\qquad\qquad\qquad\qquad\qquad\qquad\qquad \text{(A.0.90)}\\ := \inf\Big\{ \sum_{j=1}^{\infty} |\lambda_j| : f = \sum_{j=1}^{\infty} \lambda_j m_j & \text{ in } \big(\mathrm{Lip}_c(\Sigma)\big)' \text{ with } \{\lambda_j\}_{j\in\mathbb{N}} \in \ell^1(\mathbb{N})\\ & \text{and each } m_j \text{ a } (q,\lambda)\text{-dull molecule on } \Sigma \Big\}\end{aligned}$$

M_b, the operator of pointwise multiplication by the function $b \in L^\infty(\Sigma,\sigma)$

$[M_b, T]$, the commutator of M_b (the pointwise multiplication by b) with the operator T, defined by $[M_b,T]f := b(Tf) - T(bf)$ (2.7.4)

N

$\mathbb{N}_0 := \mathbb{N} \cup \{0\} = \{0, 1, 2, \dots\}$

NTA domain: a nonempty open subset Ω of $\mathbb{R}^n$ satisfying a two-sided corkscrew condition as well as a Harnack chain condition (cf. [112, Definition 5.11.1])

two-sided NTA domain: a nonempty open subset Ω of $\mathbb{R}^n$ satisfying a two-sided corkscrew condition as well as a two-sided Harnack chain condition (cf. [112, Definition 5.11.1])

one-sided NTA domain (or interior NTA domain): a nonempty open subset Ω of $\mathbb{R}^n$ satisfying an interior corkscrew condition as well as a Harnack chain condition (cf. [112, Definition 5.11.1])

$\mathcal{N}_\kappa u$, the (κ-)nontangential maximal operator acting on the function measurable $u : \Omega \to \mathbb{R}^n$ according to (cf. [112, (8.2.1)]):

$$\mathcal{N}_\kappa u : \partial\Omega \longrightarrow [0,+\infty], \quad (\mathcal{N}_\kappa u)(x) := \|u\|_{L^\infty(\Gamma_\kappa(x),\mathcal{L}^n)} \text{ for all } x \in \partial\Omega \qquad \text{(A.0.91)}$$

$\mathcal{N}_\kappa^\varepsilon u$, the ($\kappa$-)nontangential maximal operator truncated at height $\varepsilon > 0$, acting on the function $u : \Omega \to \mathbb{R}^n$ according to (cf. [112, (1.5.5)]):

$$\mathcal{N}_\kappa^\varepsilon u := \mathcal{N}_\kappa(u \cdot \mathbf{1}_{O_\varepsilon}) \text{ where } O_\varepsilon := \{x \in \Omega : \text{dist}(x, \partial\Omega) < \varepsilon\}. \tag{A.0.92}$$

$\mathcal{N}_\kappa^E u$, the restricted nontangential maximal function of $u : \Omega \to \mathbb{R}$, relative to the set E (cf. [112, (8.2.4)]):

$$\begin{gathered} \mathcal{N}_\kappa^E u : \partial\Omega \longrightarrow [0, +\infty] \quad \text{defined as} \\ (\mathcal{N}_\kappa^E u)(x) := \|u\|_{L^\infty(\Gamma_\kappa(x)\cap E, \mathcal{L}^n)} \quad \text{for each} \quad x \in \partial\Omega \end{gathered} \tag{A.0.93}$$

ν, the geometric measure theoretic outward unit normal to a set $\Omega \subseteq \mathbb{R}^n$ of locally finite perimeter, defined via the requirement that (cf. [112, (5.6.2)-(5.6.3)]):

> there exist a locally finite Borel-regular measure σ_* in $\mathbb{R}^n$ and a vector-valued function $\nu \in \left[L^\infty(\mathbb{R}^n, \sigma_*)\right]^n$ satisfying $|\nu(x)| = 1$ at σ_*-a.e. $x \in \mathbb{R}^n$ and with such that $\nabla \mathbf{1}_\Omega = -\nu\sigma_*$ in $\left[\mathcal{D}'(\mathbb{R}^n)\right]^n$ (A.0.94)

ν_g, the geometric measure theoretic outward unit normal induced by the metric tensor g

ν^{E}, the geometric measure theoretic outward unit normal induced by the standard Euclidean metric

$\nu \bullet \vec{F}$, the "bullet" product involving a vector field $\vec{F} \in \left[L^1_{\text{bdd}}(\Omega, \mathcal{L}^n)\right]^n$ (where Ω is an arbitrary open subset of $\mathbb{R}^n$) whose divergence, considered in the sense of distributions in Ω, satisfies $\text{div}\vec{F} \in L^1_{\text{bdd}}(\Omega, \mathcal{L}^n)$, defined as a functional acting on each $\psi \in \text{Lip}_c(\partial\Omega)$ according to (cf. [112, Proposition 4.2.3]):

$$\langle \nu \bullet \vec{F}, \psi \rangle := \int_\Omega \vec{F} \cdot \nabla\Psi \, d\mathcal{L}^n + \int_\Omega (\text{div}\vec{F})\Psi \, d\mathcal{L}^n, \tag{A.0.95}$$

where Ψ is any complex-valued function satisfying

$$\begin{gathered} \Psi \in \text{Lip}(\overline{\Omega}),\ \Psi\big|_{\partial\Omega} = \psi, \text{ and } \Psi \equiv 0 \text{ out-} \\ \text{side of some compact subset of } \overline{\Omega} \end{gathered} \tag{A.0.96}$$

$\nu \circledbullet u$, the Clifford bullet product of ν with u (cf. [113, (10.2.100)]):

$$\nu \circledbullet u := (-\mathrm{i})\text{Sym}(D; \nu) \bullet u \tag{A.0.97}$$

$N_\kappa^p(\Omega; \mu)$, the space of measurable functions in Ω with a p-th power integrable non-tangential maximal function on $\partial\Omega$ with respect to the measure μ (cf. [112, (8.3.31)]):

$$N^p_\kappa(\Omega;\mu) := \big\{u : \Omega \to \mathbb{C} : u \text{ is } \mathcal{L}^n\text{-measurable, and} \quad \|u\|_{N^p_\kappa(\Omega;\mu)} := \|\mathcal{N}_\kappa u\|_{L^p(\partial\Omega,\mu)} < +\infty\big\} \tag{A.0.98}$$

$\big(u\big|^{\kappa-\mathrm{n.t.}}_{\partial\Omega}\big)(x)$, the nontangential trace of the function $u : \Omega \to \mathbb{R}$ at the point $x \in \partial\Omega$ such that $x \in \overline{\Gamma_\kappa(x)}$, defined as (cf. [112, Definition 8.9.1]):

$\big(u\big|^{\kappa-\mathrm{n.t.}}_{\partial\Omega}\big)(x)$ is the number $a \in \mathbb{R}$ with the property that for every $\varepsilon > 0$ there exists some $r > 0$ such that $|u(y) - a| < \varepsilon$ for $\mathcal{L}^n$-a.e. point $y \in \Gamma_\kappa(x) \cap B(x, r)$ (A.0.99)

$\|T\|_{X\to Y}$, the operator "norm" of a positively homogeneous mapping T acting from the quasi-normed vector space $\big(X, \|\cdot\|_X\big)$ with values in the quasi-normed vector space $\big(Y, \|\cdot\|_Y\big)$

$$\|T\|_{X\to Y} := \sup\big\{\|Tu\|_Y : u \in X,\ \|u\|_X = 1\big\} \in [0, +\infty] \tag{A.0.100}$$

$\|T\|_{\mathrm{Bd}(X\to Y)} := \|T\|_{X\to Y}$

$\|T\|_{\mathrm{Bd}(X)}$, the "norm" of a positively homogeneous mapping T acting from the quasi-normed vector space $\big(X, \|\cdot\|_X\big)$ into itself

$$\|T\|_{\mathrm{Bd}(X)} := \|T\|_{\mathrm{Bd}(X\to X)} = \sup_{x\in X,\ \|x\|_X=1} \|Tx\|_X \tag{A.0.101}$$

$\|T\|^{\mathrm{ess}}_{X\to Y}$, the essential norm of the operator $T \in \mathrm{Bd}(X \to Y)$, where X, Y are quasi-normed spaces (cf. [113, §1.2]):

$$\begin{aligned}\|T\|^{\mathrm{ess}}_{X\to Y} &:= \mathrm{dist}\big(T, \mathrm{Cp}(X \to Y)\big)\\ &= \inf\big\{\|T - K\|_{X\to Y} : K \in \mathrm{Cp}(X \to Y)\big\}\end{aligned} \tag{A.0.102}$$

O

$\mathbf{1}_E$, the characteristic function of a given set E

$\omega_{n-1} := \mathcal{H}^{n-1}(S^{n-1})$, the surface area of S^{n-1} (the $(n-1)$-dimensional sphere in $\mathbb{R}^n$)

$O_\varepsilon := \big\{x \in \Omega : \delta_{\partial\Omega}(x) < \varepsilon\big\}$, the one-sided collar neighborhood of $\partial\Omega$ of "width" $\varepsilon > 0$

ω^x, the harmonic measure on $\partial\Omega$ with pole at x

P

$\mathfrak{P}$, the $\mathfrak{P}$-maximal operator acting on a Lebesgue measurable function $u : \Omega \to \mathbb{R}^n$ at the point $x \in \partial\Omega$ according to (cf. [113, §10.1]):

$$(\mathfrak{P}u)(x) := \sup_{0<r<2\,\mathrm{diam}(\partial\Omega)} \left\{ \frac{1}{\sigma\big(\partial\Omega \cap B(x,r)\big)} \int_{\Omega\cap B(x,r)} |u| \, \mathrm{d}\mathcal{L}^n \right\} \tag{A.0.103}$$

$\langle \cdot,\cdot \rangle_{\mathcal{E}}$, the pointwise (real) pairing in the fibers of the Hermitian vector bundle $\mathcal{E}$

${}_{\mathscr{E}'(\Omega)}\langle\!\langle \cdot,\cdot \rangle\!\rangle_{\mathscr{E}(\Omega)}$, the pairing between a compactly supported distribution u in Ω and a smooth function $f \in \mathscr{C}^\infty(\operatorname{supp} u)$, say $f \in \mathscr{C}^\infty(O)$ with $O \subseteq \Omega$ open set containing $\operatorname{supp} u$, defined for each $F \in \mathscr{C}^\infty(\Omega)$ with the property that $F = f$ near $\operatorname{supp} u$ as (cf. [112, (2.2.33)]):

$${}_{\mathscr{E}'(\Omega)}\langle\!\langle u, f \rangle\!\rangle_{\mathscr{E}(\Omega)} := {}_{\mathscr{E}'(\Omega)}\langle u, F \rangle_{\mathscr{E}(\Omega)} \tag{A.0.104}$$

$\langle \cdot,\cdot \rangle_{\Lambda^\ell TM}$, the (real) pointwise pairing on $\Lambda^\ell TM$

$\wedge$, the exterior product of differential forms

$\vee$, the interior product of differential forms

$\partial_{\tau_{XY}}$, the tangential derivative operator on the boundary of a subset Ω (of locally finite perimeter) of a Riemannian manifold (M,g), associated with two vector fields $X,Y \in TM$, acting on any function $f \in L^p_1(\partial_*\Omega,\sigma_g) \otimes \mathcal{E}$ according to (cf. [112, Definition 1.12.7]):

$$\partial_{\tau_{XY}} f := -h_{XY} - \big(\nu_g(X)\,\mathrm{div}_g Y - \nu_g(Y)\,\mathrm{div}_g X + \nu_g([X,Y])\big) f \tag{A.0.105}$$

at σ_g-a.e. point on $\partial_*\Omega$, where $h_{XY} \in L^p(\partial_*\Omega,\sigma_g) \otimes \mathcal{E}$ is the unique function with the property that for each section $\varphi \in \mathscr{C}^1_c(M,\mathcal{E})$ one has

$$\int_{\partial_*\Omega} \langle f, \partial_{\tau_{XY}}\varphi \rangle_{\mathcal{E}} \, \mathrm{d}\sigma_g = \int_{\partial_*\Omega} \langle h_{XY}, \varphi|_{\partial_*\Omega} \rangle_{\mathcal{E}} \, \mathrm{d}\sigma_g \tag{A.0.106}$$

$\overset{\bullet}{\partial}_{\tau_{jk}} u$, the weak tangential derivative of a given function $u \in L^1_{\mathrm{loc}}(\Omega,\mathcal{L}^n)$ with the property that $\nabla u \in \big[L^1_{\mathrm{bdd}}(\Omega,\mathcal{L}^n)\big]^n$, defined as (cf. [112, Example 4.2.4]):

$$\overset{\bullet}{\partial}_{\tau_{jk}} u := \nu \bullet \big((\partial_k u)\mathbf{e}_j - (\partial_j u)\mathbf{e}_k\big), \tag{A.0.107}$$

i.e., the "bullet" product $\nu \bullet \vec{F}^u_{jk}$, where

$$\vec{F}^u_{jk} := (\partial_k u)\mathbf{e}_j - (\partial_j u)\mathbf{e}_k \tag{A.0.108}$$

$\partial_* E$, the measure theoretic boundary of a Lebesgue measurable set $E \subseteq \mathbb{R}^n$, defined as (cf. [112, (5.2.1)]):

$$\partial_* E := \Big\{x \in \mathbb{R}^n : \limsup_{r\to 0^+} \frac{\mathcal{L}^n(B(x,r)\cap E)}{r^n} > 0 \text{ and } \limsup_{r\to 0^+} \frac{\mathcal{L}^n(B(x,r)\setminus E)}{r^n} > 0\Big\} \tag{A.0.109}$$

$\partial^* E$, the reduced boundary of a set $E \subseteq \mathbb{R}^n$ of locally finite perimeter, defined as (cf. [112, (5.6.13)]):

$\partial^* E$ consists of all points $x \in \partial E$ satisfying the following three properties: $0 < \sigma_*(B(x,r)) < +\infty$ for each $r \in (0,\infty)$, formula $\lim_{r\to 0^+} ⨍_{B(x,r)} \nu \, d\sigma_* = \nu(x)$ is valid, and $|\nu(x)| = 1$. (A.0.110)

$\partial_{\rm nta}\Omega$, the nontangentially accessible boundary of an open proper subset Ω of $\mathbb{R}^n$, defined as (cf. [112, Definition 8.8.5]):

$$\partial_{\rm nta}\Omega := \bigcap_{\kappa>0} A_\kappa(\partial\Omega) = \big\{x \in \partial\Omega : x \in \overline{\Gamma_\kappa(x)} \text{ for each } \kappa > 0\big\} \tag{A.0.111}$$

$(a)_+ := \max\{a, 0\}$, the positive part of the number $a \in \mathbb{R}$

$\pi_\kappa(E) = \pi_{\Omega,\kappa}(E)$, the "shadow" (or projection) of a given set $E \subseteq \Omega$ onto $\partial\Omega$, defined as (cf. [112, (8.1.15)]):

$$\pi_\kappa(E) = \pi_{\Omega,\kappa}(E) := \big\{x \in \partial\Omega : \Gamma_\kappa(x) \cap E \neq \varnothing\big\} \tag{A.0.112}$$

$p_{\mathbb{X}}$, the lower Boyd index of a rearrangement invariant Banach function space $\mathbb{X}$ on a non-atomic sigma-finite measure space $(X, \mathfrak{M}, \mu)$

$\partial_\nu^A(\cdot,\cdot)$, the conormal derivative operator with respect to the coefficient tensor A acting from weighted Sobolev spaces (cf. [113, Proposition 8.5.3]), and from Besov/Triebel-Lizorkin spaces (cf. [113, Proposition 9.5.2])

$\dot{\partial}_\nu^A u$, the weak conormal derivative of u with respect to the coefficient tensor A defined (cf. [113, Definition 10.2.18]) by first introducing

$$\vec{F}^\alpha := (A\nabla u)_\alpha = \big(a_{rs}^{\alpha\beta}\partial_s u_\beta\big)_{1\le r\le n} \text{ for each } \alpha \in \{1,\dots,M\} \tag{A.0.113}$$

and then setting

$$\dot{\partial}_\nu^A u := \big(\nu \bullet \vec{F}^\alpha\big)_{1\le\alpha\le M} \in \Big[\big(\mathrm{Lip}_c(\partial\Omega)\big)'\Big]^M \tag{A.0.114}$$

$\partial_{\tau_{jk}}\varphi$ with $\varphi \in \mathscr{C}^1(O)$, the pointwise tangential derivative operator (cf. [113, §11.1]):

$$\partial_{\tau_{jk}}\varphi := \nu_j\big[\partial_k\varphi\big]\big|_{O\cap\partial_*\Omega} - \nu_k\big[\partial_j\varphi\big]\big|_{O\cap\partial_*\Omega} \text{ at } \sigma_*\text{-a.e. point on } O\cap\partial_*\Omega \tag{A.0.115}$$

$\partial_{\tau_{jk}}$, the tangential derivative operator defined in a weak, or distributional, sense (cf. [113, (11.1.17) and Definition 11.1.2], and [113, Definition 11.2.1])

$\overline{\partial} := \partial_{\bar{z}} := \frac{1}{2}(\partial_x + i\partial_y)$, the Cauchy-Riemann operator (1.1.2)

$\partial := \partial_z := \frac{1}{2}(\partial_x - i\partial_y)$, the (complex) conjugate of the Cauchy-Riemann operator (1.1.2)

∂_ν^A pointwise conormal derivative operator with respect to the coefficient tensor $A := \big(a_{rs}^{\alpha\beta}\big)_{\substack{1\le\alpha,\beta\le M\\ 1\le r,s\le n}}$ whose action on $u = (u_\beta)_{1\le\beta\le N} \in \big[\mathcal{D}'(\Omega)\big]^N$ is defined as $\partial_\nu^A u := \Big(\nu_r\big(a_{rs}^{\alpha\beta}\partial_s u_\beta\big)\Big|_{\partial\Omega}^{\kappa-\mathrm{n.t.}}\Big)_{1\le\alpha\le M}$ at $\mathcal{H}^{n-1}$-a.e. point on $\partial_*\Omega$ (1.7.9)

$\partial_\nu^{\widetilde{D},D}$, the conormal derivative operator associated with the factorization of the second-order system L as $\widetilde{D}D$, where $\widetilde{D}$ and D are homogeneous, constant complex coefficient, first-order systems in $\mathbb{R}^n$, defined as $\partial_\nu^{\widetilde{D},D}u := (-i)\mathrm{Sym}(\widetilde{D};\nu)(Du)\big|_{\partial\Omega}^{\kappa-\mathrm{n.t.}}$ for $u \in \big[\mathcal{D}'(\Omega)\big]^N$ (1.7.36)

$\partial_\xi^A u := \Big(\sum_{I=1}^{M}\sum_{r,s=1}^{n} a_{rs}^{JI}\xi_r\partial_s u_I\Big)_{1\le J\le M}$, the conormal derivative of $u = (u_I)_I$, associated with the coefficient tensor $A := \big(a_{rs}^{JI}\big)_{\substack{1\le J,I\le M\\ 1\le r,s\le n}}$ used in the writing of the vector Laplacian as $\Delta = \mathrm{div}A\nabla$ and along $\xi = (\xi_r)_r \in \mathbb{C}^n$ (6.2.21)

p_Ω, the Hölder conjugate exponent of q_Ω (5.7.46)

$\mathscr{P}_{\delta,b}(x,h) := \big\{y \in B(x,R) : |y-x|\delta(|y-x|) < h\cdot(y-x) < b\big\} \subseteq \mathbb{R}^n$, the pseudo-ball in $\mathbb{R}^n$ with apex at $x \in \mathbb{R}^n$, symmetry axis along $h \in S^{n-1}$, truncated at height $b \in (0,\infty)$, whose profile is modeled by some "shape function" $\delta : [0,R] \to [0,\infty)$ (5.8.20)

PI, the Poisson integral operator (5.2.6)

$\mathrm{P.V.}\big(b\,k(x-\cdot)\big|_\Sigma$, the principal-value distribution associated with the smooth odd kernel k and the bounded function b on the countably rectifiable upper Ahlfors regular set $\Sigma \subseteq \mathbb{R}^n$, acting on each test function $\phi \in \mathrm{Lip}_c(\Sigma)$ as (cf. [113, Proposition 11.9.1]):

$$\big\langle \mathrm{P.V.}\big(b\,k(x-\cdot)\big|_\Sigma\big),\phi\big\rangle := \lim_{\varepsilon\to 0^+}\int_{\substack{y\in\Sigma\\ |y-x|>\varepsilon}} b(y)k(x-y)\phi(y)\,d\sigma(y)$$

Q

X/Y, the quotient space of a vector space X and a linear subspace Y of X

$q_{\mathbb{X}}$, the upper Boyd index of a rearrangement invariant Banach function space $\mathbb{X}$ on a non-atomic sigma-finite measure space $(X, \mathfrak{M}, \mu)$

$q_\Omega := \sup\big\{q \in (1,\infty) : \omega^{x_0} \in B_q(\sigma)\big\} \in (1,\infty]$ where ω^{x_0} is the harmonic measure with pole at $x_0 \in \Omega$ and $B_q(\sigma)$ consisting of all locally finite Borel measures

λ on $\partial\Omega$ satisfying $\lambda \ll \sigma$ and for which $d\lambda/d\sigma$ belongs to the reverse Hölder class RH_q with respect to σ (5.7.45)

R

$\mathbb{R}^n_+$, the (open) upper half-space in $\mathbb{R}^n$

$\mathbb{R}^n_-$, the (open) lower half-space in $\mathbb{R}^n$

$\mathrm{Res}(f, z)$, the residue of the function f at the point z (1.1.77)

R_j, the j-th boundary-to-boundary Riesz transform, with $j \in \{1, \dots, n\}$, acting on $f \in L^1\big(\Sigma, \frac{\sigma}{1+|\cdot|^{n-1}}\big)$ at σ-a.e. $x \in \Sigma$ as in (2.5.336):

$$R_j f(x) := \lim_{\varepsilon \to 0^+} \frac{2}{\omega_{n-1}} \int_{\substack{|x-y|>\varepsilon \\ y \in \Sigma}} \frac{x_j - y_j}{|x-y|^n} f(y)\, d\sigma(y)$$

$\mathcal{R}_j$, the j-th boundary-to-domain Riesz transform, with $j \in \{1, \dots, n\}$, acting on each $f \in L^1\big(\Sigma, \frac{\sigma}{1+|\cdot|^{n-1}}\big)$ at every point $x \in \mathbb{R}^n \setminus \Sigma$ as in (2.5.335):

$$\mathcal{R}_j f(x) := \frac{2}{\omega_{n-1}} \int_{\Sigma} \frac{x_j - y_j}{|x-y|^n} f(y)\, d\sigma(y)$$

$\mathcal{R}_{jk}$, the boundary-to-domain integral operator, with $j, k \in \{1, \dots, n\}$, acting on each $f \in L^1\big(\partial_*\Omega, \frac{\sigma(x)}{1+|x|^{n-1}}\big)$ at any point $x \in \Omega$ as in (2.5.256):

$$\mathcal{R}_{jk} f(x) := \frac{-1}{\omega_{n-1}} \int_{\partial_*\Omega} \frac{\nu_j(y)(x_k - y_k) - \nu_k(y)(x_j - y_j)}{|x-y|^n} f(y)\, d\sigma_*(y)$$

R_{jk}, the principal-value singular integral operator, with $j, k \in \{1, \dots, n\}$, acting on each $f \in L^1\big(\partial_*\Omega, \frac{\sigma(x)}{1+|x|^{n-1}}\big)$ at σ_*-a.e. point $x \in \partial_*\Omega$ as in (2.5.268):

$$R_{jk} f(x) := \lim_{\varepsilon \to 0^+} \int_{\substack{y \in \partial_*\Omega \\ |x-y|>\varepsilon}} \Big[\nu_j(y)(\partial_k E_\Delta)(y-x) - \nu_k(y)(\partial_j E_\Delta)(y-x)\Big] f(y)\, d\sigma_*(y)$$

$(R_{jk})_{\max}$, the maximal operator, with $j, k \in \{1, \dots, n\}$, acting on each function $f \in L^1\big(\partial_*\Omega, \frac{\sigma(x)}{1+|x|^{n-1}}\big)$ at any point $x \in \partial_*\Omega$ as in (2.5.270):

$$\begin{aligned} &(R_{jk})_{\max} f(x) \\ &:= \sup_{\varepsilon>0} \Big| \int_{\substack{y \in \partial_*\Omega \\ |x-y|>\varepsilon}} \big[\nu_j(y)(\partial_k E_\Delta)(y-x) - \nu_k(y)(\partial_j E_\Delta)(y-x)\big] f(y)\, d\sigma_*(y) \Big| \end{aligned}$$

$\mathcal{R}_{C\ell}$, the boundary-to-domain Clifford-Riesz transform, acting on any given Clifford algebra-valued function $f \in L^1\big(\partial\Omega, \frac{\sigma(x)}{1+|x|^{n-1}}\big) \otimes C\ell_n$ at each point $x \in \Omega$ as

$$\mathcal{R}_{C\ell} f(x) := \frac{2}{\omega_{n-1}} \int_{\partial\Omega} \frac{x-y}{|x-y|^n} \odot f(y)\, d\sigma(y) \tag{A.0.116}$$

$R_j^{\rm mod}$, the j-th modified Riesz transform (with $j \in \{1, \dots, n-1\}$) in the Euclidean space $\mathbb{R}^{n-1}$, acting on any given function $f \in L^1\big(\mathbb{R}^{n-1}, \frac{dx'}{1+|x'|^n}\big)$ at $\mathcal{L}^{n-1}$-a.e. point $x' \in \mathbb{R}^{n-1}$ as

$$\begin{aligned}(R_j^{\rm mod} f)(x) := \lim_{\varepsilon\to 0^+} \frac{2}{\omega_{n-1}} \int\limits_{\mathbb{R}^{n-1}} \Big\{ &\tfrac{x_j-y_j}{|x'-y'|^n} \mathbf{1}_{\mathbb{R}^{n-1}\setminus \overline{B_{n-1}(x',\varepsilon)}}(y') \\ &- \tfrac{-y_j}{|-y'|^n} \mathbf{1}_{\mathbb{R}^{n-1}\setminus \overline{B_{n-1}(0',1)}}(y') \Big\} f(y')\, dy'\end{aligned} \tag{A.0.117}$$

where, generally speaking, $B_{n-1}(z', r)$ is the $(n-1)$-dimensional ball centered at $z' \in \mathbb{R}^{n-1}$ and of radius r

S

$\sigma := \mathcal{H}^{n-1} \lfloor \partial\Omega$, the surface measure on $\partial\Omega$

$\sigma_* := \mathcal{H}^{n-1} \lfloor \partial_*\Omega$, the surface measure on $\partial_*\Omega$

$\sigma := \mathcal{H}^{n-1} \lfloor \Sigma$, the surface measure on the closed Ahlfors regular set $\Sigma \subseteq \mathbb{R}^n$

σ_g, the surface measure induced by the metric tensor g

$\sigma^{\rm E}$, the surface measure induced by the standard Euclidean metric

$\mathrm{Sym}(D;\xi)$, the principal symbol of the first-order system $D = \Big(\sum\limits_{j=1}^n a_j^{\alpha\beta} \partial_j\Big)_{\substack{1\le\alpha\le N\\ 1\le\beta\le M}}$:

$$\mathrm{Sym}(D;\xi) := \mathrm{i}\Big(\sum_{j=1}^n a_j^{\alpha\beta}\xi_j\Big)_{\substack{1\le\alpha\le N\\ 1\le\beta\le M}} \tag{A.0.118}$$

$*$, the Hodge star operator

$\xi \mapsto \xi^\sharp$, $X \mapsto X^\flat$, the musical isomorphisms (metric identifications via lowering and rising indices), defined as (cf. [112, (1.12.139)]):

$$\begin{aligned} TM \ni X = X_j\partial_j &\longmapsto X^\flat := g_{jk} X_k dx_j \in T^*M, \\ T^*M \ni \xi = \xi_j dx_j &\longmapsto \xi^\sharp := g^{jk}\xi_j \partial_k \in TM \end{aligned} \tag{A.0.119}$$

$S^{n-1} := \partial B(0,1)$, the unit sphere in $\mathbb{R}^n$

$\mathscr{C}^k$-singsup u, the smallest closed set outside of which the distribution u is of class $\mathscr{C}^k$

$\operatorname{supp} f$, the support of the μ-measurable function f defined on a topological space (X, τ) (where μ is a Borel measure measure on (X, τ)), defined as (cf. [112, (3.8.7) in Definition 3.8.3]):

$$\operatorname{supp} f := \left\{ x \in X : \int_O |f| \, d\mu > 0 \text{ for each open set } O \subseteq X \text{ with } x \in O \right\} \quad \text{(A.0.120)}$$

$\mathscr{S}(\mathbb{R}^n)$, the space of (smooth, rapidly decreasing) Schwartz functions in $\mathbb{R}^n$

$\mathscr{S}'(\mathbb{R}^n)$, the space of tempered distributions in $\mathbb{R}^n$

$u_{\star,\theta}$, the solid maximal function of $u : \Omega \to \mathbb{C}$, defined at each point $x \in \Omega$ according to (cf. [112, (6.6.2)]):

$$u_{\star,\theta}(x) := \|u\|_{L^\infty(B(x,\theta\delta_{\partial\Omega}(x)),\mathcal{L}^n)} \in [0, \infty] \quad \text{(A.0.121)}$$

$\rho_\#$, the regularized version of the quasi-distance ρ for a quasi-metric space (X, ρ), defined as (cf. [112, (7.1.17) in Theorem 7.1.2]):

$$\rho_\# := (\rho_{sym})_\alpha \quad \text{for} \quad \alpha := (\log_2 C_\rho)^{-1} \in (0, \infty]. \quad \text{(A.0.122)}$$

$f_p^\#$, the L^p-based Fefferman-Stein sharp maximal function of $f \in L^1_{\text{loc}}(X, \mu)$, defined as (cf. [112, (7.4.110)]):

$$f_p^\#(x) := \sup_{r>0} \left(\fint_{B_\rho(x,r)} |f(y) - f_{B_\rho(x,r)}|^p \, d\mu(y) \right)^{1/p}, \quad \forall x \in X \quad \text{(A.0.123)}$$

T

$\mathcal{T}_\gamma(x)$ the family of "bump" (i.e., localized, and normalized in the Hölder norm) functions centered at $x \in \Sigma$ (cf. [113, §4.1])

τ_ρ, the topology induced by the quasi-distance ρ in a quasi-metric space (X, ρ), defined as (cf. [112, (7.1.7)]):

$$O \text{ is open in } \tau_\rho \overset{\text{def}}{\Longleftrightarrow} \text{ for any } x \in O \text{ there is } r > 0 \text{ with } B_\rho(x, r) \subseteq O \quad \text{(A.0.124)}$$

$\operatorname{Tr}_{\mathbb{R}^n \to \Sigma}$, the trace operator from $\mathbb{R}^n$ to Σ whose action on each given function $u \in W^{1,p}(\mathbb{R}^n, \delta_\Sigma^{ap} \mathcal{L}^n)$ is the limit (cf. [113, Theorem 8.1.1]):

$$(\operatorname{Tr}_{\mathbb{R}^n \to \Sigma} u)(x) := [u]_{\mathbb{R}^n}(x) := \lim_{r \to 0^+} \fint_{B(x,r)} u(y) \, dy \quad \text{(A.0.125)}$$

$\operatorname{Tr}_{\Omega \to \partial\Omega}$, the trace operator from $\mathbb{R}^n$ to $\partial\Omega$ whose action on each given function $u \in W_a^{1,p}(\Omega) := W^{1,p}(\Omega, \delta_{\partial\Omega}^{ap} \mathcal{L}^n)$ is the limit (cf. [113, Theorem 8.3.6]):

$$(\operatorname{Tr}_{\Omega \to \partial\Omega} u)(x) := [u]_{\overline{\Omega}}(x) := \lim_{r \to 0^+} \fint_{B(x,r) \cap \Omega} u(y) \, dy \quad \text{(A.0.126)}$$

$\mathrm{Tr}_{\Omega\to\partial\Omega}$, the trace operator from the open set $\Omega \subseteq \mathbb{R}^n$ to its boundary $\partial\Omega$, acting on each given $u \in A^{p,q}_{\alpha}(\Omega)$ (and with $w \in A^{p,q}_{\alpha}(\mathbb{R}^n)$ such that $w\big|_{\Omega} = u$) according to (cf. [113, Theorem 9.4.5]):

$$\big(\mathrm{Tr}_{\Omega\to\partial\Omega}\, u\big)(x) := \lim_{r\to 0^+} \fint_{B(x,r)} w \, d\mathcal{L}^n \quad \text{at } \sigma\text{-a.e. } x \in \partial\Omega \tag{A.0.127}$$

T_{ε}, the truncated singular integral operator (at the threshold $\varepsilon > 0$), defined as in (2.3.6):

$$(T_{\varepsilon} f)(x) := \int_{y\in\Sigma,\ |x-y|>\varepsilon} k(x-y) f(y)\, d\sigma(y)$$

for each function $f \in L^1\big(\Sigma, \frac{\sigma(x)}{1+|x|^{n-1}}\big)$ and each point $x \in \Sigma$

$T_{\max}$, the maximal operator defined by $(T_{\max} f)(x) := \sup_{\varepsilon>0} \big|(T_{\varepsilon} f)(x)\big|$ for all $f \in L^1\big(\Sigma, \frac{\sigma(x)}{1+|x|^{n-1}}\big)$ and all $x \in \Sigma$ (2.3.5)

$T_{\rm mod}$, the modified principal-value singular integral operator acting on each function $f \in L^1\big(\Sigma, \frac{\sigma(x)}{1+|x|^{n}}\big)$ at σ-a.e. point in Σ as in (2.3.31):

$$T_{\rm mod} f := \lim_{\varepsilon\to 0^+} \int_{\Sigma} \big\{k_{\varepsilon}(\cdot - y) - k_1(-y)\big\} f(y)\, d\sigma(y)$$

where $k_{\varepsilon} := k \cdot \mathbf{1}_{\mathbb{R}^n \setminus \overline{B(0,\varepsilon)}}$ for each $\varepsilon > 0$

$\mathcal{T}_{\rm mod}$, the boundary-to-domain modified integral operator acting on each function $f \in L^1\big(\Sigma, \frac{\sigma(x)}{1+|x|^{n}}\big)$ at every $x \in \Omega$ as in (2.5.29):

$$(\mathcal{T}_{\rm mod} f)(x) := \int_{\partial\Omega} \big\{k(x-y) - k_1(-y)\big\} f(y)\, d\sigma(y)$$

where $k_1 := k \cdot \mathbf{1}_{\mathbb{R}^n \setminus \overline{B(1,0)}}$

U

UR set (cf. [112, Definition 5.10.1]): a closed set $\Sigma \subset \mathbb{R}^n$ which is (upper) Ahlfors regular and has Big Pieces of Lipschitz Images (in a uniform, quantitative, scale-invariant fashion)

UR domain (cf. [112, Definition 5.10.6]): a nonempty open subset Ω of $\mathbb{R}^n$ such that $\partial\Omega$ is a UR set and

$$\mathcal{H}^{n-1}(\partial\Omega \setminus \partial_*\Omega) = 0 \tag{A.0.128}$$

$U \sqcup V$, the union of two disjoint sets U, V

$\mathrm{UC}(X, \rho)$, the space of uniformly continuous functions on the metric space (X, ρ)

$[u]^{A}_{\infty}$, the contribution at infinity of a null-solution u for the vector-Helmholtz operator, associated with the coefficient tensor $A := \big(a^{JI}_{rs}\big)_{\substack{1\leq J,I\leq M\\ 1\leq r,s\leq n}}$ in the writing of the vector Laplacian as $\Delta = \mathrm{div} A\nabla$, defined at each $x \in \mathbb{R}^n$ as in (6.3.34):

$$\lim_{R\to\infty}\Big(\int_{|y|=R}\widehat{y}_s\{a^{IJ}_{rs}(\partial_r\Phi_k)(x-y)u_J(y) + a^{IJ}_{sr}\Phi_k(x-y)(\partial_r u_J)(y)\}\,d\mathcal{H}^{n-1}(x)\Big)_{1\le I\le M}$$

where Φ_k is the radiating fundamental solution of the Helmholtz operator $\Delta + k^2$

V

$dV_g := \sqrt{g}\,dx_1 \wedge \cdots \wedge dx_n$, the volume element induced by the metric tensor g

$\mathrm{VMO}(X,\mu)$, the Sarason space of functions of vanishing mean oscillations on the measure metric space (X,ρ,μ) defined (with $\mathrm{UC}(X,\rho)$ denoting the space of uniformly continuous functions on (X,ρ)) as the space (cf. [113, §3.1]):

$$\mathrm{VMO}(X,\mu) := \text{the closure of } \mathrm{UC}(X,\rho)\cap\mathrm{BMO}(X,\mu) \text{ in } \mathrm{BMO}(X,\mu) \qquad \text{(A.0.129)}$$

W

weakly elliptic system: a system L whose characteristic matrix satisfies the condition $\det[L(\xi)] \neq 0$ for each $\xi \in \mathbb{R}^n \setminus \{0\}$

weakly elliptic coefficient tensor: a coefficient tensor A with the property that the canonically associated second-order system L_A is weakly elliptic, i.e., we have $\det[L_A(\xi)] \neq 0$ for each $\xi \in \mathbb{R}^n \setminus \{0\}$

$W^{k,p}(\Omega)$, the L^p-based Sobolev space of order k in Ω (intrinsically defined)

$W^{k,p}_{\mathrm{loc}}(\Omega)$, the local L^p-based Sobolev space of order k in Ω

$W^{k,p}_{\mathrm{bdd}}(\Omega)$, the space of Sobolev functions on any bounded measurable subset of Ω (cf. [112, (3.0.4)]):

$W^{k,p}_{\mathrm{bdd}}(\Omega)$ denote the space of functions $u \in W^{k,p}_{\mathrm{loc}}(\Omega)$ with the property that $\partial^\alpha u \in L^p(O,\mathcal{L}^n)$ for each $\alpha \in \mathbb{N}_0^n$ with $|\alpha| \le k$ and each bounded Lebesgue measurable subset O of Ω. (A.0.130)

$W^{k,p}_a(\Omega)$, the weighted Sobolev space in Ω defined as in [113, (8.3.5)] for the weight $w := \delta^{ap}_{\partial\Omega}$, with $k \in \mathbb{N}_0$, $p \in (0,\infty)$, $a \in \mathbb{R}$, and equipped with the quasi-norm (cf. [113, Definition 8.3.4]):

$$\|u\|_{W^{k,p}_a(\Omega)} := \sum_{|\alpha|\le k}\Big(\int_\Omega |(\partial^\alpha u)(x)|^p \delta_{\partial\Omega}(x)^{ap}\,dx\Big)^{1/p} \qquad \text{(A.0.131)}$$

X

${}_{X^*}(\cdot,\cdot)_X$, the duality pairing between a vector space X and its algebraic dual X^*

$\mathbb{X}'$, the associated space (aka Köthe dual) of the Generalized Banach Function Space $\mathbb{X}$, equipped with the norm $\|\cdot\|_{\mathbb{X}'}$ (cf. [113, Definition 5.1.11])

$\|x\|_p := \|\cdot\|_{X,p} := \inf\left\{\lambda > 0 : \lambda^{-1}x \in \left\langle B_X(0,1)\right\rangle_p\right\}$ for all $x \in X$, the Minkowski functional associated with the absolutely p-convex hull of the unit ball in X (cf. [113, (7.8.6)])

$\mathscr{X}_{\rm bdd}(\Omega)$, or $\mathscr{X}(\Omega)_{\rm bdd}$, the space of distributions in the open set $\Omega \subseteq \mathbb{R}^n$ defined as (cf. [113, Convention 8.3.7]):

$$\left\{u \in \mathcal{D}'(\Omega) : \left(\psi\big|_{\Omega}\right)u \in \mathscr{X}(\Omega) \text{ for each } \psi \in \mathscr{C}_c^\infty(\mathbb{R}^n)\right\} \tag{A.0.132}$$

$\widehat{x} := x/|x|$ for each $x \in \mathbb{R}^n \setminus \{0\}$ (6.1.1)

References

1. M. Abramowitz and I.A. Stegun, *Handbook of Mathematical Functions*, Dover, New York, 1972.
2. S. Agmon, A. Douglis, and L. Nirenberg, *Estimates near the boundary for solutions of elliptic partial differential equations satisfying general boundary conditions, I*, Comm. Pure Appl. Math., 12 (1959), 623–727.
3. S. Agmon, A. Douglis, and L. Nirenberg, *Estimates near the boundary for solutions of elliptic partial differential equations satisfying general boundary conditions, II*, Comm. Pure Appl. Math., 17 (1964), 35–92.
4. L.V. Ahlfors, *Remarks on the Neumann-Poincaré integral equation*, Pacific J. Math., 2 (1952), no. 3, 271–280.
5. R. Alvarado, D. Brigham, V. Maz'ya, M. Mitrea, and E. Ziadé, *On the regularity of domains satisfying a uniform hour-glass condition and a sharp version of the Hopf-Oleinik Boundary Point Principle*, Journal of Mathematical Sciences, 176 (2011), no. 3, 281–360.
6. R. Alvarado, I. Mitrea, and M. Mitrea, *Whitney-type extensions in geometrically doubling quasi-metric spaces*, Communications in Pure and Applied Analysis, 12 (2013), no. 1, 59–88.
7. R. Alvarado and M. Mitrea, *Hardy Spaces on Ahlfors-Regular Quasi-Metric Spaces. A Sharp Theory*, Lecture Notes in Mathematics, Vol. 2142, Springer, 2015.
8. P. Auscher and T. Hytönen, *Orthonormal bases of regular wavelets in spaces of homogeneous type*, Applied and Computational Harmonic Analysis, 34 (2013), no. 2, 266–296.
9. J. Azzam, *Semi-uniform domains and a characterization of the A_∞ property for harmonic measure*, preprint, (2017). arXiv:1711.03088v2
10. J. Azzam, S. Hofmann, J.M. Martell, K. Nyström, and T. Toro, *A new characterization of chord-arc domains*, J. Eur. Math. Soc., 19 (2017), no. 4, 967–981.
11. B. Bennewitz and J.L. Lewis, *On weak reverse Hölder inequalities for nondoubling harmonic measures*, Complex Var. Theory Appl., 49 (2004), no. 7-9, 571–582.
12. C.A. Berenstein, *An inverse spectral theorem and its relation to the Pompeiu problem*, J. Anal. Math., 37 (1980), 128–144.
13. H. Brezis and L. Nirenberg, *Positive solutions of nonlinear elliptic equations involving critical Sobolev exponents*, Comm. Pure Appl. Math., 36 (1983), 437–477.
14. A.P. Calderón, *Cauchy integrals on Lipschitz curves and related operators*, Proc. Nat. Acad. Sci. USA., 74 (1977), no. 4, 1324–1327.
15. A.P. Calderón, *Commutators, singular integrals on Lipschitz curves and applications*, pp. 85–96 in "Proceedings of the International Congress of Mathematicians" (Helsinki, 1978), Acad. Sci. Fennica, Helsinki, 1980.

D. Mitrea et al., *Geometric Harmonic Analysis II*, Developments in Mathematics 74,
https://doi.org/10.1007/978-3-031-22735-6

16. L. Capogna, C.E. Kenig, and L. Lanzani, *Harmonic Measure: Geometric and Analytic Points of View*, University Lecture Series, Vol. 35, Amer. Math. Soc., Providence, Rhode Island, 2005.
17. A.L. Cauchy, *Sur la mécanique céleste et sur un nouveau calcul appelé calcul des limites*, Turin, 1831.
18. M. Christ, *A $T(b)$ theorem with remarks on analytic capacity and the Cauchy integral*, Colloq. Math., LX/LXI (1990), 601–628.
19. R.R. Coifman and C. Fefferman, *Weighted norm inequalities for maximal functions and singular integrals*, Studia Math., 51 (1974), 241–250.
20. R. Coifman, A. McIntosh, and Y. Meyer, *L'intégrale de Cauchy definit un opérateur borné sur L^2 pour les courbes lipschitziennes*, Ann. Math., 116 (1982), 361–388.
21. R.R. Coifman, R. Rochberg and G. Weiss, *Factorization theorems for Hardy spaces in several variables*, Ann. of Math., 103 (1976), no. 3, 611–635.
22. R.R. Coifman and G. Weiss, *Analyse Harmonique Non-Commutative sur Certains Espaces Homogènes*, Lecture Notes in Mathematics No. 242, Springer-Verlag, 1971.
23. R. Courant and D. Hilbert, *Methods of Mathematical Physics*, Vol. I-II, New York, Interscience, 1962.
24. D. Cruz-Uribe, J.M. Martell, and C. Pérez, *Weights, extrapolation and the theory of Rubio de Francia*, Operator Theory: Advances and Applications, Vol. 215, Birkhäuser/Springer Basel AG, Basel, 2011.
25. B.E.J. Dahlberg, *Estimates of harmonic measure*, Arch. Rational Mech. Anal., 65 (1977), no. 3, 275–288.
26. B.E.J. Dahlberg, *On the Poisson integral for Lipschitz and C^1 domains*, Studia Math., 66 (1979), 13–24.
27. B.E.J. Dahlberg, D.S. Jerison and C.E. Kenig, *Area integral estimates for elliptic differential operators with non-smooth coefficients*, Ark. Mat., 22 (1984), no. 1-2, 97–108.
28. B.E.J. Dahlberg and C.E. Kenig, *Hardy spaces and the Neumann problem in L^p for Laplace's equation in a Lipschitz domain*, Ann. of Math., 125 (1987), 437–465.
29. B.E.J. Dahlberg, C.E. Kenig, and G.C. Verchota, *Boundary value problems for the systems of elastostatics in Lipschitz domains*, Duke Math. J., 57 (1988), no. 3, 795–818.
30. G. David, *Opérateurs intégraux singulière sur certaines courbes du plan complexe*, Ann. Scient. École Norm. Sup. (4) 17 (1984), 157–189.
31. G. David, *Opérateurs d'intégrale singulière sur les surfaces régulières*, Ann. Scient. Ecole Norm. Sup., 21 (1988), 225–258.
32. G. David and D. Jerison, *Lipschitz approximation to hypersurfaces, harmonic measure, and singular integrals*, Indiana Univ. Math. J., 39 (1990), no. 3, 831–845.
33. G. David and J.L. Journé, *A boundedness criterion for generalized Calderón-Zygmund operators*, Annals of Math., 120 (1984), 371–397.
34. G. David, J.L. Journé, and S. Semmes, *Opérateurs de Calderón-Zygmund, fonctions para-accrétives et interpolation*, Rev. Mat. Iberoamericana, 1 (1985), 1–56.
35. M. Dindoš, D. Mitrea, I. Mitrea, and M. Mitrea, *Green function estimates for elliptic systems in the upper half-space*, preprint, (2018).
36. J. Duoandikoetxea, *Forty years of Muckenhoupt weights*, pp. 23–75 in "Function Spaces and Inequalities," Lecture Notes Paseky nad Jizerou 2013, J. Lukes and L. Pick ed., Matfyzpress, Praga, 2013.
37. X.T. Duong, S. Hofmann, D. Mitrea, M. Mitrea, and L. Yan, *Hardy spaces and regularity for the inhomogeneous Dirichlet and Neumann problems*, Revista Matemática Iberoamericana, 29 (2013), no. 1, 183–236.
38. L. Escauriaza and M. Mitrea, *Transmission problems and spectral theory for singular integral operators on Lipschitz domains*, J. Funct. Anal., 216 (2004), no. 1, 141–171.
39. L.C. Evans and R.F. Gariepy, *Measure Theory and Fine Properties of Functions*, Studies in Advanced Mathematics, CRC Press, Boca Raton, FL, 1992.
40. E.B. Fabes, *Layer potential methods for boundary value problems on Lipschitz domains*, pp. 55—80, Lecture Notes in Mathematics, Vol. 1344, Springer, Berlin-Heidelberg-New York, 1988.

41. E.B. Fabes, M. Jodeit Jr., and N.M. Rivière, *Potential techniques for boundary value problems on C^1-domains*, Acta Math., 141 (1978), no. 3–4, 165–186.
42. E. Fabes, C. Kenig, and G.C. Verchota, *The Dirichlet problem for the Stokes system on Lipschitz domains*, Duke Math. J., 57 (1988), no. 3, 769–793.
43. E. Fabes, O. Mendez, and M. Mitrea, *Boundary layers on Sobolev-Besov spaces and Poisson's equation for the Laplacian in Lipschitz domains*, J. Funct. Anal., 159 (1998), no. 2, 323–368.
44. H. Federer, *Curvature measures*, Trans. Amer. Math. Soc., 93 (1959), 418–491.
45. C. Fefferman and E.M. Stein, *H^p spaces of several variables*, Acta Math., 129 (1972), no. 3-4, 137–193.
46. G.B. Folland, *Real Analysis, Modern Techniques and Their Applications*, 2-nd edition, John Wiley and Sons, Inc., New York, 1999.
47. L. Friedlander, *Some inequalities between Dirichlet and Neumann eigenvalues*, Arch. Ration. Mech. Anal., 116 (1991), 153–160.
48. J. García-Cuerva and J.L. Rubio de Francia, *Weighted Norm Inequalities*, North Holland, Mathematics Studies Vol. 116, 1985.
49. J. Garnett and D.E. Marshall, *Harmonic Measure*, Cambridge University Press, 2005.
50. F. Gesztesy and M. Mitrea, *Generalized Robin boundary conditions, Robin-to-Dirichlet maps, and Krein-type resolvent formulas for Schrödinger operators on bounded Lipschitz domains*, pp. 105–173 in the Proceedings of Symposia in Pure Mathematics, Vol. 79, Amer. Math. Soc., 2008.
51. D. Gilbarg and N. Trudinger, *Elliptic Partial Differential Equations of Second Order*, 2nd edition, Springer-Verlag, Berlin, Heidelberg, 1983.
52. I. Gohberg and N. Krupnik, *One-Dimensional Linear Singular Integral Equations*, Vol. I-II, Operator Theory: Advances and Applications Vol. 53-54, Birkhäuser Verlag Basel, 1992.
53. G. Green, *An Essay on the Application of Mathematical Analysis to the Theories of Electricity and Magnetism*, Nottingham, England, T. Wheelhouse, 1828.
54. P. Grisvard, *Elliptic problems in nonsmooth domains*, Monographs and Studies in Mathematics, Vol. 24 Pitman Boston, MA, 1985.
55. M. Grüter and K.-O., Widman, *The Green function for uniformly elliptic equations*, Manuscripta Math., 37 (1982), no. 3, 303–342.
56. N.M. Günter, *Potential Theory and its Applications to Basic Problems of Mathematical Physics*, F. Ungar, 1967.
57. A. Harnack, *Beiträge zur Theorie des Cauchy'schen Integral*, Berichte d. k. Sächs. Ges. d. Wiss., Math.-Phys. Classe, 37 (1885), 379–398.
58. L.L. Helms, *Potential Theory*, 2nd edition, Universitext, Springer, London, 2014.
59. K. Hoffman, *Banach Spaces of Analytic Functions*, Dover Publications Inc., New York, 1962.
60. S. Hofmann and S. Kim, *The Green function estimates for strongly elliptic systems of second order*, Manuscripta Math., 124 (2007) no. 2, 139–172.
61. S. Hofmann, D. Mitrea, M. Mitrea, and A.J. Morris, *L^p-Square Function Estimates on Spaces of Homogeneous Type and on Uniformly Rectifiable Sets*, Memoirs of the American Mathematical Society, Vol. 245, No. 1159, 2017.
62. S. Hofmann, M. Mitrea, and M. Taylor, *Geometric and transformational properties of Lipschitz domains, Semmes-Kenig-Toro domains, and other classes of finite perimeter domains*, J. Geom. Anal., 17 (2007), no. 4, 593–647.
63. S. Hofmann, M. Mitrea, and M. Taylor, *Singular Integrals and Elliptic Boundary Problems on Regular Semmes-Kenig-Toro Domains*, Int. Math. Res. Not. IMRN (2010), no. 14, 2567–2865.
64. S. Hofmann, M. Mitrea, and M. Taylor, *Symbol calculus for operators of layer potential type on Lipschitz surfaces with VMO normals, and related pseudodifferential operator calculus*, Analysis and PDE, 8 (2015), no. 1, 115–181.
65. L Hörmander, *The Analysis of Linear Partial Differential Operators*, Vol. 3, Springer-Verlag, New York, 1985.
66. L. Hörmander, *The Analysis of Linear Partial Differential Operators. I. Distribution Theory and Fourier Analysis*, reprint of the second (1990) edition, Classics in Mathematics, Springer-Verlag, Berlin, 2003.

67. T. Jakab, I. Mitrea, and M. Mitrea, *Traces of functions in Hardy and Besov spaces on Lipschitz domains with applications to compensated compactness and the theory of Hardy and Bergman type spaces*, J. Funct. Anal., 246 (2007), no. 1, 50–112.
68. B. Jawerth and M. Mitrea, *Higher dimensional scattering theory on C^1 and Lipschitz domains*, Amer. J. Math., 117 (1995), no. 4, 929–963.
69. D. Jerison and C. Kenig, *An identity with applications to harmonic measure*, Bull. Amer. Math. Soc., 2 (1980), 3, 447–451.
70. D. Jerison and C. Kenig, *The Neumann problem on Lipschitz domains*, Bull. Amer. Math. Soc., 4 (1981), 2, 203–207.
71. D.S. Jerison and C.E. Kenig, *Boundary behavior of harmonic functions in nontangentially accessible domains*, Advances in Mathematics, 46 (1982), no. 1, 80–147.
72. D.S. Jerison and C.E. Kenig, *Hardy spaces, A_∞, and singular integrals on chord-arc domains*, Math. Scand., 50 (1982), 221–247.
73. D. Jerison and C.E. Kenig, *The inhomogeneous Dirichlet problem in Lipschitz domains*, J. Funct. Anal., 130 (1995), 161–219.
74. F. John, *Plane Waves and Spherical Means Applied to Partial Differential Equations*, Interscience Publishers, New York-London, 1955.
75. J.-L. Journé, *Calderón-Zygmund Operators, Pseudodifferential Operators, and the Cauchy Integral of Calderón*, Lecture Notes in Mathematics, Vol. 994, Springer-Verlag, 1983.
76. N. Kalton, S. Mayboroda, and M. Mitrea, *Interpolation of Hardy-Sobolev-Besov-Triebel-Lizorkin spaces and applications to problems in partial differential equations*, pp. 121–177 in "Interpolation Theory and Applications," AMS Contemporary Mathematics, Vol. 445, 2007.
77. N. Kalton and M. Mitrea, *Stability of Fredholm properties on interpolation scales of quasi-Banach spaces and applications*, Trans. Amer. Math. Soc., 350, (1998), 3837–3901.
78. C.E. Kenig, *Weighted H^p spaces on Lipschitz domains*, Amer. J. Math., 102 (1980), 129–163.
79. C.E. Kenig, *Harmonic Analysis Techniques for Second Order Elliptic Boundary Value Problems*, CBMS Regional Conference Series in Mathematics, Vol. 83, AMS, Providence, RI, 1994.
80. C. Kenig and T. Toro, *Free boundary regularity for harmonic measures and Poisson kernels*, Annals of Mathematics, 150 (1999), 369–454.
81. C. Kenig and T. Toro, *Poisson kernel characterization of Reifenberg flat chord arc domains*, Ann. Sci. École Norm. Sup. (4), Vol. 36 (2003), 323–401.
82. C. Kenig and T. Toro, *Free boundary regularity below the continuous threshold: 2-phase problems*, J. Reine Angew. Math., 596 (2006), 1–44.
83. B.V. Khvedelidze, *Linear discontinuous boundary value problems of function theory, singular integral equations, and some of their applications*, (Russian), Trudy Tbil. Matem. Instituta Akademii Nauk GSSR., 23 (1956), 3–158.
84. P. Koosis, *Introduction to H_p Spaces*, Cambridge University Press, 1998.
85. V.A. Kozlov, V.G. Maz'ya and J. Rossmann, *Elliptic Boundary Value Problems in Domains with Point Singularities*, Mathematical Surveys and Monographs, Vol. 52, Amer. Math. Soc., Providence, RI, 1997.
86. V.D. Kupradze, *Potential Methods in the Theory of Elasticity*, Daniel Davey & Co., Inc., New York 1965.
87. O.A. Ladyzhenskaya, *The Mathematical Theory of Viscous Incompressible Flow*, Gordon and Breach Science Publishers, New York-London, 1963.
88. J.-L. Lions and E. Magenes, *Non-Homogeneous Boundary Value Problems and Applications*, Vol. 1, Springer-Verlag, Berlin, Heidelberg, New York, 1972.
89. W. Littman, G. Stampacchia, and H.F. Weinberger, *Regular points for elliptic equations with discontinuous coefficients*, Ann. Scuola Norm. Sup. Pisa, (3) 17 (1963), 43–77.
90. A.M. Lyapunov, *Papers on potential theory*, Gostekhizdat, Moscow-Leningrad (1949).
91. E. Marmolejo-Olea, D. Mitrea, I. Mitrea, and M. Mitrea, *Radiation conditions and integral representations for Clifford algebra-valued null-solutions of the Helmholtz operator*, Journal of Mathematical Sciences, 231, (2018), no. 3, 367–472.

92. E. Marmolejo-Olea and M. Mitrea, *Harmonic analysis for general first order differential operators in Lipschitz domains*, pp. 91–114 in "Clifford Algebras: Application to Mathematics, Physics, and Engineering", Birkhäuser Progress in Mathematical Physics Series, 2003.
93. J.M. Martell, *Sharp maximal functions associated with approximations of the identity in spaces of homogeneous type and applications*, Studia Mathematica, 161 (2004), no. 2, 113–145.
94. J. Martell, D. Mitrea, I. Mitrea, and M. Mitrea, *The higher order regularity Dirichlet problem for elliptic systems in the upper half-space*, pp. 123–141 in "Harmonic Analysis and Partial Differential Equations," Contemporary Mathematics, Vol. 612, Amer. Math. Soc., Providence, RI, 2014.
95. J.M. Martell, D. Mitrea, I. Mitrea, and M. Mitrea, *The Dirichlet problem for elliptic systems with data in Köthe function spaces*, Revista Matemática Iberoamericana, 32 (2016), no. 3, 913–970.
96. S. Mayboroda and M. Mitrea, *The solution of the Chang-Krein-Stein conjecture*, pp. 61-154 in the Proceedings of the Conference in Harmonic Analysis and its Applications, Tokyo Woman's Christian University, Akihiko Miyachi editor, March 24-26, 2007, Tokyo, Japan.
97. S. Mayboroda and M. Mitrea, *Sharp estimates for Green potentials on non-smooth domains*, Math. Res. Lett., 11 (2004), no. 4, 481–492.
98. V. Maz'ya, *Differential Equations of My Young Years*, Birkhüser, 2014.
99. V. Maz'ya, M. Mitrea, and T. Shaposhnikova, *The Dirichlet problem in Lipschitz domains for higher order elliptic systems with rough coefficients*, Journal d'Analyse Mathématique, 110 (2010), no. 1, 167–239.
100. V. G. Maz'ya and T. O. Shaposhnikova, *Theory of Multipliers in Spaces of Differentiable Functions*, Monographs and Studies in Mathematics, Vol. 23, Pitman, Advanced Publishing Program, Boston, MA, 1985.
101. R. Mazzeo, *Remarks on a paper of Friedlander concerning inequalities between Neumann and Dirichlet eigenvalues*, Int. Math. Res. Not., 4 (1991), 41–48.
102. A. McIntosh and M. Mitrea, *Clifford algebras and Maxwell's equations in Lipschitz domains*, Math. Meth. Appl. Sci., 22 (1999) 1599–1620.
103. W. McLean, *Strongly Elliptic Systems and Boundary Integral Equations*, Cambridge University Press, 2000.
104. Y. Meyer, *Ondelettes et Opérateurs II. Opérateurs de Calderón-Zygmund*, Hermann, Paris, 1990.
105. Y. Meyer, *Ondelettes et Opérateurs III. Opérateurs multilinéarires*, Hermann, Paris, 1990.
106. S.G. Mikhlin, *Singular integral equations*, Uspekhi Matem. Nauk. III, 25 (1948), no. 3, 29–112.
107. C. Miranda, *Partial Differential Equations of Elliptic Type*, second edition, Springer-Verlag, Berlin, Heidelberg, New York, 1970.
108. D. Mitrea, *Distributions, Partial Differential Equations, and Harmonic Analysis*, Springer, New York, 2013.
109. D. Mitrea, *Distributions, Partial Differential Equations, and Harmonic Analysis*, Second edition, Springer Nature Switzerland, 2018.
110. D. Mitrea, I. Mitrea, and M. Mitrea, *On the Geometry of Sets Satisfying Uniform Ball Conditions*, preprint, (2014).
111. D. Mitrea, I. Mitrea, and M. Mitrea, *A sharp divergence theorem with nontangential traces*, Notices of the American Mathematical Society, 67 (2020), no. 9, 1295–1305.
112. D. Mitrea, I. Mitrea, and M. Mitrea, *Geometric Harmonic Analysis I: A Sharp Divergence Theorem with Nontangential Pointwise Traces*, Developments in Mathematics Vol. 72, Springer Nature, Switzerland, 2022.
113. D. Mitrea, I. Mitrea, and M. Mitrea, *Geometric Harmonic Analysis II: Function Spaces Measuring Size and Smoothness on Rough Sets*, Developments in Mathematics Vol. 73, Springer Nature, Switzerland, 2022.
114. D. Mitrea, I. Mitrea, and M. Mitrea, *Geometric Harmonic Analysis III: Integral Representations, Calderón-Zygmund Theory, Fatou Theorems, and Applications to Scattering*, Developments in Mathematics Vol. 74, Springer Nature, Switzerland, 2022.

115. D. Mitrea, I. Mitrea, and M. Mitrea, *Geometric Harmonic Analysis IV: Boundary Layer Potentials in Uniformly Rectifiable Domains, and Applications to Complex Analysis*, Developments in Mathematics Vol. 75, Springer Nature, Switzerland, 2022.
116. D. Mitrea, I. Mitrea, and M. Mitrea, *Geometric Harmonic Analysis V: Fredholm Theory and Finer Estimates for Integral Operators, with Applications to Boundary Problems*, Developments in Mathematics Vol. 76, Springer Nature, Switzerland, 2022.
117. D. Mitrea, I. Mitrea, M. Mitrea, and S. Monniaux, *Groupoid Metrization Theory with Applications to Analysis on Quasi-Metric Spaces and Functional Analysis*, Birkhäuser, 2013.
118. D. Mitrea, I. Mitrea, M. Mitrea, and M. Taylor, *The Hodge-Laplacian: Boundary Value Problems on Riemannian Manifolds*, Studies in Mathematics, Vol. 64, De Gruyter, 2016.
119. D. Mitrea, M. Mitrea, and M. Taylor, *Layer Potentials, the Hodge Laplacian and Global Boundary Problems in Non-Smooth Riemannian Manifolds*, Memoirs of the American Mathematical Society, Vol. 150, No. 713, Providence RI, 2001.
120. D. Mitrea, M. Mitrea, and J. Verdera, *Characterizing regularity of domains via the Riesz transforms on their boundaries*, Analysis and PDE, 9 (2016), no. 4, 955–1018.
121. D. Mitrea, M. Mitrea, and L. Yan, *Boundary value problems for the Laplacian in convex domains and semiconvex domains*, Journal of Functional Analysis, 258 (2010), no. 8, 2507–2585.
122. I. Mitrea and M. Mitrea, *Multi-Layer Potentials and Boundary Problems for Higher-Order Elliptic Systems in Lipschitz Domains*, Lecture Notes in Mathematics, Vol. 2063, Springer, Berlin, 2013.
123. I. Mitrea, M. Mitrea, and M. Taylor, *Cauchy integrals, Calderon projectors, and Toeplitz operators on uniformly rectifiable domains*, Adv. in Math., 268 (2015), 666–757.
124. I. Mitrea, M. Mitrea, and M. Taylor, *Riemann-Hilbert Problems, Cauchy Integrals, and Toeplitz Operators on Uniformly Rectifiable Domains*, book manuscript, in preparation (2017).
125. M. Mitrea, *Clifford Wavelets, Singular Integrals, and Hardy Spaces*, Lecture Notes in Mathematics, Vol. 1575, Springer-Verlag, Berlin, 1994.
126. M. Mitrea, *The method of layer potentials in electro-magnetic scattering theory on non-smooth domains*, Duke Math. J., 77 (1995), no. 1, 111–133.
127. M. Mitrea, *Generalized Dirac operators on non-smooth manifolds and Maxwell's equations*, J. Fourier Anal. Appl., 7 (2001), 207–256.
128. M. Mitrea and B. Schmutzler, *The regularity problem in rough subdomains of Riemannian manifolds*, Chapter 36, pp. 427–440 in "Integral Methods in Science and Engineering: Theoretical and Computational Advances", C. Constanda and A. Kirsch editors, Birkhäuser, 2015.
129. M. Mitrea and M. Taylor, *Boundary layer methods for Lipschitz domains in Riemannian manifolds*, J. Funct. Anal., 163 (1999), no. 2, 181–251.
130. M. Mitrea and M. Taylor, *Potential theory on Lipschitz domains in Riemannian manifolds: Hölder continuous metric tensors*, Communications in Partial Differential Equations, 25 (2000), no. 7-8, 148
131. M. Mitrea and M. Taylor, *Potential theory on Lipschitz domains in Riemannian manifolds: Sobolev-Besov space results and the Poisson problem*, J. Funct. Anal., 176 (2000), no. 1, 1–79.
132. M. Mitrea and M. Taylor, *Potential theory on Lipschitz domains in Riemannian manifolds: L^p, Hardy, and Hölder space results*, Communications in Analysis and Geometry, 9 (2001), no. 2, 369–421.
133. M. Mitrea and M. Taylor, *Boundary problems for the Navier-Stokes system on non-smooth Riemannian manifolds*, Math. Annalen, 321 (2001), no. 4, 955–987.
134. M. Mitrea and M. Taylor, *Potential theory for Lipschitz subdomains of Riemannian manifolds equipped with Dini continuous metric tensors*, Transactions of the American Mathematical Society, 355 (2003), no. 5, 1961–1985.
135. M. Mitrea and M. Wright, *Boundary Value Problems for the Stokes System in Arbitrary Lipschitz Domains*, Astérisque, Soc. Math. de France, Vol. 344, 2012.

136. G. Moisil, *Sur la généralisation des fonctions conjuguées*, Rendiconti della Reale Accad. Naz. dei Lincei, 14 (1931), 401–408.
137. Gr.C. Moisil, *Les fonctions monogènes dans les espaces à plusieurs dimensions*, p. 129 in Comptes Rendus LXIV Congrès des Soc. Savantes Clermont-Ferrant, 1931.
138. G.C. Moisil and N. Teodorescu, *Fonctions holomorphes dans l'espace*, Mathematica (Cluj), 5 (1931), 142–159.
139. C.B. Morrey, *Second order elliptic systems of differential equations. Contributions to the theory of partial differential equations*, Ann. Math. Studies, 33 (1954), 101–159.
140. C. Müller, *Zur mathematischen Theorie elektromagnetischer Schwingungen*, Abh. deutsch. Akad. Wiss. Berlin, 3 (1945/1946), 5–56.
141. J. Nečas, *Direct Methods in the Theory of Elliptic Equations*, corrected 2nd printing, Springer Monographs in Mathematics, Springer, Heidelberg, Dordrecht, London, New York, 2012.
142. N. Ortner and P. Wagner, *Fundamental Solutions of Linear Partial Differential Operators. Theory and Practice*, Springer, 2015.
143. V.A. Paatashvili and G.A. Khuskivadze, *On the boundedness of the singular Cauchy integral on Lebesgue spaces in the case on non-smooth contours* (Russian), Trudy Tbilisk. Matem. Inst. AN GSSR., 69 (1982), 93–107.
144. L.E. Payne and H.F. Weinberger, *New bounds in harmonic and biharmonic problems*, J. Math. Phys., 33 (1954), 291–307.
145. J. Plemelj, *Ein Ergänzungssatz zur Cauchyschen Integraldarstellung analytischer Funktionen, Randwerte betreffend*, Monatshefte für Mathematik und Physik, 19, (1908), 205–210.
146. J. Plemelj, *Riemannsche Funktionenscharen mit gegebener Monodromiegruppe*, Monatshefte für Mathematik und Physik, 19, (1908), 211–246.
147. H. Poincaré, *Sur les equations aux dérivée partielles de la physique mathématiques*, Amer. J. Math., 12 (1890), 211–294.
148. D. Pompeiu, *Sur une classe d'intégrales doubles*, Bull. de la Sect. Scient. de l'Académie Roumaine, 1 (1912), 128–131.
149. D. Pompeiu, *Sur une classe d'intégrales doubles* (deuxième note), Bull. de la Sect. Scient. de l'Académie Roumaine, 1 (1912), 265–271.
150. D. Pompeiu, *Sur une classe d'intégrales doubles* (troisième note), Bull. de la Sect. Scient. de l'Académie Roumaine, 1 (1912), 289–296.
151. D. Pompeiu, *Sur une classe de fonctions d'une variable complexe et sur certaines équations intégrales*, Rendiconti del Circolo Matematico di Palermo, 35 (1913), 277–281.
152. P. Pucci and J. Serrin, *Critical exponents and critical dimensions for polyharmonic operators*, J. Math. Pures Appl., 69 (1990), 55–83.
153. J. Rauch and M. Taylor, *Regularity of functions smooth along foliations, and elliptic regularity*, J. Funct. Anal., 225 (2005), 74–93.
154. F. Rellich, *Darstellung der Eigenwerte von $\Delta u + \lambda u = 0$ durch ein Randintegral*, Math. Z., 46 (1940), 635–636.
155. W. Rudin, *Functions Theory in the Unit Ball of $\mathbb{C}^n$*, Springer-Verlag, New York, Heidelberg, Berlin, 1980.
156. W. Rudin, *Real and Complex Analysis*, 3-rd edition, McGraw-Hill, Boston, Massachusetts, 1987.
157. H.A. Schwarz, *Zur integration der partiellen differentialgleichung $\partial^2 u/\partial x^2 + \partial^2 u/\partial y^2 = 0$*, J. Reine Angew. Math., 74 (1872), 218–253.
158. A. Seeger, *A note on Triebel-Lizorkin spaces*, Approximation and Function Spaces, pp. 391–400, Banach Center Publ., No. 22, PWN, Warsaw, 1989.
159. S.W. Semmes, *A criterion for the boundedness of singular integrals on hypersurfaces*, Trans. Amer. Math. Soc., 311 (1989), no. 2, 501–513.
160. S. Semmes, *Analysis vs. geometry on a class of rectifiable hypersurfaces in $\mathbb{R}^n$*, Indiana Univ. Math. J., 39 (1990), no. 4, 1005–1035.
161. Z. Shapiro, *On elliptical systems of partial differential equations*, C.R. (Doklady) Acad. Sci. URSS (N. S.), 46 (1945).
162. S. Silver, *Microwave Antenna Theory and Design*, M.I.T. Radiation Laboratory Series, Vol. 12, McGraw-Hill, New York, 1949.

163. Yu.V. Sokhotski, *On definite integrals and functions employed in expansions into series*, St. Petersburg, 1873.
164. A. Sommerfeld, *Die Greensche Funktion der Schwingungsgleichung*, Jahresbericht der Deutschen Mathematiker-Vereinigung, 21 (1912), 309–353. Reprinted in Gesammelte Schriften, Vol. 1, pp. 272–316.
165. E.M. Stein, *Conjugate harmonic functions in several variables*, pp. 414–420 in International Congress of Mathematicians, Stockholm, 1962.
166. E.M. Stein, *Singular Integrals and Differentiability Properties of Functions*, Princeton Mathematical Series, No. 30, Princeton University Press, Princeton, NJ, 1970.
167. E.M. Stein, *Harmonic Analysis: Real-Variable Methods, Orthogonality, and Oscillatory Integrals*, Princeton Mathematical Series, Vol. 43, Monographs in Harmonic Analysis, III, Princeton University Press, Princeton, NJ, 1993.
168. E.M. Stein, *Singular Integrals: The Roles of Calderón and Zygmund*, Notices of AMS, 45 (1998), no. 9, 1130–1140.
169. E.M. Stein and G. Weiss, *On the theory of harmonic functions of several variables, I. The theory of H^p spaces*, Acta Math., 103 (1960), 25–62.
170. E.M. Stein and G. Weiss, *Generalization of the Cauchy-Riemann equations and representations of the rotation group*, Amer. J. Math., 90 (1968), no. 1., 163–196.
171. E.M. Stein and G. Weiss, *Introduction to Fourier Analysis on Euclidean Spaces*, Princeton University Press, Princeton, N.J., 1971.
172. J.A. Stratton and L.J. Chu, *Diffraction theory of electromagnetic waves*, Physical Review, 56 (1939), 99–107.
173. Ş. Strătilă and L. Zsidó *Lectures on von Neumann Algebras*, Editura Academiei, Bucureşti, Romania/Abacus Press, Tunbridge Wells, Kent, England, 1979.
174. M.E. Taylor, *Pseudodifferential Operators and Nonlinear PDE*, Birkhäuser, Boston, 1991.
175. M. Taylor, *Partial Differential Equations*, Vol. I, Springer-Verlag, 1996 (2nd ed., 2011).
176. M. Taylor, *Partial Differential Equations*, Vol. II, Springer-Verlag, 1996 (2nd ed., 2011).
177. M. Taylor, *Regularity for a class of elliptic operators with Dini continuous coefficients*, J. Geom. Anal., 21 (2011), 174–194.
178. N. Théodoresco, *La Dérivée Areolaire*, Ann. Roumanine des Math., Cahier 3, Bucharest, 1936.
179. G. Verchota, *Layer potentials and regularity for the Dirichlet problem for Laplace's equation in Lipschitz domains*, J. Funct. Anal., 59 (1984), no. 3, 572–611.
180. N. Wiener, *The Dirichlet problem*, J. Math. Phys., 3 (1924), 127–146.
181. T.H. Wolff, *Counterexamples with harmonic gradients in $\mathbb{R}^3$*, pp. 321–384 in "Essays on Fourier Analysis in Honor of Elias M. Stein," Princeton Math. Ser., Vol. 42, Princeton Univ. Press, Princeton, NJ, 1995.

Subject Index

D. Mitrea et al., *Geometric Harmonic Analysis II*, Developments in Mathematics 74, https://doi.org/10.1007/978-3-031-22735-6

Symbol Index

D. Mitrea et al., *Geometric Harmonic Analysis II*, Developments in Mathematics 74,
https://doi.org/10.1007/978-3-031-22735-6